# CONTRIBUTIONS
# TO STATISTICS

**WILLIAM G. COCHRAN**

# CONTRIBUTIONS TO STATISTICS

William G. Cochran

**JOHN WILEY & SONS**

New York   Chichester   Brisbane   Toronto   Singapore

*Library of Congress Cataloging in Publication Data:*

Cochran, William Gemmell, 1909–
   Contributions to statistics.

  (Wiley series in probability and mathematical
statistics, ISSN 0271-6232. Probability and
statistics section)
   Includes bibliographical references.
   1. Mathematical statistics—Collected works.
I. Title.    II. Series.

| QA276.A1C6 | 519.5 | 81-13077 |
| ISBN 0-417-09786-1 | | AACR2 |

Printed in the United States of America

10  9  8  7  6  5  4  3  2  1

# COMPILER'S NOTE

My husband William Cochran had almost completed the collection of his papers when he died in March 1980. He left a list of the publications, and I was able to finish the collection. If there exist papers that I did not find, I must take the responsibility for their omission. Conversely, it has been my decision to include, in a few instances, discussions that followed the presentation of his papers. In one case I have included all three papers that were part of a unit discussion of agricultural meterology in which my husband took part.

I have enjoyed the help and goodwill of many people in this project. For initial encouragement I am indebted to my lifelong friend Dr. Mildred M. (Barnard) Prentice of Brisbane, who but for distance would have been the ideal compositor. For staunch support and advice I am graeful to Dr. Frederick Mosteller and to my daughter Dr. Elizabeth C. Welsh.

Dr. Francis Dressel of the U.S. Army Research Office found a paper that my husband had not listed. Dr. Theodore Colton of Boston University School of Public Health and Dr. Philip Sartwell, retired, of Johns Hopkins School of Hygiene helped with medical references.

Where I was able to contact co-authors of papers, I have received generous acquiescence for the inclusion of the joint papers. Finding the whereabouts of some co-authors has proved impossible, and some predeceased my husband.

I have received generous help from the office of the Department of Statistics at Harvard, and my editor in Wiley's Interscience Division, Miss Beatrice Shube, and her colleagues patiently helped me to bring the whole project to completion.

My thanks are due, too, to the holders of the copyrights to these papers who without exception agreed to their reproduction in this book.

The volume is intended to be a single source of my husband's work for researchers in many fields and includes some minor pieces that may have significance in the history of statistics. It leaves the evaluation and discussion of his work to others. I will end by saying that William was devoted to his work.

Betty I. M. Cochran

*September 1981*

v

# FOREWORD

William Gemmell Cochran was born in the Royal Burgh of Rutherglen, Scotland, and later lived in Gourock and Glasgow. Bill's brother, Oliver, says that they had a fairly normal boyhood, the usual illnesses, school, and games for a family in extremely modest circumstances. Their father was a railway employee all his life. At home Bill was called Willie (pronounced Wully), though when wife-to-be Betty appeared, she preferred William. Bill was always good at academic subjects but, Oliver says, not at art. Were it not for Oliver, Bill's friends would probably not have appreciated his intense desire to be first in examinations. Oliver says that Bill wanted to be first and, more or less coldbloodedly, set out to study just enough to achieve this. His repeated successes won him many prizes and finally scholarships at Glasgow and later Cambridge University, making his academic career possible.

At Cambridge he studied mathematics, applied mathematics, and statistics with Wishart. Yates persuaded Cochran to go to Rothamsted without a Ph.D., and there he began his practical work. Bill loved to tell about a complicated analysis he had done of a field experiment of crop fertilizers with many missing plots. When he gave Yates his detailed results, Yates scrawled "Crop failed" across the top with no further comment. I think Bill liked to emphasize to his students that some data are not worth analyzing, even when we have the big guns of a mathematical arsenal ready for attack. While working at Rothamsted, Cochran attended lectures by Fisher at University College.

He married Betty I. M. Mitchell, a plant pathologist, in London, July 17, 1937.

In addition to extensive consulting at Rothamsted, Cochran carried out major analyses of long-term agricultural experiments, attending to issues of weather, differential fertility of plots, and lack of replication. When he left for America, D. J. Finney took his place. In his six years at Rothamsted, Cochran completed 18 papers. Geoffrey Watson reports from communications with Yates that Cochran had often worked on sample surveys while at Rothamsted, including the 1938/39 census of woodlands.

Cochran visited Ames, Iowa, in 1938 and agreed to return as professor at Iowa State College (now Iowa State University) in 1939. George Snedecor had already brought Fisher's important ideas to Ames, and Cochran's move brought great research strength as well. At Ames, he was acquainted with Raymond Jessen who led sampling developments there. At various times later, Cochran served on and chaired an advisory panel to the U.S. Census.

During World War II, 1943–44, he worked with the Statistical Research Group of Princeton (headed by Samuel S. Wilks), and I became much better acquainted with him at that time. In Princeton, Betty and Bill lived in a large house with Harriet and Alexander Mood. Alex, a fine statistics scholar who took his doctorate at Princeton, returned to Iowa with Cochran, and in a letter (Watson, 1981) describes Cochran's breadth of knowledge of theory and applications and his marvelous consulting skills.

Gertrude Cox, who was organizing the Institute of Statistics in North Carolina, persuaded Cochran to head the graduate program in experimental statistics at North Carolina State College at Raleigh. Harold Hotelling headed the Mathematical Statistics program at University of North Carolina at Chapel Hill.

In 1949, Bill moved to The Johns Hopkins University where he chaired the Department of Biostatistics in the School of Hygiene and Public Health. His books *Sampling Techniques* and, with Gertrude Cox, *Experimental Designs* came out while he was at Hopkins.

Bill and I engaged in our longest period of joint scholarly work at this time. Kinsey, Pomeroy, and Martin at Indiana University had brought out their pathbreaking and widely discussed book *Sexual Behavior in the Human Male*. Although the media treated it extensively and enthusiastically, reviewers heavily criticized its statistical methodology. The sponsoring organization was the Committee for Research in Problems of Sex of the National Research Council, the working, rather than honorific, arm of the National Academy of Sciences. In view of the criticisms, the Committee asked Samuel Wilks, then President of the American Statistical Association, to appoint a committee to review the work. Wilks, who had been my thesis director, appointed Cochran chairman, a natural choice especially with his work in health and medicine at Hopkins, John W. Tukey from the Department of Mathematics at Princeton, and me from the Department of Social Relations at Harvard. By then, the three of us had known each other in several capacities.

Among the attractions of working with Cochran were his fairness and his down-to-earth attitude. When Tukey and I would outline a scheme for appraising the report that included a great deal of research and writing, Cochran, as a responsible chairman, would give us some notion of the timetable and effort required by our plans. Since we had no time off from our regular jobs for this work, he pointed out that our proposal could not be completed. He helped us set up a solid core of work and writing. Then, he said, we can add anything else we have time for. Under pressure from Tukey, we added a great deal.

One core effort dealt with reviews of Kinsey, Pomeroy, and Martin's book. We chose six reviews of considerable stature and length, and systematically dealt with *every* comment by these reviewers agreeing, explaining, disagreeing, and so on. The misery of labeling and losing small slips of paper taught us new facets of the research process. Cochran made us take care to think about, discuss, and review every comment. He never wanted to make a clever remark, but he did like to say things clearly, elegantly, and briefly. He had no interest in scoring points against anyone. That experience helped me in later years because I learned to look more carefully at my writing and to be more sensitive to how others might interpret it.

The idea of being satisfied with a core job plus what extras one has time for seemed to me a valuable one for many of us, protecting us from overweening ambition. It helps in advising students and planning programs. Cochran also used this idea in grading examinations. Sometimes a student will miss the point of a question and do a good job on a different question that wasn't asked. How should the answer be graded? Cochran's view was that the core job—answer the question—came first and extra points could be added for embellishments.

Bill knew a lot of Gilbert and Sullivan and liked to sing. Tukey knew a lot of it too, though I had just learned a few verses. While walking on the campus at Indiana University where Kinsey's headquarters were, we began singing together. It was fun, but we were just getting the hang of it. One morning after a long meeting with Kinsey, Pomeroy, and Martin, we went to our assigned office and for some reason decided to try a verse during coffee break. In the middle of "I was the ruler of the Queen's Navy," Kinsey burst into the office. We were embarrassed to learn that air conditioning ducts carried our little song to every office, thus disturbing everyone. I'm afraid there was some feeling that we weren't taking sexual research seriously enough

In all, we prepared a book, but Cochran said he usually spent more time writing a sentence than the three of us had spent on a page. When at some stage we found that we hadn't enough funds to do all the things we wanted to, Cochran was philosophical about it. "Perhaps this amount of money represents the amount of advice the client wants to buy," he said.

Betty Cochran contributed a great deal to this work by putting up with the team and the mess it made of her home, and by her cordiality in welcoming, feeding, and entertaining us.

When the Department of Statistics was established at Harvard University in 1957, we needed an additional senior person to lend strength to the department. A national search was instituted, and we were delighted to find that Bill Cochran was willing to come to Harvard.

He, John Pratt, Howard Raiffa, and I and later Arthur Dempster created the curriculum, set up the undergraduate and graduate requirements, and generally laid the foundation for the department. Bill's many students, both undergraduates and graduates, have gone on to become leaders in the field.

Perhaps the spirit that Cochran created in both his students and in young colleagues might be communicated by a few quotations from letters written by some of these distinguished scholars:

"Despite his many commitments outside of teaching and research, Professor Cochran always made time for his students, and carefully read and commented on all written material given to him. He supervised the research that I did for my doctoral thesis at a particularly difficult time for him, as he was quite a sick man." Robert W. Mellor.

"Whenever you went in to see Cochran about a problem, he had an intriguing personal habit that perhaps others may have noticed. He would get up off the chair, pace around the room, stroke his lips with one or two fingers, and either mumble to himself or hum a tune. And then, wham, out came the solution!" Bernard A. Greenberg.

"In addition, he was a fountain of information on how to tackle seemingly insolvable problems. . . . I presented my up-to-then results to a seminar . . . the distribution theory . . . was so complicated that closed-form solutions seemed intractable for $N > 9$. The next morning I found a short note from W. G. C. . . .

'I believe you might find my article on the distribution of quadratic forms help-ful.' . . . [It] was the solution to all my problems." R. L. Anderson.

"Cochran was the perfect vehicle to carry English statistics—its approach and its results—across the Atlantic. He knew every aspect of it, practically and theo-retically. He was absolutely impartial. In a subject with wide differences of opinion and strong clashing characters, it is no mean feat to be friendly to every-body and sympathetic to all reasonable ideas." Geoffrey Watson.

"He had a keen sense of what a professional should be able to do and should do. Whenever I . . . overstepped and asked for help inappropriately he would gently remind me 'You're a professional; that is your problem, you work it out.' " Carl E. Hopkins.

"I was always amazed at how he remembered just where I stood on a problem; it was as if he'd been spending all of his time since our previous meeting thinking about it. . . . At several points after solving some theoretical problem on my thesis topic he would ask, 'but what will this mean for the practical man?' " Joel Kleinman.

"I had various opportunities to talk to Fisher between [1946] and when he died in 1963, and have the impression that he considered Cochran the greatest of his many distinguished students and followers." Nathan Keyfitz.

Bill's research was largely programmatic. By that I mean that he had special fields that he worked in, sample surveys and experimental design and analytic techniques, including the analysis of counted data, and he had these fields laid out completely in his mind. One of our statistics journals, *Statistical Theory and Method Abstracts,* produces abstracts in about 13 different colors corresponding to the field of work such as pink for probability, yellow for estimation, and blue for design of experiments. Cochran used to joke with me saying "Why do you look at all those colors? I find two do me very well." On the one hand, there was something to it. He kept up with sample surveys and experimental design religiously. On the other hand, he was a closet reader, and I think he studied nearly every color.

Some spots in his special fields had holes or gaps that needed to be filled. These gaps provided a continuous source of research problems for students and for him-self. He knew that the problems were important because they fitted into the scheme of needs of researchers in the field. And he had a clear notion of what sort of effort it might take to solve a problem, and therefore students who did not bring their own problems could expect to find with him a thesis of manageable size and one that was reasonably sure to have a satisfactory outcome. Such pro-grammatic work advances the field and also is invaluable for an academic de-partment because it keeps students moving smoothly.

Although Cochran devoted much of his life to helping thousands with their statistical problems, he rarely asked for help himself. As others have noted, one reason was that he scoured the literature, and so had less need than others might. In revising Snedecor's *Statistical Methods,* he felt a special responsibility to the research workers who use it almost as a bible, and so I can recall his asking me questions a few times about new techniques in areas where he knew I had been specializing.

Although Cochran didn't like to talk with people about their general troubles, he did enjoy talking about their illnesses and his own. He questioned people closely about their sicknesses and went into concrete detail about his own and others' illnesses. Colds, sprained ankles, strokes seemed all the same to him. Although he talked about these matters, he didn't complain, rather he regarded them as a useful conversation piece at meals. He didn't like to talk about work at lunch. He liked to talk about daily events, to mention Betty and his children and their families in the conversation.

He was remarkably prompt at nearly every activity. When money was due—coffee fund contribution, repayment of a lunch money loan, presents for fellow workers—he was terribly on time. When he was asked to prepare a report for the department, he hurried to it and one could expect a thorough treatment quickly. Similarly, when he agreed to work jointly on developing a course, he went right at it and pursued it until it was complete. When he promised to give a paper at a meeting or to contribute a paper to a book or an encyclopedia, he was usually the first one finished.

He was not especially interested in administration, but he cheerfully took on the chairmanship of the department when others went on sabbatical, expressing neither pleasure nor pain, but not pushing either for expansion.

Had he other interests? Yes, he played badminton in his younger days. He liked square dancing and often acted as the caller. He liked serious literature. Betty and he kept up with movies and plays. And he enjoyed sports events on the telly. Betty and he put on annual parties, often a barbecue in their lovely yard at Winchester, Massachusetts. Betty and Bill have always been most gracious hosts, mixing statisticians with other fascinating people from the community. They led a very full life with sailing and swimming at their summer home in Orleans.

Bill loved to travel, and as often as he could he went abroad. Even after his strokes, he still wanted to go, and I remember riding with him in a bus on a grueling trip from New Delhi to the Taj Mahal. He was enjoying every minute of it. We talked statistics most of the way.

Cochran didn't mourn much over the time and work spent on hard jobs. I recall that when he chaired a committee to redo the constitution of a national statistics society, in his oral report, he claimed that the item that had held them up so long—the work took well over a year and a half—was the one that stated the qualifications for membership. On the one hand, we had an elite society of mathematical scholars and, on the other hand, no one could seem to come up with a graceful way of saying that there were just two qualifications for membership—first you had to have $5 (those were the days!), second you had to be willing to give it to the society. Cochran had a very chuckly sense of humor.

The nation was well aware of his ability and he was asked to serve on many committees. He was careful about his total commitments. He served as chairman of the Advisory Committee to the Bureau of the Census. He served on the Committee to Consider the Effect of Battery Additives on the Life of Batteries. Although a smoker, he served on the Surgeon General's Committee on Smoking and Health which came out strongly against smoking. That was a report that rocked the stock market!

He was a member of the Advisory Committee to the Atomic Bomb Casualty Commission, a member of the Committee on Epidemiology and Biometry at National Institutes of Health, and chairman of a World Health Organization Advisory Committee on Health Statistics.

He was a sought-after consultant in health, medicine, and other areas. Beyond this he was a much sought-after teacher for summer schools, and he enjoyed doing such teaching.

He was a great believer in the fellowship of man and made close friends of people from any part of the globe. If anyone seemed to speak against a national or racial group, he was quick to defend. Indeed, these were the only times I recall him speaking sharply. He epitomized Will Rogers' maxim "I never met a man I didn't like." I cannot recall him speaking ill of anyone. Everyone liked him. Hundreds of statisticians from all over the world turned out for his retirement party.

The first national statistics meetings I ever attended were joint between the Institute of Mathematical Statistics and the American Statistical Association in Detroit during the Christmas period in 1938. Naturally, being a graduate student, I was most excited because I met distinguished people whose papers I had been reading: A. T. Craig, C. C. Craig, W. E. Deming, Paul Dwyer, Harold Hotelling, J. Neyman, Paul Rider, Walter Shewhart, and S. S. Wilks, to name a few.

Before the first session, my teacher, E. G. Olds, told me that we were to hear a brilliant young man from Rothamsted. The first paper I heard was presented by William G. Cochran, and it was easy to follow. I had a little trouble with his Scottish accent but not with his exposition. I was totally unprepared for the discussion, including a ferocious attack made on the paper by Joseph Berkson. Bill handled Joe's complaints cooly without being at all flurried, and so I came away much impressed both with how easy it was to follow presentations at professional meetings and with the vigorous style discussions seemed to take. But it only took one or two more sessions to disabuse me of both my ideas. You usually can't follow papers at meetings, and few discussants are ferocious. As the years went by, I became friends with both these men, and I gradually learned that Joe just loved a good fight and didn't like to discuss in any other way.

After Bill died, Berkson wrote "I was the unmathematical David challenging the mathematical Goliath. Bill Cochran was the unflappable thorough one, to keep us on the solid ground of established principle. That is my recalled impression of him, coupled with a sense of his unswerving integrity. He was loyal to his teacher R. A. Fisher, but did not conceal his difference of opinion when it became a matter of public discussion. He exerted a wide influence in dissemination of sound statistical practice. We wanted him for the Mayo Clinic but he was unavailable. I was deeply indebted to him personally for he unfailingly answered my questions, those too elementary for the usual professional mathematician or those requiring considerable mathematical endeavor. Asked by someone to evaluate me, he wrote in great detail. I was pleased with his occasional praise, but overwhelmed by his thoroughness. He had apparently read all my papers. Nothing superficial with Bill Cochran."

The many honors Bill had showed in what high esteem he was held by the profession. He was President of the International Statistical Institute, itself an honorary body, the American Statistical Association, the Institute of Mathematical Statistics, and the Biometric Society. He was elected to the American Academy of Arts and Sciences and to the National Academy of Sciences, and was an honorary fellow of the Royal Statistical Society. He was awarded the S. S. Wilks Medal by the American Statistical Association for his many contributions to the advancement of the design and analysis of experiment, the Guy Medal of the Royal Statistical Society for his analysis of field crop data, and the Outstanding Statistician Award of the Chicago Chapter of the American Statistical Association. He held a Guggenheim fellowship and had honorary doctorates from Glasgow University and from The Johns Hopkins University.

William Gemmell Cochran died at the age of 70 at Orleans, Massachusetts on March 29, 1980. He had had repeated strokes and recoveries to various degrees over a period of several years. He continued active scholarship to the end in spite of serious infirmities.

One of his posthumously published papers beautifully appreciates the early work in the design and analysis of variance by R. A. Fisher. He cooly explains Fisher's mistakes and successes. What was important about Cochran's historical research was that he returned to the foundations again and again and thought about them deeply each time with new insights about the meaning and importance of the work. It is essential now that someone take up the same searching approach to an appreciation of Bill's contribution to the advancement of statistics.

FREDERICK MOSTELLER

*Harvard University*

## REFERENCES

Anderson, R. L. (December 1980), William Gemmell Cochran, 1909–1980, A Personal Tribute, *Biometrics,* **36,** 574–578.

Anderson, T. W. (1965), Samuel Stanley Wilks, 1906–1964, *Annals of Mathematical Statistics,* **36.**

Watson, G. S. (1981), William Gemmell Cochran 1909–1980, *Annals of Statistics,* 1982.

# CONTENTS

CONTRIBUTIONS
TO STATISTICS

# 1

*The distribution of quadratic forms in a normal system, with applications to the analysis of covariance.* By W. G. Cochran, B.A., St John's College. (Communicated by Mr J. Wishart.)

[*Received* 17 February, *read* 5 March 1934.]

1. Many of the most frequently used applications of the theory of statistics, such for example as the methods of analysis of variance and covariance, the general test of multiple regression and the test of a regression coefficient, depend essentially on the joint distribution of several quadratic forms in a univariate normal system. The object of this paper is to prove the main relevant results about this distribution. As an application of these results, the theory involved in the method of analysis of covariance will be investigated.

2. Let $x_1, x_2, \ldots, x_n$ be normally and independently distributed with unit standard deviation and means zero. We consider the distribution of any quadratic form $\Sigma_{ij} a_{ij} x_i x_j$ in the variates $x_1, \ldots, x_n$. Let $A$ denote the matrix $\{a_{ij}\}$ and $r(A)$ the rank of $A$, which will be called alternatively the number of degrees of freedom (d.f.) of the quadratic form, this being the term used in statistics. The frequency distribution of $\Sigma_{ij} a_{ij} x_i x_j$ is given by

$$f(Q)\, dQ = \frac{1}{(2\pi)^{\frac{1}{2}n}} \int \ldots \int e^{-\frac{1}{2}\sum_1^n x_i^2} \prod_{i=1}^n dx_i$$

over all values of $x_1, \ldots, x_n$ for which $\Sigma_{ij} a_{ij} x_i x_j$ lies between $Q$ and $Q + dQ$. Now there is an orthogonal matrix $L$ such that $L'AL = \Lambda_r$, where $\Lambda_r$ is a diagonal matrix, whose elements in the leading diagonal are the latent roots $\lambda_i$ of $A$, and we may further suppose that $\lambda_1, \ldots, \lambda_r$ are the non-zero latent roots, where of course $r = r(A)$.

Let $L = \{l_{ij}\}$, and make the transformation $y_i = \sum_{j=1}^n l_{ij} x_j$. This gives

$$f(Q)\, dQ = \frac{1}{(2\pi)^{\frac{1}{2}n}} \int \ldots \int e^{-\frac{1}{2}\sum_1^n y_i^2} \prod_{i=1}^n dy_i$$

over all values of $y_1, \ldots, y_n$ for which $\sum_1^r \lambda_i y_i^2$ lies between $Q$ and $Q + dQ$, i.e.

$$f(Q)\, dQ = \frac{1}{(2\pi)^{\frac{1}{2}r}} \int \ldots \int e^{-\frac{1}{2}\sum_1^r y_i^2} dy_1 \ldots dy_r$$

over the same range. But this is the distribution of $\sum_1^r \lambda_i y_i^2$, where

$y_1, \ldots, y_r$ are normally and independently distributed with unit standard deviation and zero means. Thus we get the result

I. With the conditions stated above, every quadratic form $\Sigma_{ij} a_{ij} x_i x_j$ is distributed as is the linear form $\overset{r}{\underset{1}{\Sigma}} \lambda_i z_i$, where the $z_i$ are independent and each follows the $\chi^2$ distribution with 1 d.f., and $\lambda_i$ are the non-zero latent roots of $A$.

*Corollary* 1. A necessary and sufficient condition that $\Sigma_{ij} a_{ij} x_i x_j$ follows a $\chi^2$ distribution is that the non-zero latent roots of $A$ are all 1.

*Corollary* 2. If $\Sigma_{ij} a_{ij} x_i x_j$ is distributed as $\chi^2$, then

$$r(A) + r(1 - A) = n. \tag{1}$$

For, by Corollary 1, the transformation $y_i = \Sigma_j l_{ij} x_j$ above changes $\Sigma_{ij} a_{ij} x_i x_j$ into $\overset{r}{\underset{1}{\Sigma}} y_i^2$ and $\overset{n}{\underset{1}{\Sigma}} x_i^2$ into $\overset{n}{\underset{1}{\Sigma}} y_i^2$, therefore it changes

$$\Sigma_{ij} (\delta_{ij} - a_{ij}) x_i x_j \text{ into } \overset{n}{\underset{r+1}{\Sigma}} y_i^2,$$

so that $\qquad\qquad r(1 - A) = n - r = n - r(A).$

We shall denote the distribution of $\overset{r}{\underset{1}{\Sigma}} \lambda_i z_i$ in I above by the symbol $P(k, \lambda_1, \lambda_2, \ldots, \lambda_q)$, where $k$ is the number of unit latent roots and $\lambda_1, \ldots, \lambda_q$ are latent roots which differ from both 1 and 0. Thus a $P(k)$ distribution is a $\chi^2$ distribution with $k$ d.f. and in general $k + q = r$.

II. With the same initial conditions, let

$$\overset{n}{\underset{1}{\Sigma}} x_i^2 = q_1 + q_2 + \ldots + q_k,$$

where $q_1, \ldots, q_k$ are quadratic forms in the $x_i$ with d.f. $n_1, n_2, \ldots, n_k$ respectively.

Then a necessary and sufficient condition that $q_1, \ldots, q_k$ are independently distributed in $\chi^2$ distributions with d.f. $n_1, \ldots, n_k$ respectively is

$$n_1 + n_2 + \ldots + n_k = n. \tag{2}$$

*Necessity.*

If $q_1, \ldots, q_k$ are independently distributed in $\chi^2$ distributions with d.f. $n_1, \ldots, n_k$, then $q_1 + q_2 + \ldots + q_k$ is distributed in a $\chi^2$ distribution with d.f. $n_1 + n_2 + \ldots + n_k$, by the additive property of the $\chi^2$ distribution. But $q_1 + \ldots + q_k = x_1^2 + \ldots + x_n^2$ and hence has a $\chi^2$ distribution with $n$ d.f., so that

$$n = n_1 + n_2 + \ldots + n_k.$$

*Sufficiency.*

Let $n_1 + n_2 + \ldots + n_k = n$. Then we can find $z_1, \ldots, z_{n_1}$, so that $q_1$ can be expressed as $\sum_{1}^{n_1} c_i z_i^2$ by a real linear transformation, where $c_i$ are all either $\pm 1$. Similarly we can find $z_{n_1+1}, \ldots, z_{n_1+n_2}$ so that $q_2$ is expressed as $\sum_{n_1+1}^{n_2} c_i z_i^2, \ldots$, and finally $z_{n_1+\ldots+n_{k-1}+1}, \ldots, z_n$ so that $q_k$ is expressed as $\sum_{n_1+\ldots+n_{k-1}+1}^{n} c_i z_i^2$. Then the transformation to $z_1, \ldots, z_n$ changes $\sum_{1}^{n} x_i^2$ into $\sum_{1}^{n} c_i z_i^2$, where $c_i$ are all $\pm 1$. Also the transformation is non-singular, for, if $Z$ is its matrix,

$$| Z' 1 Z | = | Z |^2 = \pm 1.$$

Hence the $z_i$ must be all linearly independent, otherwise the number of d.f. of $\sum_{1}^{n} x_i^2$ would be less than $n$, and

$$c_i = + 1, \; i = 1, \ldots, n,$$

since $\sum x_i^2$ is positive definite, so that the matrix $Z$ is orthogonal.

Now the joint distribution of $q_1, \ldots, q_k$ is

$$\Phi(\Sigma_1, \ldots, \Sigma_k) \, d\Sigma_1 \ldots d\Sigma_k = \frac{1}{(2\pi)^{\frac{1}{2}n}} \int \ldots \int e^{-\frac{1}{2} \sum_{1}^{n} x_i^2} \prod_{i=1}^{n} dx_i$$

over all values of $x_1, \ldots, x_n$ for which $q_i$ lies between $\Sigma_i$ and $\Sigma_i + d\Sigma_i$ for $i = 1, 2, \ldots, k$. Transforming to the variables $z_1, \ldots, z_n$, we get

$$\Phi(\Sigma_1, \ldots, \Sigma_k) \, d\Sigma_1 \ldots d\Sigma_k$$
$$= \frac{1}{(2\pi)^{\frac{1}{2}n}} e^{-\frac{1}{2}(\Sigma_1 + \Sigma_2 + \ldots + \Sigma_k)} \int \ldots \int dz_1 \ldots dz_n$$

over all values for which

$$z_1^2 + \ldots + z_{n_1}^2 \text{ lies between } \Sigma_1 \text{ and } \Sigma_1 + d\Sigma_1,$$
$$z_{n_1+1}^2 + \ldots + z_{n_1+n_2}^2 \text{ lies between } \Sigma_2 \text{ and } \Sigma_2 + d\Sigma_2,$$

and so on. But this integral is the product of the increments of volume of spheres of radii $\sqrt{\Sigma_1}, \sqrt{\Sigma_2}, \ldots$, in $n_1, n_2, \ldots$, dimensions respectively. Therefore

$$\Phi(\Sigma_1, \ldots, \Sigma_k) \, d\Sigma_1 \ldots d\Sigma_k$$
$$= c \, (\Sigma_1^{\frac{1}{2}(n_1-2)} e^{-\frac{1}{2}\Sigma_1} d\Sigma_1)(\Sigma_2^{\frac{1}{2}(n_2-2)} e^{-\frac{1}{2}\Sigma_2} d\Sigma_2) \ldots.$$

This proves the result.

*Corollary.* With the same initial conditions, if $\Sigma_{ij} a_{ij} x_i x_j$ is distributed as $\chi^2$ with $r$ d.f., then $\Sigma_{ij} (\delta_{ij} - a_{ij}) x_i x_j$ is distributed as $\chi^2$ with $n - r$ d.f., and the two distributions are independent.

This follows at once from I, Corollary 2, and II above (cf. Fisher (4)).

*Example.* $n\bar{x}^2$ is clearly distributed as $\chi^2$ with 1 d.f.

Thus $\overset{n}{\underset{1}{\Sigma}} x_i^2 - n\bar{x}^2 = \overset{n}{\underset{1}{\Sigma}} (x_i - \bar{x})^2$ is distributed as $\chi^2$ with $n - 1$ d.f., and the two distributions are independent.

III. With the same initial conditions, a necessary and sufficient condition that the two quadratic forms $\Sigma_{jk} a_{jk} x_j x_k$ and $\Sigma_{jk} b_{jk} x_j x_k$ are independently distributed is

$$| 1 - it_1 A - it_2 B | = | 1 - it_1 A | \, | 1 - it_2 B | \qquad (3)$$

for all real $t_1, t_2$.

For if $M_A$, $M_B$, $M_{AB}$ are the moment-generating functions (M.G.F.) of $\Sigma_{jk} a_{jk} x_j x_k$, $\Sigma_{jk} b_{jk} x_j x_k$, and the joint distribution of the pair, respectively, then a necessary and sufficient condition for independence is

$$M_{AB} = M_A M_B.$$

Now the joint M.G.F. may be written (with $\tfrac{1}{2}t_1$, $\tfrac{1}{2}t_2$ in place of the usual $t_1, t_2$),

$$\left(\frac{1}{2\pi}\right)^{\frac{1}{2}n} \int_0^\infty \cdots \int_0^\infty e^{-\frac{1}{2}\Sigma_{jk} (\delta_{jk} - it_1 a_{jk} - it_2 b_{jk}) x_j x_k} \overset{n}{\underset{1}{\prod}} dx_j.$$

In this, the real part of $1 - it_1 A - it_2 B$ is positive definite for all real $t_1, t_2$. Therefore (cf. Wishart and Bartlett (1))

$$M_{AB} = | 1 - it_1 A - it_2 B |^{-\frac{1}{2}}$$

and similarly

$$M_A = | 1 - it_1 A |^{-\frac{1}{2}}, \quad M_B = | 1 - it_2 B |^{-\frac{1}{2}}.$$

This proves the result.

We may note that from the theorems proved above, alternative sufficient conditions are that (i) $\Sigma_{jk} (a_{jk} + b_{jk}) x_j x_k$ is distributed as $\chi^2$ and (ii) $r(A) + r(B) = r(A + B)$. In general, however, the condition $r(A) + r(B) = r(A + B)$, while necessary, is not sufficient.

*Applications to the analysis of covariance.*

3. We first prove certain results and then go on to discuss their significance. Since the use of the method has hitherto for the most part been confined to agricultural experimentation, the applications mentioned will be couched in the language of that science.

Suppose that we have a sample of $n$ pairs of variates $(x, y)$, which are distributed in such a way that the regression of $y$ on $x$

12-2

is linear and the variate $y - \beta x$ is normally distributed about a mean zero with unit standard deviation, $\beta$ being the population regression coefficient of $y$ on $x$. This implies that the joint distribution of the $n$ pairs $(x_i, y_i)$ is of the form

$$f(x_1, x_2, \ldots, x_n) e^{-\frac{1}{2}\sum\limits_{1}^{n}(y_i - \beta x_i)^2} \prod_{i=1}^{n} dx_i dy_i. \qquad (4)$$

Let the $n$ values of $(x, y)$ be arranged in a two-way array $(x_{uv}, y_{uv})$, with $p$ columns and $q$ rows, and let

$$\bar{x}_{u\bullet} = (x_{u1} + x_{u2} + \ldots + x_{up})/p, \quad \bar{x}_{\bullet v} = (x_{1v} + x_{2v} + \ldots + x_{qv})/q,$$

with similar definitions for $\bar{y}_{u\bullet}$ and $\bar{y}_{\bullet v}$. Let $\Sigma$ refer always to summation over the $n$ values in the sample and define

$$b = \frac{\Sigma\,(x_{uv} - \bar{x})\,(y_{uv} - \bar{y})}{\Sigma\,(x_{uv} - \bar{x})^2}, \quad b_R = \frac{\Sigma\,(\bar{x}_{u\bullet} - \bar{x})\,(\bar{y}_{u\bullet} - \bar{y})}{\Sigma\,(\bar{x}_{u\bullet} - \bar{x})^2},$$

$$b_C = \frac{\Sigma\,(\bar{x}_{\bullet v} - \bar{x})\,(\bar{y}_{\bullet v} - \bar{y})}{\Sigma\,(\bar{x}_{\bullet v} - \bar{x})^2}, \quad b_1 = \frac{\Sigma\,(x_{uv} - \bar{x}_{u\bullet})\,(y_{uv} - \bar{y}_{u\bullet})}{\Sigma\,(x_{uv} - \bar{x}_{u\bullet})^2},$$

$$b_2 = \frac{\Sigma\,(x_{uv} - \bar{x}_{u\bullet} - \bar{x}_{\bullet v} + \bar{x})\,(y_{uv} - \bar{y}_{u\bullet} - \bar{y}_{\bullet v} + \bar{y})}{\Sigma\,(x_{uv} - \bar{x}_{u\bullet} - \bar{x}_{\bullet v} + \bar{x})^2},$$

these being all in a sense sample estimates of $\beta$.

We now proceed to consider the analysis of $\Sigma\,(y_{uv} - \beta x_{uv})^2$. By making the substitutions $w_{uv} = y_{uv} - \beta x_{uv}$, $x'_{uv} = x_{uv}$, and integrating out for the $x'_{uv}$, we see from (4) that the results proved in section 2 all apply to any division of $\Sigma\,(y_{uv} - \beta x_{uv})^2$ into quadratic forms in $y_{uv} - \beta x_{uv}$, provided that the latent roots are independent of the $x_{uv}$.

$$\Sigma\,(y_{uv} - \beta x_{uv})^2 = \Sigma\,\{(y_{uv} - \bar{y}) - \beta\,(x_{uv} - \bar{x})\}^2 + \Sigma\,(\bar{y} - \beta \bar{x})^2 \qquad (5)$$

$$= \Sigma\,\{(y_{uv} - \bar{y}) - b\,(x_{uv} - \bar{x})\}^2$$
$$+ (b - \beta)^2\,\Sigma\,(x_{uv} - \bar{x})^2 + \Sigma\,(\bar{y} - \beta\bar{x})^2, \qquad (6)$$

by definition of $b$.

Now the parts on the right-hand side of (6) are all quadratic forms in $y_{uv} - \beta x_{uv}$, and $\Sigma\,(\bar{y} - \beta\bar{x})^2$, $(b - \beta)^2\,\Sigma\,(x_{uv} - \bar{x})^2$ have each 1 d.f., each being the square of a single linear form. Also

$$\Sigma\,\{(y_{uv} - \bar{y}) - b\,(x_{uv} - \bar{x})\} = 0,$$
$$\Sigma\,[(x_{uv} - \bar{x})\,\{(y_{uv} - \bar{y}) - b\,(x_{uv} - \bar{x})\}] = 0,$$

by definition of $b$. Therefore

$$\Sigma\,\{(y_{uv} - \bar{y}) - b\,(x_{uv} - \bar{x})\}^2$$

has at most $n - 2$ d.f.

But
$$(n - 2) + 1 + 1 = n$$

and
$$r\,(A) + r\,(B) + r\,(C) \geqslant r\,(A + B + C).$$

Hence the conditions of II are satisfied and the three parts on the right-hand side of (6) are independently distributed as $\chi^2$ with d.f. $n-2, 1, 1$, respectively.

Further, from (5),

$$\Sigma\,(y_{uv} - \beta x_{uv})^2 = \Sigma\,\{(y_{uv} - \bar{y}_{u\cdot}) - \beta\,(x_{uv} - \bar{x}_{u\cdot})\}^2$$
$$+ \Sigma\,\{(\bar{y}_{u\cdot} - \bar{y}) - \beta\,(\bar{x}_{u\cdot} - \bar{x})\}^2 + \Sigma\,(\bar{y} - \beta\bar{x})^2 \quad (7)$$
$$= \Sigma\,\{(y_{uv} - \bar{y}_{u\cdot}) - b_1\,(x_{uv} - \bar{x}_{u\cdot})\}^2 + (b_1 - \beta)^2\,\Sigma\,(x_{uv} - \bar{x}_{u\cdot})^2$$
$$+ \Sigma\,\{(\bar{y}_{u\cdot} - \bar{y}) - b_R\,(\bar{x}_{u\cdot} - \bar{x})\}^2$$
$$+ (b_R - \beta)^2\,\Sigma\,(\bar{x}_{u\cdot} - \bar{x})^2 + \Sigma\,(\bar{y} - \beta\bar{x})^2. \quad (8)$$

Now

$$\Sigma\,(\bar{y} - \beta\bar{x})^2, \quad (b_1 - \beta)^2\,\Sigma\,(x_{uv} - \bar{x}_{u\cdot})^2, \quad (b_R - \beta)^2\,\Sigma\,(\bar{x}_{u\cdot} - \bar{x})^2$$

have each 1 d.f.

Also
$$\Sigma\,\{(\bar{y}_{u\cdot} - \bar{y}) - b_R\,(\bar{x}_{u\cdot} - \bar{x})\} = 0$$

and
$$\Sigma\,[(\bar{x}_{u\cdot} - \bar{x})\,\{(\bar{y}_{u\cdot} - \bar{y}) - b_R\,(\bar{x}_{u\cdot} - \bar{x})\}] = 0,$$

by definition of $b_R$. Hence

$$\Sigma\,\{(\bar{y}_{u\cdot} - \bar{y}) - b_R\,(\bar{x}_{u\cdot} - \bar{x})\}^2$$

has at most $q-2$ d.f. And from the $q+1$ linearly independent relations

$$\overset{p}{\underset{v=1}{\Sigma}}\,\{(y_{uv} - \bar{y}_{u\cdot}) - b_1\,(x_{uv} - \bar{x}_{u\cdot})\} = 0,$$

for $u = 1, 2, \ldots, q$,

$$\Sigma\,(x_{uv} - \bar{x}_{u\cdot})\,\{(y_{uv} - \bar{y}_{u\cdot}) - b_1\,(x_{uv} - \bar{x}_{u\cdot})\} = 0,$$

by definition of $b_1$, it follows that

$$\Sigma\,\{(y_{uv} - \bar{y}_{u\cdot}) - b_1\,(x_{uv} - \bar{x}_{u\cdot})\}^2$$

has at most $n-q-1$ d.f. But

$$(n-q-1) + 1 + (q-2) + 1 + 1 = n.$$

Thus, by II, the parts on the right-hand side of (8) are independently distributed as $\chi^2$ with d.f. $n-q-1, 1, q-2, 1, 1$, respectively. Notice that in each sum of squares we have used the regression coefficient appropriate to the sum.

The next analysis, in which we take out the variance between both rows and columns, is that used in the method of *randomised blocks* (cf. Fisher(2), pp. 243–246). We have

$$\Sigma\,(y_{uv} - \beta x_{uv})^2 = \Sigma\,\{(\bar{y}_{u\cdot} - \bar{y}) - b_R\,(\bar{x}_{u\cdot} - \bar{x})\}^2$$
$$+ (b_R - \beta)^2\,\Sigma\,(\bar{x}_{u\cdot} - \bar{x})^2 + \Sigma\,\{(\bar{y}_{\cdot v} - \bar{y}) - b_C\,(\bar{x}_{\cdot v} - \bar{x})\}^2$$
$$+ (b_C - \beta)^2\,\Sigma\,(\bar{x}_{\cdot v} - \bar{x})^2 + \Sigma\,\{(y_{uv} - \bar{y}_{u\cdot} - \bar{y}_{\cdot v} + \bar{y})$$
$$- b_2\,(x_{uv} - \bar{x}_{u\cdot} - \bar{x}_{\cdot v} + \bar{x})\}^2 + (b_2 - \beta)^2$$
$$\times \Sigma\,(x_{uv} - \bar{x}_{u\cdot} - \bar{x}_{\cdot v} + \bar{x})^2 + \Sigma\,(\bar{y} - \beta\bar{x})^2. \quad (9)$$

The d.f. on the right-hand side are $q-2, 1, p-2, 1, n-p-q, 1, 1$ respectively. For, by analogy with rows, $\Sigma\left\{(\bar{y}_{\cdot v}-\bar{y})-b_C(\bar{x}_{\cdot v}-\bar{x})\right\}^2$ has at most $p-2$ d.f. and we need only consider

$$\Sigma\left\{(y_{uv}-\bar{y}_{u\cdot}-\bar{y}_{\cdot v}+\bar{y})-b_2(x_{uv}-\bar{x}_{u\cdot}-\bar{x}_{\cdot v}+\bar{x})\right\}^2,$$

say $\Sigma z_{uv}^2$. Then the $z_{uv}$ are connected by the $p+q$ linearly independent relations

$$\overset{p}{\underset{v=1}{\Sigma}}\; z_{uv}=0 \quad\text{for each } u=1, 2, \ldots, q, \qquad q$$

$$\overset{q}{\underset{u=1}{\Sigma}}\; z_{uv}=0 \quad\text{for } v=1, 2, \ldots, p-1, \qquad p-1$$

(the relation $\overset{q}{\underset{u=1}{\Sigma}}\, z_{up}=0$ can be deduced from the above) and

$$\Sigma\left(x_{uv}-\bar{x}_{u\cdot}-\bar{x}_{\cdot v}+\bar{x}\right)z_{uv}=0,$$

by definition of $b_2$, 1. Thus $\Sigma z_{uv}^2$ has at most $n-p-q$ d.f. and the analysis is as stated above, since

$$(q-2)+1+(p-2)+1+(n-p-q)+1+1=n.$$

The Latin square (cf. Fisher (2), pp. 246–249) gives the last subdivision that is possible with a two-way array. In this $p=q$ and, in addition to taking out the variance between rows and columns, we take it out between treatments, the experiment being arranged so that each row and each column contains one and only one member with a given treatment. This is essential to the proof. Let $\bar{x}_T$ be the mean of the variates in a typical treatment and let

$$b_T=\frac{\Sigma(\bar{x}_T-\bar{x})(\bar{y}_T-\bar{y})}{\Sigma(\bar{x}_T-\bar{x})^2},$$

and

$$b_3=\frac{\Sigma(x_{uv}-\bar{x}_{u\cdot}-\bar{x}_{\cdot v}-\bar{x}_T+2\bar{x})(y_{uv}-\bar{y}_{u\cdot}-\bar{y}_{\cdot v}-\bar{y}_T+2\bar{y})}{\Sigma(x_{uv}-\bar{x}_{u\cdot}-\bar{x}_{\cdot v}-\bar{x}_T+2\bar{x})^2}.$$

Then

$$\begin{aligned}
\Sigma(y_{uv}-\beta x_{uv})^2=\;&\Sigma\left\{(\bar{y}_{u\cdot}-\bar{y})-b_R(\bar{x}_{u\cdot}-\bar{x})\right\}^2+(b_R-\beta)^2\,\Sigma(\bar{x}_{u\cdot}-\bar{x})^2\\
&+\Sigma\left\{(\bar{y}_{\cdot v}-\bar{y})-b_C(\bar{x}_{\cdot v}-\bar{x})\right\}^2+(b_C-\beta)^2\,\Sigma(\bar{x}_{\cdot v}-\bar{x})^2\\
&+\Sigma\left\{(\bar{y}_T-\bar{y})-b_T(\bar{x}_T-\bar{x})\right\}^2+(b_T-\beta)^2\,\Sigma(\bar{x}_T-\bar{x})^2\\
&+\Sigma\left\{(y_{uv}-\bar{y}_{u\cdot}-\bar{y}_{\cdot v}-\bar{y}_T+2\bar{y})-b_3(x_{uv}-\bar{x}_{u\cdot}-\bar{x}_{\cdot v}-\bar{x}_T+2\bar{x})\right\}^2\\
&+(b_3-\beta)^2\,\Sigma(x_{uv}-\bar{x}_{u\cdot}-\bar{x}_{\cdot v}-\bar{x}_T+2\bar{x})^2+\Sigma(\bar{y}-\beta\bar{x})^2, \quad(10)
\end{aligned}$$

the expansion holding since each row and each column contains one and only one member of each treatment. The only term whose d.f. are not known is the third from the right, say $\Sigma\, z_{uv}{}^2$. Then the $z_{uv}$ are connected by $3p-1$ independent relations

$$\sum_{v=1}^{p} z_{uv}=0, \quad \text{for } u=1,2,\ldots,p, \qquad\qquad p$$

$$\sum_{u=1}^{p} z_{uv}=0, \quad \text{for } v=1,2,\ldots,p-1, \qquad p-1$$

$$\underset{\text{over } i\text{th treatment}}{\Sigma}\, z_{uv}=0, \quad \text{for } i=1,2,\ldots,p-1, \qquad p-1$$

and $\qquad \Sigma\,(x_{uv}-\bar{x}_{u\bullet}-\bar{x}_{\bullet v}-\bar{x}_T+2\bar{x})\,z_{uv}=0,$ \qquad 1.

Thus the parts on the right-hand side of (10) are independently distributed as $\chi^2$ with d.f. $p-2,\,1,\,p-2,\,1,\,p-2,\,1,\,n-3p+1,1,1$ respectively.

The successive analyses of equations (7), (8), (9) and (10) are exactly parallel to the corresponding analyses of variance (for which cf. Irwin (3), where the same notation is used), except that in each sum of squares we take out the regression coefficient appropriate to the sum and deduct one from the number of d.f. For the case of $p$ variates, in which we take out the multiple linear regression of $x_p$ on $x_1, x_2, \ldots, x_{p-1}$, the analysis of

$$\Sigma\,(x_p - \beta_{p1.2\ldots p-1}\,x_1 - \beta_{p2.13\ldots p-1}\,x_2 - \ldots$$
$$- \beta_{pp-1,12\ldots p-2}\,x_{p-1})^2 = \Sigma\, x^2{}_{p.12\ldots p-1}$$

is exactly similar, assuming $x_{p.12\ldots p-1}$ normally distributed about mean zero with variance one in all arrays, except that $p-1$ d.f. are subtracted for the $p-1$ regression coefficients taken out. The variates $x_1,\ldots,x_{p-1}$ need not, of course, be in any sense independent, for instance, $x_2$ could be $x_1{}^2$ or $\log x_1$.

4. Briefly, the circumstances to which the method applies are as follows. Suppose that in an experiment we are interested in the effect of a number of treatments on the value of the variate $x_p$, and that this is known to be affected also by the values of variates $x_1,\ldots,x_{p-1}$ which can themselves be measured during the experiment. Now it may be the case that either the values obtained for $x_1,\ldots,x_{p-1}$ during the experiment cannot to any extent be controlled, or, even if they can, any significant difference found in $x_p$ due to treatments may be of very restricted applicability unless it holds over a wide range of variation of $x_1,\ldots,x_{p-1}$. In such cases, if the relation between $x_p$ and $x_1,\ldots,x_{p-1}$ can be expressed as a regression, we take out the regression before making comparisons, so that any treatment difference found will be free from the effects of the other variates and valid over the range of those variates in the experiment. The results of section 3 show

that in an analysis of variance, comparisons can be made as usual by the $z$-test, except that the regression coefficient used for each sum of squares must be that appropriate to the sum, and the correct number of d.f. is to be subtracted from each sum (cf. also Fisher(2), pp. 249–262).

As an example in which the purpose of using the method is slightly different from the above, suppose that we have a randomised blocks lay-out to test the effect of treatments on yield. Now the yield of a plot is affected by the fertility of the soil as well as by the treatment applied, and while we eliminate most of the effect of differences of fertility between whole blocks, differences of fertility between different plots in the same block would tend to vitiate the results of the experiment. These differences show themselves partly in the differences between the plant numbers of different plots, which can be counted. Thus, if the effects of the treatments on plant number are negligible, the value of the comparisons would be greatly increased by taking out the regression of yield on plant number. It is, of course, doubtful whether the increase would be worth the extra labour.

We may mention here a point dealt with by Irwin(3) for the analysis of variance. The analysis of equation (9) does not apply, as it stands, to the case of randomised blocks. For the point of taking out the variance between blocks and sacrificing $q-1$ d.f. from the error estimate is that we believe that, owing to soil heterogeneity, $\Sigma\{(\bar{y}_{u\bullet}-\bar{y})-b_R(\bar{x}_{u\bullet}-\bar{x})\}^2$ will not be distributed as $\chi^2$, and hence that (4) does not hold. But for the comparison of treatment and error, it is only necessary that the analysis between the terms

$$\Sigma\{(\bar{y}_{\bullet v}-\bar{y})-b_C(\bar{x}_{\bullet v}-\bar{x})\}^2$$

and

$$(b_2-\beta)^2\,\Sigma\,(x_{uv}-\bar{x}_{u\bullet}-\bar{x}_{\bullet v}+\bar{x})^2$$

on the right-hand side of (9) should be valid, in the absence of treatment effect. And for this to be so, it is sufficient that the general joint distribution, instead of (4), should be of the form

$$\phi\,(\overline{w}_{u\bullet})\,f\,(x_{uv})\,e^{-\frac{1}{2}\Sigma\,(w_{uv}-\overline{w}_{u\bullet})^2}\,\Pi\,dx_{uv}\,dw_{uv},\qquad(11)$$

where

$$w_{uv}=y_{uv}-\beta x_{uv},$$

$$\phi\,(\overline{w}_{u\bullet})\equiv\phi\,(\overline{w}_1{}_\bullet,\overline{w}_2{}_\bullet,\ldots,\overline{w}_q{}_\bullet),$$

$$f\,(x_{uv})\equiv f\,(x_{11},\ldots,x_{qp}),$$

so that the joint distribution of the $w_{uv}$ is of the form

$$g\,(\overline{w}_{u\bullet})\,e^{-\frac{1}{2}\Sigma\,(w_{uv}-\overline{w}_{u\bullet})^2}\,\Pi\,dw_{uv}.\qquad(12)$$

For, from the equation

$$\Sigma\, w_{uv}{}^2 = \Sigma\, (w_{uv} - \overline{w}_{u\cdot})^2 + p\overline{w}_1\cdot{}^2 + p\overline{w}_2\cdot{}^2 + \dots + p\overline{w}_q\cdot{}^2, \quad (13)$$

with d.f. $n - q, 1, 1, \dots, 1$, on the right-hand side, and the proof of II, it follows that we can find an orthogonal transformation to variates $\zeta_1, \dots, \zeta_n$, where

$$\zeta_i = p\overline{w}_i\cdot \ (i = 1, 2, \dots, q) \ \text{ and } \ \Sigma\, (w_{uv} - \overline{w}_{u\cdot})^2 = \sum_{q+1}^{n} \zeta_i{}^2.$$

Noting that the analysis required is an analysis of

$$\Sigma\, (w_{uv} - \overline{w}_{u\cdot})^2,$$

we make the transformation to $\zeta_1, \dots, \zeta_n$, and integrate with regard to $\zeta_1, \dots, \zeta_q$, and the analysis required follows at once, since the parts into which $\Sigma\, (w_{uv} - \overline{w}_{u\cdot})^2$ is divided are quadratic forms in the $\zeta_i \ (i = q + 1, \dots, n)$ with the same number of d.f. as before.

If we suppose that the $w_{uv}$ in (12) are independently distributed, we can prove that they are normally distributed about means varying from row to row. For since the rows are now independent, the joint distribution of the $p$ members of the $u$th row will be of the form

$$\psi\,(\overline{w}_{u\cdot})\, e^{-\frac{1}{2}\sum\limits_{v=1}^{p}(w_{uv}-\overline{w}_{u\cdot})^2}\, \prod_{v=1}^{p} dw_{uv}, \qquad (14)$$

and if this is to be of the form $\prod\limits_{v=1}^{p} F_v\,(w_{uv})\, dw_{uv}$, we have

$$\psi\,(\overline{w}_{u\cdot})\, e^{+\frac{1}{2}p\overline{w}_u\cdot{}^2} = \prod_{v=1}^{p} \left\{ F_v\,(w_{uv})\, e^{+\frac{1}{2}w_{uv}{}^2} \right\},$$

so that
$$\frac{1}{p}\frac{\psi'\,(\overline{w}_{u\cdot})}{\psi\,(\overline{w}_{u\cdot})} + \overline{w}_{u\cdot} = \frac{F_v{}'\,(w_{uv})}{F_v\,(w_{uv})} + w_{uv}$$

for each $v = 1, 2, \dots, p$ and must therefore be constant, and equal to $m_u$ (say).

Hence
$$F_v\,(w_{uv}) = c\,e^{-\frac{1}{2}(w_{uv} - m_u)^2}.$$

The general question of the minimum conditions necessary for the tests to hold has not yet received full consideration.

5. In any test of significance in which we look up the 5 or 1 per cent. points of a table, we automatically keep down the relative frequency of cases in which we find a significant effect where in reality there is none. The superiority of one test of this type over another therefore lies in its greater power to show significant effects where they do exist. On this score, there are two important drawbacks to the method outlined above. For the d.f. subtracted from the sum of squares for treatment, when we

take out the regression, involves generally a considerable loss in precision because as a rule the number of different treatments is small. In the quite common case where there are two treatments only, the method breaks down, since the residual sum of squares vanishes identically. Secondly, the method may not be very sensitive in detecting significant effects because the treatment regression coefficient itself changes when there is a significant treatment effect, e.g. in the limiting case of two treatments, no matter how great their difference, the treatment regression coefficient automatically alters so that the residual sum of squares still vanishes identically.

These considerations naturally lead us to examine the possibility of taking the error coefficient out of all sums of squares, e.g., in the case of randomised blocks, of examining the analysis

$$\Sigma \{(y_{uv} - \bar{y}) - b_2 (x_{uv} - \bar{x})\}^2$$
$$= \Sigma \{(\bar{y}_{u\bullet} - \bar{y}) - b_2 (\bar{x}_{u\bullet} - \bar{x})\}^2 + \Sigma \{(\bar{y}_{\bullet v} - \bar{y}) - b_2 (\bar{x}_{\bullet v} - \bar{x})\}^2$$
$$+ \Sigma \{(y_{uv} - \bar{y}_{u\bullet} - \bar{y}_{\bullet v} + \bar{y}) - b_2 (x_{uv} - \bar{x}_{u\bullet} - \bar{x}_{\bullet v} + \bar{x})\}^2. \quad (15)$$

The numbers of d.f. on the right-hand side are now $q-1$, $p-1$ and $n-p-q$, since the relations

$$\Sigma (\bar{x}_{u\bullet} - \bar{x}) \{(\bar{y}_{u\bullet} - \bar{y}) - b_2 (\bar{x}_{u\bullet} - \bar{x})\} = 0,$$
$$\Sigma (\bar{x}_{\bullet v} - \bar{x}) \{(\bar{y}_{\bullet v} - \bar{y}) - b_2 (\bar{x}_{\bullet v} - \bar{x})\} = 0$$

no longer hold.

Thus we do not lose a d.f. by taking out the error regression coefficient from rows and columns: further, the d.f. add up to $n-2$, which is the number of d.f. on the left-hand side, for, if

$$z_{uv} \equiv (y_{uv} - \bar{y}) - b_2 (x_{uv} - \bar{x}),$$

we have

$$\Sigma z_{uv} = 0$$

and

$$\Sigma (x_{uv} - \bar{x}_{u\bullet} - \bar{x}_{\bullet v} + \bar{x}) z_{uv} = 0,$$

by definition of $b_2$, and these are the only linear relations connecting the $z_{uv}$. Therefore the analysis will follow if we can show that

$$\Sigma \{(y_{uv} - \bar{y}) - b_2 (x_{uv} - \bar{x})\}^2$$

has a $\chi^2$ distribution with $n-2$ d.f.

Now let $B$ be any linear function of $y_{uv} - \beta x_{uv}$. Then

$$\Sigma (y_{uv} - \beta x_{uv})^2 = \Sigma \{(y_{uv} - \bar{y}) - \beta (x_{uv} - \bar{x})\}^2 + \Sigma (\bar{y} - \beta \bar{x})^2$$
$$= \Sigma (\bar{y} - \beta \bar{x})^2 + \Sigma \{(y_{uv} - \bar{y}) - B (x_{uv} - \bar{x})\}^2 + (B - \beta)^2 \Sigma (x_{uv} - \bar{x})^2$$
$$+ 2 (B - \beta) \Sigma [(x_{uv} - \bar{x}) \{(y_{uv} - \bar{y}) - B (x_{uv} - \bar{x})\}].$$

Therefore,

$$\Sigma\,(y_{uv}-\beta x_{uv})^2 - \Sigma\,\{(y_{uv}-\bar{y}) - B\,(x_{uv}-\bar{x})\}^2$$
$$= \Sigma\,(\bar{y}-\beta\bar{x})^2 - (B-\beta)^2\,\Sigma\,(x_{uv}-\bar{x})^2$$
$$\quad + 2\,(B-\beta)\,\Sigma\,[(x_{uv}-\bar{x})\,\{(y_{uv}-\bar{y}) - \beta\,(x_{uv}-\bar{x})\}]$$
$$= n\,(\bar{y}-\beta\bar{x})^2 - (B-b)^2\,\Sigma\,(x_{uv}-\bar{x})^2 + (b-\beta)^2\,\Sigma\,(x_{uv}-\bar{x})^2 \quad (16)$$
$$= z_1{}^2 - z_3{}^2 + z_2{}^2,$$

where 
$$z_1 = \sqrt{n}\,(\bar{y}-\beta\bar{x}), \quad z_2 = \sqrt{\{\Sigma\,(x_{uv}-\bar{x})^2\}}\,(b-\beta),$$
$$z_3 = \sqrt{\{\Sigma\,(x_{uv}-\bar{x})^2\}}\,(B-b).$$

Now, if $B = b$, this is of rank 2 and non-negative definite. But if $B = b_1$, $b_2$ or $b_3$, $z_1$, $z_2$ and $z_3$ are linearly independent, therefore the left-hand side of (16) is of rank 3 and signature 1.

It follows that $\Sigma\,\{(y_{uv}-\bar{y}) - b_i\,(x_{uv}-\bar{x})\}^2$, $i = 1,\,2,\,3$, does not follow a $\chi^2$ distribution, for if it did, then by I, Corollary 2, the left-hand side would be of rank 2 and non-negative definite.

Further, if $B$ is $b_1$, $b_2$ or $b_3$, the transformation to $z_i$ can be made normal and orthogonal. For instance, if $B = b_2$, the coefficients of the term in $y_{uv} - \beta x_{uv}$ in the expressions for $z_1$, $z_2$, $z_3$ are

$$\frac{1}{\sqrt{n}}, \quad \frac{x_{uv}-\bar{x}}{\sqrt{\{\Sigma\,(x_{uv}-\bar{x})^2\}}}$$

and 
$$\left( \frac{x_{uv}-\bar{x}_{u\cdot}-\bar{x}_{\cdot v}+\bar{x}}{\Sigma\,(x_{uv}-\bar{x}_{u\cdot}-\bar{x}_{\cdot v}+\bar{x})^2} - \frac{x_{uv}-\bar{x}}{\Sigma\,(x_{uv}-\bar{x})^2} \right) \sqrt{\{\Sigma\,(x_{uv}-\bar{x})^2\}}$$

respectively.

It is obvious that $z_1$ is orthogonal to $z_2$ and $z_3$. Also

$$\Sigma\,[(x_{uv}-\bar{x})\,(x_{uv}-\bar{x}_{u\cdot}-\bar{x}_{\cdot v}+\bar{x})]$$
$$= \Sigma\,[(x_{uv}-\bar{x})\,\{(x_{uv}-\bar{x}) - (\bar{x}_{u\cdot}-\bar{x}) - (\bar{x}_{\cdot v}-\bar{x})\}]$$
$$= \Sigma\,(x_{uv}-\bar{x})^2 - \Sigma\,(\bar{x}_{u\cdot}-\bar{x})^2 - \Sigma\,(\bar{x}_{\cdot v}-\bar{x})^2$$
$$= \Sigma\,(x_{uv}-\bar{x}_{u\cdot}-\bar{x}_{\cdot v}+\bar{x})^2,$$

from which it follows that $z_2$ and $z_3$ are orthogonal.

Thus if we write $l_1 = z_1$, $l_2 = z_2$,

$$l_3 = \sqrt{\left\{ \frac{\Sigma\,(x_{uv}-\bar{x}_{u\cdot}-\bar{x}_{\cdot v}+\bar{x})^2}{\Sigma\,(\bar{x}_{u\cdot}-\bar{x})^2+\Sigma\,(\bar{x}_{\cdot v}-\bar{x})^2} \right\}}\,z_3,$$

the transformation to $l_1$, $l_2$, $l_3$ is normal and orthogonal; i.e. there is an orthogonal transformation which changes

$$\Sigma\,(y_{uv}-\beta x_{uv})^2 - \Sigma\,\{(y_{uv}-\bar{y}) - b_2\,(x_{uv}-\bar{x})\}^2 \text{ into } l_1{}^2 + l_2{}^2 - kl_3{}^2,$$

where 
$$k = \frac{\Sigma\,(\bar{x}_{u\cdot}-\bar{x})^2 + \Sigma\,(\bar{x}_{\cdot v}-\bar{x})^2}{\Sigma\,(x_{uv}-\bar{x}_{u\cdot}-\bar{x}_{\cdot v}+\bar{x})^2}.$$

Hence if $\phi_A(\lambda)$ is the characteristic function of this form,

$$\phi_A(\lambda) = (-\lambda)^{n-3}(1-\lambda)(1-\lambda)(-k-\lambda).$$

Now

$$\phi_{1-A}(\lambda) = |1 - A - \lambda 1| = (-)^n |A - (1-\lambda)1| = (-)^n \phi_A(1-\lambda)$$
$$= (-)^n(\lambda - 1)^{n-3}\lambda^2(\lambda - 1 - k).$$

Thus $\Sigma = \Sigma\{(y_{uv} - \bar{y}) - b_2(x_{uv} - \bar{x})\}^2$ is of rank $n-2$ (which we know already), and its non-zero latent roots are

$$1 \ (n-3 \text{ times}), \text{ and } 1 + k = \frac{\Sigma(x_{uv} - \bar{x})^2}{\Sigma(x_{uv} - \bar{x}_{u\cdot} - \bar{x}_{\cdot v} + \bar{x})^2}.$$

Hence, by I, $\Sigma$ is distributed as the sum of two independent parts, one following a $\chi^2$ distribution with $n-3$ d.f., the other being

$$\frac{y^2 \Sigma(x_{uv} - \bar{x})^2}{\Sigma(x_{uv} - \bar{x}_{u\cdot} - \bar{x}_{\cdot v} + \bar{x})^2},$$

where $y$ is normally distributed. We can get no further unless we assume some particular form for the function $f(x_{11}, \ldots, x_{qp})$ in (4), so that the generality of the results no longer holds.

To take the simplest case, assume the $x_{uv}$ to be fixed from sample to sample. Then $\Sigma$ has a $P(n-3, \lambda)$ distribution, where

$$\lambda = \frac{\Sigma(x_{uv} - \bar{x})^2}{\Sigma(x_{uv} - \bar{x}_{u\cdot} - \bar{x}_{\cdot v} + \bar{x})^2}.$$

It is easy to show by integration that this is of the form

$$z^{\frac{1}{2}(n-4)} e^{-\frac{1}{2}z/\lambda} F\{\tfrac{1}{2}(n-3), \ \tfrac{1}{2}(n-2), \ -\tfrac{1}{2}(1-\lambda^{-1})z\},$$

where $F(\alpha, \gamma, x)$ is the confluent hypergeometric function.

Since $\Sigma$ does not follow a $\chi^2$ distribution, it is, by III, no longer a sufficient condition for independence of its parts that the numbers of d.f. add up to those of $\Sigma$, and the question of independence would require further consideration. We may just note here that, if these parts are independent, we can find their distributions. For by I they will all follow $P$ distributions. Also if $x$ and $y$ are independent, and $x$ follows a $P(m, \lambda_1, \ldots, \lambda_k)$, $y$ a $P(n, \lambda_{k+1}, \ldots, \lambda_s)$ distribution, then $x + y$ follows a $P(m+n, \lambda_1, \ldots, \lambda_s)$ distribution, by the additive property of semi-invariants. Now in (15) $\Sigma$ follows a $P(n-3, \lambda)$ distribution and

$$\Sigma\{(y_{uv} - \bar{y}_{u\cdot} - \bar{y}_{\cdot v} + \bar{y}) - b_2(x_{uv} - \bar{x}_{u\cdot} - \bar{x}_{\cdot v} + \bar{x})\}^2$$

a $P(n-p-q)$ distribution. Therefore if the parts are independent, (rows + columns) follows a $P(p+q-3, \lambda)$ distribution, which means that one of rows and columns follows a $\chi^2$ distribution and the other does not. The unsymmetrical nature of this result tends to indicate that rows, columns and error are not all independent.

It seems therefore that the chance of getting a test valid under general conditions by taking out the error coefficient is not very promising, though the point calls for further investigation, particularly of the error involved in using the ordinary $z$-test *.

6. While most of the space above has been taken up with the applications to the analysis of covariance, it is hoped that the results proved in section 2 are sufficient to deal with all the applications in common use. This work was undertaken at the suggestion of Dr J. Wishart, whom I have to thank for advice and criticism at all stages.

## REFERENCES.

(1) J. WISHART and M. S. BARTLETT, *Proc. Camb. Phil. Soc.* 29 (1932), 260–270 (261).

(2) R. A. FISHER, *Statistical Methods for Research Workers*, 4th edition (1932).

(3) J. O. IRWIN, *Journal Royal Statistical Soc.* 94 (1931), 284–300.

(4) R. A. FISHER, *Metron* (5) (1926), 90–104 (97).

* This is at present under consideration. It can be shown that, if the error regression coefficient is taken out of all sums of squares, treatments and error remain independently distributed, even in the general case represented by equation (4).

*The flow due to a rotating disc.* By W. G. Cochran, B.A., St John's College. (Communicated by Dr S. Goldstein.)

[*Received* 21 May, *read* 28 May 1934.]

1. The steady motion of an incompressible viscous fluid, due to an infinite rotating plane lamina, has been considered by Kármán *. If $r$, $\theta$, $z$ are cylindrical polar coordinates, the plane lamina is taken to be $z = 0$; it is rotating with constant angular velocity $\omega$ about the axis $r = 0$. We consider the motion of the fluid on the side of the plane for which $z$ is positive; the fluid is infinite in extent and $z = 0$ is the only boundary. If $u$, $v$, $w$ are the components of the velocity of the fluid in the directions of $r$, $\theta$ and $z$ increasing, respectively, and $p$ is the pressure, then Kármán shows that the equations of motion and continuity are satisfied by taking

$$u = rf(z), \quad v = rg(z), \quad w = h(z), \quad p = p(z). \qquad (1)$$

The boundary conditions are $u = 0$, $v = \omega r$, $w = 0$ at $z = 0$, and $u = 0$, $v = 0$ at $z = \infty$. $w$ does not vanish at $z = \infty$, but must tend to a finite negative limit. For the rotating lamina acts as a kind of centrifugal fan; the fluid moves radially outwards, especially near the lamina, and there must, therefore, be an axial motion towards the lamina in order to preserve continuity.

If $\mathbf{v}$ is the velocity of the fluid, $\rho$ its density and $\nu$ its kinematic viscosity, the vector equation of motion of the incompressible fluid is ($\rho$ being constant)

$$\text{grad } \tfrac{1}{2}\mathbf{v}^2 - \mathbf{v} \wedge \text{curl } \mathbf{v} = - \text{ grad } p/\rho - \nu \text{ curl curl } \mathbf{v}, \qquad (2)$$

and the equation of continuity is

$$\text{div } \mathbf{v} = 0. \qquad (3)$$

With components as in equations (1) these give the four equations

$$\left. \begin{aligned}
f^2 - g^2 + h \frac{df}{dz} &= \nu \frac{d^2 f}{dz^2}, \\[2mm]
2fg + h \frac{dg}{dz} &= \nu \frac{d^2 g}{dz^2}, \\[2mm]
h \frac{dh}{dz} &= -\frac{1}{\rho} \frac{dp}{dz} - 2\nu \frac{df}{dz}, \\[2mm]
2f + \frac{dh}{dz} &= 0,
\end{aligned} \right\} \qquad (4)$$

* *Zeitschrift für angewandte Mathematik u. Mechanik*, **1** (1921), 244–7.

and the boundary conditions are

$$f(0) = 0, \quad g(0) = \omega, \quad h(0) = 0, \quad f(\infty) = 0, \quad g(\infty) = 0. \quad (5)$$

To obtain the equations in a non-dimensional form, put

$$f = \omega F, \quad g = \omega G, \quad h = (\nu\omega)^{\frac{1}{2}} H, \quad p = \rho\nu\omega P, \quad z = (\nu/\omega)^{\frac{1}{2}} \zeta. \quad (6)$$

The equations become

$$\left. \begin{aligned} F^2 - G^2 + HF' &= F'', \\ 2FG + HG' &= G'', \\ HH' + 2F' &= P', \\ 2F + H' &= 0, \end{aligned} \right\} \quad (7)$$

and the boundary conditions become

$$F(0) = 0, \quad G(0) = 1, \quad H(0) = 0, \quad F(\infty) = 0, \quad G(\infty) = 0. \quad (8)$$

$H$ must tend to a finite negative limit, say $-c$, as $\zeta$ tends to infinity. The first two and the fourth of equations (7) give $F, G$ and $H$. The third gives $P$.

If $F, G \to 0$ at infinity, and $H \to -c$, then, to a first approximation, we have, for large values of $\zeta$,

$$-c F' = F'', \quad -c G' = G'', \quad (9)$$

giving $\quad\quad F = A e^{-c\zeta}, \quad G = B e^{-c\zeta}, \quad H = -c + \dfrac{2A}{c} e^{-c\zeta}. \quad (10)$

There are, in fact, formal expansions of $F, G, H$ in powers of $e^{-c\zeta}$ (see below). Thus $F$ and $G$ tend to zero exponentially, and are practically zero for some finite value of $\zeta$. Hence, if $\nu/\omega$ is small, $u/\omega r$ and $v/\omega r$ are appreciable only in a thin layer near the lamina whose thickness is of order $(\nu/\omega)^{\frac{1}{2}}$. Also, from the third of equations (7), if $P_0$ is the value of $P$ at the plate,

$$P - P_0 = \tfrac{1}{2}H^2 + 2F, \quad (11)$$

and the differences of the pressure $p$, throughout the layer in which $u/\omega r$ and $v/\omega r$ are sensible, are of order $\rho\nu\omega$. These results agree with the general results of the boundary layer theory for large Reynolds numbers, but for this particular case have been obtained from the equation of motion without any approximation. It is in this that the importance of the solution lies. We may note also that the axial inflow velocity is of the order $(\nu\omega)^{\frac{1}{2}}$.

If we neglect edge effect, we can find the frictional moment on a rotating disc of radius $a$. The shearing stress is given by

$$p_{z\theta} = \rho\nu \frac{dv}{dz} = \rho (\nu\omega^3)^{\frac{1}{2}} r G'(\zeta), \quad (12)$$

and hence the moment is

$$-\int_0^a 2\pi r^2 p_{z\theta}\, dr = -\tfrac{1}{2}\pi a^4 \rho\, (\nu\omega^3)^{\frac{1}{2}}\, G'\,(0). \qquad (13)$$

This is the moment for one side only. For both sides the result must be doubled. Hence, in terms of the Reynolds number

$$R = a^2\,\omega/\nu, \qquad (14)$$

we have for the moment

$$M = -\pi G'\,(0)\,\rho a^5\,\omega^2/R^{\frac{1}{2}}, \qquad (15)$$

and for the non-dimensional moment coefficient, with $S = \pi a^2$,

$$k_M = \frac{M}{\rho a^3 \omega^2 S} = -G'\,(0)/R^{\frac{1}{2}}. \qquad (16)$$

The neglect of edge effect is probably justified if the radius is large compared with the thickness of the boundary layer, or with $(\nu/\omega)^{\frac{1}{2}}$. The results of this section were all obtained by Kármán.

2. *Kármán's method of approximate solution.* The system of equations (7) and (8) does not appear to have been integrated accurately. Kármán uses an approximate process which he had invented for the integration of boundary layer equations (cf. Kármán, *loc. cit.* pp. 235, 236; Pohlhausen, *loc. cit.* pp. 252–68). Integrating the first two of the equations (7) between 0 and $\infty$, and using the relations

$$\left.\begin{aligned}
\int_0^\infty HF'd\zeta = \left[HF\right]_0^\infty - \int_0^\infty H'F\,d\zeta = 2\int_0^\infty F^2 d\zeta, \\
\int_0^\infty HG'd\zeta = \left[HG\right]_0^\infty - \int_0^\infty H'G\,d\zeta = 2\int_0^\infty FG\,d\zeta,
\end{aligned}\right\} \qquad (17)$$

together with $F'\,(\infty) = 0$, $G'\,(\infty) = 0$, we get

$$\left.\begin{aligned}
-F'\,(0) = \int_0^\infty (3F^2 - G^2)\,d\zeta, \\
-G'\,(0) = 4\int_0^\infty FG\,d\zeta.
\end{aligned}\right\} \qquad (18)$$

Kármán now assumes that $F^2$, $G^2$ and $FG$ fall off sufficiently rapidly so that a good approximation, particularly to $G'\,(0)$, which occurs in the expression for the moment, can be found by replacing the upper limit in the integrals by some finite value $\zeta_0$ of $\zeta$, or rather by supposing $F$ and $G$ zero for values of $\zeta$ greater than $\zeta_0$. We must then have

$$F\,(\zeta_0) = 0, \quad F'\,(\zeta_0) = 0, \quad G\,(\zeta_0) = 0, \quad G'\,(\zeta_0) = 0. \qquad (19)$$

$F$ and $G$ must also satisfy the conditions at the plate, namely

$$F(0) = 0, \quad G(0) = 1, \tag{20}$$

and by putting $\zeta = 0$ in (7), we obtain the conditions

$$F''(0) = -1, \quad G''(0) = 0. \tag{21}$$

Further boundary conditions, obtained by differentiating the equations (7) before putting $\zeta = 0$, are ignored, and so are the conditions that the second and higher derivatives of $F$ and $G$ must vanish at $\zeta = \zeta_0$. Then if $F'(0) = a \ (= \alpha/\xi_0$ in Kármán's notation), expressions satisfying (19), (20) and (21) are

$$
\left.
\begin{aligned}
F &= \left(1 - \frac{\zeta}{\zeta_0}\right)^2 \left[a\zeta + \left(\frac{2a}{\zeta_0} - \tfrac{1}{2}\right)\zeta^2\right], \\
G &= \left(1 - \frac{\zeta}{\zeta_0}\right)^2 \left(1 + \frac{\zeta}{2\zeta_0}\right),
\end{aligned}
\right\} \tag{22}
$$

so that $G'(0) = -\dfrac{3}{2\zeta_0}$. These expressions are then substituted in (18), with the upper limits in the integrals replaced by $\zeta_0$. Thus we get a pair of simultaneous equations for $a$ and $\zeta_0$, which are solved to give the approximate solution.

There is, however, an error in Kármán's working. In the first of equations (22) he writes the last term

$$\left(\frac{2a}{\zeta_0} - \frac{1}{2\zeta_0{}^2}\right)\zeta^2$$

instead of

$$\left(\frac{2a}{\zeta_0} - \tfrac{1}{2}\right)\zeta^2.$$

[In addition, the first of Kármán's equations (28) (*loc. cit.* p. 246) should read, in Kármán's notation, even without making the change noticed above,

$$\int_0^{\xi_0} f^2 d\xi = \xi_0 \, [0 \cdot 0301\alpha^2 - 0 \cdot 00675\alpha + 0 \cdot 00040],$$

i.e. the numbers $0 \cdot 00326$ and $0 \cdot 00159$ as given by Kármán should be $0 \cdot 00675$ and $0 \cdot 00040$ respectively.]

Thus the results at which Kármán arrived,

$$\zeta_0 = 2 \cdot 58, \quad F'(0) = 0 \cdot 40, \quad G'(0) = -0 \cdot 58, \quad H(\infty) = -0 \cdot 71,$$

are inaccurate even according to the approximate method*. In fact,

---

* These results are given, as Kármán stated them, in the *Handbuch der Physik*, 7 (1927), 158, 159; the *Handbuch der Experimental-Physik*, 4 (1931), Part I, 255-7; *Bulletin No. 84* of the National Research Council; *Hydrodynamics* (1932), 280-2; Müller, *Einführung in die Theorie der zähen Flüssigkeiten* (1932), 226-9. A comparison with experiment was given by Kempf, *Vorträge aus dem Gebiete der Hydro- und Aerodynamik* (Innsbruck, 1922) edited by Kármán and Levi-Cività (1924), 168-70.

If $\omega a^2/\nu$ is too large (greater than about $5 \times 10^5$), the motion is turbulent.

Kármán's equations (29) should read

$$0\cdot09048a^2 - 0\cdot02024a + 0\cdot00119 = \frac{1}{\zeta_0{}^4}(0\cdot23571 - a),$$

$$0\cdot24286a - 0\cdot02262 = \frac{3}{2\zeta_0{}^4}.$$

$$(23)$$

These give a quadratic equation for $a$, whose roots are $+0\cdot09648$ and $+0\cdot19466$. The larger root gives

$$\zeta_0 = 2\cdot79, \quad F'(0) = 0\cdot54, \quad G'(0) = -0\cdot54, \quad H(\infty) = -0\cdot55. \quad (24)$$

The smaller root need not be considered since it leads to a function $F$ which in $(0, \zeta_0)$ rises to a maximum, decreases to zero and crosses the $\zeta$-axis to a negative minimum, increasing again to zero at $\zeta_0$. Such a graph is inadmissible as an approximation. The graphs of $F$, $G$, $H$, as obtained from the larger root, are shown by the broken lines in Fig. 1, and those of $F'$, $G'$ and $P$ by the broken lines in Fig. 2.

If we write

$$F = \left(1 - \frac{\zeta}{\zeta_0}\right)^2 F_1, \quad G = \left(1 - \frac{\zeta}{\zeta_0}\right)^2 G_1,$$

so that $F$ and $G$ satisfy the conditions

$$F(\zeta_0) = F'(\zeta_0) = G(\zeta_0) = G'(\zeta_0) = 0,$$

then the above approximations have been obtained by making $F_1$ quadratic and $G_1$ linear in $\zeta$, and satisfying the additional conditions

$$F(0) = 0, \quad F'(0) = a, \quad F''(0) = -1, \quad G(0) = 1, \quad G''(0) = 0.$$

The following alternative assumptions were also investigated:

(i) $F_1$ linear, $G_1$ quadratic, chosen to satisfy

$$F(0) = 0, \quad F''(0) = -1, \quad G(0) = 0, \quad G'(0) = b, \quad G''(0) = 0.$$

These give

$$F = \frac{\zeta_0 \zeta}{4}\left(1 - \frac{\zeta}{\zeta_0}\right)^2,$$

$$G = \left(1 - \frac{\zeta}{\zeta_0}\right)^2 \left[1 + \left(b + \frac{2}{\zeta_0}\right)\zeta + \left(\frac{2b}{\zeta_0} + \frac{3}{\zeta_0{}^2}\right)\zeta^2\right].$$

$$(25)$$

and the equations corresponding to (22) are, if $y = b\zeta_0$,

$$0\cdot03016y^2 + 0\cdot21190y + 0\cdot23571 = 0\cdot00179\zeta_0{}^4,$$

$$0\cdot01668y + 0\cdot06308 = -\frac{y}{\zeta_0{}^4}.$$

$$(26)$$

We thus obtain a cubic equation in $y$, of which only one root, $y = -1\cdot074$, gives a real value of $\zeta_0$. From this we get

$$\zeta_0 = 2\cdot21, \quad F'(0) = 0\cdot55, \quad G'(0) = -0\cdot49, \quad H(\infty) = -0\cdot45.$$

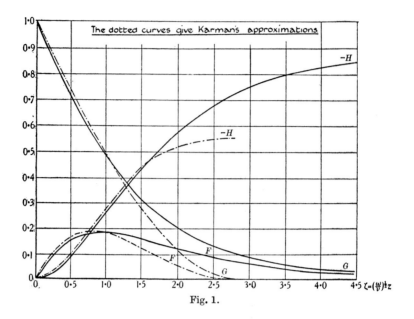

Fig. 1.

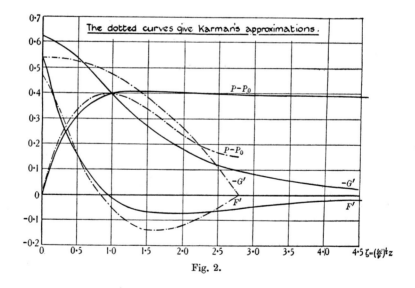

Fig. 2.

(ii) $F_1$, $G_1$ quadratic, chosen to satisfy
$$F(0) = 0, \quad F''(0) = -1, \quad F'''(0) = -2b,$$
$$G(0) = 1, \quad G'(0) = b, \quad G''(0) = 0,$$
the third of these equations being obtained by differentiating the first of equations (7) and putting $\zeta = 0$. This leads by the same process to a quartic equation in $y = b\zeta_0$. The equation has two real roots, but the graphs for $F$ corresponding to these are both inadmissible, being similar in shape to the inadmissible graph already mentioned.

(iii) $F_1$, $G_1$ quadratic, chosen to satisfy
$$F(0) = 0, \quad F'(0) = a, \quad F''(0) = -1,$$
$$G(0) = 1, \quad G''(0) = 0, \quad G'''(0) = 2a,$$
the last condition being obtained by differentiating the second of equations (7), and putting $\zeta = 0$. These lead eventually to a quintic equation in $(a\zeta_0)^3$, only one of whose roots, $(a\zeta_0)^3 = 31 \cdot 073$, gives $\zeta_0$ real. The corresponding $F$, however, has again a negative minimum in $(0, \zeta_0)$ and is therefore inadmissible.

One might perhaps have expected that by satisfying the extra conditions in (ii) and (iii) above, better approximations would be obtained, whereas Kármán's method does not yield an approximation at all in either case. As a check, the results of this section were also obtained independently by Mr L. Howarth, B.A., of Gonville and Caius College, to whom my thanks are due.

3. *Numerical integration of the equations.* By substituting in equations (7) and equating coefficients, we obtain formal expressions for $F$, $G$ and $H$ in powers of $e^{-c\zeta}$, as mentioned above. The first few terms are

$$
\left.
\begin{aligned}
F &= A e^{-c\zeta} - \frac{(A^2 + B^2)}{2c^2} e^{-2c\zeta} + \frac{A(A^2 + B^2)}{4c^4} e^{-3c\zeta} \\
&\quad - \frac{1}{144c^6} (A^2 + B^2)(17A^2 + B^2) e^{-4c\zeta} + \dots, \\
G &= B e^{-c\zeta} - \frac{B(A^2 + B^2)}{12c^4} e^{-3c\zeta} + \frac{1}{18c^6} AB(A^2 + B^2) e^{-4c\zeta} + \dots, \\
H &= -c + \frac{2A}{c} e^{-c\zeta} - \frac{(A^2 + B^2)}{2c^3} e^{-2c\zeta} + \frac{A(A^2 + B^2)}{6c^5} e^{-3c\zeta} \\
&\quad - \frac{1}{288c^7} (A^2 + B^2)(17A^2 + B^2) e^{-4c\zeta} + \dots, \\
P &= \text{constant} + \frac{(3A^2 - B^2)}{2c^2} e^{-2c\zeta} - \frac{2A(A^2 + B^2)}{3c^4} e^{-3c\zeta} \\
&\quad + \frac{1}{96c^6} (A^2 + B^2)(27A^2 + 11B^2) e^{-4c\zeta} + \dots,
\end{aligned}
\right\} \quad (27)
$$

the last being obtained from (11). These give solutions, valid for large $\zeta$, of a differential system I, consisting of equations (7) together with the boundary conditions

$$F(\infty) = 0, \quad G(\infty) = 0, \quad H(\infty) = -c. \tag{28}$$

Now by numerical integration in the direction of $\zeta$ increasing, we can obtain a solution of a differential system II consisting of equations (7) together with the boundary conditions

$$F(0) = 0, \quad G(0) = 1, \quad H(0) = 0, \quad F'(0) = a, \quad G'(0) = b. \tag{29}$$

Our aim is to find $a, b, A, B, c$ so that systems I and II have a common solution, for such a solution will satisfy the equations (7) and (8) and therefore be a solution of the physical problem which we are considering.

I began with $a = 0.54$, $b = -0.54$, the approximations given by equations (24), and integrated the first, second and last of equations (7), with a view to joining $F, G, H, F'$ and $G'$ at a suitable point to the asymptotic expressions (27). It was soon clear that these were not sufficiently good approximations to make a join. $F$ and $G$ appeared to be tending to $-\infty$ and $H$ to $+\infty$. By trial and error I was led to obtain the sets of solutions corresponding to

$$a = 0.54, b = -0.62; \quad a = 0.50, b = -0.62; \quad a = 0.54, b = -0.60.$$

From these three sets it appeared likely on inspection that the correct value of $a$ lay between $0.50$ and $0.54$, and that the correct $b$ was not far from $-0.62$. I made a join with the set $a = 0.54$, $b = -0.62$ at $\zeta = 2.5$, and by an application of Newton's rule, in a form suitable for five variables, obtained the values $a = 0.513$, $b = -0.618$, from the three sets.

The method was as follows. Let $F_0, G_0, H_0, F_0', G_0'$ be values of $F, G, H, F', G'$ for some value of $\zeta$ obtained from the numerical integration of the system II and let $F_1, G_1, H_1, F_1', G_1'$ be values obtained from the asymptotic series (27) for the same value of $\zeta$, and let

$$\lambda_1 = F_0 - F_1, \quad \lambda_2 = G_0 - G_1, \quad \lambda_3 = H_0 - H_1,$$

$$\lambda_4 = F_0' - F_1', \quad \lambda_5 = G_0' - G_1',$$

so that, for fixed $\zeta$, $\lambda_i$ is a function of $a, b, c, A$ and $B$ ($i = 1, 2, \ldots, 5$). We want to find $a, b, c, A$ and $B$ so that $\lambda_i$ vanish over a range of $\zeta$. For $a = 0.54$, $b = -0.62$, $\zeta = 2.5$, $A_1, B_1$ and $c_1$ were chosen to give the smallest values of the set $\lambda_i$. The results from the integrations for $a = 0.50$, $b = -0.62$; $a = 0.54$, $b = -0.60$ enable us to get estimates of

$$\frac{\partial \lambda_i}{\partial a} \quad \text{and} \quad \frac{\partial \lambda_i}{\partial b},$$

($i = 1, \ldots, 5$), for $a = 0.54$, $b = -0.62$, $A = A_1$, $B = B_1$, $c = c_1$, $\zeta = 2.5$. From the asymptotic series we can find

$$\frac{\partial \lambda_i}{\partial A}, \quad \frac{\partial \lambda_i}{\partial B}, \quad \frac{\partial \lambda_i}{\partial c},$$

for $a = 0.54$, $b = -0.62$, $A = A_1$, $B = B_1$, $c = c_1$, $\zeta = 2.5$. We then have sufficient material to apply Newton's rule, which gives the neighbouring values of $a$, $b$, $A$, $B$ and $c$ so that $\lambda_1, \ldots, \lambda_5$ vanish at $\zeta = 2.5$.

The join for $a = 0.513$, $b = -0.618$ was still not satisfactory, $H$ and $F''$ being about 10 per cent. out, but it was quite clear in what direction variation was necessary. With $a = 0.510$, $b = -0.616$, $c = 0.886$, $A = 0.934$, $B = 1.208$, the values of $F_0$, $G_0$, $H_0$, $F_0'$, $G_0'$, and $F_1$, $G_1$, $H_1$, $F_1'$, $G_1'$ are compared in the range $\zeta = 1.9$ to $\zeta = 2.5$ in the tables below.

| $\zeta$ | $F_0$ | $F_1$ | $G_0$ | $G_1$ | $H_0$ | $H_1$ |
|---|---|---|---|---|---|---|
| 1·9 | 0·126 | 0·127 | 0·222 | 0·222 | −0·548 | −0·548 |
| 2·1 | 0·111 | 0·112 | 0·186 | 0·187 | −0·596 | −0·596 |
| 2·3 | 0·097 | 0·098 | 0·156 | 0·157 | −0·637 | −0·638 |
| 2·5 | 0·084 | 0·085 | 0·131 | 0·131 | −0·674 | −0·675 |

| $\zeta$ | $F_0'$ | $F_1'$ | $G_0'$ | $G_1'$ |
|---|---|---|---|---|
| 1·9 | − 0·075 | − 0·075 | −0·194 | −0·194 |
| 2·1 | − 0·072 | − 0·073 | −0·164 | −0·163 |
| 2·3 | − 0·067 | − 0·068 | −0·138 | −0·138 |
| 2·5 | − 0·061 | − 0·062 | −0·116 | −0·116 |

The graphs of $F$, $G$ and $H$ for these values of the constants are shown by the continuous curves in Fig. 1 from $\zeta = 0$ to $\zeta = 4.5$, and similarly for $F'$, $G'$ and $P$ in Fig. 2; in both cases they are compared with Kármán's approximations, shown by the dotted curves. The approximate value, $-0.54$, of $G'(0)$ given by the correct application of Kármán's method is about 12 per cent. out, though, owing to errors, the value given by Kármán himself was about 6 per cent. out. The graph of $P - P_0$ rises to a maximum at about $\zeta = 1.45$, and then decreases slowly and steadily to $\frac{1}{2}c^2 = 0.3925$. This maximum at a finite distance from the plate would persist even with absolutely correct values of the constants $a$, $b$, $c$, $A$, $B$, for it can be seen from the fourth of equations (27) that $P$ will increase steadily from 0 to infinity only if $B > \sqrt{3}\,A$. Now for the values above $B/A = 1.29$, and for absolutely correct values the above condition, $B > \sqrt{3}\,A$, is certainly not satisfied.

The first, third and fourth of equations (7), which were those integrated numerically, are equivalent to five first order linear equations. By taking $H$ as independent variable, the system can be reduced to four first order linear equations, but these are more complicated and no time would be saved by the transformation. The equations were therefore integrated numerically as they stand. The method of Adams* was used, proceeding from $\zeta = 0$ by intervals of $0.1$ and working to third differences. The four values of $F$, $G$, $H$, $F'$ and $G'$ required to start the process were obtained by finding, by means of substitution in equations (7), series for $F$, $G$, $H$, $F'$ and $G'$ in ascending powers of $\zeta$. The terms necessary to give these values to four decimal places, which was the figure used, for $\zeta = 0.1, 0.2, 0.3$ are

$$F = a\zeta - \tfrac{1}{2}\zeta^2 - \tfrac{1}{3}b\zeta^3 - \tfrac{1}{12}b^2\zeta^4 - \tfrac{1}{60}a\zeta^5$$

$$+ \left(\frac{1}{360} - \frac{ab}{90}\right)\zeta^6 + \left(\frac{b}{315} + \frac{ab^2}{1260}\right)\zeta^7 + \cdots,$$

$$G = 1 + b\zeta + \tfrac{1}{3}a\zeta^3 + \tfrac{1}{12}(ab - 1)\zeta^4 - \tfrac{1}{15}b\zeta^5$$

$$- \left(\frac{a^2}{90} + \frac{b^2}{45}\right)\zeta^6 + \left(\frac{a}{315} - \frac{b^3}{315} - \frac{a^2 b}{252}\right)\zeta^7 + \cdots,$$

$$H = -\left[a\zeta^2 - \tfrac{1}{3}\zeta^3 - \tfrac{1}{6}b\zeta^4 - \tfrac{1}{30}b^2\zeta^5 - \tfrac{1}{180}a\zeta^6 + \cdots\right].$$

$F'$ and $G'$ are obtained, up to the term in $\zeta^6$, by differentiation.

The table below gives $F$, $G$, $H$, $F'$, $G'$ and $P$ to three decimal places from zero to $2.6$ by intervals of $0.1$, and from $2.6$ to $4.4$ by intervals of $0.2$. Above $\zeta = 4.4$, all are given to within one unit in the third decimal place by the term in $e^{-c\zeta}$ in the asymptotic series.

* Cf. Whittaker and Robinson, *Calculus of Observations*, 2nd ed., 363–7.

| $\zeta$ | $F$ | $G$ | $H$ | $F'$ | $G'$ | $P - P_0$ |
|---|---|---|---|---|---|---|
| 0 | 0 | 1·000 | 0 | 0·510 | −0·616 | 0 |
| 0·1 | 0·046 | 0·939 | −0·005 | 0·416 | −0·611 | 0·092 |
| 0·2 | 0·084 | 0·878 | −0·018 | 0·334 | −0·599 | 0·167 |
| 0·3 | 0·114 | 0·819 | −0·038 | 0·262 | −0·580 | 0·228 |
| 0·4 | 0·136 | 0·762 | −0·063 | 0·200 | −0·558 | 0·275 |
| 0·5 | 0·154 | 0·708 | −0·092 | 0·147 | −0·532 | 0·312 |
| 0·6 | 0·166 | 0·656 | −0·124 | 0·102 | −0·505 | 0·340 |
| 0·7 | 0·174 | 0·607 | −0·158 | 0·063 | −0·476 | 0·361 |
| 0·8 | 0·179 | 0·561 | −0·193 | 0·032 | −0·448 | 0·377 |
| 0·9 | 0·181 | 0·517 | −0·230 | 0·006 | −0·419 | 0·388 |
| 1·0 | 0·180 | 0·468 | −0·266 | −0·016 | −0·391 | 0·395 |
| 1·1 | 0·177 | 0·439 | −0·301 | −0·033 | −0·364 | 0·400 |
| 1·2 | 0·173 | 0·404 | −0·336 | −0·046 | −0·338 | 0·403 |
| 1·3 | 0·168 | 0·371 | −0·371 | −0·057 | −0·313 | 0·405 |
| 1·4 | 0·162 | 0·341 | −0·404 | −0·064 | −0·290 | 0·406 |
| 1·5 | 0·156 | 0·313 | −0·435 | −0·070 | −0·268 | 0·406 |
| 1·6 | 0·148 | 0·288 | −0·466 | −0·073 | −0·247 | 0·405 |
| 1·7 | 0·141 | 0·264 | −0·495 | −0·075 | −0·228 | 0·404 |
| 1·8 | 0·133 | 0·242 | −0·522 | −0·076 | −0·210 | 0·403 |
| 1·9 | 0·126 | 0·222 | −0·548 | −0·075 | −0·193 | 0·402 |
| 2·0 | 0·118 | 0·203 | −0·572 | −0·074 | −0·177 | 0·401 |
| 2·1 | 0·111 | 0·186 | −0·596 | −0·072 | −0·163 | 0·399 |
| 2·2 | 0·104 | 0·171 | −0·617 | −0·070 | −0·150 | 0·398 |
| 2·3 | 0·097 | 0·156 | −0·637 | −0·067 | −0·137 | 0·397 |
| 2·4 | 0·091 | 0·143 | −0·656 | −0·065 | −0·126 | 0·396 |
| 2·5 | 0·084 | 0·131 | −0·674 | −0·061 | −0·116 | 0·395 |
| 2·6 | 0·078 | 0·120 | −0·690 | −0·058 | −0·106 | 0·395 |
| 2·8 | 0·068 | 0·101 | −0·721 | −0·052 | −0·089 | 0·395 |
| 3·0 | 0·058 | 0·083 | −0·746 | −0·046 | −0·075 | 0·395 |
| 3·2 | 0·050 | 0·071 | −0·768 | −0·040 | −0·063 | 0·395 |
| 3·4 | 0·042 | 0·059 | −0·786 | −0·035 | −0·053 | 0·394 |
| 3·6 | 0·036 | 0·050 | −0·802 | −0·030 | −0·044 | 0·394 |
| 3·8 | 0·031 | 0·042 | −0·815 | −0·025 | −0·037 | 0·393 |
| 4·0 | 0·026 | 0·035 | −0·826 | −0·022 | −0·031 | 0·393 |
| 4·2 | 0·022 | 0·029 | −0·836 | −0·019 | −0·026 | 0·393 |
| 4·4 | 0·018 | 0·024 | −0·844 | −0·016 | −0·022 | 0·393 |

This work was suggested to me by Dr S. Goldstein, who has helped me at every stage. It is a pleasure to express my thanks to him.

# 3

## A NOTE ON THE INFLUENCE OF RAINFALL ON THE YIELD OF CEREALS IN RELATION TO MANURIAL TREATMENT.

By W. G. COCHRAN, B.A.

(*Statistical Department, Rothamsted Experimental Station, Harpenden.*)

(With One Text-figure.)

### INTRODUCTION.

EXPERIMENTS on the continuous growth of wheat and barley under the same manurial condition year after year have been in progress at Rothamsted, on Broadbalk and Hoosfields respectively, since 1844, and at Woburn, on Stackyard field, since 1877. Each field contained a number of plots receiving different manurial treatments. The systems of manuring were not identical in the four series, but they had certain common features, each being designed to cover a certain range of nitrogenous and mineral manuring.

In view of the relatively long series of years available, these experiments provide valuable data for the quantitative study of the effect of meteorological factors on the yield of wheat and barley. Progress in agricultural meteorology has, in fact, been greatly hampered by the scarcity of data sufficiently long that valid conclusions might be drawn from them. In 1923 R. A. Fisher[1] made an examination of the effect of rainfall on the annual yields of wheat on thirteen plots at Broadbalk which had received continuous manuring over the period 1852–1918. The other three series of data were later examined by the same method for rainfall effects, the barley yields from Hoosfield by J. Wishart and W. A. Mackenzie[2], and the wheat and barley yields from Woburn by A. M. Webster. Thirteen of the Hoos plots were chosen for study, the period being 1853–1921; at Woburn ten plots were taken for each crop over the much shorter period 1877–1906. The same method of analysis was also applied by R. J. Kalamkar[3] to the continuous experiments with mangolds at Rothamsted.

Probably the most striking of the manurial effects in these experiments, apart from that on mean yields, was the difference which they produced in the rates of deterioration of the yields throughout the

---

experimental period. The primary object of the present note, however, is to show from the data at Rothamsted and Woburn that there is a close relation between manurial treatment and the variation in yield from year to year even after eliminating the effect on variation of such factors as deterioration of the soil. In order to appreciate the bearing of this point on previous studies of the Rothamsted and Woburn experiments, it is necessary to recapitulate briefly some of the methods and results in the papers mentioned above, beginning with Fisher's examination[1] of the effect of rainfall on wheat at Broadbalk.

Before investigating the rainfall effect it was necessary to eliminate from the series of annual yields the slow changes throughout the period, as these were considered to be due to deterioration of the soil, influence of weeds, etc., but not to any marked extent to the effects of the weather. To do this Fisher fitted a smooth curve, a polynomial of the fifth degree, to the annual yield of each of the thirteen plots and used the deviation in yield each year from the value given by the curve for the examination of rainfall effects. He also developed a technique for finding the linear regression of the annual yields on the rainfall throughout the year which leads to curves showing the change in yield expected from an inch of rain above the average falling at any time of year. A separate curve was fitted to each of the thirteen plots.

The curves were definitely significant on most of the plots, and on four plots more than 30 per cent. of the total variation in yield from year to year could be ascribed to rainfall effects. Further, when considered in relation to the manurial treatments of the different plots, they were capable of a simple and consistent interpretation. To quote Fisher (*loc. cit.* p. 125): "An examination of these diagrams shows how intimately the response of the crop to weather is connected with the manurial conditions of the soil. Classing the plots solely by inspection of the curves of response to rainfall we shall put together every case in which the manurial treatment is alike, and indeed the whole series of curves arrange themselves in sequence of order of increasing abundance of nitrogenous fertilisers."

In marked contrast to the results obtained on Broadbalk, the regression function was not nearly significant on any of the thirty-three plots at Rothamsted on Hoosfield and at Woburn. At Woburn the number of years involved in the test was unfortunately rather small, but for each crop three plots received continuous manuring for 50 years, and the rainfall curves for these showed the same lack of significance as the 30-year curves.

33–2

While not individually significant, the curves in each series grouped themselves by position and shape according to the type of manurial treatment they received just as on Broadbalk. This at first sight seemed considerably to increase the confidence to be put in their interpretation, for if the curves did not represent a real effect their grouping might be expected to be entirely fortuitous. Further, it was possible that while none of the individual curves was significant, the differences between the curves on two plots with different types of manurial treatment might be significant. At Woburn, the significance of the difference between the rainfall curves was tested for the barley plots which received mineral manures and mineral manures plus nitrate of soda respectively. The method used was to fit a rainfall curve to the differences between the deviations from the long-term polynomials of the two plots for each year, in exactly the same way as rainfall curves were fitted to the deviations of the individual plots. The curve can, of course, be obtained by taking at every point the difference between the two rainfall curves already fitted. A test of the significance of this curve against the deviations of the differences from the curve is a test of the difference between the curves for the two plots. This test is not the only one which suggests itself, and a discussion of its appropriateness will be given after the main results of the paper.

The actual difference tested at Woburn showed no sign of significance, and it is expected that if any other two plots had been taken the differences would not have approached significance either. In view of this and of the lack of significance of the rainfall curves themselves, it cannot be asserted with any confidence that they represent a real effect at all, despite the apparent grouping.

Considerable further search was made for particular rainfall variates which might significantly affect the yields of wheat and barley at Woburn, account being taken of quadratic and cubic terms and of variates related to the sowing date instead of to the calendar date. A number of possible factors were examined, but for wheat not one was found which had a significant influence. For barley, however, the combination March–April rainfall, its square and cube, and rainfall 60–90 days after sowing and its square gave regression curves significant at the 5 per cent. point on six of the ten plots.

### The association of seasonal variations in yield with manurial treatment.

It has already been mentioned that, before examining regressions on rainfall, the effects of deterioration and slow changes were eliminated from yield by fitting a smooth curve to the annual yields. The deviations of the actual yield from the value given by the curve are considered to represent that part of yield which is due to the effects of the season. High correlations are to be anticipated, and are actually found to exist between the deviations for different plots on the same field, for in a favourable season yields are in general higher than usual whatever the system of manuring. It was noticed by Yates, however, that the correlations between deviations appear to be much higher for plots receiving the same type of manuring than for those receiving different types. In other words, the deviations show the same natural grouping according to manurial treatment which has been evident in all the rainfall curves discussed above.

The object of this note is to draw attention to this fact and to show by a numerical test that the grouping is statistically significant. The circumstance itself is of considerable importance in the study of the effects of weather on yield, particularly in view of the comparative failure of the investigations at Woburn to discover any clear weather influences. For it points definitely to the existence of some combination of factors, possibly meteorological, whose influence on yield is closely related to the manurial treatment of the soil. Incidentally, it explains why the rainfall curves at Woburn grouped themselves according to manurial treatment, although there was no clear evidence of rainfall effect of the type examined. The grouping, if not a true rainfall effect, was apparently simply a reflection of that in the deviations to which the curves were fitted.

The search for such a combination of variates by the usual regression method would involve in any particular case tests on two points: (1) the significance of the regressions, (2) whether the high correlations between plots receiving the same type of manure remained after taking account of the effect of the particular factors studied.

It seems clear that the rainfall curves for wheat and barley at Woburn, as found by Fisher's method, satisfy neither of the two tests: they are not nearly significant, and it is quite certain that the grouping would remain if deviations from the rainfall curves were taken instead of deviations from the long-term polynomials. With wheat, indeed, since no meteorological factor was found which significantly affected yields, the precise explanation of the grouping has not been found. With barley, however, it is worth while applying the second test to the factors March–April rainfall, its square and cube and rainfall 60–90 days after sowing and its square, in view of the fact that they were found to have a significant effect on yield.

The regressions on March–April rainfall, etc., were worked out for the period 1878–1906 on deviations from six-year means instead of from the long-term polynomials. The former will therefore also be used in testing the significance of the grouping before taking account of rainfall effects, though it would show equally well in the deviations from the long-term polynomials. Six plots were taken, and these were found to be grouped according to manurial treatment into three groups as follows: (1) and (7) (no manure) and (4) (minerals); (3) (nitrate of soda) and (6) (nitrate of soda and minerals); (11$b$) (farmyard manure). Let (1), (7), (4), (3), (6) and (11$b$) denote the deviations of the corresponding plots from the six-year means in any year, so that $\Sigma\{(1)-(7)\}^2$, for instance, is the sum of squares of the differences between the deviations for plot 1 and those for plot 7, over the 29 years 1878–1906. This sum has 24 degrees of freedom. The differences in degree of association will be shown by comparing the sums $\frac{1}{2}\Sigma\{(1)-(7)\}^2$, $\frac{1}{6}\Sigma\{(1)+(7)-2(4)\}^2$, $\frac{1}{2}\Sigma\{(3)-(6)\}^2$, $\frac{1}{6}\Sigma\{(3)+(6)-2(11b)\}^2$ and $\frac{1}{6}\Sigma\{(1)+(7)+(4)-(3)-(6)-(11b)\}^2$, each of which has 24 degrees of freedom. The theory of the method may be explained briefly for the comparison of

$$\Sigma_1 = \tfrac{1}{2}\Sigma\{(1)-(7)\}^2 \quad \text{and} \quad \Sigma_2 = \tfrac{1}{6}\Sigma\{(1)+(7)-2(4)\}^2.$$

Examination of $\Sigma(1)^2$, $\Sigma(7)^2$ and $\Sigma(4)^2$ shows that it is legitimate to assume that (1), (7) and (4) have the same standard deviation. If the correlation between any pair of (1), (7) and (4) is the same, it is easy to see that $\frac{1}{24}\Sigma_1$ and $\frac{1}{24}\Sigma_2$ will be independent estimates of the same variance. If, on the other hand, (1) and (7) agree more closely in their variation from year to year than does either with (4), $\Sigma_2$ may be expected to be larger than $\Sigma_1$. Thus a test of the ratio $\Sigma_2/\Sigma_1$ by Fisher's $z$ distribution is a test whether the yield of the minerals plot 4 shows a difference in its manner of variation from year to year from that of plots 1 and 7 which receive no manures. Similarly any other pair of the seven sums given above may be tested by the $z$ test. The test assumes that (1), (7), (4), (3), (6) and (11$b$) have all the same standard deviation. This is not quite the case. For both wheat and barley the standard deviations of

(1), (7) and (4) may be taken to be the same, and so may those of (3), (6) and (11b), but the latter are much higher than the former. For barley, for instance, the mean squares $\frac{1}{24}\Sigma\{(1)\}^2$ etc., are: plot 1, 43·5; plot 7, 38·6; plot 4, 59·3; plot 3, 100·3; plot 6, 95·0; plot 11b, 107·6. These estimates were pooled, and the variance of (1), (7) and (4) was taken to be 47·1 and that of (3), (6) and (11b) 101·0. To make the standard deviations equal, each yield was divided by its standard deviation, thus reducing (1), (7), (4), (3), (6) and (11b) to a set of quantities each having unit standard deviation. The mean squares obtained from the deviations from six-year means for wheat and barley are shown below.

|  |  | Mean squares (24 degrees of freedom) | |
| --- | --- | --- | --- |
|  | Manurial comparisons | Barley | Wheat |
| Within groups: |  |  |  |
| (1) − (7) | Between unmanured plots | 0·101 | 0·134 |
| (1) + (7) − 2 (4) | No manure v. minerals | 0·107 | 0·145 |
| (3) − (6) | N/S v. N/S + minerals | 0·061 | 0·131 |
| Between groups: |  |  |  |
| (3) + (6) − 2 (11b) | N/S v. F.Y.M. | 0·228 | 0·307 |
| (1) + (4) + (7) − (3) − (6) − (11b) | No N v. N | 0·372 | 0·414 |

The first point which the analysis reveals is the high correlations which exist between the deviations for all plots compared. For owing to the method of weighting adopted, the mean squares should, in the absence of any correlation between the plots compared, have an average value 1, and when compared with 1 by the $\chi^2$ distribution they are all significantly lower at the 1 per cent. point.

As regards the comparison of mean squares between and within groups, the results are similar for the two crops. The three sets of 24 degrees of freedom representing the variance within groups agree satisfactorily, whereas the two variances between groups are both significantly higher than any of the variances within groups. This demonstrates clearly the grouping of the deviations.

The next step is to test whether the grouping persists for barley after eliminating the effect of the significant weather variates which were found, viz. March–April rainfall, its square and cube, and rainfall 60–90 days after sowing and its square. To make the test, exactly the same procedure is carried out on deviations from the rainfall regression curves. In this case the mean squares $\frac{1}{19}\Sigma(1)^2$, etc., were (1) 35·5, (7) 25·4, (4) 38·4, (3) 67·1, (6) 58·8, (11b) 78·8, so that the estimated variances for the two groups were 33·1 and 68·2 respectively. Each yield was divided by the corresponding estimate of its standard error as before. The results in this case were:

| | Mean squares (19 degrees of freedom) |
|---|---|
| Within groups: | |
| $(1) - (7)$ | 0·163 |
| $(1) + (7) - 2(4)$ | 0·118 |
| $(3) - (6)$ | 0·081 |
| Between groups: | |
| $(3) + (6) - 2(11b)$ | 0·265 |
| $(1) + (4) + (7) - (3) - (6) - (11b)$ | 0·602 |

If the high correlations between the deviations from six-year means of plots receiving the same type of manuring were entirely due to the different effects of the weather variates we are considering, the mean squares between and within groups would be of the same size. The results above, however, are not substantially different from those obtained for barley before eliminating the effects of March–April rainfall, etc. There is some sign of a removal of part of the difference between the nitrate of soda and farmyard manure plots by the rainfall regressions, since the mean square of $[(3) + (6) - 2(11b)]$ is no longer significantly higher than that of $[(1) - (7)]$ or $[(1) + (7) - 2(4)]$. On the other hand, it is still significantly higher than $[(3) - (6)]$, with which it is more strictly comparable; and the mean square representing the difference between the no-nitrogen groups is still significantly above all mean squares within groups. It will be noticed that all mean squares rose after taking account of the rainfall effect. Such a rise means that the correlation between the sets of plots which we are comparing is less than before. This is what we should expect, since general similarity in rainfall response is presumably one of the causes of the high correlations which exist between all the plots. If the grouping of plots were due in part to a differential effect of rainfall, however, the percentage increase in mean square would be greater for the comparisons within groups than for those between groups.

The general conclusion from the above investigation is that for barley at Woburn, as for wheat, the explanation of the grouping has not been found. It must be remembered, however, that while the statistical methods employed in the study of rainfall effects constitute an advance on any used in similar studies previous to Fisher's paper[1], the particular types of weather effect studied are extremely simple and crude from the point of view of plant growth. It seems likely that little further progress towards the elucidation of weather effects on the yields at Woburn will be made by the statistician alone. For any further work on the present lines must resolve itself into an examination of correlations suggested by the data themselves. This method must be used with caution, particularly in view of the small number of years available for study, and it is unlikely to lead to fruitful results. The data will, however, remain of value for testing any theories which may arise as to the quantitative influence of weather on yields.

It is of interest to apply the two tests made on the Woburn data to some of the rainfall regressions found by Fisher's methods on Broadbalk. Here, as has been said, the rainfall curves were definitely significant and the differences between curves of plots in different manurial groups appear to be also significant. There seems, therefore, reason to expect that some of the difference between plots with different manurial treatments will disappear when the difference in rainfall effects is eliminated. Five plots were chosen which can be divided into three groups as follows:

(3) (no manure) and (5) (minerals).

(10) (ammonium salts, 200 lb. per acre) and (11) (ammonium salts and superphosphate).

(2b) (farmyard manure, 14 tons per acre).

The deviations on which the calculations are made are in this case measured from the long-term polynomials. As with Woburn, the variances of (3) and (5), and likewise those of (10) and (11), may be assumed the same, but that of (2b) differs from both of the others and the estimates, based on 54 and 48 degrees of freedom before and after eliminating rainfall effects respectively, have to be used as the true standard deviation. It may be mentioned that, while making the necessary calculations, the differences in rainfall curves corresponding to the two comparisons between groups made below were tested by the test mentioned above, both being found to be significant. Thus there is no doubt that on Broadbalk the differences in rainfall response between plots with different manuring are real. The mean squares obtained are shown below.

|  |  | Mean square of deviations from | |
|---|---|---|---|
|  |  | Poly-nomials (54 degrees of freedom) | Rainfall curves (48 degrees of freedom) |
| Within groups: | Manurial comparisons |  |  |
| (3) − (5) | No manure v. minerals | 0·091 | 0·126 |
| (10 − (11) | Ammonium salts v. ammonium salts + superphosphate | 0·150 | 0·191 |
| Between groups: |  |  |  |
| (3) + (5) − (10) − (11) | No N v. ammonium salts | 0·607 | 0·709 |
| (3) + (5) + (10) + (11) − 4 (2b) | F.Y.M. v. rest | 0·448 | 0·483 |

The mean square has again risen in all cases after eliminating rainfall effect. Further, the percentage increases are considerably greater for the comparisons within groups than for those between groups, indicating that part of the grouping is due to a differential response to rainfall. The correlations within groups, however, remain significantly higher than those between groups after eliminating the linear effect of rainfall, just as for Woburn barley. Thus on Broadbalk also the close relation between seasonal variation in yield and manurial treatment remains after eliminating that part of the variation, in some cases quite considerable, which may be ascribed to the linear effect of rainfall.

It is of course to be expected that when the same manuring is applied year after year to a plot it will have its effect on the way in which the crop responds to the varying succession of seasons; and as the principal result of this paper is merely a confirmation of this expectation it might not, if standing by itself, be considered of much interest. The interest of the result lies in its bearing on the interpretation of the studies of rainfall effect on Hoosfield, Rothamsted and on Stackyard, Woburn, for it shows why the rainfall curves could be classed according to the type of manurial treatment despite their lack of significance, and suggests that the failure to obtain significant effects of rainfall was due to the complexity and not to the relative unimportance of rainfall effect.

### TESTS OF SIGNIFICANCE OF THE DIFFERENCE BETWEEN RAINFALL CURVES.

It has been mentioned above that one method of testing the significance of the difference between the rainfall curves of two plots is to test in the usual way the significance of the rainfall curve fitted to their difference. Given replicates of the same treatment on the same field, a second test suggests itself. In this we fit a rainfall curve to each replicate, and test the difference between the mean curves for each treatment against the difference between the rainfall curves for replicates of the same treatment. In this case, of course, the first test can be made on the mean curves for the two treatments.

The two methods are not, in general, equivalent. An obvious difference between them is that a significant difference found by the first method is not necessarily attributable to the different manurial treatments, but may be due to soil heterogeneity, since no account has been taken in the test of the differences between replicates. There is, however, another important difference between the tests which is best illustrated by examining in more detail a simple particular case.

Suppose that we are examining the linear regression of yield on total rainfall for a particular treatment of which we have several replicates, and that the true response curve of yield to rainfall for one of the replicates is as shown in Fig. 1. We may assume further for simplicity that the points to which the regression is fitted lie closely along the true response curve. If the number of years taken is sufficiently large that the total range of variation of rainfall is reasonably well sampled, the regression line obtained for the mean yield might be somewhat as shown by the line $R_1$ in the figure. Let its slope be $b$, with an estimated standard error $s_1$, obtained from the deviations of the yields from the regression line. Then since $s_1$ includes contributions from the deviations of the

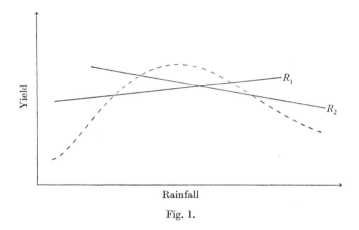

Fig. 1.

annual yields from the line $R_1$ in an approximately random sample of rainfall totals, the limits $(b \pm 2s_1)$ may be expected to include the slope which would have been obtained had the rainfall totals been any other random sample of the same size from the whole population of rainfall totals. As the line $R_1$ has been drawn, its gradient would probably not be significant on the first test, and this may be said to represent the fact that with another sample of rainfall totals, slightly wetter than the one obtained, the gradient would be very small and might even be reversed as shown by the line $R_2$. The slope of $R_2$ would probably not be significant either on the first test. Since there are replicates, we can here apply the second test also by fitting a regression line on rainfall to each replicate and testing the deviation of the average gradient from zero by comparison with the differences between the gradients obtained for the various replicates. Each of these gradients, is, however, obtained from the same set of rainfall totals, and if the difference in rainfall response due to soil heterogeneity is small, they would probably all slope in the same direction as $R_1$. Thus $R_1$ might well be significant on the second test. Similarly, with the set of rainfall totals which lead to $R_2$, it might also be found significant on the second test, though the gradients of $R_1$ and $R_2$ are opposite in sign. In fact, if $s_2$ is the estimated standard error of $b$ as found by the second test, the limits $(b \pm 2s_2)$ give no indication of the gradient which would have been obtained had the set of rainfall totals been different. Any deductions made from the second test are, therefore, valid only for the particular set of rainfall totals which happened to be present in the data.

The question as to which is the better test does not really arise. What is important is to realise the correct interpretation to be put on the results of each test, and for this the differences between the tests should be kept in mind. If there is no real difference in rainfall curves due to soil heterogeneity, it is clear that, when the tests are not equivalent, the first will be the more stringent. For it takes account of all sources of variation of which account is taken in the second test other than soil heterogeneity.

MULTIPLE FIELD TRIALS.

Another instance in which two alternative tests of significance arise occurs when we are considering the results of a number of replicated experiments with the same system of treatments carried out at different places and wish to test the significance of the average of some treatment effect over the whole set of experiments. It is only by such series of experiments that results which are reliable and have a wide range of application can be obtained, and the question of the analyses of the results of such a series is, therefore, of considerable importance. Two errors are possible, the first obtained from the interaction of the treatment effect with places, and the second by combining the estimates of error obtained by the analyses of variance at the separate places. Comparison of the two errors by Fisher's $z$ test should always be made, and tells whether the treatment effect differs significantly from place to place. If it does so, in which case the first error is significantly greater than the second, the former, and not the latter, is the appropriate error for testing the significance of the average treatment effect. For the second test takes no account of the variation in treatment effect from place to place. Thus it may, for instance, announce as a significant average response what is really a large response at a few individual places; further it gives no information as to limits within which the average response might be expected to lie if a different set of places were taken. The second test is not in this case either wrong or misleading, but the information which it supplies has ceased to be of much interest.

If it is decided to test the average treatment effects over the whole set of experiments against the interaction with places as error, there is a further point to be borne in mind. Suppose, for example, that there are ten places and that the treatments in each experiment are no manure (0), nitrogen (N), phosphate (P) and the two dressings combined (N×P), so that the three degrees of freedom for treatments split up into three single degrees representing respectively the response to N, that to P and their interaction N × P. The appropriate analysis of variance for examining the treatment responses is

|  | Degrees of freedom |  |  |  |
|---|---|---|---|---|
| Places | 9 |  |  |  |
| Treatments | 3 ← | N | | 1 |
|  |  | P | | 1 |
|  |  | N × P | | 1 |
| Error 1. Interaction: Treatments × Places | 27 ← | N | × Places | 9 |
|  |  | P | × Places | 9 |
|  |  | (N × P) × Places | | 9 |
| Error 2. Within places. |  |  |  |  |

The 27 degrees of freedom for the interaction of treatments with places divide as indicated above into sets of 9 degrees representing the separate interactions of the three treatment effects, N, P and N × P, which have been distinguished. To test the average response to N, however, it is not in general correct to use the whole 27 degrees of freedom as error. For there is no reason to suppose that the variation in response from place to place will be the same for N, P and N × P. The 27 degrees of freedom will therefore not, in general, be homogeneous, and in using the whole set to test individual treatment degrees of freedom we would obtain biased estimates of error. Each treatment effect should on this account be tested against its own interactions with places, so that in the above experiment we would use separate errors for N, P and N × P, each based on 9 degrees of freedom.

## Summary.

The study of the effect of rainfall on the yields of wheat from the continuous experiments on Broadbalk, Rothamsted, gave clear evidence of a close relation between the response in yield to rainfall and the manurial treatment of the soil. In later investigations of a similar nature on barley at Rothamsted and on wheat and barley at Woburn, however, the evidence did not point to a significant effect of rainfall on yields, and on this account little can be said with confidence from these investigations. The present note shows that the relation between seasonal variations in yield and manurial treatment is just as clear at Woburn as on Broadbalk, the difference between the two centres being that similar studies on rainfall effects have had more definite and successful results on Broadbalk. At Woburn, indeed, little progress has been made towards elucidating the particular weather factors whose quantitative influence is important.

For both barley at Woburn and wheat at Rothamsted, which were the cases examined in detail, the grouping of yields according to manurial treatment remained after eliminating the effect of the significant weather factors which were found. This shows that at both centres there are influences, other than rainfall effects of the type examined, whose effect on the seasonal variations in yield is closely associated with manurial treatment.

Some discussion is given of the appropriate test of significance of the difference between two rainfall curves and of a somewhat analogous case which arises in the interpretation of the results of a series of replicated experiments at different centres.

This investigation, and the method of attack, were suggested to me by Mr F. Yates. I am also indebted to Mr Yates for considerable help received during discussion.

REFERENCES.

(1) FISHER, R. A. *Philos. Trans.* (1924) B, **213**, 89.
(2) WISHART, J. and MACKENZIE, W. A. *J. agric. Sci.* (1930), **20**, 413.
(3) KALAMKAR, R. J. *J. agric. Sci.* (1933), **23**, 571.

(*Received May 25th*, 1935.)

The Statistical Analysis of Field Counts of Diseased
Plants.

By W. G. Cochran, B.A.

## Introduction.

In the study of the propagation of plant diseases, a common method
of obtaining data is to examine every plant in a field or greenhouse for
symptoms of a particular disease at certain intervals. Thus after
each examination a field map can be prepared showing for each
diseased plant in the field the earliest count at which the disease was
noticed. The object of this paper is to discuss the statistical analysis
of such data. The interest to the plant pathologist of those aspects
of the distribution of diseased plants which are discussed below
will vary with the particular disease and the state of knowledge of
the mechanism by which it is propagated, but it is hoped that the
questions considered will be of fairly general interest. In the discussion two questions will be considered : (1) the distribution of the
plants which have become diseased in a given interval, (2) the relation
of that distribution to the distribution of plants previously diseased.

The discussion arose out of an examination by Bald of the spread
of spotted wilt of tomatoes, a virus disease which is carried by a
species of thrips, his data being obtained from field trials made at
the Waite Institute, Australia. The map from which numerical
examples will be taken was of a field of 16 plots, each of 6 rows
containing 15 plants each, so that there were 1,440 plants in the
field. The tomatoes were planted out in November 26, 1929, and
counts were made on December 18 and 31 and January 7, 15, and 22.
There were four treatments, arranged in a 4 × 4 Latin square, but
as these produced no appreciable effect, and as the figures are being
used merely to illustrate the methods of applying the tests, the
differences in treatment will be ignored. It may be stated that,
owing to previous knowledge of the plant pathologist about the
mechanisms by which the plants may become diseased, this case is a
relatively simple one, and in it certain of the tests of significance
discussed below are not really required.

The field map at the end of the second count is shown below, a ×
representing a plant which was diseased at the first count (December
18) and a + one which was diseased at the second count (December
31). Plants which were found to be diseased at later counts are
marked by a ● , but have not been used as examples.

## *The Areal Distribution of Diseased Plants.*

We consider at present the results of the first field count, showing the position of each diseased plant in the field. If every plant in the field has an equal and independent chance of becoming infected, the resulting distribution of infected plants over the area will be called a random one. The actual distribution may, however, deviate from a random one, owing to groups of diseased plants coming together more often than would occur by chance. The groups them-

Field Map of Diseased Plants.

$\times$ = diseased at first count.    $+$ = diseased at second count.
$\bullet$ = diseased at a later count.

selves may be scattered irregularly over the area, such as might happen if an insect carrying the disease had equal access to all plants in the field, but when feeding was able to infect several neighbouring plants at the same time. On the other hand, the deviation from randomness may be of a more regular type, infection being higher, for instance, near the borders than in the interior, or on one side, owing to a source of infective insects near by.

An examination of whether the distribution of infected plants may be regarded as random, and of the type of departure which it

shows, if any, may be made by dividing the area into small groups
containing the same number of plants, say from 6 to 12 per group.
If there are $N$ groups and $n$ plants per group, and every plant in the
field has an equal chance $p$ $(= 1 - q)$ of becoming infected, the
distribution of numbers of diseased plants per acre is the binomial
one, the expected number of areas with $r$ diseased plants being

$$N \, {}^{n}C_r p^r q^{n-r} \qquad . \quad . \quad . \quad . \quad . \quad . \quad (1)$$

By counting the number $N_r$ of groups observed with $r$ diseased
plants, we may compare the observed and expected series by the $\chi^2$
test.   Since, however, we expect the observed series to deviate from
the expected one by having too many groups with many diseased
plants, too many with few diseased plants and too few with numbers
of diseased plants about the average, the estimated variance would
appear to be a more sensitive quantity to use than $\chi^2$, which is a
general test for all types of deviation.   A test of significance based on
the estimated variance in a binomial distribution has been given by
Fisher,[1] and may be obtained by noting that if $\bar{r} = \dfrac{\Sigma r N_r}{N}$, where, of

course, $N = \overset{n}{\underset{0}{\Sigma}} N_r$, then

$$\frac{n \overset{n}{\underset{r=0}{\Sigma}} N_r (r - \bar{r})^2}{\bar{r}(n - \bar{r})}$$

is distributed as $\chi^2$ with $N - 1$ degrees of freedom.

It was pointed out to me by Dr. Fisher that a simple proof of this
may be obtained by considering the data as arranged in a contingency table.   The proof will be given for the more general case in
which the numbers of plants in the small groups vary.   In the $k$th
group $(k = 1, 2 \ldots N)$ let there be $n_k$ plants, $r_k$ of which are
diseased.   Then our estimate of $p$ is $\hat{p} = \Sigma r_k / \Sigma n_k$.   For each
group the plants are divided into two classes, diseased $(r_k)$ and
healthy $(n_k - r_k)$, the expectations in the two classes being $n_k \hat{p}$ and
$n_k \hat{q}$ respectively.   Thus the value of $\chi^2$ obtained by the usual
formula for testing the departure from independence in a contingency
table is

$$\chi^2 = \overset{N}{\underset{k=1}{\Sigma}} \left\{ \frac{(r_k - n_k \hat{p})^2}{n_k \hat{p}} + \frac{(n_k - r_k - n_k \hat{q})^2}{n_k \hat{q}} \right\} \quad . \quad . \quad (2)$$

$$= \overset{N}{\underset{k=1}{\Sigma}} \frac{(r_k - n_k \hat{p})^2}{n_k \hat{p}\hat{q}} \qquad . \quad . \quad . \quad . \quad . \quad . \quad (3)$$

and has $(N - 1)$ degrees of freedom.

If $n_k = n$, for all $k$, this reduces to

$$\chi^2 = \sum_{k=1}^{N} \frac{(r_k - \bar{r})^2}{n\hat{p}\hat{q}} = n \sum_{k=1}^{n} \frac{(r - \bar{r})^2}{\bar{r}(n - \bar{r})} \quad . \quad . \quad . \quad (4)$$

since          $\hat{p} = \dfrac{\bar{r}}{n}.$

If there are $N_r$ groups with $r$ diseased plants, this may be written.

$$\chi^2 = n \sum_{r=0}^{n} N_r \frac{(r - \bar{r})^2}{\bar{r}(n - \bar{r})} \quad . \quad . \quad . \quad . \quad (4a)$$

In the proof quoted above of the $\chi^2$ distribution for a contingency table, it is assumed that the expectation in any class—in this case $np$ or $nq$—is sufficiently large for the actual values obtained to be regarded as normally distributed about the expectation. In this work it may, however, be desirable to make tests where either $np$, $nq$, or both, may be small. In the example shown below, for instance, $np$ is 1·63 and $nq$ is 7·37 both small numbers. No examination has yet appeared in print of the disturbance to the $\chi^2$ distribution for the binomial series when the expectations are small, though the question merits consideration. It is in many cases possible to find the exact distribution of $\chi^2$—which is discontinuous—without undue labour. The exact distribution of $\chi^2$ is given below for the case $n = 9$, $p = 0·2$, $N = 10$, for that region which is of importance in testing for significance. The expectations in the classes are in this case 1·8 and 7·2. The ordinary $\chi^2$ distribution for 9 degrees of freedom is also shown and gives an idea of the discrepancy from the true probability.

$\chi^2$ distribution for the binomial series $n = 9$, $p = 0·2$, $N = 10$.

*Probability of a value* $\gg \chi^2$.

| Value of $\chi^2$. | True Probability. | Probability for ordinary $\chi^2$ distribution. | Discrepancy. |
|---|---|---|---|
| 12·222 | 0·2397 | 0·2006 | −0·0391 |
| 13·611 | 0·1630 | 0·1366 | −0·0264 |
| 15·000 | 0·1032 | 0·0908 | −0·0124 |
| 16·389 | 0·0655 | 0·0591 | −0·0064 |
| 17·778 | 0·0405 | 0·0378 | −0·0027 |
| 19·167 | 0·0246 | 0·0238 | −0·0008 |
| 20·556 | 0·0141 | 0·0148 | +0·0007 |
| 21·944 | 0·0089 | 0·0091 | +0·0002 |
| 23·333 | 0·0048 | 0·0055 | +0·0005 |

The agreement here is reasonably good, the highest percentage discrepancy being about 16 per cent. Since the values of $\chi^2$ are evenly spaced, a correction for continuity could be made, but would not improve the agreement near the significance points. In general, it may be expected that with higher values of $n$ or $p$ the agreement will be better than that shown above, but with lower values it will be worse.

As an example of the application of the $\chi^2$ test for the binomial series, the field was divided into areas, each containing 3 plants from each of 3 rows, making 160 areas of 9 plants each. The distribution of numbers of diseased plants obtained at the first count is shown below.

| $r$. | Observed $x_r$. | Expected. |
|---|---|---|
| 0 | 36 | 26·45 |
| 1 | 48 | 52·70 |
| 2 | 38 | 46·67 |
| 3 | 23 | 24·11 |
| 4 | 10 | 8·00 |
| 5 | 3 | 1·77 |
| 6 | 1 | 0·25 |
| 7 | 1 | 0·03 |
| 8 | 0 | 0·00 |
| $N = 160$ | | 159·98 |

The values obtained by fitting a binomial series are also shown. It will be noticed that the observed series differs from the binomial in having too many groups with 4 or more diseased plants and no diseased plants, and too few with numbers of diseased plants from 1 to 3. In this case $\chi^2$ as calculated from (4a) above, is 225·55 with 159 $(= N - 1)$ degrees of freedom. Hence $\sqrt{2\chi^2} = 21\cdot24$ and $\sqrt{2N - 3} = 17\cdot86$, the difference being 3·38 with unit standard error. The difference is definitely significant, the 1 per cent. point being 2·326, as only one tail of the normal distribution is being considered. The ordinary $\chi^2$ test on this data gives $\chi^2 = 7\cdot967$ with 3 degrees of freedom, which is just significant at the 5 per cent. point.

The binomial series test as applied above takes no account of the relative positions in the field of the groups of diseased plants. It is on this account advisable to supplement the test by performing, where possible, an analysis of variance on the numbers of diseased plants in the groups. In a rectangular field the variation may be analysed into rows, columns and remainder, which is taken as error. This enables the significance to be tested of any type of gradient of infectivity which can be expressed as a component of row or column variation, such as, for instance, a steady increase in infection from one side of the field to the other. Similarly, a test whether the incidence of infection is higher in border plots than in interior plots may be made by dividing the plots into border and interior plots, and testing the difference between the mean infection per plot of the two sections against the pooled variation within sections. In these tests the exact $z$-distribution will not be followed, since we are dealing with data from an approximately binomial, and not a normal, distribution, but investigations have shown that the analysis of variance is applicable over a fairly wide range of non-normality.

After removing a component in the analysis of variance, such as the variation between rows or that between border and interior plots, it is possible to make a $\chi^2$ test on the residual, or intra-class, variation. If there are $h$ rows and $k$ columns and $r_{uv}$ is the total number of diseased plants out of $n$ in the plot in the $u$th row and $v$th column, the quantity

$$n \sum_{uv} \frac{(r_{uv} - \bar{r}_u.)^2}{\bar{r}_u.(n - \bar{r}_u.)}$$

where

$$\bar{r}_u. = \frac{1}{k} \sum_{v=1}^{k} r_{uv}$$

will be distributed as $\chi^2$ with $h(k-1)$ degrees of freedom, provided that each plant in a given row has an equal and independent chance of becoming infected. The use of this type of test in conjunction with the analysis of variance enables the experimenter to distinguish whether, in addition to some regular variation in percentage of infection over the area, there is a tendency for diseased plants to congregate in patches.

Both types of test were made on the data used in the example above. The results of the analysis of variance are shown below :

|  | d.f. | Sums of squares. | Mean squares. |
|---|---|---|---|
| Rows ... ... ... ... | 7 | 21·00 | 3·000 |
| Columns ... ... ... ... | 19 | 36·62 | 1·927 |
| Remainder ... ... ... | 133 | 243·63 | 1·832 |

The mean square for rows is considerably, though not significantly, above the error mean square. There is, however, some suggestion in the data of a regular increase in percentage of infection from the upper to the lower half of the field, and the row regression degree of freedom corresponding to this effect has a mean square 8·702, and is therefore significant. There seems no evidence of any change in the percentage of infection from column to column.

A value of $\chi^2$ was calculated from the variation within rows for each row, to see whether the deviation from a random distribution could be accounted for by a variation in percentage of infection from row to row. The values obtained, each with 19 degrees of freedom, were, 15·902, 28·077, 29·493, 18·348, 27·887, 18·219, 42·500, and 24·156. Since the 5 per cent. point is 30·144, only one of these is individually significant. The sum of the values is 204·582 with 152 degrees of freedom and is itself significant. The conclusion from these tests is that, in addition to a gradual increase in the degree of infectivity from the top to the bottom of the field, there is a tendency for diseased plants to congregate in small patches.

### The Distribution of Groups of Diseased Plants in a Row.

Instead of examining whether groups of diseased plants in a small area tend to occur together more often than on the hypothesis of random distribution, it may be of more interest to consider groups of diseased plants taken along or across the rows. This would be the case, for instance, where the disease might be spreading from a plant to its neighbour in the row by mechanical or root contact, or by an infective insect crawling from one plant to the next in the row and retaining its infective power. The $\chi^2$ test by means of the binomial series may be carried out as above. Where it is desired to test, however, whether groups of contiguous, and not merely neighbouring, diseased plants in a row, occur together more often than on the hypothesis of random distribution, a more sensitive test may be obtained by considering the distribution of runs of diseased plants in a row, $r$ consecutive diseased plants which separate two healthy plants constituting a run of length $r$.

If there are $n$ plants in a row and these have an equal and independent chance $p$ of becoming infected, the expected number of runs of $r$ diseased plants is

$$f_{n,r} = \left\{ \begin{array}{ll} 2p^r q + (n - r - 1)p^r q^2 \ldots 1 \leq r \leq n - 1 \\ p^r \qquad\qquad\qquad\qquad\qquad r = n \end{array} \right\} . \quad (5)$$

This formula was given by Marbe,[2] who used it to examine runs of births of the same sex in records of vital statistics, runs of heads and tails in tosses of a coin and runs of red and black in roulette. A proof of the distribution was given by Marbe [3] (p. 9), but this appears to omit essential steps. An alternative proof may, however, be obtained by induction. For it is easy to verify that the formula holds for $n = 2$ or 3, and by noting that the $2^{n+1}$ possible configurations in a row of length $(n + 1)$ may be derived from the $2^n$ in a row of length $n$ by prefixing to each configuration a $p$ and a $q$ the following relation may be derived :

$$f_{n+1,r} = f_{n,r} + p^r q^2 \ldots 0 \leq r \leq n - 1$$

from which the induction follows at once.

If all runs are considered, irrespective of whether they are healthy or diseased, the corresponding formula is

$$f_{n,r} = \begin{cases} 2(p^r q + pq^r) + (n - r - 1)(p^r q^2 + p^2 q^r) & 1 \leq r \leq n-1 \\ p^r + q^r & r = n \end{cases}. \quad (6)$$

Marbe used the series in testing a theory of his that long runs occur in practice less frequently than is to be expected on the hypothesis of independent and equal chances at each trial. His method was to place the observed and expected series side by side, and decide by inspection whether the data supported his hypothesis—that is, he made no objective test, such as the $\chi^2$ test. It is interesting to notice, however, that if in fitting the expected series we take $p$, the only undetermined parameter, as the observed fraction of diseased plants in the field, the totals of the observed and expected sets of frequencies will not in general coincide, since the total number of runs is in this case a variable quantity whose sampling distribution may be found on the hypothesis we are considering. Thus the first condition for the application of the $\chi^2$ test—namely, that the observed and expected frequencies should have the same total—is not here satisfied, and it is not at first sight clear whether the $\chi^2$ distribution will be followed. It is, in fact, possible to show that $\chi^2$, as calculated in the usual way from a set of rows each of length $n$, is distributed as

$$\lambda_1 x_1^2 + \lambda_2 x_2^2 + \ldots + \lambda_r x_r^2$$

where $x_1, x_2 \ldots$ are independently and normally distributed with unit standard deviation, but the $\lambda_i$ are not all equal to unity. This may be illustrated in the simplest case, in which the observed data consist of $N$ rows of length 2. It will be assumed that no distinction is being made between healthy and diseased plants. If $\times$ denotes a diseased plant and $\bullet$ a healthy plant, the data may be divided into two mutually exclusive configurations :

    (1)  $\times$ $\bullet$ or $\bullet$ $\times$ with probability $2pq = \dfrac{m_1}{N}$.

    (2)  $\times$ $\times$ or $\bullet$ $\bullet$ with probability $p^2 + q^2 = \dfrac{m_2}{N}$.

The probability that of $N$ rows $x_1$ are of type (1) and $x_2$ of type (2) is given by the binomial term

$$f = \frac{N!}{x_1! \, x_2!} \left(\frac{m_1}{N}\right)^{x_1} \left(\frac{m_2}{N}\right)^{x_2}$$

If $N$ is large, we may replace this (cf. [4]) by the corresponding normal distribution

$$f = ce^{-\frac{1}{2}\left\{\frac{(x_1 - m_1)^2}{m_1} + \frac{(x_2 - m_2)^2}{m_2}\right\}} \, d\tau$$

where $x_1 - m_1 + x_2 - m_2 = 0$ and $d\tau$ is the element of area. Each configuration of type (1) gives two runs of length 1, and each configuration of type (2) gives one run of length 2. Thus there are $2x_1$ runs of length 1 and $x_2$ of length 2, so that

$$\chi'^2 = \frac{2(x_1 - m_1)^2}{m_1} + \frac{(x_2 - m_2)^2}{m_2} = (x_1 - m_1)^2\left(\frac{2}{m_1} + \frac{1}{m_2}\right)$$

But $f$ may be written :

$$C'e^{-\frac{1}{2}(x_1 - m_1)^2\left(\frac{1}{m_1} + \frac{1}{m_2}\right)} dx.$$

Hence
$$\chi'^2 = \chi^2\frac{(m_1 + 2m_2)}{m_1 + m_2} = 2\chi^2(1 - pq)$$

Thus $\chi'^2$ varies between $\frac{3}{2}\chi^2$ and $2\chi^2$, so that the use of the $\chi^2$ distribution as a test would considerably over-estimate the significance of any departure.

The disturbance is due to the fact that the runs of length 1 are only independent when grouped in pairs, since to any run of length 1 there must correspond another of length 1.

The same method of attack may be used in rows of length 3, 4, 5, etc., and also when diseased plants alone are being counted, but the evaluation of the values of the $\lambda_i$ rapidly becomes laborious. With long runs it is possible to show that the significance levels of $\chi'^2$ are higher than those of $\chi^2$.

The data considered by Marbe differed from those with which we are concerned here in that, instead of having, say, 96 rows each of length 15, he had in all his examples one single long row. In this latter case a simple and satisfactory solution of the problem may be found by a slight change in the specification. Suppose that the method of obtaining a distribution of runs is to proceed along the row until a fixed number $N$ of runs has been secured, and not, as in the $f_{n,r}$ series, until a fixed length of row has been covered. This is equivalent to regarding the observed total number of runs as ancillary information. Suppose that the long row consists of diseased and healthy plants, and that runs of diseased plants are being taken. To obtain the frequency distribution we proceed along the row until a diseased plant is reached. The probability that the next $(r - 1)$ plants are diseased and the next after is healthy is $p^{r-1}q$, and this gives a run of $r$ diseased plants. Hence

$$f_r = Np^{r-1}q \qquad\qquad r = 1, 2, \ldots$$

so that the probabilities follow a geometric series law.

If both diseased and healthy plants are being counted, the corresponding expression is

$$f_r = N(p^rq + pq^r) \qquad\qquad r = 1, 2, \ldots$$

In dealing with a single long row, the difference between Marbe's series and the geometric series is of academic interest only, since the latter is the appropriate series to use and is easily handled. In the present case, however, the geometric series can be applied only if the rows, which contain 15 plants each, are considered as joined end to end to form one continuous line. Since, however, no test has been found from Marbe's series which would be arithmetically simple to apply, this course has had to be taken.

A test of departure from independence in the geometric series may be obtained by estimating $p$ from the distribution of runs of diseased plants (which, it will be noted, takes no account of the number of healthy plants), and comparing this with the observed fraction of diseased plants. The probability of obtaining $x_r$ runs of length $r$, $r = 1, 2, \ldots n$, from $N$ runs is

$$P = \frac{N!}{x_1!\, x_2!\, \ldots\ldots}\, p^{\Sigma(r-1)x_r}\, \frac{N}{q}$$

Hence
$$L = \Sigma(r-1)x_r \log p + N \log q$$

$$\frac{\partial L}{\partial p} = \frac{\Sigma(r-1)x_r}{p} - \frac{N}{q}$$

so that the maximum likelihood estimate of $p$ is

$$\hat{p} = \frac{\Sigma(r-1)x_r}{\Sigma r x_r}$$

That this quantity is an estimate of $p$ may be seen alternatively by considering that a run of length $r$ occurs if the first $(r-1)$ independent trials following the diseased plant at the beginning of the run all give diseased plants, and the next trial gives a healthy plant. The whole distribution or runs of diseased plants thus represents $\Sigma x_r$ independent trials, $\Sigma(r-1)x_r$ of which have each given a diseased plant. It is also clear that, for a given total number $\Sigma r x_r$ of diseased plants, $\hat{p}$ is greatest when there are many long runs. Thus the amount by which $\hat{p}$ exceeds $p$, the observed fraction of diseased plants, is a test of the association of diseased plants in groups. The variance of $\hat{p}$, obtained from the average value of $\frac{\partial^2 L}{\partial p^2}$, is $\frac{pq^2}{N}$, so that we may compare $\hat{p} - p$ with its standard error $\sqrt{\frac{pq^2}{N}}$.

In applying the test in practice, the effect of the assumption that the rows may be placed end to end must be considered. In general, the effect will be slightly to diminish the sensitiveness of the test, since we regard certain plants as contiguous which are not really so. If, however, there is higher infection on the edges of the field, it is important that these be kept aside when making the test, otherwise long runs may occur in joining one row to another, and yet be due entirely to the higher incidence of infection near the ends of the rows. A variation in the incidence of infection would, of course, affect any such test, whether it involved joining rows together or not.

The binomial series and geometric series tests, both made along the rows, would be far from independent, but there are differences in the types of departure from randomness which they detect. The geometric series test is most appropriate where the hypothesis is being studied that departure from randomness arises through spread or carriage of the disease from a plant to its direct neighbours in the row, and not merely to a nearby plant in the row. As an illustration of this, if $\times$ represents a diseased plant, $\bullet$ a healthy one, the two configurations $\bullet \times \times \bullet \times \bullet \times$ and $\bullet \bullet \times \times \times \times \bullet$ would count as the same in a binomial series test (if lying in the same set of $n$ plants), but the latter would give much more weight to the hypothesis of departure from randomness in the geometric series test. Where it is thought that the disease is being carried from a plant to the next in the row, or that groups of contiguous plants are becoming diseased at the same time, the geometric series will be more sensitive than the binomial, and is the more natural and appropriate test. Otherwise, however, the binomial series is to be preferred, since the data for its application can be abstracted much more quickly.

In the data used in the example above, the binomial series test made on groups of 9 plants showed that the diseased plants were scattered in patches over the area. The significant result when making this test might have risen from the fact that groups of diseased plants in a row tended to occur together, *i.e.* the grouping might be associated with rows and not with areas. The geometric series will therefore be applied to groups of diseased plants taken along the rows. The distribution obtained is given below, the expected geometric series distribution being included for comparison.

| $r$. | Observed $x_r$. | Expected $f_r$. |
|---|---|---|
| 1 | 164 | 169·5 |
| 2 | 33 | 30·7 |
| 3 | 9 | 5·6 |
| 4 | 1 | 1·0 |
| 5 | 0 | 0·2 |
| | 207 | 207·0 |

In this case there were 261 diseased plants in the field out of 1440, so that $p = \dfrac{261}{1440} = 0\cdot18125$.

$$\hat{p} = \frac{54}{261} = 0\cdot2069$$

$$\hat{p} - p = 0\cdot0257 \pm 0\cdot0244$$

The difference is not significant, so that grouping would appear to be associated with areas, and not with rows. This opinion is strengthened by making the geometric series test on lines taken at right angles to the rows. In this case $\hat{p}$ also exceeds $p$, but not significantly so.

### *A Test of Significance of Neighbour Infection.*

The geometric series test, as presented above, is intended as a general test to detect the occurrence of runs of contiguous diseased plants and is suggested on common-sense grounds only. A discussion of the appropriateness of the test in particular cases would, however, appear to be unnecessary here, since if it is possible to specify mathematically the type of departure from the geometric series distribution produced by any mechanism, an efficient test for this type of departure may be made by the method of maximum likelihood. A common method by which groups of contiguous diseased plants may occur is by spreading of disease from a plant to its direct neighbours, either by mechanical or root contact or through the medium of the soil; and it is worth while attempting to derive a sensitive test for this type of spread.

To obtain such a test, it is necessary to examine the effect on the geometric series distribution of a spreading of disease. We suppose that in the first interval each plant in the field has an equal and independent chance $p$ of becoming infected, and that in the second interval a plant which is next to a diseased plant has a probability $s$ of being infected by it, healthy plants which are not next to a diseased plant remaining healthy. It follows that if a healthy plant is between two diseased plants, its chance of remaining healthy is $(1 - s)^2$, so that its chance of becoming diseased is $2s - s^2$. The probability $F_r$ of a run of $r$ diseased plants will now be a function of $n$, $p$, and $s$. The exact solution has not been found, but approximate solutions neglecting any power of $s$ may be obtained. A proof of the first approximation, which neglects $s^2$, will be given. Further approximations may be obtained by the same method, but are rather cumbersome.

As in the proof of the geometric series already given, we suppose that the plants are arranged in an endless line, and that we proceed along the line until $N$ runs have been obtained. It will be convenient to include conventionally runs of length zero, each healthy plant, except those which conclude a run of diseased plants, being considered as a run of length zero. This convention has the advantage that, when we are considering the probability of any run, it may be assumed that the plant immediately preceding the run is healthy. Thus at any point in the line there will be a run of length zero if the plant following was healthy at the end of the first interval and has not since been infected by a diseased plant following it. The probability is

$$q\{q + p(1 - s)\} = q(1 - ps)$$

If $x$ represents a plant originally diseased, ● a healthy plant and $+$ a plant which has been infected by a neighbour, these cases may be represented as

|  ● ● or ● $+$ | ● $\times$ |
|:---:|:---:|
| Probability $\quad q^2$ | $qp(1 - s)$ |

For a run of length 1 there are two cases

|  $\times$ ● ● or $\times$ ● $+$ | $\times$ ● $\times$ |
|:---:|:---:|
| Probability $\quad pq^2(1 - s)$ | $p^2q(1 - 2s)$. |

The sum is $\qquad pq\{1 - s(1 + p)\}$

For a run of length $r$ there are two similar cases, with a total probability $\qquad p^rq\{1 - s(1 + p)\}$

A run of length $r$ may also arise in the two following ways

$+\times\times$ ●–●–● $\times$ ● or $\times\times$ ●–●–● $\times+$ ● $\qquad$ $\times\times$ ●–●–●$+$ ●–●$\times$ ●

| Probability $\quad 2p^{r-1}q^2s$ | $2(r - 2)p^{r-1}q^2s$. |
|:---:|:---:|

The factor $(r - 2)$ in the last term arises because the $+$ may occupy any of the $(r - 2)$ interior positions in the run, and the probability of infection is here $2s$ (more accurately $2s - s^2$), since there is a diseased plant on both sides. Any run which has two or more $+$ terms may be ignored, since its probability will contain a factor $s^2$. The total probability is therefore

$$F_r = p^rq\{1 - s(1 + p)\} + 2(r - 1)p^{r-1}q^2s \qquad\qquad r \geqslant 1$$
$$F_0 = q(1 - sp).$$

A test of significance of the deviation of $s$ from zero may be obtained by estimating $p$ and $s$ by the method of maximum likelihood, which also provides the sampling variance of the estimate of $s$. The likelihood function is

$$L = x_0 \log q + x_0 \log (1 - ps) + \sum_1^\infty rx_r \log p + \sum_1^\infty x_r \log q$$

$$+ \sum_1^\infty x_r \log \left\{1 - s(1 + p) + 2(r - 1)q\frac{s}{p}\right\}$$

Hence $\qquad\qquad \dfrac{\partial L}{\partial p} = \dfrac{X}{p} - \dfrac{N}{q} - Ns - \dfrac{2X_1s}{p^2}$

where $X = \sum_1^\infty rx_r$ is the total number of diseased plants, $N$ is the total number of runs and $X_1 = \overset{\infty}{\underset{1}{\Sigma}}(r-1)x_r$.  The equation of estimation of $p$ is

$$\left(\frac{X}{p} - \frac{N}{q}\right) = s\left(N + \frac{2X_1}{p^2}\right) \quad . \quad . \quad . \quad . \quad (8)$$

If $s = 0$, this reduces to $\dfrac{p}{q} = \dfrac{X}{N}$ or $p = \dfrac{X}{N+X}$, the observed fraction of diseased plants in the field.

$$\frac{\partial L}{\partial s} = - px_0(1 + ps) - N_1(1 + p) - N_1 s(1 + p)^2 + 4X_1\frac{qs}{p}(1 + p)$$
$$+ \frac{2X_1 q}{p} - 4X_2\frac{q^2}{p^2}s$$

where $\qquad N_1 = \overset{\infty}{\underset{1}{\Sigma}}x_r, \; X_2 = \overset{\infty}{\underset{1}{\Sigma}}(r-1)^2 x_r.$

This gives for $s$

$$\left(2X_1\frac{q}{p} - N_1 - pN\right)$$
$$= s\left\{4X_2\frac{q^2}{p^2} - 4X_1\frac{q}{p}(1 + p) + N_1(1 + p)^2 + p^2 x_0\right\} \quad . \quad . \quad (9)$$

The two equations (8) and (9) may be solved without much labour by interpolation, substituting trial values for $p$ in both equations, calculating the two values of $s$ and interpolating for their difference. The solutions will be denoted by $\hat{p}$ and $\hat{s}$.

Further, when $s = 0$

$$\frac{\partial^2 L}{\partial p^2} = -\frac{X}{p^2} - \frac{N^i}{q^2}, \; \frac{\partial^2 L}{\partial s \partial p} = -\frac{2X_1}{p^2} - N,$$
$$\frac{\partial^2 L}{\partial s^2} = - p^2 x_0 - N_1(1 + p)^2 + 4X_1\frac{q}{p}(1 + p) - 4X_2\frac{q^2}{p^2}$$

and these have respectively the mean values

$$-\frac{N}{pq^2}, \; -\frac{N}{q}(3 - p) \text{ and } - N(4 + pq).$$

Hence $\qquad V(\hat{s}) = \dfrac{1}{N\{4 + pq - p(3 - p)^2\}} = \dfrac{1}{Nq(1 + q)^2},$

As an example of the application of the test, it is made below on groups of diseased plants at the first count, taken along the rows. The geometric series test has already been made on this data, and the observed distribution is shown on p. 59.  For this data

$$X_1 = 54, X_2 = 78, N_1 = 207, X = 261,$$

and the equations of estimation may be written :

$$261 - 1440p = q/ps(108 + 1179p^2)$$

$$108\frac{q}{p} - 207 - 1179p$$

$$= s\{312(q/p)^2 - 216q/p(1 + p) + 207(1 + p)^2 + 972p^2\}$$

The solutions are $\hat{p} = 0.1733$ $\hat{s} = 0.0167$.
The standard error of $\hat{s}$ when $s = 0$, is $0.0178$, so that the value of $\hat{s}$ is not significant.

### The Analysis of Later Counts.

In considering the second and later counts, the experimenter may wish to study the distribution of plants which have become diseased since the last count, and its relation to the distribution of diseased plants at the beginning of the interval. In considering the first of these questions, plants which were previously diseased are ignored.

The binomial series test and the analysis of variance may be applied to the same small areas as were used for the first count, if these are considered suitable. In this case the total number $n$ of plants in the small areas, ignoring plants diseased at the beginning of the interval, will no longer be constant. If there are $n_k$ plants remaining in the $k$th area, $k = 1, 2, \ldots, N$, of which $r_k$ become diseased during the interval, $\chi^2$ is calculated from the formula (3) above :

$$\chi^2 = \sum_{k=1}^{N} \frac{(r_k - n_k p)^2}{n_k pq}$$

As will be expected, the test takes considerably longer to apply in practice than the usual test with equal values of $n$.

The analysis of variance should in this case be made on the percentages of infection in the area, giving to each percentage a weight $n_k$. Where the data are considered as classified in only one way, such as, for example, when we wish to compare the variance between rows with the remainder, or the difference between the percentage infections of border and interior plots, the analysis is easily made. For if $n_{ij}, r_{ij}$ are the totals of all plants and of diseased plants respectively in the $ij$th area, and $N_i = \sum_j n_{ij}$, $R_i = \sum_j r_{ij}$, summed over the $i$th row, and $N = \sum_i N_i$, $R = \sum_i R_i$, we have the identity

$$\sum n_{ij}\left(\frac{r_{ij}}{n_{ij}} - \frac{R}{N}\right)^2 = \sum n_{ij}\left(\frac{r_{ij}}{n_{ij}} - \frac{R_i}{N_i}\right)^2 + \sum N_i\left(\frac{R_i}{N_i} - \frac{R}{N}\right)^2$$

where the summation extends in each case over all areas. From this it follows that the variance between rows may be calculated as

$$\sum N_i\left(\frac{R_i}{N_i} - \frac{R}{N}\right)^2.$$

If, however, there are two criteria of classification, such as into rows and columns, and we wish to take out the variance due to each, a rigorous solution can be made only by fitting constants. For a discussion of the problem and of some useful approximate methods, reference should be made to Yates [5] and Snedecor.[6]

An approximate analysis may, of course, be made by ignoring the differences in weight of the percentages of infection in the different areas. This is satisfactory only if the variations in $n_k$ are small.

The binomial series test and the analysis of variance were made on the data at the second count on the same set of small areas as used before. The value of $\chi^2$ for the binomial series exceeded its mean value, but not significantly so, and the analysis of variance showed no sign of any regular gradient of infection.

Similar considerations apply to the binomial series test made along the rows. The geometric series test, however, remains unchanged when we regard the plants which have not been ignored as lying in one long row. Two plants which are separated only by plants which have been ignored, have, however, to be considered as contiguous, and for this reason the test loses much of its appropriateness and does not appear to be of much interest.

To pass to the relation between the above distribution and that of plants previously diseased, the chief question of interest will be whether the incidence of infection in the last interval is higher in the neighbourhood of a plant previously diseased. Where the binomial series test has been made as described above on plants diseased during the last interval, information on this question may be gained by comparing the percentage of infection in each small area during the last interval with the percentage previous to the beginning of the interval. Since these sets of figures will both be available, it is easy to examine by an analysis of covariance whether there is any apparent connection between them. In the data analysed above no relationship was found.

A simple and effective test on this point may, however, be made by means of a $2 \times 2$ contingency table. This depends on a classification of all plants which were healthy at the beginning of the last interval, into those which were in the neighbourhood of a plant previously diseased and those which were not so. The actual choice of a " neighbourhood " depends on the nature and extent of the spread to be expected. If, for instance, the disease is considered to be spread by a prevailing wind sweeping along the rows from left to right, and it is thought that the disease might have spread from a plant to any of the next three on its right in the row, the classification would be made at this basis. If, on the other hand, it is considered that the disease may be spreading by mechanical contact, the " neighbourhood " of a diseased plant would be taken to include all plants to which the infection from that plant might have spread by mechanical means. The classification having been made, the plants in each class are further divided into those which were diseased at the end of the interval and those which were healthy then, the usual test (cf. Fisher[1], §§ 21–21.02) for a 2 × 2 table being made to determine whether the percentage of infection is significantly higher near a diseased plant.

As an example of the application of this test, the data at the second count, December 31, were classified into those from which a diseased plant could be reached either by moving one place along the row or to the corresponding place in the next row on either side, and the remainder, the numbers of diseased and healthy plants in the two groups being recorded. The following table was obtained.

|  | Diseased. | Healthy. | Total. |
|---|---|---|---|
| Near a diseased plant     ...     ... | 111 | 441 | 552 |
| Not near diseased plant     ...     ... | 115 | 512 | 627 |
| Total     ...     ...     ... | 226 | 953 | 1,179 |

It will be observed that the incidence of disease is slightly higher in the neighbourhood of previously diseased plants. To test for significance, we calculate

$$\chi^2 = \frac{(110\frac{1}{2} \times 512\frac{1}{2} - 441\frac{1}{2} \times 115\frac{1}{2})^2 \times 1179}{226 \times 953 \times 552 \times 627} = 0.50.$$

The value is not nearly significant. It should be noted that if in this case the only type of departure from independence which is being taken into account is that in which the percentage of infection is higher near a diseased plant, the 5 per cent. value of $\chi^2$ is 2·706, and not 3·841. This can be verified at once if we remember that the quantity $\chi$, defined as positive when the proportion of diseased plants is higher near a previously diseased plant and negative when it is lower, is distributed normally about zero with unit standard error. If $\chi^2 = 3.841$, then $\chi = \pm 1.960$, and this is the 5 per cent. point of the distribution of $\chi$ when both tails are being taken into account. Thus the 5 per cent. point of $\chi^2$ is 3·841 only if we are prepared to consider that the presence of a neighbouring diseased plant might either increase or decrease the chance of infection in the second interval. In the hypothesis mentioned at the beginning of this paragraph, the 5 per cent. point value of $\chi^2$ is 2·706, corresponding to the 5 per cent. value 1·645 of the positive tail of the distribution of $\chi$.

An advantage of this table is that if it shows a significant departure from the hypothesis that the chance of infection of a plant is independent of its position relative to previously diseased plants, an estimate can at once be made of the chance of infection in the neighbourhood of a plant previously diseased.

### Discussion of Results.

In the analysis given above, each test has been made in turn on the data, without reference to any previous knowledge of the plant pathologist about the spread of this particular disease. It is, however, worth noting in conclusion how the results support such knowledge in this case. The analysis on the data at the first count shows that there are signs of a gradient of infection across the field and that infection tends to congregate in small patches. These results are both compatible with the fact that the disease is carried by insects. The patchiness could be explained by certain sections of the field being more attractive or more easily accessible to the insects, or, alternatively, if an insect retained its infective power after feeding on a plant and tended to crawl or make a short flight to a nearby plant to feed again. Further, in this case it is known that the disease is not likely to spread from plant to plant along the rows unless carried by the insects or unless infective juice were transferred, as might happen during pruning. The latter possibility may be excluded for the first two counts, and non-significance in the geometric series test along the rows and in the direct test of neighbour infection is to be expected.

In the second count the indications of patchiness were very slight, and there was no marked irregularity in the distribution of the percentage of infection over the field. No connection could be established between the incidence of infection in the second interval and the presence or absence nearby of plants previously diseased, which is not surprising in view of the considerations above. In this case, however, even if a relation had been found, it would not necessarily mean that disease was spreading from plants previously diseased, since the results could be explained on the hypothesis that certain small areas attracted the infective insects and were continuing to do so.

### Summary.

The statistical analysis of counts of diseased plants in a field or greenhouse is discussed. Tests of significance are presented to examine (1) whether diseased plants tend to congregate in patches scattered over the area or in groups along or across the rows, (2) where more than one disease count has been made, whether the distribution of plants recently infected is related to that of plants previously infected.

A test is also given which is designed to detect the spreading of infection from neighbour to neighbour in a row.

It is a pleasure to thank Mr. J. G. Bald and Mr. F. Yates for considerable help in discussion, and the former for permission to use the data as examples.

### REFERENCES.

[1] Fisher, R. A., *Statistical Methods for Research Workers.*  Edinburgh, Oliver and Boyd (5th Edition, 1934), § 19.

[2] Marbe, K., *Grundfragen der angewandten Wahrscheinlichkeitsrechnung und theoretischen Statistik.*  München, C. H. Beck, 1934, p. 26.

[3] Marbe, K., *Mathematische Bemerkungen.*  München, C. H. Beck, 1916, pp. 8–9.

[4] Fisher, R. A., *J. Roy. Stat. Soc.*, Vol. 85, Part 1 (1922), p. 89.

[5] Yates, F., *J. Amer. Stat. Soc.*, March 1934, pp. 51–66.

[6] Snedecor, G. W., *J. Amer. Stat. Soc.*, December 1934, pp. 389–93.

# 5

## AN EXPERIMENT ON OBSERVER'S BIAS IN THE SELECTION OF SHOOT-HEIGHTS

W. G. COCHRAN AND D. J. WATSON

*(Rothamsted Experimental Station)*

### *Introduction*

SAMPLING is now of common use in the study of agricultural problems. A recent paper by Yates [1] contains some discussion of the principles which must be followed if a truly representative sample is to be obtained, the most important of these being, to quote the author's own words, that 'the selection of the samples must be determined by some process un-influenced by the qualities of the objects sampled and free from any element of choice on the part of the observer'. As an example of the biases which may arise in the sampling of agricultural material in cases where the observer's choice has influenced the selection of the sample, Yates quotes a case from the observations of the Crop-Weather scheme of the Ministry of Agriculture. Under this scheme, records are taken by sampling of the growth of the wheat-plant at ten stations. The sampling-unit consists of a quarter-metre of each of four consecutive rows and is marked off for observation by inserting a U-shaped rod, whose arms are parallel and a quarter of a metre apart, close to the ground and at right angles to the rows.

Measurements of shoot-height are made only on the two shoots nearest to each end of each row of a sampling-unit, so that there are eight measurements per sampling-unit. At one of the centres, only three rows were available for sampling, and in order to make up the full eight measurements the observer selected the two additional shoots 'at random' by eye. Comparison of the two additional observations with the six regular ones showed relatively large biases in the former. Of these Yates writes, 'On May 31, when the wheat was under 2 ft. high, there is apparently a simple tendency to select the taller shoots, with perhaps a special preference for the very tall shoots. On June 14, this tendency, though much less, is still noticeable, but there is now a tendency to avoid very high values. On June 28, when the wheat was about 4 ft. high, there is a marked avoidance of the tallest shoots, and also to a less extent of the shortest shoots. The general bias is now strongly negative.' And later, 'The biases are of course of the type that might be expected. When the shoots are low and there is nothing much to be seen except the top leaves, there will be a tendency to pick the higher shoots, but when they have come into ear the observer can see the shoots of all heights, and is more likely to select shoots somewhere near the average, omitting both very high and very low values. The strong negative biases of the last set of measurements show that this selection was not particularly effective in improving the accuracy of the sample.'

At a conference of the observers of the Crop-Weather Scheme, held at Rothamsted in June 1935, the opportunity was taken of carrying out an experiment to see to what extent these biases are common to all observers who are making deliberate selection. The object of this note is to describe the experiment and its results.

For the experiment, six sampling-units were marked out, and the height of every shoot in each was measured. The observers were told to measure the height of two shoots in each row as usual, but instead of picking the end shoots, they were to select any two in the row which they pleased, so as to give what they considered to be a random sample of the shoot-heights in the sampling-units. There were twelve observers, who were divided into two groups of six. The members of the same group took their observations at the same time, but the order in which they made their selections from the various sampling-units was fixed by means of a $6 \times 6$ Latin-square arrangement, so that no two observers were measuring the same sampling-unit together. There were about 20 shoots per row within a sampling-unit, and the experiment was made about a week before ear-emergence, the wheat being 70 cm. high.

### The Biases in Mean Height

The means of the 48 measurements on shoot-height for all the observers are shown in Table 1 below:

TABLE 1. *Observers' Estimates of Mean Shoot-height*

| Group I | | Group II | |
|---|---|---|---|
| Observer | Height | Observer | Height |
| | cm. | | cm. |
| R. W. K. . | 76·79 | R. M. H. . | 77·87 |
| G. A. T. . | 76·31 | A. J. G. B. | 76·14 |
| W. R. . | 76·87 | F. H. . . | 77·19 |
| M. E. B. N. . | 77·62 | A. J. M. . | 76·08 |
| A. F. H. . | 73·42 | B. J. T. . | 77·77 |
| H. M. . . | 71·96 | I. Z. . . | 72·77 |

True mean height = 70·75 cm.

In every case the estimate was higher than the true value and in all but three cases the bias was considerable, amounting to about 8 per cent.

The difference of each shoot-height selected by an observer from the mean height of the row in which the shoot lay was calculated. The mean differences per sampling-unit for the members of the first group are shown in Table 2:

TABLE 2. *Mean Differences in Shoot-heights*: Group I

| Observer | Sampling-unit | | | | | | Total |
|---|---|---|---|---|---|---|---|
| | *1* | *2* | *3* | *4* | *5* | *6* | |
| R. W. K. . | +5·88[6] | +7·18[5] | +6·68[1] | +4·48[2] | +5·10[4] | +6·95[3] | +36·27 |
| G. A. T. . | +8·88[2] | +4·18[1] | +4·55[4] | +3·72[3] | +3·48[5] | +8·58[6] | +33·39 |
| W. R. . | +9·62[3] | +10·18[4] | +4·05[5] | +2·98[6] | +4·10[1] | +5·82[2] | +36·75 |
| M. E. B. N. | +10·50[4] | +7·55[6] | +5·82[2] | +6·60[1] | +5·98[3] | +4·95[5] | +41·40 |
| A. F. H. . | +3·12[5] | +3·68[3] | −0·70[6] | +3·72[4] | +2·36[2] | +3·82[1] | +16·00 |
| H. M. . | +3·50[1] | +1·92[2] | −2·70[3] | −3·28[5] | +3·98[6] | +3·82[4] | +7·24 |
| Mean | +6·92 | +5·78 | +2·95 | +3·04 | +4·17 | +5·66 | |

The small figures indicate the order of the observations.

*Time Means*

| Time: | 1 | 2 | 3 | 4 | 5 | 6 |
|-------|-----|-----|-----|-----|-----|-----|
| Mean: | +4·81 | +4·88 | +4·54 | +6·31 | +3·25 | +4·71 |

There appear to be fairly large differences in the biases from observer to observer and from sampling-unit to sampling-unit. This is brought out by the analysis of variance on mean differences, which is shown below for the members of Group I. The analysis for Group II is in all respects similar.

*Analysis of variance: mean differences per sampling-unit*

|  | Degrees of freedom | Sum of squares | Mean square |
|---|---|---|---|
| Observers . . . . | 5 | 154·4 | 30·87 |
| Times . . . . . | 5 | 28·54 | 5·71 |
| Sampling-units: |  |  |  |
| Regression on mean shoot-height | 1 | 59·91 | 59·91 |
| Deviations from regression . | 4 | 18·66 | 4·66 |
| Remainder . . . . | 20 | 66·61 | 3·22 |
| Error |  |  |  |
| Random sampling variation . | 288 | 1,684·44 | 5·85 |

It will be noticed that the 20 degrees of freedom which remain after removing the sums of squares due to observers, times, sampling-units, and the mean have been called 'remainder' and not 'error'. The reason for this is that although these 20 degrees of freedom contain a component representing the sampling variation of the observers, which is the true error, they are also influenced by interactions between observers and sampling-units and between observers and times, if these exist. Thus if the difference in bias between observers were much greater in some sampling-units than in others, this would inflate the 'remainder'. In the absence of anything better, the 'remainder' would have been used as error, but as it happens, we can get a proper estimate of the sampling variation of observers from the differences of the two samples taken per row. This is shown in the last line of the table and has 288 degrees of freedom. It will be seen that the 'remainder' mean square is actually lower than the error mean square, so that the interactions mentioned above appear to be negligible.

Comparing the mean squares of the various items with the error mean square, we see that the differences in the sizes of the biases of different observers are significant. This is quite to be expected from previous experience. The size of the bias was not, however, affected by the order in which the sampling-units were measured, for time-effects are not significant, nor is there any sign in the times-means given in Table 2 of a regular change in the bias from the first sampling-unit measured to the last.

The size of the bias varies significantly from sampling-unit to sampling-unit, and the question arises, to what is this due? In the example by

Yates quoted above, it was shown that the bias in selection was correlated with the true mean shoot-height, being positive when the shoots were low and negative when they were high. A comparison of the mean biases per sampling-unit with the true mean shoot-heights (shown in Table 3) indicates that a similar relation is present here also.

TABLE 3. *Mean Biases and Mean Shoot-heights*

| Sampling-unit | Mean bias | | True mean Shoot-height |
|---|---|---|---|
| | Group I | Group II | |
| 1 | 6·92 | 8·12 | 59·0 |
| 2 | 5·78 | 7·63 | 66·2 |
| 3 | 2·95 | 2·59 | 76·4 |
| 4 | 3·04 | 5·14 | 74·5 |
| 5 | 4·17 | 4·45 | 76·0 |
| 6 | 5·16 | 5·20 | 72·3 |

In the analysis of variance, the linear regression of the mean differences on the true mean shoot-height was taken out. The regression is highly significant, and the mean square of the deviations from the regression line is slightly below the error mean square. The variation between sampling-units, apart from that due to error, may thus be accounted for by the linear regression of the bias on mean shoot-height. The positive bias in the height of a shoot increases by about 0·2 cm. for each cm. decrease in the true shoot-height. This provides a striking illustration of the errors to which human judgement may be subject.

### The Distribution of the Sampled Shoots

The distributions of the differences of the heights of the sampled shoots and of all shoots from the means of the rows in which they lay are shown on the same scale by the unshaded and shaded portions respectively of the histogram in Fig. 1. The positive bias already referred to in the sampled observations is clearly shown. A striking feature of the comparison is the much smaller variability in the differences of the sampled shoots, the sampled shoot-heights being much more closely grouped about the mean. Shoots which were smaller than the mean by more than 20 cm. are scarcely represented at all in the samples. This confirms the result found in previous investigations that observers, when given freedom of choice, select samples which conform much too closely to the average, even when they are instructed to make their samples representative of all the material which is being sampled.

The distributions of the differences of the sampled shoot-heights were calculated separately for the centre and outside rows, to see whether the easier accessibility of the two outside rows made any difference to the type of sample which was taken. The results are shown below in Table 4, the two distributions together forming that represented in the unshaded histogram in Fig. 1.

There is no marked difference between the two distributions. That from the outside rows shows, however, a slightly greater variability, having rather more observations in the two extreme groups at either end of the scale. The fraction of observations in these four groups was not, however, significantly higher for the outside rows.

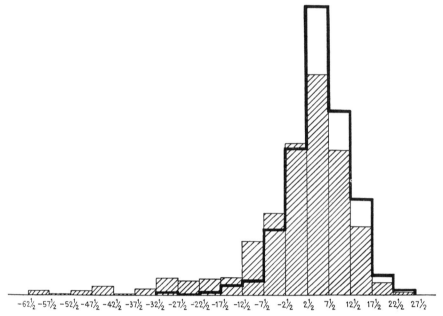

$-62\frac{1}{2}$ $-57\frac{1}{2}$ $-52\frac{1}{2}$ $-47\frac{1}{2}$ $-42\frac{1}{2}$ $-37\frac{1}{2}$ $-32\frac{1}{2}$ $-27\frac{1}{2}$ $-22\frac{1}{2}$ $-17\frac{1}{2}$ $-12\frac{1}{2}$ $-7\frac{1}{2}$ $-2\frac{1}{2}$ $2\frac{1}{2}$ $7\frac{1}{2}$ $12\frac{1}{2}$ $17\frac{1}{2}$ $22\frac{1}{2}$ $27\frac{1}{2}$

FIG. 1. Distribution of sampled shoots (unshaded) and of all shoots (shaded).

It has already been pointed out that the observers' samples showed a much smaller amount of variation than the population from which they were sampling. A more detailed examination was made of the variation in the observers' samples. For each observer there are eight differences per sampling-unit and an estimate of the variation between these may be made by taking the sum of the squares of the deviations

TABLE 4. *Distribution of the Observers' Differences in Centre and Outside Rows*

| Group | $-32\frac{1}{2}-27\frac{1}{2}$ | $-27\frac{1}{2}-22\frac{1}{2}$ | $-22\frac{1}{2}-17\frac{1}{2}$ | $-17\frac{1}{2}-12\frac{1}{2}$ | $-12\frac{1}{2}-7\frac{1}{2}$ | $-7\frac{1}{2}-2\frac{1}{2}$ |
|---|---|---|---|---|---|---|
| Number of observations: | | | | | | |
| Centre | 0 | 0 | 2 | 5 | 3 | 25 |
| Outside | 2 | 0 | 0 | 2 | 7 | 20 |

| Group | $-2\frac{1}{2}+2\frac{1}{2}$ | $+2\frac{1}{2}+7\frac{1}{2}$ | $+7\frac{1}{2}+12\frac{1}{2}$ | $+12\frac{1}{2}+17\frac{1}{2}$ | $+17\frac{1}{2}+22\frac{1}{2}$ | $+22\frac{1}{2}+27\frac{1}{2}$ | Total |
|---|---|---|---|---|---|---|---|
| Number of observations: | | | | | | | |
| Centre | 43 | 110 | 60 | 35 | 5 | 0 | 288 |
| Outside | 58 | 89 | 67 | 31 | 9 | 3 | 288 |

of these differences from the mean differences and dividing by seven. An analysis of variance was performed on these estimates of variance in the same way as that already described on mean differences. In both groups of observers the differences in variability between different observers were significant. The variability did not, however, differ significantly, or indeed at all abnormally, from sampling-unit to sampling-unit or from time to time. The observers' estimates of the variability within a sampling-unit lay between 16·0 and 93·4. The correct value from the whole set of shoots was 187·6. Thus even the best of the observers in this respect estimated the variance at less than half its true value, whereas the worst obtained only about a twelfth of the true value. It is worth noting that the three observers who made the best estimates of mean shoot-height were amongst those who made the best estimates of the variability.

## The Validity of the Sampling Process actually Used

The sampling process actually used consists, as mentioned above, in taking the two shoots nearest to the ends of the rows. The location within the sampling-unit of the shoots which are measured is certainly free from any element of choice on the part of the observer. The eight shoots which are measured do, however, always occupy the same positions within the sampling-unit, and it is of interest to see whether in this case they formed a representative sample of the shoot-heights in the six sampling-units. One reason on general grounds for doubting this is that with a patchy growth the outside shoots of a plant have a greater chance of being selected than the inner shoots, owing to the presence of a blank space in the row next to the plant. In the present case the growth was fairly even.

We may examine the heights of the 48 end shoots in the same way as the observers' samples were examined, and consider whether the former give an unbiased estimate of the true mean shoot-height and a reasonably good representation of the distribution of shoot-heights.

The mean of the 48 end shoot-heights was 71·375 cm., the true mean being 70·75 cm. The difference is +0·625 cm., its standard error, calculated from the 48 deviations, being ±1·987 cm. The difference is well within its standard error, so that there is no sign of bias in the estimate of mean shoot-height.

The distributions of the differences of the end shoots and of all shoots from the row means are shown by the unshaded and shaded histograms in Fig. 2. The unshaded histogram shows no sign of abnormality and could reasonably be regarded as a random sample from the whole distribution. In particular, the estimate from the end shoots of the variance with a sampling-unit, calculated in the same way as for the observers' samples, comes to 174·4, which agrees well with the true value of 187·6.

The evidence of this experiment thus provides no reason to doubt the validity of the sampling process which is being used for shoot-heights in the Wheat Sampling Observations.

## Conclusions

It is obvious that samples that are picked by a process of randomization which gives every sample in the population an equal chance of being picked, must be representative of the population from which they are drawn and give an unbiased estimate of the quantity which it is desired to measure. Those who have little experience of the technique of sampling might, however, be unwilling to admit that they could not do as well, or better, by choosing the samples themselves. In this experiment, out of twelve observers, all of whom have had some train-

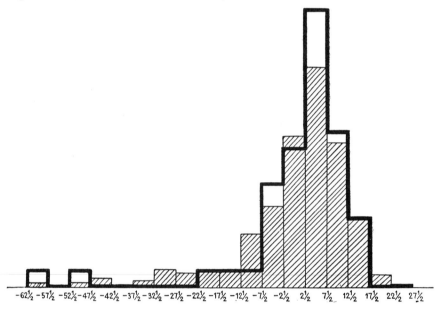

FIG. 2. Distribution of end shoots (unshaded) and of all shoots (shaded).

ing in sampling, not one managed to pick a sample that could be called representative of the material from which they were sampling, and all except three obtained relatively large biases in their estimate of the quantity, mean shoot-height, which was being measured. Further, the biases, both in mean shoot-height and sampling variance, showed large differences from observer to observer. What is even more serious and striking is that the individual observers were not consistent throughout the experiment; the positive bias in selection increased regularly as the mean height of the sampling-unit decreased.

This experiment, in short, very strongly supports the evidence from other investigations that the only sure method of avoiding bias is for the sampling to be random. The plea has sometimes been made that observer's bias is not important provided that the same observer does all the sampling. The answer to that in the present case is that there is no excuse for bias at all in such a simple problem as the estimation of the shoot-height of a field. In many sampling problems, however, particularly in sampling from bulk, it is much more difficult or troublesome to obtain a proper random sample, and for this reason the plea merits some investigation.

In the first place the plea is based on the assumption that the observer's bias remains constant. Neither the present example nor the one discussed by Yates supports this view; in both, the positive bias in the estimate of mean shoot-height decreased regularly as the true mean height increased. Thus the observer's estimates of differences between the mean shoot-heights of different varieties in the field would be biased. The true mean shoot-heights on this date (June 17) and the observer's estimates are shown below for two of the varieties grown at Rothamsted:

| Variety | True values cm. | Observer's estimates cm. |
|---------|-----------------|--------------------------|
| Victor . . | 70·7 | 75·8 |
| Yeoman . . | 63·7 | 70·5 |
| Difference . | 7·0 | 5·3 |

Victor was the variety used in the experiment. The observer's estimate for Yeoman was calculated from the regression of observers' bias on true mean shoot-height obtained above. The true difference in height, 7·0 cm., would have been under-estimated by 24 per cent. A similar case occurs in the example discussed by Yates [1], in which the growth-rate would have been under-estimated by 9 per cent.

Even granting that an observer might have a constant bias, the uses to which his results can be put are very limited. A constant bias would give a correct estimate of differences, but almost all other estimates based on his figures, such as for example percentage differences or regression coefficients, would be biased in different ways. No one can foresee, when making observations, all the estimates or comparisons which may be made from them at some future date. Further, the comparison of one observer's work with that of another will be vitiated by the fact that they will have different personal biases; and this difficulty is almost certain to arise in any work carried on at more than one place, or for any length of time at the same place. In short, the presence of observers' bias in sampling results greatly detracts from the value of the results, and one of the most important problems in the application of statistical methods to agriculture and industry is to devise reasonably quick methods of taking a proper random sample in cases where the material sampled is difficult to demarcate or handle.

### REFERENCE

1. F. YATES. Some Examples of Biased Sampling. Annals of Eugenics, 1935, 6, Pt. II.

(*Received November 5, 1935*)

# 6

# THE $\chi^2$ DISTRIBUTION FOR THE BINOMIAL AND POISSON SERIES, WITH SMALL EXPECTATIONS

By W. G. COCHRAN, B.A.

*Rothamsted Experimental Station*

## § 1. Introduction

ONE of the more common uses of the $\chi^2$ distribution is to compare the variance in a sample of $N$ from the binomial series distribution $(p+q)^n$, with the expected population variance $n\hat{p}\hat{q}$, where $\hat{p}$ and $\hat{q}$ are estimated from the sample. A single member of the binomial series distribution (i.e. a sample of one) is taken to mean the number of successes in $n$ independent trials, in each of which the chance of success is $p$. A sample of $N$ members may be written $x_1, x_2, ..., x_N$, where any $x$ is an integer between 0 and $n$. For large expectations it is known, cf. (1), that

$$\Sigma \frac{(x-\bar{x})^2}{n\hat{p}\hat{q}} = \Sigma \frac{n\,(x-\bar{x})^2}{\bar{x}\,(n-\bar{x})}$$

is distributed as $\chi^2$ with $(N-1)$ degrees of freedom.

The quantity $\dfrac{\Sigma n\,(x-\bar{x})^2}{\bar{x}\,(n-\bar{x})}$ has been called the index of dispersion for the binomial series. The following is an example of its use in tests of significance. In a study of the distribution of diseased plants in a field, it is of interest to know whether diseased plants tend to congregate in patches, an effect which might occur, for instance, if the disease were liable to spread from one diseased plant to its neighbours. This point may be studied by dividing the field into small areas, each containing say twenty plants, and calculating the total number $x$ of diseased plants in each area. If diseased plants are grouped together there will be more areas with many diseased plants or with very few diseased plants than would occur in random sampling from the binomial series. The index of dispersion will thus be larger than expectation and the $\chi^2$ test may be used to detect the grouping of diseased plants.

I recently recommended(2) the use of this test in the study of the distribution of diseased plants in a field or greenhouse in cases in which the expectations were likely to be small, perhaps as low as 1 or 2. It is not known in these cases how satisfactory a representation of the distribution of

$$\Sigma \frac{(x-\bar{x})^2}{n\hat{p}\hat{q}}$$

can be obtained from the ordinary $\chi^2$ distribution, and this paper was originally intended as a discussion, with some numerical examples, of the agreement between the true and the $\chi^2$ distribution in this case. Some notes have also been added on the general question of the use of $\chi^2$ as a test of discrepancies between observation and hypothesis with small expectations, and, while no essentially new points have arisen, the importance of the subject perhaps justifies a brief discussion.

## § 2. Some examples of the exact distribution of $\chi^2$

The distribution of $\chi_s^2$, defined as $\dfrac{\Sigma n\,(x-\bar{x})^2}{\bar{x}\,(n-\bar{x})}$ for the binomial series, depends on three variables, the size $N$ of the sample, the index $n$ of the binomial, and the expectation $m=np$ of success. Of these $m$ and $N$ are the most important, and for fixed $m$ and $N$ the distribution changes slowly and regularly as $n$ increases from 1 to $\infty$, in which case the distribution becomes the Poisson series with mean $m$. This property was used by Yates[3] in his discussion of the $2\times 2$ contingency table and appears to hold quite generally. When $N$ is large, $m$ and $n$ remaining small, the distribution of $\chi_s^2$, like that of the tabular $\chi^2$, tends to normality, though not with quite the same mean or variance as $\chi^2$. The normal approximation to $\chi_s^2$ will be given in the next section and can be used in tests of significance when appropriate.

When $N$ and $m$ are small, the exact distribution of $\chi^2$ may be found in any particular case without much computation. It is first necessary to find the probability of obtaining any given configuration.

Let $x_1, x_2, \ldots, x_N$ be the numbers of successes out of $n$ in that order in a sample of $N$ values of the binomial series distribution, where

$$T=x_1+x_2+\ldots+x_N$$

The probability of obtaining such a sample is

$$P=\prod_{i=1}^{N} \tbinom{n}{i}\, p^{x_i}\, q^{n-x_i}$$

$$=\frac{(n\,!)^N\, p^T\, q^{nN-T}}{\Pi x_i!\,(n-x_i)!}\,,$$

and $\Sigma P=1$, taken over all possible samples.

In estimating $p$ we make the population total coincide with the sample total $T$, and the frequency distribution of $\chi^2$ is accordingly to be taken only over the set of samples with total $T$. To find the probabilities in this set it is necessary to divide $P$ by $\Sigma P$ taken over the members of the set. But this is the probability of obtaining $T$ successes in a single trial of the binomial $(p+q)^{nN}$, and is therefore

$$\frac{(nN)\,!}{T!\,(nN-T)!}\, p^T q^{nN-T}.$$

Hence the probability of drawing the sample $x_1, x_2, \ldots, x_N$ in that order out of the set with total $T$ is

$$P'=\frac{(n\,!)^N\, T!\,(Nn-T)!}{(Nn)!\,\Pi x_i!\,(n-x_i)!}.$$

If the sample is regarded as representing a $2 \times N$ contingency table, with marginal totals $T$, $Nn - T$, and $N$ $n$'s, the probability is seen to be the product of the factorials of the marginal totals divided by the product of the factorials of the grand total and the individual cell numbers. This result holds generally for contingency tables, as was pointed out by Yates ((3), p. 233).

The corresponding result for a sample of $N$ from the Poisson series is obtained by making $n$ tend to infinity. This gives

$$P' = \frac{T!}{\Pi x_i!}.$$

In practice what is wanted is the probability of the configuration $(x_1, \ldots, x_N)$, irrespective of the order in which they come. If $x_1, \ldots, x_N$ were all distinct, this would be $N! P'$, and if there are $a_1$ 0's, $a_2$ 1's, etc., the probability is

$$\frac{N!}{\Pi a_i!} P'.$$

Thus the probability of obtaining $a_1$ $x_1$'s, $a_2$ $x_2$'s, etc., in a sample of $N$ is

$$\frac{N! \, (n!)^N \, T! \, (Nn - T)!}{(Nn)!} \times \frac{1}{\Pi \left[ \{(x_i)! \, (n - x_i)!\}^{a_i} (a_i)! \right]}.$$

In any particular case in which the exact distribution of $\chi_s^2$ is required, only the second part is variable. The practical procedure is first to enumerate all possible configurations. For each configuration there is a probability and a value of $\chi_s^2$. By forming a table of the values of $\{(x_i)! \, (n - x_i)!\}$ for $x_i = 0, 1, 2, \ldots, n$ and using logs the work may be reduced to a minimum.

In cases where the expectation $m$ ($= np$) is not too small, so that the agreement between the $\chi_s^2$ and the $\chi^2$ distribution is fairly close, the chief source of discrepancy is that the former is a discontinuous distribution and the latter continuous. The true values of $\chi_s^2$ are, however, always evenly spaced, apart from values with zero probability, since the values of $\Sigma (x - \bar{x})^2$ change by 2's when arranged in increasing order, and a correction for continuity may easily be made by taking $\chi^2$ less by half an interval than the value which is being tested. For low values of $m$, however, other sources of disturbance have become important, and the continuity correction does not necessarily improve the agreement in the region in which the probability lies between 0·1 and 0·01. For instance, the $\chi_s^2$ distribution has a finite range, while that of $\chi^2$ is from 0 to $\infty$, and this introduces a tendency for $\chi^2$, when corrected for continuity, to overestimate the true probabilities near the end of the tail. At the other end of the distribution, which is of less practical importance, the $\chi^2$ distribution seems always to overestimate the probability of obtaining so good an agreement by chance.

The cases $m = 2$, $N = 4$ are shown in Table I for $n = 8$ and $n = \infty$. There are only 15 distinct configurations, each of which gives a separate value of $\chi_s^2$. The probabilities shown are those of obtaining a value of $\chi_s^2$ at least as great as that given.

Table I

| $\chi_s{}^2$ | $P$ | $P'$ tabular $\chi^2$ | $P'$ corrected for continuity | $\chi_s{}^2$ | $P$ | $P'$ tabular $\chi^2$ | $P'$ corrected for continuity |
|---|---|---|---|---|---|---|---|
| \multicolumn{4}{}{$n=8$} | | | | \multicolumn{4}{}{$n=\infty$ (Poisson)} | | | |
| 0 | 1·000 | 1·000 | 1·000 | 0 | 1·000 | 1·000 | 1·000 |
| 1·3 | 0·942 | 0·722 | 0·881 | 1 | 0·962 | 0·801 | 0·919 |
| 2·6 | 0·541 | 0·446 | 0·573 | 2 | 0·654 | 0·573 | 0·682 |
| 4·000 | 0·426 | 0·262 | 0·343 | 3 | 0·551 | 0·392 | 0·476 |
| 5·3 | 0·183 | 0·149 | 0·198 | 4 | 0·295 | 0·262 | 0·321 |
| 6·6 | 0·120 | 0·0834 | 0·112 | 5 | 0·218 | 0·172 | 0·213 |
| 8·000 | 0·0489 | 0·0461 | 0·0620 | 6 | 0·116 | 0·112 | 0·139 |
| 9·3 | 0·0380 | 0·0252 | 0·0341 | 7 | 0·0951 | 0·0720 | 0·0898 |
| 10·6 | 0·0094 | 0·0137 | 0·0186 | 8 | 0·0336 | 0·0461 | 0·0576 |
| 12·000 | 0·0066 | 0·0074 | 0·0101 | 9 | 0·0272 | 0·0293 | 0·0368 |
| 14·6 | 0·0030 | 0·0021 | 0·0040 | 11 | 0·0169 | 0·0117 | 0·0186 |
| 16·0 | 0·0010 | 0·0011 | 0·0016 | 12 | 0·0067 | 0·0074 | 0·0093 |
| 22·6 | 0·0001 | 0·0001 | — | 17 | 0·0015 | 0·0007 | 0·0023 |

The two highest values of $\chi^2$ have been omitted in both distributions. In the region below 0·1, the ordinary distribution shows no consistent tendency either to overestimate or underestimate, while the corrected distribution generally overestimates the probability and gives no better agreement on the whole. The maximum discrepancies in the probabilities in the region used in tests of significance are of the order of 40 per cent; whether this is considered satisfactory will depend on the standard of accuracy required.

Comparing the Poisson distribution with the binomial $n=8$, it will be seen that the interval between successive values of $\chi_s^2$ has shortened by one-third in the former case, while the probabilities of the values themselves are less for the Poisson series than the binomial with low values of $\chi_s^2$ and greater with high values. This appears to be the general type of change in the distribution as $n$ tends to infinity. The agreement between the $\chi_s^2$ and ordinary $\chi^2$ distributions is of the same order in both cases.

Table II below gives a summary of the results obtained in this and four other examples. The maximum percentage discrepancies, both positive and negative, in the probabilities in the range $0·1 \geqslant p \geqslant 0·005$ are shown for the uncorrected and the corrected distributions.

Table II

| $m$ | $N$ | $n$ | Number of values of $\chi_s{}^2$ | Ordinary $\chi^2$ + | Ordinary $\chi^2$ − | Corrected $\chi^2$ + | Corrected $\chi^2$ − |
|---|---|---|---|---|---|---|---|
| \multicolumn{4}{}{} | | | | \multicolumn{4}{}{Maximum percentage discrepancy} | | | |
| 3·0 | 6 | 6 | 17 | 12 | 21 | 48 | — |
| ⎧2·0 | 4 | 8 | 13 | 46 | 34 | 98 | 10 |
| ⎩2·0 | 4 | ∞ | 13 | 37 | 31 | 71 | 6 |
| 1·8 | 10 | 9 | 27 | 4 | 9 | 33 | — |
| 1·8 | 5 | 9 | 14 | 9 | 21 | 44 | — |
| 0·9 | 10 | 9 | 17 | 11 | 30 | 70 | — |

The numbers of values of $\chi_s^2$ given are those which account for all but 0·0001 of the total probability. The examples taken are all extreme cases, the total number of values of $\chi_s^2$ being small, and were all computed without much labour.

For the uncorrected $\chi^2$ distribution the largest absolute percentage deviation is negative in four out of the five cases, indicating that the uncorrected $\chi^2$ distribution tends to underestimate the probability of a discrepancy in this range. The corrected distribution, on the other hand, definitely overestimates; in only one case, $m = 2$, $N = 4$, was there any underestimation at all. The absolute percentage discrepancies are in all cases considerably larger for the corrected than for the ordinary $\chi^2$ distribution. The agreement between the true and the uncorrected $\chi^2$ distribution is, in fact, surprisingly good; the maximum discrepancies are under 25 per cent except in the cases in which $m = 0·9$ or $N = 4$.

The conclusion from this very limited examination of the possible range of variation of $m$, $N$ and $n$ is the reassuring one that the tabular $\chi^2$ distribution will give a satisfactory approximation to the true distribution in the region used in tests of significance, except in those cases in which the number of possible configurations is so small that it is easy to calculate the true distribution. The only exception to this rule seems to be the case in which $m$ and $n$ are small but $N$ is large; this will be dealt with in the next section.

## § 3. THE NORMAL APPROXIMATION TO THE $\chi_s^2$ DISTRIBUTION

The notation used in this section will be that adopted by Fisher[4]. In this $k_r$ is the consistent estimate from the sample of the $r$th semi-invariant $\kappa_r$, and $\kappa$ ($2^2 1^3$), for instance, is $\kappa_{23}$ of the joint distribution of $k_2$ and $k_1$. In this notation $\chi_s^2 = \dfrac{(N-1)\, k_2}{\kappa_2}$.

When $N$ is large, the distribution of $\chi_s^2$ tends to normality, and tests of significance may be made when the mean and variance are known. If we are sampling from a binomial distribution with known $p$,

$$\text{Mean } (\chi_s^2) = (N-1),$$

$$V(\chi_s^2) = \frac{(N-1)^2}{\kappa_2^2}\, V(k_2) = \frac{(N-1)^2}{\kappa_2^2}\, \kappa\,(2^2)$$

$$= \frac{(N-1)^2}{\kappa_2^2} \left\{ \frac{2\kappa_2^2}{N-1} + \frac{\kappa_4}{N} \right\}$$

$$= 2\,(N-1) \left\{ 1 + \frac{(N-1)}{2N}\, \frac{\kappa_4}{\kappa_2^2} \right\} \qquad\qquad \text{......(1)}$$

$$= 2\,(N-1) \left\{ 1 + \frac{(N-1)\,(1-6pq)}{2N}\, \frac{}{npq} \right\} \qquad\qquad \text{......(2).}$$

This is less than the variance $2\,(N-1)$ of the ordinary $\chi^2$ distribution if $1 - 6pq < 0$, i.e. if $0·21 \leqslant p \leqslant 0·79$. It is the appropriate variance to use in the normal approximation where the value of $p$ in the binomial series is known and does not have to be estimated.

Where the value of $p$ has to be estimated from the sample, as is usually the case, the variance and the mean value of $\chi_s^2$ are required in that array in which the mean of the sample is equal to the true mean of the population. The restriction on the mean would not affect the distribution of $\chi_s^2$ if the latter were independent of the distribution of the mean. This is the case for the normal distribution but not for the binomial series.

The moments of the distribution of $\chi_s^2$ in arrays in which the mean is fixed may be obtained to any order of approximation by a method similar to that used by K. Pearson [5], which is of wide application.

If $x$ and $y$ are two correlated variates referred to the means of their distributions, their moment generating function

$$M = \int\int e^{t_1 y + t_2 x} f(xy) \, dx \, dy$$

$$= 1 + \frac{t_1^2}{2!} \mu_{20} + \frac{t_1}{1!} \frac{t_2}{1!} \mu_{11} + \frac{t_2^2}{2!} \mu_{02} + \dots \qquad \dots\dots(3).$$

Integrating first with respect to $y$ we have

$$M = \int e^{t_2 x} \, dx \, g_0(x) \left[ 1 + g_1(x) t_1 + g_2(x) \frac{t_1^2}{2!} + \dots \right],$$

where $g_1(x)$, $g_2(x)$, ... are the moments of the distribution of $y$ in arrays in which $x$ is fixed. Expanding $e^{t_2 x}$ we get

$$M = \int g_0(x) \, dx \left[ 1 + x t_2 + x^2 \frac{t_2^2}{2!} + \dots \right] \left[ 1 + g_1(x) t_1 + g_2(x) \frac{t_1^2}{2!} + \dots \right] \qquad \dots\dots(4).$$

By comparing coefficients of powers of the $t$'s in (3) and (4) the moments of the functions $g_0(x) \, g_r(x)$ may be obtained. The general expression is

$$\mu_{rs} = \int x^s \, g_0(x) \, g_r(x) \, dx \qquad \dots\dots(5).$$

Approximations will be obtained to $g_1(x)$ and $g_2(x)$ which are polynomials in $x$. Let $\psi_0(x)$, $\psi_1(x)$, ... be a set of polynomials in $x$ of degree 0, 1, ..., etc., which are orthogonal to $g_0(x)$, i.e. such that

$$\int g_0(x) \, \psi_r(x) \, \psi_s(x) \, dx = 0 \quad \text{if } r \neq s.$$

It is easy to see that if the coefficient of $x^r$ in $\psi_r$ is taken to be unity

$$\psi_0 = 1, \quad \psi_1 = x, \quad \psi_2 = x^2 - \frac{\mu_{03}}{\mu_{02}} x - \mu_{02}.$$

Write
$$g_1(x) = a_0 \psi_0 + a_1 \psi_1 + a_2 \psi_2.$$

Then
$$0 = \int g_0 g_1 \psi_0 \, dx = a_0 \int g_0 \psi_0^2 \, dx = a_0.$$

Since we want $g_1(0)$, the value of $a_1$ need not be found.

For $a_2$, $$\int g_0 g_1 \left( x^2 - \frac{\mu_{03}}{\mu_{02}} x - \mu_{02} \right) dx = a_2 \int \left( x^2 - \frac{\mu_{03}}{\mu_{02}} x - \mu_{02} \right)^2 dx,$$

i.e. $$\mu_{12} - \frac{\mu_{03}}{\mu_{02}} \mu_{11} = a_2 \left( \mu_{04} - \frac{\mu_{03}^2}{\mu_{02}} - \mu_{02}^2 \right) = a_2 \mu_{02}^2 \left( 2 + \gamma_2 - \gamma_1^2 \right),$$

where $\gamma_1$, $\gamma_2$ are Fisher's measures of departure from normality for the distribution of $x$. If $y = k_2$ and $x = k_1$, and all symbols refer to the original distribution, this becomes

$$\kappa \left( 21^2 \right) - \frac{\kappa \left( 1^3 \right)}{\kappa (1^2)} \kappa (21) = a_2 \kappa (1^2) \left( 2 + \frac{\gamma_2 - \gamma_1^2}{N} \right),$$

i.e. $$\frac{1}{N} \left( \kappa_4 - \frac{\kappa_3^2}{\kappa_2} \right) = a_2 \kappa_2 \left( 2 + \frac{\gamma_2 - \gamma_1^2}{N} \right).$$

The deviation of the mean of the array $x = 0$ from the true mean of the whole distribution is to this order of approximation

$$a_2 \psi_2 (0) = - \mu_{02} a_2$$

$$= \frac{\kappa_2}{N} \frac{(\gamma_1^2 - \gamma_2)}{\left( 2 + \frac{\gamma_2 - \gamma_1^2}{N} \right)}.$$

Hence the deviation in $\chi_s^2$ is $\left( \text{neglecting terms in } \dfrac{1}{N^2} \right)$

$$\tfrac{1}{2} (\gamma_1^2 - \gamma_2) \left[ 1 - \frac{1}{N} \left\{ 1 - \frac{\gamma_1^2 - \gamma_2}{2} \right\} \right] \qquad \dots\dots(6).$$

For the binomial series, $$\gamma_1^2 = \frac{(q-p)^2}{npq}, \qquad \gamma_2 = \frac{1-6pq}{npq}.$$

$$\gamma_1^2 - \gamma_2 = \frac{2}{n}.$$

$$\therefore \text{ Mean} = (N-1) + \frac{1}{n} \left[ 1 - \frac{1}{N} \left( 1 - \frac{1}{n} \right) \right] \qquad \dots\dots(7).$$

This formula has been found to give a good approximation with values of $N$ as low as 4. The large sample approximation is

$$(N-1) + \frac{1}{n} \qquad \dots\dots(8).$$

For the Poisson series the mean is $(N-1)$ for all values of $N$.

The same method may be used to obtain an approximation to $g_2 (x)$. The expression is rather complicated, if terms of order $N^0$ are retained, and involves semi-invariants up to $\kappa_6$. The approximation to order $N$ for the variance is

$$V = 2 (N-1) \left[ 1 - \frac{\gamma_1^2 - \gamma_2}{2} \right] \qquad \dots\dots(9).$$

This expression could be obtained alternatively by noting that for the bivariate normal distribution the variance of $y$ in arrays in which $x$ is fixed is $\sigma_y^2 (1 - \rho^2)$. For the binomial series the variance is

$$V = 2 (N-1) \left( 1 - \frac{1}{n} \right) \qquad \dots\dots(10).$$

Thus for the Poisson series, the variance of the $\chi_s^2$ distribution is equal to that of the ordinary $\chi^2$ distribution. This result, however, unlike that for the mean, holds only when $N$ is large.

As an example, the exact $\chi_s^2$ distribution for the case $n=2$, $N=160$, $m=0.4$ was worked out. There are only 24 possible configurations. The formula above gives $M=159.5$, $V=159$, the tabular $\chi^2$ distribution gives $M=159.0$, $V=318$, and the exact values were $M=159.49$, $V=159.28$. The normal approximation, corrected for continuity, gives results correct to

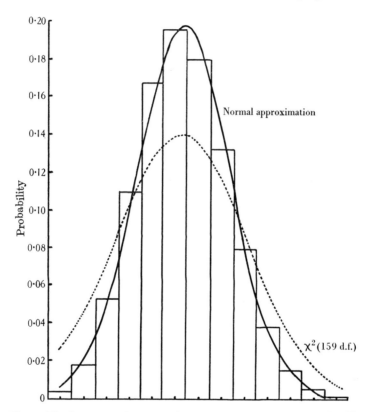

Fig. 1. The histogram shows the $\chi_s^2$ distribution for $m=0.4$, $n=2$, $N=160$.

within 15 per cent in the region used for tests of significance. The three distributions are shown in Fig. 1, the exact $\chi_s^2$ distribution being given as a histogram. It will be seen that the chief cause of discrepancy between the $\chi_s^2$ distribution and the normal approximation is the slight skewness of the former. The tabular $\chi^2$ distribution is a very bad approximation.

For any distribution with finite semi-invariants the distribution of $k_2$ from a sample of $N$ must tend to normality as $N$ increases, and expressions (1), (6) and (9) may be used to determine the mean and variance, which may be used if the ordinary $\chi^2$ approximation is found to be inapplicable.

## § 4. The general $\chi^2$ problem with small expectations

The cases in which the $\chi^2$ distribution is commonly used in tests of significance may be divided into three broad groups, in each of which the theory has been for some time reasonably complete where the expectations involved are large: (1) the comparison of a sample of grouped observations with a hypothetical frequency distribution, which may in the process of fitting be made to agree with the sample in one or more respects; (2) tests of departure from independence in contingency tables; (3) the comparison of the variance calculated from a sample in a discontinuous frequency distribution with the population variance. This is the case which we have been considering.

Little is known about the $\chi^2$ distribution in cases (1) and (2) when the expectations are not large. The only case which has been considered in detail is the $2 \times 2$ contingency table, for which Yates[3] has shown how to obtain the exact probability distribution. This paper also contains a table which greatly facilitates the use of exact tests of significance with expectations as low as unity; and this particular case, very important in practice, may be regarded as solved.

There are two points involved. It is not known to what extent the true $\chi^2$ distribution deviates from that given in the tables of Fisher and Elderton, and it is not clear that $\chi^2$, as ordinarily calculated, is the most appropriate general test of discrepancies between observation and hypothesis. Fisher suggested in 1922[6] that $\chi^2$ derives its validity from the fact that with large expectations it is approximately equal to twice the logarithm ($L$) of the likelihood of the sample, with its sign changed, and that where this equivalence does not hold, $L$ itself should be used instead of $\chi^2$. If $m$ is the expectation in any cell and $x$ the value observed,

$$L = Sx \log \frac{x}{m}.$$

The same suggestion was made by Neyman & Pearson[7] in 1928 as a result of applying their criterion of likelihood to this case. They also calculated[8] the exact distribution of $L$ and compared it with the tabular $\chi^2$ distribution in samples of 10 from a population divided into three classes with expectations 3, 5 and 2 respectively. They expressed themselves as agreeably surprised with the closeness of the agreement between the $L$ and $\chi^2$ tests in this case. The agreement between the exact and the ordinary $\chi^2$ distributions in the region used in tests of significance was excellent, and the discrepancies between $L$ and $\chi^2$ arose because $L$ arranges the observations in a different order from $\chi^2$.

In further work on this point, it would seem advisable to concentrate on the distribution of $L$ itself rather than that of $\chi^2$. For it is just as easy to work out the exact distribution of $L$ as that of $\chi^2$ in particular cases. Further, in contingency tables it is quite common for several distinct configurations to give the same value of $\chi^2$, whereas in general no two distinct configurations give the same value of $L$, so that a better representation of the distribution by a continuous curve should be obtainable for $L$ than for $\chi^2$. The first approximation to the distribution of $2L = 2Sx \log \dfrac{x}{m}$ is, of course, the tabular $\chi^2$ distribution. The way in which they appear to diverge with small expectations is shown in Fig. 2, in which the general forms of the $\chi^2$ and ordinary smoothed $L$ distribution are shown.

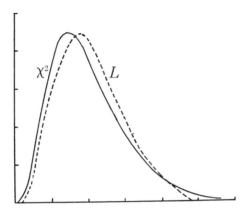

Fig. 2. Comparison of $\chi^2$ and smoothed $L$ distributions.

The two distributions are compared for the Poisson series $m = 2$, $N = 4$, already considered, in Table III below. The values given in the column headed $P\chi^2$ are obtained by fitting a tabular $\chi^2$ distribution to the distribution of $2L$, just as in Table I we fitted a tabular $\chi^2$ distribution to that of $\chi_0^2$. The $\chi^2$ distribution has been corrected for continuity as this considerably improves the fit.

Table III

| $2L$ | $P \geqslant L$ | $P\chi^2$ (corrected for continuity) |
|---|---|---|
| 6·595 | 0·1976 | 0·1084 |
| 7·778 | 0·0951 | 0·0663 |
| 10·412 | 0·0336 | 0·0281 |
| 11·094 | 0·0233 | 0·0131 |
| 11·596 | 0·0169 | 0·0100 |
| 13·185 | 0·0067 | 0·0062 |
| 16·155 | 0·0015 | 0·0020 |

Except at the end of the scale, the $\chi^2$ distribution considerably underestimates the probability, even after correction for continuity. This appeared in all cases examined and in particular in Neyman & Pearson's example. The agreement is far from satisfactory, though it is not sensibly worse than that between the exact $\chi_s^2$ and the corrected tabular $\chi^2$ in Table I above. It is to be hoped that some frequency distribution will be found which will adequately represent that of $L$ in this region, though no simple adjustment of the tabular $\chi^2$ distribution appears to meet the case.

## Summary

Some examples are given of the agreement between the exact and the tabular $\chi^2$ distribution in samples from the binomial and Poisson series with small expectations. The ordinary $\chi^2$ distribution tends slightly to underestimate the probability of discrepancies in the region used in tests of significance, but appears to give a satisfactorily close agreement except in very extreme cases (e.g. with expectations less than unity). Correction for continuity does not improve the agreement.

A method is given for obtaining for any population approximations to any given order for the mean and variance of $\chi^2$ in samples in which the mean of the sample is fixed, and from this the exact normal approximation to the $\chi^2$ distribution for the binomial series is obtained. Except for the Poisson series, this is not the same as the normal approximation to the ordinary $\chi^2$ distribution.

A brief discussion is given of the general problem of testing discrepancies between observation and hypothesis, in which it has been suggested that the likelihood, as defined by Fisher, is more appropriate than $\chi^2$ as a test criterion.

## References

(1) FISHER, R. A. *Statistical Methods for Research Workers*, § 19.

(2) COCHRAN, W. G. (1936). "The statistical analysis of field counts of diseased plants." *J. roy. Statist. Soc.*, Suppl., **3**, No. 1.

(3) YATES, F. (1934). "Contingency tables involving small numbers and the $\chi^2$ test." *J. roy. Statist. Soc.*, Suppl., **1**, No. 2.

(4) FISHER, R. A. (1928). "Moments and product moments of sampling distributions." *Proc. London math. Soc.* Ser. 2, **30**, part 3.

(5) PEARSON, K. (1921). "On a general method of determining the successive terms in a skew regression line." *Biometrika*, **13**.

(6) FISHER, R. A. (1922). "On the mathematical foundations of theoretical statistics." *Philos. Trans.* A, **222**.

(7) NEYMAN, J. & PEARSON, E. S. (1928). "On the use and interpretation of certain test criteria for purposes of statistical inference. Part II." *Biometrika*, **20**A.

(8) —— & —— (1931). "Further notes on the $\chi^2$ distribution." *Biometrika*, **22**.

# Note on J. B. S. Haldane's Paper: "The Exact Value of the Moments of the Distribution of $\chi^2$." (Biometrika, XXIX, 133–143.)

## By W. G. COCHRAN

In this paper Haldane points out (p. 142) a difference between his results for the mean and variance of $\chi^2$ in a $2 \times n$-fold contingency table when the expectation $p$ is fixed and the results obtained by me in my paper (*Annals of Eugenics*, Vol. VII, part III, p. 211). The difference is that I have $(n-1)$ throughout where Haldane has $n$. Haldane writes, "my own results would appear to be slightly more accurate than Cochran's", which might, I think, give the impression that both Haldane's results and mine are only approximations. In fact, both results are mathematically exact, the difference between them being one of definition of $\chi^2$. My paper is almost entirely concerned with the distribution of $\chi^2$ when the expectation $p$ is not known. In the results which I gave for the distribution of $\chi^2$ when $p$ is known, I retained the term $S(x-\bar{x})^2$ in the numerator of $\chi^2$ instead of $S(x-np)^2$, to facilitate comparison between this and my other results. Thus my $\chi^2$ has $(n-1)$ degrees of freedom, whereas Haldane's $\chi^2$ has $n$ degrees of freedom.

Unfortunately I did not emphasize this point in the passage concerned, and as it may have appeared misleading to others besides Haldane, I welcome this opportunity of drawing attention to it. Haldane's $\chi^2$ is, of course, the one which is normally appropriate in testing the departure from independence when the expectation is known.

The reader may perhaps also wonder why in § 3 I have replaced $\dfrac{\bar{x}}{n}(n-\bar{x})$ by $\kappa_2$ in the denominator of $\chi^2$. This was done because I am discussing the distribution of $\chi^2$ in arrays in which the mean of the sample is equal to the mean of the population, so that in this paragraph the two expressions can be regarded as equivalent.

# 8

## Problems Arising in the Analysis of a Series of Similar Experiments.

### By W. G. Cochran, B.A.

#### § 1. *Introduction.*

An efficient type of modern field experiment is that in which a replicated trial is laid down in the same year at a number of centres, or carried out at the same centre independently throughout a number of years. The statistical problems which arise in the interpretation of the results of such a set of data are of wide generality. For any treatment effect, we obtain at each centre an estimate $x$ and an estimate $s$ of its standard error, based on $n$ degrees of freedom. As a preliminary to more detailed examination, the experimenter wants to estimate and test the significance of the mean treatment effect and to find whether it has varied from centre to centre.

If the individual experiments are assumed to be equally accurate and the estimated standard errors do not contradict this assumption, the statistical treatment is easy and familiar. If there are $k$ centres, the data for any particular treatment response may be analysed into

|  | D.F. |
|---|---|
| Mean response ... ... ... ... ... | 1 |
| Interaction of response with centres ... ... | $(k-1)$ |
| Local experimental error ... ... ... | $nk$ |

The interaction of the treatment response may be tested against the combined experimental errors, and the mean response may be tested against either the interaction or the experimental error. The interpretation of these two tests has been discussed by Fisher (1).

It will, however, be the exception to find the individual experiments all of the same precision. The object of this paper is to discuss the estimation of the mean response and the test of significance of the interaction of the response with centres, where we do not wish to assume that the standard errors are all equal.

#### § 2. *The Equations of Estimation.*

The most general hypothesis to be considered is that the treatment response at any centre is the sum of two parts, each normally and independently distributed; one, representing the contribution of local experimental errors, varies about zero with standard deviation $\sigma_i$, while the other, which represents the responsiveness of the centre to the treatment, varies about a general mean $\mu$ with standard deviation $\sigma$. The parameters $\mu$ and $\sigma$ are the same for all centres, but

$\sigma_i$ varies from centre to centre. An estimate $s_i$ of $\sigma_i$, based on $n$ degrees of freedom, is available from the local analysis of variance.

For a single centre, the joint sampling distribution of $s_i$ and $x_i$ may be written, apart from the constant of integration

$$\frac{s_i^{n-1}}{\sigma_i^n \sqrt{\sigma^2 + \sigma_i^2}} e^{-\frac{1}{2}\left\{\frac{ns_i^2}{\sigma_i^2} + \frac{(x_i - \mu)^2}{\sigma^2 + \sigma_i^2}\right\}} dx_i ds_i$$

Hence the logarithm of the likelihood for all centres is

$$L = -nS \log \sigma_i - \tfrac{1}{2}S \log (\sigma^2 + \sigma_i^2) - \tfrac{1}{2}S\left\{\frac{ns_i^2}{\sigma_i^2} + \frac{(x_i - \mu)^2}{\sigma^2 + \sigma_i^2}\right\}$$

the sum being taken over all $k$ centres.

The equations of estimation of $\mu$, $\sigma_i$, $\sigma$ are

$$\frac{\partial L}{\partial \mu} = S \frac{(x_i - \mu)}{\sigma^2 + \sigma_i^2} = 0 \quad \ldots \quad \ldots \quad \ldots \quad (1)$$

$$\frac{\partial L}{\partial \sigma_i} = -\frac{n}{\sigma_i} - \frac{\sigma_i}{\sigma^2 + \sigma_i^2} + \frac{ns_i^2}{\sigma_i^3} + \frac{\sigma_i(x_i - \mu)^2}{(\sigma^2 + \sigma_i^2)^2} = 0 \quad . \quad (2)$$

$$\frac{\partial L}{\partial \sigma} = -S \frac{\sigma}{\sigma^2 + \sigma_i^2} + S \frac{\sigma(x_i - \mu)^2}{(\sigma^2 + \sigma_i^2)^2} = 0. \quad \ldots \quad (3)$$

The equations are complicated and have no simple general solution. The complication is mainly due to the fact that the value of $x_i$ provides information about all three parameters $\mu$, $\sigma$, and $\sigma_i$.

### § 3. *The Estimation of the Response when it is Assumed Constant at all Centres.*

If we assume $\sigma = 0$, the equations of estimation of $\mu$ and $\sigma_i$ give

$$S \frac{(x_i - \mu)}{\sigma_i^2} = 0 \quad \ldots \quad \ldots \quad (4)$$

$$\sigma_i^2 = \frac{ns^2 + (x_i - \mu)^2}{n + 1} \quad \ldots \quad \ldots \quad (5)$$

Thus the equation of estimation of the mean response is

$$S \frac{x_i - \mu}{s_i^2 + \frac{(x_i - \mu)^2}{n}} = 0 \quad \ldots \quad \ldots \quad (6)$$

If the values of $\sigma_i$ were known exactly, the sufficient estimate of the mean would be given by the solution of

$$S \frac{(x_i - \mu)}{\sigma_i^2} = 0 \quad \ldots \quad \ldots \quad (7)$$

*i.e.* by the weighted mean. Where the $\sigma_i$ have also to be estimated, we do not simply replace $\sigma_i$ by $s_i$ in equation (7) to obtain an efficient estimate, but make use of the extra information about $\sigma_i$ contained in

$x_i$. Equation (6) may be solved fairly quickly by successive approximation, starting with the unweighted mean as a first estimate.

To find the amount of information in the estimated mean and perform tests of significance, it is necessary to calculate the sampling variance of the solution, $\hat{\mu}$ say, of equation (6). For a given set of $\sigma_i$ and $\mu$, this would at first sight appear to depend on all these unknown parameters, but Bartlett (2) has shown how to use the information available about the unknown $\sigma_i$ to obtain the sampling variance of $\hat{\mu}$ in terms of the single unknown $\mu$. In the joint distribution of $x$ and $s$ at any centre

$$Cs^{n-1} e^{-\frac{1}{2\sigma^2}\{ns^2 + (x-\mu)^2\}} \, dx ds \quad . \quad . \quad . \quad (8)$$

he writes $\Sigma = ns^2 + (x - \mu)^2$ and substitutes for $s$ in terms of $\Sigma$. This gives for the joint distribution of $x$ and $\Sigma$

$$C\Sigma^{\frac{n}{2}-1}\{1 - (x-\mu)^2/\Sigma\}^{\left(\frac{n}{2}-1\right)} e^{-\frac{1}{2}\frac{\Sigma}{\sigma_i}} dx d\Sigma \quad . \quad . \quad (9)$$

It will be noted that the distribution of $x$ for fixed $\Sigma$ depends only on the unknown mean $\mu$. Thus the variance of $\hat{\mu}$ for a fixed $\Sigma$, which is known if $\mu$ is known, will depend only on the unknown mean $\mu$, and not on the $\sigma_i$.

To find the variance of $\hat{\mu}$, write

$$X(\hat{\mu}) = S\frac{(x-\hat{\mu})}{ns^2 + (x-\hat{\mu})^2} = 0 \quad . \quad . \quad . \quad (10)$$

Now

$$0 = X(\hat{\mu}) = X(\mu) + (\hat{\mu} - \mu)\frac{\partial X}{\partial \mu} + . \quad . \quad . \quad (11)$$

Thus if the number of centres is large,

$$E(X^2) = \sigma_{\hat{\mu}}^2 E^2\left(\frac{\partial X}{\partial \mu}\right) . \quad . \quad . \quad . \quad (12)$$

From equation (9), the variance of $x$ for fixed $\Sigma$ is $\Sigma/n + 1$. Thus

$$E(X^2) = S\left\{E\frac{(x-\mu)^2}{\Sigma^2}\right\} = \frac{1}{(n+1)}S\left(\frac{1}{\Sigma}\right) \quad . \quad . \quad (13)$$

and

$$E\left(\frac{\partial X}{\partial \mu}\right) = S\left\{E\left(\frac{(x-\mu)^2}{\Sigma^2} - \frac{1}{\Sigma}\right)\right\} = -\left(\frac{n-1}{n+1}\right)S\left(\frac{1}{\Sigma}\right). \quad (14)$$

so that

$$\sigma_{\hat{\mu}}^2 = \frac{(n+1)}{(n-1)^2}1/S\left(\frac{1}{\Sigma}\right) . \quad . \quad . \quad . \quad (15)$$

The average amount of information for a fixed set of $\sigma_i$ is $E\left(\frac{1}{\sigma_{\hat{\mu}}^2}\right)$ and is easily found to be

$$\frac{(n-1)}{(n+1)}S\left(\frac{1}{\sigma^2}\right) . \quad . \quad . \quad . \quad . \quad (16)$$

Thus the average fraction of information lost through the inaccuracy of the weights is $\dfrac{2}{n+1}$.

An alternative which suggests itself to the use of the maximum likelihood solution is the weighted mean

$$\mu_w = S\left(\frac{x}{s^2}\right) \Big/ S\left(\frac{1}{s^2}\right) \quad . \quad . \quad . \quad . \quad (17)$$

This has the advantage that it can be calculated directly and is a familiar type of mean in statistical work, and it is worth while estimating its efficiency. For a fixed set of $s_i$

$$V(\mu_w) = S\left(\frac{\sigma^2}{s^4}\right) \Big/ S^2\left(\frac{1}{s^2}\right) \quad . \quad . \quad . \quad (18)$$

so that a knowledge of the $s_i$ does not, in this case, enable us to dispense with the knowledge of the $\sigma_i$, though it has the advantage that for fixed $s_i$, $\mu_w$ is normally distributed. Since

$$E\left(\frac{1}{s^2}\right) = \frac{n}{n-2}\frac{1}{\sigma^2} \text{ and } E\left(\frac{1}{s^4}\right) = \frac{n^2}{(n-2)(n-4)}\frac{1}{\sigma^4}$$

the average variance of $\mu_w$ for a fixed set of $\sigma_i$ is, provided $n>4$,

$$\left(\frac{n-2}{n-4}\right) 1\Big/S\left(\frac{1}{\sigma^2}\right) \quad . \quad . \quad . \quad (19)$$

Thus the amount of information in the weighted mean is

$$\left(\frac{n-4}{n-2}\right) S\left(\frac{1}{\sigma^2}\right) \quad . \quad . \quad . \quad . \quad (20)$$

Comparison with (16) shows that the superior efficiency of the maximum likelihood solution is equivalent to having 3 extra degrees of freedom in the estimates $s_i$.

For $n=4$, expression (19) gives an infinite value for the variance of the weighted mean $\mu_w$. This is not correct, since (18) shows that for a fixed set of $\sigma_i$ the variance of $\mu_w$ cannot exceed the greatest of the variances $\sigma_i^2$, *i.e.* the weighted mean cannot do worse than give as the estimate of $\mu$ the most variable single value $x_i$. For $n=4$, the average variance of $\mu_w$ can be shown to be

$$S\left(\frac{1}{\sigma_i^2}\right) \Big/ S\left(\frac{1}{\sigma_i^4}\right) \quad . \quad . \quad . \quad . \quad (21)$$

This value lies between the most and least accurate of the individual estimates at the various centres. What is happening is that individual low values of $s_i$ are turning up so frequently that usually all the information about $\mu$ is being derived from a single centre. Thus the percentage information retained tends to zero as the number of centres increases.

E 2

The percentage efficiencies of $\hat{\mu}$ and $\mu$ are shown for small values of $\mu$ in the table below.

TABLE I.

*Percentage Efficiencies of $\hat{\mu}$ and $\mu_w$.*

| $n$ ... ... | 2 | 3 | 4 | 5 | 6 | 8 | 10 | 15 | 20 |
|---|---|---|---|---|---|---|---|---|---|
| $\hat{\mu}$ ... ... | 33 | 50 | 60 | 67 | 71 | 78 | 82 | 88 | 90 |
| $\mu_w$ ... ... | 0 | 0 | 0 | 33 | 50 | 67 | 75 | 85 | 89 |

For $n$ greater than 15 there is little to choose between the two estimates, but for $n$ less than 10 the increase in efficiency of the maximum likelihood solution is worth the extra labour. For values of $n$ between 2 and 6 a good deal of information is being lost in the process of estimation of the weights. As these cases may be of practical importance (*e.g.* $n = 2$ might represent a set of $3 \times 3$ Latin squares), it is worth considering the relative efficiency of two other types of mean which suggest themselves.

One is the unweighted mean. This always retains a finite fraction of the information, the fraction decreasing as the true accuracies of the individual experiments diverge, and is not subject to any loss due to estimation of weights. The other method is to fix arbitrarily an upper limit to the weights, and below that to weight inversely as the estimated variance. This is equivalent to recognizing that in practice there is a limit to the accuracy with which an individual experiment may be carried out, and that very low values of $s_i$ are likely to be under-estimates of the corresponding $\sigma_i$. The method has the advantage that no single experiment exerts too predominating an influence on the mean, while bad experiments are properly scaled down in weight; on the other hand, it has an element of arbitrariness in the choice of the upper limit. A comparison of the four types of mean for low values of $n$ is given in the next section, the number of centres being assumed to be large.

### § 4. *The Relative Efficiencies of Four Types of Mean.*

The relative efficiencies of the weighted mean and the maximum likelihood solution have been shown to be $\left(\dfrac{n-4}{n-2}\right)$ and $\left(\dfrac{n-1}{n+1}\right)$ respectively, where $n$ is the number of degrees of freedom in the estimates $s_i$. The variance of the unweighted mean is

$$\frac{1}{k^2} S \sigma_i^2$$

where $k$ is the number of centres, and can be calculated for any given set of $\sigma_i$.

To calculate the efficiency of the weighted mean with an arbitrary upper limit, let the minimum true error variance be guessed as $\sigma_0^2$. Then in sampling from a set of experiments in which the true variance is $\sigma^2$, we take the weight $w$ as $\dfrac{1}{\sigma_0^2}$ whenever $s^2 \leq \sigma_0^2$ and as $\dfrac{1}{s^2}$ whenever $s^2 > \sigma_0^2$. The variance of the mean $S(wx)/S(w)$ for any given set of $\sigma_i^2$ and fixed weights is

$$S(w^2\sigma^2)/S^2(w)$$

and the average variance for a given set of $\sigma_i^2$ is

$$S(\overline{w^2\sigma^2})/S^2(\overline{w})$$

For $n = 6$, for example, the probability that $s^2 \leq \sigma_0^2$ is, (cf. (3)),

$$P(s^2 \leq \sigma_0^2) = 1 - e^{-\frac{1}{2}\chi^2}\left\{1 + \frac{1}{2}\chi^2 + \frac{1}{2!}\left(\frac{1}{2}\chi^2\right)^2\right\}$$

where $\frac{1}{2}\chi^2 = 3\sigma_0^2/\sigma_i^2$.

We require also the mean value of $\dfrac{1}{s^2}$ in the range $(\sigma_0, \infty)$. This is found to be

$$\frac{3}{2\sigma_i^2}e^{-\frac{1}{2}\chi^2}\{1 + \tfrac{1}{2}\chi^2\}$$

Thus

$$\overline{w} = \left[1 - e^{-\frac{1}{2}\chi^2}\left\{1 + \frac{1}{2}\chi^2 + \frac{1}{2!}\left(\frac{1}{2}\chi^2\right)^2\right\}\right]\Big/\sigma_0^2 + \frac{3}{2\sigma^2}e^{-\frac{1}{2}\chi^2}\left\{1 + \frac{1}{2}\chi^2\right\}$$

Similarly

$$\overline{w^2} = \left[1 - e^{-\frac{1}{2}\chi^2}\left\{1 + \frac{1}{2}\chi^2 + \frac{1}{2!}\left(\frac{1}{2}\chi^2\right)^2\right\}\right]\Big/\sigma_0^4 + \frac{9}{2\sigma^4}e^{-\frac{1}{2}\chi^2}$$

The relative efficiencies of these four types of mean will depend on the distribution of experimental errors $\sigma_i^2$. To obtain some actual figures, the efficiencies have been calculated for some sets of hypothetical values of $\sigma_i^2$ which are intended to cover the range likely to occur in practice. It is assumed that the $k$ centres are divided into three groups as regards accuracy : a number $\lambda k$ have the same experimental variance $lv$, $\mu k$ have variance $mv$, while the remaining $\nu k$ have variance $nv$. We have

$$\lambda + \mu + \nu = 1 \text{ and } l < m < n$$

The sets of values assigned to $(\lambda, \mu, \nu)$ are : $(0\cdot2, 0\cdot6, 0\cdot2)$ and $(0\cdot3, 0\cdot4, 0\cdot3)$, the first set representing, for instance, the case in which 60 per cent. of the experiments have the same accuracy, while 20 per cent. are less accurate and 20 per cent. more accurate. For each of these sets, $(l, m, n)$ have been given the values $(\frac{1}{2}, 1, 2)$ and $(\frac{1}{4}, 1, 4)$ respectively. The case $(\lambda, \mu, \nu) = (0\cdot2, 0\cdot6, 0\cdot2)$, $(l, m, n) = (\frac{1}{2}, 1, 2)$ means, for instance, that $\frac{1}{5}$ of the centres have experimental variances $\frac{1}{2}v$, $\frac{3}{5}$ have variances $v$ and the remaining $\frac{1}{5}$ have variances $2v$. The four cases resulting cover a fairly wide range of variation in the accuracy of individual experiments, the relative accuracy of the best and the worst experiments ranging from 4 to 16.

Comparison of the four types of mean suggested is made in Table II for $n = 6$, 4 and 2.

Table II.

*Relative Efficiencies of Four Types of Mean.*

| $n$. | Relative Accuracy of the Groups $(l, m, n)$. | Proportion in Groups $(\lambda, \mu, \nu)$. | Maximum Likelihood Solution. | Weighted Mean. | Weighted Mean with Upper Limit. | | | | Unweighted Mean. |
|---|---|---|---|---|---|---|---|---|---|
| | | | | | $\frac{2}{v}$ | $\frac{1}{v}$ | $\frac{2}{3v}$ | $\frac{1}{2v}$ | |
| 6 | $\frac{1}{2}, 1, 2$ | 0·2, 0·6, 0·2 | 71 | 50 | 64 | 82 | 85 | 86 | 83 |
| | | 0·3, 0·4, 0·3 | 71 | 50 | 62 | 81 | 82 | 82 | 76 |
| 4 | $\frac{1}{2}, 1, 2$ | 0·2, 0·6, 0·2 | 60 | 0 | 67 | 78 | 83 | 83 | 83 |
| | | 0·3, 0·4, 0·3 | 60 | 0 | 68 | 76 | 80 | 79 | 76 |
| 2 | $\frac{1}{2}, 1, 2$ | 0·2, 0·6, 0·2 | 33 | 0 | 60 | 74 | 78 | 82 | 83 |
| | | 0·3, 0·4, 0·3 | 33 | 0 | 60 | 70 | 74 | 77 | 76 |
| 6 | $\frac{1}{4}, 1, 4$ | 0·2, 0·6, 0·2 | 71 | 50 | 69 | 72 | 71 | 69 | 48 |
| | | 0·3, 0·4, 0·3 | 71 | 50 | 73 | 75 | 71 | 67 | 36 |
| 4 | $\frac{1}{4}, 1, 4$ | 0·2, 0·6, 0·2 | 60 | 0 | 60 | 63 | 65 | 65 | 48 |
| | | 0·3, 0·4, 0·3 | 60 | 0 | 63 | 63 | 62 | 60 | 36 |
| 2 | $\frac{1}{4}, 1, 4$ | 0·2, 0·6, 0·2 | 33 | 0 | 44 | 49 | 53 | 56 | 48 |
| | | 0·3, 0·4, 0·3 | 33 | 0 | 42 | 45 | 47 | 48 | 36 |

As the upper limit to the possible weights decreases from infinity (no adjustment) to zero, the efficiency increases from the value given by the weighted mean to a maximum, and thereafter decreases to the value given by the unweighted mean. Owing to the variability of the estimated weights, the maximum efficiency does not coincide with the case in which the upper limit is guessed correctly, but with a somewhat lower value. The efficiency is not very sensitive to variation in the point at which the upper limit is fixed, provided that it is not fixed too high.

Where $(l, m, n) = (\frac{1}{2}, 1, 2)$, so that the experiments do not vary widely in accuracy, there is little to choose between the unweighted mean and the weighted mean with fixed upper limit, the maximum likelihood solution being definitely inferior. With the wider variations in precision, the weighted mean with fixed upper limit is the most accurate, and for $n = 4$ or 6 the maximum likelihood solution comes next.

To sum up the question of estimation of a mean response where it is assumed that there is no variation in response from experiment to experiment, the weighted mean, weighting inversely as the estimated variances, may be recommended if 15 or more degrees of freedom are available in the estimates of the weights. With less than 15 degrees of freedom one of the other means is advisable, and each has something in its favour.

The maximum likelihood solution is satisfactory from the point of view of information for values of $n$ as low as 6, though it is slightly more tedious to calculate than the other means. The unweighted mean has simplicity to commend it, and is particularly suitable with sets of experiments which do not vary widely in precision and with low values of $n$. For values of $n$ below 6 the weighted mean with a fixed upper limit is the most accurate of the three, if the limit can be chosen to represent the accuracy of the best group of experiments. The difficulty of assigning a standard error to this mean, is, however, a serious disadvantage.

## § 5. *The Test of Significance of the Mean Response.*

The tests given here are strictly appropriate to a large number of centres; the mean is assumed to be normally distributed, and is compared in the normal probability table with an unbiased estimate of its variance. In the analogous case of equal precision, the approximation is equivalent to replacing the $t$-distribution for $nk$ degrees of freedom by the normal distribution. The agreement between the exact and approximate tests may not be as good with unequal as with equal variances, but may be expected to be satisfactory unless $k$ is small. The case $k = 2$ is being investigated.

The estimated variances of the weighted mean and the maximum likelihood solution have already been found. The average variance of the weighted mean is $\left(\dfrac{n-2}{n-4}\right) \Big/ S\left(\dfrac{1}{\sigma^2}\right)$. Replacing $S\left(\dfrac{1}{\sigma^2}\right)$ by an unbiased estimate in terms of $s^2$, we get for the standard error

$$\sqrt{\frac{n}{n-4}} \Big/ \sqrt{S\left(\frac{1}{s^2}\right)}$$

In the maximum likelihood solution, the estimate of the experimental variance $\sigma_i^2$ at any centre is taken as

$$\{ns_i^2 + (x_i - \dot{\mu})^2\}/(n+1)$$

and the standard error of the mean may be written

$$\left(\frac{n+1}{n-1}\right) \Big/ \sqrt{S\left(\frac{n+1}{ns^2 + (x - \dot{\mu})^2}\right)}$$

where $\dot{\mu}$ is the estimate of the mean.

The estimated standard error of the unweighted mean $\bar{x}$ is

$$s_{\bar{x}} = \sqrt{S(s^2)}/k$$

It should be noted that the distribution of $\dfrac{\bar{x}}{s_{\bar{x}}}$ depends on the ratios of the unknown $\sigma^2$, and is not that of Student's $t$ unless the $\sigma^2$ are all equal. Where the product $nk$ is sufficiently small that a $t$-test is indicated, a first approximation to the exact test may be found by a device which has been used by Fairfield Smith (4). The variance of any individual estimate $s^2$ is

$$V(s^2) = 2\sigma^4/n$$

Hence an estimate of the variance of $s_x^2$ is

$$V(s_x^2) = \frac{2}{nk^2} S(s^4)$$

But if $n_{\bar{x}}$ is the number of degrees of freedom appropriate to $s_{\bar{x}}$, an estimate of $V(s_{\bar{x}}^2)$ is

$$\frac{2}{n_{\bar{x}}} s_x^4 = 2S^2(s^2)/n_{\bar{x}}k^2$$

Thus the relative precision of $s_{\bar{x}}$ is estimated by assigning to it a number of degrees of freedom equal to

$$n_{\bar{x}} = nS^2(s^2)/S(s^4)$$

This number only attains its maximum, $nk$, if all experiments have the same $s^2$. It reaches its minimum, $n$, if a single centre is much less accurate than any of the others, and in this case indicates, quite correctly, a $t$-test against $n$ instead of $nk$ degrees of freedom. In general, the integral part of $n_{\bar{x}}$ may be taken as the number of degrees of freedom in $s_{\bar{x}}$. No examination of the closeness of this appoximation has yet appeared in print, but consideration of the case with two centres only indicates that the approximation may over-estimate on the average the probability of a deviation arising by chance.

I have been unable to find any method of obtaining a simple estimate of the standard error of the weighted mean with fixed upper limit. Even if a method were found, a weighted mean with a badly chosen upper limit would be assigned an under-estimate of its standard error. Unless this difficulty can be satisfactorily over-come, this mean is ruled out where an exact test of significance is required.

## § 6. *The Test of Significance of the Variation in Response from Centre to Centre.*

A test could be obtained by solving equations (1), (2) and (3) for $\sigma$ and using the solution, $\hat{\sigma}$, as a test criterion, the significance levels being obtained from the frequency distribution of $\hat{\sigma}$ when $\sigma = 0$. The test would be efficient if the mathematical specification of the problem set up in § 2 conformed to practice, but in any case the method cannot be used owing to the complexity of the equations.

If the mean response $\mu$ were known, the values $(x - \mu)/s$ would, in the absence of any variation in response, be distributed as $t$ with $n$ degrees of freedom, and a sensitive test could presumably be based on the value of

$$Q = S \frac{1}{s^2} (x - \mu)^2$$

Where the value of $\mu$ is unknown, analogy with the analysis of variance suggests that the appropriate estimate of it for this purpose is the weighted mean $\mu_w = S \frac{x}{s^2} \bigg/ S \frac{1}{s^2}$, which is the value of $\mu$ which minimizes $Q$. With this value inserted, $Q$ may be written

$$Q = S \frac{(x - \mu)^2}{s^2} - (\mu_w - \mu)^2 S \frac{1}{s^2}$$

The efficiency of this quantity as a test criterion will depend on the type of variation in response which occurs in practical applications, but it seems reasonable on common-sense grounds. It may be noted that the same type of expression is used to test the departure from independence in a $2 \times N$ contingency table (cf. (5), § 21).

If the weights were known exactly, $Q$ would follow the $\chi^2$ distribution with $(k-1)$ degrees of freedom, $k$ being the number of centres. In general, $Q$ may be written

$$Q = S(t^2) - S^2 \left(\frac{t}{s}\right) \bigg/ S \left(\frac{1}{s^2}\right)$$

so that it is distributed as the sum of the squares of $k$ values of $t$, less a correction term which is a weighted mean of the values of $t$. A good approximation to the distribution of $Q$ should be obtained by replacing the weighted mean in the correction term by an unweighted mean. In particular

$$\text{Mean } (Q) = \left(\frac{n}{n-2}\right) \left\{ k - 1 - \frac{4}{n(n-4)} \right\}$$

$$< \text{Mean } \{S(t-\bar{t})^2\} = \left(\frac{n}{n-2}\right)(k-1)$$

so that in replacing $Q$ by $S(t-\bar{t})^2$ we are probably tending to overestimate slightly the probability of a discrepancy arising by chance.

Further

$$V\{S(t-\bar{t})^2\} = 2 \left(\frac{n}{n-2}\right)^2 (k-1) \left(\frac{n-1}{n-4}\right) \left\{ 1 - \frac{3}{k(n-1)} \right\}$$

Thus $\left(\dfrac{n-2}{n}\right) S(t-\bar t)^2$ has the same mean as $\chi^2$ with $(k-1)$ degrees of freedom, but its variance is too large, approximately in the ratio $\left(\dfrac{n-1}{n-4}\right)$. A transformation which leads to the same mean and variance as $\chi^2$ is obtained by putting

$$\chi_w^2 = (k-1) + \sqrt{\frac{n-4}{n-1}}\left\{\left(\frac{n-2}{n}\right) S(t-\bar t)^2 - (k-1)\right\}$$

The distribution of $\chi_w^2$ and the tabular $\chi^2$ tend to the same normal distribution, for any value of $n$, as $k$ tends to infinity; they also tend to coincidence, for any $k$, as $n$ tends to infinity, and have the same mean and variance for all values of $n$ and $k$, except for very small values of both, in which the additional factor $\left(1 - \dfrac{3}{k(n-1)}\right)$ may be brought into the transformation. The agreement between the distribution of $\chi_w^2$ and $\chi^2$ may be expected to be at its worst for low values of both, since $\chi_w^2$ has a lower limit $(k-1)\left(1 - \sqrt{\dfrac{n-4}{n-1}}\right)$ instead of zero. The difference even here is likely to be small for moderate values of $n$ and $k$; for $n = 10$, $k = 11$, for instance, the lower limit of $\chi_w^2$ is $1\cdot835$ and the probability of getting a value of $\chi^2$ lower than this for 10 degrees of freedom is only $0\cdot0025$.

It is therefore suggested that the transformation

$$\chi^2 = (k-1) + \sqrt{\frac{n-4}{n-1}}\left\{\left(\frac{n-2}{n}\right) Q - (k-1)\right\}$$

may be used in testing the significance of the variation in response from centre to centre.

For a given value of $n$, this approximation will be worst when there are only two centres. The difference is, however, still on the side of declaring too few significant results, for $Q = \dfrac{(x_1 - x_2)^2}{s_1^2 + s_2^2}$ and the $Q,\chi^2$ transformation is based on the $t^2$ distribution with $n$ degrees of freedom, whereas the exact distribution of $Q$ is probably better approximated by using in $t^2$ a number of degrees of freedom lying between $n$ and $2n$, as indicated by the method suggested in § 5. Even with only two centres the $Q,\chi^2$ transformation will thus avoid the danger of obtaining too many significant results, though the method of determining the equivalent number of degrees of freedom is to be recommended as more sensitive, and further work on this important particular case is needed.

In general, the use of $Q$ as a test criterion is inadvisable for values of $n$ below 6; the $Q,\chi^2$ transformation breaks down when $n = 4$, and the variance of the $t$-distribution itself is infinite when $n = 2$.

In these cases it is best to obtain the individual values of $t = (x - \mu)/s$ and compare these with the tabulated $t$-distribution. The estimate of $\mu$ used should be one of the three other types of mean suggested, but not the weighted mean. The transformation from $Q$ to $\chi^2$ cannot, however, be used with these means, since they will always give higher values of $Q$ than the weighted mean.

### § 7. *Estimation of the Mean Response when it Varies from Centre to Centre.*

If it cannot be assumed that the interactions do not exist, the question of estimation of the mean response is more difficult. Equation (1)

$$S \frac{(x_i - \mu)}{\sigma^2 + \sigma_i^2} = 0$$

indicates that a kind of semi-weighted mean is appropriate, but the complete solution of equations (1), (2) and (3) would be very tedious. If a fairly efficient solution is required in a particular case, there will probably be very little information lost if the $s_i^2$ are used as estimates of $\sigma_i^2$ and $\mu$ and $\sigma^2$ are estimated from the simultaneous equations

$$S \frac{x_i - \mu}{\sigma^2 + s_i^2} = 0$$

$$S \frac{1}{\sigma^2 + s_i^2} = S \frac{(x_i - \mu)^2}{(\sigma^2 + s_i^2)^2}$$

The solution of these equations is as a rule quite rapid.

In general, a simpler solution will be wanted, and it is worth comparing the efficiencies of the unweighted and weighted means. Consider first the case in which the values of $\sigma^2$ and $\sigma_i^2$ are known exactly. The variance of the semi-weighted mean

$$S \frac{x_i}{\sigma^2 + \sigma_i^2} \bigg/ S \frac{1}{\sigma^2 + \sigma_i^2}$$

is

$$1 \bigg/ S \frac{1}{\sigma^2 + \sigma_i^2}$$

The variance of the unweighted mean is

$$\frac{1}{k} \left( \sigma^2 + \frac{1}{k} S \sigma_i^2 \right)$$

while that of the weighted mean

$$S \frac{x_i}{\sigma_i^2} \bigg/ S \frac{1}{\sigma_i^2}$$

is

$$S \frac{\sigma^2 + \sigma_i^2}{\sigma_i^4} \bigg/ S^2 \frac{1}{\sigma_i^2} = \frac{1}{S \dfrac{1}{\sigma_i^2}} \left[ 1 + \sigma^2 \frac{S \dfrac{1}{\sigma_i^4}}{S \dfrac{1}{\sigma_i^2}} \right]$$

The relative efficiencies of these three types of mean will depend on the distribution of experimental errors $\sigma_i^2$ and on the interaction $\sigma^2$. To obtain some actual figures, the efficiencies have been calculated for the set of experimental variances used in Table II, giving to $(\lambda, \mu, \nu)$ the additional values $(0\cdot1, 0\cdot8, 0\cdot1)$ and to $(l, m, n)$ the additional values $(\frac{1}{3}, 1, 3)$. This provides nine instead of four examples of variation in experimental errors.

The interaction variance has now to be considered. For each of the nine selected cases, the efficiencies are continuous functions of the interaction variance $rv$, say. The mean experimental variance for any set $(\lambda, \mu, \nu)$, $(l, m, n)$ is

$$(\lambda l + \mu m + \nu n)v = \kappa v \text{ (say)}$$

and for each of the nine cases, the efficiencies have been calculated for $r = 0, \frac{1}{2}\kappa, \kappa, 2\kappa$, so that the interaction variance is respectively $0, \frac{1}{2}, 1, 2$ times the mean experimental variance. This gives a $3 \times 3 \times 4$ table of 36 pairs of entries.

TABLE III.

*Efficiencies of the Unweighted and Weighted Means.*

| Proportions in Groups. | Relative Accuracies of the Three Groups of Centres. | | | | | | | | | | | |
|---|---|---|---|---|---|---|---|---|---|---|---|---|
| | $(\frac{1}{2}, 1, 2)$. | | | | $(\frac{1}{3}, 1, 3)$. | | | | $(\frac{1}{4}, 1, 4)$. | | | |
| | $i=0$. | $i=0\cdot5$. | $i=1$. | $i=2$. | $i=0$. | $i=0\cdot5$. | $i=1$. | $i=2$. | $i=0$. | $i=0\cdot5$. | $i=1$. | $i=2$. |
| $(0\cdot1, 0\cdot8, 0\cdot1)$ | | | | | | | | | | | | |
| U | 91 | 96 | 98 | 99 | 78 | 91 | 94 | 97 | 67 | 86 | 91 | 95 |
| W | 100 | 98 | 97 | 95 | 100 | 94 | 89 | 85 | 100 | 88 | 81 | 75 |
| $(0\cdot2, 0\cdot6, 0\cdot2)$ | | | | | | | | | | | | |
| U | 83 | 92 | 96 | 98 | 62 | 84 | 90 | 95 | 48 | 77 | 86 | 93 |
| W | 100 | 97 | 94 | 91 | 100 | 91 | 85 | 79 | 100 | 86 | 78 | 71 |
| $(0\cdot3, 0\cdot4, 0\cdot3)$ | | | | | | | | | | | | |
| U | 76 | 89 | 94 | 97 | 51 | 78 | 88 | 94 | 36 | 72 | 83 | 92 |
| W | 100 | 96 | 93 | 89 | 100 | 90 | 83 | 76 | 100 | 84 | 75 | 68 |

U = Unweighted mean.          W = Weighted mean.

$i$ = Ratio of interaction variance to the mean local error variance.

Several features of the table are obvious. If there is no interaction $(i = 0)$, the weighted mean is the same as the efficient solution. This is also the most unfavourable case for the unweighted mean; its efficiency decreases, as the individual experiments diverge more widely in precision, from 91 per cent. to 36 per cent. As the interaction variance increases, the efficiency of the weighted mean falls steadily, while that of the unweighted mean rises steadily. For $i = 0\cdot5$, the weighted mean is still superior to the unweighted, but for $i = 1$, the unweighted mean is superior in all cases, the maximum loss of information being 17 per cent. in the worst case. For $i = 2$ the loss of information with the unweighted mean is small in all cases.

In practice the situation is much more favourable to the unweighted mean than Table III indicates. For the weights in the semi-weighted mean and in the weighted mean have to be estimated, and the estimation results in a loss of information on these means to which there is no corresponding loss on the unweighted mean. In particular, the information in the weighted mean has been shown above to be decreased in the ratio $\left(\dfrac{n-4}{n-2}\right)$, where $n$ is the number of degrees of freedom in the local experimental errors. With $n = 16$, for instance, the efficiencies of the weighted means in Table III have to be multiplied by $\frac{6}{7}$. This would make the unweighted mean superior to the mean throughout Table III, except in a few cases in which there was no interaction.

These results indicate that the unweighted mean may safely be recommended where we do not assume that interactions are non-existent, particularly since, with a large number of centres, it is usually necessary to keep the individual experiments small, so that there will rarely be as many as 20 degrees of freedom in the estimates of the local experimental errors.

Where the response varies from centre to centre, it is usually appropriate to test the mean response by comparing it with the variation from centre to centre, especially if the centres constitute a random sample from all possible centres. The usual expression $s_{\bar{x}}^2 = S(x - \bar{x})^2/k(k-1)$ taken over all centres, is an unbiased estimate of the variance of the unweighted mean $\bar{x}$. The $t$-test with $(k-1)$ degrees of freedom will, however, lead to too many significant results unless the interaction variance is large compared with the local experimental variances, since with one very inaccurate experiment, for instance, the estimate $s_{\bar{x}}^2$ might have a precision based on only one instead of $(k-1)$ degrees of freedom. The $t$-test is, however, known to be relatively insensitive to most types of departure from normality in the original data and may be recommended in the great majority of cases, except where the probabilities of a number of tests are being combined, or where the test gives a result very near one of the significance levels and an exact verdict is wanted.

Two alternative tests may prove useful. In many cases it may be sufficiently precise to take account of the signs of the responses only. The efficiency of this test, where the variances are equal, has been shown to be $\dfrac{2}{\pi}$ or 64 per cent. in (6), where a table of the

5 per cent. points is given. In doubtful cases, with a small number of centres, an exact test of significance may be made by assuming that in the absence of a true mean response, the responses observed at the centres would have occurred with positive or negative signs equally frequently. The complete distribution of the mean may be worked out, as exemplified in (1), pp. 50–4. Bartlett (7) has considered the approximation to this distribution by a continuous frequency distribution, and further work on his lines may enable us to assign an appropriate number of degrees of freedom to $s_x^2$ in any particular case by use of some statistic such as Fisher's $\zeta_2$ ((5), § 14). In a few examples I have worked out, the $t$-test gives a good approximation.

### § 8. *A Test of Significance of the Variation in the Local Experimental Errors.*

It sometimes happens that the individual experiments may all be regarded as having the same precision, and in this case, as pointed out in § 1, the analysis is much simplified. It is on this account worth having an idea of the amount of purely random sampling variation to be expected in the experimental errors. To obtain this, the experimental variances $s_i^2$ are each divided by their mean $\overline{s^2}$ over all $k$ centres. The corresponding values, multiplied by $n$, should be distributed approximately as $\chi^2$ with $n$ degrees of freedom, and their range may be compared with the published table (5).

A general test of significances of departure from the hypothesis that the variances are all estimates of the same quantity has been given by Neyman and Pearson (8). The test function which they recommend is

$$\prod_{i=1}^{k} \left( \frac{s_i^2}{\overline{s^2}} \right)^{n/2}$$

Tables have been given by Nayer (9).

In conclusion, while the above text has referred verbally to agricultural field experiments, the problems discussed are likely to turn up in any large-scale co-operative experiment. In particular the data considered by Neyman and Pearson (8), which arose in a factory experiment on the control of uniformity of product, are of exactly the same form as those discussed above. Their figures give the breaking strength under tension of small briquettes of cement-mortar. The cement was mixed on each of 10 different days, and 5 briquettes were tested each day. Thus the results provide a mean breaking strength $x_i$ for each day and an estimate $s_i$, based on 4 degrees of freedom, of the variability in the strengths within days. The questions which are of interest in this and similar factory experiments on control of quality may be somewhat different from those in agricultural field experiments. The question whether the standard errors $s_i$ vary from day to day, which in agricultural field experiments is not of practical importance, except in so far as it affects the efficiency of the experiments, and indeed is usually taken for granted, is one of the prime factors to be tested in a manufacturing experiment. On the other hand, the estimation of the mean of the $x_i$ and the question whether the $x_i$ have varied from day to day or from centre to centre is usually of common interest in both problems, and the discussion given above of the tests of significance may be of use in factory as well as in field experiments.

### Summary.

This paper considers the statistical analysis appropriate to experiments which yield, at each of a number of centres or times, an estimate $x_i$ of a treatment effect and an estimate $s_i$ of its standard error, based on $n$ degrees of freedom. This type of data may arise in many modern types of research, as, for instance, series of agricultural field experiments, or factory experiments on the control of quality. The problems considered are the estimation and test of significance of the mean treatment effect and of its variation from centre to centre, these being the most important preliminary questions in agricultural experiments of this type.

If the estimates $x_i$ may be considered equally accurate, *i.e.* if the quantities $s_i$ are all estimates of the same $\sigma$, the analysis of variance gives a convenient and familiar method of treatment. Where this is not so, the question is more difficult.

In the absence of any variation in treatment effect from centre to centre, the weighted mean $S\left(\dfrac{x}{s^2}\right)\bigg/ S\left(\dfrac{1}{s^2}\right)$ is suggested as a suitable estimate of the average treatment effect if at least 15 degrees of freedom are available in the estimates $s_i$. With fewer than 15 degrees of freedom, the weighted mean is not very efficient, and the maximum likelihood estimate, the solution of

$$S\,\frac{x-\mu}{ns^2+(x-\mu)^2}=0$$

is preferable, since its increased precision, which is equivalent to having three extra degrees of freedom for the estimation of the weights, is well worth the extra labour it involves. With fewer than 6 degrees of freedom in $s_i$, estimation of the weights involves a considerable loss of information. A comparison is made in this case of the relative efficiencies of the maximum likelihood solution, the unweighted mean, and the weighted mean with an arbitrarily chosen upper limit to the possible weights, in a set of hypothetical examples designed to cover the variation in experimental errors likely to occur in practice. The weighted mean with fixed upper limit is very satisfactory from the point of view of precision, but the difficulty of assigning a standard error to it is a serious disadvantage. Tests of significance of the ordinary weighted mean, the maximum likelihood solution and the unweighted mean are given.

Where the variation in treatment effect is not assumed non-existent, the unweighted mean should be used. The question of testing its significance by comparison with the variation in response from centre to centre is discussed.

The weighted sum of squares of deviations $Q = S \dfrac{1}{s^2} (x - \bar{x})^2$ is recommended to test the significance of the variation in treatment effect from centre to centre. An investigation of the frequency distribution of $Q$ is made and a transformation given by which it may be referred to the published table of $\chi^2$.

Further work is needed to determine more precise tests for the case of a few centres only.

I have to thank Mr. F. Yates for some useful suggestions.

*References.*

[1] R. A. Fisher, *The Design of Experiments.* Edinburgh : Oliver and Boyd, 1935, pp. 211–5.

[2] M. S. Bartlett, " The Information Available in Small Samples," *Proc. Camb. Phil. Soc.,* 1936, Vol. XXXII, p. 562.

[3] R. A. Fisher, " The Mathematical Distributions Used in the Common Tests of Significance," *Econometrika,* 1935, Vol. 3, No. 4, p. 356.

[4] H. Fairfield Smith, " The Problem of Comparing the Results of Two Experiments with Unequal Errors," *C.S.I.R. Journal,* 1936, Vol. 9, No. 3, pp. 211–2.

[5] R. A. Fisher, *Statistical Methods for Research Workers.* Edinburgh : Oliver and Boyd, 6th ed., 1936.

[6] W. G. Cochran, " The Efficiencies of the Binomial Series Tests of Significance of a Mean and of a Correlation Coefficient," *J. Roy. Stat. Soc.,* 1937, Vol. C, Part I, pp. 69–73.

[7] M. S. Bartlett, " The Effect of Non-Normality on the *t*-Distribution," *Proc. Camb. Phil. Soc.,* 1935, Vol. XXXI, p. 228.

[8] J. Neyman and E. S. Pearson, " On the Problem of $k$ Samples," *Bull. de l'Acad. Polonaise des Sciences et des Lettres,* A, 1931, pp. 460–81.

[9] P. P. N. Nayer, " An Investigation into the Application of the $L_1$-test, with Tables of Percentage Limits," *Statistical Research Memoirs,* 1936, Vol. I.

## THE EFFICIENCIES OF THE BINOMIAL SERIES TESTS OF SIGNIFICANCE OF A MEAN AND OF A CORRELATION COEFFICIENT.

### By W. G. COCHRAN, B.A.

#### 1. *Introduction.*

IN a preliminary survey of a set of data, one sometimes wishes to decide rapidly, without any elaborate calculation, whether there is any indication of a correlation between two sets of figures, or whether a set of differences appears to indicate a real effect. In such a case a rough test of significance is wanted, by means of which the data can be separated quickly into those effects which require further examination by more sensitive methods and those which can at once be set aside as either proved beyond dispute as real, or showing no signs of being real.

A convenient and familiar test of this type is provided by the binomial series distribution. In testing a set of differences, the numbers of positive and negative differences are counted, the sizes of the differences being ignored. If there is no real difference, and the errors are symmetrically distributed, positive and negative values should occur with equal frequency. The probability of obtaining $r$ differences of the same sign out of $n$ is

$$ {}^nC_r \frac{1}{2^n} $$

and the exact probability of obtaining $r$ or more than $r$ differences of the same sign is easily calculated. Table I shows, for values of

### TABLE I.

*Minimum Number (r) of Like Signs out of n Required to Reach 5 per cent. Significance.*

| $n$. | $r$. | $n$. | $r$. | $n$. | $r$. | $n$. | $r$. | $n$. | $r$. |
|---|---|---|---|---|---|---|---|---|---|
| — | — | 11 | 10 | 21 | 16 | 31 | 22 | 41 | 28 |
| — | — | 12 | 10 | 22 | 17 | 32 | 22 | 42 | 28 |
| — | — | 13 | 11 | 23 | 17 | 33 | 23 | 43 | 29 |
| — | — | 14 | 12 | 24 | 18 | 34 | 24 | 44 | 29 |
| — | — | 15 | 12 | 25 | 18 | 35 | 24 | 45 | 30 |
| 6 | 6 | 16 | 13 | 26 | 19 | 36 | 25 | 46 | 30 |
| 7 | 7 | 17 | 13 | 27 | 20 | 37 | 25 | 47 | 31 |
| 8 | 8 | 18 | 14 | 28 | 20 | 38 | 26 | 48 | 32 |
| 9 | 8 | 19 | 15 | 29 | 21 | 39 | 26 | 49 | 32 |
| 10 | 9 | 20 | 15 | 30 | 21 | 40 | 27 | 50 | 33 |

Above $n = 50$, $r$ may be taken as the smallest integer greater than $\left(\frac{n}{2} + \sqrt{n}\right)$.

$n$ up to 50, the number of differences of like sign, either positive or negative, required to reach the 5 per cent. level of significance. These numbers do not, of course, correspond exactly to 5 per cent. probabilities, since the binomial distribution is discrete. The probability of obtaining 9 or more like signs out of 10 is, for example, 0·0215, but the probability of obtaining 8 or more is 0·109, which does not reach the 5 per cent. level of significance. The 5 per cent. level of significance is not reached until $n = 6$, even if all signs are the same.

The test may also be used in examining a possible correlation between two sets of figures. The method consists in finding the mean of each set, and noting the number of pairs of values whose deviations from the mean are of the same sign. If this is unusually large, a positive correlation is indicated; if it is unusually small, a negative one. This test is perhaps more useful than the corresponding test on differences, in that the exact calculation of the correlation coefficient is more laborious to make than the test of a mean difference.

An example of the use of the binomial series distribution in testing a set of differences has been given by Fisher,* and may also be used to illustrate the test of a correlation. Table II is taken from the data given by Fisher.

TABLE II.

*Additional Hours of Sleep Gained by the Use of Hyoscyamine Hydrobromide.*

| Patient. | 1 (dextro-). | 2 (lævo-). | Difference (2 − 1). |
|---|---|---|---|
| 1 | + 0·7 | + 1·9 | + 1·2 |
| 2 | − 1·6 | + 0·8 | + 2·4 |
| 3 | − 0·2 | + 1·1 | + 1·3 |
| 4 | − 1·2 | + 0·1 | + 1·3 |
| 5 | − 0·1 | − 0·1 | 0·0 |
| 6 | + 3·4 | + 4·4 | + 1·0 |
| 7 | + 3·7 | + 5·5 | + 1·8 |
| 8 | + 0·8 | + 1·6 | + 0·8 |
| 9 | 0·0 | + 4·6 | + 4·6 |
| 10 | + 2·0 | + 3·4 | + 1·4 |
| Mean ($\bar{x}$) | 0·75 | + 2·33 | + 1·58 |

One of the differences is zero, but the remaining nine are all positive, which points strongly to a real superiority of Lævo- over Dextro-. The significance level indicated by this test is $2/2^9$ or 1 in 256, whereas the exact level given by the $t$ test is about 1 in 700. We might also wish to test whether the patients' responses to the two types of drug were correlated. It is easily verified from Table II that the responses to the two drugs lie on the same sides of their

* R. A. Fisher, *Statistical Methods for Research Workers*, § 24. Edinburgh: Oliver and Boyd.

respective mean responses, except for patients 8 and 9. Thus there are 8 similar responses out of 10. This does not reach the 5 per cent. significance level, as reference to Table I shows, but it is sufficiently near to call for further examination by the exact test of significance. We find $r = 0\cdot795$, which is significant at the 1 per cent. point.

The above is not a good illustration of the practical circumstances in which the tests would be useful. The number of pairs of observations (10) is small, and one would be reluctant to discard any information; further, for the same reason, the exact tests of significance can be made without labour. It is when the number of observations is large, say over 40, so that some information may be spared and the exact tests are more laborious, that the approximate tests will be useful in a preliminary analysis.

The object of this note is to compare the efficiencies of these tests with those of the exact tests, *i.e.* to determine how much information is thrown away by ignoring the sizes of the differences or of the deviations from the means. The point is of some importance, because if the tests prove to be of low efficiency, further examination may still be required when they give a result of non-significance.

### 2. *The Efficiency of the Binomial Series Test of a Difference.*

If the differences vary normally and independently about a mean $\mu$, with standard deviation $\sigma$, the probability $p$ of obtaining a positive difference is

$$p = \frac{1}{\sigma\sqrt{2\pi}} \int_0^\infty e^{-\frac{(x-\mu)^2}{2\sigma^2}} dx$$

$$= \frac{1}{\sqrt{2\pi}} \int_{-\frac{\mu}{\sigma}}^\infty e^{-y^2/2} dy$$

Hence, by estimating $p$ from the observed distribution of positive and negative signs, we obtain an estimate of $\frac{\mu}{\sigma} = \tau$ (say)

Now,     $\delta p = + \frac{1}{\sqrt{2\pi}} e^{\frac{-\tau^2}{2}} \delta\tau$

Hence     $V(\tau) = 2\pi e^{\tau^2} V(p) = 2\pi e^{\tau^2} pq/n.$

An efficient estimate of $\frac{\mu}{\sigma}$ is

$$\hat{\tau} = \frac{\bar{x}}{\sqrt{\frac{S(x-\bar{x})^2}{n-1}}} = \frac{t}{\sqrt{n}}, \text{ where } t \text{ is Student's } t.$$

And     $V(\hat{\tau}) = \frac{1}{n-2}.$

The efficiency of estimation by the binomial series is given by the ratio of the variances of $\hat{\tau}$ and $\tau$ when $n$ is large.

$$\text{Efficiency} = \frac{V(\hat{\tau})}{V(\tau)} = \frac{1}{2\pi pq e^{\tau^2}}.$$

Table III shows the percentage efficiency of estimation for a series of values of $\frac{\mu}{\sigma}$.

TABLE III.

*Percentage Efficiency of Estimation of $\frac{\mu}{\sigma}$ by Taking Account of Signs Only.*

| $\frac{\mu}{\sigma}.$ | 0 | $\pm \frac{1}{2}.$ | $\pm 1.$ | $\pm 1\frac{1}{2}.$ | $\pm 2.$ | $\pm 2\frac{1}{2}.$ | $\pm 3.$ |
|---|---|---|---|---|---|---|---|
| | 63·7 | 58·1 | 43·9 | 26·9 | 13·1 | 5·0 | 1·5 |

The efficiency is greatest when $\frac{\mu}{\sigma} = 0$ at which it has the value $\frac{2}{\pi}$. Thereafter it decreases steadily to zero. This is to be expected, since when $\frac{\mu}{\sigma}$ is large, the signs of the differences will all be positive except very rarely, and the estimate of $\frac{\mu}{\sigma}$ will almost always be $+\infty$.

The efficiency of the test as a test of significance is its efficiency of estimation when $\frac{\mu}{\sigma} = 0$, and is $\frac{2}{\pi}$, *i.e.* just under $\frac{2}{3}$. Thus about $\frac{1}{3}$ of the information is discarded by taking account of signs only.

### 3. *The Efficiency of the Binomial Series Test of a Correlation Coefficient.*

We may assume that the pairs of values, $x$ and $y$, are distributed about zero in a bivariate normal distribution. The probability of obtaining two deviations of like sign is

$$p = \frac{1}{2\pi\sigma_1\sigma_2\sqrt{1-\rho^2}} \int_{-\infty}^{0}\int_{-\infty}^{0} + \int_{0}^{\infty}\int_{0}^{\infty} e^{-\frac{1}{2(1-\varrho^2)}\left(\frac{x^2}{\sigma_1^2} - \frac{2\varrho xy}{\sigma_2\sigma_2} + \frac{y^2}{\sigma_2^2}\right)} dxdy$$

This may be written

$$p = \frac{1}{\pi\sqrt{1-\rho^2}} \int_{0}^{\infty}\int_{0}^{\infty} e^{-\frac{1}{2(1-\varrho^2)}(x^2 - 2\varrho xy + y^2)} dxdy$$

so that the value of $p$ gives an estimate of the correlation coefficient $\rho$.

$$p = \frac{1}{\pi\sqrt{1-\rho^2}} \int_0^\infty \int_0^\infty e^{-\frac{1}{2}y^2} e^{-\frac{1}{2}\frac{(x-\rho y)^2}{1-\rho^2}} \, dx\, dy$$

$$= \frac{1}{\pi} \int_0^\infty e^{-\frac{1}{2}y^2} dy \int_{\frac{-\rho y}{\sqrt{1-\rho^2}}}^\infty e^{-\frac{1}{2}u^2} du$$

$$= \tfrac{1}{2} + \frac{1}{\pi} \int_0^\infty e^{-\frac{1}{2}y^2} dy \int_0^{\frac{\rho y}{\sqrt{1-\rho^2}}} e^{-\frac{1}{2}u^2} \, du.$$

This may be written

$$\tfrac{1}{2} + \frac{1}{\pi} \int\int e^{-\frac{1}{2}(x^2 + y^2)} \, dx\, dy$$

over the sector of the $x$, $y$ plane in which $\theta$ lies between 0 and $\sin^{-1}\rho$. A change to polar co-ordinates gives at once

$$p = \tfrac{1}{2} + \frac{\sin^{-1}\rho}{\pi}$$

Hence

$$\frac{dp}{d\rho} = \frac{1}{\pi\sqrt{1-\rho^2}}$$

Hence if $\rho_b$ is the estimate of $\rho$

$$V(\rho_b) = \pi^2(1-\rho^2) \; V(p) = \frac{\pi^2(1-\rho)^2 pq}{n}$$

The variance of an efficient estimate of $\rho$ is $\dfrac{(1-\rho^2)^2}{n-1}$, so that the efficiency of estimation of $\rho$ by the binomial series distribution is

$$E = \frac{(1-\rho^2)}{\pi^2 pq}$$

The percentage efficiencies for a series of values of $\rho$ are shown in Table IV.

TABLE IV.

*Percentage Efficiency of Estimation of $\rho$ by Taking Account of Signs Only.*

| $\varrho$ | 0·0. | ±0·1. | ±0·2. | ±0·3. | ±0·4. | ±0·5. | ±0·6. | ±0·7. | ±0·8. | ±0·9. | ±1·0. |
|---|---|---|---|---|---|---|---|---|---|---|---|
| E | 40·5 | 40·3 | 39·6 | 38·3 | 36·6 | 34·2 | 31·2 | 27·5 | 22·4 | 15·6 | 0·0 |

The efficiency decreases slowly from the value $\dfrac{4}{\pi^2}$, at $\rho = 0$, to zero at $\rho = \pm 1$. The efficiency of the binomial series test of significance is 40 per cent. This rather low figure indicates that the test must be used with caution in setting aside data on the grounds that there is no apparent correlation. A glance at the sizes of the deviations from the mean will, however, generally enable one to decide whether the question requires further investigation.

# 10

## A Catalogue of Uniformity Trial Data.

### By W. G. Cochran, Rothamsted Experimental Station.

#### Some Uses of Uniformity Trial Data.

In a field uniformity trial, the area under experiment is divided into a number of plots, usually all of the same dimensions; the same variety of the crop is grown and the same manurial and cultural operations are carried out on each plot. The yield of each plot is recorded separately at harvest. In some cases other observations are made as well as yield, e.g. stand of plants. In the case of tree crops the plot may consist of a single tree or a group of trees.

The usefulness of a uniformity trial lies in the fact that neighbouring units may be amalgamated to form larger plots of various sizes and shapes. The variation in yield over the field due to soil heterogeneity, slight differences in the distribution of manures, errors in weighing, etc. (generally summed up in the term "experimental error"), may be calculated for each type of plot formed. The most obvious use of the data is to provide information on the optimum size and shape of plot, and this is the manner in which the majority of the trials given below have been used. In such studies, once the optimum size and shape have been determined, the standard error per plot and the number of replications required to reach a given degree of accuracy in the comparison of the mean treatment yields are also of interest. This type of information is not, of course, peculiar to uniformity trial data, but is supplied by every properly designed replicated experiment for the particular type of plot used.

A comprehensive study of the uniformity trial data on size and shape of plot has been made by Fairfield Smith,[1] who derives from them an empirical relation of wide applicability between variance per plot and size of plot.

Uniformity trials can also be used to compare the relative efficiencies of different types of experimental design, and, in particular, to test whether any newly proposed design seems suitable for a certain crop. For example, Yates[2] tested the efficiency of a new method of arranging variety trials on Parker and Batchelor's uniformity data with oranges (catalogue 64). Unfortunately only a small proportion of the trials given below are suitable for comparisons of this kind. For if a trial is intended to provide information on the optimum size and shape of plot, as most of the trials are, the smallest unit harvested requires to be somewhat smaller than the size of plot likely to be used in practice, so that various shapes of plot may be obtained by amalgamation. In consequence, many trials contain only a few plots of the size which is finally recommended.

The further question whether differences in soil heterogeneity from plot to plot in a field persist year after year is obviously of practical importance. Several trials have been continued on the same site for a number of years, some with the same crop, *e.g.* the trials on Ragi discussed by Lehmann (catalogue 77) and some with varying crops, *e.g.* the Huntley uniform cropping experiment (catalogue 1). As a rule, the yields of the same plot in successive years have been found to be positively correlated, whether the same crop followed or a different crop, but the closeness of the correlation has varied considerably.

The next step was to consider whether these correlations might be used to improve the accuracy of field experiments. With a high correlation it might clearly be worth while to run a uniformity trial as a preliminary to a field trial. The question of how to adjust the yields of the final experiment for differences shown in the uniformity trial at first caused some difficulty. The introduction of the statistical method known as the analysis of covariance, however, provided a means of correction free from any element of arbitrariness, and gave a stimulus to studies on the value of a uniformity trial as a preliminary to field experimentation. The results of these investigations are now well known. With annual agricultural crops, uniformity trials have not in general doubled the precision of subsequent field trials, whereas they entail approximately double the labour of a field trial with no previous uniformity trial, and a year's delay in the experimental results. With perennial plants, such as rubber, for example, where each plot consists of the same trees or bushes year after year, the gain in precision is decidedly higher, and preliminary records may often be obtained without much extra labour, or may indeed be part of a standard observational programme. The case for a preliminary uniformity trial is then considerably stronger. In animal nutrition work, also, the experimental unit is the same in the uniformity trial and the actual experiment, and the covariance method has proved strikingly successful in some such cases (cf. Bartlett).[3]

Uniformity trial data have also occasionally been used as a check on the applicability to field experiments of the analysis of variance and the tests of significance based on it. The mathematical theory from which the $z$ table is derived requires the assumptions that the experimental yields are normally distributed and that their deviations from the means about which they vary are uncorrelated. These assumptions are known to be untrue for field trials. A preliminary requirement for the application of the analysis of variance to be possible is that the experimental design used should be chosen at random from a set of designs such that, in the absence of any treatment effect, the average treatment mean square over the set should equal the average error mean square. The repeated use of the same design, however excellent in itself, is condemned on these grounds, and Tedin [4] has estimated the bias in the Knut Vik square from a set of uniformity trial data. The further question arises : how good an approximation to the tabulated $z$ distribution is generated by the process of randomization used ? There again the question may be tested from uniformity trial data. One such example has been worked out by Eden and Yates,[5] and further examples would be of considerable interest.

The large number of uniformity trials which have been carried out and the applications mentioned above testify that uniformity trial data play an important part in modern research on field technique. A catalogue of the uniformity trial data at present available therefore appears likely to be of value, in order to facilitate further research in field technique, and also to bring to light unpublished material which might otherwise be lost. With this end in view, we have at Rothamsted during the last few years been constructing a card index of such trials, and we have also encouraged workers with whom we have come into contact and who have conducted uniformity trials, but who have for various reasons been unable to examine the results, or who have examined them but have not published their conclusions, to file a copy of their material at Rothamsted. The following workers have furnished us with material of this nature :

| Worker. | Crop. | No. and size of plots, etc. | Total area. | Catalogue No. |
|---------|-------|-----------------------------|-------------|---------------|
| G. H. Goulden... | Barley | 2304 plots 3′ × 3′ | $\frac{1}{2}$ acre | 8 |
| F. J. Pound   ... | Cacao | Several thousand trees since 1914 | — | 14 |
| S. M. Gilbert   ... | Coffee | 12,000 tree yields of cherry | — | 18 |
| H. C. Ducker  ... | Cotton | 490 plots 1 row × 21′ | $\frac{3}{4}$ acre | 26 |
| O. V. S. Heath... | Cotton | 3696 individual plants | — | 28 |
| D. MacDonald ... | Cotton | 1152 plots $3\frac{1}{2}$′ × 30′ or 40′ | $3\frac{1}{2}$ acres | 30 |
| A. H. McKinstry | Cotton | 480 plots 1 row × 25′ | 1 acre | 31 |
| A. R. Saunders | Maize | 250 plots 1 row × 10 plants | — | 40 |
| Huntley (Mon.) | Oats | 46 plots $23\frac{1}{2}$′ × 317′ | 8 acres | 55 |
| J. Grantham * ... | Rubber | 1000 trees for 10 years | — | 85 |
| H. Evans      ... | Sugar Cane | 710 plots 5′ × 50′ | 6 acres | 106 |
| H. F. Smith   ... | Wheat | 1080 plots 6″ × 1′ | $\frac{1}{80}$ acre | 130 |

* Grantham's data on rubber have already been utilized by Murray in the paper referred to in the catalogue.

In most cases in which uniformity trials have been used as the basis of published work, but in which the original data have not been published, we have written to the authors concerned suggesting that they might like to file copies of these data at Rothamsted, so as to make it accessible to other workers. This suggestion has been met for nineteen trials.

Finally, in seven cases the data, although not published, are known to have been filed elsewhere.

The entries in the catalogue can therefore be classified as follows :

| Material. | No of entries in catalogue. | | No. of trials. | | Average no. of plots per trial. | |
|---|---|---|---|---|---|---|
| | Field crops. | Trees. | Field crops. | Trees. | Field crops. | Trees. |
| Already published    ... | 73 | 15 | 135 | 25 | 221 | 225 |
| Not published    ...    ... | 14 | 1 | 22 | 1 | 554 | 500 |
| Not published but copy filed    ...    ...    ... | | | | | | |
| ∫ at Rothamsted    ... | 21 | 4 | 28 | 4 | 539 | 3,761 |
| ∖ elsewhere    ...    ... | 5 | 1 | 6 | 1 | 1,440 * | 50 |

* One entry contains 203 sugar-cane trials each of 36 plots. This entry has been omitted when finding the average number of plots per trial.

As is to be expected from considerations of space, the average number of plots per trial is considerably greater for trials the yields of which have not been published than for those which have. This makes the recovery of such data the more valuable.

It must not be assumed that in the 14 entries given as " not published " the data are inaccessible. In some cases we have not been able to get into touch with the author, perhaps owing to change of address, and in others replies have not yet been received, but we hope in time to reduce this, and students are meanwhile advised to write to the author concerned about such trials.

This catalogue will not have been in vain if it has rescued from oblivion the 32 uniformity trials now filed at Rothamsted. It will be more valuable if, as we hope, it encourages other workers at present unknown to us, who have carried out uniformity trials, to furnish particulars of these, and if possible to make available copies of the original data. We should also be grateful for any information on omissions from this list. Although it is, we hope, fairly comprehensive as regards English (including Empire) and American Journals, we make no pretence to have searched the Continental literature at all thoroughly. This task we commend to some other worker.

In conclusion, we must thank the workers who helped in the compilation by sending data or information.

*References.*

[1] H. Fairfield Smith (in the press), *J. Agric. Sci.*, 1937.
[2] F. Yates, *ibid.*, 1936, Vol. 26, pp. 424–55.
[3] M. S. Bartlett, *ibid.*, 1935, Vol. 25, pp. 238–44.
[4] O. Tedin, *ibid.*, 1931, Vol. 21, pp. 191–208.
[5] T. Eden and F. Yates, *ibid.*, 1933, Vol. 23, pp. 6–17.

*The Catalogue.*

The entries are arranged alphabetically under crops, and for each crop alphabetically under author's names. The information given is as follows : the size and shape of the smallest unit harvested, its approximate area as a fraction of an acre, and the approximate total area (T.A.) occupied by the trial. In some cases complete information on these points was not available.

The following symbols have been used to show where the data may be found :—

| | |
|---|---|
| Published in the paper ...    ...    ...    ... | G |
| Not published    ...    ...    ...    ...    ... | N |
| Not published, but filed : | |
| at Rothamsted    ...    ...    ...    ... | R |
| Elsewhere    ...    ...    ...    ...    ... | E |

Notes have occasionally been added to the entries in cases where several measurements were made on the crop.

Alfalfa.    1. 46 plots, each $23\frac{1}{2}' \times 317' = \frac{1}{6}$ a.  T.A. 8 a.    G.

(1) 1912–14.  Harris, J. A., and Scofield, C. S.  Permanence of differences in the plats of an experimental field. *J. Agric. Res.*, **20**, 335–56.

(2) 1922–23–24.

Further studies on the permanence of differences in the plats of an experimental field. *J. Agric. Res.*, **36**, 15–40.

2. 36 plots, each $\frac{1}{20}$ a.  T.A. 2 a.    R.

3 years 1930–31–32.

Metzger, W. H.  The relation of varying rainfall to soil heterogeneity as measured by crop production. *J. Amer. Soc. Agron.*, **27**, 274–78.

3. 175  plots,  each  $13\cdot2' \times 13\cdot2' = \frac{1}{250}$ a.  G. T.A. $\frac{3}{4}$ a.

Summerby, R.  The value of preliminary uniformity trials in increasing the precision of field experiments. *Macdonald Coll. Tech. Bull.* 15.

к 2

Apples.    4. 512 individual tree yields.                                    G.

Batchelor, L. D., and Reed, H. S.   Relation of the variability of yields of fruit trees to the accuracy of field trials.  *J. Agric. Res.*, **12**, 245–83.

5. 50 individual tree yields.                                        E.

Collison, R. C., and Harlan, J. P.   Variability and size relations in apple trees.  *New York (Geneva), Agr. Exp. Sta. Tech. Bull.* 164, 1–38.

Yields filed at the New York State Agricultural Experiment Station, Geneva, N.Y.

6. 187 individual tree yields.                                       G.

Strickland, A. G.   Error in horticultural experiments.  *J. Dept. Agric. Victoria*, 1935, **32**, 408–16.

Time in weeks for a stored apple to reach 5 per cent. waste and 5 per cent. breakdown.

Barley.    7. 390 plots, each $4' \times 4' = \frac{1}{3000}$ a.   T.A. $\frac{1}{8}$ a.   G.

Bose, R. D.   Some soil heterogeneity trials at Pusa and the size and shape of experimental plots.  *Ind. J. Agric. Sci.*, **5**, 545.

8. 2304 plots, each $3' \times 3' = \frac{1}{4840}$ a.   T.A. $\frac{1}{2}$ a.   R.

Goulden, C. H.   Unpublished data.

9. (1) 30 plots, each $\frac{1}{100}$ a.   T.A. $\frac{1}{2}$ a.                G.

(2) 128 plots, each $\frac{1}{100}$ a.   T.A. 1 a.

Hanson, N. A.   Prøvedyrkning paa Forsøgsstationen ved Aarslev.  *Tids. for Landbrugets Planteavl.*, **21**, 553.

10. 46 plots, each $23\frac{1}{2}' \times 317' = \frac{1}{6}$ a.   T.A. 8 a.   G.

(1) 1912–14, Harris, J. A., and Scofield, C. S. Permanence of differences in the plats of an experimental field.  *J. Agric. Res.*, **20**, 335–56.

(2) 1922–23–24.

Further studies on the permanence of differences in the plats of an experimental field.  *J. Agric. Res.*, **36**, 15–40.

11. 234 plots, each $24\frac{1}{2}' \times 34\frac{1}{2}' = \frac{1}{50}$ a.   T.A. $4\frac{1}{2}$ a.                                                              G.

Kristensen, R. K.   Anlaeg og Opgrelsa of Marksforsq.  *Tids. for Landbrugets Planteavl.*, **31.**

12. 96 plots, each $3 \cdot 3' \times 3 \cdot 3' = \frac{1}{4000}$ a.    T.A. $\frac{1}{40}$ a. G.
N contents given, but not yields.
Barbacki, St.  *Mémoires de l'Institut National Polonais d'Economice Rural. à Putawy*, T. XIV, No. 213.

Cacao.

13. 500 trees : yields in pods.    N.
Cheesman, E. E., and Pound, F. J.  Uniformity trials on Cacao.  *Trop. Agric.*, **9**, 277–88.

14. Pound, F. J.  Unpublished data of several thousand trees since 1914.    R.

Clover.

15. 35 plots, each $13 \cdot 2' \times 66' = \frac{1}{50}$ a.    T.A. $\frac{3}{4}$ a.    G.
Each year 1928–32 on different parts of the same field.
Summerby, R.  The value of preliminary uniformity trials in increasing the precision of field experiments.

Coconuts.

16. 60 plots, each of 6 trees.  T.A. 12 a.    G.
Joachim, A. W. R.  A uniformity trial with coconuts.  *Tropical Agriculturist*, **85**, 4, 198–207.
Yields of nuts over 8 months.

17. 44 plots, each of 25 palms.
Yields each year from 1919 to 1928.
Beckett, W. H.    R.
Randomization in Field Experiment and its application on experiment stations.  *Bull. No. 20, Dept. of Agric., Gold Coast*, number of nuts given.

Coffee.

18. 12,000 individual tree yields of cherry for each  R. of 3 years.  Gilbert, S. M.  Unpublished data.

Corn.

19. 3 trials, 2304 plots, each 1 hill $\times$ 1 row (1) 1923, (2) 1925, (3) 1925.  T.A. $\frac{3}{4}$ a.    E.
Bryan, A. A.  Factors affecting experimental error in field plot tests with corn. *Iowa Agric. Expt. Sta. Report*, 1930–31, 67. Individual yields filed with Iowa Agric. Exp. Sta.

20.  450 plots, each $21' \times 68'$ ($3\frac{1}{2}'$ discard all round) $= \frac{1}{51}$ a.  T.A. 9 a.    G.
Garber, R. J., McIlvaine, J. C., and Hoover, M. M.  A method of laying out experimental plats.  *J. Amer. Soc. Agron.*, **23**, 286–98.

21. 46 plots, each $23\frac{1}{2}' \times 317' = \frac{1}{6}$ a.    T.A. 8 a.  G. 1915–16.
    Harris, J. A., and Scofield, C. S. Permanence of differences in the plats of an experimental field. *J. Agric. Res.*, **20**, 335–56.

22. 36 plots, each $\frac{1}{20}$ a.    T.A. $1\frac{1}{2}$ a.                    R.
    Metzger, W. H.  The relation of varying rainfall to soil heterogeneity as measured by crop production. *J. Amer. Soc. Agron.*, **27**, 274–78.

23. 438 plots, each 1 row $\times$ 66' $= \frac{1}{181}$ a.  T.A. $2\frac{1}{2}$ a.                                       G.
    McClelland, C. K.  Some determinations of plot variability. *J. Amer. Soc. Agron.*, **18**, 819–23.

24. 120 plots, each $\frac{1}{10}$ a.  T.A. 12 a.            G.
    Smith, L. H.  Plot arrangement for variety experiments with corn. *Proc. Amer. Soc. Agron.*, **1**, 84–89.

Cotton.

25. 5 trials, each of about 160 plots, each 20 ridges $\times$ 7 metres $= \frac{1}{40}$ a.  T.A. 4 a.            G.
    Bailey, M. A., and Trought, T.  An account of experiments carried out to determine the experimental error of field trials with cotton in Egypt. *Min. Agric. Egypt Tech. and Sc. Service Bull.* 63.

26. 490 plots, each 1 row $\times$ 21' $= \frac{1}{700}$ a.  T.A. $\frac{3}{4}$ a.                                        R.
    Ducker, H. C.  Unpublished data.

27. (1) 200 plots, each 1' $\times$ 24' $= \frac{1}{1800}$ a. T.A. $\frac{1}{9}$ a.                                       N.
    (2) 200 plots, each $4\frac{1}{2}' \times 16' = \frac{1}{600}$ a. T.A. $\frac{1}{3}$ a.
    Fu Siao.  Uniformity trials with cotton. *J. Amer. Soc. Agron.*, **27**, 12.

28. 3696 individual plants.                    R.
    Heath, O. V. S.  Unpublished data.
    Height, node number and dry matter of individual cotton plants.

29. 1280 plots, each 4 rows $\times$ 4·8' $= \frac{1}{1900}$ a. T.A. $\frac{3}{4}$ a.                                       R.
    Hutchinson, J. B., and Panse, V. G. Studies in the technique of field experiments. I. *Indian J. Agric. Sci.*, **5**, 523–38.

30. (1) 576 plots, each $3\cdot5' \times 40' = \frac{1}{300}$ a.  T.A.
2 a.

    (2) 576 plots, each $3\cdot5' \times 30' = \frac{1}{400}$ a.
T.A. $1\frac{1}{2}$ a.                                                    R.

MacDonald, D.  Unpublished data.

31. 480 plots, each 1 row $\times 25' = \frac{1}{500}$ a.  T.A. 1 a.  R.
McKinstry, A. H.  Unpublished data.

32. (1) 300 plots, each $3' \times 48' = \frac{1}{300}$ a.  T.A. 1 a.  R.

    (2) 700 plots, each $3\frac{1}{3}' \times 47' = \frac{1}{300}$ a.  T.A.
$2\frac{1}{2}$ a.

Reynolds, E. B., Killough, D. T., and Vantine, J. T.

Size, shape and replication of plats for field experiments with cotton. *J. Amer. Soc. Agron.*, **26**, 725–34.

**Fodder Corn.**

33. 63 plots, each $15' \times 112\frac{1}{2}' = \frac{1}{26}$ a.  T.A. $2\frac{1}{2}$ a.  G.
Morgan, J. O.  Some experiments to determine the uniformity of certain plats for field tests. *Proc. Amer. Soc. Agron.*, **1**, 58–67.

**Grapes.**

34. 200 vines, 8′ apart in rows 10′ apart.                        G.
Strickland, A. G., Forster, H. C., and Vasey, A. J.  A vine uniformity trial. *J. Agric. of Victoria*, **30**, 584.

**Hops.**

35. 30 plots, each 1 row $\times 210'$.  Yields each year. 1909–14.                                                             G.
Stockberger, W. W.  Relative precision of formulæ for calculating normal plot yields. *J. Amer. Soc. Agron.*, **8**, 167–75.

**Lemons.**

36. 364 individual tree yields.                                    G.
Batchelor, L. D., and Reed, H. S.  Relation of the variability of yields of fruit trees to the accuracy of field trials. *J. Agric. Res.*, **12**, 245–83.

**Lentils.**

37. 390 plots, each $4' \times 4' = \frac{1}{2270}$ a.  T.A. $\frac{1}{4}$ a.  G.
Bose, R. D.  Some soil heterogeneity trials at Pusa and the size and shape of experimental plots. *Ind. J. Agric. Sci.*, **5**, 545.

**Maize.**

38. 83 plots, each $33' \times 33' = \frac{1}{40}$ a.  T.A. $2\frac{1}{8}$ a.  G.
Beckett, W. H., and Fletcher, S. R. B.  A uniformity trial with maize. *Gold Coast Dept. Agric. Bull.* **16**, 222–26.

Germination and ear number counts given.
Yields measured for 15 plots only.

39. 300 plots, each 1 row $\times$ 60'.                    R.
    Saunders, A. R.   Statistical methods with special reference to field experiments. *Union of South Africa, Dept. of Agric. and Forestry, Science Bull.* 147.

40. 250 plots, each 1 row $\times$ 10 plants.              R.
    Saunders, A. R.   Unpublished data.

41. (1) 175 plots each $13 \cdot 2' \times 13 \cdot 2' = \frac{1}{250}$ a. T.A. $\frac{3}{4}$ a.                                    G.
    Yields each year from 1922–26.
    (2) 35 plots, each $13 \cdot 2' \times 66' = \frac{1}{50}$ a. T.A. $\frac{3}{4}$ a.
    Yields each year from different ranges 1927, 1928, 1929, 1930, 1931, 1932.
    Summerby R.   The value of preliminary uniformity trials in increasing the precision of field experiments.

Mangolds. 42. 30 plots, each $\frac{1}{100}$ a.   T.A. $\frac{1}{2}$ a.        G.
    Hanson, N. A.   Prøvedyrkning paa Forsøgsstationen ved Aarslev. *Tids. for Landbrugets Planteavl.*, **21**, 553.

43. 200 plots, each 3 rows $\times$ $30\frac{1}{4}' = \frac{1}{200}$ a.   T.A. 1 a.                                    G.
    Mercer, W. B., and Hall, A. D.   The experimental error of field trials. *J. Agric. Sci.*, **4**, 107–132.

44. (1) 175 plots, each $13 \cdot 2' \times 13 \cdot 2' = \frac{1}{250}$ a. T.A. $\frac{3}{4}$ a.                                    G.
    (2) 150 plots, each $13 \cdot 2' \times 13 \cdot 2' = \frac{1}{250}$ a. T.A. $\frac{1}{2}$ a.
    Summerby, R.   The value of preliminary uniformity trials in increasing the precision of field experiments.

45. 1050 plots, each $\frac{1}{1000}$ a.   T.A. 1 a.        N.
    Wood, T. B., and Stratton, F. J. M.   The interpretation of experimental results. *J. Agric. Sci.*, **3**, 417–40.

Millet. 46. 105 plots, each $\frac{1}{200}$ a.   T.A. $\frac{1}{2}$ a.        G.
    Lehmann, A. *Report of Agricultural Chemist. Dept. of Agric. Mysore State,* 1900–7. Roemer, Th. Der Feldversuch. *Arbeiten der Deutschen Landw. Gesellschaft.,* 302.

47. 600 plots, each $1' \times 15' = \frac{1}{3000}$ a.   T.A. $\frac{1}{4}$ a.   G.
   Li, H. W., Meng, C. J., and Liu, T. N.
   Field results in a millet-breeding experiment.

Mushrooms.   48. (1) 50 plots each $2' \times 5' = \frac{1}{4840}$ a.   T.A. $\frac{1}{100}$ a.   G.
   (2) 50 plots each $4' \times 6' = \frac{1}{1860}$ a.   T.A. $\frac{1}{40}$ a.
   (3) 40 plots each $4' \times 6' = \frac{1}{1860}$ a.   T.A. $\frac{1}{50}$ a.
   Lambert, E. B.   Size and arrangement of plots for yield tests with cultivated mushrooms.   *J. Agric. Res.*, **48**, 1971–80.

Oats.   49. (1) 66 plots, each $\frac{1}{4}$ a.   T.A. 17 a.   R.
   (2) 68 plots, each $\frac{1}{4}$ a.   T.A. 17 a.
   Farrell, F. D.   Interpreting the variability of plat yields.   *U.S. Dept. of Agric. Bureau of Plant Industry Circular No.* 109, 27–32.

50. 295 plots, each $21' \times 68' = \frac{1}{51}$ a.   T.A. 6 a.   G.
   Garber, R. J., McIlvaine, T. C., and Hoover, M. M.   A study in soil heterogeneity in experiment plots.   *J. Agric. Res.*, **33**, 255–68.

51. 450 plots, each $21' \times 68'$ ($3\frac{1}{2}$ feet discard all round) $= \frac{1}{51}$ a.   T.A. 9 a.   G.
   Garber, R. J., McIlvaine, T. C., and Hoover, M. M.   A method of laying out experimental plots.   *J. Amer. Soc. Agron.*, **23**, 286–98.

52. (1) 200 plots (3 yields missing) each $= \frac{1}{450}$ a.   G.
   (2) 300 plots (3 yields missing) each $= \frac{1}{450}$ a.
   Gorski, M., and Stefaniow, M.   Die Anwendbarkeit der Wahrscheinlichkeitsrechnung bei Feldversuchen.   *Landw. Versuchsstationen*, **90**, 225–40.

53. (1) 30 plots, each $\frac{1}{100}$ a.   T.A. $\frac{1}{2}$ a.   G.
   (2) 128 plots, each $\frac{1}{100}$ a.   T.A. 1 a.
   Hanson, N. A.   Prøvedyrkning paa Forsøgsstationen ved Aarslev.   *Tids. for Landbrugets Planteavl.*, **21**, 553.

54. 46 plots, each $23\frac{1}{2}' \times 317' = \frac{1}{6}$ a.   T.A. 8 a.   G.
   1917.   Harris, J. A., and Scofield, C. S.   Permanence of differences in the plats of an experimental field.   *J. Agric. Res.*, **20**, 335–56.

55. 46 plots, each $23\frac{1}{2}' \times 317' = \frac{1}{6}$ a.   T.A. 8 a.   R.
   Same trial as No. 54.   1911, total produce only.

56. 207 plots, each $\frac{1}{30}$ a.  T.A. 7 a.                          G.
   Kiesselbach, T. A.  Studies concerning the elimination of experimental error in comparative crop tests.  *Res. Bull. Nebraska Agric. Stat.*, **13**, 1–95.

57. 36 plots, each $\frac{1}{20}$ a.  T.A. $1\frac{1}{2}$ a.                          R.
   Metzer, W. H.  The relation of varying rainfall to soil heterogeneity as measured by crop.  *J. Amer. Soc. Agron.*, **27**, 274.

58. 24 plots, each $33' \times 132' = \frac{1}{10}$ a.  T.A. $2\frac{1}{2}$ a.  G.
   McClelland, C. K.  Some determinations of plot variability.  *J. Amer. Soc. Agron.*, **18**, 819–23.

59. 240 plots, each $\frac{1}{1200}$ a.  T.A. $\frac{1}{4}$ a.                          G.
   Roemer, Th.  Der Feldversuch.  *Arbeiten der deutschen Landw. Gesellschaft*, 302.

60. 48 plots, each $\frac{1}{10}$ a.  T.A. 5 a.                          G.
   *Roth. Exp. Sta. Report*, 1927–28, p. 153.

61. 512 plots, each $1' \times 15' = \frac{1}{3000}$ a.  T.A. $\frac{1}{4}$ a.  G.
   Summerby, R.  A study of size of plats, numbers of replications, and the frequency and methods of using cheek plats in relation to accuracy in field experiments.  *J. Amer. Soc. Agron.*, **17**, 140–50.

62. (1) 175 plots, each $13\cdot2' \times 13\cdot2' = \frac{1}{250}$ a.  T.A. $\frac{3}{4}$ a.                          G.
   Yields each year 1922–26 and 1924–25–26.
   (2) 35 plots, each $13\cdot2' \times 66' = \frac{1}{50}$ a.  T.A. $\frac{3}{4}$ a.  Each year on different ranges from 1927 to 1932.
   Summerby, R.  The value of preliminary uniformity trials in increasing the precision of field experiments.

63. 124 plots, each $33' \times 132' = \frac{1}{10}$ a.  T.A. $12\frac{1}{2}$ a.                          G.
   Wyatt, F. A.  Variation in plot yields due to soil heterogeneity.  *Sci. Agr.*, **7**, 248–56.

Oranges.    64. (1) 1000 individual tree yields.                          G.
   (2) 495 individual tree yields.
   (3) 240 individual tree yields.
   Batchelor, L. D., and Reed, H. S.  Relation of the variability of yields of fruit trees to the accuracy of field trials.  *J. Agric. Res.*, **12**, 245–83.

65. 193 plots, each of 8 trees, yields given each year
from 1921 to 1927.                                    G.
    Parker, E. D., and Batchelor, L. D.   Varia-
tion in the yields of fruit trees in relation to the
planning of future experiments.  *Hilgardia* 7,
No. 2, 1932.

Paddy.
66. (1) 104 plots, each $6·6' \times 122' = \frac{1}{54}$ a.  T.A.
2 a.                                                  N.
    (2) 72 plots, each $6·6' \times 174' = \frac{1}{38}$ a.   T.A.
2 a.
    Lord, L.  Irrigated paddy : a contribution
to the study of field plot technique.  *Agric. J.
India*, **19**, 20–27.

Pasture.
67. 760 plots, each $6·6' \times 3·3' = \frac{1}{2000}$ a.  T.A. $\frac{1}{2}$ a. G.
    Davies, J. G.  The experimental error of the
yield from small plots of natural pasture.
*Council Sci. and Indust. Res. (Aust.) Bull.* 48.

Peaches.
68. 144 individual tree yields.                        G.
    Strickland, A. G.  Error in horticultural
experiments.  *J. Dept. Agric. Victoria*, **33**,
408–16.

Pineapples.
69. (1) 24 plots, each 4 rows $\times$ 75'.            G.
(2) 24 plots, each 4 rows $\times$ 75'.
(3) 25 plots, each 4 rows $\times$ 60'.
    Magistad, O. C., and Farden, C. A.  Experi-
mental error in field experiments with pineapples.  *J. Amer. Soc. Agron.*, **26**, 631–44.

Potatoes.
70. 750 single-row plots.                              E.
    Jakowski, Z.  Unpublished data, see Ney-
man, J.  Statistical problems in agricultural
experimentation.  *J. Roy. Stat. Soc. Suppl.*, **2**,
2, 107–54.
    Yields filed with J. Neyman.

71. 618 plots, each $2·2' \times 33·5' = \frac{1}{600}$ a.  T.A. 1 a. E.
    Justesen, S.H.  Influence of size and shape of
plots on the precision of field experiments
with potatoes.  *J. Agric. Sci.*, **22**, 366–72.
    Data filed with the N.I.A.B., Cambridge,
England.

72. 576 plots, each $3' \times 22' = \frac{1}{700}$ a.  T.A. 1 a.  G.
    Kalamkar, R. J.  Experimental error and
the field plot technique with potatoes.  *J.
Agric. Sci.*, **22**, 373–85.

73. 204 plots, each $2\frac{3}{4}' \times 72\frac{1}{2}' = \frac{1}{210}$ a.   T.A. 1 a.   G.
Lyon, T. L.   Some experiments to estimate errors in field plat tests. *Proc. Amer. Soc. Agron.*, **3**, 89–114.

74. 720 plants, every fifth hill missing.                        G.
Stewart, F.   Missing hills in potato fields : their effect upon the yield. *New York Agric. Exp. Sta. Bull.* 459, 45–69.

75. 4 sets, each $3' \times 15' = \frac{1}{1000}$ a.                        N.
(1) 1000 plots.   (2) 1560 plots.   (3) 2000 plats.   (4) 1000 plots.
Thompson, R. C.   Size, shape, etc., in sweet potatoes field-plot experiments. *J. Agric. Res.*, **48**, 379–99.

76. 51 plots, each $\frac{1}{20}$ a.   T.A. $2\frac{1}{2}$ a.                        N.
Westover, K. C.   The influence of plat size and replication on experimental errors in field trials with potatoes. *West. Virginia. Agr. Expt. Sta. Bull.* 189.

**Ragi.**

77. 34 plots, each $\frac{1}{200}$ a.   T.A. $\frac{1}{6}$ a.                        G.
Yields for 4 years 1905–8.
Lehmann, A.   *Report of Agric. Chemist. Dept. of Agric. Mysore State*, 1900–7.   See also Roemer, *Der Feldversuch.*   1st Ed.

**Rice.**

78. (1) 144 plots each $5' \times 5' = \frac{1}{1740}$ a.   T.A. $\frac{1}{12}$ a.                        N.
(2) 144 plots each $5' \times 5' = \frac{1}{1740}$ a.   T.A. $\frac{1}{12}$ a.
Plots arranged in a $12 \times 12$ Latin square.
Bose, S. S., Ganguli, P. M., and Mahalanobis, P. C.   The frequency distribution of plot yields and the optimum size of plots in a uniformity trial with rice in Assam. *Indian J. Agric. Sci.*, 1936, **6** part 5, pp. 1107–22.

79. 3 series of 100 plots, each $1\frac{1}{2}' \times 14\cdot 2' = \frac{1}{2000}$ a. T.A. $\frac{1}{20}$ a.                        N.
Chien-Liang-Pan.   Uniformity trials with rice. *J. Amer. Soc. Agron.*, **27**, 279.

80. 54 plots each $33' \times 33' = \frac{1}{40}$ a.   T.A. $1\frac{1}{4}$ a.                        G.
Coombs, G. E., and Grantham, J.   Field experiments and the interpretation of their results. *Agr. Bull. Fed. Malay States*, 4.

81. 300 plots, each $1 \cdot 5' \times 14 \cdot 25' = \frac{1}{2000}$ a.  T.A. $\frac{1}{6}$ a.  N.

Li-Ying-Shen. Statistical analysis of a blank test of rice with suggestions for field technique. *Agricultura Sinica*, 1934, **1** No. 4, pp. 107–50.

82. 560 plots, each $10' \times 10' = \frac{1}{450}$ a.  T.A. $1\frac{1}{4}$ a.  G.

Lord, L. A uniformity trial with irrigated broadcast rice. *J. Agric. Sci.*, **21**, 178–86.

83. Plots $3' \times 3' = \frac{1}{4840}$ a.  N.

Mitra, S. H., and Ganguli, P. M. A uniformity trial in rice. *Proc. 21st Annual Indian Sci. Congress Bombay*, 1934, 71.

84. 280 plots, each 2 rows $\times$ 10 plants.  N.

Parnell, F. R. Experimental error in variety tests with rice. *Agric. J. India*, **14**, 747–57.

See also No. 66.

Rubber.

85. 1000 trees yields for each of 10 years.  R.

Murray, R. K. S. The value of a uniformity trial in field experimentation with rubber. *J. Agric. Sci.*, **24**, 177–84.

86. 161 trees each year from 1921–22 to 1924–25.  G.

Taylor, R. A. The inter-relationship of yield and the various vegetative characters in *Hevea Brasilensia*. *Dept. of Agric. Ceylon Bull.* 77.

Rye.

87. (1) 30 plots, each $\frac{1}{100}$ a.  T.A. $\frac{1}{2}$ a.  G.
    (2) 128 plots, each $\frac{1}{100}$ a.  T.A. 1 a.

Hanson, N. A. Prøvedyrkning paa Forsøgsstationen ved Aarslev. *Tids. for Landbrugets Planteavl.*, **21**, 553.

Seeds.

88. 128 plots, each $\frac{1}{100}$ a.  T.A. 1 a.  G.

Hanson, N. A. Prøvedyrkning paa Forsøgsstationen ved Aarslev. *Tids. for Landbrugets Planteavl.*, **21**, 553.

Silage corn.

89. 46 plots, each $23\frac{1}{2}' \times 317' = \frac{1}{6}$ a.  T.A. 8 a.  G.

1918. Harris, J. A., and Scofield, C. S. Permanence of differences in the plats of an experimental field. *J. Agric. Res.*, **20**, 335–56. 1920, 1925.

Further studies on the permanence of differences in the plats of an experimental field. *J. Agric. Res.*, **36**, 15–40.

Sorghum.     90. 160 plots, each $\frac{1}{160}$ a. for 1930–31–32.  T.A. 1 a.  G.

Kulkarni, R. K., Bose, S. S., and Mahalanobis, P. C.  The influence of shape and size of plots on the effective precision of field experiments with sorghum.  *Indian J. Agric. Sci.*, **6**, 460–74.

91. 2000 plots, each 1 row $\times$ 1 rod $= \frac{1}{792}$ a.  T.A. $2\frac{1}{2}$ a.     G.

Stephens, J. C., and Vinall, H. N.  Experimental methods and the probable error in field experiments with sorghum.  *J. Agric. Res.*, **37**, 629–46.

92. 400 plots, each $3\cdot3' \times 33' = \frac{1}{400}$ a.  T.A. 1 a.  N.

Swanson, A. F.  Variability of grain sorghum yields as influenced by size, shape and number of plats.  *J. Amer. Soc. Agron.*, **22**, 833–38.

Sorgo.     93. 36 plots, each $\frac{1}{20}$ a.  T.A. $1\frac{3}{4}$ a.     R.

2 years, 1932–33.

Metzger, W. H.  The relation of varying rainfall to soil heterogeneity as measured by crop production.  *J. Amer. Soc. Agron.*, **27**, 274–78.

Soy beans.     94. 30 plots : artificially constructed in frames, each $4\frac{2}{3}' \times 9\frac{1}{3}' = \frac{1}{1000}$ a.  T.A. $\frac{1}{30}$ a.     G.

Garber, R. J., and Pierre, W. H.  Variation of yields obtained in small artificially constructed field plats.  *J. Amer. Soc. Agron.*, **25**, 98–105.

95. (1) 882 plots, each 1 row $\times$ 8$' = \frac{1}{2100}$ a.  T.A. $\frac{1}{2}$ a.     G.

(2) 1540 plots, each 1 row $\times$ 8$' = \frac{1}{2100}$ a.  T.A. $\frac{3}{4}$ a.

Odland, T. E., and Garber, R. J.  Size of plot and number of replications in field experiments with soy beans.  *J. Amer. Soc. Agron.*, **20**, 94–108.

Strawberries.  96. (1) 120 plots, each 4$' \times$ 68$' = \frac{1}{160}$ a.  T.A. $\frac{3}{4}$ a.  N.

(2) 80 plots, each 4$' \times$ 34$' = \frac{1}{320}$ a.  T.A. $\frac{1}{4}$ a.

Wilcox, A. N.  A study of field plot technique with strawberries.  *Scientific Agriculture*, **8**, 171–74.

Sugar-beet.  97. 46 plots, each $23\frac{1}{2}' \times 317' = \frac{1}{6}$ a.  T.A. 8 a.  G.
Harris, J. A., and Scofield, C. S.  Permanence of differences in the plats of an experimental field.  *J. Agric. Res.*, **20**, 335–56.

98. 600 plots, each 1 row $\times$ 33$' = \frac{1}{700}$ a.  T.A. 1 a. G.
Immer, F. R.  Size and shape of plots in relation to field experiments with sugar-beets.  *J. Agric. Res.*, **44**, 649–68.

99. 600 plots, each 1 row $\times$ 33$' = \frac{1}{700}$ a.  T.A. 1 a.  R.
Immer, F. R., and Raleigh, S. M.  Further studies of size and shape of plot in relation to field experiments with sugar-beet.  *J. Agric. Res.*, **47**, 591–98.

100. 416 plots, each $8' \times 135' = \frac{1}{40}$ a.  T.A. $10\frac{1}{2}$ a.  G.
Roemer, Th.  Der Feldversuch.  *Arbeiten der deutschen Landw. Gesellschaft*, **302**.

101. 96 plots, each 1 row $\times$ 55·8$'$.  G.
Two sets, 1916 and 1918.
Roemer, Th.  Der Feldversuch.  *Arbeiten der deutschen Landw. Gesellschaft*, **302**.

Sugar cane. 102. 49 plots, each $\frac{1}{50}$ a.  T.A.  1 a.  R.
Barbados, 1927.

103. 48 plots, each $30' \times 75' = \frac{1}{19}$ a.  T.A. $2\frac{1}{2}$ a.  G.
Borden, R. J.  Replications of plot treatments in field experiments.  *Hawaiian Planters' Record*, **34**, 151–55.

104. 203 trials each of 36 plots.  E.
Demandt, E.  Die Resultaten der Blanco-Proeven met 2878 PoJ van Oogstjaar 1931.  Archief voor de Suikerindustrie in Nederlandsch-Indië Deel III.  *Med. van het Proefstation voor Java Suikerindustrie Jahrgang*, 1932, 14.
Yields filed at the Proefstation voor de Java.  Suikerindustrie, Soerabaia, Java.

105. Yields of 1200 individual stools.  G.
Evans, H.  Some preliminary data concerning the best shape and size of plot for field experiments with sugar cane.  *Dept. of Agric. Mauritius.  Sugar Cane Research Station Bull.* 3.

106. 710 plots, each $5' \times 50' = \frac{1}{175}$ a.  T.A. 6 a.  R.
H. Evans.  Unpublished data.

107. (1) 960 plots, each $3' \times 30\frac{1}{4}' = \frac{1}{480}$ a.  T.A. $1\frac{3}{4}$ a.  R.
     (2) 1088 plots, each $3' \times 60' = \frac{1}{242}$ a.  T.A. $4\frac{1}{2}$ a.                                                     G.

Wynne Sayer, Vaidyanathan and Subramaria Iyer.  Ideal size and shape of sugar-cane experimental plots based upon tonnage experiments with Co 205 and Co 213 conducted in Pusa.  *Indian J. Agric. Sci.*, 1936, **6**.

108. 968 plots, each $3' \times 60' = \frac{1}{242}$ a.  T.A. 4 a.  G.

Wynne Sayer and Krishna Iyer.  On some of the factors that influence the error of field experiments with special reference to sugar cane.  *Indian J. Agric. Sci.*, 1936, **6**, 917.

Swedes.  109. 48 plots, each $\frac{1}{10}$ a.  T.A. 5 a.                        G.
Roth. Exp. Sta. Report, 1925–26.
Roots, Tops and Plant number given.

Tea.  110. 144 plots, each $\frac{1}{72}$ a.  T.A. 2 a.                          G.
Eden, T.  Studies in the yield of tea.  *J. Agric. Sci.*, **21**, 547–73.
Yields and dry matter at 94° C. given.

111. 24 plots.                                                                   G.
Vaidyanathan, M.  The method of covariance applicable to the utilization of the previous crop records for judging the improved precision of experiments.  *Ind. J. Agric. Sci.*, **4**, 327–42.

Timothy hay.  112. 240 plots, each $16\frac{1}{2}' \times 16\frac{1}{2}' = \frac{1}{160}$ a.  T.A. $1\frac{1}{2}$ a.                  G.
Holtsmark, G. U., and Larsen, B. R.  Über die Fehler, welche bei Feldversuchen durch die Ungleichartigkeit des Bodens bedingt werden.  *Landw. der Versuchsstationen*, **65**, 1–22.  See also Roemer, Th., *der Feldversuch.*  1st Ed.

113. 35 plots, each $13 \cdot 2' \times 66' = \frac{1}{50}$ a.  T.A. $\frac{3}{4}$ a.  G.
Each year 1929–32 on different parts of the same field.
Summerby, R.  The value of preliminary uniformity trials in increasing the precision of field experiments.

Tomatoes.  114. 180 plots, each of 6 plants.                                     G.
Strickland, A. G.  Error in horticultural experiments.  *J. Dept. Agric. Victoria*, **32**, 408–16.

Walnuts.   115. 320 individual seedling tree yields.   G.
Batchelor, L. D., and Reed, H. S.   Relation of the variability of yields of fruit trees to the accuracy of field trials.   *J. Agric. Res.*, **12**, 245–83.

Wheat.   116. 390 plots, each $4' \times 4' = \frac{1}{3000}$ a.   T.A. $\frac{1}{8}$ a.   G.
Bose, R. D.   Some soil heterogeneity trials at Pusa and the size and shape of experimental plots.   *Ind. J. Agric. Sci.*, **5**, 545.

117. (1) 288 plots, each $8'' \times 7\frac{1}{2}' = \frac{1}{9000}$ a.   T.A. $\frac{1}{30}$ a.   G.
(2) 288 plots, each $8'' \times 8' = \frac{1}{8000}$ a.   T.A. $\frac{1}{25}$ a.   R.
Christidis, B. G.   The importance of the shape of plots in field experimentation.   *J. Agric. Sci.*, **21**, 14–37.

118. 3100 plots, each $8'' \times 5' = \frac{1}{13000}$ a.   T.A. $\frac{1}{4}$ a.   N.
Day, J. W.   The relation of size, shape and number of replications of plots to probable error in field experimentation.   *J. Amer. Soc. Agron.*, **12**, 100–5.

119. 160 plots, each $13 \cdot 2' \times 19 \cdot 8' = \frac{1}{160}$ a.   T.A. 1 a.   G.
Forster, H. C., and Vasey, A. J.   Experimental error of field trials in Australia.   *Victoria J. Dept. Agric.*, **27**, 385–95.

120. 450 plots, each $21' \times 68'$ ($3\frac{1}{2}'$ discard all round) $= \frac{1}{51}$ a.   T.A. 9 a.   G.
Garber, R. J., McIlvaine, T. C., and Hoover, M. M.   A method of laying out experimental plots.   *J. Amer. Soc. Agron.*, **23**, 286–98.

121.  295 plots, each $21' \times 68'$ (a border of $3\frac{1}{2}'$ all round rejected) $= \frac{1}{51}$ a.   T.A. $5\frac{3}{4}$ a.   G.
Garber, R. J., McIlvaine, T. C., and Hoover, M. M.   A study in soil heterogeneity in experimental plots.   *J. Agric. Res.*, **33**, 255–68.
Yields obtained by sampling 5 rod rows.

122. 30 plots, artificially constructed in frames each $4\frac{2}{3}' \times 9\frac{1}{3}'$.   G.
Garber, R. J., and Pierre, W. H.   Variation of yields obtained in small artificially constructed field plots.   *J. Amer. Soc. Agron.*, **25**, 98–105.

123. 1280 plots, each $\frac{1}{2}' \times 1\cdot6' = \frac{1}{60000}$ a.  T.A.
$\frac{1}{40}$ a.                                                    G.
      Kalamkar, R. J.   A study in sampling tech-
nique with wheat.  *J. Agric. Sci.*, **22**, 783–96.

124. 500 plots, each 11 rows $\times$ 10·82$' = \frac{1}{500}$ a.
T.A. 1 a.                                                           G.
      Mercer, W. B., and Hall, A. D.   The experi-
mental error of field trials.  *J. Agric. Sci.*, **4**,
107–32.

125. 36 plots, each $\frac{1}{20}$ a.  T.A. 1$\frac{3}{4}$ a.                    R.
      Metzger, W. H.   The relation of varying
rainfall to soil heterogeneity as measured by
crop production.  *J. Amer. Soc. Agron.*, **27**,
274–78.

126. 224 plots, each $5\frac{1}{2}' \times 5\frac{1}{2}' = \frac{1}{1424}$ a.  T.A.
$\frac{1}{4}$ a.                                                    G.
      Montgomery, E. G.   Experiments on wheat
breeding.  *U.S. Dept. Bulletin Bureau of plant
Industry Bull.* 269.

127. 63 plots, each $15' \times 112\frac{1}{2}' = \frac{1}{26}$ a.  T.A.
2$\frac{1}{2}$ a.                                                    G.
      Morgan, J. O.   Some experiments to deter-
mine the uniformity of certain plots for field
tests.  *Proc. Amer. Soc. Agron.*, **1**, 58–67.

128. (1) Winter wheat.  240 plots, each $\frac{1}{1210}$ a.
T.A. $\frac{1}{4}$ a.                                                G.
      (2) Summer wheat.  230 plots, each $\frac{1}{1210}$ a.
T.A. $\frac{1}{4}$ a.
      Roemer, Th.   Der Feldversuch.  *Arbeiten
der deutschen Landw. Gesellschaft*, **302**.

129. 48 plots, each $\frac{1}{10}$ a.  T.A. 5 a.                              G.
      *Roth. Exp. Sta. Report*, 1925–26.

130. 1080 plots, each $6'' \times 1' = \frac{1}{87000}$ a.  T.A.
$\frac{1}{80}$ a.                                                    R.
      Smith, H. F.   Unpublished data.

131. 360 plots, each 9 rows $\times$ 1 chain $= \frac{1}{120}$ a.
T.A. 3 a.                                                           E.
      *Waite Institute (Adelaide) Report*, 1925–32.
      Yields filed at the Waite Institute.

132. 1500 plots, each 1 row $\times$ 15$' = \frac{1}{3000}$ a.  T.A.
$\frac{1}{2}$ a.                                                    G.
      Wiebe, G. A.   Variation and correlation in
grain among 1500 wheat nursery plots.  *J.
Agric. Res.*, **50**, 331–57.

133. 94 plots, each $\frac{1}{100}$ a.  T.A. 1 a.                    N.
Wiener, W. T. G., and Broadfoot, R.  The
amount of variability which may be expected
to occur in a determination of comparative
yields in small grains.  *Proc. Fifth Ann.
Meetings Western Canadian Soc. Agr.*, 17–24.

134. 124 plots, each $33' \times 132' = \frac{1}{10}$ a.  T.A. $12\frac{1}{2}$ a.  G.
Wyatt, F. A.  Variation in plot yields due to
soil heterogeneity.  *Sci. Agric.*, **7**, 248–56.

# 11

551.501.45

# AN EXTENSION OF GOLD'S METHOD OF EXAMINING THE APPARENT PERSISTENCE OF ONE TYPE OF WEATHER

By W. G. COCHRAN

(Rothamsted Experimental Station)

## 1. *Introduction*

Gold (1929) developed a formula which is helpful in deciding whether sequences of similar meteorological events, such as runs of consecutive wet months, may or may not be expected to be due to chance variation. If a meteorological event is of two types only, for instance wet or dry months, and these occur *independently* with equal probability, Gold showed that the probable number of runs of length $r$ out of $m$ events is

$$(m + 3 - r)/2^{r+1} \qquad . \qquad . \qquad . \qquad . \qquad (1)$$

This formula was used by Gold to examine the sequence of rain days and fine days for 10 years at Kew, and the sequence of two types of pressure distribution. It has been applied by Hawke (1934) to the frequency distributions of wet and dry months and of warm and cold months at Greenwich, and by Bilham (1934, 1938) to the frequency distribution of wet and dry months over England and Wales as a whole.

In each of these applications the point at issue was to see whether there was any tendency towards the persistence of the same type of weather. This was judged by a mere comparison of the observed distribution of runs with the expected distribution on Gold's formula. While eye inspection was sufficient to decide the question in most cases, Bilham's example in particular was more doubtful, there being only a slight excess of the longer runs. Further, the observed fraction of dry months was significantly greater than 0·5, so that there was some doubt whether Gold's formula could be applied. The example suggests two further results which are needed to complete the application of Gold's formula :— (i) the extension of the formula to cases in which the chances of two events are not even ; (ii) the development of an objective test of significance giving the exact probability that a given set of runs, or a more divergent one, should have occurred without any real tendency to recurrence. These points will be considered in this note.

## 2. *The extension of Gold's formula when the chances are unequal*

The mathematical problem which Gold solved has also attracted attention in studying the distribution of runs of red and black at roulette, the sex distribution in human births, and the spread of disease from plant to plant in a row. Proofs of the general formula when the chances of the two events were not equal have been given by Marbe (1916) and by the present author (Cochran, 1936). Let $p$ be the chance of the event $P$ and $q$ be the chance of the event $Q$,

Reproduced with permission from the *Quarterly Journal of the Royal Meteorological Society*, Vol. LXIV, No. 277, pages 631–634. Copyright © 1938.

11.631

and suppose that successive trials of the event are independent; of course $p + q = 1$. The expected number of runs of length $r$ *of the event P only*, out of $m$ trials is $f_{m, r}$, which, when $1 \leq r \leq (m - 1)$, is equal to $2 p^r q + (m - r - 1) p^r q^2$, and when $r = m$, is equal to $p^r$    (2)

The corresponding formula for runs of the event $Q$ is obtained by interchanging $p$ and $q$ in (2). If runs are being counted irrespective of whether they are $P$'s or $Q$'s, the formula is

$$\left. \begin{array}{l} f_{m, r} = 2 \, (p^r q + p q^r) + (m - r - 1) \, (p^r q^2 + p^2 q^r) \text{ when } 1 \leq r \leq (m - 1) \\ f_{m, r} = p^r + q^r \text{ when } r = m \, . \quad . \quad . \quad . \quad . \quad . \quad . \quad . \end{array} \right\} (3)$$

When $p = q = \frac{1}{2}$, the formula becomes

$$\left. \begin{array}{l} f_{m, r} = (m + 3 - r)/2^{r+1} \text{ when } 1 \leq r \leq (m - 1) \\ f_{m, r} = 4/2^{r+1} \text{ when } r = m \quad . \quad . \quad . \quad . \end{array} \right\} (4)$$

With this slight extension of definition, Gold's formula holds for all values of $m$ and not only for very large values, as Gold seems to suggest.

When $p$ and $q$ are unequal, the combined distribution (3) has more long runs and fewer short runs than Gold's formula. It follows that if an observed set of runs, with $p$ and $q$ unequal, is compared with Gold's formula, the observed set will tend to have more long runs and fewer short runs. The comparison will suggest the persistence of long runs in the observed set, whether there is any real persistence or not. Before making comparisons with Gold's formula, it is therefore necessary to test whether the two events can be regarded as occurring in the data with an equal chance.

### 3. *A test of significance of the persistence of long runs of the same event*

As a numerical example, we may take the data studied by Bilham (1934 and 1938). These consist of the monthly rainfalls over England and Wales from January, 1727 to April, 1934, each month being expressed as a percentage of its normal for the period 1881-1915. There were 1,341 dry months, 1,125 wet months and 22 months with percentages of exactly 100. The last may be ignored in studying persistence. The probability of a dry month, as estimated from the data, is $(1,341)/(2,466) = 0.5438$, with standard error* $1/(2 \sqrt{2,466}) = \pm 0.0101$. The probability is clearly significantly different from $\frac{1}{2}$ and the runs of wet and dry months must be kept separate.

If there is a real persistence of, say, wet weather, the probability of a wet month will be greater if the preceding month was wet than if the preceding month was dry. A test of persistence may thus be obtained by classifying the wet months according as the preceding month was wet or dry. The classification may be made directly from the original data. It can also be obtained from the distributions of runs. The total number of runs of wet months gives the number of cases in which a dry month was preceded by a wet month. By subtracting this number from the total number of dry months, we obtain the number of cases in which a dry month

---

* The standard error of a fraction $x/n$ is $\sqrt{(pq/n)}$ (Fisher, **1936**, p. 78), where $p$ is the probability of the event, $x$ the number of occurrences, and $p + q = 1$. In this case we take $p = q = \frac{1}{2}$.

was preceded by a dry month, and similarly for the wet months. The classification for Bilham's data (omitting the 22 normal months) is as follows :—

|  | Preceding month | | |
|---|---|---|---|
| Current month— | Wet | Dry | |
| Wet .   .   .   . | 542 | 582 | 1124 |
| Dry .   .   .   . | 582 | 759 | 1341 |
| Total  .   .   . | 1124 | 1341 | 2465 |
| Fraction of wet months | ·4822 | ·4340 | |
| Standard error   .   . | ±·01485 | ±·01360 | |

The total number of months included is 2,465, the first month being omitted because it is not known whether it was preceded by a wet or a dry month. The proportion of wet months, as estimated from the data, is 0·4822 if the previous month was wet and 0·4340 if the previous month was dry, thus indicating a tendency towards persistence. The standard error of the first fraction is $\sqrt{\{(0·5438)(0·4562)/1,124\}} = \pm 0·01485$. The difference between the two fractions is 0·0482 with standard error $\sqrt{\{(0·01485)^2 + (0·01360)^2\}} = \pm 0·02013$. The probability that this or a greater difference should arise by chance is about 1 in 60.

This test is identical with the $\chi^2$ test for independence (Fisher, 1936, §21). A slight adjustment to the test is necessary if any of the numbers in the cells are under 500 (Fisher, 1936, §21.01). It will be noticed that in classifying the wet months, the dry months are automatically classified also; exactly the same test is obtained from the proportions of dry months.

If the two events are known to have an even chance, a more sensitive test may be made by classifying the events (of either kind) according as the previous event was the same or different. This test is not valid for Bilham's data, which have proved that the chance cannot reasonably be regarded as even; the test will, however, be made for the sake of an example. The figures are obtained by adding opposite corners of the 2 × 2 table above.

| Previous month | | |
|---|---|---|
| Same | Different | Total |
| 1,301 | 1,164 | 2,465 |

The proportion of cases in which the weather changed is $(1,164)/(2,465) = 0·4722 \pm 0·0101$. If, however, wet and dry months occur independently with an equal chance, this fraction should be $\frac{1}{2}$. The probability of obtaining an estimate as divergent as 0·4722 from 2,465 events is about 1 in 170.

A similar test, made on the monthly sunshine totals at Rothamsted from December, 1891, to November, 1937, gave odds of about 1 in 17 that the observed runs of sunny and dull months could have arisen without association between the sunshine totals of successive months.

An alternative method of testing for association is to calculate the correlation coefficient between successive values. This is likely to be more sensitive than the above test, since it takes account of the sizes of the departures from the normal and not merely of the signs. It should, however, be noted that the distribution of such a correlation coefficient is not the same as that of the ordinary correlation coefficient (except approximately for very long series) and has not yet been obtained.

REFERENCES

| | | |
|---|---|---|
| Bilham, E. G. | 1934 | *Quart. J.R. Met. Soc.*, **60,** 514-516. |
| ———————— | 1938 | *Ibid*, **64,** 324. |
| Cochran, W. G. | 1936 | *J.R. Statist. Soc.*, suppl. 3, 49-67. |
| Fisher, R. A. | 1936 | *Statistical methods for research workers.* Oliver and Boyd, Edinburgh, 6 Ed. |
| Gold, E. | 1929 | *Quart. J.R. Met. Soc.*, **55,** 307-309. |
| Hawke, E. L. | 1934 | *Quart. J.R. Met. Soc.*, **60,** 71-73. |
| Marbe, K. | 1916 | *Mathematische Bemerkungen.* C. H. Beck, München. |

# 12

APPENDIX

THE INFORMATION SUPPLIED BY THE SAMPLING RESULTS

BY W. G. COCHRAN

*Statistical Department, Rothamsted Experimental Station*

## 1. *Introduction*

As the quotations at the beginning of Ladell's paper indicate, previous writers on the subject of wireworm control fully realized the need for estimating the wireworm population, but appeared to have no figures from which to assess the amount of work required to obtain a reasonably accurate estimate. An attempt to obtain such data was made by Jones (1937), who took samples with surface areas of 1, $\frac{1}{4}$ and $\frac{1}{16}$ sq. ft. respectively from a number of fields and compared the standard errors per sample. Jones finds, as one would expect, that the accuracy per sample increases as the size of the sample is increased; unfortunately, however, he does not balance this gain against the extra work required in taking larger samples, so as to find which size gives the best results per unit of work expended.

When a field experiment on the control of wireworms is under consideration, a preliminary sampling of the type which Ladell undertook is essential to determine the amount of work which is likely to be involved in estimating the effects of the treatments on the numbers of wireworms. The points on which a preliminary sampling may be expected to supply information are: (1) What size of treatment effect can we hope to detect with the amount of work done in the preliminary sampling? (2) If the treatment response which will be detected is considered too large, by how much must the sampling be increased to detect a treatment response of given size? If the standard of accuracy aimed at is found to involve too much labour, the postponement of the experiment must be seriously considered. (3) In a replicated field experiment of the type carried out by Ladell, the accuracy may be increased either by increasing the number of replications or by increasing the amount of sampling per plot. Which is the more profitable?

The purpose of this note is to show how the sampling technique used by Ladell enables us to answer these questions. We will consider in detail the results of the first sampling, which are given in Table I, p. 344.

When the experiment was started, it was of course unknown whether the distribution of wireworms was a random one over the whole experimental area. The experimental design used, a Latin square, was chosen to take advantage of any regular gradients of infestation which might exist throughout the site, since differences in wireworm population between whole rows or columns do not affect the treatment comparisons. Further, the six soil samples taken per plot were restricted so that three fell in the north half and three in the south half of the plot; thus differences in infestation between these halves do not influence the treatment comparisons. This type of restriction, known as local control, is always worth while with new work, since one cannot lose anything in accuracy by it, and may gain substantially. The only limitation to its use is that at least four samples per plot are required to estimate the sampling error.

## 2. *The analysis of variance*

Before discussion, a complete analysis of variance is required. Owing to the small numbers of wireworms obtained per sample, their distribution is by no means normal, and before analysis the data ought to be transformed to some scale, such as square roots, on which they are approximately normally distributed. However, to keep the example as simple as possible, the analysis will be made on the numbers themselves; the conclusions are not altered thereby in this particular case.

The variation may be divided into: (I) Between-plot variation, which consists of variation between whole rows, with 4 degrees of freedom, variation between columns, with 4 degrees of freedom and the experimental error, with 16 degrees of freedom.* (II) The variation within plots between half-plots, derived from the differences of the totals of the north and south halves of a plot. As mentioned above, this variation does not enter into the experimental error, but it is worth calculating to see how much, if anything, has been gained by the local control. (III) The variation within half-plots, which constitutes the sampling error.

The first part, (I), is the ordinary analysis of variance of a Latin square and its calculation will not be given in detail here, as it is described with full numerical working in many text-books, such as that by Fisher & Wishart (1930). The ordinary "treatments" and "error" terms should be combined, as there are no treatments. If, however, the first sampling contained different treatments, the error term alone would be used. This analysis will be on a single plot basis (total of six samples).

To obtain (II), first take the differences (ignoring sign) between totals of the south and north halves of each plot. These are shown in Table I in the columns headed (S — N). The sum of the squares of these twenty-

* Plots 103–107 form the first row, and plots 103, 108, 113, 118 and 123 form the first column, etc.

five differences is 2269 and is on a single plot basis. The sampling error
(III) may be obtained by a subtraction. Calculate the sum of the squares
of all 150 samples; this comes to 3767 and is on a single *sample* basis.
Multiplying by 6, to bring this to a single *plot* basis, gives 22,602. Sub-
tract the product of the grand total, 607, and the general mean 24·28.
This gives the total sum of squares 7864·04 with 149 degrees of freedom.
The sampling variation may now be obtained by subtracting (I) and (II)
from the total and has 100 degrees of freedom.

The complete analysis of variance is as follows:

|  | D.F. | Sums of squares | Mean square | S.E. |
|---|---|---|---|---|
| Rows | 4 | 515·44 | 128·86 | — |
| Columns | 4 | 523·44 | 130·86 | — |
| Experimental error | 16 | 712·16 | 44·51 | 6·672 |
| Between half-plots | 25 | 2269·00 | 90·76 | — |
| Sampling error | 100 | 3844·00 | 38·44 | 6·120 |
| Total | 149 | 7864·04 | | |

### 3. *The information supplied by the preliminary sampling*

The first point to notice is that the experimental design considerably
improved the accuracy of the results, since the mean squares due to
rows, columns and differences between half-plots are all substantially
above the experimental and sampling mean squares. Had the experiment
been randomized completely within the site chosen, on the ground that
the wireworm distribution was a random one, the experimental error
(with local control) would have been

$$\tfrac{1}{24}(515·44 + 523·44 + 712·16) = 72·96 \text{ instead of } 44·51.$$

Further, if no local control had been used, the sampling error would
have been

$$\tfrac{1}{125}(2269 + 3844) = 48·90 \text{ instead of } 38·44.$$

Thus an estimate of the experimental error with complete randomization
and no local control is

$$72·96 + 49·90 - 38·44 = 83·42 \text{ instead of } 44·51,$$

so that the accuracy of the experiment has been nearly doubled by the
design.

The experimental error is 6·672 per plot and the standard error of a
treatment mean (5 replicates) is $6·672/\sqrt{5} = 2·98$, which is 12·3 % of the
general mean (24·28). Thus the standard error of the difference between
two treatments is 17·4 % of the general mean. To find the percentage
difference which would be detected at the 5 % level of significance, this
must be multiplied by 2·120, the value of "$t$" for 16 degrees of freedom.

Thus an apparent difference of 37 % between two treatment means will be significant.

When we are aiming at a given standard of accuracy, a further point, slightly more subtle, must be appreciated. If the *true* difference between say one fumigant and the control were 37 %, then in a number of experiments the estimated difference would vary about this value, being above it in half the experiments and below it in half. Thus in an individual experiment, a *true* difference of 37 % has only a chance of one in two of being detected as significant. The question arises, how large must the true difference be so that it will almost certainly be detected, say in nineteen experiments out of twenty? The difference must clearly be so large that only in 5 % of cases will the observed value of $t$ fall below 2·120.

If $x$ is the observed treatment difference, $s$ its estimated standard error and $\mu$ the real treatment difference, then the tabulated $t = (x - \mu)/s$ and we want the value of $t$ which is exceeded in all but 5 % of cases. From the $t$-table for 16 degrees of freedom we see that the value of $t$ lies inside the limits $\pm 1·746$ in all but 10 % of cases. Since the $t$-distribution is symmetrical, the value must therefore exceed $-1·746$ in all but 5 % of cases. Thus the real difference must be such that when $(x - \mu)/s = -1·746$, then $x/s$, the observed $t$, is 2·120. This gives $\mu = 3·866\,s = 16·29$, which is 67 % of the general mean.

Thus if we wish to be reasonably certain of detecting a response to a treatment, the response must be at least 67 %. This answers the first question in the introduction.

To consider the second question, suppose that we wished to detect a difference of 50 % in 95 % of cases. The standard error per plot would have to be reduced to $\frac{50}{67}$ of its present value and the experimental variance reduced in the ratio $(\frac{50}{67})^2$, i.e. to 24·8.

The third question concerns the best way to do this. The sampling variance, 38·44, accounts for all but 6·07 of the experimental variance. Thus doubling the sampling per plot, but keeping the size of the experiment fixed, would reduce the experimental error by $\frac{1}{2}(38·44) = 19·22$, i.e. to 25·3. On the other hand, keeping the *total* amount of sampling fixed, but doubling the size of the experiment, would reduce the experimental error by only $\frac{1}{2}(6·07) = 3·04$, i.e. to 41·47. There is thus a much smaller return from doubling the experimental area (using say two $5 \times 5$ Latin squares) than from doubling the amount of sampling. This must, however, be balanced by the experimenter against the expenditure of time and labour in doubling the sampling and the experimental work. It is probable that with this type of work the former will be the more exacting operation and may even be a limiting factor to the size of the experiment.

In any investigation in which the major portion of the work is taken up in handling samples, the best return for the labour expended is obtained when the sampling variance accounts for all, or almost all, the experimental variance.

The only change in Ladell's technique suggested as a result of the preliminary sampling was to increase the local control at the next sampling by taking two samples per third of a plot instead of three per half-plot. The precision attained on the treatment comparisons was considered sufficient, as only large treatment effects would be of commercial interest, and in any case the experiment was regarded as one mainly on technique. As the experimental error was almost entirely due to variation between samples, no reduction in the amount of sampling could be recommended.

### 4. *The experimental and sampling errors*

The results of Ladell's experiments from the point of view of sampling technique are summarized in Table II. In most cases a high proportion of the experimental error was due to errors in sampling. As the amount of sampling was the limiting factor in these experiments, this result is gratifying and means that the labour of sampling could not have been decreased without a considerable sacrifice in accuracy. The experimental errors are high; nevertheless certain significant treatment effects were detected.

The amount of sampling varied from $\frac{1}{7}$ to $\frac{1}{2}\%$ and the areas within which the sampling variation was taken ranged from $\frac{1}{200}$ to $\frac{1}{120}$ acre.

For comparison with other workers' data the sampling errors are best expressed in terms of a single sample, as shown in the table below.

*Sampling errors per sample*

| Size of sample in. | Size of area sampled acres | $W$ Number of wire-worms per sample | $E$ Sampling error % per sample | $E\sqrt{W}$ |
|---|---|---|---|---|
| $9 \times 9$ | $\frac{1}{120}$ | 4·0 | 61 | 122 |
| $9 \times 9$ | $\frac{1}{180}$ | 2·4 | 83 | 129 |
| $6 \times 6$ | $\frac{1}{120}$ | 1·2 | 96 | 105 |
| $9 \times 9$ | $\frac{1}{140}$ | 21·0 | 38 | 174 |
| $6 \times 6$ | $\frac{1}{140}$ | 6·2 | 48 | 119 |
| $6 \times 6$ | $\frac{1}{200}$ | 7·7 | 74 | 203 |
| $6 \times 6$ | $\frac{1}{200}$ | 6·3 | 72 | 181 |

When an experiment of this type is contemplated with new material of unknown variability, it is sometimes useful to note that a lower limit to the sampling error may be obtained from theoretical considerations alone. If the wireworms were scattered at random throughout the sub-plots sampled, the numbers found per sample should be distributed in a Poisson series distribution. For this distribution the mean $W$ is equal to the variance, so that the standard error $E$ % per sample should be $100/\sqrt{W}$. Other sources of variation may, of course, increase this error.

The values of $E\sqrt{W}$ for Ladell's experiments are given in the last column of the table. They show quite a close agreement with theory in the first experiment, where the numbers per sample were very small, but with higher numbers per sample other sources of variation become important.

Jones's figures agree remarkably well with these, despite the fact that he was sampling from much larger areas. For the 25 sets of 100 sq. ft. samples in his Table III, the means ranged from 0·22 to 10·78 per sample and the values of $E\sqrt{W}$ from 85 to 237. The values of $E\sqrt{W}$ also increase as the means increase; in fact, apart from one or two widely aberrant sets, a good summary of Jones's results may be obtained by taking a linear regression of the form

$$E\sqrt{W} = a + b\sqrt{W},$$

where $a$ and $b$ are constants.

This agreement between results obtained under widely differing conditions gives one confidence in recommending Jones's and Ladell's sampling errors to workers who are planning field experiments on wire-worms and wish to obtain a preliminary idea of the scale on which they will have to sample. This scale should, of course, be reconsidered as soon as the results of their first sampling are available, by the method in-dicated in section 2. It should be noted, however, that Jones's sampling technique, which involved a systematic rather than a random distri-bution of samples, is not recommended for sampling work in field experiments.

## Summary

In any field experiment which involves sampling of a laborious nature, it is important to know as soon as possible what degree of accuracy in the treatment mean values will be reached with a given amount of work, how much work must be done to reach a given standard of accuracy and how best to distribute one's resources between the amount of sampling and the amount of replication.

The first sampling, whether it contains experimental treatments or is uniformly treated, can supply information on all these points if properly carried out. Ladell's first wireworm sampling is taken as a simple numerical example of the way in which these questions can be answered with the help of an analysis of variance.

The sampling and experimental errors of Ladell's experiments are discussed. The sampling error accounts for a large proportion of the experimental error in most cases, as it is always advisable where the labour involved in sampling is high.

Ladell's sampling errors agree well with those obtained under widely different conditions by Jones, and both may be recommended to other workers as an indication of the amount of variability to be expected in field sampling for wireworms.

## REFERENCES

JONES, A. W. (1937). Practical field methods of sampling soil for wireworms. *J. agric. Res.* **54**, 123–34.

FISHER, R. A. & WISHART, J. (1930). The arrangement of field experiments and the statistical reduction of the results. *Imp. Bur. Soil Sci.*, Tech. Comm. No. 10.

(*Received* 20 *October* 1937)

## This table is from W.R.S.Ladell's paper.

### Table I

*Knott Wood experiment. Preliminary survey: before fumigation*

No. of wireworms in soil samples $9 \times 9 \times 5$ in.

| Serial no. of plot | South half | | | North half | | | Totals S | Totals N | S + N | (S − N) |
|---|---|---|---|---|---|---|---|---|---|---|
| 103 | 0 | 2 | 0 | 2 | 2 | 0 | 2 | 4 | 6 | 2 |
| 104 | 1 | 0 | 1 | 1 | 2 | 1 | 2 | 4 | 6 | 2 |
| 105 | 1 | 1 | 0 | 4 | 4 | 10 | 2 | 18 | 20 | 16 |
| 106 | 2 | 4 | 1 | 6 | 7 | 14 | 7 | 27 | 34 | 20 |
| 107 | 0 | 5 | 1 | 2 | 3 | 8 | 6 | 13 | 19 | 7 |
| 108 | 0 | 0 | 4 | 4 | 3 | 4 | 4 | 11 | 15 | 7 |
| 109 | 5 | 2 | 5 | 3 | 0 | 5 | 12 | 8 | 20 | 4 |
| 110 | 1 | 5 | 4 | 4 | 9 | 11 | 10 | 24 | 34 | 14 |
| 111 | 1 | 1 | 5 | 9 | 8 | 1 | 7 | 18 | 25 | 11 |
| 112 | 2 | 4 | 2 | 5 | 4 | 9 | 8 | 18 | 26 | 10 |
| 113 | 2 | 8 | 0 | 4 | 5 | 1 | 10 | 10 | 20 | 0 |
| 114 | 4 | 2 | 3 | 6 | 1 | 1 | 9 | 8 | 17 | 1 |
| 115 | 4 | 8 | 3 | 5 | 2 | 2 | 15 | 9 | 24 | 6 |
| 116 | 4 | 5 | 11 | 4 | 4 | 5 | 20 | 13 | 33 | 7 |
| 117 | 0 | 2 | 2 | 7 | 8 | 3 | 4 | 18 | 22 | 14 |
| 118 | 6 | 5 | 5 | 8 | 7 | 8 | 16 | 23 | 39 | 7 |
| 119 | 2 | 0 | 3 | 9 | 3 | 7 | 5 | 19 | 24 | 14 |
| 120 | 9 | 1 | 3 | 4 | 9 | 10 | 13 | 23 | 36 | 10 |
| 121 | 5 | 2 | 5 | 4 | 4 | 4 | 12 | 12 | 24 | 0 |
| 122 | 0 | 4 | 3 | 10 | 11 | 4 | 7 | 25 | 32 | 18 |
| 123 | 3 | 4 | 3 | 12 | 2 | 2 | 10 | 16 | 26 | 6 |
| 124 | 3 | 4 | 5 | 0 | 3 | 7 | 12 | 10 | 22 | 2 |
| 125 | 1 | 5 | 9 | 6 | 6 | 8 | 15 | 20 | 35 | 5 |
| 126 | 5 | 4 | 2 | 5 | 7 | 6 | 11 | 18 | 29 | 7 |
| 127 | 4 | 0 | 4 | 5 | 4 | 2 | 8 | 11 | 19 | 3 |

Standard errors per plot.
Sampling error $\pm 6\cdot06$ or $25\%$.
Experimental error $\pm 6\cdot99$ or $29\%$.

# 13

## THE ANALYSIS OF GROUPS OF EXPERIMENTS

By F. YATES and W. G. COCHRAN

*Statistical Department, Rothamsted Experimental Station, Harpenden*

(With One Text-figure)

### 1. INTRODUCTION

AGRICULTURAL experiments on the same factor or group of factors are usually carried out at a number of places and repeated over a number of years. There are two reasons for this. First, the effect of most factors (fertilizers, varieties, etc.) varies considerably from place to place and from year to year, owing to differences of soil, agronomic practices, climatic conditions and other variations in environment. Consequently the results obtained at a single place and in a single year, however accurate in themselves, are of limited utility either for the immediate practical end of determining the most profitable variety, level of manuring, etc., or for the more fundamental task of elucidating the underlying scientific laws. Secondly, the execution of any large-scale agricultural research demands an area of land for experiment which is not usually available at a single experimental station, and consequently much experimental work is conducted co-operatively by farmers and agricultural institutions which are not themselves primarily experimental.

The agricultural experimenter is thus frequently confronted with the results of a set of experiments on the same problem, and has the task of analysing and summarizing these, and assigning standard errors to any estimates he may derive. Though at first sight the statistical problem (at least in the simpler cases) appears to be very similar to that of the analysis of a single replicated trial, the situation will usually on investigation be found to be more complex, and the uncritical application of methods appropriate to single experiments may lead to erroneous conclusions. The object of this paper is to give illustrations of the statistical procedure suitable for dealing with material of this type.

### 2. GENERAL CONSIDERATIONS

Agronomic experiments are undertaken with two different aims in view, which may roughly be termed the technical and the scientific. Their aim may be regarded as scientific in so far as the elucidation of the underlying laws is attempted, and as technical in so far as empirical rules for the conduct of practical agriculture are sought. The two aims are, of course, not in any sense mutually exclusive, and the results of most well-conducted experiments on technique serve to add to the structure of general scientific law, or at least to indicate places where the existing structure is inadequate, while experiments on questions of a more fundamental type will themselves provide the foundation of further technical advances.

In so far as the object of a set of experiments is technical, the estimation of the average response to a treatment, or the average difference between varieties, is of considerable importance even when this response varies from place to place or from year to year. For unless we both know the causes of this variation and can predict the future incidence of these causes we shall be unable to make allowance for it, and can only base future practice on the average effects. Thus, for example, if the response to a fertilizer on a certain soil type and within a certain district is governed by meteorological events subsequent to its application the question of whether or not it is profitable to apply this fertilizer, and in what amount, must (in the absence of any prediction of future meteorological events) be governed by the average response curve over a sufficiently representative sample of years. In years in which the weather turns out to be unfavourable to the fertilizer a loss will be incurred, but this will be compensated for by years which are especially favourable to the fertilizer.

Any experimental programme which is instituted to assess the value of any particular treatment or practice or to determine the optimal amount of such treatment should therefore be so designed that it is capable of furnishing an accurate and unbiased estimate of the average response to this treatment in the various combinations of circumstances in which the treatment will subsequently be applied. The simplest and indeed the only certain way of ensuring that this condition shall be fulfilled is to choose fields on which the experiments are to be conducted by random selection from all fields which are to be covered by the subsequent recommendations.

The fact that the experimental sites are a random sample of this nature does not preclude different recommendations being made for different categories included in this random sample. We may, for instance, find that the response varies according to the nature of the previous crop, in which case the recommendations may be correspondingly varied. Moreover, in a programme extending over several years, the recommendations may become more specific as more information is accumulated, and the experiments themselves may be used to determine rules for the more effective application of the treatments tested, as in fertilizer trials in which the chemical examination of soil samples may lead to the evolution of practical chemical tests for fertilizer requirements.

At present it is usually impossible to secure a set of sites selected entirely at random. An attempt can be made to see that the sites actually used are a "representative" selection, but averages of the responses from such a collection of sites cannot be accepted with the same certainty as would the averages from a random collection.

On the other hand, comparisons between the responses on different sites are not influenced by lack of randomness in the selection of sites (except in so far as an estimate of the variance of the response is required) and indeed for the purpose of determining the exact or empirical natural laws governing the responses, the deliberate inclusion of sites representing extreme conditions may be of value. Lack of randomness is then only harmful in so far as it results in the omission of sites of certain types and in the consequent arbitrary restriction of the range of conditions. In this respect scientific research is easier than technical research.

### 3. The analogy between a set of experiments and a single replicated trial

If a number of experiments containing the same varieties (or other treatments) are carried out at different places, we may set out the mean yields of each variety at each place in the form of a two-way table. The marginal means of this table will give the average differences between varieties and between places. The table bears a formal analogy to the two-way table of individual plot yields, arranged by blocks and varieties, of a randomized block experiment, and we can therefore perform an analysis of variance in the ordinary manner, obtaining a partition of the degrees of freedom (in the case of six places and eight varieties, for example) as follows:

|  | Degrees of freedom |
|---|---|
| Places | 5 |
| Varieties | 7 |
| Remainder | 35 |
| Total | 47 |

The remainder sum of squares represents that part of the sum of squares which is due to variation (real or apparent) of the varietal differences at the different places. This variation may reasonably be called the *interaction* between varieties and places. It will include a component of variation arising from the experimental errors at the different places.

If the experiments are carried out in randomized blocks (or in any other type of experiment allowing a valid estimate of error) the above analysis may be extended to include a comprehensive analysis of the yields of the individual plots. If there are five replicates at each place, for example, there will be 240 plot yields, and the partition of the degrees of freedom will then be as follows:

|  | Degrees of freedom |
|---|---|
| Places | 5 |
| Varieties | 7 |
| Varieties × places | 35 |
| Blocks | 24 |
| Experimental error | 168 |
| Total | 239 |

It should be noted that in this analysis the sums of squares for varieties and for varieties × places are together equal to the total of the sums of squares for varieties in the analyses of the separate experiments. Similarly the sums of squares for blocks and for experimental error are equal to the totals of these items in the separate analyses. If, as is usual, the comprehensive analysis is given in units of a single plot yield, the sums of squares derived from the two-way table of places and varieties must be multiplied or divided by 5 according as means or totals are there tabulated.

The first point to notice about this comprehensive analysis of variance is that the estimates of error from all six places are pooled. If the errors of all experiments are substantially the same, such pooling gives a more accurate estimate than the estimates derived from each separate experiment, since a larger number of degrees of freedom is available. If the errors are different, the pooled estimate of the error variance is in fact the estimate of the mean of the error variances of the separate experiments. It will therefore still be the correct estimate of the error affecting the mean difference (over all places) of two varieties, but it will no longer be applicable to comparisons involving some of the places only. Moreover, as will be explained in more detail below, the ordinary tests of significance, even of means over all places, will be incorrect.

If the errors of all the experiments are the same, the other mean squares in the analysis of variance table may be compared with the mean square for experimental error by means of the $z$ test. The two comparisons of chief interest are those for varieties and for varieties × places. The meaning of these will be clear if we remember that there is a separate set of varietal means at each place, and that the differences between these

36–2

means are not necessarily the same at all places. If the mean square for varieties is significant, this indicates the significance of the average differences of the varieties *over the particular set of places chosen*. If varieties × places is also significant, a significant variation from place to place in the varietal differences is indicated. In this latter case it is clear that the choice of places must affect the magnitude of the average differences between varieties: with a different set of places we might obtain a substantially different set of average differences. Even if varieties × places is not significant, this fact cannot be taken as indicating *no* variation in the varietal differences, but only that such variation is likely to be smaller than an amount which can be determined by the arguments of fiducial probability.

We may, therefore, desire to determine the variation that is likely to occur in the average differences between the varieties when different sets of places are chosen, and in particular whether the average differences actually obtained differ significantly from zero when variation from place to place is allowed for. Endless complications affect this question, and with the material ordinarily available a definite answer is usually impossible. The various points that arise will be made clear by an actual example, but first we may consider the situation in the ideal case where the chosen places are a strictly random selection from all possible places.

At first sight it would appear to be legitimate in this case to compare the mean square for varieties with that for varieties × places by means of the $z$ test. There is, however, no reason to suppose that the variation of the varietal differences from place to place is the same for each pair of varieties. Thus the eight varieties of our example might consist of two sets of four, the varieties of each set being closely similar among themselves but differing widely from those of the other set, not only in their average yield, but also in their variations in yield from place to place.

The sums of squares for varieties and for varieties × places would then have large components derived from one degree and five degrees of freedom respectively, while the remaining components might be of the order of experimental error. In the limiting case, therefore, when the experimental error is negligible in comparison with the differences of the two sets and the average difference over all possible places is zero, the $z$ derived from the two mean squares will be distributed as $z$ for 1 and 5 degrees of freedom instead of as $z$ for 7 and 35 degrees of freedom. Verdicts of significance and of subnormal variation will therefore be reached far more often than they should be.

The correct procedure in this case is to divide the sums of squares for varieties and for varieties × places into separate components, and compare each component separately. Thus we shall have:

|  | Degrees of freedom |
|---|---|
| Varieties: Sets | 1 |
| Within sets | 6 |
| Varieties × places: Sets | 5 |
| Within sets | 30 |

The 1 degree of freedom between sets can now legitimately be compared with the 5 degrees of freedom for sets × places, but the degrees of freedom within sets may require further subdivision before comparison.

It is worth noting that the test of a single degree of freedom can be made by the $t$ test, by tabulating the differences between the means of the two sets for each place separately. This test is in practice often more convenient than the $z$ test, of which it is the equivalent.

### 4. AN ACTUAL EXAMPLE OF THE ANALYSIS OF A SET OF VARIETY TRIALS

Table I, which has been reproduced by Fisher (1935) as an example for analysis by the reader, gives the results of twelve variety trials on barley conducted in the State of Minnesota and discussed by Immer *et al.* (1934). The trials were carried out at six experiment stations in each of two years, and actually included ten varieties of which only five (those selected by Fisher) are considered here.

Table I. *Yields of barley varieties in twelve independent trials.*
*Totals of three plots, in bushels per acre*

| Place and year | Manchuria | Svansota | Velvet | Trebi | Peatland | Total |
|---|---|---|---|---|---|---|
| University Farm 1931 | 81·0 | 105·4 | 119·7 | 109·7 | 98·3 | 514·1 |
| 1932 | 80·7 | 82·3 | 80·4 | 87·2 | 84·2 | 414·8 |
| Waseca 1931 | 146·6 | 142·0 | 150·7 | 191·5 | 145·7 | 776·5 |
| 1932 | 100·4 | 115·5 | 112·2 | 147·7 | 108·1 | 583·9 |
| Morris 1931 | 82·3 | 77·3 | 78·4 | 131·3 | 89·6 | 458·9 |
| 1932 | 103·1 | 105·1 | 116·5 | 139·9 | 129·6 | 594·2 |
| Crookston 1931 | 119·8 | 121·4 | 124·0 | 140·8 | 124·8 | 630·8 |
| 1932 | 98·9 | 61·9 | 96·2 | 125·5 | 75·7 | 458·2 |
| Grand Rapids 1931 | 98·9 | 89·0 | 69·1 | 89·3 | 104·1 | 450·4 |
| 1932 | 66·4 | 49·9 | 96·7 | 61·9 | 80·3 | 355·2 |
| Duluth 1931 | 86·9 | 77·1 | 78·9 | 101·8 | 96·0 | 440·7 |
| 1932 | 67·7 | 66·7 | 67·4 | 91·8 | 94·1 | 387·7 |
| Total | 1132·7 | 1093·6 | 1190·2 | 1418·4 | 1230·5 | 6065·4 |

The experiments were all arranged in randomized blocks with three replicates of each variety. When all ten varieties are included there are therefore 18 degrees of freedom for experimental error at each station.

The error mean squares for the twelve experiments were computed from the yields of the separate plots, which have been given in full by Immer. They are shown in Table II.

Table II. *Error mean squares of barley experiments*

| | Mean square | | Approximate $\chi^2$ | |
|---|---|---|---|---|
| | 1931 | 1932 | 1931 | 1932 |
| University Farm | 21·25 | 15·98 | 16·43 | 12·36 |
| Waseca | 26·11 | 25·21 | 20·19 | 19·49 |
| Morris | 18·62 | 20·03 | 14·40 | 15·49 |
| Crookston | 30·27 | 21·95 | 23·40 | 16·97 |
| Grand Rapids | 26·28 | 26·40 | 20·32 | 20·41 |
| Duluth | 27·00 | 20·28 | 20·88 | 15·68 |

If the errors at all stations are in fact the same, these error mean squares, when divided by the true error variance and multiplied by the number of degrees of freedom, 18, will be distributed as $\chi^2$ with 18 degrees of freedom. If we take the mean of the error mean squares, 23·28, as an estimate of the true error variance, the distribution obtained will not be far removed from the $\chi^2$ distribution. The actual values so obtained are shown in Table II, and their distribution is compared with the $\chi^2$ distribution in Table III. Variation in the experimental error from station to station would be indicated by this distribution having a wider dispersion than the $\chi^2$ distribution.

Table III. *Comparison with the theoretical $\chi^2$ distribution*

| $P$ | 1·0 | 0·99 | 0·98 | 0·95 | 0·90 | 0·80 | 0·70 | 0·50 | 0·30 | 0·20 | 0·10 | 0·05 | 0·02 | 0·01 | 0 |
|---|---|---|---|---|---|---|---|---|---|---|---|---|---|---|---|
| $\chi^2$ | 0 | 7·02 | 7·91 | 9·39 | 10·86 | 12·86 | 14·44 | 17·34 | 20·60 | 22·76 | 25·99 | 28·87 | 32·35 | 34·80 | ∞ |
| No. observed | 0 | 0 | 0 | 0 | 1 | 1 | 4 | 4 | 1 | 1 | 0 | 0 | 0 | 0 | |
| No. expected | 0·12 | 0·12 | 0·36 | 0·6 | 1·2 | 1·2 | 2·4 | 2·4 | 1·2 | 1·2 | 0·36 | 0·36 | 0·12 | 0·12 | |

In this case it is clear that the agreement is good, and consequently we shall be doing no violence to the data if we assume that the experimental errors are the same for all the experiments. This gives 23·28 as the general estimate (216 degrees of freedom) for the error variance of a single plot, and the standard error of the values of Table I is therefore $\sqrt{(3 \times 23 \cdot 28)}$ or $\pm 8 \cdot 36$.

The analysis of variance of the values of Table I is given in Table IV in units of a single plot. The components due to places and years have been separated in the analysis in the ordinary manner.

Every mean square, except that for varieties × years, will be found, on testing by the $z$ test, to be significantly above the error mean square. Examination of Table I indicates that variety Trebi is accounting for a good deal of the additional variation due to varieties and varieties × places, for the mean yield of this variety over all the experiments is much above that of the other four varieties, but at University Farm and Grand Rapids, two of the lowest yielding stations, it has done no better than the other varieties.

Table IV. *General analysis of variance (units of a single plot)*

| | Degrees of freedom | Sum of squares | Mean square |
|---|---|---|---|
| Places | 5 | 7073·64 | 1414·73 |
| Years | 1 | 1266·17 | 1266·17 |
| Places × years | 5 | 2297·96 | 459·59 |
| Varieties | 4 | 1769·99 | 442·50 |
| Varieties × places | 20 | 1477·67 | 73·88 |
| Varieties × years | 4 | 97·27 | 24·32 ⎫ 42·72 |
| Varieties × places × years | 20 | 928·09 | 46·40 ⎭ |
| Total | 59 | 14910·79 | |
| Experimental error | 216 | | 23·28 |

In order to separate the effect of Trebi it is necessary to calculate the difference between the yield of Trebi and the mean of the yields of the other four varieties for each of the twelve experiments, and to analyse the variance of these quantities. For purposes of computation the quantities:

$5 \times$ yield of Trebi $-$ total yield of station $= 4$ (yield of Trebi $-$ mean of other varieties)

are more convenient.

The analysis of variance is similar to that which gives places, years and places × years in the main analysis. The divisor of the square of a single quantity is $3 \times (4^2 + 1 + 1 + 1 + 1) = 60$ and the square of the total of all twelve quantities, divided by 720, gives the sum of squares representing the average difference between Trebi and the other varieties over all stations.

The four items involving varieties in the analysis of variance are thus each split up into two parts, representing the difference of Trebi and the other varieties, and the variation of these other varieties among themselves. This partition is given in Table V. The second part of each sum of

Table V. *Analysis of variance, Trebi v. Remainder*

| | Degrees of freedom | Sum of squares | Mean square |
|---|---|---|---|
| Varieties: Trebi | 1 | 1463·76 | 1463·76 |
| Remainder | 3 | 306·23 | 102·08 |
| Varieties × places: Trebi | 5 | 938·09 | 187·62 |
| Remainder | 15 | 539·58 | 35·97 |
| Varieties × years: Trebi | 1 | 7·73 | 7·73 |
| Remainder | 3 | 89·54 | 29·85 |
| Varieties × places × years: Trebi | 5 | 162·10 | 32·42 |
| Remainder | 15 | 765·97 | 51·06 |

squares can be derived by subtraction, or by calculating the sum of squares of the deviations of the four remaining varieties from their own mean.

Study of this table immediately shows that the majority of the variation in varietal differences between places is accounted for by the difference of Trebi from the other varieties. The mean square for varieties × places has been reduced from 73·88 to 35·97 by the elimination of Trebi, and this latter is in itself not significantly above the experimental error. The mean square for varieties × places × years has not been similarly reduced, however, in fact it is actually increased (though not significantly), and the last three remainder items taken together are still significantly above the experimental error. There is thus still some slight additional variation in response from year to year and place to place. The place to place variation appears to arise about equally from all the remaining three varieties, but the differences between the different years are almost wholly attributable to the anomalous behaviour at Grand Rapids of Velvet, which yielded low in 1931 and high in 1932. If the one degree of freedom from varieties × years and varieties × places × years arising from this difference is eliminated by the "missing plot" technique (Yates, 1933) we have

|  | Degrees of freedom | Sum of squares | Mean squares |
|---|---|---|---|
| Velvet at Grand Rapids | 1 | 453·19 | 453·19 |
| Remainder | 23 | 572·16 | 24·88 |

Thus the remainder is all accounted for by experimental error.

It has already been noted that Trebi yielded relatively highly at the high-yielding centres. The degree of association between varietal differences and general fertility (as indicated by the mean of all five varieties) can be further investigated by calculating the regressions of the yields of the separate varieties on the mean yields of all varieties. The deviations of the mean yields of the six stations from the general mean in the order given are nearly proportional to

$$-2, +10, +1, +2, -6, -5.$$

The sum of the squares of these numbers is 170. Multiplying the varietal totals at each place by these numbers we obtain the sums

| | |
|---|---|
| Manchuria | 1004·6 |
| Svansota | 1196·2 |
| Velvet | 1137·8 |
| Trebi | 1926·8 |
| Peatland | 736·3 |
| Total | 6001·7 |

The sum of the squares of the deviations of these sums, divided by $170 \times 6$, gives the part of the sum of squares accounted for by the differences of the regressions. This can be further subdivided as before, giving:

|  | Degrees of freedom | Sum of squares | Mean square |
|---|---|---|---|
| Varieties × Places: |  |  |  |
| Differences of regressions: Trebi | 1 | 646·75 | 646·75 |
| Remainder | 3 | 123·17 | 41·06 |
| Total | 4 | 769·92 |  |
| Deviations from regressions: Trebi | 4 | 291·34 | 72·84 |
| Remainder | 12 | 416·39 | 34·70 |
| Total | 16 | 707·73 |  |

Thus the greater part of the differences between Trebi and the remaining varieties is accounted for by a linear regression on mean yield. There is still a somewhat higher residual variation (M.S. 72·84) of Trebi from its own regression than of the other varieties from their regressions, though the difference is not significant. Of the four remaining varieties Peatland appears to have a lower regression than the others, giving significantly higher yields at the lower yielding stations only, the difference in regressions being significant when Peatland is tested against the other three varieties.

The whole situation is set out in graphical form in Fig. 1, where the stations are arranged according to their mean yields and the calculated regressions are shown. It may be mentioned that of the remaining five varieties not included in the discussion, three show similar regressions on mean yield.

This artifice of taking the regression on the mean yield of the difference of one variety from the mean of the others is frequently of use in revealing relations between general fertility and varietal differences. A similar procedure can be followed with response to fertilizers or other treatments. The object of taking the regression on the mean yield rather than on the yield of the remaining varieties is to eliminate a spurious component of regression which will otherwise be introduced by experimental errors. If the variability of each variety at each centre is the same, apart from the components of variability accounted for by the regressions, the regression so obtained will give a correct impression of the results. This is always the situation as far as experimental error is concerned (except in those rare cases in which one variety is more variable than the others). It may or may not be the case for the other components of variability. In our example, as we have seen, the deviations of Trebi from its regression are somewhat greater than those of the other varieties. In such cases we should theoretically take regressions on a weighted mean yield, but there will be little change in the results unless the additional components of variance are very large.

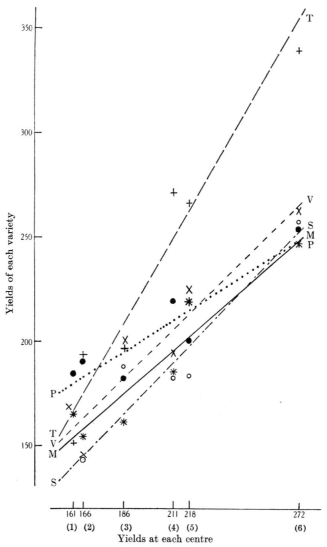

(1) Grand Rapids, (2) Duluth, (3) University Farm, (4) Morris, (5) Crookston, (6) Waseca

Trebi + ———  ——  Velvet × – – – –  Svansota ○ —·—·—
Manchuria ✳ ————— Peatland ● ·········

Fig. 1. Regressions on mean yield. The yields shown are totals of the two years. The mean yields per plot (bushels per acre) are ⅙ of these totals.

We can also examine how far it is possible to make practical recommendations from the results of these experiments.[1] The following points are of importance:

(1) How far is the superiority of Trebi over the remaining four varieties likely to recur in future years at the same set of stations?

(2) How far is Trebi likely to be superior to the other varieties at other stations in years similar to 1931 and 1932, and in particular with what degree of confidence may Trebi be recommended for general adoption in preference to the other four varieties in some or all of the districts of the state of Minnesota?

The answer to question (1) is clearly limited by the fact that only 2 years' results are available. Therefore, we cannot make any general statement as regards years which are radically different in weather conditions to 1931 and 1932. If, however, 1931 and 1932 themselves differed considerably as regards weather conditions, and if, moreover, the weather conditions varied considerably from station to station in the same year, the weather conditions over the twelve experiments might be regarded as an approximation to a random sample from all possible conditions, in which case the pooled estimate of varieties × years and varieties × places × years (these not being significantly different) might be regarded as an appropriate estimate of the variance due to weather conditions, to differences between fields at the same station, and to experimental error. In this respect Trebi is no more variable than the remaining varieties, and if the anomalous variation of Velvet at Grand Rapids be excluded the pooled estimate (23 degrees of freedom) is only slightly above experimental error. In default of any special explanation of this anomalous variation, however, it will be more reasonable not to exclude this degree of freedom, in which case we should assess the variance due to the above causes as 42·72 (24 degrees of freedom) and this agrees closely with the similar estimate (which cannot be attributed to any one outstanding difference) from the varieties which have been omitted from the present analysis.

Since three plots go to make up the total at any one place in one year, the additional variance of a single varietal mean (on a single plot basis) due to weather conditions and differences between fields at the same station, over that arising from experimental error, is

$$\tfrac{1}{3}\,(42\!\cdot\!72 - 23\!\cdot\!28) = 6\!\cdot\!48,$$

---

[1] It should be noted that in practice there are many other factors besides yield which must be taken into consideration. In this instance Dr Immer informs me that the malting quality of Trebi is poor.

and the variance of the difference of two varieties due to these causes will be double this, i.e. 12·96.

In addition to this real variation in subsequent years of the varietal differences at a place, the errors of the estimates obtained from the 2 years' experimental results must also be taken into account. These are calculated in the ordinary manner from the mean square, 42·72, for varieties × years, and varieties × places × years. Thus the error variance of the estimate of the difference of two varieties at any one place is twice $\frac{1}{6}(42\cdot72) = 14\cdot24$. (If varieties × years and varieties × places × years were different further subdivision of the components, similar to that illustrated below when considering place to place variation, would be necessary.)

In one sense, therefore, the variance of the expected difference of any two varieties, say of Trebi and Peatland at Waseca, in any subsequent year (with similar weather conditions) is $12\cdot96 + 14\cdot24 = 27\cdot20$, but it must be remembered that if a whole series of subsequent years is taken the actual differences will not be distributed about the estimated difference 14·2 with variance 27·20, but about some unknown "true" difference with variance 12·96, the unknown true difference having itself a fiducial distribution about the estimated difference given by the variance 14·24.

The answer to the second question depends on how far the actual stations may be regarded as a random sample of all stations. If this is the case, the estimate of varieties × places for Trebi will be the appropriate estimate of the variance from place to place, including one-half the variance due to differences between fields at the same station, to experimental error and to weather conditions except in so far as they are constant over all stations in each year. This is based on only five degrees of freedom and is therefore ill determined, but accepting the value 187·62, the variance of the difference of Trebi and the remaining varieties due to places only is:

$$\tfrac{1}{6}\left(1+\tfrac{1}{4}\right)(187\cdot62 - 42\cdot72) = 30\cdot19,$$

and, therefore, that due to place, field and weather conditions (but excluding experimental error) is

$$30\cdot19 + \left(1+\tfrac{1}{4}\right)6\cdot48 = 38\cdot29.$$

In addition to this variation the error of the estimated mean difference must be taken into account. The mean difference of Trebi from the mean of the other varieties is 7·1 per plot, and this has an estimated variance due to places, fields, differences in weather conditions from place to place, and experimental error, of

$$\tfrac{1}{36}\left(1+\tfrac{1}{4}\right)187\cdot62 = 6\cdot51,$$

and also an additional undetermined component of variance due to differences between years.

Hence, the difference to be expected in a single field is subject to a variation about the true mean having an estimated variance of 38·29, and the estimate of the true mean 7·1 has an error variance of 6·51. Consequently it will frequently happen in default of other information that Trebi will yield less than some other variety that might have been grown. At Grand Rapids, Trebi yielded 15–20 % less than Peatland in the 2 years of the experiment. It is poor consolation to the farmer of a farm similar to this to be told that Trebi is giving substantially higher yields on other farms.

It would be rash, however, to recommend Peatland for the whole of the Grand Rapids and Duluth districts and Trebi for the whole of the other districts, in particular the Waseca district, until we know how far variation in the varietal differences depends on factors common to the whole of a district or soil type and how far on factors exclusive to individual farms, such as variations in manuring and cultivation practices and differing crop rotations. Only parallel experiments in the same district on farms which may themselves be reasonably regarded as a random selection from all farms in the district will separate these two sources of variation, and it is therefore impossible from the general analysis of variance to say with any confidence whether Trebi is particularly suited to the district of Waseca or whether its high yield here is due to special conditions at the experimental station. As we have seen, however, the superiority of Trebi is associated with high general level of fertility. If, therefore, we know that the Waseca district is as a whole high yielding we may confidently recommend Trebi for general adoption in the district (with a reservation as to weather conditions). On the other hand, if the general yield of the district is only average, the experimental station being outstanding, then we should only be justified in recommending Trebi for farms of high fertility in this district but might also include farms of high fertility in other districts.

Immer does not report the soil types of the various stations, but it is noteworthy that Peatland, which proved the best variety at the low-yielding stations, has (as its name implies) been specially selected for peat soils, which are likely to be low yielding.

## 5. EXAMPLE OF VARIATION IN EXPERIMENTAL ERROR

If the experimental errors of the different experiments are substantially different the use of the $z$ test in conjunction with the pooled estimate of error may be misleading, in just the same way as the pooling of all the degrees of freedom from varieties × places was misleading in the set of varietal trials already considered.

The following is an example in which the $z$ test indicates an almost significant interaction between a treatment effect and places, whereas proper tests show that there is no indication of any such variation in the treatment effect. The example is particularly interesting in that on a first examination of the data the results of the $z$ test led the experimenter to draw false conclusions.

The experiments consisted of a series of thirteen $3 \times 3$ Latin squares, described by Lewis & Trevains (1934) and carried out in order to test the effectiveness of, and difference between, an ammonium phosphate mixture and an ordinary fertilizer mixture on sugar beet. Large responses to the fertilizers were shown in all the experiments. The question arose as to whether there was any significant difference between the two forms of fertilizer.

Table VI. *Experiments on sugar beet*

| | Yields of roots, tons per acre | | | | Amm. phos. − ordinary ($x$) | Error mean square per plot ($s^2$) | $\chi^2$ | $t$ |
|---|---|---|---|---|---|---|---|---|
| Centre | No fertilizer | Amm. phos. mixture | Ordinary mixture | Mean response | | | | |
| 1 | 7·44 | 15·69 | 13·75 | +7·28 | +1·94 | 0·8599 | 3·74 | +2·56 |
| 2 | 7·19 | 12·28 | 11·32 | +4·61 | +0·96 | 0·3543 | 1·54 | +1·98 |
| 3 | 10·07 | 13·93 | 13·10 | +3·44 | +0·83 | 0·5329 | 2·32 | +1·39 |
| 4 | 7·74 | 10·97 | 11·89 | +3·69 | −0·92 | 1·1528 | 5·02 | −1·05 |
| 5 | 11·88 | 13·96 | 15·06 | +2·63 | −1·10 | 0·2638 | 1·15 | −2·68 |
| 6 | 11·94 | 14·35 | 14·36 | +2·42 | −0·01 | 1·7249 | 7·51 | −0·01 |
| 7 | 6·20 | 10·27 | 10·02 | +3·94 | +0·25 | 0·4803 | 2·09 | +0·44 |
| 8 | 8·99 | 11·17 | 11·47 | +2·33 | −0·30 | 0·1107 | 0·482 | −1·10 |
| 9 | 9·46 | 12·54 | 12·46 | +3·04 | +0·08 | 0·0184 | 0·0801 | +0·72 |
| 10 | 7·42 | 10·93 | 10·79 | +3·44 | +0·14 | 0·0046 | 0·0200 | +2·53 |
| 11 | 3·70 | 5·46 | 5·38 | +1·72 | +0·08 | 0·0073 | 0·0318 | +1·15 |
| 12 | 9·62 | 12·72 | 13·01 | +3·24 | −0·29 | 0·1920 | 0·836 | −0·81 |
| 13 | 9·47 | 13·53 | 13·72 | +4·16 | −0·19 | 0·2706 | 1·18 | −0·45 |
| Mean | 8·55 | 12·14 | 12·02 | +3·53 | +0·1115 | 0·4594 | 2·00 | |

The yields are shown in Table VI. The mean yields of the two forms of fertilizer over all experiments are practically identical, indicating an absence of any consistent difference between the two forms. The analysis of variance, using a pooled estimate of error from all squares, is given in Table VII.

The $z$ between treatment × centres and error is 0·361, which is almost significant, the 5 % point being 0·382. Inspection of the differences between the two forms shows that eight of these are small ( ⩽ 0·30), while the remaining five range numerically from 0·83 to 1·94. These five are all associated with large error mean squares. The values of $t$ for the separate

Table VII. *Analysis of variance of responses to fertilizer and of difference between mixtures (single-plot basis)*

|  | Degrees of freedom | Sum of squares | Mean square | $z$ |
|---|---|---|---|---|
| Response to fertilizer: |  |  |  |  |
| Mean response | 1 | 324·7605 | 324·7605 |  |
| Response × centres | 12 | 45·8462 | 3·8205 |  |
| Differences between mixtures: |  |  |  |  |
| Mean difference | 1 | 0·2493 | 0·2493 |  |
| Difference × centres | 12 | 11·3542 | 0·9462 | 0·361 |
| Error | 26 | 11·9450 | 0·4594 |  |

experiments have therefore been calculated and are given in Table VI. These thirteen observed values are compared with the theoretical $t$ distribution for 2 degrees of freedom in Table VIII. The two distributions agree excellently, not one of the values of $t$ being below the 0·1 level of probability. We must conclude, therefore, that there is no evidence from the experiments that the two mixtures behaved differently at any centre.

Table VIII. *Comparison with theoretical $t$ and $\chi^2$ distributions*

| $P$ | 1·0 | 0·9 | 0·8 | 0·7 | 0·6 | 0·5 | 0·4 | 0·3 | 0·2 | 0·1 | 0·05 | 0·02 | 0·01 | 0 |
|---|---|---|---|---|---|---|---|---|---|---|---|---|---|---|
| $t$ | 0 | 0·14 | 0·29 | 0·44 | 0·62 | 0·82 | 1·06 | 1·39 | 1·89 | 2·92 | 4·30 | 6·96 | 9·92 | ∞ |
| Expected | | 1·3 | 1·3 | 1·3 | 1·3 | 1·3 | 1·3 | 1·3 | 1·3 | 0·65 | 0·39 | 0·13 | 0·13 | |
| Observed | | 1 | 0 | ½ | 1½ | 2 | 1 | 2½ | ½ | 4 | 0 | 0 | 0 | 0 |

| $P$ | 1·0 | 0·99 | 0·98 | 0·95 | 0·90 | 0·80 | 0·70 | 0·50 | 0·30 | 0·20 | 0·10 | 0·05 | 0·02 | 0·01 | 0 |
|---|---|---|---|---|---|---|---|---|---|---|---|---|---|---|---|
| $\chi^2$ | 0 | 0·0201 | 0·0404 | 0·103 | 0·211 | 0·446 | 0·713 | 1·39 | 2·41 | 3·22 | 4·60 | 5·99 | 7·82 | 9·21 | ∞ |
| Expected | | 0·13 | 0·13 | 0·39 | 0·65 | 1·3 | 1·3 | 2·6 | 2·6 | 1·3 | 1·3 | 0·65 | 0·39 | 0·13 | 0·13 |
| Observed | | 1 | 1 | 1 | 0 | 0 | 1 | 3 | 3 | 0 | 1 | 1 | 1 | 0 | 0 |

The above method requires modification if the true difference $\mu$ between the two forms of fertilizer is appreciably different from zero, for the quantities $t' = (x - \mu)/S$ will then conform to the $t$ distribution, instead of the $t$'s calculated as above. The quantity $\mu$ is not exactly known, but if the centres are at all numerous the use of the mean difference $\bar{x}$, or some form of weighted mean difference, such as one of those discussed in the next section, will give quantities which closely approximate to the $t$ distribution.

Although inspection of Table VIII shows quite conclusively in the present example that the observed $t$'s are in no way abnormal, border-line cases will arise in which a proper test of significance is desirable. An obvious form of test would be that based on the variance of the observed $t$'s, or of some analogous function. One such function is the "weighted sum of squares of deviations",

$$Q = Sw \, (x - \bar{x}_w)^2 = Swx^2 - \bar{x}_w \cdot Swx,$$

where the weights $w$ are the reciprocals of the estimates of the error variances of the differences $x$, and $\bar{x}_w$ is the weighted mean of the $x$'s, i.e.

$$\bar{x}_w = \frac{Swx}{Sw}.$$

If we then calculate

$$\chi'^2 = (k-1) + \sqrt{\frac{n-4}{n-1}\left\{\frac{n-2}{n}\,Q - (k-1)\right\}},$$

$\chi'^2$ will be distributed approximately as $\chi^2$ with $k-1$ degrees of freedom. The relation of $Q$ to the ordinary expression for the variance of $t'$ is shown by the alternative form

$$Q = S\,(t'^2) - \{S\,(t'\sqrt{w})\}^2/S\,(w).$$

This test should not be used if $n$ is less than 6. Actual comparison of the distribution of the $t'$'s with the $t$ distribution should then be resorted to, the unweighted mean $\bar{x}$ being used as the estimate of $\mu$.

An example of the calculation of $Q$ and $\chi'$ is given in § 7.

### 6. METHODS OF ESTIMATING THE AVERAGE RESPONSE

As has been pointed out in the second section, the average response to a treatment over a set of experiments is frequently of considerable importance, even when the response varies from experiment to experiment. The problem of how it may best be estimated from the results of the separate experiments must therefore be considered.

If the experiments are all of equal precision the efficient estimate is clearly the ordinary mean of the apparent responses in each experiment, whether the true responses are the same or vary from experiment to experiment. If, on the other hand, some of the experiments are more precise than others, the ordinary mean, by giving equal weight to both the less and the more accurate results, may appear at first sight to furnish a considerably less precise estimate than might be obtained by more refined statistical processes. As will appear from what follows, however, there are several factors which increase the advantages of the ordinary mean in relation to other possible estimates, so that unless the experiments differ widely in accuracy, or the conditions are somewhat different from those ordinarily met with in agriculture, the ordinary mean is in practice the most satisfactory as well as the most straightforward estimate to adopt.

The simplest alternative to the ordinary mean is the weighted mean mentioned at the end of the last section, in which the weights are inversely proportional to the error variances of the estimates derived from the various experiments. This weighted mean would be the efficient estimate if there were no variation in the true response from experiment to experiment, and if, moreover, the error variances of the experiments were accurately known. If the error variances are only estimated from a small number of degrees of freedom, however, the weighted mean loses greatly in efficiency and is frequently less efficient than the unweighted mean.

If the true response varies from experiment to experiment, having a variance of $\sigma_0^2$, and the error variances are accurately known, the efficient estimate of the mean response is provided by a weighted mean with weights inversely proportional to $\sigma_0^2 + \sigma_1^2$, $\sigma_0^2 + \sigma_2^2 \ldots$, where $\sigma_1^2, \sigma_2^2, \ldots$, are the error variances of the estimates from the various experiments. This has been called the *semi-weighted mean*, since the weights are intermediate between those of the weighted mean and the equal weights of the ordinary mean.

If the response does not vary from experiment to experiment, but the error variances are not accurately known, being estimated from $n_1$, $n_2$, ..., degrees of freedom, the efficient estimate is obtained by the solution of the maximum likelihood equation:

$$\mu = S \frac{(n_i + 1)\, x_i}{n_i s_i^2 + (x_i - \mu)^2} \bigg/ S \frac{(n_i + 1)}{n_i s_i^2 + (x_i - \mu)^2}\,.$$

This solution has the effect of giving lower weights to the more discrepant values than would be given by the ordinary weighted mean. It is not difficult to solve the equation by successive approximation, starting with a value of $\mu$ equal to the unweighted mean, but since in agricultural experiments cases in which the response can confidently be asserted not to vary are rare, the additional numerical work is not ordinarily justifiable, except when exact tests of a significance are required and when the $n_i$ are small.

Thus the available rigorous methods of weighting are not of much use in the reduction of the results of the type ordinarily met with. On the other hand, when a set of experiments of widely varying accuracy is encountered, some method of discounting the results of the less accurate experiments is required. The simplest method would be to reject the results of the less accurate experiments entirely, but this involves the drawing of an arbitrary line of division. Anyone who has attempted this will know how easy it is in certain cases to produce substantial changes in the mean response by the inclusion or exclusion of certain border-line experiments.

An alternative procedure is that of fixing an upper limit to the weight assignable to any one experiment. All experiments having error variances which give apparent weights greater than this upper limit are treated as of equal weight. Experiments having a lesser accuracy are weighted inversely as their error variances. The efficiency of this procedure is discussed in Cochran (1937), where it is shown to be substantially more efficient than the use of the ordinary weighted mean if the numbers of degrees of freedom for error are small. Quite large changes in the choice of the upper limit do not seriously affect the efficiency, and equally will not produce any great changes in the resultant estimate. In most agricultural field experiments the upper limit given by an error variance corresponding to a standard error of 5–7 % per plot would seem appropriate in cases in which there is no evidence of variation in the response from experiment to experiment.

A further alternative procedure which produces much the same effect is provided by the use of the semi-weighted mean, assigning some arbitrary value to $\sigma_0^2$. This procedure has the advantage of being easily adaptable to cases in which there is evidence of variation in response from experiment to experiment. If, for instance, there are eight replicates of each treatment, the value of $\sigma_0^2$ corresponding to 4 % per plot will be $(0.04)^2 (\frac{1}{8} + \frac{1}{8})$, i.e. 0·0004 times the square of the mean yield. This will produce about the same effect as taking a lower limit corresponding to a standard error of 5 % per plot. If in addition the estimated variance of the response from centre to centre is 0·0006 times the square of the mean yield, then we might reasonably take a value of $\sigma_0^2$ corresponding to 0·0010 times the square of the mean yield.

If the error variances of the various experiments are accurately known the error variance of any form of weighted mean is given by

$$\frac{w_1'^2 \sigma_1^2 + w_2'^2 \sigma_2^2 + \ldots}{(w_1' + w_2' + \ldots)^2},$$

where $w_1'$, $w_2'$, ... represent the weights actually adopted. If $w_1'$, $w_2'$, ... are equal to $1/\sigma_1^2$, $1/\sigma_2^2$, ..., this expression reduces to the expression for the error variance of the fully weighted mean, namely $1/(w_1' + w_2' + \ldots)$, and if all the weights are equal the error variance of the unweighted mean of $k$ estimates

$$\frac{1}{k^2} (\sigma_1^2 + \sigma_2^2 + \ldots)$$

is obtained.

If, however, the error variances are estimated, and the weights depend on these estimates, the above expression will not be correct. In particular the estimated error variance of the fully weighted mean in a group of experiments each with $n$ degrees of freedom for error will be $n/(n-4)$ times the expression given above (the variances being replaced by their estimates throughout). The error variance of any semi-weighted mean, or weighted mean with upper limit, will have to be similarly increased. No exact expressions are available, but in general the additional factor must lie between $n/(n-4)$ and unity. In the case of the weighted mean with upper limit to the weights the inclusion of the factor $n/(n-4)$ in the terms which have weights below the upper limit is likely to give a reasonable approximation.

The mean (weighted or unweighted) may be tested for significance by means of the $t$ test, using the estimated standard error. The test is not exact, since the number of degrees of freedom is not properly defined, but if a number somewhat less than the total number of degrees of freedom for error in the whole set of experiments is chosen the test will be quite satisfactory.

There is one further point that must be examined before using any form of mean in which the weights depend on the relative precision of the various experiments. If the precision is associated in any way with the magnitude of the response, such a weighted mean will produce biased estimates and must not be used. Thus, for example, the response to a fertilizer might be greater on poor land, and this land might be more irregular than good land, so that experiments on poor land would give results of lower precision. In such a case any of the above weighted means would lead to an estimate of the average response which would be smaller than it should be.

To see whether association of this type exists the experiments may be divided into two or more classes according to accuracy, and the differences between the mean response in each class examined. Alternatively the regression of the responses on the standard errors of the experiments may be calculated.

## 7. Example of the analysis of a set of experiments of unequal precision

Table IX gives the responses (in yield of roots) to the three standard fertilizers in a set of $3 \times 3 \times 3$ experiments on sugar beet. These experiments were conducted in various beet growing districts in England. The results shown are those of the year 1934 and are reported in full in the Rothamsted Report (1934). It cannot be claimed that the sites were selected at random (practical considerations precluded this course) and consequently any values obtained for the average responses must be accepted with caution, but the results will serve to illustrate the statistical points involved.

Table IX. *Responses to fertilizers in a series of experiments on sugar beet*

Washed roots (tons per acre)

| Station | Mean yield | Linear response to | | | Standard error | Wt. | Degrees of freedom |
|---|---|---|---|---|---|---|---|
| | | N | P | K | | | |
| Allscott | 10·97 | −0·24 | +0·63 | +0·57 | ±0·519 | 3·7 | 15 |
| Bardney | 11·44 | +1·23† | +0·35 | +0·01 | ±0·285 | 12·3 | 22 |
| Brigg | 13·42 | +0·11 | −0·38 | −0·21 | ±0·603 | 2·8 | 22 |
| Bury | 13·83 | +2·08† | −0·05 | −0·22 | ±0·351 | | 15 |
| Cantley | 12·90 | +0·20 | +0·32 | +0·14 | ±0·453 | 4·9 | 15 |
| Colwick | 10·12 | +1·05† | +0·87† | −0·07 | ±0·287 | 12·2 | 15 |
| Ely | 12·46 | −1·14 | +0·80 | −0·08 | ±0·886 | 1·3 | 15 |
| Felstead | 11·28 | +3·34† | +0·11 | +0·23 | ±0·356 | 7·9 | 15 |
| Ipswich | 12·45 | +1·64† | +0·57 | +0·34 | ±0·344 | 8·5 | 15 |
| King's Lynn | 19·54 | +0·52 | +0·12 | −0·57 | ±0·481 | 4·3 | 15 |
| Newark | 14·10 | +1·37† | +0·54* | −0·33 | ±0·198 | 25·5 | 15 |
| Oaklands | 12·84 | 0·00 | −0·14 | +0·40 | ±0·622 | 2·6 | 15 |
| Peterborough | 17·99 | −0·14 | +1·02 | −1·34* | ±0·618 | 2·6 | 15 |
| Poppleton | 14·21 | +2·72† | −0·21 | −0·18 | ±0·357 | 7·8 | 22 |
| Wissington | 14·55 | +3·32† | +0·19 | +0·38 | ±0·443 | 5·1 | 15 |
| Mean | 13·47 | +1·07 | +0·32 | −0·06 | ±0·125 | 109·6 | 246 |

*  5 % significance.    †  1 % significance.

At Bardney, Brigg and Poppleton there were two complete replications, i.e. fifty-four plots, while at each of the remaining centres there was a single replication, twenty-seven plots, only, the error being estimated from the interactions of the quadratic components of the responses and from the unconfounded second order interactions. At all centres the experiments were arranged in blocks of nine plots.

The size of the plot varied. It is immediately apparent, from inspection, or by application of the process described in § 4, that the experiments are of very varying precision. In general the larger plots, as might be expected, gave the more accurate results, though the gain in precision was not proportional to the increase in area.

(a) *The response to nitrogen.*

The response to nitrogen clearly varies significantly from centre to centre, this variation being large in comparison with experimental error. The ordinary analysis of variance of these responses is given in Table X.

The pooled estimate of error is equal to the mean of the squares of the standard errors given in Table IX. The estimate of the standard error of the average response is therefore

$$\sqrt{(0·2345/15)} = ±0·125.$$

Table X. *Analysis of variance of response to nitrogen*

|  | Degrees of freedom | Sum of squares | Mean square |
|---|---|---|---|
| Average response | 1 | 17·1949 | 17·1949 |
| Response × centres | 14 | 25·5891 | 1·8278 |
| Pooled estimate of error |  |  | 0·2345 |

Since the errors are unequal the $t$ distribution will not be exactly followed, the actual 5 % point being subject to slight uncertainty, but in any case intermediate between those given by $t$ for 15 and for 246 degrees of freedom.

The variation of the response from centre to centre has therefore an estimated variance of

$$1·8278 - 0·2345 = 1·5933,$$

excluding variance due to error. This method of estimation is not fully efficient, but may be used in cases such as the present in which the variation is large in comparison with the experimental errors.

It is clear that even were the precision of the experiments known with exactitude, the standard errors being those of Table IX, the semi-weighted mean of § 6, which could then be used, would differ little from the unweighted mean, since the weights would only range from

$$\frac{1}{1·5933 + 0·0392} \quad \text{to} \quad \frac{1}{1·5933 + 0·7850},$$

i.e. from 0·61 to 0·42. The unweighted mean is therefore the only estimate of the average response to nitrogen that need be considered. It may be noted that in this set of experiments there appears to be some association between degree of accuracy and magnitude of response to nitrogen. This is an additional reason for not using any form of weighted mean.

*(b) The response to superphosphate.*

The responses to superphosphate are of much smaller magnitude than those to nitrogen. Only two, those of Colwick and Newark, are significant, but eleven out of the fifteen are positive, and consequently there is some evidence for a general response.

The unweighted mean of the responses is $+0·32$ and the standard error of the quantity is, as before, $\pm 0·125$. The unweighted mean is therefore significant.

The weights corresponding to the estimated standard errors are given in Table IX. The sum of the products of these weights and the responses to phosphate, divided by the sum of the weights, gives the weighted mean, $+0·365$. This differs somewhat from the unweighted mean, and inspection shows that the difference is largely due to the fact that the two stations which gave significant results received high weights. The weight assigned to Newark is nearly $\frac{1}{4}$ of the total weight. The estimated error of this experiment is 3 % per plot, which would appear to be lower than is likely to be attained in practice. Fixing a lower limit of error at 5 % per plot, which is equivalent to a weight of 10·0 for the experiments of twenty-seven plots and of 20·0 to experiments with fifty-four plots, we obtain the weighted mean with upper limit, 0·323.

Following the rule given in § 6, the estimated standard error of the weighted mean will be given by the square root of

$$\frac{3\cdot7 \times \dfrac{15}{11} + 12\cdot3 \times \dfrac{22}{18} + \ldots}{(3\cdot7 + 12\cdot3 + \ldots)^2}.$$

This gives the value $\pm 0\cdot111$. Similarly the estimated standard error of the weighted mean with upper limit to the weights is given by the square root of

$$\frac{3\cdot7 \times \dfrac{15}{11} + 12\cdot3 \times \dfrac{22}{18} + \ldots + \dfrac{10^2}{12\cdot2} + \ldots}{(3\cdot7 + 12\cdot3 + \ldots + 10 + \ldots)^2}.$$

This gives a value of $\pm 0\cdot112$. This is presumably somewhat of an over-estimate, as this mean is likely to be somewhat more accurate than the weighted mean.

In order to test whether there is any evidence of variation in the phosphate response from centre to centre the weighted sum of squares of deviations $Q$ may be calculated by the formula given at the end of § 5. For convenience in this calculation it is best to tabulate the products $wx$ of the weights and the responses separately for each centre. The values obtained are

$$Swx^2 = 27\cdot6068$$
$$\bar{x}_w \, Swx = 14\cdot6044$$
$$Q = 13\cdot0024$$

The value of $\chi'^2$ may now be calculated. In the present case the number of degrees of freedom for error varies from experiment to experiment. We will take $n$ to be equal to the mean 16·4 of these numbers. This procedure will be satisfactory if the numbers do not differ too widely and are reasonably large in all experiments. Using this value we have

$$\chi'^2 = 14 + \sqrt{\frac{12\cdot4}{15\cdot4} \left( \frac{14\cdot4}{16\cdot4} \, 13\cdot0024 - 14 \right)} = 11\cdot7.$$

Clearly there is no evidence of any variation in response from centre to centre.

It may, however, be considered that although there is no evidence of variation in response, such variation should not be precluded, and that consequently the upper limit of the weights should be lower than the values of 10 and 20 taken above. Fixing the limit at 7·8, so as to give seven of the fifteen experiments equal weight, we obtain a mean of $+0·301$, with an estimated standard error of $\pm 0·109$. In fertilizer experiments such as the present, where from the nature of the treatments, constancy of response would seem unlikely, this last estimate of the mean response appears to be the most satisfactory, since it gives equal weight to all the more accurate experiments and at the same time prevents the less accurate experiments from unduly influencing the results.

(c) *The response to potash.*

The effect of potash shows no significance, either in mean response or in variation in response from centre to centre. The significant depression at Peterborough can consequently be reasonably attributed to chance.

The analysis follows the lines already given and need not be set out in detail here. The weighted mean with upper limit 7·8 has the value $-0·03 \pm 0·109$. The value of $\chi'^2$ for testing the significance of the variation in response is 11·6.

The results discussed here are, of course, only a part of the full results of the experiments. No consideration has been given to the curvature of the response curves, or to the interactions of the different fertilizers. The whole set of experiments provides an excellent illustration of the power of factorial design to provide accurate and comprehensive information. It will be noted, among other things, that the mean responses to the three fertilizers are determined with a standard error of less than 1 % of the mean yield.

## SUMMARY

When a set of experiments involving the same or similar treatments is carried out at a number of places, or in a number of years, the results usually require comprehensive examination and summary. In general, each set of results must be considered on its merits, and it is not possible to lay down rules of procedure that will be applicable in all cases, but there are certain preliminary steps in the analysis which can be dealt with in general terms. These are discussed in the present paper and illustrated by actual examples. It is pointed out that the ordinary analysis of variance procedure suitable for dealing with the results of a single experiment may require modification, owing to lack of equality in the errors of the different experiments, and owing to non-homogeneity of the components of the interaction of treatments with places and times.

## REFERENCES

COCHRAN, W. G. (1937). *J. R. statist. Soc.*, Suppl. **4**, 102–18.

FISHER, R. A. (1935). *The Design of Experiments*. Edinburgh.

IMMER, F. R., HAYES, H. K. & POWERS, LE ROY (1934). *J. Amer. Soc. Agron.* **26**, 403–19.

LEWIS, A. H. & TREVAINS, D. (1934). *Emp. J. exp. Agric.* **2**, 244.

*Rep. Rothamst. exp. Sta.* (1934), p. 222.

YATES, F. (1933). *Emp. J. exp. Agric.* **1**, 129–42.

(*Received* 15 *February* 1938)

## The Omission or Addition of an Independent Variate in Multiple Linear Regression

### By W. G. Cochran

#### § 1. *Introduction*

If $y$ is the dependent variate and $x_1, x_2, \ldots x_r$ are the independent variates, the equations to determine the linear regression coefficients $b_1, b_2, \ldots b_r$ of $y$ on $x_1, x_2, \ldots x_r$ are

$$\left. \begin{aligned} b_1 S(x_1{}^2) + b_2 S(x_1 x_2) + \ldots + b_r S(x_1 x_r) &= S(x_1 y) \\ \cdot \quad \cdot \quad \cdot \quad \cdot \quad \cdot \quad \cdot \quad \cdot \quad \cdot \quad \cdot \\ b_1 S(x_r x_1) + b_2 S(x_r x_2) + \ldots + b_r S(x_r{}^2) &= S(x_r y) \end{aligned} \right\} \quad . \quad (1)$$

In solving these equations, Fisher (1) has suggested that a set of auxiliary quantities $c_{pq}(p, q = 1, 2, \ldots r)$ should first be obtained. The quantities $c_{p1}, c_{p2}, \ldots c_{pr}$ are the solutions of the above equations with the right-hand side of the $p^{th}$ equation replaced by 1, and the right-hand sides of the other equations by 0. The regression coefficients are obtained from the $c$'s by means of the relations

$$b_i = \sum_{q=1}^{r} c_{iq} S(x_q y) \qquad i = 1, 2, \ldots r \quad . \quad . \quad (2)$$

To students carrying out a regression analysis for the first time, this procedure has sometimes seemed, as indeed it is, a somewhat roundabout method of determining the regression coefficients. The values of the $b$'s alone, however, provide a very incomplete picture of the relationship between $y$ and $x_1, \ldots x_r$; they do not show which of the independent variates are significantly related to the dependent variate, nor can limits be assigned from them within which the true values of the regression coefficients are likely to lie. When these points are realized, the convenience of Fisher's method may be appreciated, for the estimated standard error of $b_i$ has been shown to be $s\sqrt{c_{ii}}$ (where $s$ is the estimated standard error of a single observation), and is readily obtainable if the $c$'s have been found.

Other properties of the $c$'s which may sometimes be useful have been pointed out by Fisher. (1) The mean covariance of $b_1$ and $b_2$ is $s^2 c_{12}$. Thus the standard error of the sum or difference of two regression coefficients may be obtained. This will be required if, for instance, independent variates such as maximum and minimum temperature are being replaced by mean temperature and range of temperature after the regression equations have been solved. (2) If the regressions of a number of dependent variates on the same set of independent variates are being examined, the $c$'s remain the same throughout and serve for the determination of all regression coefficients. (3) It frequently happens that no apparent relation is found between the dependent variate and one or more of the independent variates. When this is the case, it is sometimes desirable to omit such variates from the regression equations. Knowing the $c$'s, this may be done without the labour of re-solving the regression equations with the superfluous variates omitted. Fisher (1) has given formulæ for the adjustments required in the regression coefficients, and the corresponding adjustments in the $c$'s are easily found. In this note the process will be reversed to show how to add a new independent variate to the equations without re-solving them.

### § 2. *The Omission of an Independent Variate*

The new regression coefficients, with the variate $x_r$ omitted from the regression, will be denoted by $b'_1, b'_2, \ldots b'_{r-1}$.

The $(r-1)$ equations satisfied by these are

$$
\left.
\begin{aligned}
b'_1 S(x_1{}^2) + b'_2 S(x_1 x_2) + \ldots + b'_{r-1} S(x_1 x_{r-1}) &= S(x_1 y) \\
\cdot \quad \cdot \quad \cdot \quad \cdot \quad \cdot \quad \cdot \quad \cdot \quad \cdot \quad \cdot \quad \cdot \quad \cdot \quad \cdot \\
b'_1 S(x_{r-1} x_1) + b'_2 S(x_{r-1} x_2) + \ldots + b'_{r-1} S(x_{r-1}{}^2) &= S(x_{r-1} y)
\end{aligned}
\right\} \quad (3)
$$

By subtracting the corresponding equation of set (1) from each of the above equations, the following equations are obtained:

$$
\left.
\begin{aligned}
\delta b_1 S(x_1{}^2) + \delta b_2 S(x_1 x_2) + \ldots + \delta b_{r-1} S(x_1 x_{r-1}) - b_r S(x_1 x_r) &= 0 \\
\cdot \quad \cdot \quad \cdot \quad \cdot \quad \cdot \quad \cdot \quad \cdot \quad \cdot \quad \cdot \quad \cdot \quad \cdot \\
\delta b_1 S(x_{r-1} x_1) + \delta b_2 S(x_{r-1} x_2) + \ldots \qquad\qquad \\
+ \delta b_{r-1} S(x_{r-1}{}^2) - b_r S(x_{r-1} x_r) &= 0
\end{aligned}
\right\} \quad (4)
$$

where $\delta b_1 = b'_1 - b_1$ is the adjustment in $b_1$ produced by the elimination of $x_r$. From these equations we may determine the ratios $\delta b_1/b_r, \delta b_2/b_r \ldots$ The equations are, however, the same as the first $(r-1)$ equations satisfied by $c_{1r}, c_{2r}, \ldots c_{rr}$, with $\delta b_1$ in place of $c_{1r}$, etc. and $-b_r$ in place of $c_{rr}$.

Hence
$$\delta b_1/(-b_r) = c_{1r}/c_{rr} \quad . \quad . \quad . \quad . \quad (5)$$

that is
$$\delta b_1 = b'_1 - b_1 = -(c_{1r}/c_{rr})b_r \quad . \quad . \quad . \quad (6)$$

as given by Fisher (1).

A similar treatment of the equations for the $c'$'s and $c$'s gives the results

$$\delta c_{11} = c'_{11} - c_{11} = -(c_{1r}{}^2/c_{rr}) \quad . \quad . \quad . \quad (7)$$

$$\delta c_{12} = c'_{12} - c_{12} = -c_{1r} c_{2r}/c_{rr} \quad . \quad . \quad . \quad (8)$$

Equations (6), (7) and (8) provide all the necessary adjustments to form the new $b$'s and $c$'s. If only the $b'$'s and their standard errors are required, the non-diagonal $c'$'s need not be found. The elimination of two variates is best carried out in two stages.

Where only a single independent variate is eliminated, this method is quicker than re-solving the regression equations, except when there are only two independent variates in the first instance. If two variates are being eliminated, the method is quicker if the original number of variates is six or more, and probably also with five variates.

### § 3. *The Addition of an Independent Variate*

If $x_n$ is the new variate, the first step is to calculate $S(x_1 x_n)$, ... $S(x_n^2)$ and $S(x_n y)$. Let dashes again denote the *new* coefficients. Regarding the original equations as obtained by eliminating $x_n$ from the new equations, the adjustment equations (6), (7) and (8) become respectively

$$\delta b_1 = b'_1 - b_1 = + (c'_{1n}/c'_{nn})b'_n \quad . \quad . \quad . \quad . \quad . \quad (9)$$

$$\delta c_{11} = c'_{11} - c_{11} = + c'^2_{1n}/c'_{nn} \quad . \quad . \quad . \quad . \quad . \quad (10)$$

$$\delta c_{12} = c'_{12} - c_{12} = + (c'_{1n} c'_{2n})/c'_{nn}, \text{ etc.} \quad . \quad . \quad . \quad (11)$$

These equations may be used to adjust all the existing coefficients; it is, however, first necessary to know the values of $c'_{1n} \ldots c'_{nn}$ and $b'_n$.

By writing down the equations satisfied by $c_{11}, c_{12}, \ldots c_{1r}$ and subtracting from each the corresponding equation satisfied by $c'_{11}, c'_{12} \ldots c'_{1n}$, we obtain the equations

$$\left. \begin{aligned} \delta c_{11} S(x_1^2) + \delta c_{12} S(x_1 x_2) + \ldots \\ + \delta c_{1r} S(x_1 x_r) = - c'_{1n} S(x_1 x_n) \\ . \quad . \quad . \quad . \quad . \quad . \quad . \quad . \quad . \quad . \\ \delta c_{11} S(x_r x_1) + \delta c_{12} S(x_r x_2) + \ldots \\ + \delta c_{1r} S(x_r^2) = - c'_{1n} S(x_r x_n) \end{aligned} \right\} \quad (12)$$

These equations are, however, the same as the original equations for $b_1, \ldots b_r$ with $- c'_{1n} S(x_q x_n)$ in place of $S(x_q y)$ on the right-hand side. Hence

$$\delta c_{1p} = - c'_{1n} \sum_{q=1}^{r} c_{pq} S(x_q x_n) \qquad p = 1, 2, \ldots r . \quad . \quad (13)$$

Hence by equations (11)

$$c'_{pn}/c'_{nn} = - \sum_{q=1}^{r} c_{pq} S(x_q x_n). \quad . \quad . \quad . \quad (14)$$

The last of the equations satisfied by $c'_{nn}$ is

$$c'_{1n} S(x_n x_1) + \ldots + c'_{rn} S(x_n x_r) + c'_{nn} S(x_n^2) = 1 \quad . \quad . \quad (15)$$

By substituting from (14) for $c'_{1n} \ldots$ in terms of $c'_{nn}$, we get

$$c'_{nn}[S(x_n^2) - \sum_{p,q=1}^{r} c_{pq} S(x_p x_n) S(x_q x_n)] = 1 \quad . \quad . \quad . \quad (16)$$

Equations (14) and (16) give $c'_{1n} \ldots c'_{nn}$. We may then find $b'_n$ from the usual relations between the $b''$s and the $c''$s, and hence adjust all the other $b$'s and $c$'s.

This process is in all cases more expeditious than re-solving the equations.   The arrangement of the computations is best illustrated by a numerical example.

### § 4. *Example of the Addition of an Independent Variate*

In a study of the effects of weather factors on the numbers of noctuid moths per night caught in a light trap, regressions were worked out on the minimum night temperature, the maximum temperature of the previous day, the average speed of the wind during the night and the amount of rain during the night.   The dependent variable was log (number of moths $+$ 1).   This was found to be roughly normally distributed, whereas the numbers themselves had an extremely skew distribution.   Further, a change in one of the weather factors was likely to produce the same *percentage* change at different times in the numbers of moths rather than the same *actual* change.   Three years' data were included.   These were grouped in blocks of nine consecutive days, so as to eliminate as far as possible the effects of the lunar cycle.   After the removal of differences between blocks, 72 degrees of freedom remained for the regressions.

The regression coefficients and their standard errors in convenient working units are as follows:

| Min. Temp. | Max. Temp. | Wind | Rain |
|---|---|---|---|
| 0·1981407 $\pm$0·0650 | 0·0385284 $\pm$0·0588 | $-$0·5086492 $\pm$0·1515 | $+$0·0318482 $\pm$0·0499 |

The analysis of variance is shown below:

Table I

|  |  |  | D.F. | Sums of Squares | Mean Squares |
|---|---|---|---|---|---|
| Regression | ... | ... | 4 | 0·8274 | 0·2068 |
| Deviations | ... | ... | 68 | 2·7245 | 0·04007 |
| Total | ... | ... | 72 | 3·5519 | 0·04933 |

It was subsequently decided to investigate the effect of cloudiness, measured on a conventional scale as the percentage of starlight obscured by clouds in a night sky camera.

The calculations are shown in Table II.   The original $c$'s are first written down, and the corresponding sums of products of each variate with the new variate are placed in the right-hand column. The sum of products of each column with the right-hand column is placed at the foot of the column, *with the signs reversed*.   By equations (14), these values are $c_{15}'/c_{55}'$ . . . .

The sum of the products of these numbers with the corresponding numbers in the right-hand column is then calculated.   The sum of

TABLE II

*Addition of an Independent Variate*

| | Min. Temp. (1) | Max. Temp. (2) | Wind (3) | Rain (4) | Sums of Products with Cloud |
|---|---|---|---|---|---|
| | | $c_{pq}$ | | | $S(x_p x_5)$ |
| (1) | +0·10542356 | -0·04194620 | -0·09606709 | -0·01849096 | -4·867 |
| (2) | -0·04194620 | +0·08603869 | +0·03317271 | +0·01290358 | +0·206 |
| (3) | -0·09606709 | +0·03317271 | +0·57265201 | +0·00811662 | -0·5446 |
| (4) | -0·01849096 | +0·01290358 | +0·00811662 | +0·06227532 | -5·42 |
| (5) | — | | | | +7·87 |
| | | | $c_{p5}'/c_{55}' = -\Sigma c_{pq}S(x_q x_5)$ | | $c_{55}'$ |
| | +0·36919824 | -0·13387286 | -0·11853374 | +0·24929891 | +0·21013314 |
| | | | $S(x_{py})$ | | |
| | +2·0744 | +1·5747 | -0·6440 | +0·885 | -1·933 |
| | $b_1$ | $b_2$ | $b_3$ | $b_4$ | |
| | +0·1981407 | +0·0385284 | -0·5086492 | +0·0318482 | — |
| | $b_1'$ | $b_2'$ | $b_3'$ | $b_4'$ | $b_5'$ |
| | +0·1142775 | +0·0689376 | -0·4817243 | -0·0247799 | -0·2271496 |
| | ±0·0704 | ±0·0576 | ±0·1459 | ±0·0528 | ±0·0882 |
| | | $c_{pq}'$ | | | |
| (1) | +0·13406625 | -0·05233216 | -0·10526303 | +0·00084984 | +0·07758079 |
| (2) | — | +0·08980468 | +0·03650720 | +0·00589052 | -0·02813112 |
| (3) | — | — | +0·57560443 | +0·00190712 | -0·02490787 |
| (4) | — | — | — | +0·07533508 | +0·05238596 |
| (5) | — | — | — | — | +0·21013314 |

squares of the new variate (7·87) is added on the calculating machine. By equation (16) the reciprocal of the total is $c_{55}'$ (0·21013314).

The regression coefficient $b_5'$ may now be found. Since

$$b_5' = c_{15}'S(x_1y) + \ldots + c_{55}'S(x_5y) \quad . \quad . \quad (17)$$

$$b_5' = \{(c_{15}'/c_{55}')S(x_1y) + \ldots + (c_{45}'/c_{55}')S(x_4y) + S(x_5y)\} \times c_{55}' \quad (18)$$

$$= \{0·36919824 \times 2·0744 + \ldots \quad -1·933\} \times (0·21013314)$$

$$= -·2271496$$

which is obtained on the machine without any intermediate writing down.

At this stage the significance of the coefficient $b_5'$ may be tested; if the new variate has no apparent effect, it may not be worth while to complete the calculations. The reduction in the sum of squares due to cloud is $b_5'^2/c_{55}' = 0·2455$. From Table I the residual mean square (67 degrees of freedom) is found to be 0·03700, so that $b_5'$ is definitely significant.

The calculations are completed by means of the adjustment equations (9), (10) and (11). In particular

$b_1' = 0·1981407 + (0·36919824) \times (-0·2271496) = 0·1142775.$

$c_{15}' = (0·36919824) \times (0·21013314) = 0·07758079.$

$c_{11}' = 0·10542356 + (0·36919824) \times (0·07758079) = 0·13406625.$

$c_{12}' = -0·04194620 + (0·36919824) \times (-0·02813112) =$
$$-0·05233216$$

In the last two cases the combined use of the ratios $c_{15}'/c_{55}'$ . . . and the values $c_{15}'$ . . . gives the adjustment terms in a single multiplication.

As a final check on the calculations, $b_1'$, . . . $b_5'$ should be substituted in the regression equations. The $c''$s may then be checked by verifying that the $b''$s obtained from the $c''$s in the usual way agree with the values already found. An intermediate check on the values $c_{15}'/c_{55}'$ . . . may also be obtained by adding the four $c$'s in each row and calculating the sum of the products of the totals with the values $S(x_1x_5)$. . . . This, with its sign reversed, is equal to

$$0·36919824 - 0·13387286 - 0·11853374 + 0·24929891.$$

The number of decimal places carried in the above calculation is excessive, though it facilitates the detection of errors when the final substitution in the regression equations is made. Six decimal places would have been sufficient in ordinary work.

*Reference*

Fisher, R. A., " Statistical Methods for Research Workers." Oliver and Boyd, Edinburgh, 6th Ed., 1936, §§ 29, 29.1.

# 15

SUPPLEMENT

TO THE

JOURNAL OF THE ROYAL STATISTICAL SOCIETY

Vol. V., No. 1, 1938.

CROP ESTIMATION AND ITS RELATION TO AGRICULTURAL
METEOROLOGY

[A Discussion before the INDUSTRIAL AND AGRICULTURAL RESEARCH SECTION
of the ROYAL STATISTICAL SOCIETY, November 18th, 1937, MR. H. D.
VIGOR in the Chair.]

DR. J. O. IRWIN

## I. *Introductory*

I CONSIDER it a great honour to be asked to open a discussion on
a subject in which I have taken no active part for nearly seven years.
Between 1928 and 1931 it was my duty and privilege to examine
the Crop–Weather Scheme of the Ministry of Agriculture and to
make recommendations for its improvement. Feeling that the
importance of agricultural meteorology from a practical point of
view lay in the possibility of using that science for forecasting or
estimating in advance the acreage, yield or quality of crops, I was
led to do two things,

(i) To examine what had been done, up to that time, in
making forecasts of crops, in different parts of the world;

(ii) To examine how far what was being done by the various
research stations co-operating in the Crop–Weather Scheme
was adequate, and if not adequate, how it could be improved.

The results of the first enquiry were published in a paper read
to the Conference of Empire Meteorologists in 1929, (1), the results
of the second enquiry were only published in part (2 and 3), but I
believe some of the recommendations made in it have since been
adopted.

Thus I can only tell you what was the position—how much had
been achieved—by the end of 1930. I must leave it to later
speakers to tell us of the progress that has been made since.

## II. *Crop forecasting up to* 1929

(i) *The problem and methods.* There are two main methods of
forecasting crops : subjective and objective. In examining the

SUPP. VOL. V. NO. 1.                  B

state of affairs in 1929, it appeared that official forecasts, which had been practised for many years by Government Departments throughout the world, were mainly of the former kind. The Departments secured, from large numbers of crop reporters or estimators distributed throughout the agricultural area, their estimates of acreage or yield expressed in some numerical form. These estimates were based on the crop reporters' *opinion* of the prospects of the growing crops after inspection of them, no doubt taking into account their general experience of the effect on them of previous weather conditions. There was already a tendency to supplement these estimates by more objective methods.

The latter methods have always looked to the weather as providing the objective basis. In the examination of the effect of weather on crops, and in the search for relations between agricultural and meteorological phenomena by which yields or other properties of the crop might be predicted, scientific workers have followed two main lines of research. First, they have looked for cycles, or regular sequences of a definite number of years at the end of which similar meteorological and, as a consequence, agricultural conditions were reproduced; secondly, they have ascertained the effects of different types of weather during or shortly before the growing period of the crop.

Whatever method of forecasting the crop be adopted, it is impossible to over-emphasize the importance of accurate final estimates of acreage and yield. Without such estimates, the assessment or the accuracy of a forecast becomes an impossibility.

(ii) *Official methods in use up to* 1929. The official methods in use in England and Wales, in the United States and in India were described in detail in the paper already mentioned. Here they can only be rapidly summarized.

(a) *England and Wales.* In 1929 there were rather more than 300 crop reporters from whom the Ministry of Agriculture and Fisheries obtained monthly crop reports, and I presume there are about the same number now. In the earlier of these reports forecasts of yield are made as a percentage of the last ten years' average, in the later ones in precise numerical terms, *e.g.*, in cwts. per acre for cereals and tons per acre for root crops. There is no difficulty about obtaining precise figures of acreage, since returns, compulsory under the Agricultural Returns Act, 1925, for all holdings exceeding one imperial acre of land used for the production of crops or of grazing are made to the crop reporters. From the information received from the crop reporters, the Ministry compiles its own monthly forecasts. Final estimates are based on the crop reporters' final figures sent to the Ministry on November 15th. Table I throws some light on the accuracy of the Ministry's forecasts. The tendency for forecasts to be below the final estimates is obvious. Not much information was available about the accuracy of the final estimates, but after an examination of 30 years' estimates (1893–1922) for Great Britain and Hertfordshire and of the yields (ascertained by actual weighings) on the farmyard-manure plot on Broadbalk field at Rothamsted, I suggested that it was possible that the official estimates under-estimated the variability in yield from year to year. I should be glad to know if this point has been further investigated since then.

TABLE I

*Mean and standard deviation of percentage difference between forecast and final estimate for the yield of certain crops in England and Wales (1923–8)*

| | | | July 1st | Aug. 1st | Sept. 1st | Oct. 1st | Nov. 1st |
|---|---|---|---|---|---|---|---|
| Wheat | Mean | ... | — | − 0·3 | − 0·4 | − 1·2 | — |
| | S.D.* | ... | — | 5·5 | 4·4 | 1·8 | — |
| Barley | Mean | ... | — | − 4·1 | − 2·4 | − 2·5 | — |
| | S.D. | ... | — | 3·8 | 3·5 | 2·8 | — |
| Oats | Mean | ... | — | − 4·4 | − 3·9 | − 4·1 | — |
| | S.D. | ... | — | 3·4 | 3·1 | 2·9 | — |
| Beans | Mean | ... | — | − 3·3 | − 4·2 | − 2·8 | — |
| | S.D. | ... | — | 2·0 | 1·8 | 1·8 | — |
| Peas | Mean | ... | — | + 1·0 † | − 1·4 | − 0·4 | — |
| | S.D. | ... | — | 7·5 | 4·8 | 4·1 | — |
| Seeds Hay | Mean | ... | − 4·1 | 0·0 | 0·0 ‡ | — | — |
| | S.D. | ... | 3·7 | 3·2 | 3·8 | — | — |
| Meadow Hay | Mean | ... | − 7·1 | − 2·8 | − 2·7 ‡ | — | — |
| | S.D. | ... | 4·5 | 2·6 | 2·9 | — | — |
| Potatoes | Mean | ... | — | — | − 5·4 | − 4·0 | − 3·5 |
| | S.D. | ... | — | — | 6·7 | 3·8 | 1·7 |
| Turnips & Swedes | Mean | ... | — | — | — | − 1·8 | − 0·7 |
| | S.D. | ... | — | — | — | 1·4 | 1·6 |
| Mangolds | Mean | ... | — | — | — | − 3·2 | − 1·8 |
| | S.D. | ... | — | — | — | 1·4 | 1·0 |

*A negative sign denotes that the forecast is below the final estimate.*

* Calculated from $\sqrt{S(x - \bar{x})^2/(n - 1)}$.
† 5 years only.
‡ 4 years only.

(b) *The United States.* I shall not describe here the personnel of the United States crop-reporting system, but refer you to the

paper already cited.(1)   Forecasts of acreage are made partly subjectively by asking reporters to express the acreage as a percentage
of the usual acreage, partly by a sampling process of securing reports
from individual farmers who report the acreages on their farms
under each crop in the current and previous years, and partly by
"field count methods," such as counting by means of a special
speedometer the number of feet along the road under each kind of
crop.   Forecasts of yield are made by asking reporters for estimates
of the condition of the crop in terms of the normal as 100.   If these
estimates are interpreted after a statistical examination of the
relation between similar estimates in previous years and the final
estimate of the crop after harvest, much information can be obtained
from them.   They are parallel to the early estimates, made by the
crop reporters in England and Wales, of yield as a percentage of
the last ten years average.   Table II gives an instance of the sort
of accuracy obtained.

TABLE  II

*Mean and standard deviation of percentage difference between forecast
and final estimate for the yield of corn in Iowa* (1914–22)

|  | July | Aug. | Sept. | Oct. |
|---|---|---|---|---|
| Mean       ...      ... | − 4·14 | − 1·41 | − 1·98 | − 1·49 |
| S.D. ...      ...      ... | 8·73 | 5·47 | 3·60 | 3·39 |

*A negative sign denotes that the forecast is below the final estimate.*

Even before 1929, the Division of Crop and Livestock Estimates,
in conjunction with the United States Weather Bureau, was experimenting with objective correlation methods, having as their
object the forecasting of crops from weather.   Doubtless much
progress has been made since then.

(c) *India*.   The official methods of crop forecasting in India
are in some respects similar to those in the United States, but again
I must refer you elsewhere (1) for details.   In India the land forms
the basis of taxation; forecasts and estimates of acreage are in
consequence usually good, but probably better in districts which are
not permanently settled.   Permanently settled districts consist of
estates which are taxed on a fixed assessment under a deed of
permanent ownership granted by the state.   In many parts of
India the crop reporter is the *patwari*, or village accountant.   He
does not give his estimate of yield as a percentage of the normal
crop, but he estimates in terms of annas.   In some parts of India
a 12-anna crop and in others a 16-anna crop is taken as a standard.
There is no objection to this, if the relation of such estimates to final
yield is properly determined.   In 1929 I came to the conclusion

that, granted the necessity of the subjective method of forecasting and estimation, the acreage figures in India were extremely good, but the actual yields and condition figures left room for considerable improvement.

(iii) *Scientific work on the relation between weather and crops up to* 1929.  My next step was to review the scientific work on forecasting crops from the weather that had already been done.  I shall not repeat that review here.  I think one was justified in concluding that the work done on cycles was disappointing, in accepting Hooker's conclusion that,

> " the chief difficulty in applying the undoubtedly significant results that had been attained, to anticipate the probable course of events in the future, seems to lie in the bewildering number of cycles actually or apparently determined."

It was the other line of approach that consisted in comparing deviations of the crop from the average with deviations of various meteorological phenomena from the average at all periods of the year, that led to the most fruitful results.  The principal handicap was the absence of accurately determined quantitative estimates of final yield.  I wonder how far that handicap still persists to-day.

It is unnecessary to repeat that review again, because regression and correlation methods and analysis of variance must be familiar in many places where they were strange seven years ago.  Further, the early work of Lawes and Gilbert on wheat yield and winter rainfall, the later work of Napier Shaw, R. H. Hooker's classical study on the relation between weather and crops, the comprehensive study of Wallen in Sweden, that of Walter in Mauritius and Jacob in India and R. A. Fisher's study of the influence of rainfall on the yield of wheat at Rothamsted—which introduced a new and powerful statistical method—are, I take it, known to all agricultural statisticians.  These studies and many others established and tested the technique necessary for obtaining crop forecasts in terms of weather.  But I do not know how far such methods have yet been used for official forecasts.  As early as 1917, H. L. Moore claimed to be able to forecast the United States cotton crop more accurately from the weather than was being done by official estimates.  Perhaps subsequent speakers may be able to tell us what progress in this direction has been made since 1929.

### III. *The crop–weather scheme and the report of* 1930

Previous work on forecasting had, I found, almost all been concerned with acreage or yield of crops over wide or fairly wide areas.  Investigating the influence of weather on the quality of

crops, confining our attention to Great Britain, was going to involve
rather intensive studies on small areas well distributed over the
country.   Accurate observations of many phases of the crop and
its environment other than final yield were going to be necessary
if the influence of weather was to be distinguished from other
influences such as soil, manuring, cultivation, previous cropping
and previous growth of the crop itself, the latter not an ultimate
influence but one very important to study if the nature of meteorological and other influences was to be understood.   This seems to
have been realized by the Agricultural Meteorological Committee
which drew up the scheme in 1924 and recommended,

> " the organization by the Ministry of Agriculture *on a uniform
> plan* of the collection of observations on the state of the crops
> (autumn and spring sown cereals, root crops, potatoes, grass
> and horticultural crops), and the incidence of insects and fungi,
> accompanied by meteorological observations of agricultural
> experiment stations in all parts of the country."

After visiting many of the research stations which were co-
operating in the scheme, it became apparent to me that, while a
good body of meteorological data was being collected the crop
data were in need of considerable improvement.   Crop yield and
quality were functions of weather and previous growth, itself a
complicated function of soil, manuring, cultivation and weather
factors.   If the maximum use was to be made of the results in
improving crop yields, as well as in predicting them, it was necessary
to know not only the effects of weather and previous growth, but
the reasons for these effects, in order that yields could be improved
as far as possible by the adjustment of controllable conditions.
It followed therefore that the observations collected under the
scheme should be as precise as possible, and that the same kind of
observations should be comparable as between different observers
in different places.   I did not find that these conditions were being
fulfilled, though I cannot review here all the evidence that led me
to that conclusion.   There was, however, an exception to this, in
the " Precision Records " scheme for observations on wheat, the
invention of Mr. A. R. Clapham, which had been tried out at
Rothamsted in the season 1927–8, and at Wye, Long Sutton and
Rothamsted in 1928–9.

It would be impossible to tell you in detail what information
was being collected up to 1929 for all the agricultural and horticultural crops considered in the crop–weather scheme.   I shall
therefore illustrate the position by considering one crop only—
wheat.   The information asked for was as follows : Variety, soil

characteristics, previous cropping, manuring, cultural operations, date of sowing, date of appearance above ground, date of breaking into ear, date of flowering, date of harvest, yield per acre of grain and straw, the weight per bushel of grain and attacks of diseases and pests. Here are some details noted at the time of the information which was actually given.

(i) *Variety.* All those stations which were participating in the National Institute of Agricultural Botany's variety trials, reported on those variety trials to the Ministry for the purposes of the Agricultural Meteorological Scheme. In this case the varieties to be grown and the size of the plots on which they were grown were in accordance with the scheme of experimentation of the National Institute of Agricultural Botany, and the varieties were therefore being grown under, as far as possible, comparable conditions. But apart from this element of uniformity, stations seem to have decided for themselves, among all the varieties they were growing, what varieties to report on, consequently the data obtained referred to a somewhat arbitrary set of varieties grown under very diverse conditions, sometimes on a commercial field, sometimes on an area of experimental field size, and in one case on an area five-foot square in the phenological garden.

(ii) *Soil characteristics.* A description of the soil in very general terms was given. There were no mechanical or chemical analyses.

(iii) *Previous cropping.* This had sometimes been given for one, sometimes for two or three and occasionally for four years.

(iv) *Manuring.* This had only been recorded for the year in question, so no information was available about manurial residues.

(v) *Cultural operations.* The description given of these was adequate.

(vi) *Date of sowing.* This was a straightforward observation which called for no comment.

(vii) *Date of appearance above ground.* Even a casual glance at the summaries of data showed the strong tendency there was for varieties in the same place, sown on the same day, to be reported as appearing above ground on the same day. At a meeting of the Agricultural Meteorological Committee held on September 13th, 1926, it was agreed that a paper in which various definitions of brairding had been given by various crop observers should be circulated to the stations with an indication that where seed had been drilled " Brairding " should be defined as when the rows can just be seen (roughly equivalent to when about half the plants had appeared above ground) and where seed had been broadcast when about half of the plants have been seen above ground.

Even if daily observations were strictly adhered to and the

recorders as careful as possible, I did not think that such a criterion would secure a date that was precise and comparable as between different observers in different places.   As regards the first definition, difficulties might arise from the crop being more advanced at one end of the field than the other, or from the variation due to conditions of illumination and the aspect from which the field was viewed. In regard to the second, the observer had obviously got to make a mental estimate of how many plants there were ultimately going to be, in order to make up his mind whether 50 per cent. were up or not.   It was obvious that personal equation would enter to a very considerable extent into a determination by either of these methods.

(viii) *Date of breaking into ear and date of flowering.*   The criticisms already made of the manner of determination of date of appearance above ground would apply here also.   I did not think that merely visual observations could determine these dates in a manner which would be precise and comparable as between different observers in different places.   This had already been recognized by the introduction of Mr. Clapham's precision records scheme where counts of numbers of emerged ears were made on 128 foot-lengths, for each variety, a procedure which enabled us to determine with precision and in a uniform manner from place to place the date when half the ears had emerged.

(ix) *Date of harvest.* This was a perfectly straightforward observation.

(x) *Yield per acre of grain and straw.*   For the National Institute of Agricultural Botany's field trials this had been determined for the whole produce of the experimental plots, but in other cases it had been estimated or not given at all, and as far as I knew, nothing had been done to secure uniformity of estimation.

(xi) *Weight per bushel of grain.*   This was not always given.

These were the observations for improving which recommendations had to be made.   Mr. Clapham's precision records scheme consisted of an experiment on wheat in which two varieties, Yeoman and Squarehead's Master, were used.   There were 16 half-drilled strips, 30 paces long, each divided into two transversely, giving 32 half plots.   A rod 10 feet long, divided into 4-foot-lengths separated by three 2-foot-lengths, was placed twice at a random distance down a random row of each half-plot.   This yielded 128 foot-lengths for each variety in which observations were made. Counts were made from time to time of plant number, shoot number, shoot height, and ear number in each foot-length separately.   Grain and straw from each set of 4 foot-lengths were weighed separately. By means of an analysis of variance, the data at each count provided a standard error for the mean count of each variety and also an estimate of the error of sampling. The final data provided a standard error of the varietal yield. These standard errors were reduced to satisfactory dimensions by the elimination of block differences which the arrangement of the experiment rendered possible.

In Mr. Clapham's scheme I thought I saw the kind of observations which were required. Accordingly, a scheme was drawn up for treating all the agricultural crops on a basis similar in principle, though here we confine our attention to wheat. In the first place it was suggested that observations should be limited to two varieties (Yeoman and Squarehead's Master), and that observations (ii) to (v) should be improved by stating the condition of soil at time of sowing, whether a fine tilth or cloddy, and whether wet or dry; by mechanical and chemical analyses of soil samples; by stating previous cropping for three years previously (except where a rotation longer than a four-course had been used when the whole rotation should be stated); by stating the nature and amount of manuring supplied if possible for three previous years as well as the current year, with dates of dressings.

The adoption of Mr. Clapham's experimental arrangement was suggested. The pairs of half-drilled strips were, however, replaced by blocks of two strips to which the two varieties were assigned at random, and foot-lengths were replaced by quarter-metres. It was provisionally recommended that date of appearance above ground should be determined as follows :—Transversely across each end of the experimental area, and at such a distanc outside the boundary of that area as to enclose two square metres for each half-plot, a string should be stretched. The date of appearance above ground for each half-plot should be considered to be when twenty plants were visible in the two-square-metre area marked off. This would necessitate daily counts starting as soon as any plants were visible at all. The following counts and measurements were recommended at appropriate times in each quarter-metre : number of plants and shoots, heights of two shoots, number of leaves on each tiller of two plants, number of emerged ears, length of two ears just before harvest, number of emerged ears which had and had not reached the flowering stage. Where two plants, shoots or ears had to be chosen, they were to be the ones nearest each end of each quarter-metre. The appropriate times for the observations were indicated, an exact botanical description of each required measurement was given and the accuracy of measurement required was stated. Diseases and pests were to be noted with their character and date of appearance, also any non-measurable

B 2

observations of importance such as lodging or animal damage. The following harvest observations were suggested :—

" Cut each quarter metre-length close to the ground when the crop is judged fit for harvest. The produce from the four quarter-metre lengths cut at each placing of the rod may be tied into one sheaf. There will therefore be two sheaves from each half-plot, thirty-two from each variety, and sixty-four in all. Each sheaf should be labelled with its half-plot symbol and should be stored in a dry place at least one week. Weigh each sheaf (in grammes). Count the fertile and sterile flowers in ten ears selected at random from each half-plot. Diseased grains should also be noted. (Replace the ears and thresh out the grains from each sheaf and weigh in grammes.) Obtain a 1000 corn-weight, for each half-plot, after mixing the grain of the two sheaves. (The 1000 corn-weights should be obtained to the nearest tenth of a gramme; other weights should be to a similar percentage accuracy.) Obtain the percentage of moisture in a sample of the grain from each half-plot. Date of harvest should be given."

This was the plan for experiments on wheat, in a number of different centres throughout Great Britain, by which it was felt the defects of the earlier observations could be remedied. It obviously provided means by which dates of crucial phenomena could be estimated precisely and in a manner comparable from one place to another. Date of emergence of ear, for instance, would be the date when half the observed ears had emerged. Schemes similar in principle were drawn up for winter and spring oats, spring barley, swedes and meadow hay. The only results obtained from the earlier data, other than Mr. Clapham's precision record experiments, were embodied in a short study of the influence of soil temperature on the germination interval, or interval between sowing and appearance above ground of wheat, winter oats, spring oats, spring barley and swedes (2).

In the horticultural portion of the crop–weather scheme, observations were taken on apples, plums, black-currants and peas. Difficulties similar to those in the agricultural section of the scheme were encountered. Nevertheless, the data existing in 1929 were sufficient for making a study of the relation between temperature and duration of flowering, and of the influence of season and geographical position on the growth rate of apple trees (3). I realized that dates of flowering, of fruit set, of the June drop of apples and plums, and of defoliation also required a " precision record " technique. To show that this was not impracticable, I carried out an experiment on eight young apple trees at East Malling. Four of these trees were Bramley's Seedling and four Worcester Pearmain, all six-year-old trees on Doucin (No. II) stock.   Four samples of 200 cms. of wood were measured and marked off on each tree. Every day throughout the flowering season counts were made on each sample of each tree of the number of flowers open (all stamens visible), and dropped (all petals dropped).   The date of flowering would then be taken as the date when half the flowers had opened, and the date of drop could also be similarly determined (see 3 for details).   Schemes were then drawn up for the horticultural crops on lines similar in principle to those for the agricultural crops.

I expressed the opinion at the time that if these plans for a revised crop–weather scheme were adopted, we should know more about the relation of weather and crop after 10 years than we did then, but that 40 or 50 years were needed to get the full benefit from such a programme.   Looking back seven years later, I am not surprised that this was not received with much enthusiasm.   Fifty years is not nearly as long a period as the period for which records are already available at Rothamsted, and in 1930 it seemed to me reasonable for administrators to take a long view.   To-day we are happy if we find the peace has been preserved for another six months.   Nevertheless, it is the business of scientific workers to take long views.   In a lecture given as recently as October 22nd last, the late Mr. Ramsay MacDonald expressed the view that though there was a feeling among many scientists that the ease with which their labours might be misused should make them, as citizens sharers in creating and upholding a public opinion of the nations to which they belonged, interested in protecting their work from being outrageously abused, yet if the communities were so blind as to follow the aggressive actions of their rulers, democratic or dictatorial, the responsibility was theirs, not the scientists'.   We must go on presenting the truth as we see it, and leave the community to apply our results if they will.

Here I must close my account of the position of our subject in 1929 : I trust I have provided enough material open to argument to prevent the discussion from languishing.

*References.*

(1) '' Crop-forecasting and the use of meteorological centres in its improvement,'' *Conference of Empire Meteorologists,* 1929.

(2) '' On the influence of soil temperature on the germination interval of crops,'' *J. Agr. Sci.,* Vol. XXI, pp. 241–50, 1931.

(3) '' Precision records in horticulture,'' *J. Pomology and Horticultural Science,* Vol. IX, pp. 149–94, 1931.

MR. W. G. COCHRAN.

My task is to describe the later work in the sampling observations
on wheat, in so far as it has a bearing on crop estimation. I hope
to show that these experiments throw some light on the question :
how difficult is it to estimate the mean yield of a crop by methods
which are based on measurement alone and not on opinions?
As Dr. Irwin has mentioned, most published estimates are at present
derived from eye inspections of the state of the crop, and not from
any actual measurements. The merits and defects of such estimates
and the biases to which they may be subject have been widely
debated, both in this Society and elsewhere. I do not propose to
add to this discussion, but in assessing the value of one method of
crop estimation it is important to know what alternatives are
available. For this reason the problem of obtaining purely *objective*
estimates of crop yields is of direct practical interest, and I want
to make a few general remarks about it.

The problem divides itself into two parts, according as the
estimates are required before harvest or at harvest. The former is
much the more difficult problem. To solve it, one must find measurable characteristics of the crop or the weather (or a combination
of the two), which are consistently correlated with yield. This
demands a preliminary programme of research extending over
several years and covering the principal districts and the various
varietal, manurial and cultural conditions in which the crop is
grown. Further, the success of such a programme depends on the
growth habits of the crop, and may be in no way commensurate
with the efforts spent on it.

Most of the previous work on yield prediction formulæ has been
based on the correlation of yield with meteorological factors. In
countries like our own, in which the weather is not a limiting factor
to the growth of the crop, these studies have given somewhat meagre
results. This is not surprising, since the influence of weather
factors is fairly complex, the effect of an inch of rain, for instance,
depending both on the soil type and on the amount of rain which has
preceded it.

However, even if the variations from year to year in the yield
of a single field could be predicted with reasonable accuracy from
the weather, meteorological factors alone are not likely to be
sufficient for predictions extending over a whole country, since they
cannot take into account many important sources of variation in
the average yield of a country, such as changes in agricultural
technique or in the quality of land on which the crop is grown.
In this respect the use of measurements on the growing crop itself
is more promising, since changes in soil fertility, amount of manuring

and cultivation may all reflect themselves in the condition of the crop at a given stage and be taken into account by a single prediction formula.  Any such formula would of course have to be tested for all the principal commercial varieties of the crop.

The technique of prediction of cereal crop yields is likely to be quite different from that of potatoes and root crops, and it is not clear which is the easier problem.  With cereals the principal constituent of the final crop—the grain—cannot be threshed until harvest and indices of the crop's progress have to be obtained at second hand; however, apart from the ripening of the ears, growth has practically ceased some time before harvest, so that any indices which may be chosen have attained their final values some time before harvest.  On the other hand, with potatoes and root crops, the principal constituent of the crop, *e.g.* the tuber in potatoes, can be lifted and weighed some three months before harvest, but the weights often increase right up to the final time of lifting, and these changes in weight may not be consistent from field to field or from year to year.

The second problem which I mentioned—the estimation of the mean yield of a country by weighings at harvest—is mainly a question of organization.  Here the use of sampling promises to be helpful, indeed essential, both for determining the yield of a field and of a district.  Preliminary research would be needed to develop a good sampling technique, to train observers in its use and to assess the amount of time and labour required to estimate the mean yield of a country with a given degree of accuracy.

The provision of unbiased estimates of the yield of a crop before harvest and at harvest does not settle the whole problem of crop estimation.  Losses occur in the carting and storage of crops, and in some cases these are quite large, so that the yield at harvest is an over-estimation of the total crop which ultimately reaches the market.  Allowance for losses after harvest is a separate problem, but it should not prove insuperable.

### The prediction of wheat yields

As Dr. Irwin has mentioned, the wheat-sampling observations under the Agricultural Meteorological Scheme were intended for the study of the effects of weather on crop growth.  They also happen to provide most of the material for judging to what extent wheat yields can be predicted before harvest.  The main features of the scheme have already been described.  The relevant facts for the present purpose are as follows.  Two varieties, Squarehead's Master and Yeoman, are grown at each of ten stations, one in Scotland, and the others in different counties in England.  The changes in plant, shoot and ear number and in shoot and ear height are studied from germination until harvest by means of actual counts. The full scheme has now completed five seasons.

At the end of the first three years Dr. Barnard examined the correlation between the grain yields and four weather factors, winter and summer rainfall and winter and summer temperature. No significant effects were found. The seasons were, however, very similar in many respects. The past two seasons have brought a marked change in weather, and it is hoped that a re-examination of the weather effects will be undertaken shortly.

Dr. Barnard also examined the relation of grain yields to previous measurements on the crop, and this work has recently been re-calculated to include all five years. Three factors were found to be correlated with the yield of grain : shoot height at ear emergence, plant number in early spring and ear number. Of these the most important factor is shoot height. The correlation of yield with plant number is actually *negative*, indicating that inter-plant competition plays an important part in a normal crop. The prediction formula reads :—

$$\text{Yield} = 28 \cdot 5 + 0 \cdot 034\ (h - 712) - 0 \cdot 0060\ (p - 1250)$$
$$+ 0 \cdot 0073\ (e - 1710)$$

where the yield is in cwt. of total grain per acre, $h$ is the shoot height in mms. and $p$ and $e$ are the plant and ear numbers per 32 metres of row length. The standard errors of the regression coefficients are : $h \pm 0 \cdot 0120$; $p \pm 0 \cdot 00285$; $e \pm 0 \cdot 00411$.

This formula is based on the means of the two standard varieties. There is no reason to expect that the same formula will fit both varieties, since they differ markedly in their habits of growth. In this particular case, however, there was little difference between the formulæ calculated separately for each variety.

The formula gives a prediction of the harvest yield at the end of ear emergence; that is, about the end of June or five to six weeks before harvest. The regression accounts for 60 per cent. of the variation in grain yields, and the standard error of the deviations from the formula is $16\frac{1}{2}$ per cent. Whether this is satisfactory depends on the magnitude of other sources of variation in the grain yields, such as variation from field to field in a district.

The observed yields and the deviations of the predicted yields are shown in Table III to the nearest cwt.

Three stations had to be omitted from the calculation of the prediction formula, owing to incomplete observations. The differences between years were eliminated by the analysis of variance before fitting the formula, so that the ability of the formula to predict differences between years is an independent test of its efficacy. The formula has been reasonably successful in predicting the mean yields of the stations in 1933, 1935 and 1936. In 1934, however, in which the actual yields were high, the estimated yields

TABLE III

*Observed and predicted yields* (cwt. per acre)

| County | Station | 1933 | | 1934 | | 1935 | | 1936 | | 1937 | |
|---|---|---|---|---|---|---|---|---|---|---|---|
| | | Yield | Discre-pancy | Yield | Discre-pancy | Yield | Discre-pancy | Yield | Discre-pancy | Yield | Discre-pancy |
| Devon | Sealehayne... | 21 | + 5 | 32 | − 7 | 26 | + 1 | 23 | + 2 | 22 | + 3 |
| Sussex | Plumpton ... | — | — | 35 | − 10 | 47 | − 9 | 23 | + 4 | 25 | + 7 |
| Kent | Wye... ... | 11 | — | 48 | − 6 | 15 | + 10 | 21 | − 1 | 19 | + 10 |
| Herts | Rothamsted | 22 | + 3 | 32 | − 4 | 35 | − 1 | 32 | − 4 | 10 | + 6 |
| Norfolk | Sprowston ... | 25 | + 5 | 28 | − 2 | 21 | + 2 | 21 | + 5 | 21 | + 2 |
| Salop | Newport ... | 35 | − 1 | 44 | − 5 | 40 | − 6 | 32 | + 5 | 12 | — |
| Lothian | Boghall ... | 33 | 0 | 36 | − 3 | 30 | − 3 | 28 | − 9 | 33 | − 1 |
| | Mean ... | 27·2 | + 2·4 | 36·4 | − 5·3 | 30·6 | − 0·9 | 25·7 | + 0·3 | 21·7 | + 4·5 |
| Rainfall (1st 20 weeks) | | | 9·8 | | 8·1 | | 9·9 | | 10·3 | | 17·9 |

A + discrepancy indicates that the predicted yield is higher than the observed.

were too low at all stations, being 5 cwt. too low in the estimate of the mean. In 1937, a low-yielding season at the stations concerned, the formula over-estimated at all stations except Boghall, and was $4\frac{1}{2}$ cwt. too high on the average. Thus, in some seasons at least, there are factors influencing the mean yield which are not taken into account in the predictions. To this extent the formula is failing in its main purpose, which is to detect changes in yield from year to year.

In seeking a cause of the failure of the formula in 1934 and 1937, we may note that the outstanding feature of the 1937 season was the continuous heavy rainfall in winter and early spring. The total rainfall for the first 20 weeks of each year is shown in the last row of Table III. It will be seen that while 1937 had much the wettest spring, 1934 had the driest spring. This comparison by no means proves that rainfall was the cause of the discrepancy in the predictions for these two years, but it gives, I think, an illustration of the way in which meteorological factors and measurements on the crop's growth may have to be combined to obtain a successful prediction of the changes in yield from year to year.

These results were obtained from crops grown in small plots at stations with some knowledge of experimental technique. This is probably the most convenient way in which to start experiments on the possibility of predicting yields. Once a likely formula has been found, however, it will have to be applied to ordinary commercial crops for the purpose of obtaining the mean yield of a country.

Sampling observations similar to those on wheat have been carried out on barley and oats at one centre and on potatoes and sugar-beet at two centres. The results are as yet too few for discussion.

*The estimation of commercial wheat yields at harvest by sampling*

Eight of the stations which participate in the wheat-sampling observations also undertook voluntarily to sample some fields of commercial wheat in their neighbourhood. Samples have been taken at each of the last four harvests.

The selection of farms is not at random, since the stations have to obtain permission from the farmer to take samples. However, apart from the first year, two fields have been sampled on each farm wherever possible, and these are selected at random from all fields growing wheat on the farm. The stations co-operating in the scheme are asked to sample three farms each, but the number

Plan of Field.

← Drilling →

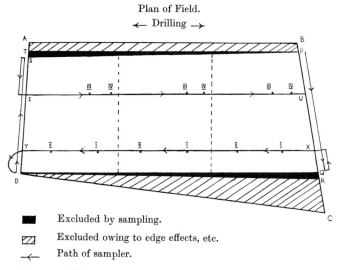

■ Excluded by sampling.

▨ Excluded owing to edge effects, etc.

← Path of sampler.

actually sampled has varied according to the labour available and the enthusiasm for this type of work. It is gratifying to note that though the samples have to be taken at a very busy time on the stations concerned, some have managed to sample as many as eight or nine farms on occasion.

In view of the voluntary nature of the scheme, a method of sampling had to be found which was quick and convenient to carry

out and involved little trampling of the crop.   It is therefore worth
giving some indication of the method used.

The sampler on entering the field paces one side PQ (perpendicular to the drill rows).   The starting point X on that side is selected
at random from the number of paces in the side.   After walking to
the starting point, the sampler crosses the field XY in the direction of
the rows, and on his way cuts two sets I and II of about three samples
each—that is, about six samples in all.   Each sample is six rows wide
and $\frac{1}{4}$ metre long.   The location of the samples is determined as
follows.   Before starting to cross the field, the sampler estimates its
width in paces by eye.   Suppose this is 350 paces.   One-third of this
is roughly 120 paces.   Two random numbers are chosen between
0 and 120, say 35 and 86.   The three samples of the first set are
taken at 35, 155 and 275 paces across the field and the samples of
the second set at 86, 206 and 326 paces : that is, there is one sample
of each set in each third of the way across.   After cutting each
sample, the sampler steps three paces to the left and continues
along the drill row, so that his path across the field is slightly stepped.
If he has misjudged the width of the field he may have to take more
or less than three samples per set.   On reaching the other side at Y
he paces it and repeats the same process on the way back, collecting
a further six samples.   Thus the field is crossed only twice, and the
sampler finishes at U on the same side as he started.

A small wedge-shaped area at each end of the field PTS and
DRQ is automatically excluded from the area sampled by the
stepping process.   The sampler is also asked to exclude irregularly
shaped areas and portions at the edges which may be affected by the
shading of trees or hedges or other edge effects.   ABPT and DRC,
the total area of such portions, and their estimated yield as a
percentage of the yield of the sampled area are noted, so that the
sampling yield may be adjusted if necessary to give an unbiased
estimate of the yield of the field per acre.   Such corrections have
been small in nearly all cases.

Two observers working together can cut, tie and label the
samples in a ten-acre field in an hour.   The weights of grain can be
determined by threshing on a small machine in a further twenty
minutes.

The analyses of variance of these results are shown in Table IV
and are of considerable interest.

The differences between the two sets in each line across the
field enable us to estimate whether the number of samples taken
during the sampler's journey across the field is sufficient.   The
differences between the totals of the two sets of six samples provide
a single degree of freedom in each field for estimating the sampling

error. Except in 1937, the variation between sets has been quite
a small proportion of the sampling variation, so that there is little
point in increasing the number of samples per line. The sampling

Table IV

*Commercial wheat sampling yields of grain (cwt. per acre). Analysis
of variance per field (total of four sets).*

Mean Squares

|  | 1934 | | 1935 | | 1936 | | 1937 | |
|---|---|---|---|---|---|---|---|---|
|  | d.f. | m.s. | d.f. | m.s. | d.f. | m.s. | d.f. | m.s. |
| Between districts ... | 4 | 66·5 | 6 | 318·4 | 4 | 79·4 | 5 | 81·7 |
| Within districts between farms ... | 11 | 38·9 | 12 | 27·1 | 7 | 62·2 | 19 | 52·9 |
| Within farms between fields ... | — | — | 15 | 22·8 | 8 | 31·2 | 11 | 24·1 |
| Within fields: Sampling variation ... ... | 16 | 5·33 | 40 | 6·20 | 22 | 11·39 | 39 | 6·58 |
| Between sets ... | 32 | 2·11 | 80 | 2·18 | 45 | 2·52 | 78 | 5·09 |

Standard Deviation per cent. 1934–37

| | | | | | | | |
|---|---|---|---|---|---|---|
| Between fields | ... | ... | ... | — | 22 | 24 | 19 |
| Sampling error | ... | ... | ... | 8 | 12 | 14 | 10 |

errors per cent. are shown in the last line of the Table, and range
from 8 to 14. This is remarkably low in view of the fact that in
a 10-acre field only $\frac{1}{15,000}$ of the total crop is sampled. Further, the
sampling variation only contributes between a quarter and a third
of the total variation between fields on the same farm. Thus any
increase in accuracy in the estimate of the mean yield of a district
is much more profitably attained by sampling more fields than by
increasing the amount of sampling per field.

The variation between farms in the same district is not correctly
estimated, owing to the non-random selection of farms. The mean
squares indicate, however, as one would expect, that there are some
differences between the levels of yield on different farms. The
variation between districts is somewhat greater than that between
farms in the same district in all four years, but the differences were
small except in 1935.

The figures also enable us to calculate the amount of sampling
which would be necessary in estimating the mean yield of the country
with a given degree of accuracy. The variation between districts
need not concern us, as it could be eliminated from the estimate
by sampling separately within each district. The variation between

farms in the same district could also be eliminated, by sampling every year from a fixed set of farms, initially selected at random. This would give each year an estimate, not of the mean yield of the whole country, but of the mean yield of a selected group of favour in the country and might result in a slight bias, but it would have many advantages from the point of view of organization.  Thus we have only to consider the variation between fields on the same farm (including its component of sampling variation).  This has been re-markably consistent, the standard deviation ranging from 19 to 24 per cent.  Taking 22 per cent. as an average figure, an estimate of the mean yield of the country with a standard error of only 1 per cent. could be obtained by sampling 500 fields.  Since harvest in most districts is spread over a fortnight at least, the estimate of the mean yield of a district would probably be available immediately after the last of the selected farms in the district was ripe for cutting.

These results show, I think, that the estimation of wheat yields by actual weighings at harvest is not an insuperable problem.  I will mention one further difficulty.  Cereal samples do not suffer any appreciable losses in cutting, transport and small-scale threshing, whereas commercial cereal crops, as ordinarily handled, may be subject to substantial losses.  Thus the sampling yields are likely to be higher than the threshed yields, and some allowance will have to be made for this.

Another feature of this scheme is worth mention.  The sampler is asked to estimate the yield per acre of each field by eye before he takes his samples.  The farmer supplies an independent eye estimate, and is asked to send the threshed yield of the field if this is obtain-able.  Thus a body of material is being accumulated by which eye-estimates, sampling yields and threshed yields of the whole field may be compared.  Only eighteen threshed yields have been obtained to date, so that the material is as yet insufficient for a detailed examination.

Two points have emerged.  The sampling yields are fairly consistently higher than the threshed yields, as I indicated in the last paragraph.  The figures also support Dr. Irwin's opinion that eye-estimates tend to under-estimate the variability in yield.  The standard deviation of the set of eighteen threshed yields is 32·3 per cent., which is of the order of magnitude indicated by the sampling results, since it contains a mixture of field, district and year differ-ences.  The sampling standard deviation is 37·1 per cent., somewhat larger than the previous figure, owing to its component of sampling variation.  The farmer's eye-estimates, however, give a standard deviation of only 23·2 per cent., considerably lower than that of the threshed yields themselves.  If the eye-estimates were unbiased, their standard deviation should be somewhat higher than that of the threshed yields, because of the errors of eye-estimation. The observers appear to be under-estimating high yields and over-estimating low yields.

In conclusion, reference should be made to previous papers by F. Yates (4), (5) and (6), describing the wheat sampling observations from the point of view of crop estimation.

*References.*

(4) Yates, F. "The place of quantitative measurements on plant growth in agricultural meteorology and crop forecasting." *Conference of Empire Meteorologists*, 1935, pp. 169–172.

(5) Yates, F. "Crop estimation and forecasting : Indications of the sampling observations on wheat." *Journ. Min. of Agric.*, 1935, Vol. XLIII, pp. 545–577.

(6) Yates, F. "Applications of the sampling technique to crop estimation and forecasting." *Man. Stat. Soc.*, 1936.

## Dr. J. Wishart.

It is the custom in the Japanese cinema, and to some extent also, I believe, on the legitimate stage, for a commentator to narrate the story in simple language, and in a loud voice, for the benefit of the audience. The custom is informative, but somewhat disturbing in the case of the average " talkie." Intervening, as I do, at this stage in the present discussion, I can only take the part of commentator, since it has never fallen to my lot to be associated officially with problems of crop forecasting, or with the developments of agricultural meteorology. There is this point of difference, however, in that the principal actors must be silent while I am on my feet; also that they know the story better than I do. I trust that the resultant disturbance will not be inflicted on you for very long. In the time allotted I shall try to suggest one or two points on which light may be thrown by subsequent speakers.

Working at Rothamsted, as I did, in close association with Dr. Irwin during the years in which he was engaged on those investigations which he has summarised to-day, I am in a position to pay a tribute to the thoroughness with which he examined a complex and difficult problem, and made recommendations as to the future for the attention of the responsible authorities. With him I would associate Dr. Clapham, who did excellent pioneer work on sampling, and to whom is due the Precision Record Scheme, forerunner of the scheme now in being which has been dealt with by Mr. Cochran. The main points on which Dr. Irwin has touched, and on which Mr. Cochran has thrown further light, are three :

(1) The necessity for obtaining accurate estimates of acreage

and yield for the country as a whole.   This is an important problem for the administrator, which accounts for the fact that a service for obtaining this information has grown up in a number of countries. Such a survey must of necessity be an extensive one, so that even where objective methods are used the accuracy of the results must still remain in doubt.   Mr. Cochran, however, has described a method whereby the sampling methods developed for the study of single plots in a field experiment are being utilized in an experimental way on a larger scale to obtain, not only estimates of the yields of crops, but also estimates of the errors involved.   The analysis of results that he gives can only, I imagine, be treated as preliminary in character, but it promises well.   May I say a word or two in connection with Mr. Cochran's Table IV?   He might with advantage have given greater details concerning the composition of his data, as it is impossible to reconstruct these merely from a knowledge of the number of degrees of freedom.   All one can be sure of is that only one field was sampled per farm in 1934, and that in subsequent years some farms were still only furnishing data from a single field. Since the variation between fields on the same farm is used to determine the error of the estimate of the country's mean yield, it is important to notice that not only do the data consist of only a small number of farms in all, but some of them are not utilized in estimating the error.   The statement that the mean yield of the country could be estimated with a standard error of only 1 per cent. by sampling 500 fields is remarkable, if true.   The yield is estimated as the mean of a number of sample determinations, and, while the process outlined may secure a mean reasonably free from bias, it seems to me that even if each district is sampled separately, the standard error of the mean must incorporate the variation between farms as well as between fields on the same farm.   The proposal to sample from a fixed set of farms certainly enables one to estimate the error of the mean yield for those farms from the variation between fields, but I am in some doubt as to how far this figure can be taken as applying to the country as a whole, while difficulties are likely to be experienced in the way in which farms are to be selected.

A parallel problem is the determination of soil characteristics in various parts of the country, and the extent to which accurate estimates may be possible of attainment by similar sampling methods. I can remember taking part in discussions of this problem some six years ago; detailed schemes were even then being worked out, and I shall be interested to hear from anyone competent to inform me whether they have proceeded beyond the embryonic stage.

(2) Scientific work on the relation between weather and crop yields.   Since weather is one of the most important factors it is

important that such scientific work should go on, and be crystallised out in due course in the shape of official schemes designed to build up in time a body of accurate data which can be made the basis of forecasts. The most important recent work along these lines has consisted in the application of Professor Fisher's ingenious correlation method. It must be a source of gratification to Professor Fisher that his method has been copied so often, and down to the last detail, even to the order of the polynomials fitted to the weather data. It will not, I think, detract from the value of the method if I say that it may not always be necessary to fit fifth-degree polynomials, or to consider the whole twelve-month period of rainfall distribution. There is one aspect, however, in which Professor Fisher's original study stands alone, and that is in the length of the period available for study. If it is desirable to have more detailed observations on the crop than are given by yield alone, and to extend the survey to cover the whole country, then it is important that an official scheme be adopted as soon as possible. Even if it is a somewhat melancholy thought that a long period of patient accumulation of data must be gone through before they can be of value for forecasting purposes, nevertheless it should be begun sometime on a sufficiently comprehensive scale to be of real value.

The practical adoption of methods for collecting such information brings us to Dr. Irwin's third main point :

(3) The Crop Weather Scheme, and its possible improvement by the taking of observations of the nature of the Precision Record Scheme. Mr. Cochran has described the more recent developments in this connection. While it is of interest to hear of what has been done by Miss Barnard on the correlation of such observations as have been recorded with weather factors, and of the subsequent examination of the correlations that exist between yield and previous measurements on the growth of the crop, it will be agreed, I think, that a five-year period is much too short for information of real value to be obtained. In this connection I should like to ask whether there are any data from Rothamsted or elsewhere on the correlation between yield obtained from full harvesting of plots and previous measurements obtained by sampling only. It is not clear from the description given of the Precision Record scheme whether the same sample is adhered to throughout or not. If I remember rightly, work carried out at Cambridge some years ago showed a similar degree of relationship between yield and previous measurements in the samples, but it proved to be very difficult to find significant results when the yield variable was the result of full-scale harvesting. This result threw some doubt at the time on the value of the sample determinations which had been made of plant numbe

shoot height, etc., measurements which are only feasible by a sampling process, whereas it is usually possible to harvest entirely the plots of a field experiment.

It will be clear, also, from the figures that Mr. Cochran has submitted showing the value of previous measurements for prediction purposes, that, as he said, meteorological factors and crop measurements will have to be combined to obtain a successful prediction of the changes in yield from year to year.

Seven of Dr. Irwin's ten years have come and gone, and while it is true to say that progress has been made, it is also true to say that we are very unlikely by 1940 to be in a position to say anything very conclusive on a nation-wide scale about the relation between weather and crop yields. There is no doubt that a reasonably satisfactory scheme of observations could be quickly worked out; it existed in fact, in embryo form as long ago as 1930. Is it rash of me to commend to the competent authorities the adoption of a more or less comprehensive scheme within the next year or two?

Perhaps I may be permitted a word or two in relation to what has been done on the topic under discussion in two countries, both considerably larger than our own. I hope that following speakers will be able to amplify what I say, and also deal with other countries. Dr. Irwin has touched on methods of crop-reporting in India, and I trust that Mr. Vaidyanathan, whom I am glad to see present, will tell us about present day methods in that country and their value. Since the ideal to be aimed at is the collection of agricultural records of a precision equivalent to those of meteorological data, and at the same station, it is of interest to note the establishment of the Agricultural Meteorological Section of the Meteorological Office at Poona. This work has been functioning for some five years, under the capable directorship of Dr. Ramadas, and falls under two heads (a) statistical and (b) experimental or biological. Under the first head a critical enquiry is made of the available data on the area and yield of crops for the various presidencies and districts in India, and it is sought to find correlations with the accumulated meteorological data. Help to the farmer is envisaged by the adoption of a sufficiently elaborate organization to provide quick dissemination of information. In this connection it is of interest to record that selected experimental farms throughout the country are being furnished with standard meteorological equipment in order that regular weather records may be furnished side by side with the crop data. India is to be congratulated on the pioneer work being done in this connection. Already a number of interesting research papers have been produced from the new organization.

The other country which I shall mention is China.   This country has not gone as far as India, in that agriculture and meteorology have not yet come together.   Nevertheless there are good possibilities there.   It falls outside our scope to touch on meteorological services pure and simple, yet it should be known that there is a very satisfactory meteorological service operating throughout China. The distances are tremendous, but modern methods of communication have speeded things up considerably.   For example, Nanking's complaint is that morning weather reports from the remote interior do not reach the capital until the afternoon !   On the crop-reporting side a nation-wide organization has existed since 1929.   Sponsored at first by the Directorate of Statistics of the National Government, and later by the Department of Agricultural Economics of the University of Nanking, crop reporting is now a major activity of the National Agricultural Research Bureau.   The Bureau collects information on (1) the production of crops that China has, or is likely to have, in a given year as compared with previous years, and (2) the changes that have taken place in the rural districts that will affect the economic situation of the future.   In the first case, data on the acreage and yield of the important crops, prices, etc., are furnished by volunteer crop reporters, who numbered 6,369 at the end of 1935, located in 1,241 different *hsiens* (counties), in 23 provinces.   Each reporter fills in a monthly questionnaire, and in return for his services receives a copy of an agricultural newspaper, which is sent out every ten days from the Bureau, and is a means of disseminating far and wide to farmers the results of scientific work done at headquarters and at the various co-operating stations in the provinces.   The crop-reporters are farmers, rural workers and school teachers, educationalists and preachers, and there is no doubt that they do their work well, a stimulus to exertion being, no doubt, the awards given at the end of the year to those who have reported regularly. The information provided is analysed at headquarters and published in a series of comprehensive monthly bulletins, which are a mine of information in relation to the agricultural statistics of China, and also in annual summaries.   It would take too long to enumerate all the information that is provided.   Suffice to say that crop-reporting appears to be on a very satisfactory basis, and the time is ripe for consideration of the related meteorological data.   When it is considered that the Bureau works in close association with upwards of twenty provincial colleges and experiment stations it would not be difficult to make a beginning with these in the simultaneous collection of accurate meteorological and crop data for subsequent correlational studies.   The present, unfortunately, is not a happy time for this collaboration to continue to its full past extent, for communications between North and South China are no doubt broken at present, but as far as possible the order of the day in scientific agriculture is " business as usual."

MR. BATTLE : After reading the proofs of these excellent papers, I feared that anything I might say about sugar beet, on which I had been asked to make a contribution, would not be technical enough for the meeting. However, as the subject is of great importance to beet-sugar factories I venture to give some account of the position to-day. I am indebted to my colleague, Mr. Swannack, for the preparation of a large part of these notes.

The sugar beet, which is the raw material, is the basis of all calculations in the manufacturing process. The purchase of coal, limestone, sugar and pulp bags, involving large sums of money, are all dependent on the quantity of beet to be delivered. It is not practicable to rely upon ordering these supplies from hand to mouth, they must be purchased in advance ; nor would it be economical to make a regular habit of over-estimating when buying these supplies in advance, by basing requirements upon the highest probable yield of the contracted acreage.

Sales of sugar and pulp must also be based on the raw material available. Supply and demand must be watched if the best prices are to be realized.

Owing to our short manufacturing season, to supply the trade throughout the year a certain amount of sugar must be stored ; but storage accommodation is limited, and incorrect estimates of the raw material may incur considerable expense in taking " outside " storage. Further, it is important to the growers that the manufacturing period should be so arranged as to enable them to lift their beet with as little inconvenience as possible, and to the best advantage. This can only be done if a reliable estimate is available of the quantity to be delivered. It is important, therefore, to find a reliable and practical method of estimating the yield during the month of August. As far as my limited knowledge goes, little scientific work has been done in this country on the subject. On the Continent the several large seed-breeding stations, the majority of which have considerable acreages of commercial beet directly or indirectly under their control, have evolved more or less reliable methods of estimating yields.

As there are no large sugar-beet seed-breeders and no sugar-beet research stations in England, the only estimates attempted are those of the agricultural departments of the various factories.

*Factory methods*

Until eighteen months ago our factories were divided into five different groups, no two of which had identically the same agricultural organization, and in consequence the methods employed in estimating yields differed widely. One group of factories based their individual crop estimates on the previous three years' average yield of that particular grower; another group estimated individual crops during the month of September, and a third made individual estimates during August.

In one particular group, for example, individual estimating was done with considerable care, the fieldmen having available full details of the previous cropping, cultivations and manuring. Speaking generally, the results attained a reasonable degree of accuracy, but this accuracy depended almost entirely on the fieldman's practical knowledge and experience.

The following give percentage errors, for three factories representing different groups, during the past three years :—

| Factory | | | 1934 Estimated | | 1935 Estimated | | 1936 Estimated | |
|---|---|---|---|---|---|---|---|---|
| | | | Under | Over | Under | Over | Under | Over |
| A | ... | ... | — | 1·46% | — | 12·44% | 1·07% | — |
| B | ... | ... | 16·37% | — | — | 5·10 | — | 3·1% |
| C | ... | ... | 4 | — | 10% | — | 3 | — |

The fieldmen have had to assume that the climatic conditions in the interim, between August and September and when the crop is lifted, would be normal, and generally speaking the nearer to normal the weather proved to be, the nearer were the estimated yields to the actual delivered tonnages.

## Field samples

For several years past, beginning the last week in August, at the instigation of the Ministry of Agriculture, each factory, through its agricultural department, selected some twenty crops from which to take field samples. For reasons previously given, there was little uniformity in this work. At the beginning *no* detailed instructions were given. Some factories took three beets from each field, others six, one or two as many as ten. No consideration was given to the types of soil, nor care taken that the crops selected were representative of the average of the crops in the area.

Later instructions were given that the fields selected must be representative of their soils, and that *average* crops and roots must be selected.

This attempt to obtain uniformity did effect some slight improvement, but the results of the field samples are not a true indication of the actual deliveries.

## Investigatory work

It is now recognized that there is room for considerable improvement. The work of estimating the yield of sugar beet differs from that of estimating wheat yields, in that the number of plants per acre can be used as an important factor in the calculations. To use this factor accurate figures of the following are essential :—

(1) Crop acreage.
(2) Number of plants per acre.
(3) Average weight of individual plants.

(1) *Crop acreage.*   There is considerable difficulty in ascertaining this figure.   In contracting acreage, growers invariably quote Ordnance Survey figures, which of course include boundary hedges or dykes, and in some cases half the adjoining road.   Often there are further complications, as for instance when the crop has failed partially or completely in some parts of the field.

However, with proper care a reasonable degree of accuracy in recording crop acreage should be achieved.

(2) *Plant population.*   This presents greater difficulty owing to the wide variation over the field.   It is impossible to count all the plants, and only sections of rows of a given length can be taken as samples.   By the old method the selection of such part rows was made by the eye, and undoubtedly the result was influenced by human bias.

(3) *Average weight of individual plants.*   Here it is necessary to select beet truly representative of the whole of the roots comprising the crop, if the result is to be accurate.   By the old method such beet were selected by the eye, and results showing bias were obtained.

The crucial difficulty appears to be to find an accurate method, which is also practicable, of selecting average places in crops, (*a*) to take the count for the plant population, and (*b*) from which to take the sample roots ; and further the method of selecting average roots from (*b*).

In the summer of 1936, in conjunction with Dr. Crowther and Mr. Yates of Rothamsted, the Bardney and Brigg Factories attempted to improve the existing methods.   It was suggested, as an alternative to relying on the fieldman's judgment, that the method of pre-determined selection should be used.

*Pre-determined selection*

This method was adopted in 1936 by the Bardney and Brigg factories for ascertaining the plant population, and later was used for taking field samples.

Briefly, this method consisted of :—

1. Selecting, by judgment, in each fieldman's district, average yielding crops representative of the soils therein.   Each fieldman has an average district.

2. Selecting, again by judgment, an average place in such crops of the size required to embrace as many *more* rows as there were weeks in the sampling period, and number of beet in the rows ascertained as follows :—

Number of sampling weeks, multiplied by number of beet to be taken each week.

3. Every alternate row was staked and numbered.

4. By cutting numbered cards in a prescribed manner, the actual row and the ten consecutive beet in the row were selected for sampling.   This selection was done at the factory office.

5. Each fieldman was supplied with a separate card for each week appropriately ruled and indicating the particular

row from which the beet were to be taken, and the particular numbered beet in the row to be selected.

An attempt was made to check the estimates reached by this method, and the results were analysed by Mr. Yates, who, in concluding his report, expressed disappointment at the general inaccuracy found.

He stated, however, that the technique of the average weight method could be improved in many respects, and recommended that further work be undertaken the next year (*i.e.* this year) using an improved technique. I understand that the work has been continued in the Bardney and Brigg areas, and that the figures will again be submitted to Mr. Yates for his analysis. The change in factory organization caused by the merger of all factories into British Sugar Corporation Limited may facilitate further experimentation, as so far only the fringe of the problem has been touched.

## *Delivery samples*

During the delivery period, from every consignment of beet delivered (average weight of consignments—road $4$–$4\frac{1}{2}$ tons, rail $7$ tons—both clean beet), a sample weighing 28 lbs. dirty beet is taken. The clean topped weight of the individual beets comprising this sample is recorded.

This figure should give a very reliable indication of the final yield per acre, if a practical method of ascertaining plant population accurately could be found.

## *Following weather*

One vitally important factor is the effect of the weather experienced between the end of August—the date estimating is concluded—and the date the crop is lifted, say between the end of October and the end of November.

Experience shows that in this country, in almost every season, the ultimate yield is influenced very appreciably by climatic conditions during this period. I must apologize for merely outlining the practical difficulties which have to be contended with in yields of sugar beet, and for contributing very little to the scientific side of the problem of crop estimating.

We are still seeking the answer to the questions :—

1. What is the most reliable method, if such can be found practicable in application, which will give us accurate estimates, made during the growing period, of the ultimate yield ? and

2. Can the effect of the rainfall, temperatures, and atmospheric conditions, be measured progressively during the period September–November, *i.e.* the interim between the conclusion of the estimating and the time of lifting ?

A satisfactory answer to these would render an invaluable service to the sugar-beet industry.

Whilst the difficulties of the problem are appreciated, now that experienced fieldmen working on similar lines are available in most of the factory areas the work should be facilitated.

Rao Bahadur M. Vaidyanathan said he felt it a great privilege and honour to have been invited to take part in the discussion. For an Indian statistician on leave in England, it was a rare opportunity to attend a meeting of the Royal Statistical Society.

Mr. Vaidyanathan thanked Dr. Wishart for his kind reference to him and to the work done in India on the subject of crop estimation, particularly to the able work done by his friend Dr. Ramdas. The Agricultural Meteorological Section of the Meteorological Office at Poona had been functioning for only five years, but he hoped that in a few years they would be able to get some valuable results, useful for the cultivator, for whose benefit the work was being done.

The method of crop estimation was as old as agriculture itself, but as in late years, especially since the War, every country had been anxious to practise self-sufficiency, the subject was looming large in the eyes of the public, and was now engaging greater attention than before, when proper methods of crop estimation had not been built up. It provided, however, a very good platform, where the agriculturist, the meteorologist, the economist and the statistician each had a place.

From a purely economic standpoint, it would be valuable if the economist could estimate consumption figures for each commodity consumed in the country, but so far in all the literature he had seen on the subject the economist was in the background. The simple formula that the *total produce* of a crop was equal to *consumption* minus *net imports* would be of help in checking the *final estimates* arrived at independently. Estimates of consumption, at least in the case of non-food crops (*e.g.*, cotton) were simpler, as factory figures might mainly account for them. It had been found possible in America (and to a certain extent in India, for cotton) to compare the final estimates by independent methods. If such comparisons were possible for a series of years, they might help in suggesting improvements in the methods adopted, provided, however, the error of each estimate was known.

Apart from the economic standpoint, the problem of estimation of crops at different stages of plant growth (*i.e.* forecasts) was very complex, owing to the several factors, some unknown, which were involved in such estimation. A great deal had been said with regard to subjective and objective crop estimation. Both seemed important since they could be used for mutual checking of the data. In his own country, where every village had a headman who was expected to report the condition of crops several times in the course of a year, such estimates for a series of years for each village would provide a great deal of material upon which to work, and if properly analysed would throw a flood of light, and show the extent of bias in those estimates. Besides, forecasts for *neighbouring* areas would provide a useful mutual check.

Next, with regard to the objective aspect of crop estimation, the whole problem was one of proper *sampling* with regard to *soil, climate, crop* which involves the particular variety grown, and the *cultural practice* adopted. The tract under examination should be so subdivided as to involve homogeneity with regard to as many factors as possible, and the number of fields to be sampled in each subdivision would then depend upon the extent of its variability. The work done at Rothamsted provided an excellent example of how several factors under observation at least a few weeks before harvest could be correlated to yield. Mr. Cochran had shown that so far as his small experimental plots were concerned, in three years out of five years in which wheat yields were under examination, the partial regression formula had worked well. This only emphasized the need for a long trend of observations, and as data were accumulated each station would have a separate analysis. The factors to be correlated to yield had also been studied thoroughly at Rothamsted. With regard to commercial farms, the administrator need not be frightened at the number of fields to be sampled, and the statistician who based his estimates on $n$ need not shirk his responsibility of stating the *size* to be sampled for any given degree of accuracy. A fair compromise was possible provided it was recognized that without a sufficient number of samples to attain the needed accuracy, a satisfactory mean *yield* was not calculable.

Again, the same problem of sampling arose in the case of unsurveyed lands whose area under crop was not known. A lot of work had been done in America in connection with forest surveys, and in any sampling method that was adopted it was necessary to know the errors that should be attached to such estimates.

Now, with regard to meteorological observations themselves, *rainfall* was one factor which had now been studied satisfactorily and correlated to yield in most of the experimental stations. Mr. Vaidyanathan said he would like to pay his humble tribute to Dr. R. A. Fisher for his notable work in this connection, and the lead he has always given, as his method of partial regression on the yield of the coefficients of distribution of rainfall in the course of the year was found generally to be very effective. But as Dr. Wishart had pointed out, it was not always necessary to fit in a fifth-degree polynomial for, as in the case of India where rainfall during the season has generally a single maximum and a single minimum, a third degree polynomial would answer very well. The problem however of meteorological effects on yield is very complex, and a joint discussion between the Royal Statistical Society and the Royal Meteorological Society would help to settle most of the issues connected with the problem.

Mr. M. G. KENDALL said that in view of the limited time available he proposed to prune his remarks severely.

On looking through the paper, it seemed to him that the fundamental difficulty in crop estimation had been glossed over by Mr. Cochran. In Dr. Irwin's paper there was a detailed account of how to take a random sample from the crop when once the field had been chosen. He looked in vain in Mr. Cochran's contribution for any suggestion as to how to take a random sample of farms as a whole; and that, for a very good reason. To obtain such a sample in present circumstances was not only difficult, but impossible.

Mr. Kendall thought the agricultural economist had just cause for reproaching the theoretical statistician at the present time, in that theory had outrun practice in this direction. For economic purposes they knew as much as they need know about random sampling, except how to obtain a random sample to which the theory could be applied. This was a practical difficulty but was none the less real on that account.

Mr. Cochran had specifically mentioned a random sample of farms, but imagine what would happen if attempts were made to take such a sample in this country. Even if those engaged in the research were very precise in choosing the farms, they would then have to go to the farmers concerned to secure their co-operation, and probably in 30 per cent. of cases this co-operation would not be forthcoming. At once the randomness of the sample and the objective nature of the inferences desired to be drawn from it would be lost. Occasionally the Ministry of Agriculture undertook voluntary surveys, and the normal return was in the neighbourhood of 70 per cent., but in this case the proportion of refusals might well prove to be higher, because the farmer would not only have to fill up forms, but would have to suffer people to walk all over his farm. Mr. Kendall saw no way of overcoming this difficulty except by passing an Act of Parliament making it compulsory on a farmer to co-operate in a crop scheme.

Assuming, however, that a random sample could be taken and a panel of farmers compiled in the manner contemplated by Mr. Cochran, the difficulties were by no means at an end. He understood Mr. Cochran's suggestion to be that that sample should be preserved from year to year. It would be found in the course of five or ten years that something like 50 per cent. of the farms had changed hands for one reason or another. The original sample would thus melt away. There was an illustration of this in an enquiry into the financial aspects of milk management financed by the Milk Marketing Board and carried out by the Advisory Economists in England and Wales. The total sample for the first year was between 700 and 800 farms. In the first year there was a mortality of 30 per cent., and in the second of about 25 per cent. What the figures were for the third year he did not know, but the fact remained that two years after the scheme had started they were working with only a little more than half the original sample. Of course the total number was filled up to the original complement by fresh recruitment; but such a procedure, as he understood it, would not meet Mr. Cochran's requirements.

Passing to the question of crop forecasting, Mr. Kendall proposed to confine his remarks to one point. Apart from the question of the relationship between weather and crops, which itself was important, a fundamental assumption behind crop forecasting from meteorological data was that the greater part of variations in crop yields from year to year were due to meteorological influences. That was probably true before the War, but now they seemed to be approaching the time when it might be true to a smaller extent. The majority of the factors affecting yield were what might be called long-term factors, in that variation in their effect from year to year was not great, or was practically negligible when compared with climatic influences. However, new short-term factors were appearing. The present year was an instance in point. The Government had passed several Acts, such as the Agriculture Act, 1937, which did all sorts of things to affect the possible crop yield next year. It had raised the limit of the total amount which could qualify for full deficiency payment under the Wheat Act; it had initiated subsidies for barley and oats; it had provided funds to subsidise the use of lime and basic slag, which, though intended primarily for grassland, would have some effect on arable products; it had extended the period to which assistance was to be given to the live-stock industry, and that, in so far as it helped to build up live-stock herds, would have an influence on agricultural productivity, inasmuch as farmyard manure was an important fertilizing agent for arable soils. In fact, Mr. Kendall did not think it would be possible to forecast from any available data at the present time what next year's crop was to be. Although this year was exceptional, it could not be contemplated that in future, however long a series of years were taken, there would not appear peculiar breaks in the series due to factors imposed by the Government or other human agencies.

Finally, Mr. Kendall felt they were still in the " catswhisker " stage in this question of crop estimation and forecasting. Sometimes they got good, and sometimes bad results, and sometimes the results faded out under their very eyes. Until someone devised an analytical instrument comparable in power to the thermionic valve, he doubted whether they would get much further with this problem; the question was, where was one to look for that instrument? He could not suggest an answer to that question, but hoped no one would leap at his throat if he suggested the possibility of the extension of the Spearman factor analysis being worth inquiring into. In many cases it was not known whether the factors contributing to crop yields were capable of measurement, but it might emerge as an empirical fact that crop variation from year to year could be represented as a linear function of two or three simple variates. In that case they would have discovered something important which might provide a basis for forecasting and estimation.

Mr. T. N. Hoblyn had been asked to say something about the possibilities of forecasting the yields of horticultural crops. He feared he could do no more than point out some of the difficulties.

The first point to be observed was the necessity for an accurate estimation of acreage on which to base a forecast. This had already been insisted upon by previous speakers, and was no less important for horticultural crops. This estimate must be infinitely more detailed than any at present available. For small fruits the various kinds, strawberries, raspberries, currants, etc., must be given separately; and to be of any value at present, some information as to the amount of disease must be available. With tree fruits, not only must the kind of fruit be stated, viz., apples, pears, plums, etc. but it was essential to know the areas devoted to different varieties. Thus in England at the present time there were three outstanding varieties of apple : Worcester Pearmain, Cox's Orange Pippin and Bramley's Seedling. This year there had been a very heavy crop of the first, with only medium to light crops of the other two, and the effect on the market was quite different from that, say, of a big crop of Bramley's Seedling.

Next it was necessary to know the relative acreages of trees of different age. Recently there had been heavy plantings of Cox's Orange Pippin, but at the moment they were in no way differentiated in the returns from twenty- or thirty-year-old trees in full bearing. All that was known was the approximate area under apples. On visiting a fruit-growing area in the United States or in Canada, this information was available. Mr. Hoblyn had with him two such summaries for the Hudson Valley, New York State and the Okanagan Valley in British Columbia. The acreages were categorized under kinds of fruits, districts and varieties, and each category contained three or more age groups. Such surveys as these could, if kept up to date, provide a reasonable basis for a forecast; but this must be done on a national scale, to be of great value. Having obtained this basis to work on, one's troubles were about to begin; for in the fruit-grower's year there were so many critical periods.

For small fruits, *e.g.* strawberries and raspberries, he doubted whether any more accurate figure could be obtained than the acreage cropped. For the crop depended to a very large extent upon the weather conditions during the actual period of fruiting.

There were, however, certain outstanding periods after which they could obtain some guidance as to the probable apple, pear and plum crops. Firstly, the season before had its influence. Other things being equal, a large crop in one season was often followed by a light one in the next, and vice versa. Further, weather conditions during the previous autumn might materially affect the " strength " of the blossom the following spring, though they had but the haziest idea at present of the actual relationship. Fortunately, crops were not so susceptible to winter conditions in this country as on the other side of the Atlantic. None the less, a very wet winter, such as the last, probably greatly affected the crops this season.

The next, and perhaps most critical period, was the blossoming season, when literally anything might happen. And it was after this that the first tentative forecast could be made. This, however, was followed by another anxious period, known as the " June drop." At this time, usually early July, the tree shed a large proportion of its crop. The proportion, however, which was shed would vary from year to year. The causes were probably very complex, combining the effects of strength of blossom, previously alluded to, and conditions affecting pollination.

After this the main forecast should be made. But even then there was a long period during which hailstorms and gales might alter the value, if not the amount, of his crop very considerably.

Finally, where fruit was normally stored, conditions during the previous growing season might affect the storage qualities of the fruit, and thus the quantity ultimately marketed and the time at which it was sold.

Mr. Hoblyn said he had pointed out some of the difficulties in forecasting horticultural crops, and in doing so he had betrayed the lamentable lack of knowledge of the effect of the weather on the fruit crops. Some of these effects were already being investigated, but the comprehensive crop-weather scheme, with the definite object of elucidating some of these points, of which some dreamt, was still to come. Dr. Irwin had told of the original horticultural crop-weather scheme. The reasons why it came to an end were complicated, but in the main were :—

    (1) The material which was being used.
    (2) The impossibility of separating soil and weather effects.
    (3) The lack of a definition in the experimental design.

Although the original scheme had been dropped, the information gained therefrom would place them in a stronger position for the time when a new scheme could be started. The importance of this and of accurate acreage surveys was unquestioned. Perhaps in time these might be obtained, but until that time he feared they would not advance very much nearer to an accurate forecast than they were at present.

Miss Cockburn Millar was exceedingly grateful to have been given the opportunity of attending the meeting. The approach to the problem of weather and yields which she had taken was rather different from those which had been discussed that afternoon. It was in general much more rough and ready, and possibly did not possess that statistical purity which was so desirable. She felt, however, that a brief description of what she was doing might be of general interest.

Her interest in this problem was mainly, of course, economic. Organized marketing, which was primarily designed for supply control and price-fixing, had brought into the forefront the question of variability in yield. This uncontrollable fluctuation in supplies was particularly significant from the point of view of certain crops which had an inelastic demand such as potatoes, which she had chosen: in other words, a crop where a surplus had a proportionately greater effect on the price. Where the total output was subject to excessive fluctuation, the problem of controlling the market supply became necessarily more acute. As was well known, supply control was now possible for potato producers, and if the Board wished to maximize its returns over the season, it was extremely important that it should have all possible information about prospective output in framing its price policy for the season. It was with this in view that she had started to examine the forecasts and actual yields of potatoes as given by the Ministry of Agriculture and *The Times*. As had already been indicated in the papers that afternoon, there were very considerable discrepancies between the forecasts and the final yields, and she wanted, if possible, assuming the final estimate to be the correct one, to devise some means of getting a more accurate forecast at an earlier date in the season.

The total supply of potatoes was much more a function of the

yield in any one year than of the acreage, and it appeared in the subsequent analysis that the yield was largely dependent upon weather
factors.   After deciding which were the principal potato-producing
areas in Great Britain, she took the weather statistics of the nearest
meteorological station.   This was not entirely satisfactory, since
the conditions, particularly for rainfall, varied considerably within
quite small areas, and it was not always possible to get a station
very close to the producing area.   For wheat the season under
review was in fact a whole calendar year, the seed being sown in
autumn, whereas for potatoes the growing season was much shorter,
as they were not planted until April.   Therefore, weather conditions were observed only from April to the end of September;
though it must be admitted that earlier rainfall would affect the
conditions of the soil for planting, this fact did not appear to be very
significant.

Rainfall and hours of sunshine, on a weekly basis, were taken
for the years 1922–23, the averages for this period for each week
being taken as normal.   A comparison of actual rainfall and sunshine and these " normals " showed fairly clearly, when considered
in relation to the potato yields for these areas, that there were very
definite critical periods during the season when sub-normal or
super-normal rainfall or sunshine had a marked effect on the final
yield.

By means of the graphical curvilinear correlation analysis as
worked out by L. H. Bean of the United States Department of
Agriculture, an estimate of the yield based on weather factors was
obtained for each of the areas.   The results were, on the whole,
encouraging, considering the roughness of the data.   By weighting
the estimates thus obtained according to the production in the area
in question, estimates were made for England and for Scotland.
These estimates were further improved when they were corrected
to allow for the fact that the areas chosen were the best potato
areas.   The index of correlation obtained for England, when corrected for degrees of freedom, was 0·85, and for Scotland 0·81.
The rain and sunshine factors used in obtaining this degree of correlation included, in certain areas, weather conditions up to the time
of lifting the crop.   It was obvious that this showed little advance
for practical purposes on the October estimates of the Ministry
of Agriculture, though her (Miss Millar's) calculations were more in
line with the final estimate than their October forecast.   It was
possible, however, to cut out the independent variables for September
and the last half of August without materially affecting the index of
correlation.   For example, the relationship between weather and
yield for mid-August was, for England, 0·80, compared with 0·85
where all factors up to October were included.   The yield based on
the mid-August data thus gave a very fair estimate of the final
output; in addition, it was 3 months in advance of the Ministry's
final estimate and was more accurate than their mid-October
estimate, their mean error being 0·325 as against hers of 0·225.

This approach was admittedly rather rough and ready, but her own
results suggested that it might be worth while to consider this wider

approach in greater detail. For example, a more careful choice of weather statistics or the special collection of rainfall and sunshine data actually in the main producing areas might provide a better basis for the estimation of output. The point she would like to make was, however, that this relationship between weather and yield had become more important than ever since the inauguration of organized marketing, for powers had been given to producers to control their market supply.

MR. HOUGHTON thought that as this question under discussion was of such great interest to the Ministry of Agriculture, it was perhaps only proper that another member of the Ministry should offer a few remarks. In doing so he proposed to speak from the point of view of the administrator rather than the scientist. The subject could be divided into two parts, the scientific and the administrative, but nearly all the discussion had been devoted to the former. The main function of the scientist was the pursuit of truth, and while that pursuit continued, the administrator could do little more than watch the pursuit with a benevolent and perhaps expectant interest. It was only when the pursuit was finished and the new truth established that the administrator was called upon to say what he was going to do about it.

So far as the scientific work was concerned, a certain amount of bait had been thrown out that evening in the hope of ascertaining what the Ministry were likely to do in the way of encouraging this work in the future. He felt sure he would not be expected to make any pronouncement on that subject; this was not exactly the right place for it. He could only say that the work on the subject was continuing, and if the persons concerned considered that the Ministry's assistance was not sufficiently generous, then the usual channels were open to them for their representations.

So far as the administrative aspects were concerned, it seemed to him that the methods of forecasting yields of crops from meteorological data or from measurements of crop growth had not yet passed the experimental stage. Certainly with regard to the relation between meteorological data and crop yield, most of the speakers had been rather cautious as to the probable outcome of those investigations. He would put Miss Millar amongst the more optimistic of the workers in this connection. Before she spoke he had gathered that it would be a very long time before this question reached the administrative stage; indeed, he had doubted whether it would ever do so. Dr. Irwin himself said it would be 40 or 50 years before the full benefit of this programme could be reaped, and he (Mr. Houghton) had felt that it would not impinge upon the administrative field during his own official lifetime. At all events he was not sure that he could now be so confident, in view of what Miss Millar had said; but she had admitted that there was room for further investigation on the subject, so that he could remain in the position of observer for some time longer.

With regard to observations of crop growth, most speakers had admitted that that was still in the experimental stage.

On the question of sampling for yield at harvest, Mr. Kendall had mentioned some of the administrative difficulties of the problem, and when he said that an Act of Parliament would be necessary in order to obtain a random sample, he (Mr. Houghton) thought he could answer confidently that it would be very difficult to secure the passage of an Act of Parliament especially for that purpose. A further difficulty of the sampling method that concerned the administrator was the actual field work. Dr. Wishart had questioned Mr. Cochran's statement that samples from 500 fields would be sufficient to yield a result having an error of not more than 1 per cent. But even if Mr. Cochran were right, the actual field work in itself would be a formidable task. If samples from 500 fields had to be taken for each crop, and this method were used for the ten crops covered by the Ministry's estimates of yields, samples would have to be taken from 5,000 fields in all, and if they had to take a random sample of each crop, this might involve visits to about 5,000 farms. He estimated that each pair of samplers could not do more than an average of four fields per day.

There was the further inherent difficulty in this method mentioned by Mr. Cochran, viz. the difference between the results of small-scale threshing in the case of the sample and the results of commercial threshing; some allowance would have to be made for this; the actual amount could only be determined after much more information had been obtained.

During the course of the discussion that had taken place, he could not help feeling that the Englishman had lived up to his usual reputation for self-depreciation; for whereas he noticed that such criticism as there was of existing methods was directed against the system of the Ministry of Agriculture, the systems in India and China were regarded as admirable. In the course of that criticism they were told that the Ministry's estimates tended to under-state the variation in the yields from year to year. The crop reporters were said to be too conservative; this was surprising in view of the source of their information and the method they adopted. For an exhaustive examination of this particular criticism, which was not new, he would refer them to the paper read to the Society by Mr. Vigor a few years ago.

To sum up, he would say that he thought the criticisms that had been made of the Ministry of Agriculture's methods and the possible alternatives that had been referred to had not up to the present been of such a nature as would call for any appreciable alteration being made.

[Mr. Houghton later sent in the tables given on pp. 38–42 relating to Forecast of Yield in England and Wales, to which he had no time to refer at the meeting.]

THE CHAIRMAN said that before the meeting closed he would like, on behalf of all present, to express thanks and appreciation to them for contributions which were interesting, fascinating and provocative of thought.

TABLE A

*Wheat*

| Year | Forecast of Yield (cwt. per acre) | | | | | | Final Esti-mate |
| | August 1st | | September 1st | | October 1st | | |
| | Yield | % Diff. from Final | Yield | % Diff. from Final | Yield | % Diff. from Final | |
|---|---|---|---|---|---|---|---|
| 1929 ... ... | 16·8 | —12·0 | 17·7 | —7·3 | 18·3 | —4·2 | 19·1 |
| 1930 ... ... | 17·3 | + 8·8 | 16·6 | +4·4 | 15·8 | —0·6 | 15·9 |
| 1931 ... ... | 16·9 | + 5·0 | 16·3 | +1·2 | 16·1 | 0·0 | 16·1 |
| 1932 ... ... | 17·2 | 0·0 | 17·0 | —1·2 | 16·9 | —1·7 | 17·2 |
| 1933 ... ... | 18·4 | — 3·2 | 18·8 | —1·1 | 18·9 | —0·5 | 19·0 |
| 1934 ... ... | 18·2 | — 8·5 | 18·7 | —6·0 | 19·0 | —4·5 | 19·9 |
| 1935 ... ... | 17·8 | — 2·7 | 17·7 | —3·3 | 17·6 | —3·8 | 18·3 |
| 1936 ... ... | 16·6 | + 2·5 | 16·7 | +3·1 | 15·7 | —3·1 | 16·2 |
| Mean Diff., % | — | — 1·3 | — | —1·3 | — | —2·3 | — |
| Estimated S.D. | — | 7·8 | — | 4·1 | — | 1·8 | — |

TABLE B

*Barley*

| Year | Forecast of Yield (cwt. per acre) | | | | | | Final Esti-mate |
| | August 1st | | September 1st | | October 1st | | |
| | Yield | % Diff. from Final | Yield | % Diff. from Final | Yield | % Diff. from Final | |
|---|---|---|---|---|---|---|---|
| 1929 ... ... | 15·8 | —11·2 | 16·5 | —7·3 | 17·0 | —4·5 | 17·8 |
| 1930 ... ... | 15·6 | + 8·3 | 14·8 | +2·8 | 14·2 | —1·4 | 14·4 |
| 1931 ... ... | 15·3 | + 2·0 | 15·0 | 0·0 | 14·8 | —1·3 | 15·0 |
| 1932 ... ... | 15·5 | — 3·1 | 15·6 | —2·5 | 15·5 | —3·1 | 16·0 |
| 1933 ... ... | 16·4 | — 2·4 | 16·7 | —0·6 | 16·6 | —1·2 | 16·8 |
| 1934 ... ... | 15·4 | — 8·9 | 15·7 | —7·1 | 16·0 | —5·3 | 16·9 |
| 1935 ... ... | 15·7 | — 5·4 | 15·6 | —6·0 | 15·7 | —5·4 | 16·6 |
| 1936 ... ... | 16·2 | 0·0 | 16·1 | —0·6 | 15·8 | —2·5 | 16·2 |
| Mean Diff., % | — | — 2·6 | — | —2·7 | — | —3·1 | — |
| Estimated S.D. | — | 6·2 | — | 3·7 | — | 1·8 | — |

TABLE C

*Oats*

| Year | Forecast of Yield (cwt. per acre) | | | | | | | Final Esti- mate |
| | August 1st | | September 1st | | October 1st | | | |
| | Yield | % Diff. from Final | Yield | % Diff. from Final | Yield | % Diff. from Final | | |
|---|---|---|---|---|---|---|---|---|
| 1929 ... ... | 15·0 | —8·5 | 15·5 | —5·5 | 15·7 | —4·3 | | 16·4 |
| 1930 ... ... | 15·2 | +0·7 | 15·1 | 0·0 | 14·7 | —2·6 | | 15·1 |
| 1931 ... ... | 15·4 | +2·7 | 15·2 | +1·3 | 15·0 | 0·0 | | 15·0 |
| 1932 ... ... | 15·1 | —4·4 | 15·5 | —1·9 | 15·5 | —1·9 | | 15·8 |
| 1933 ... ... | 15·8 | —3·7 | 16·1 | —1·8 | 16·1 | —1·8 | | 16·4 |
| 1934 ... ... | 15·0 | —5·7 | 15·1 | —5·0 | 15·3 | —3·8 | | 15·9 |
| 1935 ... ... | 15·5 | —3·1 | 15·5 | —3·1 | 15·3 | —4·4 | | 16·0 |
| 1936 ... ... | 14·9 | —2·0 | 14·9 | —2·0 | 14·6 | —3·9 | | 15·2 |
| Mean Diff., % | — | —3·0 | — | —2·3 | — | —2·8 | | — |
| Estimated S.D. | — | 3·5+ | — | 2·5— | — | 1·5+ | | — |

TABLE D

*Potatoes*

| Year | Forecast of Yield (tons per acre) | | | | | | | Final Esti- mate |
| | September 1st | | October 1st | | November 1st | | | |
| | Yield | % Diff. from Final | Yield | % Diff. from Final | Yield | % Diff. from Final | | |
|---|---|---|---|---|---|---|---|---|
| 1929 ... ... | 5·7 | —17·4 | 6·2 | —10·1 | 6·5 | —5·8 | | 6·9 |
| 1930 ... ... | 6·0 | — 7·7 | 6·1 | — 6·2 | 6·1 | —6·2 | | 6·5 |
| 1931 ... ... | 5·5 | 0·0 | 5·3 | — 3·6 | 5·3 | —3·6 | | 5·5 |
| 1932 ... ... | 6·4 | — 3·0 | 6·2 | — 6·1 | 6·3 | —4·5 | | 6·6 |
| 1933 ... ... | 6·4 | — 4·5 | 6·2 | — 7·5 | 6·4 | —4·5 | | 6·7 |
| 1934 ... ... | 6·0 | —15·5 | 6·3 | —11·3 | 6·7 | —5·6 | | 7·1 |
| 1935 ... ... | 5·6 | — 9·7 | 5·7 | — 8·1 | 6·0 | —3·2 | | 6·2 |
| 1936 ... ... | 6·0 | — 3·2 | 5·9 | — 4·8 | 5·8 | —6·5 | | 6·2 |
| Mean Diff., % | — | — 7·6 | — | — 7·2 | — | —5·0 | | — |
| Estimated S.D. | — | 6·2 | — | 2·6 | — | 1·2 | | — |

TABLE E

*Turnips and Swedes*

| Year | Forecast of Yield (tons per acre) | | | | Final Estimate |
| | October 1st | | November 1st | | |
| | Yield | % Diff. from Final | Yield | % Diff. from Final | |
|---|---|---|---|---|---|
| 1929 ... ... | 11·1 | −6·7 | 11·5 | −3·4 | 11·9 |
| 1930 ... ... | 11·2 | −5·1 | 11·5 | −2·5 | 11·8 |
| 1931 ... ... | 11·2 | −0·9 | 11·1 | −1·8 | 11·3 |
| 1932 ... ... | 12·6 | −3·1 | 13·0 | 0·0 | 13·0 |
| 1933 ... ... | 10·1 | −6·5 | 10·6 | −1·9 | 10·8 |
| 1934 ... ... | 8·4 | −6·7 | 8·7 | −3·3 | 9·0 |
| 1935 ... ... | 9·5 | +2·1 | 9·5 | +2·1 | 9·3 |
| 1936 ... ... | 12·0 | −4·0 | 12·4 | −0·8 | 12·5 |
| Mean Diff., % ... | — | −3·9 | — | −1·4 | — |
| Estimated S.D. ... | — | 3·1 | — | 1·8 | — |

TABLE F

*Mangolds*

| Year | Forecast of Yield (tons per acre) | | | | Final Estimate |
| | October 1st | | November 1st | | |
| | Yield | % Diff. from Final | Yield | % Diff. from Final | |
|---|---|---|---|---|---|
| 1929 ... ... | 17·7 | − 6·8 | 18·5 | − 2·6 | 19·0 |
| 1930 ... ... | 17·9 | − 5·3 | 18·4 | − 2·6 | 18·9 |
| 1931 ... ... | 16·6 | − 1·2 | 18·5 | +10·1 | 16·8 |
| 1932 ... ... | 17·6 | − 7·0 | 18·2 | − 3·7 | 18·9 |
| 1933 ... ... | 16·2 | − 7·0 | 17·1 | − 1·7 | 17·4 |
| 1934 ... ... | 17·2 | −10·4 | 18·3 | − 4·7 | 19·2 |
| 1935 ... ... | 17·2 | − 6·0 | 17·8 | − 2·7 | 18·3 |
| 1936 ... ... | 17·8 | − 6·8 | 18·6 | − 2·6 | 19·1 |
| Mean Diff., % ... | — | − 6·3 | — | − 1·3 | — |
| Estimated S.D. ... | — | 2·5 + | — | 4·7 | — |

## TABLE G

*Beans*

| Year | Forecast of Yield (cwt. per acre) | | | | | | Final Esti-mate |
| | August 1st | | September 1st | | October 1st | | |
| | Yield | % Diff. from Final | Yield | % Diff. from Final | Yield | % Diff. from Final | |
|---|---|---|---|---|---|---|---|
| 1929 ... ... | 14·2 | —6·5 | 14·1 | —7·2 | 14·5 | —4·6 | 15·2 |
| 1930 ... ... | 17·0 | —1·2 | 16·6 | —3·5 | 16·3 | —5·2 | 17·2 |
| 1931 ... ... | 16·5 | —0·6 | 16·0 | —3·6 | 16·0 | —3·6 | 16·6 |
| 1932 ... ... | 17·1 | 0·0 | 16·7 | —2·3 | 16·7 | —2·3 | 17·1 |
| 1933 ... ... | 16·9 | 0·0 | 16·5 | —2·4 | 16·7 | —1·2 | 16·9 |
| 1934 ... ... | 16·7 | —2·9 | 16·4 | —4·7 | 16·4 | —4·7 | 17·2 |
| 1935 ... ... | 13·8 | —0·7 | 13·3 | —4·3 | — | — | 13·9 |
| 1936 ... ... | 15·7 | +2·6 | 15·4 | +0·7 | — | — | 15·3 |
| Mean Diff., % | — | —1·2 | — | —3·4 | — | —3·6* | — |
| Estimated S.D. | — | 2·6 | — | 2·3 | — | 1·6 | — |

\* For 6 years.

## TABLE H

*Peas*

| Year | Forecast of Yield (cwt. per acre) | | | | | | Final Esti-mate |
| | August 1st | | September 1st | | October 1st | | |
| | Yield | % Diff. from Final | Yield | % Diff. from Final | Yield | % Diff. from Final | |
|---|---|---|---|---|---|---|---|
| 1929 ... ... | 15·9 | — 2·4 | 15·9 | —2·4 | 16·2 | —0·6 | 16·3 |
| 1930 ... ... | 15·6 | + 5·4 | 14·8 | 0·0 | 14·3 | —3·4 | 14·8 |
| 1931 ... ... | 15·0 | + 7·1 | 14·2 | +1·4 | 14·0 | 0·0 | 14·0 |
| 1932 ... ... | 13·7 | + 0·7 | 13·6 | 0·0 | 13·6 | 0·0 | 13·6 |
| 1933 ... ... | 14·8 | + 2·8 | 14·8 | +2·8 | 15·4 | +6·9 | 14·4 |
| 1934 ... ... | 14·8 | — 9·2 | 14·9 | —8·6 | 14·9 | —8·6 | 16·3 |
| 1935 ... ... | 13·3 | —10·1 | 13·6 | —8·1 | — | — | 14·8 |
| 1936 ... ... | 13·4 | 0·0 | 13·1 | —2·2 | — | — | 13·4 |
| Mean Diff., % | — | — 0·7 | — | —2·1 | — | —0·9* | — |
| Estimated S.D. | — | 6·3 | — | 4·2 | — | 5·1 | — |

\* 6 Years only.

c 2

## TABLE I
### Seeds Hay

| Year | Forecast of Yield (cwt. per acre) | | | | | | Final Estimate |
|------|------|------|------|------|------|------|------|
| | July 1st | | August 1st | | September 1st | | |
| | Yield | % Diff. from Final | Yield | % Diff. from Final | Yield | % Diff. from Final | |
| 1929 ... ... | 22·3 | —2·6 | 22·1 | —3·5 | 21·5 | —6·1 | 22·9 |
| 1930 ... ... | 28·5 | —2·4 | 29·2 | 0·0 | 29·2 | 0·0 | 29·2 |
| 1931 ... ... | 31·5 | +4·3 | 31·4 | +4·0 | 31·1 | +3·0 | 30·2 |
| 1932 ... ... | 27·8 | —1·8 | 28·4 | +0·4 | 28·6 | +1·1 | 28·3 |
| 1933 ... ... | 24·9 | +1·2 | 24·9 | +1·2 | 24·6 | 0·0 | 24·6 |
| 1934 ... ... | 23·5 | —2·5 | 23·7 | —1·7 | 23·5 | —2·5 | 24·1 |
| 1935 ... ... | 25·2 | —2·7 | 26·0 | +0·4 | 26·1 | +0·8 | 25·9 |
| 1926 ... ... | 24·5 | —2·0 | 23·8 | —4·8 | 24·2 | —3·2 | 25·0 |
| Mean Diff., % | — | —1·1 | — | —0·5 | — | —0·9 | — |
| Estimated S.D. | — | 2·5+ | — | 2·8 | — | 2·5 | — |

## TABLE J
### Meadow Hay

| Year | Forecast of Yield (cwt. per acre) | | | | | | Final Estimate |
|------|------|------|------|------|------|------|------|
| | July 1st | | August 1st | | September 1st | | |
| | Yield | % Diff. from Final | Yield | % Diff. from Final | Yield | % Diff. from Final | |
| 1929 ... ... | 15·8 | + 3·3 | 15·0 | —2·0 | 15·3 | 0·0 | 15·8 |
| 1930 ... ... | 22·2 | + 0·5 | 22·4 | +1·4 | 22·1 | 0·0 | 22·1 |
| 1931 ... ... | 22·3 | 0·0 | 22·5 | +0·9 | 22·1 | —0·9 | 22·3 |
| 1932 ... ... | 19·0 | — 7·3 | 19·6 | —4·4 | 19·7 | —3·9 | 20·5 |
| 1933 ... ... | 17·8 | — 1·1 | 17·8 | —1·1 | 17·7 | —1·7 | 18·0 |
| 1934 ... ... | 16·7 | — 2·9 | 16·6 | —3·5 | 16·8 | —2·3 | 17·2 |
| 1935 ... ... | 18·7 | —10·5 | 20·4 | —2·4 | 20·8 | —0·5 | 20·9 |
| 1936 ... ... | 18·7 | — 7·9 | 19·2 | —5·4 | 19·8 | —2·5 | 20·3 |
| Mean Diff., % | — | — 3·2 | — | —2·1 | — | —1·5— | — |
| Estimated S.D. | — | 4·8 | — | 2·4 | — | 1·4 | — |

DR. ISSERLIS said he thought, as this was the first meeting of the Session, they ought to put on record the very great loss that the parent Society, and the Section, had suffered through the death of Mr. W. S. Gosset. If he had been present at the meeting he would have done a few little sums in the train on the way to the meeting, on the figures provided by Dr. Irwin and Mr. Cochran, and would have said something very illuminating. His loss would be greatly felt, and should be recorded in the minutes.

The following contribution was received from Mr. H. G. Hudson after the meeting :—

An integral part of Mr. Cochran's scheme for estimating the grain yield from a given selection of farms in this country is the adequacy of his sampling technique. Mr. Cochran proposes to take 1/15,000, or 0·0067 per cent. of a field of 10 acres, a percentage that seems, at first sight, so small that any estimate based on it must be doubtful. At Cambridge I have been investigating the sampling percentage necessary to give a 5 per cent. sampling error in actual field plots varying from 1/1200 to 1/20 acre, and found that the sampling percentage ($y$) appears to decrease as the plot size ($x$) increases. The relationship may be represented fairly accurately by an equation of the form $1/v = a + bx$. Stimulated by Mr. Cochran's figure, I calculated the sampling percentage for plot sizes many times greater than those within my observed range—an extrapolation procedure which is, of course, of very doubtful validity—and found that, to obtain a sampling error of 10 per cent. in a 10-acre field, it would be necessary to take 0·0062 per cent. The agreement with Mr. Cochran, whose figures were based on sampling errors of from 8 to 14 per cent., was close—a result which is no doubt fortuitous—but the figures are of interest, in that they appear to confirm that the degree of sampling practised, though it seems extraordinarily low, is adequate to the degree of accuracy aimed at.

Mr. Cochran sent a further contribution as follows :—

It was gratifying to see that most of the speakers at the meeting were convinced of the importance of accurate crop forecasts and estimates. The opening paragraphs of Mr. Battle's remarks draw attention to this need for the sugar-beet industry in particular. The problem is being tackled by the factories, and it is interesting to note that forecasts are being based on early lifting of some of the roots, a method which, as I indicated in my remarks, is available for root crops and potatoes, but not for cereals.

The Potato Marketing Board also realize the necessity of improved forecasts, as we may judge by a quotation from their 1936 Report (pp. 6–7) : " The most difficult period during the whole of the year is the period August–October. Although the acreage is known, the national yield of potatoes is still obscure as is shown by the following figures of the comparative estimates of crop arrived at from month to month by the Ministry of Agriculture and Fisheries.

" A yield of one ton per acre is equivalent to a total production of 600,000 tons. The Board are accordingly faced with a difficult task in shaping their policy early in the season, since they are then in possession of but meagre knowledge of the yield per acre and the prevalence of disease and extent of waste, any or all of which may have a great ultimate bearing on the market situation during the ensuing months."

The Board are testing the method of obtaining forecasts of yield by lifting and weighing samples from the fields. Some 200 samples were taken this year, and the number may be considerably increased in future.

That the problems of crop forecasting and estimation raise difficulties has, I think, been sufficiently stressed in Mr. Yates's previous

ENGLAND AND WALES

| Dates of Estimates | Comparative Estimates of Crop | | |
|---|---|---|---|
| | 1935 Tons per acre | 1934 Tons per acre | 1933 Tons per acre |
| September 1st       ...    ... | 5·6 | 6·0 | 6·4 |
| October 1st       ...    ... | 5·7 | 6·3 | 6·2 |
| November 1st       ...    ... | 6·0 | 6·7 | 6·4 |
| December 1st       ...    ... | 6·3 | 7·1 | 6·7 |
| Difference between first and final estimate       ...    ... | 0·7 | 1·1 | 0·3 |

accounts of the Rothamsted work, and in the papers and discussions
at this meeting. As regards forecasting by the use of correlations
of yields with weather and with earlier measurements on the crop,
much further research appears to be needed. Miss Cockburn Millar's
results with potatoes sound very promising, and it will be interesting
to see a full report on these.

To turn to the estimation of yield by sampling at harvest, the
required number of fields, namely 500, indicated by the Rothamsted
results is by no means alarmingly large. This number is, of course,
itself subject to a sampling error, and would presumably be increased
if a random selection of farms were made *each year*, since in that case
the variation between farms in a district would enter into the standard
error of the mean yield of the country, as Dr. Wishart has indicated.
The increase required might not, however, be great if the country
were first divided into a number of small districts.

Mr. Kendall objects that it would be impossible to take anything
approaching a random sample of the farms in the country, because
some 30 per cent. or more of the farmers would refuse to co-operate.
The force of this objection can only be judged when the selection of
random samples has been tried, but one can point to investigations
which give a much more optimistic picture of the willingness of the
farmers to co-operate. A study of the blackening of potatoes was
recently made by the Potato Marketing Board. At one stage of the
work the supervisors wished to take samples of potatoes from any
farm at which they noticed potatoes being packed for market.
This type of selection is not, of course, random in the statistical
sense, but it is a selection of farms made *before* consulting the farmer.
Out of 500 farms approached, only one refusal was received, while
one farmer requested payment for the sample. If this measure of
co-operation could be obtained in crop estimation, the taking of
random samples would be quite feasible. Further, in the crop
estimation work carried out by the United States Department of
Agriculture, the voluntary co-operation of the farmers has been
secured, and the taking of random samples is coming to be regarded
as a routine practice.

When the initial panel of farms had been selected, a farm would remain on the panel even if it subsequently changed hands. The new tenant might reasonably be expected to continue a co-operation which was already in existence on the farm. Thus Mr. Kendall's objection that the samples would gradually melt away does not arise. A further advantage of having a fixed panel of farms is that opportunities would be available to study other important aspects of farming (*e.g.*, quantity of livestock kept) on which information at present is unreliable or scarce. An alternative solution might be to have a number of separate bodies, each responsible for the estimation, etc., of a single crop, with separate panels of farms if desirable.

Mr. Kendall's suggestion of the use of the Spearman factor analysis seems to me inappropriate. Factor analysis deals with problems in which there are a number of variates, no one of which can be considered as the dependent variate. In crop estimation and forecasting, however, yield is always the dependent variate, and the familiar method of partial regression is well suited to the problem. It cannot be too strongly emphasized that the difficulties in crop estimation and forecasting are not due to weaknesses in statistical theory, as Mr. Kendall suggests, but to lack of data. Unless more data can be obtained from properly conducted sampling investigations the problems are likely to remain indefinitely " in the experimental stage," as Mr. Houghton puts it.

Mr. Houghton states that the field work involved in obtaining crop estimates at harvest would be a formidable task. If his figure of 5,000 fields or 2,500 man-days for the ten principal crops were to turn out to be correct, this would, in my opinion, be a very economical piece of work. It would be interesting to know how much labour is spent in collecting the Ministry's present crop estimates. If there are still 300 crop reporters, the number of man-days must be considerable.

Mr. Houghton's tables showing the discrepancies between the Ministry's earlier and final estimates are of interest. I have already quoted the Potato Marketing Board's opinion of the early potato estimates. The tables do not, however, enable one to assess the accuracy of the Ministry's estimates, since there is no information about the accuracy of the final estimates themselves.

The positive bias in the wheat-sampling results as compared with the harvested yields will require further study. There are indications of a similar bias in the sampling results in the sugar-beet work to which Mr. Battle referred, though it is not yet certain to what causes the bias is due in this case. While the bias may turn out to be substantially constant from year to year, and thus capable of adjustment, its existence sets a limit to the accuracy which is desirable in the estimates from sampling yields.

In conclusion, the accounts which have been given of the work done by the sugar-beet factories, by the Potato Marketing Board and under the Ministry's crop weather scheme are only interim reports. The work continues, and it will be interesting to see what progress is made in the next few years.

# 16a

## I. *Moments and Semi-Invariants of Sampling Distributions.*

### By W. G. COCHRAN.

From a sample of size $n$ drawn at random from an infinite population, specified either by moments $\mu_r$ about the mean or by semi-invariants $\kappa_r$, estimates may be derived of $\mu_r$ and $\kappa_r$. The estimate of $\mu_r$ generally taken is

$$m_r = \frac{1}{n} \sum_{i=1}^{n} (x_i - \bar{x})^r$$

and Fisher (B 2) has shown that a suitable estimate $k_r$ of $\kappa_r$ may be obtained from the condition that the mean value of $k_r$ over all possible samples shall be $\kappa_r$. The problem of finding the joint frequency distributions of the $m_r$ has been attacked with some success by Craig (B 1) in 1928 and alternatively by St. Georgescu (B 5) in 1932, who gave methods for finding the semi-invariants of the distribution of the $m_r$ in terms of the $\kappa_r$; while Fisher (B 2) in 1928 gave combinatorial rules for finding the semi-invariants of the $k_r$ in terms of the $\kappa_r$.

In the year under review, Wishart (A 113) gives an interesting comparison of the results of the two methods of approach. He shows, in particular, how to obtain the Craig–St. Georgescu results from those of Fisher, working out $S(4^2)$ and $S(24)$ as examples of the general procedure. Here $S(4^2)$ denotes the 2nd semi-invariant of the distribution of $m_4$ and $S(24)$, the (1,1) semi-invariant of the simultaneous distribution of $m_2$ and $m_4$. To obtain $S(24)$, for instance, we first express $m_2$ and $m_4$ as functions of the $k_r$, in this case of $k_2$ and $k_4$. Knowing these functions, we can, by an operational process due to Fisher (B 2), find the joint moment generating function of $m_2$ and $m_4$ in terms of the moments of $k_2$ and $k_4$. From this the term required—the $\kappa_{11}$ of the distribution of $m_2$ and $m_4$—is obtained by taking logarithms. Thus we have $S(24)$ in terms of the moments of $k_2$ and $k_4$. By the known relations connecting moments and semi-invariants, we derive $S(24)$ in terms of the semi-invariants of $k_2$ and $k_4$, *i.e.* of the Fisher results. As a further step we can clearly find the moments of the distributions of the $m_r$ in terms either of the $\kappa_r$ or the $\mu_r$. Since the results in any desired form can be deduced from Fisher's results, the latter, as

Wishart points out, are the most suitable for reference purposes in view of their relative simplicity.

A short paper by Geary (A 34) also refers to previous work of Fisher and Craig. In studying the problem of obtaining tests of departure from normality, Fisher (B 3) showed how to calculate the moments of functions of the type $\dfrac{k_p}{k_2^{\frac{p}{2}}}$ in sampling from a normal population, and gave the first six moments of $\dfrac{k_3}{k_2^{\frac{3}{2}}}, \dfrac{k_4}{k_2^{2}}$. Geary finds a general expression for the $r$th moment about the origin of the corresponding quantity $\dfrac{m_p}{m_2^{\frac{p}{2}}}$ ($rp$ even). To obtain this, he first finds, using Craig's method of attack, a general expression for the $r$th moment of $m_p$ about the origin in a normal universe. He then shows that $m_2$ is distributed independently of $\dfrac{m_p}{m_2^{\frac{p}{2}}}$ for $p > 2$, so that

$$\mu_r{}'(m_p) = \mu_r{}'\!\left(\frac{m_p}{m_2^{\frac{p}{2}}}\right) \times \mu_{\frac{rp}{2}}{}'(m_2) . \quad . \quad . \quad . \quad (1)$$

and thus obtains $\mu_r{}'\!\left(\dfrac{m_p}{m_2^{\frac{p}{2}}}\right)$, knowing $\mu_{\frac{rp}{2}}{}'(m_2)$. The relation corresponding to (1) for the $k_r$ has already been given by Fisher (B 3). Geary gives no discussion of the general formula or of the simpler cases. By using the independence of $m_2$ and $\dfrac{m_p}{m_2^{\frac{p}{2}}}$, he obtains the following relations for the $(kl)$ semi-invariants of $m_2$ and $m_p$ in a normal system :

$$\left. \begin{aligned} S_{ko} &= \frac{(k-1)\,!}{2}\,(n-1)\left(\frac{2}{n}\right)^{k} \\ S_{kl} &= S_{ol}\,\frac{pl(pl+2)\,\ldots\,(pl+2k-2)}{n^{k}} \quad l \neq 0 \end{aligned} \right\} \quad . \quad . \quad (2)$$

Craig (B 1) surmised the existence of these relations, but was unable to prove them.

Two papers have appeared dealing with the sampling problem in the case of a compound population. Brown (A 11) distinguishes two cases : (i) in which we are sampling from a single population whose frequency distribution is of the form

$$p\phi_1(x) + q\phi_2(x) . \quad . \quad . \quad . \quad . \quad (3)$$

so that it is made up of two components $\phi_1(x)$, $\phi_2(x)$; (ii) in which a sample of size $N$ of the compound population is assembled by picking a random sample of $r$ from the population $\phi_1(x)$, one of $s(= N - r)$ from $\phi_2(x)$, and combining the two samples. While in both (i) and (ii) the frequency distribution of a variate drawn at random from the compound population is of the form (3), it is clear that the sampling distribution of any statistic in repeated samples of size $N$ will be different in the two cases, owing to the different ways of building up the sample. The sampling problem thus falls into two parts.

(i) Since we are here dealing with a single population, any previously obtained results in the sampling problem (such as those of Fisher (B 2) and Craig (B 1), can be used if expressions are found for the semi-invariants of the compound population in terms of those of $\phi_1(x)$ and $\phi_2(x)$. Brown considers first the case in which $\phi_1(x)$ and $\phi_2(x)$ are normal, with means $m_1$, $m_2$ and standard deviations $\sigma_1$, $\sigma_2$. He obtains explicit expressions for the semi-invariants $L_n$ of the compound frequency function in terms of $a = m_2 - m_1$, $b = \sigma_2{}^2 - \sigma_1{}^2$, $p$ and $q$. In fact, if $n > 4$,

$$L_n = \sum_{\kappa=0}^{\left[\frac{n-1}{2}\right]} \frac{n^{(2\kappa)}}{2^\kappa \kappa!} b^\kappa a^{n-2\kappa} q'_{n-\kappa} * . \quad . \quad . \quad . \quad (4)$$

where $q'_{n-\kappa}$ is a known polynomial in $q$. To facilitate calculation of the $L_n$ in particular cases, he tabulates $q_r'$ from $q = \cdot 01$ to $\cdot 50$ at intervals of $\cdot 01$, for $r = 2, 3, \ldots 9$.

Craig ((B 1) pp. 45–56) obtained expressions for the first four semi-invariants of the sampling distributions of $\alpha_3 = \dfrac{m_3}{m_2{}^{\frac{3}{2}}}$, $\alpha_4 = \dfrac{m_4}{m_2{}^2}$, $\sigma_x = \sqrt{m_2}$, in terms of the semi-invariants of the sampling distributions of the $m_r$. Brown takes the calculation a stage further, obtaining the same semi-invariants in terms of the $L_n$ (neglecting terms of order $N^{-3}$ or higher). They can then be expressed in terms of the semi-invariants of the component populations, by means of the equations (4) above.

Brown goes on to consider the case in which the third and higher semi-invariants of $\phi_1(x)$, $\phi_2(x)$ do not vanish, and shows that in the general case the actual writing down of the expressions for the $L_n$ in terms of the semi-invariants of $\phi_1(x)$ and $\phi_2(x)$ may be reduced to a partition process and a taking of derivatives.

* Here $\left[\dfrac{n-1}{2}\right]$ = integral part of $\dfrac{n-1}{2}$

$$n^{(2\kappa)} = n(n-1) \ldots (n - 2\kappa + 1).$$

No study is made of compound populations with more than two components, except when the compound frequency function is of the form

$$\sum_{i=1}^{s} p\phi(x - M_i)$$

and $s \to \infty$, $(p \to 0)$. In this case the variate of the limiting compound frequency function is distributed as the sum of two independent variates, one following the distribution $\phi(x)$ and the other the frequency distribution of the means $M_i$.

(ii) In this case the sampling problem is essentially new. For semi-invariants $S_\kappa$ of moments $m_n'$ about a fixed point, Brown shows easily that

$$S_\kappa(m_n') = \frac{r^\kappa S_\kappa'(m_n') + s^\kappa S_\kappa''(m_n')}{(r + s)^\kappa} \qquad . \quad . \quad . \quad (5)$$

where $S_\kappa'(m_n')$, $S_\kappa''(m_n')$ are the semi-invariants of $m_n'$ in samples of sizes $r, s$ from $\phi_1(x)$, $\phi_2(x)$ respectively. The problem of semi-invariants $S_\kappa$ of moments $m_r$ about the sample mean is more difficult. Brown follows Craig's method of attack very closely, making the necessary modifications, and, when $\phi_1(x)$, $\phi_2(x)$ are normal, derives expressions for $S_1(m_2)$, $S_1(m_3)$, $S_1(m_4)$, $S_2(m_2)$, $S_2(m_3)$, $S_2(m_4)$, $S_{11}(m_2,m_3)$, $S_{11}(m_2,m_4)$, and $S_{11}(m_3,m_4)$ in terms of $N$, $r$, $s$ and the semi-invariants of $\phi_1(x)$ and $\phi_2(x)$. The method of attack does not in any way depend on the normality of $\phi_1(x)$ and $\phi_2(x)$, this condition being introduced only when writing down the results.

A problem somewhat similar to that arising in (i) above is studied by Baten (A 7). Out of two parent populations A and B, consisting of $n$ and $m$ variates respectively, $r$ variates are drawn at random from A and $t$ from B, to form a sample of $N = r + t$. The problem is to find the moments of the distribution of the sample mean in terms of the moments of the populations A and B. Baten finds expressions for the first eight moments, and hence by induction a general expression for $\alpha_\kappa = \dfrac{\mu_\kappa}{\mu_2^{\frac{\kappa}{2}}}$, making free use of results obtained by other writers. In the case of many parent populations he writes down expressions for the first five moments.

In 1932 a paper by T. Kondo (B 4) appeared with the title " A new method of finding moments of moments." The method developed, however, while it may not previously have been applied to finding moments of moments, is simply the algebraic process used by Fisher (B 2, p. 207) for finding the semi-invariants of his $k_r$ functions. The essence of the method consists in expressing the function whose mean is to be found, say $m_3{}^2m_2$ or $k_3{}^2k_2$, in terms of the $s_r$ functions, where

$$s_r = \sum_{i=1}^{N} x_i{}^r, \quad x_1, x_2, \ldots, x_N$$

being the $N$ members of a sample. The problem thus reduces to finding the mean value of an expression of the type

$$s_{p_1}{}^{l_1}s_{p_2}{}^{l_2} \ldots s_{p_i}{}^{l_i}$$

Fisher (B 2, p. 207) has indicated a rule by which the mean value of any such expression may be written down at once, but Kondo does not appear to have been aware of this.

16a.88

### III.  *Orthogonal Polynomial Theory.*

### By W. G. Cochran.

Aitken (A 2 and A 3) completes his valuable work of simplifying and systematizing the numerical procedure of fitting polynomials to data by least squares in two papers, one dealing with weighted data, the other with data having weighted and correlated errors. At the same time a paper by Lidstone (A 66) has appeared which covers most of the ground of the first of Aitken's three papers on the subject (D 1). Lidstone's aim is to present the proofs in Aitken's first paper in a form suitable for actuaries. Aitken's own paper, though concise, makes very clear reading; and Lidstone differs from him mainly in giving the steps of the proofs in greater detail and using as far as possible auxiliary formulæ contained in, or derived from Freeman's *Actuarial Mathematics*. Lidstone includes also a modification of the proof given by Allan (D 2) of the formulæ for the Tchebycheff polynomial $T_r(x)$ and for $\Sigma T_r{}^2(x)$. He includes a numerical example of the method of fitting, but, unfortunately, from the point of view of completeness, he gives no example of the use of central factorial moments in fitting, this being much the shorter method.

We now turn to Aitken's paper on weighted data (A 2). Owing to the arbitrary nature of the weights, orthogonal polynomials are of little use, and the methods developed enable us to express the solution either in terms of powers of $x$ or of factorials referred to the origin or to some central value. Let $u_x$, $x = 0, 1, 2, \ldots (n-1)$ be the $n$ data, $w_x$ the weights, let $x^{(r)} = x(x-1) \ldots (x-r+1)$ and

$$U_k(x) = a_0 + a_1 x^{(1)} + a_2 x^{(2)} + \ldots + a_k x^{(k)}$$

the polynomial expressed in factorials. Then if

$$\sum_{x=0}^{n-1} (x+p)^{(r)} w_x u_x = M_{(r, p)}, \quad \sum_{x=0}^{n-1} (x+p)^{(r)} w_x = W_{(r, p)}$$

Aitken shows that the least squares conditions lead to the following linear equations for the coefficients $a_i$:

$$\sum_{r=0}^{k} a_r W_{(r+p, p)} = M_{(p, p)} \quad . \quad . \quad . \quad . \quad (19)$$

$$p = 0, 1, 2, \ldots k.$$

By introducing a set of orthogonal polynomials, he proves that the reduction in the sum of squares in passing from the $(k-1)$th to the $k$th degree is

$$S^2 = \frac{D^2}{W W_{2kk}}$$

where $W$ is the determinant of the coefficients on the left-hand side of (19), $W_{2kk}$ the minor of $W_{(2k, k)}$ in $W$, and $D$ the determinant found by replacing the last column in $W$ by the column on the right-hand side of (19).

In the working, the $W_{(r, p)}$ and $M_{(pp)}$ are found by successive summation; for instance, $W_{(r, p)}$ is $r$! times the $(r-p+1)$th element from the top of the column $\Sigma^{r+1}$ of summations of $w_x$, the summations being, as usual in these cases, from the foot of each column.

The equations (19) are solved by an ingenious method which has two advantages: (i) it gives not only $U_k(x)$ but all corresponding polynomials of lower degree at intermediate stages; (ii) the reduction in the sum of squares at any previous stage is known before that stage is fitted. The solving scheme is the same whether powers, factorials or factorials referred to a new origin are being used; the methods differ only in the early stages.

Aitken points out that these methods can, of course, be applied in the particular case when the data are equally weighted; and if the use of tables is not desired, the methods are to be heartily recommended for this purpose. The paper deals mainly with the case in which the values of $x$ are equidistant; but the solving method applies equally well when this is not so, except that it is quicker in that case to use power moments and express $U_k(x)$ in powers of $x$.

In the third paper (A 3) the errors $u_x - U_k(x)$ are correlated in a known manner, and we have to minimize a quadratic form to satisfy the least squares condition. In this case the coefficients are found independently of each other, and the method again gives estimates of the reduction in the sum of squares at each stage.

Davis (A 22) also makes a contribution to the useful work of reducing the labour of fitting polynomials to data by least squares, by the use of tables. He considers the case of $m$ equidistant values of $x$. For $m$ odd, $(= 2p + 1)$, these are taken to be $- p, - p + 1$, . . ., $p$ and for $m$ even $(= 2p)$, $\dfrac{- (2p - 1)}{2}$, $\dfrac{- (2p - 3)}{2}$, . . ., $\dfrac{(2p - 1)}{2}$. In fitting a polynomial $\Sigma a_r x^r$ to the data $y_x$ by least squares, the coefficients $a_r$ are linear functions of the moments $M_r$ of the data, and in a previous paper (D 3) Davis and Latshaw determined the coefficients of the $M_r$ in the linear functions for equations up to the 7th degree. These coefficients are tabulated to 10 significant figures in the paper by Davis, for values of $p$ up to 25. Thus in fitting the polynomial, we calculate the moments, and look up the tables for the appropriate coefficients to obtain the $a_r$. The tables enable us to fit for values of $m$ up to 51.

Two papers on graduation deserve mention. In the second part of the paper to which reference has already been made, Lidstone applies the Tchebycheff polynomials to the problem of graduation by weighted means. The paper consists mainly of a new and short proof of the equivalence of the methods of graduation by least squares and by reduction of probable error. Vaughan (A 109) has written a most interesting paper on summation formulæ. He considers the usual operator $[p][q][r]$ in the case in which the longest range is the sum of the two shorter ranges, and obtains by its use six new and powerful summation formulæ. The paper also considers the choice of operand, the method being to find first the theoretically best operand and then substitute more convenient coefficients by inspection.

## VIII. *Bibliography and References.*

### A. Current Literature.

(1) AITKEN, A. C. On Factorial Nomenclature and Notation.
   *J. Inst. Actuaries*, Vol. XLIV, Part III, pp. 449–53. 1933.
(2) AITKEN, A. C. On Fitting Polynomials to Weighted Data by Least Squares.
   *Proc. Roy. Soc. Edin.*, Vol. LIV, pp. 1–11. 1933.
(3) AITKEN, A. C. On Fitting Polynomials to Data with Weighted and Correlated Errors.
   *Proc. Roy. Soc. Edin.*, Vol. LIV, pp. 12–16. 1933.

(4) Bailey, V. A.   On the Interaction between Several Species of
       Hosts and Parasites.
           *Proc. R.S.,* A. 143, pp. 75–88.   1933.
(5) Bailey, V. A.   Non-continuous Interaction between Hosts and
       Parasites.
           *Proc. Camb. Phil. Soc.,* Vol. 29, Part 4, pp. 487–91.   1933.
(6) Bartlett, M. S.   On the Theory of Statistical Regression.
           *Proc. Roy. Soc. Edin.,* Vol. LIII, Part III, pp. 260–83.
           1933.
(7) Baten, W. D.   Sampling from Many Parent Populations.
           *Tôhoku Math. Journal,* Vol. 36, pp. 206–22.   1933.
(8) Bonferroni, C.   *Sulla Probabilità Massima nello Schema di
       Poisson.*
           *Giorn. Ist. Ital. Attuari,* Vol. IV, No. 1, pp. 109–15.
           1933.
(9) Brandner, F. A.   A Test of the Significance of the Difference
       of the Correlation Coefficients in Normal Samples.
           *Biometrika,* Vol. XXV, pp. 102–9.   1933.
(10) Brandt, A. E.   The Analysis of Variance in a $2 \times s$ Table
       with Disproportionate Frequencies.
           *J. Amer. Stat. Ass.,* Vol. XXVIII, pp. 164–73.   1933.
(11) Brown, G. M.   On Sampling from Compound Populations.
           *Annals Math. Statistics,* Vol. IV, No. 4, pp. 288–342.   1933.
(12) Bruen, C.   Five Variable Straight Line Diagrams.
           *Metron,* Vol. XI, N. 2, pp. 137–49.   1933.
(13) Burks, B. S.   A Statistical Method for Estimating the Dis-
       tribution of Sizes of Completed Fraternities in a Population
       Represented by a Random Sampling of Individuals.
           *J. Amer. Stat. Ass.,* Vol. XXVIII, pp. 388–94.   1933.
(14) Cantelli, F. P.   *Considerazioni sulla Legge Uniforme dei
       Grandi Numeri e sulla Generalizzazione di un Fondamentale
       Teorema del Sig. Paul Levy.*
           *Giorn. Ist. Ital. Attuari,* Vol. IV, No. 3, pp. 327–50.   1933.
(15) Cantelli, F. P.   *Sulla Determinazione Empirica della Leggi di
       Probabilità.*
           *Giorn. Ist. Ital. Attuari,* Vol. IV, No. 3, pp. 421–24.   1933.
(16) Carver, H. C.   Note on the Computation and Modification of
       Moments.
           *Annals Math. Statistics,* Vol. IV, No. 3, pp. 229–39.   1933.
(17) Castellano, V.   *Sulle Relazioni tra Curve di Frequenza e
       Curve di Concentrazione e sui Rapporti di Concentrazione
       Corrispondenti a Determinate Distribuzioni.*
           *Metron,* Vol. X, N. 4, pp. 1–60.   1933.
(18) Castellano, V.   *Sulla Interpretazione Dinamica del Rapporto
       di Concentrazione.*
           *Giorn. Ist. Ital. Attuari,* Vol. IV, No. 2, pp. 268–74.   1933.
(19) Craig, A. T.   On the Correlation between Certain Averages
       from Small Samples.
           *Annals Math. Statistics,* Vol. IV, No. 2, pp. 127–42.   1933.
(20) Craig, C. C.   On the Tchebycheff Inequality of Bernstein.
           *Annals Math. Statistics,* Vol. IV, No. 2, pp. 94–102.   1933.

(21) DAVIES, O. L.   On Asymptotic Formulæ for the Hypergeo-
        metric Series.
            I. Hypergeometric Series, in which the Fourth Element $x$
        is Unity.
            *Biometrika*, Vol. XXV, pp. 295–322.   1933.
(22) DAVIS, H. T.   Polynomial Approximation by the Method of
        Least Squares.
            *Annals Math. Statistics*, Vol. IV, No. 3, pp. 155–95.   1933.
(23) DE FINETTI, B.   *A Proposito di un Caso Limite delle Legge di
        Makeham.*
            *Giorn. Ist. Ital. Attuari*, Vol. IV, No. 1, pp. 129–30.   1933.
(24) DE FINETTI, B.   *Sull' Approssimazione Empirica di una Legge
        di Probabilità.*
            *Giorn. Ist. Ital. Attuari*, Vol. IV, No. 3, pp. 415–24.   1933.
(25) DEL VECCHIO, E.   *Sulla Dipendenza Statistica.*
            *Giorn. Ist. Ital. Attuari*, Vol. IV, No. 2, pp. 235–44.   1933.
(26) EDEN, T., and YATES, F.   On the Validity of Fisher's z Test
        when Applied to an Actual Example of Non-Normal Data.
            *J. Agr. Sci.*, Vol. XXIII, Part I, pp. 6–16.   1933.
(27) ELDERTON, W. P.   Adjustment for the Moments of J-Shaped
        Curves.
            *Biometrika*, Vol. XXV, pp. 179–80.   1933.
(28) FISCHER, C. H.   On Correlation Surfaces of Sums with a Certain
        Number of Random Elements in Common.
            *Annals Math. Statistics,* Vol. IV, No. 2, pp. 103–26.   1933.
(29) FISCHER, C. H.   On Multiple and Partial Correlation Coefficients
        of a Certain Sequence of Sums.
            *Annals Math. Statistics*, Vol. IV, No. 4, pp. 278–84.   1933.
(30) FUHRICH, J.   *Über die Numerische Ermittelung von Periodizi-
        täten und Ihre Beziehungen zum Zufallsgesetz.*
            *Revue Statistique Tchécoslovaque*, Sesit. 8–10, pp. 471–81.
        1933.
(31) GALVANI, L.   *Punti di Contatto e Scambi di Concetti tra la
        Statistica e la Matematica.*
            *Giorn. Ist. Ital. Attuari*, Vol. IV, No. 3, pp. 402–14.   1933.
(32) GALVANI, L.   *Sulla Determinazione del Centro di Gravità e del
        Centro Mediano di una Popolazione, con Applicazione alla
        Popolazione Italiana Censita il 1° Decembre 1921.*
            *Metron*, Vol. XI, N. 1, pp. 15–47.   1933.
(33) GARWOOD, F.   The Probability Integral of the Correlation
        Coefficient in Samples from a Normal Bivariate Population.
            *Biometrika*, Vol. XXV, pp. 71–78.   1933.
(34) GEARY, R. C.   A General Expression for the Moments of
        Certain Symmetrical Functions of Normal Samples.
            *Biometrika*, Vol. XXV, pp. 184–6.   1933.
(35) GINI, C., BOLDRINI, L., GALVANI, L., VENERE, A.   *Sui Centri
        della Popolazione e Sulle Loro Applicazioni.*
            *Metron*, Vol. XI, N. 2, pp. 3–102.   1933.
(36) GLIVENKO, V.   *Sulla Determinazione Empirica delle Leggi di
        Probabilità.*
            *Giorn. Ist. Ital. Attuari*, Vol. IV, No. 1, pp. 92–99.   1933.

(37) GOLDZIHER, R.   *Contributi alla Teoria della Funzione Logistica.*
       *Giorn. Ist. Ital. Attuari*, Vol. IV, No. 4, pp. 530–61.  1933.
(38) GRIFFIN, F. L.   The Centre of Population for Various Continuous
       Distributions of Population over Areas of Various Shapes.
       *Metron*, Vol. XI, N. 1, pp. 1–15.  1933.
(39) GRUZEWSKA, H. M.   The Precision of the Weighted Average.
       *Annals Math. Statistics*, Vol. IV, No. 3, pp. 196–215.
       1933.
(40) GULDBERG, A.   " *Ist die Normale Stabilität Empirisch Nach-
       weisbar.*"
       *Tôhoku Math. Journal*, Vol. 37, pp. 127–32.  1933.
(41) GUMBEL, E. J.   *Die Gaussche Verteilung der Gestorbenen.*
       *Jahrbücher für Nationalökonomie und Statistik*, 138 Band—
       III Folge, Band 83, pp. 365–89.  1933.
(42) GUMBEL, E. J.   *Die Verteilung der Gestorbenen und das Normal-
       alter.*
       *Aktuárské Védy* IV, 2, pp. 65–96.  1933.
(43) GUMBEL, E. J.   *La signification des constantes dans la formule
       de Gompertz-Makeham.*
       *Comptes Rendus*, t. 196, p. 592.  1933.
(44) GUMBEL, E. J.   *Erreur moyenne et moyenne arithmétique.  Age
       moyen des vivants et age moyen de la mort.*
       *Comptes Rendus*, t. 196, pp. 1710–12.  1933.
(45) GUMBEL, E. J.   *La plus petite valeur parmi les plus grandes.*
       *Comptes Rendus*, t. 196, pp. 1857–8.  1933.
(46) GUMBEL, E. J.   *La plus petite valeur parmi les plus grandes, et
       la plus grande valeur parmi les plus petites.*
       *Comptes Rendus*, t. 197, pp. 965–7.  1933.
(47) GUMBEL, E. J.   *La distribution limite de la plus petite valeur
       parmi les plus grandes.*
       *Comptes Rendus*, t. 197, pp. 1082–4.  1933.
(48) GUMBEL, E. J.   *La distribution limite de la plus grande valeur
       parmi les plus petites.*
       *Comptes Rendus*, t. 197, pp. 1381–2.  1933.
(49) GUMBEL, E. J.   *L'espérance mathématique de la $m^{ieme}$ valeur.*
       *Comptes Rendus*, t. 198, pp. 33–5.  1934.
(50) HARZER, P.   *Tabellen für alle Statistischen Zwecke.*
       *Abhandlungen der Bayerischen Akademie der Wissenschaften,
       Mathematisch-naturwissenschaftliche Abteilung, Neue Folge,*
       Heft 21, 1933.
(51) HOGBEN, L.   The Effect of Consanguineous Marriage on the
       Metrical Character of the Offspring.
       *Proc. Roy. Soc. Edin.*, Vol. LIII, Part III, pp. 239–51.
       1933.
(52) HOJO, T.   A Further Note on the Relation between the Median
       and the Quartiles in Small Samples from a Normal Population.
       *Biometrika*, Vol. XXV, pp. 79–90.  1933.
(53) HOTELLING, H.   Analysis of a Complex of Statistical Variables
       into Principal Components.
       Reprinted from *Journal of Educational Psychology*, Sept.
       Oct. 1933, *Warwick and York Inc. Baltimore*, 1933.

(54) INGHAM, A. E. An Integral which occurs in Statistics.
*Proc. Camb. Phil. Soc.*, Vol. XXIX, Part 2, pp. 271–6. 1933.

(55) JACOB, M. *Sullo Sviluppo di una Curva di Frequenza in Serie di Charlier Tipo B.*
*Giorn. Ist. Ital. Attuari*, Vol. IV, No. 2, pp. 221–34. 1933.

(56) JORDAN, C. *Problema delle Prove Ripetute a Più Variabili Indipendenti.*
*Giorn. Ist. Ital. Attuari*, Vol. IV, No. 3, pp. 351–68. 1933.

(57) JORDAN, C. *Inversione della Formula di Bernouilli Relativa al Problema delle Prove Ripetute a Più Variabili.*
*Giorn. Ist. Ital. Attuari*, Vol. IV, No. 4, pp. 505–13. 1933.

(58) JOSEPH, A. W. The Sum and Integral of the Product of Two Functions.
*J. Inst. Act.*, Vol. LXIV, Part II, pp. 329–49. 1933.

(59) KARN, M. N. A Further Study of Methods of Constructing Life Tables when Certain Causes of Death are Eliminated.
*Biometrika*, Vol. XXV, pp. 91–101. 1933.

(60) KERMACK, W. O., and McKENDRICK, A. G. Further Studies of the Problem of Endemicity.
*Proc. R.S.*, A. 141, pp. 94–122. 1933.

(61) KOEPPLER, H. *Equazioni alle Derivative Parziali della Teoria delle Probabilità che Intervengono anche nella Teoria del Calore.*
*Giorn. Ist. Ital. Attuari*, Vol. IV, No. 2, pp. 245–67. 1933.

(62) KOLMOGOROFF. *Sulla Determinazione Empirica di una Legge di Distribuzione.*
*Giorn. Ist. Ital. Attuari*, Vol. IV, No. 1, pp. 83–91. 1933.

(63) KOLODZIEJCZYK, S. *Sur l'erreur de la seconde catégorie dans le problème de M. Student.*
*Comptes Rendus*, t. 197, pp. 814–16. 1933.

(64) KOSHAL, R. S. Application of the Method of Maximum Likelihood in the Improvement of Curves Fitted by the Method of Moments.
*J. Roy. Stat. Soc.*, Vol. XCVI, pp. 303–13. 1933.

(65) LESSER, P. C. V. Note on the Shrinkage of Physical Characters in Men and Women with Age, as an Illustration of the Use of $\chi^2 - P$ Methods.
*Biometrika*, Vol. XXV, pp. 197–202. 1933.

(66) LIDSTONE, G. J. Note on Orthogonal Polynomials and their Application to Least Square Methods,
(1) Of Fitting Polynomial Curves to Data.
(2) Graduation by Weighted Means.
*J. Inst. Actuaries*, Vol. LXIV, Part II, pp. 128–59. 1933.

(67) LINDERS, E. J. *Über die Berechnung des Schwerpunkts und der Trägheitsellipse einer Bevölkerung.*
*Metron*, Vol. XI, N. 1, pp. 1–10. 1933.

(68) LORENZ, P. *Über Näherungs-parabeln hohen Grades und ihre Aufgabe in der Konjunkturforschung.*
*Metron*, Vol. X, N. 4, pp. 61–78. 1933.

(69) LOTKA, A. J. *Applications de l'analyse au phenomène démographique.*
*J. Soc. Stat. de Paris*, Nov. 1933.

(70) LOTKA, A. J.   Industrial Replacement.
      *Skandinavisk Aktuarietidskrift*, pp. 51–63.   1933.
(71) LOWTHER, H. P., Jr.   The Extended Probability Theory for
      the Continuous Variable with Particular Applications to the
      Linear Distribution.
      *Annals Math. Statistics*, Vol. IV, No. 4, pp. 241–62.   1933.
(72) McKAY, A. T.   The Distribution of $\sqrt{\beta_1}$ in Samples of Four
      from a Normal Universe.
      *Biometrika*, Vol. XXV, pp. 204–10.   1933.
(73) McKAY, A. T.   Distribution of $\beta_2$ in Samples of 4 from a Normal
      Universe.
      *Biometrika*, Vol. XXV, pp. 411–15.   1933.
(74) McKAY, A. T., and PEARSON, E. S.   A Note on the Distribution
      of Range in Samples of $n$.
      *Biometrika*, Vol. XXV, pp. 415–20.   1933.
(75) MAZZONI, P.   *Sulle Aree Moltiplicabili del Cantelli.*
      *Giorn. Ist. Ital. Attuari*, Vol. IV, No. 1, pp. 100–108.   1933.
(76) MERRELL, M.   On Certain Relationships between $\beta_1$ and $\beta_2$ for
      the Point Binomial.
      *Annals Math. Statistics*, Vol. IV, No. 3, pp. 196–214.   1933.
(77) MERZRATH, E.   *Anpassung von Flächen an zweidimensionale
      Kollektivgegenstände und ihre Auswerkung für die Korrelations
      theorie.*
      *Metron*, Vol. XI, N. 2, pp. 103–36.   1933.
(78) MESSINA, I.   *Un Teorema sulla Legge Uniforme dei Grandi
      Numeri.*
      *Giorn. Ist. Ital. Attuari*, Vol. IV, No. 1, pp. 116–28.   1933.
(79) NEYMAN, J.   An Outline of the Theory and Practice of
      Representative Method Applied in Social Research.   *Polish
      Institute of Social Problems.   Actuarial Series*, No. 1, pp.
      1–123.   Warsaw, 1933.
(80) NEYMARCK, P.   *"Applications de l'analyse au phenomène démo-
      graphique."*
      *J. Soc. Stat. de Paris*, Vol. 74, pp. 336–45.   1933.
(81) O'TOOLE, A. L.   On the System of Curves for which the Method
      of Moments is the Best Method of Fitting.
      *Annals Math. Statistics*, Vol. IV, No. 1, pp. 1–29.   1933.
(82) O'TOOLE, A. L.   A Method of Determining the Constants in the
      Bimodal, Fourth Degree, Exponential Functions.
      *Annals Math. Statistics*, Vol. IV, No. 2, pp. 79–93.   1933.
(83) O'TOOLE, A. L.   On the Degree of Approximation of Certain
      Quadrature Formulas.
      *Annals Math. Statistics*, Vol. IV, No. 2, pp. 143–53.   1933.
(84) PAE-TSI-YUAN.   On the Logarithmic Frequency Distribution
      and the Semi-Logarithmic Correlation Surface.
      *Annals Math. Statistics*, Vol. IV, No. 1, pp. 30–74.   1933.
(85) PANKRAZ, O.   *Sui Gruppi Statistici.*
      *Giorn. Ist. Ital. Attuari*, Vol. IV, No. 2, pp. 215–20.   1933.
(86) PEARSON, E. S.   Statistical Method in the Control and Standard-
      isation of the Quality of Manufactured Products.
      *J. Roy. Stat. Soc.*, Vol. XCVI, pp. 21–60.   1933.

(87) PEARSON, E. S., and WILKS, S. S.  Methods of Statistical
     Analysis Appropriate for $k$ Samples of Two Variables.
        *Biometrika*, Vol. XXV, pp. 353–78.  1933.
(88) PEARSON, K.  Note on the Fitting of Frequency Curves.
        *Biometrika*, Vol. XXV, pp. 13–16.  1933.
(89) PEARSON, K.  " On the Parent Population with Independent
     Variates which gives the Minimum Value of $\phi^2$ for a Given
     Sample."
        *Biometrika*, Vol. XXV, pp. 134–46.  1933.
(90) PEARSON, K.  " On the Application of the Double Bessel
     Function $K_{\tau_1, \tau_2}(x)$ to Statistical Problems."
        *Biometrika*, Vol. XXV, pp. 158–78.  1933.
(91) PEARSON, K.  Note on Mr. Palin Elderton's Corrections to the
     Moments of J-Curves.
        *Biometrika*, Vol. XXV, pp. 180–84.  1933.
(92) PEARSON, K.  Note on McKay's Paper.
        *Biometrika*, Vol. XXV, pp. 210–13.  1933.
(93) PEARSON, K.  On a Method of Determining whether a Sample
     of Size $n$ supposed to have been drawn from a Parent Popula-
     tion having a known Probability Integral has probably been
     drawn at Random.
        *Biometrika*, Vol. XXV, pp. 379–410.  1933.
(94) PERLO, V.  On the Distribution of Student's Ratio for Samples
     of Three drawn from a Rectangular Population.
        *Biometrika*, Vol. XXV, pp. 203–4.  1933.
(95) PIAGGIO, H. T. H.  Three Sets of Conditions Necessary for the
     Existence of a " $g$ " that is Real and Unique except in Sign.
        *Brit. J. Psych.*, Vol. XXIV, Part I, pp. 88–105.  1933.
(96) PIAGGIO, H. T. H.  Mathematics and Psychology.
        *Mathematical Gazette*, Vol. XVII, No. 222, pp. 36–42.
     1933.
(97) RIDER, P. R.  Criteria for Rejection of Observations.
     Washington University Studies—New Series.  *Science and
     Technology*, No. 8, Oct. 1933.
(98) RIGBY, C. M.  On a Recurrence Relation Connected with the
     Double Bessel Functions, $K_{\tau_1, \tau_2}(x)$ and $T_{\tau_1, \tau_2}(x)$.
        *Biometrika*, Vol. XXV, pp. 420–1.  1933.
(99) ROBINSON, S.  An Experiment Regarding the $\chi^2$ Test.
        *Annals Math. Statistics*, Vol. IV, No. 4, pp. 285–87.  1933.
(100) ROMANOVSKY, V.  On a Property of the Mean Ranges in
     Samples from a Normal Population and on Some Integrals
     of Professor T. Hojo.
        *Biometrika*, Vol. XXV, pp. 195–97.  1933.
(101) ROYER, E. B.  A Simple Method for Calculating Mean Square
     Contingency.
        *Annals Math. Statistics*, Vol. IV, No. 1, pp. 75–78.  1933.
(102) SANSONE, G.  *La Chiusura dei Sistemi Ortogonali di Legendre, de
     Laguerre e di Hermite rispetto alle Funzioni di Quadrato
     Sommabile.*
        *Giorn. Ist. Ital. Attuari*, Vol. IV, No. 1, pp. 71–82.
     1933.

(103) SCATES, D. E.   Locating the Median of the Population in the United States.
    *Metron*, Vol. XI, N. 1, pp. 49–65.   1933.

(104) SCHULTZ, H.   The Standard Error of the Coefficient of Elasticity of Demand.
    *J. Amer. Stat. Ass.*, Vol. XXVIII, pp. 64–69.   1933.

(105) SCHULTZ, T. W., and SNEDECOR, E.   Analysis of Variance as an Effective Method of Handling the Time Element in Certain Economic Statistics.
    *J. Amer. Stat. Ass.*, Vol. XXVIII, pp. 14–30.   1933.

(106) SIBIRANI, F.   *Intorno ai Problemi sulle Prove Ripetute.*
    *Giorn. Ist. Ital. Attuari*, Vol. IV, No. 2, pp. 207–10.   1933.

(107) STOUFFER, S., and TIBBITS, C.   Tests of Significance in Applying Westergaard's Method of Expected Cases to Sociological Data.
    *J. Amer. Stat. Ass.*, Vol. XXVIII, No. 183, pp. 293–302. 1933.

(108) THOMPSON, W. R.   On the Likelihood that one Unknown Probability Exceeds Another in View of the Evidence of Two Samples.
    *Biometrika*, Vol. XXV, pp. 286–94.   1933.

(109) VAUGHAN, H.   Summation Formulas of Graduation with a Special Type of Operator.
    *J. Inst. Actuaries*, Vol. XLIV, Part III, pp. 428–48.   1933.

(110) WATKINS, G. P.   An Ordinal Index of Correlation.
    *J. Amer. Stat. Ass.*, Vol. XXVIII, pp. 139–51.   1933.

(111) WICKSELL, S. D.   On Correlation Functions of Type III.
    *Biometrika*, Vol. XXV, pp. 121–33.   1933.

(112) WILSON, E. B., and PUFFER, R. R.   Least Squares and Population Growth.
    *Proc. Amer. Acad. Arts and Science*, Vol. 68, No. 9.   1933.

(113) WISHART, J.   A Comparison of the Semi-Invariants of the Distributions of Moment and Semi-Invariant Estimates in Samples from an Infinite Population.
    *Biometrika*, Vol. XXV, pp. 52–60.   1933.

(114) WISHART, J., and BARTLETT, M. S.   The Generalised Product Moment Distribution in a Normal System.
    *Proc. Camb. Phil. Soc.*, Vol. XXIX, Part 2, pp. 260–70. 1933.

(115) YATES, F.   The Principles of Orthogonality and Confounding in Replicated Experiments.
    *J. Agr. Sci.*, Vol. XXIII, Part I, pp. 108–45.   1933.

(116) YATES, F.   The Analysis of Replicated Experiments when the Field Results are Incomplete.
    *Empire Journal of Experimental Agriculture*, Vol. I, No. 2, pp. 129–42.   1933.

(117) YATES, F.   The Formation of Latin Squares for Use in Field Experiments.
    *Empire Journal of Experimental Agriculture*, Vol. I, No. 3. Sept. 1933.

OTHER REFERENCES.

B.—Section I.

(1) CRAIG, C. C.  An Application of Thiele's Semi-Invariants to the Sampling Problem.
*Metron*, Vol. 7, N. 4, pp. 3–74.  1928.
(2) FISHER, R. A.  Moments and Product Moments of Sampling Distributions.
*Proc. Lond. Math. Soc.*, Vol. 30, pp. 199–238.  1928.
(3) FISHER, R. A.  The Moments of the Distribution for Normal Samples of Measures of Departure from Normality.
*Proc. Roy. Soc.*, A, Vol. 130, pp. 16–28.  1930.
(4) KONDO, T.  A New Method of Finding Moments of Moments.
*Tôhoku Math. Journal*, Vol. 355, pp. 142–70.  1932.
(5) ST. GEORGESCU, N.  Contributions to the Sampling Problem.
*Biometrika*, Vol. 24, pp. 65–107.  1932.

C.—Section II.

(1) FISHER, R. A.  The General Sampling Distribution of the Multiple Correlation Coefficient.
*Proc. Roy. Soc.*, A, Vol. 121, pp. 669–70.  1928.
(2) ROMANOVSKY, V.  Generalisation of Some Types of the Frequency Curves of Professor Pearson.
*Biometrika*, Vol. XVI, pp. 114–16.  1924.

D.—Section III.

(1) AITKEN, A. C.  On the Graduation of Data by the Orthogonal Polynomials of Least Squares.
*Proc. Roy. Soc. Edin.*, Vol. 53, pp. 54–78.  1933.
(2) ALLAN, F. E.  The General Form of the Orthogonal Polynomials for Simple Series, with Proofs of their Simple Properties.
*Proc. Roy. Soc. Edin.*, Vol. 50, pp. 310–20.  1930.
(3) DAVIS, H. T., and LATSHAW, V. V.  Formulas for the Fitting of Polynomials to Data by the Method of Least Squares.
*Annals of Mathematics*, 2nd Series, Vol. 31, No. 1, pp. 52–78.  1930.

E.—Section IV.

(1) FISHER, R. A.  *Statistical Methods for Research Workers*, 3rd Edn., p. 163.  1920.
(2) NEYMAN, J., and PEARSON, E. S.  On the Problem of $k$ Samples.
*Bulletin de l'Académie Polonaise des Sciences et des Lettres*, A, pp. 960–81.  1931.
(3) WILKS, S. S.  Certain Generalizations in the Analysis of Variance.
*Biometrika*, Vol. XXIV, p. 476.  1932.

F.—Section V.

(1) BAILEY, V. A.  The Interaction between Hosts and Parasites.
*Quarterly Journal of Mathematics*, Vol. 2, No. 5, pp. 68–77.  1931.

(2) GINI, C., and GALVANI, L.   *Di Talune Estensioni dei Concetti di Media ai Caratteri Qualitativi.*
   *Metron*, Vol. VIII, N. 1-2, p. 136.   1929.
(3) KARN, M. N.   An Enquiry into Various Death Rates and the Comparative Influence of Certain Diseases on the Duration of Life.
   *Annals of Eugenics*, Vol. IV, pp. 279–326.   1931.
(4) KERMACK, W. O., and McKENDRICK, A. G.   Contributions to the Mathematical Theory of Epidemics.   II. The Problem of Endemicity.
   *Proc. Roy. Soc.*, A, Vol. 138, pp. 55–83.   1932.
(5) TODHUNTER, I.   *History of the Theory of Probability*, pp. 423–24.   1865.
(6) TORRICELLI, E.   *Stabilimento tipo-litografico G. Montanari*, p. 90 *et seq.*   Faenza, 1919.

G.—Section VI.

(1) BOREL, E.   *Principes et formules classique du calcul des probabilités (Traité du calcul des probabilités et de ses applications, Tome 1, fascicule 1)*, pp. 1–3.   Paris, 1925.
(2) BURNSIDE, W.   *Theory of Probability*, p. 11.   Cambridge, 1928.
(3) CANTELLI, F. P.   *Una Teoria Astratta del Calcolo delle Probabilità.*
   *Giornale dell' Istituto Italiano degli Attuari*, III, pp. 257–65.   1932.
(4) FRÉCHET, M.   *L'Arithmétique de l'Infini*, p. 12.   Paris, 1934.
(5) LOMNICKI, A.   *Nouveaux Fondements du Calcul des Probabilités.*
   *Fundamenta Mathematicae*, IV, pp. 34–71.   1923.

H.—Section VII.

(1) FISHER, R. A.   The Mathematical Foundations of Theoretical Statistics.
   *Proc. Roy. Soc.*, A, Vol. 222, pp. 355–56.   1922.

# 16b

II. *Moments and Semi-Invariants of Sampling Distributions.*

By W. G. Cochran.

The work in 1934 which comes under this heading is small in volume. In particular no paper has come to hand dealing with the general problem of finding the sampling distributions of the usual estimates from a sample of moments and semi-invariants, a subject on which several important ones have appeared in the past few years.

In a paper (C 3) which appeared in 1932, and was reviewed in our article covering that year, Wilks considered the generalization for samples from a multivariate normal population of some well-known statistics, such as, for example, the correlation ratio, which had previously been defined only for bivariate or univariate populations. The problem of finding the sampling distributions of these statistics, when generalized, involves some rather complicated algebra, and in the paper mentioned their moments were found by a method involving variation of the sample number. An alternative method of solving this and similar problems under rather more general conditions is presented in a more recent paper (A 120). To follow the argument developed there in any detail would require too much space, but a general idea of the method of attack will be given.

For convenience the suffices $ij$ will be assumed to take the values $(1, 2 \ldots, t)$, the suffices $pq$ the values $(t + 1, t + 2, \ldots n)$ and $rs$ the values $(1, 2, \ldots, n)$. The simultaneous distribution of the variates $x_i$ is taken to depend on an additional set of variates $x_p$ in such a way that the probability law of the variates $x_i$ is a multivariate normal law of the form

$$f(\bar{x}_i) = ke^{-\Sigma A_{rs}x_r x_s}d\bar{x}_i \quad . \quad . \quad . \quad . \quad (45)$$

where $A_{rs}$ is symmetric and $A_{ij}$ positive definite, $d\bar{x}_i$ standing for the product of the differentials $dx_i$. The variates $x_p$ are introduced so that the results cover the case, which may arise in regression problems, where some of the independent variates are fixed. Let a sample of $N$ individuals be drawn from the population (45) and in the $\alpha$th drawing let $x_r$ take the value $x_{ra}$. Then the probability law of the sample is

$$e^{-\Sigma A_{rs}x_{ra}x_{sa}}d\bar{x}_{ia} \quad . \quad . \quad . \quad . \quad (46)$$

except for a constant factor. Write $a_{rs} = \overset{N}{\underset{a=1}{\Sigma}} x_{ra}x_{sa}$. By integrating the expression (46) over all values of $x_{ia}$ ($\alpha = 1, 2, \ldots N$), Wilks obtains the relation

$$\int e^{-\Sigma A_{rs}a_{rs}}d\bar{x}_{ia} = \Pi^{\frac{Nt}{2}}A^{-\frac{N}{2}}e^{-\Sigma B_{pq}a_{pq}} \quad . \quad . \quad . \quad (47)$$

where $A = |A_{ij}|$ and $B_{pq} = A_{pq} - \underset{ij}{\Sigma}A_{ip}A_{qj}A^{ij}$, $A^{ij}$ being the inverse of $A_{ij}$. The relation (47) is of course simply an expression giving the value of the constant factor in (46). The generalizations with which Wilks is concerned can be expressed as functions of determinants similar in structure to $|a_{rs}|$ and its principal minors. Hence the mathematical problem to be solved is to find the mean values of functions of the form $\overset{m}{\underset{\beta=1}{\Pi}}|a_{\mu_\beta \nu_\beta}|^{k_\beta}$ where the set of determinants is taken from $|a_{rs}|$ and its principal minors. This is equivalent to finding

$$\int \overset{m}{\underset{\beta=1}{\Pi}}|a_{\mu_\beta \nu_\beta}|^{k_\beta}e^{-\Sigma A_{rs}a_{rs}}d\bar{x}_{ia}. \quad . \quad . \quad . \quad (48)$$

The method of attack is as follows : the equation (47) holds provided $A_{ij}$ is positive definite. Wilks makes a transformation of the matrix $A_{rs}$ such that (i) $A_{ij}$ when transformed remains positive definite, (ii) a set of parameters $\xi$ is introduced. Both sides of (47) are integrated with respect to the $\xi$'s from $-\infty$ to $+\infty$, and by suitable choice of the transformation this yields on the left-hand side of (47) an expression of the form (48) for $m = 1$ and $k_\beta = -\frac{1}{2}$ and on the right-hand side of (47) an integral which can be evaluated. The operator which produces this transformation is called by Wilks a moment generating operator. The case $m = 1$, $k_\beta = -(\lambda/2)$ (where $\lambda$ is a positive integer) is obtained by repeated applications of the same operator and the case $m = 1$, $k_\beta = +(\lambda/2)$ by definition of an inverse operator. The general case $m > 1$ is solved by applying a set of operators in succession.

In the sections which deal with applications of the method, Wilks obtains the moments of (i) a generalized variance of deviations from regression functions, (ii) the generalization of "Student's" ratio, (iii) the multiple correlation coefficient, and (iv) the generalized Neyman–Pearson $\lambda_H$ criterion for $k$ samples.

In two short notes Ayyangar (A 3), (A 4) provides easy proofs of the recurrence formulæ obtained by Frisch (C 1) and K. Pearson (C 2) for the moments of the point binomial and of the hypergeometric series respectively, deducing in the process one or two results which are new. The methods of attack used have the advantage of being applicable to complete and incomplete moments alike.

E. S. Martin (A 73) considers the problem of fitting by the method of moments a curve to a set of grouped data in which the start of the frequency distribution is not known. An example occurs, for instance, in the frequency distribution of the value of houses for income-tax purposes. As officially published this is given in the frequency groups £0–20, £20–40, etc., but no house is valued at nothing per annum. If the data are grouped in intervals of size $h$, the method assumes that the range of the first group is actually $\lambda_h$. Three methods are suggested for finding $\lambda$. In the first two, a curve of the fifth degree is fitted to the first five ordinates of $Z$, the integral of the frequency distributions from $x$ upwards. This determines the constants of the curve in terms of $\lambda$, and the value of $\lambda$ is obtained either by making the observed and calculated sixth ordinates agree as closely as possible, or by using the fact that the original frequency distribution starts with a zero ordinate. In the third method $\lambda$ is found by fitting a curve of the form

$$Z = N + Ax^q e^{ax + bx^2 +} \ldots q > 0 \quad . \quad . \quad . \quad (49)$$

where the number of constants fitted is varied according as 3, 4, 5, 6 or 7 observed frequencies are used. Tables are given to facilitate the numerical computation of $\lambda$ in each case. None of the methods is in all cases superior to the others but the third appears to give generally the best results, and can be used for either asymptotic or non-asymptotic frequency distributions. The value of $\lambda$ having been estimated, the moments of the observed distribution are corrected for the fact that the first group is of length $\lambda h$ instead of $h$ and the curve fitting proceeds. The examples given are numerous and well chosen.

IV. *Orthogonal Polynomial Theory and Least Squares.*

By W. G. Cochran and J. O. Irwin.

Papers by W. Andersson (A 1) and Dieulefait (A 40) both deal with the subject of non-linear regression. Several writers, amongst whom may be mentioned K. Pearson (E 2) and Neyman (E 1) have previously given a mathematical presentation of general formulæ for estimates of the unknown parameters in a polynomial fitted as a regression line by least squares. In fitting a polynomial of degree $n$, estimated moments of order $2n$ in the independent variates $x$ are required. When the problem is looked at from the standpoint of the bivariate frequency distribution, these estimated high order moments are known to have large sampling variances. Accordingly Wicksell (E 3) suggested in 1930 that the moments of a suitably chosen frequency function might be inserted in the least square expressions instead of the estimated higher marginal or $x$ moments, that is, before fitting the regression some simple form of distribution should be fitted to the observed values of $x$. The solution obtained by this method does not give a strict least-squares adjustment of the observed array means of $y$. Wicksell suggested, however, that by a slight adjustment the method might be turned into a strict least-square solution.

The paper by Andersson (A 1) shows how this adjustment may be made, working out the normal and Pearson type III cases as an example. He gives three numerical examples, and from the results of these considers that there is little, if anything, to be gained by making the adjustment.

The paper by Dieulefait (A 40) first obtains, by means of the theory of orthogonal functions, some results by previous writers on the expansion of a univariate frequency function in series, of which the Gram–Charlier series are the best known. In the part which claims to be new, he considers an expansion of the bivariate frequency function $f(x, y)$ in a series of the form

$$f(x, y) = \psi(x)\phi(y)[1 + \Sigma_{sj} w_{sj} X_s(x) Y_j(y)] \quad . \quad . \quad (58)$$

where $\psi(x)$, $\phi(y)$ are the marginal frequency distributions of $x$ and $y$ respectively, and $X_s(x)$, $Y_j(y)$ are sets of functions orthogonal to $\psi(x)$ and $\phi(y)$ respectively. The idea behind this expansion is that by specifying particular forms for $\psi(x)$ and $\phi(y)$ a development in series of a bivariate frequency function will be obtained analogous to the well-known developments of a univariate function. From this expansion he deduces formal expressions for the regression line of $y$ on $x$, the correlation ratio $\eta_{yx}$, $\sqrt{\beta_1(y_x)}$ and $\beta_2(y_x)$. The paper as it stands belongs rather to the subject of pure mathematics than to statistics, but a further paper on applications is promised. Dieulefait admits, however, that the advantages of expansions of the form (58) from the practical point of view of curve-fitting are open to dispute.

One of the most general discussions of the method of least squares that we have seen, is given by Deming (A 34–36). He supposes that there are $n$ observations on $q$ different kinds of quantities, $x_1, x_2 \ldots x_n$; $x_{n+1}, x_{n+2} \ldots x_{2n}$; $\ldots$; $\ldots x_{qn}$. For example, $x_1 \ldots x_n$ might be pressure observations, $x_{n+1} \ldots x_{2n}$ might be volume observations, $x_{2n+1} \ldots x_{3n}$ might be temperature observations or $x_1 \ldots x_n$ might be $n$ observations of height, $x_{n+1} \ldots x_{2n}$ might be $n$ observations of body weight. $x_i$ is supposed to have the weight $w_i$ and $r_i$ to be its adjusted value. Then the general problem of the method of least squares is to minimize

$$\Sigma_i w_i(r_i - x_i)^2 \quad . \quad . \quad . \quad . \quad . \quad (59)$$

where there may be a number of equations of condition between the $r$'s and (say) $p$ unknown parameters $a, b \ldots p$, and if the parameters are not all independent, there will be equations of condition between these also. The equations of condition are expressed by the relations

$$F^h(r_1, r_2 \ldots r_{qn}; a, b \ldots, p) = 0 \quad h = 1, 2 \ldots m \quad . \quad . \quad (60)$$

$$F^h(a, b \ldots p) = 0 \quad h = m + 1, m + 2 \ldots m + l \quad . \quad (60 \, bis)$$

By the use of approximate values of the parameters the equations of condition may be made linear in form and can then be minimized in the usual way by the use of indeterminate multipliers, leading to a series of linear equations in the $r$'s, the parameters and the indeterminate multipliers. If some of the observations are without error, the situation can be met by making their weights infinite and $r_i - x_i$ zero. The corresponding terms will disappear from (59).

As an example we may take the well-known case of fitting a straight line $y = a + bx$ to $n$ points. In this case $y_h = x_{n+h}$ $(h = 1, 2 \ldots n)$ and there are $n$ equations of condition

$$r_{n+h} = a + br_h \quad (h = 1, 2 \ldots n) \quad . \quad . \quad . \quad (61)$$

If the $x$'s are not subject to error

$$\left. \begin{array}{c} w_h = \infty \\ r_h - x_h = 0 \end{array} \right\} \quad h = (1, 2 \ldots n) \quad . \quad . \quad . \quad (62)$$

and we are led to the ordinary expression for the regression line.

Deming gives a number of good examples. The effect of weighting is illustrated by fitting the circle

$$(x - a)^2 + (y - b)^2 = c^2$$

to five points (i) when the $x$'s and $y$'s have equal weights, (ii) when the $y$'s only are subject to error, (iii) when the $x$'s only are subject to error.

Particularly interesting, especially to physicists, is the treatment of the laws represented by

$$ya^x = b, \quad yz^x = b, \quad yz^x = w$$

where $a, b$ are parameters and $x, y, z, w$ are observed co-ordinates. The last three equations are the ones needed in the determination of $e$ and $h$ (the charge on an electron and Planck's constant) (i) when neither $e$ nor $h$ are directly observed (here $a = e, b = h$), (ii) when direct observations on $e$ are included (here $z = e, b = h$), (iii) when direct observations on both $e$ and $h$ are included (here $z = e, w = h$). The normal equations are set up and illustrated with and without forcing the satisfaction of the condition

$$\frac{hc}{2\pi e^2} = 137 \quad . \quad . \quad . \quad . \quad . \quad . \quad (63)$$

Deming also treats the problem of curve fitting, and shows that if a curve has been fitted to $n$ points, $V_x$, $V_y$ being the residuals, $n_x$, $n_y$ the number of observations made on each co-ordinate, $w_x$, $w_y$ the weights at the point $x$, $y$, then

$$\frac{1}{\sigma^2}\Sigma(w_x V_x{}^2 + w_y V_y{}^2) \quad . \quad . \quad . \quad . \quad (64)$$

is distributed as $\chi^2$ with $(n - p)$ degrees of freedom, $p$ being the number of parameters in the empirical formula for the curve and $\sigma$ the root mean square error of an observation of unit weight.

If $\sigma$ is known, the $\chi^2$ test provides a test of goodness of fit for the empirical formula, if $\sigma$ is not known by assuming that $\chi^2$ has its average value $(n - p)$, we obtain the estimate of $\sigma$

$$\frac{1}{(n - p)}\Sigma(w_x V_x{}^2 + w_y V_y{}^2) \quad . \quad . \quad . \quad (65)$$

This is in fact what is usually done in text-books on least squares. It assumes that the empirical formula selected does in fact fit. In the course of the investigation it is shown that if $U_x$, $U_y$ are the true errors at the point $x$, $y$ then

$$\Sigma(w_x U_x V_x + w_y U_y V_y) = 0 \quad . \quad . \quad . \quad . \quad (66)$$

The remaining papers on this subject, which have come to our notice, deal with points of an elementary character. One may be mentioned.

Smith (A 103) considers the problem of fitting a polynomial to a set of data, where the criterion to be used is that the greatest deviation between any observation and the curve shall be as small as possible. The least square solution does not in general satisfy this criterion. The data are first graphed, and the points which appear to lie farthest from a curve of the order which is to be fitted are selected by eye. The number of outlying points chosen is two more than the degree of the curve, and they are taken to lie alternately on opposite sides of the curve. If $y_1, \ldots, y_{n+1}$ are the points chosen and $f(x)$ is the polynomial, of degree $(n - 1)$, Smith shows that the polynomial for which the maximum deviation is least is that which makes

$$y_1 - f(x_1) = f(x_2) - y_2 = \ldots = \varepsilon \quad . \quad . \quad . \quad (67)$$

This polynomial may be easily found. For if $F(y_1, y_2 \ldots y_{n+1}, x)$ is the Lagrangian polynomial of order $n$ through the outlying points,

$$f(x) = F(y_1, y_2 \ldots y_{n+1}, x) - \varepsilon F(1, -1, 1 \ldots) \quad . \quad . \quad (68)$$

where $\varepsilon$ is so chosen that the coefficient of $x^n$ in $f(x)$ vanishes.

Explicit expressions for $\varepsilon$ and the coefficients in $f(x)$ are given.

When the curve has been fitted it may be verified whether the correct set of outlying points has been chosen, and if the maximum deviation is considered too large a curve of higher order may be fitted by the same method.

VI. *Analysis of Variance.*

By W. G. COCHRAN.

A subject which attracted considerable attention in 1934 was the use of the analysis of covariance, first introduced by Fisher in the fourth edition of his book *Statistical Methods for Research Workers.* Of the eight papers in this section which have come to our notice during 1934, four contain references to this topic, two on the theoretical side and two illustrating practical applications; and several papers which are not reviewed below also exhibited the use of this new and powerful weapon.

Fisher and Yates (A 49) give the details of their method of enumerating the 6 × 6 Latin Squares, a task which had not previously been correctly performed. This is done by first finding the number of *reduced squares,* a *reduced square* being a square in which the first row and column have their letters in the order *ABCDEF.* Since each reduced square generates 6 ! 5 ! distinct squares by permuting all rows except the first and all columns, the number of reduced squares has only to be multiplied by 86,400 to give the total number of 6 × 6 Latin squares. The enumeration of the number of reduced squares without excessive labour was carried out by making use of the properties of a type of transformation called intramutation, which consists of permuting the letters other than *A* in a reduced square and then rearranging the rows and columns so as to give another reduced square. The enumeration was carried out in three stages : (i) the exhaustive enumeration of all possible types of leading diagonal, (ii) the determination of the number of distinct diagonals which can be generated by intramutation from each typical diagonal, (iii) the enumeration by trial of all possible reduced squares having the given typical diagonals. The same number of reduced squares will be derivable from all diagonals which can be generated by intramutation from the typical diagonal under consideration, so that these operations enable the number of reduced squares to be obtained. The number is 9,408, arranged in 111 intramutation sets.

In presenting the 812,851,200 6 × 6 Latin squares, a more

16b.737

16b.749

general type of transformation was used, all the members of a transformation set being derivable from the typical member shown by the permutation of all rows, all columns and all letters. This enabled the presentation to be made by means of 17 examples, involving only 12 distinct types of square. This presentation, together with a method of picking a square at random from the set of all squares, had already been given by one of the authors (G 5).

The problem of the enumeration of $6 \times 6$ Latin squares was first discussed by Euler (G 3) in 1782, while he was attempting to show that no $6 \times 6$ Græco-Latin square exists. It is easy to verify from Fisher and Yates's enumeration that this is so, but, remarkably enough, this is the first rigorous proof of the fact which has appeared.

Bartlett (A 7) gives a review in vector notation of all cases of the analysis of variance of one dependent variate. A sample of $n$ observations of a variate $x$ is written as the vector

$$S = (x^{(1)} \ldots x^{(n)}) \quad . \quad . \quad . \quad (116)$$

where
$$S^2 = SS' = \Sigma x^2 . \quad . \quad . \quad . \quad (117)$$

The operation of fitting a linear regression equation may be represented by regarding $S$ as being related to a matrix $Z$ of vectors $Z_1, \ldots, Z_p$ by the equation

$$S = BZ + V$$

where $B$ is a single row matrix of coefficients $b_i$ ($i = 1, \ldots p$). In geometric terms, $BZ = U$ say, is the component of $S$ in the " plane " of $Z_1, \ldots Z_p$, and $V$ the residual component. The most general case is obtained by writing $U = X + Y$, where $X$ is the component in the restricted " plane " $Z_1, \ldots Z_q$ ($q \leq p$) and is to be eliminated, and $Y$ is the remaining component. Thus

$$S = X + Y + V \quad . \quad . \quad . \quad . \quad (118)$$

and it follows at once that

$$S^2 = X^2 + Y^2 + V^2 . \quad . \quad . \quad (119)$$

Up to this point the paper has been concerned with establishing the *algebraic* independence of the components in an analysis. Bartlett next proves the necessary and sufficient vector condition for the normal law, which is that the chance of a vector $S$ depends only on its length and not on its direction. It follows at once that algebraic independence implies independence of frequency distributions in these cases. As an example of the interpretation of equation (119) Bartlett derives the exact test of significance for the treatment sum of squares in an analysis of covariance.

The results are then extended to the case of two or more correlated variates, in which $S$ becomes a matrix representing the set of vectors $S_1$, $S_2$. . . . The general equation remains in the form (118) but instead of (119) we have

$$SS' = XX' + YY' + VV' \quad . \quad . \quad . \quad (120)$$

If, for example, there were two variates $x$ and $y$, (120) stands for three equations, representing respectively the analysis of variance of $x$, that of the covariance $xy$ and that of the variance of $y$. The question of a single test of significance of $Y$, in cases where it may be of interest, is considered, the function

$$\Lambda = V^2/(Y + V)^2 . \quad . \quad . \quad . \quad (121)$$

being suggested.

Papers by Cochran (A 24) and Irwin (A 63) deal essentially with equation (119) above expressed in the ordinary algebraic form, for the particular cases of randomized blocks and the Latin square, both providing in these cases proofs of the mathematical theory underlying the analysis of variance. It is a well-known result that in an analysis of variance any set of degrees of freedom may be expressed as independent single degrees of freedom in an infinite number of ways. Irwin finds for the randomized blocks design a simple set of linear functions of the plot yields which are independent and make up the block, treatment and error degrees of freedom. For the Latin square there is in the general case no obvious simple set of single degrees of freedom for error, but one can be found easily when the size of the square is a prime, and Irwin obtains this for the cases $n = 3$ and $n = 5$. The second part of Cochran's paper considers equation (119), also from the algebraic point of view, in the case of the analysis of covariance, though the exact test of significance for the treatment sum of squares is not reached.

The question of the analysis of variance of two-way classifications with unequal numbers of objects in the different classes is of considerable practical importance, and papers on this subject by Brandt (G 2) and Yates (A 130) were reviewed previously (G 1). Where it is desired to test both main effects and their interaction against the variation within classes, a rigorous test is in general only possible by the rather laborious method of fitting constants. Snedecor (A 105) has suggested a method which is easier to apply and gives in most cases results practically equivalent to those obtained by fitting constants. If $n_{rs}$ is the number of objects in the $r$ throw and $s$th column, and $n_r . = \Sigma_s n_{rs}$, $n ._s = \Sigma_r n_{rs}$ and $n = \Sigma_{rs} n_{rs}$ are the totals in the $r$th row and $s$th column and the general total respectively, he replaces $n_{rs}$ by an expected number $\frac{n_r \cdot n_s \cdot}{n}$. The numbers of objects in the classes are still unequal, but have now the property of proportionality. As a result of this, the main effects may be obtained by a weighted sum of squares of totals and the interaction by subtraction. Further, in this case the additive property of sums of squares holds, so that the test for main effects is the same whether the interactions are assumed to exist or not. There are cases in which the process of replacing the actual numbers in the sub-classes by expected numbers has some theoretical justification, but in general the adjustment will be a numerical convenience and stand or fall by the closeness with which the results agree with those given by the method of fitting constants. An investigation on this point was made by Snedecor and Cox (A 106), who found good agreement except in one case, and in that the discrepancy between observed and expected numbers, as measured by the $\chi^2$ test, was highly significant.

A paper by Yates (A 132), describes the design, conduct and statistical analysis of a feeding experiment of the factorial type on pigs and provides a good example of how the introduction of analysis of covariance has enabled the design of such an experiment to be improved.

Sterne (A 108) gives a proof of the $t$ distribution for the ratio of the deviation of a fitted constant from its true value to the estimated residual standard deviation in a least squares solution. He claims to have established the $t$ distribution under more general conditions than those assumed in the proof by Fisher (G 4), but this does not appear to the reviewer to be the case.

## VIII. *Bibliography and References.*

### A. Current Literature.

(1) ANDERSSON, W.  On a new method of computing non-linear regression curves.
*Annals Math. Statistics*, Vol. 5, No. 2, pp. 81–106.  1934.

(2) AWBERY, J. H.  The determination of a parabolic formula to represent a series of observations.
*Proc. Physical Soc.*, Vol. 46, pp. 574–82.  1934.

(3) AYYANGAR, A. A. K.  Note on the recurrence formulæ for the moments of the point binomial.
*Biometrika*, Vol. XXVI, pp. 262–3.  1934.

(4) AYYANGAR, A. A. K.  Note on the incomplete moments of the hypergeometrical series.
*Biometrika*, Vol. XXVI, pp. 264–5.  1934.

(5) BAKER, G. A.  Transformation of non-normal frequency distributions into normal distributions.
*Annals Math. Statistics*, Vol. 5, No. 2, pp. 113–23.  1934.

(6) BARTLETT, M. S.  The problem in statistics of testing several variances.
*Proc. Camb. Phil. Soc.*, Vol. XXX, pp. 164–69.  1934.

(7) BARTLETT, M. S.  The vector representation of a sample.
*Proc. Camb. Phil. Soc.*, Vol. XXX, pp. 327–40.  1934.

(8) BARTLETT, M. S., and HALDANE, J. B. S.  The theory of inbreeding in autotetraploids.
*J. Genetics*, Vol. XXIX, No. 2, pp. 175–80.  1934.

(9) BATEN, W. D.  Combining two probability functions.
*Annals Math. Statistics*, Vol. 5, No. 1, pp. 13–20.  1934.

(10) BOLDYREFF, J. W.  Mathematical foundation for a method of statistical analysis of household budgets.
*Annals Math. Statistics*, Vol. 5, No. 3, pp. 216–26.  1934.

(11) BOSE, R. C.  On the application of hyperspace geometry to the theory of multiple correlation.
*Sankhyā*, Vol. I, pp. 338–44.  1934.

(12) BOSE, S.  Tables for testing the significance of linear regression in the case of time series and other single-valued samples.
*Sankhyā*, Vol. I, pp. 277–88.  1934.

(13) BOSE, S.  A note on the mathematical expectation of the value of the regression coefficient.
*Sankhyā*, Vol. I, pp. 432–4.  1934.

(14) BOYD, W. C.  Nomogram for rapid computation of the frequency of the blood grouping genes in populations.
*Human Biology*, Vol. 6, No. 3, pp. 558–61.  1934.

(15) BROWN, W.  The theory of two factors versus the sampling theory of mental ability.
*Nature*, Vol. 133, p. 724.  May 12, 1934.

(16) BURRAU, C.  Contribution to the problem of dissection of a given frequency curve.
*Nordic Statistical Journal*, Vol. 5, pp. 43–64.  1934.

(17) CAMP, B. H.  Spearman's general factor again.
*Biometrika*, Vol. XXVI, pp. 260–1.  1934.

(18) CANTELLI, F. P.  *Considérations sur la convergence dans le calcul des probabilités.*
*Annales de l'Institut Henri Poincaré*, Vol. V, pp. 1–50.  1935.

(19) CARVER, H. C.  Punched cards systems and statistics.
*Annals Math. Statistics*, Vol. 5, No. 2, pp. 153–60.  1934.

(20) CARVER, H. C.  A new type of average for security prices.
*Annals Math. Statistics*, Vol. 5, No. 1, pp. 73–80.  1934.

(21) CASTELLANO, V.  *Sullo Scarto quadratico medio della probabilità di transvariazione.*
*Metron*, Vol. XI, No. 4, pp. 19–75.  1934.

(22) CHEN-NAN LI.  Summation method of fitting parabolic curves and calculating linear and curvilinear correlation coefficients on a scatter-diagram.
*J. Amer. Stat. Ass.*, Vol. XXIX, pp. 405–9.  1934.

(23) CLOPPER, C. J., and PEARSON, E. S.  The use of confidence or fiducial limits illustrated in the case of the binomial.
*Biometrika*, Vol. XXVI, pp. 404–13.  1934.

(24) COCHRAN, W. G.  The distribution of quadratic forms in a normal system, with applications to the analysis of covariance.
*Proc. Camb. Phil. Soc.*, Vol. 30, pp. 178–91.  1934.

(25) ·CRAMÉR, H.  *Su un teorema relativo alla legge uniforme dei grandi numeri.*
*Giorn. Ist. Ital. Attuari*, Vol. V, No. 1, pp. 1–13.  1934.

(26) CRAMÉR, H.  Prime numbers and probability.
*Åttonde Skandinaviska Matematikerkongressen*, Stockholm. 1934.

(27) CRATHORNE, A. R.  *Moments de la binomiale par rapport à l'origine.*
   *Comptes Rendus,* t. 198, p. 1202.   1934.

(28) DARMOIS, G.   *Sur la théorie des deux facteurs de Spearman.*
   *Comptes Rendus,* t. 199, pp. 1176–8.   1934.

(29) DARMOIS, G.   *Sur la théorie des deux facteurs de Spearman.*
   *Comptes Rendus,* t. 199, pp. 1358–60.   1934.

(30) DARMOIS, G.   *Développements récents de la technique statistique.*
   *Econometrica,* Vol. II, No. 3, pp. 238–48.   1934.

(31) DAVID, F. N.   On the $P\lambda_n$ test for randomness; remarks, further illustration, and table for $P\lambda_n$.
   *Biometrika,* Vol. XXVI, pp. 1–11.   1934.

(32) DAVIES, O. L.   On asymptotic formulæ for the hypergeometric series.
   *Biometrika,* Vol. XXVI, pp. 59–107.   1934.

(33) DAVIES, O. L., and PEARSON, E. S.   Methods of estimating from samples the population standard deviation.
   *J. Roy. Stat. Soc.,* Suppl., Vol. I, pp. 76–93.   1934.

(34) DEMING, W. E.   On the application of least squares.—I.
   *Phil. Mag. Ser.,* 7, Vol. XI, pp. 146–58.   1931.

(35) DEMING, W. E.   On the application of least squares.—II.
   *Phil. Mag. Ser.,* 7. Vol. XVII, pp. 804–29.   1934.

(36) DEMING, W. E.   On the application of least squares.—III.
   *Phil. Mag. Ser.,* 7, Vol. XIX, pp. 389–402.   1935.

(37) DEMING, W. E.   The chi-test and curve fitting.
   *J. Amer. Stat. Ass.,* Vol. XXIX, pp. 372–82.   1934.

(38) DEMING, W. E., and BIRGE, R. T.   On the statistical theory of errors.
   *Reviews of Modern Physics,* Vol. 6, No. 3, pp. 120–61. 1934.

(39) DE MONTESSUS DE BALLORE.   *Determination de la médiane dans la fonction binomiale.*
   *Comptes Rendus,* t. 198, pp. 784–6.   1934.

(40) DIEULEFAIT, C. E.   *Sur les développements des fonctions des fréquences en séries de fonctions orthogonales.*
   *Metron,* Vol. XI, N. 4, pp. 77–81.   1934.

(41) DIEULEFAIT, C. E.   *Contribution à l'étude de la théorie de la corrélation.*
   *Biometrika,* Vol. XXVI, pp. 379–403.   1934.

(42) DOOB, J. L.   Stochastic processes and statistics.
   *Proc. Nat. Acad. Sci.,* Vol. 20, pp. 376–7.   1934.

(43) ELDERTON, W. P., and HANSMANN, G. H.   Improvement of curves fitted by the method of moments.
   *J. Roy. Stat. Soc.,* Vol. XCVII, pp. 331–3.   1934.

(44) ENLOW, E. R.   Quadrature of the normal curve.
   *Annals Math. Statistics,* Vol. V, No. 2, pp. 136–45.   1934.

(45) FISHER, R. A.   Two new properties of mathematical likelihood.
   *Proc. Roy. Soc.,* A. 144, pp. 285–307.   1934.

(46) FISHER, R. A.   Probability, likelihood and quantity of information in the logic of uncertain inference.
   *Proc. Roy. Soc.,* A. 146, pp. 1–8.   1934.

(47) FISHER, R. A.  The effect of methods of ascertainment upon the estimation of frequencies.
*Annals of Eugenics*, Vol. VI, pp. 13–25.  1934.

(48) FISHER, R. A.  The amount of information supplied by records of families as a function of the linkage in the population sampled.
*Annals of Eugenics*, Vol. VI, pp. 66–76.  1934.

(49) FISHER, R. A., and YATES, F.  The $6 \times 6$ Latin squares.
*Proc. Camb. Phil. Soc.*, Vol. 30, pp. 492–507.  1934.

(50) GEIRINGER, H.  *Une méthode générale de statistique théorique.*
*Comptes Rendus*, t. 198, pp. 420–2.  1934.

(51) GEIRINGER, H.  *Applications d'une méthode générale de statistique théorique.*
*Comptes Rendus*, t. 198, pp. 696–8.  1934.

(52) GOLDZIHER, K.  *Über Mittelwertinterpolation in der dynamischen Statistik.*
*Versicherungsarchiv*, 9, pp. 781–96.  1934.

(53) GUMBEL, E. J.  *L'espérance mathématique de la $m^{ième}$ valeur.*
*Comptes Rendus*, t. 198, pp. 33–5.  1934.

(54) GUMBEL, E. J.  *Les moments des distributions finales de la première et de la dernière valeur.*
*Comptes Rendus*, t. 198, pp. 141–3.  1934.

(55) GUMBEL, E. J.  *Les moments des distributions finales de la $m^{ième}$ valeur.*
*Comptes Rendus*, t. 198, pp. 313–5.  1934.

(56) GUMBEL, E. J.  *Le paradoxe de l'âge limite.*
*Comptes Rendus*, t. 199, pp. 918–9.  1934.

(57) GUMBEL, E. J.  *La distribution finale des valeurs voisines de la médiane.*
*Comptes Rendus*, t. 199, pp. 1174–6.  1934.

(58) GUMBEL, E. J.  *Les valeurs extrêmes des distributions statistiques.*
*Annales de l'Institut Henri Poincaré*, Vol. V, pp. 115–58. 1934.

(59) GUMBEL, E. J.  *L'età limite.*
*Giorn. Ist. Ital. Attuari*, Vol. V, N. I, pp. 52–80.  1934.

(60) HANSMANN, G. H.  On certain non-normal symmetrical frequency distributions.
*Biometrika*, Vol. XXVI, pp. 129–95.  1934.

(61) HENDRICKS, W. A.  The standard error of any analytic function of a set of parameters evaluated by the method of least squares.
*Annals Math. Statistics*, Vol. 5, No. 2, pp. 107–12.  1934.

(62) HERSCH, L.  *Essai sur les variations périodiques et leur mensuration.*
*Metron*, Vol. XII, N. 1, pp. 3–184.  1934.

(63) IRWIN, J. O.  On the independence of the constituent items in the analysis of variance.
*J. Roy. Stat. Soc.*, Suppl. Vol. I, pp. 236–55.  1934.

(64) JEFFREYS, H.  Probability and scientific method.
*Proc. Roy. Soc.*, A. 146, pp. 9–16.  1934.

(65) JORDAN, C.  *Teoria della perequazione e d'ell approssimazione.*
*Giorn. Ist. Ital. Attuari*, Vol. V, N. 1, pp. 81–107.  1934.

(66) JOSEPH, A. W.   Further notes on the sum and integral of the product of two functions.
  *J. Inst. Actuaries*, Vol. LXV, pp. 277–309.   1934.

(67) KEMP, W. B.   Some methods for statistical analysis.
  *J. Amer. Stat. Ass.*, Vol. XXIX, pp. 147–58.   1934.

(68) KULLBACK, S.   An application of characteristic functions to the distribution problem of statistics.
  *Annals Math. Statistics*, Vol. V, No. 4, pp. 264–307. 1934.

(69) LEVY, P.   *Généralisation de l'espace différentiel de N. Wiener.*
  *Comptes Rendus*, t. 198, pp. 786–8.   1934.

(70) LEVY, P.   *Sur les espaces V et W.*
  *Comptes Rendus*, t. 198, pp. 1203–5.   1934.

(71) LEVY, P.   *Complément à l'étude des espaces V et W.*
  *Comptes Rendus*, t. 198, pp. 1661–2.   1934.

(72) LÜDERS, R.   *Die Statistik der seltenen Ereignisse.*
  *Biometrika*, Vol. XXVI, pp. 108–28.   1934.

(73) MARTIN, E. S.   On the corrections for the moment coefficients of frequency distributions when the start of the frequency is one of the characteristics to be determined.
  *Biometrika*, Vol. XXVI, pp. 12–58.   1934.

(74) MAZZONI, P.   *Su un'origine geometrica di tipi di distribazioni di frequenze.*
  *Giorn. Ist. Ital. Attuari*, Vol. V, N. 2–3, pp. 219–31.   1934.

(75) MAHALANOBIS, P. C.   Tables for *L* tests.
  *Sankhyā*, Vol. I, pp. 109–22.   1934.

(76) MAHALANOBIS, P. C., BOSE, S. S., RAY, P. R., and BANERJI, S. K.   Table of random samples from a normal population.
  *Sankhyā*, Vol. I, pp. 289–328.   1934.

(77) McCORMICK, T.   A coefficient of independent determination.
  *J. Amer. Stat. Ass.*, Vol. XXIX, pp. 76–8.   1934.

(78) McKAY, A. T.   Sampling from batches.
  *J. Roy. Stat. Soc.*, Suppl. Vol. I, pp. 207–16.   1934.

(79) MIHOC, M. G.   *Sur les chaînes multiples discontinues.*
  *Comptes Rendus*, t. 198, pp. 2135–6.   1934.

(80) MILLER, J. C. P.   On a special case in the determination of probable errors.
  *Monthly Notices R. Ast. Soc.*, Vol. 94, pp. 860–6.   1934.

(81) MINER, J.   The variance of grouped frequency distribution when the mean of each class is taken as a centering point.
  *Human Biology*, Vol. 6, No. 3, pp. 561–3.   1934.

(82) MORRISON, J. T.   Note on the correlation of time series.
  *Phil. Mag. Ser.*, 7, Vol. XVIII, pp. 545–53.   1934.

(83) MYERS, R. J.   Note on Koshal's method of improving the parameters of curves by the use of maximum likelihood.
  *Annals Math. Statistics*, Vol. V, No. 4, pp. 320–3. 1934.

(84) NEYMAN, J.   On the two different aspects of the representative method; the method of stratified sampling and the method of purposive selection.
  *J. Roy. Stat. Soc.*, Vol. XCVIII, pp. 558–606.   1934.

D D 2

(85) O'TOOLE, A. L.   On a best values of *r* in samples of *R* from a finite population of $\dot{N}$.
    *Annals. Math. Statistics*, Vol. 5, No. 2, pp. 146–52.   1934.

(86) PALMER, C. E.   Note on the statistical significance of the difference of two series of comparable means.
    *Human Biology*, Vol. 6, No. 2, pp. 402–5.   1934.

(87) PEARSON, E. S.   Sampling problems in industry.
    *J. Roy. Stat. Soc.*, Suppl., Vol. I, pp. 107–36.   1934.

(88) PEARSON, K.   Remarks on Professor Steffensen's measure of contingency.
    *Biometrika*, Vol. XXVI, pp. 255–9.   1934.

(89) PEARSON, K.   On a new method of determining goodness of fit.
    *Biometrika*, Vol. XXVI, pp. 425–42.   1934.

(90) PICONE, M.   *Trattazione elementaire d'ell' approzimazione lineare in insiemi non limitati.*
    *Giorn. Ist. Ital. Attuari*, Vol. V, N. 2–3, pp. 155–95.   1934.

(91) POLLARD, H. S.   On the relative stability of the median and arithmetic mean, with particular reference to certain frequency distributions which can be dissected into normal distributions.
    *Annals Math. Statistics*, Vol. 5, No. 3, pp. 227–62.   1934.

(92) RICH, C. D.   The measurement of the rate of population growth.
    *J. Inst. Actuaries*, Vol. LXV, pp. 38–77.   1934.

(93) RIDER, P. R.   Recent progress in statistical method.
    *J. Amer. Stat. Ass.*, Vol. XXX, pp. 58–88.   1935.

(94) RIDER, P. R.   The third and fourth moments of the generalized Lexis theory.
    *Metron*, Vol. XII, N. 1, pp. 185–200.   1934.

(95) ROMANOVSKY, V.   *Su due problemi di distribuzione casuale.*
    *Giorn. Ist. Ital. Attuari*, Vol. 5, N. 2–3, pp. 196–218.   1934.

(96) ROMANOVSKY, V.   On the Tchebycheff's inequality for the two-dimensional case.
    *Acta Universitatis Asiæ Mediæ*, Series V–a Mathematica Fasc. 11–13.   1934.

(97) SALVEMINI, T.   *Ricerca sperimentale sull' interpolazione grafica di istogrammi.*
    *Metron*, Vol. XI, N. 4, pp. 83–197.   1934.

(98) SCHMIDT, R.   Statistical analysis of one dimensional distributions.
    *Annals Math. Statistics*, Vol. 5, No. 1, pp. 30–72.   1934.

(99) SILBERSTEIN, L.   Probability problem in integers.
    *Phil. Mag. Ser.* 7, Vol. XVIII, pp. 1132–4.   1934.

(100) SMART, W. M.   Some theorems in the statistical treatment of stellar motions.
    *Monthly Notices, R. Ast. Soc.*, Vol. 95, pp. 116–31.   1934.

(101) SMITH, E. D.   The exponential equation fitted by the mean value method.
    *J. Amer. Stat. Ass.*, Vol. XXIX, pp. 184–7.   1934.

(102) SMITH, T.   Note on integrals of products of experimentally determined magnitudes.
    *Proc. Physical Soc.*, Vol. 46, pp. 365–71.   1934.

(103) SMITH, T.  The mid-course method of fitting a parabolic formula of any order to a set of observations.
*Proc. Physical Soc.*, Vol. 46, pp. 560–73.  1934.

(104) SNEDECOR, G. W.  Calculation and interpretation of analysis of variance and covariance.
Iowa.  1934.

(105) SNEDECOR, G. W.  The method of expected numbers for tables of multiple classification with disproportionate sub-class numbers.
*J. Amer. Stat. Ass.*, Vol. 29, pp. 389–93.  1934.

(106) SNEDECOR, G. W., and COX, G. M.  Disproportionate sub-class numbers in tables of multiple classification.
*Iowa Agricultural Experiment Station Research Bulletin.* No. 180.  1934.

(107) STEFFENSEN, J. F.  On certain measures of dependence between statistical variables.
*Biometrika*, Vol. XXVI, pp. 251–5.  1934.

(108) STERNE, T. E.  The accuracy of least square solutions.
*Proc. Nat. Acad. Sci.*, Vol. 20, pp. 565–71 and 601–3.  1934.

(109) "STUDENT."  A calculation of the minimum number of genes in Winter's selection experiment.
*Annals of Eugenics*, Vol. VI, pp. 77–82.  1934.

(110) TEDESCHI, B.  *Nuovo contributo al problema dell' interpolazione lineare.*
*Giorn. Ist. Ital. Attuari*, Vol. 5, N. 2–3, pp. 232–49.  1934.

(111) THOMSON, G. H.  Hotelling's method modified to give Spearman's g.
*J. Educational Psychology*, Vol. 25, pp. 366–74.  1934.

(112) THOMSON, G. H.  The orthogonal matrix transforming Spearman's two-factor equations into Thomson's sampling equations in the theory of ability.
*Nature*, Vol. 134, p. 700, November 3, 1934.

(113) TOOPS, A. H.  On the systematic fitting of straight line trends by stencil and calculating machine.
*Annals Math. Statistics*, Vol. 5, No. 1, pp. 21–29.  1934.

(114) TRELOAR, A. E., and WILDER, M. A.  The adequacy of Student's criterion of deviations in small sample means.
*Annals Math. Statistics*, Vol. V, No. 4, pp. 324–41.  1934.

(115) TSCHUPROW, A. A.  The mathematical foundations of the methods to be used in statistical investigation of the dependence between two chance variables.
*Nordic Statistical Journal*, Vol. 5, pp. 34–42.  1934.

(116) VAUGHAN, H.  Selecting an operand for a summation formula of graduation.
*J. Inst. Actuaries*, Vol. LXV, pp. 86–93.  1934.

(117) VON SCHELLING, H.  *Die Konzentration einer Verteilung, und ihre Abhängigkeit von den Grenzen des Variationsbereiches.*
*Metron*, Vol. XI, N. 4, pp. 3–17.  1934.

(118) WALKER, H. M., and SANFORD, V.  The accuracy of computation with approximate numbers.
*Annals Math. Statistics*, Vol. V, No. I, pp. 1–12.  1934.

(119) WEIDA, F. M.   On measures of contingency.
    *Annals Math. Statistics*, Vol. V, No. 4, pp. 308–19.   1934.
(120) WILKS, S. S.   Moment-generating operators for determinants
    of product moments in samples from a normal system
    *Annals of Mathematics*, Vol. 35, No. 2, pp. 312–39.   1934.
(121) WILSDON, B. H.   Discrimination by specification statistically
    considered and illustrated by the standard specification for
    Portland cement.
    *J. Roy. Stat. Soc.*, Suppl., Vol. I, pp. 152–92   1934.
(122) WILSON, E. B., and WORCESTER, J.   The resolution of four
    tests.
    *Proc. Nat. Acad. Sci.*, Vol. 20, pp. 189–92.   1934.
(123) WILSON, E. B.   On resolution into generals and specifics.
    *Proc. Nat. Acad. Sci.*, Vol. 20, pp. 193–6.   1934.
(124) WILSON, E. B.   Boole's challenge problem.
    *J. Amer. Stat. Ass.*, Vol. XXIX, pp. 301–4.   1934.
(125) WISHART, J.   Statistics in agricultural research.
    *J. Roy. Stat. Soc.*, Suppl., Vol. I, pp. 26–51.   1934.
(126) WOLD, H.   *Sulla correzione di Sheppard.*
    *Giorn. Ist. Ital. Attuari*, Vol. V, No. 2–3, pp. 304–14.   1934.
(127) WOLD, H.   Sheppard's correction formulæ in several variables.
    *Skandinavisk Aktuarietidskrift*, Vol. XVII, pp. 248–55.
    1934.
(128) WRIGHT, S.   The method of path coefficients.
    *Annals Math. Statistics*, Vol. 5, No. 3, pp. 161–215.   1934.
(129) YASUKAWA, K.   On the deviation from normality of the
    frequency distributions of functions of normally distributed
    variates.
    *Tohoku Math. J.*, Vol. 38, pp. 465–79.   1934.
(130) YATES, F.   The analysis of multiple classifications with
    unequal numbers in the different classes.
    *J. Amer. Stat. Ass.*, Vol. XXIX, pp. 51–66.   1934.
(131) YATES, F.   Contingency tables involving small numbers and
    the $\chi^2$ test.
    *J. Roy. Stat. Soc.*, Suppl., Vol. I, pp. 217–35.   1934.
(132) YATES, F.   A complex pig-feeding experiment.
    *J. Agric. Sci.*, Vol. XXIV, pp. 511–31.   1934.
(133) ZOCH, R. T.   Invariants and covariants of certain frequency
    curves.
    *Annals Math. Statistics*, Vol. 5, No. 2, pp. 124–35.   1934.

OTHER REFERENCES.

B. SECTION I. *

(1) CASTELNUOVO, G.
    *Calcolo delle probabilità*, 2nd Edition.   Bologna 1926–8.
    Chapters III–VII and Appendix.

* The references in this section are given in the order in which they occur in
the text.

(2) CANTELLI, F. P.  *La tendenza ad un limite nel senso del calcolo delle probabilità.*
Rendiconti del circolo matematico di Palermo, XLI, pp. 191–201.  1916.

(3) CANTELLI, F. P.  *Sulla probabilità come limite delle frequenza.*
Rendiconti della R. Accademia dei Lincei, Series 5, XXVI, pp. 39–45.  1917.

(4) WRINCH, D., and JEFFREYS, H.  On some aspects of the theory of probability.
Phil. Mag., 6th Series, 38, pp. 715–31.  1919.

(5) BARTLETT, M. S.  Probability and chance in the theory of statistics.
Proc. Roy. Soc., A., 141, pp. 518–43.  1933.

(6) KOLMOGOROFF, A.  *Sur la loi forte des grands nombres.*
Comptes Rendus, t. 191, pp. 910–12.  1930.

(7) DE FINETTI, B.  *Classi di numeri aleatori equivalenti.*
Rendiconti della R. Accademia dei Lincei (6th Series), XVIII, pp. 107–10.  1933.

(8) DE FINETTI, B.  *La legge dei grandi numeri nel caso dei numeri aleatori equivalenti.*
Rendiconti della R. Accademia dei Lincei (6th Series), XVIII, pp. 203–07.  1933.

(9) MESSINA, I.  *Intorno a un nuovo teorema di calcolo delle probabilità.*
Giorn. di Matematiche di Battaglini, LVI, pp. 191–208. 1918.

(10) MESSINA, I.  *Un teorema sulla legga uniforme dei grandi numeri.*
Giorn. Ist. Ital. Attuari, IV, pp. 116–30.  1933.

(11) CLOPPER, C. J., and PEARSON, E. S.  The Use of confidence or fiducial limits illustrated in the case of the binomial.
Biometrika, XXVI, pp. 404–13.  1934.

(12) SNEDECOR, G. W.  Calculation and interpretation of analysis of variance and covariance.
Iowa.  1934.

(13) KHINTCHINE, A.  *Sur un théorème général relatif aux probabilités dénombrables.*
Comptes Rendus, t. 178, pp. 617–18.  1924.

(14) KHINTCHINE, A.  *Ein Satz der Wahrscheinlichkeitsrechnung :*
Fundamenta Mathematicæ, VI, pp. 9–20.  1924.

(15) KHINTCHINE, A.  *Über das Gesetz der grossen Zahlen.*
Mathematische Annalen, 96, pp. 152–68.  1927.

(16) KOLMOGOROFF, A.  *Über das Gesetz des iterierten Logarithmus.*
Mathematische Annalen, 101, pp. 126–35.  1929.

(17) LÉVY, P.  *Quelques théorèmes sur les probabilités dénombrables.*
Comptes Rendus, t. 192, pp. 658–9, 1931.

(18) LÉVY, P.  *Sulla legge forte dei grandi numeri.*
Giorn. Ist. Ital. Attuari, IV, pp. 1–21.  1931.

(19) LÉVY, P.  *Nuovo formulo relative al giuoco di testa e croce.*
Giorn. Ist. Ital. Attuari, IV, pp. 127–60.  1931.

(20) LÉVY, P.  *Sur un théorème de M. Khintchine.*
Bulletin des Sciences Mathématiques, 2nd Series, LV, pp. 145–60.  1931.

(21) CANTELLI, F. P. *Considerazione sulla legge uniforme dei grandi numeri e sulla generalizzazione di un fondamentalo teorema del Sig. Paul Lévy.*
Giorn. Ist. Ital. Attuari, IV, pp. 327–50. 1933.

(22) CRAMÉR, H. *Su un teorema relativo alle legge uniforme dei grandi numeri.*
Giorn. Ist. Ital. Attuari, V, pp. 1–13. 1934.

(23) GLIVENKO, V. *Sulla determinazione empirica delle leggi di probabilità.*
Giorn. Ist. Ital. Attuari, IV, pp. 92–9. 1933.

(24) KOLMOGOROFF, A. *Sulla determinazione empirica delle leggi di probabilità.*
Giorn. Ist. Ital. Attuari, IV, pp. 83–91. 1933.

(25) DE FINETTI, B. *Sull' approssimazione empirica di una legge di probabilità.*
Giorn. Ist. Ital. Attuari, IV, pp. 415–20. 1933.

(26) CANTELLI, F. P. *Sulla determinazione empirica delle leggi di probabilità.*
Giorn. Ist. Ital. Attuari, IV, pp. 421–24. 1933.

(27) CANTELLI, F. P. *Considérations sur la convergence dans le calcul des probabilités.*
Annales de l'Institut Henri Poincaré, Vol. V, pp. 1–50. 1935.

(28) DE FINETTI, B. *Sulla legge di distribuzione dei valori in une successione di numeri aleatori equivalenti.*
Rendiconti della R. Accademia dei Lincei (6th Series), XVIII, pp. 279–84. 1933.

(29) BELARDINELLI, G. *Su una teoria astratta del calcolo delle probabilità.*
Giorn. Ist. Ital. Attuari, V, pp. 418–34, 1934.

### C.—Section II.

(1) FRISCH, R. Recurrence formulæ for the moments of the point binomial.
Biometrika, Vol. 17, pp. 170. 1925.

(2) PEARSON, K. On the moments of the hypergeometrical series.
Biometrika, Vol. 16, pp. 159. 1924.

(3) WILKS, S. S. Certain generalizations in the analysis of variance.
Biometrika, Vol. 24, pp. 471–94. 1932.

### D.—Section III.

(1) FISHER, R. A., and TIPPETT, L. H. C. Limiting forms of the frequency distribution of the largest or smallest member of a sample.
Proc. Camb. Phil. Soc., Vol. XXIV, pp. 180–90. 1928.

(2) GREENWOOD, M., and YULE, G. U. An enquiry into the nature of frequency distributions representative of multiple happenings with particular reference to the occurrence of multiple attacks of disease or of repeated accidents.
J. Roy. Stat. Soc., Vol. LXXXIII, pp. 255–79. 1920.

## E.—Section IV.

(1) NEYMAN, J.  Further notes on non-linear regression.
*Biometrika*, Vol. XVIII, pp. 256–62.  1926.
(2) PEARSON, K.  On a general method of determining the successive terms in a skew regression line.
*Biometrika*, Vol. XIII, pp. 296–300.  1920–1.
(3) WICKSELL, S. D.  Remarks on regression.
*Annals Math. Statistics*, Vol. I, pp. 3–13.  1930.

## F.—Section V.

(1) FISHER, R. A.  The mathematical foundations of theoretical statistics.
*Phil. Trans.*, A., Vol. 222, pp. 309–68.  1922.
(2) FISHER, R. A.  The theory of statistical estimation.
*Proc. Camb. Phil. Soc.*, Vol. XXII, pp. 700–25, 1925.
(3) FISHER, R. A.  *Statistical Methods for Research Workers.* 5th Edn., p. 103.
(4) NEYMAN, J., and PEARSON, E. S.  Sufficient statistics and uniformly most powerful tests of statistical hypotheses.
*Statistical Research Memoirs of Department of Statistics, University College, London*, Vol. I, pp. 113–37, 1936.
(5) WEINBERG, W.  *Auslesewirkungen bei biologisch-statistichen Problemen.*
*Archiv für Rassen- und Gesellschafts-Biologie*, pp. 418–581.  1913.

## G.—Section VI.

(1) BARTLETT, M. S.  Recent work on the analysis of variance.
*J. Roy. Stat. Soc. Suppl.*, Vol. 1, pp. 252–5.  1934.
(2) BRANDT, A. E.  The analysis of variance in a $2 \times 5$ table with disproportionate frequencies.
*J. Amer. Stat. Ass.*, Vol. 28, pp. 164–73.  1933.
(3) EULER, L.  *Recherches sur une nouvelle espèce de quarrés magiques.*
*Verh. v. h. Zeeuwsch Genootsch. der Wetensch.*, Vlissingen 9, pp. 85–239.  1782.
(4) FISHER, R. A.  Applications of " Student's " distribution.
*Metron.*, Vol. 5, No. 3, pp. 90–104.  1925.
(5) YATES, F.  The formation of Latin squares, for use in field experiments.
*Empire Journal of Experimental Agriculture*, Vol. I, pp. 235–44.  1933.

# 16c

Recent Advances in Mathematical Statistics

*Recent Work on the Analysis of Variance*

By W. G. Cochran

Papers on the analysis of variance which have appeared recently may be summarized in five sections. (1) *Design of experiments*. The principles underlying the planning and interpretation of experiments with variable material are discussed in Fisher's book *The Design of Experiments*, which has recently reached its second edition. The lay-out of factorial experiments has been comprehensively studied in a number of papers by Yates, who has also introduced new designs for experiments involving a large number of treatments. (2) The *discriminant function method*, an analytical method, due to Fisher, of picking out that linear function of a set of characters which best distinguishes between several groups in which the characters have been measured. This method has already found applications in craniometry, in taxonomy, in plant selection, and in the selection of candidates by intelligence tests. (3) Papers giving advice on the use and extension of the analysis to problems where the simple method is not immediately applicable or where considerable discretion is required. (4) Further applications of the analysis of covariance. This comparatively recent technique attracted a good deal of attention in previous summaries, and papers are still appearing whose chief interest is as an illustration of the wide usefulness of the method. (5) Other applications.

## I. *Design of Experiments*

The first comprehensive examination of designs employing confounding which are likely to be useful is given by Yates in (11), and a thorough study of this in conjunction with *The Design of Experiments* will bring the reader up to date on this subject. The results obtained are summarized below.

### (a) *Confounding of Factorial Experiments in Randomized Blocks.*

In agricultural field experiments the size of block is flexible and the argument for or against confounding is determined mainly by a consideration of the relative efficiencies of experiments with and without confounding, which in turn depends upon the increase in standard error per plot with increasing size of block. A *résumé* of the information on this question is, therefore, relevant. Very little work has, however, been done. Information on this point can be obtained readily from experiments which have been carried out in randomized blocks by finding, as described by Yates in (6), the information which would have been obtained from these experiments had they been completely randomized. For example, an experiment in 8 randomized blocks of 4 plots each gives an estimate of the gain in information by reducing block size from 32 plots to 4 plots, and is relevant to the question of confounding a $2^5$ factorial design in blocks of 4 plots. The average information per experiment obtained in the randomized blocks experiments carried out by Rothamsted from 1927 to 1934 is 1·67 as compared with 1 for complete randomization. Pending more data and more detailed examination, this figure may be taken as representative of the average gain due to the reduction of block size by confounding. It indicates that confounding is well worth while.

| Factors ABC ... | Number of Treatments | Size of Block | Interactions Confounded in a Single Replication | Number * |
|---|---|---|---|---|
| $2^3$ | 8 | 4 | ABC | 1 |
| $2^4$ | 16 | 8 | ABCD | 1 |
| $2^4$ | 16 | 4 | { AB, ACD, BCD or | 6 |
| | | | { AB, CD, ABCD | 4 |
| $2^5$ | 32 | 8 | ABC, ADE, BCDE | 5 |
| $2^5$ | 32 | 4 | BD, CE, ABC, ADE, ACD, ABE, BCDE | 5 |
| $2^6$ | 64 | 16 | ABCD, ABEF, CDEF | 5 |
| $2^6$ | 64 | 8 | ACE, BDE, BCF, ADF, ABCD, ABEF, CDEF | 10 |
| $2^6$ | 64 | 4 | AB, CD, EF, ACE, BCE, ACF, BCF, ADE, BDE, ADF, BDF, ABCD, ABEF, CDEF, ABCDEF | 5 |
| $3^3$ | 27 | 9 | ABC | 4 |
| $3^4$ | 81 | 9 | ABC, ABD, ACD, BCD | 4 |
| $3 \times 2^2$ | 12 | 6 | BC, ABC | 3 |
| $3 \times 2^3$ | 24 | 6 | BC, BD, CD, ABC, ABD, ACD, ABCD | 3 |
| $3^2 \times 2$ | 18 | 6 | AB, ABC | 4 |
| $3^3 \times 2$ | 54 | 6 | AB, AC, BC, ABD, ACD, BCD, ABCD | 4 |
| $4^2$ | 16 | 8 | AB | 9 |
| $4^2$ | 16 | 4 | AB | 3 |

* Number of replications for a balanced arrangement.

The Table summarizes most of the simpler designs obtained by Yates for factorial experiments with factors at 2, 3 or 4 levels, the aim being, of course, to confound two- and three-factor interactions as little as possible. The letters $A$, $B$, $C$ ... representing factors, are to be read from the left; *e.g.* the $AB$ interaction in the $3^2 \times 2$ design is the first order interaction (4 degrees of freedom) between the two factors which occur at three levels each. The results given apply to a single replication only; with several replications different degrees of freedom may be confounded in the different replications, and with sufficient replication we can always arrange that all interactions of a given order and type are confounded equally. This type of arrangement is called a *balanced* one, and has the advantages that the loss of information on the partially confounded interactions is at a minimum and that the computations are easier. For example, with a $2^5$ design in blocks of 8 plots, the following arrangement with five replications clearly confounds all second- and third-order interactions equally.

| Replications | Interactions confounded |
|:---:|:---:|
| 1 | *ABC, ADE, BCDE* |
| 2 | *ABD, BCE, ACDE* |
| 3 | *ACE, BCD, ABDE* |
| 4 | *ACD, BDE, ABCE* |
| 5 | *ABE, CDE, ABCD* |

Since each of the interactions concerned is completely confounded in one out of five replications, one-fifth of the information relative to unconfounded effects has been sacrificed. The minimum number of replications required to obtain complete balance is shown in the last column of the table.

The following example shows the way in which treatments are assigned (before randomization) to blocks in the first replication above :

Block

| I | II | III | IV |
|:---:|:---:|:---:|:---:|
| (1) | *ab* | *b* | *a* |
| *bc* | *ac* | *c* | *abc* |
| *abd* | *d* | *ad* | *bd* |
| *acd* | *bcd* | *abcd* | *cd* |
| *abe* | *e* | *ae* | *be* |
| *ace* | *bce* | *abce* | *ce* |
| *de* | *abde* | *bde* | *ade* |
| *bcde* | *acde* | *cde* | *abcde* |

The interaction *ABC*, for instance, is obtained from the differences between the total of blocks III and IV and the total of blocks I and II. To construct this design, the 32 treatments are first divided into two groups of 16 treatments each, such that the difference between the totals of the two groups represents the interaction *ABC*. Thus the first group contains 16 treatments (1), *ab, ac, bc, d, abd, acd, bcd, e, abe, ace, bce, de, abde, acde, bcde*. These 16 treatments are then subdivided into two groups of 8 treatments, such that the difference between the totals of the two groups is a part of the *ADE* interaction. To do this, we assign to the first eight all treatments with *none* or *two* of the factors *a, d* and *e*, and to the second eight all treatments with *one* or *three* of the factors *a, d* and *e*. These groups form respectively blocks I and II in the above plan. Blocks III and IV are constructed similarly.

(b) *Confounding of High-order Interactions in Latin Squares.*

Yates also investigates in (11) the possibilities of arranging confounded designs in Latin squares. In the Latin square experiments carried out by Rothamsted from 1927 to 1934 the average information per plot was $2 \cdot 22$ times that which would have been obtained by complete randomization. The corresponding figure for randomized blocks was $1 \cdot 67$, so that on this comparison the increase in efficiency over complete randomization is nearly twice as great for the Latin square as for randomized blocks.

As a simple example, consider the confounding of the $2^3$ design with treatments (1), $n$, $p$, $k$, $np$, $nk$, $pk$, $npk$ in $4 \times 4$ Latin squares.

| (1) | $np$ | $nk$ | $pk$ |
|-----|------|------|------|
| $n$ | $p$ | $k$ | $npk$ |
| $pk$ | $nk$ | $np$ | (1) |
| $npk$ | $k$ | $p$ | $n$ |

The square shown, in which the first two rows or columns are a complete replication, clearly confounds $NPK$ completely with rows and $PK$ completely with columns. By interchange of letters, squares can be constructed which confound $NPK$ and $NK$ or $NPK$ and $NP$. Thus with six replications there is a design in three $4 \times 4$ Latin squares in which $\frac{2}{3}$ of the information is retained on each of the two-factor interactions. It should be noted that the rows and columns of any Latin square must be randomized to obtain an unbiased error, so that the replications within a square in the above example cannot be kept apart from each other in the field.

Arrangements in Latin squares which avoid confounding any first-order interactions are :—

$2^3$ in two interlaced $4 \times 4$ Latin squares
$2^4$, $2^5$ or $2^6$ in       an       $8 \times 8$ Latin square
$3^3$ or $3^4$ in       a       $9 \times 9$   ,,       ,,

(c) *Varietal Trials Involving a Large Number of Varieties.*

For work on plant selection a technique is required to cope with experiments involving, say, 25, or perhaps many more, varieties in which all comparisons between pairs of varieties may be of equal interest. The ordinary Latin square is clearly out of the question, and randomized blocks comparing as many as 25 plots may be expected to be relatively inefficient in eliminating the effects of the major soil irregularities. These difficulties have been partly overcome in the past by selecting one variety as control and dividing the varieties into small groups, each group with the control being laid out as a separate experiment in randomized blocks or a Latin square. Comparisons between varieties not in the same group are possible by comparing each with the control variety in its group, though they are consideraby less accurate than comparisons between varieties in the same group. Alternatively, the control plots may be arranged systematically, *e.g.* by repetition at fixed intervals along a line of plots. In both these cases the efficiency is reduced by the fact that more plots are allotted to the control variety than to any other.

To obtain a reduction of block size, the varieties in any replication must be arranged in a number of small groups, differences between groups being confounded with block differences. The defects of the above methods may, however, be partly avoided by taking different small groups in different replications. This type of design may be described as quasi-factorial, because its structure and analysis are most easily understood by pretending for convenience that the experiment is a factorial one. With 25 varieties, for instance, by re-numbering the varieties by means of two suffixes, 11, 12, 13, ... 55, we may regard the experiment as a 5 × 5 factorial design which we wish to confound, using blocks of 5 plots. Since main effects are now of no more importance than interactions, we may confound whichever we choose, and it is more convenient to confound the former. With two replications the main effects of each factor may be confounded in a separate replication. Before randomization within blocks the design is as follows (10 blocks of 5 plots each) :

| 1st Replication | | | | | | 2nd Replication | | | | |
|---|---|---|---|---|---|---|---|---|---|---|
| 11 | 21 | 31 | 41 | 51 | | 11 | 12 | 13 | 14 | 15 |
| 12 | 22 | 32 | 42 | 52 | | 21 | 22 | 23 | 24 | 25 |
| 13 | 23 | 33 | 43 | 53 | | 31 | 32 | 33 | 34 | 35 |
| 14 | 24 | 34 | 44 | 54 | | 41 | 42 | 43 | 44 | 45 |
| 15 | 25 | 35 | 45 | 55 | | 51 | 52 | 53 | 54 | 55 |
| 1 | 2 | 3 | 4 | 5 | | 6 | 7 | 8 | 9 | 10 |

Blocks               Blocks

With this design, the reader may verify that the 16 degrees of freedom representing interactions between the two factors are not in any way confounded with block differences. The 4 degrees of freedom representing the main effects of the first factor are, however, completely confounded with blocks in the first replication, and are estimated from the second replication only. Similarly the main effects of the second factor are obtained from the first replication only. It follows that if the reduction in block size brings no decrease in standard error per plot, this design is not as accurate as randomized blocks of 25 plots, since the standard error of any of the 8 degrees of freedom representing main effects is $\sqrt{2}$ times as great as it would be with randomized blocks. The average variance of a varietal comparison is thus : $(16 + 2 \times 8)/24 = \frac{4}{3}$ times as great as with randomized blocks of 25 plots. The reciprocal factor $\frac{3}{4}$, or in general $\frac{p+1}{p+3}$ with $p^2$ varieties, is called the " efficiency factor " of the arrangement, and tells by how much the variance has to be reduced by the new blocks to make the new arrangement as efficient as randomized blocks.* The various possible designs of this type, including the extension to three replications and to cases in which the number of varieties is not a complete square, will be found in (9). In general, the disadvantage remains that some varietal comparisons are more accurate than others, but the differences are slight.

Arrangements of this type in quasi-Latin squares are discussed in (10). As an example of the very elegant designs which result, consider the following experiment with 3 replications of 25 varieties, arranged in $5 \times 5$ quasi-Latin squares (before randomization). The varieties are numbered from 1 to 25.

|  | I |  |  |  |  | II |  |  |  |  | III |  |  |  |
|---|---|---|---|---|---|---|---|---|---|---|---|---|---|---|
| 1 | 2 | 3 | 4 | 5 | 1 | 13 | 25 | 7 | 19 | 1 | 15 | 24 | 8 | 17 |
| 6 | 7 | 8 | 9 | 10 | 20 | 2 | 14 | 21 | 8 | 18 | 2 | 11 | 25 | 9 |
| 11 | 12 | 13 | 14 | 15 | 9 | 16 | 3 | 15 | 22 | 10 | 19 | 3 | 12 | 21 |
| 16 | 17 | 18 | 19 | 20 | 23 | 10 | 17 | 4 | 11 | 22 | 6 | 20 | 4 | 13 |
| 21 | 22 | 23 | 24 | 25 | 12 | 24 | 6 | 18 | 5 | 14 | 23 | 7 | 16 | 5 |

The important point about this design is that any pair of varieties occurs once, and only once, in the same row or column. This can be verified at once for variety 1 and any other variety, by examining the composition of the first row and column of each square. It follows that all varietal comparisons must be of equal accuracy; further, the statistical analysis is very easy. This design depends on the existence of a completely orthogonal set of $5 \times 5$ Latin squares. Such sets exist for all prime numbers and have also been found for 4, 8 and 9, but not for 6.

(d) In the designs described above, the block size has always been a factor of the number of treatments (e.g. $3^4$ in a $9 \times 9$ Latin square). In agricultural field experiments this restriction is usually of minor importance, except that with small fields it may prevent the most efficient *shape* of block being used. In other branches of biological research, however, blocks of a particular size may be the natural unit, or there may be a limit to the possible size. Examples are monozygotic twins, the halves of a leaf and the numbers pro-

* In further work to be published shortly, Yates considers the possibility of recovering some of the loss of information represented in the efficiency factor, by making use of the information about treatment effects contained in the block totals. He has found that where there are no block effects, all but a small percentage of the lost information can usually be recovered, so that even in its most unfavourable case the quasi-factorial design can be made only slightly less accurate than randomized blocks.

duced in a litter by animals. Yates has therefore examined in
(8) the design of experiments in blocks of any given size. The
principle involved is similar to that in the Latin-square design just
described—namely, that each pair of treatments occurs equally
frequently in the same block. This ensures that all treatment
comparisons are of equal accuracy and that the analysis and presentation of results are simple. The main relevant properties of the
design are the minimum number of replications for which a design
exists, which is sometimes inconveniently large, and the " efficiency
factor," and are studied by Yates. Goulden (5) compares the
efficiencies of incomplete blocks and ordinary randomized blocks for
eight different crops from uniformity trial data. He concludes that
the increases in efficiency due to the use of incomplete blocks vary
from 20 to 50 per cent. Considering this gain and the greater
adaptability of incomplete blocks to irregularly-shaped fields, Goulden
recommends their use in field trials.

Some extensions of the split-plot design deserve mention (11).
(1) The arrangement of the sub-plots in a Latin-square formation is
possible in certain restricted cases. This may be expected to increase
further the accuracy of split-plot comparisons. (2) The use of
split-plots for varietal trials with, say, from 8 to 25 varieties. With
this number of varieties neither randomized blocks nor the Latin
square is particularly suitable, and the special methods available
for a large number of varieties have rather low efficiency factors.
(3) The superposition of treatments on whole rows and columns of
a Latin square. This type of design has been tried in the past, but
the principles involved have not previously been clearly stated.

One minor advantage of the randomized block design over the
Latin square has been that if for any reason the yields from one
block or treatment are missing, or have to be rejected, the statistical
analysis takes exactly the same simple form as before. The treatment of the corresponding case with Latin squares is discussed by
Yates in (7). When a single row, column or treatment is missing,
the remaining pair of factors become partially confounded, but the
statistical analysis remains comparatively simple. The methods also
apply when a row and a treatment are missing, but no simple analysis
has been found when two rows are missing. It is worth noting that
a Latin square with one treatment missing, or one row and one
treatment missing, is a valid experimental arrangement, and if
suitable could be deliberately laid down.

An important task in introducing new designs which involve
an unfamiliar process of analysis is to systematize the numerical
working in the best possible manner. Full attention has been paid
to this in (11).

## II. *The Discriminant Function Method*

The problem of classifying, say, a set of skulls, according to period or sex, by means of measurements made on parts of the skulls, is familiar in anthropology. Where a set of skulls have been correctly assigned to, say, two groups by other means, we may expect to gain some information from this classification as to which measurements are likely to be useful in classifying future material of this kind. Two relevant points will be appreciated : (1) some measurements may have practically the same values in both groups, and so be of little use in classifying new material into one or the other ; (2) the information supplied by different measurements will not necessarily be independent. The discriminant function method shows how to combine the measurements to form a single index which will best discriminate between the two groups. The index is taken to be a linear function of the measurements, but more complicated functions can be used in exactly the same way as the method of "linear" regression is extended to formulæ non-linear in the independent variate.

The standard method of discriminating between two groups in statistics is to compare the variation between groups with that within groups by means of a $t$-test. Thus, given a set of measurements $x_1, x_2, \ldots x_r$, we wish to choose multipliers $\lambda_1, \lambda_2 \ldots \lambda_r$ so that the index $\lambda_1 x_1 + \ldots - + \lambda_r x_r$ maximizes the ratio of the variance between groups to that within groups. If there are $n$ objects in each group, the variance between groups is

$$\frac{n}{2} D^2 = \frac{n}{2} (\lambda_1 d_1 + \lambda_2 d_2 + \ldots + \lambda_r d_r)^2$$

where $d_1$ is the difference between the means of $x_1$ in the two groups. The variance within groups is

$$S = \sum_{p=1}^{r} \sum_{q=1}^{r} \lambda_p \lambda_q S_{pq}$$

where $S_{pq}$ is the co-variance within groups of $x_p$ and $x_q$. Differentiating $D^2/S$ with respect to $\lambda_i$ we have :

$$\frac{2D}{S} \frac{\partial D}{\partial \lambda_i} - \left(\frac{D}{S}\right)^2 \frac{\partial S}{\partial \lambda_i} = 0$$

*i.e.*
$$\frac{1}{2} \frac{\partial S}{\partial \lambda_i} = \frac{S}{D} \frac{\partial D}{\partial \lambda_i} \quad i = 1, 2, \ldots r.$$

The factor $\dfrac{S}{D}$ occurs in all equations; thus the solutions $\lambda_{1p} \ldots \lambda_r$ are proportional to the solutions of the equations

$$\frac{1}{2} \frac{\partial S}{\partial \lambda_i} = \frac{\partial D}{\partial \lambda_i}$$

*i.e.*     $S_{i1} \lambda_1 + S_{i2} \lambda_2 + \ldots + S_{ir} \lambda_r = d_i \quad i = 1, 2, \ldots r$

These equations are formally the same as occur in fitting a multiple linear regression, and may be solved in the same way. Having found the relative values of the $\lambda_i$, we may construct the index by putting $\lambda_1 = 1$.

To classify a new specimen, we calculate the index for that specimen and assign it to that group whose mean index is nearest. The question now arises : how often will a mistake be made?; that is, how often will the index of a member of one group appear to lie nearer to the mean value in the other group? To answer this, we require the standard deviation of the index, calculated from the variation of the $x_i$ within groups. For any linear function $\lambda_1 x_1 + \ldots + \lambda_r x_r$ with *constant* coefficients, the variance within groups could be estimated by assigning $2(n-1)d.f.$ to the total variation of the function within groups. This would not, however, be correct for the index chosen, since the $\lambda_i$ are functions of the $x$'s, and have been chosen so as to minimize the relative effect of the variation within groups. The correct procedure is, however, equally simple. We can construct a quantity $y$ which is completely determined by the group to which it belongs, by giving it the value of $+\frac{1}{2}$ in the first group and $-\frac{1}{2}$ in the second. It may be shown that the equations for finding the $\lambda_i$ are the usual equations for determining the partial regression coefficients of $y$ on $x_1 x_2, \ldots x_r$, for both species combined; further, that the variance between groups of the index is that part of the variance of $y$ accounted for by its regression on $x_1, \ldots x_r$ and has $r$ $d.f.$ Hence the variation of the index *within* groups has $(2n - r - 1)d.f.$ Knowing the standard deviation within groups, we may at once calculate how frequently an index belonging to the first group would appear to lie nearer to the mean of the second group owing to chance variations in the $x_i$ within groups.

A straightforward application of the above method to discrimination between two species of Iris plants has been given by Fisher (13). In a totally different sphere, Travers (15) has applied the method to the selection of managers from salesmen by intelligence tests in a business organization. A variety of tests was given to a group of salesmen and to a group of competent managers, and from the tests which showed differences between the groups an index was constructed for use in the future selection of managers from salesmen. As the managers in the original test were naturally somewhat older than the salesmen, part of the difference between groups might be due to age differences, and the regression on age was therefore removed before finding the index.

Where the material is classified into more than two groups, the method may be applied to a single degree of freedom from the variation between groups. Thus Barnard (12), considering four series of Egyptian skulls, chooses the index by maximizing the regression of the four series on time. Fisher (13) tests the hypothesis that the index $i_3$ from a third species of Iris divides the difference between the indices $i_1$, $i_2$ of the first two species in the ratio $2:1$. If this were true, the appropriate method of choosing the index would be to maximize the single square $(4i_1 + i_3 - 5i_2)^2$.

An interesting application to plant selection has been indicated by Fairfield Smith (14). There are a number of desirable characters in a variety of wheat, *e.g.* high yield, resistance to disease, baking quality. In an experimental programme for the selection of varieties, it is essential to decide the relative importance to be attached to each of these characters, or, in strictly quantitative terms, to evaluate each in terms of a standard—say $x_1$. We might perhaps consider, for example, that a consistent increase of 10 in baking score was worth an extra bushel of yield. In interpreting the results of a replicated varietal trial, however, it would not necessarily be correct to score a difference of 10 in the mean baking score of two varieties as worth 1 unit in bushels of yield, because the observed difference is influenced by a large number of non-heritable factors, *e.g.* soil heterogeneity. In short, instead of observing $\xi$, the true genotypic value of a character, we can only observe

$$x = \xi + t$$

where $t$ represents the non-heritable part of $x$. We can, however, estimate the variance of the two parts of $x$, that of $t$ by the intra-varietal variance and that of $\xi$ by subtracting the intra-varietal from the inter-varietal variance. The problem is to find the linear function $\gamma = \lambda_1 x_1 + \lambda_2 x_2 + \ldots + \lambda_r x_r$ of the known values $x$, which best represents the score

$$\psi = a_1 \xi_1 + a_2 \xi_2 \ldots + a_r \xi_r$$

*i.e.* we choose the $\lambda_i$ to maximize the correlation between $\gamma$ and $\psi$. This leads to the type of mathematical problem considered above.

### III. *Applications*

A number of papers have appeared giving advice on the application of the analysis of variance to practical problems and dealing with some of the difficulties involved in the newer applications. One of the most common of these is that for many types of data, such as, for example, percentages, the variance is not independent of the mean, so that a straightforward analysis might be misleading. This difficulty can often be overcome by a change of scale, the

Q 2

commonest transformations in use being to replace $x$ by $\sqrt{x}$, $\log x$ or $\sin^{-1} \sqrt{x}$, which theoretically equalize the variance when it is proportional to $x$, $x^2$ or $x(1 - x)$ respectively. Included in these categories are data which might be expected to follow the binomial or Poisson distributions, but in which the appropriate $\chi^2$ tests are regarded with suspicion because of possible heterogeneity. The square root and inverse sine transformations and the general difficulties encountered in interpreting results on the new scale are discussed by Bartlett (16), who also gives some examples of the practical utility of these transformations. He points out a further complication with factorial designs; the scale in which the main effects of different factors may reasonably be regarded as additive may not be either the scale in which the data are originally recorded or the transformed scale.

The difficulties caused by unequal variances also appear when the results of a number of experiments with the same treatments are being summarized. When the individual experiments have different precision, as will commonly be the case, the problem of estimating and testing the mean response to a treatment and of testing its variation from experiment to experiment can no longer be solved by a simple analysis of variance. Cochran (19) discusses the extension of the technique to these cases. In particular, it may be noted that while the efficient estimate of the mean from a number of quantities $x_i$ with known standard deviations $\sigma_i$ is

$$\hat{\mu} = \sum \frac{x_i}{\sigma_i^2} \Big/ \sum \left(\frac{1}{\sigma_i^2}\right) \quad . \quad . \quad . \quad . \quad (1)$$

the efficient estimate when we have only estimates $s_i(n_i\ d.f.)$ of the $\sigma_i$ is the solution of

$$\sum \frac{x_i - \hat{\mu}}{n_i s_i^2 + (x_i - \hat{\mu})^2} = 0 \quad . \quad . \quad . \quad (2)$$

*i.e.* we cannot simply replace the weights in (1) by their estimate.

A wide selection of the practical problems which confront the statistician and a wealth of good advice will be found in a paper by Bartlett (17).

## IV. *Analysis of Covariance*

The number of recent papers which have been devoted to the analysis of covariance indicates the growing usefulness and popularity of this method. Most of these papers fulfil two objects: they demonstrate the arithmetical calculations necessary to carry out an analysis of covariance, and illustrate the varied uses to which the method may be put.

We will consider, first, papers which the reader may consult for instruction in the calculations involved. Brady (27) gives the details of the analysis with *two* independent variates, and shows how to obtain an exact z-test of the significance of the treatment differences as a whole after eliminating the linear regression on these two variates.* He does not consider the test of significance of the difference between a pair of treatments, but this has been given by Wishart (35). Wishart first presents the calculations for one and two independent variates, illustrating the latter case on Brady's data, and goes on to establish the general formula with any number of independent variates.

In the paper already mentioned, Bartlett (17) discusses the use of covariance in an experiment of the split-plot design. Here it may be necessary to make separate corrections in the main- and sub-plot analyses, but, as Bartlett points out, there is a certain convenience in using the sub-plot regression coefficient for the main-plot adjustments, and this procedure is quite efficient if the two regression coefficients differ little, a point which can, of course, be tested. Bartlett (17) also shows how to obtain appropriate and exact analyses of variance in experiments in which a missing or rejected value occurs in the dependent or independent variate (or in both). Where the value is missing in one variate only, the corresponding value in the other variate must also be estimated, even though a figure is already available for it. Further, Bartlett points out that the estimation of a missing value by the usual method of least squares may itself be regarded as an analysis of covariance in which the independent variate has a value $+1$ for the missing plot and zero everywhere else. Thus the combination of the estimation of a missing value with an analysis of covariance containing *one* independent variate may be treated as an analysis of covariance with *two* independent variates, and the exact test of significance is obtained by the usual methods of analysis of covariance.

The computations required for an analysis of covariance with *three* independent variates are given in detail by Day and Fisher (29) in a paper which provides an admirable text-book example of the handling of a complicated statistical analysis. The original data were supplied by J. W. Gregor, who collected samples of the seed of *Plantago maritima* from 29 different localities in Scotland and grew the plants together in an experimental garden. The 29 localities obtained are classified into three regions, Inland, Coastal

---

* A similar numerical example was previously given by Crampton and Hopkins, *Journal of Nutrition*, Vol. 8, 329–39, 1934. This paper does not, however, contain the exact test of significance of the adjusted treatment effects.

and Island, and the coastal region is further subdivided into four types of habitat. Gregor measured, amongst other factors, the length, breadth and thickness of 100 leaves from each locality. The estimated standard deviations within localities of each of these three factors form the raw material for this study, which, it will be noted, is an analysis of *variability*, since it is concerned with estimated standard deviations, and not with individual measurements.

Two points of difficulty arise at once : (1) while the distribution of an estimated standard deviation $s$ based on 99 degrees of freedom will not be far from normal, the variance of $s$ is proportional to $\sigma^2$, so that the analysis can hardly be performed directly on the values of $s$ ; (2) in natural populations in which the means vary, the standard deviation often shows an association with the mean, and if the association is at all marked, an analysis of standard deviations will merely be repeating the information which is obtained from an analysis of means. In earlier studies of variability an attempt was sometimes made to allow for this correlation by working on the ratio of the standard deviation to the mean. This, the authors point out, assumes, without logical justification, a particular form of relationship, and is analogous to the type of correction of yield for variation in plant density which was used before the introduction of the analysis of covariance.

It was noted in section III of this review that the use of log $s$ in the analysis solves the first type of difficulty. The use of log $s$, correctly adjusted for its linear regression on log (mean), also provides the correct allowance for the association of standard deviation and mean. The authors go a step further, and adjust the values of log $s$ for its association with the log (mean) of each of the three measurements, length, breadth and thickness, by a partial regression on three independent variates. The other new point in the analysis is that whilst there are three independent variates, there are also three dependent variates (the logs. of the estimated standard deviation of length, breadth and thickness). The analysis is carried out for all the variances and covariances of the dependent variates.

To turn to the applications which can be made of the analysis of covariance, Bartlett (26) gives a brief discussion of these. He distinguishes two main types. In the first, covariance is used to increase the accuracy of treatment comparisons in replicated experiments. Examples of this use are made by Bartlett (25) to experiments on the nutrition of cows, by Miles and Bryan (31) to the elimination of the effects of variations in plant density on the yields of cereals, and by Hutchinson and Panse (30) to experiments on the selection of cotton varieties for resistance to disease.

Hutchinson and Panse's method is an ingenious one; they sow alternate rows of their plots with a susceptible control variety, which is used as an indicator of the liability of the plot to attack. This method must, however, be used with discretion, as the authors point out.

In the second type of application one is interested in examining the dependence of different sets of observations on each other, and possibly also in predicting one set of observations from previous observations. Instances of this type are Brady's data (27), Snedecor's paper (32) on the control of the marking systems in University examinations in mathematics, and Snedecor and Cox's analysis (33) of the relation between the yields of different varieties of maize and the time taken to the first appearance of silks in each variety.

## V. *Other Papers*

A paper by Welch (39) concerns the validity of the application of the $z$-test to randomized blocks and the Latin squares. Fisher has pointed out that a test of significance can be made, free from any assumption of normality in the data, by calculating the value of $z$ for each of the particular set of designs from which the design used was drawn at random. A comparison on these lines of the actual and theoretical $z$-distribution was made by Eden and Yates,[*] who found good agreement. The calculation of the value of $z$ for every possible arrangement is, however, usually a lengthy process, and Welch proposes to approximate to the discontinuous $z$-distribution by a continuous curve. An approximation on the same lines for the $t$-distribution was previously given by Bartlett.[†]

Instead of $z$, Welch considers for convenience the ratio of the treatment sum of squares to the total of the treatment and error sums of squares. In normal theory, this is distributed as

$$CU^{t/2-1} (1 - U)^{e/2-1} dU \quad . \quad . \quad . \quad (3)$$

where $t$ and $e$ are the numbers of degrees of freedom attributed to treatments and error respectively. In the $U$-distribution generated by the randomization process, the mean value of $U$ is the same as that given by normal theory, but the variance is not the same for either randomized blocks or the Latin square. Welch approximates to the discontinuous $U$-distribution by choosing $t$ and $e$ in (3) to give the correct mean and variance. In actual examples the limits of the discontinuous $U$-distribution will not, except very rarely, be either 0 or 1. The effects of this discrepancy are not taken into account

[*] Eden, T., and Yates, F. *Journ. Agric. Sci.*, Vol. XXIII, 6–16, 1933.
[†] Bartlett, M. S. *Proc. Camb. Phil. Soc.*, Vol. XXXI, 228, 1935.

in the approximation, and will unfortunately be most disturbing at the tails, from which the 5 and 1 per cent. values are calculated. Welch compares the results given by normal theory and by his *U*-distribution in a number of examples, some of them uniformity trial data and some artificially constructed. He finds good agreement in all four cases of randomized blocks, but a striking difference in the significance levels in two of his six Latin squares. These results are insufficient to prove whether there is any consistent difference between normal theory and the randomization test. Welch suggests that an examination of further uniformity trial data would be useful.

BIBLIOGRAPHY

I. *Design of Experiments*

(1) Barbacki, S., and Fisher, R. A. A test of the supposed precision of systematic arrangements. *Annals of Eugenics*, Vol. VII, 189–93, 1936.
(2) Barnard, M. M. An enumeration of the confounded arrangements in the $2 \times 2 \times 2 \ldots$ factorial designs. *J. Roy. Stat. Soc., Suppl.*, Vol. III, 195–202, 1936.
(3) Cornish, E. A. Non-replicated factorial experiments. *J. Aust. Inst. of Agric. Sci.*, Vol. 2, 79–82, 1936.
(4) Fisher, R. A. The independence of experimental evidence in agricultural research. *Trans. Third International Congress of Soil Sci.*, Vol. II, 112–19, 1935.
(5) Goulden, C. H. Efficiency in field trials of pseudo-factorial and incomplete randomized block methods. *Canadian J. Res.*, C 15, 231–41, 1937.
(6) Yates, F. Complex experiments. *J. Roy. Stat. Soc., Suppl.*, Vol. II, 181–247, 1935.
(7) Yates, F. Incomplete Latin squares. *J. Agric. Sci.*, Vol. XXVI, 301–15, 1936.
(8) Yates, F. Incomplete randomized blocks. *Annals of Eugenics*, Vol. VII, 121–40, 1936.
(9) Yates, F. A new method of arranging variety trials involving a large number of varieties. *J. Agric. Sci.*, Vol. XXVI, 424–55, 1936.
(10) Yates, F. A further note on the arrangement of variety trials, quasi-Latin squares. *Annals of Eugenics*, Vol. VII, 319–32, 1937.
(11) Yates, F. The design and analysis of factorial experiments. *Imp. Bur. Soil Sci. Tech. Comm.*, No. 35, 1937.

II. *The Discriminant Function*

(12) Barnard, M. M. The secular variations of skull characters in four series of Egyptian skulls. *Annals of Eugenics*, Vol. VI, 352–71, 1935.
(13) Fisher, R. A. The use of multiple measurements in taxonomic problems. *Annals of Eugenics*, Vol. VII, 179–88, 1936.
(14) Smith, H. Fairfield. A discriminant function for plant selection. *Annals of Eugenics*, Vol. VII, 240–50, 1936.
(15) Travers, R. *British Management Congress Proc.* (in the press), 1937.

III. *Applications*

(16) Bartlett, M. S. Square root transformations in the analysis of variance. *J. Roy. Stat. Soc., Suppl.*, Vol. III, 68–78, 1936.
(17) Bartlett, M. S. Some examples of statistical methods of research in agriculture and applied biology. *J. Roy. Stat. Soc., Suppl.*, Vol. IV, 137–83, 1937.
(18) Cochran, W. G. Statistical analysis of field counts of diseased plants. *J. Roy. Stat. Soc., Suppl.*, Vol. III, 49–67, 1936.

(19) Cochran, W. G.   Problems arising in the analysis of a series of similar experiments.  *J. Roy. Stat. Soc., Suppl.*, Vol. IV, 102–18, 1937.

(20) Immer, F. R., and Le Clerg, E. L.   Errors of routine analysis for percentage of sucrose and apparent purity coefficient with sugar beets taken from field experiments.  *J. Agric. Res.*, Vol. 52, 505–15, 1936.

(21) McCleery, F. C.   Applications of the analysis of variance in experimental arrangement.  *J. Aust. Inst. of Agric. Sci.*, Vol. 1, 96–105, 1935.

(22) Saunders, A. R.   Statistical methods with special reference to field experiments.  *Union of S.A. Dept. of Agric. and Forestry Sci. Bull.*, No. 147, 1935.

(23) Smith, H. Fairfield.   The problem of comparing the results of two experiments with unequal errors.  *J. Council Sci. Indus. Res.*, Vol. 9, 211–12, 1936.

(24) Summerby, R.   The use of the analysis of variance in soil and fertilizer experiments with particular reference to interactions.  *Sci. Agric.*, Vol. XVII, 303–11, 1937.

### IV.*Analysis of Covariance*

(25) Bartlett, M. S.   An examination of the value of covariance in dairy-cow nutrition experiments.  *J. Agric. Sci.*, Vol. XXV, 238–44, 1935.

(26) Bartlett, M. S.   A note on the analysis of covariance.  *J. Agric. Sci.*, Vol. XXVI, 488–91, 1936.

(27) Brady, J.   A biological application of the analysis of covariance.  *J. Roy. Stat. Soc., Suppl.*, Vol. II, 99–106, 1935.

(28) Cox, G. M., and Snedecor, G. W.   Covariance used to analyse the relation between corn yield and average.  *Journ. Farm. Econ.*, Vol. 18, 597–607, 1936.

(29) Day, B., and Fisher, R. A.   The comparison of variability in populations having unequal means : an example of the analysis of covariance with multiple dependent and independent variates.  *Annals of Eugenics*, Vol. VII, 333–48, 1937.

(30) Hutchinson, J. B., and Panse, V. G.   An application of the method of covariance to selection for disease and resistance in cotton.  *Indian J. Agric. Sci.*, Vol. 5, 554–58, 1935.

(31) Miles, L. G., and Bryan, W. W.   The analysis of covariance and its use in correcting for irregularities of stand in agricultural trials for yield.  *Proc. Roy. Soc. Queensland*, Vol. XLVIII, No. 4, 1936.

(32) Snedecor, G. W.   Analysis of covariance of statistically controlled grades.  *Proc. Amer. Stat. Ass.*, Vol. 30, 263–68, 1935.

(33) Snedecor, G. W., and Cox, G. M.   Analysis of covariance of yield and time to first silks in maize.  *J. Agric. Res.*, Vol. 54, 449–59, 1937.

(34) Watson, S. J., and Ferguson, W. S.   Foodstuffs and quality of milk.  *J. Agric. Sci.*, Vol. III, 79–82, 1936.

(35) Wishart, J.   Tests of significance in analysis of covariance.  *J. Roy. Stat. Soc., Suppl.*, Vol. III, 79–82, 1936.

### V. *Other Papers*

(36) Hendricks, W. A.   Analysis of variance considered as an application of simple error theory.  *Annals Math. Stat.*, Vol. VI, 117–26, 1935.

(37) Kullback, S.   A note on the analysis of variance.  *Annals Math. Stat.*, Vol. VI, 76–77, 1935.

(38) Steffenson, J. F.   Free functions and the Student–Fisher Theorem.  *Skandinavisk Actuarietidskrift*, Heft 1–2, 108–28, 1936.

(39) Welch, B. L.   On the z-test in randomized blocks and Latin squares.  *Biometrika*, Vol. XXIX, 21–52, 1937.

## SOME DIFFICULTIES IN THE STATISTICAL ANALYSIS OF REPLICATED EXPERIMENTS

### W. G. COCHRAN

*(Rothamsted Experimental Station, Harpenden)*

1. *Introduction.*—The designs known as *randomized blocks* and the *Latin square*, first introduced by Fisher for experiments on English farm crops, have proved so successful that they are now in common use in most parts of the world. *The analysis of variance*, as the arithmetical procedure to be followed in working out the results is called, is not difficult to grasp or carry out, and practical workers with little mathematical knowledge have been able to plan and interpret their experiments successfully without recourse to expert advice.

In text-book examples of the analysis of variance, the data analysed are usually the weights of some crop, measured to three- or four-figure accuracy, and the treatment responses are usually quite small, of the order of 10–20 per cent. Experimenters often find, however, that the data they have to handle are quite different, qualitatively and quantitatively, from these. The differences which are most commonly met in practice may be grouped under three headings: (1) certain treatment differences may be of the order of several hundred per cent.; (2) the data to be analysed may take the form of whole numbers (e.g. numbers of plants in a plot or of wireworms in a sample of soil from the plot) or of percentages (e.g. the percentage of healthy plants in a plot, the percentage of potatoes standing on a riddle of a certain size); and (3) there may be a group of related figures to be analysed, instead of a single figure (e.g. the number of plants classed in each of four grades according to the severity of an attack by an insect, or the weights of saleable brussel sprouts picked on each of a number of occasions).

In these cases the experimenter is often in doubt whether the straightforward analysis of variance will apply, and if not, how to handle the data otherwise. This paper is an attempt to give, as simply as possible, advice on these questions.

It may be said at once that even with such data special methods are the exception rather than the rule. So long as treatment responses are the order of 50 per cent. or under and the standard error per plot is under 12 per cent., there can be little wrong with the use of the analysis of variance, no matter what form the data analysed may take.

2. *Discussion of an example.*—The first thing which the experimenter presumably wants to know is how to tell, from an inspection of the data to be analysed, whether the ordinary analysis of variance can be used. Before discussing this, it will be helpful to consider an example showing how and when the straightforward analysis of variance gives misleading results. Table 1 shows the results of counts of the numbers of poppies in different plots of an experiment on oats, the experiment consisting of four randomized blocks of six treatments.[1]

[1] The data are taken from a paper [1] by Bartlett, who discusses briefly their analysis. The treatments have been renumbered for convenience.

Whether he carries out the computations himself or not, the experimenter should make a point of looking at the data before analysis and judging the results by inspection. If the results of the analysis of variance do not agree with his original expectations, he should ask himself why this is so. Sometimes an inspection of the individual figures will show that treatment differences were not so consistent as he had at first supposed, and sometimes the comparison leads to the detection of errors in the analysis itself, particularly with experiments of more complicated design. Practice in the inspection of variable material is invaluable to the experimenter and will richly repay the efforts expended on it.

TABLE 1. *Numbers of Poppies in Oats*

| Treatment | (1) | (2) | (3) | (4) | (5) | (6) | Mean |
|---|---|---|---|---|---|---|---|
| Block A . | 538 | 438 | 77 | 115 | 17 | 18 | 200·5 |
| „   B . | 422 | 442 | 61 | 57 | 31 | 26 | 173·2 |
| „   C . | 377 | 319 | 157 | 100 | 87 | 77 | 186·2 |
| „   D . | 315 | 380 | 52 | 45 | 16 | 20 | 138·0 |
| Mean | 413·0 | 394·8 | 86·8 | 79·2 | 37·8 | 35·2 | |

The results of this experiment are fairly clear on inspection. Treatments (1) and (2) are much higher than any of the others; further, (3) and (4) are clearly significantly above (5) and (6), being well above them in every block. On the other hand, there is nothing to choose between the members of the three pairs (1) and (2), (3) and (4), or (5) and (6). Such an experiment, if standing by itself, need cause the experimenter no trouble, and an analysis of variance is hardly necessary. Standard errors might, however, be required for comparison or combination with a number of similar experiments. The analysis of variance is as follows:

| | D.F. | Sum of squares | Mean square |
|---|---|---|---|
| Blocks   . | 3 | 12,877 | |
| Treatments   . | 5 | 641,024 | 128,205 |
| Error   . | 15 | 39,799 | 2,653 |
| Total | 23 | 693,700 | |

The standard error per plot is $\sqrt{(2,653)} = 51\cdot51$, so that the standard error of a treatment mean (4 blocks) is $25\cdot76$, and since there are 15 degrees of freedom for error, the significant difference between two treatments is

$$25\cdot76 \times 1\cdot414 \times 2\cdot131^{1} = 77\cdot6$$

The experimenter will notice with some surprise that neither (3) nor (4) is significantly above (5) or (6), though by inspection it seemed obvious that they ought to be. To discover what is wrong, the 15 degrees of freedom which make up the error must be examined further. The subdivision of error degrees of freedom in a randomized-blocks experiment is not difficult and should be familiar to every worker. The 15 degrees of freedom may be divided into 3 which are strictly appropriate to the

---

[1] The value of *t* for 15 degrees of freedom.

comparison of treatments (1) and (2), a further 3 which are the error of the comparison of the mean of (1) and (2) with the mean of (3), (4), (5), and (6), and a further 9 which refer to the comparisons of (3), (4), (5), and (6) with each other. To obtain these, first calculate

$$(1)-(2), \text{ and } 2(1)+2(2)-(3)-(4)-(5)-(6)$$

as shown in Table 2.

TABLE 2. *Partition of Error Sum of Squares*

| Block | (1)−(2) | $2[(1)+(2)]$ $-[(3)+(4)+(5)+(6)]$ |
|-------|---------|-----------------------------------|
| A | − 100 | 1725 |
| B | + 20 | 1553 |
| C | − 58 | 971 |
| D | + 65 | 1257 |
| Total | − 73 | 5506 |

Sums of squares of deviations (3 d.f.)

| 16,657 | 331,315 |
|--------|---------|
| 8,328* | 27,610* |

* On a single-plot basis.

Thus, for example, $1,725 = 876+1076-77-115-17-18$, and is four times the difference between the mean of (1) and (2) and the mean of the other four treatments in block A. The sums of squares of deviations from the mean, shown at the foot of each column, must be divided by 2 and $2^2+2^2+1+1+1+1 = 12$ respectively [2], before putting them in the analysis of variance, which is based on the yield of a single plot. The resulting sums of squares are the figures with an asterisk in Table 2.

The remaining 9 degrees of freedom may be found by analysing the part of the experiment containing treatments (3) to (6) only, as 4 randomized blocks of 4 treatments each. This gives, on a single-plot basis, 3,861. Since

$$39,799 = 8,328+27,610+3,861,$$

this proves that the subdivision made is a correct one and at the same time checks all the arithmetic involved. Thus we have the following subdivision of the error degrees of freedom:

| Comparison | D.F. | Sum of squares | Mean square | Standard error |
|------------|------|----------------|-------------|----------------|
| (1)−(2)  .    . | 3 | 8,328 | 2,776 | 52·7 |
| $\frac{1}{2}[(1)+(2)]$−rest  . | 3 | 27,610 | 9,203 | 95·9 |
| rest  .    .    . | 9 | 3,861 | 429 | 20·7 |

The mean squares of the two comparisons involving the first two treatments are respectively 6·47 and 21·5 times the mean square involving comparisons of treatments (3) to (6) only. Both these ratios are significant, as may be verified from the *z*-table or the variance-ratio table.

Thus comparisons involving the two treatments with high numbers of poppies have a higher standard error than comparisons involving treatments (3) to (6), which gave much smaller numbers of poppies. The pooling of these errors into one estimate is therefore quite unjustified and has the effect of under-estimating the standard error of some treatment comparisons and over-estimating that of others. In particular the standard error per plot appropriate to comparisons between treatments (3) to (6) appears to be 20·5, instead of 51·5 as given by a combined analysis.

At this stage we may feel that the 9 d.f. error also merits further examination, since it too is composed of the errors of a large treatment difference [(3)+(4)—(5)—(6)], and two small ones. As an exercise, the reader may divide the 9 degrees of freedom into three sets of 3 degrees of freedom as follows:

| Comparison | D.F. | Sum of squares | Mean square | Standard error |
|---|---|---|---|---|
| (3)+(4)—(5)—(6) . | 3 | 1536·0 | 512·0 | 22·6 |
| (3)—(4)    .    . | 3 | 2266·5 | 755·5 | |
| (5)—(6)    .    . | 3 | 58·5 | 19·5 | |
| Total | 9 | 3861·0 | | |

The 5 per cent. and 1 per cent. values of the ratios of the mean squares for two sets of 3 degrees of freedom are 9·28 and 29·5 respectively, so that the errors of comparisons involving treatments (3) and (4) are significantly higher than that involving only the difference of the treatments (5) and (6). The first two mean squares are not significantly different, but considering how variable the errors in this experiment have proved, it would not be fair to combine the two. Thus for an error for the comparison of the mean (3) and (4) with that of (5) and (6), we are forced to the value 22·6, based on only 3 d.f., so that the value previously reached, 20·5, is not only too low, but based on too many degrees of freedom. With the new error, the 5 per cent. significant difference between the mean of (3) and (4) and that of (5) and (6) is found to be 36·0, and the mean difference itself is 46·5.

This example illustrates the most important condition for the application of a combined analysis of variance: *the treatment comparisons must be such that they may all be reasonably expected to have the same variance.* When this is not so, the error found by the analysis is a weighted mean of a number of different errors and is not really applicable to any of the treatment comparisons.

The question of differential variability may be looked at from another aspect, which is useful for detecting cases that may require special methods. An idea of the variability of a treatment in this experiment may be obtained by subtracting the smallest plot bearing that treatment from the largest. Thus for treatment (1) the difference, which may be called the *range*, is

$$538 - 315 = 223.$$

The ranges and the mean yields are:

*Treatment*

|          | (1) | (2) | (3) | (4) | (5) | (6) |
|----------|-----|-----|-----|-----|-----|-----|
| Range .    . | 223 | 123 | 105 | 70 | 71 | 59 |
| Mean .    . | 413 | 395 | 87 | 79 | 38 | 35 |

The range for each treatment shows quite a close relation to the mean, increasing as the mean increases. This relation of the range to the mean is a warning that a combined analysis of variance is not justified; conversely, if the range shows no relation to the mean, a combined analysis is usually quite correct, though exceptions to this rule occasionally occur. The use of range in this respect is only satisfactory where treatment differences are large compared with block differences, as in this experiment, or where block differences are large compared with treatment differences.

3. *Types of data which may require special treatment.*—The theoretical conditions for the application of the analysis of variance are that the experimental errors to which the data are subject shall be independently and normally distributed with the same variance. Of these restrictions the last is the most important, as illustrated by the preceding example. Indeed, as every experimenter knows, the analysis of variance is widely used in practice without any guarantee that the data are strictly normal in distribution.

With certain types of non-normal data, however, there are theoretical reasons for expecting that the standard error of a treatment yield will depend on the mean value of the treatment yield. Thus different treatments (assuming that their means differ) will have different variances, and the pooling of the errors which is implicit in the simple analysis of variance is unjustified. The most important examples of this class will be discussed later. Their common feature is that the experimental errors are markedly *skew* in distribution.

The advantage of pooling the errors is that the resultant error is more accurately determined and shows itself in every use to which the results of the experiment may be put. For instance, in the result of the last section, the value of $t$ for 15 d.f. (the original error) is 2·131, as against 3·182 for 3 d.f., so that a treatment difference must be one and a half times as large in the second case to be detected at the same level of significance. Thus there is a real loss in splitting the combined error, and with markedly skew data the principal method of attack is to transform them at once to a scale on which the data are sufficiently normal that a combined analysis may be made.

The cases in which the experimenter must think twice before going ahead with an analysis may be divided into two types: (1) Material in which there is no particular reason to suspect violent skewness or departure from normality, but where, in a particular experiment, the treatments appear on inspection to have different variances. These cases include, generally speaking, all crop yields and all whole-number counts which are over 100 on every plot. (2) Material in which there are theoretical grounds for suspecting non-normality and a consequent relation between the standard error and the mean. This type includes small whole numbers and percentages based on small numbers, and also occurs occasionally in new investigations.

The first type should be noted by examination of the yields before analysis. No hard and fast rules can be given, but a combined analysis should not be used when some treatment differences are much larger than others. There is no reason to suppose that the variations caused in plots carrying a 30-cwt. crop of wheat grain by fertility differences, bird damage, lodging, variations in seeding and germination, &c., will be of the same size as those on plots carrying a 10-cwt. crop, and if most of the treatments in an experiment give yields varying round 25–35 cwt., but one or two vary round 8–12 cwt., not only are differences between the sets clearly significant, but we have no grounds for using the variations in one set as part of the estimate of the standard error of the other set. A common example of this type is when there is a partial failure of one or two treatments.

It is important that with such cases the experimenter should have an *objective* rule to decide when to do a combined analysis and when to omit some treatments, because there is no certainty that the omission of some treatments will reduce the apparent standard error of the comparisons in which he is most interested, and in the absence of an objective rule there is a temptation to take whichever method gives the lower error. A rule which appears to work successfully for routine analysis on English farm crops is to omit from the main analysis treatments which yield consistently more than double or less than half the main group of treatments.

Sometimes, of course, the treatment variances may be different even when their mean yields do not differ greatly. It is here that inspection of the data before analysis is useful. As an example, consider a liming experiment on mangolds grown on acid soil. The treatments consisted of three levels of two types of lime—limestone and chalk—with two controls, applied in four randomized blocks of eight plots each. The plant numbers are shown in Table 3.

TABLE 3. *Mangolds, Plant Numbers*

| Block | Control | | Chalk | | | Limestone | | | Total |
|---|---|---|---|---|---|---|---|---|---|
| | 0 | 0 | 1 | 2 | 3 | 1 | 2 | 3 | |
| I | 140 | 49 | 98 | 135 | 117 | 81 | 147 | 130 | 897 |
| II | 142 | 37 | 132 | 151 | 137 | 129 | 131 | 112 | 971 |
| III | 36 | 114 | 130 | 143 | 137 | 135 | 103 | 130 | 928 |
| IV | 129 | 125 | 153 | 146 | 143 | 104 | 147 | 121 | 1068 |
| Total | 447 | 325 | 513 | 575 | 534 | 449 | 528 | 493 | |
| Range | 106 | 88 | 55 | 16 | 26 | 54 | 44 | 18 | |

Here the plant numbers on some of the control plots are as good as those given by the best treatments, but the controls are much more variable than the treated plots, owing to partial failures on some of the former—a common effect of acidity. The same effect is also apparent, to a smaller extent, on the single dressings. It is already clear that the liming has counteracted one of the harmful effects of acidity, and in examining the further questions of the optimum dressing to apply and the relative effectiveness of chalk and limestone, the controls are best left out. The reader may verify that the standard error per plot is reduced from 28·5 to 17·1 by the omission of the controls. The question whether the standard error is reduced or increased is, however, beside the point, since the aim of the analysis of variance is to obtain, not the lowest possible apparent error, but the most accurate estimate of the actual errors to which the treatment comparisons are subject.

As regards the general treatment of such cases, there are several possibilities. (1) Sometimes unequal variability is caused by a comparative failure of one or two plots, which may be traced on inquiry to damage or error which was not in any way caused by the treatments on these plots and which the other plots completely escaped. In these cases new yields may be substituted on these plots by methods described by Yates [3], and the complete analysis may then be carried out. The reader is advised to study Yates's paper in detail, as the missing plot technique is a valuable weapon and its correct use requires much discretion. (2) Often the analysis can be carried out by omitting one or more treatments (or blocks) from the main analysis, as suggested in the last example. With randomized blocks this causes no computational difficulties. With Latin squares one treatment (or one row or column), or one treatment and one row or column, or one row and one column may be omitted without complicating the work much. The alterations required in the analysis are described by Yates [4], who gives two worked examples, one with a treatment missing and one with a column and a treatment missing, which the reader is recommended to work out for himself.

If two treatments (or rows or columns) of a Latin square have to be omitted or are missing, the exact analysis is rather complicated. There is an approximate solution which is not difficult to carry out and gives good results; but in my experience the Latin square with two treatments missing is too rare to merit its inclusion here.

In the next section methods will be developed for making a combined analysis by first transforming the data to a scale in which the variances of the different treatments are approximately equal. These methods should not be used with the type of data we have been considering in this section. The function of the transformation is to turn skew distributions into distributions which are approximately normal with the same variance. The present type of data is, however, as a rule normally distributed in the original scale (though perhaps with different variances), and although an empirical transformation might be found which apparently equalized the variances, the transformed data would probably be skew.

Finally, the standard error per plot as a percentage of the mean yield (sometimes called the *coefficient of variation*) may be noted. Where a number of experiments on the same crop are made, the experimenter soon learns the limits of accuracy which are to be expected. This encourages him to look for improvements in the technique if the standard of accuracy attained is not satisfactory, and it also facilitates the detection of bad sites, careless field-work, and copying or arithmetical errors.

4. *Small whole numbers.*—There are two common distributions involving whole numbers—the *binomial* and the *Poisson*. The reader who is unacquainted with these should read §§ 15–19 of Fisher's *Statistical Methods*. If all plants in a field had an equal chance (say 1 in 5) of being infected with a certain disease, and we divided the field into plots each containing twenty plants and counted the total numbers of diseased plants in each plot, these numbers would follow a binomial distribution. The distribution of a *percentage* is therefore obtained directly from the binomial distribution. The Poisson distribution is also a special case of the binomial—when the chance of the event in question is very small, but we survey the result of a large number of trials.

The properties of these distributions which concern us are that they are both skew (except the binomial for an even chance) and that there is a known relation between the standard error and the mean.

For a binomial distribution with plots of $n$ units, in each of which the chance of the feature which we are counting (e.g. disease) is $p$, the standard error of a plot total $x$ is

$$\sqrt{\{np(1-p)\}}$$

The standard error of the observed *percentage* $\dfrac{100x}{n}$ of diseased plants in a plot is

$$\frac{100}{n}\sqrt{\{np(1-p)\}} = \sqrt{\left\{\frac{P(100-P)}{n}\right\}},$$

where $P = 100p$ is the true percentage of diseased plants. In both these cases the standard error is greatest at 50 per cent., about which value it is symmetrical, decreasing to zero at 0 or 100 per cent. For a Poisson distribution with mean $m$ the standard is $\sqrt{m}$. Thus if the treatments in an experiment produce differences in the values of $p$ and $m$, they have different variances.

Tests of significance for the Poisson and binomial series are given in *Statistical Methods* (§§ 16–19), by calculating a value which may be looked up in the $\chi^2$ Table. These tests apply to conditions which are much more uniform than those in replicated field trials, and should not be used here. They require, for instance, that the chance of a plant being diseased shall be the same in all replicates of a given treatment, and thus take no account of differences in the incidence of disease on different plots. In fact, the variance with small whole numbers may be divided into two components, one being the variation of the binomial or Poisson type and the other the usual plot-to-plot variation.

With yield data and counts running into three figures per plot, a standard error per plot of 12 per cent. or under is commonly attained. With small whole numbers, however, consideration of the binomial or Poisson variation alone shows that this standard of accuracy is not to be expected. For the binomial series the standard error of a plot total $x$

is $\sqrt{\{np(1-p)\}}$, and the mean value of $x$ is $np = m$ (say). Thus the standard error per cent. per plot is

$$\frac{100\sqrt{\{np(1-p)\}}}{np} = 100\sqrt{\left(\frac{1-p}{m}\right)}.$$

This figure also applies of course to fractions and percentages. The value for the Poisson series with a mean plot total $m$ is similarly found to be $100/\sqrt{m}$. For plot totals of the order of 20, 10, and 5 the standard errors per plot are thus of the order of 22, 31, and 44 per cent., plus a component due to inter-plot variation. It follows that with small whole numbers, treatment differences have to be large before they have a chance of being significant, and, of course, the larger the treatment differences the more unequal their variances are likely to be. Thus with small whole numbers we cannot take the line used in previous sections, that if treatment differences are large they are bound to be clearly significant.

5. *Transformations.*—With a distribution in which the relation between the variance and the mean is known, a transformation can always be found to a new scale in which the variance is, to a first approximation, independent of the mean. The three transformations which have been found most useful in practice are shown in Table 4.

TABLE 4. *Transformations*

| Distribution | Data | Relation between variance and mean | Transformation | Variance in new scale |
|---|---|---|---|---|
| Poisson | small whole numbers ($x$) | $V = x$ | $\sqrt{x}$ | $\frac{1}{4}$ |
| Binomial | fractions ($p$) percentages ($P$) | $V = p(1-p)/n$ <br> $V = P(100-P)/n$ | $\sin^{-1}\sqrt{p}$ <br> $\sin^{-1}\sqrt{(P/100)}$ | $821/n$ <br> $821/n$ |
| | numbers ($x$) | $V = \lambda x^2$ | $\log_{10} x$ | $0.189\lambda$ |

The relation between the variance and the mean for the Poisson and binomial distributions has already been discussed. For data of the Poisson type, the transformation suggested is to carry out the analysis on the *square roots* of the original data. A table of square roots will be found in almost every book of four-figure tables. The transformation to be used with fractions or percentages is called the *inverse sine* and looks much more complicated; we use the angle (between 0° and 90°) whose sine is the square root of the fraction. In fact, however, once a table has been prepared, this transformation is just as easy to use as the square-root one. A table will be found in Fisher and Yates's Tables [5]. The values can also be read off directly on any slide-rule which has a sine scale on the back, by setting the sine scale in position at the front. To transform 20·4 per cent., place the slide at 20·4 on the upper (square) scale and the value in degrees is read off at once as 26·9 on the sine scale.

The third type of data shown in the Table arises when the errors to which any plot-'yield' is subject are proportional to the value of the 'yield'. In this case the analysis is performed on the logs. of the original data.

The last column of the Table shows the theoretical values of the

3988.22                                    M

variance in the new units. These figures are not of any great practical importance, because other sources of variation tend to increase the variance in an actual experiment. With data of a descriptive or qualitative nature, however, the observer who is scoring the plots may be influenced, perhaps unconsciously, if he knows the treatments, and this is sometimes detected by noticing that the errors are consistently lower than their theoretical values.

The way in which the transformations are derived is not difficult to follow and will be exemplified for the square-root transformation. If $m$ is the mean value of $x$ and $\epsilon = x - m$ is the error of $x$, then

$$\sqrt{x} = \sqrt{(m+\epsilon)} = \sqrt{m} \sqrt{(1+\epsilon/m)}$$
$$= \sqrt{m}(1+\epsilon/2m), \tag{1}$$

approximately, if $\epsilon$ is small compared with $m$.

For instance, if $m = 9$, $\epsilon = 1$, the left-hand side is $\sqrt{10} = 3\cdot162$ and the right-hand side is $3(1+\frac{1}{18}) = 3\cdot167$.

Thus
$$\sqrt{x} = \sqrt{m} + \epsilon/2\sqrt{m}. \tag{2}$$

Since the mean of $\epsilon$ is zero, it follows that, to a first approximation, the mean value of $\sqrt{x}$ is $\sqrt{m}$. The variance of $\sqrt{x}$ is by definition the mean value of $(\sqrt{x} - \sqrt{m})^2$, and from equation (2) is equal to

$$\text{Mean } \epsilon^2/4m = \frac{\text{Variance of } x}{4m}.$$

Thus if the variance of $x$ is $m$ (as it is for the Poisson series), the variance of $\sqrt{x}$ is independent of $m$ and is $\frac{1}{4}$.

The reader with an elementary knowledge of the integral calculus can verify from this example that, in general, if the variance of $x$ is $f(m)$, the transformation which equalizes the variance is the integral of $1/\sqrt{\{f(x)\}}$. The condition under which the transformation holds is that the standard error of $x$ must be small compared with the mean value of $x$.

For the Poisson distribution, the ratio of the standard error to the mean is $1/\sqrt{m}$, which is too large for the conditions of the transformation when $m$ is below 10. Bartlett [1] has shown that with $m$ below 10 the square-root transformation tends to over-correct, so that the treatments with the smallest means have the largest ranges in square roots. He has shown that by adding $\frac{1}{2}$ to each observation before taking the square root, the variances are much better equalized, and with experiments in which the majority of the numbers are below 10, this transformation should be used instead of the simple square roots.

For individual experiments, in my own experience, the inverse-sine transformation is rarely needed. The relation between the variance and the mean for percentages is shown below (the variance is symmetrical about 50 per cent.):

TABLE 5.

| Percentage | 1 | 5 | 10 | 20 | 30 | 40 | 50 | 60 | 70 | 80 | 90 |
|---|---|---|---|---|---|---|---|---|---|---|---|
| $\dfrac{n \text{ Variance}}{100}$ | 1 | 4·8 | 9 | 16 | 21 | 24 | 25 | 24 | 21 | 16 | 9 |

For percentages between o and 20, the variance increases roughly as the mean, so that a square-root transformation can be used. For percentages between 30 and 70, on the other hand, the variance is more or less independent of the mean, so that the analysis can be performed on the original data without transformation. For percentages between 80 and 100, the square root of $(100-P)$ may be used. Most experiments with percentages can be dealt with by one of these methods if desired. If, however, experiments involving percentages are regularly met with, or with a group of experiments on the same point, it is worth having a uniform method of analysis, and the inverse-sine transformation should be used throughout.

These methods are only necessary with percentages in which the *numerators* are under 100; with numerators over 100 a direct analysis can usually be made. With numerators between 50 and 100 per plot the case is more doubtful, but a direct analysis can often be made, provided that treatment and block differences are not large and that there is no clear correlation between the range and the mean.

Numerical examples of the uses of the three transformations are given in the next three sections.

6. *The square-root transformation.*—The plan below shows the results of an experiment with a $5 \times 5$ Latin square, the treatments being a control (o) and four different soil fumigations applied in the previous year [6]. The data shown directly under the treatment symbols are the total numbers of wireworms in four samples per plot. The 'yields' are, with one exception, all under 10, so that the $\sqrt{(x+\frac{1}{2})}$ transformation is appropriate. The actual square roots are, however, given for comparison in the second line of figures, and the square roots of $(x+\frac{1}{2})$ in the third line.

TABLE 6. *Plan and Numbers of Wireworms*

| P | O | N | K | M |
|---|---|---|---|---|
| 3* | 2 | 5 | 1 | 4 |
| 1·73† | 1·41 | 2·24 | 1·00 | 2·00 |
| 1·87‡ | 1·58 | 2·34 | 1·22 | 2·12 |
| M | K | O | N | P |
| 6 | o | 6 | 4 | 4 |
| 2·45 | 0·00 | 2·45 | 2·00 | 2·00 |
| 2·55 | 0·71 | 2·55 | 2·12 | 2·12 |
| O | M | K | P | N |
| 4 | 9 | 1 | 6 | 5 |
| 2·00 | 3·00 | 1·00 | 2·45 | 2·24 |
| 2·12 | 3·08 | 1·22 | 2·55 | 2·34 |
| N | P | M | O | K |
| 17 | 8 | 8 | 9 | o |
| 4·12 | 2·83 | 2·83 | 3·00 | 0·00 |
| 4·18 | 2·92 | 2·92 | 3·08 | 0·71 |
| K | N | P | M | O |
| 4 | 4 | 2 | 4 | 8 |
| 2·00 | 2·00 | 1·41 | 2·00 | 2·83 |
| 2·12 | 2·12 | 1·58 | 2·12 | 2·92 |

* Number of wireworms ($x$).      † $\sqrt{x}$.      ‡ $\sqrt{(x+\frac{1}{2})}$.

The treatment totals and the ranges in each set of units are shown below:

| Scale | | K | P | O | M | N | S.E. |
|---|---|---|---|---|---|---|---|
| Original $x$ . | Total | 6 | 23 | 29 | 31 | 35 | |
| | Range | 4 | 6 | 7 | 5 | 13 | |
| $\sqrt{x}$ . . | Total | 4·00 | 10·42 | 11·69 | 12·28 | 12·60 | |
| | Range | 2·00 | 1·42 | 1·59 | 1·00 | 2·12 | |
| $\sqrt{(x+\frac{1}{2})}$ . | Total | 5·98 | 11·04 | 12·25 | 12·79 | 13·10 | ±1·36 |
| | Range | 1·41 | 1·34 | 1·50 | 0·96 | 2·06 | |

The correlation between the range and the mean is clearly seen in the original data. The tendency of square roots to over-correct with such small numbers is also apparent, as the range of $K$ is now larger than that of the next three treatments in order of increasing means, and indeed is only less than that of $N$ because the latter contained one value, 17, which was double the number on any other plot on the field. The analysis of variance of $\sqrt{(x+\frac{1}{2})}$ is as follows:

| | D.F. | S.S. | M.S. |
|---|---|---|---|
| Rows . . . | 4 | 2·4812 | |
| Columns . . | 4 | 0·9026 | |
| Treatments . . | 4 | 6·8747 | 1·7187 |
| Error . . . | 12 | 4·4551 | 0·3712 |

The mean-square error is not greatly above the theoretical value 0·25. The standard error is 0·609, so that the standard error of a treatment total is 1·36. The reduction in wireworm numbers due to $K$ is clearly significant, but there are no other treatment effects. All tests of significance of treatment differences must be made in the transformed scale, using the standard error, 1·36.

In presenting a summary of the results of this experiment, the data should first be re-converted to actual numbers, since it is the numbers themselves in which we are interested. The transformation is made by squaring each treatment *mean* per plot and subtracting one-half. The numbers are:

*Treatment Means per Plot*

| K | P | O | M | N |
|---|---|---|---|---|
| 0·94 | 4·38 | 5·50 | 6·04 | 6·36 |

The values are slightly lower than those obtained by taking the means of the original numbers; this is a property of transformations of the square-root type. In presenting these means, it should be stated that no single standard error can be applied to them, but that the standard error of $\sqrt{(x+\frac{1}{2})}$ is 0·272. This is important, because a reader of the report may wish to use the figures for a comparison quite different from those in which the experimenter himself was interested and must be given sufficient data to do so. One of the most valuable features of the modern replicated experiment is that the information it provides is objective and open to all, and summaries of results should be presented with this fact in mind.

7. *The inverse-sine transformation.*—The data in Table 7 are the results of an experiment with $5 \times 5$ randomized blocks on the control of virus infection of the hyoscyamus plant, which is grown for the manufacture of pharmaceutical preparations [7]. The treatments were a control (A) and different times of spraying. The plants were graded into three classes: clean, moderately infected, and severely infected.

TABLE 7. *Hyoscyamus: Healthy and Infected Plants.*

| Treat-ments | | *I* | *II* | *III* | *IV* | *V* | *Total* |
|---|---|---|---|---|---|---|---|
| | | | | *Blocks* | | | |
| A | No. of plants . . | 147 | 128 | 170 | 143 | 149 | |
| | Moderate infection . | 19 | 30 | 28 | 17 | 4 | |
| | 'Severe' infection . | 11 | 5 | 2 | 5 | 7 | |
| | Per cent. infection . | 20·4 | 27·3 | 17·7 | 15·4 | 7·4 | 88·2 |
| | Degrees . . . | 26·9 | 31·5 | 24·9 | 23·1 | 15·7 | 122·1 |
| B | No. of plants . . | 140 | 127 | 156 | 158 | 201 | |
| | Moderate infection . | 30 | 25 | 23 | 10 | 4 | |
| | 'Severe' infection . | 2 | 1 | 4 | 4 | 6 | |
| | Per cent. infection . | 22·9 | 20·5 | 17·3 | 8·9 | 5·0 | 74·6 |
| | Degrees . . . | 28·6 | 27·0 | 24·6 | 17·4 | 12 9 | 110·5 |
| C | No. of plants . . | 140 | 186 | 135 | 107 | 216 | |
| | Moderate infection . | 20 | 20 | 19 | 15 | 3 | |
| | 'Severe' infection . | 6 | 4 | 0 | 3 | 10 | |
| | Per cent. infection . | 18·6 | 12·9 | 14·1 | 16·8 | 6·0 | 68·4 |
| | Degrees . . . | 25·5 | 21·0 | 22·1 | 24·2 | 14·2 | 107·0 |
| D | No. of plants . . | 148 | 142 | 158 | 126 | 147 | |
| | Moderate infection . | 21 | 18 | 11 | 13 | 10 | |
| | 'Severe' infection . | 5 | 7 | 6 | 1 | 0 | |
| | Per cent. infection . | 17·6 | 17·6 | 10·8 | 11·1 | 6·8 | 63·9 |
| | Degrees . . . | 24·8 | 24·8 | 19·2 | 19·5 | 15·1 | 103·4 |
| E | No. of plants . . | 130 | 136 | 202 | 147 | 109 | |
| | Moderate infection . | 16 | 0 | 13 | 14 | 2 | |
| | 'Severe' infection . | 0 | 6 | 6 | 2 | 0 | |
| | Per cent. infection . | 12·3 | 4·4 | 9·4 | 10·9 | 1·8 | 38·8 |
| | Degrees . . . | 20·5 | 12·1 | 17·9 | 19·3 | 7·6 | 77·4 |
| | Total per cent. infection . . | 91·8 | 82·7 | 69·3 | 63·1 | 27·0 | 333·9 |
| | Total degrees . . | 126·3 | 116·4 | 108·7 | 103·5 | 65·5 | 520·4 |

Two of the figures given in [7], Table IV, were in error and have been corrected in the above Table.

Consider first the percentage of infected plants (shown in the fourth row of the Table for each plot). These percentages are based on numerators of about 20 per plot, which is sufficiently small for a transformation to be advisable. The percentages are nearly all under 20, so that a square-root transformation could be used; the data will, however, be taken as a numerical example of the use of the inverse sine. The transformation to degrees was made by Fisher and Yates's table [5], proportional parts being used mentally for the decimal place. The analysis of variance in degrees is:

|  | D.F. | S.S. | M.S. |
|---|---|---|---|
| Blocks . . . | 4 | 431·12 | .. |
| Treatments . . | 4 | 217·35 | 54·34 |
| Error . . . | 16 | 147·38 | 9·21 |

The summary of results in degrees and in percentages is shown below.

TABLE 8. *Summary of Results*

|  | Treatment | | | | | |
|---|---|---|---|---|---|---|
|  | A | B | C | D | E | S.E. |
| Degrees . . . | 24·4 | 22·1 | 21·4 | 20·7 | 15·5 | ±1·36 |
| Percentages . . . | 17·0 | 14·1 | 13·3 | 12·5 | 7·2 | |

The percentages are obtained by transferring back from degrees, using the same table or a slide-rule. For instance, to transform $D$ $(20\cdot7°)$ by the table, we note that the readings for 12 per cent. and 13 per cent. are $20\cdot3°$ and $21\cdot1°$ respectively. Thus $20\cdot7° = (12+\frac{4}{8})$ per cent. $= 12\cdot5$ per cent.

All spraying treatments have reduced the percentage of diseased plants, and treatment $E$, which was the most intensive spraying, has proved significantly better than any of the other treatments. The field results in Table 7 also give the numbers of moderately and severely infected plants. This raises the general question of the analysis of graded data. If a single index of the effect of the treatments is required, the cash value of the plot-yield has an immediate interest if it can be obtained, as is generally the case with data which are graded for the market. Where the cash value is not obtainable, a single index which is quite effective in practice may be obtained by assigning conventional scores on common-sense grounds to the various grades. With the present data we might, for instance, assign a score of 3 to a healthy plant, 1 to a moderately diseased plant, and 0 to a severely diseased plant. The 'value' of the first plot is then $3(117)+19 = 370$.

Where experiments are carried out for the increase of knowledge and not principally for the immediate commercial interest, the experimenter should know beforehand what sort of information he expects to derive from any set of observations which he decides to make. The position of the experimenter who has amassed a pile of field data and is somewhat embarrassed to know what to do with them is not one to be commended. With the type of data in the present example, the information might be divided into three parts:

1. The total plant-number (in this particular case the treatments were applied too late to have any possible effect).
2. The percentage of diseased plants, which has been analysed.
3. The severity of the disease, which might be measured by assigning a score of $+1$ to a plant moderately diseased and $-1$ to a plant severely diseased.

The mean score for plot 1 is $(19-11)/30 = +0.267$.[1] These data could also be analysed. It is, however, clear on inspection that the treatments have had no effect on severity of disease. With diseased plants in three or four grades, convenient scores are 1, 0, −1, and 3, 1, −1, −3, respectively.

The above division is a useful one from which to summarize the results. The three factors do not, however, contain the whole of the information unless they are independent. With sugar-beet, for instance, there is commonly a negative correlation between yield of roots and sugar percentage, so that the standard errors of the total *weight* of sugar (their product) cannot be obtained from individual analyses of the roots and sugar percentage, but demands a knowledge of their covariance.

8. *The logarithmic transformation.*—The transformation to logs. equalizes the variance when it is proportional to the square of the mean; it is thus a much more powerful transformation than the square root or the inverse sine. An important case of this type of variation occurs when we are assessing the effect of a treatment on the *consistency* of the experimental material, as measured by the range or standard deviation of the material within a plot. An example of the use of logs. in this connexion, with full numerical working, is given in [8].

The log. transformation also proves useful in dealing with new material of unknown distribution. A good example occurs in an experiment on the times which a person takes to respond to each of a list of words read out to him. A list of 100 words was divided into four groups (A, B, C, D) of 25 words and each group was read out in random order to the subject. This was repeated for six sittings. Table 9 shows in descending order the total time taken to respond to 25 words at each sitting and the sum of squares of the deviations of these 25 times from their mean. The distribution of the reaction times is a very skew one, as the relation between the sum of squares and the mean indicates. The third line of the table shows the ratio of the square root of the sum of squares to the mean. The ratios are much less variable and show little relation to the means. Thus the standard errors of the original data are proportional to the means, and a logarithmic transformation is suggested.

The principal objection to transformations—that we are really interested in the original and not the transformed data—scarcely applies here, because there is nothing fundamental about the reaction times themselves; the observer might, for instance, with equal validity have measured the rates of reaction.

Logarithms may also be of use with whole-number data which cover a wide range. An example of this occurred in the analysis of the number of insects caught in a light-trap on different nights. Here the catches varied from 0 to 72,000 per night, and the resulting distribution for any species was markedly skew. The log. transformation cannot be used directly for zero values, and indeed with such a wide range the

[1] Notice that this score, being a measure of the severity, and not of the amount, of disease, is unaltered if the numbers of moderately and severely diseased plants are halved by a treatment.

transformation wanted was one which acted like the square-root transformation for small numbers, but like the log. transformation for large
numbers. Such a transformation may be found by adding 1 to each
number before taking the log. This behaves like the square root for
numbers up to 10 and thereafter approaches the direct log.

A further justification of the use of logs. in this case is that the effects
of weather factors such as temperature, amount of cloud, &c., may be
presumed to be proportional to the number of insects and hence additive
on a log. scale. A discussion of this example is given by Williams [9].

TABLE 9. *Reaction Times*

| Sittings | Sets of 25 words | | | |
|---|---|---|---|---|
| | A | B | C | D |
| I | 435* | 523 | 349 | 481 |
| | 1,640† | 3,350 | 1,089 | 2,223 |
| | 0·093‡ | 0·111 | 0·095 | 0·098 |
| II | 226 | 310 | 220 | 301 |
| | 415 | 908 | 438 | 809 |
| | 0·090 | 0·097 | 0·095 | 0·094 |
| III | 275 | 292 | 311 | 257 |
| | 684 | 789 | 1,100 | 651 |
| | 0·095 | 0·096 | 0·107 | 0·099 |
| IV | 207 | 204 | 228 | 223 |
| | 377 | 329 | 479 | 398 |
| | 0·094 | 0·089 | 0·096 | 0·089 |
| V | 188 | 214 | 202 | 283 |
| | 194 | 212 | 296 | 823 |
| | 0·074 | 0·068 | 0·085 | 0·101 |
| VI | 209 | 263 | 201 | 282 |
| | 354 | 646 | 321 | 915 |
| | 0·090 | 0·097 | 0·089 | 0·107 |

Whately Carington's data.

\* Total time $S(x)$.　　† Sum of squares of deviations $S(x-\bar{x})^2$.
‡ $\sqrt{\{S(x-\bar{x})^2\}/S(x)}$.

9. *The use of transformations in factorial experiments.*—The advantages
of combining several different factors in the same experiment have been
pointed out many times [2, § 1 a]. In addition to giving full information
on the effects of the different factors, the experiments also enable us to
test whether the factors are acting independently of each other. In
agricultural field trials, the factors are usually taken to be independent
when their effects are additive in the scale in which the data are measured;
for instance, in an experiment on wheat involving sulphate of ammonia
and potash, a consistent increase of say 5 cwt. per acre to sulphate of
ammonia at all levels of potash is taken to indicate absence of interaction.
This definition has great practical convenience, since interactions can
easily be isolated and tested in an analysis of variance. It would, however, probably be more natural to regard sulphate of ammonia and
potash as acting independently if sulphate of ammonia produced the

same *percentage* increase at all levels of potash; that is, to regard the main effects as independent when they are additive on a log. scale. This was pointed out in 1923, in the earliest paper dealing with the analysis of a factorial experiment, by Fisher and Mackenzie [10], who also showed how to test the significance of the interaction on this definition. This test is, however, more complicated than the usual test. Further, the two definitions differ little unless the responses to both factors are large, as the reader may verify by constructing a few examples showing absence of interaction on the two scales. For these reasons the first definition given above is the one which has been commonly used in practice.

With the types of data for which the transformations have to be used, however, it has been shown that treatment differences generally have to be large to be significant. When this is so, different definitions of interactions cannot be regarded as roughly equivalent. The experimenter must decide for himself what is the most natural definition of the independence of two factors, and if necessary treat the test of an interaction as a separate problem. The following example illustrates this point. Table 10[1] shows part of the results of a factorial experiment on the transmission of a virus disease of tobacco by an insect. The treatments are all combinations of different lengths of starvation periods of the transmitting insects, before and after access to an infected plant. The data are the numbers of diseased plants out of 50 and are the totals of ten replications.

TABLE 10. *Numbers of Diseased Plants (out of 50)*

| Preliminary starving | Post-infection starving | | | Total | Rate of fall | 0 min.—1 hour Original units | Degrees |
|---|---|---|---|---|---|---|---|
| | *0* | *15 min.* | *1 hour* | | | | |
| 0 . . | 6 | 2 | 1 | 9 | 1·67 | 5 | 12·2 |
| 15 min. . . | 18 | 11 | 6 | 35 | 1·03 | 12 | 16·6 |
| 1 hour . . | 27 | 15 | 7 | 49 | 1·22 | 20 | 25·3 |
| 6 hours . | 32 | 22 | 9 | 63 | 1·10 | 23 | 28·0 |
| Total | 83 | 50 | 23 | | | | |

It is clear that the transmission of the disease increases as the preliminary starving-time is increased, and decreases as the starving-time after access to an infected plant is increased. The natural way in which to interpret interaction in this case is to regard these factors as acting independently if the percentage rate of fall from 0 to 1 hour's post-infection starving is the same for all times of preliminary starving. This rate of fall is probably best measured by the ratio of the fall from 0 to 1 hour to the mean of the three values 0, 15 min., and 1 hour. These ratios are given in the Table. There is little indication of an interaction, except possibly that the rate of fall may be higher with no preliminary starving, though this value is subject to a large error on account of the small numbers involved.

The conventional interaction-term in the original scale is the comparison of the four differences between the values for 0 and 1 hour's

[1] M. A. Watson's data, to be published shortly.

post-infection starving. These figures are shown in the Table and indicate a consistent positive interaction. The data, being essentially percentages, might be converted to degrees for analysis. The interaction-term in degrees (see Table 10) also indicates a positive interaction. Thus the conventional interaction-term is of little interest in either the original or the transformed scale, and whilst the interaction should of course be separated from the experimental error in an analysis of variance, it cannot be regarded in this case as throwing any light on the question of the independence of the two factors.

It will often be found, when dealing with percentages and small whole numbers, that main effects are most naturally interpreted as additive on a log. scale. Transformation to degrees or square roots brings the data somewhat nearer to this scale, but not sufficiently near that the conventional test of an interaction in the transformed scale may be regarded as roughly equivalent to the real test of departure from independence. In such cases the true interaction must be tested separately by examining its consistency in the different replications. Where the transformation is itself a logarithmic one, it often equalizes the variances and provides the sensible test of interaction at the same time.

In conclusion, I have omitted for simplicity the discussion of the efficiency of estimation of main effects in the transformed scale. For this, reference should be made to Bartlett's 'Square-Root Transformations in the Analysis of Variance' [1], to which the present paper owes a great deal. It is, however, worth noting, that in the examples presented above of the square-root and inverse-sine transformations, the ratio of the treatments mean square to the error mean square is considerably increased by the transformation; the same is true if the example of § 2 is analysed in square roots, as Bartlett suggests. In my experience this is a common effect of a suitable transformation.

## Summary

The analysis of variance is now widely applied in interpreting the results of replicated experiments. Sometimes, however, a combined analysis on the original data has little meaning and gives misleading results, because the treatments have different variances. A numerical example is given to illustrate such a case.

These cases may be divided into two groups. (1) With yield-data, or whole-number counts of over 100 per plot, they occur very rarely, but may do so if some treatment differences are of the order of several hundred per cent., or if there is a partial failure of certain treatments or plots. The analysis is best carried out by omitting some treatments or plots. (2) With small whole numbers or percentages, the distributions tend to follow the Poisson and binomial types, respectively, and there is a known relation between the variance and the mean. Data of this type should be transformed before an analysis to a scale on which the variances are equal.

Three transformations have proved particularly useful in practice. (a) The square root, for whole numbers per plot between 10 and 100. If the majority of the plot-yields are under 10, one-half should be added to each plot-yield before taking the square root. (*b*) The inverse sine, i.e., the angle whose sine is the square root of the fraction, for percentages and fractions based on the ratio of small numbers. Percentages can, however, often be dealt with either by square roots, for small percentages, or by a direct analysis, for percentages from 30 to 70. (*c*) The logarithm, for distributions in which the standard error is proportional to the mean.

Numerical examples are worked, illustrating the use of each of these transformations and the way in which to present the results of the experiment.

A brief discussion is given of the analysis when the results consist of the number of plants in each of a number of grades (e.g. healthy, slightly diseased, severely diseased).

With factorial experiments in which the main effects produce large differences, the experimenter must consider what is the most natural definition of the independence of two factors, since the conventional test of interactions in either the original or the transformed scale may have little relation to this. A numerical example is given illustrating this point.

## REFERENCES

1. M. S. Bartlett, Square-root Transformations in the Analysis of Variance. J. Roy. Stat. Soc. Suppl. 1936, **3**, No. 1, 68–78.
2. F. Yates, The Design and Analysis of Factorial Experiments. Imp. Bur. Soil Sci., Tech. Comm. No. 35, 1937, p. 91.
3. ——, The Analysis of Replicated Experiments when the Field Results are Incomplete. Empire J. Expt. Agric. 1933, **1**, 129–42.
4. ——, Incomplete Latin Squares. J. Agric. Sci., 1936. **26**, 301–15.
5. R. A. Fisher and F. Yates, Statistical Tables for Biological, Medical, and Agricultural Research. 1938. Edinburgh: Oliver and Boyd (in the press).
6. Rothamsted Expt. Stat. Rept. 1936, p. 209.
7. M. A. Watson, Field Experiments on the Control of Aphis-transmitted Virus Diseases of *Hyoscyamus Niger*. Ann. Appl. Biol., 1937, **24**, 557–73.
8. B. Day and R. A. Fisher, The Comparison of Variability in Populations having Unequal Means. Ann. Eugenics, 1937, **7**, 333–48.
9. C. B. Williams, The Use of Logarithms in the Interpretation of Certain Entomological Problems. Ann. Appl. Biol., 1937, **24**, 404–14.
10. R. A. Fisher and W. Mackenzie, The Manurial Response of Different Potato Varieties. J. Agric. Sci., 1923, **13**, 311–20.

(*Received November 8, 1937*)

## Long-Term Agricultural Experiments

### By W. G. Cochran

Read before the Industrial and Agricultural Research Section of the Royal Statistical Society, May 25th, 1939, Sir John Russell, F.R.S., in the Chair.]

*Introduction*

In recent years this Society has received a number of papers dealing wholly or partly with the subject of experimental design. On the agricultural side, the design and statistical analysis of single re-plicated experiments have been discussed (1), (2), (3), while a further paper (4) dealt with the difficulties involved in carrying out a series of experiments on the same problem at a number of centres and possibly repeated on different sites over a number of years. To-night I propose to discuss the design of experiments which are planned to persist *on the same site* for a number of years.

So far as I am aware, only two types of experiment falling into this class have hitherto received any adequate discussion in print. The first of these are the so-called " classical " experiments at Rothamsted and Woburn. In these experiments the same crop was grown every year, while the manurial treatments on the different plots covered a variety of combinations of nitrogenous and mineral fertilizers. Each plot received the same manurial treatment every year. The object of the experiments was to determine which of the chemical constituents of the fertilizers were essential to the growth of the crop. The experiment on wheat has been running since 1843, that on barley since 1852, and that on mangolds since 1876. The statistical analysis of these experiments formed one of the first tasks of the Statistical Department at Rothamsted on its inception in 1919, and resulted in a series of papers on the subject (5)–(12).

The second class consists of experiments on perennial and tree crops, which from the nature of the crop must be extended over a number of years to enable the estimation of the full effect of any treatments. This class constitutes a very important part of the field of agricultural research, particularly in tropical and sub-tropical countries. Amongst the many questions that have been discussed are methods of raising homogeneous material for experimental purposes, the best size and shape of plot, the minimum number of replications necessary, the best types of design to use, and the importance of subsidiary and quality measurements.

Either of these classes could well form the subject of a whole paper, but in view of the work which has already been done, I propose to restrict my remarks upon them in order to leave more space for the consideration of long-term rotation experiments.

It is not difficult to find reasons for the absence of previous discussion of rotation experiments. Owing to their considerable cost and to the necessary ear-marking of part of the resources of an experiment station for many years, the number of these experiments which have been undertaken in this country is small. Consequently the statistician's opportunities for gaining experience in the problems involved have been very restricted, though the subject is one which calls essentially for first-hand knowledge of the actual problems. Further, the wide diversity of types of rotation experiment makes it difficult to find rules of procedure, and while some useful general advice may be given, each experiment must be considered on its own merits.

The possible variations in the types of long-term experiment are illustrated in the diagram below.

### Types of Long-term Experiment

| | Treatments | | Information supplied on | | Crops | |
|---|---|---|---|---|---|---|
| Fixed | Applied on the same plots | Every year | Cumulative effects | | Single crop | annual |
| | | First year only | Residual effects | ×  | | perennial |
| | | At fixed intervals | Direct and Residual effects | | Fixed rotation | |
| Rotating | Applied on different plots in successive years | | Direct and Residual effects | | Effects of different crops | |

Fig. 1.

The treatments may either remain on the same plots throughout (fixed) or rotate from plot to plot in successive years. The type of information supplied by various alternative methods of applying the treatments is indicated in the second column. For instance, if treatments rotate from plot to plot in successive years, the yield on a plot in any given year is a measure of the direct effect of the treatment on that plot, plus the residual effects (if any) of the other treatments applied to that plot in previous years. As will be seen later, the design may sometimes be arranged to permit a separation of the direct effect and of the residual effects of at least the two previous years.

In a fixed rotation of crops, a definite order of succession, such as swedes, barley, rye-grass, wheat, is maintained in successive years, the wheat crop being followed by swedes in the fifth year and so on. Finally, the experiment may involve several different successions of crops whose effects are being compared. For instance, one may wish to know what series of crops will best restore poor land to a state of moderate or high fertility, or how often a farmer may include his best cash crops in his system of rotations without serious soil deterioration.

In a given experiment, one or more of the different methods of applying the treatments may be combined with any of the crop classes, so that a wide variation in types is obtained.

The responsibility of the statistician who is advising on the design of long-term experiments is a heavy one. With annual experiments the defects of a faulty design, if recognized, can be corrected in any repetitions of the experiment on another site in subsequent years, but with long-term experiments the results of years of labour may be largely vitiated by a poor design at the outset. Further, since this type of experimentation is costly, there is a strong economic argument in favour of keeping down the size of the experiment to a bare minimum, and the statistician may have to face considerable pressure on this point. It is not sufficient for him to provide the best possible design to suit the size of the experiment; it is also his duty to advise whether he thinks the experiment as designed is worth doing, or whether it should be postponed until more resources are available. The standards of amount of replication are of course not necessarily the same as in single annual experiments, because in most experiments, as we shall see, there is a certain amount of replication provided by the results in different years.

This paper will also be concerned to some extent with methods of analysing the results of long-term experiments. The experiments in which advantage has been taken of modern research on the question of design were naturally started only within the last few years, and the problem of analysing the complete results has not yet arisen. It is, however, a good plan to have at least the main outlines of the method of analysis in mind when the experiment is being designed, as this forms a useful check on the soundness of the design. Further, it is neither necessary nor desirable to wait perhaps some twenty years before examining any of the results of such experiments. The inspection of the mean yields of the various treatments should be undertaken every year, and an *interim* analysis after a few years will often provide useful knowledge, and perhaps anticipate the final results of the experiment.

In this respect I have assumed that the reader has a fair working knowledge of the statistical analysis of the commoner types of design, and of the principles of the analysis of variance. This assumption is necessary to keep the length of the paper within reasonable limits and is, I hope, justified by the excellent instruction on these points which has been given in previous papers to this Society (1)–(3) and in some of the text-books on the subject.

Classification of the various types of long-term experiments for descriptive purposes is by no means easy.   The order which I have adopted is roughly that of increasing complexity from the point of view of design and is as follows :  experiments on a single crop with fixed treatments;  rotation experiments with fixed treatments; single-crop and rotation experiments including rotating treatments; and, lastly, experiments in which different crops are compared.   This order does not necessarily coincide with that of increasing complexity from the point of view of analysis, for where the *design* permits the use of considerable ingenuity on the part of the statistician, he usually takes the opportunity of simplifying, as far as possible, the future problem of *statistical analysis*.

*Single-crop experiments with fixed treatments*

With long-term experiments on a single crop in which the treatments remain on the same plots throughout, no essentially new problem in design is raised.   Preliminary questions, such as the raising of uniform material for experimental purposes, the size and shape of plot and the number of guard rows necessary between plots, often involve difficult problems with perennial or tree crops, but when some decision has been reached on these points, the experiment in plan will look the same as an annual experiment.   The results of research on the latter can therefore be utilized in planning the layout of the experiment.   In particular, the accuracy and convenience of randomized blocks and Latin squares, the efficiency of factorial design, and the possibility of confounding wholly or partly some of the less important treatment comparisons should be borne in mind. With varietal trials involving a large number of varieties, the *lattice* designs recently introduced by Yates (13) may prove useful, provided that the necessary number of replications can be faced.   As Yates has shown (13), these designs can be analysed either as randomized blocks, or with the elimination of the effects of differences between rows and columns of the squares.   Thus advantage may be taken of any increase in accuracy due to the use of the Latin square design, but if the increase in accuracy is small or non-existent, or if certain subsidiary measurements do not warrant more than an approximate analysis, the experiment may be treated as if it were in randomized blocks.

The possibility of running the experiment as a uniformity trial for the first year or two should be considered.   The case in favour of this course is clearly stronger than with an experiment which has to run on the site for one year only, because it is particularly important in a long-term experiment to lay out the plots to the best advantage and to avoid the use of highly variable sites, while a delay of one year in, say, a ten-year experiment is of no great account.   The issue depends greatly on the extent to which the first few years' results are an indication of the relative fertility of the different plots or of the variability of the site in future years. This question can only be settled by an extensive examination of the appropriate data from uniformity trials, and, as in most questions of this type, the amount of uniformity trial data available for research and the amount of research done are alike scanty. As far as the results go, they indicate that with perennial or tree crops the correlation between yields in successive years is sufficiently high that an initial uniformity trial of one or two years is to be recommended (14), (15).   With annual crops, as would be expected, the correlation between yields in successive years is smaller, and in experimentation on common farm crops it is doubtful whether any substantial increase in accuracy can be made by using the uniformity trial results. In experiments on a new crop, on the other hand, a preliminary uniformity trial would give valuable experience on handling the crop under experimental conditions and safeguard against the use of an inaccurate design.

The way in which the uniformity trial results should be used in planning the design of the experiment also requires consideration. The principal alternatives in a design in randomized blocks appear to be :

(1) To group adjacent plots into blocks as usual. The uniformity trial results are used to decide the shape of the blocks and to adjust subsequent years' yields by covariance.

(2) To group plots of equal fertility (not necessarily adjacent) into blocks on the basis of the uniformity trial results. Thus in an experiment with six treatments and thirty plots, the plots would be grouped into five blocks in order of increasing fertility, the six poorest plots in the uniformity trial being assigned to the first block. Treatments would then be assigned at random to the members of each " block." As in (1) the uniformity trial data can also be used to adjust subsequent years' results, though if this method of laying-out the plots is successful, most of the increase in accuracy has probably been secured without using covariance.

(3) With, say, six treatments, the plots are divided into six groups whose mean yields are as close as possible to each other. All plots in one group are then assigned to the same treatment, the object being to make the comparison of treatment *means* as accurate as possible. In this method, as in the most accurate types of systematic designs, the variation between plots with the same treatments is likely to be an over-estimate of the true error of the treatment means, so that no proper test of significance of the *unadjusted* treatment means is available in future years. The use of a covariance on the uniformity trial results *might* give a correct test of significance and an

unbiased estimate of the errors of the *adjusted* treatment means, but this point does not appear to have been investigated.

Forster (16) compared the first two methods in a rotation experiment with two years of maize followed by one year of oats. He found that the second method gave slightly lower standard errors for the *unadjusted* yields in subsequent years. When covariance was used, however, the first method proved definitely superior, there being, in fact, no reduction in error by the use of covariance in the second method. He points out that the first method, with covariance, takes advantage both of the tendency of adjacent areas of land to yield alike and of the tendency of plots initially similar in yield to remain similar. For this reason, it seems likely that the first method will prove the best on other crops as well, since the last two methods depend for their success on the existence of a high correlation between yields on the same plots in adjacent years, and largely ignore the additional correlation between neighbouring plots. The third method in particular has little to recommend it.

I do not think that it is possible to give useful advice on the important question of the amount of replication, except in very general terms. So much depends on the variability of the material, the size of treatment difference which is expected or which it is considered important to detect, and the duration of the experiment. With experiments which are intended to last for only a few years, the amount of replication should not be appreciably less than that in single annual experiments, but with experiments planned for many years, in which the cumulative effects of the treatments may produce large differences in the long run, some relaxation of the usual standards of replication may be allowed. The classical experiments at Rothamsted and Woburn were neither replicated nor randomized, yet they have yielded strikingly clear results on many points, and it is certain that the absence of randomization in these experiments was a much more serious defect than the absence of replication.

It should, however, be noted that since the treatments remain on the same plots, any persistent differences between plots will not be " smoothed out " by averaging over a number of years. Thus the accuracy with which the differences between treatments are assessed will not necessarily be increased to any great extent by averaging the results of a number of years. This point has an important bearing on the method of analysing the results of such experiments, which will be illustrated by a numerical example.

*Numerical example.*

In a sugar-beet experiment on acid land dressings of 0, 1, 2, 3, and 4 tons of chalk per acre were applied in 1932 in a 5 × 5 Latin

E 2

square. The same crop was grown *without further treatment*, for four years. The weights of clean roots in lbs. per plot are shown in Table I. The plots receiving no chalk gave only a small

TABLE I

*Sugar-beet : Tunstall.   Plan and yields of clean roots*
(lb. per plot)

(1/56 acre)

|  |  |  | 3 | 1 | 0 | 4 | 2 |
|---|---|---|---|---|---|---|---|
| 1932 | ... | ... | 599 | 487 | 76 | 616 | 606 |
| 1933 | ... | ... | 536 | 503 | 147 | 565 | 585 |
| 1934 | ... | ... | 617 | 491 | 19 | 636 | 604 |
| 1935 | ... | ... | 622 | 568 | — | 591 | 597 |
|  | Total | ... | 2,374 | 2,049 | — | 2,408 | 2,392 |
|  |  |  | 0 | 2 | 4 | 1 | 3 |
| 1932 | ... | ... | 70 | 570 | 569 | 466 | 577 |
| 1933 | ... | ... | 117 | 565 | 587 | 431 | 589 |
| 1934 | ... | ... | 9 | 609 | 589 | 475 | 611 |
| 1935 | ... | ... | — | 606 | 571 | 499 | 594 |
|  | Total | ... | — | 2,350 | 2,316 | 1,871 | 2,371 |
|  |  |  | 4 | 3 | 1 | 2 | 0 |
| 1932 | ... | ... | 576 | 522 | 473 | 491 | 107 |
| 1933 | ... | ... | 584 | 555 | 457 | 461 | 152 |
| 1934 | ... | ... | 641 | 602 | 450 | 526 | 33 |
| 1935 | ... | ... | 639 | 553 | 539 | 533 | — |
|  | Total | ... | 2,440 | 2,232 | 1,919 | 2,011 | — |
|  |  |  | 1 | 0 | 2 | 3 | 4 |
| 1932 | ... | ... | 543 | 45 | 502 | 489 | 497 |
| 1933 | ... | ... | 518 | 97 | 519 | 464 | 492 |
| 1934 | ... | ... | 526 | 24 | 561 | 578 | 561 |
| 1935 | ... | ... | 597 | — | 616 | 555 | 578 |
|  | Total | ... | 2,184 | — | 2,198 | 2,086 | 2,128 |
|  |  |  | 2 | 4 | 3 | 0 | 1 |
| 1932 | ... | ... | 523 | 517 | 500 | 44 | 405 |
| 1933 | ... | ... | 517 | 554 | 508 | 75 | 370 |
| 1934 | ... | ... | 550 | 581 | 521 | 27 | 387 |
| 1935 | ... | ... | 607 | 592 | 546 | — | 520 |
|  | Total | ... | 2,197 | 2,244 | 2,075 | — | 1,682 |

The treatment symbols refer to the number of tons of chalk per acre applied in 1932.   In 1935 no yields were recorded for the plots without chalk.

fraction of the yields of the other plots, and must clearly be omitted from the statistical analysis. The other treatment totals are shown in Table II for each of the four years.

In all four years the higher dressings of chalk gave increased yields over the 1-ton dressing, with a regular falling-off in the amount

Table II

*Sugar-beet : Tunstall.   Totals of Five Plots* (in lb.)

|  | Amount of chalk per acre in 1932 | | | |
|---|---|---|---|---|
|  | 1 ton | 2 tons | 3 tons | 4 tons |
| 1932 | 2374 | 2692 | 2687 | 2775 |
| 1933 | 2279 | 2647 | 2652 | 2782 |
| 1934 | 2329 | 2850 | 2929 | 3008 |
| 1935 | 2723 | 2959 | 2870 | 2971 |

of response to the 3- and 4-ton dressings.   The analyses of variance for the separate years are shown in Table III.

Table III

*Sugar-beet : Tunstall.   Analyses of variance of yields of roots* (lb. per plot)

|  | Degrees of freedom | 1932 | | 1933 | | 1934 | | 1935 | |
|---|---|---|---|---|---|---|---|---|---|
|  |  | Sums of Squares | Mean Squares | Sums of Squares | Mean Squares | Sums of Squares | Mean Squares | Sums of Squares | Mean Squares |
| Rows      ...    ... | 4 | 22,624 | — | 15,309 | — | 14,149 | — | 3,565 | — |
| Columns ...   ... | 4 | 6,136 | — | 10,503 | — | 7,497 | — | 10,817 | — |
| Treatments : |  |  |  |  |  |  |  |  |  |
| Linear    ... | 1 | 14,352 | 14,352 | 22,922 | 22,922 | 44,775 | 44,775 | 4,290 | 4,290 |
| Quadratic  ... | 1 | 2,645 | 2,645 | 2,832 | 2,832 | 9,768 | 9,768 | 911 | 911 |
| Cubic ...   ... | 1 | 1,731 | 1,731 | 2,381 | 2,381 | 1,954 | 1,954 | 2,652 | 2,652 |
| Error   ...   ... | 8 | 8,876 | 1,110 | 12,468 | 1,558 | 5,349 | 669 | 3,242 | 405 |
| Total    ... | 19 | 56,365 | — | 66,416 | — | 83,491 | — | 25,479 | — |

The analysis is unfortunately slightly complicated by the omission of the plots without chalk.   The method of dealing with this case has, however, been described by Yates in (17), where the analysis of the yields of dirty roots from this experiment in 1932 is given as an example, to which the reader who is unfamiliar with the method is referred.   The four treatments are clearly orthogonal with rows and columns, and the " treatments " term in the analysis is found in the usual way.   Rows and columns are, however, not orthogonal, since one column is unrepresented in each row and vice versa.   In the analysis in Table III the " column " terms have been found, without any adjustment, in the usual way, while the " row " terms represent the differences between rows after adjusting for columns as described by Yates.   These two terms add up to the correct sum of squares for rows and columns combined.

The degrees of freedom for treatments have been subdivided into their linear, quadratic, and cubic components, the linear term forming an estimate of the average increase to higher dressings of chalk over the 1-ton dressing, while the quadratic term may be used to test the falling-off in responsiveness at the highest levels of application.

Considering the experiment as a whole, it will be realized that it provides information on the average differences between treatments in the four years and on the variation in treatment differences from year to year. The analyses of variance in Table III give valid tests of significance of the treatment effects in individual years, but owing to the persistence of treatments on the same plots they cannot be regarded as independent in testing the average treatment effect or the interaction of treatments with seasons. That the correlation between plots in different years was high in this experiment is evident from inspection of Table I. The plots with the 1-ton dressing of chalk, for instance, ranged themselves in practically the same order of yield every year.

TABLE IV

*Sugar-beet : Tunstall.    Combined analysis of variance of roots*
(lb. per plot)

Average effects of treatments

| | Degrees of freedom | Sums of Squares | Mean Squares |
|---|---|---|---|
| Rows        ...    ...    ...    ...    ... | 4 | 46,387 | — |
| Columns     ...    ...    ...    ...    ... | 4 | 27,092 | — |
| Treatments : | | | |
| Linear Response ...    ...    ...    ... | 1 | 75,158 | 75,158 |
| Quadratic Response    ...    ...    ... | 1 | 13,650 | 13,650 |
| Cubic Response  ...    ...    ...    ... | 1 | 8,658 | 8,658 |
| Error       ...    ...    ...    ...    ... | 8 | 22,328 | 2,791 |
| Total    ...    ...    ... | 19 | 193,274 | — |

Interaction of treatments with years

| | Degrees of freedom | Sums of Squares | Mean Squares |
|---|---|---|---|
| Years       ...    ...    ...    ...    ... | 3 | 43,172 | — |
| Rows × years    ...    ...    ...    ... | 12 | 9,260 | — |
| Columns × years   ...    ...    ...    ... | 12 | 7,862 | — |
| Treatments × years : | | | |
| { Linear × years    ...    ...    ... | { 3 | { 11,181 | 3,727 |
| { Remainder × years    ...    ...    ... | { 6 | { 2,566 | 428 |
| Total    ...    ...    ...    ... | 9 | 13,747 | 1,527 |
| Error       ...    ...    ...    ...    ... | 24 | 7,608 | 317 |
| Total    ...    ...    ... | 79 | 274,923 | — |

The analysis of variance appropriate to a summary of the experiment as a whole is shown in Table IV (in the same units as in

Table III).  The average differences between treatments are tested by analysing the totals over the four years on each plot, which are shown in Table I.  The error mean square for the average treatment effects (2791) is about three times the average of the error mean squares in the individual years (935·5).  If these individual errors were independent, the error mean square for the average treatment effects (in the units given) would be about the same size as their average (apart from its sampling variation), while if the individual errors were completely correlated it would be four times as large. In this experiment the average treatment effects are only slightly more accurately determined than the effects in a single year.

The remainder of the analysis may be carried out by noting the analogy with the analysis of a split-plot experiment, the four years' results on any one plot corresponding to the yields on the sub-plots. The analogy should not, however, be carried too far, since certain components of the interaction of treatment with years (the " sub-plot " treatments) may have different errors.

In the present example the interaction mean square (1,527) is significantly above the appropriate error (317).  Inspection of the treatment totals in Table II shows that this interaction is mainly due to variation from year to year in the linear response to higher dressings of chalk over the 1-ton dressing, and this term has been isolated in Table IV.

Frequently in such experiments the most interesting term in the interaction of treatments with years is the comparison of the linear regression of yield on years for the different treatments, since this term tests whether the treatment differences are becoming more or less pronounced as the experiment proceeds.  In testing these linear regressions we must allow for the possibility of systematic changes in the relative fertility of different plots as the experiment proceeds, just as in testing the average effects of treatments we allowed for the possibility of *consistent* differences between the fertilities of different plots.  A separate error term appropriate to the linear regression may be obtained by calculating the regression separately for each plot and analysing these figures in the same way as the plot totals were analysed.  The quadratic and higher terms of the regression of yield on years may be dealt with similarly if necessary.  In the present example there is, however, no evidence that the residual effectiveness of the different levels of manuring was changing systematically.

With experiments which last for many years, an analysis of the above type should be only a preliminary to a more detailed investigation of the effect of the season on the responses to treatments.  The researches of Fisher and his co-workers at Rothamsted on the influence of rainfall on yields may be cited as an example of further investigations of this type, (5)–(12).

In conclusion, the existence of *negative* correlations between the yields of the same plots in successive years has also been observed, particularly with perennial or tree crops. Hoblyn (18) gives an example in a variety of apple trees, and there are indications of the phenomenon in experiments on pyrethrum at Woburn (19). This effect may be anticipated even with annual or rotating crops, since a crop which does particularly well in a favourable season may temporarily deplete the soil nutrients, so that the crop in the next year tends to be poor. In this case the results for individual years would be highly variable, whereas the averages over a number of years would be more accurate than if the analyses for the individual years were independent. The effect can be studied by an analysis of covariance on the previous year's yield. The adjusted analysis is not relevant to the comparison of the average effects of treatments in experiments in which the treatments remain throughout on the same plots. It may be of service in experiments in which the treatments on a given plot change from year to year, since a treatment might be penalized merely because the particular plots on which it was applied happened to give high yields in the previous year. The question is, however, complicated if residual effects of the previous year's treatment are also present, and further investigation is needed before the utility of a covariance analysis can be assessed.

*Rotation experiments with fixed treatments*

*General considerations*

The most important rule about rotation experiments is that each crop in the rotation must be grown every year. Suppose, for example, that the choice in a four-course rotation lies between growing each crop every year in single replication or each crop every four years in a four-fold replication. In the latter case there is increased accuracy in the results of a single year on the particular crop which is grown, but the experiment will have to last longer to obtain equal information on the long-term effects of the treatments, and if seasonal variations are large compared with variations in a single year, it will have to last almost four times as long. By the time such an experiment had finished, the treatments compared might have lost most of their agricultural interest. Moreover, the effects of the treatments on the separate crops are obtained under different seasonal conditions, and a compact summary of the results of the experiment as a whole is made exceedingly difficult.

This rule automatically imposes a lower limit to the size of the experiment. For example, in a four-course rotation with 12 treatments, a single replication requires 48 plots.  As with single-crop experiments, the question of replication can be discussed only in general terms.  The principal advantage of replication is that it increases the accuracy with which the treatment means for any crop in any year are assessed, though if the experiment is planned to last for many years, or if seasonal variations are large, high precision in individual years is of less importance.  A further advantage of replication is that it provides a test of significance of the effects of treatments on every crop in every year.  If, however, the treatments form a factorial system, there will be internal replication on at least the main effects, and an estimate of error may be available from certain of the interaction terms.  The principal factors to be considered in reaching a decision are the number of crops and treatments (which for economic reasons alone may decide the issue), the expected duration of the experiment and the accuracy which is aimed at.

It should, however, be realized that replication is provided by the results in different years, since the crops are grown on different pieces of land in successive years and return to the same plots only after a complete cycle of the rotation.  This replication decreases to some extent the influence of permanent differences between plots on the average results, though if treatments are applied to the same plots every year, there is no further " smoothing out " of these differences after one complete rotation has been grown.

In laying-out an experiment in single replication each crop should be assigned to a compact group of plots, which is usually called a *series*.  This facilitates the farming operations and increases the accuracy of comparisons between treatments on a given crop. If there is replication, replicates of the same crop should be kept near each other; *i.e.* each crop should be laid out as a separate randomized blocks or Latin square experiment.  Under this arrangement, the comparison of yields of a crop in successive years is affected by differences between series as well as by differences between seasons.

*Numerical example.*

The replication between years in these experiments also permits tests of significance of the effects of treatments, even when there is no replication within years.  The test will be illustrated by an example.  The plan and yields of barley grain (on half-plots) from part of a four-course rotation (roots, barley, seeds, wheat) are shown for twelve years (three complete cycles) in Table V.  During this period the whole of the rotation was grown without manures on these half-plots.  In the previous two cycles of the rotation, four treatments were compared : a high and low level of organic manuring (made from cotton cake and maize meal respectively) and artificial manures calculated as equivalent in nitrogen, potash, and phosphate to the respective organic manures (for details see (20)).   The yields of the crops in the first two cycles were considered to be too high to give any chance of detecting differences between the manures.   The plots were accordingly halved, and the succeeding three courses were grown without manures on half the plots in an attempt to reduce the fertility of the soil for a continuation of the experiment.   The reader is incidentally referred to this experiment (20) as an example of the difficulties which are created in attempting to improve an experiment by changes while it is in progress.

TABLE  V

*Woburn Rotation Experiment.   Yields of barley grain*
(lb. per half-plot)

| Year | Series | Previous treatment | | | |
|---|---|---|---|---|---|
| | | Cotton Cake | Maize Meal | Artificial manures equivalent to | |
| | | | | Cake | Meal |
| 1886 ... ... | II | 207 | 215 | 229 | 210 |
| 1887 ... ... | IV | 149 | 156 | 184 | 180 |
| 1888 ... ... | I | 155 | 154 | 170 | 153 |
| 1889 ... ... | III | 141 | 148 | 169 | 164 |
| 1890 ... ... | II | 214 | 193 | 201 | 181 |
| 1891 ... ... | IV | 128 | 116 | 136 | 155 |
| 1892 ... ... | I | 142 | 126 | 123 | 130 |
| 1893 ... ... | III | 131 | 125 | 139 | 198 |
| 1894 ... ... | II | 167 | 180 | 150 | 137 |
| 1895 ... ... | IV | 76 | 98 | 102 | 104 |
| 1896 ... ... | I | 82 | 80 | 68 | 72 |
| 1897 ... ... | III | 92 | 93 | 101 | 85 |

The rotation was a four-course one, each crop being grown every year on a separate block.   The block on which the barley crop was being grown in any given year is shown in column 2 above.

*Woburn Rotation Experiment :  Plan* (1886)

Series I.          Series II.          Series III.          Series IV.

| | Wheat | | | | Barley | | | | Peas | | | | Mangolds | | |
|---|---|---|---|---|---|---|---|---|---|---|---|---|---|---|---|
| 4 | 3 | 2 | 1 | 4 | 3 | 2 | 1 | 4 | 3 | 2 | 1 | 4 | 3 | 2 | 1 |

The numbers refer to the treatments.
Area of each plot = ½ acre.

Inspection of Table  V  shows that yields were clearly falling throughout the period.   There is also a suggestion that the yields following inorganic manures were initially higher than the yields following organic manures, but fell more rapidly throughout the period. These differences in the rates of deterioration following different treatments will be tested. The treatments were not randomized within series, so that any estimate of error is open to doubt, but the experiment may be taken as a convenient example of the method of analysis.

Since the four series constitute the replications, the deterioration must be calculated separately for each treatment in each series. Each treatment appears three times in each series, at four-yearly intervals, so that the best estimate of deterioration from the results of a single series is the difference between the yield in the third cycle and the yield in the first cycle. These differences (with signs reversed) are shown in Table VI.

TABLE VI

*Woburn Barley* (lb. per half plot)

Differences between yields on the same plots in the first and third cycles

| Series | Previous treatment | | | | Total |
|---|---|---|---|---|---|
| | Cotton Cake | Maize Meal | Artificials equivalent to cake | Artificials equivalent to meal | |
| II | 40 | 35 | 79 | 73 | 227 |
| IV | 73 | 58 | 82 | 76 | 289 |
| I | 73 | 74 | 102 | 81 | 330 |
| III | 49 | 55 | 68 | 79 | 251 |
| Total | 235 | 222 | 331 | 309 | 1097 |
| | | | $\pm$ 18·09 | | |

The analysis of variance of these figures is shown below.

*Analysis of Variance of Decreases in Yield*

| | Degrees of freedom | Sums of squares | Mean squares |
|---|---|---|---|
| Series...    ...    ... | 3 | 1524·7 | 508·2 |
| Treatments  ...    ... | 3 | 2174·7 | 724·9 |
| Error       ...    ... | 9 | 736·5 | 81·83 |
| Total    ...    ... | 15 | 4435·9 | — |

The standard error of an individual figure in Table VI is $\pm$ 9·046, and of a treatment total $\pm$ 18·09. The yields following the inorganic manures deteriorated significantly more rapidly than those following the organic manures. It will be noted that the differences between series are also partly differences between seasons.

The analysis of variance of these results may be completed by

## Table VII

*Woburn Rotation Experiment.    Complete analyses of variance of barley yields per half plot*

### Average effects of treatments

|  | Degrees of freedom | Sums of squares | Mean squares |
|---|---|---|---|
| Series ... ... ... | 3 | 38,005·2 | — |
| Treatments ... ... | 3 | 503·9 | 168·0 |
| Error ... ... ... | 9 | 2,685·0 | 298·3 |
| Total ... ... | 15 | 41,194·1 | — |

### Linear regression on years

|  | Degrees of freedom | Sums of squares | Mean squares |
|---|---|---|---|
| Series ... ... ... | 3 | 762·4 | — |
| Treatments ... ... | 3 | 1,087·4 | 362·5 |
| Error ... ... ... | 9 | 368·2 | 40·9 |
| Total ... ... | 15 | 2,218·0 | — |

### Quadratic regression on years

|  | Degrees of freedom | Sums of squares | Mean squares |
|---|---|---|---|
| Series ... ... ... | 3 | 271·1 | — |
| Treatments ... ... | 3 | 656·4 | 218·8 |
| Error ... ... ... | 9 | 539·3 | 59·9 |
| Total ... ... | 15 | 1,466·8 | — |

analysing (1) the totals of the yields on any plot in the first, second, and third cycles and (2) the quadratic component of the regression of yield on years, which is obtained by multiplying the yields on any plot in the first, second, and third cycles by $-1$, $+2$, and $-1$ respectively, and adding. The complete analysis is shown in Table VII, the unit being the yield of a single half-plot in a single year. (The analysis of variance of the totals is divided by 3, that of the linear component by 2, and that of the quadratic component by 6 to obtain the results in these units.) As in the previous sugar-beet example, there is clear evidence of permanent differences between plots, the error mean square for the totals being 298·3 as compared with 40·9 for the linear component. On the other hand, the error mean square for the linear component is no larger than that for the quadratic component, so that there do not appear to be any consistent changes throughout the period in the relative fertility of different plots.

As is evident from inspection of the yields, there is no indication of any real difference between the average yields under different previous treatments. The quadratic component shows almost significant differences between treatments. The explanation is that the yields following cotton cake were particularly well maintained in the second cycle.

An analysis of the above type is likely to be of general utility as a first step to the interpretation of the results of rotation experiments.

It is instructive to consider whether a more accurate estimate of the deterioration can be obtained from these results.* Let (1), (2) . . . denote the yields on the plots receiving a given treatment in the first, second . . . years. If the plot errors in the twelve years are independent and have equal variance $\sigma^2$, the best estimate of the average fall in yield *per year* is obtained from the linear regression of yields on years and is given by

$$l = \tfrac{1}{286}[11\,(12) + 9\,(11) + 7\,(10) + \ldots -7\,(3) - 9\,(2) - 11\,(1)]$$

the variance of $l$ being $\dfrac{\sigma^2}{143}$.

Write

| Totals | Linear components | Quadratic components |
|---|---|---|
| $a_0 = (1) + (5) + (9)$ | $a_1 = (9) - (1)$ | $a_2 = (9) - 2\,(5) + (1)$ |
| $b_0 = (2) + (6) + (10)$ | $b_1 = (10) - (2)$ | $b_2 = (10) - 2\,(6) + (2)$ |
| $c_0 = (3) + (7) + (11)$ | $c_1 = (11) - (3)$ | $c_2 = (11) - 2\,(7) + (3)$ |
| $d_0 = (4) + (8) + (12)$ | $d_1 = (12) - (4)$ | $d_2 = (12) - 2\,(8) + (4)$ |

Then

$$l = \tfrac{1}{286}[(3d_0 + c_0 - b_0 - 3a_0) + 8(a_1 + b_1 + c_1 + d_1)].$$

In the previous section the deterioration was estimated from $a_1, b_1, c_1, d_1$ alone. The estimated fall in yield per year from these quantities is $\tfrac{1}{32}[a_1 + b_1 + c_1 + d_1]$ and has variance $\dfrac{\sigma^2}{128}$. Thus, even if the plot errors are independent, the fraction of the information lost by ignoring $a_0, b_0, c_0, d_0$ is only $\tfrac{15}{143} = 10 \cdot 5$ per cent.

In the present experiment, the crop returns in years 5 and 9 to the same plots as in year 1, and so on. Thus if there are *permanent* differences between plots, the variance of $a_0, b_0, c_0, d_0$ is increased relatively to that of $a_1, b_1, c_1, d_1$. It follows that the linear regression is no longer the best estimate of the deterioration and also that the fraction of information lost by ignoring $a_0, b_0, c_0, d_0$ is less than $10 \cdot 5$ per cent. If the error variances of $a_0, b_0, c_0, d_0$ and $a_1, b_1, c_1, d_1$

* This method of approach was suggested to me by Dr. F. Yates.

*on the same basis* are $\sigma_0{}^2$ and $\sigma_1{}^2$, the best estimate of the decrease in yield per year is

$$\frac{\frac{15}{\sigma_0{}^2}[\frac{1}{30}(3d_0 + c_0 - b_0 - 3a_0)] + \frac{128}{\sigma_1{}^2}[\frac{1}{32}(a_1 + b_1 + c_1 + d_1)]}{\frac{15}{\sigma_0{}^2} + \frac{128}{\sigma_1{}^2}}$$

In the present example, the error variances of these quantities are estimated from only 9 degrees of freedom, so that the weights are not determined very accurately. It is, however, clear from Table VII that there is very little additional information available from $a_0$, $b_0$, $c_0$, $d_0$, since their error variance is 298·3 as compared with 40·9 for $a_1$, $b_1$, $c_1$, $d_1$.

If a reasonable number of degrees of freedom, say over 30, are available, it will probably be fairly safe to weight the two estimates according to the inverse square of their estimated variances. This method can be extended to the term of any degree in the regression of yield on years. In the present example the quadratic component of the regression of the twelve individual years on yield is (apart from a constant multiplier)

$$q = [55\,(12) + 25\,(11) + 1\,(10) - 17\,(9) - 29\,(8) - 35\,(7) - 35\,(6)$$
$$- 29\,(5) - 17\,(4) + 1\,(3) + 25\,(2) + 55\,(1)] = [3(d_0 - c_0 - b_0$$
$$+ a_0) + 12(3d_1 + c_1 - b_1 - 3a_1) + 16(d_2 + c_2 + b_2 + a_2)]$$

The general method of separation is clear from this case. The quadratic component is expressible in terms of :

(1) the quadratic component of the four totals $a_0$, $b_0$, $c_0$, $d_0$, *i.e.* of the terms of degree zero,

(2) the linear component of the linear terms $a_1$, $b_1$, $c_1$, $d_1$ and

(3) the total, *i.e.* the term of degree zero, of the quadratic terms $a_2$, $b_2$, $c_2$, $d_2$.

Estimates of the error variances of each of these terms can be obtained from an analysis of the type shown in Table VII. Hence the standard error of the term of any degree in the regression on years of the individual annual yields under any treatment can be calculated. In studying the variation in yields from year to year, we may therefore use (1) the regression coefficient calculated entirely from within series, (2) the regression coefficient calculated from the individual annual yields or, if there are sufficient numbers of degrees of freedom for error, (3) the most accurate estimate of the regression, obtained by weighting the separate estimates from the terms of degree 0, 1, 2 . . . One of the first two will perhaps usually be preferred, on account of their simplicity and familiarity, but the possibility of obtaining a still more accurate estimate should be realized.

*Treatments applied at intervals only*

It is clearly of considerable importance in practical farming to know whether the effect of any treatment is confined to the year in which it is applied, or whether it persists in some degree for several years thereafter. In experiments in which treatments are applied every year, any such residual effects are present in the cumulative differences between treatments, but it is not possible to determine how large these effects are relatively to the direct effects or how long they persist. The direct and residual effects may, however, be separated by applying the treatments only at fixed intervals. The yields in the year in which treatments are applied provide a measure of the direct effect, the yields in the year after the treatments are applied provide a measure of the first-year residual effects, and so on.

An important rule governing these experiments is that all phases of the treatment cycle must be present. If, for instance, the treatments are applied every third year, some plots must receive the treatments in the first, fourth . . . years, some in the second, fifth . . . years, and some in the third, sixth . . . years. If this is not done, the differences between the direct and residual effects are mixed up with differences between seasons. If, however, all phases are present, then in any year there are plots which receive the treatments, plots which received the treatments last year, and so on. Thus we can measure the difference between the direct response and the first-year residual response, the difference between the first- and second-year residual responses, and so on, and so determine the extent to which treatment effects persist.

This rule increases the minimum possible size of the experiment. With eight treatments, applied every three years, and a four-course rotation, a single replication requires 96 plots. This is, however, to some extent compensated for by increased replication in years on certain of the treatment comparisons. In the above example, the sequence of crops ($a$, $b$, $c$, $d$) and treatments ($t =$ treated, $u =$ untreated) in the first twelve years on a plot which receives treatment in the first year is : $1\,a\,t$; $2\,b\,u$; $3\,c\,u$; $4\,d\,t$; $5\,a\,u$; $6\,b\,u$; $7\,c\,t$; $8\,d\,u$; $9\,a\,u$; $10\,b\,t$; $11\,c\,u$; $12\,d\,u$. To assess the *average* effects of a three-yearly dressing of a treatment, we calculate the average of all plots which received that treatment either in the present, previous, or second previous year. On this comparison there is only four-fold replication in years for a given crop, since the crops return to the same plots every four years. If, however, we are comparing treatments in the year of application only, or any fixed number of years after application, the replication is twelve-fold, since it will be observed that with crop $a$, for instance, the plot is treated only in the first year.

If we are comparing the direct response with the residual responses, the accuracy is still greater.   In the first year the yield of crop $a$ on the above plot represents the direct effect, in the fifth year it represents the first-year residual effect, and in the ninth year the second-year residual effect.   Thus permanent differences between plots are entirely eliminated from this comparison over three complete rotations.   It is important to realize that this happens only because *the periods of the crop and treatment cycles are different.* Had the treatments been applied every fourth year, crop $a$ would have been treated in the fifth and ninth years and there would have been no " smoothing out " of permanent differences between plots. In experiments in which the " residual response curve " is of interest, it is worth while choosing the interval between treatments to obtain increased accuracy in this way.

Two examples of experiments of this type will be given.

*Three-course rotation experiment*

This experiment (21) started at Rothamsted in 1933 on a three-course rotation of barley, sugar-beet, potatoes.   There were four manurial treatments as follows :

  1. Artificials applied in spring.
  2. Straw applied in autumn, artificials in spring.
  3. Straw applied in autumn, artificials applied half in autumn and half in spring.
  4. Straw made into Adco compost applied in autumn.

The amount of straw rotted in the compost treatment ($53\frac{1}{3}$ cwt. per acre) is equal to the straw applied in the second and third treatments.   The amounts of N, $P_2O_5$, and $K_2O$ used in making the Adco compost are equal to those applied in the other treatments, but the compost suffers some loss in organic matter while it is being made. One of the main objects is to study whether yields can be maintained by the use of artificials, with or without the addition of straw, so that the experiment is planned to last for a considerable period.

The treatments are applied to the same plots every alternate year. Following the rule stated in the previous section, half the plots receive the manures in 1933 and alternate years, while the other half receive the manures in 1934 and alternate years.

The experiment also contains a study of the effects of growing vetches and rye as green manure crops between each crop of the rotation.   The three treatments, no green manure, vetches, and rye, were combined factorially with the manurial treatments, making 24 treatments in all, which were laid out in three randomized blocks (or series) of 24 plots each, one crop to each block.   The size of the block could have been reduced to 12 plots by putting the manures applied in odd and even years in separate blocks. This, however, was not advisable, since the comparison of the two groups of twelve treatments is necessary to establish the relative sizes of the direct and residual responses in any year.

Since half the plots do not receive any treatment in the first year, the experiment proper does not begin until the second year. Five years' complete results are available to date. The mean yields of the potato crop under the different manurial treatments will be given as an example of a brief *interim* inspection of the results.

Table VIII

*Three-course rotation experiment, Rothamsted Potatoes* (tons per acre)

| Year | Manured in current year | | | | | Manured in previous year | | | | |
|------|-------------------|---------------|--------------------------------------|------|------|-------------------|---------------|--------------------------------------|------|------|
|  | Arts. in Spring | Arts. in Spring | Arts. ½ in autumn ½ in spring | Adco | Mean | Arts. in Spring | Arts. in Spring | Arts. ½ in autumn ½ in spring | Adco | Mean |
|  |  | | Straw in autumn | | |  | | Straw in autumn | | |
| 1934 | 7·71 | 8·14 | 7·12 | 5·59 | 7·14 | 5·56 | 6·38 | 5·51 | 6·00 | 5·86 |
| 1935 | 7·22 | 8·24 | 7·16 | 6·37 | 7·25 | 4·90 | 5·69 | 6·02 | 6·15 | 5·69 |
| 1936 | 10·84 | 9·62 | 9·69 | 8·71 | 9·72 | 8·30 | 9·19 | 8·49 | 8·69 | 8·67 |
| 1937 | 5·11 | 5·67 | 4·27 | 3·87 | 4·73 | 3·81 | 3·93 | 4·06 | 3·52 | 3·83 |
| 1938 | 9·52 | 10·10 | 9·96 | 7·69 | 9·32 | 6·91 | 7·71 | 8·15 | 7·56 | 7·58 |
| Mean | 8·08 | 8·35 | 7·64 | 6·45 | 7·63 | 5·90 | 6·58 | 6·45 | 6·38 | 6·33 |
|  | | ±0·217 | | | | | ±0·158 | | | |

There were clear responses every year to treatments in the year of application. Since the experiment has run for only five years, each year constitutes a separate replication both of seasons and plots for the comparison of the treatment responses either in the year of application or in the year after application. The standard errors of the treatment means, derived from the interaction of treatments with seasons, are shown at the foot of the table. As would be expected, the residual effects were less variable than the direct effects.

In the year of application, artificials applied in spring, with or without straw, gave the highest yields. The addition of straw did not produce a significant increase in yields, but gave the higher yield in four seasons out of five. Comparing the two straw treatments, artificials applied in spring were significantly more effective than artificials applied half in autumn and half in spring. The Adco compost was distinctly inferior to the other treatments.

The results in the year after application tell a different story. The two straw treatments and the Adco gave significantly higher yields than artificials applied alone, indicating a residual effect of the former treatments. From this result one might expect in time a progressive improvement in the direct effects of these treatments relative to artificials alone. There is, however, no indication of this so far.

To compare the average effects of two-yearly dressings of the different manures, corresponding figures in the right- and left-hand sides of Table VIII are averaged. On this comparison, there is, as stated earlier, only three-fold replication. An analysis of variance, if required, would be based on the mean of 1934 and 1937, the mean of 1935 and 1938, and the 1936 figures as three replicates. The first two replicates are somewhat more accurately determined than the third, but as a first approximation the three should be given equal weight in carrying out an analysis. This analysis will be simpler after the 1939 results, when all plots have been represented twice.

The green manure crops gave very poor yields, and were discontinued in 1938. In future the experiment will provide three-fold replication within years on the manurial treatment.

*Four-course rotation experiment (Rothamsted)*

This is a more complicated example of the same type of experiment (22). There are four crops, swedes, barley, seeds, and wheat, and five manurial treatments, dung, Adco compost, straw and artificials, superphosphate, and rock phosphate. As in the three-course rotation experiment just described, the treatments remain on the same plots throughout and are applied only at intervals. An interval of five years was chosen, in order that the period of the treatment cycle should be different from that of the crop cycle. By this device twenty years elapse before a treatment is applied again on the same plot to the same crop, and the differences between direct and successive residual effects are entirely freed from permanent differences between plots over a period of twenty years. The experiment will thus provide highly accurate information on the magnitude and duration of the residual effects.

The experiment requires four series of 25 plots each, one crop to each series. Within each series the treatments could have been randomized completely, but confounding was introduced in order to reduce the size of the block to five plots. In any year each block of five plots contains one plot receiving each of the five treatments and one plot which received its treatment in the present, previous, and second-, third-, and fourth-previous years respectively. The plan of the group of plots growing seeds in the first year is shown below. Thus treatment 5 is applied to the appropriate plot in block a in the fourth year.

Table IX

*Plan of plots 1–25 in the four-course rotation experiment,*
*Rothamsted*

|       |   |        |       |       |        |        |
|-------|---|--------|-------|-------|--------|--------|
|       | a | 5  IV  | 2  I  | 1 III | 3  V   | 4  II  |
|       | b | 5  II  | 1  V  | 3  I  | 4  IV  | 2 III  |
| Block | c | 3  II  | 2 IV  | 5  V  | 4 III  | 1  I   |
|       | d | 1  II  | 3 IV  | 4  I  | 5 III  | 2  V   |
|       | e | 4  V   | 1 IV  | 5  I  | 3 III  | 2  II  |

1, 2, 3, 4, 5 refer to the treatments
I, II, III, IV, V refer to the year of application

The above arrangement was reached by first assigning the treatments 1, 2, 3, 4, 5 at random within each block, and using a $5 \times 5$ Latin square to fix the arrangement between blocks, treatments, and year of application. Different Latin squares were used for the other three crops.

This experiment has now been in progress nine years, of which four are preliminary. I do not propose to discuss the complete statistical analysis, which will require careful consideration, but one or two observations may be made. The comparison of the average effects of five-yearly dressings of the various treatments presents no difficulty, since the effect of blocks is automatically eliminated in averaging the five plots receiving any one treatment. These effects have four-fold replication on the four groups of 25 plots and the statistical analysis will follow the general lines indicated in discussing the experiment on p. 122.

The average " residual-response curve " for the five treatments is similarly free from block differences, and may also be analysed in four-fold replication. It is, however, already clear that this will not be of great interest, since the residual effects for the two phosphatic treatments are distinctly different from those for the three organic treatments, owing to the fact that the former receive a dressing of nitrogen every year.

Each year four degrees of freedom from the interaction of treatments with year of application are completely confounded with block differences. It is, however, easy to see that a different set of four degrees of freedom are confounded each time the same crop returns to the plots, until all sixteen degrees of freedom representing the interaction have been equally confounded. This property will probably facilitate the adjustment for block differences when a complete analysis is attempted.

*Experiments containing non-cumulative treatments*

*Introduction*

Up to the present we have been concerned with experiments in which the cumulative effects of treatments, applied every year or at intervals, are being studied. The majority of long-term experiments are likely to be of this type, for where the effects of treatments in the year of application only are required, the tendency is to change the site of the experiment in different years. There are, however, a number of cases in which non-cumulative treatments are included in long-term experiments. If we wish to accumulate material for a study of the influence of weather on the direct effects of treatments, many disturbing factors which might increase the experimental errors can be minimized or eliminated by keeping the experiment on the same site, *e.g.* variations in the fertility of different sites or in the exposure to wind and rain. Again, some organic manures, which compare unfavourably with artificial manures in their direct responses, are frequently recommended in farming practice because of their presumed cumulative effects and because they are thought to keep the land in good condition. An experiment to study this question might involve plots on which the treatments changed from year to year, to measure the direct effects, as well as plots on which treatments were applied every year. A similar type of experiment might be suggested for certain of the newer implements for breaking up the soil before planting, which may be suitable if used occasionally, but deleterious if used on the same piece of land year after year.

In planning an experiment on direct effects, it might at first be thought best to re-randomize the treatments each year, since any residual effects, if present, may be assumed to be allowed for in the experimental error. This method, however, is inaccurate in two senses. At the conclusion of the experiment it will be found, apart from very exceptional chance cases, that some treatments occurred more often on certain plots than others. Thus the influence of persistent plot differences has not been eliminated so well as if we had ensured that every treatment occurred on every plot an equal number of times. Secondly, if residual effects are present, it is easier to measure them, and to allow for them, if treatments succeed one another in some orderly fashion in successive years.

These two conditions imply that the treatments on a given plot should rotate cyclically in successive years, and that the cycles should be chosen so as to facilitate the evaluation of residual effects. This raises an interesting subject.

*Qualitative effects of treatments*

Consider first the case in which treatments are all of the same kind, such as different types of nitrogenous fertilizer or different cultivation implements. With three treatments, $a$, $b$, $c$, there are only two possible cycles, as follows :

| Year | Plot | | | Plot | | |
|---|---|---|---|---|---|---|
| | 1 | 2 | 3 | 4 | 5 | 6 |
| 1 | $a$ | $b$ | $c$ | $a$ | $b$ | $c$ |
| 2 | $b$ | $c$ | $a$ | $c$ | $a$ | $b$ |
| 3 | $c$ | $a$ | $b$ | $b$ | $c$ | $a$ |

FIG. 2.

If direct and residual effects are assumed to be *additive*, the yield of any plot in any year may be represented by a prediction formula. For example, at the end of a complete cycle (*i.e.* in the third, sixth . . . years), the formula for plot 1 is

$$y = m + c_0 + b_{-1} + a_{-2} + c_{-3} + \ldots$$

where the suffix $_0$ refers to the direct response, $_{-1}$ to the first year residual response, and so on, and the terms in the equation extend back to the first year of the experiment. Since we are assessing *differences* between $a$, $b$, and $c$, we may put $a_0 + b_0 + c_0 = 0$, and similarly for residual effects.

There are five degrees of freedom between the six plots, and of these the difference between the totals of plots 1, 2, 3 and plots 4, 5, 6. is not a treatment effect in the above scheme. Thus there are only four degrees of freedom available to estimate the constants. Hence we must ignore the second-year and more remote residual effects in order to estimate the direct and first-year residual effects.

The least-square normal equations can easily be solved, but it is more instructive to obtain the solutions directly. Consider $(a_0 - b_0)$, ignoring the second-year and more remote residual effects. The difference between plots 2 and 4 is a direct estimate of this, since both plots received $c$ in the previous year. A second estimate may be obtained by estimating $(a - c)$ from plots 6 and 1 (which had $b$ in the previous year) and $(c - b)$ from plots 5 and 3 (which had $a$ in the previous year). In combining these two estimates, the first is given double weight, since its variance is half that of the second. We obtain :

$$3(a_0 - b_0) = 2(acb - bca) + (abc - cba) + (cab - bac) \quad . \quad (1)$$

where the letters, from left to right, represent treatments in the current, previous and second-previous years respectively. Similarly

$$3(a_0 - c_0) = 2(abc - cba) + (acb - bca) + (bac - cab) \quad . \quad (2)$$

Adding, we obtain, since $a_0 + b_0 + c_0 = 0$,

$$3a_0 = 2(acb + abc) + (bac + cab) - 3 \text{ (mean)} \quad . \quad (3)$$

This is clearly free from first-year residual effects; if other residual effects are present, the right-hand side is actually an estimate of

$$3(a_0 - a_{-2} + a_{-3} - a_{-5} + \ldots)$$

Equation (3) may be written, more succinctly :

$$3a_0 = 2A_0 + A_{-1} - 3 \text{ (mean)}. \qquad . \qquad . \quad (4)$$

where $A_0$ is the sum of the yields of plots having $a$ in the present year and $A_{-1}$ the sum of the yields of plots having $a$ in the previous year. We find similarly :

$$3a_{-1} = A_0 + 2A_{-1} - 3 \text{ (mean)} \qquad . \qquad . \quad (5)$$

and, if other residual effects are present, this is really an estimate of

$$3(a_{-1} - a_{-3} + a_{-4} - a_{-6} + \ldots)$$

In analysing the results of experiments of this type, it is important to consider carefully whether residual effects can be ignored, since the adjustment of direct effects to correct for non-existent residual effects results in a loss of information. As would be expected, the constants $a_0$ and $a_{-1}$ are not orthogonal. The sum of squares taken out by the residual effects (after fitting the direct effects) is the sum of squares of *deviations* of the three quantities $(acb - abc)$, $(bac - bca)$, $(cba - cab)$ from their mean. A test of significance of the residual effects in a given year is obtained only if the six plots are replicated for each crop. Further, the fact that the mean square for residual effects does not reach significance in a given year does not prove that residual effects are non-existent. However, a fairly sound decision on this question should be possible by inspecting the results of a number of years.

In planning an experiment of this kind, there is no objection to reducing the size of block by putting the different cycles in different blocks, since the treatment effects which have been extracted are orthogonal with blocks.

With four treatments, there are many possible variations. The most convenient appears to be to arrange three replications which, if superimposed, would form a completely orthogonal set of $4 \times 4$ Latin squares. An example of this arrangement is :

| Year | Plot | | | | | | | | | | | |
|---|---|---|---|---|---|---|---|---|---|---|---|---|
|  | 1 | 2 | 3 | 4 | 5 | 6 | 7 | 8 | 9 | 10 | 11 | 12 |
| 1 | a | b | c | d | a | b | c | d | a | b | c | d |
| 2 | b | a | d | c | d | c | b | a | c | d | a | b |
| 3 | c | d | a | b | b | a | d | c | d | c | b | a |
| 4 | d | c | b | a | c | d | a | b | b | a | d | c |

Fig. 3.

Of the three plots receiving treatment $a$ in any year, one had $b$ in the previous year, one $c$ and one $d$, and the same is true of the second previous year, and of the other treatments. This property facilitates the separation of direct, residual, and second-year residual effects, which can all be compared for the four treatments, since there are now nine degrees of freedom available. The direct effect of $a$ (assuming that first- and second-year residual effects exist) is given by :

$$4a_0 = 2A_0 + A_{-1} + A_{-2} - 4 \text{ (mean)}$$

With greater numbers of treatments, designs having these properties can always be obtained from completely orthogonal sets of squares (where they exist), though the number of plots required becomes large. I have not investigated the possibility of obtaining convenient estimates of residual effects from fewer numbers of replications, though a demand for this might arise if the treatments are numerous.

In these experiments the effects of permanent differences between plots can be eliminated from the treatment comparisons, provided that the number of treatments is not a factor of the number of crops. With a three-course rotation and four treatments, for example, no treatment reappears on the same plots with the same crop until after twelve years, in which period every treatment has appeared once with every crop on any plot. With a four-course rotation and six treatments, the combined crop and treatment cycle on any plot begins to repeat itself after 12 years, though in this period each crop has appeared with only *three* of the treatments. If the experiment is to be continued further, it may be advisable, if practicable agriculturally, to omit one crop of the rotation in the thirteenth year, *i.e.* to grow the crops which should have been grown in the fourteenth year, thereafter continuing the four-course rotation as before. It may be verified that with this device the combined crop and treatment cycle does not begin to repeat until after 24 years, by which time every crop has appeared once with every treatment on every plot.

*Cultivation experiment, Rothamsted* 1934–1939

In this experiment, both direct and cumulative treatment effects are being studied (23). The rotation is a three-course one of wheat, mangolds, barley. The principal object is to compare the effects of three methods of breaking up the soil before sowing—ploughing ($P$), rotary cultivation with the simar implement ($S$), and stirring the soil with a cultivator ($C$). These are compared at two depths of cultivation 4 inches (shallow) and 8 inches (deep). Further, since weeds may accumulate under certain cultivation treatments, two nitrogenous fertilizers are included—nitro-chalk ($N$) and cyanamide

(*Cy*), the latter of which is sometimes considered a useful weed-killer. All combinations of the three factors are included, making 12 primary treatments.

In one half of the experiment the treatments remain unchanged throughout on each plot, each crop being arranged in two randomized blocks of twelve treatments each, giving two-fold replication. On these plots, cumulative differences introduced by the continuous use of one type of cultivation are built up. In the other half of the experiment, also consisting of two randomized blocks of twelve treatments each for each crop, the treatments rotate on a given plot in successive years so as to provide estimates of the direct effects of the treatments in any year. In one block the cultivation treatments follow the cycle described on plots 1–3 in Fig. 2 above, and in the other block, the cycle on plots 4–6. On each of these blocks the deep and shallow cultivations alternate in successive years and the nitrogenous fertilizers alternate at two-yearly intervals. The effects of this scheme on the accuracy of the experiment may be seen by considering the sequence of treatments on a given plot in successive years. For the plot in the first block which is ploughed deep and receives nitro-chalk in the first year, the sequence is 1 *PDN*; 2 *SShN*; 3 *CDCy*; 4 *PShCy*; 5 *SDN*; 6 *CShN*; 7 *PDCy*; 8 *SShCy*; 9 *CDN*; 10 *PShN*; 11 *SDCy*; 12 *CShCy*. At the end of twelve years each treatment has appeared once on every plot. However, on the comparisons of the average effects of ploughing, simaring, and cultivating, the replication in years is only three-fold, since the crop and treatment cycles have the same period. The comparison of deep and shallow ploughing is more accurate, since plots ploughed deep in any year are ploughed shallow when the same crop returns three years later, so that over a six-yearly period (the intended duration of the experiment) the effect of systematic plot errors is entirely eliminated. This is also true for nitro-chalk versus cyanamide over a twelve-yearly period, though not over a six-yearly period (the plot given above, for example, receives nitro-chalk in the second and fifth years and cyanamide in the eighth and eleventh years).

The experiment has now completed five years, and will finish at the end of the sixth year. In discussing the results we will confine our attention to the average effects of the three cultivation implements, which are shown for each crop in Table X. Ploughing gave higher yields than simaring or cultivating, and simaring proved somewhat better than cultivating on the continuous plots. It should, however, be noted that except in 1938 all three types of cultivation were unfortunately carried out at the same time, and not at different times chosen as most suitable for each separate implement.

## Table X

*Cultivation Experiment Rothamsted, 1933–8*

| Last year / This year | Continuous | | | | Cycle A | | | Cycle B | | | |
|---|---|---|---|---|---|---|---|---|---|---|---|
| | P P | S S | C C | Mean | C P | P S | S C | S P | C S | P C | Mean |
| **Year** | Wheat grain, cwt. per acre | | | | | | | | | | |
| 1934 ... ... | 25·8 | 21·0 | 23·2 | 23·4 | 23·8 | 20·6 | 24·4 | 24·8 | 22·3 | 25·4 | 23·6 |
| 1935 ... ... | 23·1 | 20·6 | 20·1 | 21·3 | 20·4 | 19·9 | 18·3 | 24·5 | 22·2 | 18·0 | 20·6 |
| 1936 ... ... | 22·6 | 21·3 | 20·9 | 21·6 | 21·4 | 21·7 | 20·4 | 20·4 | 19·4 | 19·4 | 20·4 |
| 1937 ... ... | 20·8 | 12·1 | 9·5 | 14·1 | 21·6 | 18·1 | 11·8 | 17·4 | 9·8 | 11·8 | 15·1 |
| 1938 ... ... | 15·4 | 8·4 | 11·0 | 11·6 | 16·4 | 9·9 | 11·0 | 17·6 | 9·0 | 11·8 | 12·6 |
| Mean ... | 21·5 | 16·7 | 16·9 | 18·4 | 20·7 | 18·0 | 17·2 | 20·9 | 16·5 | 17·3 | 18·5 |
| **Year** | Mangolds roots, tons per acre | | | | | | | | | | |
| 1934 ... ... | 37·1 | 34·2 | 36·7 | 36·0 | 36·5 | 35·5 | 37·1 | 37·4 | 32·4 | 36·4 | 35·9 |
| 1935 ... ... | 24·1 | 20·2 | 17·4 | 20·6 | 22·3 | 21·8 | 21·1 | 19·7 | 18·4 | 17·8 | 20·2 |
| 1936 ... ... | 22·4 | 19·7 | 18·0 | 20·0 | 22·1 | 20·8 | 19·1 | 23·1 | 20·1 | 22·4 | 21·3 |
| 1937 ... ... | 20·6 | 21·1 | 17·6 | 19·8 | 19·9 | 19·6 | 17·3 | 21·6 | 16·9 | 19·2 | 19·1 |
| 1938 ... ... | 15·7 | 14·2 | 13·4 | 14·4 | 13·3 | 14·0 | 13·3 | 12·3 | 12·8 | 12·2 | 13·0 |
| Mean ... | 24·0 | 21·9 | 20·6 | 22·2 | 22·8 | 22·3 | 21·6 | 22·8 | 20·1 | 21·6 | 21·9 |
| **Year** | Barley grain, cwt. per acre | | | | | | | | | | |
| 1934 ... ... | 26·2 | 25·8 | 26·4 | 26·1 | 25·3 | 26·1 | 24·6 | 28·4 | 27·2 | 28·5 | 26·7 |
| 1935 ... ... | 36·4 | 36·0 | 34·6 | 35·7 | 34·8 | 34·6 | 34·8 | 33·4 | 33·4 | 33·8 | 34·1 |
| 1936 ... ... | 27·8 | 25·6 | 21·5 | 25·0 | 29·8 | 28·2 | 28·1 | 28·2 | 28·2 | 25·4 | 28·0 |
| 1937 ... ... | 16·0 | 14·2 | 14·8 | 15·0 | 16·2 | 17·2 | 14·3 | 15·9 | 13·0 | 13·1 | 15·0 |
| 1938 ... ... | 16·8 | 18·6 | 16·8 | 17·4 | 10·3 | 15·0 | 12·2 | 15·1 | 18·5 | 20·2 | 15·2 |
| Mean ... | 24·6 | 24·0 | 22·8 | 23·8 | 23·3 | 24·2 | 22·8 | 24·2 | 24·1 | 24·2 | 23·8 |

The direct and first-year residual effects can be estimated from the non-continuous part of the experiment, as described previously. If more remote residual effects are present, the sum of the estimated direct and first-year residual effects is actually an estimate of

$$a_0 + a_{-1} - a_{-2} + a_{-4} - a_{-5} + \ldots$$

If $a_{-1}$ is small, the terms $a_{-2}$, $a_{-4} \ldots$ may be assumed negligible, but, if necessary, a rough estimate of these terms could be obtained graphically from the values of the $a_0$ and $a_{-1}$. The sum of the direct and first-year residual effects could thus be corrected to represent the sum of the direct effect and all residual effects which are not negligible. Thus we can predict what the results of the continuous part of the experiment should be, if the continuous effects are nothing more than the direct effect plus the residual effects. If the results on the continuous blocks diverge *systematically* from these predictions as the experiment proceeds, this shows that there is a real cumulative effect which is not explained simply in terms of direct and residual effects. The test of this point is one of the main objects of the experiment.

In Table X there are fairly clear indications of a beneficial residual effect of ploughing on mangolds in 1936, and on all three crops in 1937, so that first-year residual effects cannot be ignored. The direct and first-year residual differences between the two extreme treatments, ploughing and cultivating, are shown in Table XI, and their sum is compared with the corresponding difference on the continuous plots. Second-year residual effects were ignored.

TABLE XI

*Cultivation Experiment Rothamsted, 1933–8.   Difference between ploughing and cultivating*

| | Direct | First-year Residual | Direct + Residual | Continuous | Continuous minus (Direct + Residual) | Block |
|---|---|---|---|---|---|---|
| Year | Wheat grain, cwt. per acre | | | | | |
| 1934 | −0·6 | — | − 0·6 | + 2·6 | +3·2 | II |
| 1935 | +4·2 | −0·3 | + 3·9 | + 3·0 | −0·9 | III |
| 1936 | +1·5 | +0·8 | + 2·3 | + 1·7 | −0·6 | I |
| 1937 | +9·8 | +4·1 | +13·9 | +11·3 | −2·6 | II |
| 1938 | +6·2 | +1·3 | + 7·5 | + 4·4 | −3·1 | III |
| Year | Mangolds roots, tons per acre | | | | | |
| 1934 | +0·2 | — | + 0·2 | + 0·4 | +0·2 | I |
| 1935 | +1·7 | +0·3 | + 2·0 | + 6·7 | +4·7 | II |
| 1936 | +2·8 | +1·9 | + 4·7 | + 4·5 | −0·2 | III |
| 1937 | +4·0 | +3·0 | + 7·0 | + 3·0 | −4·0 | I |
| 1938 | +0·1 | +0·1 | + 0·2 | + 2·3 | +2·1 | II |
| Year | Barley grain, cwt. per acre | | | | | |
| 1934 | +0·3 | — | + 0·3 | − 0·2 | −0·5 | III |
| 1935 | −0·2 | 0·0 | − 0·2 | + 1·8 | +2·0 | I |
| 1936 | +1·5 | −1·4 | + 0·1 | + 6·2 | +6·1 | II |
| 1937 | +3·5 | +2·3 | + 5·8 | + 1·2 | −4·6 | III |
| 1938 | −2·5 | +1·9 | − 0·6 | 0·0 | +0·6 | I |

Rate of change of " continuous minus (direct + residual) "

| | Wheat | | Mangolds | | Barley | |
|---|---|---|---|---|---|---|
| | Block | Difference | Block | Difference | Block | Difference |
| 1937 − 1934  ... | II | −5·8 | I | −4·2 | III | −4·1 |
| 1938 − 1935  ... | III | −2·2 | II | −2·6 | I | −1·4 |

It must be borne in mind that for each crop the treatments return to the same plots every three years. The method of procedure is thus similar to that which was used in examining the rate of deterioration of yields in the four-course rotation experiment on p. 124. For each block we take the difference between the values of " continuous minus (direct plus residual) " for the two years in which any crop was grown upon it. These are shown at the foot of Table XI. At present, there is only two-fold replication, but a third replication will be added when the 1939 results are known.

The results are somewhat surprising, and appear to indicate that as the experiment proceeds the differences between continuous treatments are becoming progressively *smaller* in relation to the direct and first-year residual differences. Too much stress cannot be laid upon the small number of figures available, and moreover the results from different crops on the same blocks are not independent, but they may at least be regarded as showing that no additional differences between treatments have been introduced by applying them continuously on the same plots.

It will be observed from Table X that the yields fell drastically in the last two years. In particular, 1938 was an excellent year for wheat, yet the yields on these plots are poor. The cause of this failure, which considerably decreases the value of any results of the experiment, is not certain, but contributory factors may have been the lack of provision for cross-cultivation in the layout of the experiment, the absence of dung, and the rather exhausting nature of the rotation itself.

*Quantitative effects of treatments*

With experiments containing different quantities of the same treatment, the choice of a suitable treatment cycle appears to be more difficult, unless residual effects can be ignored. To take a simple case, consider a factorial experiment with four treatments $o, n, p, np$. The following is a typical cycle :

| Year | Treatment | | | |
|------|------|------|------|------|
| 1 | $o$ | $p$ | $n$ | $np$ |
| 2 | $p$ | $n$ | $np$ | $o$ |
| 3 | $n$ | $np$ | $o$ | $p$ |
| 4 | $np$ | $o$ | $p$ | $n$ |

In any year the direct effect of $n$ is confounded with the residual interaction, which can probably be assumed negligible, but the direct effect of $p$ is confounded with the residual effect of $p$, and the direct interaction is confounded with the residual effect of $n$. By having different cycles present, it is possible to disentangle the direct and residual effects, particularly if some of the latter can be ignored. The reader is invited to consider the investigation of convenient designs of this type.

*Experiments on the effects of different crops*

In the previous sections the experiments described were concerned with the effects of manurial and cultivation treatments on a particular crop or succession of crops.   A certain amount of information was also obtained incidentally on the effects of crops on the soil. For instance, the classical wheat experiments at Rothamsted showed that after a time the yields of wheat grown continuously deteriorated under every one of the manurial treatments which were applied, though certain treatments mitigated the rate of fall considerably. Information of this kind is, however, limited to the particular crop, or succession of crops, which were grown.

The importance of studying the effects of crops, as well as of manures, varieties, and cultivations, need not be stressed.   In the United States, for example, it is generally admitted that the disastrous effects of soil erosion have been accentuated on some farms by the growth of a succession of crops which gave no protection to the soil. In these cases the problem of recommending a series of crops which will help to conserve the soil, or to restore badly depleted land, has now arisen in an acute form.

Few long-term experiments on the effects of different crops have been attempted in this country, except in a modest way, such as the inclusion of vetches and rye as green manures in the three-course rotation described on p. 122.   The practical difficulties are considerable; in addition to those which present themselves in any long-term experiment, they involve the growing of different crops on contiguous small plots in the same block.   However, with increasing experience these experiments will perhaps lose some of their terrors, and more attention may be devoted in future to a branch of experimental agriculture which has hitherto been much neglected.

I shall not attempt to discuss in detail the different types of experiment which may arise.   The general statistical principles of design may be brought out by an example.

*Ley experiment at Woburn*

This experiment was started in 1937, on a site which was considered to be in a poor state of fertility.   There are four crop " treatments," each of which lasts for three years, as follows :

|  |  | 1st | 2nd | Crops grown in 3rd Year | 4th | 5th |
|---|---|---|---|---|---|---|
| 1. | Three-year ley (clover–rye-grass mixture) ... | $L_1$ | $L_2$ | $L_3$ | $P$ | $B$ |
| 2. | Three-year lucerne    ...    ...    ...    ... | $Lu_1$ | $Lu_2$ | $Lu_3$ | $P$ | $B$ |
| 3. | Arable with one-year ley (potatoes, wheat, hay) ...    ...    ...    ...    ...    ... | $P$ | $W$ | $H$ | $P$ | $B$ |
| 4. | Arable without ley (potatoes, wheat, kale)... | $P$ | $W$ | $K$ | $P$ | $B$ |

Each of these cycles is followed by potatoes on all plots in the fourth year and barley in the fifth year.  These two crops will be used to assess differences in the fertility of the soil introduced by the previous crop treatments.  It is expected that the arable treatments, and particularly the one without any ley, will prove more exhausting than the other treatments.

Thus far, the experiment, in single replication, requires only four plots, and provides comparisons of the effects of the crop cycles only at five-yearly intervals.  In order to obtain these comparisons every year, the same rule was adopted as in previous rotation experiments, *each phase of the five-year cycle being represented every year*.  This requires twenty plots.

When the experiment reaches its fifth year, two courses are possible.  We may either retain the same crop treatments on each plot, to produce cumulative differences between the crop cycles, or rotate the cycles from plot to plot, to measure direct effects in subsequent years.  In order to compare the cycles in both conditions, the experiment was increased to forty plots, the cycles remaining the same throughout on one half and rotating from plot to plot on the other half.

With four treatments, there are six possible ways of rotating the treatments on a plot cyclically.  It was, however, decided to exclude rotations in which one arable treatment was succeeded by the other, or ley by lucerne, so as to keep the comparisons on the " direct " part of the experiment at a steady fertility level.  This leaves only two possible cycles :  1, 3, 2, 4 or 1, 4, 2, 3.  If only one cycle is used, each treatment always follows the same previous treatment and the comparisons may be disturbed by residual effects.  With only twenty plots available, we cannot balance out these residual effects on a given crop in a single year, but we can balance them every two years, by ensuring that if treatment 1 following 3 is compared with 2 following 4 in one year, then treatment 1 following 4 is compared with 2 following 3 in the next year.  This was arranged by having different cycles for the plots entering the experiment in odd and even years.  The cycles used are shown below :

| Years from year of inclusion in experiment | Odd years.  Treatment | Even years.  Treatment |
|---|---|---|
| 1–3 | 1  2  3  4 | 1  2  3  4 |
| 6–8 | 3  4  1  2 | 4  3  2  1 |
| 11–13 | 2  1  4  3 | 2  1  4  2 |
| 16–18 | 4  3  2  1 | 3  4  1  3 |

Thus in years 6–8, 1 follows 3; in years 7–9, 1 follows 4, and so on.  Notice that the cycles 1, 3, 2, 4 and 1, 4, 2, 3 are represented in

## Table XII

### *Woburn Ley Experiment*

Sequence of crops for the first twenty years on the "rotating" part of the experiment

| Year | Block 1 | | | | Block 2 | | | | Block 3 | | | | Block 4 | | | | Block 5 | | | |
|---|---|---|---|---|---|---|---|---|---|---|---|---|---|---|---|---|---|---|---|---|
| | 1 | 2 | 3 | 4 | 5 | 6 | 7 | 8 | 9 | 10 | 11 | 12 | 13 | 14 | 15 | 16 | 17 | 18 | 19 | 20 |
| 1 | $L_1$ | $Lu_1$ | P | P | $L_2$ | $Lu_2$ | W | W | $L_3$ | $Lu_3$ | H | K | Potatoes | Potatoes | Potatoes | Potatoes | Barley | Barley | Barley | Barley |
| 2 | $L_2$ | $Lu_2$ | W | W | $L_3$ | $Lu_3$ | H | K | Potatoes | Potatoes | Potatoes | Potatoes | Barley | Barley | Barley | Barley | P | P | $Lu_1$ | $L_1$ |
| 3 | $L_3$ | $Lu_3$ | H | K | Potatoes | Potatoes | Potatoes | Potatoes | Barley | Barley | Barley | Barley | P | P | $Lu_1$ | $L_1$ | W | W | $Lu_2$ | $L_2$ |
| 4 | Potatoes | Potatoes | Potatoes | Potatoes | Barley | Barley | Barley | Barley | P | P | $Lu_1$ | $L_1$ | W | W | $Lu_2$ | $L_2$ | H | K | $Lu_3$ | $L_3$ |
| 5 | Barley | Barley | Barley | Barley | P | P | $Lu_1$ | $L_1$ | W | W | $Lu_2$ | $L_2$ | H | K | $Lu_3$ | $L_3$ | Potatoes | Potatoes | Potatoes | Potatoes |
| 6 | P | P | $Lu_1$ | $L_1$ | W | W | $Lu_2$ | $L_2$ | H | K | $Lu_3$ | $L_3$ | Potatoes | Potatoes | Potatoes | Potatoes | Barley | Barley | Barley | Barley |
| 7 | W | W | $Lu_2$ | $L_2$ | H | K | $Lu_3$ | $L_3$ | Potatoes | Potatoes | Potatoes | Potatoes | Barley | Barley | Barley | Barley | $Lu_1$ | $L_1$ | P | P |
| 8 | H | K | $Lu_3$ | $L_3$ | Potatoes | Potatoes | Potatoes | Potatoes | Barley | Barley | Barley | Barley | $Lu_1$ | $L_1$ | P | P | $Lu_2$ | $L_2$ | W | W |
| 9 | Potatoes | Potatoes | Potatoes | Potatoes | Barley | Barley | Barley | Barley | $Lu_1$ | $L_1$ | P | P | $Lu_2$ | $L_2$ | W | W | $Lu_3$ | $L_3$ | K | H |
| 10 | Barley | Barley | Barley | Barley | $Lu_1$ | $L_1$ | P | P | $Lu_2$ | $L_2$ | W | W | $Lu_3$ | $L_3$ | K | H | Potatoes | Potatoes | Potatoes | Potatoes |
| 11 | $Lu_1$ | $L_1$ | P | P | $Lu_2$ | $L_2$ | W | W | $Lu_3$ | $L_3$ | K | H | Potatoes | Potatoes | Potatoes | Potatoes | Barley | Barley | Barley | Barley |
| 12 | $Lu_2$ | $L_2$ | W | W | $Lu_3$ | $L_3$ | K | H | Potatoes | Potatoes | Potatoes | Potatoes | Barley | Barley | Barley | Barley | P | P | $L_1$ | $Lu_1$ |
| 13 | $Lu_3$ | $L_3$ | K | H | Potatoes | Potatoes | Potatoes | Potatoes | Barley | Barley | Barley | Barley | P | P | $L_1$ | $Lu_1$ | W | W | $L_2$ | $Lu_2$ |
| 14 | Potatoes | Potatoes | Potatoes | Potatoes | Barley | Barley | Barley | Barley | P | P | $L_1$ | $Lu_1$ | W | W | $L_2$ | $Lu_2$ | K | H | $L_3$ | $Lu_3$ |
| 15 | Barley | Barley | Barley | Barley | P | P | $L_1$ | $Lu_1$ | W | W | $L_2$ | $Lu_2$ | K | H | $L_3$ | $Lu_3$ | Potatoes | Potatoes | Potatoes | Potatoes |
| 16 | P | P | $L_1$ | $Lu_1$ | W | W | $L_2$ | $Lu_2$ | K | H | $L_3$ | $Lu_3$ | Potatoes | Potatoes | Potatoes | Potatoes | Barley | Barley | Barley | Barley |
| 17 | W | W | $L_2$ | $Lu_2$ | K | H | $L_3$ | $Lu_3$ | Potatoes | Potatoes | Potatoes | Potatoes | Barley | Barley | Barley | Barley | $L_1$ | $Lu_1$ | P | P |
| 18 | K | H | $L_3$ | $Lu_3$ | Potatoes | Potatoes | Potatoes | Potatoes | Barley | Barley | Barley | Barley | $L_1$ | $Lu_1$ | P | P | $L_2$ | $Lu_2$ | W | W |
| 19 | Potatoes | Potatoes | Potatoes | Potatoes | Barley | Barley | Barley | Barley | $L_1$ | $Lu_1$ | P | P | $L_2$ | $Lu_2$ | W | W | $L_3$ | $Lu_3$ | H | K |
| 20 | Barley | Barley | Barley | Barley | $L_1$ | $Lu_1$ | P | P | $L_2$ | $Lu_2$ | W | W | $L_3$ | $Lu_3$ | H | K | Potatoes | Potatoes | Potatoes | Potatoes |

$L_1$, $L_2$, $L_3$ = first, second and third year's ley.
$Lu_1$, $Lu_2$, $Lu_3$ = first, second and third year's lucerne.

P = potatoes.  
W = wheat.

H = hay.  
K = kale.

both the odd years and the even years. This is necessary to ensure that 1 follows 4 in years 11–13 and that 1 follows 3 in years 16–19.

The whole scheme of treatments for the rotating part of the experiment is shown (before randomization) in Table XII for the first 20 years. The cropping on plots which have not yet entered the experiment was chosen to give supplementary information on the uniformity of the plots and, in block 5, on the comparison of hay with kale.

The arrangement of the experiment on its site is worth consideration. The five phases may be put in different blocks, since the indicator crops are grown in different years on these blocks. The different crop treatments in any phase must, however, be grown in the same block, since the comparison of their effects must be made as accurate as possible. Within each block, which will have eight plots, we may either randomize the treatments completely, or split the block in two, putting the continuous treatments in one half-block and the rotating treatments in the other. There is no objection to the latter course as far as the indicator crops are concerned, but we may, for instance, wish to compare the yield of wheat from an arable sequence following leys with the yield of wheat in the corresponding continuous arable sequence, and these plots would be in different half-blocks. It was therefore considered best to randomize the eight treatments completely within blocks.

In the above example the different crop treatments are compared by including two indicator crops. It is an interesting question to consider how frequently indicator crops should be grown and whether it is possible to dispense with them. If the crop treatments can act as their own indicators, some time will clearly be saved. The question naturally depends on the types of crop sequence which are being compared. For example, with *continuous* leys, $L\,L\,L\,L\,\ldots$, versus a *continuous* arable sequence without leys, $A\,A\,A\,A\,\ldots$, indicator crops can be avoided only if both phases $L\,A\,L\,A\,\ldots$ and $A\,L\,A\,L\,\ldots$ of the corresponding rotating sequence are introduced as extra treatments, and even with this addition the comparison is rather indirect. Where the crop sequences contain common crops, less difficulty arises.

| Year | Continuous Maize | Two-course Rotation | | Three-course Rotation | | |
|---|---|---|---|---|---|---|
| 1 | *M* | *M* | *O* | *M* | *M* | *O* |
| 2 | *M* | *O* | *M* | *M* | *O* | *M* |
| 3 | *M* | *M* | *O* | *O* | *M* | *M* |
| 4 | *M* | *O* | *M* | *M* | *M* | *O* |
| 5 | *M* | *M* | *O* | *M* | *O* | *M* |
| 6 | *M* | *O* | *M* | *O* | *M* | *M* |

Consider an experiment in which continuous maize is being compared with a two-course rotation of maize and oats and a three-course rotation of maize, maize, oats. If all phases of each replication are represented, a single replication requires six plots, which are cropped as follows in successive years.

In any year four of the six plots carry maize and provide a comparison of continuous maize with maize grown in two-course rotation and with both maize crops in the three-course rotation. On the remaining plots the oats crop in the two-course rotation is compared with the oats crop in the three-course rotation. From these comparisons it should be possible to assess the relative effects of the crop sequences on the fertility of the soil, though it would be wise to grow an indicator crop at the end of the experiment. It would also be advisable to duplicate or perhaps triplicate the continuous plots in every replication, as these enter into the comparisons on maize every year.

*Concluding remarks*

In the preceding sections I have endeavoured to outline the statistical principles governing the design of some of the more important types of long-term experiments, and to indicate methods of approach which may be useful in analysing the results. There are, however, many aspects of the subject which have not been discussed, and I wish to mention briefly one or two of these.

No matter how elegant the design may appear on paper, the experiment risks the possibility of failure unless the operations essential to good farming are carried out. Adequate turning headlands should be provided to permit cultivations both along and across the rows. If the whole of each series carries the same crop and is being cultivated in the same way, it will only be necessary to provide headlands between series. If, however, different cultivation treatments are being applied to different plots within the same series, headlands must be allowed between neighbouring plots, even though the headlands may occupy a fair proportion of the total area of the experiment. The crops should be protected from animal and bird damage. Unless exhaustion of the soil is being deliberately produced, a basal manuring should be given to all plots sufficient to maintain yields at a reasonable level. The succession of crops in any crop cycle should be chosen so that each crop can be sown and harvested at the most suitable time, and crops which are likely to fail repeatedly should be avoided. If leys are included which are to be grazed by sheep, the plot size must be sufficiently large so that this can be done effectively, though the other crops in the experiment could perhaps have been grown on smaller plots. Practical points such as these are liable to be overlooked, but are vital to the success of the experiment.

This leads to a further point. Long-term experiments are probably more liable to accidents and misfortunes than annual experiments. If, for example, certain crops fail repeatedly, or if the results in the first few years are highly inaccurate, perhaps owing to a variable site or to insufficient replication, should the experiment be abandoned, or should changes be introduced in the hope of improving it? Some of the older rotation experiments provide eloquent tribute to the danger of making repeated changes, but clearly each case must be considered on its own merits.

I have not considered to what extent different crops can be combined in analysing the results of rotation experiments. In general it is best to consider each crop separately, as treatment effects are likely to be different on different crops, and it is difficult to decide what weight of one crop is equivalent to unit weight of another crop. There may, however, be cases in which certain treatment effects are similar on different crops, but do not reach significance on any individual crop, though a combined analysis might establish the significance of the average treatment effects over all crops. The examination of the " cumulative " effects in the cultivation experiment on p. 129 is perhaps an example of this type, and the possibility of a combined analysis is being investigated. It should be remembered that if all crops are combined in the same analysis, each series is represented *every year*, and the analysis will follow the lines indicated on p. 115 for replicated single-crop experiments, the series constituting the replications.

In conclusion, long-term experiments are a valuable weapon for increasing our scientific knowledge of agriculture. Any experiment station which undertakes them seriously, commits itself to a considerable expenditure of time, labour, and money and is, I think, entitled to expect whole-hearted co-operation from the statistician. I hope I have shown that this co-operation provides the statistician with some stimulating problems.

In writing this paper I have naturally drawn freely on the experience of previous members of the Statistical Department at Rothamsted. In particular, I wish to thank Dr. F. Yates for many useful suggestions and criticisms.

### References

(1) J. Wishart, " Statistics in agricultural research," *J. Roy. Stat. Soc. Suppl.*, 1934, **1**, 26–61.
(2) F. Yates, " Complex experiments," *ibid.*, 1935, **2**, 181–247.
(3) M. S. Bartlett, " Some examples of statistical methods of research in agriculture and applied biology," *ibid.*, 1937, **4**, 137–83.

(4) W. S. Gosset, "Co-operation in large-scale experiments," *ibid.*, 1936, **3**, 115–36.

(5) R. A. Fisher, "An examination of the yield of dressed grain from Broadbalk," *J. Agric. Sci.*, 1921, **11**, 107–35.

(6) *Idem*, "The influence of rainfall on the yield of wheat at Rothamsted," *Phil. Trans. Roy. Soc.*, 1924, **213**, 89–142.

(7) W. A. Mackenzie, "An examination of the yield of dressed grain from Hoosfield," *J. Agric. Sci.*, 1924, **14**, 434–60.

(8) J. Wishart and W. A. Mackenzie, "The influence of rainfall on the yield of barley at Rothamsted," *ibid.*, 1930, **20**, 417–39.

(9) R. J. Kalamkar, "A statistical examination of the yield of mangolds from Barnfield at Rothamsted," *ibid.*, 1933, **23**, 161–75.

(10) *Idem*, "The influence of rainfall on the yield of mangolds at Rothamsted," *ibid.*, 1933, **23**, 571–9.

(11) L. H. C. Tippett, "On the effect of sunshine on wheat yield at Rothamsted," *ibid.*, 1926, **16**, 159–65.

(12) W. G. Cochran, "A note on the influence of rainfall on the yield of cereals in relation to manurial treatment," *ibid.*, 1935, **25**, 510–22.

(13) F. Yates, "The recovery of inter-block information in variety trials arranged in three-dimensional lattices," *Ann. Eug.* (In the press.)

(14) R. K. S. Murray, "The value of a uniformity trial in field experimentation with rubber," *J. Agric. Sci.*, 1934, **24**, 177–84.

(15) T. Eden, "The experimental errors of field experiments with tea," *ibid.*, 1931, **21**, 547–73.

(16) H. C. Forster, "Design of agronomic experiments for plots differentiated in fertility by past treatments," *Iowa State College Research Bulletin* 226, 1937.

(17) F. Yates, "Incomplete Latin squares," *J. Agric. Sci.*, 1936, **26**, 301–15.

(18) T. N. Hoblyn, "Field experiments in horticulture," Imp. Bur. of Fruit Production, Tech. Comm. No. 2, 1931.

(19) J. T. Martin, H. H. Mann, and F. Tattersfield, "The manurial requirements of pyrethrum," *Ann. App. Biol.*, 1939, **26**, 21.

(20) Sir E. J. Russell and J. A. Voelcker, *Fifty years of field experiments at the Woburn Experimental Station.* Longmans Green and Co., London, 1936, p. 196.

(21) *Rothamsted Experimental Station Annual Report*, 1933, p. 118, and subsequent years.

(22) *Ibid.*, 1930, p. 125, and subsequent years.

(23) *Ibid.*, 1934, p. 175, and subsequent years.

## Discussion on Mr. Cochran's Paper

The CHAIRMAN said that this subject was one of particular importance to all agricultural workers and the author was to be sincerely thanked for the trouble he had taken in assembling the material and presenting it in so lucid a form. In listening to this paper he could not help recalling a conversation he had had at Rothamsted almost exactly twenty years ago with a young mathematician about the enormous mass of data accumulated there in the course of the field experiments that had then been going on for nearly eighty years. Yield and other measurements were made annually, five- and ten-year averages had been taken out and found to furnish some useful information, but it seemed highly improbable that these elementary methods could extract all the information the data contained, and it was equally improbable that the methods employed at the time of which he was speaking would carry matters much further. Doubts had arisen as to whether it was worth while piling up records of this kind and simply putting them away, and the question which he put to the young mathematician was this : " Can you devise methods by which you can extract more information from these masses of figures than can be obtained by mere inspection ? " Fortunately the mathematician was R. A. Fisher : he came to Rothamsted to examine the data, and soon decided that something could be done : he put in fourteen fruitful years of work, which was continued by J. Wishart, F. Yates, and Mr. Cochran.

The application of these modern methods to the examination of the data so laboriously accumulated had yielded very valuable results. The new methods extracted far more information than was possible for the older ones. The relations between yields and meteorological conditions, which previously had been examined only very crudely by comparing what Lawes and Gilbert and afterwards Hall called a " typical dry season " with a " typical wet season," were worked out on a sound basis. The old methods had certainly given useful information in spite of their serious defects, but they remained at best a kind of glorified guesswork, and the new methods allowed far better comparisons to be made. They soon showed the inadequacy of the present methods of recording meteorological data in relation to the growth of crops, and they provided methods of examining the relations between meteorological conditions and crop growth which had opened up the possibility of developing a science of agricultural meteorology—which would be of profound importance if only we could get well on with it. Further, the new methods opened up the possibilities of crop forecasting, one of the most important problems of our time, in view of the necessity for producing adequate amounts of foodstuffs and raw materials, while at the same time avoiding the waste resulting from production greatly in excess of requirements. Methods of crop forecasting were obviously needed for organizing large-scale agricultural production of the world, and already at Rothamsted considerable steps had been taken to develop that side of the work.

Further, these new methods showed how the field data could be improved. The first method of agricultural field experiment was to select what appeared to be a suitable site and on this to lay out an experiment. When Lawes and Gilbert began their experiments at Rothamsted they introduced the great improvement of setting up permanent plots, growing the same crops year after year, with the same fertilizer treatment, and as nearly as possible the same cultivation treatment. That constituted an enormous advance on anything that had gone before. The plots were large, neatly and systematically arranged, and they furnished a mass of valuable information which laid the foundations of agricultural chemistry and of scientific crop production and, further, served as useful demonstrations. One just walked along the end of the plots and could see the results; the whole story was perfectly easy to follow, and from the point of demonstration it was difficult to beat those old plots laid out by Lawes and Gilbert.

Later on modifications became necessary, because when the

F 2

experiments were repeated elsewhere, it was not always possible to obtain the clear-cut results shown in the Rothamsted fields. Not infrequently the results did not accord with expectations, and agricultural experiments were usually started in the expectation that "You never can tell". Much turned on soil and season. A rather pernicious habit developed of explaining away the unexpected results and holding fast to those that accorded with anticipation—which of course is not the way of progress in any science. To avoid these troubles the plots were duplicated or sometimes triplicated, but always were kept in a neat sequence, always systematically arranged. Further, it was always assumed that an experiment should be regarded as a question put to nature, and that it was far simpler for everybody concerned if the questions were put one at a time, and each experiment dealt with only one problem.

The general method of the old field experiment thus involved large plots duplicated but systematically arranged in some simple design, one experiment dealing with only one question.

The new methods soon changed all that. In the first instance, as Mr. Cochran had emphasized in his paper, the orderly arrangement had to be replaced by randomization. Secondly, instead of having separate simple experiments to deal with separate problems, they were combined into a much more complex experiment in which half-a-dozen questions might be put simultaneously. That, of course, meant a large number of plots, and before long special practical difficulties arose which necessitated close collaboration between the statisticians and the agricultural workers. Fortunately for the development of the subject, Dr. Fisher and his colleagues were all deeply interested in the agricultural implications of their work, and were not content simply to take the figures as raw material to be handled by statistical methods. In that way they secured from the outset the interest of the agricultural workers, and a succession of field experimenters, like T. Eden, H. J. Hines, H. V. Garner, D. J. Watson and others, had played an important part in working out the practical details of ways by which it was possible to satisfy the rather difficult requirements imposed by the statistical workers. In the end, through the happy collaboration of the statistical and the field workers, a series of practicable methods had been elaborated capable of producing results much in advance of anything that was done twenty years ago. He need only mention the extensive series of field experiments carried out by Imperial Chemical Industries and the really remarkable experiments of Mr. Frank Crowther in Egypt, each involving 270 plots, which had yielded results unobtainable in any other way.

At the outset the older experimenters did not like the new methods, because they involved so many little plots, arranged in so disorderly a fashion. But the methods had now secured widespread adoption, and were used all over the Empire. During the last few years he had had occasion to travel over a great part of the Empire, and it had been extremely interesting to go into some remote part of India and find there an experiment on sugar cane being dealt with on the new lines laid down by Dr. Fisher. The same was true in Ceylon with experiments on tea and in Malaya with experiments on rubber, and the Sudan on cotton : all were laid down on careful statistical lines, and carried out in spite of the labour involved because they furnished much more useful results than the old ones.

Altogether it was a very fine example of the way in which a piece of scientific work well done would spread out in unexpected directions : one might know where it started, but one never knew where it was going to end. The subject was still young, and had great possibilities.

In conclusion he expressed gratitude to the Royal Statistical Society for organizing its Industrial and Agricultural Research Section, where agricultural problems could be taken up. He felt sure that they were only at the beginning of the development of this subject.

PROFESSOR R. A. FISHER said that Mr. Cochran's paper was clearly the outcome of practical and competent experience in all the subjects with which it dealt. One could scarcely read any paragraph of this quite long paper without realizing that the author had been considering carefully the aims in view in carrying out the experiment as well as, what naturally concerned him most, the practical handling of the statistical data which at the end of the long and laborious process came to be analysed. In contrasting the methods here described with those of the earlier experimenters, one of the things one always noticed, if one tried to get back to the thought of the pioneers in any subject, was how far forward their thought reached. For example, there was no doubt that Lawes and Gilbert were deliberately aiming at replication in different years. They were trying to get the advantages of replication out of this long series of experiments. The method was inadequate in respect of permanent differences in fertility between plots, and that they could scarcely have foreseen—at any rate they did not foresee it. But in respect of the interaction between season and treatment their replication in ten or fifteen years was probably quite adequate. They used replication in the second way, too, in an estimate of error, or at least in judging the confidence with which a result could be accepted. The simple fact that in an experience of twenty years one treatment had beaten another nineteen times appeared to be convincing until the consideration of a possible permanent difference in fertility showed the argument to be invalid.

With regard to rotations these workers were certainly less fortunate, and he thought the reason was that the rule on which Mr. Cochran had insisted throughout the present paper was not followed, namely, that of maintaining every phase of the rotation in every year. With regard to replication in different years, it was worth pointing out that the method which Mr. Cochran had applied to the Tunstall experiment—that is to say, of comparing linear regressions over a series of years on the different plots—gave a comparison between the rates of improvement or deterioration induced by the variation in treatments which in parallel plots was not in the speaker's experience appreciable. Although in theory he agreed that it was a safeguard, if not a necessity, to have sufficient replication to make sure that the different plots were not spontaneously deteriorating or improving in fertility at unequal rates, yet in fact the earlier experiments, when analysed in that way, did not give misleading results, even where there was no replication as a safeguard.

He desired to qualify one remark in the early part of the paper. Mr. Cochran wrote, " With varietal trials involving a large number of varieties, the *lattice* designs recently introduced by Yates may prove useful, provided that the necessary number of replications can be faced." In his opinion what was really remarkable about the method Dr. Yates had put forward for handling either a number of varieties or a number of combinations in a factorial experiment was that he had succeeded in diminishing the number of replications which otherwise would be absolutely necessary. Perhaps the phrase in the paper might be read by someone unfamiliar with what Dr. Yates had done, and give the impression that Yates's procedure required more replications than some available alternative. The reverse, of course, was the truth.

Mr. Cochran a little later in his paper had raised a point on which he would like to hazard a conjecture : where he compared three methods of using uniformity trial results. The first of these was to group adjacent plots into blocks as usual and to use the uniformity trial as concomitant observations for the correction of the final results by a method of covariance. The second was to group plots of equal fertility into blocks on the basis of the uniformity trial results. The third method described by Mr. Cochran was one which had often been mentioned, but was seldom used— the method, namely, of dividing the plots into six groups whose mean yields were as close as possible to each other. Mr. Cochran raised the interesting question whether the use of covariance could eliminate the bias introduced by that third method. He himself would hazard the suggestion that there was no reason why it should not, and he would certainly expect it to be entirely effective in doing so. But it would leave the third method in exactly the same position as the second method, that is to say, with lower precision, as Forster had shown, than the method which used grouping adjacent blocks as a means of eliminating error.

Dr. J. Wishart said that he remembered Professor Karl Pearson once telling him how he first began to feel old when he heard of former pupils retiring. In one sense Mr. Cochran was retiring, for he was giving up his post at Rothamsted to take up work in another country. His sense of regret at this event was mingled with satisfaction at the real worth of his work at Rothamsted as exemplified by the present paper. The subject was of very great importance to agriculturists, and the particular paper now presented carried them a long way, and was a worthy successor to previous papers of the kind given to the Section. The Chairman had recalled the circumstances of twenty years ago. He thought he was right in saying that the remarkable foresight which Sir John Russell showed at that time had been amply vindicated by the great developments that had taken place since. First there came the stage where the results of the classical experiments were worked out, and related to meteorological and other factors; secondly, the ·stage where new methods for single experiments were designed by Professor Fisher and carried out at Rothamsted, and thirdly, the stage where experiments of a more complex character were devised, though still with the idea of the single experiment as a basis. It was a natural extension of work of this kind to apply it to crops grown at different places, and at different times, and this led to series of multiple experiments being carried out. An interesting stage came when the experimental design was applied, not to the individual crop of a rotation cycle, but to the whole cycle, and a beginning made of a new series of experiments of a long-term character, involving a combination of crop and manurial treatment, designed to cover a number of rotations, both with fixed and with moving treatments. The first experiments of this kind were begun some nine years ago, before he (Dr. Wishart) left Rothamsted, and it was interesting to hear how they had worked out. The high water mark in this sort of work was to be found in the exceedingly interesting experiments described by the present author at the end of his paper.

In conclusion he wished to touch once more on the personal note, and to say that not only was the paper a very good one, but also the author was to be commended on the admirable way in which he had presented to the meeting his summary of a very long paper. No agricultural experimental station or college need have the slightest qualms at employing a mathematician who could begin this subject as the author had begun it, treat it throughout as he had done, and end on the exceedingly practical note on which he had ended. He had maintained the biological point of view throughout, and that was of very great importance.

Dr. F. YATES said that he did not propose to add much to what Mr. Cochran had already said and written in his paper. He had had opportunities of thorough discussion with Mr. Cochran, and found himself in substantial agreement with him on all points. He would, however, strongly recommend those present at the meeting to give the paper considerable further study. The subject was a very intricate one, and had great theoretical as well as practical interest to all those concerned with the design of experiments and with combinatorial problems.

To those concerned with the design of rotation experiments, he would only offer one word of advice, that was to examine all the various alternatives very thoroughly, and not to undertake an experiment until the whole matter had been very fully discussed with all those interested. A rotation experiment, once it was laid down, involved an experimental station in continuous work for many years, and in general no alterations in the original design could be made. Consequently a mistake at the start was likely to be very costly. As an example of the benefit of such discussion he might mention that the ley experiment at Woburn, which Mr.

Cochran described, was the result of two years of meetings and discussions between four or five people. The design of that experiment had been very considerably modified in the course of these discussions, and had certainly been greatly improved.

The problems that arose in the design of rotation experiments were also of interest in other branches of agriculture and in other biological sciences. In particular, in nutrition experiments on animals the different diets could be rotated in just the same way as treatments could be rotated on agricultural field plots. Those concerned with such experiments would find it well worth while to study what had been done so far in agricultural rotation experiments.

Finally, he would emphasize that the design of rotation experiments was a branch of experimental design that was only just being developed. The problems and difficulties involved had only recently been fully realized, and he had every confidence that very considerable further progress in the technique of design would be made in the coming years. He was sure that Mr. Cochran would agree with him that his paper was not intended as a final or ultimate exposition of the methods to be followed, but rather as a description of the progress that had been made. The paper was full of stimulating suggestions which were ripe for further development.

Dr. M. S. Bartlett said that Rothamsted had been the pioneer in long-term experiments, and so far as these particular types of experiment were concerned, other stations were still lagging behind, so that those people who had not had the fortunate experience of being at Rothamsted for any length of time would find it difficult to criticize or comment on this type of experiment. He wanted first of all simply to emphasize certain remarks of the author at the beginning of his paper with regard to what was perhaps the most simple type of long-term experiment—that is, where the treatment was going on year after year. It was a matter of psychological experience that some agriculturists, if they were not too expert in statistics, did feel when they took observations year after year, even if there was very inadequate replication, that from all the figures so obtained the statistician must be able to get something of value. Such people were a little surprised when told that the replication was rather poor. It was true that some of these experiments—he was thinking of experiments he had come across on rubber in Malaya—with reasonable replication might yield useful results, taking the first year or so as a pseudo-uniformity trial, and the covariance method then used to get some sort of analysis. In a way he supposed that was crudely equivalent to the method mentioned by the author of studying the linear regression of treatments with time.

With regard to the point mentioned by Professor Fisher concerning the third of the three principal alternatives in a design in randomized blocks, that particular method of using covariance might not have been much used in agriculture, but it had been used in work on animals to equalize up the initial mean yields of cows in nutrition experiments. The experimenters had merely compared the cows on the basis of group yield, and if such results, obtained by non-statisticians, were submitted to a statistician, he could sometimes do something with them by using covariance to get an idea of the error involved.

MISS H. N. TURNER said that they were often faced in agricultural processes with the problem of analysing wool yields month by month from the same sheep and milk yields month by month from the same cow, and the methods which Dr. Yates had just mentioned would be very useful in that connexion.

MR. KENDALL said that there was one question he would like to put to the author. It seemed to him that the long-term experiments might introduce a new principle which apparently had not been considered before, in that one plot might—using the word in the colloquial sense—interact on another. For example, it was known that if beet were grown on the same land for a considerable number of years, there was a tendency to encourage eel-worm which would tend to spread to other plots. Thus one not only exhausted the fertility on the particular plot dealt with, but also affected adjacent plots. The same kind of thing would happen in producing fruit on trees which were subject to pests which might spread to other trees. The effect was, perhaps, unimportant in short-term experiments, but would possibly become serious in long-term experiments on the same piece of land. Had factors like that been considered? Were they supposed to be important over a long period of years? And if so, what did the author propose to do about them?

DR. HARTLEY said that it would be useful to apply the statistical technique developed by Mr. Cochran to animal-experimentation, because such experiments were often carried out in a shorter time than those on crops. There were certain " long-term " experiments on animals which could be carried out within a year, and would thus readily lend themselves to testing the technique described in practical work within a comparatively short period. As an example, in poultry-breeding experiments, where hatches were produced at intervals of three weeks or less, so that 10 hatches could easily be hatched within a year, whereas a corresponding long-term experiment on crops would last 10 years.

There were, of course, various ways in which experiments on crops differed from those on animals and the experience gained with the one did not necessarily apply to the other. For example, from his experience with poultry experiments he would say that the frequent occurrence of " missing plots " would certainly be a difficulty when applying long-term designs to such experiments. Although it was known that with crops the occurrence of missing plots was less frequent, he would like to know about Mr. Cochran's experience on this point; for it was obvious that with long-term experiments a missing plot was more likely to occur than with annual experiments, as one had to rely on the experimental site for a longer period.

MR. COCHRAN, in reply, said that he fully appreciated Professor Fisher's point about the ambiguity of the phrase he had used in describing Dr. Yates's method. It was an interesting example of how the same words could convey quite different impressions to different minds. What he had in mind was that, for example, a $7 \times 7$ lattice square required four replications in 196 plots. It might be impossible for economic reasons—or the experimental station might maintain that it was impossible—to carry out more than two replications of a $7 \times 7$ square, and such a design would then be out of the question. But he would be sorry to give the impression that he regarded these designs as lavish in replication compared with other designs.

He had been interested to hear Professor Fisher's remark on the way in which covariance might get rid of the possible bias in error in the third method using the uniformity trial results. That was their own impression, he thought, at Rothamsted, but he had not been sure that there was not a snag somewhere.

With regard to Dr. Bartlett's remarks on the use of the first year of the experiment as a kind of uniformity trial, he did not think that one should recommend that method without discretion to inexpert statisticians. If the treatments had really been producing some effect in the first year it would invalidate the corrections.

On the point raised by Mr. Kendall, no examples of damage spreading from one plot to another had been noticed in these long-term experiments. If such were noticed, one would try to get an estimate of the damage and correct for it by covariance. Where an experiment was particularly liable to damage spreading in this way from plot to plot, the remedy was to provide adequate guards between plots in the design, though with a flying insect this was often difficult. In reply to Dr. Hartley, the long-term experiments at Rothamsted and Woburn had been remarkably free from mishaps to single plots, though a number of years had to be omitted for certain crops, owing to failure to obtain a stand or to a decision to fallow the whole experiment. He did not think that the general method of analysis would be changed by the presence of a number of individual missing values, though the least squares solutions then became somewhat more involved. In conclusion, he wished to associate himself strongly with what Dr. Yates had said. It would be very unfortunate if the impression were gained that because this paper was bulky it dealt finally and exhaustively with the question of long-term experiments. There were many more points involved in the design and analysis than he had had time to discuss. The subject was still a young one and the conclusions were tentative.

# 19

## THE USE OF THE ANALYSIS OF VARIANCE IN ENUMERATION BY SAMPLING*

By W. G. Cochran

*Rothamsted Experimental Station, England*

### 1 INTRODUCTION

In sampling, two considerations must be kept in mind—representativeness and accuracy. In the popular sense, a sample is representative if any measurements made on it are equivalent to the same measurements made on the whole population, apart from the inaccuracy produced by the restricted size of the sample. Much has been learned in recent years about the methods of selection necessary to secure such samples. In particular, selection on the judgment of the sampler, where the population can be inspected, has been found to give samples which are usually biased (Yates 1935, Cochran and Watson 1936). Kiser (1934) has shown that biases arose in a house to house study through failure to revisit houses which were empty at the first call. Further, Neyman (1934) examined the method of making the sample means coincide with the population means in a number of features (called *controls*) and was of the opinion that this would rarely give a representative sample.

A representative sample (in the above sense) can clearly be obtained by giving every unit in the population an equal chance of being included in the sample. Many variations are, however, possible in the method of selecting the samples, and a brief account of some of the most useful devices for increasing the accuracy of sampling will be given later.

From considerations of cost, it is rarely possible in practice to test the accuracy of two different methods of sampling by trying both on the same population. If the sampling is suitably planned, however, it will usually provide information on the relative precision of various alternative methods of sampling which might have been adopted. This information can be extracted without difficulty from the analysis of variance of the sampling results and is of considerable service in improving the accuracy of future samplings on the same type of material. The use of the analysis of variance in this respect for sampling in field experiments has been studied in a number of papers, of which the most comprehensive is by Yates and Zacopanay (1935). Their results do not, however, appear to be sufficiently familiar to workers engaged in

* Revision of a paper presented at the One-hundredth Annual Meeting of the American Statistical Association, Detroit, Michigan, December 29, 1938.

sampling censuses; and the object of this paper is to give an example of the use of the technique.

## 2  SOME ASPECTS OF THE TECHNIQUE OF SAMPLING

The accuracy of a sample may often be greatly increased by careful planning. The most common devices for this purpose may be classed under four heads: (1) subdivision, (2) sub-sampling, (3) choice of sampling unit, (4) double sampling. They are not, of course, in any sense mutually exclusive—a sampling study may involve some or all of them.

*Subdivision:* The population is sub-divided into a number of groups, chosen if possible so that the quantity to be studied varies little within each group. The sampling is subjected to the restriction that a certain number of samples are to be drawn at random from each group. The mean of the sample in any group is an estimate of the population mean of the group and its standard error clearly depends only on the amount of variation within the group. Thus if the relative sizes of the different groups are known, these estimates can be combined into an average for the whole population whose standard error depends only on the amount of variation within groups. If the group means differ widely and the number of samples in each group is small, there will be a considerable increase in precision. A further advantage of this method is that often it is convenient for administrative purposes, quite apart from its effect on the accuracy of sampling.

When a sampling investigation involving sub-division has been completed, the analysis of variance of the data provides estimates of the mean square *between groups* and *within groups*. It is, of course, necessary to have at least *two* samples at random within each group. From these it is possible to estimate the sampling error to be expected from a sample of the same size if the population had not been sub-divided. Thus the increase in precision which resulted from the sub-division may be estimated.

The term *stratification*, borrowed from geology, has sometimes been applied to describe this process. While this is possibly a fairly apt description in economic studies, where the population may be arranged for example in increasing income levels, it is less appropriate where the grouping is on a geographical basis, whereas the term *subdivision* appears to cover all cases.

*Subsampling:* The sampling unit need not be measured completely; it may itself be enumerated by sub-sampling. A common example of the method occurs in the estimation of the yields of field crops. Here the sampling-unit is usually a single field, the yield of the field being estimated by taking several small samples from the field, instead of by harvesting the entire crop in the field. From the point of view of the analysis of variance, this method might be regarded as a case of incomplete sub-division; the material is grouped by fields, but not all fields are sampled, so that the sampling error consists partly of variation between fields and partly of variation between sub-samples within fields.

An important practical consideration in sub-sampling is the division of resources between the amount of sub-sampling per sampling-unit and the number of sampling-units chosen. For a given expenditure, one can only be increased at the expense of the other. The best compromise will depend on the relative costs of increasing the number of subsamples per sampling-unit and of increasing the number of samplingunits, and on the relative increases in accuracy obtained by such changes. When a particular scheme has been used in a sampling investigation, the analysis of variance provides estimates of the sampling variance to be expected, in similar material, from any change in the amount of sub-sampling per sampling-unit or in the number of sampling-units. In planning future sampling work, these changes may be balanced against the estimated costs of making them.

*Choice of Sampling Unit:* The size and structure of the sampling unit plays an important part in determining the accuracy of sampling. In an areal survey, for instance, the choice may lie between a section, a square block of four sections, or a township. Here again the problem is to strike the most effective balance between the amount of work and the statistical efficiency, since, for a given percentage sampled, a few large sampling-units are usually less expensive to collect than a large number of more widely scattered small sampling-units.

A thorough study of the relative effectiveness of different sizes of unit cannot as a rule be made without a special investigation for that purpose. Some useful information can, however, often be gained from the analysis of variance of a particular sampling study. If, for instance, a four-section block has been chosen as the unit, the data may perhaps conveniently be recorded by sections. By using the variance between sections in the same four-section block, an estimate may be obtained of the sampling error to be expected if a single section had been taken as the sampling unit.

*Double Sampling:* The possibility of increasing accuracy by using a second character which is correlated with the character to be enumerated and which can be easily measured has attracted some attention in recent years. The method has been named *double sampling* because it involves two sampling investigations. The first is a large sample, in which the second character alone is enumerated, and the second a small sample, in which both characters are usually enumerated. There are several possible ways of using the results. In one method, which has been studied by Neyman (1938), the first sample is used to sub-divide the population into groups within which the second character varies little. If the correlation between the two factors is high, this will also be an effective sub-division for the first factor, and the second sample is planned accordingly. In order to obtain the increased accuracy of sub-divided sampling, the relative total numbers of sampling units in the different groups must be known. This will not be the case, but if the first sampling is sufficiently large it provides fairly accurate estimates of these numbers.

Another possibility is to use the small sample to determine the regression of the first character on the second. The mean value of the first character for the whole population is then estimated by the predicted value in the regression equation which corresponds to the mean value of the second character in the large sample. This method was used by Watson (1937) to estimate the mean leaf area of a large batch of leaves. He weighed the whole batch of leaves and selected a small sample from which to estimate the regression of leaf area on leaf weight.

An interesting application of the regression method to the estimation of the total volume of certain species of timber in the counties of Nottingham and Lincoln in Great Britain has recently been made by Yates.*

In this study a complete survey of the area was made. Each block of woodland was visited and divided by eye inspection into areas uniform for descriptive purposes. These areas, called "stands," were numbered and demarcated on the map, and eye estimates were made of the volume of each species of timber in each stand. Independent estimates of volume were obtained by actual measurement of the trees in a number of small sample plots located on a grid pattern across the area. On comparing the estimates of volume obtained by measurement of the sample-plots with the eye estimates for the stands in which the sample-plots lay, the eye-estimates were found to have a considerable negative bias.

We need only consider the estimation of the mean volume of any class of timber *per unit area*, since the total acreage of the class is

* I am indebted to the Forestry Commission and to Mr. Yates for permission to use this example, as yet unpublished.

known from the complete survey. There are two obvious methods of procedure (1) to ignore the eye estimates. The estimate chosen is the mean volume $\bar{y}_s$ per unit area of the class of timber in the sample plots, (2) to use the eye estimates, corrected for bias. The mean difference, $(\bar{y}_s - \bar{x}_s)$, between the sample plots and the corresponding eye estimates is a measurement of the bias, and is added to the eye estimate $\bar{x}_a$ for the whole area, the final estimate being $\bar{x}_a + (\bar{y}_s - \bar{x}_s) = \bar{y}_s + (\bar{x}_a - \bar{x}_s)$.

There is, however, a more accurate method. The regression of the sample-plot estimates $y$ on the corresponding eye estimates $x$ is calculated, and this equation is used to predict the average value of $y$ for the whole area from the known average $\bar{x}_a$ of $x$. If $b$ is the regression coefficient, the estimate is

$$\bar{y}_s + b(\bar{x}_a - \bar{x}_s).$$

The first method above is obtained by assuming $b=0$ in this equation, and is the most accurate only if the eye estimates provide no information about the volumes. The second method assumes $b=1$, and is appropriate only if the amount of bias in the eye estimates is independent of the sample-plot values. The third method makes the most accurate use possible of the information contained in the eye estimates in addition to that already provided by the sample-plot values. This example differs from Watson's in that the second variable is not a separate character correlated with the first, but an alternative estimate of the first character.

The mean squares between stands for certain classes of timber are shown below for the three methods of estimation.

| Estimate | Mean square | Relative information |
|---|---|---|
| Sample-plot value | 7829 | 74 |
| Eye estimate, corrected for bias | 7132 | 81 |
| Sample-plot value, adjusted for regression on eye estimate ($b=.55$) | 5782 | 100 |

The regression method gave about $\frac{1}{3}$ more information than the sample plot estimate and about $\frac{1}{4}$ more information than the eye estimate corrected for bias.

Before proceeding to discuss an illustrative example, the way in which the analysis of variance facilitates the separation of different sources of variation will be indicated. Let the population consist of $MN$ sampling units, which are divided into $M$ groups of $N$ elements each. In addition to the variation between sampling units in the same group, there may be a real variation in the means from group to group. To describe this situation mathematically, any sampling unit, for instance the $i^{th}$ unit in the $j^{th}$ group, is considered to be the sum of two independent parts, $x_j + x_{ji}$, where $x_j$ is common to all members of the $j^{th}$ group, but varies from group to group with variance $\sigma_g^2$, while $x_{ji}$ varies within the $j^{th}$ group with variance $\sigma^2$. Thus the total of the $j^{th}$ group is

$$Nx_j + \sum_{i=1}^{N} x_{ji}$$

and has a variance $(N^2\sigma_g^2 + N\sigma^2)$ or, if reduced to a single unit basis by dividing by $N$, $(N\sigma_g^2 + \sigma^2)$. The mean square between groups will be an estimate of this quantity, based on $(M-1)$ degrees of freedom, while the mean square within groups is an estimate of $\sigma^2$, based on $M(N-1)$ degrees of freedom. Hence an estimate of $\sigma_g^2$ can be obtained for use in any further calculations. All subsequent applications of the analysis of variance in this paper are based on this principle.

### 3 NUMERICAL EXAMPLE

During the past five years, a number of commercial fields of wheat have been sampled for yield of grain in each of several districts in Great Britain. The choice of farms was not randomly made. Samplers were instructed to sample wherever possible two fields drawn at random from all fields growing wheat on the farm chosen, though in many cases only a single field per farm was sampled. The method of sampling a field has been described in detail elsewhere (Irwin et al., 1938, p. 17). The sampler on entering the field selects a random starting-point on one of the two sides which are perpendicular or oblique to the rows. From the starting-point he estimates the width of the field by eye. Suppose this is 400 paces. Dividing by three gives roughly 130 paces. Two random numbers are selected between 1 and 130, say 23 and 83. Crossing the field in the direction of the rows to minimize trampling, the sampler cuts his first sample at 23 paces and his second at 83 paces. Two further samples are taken in the middle third of the way across, at $(130+23) = 153$ paces and 213 paces respectively, while two further samples are taken at $(260+23) = 283$ paces and 343 paces respectively in the last third. Each sample consists of $\frac{1}{4}$ metre of each of 6 neighboring rows, and the sampler steps three paces to the left after taking each sample, so that his path across the field is zig-zagged. The six samples are bulked into two sets before threshing, the samples taken at 23, 153, and 283 paces going into the same set. A new random starting-point is selected on the other side of the field, and a further six samples are collected on the way back.

In this method the sampling unit is the set of six samples which are taken in one path across the field. The difference between the two sets of six samples per field gives a single degree of freedom per field to determine the sampling error. The difference between the two sets of three samples in the same line across the field does not provide a true estimate of sampling error, since the sets lie on the same line, but is worth calculating for the information it gives on the efficiency of the sampling unit.

The 1937 results are shown in Table I. The individual figures in the table are the mean grain yields per sample for each set in grams. Since the width between rows varied, the figures have been adjusted to 6-inch rows and represent the yield of an area of $1\frac{1}{2}$ metres$\times$6 inches. Within each group of four sets the sets are arranged in the order in which they were cut, so that in descending order the first two sets constitute the first sampling-unit. There are 39 groups of four sets each, which come from 36 fields, since in district III two varieties were growing on each of the fields sampled and these varieties were sampled separately. The data provide, without undue labour, a convenient example of the working out and interpretation of the analysis of variance.

The total sum of squares of deviations from the mean (155 degrees of freedom) is found in the usual manner. This figure refers to the variation of a single set in grams per $1\frac{1}{2}$ metres$\times$6 inches. Since it is convenient for later discussion to have estimates of variance per field of four sets, the figure should be multiplied by four. The variation between sets in the same line is found by tabulating the 78 differences between the two sets ignoring sign (the first of these in Table I being $63-47=16$). The sum of squares of these quantities is 26,184 and is multiplied by 2 to give the sum of squares for a four set total. The sum of squares for the sampling error is obtained by first writing down the 39 differences between the totals of the two lines (the first of these in Table I being $67+55-47-63=12$). The sum of squares of these quantities is 33,924 and is already on a four set basis.

Two fields were sampled from the same farm in nine cases, and three fields in one case (District III). For the first nine cases, the sum of squares of the differences between the totals of the two fields is 55,404, while the sum of the squares of the deviations of the three fields in district III from their mean is 14,802. To obtain the corresponding sums of squares for the total of four sets, each figure must be divided by 2. Their sum gives 35,103 with 11 degrees of freedom.

Table I. Commercial wheat samples 1937
Mean yields of grain per sample in gms. (on a 6″ basis)

| | District I Farm | | | | District II Farm | | | District III Farm | | | | | |
|---|---|---|---|---|---|---|---|---|---|---|---|---|---|
| | 1 | | 2 | | 1 | | | 1 | | | | | |
| 1st line | 47 48 | | 75 105 | | 93 58 76 | | | 92 89 | 89 75 | | 70 80 | | |
| | 63 51 | | 71 82 | | 84 78 57 | | | 83 111 | 58 72 | | 85 97 | | |
| 2nd line | 67 45 | | 75 97 | | 75 68 79 | | | 93 90 | 70 82 | | 76 111 | | |
| | 55 46 | | 85 86 | | 80 83 78 | | | 96 115 | 70 81 | | 102 66 | | |
| | 232 190 | | 306 370 | | 332 287 290 | | | 364 405 | 287 310 | | 333 354 | | |
| | 422 | | 676 | | 619 | | | 769 | 597 | | 687 | | |
| | (4) 1098 | | | | (3) 909 | | | (6) 2053) | | | | | |

| | District IV Farm | | | | | | | | | | | | | | District V Farm |
|---|---|---|---|---|---|---|---|---|---|---|---|---|---|---|---|
| | 1 | | 2 | | 3 | 4 | 5 | 6 | 7 | 8 | 9 | | | | 1 |
| 1st line | 29 45 | | 57 69 | | 78 | 59 | 68 | 97 | 60 | 65 | 81 | | | | 77 60 |
| | 21 39 | | 63 55 | | 109 | 59 | 56 | 88 | 53 | 59 | 94 | | | | 74 88 |
| 2nd line | 29 69 | | 46 21 | | 90 | 58 | 74 | 109 | 43 | 49 | 93 | | | | 44 84 |
| | 31 57 | | 66 40 | | 51 | 53 | 61 | 95 | 48 | 71 | 92 | | | | 57 97 |
| | 110 210 | | 232 185 | | 328 | 229 | 259 | 389 | 204 | 244 | 360 | | | | 252 329 |
| | 320 | | 417 | | | | | | | | | | | | 581 |
| | (11) 2750 | | | | | | | | | | | | | | (2) 581 |

| | District VI Farm | | | | | | | | | | | | | | | |
|---|---|---|---|---|---|---|---|---|---|---|---|---|---|---|---|---|
| | 1 | | 2 | | 3 | | 4 | 5 | 6 | 7 | 8 | 9 | 10 | | | |
| 1st line | 66 93 | | 55 127 | | 84 80 | | 81 | 93 | 21 | 84 | 87 | 79 | 90 | | | |
| | 73 70 | | 56 106 | | 80 86 | | 107 | 106 | 63 | 51 | 67 | 79 | 117 | | | |
| 2nd line | 64 80 | | 83 84 | | 63 88 | | 135 | 71 | 50 | 82 | 135 | 71 | 112 | | | |
| | 73 67 | | 60 98 | | 89 110 | | 82 | 83 | 29 | 80 | 114 | 89 | 122 | | | |
| | 276 310 | | 254 415 | | 316 364 | | 405 | 353 | 163 | 297 | 403 | 318 | 441 | | | |
| | 586 | | 669 | | 680 | | | | | | | | | | | |
| | (13) 4315 | | | | | | | | | | | | | | | |

The 19 degrees of freedom for the variation between farms in the same district are best calculated separately for each district. The differences in the numbers of sets per farm must be taken into account. For district IV, for instance, the sum of squares is

$$\tfrac{1}{2}(320^2 + 417^2) + 328^2 + 229^2 + 259^2 + 389^2 + 204^2 + 244^2 + 360^2$$
$$- (2750)^2/11 = 59,824.$$

Finally, the district totals and means on a four set basis should be tabulated separately, the latter being carried to 4 decimal places. The sum of products of the totals and means, minus the product of the grand total and its mean gives the sum of squares between districts.

Table II. Analysis of variance of the total of four sets
Unit: gms. per 6 metres $\times$ 6 inches ( = .000226 acre)

|  | Degrees of freedom | Sums of squares gms. | Mean squares | |
|---|---|---|---|---|
|  |  |  | gms. | cwt. per acre |
| Between districts | 5 | 54,224 | 10,845 | 82.27 |
| Between farms within districts | 19 | 132,062 | 6,951 | 52.73 |
| Between fields within farms | 11 | 35,103 | 3,191 | 24.21 |
| Sampling error | 39 | 33,924 | 869.8 | 6.598 |
| Between sets within lines | 78 | 52,368 | 671.4 | 5.093 |
| Between varieties (District III) | 3 | 1,326 | — | — |
| Total | 155 | 309,007 | — | — |

The complete analysis of variance is shown in Table II. To make the total sum of squares provide a check on the other calculations, the three degrees of freedom representing differences between varieties in the same field in district III are included, though they are not required in the discussion. The mean squares have been converted to cwt. per acre by multiplying by .0075861.

Two of the purposes of this sampling investigation were to develop a practical technique for the sampling of wheat fields and to study the feasibility of carrying out a crop estimation scheme by sampling methods. It will be realised that a small sample such as the above, taken in a single year, cannot give results of any great precision. The sampling scheme has, however, been carried out for five years. The corresponding analyses of variance for each of the five years are shown in Table III below. The results from different years are not entirely independent, since in a number of cases the same farms were included in the samples, but the samples from different years are much more valuable than a single sample five times as large taken in a single year. Reference to the results in other years will be made when discussing the 1937 analysis.

Table III. Analysis of variance per field of yields of wheat grain
(cwt. per acre)

|  | 1934 | | 1935 | | 1936 | | 1937 | | 1938 | |
|---|---|---|---|---|---|---|---|---|---|---|
|  | d.f. | m.s. | d.f. | m.s. | d.f. | m.s. | d.f. | m.s. | d.f. | m.s. |
| Between districts | 4 | 66.5 | 6 | 318.4 | 4 | 79.4 | 5 | 82.3 | 4 | 206.8 |
| Within districts between farms | 11 | 38.9 | 12 | 27.1 | 7 | 62.2 | 19 | 52.7 | 14 | 65.3 |
| Within farms between fields | — | — | 15 | 22.8 | 8 | 31.2 | 11 | 24.2 | 8 | 12.1 |
| Within fields: | | | | | | | | | | |
|   Sampling error | 16 | 5.33 | 40 | 6.20 | 22 | 11.39 | 39 | 6.60 | 28 | 9.80 |
|   Between sets | 32 | 2.11 | 80 | 2.18 | 45 | 2.52 | 78 | 5.09 | 55 | 4.78 |
| Mean yield | 29.1 | | 23.3 | | 24.3 | | 26.2 | | 30.7 | |

In discussing the efficiency of the sampling unit and the adequacy of the amount of sampling per field, these must be related not only to the sampling error, but also to other sources of variation which affect the estimate of the mean yield of the country. The sampling error can always be halved by doubling the amount of sampling per field, but if this produces only a 10 per cent reduction in the variance of the mean yield of the country, it is doubtful whether the result justifies the extra labour. The appropriate variance for the mean yield of the country will be discussed later, but it is clear that it must at least involve the variation between fields on the same farm.

*Variation between sets*

As pointed out in §2, the alternative types of sampling unit with which the present one can be compared are somewhat restricted. We may, however, consider variations in the number of sets per line. In 1937 the variation between sets was almost as large as the variation between lines (i.e. there was little true variation between lines), but this was not the case in most other years. If the number of sets per line were doubled, the contribution to the sampling error of a field from variations between sets would be halved. This would have reduced the sampling variance by 20%, 18%, 11%, 38% and 24% respectively in the five years. Even if the sampling error were the only error to be considered, the increase in accuracy would scarcely be worth the labour of handling twice the number of samples. Further, the percentage decreases in the estimated variances *between fields* are only about a third of the above figures. It is clear that no increase in the number of sets per line can be recommended. On the other hand, a reduction of the number of sets to one per line might be advocated in a crop estimation scheme, although in the present study it is useful to have two sets per line to estimate the variance between sets. It may be verified that the percentage increases in the variances between fields with one set per line would be 10%, 8%, 21% and 40% respectively in the years 1935–8, these losses in precision being compensated for by a halving of the number of samples to be handled and a slight decrease in the labour of locating the samples.

### Variation between lines

Except in 1938, the sampling error per field lies between about $\frac{1}{4}$ and $\frac{1}{3}$ of the variation between fields on the same farm. It is clear that increased accuracy is to be sought in the sampling of more fields rather than in an increase in the number of lines sampled per field. In 1937, for example, twice the amount of sampling per field would reduce the variation between fields by 3.30, i.e. to 20.9, whereas doubling the number of fields sampled would give a corresponding variance of only 12.1.

### Variation between fields on the same farm

If a fixed panel of farms were selected for crop estimation and adhered to each year, the sampling error appropriate to the mean yield of these farms would be derived from the variance between fields on the same farm (plus its component of sampling error of a field). With four sets per field, the corresponding standard errors per field were respectively 21%, 23%, 19% and 11% of the mean yield in the last four years. A sample from about 500 fields would give on the average a standard error of about or under 1% for the mean. If only one set per line were taken, the number of fields would have to be increased by about 20 per cent as indicated before. This scheme would have considerable practical convenience and requires a surprisingly small amount of sampling to obtain an accurate mean yield. Against this is the possibility of biases in the selected panel as compared with the country as a whole, though if these remained constant from year to year the *changes* in mean yield would still be estimated without bias.

It should be noted that it would not be necessary to sample two fields from every farm, provided that sufficient farms, say 150, were sampled twice to give a reliable estimate of the variation between fields.

*Variation between farms in the same district*

If on the other hand the farms to be sampled were selected at random *each year*, the variance of the estimated mean yield of the country would contain a portion representing variation between farms, in addition to sampling and field variation. The estimation of the true variation between farms is slightly complicated by the fact that the number of fields per farm was not constant. In general, if there are $m$ groups and $n_i$ sampling units are taken from the $i^{th}$ group, the mean square variance between groups is an estimate of

$$\sigma_w^2 + \frac{1}{m-1} \left[ \sum n_i - \sum n_i^2 \Big/ \sum n_i \right] \sigma_g^2$$

where $\sigma_w^2$ is the variance within groups and $\sigma_g^2$ the true variance between groups. If $n_i = n$ in all groups, the coefficient of $\sigma_g^2$ reduces to $n$; otherwise the coefficient is somewhat smaller than the average number of sampling units per group.

Table IV. Estimation of the true variance between farms within districts
(for the total of four sets)

| District | Number of farms with | | Number of d.f. between farms | Mean square is an estimate of |
|---|---|---|---|---|
| | 1 field | 2 fields | | |
| I | 0 | 2 | 1 | $\sigma_f^2 + 2\sigma_F^2$ |
| II | 1 | 1 | 1 | $\sigma_f^2 + 1.33\sigma_F^2$ |
| IV | 7 | 2 | 8 | $\sigma_f^2 + 1.20\sigma_F^2$ |
| VI | 7 | 3 | 9 | $\sigma_f^2 + 1.28\sigma_F^2$ |
| Total | | | 19 | $\sigma_f^2 + 1.29\sigma_F^2$ |

$\sigma_F^2 =$ true variance within districts between farms.
$\sigma_f^2 =$ variance between fields within farms (including its component of sampling error).

The details of the calculation for the 1937 results are shown in Table IV. Each district must be considered separately. Districts III and V have only one farm each and do not enter into the calculation.

In district IV, for example, there are 7 farms with one field and 2 farms with two fields. The coefficient of $\sigma_F^2$ is $\frac{1}{8}\left[11 - (7 \times 1^2 + 2 \times 2^2)/11\right]$

$=1.20$. The average coefficient, 1.29, is found by weighting the districts according to the numbers of degrees of freedom. The mean number of fields per farm in the districts which contributed to the variation between farms was 1.35. This is only slightly above the correct coefficient, 1.29, and could be used in a rapid survey of the data. Thus an estimate from Table II of the true variance per field between farms in the same district is $[52.7-24.2]/1.29=22.1$. This figure is, of course, open to suspicion since the farms were not selected at random, but for the present example it will be assumed to be unbiased. The corresponding figures for 1936 and 1938 were of the same order of magnitude, but the 1935 figure was much smaller.

With one field per farm, the variance per field between farms in 1937 would be $24.2+22.1=46.3$, whereas with two fields per farm it would be 68.4. With a random selection of farms one would have to sample $\frac{2}{3}$ of the number of fields, though $\frac{1}{3}$ more farms, to obtain equal accuracy with one field per farm as with two. On these figures it is probable that only a single field would be sampled per farm. In this case the standard errors per cent per field from variation between farms are found to be 23%, 22%, 29%, 26% and 23%. It appears that about 750 fields would be required to reduce the standard error of the mean to 1 per cent. This figure may, of course, be an underestimate, since the farms were not selected at random and since districts which were chosen for administrative purposes might be considerably larger than the districts covered by the present scheme.

*Variation between districts*

For this reason it is rather artificial to estimate the gain due to subdivision into districts from the analysis of variance. The calculation will, however, be made from the 1937 figures to illustrate the method. The estimated true variation between districts will be found to be $[82.3-52.7]/5.98=4.9$. To find the variation between farms without subdivision into districts, the average variation between farms in the whole population must be calculated. This may be done by reconstructing the analysis of variance for the whole population. The calculation involves the total number of fields in each district; this may for simplicity be assumed to be $n$ where $n$ will be large. The analysis of variance of a field for the whole population is shown in Table V. The arrows indicate the order in which the various estimates are found. It is assumed that one field per farm is being sampled.

Here $\sigma^2$ is the variance between farms in the same district, including the components from variation between fields and from sampling errors. Since estimates of $\sigma^2$ and $\sigma_d^2$ are available, the estimated mean squares for the whole population may be obtained. From these the corresponding sums of squares are calculated and by addition the total sum of squares of deviations from the general mean is found. Division by $(6n-1)$ gives the required mean square. Since $n$ is large, this is approximately equal to $46.3+\frac{5}{6}\times4.9=50.4$. The increase in precision by sub-dividing into districts is only 9 per cent. The gain was of the

Table V. Analysis of variance for the whole population
(on a single field basis)

| | d.f. | Expected mean squares | Estimated mean squares | Estimated sums of squares |
|---|---|---|---|---|
| Between districts | 5 | $\sigma^2+n\sigma_d^2\rightarrow$ | $46.3+n\times4.9$ $\rightarrow$ | $5\times46.3+5n\times4.9$ |
| Between farms within districts | $6(n-1)$ | $\sigma^2$ $\rightarrow$ | $46.3$ $\rightarrow$ | $6(n-1)\times46.3$ $\downarrow$ |
| Total | $(6n-1)$ | | $46.3+\dfrac{5n}{6n-1}\times4.9$ | $(6n-1)\times46.3$ $\leftarrow$ $+5n\times4.9$ |

same order of magnitude in 1936 and 1934, but reached 44 per cent in 1938 and was still higher in 1935. In order to attain these gains in practice it is of course necessary to know accurately the total acreage of wheat in each district.

It is hoped that the above discussion will give some idea of the type of information which may be derived from a study of the results of a sampling scheme.

## 4 THE SAMPLING ERROR OF THE MEAN WHEN AN APPRECIABLE FRACTION OF THE POPULATION IS SAMPLED

In sampling for enumeration the population is usually large and it will rarely be expedient to sample more than a small fraction of the total. If, for instance, the sampling units in a variable population have a standard error of 100 per cent, the standard error of the mean of a sample of 2,500 taken at random is only 2 per cent and competent use of the opportunities for greater accuracy described in §2 may reduce this still further. Yet 2,500 is a negligible fraction of a population of say 1 million. It follows that discussion of the case in which the fraction sampled is appreciable is unlikely to be of frequent practical interest. The method of dealing with this case will, however, be indicated.

The finite population should itself be regarded as a random sample from some infinite population; thus the sample which is taken for enumeration is regarded as a subsample from a larger sample of the same infinite population. Further, in so far as the sampling is carried out for the purpose of estimating the mean of some character in the finite population (i.e. in the larger sample), sampling errors must be measured about the mean of the larger sample. With these two points in mind, the ordinary rules of the analysis of variance may be applied. Consider the variance of the mean $\bar{x}_n$ of a random sample drawn from a larger sample with $N$ units and mean $\bar{x}_N$. Let $\sigma^2$ be the variance in the infinite population. Then

$$\bar{x}_n - \bar{x}_N = \left(\frac{1}{n} - \frac{1}{N}\right)(x_1 + x_2 + \cdots + x_n)$$
$$-\frac{1}{N}(x_{n+1} + \cdots + x_N).$$

Hence
$$V(\bar{x}_n - \bar{x}_N) = \left\{n\left(\frac{1}{n} - \frac{1}{N}\right)^2 + \frac{N-n}{N^2}\right\}\sigma^2 \tag{1}$$
$$= \frac{N-n}{N}\frac{\sigma^2}{n}.$$

This differs from the usual expression $\sigma^2/n$, by the factor $(N-n)/N$. The value of $\sigma^2$ is, of course, unknown, but the mean square within the sample of $n$ is an unbiased estimate of it.

It follows for example that the variance of the mean $\sum N_i \bar{x}_i / \sum N_i$ of a sample sub-divided into groups of different sizes $N_i$ is

$$\sum \left\{\frac{N_i}{n_i}(N_i - n_i)\sigma_i^2\right\} / \sum{}^2 N_i \tag{2}$$

where $\sigma_i^2$ is the variance and $n_i$ the size of the sample in the $i^{\text{th}}$ group. If the variances within groups are all equal, the mean is most accurate when $n_i$ is chosen proportional to $N_i$. It sometimes happens that the total numbers $N_i$ are not themselves known accurately, but are estimated from other data or are themselves the subject of a sampling investigation. In this case the variance of the weighted mean $\bar{x}_w$ is increased, to a first approximation, by

$$\sum_i \left\{(\bar{x}_i - \bar{x}_w)^2 V(N_i)\right\} / \sum{}^2 N_i. \tag{3}$$

As is to be expected, the effect of inaccuracy in the group total $N_i$ on the weighted mean is greatest when the group mean deviates greatly from the general mean. It should also be remembered that with a given sampling error in the estimates of the group *means* a point is reached beyond which increased accuracy in the determination of the group totals is not worth while.

The mathematical theory of sampling from a fixed batch of known numbers has been extensively studied, usually under the title "sampling from a finite universe." Dwyer (1938) has given an extensive bibliography. It might be thought that this large body of results would be pre-eminently suitable for application to the problem in hand, but this is not the case. Where the population consists of a single group, the results obtained by "finite sampling theory" agree with those obtained by the analysis of variance. The former is, however, not easily extended to the case in which the population is sub-divided into groups, at least so far as the situations arising in practice are concerned. Further, it is far removed from reality to regard the population as a fixed batch of known numbers. In economic and sociological studies the population is changing from day to day. The population at any one time is often conventional, as for example with a population of farms or carpenters, owing to the difficulty in defining a member of the population. Errors in counting are bound to occur in any large-scale investigation and though they are not usually differentiated from the sampling errors, they will contribute to inaccuracy in any means which are calculated.

### 5  SAMPLING ERRORS OF THE RELATIVE EFFICIENCY OF TWO METHODS OF SAMPLING

When the efficiencies of two alternative types of sampling are compared by the methods outlined above, the estimates of the two sampling errors concerned are themselves subject to sampling variation. The estimated variance of the mean of the sample which was actually taken presents no difficulty, since it is almost always a multiple of a single mean square in the analysis of variance and hence its sampling variation may be found from the $\chi^2$ tables. The estimated sampling variance of the alternative method is generally a weighted mean of two estimated variances. Distributions of this type have not yet been reduced to a form in which they can be conveniently handled in practice. The omission is not, however, serious, since we are more interested in the *relative* efficiency of the new method to the method which was previously used. This is usually a simple function of the ratio of two variances and its sampling limits can be deduced from the known distribution of that quantity.

For example, with a fixed panel of farms, the percentage increase in information due to doubling the amount of sampling per field in the example in §3 may be written as $15.8\% = 100/\{2(24.2/6.6)-1\}$ where the estimate 24.2 is based on 11 degrees of freedom and 6.6 is based on 39 degrees of freedom. The 5 per cent value of the variance ratio for 11 and 39 degrees of freedom is 2.046. If the true variance ratio is $(1/2.046)\times(24.2/6.6)$, the observed ratio, or a higher value, would occur only once in twenty trials. Thus the 5 per cent upper fiducial limit to the increase in information is

$$\frac{100}{\dfrac{2}{2.046}\left(\dfrac{24.2}{6.6}\right)-1} = 38.7\%.$$

Similarly the lower 5 per cent fiducial limit is found to be 5.7 per cent.

The gain due to sub-division will also be usually of this form, though in the example considered above it will be found to contain three variances, owing to our consideration of the case in which there was one field per farm.

### 6 THE STATISTICAL ANALYSIS OF LARGE SAMPLES

In large scale surveys, where the number of samples taken may run into thousands, the task of performing the statistical analysis without undue labour raises special problems which are outside the scope of this paper. One or two points will, however, be discussed.

Table VI. Standard errors % per plot: roots

| Analysis of yield | Sowing date | |
|---|---|---|
| | April 30 | May 25 |
| (1) Per plant | 18.9 | 15.0 |
| (2) Per unit area | 15.4 | 9.6 |
| (3) Per plant (covariance) | 14.8 | 9.6 |

The choice of variable to be analysed is worth consideration. For example, in a sampling study on sugar-beet similar to that on wheat discussed above, the sampling unit consisted of five consecutive beet in the same row. The distance between the extreme beet of the sample was also measured. Here there are three possible variables: (1) the total weight of five plants, (2) the weight per unit length of row, and (3) the total weight adjusted by its covariance on the distance apart of the extreme beet. If there is no competition between plants in the same row, (1) will be the most convenient and accurate variable to use. If there is competition, (2) will be more stable than (1), and (3) gives the smallest possible sampling error, though it has the disadvantage of requiring extra computation. The percentage standard errors per plot of 15 beet are given below for the final sampling date at Rothamsted in 1937. There were two sowing dates, which were analysed separately.

The standard error was substantially reduced by the choice of either (2) or (3), while (2) was almost as effective as (3) and would be recommended on account of the smaller amount of work involved. In the earlier sampling dates, on the other hand, there was little evidence of competitition, and (1) is the most suitable variable.

The choice of variable to be analysed must, of course, be related to the purpose of the investigation. Particular care is needed where the original sample is taken to obtain information on the practicability of sampling for the purpose in hand. In sampling to estimate the total yield of a crop, for instance, an analysis based on the total yield *per farm* would indicate high variability owing to variations in the sizes of the farms. If opinions as to practicability were based on this analysis, such large samples would appear to be needed to obtain accurate information that the whole scheme might be rejected. If, however, the total acreage of the crop is known even approximately, this information should be utilized by analysing the yield *per unit area*. The effect of inaccuracy in the estimate of the acreage can be taken into account, since the *percentage* variance of the estimate of total yield is approximately the sum of the *percentage* variances of the estimated mean yield per acre and of the estimated acreage. For example, if the acreage estimate is fairly certain to be correct to within $\pm 6$ per cent, so that it may be assigned a standard error of 3 per cent, the standard error per cent of the estimate of the total yield is $\sqrt{9+p^2}$, where $p$ is the standard error per cent of the estimate of mean yield per acre. In this case it would not be worth while to reduce $p$ below $1\frac{1}{2}$ to 2 per cent in sampling for mean yield per acre, and if very high precision is necessary in the estimate of total yield, the acreage estimates must be improved.

If only the total acreage of all crops is known reasonably accurately, the position is more complicated. The relevant factors for a particular crop are the mean yield per acre, and the acreage as a fraction of the acreage of all crops. Different amounts of sampling might be required for these two variables to carry out the estimation most efficiently.

In performing the statistical analysis, the chief difficulty is likely to arise from the large numbers of samples to be handled. The material will usually present itself in a series of successive sub-classifications, with probably different numbers in the various classes at any stage. Estimates of variance can be obtained as a sum of products of totals and means, as in the example discussed above. The variances may differ from class to class, for instance in a sample subdivided into groups whose means are widely different. Even in this case an unbiased estimate of the average variance can be obtained without difficulty, though this quantity will not be distributed as a multiple of $\chi^2$ and will be more variable than its number of degrees of freedom would indicate.

It may be sufficient, at certain stages, to analyse only a small portion of the data. With a sample of 2000 fields, each providing one degree of freedom for sampling error, the evaluation of this variance from one field picked at random in the first ten, and every subsequent tenth field, will give satisfactory precision. It may even be advisable to carry out the whole of the statistical analysis on a small sub-sample of the data. Since, however, the whole of the data will at least have to be added to calculate the desired averages, the necessary figures for the analysis of the classifications with fewer classes can be obtained by noting sub-totals at various stages.

### SUMMARY

The results of a properly planned sampling investigation, in addition to providing an estimate of the accuracy of the sample, often provide estimates of the accuracy of various alternative methods of sampling which might have been used. These estimates are helpful in increasing the efficiency of sampling in future studies on similar material. The use of the analysis of variance of the sampling results for this purpose is discussed and illustrated by a numerical example. The case in which an appreciable fraction, say more than 10%, of the total population is sampled is discussed briefly. The estimate of the relative accuracy of two methods of sampling is shown to be in most cases a simple function of the variance-ratio, so that its sampling limits are easily obtainable. Some advice is given on the problem of analysing the results of large samples without excessive labour.

### REFERENCES

Cochran and Watson (1936). *Emp. Journ. Exp. Agric.*, IV, 69.

Dwyer (1938). *Ann. Math. Stat.*, IX, 1.

Irwin (1938). *J. R. Statist. Soc. Suppl.*, V, 1.

Kiser (1934). *J. Amer. Stat. Assn.*, 29, 250.

Neyman (1934). *J. R. Statist. Soc.*, 97, 558.

Neyman (1938). *J. Amer. Stat. Assn.*, 33, 101.

Watson (1937). *J. Agric. Sci.*, 27, 474.

Yates (1935). *Ann. Eugen.*, VI, 202.

Yates and Zacopanay (1935). *J. Agric. Sci.*, 25, 545.

# 20

September, 1940                    Research Bulletin 281

# The Analysis of Lattice and Triple Lattice Experiments in Corn Varietal Tests

## I   Construction and Numerical Analysis

By Gertrude M. Cox and Robert C. Eckhardt

## II   Mathematical Theory

By W. G. Cochran

AGRICULTURAL EXPERIMENT STATION
IOWA STATE COLLEGE OF AGRICULTURE
AND MECHANIC ARTS

STATISTICAL SECTION
FARM CROPS SUBSECTION

DIVISION OF AGRICULTURAL STATISTICS, AGRICULTURAL MARKETING
SERVICE, DIVISION OF CEREAL CROPS AND DISEASES, BUREAU
OF PLANT INDUSTRY, UNITED STATES DEPARTMENT
OF AGRICULTURE
Cooperating

AMES, IOWA

# CONTENTS

# SUMMARY

Two incomplete block designs, the lattice and triple lattice, are discussed. Their construction, the field plans, experimental results, the statistical analysis and new features of the mathematical theory are included.

The two experiments consist of yield tests of 81 double-crosses of corn. They are used to illustrate a new method of analysis in which the inter-block information is recovered. The analysis is presented in such a way that it can be adopted as the standard method of analyzing lattice and triple lattice experiments.

The recovery of inter-block information and the reduction of block size from 81 to 9 plots per block resulted in a notable increase in precision when compared with the randomized complete block designs. The gain was 85 percent for the lattice experiment and 73 percent for the triple lattice experiment.

Using inter-block information, the experiment can in no event be appreciably less accurate than arrangements in randomized complete blocks, and will be considerably more precise in the fields where the smaller blocks eliminate a considerable portion of the variance ascribed to soil heterogeneity.

These designs are well adapted to the testing of differences among large numbers of varieties. They are especially desirable when little is known about the variability of the experimental field.

For the triple lattice experiment, the use of covariance analysis is illustrated.

For each design the mathematical basis is given for the estimation of the varietal means together with their standard errors, the weights assigned to intra- and inter-block error variance and the tests of significance. Tables are included indicating the efficiencies to be expected from the use of these designs instead of randomized complete blocks.

# The Analysis of Lattice and Triple Lattice Experiments in Corn Varietal Tests[1]

## I. Construction and Numerical Analysis

By

GERTRUDE M. COX[2] AND ROBERT C. ECKHARDT[3]

Well-known methods of design, such as the Latin square or randomized complete block, are available and efficient when the number of varieties to be compared is small. As the number of varieties increases, these designs may become less efficient through failure to eliminate soil heterogeneity. Furthermore, the Latin square design becomes cumbersome, because it requires replicates equal in number to the varieties. Several methods exist for making comparisons among a large number of varieties. 1. The classical way of arranging such trials is by the use of controls or check plots. These are employed in one of two ways: (a) Small groups of varieties, together with one or more controls, are arranged in either randomized blocks or Latin squares. Comparing each variety with the control variety in its group makes possible the comparison of varieties not in the same group, such comparisons being less accurate than comparisons between varieties in the same group; (b) controls are inserted systematically in a randomized complete block arrangement of all the varieties. These controls furnish an index of soil fertility variations within the blocks. In both cases, the precision of the experiment is reduced by the fact that more plots are allotted to the control variety than to any other. 2. The semi-Latin square devised by "Student" [9] has been shown to give a biased estimate of error [11]. 3. Richey [7] devised an ingenious method by which variety yields are adjusted to their regression on a moving average. This method, how-

[1] Project 514 of the Statistical Section of the Iowa Agricultural Experiment Station in cooperation with the Division of Agricultural Statistics, Agricultural Marketing Service, United States Department of Agriculture, and project 163 of the Farm Crops Subsection of the Iowa Agricultural Experiment Station in cooperation with the Division of Cereal Crops and Diseases, Bureau of Plant Industry, United States Department of Agriculture.

[2] Research Assistant Professor, Statistical Section, Iowa Agricultural Experiment Station; also Agent, Division of Agricultural Statistics, Agricultural Marketing Service, United States Department of Agriculture.

[3] Agent, Division of Cereal Crops and Diseases, Bureau of Plant Industry, United States Department of Agriculture; also Collaborator, Agronomy Section, Farm Crops Subsection, Iowa Agricultural Experiment Station.

ever, requires involved computation, calls for an arbitrary decision as to the number of varieties which appear in each moving average group and affords no exact test of significance.  4. A similar method was later suggested by Papadakis [6]. He uses covariance with the yields of neighboring plots to reduce the error of replicated experiments.  A review and comments on this method are given by Bartlett [1]. He says, "The calculations are rather laborious, and would compare unfavorably with the simple modifications necessary in the analysis of well-designed experiments in which confounding has been introduced."  Although  these  four methods were devised to test large numbers of varieties, their limitations constitute a serious bar to their use.

Because of the limitations of the above methods, several modifications of the complete block design have been devised by Yates.  These new designs are more efficient than complete block designs when there are appreciable inequalities in fertility within replications.  They also have the common characteristic that a block contains fewer varieties than the total number of varieties to be compared.  These more homogeneous small blocks are referred to as incompete blocks. [13].

The incomplete block designs were first suggested for experiments on heterogeneous soil.  The method of analysis appropriate to this case has been given by Yates [12] and Goulden [4, 5].  It sometimes happened that the soil was relatively homogeneous; that is, differences between blocks were small.  In such experiments the efficiency of the incomplete block designs was less than that in the complete block designs.  Yates [15] and Fisher and Yates [3] proposed a method for the recovery of the information about variety differences contained in the blocks (groups of varieties), provided that the experiment has sufficient replications to give an adequate estimate of error for the inter-block as well as the intra-block comparisons.  In a recent article, Yates [16] has described this method for variety trials arranged in three-dimensional lattice (cubic lattice) design.  He states, "The recovery of the inter-block information greatly increases the attractiveness of these experimental designs, for when this information is utilized they can in no event be appreciably less accurate than arrangements in ordinary randomized blocks containing all the varieties, and they will of course be considerably more accurate if there is any appreciable reduction in variance resulting from the use of smaller blocks."

The object of this bulletin is to present the field plans, statistical methods, experimental results and mathematical theory of two incomplete block designs, the lattice and triple lattice, which in addition to the lattice square designs [10] are particularly well adapted to testing large numbers of varieties. A new method of analysis is presented in such a way that it can be adopted as the standard method of procedure for lattice and triple lattice experiments. Also, the variety means are computed so as to recover the information about varieties contained in the blocks. The standard errors used to test the significance of the differences between variety means are given for each example. For the triple lattice experiment, the use of covariance analysis is illustrated.

## LATTICE DESIGNS[4]

The construction of lattice designs has been presented by Yates [12] and by Goulden [4, 5]. In the lattice designs the number of varieties is a perfect square, $v = k^2$. To illustrate the construction let $v = 81$, and therefore, $k = 9$. The 81 variety identification numbers are arranged in a square:

```
 1  2  3  4  5  6  7  8  9
10 11 12 13 14 15 16 17 18
19 20 21 22 23 24 25 26 27
28 29 30 31 32 33 34 35 36
37 38 39 40 41 42 43 44 45
46 47 48 49 50 51 52 53 54
55 56 57 58 59 60 61 62 63
64 65 66 67 68 69 70 71 72
73 74 75 76 77 78 79 80 81
```

These 81 varieties are next arranged in incomplete blocks, each block containing nine varieties. There are two groups of these blocks, the blocks of one group being so arranged that they cut across those of the other group. In one group, designated as X, the varieties that are together in a row in the square above are put into a block. In group Y, the varieties that are together in columns of the square make up the blocks. Thus there are nine blocks in a group, each block containing nine varieties. The arrangement is as follows:

| Block | Group X | | | | | | | | | Block | Group Y | | | | | | | | |
|---|---|---|---|---|---|---|---|---|---|---|---|---|---|---|---|---|---|---|---|
| (a) | 1 | 2 | 3 | 4 | 5 | 6 | 7 | 8 | 9 | (a) | 1 | 10 | 19 | 28 | 37 | 46 | 55 | 64 | 73 |
| (b) | 10 | 11 | 12 | 13 | 14 | 15 | 16 | 17 | 18 | (b) | 2 | 11 | 20 | 29 | 38 | 47 | 56 | 65 | 74 |
| (c) | 19 | 20 | 21 | 22 | 23 | 24 | 25 | 26 | 27 | (c) | 3 | 12 | 21 | 30 | 39 | 48 | 57 | 66 | 75 |
| (d) | 28 | 29 | 30 | 31 | 32 | 33 | 34 | 35 | 36 | (d) | 4 | 13 | 22 | 31 | 40 | 49 | 58 | 67 | 76 |
| (e) | 37 | 38 | 39 | 40 | 41 | 42 | 43 | 44 | 45 | (e) | 5 | 14 | 23 | 32 | 41 | 50 | 59 | 68 | 77 |
| (f) | 46 | 47 | 48 | 49 | 50 | 51 | 52 | 53 | 54 | (f) | 6 | 15 | 24 | 33 | 42 | 51 | 60 | 69 | 78 |
| (g) | 55 | 56 | 57 | 58 | 59 | 60 | 61 | 62 | 63 | (g) | 7 | 16 | 25 | 34 | 43 | 52 | 61 | 70 | 79 |
| (h) | 64 | 65 | 66 | 67 | 68 | 69 | 70 | 71 | 72 | (h) | 8 | 17 | 26 | 35 | 44 | 53 | 62 | 71 | 80 |
| (i) | 73 | 74 | 75 | 76 | 77 | 78 | 79 | 80 | 81 | (i) | 9 | 18 | 27 | 36 | 45 | 54 | 63 | 72 | 81 |

[4] Lattice designs have been referred to as pseudo-factorial arrangements in two equal groups of sets [12], as two dimensional pseudo-factorial experiments with two equal groups of sets [4] and as two dimensional quasi-factorial designs in randomized blocks in two equal groups of sets (lattice) [14].

For field plans the varieties within the blocks and the blocks in the group as well as the groups themselves are randomized. Below is given one random arrangement of groups X and Y.

RANDOM ARRANGEMENT

| Block | Group X | | | | | | | | | Group Y | | | | | | | | |
|---|---|---|---|---|---|---|---|---|---|---|---|---|---|---|---|---|---|---|
| (c)(1) | 27 | 20 | 22 | 26 | 25 | 21 | 23 | 24 | 19 | (f)(1) | 69 | 15 | 78 | 33 | 51 | 24 | 60 | 42 | 6 |
| (h)(2) | 67 | 64 | 72 | 71 | 65 | 68 | 70 | 66 | 69 | (d)(2) | 58 | 22 | 13 | 76 | 31 | 67 | 4 | 49 | 40 |
| (g)(3) | 56 | 59 | 62 | 61 | 58 | 55 | 63 | 57 | 60 | (b)(3) | 29 | 11 | 20 | 2 | 47 | 74 | 56 | 65 | 38 |
| (a)(4) | 4 | 9 | 3 | 7 | 6 | 1 | 5 | 2 | 8 | (e)(4) | 23 | 77 | 5 | 41 | 14 | 68 | 32 | 50 | 59 |
| (f)(5) | 53 | 51 | 50 | 47 | 46 | 48 | 49 | 54 | 52 | (c)(5) | 3 | 75 | 21 | 12 | 66 | 48 | 39 | 57 | 30 |
| (i)(6) | 78 | 73 | 80 | 75 | 74 | 77 | 79 | 81 | 76 | (i)(6) | 36 | 63 | 54 | 81 | 45 | 72 | 27 | 9 | 18 |
| (b)(7) | 15 | 18 | 17 | 14 | 11 | 12 | 10 | 16 | 13 | (a)(7) | 64 | 73 | 10 | 37 | 1 | 46 | 55 | 28 | 19 |
| (d)(8) | 28 | 35 | 36 | 29 | 33 | 34 | 30 | 31 | 32 | (h)(8) | 26 | 53 | 35 | 80 | 44 | 17 | 8 | 71 | 62 |
| (e)(9) | 41 | 38 | 42 | 37 | 40 | 44 | 39 | 43 | 45 | (g)(9) | 70 | 61 | 25 | 43 | 34 | 16 | 52 | 7 | 79 |

Block (c) in group X has by random selection become block (1). Notice that block (1) contains the same varieties as block (c) but in a random order. This random arrangement within and between the blocks in a group, insuring as it does that no variety shall be favored, provides an unbiased estimate of error. For a discussion of the fundamental theory and assumptions back of randomization and replications see [2]. Groups X and Y are replicated *(r)* the same number of times according to the degree of precision required, each replication having a different random arrangement. The nine blocks in group X should be kept together in the field as a distinct replication. Similarly, the nine blocks in group Y should be kept together. By having the varieties in groups or complete replications,

(a) The inter-block comparisons are made as accurate as possible.

(b) The results can be analyzed as a randomized complete block design. This allows one or more varieties to be omitted if the experiment has large variety differences; the components of error can be isolated if the experiment has unequal variances and the unadjusted varietal means can be used for subsidiary measurements.

(c) The gain in precision of the design over a complete block design can easily be tested.

FIELD PLANS

A brief statement as to the placing of lattice experiments in the field follows. For corn where 4 × 5 hill plots are used, the field plan for a plot and for block (1) of the random arrangement of group X is illustrated.

Plot

```
x x x x
x
x
x
x
```

Block (1)

| x x x x x | x x x x | x x x x |
|-----------|---------|---------|
| x         |         |         |
| x 27      | 20      | 22      |
| x         |         |         |
| x         |         |         |
| x         |         |         |
| x         |         |         |
| x 26      | 25      | 21      |
| x         |         |         |
| x         |         |         |
| x         |         |         |
| x         |         |         |
| x 23      | 24      | 19      |
| x         |         |         |
| x         |         |         |

When $2 \times 10$ hill plots are used, the field plan for a plot and for block (1) is illustrated.

Plot

```
x x
x x
x x
x x
x x
x x
x x
x x
x x
x x
```

Block (1)

| x x | x x | x x | x x | x x | x x | x x | x x | x x |
|-----|-----|-----|-----|-----|-----|-----|-----|-----|
| x   |     |     |     |     |     |     |     |     |
| x   |     |     |     |     |     |     |     |     |
| x   |     |     |     |     |     |     |     |     |
| 27  | 20  | 22  | 26  | 25  | 21  | 23  | 24  | 19  |
| x   |     |     |     |     |     |     |     |     |
| x   |     |     |     |     |     |     |     |     |
| x   |     |     |     |     |     |     |     |     |
| x   |     |     |     |     |     |     |     |     |
| x   |     |     |     |     |     |     |     |     |

Plot and block size vary with the experiment, no one size or shape being best for all crops on all soils.

Blocks are placed in the field in succession or in groups on land representative of the district where the varieties will ultimately be planted. It may be advisable to leave a space between two blocks in order to avoid an undesirable spot in the field; this is permissible when using lattice arrangements. In this respect, the lattice design is more flexible than the lattice square design [10] where the plots must be kept in a fixed arrangement. If at all possible, keep the blocks of a replication together. A field plan for group X using $4 \times 5$ hill plots is given in fig. 1. This gives not only compact blocks, but also a replication which is almost square. The field plan which has been commonly used for the corn experiments at Iowa State College is given in fig. 2. This field plan has the desirable characteristic that the blocks can be placed in succession across the field. However, the blocks in the same replication may be rather far apart. This reduces

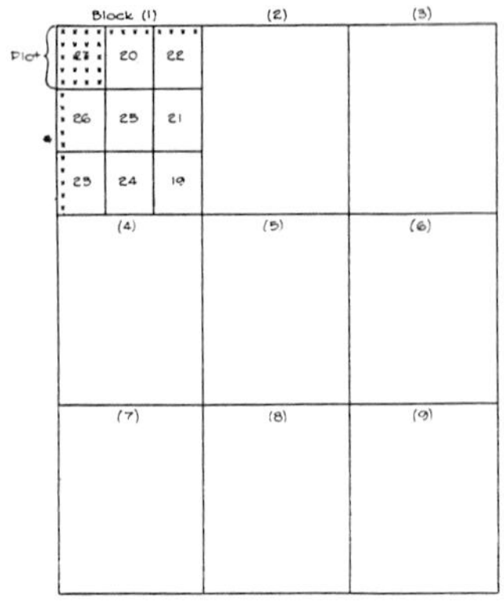

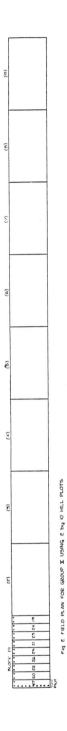

Fig 2 FIELD PLAN FOR GROUP II USING 2 by 10 HILL PLOTS

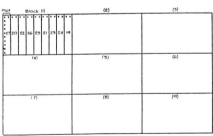

Fig 1 FIELD PLAN FOR GROUP II USING 4 by 5 HILL PLOTS

Fig 3 FIELD PLAN FOR GROUP II USING 2 by 10 HILL PLOTS

the value of inter-block information and also reduces the accuracy of the design if it is analyzed as a randomized complete block arrangement. Therefore, fig. 3 is presented as perhaps the best field arrangement when using $2 \times 10$ hill plots. In some of the other lattice designs the blocks will not fit into as compact a replication as given in fig. 3.

It should be noted that the varieties are apportioned into blocks in such a way that the comparisons between pairs of varieties cannot all be made with equal precision. Comparisons between varieties which appear in a block together are slightly more accurate than comparisons between varieties which have no block in common.

## EXAMPLE OF LATTICE EXPERIMENT

In order to illustrate the lattice type of incomplete block design which has just been discussed, an experiment used to test the yields of 81 double-crosses of corn is presented. All of these double-crosses were the hybrids retained from a larger number of double-crosses previously tested for 2 or more years. Because of this selection, they did not differ greatly in yield. The experiment consists of four replications; replications 1 and 2 are two random arrangements of group X, while replications 3 and 4 are two random arrangements of group Y. This gives 36 blocks of 9 plots each; the 36 blocks are in four replication groups of nine blocks each. This lattice design was used in order to place one replication of groups X and Y on each of two fertility levels. Replication 1 (group X) and replication 3 (group Y) were grown on soil of medium fertility, while replications 2 and 4 were grown in a neighboring field on soil of high fertility. A preliminary analysis of each field separately gave results which justify using all the data in one analysis. A balanced lattice square design [10, 14] might have been better if there had been enough land of approximately the same fertility available for the five replications required to secure balance.

The double-crosses were arranged in the field according to the plan (fig. 2) for $2 \times 10$ hill plots. The field plan is for replication 1, with varieties 27, 20, 22, 26, 25, 21, 23, 24 and 19 in block (1). Block (2) followed adjacent to block (1). The blocks of replication 3 were placed directly after the blocks of replication 1. Table 1 indicates the random arrangement of the varieties in each block. The varieties are indicated by the numbers 1, 2, 3, . . . , 81, given in boldface type. Since the results are for a single year, the varieties are designated by numbers and not by name. Following the variety identification, the individual plot yields are presented. Block totals are also given in table 1.

The variety yields from replications 1 and 2 are collected into group X of table 2. That is, for variety 1, 19.5 + 26.4 = 45.9. Replications 3 and 4 have their variety yields summed in group Y. The totals for rows and columns are given for both groups. The combination of the variety yields from all four replications appears in table 3.

### TABLE 1. PLOT YIELDS OF THE LATTICE EXPERIMENT.

(Variety identification in boldface type.)

#### Replication 1 (Group X).

| Block | | | | | | | | | | Block totals |
|---|---|---|---|---|---|---|---|---|---|---|
| (1) | **27** 27.6 | **20** 24.3 | **22** 24.9 | **26** 25.9 | **25** 25.2 | **21** 22.8 | **23** 24.5 | **24** 22.9 | **19** 24.4 | 222.5 |
| (2) | **67** 28.8 | **64** 29.6 | **72** 25.2 | **71** 26.6 | **65** 27.9 | **68** 24.0 | **70** 26.8 | **66** 26.8 | **69** 25.2 | 240.9 |
| (3) | **56** 27.0 | **59** 28.4 | **62** 30.9 | **61** 31.4 | **58** 33.7 | **55** 29.0 | **63** 28.7 | **57** 29.1 | **60** 27.0 | 265.2 |
| (4) | **4** 23.7 | **9** 21.2 | **3** 24.1 | **7** 24.5 | **6** 25.8 | **1** 19.5 | **5** 24.0 | **2** 23.3 | **8** 27.1 | 213.2 |
| (5) | **53** 27.7 | **51** 27.3 | **50** 36.0 | **47** 31.3 | **46** 26.8 | **48** 30.5 | **49** 25.6 | **54** 27.8 | **52** 24.2 | 257.2 |
| (6) | **78** 31.1 | **73** 33.7 | **80** 31.7 | **75** 29.2 | **74** 31.3 | **77** 26.8 | **79** 30.8 | **81** 31.6 | **76** 23.7 | 269.9 |
| (7) | **15** 28.7 | **18** 28.2 | **17** 25.9 | **14** 28.9 | **11** 26.5 | **12** 26.4 | **10** 27.3 | **16** 25.6 | **13** 26.2 | 243.7 |
| (8) | **28** 29.4 | **35** 26.3 | **36** 23.3 | **29** 30.0 | **33** 25.5 | **34** 24.3 | **30** 25.7 | **31** 24.8 | **32** 28.1 | 237.4 |
| (9) | **41** 25.0 | **38** 25.4 | **42** 25.3 | **37** 24.1 | **40** 25.0 | **44** 24.8 | **39** 24.3 | **43** 25.7 | **45** 23.9 | 223.5 |
| | | | | | | | | | | 2173.5 |

#### Replication 2 (Group X).

| Block | | | | | | | | | | Block totals |
|---|---|---|---|---|---|---|---|---|---|---|
| (1) | **1** 26.4 | **8** 32.8 | **9** 32.2 | **2** 31.0 | **5** 34.9 | **3** 27.1 | **6** 32.2 | **4** 28.0 | **7** 27.8 | 272.4 |
| (2) | **38** 29.6 | **45** 32.1 | **37** 32.7 | **44** 33.0 | **43** 33.0 | **40** 33.5 | **42** 31.3 | **41** 32.5 | **39** 35.8 | 293.5 |
| (3) | **66** 36.7 | **68** 32.0 | **71** 34.2 | **69** 38.6 | **65** 35.4 | **64** 38.2 | **72** 31.6 | **67** 36.3 | **70** 38.5 | 321.5 |
| (4) | **75** 37.2 | **80** 34.1 | **78** 36.1 | **79** 41.1 | **74** 39.4 | **81** 37.3 | **77** 32.8 | **73** 36.2 | **76** 33.1 | 327.3 |
| (5) | **34** 34.0 | **32** 36.6 | **28** 36.6 | **36** 30.6 | **33** 31.3 | **31** 34.8 | **30** 36.6 | **35** 35.5 | **29** 33.8 | 309.8 |
| (6) | **55** 33.7 | **58** 37.0 | **56** 36.8 | **57** 35.0 | **60** 33.5 | **61** 35.5 | **62** 38.7 | **59** 35.1 | **63** 39.1 | 324.4 |
| (7) | **16** 34.1 | **11** 25.3 | **18** 30.9 | **14** 29.7 | **13** 31.9 | **12** 29.7 | **10** 29.3 | **17** 25.0 | **15** 27.3 | 263.2 |
| (8) | **47** 25.5 | **50** 26.4 | **53** 24.0 | **46** 22.7 | **48** 23.9 | **54** 28.3 | **51** 26.9 | **49** 23.6 | **52** 28.1 | 229.4 |
| (9) | **21** 29.0 | **20** 28.4 | **26** 29.8 | **24** 33.1 | **22** 33.2 | **27** 34.5 | **19** 28.9 | **25** 34.4 | **23** 30.3 | 281.6 |
| | | | | | | | | | | 2623.1 |

TABLE 1. Continued.

Replication 3 (Group Y).

| Block | | | | | | | | | | |
|---|---|---|---|---|---|---|---|---|---|---|
| (1) | 69<br>29.4 | 15<br>24.9 | 78<br>26.2 | 33<br>28.2 | 51<br>24.5 | 24<br>27.9 | 60<br>24.3 | 42<br>21.8 | 6<br>28.4 | 235.6 |
| (2) | 58<br>31.9 | 22<br>27.7 | 13<br>29.2 | 76<br>25.3 | 31<br>29.9 | 67<br>32.6 | 4<br>28.5 | 49<br>26.2 | 40<br>29.6 | 260.9 |
| (3) | 29<br>30.5 | 11<br>27.3 | 20<br>28.3 | 2<br>25.6 | 47<br>27.9 | 74<br>30.0 | 56<br>31.2 | 65<br>30.3 | 38<br>25.6 | 256.7 |
| (4) | 23<br>28.1 | 77<br>26.5 | 5<br>25.2 | 41<br>26.2 | 14<br>27.7 | 68<br>27.9 | 32<br>27.7 | 50<br>29.5 | 59<br>28.4 | 247.2 |
| (5) | 3<br>25.6 | 75<br>28.3 | 21<br>28.9 | 12<br>28.9 | 66<br>33.1 | 48<br>29.8 | 39<br>29.3 | 57<br>32.9 | 30<br>30.0 | 266.8 |
| (6) | 36<br>24.4 | 63<br>26.9 | 54<br>27.8 | 81<br>27.3 | 45<br>22.3 | 72<br>27.9 | 27<br>26.6 | 9<br>25.7 | 18<br>26.6 | 235.5 |
| (7) | 64<br>27.4 | 73<br>29.0 | 10<br>29.5 | 37<br>24.4 | 1<br>20.0 | 46<br>24.5 | 55<br>28.7 | 28<br>25.0 | 19<br>24.8 | 233.3 |
| (8) | 26<br>26.6 | 53<br>29.0 | 35<br>34.5 | 80<br>32.2 | 44<br>33.9 | 17<br>28.3 | 8<br>32.4 | 71<br>33.8 | 62<br>38.1 | 288.8 |
| (9) | 70<br>30.6 | 61<br>36.0 | 25<br>36.4 | 43<br>33.6 | 34<br>33.1 | 16<br>33.8 | 52<br>32.1 | 7<br>30.4 | 79<br>39.9 | 305.9 |
| | | | | | | | | | | 2330.7 |

Replication 4 (Group Y).

| Block | | | | | | | | | | |
|---|---|---|---|---|---|---|---|---|---|---|
| (1) | 76<br>29.1 | 13<br>29.4 | 40<br>28.5 | 4<br>29.3 | 49<br>24.3 | 67<br>27.5 | 58<br>31.3 | 31<br>28.0 | 22<br>28.0 | 255.4 |
| (2) | 2<br>25.1 | 11<br>29.8 | 47<br>29.6 | 65<br>32.6 | 56<br>33.5 | 38<br>30.8 | 29<br>33.4 | 20<br>30.0 | 74<br>39.3 | 284.1 |
| (3) | 6<br>34.8 | 42<br>35.1 | 69<br>38.4 | 51<br>34.2 | 33<br>33.4 | 24<br>40.4 | 15<br>30.8 | 60<br>36.5 | 78<br>32.4 | 316.0 |
| (4) | 25<br>35.5 | 70<br>40.5 | 43<br>25.4 | 61<br>36.1 | 34<br>37.4 | 7<br>30.0 | 52<br>34.1 | 16<br>33.3 | 79<br>39.7 | 312.0 |
| (5) | 3<br>34.5 | 21<br>35.6 | 30<br>36.8 | 48<br>34.3 | 39<br>26.6 | 12<br>39.2 | 57<br>34.1 | 66<br>40.5 | 75<br>33.9 | 315.5 |
| (6) | 5<br>31.8 | 77<br>30.1 | 59<br>34.0 | 32<br>36.4 | 68<br>32.0 | 23<br>33.9 | 41<br>34.5 | 50<br>41.6 | 14<br>35.7 | 310.0 |
| (7) | 62<br>38.7 | 71<br>35.0 | 80<br>32.5 | 26<br>34.8 | 44<br>34.1 | 35<br>36.3 | 53<br>33.4 | 17<br>31.6 | 8<br>32.6 | 309.0 |
| (8) | 46<br>29.6 | 10<br>30.9 | 28<br>38.2 | 55<br>34.6 | 37<br>32.0 | 19<br>34.8 | 1<br>28.3 | 64<br>40.3 | 73<br>36.8 | 305.5 |
| (9) | 45<br>35.9 | 54<br>38.1 | 72<br>32.1 | 81<br>35.9 | 27<br>34.4 | 9<br>32.9 | 63<br>38.0 | 18<br>34.7 | 36<br>29.4 | 311.4 |
| | | | | | | | | | | 2718.9 |

It is assumed that the experimenter is familiar with analysis of variance and the associated tests of significance. Excellent presentations are given by Fisher [2], by Snedecor [8] and by Goulden [5]. The computational details, however, are included here so that they can be followed by a computer.

In the new method of analysis which we present, an estimate of the inter-block variance is made, and adjustments to the variety plot means are calculated so that the inter- and intra-block comparisons are weighted according to their relative precision.

TABLE 2. COMBINATION OF REPLICATIONS.

Group X (Replication 1 + Replication 2).

| 1 | 2 | 3 | 4 | 5 | 6 | 7 | 8 | 9 | Row totals |
|---|---|---|---|---|---|---|---|---|---|
| 45.9 | 54.3 | 51.2 | 51.7 | 58.9 | 58.0 | 52.3 | 59.9 | 53.4 | 485.6 |
| 10 56.6 | 11 51.8 | 12 56.1 | 13 58.1 | 14 58.6 | 15 56.0 | 16 59.7 | 17 50.9 | 18 59.1 | 506.9 |
| 19 53.3 | 20 52.7 | 21 51.8 | 22 58.1 | 23 54.8 | 24 56.0 | 25 59.6 | 26 55.7 | 27 62.1 | 504.1 |
| 28 66.0 | 29 63.8 | 30 62.3 | 31 59.6 | 32 64.7 | 33 56.8 | 34 58.3 | 35 61.8 | 36 53.9 | 547.2 |
| 37 56.8 | 38 55.0 | 39 60.1 | 40 58.5 | 41 57.5 | 42 56.6 | 43 58.7 | 44 57.8 | 45 56.0 | 517.0 |
| 46 49.5 | 47 56.8 | 48 54.4 | 49 49.2 | 50 62.4 | 51 54.2 | 52 52.3 | 53 51.7 | 54 56.1 | 486.6 |
| 55 62.7 | 56 63.8 | 57 64.1 | 58 70.7 | 59 63.5 | 60 60.5 | 61 66.9 | 62 69.6 | 63 67.8 | 589.6 |
| 64 67.8 | 65 63.3 | 66 63.5 | 67 65.1 | 68 56.0 | 69 63.8 | 70 65.3 | 71 60.8 | 72 56.8 | 562.4 |
| 73 69.9 | 74 70.7 | 75 66.4 | 76 56.8 | 77 59.6 | 78 67.2 | 79 71.9 | 80 65.8 | 81 68.9 | 597.2 |
| Column totals 528.5 | 532.2 | 529.9 | 527.8 | 536.0 | 529.1 | 545.0 | 534.0 | 534.1 | 4796.6 |

Group Y (Replication 3 + Replication 4).

| 1 | 10 | 19 | 28 | 37 | 46 | 55 | 64 | 73 | Row totals |
|---|---|---|---|---|---|---|---|---|---|
| 48.3 | 60.4 | 59.6 | 63.2 | 56.4 | 54.1 | 63.3 | 67.7 | 65.8 | 538.8 |
| 2 50.7 | 11 57.1 | 20 58.3 | 29 63.9 | 38 56.4 | 47 57.5 | 56 64.7 | 65 62.9 | 74 69.3 | 540.8 |
| 3 60.1 | 12 68.1 | 21 64.5 | 30 66.8 | 39 55.9 | 48 64.1 | 57 67.0 | 66 73.6 | 75 62.2 | 582.3 |
| 4 57.8 | 13 58.6 | 22 55.7 | 31 57.9 | 40 58.1 | 49 50.5 | 58 63.2 | 67 60.1 | 76 54.4 | 516.3 |
| 5 57.0 | 14 63.4 | 23 62.0 | 32 64.1 | 41 60.7 | 50 71.1 | 59 62.4 | 68 59.9 | 77 56.6 | 557.2 |
| 6 63.2 | 15 55.7 | 24 68.3 | 33 61.6 | 42 56.9 | 51 58.7 | 60 60.8 | 69 67.8 | 78 58.6 | 551.6 |
| 7 60.4 | 16 67.1 | 25 71.9 | 34 70.5 | 43 59.0 | 52 66.2 | 61 72.1 | 70 71.1 | 79 79.6 | 617.9 |
| 8 65.0 | 17 59.9 | 26 61.4 | 35 70.8 | 44 68.0 | 53 62.4 | 62 76.8 | 71 68.8 | 80 64.7 | 597.8 |
| 9 58.6 | 18 61.3 | 27 61.0 | 36 53.8 | 45 58.2 | 54 65.9 | 63 64.9 | 72 60.0 | 81 63.2 | 546.9 |
| Column totals 521.1 | 551.6 | 562.7 | 572.6 | 529.6 | 550.5 | 595.2 | 591.9 | 574.4 | 5049.6 |

CALCULATIONS FOR THE ANALYSIS OF VARIANCE.

The calculations necessary for the analysis of variance and adjustments to the varietal means are as follows.

The correction term $(c)$ is the square of the total divided by the total number of plots:

$$c = \frac{(9846 \; 2)^2}{324} = 299{,}221.16$$

The total sum of squares is obtained by adding the squares of the individual plot yields, table 1, and subtracting the correction term:

$$(27.6)^2 + (24.3)^2 + \ldots + (29.4)^2 - 299,221.16 = 6708.90$$

The sum of squares for replications is

$$\frac{(2173.5)^2 + (2623.1)^2 + (2330.7)^2 + (2718.9)^2}{81} - c =$$

$$301,596.74 - 299,221.16 = 2375.58$$

The sum of squares for varieties (ignoring blocks) is obtained by adding the squares of the variety totals in the cells in table 3, dividing by the number of plots added to give the variety totals, and subtracting the correction term:

$$\frac{(94.2)^2 + (105.0)^2 + \ldots + (132.1)^2}{4} - c =$$

$$301,375.83 - 299,221.16 = 2154.67$$

TABLE 3. VARIETY TOTAL YIELDS.

| 1 | 2 | 3 | 4 | 5 | 6 | 7 | 8 | 9 | Row totals |
|---|---|---|---|---|---|---|---|---|---|
| 94.2 | 105.0 | 111.3 | 109.5 | 115.9 | 121.2 | 112.7 | 124.9 | 112.0 | 1006.7 |
| 10 | 11 | 12 | 13 | 14 | 15 | 16 | 17 | 18 | |
| 117.0 | 108.9 | 124.2 | 116.7 | 122.0 | 111.7 | 126.8 | 110.8 | 120.4 | 1058.5 |
| 19 | 20 | 21 | 22 | 23 | 24 | 25 | 26 | 27 | |
| 112.9 | 111.0 | 116.3 | 113.8 | 116.8 | 124.3 | 131.5 | 117.1 | 123.1 | 1066.8 |
| 28 | 29 | 30 | 31 | 32 | 33 | 34 | 35 | 36 | |
| 129.2 | 127.7 | 129.1 | 117.5 | 128.8 | 118.4 | 128.8 | 132.6 | 107.7 | 1119.8 |
| 37 | 38 | 39 | 40 | 41 | 42 | 43 | 44 | 45 | |
| 113.2 | 111.4 | 116.0 | 116.6 | 118.2 | 113.5 | 117.7 | 125.8 | 114.2 | 1046.6 |
| 46 | 47 | 48 | 49 | 50 | 51 | 52 | 53 | 54 | |
| 103.6 | 114.3 | 118.5 | 99.7 | 133.5 | 112.9 | 118.5 | 114.1 | 122.0 | 1037.1 |
| 55 | 56 | 57 | 58 | 59 | 60 | 61 | 62 | 63 | |
| 126.0 | 128.5 | 131.1 | 133.9 | 125.9 | 121.3 | 139.0 | 146.4 | 132.7 | 1184.8 |
| 64 | 65 | 66 | 67 | 68 | 69 | 70 | 71 | 72 | |
| 135.5 | 126.2 | 137.1 | 125.2 | 115.9 | 131.6 | 136.4 | 129.6 | 116.8 | 1154.3 |
| 73 | 74 | 75 | 76 | 77 | 78 | 79 | 80 | 81 | |
| 135.7 | 140.0 | 128.6 | 111.2 | 116.2 | 125.8 | 151.5 | 130.5 | 132.1 | 1171.6 |
| Column totals | | | | | | | | | |
| 1067.3 | 1073.0 | 1112.2 | 1044.1 | 1093.2 | 1080.7 | 1162.9 | 1131.8 | 1081.0 | 9846.2 |

The sum of squares for blocks (eliminating varieties) is made up of two components as follows:

*Component (a)* consists of two sets of differences in yield between paired blocks,

| Block totals | | Set X | Block totals | | Set Y |
|---|---|---|---|---|---|
| Rep. 1 | Rep. 2 | Differences | Rep. 3 | Rep. 4 | Differences |
| 222.5 | 281.6 | —59.1 | 235.6 | 316.0 | —80.4 |
| 240.9 | 321.5 | —80.6 | 260.9 | 255.4 | 5.5 |
| 265.2 | 324.4 | —59.2 | 256.7 | 284.1 | —27.4 |
| 213.2 | 272.4 | —59.2 | 247.2 | 310.0 | —62.8 |
| 257.2 | 229.4 | 27.8 | 266.8 | 315.5 | —48.7 |
| 269.9 | 327.3 | —57.4 | 235.5 | 311.4 | —75.9 |
| 243.7 | 263.2 | —19.5 | 233.3 | 305.5 | —72.2 |
| 237.4 | 309.8 | —72.4 | 288.8 | 309.0 | —20.2 |
| 223.5 | 293.5 | —70.0 | 305.9 | 312.0 | — 6.1 |
| 2173.5 | 2623.1 | —449.6 | 2330.7 | 2718.9 | —388.2 |

These two sets of differences are computed from the sums of the blocks containing the same group of nine varieties. In replication 1, block (1) contains varieties 27, 20, 22, 26, 25, 21, 23, 24 and 19. These same nine varieties appear in another random order in block (9) of replication 2. It is clear that such sets of differences can be computed only in those lattice experiments which have four or more replications.

The sums of squares of the deviations of these two sets of differences give the variance for paired blocks. The divisors are $2k$ (18) and $2k^2$ (162).[5]

For set X

$$\frac{(59.1)^2 + (80.6)^2 + \ldots + (70.0)^2}{18} - \frac{(449.6)^2}{162} =$$

$$1754.89 - 1247.78 = 507.11$$

For set Y

$$\frac{(80.4)^2 + (5.5)^2 + \ldots + (6.1)^2}{18} - \frac{(388.2)^2}{162} =$$

$$1387.76 - 930.24 = 457.52$$

Summary of component (a)

|  | Degrees of freedom | Sum of squares |
|---|---|---|
| Set X | 8 | 507.11 |
| Set Y | 8 | 457.52 |
| Between paired blocks | 16 | 964.63 |

The total sums of squares of set X and set Y is the sum of squares between paired blocks.

Component (a) is secured with fewer operations by summing the squares of the differences in sets X and Y, then subtracting the two corrections. Thus:

$$\frac{(59.1)^2 + (80.6)^2 + \ldots + (6.1)^2}{18} - \frac{(449.6)^2 + (388.2)^2}{162} =$$

$$3142.65 - 2178.02 = 964.63$$

---

[5] For six replications there are three columns of block totals for each set. Component (a) for set X is calculated by analysis of variance, as follows:

| Replications | 2 |
| Set X totals | $(k-1)$ |
| Component (a) or interaction | $2(k-1)$ |

*Component (b)* consists of two sets of differences giving an estimate of block yield freed of varietal effects. In table 2 in both group X and group Y the row totals are combinations of two block totals from table 1. That is, for group X the first row total comes from block (4) in replication 1 and from block (1) in replication 2.

$$213.2 + 272.4 = 485.6$$

These totals cannot be used directly to compute block sum of squares because they also contain varietal effects. The above total (485.6) contains only varieties 1, 2, ..., 9. However, an estimate of the sum of these nine varieties is given in the first column total of group Y. This estimate is unconfounded with blocks, since each block is equally represented in the total. Hence, an estimate of block effect freed from varietal differences is given by,

$$485.6 - 521.1 = -35.5$$

Later, these values are to be used to calculate the adjustments necessary to evaluate the varietal means. In making these adjustments, the block effects must clearly be *subtracted*, since a variety which occurs in a set of good blocks should have its mean yield reduced to make it comparable to the other variety means. However, since it is more convenient in computation to *add* the block effect correction terms, we compute their negative values, designated in table 4 as $rkc_X$ and $rkc_Y$, $r$ and $k$ being respectively the number of replications and number of plots per block. The reversed signs of the differences do not affect the sum of squares.

The $rkc_X$ and $rkc_Y$ values may be calculated in either of two ways,

$rkc_X$ = column total of group Y — row total of group X
or $rkc_X$ = row total of table 3 — 2(row total of group X)
$rkc_Y$ = column total of group X — row total of group Y
or $rkc_Y$ = column total of table 3 — 2(row total of group Y)

As an illustration, the first row total (table 3) is 1006.7 and the first row total of group X (table 2) is 485.6. The first $rkc_X$ value is, therefore,

$$1006.7 - (2)(485.6) = 35.5$$

The sum of the eighteen $rkc_X$ and $rkc_Y$ values is equal to 0.

TABLE 4.  rkc VALUES.

| rkc$_x$ | rkc$_Y$ |
|---|---|
| 35.5 | —10.3 |
| 44.7 | — 8.6 |
| 58.6 | —52.4 |
| 25.4 | 11.5 |
| 12.6 | —21.2 |
| 63.9 | —22.5 |
| 5.6 | —72.9 |
| 29.5 | —63.8 |
| —22.8 | —12.8 |
| 253.0 | —253.0 |

The sums of squares of the deviations of these two sets of *rkc* values give an estimate of the variance between blocks (eliminating varieties). The divisors are *rk* (36) and $rk^2$ (324).

The sum of squares for component (b) is, therefore,

$$\frac{(35.5)^2 + (44.7)^2 + \ldots + (12.8)^2}{36} - \frac{(253.0)^2 + (253.0)^2}{324} =$$

$$737.87 - 395.12 = 342.75$$

The foregoing results are brought together in table 5, the analysis of variance of the lattice experiment.

TABLE 5.  ANALYSIS OF VARIANCE OF LATTICE EXPERIMENT.

| Source of variation | Degrees of freedom | Sum of squares | Mean square |
|---|---|---|---|
| Replications | 3 | 2375.58 | 791.860 |
| Component (a) | 16 | 964.63 | |
| Component (b) | 16 | 342.75 | |
| Blocks (eliminating varieties) | 32 | 1307.38 | 40.856 |
| Varieties (ignoring blocks) | 80 | 2154.67 | |
| Error (intra-block) | 208 | 871.27 | 4.189 |
| Total | 323 | 6708.90 | |

TEST OF SIGNIFICANCE.

If there is any question about the significance of the differences among the variety means, a test should be made. The test appropriate to the inter-block analysis is given in the mathematical theory for lattice designs, section 6. However, a less precise test, easily made, is usually adequate. The lattice experiment is treated as a randomized complete block design, the variation among the smaller, incomplete blocks being included in experimental error. Clearly, if the varieties are adjudged significantly different with this estimate of error, the testing need be carried no further. Table 5 contains all the data needed for the analysis of variance given in table 6; the sums of squares and degrees of freedom for blocks and error are added, and the sums of squares and degrees of freedom for replications and varieties remain the same.

TABLE 6. ANALYSIS OF VARIANCE AS RANDOMIZED COMPLETE BLOCKS.

| Source of variation | Degrees of freedom | Sum of squares | Mean square |
|---|---|---|---|
| Replications | 3 | 2375.58 | 791.860 |
| Varieties (ignoring blocks) | 80 | 2154.67 | 26.933 |
| Error | 240 | 2178.65 | 9.078 |
| Total | 323 | 6708.90 | |

The test of significance of variety differences:
$$F = 26.933/9.078 = 2.97**$$

Since the value of F indicates highly significant differences among the mean yields of the varieties in this experiment, no further test is necessary. It is proper to interpret the differences among the variety yields as due in part to some inherent qualities of the varieties. If the yields of the varieties in an experiment do not give a significant F-value, caution should be used before claiming significance for individual comparisons.

### VARIETY MEANS.

In the method of analysis being presented, the variety means are calculated by using both the inter- and intra-block variance. To obtain the variety mean yields by the old method, two sets of corrections, $c_X$ and $c_Y$, were calculated. Having recovered the inter-block information by the new method of analysis, better estimates of variety means are secured by weighting $c_X$ and $c_Y$ giving $c_X'$ and $c_Y'$. These corrections are added to the averages computed from table 3 to get the variety means free from block effects.

The weighting factor is $\dfrac{w - w'}{w + w'}$, where

$$w = 1/E, \quad w' = 3/(4B - E)$$

E, B[6] being respectively the error and block mean squares in table 5. This expression is true only for four replications, the formulas for two and six replications being given in the mathematical theory of lattice designs, section 2. For the present example,

$$w = \frac{1}{E} = \frac{1}{4.189} = .23872$$

$$\text{and } w' = \frac{1}{4B - E} = \frac{3}{4(40.856) - 4.189} = .01884$$

---

[6] If B is less than or equal to E, it is assumed that there are no real differences between blocks and no adjustments for blocks are necessary. In this case the averages in table 7 are the correct variety means.

The value of the weighting factor is

$$\frac{w - w'}{w + w'} = \frac{.21988}{.25756} = .8537$$

The $rkc_X$ and $rkc_Y$ values given in table 4 are weighted to secure the $c_X'$ and $c_Y'$ corrections.

$$c_X' = \frac{w - w'}{rk(w + w')}\ rkc_X = .023714\ rkc_X$$

$$c_Y' = .023714\ rkc_Y$$

The first $c_X'$ is

$$(.023714)(35.5) = .8418$$

In table 7 are given the averages corresponding to the variety totals of table 3. That is, each cell total of table 3 is divided by 4 and the resulting value is entered in table 7. For example, $(94.2)/4 = 23.550$.

The $c_X'$ values are placed at the side of table 7 and the $c_Y'$ values at the bottom. Each unconfounded variety mean is secured by adding to the variety average the proper $c_X'$ and $c_Y'$ values. Thus, the mean for variety 1 is $23.550 + 0.8418 — 0.2443 = 24.15$. The variety means are given in table 8.

### STANDARD ERRORS OF THE DIFFERENCES BETWEEN VARIETY MEANS.

The intra-block error mean square in table 5 is the estimate of the uncontrolled variance ($s^2$) of a single plot. Using $s^2$, we can find out which of the differences between variety yield values are large enough to be regarded as real.

1. The standard error of the differences between the means of two varieties occurring in the same block is

$$\sqrt{\frac{2s^2}{rk}\left[\frac{2w}{w + w'} + (k - 1)\right]} = \sqrt{\frac{(2)(4.189)}{(4)(9)}\left[\frac{(2)(.23872)}{.23872 + .01884} + (9 - 1)\right]}$$

$$= \sqrt{2.2930} = 1.51$$

2. Similarly, the standard error of the differences between the means of two varieties which do not occur in the same block is

$$\sqrt{\frac{2s^2}{rk}\left[\frac{4w}{w + w'} + (k - 2)\right]} = \sqrt{\frac{(2)(4.189)}{(4)(9)}\left[\frac{(4)(.23872)}{.23872 + .01884} + (9 - 2)\right]}$$

$$= \sqrt{2.4917} = 1.58$$

## TABLE 7. AVERAGES AND c' VALUES.

| | | | | | | | | | cx' |
|---|---|---|---|---|---|---|---|---|---|
| 1:23.550 | 2:26.250 | 3:27.825 | 4:27.375 | 5:28.975 | 6:30.300 | 7:28.175 | 8:31.225 | 9:28.000 | 0.8418 |
| 10:29.250 | 11:27.225 | 12:31.050 | 13:29.175 | 14:30.500 | 15:27.925 | 16:31.700 | 17:27.700 | 18:30.100 | 1.0600 |
| 19:28.225 | 20:27.750 | 21:29.075 | 22:28.450 | 23:29.200 | 24:31.075 | 25:32.875 | 26:29.275 | 27:30.775 | 1.3896 |
| 28:32.300 | 29:31.925 | 30:32.275 | 31:29.375 | 32:32.200 | 33:29.600 | 34:32.200 | 35:33.150 | 36:26.925 | 0.6023 |
| 37:28.300 | 38:27.850 | 39:29.000 | 40:29.150 | 41:29.550 | 42:28.375 | 43:29.425 | 44:31.450 | 45:28.550 | 0.2988 |
| 46:25.900 | 47:28.575 | 48:29.625 | 49:24.925 | 50:33.375 | 51:28.225 | 52:29.625 | 53:28.525 | 54:30.500 | 1.5153 |
| 55:31.500 | 56:32.125 | 57:32.775 | 58:33.475 | 59:31.475 | 60:30.325 | 61:34.750 | 62:36.600 | 63:33.175 | 0.1328 |
| 64:33.875 | 65:31.550 | 66:34.275 | 67:31.300 | 68:28.975 | 69:32.900 | 70:34.100 | 71:32.400 | 72:29.200 | 0.6996 |
| 73:33.925 | 74:35.000 | 75:32.150 | 76:27.800 | 77:29.050 | 78:31.450 | 79:37.875 | 80:32.625 | 81:33.025 | −0.5407 |
| cy' −0.2443 | −0.2039 | −1.2426 | 0.2727 | −0.5027 | −0.5336 | −1.7288 | −1.5130 | −0.3035 | |

## TABLE 8. VARIETY MEANS.

(pounds per plot)

| | | | | | | | | |
|---|---|---|---|---|---|---|---|---|
| 1:24.15 | 2:26.89 | 3:27.42 | 4:28.49 | 5:29.31 | 6:30.61 | 7:27.29 | 8:30.55 | 9:28.54 |
| 10:30.07 | 11:28.08 | 12:30.87 | 13:30.51 | 14:31.06 | 15:28.45 | 16:31.03 | 17:27.25 | 18:30.86 |
| 19:29.37 | 20:28.94 | 21:29.22 | 22:30.11 | 23:30.09 | 24:31.93 | 25:32.54 | 26:29.15 | 27:31.86 |
| 28:32.66 | 29:32.32 | 30:31.63 | 31:30.25 | 32:32.30 | 33:29.67 | 34:31.07 | 35:32.24 | 36:27.22 |
| 37:28.35 | 38:27.94 | 39:28.06 | 40:29.72 | 41:29.35 | 42:28.14 | 43:28.00 | 44:30.24 | 45:28.55 |
| 46:27.17 | 47:29.89 | 48:29.90 | 49:26.71 | 50:34.39 | 51:29.21 | 52:29.41 | 53:28.53 | 54:31.71 |
| 55:31.39 | 56:32.05 | 57:31.67 | 58:33.88 | 59:31.17 | 60:29.92 | 61:33.15 | 62:35.22 | 63:33.00 |
| 64:34.33 | 65:32.05 | 66:33.73 | 67:32.27 | 68:29.17 | 69:33.07 | 70:33.07 | 71:31.59 | 72:29.60 |
| 73:33.14 | 74:34.26 | 75:30.37 | 76:27.53 | 77:28.01 | 78:30.38 | 79:35.61 | 80:30.57 | 81:32.18 |

3. The mean standard error of all comparisons is

$$\sqrt{\frac{2s^2}{r(k+1)}\left[\frac{4w}{w+w'}+(k-1)\right]} =$$

$$\sqrt{\frac{(2)(4.189)}{(4)(10)}\left[\frac{(4)(.23872)}{.23872+.01884}+(9-1)\right]} = \sqrt{2.4522} = 1.57$$

Ordinarily, this latter variance may be used for all comparisons without appreciable error.

Variety 8 with a yield of 30.55 lbs. per plot is a standard variety. To test the significance of the difference between the means of varieties 8 and 81 which do not occur together in an incomplete block,

$$\overline{d} = 32.18 - 30.55 = 1.63$$

$$t = \frac{1.63}{1.58} = 1.03,$$

not a significant difference.

### PRECISION OF LATTICE EXPERIMENT.

(a)   Without recovery of inter-block information.

The error variance which resulted by analyzing our example as a randomized complete block arrangement was 9.078 (table 6), whereas the intra-block error variance was 4.189 (table 5). These figures cannot be directly compared to assess the relative accuracy of the two types of designs, since some replication was sacrificed in the later to obtain the reduction of block size. We may, however, legitimately compare the average variance of the difference between two varietal means for the two designs. For the randomized complete block design in four replications, this is $2v/r = 2(9.078)/4$, and for the lattice design, it is $\left(\frac{2}{4}\right)\frac{(k+3)}{(k+1)}$ (4.189). An explanation of this factor $(k+3)/(k+1)$ is given by Yates [12]. The ratio of these two variances is

$$\frac{(2/4)(9.078)}{(2/4)(12/10)(4.189)} = \frac{4.5390}{2.5134} = 181 \text{ percent,}$$

an 81 percent gain in precision.

(b)   With recovery of inter-block information.

The average variance of the difference between two varietal means for the lattice design given in 3 of the preceding section is 2.4522. For the randomized complete block design in four replications, this is 4.5390. Assuming the precision obtainable in the complete block design as 100 percent, $\dfrac{4.5390}{2.4522} = 185$ percent was obtained with the lattice design, representing a gain of 85 percent.

The reduction of block size from 81 to 9 plots per block has resulted in a notable increase in precision, a gain of 81 percent when inter-block information is ignored and a gain of 85 percent with the recovery of inter-block information. So small a percent increase obtained by using inter-block variance is due to the fact that in this experiment the variation between blocks is very large.

## TRIPLE LATTICE DESIGNS[7]

If a group (Z) is added to the groups X and Y of the lattice design, this new design is called a triple lattice. When $k$ is a prime number or a power of a prime, this process of adding more orthogonal groups can be continued until all possible groups are used, the result being a balanced set of groups [10, 14].

If the number of varieties is a perfect square, it is always possible to superimpose a Latin square arrangement on the square of variety numbers. Given below is a Latin square for the nine Latin letters, A, B, . . . , I.

```
A I H G F E D C B
B A I H G F E D C
C B A I H G F E D
D C B A I H G F E
E D C B A I H G F
F E D C B A I H G
G F E D C B A I H
H G F E D C B A I
I H G F E D C B A
```

Superimpose this Latin square on the square of variety numbers which was used when constructing the lattice design. Put the numbers corresponding to the Latin letter A in the first of a new group of blocks, and the numbers corresponding to B in the second. This process gives the nine blocks of group Z.

| Block | | Group Z | | | | | | | | |
|---|---|---|---|---|---|---|---|---|---|---|
| | (a) | 1 | 11 | 21 | 31 | 41 | 51 | 61 | 71 | 81 |
| | (b) | 10 | 20 | 30 | 40 | 50 | 60 | 70 | 80 | 9 |
| | (c) | 19 | 29 | 39 | 49 | 59 | 69 | 79 | 8 | 18 |
| | (d) | 28 | 38 | 48 | 58 | 68 | 78 | 7 | 17 | 27 |
| | (e) | 37 | 47 | 57 | 67 | 77 | 6 | 16 | 26 | 36 |
| | (f) | 46 | 56 | 66 | 76 | 5 | 15 | 25 | 35 | 45 |
| | (g) | 55 | 65 | 75 | 4 | 14 | 24 | 34 | 44 | 54 |
| | (h) | 64 | 74 | 3 | 13 | 23 | 33 | 43 | 53 | 63 |
| | (i) | 73 | 2 | 12 | 22 | 32 | 42 | 52 | 62 | 72 |

[7] The triple lattice designs have been called two-dimensional pseudo-factorial arrangements in three groups of sets forming a Latin square [12], two-dimensional pseudo-factorial experiments with three groups of sets [4] and two-dimensional quasi-factorial designs in randomized blocks in three equal groups of sets forming a Latin square (triple lattice) [14].

This group Z together with groups X and Y make up the triple-lattice arrangement. The three groups of blocks bear the same orthogonal relationship to one another, every block of each group containing only one variety from every block of each of the other two groups. The randomization is the same as that for the lattice design; that is, the plots in each block, the blocks in each group and the groups are arranged at random. Groups X, Y and Z comprise three replications. This design can be arranged in six or nine replications by using other random arrangements of the three groups. However, if there is sufficient land in one field to allow five replications the balanced arrangement should be used.

### EXAMPLE OF TRIPLE LATTICE EXPERIMENT

The triple lattice design is illustrated by a yield test of 81 double-crosses of corn. These double-crosses had been previously tested for 2 or more years and represent the best of a large number of experimental combinations. In this experiment, groups X, Y and Z together were put in each of two neighboring fields of different productivity, making a total of 54 blocks with 9 plots in each block. Each group of nine blocks was kept together in a complete replication. The plots consisted of 20 hills arranged in 2 rows of 10 hills each. The arrangement in each field (fig. 2), was similar to the one used for the lattice experiment just illustrated, except for the addition of the Z group of blocks.

The yield and number of missing hills for the individual plots and for the blocks are presented in table 9. The table indicates the random order of the varieties in each block. The varieties are identified by the numbers 1, 2, 3, . . . , 81, given in boldface type. The numbers of missing hills are indicated in italics. When there were no missing hills in the individual plots, the space is left blank.

The experimental results for each field were first analyzed separately, giving results which justify using all the data in one analysis. Therefore, the sum of the variety yields and the sum of the number of missing hills from replications 1 and 2 are recorded in order by rows in group X of table 10. In replication 1, block (4) contains varieties 4, 9, 3, 7, 6, 1, 5, 2 and 8. Their yield values and their numbers of missing hills are added to those in block (1) of replication 2, which contains the same varieties but in another random order. For variety 1 the yield is 19.5 in replication 1 and 26.4 in replication 2, giving 45.9, the first value in group X of table 10. The sums of replications 3 and 4 are recorded by columns in group Y. Replications 5 and 6 give group Z. The row totals, which are the sums of two block totals, are given at the right-hand side for each group. The sums of the variety yields and the sums of the number of missing hills for all

## TABLE 9. PLOT YIELDS AND MISSING HILLS OF TRIPLE LATTICE EXPERIMENT.
### (Variety identification in boldface type.)

**Replication 1 (Group X).**

| Block | | | | | | | | | | Block totals | Missing hills Yield |
|---|---|---|---|---|---|---|---|---|---|---|---|
| (1) | **27** | **20** | **22** | **26** | **25-2** | **21-1** | **23** | **24** | **19** | *3* | Yield |
|  | 27.6 | 24.3 | 24.9 | 25.9 | 24.4 | 22.5 | 24.5 | 22.9 | 24.4 | 221.4 | |
| (2) | **67** | **64** | **72** | **71** | **65** | **68** | **70** | **66** | **69** | *0* | |
|  | 28.8 | 29.6 | 25.2 | 26.6 | 27.9 | 24.0 | 26.8 | 26.8 | 25.2 | 240.9 | |
| (3) | **56** | **59** | **62** | **61** | **58** | **55** | **63** | **57** | **60** | *0* | |
|  | 27.0 | 28.4 | 30.9 | 31.4 | 33.7 | 29.0 | 28.7 | 29.1 | 27.0 | 265.2 | |
| (4) | **4-1** | **9** | **3** | **7** | **6** | **1** | **5** | **2** | **8** | *1* | |
|  | 23.3 | 21.2 | 24.1 | 24.5 | 25.8 | 19.5 | 24.0 | 23.3 | 27.1 | 212.8 | |
| (5) | **53** | **51** | **50** | **47** | **46** | **48** | **49-1** | **54** | **52** | *1* | |
|  | 27.7 | 27.3 | 36.0 | 31.3 | 26.8 | 30.5 | 25.2 | 27.8 | 24.2 | 256.8 | |
| (6) | **78** | **73** | **80** | **75** | **74** | **77** | **79** | **81** | **76-3** | *3* | |
|  | 31.1 | 33.7 | 31.7 | 29.2 | 31.3 | 26.8 | 30.8 | 31.6 | 22.6 | 268.8 | |
| (7) | **15** | **18** | **17-1** | **14-1** | **11** | **12** | **10** | **16-1** | **13** | *3* | |
|  | 28.7 | 28.2 | 25.5 | 28.5 | 26.5 | 26.4 | 27.3 | 25.2 | 26.2 | 242.5 | |
| (8) | **28** | **35** | **36** | **29** | **33** | **34** | **30** | **31** | **32** | *0* | |
|  | 29.4 | 26.3 | 23.3 | 30.0 | 25.5 | 24.3 | 25.7 | 24.8 | 28.1 | 237.4 | |
| (9) | **41** | **38** | **42** | **37** | **40·** | **44** | **39** | **43** | **45** | *0* | *11* |
|  | 25.0 | 25.4 | 25.3 | 24.1 | 25.0 | 24.8 | 24.3 | 25.7 | 23.9 | 223.5 | 2169.3 |

**Replication 2 (Group X)**

| Block | | | | | | | | | | Block totals | Missing hills Yield |
|---|---|---|---|---|---|---|---|---|---|---|---|
| (1) | **1** | **8** | **9** | **2** | **5** | **3-2** | **6** | **4** | **7** | *2* | |
|  | 26.4 | 32.8 | 32.2 | 31.0 | 34.9 | 26.3 | 32.2 | 28.0 | 27.8 | 271.6 | |
| (2) | **38** | **45-1** | **37-1** | **44** | **43** | **40** | **42** | **41** | **39** | *2* | |
|  | 29.6 | 31.6 | 32.2 | 33.0 | 33.0 | 33.5 | 31.3 | 32.5 | 35.8 | 292.5 | |
| (3) | **66** | **68-1** | **71-1** | **69** | **65-3** | **64-1** | **72-2** | **67** | **70** | *8* | |
|  | 36.7 | 31.5 | 33.7 | 38.6 | 33.8 | 37.6 | 30.7 | 36.3 | 38.5 | 317.4 | |
| (4) | **75-1** | **80-1** | **78** | **79** | **74** | **81-2** | **77-2** | **73-2** | **76** | *8* | |
|  | 36.6 | 33.6 | 36.1 | 41.1 | 39.4 | 36.2 | 31.8 | 35.1 | 33.1 | 323.0 | |
| (5) | **34** | **32** | **28** | **36-1** | **33** | **31** | **30** | **35-1** | **29** | *2* | |
|  | 34.0 | 36.6 | 36.6 | 30.1 | 31.3 | 34.8 | 36.6 | 35.0 | 33.8 | 308.8 | |
| (6) | **55** | **58** | **56** | **57** | **60** | **61** | **62** | **59-1** | **63** | *1* | |
|  | 33.7 | 37.0 | 36.8 | 35.0 | 33.5 | 35.5 | 38.7 | 34.6 | 39.1 | 323.9 | |
| (7) | **16-2** | **11-5** | **18-6** | **14-3** | **13-5** | **12** | **10** | **17** | **15** | *21* | |
|  | 33.0 | 23.4 | 28.1 | 28.4 | 29.5 | 29.7 | 29.3 | 25.0 | 27.3 | 253.7 | |
| (8) | **47-1** | **50** | **53** | **46** | **48** | **54-1** | **51-1** | **49-3** | **52-2** | *8* | |
|  | 25.1 | 26.4 | 24.0 | 22.7 | 23.9 | 27.9 | 26.5 | 22.5 | 27.3 | 226.3 | |
| (9) | **21** | **20** | **26** | **24** | **22** | **27** | **19-2** | **25** | **23-2** | *4* | *56* |
|  | 29.0 | 28.4 | 29.8 | 33.1 | 33.2 | 34.5 | 28.0 | 34.4 | 29.4 | 279.8 | 2597.0 |

**Replication 3 (Group Y)**

| Block | | | | | | | | | | Block totals | Missing hills Yield |
|---|---|---|---|---|---|---|---|---|---|---|---|
| (1) | **69** | **15-3** | **78-3** | **33** | **51** | **24** | **60-1** | **42-1** | **6** | *8* | |
|  | 29.4 | 23.8 | 25.0 | 28.2 | 24.5 | 27.9 | 23.9 | 21.5 | 28.4 | 232.6 | |
| (2) | **58** | **22** | **13** | **76-2** | **31-1** | **67** | **4** | **49** | **40** | *3* | |
|  | 31.9 | 27.7 | 29.2 | 24.5 | 29.4 | 32.6 | 28.5 | 26.2 | 29.6 | 259.7 | |
| (3) | **29** | **11** | **20** | **2** | **47** | **74** | **56-3** | **65-4** | **38-1** | *8* | |
|  | 30.5 | 27.3 | 28.3 | 25.6 | 27.9 | 30.0 | 29.8 | 28.5 | 25.2 | 253.1 | |
| (4) | **23** | **77** | **5** | **41** | **14** | **68** | **32** | **50** | **59-1** | *1* | |
|  | 28.1 | 26.5 | 25.2 | 26.2 | 27.7 | 27.9 | 27.7 | 29.5 | 28.0 | 246.8 | |
| (5) | **3** | **75-1** | **21** | **12** | **66** | **48-3** | **39** | **57-1** | **30** | *5* | |
|  | 25.6 | 27.9 | 28.9 | 28.9 | 33.1 | 28.5 | 29.3 | 32.4 | 30.0 | 264.6 | |
| (6) | **36** | **63** | **54** | **81** | **45-3** | **72** | **27** | **9** | **18** | *3* | |
|  | 24.4 | 26.9 | 27.8 | 27.3 | 21.3 | 27.9 | 26.6 | 25.7 | 26.6 | 234.5 | |
| (7) | **64-3** | **73** | **10** | **37** | **1** | **46** | **55-1** | **28-1** | **19** | *5* | |
|  | 26.2 | 29.0 | 29.5 | 24.4 | 20.0 | 24.5 | 28.3 | 24.6 | 24.8 | 231.3 | |
| (8) | **26-1** | **53-2** | **35** | **80** | **44** | **17-4** | **8** | **71** | **62** | *7* | |
|  | 26.2 | 28.1 | 34.5 | 32.2 | 33.9 | 26.6 | 32.4 | 33.8 | 38.1 | 285.8 | |
| (9) | **70-2** | **61** | **25** | **43** | **34** | **16** | **52** | **7** | **79** | *2* | *42* |
|  | 29.7 | 36.0 | 36.4 | 33.6 | 33.1 | 33.8 | 32.1 | 30.4 | 39.9 | 305.0 | 2313.4 |

## TABLE 9. Continued.

### Replication 4 (Group Y)

| Block | | | | | | | | | | Block totals | Missing hills Yield |
|---|---|---|---|---|---|---|---|---|---|---|---|
| (1) | 76-2<br>28.2 | 13<br>29.4 | 40<br>28.5 | 4<br>29.3 | 49<br>24.3 | 67<br>27.5 | 58<br>31.3 | 31<br>28.0 | 22<br>28.0 | 254.5 | *2* |
| (2) | 2-4<br>23.6 | 11<br>29.8 | 47-1<br>29.2 | 65<br>32.6 | 56<br>33.5 | 38-1<br>30.3 | 29-3<br>31.9 | 20-4<br>28.2 | 74<br>39.3 | 278.4 | *13* |
| (3) | 6<br>34.8 | 42<br>35.1 | 69<br>38.4 | 51<br>34.2 | 33<br>33.4 | 24<br>40.4 | 15<br>30.8 | 60<br>36.5 | 78-1<br>31.9 | 315.5 | *1* |
| (4) | 25<br>35.5 | 70<br>40.5 | 43-3<br>24.3 | 61<br>36.1 | 34<br>37.4 | 7<br>30.0 | 52<br>34.1 | 16-2<br>32.3 | 79<br>39.7 | 309.9 | *5* |
| (5) | 3-1<br>34.0 | 21-1<br>35.1 | 30-1<br>36.2 | 48-4<br>32.2 | 39-7<br>23.8 | 12-1<br>38.6 | 57-4<br>32.1 | 66-3<br>38.7 | 75-3<br>32.4 | 303.1 | *25* |
| (6) | 5-1<br>31.3 | 77-2<br>29.2 | 59-3<br>32.5 | 32-2<br>35.3 | 68<br>32.0 | 23-1<br>33.4 | 41<br>34.5 | 50<br>41.6 | 14<br>35.7 | 305.5 | *9* |
| (7) | 62-1<br>38.1 | 71<br>35.0 | 80<br>32.5 | 26<br>34.8 | 44<br>34.1 | 35-2<br>35.2 | 53<br>33.4 | 17<br>31.6 | 8-1<br>32.1 | 306.8 | *4* |
| (8) | 46-2<br>28.7 | 10-4<br>29.0 | 28-2<br>37.1 | 55-5<br>32.0 | 37<br>32.0 | 19<br>34.8 | 1-1<br>27.9 | 64<br>40.3 | 73<br>36.8 | 298.6 | *14* |
| (9) | 45<br>35.9 | 54-1<br>37.5 | 72-1<br>31.6 | 81<br>35.9 | 27<br>34.4 | 9<br>32.9 | 63<br>38.0 | 18<br>34.7 | 36-1<br>29.0 | 309.9 | *3*    *76*  2682.2 |

### Replication 5 (Group Z).

| Block | | | | | | | | | | Block totals | Missing hills |
|---|---|---|---|---|---|---|---|---|---|---|---|
| (1) | 55<br>39.1 | 34<br>32.5 | 75<br>32.2 | 65<br>36.0 | 24<br>35.1 | 4<br>32.0 | 54<br>35.3 | 44<br>33.5 | 14<br>31.0 | 306.7 | *0* |
| (2) | 29-1<br>31.0 | 18<br>30.6 | 59<br>32.9 | 39<br>29.6 | 49<br>29.8 | 79<br>34.6 | 69<br>31.6 | 8<br>28.0 | 19-1<br>29.3 | 277.4 | *2* |
| (3) | 11<br>31.7 | 81<br>30.2 | 1<br>23.4 | 31<br>31.7 | 21<br>29.4 | 61<br>32.2 | 51<br>28.4 | 71-3<br>28.9 | 41-1<br>27.9 | 263.8 | *4* |
| (4) | 78<br>30.6 | 68<br>28.1 | 58<br>30.0 | 48<br>29.0 | 27<br>27.9 | 7<br>24.3 | 38<br>25.1 | 28<br>29.5 | 17<br>27.6 | 252.1 | *0* |
| (5) | 35<br>27.0 | 66<br>30.2 | 15<br>25.9 | 5-2<br>23.5 | 45-3<br>22.3 | 46-3<br>23.2 | 76-1<br>23.8 | 25<br>29.0 | 56<br>26.9 | 231.8 | *9* |
| (6) | 43<br>23.8 | 63<br>27.0 | 13<br>20.7 | 3<br>20.1 | 74<br>22.8 | 23<br>27.2 | 33<br>24.9 | 53<br>30.1 | 64<br>34.0 | 230.6 | *0* |
| (7) | 60<br>27.6 | 40<br>27.4 | 80<br>27.7 | 10<br>25.9 | 20<br>25.0 | 70-1<br>28.3 | 9<br>24.8 | 50<br>31.4 | 30<br>25.0 | 243.1 | *1* |
| (8) | 67<br>28.4 | 6<br>23.8 | 36<br>25.7 | 47<br>29.2 | 16<br>24.9 | 26<br>30.2 | 37<br>27.1 | 77<br>29.7 | 57<br>30.4 | 249.4 | *0* |
| (9) | 22<br>27.4 | 12<br>28.9 | 42<br>26.7 | 52<br>27.9 | 32<br>29.0 | 72<br>30.4 | 73<br>30.4 | 2-1<br>19.4 | 62<br>31.4 | 251.5 | *1*    *17*  2306.4 |

### Replication 6 (Group Z)

| Block | | | | | | | | | | Block totals | Missing hills |
|---|---|---|---|---|---|---|---|---|---|---|---|
| (1) | 31<br>34.2 | 1<br>30.0 | 81<br>33.3 | 41<br>30.6 | 61<br>34.2 | 11<br>32.4 | 51<br>34.2 | 71<br>38.5 | 21<br>33.5 | 300.9 | *0* |
| (2) | 74<br>40.0 | 23<br>35.0 | 13-2<br>34.9 | 33-2<br>31.1 | 53-2<br>31.5 | 3<br>31.5 | 43<br>36.1 | 64<br>39.7 | 63-4<br>32.6 | 312.4 | *10* |
| (3) | 49<br>31.9 | 29<br>38.2 | 69<br>35.2 | 59<br>37.5 | 39<br>31.8 | 79<br>39.5 | 8<br>32.5 | 19<br>32.4 | 18<br>34.5 | 313.5 | *0* |
| (4) | 77<br>33.4 | 16<br>33.4 | 37<br>32.5 | 67<br>36.2 | 36<br>31.5 | 47<br>31.4 | 26<br>36.4 | 57<br>38.0 | 6<br>29.7 | 302.5 | *0* |
| (5) | 34<br>32.4 | 55<br>37.6 | 44<br>33.6 | 75<br>32.0 | 14<br>32.0 | 24<br>33.4 | 54<br>34.5 | 4<br>28.0 | 65<br>36.1 | 299.6 | *0* |
| (6) | 22-2<br>29.9 | 72-2<br>29.6 | 62-2<br>32.4 | 12-2<br>30.0 | 2-1<br>27.5 | 32<br>32.5 | 73<br>37.8 | 52<br>28.3 | 42<br>31.5 | 279.5 | *9* |
| (7) | 20<br>32.1 | 40<br>31.0 | 9<br>26.0 | 70<br>32.0 | 80-3<br>20.4 | 60<br>17.5 | 50<br>33.0 | 30<br>28.6 | 10<br>31.5 | 252.1 | *3* |
| (8) | 25<br>30.1 | 45<br>30.5 | 76<br>33.0 | 35<br>32.5 | 66<br>35.0 | 56<br>35.4 | 5<br>32.0 | 15-1<br>31.3 | 46<br>26.8 | 286.6 | *1* |
| (9) | 27<br>35.6 | 28<br>33.5 | 68<br>33.0 | 78<br>32.5 | 17<br>31.4 | 58<br>37.7 | 7<br>33.1 | 38<br>35.0 | 48<br>35.5 | 307.3 | *0*    *23*  2654.4 |

## TABLE 10.  COMBINATION OF REPLICATIONS.

### Group X (Replications 1 and 2)

| 1 | 2 | 3 | 4 | 5 | 6 | 7 | 8 | 9 | Row totals |  |
|---|---|---|---|---|---|---|---|---|---|---|
| **1** 45.9 | **2** 54.3 | **3-2** 50.4 | **4-1** 51.3 | **5** 58.9 | **6** 58.0 | **7** 52.3 | **8** 59.9 | **9** 53.4 | *3* 484.4 |  |
| **10** 56.6 | **11-5** 49.9 | **12** 56.1 | **13-5** 55.7 | **14-4** 56.9 | **15** 56.0 | **16-3** 58.2 | **17-1** 50.5 | **18-6** 56.3 | *24* 496.2 |  |
| **19-2** 52.4 | **20** 52.7 | **21-1** 51.5 | **22** 58.1 | **23-2** 53.9 | **24** 56.0 | **25-2** 58.8 | **26** 55.7 | **27** 62.1 | *7* 501.2 |  |
| **28** 66.0 | **29** 63.8 | **30** 62.3 | **31** 59.6 | **32** 64.7 | **33** 56.8 | **34** 58.3 | **35-1** 61.3 | **36-1** 53.4 | *2* 546.2 |  |
| **37-1** 56.3 | **38** 55.0 | **39** 60.1 | **40** 58.5 | **41** 57.5 | **42** 56.6 | **43** 58.7 | **44** 57.8 | **45-1** 55.5 | *2* 516.0 |  |
| **46** 49.5 | **47-1** 56.4 | **48** 54.4 | **49-4** 47.7 | **50** 62.4 | **51-1** 53.8 | **52-2** 51.5 | **53** 51.7 | **54-1** 55.7 | *9* 483.1 |  |
| **55** 62.7 | **56** 63.8 | **57** 64.1 | **58** 70.7 | **59-1** 63.0 | **60** 60.5 | **61** 66.9 | **62** 69.6 | **63** 67.8 | *1* 589.1 |  |
| **64-1** 67.2 | **65-3** 61.7 | **66** 63.5 | **67** 65.1 | **68-1** 55.5 | **69** 63.8 | **70** 65.3 | **71-1** 60.3 | **72-2** 55.9 | *8* 558.3 |  |
| **73-2** 68.8 | **74** 70.7 | **75-1** 65.8 | **76-3** 55.7 | **77-2** 58.6 | **78** 67.2 | **79** 71.9 | **80-1** 65.3 | **81-2** 67.8 | *11* 591.8 | *67* 4766.3 |

### Group Y (Replications 3 and 4).

| 1 | 2 | 3 | 4 | 5 | 6 | 7 | 8 | 9 | Row totals |  |
|---|---|---|---|---|---|---|---|---|---|---|
| **1-1** 47.9 | **10-4** 58.5 | **19** 59.6 | **28-3** 61.7 | **37** 56.4 | **46-2** 53.2 | **55-6** 60.3 | **64-3** 66.5 | **73** 65.8 | *19* 529.9 |  |
| **2-4** 49.2 | **11** 57.1 | **20-4** 56.5 | **29-3** 62.4 | **38-2** 55.5 | **47-1** 57.1 | **56-3** 63.3 | **65-4** 61.1 | **74** 69.3 | *21* 531.5 |  |
| **3-1** 59.6 | **12-1** 67.5 | **21-1** 64.0 | **30-1** 66.2 | **39-7** 53.1 | **48-7** 60.7 | **57-5** 64.5 | **66-3** 71.8 | **75-4** 60.3 | *30* 567.7 |  |
| **4** 57.8 | **13** 58.6 | **22** 55.7 | **31-1** 57.5 | **40** 58.1 | **49** 50.5 | **58** 63.2 | **67** 60.1 | **76-4** 52.7 | *5* 514.2 |  |
| **5-1** 56.5 | **14** 63.4 | **23-1** 61.5 | **32-2** 63.0 | **41** 60.7 | **50** 71.1 | **59-4** 60.5 | **68** 59.9 | **77-2** 55.7 | *10* 552.3 |  |
| **6** 63.2 | **15-3** 54.6 | **24** 68.3 | **33** 61.6 | **42-1** 56.6 | **51** 58.7 | **60-1** 60.4 | **69** 67.8 | **78-4** 56.9 | *9* 548.1 |  |
| **7** 60.4 | **16-2** 66.1 | **25** 71.9 | **34** 70.5 | **43-3** 57.9 | **52** 66.2 | **61** 72.1 | **70-2** 70.2 | **79** 79.6 | *7* 614.9 |  |
| **8-1** 64.5 | **17-4** 58.2 | **26-1** 61.0 | **35-2** 69.7 | **44** 68.0 | **53-2** 61.5 | **62-1** 76.2 | **71** 68.8 | **80** 64.7 | *11* 592.6 |  |
| **9** 58.6 | **18** 61.3 | **27** 61.0 | **36-1** 53.4 | **45-3** 57.2 | **54-1** 65.3 | **63** 64.9 | **72-1** 59.5 | **81** 63.2 | *6* 544.4 | *118* 4995.6 |

### Group Z (Replications 5 and 6)

| 1 | 2 | 3 | 4 | 5 | 6 | 7 | 8 | 9 | Row totals |  |
|---|---|---|---|---|---|---|---|---|---|---|
| **1** 53.4 | **11** 64.1 | **21** 62.9 | **31** 65.9 | **41-1** 58.5 | **51** 62.6 | **61** 66.4 | **71-3** 67.4 | **81** 63.5 | *4* 564.7 |  |
| **10** 57.4 | **20** 57.1 | **30** 53.6 | **40** 58.4 | **50** 64.4 | **60** 45.1 | **70-1** 60.3 | **80-3** 48.1 | **9** 50.8 | *4* 495.2 |  |
| **19-1** 61.7 | **29-1** 69.2 | **39** 61.4 | **49** 61.7 | **59** 70.4 | **69** 66.8 | **79** 74.1 | **8** 60.5 | **18** 65.1 | *2* 590.9 |  |
| **28** 63.0 | **38** 60.1 | **48** 64.5 | **58** 67.7 | **68** 61.1 | **78** 63.1 | **7** 57.4 | **17** 59.0 | **27** 63.5 | *0* 559.4 |  |
| **37** 59.6 | **47** 60.6 | **57** 68.4 | **67** 64.6 | **77** 63.1 | **6** 53.5 | **16** 58.3 | **26** 66.6 | **36** 57.2 | *0* 551.9 |  |
| **46-3** 50.0 | **56** 62.3 | **66** 65.2 | **76-1** 56.8 | **5-2** 55.5 | **15-1** 57.2 | **25** 59.1 | **35** 59.5 | **45-3** 52.8 | *10* 518.4 |  |
| **55** 76.7 | **65** 72.1 | **75** 64.2 | **4** 60.0 | **14** 63.0 | **24** 68.5 | **34** 64.9 | **44** 67.1 | **54** 69.8 | *0* 606.3 |  |
| **64** 73.7 | **74** 62.8 | **3** 51.6 | **13-2** 55.6 | **23** 62.2 | **33-2** 56.0 | **43** 59.9 | **53-2** 61.6 | **63-4** 59.6 | *10* 543.0 |  |
| **73** 68.2 | **2-2** 46.9 | **12-2** 58.9 | **22-2** 57.3 | **32** 61.5 | **42** 58.2 | **52** 56.2 | **62-2** 63.8 | **72-2** 60.0 | *10* 531.0 | *40* 4960.8 |

## TABLE 11. VARIETY TOTAL YIELDS AND TOTAL NUMBER OF MISSING HILLS.

|  |  |  |  |  |  |  |  |  | Row totals |
|---|---|---|---|---|---|---|---|---|---|
| 1-1 | 2-6 | 3-3 | 4-1 | 5-3 | 6 | 7 | 8-1 | 9 | 15 |
| 147.2 | 150.4 | 161.6 | 169.1 | 170.9 | 174.7 | 170.1 | 184.9 | 162.8 | 1491.7 |
| 10-4 | 11-5 | 12-3 | 13-7 | 14-4 | 15-4 | 16-5 | 17-5 | 18-6 | 43 |
| 172.5 | 171.1 | 182.5 | 169.9 | 183.3 | 167.8 | 182.6 | 167.7 | 182.7 | 1580.1 |
| 19-3 | 20-4 | 21-2 | 22-2 | 23-3 | 24 | 25-2 | 26-1 | 27 | 17 |
| 173.7 | 166.3 | 178.4 | 171.1 | 177.6 | 192.8 | 189.8 | 183.3 | 186.6 | 1619.6 |
| 28-3 | 29-4 | 30-1 | 31-1 | 32-2 | 33-2 | 34 | 35-3 | 36-2 | 18 |
| 190.7 | 195.4 | 182.1 | 183.0 | 189.2 | 174.4 | 193.7 | 190.5 | 164.0 | 1663.0 |
| 37-1 | 38-2 | 39-7 | 40 | 41-1 | 42-1 | 43-3 | 44 | 45-7 | 22 |
| 172.3 | 170.6 | 174.6 | 175.0 | 176.7 | 171.4 | 176.5 | 192.9 | 165.5 | 1575.5 |
| 46-5 | 47-2 | 48-7 | 49-4 | 50 | 51-1 | 52-2 | 53-4 | 54-2 | 27 |
| 152.7 | 174.1 | 179.6 | 159.9 | 197.9 | 175.1 | 173.9 | 174.8 | 190.8 | 1578.8 |
| 55-6 | 56-3 | 57-5 | 58 | 59-5 | 60-1 | 61 | 62-3 | 63-4 | 27 |
| 199.7 | 189.4 | 197.0 | 201.6 | 193.9 | 166.0 | 205.4 | 209.6 | 192.3 | 1754.9 |
| 64-4 | 65-7 | 66-3 | 67 | 68-1 | 69 | 70-3 | 71-4 | 72-5 | 27 |
| 207.4 | 194.9 | 200.5 | 189.8 | 176.5 | 198.4 | 195.8 | 196.5 | 175.4 | 1735.2 |
| 73-2 | 74 | 75-5 | 76-8 | 77-4 | 78-4 | 79 | 80-4 | 81-2 | 29 |
| 202.8 | 202.8 | 190.3 | 165.2 | 177.4 | 187.2 | 225.6 | 178.1 | 194.5 | 1723.9 |

Column totals

| 29 | 33 | 36 | 23 | 23 | 13 | 15 | 25 | 28 | 225 |
|---|---|---|---|---|---|---|---|---|---|
| 1619.0 | 1615.0 | 1646.6 | 1584.6 | 1643.4 | 1607.8 | 1713.4 | 1678.3 | 1614.6 | 14722.7 |

|  |  |  |  |  |  |  |  |  | Latin letter totals |
|---|---|---|---|---|---|---|---|---|---|
| 1-1 | 11-5 | 21-2 | 31-1 | 41-1 | 51 | 61 | 71-4 | 81-2 | 17 |
| 147.2 | 171.1 | 178.4 | 183.0 | 176.7 | 175.1 | 205.4 | 196.5 | 194.5 | 1627.9 |
| 10-4 | 20-4 | 30-1 | 40 | 50 | 60-1 | 70-3 | 80-4 | 9 | 17 |
| 172.5 | 166.3 | 182.1 | 175.0 | 197.9 | 166.0 | 195.8 | 178.1 | 162.8 | 1596.5 |
| 19-3 | 29-4 | 39-7 | 49-4 | 59-5 | 69 | 79 | 8-1 | 18-6 | 30 |
| 173.7 | 195.4 | 174.6 | 159.9 | 193.9 | 198.4 | 225.6 | 184.9 | 182.7 | 1689.1 |
| 28-3 | 38-2 | 48-7 | 58 | 68-1 | 78-4 | 7 | 17-5 | 27 | 22 |
| 190.7 | 170.6 | 179.6 | 201.6 | 176.5 | 187.2 | 170.1 | 167.7 | 186.6 | 1630.6 |
| 37-1 | 47-2 | 57-5 | 67 | 77-4 | 6 | 16-5 | 26-1 | 36-2 | 20 |
| 172.3 | 174.1 | 197.0 | 189.8 | 177.4 | 174.7 | 182.6 | 183.3 | 164.0 | 1615.2 |
| 46-5 | 56-3 | 66-3 | 76-8 | 5-3 | 15-4 | 25-2 | 35-3 | 45-7 | 38 |
| 152.7 | 189.4 | 200.5 | 165.2 | 170.9 | 167.8 | 189.8 | 190.5 | 165.5 | 1592.3 |
| 55-6 | 65-7 | 75-5 | 4-1 | 14-4 | 24 | 34 | 44 | 54-2 | 25 |
| 199.7 | 194.9 | 190.3 | 169.1 | 183.3 | 192.8 | 193.7 | 192.9 | 190.8 | 1707.5 |
| 64-4 | 74 | 3-3 | 13-7 | 23-3 | 33-2 | 43-3 | 53-4 | 63-4 | 30 |
| 207.4 | 202.8 | 161.6 | 169.9 | 177.6 | 174.4 | 176.5 | 174.8 | 192.3 | 1637.3 |
| 73-2 | 2-6 | 12-3 | 22-2 | 32-2 | 42-1 | 52-2 | 62-3 | 72-5 | 26 |
| 202.8 | 150.4 | 182.5 | 171.1 | 189.2 | 171.4 | 173.9 | 209.6 | 175.4 | 1626.3 |
|  |  |  |  |  |  |  |  |  | 225 |
|  |  |  |  |  |  |  |  |  | 14722.7 |

six replications appear in table 11. These are set up in two ways in order to give the three sets of marginal totals, those for rows, columns and Latin letters.

An examination of the original field data, as given in table 9, shows that the missing hills appeared in spots. In no case is the damage so severe as to make the data of questionable value, but some allowance should be made for the missing hills, provided these are accidental and not varietal effects. This difficulty may be handled by covariance analysis of number of missing hills on yield or by using on the original plot yields an arbitrary correction factor (suggested by previous experience) for the number of missing hills. The yield values are first analyzed without any correction for missing hills to illustrate the new method of analysis when only a single variable is used. Later, a covariance analysis is given.

CALCULATIONS FOR THE ANALYSIS OF VARIANCE.

The calculations necessary for the analysis of variance and the varietal means are given below.

The correction term is

$$c = \frac{(14,722.7)^2}{486} = 446,003.90$$

The total sum of squares is

$$(27.6)^2 + (24.3)^2 + \ldots + (35.5)^2 - c =$$
$$455,654.05 - 446,003.90 = 9650.15$$

The sum of squares for replications is

$$\frac{(2169.3)^2 + (2597.0)^2 + \ldots + (2654.4)^2}{81} - c =$$
$$448,908.74 - 446,003.90 = 2904.84$$

The sum of squares for varieties (ignoring blocks) is obtained by adding the squares of the sub-totals in table 11, dividing by the number of replications, and subtracting the correction term.

$$\frac{(147.2)^2 + (150.4)^2 + \ldots + (194.5)^2}{6} - c =$$
$$448,786.95 - 446,003.90 = 2,783.05$$

In a triple lattice experiment having six replications, the sum of squares for blocks (eliminating varieties) is made up of two components as follows:

*Component (a)* consists of three sets of differences in yield between paired blocks,

(Yield)

| Block totals Rep. 1 | Rep. 2 | Set X Differ- ences | Block totals Rep. 3 | Rep. 4 | Set Y Differ- ences | Block totals Rep. 5 | Rep. 6 | Set Z Differ- ences |
|---|---|---|---|---|---|---|---|---|
| 221.4 | 279.8 | —58.4 | 232.6 | 315.5 | —82.9 | 306.7 | 299.6 | 7.1 |
| 240.9 | 317.4 | —76.5 | 259.7 | 254.5 | 5.2 | 277.4 | 313.5 | —36.1 |
| 265.2 | 323.9 | —58.7 | 253.1 | 278.4 | —25.3 | 263.8 | 300.9 | —37.1 |
| 212.8 | 271.6 | —58.8 | 246.8 | 305.5 | —58.7 | 252.1 | 307.3 | —55.2 |
| 256.8 | 226.3 | 30.5 | 264.6 | 303.1 | —38.5 | 231.8 | 286.6 | —54.8 |
| 268.8 | 323.0 | —54.2 | 234.5 | 309.9 | —75.4 | 230.6 | 312.4 | —81.8 |
| 242.5 | 253.7 | —11.2 | 231.3 | 298.6 | —67.3 | 243.1 | 252.1 | — 9.0 |
| 237.4 | 308.8 | —71.4 | 285.8 | 306.8 | —21.0 | 249.4 | 302.5 | —53.1 |
| 223.5 | 292.5 | —69.0 | 305.0 | 309.9 | — 4.9 | 251.5 | 279.5 | —28.0 |
| 2169.3 | 2597.0 | —427.7 | 2313.4 | 2682.2 | —368.8 | 2306.4 | 2654.4 | —348.0 |

These three sets of differences are from the sums of the blocks containing the same group of nine varieties. Since block (1) in replication 1 contains varieties 27, 20, 22, 26, 25, 21, 23, 24 and 19 with a total yield of 221.4 lbs., and since these same nine varieties are in another random order in block (9) of replication 2, with a total of 279.8 lbs., the first difference in set X is $221.4 — 279.8 = — 58.4$. Such sets of differences can only be obtained in those triple lattice experiments which have one or more complete replications of groups X, Y and Z.

The sums of squares of the deviations of these three sets of differences give the variance for paired blocks. The divisors are $2k$ (18) and $2k^2$ (162)[8].

The sum of squares for component (a) is

$$\frac{(58.4)^2 + (76.5)^2 + \ldots + (28.0)^2}{18} - \frac{(427.7)^2 + (368.8)^2 + (348.0)^2}{162} =$$

$$4017.84 — 2716.33 = 1301.51$$

*Component (b)* consists of three sets of values used to give an estimate of block differences freed from varietal effects. In group X (table 10) the row totals are combinations of two block totals from table 9. The first row total of group X is

$$212.8 + 271.6 = 484.4$$

Since these totals contain varietal effect as well as block

---

[8] For nine replications there are three columns of block totals for each set. Component (a) for set X is calculated by analysis of variance, as follows:

| | |
|---|---|
| Replications | 2 |
| Set X totals | (k — 1) |
| Component (a) or interaction | 2(k — 1) |

effect, they are not used to compute block sum of squares. The above total (484.4) contains only the plot yields for varieties 1, 2, . . . , 9. An unconfounded estimate of the sum of these nine varieties is given in the total (517.7) for the first column of group Y and in the total (489.6) for these nine variety yields in group Z. An estimate of block effect freed from varietal differences is given by

$$2(484.4) - (517.7) - (489.6) = -38.5$$

This same value can more conveniently be secured by taking three times the group X row totals (table 10) minus the row totals of table 11.

$$3(484.4) - 1491.7 = -38.5$$

These values are also to be used to calculate the adjustments necessary to evaluate the varietal means. For reasons given in analysis of lattice experiment (page 17) their negative values are computed and designated as $2rkc_X$, $2rkc_Y$ and $2rkc_Z$ in table 12. The change in sign does not affect the sum of squares.

The $2rkc_X$, $2rkc_Y$ and $2rkc_Z$ values are calculated as follows:

$2rkc_X$ = row total of table 11 — 3(row total of group X)
$2rkc_Y$ = column total of table 11 — 3(row total of group Y)
$2rkc_Z$ = Latin letter total of table 11 — 3(row total of group Z)

TABLE 12. 2rkc VALUES FOR YIELD.

| $2rkc_X$ | $2rkc_Y$ | $2rkc_z$ |
|---|---|---|
| 38.5 | 29.3 | — 66.2 |
| 91.5 | 20.5 | 110.9 |
| 116.0 | — 56.5 | — 83.6 |
| 24.4 | 42.0 | — 47.6 |
| 27.5 | — 13.5 | — 40.5 |
| 129.5 | — 36.5 | 37.1 |
| — 12.4 | —131.3 | —111.4 |
| 60.3 | — 99.5 | 8.3 |
| — 51.5 | — 18.6 | 33.3 |
| 423.8 | —264.1 | —159.7 |

The sum of all the above 2rkc values is equal to 0,

$$423.8 + (-264.1) + (-159.7) = 0$$

The sums of squares of the deviations of these three sets of values give an estimate of the variance between blocks (eliminating varieties). The divisors are $2rk$ (108) and $2rk^2$ (972).

The sum of squares for component (b) is

$$\frac{(38.5)^2 + (91.5)^2 + \ldots + (33.3)^2}{108} - \frac{(423.8)^2 + (264.1)^2 + (159.7)^2}{972} =$$

$$1163.43 - 282.78 = 880.65$$

The results are now summarized in table 13, the analysis of variance of the triple lattice experiment.

TABLE 13. ANALYSIS OF VARIANCE OF TRIPLE LATTICE EXPERIMENT.

| Source of variation | Degrees of freedom | Sum of squares | | Mean square |
|---|---|---|---|---|
| (Yield). | | | | |
| Replications | 5 | | 2904.84 | 580.968 |
| Component (a) | 24 | 1301.51 | | |
| Component (b) | 24 | 880.65 | | |
| Block (eliminating varieties) | 48 | | 2182.16 | 45.462 |
| Varieties (ignoring blocks) | 80 | | 2783.05 | |
| Error (intra-block) | 352 | | 1780.10 | 5.057 |
| Total | 485 | | 9650.15 | |

TEST OF SIGNIFICANCE.

As in the lattice design, the test appropriate to the interblock analysis involves additional computation. If the sums of squares for blocks and error in table 13 are added, the resulting mean square, based on 400 degrees of freedom, may be used to test the variety mean square in table 13. This test is equivalent to analyzing the experiment as randomized complete blocks. This gives a close approximation to the correct test if the block mean square in table 13 is not much greater than the error mean square.

TABLE 14. ANALYSIS OF VARIANCE AS RANDOMIZED COMPLETE BLOCKS.

| Source of variation | Degrees of freedom | Sum of squares | Mean square |
|---|---|---|---|
| Replications | 5 | 2904.84 | 580.968 |
| Varieties (ignoring blocks) | 80 | 2783.05 | 34.788 |
| Error | 400 | 3962.26 | 9.906 |
| Total | 485 | 9650.15 | |

The test of significance of variety differences:
$F = 34.788/9.906 = 3.51^{**}$

Since there are highly significant differences among the mean yields of these varieties, no further test is necessary for this experiment. If, however, the F-value does not quite reach significance, and if the block mean square in table 13 is above the error mean square, the precise test, as given in the mathematical theory, is needed.

VARIETY MEANS.

To obtain the variety means, three sets of corrections, $c_X'$, $c_Y'$ and $c_Z'$ are calculated. These are obtained from the $c_X$, $c_Y$ and $c_Z$ values (secured from table 12), by multiplying them by a weighting factor.

The weighting factor is $\dfrac{2(w - w')}{2w + w'}$, where

$$w = 1/E, \quad w' = 5/(6B - E)$$

E and B being the error and block mean squares in table 13. The formulas for 3 and 9 replications are given in the mathematical theory of triple lattice designs, section 2. For this example,

$$w = \frac{1}{E} = \frac{1}{5.057} = .19775$$

and $w' = \dfrac{5}{6B - E} = \dfrac{5}{6(45.462) - 5.057} = .01868$

The value of the weighting factor is

$$\frac{2(w - w')}{2w + w'} = .8647$$

To secure the variety mean yields, solve for the weighted correction terms.

$$c_X' = \frac{1}{2rk} \frac{2(w - w')}{(2w + w')} (2rkc_X) = .0080065 \ (2rkc_X)$$

$$c_Y' = .0080065(2rkc_Y)$$

and $c_Z' = .0080065(2rkc_Z)$

The first $c_X'$ is

$$(.0080065)(38.5) = .3082$$

The variety mean yields can now be secured by using the totals of the yield values in table 11. These values are divided by six, the number of replications, and the averages entered into table 15. That is, for variety 1, $\dfrac{147.2}{6} =$ 24.5333, the average yield per plot. The weighted corrections ($c'$) are placed with table 15, the $c_X'$ values at the end of the rows, the $c_Y'$ values at the bottom of the columns, and the $c_Z'$ values are tabulated for convenience along the bottom of the table with their identification letters. Following the variety identifications, which are in boldface type, enter the Latin letters from the Latin square used in making group Z of the triple lattice designs. The variety mean yields are now secured by adding to each value in table 15 its appropriate $c_X'$, $c_Y'$ and $c_Z'$ values.

Mean yield of variety 1 = 24.5333 + .3082 + .2346 − .5300 = 24.5461
Mean yield of variety 2 = 25.0667 + .3082 + .1641 + .2666 = 25.8056

The variety mean yields are given in table 16. When the variety means have all been computed, they can be checked thus:

$$6(\text{sum of means}) = \text{grand total}.$$

TABLE 15.  AVERAGES AND c' VALUES.

(Yield)

| | | | | | | | | | cx' |
|---|---|---|---|---|---|---|---|---|---|
| 1 A 24.5333 | 2 I 25.0667 | 3 H 26.9333 | 4 G 28.1833 | 5 F 28.4833 | 6 E 29.1167 | 7 D 28.3500 | 8 C 30.8167 | 9 B 27.1333 | .3082 |
| 10 B 28.7500 | 11 A 28.5167 | 12 I 30.4167 | 13 H 28.3167 | 14 G 30.5500 | 15 F 27.9667 | 16 E 30.4333 | 17 D 27.9500 | 18 C 30.4500 | .7326 |
| 19 C 28.9500 | 20 B 27.7167 | 21 A 29.7333 | 22 I 28.5167 | 23 H 29.6000 | 24 G 32.1333 | 25 F 31.6333 | 26 E 30.5500 | 27 D 31.1000 | .9288 |
| 28 D 31.7833 | 29 C 32.5667 | 30 B 30.3500 | 31 A 30.5000 | 32 I 31.5333 | 33 H 29.0667 | 34 G 32.2833 | 35 F 31.7500 | 36 E 27.3333 | .1954 |
| 37 E 28.7167 | 38 D 28.4333 | 39 C 29.1000 | 40 B 29.1667 | 41 A 29.4500 | 42 I 28.5667 | 43 H 29.4167 | 44 G 32.1500 | 45 F 27.5833 | .2202 |
| 46 F 25.4500 | 47 E 29.0167 | 48 D 29.9333 | 49 C 26.6500 | 50 B 32.9833 | 51 A 29.1833 | 52 I 28.9833 | 53 H 29.1333 | 54 G 31.8000 | 1.0368 |
| 55 G 33.2833 | 56 F 31.5667 | 57 E 32.8333 | 58 D 33.6000 | 59 C 32.3167 | 60 B 27.6667 | 61 A 34.2333 | 62 I 34.9333 | 63 H 32.0500 | —.0993 |
| 64 H 34.5667 | 65 G 32.4833 | 66 F 33.4167 | 67 E 31.6333 | 68 D 29.4167 | 69 C 33.0667 | 70 B 32.6333 | 71 A 32.7500 | 72 I 29.2333 | .4828 |
| 73 I 33.8000 | 74 H 33.8000 | 75 G 31.7167 | 76 F 27.5833 | 77 E 29.5667 | 78 D 31.2000 | 79 C 37.6000 | 80 B 29.6833 | 81 A 32.4167 | —.4123 |
| cy' .2346 | .1641 | —.4524 | .3363 | —.1081 | —.2922 | —1.0513 | —.7966 | —.1489 | |
| cz' A —.5300 | B .8879 | C —.6693 | D —.3811 | E —.3243 | F .2970 | G —.8919 | H .0665 | I .2666 | |

TABLE 16.  VARIETY MEANS.

(pounds per plot).

| | | | | | | | | |
|---|---|---|---|---|---|---|---|---|
| 1:24.55 | 2:25.81 | 3:26.86 | 4:27.94 | 5:28.98 | 6:28.81 | 7:27.23 | 8:29.66 | 9:28.18 |
| 10:30.61 | 11:28.88 | 12:30.96 | 13:29.45 | 14:30.28 | 15:28.70 | 16:29.79 | 17:27.50 | 18:30.36 |
| 19:29.44 | 20:29.70 | 21:29.68 | 22:30.05 | 23:30.49 | 24:31.88 | 25:31.81 | 26:30.36 | 27:31.50 |
| 28:31.83 | 29:32.26 | 30:30.98 | 31:30.50 | 32:31.89 | 33:29.04 | 34:30.54 | 35:31.45 | 36:27.06 |
| 37:28.85 | 38:28.44 | 39:28.20 | 40:30.61 | 41:29.03 | 42:28.76 | 43:28.65 | 44:30.68 | 45:27.95 |
| 46:27.02 | 47:29.89 | 48:30.14 | 49:27.35 | 50:34.80 | 51:29.40 | 52:29.24 | 53:29.44 | 54:31.87 |
| 55:32.53 | 56:31.93 | 57:31.96 | 58:33.46 | 59:31.44 | 60:28.16 | 61:32.55 | 62:34.30 | 63:31.87 |
| 64:35.35 | 65:32.24 | 66:33.74 | 67:32.13 | 68:29.41 | 69:32.59 | 70:32.95 | 71:31.91 | 72:29.83 |
| 73:33.89 | 74:33.62 | 75:29.96 | 76:27.75 | 77:28.72 | 78:30.11 | 79:35.47 | 80:29.36 | 81:31.33 |

STANDARD ERRORS OF THE DIFFERENCES BETWEEN VARIETY
MEANS.

Using $s^2 (5.057)$, table 13, the standard errors for variety mean differences are calculated.

1. The standard error of the differences between the mean yields of two varieties occurring in the same block of one of the groups is

$$\sqrt{\frac{2s^2}{rk}\left[\frac{6w}{2w+w'}+(k-2)\right]} = \sqrt{\frac{2(5.057)}{(6)(9)}\left[\frac{(6)(.19775)}{(2)(.19775)+.01868}+7\right]}$$

$$= \sqrt{1.8477} = 1.36$$

2. The standard error of the differences between the mean yields of two varieties which do not occur in the same block in any group is

$$\sqrt{\frac{2s^2}{rk}\left[\frac{9w}{2w+w'}+(k-3)\right]} = \sqrt{\frac{2(5.057)}{(6)(9)}\left[\frac{(9)(.19775)}{(2)(.19775)+.01868}+6\right]}$$

$$= \sqrt{1.9286} = 1.39$$

3. The mean standard error of all comparisons is

$$\sqrt{\frac{2s^2}{r(k+1)}\left[\frac{9w}{2w+w'}+(k-2)\right]} =$$

$$\sqrt{\frac{2(5.057)}{(6)(10)}\left[\frac{(9)(.19775)}{(2)(.19775)+.01868}+7\right]} = \sqrt{1.9043} = 1.38$$

Variety 10 with a yield of 30.61 lbs. per plot is considered a standard variety. To test the significance of the difference between the means of varieties 10 and 50,

$$\overline{d} = 34.80 - 30.61 = 4.19$$

$$t = \frac{4.19}{1.36} = 3.08**$$

a highly significant difference.

### PRECISION OF TRIPLE LATTICE EXPERIMENT.

In field plot experimental work the size of the block is flexible. Therefore, it is quite important to have some idea of the relative precision of experiments using blocks of different sizes.

The error variance which resulted by analyzing the triple lattice design as a randomized complete block arrangement was 9.906 (table 14). The average variance of the difference between two varietal means is $2v/r = 2(9.906)/6 = 3.302$ for the randomized complete block design and 1.9043 for the triple lattice design. This gives $\frac{3.302}{1.9043} = 173$ percent, a gain of 73 percent by the recovery of inter-block variance and by reducing the block size from 81 to 9 plots per block.

## COVARIANCE ANALYSIS OF TRIPLE LATTICE EXPERIMENT

The covariance technique which offers a method of adjusting yields on the basis of number of missing hills is illustrated by use of data in tables 9, 10 and 11. The sum of squares for number of missing hills and the cross products of yield by number of missing hills are now computed.

The correction terms are

$$\text{Missing hills } c_1 = \frac{(225)^2}{486} = 104.1667$$

$$\text{Yield} \times \text{missing hills } c_2 = \frac{(14,722.7)(225)}{486} = 6816.06$$

The total sums of squares and cross products are

Missing hills

$$(2)^2 + (1)^2 + \ldots + (1)^2 - c_1 = 623.0000 - 104.1667 = 518.8333$$

Yield $\times$ missing hills

$$(24.4)(2) + (22.5)(1) + \ldots + (31.3)(1) - c_2 =$$
$$6495.6000 - 6816.0648 = -320.4648$$

Note that most of the plots, those in which there are no italic numbers, have no missing hills.

The sums of squares and cross products for replications in table 9 are

Missing hills

$$\frac{(11)^2 + (56)^2 + \ldots + (23)^2}{81} - c_1 =$$
$$143.3950 - 104.1667 = 39.2283$$

Yield $\times$ missing hills

$$\frac{(2169.3)(11) + (2597.0)(56) + \ldots + (2654.4)(23)}{81} - c_2 =$$
$$7,044.0037 - 6816.0648 = 227.9389$$

The sums of squares and cross products for varieties (ignoring blocks) are obtained by using the sub-totals in table 11:

Missing hills

$$\frac{(1)^2 + (6)^2 + \ldots + (2)^2}{6} - c_1 =$$

$$165.5000 - 104.1667 = 61.3333$$

Yield $\times$ missing hills

$$\frac{(147.2)(1) + (150.4)(6) + \ldots + (194.5)(2)}{6} - c_2 =$$

$$6731.9500 - 6816.0648 = -84.1148$$

*Component (a)*, the three sets of differences in number of missing hills between paired blocks,

(Number of missing hills)

| Block totals | | Set X Differences | Block totals | | Set Y Differences | Block totals | | Set Z Differences |
|---|---|---|---|---|---|---|---|---|
| Rep. 1 | Rep. 2 | | Rep. 3 | Rep. 4 | | Rep. 5 | Rep. 6 | |
| 3 | 4 | — 1 | 8 | 1 | 7 | 0 | 0 | 0 |
| 0 | 8 | — 8 | 3 | 2 | 1 | 2 | 0 | 2 |
| 0 | 1 | — 1 | 8 | 13 | — 5 | 4 | 0 | 4 |
| 1 | 2 | — 1 | 1 | 9 | — 8 | 0 | 0 | 0 |
| 1 | 8 | — 7 | 5 | 25 | —20 | 9 | 1 | 8 |
| 3 | 8 | — 5 | 3 | 3 | 0 | 0 | 10 | —10 |
| 3 | 21 | —18 | 5 | 14 | — 9 | 1 | 3 | — 2 |
| 0 | 2 | — 2 | 7 | 4 | 3 | 0 | 0 | 0 |
| 0 | 2 | — 2 | 2 | 5 | — 3 | 1 | 9 | — 8 |
| 11 | 56 | —45 | 42 | 76 | —34 | 17 | 23 | — 6 |

These three sets of differences were computed in the same manner as the yield differences.

The sums of squares and cross products for component (a) are

Missing hills

$$\frac{(1)^2 + (8)^2 + \ldots + (8)^2}{18} - \frac{(45)^2 + (34)^2 + (6)^2}{162} = 55.8642$$

Yield $\times$ missing hills

$$\frac{(-58.4)(-1) + (-76.5)(-8) + \ldots + (-28.0)(-8)}{18} -$$

$$\frac{(-427.7)(-45) + (-368.8)(-34) + (-348.0)(-6)}{162} = -38.1414$$

*Component (b)* consists of three sets of differences between totals giving an estimate of block differences freed from varietal effects. A discussion of the methods for securing these differences is given in the corresponding section of the analysis of variance of yield.

The $2rke_X$, $2rke_Y$ and $2rke_Z$ values (given in table 17) for number of missing hills are calculated as follows:

$2rke_x$ = row total of table 11 — 3(row total of group X)

$2rke_Y$ = column total of table 11 — 3(row total of group Y)

$2rke_Z$ = Latin letter total of table 11 — 3(row total of group Z)

TABLE 17.   2rke VALUES FOR NUMBER OF MISSING HILLS.

| $2rke_x$ | $2rke_Y$ | $2rke_z$ |
|---|---|---|
| 6 | —28 | 5 |
| —29 | —30 | 5 |
| — 4 | —54 | 24 |
| 12 | 8 | 22 |
| 16 | — 7 | 20 |
| 0 | —14 | 8 |
| 24 | — 6 | 25 |
| 3 | — 8 | 0 |
| — 4 | 10 | — 4 |
| 24 | —129 | 105 |

The sum of the $2rke_X$, $2rke_Y$ and $2rke_Z$ values is equal to zero,

$$24 + (—129) + 105 = 0$$

The sums of squares and cross products of the deviations of these three sets of values give an estimate of variance between blocks (eliminating varieties). The divisors are $2rk$ (108) and $2rk^2$ (972).

The sums of squares and cross products for component (b) are

Missing hills
$$\frac{(6)^2 + (29)^2 + \ldots + (4)^2}{108} - \frac{(24)^2 + (129)^2 + (105)^2}{972} = 56.2963$$

Yield × missing hills
$$\frac{(38.5)(6) + (91.5)(—29) + \ldots + (33.3)(—4)}{108} -$$

$$\frac{(423.8)(24) + (—264.1)(—129) + (—159.7)(105)}{972} = —68.7353$$

ANALYSIS OF VARIANCE OF NUMBER OF MISSING HILLS.

The analysis of variance for number of missing hills is summarized in table 18.

TABLE 18.  ANALYSIS OF VARIANCE OF TRIPLE LATTICE
EXPERIMENT.

| Source of variation | Degrees of freedom | Sum of squares | Mean square |
|---|---|---|---|
| (Number of missing hills) | | | |
| Replications | 5 | 39.2283 | |
| Component (a) | 24 | 55.8642 | |
| Component (b) | 24 | 56.2963 | |
| Blocks (eliminating varieties) | 48 | 112.1605 | 2.3367 |
| Varieties (ignoring blocks) | 80 | 61.3333 | |
| Error (intra-block) | 352 | 306.1112 | .8696 |
| Total | 485 | 518.8333 | |

In this case, a precise test of the significance of the differences among the variety means is required.  The test appropriate for inter-block analysis is given in the mathematical theory of triple lattice designs, section 5.  The mean square for varieties is 0.6431 as compared to the intra-block error mean square of .8696, which indicates that the number of missing hills is not a variety characteristic.

ANALYSIS OF COVARIANCE.

The covariance results are brought together in table 19.

TABLE 19.  ANALYSIS OF COVARIANCE OF TRIPLE LATTICE
EXPERIMENT.

| Source of variation | Degrees of freedom | Sums of squares and products | | | Errors of estimate | | |
|---|---|---|---|---|---|---|---|
| | | No. of missing hills | | Yield | Sum of squares | Degrees of freedom | Mean square |
| | | $Sx^2$ | Sxy | $Sy^2$ | | | |
| Total | 485 | 518.8333 | —320.4648 | 9,650.15 | | | |
| Replications | 5 | 39.2283 | 227.9389 | 2,904.84 | | | |
| Component (a) | 24 | (55.8642) | (—38.1414) | (1,301.51) | | | |
| Component (b) (eliminating varieties) | 24 | (56.2963) | (—68.7353) | (880.65) | | | |
| Blocks (eliminating varieties) | 48 | 112.1605 | —106.8767 | 2,182.16 | | | |
| Varieties (ignoring blocks) | 80 | 61.3333 | — 84.1148 | 2,783.05 | | | |
| Error (intra-block) | 352 | 306.1112 | —357.4122 | 1,780.10 | 1362.79 | 351 | ·3.883 |
| Block + error | 400 | 418.2717 | —464.2889 | 3,962.26 | 3446.89 | 399 | |
| Regression block means | | | | | 2084.10 | 48 | 43.419 |

To find the mean square for blocks adjusted to an average number of missing hills, compute the residual sum of squares for error + block, and from this subtract the residual sum of squares for error.  For a detailed description of this procedure see [8].

The residual sum of squares for error is

$$Sy^2 - \frac{(Sxy)^2}{Sx^2} = 1{,}780.10 - \frac{(-357.4122)^2}{306.1112} = 1{,}362.79$$

This value is entered in table 19 under the errors of estimate sum of squares for error.

The residual sum of squares for block + error is

$$3{,}962.26 - \frac{(-464.2889)^2}{418.2717} = 3{,}446.89$$

The mean square for blocks adjusted to an average number of missing hills is 43.419 as compared with 45.462 (table 13) before adjusting. The error variance of yield has been reduced from 5.057 to 3.883.

VARIETY MEANS.

To compute the variety means adjusted for missing hills, the regression coefficients for both blocks and intra-block error are to be considered.

The block regression coefficient is

$$b = \frac{Sxy}{Sx^2} = \frac{-106.8767}{112.1605} = -.9529$$

while that for error is

$$\frac{-357.4122}{306.1112} = -1.1676$$

Since the two coefficients are quite similar, the error regression coefficient is used to adjust both intra- and inter-block estimates to obtain yield means adjusted for number of missing hills.

The equation for yield means adjusted to the mean number of missing hills is

$$\text{Yield mean (adjusted)} = \overline{y}_i - b(\overline{x}_i - \overline{x})$$
$$= \overline{y}_i + 1.1676(\overline{x}_i - .4630)$$
$$= \overline{y}_i + 1.1676\,\overline{x}_i - .5406$$

where $\overline{y}_i$ is the average yield (table 15) and $x_i$ is the average number of missing hills (table 20) for each variety. $\overline{x}$ is the mean number of missing hills for all plots.

TABLE 20. AVERAGE NUMBER OF MISSING HILLS.

| | | | | | | | | |
|---|---|---|---|---|---|---|---|---|
| 1:0.1667 | 2:1.0000 | 3:0.5000 | 4:0.1667 | 5:0.5000 | 6:0.0 | 7:0.0 | 8:0.1667 | 9:0.0 |
| 10:0.6667 | 11:0.8333 | 12:0.5000 | 13:1.1667 | 14:0.6667 | 15:0.6667 | 16:0.8333 | 17:0.8333 | 18:1.0000 |
| 19:0.5000 | 20:0.6667 | 21:0.3333 | 22:0.3333 | 23:0.5000 | 24:0.0 | 25:0.3333 | 26:0.1667 | 27:0.0 |
| 28:0.5000 | 29:0.6667 | 30:0.1667 | 31:0.1667 | 32:0.3333 | 33:0.3333 | 34:0.0 | 35:0.5000 | 36:0.3333 |
| 37:0.1667 | 38:0.3333 | 39:1.1667 | 40:0.0 | 41:0.1667 | 42:0.1667 | 43:0.5000 | 44:0.0 | 45:1.1667 |
| 46:0.8333 | 47:0.3333 | 48:1.1667 | 49:0.6667 | 50:0.0 | 51:0.1667 | 52:0.3333 | 53:0.6667 | 54:0.3333 |
| 55:1.0000 | 56:0.5000 | 57:0.8333 | 58:0.0 | 59:0.8333 | 60:0.1667 | 61:0.0 | 62:0.5000 | 63:0.6667 |
| 64:0.6667 | 65:1.1667 | 66:0.5000 | 67:0.0 | 68:0.1667 | 69:0.0 | 70:0.5000 | 71:0.6667 | 72:0.8333 |
| 73:0.3333 | 74:0.0 | 75:0.8333 | 76:1.3333 | 77:0.6667 | 78:0.6667 | 79:0.0 | 80:0.6667 | 81:0.3333 |

The variety average yield and number of missing hills are substituted in the equation to give the variety yield means adjusted for number of missing hills. For example, the yield mean adjusted for number of missing hills for variety 1 is

$$\text{Yield mean (adjusted) variety } 1 = 24.5333 + 1.1676(.1667) - .5406$$
$$= 24.1873$$

These variety means are recorded in table 21.

To obtain the weighted variety means adjusted for number of missing hills, three sets of corrections, $v_X$, $v_Y$ and $v_Z$ are calculated. These corrections are obtained from the $c_X$, $c_Y$, $c_Z$ and the $e_X$, $e_Y$, $e_Z$ values, by using the regression co-efficient and a weighting factor. The $v_X$ values are

$$v_x = \frac{1}{2rk} \frac{2(u - u')}{2u + u'} (2rkc_x - b2rke_x)$$

where $\dfrac{2(u - u')}{2u + u'}$ is the weighting factor using the inter-block information

$$u = \frac{1}{E'} = \frac{1}{3.883} = .25753$$

$$\text{and } u' = \frac{5}{6B' - E'} = \frac{5}{6(43.419) - 3.883} = .01948$$

$E'$ and $B'$ are the error and block mean squares for yield after they have been adjusted to an average number of missing hills (table 19).

The value of the weighting factor is

$$\frac{2(u - u')}{2u + u'} = .8907$$

This method of weighting ignores the sampling variance of the regression coefficient, which would require consideration when using smaller designs.

The $v$ values are computed using the 2rkc values of table 12 and the 2rke values of table 17.

$$v_x = \frac{1}{2rk} \frac{2(u - u')}{2u + u'} (2rkc_x + 1.1676 \; 2rke_x)$$

$$= .008247(2rkc_x + 1.1676 \; 2rke_x)$$

The first $v_X$ value is

$$(.008247) [38.5 + (1.1676)6] = .3753$$

## TABLE 21. YIELD MEANS ADJUSTED FOR NUMBER OF MISSING HILLS.

| | | | | | | | | | vz |
|---|---|---|---|---|---|---|---|---|---|
| **1 A** 24.1873 | **2 I** 25.6937 | **3 H** 26.9765 | **4 G** 27.8373 | **5 F** 28.5265 | **6 E** 28.5761 | **7 D** 27.8094 | **8 C** 30.4707 | **9 B** 26.5927 | .3753 |
| **10 B** 28.9878 | **11 A** 28.9491 | **12 I** 30.4599 | **13 H** 29.1383 | **14 G** 30.7878 | **15 F** 28.2045 | **16 E** 30.8657 | **17 D** 28.3824 | **18 C** 31.0770 | .4754 |
| **19 C** 28.9932 | **20 B** 27.9545 | **21 A** 29.5819 | **22 I** 28.3653 | **23 H** 29.6432 | **24 G** 31.5927 | **25 F** 31.4819 | **26 E** 30.2040 | **27 D** 30.5594 | .9181 |
| **28 D** 31.8265 | **29 C** 32.8045 | **30 B** 30.0040 | **31 A** 30.1540 | **32 I** 31.3819 | **33 H** 28.9153 | **34 G** 31.7427 | **35 F** 31.7932 | **36 E** 27.1819 | .3168 |
| **37 E** 28.3707 | **38 D** 28.2819 | **39 C** 29.9216 | **40 B** 28.6261 | **41 A** 29.1040 | **42 I** 28.2207 | **43 H** 29.4599 | **44 G** 31.6094 | **45 F** 28.4049 | .3809 |
| **46 F** 25.8824 | **47 E** 28.8653 | **48 D** 30.7549 | **49 C** 26.8878 | **50 B** 32.4427 | **51 A** 28.8373 | **52 I** 28.8319 | **53 H** 29.3711 | **54 G** 31.6486 | 1.0680 |
| **55 G** 33.9103 | **56 F** 31.6099 | **57 E** 33.2657 | **58 D** 33.0594 | **59 C** 32.7491 | **60 B** 27.3207 | **61 A** 33.6927 | **62 I** 34.9765 | **63 H** 32.2878 | .1288 |
| **64 H** 34.8045 | **65 G** 33.3049 | **66 F** 33.4599 | **67 E** 31.0927 | **68 D** 29.0707 | **69 C** 32.5261 | **70 B** 32.6765 | **71 A** 32.9878 | **72 I** 29.6657 | .5262 |
| **73 I** 33.6486 | **74 H** 33.2594 | **75 G** 32.1491 | **76 F** 28.5495 | **77 E** 29.8045 | **78 D** 31.4378 | **79 C** 37.0594 | **80 B** 29.9211 | **81 A** 32.2653 | —.4632 |
| vy —.0280 | —.1198 | —.9859 | .4234 | —.1787 | —.4358 | —1.1406 | —.8976 | —.0571 | |
| vz **A** —.4978 | **B** .9627 | **C** —.4583 | **D** —.1807 | **E** —.1414 | **F** .3830 | **G** —.6780 | **H** .0684 | **I** .2361 | |

## TABLE 22. VARIETY MEANS.

(Pounds per plot).

| | | | | | | | | |
|---|---|---|---|---|---|---|---|---|
| 1:24.04 | 2:26.19 | 3:26.43 | 4:27.96 | 5:29.11 | 6:28.37 | 7:26.86 | 8:29.49 | 9:27.87 |
| 10:30.40 | 11:28.81 | 12:30.19 | 13:30.11 | 14:30.41 | 15:28.63 | 16:30.06 | 17:27.78 | 18:31.04 |
| 19:29.42 | 20:29.72 | 21:29.02 | 22:29.94 | 23:30.45 | 24:31.40 | 25:31.64 | 26:30.08 | 27:31.24 |
| 28:31.93 | 29:32.54 | 30:30.30 | 31:30.40 | 32:31.76 | 33:28.86 | 34:30.24 | 35:31.60 | 36:27.30 |
| 37:28.58 | 38:28.36 | 39:28.86 | 40:30.39 | 41:28.81 | 42:28.40 | 43:28.77 | 44:30.41 | 45:29.11 |
| 46:27.31 | 47:29.67 | 48:30.66 | 49:27.92 | 50:34.29 | 51:28.97 | 52:29.00 | 53:29.61 | 54:31.98 |
| 55:33.33 | 56:32.00 | 57:32.27 | 58:33.43 | 59:32.24 | 60:27.98 | 61:32.18 | 62:34.44 | 63:32.43 |
| 64:35.37 | 65:33.03 | 66:33.38 | 67:31.90 | 68:29.24 | 69:32.16 | 70:33.02 | 71:32.12 | 72:30.37 |
| 73:33.39 | 74:32.74 | 75:30.02 | 76:28.89 | 77:29.02 | 78:30.36 | 79:35.00 | 80:29.52 | 81:31.25 |

20.43

The $v_X$ weighted corrections are placed at the end of the rows of table 21, the $v_Y$ values at the bottom of the columns, and the $v_Z$ values are tabulated for convenience along the bottom of the table with their identification letters. Following the variety identifications, enter the Latin letters from the Latin square used in making group Z of the triple lattice designs. The weighted variety mean yields are now secured by adding to each value in table 21 its appropriate $v_X$, $v_Y$ and $v_Z$ values. The variety mean yields are given in table 22.

STANDARD ERRORS OF THE DIFFERENCES BETWEEN VARIETY MEANS.

1. The standard error of the differences between the means of two varieties occurring in the same block of one of the groups is

$$\sqrt{\frac{2s^2}{rk}\left[\frac{6u}{2u+u'}+(k-2)\right]} = \sqrt{\frac{2(3.883)}{(6)(9)}\left[\frac{(6)(.25753)}{2(.25753)+.01948}+7\right]}$$
$$= 1.19$$

2. The standard error of the differences between the mean yields of two varieties not having a block in common is

$$\sqrt{\frac{2s^2}{rk}\left[\frac{9u}{2u+u'}+(k-3)\right]} = \sqrt{\frac{2(3.883)}{(6)(9)}\left[\frac{(9)(.25753)}{2(.25753)+.01948}+6\right]}$$
$$= 1.22$$

3. The mean standard error of all comparisons is

$$\sqrt{\frac{2s^2}{r(k+1)}\left[\frac{9u}{2u+u'}+(k-2)\right]} =$$
$$\sqrt{\frac{2(3.883)}{(6)(10)}\left[\frac{(9)(.25753)}{2(.25753)+.01948}+7\right]} = 1.21$$

Variety 10 with a yield of 30.40 lbs. per plot is the standard variety. To test the significance of the difference between the means of varieties 10 and 36,

$$\overline{d} = 30.40 - 27.30 = 3.10$$

$$t = \frac{3.10}{1.22} = 2.54*$$

# II. Mathematical Theory

By W. G. Cochran[9]

The object of this portion of the bulletin is to give some indication of the mathematical basis for the methods used in the previous paper. The mathematical analysis follows closely that given by Yates[10] [16] for the more difficult case of the cubic lattice. In reading this part, frequent reference to the numerical examples will be necessary.

## LATTICE DESIGNS

### 1. CALCULATION OF THE ADJUSTED VARIETAL MEANS

The analysis is most easily followed by regarding the design as factorial. This device was used in the paper [12] which introduced these designs and is familiar to those who have studied the literature. If the varieties are renumbered so that $v_1$ becomes $v_{11}$, $v_2$ becomes $v_{12}$, . . . $v_9$ becomes $v_{19}$, $v_{10}$ becomes $v_{21}$ and $v_{81}$ becomes $v_{99}$, they may be regarded as given by all combinations of two factors, X and Y, each as nine levels. In this representation the first level of the main effect of X is the mean yield of varieties $v_1$, $v_2$, . . . $v_9$, while the first level of the main effect of Y is the mean yield of varieties $v_1$, $v_{10}$, $v_{19}$, . . . $v_{73}$. In the numerical example worked above, the main effects of X are completely confounded with blocks in replications 1 and 2, and those of Y in replications 3 and 4. These effects account for 16, or in general $2(k-1)$ of the degrees of freedom between varieties. The remaining 64, or $(k-1)^2$ degrees of freedom, representing the XY interactions, are not confounded in any replication, and the new method of analysis does not affect them.

The estimates of the main effects X and Y were previously taken only from those replications in which they were unconfounded with blocks. These estimates are denoted by $X_i$ and $Y_i$ (i for intra-blocks). Estimates can, however, also be made from the replications in which X and Y are confounded with blocks; these will be denoted by $X_b$ and $Y_b$ (b for blocks). In the numerical example, the values of $X_i$ and $X_b$ for the first level of X are (table 2)

$$X_i = \frac{521.1}{18} = 28.95, \qquad X_b = \frac{485.6}{18} = 26.98$$

---

[9] Research Professor, Statistical Section, Iowa Agricultural Experiment Station.

[10] Dr. Yates has worked out the corresponding analysis for all types of lattice designs.

The new method of analysis was developed from the discovery that the accuracy of the block estimates $X_b$ and $Y_b$ could be determined, or in other words that estimated standard errors could be found for $X_b$ and $Y_b$ as well as for $X_i$ and $Y_i$. If the blocks vary in fertility, the standard errors of $X_b$ and $Y_b$ are higher than those of $X_i$ and $Y_i$. Nevertheless, knowing the relative accuracy of the two types of estimate, combined estimates may be made which are more accurate than $X_i$ or $Y_i$. If $\sigma^2$ and $\sigma'^2$ are the true error variances *per plot* of the estimates $X_i$ and $X_b$, it is known that the most accurate combined estimate is

$$X = \frac{wX_i + w'X_b}{w + w'} \tag{1}$$

where $w = 1/\sigma^2$ and $w' = 1/\sigma'^2$.

In practice, $w$ and $w'$ are not known exactly, but if they are based on sufficient numbers of degrees of freedom, the disturbance from this point is negligible. Discussion of this question and of the method of estimating $w'$ is given later.

The adjusted varietal means, using the block estimates $X_b$ and $Y_b$, are more quickly computed if (1) is expressed in terms of the unweighted varietal means $X_o$ and $Y_o$, *taken over all replications*. Since $X_i$ and $X_b$ are based on the same number of replications,

$$X_o = \frac{X_i + X_b}{2} \tag{2}$$

Hence $$X_i = 2X_o - X_b \tag{3}$$

Eliminating $X_i$ from (1) and (2), we find

$$X = X_o + \frac{w - w'}{w + w'}(X_o - X_b) \tag{4}$$

Inspection of the numerical example shows that the quantities $rkc_X$, calculated in table 4, are simply

$$\frac{rk}{2}(X_i - X_b) = rk(X_o - X_b) \tag{5}$$

Hence, the adjustment per plot to $X_o$, as calculated from the mean over all replications, is

$$c_x' = \lambda\, c_x \text{ where } \lambda = \frac{w - w'}{w + w'}$$

Similarly the quantities $c_Y'$ are the adjustments per plot to $Y_o$. An interaction term XY does not receive any adjustment.

In practice it is more convenient to adjust each varietal mean directly. The necessary adjustments may be obtained by expressing the varietal means in terms of main effects and interactions. If $v_o$ is an unadjusted varietal mean, and M is the general mean, we may write

$$v_o - M = (X_o - M) + (Y_o - M) + (v_o - X_o - Y_o + M) \qquad (6)$$

where $X_o$, $Y_o$ are the corresponding row and column means. Similarly for an adjusted mean v we must have

$$v - M = (X - M) + (Y - M) + (v - X - Y + M) \qquad (7)$$

M remains unchanged, since the general mean is unaffected by the adjustments. Also the interaction terms (the extreme right-hand brackets) must be the same in both cases, since the interactions are unconfounded. Hence,

$$v - v_o = (X - X_o) + (Y - Y_o) \qquad (8)$$

i. e., each variety is adjusted by adding the corresponding $c_X'$ and $c_Y'$ terms.

## 2. ESTIMATION OF w AND w′

An unbiased estimate of the variance *per plot* of $X_i$ and $Y_i$ is given by the intra-block error mean square. Hence w is taken as the reciprocal of this mean square.

The method of estimating the error mean square between blocks depends on the number of replications. In the present example, with four replications, a direct estimate may be made from the *differences* between the totals of pairs of blocks containing the same set of nine varieties. These differences are obviously clear of varietal effects. This component is described as component (a) in the numerical example and is based on 16 degrees of freedom. It is interesting to note that the remaining 16 degrees of freedom between blocks (unadjusted) represent the variance between the $X_b$ and $Y_b$ estimates of the varieties. Thus the analysis of variance might be written as follows, by an extension of table 5:

TABLE 23.

| Source of variation | Degrees of freedom | Sums of squares | Mean square |
|---|---|---|---|
| Replications | 3 | 2375.58 | |
| $X_b$ and $Y_b$ | 16 | 1267.19 | 79.199 |
| Error (a) or Component (a) | 16 | 964.63 | 60.289 |
| Varieties (eliminating blocks) | 80 | 1230.23 | 15.378 |
| Error (intra-block) | 208 | 871.27 | 4.189 |

In this form, the analogy with a split-plot design is easily seen. Components $X_b$ and $Y_b$, the main-plot treatments, are both tested in two randomized blocks of nine "plots" each, the "plots" being nine times as large as the unit plots. The principal difference from the ordinary split-plot design is that the "sub-plot" treatments also contain the X and Y components of the varietal comparisons, i. e., the main-plot treatments are also included as part of the sub-plot treatments. From this analysis we obtain estimates,

$$w = \frac{1}{4.189} = .23872 \qquad w' = \frac{1}{60.289} = .01659$$

This estimate of $w'$, based on 16, or $2(k-1)$, degrees of freedom, is probably sufficiently precise for our purpose. However, in corresponding experiments with fewer numbers of varieties, $w'$ would be based on rather few degrees of freedom. Moreover, this estimate cannot be made if there are only two replications.

An alternative estimate is obtained from the *sums* of pairs of blocks containing the same set of varieties. These sums are partially confounded with varietal effects, but they may be corrected for varietal effects in the manner described in the paper. The corrected sums yield component (b), also based on 16 degrees of freedom. It will be observed from table 5 that the mean square for component (a) is significantly greater than that for component (b). This is probably not a chance result; component (b) actually has a smaller expectation than component (a), owing to the adjustments made to free the former from varietal effects. Thus component (b) cannot be used directly to estimate the mean square between blocks.

This may be proved as follows. The error of any plot may be regarded as the sum of two independent parts, (1) a part $e$ which varies independently from plot to plot with variance $\sigma_i^2$, (2) a part $b$, constant for all plots in the same block, and varying from block to block with variance $\sigma_b^2$. Since the error of the difference between two plots in the same block is $(e_{11} - e_{12})$, $\sigma_i^2$ is the intra-block error variance. With this scheme, the error of a block total may be written,

$$E_1 = e_{11} + e_{12} + \ldots + e_{19} + 9b_1 \qquad (9)$$

This has a variance $(9\,\sigma_i^2 + 81\,\sigma_b^2)$. Hence the variance of the sum or difference of two block totals is

$$18\,\sigma_i^2 + 162\,\sigma_b^2 \qquad (10)$$

In obtaining component (a), the squares of the differences

between two block totals were divided by 18. Thus the mean square for component (a) is an estimate of

$$\sigma_i^2 + 9 \, \sigma_b^2 \tag{11}$$

Component (b) is based on the *sum* of two block totals, plus a correction term which is derived from a different pair of replications, and is therefore independent. The correction term is the total of 18 plots, one from every block in the pair of replications, and has a variance $(18 \, \sigma_i^2 + 18 \, \sigma_b^2)$. There is, however, one additional complication. All correction terms from the same pair of replications contain the same set of 18 b's and are therefore not independent of each other. Since, however, component (b) is a sum of squares of *deviations*, it involves only differences between correction terms, in which the b's cancel. Thus the appropriate error variance for the totals from which component (b) is derived is

$$36 \, \sigma_i^2 + 162 \, \sigma_b^2 \tag{12}$$

In the analysis of variance, these squares were divided by 36, giving

$$\sigma_i^2 + \tfrac{1}{2} 9 \, \sigma_b^2 \tag{13}$$

The reader may verify that the same expression holds for component (b) when there are only two replications, and that the general formula is $(\sigma_i^2 + \tfrac{1}{2}k \, \sigma_b^2)$.

Since an estimate of $\sigma_i^2$ is available, component (b) can be adjusted so as to provide an estimate of $(\sigma_i^2 + 9 \, \sigma_b^2)$. In combining this with the estimate from component (a), more weight should be given to the latter. The exact method of weighting is, however, complicated.

Yates [16], discussed this problem and suggested a simple method of weighting which is sufficiently accurate. The average mean square B of components (a) and (b) is an estimate of $\sigma_i^2 + \tfrac{3}{4} 9 \, \sigma_b^2$. Since the intra-block error mean square E is an estimate of $\sigma_i^2$, an estimate of $(\sigma_i^2 + 9 \, \sigma_b^2)$, the true variance between blocks, is given by $E + 4(B - E)/3 = (4B - E)/3$. Hence w' may be taken as $3/(4B - E)$.

The general formulas for estimating w and w' are given below for application in other examples.

*Two replications:*
   E = intra-block error mean square,
   B = mean square for component (b), based on $2(k - 1)$
      degrees of freedom,

$$w = 1/E, \quad w' = 1/(2B - E) \tag{14}$$

*Four replications:*

Here B = average mean square for components (a) and (b) based on $4(k-1)$ degrees of freedom,

$$w = 1/E, \quad w' = 3/(4B - E) \tag{15}$$

*Six replications:*

Component (a), based on $4(k-1)$ degrees of freedom, is sufficient. If B is the mean square for component (a),

$$w = 1/E, \quad w' = 1/B \tag{16}$$

Component (b) need not be used.

If, in any of these cases, B is less than or equal to E, w and w' should be taken as equal, i. e., the randomized complete blocks analysis should be used.

### 3. STANDARD ERRORS OF ADJUSTED VARIETAL MEANS

The expressions for the error variance of the difference between two adjusted varietal means are most easily obtained by expressing each mean in terms of X, Y effects and their interactions.

#### VARIETIES IN THE SAME BLOCK.

Let $v_{11}$, $v_{21}$ be the adjusted mean yields of two varieties which appear in the same block. The suffixes now refer to the *levels* of X and Y. Since the varieties appear in the same block, they have one suffix in common, which has been taken as the Y suffix.

$$v_{11} - M = (X_1 - M) + (Y_1 - M) + (v_{11} - X_1 - Y_1 + M) \tag{17}$$
$$v_{21} - M = (X_2 - M) + (Y_1 - M) + (v_{21} - X_2 - Y_1 + M) \tag{18}$$

Hence

$$v_{11} - v_{21} = (X_1 - X_2) + (v_{11} - X_1 - v_{21} + X_2) \tag{19}$$

Now

$$x_1 = \frac{wX_{11} + w'X_{b1}}{w + w'},$$

and if there are r replications, the variance of $X_{i1}$ is $2/rkw$ and that of $X_{b1}$ is $2/rkw'$, $X_{i1}$ and $X_{b1}$ being the means of $rk/2$ quantities. Hence

$$\text{variance } (X_1) = \frac{2}{rk}\left(\frac{1}{w + w'}\right) \tag{20}$$

Since $X_1$ and $X_2$ are independent, being derived from completely different sets of plots,

$$\text{variance } (X_1 - X_2) = \frac{4}{rk}\left(\frac{1}{w + w'}\right) \tag{21}$$

For the interaction term it is useful to note that this is unconfounded, and therefore must give the same result if calculated from unadjusted values of $v_{11}$, $v_{21}$, $X_1$ and $X_2$. The variance of this term is most easily found by expressing it as a linear function of single-plot yields. Since there are $r$ replicates, it is sufficient to do this for one replication, dividing the resulting variance by $r$.

In each replication, one plot of $X_1$ carries the variety $v_{11}$, and one plot of $X_2$ carries $v_{21}$. Hence the interaction term for a single replication contains 2 plot yields with coefficient $\pm\left(\dfrac{k-1}{k}\right)$ and $2(k-1)$ plot yields with coefficient $\left(\pm\dfrac{1}{k}\right)$. Thus the variance of the interaction term is

$$\frac{2}{rw}\left[\left(\frac{k-1}{k}\right)^2 + \frac{(k-1)}{k^2}\right] = \frac{2(k-1)}{rwk} \tag{22}$$

The main effect and interaction terms are independent. Hence

$$\text{variance } (v_{11} - v_{21}) = \frac{2(k-1)}{rwk} + \frac{4}{rk}\left(\frac{1}{w + w'}\right) \tag{23}$$

$$= \frac{2}{rwk}\left[(k-1) + \frac{2w}{w + w'}\right] \tag{24}$$

If $w' = 0$, this reduces to $\dfrac{2}{rw}\dfrac{(k+1)}{k}$, the value for the intra-block analysis. If $w' = w$, the variance becomes $2/rw$, the formula for randomized complete blocks.

VARIETIES NOT IN THE SAME BLOCK.

If $v_{11}$, $v_{22}$ are the adjusted mean yields of two varieties which do not appear in the same block,

$$v_{11} - v_{22} = (X_1 - X_2) + (Y_1 - Y_2) +$$

$$(v_{11} - X_1 - Y_1 - v_{22} + X_2 + Y_2) \tag{25}$$

By the previous analysis, the variance of the main-effect component is

$$\frac{8}{rk}\left(\frac{1}{w + w'}\right)$$

The expression of the interaction term as a linear function of independent plot yields requires a little more care and is facilitated by constructing a diagram representing a typical replication. It will be found that in each replication the function contains two plots with coefficient $\pm\left(\frac{k-2}{k}\right)$, two plots with coefficient 0, and $4(k-2)$ plots with coefficient $\pm\frac{1}{k}$. Hence the variance of the interaction term is

$$\frac{1}{rw}\left[2\left(\frac{k-2}{k}\right)^2 + 4\frac{(k-2)}{k^2}\right] = \frac{2}{rw}\left(\frac{k-2}{k}\right) \quad (26)$$

This gives

$$\text{variance } (v_{11} - v_{22}) = \frac{2}{rw}\left(\frac{k-2}{k}\right) + \frac{8}{rk}\left(\frac{1}{w+w'}\right) \quad (27)$$

$$= \frac{2}{rkw}\left[(k-2) + \frac{4w}{w+w'}\right] \quad (28)$$

This reduces to the correct values when $w' = 0$, or $w' = w$.

### 4. AVERAGE ERROR VARIANCE OF ALL VARIETAL COMPARISONS

This may be calculated either from the above results, or from the analysis of variance. There are $\frac{1}{2}k^2(k^2-1)$ comparisons between pairs of varieties, of which $k^2(k-1)$ are between varieties in the same block, and the remaining $\frac{1}{2}k^2(k-1)^2$ are between varieties which do not appear in the same block. The appropriate weighted mean of (24) and (28) is found to be

$$\frac{2}{rw(k+1)}\left[(k-1) + \frac{4w}{w+w'}\right] \quad (29)$$

Alternatively we may note that on a single plot basis, the $2(k-1)$ degrees of freedom representing main effects have variance $2/(w+w')$, while the $(k-1)^2$ degrees of freedom representing interactions have variance $(1/w)$. Taking the corresponding weighted mean, and dividing by $r/2$ to give the average variance of the difference between two varietal means, we verify (29).

## 5. EFFICIENCY OF THE LATTICE DESIGN RELATIVE TO RANDOMIZED BLOCKS

If the experiment were arranged in randomized complete blocks of $k^2$ plots, the error variance for any replication would be a weighted mean of the $(k - 1)$ degrees of freedom between incomplete blocks, with variance $1/w'$, and the $k(k - 1)$ degrees of freedom within incomplete blocks, with variance $1/w$. Thus the average error variance per plot is $\left(\dfrac{k}{w} + \dfrac{1}{w'}\right)\Big/(k + 1)$. This is comparable with $r/2$ times the average variance of the difference between two varieties in the previous section. Hence, with recovery of inter-block information, the efficiency of the lattice design relative to randomized complete blocks is measured by

$$\frac{k + w/w'}{(k - 1) + 4w/(w + w')} \tag{30}$$

In this ratio the loss of information due to inaccurate weighting is ignored.

The percentage efficiency is shown in table 24 for various values of $w/w'$, with and without recovery of inter-block information.

TABLE 24. PERCENTAGE EFFICIENCIES OF THE LATTICE DESIGN RELATIVE TO RANDOMIZED COMPLETE BLOCKS.

| k | w/w' | | | | | | |
|---|---|---|---|---|---|---|---|
| | 1 | 2 | 3 | 4 | 6 | 8 | 10 |
| 5 | 100* | 105.0 | 114.3 | 125.0 | 148.1 | 172.0 | 196.4 |
| | 75.0* | 87.5 | 100 | 112.5 | 137.5 | 162.5 | 187.5 |
| 6 | 100 | 104.3 | 112.5 | 122.0 | 142.4 | 163.6 | 185.3 |
| | 77.8 | 88.9 | 100 | 111.1 | 133.3 | 155.6 | 177.8 |
| 7 | 100 | 103.8 | 111.1 | 119.6 | 137.9 | 157.0 | 176.4 |
| | 80.0 | 90.0 | 100 | 110.0 | 130.0 | 150.0 | 170.0 |
| 8 | 100 | 103.4 | 110.0 | 117.6 | 134.2 | 151.6 | 169.2 |
| | 81.8 | 90.9 | 100 | 109.1 | 127.3 | 145.5 | 163.6 |
| 9 | 100 | 103.1 | 109.1 | 116.1 | 131.2 | 147.1 | 163.3 |
| | 83.3 | 91.7 | 100 | 108.3 | 125.0 | 141.7 | 158.3 |
| 10 | 100 | 102.9 | 108.3 | 114.8 | 128.7 | 143.4 | 158.3 |
| | 84.6 | 92.3 | 100 | 107.7 | 123.1 | 138.5 | 153.8 |
| 11 | 100 | 102.6 | 107.7 | 113.6 | 126.6 | 140.2 | 154.0 |
| | 85.7 | 92.9 | 100 | 107.1 | 121.4 | 135.7 | 150.0 |
| 12 | 100 | 102.4 | 107.1 | 112.7 | 124.8 | 137.4 | 150.3 |
| | 86.7 | 93.3 | 100 | 106.7 | 120.0 | 133.3 | 146.7 |
| 13 | 100 | 102.3 | 106.7 | 111.8 | 123.1 | 135.0 | 147.1 |
| | 87.5 | 93.8 | 100 | 106.2 | 118.8 | 131.2 | 143.8 |

* The upper and lower figures show the percentage efficiency with and without recovery of inter-block information respectively. The loss of information due to inaccuracy in the weights is not taken into account.

This table gives some indication of the gains in efficiency to be anticipated from the use of lattice designs instead of randomized complete blocks. The recovery of inter-block information is naturally most valuable when w/w' is small.

In individual experiments, the percentage efficiency may be estimated by substituting the observed value of w/w' in (30). For the numerical example, w/w' = 12.67, giving a percentage efficiency of 185, which checks with the figure on page 23.

### 6. THE F-TEST OF THE VARIETAL DIFFERENCES

In most experiments involving large numbers of varieties, it is reasonable to expect that there will be real differences in yield among the varieties, and a precise test of significance may seldom be required. The approximate test described in the first part of this bulletin should be sufficient for practically all cases; the test appropriate to the adjusted yields using inter-block information is, however, given here for completeness.

The 64 degrees of freedom representing interactions are unaffected by confounding and are tested against the intra-block error. It is, however, necessary to recompute the 16 degrees of freedom for main effects, since these estimates are changed by the use of the inter-block information.

If $X_i' = \dfrac{rk}{2} X_i$, $X_b' = \dfrac{rk}{2} X_b$, so that $X_i'$ and $X_b'$ represent *totals*, the adjusted main effects are derived from $(wX_i' + w'X_b')$. Assuming that the weights are known exactly, the variance of this quantity is $\dfrac{rk}{2}(w + w')$. Thus the sum of squares representing X effects must be divided by $\dfrac{rkw}{2}(w + w')$, to make the resulting mean square comparable with the intra-block mean square $1/w$. In replacing the sum of squares of unadjusted X effects by the sum of squares of adjusted X effects, the change in the total unadjusted varietal sum of squares is therefore

$$\frac{2}{rkw(w + w')} S(wX_1' + w'X_b')^2 - \frac{1}{rk} S(X_1' + X_b')^2 \qquad (31)$$

By collecting the coefficients of $X_i'^2$, $X_i'X_b'$ and $X_b'^2$, this may be written

$$\frac{(w-w')}{rkw(w+w')} \; S \; [wX_1'^2 - 2wX_1'X_b' - (w+2w')X_b'^2] \qquad (32)$$

$$= \frac{(w-w')}{rkw(w+w')} \; S \; [w(X_1'-X_b')^2 - 2(w+w')X_b'^2] \qquad (32a)$$

Hence the total change in the unadjusted varietal sum of squares to adjust for X and Y effects is

$$- \lambda \left[ \left( 1 + \frac{w'}{w} \right) B_u - B_a \right] \qquad (33)$$

where $\lambda = (w-w')/(w+w')$, and $B_u$ and $B_a$ are respectively the unadjusted and adjusted sums of squares for component (b) of the blocks. In this expression $\lambda$, w, w' and $B_a$ have already been calculated in the standard analysis of the experiment, so that the only extra computation required is to find $B_u$, which is the sum of squares of deviations of the row totals in table 2, divided by 18 (or in general rk/2). Actually, $B_u$ was calculated in table 23 and was found to be 1267.19. Also $B_a = 342.75$ (table 5), w = .23872, w' = .01884 and $\lambda = .8537$ (p. 20). With these values, the adjustment to the varietal sum of squares in table 5 is —874.57, giving 1280.10 for the adjusted sum of squares of varieties. The F-value is 3.82, as against 2.97 for the approximate test in table 6.

The above test gives in general an F-value which is slightly too high, because the true variance of $(wX_1' + w'X_b')$ is in general somewhat larger than $\frac{rk}{2} (w+w')$, owing to inaccuracies in the weighting. The over-estimation is negligible in an experiment of this size.

### 7. LOSS OF INFORMATION DUE TO THE USE OF INACCURATE WEIGHTS.

The use of estimated weights, w and w', instead of the unknown true weights, results in a loss of precision in the X and Y estimates. Upon investigation, this loss was found to be negligible for the experiment discussed in the numerical example. It is, however, important to investigate the loss for smaller lattice designs with only two replicates, in order to form some judgment on the minimum size of experiment for which the recovery of inter-block information is worthwhile.

In the notation of section 2, the true intra- and inter-block error variances per plot are $\sigma_i^2$ and $\sigma'^2 = \sigma_i^2 + k\sigma_b^2$. The true variance of $(wX_i + w'X_b)/(w + w')$, from two replications is therefore

$$\frac{1}{k} \frac{w^2\sigma_i^2 + w'^2\sigma'^2}{(w + w')^2} = \frac{\frac{1}{k}\left(\frac{w^2}{W} + \frac{w'^2}{W'}\right)}{(w + w')^2} \tag{34}$$

where $W = 1/\sigma_i^2$, $W' = 1/(\sigma_i^2 + k\sigma_b^2)$.

The variance of the correctly-weighted mean,

$$(WX_i + W'X_b)/(W + W')$$

is $1/k \ (W + W')$. Hence the increase in variance due to inaccurate weighting is

$$\frac{1}{k}\left[\frac{\left(\frac{w^2}{W} + \frac{w'^2}{W'}\right)}{(w + w')^2} - \frac{1}{(W + W')}\right]$$

$$= \frac{\left(\frac{w}{w'}\cdot\frac{W'}{W} - 1\right)^2}{kW'(1 + W'/W)(1 + w/w')^2} \tag{35}$$

after simplification.

From (14), $w/w' = (2B/E - 1)$. The mean square B is an estimate of $(\sigma_i^2 + \frac{1}{2}k\sigma_b^2)$, based on $2(k - 1)$ degrees of freedom, while E is an estimate of $\sigma_i^2$, based on $(k - 1)^2$ degrees of freedom. Thus B/E may be written

$$F\left(1 + \frac{k\sigma_b^2}{2\sigma_i^2}\right)$$

where $F = e^{2z}$. Making the necessary substitutions, (35) can be expressed as

$$\frac{1}{k}\frac{W'}{W(W + W')}\left(\frac{F - 1}{F}\right)^2 \tag{36}$$

where F is based on $2(k - 1)$ and $(k - 1)^2$ degrees of freedom respectively.

This increase in variance applies to each of the $2(k - 1)$ degrees of freedom for main effects. Hence the average increase in variance over all $(k^2 - 1)$ degrees of freedom between varieties is

$$\frac{2}{k(k + 1)}\frac{W'}{W(W + W')}\left(\frac{F - 1}{F}\right)^2 \tag{37}$$

The average variance with correct weights is

$$\frac{1}{k(k^2-1)}\left[\frac{(k-1)^2}{2W}+\frac{2(k-1)}{W+W'}\right]=\frac{(k+3)W+(k-1)W'}{k(k+1)2W(W+W')} \tag{38}$$

Hence the fractional increase in the average variance due to the use of inaccurate weights is

$$Q=\frac{4}{(k+3)\dfrac{W}{W'}+(k-1)}\left(\frac{F-1}{F}\right)^2 \tag{39}$$

and the corresponding fractional loss of information is

$$Q/(1+Q)$$

This loss is a function of the unknown ratio of the true weights and of the F-value which is obtained in any experiment. The loss increases without bound as F decreases towards zero. However, by the rule given above that equal weights are used when $B \leqq E$, a lower limit is set to this portion of the loss.

The average loss may be calculated, by numerical integration, for any lattice design with two replicates, and for any given ratio $W/W'$. The $5 \times 5$ design was selected for detailed investigation. The F-ratio is based on only 8 and 16 degrees of freedom. The average percentage losses of information are tabulated below for a set of values of $W/W'$.

| $W/W'$ | 1 | 2 | 4 | 6 | 8 |
|---|---|---|---|---|---|
| Average % loss of information | 2.21 | 3.07 | 4.54 | 4.37 | 3.91 |

The loss of information increases to a maximum of presumably about 5 percent and thereafter decreases. The losses are surprisingly small, considering the small numbers of degrees of freedom. However, the efficiency factor of this design is considerably higher than that of the $3 \times 3 \times 3$ cubic lattice considered by Yates [16].

It seems clear that for all lattice designs of size larger than the $5 \times 5$, the use of estimated weights will result in only a slight loss of precision. Even with the $5 \times 5$, the loss is not sufficiently great to invalidate the recovery of inter-block information.

## TRIPLE LATTICE DESIGNS

The mathematical analysis is an easy extension of that for the lattice and will be given in less detail. The varietal degrees of freedom are divided into X, Y and Z effects, each based on $(k-1)$ degrees of freedom, and "interactions", based on $(k-1)(k-2)$ degrees of freedom. The X, Y, and Z effects are confounded in one of each set of three replications, while the interactions are clear of blocks. Unless otherwise mentioned, the notation is the same as for the lattice.

### 1. CALCULATION OF THE ADJUSTED VARIETAL MEANS.

Since the number of replications for $X_i$ is twice that for $X_b$,

$$X = \frac{2wX_i + w'X_b}{2w + w'} \tag{40}$$

$$X_o = \frac{2X_i + X_b}{3} \tag{41}$$

Hence

$$X = X_o + \left(\frac{w - w'}{2w + w'}\right)(X_o - X_b) \tag{42}$$

The $2rkc_X$ values in table 12 are equal to $rk(X_o - X_b)$, since the row totals of table 11 are $rkX_o$ and the row totals of group X are $\frac{rk}{3}X_b$. Thus the adjustment per plot to $X_o$ is

$$c_X' = 2\left(\frac{w - w'}{2w + w'}\right)c_X$$

as given on page 33.

It follows as in the lattice design that individual varietal means are adjusted by adding the appropriate values of $c_X'$, $c_Y'$ and $c_Z'$.

### 2. ESTIMATION OF w AND w'

The mean square for component (a) gives a direct estimate of the variance per plot of the $X_b$, $Y_b$ and $Z_b$ estimates. This is based on 24, or $3(k-1)$ degrees of freedom, and is sufficiently accurate to determine the weight $w'$. However, this estimate can be made only when there are six or more replications, and even with six replications it is based on rather few degrees of freedom for the smaller lattice designs.

By the method of analysis used in the lattice, it may be shown that the mean square for component (b) is an estimate of $(\sigma_i^2 + \frac{2}{3} 9 \sigma_b^2)$. Thus the average mean square B for components (a) and (b) is an estimate of $(\sigma_i^2 + \frac{5}{6} \, 9 \, \sigma_b^2)$. Hence we may take

$$w' = \frac{1}{6/5 (B - E) + E} = \frac{5}{6B - E} \tag{43}$$

General rules for the estimation of w and w' in the k × k triple lattice are given below.

*Three replications:*

$$w = 1/E, \; w' = 2/(3B - E) \tag{44}$$

In this case, B is the mean square for component (b) only, since component (a) does not exist.

*Six replications:*

$$w = 1/E, \quad w' = 5/(6B - E) \tag{45}$$

B is the average mean square for components (a) and (b).

*Nine replications:*

$$w = 1/E, \quad w' = 1/B$$

where B is the mean square for component (a). Component (b) need not be used.

If B is less than or equal to E, w and w' are regarded as equal and the randomized complete blocks solution is used. In this case no adjustments are required for the unweighted varietal means.

### 3. STANDARD ERRORS OF ADJUSTED VARIETAL MEANS

Numerical suffixes will be used to denote the *levels* of the X, Y and Z factors.

#### VARIETIES IN THE SAME BLOCK.

The adjusted means of two typical varieties which appear in the same block may be represented by $v_{111}$ and $v_{221}$.

$$v_{111} - M = (X_1 - M) + (Y_1 - M) + (Z_1 - M) + \\ (v_{111} - X_1 - Y_1 - Z_1 + 2M) \tag{46}$$

$$v_{221} - M = (X_2 - M) + (Y_2 - M) + (Z_1 - M) + \\ (v_{221} - X_2 - Y_2 - Z_1 + 2M) \tag{47}$$

Hence

$$v_{221} - v_{111} = (X_2 - X_1) + (Y_2 - Y_1) + \\ (v_{221} - v_{111} - X_2 + X_1 - Y_2 + Y_1) \tag{48}$$

Now $X_1 = \dfrac{2wX_{i1} + w'X_{b1}}{2w + w'}$. With $r$ replications, the variance

of $X_{i1}$ is $3/(2rkw)$ and that of $X_{b1}$ is $3/(rkw')$.  Hence

$$\text{variance } (X_1) = \frac{3}{rk}\left(\frac{1}{2w + w'}\right) \tag{49}$$

Thus the variance of the main effect component in (48) is

$$\frac{12}{rk}\left(\frac{1}{2w + w'}\right) \tag{50}$$

The variance of the interaction component is most easily obtained by noting that this component is unchanged if adjusted values of v, X and Y are replaced by unadjusted values and expressing the component as a linear function of individual plot yields.  For a single replication (apart from a common divisor $r$) the multipliers of the plot yields are as follows:

| Number of plots | Multiplier | Contribution to variance |
|---|---|---|
| 2 | $\pm\left(\dfrac{k-2}{k}\right)$ | $\dfrac{2}{w}\left(\dfrac{k-2}{k}\right)^2$ |
| 2 | 0 | 0 |
| $4(k-2)$ | $\pm\dfrac{1}{k}$ | $\dfrac{4(k-2)}{wk^2}$ |
| Total: $4k-4$ | —— | $\dfrac{2(k-2)}{wk}$ |

Hence the variance of the interaction term is $\dfrac{2}{rw}\left(\dfrac{k-2}{k}\right)$.
By adding this expression to (50), the variance of the difference between two varietal means which occur in the same block is found to be

$$\frac{2}{rw}\left(\frac{k-2}{k}\right) + \frac{12}{rk}\left(\frac{1}{2w + w'}\right)$$

$$= \frac{2}{rkw}\left[(k-2) + \frac{6w}{2w + w'}\right] \tag{51}$$

VARIETIES NOT IN THE SAME BLOCK.

The adjusted means may be represented by $v_{111}$ and $v_{222}$. The main effect component of the difference now contains $\overline{X}$, Y and Z terms, and its variance from (49) is $\dfrac{18}{rk}\left(\dfrac{1}{2w + w'}\right)$.

The interaction component for a single replication is found to involve $(6k - 10)$ plots, of which two have coefficient $(k - 3)/k$, six have coefficient 0, and 6 $(k - 3)$ have coefficient $\underline{+}1/k$. Hence the variance of the interaction term is

$$\frac{1}{rw}\left[\frac{2(k - 3)^2}{k^2} + \frac{6(k - 3)}{k^2}\right] = \frac{2}{rw}\left(\frac{k - 3}{k}\right) \tag{52}$$

The variance of the difference between the adjusted means of two varieties not in the same block is therefore

$$\frac{2}{rw}\left(\frac{k - 3}{k}\right) + \frac{18}{rk}\left(\frac{1}{2w + w'}\right) = \frac{2}{rkw}\left[(k - 3) + \frac{9w}{2w + w'}\right] \tag{53}$$

By an appropriate weighting of (51) and (53), or from the analysis of variance, the average variance of the difference between two varietal means is found to be

$$\frac{2}{rw(k + 1)}\left[(k - 2) + \frac{9w}{2w + w'}\right] \tag{54}$$

The reader may verify that these expressions check with the values given by the intra-block analysis when $w' = 0$, and with the values for randomized complete blocks when $w = w'$.

### 4. EFFICIENCY OF THE TRIPLE LATTICE DESIGN RELATIVE TO RANDOMIZED BLOCKS

If the experiment were arranged in randomized complete blocks of $k^2$ plots, the average error variance per plot would be $\left(\dfrac{k}{w} + \dfrac{1}{w'}\right)\Big/(k + 1)$, as proved in the corresponding section for the lattice design. Thus the efficiency of the triple lattice design relative to randomized complete blocks, is

$$\frac{k + \dfrac{w}{w'}}{(k - 2) + \dfrac{9w}{2w + w'}}$$

TABLE 25.   PERCENTAGE EFFICIENCIES OF THE TRIPLE LATTICE
DESIGN RELATIVE TO RANDOMIZED COMPLETE BLOCKS.

| k | w/w' | | | | | | |
|---|---|---|---|---|---|---|---|
|  | 1 | 2 | 3 | 4 | 6 | 8 | 10 |
| 4 | 100* | 107.1 | 119.5 | 133.3 | 162.5 | 192.5 | 222.7 |
|  | 76.9* | 92.3 | 107.7 | 123.1 | 153.8 | 184.6 | 215.4 |
| 5 | 100 | 106.1 | 116.7 | 128.6 | 153.8 | 179.7 | 205.9 |
|  | 80.0 | 93.3 | 106.7 | 120.0 | 146.7 | 173.3 | 200.0 |
| 6 | 100 | 105.3 | 114.5 | 125.0 | 147.2 | 169.9 | 193.1 |
|  | 82.4 | 94.1 | 105.9 | 117.6 | 141.2 | 164.7 | 188.2 |
| 7 | 100 | 104.7 | 112.9 | 122.2 | 142.0 | 162.4 | 183.1 |
|  | 84.2 | 94.7 | 105.3 | 115.8 | 136.8 | 157.9 | 178.9 |
| 8 | 100 | 104.2 | 111.6 | 120.0 | 137.9 | 156.3 | 175.0 |
|  | 85.7 | 95.2 | 104.8 | 114.3 | 133.3 | 152.4 | 171.4 |
| 9 | 100 | 103.8 | 110.5 | 118.2 | 134.5 | 151.3 | 168.4 |
|  | 87.0 | 95.7 | 104.3 | 113.0 | 130.4 | 147.8 | 165.2 |
| 10 | 100 | 103.4 | 109.6 | 116.7 | 131.6 | 147.1 | 162.8 |
|  | 88.0 | 96.0 | 104.0 | 112.0 | 128.0 | 144.0 | 160.0 |
| 11 | 100 | 103.2 | 108.9 | 115.4 | 129.2 | 143.6 | 158.1 |
|  | 88.9 | 96.3 | 103.7 | 111.1 | 125.9 | 140.7 | 155.6 |
| 12 | 100 | 102.9 | 108.2 | 114.3 | 127.2 | 140.5 | 154.0 |
|  | 89.7 | 96.6 | 103.4 | 110.3 | 124.1 | 137.9 | 151.7 |
| 13 | 100 | 102.7 | 107.7 | 113.3 | 125.4 | 137.8 | 150.5 |
|  | 90.3 | 96.8 | 103.2 | 109.7 | 122.6 | 135.5 | 148.4 |

* The upper and lower figures give the percentage efficiencies with and without recovery of inter-block information respectively.   The loss of information through inaccuracy in the weights is ignored.

Table 25 shows these efficiencies in a similar manner to
table 24 and Yates' table VI [16].   For given values of k
and w/w', the percentage efficiency of the triple lattice is
somewhat higher than that of the lattice.   This is the result
of spreading the confounding more evenly over all varietal
comparisons.   On the other hand, the *extra* gain due to the
new method of analysis over the intra-block analysis is
smaller with the triple lattice than with the lattice, since
the intra-block analysis rejects a smaller fraction of the total
information in the former design.

### 5.   THE F-TEST OF THE VARIETAL DIFFERENCES

To obtain the varietal sum of squares appropriate to the
analysis with the recovery of inter-block information, the
unadjusted varietal sum of squares is *diminished* by the
amount,

$$\lambda \left[ \left( 1 + \frac{w'}{2w} \right) B_u - B_a \right] \tag{56}$$

where $\lambda = 2(w - w')/(2w + w')$, and $B_u$ and $B_a$ are as before the unadjusted and adjusted sums of squares for component (b) of the blocks respectively.

As an example, this test will be applied to the numbers of missing hills. A precise test is required in this case, since it is important to know whether the varieties show differences in the numbers of missing hills before adjusting the mean yields for missing hills. From table 18 we find $w = 1/E = 1/.8696 = 1.1500$, $w' = 5/(6B - E) = 0.3802$, $\lambda = .5744$, and $B_a = 56.2963$. From the row totals in table 10, $B_u$ is found to be 63.074. On substituting in (56), the adjustment to the varietal sum of squares amounts to —9.883, giving 51.451 for the new varietal sum of squares. The mean square, 0.6431, is below the intra-block error mean square, so that there is no evidence of real differences between the varieties in the number of missing hills.

6. LOSS OF INFORMATION DUE TO THE USE OF INACCURATE WEIGHTS

For a given value of $k$, the weights are least accurately determined when there are only three replications. The investigation of the loss has been confined to this case.

By an analysis similar to that given in section 7 for the lattice design, the fractional increase in the average variance of a varietal mean due to the use of inaccurate weights is found to be

$$Q = \frac{9}{2} \frac{1}{(2k + 5)\frac{W}{W'} + (k - 2)} \left(\frac{F - 1}{F}\right)^2 \tag{57}$$

where $W$, $W'$ are the true weights, and $F = e^{2z}$, is based on $3(k - 1)$ and $(k - 1)(2k - 1)$ degrees of freedom. The corresponding fractional loss of information is $Q/(1 + Q)$. Comparison of (39) and (57) shows that for given values of k and $W/W'$, the average loss of information is always smaller for the triple lattice than for the lattice. This is to be expected, since the former has a higher efficiency factor and gives more degrees of freedom for the estimation of w and w'. Thus the $5 \times 5$ triple lattice, and all larger triple lattices, give a smaller average loss of information than that found above for the $5 \times 5$ lattice.

The average losses are shown below for the $4 \times 4$ triple lattice, this being about the smallest size likely to be used in practice. The numbers of degrees of freedom in the F-ratio are 9 and 21.

| W/W' | 1 | 2 | 4 | 6 |
|---|---|---|---|---|
| Average % loss of information | 1.73 | 3.00 | 3.73 | 3.19 |

The maximum loss appears to be about 4-5 percent.

## 7. THE ANALYSIS OF COVARIANCE

In this case, the relative weights to be attached to the block and intra-block estimates must be determined after the yields have been adjusted, since, for example, the adjustments might reduce block variability very considerably if the numbers of missing hills varied widely between different blocks. The first question which arises is whether to use a single regression coefficient for the adjustments, or separate coefficients for blocks and the intra-block error. In the numerical example, the two coefficients agree closely, and it was decided to make all adjustments by means of the intra-block regression coefficient. A weighted mean of the two might have been employed, but the relative information in the block estimate was negligible. The use of a single regression coefficient makes the calculation of adjusted means somewhat easier. On the other hand, it would not be an efficient procedure if the two coefficients differed widely, and it introduces a correlation between the adjusted block estimates and the adjusted intra-block estimates.

The methods given in the paper for calculating the relative weights of the adjusted inter- and intra-block estimates and the standard errors of the final varietal means are not strictly correct, owing to certain difficulties introduced by the sampling variance of the regression coefficient. For this design, and probably also for similar designs of the same size, they are sufficiently precise, but the complications are mentioned below, as their effect becomes more important with smaller designs. The complications arise from several sources: 1. The adjusted mean square for blocks (43.419 in table 19) is not an estimate of $\sigma_i^2 + 5/6 \, k\sigma_b^2$, the coefficient 5/6 being slightly reduced because the block degrees of freedom are not orthogonal with the regression degree of freedom. The correct coefficient can be calculated without difficulty. 2. In a covariance analysis, the variance of the difference between two adjusted treatment means, $\bar{y}_1 - \bar{y}_2 - b(\bar{x}_1 - \bar{x}_2)$ is

$$\frac{2\sigma^2}{r} \left( 1 + \frac{(\bar{x}_1 - \bar{x}_2)^2}{2E_{xx}} \right) \tag{57}$$

where $\sigma^2$ is the reduced error variance, $r$ is the number of replicates, and $E_{xx}$ the error sum of squares for the independent variable. Thus the variance of the difference between two adjusted X, Y and Z components of the variety sum of squares depends upon the corresponding difference in missing hills as well as on the appropriate $\sigma^2$. As a compromise, the relative weights of the block and intra-block estimates may be based on the average variance of the difference between two X, Y or Z components. For the intra-block estimate, this is

$$\frac{2\sigma_i^2}{r}\left( 1 + F/n_i \right)$$

(58)

where F is the variance ratio for the intra-block X, Y and Z components of the missing hills against the intra-block error, and $n_i$ is the number of degrees of freedom in the intra-block error. Since the covariance is unlikely to be used unless F is near unity, the additional factor is of the order of $(1 + 1/n_i)$, being about 1.003 in the numerical example. The corresponding factor for the inter-block estimate is also very close to unity. 3. As mentioned above, the use of a common regression coefficient introduces a correlation, usually positive, between the adjusted inter- and intra-block estimates of the X, Y and Z components. In this situation, the most accurate weights and the resulting standard errors of the varietal means depend on the value of the correlation coefficient, the standard errors being somewhat higher than the values obtained by ignoring the correlation. In the present example, the effects of the correlation are negligible owing to the low sampling variance of the regression coefficient.

If separate regression coefficients are used for the block and intra-block estimates, the difficulties mentioned in (1) and (3) above disappear, but the calculation of adjusted means becomes slightly more laborious. With the smaller designs, allowance may be made for the extra complications when the need arises.

# REFERENCES CITED

[1] Bartlett, M. S. The approximate recovery of information from replicated field experiments with large blocks. Jour. Agr. Sci., 28:418-427. 1938.

[2] Fisher, R. A. The design of experiments. Oliver and Boyd, Edinburgh. 2nd Edition. 1937.

[3] Fisher, R. A. and Yates, F. Statistical tables for biological, agricultural, and medical research. Oliver and Boyd, Edinburgh. 1938.

[4] Goulden, C. H. Modern methods for testing a large number of varieties. Dominion of Canada, Dept. of Agr., Tech. Bul. 9. 1937.

[5] Goulden, C. H. Methods of statistical analysis. John Wiley and Sons, New York. 1939.

[6] Papadakis, J. S. Bul. Inst. Amél. Plantes i Salonique, No. 23.

[7] Richey, F. D. Adjusting yields to their regression on a moving average, as a means of correcting for soil heterogeneity. Jour. Agr. Res., 27:79-118. 1924.

[8] Snedecor, G. W. Statistical methods. Collegiate Press, Ames, Iowa. 1938.

[9] "Student." Yield trials. Encyclopedia of Scientific Agriculture, H. Hunter. 2:1342-1361. 1931.

[10] Weiss, M. G., and Cox, G. M. Balanced incomplete block and lattice square designs for testing yield differences among large numbers of soybean varieties. Iowa Agr. Exp. Sta., Res. Bul. 257:291-316. 1939.

[11] Yates, F. Complex experiments. Supp. Jour. Roy. Statis. Soc., 2:181-247. 1935.

[12] Yates, F. A new method of arranging variety trials involving a large number of varieties. Jour. Agri. Sci., 26:424-455. 1936.

[13] Yates, F. Incomplete randomized blocks. Ann. Eugenics, 7:121-140. 1936.

[14] Yates, F. A further note on the arrangement of variety trials: Quasi-Latin squares. Ann. Eugenics, 7:319-332. 1937.

[15] Yates, F. The design and analysis of factorial experiments. Imp. Bur. Soil Sci., Tech. Comm. No. 35. 1937.

[16] Yates, F. The recovery of inter-block information in variety trials arranged in three-dimensional lattices. Ann. Eugenics. 9:136-156. 1939.

## NOTE ON AN APPROXIMATE FORMULA FOR THE SIGNIFICANCE LEVELS OF $Z$

### By W. G. Cochran

1. **Introduction.** An important part has been played in modern statistical analysis by the distribution of $z = \frac{1}{2} \log \frac{s_1^2}{s_2^2}$, when $s_1^2$ and $s_2^2$ are two independent estimates of the same variance. In particular, all tests of significance in the analysis of variance and in multiple regression problems are based on this distribution. Complete tabulation of the frequency distribution of $z$ is a heavy task, because the distribution is a two-parameter one, the parameters being the number of degrees of freedom, $n_1$ and $n_2$ in the estimates $s_1^2$ and $s_2^2$. Thus each significance level of $z$ requires a separate two-way table. Fisher constructed a table of the 5 percent points in 1925 [1], and this has since been extended by several workers [2] to the 20, 1, and 0.1 percent level for a somewhat wider range of values of $n_1$ and $n_2$.

With his original table, Fisher gave an approximate formula for the 5 percent values of $z$, for high values of $n_1$ and $n_2$ outside the limits of his table. The formula reads:

(1)
$$z \, (5 \text{ percent}) = \frac{1.6449}{\sqrt{h - 1}} - 0.7843 \left( \frac{1}{n_1} - \frac{1}{n_2} \right),$$

$$\text{where } \frac{2}{h} = \frac{1}{n_1} + \frac{1}{n_2}.$$

The constant 1.6449 is the 5 percent significance level for *a single tail* of the normal distribution, and the constant 0.7843 will be found to be $\frac{1}{6}\{2 + (1.6449)^2\}$. Thus the general formula for the significance levels of $z$ derivable from (1) is

$$z = \frac{x}{\sqrt{h - 1}} - \left( \frac{x^2 + 2}{6} \right)\left( \frac{1}{n_1} - \frac{1}{n_2} \right),$$

where $x$ is a normal deviate with unit standard error. By inserting the appropriate significance level of $x$, this formula has been extended [2] to the tables of the 20, 1, and 0.1 percent levels of $z$ and commonly appears with all published tables of $z$. The objects of this note are to indicate the derivation of the formula and to suggest an improvement upon it in the latter cases.

Reproduced with permission from the *Annals of Mathematical Statistics,* Vol. 11, No. 1, pages 93–95. Copyright © 1940, The Institute of Mathematical Statistics.

**2. The transformation of the $z$-distribution to normality.** For high values of $n_1$ and $n_2$, the distribution of $z$ approaches the normal distribution, the principal deviation being a slight skewness introduced by the inequality of $n_1$ and $n_2$. It is therefore natural to seek an approximate formula for the distribution of $z$ by examining its relation to the normal distribution. For the $z$-distribution the ratio $\kappa_r/\kappa_2^{r/2}$, where $\kappa_r$ is the $r^{\text{th}}$ cumulant, is of the order $n^{-(\frac{1}{2}r-1)}$, where $n$ is the smaller of $n_1$ and $n_2$. This property is common to a large number of distributions which tend to normality; for example, the distribution of the mean of a sample of size $n$ from any distribution with finite cumulants. Fisher and Cornish [3] have recently given a method, applicable to all distributions with this property, for transforming the distribution to a normal distribution to any desired order of approximation. They also obtained explicit expressions for the significance levels of the original distribution in terms of the significance levels of the normal distribution, discussing the $z$-distribution as a particular example. The relation between $z$ and the normal deviate $x$ at the same level of probability was found to be

$$(2) \quad z = \frac{x}{\sqrt{h}} - \tfrac{1}{6}(x^2+2)\left(\frac{1}{n_1}-\frac{1}{n_2}\right) + \frac{1}{\sqrt{h}}\left\{\frac{x^3+3x}{12\,h} + \frac{x^3+11x}{144}h\left(\frac{1}{n_1}-\frac{1}{n_2}\right)^2\right\},$$

the three terms on the right hand side being respectively of order $n^{-\frac{1}{2}}$, $n^{-1}$, and $n^{-\frac{3}{2}}$, so that terms of order $n^{-2}$ are neglected.[1]

If this equation is compared with equation (1), the latter appears at first sight to be the approximation of order $n^{-1}$ to the $z$-distribution, except that the divisor of $x$ is $\sqrt{h-1}$ in (1) and $\sqrt{h}$ in (2). Computation of a few values shows that at the 5 percent level, equation (1) is the better approximation. For example, for $n_1 = 40$, $n_2 = 60$, (1) gives $z$ (5 percent) $= .2334$, (2) gives $.2309$, and the exact value is $.2332$.

Since

$$\frac{x}{\sqrt{h-1}} = \frac{x}{\sqrt{h}} + \frac{x}{2h\sqrt{h}} + \text{terms of order } n^{-2},$$

Fisher's approximation differs from (2) by including a correction term of order $n^{-\frac{3}{2}}$. Inspection of the true correction terms of this order in equation (2) shows that for finite values of $n_1$ and $n_2$ the term $\dfrac{x^3+11x}{144}\sqrt{h}\left(\dfrac{1}{n_1}-\dfrac{1}{n_2}\right)^2$ is considerably smaller than the term $\dfrac{x^3+3x}{12h\sqrt{h}}$, since the former has a smaller numerical coefficient and involves the difference between $\dfrac{1}{n_1}$ and $\dfrac{1}{n_2}$. Thus Fisher's formula gives a close approximation to the true formula of order $n^{-\frac{3}{2}}$, provided that $\dfrac{x}{2}$ is approximately equal to $\dfrac{x^3+3x}{12}$; i.e. if $\dfrac{x^2+3}{6}$ is approximately equal

---

[1] Fisher and Cornish also gave the two succeeding terms.

to 1. For the 5 percent level, $x = 1.6449$, and $\dfrac{x^2 + 3}{6} = 0.951$. Thus at the 5 percent level the use of $\sqrt{h - 1}$ in (1) instead of $\sqrt{h}$ extends the validity of Fisher's approximation from order $n^{-1}$ to order $n^{-\frac{3}{2}}$.

This ingenious device, however, requires adjustment at other levels of significance. The values of $(x^2 + 3)/6$ at the principal significance levels are shown below.

| Significance level—% | 40 | 30 | 20 | 10 | 5 | 1 | 0.1 |
|---|---|---|---|---|---|---|---|
| $\lambda = (x^2 + 3)/6$ | 0.51 | 0.55 | 0.62 | 0.77 | 0.95 | 1.40 | 2.09 |

If $\sqrt{h - 1}$ in formula (1) is replaced by $\sqrt{h - \lambda}$, with the above values of $\lambda$, Fisher's formula will be approximately valid to order $n^{-\frac{3}{2}}$ at all levels of significance. In particular, for the tables already published of the 20, 1 and 0.1 percent points, $\lambda$ may be taken as 0.6, 1.4 and 2.1 respectively. The values of $z$ given by the use of $\sqrt{h - 1}$ and $\sqrt{h - \lambda}$ are compared below for $n_1 = 24$, $n_2 = 60$.[2]

| Significance Level | Approximate formula | | Exact value |
|---|---|---|---|
| | $\sqrt{h - 1}$ | $\sqrt{h - \lambda}$ | |
| 20% | .1346 | .1337 | .1338 |
| 1% | .3723 | .3748 | .3746 |
| 0.1% | .4875 | .4966 | .4955 |

The use of $\sqrt{h - \lambda}$ gives values practically correct to 4 decimal places, except for the 0.1 level of significance, at which the higher terms become more important.

With the aid of this formula, complete tabulation of the $z$-distribution for a given pair of high values of $n_1$ and $n_2$ is relatively simple. If very low probabilities at the tails are required, the further approximations given by Fisher and Cornish [3] may be used.

## REFERENCES

[1] R. A. FISHER. *Statistical Methods for Research Workers*. Edinburgh, Oliver and Boyd. 1st Ed. 1925.

[2] R. A. FISHER AND F. YATES. *Statistical Tables*. Edinburgh, Oliver and Boyd. 1938.

[3] E. A. CORNISH AND R. A. FISHER. "Moments and Cumulants in the Specification of Distributions," *Revue de l'Institut International de Statistique*, Vol. 4 (1937).

IOWA STATE COLLEGE,
AMES, IOWA.

[2] The numerical terms in the approximate formula given for the 20 percent points on p. 28 of Fisher and Yates' *Statistical Tables* are in error. Their formula should read:

$$z = \frac{0.8416}{\sqrt{h - 1}} - 0.4514 \left( \frac{1}{n_1} - \frac{1}{n_2} \right)$$

## THE ANALYSIS OF VARIANCE WHEN EXPERIMENTAL ERRORS FOLLOW THE POISSON OR BINOMIAL LAWS

### By W. G. Cochran

**1. Introduction.** The use of transformations has recently been discussed by several writers [1], [2], [3], [4], in applying the analysis of variance to experimental data where there is reason to suspect that the experimental errors are not normally distributed. Two types of transformations appear to be coming into fairly common use: $\sqrt{x}$ and $\sin^{-1}\sqrt{x}$. The former is considered appropriate where the data are small integers whose experimental errors follow the Poisson law, while the latter applies to fractions or percentages derived from the ratio of two small integers, where the experimental errors follow the binomial frequency distribution. In each case the object of the transformation is to put the data on a scale in which the experimental variance is approximately the same on all plots, so that all plots may be used in estimating the standard error of any treatment comparison. The extent to which these transformations are likely to succeed in so doing has been examined by Bartlett [2]. The object of the present paper is to discuss the theoretical basis for these transformations in more detail, and in particular to examine their relation to a more exact analysis.

**2. Experimental variation of the Poisson type.** The first step in an exact statistical analysis of the results of any field experiment, is to specify in mathematical terms (1) how the expected values on each plot are obtained in terms of unknown parameters representing the treatment and block (or row and column) effects (2) how the observed values on the plots vary about the expected values. In this section, the variation is assumed to follow the Poisson law.

The specification of the expected values requires some consideration. In the standard theory of the analysis of variance, treatment and block (or row and column) effects are assumed to be additive. In the case of a Latin square, for example, the expected yield $m_i$ of the $i$th plot, which receives the $t$th treatment and occurs in the $r$th row and the $c$th column is written

$$(1) \qquad m_i = G + T_t + R_r + C_c$$

where $G$ is a parameter representing the average level of yield in the experiment, and $T_t$, $R_r$ and $C_c$ represent the respective effects of the treatment, row and column to which the plot corresponds. Since the $T$, $R$ and $C$ constants are required only to measure differences between different treatments, rows and columns, we may put

$$(2) \qquad \sum_t T_t = \sum_r R_r = \sum_c C_c = 0.$$

Reproduced with permission from the *Annals of Mathematical Statistics*, Vol. 11, No. 3, pages 335–347. Copyright © 1940, The Institute of Mathematical Statistics.

If the experimental errors are normally and independently distributed with equal variance, this specification leads to very simple equations of estimation for the unknown parameters, the maximum likelihood estimate of $T_t$, for example, being the difference between the mean yield of all plots receiving that treatment and the general mean. In addition to its simplicity, this type of prediction formula is fairly suitable for general use, because it gives a good approximation to most types of law which might be envisaged, provided that row and column differences are small in relation to the mean yield. However, in considering an exact analysis with Poisson variation, the prediction formula is assumed chosen, without reference to computational simplicity, as being the most suitable to describe the combined actions of treatment and soil effects.

The probability of obtaining a given set of plot yields $x_i$ with expectations $m_i$ may be written

$$\prod_i \frac{e^{-m_i} m_i^{x_i}}{x_i!}.$$

Thus $L$, the logarithm of the likelihood, is given by

$$(3) \qquad L = \sum_i (x_i \log m_i - m_i) - \sum_i \log x_i!.$$

Hence the maximum likelihood equation of estimation for any parameter $\theta$ assumes the form

$$(4) \qquad \sum \frac{(x_i - m_i)}{m_i} \frac{\partial m_i}{\partial \theta} = 0$$

where the summation extends over all plots whose expectations involve $\theta$. The function $\dfrac{\partial m_i}{\partial \theta}$ will usually involve a number of parameters. Since the specification of row, column and treatment effects in a 6 x 6 Latin square requires 16 independent parameters, the solution of these equations may be expected to be laborious, though it may be shortened by the intelligent use of iterative methods. The problem of obtaining exact tests of significance is also difficult. The method of maximum likelihood provides estimates of the variances and co-variances of the treatment constants, which under certain conditions can be assumed to be normally distributed if there is sufficient replication, but this can hardly be considered an exact "small sample" solution.

These remarks show that the exact solution is somewhat too complicated for frequent use. The difficulty arises principally because the typical equation of estimation consists of a *weighted* sum of the deviations of the observed from the expected values, the weights being $\dfrac{1}{m_i} \dfrac{\partial m_i}{\partial \theta}$. The factor $\dfrac{1}{m_i}$ was introduced into the weight by the Poisson variation of the experimental errors, and must be retained in any theory which claims to apply to Poisson variation. It is, however, worth considering whether some simplification cannot be introduced into the equations by assuming some particular form for the prediction formula. This line of approach seems promising when one considers the simplification introduced into the "normal theory" case by assuming the prediction formula to be linear.

For Poisson variation, the linear law does not appear to be particularly suitable, since it may give negative expectations on some plots (as happens in the numerical example considered in the next section). Further, while $\dfrac{\partial m_i}{\partial \theta}$ becomes a constant, the factor $\dfrac{1}{m_i}$ remains in the weight.

The entire weight can be made constant by assuming a linear prediction formula in the square roots and transforming the data to square roots. For a Latin square, this prediction formula is written

$$(5) \qquad \sqrt{m_i} = \alpha_i = G + T_t + R_r + C_c,$$

where

$$(6) \qquad \sum_t T_t = \sum_r R_r = \sum_c C_c = 0.$$

To find the maximum value of (3) subject to the restrictions (6), we may use the method of undetermined multipliers, maximizing

$$(7) \qquad L + \lambda(\sum_t T_t) + \mu(\sum_r R_r) + \nu(\sum_c C_c).$$

The equation of estimation for a typical treatment constant $T_t$ becomes

$$(8) \qquad \Sigma\left(\frac{x_i - m_i}{m_i}\right)\frac{dm_i}{d\alpha_i}\frac{\partial\alpha_i}{\partial T_t} + \lambda = 0, \quad \text{i.e.,} \quad \Sigma\frac{2(x_i - m_i)}{\sqrt{m_i}} + \lambda = 0,$$

the summation being extended over all plots receiving the treatment. If $a_i = \sqrt{x_i}$, then by Taylor's theorem

$$(9) \qquad x_i - m_i = (a_i - \alpha_i)\frac{dm_i}{d\alpha_i} + \frac{1}{2!}(a_i - \alpha_i)^2\frac{d^2m_i}{d\alpha_i^2} + \cdots.$$

If $m_i$ is reasonably large, only the first term on the right-hand side need be retained. When $m_i$ is small, we may use, instead of the exact square root, a quantity $a_i'$ defined so that

$$(10) \qquad x_i - m_i = (a_i' - \alpha_i)\frac{dm_i}{d\alpha_i} = 2\sqrt{m_i}(a_i' - \alpha_i).$$

Thus if the analysis is performed on the quantities $a_i'$ instead of on the original data, equation (8) becomes

$$(11) \qquad \sum_{T_t} 4(a_i' - \alpha_i) + \lambda = 0.$$

On substituting the expectations for $\alpha_i$ from (5), and using (6), we obtain

$$(12) \qquad \sum_{T_t} 4(a'_i - G - T_t) + \lambda = 0.$$

The corresponding equation for $G$ is

$$(13) \qquad \sum_i 4(a'_i - G) = 0,$$

so that $G$ is the general mean of the quantities $a'$. By adding equations (12) over all treatments, and comparing the total with (13), we find $\lambda = 0$ Hence $T_t$ is the difference between the mean yield of $a'$ over all plots receiving $T_t$ and the general mean of $a'$. In this scale the simplicity of the "normal theory" equations has apparently been recovered. Actually, the quantities $a'$ are not known exactly, since

$$(14) \qquad a' = \alpha + \frac{(x - m)}{2\sqrt{m}} = \frac{1}{2}\left(\alpha + \frac{x}{\alpha}\right)$$

where $\alpha$ is the expected value of $\sqrt{x}$. However, this process provides a means of successively approximating the maximum likelihood solution, by choosing first approximations to the quantities $\alpha$, constructing the $a''$s, solving for the unknown constants and hence obtaining second approximations to the expected values. The close relation of $a'$ to $\sqrt{x}$ is seen by remembering one of the common rules for finding square roots. This consists in guessing an approximate root ($\alpha$), dividing $x$ by the approximate root, and taking the mean of the approximate root ($\alpha$) and the resulting quotient ($x/\alpha$).

The suitability of the linear prediction formula in square roots must be considered in any example in which the above analysis is being employed. The law is intermediate in its effects between the linear law and the product law in the original data. My experience is that it is fairly satisfactory for general use, (cf. [2], p. 72) An exception may occur when it is desired to test the interaction between two treatments, both of which produce large effects. In this case the definition chosen for absence of interaction may not coincide at all closely with the definition implied in using the linear law in square roots. An example of this case was given in a previous paper [1].

In this connection it should be noted that an approximate "goodness of fit" test may be obtained of the validity of the assumptions made. Since the quantities $a'_i$ enter into the equations of estimation with weight 4, the quantity $4\sum_i (a'_i - \alpha_i)^2$ is distributed approximately as $\chi^2$ with the number of degrees of freedom in the error term of the analysis of variance. Some idea of the closeness of the approximation may be gathered by considering the simplest case in which only the mean yield is being estimated. In this case the observed values $x$ are assumed to be drawn from the same Poisson distribution, and the sufficient statistic for the mean $G$ is known to be $\Sigma(x_i)/n$. Since, however, the prediction formula is here the same in square roots as in the original scale, and since the maximum likelihood solution is invariant to change of scale, the mean value $\alpha$ of $a'$ must be *exactly* $\sqrt{\Sigma(x)/n}$, as the reader may verify by working any particular example. Thus $\Sigma 4(a' - \alpha)^2$ is found to be $\Sigma(x - \bar{x})^2/\bar{x}$, the usual $\chi^2$ test for examining whether a set of values $x$ may reasonably be assumed to come from the same Poisson distribution. By working out the exact distribution of $\Sigma(x_i - \bar{x})^2/\bar{x}$ in a number of cases [5], I previously expressed the opinion that this quantity followed the $\chi^2$ distribution sufficiently closely for most practical uses, even for values of the mean as low as 2. This opinion has since been substantiated by Sukhatme, [6] who sampled this distribution for $m = 1, 2, 3, 4$, and 5.

A high value of $\chi^2$ means either that the prediction formula is not satisfactory or that the experimental errors are higher than the Poisson distribution indicates, or that both causes are operating. These effects can sometimes be separated by examining whether the observed yields deviate from the expected yields in a systematic or a random manner. If the deviation is systematic, the prediction formula is probably unsatisfactory.

The type of approach used above resembles in many features the "exact" analysis for the probit transformation [7]. The principal difference is that in the case of probits the transformation is made to suit the *a priori* prediction formula, which postulates that the probits are a linear function of the dosage, or of the log (dosage). Thus with probits the equations of estimation still involve weights in the transformed scale. These do not seriously complicate the analysis, since only two parameters require to be estimated for a given poison. With, however, the much greater number of parameters usually involved in specifying the results of a field experiment, the attractiveness of a solution which does not involve weighting is greatly increased.

### 3. Numerical example of the square root transformation.

A $5 \times 5$ Latin square experiment on the effects of different soil fumigants in controlling wireworms was selected as an example. The average number of wireworms per plot (total of four soil samples) was just under five. Previous studies [8], [9] have indicated that with small numbers per sample, the distribution of numbers of wireworms tends to follow the Poisson law.

The plan and yields are shown in Table I. The first two figures under the treatment symbols are the numbers of wireworms and their square roots respectively, the latter being regarded as first approximations to the values $a'$. Two of the plots receiving treatment $K$ gave no wireworms. Since these plots are likely to be changed most in the transition from square roots to $a'$, better approximations were estimated for them before proceeding with the calculations. The best simple approximations appeared to be obtained from the square roots of the means in the original units. For the plot in the second row and second column, the square roots of the row, column and treatment means in the original

## TABLE I

*Plan and number of wireworms per plot*

| P | O | N | K | M | Mean |
|---|---|---|---|---|------|
| 3[1] | 2 | 5 | 1 | 4 | |
| 1.73[2] | 1.41 | 2.24 | 1.00 | 2.00 | 1.676[2] |
| 1.76[3] | 1.45 | 2.25 | 1.11 | 2.00 | 1.714[3] |
| 1.77[4] | 1.46 | 2.25 | 1.10 | 2.00 | 1.716[4] |

| M | K | O | N | P | |
|---|---|---|---|---|---|
| 6 | 0 | 6 | 4 | 4 | |
| 2.45 | (0.39) | 2.45 | 2.00 | 2.00 | 1.858 |
| 2.45 | 0.32 | 2.50 | 2.02 | 2.02 | 1.862 |
| 2.46 | 0.32 | 2.49 | 2.02 | 2.02 | 1.862 |

| O | M | K | P | N | |
|---|---|---|---|---|---|
| 4 | 9 | 1 | 6 | 5 | |
| 2.00 | 3.00 | 1.00 | 2.45 | 2.24 | 2.138 |
| 2.10 | 3.09 | 1.00 | 2.47 | 2.25 | 2.182 |
| 2.13 | 3.08 | 1.00 | 2.46 | 2.25 | 2.184 |

| N | P | M | O | K | |
|---|---|---|---|---|---|
| 17 | 8 | 8 | 9 | 0 | |
| 4.12 | 2.83 | 2.83 | 3.00 | (0.79) | 2.714 |
| 4.18 | 2.84 | 2.83 | 3.00 | 0.77 | 2.724 |
| 4.17 | 2.84 | 2.83 | 3.00 | 0.77 | 2.722 |

| K | N | P | M | O | |
|---|---|---|---|---|---|
| 4 | 4 | 2 | 4 | 8 | |
| 2.00 | 2.00 | 1.41 | 2.00 | 2.83 | 2.048 |
| 2.14 | 2.02 | 1.49 | 2.04 | 2.92 | 2.122 |
| 2.10 | 2.03 | 1.50 | 2.05 | 2.90 | 2.116 |

| Mean | | | | | |
|------|---|---|---|---|---|
| 2.460[2] | 1.926 | 1.986 | 2.090 | 1.972 | 2.087[2] |
| 2.526[3] | 1.944 | 2.014 | 2.128 | 1.992 | 2.121[3] |
| 2.526[4] | 1.946 | 2.014 | 2.126 | 1.988 | |

### Treatment Means

| K | P | O | M | N |
|---|---|---|---|---|
| 1.036[2] | 2.084 | 2.338 | 2.456 | 2.520 |
| 1.068[3] | 2.116 | 2.394 | 2.482 | 2.544 |
| 1.058[4] | 2.118 | 2.396 | 2.484 | 2.544 |

[1] Original numbers.    [2] Square roots.    [3] Second approximations.    [4] Third approximations.

units are respectively 2.000, 2.145 and 1.095, and the square root of the general mean is 2.227. Hence

$$a' = \tfrac{1}{2}[2.000 + 2.145 + 1.095 - 2(2.227)] = 0.39.$$

The other zero value was similarly found to give $a' = 0.79$. The corresponding estimates from the means of the square roots were considerably too low, since the $a'$ values tend to be higher than the square roots. The use of "missing plot" technique gave very poor approximations, because it ignores the fact that the plots in question had zero yields.

With the estimated values inserted, the row, column, and treatment means of the square roots are as shown in Table I. A second approximation to $a'$ was calculated for each plot. For the plot in the first row and the first column, the expected yield is

$$\alpha = 1.676 + 2.460 + 2.084 - 2(2.087) = 2.046.$$

Hence $a' = \tfrac{1}{2}(2.046 + 3/2.046) = 1.76$. These values constitute the third set of figures in Table I. Theoretically, it is advisable to readjust the row, column, and treatment means after each new value of $a'$ has been obtained, in order to secure rapid convergence. This is rather laborious in practice, and a complete set of new plot values was obtained before readjusting the means. The third approximations obtained by this method are shown in the fourth lines in Table I and are correct to two decimal places.

It is noteworthy how closely the square roots agree with the third approximations on all plots except those which originally gave zero yields. The differences between the second and third approximations are trivial.

The next step is to make a $\chi^2$ test by means of the quantity $4\Sigma(a' - \alpha)^2$. From the manner in which the values $\alpha$ are constructed from the $a''$s, it follows that $\Sigma(a' - \alpha)^2$ is simply the error sum of squares in the conventional analysis of variance of the values $a'$. The analysis of variance of the third approximations is shown in Table II.

TABLE II

*Analysis of variance of adjusted square roots*

|            | Degrees of freedom | Sum of squares | Mean square |
|------------|:---:|:---:|:---:|
| Rows       | 4  | 2.9815 |        |
| Columns    | 4  | 1.1190 |        |
| Treatments | 4  | 7.5815 | 1.8954 |
| Error      | 12 | 4.5970 | 0.3831 |

The value of $\chi^2$ is $4 \times 4.597 = 18.39$, with 12 degrees of freedom, which is just about the 10 percent level. If the hypothesis is regarded as disproved only when $\chi^2$ exceeds the 5 percent level, the treatment means may be tested by regarding them as approximately normally distributed with variance

$1/5 \times 0.25 = 0.05$.   It is, however, more prudent to use the actual error mean square as an estimate of the experimental error variance, performing the usual tests associated with the analysis of variance.   This may be justified on the grounds that the calculations have produced a set of plot values $a'$ of equal weight.   On this basis the standard error of a treatment mean is $\sqrt{0.3831/5} = 0.2768$.   Treatment $K$ reduced the number of wireworms significantly below all other treatments, but there is no indication of any difference between the other treatments.   The treatment means may be reconverted to the original units by squaring.

**4. Experimental variation of the binomial type.**   In this case the yields are obtained by examining a constant number $n$ units per plot and noting those which possess a certain attribute (e.g., plants which are diseased).   Experimental variation is presumed to arise solely from the binomial variation of the observed fraction $p$ possessing the attribute about the expected fraction $P$, which is specified in terms of unknown parameters representing the treatment and soil effects.

If $r_i$ is the number possessing the attribute on a typical plot, so that $p_i = r_i/n$ the likelihood function takes the form

$$\prod_i \frac{n!}{r_i!(n-r_i)!}\, P_i^{r_i} Q_i^{n-r_i}.$$

Hence the terms in the logarithm which involve the unknown parameters are given by

(15) $$L = \sum_i \{r_i \log P_i + (n - r_i) \log Q_i\}.$$

The equation of estimation for a typical constant $\theta$ is

(16) $$\Sigma \frac{n}{P_i Q_i}\,(p_i - P_i)\,\frac{\partial P_i}{\partial \theta} = 0$$

where the summation is over all plots whose expectations involve $\theta$.

As in the Poisson case, an exact solution is laborious because of the weights $\frac{n}{P_i Q_i} \cdot \frac{\partial P_i}{\partial \theta}$.   The unequal weighting may be removed by transforming to the variate $\alpha_i = \sin^{-1} \sqrt{P_i}$, and assuming that the prediction formula is linear in the transformed scale.   For a Latin square the prediction formula is assumed to be

(17) $$\alpha_i = G + T_t + R_r + C_c$$

where the $i$th plot receives treatment $t$ and lies in the $r$th row and $c$th column. Further

(18) $$\sum_t T_t = \sum_r R_r = \sum_c C_c = 0.$$

Since $P_i = \sin^2 \alpha_i$, $\dfrac{dP_i}{d\alpha_i} = 2\sqrt{P_i Q_i}$. A set of variates $a_i'$ is defined so that on each plot

$$(19) \qquad p_i - P_i = (a_i' - \alpha_i) \frac{dP_i}{d\alpha_i} = 2\sqrt{P_i Q_i}\,(a_i' - \alpha_i).$$

With these substitutions, the equation of estimation for $T_t$, for instance, becomes

$$(20) \qquad \sum_{T_t} 4n(a_i' - \alpha_i) + \lambda = 0$$

where, as before, $\lambda$ is an undetermined multiplier. The remainder of the solution proceeds exactly as in the Poisson case, $T_t$ being found to be the difference between the mean value of $a_i'$ over all plots receiving this treatment and the general mean of $a_i'$. A $\chi^2$ test may be made with $\sum_i 4n(a_i' - \alpha_i)^2$.

From (19)

$$(21) \qquad a_i' = \alpha_i + \frac{1}{2\sqrt{P_i Q_i}}\,(p_i - P_i) = \alpha_i + \frac{1}{2\sqrt{P_i Q_i}}\,(Q_i - q_i)$$

$$(22) \qquad = \alpha_i + \tfrac{1}{2}\cot \alpha_i - q_i \operatorname{cosec}(2\alpha_i)$$

where $q_i$ is the observed fraction which does not possess the attribute. The calculation of approximations to $a_i'$ thus involves finding a predicted value $\alpha_i$ from the treatment and block (or row and column) means, and using equation (22). Tables [10] of the values of $\sin^{-1} \sqrt{P_i}$, $\alpha_i + \tfrac{1}{2}\cot \alpha_i$, and $\operatorname{cosec}(2\alpha_i)$ have been prepared to facilitate the computations. It should be noted that these tables are in degrees, whereas the above equations assume that $\alpha_i$ is measured in radians. In degrees, equation (20) above becomes

$$(23) \qquad \sum_{T_t} \frac{\pi^2 n}{8100}\,(a_i' - \alpha_i) = 0$$

while

$$(24) \qquad a_i' = \alpha_i + \frac{180}{\pi}\,\{\tfrac{1}{2}\cot \alpha_i - q_i \operatorname{cosec}(2\alpha_i)\}.$$

As in the Poisson case, the appropriateness of the linearly additive law in equivalent angles depends on the way in which treatment and soil effects operate. As Bliss has shown [11], the effect of the transformation is to flatten out the cumulative normal frequency distribution, extending the range over which it can be approximated by a straight line.

**5. Numerical example of the angular transformation.** The data were selected from a randomized blocks experiment by Carruth [12] on the control by mechanical and insecticidal methods of damage due to corn ear worm larvae.

The control and the six types of mechanical protection were chosen for analysis, the "yields" being the percentages of ears unfit for sale. The numbers of ears varied somewhat from plot to plot, the average being 36.5, but the variations were fairly small and appeared to be random. It was considered that variations in the weight ($4n$) could be ignored in solving the equations of estimation.

## TABLE III
### *Percentages of unfit ears of corn*

| Treatments | Blocks | | | | | | Means | |
|---|---|---|---|---|---|---|---|---|
| | I | II | III | IV | V | VI | | |
| | 42.4[1] | 34.3 | 24.1 | 39.5 | 55.5 | 49.1 | | |
| 1 | 40.6[2] | 35.8 | 29.4 | 38.9 | 48.2 | 44.5 | 39.57[2] | |
| | 40.7[3] | 36.0 | 29.4 | 38.9 | 48.6 | 44.6 | 39.70[3] | |
| | 23.5 | 15.1 | 11.8 | 9.4 | 31.7 | 15.9 | | |
| 2 | 29.0 | 22.9 | 20.1 | 17.9 | 34.3 | 23.5 | 24.62 | |
| | 29.1 | 23.1 | 20.3 | 18.2 | 34.3 | 23.5 | 24.75 | |
| | 33.3 | 33.3 | 5.0 | 26.3 | 30.2 | 28.6 | | |
| 3 | 35.2 | 35.2 | 12.9 | 30.9 | 33.3 | 32.3 | 29.97 | |
| | 35.5 | 35.3 | 14.5 | 31.0 | 33.4 | 32.4 | 30.35 | |
| | 11.4 | 13.5 | 2.5 | 16.6 | 39.4 | 11.1 | | |
| 4 | 19.7 | 21.6 | 9.1 | 24.0 | 38.9 | 19.5 | 22.13 | |
| | 19.8 | 21.7 | 10.0 | 24.4 | 39.9 | 19.6 | 22.57 | |
| | 14.3 | 29.0 | 10.8 | 21.9 | 30.8 | 15.0 | | |
| 5 | 22.2 | 32.6 | 19.2 | 27.9 | 33.7 | 22.8 | 26.40 | |
| | 22.6 | 32.7 | 19.2 | 28.0 | 33.7 | 22.9 | 26.52 | |
| | 8.5 | 21.9 | 6.2 | 16.0 | 13.5 | 15.4 | | |
| 6 | 17.0 | 27.9 | 14.4 | 23.6 | 21.6 | 23.1 | 21.27 | |
| | 17.4 | 28.2 | 14.5 | 24.0 | 22.1 | 23.2 | 21.57 | |
| | 16.6 | 19.3 | 16.6 | 2.1 | 11.1 | 11.1 | | |
| 7 | 24.0 | 26.1 | 24.0 | 8.3 | 19.5 | 19.5 | 20.23 | |
| | 24.3 | 26.2 | 28.8 | 10.9 | 20.1 | 19.5 | 21.63 | |
| Means | 26.81[2] | 28.87 | 18.44 | 24.50 | 32.79 | 26.46 | 26.31 | |

[1] Percentage.  [2] Equivalent angle.  [3] Second approximation.

The percentages of unfit ears, the equivalent angles and the second approximations to $a'$ are shown in descending order in Table III. The percentages on individual plots vary from 2.1 to 55.5. The second approximations were calculated from the block and treatment means of the angles. For the control plot (treatment 1) in block I, for example, the expected value is

$$39.57 + 26.81 - 26.31 = 40.07.$$

Since Fisher and Yates's tables of $\alpha + \frac{1}{2} \cot \alpha$ and cosec $(2\alpha)$ are given for values of $\alpha$ from 45° to 90°, we take the complement of the expected value, which is 49.93. Interpolating mentally from the table, we find

$$\alpha + \tfrac{1}{2} \cot \alpha = 74.0, \text{ cosec } (2\alpha) = 58.3.$$

Thus the second approximation to the complement of the angle is

$$74.0 - 0.424 \times 58.3 = 49.3.$$

Hence the second approximation to $a'$ is 40.7, which agrees very closely with the equivalent angle.

On the majority of the plots, the second approximation differs by only a trivial amount from the equivalent angle. The plots with the three lowest percentages (2.1, 2.5, and 5.0) have increased somewhat more, and also one or two other plots where the angles deviated considerably from the expected values. A third set of approximations was not considered necessary.

The analysis of variance of the second approximations is given in Table IV.

TABLE IV

|            | Degrees of freedom | Sum of squares | Mean squares |
|------------|--------------------|----------------|--------------|
| Blocks     | 5                  | 709.79         |              |
| Treatments | 6                  | 1,531.56       | 255.26       |
| Error      | 30                 | 982.67         | 32.76        |

Taking $n$ as 36.5, the expected value of the error mean square is $820.7/36.5 = 22.48$. Thus $\chi^2 = 982.67/22.48 = 43.71$, with 30 degrees of freedom, which is almost exactly at the 5 percent level. This, together with the appreciable amount of the variance removed by blocks, indicates that the experimental error probably contains some element other than binomial variation. As in the preceding case, it would be wise to make the usual analysis of variance tests with the actual error mean square.

**6. Discussion.** It must be emphasized that the solutions given above apply to the case where the whole of the experimental error variation is of the Poisson or binomial type. The methods are therefore likely to be useful in practice only where the experimental conditions have been carefully controlled, or where the data are derived from such small numbers that the Poisson or binomial variation is much larger than any extraneous variation. The $\chi^2$ test is helpful in deciding whether this assumption is justified. Further, the examples worked above indicate that the transformed values form very good approximations on most plots. It will often be sufficient to adjust only those plots which give zero or very small values in the Poisson case, or zero or 100 percent values in the binomial case. In this connection the method of adjustment given above may perhaps be considered as an improvement on the empirical rule given by Bartlett [13] of counting $n$ out of $n$ as $(n - 1/4)$ out of $n$.

Where extraneous variation becomes important, as is probably the normal case with data derived from field experiments, there seem to be no theoretical grounds for using the adjusted values. If we were prepared to describe accurately the nature of the variation other than that of the Poisson or binomial type, a new set of maximum likelihood equations could be developed. These would, however, lead to a different type of adjustment.

The justification for the use of transformations has no direct relation to the Poisson or binomial laws in this case, or in cases where percentages are derived from the ratios of two *weights* or volumes, as in chemical analyses, or from an arbitrary observational scoring With percentages, for example, it may be said, without describing the experimental variation in detail, that the variance must vanish at zero and 100 percent and is likely to be greatest in the middle. The formula $V = \lambda PQ$ is at least a first approximation to this situation. The angular transformation will approximately equalize a distribution of variances of this type, provided that $\lambda$ is sufficiently small. We have, of course, returned to an "approximate" type of argument. It follows that the original data should be scrutinized carefully before deciding that a transformation is necessary and that any presumed opinions about the nature of the experimental variation should be verified as far as possible.

**7. Summary.** This paper discusses the theoretical basis for the use of the square root and inverse sine transformations in analyzing data whose experimental errors follow the Poisson and binomial frequency laws respectively.

The maximum likelihood equations of estimation are developed for each case, but are in general too complicated for frequent use. If, however, the expected yield of any plot is assumed to be an additive function of the treatment and soil effects in the transformed scale, a transformation can be found so that the equations of estimation assume the simple "normal theory" form. The transforms are closely related to the square roots and inverse sines respectively.

The nature of the assumed formula for the expected values is briefly discussed, and a $\chi^2$ test is developed for the combined hypotheses that the prediction formula is satisfactory and that the experimental errors follow the assumed law.

Numerical examples are worked for both types of transformation. These indicate that even for data derived from small numbers, the square roots or inverse sines are good estimates of the correct transforms on almost all plots, except those which give zero yields in the Poisson case, or percentages near zero or 100 in the binomial case.

In practice, these new methods are not recommended to supplant the simple transformations for general use, because it can seldom be assumed that the whole of the experimental error variation follows the Poisson or binomial laws. The more exact analysis may, however, be useful (*i*) for cases in which the plot yields are very small integers or the ratios of very small integers (*ii*) in showing how to give proper weight to an occasional zero plot yield.

## REFERENCES

[1] W. G. Cochran, "Some difficulties in the statistical analysis of replicated experiments," *Empire J. Expt. Agric.*, Vol. 6 (1938), pp. 157–75.

[2] M. S. Bartlett, "The square root transformation in the analysis of variance," *J. Roy. Stat. Soc. Suppl.*, Vol. 3 (1936), pp. 68–78.

[3] C. I. Bliss, "The transformation of percentages for use in the analysis of variance," *Ohio J. Sci.*, Vol. 38 (1938), pp. 9–12.

[4] A. Clark and W. H. Leonard, "The analysis of variance with special reference to data expressed as percentages," *J. Amer. Soc. Agron.*, Vol. 31 (1939), pp. 55–56.

[5] W. G. Cochran, "The $\chi^2$ distribution for the Binomial and Poisson series, with small expectations," *Ann. Eugen.*, Vol. 7 (1936), pp. 207–17.

[6] P. V. Sukhatme, "On the distribution of $\chi^2$ in samples of the Poisson series," *J. Roy. Stat. Soc. Suppl.*, Vol. 5 (1938), pp. 75–9.

[7] C. I. Bliss, "The determination of the dosage-mortality curve from small numbers," *Quart. J. Pharmacy and Pharmacology*, Vol. 11 (1938), pp. 192–216.

[8] A. W. Jones, "Practical field methods of sampling soil for wireworms," *J. Agric. Res.*, Vol. 54 (1937), pp. 123–34.

[9] W. G. Cochran, "The information supplied by the sampling results," *Ann. App. Biol.*, Vol. 25 (1938), pp. 383–9.

[10] R. A. Fisher and F. Yates, *Statistical tables for agricultural, biological and medical research*, Edinburgh, Oliver and Boyd, 1938.

[11] C. I. Bliss, "The analysis of field experimental data expressed in percentages," *Plant Protection* (Leningrad), 1937, pp. 67–77.

[12] L. A. Carruth, "Experiments for the control of larvae of *Heliothis Obsoleta Fabr*," *J. Econ. Ent.*, Vol. 29 (1936), pp. 205–9.

[13] M. S. Bartlett, "Some examples of statistical methods of research in agriculture and applied biology," *J. Roy. Stat. Soc. Suppl.*, Vol. 4 (1937), p. 168, footnote.

Iowa State College,
Ames, Iowa

# 23

## THE ESTIMATION OF THE YIELDS OF CEREAL EXPERIMENTS BY SAMPLING FOR THE RATIO OF GRAIN TO TOTAL PRODUCE

By W. G. COCHRAN

*Statistical Department, Rothamsted Experimental Station, Harpenden, Herts*

### 1. Introduction

THRESHING difficulties constitute a formidable barrier in planning an extensive series of small-plot cereal experiments of modern design, or in studying the residual effects of treatments on the succeeding cereals in a series of experiments on root crops. Failing the provision of a small threshing machine, sampling is at present the only practicable method for obtaining the grain yields of small plots located on commercial farms.

Yates & Zacopanay (1935) summarized the work carried out at Rothamsted and its associated outside centres on the estimation of the yields of cereal crop experiments by sampling. In these experiments a number of small areas (e.g. $\frac{1}{2}$ m. of each of four contiguous rows) were selected at random in each plot or subplot. The standing crop in each of these areas was cut close to the ground, bagged, and transported to Rothamsted for threshing, the yields of grain and straw per unit area being estimated entirely from the samples.

The authors suggested that, if the total produce of each plot were weighed on the field, the samples need be used only to determine the ratio of the weight of grain to the weight of total produce. In view of the high correlation which normally exists between grain and straw yields, the sampling errors of this ratio might be expected to be considerably smaller than those of the yield of grain itself, so that less sampling would be required to obtain results of equal precision. They found from the average of nine experiments that the sampling error per metre of row length was 7·14 % for the ratio of grain to total produce, compared with 23·9 % for the yield of grain. Judging from these figures, only about one-tenth of the number of samples is required to obtain equal information if the total produce is weighed. The estimation of yields by this method has been tried in a number of experiments since 1935. The present paper reviews their results from the point of view of sampling technique.

Since total produce, when weighed on the field, usually contains some moisture, the samples must be weighed on the field as well as before threshing, to enable a correction to be made in the grain and straw figures for the loss of moisture. If the samples are taken from the standing crop, this should be done immediately before the crop is reaped, in order that the samples and the total produce may be weighed on the field at the same time. This may not always be convenient, but with this method the samples can alternatively be taken from the crop while it is lying in the stooks, since there is no need to know the area of the land from which a sample was taken. As the crops usually lie in the stooks for some days, this gives a wider choice in the time during which the sampling must be carried out. In most of the experiments discussed below, the samples were taken from the stooks.

## 2. Method of sampling from the stooks

Total produce is first weighed on each plot. A spring balance may be used, weighing the sheaves one at a time. This method is rather tedious unless the plot size is small or the crop is a poor one, since a 1/40 acre plot may contain thirty sheaves. If a portable tripod is available with a platform which can hold all the sheaves in one stook, some time will be saved. After weighing, the sheaves should be laid separately on the ground, to facilitate the sampling operations.

The next step is to select the samples. A method which gives reasonably random samples is as follows: Suppose that there are eighteen sheaves on a plot and that each sample is to be approximately 1 % of the produce of the plot. A sheaf is first selected at random. The binding tape is cut and the sheaf is divided into six portions of about equal weight. One of these is selected at random and constitutes the sample. The division of the sheaf is usually most quickly done by successive subdivision into halves, selecting one-half at random at each stage for further subdivision, until a sample of about the required size is reached. This method also has the advantage that it reduces to a minimum the number of small bundles which are scattered about the plot. For selecting the halves at random, a piece of paper bearing a selection of odd and even numbers drawn from a book of random numbers may be used; alternatively, a set of disks containing an equal number of two different colours may be carried in the pocket.

When the samples have been selected, labelled and bagged, they are weighed. For the calculation of the grain yields on each plot, it is necessary to know only the *total* weight of the samples from the plot, but if a full investigation of sampling errors is required, each sample must be weighed individually. As the samples may weigh less than one pound each, a fairly accurate balance is required, and the weighings should if possible be done indoors whenever there is any appreciable wind. The average weight of a bag, with its label and string, must also be recorded.

This completes the experimental operations on the field. The sheaves should be restooked unless they are being carted off immediately.

The taking of random samples from the stooks is a lengthy process. Following a suggestion made by Yates & Zacopanay, samples were also taken by picking a few shoots from each of several sheaves until a sample of about the agreed size had been amassed. These samples, which will be called grab samples, can be taken in about one-third of the time required for random samples, since the sheaves need not even be opened unless they are very tightly bound. It is, however, not clear *a priori* whether grab samples give unbiased estimates of the grain/total produce ratios or how they compare in accuracy with random samples. In grabbing, no attempt was made to select representative shoots, as this method is known to be likely to introduce bias. One might, however, expect a tendency to miss the shorter and less vigorous shoots, and possibly also to free the shoots of weeds in pulling them from the sheaves. Both factors would tend to increase the apparent grain/total produce ratio. A comparison of the results with random and grab samples will be given later in this paper.

### 3. MATERIAL

A list of the experiments discussed is given in Table I below. The plots were not subdivided for sampling, so that the plot area given is in all cases the area to which the sampling errors apply.

The random samples were taken from the stooks in five experiments. The grab samples were taken from the stooks in all cases except in exp. 5, where they were taken from the crop as it lay on the ground immediately after scything.

## Table I. *List of experiments*

| No. | Year | Place | Type | Size of plot acres | Random samples taken from | No. of samples per plot | |
|-----|------|-------|------|------|------|------|------|
| | | | | | | Random | Grab |
| | | | Wheat | | | | |
| 1 | 1935 | Rothamsted | 6 × 6 L.S.* | 1/40 | Stooks | 2 | 1 |
| 2 | 1935 | Woburn | 6 × 6 L.S. | 1/100 | Stooks | 2 | 1 |
| 3 | 1936 | Woburn | 6 × 6 L.S. | 1/100 | Standing crop | 2 | 1 |
| | | | Barley | | | | |
| 4 | 1936 | Rothamsted | 4, 16 R.B.† | 1/40 | Stooks | 2 | 1 |
| 5 | 1937 | Wye | 6 × 6 L.S. | 1/120 | Standing crop | 2 | 2 |
| 6 | 1937 | Tunstall | 3, 9 R.B. | 1/40 | Stooks | 2 | 2 |
| | | | Oats | | | | |
| 7 | 1938 | Rothamsted | 4, 18 R.B. | 1/60 | Stooks | 3 | 2 |

* Latin square.
† I.e. four randomized blocks of sixteen plots each.

### 4. Sampling errors per cent per metre

The sampling errors per cent per metre of row length for the ratio $r$ of grain to total produce are shown in Table II. Where the samples were taken from the stooks, the average number of metres sampled was estimated from the ratio of the weight of the sample to the weight of the whole crop on the plot. The sampling errors in all cases refer to the ratios of grain to *dry* total produce, as these were the figures with which Yates & Zacopanay dealt.

### Table II. *Sampling errors of the ratio of grain to total produce*

| Exp. | Method of sampling | Mean yield of grain cwt. per acre | Area of plot acres | Size of sampling unit metres | Sampling error % per metre |
|------|------|------|------|------|------|
| 1 | R. | 32·3 | 1/40 | 5·4 | 16·8 |
| 2 | R. | 29·9 | 1/100 | 1·9 | 10·0 |
| 3 | R. | 20·8 | 1/100 | 2·0 | 15·0 |
| 4 | R. | 25·1 | 1/40 | [5·9] | [29·1] |
| 5 | { R. | 14·7 | 1/120 | { 4·0 | { 13·8 |
| | { G. | 15·1 | | { 3·0 | { 12·5 |
| 6 | G. | 5·6 | 1/40 | 5·6 | 13·7 |
| 7 | { R. | 33·5 | 1/40 | { 2·6 | { 7·0 |
| | { G. | 33·6 | | { 3·1 | { 7·3 |
| | | | Mean | 3·5 | 12·6 |

In the first four experiments sampling errors are obtainable only for the random samples, since only one grab sample was taken per plot. In exp. 6 the random samples were unfortunately bulked for threshing.

The sampling errors per cent per metre are considerably higher than Yates & Zacopanay's figure of 7·14 %. Exp. 4 may perhaps be omitted in reaching an average figure, since 12 % of the samples were reported as damaged by mice during storage. These samples were excluded from the statistical analysis, but five other samples also showed an anomalously low ratio of dry total produce to wet total produce, as well as an anomalously low grain/total produce ratio. These samples might perhaps be regarded as affected by damage which was not reported. The experiment was, however, one in which different leys were growing under barley and the samples in question all came from plots growing a clover-ryegrass mixture, so that they may have contained a substantial amount of the undergrowth. In any case it is clear that if there is a vigorous and variable undergrowth of ley or weeds, this method is likely to give high sampling errors.

Excluding exp. 4, the average value for the sampling error is 12·6 % per metre. There are several reasons which might account, in part at least, for the higher value obtained.

### Size of plot

The criterion used, sampling error per cent per metre, is likely to increase as the size of the plot increases. While no correlation is evident in Table II between sampling error and size of plot, the average plot size in these experiments was considerably larger than in Yates & Zacopanay's experiments, in which most of the sampling subplots were only 1/200 acre. Since, however, Yates & Zacopanay used only a fraction of their data for this particular calculation, some additional information on the effect of plot size was obtained by calculating the sampling error of $r$ for six of their experiments in which the plots were 1/80 acre. The results are shown in Table III.

Table III. *Sampling errors of the ratio of grain to total produce (from 1/80 acre plots)*

| Exp.* | Crop | Size of sampling unit metres | Sampling error % per metre | |
|---|---|---|---|---|
| | | | $r$† | Grain |
| 4 | Barley | 5 | 5·98 | 23·7 |
| 7 | Barley | 1 | 9·06 | 30·3 |
| 10 | Wheat | 1 | 8·89 | 28·8 |
| 10 | Barley | 1 | 10·57 | 32·5 |
| 11 | Wheat | 1 | 9·42 | 22·8 |
| 11 | Barley | 1 | 6·76 | 33·5 |
| | | Mean | 8·45 | 28·6 |

\* In Yates & Zacopanay's notation.
† The method by which these figures were obtained is discussed in the Appendix.

The average value, 8·45, is somewhat larger than the previous figure of 7·14 for smaller plots, but is still considerably below 12·6. The average sampling error for the yields of grain in the same experiments was 28·6 %, so that the relative efficiency of the two methods works out at almost the same figure as Yates & Zacopanay obtained. It does not appear as if the difference in the size of the plots can account for more than a small part of the increase from 7·1 to 12·6 %.

### Size and type of sampling unit

The sampling error of $r$ will also depend to some extent on the size and shape of the sampling unit. As a rule, it is to be expected that for the same total percentage sampled, a few large sampling units will be less efficient than a larger number of small sampling units. In the present experiments the average size of the sampling unit was 3·5 m. as against 2·0 m. in Yates & Zacopanay's experiments, and this difference might partly account for the higher sampling error. In this connexion it would have been instructive to compare the variation in $r$ between samples taken from the same sheaf with that between samples taken from different sheaves, but this is not possible from the way in which the samples were selected. It is also possible that the reaper or scythe gives a less even cut than is obtained when small samples are cut by hand from the standing crop.

### Presence of weeds or undergrowth

This point has already been mentioned in discussing exp. 4 in Table II, but it applies, to a less extent, to all experiments. In Yates & Zacopanay's experiments, the samples were cleared of weeds before determining the weights of grain and straw, whereas in sampling for the grain/total produce ratio it is essential that the sample should not be cleaned of weeds. Thus the presence of weeds, from which few experiments are entirely free, adds to the variability of $r$, particularly so as weeds compete with the crop and are more likely to abound in poorer patches, where the value of $r$ is already low.

### 5. THE CORRECTION FOR LOSS OF MOISTURE

No discussion has so far been given for the correction which must be made for the amount of moisture in the total produce as weighed on the field. Since this correction is made from the samples, it will involve some loss of information, so that the sampling errors given in the preceding section for the ratio of grain to *dry* total produce do not represent the whole of the sampling error involved in this method.

The yield of dry grain of any plot is most simply obtained by multiplying the yield of wet total produce by the ratio, in samples from that plot, of the total yield of dry grain to the total yield of wet total produce. The percentage sampling variance per plot of the yield of dry grain will be given (with all necessary accuracy) by the percentage sampling variance of the ratio of dry grain to wet total produce, divided by the number of samples taken per plot. This can be calculated if the samples were weighed individually on the field and threshed individually.

Since the sampling errors of the ratio of dry grain to dry total produce have already been discussed, it will be more convenient to discuss here the sampling errors of the ratio of dry total produce to wet total produce, assuming these ratios to be independent. In general, however, the more direct approach is preferable, since the assumption of independence is not likely always to hold.

Unfortunately, little evidence on the dry/wet ratio is obtainable from these experiments. The samples were weighed individually on the field in only three experiments, nos. 3, 4 and 7, mainly because the accuracy of the spring balance and the external conditions did not appear to justify weighing each sample. Of these experiments, no. 4 has already been noted as exceptionally variable, while in no. 7 there appears to have been a zero error in the spring balance, since almost all the dry weights of total produce were slightly higher than the wet weights.

In exp. 3, the sampling error per cent *per plot* for the ratio of dry to wet total produce was 7·03, as compared with 7·50 for the ratio $r$ of dry grain to dry total produce. The corresponding figures in exp. 4, omitting the plots undersown with the clover-ryegrass mixture, were 7·45 and 8·50. These figures suggest that almost as much information is being lost in estimating the correction for drying as in estimating the ratio of dry grain to dry total produce. If this is true, the accuracy of the method is only half that indicated by the figures in the last section. There is, however, reason to believe that these results are not representative, since rain fell during the sampling of exp. 3, some samples being wet when weighed, and in both experiments there was an unusual amount of drying-out, the mean values of the ratio of dry to wet total produce being 0·628 and 0·673 respectively.

For the remaining experiments, the experimental error between plots for the ratio of dry to wet total produce of the samples may be used as an upper limit to the corresponding sampling error within plots. It may be mentioned that in exp. 3, the sampling variance of the dry/wet ratio was practically equal to the experimental variance, though there were significant differences between rows, columns and treatments, while in exp. 4 the sampling variance was less than half the experimental variance. The results *per plot* for the other experiments are shown in Table IV.

Table IV. *Experimental errors per cent per plot of the ratio of dry to wet total produce*

| Exp. | Mean ratio dry/wet | Experimental error % of dry/wet | Sampling error % of $r$ |
|------|------|------|------|
| 1 | 0·849 | 2·91 | 5·17 |
| 2 | 0·707 | 4·44 | 5·17 |
| 5 | 0·878 | 2·26 | 4·87 |
| 6 | 0·859 | 2·59 | 4·07 |

In exps. 1, 5 and 6 the percentage sampling variance of the dry/wet ratio cannot exceed about one-third of the percentage sampling variance of $r$, and may be substantially less. In exp. 2, in which the amount of drying was much greater, the additional loss of information was probably also greater.

If the dry/wet ratios are very variable the question arises whether the use of some average correction figure will improve matters. Clearly such an average can only be properly employed if the dry/wet ratios are unaffected by the treatments, for if they are so affected the use of an average will distort the treatment differences. Actually four of the six experiments considered showed clear differences between treatments, and exp. 3 also falls into this category if the clover-ryegrass plots are included. The use of an average correction figure is therefore inadvisable.

Such distortion can of course be avoided by using a separate correction factor for each treatment, based on the average dry/wet ratio for all replicates of that treatment. There is no point in following this course, however, since the results will be almost the same as if each plot is corrected separately. The only effect will be to give a spuriously low estimate of experimental error.

## 6. Comparison of grab with random samples

Direct comparison of the sampling errors of $r$ for random and grab samples can be made in only two of the experiments in Table II, nos. 5 and 7.

In exp. 5 grab sampling was somewhat more accurate, though not significantly so, while in exp. 7 there was little to choose between the two methods.

A less direct comparison may be obtained by calculating the between-plot errors of the yields of grain given by the two methods (after elimination of treatment and block effects). Some allowance must be made for the difference in the amounts which were sampled under the two methods. In exp. 1, for example, two random samples each of about 954 g. total produce were taken, as against one grab sample of 794 g. The sampling and experimental errors per cent *per plot* for the random samples were 5·15 and 8·67 respectively. The estimated experimental error per cent, if only one random sample of 794 g. had been taken is

$$\left\{(8\!\cdot\!67)^2 + \left(\frac{2 \times 954}{794} - 1\right)(5\!\cdot\!15)^2\right\}^{\frac{1}{2}} = 10\!\cdot\!8,$$

and this figure is comparable with the experimental error per cent for grab sampling. The adjustment for the size of the individual sample in the above formula is open to question, since a sample of twice the size, *taken from the same sheaf*, would probably not be twice as accurate. Since the grab samples were usually the larger, the adjustment possibly favours the random samples slightly.

Table V. *Experimental errors per cent per plot of the yields of grain*

| Exp. | Method of sampling | |
|------|--------|------|
|      | Random | Grab |
| 1    | 10·8   | 7·9  |
| 2    | 7·6    | 10·4 |
| 3    | 15·5   | 12·8 |
| 4    | 17·2   | 17·1 |
| 5    | 8·3    | 9·3  |
| 6    | 14·7   | 14·0 |
| 7    | 6·8    | 6·9  |
| Mean | 11·6   | 11·2 |

Random samples gave a smaller experimental error in two experiments, grab samples in three, while the remaining two experiments showed practically identical results. Thus grab sampling appears to be no less accurate than random sampling.

The mean yields of grain obtained by random and grab sampling are shown in Table VI. The right-hand column shows the difference between the yields from grab and random sampling as a percentage of the yield given by random sampling.

Except in exp. 4 the grab samples gave slightly higher yields of grain than the random samples. The biases are, however, in no case large.

Both random and grab sampling gave a positive bias in yields as compared with full harvesting in exps. 1 and 3. In exp. 3 the difference is due almost entirely to a greater drying out of the total produce than of the samples.

Table VI. *Comparison of mean yields by random and grab sampling*

Grain: cwt. per acre

| Exp. | Crop | Full harvesting | Random sampling | Grab sampling | % bias in grab |
|------|-------|------|------|------|------|
| 1 | Wheat | 30·59 | 32·36 | 34·33 | +6·1 |
| 2 | Wheat | — | 29·88 | 31·55 | +5·6 |
| 3 | Wheat | 18·57 | 20·83 | 21·45 | +3·0 |
| 4 | Barley | — | 25·07 | 23·95 | −4·5 |
| 5 | Barley | — | 14·74 | 15·14 | +2·7 |
| 6 | Barley | — | 5·44 | 5·57 | +2·4 |
| 7 | Oats | — | 33·50 | 33·60 | +0·3 |

A more detailed examination of the treatment means in these experiments shows close agreement between results from random and grab sampling.

## 7. Discussion of results

Owing to the uncertainty about the sampling variance of the ratio of dry to wet total produce, the total sampling error involved in sampling for the ratio of grain to total produce cannot be fixed definitely for these experiments. An average figure of 14·5 % per metre of row length for the ratio of dry grain to wet total produce is probably not far wrong. (This represents an increase in the average sampling variance in Table II by one-third to allow for the sampling variance of the dry/wet ratio.) With this figure, a sample of 25 m. per plot gives a sampling error of 2·9 % per plot. With an experimental error of 7·5 % per plot, the loss of information is about 13 %, i.e. an amount which could be more than offset by adding an extra replication to an experiment with between four and seven replications. This amount of sampling represents about 5 % of the total produce in a 1/40 acre plot.

This figure is subject to qualification according to the conditions of the experiment. If the crop is fairly dry and free from weeds or undergrowth when it is being sampled, or if the plot size is only 1/100 acre, some reduction may perhaps be allowed in the number of metres sampled, though it would be advisable to collect more experimental data on this point.

The size of the samples taken in these experiments was probably too large. It might be better to take not more than 2 m. of row length for each sample. It is easy to calculate, for any particular experiment, the fraction of a sheaf necessary to secure such samples. For example, in a 1/60 acre plot, sown at 7 in., there are about 380 m. of row, and if there are twenty sheaves per plot, each sample should be $\frac{2 \times 20}{380}$ of a sheaf, i.e. about one-tenth.

Apart from the small positive bias in the yield of grain, there appears to be no objection to grab sampling as carried out in these experiments. In practice, one might take first two random samples of about 2 m. each, and then a further eight grab samples of about the same size. The random samples would serve as a check on the others, and the whole process would require considerably less time than ten random samples.

Although this method has not proved as accurate as was anticipated, considerable time and labour still appears to be saved as compared with the previous method of random sampling from the standing crop without weighing total produce. If the figure of 28·6 % per metre (from Table III) is taken as a comparable value for the sampling error of the yield of grain by the latter method, about one-quarter of the number of samples is needed if total produce is weighed and the samples are taken from the sheaves. If most of the sampling is done by grabbing, this can be done in little more than one-eighth of the time (in exp. 5, for example, seventy-two random samples from the standing crop required 10 man-hours, including bagging and labelling, while an equal number of grab samples took 5 man-hours). As far as can be judged, the time taken to weigh the samples and total produce is not more than twice the time required to select, bag and label an equal number of grab samples. Thus the field operations require only about three-eighths of the time taken by the previous method. There is also a considerable gain in time during threshing, which is also of importance, as with the machines at present available threshing occupies a large proportion of the total time.

The new method is somewhat more exposed to weather hazards at the time of sampling. For instance, if rain falls after total produce has been weighed and while the samples are being taken, the samples which have already been drawn must be protected from the rain, while the total produce may have to be weighed again on plots which have not yet been sampled. On the other hand, it is extremely difficult to sample from the standing crop if it is badly lodged, whereas such a crop presents no special difficulty once it is in the sheaves.

## Summary

In a number of cereal experiments, three on wheat, three on barley and one on oats, the yields of grain and straw per plot were estimated by weighing the total produce on each plot and taking samples, usually from the sheaves, to estimate the ratio of grain to total produce. This paper discusses the sampling errors of this method. The method proved considerably less accurate than was anticipated from previous calculations made by Yates & Zacopanay. Amongst the reasons which are suggested to account for this are the larger sizes of plot and sampling unit used in these experiments and the additional variability introduced by the presence of weeds, undergrowth and moisture.

Nevertheless, the method appears to be substantially superior to the older method of cutting small areas from the standing crop, without weighing total produce, only about one-quarter of the number of samples being required to obtain results of equal precision.

The samples were taken both by an approximately random process and by grabbing a few shoots haphazardly from each of several sheaves. The grab samples gave on the whole a slightly higher yield of grain, the greatest positive bias being 6 %, but were otherwise just as accurate as the random samples. Since the grab samples can be selected and bagged in about one-third of the time required for random samples, their use is recommended for the majority of the samples required in any experiment.

The validity of an approximate formula for calculating the variance of a ratio (in the present instance the ratio of grain to total produce) is discussed briefly in an appendix.

## APPENDIX

### *The validity of an approximate formula for the variance of a ratio*

To avoid the labour of calculating the actual ratios of grain to total produce, Yates & Zacopanay used an approximate formula expressing the variance of the ratio in terms of the variances and covariance of grain and total produce, most of which they had already calculated for the earlier part of their paper.

Let $g$, $t$ denote the grain and total produce yields of a sample and $r$ their ratio, and let $\bar{g}$, $\bar{t}$ and $\bar{r}$ be the corresponding means over all samples taken. Then as a first approximation

$$\frac{1}{\bar{r}^2}\, V(r) = \frac{1}{\bar{g}^2}\, V(g) + \frac{1}{\bar{t}^2}\, V(t) - \frac{2}{\bar{g}\bar{t}}\, \text{Cov.}\,(gt).$$

The most important condition required for this approximation to be satisfactory is that the standard errors of $g$ and $t$ should be small relative to their mean values, though so far as I am aware, the limits within which the formula applies have never been investigated. The standard errors of $g$ and $t$ were nearly all under $20\%$ in the experiments which Yates & Zacopanay used, but in the 1/80 acre plots given in Table III, they were sometimes as high as $30\%$, so that some investigation is needed of the accuracy of the approximation under these conditions.

To proceed to a second approximation, the form of the joint distribution of $g$ and $t$ must be specified. If we assume that they follow the bivariate normal distribution, and write

$$c_{11} = \frac{1}{\bar{g}^2} V(g), \quad c_{12} = \frac{1}{\bar{g}\bar{t}} \operatorname{Cov.}(gt), \quad c_{22} = \frac{1}{\bar{t}^2} V(t),$$

the second approximation is

$$\frac{1}{\bar{r}^2} V(r) = (c_{11} + c_{22} - 2c_{12}) + (6c_{22}^2 + c_{11}c_{22} + c_{12}^2 + 2c_{11}c_{12} - 10c_{12}c_{22}).$$

In the present examples, $c_{11}$ and $c_{22}$ are always nearly equal. If we write $\sigma^2 = \frac{1}{2}(c_{11} + c_{22})$, $\rho\sigma^2 = c_{12}$, the second approximation may be written, with sufficient accuracy,

$$\frac{1}{\bar{r}^2} V(r) = (c_{11} + c_{22} - 2c_{12})\{1 + \frac{1}{2}(7 - \rho)\sigma^2\},$$

or, since the correlation coefficient $\rho$ is usually near 1,

$$\frac{1}{\bar{r}^2} V(r) = (c_{11} + c_{22} - 2c_{12})(1 + 3\sigma^2).$$

This formula shows that if the standard errors of $g$ and $t$ are about $10\%$, the first approximation underestimates by about $3\%$ relative to the second approximation; if the standard errors are $30\%$, the first approximation underestimates by $27\%$.

The errors involved in the use of 'the second approximation arise from (1) the assumption that $c_{11}$, $c_{12}$ and $c_{22}$ are equal to the corresponding population values. This error is unlikely to be serious in these experiments since the estimates were derived from a large number of degrees of freedom; (2) the assumption that $g$ and $t$ follow the bivariate normal law. It is not possible to assess the magnitude of the error from this source without further computation, although the marginal distributions of $g$ and $t$ tend to be slightly positively skew; (3) neglect of the higher terms in the approximation. These terms are likely to be important only when the standard errors of $g$ and $t$ are over $25\%$.

As a check on these approximations, the actual sampling errors of grain/total produce were computed for about half the data from each plot in four of Yates & Zacopanay's experiments. These are compared with the first and second approximations in Table VII below.

Table VII. *Check on the approximate formula for the sampling variance of r*

| | | Sampling variance of $r$ | | |
|---|---|---|---|---|
| Exp.* | Crop | First approximation | Second approximation | Actual |
| 7 | Barley | 62·9 | 79·8 | 82·1 |
| 11 | Wheat | 44·9 | 51·7 | 45·7 |
| 11 | Barley | 76·5 | 100·1 | 88·8 |
| 17 | Barley | 18·6 | 20·0 | 22·5 |

\* In Yates & Zacopanay's notation.

The first approximations are all less than the actual values, but the second approximations do not appear to underestimate on the whole. In exp. 11 the second approximations are rather poor, but in view of the large variation in the sampling variance of $r$, they are probably sufficiently close for the present purpose.

The substitution of the second for the first approximation makes little difference to Yates & Zacopanay's results, merely increasing the average standard error of $r$ % per metre from 7·14 to 7·50. The figures given in Table III for the 1/80 acre plots are the actual values of the sampling errors of $r$ in the exps. 11 and 17 and the second approximations in the other experiments.

REFERENCE

Yates, F. & Zacopanay, I. (1935). *J. agric. Sci.* **25**, 545–77.

(*Received* 11 *November* 1939)

## LATTICE DESIGNS FOR WHEAT VARIETY TRIALS[1]

### W. G. Cochran[2]

LATTICE designs were first described by Yates (8).[3] Their object is to provide a more accurate design than randomized blocks for varietal trials with a large number of varieties, while retaining as far as possible equal accuracy of comparison between every pair of varieties in the experiment. Although Goulden (3) recommended these designs for experiments on wheat varieties, so far they have not received an extensive trial in this country. In contacts with several wheat breeders who have used the designs, it appeared that the field operations presented no new difficulties; however, some questions were raised concerning the statistical analysis. It is hoped that a brief discussion of the nature of the designs and their analysis may be useful to those who contemplate using them. In particular, the method of adjusting yields will be considered in some detail, since this appears to be the least familiar feature.

To avoid confusion, it should be noted that lattice designs were called pseudo-factorial or quasi-factorial designs in the earlier papers on the subject. Further, the most accurate method of analyzing these designs was not discovered until some time after the designs were first published. Yates (9, 12) and Cox, et al. (2) present the new method of analysis.

### METHODS OF REDUCING BLOCK SIZE

With experiments containing a large number of varieties, two devices have been used in attempting to obtain more accurate designs than randomized blocks. One is to insert a control variety at regular intervals within the block to serve as an indicator of soil fertility variations inside the block. The second is to reduce the block size, using blocks that do not contain all the varieties. The lattice designs belong to this class, and it may be instructive to examine their relation to earlier designs of the same type.

A well-known method is to divide the varieties into groups, inserting one or more common control varieties with each group, the groups being laid out in *separate* randomized blocks or Latin square experiments. This design provides accurate comparisons between varieties which are in the same group and is very convenient for making field observations since a control variety is always near at hand for comparison. It is much less satisfactory for comparing varieties in different groups, because no estimate of experimental error is available for the difference between the means of two such varieties. This difficulty can be overcome by comparing the varieties indirectly, calculating $(v_1 - c_1) - (v_2 - c_2)$, where $c_1$, $c_2$ are the means of the common control varieties in the two groups. However, these

[1]Journal Paper No. 867 of the Iowa Agricultural Experiment Station, Ames, Iowa. Project No. 514. Also presented at the annual meeting of the Society held in Chicago, Ill., December 4 to 6, 1940. Received for publication January 15, 1941.
[2]Research Professor.
[3]Figures in parenthesis refer to "Literature Cited", p. 360.

comparisons have higher experimental errors than comparisons between varieties in the same group, an objectionable feature if all comparisons are wanted with equal accuracy. Moreover, the controls occupy a disproportionately large part of the experimental site.

Inter-group comparisons may be made directly, thus dispensing with the need for extra controls, by combining the separate experiments on the same site, as illustrated in Fig. 1a. Here the six groups are themselves arranged in a $6 \times 3$ randomized blocks design. The analogy with a split-plot design is evident. The analysis of variance of this design runs as follows:

|  | Degrees of freedom | Mean squares |
|---|---|---|
| Replications.................... | 2 | |
| Between groups.................. | 5 | |
| Inter-group error............... | 10 | a |
| Between varieties in the same group.. | 30 | |
| Intra-group error............... | 60 | b |
| Total........................ | 107 | |

Since there are three replicates, the error variance of the difference between two varietal means in the same group is $\frac{2}{3}$ b. For varieties in different groups, the corresponding error is a weighted mean of a and b, being in this case $\frac{2}{3}\left(\frac{5}{6} b + \frac{1}{6} a\right)$.

This design has many features in common with lattice designs. In particular, it utilizes both inter- and intra-group comparisons and requires no more land than randomized blocks from which it differs only in a rearrangement of the varieties within each replication. However, if the mean square a is much larger than b, indicating that the small blocks have been successful in avoiding soil heterogeneity within the replications, comparisons between varieties in different groups have a higher error than comparisons between varieties in the same group. The former are much more numerous than the latter. On the other hand, if a is about the same size as b, a randomized blocks design would have given about the same results. In fact, this design seems likely to prove more useful than randomized blocks only when comparisons within groups *are* required with higher accuracy, or when the number of groups is the same as the number of replications, permitting the use of a Latin square for inter-group comparisons.

Where all comparisons are desired with equal accuracy, the design above is at fault in keeping the same groups of varieties together in all replications, a pair of varieties either appearing *always* in the same group or *never* in the same group. It is clearly better to make the opposite rule, that a pair of varieties which are in the same group in the first replication shall not appear in the same group in any subsequent replication. This rule leads to the construction of lattice designs. Investigation was required to find whether designs obeying this rule could be constructed. Where the number of varieties is an exact square, the first three replications can always be constructed by means of an auxiliary Latin square, as shown in Fig. 1b. Varieties in the same row of the square are put in the same group in replication 1, varieties in the same column in replication 2, and varieties having the same Latin letter in replication 3. With 36 varieties, no fourth replication of this type can be found, but one exists for all other squares from 16 up to 169, except 100.

Fig. 1a
*Extension of split-plot design*
Groups

| IV | V | III | I | II | VI |
|----|----|-----|----|----|----|
| 22 | 27 | 14 | 2 | 7 | 34 |
| 21 | 30 | 15 | 3 | 8 | 36 |
| 24 | 25 | 16 | 6 | 9 | 31 |
| 19 | 29 | 18 | 5 | 12 | 33 |
| 20 | 26 | 13 | 4 | 11 | 35 |
| 23 | 28 | 17 | 1 | 10 | 32 |

| III | IV | I | VI | V | II |
|-----|----|----|----|----|----|
| 13 | 21 | 4 | 35 | 27 | 9 |
| 14 | 19 | 3 | 33 | 26 | 10 |
| 18 | 24 | 6 | 36 | 29 | 11 |
| 17 | 23 | 1 | 32 | 30 | 12 |
| 16 | 22 | 2 | 34 | 28 | 8 |
| 15 | 20 | 5 | 31 | 25 | 7 |

| II | I | IV | V | VI | III |
|----|----|----|----|----|-----|
| 11 | 2 | 21 | 30 | 31 | 14 |
| 12 | 1 | 19 | 28 | 35 | 16 |
| 7 | 6 | 22 | 25 | 34 | 18 |
| 9 | 5 | 23 | 27 | 36 | 15 |
| 10 | 3 | 24 | 26 | 32 | 13 |
| 8 | 4 | 20 | 29 | 33 | 17 |

Design: 36 varieties in
3-fold replication

The numbers 1–36 denote varieties

Fig. 1b
*Lattice design in three replications*
(Before randomization)

| 1 | 7 | 13 | 19 | 25 | 31 |
|----|----|----|----|----|----|
| 2 | 8 | 14 | 20 | 26 | 32 |
| 3 | 9 | 15 | 21 | 27 | 33 |
| 4 | 10 | 16 | 22 | 28 | 34 |
| 5 | 11 | 17 | 23 | 29 | 35 |
| 6 | 12 | 18 | 24 | 30 | 36 |

| 1 | 2 | 3 | 4 | 5 | 6 |
|----|----|----|----|----|----|
| 7 | 8 | 9 | 10 | 11 | 12 |
| 13 | 14 | 15 | 16 | 17 | 18 |
| 19 | 20 | 21 | 22 | 23 | 24 |
| 25 | 26 | 27 | 28 | 29 | 30 |
| 31 | 32 | 33 | 34 | 35 | 36 |

| 1 | 2 | 3 | 4 | 5 | 6 |
|----|----|----|----|----|----|
| 10 | 9 | 7 | 11 | 12 | 8 |
| 17 | 18 | 14 | 15 | 16 | 13 |
| 20 | 19 | 23 | 24 | 21 | 22 |
| 27 | 28 | 30 | 25 | 26 | 29 |
| 36 | 35 | 34 | 32 | 31 | 33 |

Auxiliary
Latin Square

| 1A | 2B | 3C | 4D | 5E | 6F |
|----|----|----|----|----|----|
| 7C | 8F | 9B | 10A | 11D | 12E |
| 13F | 14C | 15D | 16E | 17A | 18B |
| 19B | 20A | 21E | 22F | 23C | 24D |
| 25D | 26E | 27A | 28B | 29F | 30C |
| 31E | 32D | 33F | 34C | 35B | 36A |

Fig. 1.—Comparison of "split-plot" design and lattice design.

## COMPARISON OF LATTICE DESIGNS AND RANDOMIZED BLOCKS

### FIELD OPERATIONS

There is no single measure of the relative merits of two designs. However, probably the most important aim in experimental design is to obtain a given degree of accuracy at the least cost. As regards cost of field operations, the lattice designs, like the split-plot designs, differ from randomized blocks only in a different grouping of the varieties within each replication. Thus the cost of field operations is the same for both designs, the principal difficulty common to both being that of harvesting when the varieties mature at different times. In one respect the lattice designs are more flexible under field conditions, as Goulden (3) has pointed out. If the experiment is laid out in an irregularly shaped field, or portion of a field, it may be difficult to plan the design so that each replication is compact. The accuracy of a randomized blocks design may suffer considerably under such conditions. However, if the *small* blocks are kept compact, which is usually easy, only the inter-block comparisons in the lattice design are affected by the manner in which the small blocks are grouped to form a replication. Since the inter-block comparisons contribute only a small fraction of the total information, the accuracy of lattice designs suffers much less in these circumstances.

These remarks were strikingly illustrated on Wiebe's (7) wheat uniformity trial which appears to provide the only wheat data in the literature copious enough to test designs for many varieties. A design for 81 varieties in three replications was superimposed on the data, the plots being three rows (3 feet) $\times$ 15 feet, with only the centre row harvested. Using exactly the same plots and site, a "good" and a "bad" grouping of the plots into replications were examined. For randomized blocks, the standard errors of a varietal mean were 8.0% and 10.4% in the two cases. For the lattice design, the errors were 6.60% and 6.63%.

### NUMBERS OF VARIETIES AND REPLICATIONS

While randomized blocks designs can be constructed for any number of varieties, the lattice designs are more restricted in this respect. The type described above exists only when the number of varieties is an exact square, of which the most useful are those from 25 to 169. With numbers in this group other than 36, 100, and 144, any number of replications may be constructed until every variety has appeared once with every other variety in the same block, this requiring $(n + 1)$ replications for $n^2$ varieties. With 36 or 100 varieties, four and six replications may be secured by duplicating the designs for two and three replications, respectively. So far as the writer is aware, the 12 $\times$ 12 design has been constructed only up to four replicates and requires the same device to obtain six replicates. In fact, this device may also be used for the other numbers of varieties, with a slight loss in accuracy but some compensation in simplicity of analysis.

When the number of varieties is the product of two approximately equal numbers, e.g., 72 or 90 varieties, similar designs are available for an even number of replications (Yates, 8). These, however, require a slightly more elaborate analysis.

### STATISTICAL ANALYSIS

Since field operations are essentially the same for both designs, their relative efficiency depends primarily on the labor involved in the statistical analysis and on the accuracy attained with a given number of replications. While the complete analysis of the lattice designs is more complicated than that of randomized blocks, the former possess the useful property that they can be analyzed alternatively as an ordinary randomized blocks experiment. This statement implies two things, *viz.*, (a) the error mean square found by the usual randomized blocks analysis is an unbiased error with which to compare the varieties mean square, and (b) what is more important, this error may safely be used to test the difference between any pair of varietal means. Of course, the *true* error is not exactly the same for all pairs of varieties, since by the nature of the design some pairs occur more closely together than others. However, the same phenomenon occurs in randomized blocks, as can be seen by examining any design after it has been randomized.

Investigation by Yates (9) showed that the use of a common error was about equally valid in the two designs. It may be noted that the first property above also holds for the "split-plot" design previously discussed, but the disparity in the true errors for different pairs of varieties is usually too great for the second property to hold.

This property of lattice designs is valuable in many cases. Where a rapid preliminary inspection of the results is wanted, the randomized blocks analysis may first be calculated. Little labor is wasted by doing so even if it is intended later to complete the full lattice analysis, since all calculations already made for the randomized blocks analysis are required in the full analysis. Further, a randomized blocks analysis is adequate for any field measurements which are little affected by soil fertility variations.

The extra numerical work in the full lattice analysis consists in calculating the sum of squares between blocks and adjusting the varietal means. Practically all the extra operations are simple additions which are self-checking. The times required for the two types of analyses were compared for a lattice design with 81 varieties and four replications formed by duplicating the design for two replications. The machine used was an electric Monroe without automatic multiplication or division. With no mistakes in computation, it took 2 hours 45 minutes to form the varietal totals and the randomized blocks analysis of variance, with checks by re-computation where necessary. The lattice analysis leading to the adjusted varietal totals required 4 hours 30 minutes. In the latter, it is easier to locate mistakes in addition, since more sub-totals and subsidiary checks are available. Of course, with someone unfamiliar with the lattice design, much more time would be consumed at a first trial.

### NATURE OF THE ADJUSTMENTS

Extra complication in the statistical analysis may be a drawback to the widespread use of a design in other respects. If the experimenter does not clearly understand the assumptions involved in the statistical manipulations, or the reasons for them, he loses confidence in the final results of the calculations. The principal difference between the randomized blocks analysis and the lattice analysis is that the ordinary varietal mean yields are adjusted in the latter to what is considered a more accurate estimate. Examining the nature of the adjustment in some detail in the hope of clarifying its common-sense interpretation, the lattice design with two replications is chosen for simplicity, the nature of the adjustment being essentially the same for all lattice designs. The formula for adjusting the mean yield of a typical variety $v$ reads as follows:

$$\text{Adjusted yield} = \text{unadjusted yield} - \left(1 - \frac{E}{B}\right)(M_B - M_V) \qquad 1$$

where $E$ = error mean square, $B$ = blocks mean square
$M_B$ = mean yield of the small blocks containing $v$,
$M_V$ = mean yield, in the rest of the experiment, of the varieties appearing in those blocks.

The adjustment is the product of two factors. Consider the second factor. With two replications, every variety occurs in two of the small blocks, but no two varieties appear in the same pair of small blocks. Thus, in taking the unweighted mean, a variety which happens to be grown in a pair of good blocks is favored, and one which is grown in a pair of poor blocks is handicapped. The lattice analysis endeavors to eliminate this source of error from the varietal comparisons. In order to correct varietal means for block differences, we must have an estimate of the relative fertility of the different pairs of blocks. At first sight, it might seem sufficient to calculate the difference between the mean yield of any pair of blocks and the mean yield of the whole experiment. This is not satisfactory, however, since different blocks contain different varieties. The best available estimate is obtained by taking the mean yield of the blocks minus the mean yield *in the rest of the experiment* of those varieties which appear in the blocks. On inspecting the design, it will be found that the latter mean yield contains one plot from every block in the experiment, so that the comparison is a fair one. To make the adjustment, the above quantity must, of course, be *subtracted* from the mean yield of a variety.

The adjustments described in the preceding paragraph are based on the assumption that there *are* real differences in fertility between blocks. However, in an experiment where there are no such differences, yields adjusted by this method are less accurate than unadjusted yields, because the adjustments perform no useful purpose, merely adding to the experimental error of the mean yields. Thus it would be unwise to apply the method of adjustment automatically in all experiments. Further examination of the differences between blocks showed that the most accurate estimates of mean yield were obtained by reducing the adjustments to $\left(1 - \frac{E}{B}\right)$ of their full value.

The factor $\left(1 - \frac{E}{B}\right)$ reduces to zero (no adjustment) when $B$ is no larger than $E$, i.e., when there are no real differences between blocks. The factor approaches unity (full adjustment) only when $E/B$ is very small, i.e., when differences between blocks are large. In intermediate cases, the partially adjusted yields, as Yates (10) has called them, are more accurate than either the fully adjusted yields or the unadjusted yields.

Agronomists may feel some diffidence in adjusting the yield of a variety by using the performance of other varieties which happen to appear with it in the same block, since some varieties may be accompanied by high-yielding varieties and some by low-yielding varieties. However, it should be noted that the adjustment does not depend on the average yields of the varieties which are associated with a given variety. The addition of say, 100, to all the plots of a given variety leaves the blocks and error mean squares unaltered, while if this variety appears in the second factor, the plus and minus terms are increased to exactly the same extent. There is the further point that different varieties may not respond in the same way to variations in the fertility of the blocks. This may have been the reason for the statement by Weiss and Cox (6) quoted by Salmon (5), that, "The partial confounding of variety differences with block effects makes it unwise to employ this type of design when comparing varieties which have an extremely large range in yields".[4] However, Weiss and Cox used an earlier method of analysis in which the full adjustments were made, the factor $\left(1 - \dfrac{E}{B}\right)$ being set equal to unity. With the type of adjustment made in equation 1 above, the possible danger from this source is much smaller, for if different varieties do not respond in the same way to the variations in fertility from block to block, the blocks mean square will tend to be no larger than the error mean square, and the lattice analysis will automatically lead to unadjusted, or only slightly adjusted yields. In fact, there is no danger in using the lattice *design* instead of randomized blocks, since the results can be analyzed as randomized blocks if certain varieties give such poor yields that it is considered unsafe to use them to adjust the yields of good varieties.

### MISSING PLOTS AND VARIETIES

In trials with many varieties it is frequently necessary to analyze the results of experiments in which several plot yields, or the entire yields from several varieties, are missing. The randomized blocks analysis is extremely useful in dealing with such cases. The formulae for the estimation of missing values are fairly simple, while any number of varieties may be omitted from the analysis without any additional complications in the numerical work.

Where it is desired to use the lattice analysis, formulae for the estimation of missing plots have been given by Cornish (1). These formulae are appropriate to the earlier method of analysis in which the *full* adjustments for block differences were made. The corresponding formulae for partially-adjusted yields are at present being in-

---

[4]This statement referred to the incomplete randomized blocks design which differs somewhat from lattice designs. However, the authors later apply a similar warning to the lattice square designs.

vestigated by Cornish. It seems unlikely that much change will be necessary if the blocks mean square is appreciably above the error mean square.

No publication has yet appeared explaining how to make the full lattice analysis when several varieties are missing. From preliminary inspection of the problem, it appears that the calculations will become considerably more complicated. Further investigation is needed to set up a methodical method of computation and to consider in what circumstances the extra computational labor is justified.

### RELATIVE ACCURACY OF COMPARISONS BETWEEN VARIETIES

It is usually safe in the lattice analysis to use a common standard error for all comparisons between pairs of varieties. Strictly speaking, two standard errors are required, one for a pair of varieties which do not appear in the same block, and one, slightly lower, for a pair which appear in the same block. However, even with only two replications—the most extreme case—the ratio of the former to the latter standard error does not exceed $\sqrt{\dfrac{p+2}{p+1}}$, where $p^2$ is the number of varieties, and approaches this limit only when the differences between small blocks are large. Separate standard errors should first be calculated for the two types of comparison before deciding whether to use a common error.

Sometimes the experimenter may wish to replicate a standard variety more frequently than the new varieties, either to secure more accurate comparisons between the latter and the standard or to assist in taking field notes. In the lattice designs, this extra replication may be secured simply by choosing two or more of the $p^2$ varieties to be a common control variety. While the controls should be randomized in the same way as the new varieties, the field plan can usually be arranged so that two controls never appear in the same block, thus giving a more even distribution of the controls through each replication. In analyzing the results, it is simplest to regard the controls at first as separate varieties, so that the usual method of analysis applies. The average of the controls should be taken when the adjusted mean yields have been computed, the standard error of this average being, of course, $1/\sqrt{r}$ times the standard error of a single varietal mean, where $r$ is the number of controls averaged.

### COMPARISON OF RESULTS FROM DIFFERENT YEARS OR DIFFERENT CENTRES

In most breeding programs, promising varieties are tested in several different localities and over several seasons. A combined analysis may be wanted to discover those varieties which are consistently superior and those which are specially adapted for certain localities. Even with the simplest of designs, a satisfactory analysis is more difficult than is usually realized (cf. Yates and Cochran, 11). In particular, special methods are necessary if the experimental errors differ widely from experiment to experiment.

With lattice designs, the only additional complication arises from the slight inequality mentioned above in the relative accuracy of comparisons between different pairs of varieties. If the same design (apart from the randomization) is used at a number of places, little error is introduced by calculating the interaction of varieties with places in the usual way, using adjusted totals or means.

### GAIN IN ACCURACY

Information is obtained on the relative accuracy of randomized blocks and lattice designs with the same number of replications (a) by superimposing both designs on uniformity-trial data and (b) by analyzing lattice experiments both by the lattice analysis and by the randomized blocks analysis. To make a fair comparison, each design should be laid out in what is considered from previous experience the most accurate arrangement on the particular site used. This condition can be fulfilled with uniformity-trial data, but in a lattice experiment, where the shape of the replication is fixed when the lattice is laid down, we must consider whether the same shape would have been used had the experiment originally been planned as randomized blocks. As a rule, this should be so, because the optimum shape of replication for the lattice is usually also the optimum for randomized blocks.

The measure adopted for the statistical accuracy of a design is the inverse of the average variance of the difference between two varietal means. For the relative accuracy (or efficiency) of two designs, we use the inverse ratio of these variances. Variances are used instead of standard errors because the variance of the mean of $r$ replications is $1/r$ times the variance of a single replication. Thus if the lattice design in two replications has an efficiency of 1.5 (150%) relative to randomized blocks, this implies that a randomized blocks design with three replications would have given about the same accuracy as the lattice design with two replications.

Information on wheat is scanty. Of six experiments carried out by Dr. L. R. Waldron, with numbers of varieties ranging from 49 to 169, two showed no gain, one a negligible gain (2%), while the other three showed gains of 39, 45, and 156%.[5] With 36 varieties, Goulden (4) obtained increases of about 70 and 90% on Wiebe's data. For other crops for which comparisons are available, principally corn, the average gain appears to be about 30%. With this figure, a lattice design in three replicates is almost as accurate as a randomized blocks design in four replicates. Valuable information can be obtained by calculating the relative efficiency as a routine matter whenever lattice designs are used which provide a fair comparison. If the experiences of different workers are pooled, it should be possible to assess the usefulness of these designs for wheat trials in a short time. Results of a similar investigation on corn will be published shortly, covering over 70 lattice experiments carried out or supervised by the Iowa Experiment Station.

---

[5]The writer is indebted to Dr. Waldron for permission to use his results in calculating these gains.

## OTHER LATTICE DESIGNS

If a higher degree of replication is desired, it may be possible to use lattice designs in which *every* pair of varieties occurs once in the same block. For 25, 49, 64, 81, and 121 varieties, these designs require 6, 8, 9, 10, and 12 replications, respectively.

Certain lattice designs can be laid out in Latin squares. For 25, 49, 81, and 121 varieties, 3, 4, 5, and 6 replications are required, respectively. For 64 varieties, 9 replications are required, while no design exists for 36 and 100 varieties. The writer does not possess the data necessary to give an opinion on the advisability of Latin square designs for wheat, where the plots are usually 1 rod long and one, two, or three rows wide. Where experimenters have found ordinary Latin squares superior to randomized blocks in small varietal trials, the same result would be expected to hold with lattice designs. On Wiebe's data, lattice designs arranged in Latin squares proved substantially more accurate than lattice designs arranged in small blocks, but this result may not be typical.

For numbers of varieties over 200, cubic lattice designs arrange $p^3$ varieties in blocks of size p (9). The number of replications must be a multiple of three.

Goulden (3) also recommends the *incomplete randomized blocks designs* in which the number of varieties need not be a complete square. However, these designs cannot as a rule be arranged in separate replications or analyzed alternatively by the randomized blocks analysis.

## LITERATURE CITED

1. CORNISH, E. A. The estimation of missing values in quasi-factorial designs. Ann. Eugen. (Cambridge), 10:137–143. 1940.
2. COX, G. M., ECKHARDT, R. C., and COCHRAN, W. G. The analysis of lattice and triple lattice experiments in corn varietal tests. Iowa Agr. Exp. Sta. Res. Bul. 281. 1940.
3. GOULDEN, C. H. Modern methods for testing a large number of varieties. Can. Dept. Agr. Tech. Bul. 9. 1937.
4. ————. Efficiency in field trials of pseudo-factorial and incomplete randomized blocks methods. Can. Jour. Res. C, 15:231–241. 1937.
5. SALMON, S. C. The use of modern statistical methods in field experiments. Jour. Amer. Soc. Agron., 32:308–320. 1940.
6. WEISS, M. G., and COX, G. M. Balanced incomplete block and lattice square designs for testing yield differences among large numbers of soybean varieties. Iowa Agr. Exp. Sta. Res. Bul. 257. 1939.
7. WIEBE, G. A. Variation and correlation in grain yields among 1,500 wheat nursery plots. Jour. Agr. Res., 50:331–357. 1935.
8. YATES, F. A new method of arranging variety trials involving a large number of varieties. Jour. Agr. Sci., 26:424–455. 1936.
9. ————. The recovery of inter-block information in variety trials arranged in three-dimensional lattices. Ann. Eugen. (Cambridge), 9:136–156. 1939.
10. ————. Modern experimental design and its functions in plant selection. Emp. Jour. Exp. Agr., 8:223–230. 1940.
11. ————, and COCHRAN, W. G. The analysis of groups of experiments. Jour. Agr. Sci., 28:556–580. 1938.
12. ————. Lattice squares. Jour. Agr. Sci., 30:672–687. 1940.

# 25

## A DOUBLE CHANGE-OVER DESIGN FOR DAIRY CATTLE FEEDING EXPERIMENTS*

W. G. COCHRAN, K. M. AUTREY AND C. Y. CANNON

*Iowa Agricultural Experiment Station, Ames, Iowa*

### INTRODUCTION

During the winter of 1939–40, a feeding trial was carried out by the Iowa Agricultural Experiment Station with eighteen Holstein cows, on three planes of feeding. The three rations tested, and hereinafter designated as A, B and C, respectively, were as follows:

A. Roughage alone (alfalfa hay and corn silage).

B. Limited grain (alfalfa hay, corn silage and grain fed at the rate of 1 pound for each 7 pounds of milk produced).

C. Full grain (alfalfa hay, corn silage and grain fed at the rate of 1 pound for each 3.5 pounds of milk produced).

The grain mixture consisted of the following constituents in parts by weight: corn, 4; oats, 4; cracked soybeans, 1; bone meal, 0.25; and salt, 0.15.

The experiment was confined to a single lactation, and was of the short-time switch-over type, each cow receiving each ration for a period of six weeks. The layout of the experiment was designed so as to avoid as far as possible the difficulties that commonly arise in interpreting the results of switch-over trials, and to permit tests of significance of the differences between the effects of the rations. It is hoped that a brief description of the layout and of the statistical analysis may be of interest to those engaged in dairy work.

### DESCRIPTION OF THE DESIGN

In the ordinary group trial, where each cow receives only a single ration, the average milk-yield obtained for a given ration depends on the yielding ability of the cows receiving the ration. Since cows are highly variable in this respect, the experimental error from this source is often large, though it may be reduced by skillful grouping of the cows. The switch-over trial represents an attempt to eliminate this source of variation entirely from the comparisons between rations, by feeding every cow in turn on all rations. However, in designing switch-over trials, other sources of experimental error must be taken into account if the maximum gain in accuracy is to be realized. One arises from the characteristic drop in milk yields towards the end of the lactation period. Thus if a cow receives rations A, B and C in succession, her total milk yields under the three rations are scarcely comparable, since A was tested during the most productive period and C during the least. In

Received for publication June 16, 1941.

* Journal Paper No. J-199 of the Iowa Agricultural Experiment Station, Ames, Iowa. Project Numbers 416 and 514.

most switch-over trials whose results we have examined, this drop in yield is a preponderating factor, the yields falling even when the poorest ration is given first and the best ration last.

This difficulty can be largely overcome by designing the experiment so that in any experimental period one-third of the cows are receiving each ration. For instance, one-third of the cows might receive the sequence ABC, one-third the sequence BCA, and one-third the sequence CAB. With this arrangement the three experimental periods are equally represented in the average milk yield for any ration. However, the effects of the lactation curve are not eliminated entirely from the experimental error of these averages unless the natural rate of fall in yield from period to period is the same for the three groups of cows. For example, if the sequence ABC is given to a group of cows whose lactation curves show high yields in the first period, thereafter dropping rapidly, whereas the other two sequences are given to cows whose lactation curves have a much smaller rate of fall, ration A is favored relatively to B and C when the average yields over all three groups are calculated. Thus the division of the available cows into three groups to receive the three sequences should not be made simply at random; instead, the object should be to obtain groups which are as similar as possible in the rates of fall of their lactation curves.

In the present experiment, the eighteen cows were first divided into six groups of three on the basis of expected yielding ability, the three cows within each group being chosen as similar as possible. The three sequences were then allotted at random to the three cows of each group. The reasoning was that cows with the same yielding ability would be likely to show a similar rate of fall in yields. This assumption appeared to be justified by the results of the trial in which this grouping was used, the total rate of fall per group decreasing markedly from the high-yielding to the low-yielding groups, as shown in table 1. In this table, as in subsequent tables, all the milk yields were converted to four per cent, fat-corrected milk (2).

The decreases in yield from period I to period II and from period II to period III were almost twice as great for the high-yielding groups as for the low-yielding groups.

TABLE 1

*Total yield and rate of fall in yield of fat-corrected milk*

|  | Groups* of three cows | | | | |
|---|---|---|---|---|---|
|  | 1 | 2 | 4 | 5 | 6 |
| Total milk yield (lbs). | 14,351 | 14,168 | 12,735 | 11,391 | 10,603 |
| Decrease in yield (lbs.) |  |  |  |  |  |
| Period I–II | 868 | 889 | 385 | 521 | 472 |
| Period II–III | 1,019 | 923 | 598 | 506 | 495 |

* Group 3 omitted because one cow became sick and had to be rejected.

With this design, each group of three cows constitutes an independent experiment, of a type known as the $3 \times 3$ Latin square (fig. 1) (1).

| Period | Cow | | |
|--------|-----|-----|-----|
| | 1 | 2 | 3 |
| I | A | B | C |
| II | B | C | A |
| III | C | A | B |

FIG. 1. The $3 \times 3$ Latin Square Design (No. 1).

A valid estimate of the experimental error may be obtained by an analysis of variance, as follows:

| | Degrees of freedom |
|---|---|
| Between cows | 2 |
| Between periods | 2 |
| Between rations | 2 |
| Error | 2 |
| Total | 8 |

Differences between cows and between periods are not included in the estimate of error, since by the nature of the design they are excluded from the true errors of the ration means. Each of the six groups provides a separate analysis of variance of the above type. On combining the six analyses, the following subdivision of degrees of freedom is obtained for the whole experiment:

| | Degrees of freedom |
|---|---|
| Between groups | 5 |
| Between cows within groups | 12 |
| Between periods within groups | 12 |
| Between rations | 2 |
| Interaction of rations and groups | 10 |
| Error | 12 |
| Total | 53 |

The 12 degrees of freedom for *between cows between periods* and *error* are the totals of the corresponding terms in the individual groups. A similar set of 12 degrees of freedom is obtained for *between rations*. This may be divided into two parts: 2 degrees of freedom representing the average differences between rations over the whole experiment, and 10 degrees of freedom representing variations from group to group in the differences between the rations. If the differences between the rations are the same in all groups (apart from experimental errors), the mean square for this term should be no larger than the error mean square, and may legitimately be included in the error. However, the term should be calculated separately to test this point, particularly where the groups differ widely in yielding ability or in other respects, *e.g.*, breed. The remaining 5 degrees of freedom, corresponding to differences between the group totals, are not included in the individual analyses, being given here for completeness.

For reasons to be explained later, three of the groups were assigned to the Latin square given above, and the remaining three to the complementary square as shown in figure 2. This feature introduces no change in the method of analysis.

$$
\begin{array}{ccc}
A & B & C \\
C & A & B \\
B & C & A
\end{array}
$$

FIG. 2. The $3 \times 3$ Latin Square (No. 2).

### ANALYSIS OF VARIANCE OF THE TOTAL DIGESTIBLE NUTRIENT CONSUMPTION

The consumption of digestible nutrients per cow for each period is given in table 2. Each of the groups in this table was analyzed separately and the degrees of freedom, as well as the sums of squares, were combined (table 3) according to the procedure previously described. The data on group 3 were omitted from table 2 due to the loss of one of the cows in it, thus causing a loss of nine degrees of freedom in the combined analysis.

TABLE 2

*Individual total digestible nutrient consumption units: lbs. per cow per period*
*(total of six weeks)*

| | Group 1 | | | | | | | Group 2 | | | | | |
|---|---|---|---|---|---|---|---|---|---|---|---|---|---|
| Period | Cow | | | | | | Period total | Cow | | | | | Period total |
| | 1 | | 2 | | 3 | | | 4 | | 5 | | 6 | |
| I | A | 608 | B | 885 | C | 940 | 2433 | A 527 | B | 696 | C | 989 | 2212 |
| II | B | 715 | C | 1087 | A | 766 | 2568 | C 883 | A | 635 | B | 899 | 2417 |
| III | C | 844 | A | 711 | B | 832 | 2387 | B 785 | C | 901 | A | 657 | 2343 |
| Total | | 2167 | | 2683 | | 2538 | 7388 | 2195 | | 2232 | | 2545 | 6972 |

| | Group 3 | | | | | | Group 4 | | | | |
|---|---|---|---|---|---|---|---|---|---|---|---|
| | Cow | | | | | | Cow | | | | Period total |
| | 7 | | 8 | | 9 | | 10 | | 11 | | 12 |
| This group omitted because cow No. 7 was removed from the experiment in Period I, owing to kidney cancer. | | | | | | A | 472 | B 734 | C 897 | 2103 |
| | | | | | | C | 819 | A 644 | B 766 | 2229 |
| | | | | | | B | 778 | C 953 | A 706 | 2437 |
| | | | | | | | 2069 | 2331 | 2369 | 6769 |

| | Group 5 | | | | | | | Group 6 | | | | | |
|---|---|---|---|---|---|---|---|---|---|---|---|---|---|
| Period | Cow | | | | | | Period total | Cow | | | | | Period total |
| | 13 | | 14 | | 15 | | | 16 | | 17 | | 18 | |
| I | A | 586 | B | 635 | C | 805 | 2026 | A 489 | B | 593 | C | 788 | 1870 |
| II | B | 723 | C | 799 | A | 542 | 2064 | C 730 | A | 536 | B | 695 | 1961 |
| III | C | 892 | A | 595 | B | 681 | 2168 | B 674 | C | 758 | A | 609 | 2041 |
| Total | | 2201 | | 2029 | | 2028 | 6258 | 1893 | | 1887 | | 2092 | 5872 |

A = roughage, B = limited grain, C = full grain.

TABLE 3

*Analysis of variance (3) of t.d.n. consumption units: total consumption per cow per period, in lbs.*

| | Degrees of freedom | Sums of squares | Mean squares |
|---|---|---|---|
| 1. Between groups | 4 | 157,936 | 39,484† |
| 2. Between cows within groups | 10 | 105,336 | 10,534† |
| 3. Between periods within groups | 10 | 40,534 | 4,053* |
| 4. Between rations | 2 | 533,869 | 266,934† |
| 5. Ration × group interactions | 8 ⎱ 18 | 12,021 ⎱ 26,121 | 1,503 ⎱ 1,451 |
| 6. Error | 10 ⎰ | 14,100 ⎰ | 1,410 ⎰ |
| 7. Total | 44 | 863,796 | |

\* Significant.
† Highly significant.

Since the analysis of variance in table 3 shows the ''ration × group interaction'' to be insignificant, this term was included in the experimental error. The mean square of the ration effects (266,948) is significantly greater than the error mean square (1450), signifying that the consumption of nutrients was appreciably increased by the addition of grain to the ration. The F-test (3) reveals that the differences between groups are also highly significant, the mean square for that term being 39,490 as compared with only 1450 for the error. This shows that the arrangement of cows into outcome groups was well justified.

TABLE 4

*Digestible nutrient consumption-totals and means units: lbs. for six weeks periods*

| Rations | Ration totals (15 cows) | Ration means | Standard error |
|---|---|---|---|
| A | 9,083 | 605.5 | |
| B | 11,091 | 739.4 | ± 9.83 |
| C | 13,085 | 872.3 | |
| Grand total = | 33,259 | General mean = 739.1 | |

Table 4 shows the mean digestible nutrient consumption for each ration, and the standard error per cow. The standard error, $\sqrt{1451} = 38.08$ is 5.15 per cent of the general mean, 739.1. The standard error for the mean of 15 cows is $38.08/\sqrt{15} = 9.83$ pounds, or only 1.33 per cent of the mean. The relative performance of the groups was in the order anticipated, all differences between pairs of rations being highly significant.

#### ADJUSTMENTS FOR CARRY-OVER EFFECTS

In the case of total nutrient consumption the amount eaten in one period did not appear to be influenced by the ration given in the previous period. However, with other factors, particularly total milk yield, a carry-over effect of the ration given in the previous period may be anticipated, owing to the shortness of the change-over period from one ration to another.

If such residual effects are present, simple averages do not give unbiased estimates of the effects of the rations. For instance, with the layout shown in figure 1, ration A is preceded by ration C in both the second and third periods, and likewise B by A and C by B. If A is the poorest ration and C the best, the milk yields for ration A may be increased by the beneficial carry-over effect of C, while those for ration C may be depressed by the carry-over effect of A. In these circumstances, the simple averages would *underestimate* the differences between the direct effects of the rations.

Thus when carry-over effects are present, the average milk yields under each ration must be adjusted in some way to avoid bias. These adjustments can clearly be made only if the design of the experiment enables us to esti- mate the sizes of the carry-over effects. This cannot be done with the design shown in figure 1, since each ration is always followed by the same ration. Thus, if this design is used for all six groups, no corrections for carry-over effects are possible, except perhaps by inspecting the daily yields, omitting from the results those parts of the experimental periods which appear to show carry-over effects. This method is at best somewhat arbitrary, and would be unsatisfactory if carry-over effects persisted through most of the succeeding periods.

However, by using both the cycles shown in figures 1 and 2, a direct evaluation of the carry-over effects is possible. This may be seen by examin- ing the total milk yields for each of the six cycles, shown in table 5.

TABLE 5

*Total milk yields for each of the six ration cycles*
(Totals for 3 cows (18 weeks) in lbs.)

| Period | | Sets* | | | | | | | | | | |
|---|---|---|---|---|---|---|---|---|---|---|---|---|
| | | 1 | | 2 | | 3 | | 4 | | 5 | | 6 |
| I ...... | A | 4,383 | B | 5,370 | C | 5,720 | A | 4,427 | B | 4,675 | C | 5,236 |
| II ...... | B | 4,057 | C | 4,980 | A | 4,208 | C | 4,535 | A | 3,637 | B | 4,420 |
| III ...... | C | 3,834 | A | 3,264 | B | 3,571 | B | 3,833 | C | 3,787 | A | 2,956 |
| Set totals | | 12,274 | | 13,614 | | 13,499 | | 12,795 | | 12,099 | | 12,612 |
| Grand total | | 76,893 | | | | | | | | | | |
| Ration totals: | | A = Roughage | | | B = Limited grain | | | C = Full grain | | | | |
| | | A = 22,875 | | | B = 25,926 | | | C = 28,092 | | | | |

* To complete set 1, in which a cow was rejected, values were inserted for the missing cow by a method described below.

The results during the first period do not require adjustment, since all cows received the same preceding ration. In the second period, two sets of cows, 3 and 5, received roughage. Of these, set 3, having had full grain in the previous period, gave a total of 4208 lbs., while set 5, having had limited grain in the previous period, gave a total of 3637 lbs. The differ- ence, 571 lbs., is an estimate of the difference between the carry-over effect of full grain and that of limited grain. Similarly, a comparison of the carry-over effects of full grain and roughage is obtained from sets 1 and 6 in the second period, and a comparison between limited grain and roughage from sets 2 and 4. The same comparisons are repeated with the results in the third period. In all six comparisons, the carry-over effect was in the same direction as the direct effect (*i.e.*, full grain > limited grain > roughage), indicating strongly that real carry-over effects were present.

We may now consider how to obtain estimates of the direct effects of the rations, free from the disturbance due to carry-over effects. This may be done in many ways. For instance, an unbiased comparison of the direct effects of roughage and limited grain in each period is given by the mean milk yields of the following groups of cows:

| Period | Roughage | Limited grain |
|--------|----------|---------------|
| I | Sets 1 and 4 | Sets 2 and 5 |
| II | Set 3 | Set 6 |
| III | Set 2 | Set 4 |

Since sets 3 and 6 both had full grain in the first period, their yields in the second period, under roughage and limited grain respectively, are subject to the same carry-over effect, and similarly for sets 2 and 4 in the third period. An estimate of the difference between the two rations for the whole experiment could be obtained by taking some simple average of the differences in the individual periods. Such an estimate is, however, open to criticism on the grounds that only half of the available cows are used in the second and third periods. A further objection, probably more serious, is that the estimate would be partly a group comparison, since only the cows in sets 2 and 4 are common to both roughage and limited grain. The use of such an estimate would therefore defeat the purpose of the switch-over design, which is to avoid group comparisons.

To retain the full advantage of the switch-over design, the estimate of the difference between the effects of two rations must clearly be unaffected by differences in the yielding ability of the cows or by differences between the average yields for different periods. The most accurate estimates, subject to these restrictions, are given by a technique known as the method of least squares, which is frequently used for dealing with similar problems arising in field experiments. In this procedure, the observed milk yield for any cow in any period is expressed as a linear function of the effects of the cow's yielding ability, the period, the current and previous rations, and the experimental error. For example, consider a cow receiving in succession limited grain, full grain and roughage. Its milk yields $y_1$, $y_2$, $y_3$ in the three periods are expressed as follows:

$$y_1 = m + p_1 + d_b + e_1$$
$$y_2 = m + p_2 + d_c + r_b + e_2$$
$$y_3 = m + p_3 + d_a + r_c + e_3$$

where m = mean yield for the cow

$p_1$, $p_2$, $p_3$ = effects of the three periods

$d_a$, $d_b$, $d_c$ = direct effects of the rations

$r_b$, $r_c$ = residual effects of the rations given in the previous period

$e_1$, $e_2$, $e_3$ = experimental errors.

No parameter is required to represent residual effects in the first period. The constants $p_1$, $p_2$, $p_3$ remain unchanged for all three cows in the same group, since only the group averages of the differences between periods are eliminated from the experimental errors. After setting up equations of this type for every cow, the constants are estimated by minimizing the sum of squares of the experimental errors. The details of the mathematical analysis will not be given here. To calculate the adjusted yields given by the least squares solution, the following quantities (Table 6) are required, all being easily obtainable from table 5.

TABLE 6

Data required for calculating the adjusted yields for each ration

| A | 22,875 | a | 15,950 | $s_2 + s_6$ | 26,226 |   |   |
|---|--------|---|--------|-------------|--------|---|---|
| B | 25,926 | b | 15,407 | $s_3 + s_4$ | 26,294 | T | 76,893 |
| C | 28,092 | c | 15,725 | $s_1 + s_5$ | 24,373 | (2/3) T | 51,262 |

Here A, B, C are simply the total yields under the three rations, as given at the foot of table 5. The quantities a, b, c are the total yields of all cows receiving these respective rations *in the previous period.*

$$e.g., \ a = 4057 + 3571 + 4535 + 3787 = 15,950, \ etc.$$

Also $s_2 + s_6$ = total of sets 2 and 6 = 13,614 + 12,612 = 26,226, etc.[1] Finally, we required two-thirds of the total of all yields in table 6 [ $(\frac{2}{3})$ T = 51,262]. The adjusted mean yields per cow per period (total of six weeks) are then given by the following equations:

*Roughage*
$$72\bar{y}_a = 5A + (2a - b - c) + s_2 + s_6 - (\tfrac{2}{3}) \, T$$
$$= 5 \times 22,875 + (2 \times 15,950 - 15,407 - 15,725) + 26,226 - 51,262$$
$$= 90,107$$

Hence $\bar{y}_a = 1251.5$

*Limited grain*
$$72\bar{y}_b = 5B + (2b - a - c) + s_3 + s_4 - (\tfrac{2}{3})T = 103,801 \text{ giving } \bar{y}_b = 1441.7$$

*Full grain*
$$72\bar{y}_c = 5C + (2c - a - b) + s_1 + s_5 - (\tfrac{2}{3})T = 113,664 \text{ giving } \bar{y}_c = 1578.7$$

The corresponding unadjusted mean yields per cow per period are the totals A, B and C, divided by 18 (the number of cows in each total).

The adjustments reduced the mean yield under roughage by 19 lbs., and increased the mean yield under full grain by about the same amount, these

[1] Sets 2 and 6 received roughage during period III; Sets 3 and 4 received limited grain, and Sets 1 and 5 full grain.

TABLE 7

*Mean yields of fat corrected milk per cow per period of six weeks*

| Ration | Unadjusted | | | Adjusted for carry-over effects | | |
|--------|------|-------------------------|---|------|-------------------------|---|
| | Mean | Increase over roughage | | Mean | Increase over roughage | |
| | lbs. | lbs. | % | lbs. | lbs. | % |
| A | 1270.8 | ............ | ...... ...... | 1251.5 | ............ | ............ |
| B | 1440.3 | + 169.5 | + 13.3 | 1441.7 | + 190.2 | + 15.2 |
| C | 1560.7 | + 289.9 | + 22.8 | 1578.7 | + 327.2 | + 26.1 |
| | | | Standard error ± 21.4 | | ± 30.3 | ± 2.4 |

effects being in the direction anticipated by the previous discussion. It is interesting to observe that the mean yield is practically unaltered for the cows receiving limited grain, for which the beneficial carry-over effect from full grain and the detrimental effect from roughage appeared to cancel. By failure to adjust for carry-over effects, the differences between the rations would have been underestimated by about 11 per cent.

To verify that the yields have been freed from carry-over effects, we may create artificially an additional carry-over effect of say, the full grain ration, by adding 1000 to the yields in table 5, wherever full grain was fed in the previous period. The yields affected are 3264 (set 2), 4208 (set 3), 3833 (set 4) and 4420 (set 6). With this change, the reader may verify that the adjusted mean yields become:

Roughage 1325.6 lbs., Limited grain 1515.8 lbs., Full grain 1652.7 lbs.

The effect of the artificial carry-over effect is simply to increase all three means by the same amount, 74.1 lbs., the differences between the ration means remaining exactly the same as before. With unadjusted means, on the other hand, the yields for roughage and limited grain would both be increased by 111.1 lbs., that for full grain being unchanged. The same property holds for the adjusted mean yields if a constant amount is added to all yields following any one of the three rations.

It may be of interest to examine the differences between the carry-over effects of the three rations. For a single cow (total of six weeks), these are given by the deviations of the following quantities (Table 8) from their mean.

TABLE 8

*Estimates of the carry-over effects per cow (total for 6 weeks)*

| | | Deviation (lbs.) |
|--------|--------|-----|
| Roughage: ............ | $(3a + A + s_2 + s_6)/24 = 96,951/24 = 4,040$ | − 58 |
| Limited grain: ............ | $(3b + B + s_3 + s_4)/24 = 98,441/24 = 4,102$ | + 4 |
| Full grain: ............ | $(3c + C + s_1 + s_5)/24 = 99,640/24 = 4,152$ | + 54 |

The difference in carry-over effect between roughage and full grain was 112 lbs., about one-third of the corresponding difference (327 lbs.) between the direct effects.   The same ratio holds approximately for the other two differences between the rations.   The carry-over effects seem strikingly large in relation to the direct effects; however, the figures given above agree well with the rough estimates which may be made from table 5.   The six cows receiving limited grain during the second and third periods and full grain in the previous period gave a total yield of 4420 + 3833 = 8253 lbs., as compared with 7628 lbs. for the six cows having limited grain preceded by roughage.   The difference, 625 lbs., corresponds to 104 lbs. per cow, which checks closely with the estimate of 112 lbs. given above.

The adjustments required for the direct effects may be obtained from the above estimates for the carry-over effects.   For example, of the eighteen yields from cows receiving roughage, six followed a previous ration of limited grain, six followed full grain, while none followed roughage.   Thus the total yield of the eighteen cows should be reduced by six times the sum of the carry-over effects for limited grain and roughage, i.e., the mean should be reduced by

$$(6/18)\ [54 + 4] = 58/3 = 19.3\ \text{lbs.};$$

in agreement with the reduction already given in table 7.

### THE ANALYSIS OF VARIANCE WHEN CARRY-OVER EFFECTS ARE PRESENT

In analyzing the results for total nutrient consumption, we discarded results for the group containing the cow which was removed from the experiment.   While this is the simplest procedure, it results in an appreciable loss of information.   By a technique described by Yates (4) it is possible to include in the analysis the two healthy cows in group 3.   The first step in this technique is to estimate results for the missing cow by the method of least squares.   The completed data can then be analyzed by the usual procedure. with certain slight changes mentioned below.

The individual yields of fat corrected milk are shown in table 9, the values inserted for the missing cow being denoted by asterisks.   In the analysis of variance, the total sum of squares and the sums of squares for groups, cows within groups, periods and the periods by groups interaction are found in the same way as for total nutrient consumption.   Only the sums of squares between rations require special consideration.   Of these, two degrees of freedom measure differences between the direct effects of the three rations, and two measure differences between the carry-over effects. However, the two sets of two degrees of freedom do not add to the correct total sum of squares for direct and carry-over effects, because the two effects are entangled by the nature of the design.   This total sum of squares (4 degrees of freedom) may be obtained either as direct effects (ignoring carry-over effects) + carry-over effects, or as carry-over effects (ignoring direct effects) + direct effects.

TABLE 9

*Individual yields of fat corrected milk units: lbs. per cow per period (total of six weeks)*

| Group 1 | | | | | | Group 2 | | | | |
|---|---|---|---|---|---|---|---|---|---|---|
| Period | Cow | | | Totals | | Cow | | | Totals | |
| | 1 | 2 | 3 | | | 4 | 5 | 6 | | |
| I | A 1376 | B 2088 | C 2238 | 5702 | | A 1863 | B 1748 | C 2012 | 5623 | |
| II | B 1246 | C 1864 | A 1724 | 4834 | | C 1755 | A 1353 | B 1626 | 4734 | |
| III | C 1151 | A 1392 | B 1272 | 3815 | | B 1462 | C 1339 | A 1010 | 3811 | |
| Totals | 3773 | 5344 | 5234 | 14351 | | 5080 | 4440 | 4648 | 14168 | |

| Group 3 | | | | | | Group 4 | | | | |
|---|---|---|---|---|---|---|---|---|---|---|
| Period | Cow | | | Totals | | Cow | | | Totals | |
| | 7 | 8 | 9 | | | 10 | 11 | 12 | | |
| I | A 1665* | B 1938 | C 1855 | 5458 | | A 1384 | B 1640 | C 1677 | 4701 | |
| II | B 1517* | C 1804 | A 1298 | 4619 | | C 1535 | A 1284 | B 1497 | 4316 | |
| III | C 1366* | A 969 | B 1233 | 3568 | | B 1289 | C 1370 | A 1059 | 3718 | |
| Totals | 4548* | 4711 | 4386 | 13645 | | 4208 | 4294 | 4233 | 12735 | |

| Group 5 | | | | | | Group 6 | | | | |
|---|---|---|---|---|---|---|---|---|---|---|
| Period | Cow | | | Totals | | Cow | | | Totals | |
| | 13 | 14 | 15 | | | 16 | 17 | 18 | | |
| I | A 1342 | B 1344 | C 1627 | 4313 | | A 1180 | B 1287 | C 1547 | 4014 | |
| II | B 1294 | C 1312 | A 1186 | 3792 | | C 1245 | A 1000 | B 1297 | 3542 | |
| III | C 1317 | A 903 | B 1066 | 3286 | | B 1082 | C 1078 | A 887 | 3047 | |
| Totals | 3953 | 3559 | 3879 | 11391 | | 3507 | 3365 | 3731 | 10603 | |

A = Roughage, B = Limited grain, C = Full grain.
* Values inserted for cow which was rejected during the first period.

The sum of squares for direct effects (ignoring carry-over effects) is found in the same way as for total nutrient consumption, being:

$(1/18)$ $[(22{,}875)^2 + (25{,}926)^2 + (28{,}092)^2 - 1/3 \ (76{,}893)^2] = 763{,}282.$

The sum of squares for direct effects (adjusted for carry-over effects) is obtained from the adjusted ration totals given above (p. 944), as follows:

$(1/360)$ $[(90{,}107)^2 + (103{,}801)^2 + (113{,}664)^2 - 1/3 \ (307{,}572)^2] = 777{,}534.$

To calculate the two sums of squares for carry-over effects, we require the totals for the three rations, with and without adjustment for direct effects. The former have already been obtained (Table 8).

The adjusted sum of squares is:

$(1/72)$ $[(96{,}951)^2 + (98{,}441)^2 + (99{,}640)^2 - 1/3 \ (295{,}032)^2] = 50{,}409.$

The unadjusted totals are given by subtracting A, B and C, respectively from the corresponding adjusted totals. The sum of squares is:

$(1/90)$ $[(74{,}076)^2 + (72{,}515)^2 + (71{,}548)^2 - 1/3 \ (218{,}139)^2)] = 36{,}158.$

As a check, we may note that:

$$763{,}282 + 50{,}409 = 813{,}691 : 777{,}534 + 36{,}158 = 813{,}692;$$

this figure being the total sum of squares (4 degrees of freedom) for direct and residual effects.   The error sum of squares is found by subtraction.

TABLE 10

*Analysis of variance of fat corrected milk units: total yield per cow per period, in lbs.*

|  | Degrees of freedom | Sum of squares | Mean squares |
|---|---|---|---|
| Total | 50 | 5,124,267 | |
| Periods | 2 | 2,041,769 | |
| Interaction of periods with groups | 10 | 193,341 | 19,334 |
| Groups | 5 | 1,311,769 | 262,354 |
| Cows within groups | 11* | 654,638 | 59,513 |
| Rotations { Residual (ignoring direct) | 2 | 36,158 } | |
| Direct | 2 | 777,534 } | 388,767 |
| Direct (ignoring residual) | 2 | 763,282 } | |
| Residual | 2 | 50,409 } | 25,204 |
| Error | 18† | 109,058 | 6,059 |

\* One degree of freedom subtracted for missing cow.
† Two degrees of freedom subtracted for missing cow.

The complete analysis of variance is shown in table 10, both sets of 2 degrees of freedom for direct and carry-over effects being included.   While only one set need be calculated to obtain the error sum of squares by subtraction, both are required if tests of significance of the direct and carry-over effects are desired.   The mean squares for direct effects (388,767) and carry-over effects (25,204), are both significant, the 5 per cent F-value for 2 and 18 degrees of freedom being 3.55.   The estimated standard error per cow (total of six weeks) is 77.84 lbs., or 5.5 per cent of the mean yield of 1424 lbs.   The corresponding standard error for the mean of 18 cows would be $77.84/\sqrt{18}$.   However, this must be increased to allow for the missing cow, and for the adjustments necessary to correct for carry-over effects.   The effect of the missing cow is to decrease the effective replication from 18 to nearly 16.5, while the adjustments increase the standard error by the factor $\sqrt{1.25}$.   Thus the standard error of the adjusted mean yields in table 7 is $77.84 \times \sqrt{1.25}/\sqrt{16.5} = 21.4$ lbs., or 1.5 per cent of the mean.   The design attained a satisfactory degree of precision; an observed increase of 5 per cent would have been detected as statistically significant, while the actual differences between the ration means were all highly significant.

It has already been noticed (table 1) that the rates of fall in yield from period I to period III differed greatly for the different groups of cows.   This fact is reflected in the analysis of variance, the mean square for the interaction of periods with groups being 19,334 as against 6,059 for the error mean square.   If the six cycles had been assigned to the eighteen cows at random, with three cows to each cycle, the estimated error mean square would have been approximately $(24 \times 6,059 + 10 \times 19,334)/34 = 9,963$.   The device of first dividing the cows into six groups of three on the basis of expected yielding ability therefore resulted in a marked increase in precision.

Similarly, the mean square between cows in the same group (59,513) is much larger than the error mean square. This indicates that any estimates based on group comparisons would have been subject to much higher experimental errors than those avoiding the use of group comparisons.

There is, of course, nothing to be gained by adjusting the means in cases where there are no carry-over effects. The important feature of the design is that it enables corrections to be made where these are considered necessary. An examination of the records, presented as in table 5 will usually be sufficient to decide the issue. In more doubtful cases, the direct and residual effects should be calculated as above, comparing the mean square for residual effects with the error mean square in the analysis of variance. If the former mean square is noticeably higher than the latter, it is safer to make the adjustments.

## OTHER NUMBERS OF REPLICATIONS AND TREATMENTS

In the above design, the experimental unit consists of six cows, one receiving each cycle. To retain the full balance of the design, the number of cows used must be a multiple of six. In an experiment with 6k cows, the calculations are exactly as described above, except that all divisors should be multiplied by k/3. Thus if 24 cows had been used, the expressions $72\bar{y}_a$, $72\bar{y}_b$, $72\bar{y}_c$ for the adjusted totals would be replaced by $96\bar{y}_a$, $96\bar{y}_b$, $96\bar{y}_c$ respectively, the righthand sides of the equations remaining unchanged. The same rule applies to the divisors in the analysis of variance. Where there are no missing cows, the standard error of the adjusted means is

$\sqrt{\dfrac{1.25s^2}{6k}}$ where $s^2$ is the error mean square in the analysis of variance.

Adjustments of this type cannot be made when only two rations are being compared in a switch-over design, since each ration must always precede the other. To obtain direct information on carry-over effects, say in a single reversal trial, it would be necessary to include cows receiving the sequences AA and BB, as well as cows receiving the switch-over sequences AB and BA.

With four treatments there are twenty-four possible cycles. However, the important features of the design can be obtained with only twelve cows, by the use of the following trio of $4 \times 4$ Latin squares:

| Period | Cows | | | | Cows | | | | Cows | | | |
|---|---|---|---|---|---|---|---|---|---|---|---|---|
| | 1 | 2 | 3 | 4 | 5 | 6 | 7 | 8 | 9 | 10 | 11 | 12 |
| I | A | B | C | D | A | B | C | D | A | B | C | D |
| II | B | A | D | C | D | C | B | A | C | D | A | B |
| III | C | D | A | B | B | A | D | C | D | C | B | A |
| IV | D | C | B | A | C | D | A | B | B | A | D | C |

FIG. 3.  Change-over design for four rations.

Of the three A's in period II, one follows B, one follows C and one follows D. The same is true for any ration in any period after the first. The quantities required for estimating the direct and residual effects of ration A (with and without adjustment) are shown below.

TABLE 11

*Calculation of the direct and residual effects in an experiment with four rations*

| Effect | Totals | Divisors | |
|---|---|---|---|
| | | For means | For analysis of variance |
| Direct (unadjusted) ... | A | 12 | 12 |
| Direct (adjusted) ...... | $11A + (3a - b - c - d) + s_4 + s_7 + s_{10} - \frac{1}{2} T$ | 120 | 1320 |
| Residual (unadjusted) | $4a + s_4 + s_7 + s_{10}$ | ...... | 132 |
| Residual (adjusted) ... | $4a + A + s_4 + s_7 + s_{10}$ | 30 | 120 |

The symbols here have the same definitions as on pages 14 and 15. The sets $s_4$, $s_7$, $s_{10}$ are those in which A is given during the fourth period, the same rule applying to rations B, C and D. The first column of divisors gives the numbers by which the corresponding totals must be divided to obtain estimates per cow per period. The second column shows the divisors required in the analysis of variance; in each case these apply to the sums of squares of deviations of the totals for the four rations from their mean. Since the divisors are for a single unit of 12 cows, they must be multiplied by k where k units are used in the experiment. The formula for the standard error of the adjusted means per cow per period is $\sqrt{\frac{1.1s^2}{12k}}$, where $s^2$ is the error mean square.

Since four is probably the maximum number of rations that could be given in a single lactation, no designs are presented for higher numbers of rations.

It should be understood that the designs described above apply particularly to short-time trials. Since it is difficult to keep a group of cows going for several lactations without some losses, such designs would require modification for use in long-time trials (of 3–4 years). Animal losses in experiments of this type destroy the balance of the design, thus making a thorough and efficient analysis of the results difficult.

### SUMMARY

The design and statistical analysis of a short-time switch-over trial comparing three rations are discussed in this paper. The principal objects of the design are to secure accurate comparisons of the effects of the rations and unbiased estimates of the experimental errors. The design also makes it possible to estimate and adjust for carry-over effects, which have usually been ignored in switch-over trials. The appropriate statistical analyses are illustrated by numerical examples of the case in which carry-over effects are negligible (nutrient consumption) and the case in which they are not negligible (milk production). A corresponding design for four rations is briefly described.

A detailed report of the results of this feeding trial will be given in a future paper.

## REFERENCES

(1) FISHER, R. A. The Design of Experiments. Chap. 5. Oliver and Boyd, London, Eng. 1935.

(2) GAINES, W. L., AND DAVIDSON, F. A. Correction of Milk Yield for Fat Content. Ill. Agr. Exp. Sta. Bul. 245. 1923.

(3) SNEDECOR, G. W. Statistical Methods. 3rd ed. Chap. 10 and 11. Collegiate Press, Inc., Ames, Iowa. 1940.

(4) YATES, F. The Analysis of Replicated Experiments when the Field Results are Incomplete. Empire Jour. Expt. Agr., 1: 129–142. 1933.

# 26

## THE DISTRIBUTION OF THE LARGEST OF A SET OF ESTIMATED VARIANCES AS A FRACTION OF THEIR TOTAL

### By W. G. COCHRAN

### 1. INTRODUCTION

FOR a set of quantities $u_1$, $u_2$, ..., $u_n$, each distributed independently as $\chi^2 \sigma^2$ with two degrees of freedom, Fisher (1929) obtained the distribution of the ratio of the largest of the $u$'s to their total. He found the probability that this ratio exceeds the value $g$ to be

$$P = n(1-g)^{n-1} - \frac{n(n-1)}{2!}(1-2g)^{n-1} + \dots + (-)^{k-1}\frac{n!}{k!(n-k)!}(1-kg)^{n-1}, \qquad (1)$$

where $k$ is the greatest integer less than $1/g$. This distribution is used in the harmonic analysis of a series, in testing a particular term which is picked out on inspection because of its exceptional magnitude. It has also been employed (Fisher, 1939) to illustrate the disturbance introduced into the $z$-test of a regression equation, when the equation is 'non-linear', in the sense that the eliminant of the regression coefficients is a non-linear function of the observations which satisfy the regression equation exactly.

The object of this note is to give the result corresponding to (1) for $n$ sets of $r$ degrees of freedom each. The case $r = 2$ appears to be the most useful one practically, as well as the simplest mathematically. The general result is, however, of some mathematical interest and is occasionally helpful in testing one of a group of estimates of variance which appears to be anomalously large.

### 2. A THEOREM IN PROBABILITY

The principal difficulty in the proof arises from the fact that the ratios $u_i/(u_1 + u_2 + \dots + u_n)$, $i = 1, 2, ..., n$, are not independently distributed, since their total is unity. The distribution of the largest ratio will be obtained from a theorem in probability which is of fairly general application to problems of this type.

*Theorem.* Let the joint distribution of a set of quantities $x_1$, $x_2$, ..., $x_n$ be a symmetrical function of the variates $x_1$, ..., $x_n$. If $P_1(g)$ is the probability that a specified variate exceeds the value $g$, $P_2(g)$ the probability that a specified pair of the variates both exceed $g$, and so on, the probability that the largest of the variates exceeds $g$ is

$$nP_1(g) - \frac{n(n-1)}{2!}P_2(g) + \frac{n(n-1)(n-2)}{3!}P_3(g) - \dots + (-)^{n-1}P_n(g). \qquad (2)$$

This theorem is the analogue of a familiar result in the theory of derangements (cf. Whitworth, *Choice and Chance*, Prop. XIV).

*Proof.* The variates are considered in a definite order, $x_1, x_2, ..., x_n$. The probability, $p(x_1 \leqslant g)$, that $x_1$ does not exceed $g$, is $1 - P_1(g)$. But

$$p(x_1 \leqslant g) = p(x_1 \leqslant g, x_2 \leqslant g) + p(x_1 \leqslant g, x_2 > g)$$

and

$$p(x_1 \leqslant g, x_2 > g) = P_1(g) - P_2(g).$$

Hence

$$p(x_1 \leqslant g, x_2 \leqslant g) = 1 - 2P_1(g) + P_2(g). \tag{3}$$

Repeating the above argument with $P_1(g)$, $P_2(g)$, $P_3(g)$ in place of $1$, $P_1(g)$, $P_2(g)$ respectively, we find

$$p(x_1 \leqslant g, x_2 \leqslant g, x_3 > g) = P_1(g) - 2P_2(g) + P_3(g).$$

Hence

$$p(x_1 \leqslant g, x_2 \leqslant g, x_3 \leqslant g) = 1 - 3P_1(g) + 3P_2(g) - P_3(g). \tag{4}$$

The rule of formation of successive terms is clear from (3) and (4), the coefficients being the coefficients in the expansion of $(1-1)^r$. Thus the probability that none of the $n$ variates exceeds $g$ is found to be

$$1 - nP_1(g) + \frac{n(n-1)}{2!} P_2(g) - ... + (-)^n P_n(g). \tag{5}$$

Hence the probability that the largest of the variates exceeds $g$ is as stated in the theorem.

If the variates $x_1, x_2, ..., x_n$ are all independently distributed, $P_2(g) = P_1^2(g)$, $P_3(g) = P_1^3(g)$, .... In this case, (5) reduces to $[1 - P_1(g)]^n$ and (2) to $1 - [1 - P_1(g)]^n$. Thus the theorem may be regarded as the extension of this well-known result to the case where the variates are symmetrically, but not independently, distributed.

## 3. THE JOINT DISTRIBUTIONS OF THE RATIOS

To apply the theorem, it is necessary to study the joint distributions of groups of the ratios $x_i = u_i/(u_1 + ... + u_n)$. These may easily be obtained, since the quantities

$$y_1 = u_1/(u_1 + u_2 + ... + u_n), \quad y_2 = u_2/(u_2 + u_3 + ... + u_n),$$

$$y_3 = u_3/(u_3 + ... + u_n) ... y_{n-1} = u_{n-1}/(u_{n-1} + u_n)$$

are independently distributed. Each of the quantities $y_i$ is of the form $1 \Big/ \Big(1 + \dfrac{n_1 e^{2z}}{n_2}\Big)$, where $z$ is Fisher's $z$ for $n_1$ and $n_2$ degrees of freedom, $n_2$ being equal to $r$ throughout, while the successive values of $n_1$ are $r(n-1)$, $r(n-2)$, $r(n-3)$, ... respectively.

The distribution law of $v = e^{2z}$ for $n_1$ and $n_2$ degrees of freedom is

$$\frac{n_1^{\frac{1}{2}n_1} n_2^{\frac{1}{2}n_2}}{B(\frac{1}{2}n_1, \frac{1}{2}n_2)} \frac{v^{\frac{1}{2}n_1 - 1} dv}{(n_2 + n_1 v)^{\frac{1}{2}(n_1 + n_2)}}. \tag{6}$$

Hence the distribution of any $y = 1 \Big/ \Big(1 + \dfrac{n_1 v}{n_2}\Big)$ is

$$\frac{1}{B(\frac{1}{2}n_1, \frac{1}{2}n_2)} y^{\frac{1}{2}n_2 - 1} (1 - y)^{\frac{1}{2}n_1 - 1} dy, \quad 0 \leqslant y \leqslant 1 \tag{7}$$

with the appropriate values of $n_1$ and $n_2$.

Now $y_1 = x_1$, $y_2 = x_2/(1-x_1)$, $y_3 = x_3/(1-x_1-x_2)$, etc., so that the joint distribution of any group of the ratios $x_i$ may be found by making the necessary substitutions in the joint distributions of the $y$'s. This gives, for a single specified ratio,

$$f_1(x_1) = B_{r,n} x_1^{\frac{1}{2}r-1}(1-x_1)^{\frac{1}{2}\{r(n-1)\}-1}, \quad 0 \leqslant x_1 \leqslant 1 \tag{8}$$

For two specified ratios,

$$f_2(x_1, x_2) = \begin{cases} B_{r,n} B_{r,n-1}(x_1 x_2)^{\frac{1}{2}r-1}(1-x_1-x_2)^{\frac{1}{2}\{r(n-2)\}-1}, & x_1+x_2 \leqslant 1 \\ 0. & x_1+x_2 > 1 \end{cases} \tag{9}$$

For three specified ratios,

$$f_3(x_1, x_2, x_3) = \begin{cases} B_{r,n} B_{r,n-1} B_{r,n-2}(x_1 x_2 x_3)^{\frac{1}{2}r-1}(1-x_1-x_2-x_3)^{\frac{1}{2}\{r(n-3)\}-1}, & x_1+x_2+x_3 \leqslant 1 \\ 0, & x_1+x_2+x_3 > 1 \end{cases} \tag{10}$$

and so on, where $B_{r,n} = 1/B\left(\dfrac{r}{2}, \dfrac{r(n-1)}{2}\right)$.

## 4. THE DISTRIBUTION OF THE LARGEST RATIO

In the notation of the theorem,

$$P_1(g) = \int_g^1 f_1(x_1)\, dx_1, \tag{11}$$

$$P_2(g) = \begin{cases} 0, & \text{if } g > \frac{1}{2}, \\ \int_g^{1-g} dx_2 \int_g^{1-x_2} f_2(x_1, x_2)\, dx_1, & \text{if } g \leqslant \frac{1}{2}, \end{cases} \tag{12}$$

$$P_3(g) = \begin{cases} 0, & \text{if } g > \frac{1}{3}, \\ \int_g^{1-2g} dx_3 \int_g^{1-x_3-g} dx_2 \int_g^{1-x_2-x_3} f_3(x_1, x_2, x_3)\, dx_1, & \text{if } g \leqslant \frac{1}{3} \end{cases} \tag{13}$$

and so on.

Hence the probability that the largest ratio exceeds $g$ is

$$nP_1(g) = \frac{n(n-1)}{2!} P_2(g) + \ldots + (-)^{n-1} P_n(g), \tag{14}$$

where the number of non-zero terms appearing in the expression is the greatest integer less than $1/g$. The value of $P_n(g)$ is not required in practice, since $P_n(g) = 0$ if $g > 1/n$, and the largest ratio is certain to be as great as $1/n$.

For $r = 2$, it may easily be verified that $P_1(g) = (1-g)^{n-1}$, $P_2(g) = (1-2g)^{n-1}$, etc., in agreement with Fisher's result. For all other values of $r$ the integrals are more troublesome. The first term, $P_1(g)$, is simply the incomplete integral of a Beta-function distribution, and may be obtained from the published tables (1934), or by direct calculation outside the limits of the tables. The second and all subsequent terms require repeated integration of the Beta-function distribution. With even values of $r$, all terms reduce to a polynomial in $g$, but the expressions rapidly become complicated. For instance, $r = 4$, gives

$$P_1(g) = (1-g)^{2n-2}[1 + 2(n-1)g],$$
$$P_2(g) = (1-2g)^{2n-3}[(1-2g)^2 + 2(2n-1)g(1-2g) + (2n-1)(2n-2)g^2].$$

As shown by Fisher (1929) for the case $r = 2$, the upper 5 and 1 % significance levels of $g$ can be calculated without difficulty, with accuracy sufficient for most purposes. It was pointed out above that if the $x$-variates were independent, $P_2(g)$ would equal $P_1^2(g)$. Since, however, the probability of any other $x$ being high is decreased as $x_1$ increases, $P_2(g)$ must be less than $P_1^2(g)$ for all values of $n$ and $r$. At the 5 % level, a first approximation to $g$ is obtained by ignoring all terms except $P_1(g)$, choosing $g$ so that $P_1(g) = 0.05/n$. In this case $P_2(g)$ is less than $0.0025/n^2$, and the term $n(n-1)P_2(g)/2$ is less than $0.00125(n-1)/n$. Thus the value of $g$ obtained by using the first term only, while always slightly too high, gives a probability which is between 0.04875 and 0.05. For small values of $n$, $P_2(g)$ is considerably less than $P_1^2(g)$ at the 5 % level, and the approximation is much closer. Similarly, the use of the first term alone at the 1 % level gives a probability between 0.00995 and 0.01. A lower limit to the significance level may also be obtained, by treating all the $x$-variates as independent, and using $[1 - P_1(g)]^n = 0.95$ or $0.99$. As an example, with five sets of six degrees of freedom each, the first term alone gives the 5 % level of $g$ as 0.4783, which is correct to a unit in the last place, while the assumption of independence gives the lower limit $g = 0.4772$. For higher values of $n$, the lower limit is still less in error.

A table of the 5 % points of $g$ for small values of $n$ and $r$ is given below.

*Table of 5 % levels of the largest ratio $u/\Sigma u$*

| $n =$ number of sets | $r =$ number of degrees of freedom in each set | | | | | | | |
|---|---|---|---|---|---|---|---|---|
| | 1 | 2 | 3 | 4 | 5 | 6 | 8 | 10 |
| 3 | 0.9669 | 0.8709 | 0.7977 | 0.7457 | 0.7071 | 0.6771 | 0.6333 | 0.6025 |
| 4 | 0.9065 | 0.7679 | 0.6841 | 0.6287 | 0.5895 | 0.5598 | 0.5175 | 0.4884 |
| 5 | 0.8412 | 0.6838 | 0.5981 | 0.5441 | 0.5065 | 0.4783 | 0.4387 | 0.4118 |
| 6 | 0.7808 | 0.6161 | 0.5321 | 0.4803 | 0.4447 | 0.4184 | 0.3817 | 0.3568 |
| 7 | 0.7271 | 0.5612 | 0.4800 | 0.4307 | 0.3974 | 0.3726 | 0.3384 | 0.3154 |
| 8 | 0.6798 | 0.5157 | 0.4377 | 0.3910 | 0.3595 | 0.3362 | 0.3043 | 0.2829 |
| 9 | 0.6385 | 0.4775 | 0.4027 | 0.3584 | 0.3286 | 0.3067 | 0.2768 | 0.2568 |
| 10 | 0.6020 | 0.4450 | 0.3733 | 0.3311 | 0.3029 | 0.2823 | 0.2541 | 0.2353 |

Within the limits of the table, only the first term need be considered to give four-figure accuracy. As a check, the second term $P_2(g)$ was evaluated by a series of the form

$$\frac{(\frac{1}{2}rn - 1)!}{[(\frac{1}{2}r-1)!]^2 \{\frac{1}{2}[r(n-2)]+1\}!} g^{r-2}(1-2g)^{\frac{1}{2}\{r(n-2)\}+1}$$
$$\times \left\{ 1 + \frac{2(r-2)}{r(n-2)+4}x + \frac{(r-2)(3r-10)}{[r(n-2)+4][r(n-2)+6]}x^2 + \ldots \right\},$$

where $x = (1-2g)/g$.

This series terminates if $r$ is even, and the terms diminish fairly rapidly. For $r = 10$, $n = 10$, the value of the second term, $\frac{n(n-1)}{2}P_2(g)$ is 0.0000988, while a change of one unit in the fourth decimal place of the tabulated value alters $nP_1(g)$ by about 0.0002. Thus at this stage the value given by the first term only is about five units out in the fifth decimal place.

### 5. The limiting distribution when $n$ is large

For groups of two degrees of freedom, Fisher (1939) showed that the mean value of $g$ was

$$\bar{g} = \frac{1}{n}\left(1 + \frac{1}{2} + \frac{1}{3} + \dots + \frac{1}{n}\right). \tag{15}$$

Thus when $n$ is large, $\bar{g}$ is approximately equal to $(\gamma + \log n)/n$, where $\gamma$ is Euler's constant. The corresponding result for sets of $r$ degrees of freedom is investigated in this section.

Since the whole range of the distribution of the largest ratio must be considered in finding its mean value, it is not sufficient to consider only the first term in the probability integral of $g$. However, when $n$ is large, the ratios may be regarded as independently distributed, and the formula $1 - [1 - P(g)]^n$ used as the probability that the largest ratio exceeds $g$.

From (8) $\qquad\qquad f_1(x_1) = \dfrac{1}{B\{\frac{1}{2}r, \frac{1}{2}[r(n-1)]\}} x_1^{\frac{1}{2}r-1} (1-x_1)^{\frac{1}{2}\{r(n-1)\}-1}. \tag{16}$

The mean value of $x_1$ is $1/n$ and its variance is $0(1/n^2)$. Hence, for large $n$, $f_1(x_1)$ may be written approximately $\qquad f_1(x_1) \sim c x_1^{\frac{1}{2}r-1} e^{-\frac{1}{2}nrx_1} \tag{17}$

so that $nrx_1$ is distributed approximately as $\chi^2$ with $r$ degrees of freedom.

The integral of $f_1(x_1)$ may be taken from $g$ to $\infty$ instead of from $g$ to 1, and in the series for the integral (Fisher, 1935) only the first term need be retained. This gives

$$P_1(g) \sim e^{-\frac{1}{2}nrg}. \tag{18}$$

Hence the probability that all $n$ are less than $g$ is approximately

$$(1 - e^{-\frac{1}{2}nrg})^n. \tag{19}$$

Differentiating, we find for the frequency distribution of $g$,

$$\phi(g)\,dg \sim \tfrac{1}{2}n^2 r\, e^{-\frac{1}{2}nrg}(1 - e^{-\frac{1}{2}nrg})^{n-1}\,dg. \tag{20}$$

Put $\frac{1}{2}nrg = \log n + u$. The limits of variation for $u$ are $-\log n$ and $\frac{1}{2}nr - \log n$, which may be taken as $-\infty$ and $+\infty$. Hence the frequency distribution of $u$ is

$$\psi(u)\,du \sim e^{-u}\left(1 - \frac{e^{-u}}{n}\right)^{n-1}du \sim e^{-u - e^{-u}}\,du. \tag{21}$$

The moment-generating function of $\psi(u)$ is

$$M = \int_{-\infty}^{\infty} e^{tu} e^{-u - e^{-u}}\,du = \int_0^{\infty} y^{-t} e^{-y}\,dy = (-t)!.$$

Using the expansion formula for $\log \Gamma(x+1)$, we find

$$K = \log M = \gamma t + \frac{\pi^2 t^2}{6\ 2} + \dots. \tag{22}$$

Thus the mean value of $g$ for $n$ large, is approximately

$$\frac{2}{nr}(\gamma + \log n). \tag{23}$$

This result could alternatively have been obtained from the study by Fisher & Tippett (1928) of the general form of the limiting distribution of the largest member of a sample.

The case $r = 1$ may be related to the fitting of a sine series $A \sin \theta x$, $x = 1, 2, ..., n$, to a set of $n$ values of $y$ normally distributed about a zero mean. If

$$a_{rp} = \sqrt{\frac{2}{n+1}} \sin\left(\frac{p\pi r}{n+1}\right), \quad r = 1, 2, ..., n,$$

and $p$ takes the values $1, 2, ..., n$,

$$\sum_{r=1}^{n} a_{rp} a_{rq} = \begin{cases} 0 & p \neq q, \\ 1 & p = q. \end{cases}$$

Thus the linear functions $\sum_{r=1}^{n} a_{rp} y_r$ provide a subdivision of $\sum_{r=1}^{n} y_r^2$ into $n$ independent single degrees of freedom. The reduction in the sum of squares of $y$ given by the least-squares values of $A$ and $\theta$ cannot be less than the largest of the terms $(\Sigma a_{rp} y_r)^2$. Hence the fraction of the sum of squares accounted for by the two constants $A$ and $\theta$ tends to $2(\gamma + \log n)/n$, or a greater value, as $n$ increases, instead of $2/n$.

## SUMMARY

For a set of $n$ independent estimates of the same variance, each based on two degrees of freedom, Fisher (1929) obtained the distribution of the largest estimate as a fraction of the total. By means of a theorem in probability which may be frequently useful in problems of this type, this result is extended to a set of estimates each based on $r$ degrees of freedom. The distribution has the same general form as Fisher's result, but is simplest in the case that he considered. A table is given of the $5\%$ significance levels of the largest ratio, for small values of $r$ and $n$. The limiting distribution when $n$ is large is briefly discussed.

## REFERENCES

R. A. FISHER & L. H. C. TIPPETT (1928). 'Limiting forms of the frequency distribution of the largest or smallest member of a sample.' *Proc. Camb. Phil. Soc.* **24**, 180–90.

R. A. FISHER (1929). 'Tests of significance in harmonic analysis.' *Proc. Roy. Soc.* A, **125**, 54–59.

—— (1935). 'The mathematical distributions used in the common tests of significance.' *Econometrica*, **3**, 353–65.

—— (1939). 'The sampling distribution of some statistics obtained from non-linear equations.' *Ann. Eugen. Lond.*, **9**, 238–49.

*Tables of the incomplete Beta-function* (1934). London: Biometrika Office.

# 27

August, 1941

Research Bulletin 289

# An Examination of the Accuracy of Lattice and Lattice Square Experiments On Corn

By W. G. Cochran

AGRICULTURAL EXPERIMENT STATION
IOWA STATE COLLEGE OF AGRICULTURE
AND MECHANIC ARTS

STATISTICAL SECTION

DIVISION OF AGRICULTURAL STATISTICS, AGRICULTURAL MARKETING
SERVICE, UNITED STATES DEPARTMENT OF AGRICULTURE

Cooperating

AMES, IOWA

CONTENTS

# SUMMARY

This bulletin consists of an examination from the point of view of accuracy of the results of 93 lattice or lattice square designs used in corn varietal tests during the period 1938-40, inclusive. For the triple lattice designs at Iowa State College, three replications were on the average somewhat more accurate than five replications of the type of randomized blocks design previously used. Since part of this increase in accuracy was presumably due to the long and narrow shape of replication in the randomized blocks designs, somewhat smaller increases would be expected over a randomized blocks design with a more compact replication. For the lattice square designs, the increase in accuracy over randomized blocks represents a saving of about one replication in six with 25 varieties, one replication in five with 49 or 81 varieties and one replication in three with 121 varieties.

The average standard error per plot of 20 hills (1/200 acre) was about 8½ percent and did not vary markedly among the 3 years. The standard error increased only slightly with increasing numbers of varieties in the test, indicating the value of these designs in providing accurate comparisons for tests with many varieties.

While a large number of experiments would be required to obtain a precise comparison, the lattice square designs with 4 x 5 hill plots appeared to be no more accurate than the lattice designs with 2 x 10 hill plots, as judged by the standard errors per plot. Further evidence in support of this result was obtained by a comparison of the triple lattice and lattice square on three corn-uniformity trials. A possible explanation is that the incomplete block in the lattice designs, where 2 x 10 hill plots were used, was much more compact than the row or column in the lattice square, where 4 x 5 hill plots were used.

On the three uniformity trials mentioned above, a lattice square with 2 x 10 hill plots gave a 15 percent gain in accuracy over either a triple lattice with 2 x 10 hill plots or a lattice square with 4 x 5 hill plots. This result suggests that lattice square designs may be used profitably in corn experiments with 2 x 10 hill plots and may also prove serviceable for crops, such as the small grains and soybeans, where the plot is long and narrow.

# An Examination of the Accuracy of Lattice and Lattice Square Experiments on Corn[1]

BY W. G. COCHRAN

During the past 3 years the lattice and lattice square designs have been extensively used in corn varietal trials carried out by the Iowa Agricultural Experiment Station. These designs were introduced by Yates (1) (2) (3) (4) as an improvement on the randomized blocks design. Their layout resembles that of randomized blocks, the plots being grouped in separate replications, which should be as compact in shape as possible. Within each replication the varieties, whose number must be an exact square, are arranged either in k blocks of k varieties each, as in the simple and triple lattice designs, or in a square with k rows and k columns, as in the lattice square designs. By this device and by a suitable choice of the groupings used in successive replications, the mean yield of any variety may be adjusted for the fertility levels of the blocks (or rows and columns) in which it lies, thus permitting a more accurate comparison of the differences among varieties. Since the field operations are essentially the same as for randomized blocks, any increase in accuracy is gained at the expense of only a somewhat more elaborate statistical analysis, the calculation of the adjusted varietal mean yields requiring some extra time in computation.

When the designs were first introduced (1) some idea of the expected gain in accuracy was obtained by superimposing various lattice designs and corresponding randomized blocks designs on a set of uniformity trial data on oranges. Similar comparisons were later made by Goulden (5) on oranges, sugar cane, sugar beets, potatoes, barley and wheat, and by Zuber (6) on corn. While this method of comparison is flexible, since the size and shape of the plots and the dimensions of the replications can be varied, its scope is limited by the small amount of uniformity trial data available for any particular crop, particularly so as lattice designs require a relatively large number of plots. Moreover, fields selected for uniformity trials may not always be representative of the fields on which experiments are normally carried out.

---

[1]Project 514 of the Iowa Agricultural Experiment Station.

The results of any lattice or lattice square experiment, however, make it possible to compare the accuracy of the experiment with that of a randomized blocks design occupying the same plots and the same shape of replication (4), (7). This comparison has the advantage of being made under actual experimental conditions. Thus, when a number of lattice or lattice square designs have been carried out with any crop, the experimental results may be reviewed to discover how much has been gained by the new designs and what types of design appear to be most suitable. The purpose of this bulletin is to present the results of such an examination of the lattice and lattice square experiments on corn, of which 93 are available to date.

## MATERIAL

Access was readily obtained to the results of all lattice or lattice square experiments on corn carried out by the Iowa Agricultural Experiment Station, in cooperation with the Division of Cereal Crops and Diseases, Bureau of Plant Industry, U.S.D.A. The sources of the data are shown below. In parentheses are given the names of the persons to whom the author is indebted for permission to use the data and from whom much further help and information were received: 1. One lattice, 20 triple lattice and 56 lattice square experiments carried out under the corn-breeding program (Dr. G. F. Sprague, Mr. R. C. Eckhardt, Mr. L. A. Tatum); 2. Eleven lattice square experiments from the Iowa Corn Yield Test (Mr. M. S. Zuber); 3. Three lattice square experiments on seed treatment (Dr. C. S. Reddy); and 4. Two lattice square experiments, also on seed treatment (Dr. R. H. Porter). Of the 93 experiments, 14 were conducted in 1938, 24 in 1939 and 55 in 1940. About two-fifths of the experiments were located at or near Ames, the remainder being scattered throughout the state. The following table shows the number of experiments in each of the 12 districts into which the state is divided for the corn-yield test.

TABLE 1. DISTRIBUTION OF THE LATTICE AND LATTICE SQUARE EXPERIMENTS ON CORN

| | Section | | | | | | | | | | | |
|---|---|---|---|---|---|---|---|---|---|---|---|---|
| | Northern | | | North-central | | | South-central | | | Southern | | |
| District | 1 | 2 | 3 | 4 | 5 | 6 | 7 | 8 | 9 | 10 | 11 | 12 |
| Number of experiments | 8 | 9 | 4 | 8 | 3 | 4 | 3 | 40 | 5 | 5 | 1 | 3 |

In this scheme the state is partitioned into four sections from north to south, each section being subdivided into three districts from east to west (1, 4, 7 and 10 being west).

Apart from a few exceptions, the plots consisted of 4 x 5 hills in the lattice square experiments. For the lattice and triple lattice experiments, 2 x10 hills were used at Ames (15 experiments) and 4 x 5 hills elsewhere (6 experiments). Since the hills were usually spaced 3'4" by 3'4", the plot size was thus about 1/200 acre.

## VALIDITY OF THE COMPARISONS

To obtain an equitable comparison between two designs, each should be placed in what is considered the most accurate arrangement according to previous experience and general information about the site. With suitable uniformity trial data, this can be done, since, for example, a different shape of plot can be used for the two designs if considered appropriate. In the present method, however, the lattice or lattice square design can be compared only with a randomized blocks design which employs the same plots and the same replications. It is necessary to examine whether the comparison obtained is fair to the randomized blocks design.

We shall first describe what is considered the optimum arrangement of plots for each type of design, from the point of view of accuracy *per replication*. For the lattice designs, 2 x 10 hill plots should be used, plots in the same incomplete block lying in one line, with their longer sides contiguous. This arrangement gives a compact shape of incomplete block, the dimensions of the block being for example 14 x 10 hills with 49 varieties. The incomplete blocks should be grouped to form a replication as nearly square as possible. The lattice square designs should be laid down in 4 x 5 hill plots, the plots in each replication forming a square. For the randomized blocks designs, 2 x 10 hill plots would be preferred, while the replication should be approximately square in shape.

In the lattice and triple lattice designs at Ames, the plots in the same replication lay in one continuous line, the longer sides of the plots being contiguous. Thus, with 49 varieties, for example, the dimensions of the replication were 10 x 98 hills. The second replication continued in the same line, and so on until an edge of the field was reached. If this occurred in the middle of a replication, the remainder of the replication was placed in the next line, starting back from the edge just reached. A plan illustrating this layout was given by Cox and Eckhardt (7). While the shape of replication obtained by this method was not considered good, the method was normally used for randomized blocks designs at Ames and was continued when the new designs were introduced. Its advantages are that the whole of the experimental site is utilized without waste and that field notes are more easily taken when the plots are in one line.

Thus with the above layout the randomized blocks designs were poorly placed from the point of view of accuracy per replication. In the lattice designs, information is supplied by two types of comparison among varieties, intra-block and inter-block. For the former the layout follows the recommendations given above, but the accuracy of the inter-block information is decreased by the way in which the blocks are arranged in a replication. Since, however, most of the information is contained in the intra-block comparisons, the lattice designs probably suffer less in accuracy per replication than the randomized blocks designs. Nevertheless it is at least of local interest to calculate the relative accuracies in this case, since we are comparing the layout now used with the layout used before the new designs were introduced. The gains in accuracy obtained with the new designs may be applicable to other experiment stations which at present use a similar layout of the randomized blocks design, though they probably overestimate the gains that would be achieved at a station which now uses randomized blocks with a more compact shape of replication.

In the lattice square experiments and in the six lattice experiments carried out in the other districts, the plots (4 x 5 hills) were arranged in a square. As mentioned above this was considered the most accurate arrangement for lattice squares. The lattice designs were not placed to the best advantage, for while the replications were almost square in shape, the incomplete blocks were long and narrow, measuring for instance 28 x 5 hills in a design with 49 varieties. From the point of view of randomized blocks, the shape of replication was also good, but it might be contended that the long and narrow plots would have been used instead of square plots. To examine the effect of shape of plot with a fixed shape of replication, some calculations were made on Bryan's uniformity trial data (8).

These data comprise 2,304 hills of each of three strains, Krug (1923), Iodent (1925) and McCulloch (1925), each site being arranged in a square, 48 x 48 hills. The plot shapes compared were 2 x 8 hills (in both directions) and 4 x 4 hills. Plots 1 x 16 hills were not examined, since they would not be used in actual field tests because of the possibility of inter-strain competition. For each strain the plots were grouped to form six "replications" of 24 plots each, the dimensions of the replications being 16 x 24 hills. With this shape of replication all three types of plot can be arranged in exactly the same set of six replications, thus permitting a comparison of the effect of plot shape with a given shape of replication. The mean squares per plot within replications are shown in table 2.

TABLE 2. MEAN SQUARES PER PLOT (LBS. PER 16 HILLS) FOR THREE SHAPES OF PLOT.

| Shape of plot | Strain | | |
|---|---|---|---|
| | Krug 1923 | Iodent 1925 | McCulloch 1925 |
| 4 x 4 | 2.75 | 4.17 | 7.61 |
| 2 x 8* | 2.94 | 3.46 | 6.66 |
| 8 x 2 | 2.88 | 4.31 | 7.82 |
| Mean of 2 x 8 and 8 x 2 | 2 91 | 3.88 | 7.24 |

*Longer side east-west

Comparing the 4 x 4 plot with the average of the 2 x 8 plots, there appears to be no marked difference in variability. The 4 x 4 plots gave a slightly lower mean square with Krug 1923, and a slightly higher mean square with the other two strains, the differences being about 6 percent in each trial. From an examination of the total variation between plots on the same data, Bryan (8) concluded that "with plots as small as 16 hills, either single, two- or four-row plots may be expected to give rather similar results." We may conclude that for the lattice square designs the comparisons with randomized blocks are reasonably fair to both types of design, while for the six lattice designs the comparisons possibly favor randomized blocks.

## METHOD OF CALCULATING THE RELATIVE ACCURACY

The calculations necessary to compare the relative accuracy of the two types of design have been described by Yates (4) for the lattice square designs and by Cox, Eckhardt and Cochran (7) for the lattice and triple lattice designs. The method depends on the fact, discovered by Yates (3), that these designs can be analyzed either as ordinary randomized blocks designs or by the full lattice analyses. In the former case the unadjusted means over all replications are used as estimates of the varietal yields, the error mean square found by an ordinary randomized blocks analysis of variance providing an unbiased estimate of error for these means. In the latter case the varietal means are adjusted to correct for variations in the fertility of the incomplete blocks (or rows and columns in a lattice square). Thus by analyzing the results of an experiment in both ways, a comparison is obtained between the new design and a randomized blocks design occupying the same plots and the same replications.

As a measure of the accuracy of a design, we shall use the reciprocal of the average variance of the difference between two varietal means. This definition allows a fairly concrete meaning to be given to the term *relative accuracy*. For example, if the above variance is 1.23 for a lattice design and 1.65 for randomized blocks, the accuracy of the new design relative to randomized blocks is measured by the ratio $(1/1.23) \div (1/1.65) = 1.65/1.23 = 1.34$. Since, however, the variance of the varietal means decreases in proportion to the number of replications, the number of replications in the randomized blocks design would have to be increased in the ratio $1.65/1.23$ to make the two designs of equal accuracy. Thus with a relative accuracy of 1.34, or about 4 to 3, we may say that three replications of the lattice design are approximately as accurate as four replications of the randomized blocks design.

To illustrate the computations, an example will be worked for a 7 x 7 triple lattice, with three replications. The analysis of variance required for calculating the adjusted mean yields is shown below, the units being a single plot yield in pounds.

|  | Degrees of freedom | Sums of squares | Mean square |
|---|---|---|---|
| Replications | 2 | 77.08 | |
| Blocks (adjusted) | 18 | 214.90 | 11.939 |
| Varieties (unadjusted) | 48 | 1,558.36 | 32.466 |
| Error (intra-block) | 78 | 347.41 | 4.454 |

The sums of squares for Replications and Varieties are exactly the same as would be obtained in a randomized blocks analysis. Hence the randomized blocks error mean square per plot is $(214.90 + 347.41)/(18 + 78) = 5.857$. Since there are three replications, the error variance of the difference between two unadjusted varietal means is $2(5.857)/3$. With the complete analysis the corresponding variance is $2(4.454)f/3$, the factor f measuring the additional variability that arises from the experimental errors of the adjustments to the varietal means. For a k x k triple lattice, f has the value $1 + \dfrac{3}{2(k+1)} \left( \dfrac{B-E}{B} \right)$, where B, E are the blocks and error mean squares respectively in the analysis of variance above. In this experiment

$$f = 1 + \frac{3}{16} \times \frac{(11.939 - 4.454)}{11.939} = 1.1176$$

Omitting the common factor $\frac{2}{3}$, the accuracy of the triple lattice relative to randomized blocks is $5.857/4.454f = 5.857/4.978 = 1.18$ or 118 percent. The formula for calculating the factor f is given for each type of design at the end of this bulletin. If B is less than or equal to E, indicating that there were no real differences in fertility between the incomplete blocks, the complete analysis reduces to the randomized blocks analysis.

In calculating the factor f, the effects of sampling errors of the relative weights assigned to inter- and intra-block information were ignored. It has been shown (9) that these effects are negligible for the designs presented here, except perhaps for 5 x 5 lattice squares, for which the value of f should be increased by about 3 percent.

When these designs were first introduced by Yates (1), he used a different method of analysis, in which the information contained in the incomplete block totals was ignored. Later he discovered how this information could be utilized to improve the accuracy of the adjustments and developed the method of analysis described in references (4) and (7), which is now recommended as a standard procedure. To examine the superiority of the newer method, to which the above calculations apply, a study was made of the relative accuracies of the designs, when analyzed also by the earlier method. For this it is only necessary to replace f by the reciprocal of the efficiency factor, the latter being $(k + 1)/(k + 2\frac{1}{2})$ for the triple lattice. Hence the relative accuracy, ignoring inter-block information is

$$5.857/(4.454) \times \frac{19}{16} = 1.11$$

## RESULTS

### RELATIVE ACCURACY OF THE LATTICE DESIGNS

The relative accuracies obtained for the 21 lattice designs (20 triple lattice and 1 simple lattice) are shown in percentages in table 3, arranged in increasing order.

The results are highly variable, ranging from 101 to 265 percent where inter-block information was recovered. The most striking feature is the great increase in the accuracy of the experiments at Ames by the use of the new designs. Nine of the 15 experiments showed gains of over 50 percent, while five produced gains of over 100 percent, the mean gain over the whole set being 92 percent. Since, however, the distribution is somewhat skew, the median gain of 79 percent gives perhaps a better estimate of the center of the distribution. With this figure, three replications of the triple lattice would be more accurate than five replications of the type of randomized blocks design previously used.

TABLE 3. ESTIMATED RELATIVE ACCURACIES OF THE LATTICE DESIGNS
(IN PERCENTS).

| Number of varieties | 2 x 10 plots (Ames) Inter-block information | | Number of varieties | 4 x 5 plots Inter-block information | |
|---|---|---|---|---|---|
| | Recovered | Ignored | | Recovered | Ignored |
| 25 | 114 | 103 | 25 | 101 | 85 |
| 49 | 118 | 111 | 49 | 110 | 101 |
| 64 | 124 | 118 | 100 | 116 | 112 |
| 64* | 137 | 130 | 169 | 116 | 113 |
| 49 | 140 | 134 | 100 | 174 | 171 |
| 81 | 145 | 142 | 64 | 190 | 186 |
| 81 | 157 | 154 | | | |
| 49 | 179 | 175 | | | |
| 100 | 184 | 180 | | | |
| 64 | 190 | 196 | | | |
| 64 | 231 | 228 | | | |
| 64 | 238 | 235 | | | |
| 64 | 255 | 252 | | | |
| 169 | 291 | 290 | | | |
| 81 | 365 | 363 | | | |
| Mean | 192 | 187 | Mean | 135 | 128 |

*Simple lattice design

As anticipated earlier, the gains were smaller with 4 x 5 plots; only two of the six experiments showed large increases in accuracy, while the average gain was 35 percent.

The number of experiments is too small to provide more than a crude estimate of the relative accuracies for different numbers of varieties. As would be expected, the gains are smaller with 25 and 49 varieties than with the larger numbers. This point must be borne in mind when interpreting any average figure from the whole group of experiments.

TABLE 4. MEAN RELATIVE ACCURACIES FOR DIFFERENT NUMBERS
OF VARIETIES (WITH RECOVERY OF INTER-BLOCK INFORMATION).

| Shape of plot | Number of varieties | | | | | |
|---|---|---|---|---|---|---|
| | 25 | 49 | 64 | 81 | 100 | 169 |
| 2 x 10 | 114[1] | 146[3] | 197[6] | 222[3] | 184[1] | 291[1] |
| 4 x 5 | 101[1] | 110[1] | 190[1] | ............ | 145[2] | 116[1] |

The figures in brackets denote the number of experiments.

The recovery of inter-block information increased the relative accuracy by 14 percent in the two experiments with 25 varieties and by 6 percent in the four experiments with 49 varieties. With higher numbers of varieties the increase in relative accuracy was only about 3 percent.

While the results for the 2 x 10 plots provide valid estimates of the gains in accuracy achieved at Ames, little information is available on the gain to be expected by changing from randomized blocks with a compact shape of replication to a lattice design with a compact shape of incomplete block, since neither the 2 x 10 nor the 4 x 5 hill plots provides this type of comparison. The expected gain would presumably lie between the gains obtained with these two shapes of plot.

## RELATIVE ACCURACY OF THE LATTICE SQUARE DESIGNS

The results for the 72 lattice square designs are shown in table 5, arranged by number of varieties.

For the 5 x 5 designs the gains in accuracy were mostly small, though only two experiments produced no gain. The average gain was about 20 percent. It is interesting to note that this increase in precision was almost entirely due to the recovery of inter-block information; with the earlier method of analysis nine of the experiments were less accurate than randomized blocks, while the group as a whole was about equal in accuracy to randomized blocks. The gains increased to about 25 percent for the 7 x 7 and 9 x 9 designs with a further increase to over 50 percent for the 11 x 11 designs, for which every experiment showed a substantial gain. The highest individual gain, 362 percent, was obtained with a 7 x 7 design on a soil of low fertility. By using the new method of analysis instead of the original method, the relative accuracy was increased by 22 percent with 25 varieties, 15 percent with 49 varieties, 10 percent with 81 varieties and 6 percent with 121 varieties.

TABLE 5. ESTIMATED RELATIVE ACCURACIES OF THE LATTICE SQUARE
DESIGNS (IN PERCENTS).

| Number of varieties in the experiment | | | | | | | |
|---|---|---|---|---|---|---|---|
| 25 | | 49 | | 81 | | 121 | |
| R* | I† | R | I | R | I | R | I |
| 98 | 71 | 104 | 86 | 111 | 99 | 122 | 115 |
| 100 | 68 | 112 | 97 | 112 | 101 | 134 | 127 |
| 105 | 78 | 114 | 100 | 114 | 102 | 137 | 131 |
| 109 | 81 | 114 | 100 | 114 | 103 | 144 | 138 |
| 110 | 87 | 115 | 96 | 116 | 105 | 145 | 139 |
| 111 | 88 | 116 | 100 | 120 | 106 | 151 | 146 |
| 116 | 94 | 116 | 101 | 121 | 110 | 156 | 150 |
| 119 | 98 | 117 | 103 | 122 | 110 | 164 | 160 |
| 124 | 102 | 120 | 106 | 123 | 114 | 176 | 171 |
| 136 | 117 | 120 | 107 | 124 | 114 | 192 | 185 |
| 137 | 108 | 120 | 107 | 125 | 115 | 194 | 190 |
| 140 | 130 | 124 | 109 | 128 | 119 | 257 | 254 |
| 151 | 134 | 126 | 114 | 132 | 123 | 329 | 316 |
| 195 | 180 | 129 | 113 | 135 | 126 | | |
| | | 133 | 121 | 140 | 124 | | |
| | | 134 | 121 | 140 | 130 | | |
| | | 142 | 122 | 152 | 140 | | |
| | | 148 | 136 | 164 | 156 | | |
| | | 158 | 144 | 178 | 168 | | |
| | | 163 | 152 | 212 | 190 | | |
| | | 173 | 163 | 223 | 213 | | |
| | | 210 | 197 | | | | |
| | | 462 | 455 | | | | |

| Number of experiments | | | | | | | |
|---|---|---|---|---|---|---|---|
| 14 | | 23 | | 21 | | 13 | |

| Mean relative accuracy | | | | | | | |
|---|---|---|---|---|---|---|---|
| 125 | 103 | 147 | 133 | 138 | 127 | 177 | 171 |

| Median relative accuracy | | | | | | | |
|---|---|---|---|---|---|---|---|
| 118 | 96 | 124 | 109 | 125 | 115 | 156 | 150 |

R* = Inter-block information recovered. I† = Inter-block information ignored.
An experiment with 64 varieties (not shown above) gave relative accuracies of 108 (R) and 94 (I) percents.

The results as a whole represent a saving of about one replication in six with 25 varieties, one in five with 49 and 81 varieties and one in three with 121 varieties.

It will be noticed that the gains were considerably larger for the lattices with 2 x 10 hill plots than for the lattice squares. This is presumably due to the difference already noted in the shapes of replication in the two cases. On uniformity trial data, with the same range in numbers of varieties, Zuber (6) obtained average relative accuracies of 144 percent for lattice squares, 131 percent for triple lattices and 133 percent for simple lattices. Since Zuber used the arithmetic mean to combine results from different experiments, a comparable average for the lattice squares in the above data is $(125 + 147 + 138 + 177)/4 = 147$ percent, which agrees closely with his figure.

## EXAMINATION OF THE STANDARD ERRORS PER PLOT

From an examination of the standard errors per plot, some information is obtained about the relative accuracy of the lattice and lattice square designs, and of the designs with different numbers of varieties. The standard errors used were those given by the new method of analysis, being $\sqrt{Ef}$, where E is the intra-block (or intra-row and column) error mean square and f is the factor described on page 406. This quantity measures the *effective* standard error per plot, since it provides the average standard error of a varietal mean on division by the square root of the number of replicates.

The regression coefficients of the standard errors per plot on the mean yields of the experiments were calculated separately for each year. The coefficients were negative in 1938 and 1939 and positive in 1940 and were all small, the largest being -0.11, though two of the values were statistically significant. It was accordingly decided to use the standard errors themselves rather than the coefficients of variation in calculating averages over a number of experiments. In presenting averages, however, the standard errors are expressed as a percentage of the mean yield of the whole group of experiments.

The average standard error percent per plot and the mean yield (in pounds per plot of 20 hills) for each of the 3 years are shown at the top of the following page.

The larger value in 1940 was due mainly to a single highly variable field, which contained four experiments with an average standard error of 22 percent. If this field is omitted, the average standard error percent for 1940 falls to 8.21, while the average over all 3 years is 8.41.

| 1938 | | | 1939 | | | 1940 | | |
|---|---|---|---|---|---|---|---|---|
| No. of experiments | Standard error % | Mean yield | No. of experiments | Standard error % | Mean yield | No. of experiments | Standard error % | Mean yield |
| 14 | 8.68 | 29.2 | 24 | 7.67 | 31.1 | 55 | 9.07 | 27.5 |

The standard errors are arranged by type of design and by number of varieties in table 6.

The results indicate no marked difference in the accuracy attained in lattice square and in triple lattice experiments with 2 x 10 hill plots. Taking the weighted means of the figures for 49, 64, and 81 varieties, we obtain an average of 7.96 percent on 44 experiments for the lattice squares, as against 7.99 on 12 experiments for the triple lattices. The remaining comparisons, though based on very small numbers, appear to be in favor of the triple lattices. Since, as would be expected, there are large differences in the standard errors from field to field, comparisons made between experiments on the same field would be of much higher accuracy. Unfortunately the data provide only a few comparisons of this type. These tend to support the conclusion suggested above, which must be accepted tentatively pending the accumulation of more data in subsequent years. The result is perhaps not so surprising, for while the effects of fertility gradients in two directions are eliminated from the experimental errors in the lattice square, the triple lattice, with 2 x 10 plots, gives an incomplete block that is much more compact than either the row or column in the lattice square.

Since two of the triple lattices with 4 x 5 plots were located on the highly variable field which was omitted, only four are included in the above table. The number is too small for comparison with the other designs.

An increase in the standard errors per plot would be expected with increasing numbers of varieties. This does not appear consistently in table 4, except that the 5 x 5 designs gave the lowest average. By making a within-field com-

TABLE 6. AVERAGE STANDARD ERRORS PERCENT PER PLOT.
(The figures in brackets denote the number of experiments.)

| Design | Number of varieties | | | | | | |
|---|---|---|---|---|---|---|---|
| | 25 | 49 | 64 | 81 | 100 | 121 | 169 |
| Lattice square | 7.88 (14) | 8.02 (23) | 8.33 (1) | 7.88 (20) | ............... | 8.12 (12) | ............... |
| Triple lattice (2 x 10 plots) | 5.31 (1) | 7.88 (3) | 8.02 (6)* | 8.02 (3) | 8.02 (1) | ............... | 7.43 (1) |
| Triple lattice (4 x 5 plots) | 8.75 (1) | 7.95 (1) | 9.51 (1) | ............... | 7.95 (1) | ............... | ............... |

*Includes one simple lattice design

parison between the 5 x 5, 7 x 7, 9 x 9 and 11 x 11 experiments, the following standard errors percent were obtained: 5 x 5, 7.47; 7 x 7, 7.92; 9 x 9, 8.47; 11 x 11, 8.09. While these figures emphasize the increase from the 5 x 5 to the 7 x 7, the change thereafter is somewhat irregular. The fact that the increases are apparently small emphasizes the value of these designs for testing large numbers of varieties. With randomized blocks designs the increases would be much greater, as evidenced by the results obtained in table 5 for the gains in accuracy with different numbers of varieties.

## LATTICE SQUARE DESIGNS WITH OBLONG PLOTS

If the lattice designs with 2 x 10 hill plots are not inferior in accuracy to the lattice square designs with 4 x 5 hill plots, as tentatively suggested above, lattice square designs with 2 x 10 hill plots may prove superior to both types of design. While 4 x 5 and 2 x 10 hill plots are equally easy to lay out with corn, the latter are more convenient at harvest, since all the ears from a plot can be picked by traversing the plot between its two rows. Moreover, the problem whether to recommend lattice square or lattice designs when the plots are long and narrow is an important one for crops such as soybeans and the small grains, where such plots are normally used.

The question was examined by selecting an area measuring 40 x 40 hills from each of Bryan's three uniformity trials (8). Each area was divided into four 5 x 5 squares using [1] 4 x 4 hill plots, [2] 2 x 8 hill plots, with the longer side east-west, [3] 8 x 2 hill plots, with the longer side north-south. With each shape of plot and each trial, the effective standard error per plot for a 5 x 5 lattice square was calculated. For the oblong plots, the standard errors were also computed for 5 x 5 triple lattices, with the longer side of the plot extending the whole length of the incomplete block. While these designs require only three replicates, the results from all four replicates were used in order to compare the shapes of plot on the same total area. The results are given in table 7, the figures for 2 x 8 plots being the means of the results from the two directions of the longer side.

On the average the triple lattice designs with 2 x 8 plots were no less accurate than the lattice square designs with 4 x 4 plots, a result which supports the conclusion suggested in the previous section. The lattice square with 2 x 8 plots proved superior to each of the other designs in all three experiments, showing an average gain in accuracy of about 15 percent. Similar results were obtained by Zuber (6) on

413

TABLE 7. RELATIVE ACCURACY OF LATTICE SQUARE DESIGNS WITH
4 x 4 AND 2 x 8 HILL PLOTS.

| Field | Effective error mean squares (lbs). | | | Relative accuracy* | |
|---|---|---|---|---|---|
| | Lattice square 4 x 4 | Lattice square 2 x 8 | Triple lattice 2 x 8 | Lattice square 2 x 8 | Triple lattice 2 x 8 |
| Krug, 1923 | 2.213 | 2.138 | 2.324 | 103.5 | 95.2 |
| Iodent, 1925 | 2.579 | 2.238 | 2.678 | 115.2 | 96.3 |
| McCulloch, 1925 | 2.665 | 1.958 | 2.136 | 136.1 | 124.8 |
| Mean | 2.486 | 2.111 | 2.379 | 117.8† | 104.5† |

*Relative to the lattice square with 4 x 4 plots
†Calculated from the preceding columns

uniformity-trial data. The results indicate that lattice square designs may be used to advantage with long and narrow plots.

# APPENDIX

## CALCULATION OF THE EFFECTIVE ERROR MEAN SQUARE PER PLOT

As indicated on page 406, the effective error mean square per plot is given by Ef, where E is the intra-block (or intra-row and column) error mean square and f is a multiplying factor. The formulas or references necessary for calculating f (with recovery of inter-block information) are given below for the designs discussed in this paper.

| Type of design | Number of replications | f |
|---|---|---|
| k x k simple lattice | 2 | $1 + \frac{2}{(k+1)}\left(\frac{B-E}{E}\right)$ |
| | 4 | $1 + \frac{4}{(k+1)}\left(\frac{B-E}{2B+E}\right)$ |
| k x k triple lattice | 3 | $1 + \frac{3}{2(k+1)}\left(\frac{B-E}{B}\right)$ |
| | 6 | $1 + \frac{6}{(k+1)}\left(\frac{B-E}{4B+E}\right)$ |
| k x k lattice square (k odd) | (k+1)/2 | $1 + \frac{1}{(k-1)}\left(\frac{R-E}{R} + \frac{C-E}{C}\right)$ |
| | (k+1)* | See (4) |
| k x k lattice square (k even) | (k+1) | See (4) |

*Obtained by turning the design through one right angle, interchanging rows and columns.

B = blocks mean square (adjusted)
R = rows mean square (adjusted)
C = columns mean square (adjusted)
Where B, R or C is less than E, the corresponding term (B — E), (C — E) or (R — E) is taken as zero.

# REFERENCES CITED

(1) Yates, F.  A new method of arranging variety trials involving a large number of varieties. Jour. Agr. Sci., 26:424-455. 1936.

(2) Yates, F.  A further note on the arrangement of variety trials: Quasi-Latin squares.  Ann. Eugenics, 7:319-332. 1937.

(3) Yates, F.  The recovery of inter-block information in variety trials arranged in three-dimensional lattices.  Ann. Eugenics, 9:136-156. 1939.

(4) Yates, F. Lattice squares. Jour. Agr. Sci., 30:672-687. 1940.

(5) Goulden, C. H.  Efficiency in field trials of pseudo-factorial and incomplete randomized blocks methods.  Canadian Jour. Res., C 15:231-241. 1937.

(6) Zuber, M. S.  Relative efficiency of incomplete block designs using corn uniformity trial data.  (In press)

(7) Cox, G. M., Eckhardt, R. C. and Cochran, W. G.  The analysis of lattice and triple lattice experiments in corn varietal tests. Iowa Agr. Exp. Sta., Res. Bul. 281:166. 1939.

(8) Bryan, A. A.  Factors affecting experimental error in field plot tests with corn.  Iowa Agr. Exp. Sta., Res. Bul. 163:243-280. 1933.

(9) Yates, F.  The recovery of inter-block information in balanced incomplete block designs.  Ann. Eugenics, 10:317-325. 1940.

## SAMPLING THEORY WHEN THE SAMPLING-UNITS ARE OF UNEQUAL SIZES*

By W. G. Cochran
*Iowa State College*

IN SAMPLING, the sampling-units are usually chosen so as to be similar in size and structure. With some types of population, however, it is convenient or necessary to use sampling-units that differ in size. Thus the farm is often the sampling-unit for collecting agricultural data, though farms in the same county may vary in land acreage from a few acres to over 1,000 acres. Similarly, when obtaining information about sales or prices, the sampling-unit may be a dealer or store, these ranging from small to large concerns.

In such cases the question arises: Should differences between the sizes of the sampling-units be ignored or taken into account in selecting the sample and in making estimates from the results of the sample? This paper contains a preliminary discussion of the problem, though further research is needed, many of the results given below being only large-sample approximations. It is convenient to consider first the problem of estimation, since it appears that the best method of distributing the sample depends on the process of estimation that is to be used.

### THE PROBLEM OF ESTIMATION

To state the problem of estimation in mathematical terms, we assume that sampling units are drawn at random without regard to their sizes, and consider how to estimate the population total of some quantity $y$ which can be measured on each sampling-unit. Associated with each sampling-unit is also a quantity $x$, which is called its *area* rather than its *size*, to avoid possible confusion between the terms "size of sample" and "size of sampling-unit." Some knowledge is assumed to be available about the values of $x$ in the sample, and possibly also in the population.[1] In order to apply results from the statistical theory of estimation, it is also assumed that the number of sampling-units in the population may be considered infinite. Formulae applicable to the

---

* A paper presented at the 103rd Annual Meeting of the American Statistical Association in joint session with the Institute of Mathematical Statistics, New York, December 30, 1941.

Journal paper No. J989 of the Iowa Agricultural Experiment Station, Ames, Iowa. Project No. 611.

[1] For some populations an alternative method of specification may be more appropriate. For instance, each sampling may consist of an integral number of sub-units, as in the case of human populations where the sampling-unit is a household and the sub-unit is a single person. The specification may be made in terms of the value of $y$ per sub-unit and the number $n$ of sub-units per sampling-unit (cf. Hanson and Hurwitz, 1942). Since this approach would not apply in the examples given at the beginning of this paper, it will not be considered here.

---

practical situation of sampling from finite populations can be obtained by adding suitable correction terms.

Stated in this way, the problem of estimation is a familiar one in mathematical statistics. If the joint frequency distribution of $x$ and $y$ in the population is known, the theory of estimation provides a routine technique leading to an efficient estimate of the population total of $y$ and using to best advantage any available information about $x$. There are, however, difficulties in utilizing this method of approach. The joint frequency distribution is often known at best only vaguely from the available data, and may not appear to follow any of the few types of bivariate frequency distribution that have been studied. Further, there are strong administrative arguments for keeping the computations involved in making the estimate as simple as possible; these requirements may impose a bar on the use of estimates which, while highly efficient statistically, are rather difficult to compute.

Both difficulties can be met to some extent by restricting the estimates to those derived from the regression of $y$ on $x$. For the calculation of regression equations, it is not necessary to describe completely the joint frequency distribution of $x$ and $y$; we need only know how the mean value and the variance of $y$ change as $x$ changes. These can be examined from a graph or two-way table of the pairs of values of $x$ and $y$ constructed from any available data. If the form of the regression line and the relative weights assigned to different values of $y$ are correct, the regression estimate is a *best unbiased linear estimate* as defined by David and Neyman (1938), though it is not a maximum likelihood estimate unless in addition the values of $y$ are normally distributed within arrays in which $x$ is fixed. The computations required for the simpler types of regression line are well-known and not unduly laborious.

### ESTIMATES DERIVED FROM LINEAR REGRESSION

In the following sections it will be assumed that the quantity to be estimated is the population total of $y$; any formulae can easily be altered so as to refer to the estimation of the population mean per sampling-unit.

The simplest case occurs when the mean value of $y$ is linearly related to the area of the sampling-unit, with constant variance; i.e. $y$ is of the form $\alpha + \beta x + e$, where $e$ has mean value zero and constant variance in arrays in which $x$ is fixed. In this case, the linear regression estimate $Y_l$ for the population total of $y$ is

$$Y_l = N\{\bar{y}_s + b(\bar{x}_p - \bar{x}_s)\} \tag{1}$$

where $N$ is the number of sampling-units in the population, $b$ is the sample regression coefficient $S(y-\bar{y}_s)(x-\bar{x}_s)/S(x-\bar{x}_s)^2$, and the suffixes $p$ and $s$ refer to the population and sample respectively. It will be noted that this estimate requires a knowledge both of the total number $N$ of sampling-units and of the mean value of $x$ in the population.

In samples in which the $x$'s remain fixed, the sampling variance of $Y_l$ is

$$V(Y_l) = N^2\sigma_y^2(1-\rho^2)\left\{\frac{1}{n} + \frac{(\bar{x}_p - \bar{x}_s)^2}{S(x-\bar{x}_s)^2}\right\} \tag{2}$$

$n$ being the number of sampling-units in the sample and $\rho$ the correlation coefficient between $y$ and $x$. The distribution of $Y_l$ tends to normality as $n$ increases, being exactly normal for any size of sample if $y$ is normally distributed for fixed $x$. A sample estimate of this variance is obtained by substituting for $\sigma_y^2(1-\rho^2)$ the mean square $s_d^2$ of deviations from the sample regression line.

For comparison with other estimates we may require the average variance of the regression estimate under random sampling. From (2), this clearly depends on the form of the frequency distribution of the areas. Since the areas are essentially positive, their distribution will not in general be normal, except perhaps as an approximation. The mean value of (2) may be expanded in a series of inverse powers of $n$, the sample size. Retaining the two leading terms, we obtain

$$\overline{V}(Y_l) = \frac{N^2\sigma_y^2(1-\rho^2)}{n}\left\{1 + \frac{1}{n} + \frac{3+2\gamma_1^2}{n^2}\right\} \tag{3}$$

where $\gamma_1$ is Fisher's (1941) measure of relative skewness ($\gamma_1^2 = \kappa_3^2/\kappa_2^3$). If the areas were normally distributed, $\gamma_1$ would of course be zero, and the exact value for the term in curled brackets would be $(n-2)/(n-3)$, which agrees with the value given above to this order of approximation. With large samples the factor is close to unity.

In many problems the true regression line must pass through the origin, as for example when $y$ represents corn acreage and $x$ farm acreage. Even in such cases, it may be advisable to use the preceding type of regression, if it appears on examination that a straight-line regression not passing through the origin will provide a satisfactory fit, whereas it would be necessary to use a curvilinear regression in order to include the origin. If a straight line through the origin can be used, $y$ being of the form $(\beta\bar{x}+e)$, with constant residual variance, the regression estimate $Y_o$ ($_o$ for origin) of the population total is

$$Y_o = N \frac{S(xy)}{S(x^2)} \bar{x}_p = \{\Sigma(x)\} \frac{S(xy)}{S(x^2)} \qquad (4)$$

where $\Sigma(x)$ is the population total of the areas. The variance of $Y_o$ is

$$V(Y_o) = \{\Sigma(x)\}^2 \sigma_y^2 (1 - \rho^2)/S(x^2). \qquad (5)$$

The number of sampling-units in the population does not enter into either of these formulae, which require only the population total of the areas.

The expression for the average value of this variance, under repeated random sampling, is rather complicated. If the distribution of the areas is not far from normal, the leading terms give

$$\overline{V}(Y_o) = \frac{N^2 \sigma_y^2 (1 - \rho^2)}{n(1 + c_x)} \left\{ 1 + \frac{2c_x(2 + c_x)}{n(1 + c_x)^2} \right\} \qquad (6)$$

$c_x = \sigma_x^2/\bar{x}_p^2$ being the square of the coefficient of variation of $x$.

From formulae (3) and (6), we may compare the sampling errors of $Y_l$ and $Y_o$ with that of the estimate $Y_s$ ($s$ for sampling-unit) which is obtained by multiplying the sample mean per sampling-unit by the total number of sampling-units, and is commonly used where sampling-units are equal in size. Since the variance of $Y_s$ is $N^2\sigma_y^2/n$, the ratios of the three pairs of variances in large samples are as follows:

$$\frac{\overline{V}(Y_l)}{V(Y_s)} = (1 - \rho^2); \; \frac{\overline{V}(Y_o)}{V(Y_s)} = \frac{(1 - \rho^2)}{(1 + c_x)}; \; \frac{\overline{V}(Y_o)}{\overline{V}(Y_l)} = \frac{1}{(1 + c_x)}. \qquad (7)$$

The additional factors involving $1/n$ and $1/n^2$ have been omitted from these expressions; they should be included in practical applications unless they are negligible. In large samples, both regression estimates are more accurate than the sample-mean estimate, the gain in accuracy being considerable if $\rho$ is high. As would be expected, $Y_o$ is more accurate than $Y_l$ when the true regression line is straight and passes through the origin, the increase in accuracy depending on the coefficient of variation of the areas.

These results must be interpreted with care. They indicate that in large samples $Y_l$ can never be less accurate, on the average, than $Y_s$. This statement was proved under the assumption that the true regression is linear (whether it passes through the origin or not); in the following section it will be shown to hold substantially even if the true regression is not linear. The conclusions about $Y_o$ have a much more restricted validity, holding only if the true regression is linear and passes through the origin. If the true regression line passes through the point $y = \alpha$ when $x$ is zero, the estimate $Y_o$ is biased, the bias tending in large samples to the constant value $-N\alpha c_x/(1+c_x)$. Including this bias in the expression for the sampling error, we have, instead of (6)

$$\overline{V}(Y_o) = \frac{N^2\alpha^2 c_x^2}{(1 + c_x)^2} + \frac{N^2\sigma_y^2(1 - \rho^2)}{n(1 + c_x)}.$$ (8)

Since the component arising from the bias does not decrease as the sample size $n$ increases, a sample size is always reached beyond which both $Y_l$ and $Y_s$ are more accurate than $Y_o$, unless $\alpha$ is zero. Thus $Y_o$ cannot be recommended as an estimate unless it is known with considerable confidence that the true regression is straight and passes through the origin.

### NON-LINEAR RELATIONS BETWEEN $y$ AND $x$

It has already been pointed out that in many cases the investigator possesses only fragmentary knowledge of the true relation between the observations $y$ and the areas of the sampling-units. Since a linear regression estimate may be used without any certainty that the population regression is linear, it is worth examining how $Y_l$ is affected when the population regression is non-linear. Suppose that the true relation is of the form

$$y = \alpha + \beta x + \xi + e$$ (9)

where as before $e$ is distributed with zero mean and unit variance, independently of $x$, and $\xi$ is a non-linear function of $x$. For this reason, it may be assumed without loss of generality that $\xi$ has zero mean and zero *linear* correlation with $x$. Following the usual algebraic development of linear regression, we find that the error of estimate

$$Y_l - \Sigma(y) = N\left\{ (\bar{\xi}_s + \bar{e}_s) + (\bar{x}_p - \bar{x}_s)\frac{S(\xi + e)(x - \bar{x}_s)}{S(x - \bar{x}_s)^2} \right\}.$$ (10)

Taking the mean value over all possible samples of $n$ sampling-units, all terms become zero except the second term in $\xi$, whose mean value does not vanish on account of the non-linear correlation between $\xi$ and $x$. By a technique developed by Fisher (1929), this value can be expressed in terms of the semi-invariants $\kappa_{ij}$ of the joint distribution of $\xi$ and $x$, the first two terms of the series being

$$E\{Y_l - \Sigma(y)\} = \frac{N}{n}\left\{ \frac{-\kappa_{12}}{\sigma_x^2} + \frac{1}{n}\left( \frac{2\kappa_{12}}{\sigma_x^2} + \frac{\kappa_{14}}{\sigma_x^4} \right) \right\}.$$ (11)

Thus the regression estimate is biased, the bias however tending to zero as the sample size is increased, since $n$ appears in the denominator of (11). The numerator $\kappa_{12}$ of the largest term depends essentially on the correlation between $\xi$ and $x^2$, i.e. on the quadratic component of the regression of $y$ on $x$.

Formula (3) for the average sampling variance of $Y_l$ is also changed, but the change affects only the terms in $1/n$, $1/n^2$ etc. inside the curled brackets, the factor outside the bracket remaining $N^2\sigma_y^2 (1-\rho^2)/n$, which in this case is equal to $N^2(\sigma_\xi^2+\sigma_e^2)/n$. Since the bias in $Y_l$ changes in inverse proportion to the sample size, while the standard error of $Y_l$ changes inversely as $\sqrt{n}$, the bias ultimately becomes negligible relative to the standard error if the sample is sufficiently large.

Thus, with samples large enough so that terms in $1/n$ are negligible, the ratio of the variance of $Y_l$ to that of the sample-mean estimate $Y_s$ remains $(1-\rho^2)$ even if the population regression is non-linear. This does not of course imply that $Y_l$ is an efficient estimate in this case. If the correct form of regression line could be fitted, the variance of the regression estimate would be reduced, in large samples, to $N^2\sigma_e^2/n$, as compared with $N^2(\sigma_\xi^2+\sigma_e^2)/n$ for $Y_l$. As would be expected, the relative loss of information with $Y_l$ depends on the ratio of the variance of the "non-linear" component $\xi$ to the residual variance.

At least part of loss of accuracy could be recovered by adding terms in $x^2$, $x^3$, etc., to the regression, with a corresponding increase in the numerical computations. In order to use such regressions in constructing the estimate, however, additional population data about $x$ are required. For a quadratic regression, for example, we must know both the population mean and variance of $x$ to be able to calculate the regression adjustments to the sample mean $\bar{y}_s$. It is unlikely that these would be available without a complete frequency distribution of the population by area of sampling-unit. Where such complete information is available, there is an alternative method of estimation which will be discussed later.

It was previously remarked that when the population regression is linear, an unbiased sample estimate of the variance of $Y_l$ is obtained by substituting the residual mean square $s_d^2$ in place of $\sigma_y^2(1-\rho^2)$ in formula (2). With a non-linear population regression, $s_d^2$ is a biased estimate of $\sigma_y^2(1-\rho^2)$, but again the bias is inversely proportional to $n$, becoming negligible in large samples.

To summarize, with large samples the linear regression estimate is unbiased, and the standard formula gives an unbiased estimate of its variance even when the population regression is non-linear. There is, however, a loss of efficiency which remains fixed in large samples. While no exact small-sample theory has been reached, it appears that both the estimate itself and the estimated variance are biased in small samples.

### WEIGHTED REGRESSIONS

Thus far we have considered the case in which only the mean value of $y$ changes as $x$ changes. The variance of $y$ may also change, particularly so if there is considerable variation in the areas of the sampling-units. The theory of regression has been extended to meet this case, provided that the ratios of the variances of different values of $y$ are known exactly, a condition which rarely if ever holds in problems of this type. If the true residual variance of $y_i$ is $\sigma_i^2$, and $w_i = 1/\sigma_i^2$ the best unbiased linear estimate $Y_{wl}$ is

$$Y_{wl} = N\{\bar{y}_w + b_w(\bar{x}_p - \bar{x}_w)\} \qquad (12)$$

where $\bar{y}_w = S(w_i y_i)/S(w_i)$, $\bar{x}_w = S(w_i x_i)/S(w_i)$ are weighted sample means and $b_w = S w_i(x_i - \bar{x}_w)(y_i - \bar{y}_w)/S w_i(x_i - \bar{x}_w)^2$ is the weighted sample regression coefficient. For a fixed set of $x$'s the sampling variance of $Y_{wl}$ is

$$V(Y_{wl}) = N^2 \left\{ \frac{1}{S(w_i)} + \frac{(\bar{x}_p - \bar{x}_w)^2}{S w_i(x_i - \bar{x}_w)^2} \right\}. \qquad (13)$$

It will be noticed in (12) that $Y_{wl}$ remains unchanged if instead of the correct weights $w_i$ we use numbers $w_i' = \lambda w_i$ which are proportional to the weights; i.e. only the *relative* weights assigned to different values of $y$ need be known in order to calculate $Y_{wl}$. Formula (13) for the sampling variance cannot be used however unless the actual values of the weights are known. If only relative weights $w_i'$ are known, an unbiased sample estimate of (13) is given by

$$S^2(Y_{wl}) = N^2 \frac{S w_i'(y_i - Y_i)^2}{(n-2)} \left\{ \frac{1}{S(w_i')} + \frac{(\bar{x}_p - \bar{x}_w)^2}{S w_i'(x - \bar{x}_w)^2} \right\} \qquad (14)$$

where $S w_i'(y_i - Y_i)^2/(n-2)$ is the weighted mean square of deviations from the sample regression, using $w_i'$ as weights.

In practice, before these formulae can be used, it will be necessary to estimate the residual variances, and hence the weights, from the results of the sample and any other comparable data. Baker (1941) has recently discussed this problem for the case in which the $x$'s fall into a number of distinct groups, all $x$'s having the same value within each group. More generally, the $x$'s will show a continuous range of varia-

tion. Space does not permit a detailed investigation of the best procedure for estimating the weights in this case. If the weights are presumed
to change *continuously* as $x$ changes, the first step seems clearly to subdivide the range of variation of $x$ into a number of groups. The residual
variance of $y$ within each group can then be estimated by fitting an
unweighted linear regression of $y$ on $x$ separately for each group. From
these results, the relation between the residual variance and the area
can be studied, and a smooth curve drawn to give the variance as a
function of $x$. The weight to be assigned to any value of $y$ is then obtained by noting the area of the sampling-unit, reading the curve, and
taking the inverse of the variance.

The greater the number of groups, the more points are available for
appraising the relation between variance and area. A further advantage
of having many groups is that if the range of $x$ is small within the
groups, the within group correlation between $y$ and $x$ may be negligible,
so that the total within-group mean square of $y$ may be used as equivalent to the residual mean square, thus obviating the necessity of fitting
a regression within each group. However, as the grouping is made finer,
the number of observations within each group decreases, leading to less
accurate estimates of the within-group variances. The optimum number of groups is not clear without further examination, though at a
guess it seems advisable to have at least 20 observations in each group.

The estimated weights are, of course, subject to sampling errors.
These errors have two consequences. The estimate $Y_{wl}$ is not as accurate as it could have been made if the true weights had been known.
This loss of accuracy is unavoidable, the somewhat laborious process
described above for estimating the weights being an attempt to reduce
the loss to a minimum. Secondly, and somewhat more seriously, both
formulae (13) and (14) give biased estimates of the sampling variance
of $Y_{wl}$ *even in large samples*, i.e. even ignoring the correction terms of
order $1/n$ which have appeared in previous formulae. If $w_i'$ are the
estimated and $w_i$ the true weights, the correct sampling variance of
$Y_{wl}$ in large samples appears to be

$$N^2 S\left(\frac{w_i'^2}{w_i}\right) \Big/ (Sw_i')^2. \tag{15}$$

Substituting $w_i'$ for $w_i$, formula (13) gives for large samples

$$N^2/S(w_i') \tag{16}$$

while (14) gives, on the average

$$N^2 S\left(\frac{w_i{}'}{w_i}\right) \Big/ nS(w_i{}').\tag{17}$$

All three formulae agree if $w_i{}'=w_i$ for all $i$, this being the only case in which (16) gives the correct result, except by chance. However (17) also gives the correct result whenever $w_i{}'=\lambda w_i$ for any value of $\lambda$, and in general (17) is less subject to error than (16). Thus the process outlined above for estimating the weights should be regarded as leading merely to *relative* weights, formula (14) being used to estimate the sampling variance of $Y_{wl}$. If some idea can be formed of the probable magnitude of the errors in the estimated weights, formulae (15) and (17) can be compared to assess whether the estimated sampling error of $Y_{wl}$ is likely to be greatly or only slightly biased.

Formulae (15) and (17) also agree if all the estimated weights $w_i{}'$ are chosen equal, whether the true weights are equal or not. Thus, if we fit an *unweighted* regression, using $Y_l$ as the estimate, the formula previously given for the estimated sampling variance of $Y_l$ is still unbiased in large samples when the true weights vary. Considering how frequently the unweighted linear regression is used in statistical applications, it is reassuring to find that at least in large samples the standard formula for the estimated variance remains reliable even if the population regression is non-linear or if the true weights vary.

In view of the labor involved in estimating weights and fitting a weighted regression, it may sometimes be questioned whether the gain in accuracy is sufficient to compensate for the extra work, particularly so if the true weights do not appear to vary greatly. To obtain some idea of the gain in accuracy, we may note from either (15) or (17) that if an unweighted regression is used, the variance of $Y_l$ is approximately $N^2 S(1/w_i)/n^2$. From (13), the maximum possible accuracy attained by a weighted estimate is $N^2/S(w_i)$ to the same order of approximation. Thus the relative accuracy of $Y_l$ to $Y_{wl}$ cannot be less than approximately

$$\frac{V(Y_{wl})}{V(Y_l)} = \frac{n^2}{S(w_i)S\left(\dfrac{1}{w_i}\right)} = \frac{n^2}{S(\sigma_i{}^2)S\left(\dfrac{1}{\sigma_i{}^2}\right)}.\tag{18}$$

By inserting a series of values of $\sigma_i{}^2$ to represent the range of variation in a practical case, this formula gives some idea of the relative accuracy attained by an unweighted regression. If the true variances do not change greatly, a rough approximation to (18) is $1/(1+c_v)$, where $c_v$ is the coefficient of variation of the variances. Thus, for example, if the variance $\sigma_i^2$ appears proportional to the area of the sampling-unit, the relative accuracy is about $1/(1+c_x)$.

### ESTIMATION BY THE MEAN PER UNIT AREA

One type of weighted regression leads to an estimate which is particularly simple to calculate, and has proved serviceable in estimating crop acreages in agricultural sampling. If a weighted regression passes through the origin, the corresponding estimate $Y_{wo}$ is

$$Y_{wo} = \frac{S(w_i x_i y_i)}{S(w_i x_i^2)} \, \Sigma(x_i). \tag{19}$$

Suppose that the variance of $y_i$ increases proportionally to the area $x_i$; in this case $w_i = 1/kx_i$, and (19) reduces to

$$Y_a = \frac{S(y_i)}{S(x_i)} \, \Sigma(x_i). \tag{20}$$

Thus to calculate $Y_a$ ($a$ for area), the sample total of $y$ is divided by the total area of all sampling-units in the sample, giving a mean per unit area, which is then multiplied by the total area in the population. From the conditions mentioned above, $Y_a$ is a best unbiased linear estimate if the mean value and the variance of $y$ both change proportionally to $x$.

Goldberg (1942) has studied the sampling distribution of $Y_a$ for any type of joint frequency distribution of $y$ and $x$. He has shown that in general $Y_a$ is biased, the leading term in the bias being

$$\frac{N\bar{y}_p}{n} \, (c_x - \rho\sqrt{c_x c_y}) \tag{21}$$

while the first approximation to the variance is

$$V(Y_a) = \frac{N^2 \bar{y}_p{}^2}{n} \, (c_x + c_y - 2\rho\sqrt{c_x c_y}). \tag{22}$$

Thus in large samples the ratio of the bias to the standard deviation is proportional to $1/\sqrt{n}$. By examining the ratio as a function of $\rho$, it may be shown that the ratio cannot numerically exceed the coefficient of variation of $x$, divided by $\sqrt{n}$.

Since $Y_s$ and $Y_a$ are the two simplest estimates to calculate, it is of interest to compare their sampling variances. For samples sufficiently large so that (22) may be used as the variance of $Y_a$, Goldberg (1942) has shown that $Y_a$ has a smaller variance than $Y_s$ whenever $\rho > \frac{1}{2}\sqrt{c_x/c_y}$ and vice versa.

### ESTIMATION BY USING POPULATION WEIGHTS

If a complete tabulation of the areas of all sampling-units in the population is available, the areas can be sub-divided into groups or strata, an estimate of the total of $y$ being made for each stratum. While this procedure could be carried out with all the estimates previously discussed, this investigation will be confined to the simplest estimate $Y_s$. If $n_1, \ldots n_k, N_1, \ldots N_k$ are the numbers of sampling-units in the sample and population respectively for the $k$ groups, the estimate $Y_{gs}$ of the population total over all strata is

$$Y_{gs} = (N_1\bar{y}_1 + \cdots + N_k\bar{y}_k) \tag{23}$$

the sampling variance being

$$V(Y_{gs}) = \left(\frac{N_1^2\sigma_1^2}{n_1} + \frac{N_2^2\sigma_2^2}{n_2} + \cdots + \frac{N_k^2\sigma_k^2}{n_k}\right) \tag{24}$$

where $\sigma_1^2 \ldots \sigma_k^2$ are the within-strata variances of $y$.

As shown by Neyman (1934), this variance is smallest, for a fixed total size of sample, when the sample is distributed amongst the groups so that $n_i$ is proportional to $N_i\sigma_i$. To retain comparability with previous estimates, however, we will assume that the sample is chosen at random.

In large samples, the average value of (24) works out approximately as

$$\bar{V}(Y_{gs}) = \frac{N^2}{n}\left\{\frac{(N_1\sigma_1^2 + \cdots + N_k\sigma_k^2)}{N}\left(1 - \frac{1}{n}\right) + \frac{k}{n}\frac{(\sigma_1^2 + \cdots + \sigma_k^2)}{k}\right\} \tag{25}$$

where $n$ and $N$ are as before the total numbers of sampling-units in the sample and population respectively. The expression inside the curled brackets contains both a weighted and an unweighted mean of the within-strata variances of $y$.

If all within-strata variances are the same, this reduces to

$$\bar{V}(Y_{gs}) = \frac{N^2\sigma^2}{n}\left(1 + \frac{k-1}{n}\right). \tag{26}$$

By increasing the number of $x$-groups with a given sample, the within-group variances are presumably decreased, since that portion of the variances of $y$ which is due to variation in $x$ is decreased by cutting down the range of $x$ within each group. However, the factor involving $k$ is of course increased as $k$ increases, so that a point is reached beyond which a further increase in the number of groups will result in less accuracy. From (26) it follows that in the case of equal variances the factor involving $k$ is relatively unimportant provided that $(k-1)/n$ is less than say .05, which holds if the average number of observations per group exceeds 20.

On comparing (26) with (3), $Y_{gs}$ is found to be somewhat less accurate than the linear regression estimate $Y_l$ if the true population regression is linear with equal variances. This follows because the within-stratum variance $\sigma^2$ cannot be less than $\sigma_y^2(1-\rho^2)$, while the additional factor in $1/n$ is also larger for $Y_{gs}$ than for $Y_l$. This conclusion was to be expected, since under the conditions mentioned $Y_l$ is a best unbiased linear estimate. If however the relation between $y$ and $x$ is markedly curvilinear or discontinuous, $Y_{gs}$ may be superior to $Y_l$, since the variation in $y$ arising from *any* type of relation with $x$ can be reduced by a suitable choice of strata, whereas $Y_l$ eliminates only the effects of the linear  component of the relationship. Moreover, $Y_{gs}$ is an unbiased estimate for any type of relation between $y$ and $x$ and any size of sample. Similarly, an unbiased estimate of the variance of $Y_{gs}$ is always obtained by substituting the sample within-strata mean squares in (24).

Similar comparisons can be made between $Y_{gs}$ and the weighted linear regression estimates by means of the formulae given for the sampling errors. Goldberg (1942) has discussed briefly the properties of $Y_{ga}$, the corresponding weighted estimate derived from the sample mean per unit area within each group.

### FURTHER NOTES

Some apology is needed for presenting in the previous sections a number of large-sample approximations without guidance as to the limits within which these apply. Unfortunately these limits depend on the form of the joint frequency distribution of $x$ and $y$, and could not be specified more definitely without a classification of the types of frequency distribution. Moreover, in extensive surveys, where problems of organization are difficult, biases may arise through the method of selecting the sample, incompleteness in the returns, and errors in reporting or recording the data. Such biases, while affecting the accuracy of the estimates, may not be measured by the formula for the sampling error, so that a rough approximation to the sampling error is often sufficient for practical purposes.

If the correct form of regression is used, population estimates derived from regressions remain unbiased in non-random sampling, provided that all sampling-units *with the same area* have an equal chance of selection. Thus the large sampling-units might be allotted a greater chance of inclusion in the sample, this procedure giving a more accurate estimate whenever the variance of $y$ increases as $x$ increases. On the other hand, if the method of selection discriminates in favor of certain sampling-units amongst those of the same area, bias may arise.

The formulae in this paper will of course apply to any variable $x$ which is correlated with $y$. For example, in agricultural sampling, where the sampling-unit is sometimes a fixed area of land, $x$ may be taken as the number of farms in the area, the total farm land or the total crop land, according to which gives the highest correlation with $y$.

In developing correction terms to be applied where an appreciable fraction of the population is sampled, the initial difficulty is that of defining a regression in a finite population. Writing $y = \alpha + \beta x + e$, we may suppose that $e$ has no linear correlation with $x$ in the finite population, but if we attempt to postulate that $e$ is uncorrelated with any power of $x$, the number of conditions to be satisfied is greater than the number of values of $e$ available, so that $e$ and $x$ cannot be independently distributed in the sense in which this term is applied with infinite populations. An alternative approach is to regard the finite population as a random sample from an infinite population in which $e$ and $x$ are independent. From a preliminary investigation and from Goldberg's (1942) work, it appears that the first approximation consists in multiplying formulae (2), (3) and (22) for the sampling-variances, and formulae (11) and (21) for the biases by $(N-n)/N$, this being the same correction as in the case of the sample-mean estimate $Y_s$. In formula (24), each term is multiplied by the corresponding factor $(N_i - n_i)/N_i$. For $Y_o$, the estimate derived from a straightline regression through the origin, and $Y_{wl}$, the weighted linear regression estimate, further investigation is needed. The difficulty arises because these two estimates do not equal the true population total when the sample consists of the whole finite population; i.e. they are *inconsistent* in the sense of Fisher (1941) whereas $Y_s$, $Y_l$, $Y_a$, and $Y_{gl}$ are always *consistent*.

For sampling surveys in which the areas $x$ of the sampling-units are unequal, the properties of various estimates of the population total of some observed quantity $y$ are discussed, these estimates being mostly derived from the regression of $y$ on $x$. In order of ease of calculation, the estimates are as follows; $Y_s$, derived from the sample mean per sampling-unit; $Y_a$, derived from the sample mean per unit area; $Y_{gs}$, a weighted form of $Y_s$, using population weights; $Y_o$ and $Y_l$, based on unweighted linear regressions; and $Y_{wo}$ and $Y_{wl}$, using weighted regressions. The conditions under which each estimate is most efficient are described, with various comparisons of their relative efficiencies.

### REFERENCES

Baker (1941) this JOURNAL, 36, 500.

David and Neyman (1938) *Statistical Research Memoirs*, II, 105.

Fisher (1929) *Proceedings of the London Mathematical Society*, 30, 199.

Fisher (1941) *Statistical Methods for Research Workers*, Edinburgh, Oliver and Boyd, 8th ed.

Goldberg (1942) (Not yet published).

Hanson and Hurwitz (1942) this JOURNAL, 37, 89.

Neyman (1934) *Journal of the Royal Statistical Society*, 97, 588.

# THE $\chi^2$ CORRECTION FOR CONTINUITY[1]

W. G. COCHRAN

*From the Statistical Section, Iowa Agricultural Experiment Station*

Received September 29, 1941

## INTRODUCTION

The correction for continuity, devised by Yates (1, 2), has proved a useful device for extending the use of the $\chi^2$ test of significance to data in which the expectations are small. For the two cases which he investigated in detail—the binomial distribution and the $2 \times 2$ contingency table—Yates gave simple rules for applying the correction, and a table of the exact significance levels of the corrected $\chi^2$. He also suggested that for contingency tables larger than the $2 \times 2$ table, there appeared to be less need for a correction for continuity.

From correspondence with several research workers, it appears that there is some uncertainty about the need for a correction for continuity and about the method of making the correction, in cases not explicitly discussed by Yates. The object of this paper is to illustrate the principles involved in correcting for continuity on two of the more common applications of $\chi^2$, in the hope of providing a basis for judgment in any particular problem. Some investigation has also been made of the allied problem of the effect of a low expectation at one tail on the $\chi^2$ test for goodness of fit.

## THE CORRECTION FOR CONTINUITY

While the theoretical basis for the correction has been described by Yates (1) and Fisher (2), a brief outline will be given here for completeness. In all applications of the $\chi^2$ test to be considered, the exact distribution of $\chi^2$ is discontinuous, only a finite number of discrete values of $\chi^2$ being possible. For example, in Figure 1 the exact distribution of $\chi^2$ has been worked out for a problem (to be discussed later) in which $\chi^2$ has 10 degrees of freedom. The heights of the vertical lines give the probabilities of all possible values of $\chi^2$, which in this case are equally spaced at intervals of 4/3. The sum of these heights is, of course, unity. To calculate the exact probability of a value of $\chi^2$ as great as or greater than 20, for instance, we add the heights of all lines for which $\chi^2$ is 20 or greater.

The tabulated distribution of $\chi^2$, on the other hand, is a continuous frequency curve, $\chi^2$ taking all values from zero to infinity. The probability that $\chi^2$ exceeds a value $\chi^2_o$, as given in the tables, is the area of the tail of this frequency curve from $\chi^2_o$ to infinity. In using the continuous curve as an approximation to the exact probabilities, each vertical line must be made to correspond to some area under the curve. The most natural

---

[1] Journal Paper No. J-936 of the Iowa Agricultural Experiment Station, Ames, Iowa. Project 514.

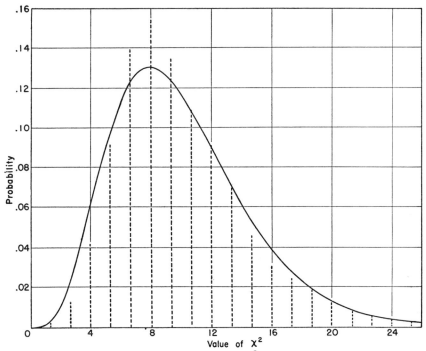

Fig. 1—Comparison of Exact and Tabular $X^2$ for 10 Degrees of Freedom

approach is to regard the exact discontinuous distribution as a *grouping* of the tabular distribution, each possible value of the exact $\chi^2$ representing all values of the continuous $\chi^2$ which are nearer to it than to any other permissible value of $\chi^2$. Thus, the exact value $\chi^2 = 20$ corresponds to all values of the continuous $\chi^2$ between $19\frac{1}{3}$ and $20\frac{2}{3}$, since the next lowest value of the exact $\chi^2$ is $18\frac{2}{3}$, while the next highest is $21\frac{1}{3}$. Consequently, the exact probability of a value of $\chi^2$ equal to 20 is approximated by the area under the tabulated distribution from $\chi^2 = 19\frac{1}{3}$ to $\chi^2 = 20\frac{2}{3}$.

It follows that the probability of a value of $\chi^2$ as great as or greater than 20 is estimated by the area under the curve from $\chi^2 = 19\frac{1}{3}$ to infinity, not from $\chi^2 = 20$ to infinity. In this example the exact probability is .0354, while the tabular probabilities are .0362 for $\chi^2 = 19\frac{1}{3}$ and .0292 for $\chi^2 = 20$, showing a marked improvement by correcting for continuity. Similarly, the probability that $\chi^2$ is less than or equal to 20 is estimated by the area under the curve from zero to $20\frac{2}{3}$, i.e., by taking one minus the tabular probability for $\chi^2 = 20\frac{2}{3}$. To estimate the probability that $\chi^2$ lies between 12 and 20 inclusive, we would calculate the area of the curve between $11\frac{1}{3}$ and $20\frac{2}{3}$.

For most tests of significance the probability wanted is that of a value of $\chi^2$ as great as or greater than the observed $\chi^2$. For this purpose, the general rule in applying corrections for continuity may be stated as follows: Calculate $\chi^2$ by the usual formula. Find the next lowest possible value of $\chi^2$ to the one to be tested, and use the tabular probability for a value of $\chi^2$ midway between the two.

If the possible values of $\chi^2$ are closely spaced together, the probabilities given by the uncorrected $\chi^2$ and the corrected $\chi^2$ may differ only by an amount that is regarded as negligible. In this case the correction may be ignored.

The rule given by Yates for correcting for continuity is to decrease the absolute difference between the observed and the expected values by one-half. It is important to trace the relation between this rule and the general rule given above. For the two cases discussed by Yates (1)—the binomial and the $2 \times 2$ contingency table—$\chi^2$ has only a single degree of freedom, the difference between the observed value $x$ and the expected value $m$ being the same for all cells. Thus, $\chi^2$ is of the form $(x - m)^2$ $(\Sigma \frac{1}{m})$, the summation being over all cells, while $\chi$ may be written $(x - m)\sqrt{\Sigma \frac{1}{m}}$. Yates actually dealt with the frequency distribution of $\chi$, for which the tabular value follows the normal law. The advantage in so doing is that a distinction may be drawn between positive and negative values of $(x - m)$, this procedure being convenient when it is desired to test the deviation in one direction only, or when the exact distribution of $\chi$ is skew. Thus, Yates was concerned with the problem of estimating the probability at one tail of the distribution of $\chi$. Since $x$ takes only a succession of integral values, successive values of $(x - m)$ must differ by unity. Hence, to correct $\chi$ for continuity by the general rule, the difference between the observed and expected values should be reduced in absolute magnitude by one-half, as Yates proposed. The point to be stressed here, however, is that this rule should be used only when it agrees with the general rule given previously.

### TESTS FOR LINKAGE

A convenient example for further illustration of the principles is provided by the use of $\chi^2$ in testing for linkage between two factors in genetical studies. If doubly heterozygous individuals $AaBb$ are inbred, where $A$ and $B$ are dominant, the expectations of the four phenotypes $AB$, $Ab$, $aB$ and $ab$ in the $F_2$ generation are as $9:3:3:1$, respectively. The results of such an experiment may therefore be put in the form:

| Type | $AB$ | $Ab$ | $aB$ | $ab$ | Total |
|---|---|---|---|---|---|
| Number observed........... | $a$ | $b$ | $c$ | $d$ | $n$ |
| Number expected........... | $\frac{9n}{16}$ | $\frac{3n}{16}$ | $\frac{3n}{16}$ | $\frac{n}{16}$ | $n$ |

It is easily seen that $\chi^2$, with 3 degrees of freedom, may be written as

$$\chi^2 = \frac{16}{n}\left(\frac{a^2}{9} + \frac{b^2}{3} + \frac{c^2}{3} + \frac{d^2}{1}\right) - n$$

As Fisher (2, § 51) has shown, this quantity may be divided into three single-degrees of freedom, as follows.

$$\chi^2_1 = (a+b-3c-3d)^2/3n$$

$$\chi^2_2 = (a+c-3b-3d)^2/3n$$

$$\chi^2_3 = (a+9d-3b-3c)^2/9n$$

The quantities $\chi^2_1$ and $\chi^2_2$ test whether the observations are segregating in a 3 : 1 ratio for the $A$ and $B$ factors, respectively. These are simply tests of a single binomial ratio, and are completely covered by Yates's discussion (1). To correct $\chi_1$ and $\chi_2$ for continuity in the above form, the quantity inside the bracket should be reduced by two in absolute magnitude,[2] since, for fixed $n$, $(a+b-3c-3d)$ and $(a+c-3b-3d)$ change by intervals of four.

The quantity $\chi^2_3$ serves as a test for linkage between the two characters. For fixed $n$, the smallest change in $(a+9d-3b-3c)$ is also $\pm 4$, produced by decreasing $a$ and increasing $b$ or $c$ by unity. Hence, to correct $\chi_3$ for continuity, $(a+9d-3b-3c)$ should also be reduced by two in absolute magnitude. The effect of the correction is smaller with $\chi^2_3$ than with $\chi^2_1$ or $\chi^2_2$, owing to the larger divisor, $9n$. Since the difference between the uncorrected and the corrected value of $\chi_3$ is $2/3\sqrt{n}$, it is easy to calculate a value of $n$ above which correction is unnecessary. If the uncorrected value is to underestimate the probability by not more than 0.5 per cent at the 5 per cent level, $n$ should be at least 300. It should be noted that $\chi_3$ cannot be interpreted as the test of a single binomial ratio, so that Yates's table of the exact significance levels of the corrected $\chi$ does not apply to this case.

Since the individual degrees of freedom provide tests of different aspects of the data, there is probably little interest in testing the combined $\chi^2$ (with three degrees of freedom), except perhaps as a preliminary indication. To correct the combined $\chi^2$, the general rule must be applied, by finding the next lowest possible value of $\chi^2$. No simple arithmetical rule appears to be available for doing this. The minimum possible change in the quantity $(a^2+3b^2+3c^2+9d^2)$ is $\pm 2$, so that the corresponding change in the combined $\chi^2$ is $\pm 32/9n$. There may, however, be gaps in the succession of values of $\chi^2$, and each case requires special inspection. An example will illustrate the relation between the corrections of the individual degrees of freedom and of the combined $\chi^2$. The data are taken from an experiment by Imai (3), quoted by Mather (4), the values being $a=47$, $b=8$, $c=11$, $d=9$.

The quantities $\chi^2_1$, $\chi^2_2$ and $\chi^2_3$ are corrected by the rule already given.

---

[2] B. L. Wade, Unpublished typescript.

TABLE 1
CORRECTION FOR CONTINUITY IN A LINKAGE TEST

|  | VALUE OF $\chi^2$ | | PROBABILITY | | PERCENTAGE ERROR IN UNCORRECTED PROBABILITY |
|---|---|---|---|---|---|
|  | Uncorrected | Corrected | Uncorrected | Corrected |  |
| $\chi_1^2$........ | 0.111 | 0.040 | .739 | .841 | −13 |
| $\chi_2^2$........ | 0.218 | 0.111 | .640 | .739 | −13 |
| $\chi_3^2$........ | 7.468 | 7.053 | .0063 | .0079 | −21 |
| Combined $\chi^2$.... | 7.797 | 7.750 | .0505 | .0515 | − 2 |

By trial, the next lowest value of the combined $\chi^2$ appears to be 7.702, given by the configuration $a=45$, $b=7$, $c=14$, $d=9$. Hence, the corrected value of the combined $\chi^2$ is $1/2\,(7.797+7.702)=7.750$. It should be noted that the corrected values of $\chi^2$ are not additive, the total of the corrected single degrees of freedom being 7.204, while the combined value, after correction, is 7.750. Further, the effect of correction is much smaller on the combined $\chi^2$ than on the single degrees of freedom. The difference between the probabilities given by the uncorrected and corrected values of the combined $\chi^2$ is negligible.

### THE ADDITION OF SINGLE DEGREES OF FREEDOM

A further example, also common in genetical work, arises when observations on the number of individuals of each of two types, $A$ and $a$, are available for several different families. For each family we may test whether the individuals are segregating in a given ratio, say $3:1$, by calculating the appropriate $\chi^2$, as described in the previous section. If all families are presumed to be segregating in the same ratio, it is customary to add the individual values of $\chi^2$ to form a "total" $\chi^2$. This is divided into two parts: $\chi^2_t$ (with 1 degree of freedom) which tests whether the totals of the $A$'s and $a$'s over all families are segregating in the proposed ratio, and $\chi^2_h$, found by subtracting $\chi^2_t$ from the "total" $\chi^2$. Of these, $\chi^2_t$ provides a sensitive test for any *consistent* departure of the families as a whole from the proposed ratio, while $\chi^2_h$ (sometimes called the "heterogeneity" $\chi^2$) tests whether the families agree mutually in their segregation ratio. The use of these tests is, of course, not confined to genetical investigations.

When testing any individual family, the $\chi^2$ should of course be corrected for continuity. The question arises: Should corrected or uncorrected values of $\chi^2$ be added when forming the "total" $\chi^2$? It is clear from the discussion in the previous section that the *uncorrected* values must be used when computing the "total" $\chi^2$, which should then be corrected, if necessary, by the general rule given. As a demonstration, we have worked out for an extreme case the exact distribution of the "total" $\chi^2$ as computed in each of the two proposed ways. The data consist of ten families, each containing only four observations, segregating 3 to 1. Thus, the "total" $\chi^2$ has ten degrees of freedom.

Since there are only four observations per family, $a$ and $A$ can take only the values 0, 1, 2, 3, and 4, with probabilities which are given by the binomial $\left(\dfrac{3}{4}+\dfrac{1}{4}\right)^4$. All possible results for a single family are shown in Table 2 below, with the corresponding probabilities and values of $\chi^2$ with and without correction for continuity.

TABLE 2

POSSIBLE VALUES OF $\chi^2$ FOR FOUR OBSERVATIONS, SEGREGATING 1:3

| $a$ | $A$ | PROBABILITY | VALUE OF $\chi^2$ | |
|---|---|---|---|---|
| | | | Uncorrected | Corrected |
| 0............... | 4 | 81 | 4/3 | 1/3 |
| 1............... | 3 | 108 | 0 | 0 |
| 2............... | 2 | 54 | 4/3 | 1/3 |
| 3............... | 1 | 12 | 16/3 | 3 |
| 4............... | 0 | 1 | 12 | 25/3 |
| | | ÷256 | | |

For $a=0$, $A=4$, for instance, the value of $\chi^2$ is, of course, $(0-1)^2/1+(4-3)^2/3=4/3$, while the corresponding value, corrected by Yates's rule, is $(1/2)^2/1+(1/2)^2/3=1/3$.

The probabilities were computed for all possible combinations of these results from ten families, and also the corresponding values of $\chi^2$ (with ten degrees of freedom) found by adding the corrected or uncorrected individual $\chi^2$'s.

In Table 3, the exact significance levels of the total of the "corrected"

TABLE 3

COMPARISON OF EXACT AND TABULAR $\chi^2$ DISTRIBUTIONS, FOR THE TOTAL OF THE "CORRECTED" SINGLE DEGREES OF FREEDOM

| $\chi^2$ | Exact $P \geqq \chi^2$ | Tabular $P \geqq \chi^2$ |
|---|---|---|
| 1............................ | .9850 | .9998 |
| 2............................ | .7143 | .9963 |
| 3............................ | .4211 | .9814 |
| 4............................ | .3926 | .9473 |
| 5............................ | .2227 | .8912 |
| 6............................ | .1114 | .8153 |
| 7............................ | .1075 | .7254 |
| 8............................ | .0627 | .6288 |
| 9............................ | .0475 | .5321 |
| 10............................ | .0375 | .4405 |
| 11............................ | .0167 | .3575 |

$\chi^2$'s are compared with these given by the tables. The possible values of the total of the *corrected* $\chi^2$'s are spaced at intervals of one-third. To save space only every third value is shown below.

The tabular $\chi^2$ overestimates the probabilities so violently that there can scarcely be said to be any relation between the two distributions. For instance, ten is the mean value of the tabular $\chi^2$, yet a value of ten or more is reached only about once in thirty times by the total of the "corrected" $\chi^2$'s. The use of the $\chi^2$ table would lead very frequently to the apparent conclusion that the data were too homogeneous. If the tabular $\chi^2$ were corrected for continuity, the agreement would be still worse.

The exact distribution of the total of the *uncorrected* values of $\chi^2$ is compared with the corresponding tabular $\chi^2$ in Figure 1 above.[3] The agreement appears remarkably good, at least in the upper tail which is relevant to tests of significance. A more exact comparison in the region between .10 and .01 is shown in Table 4.

TABLE 4

COMPARISON OF EXACT AND TABULAR $\chi^2$ DISTRIBUTIONS, FOR THE TOTAL OF THE "UNCORRECTED" SINGLE DEGREES OF FREEDOM

| $\chi^2$ | Exact $P \geqq \chi^2$ | Tabular $P \geqq \chi^2$ | Tabular, Corrected for Continuity |
|---|---|---|---|
| 16.00 | .1076 | .0996 | .1204 |
| 17.33 | .0773 | .0673 | .0821 |
| 18.67 | .0534 | .0447 | .0550 |
| 20.00 | .0354 | .0292 | .0362 |
| 21.33 | .0236 | .0189 | .0235 |
| 22.67 | .0158 | .0120 | .0151 |
| 24.00 | .0102 | .0076 | .0096 |

Since, owing to the small number of observations, the successive values of $\chi^2$ are rather far apart, the tabular $\chi^2$ requires correction for continuity, as shown in the right-hand column. The agreement with the exact probabilities in the critical region is excellent.

A similar comparison for the "heterogeneity" $\chi^2$ would show approximately the same results. Thus, the correct procedure is to form the "total" $\chi^2$ by adding the *uncorrected* single degrees of freedom. To form $\chi^2_h$, subtract $\chi^2_t$, calculated without correction. In *testing* $\chi^2_t$, however, a correction should be made, since we are testing a single binomial ratio. Theoretically, $\chi^2_h$ may also require correction after it has been computed as described above. While there appears to be no simple arithmetical rule for making this correction, its effect should be negligible unless the number of families and the expectations of the two types $a$ and $A$ are both small.

It may seem paradoxical that the uncorrected $\chi^2$, which underestimates the true probabilities with a single degree of freedom, should give good agreement when several different values are added together, while the corrected $\chi^2$ behaves in the opposite manner. The reason may be

---

[3] Since the interval between successive values of $\chi^2$ is 4/3, the ordinates of the tabular curve have been multiplied by 4/3.

exhibited more clearly by examining the two quantities further. For a single degree of freedom, the exact mean of the uncorrected $\chi^2$ is 1, while the variance is $2\left(1 + \dfrac{1-6pq}{2npq}\right)$ where $n$ is the number of observations and $p : q$ is the segregation ratio. For the total $\chi^2$ from $r$ different families, the corresponding values are, therefore,

$$\text{Mean} = r. \quad \text{Variance} = 2r\left(1 + \frac{1-6pq}{2n'pq}\right)$$

where $\dfrac{1}{n'} = \dfrac{1}{r}\left(\sum_{i=1}^{r}\dfrac{1}{n_i}\right)$ is the average value of $\dfrac{1}{n}$ for the different fam-

ilies. The corresponding mean and variance for the tabular $\chi^2$ are $r$ and $2r$, respectively. Hence, the exact and tabular $\chi^2$ have always the same mean value, and agree closely in their variances provided that $(npq)$ is reasonably large. As $r$ increases, both distributions tend to normality, being determined more and more exactly by the values of their mean and variance. Moreover, as $r$ increases, the number of possible values of the exact $\chi^2$ increases rapidly, particularly so if $n$ varies from family to family, so that the effect of a continuity correction diminishes. Hence, the agreement between the exact and tabular $\chi^2$ tends to improve as $r$ increases.

For a single degree of freedom, the correction for continuity reduces the value of $\chi^2$ in order that the tabular probability may give a better approximation to the true probability at the upper tail. The effect is, however, to reduce both the mean and variance of the corrected $\chi^2$ below these of the uncorrected $\chi^2$, considerably so if $n$ is small. If the mean of the corrected $\chi^2$, for a single degree of freedom, is $\lambda$, where $\lambda$ is less than unity, the mean of the total of $r$ values is $\lambda r$. Thus, the negative bias in the mean as compared with the tabular $\chi^2$, is $r(1-\lambda)$, which increases steadily as $r$ increases. Hence, the agreement with the tabular $\chi^2$ gets steadily worse as the number of families increases.

With a large number of families the distribution of the "heterogeneity" $\chi^2$, calculated without correction, also tends to normality with a mean value $(r-1)$ and a variance which is approximately equal to $2(r-1)$

$\left(1 + \dfrac{1-6pq}{2n'pq}\right)$, where $\dfrac{1}{n'} = \dfrac{1}{r}\sum\dfrac{1}{n_i}$. Since the discrepancy in the vari-

ance does not tend to decrease as $r$ increases, it would be somewhat more accurate, when the number of families is large, to regard $\chi^2$ as normally distributed with the above mean and variance than to use the tabular $\chi^2$. However, the latter is adequate for most cases arising in practice unless $p$ or $q$ is close to unity. If the tabular $\chi^2$ and the normal approximation are to give probabilities which agree within 0.5 per cent at the 5 per cent level, it may be shown that the average number of observations per family need only exceed 10 if $p$ is 1/4, but should exceed 140 if $p$ is 1/16, and 740 if $p$ is 1/64.

## THE EFFECT OF A SMALL EXPECTATION ON "GOODNESS OF FIT" TESTS

In using $\chi^2$ to test goodness of fit, the expectations usually become small towards one or both ends of the frequency distribution. As a working rule, it is customary to combine several classes at the ends, if necessary, so that no expectation is less than a conventional minimum, for which Fisher (2) suggests the value 5 and Aitken (5) the value 10. The reasons for this rule are to decrease the discontinuity in the exact $\chi^2$ distribution and to approximate more closely to the assumptions underlying the tabular $\chi^2$ distribution, which postulates that the observed values are normally distributed about the expected values. Of these, the second reason is probably the more important, for if the expectations in the central classes are reasonably high, the total number of values of the exact $\chi^2$ will be large even though the expectations at the tails are small.

In cases where the observed and the fitted distribution disagree most markedly at the tails, this grouping of classes diminishes the sensitiveness of the $\chi^2$ test, so that it is sometimes more appropriate to calculate $\chi^2$ without grouping the tails. Since the effects of a low expectation at one tail do not appear to have been examined, the exact distributions of $\chi^2$ were worked out for three related examples, each having three classes and a total of 20 observations. The expectations in the individual classes were as follows:

|  | EXPECTATIONS | | |
|---|---|---|---|
|  | $m_1$ | $m_2$ | $m_3$ |
| Example 1 | 11.4 | 7.6 | 1.0 |
| Example 2 | 11.7 | 7.8 | 0.5 |
| Example 3 | 11.94 | 7.96 | 0.1 |

The smallest expectation varies from 1.0 to 0.1, the other two expectations being kept in the ratio 3/2 in all three examples. Since $m_1$ and $m_2$ vary little between the three examples, the differences between the resulting $\chi^2$ distributions should be due mainly to the smallest expectation, which in all cases is well below the usually accepted minimum. The value of $\chi^2$ is,

of course, $\sum\limits_{i=1}^{3} \dfrac{(x_i - m_i)^2}{m_i}$, where $x_i$ is the observed value in the $i$th cell, and

has two degrees of freedom.

For each example, the exact significance levels and the corresponding probabilities given by the tabular $\chi^2$ (without correction for continuity) are shown for all values of $\chi^2$ giving probabilities between 0.1 and 0.01.

The examples bring out some interesting points. Except for the extreme case in which the smallest expectation is only 0.1, the successive values of $\chi^2$ are sufficiently close together so that discontinuity would not of itself introduce gross errors in the probabilities. Further, the exact and tabular probabilities agree reasonably well down to a certain significance level, this being .0271 for $m_3 = 1$, .0285 for $m_3 = 0.5$, and .1926 for $m_3 = 0.1$. Thereafter, the tabular $\chi^2$ begins rather abruptly to underestimate ser-

TABLE 5

COMPARISON OF EXACT AND TABULAR $\chi^2$ IN A GOODNESS OF FIT TEST

SMALLEST EXPECTATION ($m_3$)

| | 1.0 | | | 0.5 | | | 0.1 | |
|---|---|---|---|---|---|---|---|---|
| $\chi^2$ | Exact $P(\geq \chi^2)$ | Tabular $P(\geq \chi^2)$ | $\chi^2$ | Exact $P(\geq \chi^2)$ | Tabular $P(\geq \chi^2)$ | $\chi^2$ | Exact $P(\geq \chi^2)$ | Tabular $P(\geq \chi^2)$ |
| 4.561 | .1051 | .1022 | 4.752 | .1070 | .0929 | 1.985 | .3244 | .3706 |
| 4.640 | .0798 | .0983 | 4.923 | .0939 | .0853 | 3.451 | .1926 | .1781 |
| 4.921 | .0732 | .0854 | 5.034 | .0814 | .0807 | 5.335 | .1289 | .0694 |
| 5.272 | .0603 | .0716 | 5.308 | .0760 | .0704 | 7.638 | .1045 | .0243 |
| 5.930 | .0540 | .0516 | 5.649 | .0664 | .0593 | 8.174 | .0973 | .0168 |
| 6.105 | .0492 | .0472 | 5.855 | .0577 | .0535 | 8.216 | .0810 | .0164 |
| 6.395 | .0465 | .0409 | 6.290 | .0415 | .0431 | 8.551 | .0647 | .0139 |
| 6.456 | .0433 | .0396 | 6.744 | .0357 | .0343 | 8.677 | .0513 | .0131 |
| 6.535 | .0399 | .0381 | 7.214 | .0285 | .0271 | 9.347 | .0381 | .0093 |
| 6.877 | .0302 | .0321 | 7.701 | .0270 | .0213 | 9.556 | .0293 | .0084 |
| 6.982 | .0285 | .0305 | 8.205 | .0242 | .0165 | 10.360 | .0208 | .0056 |
| 7.377 | .0271 | .0250 | 8.385 | .0192 | .0151 | 10.561 | .0192 | .0051 |
| 8.079 | .0251 | .0176 | 9.265 | .0176 | .0098 | 10.854 | .0143 | .0044 |
| 8.430 | .0237 | .0148 | 9.539 | .0168 | .0085 | 12.194 | .0101 | .0023 |
| 8.947 | .0229 | .0114 | 9.820 | .0158 | .0074 | | | |
| 8.982 | .0200 | .0112 | 10.393 | .0155 | .0055 | | | |
| 9.263 | .0191 | .0097 | 10.983 | .0150 | .0041 | | | |
| 9.483 | .0185 | .0087 | 11.804 | .0139 | .0027 | | | |
| 9.509 | .0156 | .0086 | 12.214 | .0136 | .0022 | | | |
| 9.552 | .0156 | .0084 | 12.829 | .0134 | .0016 | | | |
| 9.553 | .0131 | .0084 | 12.855 | .0111 | .0016 | | | |
| 9.903 | .0128 | .0071 | 13.128 | .0111 | .0014 | | | |
| 10.035 | .0106 | .0066 | | | | | | |

iously. Since only a part of the underestimation in this region would be recovered by correction for continuity, the discontinuity is evidently not the principal cause of the discrepancy. For $m_3 = 0.1$ the tabular $\chi^2$ is useless as an approximation throughout the whole of the region between $p = .1$ and $p = .01$. For $m_3 = 0.5$, the tabular probability agrees almost as well as for $m_3 = 1.0$ down to $p = .028$, but thereafter the underestimation is more serious, so that at $p = .01$ the tabular $\chi^2$ is worse for $m_3 = 0.5$ than for $m_3 = 0.1$.

To investigate these results in more detail, it is necessary to isolate the contribution to $\chi^2$ from the cell with the small expectation. Let $n$ be the number of observations and $p_1, p_2, \ldots p_r$ the probabilities in the $r$ classes. The observed number $k$ in the $r$th class is distributed in the binomial series $(q_r + p_r)^n$ where $q_r = 1 - p_r$. For a fixed $k$, the observed numbers in the remaining classes are distributed in the multinomial

$$\left( \frac{p_1}{q_r} + \frac{p_2}{q_r} + \ldots + \frac{p_{r-1}}{q_r} \right)^{n-k}.$$

Consequently, if the expected values in the first $(r-1)$ classes are sufficiently large, the quantity

$$\chi^2_{r-2} = \frac{q_r}{(n-k)} \sum_{i=1}^{r-1} \left( \frac{k^2_i}{p_i} \right) - (n-k) \tag{1}$$

is distributed as $\chi^2$ with $(r-2)$ degrees of freedom. The value of $\chi^2$ for all classes may be written

$$\chi^2 = \sum_{i=1}^{r} \frac{k^2_i}{(np_i)} - n \tag{2}$$

Using (1) this becomes, after some algebraic manipulations

$$\chi^2 = \frac{(n-k)}{nq_r} \chi^2_{r-2} + \frac{(k-np_r)^2}{np_r q_r} \tag{3}$$

The first term on the right-hand side of (3) represents the combined contribution to $\chi^2$ from the first $(r-1)$ classes, while the second term is the contribution from the $r$th class, in which the expectation $np_r$ will be assumed small.

For a given value of $k$, the probability that $\chi^2$ exceeds any specified quantity depends on the value of $\chi^2_{r-2}$, and is easily found from the tabular $\chi^2$ for $(r-2)$ degrees of freedom. Since the probabilities of different values of $k$ are known from the binomial $(q_r+p_r)^n$, the total probability that $\chi^2$ exceeds any specified quantity may be found by adding the probabilities for different values of $k$. The significance levels of $\chi^2$, as defined by equation (3), clearly depend on three variables: the number of classes, $r$; the smallest expectation, $np_r$; and the total number of observations, $n$. However, for fixed $r$ and $np_r$, the significance levels vary little with $n$, provided that $n$ exceeds 20. Since in any case $n$ must exceed 20 to satisfy the condition that the expectations $np_1, \ldots np_{r-1}$ are reasonably large, the limiting value as $n$ tends to infinity was substituted in (3), giving

$$\chi^2_m = \chi^2_{r-2} + \frac{(k-m_r)^2}{m_r} \tag{4}$$

where $k$ now follows a Poisson distribution with mean $m_r$. The symbol $\chi^2_m$ is used to denote the fact that the distribution depends on the expectation $m_r$ in the smallest class.

To verify this theoretical approach, the significance levels of $\chi^2_m$ given by equation (4) were compared with the exact significance levels of $\chi^2$ for the most extreme example $m_3 = 0.1$.

Since the distribution of $\chi^2_m$ is continuous, and since the exact values of $\chi^2$ are somewhat widely spaced, the significance levels of $\chi^2_m$ are shown both with and without correction for continuity. Before correction, the probabilities are slightly underestimated by $\chi^2_m$, but the underestimation is mainly due to the discontinuity in the exact $\chi^2$. After correction, the $\chi^2_m$ probabilities give a satisfactory approximation to the correct probabilities, whereas it will be recalled that the tabular $\chi^2$ probabilities were

TABLE 6

COMPARISON OF THE EXACT $\chi^2$ DISTRIBUTION AND THE $\chi^2_m$ DISTRIBUTION

Expectations: 11.94, 7.96, 0.1

| $\chi^2$ | EXACT $P(\geqq \chi^2)$ | $P(\geqq \chi^2)$ GIVEN BY $\chi^2_m$ | |
|---|---|---|---|
| | | Uncorrected | Corrected |
| 7.638 | .1045 | .1006 | .1056 |
| 8.174 | .0973 | .0798 | .0999 |
| 8.216 | .0810 | .0756 | .0773 |
| 8.551 | .0647 | .0534 | .0621 |
| 8.677 | .0513 | .0482 | .0509 |
| 9.347 | .0381 | .0309 | .0380 |
| 9.556 | .0293 | .0272 | .0289 |
| 10.360 | .0208 | .0180 | .0218 |
| 10.561 | .0192 | .0163 | .0171 |
| 10.854 | .0143 | .0144 | .0153 |
| 12.194 | .0101 | .0093 | .0111 |

greatly in error. The $\chi^2_m$ approximation is also sufficiently accurate for practical purposes in the other two examples.

TABLE 7

PROBABILITY LEVELS OF $\chi^2_m$ ASSOCIATED WITH THE 5 AND 1 PER CENT LEVELS OF $\chi^2$

| SMALLEST EXPECTATION $m$. | NUMBER OF DEGREES OF FREEDOM IN $\chi^2$ | | | | | | | |
|---|---|---|---|---|---|---|---|---|
| | 2 | 3 | 4 | 5 | 6 | 10 | 15 | 25 |
| | 5 Per Cent Level | | | | | | | |
| 0.1 . . . . . . . . . . | .1089 | .1142 | .0909 | .0800 | .0741 | .0646 | .0605 | .0576 |
| 0.5 . . . . . . . . . . | .0486 | .0522 | .0542 | .0555 | .0564 | .0552 | 0.539 | .0528 |
| 1.0 . . . . . . . . . . | .0480 | .0537 | .0553 | .0540 | .0534 | .0527 | .0522 | .0516 |
| 2.0 . . . . . . . . . . | .0474 | .0527 | .0517 | .0516 | .0518 | .0515 | .0512 | .0509 |
| 3.0 . . . . . . . . . . | .0471 | .0507 | .0510 | .0512 | .0512 | .0510 | .0507 | .0507 |
| 5.0 . . . . . . . . . . | .0476 | .0499 | .0504 | .0506 | .0507 | .0506 | .0505 | .0504 |
| | 1 Per Cent Level | | | | | | | |
| 0.1 . . . . . . . . . . | .0334 | .0259 | .0229 | .0213 | .0203 | .0180 | .0169 | .0159 |
| 0.5 . . . . . . . . . . | .0195 | .0209 | .0197 | .0172 | .0159 | .0139 | .0129 | .0119 |
| 1.0 . . . . . . . . . . | .0181 | .0144 | .0139 | .0137 | .0136 | .0123 | .0117 | .0111 |
| 2.0 . . . . . . . . . . | .0126 | .0130 | .0128 | .0122 | .0119 | .0117 | .0109 | .0106 |
| 3.0 . . . . . . . . . . | .0119 | .0123 | .0118 | .0115 | .0115 | .0110 | .0107 | .0104 |
| 5.0 . . . . . . . . . . | .0119 | .0112 | .0112 | .0110 | .0109 | .0106 | .0104 | .0102 |

Assuming that the $\chi^2_m$ distribution takes proper account of the effect of a single small expectation $m$, we may estimate the error involved in using the tabular $\chi^2$ distribution in such cases. The error will depend on the number of degrees of freedom in $\chi^2$, since a single small class presumably contributes to a lesser degree when the number of classes is large. For $m=0.1, 0.5, 1, 2, 3$, and 5 and 2, 3, 4, 5, 6, 10, 15, and 25 degrees of freedom in $\chi^2$, the probability level of $\chi^2_m$ are shown below at the 5 and 1 per cent values of the tabular $\chi^2$. These probabilities are to be regarded as more nearly the true probabilities corresponding to apparent probabilities of 5 and 1 per cent found by using the tabular $\chi^2$.

As has generally been supposed, the tabular $\chi^2$ underestimates the probabilities, except at the 5 per cent level with two degrees of freedom (three classes). The error diminishes as the smallest expectation increases and also as the number of degrees of freedom increases, though the rate of improvement is somewhat irregular in certain parts of the table, as might be expected from the nature of the distribution. Except for the most extreme case, $m=0.1$, the errors are not alarmingly great. If a 20-per cent error is permitted in the estimation of the true probability (i.e., an error up to 1 per cent at the 5 per cent level and 0.2 per cent at the 1 per cent level), the tabular $\chi^2$ is sufficiently accurate at the 5 per cent level for all values of $m$ down to 0.5. To reach this standard of accuracy for $m=0.1$, the number of degrees of freedom in $\chi^2$ should exceed 15. At the 1 per cent level, agreement is not so good. The necessary minimum values of the number of degrees of freedom are shown below for the various values of $m$.

| Smallest expectation $m$ | 0.1 | 0.5 | 1.0 | 2.0 | 3.0 | 5.0 |
|---|---|---|---|---|---|---|
| Minimum number of degrees of freedom...... | ? | 25 | 10 | 6 | 4 | 2 |

For example, with an expectation of two in the smallest class, the expectations in the other classes being supposed large, $\chi^2$ must have at least six degrees of freedom if the error at the 1 per cent level is to be less than 0.2 per cent.

From the above investigation it appears that, *with only a single small expectation*, the conventional procedure of grouping a class with a lower limit of under 5 is on the conservative side. At the 5 per cent level, the lower limit could be set as small as 0.5, and at the 1 per cent level as low as 2 without undue error; though it should be remembered that the error is consistently in the same direction. For those desiring a better approximation to the true probabilities, a small table of the 5 and 1 per cent significance levels of $\chi^2_m$ is given in Table 8. These significance levels may be used instead of the tabular $\chi^2$ where it is desired to avoid grouping a class with an expectation of below 5. The value of $\chi^2$ is, of course, computed in the usual manner.

For values of $m$ and numbers of degrees of freedom not shown in the

TABLE 8

FIVE AND 1 PER CENT SIGNIFICANCE LEVELS OF $\chi^2_m$

| SMALLEST EXPECTATION | NUMBER OF DEGREES OF FREEDOM IN $\chi^2$ | | | | | | | |
|---|---|---|---|---|---|---|---|---|
| | 2 | 3 | 4 | 5 | 6 | 10 | 15 | 25 |
| 5 Per Cent Level | | | | | | | | |
| 0.1........... | 8.63 | 9.82 | 11.14 | 12.50 | 13.87 | 19.28 | 25.82 | 38.37 |
| 0.5........... | 5.93 | 7.94 | 9.74 | 11.41 | 13.02 | 18.67 | 25.30 | 37.91 |
| 1........... | 5.88 | 8.04 | 9.79 | 11.29 | 12.79 | 18.49 | 25.16 | 37.80 |
| 2........... | 5.87 | 7.98 | 9.57 | 11.17 | 12.69 | 18.41 | 25.09 | 37.73 |
| 3........... | 5.87 | 7.85 | 9.54 | 11.13 | 12.66 | 18.37 | 25.06 | 37.70 |
| 4........... | 6.04 | 7.80 | 9.53 | 11.11 | 12.64 | 18.36 | 25.04 | 37.69 |
| 5........... | 5.88 | 7.81 | 9.51 | 11.10 | 12.63 | 18.35 | 25.03 | 37.68 |
| $\infty$ *.......... | 5.99 | 7.82 | 9.49 | 11.07 | 12.59 | 18.31 | 25.00 | 37.65 |
| 1 Per Cent Level | | | | | | | | |
| 0.1........... | 11.85 | 14.10 | 16.04 | 17.84 | 19.56 | 25.94 | 33.34 | 47.19 |
| 0.5........... | 12.78 | 13.86 | 15.25 | 16.78 | 18.33 | 24.40 | 31.59 | 45.08 |
| 1........... | 10.31 | 12.39 | 14.34 | 16.17 | 17.80 | 23.91 | 31.13 | 44.72 |
| 2........... | 9.85 | 12.22 | 13.94 | 15.64 | 17.32 | 23.59 | 30.88 | 44.53 |
| 3........... | 9.64 | 12.00 | 13.71 | 15.47 | 17.17 | 23.47 | 30.78 | 44.46 |
| 4........... | 9.56 | 11.75 | 13.62 | 15.38 | 17.09 | 23.41 | 30.73 | 44.42 |
| 5........... | 9.82 | 11.63 | 13.56 | 15.33 | 17.04 | 23.37 | 30.70 | 44.40 |
| $\infty$ *.......... | 9.21 | 11.34 | 13.28 | 15.09 | 16.81 | 23.21 | 30.58 | 44.31 |

*Tabular $\chi^2$ significance levels.

table, interpolation will be necessary. Where the number of degrees of freedom is tabulated, linear interpolation for $m$ will probably suffice down to $m=1$, though harmonic interpolation (i. e., linear interpolation on $1/m$) is better if $m$ exceeds 2. If the appropriate number of degrees of freedom is not tabulated, interpolation should be carried out on the difference between $\chi^2_m$ and the tabular $\chi^2$. Suppose, for example, that we wish to find the 1 per cent value of $\chi^2_m$ for eight degrees of freedom, the smallest expectation being 1. For six degrees of freedom, $\chi^2_m$ is 17.80 and $\chi^2$ is 16.81, the difference being 0.99. The corresponding difference for ten degrees of freedom is 0.70. Linear interpolation gives 0.84 for eight degrees of freedom. Since the tabular 1 per cent point of $\chi^2$ is 20.09 for eight degrees of freedom, the 1 per cent point of $\chi^2_m$ is taken as 20.93. The correct value is 20.95.

With more than one small expectation, as is often the case in testing the goodness of fit to a unimodal frequency distribution, the errors in using the tabular $\chi^2$ are presumably greater. By an extension of the method above, the probability levels of $\chi^2_m$, at the 5 and 1 per cent levels of the tabular $\chi^2$ were worked out for a goodness of fit test with two expectations of 1.0, the other expectations being assumed large. The corresponding results for one and two small expectations are compared in Table 9.

TABLE 9

COMPARISON OF THE PROBABILITY LEVELS OF $\chi^2_m$ WITH ONE AND TWO SMALL EXPECTATIONS ($m = 1.0$)

| NUMBER OF SMALL EXPECTATIONS | NUMBER OF DEGREES OF FREEDOM IN $\chi^2$ | | | | | | |
|---|---|---|---|---|---|---|---|
| | 3 | 4 | 5 | 6 | 10 | 15 | 25 |
| | 5 Per Cent Level of the Tabular $\chi^2$ | | | | | | |
| 1................... | .0537 | .0553 | .0540 | .0534 | .0527 | .0522 | .0516 |
| 2................... | .0592 | .0612 | .0585 | .0569 | .0556 | .0545 | .0529 |
| | 1 Per Cent Level of the Tabular $\chi^2$ | | | | | | |
| 1................... | .0144 | .0139 | .0137 | .0136 | .0123 | .0117 | 0.111 |
| 2................... | .0177 | .0178 | .0176 | .0172 | .0147 | .0133 | .0124 |

With two small classes, the errors in using the tabular $\chi^2$ are consistently about twice as great as with one small class. At the 1 per cent level, all the tabular $\chi^2$ probabilities in Table 9 are in error by more than 20 per cent, though only one value exceeds this limit at the 5 per cent level.

In the above investigation the expectations were treated as if known exactly, whereas in practice they are usually estimated from the sample. Calculation of the exact distribution of $\chi^2$ in such cases would be illuminating, but would involve laborious computations. At first sight it appears that the errors introduced by a single small expectation are perhaps smaller than those shown in Table 7, since with one unknown parameter a $\chi^2$ with four degrees of freedom represents six classes, as against five classes where there are no unknown parameters.

I am indebted to Dr. R. L. Anderson for the greater part of the computations in Tables 7 and 8.

## SUMMARY

The $\chi^2$ correction for continuity is a device for obtaining an approximation to the sum of a number of discrete probabilities by means of a corresponding area under the tabular $\chi^2$ frequency distribution, and should be used only at the final stage of calculating the probability level attached to a value of $\chi^2$. The method of making the correction is illustrated on a common test for linkage in genetical investigations.

Where several values of $\chi^2$ (each with a single degree of freedom) are added to form a total $\chi^2$, uncorrected values must be used.

The effect of a single small expectation in goodness of fit tests is illustrated by working out the exact distribution of $\chi^2$ for three examples. The discontinuities in the exact distribution of $\chi^2$ produce only a small part of the discrepancy between the tabular and the exact $\chi^2$ distributions. By a theoretical approach taking account of the size of the smallest expectation, a more satisfactory approximation is obtained to the exact distribution of $\chi^2$ and is employed to estimate more generally the sizes of the errors involved in using the tabular $\chi^2$. With only a single small expectation, the conventional limit of five for the smallest expectation appears unduly high. At the 5 per cent level, the tabular $\chi^2$ distribution may be used without undue error with an expectation as low as 0.5, and at the 1 per cent level with an expectation as low as 2. For more exact work, a small table is given of the 5 and 1 per cent levels of $\chi^2$ as found by the new approximation.

The errors involved where two expectations are small and the effects introduced by the estimation of unknown parameters are briefly discussed.

## REFERENCES

1. YATES, F.
   Contigency tables involving small numbers and $\chi^2$ test. Jour. Roy. Stat. Soc. Suppl. Vol. 1 (1934) 217–235.

2. FISHER, R. A.
   Statistical methods for research workers. Edinburgh, Oliver and Boyd. §§ 21.01.

3. IMAI, Y.
   Linkage studies in *Pharbitis Nil* I. Genetics. Vol. 16 (1931) 26–41.

4. MATHER, K.
   The measurement of linkage in heredity. Chemical Publishing Co. of New York, Inc. (1938).

5. AITKEN, A. C.
   Statistical mathematics. Edinburgh, Oliver and Boyd. (1940).

# 30

## ROTATION EXPERIMENTS WITH COTTON
## IN THE SUDAN GEZIRA

By F. CROWTHER, *Plant Physiology Section, Agricultural Research Institute,*
*Anglo-Egyptian Sudan*

AND

W. G. COCHRAN, *Statistical Laboratory, Iowa State College, Ames, Iowa*

(With Four Text-figures)

### INTRODUCTION

Rotation of crops is based in most countries upon the accumulated experience of past years; experiments testing the relative merits of various rotations are rarely resorted to unless some serious disturbance arises, such as the inclusion of a new crop. But in the Sudan Gezira (lat. 14–15° N., long. 33–34° E.) the need for rotation experiments was urgent, since growing cotton by artificial irrigation was a recent introduction, and, moreover, the soil was known to be intractable. Any improvement in rotation which would allay seasonal fluctuation in yield and raise the general level would greatly add to the prosperity of the area.

Thus when the Gezira Research Farm was inaugurated in 1918 some unreplicated plots of various two-course rotations, and of continuous cotton, were laid down. These were supplemented in 1925–6 by a replicated trial comparing various three-course rotations, designed by E. M. Crowther of Rothamsted. Later, in 1931–2, a new 'Combined Rotation' experiment, designed by M. A. Bailey and E. M. Crowther, was added, and this comprised comparisons of one-, two-, three- and four-course rotations. These last two experiments form the subject of this paper.

The first crop of irrigated cotton was grown in 1911–12 covering 250 feddans.* From that the area has increased to the present 200,000 feddans. The water, coming from the Blue Nile, irrigates the cotton from sowing time in August to the completion of the harvest in April. The management of the Gezira Scheme† is such that the rotation can be standardized and adopted immediately. The soil is a heavy alkaline clay, and irrigation water penetrates little beyond 3–4 ft. It is deficient in available nitrogen, and the use of nitrogenous manures leads to increased yields. Details of cultural operations have been given by Crowther (1934). The rainfall of 16 in. per annum falls mainly in July and August, and before the introduction of cotton supported over much of the area annual crops of dura (*Sorghum vulgare*).

Since dura is the staple diet of the people, any rotation on the irrigated area should include dura. Because of the assured water supply a given bulk of dura can be grown on a smaller area than under rain cultivation, and land is thereby freed for other crops. Moreover, the growing of irrigated dura ensures freedom from famine when the rains fail.

---

* 1 feddan = 1·038 acres, 1 kantar = 315 rotls, 1 rotl = 0·99 lb.

† For a description of the Gezira Scheme see Bluen (1931), Crowther & Crowther (1935), Gregory, Crowther & Lambert (1932), and Norris (1934); and for studies on the soil, see Joseph (1925), Greene (1928 *a*, *b*, 1942), and Greene & Snow (1939).

---

Numerous trials have been, and still are being, made on the suitability of crops other than cotton and dura for the Gezira, but so far only lubia (*Dolichos lablab*) is both useful and easily grown. Lubia supplies food for both man and beast, and, as a legume, should benefit any rotation in which it is included.

### Description of experiments

The crops for the rotation are thus clearly defined. What is less evident is the optimum length of the fallow period between successive crops. Under the intense heat of the central Sudan, fallow improves the physical texture of the soil and facilitates weed control through cessation of irrigation.

Cotton, being the 'cash crop' of the Sudan, forms the basis of any rotation, and experiments must reveal whether the exhaustion after the cereal can be offset by the legume, and what is the position and length of fallow for maintaining and, if possible, improving the fertility of the soil.

Five rotations were included in the Three-course* Rotation Experiment started in 1925–6: (1) cotton—fallow—fallow; (2) cotton—dura—fallow; (3) cotton—dura—dura; (4) cotton—lubia—fallow; (5) cotton—dura—lubia. The rotations ranged from continuous cropping to one crop in three years, and from intensive cereal cropping to rotations without cereal.

The practical value of this experiment was limited by the inclusion of only three-course rotations. More frequent cropping might increase the gross agricultural output of the scheme; but it seemed possible also that, since land was plentiful, a three-year cycle might prove inferior to a four-year cycle.

The new 'Combined Rotations' experiment therefore provided a comparison of one-, two-, three- and four-course rotations. Ten rotations were included:

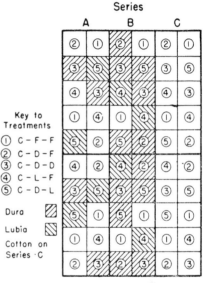

Plan of
**Three Course Rotation Experiment
1938 — 1939**

Figure 1

| (A) | (M) | (G) | (K) | (A) | (E) | (F) | (H) | (J) | (O) | (P) |
|-----|-----|-----|-----|-----|-----|-----|-----|-----|-----|-----|
| C | C | C | C | C | C | C | C | C | C | C |
|   | F | F | F |   | D/L | D/L | F | D/L | F | F |
|   |   | F | F |   | F | D/L | F | D/L | D/L | F |
|   |   |   | F |   |   |   | F | F | F | D/L |

They comprise two sets of one-, two-, three- and four-course rotations. One set has fallow only between successive cotton crops. The other set includes dura/lubia instead of one of the fallows. Dura/lubia indicates that half the plot was cropped with dura and half with lubia. When the same phase of the cycle is repeated the dura and lubia halves are reversed. This alternation, included to overcome the exhausting effect of the cereal

---

* Since all crops are sown only during the rains, a three-course rotation is a three-year one. Fallow indicates that the land was uncropped and uncultivated for a whole year, including a rain period.

and yet produce food as well as cash crops, effectively converts the rotations into four-, six- and eight-year rotations.

The same cotton variety, Sakellarides, has been grown throughout. All phases of all rotations are represented each year. In the 'Three Course' the three phases of the five rotations are replicated four times making sixty plots, each of about half a feddan. For simplicity in field operations all cotton plots are grouped together, each group being called a *series* (see Fig. 1). The same random order was used for the three series to lessen

### Plan of Combined Rotations Experiment 1938–1939

|  | Block 1 |  |  |  |  |  | Block 2 |  |  |  |  |
|---|---|---|---|---|---|---|---|---|---|---|---|
| O4 | F2 | K4 | G3 | H1 | E1 | M2 | P4 | K4 | E1 | J2 | F2 |
| H2 | J4 | M1 | P3 | K1 | M2 | P2 | A | O3 | K2 | O4 | P1 |
| P4 | K2 | E2 | O1 | J3 | F1 | J3 | H1 | G2 | F3 | E2 | J1 |
| J1 | O2 | O3 | F3 | G1 | A | G3 | M1 | J4 | H2 | K1 | O1 |
| G2 | P1 | J2 | H3 | P2 | K3 | K3 | O2 | G1 | F1 | H3 | P3 |
| J1 | O2 | M2 | K1 | K2 | H3 | J3 | O4 | J1 | O2 | M1 | P2 |
| P2 | F2 | K3 | O1 | G2 | P4 | G2 | P4 | O1 | A | E1 | H2 |
| O4 | M1 | H1 | E2 | E1 | P3 | M2 | O3 | H1 | K4 | J2 | K1 |
| J4 | G3 | F1 | G1 | P1 | A | H3 | K3 | F1 | P3 | J4 | K2 |
| K4 | H2 | O3 | J2 | F3 | J3 | F2 | G1 | F3 | G3 | P1 | E2 |
|  | **Block 3** |  |  |  |  |  | **Block 4** |  |  |  |  |

**Key to Rotations and Symbols**

|  | A | E | F | G | H | J | K | M | O | P |
|---|---|---|---|---|---|---|---|---|---|---|
| 1 | C | C | C | C | C | C | C | C | C | C |
| 2 | (D/D) L | F | F | (D) L | F | F | F | F | | |
| 3 | | F | F | (D) L | F | F | | (D) L | F | |
| 4 | | | | F | F | | F | | F | (D) L |

Dura is grown on the shaded half of the plot. Dura and Lubia alternate on each recurrence.

| P1 | G3 | M1 | K1 | O1 | E2 |
|---|---|---|---|---|---|
| A | M2 | P2 | J3 | P3 | F3 |
| H3 | K4 | O3 | F1 | K2 | J1 |
| K3 | J2 | E1 | H2 | G1 | O2 |
| O4 | F2 | J4 | G2 | H1 | P4 |

**Block 5**

Figure 2

the difficulty of field supervision. The bearing of this on the statistical analysis is discussed in the next section.

In the 'Combined Rotations' the phases of the ten rotations total 30, and here the plots are randomized without restriction in each of five blocks, making a total of 150 plots in the experiment (see Fig. 2). The halves of the plots have been picked separately since 1936–7 because of the dura/lubia cropping. The gross area is $37\frac{1}{2}$ feddans, while a subplot is $\frac{1}{4}$ feddan and a picking unit $\frac{1}{8}$ feddan, including belts.

While desirable statistically this lay-out entails much extra field work and supervision.

Small plots must be ploughed and irrigated while adjacent ones remain undisturbed. Belts between adjacent plots minimize these cultural difficulties, at the same time serving as a protection against seepage.

The fate of the crop residues is important in comparing rotations, for more nutrient may be permanently removed from the land than is contained in the portions harvested. In these experiments, after the removal of the seed cotton, the sticks are burnt outside the plots. The dura grain is removed, but the straw is eaten in situ by penned sheep in the 'Three-Course' and by cattle in the 'Combined Rotations'. Theoretically then all is eventually returned to the land except that contained in the cotton sticks and seed, and the dura heads. Practically, however, there is difficulty in keeping the cattle long enough on the plot for its residues to be returned.

These experiments are laid down and supervised by the Agricultural Section of the Agricultural Research Institute, and Messrs V. P. Walley, E. Mackinnon and E. R. John have been in turn responsible for them.

## Detailed results and statistical analysis

### The three-course rotations experiment

Since cotton is the important cash crop of the rotation, the statistical analysis is confined throughout this paper to an examination of the cotton yields. The size of the harvested plot, originally 0·50 feddan, was reduced in 1932–3 to 0·45 feddan, and varied slightly thereafter. Where necessary, the yields were adjusted to a common area of 0·50 feddan, the statistical analysis being based on these figures.

An analysis of variance was first carried out for each year. The standard errors per plot ranged from 5·9% in 1937–8 to 14·6% in 1932–3, the percentages being highest in the low-yielding years. Fig. 3 shows a graph of the mean annual yields for each rotation.

The average yields of the five rotations, from 1927–8 to 1939–40, were as follows:

|  | CLF | CFF | CDL | CDF | CDD |
|---|---|---|---|---|---|
| Kantars per feddan | 4·25 | 4·07 | 3·84 | 3·84 | 2·96 |

The most striking result is the poor yield for the rotation containing two years of dura, this yield being significantly smaller than the lowest of the other four rotations almost every year. The inclusion of one year of dura also resulted in a reduced yield. The reduction, though much smaller, was significant in most years towards the end of the period. Of the two best rotations, CLF gave generally a higher yield than CFF, but the difference in its favour was not significant in any single year.

Some further indications may be obtained from Fig. 3. In the four poorest seasons, CDL, the only rotation in which lubia immediately precedes the cotton crop, did relatively well, giving the highest mean yield of all rotations. The other rotations were in

Table 1. *Percentage increments over CDD*

| Mean yield of all rotations (kantars per feddan) | Rotation | | | |
|---|---|---|---|---|
|  | CLF | CFF | CDF | CDL |
| 2–3 (4) | 34 | 31 | 24 | 36 |
| 3–4 (5) | 46 | 43 | 30 | 30 |
| 4 + (4) | 46 | 36 | 32 | 26 |

(The figures in brackets show the numbers of years averaged.)

the same order in good and poor years, but their increments over CDD were smaller, both actually and in percentages, in the poor seasons.

### Three Course Rotations Experiment—Mean Yields

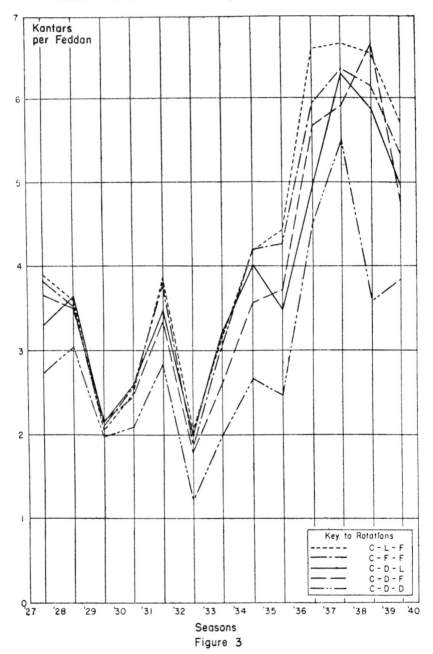

Seasons

Figure 3

Thus, apart possibly from CDL, the rotations were not successful in maintaining yields in poor seasons. Unfortunately, from the point of view of interpretation, the four years with mean yields over 4 kantars per feddan were the last four years of the period, when long-term differences might be beginning to show. This objection does not, however, apply to the yields between 3 and 4 kantars per feddan.

A preliminary statistical analysis was undertaken to test the significance of the average differences between the rotations, and to examine the variations from year to year. The analysis was made when the 1938–9 results became available, and was confined to the four complete cycles between 1927–8 and 1938–9. The year 1926–7 was omitted because the rotations were then only in their second phase.

It is not easy to find a satisfactory method of analysis. We shall describe in some detail the various factors that must be considered, as similar problems are likely to arise in other rotation experiments. The observed difference between two rotations may be regarded as the sum of the true difference and the experimental error. Since the rotations return to the same plots every three years, the experimental error in any series consists of two parts, a part which is constant for a given plot, and a part which varies independently for each plot in each year. Further, as the same randomization was used in each series, the experimental errors for the three series may not be independent. It was also evident from the analysis of the individual years that the experimental errors were different in different seasons.

The true difference between two rotations will presumably vary from year to year, according to the manner in which the individual rotations respond to a particular type of season.* As the experiment proceeds, the true difference may also be increasing or decreasing at a constant or variable rate, owing to cumulative effects of the rotations on the cotton yields. One of the principal objects of the analysis is to examine whether there are indications of such long-term differences between the rotations.

A preliminary inspection of the data threw some light on the importance of the factors mentioned above. It appeared that any cumulative differences between the rotations could be measured by a parabolic regression on years, though the interpretation of these differences is doubtful on account of the group of poor seasons in the middle and of good seasons at the end of the period. The rotations × seasons interactions were marked, and differed widely according to the particular rotations which were being compared. For studying these interactions, the most appropriate division of the degrees of freedom between rotations appeared to be: CLF $v$. CFF; CDL $v$. CDF; no dura $v$. one year of dura; and two years of dura $v$. the rest. This subdivision was not entirely satisfactory, however, since CDL behaved somewhat differently from the other rotations in poor years, as noticed above. Some information was also obtained on the relative sizes of the two components of the experimental error. If $e_p$ is the part of the error which is constant for a given plot, and $e_r$ is the random part, the total yield of the plot has an error $4e_p + e_{r_1} + e_{r_2} + e_{r_3} + e_{r_4}$, since cotton was grown on every plot in four different years. The variance of the mean, on a single-plot basis, is $\{\sigma^2_p + \frac{1}{4}(\sigma^2_{r_1} + \sigma^2_{r_2} + \sigma^2_{r_3} + \sigma^2_{r_4})\}$. On the other hand, the differences between the yields of a plot in two different years does not involve $e_p$, the average variance for the within-series comparisons being $\frac{1}{4}(\sigma^2_{r_1} + \sigma^2_{r_2} + \sigma^2_{r_3} + \sigma^2_{r_4})$. Thus we may estimate $\sigma^2_p$ for each series, by comparing the error variance of the totals with the average 'within-series' error variance. For the three series, the average error variance of the totals was 3967 per plot as compared with 2746 per plot for the within-series comparisons. Thus comparisons based on totals are less accurate than within-series comparisons.

* These variations are technically called the 'rotations × seasons interactions'.

The estimation of the linear and quadratic components of the regression of yields on years required some consideration. The linear component measures the average change in yield per year and enables us to judge whether the differences between the rotations are being accentuated, or are diminishing, as the experiment proceeds. An estimate was obtained separately for each series, by multiplying the successive yields of any plot by $-3$, $-1$, $+1$ and $+3$ respectively. These estimates are within-series comparisons, so that their experimental errors are unaffected by constant plot errors. A further estimate was available from the difference between the totals on series C (years 3, 6, 9, 12) and series A (years 1, 4, 7, 10). As mentioned earlier, this estimate was subject to a higher experimental error per plot than the within-series estimates, and appeared to contribute only about 5 % of the total information. It was accordingly ignored, the unweighted mean of the three within-series estimates being used.

For similar reasons, the quadratic term of the regression was also estimated entirely from within-series comparisons, being obtained by multiplying the individual plot yields by the following quantities:

Year

| 1 | 2 | 3 | 4 | 5 | 6 | 7 | 8 | 9 | 10 | 11 | 12 |
|---|---|---|---|---|---|---|---|---|----|----|----|
| +6 | +3 | 0 | −2 | −3 | −4 | −4 | −3 | −2 | 0 | +3 | +6 |

The rotation and error mean squares in the analysis of variance are shown in Table 2, the degrees of freedom between rotations being subdivided in the manner suggested previously. In calculating the experimental error for the total yields over the twelve years, the yields on corresponding plots were added to form a single set of twenty plot yields, which were then analysed. This procedure was necessary to obtain an unbiased estimate of error, because the same randomization was used in each series. A similar procedure was followed for the linear and quadratic estimates. The rotations × seasons interactions were divided into those derived from between-series comparisons and those derived from within-series comparisons, the former having a higher experimental error than the latter. The error from the within-series comparisons was not entirely homogeneous, but appeared to be the same for all the single degree-of-freedom comparisons.

Table 2. *Analysis of variance of the three-course rotations experiment.*
*Units: a single-plot yield in a single year, in rotls per $\frac{1}{2}$ feddan*

| | | Mean squares | | | Rotations × seasons interaction | | | |
| Rotations compared | Degrees of freedom | Totals | Linear component | Quadratic component | Degrees of freedom | Between series | Degrees of freedom | Within series |
|---|---|---|---|---|---|---|---|---|
| CFL, CFF | 1 | 15,100 | 11,544 | 2,040 | 2 | 399 | 7 | 1,434 |
| CDL, CDF | 1 | 1,512 | 1,940 | 40,280 | 2 | 7,511 | 7 | 6,516 |
| No dura, one year of dura | 1 | 97,245 | 38,292 | 1,710 | 2 | 3,636 | 7 | 10,842 |
| CDD, rest | 1 | 985,794 | 164,748 | 44,524 | 2 | 13,373 | 7 | 19,940 |
| Error | 12 | 3,734 | 2,766 | 1,663 | 24 | 4,084 | 84 | 2,898 |

The rotations × seasons interaction was significant for all comparisons except that between CFL and CFF, for which there was no indication of an interaction. Having established the existence of seasonal fluctuations in the differences between rotations, we wish to test whether these differences show in addition any real long-term trends. Even if there were no real trends, a 'random' seasonal fluctuation in the differences between rotations will inflate not only the rotations × seasons interaction mean square, but to an equal extent the totals, linear and quadratic mean squares. From this it follows that in testing for trends, the totals, linear and quadratic mean squares must be compared with their experimental errors, plus an estimate of the true rotations × seasons interaction variance. For CDL v. CDF, for example, the two estimates of the seasonal interaction variance are $7511-4084=3427$ and $6516-2898=3618$. Weighting these according to the numbers of interaction degrees of freedom, we obtain the estimate 3576. Added to the experimental error mean square of 3734, this gives an estimated mean square of 7310 with which to compare the totals mean square of 1512. A similar estimate was made for each single degree of freedom. The above method of weighting does not give the most accurate estimate of an interaction variance, but seems to be the best simple procedure.

The average yields and the linear and quadratic terms are shown for each rotation in Table 3. The linear component is the average change in yield per year, while the quadratic component is the average change per year in the linear component. Since the standard deviation (i.e. experimental error plus seasonal interaction) is different for every comparison, a subsidiary table is given for the tests of significance. The ratio of a difference to its standard deviation will not follow the $t$ distribution exactly, but conclusions should not be far in error if only 9 degrees of freedom are assigned to the error and if a slightly higher level of significance is used.

Table 3. *Long-term effects*, 1927–8 to 1938–9. *Kantars per feddan*

|  | Rotation | | | | |
|---|---|---|---|---|---|
|  | CLF | CFF | CDF | CDL | CDD |
| Mean yield | 4·128 | 3·969 | 3·788 | 3·738 | 2·889 |
| Linear component | 0·378 | 0·337 | 0·312 | 0·295 | 0·207 |
| Quadratic component | 0·157 | 0·146 | 0·183 | 0·134 | 0·114 |

Tests of significance

| Comparison | Mean yields | Standard error | Linear component | Standard error | Quadratic component | Standard error |
|---|---|---|---|---|---|---|
| CLF – CFF | +0·159* | ±0·0793 | +0·041* | ±0·0203 | +0·011 | ±0·0100 |
| CDF – CDL | +0·050 | ±0·111 | +0·017 | ±0·0308 | +0·049† | ±0·0178 |
| No dura, one year of dura | +0·286† | ±0·0908 | +0·054* | ±0·0259 | −0·007 | ±0·0154 |
| Rest, two years of dura | +1·017† | ±0·141 | +0·124† | ±0·0411 | +0·041 | ±0·0253 |

\* Difference approaches significance.    † Significant difference.

The results of the tests of significance may be summarized as follows:

*CLF v. CFF.* The rotation CLF produced a higher mean yield of cotton and a higher average rate of increase of yield than CFF, the difference being not quite significant in either case. The results suggest, however, that the superiority of CLF may become more marked in future years. It may be noted that CLF gave a higher yield than CFF in 1940.

*CDF v. CDL.* There is no indication that these rotations differed consistently in yield throughout the twelve years, or that a consistent difference is likely to appear in the near future. The two rotations gave very similar yields in 1940.

*No dura v. one year of dura.* The reduction in yield due to the inclusion of one year of dura is definitely significant. The average rate of increase of yield is also smaller with one year of dura, but not quite significantly so.

*CDD* v. *rest.* With CDD, there was a pronounced reduction in yield and a definitely significant reduction in the average rate of increase.

The only significant quadratic term is that for the comparison of CDF and CDL. In the first three years of the experiment, CDF gave slightly higher yields. This situation was reversed in the next five years, which included three of the poor seasons. Thereafter CDF has again been giving somewhat higher yields.

It must again be stressed that owing to the peculiar succession of seasons, the long-term effects could be interpreted alternatively as a correlation of the responses with the average level of yield. The mean yield of all rotations gives a correlation coefficient of $+0.745$ with the linear component, and a multiple correlation coefficient of $0.935$ with the linear and quadratic components, so that it is not possible to separate statistically the two types of effect. A crucial test of the reality of the long-term effects will not be possible until further moderate or poor seasons are obtained.

Further examination remains to be made of the rotations $\times$ seasons interactions. As a first step, this will involve the comparison of the differences in individual seasons with the characteristics of the seasons.

### The 'combined rotations' experiment

As will be seen from Fig. 2, the design of this experiment differs considerably from that of the three-course rotations experiment. To ensure accurate comparisons between the different rotations, they were grown on neighbouring plots in the same block. Every phase of each rotation was represented, there being for example four plots of each four-course rotation in every block. This rule increases the number of plots required, but has the merit of providing comparisons *each year* between the cotton yields in all rotations.

Six years' results were available when the preliminary statistical analysis was made after the 1938–9 harvest. It was pointed out earlier that the rotations involving dura and lubia are actually four-, six-, and eight-year rotations with two cotton crops in each rotation, owing to the alternation of dura and lubia on the half-plots. However, half-plot yields were not obtained during the first three years, while the whole-plot yields measure the average performance of these rotations, since they are estimates of the average yields of the two cotton crops. Accordingly, the six years' results from whole-plots were first examined, the available half-plot yields being subsequently analysed so as to separate the effects of dura and lubia. As it happened, the yields of cotton following dura and lubia were widely different, and comparisons based on whole-plot yields give a misleading picture of the effects of the individual crops.

The annual yields from 1933–4 to 1939–40 are given in Figs. 4*a*, *b*, *c*. The first graph shows the effects of one-year, two-year, and three-year fallows between successive cotton crops. Fig. 4*b* shows the effects of fallowing between successive two-year sequences: cotton, dura/lubia. The results are grouped by length of rotations in Fig. 4*c*.

In the statistical analysis the same difficulties arise as in the three-course rotations experiment. There is the additional complication that different rotations return to the same plots at different intervals, so that a separate standard deviation is required for almost every pair of rotations. Comparisons between rotations of the same length may be made by methods essentially similar to those used for the three-course rotations, analysing each length of rotation separately. For the two-course rotations, CF and C D/L, there are two 'series' of plots, one for the odd and one for the even years. The

## Combined Rotations Experiment—Mean Yields

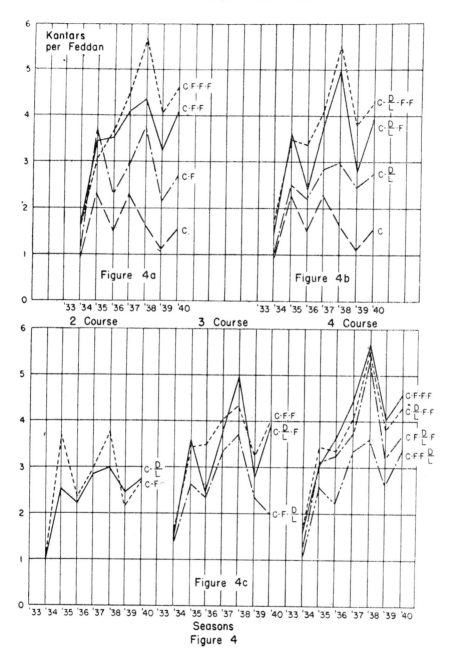

Figure 4a

Figure 4b

Figure 4c

Figure 4

Seasons

four-course rotations had repeated on only two sets of plots, and as there appeared to be no correlation between the experimental errors in the first and fifth, or the second and sixth years, constant plot errors were assumed negligible for this series.

Only the linear component of the long-term effects was considered, on account of the small number of years available. The mean yields and the average changes in yield per year are shown below. The latter were estimated by multiplying the successive annual yields by $-5$, $-3$, $-1$, $+1$, $+3$, $+5$, and dividing by 35. This was not the most accurate estimate for the two- and three-course rotations, but was chosen as a compromise to facilitate comparisons between rotations of different lengths.

Table 4. *Mean yield and average increases in yield per year, kantars per feddan*

|  | C | Two-year | | Three-year | | | Four-year | | | |
| --- | --- | --- | --- | --- | --- | --- | --- | --- | --- | --- |
|  | C | C | C | C | C | C | C | C | C | C |
|  |  | D/L | F | F | D/L | F | F | D/L | D/L | F |
|  |  |  |  | D/L | F | F | F | D/L | F | F |
|  |  |  |  |  |  |  | D/L | F | F | F |
| Mean yield | 1·64 | 2·34 | 2·68 | 2·63 | 3·16 | 3·37 | 2·58 | 3·33 | 3·66 | 3·75 |
| Significant diff.* |  | ±0·428 | | ±0·400 | | | ±0·336 | | | |
| Av. increase per year | −0·008 | +0·261 | +0·167 | +0·264 | +0·344 | +0·333 | +0·345 | +0·475 | +0·500 | +0·596 |
| Significant diff.* |  | ±0·251 | | ±0·232 | | | ±0·175 | | | |

\* At the 5 % level.

Results of the tests of significance made between rotations of the same length are given below in a condensed form. The rotations × seasons interactions were included in estimating the standard deviations.

*Two-year.\** Mean yields: CF ⩾ C D/L.
Rates of increase: No significant difference.
Rotations × seasons: Significant.

*Three-year.* Mean yields: CFF, CF D/L > C D/LF.
Rates of increase: No significant differences.
Rotations × seasons: Not significant.

*Four-year.* Mean yields: CFFF, C D/LFF ⩾ CF D/LF > CFF D/L.
Rates of increase: CFFF > CFF D/L; C D/LFF, CF D/LF ⩾ CFF D/L.
Rotations × seasons: No indication of a real interaction.

While the differences examined above were not always significant, the general trend of the results is very consistent. In every case the substitution of dura/lubia for fallow produced a decrease in yield, the decrease being greatest when the substitution immediately preceded the cotton crop. It will be seen later that the depression was mainly due to the dura. In the four-year rotations, the superiority of one-, two- and three-year fallows before cotton appears to be increasing, though only the extreme difference, CFFF − CFF D̊/L, was significant at the 5 % level.

Turning to comparisons between rotations of different length, the successive increments in yield from increasing lengths of fallow, as shown in Figs. 4 *a, b*, were tested for significance. A separate analysis was made for each pair of rotations compared, as the pooling of errors did not seem justified. In both series, all successive increments in yield and in average change in yield were significant. Up to the present, both series show a 'diminishing returns' effect, the successive increases in yield being 1·02, 0·69

\* The symbol > indicates that any rotations on the left were significantly greater than any rotations on the right at the 5 % level. The symbol ⩾ denotes that significance was attained at the 20 % level, but not at the 5 % level. This symbol denotes cases in which there was some indication of a real difference, though it did not reach the usual level of significance.

and 0·38 kantar per feddan in the C, CF, CFF, CFFF series, and 0·85, 0·46 kantar per feddan in the C D/L, C D/LF, C D/LFF series. Since, however, the successive increases to each length of fallow are becoming greater as the experiment proceeds, these figures may underestimate future gains from the longer periods of fallow. In 1940, for example, the successive increments in yield in the C ... CFFF series were 1·13, 1·39 and 0·51 kantars per feddan.

As would be expected, fallow proved less effective in maintaining cotton yields when placed after cotton and before dura/lubia, the mean yields for CF D/L and CFF D/L being 2·63 and 2·58 kantars per feddan respectively, as compared with 2·35 kantars per feddan for C D/L.

### Examination of the half-plot yields

With the inclusion of the 1940 results, four years' results from half-plots are available. The type of information obtained from half-plot yields may be exemplified by the two-course rotation, which, for a given phase, provides the two series DCLC... and LCDC.... The difference between the two cotton yields in any year represents the difference in the residual effect of dura and lubia from the previous year, *minus* the difference in residual effect from three years previous, *plus* the difference in residual effect from five years previous, and so on. Separation of the recent and more remote residual effects is not possible, since the rotations CDCD... and CLCL... were not included in the experiment. A similar difficulty arises in comparing the individual yields after dura and lubia with the CFCF... rotation.

The averages of the four years' results are shown in Table 5, arranged so as to compare cotton yields after fallow, dura and lubia. Three standard errors were required each year, one for the comparison of dura and lubia grown on the same plot, one for the comparison of dura and lubia on different plots, and one for the comparison of dura or lubia with the corresponding fallow plots.* The first is derived from the subplot error in the analysis of variance, while the last two are weighted means of the sub- and main-plot errors (cf. F. Yates, 1937).

Table 5. *Mean yields, 1936–7 to 1939–40. Kantars per feddan*

| Lubia and dura grown | Length of rotation years | Cotton after | | | Rotations compared | | |
|---|---|---|---|---|---|---|---|
| | | Fallow | Lubia | Dura | | | |
| Preceding year | 2 | 2·90 | 3·20 | 2·31 | CF | CL | CD |
| | 3 | 3·94 | 3·60 | 2·60 | CFF | CFL | CFD |
| | 4 | 4·70 | 3·83 | 2·66 | CFFF | CFFL | CFFD |
| Two years previous | 3 | 3·94 | 4·21 | 3·49 | CFF | CLF | CDF |
| | 4 | 4·70 | 4·44 | 3·56 | CFFF | CFLF | CFDF |
| Three years previous | 4 | 4·70 | 4·65 | 4·20 | CFFF | CLFF | CDFF |

The exhausting effect of dura grown immediately before the cotton crop is clearly shown in the first three lines of the table. In almost every instance the yields of cotton immediately after dura were significantly lower than those after fallow or lubia. The exhaustion appears to be little mitigated by fallows preceding the dura crop; indeed, the loss in yield due to dura increased when fallow preceded dura. The yields of cotton after dura were increased by an intervening fallow, and increased further by two years'

* The numerical values in kantars per feddan for significant differences, in the above order, were: in 1936–7, 0·71, 1·20, 1·19; in 1937–8, 0·48, 0·96, 0·93; in 1938–9, 0·42, 0·59, 0·57; and in 1939–40, 0·42, 0·61, 0·57.

intervening fallow. However, even in the latter case, the yield after dura (4·20 kantars per feddan) was significantly below the corresponding yields after fallow (4·70) or lubia (4·65).

The effectiveness of lubia relative to fallow appears to depend on the cropping *previous to the lubia*. The results are grouped below according to previous cropping, some relevant comparisons from the three-course rotations experiment being included.

Table 6. *Lubia and fallow rotations : Mean yields in kantars per feddan*

| | Previous crop: cotton or dura | | | | | | | | | |
|---|---|---|---|---|---|---|---|---|---|---|
| | CLF* | 4·25 | CDL* | 3·84 | CL | 3·20 | CLF | 4·21 | CLFF | 4·65 |
| | CFF* | 4·07 | CDF* | 3·84 | CF | 2·90 | CFF | 3·94 | CFFF | 4·70 |
| Increase to lubia | +0·18 | | 0·00 | | +0·30 | | +0·27 | | −0·05 | |

| | Previous crops: cotton and one-year fallow | | | | Previous crop: two-year fallow | |
|---|---|---|---|---|---|---|
| | CFL | 3·60 | CFLF | 4·44 | CFFL | 3·83 |
| | CFF | 3·94 | CFFF | 4·70 | CFFF | 4·70 |
| Increase to lubia | −0·34 | | −0·26 | | −0·87 | |

\* From the three-course rotations experiment.

Where lubia followed cotton or dura, it proved as effective as fallow, or possibly slightly more so. In the three-course rotations experiment, it will be recalled that the increase to CLF over CFF was almost significant. In the 'combined rotations', the increase to CL over CF was significant in 1939, and almost so in 1940, there being little difference between lubia and fallow in 1937 or 1938.\* Similarly, CLF was significantly above CFF in 1938, there being little difference in other years. The two four-year rotations agreed closely in all four years, but here any difference between lubia and fallow would be diminished by the two years' intervening fallow.

With a one-year fallow between cotton and lubia, the yields after lubia were lower than those after fallow in both comparisons, though the reductions were significant only in 1940. Where two years' fallowing preceded lubia, the reduction to lubia was large and definitely significant.

### Interpretation of results

Rotation experiments are continuous and results obtained over a period cannot be considered as final. Yet, when information is urgently required, even short periods may give broad indications as to the relative merits of the rotations compared.

Already the exhausting effect of the cereal is evident, for cotton yields of rotations including dura are consistently lower than those including lubia and fallow only. When two dura crops are interposed between successive cotton crops the reduction in yield is great, more than double that from a single dura crop. Greene (1942), describing nitrogen and moisture changes in these experiments, shows that yield differences are closely associated with corresponding differences in nitrates in the first and second feet of soil.

Surprisingly there is little benefit from the presence of a legume in the rotation, in a soil where nitrogen supply is known to be deficient. Lubia has abundant root nodules, yet a following cotton crop reaps little advantage therefrom, and in no case yields substantially more than if the land had remained fallow. Clearly, a period of fallow is of

\* It should be remembered that in this case lubia is preceded by the sequence dura, cotton, whereas the fallow is preceded by fallow, cotton.

great value to the Gezira soil and rotations must include fallow to maintain yields. But even 'Continuous Cropping' allows of several months of bastard fallowing, viz. 7, 5 and 4 months respectively for dura, lubia and cotton, since all crops are sown during the rains (July–August) and irrigations continue till November, January and late March for the three crops respectively. These bastard fallows cover the hottest weather (April-June) when desiccating conditions are most intense.

An effective fallow must last for at least a full year and include a period of rains. Moreover, the 'Combined Rotations' experiment shows that the longer the preceding fallow the higher the cotton yields, for one year is superior to none, two years to one year, and even three years to two. Possibly, if a longer rotation had been tested, the fallow period for maximum yields might have been shown to be even longer than three years. The advantageous effect of fallow has been assumed to arise primarily from the improved physical conditions of the soil through desiccation. Attention may usefully be focused in future on the behaviour of fallows during the rain period.

In the absence of experiment it might have seemed undesirable to expose uncropped land to erosion from violent tropical rain storms. It must however be remembered that soil erosion is slight in such a level plain as the Sudan Gezira, and that the irrigation ridges and banks restrict surface wash in the year in which cotton is sown. It has been shown by subsequent experiments (Crowther, 1942) that weeding the fallow land during the rains greatly increases the yield of cotton sown during the rains one year later by improving the supply of available nitrogen. Evidently the possible ill effects of soil erosion from uncropped and even clean-weeded land are offset by gains in other directions under the system of irrigation agriculture practised in the Sudan Gezira.

While it is difficult to obtain strictly comparable figures from the two experiments, the degree of superiority of one rotation over another appears to depend on the nature and previous history of the land. The mean yields of the three common rotations are shown below for the four years during which data are available from both experiments.

*Mean yields, 1936–7 to 1939–40*

|  | Three-course experiment | | Combined experiment | |
|---|---|---|---|---|
|  | Yield kantars per feddan | % of CDF | Yield kantars per feddan | % of CDF |
| CLF | 6·38 | 111 | 4·21 | 121 |
| CFF | 5·94 | 103 | 3·94 | 113 |
| CDF | 5·75 | 100 | 3·49 | 100 |

In the three-course experiment, the increases in this period were 11 and 3%, the corresponding increases for the whole period of 13 years being 11 and 6%. In the combined rotations, the percentage increases were about double these figures.

This is explained by the history of the sites. The site of the 'Combined' experiment sustained an annual crop of rain-grown dura until, in 1930, it became part of the Gezira Research Farm. The early years of the experiment therefore measure the rate of recovery of the land after continuous dura. The land of the 'Three-course' was first irrigated in 1917 and had largely recovered from such exhaustion. The 'Three-course' experiment is thus more representative of the Gezira scheme as a whole.

Though it has been established that, in general, an extensive fallow should intervene between successive cotton crops, other practical aspects must be considered. Dura provides grain for human food and straw for fodder. Moreover, a long fallow period in a scheme of artificial irrigation entails overhead charges, whether the soil is being irrigated or not. Thus it is important to consider how dura and lubia may be included in the rotation so as to interfere least with the cotton crop.

If dura alone is included, the cotton yields show least disturbance where the dura follows immediately upon cotton, i.e. in the next rain period, with one or more years' fallow before the next cotton crop. Naturally the yield of dura suffers by its following cotton, and in practice the Sudan Plantations Syndicate has found it necessary to have a year's fallow between cotton and dura to ensure a good yield of the food crop. The dura yields in the Combined Rotations Experiment abundantly confirm their experience.

The results for the effect of lubia alone on the yields of cotton are not yet conclusive. If lubia immediately follows cotton or dura, there are indications of some slight benefit as compared with fallow, though further results are necessary before these indications can be definitely established or refuted. On the other hand, lubia after fallow appears to nullify the beneficial effects of the fallow, as demonstrated strikingly in the CFFF, CFFL comparison. It may be that when fallows have built up a considerable supply of available nitrogen, the lubia may add less nitrogen to the soil than when it is grown on land exhausted by cotton or dura. Probably too the lubia is somewhat penalized by a detail in the experimental technique. On small plots the produce is consumed by cattle in less than a day and some of the nitrogen is transferred in droppings to land outside the experiment. A more likely explanation of the comparative failure of lubia to enrich the soil is that any benefit from its nitrogen fixation is more than offset by other soil changes in the year preceding cotton. Weed growth in the rains one year before cotton is sown reduce the moisture and nitrate contents of the soil during the dry season and seriously restrict the nitrogen supply and growth of the cotton (Crowther, 1942). The lubia crop seems to act in much the same way as non-leguminous weeds in spite of its nitrogen fixation.

Lubia may show to better advantage when associated with dura in the rotation, for more nitrogen is required to offset the dura requirements and the fallow period is less. Here information is scanty for lubia occurs immediately after dura in only one of the twelve rotations examined. In this, CDL, lubia immediately precedes cotton and fallow is entirely excluded. A four-course rotation is essential if cotton, dura and lubia are all to be included.*

So far, examination of the relative merits of the rotations has been confined to a consideration of cotton yields averaged over all years. Examination of the average changes in yield per year indicates that most of the differences between rotations will become larger in future years, though in the 'three-course' experiment this conclusion is at present doubtful because of the peculiar succession of seasons. The next few seasons should provide a crucial test of this and other results which are not yet established.

The preponderating influence of the season is clearly shown in Figs. 3 and 4. Indeed, in the three-course experiment, all yields improved markedly in recent years, irrespective of the rotation. Further, the better rotations appear to have little effect in maintaining yields in poor years, showing least gains over CDD in these years. There is the interesting

* A rotation practised on part of the Gezira Research Farm is three-course with both dura and lubia in the second year. Dura is sown in June–July and harvested in November. Lubia is winter sown. This rotation is impracticable on a large scale as it is unrelated to available water supply.

suggestion that cotton planted immediately after lubia holds an advantage over cotton after fallow in unfavourable seasons; but this effect, even if confirmed in future years, is numerically small.

While the order of the rotations has not changed much from year to year, there are statistically significant fluctuations from season to season. These require further investigation.

## Conclusions

Frequent fallows are the first necessity in any rotation for cotton in the Sudan Gezira. Cotton should not be sown more often than once in three years, and even that may be too frequent for maximum yields. Between cotton crops at least a year's fallow is essential and the longer the fallow the greater the benefit. Thus experimental results support fully the soundness of present rotation throughout the Gezira scheme. Up till 1932–3 the rotation was three-yearly with at least one full year's fallow per cycle. Then the rotation was changed to four-yearly with two or three years' fallow per cycle.

Inclusion of dura in the rotation invariably reduces cotton yields. Least harm is done when dura immediately follows cotton, with at least one fallow year before cotton recurs. Lubia was in no case markedly superior to fallow, but may prove slightly more beneficial than fallow when included in a short rotation following cotton or dura. On the other hand, if grown after fallow, lubia decreases the cotton yields.

In view of the large increases regularly obtained from nitrogenous fertilizers, the benefit of lubia is surprisingly small and that of fallow surprisingly great.

By contrast, fallowing on irrigated land in Egypt is rarely justified for a spell longer than two or three months, the period necessary for cultivation. The high value of the land there together with the expenses of canal digging and maintenance render fallowing uneconomical. Fortunately in the Gezira land is plentiful and cheap. If in the future its value rises and cropping has to be more intensive, lubia will replace much of the fallow; meanwhile fallowing is a simple way of controlling weed growth and allowing recovery of the land without expenses of cultivation and supervision.

## REFERENCES

Bluen, L. (1931). *Empire Cotton Grow. Rev.* **8.**
Crowther, E. M. & Crowther, F. (1935). *Proc. Roy. Soc.* B, **118**, 343–70.
Crowther, F. (1934). *Ann. Bot., Lond.,* **48**, 877–913.
Crowther, F. (1942). *Empire J. Exp. Agric.* (in the Press).
Greene, H. (1928 *a*). *J. Agric. Sci.* **18**, 518–30.
Greene, H. (1928 *b*). *J. Agric. Sci.* **18**, 531–43.
Greene, H. (1942). In the Press.
Greene, H. & Snow, O. W. (1939). *J. Agric. Sci.* **29**, 1–34.
Gregory, F. G., Crowther, F. & Lambert, A. R. (1932). *J. Agric. Sci.* **22**, 617–38.
Joseph, A. F. (1925). *J. Agric. Sci.* **15**, 407–419.
Norris, P. K. (1934). *U.S.A. Bur. Agric. Econ.* F.S. **62.**
Yates, F. (1937). *Tech. Comm. Imp. Bur. Soil Sci.* no. 35, pp. 75–6.

(*Received* 25 *April* 1941)

W. G. Cochran: **Some Developments in Statistics:**- (1) *Collection of data.*—Fisher (*Design of Experiments*, Section 66) aptly summarizes one of the most notable changes in the statistician's attitude towards his craft.

"During the period in which highly inefficient methods of estimation were commonly employed, and indeed strongly advocated by the most influential authorities, it was natural that a great deal of ingenuity should be devoted, in each type of problem as it arose, to the invention of methods of estimation, with the idea always latent, though seldom clearly expressed, of making these as accurate as possible. The attainment of a result of high accuracy was, in fact, evidence not only of the intrinsic value of the data examined, but also to some extent of the skill with which it had been treated."

In contrast, the effect of recent improvements in statistical methodology is stated as follows:

"Clearly, in any subject in which the statistical methods ordinarily employed leave little to be desired, the precision of the result obtained will depend almost entirely on the value of the data on which it is based, and it is useless to commend the statistician, if this is great, or to reproach him if it is small."

For a time there grew up amongst certain research workers the legend of the statistician as a person whose primary equipment was a copious receptacle for waste paper, to which the fruit of years of work was consigned with the pronouncement: "These data are entirely worthless". On the contrary, as a consequence of progress in methods of estimation, a steadily increasing amount of attention has been devoted in recent years to the technique of collecting data, either by controlled experiments or by extensive sampling.

(2) *Design of experiments.*—Thus, when his opinion on a proposed experiment is invited, the statistician considers whether the experiment will furnish unbiased estimates of the effects of the treatments and unbiased estimates of the experimental errors to which the treatment responses are subject. He must also inquire whether more accurate estimates of the responses might result from some change in the conduct of the experiment. During the past 20 years these questions have led to a study of the precautions which are necessary to ensure freedom from bias. Fisher stressed the importance of an element of randomization in the experimental plan. After much discussion amongst statisticians and experimenters alike, the consensus of thought seems to be that although occasionally inconvenient in practice, randomization provides a useful safeguard against many types of bias of which the experimenter might be unaware.

Methods for the reduction of experimental errors have also been investigated, particularly for errors that arise from heterogeneity in the experimental units (*e.g.*, plots of land, greenhouse pots, etc.) to which the treatments are applied. When the number of treatments is small, it is often possible to group the experimental units so that all units within the same replicate are similar in their anticipated responses. This principle forms the basis

Reproduced with permission from *Chronica Botanica,* Vol. 7, pages 383–386, 1943.

of the familiar method of randomized blocks. Recent work has concentrated mainly on experiments where the number of treatments is too large to permit the formation of homogeneous replications. In a series of designs developed by YATES (Jour. Agric. Sci. 26: 424, 1936) the treatments are arranged in groups which are smaller (usually much smaller) than a complete replication. When the appropriate statistical analysis is used, only variations within groups need contribute to the experimental errors. Insofar as the groups are more homogeneous than complete replications, more accurate comparisons of the treatments are obtained.

The designs fall into several broad types: (1) *lattices* and *balanced incomplete blocks*, devised for experiments where all treatment comparisons are of equal importance to the investigator; (2) *split-plot* designs and *confounded factorial* designs, where accuracy is sacrificed on certain treatment comparisons, either for practical convenience or to obtain the greatest accuracy for those comparisons which are of special interest; (3) *lattice* squares and *Youden* squares, which are extensions of the Latin square, available for use when the number of treatments is large.

The new designs have been used most extensively in plant breeding, where large numbers of varieties are usually tested in the early stages of the program. Corn, wheat, oats, soybeans, sugar-beets, clover, tomatoes, Kentucky bluegrass and forest nursery seedlings are among the crops represented. The increases in accuracy over randomized blocks naturally vary both with the crop and with the heterogeneity of the soil. In the great plains of the United States, for example, results to date indicate that gains are seldom substantial unless the number of varieties tested exceeds 50. In eastern states the reduction of block size has sometimes proved profitable with as few as 9 varieties.

Before a new design is introduced, the inventor can form some judgment as to its accuracy relative to existing designs from analyses on *uniformity data*, that is, data collected from experiments where no variation in treatment is introduced. Practical experience is the ultimate test. The profusion of new designs, often untried in practice, makes it difficult for the statistician to decide in many cases what design should be recommended. In this difficulty experimenters are co-operative and frequently offer to try an unfamiliar layout in order to see how it will perform. Fortunately, with most of the new designs, the accuracy relative to randomized blocks can be estimated from the results of any experiment in which the new design has been used. Thus each experiment contributes to the store of experience which will determine the field of application of the new methods.

(3) *Sampling techniques.*—Work in sampling is directed towards the same goals—freedom from bias and maximum accuracy per unit cost. While no revolutionary techniques have appeared, steady progress has been made in the exploitation of several devices. The choice of sampling-unit is important for its effect both on the accuracy of the sample and on the ease with which the sample can be taken. Subdivision of the population into more homogeneous parts (sometimes called *stratification* or *local control*) has become a standard practice. Its object is essentially to ensure a better distribution of the sample over the various parts of the population.

Where the measurements are expensive, costs can sometimes be reduced materially by the use of supplementary information on other variables which are correlated with the measurements (NEYMAN, Jour. Amer. Stat. Ass. 33: 101, 1938). For instance, in order to estimate the volume of timber in certain forest areas, a number of sample-plots were accurately measured. In addition an eye-estimate was made of the volume for *every* sample-plot in the population. The regression of the sample-plot measurements on the eye-estimates was calculated from those plots which were measured. The regression was then used to adjust the average volume of timber in the sampled plots to an average for the mean of the whole population (COCHRAN, Jour. Amer. Stat. Ass. 34:492, 1939). The technique might be used to advantage in many biological investigations.

By an ingenious application of the analysis of variance, it is often possible to estimate, from the results of a sample, the standard of accuracy which would have been attained had a different method of sampling been employed. Since the statistician is seldom permitted the luxury of taking a sample for the sole purpose of finding out something about sampling technique, this device has contributed much towards the improvement of sampling methods.

(4) *Statistical analysis.* — Of the great mass of work in the field of statistical *analysis*, perhaps the most interesting is that which deals with the simultaneous analysis of several measurements (FISHER, Ann. Eug. 7: 179, 1936). From a number of measurements on the human body, MAHALANOBIS and his collaborators in India constructed a single index which could be used for the comparison of different races and obtained its sampling distribution. Independent work by HOTELLING and FISHER reached the same problem and extended the technique in several ways. The mathematical problems involved in the theory are by no means fully solved and much additional research may be expected in the near future. The index, called a *discriminant function* by FISHER, will serve either to differentiate among several groups or to assign numerical scores to a series of qualitative grades. Within the space of nine years, applications have been made in anthropology, taxonomy, psychology, plant breeding, forestry, bacteriology, human genetics and economics.

To mention one application, FISHER (*loc. cit.*) used measurements on the petal length, petal width, sepal length and sepal width of two species of *Iris* for the development of an index which would distinguish whether a plant belonged to one or the other species. He tested also whether the measurements on a third species were in agreement with the hypothesis that this species is a polyploid hybrid of the first two species.

Work in the general theory of tests of significance has served to emphasize the vital rôle played by the *alternative hypotheses*. Suppose that we are presented with a batch of data which, according to the null hypothesis, are normally distributed with zero mean and unit variance. The criteria that are employed to test this hypothesis depend entirely on the alternative hypotheses which we are willing to entertain. If these postulate that the data may not be normal, tests of departure from normality will be applied. If normality is taken for granted in any acceptable explanation of the nature of the data, the test may be a t-test of the supposition that the mean is zero, or a chi-square test of the hypothesis of unit variance. This point is perhaps not liable to be overlooked by biologists, who are generally well aware of the various hypotheses under consideration.

From the mathematical point of view the general problem in developing tests of significance may be stated as follows: given a certain null hypothesis and a set of alternative hypotheses, can a routine procedure be discovered which leads automatically to the best test of significance? The problem naturally pre-supposes agreement as to what is meant by a "best" test. NEYMAN and PEARSON (Phil. Trans. Royal Soc. Lond. A. 231: 289, 1933) constructed an acceptable definition and a method which produces the best test, if one exists. They showed, however, that no best test exists unless the alternative hypothesis is of a very simple type. In most cases we must evidently be content with something short of perfection. Current work is concerned with the production of techniques which yield satisfactory tests when the alternative hypotheses are complex.

A number of useful additions have been made to the available tests of significance. The test for the homogeneity of a group of estimated variances (BARTLETT, Suppl. Jour. Royal Stat. Soc. 4: 158, 1937) is already well known. The z- or F-test has been extended to the case where the variances are correlated (PITMAN and MORGAN, Biometrika 31: 9, 1939). A t-test of the difference between two group means, when the groups have unequal variances, was developed independently by BEHRENS and FISHER. A table of the 5 percent levels is given by SUKHATME (Sankhyā 4: 39, 1938). This test, which involves an extension of the notion of a fiducial distribution, has provoked considerable discussion.

With the increasingly widespread use of the analysis of variance, investigations continue to appear on the validity of the z-test when the data are not normal. The verdict from a fairly substantial body of evidence is that the test is remarkably insensitive to deviations from normality. The t-tests which usually follow the z-test are more vulnerable. In particular, if the error variance is heterogeneous, the use of the pooled variance for individual t-tests may result in serious bias. Remedies which have been proposed are (i) subdivision of the error variance into homogeneous components and (ii) transformation of the data before analysis to a scale on which the variance is more homogeneous. For small numbers and percentages, as are frequently encountered in biological work, the square-root transformation (BARTLETT, Suppl. Jour. Royal Stat. Soc. 3: 68, 1936) and the angular transformation (FISHER and YATES, Statistical Tables for Agricultural, Biological and Medical Research. Edinburgh, Oliver and Boyd, 1938) are often successful. BEALE (Biometrika 32: 243, 1942) has recently suggested an inverse hyperbolic sine transformation for data where the variance is a quadratic function of the mean.

In the development of new techniques to the point where they can be used with ease by the non-mathematical research worker, an essential step is the provision of numerical tables as these become necessary. The collection of tables prepared by FISHER and YATES (*loc. cit.*) will be found exceedingly helpful by all who apply modern techniques. The laborious task of providing a comprehensive set of significance levels for the variance-ratio (F) distribution has been completed by a group of workers. A recent publication (Biometrika 32, 1942) gives the 50, 25, 10, 5, 2.5, 1 and 0.5% levels, to which may be added the previously available tables of the 20, 5, 1, and 0.1% levels. Two tables of the range have appeared recently: one gives the frequency distribution (PEARSON and HARTLEY, Biometrika 32: 301, 1942) and the other the 5 and 1% significance levels of the ratio of the range to an independent estimate of the standard deviation (NEWMAN, Biometrika 31: 20, 1939).

MOLINA (New York, D. Van Nostrand Co., 1942) has published an extensive set of tables for the Poisson distribution. Adequate tables for the binomial distribution are still lacking.

(5) *Current attitudes.*—With the rapid growth of applications of statistical techniques, specialization within the field has increased. It is felt by some that the gap between the applied statisticians and the mathematical statisticians hinders the progress of the subject, though the example cited from discriminant function analysis illustrates the rapid dissemination of a technique whose development involves difficult mathematics. The appearance in the United States of a journal comparable to Biometrika or to the Supplement to the Journal of the Royal Statistical Society would stimulate materially the applications in the field of biometry.

Considerable dissatisfaction has also been expressed with the current arrangements for the teaching of statistics. The chief complaints are that very few universities provide adequate training either for the applied or theoretical statistician; that two courses within the same university may differ by 10 or 20 years in their outlook; that the subject is too often assigned to a lecturer with no permanent interest in the field and in this connection that there are no generally approved professional qualifications for statistical posts. As far as teaching is concerned no easy solution is in sight. The core of the difficulty is that within a university we find small groups of economists, botanists, psychologists, agronomists and so on, each interested mainly in applications towards his own field and none able to devote much time for the mastery of the subject.

The opportunity of mentioning a few impressions of the consulting statistician is too good to be missed. There is still misapprehension amongst research workers about the kinds of question that the statistician should be expected to answer. Of this misapprehension a good example is the not uncommon query: "Here are my data. What can I test?" Statistical techniques are too often regarded as a set of rules designed to save the experimenter the trouble of thinking about his data—an attitude for which teachers and text-books are largely to blame. On the other hand, there is increased realization that statistical advice should be sought before the data have been collected. And it is a continual pleasure to observe how some experimenters, after brief training and some preliminary advice, advance to a confident and shrewd manipulation of their statistical equipment.

STATISTICAL LABORATORY,
IOWA STATE COLLEGE,
AMES, IOWA.                    *Spring, 1943.*

# THE COMPARISON OF DIFFERENT SCALES OF MEASUREMENT FOR EXPERIMENTAL RESULTS[1,2]

## By W. G. Cochran

### *Iowa State College*

**1. Introduction.** In some fields of research, the development of a satisfactory method for measuring the effects of experimental treatments constitutes a difficult problem. The estimation of the vitamin content of preparations of foods furnishes a good example; for most of the vitamins several years of work were required to construct a reliable method of assay. In other cases, where the ideal method for measuring treatment responses is costly or troublesome, a search may be made for a more convenient substitute. Thus in pasture or forage-crop experiments the species composition of a plot may be estimated by eye inspection as a substitute for a complete botanical separation. As a third example we may quote experiments in cookery, where the flavor and quality of the dishes are subject to the whims of human taste. Frequently a panel of judges is employed, each of whom scores the dishes independently. It is not easy to determine how the panel should be chosen, nor how representative its verdicts are of consumer preferences in general.

When such problems are investigated, experiments may be carried out specifically for the purpose of comparing two or more methods or *scales* of measurement. Where the process of measurement affects only the final stages of the experiment, as in the last two examples quoted above, all that is necessary is to score the *same* experiment by the various scales under consideration. In comparing two different methods of assaying vitamins, on the other hand, independent experiments are frequently required, the only common feature being that the same set of treatments is tested in both experiments.

In the interpretation of the results of such experiments, two types of comparison are of general interest. One concerns the *relations* between the scales. It may be summed up rather loosely in the question: Are the effects of the treatments the same in all scales? For a more exact formulation, consider the case of two scales, which is probably the most frequent in practice. Let $\xi_{1t}$, $\xi_{2t}$ be the true means of the $t$th treatment as measured on the two scales. We may wish to examine the following hypotheses:

(i) *Scales equivalent:*

$$(1) \qquad \xi_{1t} = \xi_{2t}, \qquad \text{(all } t);$$

(ii) *Scales equivalent, apart from a constant difference:*

$$(2) \qquad \xi_{1t} = \xi_{2t} + \epsilon, \qquad \text{(all } t);$$

---

[1] Paper presented at a meeting of the Institute of Mathematical Statistics, Washington, D. C., June 18, 1943.

[2] Journal Paper No. J-1136 of the Iowa Agricultural Experiment Station, Ames, Iowa. Project 514.

(iii) *Scales linearly related:*

$$(3) \qquad\qquad \alpha\xi_{1t} + \beta\xi_{2t} = \gamma, \qquad\qquad \text{(all } t\text{)};$$

(iv) *Relation monotonic, but not linear:*

$$(4) \qquad\qquad \xi_{1t} = f(\xi_{2t}, \alpha, \beta, \cdots), \qquad\qquad \text{(all } t\text{)};$$

where the function is strictly monotonic.

In this case the two scales are mutually consistent in that they place any set of treatments in the same order. The ratio of a treatment difference in one scale to the corresponding difference in the other scale is, however, not constant.

(v) *Relation not monotonic:* Here the scales do not place the treatments in the same order and consequently are not satisfactory substitutes for each other.

The second question concerns the relative *accuracy* or *sensitivity* of the two scales. For practical purposes this question may be put as follows: how many replications are required with the second scale to attain the accuracy given by $r$ replications with the first scale? It is clear that the answer depends both on the experimental errors associated with the scales and on the magnitudes of the treatment effects in the two scales. For example, Coward [1] reports that in the assay of vitamin D, male rats give a higher experimental error than females, yet provide a more accurate assay because they are more responsive. The relative accuracy may be different in different parts of the two scales. This is likely to happen whenever the relation between the scales is of type (iv) above.

This paper gives a preliminary discussion of some of the simpler questions raised above, to which recent work in multivariate analysis is applicable. A complete solution for small sample work appears to demand considerable further development in the distribution theory of multivariate analysis.

The discussion is confined to the case in which all scales measure the same experiment. The case where each scale requires a separate experiment may be expected to be somewhat simpler, but cannot conveniently be treated as a special case of the procedure for a single experiment.

**2. Assumptions.** Let $x_1, x_2, \cdots x_p$ denote measurements on the $p$ scales and let $n_1$ and $n_2$ be the numbers of degrees of freedom for treatments and error respectively. The experimental data furnish a joint analysis of variance and covariance of the $p$ variates as follows:

|  |  | d.f. | Sum of squares or products |
|---|---|---|---|
|  | Mean.................. | 1 | $m_{ij}$ |
| (5) | Treatments............. | $n_1$ | $a_{ij}$ |
|  | Error.................. | $n_2$ | $b_{ij}$ |

It will be assumed that $x_1, \cdots, x_p$ follow a multivariate normal distribution, and that for any pair of variates $x_i, x_j$ the error mean covariance $\sigma_{ij}$ is constant throughout the experiment (though it may vary as $i$ and $j$ vary). Thus the quantities $b_{ij}$ follow the standard joint distribution, Wishart [16], of sums of squares and products while the quantities $m_{ij}$ and $a_{ij}$ follow the corresponding non-central distributions and the three sets of distributions are independent.

**3. Tests for equivalence.** If there are only two scales, a test for equivalence is obtained from elementary techniques. An analysis of variance similar to (5) is computed on the *differences* between the two scales for every observation. If equations (1) hold in the population, the sums of squares for the Mean, Treatments and Error are distributed independently as $\chi^2(\sigma_{11} + \sigma_{22} - 2\sigma_{12})$. The pooled mean square for the Mean and Treatments may therefore be compared with the Error mean square in a variance-ratio test, the degrees of freedom being $(n_1 + 1)$ and $n_2$. If the scales are equivalent apart from a constant difference, the same result is valid for Treatments and Error, while the mean square for the Mean is proportional to a non-central $\chi^2$. Thus separate $z$- or $F$-tests on the Mean and Treatments assist in distinguishing between hypotheses (1) and (2).

**4. More than two scales.** Let $\xi_{it}$ be the true mean of the $t$th treatment as measured on the $i$th scale. The first two hypotheses may now be written respectively:

(1')
$$\xi_{it} = \xi_t$$

(2')
$$\xi_{it} = \xi_t + \epsilon_i$$

for $i = 1, 2, \cdots, p$. The quantities $\epsilon_i$, whose sum may be assumed zero, measure the constant differences among the scales.

If the interactions of all components with Scales are computed, the analysis of variance extends formally, with the following separation of degrees of freedom:

|  |  | d.f. |
|---|---|---|
| | Mean × Scales | $(p - 1)$ |
| (6) | Treatments × Scales | $n_1(p - 1)$ |
| | Error | $n_2(p - 1)$ |

The three lines in the analysis play the same roles as before in relation to hypotheses (1') and (2'). When $p > 2$, however, it may be shown that the three sums of squares are not distributed as multiples of $\chi^2$ unless (i) all scales have the same error variance and (ii) every pair of scales has the same correlation coefficient. Where these conditions are reasonably well satisfied, as happens possibly when experienced judges employ a similar scoring system, the above analysis supplies approximate tests. But with scales which differ widely in their experimental errors or in their degrees of interrcorrelation, the validity of variance-ratio tests is open to more serious question.

In order to obtain an exact test, we may note that hypothesis (1') is closely related to the Wilks-Lawley hypothesis (Wilks [15], Lawley [9], Hsu [7]) that the means of $k$ populations are all equal. If each treatment denotes a separate population, the Wilks-Lawley hypothesis states that

(7)
$$\xi_{it} = \xi_i \qquad (t = 1, 2, \cdots, n_1 + 1).$$

Since this differs from (1') only in the interchange of the letters $i$ and $t$, it is clear that the two hypotheses may be subjected to the same kind of test.

For the details of the procedure we first divide the $(p-1)$ comparisons among scales into $(p-1)$ single comparisons by the introduction of a set of variates $y_i$, $(i = 1, 2, \cdots, p-1)$.

$$(8) \qquad\qquad y_i = \sum_{j=1}^{p} \lambda_{ij} x_j .$$

Any set of $y$'s may be chosen, provided that they are linearly independent and that

$$(9) \qquad\qquad \sum_{j=1}^{p} \lambda_{ij} = 0, \qquad\qquad (i = 1, 2, \cdots (p-1)).$$

Thus with three scales we might use $y_1 = x_1 - x_2$, $y_2 = x_1 - x_3$ or $y_1 = 2x_1 - x_2 - x_3$, $y_2 = x_2 - x_3$.

The next step is to compute an analysis of variance and covariance of the $y$ variates, as follows:

|      |                         | d.f.  | Sum of squares or products |
|------|-------------------------|-------|----------------------------|
|      | Mean................... | 1     | $m'_{ij}$                  |
| (10) | Treatments............. | $n_1$ | $a'_{ij}$                  |
|      | Error.................. | $n_2$ | $b'_{ij}$                  |

If hypothesis (1') holds, it follows from (9) that the three sets of quantities $m'_{ij}$, $a'_{ij}$ and $b'_{ij}$ all follow the standard joint distribution for sums of squares and products. Hence Wilks' test (Wilks [15], Pearson and Wilks [11], Hsu [7]), for the equality of the means of $k$ populations may be applied. For a single test of hypothesis (1') we may use

$$(11) \qquad\qquad W = \frac{|b'_{ij}|}{|b'_{ij} + m'_{ij} + a'_{ij}|} .$$

As before, if $W$ is significant we may test whether the deviation is due to constant differences or to other types of difference among the scales by calculating

$$(12) \qquad\qquad W_m = \frac{|b'_{ij}|}{|b'_{ij} + m'_{ij}|} ,$$

and

$$(13) \qquad\qquad W_t = \frac{|b'_{ij}|}{|b'_{ij} + a'_{ij}|} .$$

The flexibility of analysis of variance tests is not sacrificed; in particular we may test any desired subgroup of the treatments or of the scales. When there are only two scales the tests reduce to those given in section 3.

The tests are invariant under homogeneous linear transformations of the $y$'s which explains why the form of the subdivision of the scale comparisons is immaterial. In fact for purposes of computation it is not necessary to introduce the $y$'s. By taking a simple transformation and expressing $a'_{ij}$ in terms of $a_{ij}$, etc., we may express $W$ directly in terms of the $x$'s, as follows:

$$(14) \qquad W = \frac{\sum_{ij} B_{ij}}{\sum_{ij} (B + M + A)_{ij}},$$

where $B_{ij}$, $(B + M + A)_{ij}$ are respectively the co-factors of the matrices $(b_{ij})$, $(b_{ij} + m_{ij} + a_{ij})$. Analogous expressions hold for $W_m$ and $W_t$. In practice it will often be preferable to compute the $y$'s in order that particular comparisons among the scale variates may be examined in detail.

The form of the frequency distribution has been worked out by Wilks [15]. For small values of $n_1$ and $p$, the test of significance can be referred to the recent tables of the significance levels of the incomplete Beta-function, Thompson [13], or to variance-ratio tables. Such cases are listed below, from Wilks [15] and Hsu [7]. In our notation, $\nu_1$ is taken as $(n_1 + 1)$ in equation (11), as 1 in equation (12) and as $n_1$ in equation (13).

$\underline{p = 3, \nu_1 > 1}$ : $f(W) \propto W^{\frac{1}{2}(n_2-3)}(1 - W^{\frac{1}{2}})^{\nu_1-1}$

$$: F\{2\nu_1, 2(n_2 - 1)\} = \frac{(n_2 - 1)(1 - W^{\frac{1}{2}})}{\nu_1 W^{\frac{1}{2}}},$$

$\underline{\nu_1 = 1}$ : $\qquad f(W) \propto W^{\frac{1}{2}(n_2-p)}(1 - W)^{\frac{1}{2}(p-3)}$

$$: F\{p - 1, n_2 - p\} = \frac{(n_2 - p)(1 - W)}{(p - 1)W}.$$

This distribution applies to all tests made on the Mean, equation (12), and all cases where a single degree of freedom is isolated from the treatment comparisons.

$\underline{\nu_1 = 2}$ : $f(W) \propto W^{\frac{1}{2}(n_2-p)}(1 - W^{\frac{1}{2}})^{p-2}$

$$: F\{2(p - 1), 2(n_2 - p + 2)\} = \frac{(n_2 - p + 2)(1 - W^{\frac{1}{2}})}{(p - 1)W^{\frac{1}{2}}}.$$

A tabulation of the distributions for four and five scales would be useful. Hsu [7] has shown that as $n_2$ becomes large, the distribution of $-n_2 \log W$ tends to that of $\chi^2$ with $\nu_1(p - 1)$ degrees of freedom. In general, this approximation does not agree very well with the exact distributions above unless $n_2$ exceeds 60.

**5. Interpretation as a problem in canonical correlations.** As an introduction to the methods that will be used in testing the hypothesis of linearity, we may note that hypotheses (1') and (2') can be described in terms of canonical correlations. Fisher [5] has pointed out that the roots $\theta$ of the equation

$$(15) \qquad |a_{ij} - \theta(a_{ij} + b_{ij})| = 0,$$

are the squares of the *sample* canonical correlations between the $x$-variates and a set of $n_1$ dummy variates which represent the $n_1$ degrees of freedom among treatments. In order to obtain the corresponding equation for the *population* correlations, we may suppose that $n_1$ and $p$ remain constant while the number of replicates $r'$ and consequently $n_2$ increase without limit. After the removal of a common factor $r'$, equation (15) becomes

$$(16) \qquad | \psi_{ij} - \rho^2(\psi_{ij} + \nu\sigma_{ij}) | = 0,$$

where

$$(17) \qquad \psi_{ij} = \sum_{t=1}^{n_1+1} (\xi_{it} - \bar{\xi}_i)(\xi_{jt} - \bar{\xi}_j).$$

The value of the coefficient $\nu$ depends on the type of experimental design. For a randomized block layout, $\nu = n_1$ and for a simple group comparison $\nu = (n_1 + 1)$.

Now if hypothesis (2') is true, i.e., $\xi_{it} = \xi_t + \epsilon_i$, it follows that $\psi_{ij}$ is independent of $i$ and $j$. In this event equation (16) has $(p - 1)$ roots $\rho^2$ which are identically zero. The remaining root corresponds to the best discriminant function, Fisher [5], and does not vanish unless the treatments have no effects on any of the $x$-variates.

Let $\Sigma\beta_i x_i$ be a population canonical variate for the scale variables. The coefficients $\beta_i$ satisfy the equations

$$(18) \qquad \sum_j \beta_j \{\psi_{ij} - \rho^2(\psi_{ij} + \nu\sigma_{ij})\} = 0. \qquad i = 1, \cdots p.$$

For a zero root $\rho^2 = 0$ we have $\psi_{ij} = $ constant. Hence if a zero root is substituted, equation (18) degenerates into

$$(19) \qquad \beta_1 + \beta_2 + \cdots + \beta_p = 0.$$

To summarize, hypothesis (2') specifies that (i) $(p - 1)$ of the population canonical correlations vanish and (ii) any variate $\Sigma\beta_i x_i$ is a canonical scale variate corresponding to a zero root, provided that equation (19) is satisfied. Analogous results hold for hypothesis (1'); in this case we replace the Treatments line of the analysis of variance by the (Treatments + Mean) line.

**6. Test for linear relationship—two scales.** We may assume $n_1 \geq 2$; otherwise no test of linearity is possible. If the values of $\alpha$, $\beta$ and $\gamma$ in equations (3) are known, the problem can be reduced to that of testing hypothesis (1) or (2). Since this case is unlikely to be encountered frequently in practice, further details are omitted.

When $\alpha$, $\beta$ and $\gamma$ are unknown, we may theoretically replace the variates $x_1$ and $x_2$ by $v_1 = \alpha x_1 + \beta x_2$ and $v_2 = \mu_1 x_1 + \mu_2 x_2$, where $\mu_1$ and $\mu_2$ are chosen so that $v_1$ and $v_2$ are independently distributed. If hypothesis (3) holds, it follows from (17) that in terms of the $v$'s, $\psi_{11} = \psi_{12} = 0$. Since in addition $\sigma_{12} = 0$, the two roots of equation (16) are

$$(20) \qquad \rho^2 = 0 \quad \text{and} \quad \rho^2 = \psi_{22}/(\psi_{22} + \nu\sigma_{22}).$$

Thus hypothesis (3) implies that one of the population canonical correlations vanishes. Unlike the previous case, however, we cannot construct the corresponding canonical variate, which requires knowledge of $\alpha$ and $\beta$.

The selection of a sample test criterion opens up some difficulties. Pending further elucidation of the problem, the natural choice seems to be the square $r_2^2$ of the lower sample canonical correlation, or the equivalent quantity $h_2 = r_2^2/(1 - r_2^2)$, where $h_2$ is the lower root of the equation:

$$(21) \qquad | a_{ij} - hb_{ij} | = 0.$$

It appears likely, however, that $r_2^2$ and $h_2$ are not sufficient estimates of the corresponding population parameters.

When $n_2$ is large, Hsu [8] has shown that the distribution of $n_2h_2$ tends to that of $\chi^2$ with $(n_1 - 1)$ degrees of freedom. A considerable advance towards the small-sample distribution is obtainable from Madow [10], who developed an expression for the exact distribution of $r_1^2$ and $r_2^2$ when one of the population correlations is different from zero. In our notation this result, which is an important generalization of the distribution found by Fisher [5] and Girshick [6] may be written as follows:

$$\frac{(n_1 + n_2 - 2)!}{4\pi(n_1 - 2)!\,(n_2 - 2)!}\,(r_1^2 r_2^2)^{\frac{n_1-3}{2}}\,\{(1 - r_1^2)(1 - r_2^2)\}^{\frac{n_2-3}{2}}\,(r_1^2 - r_2^2)\,dr_1^2\,dr_2^2$$

$$(22)$$

$$\times\,(1 - \rho_1^2)^{\frac{1}{2}(n_1+n_2)} \int_{r_2^2}^{r_1^2} \frac{F\left(\dfrac{n_1 + n_2}{2},\, \dfrac{n_1 + n_2}{2},\, \dfrac{n_1}{2},\, \rho_1^2 y\right) dy}{\sqrt{(r_1^2 - y)(y - r_2^2)}},$$

where $\rho_1$ is the non-vanishing population correlation. It is evident from the form of (22) that the distribution of $r_2^2$ or $h_2$ involves $\rho_1$. The conditional distribution of $h_2/h_1$ may be relatively insensitive to changes in $\rho_1$, though even this distribution does not seem entirely independent of $\rho_1$.

When $\rho_1$ is unity, the small-sample distribution of $h_2$ is that of the ratio of two independent sums of squares, i.e., $h_2 = (n_1 - 1)e^{2z}/n_2$, with $(n_1 - 1)$ and $n_2$ degrees of freedom. This result is a particular case of a more general result proved in section 8. From (20) it is seen that $\rho_1$ is close to unity when $\psi_{22}$ is large relative to $\sigma_{22}$, i.e., when the real differences among the treatments are large relative to the experimental errors. In the absence of a usable exact solution, the $F$-distribution may be a better approximation than the large-sample distribution of $h_2$ for data where $r_1$ is found to be close to unity, though proof of this statement is not yet available.

If it is desired to test hypothesis (3) with the additional assumption that $\gamma = 0$, we replace $a_{ij}$ by $(a_{ij} + m_{ij})$ in equation (21) for $h_2$, and $n_1$ by $(n_1 + 1)$ in the distribution theory.

## 7. Connection with the method of least squares.
The previous approach has an interesting connection with the method of least squares. We are required to test the linearity of relationship between $(n_1 + 1)$ pairs of means $(\bar{x}_{1t}, \bar{x}_{2t})$.

Both variates are subject to error and the errors are correlated; with $r'$ replications the population variances and covariance of these means are $\sigma_{11}/r'$, $\sigma_{22}/r'$ and $\sigma_{12}/r'$. For these unknown quantities we have sample estimates $b_{11}/n_2r'$, $b_{22}/n_2r'$ and $b_{12}/n_2r'$ respectively, derived from the Error line of the analysis of variance.

The procedure suggested by the method of least squares is to estimate the parameters of the line and use the deviations of the points $(\bar{x}_{1t}, \bar{x}_{2t})$ from the line for a test of linearity. If the population variances were known, the unknown quantities $\alpha$, $\beta$, $\gamma$ and $\xi_{it}$ would be estimated by minimizing the quadratic form:

$$(23) \quad \sigma^{11} \sum_{t=1}^{n_1+1} r'(\bar{x}_{1t} - \xi_{1t})^2 + 2\sigma^{12} \sum_{t=1}^{n_1+1} r'(\bar{x}_{1t} - \xi_{1t})(\bar{x}_{2t} - \xi_{2t}) + \sigma^{22} \sum_{t=1}^{n_1+1} r'(\bar{x}_{2t} - \xi_{2t})^2,$$

subject to the linear relations (3). Here $(\sigma^{ij})$ is the matrix inverse to $\sigma_{ij}$. On substitution of the estimates, expression (23), which is positive definite, would serve as a "sum of squares" of deviations from the line and therefore as a test criterion. This criterion is of course a direct generalization of the weighted sum of squares which is used when the errors are independent.

Van Uven [14] gave an elegant method by which the sum of squares of deviations can be found directly, before solving for any of the unknown quantities In our notation he showed that the sum of squares of deviations is the smaller root $H_2$ of the equation

$$(24) \qquad\qquad\qquad | a_{ij} - H\sigma_{ij} | = 0,$$

where $a_{ij}$ is as before the treatments sum of squares or products.

Suppose that in default of knowledge of the $\sigma_{ij}$ we derive the weights from the sample estimates $b_{ij}/n_2$; i.e., we minimize (23) with $b^{ij}$ in place of $\sigma^{ij}$, where $(b^{ij}) = (b_{ij}/n_2)^{-1}$. In this case the method of Van Uven shows that the sum of squares of deviations from the best-fitting line is the smaller root $H_2'$ of the equation

$$(25) \qquad\qquad\qquad \left| a_{ij} - \frac{H'}{n_2} b_{ij} \right| = 0.$$

Comparing (25) with (21) we find $H_2' = n_2h_2$. Consequently the least squares approach, with sample weights substituted in (23) for the unknown true weights, leads to $h_2$ as a test criterion. Further, Hsu's [8] proof that the distribution of $n_2h_2$ tends to $\chi^2$ with $(n_1 - 1)$ degrees of freedom establishes for this case the standard least-squares result for the distribution of the residual sum of squares: —namely that when the population weights are known, the residual sum of squares is distributed as $\chi^2$, with degrees of freedom equal to the number of points, $2(n_1 + 1)$, *minus* the number of independent unknowns, $(n_1 + 3)$. By a transformation of the $x$-variates to independent variables, this result can be obtained alternatively from a theorem by Deming [2].

**8. Test for linear relationship—more than two scales.** The extension of hypothesis (3) to the case of $p$ scales can be expressed by means of the equations

$$(3') \qquad \alpha_i \xi_{1t} + \beta_i \xi_{it} = \gamma_i : \qquad (i = 2, \cdots p)(t = 1, \cdots n_1 + 1).$$

The equations, $(p-1)(n_1+1)$ in number, postulate a linear relation between $x_1$ and every other variate and consequently imply a linear relation between any pair of variates $x_i$ and $x_j$.

Consider the variates $v_i = \alpha_i x_{1t} + \beta_i x_{it}$, $(i = 2, \cdots p)$. For $v_1$ we choose the linear function of the $x$'s which is independent of $v_2, \cdots v_p$. Thus in equation (16) for the population canonical correlations we have $\psi_{ij} = 0$, $(i, j, \geq 2)$ and $\sigma_{1j} = 0$, $(j > 1)$. It follows that all roots of equation (16) are zero except one, the non-vanishing root being $\rho^2 = \psi_{11}/(\psi_{11} + \nu \sigma_{11})$. If each treatment denotes a separate population, hypothesis (3') is therefore identical with Fisher's hypothesis [4], that the populations are *collinear*.

As a test criterion for this hypothesis Fisher has suggested the sum of the roots of equation (21), excluding the highest root, i.e., $V' = \Sigma h_i = \Sigma r_i^2/(1 - r_i^2)$. If $n_1 \geq p$ the sum extends over $(p-1)$ roots, while if $n_1 < p$ the sum extends over $(n_1 - 1)$ roots. For computational purposes it may be more expeditious to form this sum by subtraction. Hsu [7] has pointed out that the sum of all roots is given by $V = \sum_{ij} b^{ij} a_{ij}$, which is obtained readily when the inverse of $(b_{ij})$ has been calculated. The largest root of (21) is then found and subtracted from $V$.

Fisher [4] also suggested that when equations (3') hold, the distribution of $V'$ is approximately that of $\chi^2$ with $(p-1)(n_1-1)$ degrees of freedom. This result has been confirmed by Hsu [8] as the limiting form of the $V'$ distribution when $n_2$ tends to infinity. As in the case of two scales, the small-sample distribution is as yet unknown; it presumably contains $\rho_1$, the non-vanishing correlation, as a nuisance parameter.

Some progress towards the small-sample distribution can be made without difficulty in the case where $\rho_1 = 1$. For then $v_1$ must have a zero Error sum of squares in every sample from the population, i.e., $v_1$ is constant within any given treatment. Consequently (i) $b_{1i} = 0$ for $i = 1, \cdots p$, and (ii) $a_{1j}^2/a_{11}$ is a single degree of freedom from the Treatments sum of squares of $v_j$. On account of conditions (i), equation (21) reduces to

$$(26) \qquad \begin{vmatrix} a_{11} & a_{12} & \cdots & a_{1p} \\ a_{12} & a_{22} - hb_{22} & \cdots & a_{2p} - hb_{2p} \\ \cdots & \cdots & \cdots & \cdots \\ a_{1p} & a_{2p} - hb_{2p} & \cdots & a_{pp} - hb_{pp} \end{vmatrix} = 0.$$

Subtract $a_{1i}/a_{11}$ times the first row from the $i$th row, for $i = 2, \cdots p$. We see that one root is infinite; the rest are the roots of the equation

$$(27) \qquad |a_{ij}'' - hb_{ij}| = 0, \qquad i, j = 2, \cdots p,$$

where $a_{ij}'' = a_{ij} - a_{1i}a_{1j}/a_{11}$.

If hypothesis (3′) holds, the quantities $a''_{ij}$ follow the Wishart distribution [16] with $(n_1 - 1)$ degrees of freedom. Hence the joint distribution of $h_2, \cdots h_p$ or $h_{n_1}$, is that which is obtained when all the population canonical correlations vanish, with $(n_1 - 1)$ in place of $n_1$. For $n_1 \geq p$, the distribution function (apart from the constant term) is:

$$(28) \qquad \prod_{i=2}^{p} \left[ h_i^{\frac{1}{2}(n_1-p-1)} (1 + h_i)^{-\frac{1}{2}(n_1+n_2-1)} \left\{ \prod_{j=i+1}^{p} (h_i - h_j) \right\} \right].$$

For two scales, $(p = 2)$, we reach the result mentioned in section 6, that $V' = h^2$ is distributed as $(n_1 - 1)e^{2z}/n_2$. This result can also be obtained directly from (27). When $p = 3$, the distribution of $V'$ is obtainable from a result by Hsu [7].

**9. Measures of relative sensitivity.** We propose to discuss briefly the estimation of the relative sensitivity of two scales and to indicate the types of distribution that are involved. If there are only two treatments, $t$, $t'$, an appropriate definition of the true sensitivity of the $i$th scale is

$$(29) \qquad \frac{(\xi_{it'} - \xi_{it})^2}{2\sigma_{ii}},$$

or some simple function of this quantity. In justification, we may observe that for a fixed number of replicates, the power function of the $t$-test in the $i$th scale depends entirely on this quantity. An unbiased sample estimate is

$$(30) \qquad \frac{(n_2 - 2)(\bar{x}_{it'} - \bar{x}_{it})^2}{2b_{ii}} - \frac{1}{r'},$$

where $r'$ is the number of replicates. Since (30) involves a non-central variance ratio, confidence limits for the true sensitivity can be found from Fisher's Type C distribution, Fisher [3].

It follows from (3) and (29) that if two scales are linearly related (including the case of equivalence) their relative sensitivity is constant for all treatment comparisons. For scale 1 relative to scale 2 the sensitivity is measured by $\beta^2\sigma_{22}/\alpha^2\sigma_{11}$.

If the scales are equivalent, apart possibly from a constant difference, this quantity reduces to $\varphi = \sigma_{22}/\sigma_{11}$, for which $F = b_{22}/b_{11}$ serves as a sample estimate. A test of significance of the sample ratio and confidence limits for the true ratio may be obtained from Pitman [12], who showed that

$$(31) \qquad \left(\frac{F}{\varphi} - 1\right) \Big/ \sqrt{\left(\frac{F}{\varphi} + 1\right)^2 - \frac{4r_{12}^2 F}{\varphi}},$$

follows the distribution of a sample correlation coefficient from $(n_2 + 1)$ pairs of observations. In (31), $r_{12}^2 = b_{12}^2/b_{11}b_{22}$. The same procedure may be used whenever $\alpha$ and $\beta$ are known.

When $\alpha$ and $\beta$ are unknown, a sample estimate of the relative sensitivity is $b^2b_{22}/a^2b_{11}$, where $(ax_1 + bx_2)$ is the discriminant function which corresponds to the lower root of equation (21). We have not been able to reach the distribution of this estimate. Confidence limits for the relative sensitivity can, however, be obtained when $n_2$ is sufficiently large so that $\sigma_{11}$ and $\sigma_{22}$ may be assumed known. For in that case the problem reduces to that of finding confidence limits for $\beta^2/\alpha^2$. Now if $\alpha$, $\beta$ are the true coefficients, the quantity

$$(32) \qquad \frac{\alpha^2 a_{11} + 2\alpha\beta a_{12} + \beta^2 a_{22}}{\alpha^2 b_{11} + 2\alpha\beta b_{12} + \beta^2 b_{22}},$$

follows the $n_1 e^{2z}/n_2$ distribution. Any proposed values of $\alpha$ and $\beta$ which make (32) significant are rejected by the evidence of the sample. By equating (32) to the desired significance level of $n_1 e^{2z}/n_2$, we get a quadratic equation for the two limits of $\beta/\alpha$. The limits will not be narrow unless the treatment effects are large.

If the relation between the scales is non-linear, and the assumption of a constant error variance throughout an individual scale is valid, the relative sensitivity differs for different treatment comparisons. Even in this event estimates of relative sensitivity may be of interest. Attention might be restricted to a single degree of freedom from the treatment comparisons, in which case the definition for two treatments could be applied.

Alternatively an estimate might be wanted of the *average* relative sensitivity over all treatment comparisons. For a given number of replicates, the power function of the variance-ratio test of the treatment effects in the $i$th scale depends only on the quantity

$$(33) \qquad \frac{\sum\limits_{t} (\xi_{it} - \bar{\xi}_i)^2}{\sigma_{ii}}.$$

Consequently this quantity, which is an extension of (29), might be chosen as a measure of average sensitivity. The corresponding generalization of the unbiased sample estimate (20) is

$$(34) \qquad \frac{(n_2 - 2)a_{ii}}{n_1 r' b_{ii}} - \frac{1}{r'}.$$

Since the quantity $a_{ii}/b_{ii}$ is a multiple of a non-central variance ratio, the comparison of two scales involves a test of significance of the hypothesis that two non-central variance ratios are equal.

**10. Summary.** This paper discusses the analysis of data obtained when the results of a replicated experiment are measured on several different scales which we wish to compare. Recent work in multivariate analysis provides tests of the hypothesis that the treatment effects are the same in all scales, and of the hypothesis that the scales are linearly related. When the number of Error degrees of freedom is large, the significance levels of these tests are obtainable from the standard tables. For small sample tests, further investigation and tabulation of certain distributions will be needed, particularly that of the sample canonical correlations when one population correlation differs from zero.

A brief discussion is given of methods for comparing the relative sensitivity of two scales.

## REFERENCES

[1] K. H. COWARD, *The Biological Standardization of the Vitamins*, 1937.
[2] W. E. DEMING, "The chi-test and curve-fitting," *Jour. Amer. Stat. Assn.*, Vol. 29 (1934), pp. 372–382.
[3] R. A. FISHER, "The general sampling distribution of the multiple correlation coefficient," *Proc. Roy. Soc. A*, Vol. 121 (1928), pp. 654–673.
[4] R. A. FISHER, "The statistical utilization of multiple measurements," *Annals of Eugenics*, Vol. 8 (1938), pp. 376–386.
[5] R. A. FISHER, "The sampling distribution of some statistics obtained from non-linear equations," *Annals of Eugenics*, Vol. 9 (1939), pp. 238–249.
[6] M. A. GIRSHICK, "On the sampling theory of the roots of determinental equations," *Annals of Math. Stat.*, Vol. 10 (1939), pp. 203–224.
[7] P. L. HSU, "On generalized analysis of variance (I)," *Biometrika*, Vol. 31 (1940), pp. 221–237.
[8] P. L. HSU, "The problem of rank and the limiting distribution of Fisher's test function," *Annals of Eugenics*, Vol. 11 (1941), pp. 39–41.
[9] D. N. LAWLEY, "A generalization of Fisher's z-test," *Biometrika*, Vol. 30 (1938), pp. 180–187.
[10] W. G. MADOW, "Contributions to the theory of multivariate statistical analysis," *Trans. Amer. Math. Soc.*, Vol. 44 (1938), p. 490.
[11] E. S. PEARSON and S. S. WILKS, "Methods of statistical analysis appropriate for $k$ samples of two variables," *Biometrika*, Vol. 25, (1933), pp. 353–378.
[12] E. J. G. PITMAN, "A note on normal correlation," *Biometrika*, Vol. 31 (1939), pp. 9–12.
[13] C. M. THOMPSON, "Tables of percentage points of the incomplete Beta-function," *Biometrika*, Vol. 32 (1942), pp. 151–181.
[14] M. J. VAN UVEN, "Adjustment of $N$ points (in $n$-dimensional space) to the best linear $(n - 1)$ dimensional space," *Proc. Koninklizke Akad. van Wetenschappen te Amsterdam*, Vol. 33 (1930), pp. 143–157.
[15] S. S. WILKS, "Certain generalizations in the analysis of variance," *Biometrika*, Vol. 24 (1932), pp. 471–494.
[16] J. WISHART, "The generalized product moment distribution in samples from a normal multivariate population," *Biometrika*, Vol. 20A (1928), pp. 32–52.

# ANALYSIS OF VARIANCE FOR PERCENTAGES BASED ON UNEQUAL NUMBERS*

By W. G. Cochran
*Iowa State College*

SEVERAL WRITERS have discussed the use of the analysis of variance when the data are expressed as fractions or percentages. In the case where all percentages are based on the same total count (i.e., the same demoninator), the techniques which have been developed appear to be adequate for most practical purposes. According to the nature of the variations present in the data, these techniques take the form of (i) a straightforward analysis of variance of the percentages, (ii) an analysis of variance of *angles* into which the percentages have been transformed, [Bliss (1938), Cochran (1938), Clark and Leonard (1939)] or (iii) the angular transformation with some further refinements [Bartlett (1936) (1937), Cochran (1940)].

The case in which the percentages are based on different total numbers is more troublesome. As will appear later, there is a greater variety of possible methods of analysis, so that much time may be consumed in trying to decide what is a suitable method. Further, with certain types of data the efficient methods are rather tedious. The object of this discussion is to suggest approximate preliminary tests which are helpful in the choice of a method of analysis that is reasonably efficient and not unnecessarily laborious.

## NATURE OF THE VARIATION IN PERCENTAGE DATA

For convenience in this discussion fractions will be used instead of percentages. Let $n_i$ be the total count on which an observed fraction or percentage is based, and let $f_i = a_i/n_i$ be the observed *fraction*. If $p_i$ is the true fraction for this observation, the variance of $f_i$ about $p_i$ is given by the usual expression for the binomial distribution: $p_i q_i/n_i$, where $q_i = (1 - p_i)$.

In addition, the true fraction $p_i$ will frequently be found to vary from observation to observation. Where present, such variation may contribute to the experimental error on which $z$, $F$, or $t$-tests are made. Thus for the total variance of an observed fraction we may write

$$V(f_i) = \frac{p_i q_i}{n_i} + \sigma_i^2 \qquad (1)$$

* Journal paper No. J-1115 of the Iowa Agricultural Experiment Station, Ames, Iowa. Project No. 514.

where the first term represents the *binomial* variation and the second term the *extraneous* variation.

It is not clear what assumptions may best be made about $\sigma_i{}^2$. Probably no single set of assumptions is appropriate for all types of data. Since, however, $\sigma_i{}^2$ represents the variance of the *true* fraction, $\sigma_i{}^2$ will not in general depend on $n_i$. It may depend on $p_i$, for more variation would be expected when the fractions are near $\frac{1}{2}$ than when they approach 0 or 1.

Subject to the assumption, implicit in all uses of the analysis of variance with fractions, that the fractions are approximately normally distributed, equation (1) suggests that the observations should be *weighted*, where the true weight $W_i$ is

$$W_i = \frac{1}{\dfrac{p_i q_i}{n_i} + \sigma_i{}^2} = \frac{n_i}{p_i q_i + n_i \sigma_i{}^2}. \tag{2}$$

When the extraneous variation is small, the weight assigned to any fraction is $n_i/p_i q_i$; i.e., it increases directly as the total $n_i$ from which the fraction is derived. At the other extreme, when the values of $\sigma_i{}^2$ are large relative to $p_i q_i$, the weight of an observation is $1/\sigma_i{}^2$, i.e., independent of $n_i$.

The practical procedure which follows from this discussion is to estimate first the relative amounts of binomial and extraneous variation in the data and the nature of the extraneous variation. From this information an appropriate set of weights can be constructed for the subsequent analysis. But unless the data are extensive or unless *a priori* knowledge is available, it may be expected that the information for the calculation of weights will not be very accurate. Moreover the computation of weights is time-consuming. For these reasons it is worthwhile to consider the adequacy of some simple methods of weighting.

### THE EFFICIENCIES OF BINOMIAL AND EQUAL WEIGHTING

Two methods to be investigated are (i) weighting proportional to the $n_i$ (ii) weighting independent of the $n_i$. As we have seen, these methods are efficient when the ratio of the binomial to the total variation is 1 and 0 respectively.

For simplicity we will consider the estimation of the mean of a group of $s$ observations, such as the mean for one treatment over a number of replications. If weights $w_i$ are used where the true weights are $W_i$, the variance of the weighted mean is known to be

$$\sum_{i=1}^{s} \left( \frac{w_i{}^2}{W_i} \right) \bigg/ \left[ \sum_{i=1}^{s} (w_i) \right]^2. \tag{3}$$

Since the variance of a correctly-weighted mean is $1/\Sigma^s_{i-1}(W_i)$, the efficiency of the actual mean is

$$\left[\sum_{i-1}^s (w_i)\right]^2 \bigg/ \left[\sum_{i-1}^s W_i\right] \left[\sum_{i-1}^s \left(\frac{w_i^2}{W_i}\right)\right]. \qquad (4)$$

In order to obtain data which can be visualized, a number of sets of values of $n_i$ were chosen, in some of which the $n_i$ were approximately equal while in others they varied greatly. It was assumed that $(p_i q_i)$ and $\sigma_i^2$ were constant for all observations. This assumption is not considered unduly restrictive since the mean of a *single* treatment is being

EFFICIENCIES OF BINOMIAL, EQUAL AND PARTIAL WEIGHTS

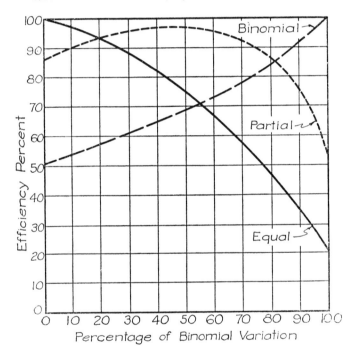

considered. The assumption might fail if, for instance, a treatment in a randomized blocks experiment gave 95 per cent in some blocks and 40 per cent in others.

With these assumptions we may take $w_i = n_i$ (binomial weighting) in method (i) and $w_i = 1$ (equal weighting) in method (ii). For each of the selected sets of $n_i$, the efficiencies of the two weightings were calculated by formula (4). These efficiencies naturally depend on the ratio of the binomial to the total variation. Since the average binomial variance is

$pq/\bar{n}_h$, where $\bar{n}_h$ is the *harmonic* mean of the $n_i$, the ratio of the binomial to the total variance was taken as

$$u = \frac{pq}{\bar{n}_h} \bigg/ \left( \frac{pq}{\bar{n}_h} + \sigma^2 \right). \tag{5}$$

Each efficiency was computed for a series of values of $u$, ranging from 0 to 1.

As would be expected, the efficiencies of both methods were always close to unity when the $n_i$ varied little. In this case either weighting is satisfactory whatever the relative amount of extraneous variation, equal weighting being preferable on account of its simplicity. For a case of extreme variation in the $n_i$, the efficiencies are shown graphically in the Chart. The third curve (partial weighting) will be explained later. The values of $n_i$ were 1, 2, 6, 8, 10, 20, 30, 40, 50 and 80, exhibiting a range which is seldom equalled or exceeded in experimental data.

In the Chart binomial weighting is seen to be more accurate when the binomial variation in the data exceeds about 55 per cent and is less accurate otherwise. Binomial weighting at its worst $(u=0)$ is superior to equal weighting at its worst $(u=1)$. Since this result is always true, binomial weighting would appear to be a better method for routine use, there being less chance of getting a very inefficient mean than with equal weights. Neither method is however satisfactory in the mid-range when the binomial variation lies between 30 and 80 per cent of the total variation.

It should be remembered that the example represents a very extreme case. For examples which are perhaps more typical, in which $n_i$ varied over a much less extreme range than used for the illustration shown in the Chart, the efficiency of binomial weighting appears to exceed 80 per cent throughout the range, while that of equal weighting may fall to about 60 per cent. Nevertheless there is need for a better method when about half the variation is binomial and the $n_i$ are far from constant.

### PARTIAL WEIGHTING

In the numerical examples which were worked, examinations of the correct weights as given by formula (2) revealed that when about half the variance is extraneous, the weights tend to change slowly for the higher values of $n_i$. This feature is illustrated in Table I, where the correct weights are compared with the values of $n_i$ in the example of the Chart in the case in which half the variation is extraneous.

All $W_i$ were multiplied by a common factor to make the true and binomial weight the same (unity) for the first observation. It will be

TABLE I
BINOMIAL WEIGHTS ($n_i$) AND CORRECT WEIGHTS WHEN $u = 0.5$

| $n_i$ | 1 | 2 | 6 | 8 | 10 | 20 | 30 | 40 | 50 | 80 |
|-------|---|---|---|---|----|----|----|----|----|----|
| $W_i$ | 1 | 1.71 | 3.26 | 3.67 | 3.98 | 4.76 | 5.10 | 5.29 | 5.41 | 5.59 |

noted that the increase in $W_i$ for a fixed increase in $n_i$ becomes steadily and rapidly smaller as $n_i$ increases. Thus when $n_i$ increases from 1 to 2, $W_i$ increases from 1 to 1.71; yet when $n_i$ increases from 50 to 80, $W_i$ increases only from 5.41 to 5.59. In fact, the extraneous variation imposes a ceiling on the weight assigned to any observation, no matter how high $n_i$ is.

In these circumstances a simple method of approximating the correct weights is to assign a fixed upper limit to the weights. Below that limit the $n_i$ are chosen as weights. This device, which may be called *partial* weighting, was previously suggested [Cochran (1937), Yates and Cochran (1938)] for an analogous problem. At first a weighting was tried in which the upper limit was chosen as the average of the $n_i$ (24.7). With this rule the values of $w_i$ are 1, 2, 6, 8, 10, 20, 25, 25, 25, 25 respectively. The examples indicated, however, that this weighting reaches its highest efficiency when there is between 80 and 90 per cent binomial variation. The best simple convention for $u$ between 30 and 80 per cent is to assign a constant weight to the upper two-thirds of the $n_i$. For the example cited these weights would be 1, 2, 6, 8, 8, 8, 8, 8, 8, 8. In computation this procedure has the additional advantage of giving fewer unequal weights to deal with.

The efficiency of this type of partial weighting is shown by the dotted curve in the Chart. When $u$ lies between 20 and 80 per cent, partial weighting is superior to both binomial and equal weighting. Its efficiency is above 90 per cent for any value of $u$ between 10 and 70 per cent.

Partial weighting has an element of arbitrariness in the choice of the upper limit. The choice appears to make relatively little difference if somewhere between one-half and two-thirds of the observations are given the same weight. The latter rule is preferable for general use.

### PRELIMINARY INVESTIGATION OF DATA TO BE ANALYZED

We now discuss certain preliminary calculations which assist in deciding how a batch of data is to be analyzed. The first question to be considered is: are the variations in the totals $n_i$ sufficiently small so that they can be ignored? Sometimes the answer, one way or the other, is obvious on inspection; but frequently with moderate variation in the totals it is doubtful whether the extra work involved in a weighted analysis is justified.

If variations in the $n_i$ are ignored when they should be taken into account as in formula (2), some loss of efficiency results (and in addition some disturbance of the significance levels for $z$, $F$, or $t$-tests). From the preceding discussion it is clear that this loss cannot properly be estimated without knowledge of the relative amounts of binomial and extraneous variation in the data.

An upper limit to the loss can however be assigned from the values of the $n_i$. For it may be shown from formula (2) that the loss is greatest when all the variation is binomial. (The Chart provides an illustration of this result.) Consequently an upper limit to the loss is obtained by calculating the efficiency obtained from equal weights on the assumption that all the variation is binomial.

The purpose of the analysis of variance is presumed to be to estimate and test certain group or treatment averages. As might be expected, the potential loss of efficiency from equal weights depends on the type of classification present in the data (e.g., single grouping, two-fold classification or multiple classification.) Only the simplest and most common types—the single grouping and the two-fold classification—will be considered here.

For the same data the loss differs also from one type of treatment comparison to another. An *average* figure may be obtained by supposing that the true treatment means are distributed about the general mean with variance $\sigma_t^2$ and that this variance is the quantity to be estimated

*Single grouping.* With $r$ treatments and $s$ observed fractions for each treatment, the totals $n_{ij}$ on which the fractions are based may be set out as in Table II.

TABLE II

| | Treatments | | |
|---|---|---|---|
| 1 | 2 | $\cdots$ | $r$ |
| $n_{11}$ | $n_{21}$ | $\cdots$ | $n_{r1}$ |
| $n_{12}$ | $n_{22}$ | $\cdots$ | $n_{r2}$ |
| $\cdots$ | $\cdots$ | $\cdots$ | $\cdots$ |
| $n_{1s}$ | $n_{2s}$ | $\cdots$ | $n_{rs}$ |
| Totals  $N_1.$ | $N_2.$ | $\cdots$ $N_r.$ | $N$ |

When reduced to a comparable basis, the expected values of the treatments mean square for the two types of analysis work as follows:

$$\text{Binomial weights}: pq + \frac{1}{(r-1)} \left[ N - \frac{N_1.^2 + \cdots + N_r.^2}{N} \right] \sigma_t^2. \quad (6)$$

The coefficient of $\sigma_t^2$ is always slightly less than the mean of the treatment totals $N_i.$ (unless these totals are identical). If the number of treatments exceeds 10, or if the $N_i.$ vary little, the mean of the $N_i.$ may be used in (6) as a sufficiently good approximation.

*Equal weights:*
$$pq + s\bar{n}_h\sigma_t^2 \tag{7}$$

where $\bar{n}_h$ is the harmonic mean of the $n_{ij}$, given by the equation

$$\frac{1}{\bar{n}_h} = \frac{1}{rs}\left(\frac{1}{n_{11}} + \cdots + \frac{1}{n_{rs}}\right). \tag{8}$$

For the estimation of $\sigma_t^2$ the lower limit to the efficiency of equal weighting is therefore

$$s(r-1)\bar{n}_h \Big/ \left[N - \frac{N_1.^2 + \cdots + N_r.^2}{N}\right]. \tag{9}$$

If this ratio exceeds 0.9 it may be concluded that variations in the $n_{ij}$ can be neglected unless the data require the best possible analysis.

*Equal weights within treatments.* It sometimes happens that the totals $n_{ij}$ are substantially constant within each treatment, but change considerably from one treatment to another. In this case, although a completely equal weighting is possibly unsatisfactory, an analysis in which each observation is weighted by the average $n_{ij}$ for that treatment may be highly efficient. The lower limit to the efficiency of this type of weighting relative to binomial weighting may be shown to be

$$(r-1)s^2/\lambda \tag{10}$$

where $\lambda$ is the rather complex expression

$$\lambda = \left[\left(N_1. - \frac{N_1.^2}{N}\right)\left(\frac{1}{n_{11}} + \cdots + \frac{1}{n_{1s}}\right) + \cdots\right.$$
$$\left. + \left(N_r. - \frac{N_r.^2}{N}\right)\left(\frac{1}{n_{r1}} + \cdots + \frac{1}{n_{rs}}\right)\right]. \tag{11}$$

A close approximation to (10) which is more quickly calculated is

$$rs^2/\mu \tag{12}$$

where

$$\mu = \left[N_1.\left(\frac{1}{n_{11}} + \cdots + \frac{1}{n_{1s}}\right) + \cdots \right.$$
$$\left. + N_r.\left(\frac{1}{n_{r1}} + \cdots + \frac{1}{n_{rs}}\right)\right]. \tag{13}$$

TABLE III

NUMBER OF SERVICES (n), NUMBER OF CONCEPTIONS (a) AND PERCENTAGE
OF CONCEPTIONS TO SERVICES IN ARTIFICIAL INSEMINATION TESTS

| Sample | Bull | | | | | | | | | | | | | | | | | |
|---|---|---|---|---|---|---|---|---|---|---|---|---|---|---|---|---|---|---|
| | 1 | | | 2 | | | 3 | | | 4 | | | 5 | | | 6 | | |
| | $n$ | $a$ | % | $n$ | $a$ | % | $n$ | $a$ | % | $n$ | $a$ | % | $n$ | $a$ | % | $n$ | $a$ | % |
| 1 | 13 | 6 | 46 | 10 | 7 | 70 | 23 | 12 | 52 | 3 | 1 | 33 | 12 | 5 | 42 | 17 | 6 | 35 |
| 2 | 13 | 4 | 31 | 8 | 3 | 38 | 16 | 7 | 44 | 15 | 7 | 47 | 11 | 7 | 64 | 22 | 15 | 68 |
| 3 | 19 | 7 | 37 | 9 | 4 | 44 | 14 | 8 | 57 | 9 | 3 | 33 | 16 | 8 | 50 | 39 | 23 | 59 |
| 4 | 6 | 3 | 50 | 8 | 5 | 62 | 15 | 6 | 40 | 14 | 3 | 21 | 16 | 11 | 69 | 16 | 6 | 38 |
| 5 | 13 | 8 | 62 | 1 | 1 | 100 | 6 | 5 | 83 | 10 | 7 | 70 | 13 | 10 | 77 | 21 | 12 | 57 |
| 6 | 9 | 3 | 33 | 2 | 0 | 0 | 15 | 10 | 67 | 5 | 4 | 80 | 16 | 13 | 81 | 7 | 6 | 86 |
| 7 | 2 | 0 | 0 | 5 | 1 | 20 | 3 | 1 | 33 | 4 | 3 | 75 | 15 | 13 | 87 | 17 | 13 | 76 |
| 8 | 9 | 5 | 56 | 1 | 0 | 0 | 6 | 3 | 50 | 8 | 4 | 50 | 8 | 5 | 62 | 14 | 8 | 57 |
| 9 | 10 | 3 | 30 | 8 | 5 | 62 | 14 | 9 | 64 | 24 | 11 | 46 | 8 | 6 | 75 | 34 | 10 | 29 |
| 10 | 7 | 3 | 43 | 17 | 10 | 59 | 23 | 16 | 70 | 21 | 3 | 14 | 2 | 2 | 100 | 30 | 18 | 60 |
| Total | 101 | 42 | 42 | 69 | 36 | 52 | 135 | 77 | 57 | 113 | 46 | 41 | 117 | 80 | 68 | 217 | 117 | 54 |
| $\Sigma$(rec.) | 1.4151 | | | 3.3449 | | | 1.0924 | | | 1.3468 | | | 1.2553 | | | 0.5758 | | |

Grand totals: $(n)$, 752; $(a)$, 398; (rec.), 9.0303

## NUMERICAL EXAMPLE

The practical use of these formulae may be illustrated by a numerical example.[1] In this investigation a number of semen samples were taken from each of a number of bulls. With each sample one or more cows were inseminated artificially. The data in Table III show the number of inseminations $(n_{ij})$ and the number of conceptions $(a_{ij})$ for every sample. The fraction $a_{ij}/n_{ij}$ measures the success of the sample in artificial insemination and is the variable to be studied. As a preliminary step in the investigation it is desired to test the differences among bulls. The original data contain the results for 21 samples for each of 17 bulls. For illustration 6 bulls were chosen, and for each bull 10 of the 21 samples were selected at random.

The numbers of services fluctuate widely both within and among bulls. If nearly all the variation is binomial, neither equal weighting nor equal weighting within treatments appears satisfactory at first sight, though the latter would be expected to be somewhat more efficient than the former.

The first step in applying the preliminary tests is to record the sum of the reciprocals of the $n_{ij}$ for each treatment (i.e. bull), as shown below the treatment totals (1.4151 etc.). With the aid of a table of

[1] I am indebted to Dr. G. W. Salisbury, Department of Animal Husbandry, Cornell University, for permission to use these data.

reciprocals these figures can be summed directly on the calculating machine. Their total over all treatments is 9.0303.

Three auxiliary quantities are now calculated as follows:

(i)
$$N - (N_1.^2 + \cdots + N_r.^2)/N$$
$$= 752 - [(101)^2 + \cdots + (217)^2]/752 = 610.07$$

(ii)          $\bar{n}_h = 60/(9.0303)$                    $= 6.644$

where $60 = rs$ is the total number of samples tested.

(iii)   $\mu = 101 \times 1.4151 + 69 \times 3.3449 + \cdots + 217 \times 0.5758 = 945.2$.

The last quantity is required for the approximate test in formula (12) of equal weights within groups and need not be calculated if the treatment totals $N_i.$ are all approximately the same.

From formula (9), the efficiency for equal weights is

$$10 \times 5 \times 6.644/610.07 = .544.$$

From formula (12), the efficiency for equal weights within groups is approximately

$$6 \times 100/945.2 = .635.$$

The exact value, by the more laborious formula (10) is found to be 0.628.

Thus if the variation within bulls is entirely binomial, the analyses by equal weights and by equal weights within treatments are equivalent to discarding nearly 50 and 40 per cent of the information respectively, a procedure which is unjustified for anything but a crude examination. This result does not however dispose of the question of weighting. Since the samples from the same bull were tested on different cows at different times and in different places, there is no *a priori* reason to suppose that the probability of securing a conception is constant for all samples from a single bull. The estimation of the percentage of extraneous variation, which requires an analysis of variance of the actual percentages of conceptions, constitutes the next step.

### ESTIMATION OF THE RELATIVE AMOUNTS OF BINOMIAL AND EXTRANEOUS VARIATION

Either equal weights or binomial weights may be used. If on inspection most of the variation appears to be extraneous the former are advisable, while if most of the variation appears to be binomial the latter are preferable unless the weighted analysis is considered too complicated. The reasons for these choices are: (i) theoretically we expect a better estimate when the weighting used is closer to the correct weighting and (ii) if the preliminary guess turns out to be approximately correct, the preliminary analysis may be satisfactory for tests of significance. In this case no further analysis will be needed.

When extraneous variation is present, the error mean square is composed of binomial variation, extraneous variation and random fluctuations. For a simple group comparison, the expected value of the error mean square of the *unweighted* fractions may be taken as

$$E(s^2) = \frac{pq}{\bar{n}_h} + \sigma_e{}^2 \tag{14}$$

where $p$ is the average of the true fractions and $\sigma_e{}^2$ is the average extraneous variance. Actually this result is not quite correct, since the expectation depends on the unknown values of the true fractions $p_{ij}$ for each observation. In equation (14) the average fraction $p$ has been substituted for these unknowns. The result should be sufficiently accurate for its purpose.

Thus if $s^2$ is the error mean square, the ratio of the binomial to the total variance is estimated as

$$\bar{f}(1 - \bar{f})/s^2 \bar{n}_h, \tag{15}$$

where $\bar{f}$ is the mean observed fraction.

If a *weighted* analysis is used, with weights $n_{ij}$, the error mean square $s_w{}^2$ is an estimate of

$$pq + \frac{\sigma_e{}^2}{r(s-1)} \left[ N - \frac{(n_{11}{}^2 + \cdots + n_{1s}{}^2)}{N_1.} - \cdots \right.$$
$$\left. - \frac{(n_{r1}{}^2 + \cdots + n_{rs}{}^2)}{N_r.} \right]. \tag{16}$$

The coefficient of $\sigma_e{}^2$ is slightly smaller than the arithmetic mean $\bar{n}$ of the $n_{ij}$. Unless the number of observations or replications per treatment is less than 4, it is usually sufficiently accurate to use $\bar{n}$ in (16), writing

$$E(s_w{}^2) = pq + \bar{n}\sigma_e{}^2. \tag{17}$$

Since the average binomial variance has been defined as $pq/\bar{n}_h$, the estimates of the two components of variance are:

$$Binomial: \; \frac{\bar{f}(1 - \bar{f})}{\bar{n}_h} \; ; \qquad Extraneous: \; \frac{s_w{}^2 - \bar{f}(1 - \bar{f})}{\bar{n}} \cdot \tag{18}$$

As a rough working rule it is suggested that if less than 30 per cent of the variation is binomial, equal weights may be employed for the analysis on which tests of significance are based. Binomial weighting is satisfactory if more than 80 per cent of the variation is binomial. In the intervening range partial weights may be used; or if the data are sufficiently extensive and important to warrant the additional work, the actual weights $W_i$ in formula (2) may be estimated from the computations which have been carried out.

It should however be remembered that unless the number of degrees of freedom for error is large, the estimate of the percentage of binomial to total variation is itself subject to a large sampling fluctuation. Even if $p$ were known exactly, the estimated percentage of binomial variation would at best be distributed as $eB/\chi^2$, where $B$ is the true percentage of binomial variation, $e$ is the number of error degrees of freedom, and $\chi^2$ has $e$ degrees of freedom. For example, if a batch of data shows 50 per cent of binomial variation and the error has 20 degrees of freedom, the 95 per cent lower and upper limits to the true percentage are at least as far apart as 27 per cent and 78 per cent. With 30 degrees of freedom for error, this range narrows to 30–73 per cent. It may be concluded that with the moderate numbers of degrees of freedom which are common in experimental work, attempts to estimate the true weights $W_i$ are scarcely justified, though the technique probably does permit the discrimination suggested between equal weights, binomial weights and partial weights.

### APPLICATION TO THE NUMERICAL EXAMPLE

In order to form an initial judgment about the extent of extraneous variation, the percentages of conceptions in Table III were inspected. Although the percentages vary widely for the same bull when the number of inseminations is small, those percentages which are based on larger numbers of inseminations seem fairly stable. It was concluded as a first guess that the extraneous variation is small relative to the binomial variation. A *weighted* analysis was therefore made.

This analysis can be carried out on the fractions by the familiar "(total)²/(number)" rule. For the sums of squares, we have

$$\text{Total:} \qquad \frac{6^2}{13} + \frac{4^2}{13} + \cdots + \frac{18^2}{30} - \frac{398^2}{752} = 24.058$$

$$\text{Among bulls:} \qquad \frac{42^2}{101} + \cdots + \frac{117^2}{217} - \frac{398^2}{752} = 6.033.$$

The error (i.e., within bulls) mean square is 0.334 with 54 degrees of freedom.

The additional figures required are:

$$\bar{n} = 752/60 = 12.53 : \bar{n}_h = 6.644 : \bar{f} = 398/752 = .5293$$

$$Binomial\ variance = \bar{f}(1 - \bar{f})/\bar{n}_h = .0375$$

$$Extraneous\ variance = [s_w{}^2 - \bar{f}(1 - \bar{f})]/\bar{n} = .0068.$$

The ratio of binomial to total variance is estimated as 375/443, or about 85 per cent. From this result binomial weighting appears adequate.

To illustrate the alternative approach, the following is the *unweighted* analysis of variance of the percentages.

|                      | d.f. | Sums of squares | Mean squares |
|----------------------|------|-----------------|--------------|
| Among bulls          | 5    | 6,274           | 1255         |
| Within bulls (error) | 54   | 24,144          | 447          |

The unweighted mean of the percentages is 52.42. From formula (15)' this gives (52.42)(47.58)/(447)(6.644), or 84 per cent of binomial variation. The agreement between the two estimates is closer than will normally be found.

Further evidence on the superiority of binomial over equal weighting is provided by the $F$-ratios, which are 3.61 for the former and 2.80 for the latter.

### THE ANGULAR TRANSFORMATION

Thus far nothing has been said about the use of the angular transformation, in which the analysis of variance is performed on the transformed variate $y = \sin^{-1}(\sqrt{f})$. For it seems best to settle the question of weighting before considering the advisability of a transformation.

The angular transformation was devised for the case in which all the variation is binomial with constant $n$. The effect of the transformation is to change the weights $n/(p_iq_i)$, which apply to $f$, into a set of approximately *constant* weights $n/821$ applicable to the angles $y$ (expressed in degrees).

When the $n_i$ vary and extraneous variation is present, the effect of the transformation is more complicated. Suppose however that the extraneous variance $\sigma_i{}^2$ is of the form $(\lambda p_iq_i)$. The correct weights for $f$ and for $y$ are then:

$$W_i(f_i) = \left(\frac{n_i}{1 + \lambda n_i}\right)\left(\frac{1}{p_iq_i}\right) : W_i(y_i) = \left(\frac{n_i}{1 + \lambda n_i}\right)\left(\frac{1}{821}\right). \quad (19)$$

These results illustrate the fact that the transformation equalizes approximately any system of variances which are proportional to

$(p_i q_i)$, but does not alter the extent to which the weights vary with the totals $n_i$. It is for this reason that we decide first how the weights shall vary with $n_i$. If an angular transformation is subsequently made, the weighting adopted should be used with the angles.

The relation $\sigma_i^2 = \lambda p_i q_i$ imposes a restrictive assumption on the nature of the extraneous variation. The assumption is perhaps not unreasonable for data where the variance is greatest at 50 per cent and smaller towards both ends of the range.

It seems therefore that the usual rules for deciding whether to use angles may be applied after the weights (equal, binomial or partial) have been chosen. If the percentages vary widely, especially as between different treatments, and if the error variance appears to be greatest around 50 per cent, the transformation may provide a more homogeneous estimate of error. In the numerical example a transformation was not considered necessary in view of the fact that the average percentage of conceptions for the different bulls ranged only from 41 per cent to 68 per cent.

### TWO-WAY CLASSIFICATIONS

The principles outlined above apply also to two-way classifications (as in a randomized block experiment), though the detailed application is more complicated. The chief practical change is an increase in the attractiveness of equal weighting. With a given amount of variation in the $n_{ij}$, the efficiency of equal weighting is usually somewhat greater than in the case of a simple grouping. In addition, it has been shown (Yates, 1934) that a weighted analysis cannot be carried out by the elementary methods of the analysis of variance, because the sums of squares for the rows and columns of the classification are in general non-orthogonal. In order to test either rows or columns against their interaction, a set of multiple regression equations must be solved. For this calculation the procedure described for a continuous variate by Yates (1934) may be followed, provided that the weights $w_{ij}$ are substituted for Yates's totals $n_{ij}$. Except in special cases, there will be no "within-classes" sum of squares.

Some modifications are required in the formulae used in the preliminary tests. According to the nature of the data, we may wish to consider one of three simple methods of weighting, each of which avoids the labor of a least squares solution: (1) equal weights; (2) equal weights within treatments; (3) equal weights within *rows* or *replications*. On the assumption that all the variation is binomial, the minimum values for the efficiencies of these methods are as follows:

*Equal weights:* $s(r - 1)\bar{n}_h/\nu$                                              (20)

*Equal weights within treatments:*

$$rs^2[N - (N_1.^2 + \cdots + N_r.^2)/N]/\mu\nu \tag{21}$$

*Equal weights within replicates:* $r(r - 1)N^2/\theta\nu$               (22)

where

$$\nu = N - \frac{(n_{11}^2 + \cdots + n_{r1}^2)}{N_{.1}} - \cdots - \frac{(n_{1s}^2 + \cdots + n_{rs}^2)}{N_{.s}} \tag{23}$$

$$\theta = N_{.1}^2 \left( \frac{1}{n_{11}} + \cdots + \frac{1}{n_{r1}} \right) + \cdots$$
$$+ N_{.s}^2 \left( \frac{1}{n_{1s}} + \cdots + \frac{1}{n_{rs}} \right). \tag{24}$$

The quantities $\nu$ and $\theta$ involve respectively sums of the squares of $n_{ij}$ and of the reciprocals of $n_{ij}$ taken along the *rows*. Formula (21) is an approximation, $\mu$ having the value defined in formula (13).

The use of any of these formulae requires a certain amount of preliminary calculation. Where data of this type are frequently encountered, the computer soon becomes experienced in estimating the efficiencies simply by inspection of the values of the $n_{ij}$, so that in course of time the calculations need seldom be made.

Should none of these methods be considered satisfactory, the relative amounts of binomial and extraneous variation must be estimated. Unless it appears fairly certain that the greater part of the variation is binomial, estimation from an equally weighted analysis of variance is preferable on account of the saving in time. With equal weights, formula (14) remains valid for the expected value of the error mean square (in this case the treatments times replicates interaction). When binomial weights are used and the least squares analysis is carried out, the expected value of the error mean square is cumbersome. The approximate formula (17) appears to be sufficiently accurate for practical use; it underestimates slightly the relative amount of binomial variation.

### SUMMARY OF SUGGESTED PROCEDURE

1. Consider whether one of the simplest methods of analysis (equal weights, equal weights within treatments, equal weights within replicates) can be used without further investigation. If in doubt apply the appropriate test.

2. If none of the methods in (1) appears satisfactory, estimate the relative amounts of binomial and extraneous variation from an analysis of variance of the fractions or percentages.

3. As a rough discriminatory rule, adopt binomial weights for the subsequent analysis if more than 80 per cent of the variation is binomial, and equal weights if less than 30 per cent of the variation is binomial. In the intervening range use partial weighting unless the data are extensive and important enough to justify the estimation of the exact weights for every observation.

With a two-way classification it may be advisable under pressure of time to weight equally when as much as 50 per cent of the variation is binomial. In the examples which I have investigated, the loss of efficiency in this case was less than half the loss when all the variation is binomial.

4. When a method of weighting has been adopted, consider whether an angular transformation is advisable.

This procedure can be compressed as experience accumulates.

A discussion of the effects of erroneous weights on the validity of $z$, $F$ or $t$-tests is beyond the scope of this paper. It appears that the more closely the weights used approach the true weights, the smaller is the disturbance in levels of significance. Consequently a procedure designed to choose weights that are approximately correct should also lead to an analysis of variance for which the tabulated significance levels are approximately valid.

### REFERENCES

Bartlett (1936). *Journal of the Royal Statistical Society*, Supplement, 3, 68.
Bartlett (1937). *Journal of the Royal Statistical Society*, Supplement, 4, 168.
Bliss (1938). *Ohio Journal of Science*, 38, 9.
Clark and Leonard (1939). *Journal of the American Society of Agronomy*, 31, 55.
Cochran (1937). *Journal of the Royal Statistical Society*, Supplement, 4, 102.
Cochran (1938). *Empire Journal of Experimental Agriculture*, 6, 157.
Cochran (1940). *Annals of Mathematical Statistics*, 11, 335.
Yates (1934). This JOURNAL, 29, 60.
Yates and Cochran (1938). *Journal of Agricultural Science*, 28, 556.

May, 1943

Research Bulletin 318

# Some Additional Lattice Square Designs

By W. G. Cochran

AGRICULTURAL EXPERIMENT STATION
IOWA STATE COLLEGE OF AGRICULTURE
AND MECHANIC ARTS

STATISTICAL SECTION

BUREAU OF AGRICULTURAL ECONOMICS,
UNITED STATES DEPARTMENT OF AGRICULTURE

Cooperating

AMES, IOWA

34.729

# CONTENTS

# Some Additional Lattice Square Designs [1]

By W. G. Cochran [2]

The *latin square* design has been extensively used in field experiments, because of its ability to eliminate from the experimental errors the effects of soil fertility variations in two directions. More recently, Yates (1, 2) developed the *lattice square* designs for varietal trials conducted under a plant breeding program. In these designs, the number of varieties must be an exact square. Within each replication, the plots are laid out in a square array, in which every row and column contains the same number of plots. (The physical dimensions of the rows and columns will, of course, be different unless the plots are square in shape.) In successive replications the groupings of the varieties into rows and columns are changed so that with the complete design every pair of varieties has appeared together once either in the same row or in the same column. This symmetry makes it possible to adjust the varietal total or average yields, by simple calculations, for variations in the fertility of different rows and columns. In this way the effects of soil fertility variations in two directions can be eliminated from the experimental errors, just as in the latin square.

These designs will be described as *balanced* lattice squares, in order to distinguish them from the designs which form the subject of this bulletin. In the useful range for most practical purposes, the available selection of balanced lattice squares is shown in table 1a.

TABLE 1a. AVAILABLE SELECTION OF BALANCED LATTICE SQUARES.

| Number of varieties | 16 | 25 | 49 | 64 | 81 | 121 | 169 |
|---|---|---|---|---|---|---|---|
| Numbers of replicates | 5, 10 | 3, 6 | 4, 8 | 9 | 5, 10 | 6 | 7 |

The selection is restricted both as to number of varieties and the number of replicates. The restricted choice of numbers of varieties does not appear to limit seriously the utility of these designs in plant breeding work, where the number of varieties introduced into an experiment can usually be

[1] Project 514 of the Iowa Agricultural Experiment Station in cooperation with the Bureau of Agricultural Economics, United States Department of Agriculture.

[2] Research Professor, Agricultural Statistics, Iowa Agricultural Experiment Station.

varied to some extent. The restrictions on the number of replicates constitute perhaps a more frequent limitation. In varietal trials it is convenient to be able to vary the number of replications anywhere between three and nine or more, according to the labor and field space available and to the state of development of the breeding program.

The purpose of the present bulletin is to describe an additional group of lattice square designs, which allow a somewhat wider choice of number of replications. In this group, the numbers of varieties and replicates which are likely to be suitable for plant breeding experiments are shown in table 1b.

TABLE 1b. SOME ADDITIONAL LATTICE SQUARE DESIGNS.

| Number of varieties | 49 | 64 | 81 | 121 | 169 |
|---|---|---|---|---|---|
| Numbers of replicates | 3 | 3, 4 | 3, 4 | 3, 4, 5 | 3, 4, 5, 6 |

It will be noted that the designs above all require fewer replicates than the corresponding balanced lattice squares in table 1a. In types of experimentation where the balanced lattice squares have proved successful, these additional designs may therefore be serviceable when it is not practicable to plant the full number of replications necessary for a balanced lattice square. In this respect, the designs in table 1b are analogous to the simple and triple lattices (3), which are used when the number of replications is insufficient for a balanced lattice.

## 1. PROPERTIES OF THE DESIGNS

The experimental plans, except that for 64 varieties, are obtained simply by selecting the desired number of replicates from the corresponding plan for a balanced lattice square. Any set of replicates may be chosen, provided that they are all different. The method of constructing balanced lattice squares has been described by Yates (1, 2) and may be followed until the required number of replicates has been obtained. It is hoped to publish a handbook showing the complete field plans for the balanced lattice squares.

If there are 64 varieties, a particular group of replicates must be taken from the balanced set. The plan for this design, with four replicates, is given in table 8. If threefold replication is desired, the first three replicates in table 8 may be used. Since the balanced lattice square with 64 varieties requires nine replications, the new designs for 64 varieties are perhaps the most useful of those in table 1b, as suggested by Yates (1).

The nature of the groupings into rows and columns may be

seen by studying table 8. It will be found that each variety appears in the same row as 28 other varieties and in the same column as a different set of 28 other varieties. There remain seven varieties which do not appear either in the same row or column as the chosen variety. Thus the additional designs lack the complete symmetry of the balanced designs.

The statistical analysis follows the same procedure as that for the balanced lattice square with $p^2$ varieties in $(p + 1)/2$ replications, except for minor changes which involve no extra work. A numerical example is given in section 3 below.

The standard error of the difference between two varietal means varies according to the relation between the varieties in the experimental plan. Theoretically, three standard errors are required, one for the difference between two varieties which appear in the same row, one for the difference between two varieties which appear in the same column, and one for the difference between two varieties which do not appear together in a row or column. Formulae are given for the three types of standard error. It is shown in section 4, however, that the use of the average standard error for all varietal comparisons is generally of sufficient accuracy for practical purposes in experiments involving three or more replicates.

With designs which are only partially balanced, there is a slight overall loss of efficiency relative to balanced designs. As Yates' results (1) indicate, this loss is negligible when at least three replicates are used (see section 5).

Lattice square designs can be constructed for numbers of replications other than those in tables 1a and 1b, e. g., for 49 varieties in five, six or seven replicates. Since the statistical analysis is more complicated, such designs will not be discussed here.

## 2. FIELD LAYOUT AND RANDOMIZATION

As we have indicated, the plots within each replication are laid out in square formation, so that differences among rows and columns represent fertility variations in two directions at right angles.

For many field crops the plots normally used for experimental purpose are long and narrow. If the narrow sides of the plots are parallel to the rows of the square, the rows may be fairly compact in shape, but the columns will be extremely long and narrow. For instance, in corn experiments with plots 2 hills by 10 hills, the dimensions of each row in a 9 x 9 lattice square are 18 hills by 10 hills, while those of each column are 2 hills by 90 hills. It may be doubted in such cases whether the extra control of variation among columns will materially reduce the experimental error.

Some information on this question was obtained by examining the results of 16 corn experiments with 2 x 10 hill plots, involving 49, 81 or 121 varieties. These experiments were carried out at various locations in Iowa during the 1941 season, as part of the corn breeding program of the Iowa Agricultural Experiment Station, in cooperation with the Division of Cereal Crops and Diseases, Bureau of Plant Industry, U. S. D. A.[3] In each of these experiments it is possible to estimate what the experimental error variance would have been if the variation among columns had not been eliminated, i. e., if lattice designs (3) had been used instead of lattice squares. The results were expressed in terms of the *relative accuracy* of the two designs, this quantity being the inverse ratio of their experimental error variances.

In 10 of the 16 experiments, no appreciable increase in precision was obtained from the elimination of column differences, the accuracy of the lattice squares relative to the lattices varying from 96 percent to 107 percent. The relative accuracies in the remaining six experiments were 120, 131, 132, 137, 193 and 199 percent. In three experiments, the variation among columns was actually greater than the variation among rows. The results indicate that although the extra control may be inoperative in the majority of cases, there are occasionally substantial fertility variations parallel to the narrow sides of the plots. For this reason the ability to adjust for variations in both directions is useful, particularly in experiments on outlying farms where the fertility contours may not be well known.

It must not be concluded from these comparisons that lattice designs would have been inferior or at best equal to lattice square designs in the experiments discussed above. With lattice designs, the best field layout, on the average, is obtained by making the incomplete blocks compact in shape, and arranging these blocks so that the replications are also as compact as is practicable. With 81 varieties, the incomplete block might measure 18 hills by 10 hills, this being the same as the row of the lattice square, while the replication might measure 54 hills by 30 hills, as compared with 18 hills by 90 hills for the lattice square. A reasonably compact replication (32 hills by 40 hills) is also possible with 64 varieties, though designs with 25, 49 or 121 varieties are less convenient in this respect. A comparison of lattice and lattice square designs, with each design arranged in the most accurate

[3] I am indebted to Dr. G. F. Sprague and Mr. L. A. Tatum for permission to use these data, and the data in the numerical example below.

layout, could be obtained only from uniformity data. Such a comparison, however, might not be entirely relevant to the conditions under which field experiments are usually laid down. In most experiment stations, the sites available for experiments in any year are fully occupied, leaving as little waste space as possible. It is frequently impractical to arrange every experiment in its most suitable layout. In these circumstances it is convenient to be able to use either a lattice or a lattice square, according to which can most appropriately be fitted into the available field space.

Some comparisons of lattice and lattice square designs have been made by Johnson and Murphy (5) on oats uniformity data, with plots 3 feet by 16 feet and 4 feet by 8 feet. When the plots within each replication were arranged in a square, Johnson and Murphy obtained a substantial reduction in the error mean square from the control of variation among columns, even in cases where a column measured 3 feet by 88 feet. Further, lattice square designs proved on the average superior in accuracy to lattice designs, although the incomplete blocks in the latter were arranged into replications as compact as possible.

While lattice squares are expected to give best results when the plots within each replicate are laid out in a square, they may occasionally be worthwhile when practical considerations necessitate a different field arrangement. For instance, varietal trials are frequently planted with the plots in one continuous line. Where there is a fertility gradient along this line, part of the resulting error will be eliminated by the replications, and a further part by grouping the plots within each replicate into incomplete blocks. If a lattice square is used with this layout, these incomplete blocks might constitute the rows of the square. The first column of the lattice square would then contain the first plot in each block, and so on. Thus variation among columns corresponds to the fertility gradient within the blocks, though only insofar as that gradient is in the same direction within all blocks of a replicate. Thus when the plots are to be laid out in a straight line, a lattice square may be advisable if there is reason to suspect a consistent fertility gradient along the line. Even if the columns prove ineffective, the loss of accuracy as compared with a lattice design is small. These remarks apply also to latin squares which have sometimes proved successful under similar conditions, for example where the plots lay side by side down a sloping field.

The designs described here are randomized in the same way as balanced lattice squares. A separate random rearrangement should be made of the rows and columns of each replication.

For reasons which will now be described, it is also advisable to assign the varieties to the variety numbers at random. The groups of variety numbers which form the rows and columns of any replication are decided by the experimental plan, and these groups are not changed by randomizing the rows and columns of each square. Further, all varieties which fall in the same row or column of a replication receive the same adjustment for that row or column, and if an adjustment is over-estimated, all varieties in the row or column are favored. Consequently, if the differences among rows are large, the variance per plot of the difference between two groups of varieties which lie almost entirely in different rows of one replicate may be considerably larger than the average error variance. In the numerical example below, where the mean square for columns happens to be relatively though not exceptionally large, the error variance of a column mean is 7.19 per plot, as against an average variance of 6.18 per plot. The additional randomization suggested above gives every group of $p$ varieties an equal chance of forming a row or column in the field layout and thus helps to ensure that the average error variance may be used for comparisons among groups of varieties, as well as between individual pairs of varieties. These considerations apply to all lattice and lattice square designs, and not merely to the designs discussed in this bulletin.

TABLE 2. PLOT YIELDS IN POUNDS.

(Variety numbers are shown in boldface type)

| | | | Replication I | | | | | Total | L | δ |
|---|---|---|---|---|---|---|---|---|---|---|
| | **17** 38.2 | **19** 32.7 | **18** 30.7 | **21** 33.3 | **15** 32.9 | **20** 33.6 | **16** 33.4 | 235.8 | −44.4 | −1.0 |
| | **31** 34.4 | **33** 35.3 | **32** 33.2 | **35** 30.5 | **29** 31.7 | **34** 33.6 | **30** 35.2 | 233.9 | −67.3 | −1.5 |
| | **45** 24.9 | **47** 27.6 | **46** 29.4 | **49** 32.6 | **43** 32.8 | **48** 29.3 | **44** 30.6 | 207.2 | −58.6 | −1.3 |
| | **38** 31.4 | **40** 33.5 | **39** 30.4 | **42** 30.1 | **36** 24.9 | **41** 30.6 | **37** 29.4 | 210.3 | −49.2 | −1.1 |
| | **3** 26.5 | **5** 34.9 | **4** 36.0 | **7** 37.6 | **1** 34.1 | **6** 31.5 | **2** 35.7 | 236.3 | −86.8 | −2.0 |
| | **24** 34.9 | **26** 35.6 | **25** 34.6 | **28** 30.5 | **22** 32.4 | **27** 37.7 | **23** 26.7 | 232.4 | −40.0 | −0.9 |
| | **10** 38.7 | **12** 33.5 | **11** 36.8 | **14** 34.5 | **8** 36.2 | **13** 31.6 | **9** 35.4 | 246.7 | −59.3 | −1.3 |
| Total | 229.0 | 234.1 | 231.1 | 229.1 | 225.0 | 227.9 | 226.4 | 1602.6 | | |
| M | −64.1 | −51.5 | −51.4 | −74.8 | −38.0 | −71.5 | −54.3 | | −405.6 | |
| ε | − 3.8 | − 3.0 | − 3.0 | − 4.4 | − 2.2 | − 4.2 | − 3.2 | | | |

TABLE 2 (continued)

| | 29 | 48 | 42 | 5 | 23 | 11 | 17 | Total | L | δ |
|---|---|---|---|---|---|---|---|---|---|---|
| Replication II | | | | | | | | | | |
| | 31.4 | 30.0 | 31.2 | 33.5 | 27.2 | 36.8 | 34.5 | 224.6 | −43.9 | −1.0 |
| | **39** | **2** | **45** | **8** | **33** | **21** | **27** | | | |
| | 33.1 | 36.0 | 28.6 | 39.2 | 34.9 | 34.8 | 36.9 | 243.5 | −67.3 | −1.5 |
| | **16** | **35** | **22** | **41** | **10** | **47** | **4** | | | |
| | 30.8 | 31.0 | 35.2 | 30.5 | 31.2 | 29.0 | 30.6 | 218.3 | −34.5 | −0.8 |
| | **49** | **12** | **6** | **18** | **36** | **24** | **30** | | | |
| | 35.1 | 32.2 | 31.1 | 33.9 | 26.9 | 32.3 | 34.5 | 226.0 | −46.2 | −1.0 |
| | **3** | **15** | **9** | **28** | **46** | **34** | **40** | | | |
| | 28.3 | 34.3 | 34.6 | 28.1 | 26.0 | 31.4 | 33.8 | 216.5 | −40.6 | −0.9 |
| | **26** | **38** | **32** | **44** | **20** | **1** | **14** | | | |
| | 36.1 | 27.2 | 28.4 | 25.5 | 26.5 | 30.0 | 33.2 | 206.9 | − 9.6 | −0.2 |
| | **13** | **25** | **19** | **31** | **7** | **37** | **43** | | | |
| | 32.3 | 35.3 | 32.5 | 30.7 | 31.1 | 29.3 | 29.7 | 220.9 | −25.8 | −0.6 |
| Total | 227.1 | 226.0 | 221.6 | 221.4 | 203.8 | 223.6 | 233.2 | 1556.7 | | |
| M | −57.3 | −46.0 | −39.1 | −34.8 | −23.1 | −43.3 | −24.3 | | −267.9 | |
| ε | − 3.4 | − 2.7 | − 2.3 | − 2.0 | − 1.4 | − 2.5 | − 1.4 | | | |

| | 1 | 48 | 28 | 10 | 30 | 19 | 39 | Total | L | δ |
|---|---|---|---|---|---|---|---|---|---|---|
| Replication III | | | | | | | | | | |
| | 29.6 | 25.4 | 27.7 | 25.7 | 29.2 | 28.7 | 21.6 | 187.9 | +75.5 | +1.7 |
| | **23** | **21** | **43** | **32** | **3** | **41** | **12** | | | |
| | 26.6 | 30.1 | 27.9 | 26.6 | 19.4 | 15.8 | 21.5 | 167.9 | +91.9 | +2.1 |
| | **40** | **31** | **11** | **49** | **20** | **2** | **22** | | | |
| | 29.5 | 32.2 | 30.2 | 23.7 | 25.2 | 24.6 | 25.4 | 190.8 | +91.5 | +2.1 |
| | **35** | **26** | **6** | **37** | **8** | **46** | **17** | | | |
| | 28.3 | 32.0 | 24.6 | 23.2 | 31.8 | 17.8 | 30.3 | 188.0 | +82.0 | +1.9 |
| | **18** | **9** | **38** | **27** | **47** | **29** | **7** | | | |
| | 30.9 | 33.8 | 23.2 | 26.7 | 18.5 | 18.2 | 9.3 | 160.6 | +135.0 | +3.1 |
| | **45** | **36** | **16** | **5** | **25** | **14** | **34** | | | |
| | 23.8 | 26.1 | 28.4 | 26.9 | 28.8 | 19.8 | 16.1 | 169.9 | +100.7 | +2.3 |
| | **13** | **4** | **33** | **15** | **42** | **24** | **44** | | | |
| | 31.8 | 30.8 | 27.6 | 26.3 | 20.0 | 26.5 | 14.8 | 177.8 | +96.9 | +2.2 |
| Total | 200.5 | 210.4 | 189.6 | 179.1 | 172.9 | 151.1 | 139.0 | 1242.9 | | |
| M | +27.8 | +31.8 | +71.1 | +109.9 | +102.0 | +149.6 | +181.3 | | +673.5 | |
| ε | + 1.6 | + 1.9 | + 4.2 | + 6.5 | + 6.0 | + 8.8 | +10.6 | | | |

Grand total 4402.2

TABLE 3. VARIETY TOTALS (UNADJUSTED).

| | 1 | 2 | 3 | 4 | 5 | 6 | 7 | Total |
|---|---|---|---|---|---|---|---|---|
| | 93.7 | 96.3 | 74.2 | 97.4 | 95.3 | 87.2 | 78.0 | 622.1 |
| | **8** | **9** | **10** | **11** | **12** | **13** | **14** | |
| | 107.2 | 103.8 | 95.6 | 103.8 | 87.2 | 95.7 | 87.5 | 680.8 |
| | **15** | **16** | **17** | **18** | **19** | **20** | **21** | |
| | 93.5 | 92.6 | 103.0 | 95.5 | 94.9 | 85.3 | 98.2 | 663.0 |
| | **22** | **23** | **24** | **25** | **26** | **27** | **28** | |
| | 93.0 | 80.5 | 93.7 | 98.7 | 103.7 | 101.3 | 86.3 | 657.2 |
| | **29** | **30** | **31** | **32** | **33** | **34** | **35** | |
| | 81.3 | 98.9 | 97.3 | 88.2 | 97.8 | 81.1 | 89.8 | 634.4 |
| | **36** | **37** | **38** | **39** | **40** | **41** | **42** | |
| | 77.9 | 81.9 | 81.8 | 85.1 | 96.8 | 76.9 | 81.3 | 581.7 |
| | **43** | **44** | **45** | **46** | **47** | **48** | **49** | |
| | 90.4 | 70.9 | 77.3 | 73.2 | 75.1 | 84.7 | 91.4 | 563.0 |
| Total | 637.0 | 624.9 | 622.9 | 641.9 | 650.8 | 612.2 | 612.5 | 4402.2 |

TABLE 4.  VARIETY TOTALS (ADJUSTED).

| 1 | 2 | 3 | 4 | 5 | 6 | 7 |
|---|---|---|---|---|---|---|
| 90.1 | 97.8 | 72.2 | 94.3 | 96.1 | 83.8 | 83.3 |

| 8 | 9 | 10 | 11 | 12 | 13 | 14 |
|---|---|---|---|---|---|---|
| 108.1 | 101.1 | 96.5 | 102.3 | 91.9 | 90.0 | 91.3 |

| 15 | 16 | 17 | 18 | 19 | 20 | 21 |
|---|---|---|---|---|---|---|
| 95.4 | 90.7 | 108.3 | 93.2 | 98.5 | 86.6 | 92.8 |

| 22 | 23 | 24 | 25 | 26 | 27 | 28 |
|---|---|---|---|---|---|---|
| 99.5 | 77.7 | 96.5 | 99.8 | 100.0 | 102.9 | 84.0 |

| 29 | 30 | 31 | 32 | 33 | 34 | 35 |
|---|---|---|---|---|---|---|
| 85.1 | 99.5 | 93.4 | 89.8 | 96.8 | 84.9 | 83.9 |

| 36 | 37 | 38 | 39 | 40 | 41 | 42 |
|---|---|---|---|---|---|---|
| 76.4 | 82.9 | 81.3 | 88.4 | 94.1 | 79.7 | 80.7 |

| 43 | 44 | 45 | 46 | 47 | 48 | 49 |
|---|---|---|---|---|---|---|
| 91.2 | 77.0 | 72.3 | 77.3 | 76.6 | 79.1 | 89.9 |

Grand total   4405.0

## 3.  STATISTICAL ANALYSIS

The statistical analysis follows the same general procedure as for a balanced lattice square with $(p + 1)/2$ replications (2). In order to present a numerical example, the fourth replication was omitted from a lattice square experiment on corn, with 49 varieties, carried out at Cresco, Iowa, in 1940. Each plot measured 4 hills by 5 hills, the hills being spaced 3 feet 6 inches apart in each direction. Individual plot yields, in pounds, are given in table 2.

In the algebraic formulae $p^2$ is the number of varieties and $r$ the number of replicates. The computing instructions are as follows:

**1.** Calculate the row and column totals for each replication, the replication totals and the grand total, inserting these in table 2 as shown. The variety totals are then obtained and inserted, in the form of a square, in table 3. It is also worthwhile to insert the row and column totals of this table as shown.

**2.** Calculate and insert the quantities L for each row of the experiment.

L = (Total yield of all varieties appearing in the row) —
r (Row total).

The sum inside the first bracket is obtained by addition from table 3. For rows of the first replication, this quantity is simply one of the row totals of table 3 and has been calculated and written down in step 1. Thus for the first row of replication I,

$$L = 663.0 - 3 (235.8) = -44.4.$$

In the other replications, time is probably saved by mentally arranging the variety symbols in each row in increasing order before summing their yields. If this is done, the first variety will be found to lie in the first row of table 3, the second variety in the second row, and so on. For the first row of replication II, for instance, the variety totals are summed in the order 5, 11, 17, 23, 29, 42, 48. This procedure facilitates the location of varieties in table 3 and reduces the likelihood of errors.

Similarly the quantities M are calculated and entered at the foot of each column, where

M = (Total yield of all varieties appearing in the column) — r (column total).

The replication totals (—405.6, —267.9, +673.5) of L and M are obtained and inserted in table 2. As a check, these numbers should be equal to

(Grand total of individual yields) — r (Replication total of individual yields)

while the total of the r numbers should be exactly zero.

**3.** The analysis of variance may now be calculated. The total sum of squares and the sums of squares for replications and varieties are obtained just as in an ordinary randomized blocks design. The sum of squares for rows, after eliminating varietal effects, is the sum of squares of deviations of the quantities L from their replication means, divided by $pr(r-1)$. In the example, $p = 7$, $r = 3$, this gives

$$\frac{1}{42} \left[ (44.4)^2 + (67.3)^2 + \ldots + (96.9)^2 - \right.$$
$$\left. \{ (405.6)^2 + (267.9)^2 + (673.5)^2 \} /7 \right] = 134.19.$$

The sum of squares for columns is calculated from the quantities M by a similar rule. The error sum of squares is found by subtraction. All results should be checked by re-calculation.

TABLE 5.   ANALYSIS OF VARIANCE: POUNDS PER PLOT.

|  | d. f. | Sums of squares | Mean square |
|---|---|---|---|
| Replications | 2 | 1,564.37 |  |
| Varieties | 48 | 1,383.11 | 28.815 |
| Rows | 18 | 134.19 | 7.455 |
| Columns | 18 | 515.65 | 28.647 |
| Error | 60 | 305.63 | 5.094 |
| Total | 146 | 3,902.95 |  |

**4.** If R, C and E are the mean squares for rows, columns and error, respectively, we now calculate the factors

$$\lambda' = \frac{(R-E)}{p(r-1)R} = \frac{(7.455 - 5.094)}{14 \times 7.455} = .02262$$

$$\mu' = \frac{(C-E)}{p(r-1)C} = \frac{(28.647 - 5.094)}{14 \times 28.647} = .05873$$

(These quantities correspond to Yates' $\lambda/p$ and $\mu/p$ for the balanced lattice square.) They provide the multipliers necessary to convert L and M into adjustments for row and column effects, applicable to the varietal totals in table 3. If either R or C is less than E, the corresponding factor is put equal to zero, no adjustment being made for the corresponding row or column effects. Similarly, should both R and C be less than E, the unadjusted varietal totals are used.

Each quantity L is multiplied by $\lambda'$, giving the quantities $\delta$ (table 2), and each quantity M by $\mu'$, giving the quantities $\epsilon$ (table 2). It is usually sufficient to carry the same number of decimal places in $\delta$ and $\epsilon$ as in the varietal totals.

The totals of all the $\delta$'s and of all the $\epsilon$'s should be zero, apart from rounding-off errors. If one or both of these totals appear suspicious, a rough check may be obtained from the result that the absolute magnitude of the rounding-off error should be less than $h(\frac{1}{2} + \sqrt{\frac{pr}{3}})$, except about once in 20 times, where $h$ is the rounding-off interval. In the present example, the $\delta$'s total $+ 0.3$ and the $\epsilon$'s total $+ 0.1$. (It is worth recording these totals, as they provide a check in step 5 below.) The rounding interval is 0.1, and the check formula above gives $(0.1)(0.5 + \sqrt{7}) = 0.31$, slightly above the higher discrepancy.

**5.** To adjust the varietal totals in table 3, the quantities $\delta$ and $\epsilon$ are added for each row and column in which the variety appears. Thus the adjusted total yield for variety 1 is

$$93.7 - 2.0 - 2.2 - 0.2 - 2.5 + 1.7 + 1.6 = 90.1$$

These values are shown in table 4 above. For a check by summation, it may be noted that the total of the adjusted yields, minus the total of the unadjusted yields, should exactly equal p times the sum of the $\delta$'s and $\epsilon$'s. Thus,

$$4405.0 - 4402.2 = 2.8 = 7 \, (+ \, 0.3 + 0.1)$$

For tests of significance, the error mean square in the analysis of variance must be increased in order to take into account the sampling error of the row and column adjustments. The average experimental error variance per plot is given by

$$E \left\{ 1 + \frac{rp}{p+1} \; (\lambda' + \mu') \right\}$$
$$= (5.094) \left\{ 1 + (2.625)(.02262 + .05873) \right\} = 6.18$$

This quantity is multiplied, as usual, by 2r to give the average variance of the difference between two varietal totals, and by 2/r to give the average variance of the difference between two varietal means.

As indicated previously, the average error will usually be sufficiently accurate for tests of significance of the difference between any pair of varieties. Individual formulae for the three types of comparisons are as follows:

Two varieties in the same row:
$$E \left\{ 1 + (r-1) \; \lambda' + r\mu' \right\} = 6.22$$

Two varieties in the same column:
$$E \left\{ 1 + r\lambda' + (r-1) \; \mu' \right\} = 6.04$$

Two varieties not appearing together:
$$E \left\{ 1 + r\lambda' + r\mu' \right\} = 6.34$$

Like the lattices and balanced lattice squares, these designs can be analyzed alternatively as if they were in randomized blocks, the rows and columns being ignored. The error sum of squares in this case is equal to the total sum of squares for rows, columns and error in table 5, and gives a mean square of 9.953. A z- or F-test of the unadjusted varietal totals can thus readily be obtained from table 5, by forming the ratio of the varieties mean square to the mean square 9.953 above. Varietal differences are highly significant. The varieties mean square must not be compared directly with the error mean square in table 5, since the former has not been adjusted for row and column effects.

A comparison of the experimental errors 9.95 and 6.18 indicates that the adjustments for row and column differences produced a marked increase in accuracy.

The randomized blocks analysis is useful where preliminary results must be summarized under pressure of time, where subsidiary measurements on the plots are little affected by local soil variations or are not required with high accuracy, or where it is necessary to omit the yields of some varieties from the results.

It may be worth noting that the analysis described above is also valid for a balanced lattice square in which one or more replications must be omitted through field damage or for other reasons.

## 4. ESTIMATION OF THE STANDARD ERRORS

The theory of the statistical analysis follows very closely that given by Yates (2) for the balanced lattice square with $(p + 1)/2$ replications. In particular, the structure of the analysis of variance and the method of calculating the row and column adjustments are exactly similar, except for variations introduced by the change from $(p + 1)/2$ to $r$ replications, and need not be reproduced here. The derivation of the varietal standard errors will, however, be indicated, since these are required in further discussion and since the balanced design contains no case in which two varieties do not appear within the same row or column.

As Yates (2) has pointed out, it is convenient, in a discussion of the theory, to estimate first the total yields of the groups of $p$ varieties which constitute the rows and columns. The relation between these yields and the yields of the individual varieties may be illustrated by considering the case of nine varieties $(p = 3)$.

TABLE 6. ORTHOGONAL SUBDIVISION OF THE EIGHT DEGREES OF FREEDOM AMONG NINE VARIETIES.

| Set I | | | Set II | | | Set III | | | Set IV | | |
|---|---|---|---|---|---|---|---|---|---|---|---|
| 1 | 4 | 7 | 1 | 2 | 3 | 1 | 2 | 3 | 1 | 2 | 3 |
| 2 | 5 | 8 | 4 | 5 | 6 | 5 | 6 | 4 | 6 | 4 | 5 |
| 3 | 6 | 9 | 7 | 8 | 9 | 9 | 7 | 8 | 8 | 9 | 7 |

The table shows 12 groups of three varieties each. Every variety belongs to four of the groups. Further, in the remaining members of these four groups, every other variety appears once. For example, in the four groups containing variety 1, its companions are 2, 3; 4, 7; 5, 9; and 6, 8. It follows that the total yield of all the groups to which a variety belongs is equal to three times the yield of the variety plus the total yield of all varieties.

Groups possessing these properties can be constructed for any value of $p$ for which a completely orthogonalized latin square (4) exists; these values include $p = 3, 4, 5, 7, 8, 9, 11, 13$. In the general case, a variety belongs to $(p + 1)$ groups, which together contain all the other varieties once. Thus the total yield of the groups is equal to $p$ times the yield of the variety plus the grand total. These relations enable the individual varietal yields to be estimated if the yields of each group total are known.

In these designs, $r$ of the groups to which a variety belongs form rows of the experiment, a further $r$ groups form columns, while the remaining $(p + 1 - 2r)$ groups are unconfounded with rows or with columns. In the case of a group which forms a row in one of the replications, two estimates of the group total are available, one from the replication in question, and one from the remaining $(r - 1)$ replications, in which the group is unconfounded with rows or columns. If the error and inter-row variances per plot are $1/w_i$ and $1/w_r$, respectively, so that $w_i$ and $w_r$ represent the error and inter-row *weights*, the variances of these two estimates are respectively $p/w_r$ and $p/(r - 1)w_i$. The factor $p$ is included because each group contains $p$ varieties. If the two estimates are combined by weighting them inversely as their variances, the variance of the weighted mean, by a well-known statistical theorem, is

$$\frac{p}{w_r + (r - 1) w_i} \tag{1}$$

For a group which constitutes a column of one of the replicates, we need only replace $w_r$ by $w_c$ in the above expression, where $w_c$ is the reciprocal of the inter-column variance per plot. The remaining group totals, which are unconfounded in all replications, are estimated each with variance $p/rw_i$.

From the discussion above of the relation between the groups and the individual varieties, the difference between the means of any two varieties is equal to the difference between the totals of their corresponding groups, divided by $p$. Every pair of varieties has, however, one group in common. Consequently, if two varieties occur together in a row, the variance of the difference between their means is

$$\frac{2p}{p^2}\left[ \frac{(r - 1)}{w_r + (r - 1)w_i} + \frac{r}{w_c + (r - 1)w_i} + \frac{(p + 1 - 2r)}{rw_i} \right] \tag{2}$$

After some algebraic manipulation, this becomes

$$\frac{2}{rw_i}\left[ 1 + (r - 1)\lambda' + r\mu' \right] \tag{3}$$

where $\lambda' = \dfrac{1}{p} \cdot \dfrac{(w_i - w_r)}{(r - 1)w_i + w_r}$, $\mu' = \dfrac{1}{p} \cdot \dfrac{(w_i - w_c)}{(r - 1)w_i + w_c}$

The corresponding result for a pair of varieties which have a column in common is found, by interchanging $w_r$ and $w_c$ in the above result, to be

$$\frac{2}{rw_i} \left[ \quad 1 + r\lambda' + (r-1)\mu' \quad \right] \tag{4}$$

For a pair of varieties having no row or column in common, the common group must be one of the $(p + 1 - 2r)$ groups which are unconfounded. The variance of the difference between the varietal means is therefore

$$\frac{2p}{p^2} \left[ \frac{r}{w_r + (r-1)\ w_i} + \frac{r}{w_c + (r-1)w_i} + \frac{(p-2r)}{rw_i} \right] \tag{5}$$

which reduces to

$$\frac{2}{rw_i} \left[ \quad (1 + r\lambda' + r\mu') \quad \right] \tag{6}$$

To find the *average* variance of the difference between two varietal means, we may note that any variety appears in the same row as $r(p-1)$ other varieties and in the same column as an equal number of other varieties. Hence, there remain $(p-1) \times (p + 1 - 2r)$ other varieties which do not occur in the same row or column anywhere in the design. The three types of variance must therefore be weighted in the ratios, $r$, $r$, $(p + 1 - 2r)$ respectively. This gives

$$\frac{2}{rw_i} \left[ 1 + \frac{rp}{p+1} \ (\lambda' + \mu') \right] \tag{7}$$

In practice, the values of $w_i$, $w_r$ and $w_c$ must be estimated from the analysis of variance. The values of $\lambda'$ and $\mu'$ given in section 3 are the maximum likelihood estimates of the theoretical values in the formulae above. In general, the use of the same symbol for the theoretical value and its estimate is not advisable, but should cause no confusion here. The substitution of estimated weights results in a slight underestimation of the standard errors. This bias is negligible in the larger designs where considerable numbers of degrees of freedom are available for estimating the weights. Even with the 7x7 design in three replicates, the underestimation of the average standard error does not appear to exceed 2 percent for any values of the true weights. With a 2 percent underestimation of the standard error, the apparent 5 percent point has a true probability of about 4.6 percent.

We may now consider the justification for the use of the

average error variance for all types of comparison between pairs of varieties. Upon investigation of the formulae given above for the three error variances, it will be found that the greatest difference between the average variance and any one of the three variances, when expressed as a fraction of the average variance, cannot exceed the larger of

$$\frac{2r}{p\{r(p+3)-(p+1)\}} \quad \text{and} \quad \frac{(p+1-r)}{p\{r(p+2)-(p+1)\}}$$

For the relevant values of p and r, the greater of these two quantities is shown below.

TABLE 7.   MAXIMUM PERCENT ERROR RESULTING FROM THE USE OF THE AVERAGE VARIANCE.

| Number of varieties | Number of replications | | | |
|---|---|---|---|---|
| | 3 | 4 | 5 | 6 |
| 169 | 1.36 | 1.23 | 1.16 | 1.13 |
| 121 | 1.82 | 1.65 | 1.57 | |
| 81 | 2.56 | 2.34 | | |
| 64 | 3.13 | 2.86 | | |
| 49 | 3.90 | | | |

The corresponding percent errors in the standard deviations are about half the values above. Thus for the 7 x 7 design, an apparent 5 percent probability obtained by using the average error would represent a true probability varying, at the most, between 4.6 and 5.4 percent. For the other designs this variation is still smaller. These maximum errors occur either in the case where both the inter-row and inter-column variations are large relative to the error variation, or in the case when one is large while the other is no larger than the error variation.

## 5.   LOSS OF EFFICIENCY DUE TO ASYMMETRY

The average error variance *per plot* applicable to the adjusted varietal yields may be obtained by deleting the factor $\frac{2}{r}$ in formula (7) above. Written in full, this becomes

$$\frac{1}{w_i}\left[1+\frac{r}{p+1}\left\{\frac{(w_i-w_r)}{(r-1)w_i+w_r}+\frac{(w_i-w_c)}{(r-1)w_i+w_c}\right\}\right] \quad (8)$$

The corresponding value for a balanced design, in $(p+1)/2$ replicates, is obtained by putting $r=(p+1)/2$ in (8). For given values of $w_i$, $w_r$ and $w_c$, i. e., for given field conditions, the variance is always slightly greater for the partially balanced design than for the balanced design. This increase in variance represents a loss of efficiency due to asymmetry of the design.

The loss of efficiency is relatively greatest when $w_r = w_c = o$, that is, when the design is highly effective in removing variation among rows and columns. In this case, the factor inside the square brackets reduces to

$$\frac{r(p + 3) - (p + 1)}{(p + 1)(r - 1)} \tag{9}$$

and to $(p + 1)/(p - 1)$ for the balanced design. Under the same field conditions, the accuracy per replication of the partially balanced design relative to the balanced design cannot therefore be less than

$$\frac{(p + 1)^2(r - 1)}{(p - 1)\{r(p + 3) - (p + 1)\}} \tag{10}$$

These ratios are shown below

| Number of varieties | Number of replicates | | | |
|---|---|---|---|---|
| | 3 | 4 | 5 | 6 |
| 169 | .960 | .980 | .990 | .996 |
| 121 | .960 | .982 | .994 | |
| 81 | .961 | .986 | | |
| 64 | .964 | .992 | | |
| 49 | .969 | | | |

The loss of accuracy does not amount to more than 4 percent in any of the above designs, and of course approaches zero as the designs become more symmetrical with increasing replication.

It may be noted that the formula (9) gives the reciprocal of the *efficiency factor* of these designs, as defined by Yates (1).

TABLE 8. LATTICE SQUARE DESIGN FOR 64 VARIETIES IN FOUR REPLICATES

| Replication 1 | | | | | | | | Replication II | | | | | | | |
|---|---|---|---|---|---|---|---|---|---|---|---|---|---|---|---|
| 1 | 2 | 3 | 4 | 5 | 6 | 7 | 8 | 1 | 44 | 62 | 56 | 27 | 39 | 18 | 13 |
| 9 | 10 | 11 | 12 | 13 | 14 | 15 | 16 | 46 | 2 | 17 | 35 | 16 | 53 | 60 | 31 |
| 17 | 18 | 19 | 20 | 21 | 22 | 23 | 24 | 64 | 23 | 3 | 25 | 54 | 12 | 45 | 34 |
| 25 | 26 | 27 | 28 | 29 | 30 | 31 | 32 | 55 | 40 | 29 | 4 | 58 | 41 | 11 | 22 |
| 33 | 34 | 35 | 36 | 37 | 38 | 39 | 40 | 28 | 9 | 50 | 63 | 5 | 24 | 38 | 43 |
| 41 | 42 | 43 | 44 | 45 | 46 | 47 | 48 | 37 | 51 | 15 | 42 | 20 | 6 | 32 | 57 |
| 49 | 50 | 51 | 52 | 53 | 54 | 55 | 56 | 19 | 61 | 48 | 14 | 33 | 26 | 7 | 52 |
| 57 | 58 | 59 | 60 | 61 | 62 | 63 | 64 | 10 | 30 | 36 | 21 | 47 | 59 | 49 | 8 |

| Replication III | | | | | | | | | Replication IV | | | | | | | |
|---|---|---|---|---|---|---|---|---|---|---|---|---|---|---|---|---|
| 1 | 11 | 20 | 30 | 34 | 48 | 53 | 63 | | 1 | 32 | 47 | 61 | 22 | 50 | 12 | 35 |
| 15 | 2 | 56 | 45 | 59 | 28 | 22 | 33 | | 29 | 2 | 14 | 49 | 39 | 64 | 43 | 20 |
| 21 | 52 | 3 | 39 | 32 | 58 | 9 | 46 | | 42 | 13 | 3 | 24 | 60 | 33 | 30 | 55 |
| 26 | 47 | 38 | 4 | 17 | 13 | 64 | 51 | | 59 | 54 | 18 | 4 | 48 | 31 | 37 | 9 |
| 40 | 62 | 31 | 19 | 5 | 49 | 42 | 12 | | 23 | 36 | 57 | 46 | 5 | 11 | 56 | 26 |
| 43 | 25 | 61 | 16 | 55 | 6 | 36 | 18 | | 52 | 63 | 40 | 27 | 10 | 6 | 17 | 45 |
| 54 | 24 | 10 | 57 | 44 | 35 | 7 | 29 | | 16 | 41 | 28 | 34 | 51 | 21 | 7 | 62 |
| 60 | 37 | 41 | 50 | 14 | 23 | 27 | 8 | | 38 | 19 | 53 | 15 | 25 | 44 | 58 | 8 |

## SUMMARY

Some additional experimental designs are described for varietal trials in which a large number of varieties are compared. The plots in each replication are usually arranged in square formation, the arrangement allowing the varietal yields to be adjusted for fertility or other differences among the rows and columns of each replicate. The designs are similar to the balanced lattice squares developed by Yates, except that a smaller number of replications are used.

The statistical analysis, which differs only in minor details from that of a balanced lattice square, is illustrated by means of a numerical example. Formulae for the standard errors of the adjusted varietal yields are derived and discussed.

Some discussion is given of the field layout and of the relative accuracy of lattice and lattice square designs.

## REFERENCES CITED

(1) Yates, F. A further note on the arrangement of variety trials: quasi—latin squares. Ann. Eugenics, 7 : 319-331. 1937.

(2) Yates, F. Lattice squares. Jour. Agr. Sci., 30 : 672-687. 1940.

(3) Cox, G. M., Eckhardt, R. C., and Cochran, W. G. The analysis of lattice and triple lattice experiments in corn varietal tests. Iowa Agr. Exp. Sta., Res. Bul. 281 : 1-66. 1940.

(4) Fisher, R. A. and Yates, F. Statistical tables for biological, agricultural and medical research. Oliver and Boyd, Edinburgh. 1938.

(5) Johnson, I. J. and Murphy, H. C. Lattice and lattice square designs with oat uniformity data and in variety trials. Jour. Amer. Soc. Agron., 35 : 291-305. 1943.

# TRAINING AT THE PROFESSIONAL LEVEL FOR STATISTICAL WORK IN AGRICULTURE AND BIOLOGY*

By W. G. COCHRAN

*Iowa State College*

### INTRODUCTION

THIS PAPER attempts to outline the kind of training that may appropriately be given by colleges and universities for statistical work at a professional level in agriculture and biology. Since the problem of training is both difficult and controversial, my views are put forward with some diffidence. Nevertheless, this opportunity for public discussion of training methods is most timely, for the future progress and usefulness of the field of applied statistics will depend greatly on the type and quality of training available to young persons who seek to enter the field.

There are three groups of students whom the colleges may serve. The first consists of those who wish to earn their living as professional statisticians, and who are looking towards posts in agriculture, biological or medical colleges or research institutes, or in the agricultural departments of federal or state agencies or of large corporations. This group is numerically small, and though an increased demand after the war seems certain, the number of careers may be expected to remain fairly small. This is, however, the group from whom future leadership is expected in the development of new statistical techniques for agriculture and biology. The second group, considerably larger in numbers, comprises students in an applied field, e.g. agronomy, agricultural economics, who want to acquire such knowledge of statistical concepts and techniques as will be useful in their future work. Finally, there are many persons, employed in posts which require varying degrees of statistical knowledge, who can profit from occasional periods spent at a college for the purpose of learning recent developments in statistical techniques and of discussing their statistical problems. The three groups will be considered in turn.

### TRAINING OF PROFESSIONAL STATISTICIANS

The decision to specialize in statistics is seldom made until the bachelor's degree has been obtained. Usually when the student presents

* A paper presented at the 104th Annual Meeting of the American Statistical Association, Washington, D. C., December 27, 1944.

himself, he has had only an introductory course or two in statistics, his undergraduate major having been either in mathematics or in the applied field. There is much variation in the amount and type of this previous training. Postgraduate training must therefore be flexible and must include, so far as needed, statistical theory, mathematics and the applied field.

Training for students who wish to acquire a mastery of statistical theory will be considered first. The following topics are basic: (i) Probability theory. Since progress in statistical research is so often hindered by inability to solve the distribution problems which are involved, a thorough training should be provided in the methods applicable to such problems, e.g., multiple integration, moment-generating functions, geometrical arguments and the use of conditional probabilities. (ii) The theory of point and interval estimation. (iii) The theory of tests of significance and (iv) the bases of the common statistical techniques and their relationship to (ii) and (iii). In this connection, regression theory and the analysis of variance are worth special emphasis by reason of their frequent utility in biological applications.

In such fundamental courses, I believe that it is important to include numerical examples in which the student applies the techniques to actual data. In these examples the questions should be stated in terms of the data and it should be left to the student to construct the appropriate mathematical model, select or develop the technique needed, carry out the computations and submit his conclusions in a nontechnical report. Similarly, the student should receive experience in handling data by such familiar devices as answering letters from workers in the applied field, submitting plans for projects which require the taking of a sample survey or the design of an experiment, and taking part in meetings where his professor is consulting with applied workers.

It is debatable how much time should be devoted in the fundamental training to this type of experience in applications. Time spent on one subject is usually stolen from another and in this case usually from the mathematical theory of statistics. Moreover, such contribution as the statistical profession is now able to make is due in no small measure to advances in mathematical statistics during the past thirty years and it is likely that in the future the value of sound fundamental training will be increasingly appreciated in applied work. Nevertheless, the applied part of the training fulfills the very useful purposes of informing the student whether he will enjoy applied work and informing both student and professor whether he is likely to be good at it. As is well known, the qualities required for success are not the same in theoretical and applied work. Aptitude for applied statistics is not easy to define; it includes an ability to distinguish what is important and what is unimportant in a statistical problem, a flair for making sound decisions from scanty data and the quality that for lack of a better name we call common sense. These are qualities that by no means every student possesses and it is doubtful whether they are developed to any great extent by academic training. A course of training that includes both theory and applications helps to direct the student towards the type of work in which he will be most happy and useful.

In a college where the usefulness of statistical techniques has been recognized, there will be other courses, given primarily to workers in the applied fields, which can profitably be attended by the would-be professional statistician. I have in mind such topics as design of experiments, methods for taking sample surveys, econometrics, psychometrics, vital statistics and applications to engineering or industry.

To turn to the training in mathematics, one might say "the more the better." In fact, it is difficult to foresee what branches of advanced mathematics will be most fruitful for future developments in statistical theory. Even to provide a background for the techniques used in present-day mathematical statistics, an extensive training in mathematics is required. Advanced calculus, particularly multiple integration, infinite series, the algebra of linear and quadratic forms, including operations with matrices, and an introduction to complex variable are essential to keep abreast of current research. Also somewhat useful are at least the elementary ideas in the theory of measure, algebraic geometry of many dimensions, finite groups, the calculus of variations and numerical methods. Since this list covers some courses which are given infrequently if at all in many mathematics departments, some special reading may be necessary in order to fill the gaps.

For the student whose undergraduate major is in mathematics, training in the applied field may present somewhat of a problem. The training must be broad, partly because it is not wise for the student to restrict himself to a narrow field of specialization and partly because of the close interrelationships among the various fields of biology. Further, while fundamentals must be emphasized, some acquaintance with current lines and methods of research is also desirable, which implies attendance at the most advanced courses that are offered. The Ph.D. requirements at some of the agricultural colleges are well adapted for this situation, since proficiency is required in one field outside of mathematics and statistics and since the student's Ph.D. committee usually contains two representatives from that field, who assist him in outlining a course of study.

The studies in statistical theory, mathematics and the applied field should culminate in the conduct of research. The objectives should be to provide experience in the planning and execution of a major piece of original work and a test of the student's aptitude for research. The research work should be selected and judged accordingly rather than by its presumed significance as a contribution to current knowledge, especially so since such significance is often beyond the control of the student. Either a problem in mathematical statistics or the interpretational analysis of complex data is suitable.

The training outlined above is designed to befit the able student for work at a high level in mathematical statistics. Students from agriculture and biology may be expected to find the program very heavy going. Only the exceptional student is likely to attain a confident mastery of the advanced mathematics. The student from the applied field, however, will continue to fill a useful place as a professional statistician. Many important applications of statistics, while employing elementary statistical techniques, demand thorough knowledge and long experience in the applied field. Moreover, such students are usually well aware of their mathematical limitations and, thanks to the numerous able and generous mathematical statisticians, help in mathematical difficulties is now easy to obtain. Thus the postgraduate training should be designed to attract students from agriculture and biology and to equip them as completely as their mathematical capabilities will allow. This objective can be secured within the framework described above by recommending for the student such parts of the program as he can profitably undertake. As Greenwood has remarked,[1] "The advantage of excluding by severe mathematical requirements many quacks is bought too dear if it shuts out a single John Graunt."

### TRAINING FOR STUDENTS IN THE APPLIED FIELD

With students in agriculture and biology for whom statistics is of secondary importance, the objectives of the teacher might be stated as follows: (i) to teach the calculations involved in the common statistical techniques and in special techniques that are appropriate for the applied field; (ii) to present as clearly as possible the reasoning behind each technique; (iii) to emphasize the assumptions on which each technique is based and its limitations. If these objectives are achieved, the students should be able to use their statistical equipment intelligently and should realize when expert advice must be sought.

On account of the scanty mathematical training given in under-

---

[1] *Journal of the Royal Statistical Society*, 102, p. 552, 1939.

graduate biology in many colleges, teaching in statistics must proceed at a low mathematical level. To many students even the elementary operations in algebra appear meaningless. With experienced teaching, however, I believe that surprisingly good results are obtained. A helpful factor is that a greater proportion of the students are familiar with the phenomenon of variability than would be the case in chemistry or engineering. The relation between sample and population can be clarified by the drawing of samples, which usually creates hearty interest. The same device may be used to illustrate the generation of the common frequency distributions and the logical process involved in a test of significance. With ingenious preparation, the assumptions and limitations involved can also be presented in concrete terms.

The most serious problem is that of finding good teachers. The problem is accentuated by the fact that demands for instruction are highly specific. Thus students in forestry benefit much more from a course where the problems are those of forestry than from one where agricultural plot yields are always under discussion. Consequently the tendency in the past has been for each department to set up its own courses of instruction, with results deplored by many statisticians because of the poor quality of instruction in many cases and the complete lack of coordination.

As a remedy, some colleges are considering or have initiated a basic course in statistics which is taken by all students from a wide group of applied fields. While this may be an improvement on the previous system and may be the best solution with present resources, the ideal solution will not be reached until each major department can offer a separate course under an instructor who is keenly interested in statistics, has the leisure and ability to follow the latest developments and can teach. With such an instructor available, there should be no objection when a group of students with special interests wish to withdraw from the general course, provided that steps can be taken to insure a continuing exchange of ideas among the various teachers of statistics within the college. With teachers who are all lovers of the subject, such coordination should not be difficult to insure. I will not attempt to predict how soon this supply of teachers is likely to appear. With regard to agriculture and biology the outlook is perhaps more hopeful than in other fields, because of the relatively high prestige which statistics now enjoys.

### TRAINING FOR PROFESSIONAL WORKERS

In view of the rapid progress of research in statistical theory, affecting even the elementary techniques, professional workers who make frequent use of statistics may find it difficult to keep in touch with developments. This difficulty can be overcome through arrangements by which such workers are assigned to a college for a period of time. The benefits are mutual, for the college statisticians receive education in the current applications of statistical techniques and information about the important statistical questions as viewed by the user of statistics. For the professional worker, not the least of the benefits is the temporary freedom from the pressures and interruptions of everyday work. The length of the period and the type of instruction or research undertaken must of course be varied according to the circumstances. Two types of arrangement which have proved successful and which perhaps represent the extremes are described briefly below.

The first is a plant science work conference conducted by the Department of Experimental Statistics, North Carolina State College, during the week of February 21–26, 1944. The conference was attended by agronomists from the southern states, all of whom had had considerable experience in handling statistical techniques. Lectures were given on the analysis of long-term experiments, rotation experiments and groups of experiments. In addition, the members presented their current and proposed experimental designs for discussion by the entire group and exchanged their experiences of the advantages and disadvantages of the new lattice designs. As a visitor during one day, I can testify to the high quality of the discussion and to the keen enjoyment of the members. A later conference of a similar type was held for workers in animal science and I understand that further conferences are planned for other groups of workers with some specialized interest.

Another example is provided by the cooperative agreement between the Bureau of Agricultural Economics and Iowa State College. In the statistical laboratory at Ames, the Bureau maintains a unit devoted to research on sampling problems, most of which have arisen in the work of the Division of Agricultural Statistics. From time to time this Division has assigned one of its members to the research unit for a period which has averaged about two years. The member's time has been divided between graduate course work in the college and research on one of the Division's problems. The length of the period has been an important factor in the success of this arrangement. In two years, sufficient time is available for the member to become experienced in the use of the newer statistical techniques and to engage in research problems which are relatively broad and complex, whereas a brief stay would permit only research of a simple and routine nature.

It is worth emphasizing this point that the length of stay at the college merits careful consideration. Persons who are already experienced statistically and who resolve many of their statistical difficulties without aid can profit greatly from a brief period of discussion or instruction. On the other hand, persons who lack statistical training cannot be taught more than a series of ill-digested working rules in a course lasting one or two weeks. In fact, there is no method for acquiring statistical competence without concentrated effort over a long period of time. In an organization which undertakes work of a statistical nature, it would therefore seem wise to insist on a fundamental training in statistical theory when recruiting employees, rather than to expect the employee to acquire the necessary knowledge in the course of his work or in short courses of instruction.

### SUMMARY

Colleges and universities may appropriately provide three types of training: (i) training for students who wish to become professional statisticians; (ii) training for students in agriculture and biology who wish to acquire an intelligent grasp of current statistical reasoning and techniques; (iii) training for professional workers who wish to keep in touch with recent developments.

For the would-be professional statisticians, the training should include mathematics, mathematical and applied statistics and some broad, fundamental courses from the applied field which the student selects. The training should be sufficiently flexible to cope with aspirants from both the mathematical and biological fields and should culminate in the conduct of research.

Although at present the teaching of statistics to agricultural and biological students must be carried out at a low mathematical level, it is possible with experience and ingenuity to make clear not only the computational processes, but also the nature of statistical reasoning and the limitations of statistical techniques. The primary obstacle to an improvement in the quality of such teaching is the dearth of teachers who are keenly interested in statistics and have the leisure and ability to follow the advances in statistical theory.

Two examples are given of the type of service which colleges can render to professional workers. It is emphasized that the amount of time to be spent at college by the professional worker should be chosen with care.

# 36

## DESIGNS OF GREENHOUSE EXPERIMENTS FOR STATISTICAL ANALYSIS[1]

GERTRUDE M. COX and W. G. COCHRAN[2]

*North Carolina State College*

Numerous greenhouse studies of soil deficiencies and plant needs are being conducted by experimenters in agronomy, plant pathology and physiology, soil science, and horticulture. Although greenhouse experiments may be performed for observational or exploratory purposes, most experiments brought to the attention of the statistician are investigations to determine small differences. Such experiments call for improved techniques, efficient methods of layout, and careful analysis of results.

In this paper some types of layout that have been applied successfully in greenhouse experiments are illustrated. The object in selecting an arrangement on the greenhouse bench, like the object of a good technique, is to obtain the maximum accuracy commensurate with the amount of time and labor available for the conduct of the experiment. In both instances, the first step toward increased accuracy is to discover the principal causes of the variation observed in the greenhouse.

### SOURCES OF VARIATION IN GREENHOUSE EXPERIMENTS

Variation within a greenhouse bench may be large, though it has been considerably reduced in the newer greenhouses built specifically for comparative tests. The major sources of variation are temperature gradients caused by such things as steam pipes under the bench, doors, and proximity to ventilators; shading effects resulting from the structures in the house, adjacent buildings or trees; and moisture differentials caused by methods of watering and air currents.

A careful appraisal of the greenhouse facilities and some experience in experimentation will usually indicate rather accurately which of these sources of variation exist. The nature and direction of environmental gradients will also be apparent. For example, steam pipes under each edge of a bench and running parallel to the bench will produce sharp gradients within 3 to 5 feet across the bench. LeClerg (6) has given a striking example of differences in the stands of sugar-beet seedlings caused by a temperature gradient from vertical radiating coils on the side walls and ends of the greenhouse. On a bench adjacent to an outside ventilator a marked temperature and possibly a moisture gradient will exist across the bench. Of course, such variations are most pronounced in greenhouses located where extreme climatic conditions prevail.

[1] Contribution from the Department of Experimental Statistics, North Carolina Agricultural Experiment Station. Published with the approval of the director as Paper No. 241 of the Journal Series.

[2] The authors are indebted to J. A. Rigney, North Carolina State College, A. G. Norman, Iowa State College, and E. L. Le Clerg, Bureau of Plant Industry, U. S. Department of Agriculture, for helpful suggestions from their experience.

Shading effects are rather universal and usually are greater along the bench than across it. Sometimes they can be avoided by not using that part of the bench near structural material. Another important source of variation arises from inequalities in watering (methods of watering are discussed elsewhere in this issue).

In addition to the effects of temperature, shading, and moisture inequalities, there remains a serious source of error which is summed up in the phrase "pot to pot variation." Often this variation is due mainly to inherent differences in the responsiveness of individual plants; because of the small numbers of plants per pot which are typical in greenhouse experiments, these plant differences are not averaged out to anything approaching the extent that occurs in a plot in a field experiment. Here again, however, much may be accomplished by good technique; for instance, by weighing the soil into the pots for fertilizer experiments and, when the plants are large, by leaving sufficient space between pots to mitigate the effects of interpot competition (or contamination in disease experiments). More experimental work on the selection of a suitable pot size might be profitable, though experience has given useful leads.

With a knowledge of the greenhouse conditions that will affect the experimental results and an appreciation of the limitations of the experimental unit to be used, the research worker is in a position to select an efficient experimental design.

### MODERN EXPERIMENTAL DESIGNS

From the point of view of their application to greenhouse experiments, modern experimental designs may be divided into two classes. In the first class, which contains randomized blocks, lattice designs, and balanced incomplete blocks, the experimental units (the pots, jars, or sections) are arranged in a series of groups in such a way that differences among the groups, whether environmental or associated with the plants, can be eliminated from the experimental errors. Such designs may be used to counteract the effects of a single gradient either along or across the bench or of differences between benches when the experiment extends over several benches. Where the treatments are applied to mature or partly mature plants, plant differences can also be eliminated to a considerable extent by placing similar plants within the same group. Naturally, the groups should be constructed so that the differences among groups represent what is considered to be the greatest source of variation. In the second class, consisting of latin squares, lattice squares, and Youden squares, the experimental units are arranged in groups in two different ways. Variations associated with both types of grouping are removed simultaneously from the errors, so that these designs can cope with gradients both along and across the bench. A brief description follows of some useful designs in each class, accompanied by an example of a greenhouse experiment.

The randomized block design is the simplest and most widely used design; indeed, there would be little need to consider other designs except that in certain circumstances they give a marked increase in accuracy over randomized blocks. In this design the experimental units are arranged in groups, each of which constitutes a complete replication of the treatments. Any number of treatments and any number of replications may be used. An example may be selected from a group of floriculture experiments. A project was planned to determine whether differences existed among seven soil types in their suitability for use in commercial greenhouses. The soil was mechanically mixed and placed on the benches, which had been divided into sections of equal size. There were three replications, each containing seven sections. The soil types were assigned to the units, a new randomization being used for each replicate. For snapdragons 18 plants, for chrysanthemums 9 plants, and for stock 36 plants were placed in each section. Some of these experiments were conducted during the winter in a greenhouse located where the outside temperature was frequently below freezing. The heating pipes were at one end of the bench, and the other end was close to outside windows. The experiments gave a pictorial demonstration of bench gradients showing open flowers, half-open flowers, and buds from the warmer to the cooler end of the bench. The arrangement of the replicates removed most of this bench variation from experimental error.

In the *latin square* design the treatments are grouped into replications in two different ways. This design has been used for virus experiments where individual leaves are inoculated. In such experiments the position of the plant on the bench has little influence on the responses of its leaves to the inoculation; the important factors are plant differences and differences in the responses of leaves of different sizes. By the latin square arrangement the treatments can be placed so that each appears on every plant and on every leaf size. For plants with five usable leaves, a layout for five treatments (A, B, C, D, E) is shown below. The symbol (a) denotes the largest leaf on a plant, (b) the second largest, and so on.

| SIZE OF LEAF | PLANT | | | | |
|---|---|---|---|---|---|
| | 1 | 2 | 3 | 4 | 5 |
| (a) | A | C | B | E | D |
| (b) | E | D | C | A | B |
| (c) | D | A | E | B | C |
| (d) | C | B | A | D | E |
| (e) | B | E | D | C | A |

For bench experiments the utility of latin squares with more than five treatments is limited by the narrowness of the bench. The larger latin squares may be useful in greenhouses that contain no bench but where large earthenware jars are used. For example, seven seed treatments were applied to a variety of soybeans. Lots of 50 seeds, one lot treated in each of the seven ways, were planted in gallon jars which were arranged in seven rows of seven jars each. Every treatment appeared in each row and in each column. The experiment covered considerable space, and the arrangement provided an opportunity to measure the environmental variations along and across that section of the greenhouse. Heat and light gradients were present. The difference produced by these gradients would have masked the treatment effects in a less appropriate design.

If the conditions of the experiment make it necessary to apply some treatments to a considerable area on the bench, while others can be applied to a single pot, the two sets of treatments can be combined in the same trial by a device known as *split-plot* Technique. An experiment was carried out to test six new selections of peas grown with three different amounts of light—light for 24 hours, light for 12 hours, and total darkness. The lighting equipment introduced the limitations which made a split-plot design suitable. Two benches (replications) were partitioned into three sections each. At random, one section on each bench was surrounded by curtains to provide a dark condition. The other two sections were provided with lighting equipment to control the length of light for 12 or 24 hours. Within each section six pots, each containing seed from one of the selections, were placed at random.

For comparisons of the effects of different amounts of light, the experimental unit (sometimes called the *whole-unit*) is a section of the bench; for comparisons among the selections the experimental unit (sometimes called the *sub-unit*) is a single pot. In the analysis of variance, separate estimates of error are calculated for the two types of unit. The whole-unit error applies to the light treatments, the sub-unit error to the selections and to the interactions between the selections and the light treatments. Except in rare cases the sub-unit error is the smaller.

Another application of this technique occurs in studies of nutrient solutions using sand cultures, where it is impractical to attach the electric pumps to small experimental units. Other treatment comparisons may be introduced within each solution. In fact, when practicable an additional set of treatments may be added to any experiment by subdividing the whole-unit. A simple example of the statistical analysis is given by Hayes and Immer (5), a more complex one by Goulden (4).

### INCOMPLETE BLOCK DESIGNS

Suppose that the width of the bench accommodates four pots and that a steep gradient has been found to exist along the bench. This gradient can be removed rather successfully by either randomized blocks or a latin square, provided that the number of treatments does not exceed four. With 16 treatments, however, there might be a substantial gradient within replicates in a randomized block layout. A similar difficulty arises when the treatments are applied to a heterogeneous batch of plants or shoots where it might be possible to assemble small groups of similar plants or shoots but not large groups. For instance, in experiments described by Morrow (8) on the comparison of self- and cross-pollination in blueberries, paired shoots of equal size and with approximately the same number of fruit buds were selected from the same plant in order to eliminate plant variation and variation in shoot vigor. If it were necessary to assemble groups of 20 "similar" shoots for the comparison of 20 treatments in randomized blocks, uniform matching within a group would have to be sacrificed to a considerable degree.

In this situation it is desirable to use an arrangement that makes possible the elimination from the error of differences among groups that are smaller than a replication. A series of designs of this type have been developed by Yates (12) since 1936. Since the small groups or blocks do not contain all the treatments, the designs are sometimes called *incomplete block designs*.

One such design, a *balanced lattice*, was used for an experiment on the effects of eight hormone treatments on the development of tobacco plants grown in each of two soil types. Small ($6\frac{1}{2}$-inch pots) were used, and the plants were thinned to six per pot. Since environmental variations across the bench were expected, the 16 treatment combinations—1, 2, . . ., 16—were placed on the bench in the following order:

| Rep. I | Rep. II | Rep. III | Rep. IV | Rep. V |
|---|---|---|---|---|
| 9 1 5 13 | 13 12 6 3 | 5 6 7 8 | 3 10 16 5 | 13 7 10 4 |
| 16 8 12 4 | 1 10 15 8 | 14 16 13 15 | 9 4 15 6 | 11 1 16 6 |
| 15 7 11 3 | 14 4 11 5 | 11 12 9 10 | 13 8 2 11 | 15 12 2 5 |
| 14 6 2 10 | 16 2 7 9 | 2 1 4 3 | 12 1 14 7 | 8 9 14 3 |

The incomplete blocks, containing four pots each, were arranged to remove differences across the bench, while the replications removed a considerable amount of the variation along the bench. The results indicated less bench effect on the roots than on the tops, perhaps because light and temperature differences affect the tops in a more direct manner; also the roots were limited in space and the tops were not.

This design belongs to the group of *lattice designs*, which have some useful properties. These designs require no more work in the greenhouse than randomized blocks with the same number of replicates. The results can be analyzed, if desired, by the usual procedure for randomized blocks, though in general more accurate estimates of the treatment effects are obtained by use of the analysis appropriate to lattice designs. The essential feature of this analysis is that the treatment means are adjusted for environmental differences among the incomplete blocks. The method of analysis is given for the balanced lattice by Wellhausen (9) and for other lattices by Cox, Eckhardt, and Cochran (1). The chief limitation to the utility of the design is that the number of treatments must be an exact square.

*Balanced incomplete block* designs are, in general, similar to balanced lattices but are not restricted to numbers of treatments which are exact squares. An example on soybeans is shown in the following diagram; glazed pots 12 inches in diameter were used with 10 plants per pot:

*Bench I*

| Lime 1 | 4 | 2 |
|---|---|---|

*Bench II*

| 3 | 1 | 2 |
|---|---|---|

*Bench III*

| 3 | 4 | 2 |
|---|---|---|

*Bench IV*

| 1 | 3 | 4 |
|---|---|---|

Four soil types (a, b, c, d), two rates of Mg, and four levels of lime were tested in all combinations, giving 32 treatments. Since this experiment was intended principally to give information on the soil types and on differences in the responsiveness of the soil types to Mg and to lime, a double split-plot arrangement was used. The soil types were tested in individual pots, the Mg rates in groups of four neighboring pots, and the lime rates in groups of eight pots.

The space available did not allow 32 pots to be placed in the same bench so as to leave sufficient space between pots. Consequently, in a randomized blocks layout of the lime rates, each replication would have extended over more than one bench. For this reason the lime rates were arranged in the balanced incomplete block design for four treatments in blocks of three units. Each block of 24 pots could be placed on a single bench. The arrangement provided three replications and occupied four benches. Differences among benches were removed from the experimental error by means of adjustments applied in the statistical analysis to the results for the lime rates. An index to balanced incomplete block designs and an account of the method of analysis are given by Fisher and Yates (3).

By a rearrangement of certain of the balanced incomplete block plans, incomplete block layouts are obtained which remove differences between two types of grouping of the pots. This rearrangement (usually called a *Youden square*) is illustrated below for the lime rates design in the previous example:

| Balanced Incomplete Blocks Blocks | | | | Youden Square Blocks | | | | |
|---|---|---|---|---|---|---|---|---|
| I | II | III | IV | Replicates | I | II | III | IV |
| 1 | 3 | 3 | 1 | I | 1 | 2 | 3 | 4 |
| 4 | 1 | 4 | 3 | II | 4 | 3 | 2 | 1 |
| 2 | 2 | 2 | 4 | III | 2 | 1 | 4 | 3 |

Without any change in the compositions of the four blocks, the rows now represent separate replicates, so that 2 degrees of freedom may be subtracted from the error for differences among replicates in addition to the 3 degrees of freedom for blocks. A single replication of this design is not recommended, since only 3 degrees of freedom remain for error. By a duplication of the design, however, six replicates provide nine error degrees of freedom, while nine replicates provide 15 error degrees of freedom. The basic plans for these designs and some interesting applications to greenhouse experiments are described by Youden (13).

Similarly, *lattice squares* (11) are constructed by a rearrangement of the balanced lattices. In addition to the removal of differences among replicates, these designs provide adjustments for differences among rows and columns within replicates, thus allowing a high degree of control over the major sources of variation.

## FACTORIAL EXPERIMENTS

Fisher (2) and Yates (10) have drawn attention to the advantages to be gained by investigating the effects of a number of different factors in the same experiment. In a *factorial* experiment, all combinations of the different factors are tested: the two split-pot experiments described above are examples. Efficient use is made of the experimental material, because every pot contributes information on all the factors under investigation. For instance, it was desired to measure the top and root growth of four species of grass on soil of high and of low fertility with four intensities of cutting. The three factors (four species, two fertility levels, four intensities of cutting) give 32 treatment combinations which were laid out in randomized blocks with four replications. There was considerable pot-to-pot variation within replicates; the uneven stands indicated variation in moisture content, crusting, and disease at planting. Nevertheless, the average effects of the factors were estimated with a satisfactory degree of accuracy because the experiment afforded 32 replications for the species and intensities comparisons and 64 replications for the comparison of the fertility levels. Also, by a study of the first-order interactions, it was possible to test whether the species responded similarly to the different intensities of cutting and to the two fertility levels and whether the effect of intensity of cutting was conditioned by the fertility level.

Factorial experiments may be arranged in randomized blocks, in latin squares, or in split-plot designs. If the total number of treatment combinations is so large that some reduction in block size is advisable, the incomplete block designs described above are suitable when the study of interactions is considered as important as that of the average effects of the factors. For many crops, however, experimentation has indicated that the high-order interactions are usually negligible or relatively small. In such cases it is preferable to reduce block size by a sacrifice of information on these interactions. Perhaps the most useful designs of this type are those for a confounded $2 \times 2 \times 2$ factorial experiment in blocks of four pots. This design was used on corn to study the root-top ratio of an inbred line (a slow-growing genetic inbred) compared with a very vigorous, uniform hybrid. These two sources of seed were grown in poor and in good soil, kept in wet and in dry condition. This gave a $2 \times 2 \times 2$ factorial consisting of two seed sources, two fertility levels, and two moisture conditions. It was known that considerable variation existed along the bench and since only four pots (6 inch) could be placed across it, the experiment was arranged in incomplete blocks of four pots across the bench. Two adjacent blocks made a replication, of which there were four. In this experiment there are three first-order interactions and one second-order interaction, each of which was confounded with blocks in one of the four replications. In effect, the experiment gives four replicates on the average effects and three replicates on the interactions. A more common variant is to confound the second-order interaction in all replicates, giving greater reliability to the measure of main effects and first-order interactions. Yates (10) gives a systematic account of confounded factorial designs, and Li (7) contributes some additional arrangements.

In the foregoing experiment, the pots within the incomplete blocks were re-randomized at intervals during the early part of the trial. It is now believed that this was unnecessary. The ability to move pots around is an advantage of greenhouse conditions over field conditions, and may be used in an attempt to equalize environmental conditions or to avoid the worst effects of inter-pot competition. Unfortunately, we have found no data on the effectiveness of the practice: its drawbacks are the labor involved, the possibility of injury to the plants, and the opportunity for unobserved biases. The use of incomplete block designs should render the practice of less value insofar as homogeneity within the blocks is attainable.

## NUMBER OF REPLICATIONS

Greenhouse experiments are less accurate than is usually realized, particularly where crop weights are concerned. Not infrequently the percentage standard errors per pot are higher than those per plot in field experiments with the same crop. When the resources of skillful technique and layout have been exhausted, the only way in which accuracy can be increased is to provide more replications.

It is not easy to give helpful advice on the desirable amount of replication. Often a compromise must be made between the conflicting claims of different experiments; moreover, the issue is complicated when it is intended to repeat the experiment several times if initial results appear promising. In the simplest case, considering only a single experiment, we require for a realistic discussion an estimate of the standard error per pot, which can often be made from previous experience, and a statement of the size of difference which the experiment is designed to detect and measure.

Suppose that the standard error per pot is 10 per cent of the average yield—

a figure not unusually high for greenhouse experiments—and consider how accurately treatment differences are measured in an experiment with four replications. The standard error of a treatment mean is $10/\sqrt{4}$ or 5 per cent and that of the difference between two treatment means is $5\sqrt{2}$ or about 7 per cent of the average yield. The difference which will be observed in the experiment varies about the true difference with this standard error. If the number of error degrees of freedom is ample, we find from the normal probability tables that four times out of five the observed difference will vary from the true difference by less than $7 \times 1.28 = 9$ per cent of the average yield. In other words, the experiment is fairly certain to determine the true difference to within $\pm 9$ per cent of the average yield.

For instance, if the true difference is 10 per cent of the average yield, the difference observed in the experiment is fairly certain to lie somewhere between $(10 - 9)$ and $(10 + 9)$ per cent; that is, between 1 per cent and 19 per cent. It is clear that this experiment cannot be expected to measure a true 10 per cent difference with any degree of precision, since the experiment might show a difference as low as 1 or 2 per cent, or a difference as large as 18 or 19 per cent. Similarly, if the true difference happens to be 30 per cent, the limits for the observed difference are $(30 \pm 9)$ per cent; that is, 21 per cent and 39 per cent. It follows that the experiment can distinguish, with reasonable certainty, between a true difference of 10 per cent and one of 30 per cent, because in the former case the observed difference is unlikely to exceed 19 per cent, whereas in the latter case the observed difference is unlikely to be lower than 21 per cent. The experiment cannot be relied upon, however, to discriminate between true differences of 10 and 20 per cent.

This kind of calculation gives some idea of the accuracy with which experiments having a given amount of replication will measure real treatment differences of specified sizes. Alternatively, in setting up a preliminary experiment, we may wish to know the number of replications necessary in order that a true difference of given size is fairly certain (say four times out of five) to be detected as significant at the 5 per cent level. In the foregoing example, with four replicates and a 10 per cent standard error per pot, it is found that a true difference of 21 per cent satisfies these conditions; that is, four times out of five the observed difference will exceed the value required for 5 per cent significance. With a true difference of 10 per cent, on the other hand, calculation shows that there is considerably less than a 50-50 chance that the experiment will indicate a significant difference. Thus it is quite likely that the existence of a true 10 per cent difference would be undiscovered in a preliminary experiment of this type.

To visualize what is involved in this sort of problem it is necessary first of all to compute the smallest difference that would be judged significant at the 5 per cent level by the $t$ test. In the present example the standard error of the difference between two treatment means is 7.1 per cent of the average yield. Assuming that there are eight treatments in randomized blocks, with four replicates per treatment, the number of degrees of freedom for estimating error is $7 \times 3 = 21$. For 21 degrees of freedom the value of $t$ required for significance is 2.080. The smallest difference that would be considered significant is, therefore, 2.08 × 7.1 = 14.8 per cent. To achieve a significant difference at least four times out of five trials, it is then necessary to get an observed difference as large as 14.8 per cent at least four times out of five trials. If the standard error of an observed difference is 7.1 per cent, and an observed difference of 14.8 per cent or more is to be attained four times in five trials, the true difference in the population must satisfy the relation $\dfrac{14.8 - d}{7.1} = -0.859$. The number $-0.859$ represents the value of $t$ which will be equalled or exceeded four times in five trials when the true difference in the population is $d$ and the standard error of the observed difference is estimated from 21 degrees of freedom. Solving this equation gives $d = 20.9$ per cent. Thus the true difference in the population must be at least 20.9 per cent before an observed difference of 14.8 per cent will be reached as often as four times in five trials, or, in other words, the true difference in the population must

TABLE 1

*Number of replications required in order to obtain a significant observed difference (5 per cent level) in four experiments out of five*

| TRUE DIFFERENCE (AS % OF AVERAGE YIELD) | STANDARD ERROR PER CENT PER UNIT (POT, SECTION, ETC.) | | | | | |
|---|---|---|---|---|---|---|
| | 14 | 12 | 10 | 8 | 6 | 4 |
| 5 | 124 | 91 | 63 | 41 | 23 | 11 |
| 10 | 31 | 23 | 16 | 11 | 6 | 3 |
| 15 | 14 | 11 | 8 | 5 | 3 | 2 |
| 20 | 8 | 6 | 5 | 3 | 2 | 2 |
| 30 | 4 | 3 | 3 | 2 | 2 | 2 |

be at least 20.9 per cent if a significant difference (5 per cent level) is to be achieved at least four times in five trials.[3]

Table 1 shows the number of replications required in order to be fairly certain (four times out of five) of obtaining a significant observed difference when the true difference is 5, 10, 15, 20, and 30 per cent, respectively. The table has been prepared for a series of standard errors per unit (pot, section, etc.) which cover the range usually encountered in greenhouse experiments.[4]

With three or four replications, it is evident that we can expect to detect only true differences of 20 per cent or more, unless the standard error per unit is un-

[3] This argument is not exactly correct. In an actual experiment, where the true standard deviation is unknown, the difference required for significance is not 14.8 per cent, but 2.08 times the estimated standard error of the difference between two treatment means. The argument, however, displays the nature of the problem and gives a very close approximation to the exact result, which requires more complicated mathematics. Values in table 1 were based on the correct argument, for which see Neyman, *Sup. Jour. Roy. Statis. Soc.*, 2, 1935.

[4] In taking into account the number of error degrees of freedom, the experiment was assumed to have eight treatments in randomized blocks; thus with $r$ replicates, there are $7(r-1)$ error degrees of freedom.

usually small. To cope with true differences of 5 or even 10 per cent, large numbers of replications are required throughout most of the range of standard errors per unit. Moreover, it should be remembered that these numbers of replications merely insure, with reasonable certainty, that the experiment will indicate the existence of a true difference. If, in addition, the experiment is to estimate the size of the true difference within narrow limits, the requirements in number of replications are considerably higher.

The table serves to emphasize that fine differences cannot be determined with few replications. The value of any reduction in standard error obtainable by good technique or layout is also apparent, as is the efficiency of factorial experiments, where there may be from 20 to 60 replications for the average effects of a factor.

### SUMMARY

For accurate experimentation under greenhouse conditions, a knowledge of the environmental variations due to the structure and location of the greenhouse and of the variability of the plant material is of prime importance. Brief descriptions are given, with illustrations from greenhouse experiments, of a number of experimental designs that enable the experimenter to utilize this knowledge so as to eliminate the major sources of variation from the experimental errors. An investigation is made of the accuracy of experiments with a given number of replicates. This study shows that under typical greenhouse conditions only large treatment differences can be detected with three or four replicates and makes evident the value of any reduction in standard error per pot obtainable from improved techniques or designs.

### REFERENCES

(1) Cox, G. M., Eckhardt, R. C. and Cochran, W. G. 1940 The analysis of lattice and triple lattice experiments in corn varietal tests. Iowa Agr. Exp. Sta. Res. Bul. 281.

(2) Fisher, R. A. 1942 The Design of Experiments, ed. 3. Oliver and Boyd, Edinburgh.

(3) Fisher, R. A., and Yates, F. 1943 Statistical Tables, ed. 2. Oliver and Boyd, Edinburgh.

(4) Goulden, C. H. 1939 Methods of Statistical Analysis. John Wiley and Sons, New York.

(5) Hayes, H. K., and Immer, F. R. 1942 Methods of Plant Breeding. McGraw-Hill, New York.

(6) Le Clerg, E. L. 1935 Factors affecting experimental error in greenhouse pot tests with sugar beets. Phytopath. 25: 1019–1025.

(7) Li, J. C. R. 1944 Design and statistical analysis of some confounded factorial experiments. Iowa Agr. Exp. Sta. Res. Bul. 333.

(8) Morrow, E. B. 1943 Some effects of cross-pollination versus self-pollination in the cultivated blueberry. Proc. Amer. Soc. Hort. Sci. 42: 469–472.

(9) Wellhausen, E. J. 1943 The accuracy of incomplete block designs in varietal trials in West Virginia. Jour. Amer. Soc. Agron. 35: 66–76.

(10) YATES, F.  1937  The design and analysis of factorial experiments.  Imp. Bur. Soil. Sci. Tech. Commun. 35.

(11) YATES, F.  1940  Lattice squares.  *Jour. Agr. Sci.* 30: 672–687.

(12) YATES, F.  1940  Modern experimental design and its function in plant selection. *Empire Jour. Exp. Agr.* 8: 223–230.

(13) YOUDEN, W. J.  1940  Experimental designs to increase accuracy of greenhouse studies.  *Boyce Thompson Inst. Contrib.* 11: 219–228.

# 37

## GRADUATE TRAINING IN STATISTICS*

### W. G. COCHRAN, Iowa State College

**1. Introduction.** The emergence of statistics as a possible field of specialization for mathematicians is rather recent. There has, of course, been some kind of a profession of statistics for many years—the professional statistical societies are over 100 years old both in this country and in England. But until about 25 years ago statistics was regarded mainly as a useful tool in economics and politics and the deliberations in the professional societies reflected this point of view.

The beginnings of research and teaching in mathematical statistics were the results either of the insight and labors of individual enthusiasts or of demands from the users of statistical methods for better guidance. In London Karl Pearson was active and prolific in research, teaching and applications from 1893 until his retirement in 1933. An inspiring leader, he attracted into the field a number of able young men. In 1911, after some years as professor of applied mathematics, he was appointed head of a department of applied statistics, a post which he held jointly with a new professorship in eugenics. Upon his retirement the department of statistics was headed by his son, E. S. Pearson, while the chair of eugenics was also occupied by a noted mathematical statistician, R. A. Fisher.

In India Mahalanobis, professor of Physics at Calcutta, started a statistical laboratory about 1927 and carried this on at his own expense. The laboratory gradually brought together a group of young Indian mathematicians and has now become one of the outstanding centers of research in mathematical statistics in the world. These developments, it may be noted, were outside of departments of mathematics. A similar case in this country occurred at Columbia, where Hotelling, appointed professor of economics in 1931, is also a distinguished mathematical statistician. In consequence, the graduate lectures and research in mathematical statistics were at first conducted almost entirely in the department of economics.

About the same time, however, some mathematics departments began to offer work in mathematical statistics. At Michigan, for example, Carver became associate professor of mathematics and statistics in 1925 and has done much to promote the welfare of mathematical statistics in this country. Presumably as a result of his presence, Michigan now has three professors of mathematics who specialize in mathematical statistics. The late Dr. Rietz exercised a similar influence at the University of Iowa. To mention a few other dates, mathematical statisticians were appointed at Cambridge in 1931, Princeton in 1933, Wisconsin in 1937, California in 1938 and Iowa State in 1939. In several of these and similar cases, at least part of the impetus came from agricultural experiment stations which appreciated the services of a consulting statistician. At Cambridge, for instance, the Readership in statistics was established in the School of Agriculture, though some lectures are given by the Reader in the Mathematical Tripos.

---

* Delivered before the twenty-ninth annual meeting of the Mathematical Association of America at Chicago on November 25, 1945.

---

Reproduced with permission from the *American Mathematics Monthly,* Vol. LIII, No. 4, pages 193–199. Copyright © 1946, American Mathematical Association.

Thus, while statistics is at present represented in mathematics departments only to a minor extent, there are a number of institutions which offer a more or less complete graduate program in mathematical statistics.

Statistics has benefited tremendously from the influx of mathematicians of good caliber. Frequently the same statistical problem occurs in different fields of application. The applied statistician is likely to solve the problem purely in terms of the specific situation in which it arose, without concerning himself with other possible applications of his result. The mathematician goes to the essence of the problem and tackles this independently of the field of application. Consequently there has been built up a central fund of statistical theory, now fairly substantial, to which all applied fields may turn for new or improved techniques. Further benefit comes simply from the greater mathematical power available. Many statistical problems, whose solution was beyond the resources of most people in the field, yield readily to treatment by advanced mathematical methods. Consequently, it is important for the future progress of statistics that able mathematicians be attracted to specialize in that field.

**2. Sources of employment.** Before considering what constitutes suitable graduate training in statistics, it is relevant to examine briefly the job opportunities for statisticians and the types of work that these jobs involve. The "Description of the profession of statistics," issued recently by the National Roster of Scientific and Specialized Personnel, contains the following account of sources of employment.

> "Statisticians are employed in the various agencies of Federal, state and local government; in educational and research institutions, both public and private; in commercial, utility and manufacturing firms; in business, labor and other types of associations; in social agencies; in insurance companies; and in advertising and market research firms."

It is evident that there is a wide range of possible sources of employment. At the moment educational institutions probably rank as the chief single source for young mathematical statisticians. A tabulation of the present employment of 17 persons known to me who recently received advanced degrees in statistics shows that 11 have gone to colleges and universities, almost all of these to posts which combine teaching and research in mathematical statistics with the provision of consulting help to research groups on the campus. Of the remainder, 4 have gone into the federal service and 2 into industry. Opportunities in these three fields seem relatively plentiful for the near future.

**3. Type of work.** A concise definition of the functions of the statistician is extremely difficult to give. The definition given in the National Roster pamphlet is as successful as any and carries some authority, having been compiled in collaboration with the two leading statistical societies in this country.

> "The statistician uses inductive reasoning, based on the mathematics of probability, to develop and apply the most effective methods for collecting, tabulating and interpreting quantitative information."

According to this definition, two characteristics distinguish the statistician. First, he is a specialist in the techniques for collecting, tabulating and interpreting data; second, he applies inductive reasoning in these functions. To illustrate, the statistician may be asked to draw a sample of the sugar-beet farms in this country, in order to provide, at minimum cost, certain information about sugar-beet growing which is needed for policy-making. He may be asked to help in the planning of an experiment to test the effectiveness of DDT as a killer of undesirable insects. He is likely to be consulted on the best machines and procedure for some complex tabulation. He may specify the conclusions which can be drawn validly from the results of the DDT experiment.

The role of inductive reasoning is vital. In practically all cases where statistical methods are used, the data which are collected or tabulated or interpreted are of interest only insofar as they represent some larger group of data which statisticians call a population. In the sugar-beet example, the information which is wanted is that which applies to the whole group of sugar-beet growers: the sample is used only as a saver of time and money. The population in this case is real and definite, assuming that one can define a sugar-beet grower without ambiguity. Similarly, in the DDT experiment, the object is likely to be to predict what will happen in future uses of DDT under certain specified conditions. The experiment is thought of as a representative of a whole series of future trials. Although the population in this example is less tangible, it nevertheless governs the statistician's thinking when he is engaged in planning the experiment.

Thus the theory of statistics deals essentially with the properties of samples drawn in a specified way from one or more specified populations, and with the inferences that can be made about the populations from a knowledge of the samples. The collection of data is regarded by the statistician as the operation of drawing from a population a sample with certain desired properties. Similarly, the first step in the interpretation of data is to consider whether the data can be assumed to be drawn from some specified population. Moreover, these populations are characterized by *variability* in that the elements which comprise the population are different from one another. Without such variability, statistical problems are either trivial or non-existent.

Within the profession of statistics, there is a wide range in the amount of theoretical knowledge that is used in a man's work. At one extreme, some statisticians confine themselves, apart from some teaching, to research in the mathematical theory. At the other, there are statisticians who are essentially administrators of action programs and who use their basic theoretical training only in that it gives them, so to speak, a statistical point of view in their work.

These considerations suggest that a well-rounded graduate program must be flexible and should make available to the student the following features:

1. A thorough training in the theory of sampling and in any supporting mathematics necessary for that study.

2. Some knowledge, at a fairly concrete level, of the applications of statistical theory.

3. For students who may undertake consulting work, some basic knowledge of an applied field in which statistics is used.

We consider now the various parts of the training program in more detail.

**4. Training in statistical theory.** Most present graduate programs start with the assumption that the student has little or no undergraduate preparation in statistics. The students come either from mathematics or from some applied field such as economics or biology. The latter students usually possess some acquaintance with the applied phases of the subject, have had their enthusiasm aroused during an undergraduate course. The mathematicians, on the other hand, may have elected statistics simply from an interest in applied rather than pure mathematics and may have attended at most a course in probability. This lack of undergraduate preparation, which limits the amount that can be achieved in the graduate training, is, I believe, temporary, since there has been a growth in the undergraduate training in statistics concurrent with the growth in graduate training. Some time will elapse, however, before good undergraduate training is available generally.

The three basic courses in statistical theory are probability and distribution theory; statistical estimation; and the testing of statistical hypotheses. Probability appropriately comes first, since it enters into all statistical techniques. The course in probability opens with a statement of the laws of probability for discrete and for continuous variables and usually proceeds to its main topic, distribution theory. The typical problem in distribution theory may be stated as follows. Given that the random variates $x_1, x_2, \cdots, x_n$ follow a known law of probability, find the law of probability followed by the known functions $\phi_i(x_1, x_2, \cdots, x_n)$, $i = 1, 2, \cdots, k \leq n$. This type of problem arises repeatedly from applications, where the $x_i$ represent the data which are to be collected or interpreted, while the functions $\phi_i$ represent the summary statistics (*e.g.*, means, totals, percentages) on which inferences about the population are based. Further, the amount and type of information which a statistic $\phi_i$ can supply about the population clearly depends on the frequency distribution of $\phi_i$.

Several methods are available for solving distribution problems. Expressions for the desired frequency distributions can always be written down either as complex summations or as multiple integrals. The problem therefore reduces to one of performing the summations or integrations so as to obtain a reasonably simple answer. Alternatively, the method of moment-generating functions or characteristic functions provides a solution by a double application of the Fourier integral theorem. In addition, a distribution problem can sometimes be interpreted as the problem of finding the volume of some solid in multi-dimensional euclidean space and resolved by metrical methods in geometry.

Many distribution problems in statistical theory are unsolved in the sense that the integrations and summations cannot be expressed in simple terms. A rather common situation is that a solution is possible if the sample is very small, *e.g.* has, two, three or four observations, and that a limiting form of the solution is easy to find when the sample is very large. In the intermediate range the solution is lacking. Accordingly, important parts of the course on distribution theory are those dealing with various forms of the law of large numbers and with methods leading to approximate solutions of distribution problems. The course should also include an account of the properties of those frequency distributions which occur commonly in the theory, such as the binomial, Poisson, Gaussian, Chi-square, Student's $t$, Fisher's $z$, and so on.

The second basic course, statistical estimation, deals with the problem of estimating the properties of the population from a knowledge of those of the sample. Given the type of population and sample and the characteristics to be estimated, the theory of estimation seeks foolproof mathematical rules which automatically lead to the best method of estimation. For certain types of population, these rules have been found. In general, the problem is much more difficult; indeed it is not entirely clear how to define a "best" estimate. Nevertheless, substantial progress has been made along productive lines.

In many uses of statistical methods, the point at issue is whether the population is of a certain specified type, or whether two or more populations are the same in certain characteristics. For example, knowing the cost of buying and applying a certain amount of artificial fertilizer to a crop, one can calculate the minimum increase in yield which must result from the fertilizer in order to make its use profitable at a given selling price for the crop. In an agricultural experiment on this question, the data would comprise two samples, one of crop yields without applied fertilizer and one with applied fertilizer. In the interpretation of these data, the interesting question might be: does the mean yield in the latter population exceed that in the former population by at least the amount required for the application to be remunerative? This is an example of a statistical hypothesis about two populations. The *test* of the hypothesis is some calculation which enables the experimenter to decide whether the data are or are not consistent with the hypothesis. Such decisions are, of course, subject to error. The theory of testing hypotheses deals with an analysis of the various types of error and attempts to discover mathematical rules which lead to a decision with the minimum risk of error.

Thus the main body of theory teaches the student how to solve distribution problems, how to estimate properties of the population and how to test statistical hypotheses. In addition, many special problems in estimation or in testing hypotheses occur so frequently in applications that the solutions to these problems should be part of the basic knowledge of the student. Sometimes these special techniques are taught as illustrations of the theory; sometimes they are given in special courses. Examples are the method of least squares, a technique for estimation of which the fundamental theorems were established by Gauss; and multivariate analysis, which deals with the relations among groups of variables.

**5. Supporting mathematics.** The essential courses in mathematics are (1) advanced calculus, particularly multiple integration and expansions in series, both of which arise repeatedly in the solution of distribution problems, and (2) the algebra of determinants and matrices and of linear and quadratic forms, which forms the mathematical groundwork for the method of least squares and for much of the development in multivariate analysis. Both are standard courses in almost any mathematics department which gives advanced lectures. With a good grasp of these topics, the student should be able to follow most of the statistical theory.

In addition, a number of other branches of mathematics are useful in that they contribute to particular parts of statistical theory, although they do not form the thread running through all parts. How the statistician rates these depends on his predilections. Personally, I would place on the "most useful" level a course in numerical methods including interpolation, numerical integration and some finite differences and a course on complex variable as far as contour integration. Numerical methods are useful because the results of theoretical investigations generally require the construction of tables before they can be handled by the non-mathematical applied worker. Complex variable is a helpful tool in dealing with distribution problems.

On the next grade of usefulness, one might mention a number of topics: combinatory analysis, useful for discrete distribution problems; vector analysis and $n$-dimensional geometry, again useful for distribution problems: point-set and measure theory, which is needed for following some of the more fundamental developments in probability; finite groups, which has recently found applications to the theory of the design of experiments: and calculus of variations, which is the basis of the general theory of testing hypotheses. This list above covers a fairly extensive range in advanced mathematics and includes some topics which may not be taught by the mathematics department. Such topics can be dealt with either by indicating a short course of reading to the student at the appropriate time or by arranging seminars.

Since it is to be anticipated that other branches of mathematics will provide fruitful applications to statistics in the future, the list of useful supplementary mathematical knowledge is not likely to be static.

**6. Training in applied statistics.** Two types of training may be offered. The first consists of courses in which the relevant body of statistical theory is applied to some subject-matter field, such as economics, psychology, engineering or biology or to some special type of activity such as the conduct of sample surveys, the design of experiments, or the setting-up of a quality-control program in an industry. Such courses are usually attended also by workers in the applied field who wish to obtain some grasp of the contributions of statistics to that field.

Further, the student may receive some experience in the independent application of his statistical knowledge by answering letters containing queries, by assisting in consulting work and by helping to construct or criticize the plans for research projects. These tasks, in which the student must himself select the appropriate statistical theory for a problem, or develop new theory if needed, often present considerable difficulty even to the student who has a good grasp of theory.

**7. Training in an applied field.** The object is to give the student a sufficient acquaintance with the field so that he can cooperate intelligently with workers in the field and can translate statistical problems arising there into theoretical counterparts which really correspond to the situation in which the problems arose. For the mathematician, in view of time limitations, such training tends to consist of some courses which deal with the fundamental theory in the subject-matter field and some which describe current research in the field.

There is some doubt amongst statisticians as to the amount of time which should be allotted to the work in applications or in the applied field at the expense of mathematics or basic theory. The answer probably depends on the student. With a student of high general intelligence and good mathematical ability who is likely to be successful in most things that he undertakes, the best procedure may be to concentrate on theory and mathematics on the grounds that youth is the time to learn mathematics and that, should applications later attract him, he can learn what is needed by his own efforts. With an average graduate student whose development is less predictable, it should be remembered that the abilities required for theoretical statistics are considerably different from those required for applied statistics. The student's training should therefore give him the opportunity to learn whether his preferences and his skill lie in theory or in applications.

**8. Concluding remarks.** The program of study described above, if supplemented by the production of a thesis, is intended to constitute an adequate Ph.D. training. Courses for a master's degree, while naturally less comprehensive, should, I believe, follow the same general plan.

The administration of such a program requires coordinated effort from a number of departments. Where a cooperative spirit is present, this coordination can be secured in several ways: it is perhaps too early to be sure which is the best way. At both Columbia and Iowa State, for instance, the programs are at present under the general direction of a committee recruited from the departments concerned, the courses being given in the appropriate departments. At London, as has been indicated, there is a separate department of statistics. With the latter arrangement, it is advisable to have all teachers of courses in statistics at least part-time members of the department of statistics.

In conclusion, I should like to re-emphasize that statistics depends primarily on mathematics and mathematicians for its future development. Anyone interested in the progress of statistics cannot but hope that able young mathematicians will continue to be attracted into this field, which offers a wide range of useful and stimulating applications. Such mathematicians need not be regarded as lost or strayed from the fold. For while statistics has as yet contributed relatively little to repay its debt to mathematics, it is to be expected that new points of view and new problems arising in mathematical statistics will in course of time enrich the body of mathematical knowledge itself.

# 38

## RELATIVE ACCURACY OF SYSTEMATIC AND STRATIFIED RANDOM SAMPLES FOR A CERTAIN CLASS OF POPULATIONS[1]

By W. G. Cochran

*Iowa State College*

**1. Summary.** A type of population frequently encountered in extensive samplings is one in which the variance within a group of elements increases steadily as the size of the group increases. This class of populations may be represented by a model in which the elements are serially correlated, the correlation between two elements being a positive and monotone decreasing function of the distance apart of the elements. For populations of this type, the relative efficiencies are compared for a systematic sample of every $k$th element, a stratified random sample with one element per stratum and a random sample.

The stratified random sample is always at least as accurate on the average as the random sample and its relative efficiency is a monotone increasing function of the size of the sample. No general result is valid for the relative efficiency of the systematic sample. In fact, there are populations in the class in which the systematic sample is more accurate than the stratified sample for one sampling rate, but is less accurate than the random sample for another sampling rate. If, however, the correlogram is in addition concave upwards, the systematic sample is on the average more accurate than the stratified sample for any size of sample.

Some numerical results are given for the cases in which the correlogram is (i) linear (ii) exponential.

**2. Introduction.** We consider a finite population consisting of the elements $x_1$, $x_2$, $\cdots$, $x_{nk}$, where $n$ and $k$ are integers. A systematic sample is drawn by choosing an element at random from the elements $x_1$, $\cdots$, $x_k$, and then selecting every $k$th consecutive element. That is, if $x_i$ is the element first chosen, the systematic sample comprises the elements $x_i$, $x_{i+k}$, $\cdots$, $x_{i+(n-1)k}$. This type of sample has found considerable use in practice, because it is often easier to select and to administer than a random or stratified random sample and because it has an intuitive appeal through spreading the sample evenly over the population. Much remains to be learned, however, about the accuracy of this systematic sample relative to that of comparable random or restricted random samples. Probably the most relevant comparison is that between the systematic sample and the stratified random sample having one element per stratum. In the latter case, the population is divided into the $n$ strata $\{x_1, \cdots, x_k\}$, $\{x_{k+1}, \cdots, x_{2k}\}$, $\cdots$, and one element is chosen independently at random from each of the strata. This type of sample is similar in many respects to the systematic

---

[1] Journal paper No. J-1341 of the Iowa Agricultural Experiment Station, Ames, Iowa. Project 891.

---

sample. Both divide the population into the same $n$ strata of $k$ elements each, with one element chosen from each stratum. Moreover, neither sample provides the data for an unbiased estimate of the sampling variance of the sample mean, at least in the sense that the estimate is unbiased whatever the form of the population of elements $x_i$.

The first thorough investigation of the properties of systematic samples was made by W. G. and L. H. Madow [1]. In particular, these authors compared the accuracies of a systematic sample and a stratified random sample of the types described above for several types of finite population. Where the elements in the population lie on the line $x_i = i$, they showed that the stratified random sample, with one element per stratum, is more accurate than the systematic sample. If the population has a periodic distribution, the stratified random sample is superior when $k$ is an integral multiple of the period, but the systematic sample is superior when $k$ is an odd multiple of the half-period. The authors also considered the more complex case where the population contains both a trend function and a periodic function.

The object of this paper is to make similar comparisons for another type of population which appears to be fairly frequently encountered in extensive samplings. The population is one in which the variance among the elements in any group of contiguous elements increases steadily as the size of the group increases. This type of population has long been regarded as applicable in field experimental work, where the variance among plots within a block is found usually to increase with the size of block. Summarizing data from 40 uniformity trials, Fairfield Smith [2] verified this notion and derived an empirical relationship from which the rate of increase may be estimated. The same type of population is also considered in several recent papers on extensive sample surveys. Thus, in a discussion of methods for sampling farm populations, Jessen [3] postulated a law in which the variance among farms within a grid is a monotone increasing function of the size of the grid and used the law for estimating the optimum number of farms which should be included in a sampling-unit. Mahalanobis [4] independently developed the same law as Fairfield Smith in a comprehensive investigation of large-scale sample surveys. Hansen and Hurwitz [5] referred to the increase in variance within a cluster with growing size of cluster as typical of many actual populations. Numerous other references could be given.

**3. Specification of the population.** Various mathematical models may be constructed to represent the situation in which the variance within any group increases with increasing size of group. For instance, we might consider that the elements $x_i$ are drawn from different populations, the population changing in some regular manner with $i$. Alternatively, the $x_i$ may be assumed to belong to the same population, but to be serially correlated. For simplicity, we assume further that the serial correlation between $x_i$ and $x_{i+u}$ is some quantity $\rho_u$ which depends only on $u$. Then if $\rho_u$ is positive and is a monotone decreasing function of $u$, it may be expected from intuition (and will be proved later) that the variance within the group of elements $x_i$, $x_{i+1}$, $\cdots$, $x_{i+k}$ is a monotone increasing function of $k$. This model seems appropriate for our purpose, since many writers refer explicitly to positive correlations between the $x$'s as the basis for the phenomenon of increasing variance.

The specification above will be qualified in one respect. To assume that the $\rho$'s are *strictly* monotone for an actual finite population of only moderate size does not seem realistic. While the correlogram may exhibit a definite downward trend, yet individual fluctuations about the trend prevent the correlogram from being strictly monotone. It is more reasonable to regard the finite population as being itself a sample from an infinite population in which the $\rho$'s are monotone. This attitude is, I believe, in accord with that of the authors referred to above, who, as I interpret their writings, regard the variance law as holding in an idealized population. Thus, comparisons between the systematic and stratified random samples will be made not for a single finite population, but for the average of finite populations drawn from an infinite population with monotone decreasing $\rho$. Results for an individual finite population will differ from the average results because the $r$'s which appear in the population fluctuate about their expectations $\rho$. As the finite population becomes larger, its results will tend to coincide with the average results.

Accordingly, the elements $x_i$, $i = 1, 2, \cdots, nk$, are assumed to be drawn from a population in which

$$E(x_i) = \mu, \; E(x_i - \mu)^2 = \sigma^2, \; E(x_i - \mu)(x_{i+u} - \mu) = \rho_u \sigma^2$$

where $\rho_u \geq \rho_v \geq 0$, whenever $u < v$.

**4. Some useful preliminary formulas.** If $\bar{x}$ is the mean of a specified finite population, the following algebraic identity, frequently useful in the analysis of variance, is easily established.

$$(1) \qquad (kn) \sum_{i=1}^{kn} (x_i - \bar{x})^2 = \sum_{i=1}^{kn} \sum_{j>i} (x_i - x_j)^2.$$

Since there are $(kn)(kn - 1)/2$ possible pairs of values $(x_i, x_j)$, this gives

$$(2) \quad \sum_{i=1}^{kn} (x_i - \bar{x})^2 = \frac{(kn - 1)}{2} E(x_i - x_j)^2 = \frac{(kn - 1)}{2} E\{(x_i - \mu) - (x_j - \mu)\}^2$$

where $E$ is taken over the finite population. Now expand the quadratic and average over all finite populations. In the $(kn)(kn - 1)/2$ combinations, there are $(kn - 1)$ in which $j$ exceeds $i$ by 1, $(kn - 2)$ in which $j$ exceeds $i$ by 2, and so on. Hence

$$(3) \quad E \sum_{i=1}^{kn} (x_i - \bar{x})^2 = (kn - 1)\sigma^2 \left\{ 1 - \frac{2}{(kn)(kn - 1)} \sum_{u=1}^{kn-1} (kn - u)\rho_u \right\}.$$

To obtain the corresponding expectation for the sum of squares within a single stratum of $k$ consecutive elements, we need only replace $(kn)$ by $k$ in (3). Since the result is the same for all $n$ strata, we obtain

$$(4) \quad E \text{ (S. S. within strata)} = n(k - 1) \sigma^2 \left\{ 1 - \frac{2}{k(k - 1)} \sum_{u=1}^{k-1} (k - u) \rho_u \right\}.$$

Formula (3) also gives the expected sum of squares within a specified systematic sample if we replace $(kn)$ by $n$ and $u$ by $(ku)$, since there are $n$ elements in the sample and since the correlations between successive elements are $\rho_k$, $\rho_{2k}$, $\cdots$ instead of $\rho_1$, $\rho_2$, $\cdots$. The result is the same for each of the $k$ systematic samples. Hence

$$(5) \quad E \text{ (S. S. within systematic samples)} = k(n - 1) \sigma^2 \left\{ 1 - \frac{2}{n(n - 1)} \right.$$
$$\left. \cdot \sum_{u=1}^{n-1} (n - u) \rho_{ku} \right\}.$$

**5. Average variance for a random sample.** The symbols $\sigma_r^2$, $\sigma_{st}^2$, $\sigma_{sy}^2$ will be used to denote the average variances of the means of the random, stratified random and systematic samples, respectively, about the mean of the finite population, this average being taken over all finite populations drawn from the infinite population specified in the previous section. Comparisons with the random sample, though not our main purpose, will be included where they are of interest.

For a single finite population, it has been shown by several writers that the variance of the mean of a random sample is

$$(6) \qquad \frac{1}{n} \cdot \frac{(kn - n)}{(kn - 1)} \cdot \frac{1}{kn} \sum_{i=1}^{kn} (x_i - \bar{x})^2$$

where $\bar{x}$ is the mean of the finite population.

From (3), we obtain

$$(7) \qquad \sigma_r^2 = \frac{\sigma^2}{n} \left( 1 - \frac{1}{k} \right) \left\{ 1 - \frac{2}{(kn)(kn - 1)} \sum_{u=1}^{kn-1} (kn - u) \rho_u \right\}.$$

**6. Average variance for a stratified random sample.** If $\bar{x}_{st}$ is the mean of a typical stratified random sample, the sampling variance of $\bar{x}_{st}$ is by definition

$$(8) \qquad\qquad\qquad E(\bar{x}_{st} - \bar{x})^2.$$

Consider first the average over a single finite population. Let $\bar{x}_1$, $\bar{x}_2$, $\cdots$, $\bar{x}_n$ be the means of the $n$ strata, respectively, and let $x_{1j}$, $x_{2j}$, $\cdots$, $x_{nj}$ be the elements selected from the respective strata. Then (8) may be written

$$(9) \qquad \frac{1}{n^2} E \left\{ (x_{1j} - \bar{x}_1) + (x_{2j} - \bar{x}_2) + \cdots + (x_{nj} - \bar{x}_n) \right\}^2$$

since

$$\sum_{i=1}^{n} x_{ij} = n\bar{x}_{st} \text{ and } \sum_{i=1}^{n} \bar{x}_i = n\bar{x}.$$

Take the average over all $k^n$ samples from the finite population. All cross-product terms vanish, since, for example, $x_{1j}$ appears equally often with $x_{21}$, $x_{22}$, $\cdots$, $x_{2k}$. This gives

$$(10) \qquad \frac{1}{kn^2} \sum_{i=1}^{n} \sum_{j=1}^{k} (x_{ij} - \bar{x}_i)^2$$

for the variance for a single finite population. The sum of squares involved is, of course, simply the sum of squares within strata. Hence, by (4)

$$(11) \qquad \sigma_{st}^2 = \frac{\sigma^2}{n}\left(1 - \frac{1}{k}\right)\left\{1 - \frac{2}{k(k-1)} \sum_{u=1}^{k-1} (k-u)\rho_u\right\}.$$

## 7. Average variance for the systematic sample.

If $\bar{x}_{sy}$ is the mean of a typical sample, the variance for a single finite population is

$$(12) \qquad E(\bar{x}_{sy} - \bar{x})^2 = \frac{1}{kn}\{n\Sigma(\bar{x}_{sy} - \bar{x})^2\}$$

where the sum is taken over the $k$ systematic samples. Since the sum of squares among samples is equal to the total sum of squares in the population *minus* the sum of squares within samples, (12) equals

$$(13) \qquad \frac{1}{kn} \sum_{i=1}^{kn} (x_i - \bar{x})^2 - \frac{1}{kn} \text{ (S. S. within systematic samples)}.$$

To obtain the average over all finite populations we substitute from (3) and (5) for the first and second terms respectively. The result is

$$(14) \quad \sigma_{sy}^2 = \frac{(kn-1)}{kn}\sigma^2\left\{1 - \frac{2}{(kn)(kn-1)} \sum_{u=1}^{kn-1} (kn-u)\rho_u\right\}$$

$$- \frac{(n-1)}{n}\sigma^2\left\{1 - \frac{2}{n(n-1)} \sum_{u=1}^{n-1} (n-u)\rho_{ku}\right\}.$$

This reduces to

$$(15) \quad \sigma_{sy}^2 = \frac{\sigma^2}{n}\left(1 - \frac{1}{k}\right)\left\{1 - \frac{2}{kn(k-1)} \sum_{u=1}^{kn-1} (kn-u)\rho_u\right.$$

$$\left. + \frac{2k}{n(k-1)} \sum_{u=1}^{n-1} (n-u)\rho_{ku}\right\}.$$

It should be noted that the formulas and notations above are different from those used by the Madows, who define $\rho$ and $\sigma^2$ with reference to a single finite population and discuss the sample variances for a single finite population.

## 8. Relative accuracies of random and stratified random samples.

First, some general comments. From (7), (11) and (15) the relative efficiencies of the three types of sample are seen to depend only on the linear functions of the $\rho$'s which appear in $\sigma_r^2$, $\sigma_{st}^2$, and $\sigma_{sy}^2$. It is easy to verify that in each case the sum of the coefficients of the $\rho$'s is unity. For the random sample, the linear function involves every serial correlation up to lag $(kn - 1)$ with coefficients which decrease linearly as the lag increases and are independent of the size of sample, depending only on $N = (kn)$, the number of elements in the finite population. For the stratified random sample, only serial correlations with lags up to $(k - 1)$ appear, $k$ being the number of elements in the stratum. As presented in (15), the formula for the systematic sample is separated into two linear functions. The first is the same function as appears in the formula for the random sample except that all coefficients are $(kn - 1)/(k - 1)$ times as large. The second, which carries a positive sign, involves correlations where the lag is a multiple of $k$.

Thus far the formulae require no restrictions on the $\rho$'s. In considering the case where the $\rho$'s are positive and monotone decreasing, the following lemma is helpful.

LEMMA. *If* $\rho_i$, $(i = 1, \cdots, m)$, *are positive and monotone decreasing, that is,* $\rho_i \geq \rho_{i+1} > 0$ *and if* $(\alpha_1 + \alpha_2 + \cdots + \alpha_m)$ *is zero, the necessary and sufficient conditions that*

$$(16) \qquad L = \alpha_1\rho_1 + \alpha_2\rho_2 + \cdots + \alpha_m\rho_m \geq 0, \qquad \text{for all admissible sets of } \rho\text{'s,}$$

$$(17) \qquad \text{are } \alpha_1 + \alpha_2 + \cdots + \alpha_i \geq 0, i = 1, 2, \cdots, (m - 1).$$

For let $\rho_i = \rho_{i+1} + \delta_i$, where by hypothesis $\delta_i \geq 0$. Then if we substitute successively for $\rho_1, \rho_2, \cdots, \rho_{m-1}$ in terms of $\delta_1, \delta_2, \cdots, \delta_{m-1}$, we find

$$(18) \qquad L = \alpha_1\delta_1 + (\alpha_1 + \alpha_2)\delta_2 + (\alpha_1 + \alpha_2 + \alpha_3)\delta_3 + \cdots$$
$$+ (\alpha_1 + \alpha_2 + \cdots + \alpha_{m-1})\delta_{m-1},$$

the final term in $\rho_m$ vanishing because $(\alpha_1 + \cdots + \alpha_m)$ is zero. Since all $\delta_i \geq 0$, the sufficiency of (17) is obvious. Also, if for any $i$ the coefficient of $\delta_i$ is negative, we can make $L$ negative by choosing that $\delta_i$ as positive and all other $\delta$'s as zero. This establishes necessity.

COROLLARY. *If* $\rho_i$ *are strongly monotone, i.e.,* $\rho_i > \rho_{i+1}$, *and if at least one of the* $\alpha_i$ *is different from zero, conditions* (17) *are sufficient to establish that L exceeds zero.* For in (18) all the $\delta$'s are greater than zero and by (17) none of the $\delta$'s has a negative coefficient. Further, the coefficient of at least one of the $\delta$'s must exceed zero, otherwise all the $\alpha$'s would be zero. Hence $L > 0$.

We now show that if the $\rho_u$ are monotone decreasing,

$$(19) \qquad L(k) = \frac{2}{k(k - 1)} \sum_{u=1}^{k-1} (k - u)\rho_u$$

is a monotone decreasing function of $k$. This is the linear function which appears in the variance of the stratified sample.

$$(20) \quad L(k) - L(k + 1) = \frac{2}{k(k - 1)} \sum_{u=1}^{k-1} (k - u)\rho_u - \frac{2}{(k + 1)k} \sum_{u=1}^{k} (k + 1 - u)\rho_u$$

$$(21) \qquad\qquad = \frac{2}{k(k^2 - 1)} \sum_{u=1}^{k} (k + 1 - 2u)\rho_u.$$

Since the sums of the coefficients of the $\rho_u$ are unity in $L(k)$ and $L(k+1)$, the sum is zero in (21). Hence the lemma may be applied. But it is obvious that the sum of the first $i$ coefficients in (21) exceeds zero, since the coefficients are all positive for $u \le (k+1)/2$ and all negative for $u > (k+1)/2$. Hence

$$(22) \qquad\qquad L(k) - L(k+1) \ge 0.$$

Further, by the corollary, if the $\rho_u$ are strongly monotone, $L(k)$ is strongly monotone. Since all $\rho_u$ are positive, this result is sufficient to prove that

$$(23) \quad 1 - \frac{2}{k(k-1)} \sum_{u=1}^{k-1} (k-u)\rho_u \le 1 - \frac{2}{(nk)(nk-1)} \sum_{u=1}^{nk-1} (nk-u)\rho_u .$$

Consequently, for any size of sample the average variance of the stratified sample cannot exceed that of the random sample. Further, the relative efficiency of the stratified sample to the random sample is monotone increasing with decreasing size of stratum, i.e. with increasing size of sample. There is, of course, nothing unexpected in these results. Equation (22) also establishes the result mentioned in the third section, that with monotone decreasing $\rho$, the average variance within strata increases steadily as the size of stratum increases. For if $n(k-1)$ degrees of freedom are assigned to the sum of squares within strata, formula (4) above shows that the average variance within strata is

$$(24) \qquad \sigma^2 \left\{ 1 - \frac{2}{k(k-1)} \sum_{u=1}^{k-1} (k-u)\rho_u \right\} = \sigma^2 \{ 1 - L(k) \}.$$

**9. Comparison of the systematic and random samples.** Upon investigation, it is soon evident that no general results can be established about the efficiency of the systematic sample relative to the random samples, unless further restrictions are made on the form of the population. In order to apply the lemma, we find the sums of the first $i$ coefficients of the linear functions of $\rho$ which appear in the variance formulae (7), (11) and (15). By elementary methods these sums are found to be

$$\sum_r = \frac{i(2nk - i - 1)}{nk(nk-1)}$$

$$(25) \qquad \sum_{st} = \frac{i(2k - i - 1)}{k(k-1)}, \qquad 1 \le i \le (k-1)$$

$$\qquad\qquad\qquad 1 \qquad\qquad , \qquad i \ge k.$$

$$\sum_{sy} = \frac{i(2nk - i - 1)}{nk(k-1)} - \frac{rk(2n - r - 1)}{n(k-1)},$$

where $r$ is the integer such that $(r+1)k > i \ge rk$.

From the lemma, in order to establish $\sigma_{sy}^2 \le \sigma_{st}^2$, it would be necessary to show that $\Sigma_{sy} \ge \Sigma_{st}$ for any $i$. Now if $i$ is less than $k$, so that $r$ is zero, clearly

(26) $$\sum_{sy} > \sum_{st} > \sum_r, \qquad i = 1, 2, \cdots, (k - 1).$$

except when $n$ is 1, in which case all three are equal.
But if $i$ is an integral multiple of $k$, say $rk$, we find

(27) $$\sum_r = \frac{r}{n}\left[1 + \frac{(n - r)k}{(nk - 1)}\right], \qquad \sum_{st} = 1, \qquad \sum_{sy} = \frac{r}{n},$$

so that

(28) $$\sum_{st} > \sum_r > \sum_{sy}.$$

Consequently the conditions of the lemma are not satisfied with regard to the systematic sample and no general theorem exists for all populations with monotone decreasing $\rho$. The result (26) and the corollary show that for any population in this class which has $\rho_u = 0$, $u > (k - 1)$, the systematic sample is more efficient than the stratified random sample. On the other hand, (28) shows that in a population with the first $k$ of the $\rho$'s equal and the rest zero, the systematic sample has a higher variance than a random sample. If these two results are collated for a population with the first $j$ of the $\rho$'s equal and the rest zero, we see that the systematic sample with stratum size $j$ is less accurate than the comparable random sample, while the systematic sample with stratum size $(j + 1)$ is more accurate than the comparable stratified random sample. Although such a population may not occur in practice, the result suggests that the graph of the variance of the mean against the size of sample is unlikely to exhibit the same regularity for the systematic as for the random samples.

**10. Populations in which the correlogram is concave upwards.** Further investigation shows that the deciding factors in determining the relative accuracies of the systematic and random samples are the second differences of the $\rho_u$ rather than the first differences. The following result will be proved.

THEOREM: *For all infinite populations in which*

$$\rho_i \geq \rho_{i+1} \geq 0, \, i = 1, 2, \cdots, (kn - 1),$$

*and*

$$\delta_i^2 = \rho_{i-1} + \rho_{i+1} - 2\rho_i \geq 0, \, i = 2, 3, \cdots, (kn - 2),$$

*then*

$$\sigma_{sy}^2 \leq \sigma_{st}^2 \leq \sigma_r^2$$

*for any size of sample. Further, $\sigma_{sy}^2 < \sigma_{st}^2$, unless $\delta_i^2 = 0, \, i = 2, 3, \cdots, (kn - 2)$.*

This result can be proved by expressing the linear functions of the $\rho_u$ in terms of second differences and establishing a new lemma applicable to second differences. An alternative approach is simpler and perhaps more instructive.

Since the $\rho_u$ are monotone decreasing, $\sigma_{st}^2 \leq \sigma_r^2$ by the results in section 8. In (13) above, the variance of the mean of a systematic sample for a specified finite population was expressed as

$$\frac{1}{kn} \sum_{i=1}^{kn} (x_i - \bar{x})^2 - \frac{1}{kn} \text{ (Total S.S. within systematic samples)}$$

(29)

$$= \frac{1}{kn} \sum_{i=1}^{kn} (x_i - \bar{x})^2 - \frac{1}{n} \text{ (Average S.S. within a systematic sample).}$$

A corresponding equation holds for stratified random samples. For if $x_{1j}$, $x_{2j}$, $\cdots$, $x_{nj}$ are the elements of any stratified random sample with mean $\bar{x}_{st}$

(30)
$$\sum_{i=1}^{n} (x_{ij} - \bar{x})^2 = \sum_{i=1}^{n} (x_{ij} - \bar{x}_{st})^2 + n(\bar{x}_{st} - \bar{x})^2.$$

Now take the average over all $k^n$ samples. This gives

(31) $\quad \frac{1}{k} \sum_{i=1}^{kn} (x_i - \bar{x})^2 = \text{(Average S.S. within samples)} + nE(\bar{x}_{st} - \bar{x})^2.$

Since the term on the extreme right is $n$ times the variance of the stratified random sample, a result analogous to (29) follows at once.

Consequently, $\sigma_{sy}^2 \leq \sigma_{st}^2$ if the average sum of squares *within* a systematic sample is greater than or equal to that *within* a stratified random sample. Now by (2), with $n$ in place of $(kn)$, each of these averages is equal to

(32)
$$\frac{(n-1)}{2} E(x_{ij} - x_{lj})^2$$

where $x_{ij}$, $x_{lj}$ are the elements in the sample from the $i$th and the $l$th strata respectively, the average being taken over all possible pairs of strata.

We consider a fixed pair of strata and let $l - i = u$. For the systematic sample, corresponding elements in the $i$th and $l$th strata are always $(ku)$ elements apart. Hence,

(33)
$$E_{sy} (x_{ij} - x_{lj})^2 = 2\sigma^2(1 - \rho_{ku}).$$

For the stratified random sample, there are $k^2$ possible pairs of elements from the two strata. One pair is $(ku - k + 1)$ elements apart, two pairs are $(ku - k + 2)$ elements apart, and so on, the numbers of pairs rising linearly to $k$ and then decreasing linearly to one for the final pair which are $(ku + k - 1)$ elements apart. This gives

(34)
$$E_{st}(x_{ij} - x_{lj})^2 = 2\sigma^2 \left\{ 1 - \frac{1}{k^2} \sum_{i=-(k-1)}^{(k-1)} (k - |i|)\rho_{ku+i} \right\}.$$

Hence, to complete the proof that $\sigma_{sy}^2 \leq \sigma_{st}^2$, it is sufficient to show that

(35)
$$\sum_{i=-(k-1)}^{(k-1)} (k - |i|)\rho_{ku+i} - k^2 \rho_{ku} \geq 0$$

for $u = 1, 2, \cdots, (n-1)$, that is, for any pair of strata. This may be written

(36)
$$\sum_{i=1}^{(k-1)} (k - i)(\rho_{ku+i} + \rho_{ku-i} - 2\rho_{ku}) \geq 0.$$

But if $\delta^2_{ku} = \rho_{ku-1} + \rho_{ku+1} - 2\rho_{ku}$ is the second central difference it is easy to show that

$$(37) \qquad \rho_{ku+i} + \rho_{ku-i} - 2\rho_{ku} = \sum_{j=-(i-1)}^{(i-1)} (i - |j|)\delta^2_{ku+j} \geq 0,$$

since by hypothesis $\delta^2_j \geq 0$, $j = 2, 3, \cdots, (kn - 2)$. This proves that the variance between the elements of the systematic sample is greater than or equal to that between the elements of the stratified random sample for any fixed pair of strata. The result for the overall average follows. Hence $\sigma^2_{sy} \leq \sigma^2_{st}$. Further, unless $\sigma^2_j = 0$, for all $j$, clearly $\sigma^2_{sy} < \sigma^2_{st}$, except for samples of one.

The essential point in the proof may be put as follows. The elements in the $i$th and $l$th strata are on the average $(ku)$ elements apart for both the systematic and the stratified random sample. When two elements in the latter sample are $(ku + i)$ elements apart, they are less correlated than on the average, since $\rho_{ku+i} \leq \rho_{ku}$, and thus provide more independent information. The variance between the elements exceeds the systematic sample variance by $2\sigma^2(\rho_{ku} - \rho_{ku+i})$. However, such cases are counterbalanced by an equal number of cases in which the elements differ by $(ku - i)$ and the variance is below the systematic sample variance by $2\sigma^2(\rho_{ku-i} - \rho_{ku})$. Because of the concavity of $\rho_u$, the losses on the average balance or outweigh the gains.

For the population discussed in section 9, in which $\rho_u = \rho$, $u = 1, 2, \cdots, j$, $\rho_u = 0$, $u > j$, we have $\delta^2_j < 0$, $\delta^2_{j+1} > 0$, and $\delta^2_u = 0$ otherwise. This reversal of the sign of the second difference is the explanation for the anomalous behavior of the systematic samples with stratum sizes $j$ and $(j + 1)$.

The theorem above does *not* prove that the relative accuracy of the systematic to the stratified random sample is a monotone function of $n$, nor even that $\sigma^2_{sy}$ decreases steadily as $n$ increases. Actually, there are populations in the class for which neither result holds, as will be illustrated in the next section.

So far as practical applications are concerned, the restriction that the $\rho_u$ should be concave upwards may not be severe. For instance, this condition is satisfied when the correlogram is linear, i.e. $\rho_u = (l - u)/l$, this being one type of correlogram which Wold [6] has considered applicable to economic data. Concavity also holds for the function $\rho_u = e^{-\lambda u}$ which Osborne [7] has suggested for forestry and land-use surveys and for the relation $\rho_u = \tanh(u^{-3/5})$ which Fisher and Mackenzie [8] used for expressing the correlation between the weekly rain at two weather stations as a function of their distance apart. In fact, if $\rho_u$ is conceived of as positive and continuous for all $u$, a concave upwards function suggests itself naturally.

**11. Linear correlograms.** It may be of interest to present some results obtained when the correlogram is (i) linear, (ii) exponential, since both types have been suggested as possible models for populations occurring in practice.

In the linear case,

$$(38) \qquad \rho_u = (L - u)/L, \, u \leq L; \quad \rho_u = 0, \, u > L.$$

If $L \geq (nk - 1)$, the correlogram is a *straight* line throughout the whole range of the finite population. Since all second differences are zero in this case, we may expect $\sigma_{sy}^2 = \sigma_{st}^2 < \sigma_r^2$. If $L < (nk - 1)$, all second differences vanish except $\delta_L^2$, which is positive. Hence we may expect $\sigma_{sy}^2 < \sigma_{st}^2 < \sigma_r^2$.

The results for these cases are found by elementary summations from the basic formulae (7), (11) and (15). Details of the summations will not be presented. For $L \geq (nk - 1)$, we find

$$(39) \quad \sigma_{sy}^2 = \sigma_{st}^2 = \frac{\sigma^2}{n}\left(1 - \frac{1}{k}\right)\frac{(k+1)}{3L}; \qquad \sigma_r^2 = \frac{\sigma^2}{n}\left(1 - \frac{1}{k}\right)\frac{(nk+1)}{3L}.$$

The ratio $\sigma_r^2/\sigma_{sy}^2$ is $(nk + 1)/(k + 1)$, which is approximately equal to $n$, the size of sample, unless the percentage sampled is large. Thus very large gains in efficiency over random sampling are obtained.

If $L < (nk - 1)$, the formulae are less simple. Consider first $k \geq L$; that is, cases where the percentage sampled is less than $100/L$. If $N = nk$,

$$(40) \qquad \sigma_r^2 = \frac{\sigma^2}{n}\left(1 - \frac{1}{k}\right)\left\{\frac{3N(N-L) + (L^2 - 1)}{3N(N-1)}\right\}$$

$$(41) \qquad \sigma_{st}^2 = \frac{\sigma^2}{n}\left(1 - \frac{1}{k}\right)\left\{\frac{3k(k-L) + (L^2 - 1)}{3k(k-1)}\right\}, \qquad\qquad k \geq L$$

$$(42) \qquad \sigma_{sy}^2 = \frac{\sigma^2}{n}\left(1 - \frac{1}{k}\right)\left\{\frac{3N(k-L) + (L^2 - 1)}{3N(k-1)}\right\}, \qquad\qquad k \geq L.$$

It is clear on inspection that $\sigma_{sy}^2 < \sigma_{st}^2$; moreover, it is easy to show that the efficiency of systematic relative to stratified random sampling increases steadily as the size of sample increases.

When the size of sample is increased further so that $k \leq L$, formula (40) remains unchanged, while $\sigma_{st}^2$ is now given by the same formula as in (39). The formula for $\sigma_{sy}^2$ is more complex. If $q$ is the integral part of the quotient when $L$ is divided by $k$ and $r$ is the remainder, so that $L = (qk + r)$, the formula may be written

$$(42') \qquad \begin{aligned} \sigma_{sy}^2 = &\frac{\sigma^2}{n}\left(1 - \frac{1}{k}\right) \\ &\cdot\left\{\frac{qk(k^2 - 1) + 3rk(n-q)(k-r) + r(r^2 - 1)}{3NL(k-1)}\right\}, \quad k \leq L. \end{aligned}$$

It is noteworthy that the last two terms in the numerator inside the curly bracket vanish whenever $L$ is exactly divisible by $k$. Further, the second term is of order $nk = N$ and, when present, exerts a much greater weight than the first term. Thus $\sigma_{sy}^2$ takes a sudden dip whenever $L$ is a multiple of $k$. In fact, for $L = qk$, (42') reduces to

$$(43) \qquad \sigma_{sy}^2 = \frac{\sigma^2}{n}\left(1 - \frac{1}{k}\right)\frac{(k+1)}{3N}, \qquad\qquad L = qk,$$

so that the variance goes to zero if $N$ is sufficiently large. By comparison with formula (39) for $\sigma_{st}^2$ we see that when $L = qk$ the relative efficiency of systematic to stratified random sampling is $N/L$, which increases beyond bound if $N$ is sufficiently large. In intermediate cases, when the remainder $r$ does not vanish, the leading term in the relative efficiency for $N$ large is $(k^2 - 1)/3r(k - r)$. This varies somewhat irregularly, depending on the relation between $L$ and $k$,

To illustrate, numerical values are given below when $L = 10$ and the finite population is large enough so that terms in $1/n$ are negligible.

The quantities $v_{st}$, $v_{sy}$ are the corresponding variances apart from a factor $\sigma^2/N$. The stratified sample variance decreases steadily with increasing percentage sampled. On the other hand the systematic sample variance goes to zero and the relative efficiency to infinity when $k$ is 2, 5 or 10. Moreover, in the intermediate cases $k = 3, 4, 6, 7, 8, 9$, the variance and the relative efficiency show no consistent relation to the percentage sampled. For samples of less than 10 per cent, including the cases outside the limits of the table, the relative efficiency decreases steadily from 4 at $k = 11$ to 1 when $k$ is large.

TABLE 1

*Variances except for a factor $\sigma^2/N$ and relative efficiency for systematic and stratified random samples for a linear correlogram*

| $k$ | 2 | 3 | 4 | 5 | 6 | 7 | 8 | 9 | 10 | 11 | 20 |
|---|---|---|---|---|---|---|---|---|---|---|---|
| % Sampled | 50 | 33 | 25 | 20 | 17 | 14 | 12 | 11 | 10 | 9 | 5 |
| $v_{st}$ | .10 | .27 | .50 | .80 | 1.17 | 1.60 | 2.10 | 2.67 | 3.30 | 4.00 | 11.65 |
| $v_{sy}$ | 0 | .20 | .40 | 0 | .80 | 1.20 | 1.20 | .80 | 0 | 1.00 | 10.00 |
| $v_{st}/v_{sy}$ | $\infty$ | 1.33 | 1.25 | $\infty$ | 1.46 | 1.33 | 1.75 | 3.33 | $\infty$ | 4.00 | 1.16 |

**12. Exponential correlograms.** For the exponential $\rho_u = e^{-\lambda u}$ the results are much more regular. Each of the linear functions of the $\rho$'s consists of a finite number of terms of an expansion of the form $(1 - x)^{-2}$. If

$$(44) \qquad f(N, \lambda) = \frac{2}{N(N - 1)} \left\{ \frac{(N - 1)e^{\lambda} - N + e^{-(N-1)\lambda}}{(e^{\lambda} - 1)^2} \right\}$$

which is the sum for $\sigma_r^2$, we find

$$(45) \qquad \sigma_r^2 = \frac{\sigma^2}{n} \left( 1 - \frac{1}{k} \right) \{1 - f(N, \lambda)\}$$

$$(46) \qquad \sigma_{st}^2 = \frac{\sigma^2}{n} \left( 1 - \frac{1}{k} \right) \{1 - f(k, \lambda)\}$$

$$(47) \qquad \sigma_{sy}^2 = \frac{\sigma^2}{n} \left( 1 - \frac{1}{k} \right) \left\{ 1 - \frac{(N - 1)}{(k - 1)} f(N, \lambda) + \frac{k(n - 1)}{(k - 1)} f(n, k\lambda) \right\}.$$

It may be shown that the variance of the systematic sample decreases steadily and its efficiency relative to stratified sampling increases steadily as the sample becomes larger.

In order to obtain some idea of the magnitude of the gain in efficiency, consider the case where $k$ and $n$ are large. For this case the relative efficiency, which actually is a function of $k$, $n$ and $\lambda$, turns out to depend almost entirely on the single quantity $(k\lambda)$; or, equally, on the correlation $e^{-k\lambda}$ between the items in successive strata in the systematic sample. If $t = (k\lambda)$, we obtain $\sigma_r^2 = \sigma^2/n$,

$$(48) \qquad \sigma_{st}^2 = \frac{\sigma^2}{n}\left\{1 - \frac{2}{t} + \frac{2}{t^2} - \frac{2e^{-t}}{t^2}\right\},$$

$$(49) \qquad \sigma_{sy}^2 = \frac{\sigma^2}{n}\left\{1 - \frac{2}{t} + \frac{2}{(e^t - 1)}\right\}.$$

The relative efficiency is given in Table 2 for a selection of values of $e^{-t}$, the correlation between the items in successive strata.

The relative efficiency has a limiting value 2 when $\rho$ tends to 1 and decreases slowly towards 1 as $\rho$ falls to zero. The gains in efficiency are quite substantial if $\rho$ exceeds 0.1.

TABLE 2

*Relative efficiency of systematic and stratified random samples for an exponential correlogram*

| $\rho$ | .9 | .8 | .7 | .6 | .5 | .4 | .3 | .2 | .1 |
|---|---|---|---|---|---|---|---|---|---|
| $\sigma_{st}^2/\sigma_{sy}^2$ | 1.96 | 1.90 | 1.84 | 1.78 | 1.71 | 1.64 | 1.55 | 1.46 | 1.33 |

It was pointed out in section 1 that no unbiased estimate of error is available from a single sample for either the systematic or the stratified random sample. This does not mean that no estimate of error can be attempted. However, any estimate must depend on certain assumptions about the form of the population which is being sampled and is likely to be vitiated insofar as these assumptions are false. If, for instance, the correlogram were assumed to be exponential, formula (47), or (49) in the particular case with $n$, $k$ large, would appear to be the appropriate basis for the estimation of error from a single systematic sample. Consider the simpler case in which (49) is valid. The correlation between successive items in the systematic sample provides an estimate of $e^{-t}$ and hence of $t$. Also, if terms in $1/n$ are negligible, the mean square within the systematic sample is found to be an unbiased estimate of $\sigma^2$. By substitution in (49) a consistent estimate of the variance of a single systematic sample would be secured, provided that the exponential assumption were correct. The gains in efficiency over stratified and random sampling could also be estimated.

REFERENCES

[1] W. G. AND L. H. MADOW, "On the theory of systematic sampling," *Annals of Math. Stat.*, Vol. 15 (1944), pp. 1–24.

[2] H. Fairfield Smith, "An empirical law governing soil heterogeneity," *Jour. Agr. Sci.*, Vol. 28 (1938), pp. 1–23.

[3] R. J. Jessen, "Statistical investigation of a sample survey for obtaining farm facts," *Iowa Agr. Exp. Sta., Res. Bull.*, No. 304, (1942).

[4] P. C. Mahalanobis, "On large scale sample surveys," *Roy. Soc. Phil. Trans.*, B231 (1944), pp. 329–451.

[5] M. H. Hansen and W. N. Hurwitz, "On the theory of sampling from finite populations," *Annals of Math. Stat.*, Vol. 14 (1943), pp. 333–362.

[6] H. Wold, "A study of the analysis of stationary time series," *Uppsala* (1938).

[7] J. G. Osborne, "Sampling errors of systematic and random surveys of cover-type areas," *Amer. Stat. Assoc. Jour.*, Vol. 37 (1942), pp. 256–270.

[8] R. A. Fisher and W. A. Mackenzie, "The correlation of weekly rainfall," *Quart. Jour. Roy. Met. Soc.*, Vol. 48 (1922), pp. 234–245.

## SOME CONSEQUENCES WHEN THE ASSUMPTIONS FOR THE ANALYSIS OF VARIANCE ARE NOT SATISFIED

W. G. Cochran

*Institute of Statistics, North Carolina State College*

1. *Purposes of the Analysis of Variance.* The main purposes are:

(i) To estimate certain treatment differences that are of interest. In this statement both the words "treatment" and "difference" are used in rather a loose sense: e.g., a treatment difference might be the difference between the mean yields of two varieties in a plant-breeding trial, or the relative toxicity of an unknown to a standard poison in a dosage-mortality experiment. We want such estimates to be *efficient.* That is, speaking roughly, we want the difference between the estimate and the true value to have as small a variance as can be attained from the data that are being analyzed.

(ii) To obtain some idea of the accuracy of our estimates, e.g., by attaching to them estimated standard errors, fiducial or confidence limits, etc. Such standard errors, etc., should be reasonably free from bias. The usual property of the analysis of variance, when all assumptions are fulfilled, is that estimated variances are unbiased.

(iii) To perform tests of significance. The most common are the *F*-test of the null hypothesis that a group of means all have the same true value, and the *t*-test of the null hypothesis that a treatment difference is zero or has some known value. We should like such tests to be *valid,* in the sense that if the table shows a significance probability of, say, 0.023, the chance of getting the observed result or a more discordant one on the null hypothesis should really be 0.023 or something near it. Further, such tests should be *sensitive* or *powerful,* meaning that they should detect the presence of real treatment differences as often as possible.

The object of this paper is to describe what happens to these desirable properties of the analysis of variance when the assumptions required for the technique do not hold. Obviously, any practical value of the paper will be increased if advice can also be given on how to detect failure of the assumptions and how to avoid the more serious consequences.

2. *Assumptions Required for the Analysis of Variance.* In setting up an analysis of variance, we generally recognize three types of effect:

(a) treatment effects—the effects of procedures deliberately introduced by the experimenter

(b) environmental effects (the term is not ideal)—these are certain features of the environment which the analysis enables us to measure. Common examples are the effects of replications in a randomized blocks experiment, or of rows and columns in a Latin square

(c) experimental errors—this term includes all elements of variation that are not taken account of in (a) or (b).

The assumptions required in the analysis of variance for the properties listed as desirable in section 1 are as follows:

(1) The treatment effects and the environmental effects must be additive. For instance, in a randomized blocks trial the observation $y_{ij}$ on the $i^{th}$ treatment in the $j^{th}$ replication is specified as

$$y_{ij} = \mu + \tau_i + \rho_j + e_{ij}$$

where $\mu$ is the general mean, $\tau_i$ is the effect of the $i^{th}$ treatment, $\rho_j$ is the effect of the $j^{th}$ replication and $e_{ij}$ is the experimental error of that observation. We may assume, without loss of generality, that the $e$'s all have zero means.

(2) The experimental errors must all be independent. That is, the probability that the error of any observation has a particular value must not depend on the values of the errors for other observations.

(3) The experimental errors must have a common variance.[1]

(4) The experimental errors should be normally distributed.

We propose to consider each assumption and to discuss the consequences when the assumption is not satisfied. The discussion will be in rather general terms, for much more research would be needed in order to make precise statements. Moreover, in practice several assumptions may fail to hold simultaneously. For example, in non-normal distributions there is usually a correlation between the variance of an observation and its mean, so that failure of condition (4) is likely to be accompanied by failure of (3) also.

3. *Previous Work on the Effects of Non-normality.* Most of the published work on the effects of failures in the assumptions has been

[1] This statement, though it applies to the simplest analyses, is an oversimplification. More generally, the analysis of variance should be divisible into parts within each of which the errors have common variance. For instance, in the split-plot design, we specify one error variance for whole-plot comparisons and a different one for subplot comparisons.

23

concerned with this item. Writing in 1938, Hey (8) gives a bibliography of 36 papers, most of which deal with non-normality, while several theoretical investigations were outside the scope of his bibliography. Although space does not permit a detailed survey of this literature, some comments on the nature of the work are relevant.

The work is almost entirely confined to a single aspect, namely the effect on what we have called the validity of tests of significance. Further, insofar as the $t$-test is discussed, this is either the test of a single mean or of the difference between the means of two groups. As will be seen later, it is important to bear this restriction in mind when evaluating the scope of the results.

Some writers, e.g., Bartlett (1), investigated by mathematical methods the theoretical frequency distribution of $F$ or $t$, assuming the null hypothesis true, when sampling from an infinite population that was non-normal. As a rule, it is extremely difficult to obtain the distributions in such cases. Others, e.g., E. S. Pearson (9), drew mechanically 500 or 1000 numerical samples from an infinite non-normal population, calculated the value of $F$ or $t$ for each sample, and thus obtained empirically some idea of their frequency distributions. Where this method was used, the number of samples was seldom large enough to allow more than a chi-square goodness of fit test of the difference between the observed and the standard distributions. A very large number of samples is needed to determine the 5 percent point, and more so the 1 percent point, accurately. A third method, of which Hey's paper contains several examples, is to take actual data from experiments and generate the $F$ or $t$ distribution by means of randomization similar to that which would be practiced in an experiment. The data are chosen, of course, because they represent some type of departure from normality.

The consensus from these investigations is that no serious error is introduced by non-normality in the significance levels of the $F$-test or of the two-tailed $t$-test. While it is difficult to generalize about the range of populations that were investigated, this appears to cover most cases encountered in practice. If a guess may be made about the limits of error, the true probability corresponding to the tabular 5 percent significance level may lie between 4 and 7 percent. For the 1 percent level, the limits might be taken as $\frac{1}{2}$ percent and 2 percent. As a rule, the tabular probability is an underestimate: that is, by using the ordinary $F$ and $t$ tables we tend to err in the direction of announcing too many significant results.

24

The one-tailed $t$-test is more vulnerable. With a markedly skew distribution of errors, where one tail is much longer than the other, the usual practice of calculating the significance probability as one-half the value read from the tables may give quite a serious over- or under-estimate.

It was pointed out that work on the validity of the $t$-test covered only the cases of a single mean or of the comparison of the means of two groups. The results would be applicable to a randomized blocks experiment if we adopted the practice of calculating a separate error for each pair of treatments to be tested, using only the data from that pair of treatments. In practice, however, it is usual to employ a pooled error for all $t$-tests in an analysis, since this procedure not only saves labor but provides more degrees of freedom for the estimation of error. It will be shown in section 6 that this use of a pooled error when non-normality is present may lead to large errors in the significance probabilities of individual $t$-tests. The same remark applies to the Latin square and more complex arrangements, where in general it is impossible to isolate a separate error appropriate to a given pair of treatments, so that pooling of errors is unavoidable.

4. *Further Effects of Non-Normality.* In addition to its effects on the validity of tests of significance, non-normality is likely to be accompanied by a loss of efficiency in the estimation of treatment effects and a corresponding loss of power in the $F$- and $t$-tests. This loss of efficiency has been calculated by theoretical methods for a number of types of non-normal distribution. While these investigations dealt with the estimation of a single mean, and thus would be strictly applicable only to a paired experiment analyzed by the method of differences, the results are probably indicative of those that would be found for more complex analyses. In an attempt to use these results for our present purpose, the missing link is that we do not know which of the theoretical non-normal distributions that have been studied are typical of the error distributions that turn up in practice. This gap makes speculation hazardous, because the efficiency of analysis of variance methods has been found to vary from 100 percent to zero. While I would not wish to express any opinion very forcibly, my impression is that in practice the loss of efficiency is not often great. For instance, in an examination of the Pearson curves, Fisher (4) has proved that for curves that exhibit only a moderate departure from normality, the efficiency remains reasonably high. Further, an analysis of the logs of the observations instead of the observations themselves has fre-

25

quently been found successful in converting data to a scale where errors are approximately normally distributed. In this connection, Finney, (3) has shown that if log $x$ is exactly normally distributed, the arithmetic mean of $x$ has an efficiency greater than 93 percent so long as the coefficient of variation of $x$ is less than 100 percent. In most lines of work a standard error as high as 100 percent per observation is rare, though not impossible.

The effect of non-normality on estimated standard errors is analogous to the effect on the $t$-test. If a standard error is calculated specifically for each pair of treatments whose means are to be compared, the error variance is unbiased. Bias may arise, however, by the application of a pooled error to a particular pair of treatments.

We now consider how to detect non-normality. It might perhaps be suggested that the standard tests for departure from normality, Fisher (5), should be applied to the errors in an analysis. This suggestion is not fruitful, however, because for experiments of the size usually conducted, the tests would detect only very violent skewness or kurtosis. Moreover, as is perhaps more important, it is not enough to detect non-normality: in order to develop an improved analysis, one must have some idea of the actual form of the distribution of errors, and for this purpose a single experiment is rarely adequate.

Examination of the distribution of errors may be helpful where an extensive uniformity trial has been carried out, or where a whole series of experiments on similar material is available. Theoretically, the best procedure would be to try to find the form of the frequency distribution of errors, using, of course, any *a priori* knowledge of the nature of the data. An improved method of estimation could then be developed by maximum likelihood. This, however, would be likely to lead to involved computations. For that reason, the usual technique in practice is to seek, from *a priori* knowledge or by trial and error, a transformation that will put the data on a scale where the errors are approximately normal. The hope is that in the transformed scale the usual analysis will be reasonably efficient. Further, we would be prepared to accept some loss in efficiency for the convenience of using a familiar method. Since a detailed account of transformations will be given by Dr. Bartlett in the following paper, this point will not be elaborated.

The above remarks are intended to apply to the handling of a rather extensive body of data. With a single experiment, standing by itself, experience has indicated two features that should be watched for:

26

(i) evidence of charges in the variance from one part of the experiment to another. This case will be discussed in section 6.

(ii) evidence of gross errors.

5. *Effects of Gross Errors.* The effects of gross errors, if undetected, are obvious. The means of the treatments that are affected will be poorly estimated, while if a pooled error is used the standard errors of other treatment means will be over-estimated. An extreme example is illustrated by the data in Table I, which come from a randomized blocks experiment with four replicates.

TABLE I

WHEAT: RATIO OF DRY TO WET GRAIN

| Block | Nitrogen applied | | | |
|-------|------|-------|--------|------|
|       | None | Early | Middle | Late |
| 1 | .718 | .732 | .734 | .792 |
| 2 | .725 | .781 | .725 | .716 |
| 3 | .704 | 1.035 | .763 | .758 |
| 4 | .726 | .765 | .738 | .781 |

As is likely to happen when the experimenter does not scrutinize his own data, the gross error was at first unnoticed when the computor carried out the analysis of variance, though the value is clearly impossible from the nature of the measurements. This fact justifies rejection of the value and substitution of another by the method of missing plots, Yates (11).

Where no explanation can be found for an anomalous observation, the case for rejection is more doubtful. Habitual rejection of outlying values leads to a marked underestimation of errors. An approximate test of significance of the contribution of the suspected observation to the error helps to guard against this bias. First calculate the error sum of squares from the actual observations. Then calculate the error when the suspected value is replaced by the missing-plot estimate: this will have one less degree of freedom and is designated the "Remainder" in the data below. The difference represents the sum of squares due to the suspect. For the data above, the results are

|  | d.f. | S.S. | M.S. |
|--|------|------|------|
| Actual error | 9 | .04729 | .00525 |
| Suspect | 1 | .04205 | .04205 |
| Remainder | 8 | .00524 | .000655 |

27

Alternatively, the contribution due to the suspected observation may be calculated directly and the remainder found by subtraction. If there are $t$ treatments and $r$ replicates, the sum of squares is $(t-1)$ $(r-1)d^2/tr$, where $d$ is the difference between the suspected observation and the value given by the missing-plot formula. In the present case $t$ and $r$ are 3 and the missing-plot value is 0.7616, so that the contribution is $9(0.2734)^2/16$, or 0.04205.[2]

The $F$ ratio for the test of the suspect against the remainder is 64.2, giving a $t$ value of 8.01, with 8 degrees of freedom. Now, assuming that the suspect had been examined simply because it appeared anomalous, with no explanation for the anomaly, account must be taken of this fact in the test of significance. What is wanted is a test appropriate to the *largest* contribution of any observation. Such a test has not as yet been developed. The following is suggested as a rough approximation. Calculate the significance probability, $p$, by the ordinary $t$ table. Then use as the correct significance probability $np$, where $n$ is the number of degrees of freedom in the actual error.[3] In the present case, with $t = 8.01$, $p$ is much less than 1 in a million, and consequently $np$ is less than 1 in 100,000. In general, it would be wise to insist on a rather low significance probability (e.g., 1 in 100) before rejecting the suspect, though a careful answer on this point requires knowledge of the particular types of error to which the experimentation is subject.

6. *Effects of Heterogeneity of Errors.* If ordinary analysis of variance methods are used when the true error variance differs from one observation to another, there will as a rule be a loss of efficiency in the estimates of treatment effects. Similarly, there will be a loss of sensitivity in tests of significance. If the changes in the error variance are large, these losses may be substantial. The validity of the $F$-test for all treatments is probably the least affected. Since, however, some treatment comparisons may have much smaller errors than others, $t$-tests from a pooled error may give a serious distortion of the significance levels. In the same way the standard errors of particular treatment comparisons, if derived from a pooled error, may be far from the true values.

[2] This formula applies only to randomized blocks. Corresponding formulas can be found for other types of arrangements. For instance, the formula for a $p \times p$ Latin square is $(p-1)(p-2)d^2/p^2$.

[3] The approximation is intended only to distinguish quickly whether the probability is low or high and must not be regarded as accurate. For a discussion of this type of test in a somewhat simpler case, see E. S. Pearson and C. Chandra Sekar, *Biometrika*, Vol. 28 (1936), pp. 308–320.

28

There is no theoretical difficulty in extending the analysis of variance so as to take account of variations in error variances. The usual analysis is replaced by a weighted analysis in which each observation is weighted in proportion to the inverse of its error variance. The extension postulates, however, a knowledge of the relative variances of any two observations and this knowledge is seldom available in practice. Nevertheless, the more exact theory can sometimes be used with profit in cases where we have good estimates of these relative variances. Suppose for instance, the situation were such that the observations could be divided into three parts, the error variances being constant within each part. If unbiased estimates of the variances within each part could be obtained and if these were each based on, say, at least 15 degrees of freedom, we could recover most of the loss in efficiency by weighting inversely as the observed variances. This device is therefore worth keeping in mind, though in complex analyses the weighted solution involves heavy computation.

TABLE II

MANGOLDS, PLANT NUMBERS PER PLOT

| Block | Control | | Chalk | | | Lime | | | Total |
|---|---|---|---|---|---|---|---|---|---|
| | 0 | 0 | 1 | 2 | 3 | 1 | 2 | 3 | |
| I | 140 | 49 | 98 | 135 | 117 | 81 | 147 | 130 | 897 |
| II | 142 | 37 | 132 | 151 | 137 | 129 | 131 | 112 | 971 |
| III | 36 | 114 | 130 | 143 | 137 | 135 | 103 | 130 | 928 |
| IV | 129 | 125 | 153 | 146 | 143 | 104 | 147 | 121 | 1068 |
| Total | 447 | 325 | 513 | 575 | 534 | 449 | 528 | 493 | 3864 |
| Range | 106 | 88 | 55 | 16 | 26 | 54 | 44 | 18 | |

Heterogeneity of errors may arise in several ways. It may be produced by mishaps or damage to some part of the experiment. It may be present in one or two replications through the use of less homogeneous material or of less carefully controlled conditions. The nature of the treatments may be such that some give more variable responses than others. An example of this type is given by the data in Table II.

The experiment investigated the effects of three levels of chalk dressing and three of lime dressing on plant numbers of mangolds. There were four randomized blocks of eight plots each, the control plots being replicated twice within each block.[4]

Since the soil was acid, high variability might be anticipated for the

[4] The same data were discussed (in much less detail) in a previous paper, Cochran (2).

29

control plots as a result of partial failures on some plots. The effect is evident on eye inspection of the data. To a smaller extent the same effect is indicated on the plots receiving the single dressing of chalk or lime. If the variance may be regarded as constant within each treatment, there will be no loss of efficiency in the treatment means in this case, contrary to the usual effect of heterogeneity. Any $t$ tests will be affected and standard errors may be biased. In amending the analysis so as to avoid such disturbances, the first step is to attempt to subdivide the error into homogeneous components. The simple analysis of variance is shown below.

TABLE III

ANALYSIS OF VARIANCE FOR MANGOLDS DATA

|  | d.f. | S.S. | M.S. |
|---|---|---|---|
| Blocks | 3 | 2,079 | ............ |
| Treatments | 6 | 8,516 | ............ |
| Error | 22 | 18,939 | 860.9 |
| Total | 31 | 29,534 | ............ |

For subdivision of the error we need the following auxiliary data.

| Block | Diff. between Controls | Total − 4 (Controls) | (C1–L1) | (C2 + L2 + C3 + L3) − 2(C1 + L1) |
|---|---|---|---|---|
| 1 | 91 | 141 | 17 | 171 |
| 2 | 105 | 255 | 3 | 9 |
| 3 | 78 | 328 | − 5 | − 17 |
| 4 | 4 | 52 | 49 | 43 |
| Total | ........ | 776 | 64 | 206 |
| Divisor for S.S. | 2 | 24 | 2 | 12 |

The first two columns are used to separate the contribution of the controls to the error. This has 7 d.f. of which 4 represent differences between the two controls in each block. The sum of squares of the first column is divided by 2 as indicated. There remain 3 d.f. which come from a comparison within each block of the total yield of the controls with the total yield of the dressings. Since there are 6 dressed plots to 2 controls per block we take

(Dressing total) − 3(Control total) = (Total) − 4(Control total)

Thus $141 = 897 - 4(140 + 49)$.

By the usual rule the divisor for the sum of squares of deviations is 24.

Two more columns are used to separate the contribution of the single dressings. There are 6 d.f. of which 3 compare chalk with line at this level while the remaining 3 compare the single level with the higher levels. The resulting partition of the error sum of squares is shown below.

TABLE IV

PARTITION OF ERROR SUM OF SQUARES

|  | d.f. | S.S. | M.S. |
|---|---|---|---|
| Total | 22 | 18,939 | 861 |
| Between controls | 4 | 12,703 | 3,176 |
| Controls v. Dressings | 3 | 1,860 | 620 |
| Chalk 1 v. Lime 1 | 3 | 850 | 283 |
| Single v. Higher Dressings | 3 | 1,738 | 579 |
| Double and Triple Dressings | 9 | 1,788 | 199 |

As an illustration of the disturbance to $t$-tests and to estimated standard errors, we may note that the pooled mean square, 861, is over four times as large as the 9 d.f. error, 199, obtained from the double and triple dressings. Consequently, the significance levels of $t$ and standard errors would be inflated by a factor of two if the pooled error were applied to comparisons within the higher dressings.

In a more realistic approach we might postulate three error variances, $\sigma_c^2$ for controls, $\sigma_1^2$ for single dressings and $\sigma_h^2$ for higher dressings. For these we have unbiased estimates of 3,176, 283 and 199 respectively from Table IV. The mean square for Controls v. Dressings (620) would be an unbiased estimate of $(9\sigma_c^2 + \sigma_1^2 + 2\sigma_h^2)/12$, while that for Single v. Higher Dressings (579) would estimate $(2\sigma_1^2 + \sigma_h^2)/3$.

What one does in handling comparisons that involve different levels depends on the amount of refinement that is desired and the amount of work that seems justifiable. The simplest process is to calculate a separate $t$-test or standard error for any comparison by obtaining the comparison separately within each block. Such errors, being based on 3 d.f., would be rather poorly determined. A more complex but more efficient approach is to estimate the three variances from the five mean squares given above. Since the error variance of any comparison will be some linear function of these three variances, it can then be estimated.

To summarize, heterogeneity of errors may affect certain treatments or certain parts of the data to an unpredictable extent. Sometimes, as in the previous example, such heterogeneity would be expected in ad-

31

vance from the nature of the experiment. In such cases the data may be inspected carefully to decide whether the actual amount of variation in the error variance seems enough to justify special methods. In fact, such inspection is worthwhile as a routine procedure and is, of course, the only method for detecting heterogeneity when it has not been anticipated. The principal weapons for dealing with this irregular type of heterogeneity are subdivision of the error variance or omission of parts of the experiment. Unfortunately, in complex analyses the computations may be laborious. For the Latin square, Yates (12) has given methods for omitting a single treatment, row or column, while Yates and Hale (14) have extended the process to a pair of treatments, rows or columns.

In addition, there is a common type of heterogeneity that is more regular. In this type, which usually arises from non-normality in the distribution of errors, the variance of an observation is some simple function of its mean value, irrespective of the treatment or block concerned. For instance, in counts whose error distribution is related to the Poisson, the variance of an observation may be proportional to its mean value. Such cases, which have been most successfully handled by means of transformations, are discussed in more detail in Dr. Bartlett's paper.

7. *Effects of Correlations Amongst the Errors.* These effects may be illustrated by a simple theoretical example. Suppose that the errors $e_1, e_2, \ldots, e_r$ of the $r$ observations on a treatment in a simple group comparison have constant variance $\sigma^2$ and that every pair has a correlation coefficient $\rho$. The error of the treatment total, $(e_1 + e_2 \ldots, + e_r)$ will have a variance

$$r\sigma^2 + r(r-1)\rho\sigma^2$$

since there are $r(r-1)/2$ cross-product terms, each of which will contribute $2\rho\sigma^2$. Hence the *true* variance of the treatment mean is

$$\sigma^2\{1 + (r-1)\rho\}/r.$$

Now in practice we would estimate this variance by means of the sum of squares of deviations within the group, divided by $r(r-1)$. But

$$\text{Mean } \Sigma(e_i - \bar{e})^2 = \text{Mean } \Sigma \ e_i{}^2 - r \ \{\text{Mean } \bar{e}^2\}$$
$$= r\sigma^2 - \sigma^2\{1 + (r-1)\rho\} = (r-1)\sigma^2(1-\rho).$$

Hence the *estimated* variance of the treatment mean is $\sigma^2(1-\rho)/r$.

Consequently, if $\rho$ is positive the treatment mean is less accurate

32

than the mean of an independent series, but is estimated to be more accurate. If $\rho$ is negative, these conditions are reversed. Substantial biases in standard errors might result, with similar impairment of $t$-tests. Moreover, in many types of data, particularly field experimentation, the observations *are* mutually correlated, though in a more intricate pattern.

Whatever the nature of the correlation system, this difficulty is largely taken care of by proper randomization. While mathematical details will not be given, the effect of randomization is, roughly speaking, that we may treat the errors as if they were independent. The reader may refer to a paper by Yates (13), which presents the nature of this argument, and to papers by Bartlett (1), Fisher (6) and Hey (8), which illustrate how randomization generates a close approximation to the $F$ and $t$ distributions.

Occasionally it may be discovered that the data have been subject to some systematic pattern of environmental variation that the randomization has been unable to cope with. If the environmental pattern obviously masks the treatment effects, resort may be had to what might be called desperate remedies in order to salvage some information.

The data in Table V provide an instance. The experiment was a $2^4$ factorial, testing the effects of lime (L), fish manure (F) and artificial fertilizers (A). Lime was applied in the first year only; the other dressings were either applied in the first year only (1) or at a half rate every year (2). Two randomized blocks were laid out, the crop being pyrethrum, which forms an ingredient in many common insecti-

TABLE V

WEIGHTS OF DRY HEADS PER PLOT
(Unit, 10 grams)

| Block 1 | | | | Block 2 | | | |
|---|---|---|---|---|---|---|---|
| LA1 | LF2 | F2 | L1 | A1 | L1 | A2 | 0 |
| 84 | 66 | 70 | 81 | 63 | 97 | 56 | 64 |
| 1 | 1 | 1 | 1 | 1 | 1 | 1 | 1 |
| LF1 | A2 | A1 | FA2 | F1 | LA2 | LA1 | LFA1 |
| 148 | 137 | 146 | 171 | 168 | 158 | 189 | 152 |
| 0 | 0 | 0 | 0 | 0 | 0 | 0 | 0 |
| LFA2 | F1 | LFA1 | LA2 | LF1 | L2 | LF2 | FA2 |
| 179 | 218 | 247 | 228 | 191 | 195 | 189 | 179 |
| 0 | 0 | 0 | 0 | 0 | 0 | 0 | 0 |
| 0 | L2 | 0 | FA1 | FA1 | LFA2 | 0 | F2 |
| 124 | 166 | 177 | 153 | 133 | 145 | 141 | 130 |
| 0 | 0 | 0 | 0 | 0 | 0 | 0 | 0 |

33

cides. The data presented are for the fourth year of the experiment, which was conducted at the Woburn Experimental Farm, England.

The weights of dry heads are shown immediately underneath the treatment symbols. It is evident that the first row of plots is of poor fertility—treatments appearing in that row have only about half the yields that they give elsewhere. Further, there are indications that every row differs in fertility, the last row being second worst and the third row best. The fertility gradients are especially troublesome in that the four untreated controls all happen to lie in outside rows. The two replications give practically identical totals and remove none of this variation.

There is clearly little hope of obtaining information about the treatment effects unless weights are adjusted for differences in fertility from row to row. The adjustment may be made by covariance.

For simplicity, adjustments for the first row only will be shown: these remove the most serious environmental disturbance. As $x$ variable we choose a variable that takes the value 1 for all plots in the first row and zero elsewhere. The $x$ values are shown under the weights in Table V. The rest of the analysis follows the usual covariance technique, Snedecor (10).

TABLE VI

SUMS OF SQUARES AND PRODUCTS
($y$ = weights, $x$ = dummy variates)

|  | d.f. | $y^2$ | $yx$ | $x^2$ |
|---|---|---|---|---|
| Blocks | 1 | 657 | 0.0 | 0.00 |
| Treatments | 13 | 33,323 | − 200.2 | 1.75 |
| Error | 17 | 46,486 | − 380.0 | 4.25 |
| Total | 31 | 80,466 | − 580.2 | 6.00 |

Note that there are only 14 distinct treatments, since L1 is the same as L2. The reduction in the error S.S. due to covariance is $(380.0)^2/$ 4.25, or 33,976. The error mean square is reduced from 2,734 to 782 by means of the covariance, i.e., to less than one-third of its original value. The regression coefficient is − 380.0/4.25, or − 89.4 units.

Treatment means are adjusted in the usual way. For L1, which was unlucky in having two plots in the first row, the unadjusted mean is 89. The mean $x$ value is 1, whereas the mean $x$ value for the whole experiment is 8/32, or ¼. Hence the adjustment increases the L1 mean by (3/4)(89.4), the adjusted value being 156. For L2, which had no plots in the first row, the $x$ mean is 0, and the adjustment reduces the mean from 180 to 158. It may be observed that the unadjusted mean

34

of L2 was double that of L1, while the two adjusted means agree closely, as is reasonable since the two treatments are in fact identical.

If it were desired to adjust separately for every row, a multiple covariance with four $x$ variables could be computed. Each $x$ would take the value 1 for all plots in the corresponding row and 0 elsewhere. It will be realized that the covariance technique, if misused, can lead to an underestimation of errors. It is, however, worth keeping in mind as an occasional weapon for difficult cases.

8. *Effects of Non-Additivity.* Suppose that in a randomized blocks experiment, with two treatments and two replicates, the treatment and block effects are multiplicative rather than additive. That is, in either replicate, treatment B exceeds treatment A by a fixed percentage, while for either treatment, replicate 2 exceeds replicate 1 by a fixed percentage. Consider treatment percentages of 20% and 100% and replicate percentages of 10% and 50%. These together provide four combinations. Taking the observation for treatment A in replicate 1 as 1.0, the other observations are shown in Table VII.

TABLE VII

HYPOTHETICAL DATA FOR FOUR CASES WHERE EFFECTS ARE MULTIPLICATIVE

| Rep. | T 20% R 10% | | T 20% R 50% | | T 100% R 10% | | T 100% R 50% | |
|---|---|---|---|---|---|---|---|---|
| | A | B | A | B | A | B | A | B |
| 1 | 1.0 | 1.2 | 1.0 | 1.2 | 1.0 | 2.0 | 1.0 | 2.0 |
| 2 | 1.1 | 1.32 | 1.5 | 1.8 | 1.1 | 2.2 | 1.5 | 3.0 |
| $d$ | .02 | | .10 | | .10 | | .50 | |
| $\sigma_{na}$ | .01 | | .05 | | .05 | | .25 | |

Thus, in the first case, 1.32 for B in replicate 2 is 1.2 times 1.1. Since no experimental error has been added, the error variance in a correct analysis should be zero. If the usual analysis of variance is applied to each little table, the calculated error in each case will have 1 d.f. If $d$ is the sum of two corners minus the other two corners, the error S.S. is $d^2/4$, so that the standard error $\sigma_{na}$ is $d/2$ (taken as positive). The values of $d$ and of $\sigma_{na}$ are shown below each table.

Consequently, in the first experiment, say, the usual analysis would lead to the statement that the average increase to B is 0.21 units $\pm$ 0.01, instead of to the correct statement that the increase to B is 20%. The standard error, although due entirely to the failure of the additive rela-

35

tionship, does perform a useful purpose. It warns us that the actual increase to B over A will vary from replication to replication and measures how much it will vary, so far as the experiment is capable of supplying information on this point. An experimenter who fails to see the correct method of analysis and uses ordinary methods will get less precise information from the experiment for predictive purposes, but if he notes the standard error he will not be misled into thinking that his information is more precise than it really is.

When experimental errors are present, the variance $\sigma_{na}^2$ will be added to the usual error variance $\sigma_e^2$. The ratio $\sigma_{na}^2/(\sigma_{na}^2 + \sigma_e^2)$ may appropriately be taken as a measure of the loss (fractional) of information due to non-additivity. In the four experiments, from left to right, the values of $\sigma_{na}$ are respectively 0.9, 3.6, 3.2, and 13.3 percent of the mean yields of the experiments. In the first case, where treatment and replicate effects are small, the loss of information due to non-additivity will be trivial unless $\sigma_e$ is very small. For example, with $\sigma_e = 5$ percent, the fractional loss is 0.81/25.81 or about 3 percent. In the two middle examples, where either the treatment or the replicate effect is substantial, the losses are beginning to be substantial. With $\sigma_e = 5$ percent in the second case, the loss would be about 30 percent. Finally, when both effects are large the loss is great.

Little study has been made in the literature of the general effects of non-additivity or of the extent to which this problem is present in the data that are usually handled by analysis of variance.[5] I believe, however, that the results from these examples are suggestive of the consequences in other cases. The principal effect is a loss of information. Unless experimental errors are low or there is a very serious departure from additivity, this loss should be negligible when treatment and replication effects do not exceed 20 percent, since within that range the additive relationship is likely to be a good approximation to most types that may arise.

Since the deviations from additivity are, as it were, amalgamated with the true error variance, the pooled error variance as calculated from the analysis of variance will take account of these deviations and should be relatively unbiased. This pooled variance may not, however, be applicable to comparisons between individual pairs of treatments. The examples above are too small to illustrate this point. But, clearly, with three treatments A, B, and C, the comparison (A–B) might be much less affected by non-additivity than the comparison (A–C).

[5] A relevant discussion of this problem for regressions in general, with some interesting results, has been given recently by Jones (7).

36

Thus non-additivity tends to produce heterogeneity of the error variance.[6]

If treatment or block effects, or both, are large, it will be worth examining whether treatment differences appear to be independent of the block means, or vice versa. There are, of course, limitations to what can be discovered from a single experiment. If relations seem non-additive, the next step is to seek a scale on which effects are additive. Again reference should be made to the paper following on transformations.

9. *Summary and Concluding Remarks.* The analysis of variance depends on the assumptions that the treatment and environmental effects are additive and that the experimental errors are independent in the probability sense, have equal variance and are normally distributed. Failure of any assumption will impair to some extent the standard properties on which the widespread utility of the technique depends. Since an experimenter could rarely, if ever, convince himself that all the assumptions were exactly satisfied in his data, the technique must be regarded as approximative rather than exact. From general knowledge of the nature of the data and from a careful scrutiny of the data before analysis, it is believed that cases where the standard analysis will give misleading results or produce a serious loss of information can be detected in advance.

In general, the factors that are liable to cause the most severe disturbances are extreme skewness, the presence of gross errors, anomalous behavior of certain treatments or parts of the experiment, marked departures from the additive relationship, and changes in the error variance, either related to the mean or to certain treatments or parts of the experiment. The principal methods for an improved analysis are the omission of certain observations, treatments, or replicates, subdivision of the error variance, and transformation to another scale before analysis. In some cases, as illustrated by the numerical examples, the more exact methods require considerable experience in the manipulation of the analysis of variance. Having diagnosed the trouble, the experimenter may frequently find it advisable to obtain the help of the mathematical statistician.

[6] It is an over-simplification to pretend, as in the discussion above, that the deviations from addivity act entirely like an additional component of random error. Discussion of the effects introduced by the systematic nature of the deviations would, however, unduly lengthen this paper.

## REFERENCES

1. Bartlett, M. S. "The Effect of Non-Normality on the *t* Distribution," *Proceedings of the Cambridge Philosophical Society* (1935), 31, 223–231.
2. Cochran, W. G. "Some Difficulties in the Statistical Analysis of Replicated Experiments," *Empire Journal of Experimental Agriculture* (1938), 6, 157–175.
3. Finney, D. J. "On the Distribution of a Variate Whose Logarithm is Normally Distributed," *Journal of The Royal Statistical Society, Suppl.* (1941), 7, 155–161.
4. Fisher, R. A. "On the Mathematical Foundations of Theoretical Statistics," *Philosophical Transactions of the Royal Society of London*, A, 222 (1922), 309–368.
5. Fisher, R. A. *Statistical Methods for Research Workers*. Oliver and Boyd, Edinburgh, § 14.
6. Fisher, R. A. *The Design of Experiments*. Oliver and Boyd, Edinburgh, § 21.
7. Jones, H. L. "Linear Regression Functions with Neglected Variablies," *Journal of the American Statistical Association* (1946), 41, 356–369.
8. Hey, G. B. "A New Method of Experimental Sampling Illustrated on Certain Non-Normal Populations," *Biometrika* (1938), 30, 68–80.
9. Pearson, E. S. "The Analysis of Variance in Cases of Non-Normal Variation," *Biometrika* (1931), 23, 114.
10. Snedecor, G. W. *Statistical Methods*. Iowa State College Press, Ames, Ia. 4th ed. (1946). Chaps. 12 and 13.
11. Yates, F. "The Analysis of Replicated Experiments When the Field Results Are Incomplete," *Empire Journal of Experimental Agriculture* (1933), 1, 129–142.
12. Yates, F. "Incomplete Latin Squares," *Journal of Agricultural Science* (1936), 26, 301–315.
13. Yates, F. "The Formation of Latin Squares for Use in Field Experiments," *Empire Journal of Experimental Agriculture* (1933), 1, 235–244.
14. Yates, F., and Hale, R. W. "The Analysis of Latin Squares When Two or More Rows, Columns or Treatments Are Missing," *Journal of the Royal Statistical Society, Suppl.* (1939), 6, 67–79.

# 40

## RECENT DEVELOPMENTS IN SAMPLING THEORY IN THE UNITED STATES

by W. G. Cochran

*University of North Carolina (United States)*

### 1. INTRODUCTION

In this country the investigation of the theory of sample surveys is of recent origin. The increased use of sampling methods during the late thirties by public opinion polls, market research agencies and government bureaus took place with relatively little reliance on the services of the mathematical statistician and did not at first give rise to extensions of the underlying theory. Thus the period under review, from about 1939 onwards, contains practically all the development in theory that has occurred here. Discoveries in theory during the period have already profoundly influenced sampling practice and have resulted in an appreciation of the utility of theoretical investigations. These consequences are due largely to the fact that the small group of persons responsible for the developments in theory are also either actively engaged in the taking of sample surveys, or are closely concected with this work.

### 2. GENERAL TRENDS IN RESEARCH

The chief guiding principle in current research work is that the object in taking a sample is to obtain estimates of maximum accuracy for a given cost of taking the sample, or alternatively to obtain estimates with a specified degree of accuracy for minimum cost. Since in practice sample estimates can usually be regarded as approximately normally distributed, maximum accuracy is identified with minimum error variance in the sample estimate. The principle is not new: it has been used, for instance, by Y a t e s and Z a c o p a n a y (1935) and by N e y m a n (1938). Nevertheless, its implications were not at first realized either by the practical sampler or by the statistician. The former was inclined to aim at getting the largest size of a sample for a given cost without appreciating fully that a large sample is itself no guarantee of accuracy. The statistician, on the other hand, tended to base his recommendations on considerations of accuracy alone, perhaps from a reluctance to become involved in cost accountancy. Indeed, knowledge of the relative costs of different sampling methods still lags behind knowledge of their relative accuracies, though a beginning has been made

If accuracy alone is taken into account, sampling units that are small and widely scattered very frequently appear to be best. A computation of costs may show, however, that a larger sampling unit gives equal accuracy more cheaply, while if subsampling is used, quite large units may be economical. Investigations on optimum size of unit will be reported later (section 6).

The principle has also brought out the importance of careful planning of a sample. In order to apply the principle one must know, before the sample is taken, the costs of the methods of sampling under consideration, the methods of estimation that will subsequently be used, and the sampling variances of the resulting estimates. These facts will not all be known exactly, but serviceable estimates of them can often be made from the results of previous surveys. This approach has led to research on cost functions (sections 5.2, 5.5, 6.2, 7.2), on the properties of different methods of estimation (section 4), and on sampling error formulae for more complex types of sampling (section 5.3, 7.1). Attempts have also been made to obtain error formulae for techniques such as systematic sampling (section 8) which though intuitively attractive have not as yet proved amenable to a treatment by standard statistical methods.

It is known that no valid sampling error formula exists unless the selection of the sample is to a certain extent determined by the laws of probability through the use of an objective method of randomization. Such random selection cannot be achieved without something equivalent to a listing of the sampling units in the population. This fact often imposes a serious obstacle to the use of theoretically sound sampling methods, particularly with human populations, where usually no listings are available. The method of *area sampling* represents a major achievement towards overcoming this difficulty. The sampling unit is a compact area of land. A listing of the units in the population is obtained by constructing these areas so that they completely cover a map that shows the population to be sampled. In the Master Sample of Agriculture, designed primarily for farm surveys (K i n g and J e s s e n, 1945), every county in the United States was divided in this way into areas. The areal sampling units average about $2\frac{1}{2}$ square miles and contain from 4 to 8 farms on the average. Important administrative problems were to devise a rule so that every farm "belongs" to one and only one area, and to provide interviewers with aids (*e.g.*, aerial photographs) to the correct identification of the boundaries of the areas. In large towns the unit is usually a city block, which can also be outlined on a map (for discussion see H a n s e n and H a u s e r, 1945). While detailed accounts of the method will not be given, some research involving area sampling is presented in sections 6 (for rural surveys) and 7 (for urban surveys).

6

In the presentation which follows, I have tried to select the easiest order for the various topics, rather than a chronological order or one of presumed importance. Further, one cannot help being struck by the similarity to developments in India and Britain, which have been reported by M a h a l a n o b i s (1944) and Yates (1946). For this reason I have not in general attempted to assign priority of discovery, especially since numerous methods and results were known for some time before publication. In order to maintain a reasonably uniform notation in this paper, I have often, when writing about some result, had to change the notation in which the author originally presented it. This will be an annoyance to the reader who wishes to study the original, but has seemed unavoidable, since sampling literature is noted for the high variance in its notation. A population variance $\sigma^2$ has been defined as the sum of squares of deviations divided by $(N-1)$, where $N$ is the number of units in the population. This will change slightly the appearance of formulae in cases where the author has used a divisor $N$.

### 3. Extension of the General Principle

On further examination, the principle of maximum accuracy for given cost or minimum cost for given accuracy, is not itself completely satisfactory. The principle assumes that in some way either the cost or the accuracy is fixed in advance. Now the specification of the desired degree of accuracy usually involves some arbitrariness. If a coefficient of variation of 1.5 percent is demanded, the sample will not be regarded as useless should the coefficient turn out to be 1.6 percent. The advance specification of a sum of money that must be spent on the sample is also open to criticism, for the accuracy obtained from this expenditure may be substantially more, or substantially less, than is needed for the use that is to be made of the estimates. Two attempts to utilize a more general principle, in which optimum cost and optimum accuracy are determined simultaneously, deserve notice.

In order to apply the principle, one must be able to answer the question: how much is a given degree of accuracy worth?. Any decisions that are based on an estimate from a sample will presumably be more fruitful if the estimate has a low error than if it has a high error. In certain cases we may be able to calculate, in monetary terms, the loss $l(z)$ that will be incurred in a decision through an error of amount $z$ in the estimate. Although the actual value of $z$ is not predictable in advance, sampling theory may enable us to predict the frequency distribution $p(z,n)$ of $z$, which for a specified method of sampling will depend on the size of sample $n$. Hence, the *expected* loss for a given size of sample is

$$L(n) = \int l(z)p(z, n)dz.$$

The purpose in taking the sample is to diminish this loss. If $C(n)$ is the cost of a sample of size $n$, clearly $n$ should be chosen so as to minimize

$$C(n) + L(n)$$

since this is the total cost involved in taking the sample and in making decisions from its results. Choice of $n$ so as to minimize this quantity will determine both the optimum amount of money to be spent on sampling and the optimum accuracy. The idea is presented here only in its simplest form: it may be extended to cover a choice between different sampling methods.

In the application described by B l y t h e (1945), the selling price of a lot of standing timber is $SV$, where $S$ is the price per unit volume, and $V$ is the volume of timber in the lot. The number $N$ of logs in the lot is counted, and the average volume per log is estimated from a sample of $n$ logs. If $\sigma$ is the standard deviation per log for the sampling method used, the standard deviation of the estimate of $V$ will be $N\sigma/\sqrt{n}$ (ignoring finite population correction).

Suppose that this estimate is made and paid for by the seller. The buyer provisionally accepts the estimate of the amount of timber which he has bought. Subsequently, however, he finds out the correct volume purchased, and the seller reimburses him if he has paid for more than was delivered. If he has paid for less than was delivered, the buyer does not mention the fact. In this situation the seller loses whenever he underestimates the volume, but does not gain when he overestimates it. The situation is artificial (my presentation is slightly different from that of Blythe). My purpose is simply to illustrate the application of the principle to a case that does not require complex mathematics.

When he underestimates the volume by an amount $z$, the seller loses an amount $Sz$. Thus we may take $l(z)$ as zero when $z$ is negative and as $Sz$ when $z$ is positive, where $z$ is the amount of underestimation. On the assumption of normal distribution of sampling errors, $p(z,n)$ is the normal distribution with mean zero and variance $N^2\sigma^2/n$. Hence

$$L(n) = \frac{\sqrt{n}}{\sqrt{2\pi}\,N\sigma} \int_0^\infty Sze^{-\frac{nz^2}{2N^2\sigma^2}} dz = \frac{SN\sigma}{\sqrt{2\pi n}}$$

If we suppose further that the cost of measuring the volume of a log is $c$, the cost function $C(n)$ is $cn$. The quantity to be minimized is therefore

$$cn + \frac{SN\sigma}{\sqrt{2\pi n}}$$

Differentiation with respect to $n$ leads to the solution

$$n = \left(\frac{SN\sigma}{2c\sqrt{2\pi}}\right)^{\frac{2}{3}}$$

In the example due to N o r d i n (1944), a manufacturer takes a sample in order to estimate the size of a market which he intends to enter. If the size is known accurately, the amount of fixed equipment and the production per unit period can be adjusted so as to maximize expected profit. Errors in the estimated size of market will result in choices of these two factors that fall short of the optimum, and lead to smaller expected profit. The sample size $n$ should therefore be such that the addition of an $(n+1)$th unit to the sample increases the profit expectation by exactly the cost of the $(n+1)$th unit.

In many cases it will be difficult to apply these ideas because no way can be found to translate the effect of a sampling error into monetary terms. Moreover, an estimate may be used by different persons for quite diverse purposes. Nevertheless, the question of the standard of accuracy needed in sample estimates has received too little attention and this type of research points in a fruitful direction.

### 4. Estimation of a Population Total

Rather naturally, persons engaged in sampling have favored methods of estimation that can be computed easily and rapidly. The potentialities of complex methods of estimation have been little explored. The gain in accuracy from a superior method of estimation may, however, often be secured fairly cheaply, since only the final computations are affected, and there are likely to be cases, with important estimates, where quite elaborate calculations would be justified if a substantial increase in accuracy resulted. In this section some sampling properties of two methods of estimation, the *ratio* and the *linear regression* methods, are presented. Both have been in use for some time: the circumstances favorable to each are now, however, better understood. For simplicity, the properties are given only for a purely random sample (that is, a sample drawn so that every group of $n$ distinct units in the population has an equal chance of being the sample).

Probably the most common method for estimating the population total $Y_p$ of some item $y$ is to multiply the sample total $Y_s$ by the inverse $N/n$ of the sampling ratio. This estimate may be described as the *mean per s.u.* estimate. Its variance is

$$(1) \qquad\qquad V_{N/n} = \frac{N(N-n)\sigma_y^2}{n}$$

In the other two methods an auxiliary variate $x$, correlated with $y$, must be obtained for each unit in the sample. In addition, the population total $X_p$ of $x$ must be known. In practice, $x$ is often the value of $y$ on some previous occasion on which a complete census was taken. The aim in both methods is to obtain increased accuracy by taking advantage of the correlation between $y$ and $x$.

**4.1.** *The ratio estimate.* This estimate, which is simple to compute, is

$$\frac{Y_s}{X_s} \cdot X_p$$

This estimate is, in the technical sense, a "best unbiased linear estimate" if two conditions are satisfied: (i) the relation between $y$ and $x$ is a straight line through the origin, and (ii) the variance of $y$ about this line is proportional to $x$ (Cochran, 1942). In exploratory work, a graph plotting the sample values of $y$ against $x$ is therefore useful in considering whether the ratio estimate is likely to be the best available.

Where these conditions are not satisfied, the distribution of the ratio estimate in small samples has not yet been expressed in convenient terms, despite numerous attempts. Unless condition (i) holds, the estimate is biased (H a s e l, 1942), though the bias is usually negligible relative to the sampling error. In large samples the distribution tends to normality with a variance which is approximately

$$(2) \qquad V_{ra} = \frac{N(N-n)}{n}\left\{\sigma_y{}^2 + R_p{}^2\,\sigma_x{}^2 - 2\,R_p\rho\sigma_y\sigma_x\right\}$$

where $R_p = Y_p/X_p$ is the population ratio of $y$ to $x$, and $\rho$ is the correlation coefficient between $y$ and $x$.

On comparison with (1) it is found that the ratio estimate has a smaller variance than the mean per s.u. estimate if

$$\rho > \tfrac{1}{2}\frac{R_p\,\sigma_x}{\sigma_y} = \tfrac{1}{2}\frac{(c.v.)_x}{(c.v.)_y}$$

where $c.v.$ denotes coefficient of variation. Thus the success of the ratio method depends not only on the size of the correlation coefficient, but also on the $c.v.$ values. In fact, if the $c.v.$ of $x$ is more than twice that of $y$, the ratio estimate always has a higher variance than the mean per s.u. If $x$ is the value of $y$ on some previous occasion, the two $c.v.$'s may be about equal. In this case the ratio estimate is superior if $\rho$ exceeds $\tfrac{1}{2}$.

*4.2. The linear regression estimate.* We first compute from the sample the least squares estimate $b$ of the linear regression of $y$ on $x$. The estimate of $Y_p$ is then taken as

$$(3) \qquad N\{y_s + b(x_p - x_s)\},$$

where $y_s$, $x_s$ are sample *means* and $x_p$ is the population *mean* of $x$. Like the ratio estimate, this estimate is slightly biased, though the bias is unimportant. For samples that are reasonably large, the variance may be taken as

$$(4) \qquad V_{\text{l.r.}} = \frac{N(N-n)}{n} \sigma_y^2(1-\rho^2)$$

The error in this formula, which arises from neglect of the sampling variance of $b$, is in relative terms about $1/n$. The expression has been shown to remain valid in large samples even if the true relation is not linear or if the variance of $y$ changes with $x$ (Cochran, 1942).

From equations (1) and (4) it is clear that in large samples the regression estimate is always more accurate than the mean per s.u. estimate, except when $\rho$ is zero. Comparison of (4) with (2) shows, after a little manipulation, that the regression estimate always has a smaller variance than the ratio estimate unless

$$(5) \qquad \rho = \frac{R_p \sigma_x}{\sigma_y} = \frac{(c.v.)_x}{(c.v.)_y}$$

in which case the two have equal variances.[1] Equation (5) holds whenever the relation between $y$ and $x$ is a straight line through the origin, so that in this event the regression and ratio estimates are equally accurate. The fact that the regression estimate is as good as the ratio estimate even when the latter is a best unbiased linear estimate is interesting.

The regression estimate is more laborious to compute, principally owing to the work in calculating $b$. However, if there is an appreciable saving in time, an inefficient estimate of $b$ can be used instead of the least squares estimate. If the estimate of $b$ has efficiency $E$, $(E<1)$, the fractional increase in the variance of the regression estimate of $Y_p$ is about $(1-E)/nE$. With large $n$, even a highly inefficient estimate of $b$ causes only a trivial increase in the variance. It should, however, be remembered that with the least squares $b$, one can get an unbiased sample estimate of $\sigma_y^2(1-\rho^2)$ very quickly, whereas with other estimates of $b$ the "short-cut" calculation of the sample residual mean square does not apply.

[1] This result was first brought to my attention by Mr. J. R. Goodman.

Another type of estimate has been used by Jessen *et al.* (1947) in a survey taken to estimate the population of Greece. The relation of $y$ to $x$ was found to be a straight line through the origin, but the variance of $y$ about the line was proportional to $x^2$ rather than to $x$. The best estimate of $Y_p$ in this case is $\bar{r}_p X_p$, where $\bar{r}_p$ is the arithmetic mean of the sample values of $y/x$. Some discussion of the properties of weighted regressions has been given by Hasel (1942) and Cochran (1942).

### 5. STRATIFIED RANDOM SAMPLING

In stratified random sampling the population is divided into a number of groups or strata, the object being to have little within-group variation in the items to be estimated. The total number of sampling units $N_h$ in the $h$th group is assumed known, and a random sample of $n_h$ is selected from the group. The principles governing stratified sampling have become familiar through the work of Neyman (1934) and the device is almost universally applied. While no major developments have occurred in the period under review, a number of useful results will be given.

*5.1. Deep stratification.* The number of strata is limited by the size of sample, since there must be at least one sampling unit in each stratum. It may happen that the size of sample must be kept small for administrative reasons, but that many useful criteria for stratification are available. A method of taking advantage of the criteria with an *incomplete* stratification has been mentioned by F r a n k e l and S t o c k (1942, p. 80).

The sample was taken as part of a monthly survey of unemployment. The population with which we are concerned contains the 447 "urban" counties in the United States. These were divided into three strata according to county population (low, medium, high). Each stratum was further stratified into three groups by geographic location of the counties (NE, SE and W), and each of the nine cells was still further stratified by the 1937 percentage of unemployment (low, medium, high). Thus there were 27 strata, which might be shown as follows:

| Location | County Population | | |
|---|---|---|---|
| | Low | Medium | High |
| NE | l, m, h [1] | l, m, h | l, m, h |
| SE | l, m, h | l, m, h | l, m, h |
| W | l, m, h | l, m, h | l, m, h |

[1] 1937 % unemployed: low (l), medium (m), high (h).

Suppose that only *nine* strata are chosen, following the latin square plan below:

3 × 3 LATIN SQUARE

|      | L | M | H |
|------|---|---|---|
| NE   | l | m | h |
| SE   | m | h | l |
| W    | h | l | m |

For instance, from the three strata which have low county population and NE location, we select only the stratum with low 1937 percent unemployed (l), and so on. Given the sample means in the nine selected cells, we can estimate from the columns of the square, the average difference between counties with low, medium, and high population, from the rows the average difference between counties with NE, SE, and W locations; and similarly from the letters in the square, the average differences caused by the third classification. If these differences are assumed to be additive, the stratum means for all 27 cells can be estimated. By use of a 3×3 Graeco-latin square, a fourth classification could have been added, giving a representation of 81 strata by only nine cells.

Not enough research has yet been conducted on this method to permit a verdict on its utility. The method will obviously be most successful when each classification shows large effects that are independent of those of other classifications. A test of the method in a survey of the city of Wilmington was made by T e p p i n g, H u r w i t z, and D e m i n g (1943). They compare four methods of deep stratification with ordinary stratified sampling and with random sampling. The results were, however, inconclusive.

5.2. *Optimum allocation with varying costs.* If $y_{hs}$ is the sample mean in the $h$th stratum, the "mean per s.u." estimate of the population total is $\Sigma N_h y_{hs}$, summed over the strata. For a given total size of sample, Neyman (1934) has shown that this estimate has minimum variance if the sample sizes $n_h$ are chosen proportional to $N_h \sigma_h$, where $\sigma_h^2$ is the variance in the stratum. This well-known principle has often resulted in marked increases in accuracy over proportional sampling, particularly in sampling business enterprises, where the variance may be immensely greater for large than for small firms.

If we apply the general principle of minimum variance for given total *cost*, the Neyman method of allocation is optimum only when the cost per unit is the same in all strata. The best allocation can be worked out for any particular case when the cost function has been obtained in terms of the $n_h$. A useful result that is easily remembered occurs when the total cost is of the form $\Sigma c_h n_h$. That is, the cost per unit varies from stratum to stratum, but within each stratum costs are proportional to the size of sample.

The variance of the estimate $\Sigma N_h y_{hs}$ is

$$(6) \qquad \sum_h \frac{N_h (N_h - n_h)}{n_h} \sigma_h^2$$

and this is to be minimized subject to the restriction that $\Sigma c_h n_h$ is fixed. Minimum variance will be found to be obtained when $n_h$ is proportional to $N_h \sigma_h / \sqrt{c_h}$. As would be expected, the effect is to diminish the amount of sampling from costly strata.

*5.3. Stratified sampling with ratio estimates.* There are several ways in which a ratio estimate of a population total $Y_p$ can be made. One is to make a separate ratio estimate of the total of each stratum, and add these totals. If $Y_{hs}$, $X_{hs}$ are the sample totals in the $h$th stratum, and $X_{hp}$ is the stratum total for $x$, this estimate is

$$\sum_h \frac{Y_{hs}}{X_{hs}} X_{hp}$$

It is clear that no assumption is made that the true ratio remains fixed from stratum to stratum: the estimate postulates, however, a knowledge of the separate $X_{hp}$.

Since sampling is independent in the different strata, the variance is found simply by summation of terms as given in formula (2).

$$(7) \qquad V_{sep.} = \sum_h \frac{N_h (N_h - n_h)}{n_h} \left\{ \sigma_{hy}^2 + R_{hp}^2 \sigma_{hx}^2 - 2 R_{hp} \rho_h \sigma_{hy} \sigma_{hx} \right\}$$

where $R_{hp} = Y_{hp} / X_{hp}$ is the true ratio for the stratum.

An alternative estimate, derived from a single *pooled* ratio, has been used by Hansen, Hurwitz and G u r n e y (1946). This is

$$\frac{\sum \frac{N_h}{n_h} Y_{hs}}{\sum \frac{N_h}{n_h} X_{hs}} \cdot X_p$$

7

Note that only the population total $X_p$ of the $x$'s need be known. The variance of this estimate works out as

$$(8) \quad V_{\text{pooled}} = \sum_h \frac{N_h (N_h - n_h)}{n_h} \left\{ \sigma_{hy}{}^2 + R_p{}^2 \, \sigma_{hx}{}^2 - 2 \, R_p \, \rho_h \, \sigma_{hy} \, \sigma_{hx} \right\}$$

This differs from (7) only in that the single ratio $R_p = Y_p/X_p$ replaces the $R_{hp}$. In order to compare (7) with (8), we may use the result (*loc. cit.* p. 179, footnote) that (8) can be written

$$(9) \quad V_{\text{pooled}} = V_{\text{sep.}}$$

$$+ \sum_h \frac{N_h(N_h - n_h)}{n_h} \left\{ (R_{hp} - R_p)^2 \, \sigma_{hx}{}^2 + 2(R_{hp} - R_p)(\rho_h \, \sigma_{hy} \, \sigma_{hx} - R_{hp} \, \sigma^2{}_{hx}) \right\}$$

The last term on the right is usually small (it vanishes if within each stratum the relation between $y$ and $x$ is a straight line through the origin). It follows that unless $R_{hp}$ is constant from stratum to stratum, the use of a separate ratio estimate in each stratum is more accurate. The advantage appears to be small, however, unless the variation in the $R_{hp}$ is violent.

For sample estimates of (7) or (8), we substitute the sample estimates of $R_{hp}$ or $R_p$ in the appropriate places. The sample mean squares $s_{hy}{}^2$, $s_{hx}{}^2$ are substituted for the corresponding variances and the sample mean covariance for the term $\rho_h \sigma_{hy} \sigma_{hx}$. It will be noted that in general the sample mean squares and covariance must be calculated separately for each stratum.

The optimum allocation of the $n_h$ may be different when a ratio estimate is used than when a mean per s. u. is used. In discussing this point, we shall use (7), on the assumption that in practice (7) and (8) will differ little. Now the quantity inside the bracket in (7) is the variance within the $h$th stratum of the variate $d = (y - R_{hp}x)$. This variance will be called $\sigma_{hd}{}^2$. If (7) is minimized subject to a total cost of the form $\Sigma c_h n_h$, it is found that the $n_h$ must be chosen proportional to $N_h \sigma_{hd}/\sqrt{c_h}$, whereas with a mean per s. u. estimate, $n_h$ is chosen proportional to $N_h \sigma_{hy}/\sqrt{c_h}$.

In the case where the ratio estimate is a best unbiased linear estimate, $\sigma_{hd}$ will be proportional to $\sqrt{x}$. The $n_h$ would then be made proportional to $N_h \sqrt{x_{hp}}/\sqrt{c_h}$, where $x_{hp}$ is the stratum mean of $x$. In other cases, *e.g.* that of estimating retail store sales discussed by Hansen, Hurwitz, and Gurney (*loc. cit.*), the variance of $d$ may be more nearly proportional to $x^2$. This leads to the allocation $n_h \propto X_{hp}/\sqrt{c_h}$, where $X_{hp}$ is the stratum total of $x$.

*5.4. Effects of errors in the strata totals.* It frequently happens in practice that for some desirable type of stratification the strata totals $N_h$ are not known exactly, being perhaps derived from a population count that is out of date, or from another sample. Definite statements about the consequences of basing a stratification upon erroneous weights cannot be made without getting down to particular cases. A few conclusions of a general nature can, however, be drawn.

For simplicity, we consider only the estimation of the population mean per unit: finite population corrections will be ignored and the cost per unit is assumed the same in all strata. If the $N_h$ were known, $n_h$ would be chosen equal to $nN_h\sigma_h \,/\, \Sigma N_h\sigma_h$. The sample estimate of the population mean would be $\Sigma N_h y_{hs}/N$, which may be written $\Sigma W_h y_{hs}$. Its variance (which may be obtained from formula (6)) simplifies to

$$(10) \qquad \frac{(\Sigma W_h\sigma_h)^2}{n}$$

Instead of the true stratum proportions $W_h$, we have estimates $w_h$. The sample estimated mean is $\Sigma w_h\, y_{hs}$. The first point to note is that this estimate is biased. Its mean value in repeated sampling is $\Sigma w_h y_{hp}$ while the true population mean is $\Sigma W_h y_{hp}$. The bias amounts to $\Sigma(w_h - W_h)y_{hp}$. Consequently, the error variance of this estimate contains two components : the variance about its own mean and the square of the bias. If optimum allocation is used (with, of course, the $N_h$ replaced by their estimates) the first component is $(\Sigma w_h\sigma_h)^2/n$. The total variance is

$$(11) \qquad \frac{(\Sigma w_h\sigma_h)^2}{n} + \{\, \Sigma(w_h - W_h)y_{hp}\}^2$$

A more general form of this expression was given by S t e p h a n (1941). He points out that the first term in (11) will usually be about the same size as (10)—they are exactly the same if the variance is the same in all strata. The loss of accuracy from incorrect weights thus depends mainly on the size of the bias, which in individual cases might either be small or large. Further, for any given set of erroneous weights, the loss varies with the size of sample taken. This is so because the bias component of the total variance is independent of the size of sample. With increasing sample size, a stage is reached where the bias term predominates, and where the stratification would be less accurate than simple random sampling.

The preceding discussion does not help much in considering whether to stratify in a survey where the weights are known to be in error, because the size of the bias term cannot be predicted. Sometimes a standard error can be attached to the estimate of each $N_h$, from knowledge of the process by which these were estimated. If the estimates of the $N_h$ are independent, and independent of the $y_{hs}$, the average value of the bias component of the total variance is roughly (Cochran, 1939)

$$(12) \qquad\qquad \Sigma(y_{hp}-y_p)^2 \; V(N_h)/N^2,$$

where $V(N_h)$ is the variance of our estimate of $N_h$. This quantity measures the expected increase in variance due to errors in the $N_h$.

King, Mc C a r t y and Mc P e e k (1942) applied this formula in research directed towards the estimation of yield per acre, protein and test weight of wheat in the wheat belt. They discuss the advisability of stratification by districts within each state. The total acreages $N_h$ for each district were themselves estimated by a sample survey, so that some knowledge of the $V(N_h)$ was available.

In sampling human populations, the use of erroneous strata totals can often be avoided by area sampling, as discussed in sections 7.2 and 7.3.

5.5. *The problem of non-response.* In many types of survey, there are certain units in the sample from which the desired information cannot be obtained at the first attempt. With human populations, this group may be persons who are not at home, or who do not reply to a mail questionnaire. Similarly, in crop surveys certain fields in the sample may not be ripe when the sampler reaches them. This "non-response" group constitutes an important practical problem. To obtain information from it may require several attempts and be costly. To ignore it may result in a sample that has a bias of unknown dimensions. An ingenious application to this problem of the idea of stratified sampling has been made by Hansen and Hurwitz (1946).

The population is envisaged as containing two strata. One, of size $N_1$, contains units that provide the information at the first try. The second, of size $N_2$, is the non-response stratum. The basic idea is that the second stratum should be sampled at a lower rate than the first, since the cost per unit is higher in that stratum. There is, however, the complication that neither the values of $N_1$ and $N_2$, nor even the units that fall in the two strata, is known in advance.

The first step, in the simplest case, is to take a random sample of $n$ units. Of these let $n_1$ be the number that provide the data sought, and $n_2$ the number in the non-response group. By repeated efforts, the data are later obtained from a random sample of $r_2$ out of the $n_2$. If $n_2 = kr_2$, the quantity $k$ is the ratio of the sampling rate in the first stratum to that in the second. The values of $n$ (initial size of sample) and $k$ are chosen so as to give a specified accuracy for the lowest cost.

The cost of taking this sample is

$$(13) \qquad c_0 n + c_1 n_1 + c_2 r_2,$$

where the $c$'s are costs per unit: $c_0$ is the cost of making the first attempt, while $c_1$ and $c_2$ are the costs of getting and processing the data in the two strata respectively. Since the values of $n_1$ and $r_2$ will not be known until the first attempt is made, the *expected* cost must be used in planning the sample. The expected values of $n_1$ and $r_2$ are respectively $W_1 n$ and $W_2 n/k$, where as previously $W_1 = N_1/N$. Thus expected cost is

$$(14) \qquad c_0 n + c_1 W_1 n + c_2 W_2 n/k.$$

The estimated population total is taken as

$$\frac{N}{n} \left\{ n_1 y_{1s} + n_2 y_{2s} \right\}$$

and its variance is shown to be

$$(15) \qquad \frac{N(N-n)}{n} \sigma^2 + \frac{(k-1)N N_2}{n} \sigma_2^2$$

where $\sigma^2$ is the variance in the whole population and $\sigma_2^2$ is that within the non-response stratum. The first component is the variance that would be obtained if all $n_2$ in the non-response group were sampled; the second is the increase from sampling only $r_2$ of the $n_2$. The quantities $n$ and $k$ are then chosen to minimize (14) for a pre-assigned value of (15).

The solutions depend on the unknowns $N_1$ and $N_2$. If fairly close estimates of these can be made from earlier experience, the estimates may be used in place of the unknowns. Even if nothing is known in advance about $N_1$ and $N_2$, the authors develop an alternative method that gives in most cases a solution close to the optimum. Extensions to stratified sampling and to ratio estimation are also presented.

## 6. Optimum Size of Sampling Unit: Cluster Sampling

In certain types of survey the size of unit may be varied over quite a wide range. In accordance with the general principle, the best size must depend on a balance between the accuracy obtained from a given size, and the cost of sampling with that size. An investigation by Jessen (1942) presents a good illustration.

The population consisted of the farms in Iowa. The unit was a compact area of land delineated on the map, or more accurately all farms defined as belonging to this area. While certain sizes of area (*e.g.*, multiples or submultiples of a square mile) have particular practical conveniences in Iowa, the size will be regarded as capable of continuous variation. For simplicity in presentation, all areas of a given size are assumed to contain the same number of farms, and the variance of any item between farms within areas of the same size is assumed constant over the state. Neither assumption seriously vitiates the conclusions in this case.

*6.1. The variance function.* For some common farm economic items, Jessen found that the variance $W$ between farms within an area which contains $M$ farms is given by the equation

$$(16) \qquad W = AM^g, \qquad g > 0,$$

where $W$ is measured on a single-farm basis, and $A$ and $g$ are constants independent of $M$. In this model $W$ increases as the size of area increases, the curve being concave upwards. This increase is familiar in farm sampling, and may be attributed to the fact that many forces, including weather and soil type, tend to exert a similar influence on farms that are close together.

Suppose that a random sample of $n$ of these areas is selected. The variance that we are interested in is that of the estimated population total or mean: Jessen considers the latter. If $B$ is the variance between areas (on a single-farm basis) the variance of the sample mean per farm is $B/Mn$, since the sample contains $Mn$ farms (finite population correction is neglected). Thus we require knowledge of $B$ as a function of $M$.

The relation between $B$ and $W$ is supplied by the following analysis of variance for *all* farms in the state. The number of farms in the state is $N$, so that the number of areas is $N/M$.

|  | d.f. | Mean square |
|---|:---:|:---:|
| Between areas with $M$ farms | $\left(\dfrac{N}{M} - 1\right)$ | $B$ |
| Within areas between farms | $\dfrac{N}{M}(M-1)$ | $W$ |
| Total between farms in the state | $(N-1)$ | $T$ |

From the additive property of sums of squares, it follows that

$$(17) \qquad \frac{(N-M)B}{M} = (N-1)T - \frac{N(M-1)W}{M}$$

It will be noted that $T$ does not depend on $M$. Substituting for $W$ from (16) and dividing by $Mn$, we express the variance of the estimated population mean per farm as shown below.

(18)     $$V(M, n) = \{ M(N-1)T - N(M-1)AM^g \}/Mn(N-M)$$

Assuming $N$ large, Jessen simplifies this to

(19)     $$V'(M, n) = \frac{1}{n} \left[ T - (M-1)AM^{g-1} \right]$$

To digress for a moment, the whole population might be regarded as a single sampling unit of size $N$. Consequently, if formula (16) still holds for such large units, we may substitute $T = AN^g$ in (18). This reduces the number of unknown constants to be determined. H e n-d r i c k s (1944) found this relation to hold satisfactorily for corn and wheat acreage data, and M c V a y (1947) has applied it to the estimation of the proportion of orchards that grow peaches. It may happen, however, that the formula for $W$, while remaining valid for the range of values of $M$ that we are likely to consider, does not hold out to the value $N$. In this case the more general relation (18) should be used. The constants $T$, $A$, and $g$ must of course be estimated from data.

*6.2. The cost function.* Jessen distinguishes two elements of cost. The element $c_1Mn$ comprises costs that vary directly with the total number of farms : thus $c_1$ contains the cost of an interview and the cost of travel from farm to farm within an area.

The second element, $c_2\sqrt{n}$, measures the cost of travel between the areas. By tests on a map it was found that this cost, for a fixed population, varied roughly with the square root of the number of areas. Total cost is therefore of the form

(20)     $$C(M, n) = c_1Mn + c_2\sqrt{n}.$$

Equation (19) for the variance is then minimized with respect to variation in $M$ and $n$, subject to the restriction (20). While there is no neat explicit solution, the numerical solution in a specific case is not difficult.

One form of the theoretical solution for $M$, obtained after some manipulation, leads to some general conclusions that are of interest.

(21)     $$\frac{M}{V'}\frac{\partial V'}{\partial M} = \left\{ 1 + \frac{4Cc_1M}{c_2^2} \right\}^{-\frac{1}{2}} \quad 1$$

The left hand side is purely a function of $M$ and the constants $T$, $A$, and $g$: it does not depend on $n$ or the cost factors. Consequently, so far as cost factors are concerned, the optimum $M$ depends only on the quantity $q = Cc_1/c_2^2$. Further, it may be shown that as $q$ increases, the optimum $M$ decreases. But $c_1$ increases as the length of interview increases, while $c_2$ decreases if travel becomes cheaper, or if farms become more dense. These facts provide the general conclusion that, other things being equal, the optimum size of sampling unit becomes smaller if (i) the length of interview increases, (ii) travel becomes cheaper, (iii) the elements (farms) become more dense, or (iv) total amount of money available ($C$) increases.

With urban sampling, the sampling units that suggest themselves are the individual, the household, and the group of households. Hansen and Hurwitz (1942) used the name *cluster* for a group of elements (*e.g.* individuals) that constitute a sampling unit. They present examples which show that for certain items in urban sampling the variance function is quite different from that used by Jessen. In estimating the sex ratio, for instance, the variance $B$ between households is only about half as large as the variance $T$ between individuals, whereas in formula (17) $B$ is always envisaged as being greater than $T$. The common presence of both husband and wife in the same house introduces a negative correlation between the sexes of members of a household. For this reason, the authors have preferred to express their mathematical model in terms of the intra-cluster correlation coefficient, which may take either positive or negative values.

## 7. SUBSAMPLING

With large sampling units, a useful device is to sample only a part of each unit that is selected. Thus if the large unit contains $M$ *sub-units*, we might obtain information from $m$ of these, chosen at random. When sub-units within the same unit give closely similar results, common sense suggests that it will not be profitable to spend time measuring a large number of them, at the expense of decreasing the number of units that can be sampled. For instance. in estimating wheat yield and quality by sampling at harvest time, King and McCarty (1941) found that it was advisable to cut only a very small quantity of wheat from each field that was sampled.

The sampling theory is simple when each unit contains the same number $M$ of sub-units, of which $m$ are chosen, and when the variance $\sigma_w^2$ between sub-units in the same unit is constant. The variance of the sample mean per sub-unit is

$$\frac{(N-n)}{N} \frac{\sigma_b^2}{n} + \frac{(M-m)}{M} \frac{\sigma_w^2}{mn}$$

where $\sigma_b{}^2$ is the variance between the *true* means of the units and $n$ units are sampled out of $N$. An unbiased estimate of this variance is obtained from an analysis of variance of the sample. If $B$ is the mean square between units and $W$ that between sub-units within units, both being on a sub-unit basis, it may be shown that

$$E(B) = m\sigma_b{}^2 + \frac{(M-m)}{M}\sigma_w{}^2, \quad E(W) = \sigma_w{}^2.$$

Consequently, the unbiased sample estimate is

$$\frac{1}{mn}\left[\frac{(N-n)}{N}B + \frac{(M-m)}{M}\frac{n}{N}W\right]$$

*7.1. Subsampling when sampling units vary in size.* This problem has arisen frequently in recent years especially with human populations, where the unit may be a city block, or a county, and the sub-unit a household or a farm. At first sight a confusing variety of methods of sampling and of estimation present themselves. The appraisal of these methods and the development of new methods is an important contribution by Hansen and Hurwitz (1943).

The methods and their properties can be illustrated for the simple population represented below.

| Unit no. (i) | $M_i$ | Sub-unit values | Unit total |
|---|---|---|---|
| 1 | 6 | 1,1,1,1,1,1 | 6 |
| 2 | 2 | 0,0 | 0 |
| | 8 | | 6 |

The population has only two units. The first has six sub-units, all of which give the value 1 for the item measured. The second has two units, each giving 0. Thus $\sigma_w{}^2$ is zero. The population total (the quantity to be estimated) is 6. It is proposed to select one unit, so that $n = 1$, $N = 2$. Two sub-units will be sampled from the unit chosen. The sample therefore comprises two of the 8 sub-units in the population.

The methods and the results for this population are summarized in Table 1.

8

TABLE 1

Subsampling Methods With Units of Unequal Sizes.

| Method | $p_i$ | $m_i$ | Estimate | Values | Bias* | $\sigma^2$ |
|---|---|---|---|---|---|---|
| Ia | equal | $m$ | $\dfrac{N}{n}\cdot\dfrac{M_i}{m}\cdot Y$ | 12, 0 | $U$ | 36 |
| Ib | equal | $m$ | $\dfrac{M}{m}\cdot Y$ | 8, 0 | $B$ | 20 |
| IIa | equal | $fM_i$ | $\dfrac{N}{n}\cdot\dfrac{M_i}{m_i}\cdot Y$ | 12, 0 | $U$ | 36 |
| IIb | equal | $fM_i$ | $\dfrac{M}{m_i}\cdot Y$ | 8, 0 | $B$ | 20 |
| IIIa | $\dfrac{M_i}{M}$ | $m$ | $\dfrac{M}{m}\cdot Y$ | $8(\tfrac{3}{4}),\ 0(\tfrac{1}{4})$ | $U$ | 12 |
| IIIb | $P_i$ | $\dfrac{tM_i}{P_i}$ | $\dfrac{1}{t}\cdot Y$ | $9(\tfrac{2}{3}),\ 0(\tfrac{1}{3})$ | $U$ | 18 |

\* $U$ = unbiased,  $B$ = biased.

*Method Ia.* Choose a unit at random and $m = 2$ sub-units from it. Multiply the sample total $Y$ by $M_i/m$ to give an unbiased estimate of the unit total, and then by $N/n$, or 2, to give an unbiased estimate of the population total. This estimate is 12 if the first unit is chosen and 0 if the second unit is chosen. The sampling variance of the estimate is 36.

On reflection, the method of estimation leaves something to be desired. The estimate is $N$(estimated unit total)/$n$. But the variation in the unit *totals* depends on the variation in the sizes $M_i$ as well as on that in the sub-unit values. Even if the latter all had the value 1, this method would give a large variance if the $M_i$ differed greatly. A technique that avoids this type of inflation of the variance is desired.

*Method Ib.* Same method of sampling. The estimate is the sample mean, multiplied by the number of sub-units in the population: *i.e.*,

$MY/m$, or in this case $4Y$. The two possible estimates are 8 and 0. Since their mean, 4, does not equal the true population total, the estimate is biased. The error variance (about the true population total) is $\frac{1}{2}(2^2 + 6^2)$, or 20. Despite the bias, this estimate is more accurate than Ia. The reason is that since the estimate is based on the sample *mean* its error is not inflated by variations in the $M_i$.

*Method IIa.* Select a unit at random as before, but take $m_i = \frac{1}{2}M_i$ (*proportional subsampling*). This procedure is derived from the idea that large units should in some way receive more representation in the sample than small units. The sample will be of size 3 if the first unit is chosen, and of size 1 if the second is chosen. While it may seem undesirable that the sample size should depend on the particular units selected, over a number of strata the actual size would be near the

expected size. The estimate is analogous to that in Ia ; *i.e.*, $\dfrac{N}{n}.\dfrac{M_i}{m_i}. Y$.

The reader will find that this method gives exactly the same estimates as Ia and hence has the same variance, 36.

*Method IIb.* Sampling method as in IIa: estimate as in Ib. $(MY/m_i)$. Results are the same as in Ib, the estimate being biased, with variance 20.

The equivalence of IIa with Ia and IIb with Ib always holds when there is no variation within units. When sub-units within the same unit differ, as will happen in practice, Hansen and Hurwitz (1943, p. 357) have shown that Ib will usually be more accurate than IIb.

*Method IIIa.* The authors propose to give greater representation to large units by drawing the units, not with equal probabilities, as in all previous methods, but with probability *proportional to their sizes*. Thus $p_i = M_i/M$. A probability $\frac{3}{4}$ is assigned to the large unit and $\frac{1}{4}$ to the small unit. As in methods I, two sub-units are sampled from the unit chosen. The estimate is $MY/m$, or $4Y$. In repeated sampling we get an estimate 8 in $\frac{3}{4}$ of the samples, and one of 0 in the remaining $\frac{1}{4}$. The estimate is unbiased. Its variance is

$$\tfrac{3}{4}(2)^2 + \tfrac{1}{4}(6)^2 = 12$$

This value is considerably smaller than that given by any previous method.

By examination of the sampling error formulae for the different techniques, the authors show that method IIIa will give the smallest between-unit contribution to the variance for many types of population that appear to be met in practice. So far as the within-unit contribution is concerned, this is not likely to differ greatly in methods IIIa and

Ib, which is usually the most accurate of the other methods described thus far (method IIIb will be discussed in section 7.2). The between-unit contribution is quite commonly, though not always, much larger than the within-unit contribution. This was the reason for neglecting the latter in the example.

*7.2. Case where the sizes are not known.* It may happen that the sizes $M_i$ of the units (*e.g.*, number of households in a block or county) are not known when the sample is drawn, though estimates of them may be obtainable from previous data. Insofar as these sizes are needed for sampling or estimation, time and money must be spent in obtaining them. It will be seen from Table 1 that with methods Ia and IIa we require the $M_i$ only *for those units that come into the sample.* With the more accurate methods Ib, IIb, and IIIa, we must count the total $M$ of *all* the units. If counting is expensive, the cheaper cost of methods Ia and IIa may go far to counterbalance their relatively large variances.

An interesting application of an extension of method IIa is presented by the sampling staff of the Bureau of the Census in *A Chapter in Population Sampling* (1947). The primary object of the sample was to estimate the total population in certain congested areas, each containing a city and the surrounding rural area. In the cities the unit was a block and the sub-unit a dwelling-place. The blocks were stratified by estimated size (number of dwelling-places) according to data obtained four years previously. This stratification would remove a considerable part of the variation of the $M_i$. The sampling ratio $n/N$ and the sub-sampling ratio $m_i/M_i$ were determined so as to minimise for a specified accuracy, the total cost of counting the $M_i$ in the sampled units, enumerating the $m_i$ and tabulating the results. Special methods were devised for trailer camps, institutions, etc., for the rural areas, and for blocks reported as having no dwelling-places in the previous data.

There is another method of sampling (IIIb) which requires a knowlege only of the $M_i$ that are being sub-sampled. It is in a sense a generalization both of methods IIIa and IIa, and is due to Hansen and Hurwitz (1943). It is an application of IIIa in the case where we have only estimates $P_i$ of the relative sizes, though mathematically the $P_i$ can be any set of positive numbers adding to unity over each stratum . We first select a *sampling rate t*, which may vary from stratum to stratum. In the example $t$ is chosen as 2/8, since a sample of size 2 is desired and $M$ is 8   A sampling unit is then chosen with a probability $P_i$ attached to the *i*th unit. The $M_i$ is counted for the unit chosen and $m_i$ is taken as $tM_i/P_i$. The estimate is $Y/t$ and therefore does not require a knowledge of $M$.

In the example, suppose that we estimated the $P_i$ as $\frac{2}{3}$ for the large unit and $\frac{1}{3}$ for the small. The sample sizes work out at 9/4 for the large unit and 3/2 for the small. We shall assume that samples of these non-integral sizes could actually be taken (they could be approximated closely if the $M_i$ were large). The estimates are 9, with probability $\frac{2}{3}$, and 0 with probability $\frac{1}{3}$. The estimate is unbiased (this is always true). The sampling variance is $\{\frac{2}{3}(3)^2 + \frac{1}{3}(6)^2\}$, or 18.

The reader may verify that if $P_i = M_i/M$, the method reduces to IIIa, while if $P_i = 1/N$, the method reduces to IIa. When the $P_i$ differ from $M_i/M$, the sampling error is increased to some extent through the variations in the $M_i$, though not greatly if the estimates of size are good.

*7.3. Further comments.* Methods IIIa and IIIb have been extended to stratified sampling and to ratio estimation (*loc. cit.*). When only one unit is sampled per stratum, the actual drawing of the unit is simple. The $P_i$, or a convenient multiple of them, are cumulated. If their grand total is $G$, a random number $r$ is chosen between 1 and $G$. The unit which contains this number in the list of cumulative totals is the unit selected. Since mathematical complications enter when more than one unit per stratum is to be drawn, it is customary to stratify to the point where one unit per stratum is taken.

With methods Ia or IIa it is highly advisable to stratify by estimated size if the $M_i$ are likely to vary considerably. With methods IIIa and IIIb, however, stratification can be by other factors, since size effects are removed by selection of units with probability proportional to size. It is worth emphasis that with methods Ia, IIa, or IIIb the stratum total numbers of sub-units need not be known exactly. Thus these methods allow stratification without risking the bias that may be introduced by the use of erroneous strata totals, as discussed in section 5.4.

When only one unit is sampled from a stratum, the subsample from that unit should represent the *stratum* rather than the unit. In some cases it is possible to stratify the sub-units in the stratum by a criterion different from that used in forming the original strata. If so, an increase in accuracy may be secured by drawing the subsample so that it is stratified according to this criterion. The sampling error formula for this technique, which is known as *area substratification*, will be found in the same reference.

### 8. Systematic Sampling

If the units in a population are numbered from 1 to $N = nk$, a common method of drawing a sample of $n$ is to take a random number between 1 and $k$, say $r$, and draw the $r$th unit and every $k$th unit thereafter. This type of sample is called a systematic sample. It can be drawn more quickly than a random sample and intuitively seems likely to be more accurate, since it spreads the sample evenly over the population. In some respects it resembles a stratified random sample in which each stratum has $k$ units, and one unit is sampled per stratum. The difference is that in the systematic sample the unit sampled always occurs at the same place in the stratum. Frequently it is used mainly for convenience, the order in which the units are numbered having no particular significance. In this case it may be predicted that the systematic sample will give about the same accuracy as a random sample of the same size. It has, however, not proved easy to make general statements about the accuracy of systematic samples.

The sampling theory was investigated by W. G. and L. H. M a d o w (1944). They give several formulas for the variance of the sample mean, of which one may be quoted. In our notation, this is

$$\frac{\sigma^2}{n} \left\{ \frac{N-1}{N} + \frac{2}{n} \sum_{d=1}^{n-1} (n-d)\, \rho'_{kd} \right\}$$

where $\sigma^2$ is the population variance and $\rho'_{kd}$ is the noncircular serial correlation coefficient with lag $kd$, defined by

$$k(n-d)\sigma^2 \rho'_{kd} = \sum_{i=1}^{k(n-d)} (y_i - y_p)(y_{i+kd} - y_p).$$

The difficulty is that the accuracy depends on the nature of the lag correlations, such as are caused by any continuous variation in the items $y_i$.

If the major variation is a linear trend, the Madows show that the systematic sample is more accurate than a random sample, but less accurate than a stratified random sample with one unit per stratum. If the major variation is periodic, the systematic sample is highly accurate if $k$ is an odd multiple of half the period, but is inaccurate if $k$ is a multiple of the period. If the major variation is such that $\rho'_i$ is a positive, monotone-decreasing, and concave upwards function of $i$, the systematic sample is always at least as accurate, on the average, as the stratified random sample (Cochran, 1946). This type of population was studied because in it the variance within a group of $i$ consecutive units increases as $i$ increases, as often seems to be the case in practice (*e.g.*, Jessen's variance formula given earlier has this property).

Investigations of this type must be supplemented by examination of the types of continuous variation that are encountered in practice. There is reason to believe that frequently the systematic sample will be substantially more accurate than either the random or stratified random samples. For instance, J o h n s o n (1943) studied 13 natural populations in which the observations were the numbers of seedlings in successive feet in a forest nursery bed. Five species were represented. In seven beds containing seedbed stock with high variability, the variance of the systematic sample was about half that of the stratified random sample. In the remaining six beds, which had rather homogeneous transplant stock, the systematic sample was about equal in accuracy to the stratified random sample and considerably more accurate than the random sample. O s b o r n e (1942) found the systematic sample twice to four times as accurate as the stratified sample in estimating the areas under different types of cover (*e.g.*, grass, woodland) from a map. In both investigations the stratified sample had a stratum of size $2k$ with two samples per stratum so as to permit estimation of the sampling error. It appears, however, that the results would be little changed if the stratum size were reduced to $k$.

For a given population, the variance of the systematic sample mean is likely to be a much less stable function of $n$ than with the random methods of sampling. This is illustrated by a silver maple bed from Johnson's data (L.H.Madow, 1946). Variances of sample means for samples from $n = 10$ to $n = 84$ are shown below : the value of $N$ was 420.

TABLE 2

Variances of Sample Means

(*from L.H.Madow*, 1946)

| $n$ | Type of sample | | |
| | Random | Stratified random* | Systematic |
|---|---|---|---|
| 10 | 10·29 | 7·21 | 4·21 |
| 21 | 4·77 | 3·00 | 3·06 |
| 28 | 3·52 | 2·09 | 2·42 |
| 30 | 3·26 | 1·90 | 0·69 |
| 42 | 2·26 | 1·29 | 1·74 |
| 60 | 1·51 | 0·82 | 0·26 |
| 84 | 1·00 | 0·51 | 1·22 |

\* One unit per stratum.

The rather erratic behavior of the systematic sample is evident. For some sample sizes ($n = 10$, 30, and 60) there are striking gains over the stratified sample; for others ($n = 21$, 28, and 42) the variance is slightly higher than for stratified sampling; while for $n = 84$ the variance is actually higher than with simple random sampling. This type of effect is to be expected. The systematic samples of size 60, for instance, are each composed of two of the samples of size 30. Thus there is a positive correlation between the variances for $n = 30$ and $n = 60$.

It is well known that from the results of a single systematic sample one cannot obtain an unbiased estimate of the sampling variance of the mean that will be valid for any type of population. It should be feasible to develop estimates that will be adequate for practical purposes, given some knowledge of the type of population that is being sampled. With his data, Osborne (1942) obtained good results from a formula $\delta^2 (1 - \bar{r}^2)/2n$, where $\delta^2$ is the mean square successive difference $\sum_{i=1}^{n-1} (y_i - y_{i+k})^2/(n-1)$, and $\bar{r}^2$ is the average of the squared correlations between a unit in the sample and all other units in the same group of $k$ units. Further study of actual populations will be profitable.

## References

*A Chapter in Population Sampling*, by the Sampling Staff, Bureau of the Census, 1947.

Blythe, R. H., "The Economics of Sample Size Applied to the Scaling of Saw Logs," *Biometrics Bulletin*, Vol. 1 (1945), pp. 67-70.

Cochran, W. G., "The Use of the Analysis of Variance in Enumeration by Sampling," *Journal of the American Statistical Association*, Vol. 34 (1939), pp. 492-510.

Cochran, W. G., "Sampling Theory When the Sampling Units are of Unequal Sizes," *Journal of the American Statistical Association*, Vol. 37 (1942), pp. 199-212.

Cochran, W. G., "Relative Accuracy of Systematic and Stratified Random Samples for a Certain Class of Populations," *Annals of Mathematical Statistics*, Vol. 17 (1946), pp. 164-177.

Frankel, L. R. and Stock, J. S., "On the Sample Survey of Unemployment," *Journal of the American Statistical Association*, Vol. 37 (1942), pp. 77-80.

Hansen, M. H. and Hauser, P. M., "Area Sampling—Some Principles of Sample Design," *Public Opinion Quarterly*, Vol. 9 (1945), pp. 183-193.

Hansen, M. H. and Hurwitz, W. N., "Relative Efficiencies of Various Sampling Units in Population Inquiries," *Journal of the American Statistical Association* Vol. 37 (1942), pp. 89-94.

Hansen, M. H. and Hurwitz, W. N., "On the Theory of Sampling from Finite Populations," *Annals of Mathematical Statistics*, Vol. 14 (1943), pp. 333-362.

Hansen, M. H. and Hurwitz, W. N., "The Problem of Non-Response in Sample Surveys," *Journal of the American Statistical Association*, Vol. 41 (1946), pp. 517-529.

Hansen, M. H., Hurwitz, W. N., and Gurney, M., "Problems and Methods of a Sample Survey of Business," *Journal of the American Statistical Association*, Vol. 41 (1946), pp. 173-189.

Hasel, A. A., "Estimation of Volume in Timber Stands by Strip Sampling," *Annals of Mathematical Statistics*, Vol. 13 (1942), pp. 179-206.

Hendricks, W. A., "The Relative Efficiencies of Groups of Farms as Sampling Units," *Journal of the American Statistical Association*, Vol. 39 (1944), pp. 367-376.

Jessen, R. J., "Statistical Investigation of a Sample Survey for Obtaining Farm Facts." *Iowa State College Research Bulletin* 304, 1942.

Jessen, R. J., Blythe, R. H., Kempthorne, O., and Deming, W. E., "On a Population Sample for Greece," *Journal of the American Statistical Association*, Vol. 42 (1947), pp. 357-383.

Johnson, F. A., "A Statistical Study of Sampling Methods for free Nursery Inventories," *Journal of Forestry*, Vol. 41 (1943), pp. 674-679.

King, A. J. and Jessen, R. J., "The Master Sample of Agriculture," *Journal of the American Statistical Association*, Vol. 40 (1945), pp. 38-56.

King, A. J., and McCarty, D. E., "Application of Sampling to Agricultural Statistics with Emphasis on Stratified Samples," *Journal of Marketing*, Vol. 5 (1941) pp. 462-474.

King, A. J., McCarty, D. E., and McPeek, M., "An Objective Method of Sampling Wheat Fields to Estimate Production and Quality of Wheat," *U. S. Department of Agriculture Technical Bulletin* 814, 1942.

Madow, L. H., "Systematic Sampling and its Relation to Other Sampling Designs," *Journal of the American Statistical Association*, Vol. 41 (1946), pp. 204-217.

Madow, W. G. and Madow, L. H., "On the Theory of Systematic Sampling I," *Annals of Mathematical Statistics*, Vol. 15 (1944), pp. 1-24.

Mahalanobis, P. C., "On Large-Scale Sample Surveys," *Philosophical Transactions of the Royal Society of London*, Vol. 231 (1944), pp. 329-451.

McVay, F. E., "Sampling Methods Applied to Estimating Numbers of Commercial Orchards in a Commercial Peach Area," *Journal of the American Statistical Association*, Vol. 42 (1947), pp, 533-540.

Neyman, J., "On Two Different Aspects of the Representative Method: The Method of Stratified Sampling and the Method of Purposive Selection," *Journal of the Royal Statistical Society*, Vol 97 (1934), pp. 558-606.

Neyman, J., "Contribution to the Theory of Sampling Human Population," *Journal of the American Statistical Association*, Vol. 33 (1938), pp. 101-116.

Nordin, J. A. "Determining Sample Size," *Journal of the American Statistical Association*, Vol. 39, (1944), pp. 497-506.

9

Osborne, J. G., "Sampling Errors of Systematic and Random Surveys of Cover-Type Areas," *Journal of the American Statistical Association*, Vol, 37 (1942), pp. 256-264.

Stephan, F. F., "Stratification in Representative Sampling," *Journal of Marketing*, Vol. 6 (1941), pp. 38-46.

Tepping, B. J., Hurwitz, W. N., and Deming, W. E., "On the Efficiency of Deep Stratification in Block Sampling," *Journal of the American Statistical Association*, Vol. 38 (1943), pp. 93-100.

Yates, F., "A Review of Recent Statistical Developments in Sampling and Sampling Surveys," *Journal of the Royal Statistical Society*, Vol. 109 (1946), pp. 12-43.

Yates, F. and Zacopanay, I., "The Estimation of the Efficiency of Sampling with Special Reference to Sampling for Yield in Cereal Experiments," *Journal of Agricultural Science*, Vol. 25 (1935), pp. 545-577.

## Résumé

Lorsqu'on effectue un sondage, on vise soit à obtenir le maximum de précision dans les estimations pour une dépense donnée, soit à minimiser le coût d'estimations d'un degré déterminé de précision. Les recherches aux Etats-Unis sont régies par ces principes. Un principe encore plus général a été introduit par *Blythe* et *Nordin*. Suivant ce principe, le coût optimum et la précision optimum sont déterminés simultanément en minimisant le coût de l'enquête et le coût de décisions erronées dûes à l'imperfection des estimations.

Dans la communication sont examinés: 1) les estimations du total de la population, 2) la méthode du sondage aléatoire stratifié, dont quelques nouvelles applications ont été dévouvertes, 3) le choix optimum de l'unité d'échantillonnage, 4) la méthode de sub-échantillonnage, 5) la méthode du sondage systématique, très souvent utilisée en pratique.

# 41

## DISCRIMINANT FUNCTIONS WITH COVARIANCE

By W. G. Cochran and C. I. Bliss

*North Carolina State College; Connecticut Agricultural Experiment Station and Yale University*

**1. Summary.** This paper discusses the extension of the discriminant function to the case where certain variates (called the covariance variates) are known to have the same means in all populations. Although such variates have no discriminating power by themselves, they may still be utilized in the discriminant function.

The first step is to adjust the discriminators by means of their 'within-sample' regressions on the covariance variates. The discriminant function is then calculated in the usual way from these adjusted variates. The standard tests of significance for the discriminant function (e.g. Hotelling's $T^2$ test) can be extended to this case without difficulty. A measure is suggested of the gain in information due to covariance and the computations are illustrated by a numerical example. The discussion is confined to the case where only a single function of the population means is being investigated.

**2. Introduction.** Discriminant function analysis is now fairly well advanced for the case where there are only two populations. The data consist of a number of measurements, called the *discriminators*, that have been made on each member of a random sample from each population. The technique has various uses. Fisher [1] used it in seeking a linear function of the measurements that could be employed to classify new observations into one or other of the two populations. He pointed out [2] that a test of significance of the difference between the two samples, developed from his discriminant, was identical with Hotelling's generalization of Student's *t* test, discovered some years earlier [3]. Mahalanobis' concept of the generalized distance between two populations [4] was also found to be closely related to the discriminant function. In any of these applications—to classification, testing significance, or estimating distance—we may also be interested in considering whether certain of the measurements really contribute anything to the purpose at hand, and helpful tests of significance are available for this purpose.

Recently the authors encountered a problem in which it seemed advisable to combine discriminant function analysis with the analysis of covariance. This case occurs whenever, in addition to the discriminators, there is a measurement whose mean is known to be the same in both populations. Suppose, for example, that the I.Q.'s of each of a sample of students are measured. The sample is then divided *at random* into two groups, each of which subsequently receives a different type of training. Measurements made at the end of the period of training would be potential discriminators, but in the case of the initial I.Q.'s we can

---

clearly assume that there is no difference in the means of the populations corresponding to the two groups.

The initial I.Q. measurements are of course of no use in themselves in studying differences introduced by the training. Nevertheless, if they are correlated with the discriminators, they may serve in some way to 'improve' the discriminant: e.g. to increase the power of Hotelling's $T^2$ test, or to reduce the number of errors in classification. This paper discusses the problem of utilizing such measurements, which will be called *covariance variates*. The problem is analogous to that which is solved by the analysis of covariance. In covariance, as applied for instance in a controlled experiment, variates that are unaffected by the experimental treatments can be used to provide more accurate estimates of the effects of the treatments or to increase the power of the $F$ test of the differences among the treatment means.

The procedure suggested is as follows. First, the multiple regression is obtained of each discriminator on all the covariance variates. These regressions are calculated from the 'within-sample' sums of squares and products: that is, from the sums of squares and products of deviations of the individual measurements from their sample means. Each discriminator is then replaced by its deviations from the multiple regression, and a new discriminant function is calculated in the usual way from these deviations. The extensions of Hotelling's $T^2$ and Mahalanobis' distance are both obtained from this discriminant, though a further adjustment factor is needed for tests of significance.

This paper is arranged in three parts. Part I presents a numerical example. The decision to place the example first was taken because most of the actual applications of the discriminant function in the literature appear to have been made by persons relatively unfamiliar with the theory of multivariate analysis. It is hoped that with the aid of the example readers in this class may be able to utilize covariance variates. For the same reason, the calculations have been presented as far as possible in terms of the operations of ordinary multiple regression, rather than in the form in which they first emerge from the theory. Actually, various equivalent methods of calculation are available, and it is not claimed that our method is necessarily the best. A mathematical statistician may prefer to follow the computing methods which come directly from theory (Part II, section 13).

The example is more complex in structure than the two-sample case. The data constitute a two-way classification, in which the row means are nuisance parameters, being of no interest, while only a single linear function of the column means is of interest. It is well known that the ordinary $t$ test can be applied not only to the difference between two sample means, but to any linear function of a number of sample means in data that are quite complex. Discriminant function technique can be extended in the same way, and readers familiar with the analysis of variance should find no great difficulty in making the appropriate extension to such data.

Part II presents the theory. The reader who is primarily interested in theory should read Part II before Part I. Since the approaches used by Mahalanobis, Hotelling and Fisher all converge, we have chosen that of Mahalanobis, mainly because the extension of his techniques to include covariance variates seems straightforward. Maximum likelihood estimation of the generalized distance is presented in full for the two-population case. The frequency distribution of the estimated distance and the extension of the $T^2$ test are worked out. An attempt is also made to obtain a quantity that will measure what has been gained by the use of covariance.

In order to illustrate how the theory applies with other types of data, the mathematical model is given for the row by column classification that occurs in the example. The major results for this model are indicated, though without proof.

In Part III it is shown that the computational methods used in the example are equivalent to those developed by theory. While this can easily be verified in a particular case, it is not intuitively obvious.

## PART I NUMERICAL EXAMPLE

**3. Description.** The data form part of an experiment on the assay of insulin of which other parts have been published [5]. Twelve rabbits were used. Each rabbit received in succession four doses of insulin, equally spaced on a log. scale. An interval of eight days or more elapsed between successive doses, and the order in which the doses were given to any rabbit was determined by randomization. Thus the experiment is of the 'randomized blocks' type, where each rabbit constitutes a block and there are 12 blocks with 4 treatments each.

The effect of insulin is usually measured by some function of the blood sugar of the rabbit in periodic bleedings after injection of the insulin. The blood sugar was measured for each rabbit at 1, 2, 3, 4, and 5 hours after injection, and also before injection. In order to simplify the arithmetic, only the initial blood sugar and the blood sugars at 3 and 4 hours after injection will be considered here. These data are shown for the first three rabbits (with totals for all 12 rabbits) in Table I.

Let $x_{iwz}$ be a typical observation of blood sugar, where $i = 3, 4$ stands for the hour after injection, $w$ for the rabbit and $d$ for the dose. The mathematical model to be used is as follows.

$$(1) \qquad x_{iwz} = \mu_i + \rho_{iw} + \gamma_{iz} + \beta_{i0}(x_{0wz} - x_{0..}) + e_{iwz}.$$

The parameters $\mu_i$, $\rho_{iw}$ and $\gamma_{iz}$ represent the true mean and the effects of rabbit and log dose respectively. The quantity $x_{0wz}$ is the initial blood sugar for the rabbit $w$ before the test at dose $z$, while $x_{0..}$ is the average initial blood sugar over the whole experiment. The blood sugar at $i$ hours has been found experimentally to be correlated with the corresponding initial blood sugar, and the relationship is represented here as a linear regression, with $\beta_{i0}$ as the regression coefficient. The residuals $e_{iwz}$ are assumed to follow a multivariate (in this case bivariate) normal distribution, with zero means. The covariance between $e_{iwz}$ and $e_{jwz}$

is taken as $\sigma_{ij\cdot0}$. The model is the standard one for the ordinary analysis of covariance, except that we have *two* measures of the effect of insulin, $x_3$ and $x_4$.

One additional assumption was made. For all post-injection readings, the blood sugar seemed linearly related to the log dose $t_z$. Since this result has been found in other experiments, we assumed that

$$\gamma_{iz} = \delta_i t_z$$

where $\delta_i$ is the regression coefficient of blood sugar on log dose.

**4. Object of the analysis.** Our object was to find the linear combination of the three blood sugar readings that would measure best the effect of the insulin. Because of the linearity of the regression on log dose, the effect of insulin on each

TABLE 1

*Sample of original data on blood sugar levels in insulin experiment*

| Rabbit No. | Log dose | | | | | | | | | | | |
|:---:|:---:|:---:|:---:|:---:|:---:|:---:|:---:|:---:|:---:|:---:|:---:|:---:|
| | Initial blood sugar $x_0$ | | | | Three hours $x_3$ | | | | Four hours $x_4$ | | | |
| | .32 | .47 | .62 | .77 | .32 | .47 | .62 | .77 | .32 | .47 | .62 | .77 |
| 1 | 75 | 94 | 107 | 94 | 95 | 76 | 67 | 56 | 96 | 95 | 115 | 91 |
| 2 | 91 | 86 | 83 | 93 | 98 | 90 | 77 | 69 | 104 | 87 | 90 | 89 |
| 3 | 97 | 99 | 90 | 91 | 84 | 76 | 59 | 48 | 93 | 102 | 85 | 90 |
| Total*........ | 1065 | 1074 | 1121 | 1070 | 932 | 872 | 731 | 591 | 1098 | 1026 | 970 | 847 |

*12 rabbits.

$x_i$ is known completely if the slope $\delta_i$ is known. It seems reasonable to choose the linear compound of the $x_i$'s which will give the maximum ratio when its estimated regression on log dose is divided by the estimated standard error of this regression. We now consider how to obtain this maximum. The argument given below is not intended to prove the validity of the method, for which reference should be made to Part II.

The true regression of the original blood sugar $x_0$ on log dose is known to be zero. Hence, it is clear that the variate $x_0$ is useful only in so far as it enables us to obtain more accurate estimates of $\delta_3$ and $\delta_4$. For this purpose we need to estimate the effect of $x_0$ upon $x_3$ and $x_4$, the blood sugar readings at 3 and 4 hours, independently of dose of insulin or of differences between rabbits. From the standard theory of covariance the best estimate is the regression coefficient $b_{i0} = E_{i0}/E_{00}$, where $E$ denotes a sum of squares or products calculated from the *error* line in the analysis of covariance; that is from the sums of squares and products of deviations of the $x_i$ from the fitted regression on row and column parameters.

The regression of the blood sugar at each hour on the log dose of insulin is calculated from totals adjusted for the regression on $x_0$. Since the 4 successive log doses ($z = 1, 2, 3, 4$) are spaced equally, they may be replaced in the computation by the coded doses $-3$, $-1$, $+1$, and $+3$. If we let $T_{iz}$ be the total blood sugar, summed over 12 rabbits, at the $i$th hour with dose $z$, the following result is well known for the analysis of covariance. The best estimate of $\delta_i (i = 3, 4)$ is

$$[(-3T_{i1} - T_{i2} + T_{i3} + 3T_{i4}) - b_{i0}(-3T_{01} - T_{02} + T_{03} + 3T_{04})]/240.$$

The divisor, 240, is $12(3^2 + 1^2 + 1^2 + 3^2)$. The expression may be written

$$\frac{d_i'}{240} = \frac{(d_i - b_{i0}d_0)}{240},$$

where

$$d_i = -3T_{i1} - T_{i2} + T_{i3} + 3T_{i4}.$$

A linear combination is formed from $d_3'$ and $d_4'$, the numerators in the best estimates of $\delta_3$ and $\delta_4$, by means of the coefficients $L_3$ and $L_4$. $L_3$ and $L_4$ are computed so as to maximize the ratio of

$$d_I = L_3 d_3' + L_4 d_4'$$

to its estimated standard error.

From the definition of $d_i'$, this requires a discriminant of the form

$$I = L_3(x_{3wz} - b_{30}x_{0wz}) + L_4(x_{4wz} - b_{40}x_{0wz}),$$

where each $x_{0wz}$ is measured from its mean.

We require next the estimated standard error of $d_I$. This depends, in turn upon the variances of $d_3'$ and $d_4'$ and their covariance. As usual in the analysis, of variance we have

$$(5) \qquad V(d_3') = V(d_3) + d_0^2 V(b_{30}) = \sigma_{33 \cdot 0}\left(240 + \frac{d_0^2}{E_{00}}\right).$$

The residual variance $\sigma_{33 \cdot 0}$ is estimated from the sums of squares and products in the error row of the analysis of covariance as

$$s_{33 \cdot 0} = E_{33 \cdot 0}/n = (E_{33} - E_{30}^2/E_{00})/n,$$

where $n$ is the degrees of freedom in each $E_{ii}$ diminished by one. Similar methods lead to the variance of $d_4'$ and to the covariance of $d_3'$ and $d_4'$. It follows that the *true* variance of $d_i$ may be written

$$(6) \qquad V(d_I) \; \alpha \; L_3^2\,\sigma_{33.0} + 2L_3 L_4\,\sigma_{34.0} + L_4^2\,\sigma_{44.0},$$

where the factor $\left(240 + \dfrac{d_0^2}{E_{00}}\right)$ in equation (5) is omitted since it does not involve the $L$'s. Similarly, the *estimated* variance of $d_I$, apart from constant factors, may be written as

(7) $$L_3^2 E_{33.0} + 2L_3 L_4 E_{34.0} + L_4^2 E_{44.0} .$$

The quantity to be maximized is therefore

$$\frac{(L_3 d_3' + L_4 d_4')}{\sqrt{L_3^2 E_{33.0} + 2L_3 L_4 E_{34.0} + L_4^2 E_{44.0}}} .$$

Formally, this is the same type of quantity that is maximized in ordinary analysis with the discriminant function. Differentiation with respect to the $L$'s leads to the equations (after omission of another constant factor)

(8) $$E_{33.0}L_3 + E_{34.0}L_4 = d_3' , \qquad E_{34.0}L_3 + E_{44.0}L_4 = d_4' .$$

The objective of the computation, therefore, is to obtain discriminant coefficients having the same ratio to each other as $L_3$ and $L_4$ in equations (8). As will be shown in the next section, this can be accomplished in practice more conveniently by substituting an alternative set of three simultaneous equations for the two in equations (8).

**5. Calculations.** The first step is to form the sums of squares and products in the analysis of covariance. With 12 rabbits and 4 doses, the conventional breakdown of each total sum is into components for rabbits (11 d.f.), doses (3 d.f.) and rabbits $\times$ doses (33 d.f.). Because of the assumed linear regression on log dose, the sum of squares for doses was further divided into two components. The first (1 d.f.) is the contribution due to this regression. For $x_i$, the sum of squares due to regression is $d_i^2/240$, or in the case of $x_3$, $(1164)^2/240$, or 5645. The remaining component, (2 d.f.) is called the *curvature*, since it measures the effect of deviations from the linear regression. The sum of squares for curvature is found by subtraction.

The following points may be noted. (i) For both $x_3$ and $x_4$, the $F$ ratio of the curvature mean square to the rabbits $\times$ doses mean square will be found to be less than 1, so that the data do not suggest rejection of the hypothesis of a linear regression on log dose. (ii) The $F$ ratios of the regression mean squares to the rabbits $\times$ doses mean squares are highly significant, being 57.8 for $x_3$ and 28.7 for $x_4$. This indicates, incidentally, that the three-hour reading may be a more responsive measure of the effect of insulin than the four-hour reading. (iii) With $x_0$, the $F$ ratio does not approach significance for either the regression or the curvature, as is to be expected.

A consequence of the assumption of linear regression on log dose is that the curvature mean squares and products are estimates of the same quantities as the rabbits $\times$ doses mean squares and products. Consequently, the lines for curvature and rabbits $\times$ doses in Table 2 could be added to give 35 d.f. for the 'error' sums of squares or products, $E_{33}$, etc. We decided, however, to estimate the error only from the 33 d.f. for rabbits × doses. This was done because it seemed to facilitate a test of the curvature of the final discriminant $I$. (This test will not be reported here.)

The $L$'s could now be obtained from equations (8). In this case the first equation would contain the terms

$$E_{33.0} = 3223 - (1259)^2/2351; \qquad E_{34.0} = 1200 - (1259)(1340)/2351;$$

$$d_3' = d_3 - b_{30}d_0 = -1164 - \left(\frac{1259}{2351}\right)62,$$

leading to the simultaneous equations

$$2548.8\,L_3 + 482.4\,L_4 = -1197.2$$
$$482.4\,L_3 + 2373.2\,L_4 = -844.3,$$

which give $L_3/L_4 = -.41848/-.27070 = 1.5459$.

TABLE 2

*Sums of squares and products*

| Component | d. f. | $x_0^2$ | $x_3^2$ | $x_4^2$ | $x_0 x_3$ | $x_0 x_4$ | $x_3 x_4$ |
|---|---|---|---|---|---|---|---|
| Between rabbits............ | 11 | 886 | 9376 | 11165 | 1952 | 2477 | 9206 |
| Between doses............. | 3 | 168 | 5806 | 2810 | −247 | −98 | 3981 |
| ⎰Reg. on log dose........ | 1 | 16 | 5645 | 2727 | −301 | −209 | 3924 |
| ⎱Curvature ............. | 2 | 152 | 161 | 83 | 54 | 111 | 57 |
| Rabbits × doses........... | 33 | 2351 | 3223 | 3137 | 1259 | 1340 | 1200 |
| Total................... | 47 | 3405 | 18405 | 17112 | 2964 | 3719 | 14387 |

Instead of using these equations, we propose to solve alternatively the set of three equations

(9)
$$S_{00}L_0 + S_{03}L_3 + S_{04}L_4 = d_0$$
$$S_{30}L_0 + S_{33}L_3 + S_{34}L_4 = d_3$$
$$S_{40}L_0 + S_{43}L_3 + S_{44}L_4 = d_4,$$

where each $S_{ij}$ ($i = 0, 3, 4$) is the sum of squares or products formed by adding the *error* line in the analysis of variance to the line for *regression on log dose*. Thus $S_{ij}$ has 34 d.f. The ratio of $L_3$ to $L_4$, as found from equations (9), is exactly the same as that found from the original equations (8), as is proved in section 18. Further, the solution of the new equations seems to be more useful for performing tests of significance, as will appear in following sections.

Accordingly, the first step after forming the analysis of variance is to set up the three equations (9).

The equations are solved by means of the inverse matrix. The values of $d_i$ on the right side of the equations are replaced successively by 1, 0, 0 by 0, 1, 0 and by 0, 0, 1 to obtain the three sets of values for $L_0$, $L_3$ and $L_4$. These results are given in the first three columns of Table 4 and are designated as $c_{ij}$.

The $L$'s follow from the $c_{ij}$ by the usual rule for regressions. For example,

$$L_3 = \{(.003209)(62) + (.227781)(-1164) + (-.199655)(-809)\} \cdot 10^{-3} =$$
$$-.103417$$

TABLE 3

*Equations for determining $L_3$ and $L_4$*

$$2367L_0 + 958L_3 + 1131L_4 = 62$$
$$958L_0 + 8868L_3 + 5124L_4 = -1164$$
$$1131L_0 + 5124L_3 + 5864L_4 = -809$$

The composite response or discriminant, adjusted for the covariance variate, is now taken as

$$I = L_3 \left( x_3 - \frac{E_{20}}{E_{00}} x_0 \right) + L_4 \left( x_4 - \frac{E_{40}}{E_{00}} x_0 \right)$$

or

$$-.103417 \left( x_3 - \frac{1259}{2351} x_0 \right) - .066883 \left( x_4 - \frac{1340}{2351} x_0 \right)$$
$$= .093503x_0 - .103417x_3 - .066883x_4 .$$

Note that the value of $L_0$ is not used at this stage and that $L_3/L_4 = 1.546$ agrees with the value found from equations (8).

TABLE 4

*Inverse matrix ($\times 10^3$) and L's*

| ($10^3 c_{ij}$) | | | $d_i$ | $L_i$ |
|---|---|---|---|---|
| .465408 | .003209 | −.092568 | 62 | .100008 |
| .003209 | .227781 | −.199655 | −1164 | −.103417 |
| −.092568 | −.199655 | .362846 | −809 | −.066883 |

A similar method may be followed when there are more discriminators or more covariance variates. With two covariance variates, $x_0$ and $x_0^1$, for instance, the adjusted discriminant would be

$$L_3(x_3 - b_{30}x_0 - b_{30}^1 x_0^1) + L_4(x_4 - b_{40}x_0 - b_{40}^1 x_0^1)$$

where $b_{30}$, $b_{30}^1$ are the partial regression coefficients of $x_3$ on $x_0$, $x_0^1$ respectively, determined from the error line, and similarly for $x_4$. Further, since any linear function of the column (dose) means may be represented as a regression on some variate $t_z$, this method may be applied to any linear function of the column means in which we are interested, provided that the mathematical model is appropriate.

**6. Test of the regression of the adjusted discriminant on log dose.** The numerator of the regression of $I$ on the coded doses is

$$d_I = L_3(d_3 - b_{30}d_0) + L_4(d_4 - b_{40}d_0).$$

Since the regressions of $x_3$ and $x_4$ on the coded doses were both significant, it may be confidently expected that the regression of $I$ will also be significant. The test of significance will, however, be given in case it may be useful in other applications. For those who are familiar with multiple regression, the test is perhaps most easily made by means of a device due to Fisher [2].

Construct a dummy variate $y_{wz}$ such that $y_{wz}$ is always equal to $t_z$, or in our case to the coded doses. That is, $y$ takes the value $-3$ for all observations at

TABLE 5

*Analysis of $y^2$ and $yx_i$*

| | d. f. | $y^2$ | $yx_i$ |
|---|---|---|---|
| Rabbits.............................. | 11 | 0 | 0 |
| Doses............................. | 3 | 240 | $d_i$ |
|   Regression on log dose................... | 1 | 240 | $d_i$ |
|   Curvature........................... | 2 | 0 | 0 |
| Rabbits × doses = error.................... | 33 | 0 | 0 |
| Sum = Error plus reg. on log dose......... | 34 | 240 | $d_i$ |
| Total............................. | 47 | 240 | $d_i$ |

the lowest dose level, and $-1$, $+1$, and $+3$ respectively for all observations at the successive higher dosage levels. We shall show that equations (9) solved in finding the $L$'s are formally the same as a set of normal equations for the linear regression of $y$ on $x_0$, $x_3$, and $x_4$.

The following analysis for $y^2$ and $yx_i$ may easily be verified.

It will be noted that the sum of products of $y$ and $x_i$ in the sum line is $d_i$. Further, $S_{ij}$ is the sum of products of $x_i$ and $x_j$ for this line. It follows that the normal equations for the regression of $y$ on the $x$'s, as calculated from the "sum" line, are

$$S_{i0}L_0 + S_{i3}L_3 + S_{i4}L_4 = d_i \qquad (i = 0, 3, 4).$$

These are just the equations solved in obtaining the $L$'s. Consequently, $L_3$ and $L_4$ are the partial regression coefficients of $y$ on $x_3$ and $x_4$. A test of the null hypothesis that the true values of $L_3$ and $L_4$ are both zero can be made by the standard method for multiple regression, as will be shown later from theory. This test is equivalent to a test of the hypothesis that the true value of $d_I$ is zero.

To apply the test, we require three items in the analysis of variance of $y$. First, the total sum of squares for the Sum line, already seen to be 240 (Table 5). Second, the reduction due to a regression on all variates (covariance variates plus discriminators). By the usual rules for regression, this is (from Table 4)

$$L_0 d_0 + L_3 d_3 + L_4 d_4 = (.100008)(62) + (-.103417)(-1164)$$

$$+ (-.066883)(-809) = 180.69.$$

Finally, we need the reduction due to a regression on the variates that are not being tested, i.e. on the covariance variates alone. From Table 4, the reduction

TABLE 6

*Analysis of variance of dummy variate y*

|  | d. f. | S. S. | M. S. |
|---|---|---|---|
| Reduction to regression on covariance variates. | 1 | 1.62 |  |
| Additional reduction to regression on discriminators................................ | 2 | 179.07 | 89.54 |
| Deviations.................................. | 31 | 59.31 | 1.913 |
| Total (from Sum line).................... | 34 | 240.00 |  |

in this case is simply $d_0^2/S_{00}$ or $(62)^2/2367$, or 1.62. The difference, $180.69 - 1.62$, represents the reduction due to the regression of $y$ on $L_3$ and $L_4$, after fitting $x_0$. The resulting analysis is given below, the degrees of freedom being apportioned by the usual rules.

The $F$ ratio, $89.54/1.913$, or 46.80, with 2 and 31 d.f., is used to test the null hypothesis that the adjusted discriminant has no real regression on log dose.

**7. Test of particular discriminators.** Another useful test is that of the null hypothesis that a particular discriminator, or group of discriminators, contribute nothing to the adjusted discriminant. In other words, this is a test of the null hypothesis that the true values of a subset of the $L$'s are all zero. The test is of interest in the present investigation, since it would be useful to know whether all five hourly readings of the blood sugar are really helpful. As might be expected by analogy with the previous section, the test is made by calculating the additional reduction due to the regression of $y$ on the particular subset of the $L$'s in question.

The test will be illustrated with respect to $L_4$. One method of making the test is to re-solve the normal equations with $L_4$ omitted. From this solution the reduction due to a regression of $y$ on $x_0$ and $x_3$ alone is obtained. The additional reduction due to a regression on $x_4$ is found by subtraction from 180.69.

However, the additional reduction can be found directly from the well-known regression theorem that it is equal to $L_4^2/c_{44}$. The $c$'s have already been found in Table 4. The result is $(.066883)^2/(.000362846)$, or 12.33. This value is tested against the residual error mean square of 1.913, $F$ having 1 and 31 d.f. The contribution is found to be significant.

In fact, by this process a kind of estimated standard error can be attached to each of the $L$'s for the discriminators, using the formula $s\sqrt{c_{ii}}$, where $s$ is the residual root mean square. Thus for $L_3$, $(-.103417)$, the 'standard error' is $\sqrt{(1.913)(.000227781)}$, or .0209. It should be stressed that at this point the analogy with regression is rather thin. The $L$'s are not normally distributed, nor do the estimated standard errors follow their usual distribution. It is, however, correct that if the true value of $L_4$ is zero, $L_4/s\sqrt{c_{44}}$ follows the $t$ distribution with 31 d.f. Thus, if omission of some discriminators seems warranted,

TABLE 7

*Analysis of variance for regression of $y$ on the discriminators*

|  | d. f. | S. S. | M. S. |
|---|---|---|---|
| Regression.............................. | 2 | 159.20 | 79.60 |
| Deviations.............................. | 32 | 80.80 | 2.525 |
| Total................................. | 34 | 240.00 | |

these $t$ ratios are relevant in deciding which variate to eliminate first. Strictly speaking, the $c$'s should be re-calculated after each elimination before deciding which other discriminators might also be discarded.

**8. Estimation of the gain due to covariance.** The tests given above enable us to state whether the discriminators contribute significantly, in the statistical sense. It is also of interest to investigate what has been gained by the use of the covariance variates. From the practical point of view, the question: "What is the gain from covariance?" might be re-phrased as: "If $x_0$ is ignored, how many rabbits must be tested in order to estimate the regression on log dose as accurately as it was estimated with the adjusted discriminant for 12 rabbits?"

The theoretical aspects of the question are discussed in section 16; the calculations are described here. The only new quantity needed is the $F$ ratio for the regression of $y$ on the discriminators alone. This can be obtained by a new solution of the normal equations, this time with the covariance variates omitted. With just one covariance variate, it is quicker to use the fact that the additional reduction to the regression of $y$ on $x_0$, after fitting $x_3$ and $x_4$, is $L_0^2/c_{00}$, or $(.100008)^2/(.000465408)$ or 21.49. Consequently, the reduction due to a regression of $y$ on $x_3$ and $x_4$ alone is $180.69 - 21.49$, or $159.20$. The $F$ ratio, $79.60/2.525$, is $31.52$, whereas the $F$ ratio with covariance is $46.80$ (from Table 6). The quantity suggested from theory for comparing the two techniques is

$$\frac{(n_2 - 2)F}{n_2} - 1$$

where $n_2$ is the number of d.f. in the denominator of $F$. These values are $\{(30 \times 31.52/32) - 1\}$ or $28.55$ with no covariance and $\{(29 \times 46.80/31) - 1\}$ or $42.78$ with covariance. The relative information is estimated as $42.78/28.55$, or $1.50$, so that the use of covariance gives 50 per cent more information. In other words about 18 rabbits would be needed if the initial blood sugars were ignored. To a slight extent this estimate favors the covariance analysis, since it ignores the increased accuracy that would accrue from the extra error d.f. if 18 rabbits were used without covariance.

## PART II Theory

**9. Notation.** The theory will be given first for the two-population case. We suppose that a random sample of size $N$ has been drawn from each population. A typical discriminator is written $x_{iw\alpha}$ and a typical covariance variate $x_{\xi w \alpha}$, where

$i, j = 1, 2, \cdots p$ denote discriminators,

$\xi, \eta = 1, 2, \cdots k$ denote covariance variates,

$w = 1, 2$ denotes the population, and

$\alpha = 1, 2, \cdots N$ denotes the order within the sample.

The population mean of $x_{iw\alpha}$ is $\mu_{iw}$, and the corresponding sample mean is $x_{iw\cdot}$. The difference $(\mu_{i2} - \mu_{i1})$ is denoted by $\delta_i$ and the corresponding estimated difference $(x_{i2\cdot} - x_{i1\cdot})$ by $d_i$.

**10. Discriminant functions and generalized distance.** Since we propose to approach the theory by means of the generalized distance, it may be well to review briefly the relation between the discriminant and the generalized distance. In the ordinary theory (with no covariance variates) it is assumed that the variates $x_{iw\alpha}$ follow a multivariate normal distribution, and that the covariance matrix $\sigma_{ij}$ between $x_{iw\alpha}$ and $x_{jw\alpha}$ is the same in both populations. The generalized distance of Mahalanobis is defined by

$$(10) \qquad p\Delta^2 = \sum_{i,j=1}^{p} \sigma^{ij} \delta_i \delta_j, \qquad \text{where} \qquad (\sigma^{ij}) = (\sigma_{ij})^{-1}.$$

In order to estimate this quantity from the sample, we first calculate the mean within-sample covariance $s_{ij}$, where

$$(11) \qquad s_{ij} = \sum_{w=1}^{2} \sum_{\alpha=1}^{N} (x_{iw\alpha} - x_{iw\cdot})(x_{jw\alpha} - x_{jw\cdot})/2(N-1),$$

The estimated distance is then taken as

$$(12) \qquad pD^2 = \sum_{i,j=1}^{p} s^{ij} d_i d_j .$$

Apart from a factor $N/(N-1)$, this is the maximum likelihood estimate.

In the discriminant function used by Fisher (1), the object is to find a linear function $I_{w\alpha} = \Sigma M_i x_{iw\alpha}$, where the $M_i$ are chosen to maximize the ratio of the sum of squares between samples to that within samples in the analysis of variance of $I$. This is equivalent to maximizing the ratio of the difference between the two sample means of $I$ to the estimated standard error of this difference. As Fisher showed (2), the $M_i$ (apart from an arbitrary multiplier) are given by

$$M_i = \sum_{j=i}^{p} s^{ij} d_j$$

Consequently, the difference between the two sample means of $I$, the discriminant function, is

$$\sum_{i=1}^{p} M_i d_i = \sum_{i,j=1}^{p} s^{ij} d_i d_j .$$

This is exactly the same as $pD^2$ in equation (12). Thus the discriminant function leads to the estimated distance, and *vice versa*.

**11. Extension to the present problem.** In our case there are $(p+k)$ variates ($p$ discriminators, $k$ covariance variates) from which to estimate the distance. All variates, $x_{iw\alpha}$ and $x_{\xi w\alpha}$, are assumed to follow a multivariate normal distribution. The covariance matrix, assumed the same in both populations, now has $(p+k)$ rows and columns, and may be denoted by

$$(13) \qquad \Lambda = \begin{pmatrix} \sigma_{ij} & \sigma_{i\eta} \\ \sigma_{\xi j} & \sigma_{\xi\eta} \end{pmatrix} .$$

For each of the covariance variates, it is known that the population means $\mu_{\xi 1}$, $\mu_{\xi 2}$ are equal, so that the difference $\delta_\xi$ is zero. This is the fact that distinguishes the problem from ordinary discriminant function analysis.

Hence, the generalized distance, as defined from all $(p+k)$ variates contains no contribution from terms in $\delta_\xi$ and is given by

$$(14) \qquad (p+k)\Delta^2 = \sum_{i,j=1}^{p} \sigma_{(p+k)}^{ij} \delta_i \delta_j .$$

The matrix $\sigma_{(p+k)}^{ij}$ is that formed by the first $p$ rows and columns of the inverse of $\Lambda$. Note that in general this will not be the same as the matrix $\sigma^{ij}$, which is the inverse of $\sigma_{ij}$.

In the next section we consider the estimation of this quantity from the sample data. By analogy with the previous section, it might be guessed that the estimate would be of the form $\Sigma s_{(p+k)}^{ij} d_i d_j$. The maximum likelihood estimate turns out to be of this form, except that instead of $d_i$ we have $d'_i$ , the difference between the two sample means of the deviations of $x_i$ from its 'within-sample' linear regression on the $x_\xi$ .

**12. Estimation of the distance.** It is known that the generalized distance is invariant under non-singular linear transformations of the variates. For convenience, we replace the $x_{iwa}$ by variates $x'_{iwa}$ , where

$$x'_{iwa} = x_{iwa} - \sum_{\xi=1}^{k} \beta_{i\xi}(x_{\xi wa} - \mu_{\xi w}).$$

Thus $x'_{iwa}$ is the deviation of $x_{iwa}$ from its population linear regression on the $x_{\xi wa}$ . The population mean of $x'_{iwa}$ is clearly $\mu_{iw}$ , and the difference between the two population means is therefore $\delta_i$ .

The covariance matrix of the $x'_{iwa}$ , $x_{\xi wa}$ may be written

$$(15) \qquad \Lambda' = \begin{pmatrix} \sigma_{ij\cdot\xi} & 0 \\ 0 & \sigma_{\xi\eta} \end{pmatrix},$$

where $\sigma_{ij\cdot\xi}$ denotes the covariance matrix of the deviations of the $x_{iwa}$ from their regressions on the $x_{\xi wa}$ . It follows that in terms of the transformed variates the generalized distance is given by

$$(16) \qquad (p + k)\Delta^2 = \sum_{i,j=1}^{p} \sigma^{ij\cdot\xi} \delta_i \delta_j$$

where $\sigma^{ij\cdot\xi}$ is the inverse of the $p \times p$ matrix $\sigma_{ij\cdot\xi}$ .

The joint distribution of the $2N$ observations on each of the $x'_{iwa}$ and $x_{\xi wa}$ is as follows:

$$(2\pi)^{-N(p+k)} \mid \sigma^{ij\cdot\xi} \mid^{+N} \mid \sigma^{\xi\eta} \mid^{+N} \Pi dx'_{iwa} dx_{\xi wo} \cdot$$

$$\exp \left\{ -\frac{1}{2} \left[ \sum_{w=1}^{2} \sum_{\alpha=1}^{N} \sum_{i,j=1}^{p} \sigma^{ij\cdot\xi}(x'_{iwa} - \mu_{iw})(x'_{jwa} - \mu_{jw}) + \right.\right.$$

$$\left.\left. \sum_{w=1}^{2} \sum_{\alpha=1}^{N} \sum_{\xi,\eta=1}^{k} \sigma^{\xi\eta}(x_{\xi wa} - \mu_{\xi w})(x_{\eta wa} - \mu_{\eta w}) \right] \right\},$$

where $\sigma^{\xi\eta}$ is the inverse of the $k \times k$ matrix $\sigma_{\xi\eta}$ .

We now proceed to estimate $\Delta^2$ in equation (16) by maximum likelihood. For this, we obviously need the sample estimates of the $\sigma^{ij\cdot\xi}$ and the $\delta_i$ , and it will appear presently that the sample estimates of the $\beta_{i\xi}$ are also required. However, it happens that the $\sigma^{\xi\eta}$ and the $\mu_{\xi w}$ are not needed. Hence the relevant part of the likelihood function is

$$(17) \qquad L = N \log \mid \sigma^{ij\cdot\xi} \mid - \frac{1}{2} \sum_{w=1}^{2} \sum_{\alpha=1}^{N} \sum_{i,j=1}^{p} \sigma^{ij\cdot\xi}(x'_{iwa} - \mu_{iw})(x'_{jwa} - \mu_{jw})$$

where

$$x'_{iw\alpha} = x_{iw\alpha} - \sum_{\xi=1}^{k} \beta_{i\xi}(x_{\xi w\alpha} - \mu_{\xi w}) .$$

Differentiating first with respect to $\mu_{iw}$ , we obtain

(18)
$$\sum_{\alpha=1}^{N} \sum_{j=1}^{p} \sigma^{ij\cdot\xi}(x'_{jw\alpha} - \hat{\mu}_{jw}) = 0.$$

Except in the case (with probability zero) where our estimate of $\sigma^{ij\cdot\xi}$ turns out to be singular, these equations have no solution except

(19)
$$\sum_{\alpha=1}^{N} (x'_{jw\alpha} - \hat{\mu}_{jw}) = 0$$

for every $j$, $w$.  Consequently

$$\hat{\mu}_{jw} = x'_{jw\cdot}.$$

so  that

$$\hat{\delta}_j = \hat{\mu}_{j2} - \hat{\mu}_{j1} = x'_{j2\cdot} - x'_{j1\cdot} = d_j - \sum_{\xi=1}^{k} \beta_{j\xi}\, d_\xi.$$

This shows that the $\beta_{i\xi}$ must also be estimated.  Now

$$\frac{\partial L}{\partial \beta_{i\xi}} = \sum_{w=1}^{2} \sum_{\alpha=1}^{N} \frac{\partial L}{\partial x'_{iw\alpha}} \frac{\partial x'_{iw\alpha}}{\partial \beta_{i\xi}} = \sum_{w=1}^{2} \sum_{\alpha=1}^{N} \sum_{j=1}^{p} \sigma^{ij\cdot\xi}(x_{\xi w\alpha} - \mu_{\xi w})(x'_{jw\alpha} - \mu_{jw}).$$

Once again, unless the estimate of $\sigma^{ij\cdot\xi}$ is singular, the only solutions of the equations formed by equating this quantity to zero are

(20)
$$\sum_{w=1}^{2} \sum_{\alpha=1}^{N} (x_{\xi w\alpha} - \mu_{\xi w})(x'_{jw\alpha} - \hat{\mu}_{jw}) = 0$$

for every $\xi$, $j$.

Since $\hat{\mu}_{jw} = x'_{jw\cdot}$ , the term in $\mu_{\xi w}$ vanishes.  Substituting for $x'$ in terms of $x$ from  (17), we  obtain

$$\sum_{w=1}^{2} \sum_{\alpha=1}^{N} x_{\xi w\alpha} \left\{ (x_{jw\alpha} - x_{jw\cdot}) - \sum_{\eta=1}^{k} b_{j\eta}(x_{\eta w\alpha} - x_{\eta w\cdot}) \right\} = 0$$

where $b_{j\eta}$ stands for the maximum likelihood estimate of $\beta_{j\eta}$ .  These equations may  be  written

(21)
$$\sum_{\eta=1}^{k} b_{j\eta} E_{\xi\eta} = E_{j\xi}$$

where $E$ denotes a sum of squares or products of deviations from the sample means, containing $2(N - 1)$ degrees of freedom.  The equations are therefore the ordinary normal equations for the 'within-sample' multiple regression of $x_{jw\alpha}$ on the $x_{\xi w\alpha}$.

Finally, differentiation of $L$ with respect to the $\sigma^{ij\cdot\xi}$ leads to

$$(22) \qquad 2N\hat{\sigma}_{ij\cdot\xi} = \sum_{w=1}^{2} \sum_{\alpha=1}^{N} (x'_{iw\alpha} - x'_{iw\cdot})(x'_{jw\alpha} - x'_{jw\cdot}).$$

This is just the 'within-samples' sum of squares or products of the variates $x'$. On substituting for the $x'$ in terms of the $x$ and using equations (21), we obtain

$$2N\hat{\sigma}_{ij\cdot\xi} = E_{ij} - \sum_{\xi=1}^{k} b_{i\xi} E_{j\xi} = E_{ij\cdot\xi} \qquad \text{(say)}.$$

To summarize, the estimated distance is given by means of the equation

$$(p+k)D^2 = \sum_{i,j=1}^{p} \hat{\sigma}^{ij\cdot\xi}\hat{\delta}_i\hat{\delta}_j = 2N \sum_{i,j=1}^{p} E^{ij\cdot\xi} d'_i d'_j,$$

where $E^{ij\cdot\xi}$ is the inverse of $E_{ij\cdot\xi}$ and

$$d'_i = d_i - \sum_{\xi=1}^{k} b_{i\xi} d_\xi.$$

This estimate was obtained by assuming *all* variates jointly normally distributed. From the form of the likelihood function (17) it can be seen that the M.L. estimate of the distance remains the same under the less restrictive assumptions that the $x_{\xi w\alpha}$ are fixed, while the deviations of the $x_{iw\alpha}$ from their regressions on the $x_{\xi w\alpha}$ are jointly normal.

**13. Computational procedure.** An orderly procedure for calculating the generalized distance will now be given. From this, the method for computing the corresponding discriminant function will be shown. The computations also lead to the generalization of Hotelling's $T^2$. The steps are as follows.

(i). First form the 'within-sample' sums of squares and products of all variates, with $2(N-1)$ degrees of freedom. These are the quantities denoted by $E_{ij}$, $E_{i\xi}$, $E_{\xi\eta}$.

(ii). Invert the matrix $E_{\xi\eta}$, giving $E^{\xi\eta}$.

(iii). The regression coefficients $b_{i\xi}$, estimates of the $\beta_{i\xi}$, are now obtainable by means of the relations

$$b_{i\xi} = \sum_{\eta=1}^{k} E_{i\eta} E^{\xi\eta},$$

as is clear from the usual matrix solution of equations (21).

(iv). The sums of squares and products of the deviations of the $x_i$ from their 'within-sample' regressions on the $x_\xi$ are now computed from equations (22)

$$2N\hat{\sigma}_{ij\cdot\xi} = E_{ij\cdot\xi} = E_{ij} - \sum_{\xi=1}^{k} b_{i\xi} E_{j\xi}.$$

(v). The final step is to invert the matrix $E_{ij\cdot\xi}$, giving $E^{ij\cdot\xi}$, and to form the product

$$(p+k)D^2 = 2N \sum_{i,j=1}^{p} E^{ij\cdot\xi} d_i' d_j', \quad \text{where} \quad d_i' = d_i - \sum_{\xi=1}^{k} b_{i\xi} d_\xi.$$

When there were no covariance variates, the discriminant function $I$ had the property that the difference between the two sample means of $I$ was equal to the estimated distance (Section 10). This relationship can be preserved when covariance variates are present by defining $I$ so that

$$I_{w\alpha} = \sum_{i=1}^{p} M_i \left( x_{iw\alpha} - \sum_{\xi=1}^{k} b_{i\xi} x_{\xi w\alpha} \right),$$

and calculating the weights $M_i$ from the equations,

$$\sum_{j=1}^{p} E_{ij\cdot\xi} M_j = d_i'.$$

For in that case,

$$M_i = \sum_{j=1}^{p} E^{ij\cdot\xi} d_j'.$$

Consequently the difference between the two sample means of $I$ is

$$\sum_{i=1}^{p} M_i d_i' = \sum_{i,j=1}^{p} E^{ij\cdot\xi} d_i' d_j',$$

which (apart from the constant $2N$) is equal to $(p+k)D^2$.

**14. Distribution of the estimated distance.** In the ordinary case, with no covariance variates, the frequency distribution of the estimated distance has been given by several authors, e.g. Hsu [6]. It will be found that in our problem the distribution is essentially the same, except that the quantity $D^2$ must be multiplied by a new factor and that one set of degrees of freedom entering into the result must be changed from $(n-p+1)$ to $(n-p-k+1)$.

Thus far we have assumed that all variates jointly follow a multivariate normal distribution. It is convenient at this stage to regard the covariance variates $x_{\xi w\alpha}$ as fixed from sample to sample, and to use the conditional distribution of the $x_{iw\alpha}$, subject to this restriction. It is well known (e.g. Cramér [7, section 24.6]) that this conditional distribution is the multivariate normal

(23)
$$(2\pi)^{-Np} |\sigma^{ij\cdot\xi}|^{+N} \prod dx_{iw\alpha}$$
$$\cdot \exp\left\{ -\tfrac{1}{2}\left[ \sum_{w=1}^{2} \sum_{\alpha=1}^{N} \sum_{i,j=1}^{p} \sigma^{ij\cdot\xi}(x_{iw\alpha} - \mu_{iw} - \gamma_{iw\alpha})(x_{jw\alpha} - \mu_{jw} - \gamma_{jw\alpha}) \right] \right\}$$

where

$$\gamma_{iw\alpha} = \sum_{\xi=1}^{k} \beta_{i\xi}(x_{\xi w\alpha} - \mu_{\xi w}).$$

Since the estimated distance is a function of the quantities $E'_{ij\cdot\xi}$, $d'_i$, we now find the joint distribution of these variates. The joint distribution of the sums of squares and products $E_{ij\cdot\xi}$ is obtained by quoting a slight extension of a result due to Bartlett [8], which may be stated as follows.

*Let the variates $x_{iw\alpha}$ follow the distribution (23) and let*

$$(i) \qquad E_{ij} = \sum_{w=1}^{2} \sum_{\alpha=1}^{N} (x_{iw\alpha} - x_{iw\cdot})(x_{jw\alpha} - x_{jw\cdot})$$

*be a typical 'within-samples' sum of squares or products,*

$$(ii) \qquad b_{i\xi} = \sum_{\eta=1}^{k} E_{i\eta} E^{\xi\eta}$$

*be the 'within-samples' partial regression coefficient of $x_i$ on $x_\xi$, and*

$$(iii) \qquad E_{ij\cdot\xi} = E_{ij} - \sum_{\xi=1}^{k} b_{i\xi} E_{j\xi}$$

*be the sum of squares or products of deviations from these regressions.* **Then**
*(a) the quantities $E_{ij\cdot\xi}$ follow the Wishart distribution*

$$c \, | E_{ij\cdot\xi} |^{\frac{1}{2}(n-k-p-1)} \exp\left\{-\tfrac{1}{2} \sum_{i,j=1}^{p} \sigma^{ij\cdot\xi} E_{ij\cdot\xi}\right\} \prod dE_{ij\cdot\xi}$$

*with $(n - k)$ d.f., where $n = 2(N - 1)$,*

*(b) this distribution is independent of that of the $b_{i\xi}$, and*
*(c) both distributions are independent of that of the means $x_{iw\cdot}$ and consequently of that of the difference $d_i = (x_{i2\cdot} - x_{i1\cdot})$.*

The result was proved by Bartlett for a sample from a single population. The extension to the case of two populations is straightforward and will not be given in detail.

From (b) and (c) it follows that the distribution of the $E_{ij\cdot\xi}$ is independent of that of the quantities

$$d'_i = d_i - \sum_{\xi=1}^{k} b_{i\xi} d_\xi.$$

Further, with the $x_\xi$ variates fixed, the $d'_i$ are linear functions of the $x_{iw\alpha}$ with constant coefficients and hence follow a multivariate normal distribution, Wilks [9]. We now find the means and the covariance matrix of this joint distribution.

From the joint distribution (23) of the $x_{iw\alpha}$, it is easily seen that

$$(24) \qquad E(d_i) = \delta_i + \sum_{\xi=1}^{k} \beta_{i\xi} d_\xi.$$

Also, since by standard regression theory the $b_{i\xi}$ are unbiased estimates of the $\beta_{i\xi}$,

$$E\left\{\sum_{\xi=1}^{k} b_{i\xi} d_\xi\right\} = \sum_{\xi=1}^{k} \beta_{i\xi} d_\xi.$$

Hence, by subtraction,

(25) $$E(d_i') = \delta_i .$$

Now

$$\text{Cov } (d_i' d_j') = \text{Cov } (d_i - \sum_{\xi=1}^{k} b_{i\xi} d_{\xi})(d_j - \sum_{\eta=1}^{k} b_{j\eta} d_{\eta}).$$

By (c) the distributions of the $d_i$, $b_{j\eta}$ are independent, so that there will be no contribution from products of the form $d_i b_{j\eta}$. Hence

(26) $$\text{Cov } (d_i' d_j') = \text{Cov } (d_i d_j) + \sum_{\xi,\eta=1}^{k} d_{\xi} d_{\eta} \text{ Cov } (b_{i\xi} b_{j\eta}).$$

Since $d_i$ is the difference between the means of two samples of size $N$, Cov $(d_i d_j)$ is $2 \sigma_{ij\cdot\xi}/N$. The covariance of $b_{i\xi}$ and $b_{j\eta}$ is more troublesome. Writing the expressions for these regression coefficients in terms of the original data, we have

$$\text{Cov } (b_{i\xi} b_{j\eta}) = \sum_{\lambda,\nu=1}^{k} E^{\lambda\xi} E^{\nu\eta} \text{ Cov } (E_{i\lambda} E_{j\nu}) =$$

$$\sum_{\lambda,\nu=1}^{k} E^{\lambda\xi} E^{\nu\eta} \sum_{w,z=1}^{2} \sum_{\alpha,\zeta=1}^{N} (x_{\lambda w\alpha} - x_{\lambda w\cdot})(x_{\nu z\zeta} - x_{\nu z\cdot}) \text{ Cov } (x_{iw\alpha} x_{jz\zeta}).$$

Since successive observations are assumed independent, the covariance term vanishes unless $w = z$ and $\alpha = \zeta$, in which case it equals $\sigma_{ij\cdot\xi}$. Thus

$$\text{Cov } (b_{i\xi} b_{j\eta}) = \sigma_{ij\cdot\xi} \sum_{\lambda,\nu=1}^{k} E^{\lambda\xi} E^{\nu\eta} E_{\lambda\nu} = \sigma_{ij\cdot\xi} E^{\xi\eta}.$$

Finally, from (26)

(27) $$\text{Cov } (d_i' d_j') = \sigma_{ij\cdot\xi} \left( \frac{2}{N} + \sum_{\xi,\eta=1}^{k} E^{\xi\eta} d_{\xi} d_{\eta} \right) = v\sigma_{ij\cdot\xi} \qquad \text{(say)}.$$

Having obtained the distributions of the $E_{ij\cdot\xi}$, $d_i'$, we may apply Hsu's result [6] for the general distribution of Hotelling's $T^2$. In our notation, this may be stated as follows.

*If the variates $d_i'/\sqrt{v}$ follow the multivariate normal distribution with means $\delta_i/\sqrt{v}$ and covariance matrix $\sigma_{ij\cdot\xi}$, and if the variates $E_{ij\cdot\xi}$ follow the Wishart distribution with $(n - k)$ d.f. and covariance matrix $\sigma_{ij\cdot\xi}$, the two distributions being independent, then*

$$y = \sum_{i,j=1}^{p} E^{ij\cdot\xi} d_i' d_j'/v, \quad .$$

*follows the distribution*

(28) $$e^{-\tau} \sum_{h=0}^{\infty} \frac{\tau^h}{h!} \frac{1}{B\{\frac{1}{2}p + h, \frac{1}{2}(n - k - p + 1)\}} \cdot y^{\frac{1}{2}p+h-1}(1 + y)^{-\frac{1}{2}(n-k+1)-h} \, dy,$$

*where*

$$\tau = \tfrac{1}{2} \sum_{i,j=1}^{p} \sigma^{ij \cdot \xi} \delta_i \, \delta_j / v,$$

$$v = \frac{2}{N} + \sum_{\xi,\eta=1}^{k} E^{\xi\eta} \, d_\xi \, d_\eta, \qquad\qquad n = 2(N - 1).$$

This distribution is, of course, the distribution of the ratio of two independent values of $\chi^2$, with $p$ and $(n - k - p + 1)$ d.f. respectively, in the case where the numerator is non-central.

**15. Tests of significance.** This result leads to the extension of Hotelling's $T^2$ test. For if $\delta_i = 0$, $(i = 1, 2, \cdots p)$, then $\tau$ is zero and

$$\sum_{i,j=1}^{p} E^{ij \cdot \xi} \, d'_i \, d'_j$$

is distributed as $vpF/(n - k - p + 1)$, with $p$ and $(n - k - p + 1)$ d.f. The distribution (28) above gives the power function of this test.

We may also wish to apply a test of this type to a subgroup $x_i$ of the discriminators $(i = 1, 2, \cdots q < p)$. Speaking popularly, this is a test of the null hypothesis that the above variates $x_i$ contribute nothing to the discrimination between the two populations, given that the remaining discriminators and the covariance variates have already been included.[1] To see what is meant more precisely, consider the following transformation:

$$x'_i = x_i - \Sigma \, \beta_{il} x_l - \Sigma \, \beta_{i\xi} x_\xi, \qquad i = 1, 2, \cdots q;$$

$$x'_l = x_l - \Sigma \, \beta_{l\xi} x_\xi, \qquad\qquad l = q + 1, \cdots p;$$

$$x'_\xi = x_\xi, \qquad\qquad\qquad\qquad \xi = 1, 2, \cdots k,$$

where the $\beta$'s are population regression coefficients. Then it is not difficult to see that the distance is now given by

$$(p + k)\Delta^2 = \sum_{i,j=1}^{q} \sigma^{ij \cdot l\xi} \delta'_i \delta'_j + \sum_{l,m=q+1}^{p} \sigma^{lm \cdot \xi} \delta_l \, \delta_m$$

where $\sigma^{ij \cdot l\xi}$ is the inverse of the covariance matrix of the deviations of the $x_i$ from their regressions on the $x_l$ plus the $x_\xi$, and

$$\delta'_i = \delta_i - \Sigma \, \beta_{il} \delta_l.$$

Consequently if $\delta'_i = 0$, $(i = 1, 2, \cdots q)$ the distance is exactly the same as it would be if the variates $x_i$ were omitted. The test in question is therefore a test of the null hypothesis that $\delta'_i = 0$, $(i = 1, 2, \cdots q)$.

If both the remaining discriminators $x_l$ and the covariance variates $x_\xi$ are regarded as fixed, the method of proof in the previous section provides an $F$ test

---

[1] The test is illustrated in section 7.

for this hypothesis also. It is found that the sums of squares or products $E^{ij \cdot l\xi}$ follow a Wishart distribution with $(n - k - p + q)$ d.f., while the quantities

$$d_i' = d_i - \sum_{l=q+1}^{p} b_{il}\, d_l - \sum_{\xi=1}^{k} b_{i\xi}\, d_\xi$$

are normally distributed, with zero means when the null hypothesis is true. This leads to the result that

$$\sum_{i,j=1}^{q} E^{ij \cdot l\xi}\, d_i'\, d_j'$$

is distributed as $v'qF/(n - k - p + 1)$, with $q$ and $(n - k - p + 1)$ d.f., and

$$v' = \frac{2}{N} + \Sigma\, E^{l\xi}\, d_l\, d_\xi,$$

the sum extending over both the covariance variates and the discriminators that are not being tested.

**16. Discussion of the gain due to covariance.** In this section we attempt to construct a measure of the amount that has been gained by the use of the covariance variates. Only a preliminary discussion will be given: a complete discussion would be rather lengthy, owing to the many different uses to which the discriminant function can be put. Perhaps the problem can most easily be seen by considering the effect on Hotelling's generalized $T^2$ test of significance.

The power function of this test, as obtained from equation (28) section 14, depends on four factors; the level of significance that is chosen, the degrees of freedom $n_1$ and $n_2$ in the numerator and denominator of $F$, and the parameter $\tau$. If the covariance variates were ignored, the usual $T^2$ test could be applied to the discriminators alone. In this case we would have

$$n_1' = p, \qquad n_2' = n - p + 1, \qquad \tau' = \tfrac{1}{2}\Sigma\sigma^{ij}\delta_i\delta_j/v', \qquad \text{where } v' = 2/N.$$

With the covariance variates, we have

$$n_1 = p, \qquad n_2 = n - p - k + 1, \qquad \tau = \tfrac{1}{2}\Sigma\sigma^{ij \cdot \xi}\delta_i\delta_j/v,$$

where

$$v = \frac{2}{N} + \Sigma\, E^{\xi\eta}\, d_\xi\, d_\eta.$$

The first point to note is that

$$\Sigma\, \sigma^{ij \cdot \xi}\delta_i\delta_j \geq \Sigma\, \sigma^{ij}\delta_i\delta_j$$

This is an instance of the general result that the addition of new variates cannot decrease the value of $p\Delta^2$. To see this, replace the covariance variates by their deviations from their regressions on the discriminators. This transformation gives

$$(29) \qquad \sum_{i,j=1}^{p} \sigma^{ij \cdot \xi} \delta_i \delta_j = \sum_{i,j=1}^{p} \sigma^{ij} \delta_i \delta_j + \sum_{\xi,\eta=1}^{k} \sigma^{\xi\eta \cdot i} \delta'_\xi \delta'_\eta ,$$

where

$$\delta'_\xi = \delta_\xi - \sum_{i=1}^{p} \beta_{\xi i} \delta_i .$$

Since the term on the right of equation (29) is a positive definite quadratic form, the result follows.

Consequently, the first effect of the covariance variates is to make the numerator of $\tau$ greater than that of $\tau'$. As a partial compensation, the denominator $v$ is also greater than $v'$, but it may be shown that the difference in the denominators will usually be trivial if $k$ is small relative to $n$. We therefore expect $\tau$ to be greater than $\tau'$. Now for fixed $n_1$, $n_2$ and significance level, it is well known that the power function (28) is monotone increasing with $\tau$. Hence, other things being equal, the increase in $\tau$ due to the covariance variates leads to a more powerful test.

The two power functions, however, differ in another respect, in that with covariance the value of $n_2$ is reduced from $(n - p + 1)$ to $(n - p - k + 1)$. This decrease in the number of degrees of freedom in the denominator of $F$ will to some extent offset the gain from an increased $\tau$. Examination of Tang's tables [10] indicates, however, that if the degrees of freedom are substantial, this effect will not be important. Moreover, in most practical applications, $k$ is likely to be only 1 or 2. Hence, as a first approximation the effect will be ignored, though to do so tends to overestimate the advantage of covariance.

Suppose now that $\tau = r\tau'$, where $r > 1$. Since $\tau'$ is proportional to $N$, the size of sample taken from each population, we could make $\tau' = \tau$ by increasing the size of sample (when covariance is not used) from $N$ to $rN$. This suggests that the ratio $\tau/\tau'$ can be used, as a first approximation, to measure the relative accuracy obtained with and without the use of covariance. This measure carries approximately the usual interpretation that the inferior method would become as good as the superior method if the sample size for the inferior method were increased by the factor $r$. A further refinement could be made to take account of the difference in the $n_2$ values. By trial and error applied to Tang's tables, one could determine $r'$ so that the two power functions would be as nearly coincident as possible.

In practice, the ratio $\tau/\tau'$ must be estimated from the data. From the power function in equation (28) it is found by integration that the mean value of $y$ is

$$(2\tau + p)/(n_2 - 2),$$

so that an unbiased estimate of $\tau$ is

$$\tfrac{1}{2}\{(n_2 - 2)y - p\} = \tfrac{1}{2}p \left\{ \frac{(n_2 - 2)}{n_2} F - 1 \right\}.$$

This suggests that the quantity

$$\frac{(n_2 - 2)}{n_2} F - 1$$

should be calculated both with and without covariance. The ratio of the two values will probably not be an unbiased estimate of $\tau/\tau'$, but may be used pending further information about its sampling distribution. This type of calculation is made for the numerical example in section 8.

**17. The case of a row by column classification.** Thus far the discussion has been confined to the case where there are only two populations. The technique may also be used when there are more than two populations. The difference $\delta_i$ between the two population means is replaced by some linear function of the population means. As an illustration we consider a row by column classification, the case that arises in the numerical example. No detailed proofs will be given, though it is hoped that the theory can be fairly easily developed from the mathematical model.

A typical variate is $x_{iwz}$, where $i = 1, 2, \cdots p$ denotes the variate, $w = 1, 2, \cdots r$ denotes the row and $z = 1, 2, \cdots c$ denotes the column, there being one observation in each cell. The variates $x_{iwz}$ follow a multivariate normal distribution, with covariance matrix $\sigma_{ij \cdot \xi}$ and means

$$E(x_{iwz}) = \mu_i + \rho_{iw} + \gamma_{iz} + \sum_{\xi=1}^{k} \beta_{i\xi}(x_{\xi wz} - x_{\xi \cdot \cdot}),$$

where $\rho_{iw}$ denotes the effect of the row and $\gamma_{iz}$ that of the column. Without loss of generality we may assume that

$$\sum_{w} \rho_{iw} = \sum_{z} \gamma_{iz} = 0.$$

In addition, there exists a *known* set of variates $t_z$ such that

$$\gamma_{iz} = \delta_i t_z, \qquad \sum_{z} t_z = 0.$$

That is, the column constants have a linear regression on a set of known numbers. The following are the maximum likelihood estimates of the relevant constants.

$$b_{i\xi} = \sum_{\eta=1}^{k} E_{i\eta} E^{\eta\xi},$$

where

$$E_{i\eta} = \sum_{w,z} x_{iwz} \left\{ x_{\eta wz} - x_{\eta w \cdot} - t_z \frac{\left( \sum t_z x_{\eta \cdot z} \right)}{\sum t_z^2} \right\}.$$

$$\hat{\delta}_1 = \frac{\sum_{w,z} t_z (x_{iwz} - \sum_{\xi} b_{i\xi} x_{\xi \cdot z})}{\sum t_z^2}.$$

In the notation used for numerical calculation,

$$\hat{\delta}_i = \frac{(d_i - \sum b_{i\xi}\, d_\xi)}{r\sum t_z^2} = \frac{d_i'}{r\sum t_z^2}, \qquad \text{where} \quad d_i = \sum_z t_z X_{i \cdot z},$$

the quantity $X_{i \cdot z}$ being the column *total*. Finally

$$rc\hat{\sigma}_{ij \cdot \xi} = E_{ij \cdot \xi} = E_{ij} - \sum_\xi b_{i\xi} E^{\xi j}.$$

The distributional properties are similar to those in the two-population case. The quantities $E_{ij \cdot \xi}$ follow a Wishart distribution ith $(rc - r - 1 - k)$ d.f. and covariance matrix $\sigma_{ij \cdot \xi}$. The variates $d_i'$ follow a multivariate normal distribution with means $r\delta_i\Sigma t_z^2$ and covariance

$$\sigma_{ij \cdot \xi}(r\Sigma t_z^2 + \Sigma E^{\xi\eta}\, d_\xi\, d_\eta) = v\sigma_{ij \cdot \xi} \qquad\qquad \text{(say)}.$$

Consequently,

$$y = \Sigma E^{ij \cdot \xi}\, d_i'\, d_j'$$

is distributed as $vpF/(rc - r - p - k)$ with $p$ and $(rc - r - p - k)$ d.f. and parameter

$$\tau = \tfrac{1}{2}(r\Sigma t_z^2)^2 \Sigma \sigma^{ij \cdot \xi}\delta_i\delta_j/v.$$

Thus in the numerical example, with $r = 12$, $c = 4$, $p = 2$, $k = 1$, this procedure would have given an $F$ test of the null hypothesis $\tau = 0$, where $F$ has 2 and 33 d.f. However, the contribution from 2 degrees of freedom was deliberately omitted from the quantities $E_{ij}$, so that $F$ actually had 2 and 31 d.f.

## PART III

**18. Justification of the 'dummy variate' approach.** It remains to show that the method of calculation used in the example (sections 5 and 6) is equivalent to that derived from theory. There are two chief points to prove. First, that the $M$'s found from the equations

$$(30) \qquad\qquad \sum_j E_{ij \cdot \xi} M_j = d_i'$$

are proportional to the corresponding $L$'s found from the equations

$$(31) \qquad\qquad \sum_a S_{ij} L_j = d_i$$

where the suffix $a$ denotes summation over both $x_i$ and $x_\xi$ variates.

Now, since $S_{ij} = E_{ij} + d_i\, d_j/240$, equations (31) are the same as

$$(32) \qquad\qquad \sum_a E_{ij} L_j = d_i(1 - \sum_a L_j\, d_j/240).$$

Hence the $L$'s in (31) are proportional to the values found from the equations

$$(33) \qquad\qquad \sum_a E_{ij} L_j' = d_i.$$

But it is well known that if the $L'_\xi$ are eliminated one by one from equations (33), we obtain .

$$\sum_j E_{ij\cdot\xi} L'_j = d'_i ,$$

which is the same as (30). This proves the first point.

The second point to establish is that the $F$ test in the example is the same as that obtained from theory. In section 15, it was shown that

(34)
$$\sum_{i,j} E^{ij\cdot\xi} d'_i d'_j/v$$

is distributed as $pF/(n - p - k + 1)$. In the analysis of variance of Table 6, section 6, the quantity following the same distribution was

(35)
$$\frac{(S_a - S_\xi)}{(240 - S_a)},$$

where

$$S_a = \sum_a S^{ij} d_i d_j , \qquad S_\xi = \sum_{\xi,\eta} S^{\xi\eta} d_\xi d_\eta .$$

Since equations (31) and (32) have the same solution, we must have

$$S^{ij} = E^{ij}\left(1 - \sum_a L_j d_j/240\right) = E^{ij}(1 - S_a/240).$$

Multiplying both sides by $d_i d_j$ and summing over all $i, j$, we obtain

$$S_a = E_a(1 - S_a/240) = E_a(1 + E_a/240),$$

where $E_a$ is defined analogously to $S_a$. Similarly

$$S_\xi = E_\xi/(1 + E_\xi/240).$$

Hence

(36)
$$\frac{S_a - S_\xi}{240 - S_a} = \frac{E_a - E_\xi}{240 + E_\xi} = \frac{E_a - E_\xi}{v}.$$

Transform the variates $x_i$, $x_\xi$ into variates $x'_i$, $x_\xi$, where $x'_i = x_i - \Sigma b_{i\xi} x_\xi$. It is easy to see that this transforms

$$\sum_a E^{ij} d_i d_j \quad \text{into} \quad \sum E^{\xi\eta} d_\xi d_\eta + \sum_{i,j} E^{ij\cdot\xi} d'_i d'_j .$$

That is,

$$E_a = E_\xi + \sum_{i,j} E^{ij\cdot\xi} d'_i d'_j ,$$

since the quantity on the left is invariant under non-singular linear transformations. Hence from (36),

$$\frac{(S_a - S_\xi)}{(240 - S_a)} = \sum_{i,j} E^{ij\cdot\xi} d'_i d'_j/v.$$

From (34) and (35), this establishes the equivalence of the $F$ tests. While the proof has been given only for the type of data encountered in the example, the same method will apply to other types of data.

In conclusion, we wish to thank the referees for many helpful suggestions in connection with the presentation of this paper.

## REFERENCES

[1] R. A. FISHER, "The use of multiple measurements in taxonomic problems," *Annals of Eugenics*, Vol. 7 (1936), pp. 179–188.

[2] R. A. FISHER, "The statistical utilization of multiple measurements," *Annals of Eugenics*, Vol 8 (1938), pp. 376–386.

[3] H. HOTELLING, "The generalization of Student's ratio," *Annals of Math. Stat.*, Vol. 2 (1931), pp. 360–378.

[4] P. C. MAHALANOBIS, "On the generalized distance in statistics," *Proc. Nat. Inst. Sci. Ind.*, Vol. 12 (1936), pp. 49–55.

[5] C. I. BLISS AND H. P. MARKS, "The biological assay of insulin," *Quart. Jour. Pharm. and Pharmacol.*, Vol. 12 (1939), pp. 82–110; 182–205.

[6] P. L. HSU, "Notes on Hotelling's generalized $T$," *Annals of Math. Stat.*, Vol. 9 (1938), pp. 231–243.

[7] H. CRAMÉR, *Mathematical methods of statistics*, Princeton University Press, 1946.

[8] M. S. BARTLETT, "On the theory of statistical regression," *Roy. Soc. Proc. Edin.*, Vol. 53 (1933), pp. 271–277.

[9] S. S. WILKS, *Mathematical Statistics*, Princeton University Press, 1943, p. 70.

[10] P. C. TANG, "The power function of the analysis of variance tests," *Stat. Res. Memoirs*, Vol. 2 (1938), pp. 126–157.

## ESTIMATION OF BACTERIAL DENSITIES BY MEANS OF THE "MOST PROBABLE NUMBER"*

WILLIAM G. COCHRAN

*School of Hygiene and Public Health*
*The Johns Hopkins University*

*Presented at a joint session of the Engineering, Laboratory and Statistics Sections of the American Public Health Association, Biometrics Section of the American Statistical Association, and the Biometric Society, New York, October 27, 1949.*

### INTRODUCTION

THIS PAPER attempts to give a simple account of the concept of the "most probable number" (m.p.n.) of organisms in the dilution method. The concept is quite old, going back to McCrady (4) in 1915, and has been discussed by various writers from time to time, so that little of what I shall present is new. In addition, some advice is given on the planning of dilution series.

The dilution method is a means for estimating, without any direct count, the density of organisms in a liquid. It is used principally for obtaining bacterial densities in water and milk. The method consists in taking samples from the liquid, incubating each sample in a suitable culture medium, and observing whether any growth of the organism has taken place. The estimation of density is based on an ingenious application of the theory of probability to certain assumptions. For a biologist, it is more important to be clear about these assumptions than about the details of the mathematics, which are rather intricate.

### ASSUMPTIONS

There are two principal assumptions. In statistical language, the first is that the organisms are distributed *randomly* throughout the liquid. This means that an organism is equally likely to be found in any part of the liquid, and that there is no tendency for pairs or groups of organisms either to cluster together or to repel one another. In practice this implies that the liquid is thoroughly mixed, and if the volume of liquid is not too great some shaking device is usually employed for this purpose.

---

*Paper 254 from the Department of Biostatistics.

The second assumption is that each sample from the liquid, when incubated in the culture medium, is certain to exhibit growth whenever the sample contains one or more organisms. If the culture medium is poor, or if there are factors which inhibit growth, or if the presence of more than one organism is necessary to initiate growth, the m.p.n. gives an underestimate of the true density.

MATHEMATICAL ANALYSIS

In the mathematical analysis we relate the probability that there will be no growth in a sample to the density of organisms in the original liquid. Suppose that the liquid contains $V$ ml., the sample contains $v$ ml., and that there are actually $b$ organisms in the liquid. By the second assumption, there will be no growth if and only if the sample contains no organisms. We will calculate the probability that none of these $b$ organisms is in the sample.

Consider a single organism. By the first assumption, the probability that it lies in the sample is simply the ratio of the volume of the sample to that of the liquid, i.e. $v/V$. The probability that it is not in the sample is therefore $(1 - v/V)$. Since there is assumed to be no kind of attracttion or repulsion between organisms, these two probabilities hold for *any* organism, irrespective of the positions of the other organisms. (Strictly, this requires the additional assumption that the space occupied by an organism is negligible relative to $v$.) Consequently, by the multiplication theorem in probability, the probability that none of the $b$ organisms is in the sample is

$$p = (1 - v/V)^b.$$

When $v/V$ is small, this is closely approximated by

$$p = e^{-vb/V}$$

where $e$, about 2.7, is the base of natural logarithms. Finally, since $b/V$ is the density $\delta$ of organisms per ml., we have

$$p = e^{-v\delta},$$

where $p$ is the probability that the sample is sterile.

THE CASE OF A SINGLE DILUTION

If $n$ samples, each of volume $v$, are taken, and if $s$ of these are found to be sterile, the proportion $s/n$ of sterile samples is an estimate of $p$. Hence we obtain an estimate $d$ of the density $\delta$ by the equation

$$\frac{s}{n} = e^{-vd}.$$

This gives

$$d = -\frac{1}{v} \ln\left(\frac{s}{n}\right) = -\frac{2.303}{v} \log\left(\frac{s}{n}\right) \tag{1}$$

where $ln$ and $\log$ stand for logarithms to base $e$ and to base 10 respectively.

The estimate $d$ is the "most probable number" of organisms per ml. The derivation given here does not reveal why this name has been ascribed to the estimate. In fact, the concept of m.p.n. is scarcely needed for this simple case. We will, however, reexamine the analysis so as to introduce the concept, which becomes useful in the more complex situation where several dilutions are used.

If $p$ is the probability that a sample is sterile, the probability that $s$ out of $n$ samples are sterile is given by the binomial distribution as

$$\frac{n!}{s!(n-s)!} p^s (1-p)^{n-s} \tag{2}$$

Since $p = e^{-v\delta}$, this expression may be written

$$= \frac{n!}{s!(n-s)!} e^{-sv\delta}(1 - e^{-v\delta})^{n-s} \tag{3}$$

If we have obtained $s$ sterile samples out of $n$, this formula enables us to plot the probability of this event against the true density $\delta$. Such curves always have a single maximum.

A curve of this type suggests a method for estimating $\delta$ which is plausible on intuitive grounds. For if we are considering two possible values of $\delta$, it seems reasonable to prefer the one which gives a higher probability to the result that was actually observed. This argument, carried to its conclusion, leads to a choice of the value of $\delta$ for which the probability of obtaining the observed result is greatest. It is this value of $\delta$ that has been called the "most probable number" of organisms. It can be shown mathematically that this is the value of $\delta$ for which $p = s/n$. Consequently the m.p.n. is the same as the estimate previously given.

In practice, more than one dilution is usually needed. The reason is that the precision of the m.p.n. is very poor when the volume $v$ in the sample is such that the samples are likely to be all fertile or all sterile. When all are fertile, the maximum on the probability curve (3) occurs when $\delta$ is infinite, so that the estimated density is infinite. When all are sterile the estimated density is zero, as may also be verified from equation (1). Thus a single dilution is successful only if $v$ happens to be chosen so that some samples are sterile and some are fertile. Such a choice of $v$ can be made only if the density $\delta$ is known fairly closely in advance.

If we possess this knowledge, it is best to select $v$ so that the expected number of organisms per sample lies somewhere between 1 and 2. For this choice the expected percentage of sterile samples will lie between 15% and 35%. In default of this knowledge, the practice is to use several dilutions (i.e. several different values of $v$) in the hope that at least one of them will give some sterile and some fertile samples.

THREE DILUTIONS

The case of three dilutions serves to illustrate the general problem. Let the suffix $i$ indicate the dilution. For the $i$th dilution the volume of the sample is $v_i$, and $s_i$ out of $n_i$ samples are found to be sterile. How do we estimate $\delta$ from these results?

From equation (1) we can obtain a separate estimate for each dilution: i.e.

$$d_i = -\frac{2.303}{v_i} \log\left(\frac{s_i}{n_i}\right).$$

However, the best way to combine the three estimates $d_i$ into a single value is not obvious. Since, as we have seen, some dilutions give very poor estimates, it is not satisfactory to take the arithmetic mean.

One solution is provided by the m.p.n. concept, which extends easily to this situation. Following the approach used in the previous section, we first write down the probability of obtaining the observed results for any hypothetical value of the true density $\delta$. The observed results are that $s_1$ samples out of $n_1$ are sterile at the first dilution, $s_2$ out of $n_2$ at the second, and $s_3$ out of $n_3$ at the third. The probability that these three events should all happen is the product of three terms, each like expression (3) in the previous section. As before, the graph of this probability against $\delta$ shows a single maximum. The value of $\delta$ at this maximum is taken as the m.p.n.

The value of the m.p.n. cannot be written down explicitly. The equation which it satisfies is as follows.

$$s_1v_1 + s_2v_2 + s_3v_3 = \frac{(n_1 - s_1)v_1e^{-v_1d}}{1 - e^{-v_1d}} + \frac{(n_2 - s_2)v_2e^{-v_2d}}{1 - e^{-v_2d}} + \frac{(n_3 - s_3)v_3e^{-v_3d}}{1 - e^{-v_3d}}$$

Methods for solving this equation by trial and error have been given by several writers: e.g. Halvorson and Ziegler (3), Barkworth and Irwin (1) and Finney (2). In laboratories where the numbers of samples $n_i$ and the dilution ratios are standardized, it is convenient to have a table which gives the m.p.n. for all sets of results that are likely to occur. A table is provided in "Standard methods for the examination of water and sewage" (5), for dilution series in which 5 samples are taken at each dilution and there are three 10-fold dilutions. A more extensive table, for dilution ratios of 2, 4, and 10 and any number of levels (except two levels with a 10-fold dilution) is given by Fisher and Yates (6). This is not a table of the m.p.n., but of a different estimate which seems to be just about as precise for series of the size usually conducted in practice. This estimate is derived from the total numbers $X$ and $Y$ of fertile and sterile samples. The quantities $x = X/n$, $y = Y/n$ are entered in the table, from which an estimate of $\log d$ is obtained.

### CRITIQUE OF THE M.P.N.

We have seen that the m.p.n. is an estimate of the density of organisms. Considered more generally, it is a *procedure for obtaining estimates*, since the same argument could be applied to other statistical problems. The only justification which I have mentioned for the procedure is that it seems intuitively reasonable. From a reading of the literature I am not certain as to the reasons which led early investigators to select this estimate, though either the intuitive approach or an appeal to a theory of inverse probability may have been responsible.

During the past 25 years the problem of making estimates from data has received much attention from statisticians. Today, most statisticians would, I believe, reject an appeal to intuition or to the theory of inverse probability as a reliable procedure for constructing estimates, since both have been found on occasion to be untrustworthy. They might also object to the name "most probable number," on the grounds that the adjective "probable" in that phrase has a different meaning from the one given to it in the theory of probability. The estimate is "most probable" only in the roundabout sense that it gives the highest probability to the observed results. But they would not reject the m.p.n. procedure itself, which has come to be regarded as a remarkably reliable tool of very wide utility. At the risk of a slight digression it is interesting to indicate the reasons for the reputation which the method has acquired.

The modern approach is to appraise any method of estimation by results. For the m.p.n. this is done, ideally, by conducting a large number of dilution series with given $v$'s and $n$'s, in circumstances where the true density is known. For each series the density is estimated by the m.p.n., so that we accumulate a large number of observations on the amounts by which the m.p.n. is in error. These observations can be summarized conveniently by plotting the frequency distribution of the m.p.n. about the true density. If this frequency distribution groups very closely about the true density, we know that the estimates are usually good. Such a set of experiments would be difficult and expensive to conduct, but if we assume that the mathematical analysis which has been applied to the dilution method is valid, we can work out the frequency distribution by purely mathematical methods.

As the numbers of samples $n_i$ become large, the frequency distribution of such an estimate (m.p.n. or other) usually tends to assume a certain limiting form—the normal distribution. An important general result has been established about these limiting distributions (7), to the effect that the limiting distribution of the m.p.n. has the smallest standard deviation that can be achieved by any method of estimation. Roughly speaking, this means that the m.p.n. gives on the average at least as precise estimates as any other method used on the same data. There is no point in seeking further for a more precise estimate. The theorem cannot be proved in general when the numbers of samples are small, but experience suggests that the m.p.n. technique is among the best methods of estimation in this case also. Consequently the m.p.n. method is now generally used in a great variety of problems of statistical estimation, though it more frequently goes by the name of the "method of maximum likelihood."

### THE PLANNING OF DILUTION SERIES

In preparation for an estimation by the dilution method, three decisions must be made: (i) what range is to be covered: i.e. what are to be the highest and lowest sample volumes; (ii) what dilution factor is to be used; and (iii) how many samples should be taken for each dilution.

Specific decisions must depend on a knowledge of the limits within which the true density is likely to lie and on the precision desired in the estimate. The way in which precision is to be measured needs some comment. Suppose that the true density is thought to lie somewhere between say 2 and 400 organisms per ml. No matter where the true density should happen to be within this range, we want to plan the series so that the estimate will have a specified "precision." This might be taken to mean that the standard error of the estimated density should be say 30 organisms. But this does not seem a reasonable definition of "equal precision," because although an estimate of $360 \pm 30$ organisms seems satisfactorily precise, an estimate of $5 \pm 30$ organisms seems very imprecise. Instead, we take "equal precision" to imply that the standard error bears a constant ratio to the true density, in other words that the coefficient of variation of the estimated density is constant. A further potent reason for adopting this concept is that in a well-designed series the m.p.n. estimates do have approximately the property that the coefficient of variation is independent of the true density. Thus in a sense we are making a virtue of necessity.

The following remarks are intended as a rough guide in the planning of dilution series. They were derived from investigations of the precision of the m.p.n.

### HIGHEST AND LOWEST SAMPLE VOLUMES

These are determined by the range of densities with which we expect to have to cope. With a single dilution it was mentioned that for the best results the expected number of organisms in the sample volume $v$ should lie between 1 and 2. It follows that in a series of dilutions the expected number of organisms in the *highest* sample volume $v_H$ should be at least 1, otherwise there is a risk that all samples will be sterile. Similarly the expected number of organisms in the *lowest* sample volume $v_L$ should not exceed 2, to avoid the risk that all samples will be fertile. This line of reasoning would lead to the rule that a dilution series is capable of estimating any density that lies between $1/v_H$ and $2/v_L$.

This rule is satisfactory if a substantial number of samples, say 20 or more, are being taken at each dilution. With very small numbers of samples per dilution, which are typical in certain lines of work, the rule is not quite stringent enough, in that it allows too much risk that all samples may be fertile. Suppose that we have three 10-fold dilutions, with sample volumes 0.01, 0.1 and 1 ml. This series should be able to estimate any true density between 1 and 200 organisms per ml. If, however, the density happens to be 200 per ml., so that the expected number of organisms per sample in the lowest sample volume is 2, then the probability of a sterile sample at this dilution is $e^{-2}$, or 0.135. The probability of a fertile sample is 0.865. If only four samples are used per dilution, the probability that all four are fertile is $(0.865)^4$, or 0.56. At the two higher concentrations, all samples are practically certain to be fertile. Thus the worker runs about a 50-50 chance that all his samples will be fertile, which usually necessitates repetition of the series. On the other hand, with 20 samples per dilution, the probability that all are fertile is $(0.865)^{20}$, or only about 0.05.

Thus in small experiments it is safer to reduce the upper density value from $2/v_L$ to $1/v_L$. In practice, we use this rule by first guessing two limits $\delta_L$ and $\delta_H$ between which we are fairly certain that the true density lies. The sample volumes are then chosen to satisfy the rules

$$v_H \geq \frac{1}{\delta_L} ; \quad v_L \leq \frac{1}{\delta_H} .$$

For example, if we are confident that the density lies between 10 and 750 per ml., the highest sample volume should be at least 1/10, or 0.1 ml. The lowest sample volume should not be more than 1/750 ml. The three 10-fold dilutions 1/10, 1/100 and 1/1000 ml., or the four 5-fold dilutions 1/10, 1/50, 1/250 and 1/1250, would amply cover this range of densities.

## THE DILUTION RATIO

As regards the selection of a dilution ratio, there are two relevant results. If the total number of samples in the whole series is kept fixed, the average precision is practically the same for any dilution ratio between 2 and 10. The advantage of a low dilution ratio, which requires more work, is that the precision is more nearly constant throughout the range of densities between $1/v_H$ and $1/v_L$. These points may be illustrated by a comparison between the dilution ratios 2 and 10, in series designed to cover the same range of densities and to use the same total number of samples, 72. The details for the two series are as follows.

| Dilution ratio | No. of samples per dilution | Volumes of samples (ml.) |
|:---:|:---:|:---|
| 2 | 9 | .01, .02, .04, .08, .16, .32, .64, 1.28 |
| 10 | 24 | .01, .10, 1.00 |

The two series should cover a range of densities from $1/v_H$ to $1/v_L$, or from about 1 to 100 organisms per ml. The dilution ratio 2 requires eight dilutions, with 9 samples per dilution, whereas the dilution ratio 10 requires only 3 dilutions and allows 24 samples per dilution.

In Figure 1 the standard error of the m.p.n., expressed as a percent of the true density, is plotted against the true density (on a log scale). With both dilution ratios the standard error per cent is fairly constant for any true density between 1 and 100 organisms per ml. Outside these limits the standard error begins to rise steeply, except that with the 10-fold series, which has 24 samples per dilution, the rise is postponed until $\delta = 200$, for reasons given in the previous section. Inside the limits the standard error shows a periodic fluctuation which is noticeable with the 10-fold dilution but negligible for the 2-fold. With a 5-fold dilution (not shown), this periodic effect would be just perceptible. It is present with the 10-fold series because practically all the information is contributed by a single dilution. When the true density is about 1.5 or 15 or 150, so that one of the dilutions has about 1.5 organisms per sample, there is a trough, with peaks in the intervening densities where no sample has a density close to this value. With the 2-fold series, several dilutions contribute information and the periodic effect is smoothed out. On the whole, the 2-fold dilution gives a slightly lower standard error over the range from 1 to 100 organisms per ml., the difference being about 7 per cent. For these reasons a low dilution ratio is preferable if the extra work involved can be accomplished easily.

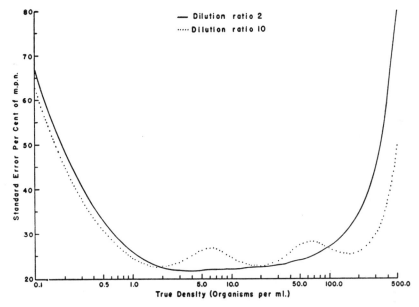

FIGURE I. COMPARISON OF DILUTION RATIOS 2 AND 10

The curves in Figure 1 were calculated by assuming that the formula which holds for the standard error in the limiting distribution, appropriate for very large samples, could be applied to this example in which the total number of samples is 72. Some unpublished work by Dr. I. J. Bross on the distribution of the m.p.n. in small samples indicates that the standard errors are higher than those obtained in this way from the limiting distribution. Further, the periodicity with the 10-fold dilution does not follow the course predicted for it. However, the two principal conclusions from Figure 1 still appear to hold in small samples, namely that the standard error is more stable with a low dilution ratio, and also tends to be slightly lower.*

*This work was carried out under contract with the Office of Naval Research.

STANDARD ERROR OF THE M.P.N.

In many types of investigation there may be only a few samples for each dilution. In this event the distribution of the estimated density $d$ is very skew, and to attach a standard error to $d$ is misleading. The distribution of $\log d$ is more nearly symmetrical, and it is recommended that tests of significance and the construction of confidence limits be performed from $\log d$ rather than from $d$. If there are $n$ samples *per dilution* (assumed the same in all dilutions), the standard error of $\log_{10} d$ may be taken as

$$0.55 \sqrt{\frac{\log_{10} a}{n}}$$

where $a$ is the dilution ratio. This formula can be used for any density which lies between $1/v_H$ and $1/v_L$, and for any dilution ratio of 5 or less. For a dilution ratio of 10, a more conservative factor of 0.58 is preferable to 0.55, to allow for the contingency that the estimation may have been made at a point where the standard error has one of its peaks. Thus for dilution ratio 10 the formula becomes simply $0.58/\sqrt{n}$. Note that the formula does not explicitly involve the number of dilutions used.

To test the significance of the difference between two estimated densities, made from independent series, we compute

$$\frac{\log d_1 - \log d_2}{0.55 \sqrt{\dfrac{\log a_1}{n_1} + \dfrac{\log a_2}{n_2}}}$$

and refer to the normal probability tables.

The construction of confidence limits may be illustrated by assuming that we have three 10-fold dilutions, with 5 samples per dilution. The standard error of $\log d$ is $0.58/\sqrt{5}$, or 0.259, so that the 95 per cent confidence limits for $\log d$ are $(\log d \pm 0.518)$. It follows that to get the upper confidence limit for $d$, we must multiply $d$ by antilog (0.518) or 3.3, and to get the lower confidence limit we must divide $d$ by 3.3.

For the common dilution ratios, 2, 4, 5, and 10, Table I shows the standard error of $\log d$ for any number of samples per dilution between 1 and 10. The table also gives the factor by which the estimated density must be multiplied and divided in order to obtain upper and lower 95 per cent confidence limits respectively. In the example presented by Fisher and Yates (6), the number of rope spore organisms per gram of potato flour was estimated to be 760. The dilution ratio was 2 and there were 5 tubes per dilution. From Table I, the factor for $n = 5$, $a = 2$ is 1.86. Hence the upper confidence limit is $760 \times 1.86$ or 1414, while the

TABLE I
STANDARD ERROR OF LOG $d$ AND FACTOR FOR CONFIDENCE LIMITS

| No. of samples per dil. | S.E.$_{(\log_{10} d)}$ Dilution ratio (a) | | | | Factor for 95% confidence limits Dilution ratio (a) | | | |
|---|---|---|---|---|---|---|---|---|
| $n$ | 2 | 4 | 5 | 10 | 2 | 4 | 5 | 10 |
| 1 | .301 | .427 | .460 | .580 | 4.00 | 7.14 | 8.32 | 14.45 |
| 2 | .213 | .302 | .325 | .410 | 2.67 | 4.00 | 4.47 | 6.61 |
| 3 | .174 | .246 | .265 | .335 | 2.23 | 3.10 | 3.39 | 4.68 |
| 4 | .150 | .214 | .230 | .290 | 2.00 | 2.68 | 2.88 | 3.80 |
| 5 | .135 | .191 | .206 | .259 | 1.86 | 2.41 | 2.58 | 3.30 |
| 6 | .123 | .174 | .188 | .237 | 1.76 | 2.23 | 2.38 | 2.98 |
| 7 | .114 | .161 | .174 | .219 | 1.69 | 2.10 | 2.23 | 2.74 |
| 8 | .107 | .151 | .163 | .205 | 1.64 | 2.00 | 2.12 | 2.57 |
| 9 | .100 | .142 | .153 | .193 | 1.58 | 1.92 | 2.02 | 2.43 |
| 10 | .095 | .135 | .145 | .183 | 1.55 | 1.86 | 1.95 | 2.32 |

lower limit is 760/1.86 or 409. This factor clearly fulfills the same general purpose as would a standard error, if it had been appropriate to attach one to $d$.

The table makes it evident that the dilution method is of low precision, as is to be expected from a method that does not use direct counts. Large numbers of samples must be taken at each dilution if a really precise result is wanted. Further, the table is likely to overestimate the accuracy of the method, since it is derived on the assumption that the mathematical analysis corresponds exactly to the practical situation. With a large volume of liquid that cannot be mixed, the distribution of organisms may be far from homogeneous. The method will determine the density in that part of the liquid from which the initial sample was taken. This might be very different from the average density over the whole liquid, and this source of error could be more important than the error in the dilution method itself.

SUMMARY OF STEPS IN PLANNING

The decisions to be made involve a choice of the dilution ratio, $a$, the number of dilutions and the actual sample volume in each dilution, and finally the number of samples $n$ to be used at each dilution. The steps may be set out as follows.

1. Decide on the limits $\delta_L$ and $\delta_H$ within which the true density appears certain to lie.

2. Calculate the lowest and highest sample volumes by means of the relations

$$v_H = \frac{1}{\delta_L}, \qquad v_L = \frac{1}{\delta_H}.$$

3. Select a dilution ratio. A low ratio is preferable whenever feasible.

4. The number of dilutions and the actual volumes for each dilution may now be chosen so as to satisfy the requirements that the highest sample volume must not be less than $v_H$ and the lowest must not exceed $v_L$.

5. The precision to be expected for any specified number $n$ of samples per dilution may be appraised from Table I, if the number of samples per dilution is less than 10, or from the formula for $S.E._{(\log d)}$. Choose the number of samples in the light of the precision that is desirable and the amount of work that it is practicable to do.

### REFERENCES

(1) Barkworth, H. and Irwin, J. O. (1938). Distribution of coliform organisms in milk and the accuracy of the presumptive coliform test. *J. Hyg.*, Cambridge 38, 446–457.

(2) Finney, D. J. (1947). The principles of biological assay. *J. Roy. Stat. Soc., Ser. B.,* 9, 46–91.

(3) Halvorson, H. O., and Ziegler, N. R. (1933). Application of statistics to problems in bacteriology. *J. Bact.* 25, 101–121.

(4) McCrady, M. H. (1915). The numerical interpretation of fermentation-tube results. *J. Infec. Dis.,* 17, 183–212.

(5) American Public Health Association (1941). *Standard Methods for the Examination of Water and Sewage.* 8th ed.

(6) Fisher, R. A. and Yates, F. (1948). *Statistical Tables for Biological, Agricultural and Medical Research.* Edinburgh, Oliver and Boyd, 3rd ed. Table VIII2.

(7) Fisher, R. A. (1921). On the mathematical foundations of theoretical statistics. *Phil. Trans. Roy. Soc.* London, A, 222, 309–368.

# 43

## THE COMPARISON OF PERCENTAGES IN MATCHED SAMPLES

By W. G. COCHRAN

*Department of Biostatistics, School of Hygiene and Public Health, Johns Hopkins University*

### 1. INTRODUCTION

The $\chi^2$ test has long been used to test the significance of differences between ratios or percentages in two or more independent samples. It sometimes happens that each member of a sample is matched with a corresponding member in every other sample, in the hope of securing a more accurate comparison among the percentages. The matching may be based either on the characteristics of the members, or on the fact that the partners in a group are subjected to some test that is the same for all members of the group but varies from one group to another.

Since the matching may introduce correlation between the results in different samples, it invalidates the ordinary $\chi^2$ test, which gives too few significant results if the matching is effective. For the case where there are two samples, an appropriate test is easily constructed. An example has been given by McNemar (1949), who presents this test. In his data, 205 soldiers were asked whether they thought that the war against Japan would last more or less than a year. They were subsequently asked the same question after a lecture on the difficulties of the war against Japan. Matching occurs because each sample contains exactly the same soldiers.

The replies may be classified in a $2 \times 2$ frequency table as shown in Table 1.

Table 1. *Comparison of ratios in two samples*

|  | After lecture | | Total |
|---|---|---|---|
|  | Less | More | |
| Before lecture: Less | 36 (a) | 34 (b) | 70 |
| More | 0 (c) | 135 (d) | 135 |
|  | 36 | 169 | 205 |

Before the lecture, 70 men out of the 205 thought that the war would last less than a year, whereas after the lecture this number has dropped to 36. The comparison which we wish to make is that between the two frequencies 70/205 and 36/205. There are several ways in which the test may be derived. Perhaps the easiest is to note that both numerators, 70 and 36, contain the 36 (a) men who persisted in thinking that the war would last less than a year. Hence, equality of the numerators would imply that the same number of men changed from 'Less' to 'More' as changed from 'More' to 'Less'. In other words, if the lecture is without effect we would expect half the persons who changed their minds to change in one direction and half in the other. Thus the test can be made by testing whether the numbers ($b$) and ($c$) are binomial successes and failures out of $n = (b+c)$ trials, with probability $\frac{1}{2}$. For this

$$\chi^2 = \frac{(b-\frac{1}{2}n)^2}{\frac{1}{2}n} + \frac{(c-\frac{1}{2}n)^2}{\frac{1}{2}n} = \frac{(b-c)^2}{b+c} = \frac{(34-0)^2}{34+0} = 34,$$

with 1 degree of freedom. A correction for continuity can be applied by subtracting 1 from the absolute value of the numerator before squaring.

The two-sample case has also been discussed in a study by Denton & Beecher (1949), where the object was to find out whether subjects reacted more frequently to an injection of a new drug than they did to one of isotonic sodium chloride, which was used as a control. They give a $\chi^2$ test, attributed to Mosteller, which differs slightly from that given above.

The object of this paper is to extend the test to the situation where there are more than two samples. An example is provided in some studies of variability among interviewers in sample surveys. Each interviewer called at a different group of houses, but any house assigned to an interviewer was matched with one of the houses assigned to each other interviewer according to the characteristics of the housewife. A test of whether the percentage of 'yes' answers to some question differed from interviewer to interviewer is a test of the type that we are considering.

In a second example, the effectiveness of a number of different media for the growth of diphtheria bacilli was investigated by the Communicable Disease Centre, U.S. Public Health Service. In one series, specimens were taken from the throats of sixty-nine suspected cases. Each specimen was grown on each of four media A, B, C, D. The probability that growth takes place will depend on the number of diphtheria bacilli present, and in a number of cases there might well be no bacilli present.

Table 2. *Method of presentation suited to more than two columns*

Diphtheria media

| | A | B | C | D | No. of cases |
|---|---|---|---|---|---|
| | 1 | 1 | 1 | 1 | 4 |
| | 1 | 1 | 0 | 1 | 2 |
| | 0 | 1 | 1 | 1 | 3 |
| | 0 | 1 | 0 | 1 | 1 |
| | 0 | 0 | 0 | 0 | 59 |
| Totals ($T_j$) | 6 | 10 | 7 | 10 | |

Soldiers' replies

| | Before | After | No. of cases |
|---|---|---|---|
| | 1 | 1 | 135 ($d$) |
| | 0 | 1 | 34 ($b$) |
| | 0 | 0 | 36 ($a$) |
| Totals | 135 | 169 | |

Results are shown in Table 2.* Where there are four media, the $2 \times 2$ table does not seem well adapted to a succinct presentation. Instead, each medium is allotted to one column of the table. A 1 denotes that growth occurred with that medium, a 0 that no growth occurred. Thus in Table 2 there were four specimens in which all four media exhibited growth, two specimens in which media A, B and D, but not C, showed growth, and so on. To illustrate

* I wish to thank Dr Martin Frobisher, Chief, Bacteriology Laboratories, Communicable Disease Centre, U.S. Public Health Service, Atlanta, for permission to use these data for illustration.

the relation to the method of presentation in a $2 \times 2$ table, McNemar's results are also shown in this form, where a 1 denotes the answer 'more than a year'.

The column totals are the total numbers of 1's. The problem is to test whether these totals differ significantly among media.

## 2. MATHEMATICAL FRAMEWORK

For a discussion of the theory of the test we shall adopt a less concise method of presentation than that given in Table 2. Each matched group will be placed in a different row of the table. Thus the table for the diphtheria data would contain 69 rows and 4 columns. The probability of a 1 is presumed to vary from row to row, usually in a manner that is known only vaguely. Nevertheless, an exact test can be developed by the familiar method in which the population is generated by randomization. The observed total number $u_i$ of successes (1's) in the $i$th row is regarded as fixed. If the null hypothesis is true, every one of the $c$ columns is considered equally likely to obtain one of these successes. The population of possible results in the $i$th row consists of the $\binom{c}{u_i}$ ways in which the $u_i$ successes can be distributed among the $c$ columns.

This specification has one consequence that might be questioned. If a row contains no successes, or $c$ successes, the population generated in that row consists only of the single case that actually occurred. As will be seen, this implies that such rows play no part in the test of significance. This is evident in the two-sample test, which makes no use of the number of cases $a$ and $d$ in the cells of Table 1 where there was no change of opinion. On the other hand, for given values of $b$ and $c$, one might feel intuitively that significance ought to be more definitely established if there are no cases in which the samples give the same result (i.e. $a$ and $d$ are zero) than if there are a large number of such cases. Whether this feeling is sound is perhaps debatable, and I do not see how weight can be given to it without losing the advantage of an exact test.

The test criterion that will be used is $\Sigma(T_j - \overline{T})^2$, where $T_j$ is the total number of successes in the $j$th column. This is the same criterion as in the ordinary $\chi^2$ test for the situation where the columns are independent. It may not be the best criterion. For the usefulness of the data from a row for the purpose of detecting differences among columns may depend on the probability of success in the row. That is, the situation may be similar to that which occurs in dosage-mortality experiments, in which, for maximum sensitivity per observation, comparisons of two drugs must be made close to the median lethal doses. This suggests that in extensive data it might be advisable to group the rows according to the value of $u_i$ and to perform some kind of weighting on the $T_j$ values for different values of $u_i$. I have occasionally used this approach, but it may be difficult to decide what form the weighting should take, particularly in a new type of experimentation. A test based on the unweighted totals will often serve our purpose.

## 3. THE LIMITING DISTRIBUTION

We consider first the limiting distribution of the test criterion when the number of rows $r$ is very large. Let the variate $x_{ij}$ take the value 1 if there is a success in the cell in the $i$th row and $j$th column, and 0 if there is a failure. By the properties of the randomization in that row, these two events occur with probabilities $u_i/c$ and $1 - u_i/c$, respectively. Hence

$$E(x_{ij}) = \frac{u_i}{c}, \quad \sigma^2(x_{ij}) = \frac{u_i}{c}\left(1 - \frac{u_i}{c}\right).$$

By symmetry, the covariance is the same for any two cells in the same row. Since the row total of the $x_{ij}$ is fixed at $u_i$ and thus has zero variance, the covariance of $x_{ij}$ and $x_{ik}$ is found to be

$$\operatorname{cov}(x_{ij}x_{ik}) = \frac{-\dfrac{u_i}{c}\left(1-\dfrac{u_i}{c}\right)}{(c-1)} \quad (j \neq k).$$

These results enable us to arrive, by non-rigorous methods, at the form of the limiting distribution. Since the randomization is independent in different rows, the means, variances and covariances of the column totals $T_j$ will be corresponding expressions above, summed over the rows. If the number of rows is large, the joint distribution of these totals may be expected to tend to the multivariate normal. Finally, if a set of $c$ variates $T_j$ follow a multivariate normal distribution with common variance $\sigma^2$ and common covariance $\rho\sigma^2$, it is well known that $\Sigma(T_j-\bar{T})^2$ is distributed as $\chi^2\sigma^2(1-\rho)$, with $(c-1)$ degrees of freedom (Walsh, 1947). In this case

$$\sigma^2 = \sum_i \frac{u_i}{c}\left(1-\frac{u_i}{c}\right), \quad \rho\sigma^2 = \frac{-\sigma^2}{(c-1)},$$

so that

$$\sigma^2(1-\rho) = \frac{1}{(c-1)}\sum_i u_i\left(1-\frac{u_i}{c}\right).$$

Hence, when the number of rows is large,

$$Q = \frac{(c-1)\sum_j (T_j-\bar{T})^2}{\sum_i u_i\left(1-\dfrac{u_i}{c}\right)} = \frac{c(c-1)\sum (T_j-\bar{T})^2}{c(\sum u_i)-(\sum u_i^2)} \tag{1}$$

is distributed as $\chi^2$ with $(c-1)$ D.F.

A rigorous proof of this result may be obtained by the method developed by Hsu (1949) and will not be given here. The only restriction needed is a rather obvious one, to guard against the possibility that as the number of rows tends to infinity, the value of $u_i$ might be $c$ or 0 in all but a finite number of rows. If this happens, the size of the population is still finite in the limit, because permutations within rows having $u_i = c$ or 0 do not generate any new cases. This situation is avoided by stipulating that for at least one intermediate value of $u_i$, the number of rows having that value must tend to infinity.

When there are only two samples ($c = 2$) the test reduces to that given in §1. If a 1 denotes a reply of 'more than a year' it will be seen from Table 1 that

$$T_1 = c+d, \quad T_2 = b+d,$$

$$\Sigma u_i = b+c+2d, \quad \Sigma u_i^2 = b+c+4d.$$

From (1) $Q$ becomes

$$Q = \frac{(2)(1)\tfrac{1}{2}(T_1-T_2)^2}{2\Sigma(u_i)-(\Sigma u_i^2)} = \frac{(b-c)^2}{b+c}.$$

## 4. Comparison with the ordinary $\chi^2$ test

In the ordinary $\chi^2$ test, valid when the samples are independent, we have

$$\chi^2 = \frac{\sum_j (T_j-\bar{T})^2}{r\dfrac{\bar{u}}{c}\left(1-\dfrac{\bar{u}}{c}\right)}, \tag{2}$$

where $\bar{u} = \Sigma u_i/r$.

Under what conditions does this test coincide with the new $(Q)$ test? It might be anticipated that this should happen when the probability of success does not change from row to row. The results are in line with this expectation.

Consider the application of both tests to a series of tables, all of which have the same set of row totals. From (1) and (2) the new test gives a greater, an equal or a smaller number of significant results than the ordinary test, according as

$$\Sigma u_i \left(1 - \frac{u_i}{c}\right) \lessgtr \frac{(c-1)\, r\bar{u}}{c} \left(1 - \frac{\bar{u}}{c}\right). \tag{3}$$

Since $\Sigma u_i = r\bar{u}$, the left-hand side may be expressed as

$$r\bar{u}\left(1 - \frac{\bar{u}}{c}\right) - \frac{\Sigma(u_i - \bar{u})^2}{c}.$$

It follows that relations (3) are equivalent to

$$r\bar{u}\left(1 - \frac{\bar{u}}{c}\right) \lessgtr \Sigma(u_i - \bar{u})^2. \tag{4}$$

If we wish to test the null hypothesis that the probability of success is the same in all rows, this could be done by an ordinary $\chi^2$ test on the row totals $u_i$. Since rows are independent, the value of $\chi^2$ would be

$$\chi_r^2 = \frac{\sum\limits_i (u_i - \bar{u})^2}{\bar{u}\left(1 - \dfrac{\bar{u}}{c}\right)},$$

with $(r-1)$ degrees of freedom. Thus relations (4) can be written

$$r \lessgtr \chi_r^2. \tag{5}$$

The expected value of $\chi_r^2$ is $(r-1)$, which is indistinguishable from $r$ if the number of rows is large. Thus the equality in relations (5) is satisfied when the value of $\chi_r^2$ in a test on the row totals is just about equal to its expectation. The analysis also shows that the $Q$ test gives more significant results when $\chi_r^2$ exceeds its expectation, and fewer significant results when $\chi_r^2$ is below expectation.

### 5. APPLICATION TO THE EXAMPLE

In the example (Table 2) we have $c = 4$,

$$\sum_j (T_j - \bar{T})^2 = 6^2 + 10^2 + 7^2 + 10^2 - (33)^2/4 = 12{\cdot}75.$$

To find the denominator of $Q$, a separate frequency distribution of the values of the row totals $u_i$ may be made.

| Value of $u_i$ | Frequency |
|:---:|:---:|
| 4 | 4 |
| 3 | 5 |
| 2 | 1 |
| 0 | 59 |

$$\sum_i u_i = \sum_j T_j = 33, \quad \sum u_i^2 = 113.$$

Hence from (1)

$$Q = \frac{(4)\,(3)\,(12{\cdot}75)}{(4)\,(33) - (113)} = 8{\cdot}05,$$

with 3 degrees of freedom, corresponding to a probability of 0·045.

It is easy to show algebraically, as may be verified in this example, that the value of $Q$ is not altered if we omit all rows in which $u_i = c$ or 0. In this respect, $Q$ behaves as we should expect a test based on the randomization process to behave.

In the example, 63 of the 69 rows have 4 or 0 successes, so that only 6 rows really contribute to the frequency distribution of $Q$. It may be doubted whether a limiting distribution which assumes a large number of such rows can be applied here. This question is discussed in §6.

### 6. The distribution of $Q$ in small samples

If there are only two columns, an exact small-sample test presents no difficulty. The $Q$ test is essentially equivalent to the *sign* test (Cochran, 1937; Dixon & Mood, 1946), for which tables are available in the references cited.* This can be seen by the argument used in §1. Apart from its divisor, $Q$ is $(T_1 - T_2)^2$, i.e. the square of the difference between the number of successes in the two columns. We may ignore all rows that contain either 2 or 0 successes, since these do not affect the value of $Q$. Consequently $(T_1 - T_2)$ is the difference between the number $r_1$ of rows in which the results are $(1 \cdot 0)$ and the number $r_2$ in which they are $(0 \cdot 1)$. If $n = r_1 + r_2$, this difference equals $(2r_1 - n)$.

For any row that has one success, the probabilities of a $(1 \cdot 0)$ and of a $(0 \cdot 1)$ on the null hypothesis are both $\frac{1}{2}$. This shows that $r_1$ is distributed in the binomial $(\frac{1}{2} + \frac{1}{2})^n$, which is the quantity that is tabulated in the *sign* test.

For an exact test when $c = 2$, the procedure is therefore as follows: (i) ignore all rows with 2 or 0 successes; (ii) count the number of rows with a single success in the first column, and refer to the tables of the *sign* test, where $n$ is the total number of rows that have one success.

In the small tables that have more than two columns, the exact distribution of $Q$ can be tabulated by enumerating all configurations generated by the randomization. Since the number of possible cases is large, a comprehensive listing of exact significance levels would be laborious to construct. As a check on the accuracy of the limiting distribution in small samples, the exact distribution of $Q$ was worked out for the following eight cases:

$$c = 3,\ r = 10;\ 2^5 1^5, \qquad c = 4,\ r = 6;\ 3^5 2.$$
$$c = 3,\ r = 10;\ 2 1^9, \qquad c = 4,\ r = 9;\ 3^3 2^3 1^3,$$
$$c = 3,\ r = 11;\ 2 1^{10}, \qquad c = 4,\ r = 10;\ 3^3 2^3 1^4,$$
$$c = 3,\ r = 16;\ 2 1^{15}, \qquad c = 5,\ r = 8;\ 4^2 3^2 2^2 1^2.$$

The figures following the semicolon are the $u_i$ values: e.g. $2^5 1^5$ means that $u_i$ is 2 in five of the rows and 1 in the remaining five. No case in which $u_i = c$ or 0 was included, since any number of such rows may be added to the basic table without affecting the value of $Q$.

Some of the cases are rather closely related in their structure. Nevertheless, it seemed best to include all of them in presenting summary comparisons. The cases were chosen as indicative of the smallest samples in which the $\chi^2$ approximation to the distribution of $Q$ is likely to be needed. Smaller samples can of course occur in practice, but in this event it is relatively easy to make an exact test of significance from the exact distribution of $Q$.

The exact distribution was compared not only with the $\chi^2$ approximation, but also with an $F$-test applied to the data by means of an analysis of variance into the components

|  | D.F. |
|---|---|
| Rows | $(r-1)$ |
| Columns | $(c-1)$ |
| Rows × columns | $(r-1)(c-1)$ |

* This has been pointed out by Mosteller (1947).

where $F$ is the ratio of the mean squares for columns and rows $\times$ columns. If the data had been measured variables that appeared normally distributed, instead of a collection of 1's and 0's, the $F$-test would be almost automatically applied as the appropriate method. Without having looked into the matter, I had once or twice suggested to research workers that the $F$-test might serve as an approximation even when the table consists of 1's and 0's. As a testimony to the modern teaching of statistics, this suggestion was received with incredulity, the objection being made that the $F$-test requires normality, and that a mixture of 1's and 0's could not by any stretch of the imagination be regarded as normally distributed. The same workers raised no objection to a $\chi^2$ test, not having realized that both tests require to some extent an assumption of normality, and that it is not obvious whether $F$ or $\chi^2$ is more sensitive to the assumption. Inclusion of the $F$-test is also worth while in view of the widespread interest in the application of the analysis of variance to non-normal data.

The total number of values in a population is sufficiently small so that correction for continuity makes an appreciable difference. Application of the correction requires a little inspection of the data. Usually the values of $\Sigma(T_j - \bar{T})^2$ increase by 2's, but with $c = 2$ or 3 the increase may be much greater, and it is necessary to discover what is the value of $Q$ immediately below the one actually obtained, and enter the table with a value midway between the two. For $\chi^2$ the results are given both with and without correction, since experience in other problems has suggested that the correction may not be helpful when there are more than two samples. For $F$ the correction was a decided improvement and only corrected values are shown.*

It is easy to build up the exact distribution row by row. Members of the first row need not be permuted, but all other rows must be. Consider the diphtheria example in Table 2. If the sixty-three rows which show either all positives or no positives are omitted, this becomes the case $c = 4$, $r = 6$; $3^5 2$. We start with the row (1110) and add successively four rows with $u_i = 3$ and one row with $u_i = 2$. Addition of the second row gives the four cases

| 1110 | 1110 | 1110 | 1110 |
| 0111 | 1011 | 1101 | 1110 |
| 1221 | 2121 | 2211 | 2220 |

At this stage the possible sets of column totals are (2220) with probability 1/4 and (2211) with probability 3/4. All permutations of the third row are now added, and so on. The total number of cases is $(4^4)(6)$, or 1536, but these combine to give only nine different values of $Q$.

## 7. RESULTS OF THE SMALL-SAMPLE TESTS

In appraising the tabular $\chi^2$ and $F$ approximations, attention was concentrated on the region in which the exact probability lies between 0·2 and 0·005. Table 3 shows the average percentage errors in the estimates of significance probabilities for each of three subdivisions of the region. The percentage error is

$$100(\text{tabular } P - \text{true } P)/(\text{true } P),$$

where the averages are taken without regard to sign. The numbers of overestimates and underestimates made by each approximation are also shown. In Table 3, $\chi^2$ denotes the uncorrected value, and $\chi'^2$ and $F'$ denote the values after correction for continuity.

* Actually, the incomplete beta function rather than $F$ was corrected for continuity, since the former was more convenient for reading significance probabilities. Results differ slightly, but not materially, from those given by correcting $F$ itself.

From the percentage errors it appears that $\chi^2$ (uncorrected) and $F'$ have performed about equally well, both being slightly better than $\chi'^2$. None of the methods is free from bias. $\chi'^2$ tends to overestimate and $F'$ to underestimate. Over the range as a whole $\chi^2$ comes off fairly well with 23 overestimates and 32 underestimates, but it appears that a negative bias in the region of 0·2 to 0·1 is being counteracted by a positive bias in the region of 0·02 to 0·005. For practical use $\chi^2$ is preferable to $F'$, since it is slightly easier to calculate, though the possible application of $F'$ to more complex tables should be borne in mind.

Table 3. *Average percentage errors in estimating significance probabilities*

| Range of exact $P$ | Percentage error | | | No. of overestimates ($+$) and underestimates ($-$) | | | | | |
| | $\chi^2$ | $\chi'^2$ | $F'$ | $\chi^2$ | | $\chi'^2$ | | $F'$ | |
| | | | | $+$ | $-$ | $+$ | $-$ | $+$ | $-$ |
| 0·2 −0·1 | 15 | 7 | 5 | 1 | 10 | 7 | 4 | 6 | 5 |
| 0·1 −0·02 | 14 | 18 | 15 | 7 | 19 | 21 | 5 | 7 | 19 |
| 0·02−0·005 | 21 | 46 | 26 | 15 | 3 | 17 | 1 | 1 | 17 |
| Average or total | 16 | 25 | 17 | 23 | 32 | 45 | 10 | 14 | 41 |

At the true 5 % level, average errors of about 14 % are to be anticipated, which means that the tabular approximations might give a value of 0·057 or 0·043 instead of 0·05. At the 1 % level, the corresponding figures are about 0·012 and 0·008. These results appear close enough for routine decisions. For true probabilities below 0·005 all methods tend to go to pieces. $F'$ may give values only one-quarter of the true probability, while the two $\chi^2$ values may be six or eight times too high. An exact assessment of a very small probability is rarely essential.

It may be of interest to note the probabilities given by the various approximations for the diphtheria example in Table 2. We have already calculated $\chi^2$, with $P = 0·045$. The exact $P$ is 81/1536, or 0·053, while $\chi'^2$ gives 0·080 and $F'$ gives 0·062. All methods agree to the extent of indicating a probability somewhere close to the region of significance.

## 8. Further notes on the small-sample case

It has been mentioned that the value of $Q$, and hence of $\chi^2$, is unaffected by any rows which contain $c$ or 0 successes. This is not so for $F$, where the degrees of freedom $(r-1)(c-1)$ in the denominator are obviously increased by the addition of rows of any kind. The value of $F$ itself is also affected. Without resort to details, what happens is that if we take a basic table containing no rows with $c$ or 0 successes, and add to it an increasing number of such rows, the probabilities given by $F'$ (corrected) or $F$ (uncorrected) increase slowly until ultimately they agree with those given by $\chi'^2$ and $\chi^2$ respectively. This implies, incidentally, that at intermediate stages $F'$ may give a better approximation than any of the methods previously presented, because for the basic table the probability given by $\chi'^2$ is in general too high and that by $F'$ too low. In the eight worked examples, this was so when half of the rows were

$c$'s or 0's. In fact, it might be possible, as a purely empirical device, to set a quota of such rows which would be included in calculating $F$ (whether they were actually present or not), so as to make $F$ or $F'$ a good approximation to the exact probability. This approach was not pursued since $\chi^2$ seems good enough for most purposes. The approach may appear slightly repugnant logically, but is no more so than the use of an empirical approximation to an exact frequency distribution.

Some investigation was undertaken in an attempt to discover why, at low values of the exact significance probability, $\chi^2$ gives an overestimate of the probability. As might be expected, the principal reason seems to be that in small samples the true variance of $Q$ is less than that ascribed to it by the $\chi^2$ approximation. The true variance of $Q$ can be obtained by the usual rather laborious methods. We find -

$$E(Q) = (c-1); \quad V(Q) = 2(c-1)\left[1 - \frac{s_2 - 2s_3 + s_4}{(s_1 - s_2)^2}\right],$$

where

$$s_k = \sum_i \left(\frac{u_i}{c}\right)^k.$$

The mean value of $Q$ agrees with that of $\chi^2$, but the variance is always slightly too low. These results provide another approximation to the exact distribution of $Q$, in which instead of the $\chi^2$ distribution we use a type III distribution with exactly the same first two moments as $Q$, and with, in general, non-integral degrees of freedom. This approximation was tested on the eight examples. It gave a substantial improvement for probabilities less than 0·005, but in the region between 0·2 and 0·005 was only slightly better than $\chi^2$. A similar elaboration of $F$ produced about the same results.

As mentioned previously, the eight examples which were worked lead to a recommendation not to use the correction for continuity with $\chi^2$. This conclusion applies only when there are more than two samples. With two samples, the argument for a continuity correction is already provided by Yates's examination of the correction when used with the binomial distribution. As a check, two exact distributions with only two columns were worked out, and both showed $\chi'^2$ superior to $\chi^2$, though $\chi'^2$ still tended to overestimate the probabilities. A subdivision of the eight worked examples into the four examples with $c = 3$ and the four examples with $c = 4$ or 5 indicated that the superiority of $\chi^2$ over $\chi'^2$ was slightly greater in the latter group. With $c = 3$, the average percentage errors were 23 for $\chi'^2$ and 18 for $\chi^2$, whereas with $c = 4$ or 5 the corresponding figures were 27 and 15 respectively.

## 9. Subdivision of $\chi^2$ into components

In the limiting distribution all totals $T_j$ have the same expectations, variances, and covariances when the null hypothesis holds. This implies that if we divide $\Sigma(T_j - \bar{T})^2$ into components by the usual rules for subdividing a sum of squares, each component, when multiplied by the factor which converts it to $Q$, will follow a $\chi^2$ distribution in large samples. This procedure requires some care in its application. The diphtheria example is not very suitable for illustration, since the total $\chi^2$ is barely significant and would probably not be subdivided into components. The artificial example in Table 4, with data similar to those in the diphtheria example but showing more significance, will be used.

In the frequency distribution of $u_i$ the rows with $u_i = 4$ or 0 have been omitted. Since

$$\Sigma(T_j - \bar{T})^2 = (6)^2 + (15)^2 + (12)^2 + (17)^2 - (50)^2/4 = 69,$$

we find

$$Q = \frac{(4)(3)(69)}{44} = \frac{(3)(69)}{11} = 18\cdot81, \quad \text{with} \quad 3 \text{ D.F.}$$

Suppose that there is some reason to expect that A may perform differently from B, C and D. We might then wish to divide $Q$ into the components A $v$. B, C, D and B $v$. C $v$. D. For the first component we calculate

$$\frac{[3(6)-44]^2}{12} = 56\cdot33, \quad Q_1 = \frac{3(56\cdot33)}{11} = 15\cdot36 \quad (1 \text{ D.F.}).$$

By subtraction from the total $Q$, 18·81, we find $Q_2$ to be 3·45 (2 D.F.). It represents a comparison of the totals of B, C and D.

Table 4. *Artificial example to illustrate subdivision of $\chi^2$*

| | A | B | C | D | No. of cases | Frequency distribution of $u$ | |
|---|---|---|---|---|---|---|---|
| | | | | | | $u_i$ | $f$ |
| | | | | | | 3 | 8 |
| | | | | | | 2 | 5 |
| | 1 | 1 | 1 | 1 | 4 | | |
| | 1 | 0 | 1 | 1 | 2 | $\Sigma u_i = 34$, $\Sigma u_i^2 = 92$. | |
| | 0 | 1 | 1 | 1 | 6 | | |
| | 0 | 1 | 0 | 1 | 5 | $c\Sigma u_i - \Sigma u_i^2 = 4(34) - 92 = 44$. | |
| | 0 | 0 | 0 | 0 | 59 | | |
| Totals ($T_j$) | 6 | 15 | 12 | 17 | | | |

The difficulty is that since $Q_1$ is definitely significant, the null hypothesis that the probability of success within a row is the same in all *four* columns can no longer be maintained for the development of a comparative test of B, C and D amongst themselves. It seems better, when $Q_1$ is significant, to recalculate $Q_2$, using only the data from the relevant columns. If we reject the first column, the B, C and D totals do not change, but the frequency distribution of $u_i$ (ignoring 3's and 0's) becomes

$$\begin{array}{cc} u_i & f \\ 2 & 7 \end{array}$$

$$\Sigma u_i = 14, \quad \Sigma u_i^2 = 28, \quad c\Sigma u_i - \Sigma u_i^2 = 14.$$

Hence
$$Q_2' = \frac{(3)(2)(12\cdot67)}{14} = 5\cdot43 \quad (2 \text{ D.F.}).$$

The difference between $Q_2$ and $Q_2'$ arises solely in the conversion factor from $\Sigma(T_j - \bar{T})^2$ to $Q$, which has been altered from 3/11 to 3/7. This changes the significance probability from 0·178 for $Q_2$ to 0·066 for $Q_2'$. The exact probability, computed from the data for B, C and D alone, is found to be 0·078.

From such cases as I have examined, the ordinary rule for the subdivision of the sum of squares, and hence of $Q$, appears good enough for a preliminary inspection. When the situation is similar to that in this example, the advisability of recomputing tests that are of special interest should be noted.

## 10. Summary

In this paper the familiar $\chi^2$ test for comparing the percentages of successes in a number of independent samples is extended to the situation in which each member of any sample is matched in some way with a member of every other sample. This problem has been encountered in the fields of psychology, pharmacology, bacteriology and sample survey design

A solution has been given by McNemar (1949) and others when there are only two samples.

In the more general case, the data are arranged in a two-way table with $r$ rows and $c$ columns, in which each column represents a sample and each row a matched group. The test criterion proposed is

$$Q = \frac{c(c-1)\Sigma(T_j - \overline{T})^2}{c(\Sigma u_i) - (\Sigma u_i^2)},$$

where $T_j$ is the total number of successes in the $j$th sample (column) and $u_i$ the total number of successes in the $i$th row. If the true probability of success is the same in all samples, the limiting distribution of $Q$, when the number of rows is large, is the $\chi^2$ distribution with $(c-1)$ degrees of freedom. The relation between this test and the ordinary $\chi^2$ test, valid when samples are independent, is discussed.

In small samples the exact distribution of $Q$ can be constructed by regarding the row totals as fixed, and by assuming that on the null hypothesis every column is equally likely to obtain one of the successes in a row. This exact distribution is worked out for eight examples in order to test the accuracy of the $\chi^2$ approximation to the distribution of $Q$ in small samples. The number of samples ranged from $c = 3$ to $c = 5$. The average error in the estimation of a significance probability was about 14% in the neighbourhood of the 5% level and about 21% in the neighbourhood of the 1% level. Correction for continuity did not improve the accuracy of the approximation although it is recommended when there are only two samples. Another approximation, obtained by scoring each success as '1' and each failure as '0' and performing an analysis of variance of the data, was also investigated. The $F$-test, corrected for continuity, performed about as well as the $\chi^2$ approximation (uncorrected), but is slightly more laborious.

The problem of subdividing $\chi^2$ into components for more detailed tests is briefly discussed.

In conclusion, my thanks are due to Miss Elizabeth O. Grant and Mrs Elizabeth S. Jamison for considerable assistance in the computations. This work was done as part of a contract with the Office of Naval Research, U.S. Navy Department.

## REFERENCES

COCHRAN, W. G. (1937). The efficiencies of the binomial series tests of significance of a mean and of a correlation coefficient. *J. R. Statist. Soc.* **100**, 69.

DENTON, J. E. & BEECHER, H. K. (1949). New analgesics. *J. Amer. Med. Ass.* **141**, 1051.

DIXON, W. J. & MOOD, A. M. (1946). The statistical sign test. *J. Amer. Statist. Ass.* **41**, 557.

HSU, P. L. (1949). The limiting distribution of functions of sample means and application to testing hypotheses. *Proceedings of the Berkeley Symposium on Mathematical Statistics and Probability*, p. 359. University of California Press.

McNEMAR, Q. (1949). *Psychological statistics.* New York: John Wiley and Sons.

MOSTELLER, F. J. (1947). Equality of margins. *Amer. Statist.* **1**, 12.

WALSH, J. E. (1947). Concerning the effect of intraclass correlation on certain significance tests. *Ann. Math. Statist.* **18**, 88.

# 44

## THE PRESENT STATUS OF BIOMETRY

by William G. Cochran

*Professor of Biostatistics, School of Hygiene and Public Health,*
*The Johns Hopkins University, Baltimore*

### I. Introduction

In the founding of the Biometric Society, a wise decision was made to regard the scope of the society as very broad. In our constitution biometry is defined, by implication, as roughly synonymous with quantitative biology and as a discipline requiring the efforts and cooperation of biologists, mathematicians and statisticians. I like this concept of biometry and shall adopt it for the purposes of my paper.

As a relevant part of a discussion of the present status of the subject, it seemed to me appropriate to present some review of the major current areas of research. The field has grown to such an extent, however, and recent literature is so voluminous and scattered, that I could not hope to cover more than a small and biased sample of it. My review inevitably reflects this selectivity and must be regarded as illustrative rather than in any sense complete. Further, it is difficult in some cases to decide what is biometric research and what is not. Techniques in biometry have long been borrowed, sometimes appropriately and sometimes not, by other branches of science, and more recently techniques developed in these other branches promise to have useful applications in biometry. Consequently it is advisable from time to time to step outside the bounds of the subject.

There are several ways in which the field can be subdivided for a discussion of recent developments. My principal subdivision is into two areas which I shall call statistical techniques and biological applications. The former are the tools which biometry places at the disposal of the biologist. They include methods for the efficient collection of data, either through controlled experiments or biological surveys, and procedures for the statistical analysis of the data so as to extract all the relevant information. The chief contributors in this area are naturally the statisticians, though the biologist as such is sometimes deprived of the credit for a brilliant idea by a tendency to regard him thereafter as a statistician. Under biological applications we shall be concerned with the uses that are made of statistical techniques to advance biological science. In some cases the uses are strictly practical, as in agricultural experimentation designed to lead to recommendations for the farmer, or in most of bioassay. In others, the object is to formulate biological laws that are much more than merely descriptive of observed phenomena. Such laws summarize the workings of the biological processes that control the phenomena, and thus give new insight into these processes.

Reproduced with permission from the *Bulletin of the International Statistical Institute*, Vol. 32, Bk. 2, pages 132–150 (1–19). Copyright © 1950.

# II. Statistical techniques in biometry

## (a) Design of experiments

Expository papers. The ability to perform experiments is an advantage enjoyed in many branches of biology. The principles of sound experimental design, set forth by Fisher in the 1920's, have gradually come to be appreciated, at first in agriculture and more recently in many other fields. Nevertheless, dissemination is very uneven. It is not hard to name lines of research where the results of different investigators, carrying out what purports to be the same experiment, appear to disagree, yet no standard errors are presented and upon investigation it is found that none can be calculated. Much ingenuity is shown by investigators in concocting possible explanations of the discrepancies among the results of different workers, although they do not even know whether any of the discrepancies are real or merely attributable to sampling variation. This state of affairs leads to a good deal of confusion. Consequently it is encouraging that the literature in recent years contains numerous papers where the principles of good design are illustrated, either by the detailed description of a well-conducted experiment or by a discussion of the problems encountered in planning experiments for a specific investigation. Examples of the topics dealt with are microbiological assays (Wood and Finney, 1946), the effect of streptomycin on pulmonary tuberculosis (British Medical Research Council, 1948), remedies for headaches (Jellinek, 1946), the assay of tetanus antitoxin (Ipsen, 1942), the assay of penicillin (Bliss, 1946), feeding trials on milk cows (Cochran et al, 1941), tests of mosquito repellents (Wadley, 1946), and the stimulation of denervated muscle by electric currents (Solandt et al, 1943).

The newer developments in the field consist mainly of specialized devices for particular purposes and do not involve any fundamental change in outlook towards the general principles of design. I shall mention three devices on which active work has been in progress, although none is yet at the stage where its utility in biological experimentation can be accurately appraised. They have the common feature that in each case the purpose is to decrease the amount of work required in an experiment. This aim is representative of much of the work in experimental design.

Fractional replication. The first is the fractional replication of large factorial experiments. Fisher (1946) has repeatedly drawn attention to the advantage of testing a number of different factors simultaneously in an experiment, rather than confining each experiment to a single factor. The experiment on denervated muscle mentioned previously, which was of the factorial type, tested the efficacy of four types of electric current, four lengths of treatment period (1, 2, 3 and 5 minutes), and three numbers of treatment periods per day (1, 3 and 6) in all combinations. Thus there were $4 \times 4 \times 3$, or 48 different treatment combinations in a single replication. Such an experiment usually gives precise estimates of the average effects of each factor because it contains internal replication. For example, each single replication has 12 subjects (in this case rats) which received a particular type of current, so that the actual amount of replication available to estimate the average effects of the different types of current is 12 times as large as would at first sight appear. The experiment also enables us to examine whether there are interactions among the effects of different factors; e. g. whether the effect of a change

2

in type of current is dependent on the number of treatment periods or on the length of a treatment period.

In an exploratory experiment in biological research there may be many factors which it would be desirable to examine. The ambitious experimenter who adopts the factorial approach may find, when he has included all such factors, that the number of treatment combinations is beyond his resources, even if only one replication is used. The subject of fractional replication (Finney, 1945, 1946) deals with the type of information that can be obtained when only a fraction, say one-half or one-quarter, of the combinations needed for a complete replication are actually tested in the experiment. It is found that no clear interpretation of the results can be made unless certain high-order interactions are negligible. Where this is so, a number of potentially useful arrangements habe been developed. One is a design for testing six factors each at two levels. A complete replication requires $2^6$ or 64 treatment combinations, but by a suitably-chosen half-replicate we can estimate (and test for significance) the average effects of all factors and all interactions between pairs of factors. The assumption required is that all interactions among three or more factors are absent. It is of interest that arrangements of this type have been studied somewhat independently from the point of view of their applications to physics (Plackett and Burman, 1946) and to the weighing of objects on a spring or a chemical balance (Mood, 1946).

The «up and down» method. A second device (Dixon and Mood, 1948), which is sometimes known as the «up and down» method, may be useful in biological assays when animals are expensive and when it is convenient to test animals one at a time. The method was originally applied to the testing of explosives. The object is to estimate the median lethal dose with the smallest possible number of animals. The test doses should be equally spaced on a scale (most commonly a log. scale) on which it is expected that the response follows a normal distribution. The first animal is tested with the dose that is thought to be nearest to the median dose. If it survives, the second animal receives a stronger dose (the next one on the scale). If on the other hand the first animal dies, the second animal receives a weaker dose, again the next one on the scale. Thus after every death the following dose is one step weaker, and after every survival it is one step stronger. The effect is to concentrate the tests in the neighborhood of the 50 per cent lethal dose. The reduction in the number of animals required for an estimate of the median dose with a specified precision may be of the order of 30 per cent as compared with the usual method in which a pre-assigned number of animals is tested at each of several doses. A similar method for the estimation of other percentage points (e. g. the 80 per cent lethal dose) has been given by Bartlett (1946).

Sequential experimentation. A more general approach, also directed at keeping the amount of experimental material at a minimum, is that of sequential experimentation (Wald, 1947). As in the «up and down» process, the method was first developed for industrial and military purposes and applies where replicates can be completed one at a time. After each replicate the results are suitably tabulated, and the object is, roughly speaking, to terminate the experiment as soon as the results are decisive. The method might prove useful, for instance, in comparing a new with a standard therapeutic treatment on patients in a hospital, when it is

3

important to continue the experiment no longer than is absolutely necessary and when the effect of each drug, if any, is speedy.

Up to the time of writing, the advancement of the technique is impeded by difficult mathematical problems. A simple procedure is available when the object of the experiment is to select the better of two treatments and the data are in the form of ratios, e. g. percentage recovered. The experiment does not supply an unbiased estimate of the difference in performance of the two treatments, being merely designed to terminate when it becomes clear which is the superior treatment. A similar method exists when the data are measurements which can be assumed normally distributed, but this requires the standard deviations of the distributions to be known, as is seldom likely to be true in biological applications.

Most biological experiments are intended to estimate the differences between the effects of different treatments rather than simply to select the «best» treatment. From this point of view there is need for a sequential method in which the experiment stops when the true difference between the effects of two treatments has been estimated with a desired degree of precision, for example by means of a confidence interval of specified width. Sequential procedures for doing this have been obtained but they are computationally difficult and it is not yet known how efficient they are. A method in which the experiment is conducted in two stages (Stein, 1945) is a step in this direction.

## (b) Sampling problems

General status. The technique of sampling has taken great strides during the past twenty years. In this progress biometric research has had a prominent part. Indeed, the principles of sampling as applied to crops and soils were fairly well established and understood before any systematic study had been made of the corresponding problems that arise in sampling human populations for economic or sociological purposes, though these problems too now receive much attention.

The present situation is that sampling is relatively simple and routine provided that (i) the population can be divided into elements (sampling units) which are easily accessible and (ii) satisfactory methods of taking the desired measurements on the sampling-units have been devised. Under these conditions a variety of sampling techniques is available, with rules for a choice of the method that will provide the most precise estimates of the properties of the population for a given expenditure of time and labor. These methods are applicable, for instance, for estimating the number of plants or the percentage of diseased plants in a field, or the number of some species of insect in the topsoil. Incidentally, one of the methods, in which the units in the sample are chosen in a systematic pattern throughout the population, has long been used by foresters, but was at first ignored or in some cases condemned by statisticians. Since it has proved both simple·and accurate in a number of tests, it is now being intensively studied, see e. g. Yates (1948) and Matérn (1947).

So far as the *use* of these methods in biological research is concerned, my impression is that the situation is less satisfactory, in that there is often a tendency to ignore the sampling problem. Suppose that the amount of some vitamin in the leaves of a nutritious plant growing in a field is to be determined. The difficult part of the work is the determination of the vitamin content. The biologist may select, in any convenient way, just enough leaf material to enable the determination to be

4

made without considering whether the material selected is a representative sample of the leaves in the field. As an indication of the «precision» of the method, duplicate readings may be taken at some stage, but these come from the same original leaf material; in fact the subdivision into duplicates may be made only towards the final stage in the chemical process. There is, of course, a natural inclination to concentrate attention on the problem of measurement, where that is difficult, but the variation in vitamin content from one leaf to another may be much greater than any error in the chemical determination.

There remain many sampling problems where a solution is not in sight. Frequently the difficulty is that no convenient way has been found to divide the population into sampling units. The sampling of a peach tree in order to estimate the total weight of fruit is an illustration. The limbs are variable in their size, number and conformation and it is hard to construct any reliable sampling method that is substantially quicker than picking all the fruit from the tree. Sampling from bulked material has received only slight attention. Sometimes such material can be mixed, with greater or less effort, before a sample is taken, but there is little accurate information about the efficacy of mixing in reducing the error due to sampling, which might serve as a guide in deciding how much effort to devote to mixing. In the study of human populations we do not know for the most part how reliably people can and will give information about their past history, though this question determines the quality and scope of the information that can be obtained in much of the research on human beings.

Mobile populations. The estimation of the size of a mobile population, e. g. of birds, fish or insects, which inhabits some natural area presents a special problem in which it is not practicable to divide the population into a number of fixed sampling units. Some ingenious methods have been developed.

One is to capture specimens by a series of traps or by traversing the area with a net. All specimens are released after marking them so that they can be distinguished if later recaptured. On a second visit a further group of specimens is captured. It is clear that the fraction of recaptures found in this group gives an indication of the size of the population. This second group is again marked and released and the procedure continues as long as is considered desirable.

If the total number $N$ in the population can be considered fixed throughout the period of sampling, the analysis is simple. Let $n_i$ be the total number of specimens caught in the ith sample, and let $r_i$ be the number of these that have previously been marked. Also let $T_i$ be the total number of specimens that were marked and released previous to the ith sampling. Then if a marked specimen is just as likely to be caught as an unmarked one, the number of recaptures $r_i$ follows a binomial distribution with $n = n_i$, $p = \dfrac{T_i}{N}$ (assuming $N$ large). Consequently from the sample on the ith day, $N$ is estimated as $\dfrac{n_i T_i}{r_i}$. The maximum likelihood estimate of $N$ from a series of samples has been obtained (Schnabel, 1938). If the total number marked is only a small fraction of $N$, this estimate is very nearly $\dfrac{\Sigma (n_i T_i)}{\Sigma (r_i)}$, where the sum is taken over all samples subsequent to the first.

5

A more elaborate method is necessary to take account of births, deaths, emigration and immigration (Jackson, 1948, Fisher and Ford, 1947). There is now no single population total, since this varies from day to day. The population $N_i$ at the ith sampling may be estimated as follows. Let $t_j$ denote those specimens that were first marked on the jth day $(j < i)$. Let $s_{ij}$ be the expected fraction of these that have survived to the ith sampling. Then in that sampling the recaptures $r_i$ will follow a binomial distribution with $n = n_i$, $p = \dfrac{\Sigma\,(t_j s_{ij})}{N_i}$ , where the sum is over all previous samplings. Thus $N_i$ is estimated as

$$\frac{n_i\,(\Sigma\,t_j s_{ij})}{r_i}\;.$$

Some method must be found for obtaining the quantities $s_{ij}$, which are basic to the argument. If the emigration of marked specimens from the area is negligible, the $s_{ij}$ can be found from a knowledge of the life-table of the species, though an accurate application would require the ages of the specimens at the time of capture to be known. Alternatively, a simple rational function may be suggested for the $s_{ij}$. For example, if the combined death and emigration rate is constant with time, $s_{ij} = d^{(i-j)}$. The parameter $d$ can be estimated from the mean interval between release and recapture, or in other ways.

A fundamental assumption is that a marked specimen has exactly the same chance of dying, emigrating, or being recaptured as an unmarked specimen. Precautions must be taken to ensure that marking does not injure or handicap the specimen after release. Subsidiary studies may have to be undertaken in order to verify this.

There is another approach (DeLury, 1947), in which specimens that are caught are removed from the population. Sufficient numbers must be caught so that the catches make appreciable inroads into the population. If this is so, the catch per unit of effort will diminish steadily with repeated catching. The form of the diminution curve gives a clue to the initial population size. By setting up reasonable assumptions about the relation of catch to effort we can estimate the initial population from data showing the catches and amounts of effort (e. g. in terms of numbers of traps set).

### (c) Techniques for the analysis of data

The literature in this area is quite voluminous and difficult to classify. Most of it is concerned with the further extension of techniques which were first put forward some years ago, rather than with the presentation of essentially new techniques. A partial exception is the study of stochastic processes. But although this subject is likely to have fruitful biological applications, research thus far has been conducted mainly outside the field of biometry.

Some general impressions may be mentioned. As is to be expected, the techniques are becoming more complex, and in a number of cases the problem of presenting them in a form that is assimilable by the biologist has not been faced. There are of course exceptions, as in the probit technique, where the solution of transcendental equations is transformed into what is hoped will be the more familiar

6

terminology of linear regression. Problems still tend to be stated and solved in the terms that are most convenient to the statistician, rather than with the assumptions that are most realistic for biological use, and the normal distribution is in no danger of losing its popularity. Here again some opposing tendencies are noticeable, as in the development of transformations for skew data, occasional investigations of the effects of non-normality and the creation of a method of bioassay based on the logistic instead of the normal curve.

I shall describe briefly some recent work in two lines of research. The first, discriminant functions, exemplifies the tendency towards more complex developments, while the second, human genetics, is especially interesting because it embodies methods of estimation from non-random samples.

Discriminant functions. Suppose that we have two distinct groups, e. g. male and female skulls, and that a number of measurements $x_1$, $x_2$, ... $x_k$ have been made on every member of each group. The discriminant function is a linear compound

$$I = w_1 x_1 + w_2 x_2 + \ldots + w_k x_k,$$

in which the weights $w_i$ are chosen so as to give the best discrimination between the two groups. More specifically, the weights satisfy the property that in an analysis of variance of $I$ the ratio of the sum of squares between groups to that within groups is a maximum.

The function was proposed by Fisher about 15 years ago, and was first applied to the dating of a series of Egyptian skulls (Barnard, 1935). It is a very flexible tool. It provides a single test of significance of the difference between the group means, based on all $k$ measurements. It serves as a criterion for assigning a new specimen to one or the other group with a minimum probability of misclassification. From it we can obtain a measure of the distance between the two groups, which turns out to be the same as Mahalanobis' generalized distance (Mahalanobis, 1936). One adaptation leads to an index, derived from the characteristics of a crop that are of economic value, which may be used for the selection or rejection of lines in a program of plant improvement (H. F. Smith, 1940). Another gives a means of assigning quantitative scores to a series of qualitative grades (Fisher, 1941). These various properties of the discriminant function have been applied to such problems as the taxonomic distinction between two forms of the black locust tree, the quality rating of sheep, the grading of potatoes, the selection of strains of sea island cotton, and the classification of hair color.

More recent work on the statistical side has extended the scope of the technique. Rao (1948) has obtained a method for classifying a specimen when the number of populations exceeds two. While a single discriminant suffices for the assignment of a specimen to one of two populations, two functions are needed when there are three populations, and in general $(p - 1)$ functions are needed with $p$ populations. Rao has also developed a method, based on the concept of generalized distance, for arranging a group of populations into clusters or constellations, such that two members of the same constellation are «closer» to one another than any two belonging to different constellations. This approach has been applied to examine the relations between 22 castes and tribes inhabiting an area in the United Provinces in India.

**7**

Another extension shows how to utilize variates which themselves have no discriminating power, but which may be advantageous in increasing the discriminating powers of other variates. In another sense this technique is an extension of the well-known analysis of covariance, in which subsidiary measurements are used for this purpose. The technique has been applied to the assay of insulin, where the problem was to discover the best combination of a number of readings on the blood sugar of rabbits, both prior and subsequent to the injection of insulin (Cochran and Bliss, 1948).

The sampling theory associated with the discriminant function is based on the assumption that all variates follow a multivariate normal distribution, with the same covariance matrix in all populations. For some applications these assumptions appear very unsuitable, as for example when the object is to divide the members of a single population into a superior and an inferior group. Practically nothing is known at present about the effects of non-normality, but a paper by C. A. B. Smith (1947) allows us to dispense with the assumption that variances and covariances are the same within each population.

Human genetics. The work described here might perhaps equally well have been treated in the previous section on sampling problems. The field of human genetics is one where experiments are seldom possible. New knowledge comes mainly from the interpretation of observational material. Certain types of information, such as the distribution of the A, B, and O blood group genes in a racial group, can be accumulated by the random sampling of the group. But much of the progress, so far as the study of gene frequencies and linkage relationships is concerned, has been derived from human characteristics that are relatively rare. With such characteristics the geneticist must seek his cases where they are to be found, and any method of «random» sampling is out of the question. It is often profitable, for example, to investigate all available near-relatives of any case that has been discovered.

The estimation of gene frequencies or linkage from such data presents problems that are absent when the samples are strictly random. It may not be clear, particularly with older records, how these were obtained: for instance, if a family shows three children who have the characteristic, were the children detected independently, or by a follow-up of relatives of a single child? Individuals from the same family group are correlated in varying degrees, so that the records cannot be treated as a series of «independent» observations. Different types of family supply differing amounts of information, a fact which must be taken into account in making the most of the data. Usually there is no lack of methods of estimation, but some which appear intuitively attractive are badly biased, or efficient only in special cases.

During the past 15 years, techniques for handling such data have been put on a sound footing by the application of the modern theory of estimation. This development was initiated by the work of Bernstein (1931), who showed that linkage can be estimated from data involving only two generations, Haldane (1934), who made a systematic study of the detection of linkage between two autosomal genes, and Fisher (1934), who emphasized the fact that the method of analysis must be governed by our knowledge of the way in which the data were collected.

The fact that the data contain internal correlations creates no insuperable obstacle to the use of maximum likelihood, because Mendelian theory enables us

8

to make full allowance for known familial relationships in the calculation of likelihood functions. The fact that cases are deliberately sought, so that case frequencies as such are meaningless, is taken account of by the use of conditional probabilities, and the same technique is found to handle records where there is only partial knowledge of the methods by which cases were collected. The chief practical difficulty is that the computations are liable to become complex and tedious. This has been met, at least in part, by an ingenious presentation of the arithmetic required, and by the adoption of simpler though less efficient methods where the maximum likelihood solution seems too formidable. Tests for linkage (Fisher, 1935, Finney, 1940) were constructed by means of a series of rules for scoring the data family by family, and in subsequent papers the application of this method to the principal modes of inheritance has been worked out. Similar methods are available for the estimation of gene frequencies (Fisher, 1940, Finney, 1948). One by-product of this research is that the amounts of information that are supplied by different types of family can be tabulated in advance, so that the geneticist is made aware of the type of data that it is most profitable to seek.

## III. Biological applications

### (a) Applied biology

This section deals with the extent to which profitable use is made of biometry in the practical applications of biology, as distinct from fundamental or pure biology. Over the past 20 years a steady and marked increase in the use of biometric methods is observable, particularly among the younger biologists, but there is great variation both between different branches of biology and within an individual branch. The uses made cannot always be said to have been profitable, since the effect of exposure to statistics is sometimes to distract the biologist's attention from the meaning of his data to the details of statistical operations.

To consider briefly the subject of plant breeding and selection, as it has developed in the United States, I have always been surprised at the readiness with which plant breeders have been willing to try new methods of layout for field experiments. The lattice designs were first introduced by Yates (1936), under a series of names which certainly carried no advertising appeal, such as «pseudo-factorial arrangements in two equal groups of sets». The statistical analysis of the data from such designs rests on rather difficult results in the theory of least squares, which few plant breeders would claim to understand. Yet the designs began to be tried very promptly, in some cases before the author had had time to publish complete methods of analysis. The chief claim made for the new methods was that they were likely to be more precise than the older methods, though the actual increase in precision would depend greatly on the heterogeneity of the soil. During the period 1940–1946 a number of papers appeared giving the gains in precision which plant breeders had secured under actual field conditions, and presenting recommendations about their use in particular areas with particular crops. At the present time lattice designs have become the standard method of layout in a number of places. In others, where the designs are not now used, they have been tested, but were not adopted either because the gain in precision was very slight or the experimentation is of a hazardous

9

type that demands simpler methods. This experience testifies both to the alertness of the plant breeder and to the confidence which the statistician enjoys.

I have mentioned plant breeding as an example of a field where biologists are ready and eager to take advantage of new biometric techniques. To some extent the same remarks could be made for bioassay, where the probit method, which cannot be too simple for a biologist to understand and use, has found wide application. Within this field, however, there is probably more variation in the willingness of biologists to follow the statistician's lead, as evidenced by the frequent pleas for an easier method, the use of biased and inefficient methods, and the occasional requests for a technique that will determine the «minimum» or the «100 %» lethal dose from a total of 10 animals per experiment. Further, official pharmacological standards for the rating of new drugs still appear to leave a good deal to be desired in many instances.

Clinical experiments with human subjects represent an area where the penetration of statistical ideas is sporadic and incomplete. It is not uncommon to find the effectiveness of a new treatment «established» by comparison with the records of some previous group of patients under the routine treatment, and in circumstances where no clear understanding has been reached as to what is meant by the «success» of any treatment. A number of factors may contribute to the slowness of the infiltration, among them the high prestige of the physician, the real and justifiable pressures against any unnecessary experiments with human beings, and the keen interest of the public press in sensational new cures. Wherever it arises, this type of «experimentation» is very unfortunate, since different groups become protagonists for different methods, and much energy is diverted to arguments and disputes which, in the absence of any sound comparison of the methods, must be mostly a matter of verbiage.

## (b) Fundamental biology

Most of us would agree, I believe, that in the elucidation of natural laws that are of wide applicability and carry within them a kind of rational explanation of some basic process, biology lags behind such fields as physics or chemistry. The reasons why this is so are more debatable. It may be because in their training physicists and chemists are more exposed to mathematics and to the logical process of reasoning that mathematics exemplifies, or because these fields have long attracted more capable research workers by the evident rewards of success and by the fact that biology is less frequently taught in high schools. It may be that biological problems are intrinsically more difficult or that often the data needed for a test of a biological hypothesis requires prolonged work and great experimental resources to collect. One hears sometimes of mathematical physicists who attain world fame at an early age without apparently having conducted any experiments. I can recall no similar instance in fundamental biology.

In a discussion of this point in his interesting book, Kostitzin (1937) makes the statement that the biologist finds mathematical reasoning repugnant. This he attributes to two factors. The first is that the biologist is accustomed to proceed one step at a time, with a constant appeal to a corroborative experiment, whereas a single chain of mathematical reasoning can cover a great many steps so that the

10

final deductions seem remote from the initial premise. The second is that in the application of mathematics one always over-simplifies the problem. Errors introduced by over-simplification may be unimportant or soon detected in a very simple deduction which is being verified by experiment, but in a long chain of reasoning they may produce results that appear ridiculous to the biologist. It is true that in the mathematical approach to biology one of the most difficult issues is to decide what can be disregarded in the process of simplification.

To illustrate the different modes of attack that are currently being employed and the progress made through them, I shall mention briefly a number of disconnected topics.

The logarithmic series distribution. It is a remarkable and helpful fact that a few frequency distributions—the normal, binomial, Poisson and more recently the negative binomial—have been found to apply, at least approximately, to data obtained in many different branches of biology. The logarithmic series distribution (Fisher et al, 1943) may be an addition to this list. It was first applied to a collection of Malayan butterflies which were classified by species after collection. Collectors of biological species have observed that most species are rare and only a few are abundant. Thus a tabulation of the number $n$ of species for which $1, 2, 3, \ldots$ specimens have been caught usually shows a mode at $n = 1$, with steadily decreasing frequencies thereafter. The data are well fitted by a distribution in which

$$N_n = \frac{\alpha\, x^n}{n}, \qquad (n \geqslant 1)$$

where $N_n$ is the expected number of species for which $n$ specimens are caught, and $\alpha$ and $x$ are parameters ($x < 1$). This discovery has encouraged the application of the distribution to plant and animal aggregations, and Williams (1947) quotes 34 series of data satisfactorily fitted.

The discovery of an empirical formula of wide applicability is no assurance that the formula represents a biological law in the sense in which I have used that term. It does, however, excite curiosity as to why the same formula should appear in many different situations, and the research so stimulated may lead to useful generalizations. Some work of this type has already been undertaken for the logarithmic series.

In deriving the distribution, Fisher pointed out that for a single species it might be reasonable to suppose that the number caught in a specified time would follow a Poisson distribution, with a mean depending on the abundance of the species. Thus if all species were equally abundant we would expect a Poisson instead of a logarithmic series distribution. Since in fact abundance varies from species to species it is more natural to anticipate a mixture of Poisson distributions. If the abundance from species to species can be assumed to follow a Pearson type III distribution, this mixture is the negative binomial distribution, in which the probability that $n$ specimens of some species are caught is

(1) $$\frac{(k + n - 1)\,!}{(k-1)\,!\, n\,!}\, x^n\, (1 - x)^k. \qquad (0 < x < 1)$$

The mean of this distribution is $m = \dfrac{k\,x}{(1 - x)}$. Its variance is $m + \dfrac{m^2}{k}$, in which

11

the second term arises from the variability in abundance. The quantity $\dfrac{1}{k}$ may be interpreted as a measure of the diversity or variability in abundance, since when $k$ is large the distribution approaches the Poisson while when $k$ is small the variance of the distribution greatly exceeds its mean.

When the negative binomial is fitted to the distribution of the number of specimens caught from the same species, the values of $k$ obtained are very close to zero. If we let $k$ tend to zero in (1), remembering that we cannot observe the number of species for which no specimens are caught, the frequencies for $n \geqslant 1$ are easily seen to be proportional to $\dfrac{x^n}{n}$, which gives the logarithmic series distribution.

This analysis shows that the logarithmic series distribution is consistent with the hypothesis that there is great diversity in the abundance of different species, and is not consistent with the hypothesis that such diversity is only slight, subject to the important practical proviso that the method of capture gives an equal probability of capture to any specimen irrespective of its species. Thus far, the analysis is of a descriptive nature. The question why species show such diversity is a more important one, of great interest in the study of evolution. D. G. Kendall (1948) has attempted to throw some light on this question by a mathematical study of the growth of populations subject to reproduction, mortality and immigration, and has shown that certain types of balance among these factors will lead to a logarithmic series distribution for the size of the population.

The Rhesus factor. Work on the rhesus factor constitutes one of the most exciting and successful applications of the biometric method. Levine (1939) found that in a case of stillbirth the mother's serum contained an antibody which would agglutinate the red cells of about 85 per cent of white Americans. It was soon discovered that a similar antibody was obtained from rabbits and guinea-pigs by injection of the blood of the rhesus monkey (Landsteiner and Wiener, 1940), a circumstance which gave the name to the genetic factor involved. In the original case the blood of the father and the child both reacted to the antibody, but that of the mother did not. The simplest hypothesis is that persons whose blood reacts are of genotypes either $RR$ or $Rr$, while those who do not react are of genotype $rr$. On this supposition the frequency of the gene $r$ can be estimated as $\sqrt{.15}$ or 39 %.

This discovery prompted the testing of mothers' sera, in cases where pregnancy involved haemolytic disease, in a number of centres of genetic research. There resulted a series of new discoveries which in the absence of a body of genetic theory would have presented a very confusing array of facts, but which with the aid of that theory have been assembled into a relatively simple picture. By 1946 it was known with reasonable certainty that there are at least eight genes, instead of the original two: six different antibodies had been obtained with which a person's phenotype can be determined by serological tests: and methods had been worked out for estimating the gene frequencies. In some instances the theoretical structure was in advance of the discoveries, since it was able to predict the existence of two of the genes, $R_y$ and $R_z$, of which $R_z$ was subsequently discovered while $R_y$, not yet discovered, can be estimated to be extremely rare. What is more important, the existence and properties of two of the antibodies could also be postulated before

12

they were discovered. The details of this development, which are rather intricate, cannot be given here: for an introductory account see Fisher (1947).

Crop-weather relations. By way of contrast I should like to mention this problem as one where, as it seems to me, relatively little progress has been made in recent years. I believe that the problem is typical of those in other fields where advances are slow.

That crop yields vary from year to year, even when grown on the same soil with similar husbandry, is very familiar. In many parts of the world, these variations are substantial, and it is clear that they must for the most part be ascribed to variations in the weather. For a given crop and soil a biometric analysis of this variation, if completely successful, would find a number of measurable weather factors which, when combined suitably in an equation, would predict most of the variation in yields that can be ascribed to weather. Moreover, this equation, which might have to be very complex, should be consistent with the known facts of plant physiology. Thus changes in crop yields predicted from changes in certain of the weather variables should be verifiable by experiment.

There are two distinct approaches towards this ambitious goal. One is by the «laboratory» method: that is, by experimentation under conditions in which weather can be controlled. This method has been workable in the greenhouse with certain individual factors, e. g. amount of water supplied or number of hours of darkness, but plant physiologists would admit that so far this work gives only some tentative suggestions as to the relationships between weather factors and yields. Further, the experimental approach is complicated by the facts that there are numerous factors which might be important, that the effect of any factor is likely to be different at different stages in the plant's growth towards maturity, that inter-relations between the factors may play a prominent role and that plants have a remarkable facility for recovery from temporary adverse conditions. Consequently one can envisage a lengthy and expensive experimental programm which is not likely to be completed in the near future.

A considerable amount of effort has also gone into a statistical approach in which crop yields are correlated with the available weather variates. The most precise data are obtained in experiments where the same crop is grown year after year in the same field, under similar fertilizer and cultural practices, so that apart from the influence of slow changes in soil fertility, which can be allowed for, the changes in yield can reasonably be regarded as caused by weather fluctuations. Many years of field work are required in order to accumulate a substantial amount of data, though the period can be shortened, with some loss in precision, by growing the crop at several places, each subject to different weather conditions. This approach, too, has met with very moderate success. Sometimes as much as 40 per cent of the variation in yields can be accounted for by a regression on weather variates, but a figure between 15 and 20 per cent is more typical. This outcome is probably not surprising to the plant physiologist, since the weather variates are usually few in number and restricted to those that can be recorded cheaply (e. g. rainfall rather than soil moisture), since they must be chosen at the start of the experiment, and since the types of regression function which have been tested, despite much ingenuity on the part of the statistician, must appear rather naive.

13

I have cited this case not as a reproach either to the plant physiologists or the statisticians, but as a problem where progress must be slow and costly, and where it is not even clear how best to allocate a large sum of money to research, if some generous donor appeared.

Mathematical investigations. Obviously, any attempt to formulate biological laws is bound to involve some mathematics, though perhaps in a very primitive way. From time to time, investigations have appeared that are primarily mathematical in nature, in which after the formulation of a mathematical model for some biological process, the mathematical deductions from the model are worked out in detail. Researches of this type tend to be restricted to processes which seem particularly adapted to mathematical description, and to be associated with relatively few individuals, since not many feel confident both in mathematics and biology. Amongst earlier investigations may be mentioned the work of Volterra and Lotka on the growth of populations or groups of populations under various assumptions as to reproduction and competition for food, that of Fisher, Wright and Haldane on evolution and that of Ross, Soper, Kermack and McKendrick on the spread of epidemic diseases.

More recently the amount of attention given to this field has increased, though not greatly. An important development is the establishment at Chicago of a group of workers, under the leadership of Rashevsky, who concentrate on problems of this type. This has produced a steady volume of work on a variety of topics. Especially noteworthy is the work on the communication of impulses or messages as applied to physiology (Householder and Landahl, 1944, Wiener and Rosenbluth, 1946, McCulloch and Pitts, 1948). As Wiener (1948) has pointed out, this work has applications in many other branches of science. Topics treated by workers elsewhere are binocular vision (Luneberg, 1947), dark adaptation (Moon and Spencer, 1945) and the Weber-Fechner law (von Schelling, 1944), in addition to further work on epidemic theory (Wilson and Worcester, 1945) and on evolution (Malécot, 1946).

On a more purely mathematical level, the recent interest in stochastic processes may have fruitful implications for biology in two respects. In the first place the great bulk, though not all, of the mathematical analysis of biology has been of a deterministic type, in that a given set of initial conditions always produces the same consequences at a specified future time. This is true, for instance, of almost all the work in epidemic theory, even though epidemics are notoriously influenced by chance factors. Increased knowledge of stochastic processes may result in more realistic mathematical models in which probabilistic influences play their proper role. Secondly, the interesting mathematical aspects of stochastic processes may induce mathematicians to make contributions which, although not in themselves biological, can be used for biological applications. Researches in mathematical biology have produced new and challenging problems in mathematics, and there is no doubt that both this subject and mathematics itself will benefit from any mathematical skills that can be enlisted. Moreover, the statistical problems which are met when stochastic processes are applied to actual data have scarcely been touched, and are likely to be of great interest, as the memoir of Lundberg (1940) on the application of stochastic processes to sickness insurance suggests.

14

# IV. Discussion

I need hardly remark that speculations in this section are both personal and debatable. There seems no doubt, however, that biometry is in a flourishing state. In addition to the vigorous output of new papers, from which I have quoted some examples, one may cite the growth of a large central body of statistical theory which furnishes efficient methods for estimation and for testing hypotheses. The establishment of new journals devoted to biometric research—for example «Population», «Heredity» and «Biometrics», and somewhat earlier the «Bulletin of Mathematical Biophysics» and «Acta Biotheoretica»—is another testimony to the greatly increased interest in the field. Courses of lectures in biometry are more widely available in universities and colleges, and while the quality of teaching is not uniformly good, it has been found that the biologist can readily grasp the notion of variability and that by the device of drawing repeated samples much of the basic theory can be presented without elaborate mathematics. Finally, the foundation of our new Biometric Society gives us the opportunity of maintaining cooperation on a worldwide basis.

To consider first statistical research, this appears to be ample in quantity, and in view of the relatively high prestige of the statistician the volume is likely to be maintained in the future. In the way of critical comment the following points occur to me. A substantial part of the work seems narrow in scope and of minor rather than major importance, as exemplified by papers which present only a slight variation on previously published work, or which claim an increase in precision that turns out in actual applications to be trivial. Since certain subjects tend to become fashionable from time to time in all branches of science, it is too much to expect that this feature will not persist, but the field would advance more rapidly if both directors of research and young statisticians would devote more time to the less «popular» problems. There is a need for more research on the effects of errors in the assumptions on which the standard techniques are based. For instance, linear regression methods are frequently used without any assurance that the true relation is linear, and of course a similar comment can be made with respect to the many techniques based on the normal distribution. Research of this kind usually leads to unattractive mathematical problems, but insofar as biometric methods are produced in order to be used the neglect of such problems is not justified.

In view of the growing complexity of statistical techniques, a trend which is unlikely to be reversed, more attention might profitably be given to simpler methods whose efficiency is satisfactory if not the maximum attainable. It is easy for the statistician, who may use his method only in the preparation of an example to be included in a publication, to persuade himself that the method is relatively simple, but his views might change if he were forced to apply it continually with poor computing equipment. I found myself that a method which I customarily recommended to biologists for fitting a catalytic curve became intolerable in the face of the problem of fitting over 300 such curves for a research study. It is interesting to note that in the field of quality control a deliberate compaign has been carried on to keep the techniques very elementary, even at a loss both in efficiency and in flexibility. While this has sometimes led to the teaching of rules of procedure rather than principles, it has contributed to the rapid dissemination of the methods and is evidence of the sound attitude that the methods are a means and not an end.

15

With regard to the extent to which biologists are receptive to and utilize biometric methods, the situation has become more favorable but is very variable from one branch of biology to another. It is not always easy to account for these variations. The amount of mathematical training which the biologist has received is sometimes thought to be important, but I rather doubt this: at least I have not found poor mathematics too much of an impediment if the biologist knows what he wants and insists on statistical advice being made clear to him in commonsense terms. A more frequent obstacle is the natural feeling of inferiority which the biologist sometimes has in discussions on statistical matters. This makes communication difficult to establish and may prevent him from questioning the advice given, even though he has a vague notion that there is something wrong with it. Another obstacle, much more disturbing, is that some biologists appear to have a very imperfect concept of the scientific method; for example, of the nature of the evidence required to decide which of two hypotheses should be accepted and which rejected. This would suggest a defect in the teaching of biology, at least in some branches, perhaps as a consequence of the increasing amount of material which it is thought must be learned and the increasing specialization within the subject. On the whole I believe that the outlook for the future is good.

So far as mathematical biology is concerned, my impression is that the accomplishments are not well known and that their influence on biology falls short of what it might be. This is likely to happen with work in the border area between two distinct sciences, since the work tends to be omitted from the main stream of teaching in both sciences. Indeed, the subject is seldom taught even in the stronger centres of biometric teaching. Further, since the mathematics is itself relatively unfamiliar, experimental biologists, even with a good grasp of standard elementary mathematics, find the material difficult to understand, so that a wide gap may exist between the mathematical and the experimental biologists within the same branch of biology. I should like to see a substantial increase in the scope and volume of work in mathematical biology; with participation by mathematicians, statisticians and especially by biologists. Even an unsuccessful attempt at a mathematical analysis may be rewarding, in that it may show where definitions and concepts have been vague, and where thinking has been confused. As to the prospects that such an increase will take place I am uncertain. In the case of the biologists this seems bound to depend on the amount of importance which is attached to mathematical training as a prerequisite for research in biology.

To turn very briefly to the role which our society may play in future developments, M. Teissier spoke some very wise words in his concluding remarks to the first Biometric Conference. He pointed out that our society is an experiment in cooperation, and one where success will be difficult to attain. If our efforts are not carefully directed, the natural tendency, it seems to me, is for our society to become primarily and perhaps almost exclusively a society of statisticians. Such a society might be useful and influential but it will not fulfill the purpose for which our society was formed. It should be realized that both the biologists and the mathematicians will require to some extent to be coaxed into cooperation, and that numerous other societies are competing for their attention. Consequently I hope that the scope of our society will continue to be conceived broadly, and that biologists and mathematicians will have their share of responsibility in determining our future activities.

16

Finally, I hope that our society will set an example in international co-operation. To mention just one aspect, I should like to see the society serve the purpose of keeping biometricians better informed of developments in countries other than their own, perhaps by means of a series of articles on recent contributions in various countries or regions. Needless to say, the desirability of such information has struck me very forcibly in the preparation of this paper.

## Résumé

### L'état actuel de la biométrie

La biométrie est considérée comme l'aspect quantitatif de la biologie. Afin de passer en revue les développements récents, la matière est divisée en deux groupes: méthodes statistiques, qui comprennent l'organisation des expériences, les méthodes de sondage et les méthodes utilisées dans l'analyse des données; et les applications biologiques, employant ces méthodes.

En ce qui concerne l'organisation des expériences, les nouvelles méthodes tendent à diminuer le travail que nécessite l'expérience. L'auteur donne trois exemples — la répétition fractionnée, la méthode «up and down» et les expériences consécutives.

La méthode des sondages s'est développée de façon à être relativement simple, pourvu que l'ensemble soit divisible en unités de sondage accessibles et qu'il soit possible de procéder aux mensurations nécessaires. Lorsque les conditions sont défavorables, par exemple pour un matériel en grands groupes hétérogènes, ou en cherchant des informations correctes d'une population, de nombreux problèmes restent à résoudre. Une question qu'il a été possible de trancher concerne l'estimation du nombre total d'un ensemble d'éléments mobiles (par exemple des insectes) dans un espace déterminé.

Les méthodes d'analyse tendent à devenir de plus en plus compliquées, et les hypothèses sur lesquelles elles se fondent ne s'accordent parfois pas avec la réalité biologique. L'analyse discriminatoire est un exemple de l'application d'une méthode compliquée. En génétique humaine, des méthodes ingénieuses ont été inventées pour l'estimation de la fréquence des gènes et du linkage lorsqu'il s'agit d'un échantillon n'ayant pas été prélevé au hasard.

Dans certaines branches de la biologie, par exemple dans le domaine de la sélection des plantes, les biologistes ont vite profité des nouvelles méthodes biométriques, mais la situation diffère fortement d'une branche à l'autre. Les essais se font souvent avec peu d'appréciation des méthodes scientifiques. Lorsque les mensurations biologiques sont difficiles, on constate une tendance à négliger les problèmes de sondage.

Les progrès en vue de formuler des lois biologiques sont moins prononcés que ceux réalisés en chimie et en physique.

Quelques exemples de recherches récentes sont discutés: la répartition dite de la série logarithmique, le facteur rhésus, et l'influence des conditions météorologiques sur les plantes domestiques.

La biologie mathématique, ou autrement dit l'élaboration et l'étude des modèles mathématiques de processus biologiques, accuse une activité croissante. A noter de nouvelles applications à la neurophysiologie et la reprise de recherches sur l'évolution et sur la théorie des épidémies. L'intérêt porté aux processus stochastiques pourrait bien porter des fruits vu que dans les modèles l'élément probabiliste fait souvent défaut, même pour des phénomènes manifestement influencés par le hasard.

En somme, la biométrie se développe de façon réjouissante. Le volume de la recherche en méthodes statistiques est grand. Afin de ne pas perdre contact avec le biologiste, il est nécessaire d'étudier des méthodes simples basées sur des hypothèses conformes aux applications biologiques, et de travailler sur des problèmes d'une plus grande diversité. Il est probable que les biologistes emploieront de plus en plus les méthodes statistiques: le manque d'attitude scientifique semble constituer un plus grand obstacle qu'une formation mathématique insuffisante.

Les progrès de la biologie mathématique ne sont pas suffisamment connus. Il y aurait profit à augmenter les efforts dans ce domaine.

Pour terminer, l'auteur met en évidence le rôle de la Société de Biométrie en vue d'activer la coopération entre biologistes, mathématiciens et statisticiens.

# 17

# References

Barnard, M. M. (1935), The secular variations of skull characters in four series of Egyptian skulls (Ann. Eugen., 6, pp. 352–371).

Bartlett, M. S. (1946), A modified probit technique for small probabilities (Jour. Roy. Stat. Soc. Supp., 8, pp. 113–117).

Bernstein, F. (1931), Zur Grundlegung der Chromosomentheorie der Vererbung beim Menschen mit besonderer Berücksichtigung der Blutgruppen (Z. indukt. Abstamm. u. Vererbungslehre, 57, pp. 113–138).

Bliss, C. I. (1946), A revised cylinder-plate assay for penicillin (Jour. Amer. Phar. Ass., 35, pp. 6–12).

British Medical Research Council (1948), Streptomycin treatment of pulmonary tuberculosis (Brit. Med. Jour., 2, pp. 769–783).

Cochran, W. G., Cannon, C. Y. and Autrey, K. M. (1941), A double change-over design for dairy cattle feeding experiments (Jour. Dairy Sci., 24, pp. 937–951).

— and Bliss, C. I. (1948), Discriminant functions with covariance (Ann. Math. Stat., 19, pp. 151–176).

DeLury, D. B. (1947), On the estimation of biological populations (Biometrics, 3, pp. 145–167).

Dixon, W. J. and Mood, A. M. (1948), A method for obtaining and testing sensitivity data (Jour. Amer. Stat. Ass., 43, pp. 109–126).

Finney, D. J. (1940), The detection of linkage (Ann. Eugen., 10, pp. 171–214).

— (1945), The fractional replication of factorial arrangements (Ann. Eugen., 12, pp. 291–301).

— (1946), Recent developments in the design of field experiments. III. Fractional replication (Jour. Agr. Sci., 36, pp. 184–191).

— (1948), The estimation of gene frequencies from family records. I. Factors without dominance (Heredity, 2, pp. 199–218).

Fisher, R. A. (1934), The effect of methods of ascertainment upon the estimation of frequencies (Ann. Eugen., 6, pp. 13–25).

— (1935), The detection of linkage with « dominant » abnormalities (Ann. Eugen., 6, pp. 187–201).

— (1940), The estimation of the proportion of recessives from tests carried out on a sample not wholly unrelated (Ann. Eugen., 10, pp. 160–170).

— (1941), Statistical methods for research workers (Oliver and Boyd, Edinburgh, 8th ed.), § 49. 2.

— Corbet, A. S. and Williams, C. B. (1943), The relation between the number of species and the number of individuals in a random sample of an animal population (Jour. An. Ecol., 12, pp. 42–58).

— (1946), The design of experiments (Oliver and Boyd, Edinburgh. 3rd ed.)

— (1947), The Rhesus factor (Amer. Scientist, 35, pp. 95–103).

— and Ford, E. B. (1947), The spread of a gene in natural conditions in a colony of the moth panaxia dominula L. (Heredity, 1, pp. 143–174).

Haldane, J. B. S. (1934), Methods for the detection of autosomal linkage in man (Ann. Eugen., 6, pp. 26–65).

Householder, A. S. and Landahl, H. D. (1944), Mathematical biophysics of the central nervous system (Principia Press, Bloomington, Indiana).

Ipsen, J. (1942), Systematische und zufällige Fehlerquellen bei Messung kleiner Antitoxinmengen (Zeitschrift für Immunitätsforschung, 102, pp. 347–368).

Jackson, C. H. N. (1948), The analysis of a tsetse-fly population. III (Ann. Eugen., 14, pp. 91–108).

Jellinek, E. M. (1946), Clinical tests on comparative effectiveness of analgesic drugs. (Biometrics, 2, pp. 87–91).

Kendall, D. G. (1948), On some modes of population growth leading to R. A. Fisher's logarithmic series distribution. (Biometrika, 35, pp. 6–15).

Kostitzin, V. A. (1937), Biologie mathématique. (Collection Armand Colin, Paris).

Landsteiner, K. and Wiener, A. S. (1940), An agglutinable factor in human blood recognized by immune sera for rhesus blood. (Proc. Soc. Exp. Biol. and Med., 43, p. 223).

Levine, P. and Stetson, R. E. (1939), An unusual case of intra-group agglutination. (Jour. Amer. Med. Ass., 113, pp. 126–127).

Lundberg, O. (1940), On random processes and their application to sickness and accident statistics. (Almquist and Wiksells, Uppsala).

Luneburg, R. K. (1947), Mathematical analysis of binocular vision. (Princeton University Press.)

18

Mahalanobis, P. C. (1936), On the generalized distance in statistics (Proc. Mat. Inst. Sci. Ind., 12, pp. 49–55).

Malécot, G. (1946), La consanguinité dans une population limitée (C. R. Acad. Sci. Paris, 222, pp. 841–843).

Matérn, B. (1947), Methods of estimating the accuracy of line and sample plot surveys (Medd. fr. Statens Skagsforsknings Inst., 36).

McCulloch, W. S. and Pitts, W. (1948), The statistical organization of nervous activity (Biometrics, 4, pp. 91–99).

Mood, A. M. (1946), On Hotelling's weighing problem (Ann. Math. Stat., 17, pp. 432–446).

Moon, P. and Spencer, D. E. (1945), Visual dark adaptation: a mathematical formulation (Jour. Math. Phys. M. I. T., 24, pp. 65–105).

Plackett, R. L. and Burman, J. P. (1946), The design of optimum multifactorial experiments (Biometrika, 33, pp. 305–325).

Rao, C. R. (1948), The utilization of multiple measurements in problems of biological classification. (Jour. Roy. Stat. Soc. Ser. B., 10, pp. 159–203).

Schnabel, Z. E. (1938), The estimation of the total fish population of a lake (Amer. Math. Monthly, 45, pp. 348–350).

Smith, C. A. B. (1947), Some examples of discrimination (Ann. Eugen., 13, pp. 272–282).

Smith, H. Fairfield (1936), A discriminant function for plant selection (Ann. Eugen., 7, pp. 240 to 250).

Solandt, D. Y., DeLury, D. B. and Hunter, J. (1943), Effect of electrical stimulation on atrophy of denervated muscle (Arch. Neur. and Psych., 49, pp. 802–807).

Stein, C. (1945), A two-sample test for a linear hypothesis whose power is independent of the variance (Ann. Math. Stat., 16, pp. 243–258).

von Schelling, H. (1944), Gedanken zum Weber-Fechnerschen Gesetz (Abh. Press. Akad. Wiss. Math.-Nat. Kl., 5).

Wadley, F. M. (1946), Incomplete-block design adapted to paired tests of mosquito repellents (Biometrics, 2, pp. 30–31).

Wald, A. (1947), Sequential Analysis (John Wiley and Sons, New York).

Wiener, N. and Rosenbluth, A. (1946), The mathematical formulation of the problem of conduction of impulses in a network of connected excitable elements, specifically in cardiac muscle (Arch. Inst. Cardiol. Mexico, 16, pp. 205–265).

— (1948), Cybernetics (John Wiley and Sons, New York).

Williams, C. B. (1947), The logarithmic series and its application to biological problems (Jour. Ecol., 34, pp. 253–272).

Wilson, E. B. and Worcester, J. (1945), The law of mass action in epidemiology (Proc. Nat. Acad. Sci., 31, pp. 109–116).

Wood, E. C. and Finney, D. J. (1946), The design and statistical analysis of microbiological assays (Quart. Jour. Phar. and Pharmacology, 19, pp. 112–127).

Yates, F. (1936), A new method of arranging variety trials involving a large number of varieties (Jour. Agr. Sci., 26, pp. 424–455).

— (1948), Systematic sampling (Phil. Trans. Roy. Soc. Lond. A, 241, pp. 345–377).

19

# TESTING A LINEAR RELATION AMONG VARIANCES*

W. G. Cochran**

*School of Hygiene and Public Health,
Johns Hopkins University.*

## 1. NATURE OF THE PROBLEM

WE HAVE A NUMBER of independent estimates $v_i$ of variances $\theta_i$ respectively, $(i = 1, 2, \cdots k)$. The estimate $v_i$ is based on $n_i$ degrees of freedom (d.f.) and follows the usual distribution of a mean square derived from normally distributed observations; namely that $n_i v_i / \theta_i$ is $\chi_i^2$ with $n_i$ d.f. We wish to test the null hypothesis that $p$ homogeneous linear relations

$$(1) \qquad c_{1j}\theta_1 + c_{2j}\theta_2 + \cdots + c_{kj}\theta_k = 0 \qquad (j = 1, 2, \cdots p)$$

hold among the $\theta_i$, where the $c_{ij}$ are known numbers.

This problem, usually involving a *single* linear relation, has been encountered occasionally in the applications of statistical methods to data. With the current increased interest in the "components of variance" technique, it is not unlikely that the problem will appear more frequently in the future. Two examples will be described briefly.

An experiment is repeated at $r$ different places on each of $c$ different occasions, because it is expected that the performance of the treatments $(t)$ will differ from one place or one time to another. The places and times are assumed to be a random sample of the population of places and times in which the results will be used. If the usual "components of variance" model is set up, the expectations of the four principal items in the analysis of variance of the results are as shown in Table I.

---

*Prepared in connection with research sponsored by the Office of Naval Research.

**Presented before the American Statistical Association annual meeting in Chicago, 1950 at sessions held jointly by the Biometrics Section of the American Statistical Association and The Biometric Society (ENAR). Department of Biostatistics paper no. 266.

TABLE I. EXPECTATIONS OF MEAN SQUARES.

| | m. s. | Expected value |
|---|---|---|
| Treatments | $v_1$ | $\theta_1 = (E) + (TPO) + r(TO) + c(TP) + rc(\Sigma(\tau_j - \tau)^2)/(t - 1)$ |
| Treatments × places | $v_2$ | $\theta_2 = (E) + (TPO) \qquad\qquad + c(TP)$ |
| Treatments × occasions | $v_3$ | $\theta_3 = (E) + (TPO) + r(TO)$ |
| Treatments × places × occasions | $v_4$ | $\theta_4 = (E) + (TPO)$ |

In this representation, all effects except the treatment means $\tau_j$ are assumed to be random variables. The variance $(TPO)$ stands for the variance contributed by the three-factor interaction, and so on. For more detailed discussion of this model, see (1) and (2).

Suppose that we wish to test the hypothesis that there are no differences in the average effects of the treatments; i.e. that all $\tau_j$ are equal. It is evident that none of the other lines in the analysis of variance supplies an appropriate denominator, or "error" mean square, for an $F$-test of the treatments mean square. In fact, the null hypothesis that all $\tau_j$ are equal implies the linear relation

$$\theta_1 - \theta_2 - \theta_3 + \theta_4 = 0.$$

Consequently a test of the null hypothesis is a test of this linear relation.

A second example occurred in a corn-breeding experiment in which $mn$ female parents from a population produced by random mating were mated $n$ to each of $m$ males from the same population. Comstock and Robinson (3) were able to show that under certain assumptions the expectations of three mean squares in the analysis of variance of the results were connected by the relation

$$c_1\theta_1 + c_2\theta_2 = \theta_3$$

In this case the coefficients $c_1$ and $c_2$ are functions of a quantity $a$ which serves as a measure of the degree of dominance, the value $a = 0$ implying no dominance, while values greater than one imply over-dominance. Thus the null hypothesis that $a$ has a specified value leads to a linear relation with known coefficients, which we may wish to test from the data.

Three tests which are already well-known are particular cases of the general problem of testing relations (1).

(i) *The F-test.* This tests the relation $\theta_1 = \theta_2$. It is an exact test.

(ii) *Bartlett's test of homogeneity of variances.* This tests the series of $(k - 1)$ relations $\theta_1 = \theta_2 = \cdots = \theta_k$. For this Bartlett (4) proposed a form of the likelihood-ratio test criterion

$$(\textstyle\sum n_i) \log_e \bar{v} - \sum (n_i \log_e v_i)$$

which is distributed approximately as $\chi^2$ with $(k - 1)$ d.f. Modifications have been suggested by Bartlett and others in order to improve the approximation to the tabular $\chi^2$ distribution.

(iii) *The two-tailed Behrens-Fisher problem.* If $x_1$ is a normally-distributed estimate of $\mu_1$ with variance $\theta_1$, with similar roles for $x_2$, $\mu_2$ and $\theta_2$, the Behrens-Fisher problem is concerned with testing the hypothesis $\mu_1 = \mu_2$. We possess estimates $v_1$, $v_2$ of $\theta_1$, $\theta_2$ respectively, but nothing is known about the relative sizes of $\theta_1$ and $\theta_2$.

The test criterion is $(x_1 - x_2)/\sqrt{v_1 + v_2}$. If, however, the test is to be two-tailed, we might equally use $(x_1 - x_2)^2$ as the basis of a test criterion. When the null hypothesis holds, $(x_1 - x_2)^2$ is an estimate of $(\theta_1 + \theta_2)$, based on 1 d.f. Thus the problem may be regarded as one of testing the relation

$$\theta_1 + \theta_2 - \theta_3 = 0$$

where the estimate $(x_1 - x_2)^2$ of $\theta_3$ happens to be derived from only 1 d.f.

This case illustrates the difficulty of testing a linear relation among variances, even when only three variances are involved. The unknown ratio $\theta_1/\theta_2$ tends to creep into the solution as a nuisance parameter, and some way must be found to get rid of it. Two approaches may be mentioned. Fisher (5) developed a test in which the ratio $v_1/v_2$ remains fixed from sample to sample, while the ratio $\theta_1/\theta_2$ varies about $v_1/v_2$ in its fiducial distribution. Tables have been constructed by Sukhatme (6). On the other hand, Welch (7) considered a population in which $\theta_1/\theta_2$ remains fixed. By successive approximation he obtained a function $h$, depending on $v_1/v_2$ and on the significance probability $P$, such that

$$Pr. \{(x_1 - x_2) > h(v_1/v_2, P)\} = P,$$

for any value of $\theta_1/\theta_2$. Aspen (8) has published tables from which the test may be made.

### 3. CONTENT OF THIS PAPER.

The subsequent discussion falls into three main parts. An approximate $F$-test of a single relation will be presented and illustrated. Relatively little is known about the closeness of this approximation, but as the test is being used in applications, it seems advisable to put some account of it on record. The test is easy to make, and is recommended, at least until a more exact test may appear. In a later section an investigation of the adequacy of the approximation and of the power of the test is given for the case where there are only three variances. Finally, in an Appendix, the large-sample limiting distribution of the test is compared with those of other tests which suggest themselves, particularly the likelihood-ratio test.

### 4. THE APPROXIMATE $F$-TEST.

Since the coefficients $c_{ij}$ in the relation are known numbers, the linear relation can always be reduced to the form

$$(2) \qquad \theta_1 + \theta_2 + \cdots + \theta_r = \theta_{r+1} + \theta_{r+2} + \cdots + \theta_k$$

If the alternative hypothesis specifies that one side of the equation is definitely greater than the other, we suppose that the left side is the greater. The test criterion suggested is

$$F' = \frac{v_1 + v_2 + \cdots + v_r}{v_{r+1} + v_{r+2} + \cdots + v_k}$$

When the null hypothesis holds, this quantity follows the $F$-distribution, approximately. The degrees of freedom $\nu_1$, $\nu_2$ are found by a rule suggested by Fairfield Smith (9) and Satterthwaite (10).

$$\nu_1 = \frac{(v_1 + v_2 + \cdots + v_r)^2}{\dfrac{v_1^2}{n_1} + \dfrac{v_2^2}{n_2} + \cdots + \dfrac{v_r^2}{n_r}} \quad : \quad \nu_2 = \frac{(v_{r+1} + \cdots + v_k)^2}{\dfrac{v_{r+1}^2}{n_{r+1}} + \cdots + \dfrac{v_k^2}{n_k}}$$

The values of $\nu_1$, $\nu_2$ are not in general integers. Interpolation in the $F$-tables is rarely necessary, since a glance at the $F$-value for the nearest integers to $\nu_1$ and $\nu_2$ usually decides the issue.

If the alternative hypothesis is two-sided, we place in the numerator of $F'$ whichever of the estimates of variance happens to be larger, so that $F' > 1$. The resulting probability is doubled.

Many different $F$-ratios could be formed for an approximate test of relation (2). Consider the first example described previously, where the null hypothesis is $(\theta_1 - \theta_2 - \theta_3 + \theta_4 = 0)$. Under the alternative hypothesis, the values of $\theta_2$, $\theta_3$ and $\theta_4$ remain unchanged but that of $\theta_1$ increases. This suggests that an $F$-ratio of the form $v_1/(v_2 + v_3 - v_4)$, instead of $(v_1 + v_4)/(v_2 + v_3)$, might be appropriate, and some workers have used this form. In the case where all d.f. $n_i$ become large, the limiting power functions of the two test criteria are the same. The form proposed was suggested intuitively on the grounds that a quantity like $(v_2 + v_3 - v_4)$, where some coefficients are negative, is not so well represented by a Type III approximation as a linear form where all coefficients are positive. Consequently, the proposed $F'$ may be distributed more like $F$, in samples of practical size, than the alternative criterion.

5. NUMERICAL EXAMPLE.

This is a test of the relation

$$\theta_1 + \theta_4 = \theta_2 + \theta_3$$

The data come from a long-term experiment on sugar-beet conducted by Rothamsted Experimental Station. Since the data are intended only to illustrate the arithmetic involved in making the test, explanatory details are omitted. A partial analysis of variance is shown in Table II.

TABLE II. PARTIAL ANALYSIS OF VARIANCE (ROOTS, TONS PER ACRE).

|  | d.f. | m.s. | Estimate of |
|---|---|---|---|
| Treatments | 3 | 32,489 | $\theta_1 = (E_e) + 5(TY) + 4(E_p) + 20(T)$ |
| Error (a) | 8 | 2,791 | $\theta_2 = (E_e) \qquad\qquad\quad + 4(E_p)$ |
| Treatments $\times$ years | 9 | 1,527 | $\theta_3 = (E_e) + 5(TY)$ |
| Error (b) | 24 | 317 | $\theta_4 = (E_e)$ |

The null hypothesis is that the component $(T)$ is zero. This is equivalent to $\theta_1 + \theta_4 = \theta_2 + \theta_3$ with a one-sided alternative. Hence

$$F' = \frac{32,489 + 317}{2,791 + 1,527} = \frac{32,806}{4,318} = 7.60$$

$$\nu_1 = \frac{(32,806)^2}{\dfrac{(32,489)^2}{3} + \dfrac{(317)^2}{24}} = \frac{1}{\dfrac{(.9903)^2}{3} + \dfrac{(.0097)^2}{24}} = 3.1$$

$$\nu_2 = \frac{(4,318)^2}{\dfrac{(2,791)^2}{8} + \dfrac{(1,527)^2}{9}} = \frac{1}{\dfrac{(.6464)^2}{8} + \dfrac{(.3536)^2}{9}} = 15.1$$

The 1% value of $F$ for 3, 15 d.f. is 5.42. The observed $F'$ is definitely significant.

6. INVESTIGATION OF THE $F'$ DISTRIBUTION.

This investigation is confined to the case of three variances, where the null hypothesis $\theta_3 = \theta_1 + \theta_2$ holds. The test criterion is $F' = v_3/(v_1 + v_2)$. The quantity computed was $P(F' > F'_{.05})$; that is, the true probability that $F'$ exceeds the 5 percent level attributed to it by the approximation used.

The approximation is the tabular $F$-distribution, with $\nu_1$, $\nu_2$ d.f. where $\nu_1 = n_3$ and

$$(3) \qquad \nu_2 = \frac{(v_1 + v_2)^2}{\dfrac{v_1^2}{n_1} + \dfrac{v_2^2}{n_2}} = \frac{(1 + u)^2}{\dfrac{u^2}{n_1} + \dfrac{1}{n_2}},$$

with $u = v_1/v_2$. Note that this distribution depends on the variance-ratio $u$, so that we do not use a *single* approximation to the exact distribution of $F'$, but a whole series of approximations, depending on the $u$ which happens to turn up. Consequently,

$$(4) \qquad P(F' > F'_{.05}) = \int_0^\infty P(F' > F'_{.05}/u)\, f(u)\, du,$$

where $f(u)$ is the frequency function of $u$. The distributions of $F'$ and $u$ are not independent, except in certain special cases, so that in (4) we must use the conditional distribution of $F'$ given $u$. This may be obtained by routine methods.

An alternative approach gives the conditional distribution more quickly. It is well known that

$$Q = \frac{v_3(n_1 + n_2)}{\theta_3\left[\dfrac{n_1 v_1}{\theta_1} + \dfrac{n_2 v_2}{\theta_2}\right]}$$

follows the $F$-distribution with $(n_3, n_1 + n_2)$ d.f. and is distributed independently of $u = v_1/v_2$. Hence the conditional distribution of $Q$ given $u$ is the same $F$-distribution. But, since $\theta_3 = \theta_1 + \theta_2$,

$$(5) \qquad F' = \frac{v_3}{v_1 + v_2} = Q\, \frac{(\theta_1 + \theta_2)\left[\dfrac{n_1 v_1}{\theta_1} + \dfrac{n_2 v_2}{\theta_2}\right]}{(v_1 + v_2)(n_1 + n_2)}$$

$$= Q\, \frac{(1 + U)\left[\dfrac{n_1 u}{U} + n_2\right]}{(1 + u)(n_1 + n_2)},$$

where $U = \theta_1/\theta_2$. Thus $F'$ is the product of $Q$ and a factor which depends solely on $u$ and the fixed parameter $U$. Hence the conditional distribution of $F'$ is transformed to an ordinary $F$-distribution, with $(n_3, n_1 + n_2)$ d.f., by multiplying by the inverse of the factor.

These results give a method for numerical evaluation of (4). The percentiles of the distribution of $u$ can be read from the $F$-table, since $u/U$ follows the $F$-distribution with $n_1$, $n_2$ d.f. For any value of $u$, $F'_{.05}$ is found by means of equation (3) and a further reference to the $F$-table with $\nu_1$, $\nu_2$ d.f. Then the conditional probability that $F'$ exceeds $F'_{.05}$ is read by transforming $F'$ to an $F$-distribution with $(n_3, n_1 + n_2)$ d.f. from (5). Finally, the integral is evaluated numerically in the form

$$(6) \qquad P(F' > F'_{.05}) = \int_0^1 P(F' > F'_{.05}/p) \, dp,$$

where $p$ is the cumulative probability obtained from $u$ so that $dp = f(u) \, du$.

Computations were made for six sets of values of the $n_i$, ranging from $n_1 = n_2 = 6$, $n_3 = 3$; to $n_1 = n_2 = 24$, $n_3 = 6$. These values were thought to be fairly representative of the numbers of d.f. available in some of the smaller applications. Four ratios for $\theta_1/\theta_2$ were included; 1, 2, 4 and 16. The true significance probabilities are shown in Table III.

TABLE III. TRUE SIGNIFICANCE PROBABILITY OF
$F'$ AT THE APPARENT 5% LEVEL.

| $n_1, n_2 =$ | $n_3 =$ | $U = \theta_1/\theta_2$ | | | |
|:---:|:---:|:---:|:---:|:---:|:---:|
| | | 1 | 2 | 4 | 16 |
| 6 | 3 | .044 | .047 | .051 | .056 |
| | 6 | .043 | .046 | .053 | .059 |
| 12 | 6 | .047 | .049 | .052 | .054 |
| | 12 | .046 | .049 | .054 | .056 |
| 24 | 6 | .050 | .050 | .052 | .053 |
| | 12 | .049 | .050 | .052 | .054 |

The primary purpose of the $F'$ test, with its varying significance levels according to the value of $\nu_1/\nu_2$, is to nullify the effect of the nuisance parameter $U = \theta_1/\theta_2$. If the approximation were fully successful, all entries in Table III would be 0.050. Looking along the rows, we see that the effect of variations in $U$ is not obliterated. When $U = 1$, the $F'$ test gives in general slightly too few significant results.

TABLE IV. PROBABILITY OF OBTAINING A SIGNIFICANT RESULT
AT THE APPARENT 5% LEVEL OF $F'$.

|  | $\varphi = \theta_3/(\theta_1 + \theta_2)$ | | |
|---|---|---|---|
|  | 2 | 4 | 8 |
| $F(6, 12)$ | .259 | .622 | .881 |
| $F' : U = 1$ | .234 (.26)* | .592 (.62) | .866 (.88) |
| $F' : U = 2$ | .236 (.24) | .583 (.59) | .857 (.86) |
| $F' : U = 4$ | .234 (.23) | .556 (.57) | .833 (.85) |
| $F' : U = 16$ | .213 (.22) | .504 (.53) | .792 (.81) |
| $F(6, 6)$ | .188 | .468 | .767 |

*Figures in ( ) denote an approximation discussed later.

As $U$ departs from 1, the proportion of significant results increases to over 1 in 20. Throughout the range of variation of $U$, however, the probability is always near enough to 0.05 so that the approximation seems adequate for practical use.

Only a few values were calculated for $U > 16$, since this case may be rare in practice. The significance probabilities appear to increase to a maximum which is not far from the probabilities for $U = 16$. As $U$ increases still further, the probabilities decline towards 0.05, which is the limiting value when $U = \infty$.

The effect of $n_3$ (number of d.f. in the numerator of the test) is seen by comparing neighboring pairs of rows. Rather surprisingly, the approximation is slightly better with the lower than with the higher of the two values of $n_3$ . As $n_1$ and $n_2$ increase, on the other hand, the approximation tends to improve, as would be expected.

This case provides little guidance as to the performance of the test when more than three variances are involved. The methods employed can be extended to the case of four variances, but the calculations are much more lengthy. We would expect the approximation to be less satisfactory with four variances, since two nuisance parameters are involved.

### 7. THE POWER FUNCTION OF THE $F'$ TEST.

If $\varphi = \theta_3/(\theta_1 + \theta_2)$, equation (5) still holds for the conditional distribution of $F'$ given $u$, except that the right side is multiplied by $\varphi$. For any known value of $\varphi$, the method of the previous section can be used to find the probability that the test gives a significant result at the 5 percent level.

Calculations were made for the case $n_i = 6$, and $\varphi = 2, 4, 8$. The probabilities are shown in Table IV. The corresponding probabilities for the ordinary $F$-test with 6 and 12 d.f. and with 6 and 6 d.f. are included for comparison. It should be noted that in the case of $F'$, computations were made for the *apparent* 5 percent levels. As shown in Table III, these levels were not actually at 5 percent, being slightly under it for $U = 1, 2$ and slightly over for $U = 4, 16$. The computations could have been adjusted to a common 5 percent level, but it seemed preferable to examine the power of the test as it actually operates.

In all cases the power lies between that of $F$ (6, 12) and that of $F$ (6, 6). This suggests the general result that the power lies between that of $F$ ($n_3$, $n_1 + n_2$) and that of $F$ ($n_3$, $n_1$), where $n_1$ is the *smaller* of $n_1$, $n_2$. I have not, however, been able to establish this result. The power tends to decline steadily as $U = \theta_1/\theta_2$ departs from unity. This again would be expected intuitively.

A rough approximation to the power function can be obtained by a method due to Welch (11). By this approximation, the $F'$ test has a power equal to that of an $F$ test with $n_3$ and $n_e$ d.f., where

$$n_e = \frac{(\theta_1 + \theta_2)^2}{\dfrac{\theta_1^2}{n_1} + \dfrac{\theta_2^2}{n_2}} = \frac{(U + 1)^2}{\dfrac{U^2}{n_1} + \dfrac{1}{n_2}}.$$

The probabilities given by this method are shown to 2 d.p. in parentheses in Table IV. The approximation overestimates the power, but not too seriously. Since $n_e$ always lies between $(n_1 + n_2)$ and the smaller of $n_1$, $n_2$, the approximation is in line with the general speculation made above.

### 8. SUMMARY COMMENTS.

So far as they go, these investigations indicate, in my opinion, that the $F'$ test is quite satisfactory when three variances are involved, though we are still in the dark for the case of four or more variances. If a more precise test turns out to be necessary, one possibility is to try to extend the approach used by Welch (7). The $F'$ test might in fact be regarded as a first approximation to the type of test developed by Welch, which will probably involve the computation of tables of significance levels. Another possibility, suggested by J. W. Tukey, is to investigate different rules for calculating the numbers of d.f. to be assigned to $F'$, in the hope of finding one under which the significance probability is less affected by the $\theta_1/\theta_2$ ratios. For example, the rule might be made to depend on the significance probability, as it does in Welch's method, whereas our rule does not take any account of this.

In addition to the $F$-test, two other tests have been considered. The first is based on the assumption that

$$(7) \qquad (v_1 + v_2 + \cdots + v_r) - (v_{r+1} + \cdots + v_k)$$

can be regarded as a normal deviate. This test might be tried if the numbers of d.f. $n_i$ are all large. The second test is the likelihood ratio test. Both tests have been used in practice. They do not appear to me as satisfactory for small-sample work as the $F'$ test, but it seems worthwhile to describe them.

As will be shown, all three tests are asymptotically equivalent. That is, if all $n_i$ tend to infinity in such a way that their ratios remain constant, the power functions of the three tests tend towards the same limiting distribution. The tests will be illustrated by applying them to the example given by Comstock and Robinson (3).

### 9. THE NORMAL DEVIATE TEST.

Henceforth, the quantity (7) above will be denoted by

$$\left( \sum_1 v_i - \sum_2 v_i \right).$$

If we divide this quantity by an estimate of its standard error, the test criterion is

$$d = \frac{\sum_1 v_i - \sum_2 v_i}{\sqrt{2 \sum \dfrac{v_i^2}{n_i}}}$$

This is regarded as a normal deviate with mean zero and unit standard deviation.

The data given by Comstock and Robinson (3) are as follows.

$$v_1 = 0.0334; \qquad v_2 = 0.0248; \qquad v_3 = 0.069$$

$$n_1 = 36; \qquad n_2 = 180; \qquad n_3 = 144.$$

The null hypothesis which corresponds to the absence of dominance is $\theta_3 = \theta_1 + \theta_2$. The alternative hypothesis (presence of some dominance) specifies that $\theta_3$ exceeds $(\theta_1 + \theta_2)$.

The value of $d$ is

$$\frac{.069 - .0334 - .0248}{\sqrt{2 \left[ \dfrac{.069^2}{144} + \dfrac{.0334^2}{36} + \dfrac{.0248^2}{180} \right]}} = 0.930$$

For a single-tailed test, this gives a probability of 0.176.

The limiting distribution of $d$ is easily found by familiar methods. Write $n_i = na_i$, where the $a_i$ remain fixed as $n$ increases. Then

$$d = \frac{\sqrt{n}\left(\sum_1 v_i - \sum_2 v_i\right)}{\sqrt{2 \sum \frac{v_i^2}{a_i}}}$$

Also let

$$\sum_1 \theta_i - \sum_2 \theta_i = \frac{\mu}{\sqrt{n}},$$

where $\mu$ remains finite, and measures the amount of divergence from the null hypothesis. The factor $\sqrt{n}$ is needed in the denominator, because in large samples the test can detect a divergence of order $1/\sqrt{n}$.

The numerator of $d$ is asymptotically normally distributed, with mean $\mu$ and variance $2 \sum \theta_i^2/a_i$. Further, since $v_i$ converges in probability to $\theta_i$, it follows, as shown by Cramér (12), that the denominator of $d$ converges in probability to

$$\sqrt{2 \sum \frac{\theta_i^2}{a_i}},$$

Hence the limiting distribution of $d$ is normal, with mean

$$\frac{\mu}{\sqrt{2 \sum \frac{\theta_i^2}{a_i}}}$$

and unit variance, (Cramér, loc. cit.)

10. THE $F'$-TEST.

For this example

$$F' = \frac{v_3}{v_1 + v_2} = \frac{.069}{.0334 + .0248} = 1.186,$$

with d.f. 144 and 98, where

$$98 = \frac{(.0334 + .0248)^2}{\frac{(.0334)^2}{36} + \frac{(.0248)^2}{180}},$$

The significance probability is found to be 0.179, in close agreement with the previous test.

For the asymptotic distribution of $F'$, we note that

$$\sqrt{n}\,(F' - 1) = \frac{\sqrt{n}\left(\sum_1 v_i - \sum_2 v_i\right)}{\sum_2 v_i} = \frac{d\sqrt{2 \sum \frac{v_i^2}{a_i}}}{\sum_2 v_i}.$$

Since the factor by which $d$ is multiplied on the right tends in probability to a constant (the corresponding function of the $\theta_i$), the limiting power functions of $F'$ and $d$ are essentially the same.

### 11. THE LIKELIHOOD RATIO TEST.

For the present, we consider that the alternative hypothesis places no restrictions on the $\theta_i$, except that they are positive: i.e. that the test is two-sided. Apart from constant factors, the log. of the likelihood is

$$L = -(1/2) \sum \frac{n_i v_i}{\theta_i} - (1/2) \sum n_i \log \theta_i$$

Under the alternative hypothesis, denoted by the suffix 2, the maximum likelihood estimates are simply $\hat{\theta}_{i2} = v_i$. This gives

$$L_2 = -(1/2) \sum n_i - (1/2) \sum n_i \log v_i$$

For the null hypothesis, we minimize $L$ subject to the restriction

$$\text{(8)} \qquad \sum_1 \hat{\theta}_i - \sum_2 \hat{\theta}_i = 0.$$

This leads to the set of equations

$$\text{(9)} \qquad \frac{n_i v_i}{\hat{\theta}_i} - n_i = \pm g\hat{\theta}_i , \qquad (i = 1, 2, \cdots k)$$

where $g$ is a Lagrange multiplier and the sign on the right is $+$ or $-$ depending on whether $i$ belongs to $\sum_1$ or $\sum_2$.
Adding the equations, we get

$$\text{(10)} \qquad \sum \frac{n_i v_i}{\hat{\theta}_i} - \sum n_i = 0.$$

Hence, the test criterion,

$$\text{(11)} \quad 2(L_2 - L_1) = - \sum n_i - \sum n_i \log v_i + \sum \frac{n_i v_i}{\hat{\theta}_i} + \sum n_i \log \hat{\theta}_i$$

$$2(L_2 - L_1) = \sum n_i \log \left[ \frac{\hat{\theta}_i}{v_i} \right],$$

using (10). It may be noted that the test criterion takes the same form as in Bartlett's test of homogeneity of variances.

Equations (9) for the maximum likelihood estimates under the null hypothesis are non-linear, and I do not know any quick means of solving them. One method is to note that

$$(12) \qquad \sum_1 v_i - \sum_2 v_i = g \sum \frac{\hat{\theta}_i^2}{n_i}$$

By substituting first approximations to the $\hat{\theta}_i$, we find an approximation to $g$. From this, a second approximation to each $\hat{\theta}_i$ is found from the equation (9) in which it appears. Convergence is rather slow.

For the numerical example, this process gives

$$\hat{\theta}_1 = .039558, \qquad \hat{\theta}_2 = .025304, \qquad \hat{\theta}_3 = .064862.$$

$$2(L_2 - L_1) = 36 \log\left(\frac{.039558}{.0334}\right) + 180 \log\left(\frac{.025304}{.0248}\right)$$

$$+ 144 \log\left(\frac{.06486}{.069}\right) = .847$$

When the null hypothesis holds, the limiting distribution is that of $\chi^2$ with 1 d.f.

As given here, the likelihood ratio test is two-sided. For comparison with the other tests, we want to make a one-sided test against the alternative $\theta_3 > (\theta_1 + \theta_2)$. To obtain a one-sided test in the general case, compute $(\sum_1 v_i - \sum_2 v_i)$. If this has a sign opposite to that specified in the alternative hypothesis, stop and declare the result non-significant. If the sign is the appropriate one, calculate $\chi^2$, and double the resulting probability. Since $\chi^2$ has 1 d.f., we obtain the probability by regarding $\sqrt{0.847} = 0.920$, as a normal deviate. This is practically the same as the normal deviate, 0.930, obtained from the $d$ test. It gives a one-tailed probability of 0.179.

12. ASYMPTOTIC DISTRIBUTION OF THE LIKELIHOOD RATIO TEST.

Investigation of this distribution presents some annoying complications relating to the existence and uniqueness of the maximum likelihood estimates under the null hypothesis. From the previous section, the equations to be solved may be written

$$(13) \qquad v_i - \hat{\theta}_i = \pm g \frac{\hat{\theta}_i^2}{n_i}, \qquad (i = 1, 2 \cdots k)$$

subject to

$$\sum_1 \hat{\theta}_i - \sum_2 \hat{\theta}_i = 0.$$

The equations are non-linear, and in finite samples there may be more than one set of solutions, though it is easy to select the "sensible" solution. However it is difficult to express the solution in a workable form. To avoid the complications, we shall use a unique and explicit set of second approximations to the maximum likelihood estimates, rather than the final estimates themselves.

These second approximations are obtained by the iterative method suggested in the previous section. We take the $v_i$ as first approximations. Then from equation (12), an approximate value of the Lagrange multiplier $g$ is

$$(14) \qquad g = \frac{\sum_1 v_i - \sum_2 v_i}{\sum \frac{v_i^2}{n_i}}$$

Hence, as second approximations $u_i$ to the $\hat{\theta}_i$ we take

$$(15) \qquad u_i = v_i \mp g \frac{v_i^2}{n_i},$$

where in the adjustment term we have substituted $v_i$ in place of the unknown $\hat{\theta}_i$.

As before, the quantities $n_i v_i / \theta_i$ are independently distributed as $\chi^2$ with $n_i$ d.f. respectively. We have $n_i = n a_i$, where the $a_i$ remain fixed as $n$ goes to infinity. Also

$$(\theta_1 + \cdots + \theta_r) - (\theta_{r+1} + \cdots + \theta_k) = \sum_1 \theta_i - \sum_2 \theta_i = \mu/\sqrt{n}.$$

Certain preliminary results which are needed follow at once from standard theorems.
(i) The limiting distribution of $g/\sqrt{n}$ is normal with

$$\text{Mean} = \frac{\mu}{\sum \frac{\theta_i^2}{a_i}} \quad : \quad \text{Variance} = \frac{2}{\sum \frac{\theta_i^2}{a_i}}$$

(ii) The quantity $g/n$ tends in probability to zero.
(iii) Since

$$u_i = v_i \mp g \frac{v_i^2}{n_i} = v_i \mp \left[\frac{g}{n}\right]\left[\frac{v_i^2}{a_i}\right],$$

it follows from (ii) that $u_i$ tends in probability to $\theta_i$.

In starting the proof, we cannot use the compact expression

$\sum n_i \log \hat{\theta}_i/v_i$ for the test criterion, since this is valid only if the correct maximum likelihood estimates have been used. This point was brought forcibly to my notice when I obtained a $\chi^2$ of *minus* 2.2 when substituting rather crude estimates into this expression. Instead, we must use the more general results derived from equation (11).

$$2(L_2 - L_1) = -\sum n_i \log \frac{v_i}{u_i} + \sum \frac{n_i(v_i - u_i)}{u_i}$$

$$= -\sum n_i \log \left[1 + \frac{v_i - u_i}{u_i}\right] + \sum \frac{n_i(v_i - u_i)}{u_i}$$

$$= -\sum n_i \log \left[1 \pm \frac{gv_i^2}{n_i u_i}\right] + \sum \left[\pm \frac{gv_i^2}{u_i}\right]$$

from (iii). Expanding the log, we obtain

$$(16) \qquad 2(L_2 - L_1) = \frac{1}{2} \sum \frac{g^2 v_i^4}{n_i u_i^2} \mp \frac{1}{3} \sum \frac{g^3 v_i^6}{n_i^2 u_i^3 \left[1 + \frac{\alpha g v_i^2}{n_i u_i}\right]^3} ,$$

where $0 < \alpha < 1$. But

$$\frac{g^3 v_i^6}{n_i^2 u_i^3} = \left[\frac{g^2}{n}\right]\left[\frac{g}{n}\right]\left[\frac{v_i^6}{a_i^2 u_i^3}\right].$$

From (i), (ii) and (iii), this tends to zero in probability, so that the sum on the extreme right of (16) also tends to zero in probability. Finally, the remaining term,

$$\frac{1}{2} \sum \frac{g^2 v_i^4}{n_i u_i^2}$$

has a limiting distribution which is the same as that of

$$\frac{1}{2} \frac{g^2}{n} \sum \frac{\theta_i^2}{a_i}$$

From (i), this is the non-central $\chi^2$ with 1 d.f. and parameter $\mu^2/\sum 2\theta_i^2/a_i$. Thus the limiting distribution is essentially the same as that for $d$ and $F'$.

In conclusion, the power functions of the three tests were first obtained by G. S. Watson: earlier, preliminary work on the likelihood ratio test had been done by Dr. R. A. Porter. I should like to thank Miss Elizabeth Grant and Miss Janice Harris, who carried out the bulk of the computations.

REFERENCES

(1) Cochran, W. G. and Cox, G. M. *Experimental Designs*. John Wiley and Sons, p. 411, 1950.

(2) Anderson, R. L. Use of variance components in the analysis of hog prices in two markets. *Jour. Amer. Stat. Ass.* 42, 627, 1947.

(3) Comstock, R. E. and Robinson, H. F. The components of genetic variance in populations of biparental progenies and their use in estimating the average degree of dominance. *Biometrics*, 4, 254–266, 1948.

(4) Bartlett, M. S. Properties of sufficiency and statistical tests. *Proc. Roy. Soc. Lond.* A, 901, 268–282, 1937.

(5) Fisher, R. A. The fiducial argument in statistical inference. *Ann. Eugen.* 6, 391–398, 1935.

(6) Fisher, R. A. and Yates, F. *Statistical Tables*. Oliver and Boyd, Edinburgh, Table V1, 3rd. ed., 1949.

(7) Welch, B. L. The generalization of Student's problem when several different population variances are involved. *Biometrika*, 34, 28–35, 1947.

(8) Aspen, A. A. Tables for use in comparisons whose accuracy involves two variances, separately estimated. *Biometrika*, 36, 290–296, 1949.

(9) Smith, H. F. The problem of comparing the results of two experiments with unequal errors. *Jour. G.S.I.R.* (Australia) 9, 211–212, 1936.

(10) Satterthwaite, F. E. An approximate distribution of estimates of variance components. *Biometrics*, 2, 110–114, 1946.

(11) Welch, B. L. The significance of the difference between two means when the population variances are unequal. *Biometrika*, 29, 350–362, 1938.

(12) Cramér, H. *Mathematical methods of statistics*. Princeton Univ. Press, 254–5, 1946.

# 46

# Modern Methods in the Sampling of Human Populations

## General Principles in the Selection of a Sample *

### WILLIAM G. COCHRAN

*Professor of Biostatistics, School of Hygiene and Public Health, Johns Hopkins University, Baltimore, Md.*

SAMPLE surveys are being employed increasingly in public health research. When successful, they provide usable information about the characteristics of a large group of people quickly and moderately cheaply. They are particularly advantageous when highly trained personnel, limited in availability, are needed for some aspect of the study.

This paper discusses some of the principles involved in the selection of the sample. This is not always the most important or difficult part of a sample survey. But it is the part where persons inexperienced in sampling may feel least sure of their ground and most open to criticism by experts in sampling theory.

### ERRORS IN SAMPLING

Any estimate made from a sample is subject to error. Although there are many sources from which error may arise, we can distinguish four broad groups.

1. Sampling errors
   a. Errors arising from the sample as selected
   b. Errors arising from non-response
2. Errors of measurement
3. Errors in the preparation of estimates
4. Errors due to the fact that the population characteristics change with time

The first group (sampling errors) is

* Presented before the Statistics Section of the American Public Health Association at the Seventy-eighth Annual Meeting in St. Louis, Mo., November 2, 1950.

always present because we measure only a part of the population or universe. This category may be subdivided into two parts—errors arising from the sample as it was selected, and errors due to failure to measure the whole of the sample that was selected. As anyone who has conducted a sample survey will agree, it is no small task to obtain answers to all the questions from all the persons in the sample. The second group consists of errors of measurement, which face us whenever, for any reason, the answers recorded for some question are incorrect. Errors are also introduced (group 3) in the editing and tabulation of results and the preparation of estimates.

The fourth group requires some explanation. Samples are often taken, not in order to describe the present or past, but as a basis for future action. In considering whether to start a health program for a chronic disease, we might measure the incidence during the past year. What we really would like to know is the incidence in future years. Thus, even if a complete census is taken and the past year's incidence is known exactly, this figure still has an error for the purpose for which we want to use it, because incidence changes from year to year.

This classification brings out the point that the way in which the sample is selected will influence only one component, and not the whole, of the error. A survey may be ruined because of poor selection of the sample, but it may also be ruined even if the sample has been very well selected. Constant attention to all stages of the operation is needed. Indeed, it is advisable to try to foresee at what point any projected survey will be most vulnerable. It may often pay to take a smaller sample than we had hoped, in order to have the resources to concentrate on the reduction of some other source of error, e.g., non-response or error of measurement. Finally, in the present paper the reader should bear in mind that my subsequent remarks refer in general to the error due to the sample as selected, rather than to the total error.

### THE GUIDING PRINCIPLE

The principle that governs modern sampling practice is the familiar economic maxim that one should get the most for one's money. When considering various ways in which the sample might be drawn, we try to select that which will give the desired degree of precision as cheaply as possible. I shall describe some of the steps involved in an attempt to apply this principle.

### THE SPECIFICATION OF PRECISION

The first question that must be answered is: how small do we want the sampling error to be? This, of course, is related to the question: how precise would we like the estimate to be? The question is usually answered by specifying the standard error $\sigma$ which we wish the estimate to have. There are good theoretical reasons for expecting that most estimates in sample surveys will be approximately normally distributed. Thus, the chances are about 19 in 20 that the estimate will not differ from the true value for the universe by more than $2\ \sigma$, so that knowledge of the standard error permits us to compute limits of error for the estimate which we can be fairly confident are valid.

The specification of the standard error should be made by the persons who pro-

pose to use the results of the survey, and should depend on the precision needed for such use. There is often some arbitrariness and perhaps indecision in the specification, but it should be made as carefully as possible, because a demand for unnecessarily high precision involves a waste of resources, while poor precision diminishes the utility of the results.

In deciding the size of sampling error at which we will aim, account should also be taken of what is known about other sources of error. We might wish to estimate a morbidity rate correctly to within 4 per cent, and therefore choose a sampling standard error of 2 per cent. But perhaps the definition of this morbidity is so difficult and so little understood by the layman that even with a complete census the rate is unlikely to be correct to closer than 10 per cent. In this situation there is a case for increasing the desired sampling standard error, on the grounds that there is no point in having a very small sampling error if other errors will make the estimate imprecise.

### SIZE OF SAMPLE

The next step is to estimate, for any sampling method that is under consideration, the size of sample that is needed to reduce the standard error of the estimate to the chosen amount. This process will be illustrated for a simple random sample. By this is meant a sample drawn by numbering all the units in the population, mixing the numbers thoroughly, and picking numbers in succession until the desired sample size is reached. If $\sigma$ is the standard deviation for some measurement in the population, the standard error of the mean $y$ from a simple random sample is known from theory to be

$$(1) \qquad \sigma_y = \frac{\sigma}{\sqrt{n}} \sqrt{1 - \frac{n}{N}}$$

where $n$ and $N$ are the sizes of the sample and population respectively. Equation (1) can be solved to provide $n$ in terms of $\sigma_y$, the desired standard error for the sample mean.

Needless to say, this example oversimplifies the problem of determining sample size. There are nearly always a number of estimates to be made, each requiring a different sample size. Separate estimates may be wanted for different segments of the population, each with its specified limits of error. Further, the formula for $\sigma_y$ changes with the method of sampling. A more extended discussion of the issues is given by Mr. Cornfield.*

Two points may be noted. If the calculation of $n$ is to be feasible, we must have a formula which expresses the standard error of the estimate as a function of the sample size. The implications of this requirement will be discussed in a later section. Secondly, as is true for all methods of sampling, the formula for $n$ involves the population standard deviation $\sigma$. This must be estimated, either from previous data, or from what it is hoped is an intelligent guess. For anyone who engages regularly in sampling, it is worth while to record any information about population standard deviations that becomes available from the results of sample surveys. Such data, which tend to be overlooked as uninteresting, may constitute the most valuable information that can be had for improving future sampling.

### COSTS

At this point we are in a position to make some estimate of the cost of taking the sample. For an unfamiliar type of sampling some guesswork is usually involved, but with experience, accurate costing becomes easier. A common fault is to be too optimistic about field costs and to allow no margin for mistakes. It

is sometimes taken for granted that the interviewer wastes no time in finding an address, and that as soon as a bell is rung, the door is opened by a housewife eager to answer questions but with no desire to make any irrelevant remarks. Costs involved in tabulating, summarizing, and interpreting the results are sometimes forgotten.

### THE COMPARISON OF SAMPLING METHODS

When the cost of obtaining the desired degree of precision has been estimated for each sampling method that is under consideration, it remains only to select the method for which the cost appears likely to be least. Space does not permit a classification of the different methods of sampling, though some account of the principal devices will be given later. The number of methods that are operationally convenient for any particular survey is usually small. Further, by means of general rules developed from sampling theory, it is often possible to select the most economical of a whole group of sampling methods without detailed examination.

### PROBABILITY SAMPLING

In the approach outlined above it is necessary to have, for any sampling method, a formula which gives the standard errors of estimates made from the sample. This formula is used to calculate sample sizes when the sample is being planned. When the survey is finished it provides standard errors for the estimates that have been made.

Sampling error formulae are available only for samples selected so that every unit in the population has a known probability of getting into the sample. The simple random sample satisfies this condition, for it gives every unit in the population an *equal* chance of being included in the sample. The phrase *probability sampling* is applied nowadays to denote this type of sample.

Some sampling techniques which have

---

* Cornfield, Jerome. The Determination of Sample Size. *A.J.P.H.* 41, 6:654-661 (June), 1951.

been widely used do not yield probability samples. This happens when the sampling is confined to an isolated segment of the population, or to those units in the population that are thought by someone to be especially "typical," or to those that happen to be particularly easy to measure. These methods often appear attractive, since they may accumulate a large body of data cheaply. They should be avoided except where their convenience definitely outweighs the disadvantage of having no clear idea of the precision of the estimates. Their use might be defended in exploratory work if the main object is to find out whether a particular study is operationally feasible at all.

### THE FRAME

How do we draw a probability sample? Without resort to details, the general procedure is as follows. We need a subdivision of the population into sampling units. These units are numbered in some fashion, and the desired probability of inclusion is decided for each one. Then, usually with the aid of a book of random numbers, the draw is made so as to conform with the system of probabilities that has been assigned.

The drawing of a probability sample thus necessitates something equivalent to a listing of the sampling-units in the population. The word *frame* has been coined by the U. N. Subcommission on Sampling to denote this subdivision of the population. One of the first tasks in planning a sample is to find out whether a frame is already available. Sometimes this presents no difficulty, because there exists a listing of the population. But experience in sampling breeds a suspicious attitude toward lists. They are often found to be incomplete, or to involve duplication, or to contain inaccurate information, despite the most authoritative assurances to the contrary. Lists are, however, so convenient for sampling that it is worth while to make

every effort to utilize them, after revision and completion, or with supplementary sampling to cover gaps in the list.

In the absence of a suitable frame, one must be constructed. This may be expensive and time-consuming, but by use of ingenuity it often proves less of a burden than had been anticipated. For instance, the sampling units into which the population is divided in the frame need not be the same as the units which we wish to measure (generally called the *units of observation*). For a sample of an urban population, the list need not consist of persons: it may be a list of households, or of city blocks. The paper by Mr. Hansen and Mr. Hurwitz * on the method of area sampling illustrates how such devices are applied to human populations.

In the following sections some of the commonest techniques used in probability sampling are briefly described.

### STRUCTURE OF THE SAMPLING UNIT

The sampling unit is often an aggregate or cluster of the units of observation. Substantial savings are sometimes made in this way. For a given bulk of sample, i.e., a given number of completed questionnaires, a large sampling unit usually gives rather imprecise results because, speaking popularly, large units provide a rather "spotty" coverage of the population. On the other hand, a large unit usually enables a larger bulk of sample to be taken for the same cost, since travel and administrative costs tend both to be reduced. In a personal interview sample designed to cover the whole of the United States, a random sample of individuals, if it could be taken, would not be efficient, because too much time and money would go into traveling among individuals and too little into securing questionnaires. The county has been found fairly satisfactory as a

---

* Hansen, Morris H., and Hurwitz, William N. Some Methods of Area Sampling in a Local Community. *A.J.P.H.* 41, 6:662–668 (June), 1951.

sampling unit in such surveys. The general objective is to obtain by trial and error the most efficient balance between cheapness in field costs and statistical precision.

The structure of the unit also plays a part. If there is a choice, the unit should be set up so that there is the maximum variability *within* the unit. This automatically decreases the variability among units, which dominates the sampling error of the estimate.

### TWO-STAGE SAMPLING

If the sampling unit consists of a cluster of the units of observation, we need not measure *all* the units of observation that belong to a chosen sampling unit. The sampling unit itself can be regarded as a little population, which is enumerated by drawing a sample from it. This process has been called *subsampling* or *two-stage* sampling, the latter name being appropriate because the sample is drawn in two steps. The first is to draw a sample of the clusters. The second is to take each cluster that comes into the sample, and draw a sample from it. This necessitates the construction of a frame within the cluster, and the definition of a secondary, or subsampling unit. The process can obviously be extended.

Thus, in a sample of the urban population of the United States, the county might be the primary unit. Within the county, the town might serve as a secondary or sub-unit, a sample of towns being drawn from each selected county. Within the town, the city block might be a sub-sub-unit, and within that the household might be the final unit.

One advantage of multi-stage sampling is that it gives more flexibility in the planning. The number of counties, the number of towns per county, the number of blocks per town, and the number of households per block in the sample are all at our disposal, and can be chosen in an attempt to minimize costs for a specified precision. A wise choice demands considerable knowledge both of the variability within the population and of the field costs. This emphasizes the value of keeping such records whenever the opportunity occurs.

### STRATIFICATION

A very common device is to divide the population into sub-populations or *strata*. There may be several reasons for this. Sometimes statistics are to be published separately for each stratum, so that it is convenient to keep the corresponding samples separate. Sometimes different strata present quite different field problems that are best handled independently, as for example a stratum comprised of prisons versus one comprised of private homes.

Even when these reasons do not apply, stratification may produce an increase in precision. This occurs when we are able to form strata such that each stratum is homogeneous internally, although one stratum may differ markedly from another. The exact size of each stratum (i.e., the number of sampling units which it contains) must also be known. A sample is taken independently in each stratum. If the stratum is homogeneous, the sample mean should be close to the corresponding stratum mean. Since the sizes of the strata are known, the sample mean from each stratum can be given its proper weight in estimating a mean for the whole population.

The sampling rate (i.e., the proportion of units that are selected) may be varied from one stratum to another, with the possibility of a further increase in precision, which is sometimes large. Formulae giving optimum rates have been worked out for a variety of situations. Suppose, for example, that the cost of taking the sample can be represented by the simple formula

$$\text{Cost} = \Sigma\, c_i\, n_i$$

where $c_i$ is the cost per sampling unit in the ith stratum and $n_i$ the number of units sampled in that stratum. The optimum rate, $n_i/N_i$ for the ith stratum, is found to be proportional to $\sigma_i/\sqrt{c_i}$, where $\sigma_i$ is the standard deviation within the stratum. In other words, it pays to increase the sampling rate within a stratum if the stratum is highly variable, or unusually cheap to sample. The result presupposes random sampling within each stratum. A sample of this kind illustrates a probability sample in which each unit does not have the same probability of inclusion, though all units within the same stratum do so.

### SIMPLE RANDOM AND SYSTEMATIC SAMPLING

For the final operation of drawing the sample from a stratum, or the subsample from a primary unit, one of two methods is commonly employed. The first is to take a simple random sample, which has already been described.

Alternatively, having numbered the units, we may sample at a rate 1 in $k$ by drawing a random number $r$ between 1 and $k$, and selecting the $r$th unit and every $k$th unit thereafter until the end of the list is reached. Thus, if a part of the population contains 33 units, to be sampled at a 20 per cent, or 1 in 5, rate, we select a random number between 1 and 5, say 2. The sample consists of the units numbered 2, 7, 12, 17, 22, 27, and 32.

This type of sample, called a *systematic* sample, is easier to select than a simple random sample, particularly if the drawing must be done in the field. It is likely to be more accurate, if there is a trend in the measurements which follows the order of numbering. Like the random sample, it gives every unit the same probability of inclusion. The chief defect of a systematic sample is that one cannot obtain from its results any formula of general validity for the sampling error of the estimate. Various

approximations which are usually over-estimates are available.

This account of the techniques which lead to economical sampling is scanty and incomplete. The drawing of units with probabilities proportional to their sizes, and the choice of an efficient method of estimation are additional devices which deserve study. Recent books by Yates[1] and Deming[2] give an extensive discussion of the subject, with numerous illustrations.

### THE PROBLEM OF NON-RESPONSE

In conclusion, it may be appropriate to mention the problem of non-response, because failure to collect the data obviously nullifies at least part of the skill and effort devoted to the selection of the sample. Moreover, in sampling human populations the problem almost invariably turns up. Some of the interviewees are difficult to contact: others are unable or unwilling to give information. This is particularly so in surveys where the data desired are technical or troublesome.

Unfortunately, any sizable percentage of non-response makes the results open to question by anyone who cares to do so. Suppose that our field methods, if applied to the complete population, would fail to get information from 20 per cent of the population. Then we really know nothing about, say, an incidence rate in that part of the population, except that it lies between zero and 100 per cent. We may believe and assert that the rate cannot differ much from the rate in the part of the population which we were able to sample, but such assertions are unsubstantiated opinions instead of facts. One sometimes reads that non-response has been "allowed for" by substituting other individuals for those who were hard to find or non-coöperative. Substitutions rarely solve the problem. They merely increase the size of sample in that part of the population which is amenable to our field methods.

Persons new to sampling often neglect to make any plans for dealing with non-response, apparently assuming that it will be negligible. There is some excuse for this, since it is difficult to foresee how much non-response will be encountered. However, ability to cope with unexpected contingencies is a very important asset in sampling, because, with all due respect to the theory of probability, some unexpected contingencies are certain to occur.

If the non-response is due to difficulty in contacting the respondent or to unwillingness to coöperate, one attack is to concentrate on a random subsample of the cases, making every effort to obtain the data from them. Methods for calculating the best size for this subsample have been given by Hansen and Hurwitz;[3] see also Birnbaum and Sirken.[4] If the trouble lies in inability to give the information, a change in the questionnaire and type of approach is indicated. Sometimes there is no remedy, the fact being that people do not possess the information that we had hoped to get. Even in this case it is well to become aware of it early rather than late.

### REFERENCES

1. Yates, F. *Sampling Methods for Censuses and Surveys.* London: Charles Griffin and Co., 1949.
2. Deming, W. E. *Some Theory of Sampling.* New York: Wiley & Sons, 1950.
3. Hansen, M. H., and Hurwitz, W. N. The Problem of Non-response in Sample Surveys. *J. Am. Stat. A.* 41:517–529, 1946.
4. Birnbaum, Z. W., and Sirken, M. G. On the Total Error Due to Non-interview and to Random Sampling. *Internat. J. Opinion & Attitude Research* 4:179–191, 1950.

# 47

# IMPROVEMENT BY MEANS
# OF SELECTION

W. G. COCHRAN

JOHNS HOPKINS UNIVERSITY

## 1. Introduction

One of the principal techniques for improving quality is to select those members of a population that appear to be of high quality and to reject those that appear to be of low quality. Usually the selection is based on a number of measurements that have been made on the available candidates. In personnel selection the measurements are sometimes obtained by competitive examination, the hope being that persons who obtain high marks will have superior ability for performing the subsequent tasks. In a program for improving hogs, the choice of a sire for breeding may be made after a study of his own characteristics, for example, weight at 180 days, plus those of his first few offspring.

A common feature in most selection problems is that at the time of selection we cannot measure directly the quantity which we wish to improve. Thus when a promotion from one type of work to another is in question, for example from salesmanship to an administrative task, success in the old occupation may not guarantee success in the new one. Stated mathematically, the problem is to improve some quantity $y$ by means of indirect selection that is made from a group of tests or measurements $x_1, x_2, \ldots, x_p$.

The mathematical foundation of most of the work that has been done thus far is Karl Pearson's memoir [1] of 1902. His primary interest was in the effects of *natural* selection on correlation and variability. On the assumption that $y$ and the $x$'s follow a multivariate normal distribution, he gave some important theorems about the means, variances and correlations of the variates after a selection based on the values of certain of the $x$'s. Various applications of these and other results are dispersed in the literature on personnel selection and on plant and animal improvement [2], [3], [4].

The object of this paper is to present the principal mathematical results that are useful for setting up a selection program. This part is mainly expository in character, though a few results are given in a more general form than hitherto. In addition, we shall discuss some of the problems that are encountered when we come to apply the theory to selection in practice. Here there appears to be need for much further research.

## 2. Statement of the problem

We shall assume that $y$ is a continuous variate. This is not always the case in practice, since the object of selection is sometimes to draw out those who possess a specific attribute. The same general approach is valid whether $y$ is continuous or discrete, though the details differ. Before selection, the variates $y, x_1, x_2, \ldots, x_p$ are assumed to follow a distribution whose frequency function is $f(y, x_1, x_2, \ldots, x_p)$. The decision to accept or reject a candidate is to be made by an objective rule that can be unambiguously applied as soon as the values of the $x$'s are known for the candidate. That is, some region $R$ in the sample space of the $x$'s is chosen as the region of selection.

When we compare different rules for selection, it is natural to make the comparison subject to the condition that all rules operate with the same intensity of selection. In other words, if any rule is applied repeatedly to the parent population, it should in the long run select a fraction $a$ and reject a fraction $(1 - a)$ of the candidates. The size of the region $R$ is therefore $a$.

It will be supposed that the specific purpose of selection is to maximize the mean value of $y$ in the selected portion of the universe. This purpose has usually been taken for granted in applications, and seems a reasonable one to adopt, though cases can be imagined where a different objective would be more appropriate. For example, we might wish to maximize the probability that $y$ exceeds some value $y_0$. This would in general require a different mathematical treatment and lead to a different rule of selection.

To simplify the notation, the symbol $x$ will be used to denote the set of $p$ variates $x_1, x_2, \ldots, x_p$, and $dx$ to denote the product of their differentials $dx_1, dx_2, \ldots, dx_p$. Given the joint frequency function $f(y, x_1, x_2, \ldots, x_p) \equiv f(y, x)$, the problem is to find a region $R$ in the sample space of the $x$'s, such that

(1)
$$\frac{1}{a} \int_{-\infty}^{\infty} dy \int_R y f(y, x) \, dx$$

is maximized subject to the restriction

(2)
$$\int_R f_1(x) \, dx = a$$

where $f_1(x)$ is the joint frequency function of the $x$'s.

### 3. The optimum rule for selection

If the regression $\eta(x)$ of $y$ on the $x$'s exists, the optimum rule is to select all members for which

$$\eta(x) \geqq k$$

where the value of $k$ is chosen so that the frequency of selection is $a$. The result requires the assumption that the cumulative distribution function of $\eta(x)$ is continuous and strongly monotone, so that for any $a$, $(0 < a < 1)$, there is one and only one $k(a)$ for which

$$P\{\eta(x) \geqq k\} = a .$$

To prove the result, we write

$$f(y, x) = \phi(y|x) f_1(x) ,$$

where $\phi(y|x)$ is the conditional frequency function of $y$, given the $x$'s. The mean value of $y$ after selection is

$$\frac{1}{a} \int_{-\infty}^{\infty} dy \int_R y\phi(y \mid x) f_1(x) dx$$

$$= \frac{1}{a} \int_R f_1(x) dx \int_{-\infty}^{\infty} y\phi(y \mid x) dy$$

(3)
$$= \frac{1}{a} \int_R \eta(x) f_1(x) dx,$$

from the usual definition of a regression function. The problem is therefore to find a region $R$ which maximizes (3) subject to (2).

This problem is analogous to that of finding the best critical region $R$ for a test of significance of a null hypothesis $H_0$ against a single specified alternative $H_1$. In the analogy, $f_1(x)$ corresponds to $p_0$, the frequency function of a sample point given that $H_0$ holds, while the product $\eta(x)f_1(x)$ corresponds to $p_1$, the frequency function of a sample point given that $H_1$ holds. Neyman and Pearson [5] have shown that the best critical region $R_0$ is defined by

$$p_1 \geqq kp_0$$

where $k$ is chosen so that the region is of size $a$. The corresponding region in our problem is

(4)
$$\eta(x) f_1(x) \geqq kf_1(x), \quad \text{or} \quad \eta(x) \geqq k .$$

Although the argument seems to apply without change, it may be worth repeating the principal part of the proof. Let $R_0$ be the region defined by (4) and let $R_1$ be any other region of size $a$ in the sample space of the $x$'s. If the two regions have a common part, denote this by $R_{01}$.

Since both regions are of size $a$, it is clear that

(5)
$$\int_{(R_0 - R_{01})} f_1(x) dx = \int_{(R_1 - R_{01})} f_1(x) dx .$$

Now

$$\int_{R_0} \eta(x) f_1(x) dx = \int_{R_{01}} \eta(x) f_1(x) dx + \int_{(R_0 - R_{01})} \eta(x) f_1(x) dx$$

$$\geqq \int_{R_{01}} \eta(x) f_1(x) dx + \int_{(R_0 - R_{01})} kf_1(x) dx$$

$$\geqq \int_{R_{01}} \eta(x) f_1(x) dx + \int_{(R_1 - R_{01})} kf_1(x) dx ,$$

using (5). But in $(R_1 - R_{01})$, we have $k > \eta(x)$, so that

$$\int_{R_0} \eta(x) f_1(x) dx \geqq \int_{R_{01}} \eta(x) f_1(x) dx + \int_{(R_1 - R_{01})} \eta(x) f_1(x) dx$$

$$\geqq \int_{R_1} \eta(x) f_1(x) dx .$$

The equality will hold only if the region $(R_1 - R_{01})$ is empty, that is, if $R_1$ and $R_0$ coincide.

This result might of course be anticipated by elementary considerations. By a selection which operates entirely on the $x$'s we cannot hope to influence the individual variations of $y$ in arrays in which the $x$'s are fixed: the most that we can hope to do is to choose arrays in which the mean value of $y$ is relatively high. The result is a convenient one, since it implies that selection can be based on a single index by which the candidates are scored. The use of the regression as an index is well known for the case where all variates follow the multivariate normal distribution, but actually it does not require normality nor linearity of regression.

## 4. The gain in $y$ due to selection

We may choose the scales so that in the original population all variates have zero means. Since $E(y) = E(\eta)$ in the unselected population, it follows that $\eta$ also has zero mean. Hence the mean values of $y$ and $\eta$ after selection are the increases or gains in these variates due to the selection.

From (3) we see that the gain in $y$, $G(y)$, is the same as that in $\eta$. This result can be put in another form that is sometimes of interest. In measuring the gain in a variate, we often express it as a fraction of the standard deviation of the variate in the original population. This device converts the gain to a type of standard scale which is invariant under any linear transformation of the units in which measurements are recorded. In standard units,

$$\frac{G(y)}{\sigma_y} = \left(\frac{\sigma_\eta}{\sigma_y}\right)\frac{G(\eta)}{\sigma_\eta}.$$

But in the original population,

$$\mathrm{cov}\,(y\eta) = \int\int y\eta f(y, x)\,dy\,dx = \int \eta f_1(x)\,dx \int y\phi(y\,|\,x)\,dy$$

$$= \int \eta^2 f_1(x)\,dx = \sigma_\eta^2.$$

Hence $\rho_{y\eta} = \sigma_\eta/\sigma_y$ and we have

(6)                    $$\frac{G(y)}{\sigma_y} = \rho_{y\eta}\frac{G(\eta)}{\sigma_\eta}.$$

In standard units, the gain in $y$ is a fraction $\rho_{y\eta}$ of that in $\eta$, where $\rho_{y\eta}$ is the correlation coefficient between $y$ and $\eta$.

It would be interesting to have a simple expression which gives the gain due to indirect selection in terms of that due to direct selection on $y$ of the same intensity, but I have been unable to discover one. It is of course easy to show that indirect selection cannot be superior to direct selection. A very simple result which connects the two gains is obtained if the variates follow the multivariate normal distribution, as discussed in the next section.

## 5. Results when all variates follow a multivariate normal distribution

In this case, which has been assumed in most applications, the results can be made more specific. In particular, $\eta$ is a linear function of the $x$'s and is normally distributed in the original population. Hence

$$\frac{G(\eta)}{\sigma_\eta} = \frac{1}{a\sqrt{2\pi\sigma_\eta^2}} \int_k^\infty \eta\, e^{-\eta^2/2\sigma_\eta^2} d\eta = \frac{1}{a\sqrt{2\pi}}\, e^{-k^2/2\sigma_\eta^2} = \frac{z(a)}{a}$$

where $z(a)$ is the ordinate of the normal frequency function at the point $k/\sigma_\eta$ at which a fraction $a$ of the total area lies above the ordinate. This gives

$$(7) \qquad \frac{G(y)}{\sigma_y} = \rho_{y\eta}\, \frac{z(a)}{a}.$$

If we were able to select directly on $y$, the gain in $y$ would be $z(a)/a$. Thus the gain due to indirect selection is a fraction $\rho_{y\eta}$ of that due to direct selection with the same intensity of selection. The correlation $\rho_{y\eta}$ is the multiple correlation coefficient between $y$ and the $x$'s.

The following are the chief properties of the distribution of $y$ after selection.

*Frequency function:* For $y_1 = y/\sigma_y$

$$(8) \qquad f(y_1) = \frac{1}{a\sqrt{2\pi}}\, e^{-y_1^2/2} \int_{\frac{t-\rho y_1}{\sqrt{1-\rho^2}}}^\infty e^{-u^2/2} du$$

where $t$ is the point on the abscissa of the normal curve above which a fraction $a$ of the area lies, and $\rho$ denotes $\rho_{y\eta}$.

*Mean:*

$$(9) \qquad G(y) = \rho\, \frac{z}{a}\, \sigma_y$$

*Variance:*

$$(10) \qquad V(y) = \sigma_y^2 \left[ 1 - \rho^2\, \frac{z}{a}\left(\frac{z}{a}-t\right) \right]$$

*Correlation between $y$ and $\eta$:*

$$(11) \qquad \rho' = \rho \sqrt{\frac{1 - \frac{z}{a}\left(\frac{z}{a}-t\right)}{1 - \rho^2\, \frac{z}{a}\left(\frac{z}{a}-t\right)}}.$$

The frequency function is positively skewed to a marked degree if $\rho$ is high and $a$ is small: otherwise skewness is only moderate and the general appearance is similar to that of a normal curve. Both the variance and the correlation between $y$ and $\eta$ are reduced by the selection.

Table I gives numerical data on some properties of the distribution of $y$ after selection, for several values of $\rho$ and $a$. Before selection, $y$ is normally distributed with mean zero and unit standard deviation.

The values shown after selection are the mean, standard deviation, and the correlation with $\eta$. As the intensity of selection increases, the mean increases, the s.d. decreases (rather slowly), and the correlation between $y$ and $\eta$ decreases. The effects are in the same direction as $\rho$ increases, except that $\rho'$ increases with $\rho$.

## 6. Selection in two stages

Sometimes the measurements that seem useful for selection, because they are thought to be correlated with $y$, become available at different times. For example, in the selection of a hog as a sire, his weight at 180 days is known before we have records on the performance of his offspring, although the latter records seem more relevant to the purpose at hand. With dairy cows that are being selected for milk yield, each successive lactation provides new data. Consequently a selection program may involve repeated selections as more measurements accumulate.

TABLE I

PROPERTIES OF THE DISTRIBUTION OF $y$ AFTER
SELECTION IMPOSED ON $\eta$

| PER CENT SELECTED 100 $a$ | MEAN $\rho$ | | | S.D. $\rho$ | | | $\rho'$ $\rho$ | | |
|---|---|---|---|---|---|---|---|---|---|
| | .5 | .8 | .95 | .5 | .8 | .95 | .5 | .8 | .95 |
| 50 | .40 | .64 | .76 | .92 | .77 | .65 | .33 | .63 | .88 |
| 25 | .64 | 1.02 | 1.21 | .89 | .72 | .56 | .27 | .55 | .83 |
| 10 | .88 | 1.40 | 1.67 | .89 | .68 | .50 | .23 | .48 | .78 |
| 5 | 1.03 | 1.65 | 1.96 | .89 | .67 | .47 | .21 | .44 | .75 |
| 1 | 1.33 | 2.13 | 2.53 | .88 | .65 | .43 | .18 | .38 | .69 |

To consider selection in two stages, suppose that the variates $x_1, x_2, \ldots, x_q$, $(q < p)$ are known when the first selection is to be made, while the remaining variates $x_{q+1}, \ldots, x_p$ do not become known until the second selection is made. If the frequencies of selection $a_1, a_2$ at the two stages have been decided in advance, the optimum rule for selection is given by the previous theory. At the first selection, we use as a selection index the regression $\eta_1(x_1, x_2, \ldots, x_q)$ of $y$ on $x_1, x_2, \ldots, x_q$. This regression is, by definition,

$$\eta_1(x_1, x_2, \ldots, x_q) = \int \ldots \int y f(y, x) \, dy \, dx_{q+1} \ldots dx_p,$$

where the integration for $y$ and $x_{q+1}, \ldots, x_p$ extends over all the sample space. We select whenever $\eta_1 \geqq k_1$, where $k_1$ is chosen so that the frequency of selection is $a_1$.

At the second stage, the optimum index is the regression $\eta_2(x)$ of $y$ on all the variables, in that fraction $a_1$ of the sample space which remains after the first selection. Since, however, the first selection operates purely on the $x$'s and does not alter the frequency distribution of $y$ in arrays in which all $x$'s are fixed, $\eta_2(x)$ is exactly the same function as $\eta(x)$, the regression in the original population. We select whenever $\eta_2 \geqq k_2$, where $k_2$ satisfies the equation

$$a_2 = \frac{1}{a_1} \int_{\substack{\eta_1 \geqq k_1 \\ \eta \geqq k_2}} f_1(x) \, dx.$$

If all variates have zero means in the original population, the gain in $y$ due to the two stage selection may be written formally as

$$G(y) = \frac{1}{a_1 a_2} \int_{\substack{\eta_1 \geq k_1 \\ \eta \geq k_2}} y f(y, x) \, dy \, dx.$$

The extension to three or more stages of selection is easily made.

This statement of the problem is not very realistic for most applications. It is more likely that only the product $a_1 a_2$, that is, the desired frequency of survivors of both selections, would be decided in advance. For given $a_1 a_2$, the question as to the optimum values of $a_1$ and $a_2$ is often asked in practice. In any specific case, this question can be answered from the equations above by trial and error, inserting various values of $a_1$ and $a_2$ to see which give the greatest gain in $y$. There does not seem to be a useful general solution in functional terms.

Even this form of the problem may not be what is wanted. For a given value of the product $a_1 a_2$, the cost of a selection program may vary according to the values of $a_1$ and $a_2$. If we decide to retain a group of hogs until information on their progeny is obtained rather than to sell them, we have to feed them in the intervening period with perhaps no compensating increase in their saleable value. The desirability of reserving judgment on a dairy cow for several lactations will depend on the amount of profit which her milk yields. Thus two stage selection problems usually have to be considered in the light of a specific cost situation, with the object of maximizing the gain in $y$ for a given outlay.

As before, results become more definite if the variates follow a multivariate normal distribution. In this event, $y$, $\eta_1$ and $\eta_2$ are all normally distributed and jointly follow a trivariate normal. If the original multivariate normal distribution is given, the covariance matrix of this trivariate normal can be found. Since the gain in $y$ is determined solely by the parameters of the trivariate distribution, there is no loss of generality in confining discussion to the trivariate distribution. In the original population, we may assume that all variates have zero means and unit variances. The parameters $\rho_1$, $\rho_2$ and $\rho$ will denote the *simple* correlations between $y$ and $\eta_1$, $y$ and $\eta_2$, and $\eta_1$ and $\eta_2$, respectively.

The points of truncation $k_1$, $k_2$ satisfy the equations

$$(12) \qquad a_1 = \frac{1}{\sqrt{2\pi}} \int_{k_1}^{\infty} e^{-\eta_1^2/2} d\eta_1;$$

$$(13) \qquad a_1 a_2 = \frac{1}{2\pi \sqrt{1-\rho^2}} \int_{k_1}^{\infty} d\eta_1 \int_{k_2}^{\infty} e^{-[1/2(1-\rho^2)]\,\{\eta_1^2 - 2\rho\eta_1\eta_2 + \eta_2^2\}} d\eta_2.$$

They can be found from the tables of the univariate and bivariate normal [6], respectively.

We now find the gain in $y$ due to selection on $\eta_1$, followed by selection on $\eta_2$. Write

$$y = \beta_1 \eta_1 + \beta_2 \eta_2 + e,$$

where $(\beta_1 \eta_1 + \beta_2 \eta_2)$ is the multiple regression of $y$ on the $\eta$'s in the original population. If $E'$ denotes a mean value in the selected part of the universe, we have

$$E'(y) = \beta_1 E'(\eta_1) + \beta_2 E'(\eta_2) + E'(e) .$$

But the distribution of $e$ is independent of that of the $\eta$'s, so that $E'(e) = 0$. Hence

(14) $$G(y) = \beta_1 G(\eta_1) + \beta_2 G(\eta_2) ,$$

so that we need only find $G(\eta_1)$ and $G(\eta_2)$. It is simpler to find $G(\eta_1 - \rho\eta_2)$ as follows.

$$a_1 a_2 G(\eta_1 - \rho\eta_2)$$

$$= \frac{1}{2\pi\sqrt{1-\rho^2}} \int_{k_2}^{\infty} d\eta_2 \int_{k_1}^{\infty} (\eta_1 - \rho\eta_2)\, e^{-[1/2(1-\rho^2)]\{\eta_1^2 - 2\rho\eta_1\eta_2 + \eta_2^2\}} d\eta_1$$

$$= \frac{\sqrt{1-\rho^2}}{2\pi} \int_{k_2}^{\infty} e^{-[1/2(1-\rho^2)]\{k_1^2 - 2\rho k_1\eta_2 + \eta_2^2\}} d\eta_2$$

$$= \frac{\sqrt{1-\rho^2}}{2\pi} \int_{k_2}^{\infty} e^{-[(\eta_2 - k_1\rho)^2/2(1-\rho^2)] - k_1^2/2} d\eta_2$$

$$= (1-\rho^2)\left(\frac{e^{-k_1^2/2}}{\sqrt{2\pi}}\right)\left(\frac{1}{\sqrt{2\pi}}\int_{\frac{k_2-\rho k_1}{\sqrt{1-\rho^2}}}^{\infty} e^{-t^2/2} dt\right)$$

$$= (1-\rho^2)\, z(k_1)\, I\left(\frac{k_2 - \rho k_1}{\sqrt{1-\rho_2}}\right) = (1-\rho^2)\, z_1 I_2, \text{ say },$$

where $z$ and $I$ denote the ordinate and the incomplete area of the normal curve, respectively. A corresponding equation holds for $G(\eta_2 - \rho\eta_1)$. Solving the two equations, we find

(14.1) $$a_1 a_2 G(\eta_1) = z_1 I_2 + \rho z_2 I_1 \qquad a_1 a_2 G(\eta_2) = \rho z_1 I_2 + z_2 I_1 .$$

Hence, from (14)

$$a_1 a_2 G(y) = \beta_1 a_1 a_2 G(\eta_1) + \beta_2 a_1 a_2 G(\eta_2)$$

$$= (\beta_1 + \rho\beta_2)\, z_1 I_2 + (\rho\beta_1 + \beta_2)\, z_2 I_1 .$$

But

$$\rho_1 = \text{cov}\,\{y\eta_1\} = \text{cov}\,\{(\beta_1\eta_1 + \beta_2\eta_2)\,\eta_1\} = \beta_1 + \rho\beta_2$$

and similarly for $\rho_2$. This gives for the final result

(15) $$G(y) = \frac{\rho_1 z_1 I_2 + \rho_2 z_2 I_1}{a_1 a_2}$$

where it is to be noted that $\rho_1$, $\rho_2$ and $\rho$ are *simple* correlation coefficients in the original population, and

$$I_1 = I\left(\frac{k_1 - \rho k_2}{\sqrt{1-\rho^2}}\right); \qquad I_2 = I\left(\frac{k_2 - \rho k_1}{\sqrt{1-\rho^2}}\right).$$

If $y$ does not have unit standard deviation, the only change needed is to multiply the right side of (15) by $\sigma_y$.

In this proof no use has been made of the fact that $\eta_2$ is the population regression of $y$ on the $x$'s. The result therefore holds for two stage selection on *any* pair of variates $\eta_1$ and $\eta_2$, provided that $\eta_1$, $\eta_2$ and $y$ follow a trivariate normal. In practice the second selection is sometimes based only on those variates that were not available in time for the first selection, presumably because it is considered that there will be little further gain in bringing into the second index variates that have already been used.

If $\eta_2$ is the population regression of $y$ on the $x$'s, the proof can be shortened a little. From equation (6) of section (4)

$$\frac{G(y)}{\sigma_y} = \rho_{y\eta_2}\frac{G(\eta_2)}{\sigma_{\eta_2}} = \rho_2\frac{G(\eta_2)}{\sigma_{\eta_2}}.$$

This equation holds for any kind of selection based on the $x$'s, so that it applies to two stage selection. But from (14.1)

$$a_1 a_2 \frac{G(\eta_2)}{\sigma_{\eta_2}} = \rho z_1 I_2 + z_2 I_1.$$

Hence

(15.1)
$$\frac{G(y)}{\sigma_y} = \frac{\rho_2(\rho z_1 I_2 + z_2 I_1)}{a_1 a_2}.$$

This result is equivalent to (15). For if $\eta_2$ is the population regression, the partial regression of $y$ on $\eta_1$, holding $\eta_2$ constant, is zero. Hence $\rho_1 = \rho_2\rho$, which makes (15) reduce to (15.1).

For specific applications, (15) can be computed from tables of the univariate normal distribution. In the case of three stage selection $G(y)$ can be expressed in terms of functions of the univariate and bivariate normals: the area of the trivariate distribution is however needed for reading $k_3$. The results (15) have been given in another form by Perotti [7].

## 7. An application to plant selection

As mentioned previously, the formulae in the preceding section are most likely to be useful in connection with specific applications. An example may help to clarify the procedure. In a program for finding superior varieties of a crop, it is quite common to start with a large number of varieties. A replicated field trial is conducted each year for several years, and at the end of each year some varieties are discarded. Suppose that we have a two year program, and at the end of two years we wish to retain only 1/24 of the original number of varieties, so that $a_1 a_2 = 1/24$. The same number of plots is available for experimentation each year. It follows that if $a_1 = 1/a$, the varieties that survive to the second year can be tested in $a$ times as many replications as the varieties in the first year. The problem is to find the best value of $a_1$ or $a$.

We will assume that we are trying to improve only a single characteristic of the crop, such as yield per acre, and are oblivious to all others, though this is somewhat of an oversimplification. Let $y$ denote the true yielding ability of a variety. From the first year's experiment, we obtain an estimate $x_1 = y + e_1$, where $x_1$ is the observed mean yield of the variety and $e_1$ is the experimental error of $x_1$. At the end of the first year, a selection is made by means of $x_1$. For any variety which survives this, we obtain a second estimate $x_2 = y + e_2$ at the end of the second year. If experimental errors per plot are the same in the two years, we will have $\sigma_{e_2}^2 = \sigma_{e_1}^2/a$.

The variates $y$, $e_1$ and $e_2$ are assumed to be normally and independently distributed, with zero means. Consequently, $y$, $x_1$ and $x_2$ follow a trivariate normal. It is easy to verify that, apart from a constant factor, the multiple regression of $y$ on the $x$'s is $(x_1 + ax_2)/(a + 1)$. This can be interpreted as the unweighted mean of all available observations on the yield, the apparent weight $a$ arising because $x_2$ is based on $a$ times as many replicates as $x_1$. From the general rule, selection at the end of the second year should be based on $(x_1 + ax_2)/(a + 1)$. Thus we may take

$$\eta_1 = x_1 = y + e_1; \qquad \eta_2 = \frac{(x_1 + ax_2)}{a + 1} = y + \frac{e_1 + ae_2}{a + 1}.$$

If $u = \sigma_{e_1}^2/\sigma_y^2$, it is found that

$$\rho_1 = \rho_{y\eta_1} = \frac{1}{\sqrt{1 + u}}; \qquad \rho_2 = \rho_{y\eta_2} = \sqrt{\frac{a + 1}{a + 1 + u}};$$

$$\rho = \rho_{\eta_1\eta_2} = \sqrt{\frac{a + 1 + u}{(a + 1)(1 + u)}}.$$

The symbol $u$ measures the ratio of the error variance to the true genetic variance of a varietal mean in the first year. Calculations were made for $u = 1, 3, 15$ and $63$. These values make the correlations $\rho_1$, between $y$ and $\eta_1$, $0.707, 0.5, 0.25$ and $0.125$ respectively. The values of $a$ were $1, 2, 3, 4, 6, 8, 12$ and $24$. The method of computation is first to find $k_1$ and $k_2$ from equations (12) and (13) of the previous section. The gain in $y$ is then computed from equation (15). Results appear in table II.

TABLE II

GAIN IN $y$ FOR VARIOUS METHODS OF TWO STAGE SELECTION

| $a_1$ | $a_2$ | $u = 1$ $\rho_1 = .707$ | $u = 3$ $\rho_1 = .5$ | $u = 15$ $\rho_1 = .25$ | $u = 63$ $\rho_1 = .125$ |
|---|---|---|---|---|---|
| $1$ | $\frac{1}{24}$ | 1.745 | 1.352 | 0.733 | 0.375 |
| $\frac{1}{2}$ | $\frac{1}{12}$ | 1.867 | 1.507 | 0.858 | 0.452· |
| $\frac{1}{3}$ | $\frac{1}{8}$ | 1.902 | 1.592 | 0.948 | 0.501 |
| $\frac{1}{4}$ | $\frac{1}{6}$ | 1.936 | 1.637 | 0.996 | 0.532 |
| $\frac{1}{6}$ | $\frac{1}{4}$ | 1.947 | 1.649 | 1.035 | 0.564 |
| $\frac{1}{8}$ | $\frac{1}{3}$ | 1.935 | 1.630 | 1.032 | 0.572 |
| $\frac{1}{12}$ | $\frac{1}{2}$ | 1.867 | 1.529 | 0.970 | 0.547 |
| $\frac{1}{24}$ | $1$ | 1.511 | 1.069 | 0.534 | 0.267 |

The best selection intensity for the first year appears to be fairly independent of the relative amounts of genetic and experimental error variation, the optimum being in the neighborhood of $a_1 = 1/6$ in all cases. Since the maxima are moderately flat, the use of equal selection rates of approximately $1/5$ in both years would be a good simple rule.

The comparison of the optimum with the last line of the table is of interest. With $a_1 = 1/24$, all the selection is made in the first year, there being no need for a trial in the second year. When genetic variance is relatively low ($u = 15$ and $u = 63$), the optimum gain is about double that when $a_1 = 1/24$. This suggests that the return per unit of work is about the same in the two years. As genetic variance increases, there is a diminishing return from the second year's work. In the limiting case $u = 0$ where there is no environmental variance (not shown in table II) there would be no return from a second year's work, since the maximum attainable gain $z(a)/a$, in this case 2.138, would be reached by a selection in the first year.

Comparison of the optimum with the top line of table II is also of interest. When $a_1 = 1$, all the selection is made at the end of the second year. The comparison is therefore one of a single versus a two stage selection, for the same total number of plots. The increase due to two stage selection is worthwhile though not spectacular. Since the gains for $a_1 = 1$ could have been obtained in the first year by growing double the number of plots, it might be questioned whether the increase from two stage selection compensates for a year's delay. On this issue the mathematical assumptions are too simple for application to plant breeding practice, where the true yielding ability of a variety would be expected to vary to some extent from year to year. Intuitively, this suggests that the advantage of a two year trial would be greater than that revealed by the comparison above.[1]

## 8. The construction of selection indices

The preceding theory assumes a knowledge of the exact form of the joint frequency distribution of $y$ and the $x$'s. In practice, such knowledge is rarely available. Instead, the general functional form of the joint distribution is assumed, with certain parameters introduced into the expression in order to give some degree of flexibility to the assumptions. The values of these parameters are estimated from some initial data. In practically all applications with which I am familiar, the functional form assumed has been the multivariate normal distribution. In this section the construction of selection indices for this distribution will be described: comment is reserved for later sections.

In some applications it is possible to obtain an initial sample in which the values of $y$ and the $x$'s are measured. For instance, if a group of tests is to be used to select personnel for some type of work, we might have available, from past data, records which show the performance of a sample of people both in the tests and in the subsequent work. In this event the procedure is the familiar one of calculating the least squares regression of $y$ on the $x$'s from this sample. The sample regression function $Y = \sum b_i x_i$ is taken as the selection index, and used to score new candidates.

In plant and animal selection, on the other hand, the value of $y$ cannot be observed directly, and a more ingenious approach is needed. The situation is that $y$ is a linear function $\sum a_i \xi_i$ of the important genetic characteristics $\xi_i$ of the candidate, where the $a_i$ are known weights. In the example cited by Fairfield Smith [8], to whom this approach is due, the $a_i$ were determined by the relative economic values of improvements in the several characteristics. Alternatively, if we are interested only in a single characteristic, we take all $a$'s zero except one. The variates $\xi_i$ are presumed to follow a multivariate normal distribution. From experiments, we can

---

[1] An interesting discussion of the problem of single stage selection for sugar beets has been given by Y. Tang [13].

observe estimates $x_i$ of the $\xi_i$, where

$$x_i = \xi_i + \epsilon_i,$$

$\epsilon_i$ being the experimental error. The variates $\epsilon_i$ are also assumed to follow a multivariate normal distribution, and the joint distributions of the $\xi_i$ and $\epsilon_i$ are taken as independent of one another. These assumptions are sufficient to ensure that $y$ and the $x$'s follow a multivariate normal. Consequently, from our previous theory, the best selection index for $y$ that can be constructed from the $x$'s is the population regression of $y$ on the $x$'s.

Let $\tau_{ij}$, $\gamma_{ij}$, and $\epsilon_{ij}$ be the population covariances of $x_i x_j$, $\xi_i \xi_j$, and $\epsilon_i \epsilon_j$ respectively, so that

$$(16) \qquad \tau_{ij} = \gamma_{ij} + \epsilon_{ij}.$$

Then if $\eta = \sum \beta_i x_i$ is the population regression of $y$ on the $x$'s, the $\beta_i$ satisfy the equations

$$(17) \qquad \sum_i \beta_i \tau_{ij} = \operatorname{cov}(x_j y) = \sum_k a_k \gamma_{jk}.$$

The theory above requires a knowledge of the parameters $\tau_{ij}$, $\gamma_{jk}$. Estimates of these are obtained from an initial experiment. In this, we take a random sample of $n$ members of the population, for example, $n$ varieties of a crop, and conduct a replicated experiment in which all $p$ measurements $x_i$ are made on every plot. The design of the experiment may take various forms, but we will suppose that it is arranged in randomized blocks, with the following analysis of covariance between $x_i$ and $x_j$.

|  | d.f. | Mean square | Unbiased estimate of |
|---|---|---|---|
| Between members (varieties) | $(n-1)$ | $l_{ij}$ | $\tau_{ij} = \gamma_{ij} + \epsilon_{ij}$ |
| Experimental error | $(n-1)(m-1)$ | $e_{ij}$ | $\epsilon_{ij}$ |

This analysis is in terms of a varietal mean over the $m$ replications. For simplicity, we have supposed that the selection is to be made from these means, so that the experimental error of a mean is the quantity $\epsilon_i$ which enters into the theoretical argument. From the analysis we can substitute unbiased estimates of $\tau_{ij}$ and $\gamma_{jk}$ into (17). For the coefficients $b_i$ in the estimated selection index $\sum b_i x_i$, this gives

$$(18) \qquad \sum_i b_i l_{ij} = \sum_k a_k (l_{jk} - e_{jk}), \qquad \text{for } j = 1, 2, \dots, p.$$

Equations (18) are linear in the weights $b_i$, and their arithmetic solution proceeds by exactly the same methods as for a set of normal equations in least squares. As Bartlett [9] has pointed out, the sampling distribution of the $b_i$ appears to be much more complicated than in least squares, because the right hand side of (18) consists of a linear function of covariances. This means that in any discussion of the sampling errors of a selection index that is computed from an initial sample, the situation in which $y$ cannot be measured will require a separate investigation from that in which $y$ can be measured.

The equations are slightly more elaborate if the means $x_i$ from which the selection is to be made are based on $m'$ replicates, whereas the means in the initial experiment by which the index was constructed are based on $m$ replicates, $(m \neq m')$. The extension to this case has been given by Nanda [10]. In other applications, particularly in animal selection, a more complex analysis of covariance may be necessary to estimate the unknown covariances: on the other hand, the exact values of some of the correlation coefficients can be predicted from Mendelian theory, so that less remains to be estimated from the initial sample. The general structure of the equations of estimation is similar to that in equations (18), though the differences mentioned above become important in any investigation of sampling error theory.

## 9. The problem of constructing a best index from an initial sample

The procedure of substituting sample values for the unknown population parameters seems natural enough that it might be taken for granted as the obvious practical method for constructing a selection index. It does, however, raise the question: in what sense is a selection index $Y = \sum b_i x_i$, constructed from a sample, the "best" index? This is, of course, a more complex problem than the construction of a selection index when the values of the $\beta_i$ are given. I have not been able to find any rule for constructing a "best" sample index, and I am doubtful whether one exists, unless a very specialized meaning of the word "best" is adopted.

In particular, in the case where the $b_i$ are obtained from a least squares regression, the index $\sum b_i x_i$ does not have any obvious optimum properties, though this does not mean that an alternative which is superior can be found. In this connection, it may be noted that in practice we tend to act as if we distrust the least squares formula to some extent. It is a common procedure, starting with, say, 7 $x$-variables, to include in the index only those whose observed partial correlations with $y$ in the initial sample are "large" enough, in some sense of this term. Some investigators retain only those $x$'s whose partial correlations with $y$ are statistically significant: others prefer a more flexible rule. Although such procedures are sometimes justified on the grounds that we want to keep the index simple, it is also thought, I believe, that the inclusion of the rejected $x$'s would actually weaken the index, because the deleterious effect of sampling errors in the weights $b_i$ would more than offset any gain that there might be in the correlation with $y$. When the $\beta_i$ are known, on the other hand, the optimum rule shows that it pays to include any $x$ whose partial correlation with $y$ in the population differs from zero.

An introduction to the problem of finding a "best" sample index will be given for the case in which the initial sample is random, of size $n$, and provides data on $y$ and the $x$'s. An underlying multivariate normal distribution for $y$ and the $x$'s is assumed. Without loss of generality, we may suppose that *in the population* the only variable that is correlated with $y$ is $x_1$, and that all $x$'s are independently distributed with unit variance. Thus

$$(19) \qquad y = \beta_1 x_1 + e_1, \quad (\beta_1 > 0); \qquad \eta = \beta_1 x_1; \qquad \rho = \rho_{y\eta} = \frac{\beta_1}{\sigma_y}.$$

Consider any linear index $I = \sum w_i x_i$, where the $w_i$ are known numbers. This index is used repeatedly to select or reject new candidates, the $w$'s remaining unchanged throughout. For these new candidates, the variates $y$ and the $x$'s are presumed to follow the same multivariate normal from which the initial sample was drawn. Hence for the new candidates the joint distribution of $y$ and $I$ is a bivariate normal in which

$$(20) \qquad \operatorname{cov}(yI) = w_1\beta_1: \qquad \sigma_I^2 = \sum_{i=1}^{p} w_i^2: \qquad \rho_{yI} = \frac{w_1\beta_1}{\sigma_y\sqrt{\sum w_i^2}}$$

since the $x$'s are independently distributed with unit variance. Thus the average increase in $y$ due to selection on $I$ is

$$(21) \qquad G(y) = \rho_{yI}\frac{z(a)}{a}\sigma_y = \frac{w_1\beta_1}{\sqrt{\sum w_i^2}} \cdot \frac{z(a)}{a},$$

where the population over which the correlation is taken consists of an infinite number of selections made by the same index $I$.

If we are seeking the values of the $w_i$ which give the best index, we might be inclined to choose them so as to maximize $G(y)$ in (21). This approach is fruitless. From the extreme right side of (21), we see that the maximizing values are $w_i = 0$, $i \geq 2$, while $w_1$ can take any positive value. This solution confirms the general theorem that $x_1$, or any positive multiple of it, is the best index. But like the general theorem it requires knowledge of the $\beta_i$, since it could not be used unless we knew that $\beta_i = 0$, $i \geq 2$.

Evidently we cannot get to the heart of the problem by considering the repeated use of a fixed index calculated from a single initial sample. It seems necessary to consider a two stage population in which (i) initial samples are repeatedly drawn, (ii) from each sample an index $I$ is calculated by some rule and (iii) each index is then used repeatedly for selection. In such a population, the correlation $\rho_{yI}$ will follow some frequency distribution. A rule for constructing $w$'s which maximize the average $\rho$ of $\rho_{yI}$ might reasonably be considered the "best" rule, since it would also maximize the average $G(y)$ in equation (21). If we are willing to confine our attention to rules for which the $w_i$ are linear functions of the $y$'s in the initial sample, the frequency distribution of $\rho_{yI}$, though complex, does not look unmanageable, but I have not been able to express it in any form in which the consequences of different rules can be studied.

## 10. The loss due to the use of a least squares index

Some insight into the nature of the distribution of $\rho_{yI}$ in the two stage population can be obtained if $I$ is the least squares index $Y = \sum b_i x_i$. Consider the conditional distribution of $\rho_{yY}$ in initial samples for which the values of the $x$'s are fixed. For any fixed set of these $x$'s, $b_i$ is normally distributed. The mean value of $b_1$ is $\beta_1$ and the mean value of any other $b_i$ is zero. Since, from the right side of (20),

$$(22) \qquad \rho_{yY}\sigma_y = \frac{b_1\beta_1}{\sqrt{\sum b_i^2}},$$

it follows that $\rho_{yY}\sigma_y$ is distributed as the ratio of a normal variate to the square root of a noncentral quadratic form in normal variables.

From standard regression theory, the covariance matrix of the $b_i$ in the conditional distribution is $\sigma_e^2 S^{ij}$, where $S^{ij}$ is the inverse of the matrix $S_{ij} = \sum x_i x_j$, taken over the initial sample. Thus in the conditional distribution the $b_i$ have somewhat different variances, and are correlated to some extent.

Consider now an averaging of these conditional distributions over all possible initial samples. In this population all $b_i$ have the same variance $\sigma_e^2/(n - p - 1)$. For by a familiar transformation the quantity $S^{ii}$ may be written $1/S_{i \cdot jk} \ldots$, where $S_{i \cdot jk} \ldots$ is the sum of squares of deviations of $x_i$ from its linear regression on the $(p - 1)$ other $x$-variables. Since the $x$'s are normally and independently distributed, $S_{i \cdot jk} \ldots$ is distributed as $\chi^2$ with $(n - p + 1)$ degrees of freedom. But if $\chi^2$ has $\nu$ degrees of freedom, the average value of $1/\chi^2$ is known to be $1/(\nu - 2)$, from which the result follows. Further, in the unconditional distribution $b_i$ and $b_j$ may easily be shown to have zero covariance. The $b$'s are not normally distributed, the distribution of $(b_i - \beta_i)$ being similar to Student's $t$-distribution.

An approximation to the distribution of $\rho_{yY}$ may be obtained by regarding the $b_i$ as normally and independently distributed with the same variance $\sigma_e^2/(n - p - 1)$, where

$$(23) \qquad \rho_{yY} = \frac{b_1 \beta_1}{\sigma_y \sqrt{\sum b_i^2}}.$$

Write $z_i = b_i/\sigma_b$, so that the $z_i$ have unit variance. The mean value of $z_1$ is $\beta_1 \sqrt{n - p - 1}/\sigma_e$. But

$$(24) \qquad \beta_1 = \rho \sigma_y, \text{ from } (19); \qquad \sigma_e^2 = \sigma_y^2 (1 - \rho^2),$$

where $\rho$ is the population multiple correlation coefficient between $y$ and the $x$'s. Hence $E(z_1) = \rho \sqrt{n - p - 1}/\sqrt{1 - \rho^2}$.

Under these assumptions, the quantity

$$t = \frac{z_1 \sqrt{p - 1}}{\sqrt{\sum_2^p z_i^2}}$$

follows the noncentral $t$-distribution, with $(p - 1)$ degrees of freedom, and parameter $\tau = \rho \sqrt{n - p - 1}/\sqrt{1 - \rho^2}$. Thus from $(23)$,

$$(25) \qquad \rho_{yY} = \left(\frac{\beta_1}{\sigma_y}\right) \frac{z_1}{\sqrt{\sum_1^p z_i^2}} = \frac{\rho t}{\sqrt{t^2 + (p - 1)}}.$$

The form of the result $(25)$ is of some interest. It shows that the correlation between $y$ and $Y$ equals the correlation between $y$ and $\eta$, multiplied by a fraction which cannot exceed unity. The average value of this fraction therefore represents that fraction of the possible gain in $y$ (possible if the $\beta$'s were known) which will actually be attained. The quantities $n$ and $p$ enter into the result only in the form

$\rho\sqrt{n-p-1}/\sqrt{1-\rho^2}$. For a given number $p$ of $x$-variates it follows that the initial sample size $n$ must be much larger when $\rho$ is small than when $\rho$ is large if the same fractional loss in $G(y)$ is to be sustained.

We now give some calculations from which the size of the fractional loss in $G(y)$ can be estimated in specific cases. For $(p-1)$ degrees of freedom, the frequency distribution of the noncentral $t$ may be written

$$(26) \qquad f(t_1)\,dt_1 = \frac{e^{-\tau^2/2}}{\sqrt{\pi}\left(\dfrac{p-3}{2}\right)!}\sum_{r=0}^{\infty}\frac{2^{r/2}\left(\dfrac{p+r-2}{2}\right)!}{r!\,(1+t_1^2)^{(p+r)/2}}(t_1\tau)^r\,dt_1,$$

where $t_1 = t/\sqrt{p-1}$, $\tau = \rho\sqrt{n-p-1}/\sqrt{1-\rho^2}$, $(p-1) > 0$.

The variate in which we are interested is $z = t_1/\sqrt{1+t_1^2}$. A routine transformation gives

$$(27) \qquad \phi(z)\,dz = \frac{e^{-\tau^2/2}}{\sqrt{\pi}\left(\dfrac{p-3}{2}\right)!}\sum_{r=0}^{\infty}\frac{2^{r/2}\left(\dfrac{p+r-2}{2}\right)!}{r!}(\tau z)^r(1-z^2)^{(p-3)/2}\,dz.$$

From term by term integration, the mean value of $z$ is found to be

$$(28) \quad E(z) = \frac{\bar{\rho}}{\rho} = \frac{\tau}{\sqrt{2}}\,e^{-\tau^2/2}\frac{\left(\dfrac{p-1}{2}\right)!}{\left(\dfrac{p}{2}\right)!}\left\{1 + \frac{(p+1)}{(p+2)}\left(\frac{\tau^2}{2}\right)\right.$$
$$\left. + \frac{(p+1)(p+3)}{(p+2)(p+4)}\frac{1}{2!}\left(\frac{\tau^2}{2}\right)^2 + \cdots\right\}.$$

In table III the values of $E(z)$ are given for $p = 2, 3, 4$ and $5$, and a series of values of the parameter $\tau^2/2$. This seems the most useful form for a succinct presentation, since by interpolation the reader can compute $\bar{\rho}/\rho$ for any case in which he is interested.

*Example* 1. Suppose that a regression on 4 $x$-variates is computed from an initial sample of size 10. If $\rho = 0.6$, what is the expected correlation between $y$ and $Y$? In this case

$$\tfrac{1}{2}\tau^2 = \frac{1}{2}\frac{0.36}{0.64}(10-4-1) = 1.4\,.$$

By interpolation in the column $p = 4$, we find $\bar{\rho} = 0.64\rho$. Hence, on the average, the correlation between $y$ and $Y$ is $(0.64)(0.6) = 0.38$.

*Example* 2. In a selection index based on 5 $x$-variates, it is confidently expected that the true multiple correlation coefficient will be at least 0.7. An initial random sample for constructing the index is to be drawn. It is desired to take this large enough so that the fraction of the potential gain in $y$ that is lost through errors in the $b$'s will not exceed 5 percent. How large must $n$ be?

We want $\bar{\rho}/\rho$ to be at least 0.95. From table III it is clear that a larger sample is needed for $\rho = 0.7$ than for any higher $\rho$. Hence we make the estimate for $\rho = 0.7$. Since the entry 0.95 lies outside the limits of the table, we use the approximation in the footnote. We want

$$-\frac{1}{1+\dfrac{4}{\tau^2}} \geqq (.95)^2 = 0.9025 .$$

This leads to

$$\tau^2 = \frac{(.49)(n-6)}{.51} \geqq \frac{4}{\dfrac{1}{.9025}-1} = 37$$

and hence to $n = 45$.

To give more concrete results, the values of $\bar{\rho}/\rho$ are shown in table IV for initial samples of·sizes 10 and 30, and for $\rho = 0.9, 0.7, 0.5$ and $0.3$.

TABLE III*

VALUES OF $\bar{\rho}/\rho$

| $\frac{1}{2}\tau^2 =$ $\dfrac{\rho^2(n-p-1)}{2(1-\rho^2)}$ | $p =$ NUMBER OF $x$-VARIATES | | | |
|---|---|---|---|---|
| | z | 3 | 4 | 5 |
| 0.2 | .377 | .324 | .288 | .262 |
| 0.4 | .510 | .441 | .394 | .360 |
| 0.6 | .597 | .521 | .469 | .430 |
| 0.8 | .661 | .581 | .526 | .484 |
| 1.0 | .710 | .629 | .572 | .528 |
| 1.5 | .793 | .714 | .656 | .611 |
| 2.0 | .844 | .770 | .714 | .670 |
| 3.0 | .900 | .838 | .788 | .747 |
| 4.0 | .928 | .876 | .833 | .796 |
| 5.0 | .944 | .900 | .863 | .830 |

\* For values of $\frac{1}{2}\tau^2$ outside these limits, use the following approximations:

$$\left(\tfrac{1}{2}\tau^2 > 5\right): \qquad \frac{\bar{\rho}}{\rho} = \frac{1}{\sqrt{1+\dfrac{(p-1)}{\tau^2}}}$$

$$\left(\tfrac{1}{2}\tau^2 < 0.2\right): \qquad \frac{\bar{\rho}}{\rho} = \frac{\left(\dfrac{p-1}{2}\right)!}{\left(\dfrac{p}{2}\right)!}\frac{\tau}{\sqrt{2}}$$

The results illustrate the fact that when an index $Y$ is computed by multiple regression from a small sample, the correlation between $y$ and $Y$ is likely to be substantially less than the multiple correlation coefficient between $y$ and $\eta$. The gain in $y$ following selection on $Y$ is reduced in the same ratio. The loss in correlation increases with the addition of each independent variate, and when measured as a fraction of $\rho$, as in table IV, it increases rapidly as $\rho$ becomes small. Since the results are derived from an approximation to the distribution of $\rho_{yY}$, not too much reliance can be placed on individual figures. But there seems little doubt that in small samples the addition of an extra $x$-variable to the index will not increase the gain due to selection unless it produces at least a moderate increase in $\rho$.

The results also suggest, as would be expected, that the decrease in the improvement in $y$ can be avoided by the choice of an initial sample which is large enough.

When $\rho$ is as high as 0.9, the initial sample may be quite small, at least if the index contains no more than 5 $x$-variables. Tables III and IV enable rough estimates to be made of the size of sample needed in specific cases. It should be noted that these tables apply to a *random* sample of size $n$. It is not uncommon to choose an initial sample in which the variation among the $x$'s is substantially larger than that in a random sample of the same size. The purpose of this device is to decrease the sampling errors of the $b$'s, and its effect is to make the appropriate size of sample for reading tables III and IV larger than the actual size.

TABLE IV

VALUES OF $\bar{\rho}/\rho$ FOR INITIAL SAMPLES OF SIZE 10, 30

| $\rho$ | $n = 10$ $p =$ No. of $x$-Variates | | | | $n = 30$ $p =$ No. of $x$-Variates | | | |
|---|---|---|---|---|---|---|---|---|
| | 2 | 3 | 4 | 5 | 2 | 3 | 4 | 5 |
| 0.9 | .98 | .96 | .93 | .90 | 1.00 | .99 | .99 | .98 |
| 0.7 | .91 | .83 | .74 | .66 | .98 | .96 | .94 | .92 |
| 0.5 | .74 | .63 | .54 | .45 | .94 | .88 | .84 | .80 |
| 0.3 | .67 | .38 | .31 | .26 | .77 | .67 | .61 | .56 |

## 11. The effect of discarding variates from the index

As has been mentioned, the practice of discarding from the index those $x$-variates which appear to have little partial correlation with $y$ may be regarded as an attempt to avoid some of the decrease in correlation which results from errors in the weights $b_i$. The practice seems obviously sound if we can be sure that the initial sample informs us correctly which variates to discard. In a small initial sample, however, the sample partial correlations may not be close to the corresponding population correlations, and the process of discarding is itself subject to errors. Precise information about the effects of such errors on $\rho_{yY}$ would be worth having. In particular, it is relevant to discover whether the reduction in $G(y)$ through errors in discarding is as great as the reduction incurred if we do not discard. Such an investigation encounters intricate mathematics. Some results will be given for a simple case in which the analysis is not difficult.

We adopt the same mathematical framework as in the previous section. That is, in the two stage population the $b_i$ are assumed as an approximation to be normally and independently distributed with variances $\sigma_e^2/(n - p - 1)$. Only $x_1$ is correlated with $y$, so that all $\beta$'s except $\beta_1$ are zero. The index $Y$ is to contain only one $x$. From any initial sample we select that $x_i$ for which the corresponding $b_i$ is greatest.

This method of discarding differs from methods that are common in practice in three respects: (i) we retain only one $x_i$, whereas all $x$'s which have significant partial correlations with $y$ are usually retained; (ii) in any specific initial sample, the conditional variances of the $b_i$ will not all be the same, and it would be more customary to retain that $x_i$ for which $b_i$ gives the highest value of Student's $t$; (iii) in our problem we either retain an index $x_1$ which is actually the optimum index, or one which is of no value at all for selection, whereas in practice the choice would probably lie among a number of variates each of which was of some value, although none was optimum. These deviations from practice were accepted in order to simplify the analysis. It does not seem that they distort the essentials of the problem unduly.

The frequency distribution of $\rho_{yY}$ has only two values, $\rho$ if $x_1$ is chosen and 0 otherwise. Hence we need consider only the probability that $x_1$ is chosen, that is, the probability that $b_1$ is the greatest (algebraically) of the $b$'s. As before, write $z_i = b_i \sqrt{n - p - 1}/\sigma_e$, so that the $z_i$ have unit variances. The mean value of $z_1$ is $\tau = \beta_1 \sqrt{n - p - 1}/\sigma_e$, or $\rho \sqrt{n - p - 1}/\sqrt{1 - \rho^2}$. All other $z$'s have zero means.

The method of calculating the probability $P$ that $z_1$ is the greatest, which was suggested by J. W. Tukey, was to use the formula

$$P = 1 - (p - 1)P_i + \frac{(p - 1)(p - 2)}{2!} P_{ij} - \frac{(p - 1)(p - 2)(p - 3)}{3!} P_{ijk} + \ldots$$

where $p$ is the number of $x$-variates from which the winner is chosen, and $P_{ijk}$, for example, is the probability that any three *specified* variables $x_u$, $u \geq 2$, will exceed $x_1$. By symmetry, this probability is the same for any choice of the three. $P_i$ is read from tables of the univariate normal distribution. $P_{ij}$ is obtained by noting that

TABLE V

MEAN VALUES OF $\bar{\rho}/\rho$ FOR TWO METHODS OF CONSTRUCTING AN INDEX

| | ONLY THE "BEST" VARIABLE RETAINED | | | | | | ALL $x$-VARIABLES RETAINED | | | | | |
| | $n' = 8$ $p$ | | | $n' = 32$ $p$ | | | $n' = 8$ $p$ | | | $n' = 32$ $p$ | | |
| $\rho$ | 2 | 3 | 4 | 2 | 3 | 4 | 2 | 3 | 4 | 2 | 3 | 4 |
|---|---|---|---|---|---|---|---|---|---|---|---|---|
| 0.9 | 1.00 | 1.00 | 1.00 | 1.00 | 1.00 | 1.00 | .98 | .97 | .96 | 1.00 | 1.00 | .99 |
| 0.7 | .98 | .95 | .94 | 1.00 | 1.00 | 1.00 | .92 | .87 | .83 | .98 | .97 | .95 |
| 0.5 | .88 | .80 | .72 | .99 | .98 | .97 | .77 | .69 | .63 | .95 | .91 | .87 |
| 0.3 | .74 | .60 | .49 | .89 | .82 | .76 | .51 | .44 | .39 | .80 | .72 | .67 |

$(z_2 - z_1)$ and $(z_3 - z_1)$ follow a bivariate normal with means $- \tau$, variances 2, and correlation $+\frac{1}{2}$. The probability that both variates exceed zero is read from tables of the bivariate normal distribution. The higher $P$'s necessitate numerical integration.

The probability $P$ is shown on the *left* side of table V for $p = 2, 3, 4$; $\rho = 0.9$, 0.7, 0.5 and 0.3; and $n' = (n - p - 1) = 8, 32$. Since $\rho_{yY}$ takes only the two values $\rho$ and 0, the quantity $P$ is also the mean value of $\bar{\rho}/\rho$, the same quantity as tabulated in tables III and IV. Fixed values of $n'$ rather than $n$ were used for convenience in calculation.

For the higher values of $\rho$, there is relatively little chance of failing to pick the best variate even with $n' = 8$, that is, with sample sizes of the order of 12. For $\rho = 0.5$ there is an appreciable loss in correlation with the smaller sample, but practically none with the larger sample. For $\rho = 0.3$ there is some loss even with the larger sample, which has about 36 observations.

The right side of table V shows the average value of $\bar{\rho}/\rho$ when all $x$-variates are retained, as calculated by the approximate method given in the previous section. So far as it goes, the comparison supports the practice of attempting to discard the $x$-variates that do not seem to contribute to the correlation, since in all cases in which there is any appreciable loss, it is greater on the right side of the table.

## 12. Further problems

There is evidently much to be learned about the properties of a least squares index computed from an initial sample, as it affects the gain in $y$ that may be expected from the use of the index for selection. The analysis should also be undertaken for indices that are computed by the use of estimated components of variance. The work of Bartlett [9] and Nanda [10] on this problem indicates its complexity and shows that the effective size of the initial sample is determined mainly by the number of varieties, rather than by the amount of replication for each variety. In this section a few additional problems are mentioned.

One concerns selection from nonnormal populations. The general theorem on optimum selection does not require normality. The two principal results which it provides are (i) $\eta(x)$ is the best selection index and (ii) the gain in $y$ is equal to that in $\eta$. Consequently, in setting up a selection program in a population that is nonnormal, we should attempt to find out the shape of the regression $\eta(x)$ of $y$ on the $x$'s, and to study the frequency distribution of $\eta(x)$ in the unselected population. Although the formula $z(a)\sigma_\eta/a$ for the gain in $\eta$ due to selection holds only if $\eta$ is normally distributed, the correct formula is easily found if the frequency distribution of $\eta$ is known.

In view of the widespread assumption of normality in applications, an investigation of the consequences of this assumption in nonnormal populations would also be worthwhile. In general, a linear index will not be the best index, and predictions of the expected gain in $y$, based on normal theory, are likely to be in error. Unfortunately it cannot be taken for granted that a moderate departure from normality will have little effect. This may be so if selection is not intense and $y$ has only a small correlation with the $x$'s, so that progress is slow. But in intense selection the gains depend primarily on the shapes of the tails of frequency distributions. As is well known, a frequency curve which looks quite similar to the normal curve may differ greatly in its tail. A combination of theoretical investigations with sampling experiments on natural populations is suggested.

Secondly, how accurately can the gain due to selection be estimated from given initial data? This question is important for policy making in plant and animal improvement, particularly at the present time, when the prospects of a steady increase in the world production of food are the subject of much study. Often, a program of selection is only one of a number of feasible means for improving quality or quantity, and its expected gains must be compared with similar estimates for other approaches. For a multivariate normal population, the expected gain in $y$ is $\rho_{y\eta}\sigma_y z(a)/a$. The standard error of the estimate of this quantity from an initial sample has been given by Nanda [10], following earlier work by Bartlett [9], for the type of estimation that arises in plant selection. For practical purposes this standard error must be regarded as a lower limit to the effective error, since disturbances due to nonnormality and to time changes in the population undergoing selection will presumably be present. Moreover, the estimated gain itself is the gain that would be attained if the true regression were known, and some reduction in this estimate to take account of sampling errors in the index may be required.

A third problem of a more specialized type arises because it is not always practicable to use a selection index in the best way. The optimum rule is to select all candidates for which $\eta > k$. When selection is made from small samples to fulfill some specific purpose, the number of candidates for which $\eta > k$ will vary from sample to sample. In some cases, however, each sample must provide a known quota of successful candidates. Consequently, we impose the restriction that the number selected from the $i$-th sample is to be $r_i$. This restriction decreases the expected gain in $y$ and changes the mathematical aspects of the problem. The changes are easily made if there is only a single stage of selection. For the multivariate normal case, with $r_i$ constant, the result is to replace the factor $z(a)/a$ by the factor $a(r, n)$, which is the average value of the largest $r$ out of a standardized normal sample of size $n$. The extension to two stage sampling presents difficulties.

In this paper we have considered that the purpose of selection is to maximize the mean value of $y$ while retaining a specified fraction $a$ of the members of the original population. Birnbaum [11] and Birnbaum and Chapman [12] investigate the related problem of maximizing the fraction of the population that is retained, subject to the condition that the mean value of $y$ in the selected universe has some preassigned value. For a multivariate normal population, they show that truncation by means of the linear regression of $y$ on the $x$'s is optimum for this problem also.

In conclusion, these problems may leave the impression, not incorrectly, that more issues have been raised than solved in this paper. The topic of selection appears to be one where the applications have run somewhat ahead of their theoretical basis, and it may be anticipated that any new advances in theory will quickly be utilized.

## REFERENCES

[1] K. Pearson, "On the influence of natural selection on the variability and correlation of organs," *Roy. Soc. London Phil. Trans.*, A, Vol. 200 (1903), pp. 1–66.

[2] J. L. Lush, "Family merit and individual merit as basis for selection," *Amer. Naturalist*, Vol. 81 (1947), pp. 241–261 and 362–379.

[3] G. E. Dickerson and L. N. Hazel, "Effectiveness of selection on progeny performance as a supplement to earlier culling in livestock," *Jour. Agr. Res.*, Vol. 69 (1944), pp. 459–476.

[4] W. T. Federer and G. F. Sprague, "A comparison of variance components in corn yield trials," *Jour. Amer. Soc. Agronomy*, Vol. 39 (1947), pp. 453–463.

[5] J. Neyman and E. S. Pearson, "On the problem of the most efficient tests of statistical hypotheses," *Roy. Soc. London Phil. Trans.*, A, Vol. 231 (1933), pp. 289–337.

[6] K. Pearson, *Tables for Statisticians and Biometricians*, Part 2, Cambridge University Press, Cambridge, 1931.

[7] J. M. Perotti, "Mean improvement in a normal variate under direct and indirect selection," M. Sc. Thesis, Iowa State College, 1943.

[8] H. Fairfield Smith, "A discriminant function for plant selection," *Annals of Eugenics*, Vol. 7 (1936), pp. 240–250.

[9] M. S. BARTLETT, "The standard errors of discriminant function coefficients," *Jour. Roy. Stat. Soc., Suppl.*, Vol. 6 (1939), pp. 169–173.

[10] D. N. NANDA, "The standard errors of discriminant function coefficients in plant breeding experiments," *Jour. Roy. Stat. Soc.*, B, Vol. 11 (1949), pp. 283–290.

[11] Z. W. BIRNBAUM, "Effect of linear truncation on a multinormal population," *Annals of Math. Stat.*, Vol. 21 (1950), pp. 272–279.

[12] Z. W. BIRNBAUM and D. G. CHAPMAN, "On optimum selections from multinormal populations," *Annals of Math. Stat.*, Vol. 21 (1950), pp. 443–447.

[13] Y. TANG, "Certain statistical problems arising in plant breeding," *Biometrika*, Vol. 30 (1938), pp. 29–56.

AN APPRAISAL OF THE REPEATED POPULATION
CENSUSES IN THE EASTERN HEALTH DISTRICT,
BALTIMORE . . . . . . . . William G. Cochran

T HE series of population censuses of the Eastern Health District of Baltimore was initiated in 1933, primarily through the interest of the late Dr. Frost. Their major purpose was to provide a background for studies on the epidemiological characteristics of chronic disease in family units. Since we now have over 15 years experience of the uses to which such data can be put, or at least have been put in the Baltimore studies, I propose to devote most of my time to an appraisal of the scope, and some limitations, of the data. A brief description of the content of the data will first be given; a more detailed account by Dr. Fales has appeared recently. (1)

## The Data

It was recognized at the outset that for analytical purposes the ideal would be a continuous register of the families in the area. This was not attempted because of the difficulties and expense in keeping such a register tolerably accurate. Instead, repeated censuses at three-year intervals were planned as a useful approximation to a continuous register. The war interrupted the series and left a gap of eight years between the 1939 and 1947 censuses.

In the 1933 and 1936 censuses, the area under study comprised Wards 6 and 7, containing about 43,000 white and 12,000 Negro persons. In the 1939 and 1947 censuses the area was enlarged, following an enlargement of the Eastern Health District, to contain slightly over 100,000 persons. In addition, a census of Ward 7 made in 1922 is available from an earlier survey by Doull.

Compared to the population of the City of Baltimore, that of the District is younger, has more foreign born, and is more homogeneous. (2) Relative to the City, the white population of the District is deficient in the professional and "white-collar" classes, and has a greater proportion of skilled and semi-skilled workers. The Negro population of the District has more unskilled workers and a slightly smaller proportion of members of the servant class than are found for the Negroes in Baltimore. On the whole, however, comparisons reveal the District as fairly representative rather than atypical of the City.

Information is collected on a household basis. The household is defined as a group of persons who live together as a single unit in the same dwelling, whose sleeping quarters are under the supervision of one person, and whose food is served at the same table from a common kitchen. For each individual in the household, the age, race, sex, marital status, relationship to the head of the household and occupation are determined, and also (since 1936) education, employment status, and date of entering the household. Characteristics pertaining to the household itself are the number of rooms, owned or rented status, bath and toilet facilities and (since 1936) date of establishment of the household and rent paid. Additional questions relating to disease and medical care are added, but the above constitute the basic demographic characteristics. The term "household" and "family" have usually been used synonymously; when desired, the family may be restricted to those having blood relationship with the head.

An important feature of the data is that each individual and each family is matched back to all previous censuses. Thus the data are in the nature of a follow-up of the individual or family, insofar as they are living in the District at the times when censuses are taken.

## The Principal Types of Use of the Data

1. *As a Source of Samples for Special Studies.* For special studies which are initiated soon after the completion of a census, the census data enable the composition of the sample to be controlled on any of the characteristics measured in the census. This helps in generalizing the sample results to the District as a whole. There are also practical advantages which come from familiarity with the population. When a new study is being planned, we can draw on past experience as to the amount of cooperativeness to be expected from

the respondents, the amount of evening work necessary to find respondents at home, and other factors which influence the cost and conduct of a study. These advantages are diminished as the interval between censuses increases, because knowledge of the District in inter-censal years is weakened by migration.

One example of a sampling procedure is the study by Gurney Clark and Turner. (3) From the 1939 census they abstracted the names of all Negroes in the age groups 20–24 and 35–39 and estimated the prevalence of syphilis in the two groups. Another example is the sample of thirty-five blocks containing white occupants, which formed the basis of the comprehensive studies of illness by monthly visits during the five year period following 1938. (4)

2. *As a Source of a "Control" Group.* Many questions of wide interest in public health revolve around the study of some particular group, e.g., those having some specific diagnosis, or method of treatment, or occupation. We are all familiar with lengthy, difficult, and expensive studies of this kind which have failed to produce any reliable conclusions because of the absence of data for a comparable group who did not possess this particular attribute. Even when the need for a control group is realized and steps are taken to secure one, it may be discovered, when the study has been in progress for some time, that the control group chosen will not serve the purpose. Sometimes this becomes apparent only when the results appear in print and are exposed to criticism from outside.

One use of the Eastern Health District material which has not been sufficiently exploited is in the construction of controls. For any special group in which we are interested in Baltimore (e.g. syphilitics), there are good resources for identifying those members who reside in the District and for obtaining the census data about them. If the study group is restricted to syphilitics who have been identified in the District, knowledge of the population of the District greatly facilitates the assembly of a control group, taken from the District, in whose comparability we can have some confidence. The relative homogeneity of the District, which is a disadvantage in attempting inferences to Baltimore as a whole, is a real advantage for this purpose. Both the control and study groups should be defined at the beginning of the investigation, and both should be followed through-out the investigation, unless there is a special and important reason why the following of the control group would distort the object of the study. This approach does not remove all the difficulty from "constructed" controls, whose comparability can never be entirely free from question, but it does give a grip on the problem.

3. *Longitudinal Demographic Studies.* These are made possible by the matching of individuals and families back to previous censuses. One direct application is to determine the percentage of families or of individuals who remain at the same address or within the District for varying lengths of time, and to compare the characteristics of permanent with migratory families. Studies of this type have been made by Luykx (5), Valaoras (6) and others. These studies are restricted to the set of time intervals provided by the censuses. In a more intensive study of migration rates, Rider and Badger (7) used the data obtained at monthly intervals in the five-year illness surveys.

For families or individuals who are identified in two or more censuses, the changes in their characteristics with time can be examined. For instance, Guralnick and Fales (8) studied occupational stability for white men identified in the 1939 and 1947 censuses. Studies of changes in the *family,* as distinct from the individual, have not been prosecuted as vigorously as might have been expected, in view of their close relation to the original purpose of the censuses. Recently, however, an investigation has been started by Taback in which it is hoped to describe the changing pattern of family structure with respect to size, composition, household density, and socioeconomic status during its natural lifetime. This analysis may throw some light on the problem of finding a relevant classification of families for demographic and epidemiological studies, although it is realized that the appropriate classification will change with the type of study.

4. *For Methodological Studies of Sampling.* Finally, any completed census of an area provides data for estimating the precision obtained from different types and intensities of sampling. The theory of sampling is well developed for single, cross-sectional

surveys, but relatively little exploration has been made of different techniques for repeated sampling. Part of the reason has been the lack of census data, at frequent intervals, on which different methods may be tried in the statistical laboratory. So far as I know, the only study of this kind within the District is Densen's examination of the consistency of reports of age on two different occasions (9), which is concerned with measurement rather than sampling as such.

*Summary.* Following their original purpose, the censuses make available a body of data about the composition, socio-economic status, and permanence of residence of families, as background material from which samples may be drawn for cross-sectional or longitudinal studies of special groups. In addition, there are two by-products. One is material for longitudinal demographic studies of the individuals or families who appear in more than one census: the other is the potential utility of the data in research on the methodology of sampling.

### SOME LIMITATIONS AND DIFFICULTIES

1. *Restriction of Long-Term Studies to a Non-Migrant Group.* In any longitudinal study which is confined to a specific geographical area, the families or persons who appear in the records over a long period of years consist of those who remain resident in the area. This is true both in demographic studies that are made from the matched census data themselves, and in special illness studies that are undertaken on a sample drawn from the District.

This property of the data may not introduce any serious difficulty in relatively short-time studies in which, say, the prevalence of chronic disease is estimated at intervals during a year. Indeed, from the first year's records in the monthly surveys of illness, Downes (4) was able to compare moving with non-moving families as to illness rates, composition, and economic status, and pointed out that comparisons of this type were one of the advantages of a sample which was composed of houses. She found that illness rates were generally higher for the moving families, the differences being greater than might be expected from their relative composition and economic status.

In a long-term study, on the other hand, the group of families who have been under observation for $T$ years shrinks steadily as $T$ increases, and becomes more and more selective for permanence of residence. This difficulty, of course, plagues any longitudinal study which does not have the resources to follow people wherever they move. In the District we have some advantages. One is that we can estimate the rate at which the sample will melt away. From experience thus far it appears that roughly two-thirds of the families remain after five years, one-half after eight years, two-fifths after fourteen years and one-fifth after twenty-five years.

In the analysis of data of this kind over a period of $T$ years, two methods have been commonly employed. One is to restrict the analysis to those families which have been under observation for all $T$ years. This method is the simplest, but results are restricted to the permanent families and the size of sample is diminished accordingly. The other method is to apply life-table techniques based on person-years of experience, so that each family is utilized for as many years as it has been under observation. This approach brings into the analysis all available data, but the sample to which it applies is a mixture of moving and permanent families in which the proportions in the mixture change steadily with time. If the initial sample is large enough, it should be possible to go a little further and compare the illness experiences of permanent and moving families. For instance, illness experience during the first three years of follow-up could be compared for families which remained under observation three, four, five . . . years. From these comparisons we could speculate about the illness experience of a group composed of the proper proportion of moving and permanent families. Some arbitrary assumptions would have to be made, but this analysis would give some idea of the magnitude of the effect of the selectivity towards permanent families inherent in the data.

In the censuses themselves, I do not think that much can be accomplished to avoid this selectivity in the data, since the three-year interval is too long to permit one to follow moving families. With more intensive studies within the District, where families are visited at shorter intervals, careful thought should be given to the possibility and the appropriateness of following a sub-sample of the migrant families. This is easier said than done. But it would be worth spending some time and money to find out at least how feasible and costly a follow-up is likely to be, and to learn something about the techniques of follow-up. This point deserves emphasis. There is a danger that the majority of longitudinal family studies will in fact be restricted mainly to families with permanence of residence, whose disease history, composition and pattern of living may differ from those of families who move frequently.

2. *Restricted Nature of the Population.* Since the results of studies are primarily of interest only if they can be generalized beyond the District, the criticism can be raised that it is unwise to devote a large amount of resources to investigations in one compact geographic area. In discussing this criticism for the District censuses, it should be remembered that there are other factors to consider. The census material is useful for teaching purposes and for giving students an initiation into research, and might be justified on these grounds even if conclusions from the studies had relatively little general interest. I should, however, like to consider the question more broadly.

The following remarks may be found rather vague and not too helpful. There are two reasons for this. The problem of deciding on the size and breadth of population which it is most profitable to study with given resources is one that requires a more extensive discussion than time permits. Secondly, although there exists a method for making decisions on this question, as will be shown, the data necessary for a practical application of the method are seldom available.

In the teaching of sampling techniques, this issue is brought into the open by the device, recently introduced, of having one term—the *sampled* population—for the population that was actually sampled in an investigation, and another term—the *target* population—for the population about which inferences are desired. Often there will be more than one target population from a given survey, but for simplicity it is assumed that there is only one. The decision whether to attempt a sampled population which closely approximates the target population or one which is much more restricted, but easier to handle, is an important one in many investigations. When the easier course is taken, a wide gap is left to be crossed, as best we can, by judgment or, from a more pessimistic point of view, by guesswork.

Authors' attitudes towards this gap when they report results vary in an interesting way. Some repeatedly warn the reader that conclusions must not be extended beyond the sample data themselves. Although this scientific caution is commendable, the attitude is a little unrealistic, because why publish data if they really are valid only for one specific group of people at one time? Other authors give an early warning not to extend the results beyond the sampled population, and then proceed to forget it, while still others generalize breezily with scarcely a mention of restrictions. Actually, when faced with a substantial difference between the sampled and the target population, the author has no comfortable way out. About the best that can be done is to state clearly what the sampled and target populations are, to present any comparative information about them that seems relevant, and to give one's opinions about the differences in results that might be anticipated for the two populations.

What we need are criteria by which we can appraise, when a study is being planned, whether the sampled population is too narrow with respect to the target population. Such criteria are hard to find and apply. One approach is to argue as follows: if we spend the same amount of money in sampling a broader population, the sample will be smaller, so that sampling errors, and probably also errors of measurement and non-response, will increase. Are the increases more than compensated for by the decrease in the gap between sampled and target population? In order to apply this reasoning to a specific situation, we would need to know how much it costs to broaden the sampled population (e.g. to migrants) and how much the differences in results between sampled and target population are thereby reduced. Although information of this type is usually scanty at present, more will become available as studies accumulate.

Pending the data for an objective examination of this question, one attitude might be, when in doubt, to bite off a little more than we can chew rather than a little less. A scientist who chooses a narrow and relatively easy population to sample is likely to be able to report a well-conducted study with low sampling errors. The human tendency, when using the figures, is to forget about the gap between sampled and target population, particularly when we note how efficiently the work was done. The scientist who attempts a broader population will probably have higher sampling errors and run into certain difficulties which raise doubts about the validity of his conclusions. The reader of his report, assuming that it is an honest one, is made more aware of the limitations of the data and the sampling errors presented are more realistic. Of course this suggestion is made from the point of view of scientific progress: for the scientist's own reputation he might be wise to take the easier job in which he can shine!

3. *More Detailed Points.* Some difficulty has been encountered in the attempt to use birth and death certificates in connection with studies in the District. At first sight these seem to offer a good opportunity, for births and deaths which occur within the District are kept in one file in the City records. By matching them back to the census files it might seem that the mortality and fertility of different subgroups within the District could be studied conveniently and at little expense. Birth and death rates found in this way tend to run too low, and although further investigation may resolve the difficulties, it appears at present as if something is slipping through our fingers. A second difficulty is that rates in other than census years require adjustment for migration in and out. Differential adjustments may be necessary for different groups, and it is not easy to satisfy oneself as to how accurate the adjustments are. This difficulty mounts as the distance between censuses increases.

Another point, perhaps of minor importance, is that the relation of the area to the city of Baltimore changes with time. For instance, the proportion of Negroes has been increasing. These changes must be watched, since they may influence the interpretation of long-term studies.

TAKING STOCK

Finally, I wish to present some points which are relevant in taking stock of the Baltimore surveys, or in considering the utility of such surveys in another area. If the primary purpose of the surveys is to provide background data for more intensive studies conducted on samples from the area, then clearly the censuses are not justified unless there exist definite plans to undertake a variety of such studies. The censuses might be carried out with or without matching. Thus far, the chief benefits from matching, which is time-consuming and moderately expensive, have been the series of demographic studies, some of which are of general sociological interest rather than of interest in public health itself. In the Baltimore situation, this has been advantageous since a School of Public Health should have interests beyond its immediate confines and certainly interests in demography and sociology. In another place it would be well to examine what will be added by matching. For repeated censuses with the general aims of the Baltimore data, the interval should definitely not be longer than three years: a census of one-third of the area every year is worth consideration. The size of the area is important. In Baltimore we will want to appraise what has been gained from the doubling of the area in the 1939 and 1947 censuses. This will depend on the sizes of the incidence rates and prevalence ratios of the diseases or conditions which it is proposed to study, and on the extent to which results require subclassification.

REFERENCES

1. Fales, W. Thurber: Matched Population Records in the Eastern Health District, Baltimore, Md.: A Base for Epidemiological Study of Chronic Disease. *The American Journal of Public Health,* August, 1951, Part 2, 41, No. 8, p. 91.

2. Reed, Lowell J.; Fales, W. Thurber; and Badger, George F.: Family Studies in the Eastern Health District. General Characteristics of the Population. *The American Journal of Hygiene,* January, 1943, 37, No. 1, p. 37.

3. Clark, E. G. and Turner, T. B.: Study of the Prevalence of Syphilis Based on Specific Age Groups of an Enumerated Population. *American Journal of Public Health,* March, 1942, 32, No. 3, p. 307.

4. Downes, Jean and Collins, Selwyn D.: A Study of Illness Among Families in the Eastern Health District of Baltimore. The Milbank Memorial Fund *Quarterly,* January, 1940, XVIII, No. 1, p. 5.

5. Luykx, H. M. C.: Permanence of Residence with Respect to Various Family Characteristics. *Human Biology,* September, 1947, 19, No. 3, p. 91.

6. Valaoras, V. G.: A Comparison of Two Groups of Urban Families Which Differ as to the Permanence of their Residence. Thesis, The Johns Hopkins University, School of Hygiene and Public Health, Baltimore, Maryland, 1936.

7. Rider, R. V. and Badger, G. F.: A Consideration of Issues Involved in Determining Migration Rates for Families. *Human Biology*, May, 1943, 15, No. 2, p. 101.

8. Guralnick, Lillian; Fales, W. Thurber: Family Studies in the Eastern Health District, Job Stability for White Men, 1939 to 1947. The Milbank Memorial Fund *Quarterly*, October, 1950, XXVIII, No. 4, p. 255.

9. Densen, P. M.: The Accuracy of Statements of Age on Census Records. *American Journal of Hygiene*, July, 1940, 32, No. 1, p. 1.

# THE $\chi^2$ TEST OF GOODNESS OF FIT [1]

## By William G. Cochran

### *Johns Hopkins University*

**1. Summary.** This paper contains an expository discussion of the chi square test of goodness of fit, intended for the student and user of statistical theory rather than for the expert. Part I describes the historical development of the distribution theory on which the test rests. Research bearing on the practical application of the test—in particular on the minimum expected number per class and the construction of classes—is discussed in Part II. Some varied opinions about the extent to which the test actually is useful to the scientist are presented in Part III. Part IV outlines a number of tests that have been proposed as substitutes for the chi square test (the $\omega^2$ test, the smooth test, the likelihood ratio test) and Part V a number of supplementary tests (the run test, tests based on low moments, subdivision of chi square into components).

**2. Introduction.** In the standard applications of the test, the $n$ observations in a random sample from a population are classified into $k$ mutually exclusive classes. There is some theory or null hypothesis which gives the probability $p_i$ that an observation falls into the $i$th class ($i = 1, 2, \cdots, k$). Sometimes the $p_i$ are completely specified by the theory as known numbers, and sometimes they are less completely specified as known functions of one or more parameters $\alpha_1, \alpha_2, \cdots$ whose actual values are unknown. The quantities $m_i = np_i$ are called the *expected* numbers, where

$$\sum_{i=1}^{k} p_i = 1, \qquad \sum_{i=1}^{k} m_i = n.$$

The starting point in the theory is the joint frequency distribution of the *observed* numbers $x_i$ falling in the respective classes. If the theory is correct, these observed numbers follow a multinomial distribution with the $p_i$ as probabilities. The joint distribution of the $x_i$ is therefore specified by the probabilities

$$(1) \qquad \frac{n!}{x_1! \, x_2! \, \cdots \, x_k!} \, p_1^{x_1} p_2^{x_2} \cdots p_k^{x_k}.$$

As a test criterion for the null hypothesis that the theory is correct, Karl Pearson [1] proposed the quantity

$$(2) \qquad X^2 = \sum_{i=1}^{k} \frac{(x_1 - m_i)^2}{m_i} = \sum_{i=1}^{k} \frac{x_i^2}{m_i} - n.$$

---

[1] Department of Biostatistics Paper No. 282.

*Editor's Note:* This paper was presented to the Boston meeting of the Institute of Mathematical Statistics, December 28, 1951, and is published in the *Annals* by invitation of the Institute Committee on Special Invited Papers.

Pearson did not mention any particular alternative hypothesis. The test has usually been regarded as applicable to situations in which the alternative hypothesis is described only in rather vague and general terms.

As with any test of a hypothesis, certain properties of the test must be worked out before it is ready for practical application. We need to know the frequency distribution of the test criterion when the null hypothesis is correct, in order that tables of significant values can be constructed. As much as possible should also be known about the performance of the test when the null hypothesis does not hold. Practically all the results in the literature deal only with the limiting distribution of $X^2$ as $n \to \infty$, the $p_i$ remaining fixed. When the null hypothesis holds, this limiting distribution is the $\chi^2$ distribution,

$$(3) \qquad \frac{1}{2^{\nu/2} \left( \frac{\nu}{2} - 1 \right)!} \; (\chi^2)^{(\nu/2)-1} e^{-\frac{1}{2}\chi^2} \; d\chi^2,$$

where $\nu$ is the number of degrees of freedom in $\chi^2$. This distribution is also known to be that followed by the quantity

$$y_1^2 + y_2^2 + \cdots + y_\nu^2,$$

where the $y_i$ are normally and independently distributed with zero means and unit variances.

To avoid confusion, the symbol $X^2$ will be used for the quantity in equation (2) which is calculated from the data when a chi square test is performed. The symbol $\chi^2$ will refer to any random variate which follows the tabular chi square distribution given in (3).

## PART I. HISTORICAL DEVELOPMENT OF THE TEST

**3. Karl Pearson's 1900 paper.** This remarkable paper is one of the foundations of modern statistics. Its style has always impressed me as unusual for a pioneering paper. Pearson writes with the air of a man who knows exactly what he is doing. The exposition, although clear, is slightly hurried and brusque, as if the reader will not wish to be troubled by elaborate details of a problem that is routine and straightforward. One misses any discussion of how Pearson came to choose the $X^2$ test criterion, and of when he first came to realize that this criterion would, under certain circumstances, follow the $\chi^2$ distribution.

The paper opens by proving that if a set of $\nu$ correlated variates $z_i$, with zero means, follow a multivariate normal distribution

$$Ce^{-\frac{1}{2}Q}dz_1 dz_2 \cdots dz_\nu,$$

then the quadratic form $Q$ is distributed as $\chi^2$ with $\nu$ degrees of freedom. This proof is accomplished, in about half a page, by a now familiar geometrical argument. Pearson points out that the ellipsoid $Q$ can be "squeezed" into a sphere. A transformation to polar coordinates is made, where $\chi$ is the radius of the sphere and $Q = \chi^2$. He then remarks that all the angles introduced in the transformation will integrate out to a constant factor, so that the probability that $Q$ exceeds $\chi_0^2$, say, reduces to

$$P = \frac{\displaystyle\int_{\chi_0}^{\infty} e^{-\frac{1}{2}\chi^2}\chi^{\nu-1}\,d\chi}{\displaystyle\int_0^{\infty} e^{-\frac{1}{2}\chi^2}\chi^{\nu-1}\,d\chi}\,.$$

This, of course, is the tabular $\chi^2$ distribution, expressed as an integral of $\chi$ rather than $\chi^2$. The result is a generalization of the result reached by Helmert in 1876, and also of the result which Student later developed in 1908 as groundwork for the $t$-distribution.

The next step is to express the probability integral in power series form, this being necessary to construct a table of the probability integral. The paper contains a table giving $P$ to 6 decimal places, for degrees of freedom from 2 to 19, and for various integral values of $\chi^2$.

Pearson now turns to the problem of testing goodness of fit. He deals first with the case in which the expectations $m_i$ are known numbers. The data have been classified as described in Section 2, so that the observations $x_i$ follow a multinomial distribution. Pearson assumes without more ado that the $x_i$ may be taken as normally distributed. It is at this point, therefore, that he is committed to the assumption that the expectations $m_i$ are large in all cells. He assigns to the $x_i$ their correct variances and covariances from the multinomial distribution, that is,

(4) $$\sigma_{ii} = np_i(1 - p_i), \qquad \sigma_{ij} = -np_ip_j \qquad\qquad (i \neq j)$$

The remainder of the proof consists in writing down the presumed multivariate normal distribution of the quantities $(x_i - m_i)$. From this comes the pleasing result that the quadratic form $Q$ in the exponent is simply

$$Q = \sum \frac{(x_i - m_i)^2}{m_i} = X^2.$$

This may be shown as follows. Since the $x_i$ are constrained to add to $n$, we must omit one of them, say $x_k$, in considering their joint distribution. If the joint frequency function of the first $(k - 1)$ of the $x$'s is $Ce^{-\frac{1}{2}Q}$, it is well known that

$$Q = \sum_{i=1}^{k-1}\sum_{j=1}^{k-i} \sigma^{ij}(x_i - m_i)(x_j - m_j),$$

where $\sigma^{ij}$ is the inverse of the matrix $\sigma_{ij}$ given in (4).

Now consider $X^2$, with $(x_k - m_k)$ replaced by $-\sum_{i=1}^{m-1}(x_i - m_i)$.

$$X^2 = \frac{(x_1 - m_1)^2}{m_1} + \cdots + \frac{(x_{k-1} - m_{k-1})^2}{m_{k-1}}$$

$$+ \frac{\{(x_1 - m_1) + \cdots + (x_{k-1} - m_{k-1})\}^2}{m_k}.$$

Hence, if we write

$$X^2 = \sum_{i=1}^{k-1} \sum_{j=1}^{k-1} a_{ij}\,(x_i - m_i)(x_j - m_j),$$

the matrix $a_{ij}$ is

(5) $$a_{ii} = \frac{1}{m_i} + \frac{1}{m_k}, \qquad a_{ij} = \frac{1}{m_k} \qquad (i \neq j).$$

The remainder of the proof consists in showing that (5) is the inverse of (4). Pearson does this by a rather complicated polar transformation, but the student who has some familiarity with the evaluation of determinants will find it a fairly easy exercise, as Hotelling [2] has pointed out. It may be helpful to write (4) as

(4') $$\sigma_{ii} = m_i \left( 1 - \frac{m_i}{n} \right), \qquad \sigma_{ij} = - \frac{m_i m_j}{n} \qquad (i \neq j),$$

and to invert (5) rather than (4), or to prove that the product of the matrices $a_{ij}$, $\sigma_{jk}$ is the unit matrix.

Hence, by the first part of Pearson's paper, we reach the result that in the limit as $n$ becomes large, $X^2$ follows the $\chi^2$ distribution with $(k - 1)$ degrees of freedom.

An approach which avoids most of the mathematical complexities in Pearson's argument has been pointed out by Fisher [3]. If the observations $x_i$ are regarded as following independent Poisson distributions, their joint frequency function is

(6) $$\prod_{i=1}^{k} \frac{e^{-m_i} m_i^{x_i}}{x_i!} = e^{-n} \prod_{i=1}^{k} \frac{m_i^{x_i}}{x_i!},$$

since $\sum m_i = n$.

Under this assumption, their total $T = \sum x_i$ also follows a Poisson distribution, with mean $\sum m_i = n$. The frequency function of $T$ is therefore

(6') $$\frac{e^{-n} n^T}{T!}.$$

Hence, on dividing (6) by (6') the *conditional* frequency function of the $x_i$, given that their total $T$ has the value $n$, is

$$\frac{n!}{x_1!\, x_2! \cdots x_k!} \left( \frac{m_1}{n} \right)^{x_1} \left( \frac{m_2}{n} \right)^{x_2} \cdots \left( \frac{m_k}{n} \right)^{x_k}.$$

This is the same as the basic multinomial (1).

This argument implies that in an investigation of the distribution of $X^2$ we may start by regarding the $x_i$ as following independent Poisson distributions, subject to the restriction that $\sum x_i = n$.

In the limit, as the $m_i$ become large, the quantities

$$y_i = \frac{x_i - m_i}{\sqrt{m_i}}$$

become normally distributed with means zero and unit standard deviations, since the Poisson distribution of $x_i$ has mean $m_i$ and standard deviation $\sqrt{m_i}$. Hence the limiting distribution of $X^2$ is that of the quantity

$$y_1^2 + y_2^2 + \cdots + y_k^2,$$

where the $y_i$ are independently distributed but are subject to the single linear restriction

$$\sum_{i=1}^{k} y_i \sqrt{m_i} = \sum_{i=1}^{k} (x_i - m_i) = 0.$$

The fact that in the limit $X^2$ follows the $\chi^2$ distribution with $(k - 1)$ degrees of freedom can now be established by integration or by quoting well known theorems on the analysis of variance. This approach also makes it clear that if further homogeneous linear restrictions are imposed on the variates $(x_i - m_i)$, either by the structure of the data or in the process of fitting, the effect will merely be to reduce the degrees of freedom in $\chi^2$.

Pearson next considers the situation in which the $m_i$ depend on parameters that have to be estimated from the sample. Denoting by $m_i'$ the expectations derived from sample estimates of these parameters, and by $m_i$ the true expectations, he discusses the difference

$$X^2 - X'^2 = \sum_{i=1}^{k} \frac{x_i^2}{m_i} - \sum_{i=1}^{k} \frac{x_i^2}{m_i'}.$$

He suggests that this difference will usually be positive, because we ought to be able to do a better job of fitting when we can adjust the estimates of the parameters to suit the vagaries of the sample. He argues, however, that the difference will be small enough so that if we regard $X'^2$ as also distributed as $\chi^2$ with $(k - 1)$ degrees of freedom, the error in this approximation will not affect practical decisions.

In this conclusion, which is reached with some sign of hesitation, he may well have been justified for many applications. We now know that the number of degrees of freedom must be reduced in order to give the correct limiting distribution. Perhaps the most common of all uses of the $X^2$ test is for the $2 \times 2$ contingency table. Unfortunately, Pearson's suggestion works rather poorly in this case, since he attributed 3 degrees of freedom to $X^2$, whereas it should receive only 1. This point caused some confusion and controversy in practical applications, and was not settled for over 20 years.

Finally, the paper contains eight numerical applications of the new technique. In two of these he pokes fun at Sir George Airy and Professor Merriman. They had both published series of observations which they claimed to be good illustrations of variates that follow the normal distribution. In the absence of any test of goodness of fit, Airy and Merriman could judge this question only by eye inspection. Pearson showed that the significance probability for Airy's data was 0.014, although the data from which Pearson calculated $X^2$ had already been smoothed by Airy. Merriman fared worse, his probability being $1\frac{1}{2}$ parts in a million. These examples show the weak position in which the scientist was placed when he had to judge goodness or badness of fit in the absence of an objective test of significance.

To summarize, Pearson established the necessary distribution theory for finding significance levels when the null hypothesis provides the exact values of the $m_i$, except that he did not show that the exact distribution of $X^2$, which is discontinuous, actually approaches $\chi^2$ as a limiting distribution. A fully rigorous proof may be given by the use of moment-generating functions [8].

**4. The distribution of $X^2$ when the expectations are estimated from the sample.** This problem is much more difficult and was not elucidated until the appearance of Fisher's 1924 paper. In the intervening period, a paper by Greenwood and Yule [4] in 1915 illustrates the perplexity which existed among critical users of the test and which led to the "degrees of freedom" battle. The authors were attempting to examine the effects of inoculation against typhoid and cholera. They present a substantial number of $2 \times 2$ tables containing subjects classified as inoculated or not, and also as to whether they contracted the disease following exposure to it. The following is an example.

*Kalain (Cholera)*

|  | Not Attacked | Attacked | Total |
|---|---|---|---|
| Inoculated..................... | 1625 | 5 | 1630 |
| Not........................... | 1022 | 11 | 1033 |
| Total........................ | 2647 | 16 | 2663 |

$X^2 = 6.08$.

Following Pearson's rule, they assign 3 degrees of freedom to $X^2$. This gives a $P$ of 0.108, whereas with 1 degree of freedom, $P$ is 0.015. They realised, however, that the hypothesis that inoculation is without effect could be tested, alternatively, by calculating the difference $(p_1 - p_2)$ between the percent ill among the inoculated and the non-inoculated. On the null hypothesis, the ratio

$$R = \frac{p_1 - p_2}{\sqrt{\dfrac{p_1 q_1}{n_1} + \dfrac{p_2 q_2}{n_2}}}$$

is approximately a normal deviate with mean zero and unit standard deviation. This test, as they found, gave more statistically significant results than Pearson's test. The quantity $R$ is exactly equal to $X$ if we use the pooled percent ill, $\bar{p}$, in estimating the two variances in the denominator, so that the "normal deviate" test and the $X^2$ test should be identical. It is not clear that Greenwood and Yule recognised this in 1915.

Although giving the impression of being somewhat in a quandary as to which test to employ, they content themselves with the decision to adopt Pearson's test, pointing out that it is the more conservative of the two, and adding that the issue deserves further theoretical investigation.

After some controversy, the matter was cleared up in theoretical papers [3], [5] by Fisher in 1922 and 1924, supported by sampling experiments which were published by Yule [6] and Brownlee [7]. Fisher's 1922 paper included a discussion of $2 \times 2$ contingency tables, and showed that for this case $X^2$ is the square of a single quantity which had a limiting normal distribution, and that the $X^2$ test and the test of $(p_1 - p_2)$ by the normal distribution are identical.

Fisher's 1924 paper is much more general. He points out that the limiting distribution of $X^2$ depends on the method of estimation. With a poor method of estimation, $X^2$ may frequently have a large value even if the theory is correct. It is therefore necessary, in a general proof of the distribution of $X^2$, to state what is to be the method of estimation. At first sight, the natural method would seem to be to choose the unknown parameters so that $X^2$ is as small as possible. Fisher shows that in the limit in large samples, this method becomes equivalent to the method of maximum likelihood. For his main proof, this result serves as an ingenious lemma, since at one point in the main proof he assumes that estimation is by maximum likelihood, while at another he assumes that it is by minimum $X^2$.

Although Fisher's main proof is not fully rigorous, it is worthwhile to outline the principal steps, because the proof does reveal the core of the problem, and a rigorous proof requires advanced methods. Fisher starts in the same way as Pearson, by considering

$$X^2 - X'^2 = \sum_{i=1}^{k} \frac{x_i^2}{m_i} - \sum_{i=1}^{k} \frac{x_i^2}{m_i'} = \sum_{i=1}^{k} x_i^2 \left( \frac{1}{m_i} - \frac{1}{m_i'} \right),$$

where $m_i$ is a specified function of a single unknown parameter $\alpha$, with $m_i = m_i(\alpha)$, $m_i' = m_i(\alpha')$. He expands in a Taylor series about the point $\alpha'$. The first two terms give

$$X^2 - X'^2 = -\sum \frac{x_i^2}{m_i'^2} \left( \frac{\partial m_i'}{\partial \alpha'} \right) \delta\alpha + \sum x_i^2 \left\{ \frac{2}{m_i'^3} \left( \frac{\partial m_i'}{\partial \alpha'} \right)^2 - \frac{1}{m_i'^2} \frac{\partial^2 m_i'}{\partial \alpha'^2} \right\} \frac{(\delta\alpha)^2}{2!},$$

where $\delta\alpha = (\alpha - \alpha')$. Since the method of estimation consists in choosing $\alpha'$ so that $X'^2$ is a minimum, we have

$$\sum \frac{x_i^2}{m_i'} \left( \frac{\partial m_i'}{\partial \alpha'} \right) = 0,$$

so that the first term on the right vanishes.

In the second term on the right, Fisher replaces $x_i$ by $m_i'$. The error introduced by this step may be shown to be of the same order as the third term in the Taylor series, which has already been ignored. Hence,

$$X^2 - X'^2 = \sum \left\{ \frac{2}{m_i'} \left( \frac{\partial m_i'}{\partial \alpha'} \right)^2 - \frac{\partial^2 m_i'}{\partial \alpha'^2} \right\} \frac{(\delta \alpha)^2}{2!} .$$

But if the identity $\sum m_i' = n$ is differentiated twice, we find

$$\sum \left( \frac{\partial^2 m_i'}{\partial \alpha'^2} \right) = 0.$$

Hence,

$$X^2 - X'^2 = \sum \left\{ \frac{1}{m_i'} \left( \frac{\partial m_i'}{\partial \alpha'} \right)^2 \right\} (\delta \alpha)^2 .$$

If $\alpha'$ is regarded as a maximum likelihood estimate of $\alpha$, we may use the standard result that the error of estimate $(\alpha' - \alpha)$ has a limiting normal distribution, with mean zero, and variance given by

$$\frac{1}{\sigma_{\alpha'}^2} = \sum \left\{ \frac{1}{m_i} \left( \frac{\partial m_i}{\partial \alpha} \right)^2 \right\} \doteq \sum \left\{ \frac{1}{m_i'} \left( \frac{\partial m_i'}{\partial \alpha'} \right)^2 \right\} .$$

This gives the neat result

$$X^2 - X'^2 = \frac{(\alpha' - \alpha)^2}{\sigma_{\alpha'}^2} .$$

Our object is to find the limiting distribution of $X'^2$. At this point the facts in our possession are: (i) $X^2$ is distributed as $\chi^2$ with $(k - 1)$ degrees of freedom (this follows from Pearson's results, since $X^2$ is calculated from the correct $m$'s); and (ii) $X^2 - X'^2$ is distributed as $\chi^2$ with 1 degree of freedom (from Fisher's argument). These facts are not sufficient to determine the distribution of $X'^2$. However, Fisher points out that the limiting distributions of $X'^2$ and $(\alpha' - \alpha)^2$ must be independent, since $X'^2$ was obtained by minimizing $X^2$ with respect to $\alpha'$. Given this additional result, it is easily shown that $X'^2$ must be distributed as $\chi^2$ with $(k - 2)$ degrees of freedom.

The argument leads to two further results. Any method of estimation that is efficient gives estimates which become, in the limit, identical with the maximum likelihood estimate. Thus the $\chi^2$ distribution, with the appropriate reduction in degrees of freedom, is valid for any efficient method of estimation. The argument also provides the limiting mean value of $X'^2$ when the estimation is inefficient. An interesting corollary is that the mean value of $X'^2$ exceeds that of $X^2$ when the efficiency is less than 50 percent.

Rigorous proofs of the general limiting distribution, when several parameters are being estimated, are scarce in the literature. For the student, one of the best is that given by Cramér [8]. The restrictions under which he proves his result are stated below. He assumes maximum likelihood estimation.

THEOREM. *Suppose that the $k$ probabilities $p_i(\alpha_1, \alpha_2, \cdots, \alpha_s)$ are known functions of $s < k$ parameters $\alpha_1, \alpha_2, \cdots, \alpha_s$. For all points of a nondegenerate interval $A$ in the $s$-dimensional space of the $\alpha_j$, the $p_i$ satisfy the following conditions:*

(a)
$$\sum_{i=1}^{k} p_i(\alpha_1, \cdots, \alpha_s) = 1;$$

(b)
$$p_i(\alpha_1, \cdots, \alpha_s) > C^2 > 0 \qquad \qquad \text{for all } i;$$

(c) *every $p_i$ has continuous derivatives $\dfrac{\partial p_i}{\partial \alpha_j}$ and $\dfrac{\partial^2 p_i}{\partial \alpha_j \, \partial \alpha_h}$ ;*

(d) *the matrix $D = \left(\dfrac{\partial p_i}{\partial \alpha_j}\right)$ is of rank $s$. Then $X^2$ is distributed in the limit, as $n \to \infty$, in a $\chi^2$ distribution with $(k - s - 1)$ degrees of freedom.*

**5. The limiting power function of the test.** The literature does not contain much discussion of the power function of the $X^2$ test. There has been little demand for this from applications, because the test is most commonly used when we do not have a clear-cut alternative in mind, and are not in a position to make computations of the power.

Suppose that we test the null hypothesis that the expectations are $m_i$ when in fact they are $m_i'$. If the values of $m_i$, $m_i'$ and the significance level are kept fixed, then as $n$ increases, it turns out, as would be expected, that the power of the test tends to 1. This has been shown by Neyman [9]. In order to examine the situation in which the power is not close to 1 in large samples, we must somehow make the task continually harder for the test as $n$ becomes larger. This can be accomplished either by making the significance probability decrease steadily as $n$ increases, thus reducing the chance of an error of type I, or by moving the alternative hypothesis steadily closer to the null hypothesis. The second method will be discussed here. Let

$$m_i' - m_i = c_i\sqrt{n}; \qquad \text{that is, } p_i' - p_i = c_i/\sqrt{n},$$

where the quantities $c_i$ remain fixed as $n$ increases.

The nature of the limiting power distribution of $X^2$ is indicated by the following argument, for which I am indebted to J. W. Tukey. We may write

(7)
$$\frac{x_i - m_i}{\sqrt{m_i}} = \frac{(x_i - m_i')}{\sqrt{m_i'}} \sqrt{\frac{m_i'}{m_i}} + \frac{m_i' - m_i}{\sqrt{m_i}}.$$

Now

$$\sqrt{\frac{m_i'}{m_i}} = \sqrt{1 + \frac{m_i' - m_i}{m_i}} = \sqrt{1 + \frac{c_i}{p_i\sqrt{n}}}.$$

This tends to 1 as $n$ becomes large, since $c_i$ and $p_i$ are presumed to remain fixed. If we adopt Fisher's approach to the distribution theory (Section 3), the quantities

$$\frac{x_i - m'_i}{\sqrt{m'_i}}$$

tend to become normally and independently distributed with means zero and unit standard deviations, as $n$ becomes large. Consequently, so do the quantities

$$y_i = \frac{(x_i - m'_i)}{\sqrt{m'_i}} \sqrt{\frac{m'_i}{m_i}}.$$

Finally, by equation (7),

$$X^2 = \sum_{i=1}^{k} \frac{(x_i - m_i)^2}{m_i} = \sum_{i=1}^{k} \left\{ y_i + \frac{m'_i - m_i}{\sqrt{m_i}} \right\}^2 = \sum_{i=1}^{k} (y_i + a_i)^2,$$

where the $y_i$ are subject to the linear restriction

$$\sum_{i=1}^{k} y_i \sqrt{m_i} = \sum_{i=1}^{k} (x_i - m'_i) = 0.$$

Thus, in the limit, $X^2$ is distributed as a sum of squares of variates $(y_i + a_i)$ independently and normally distributed with unit variances, but where the means $a_i$ are not all zero. The variates are subject to one linear restriction when the $m_i$ are completely specified.

This type of distribution is known as a noncentral $\chi^2$. It depends on two parameters—the degrees of freedom, in this case $(k - 1)$, and a parameter of non-centrality $(a_1^2 + a_2^2 + \cdots + a_k^2)$, which has the value

$$\sum_{i=1}^{k} \frac{(m'_i - m_i)^2}{m_i} = \sum_{i=1}^{k} \frac{c_i^2 n}{p_i n} = \sum_{i=1}^{k} \frac{c_i^2}{p_i}.$$

Tables of the noncentral distribution have been provided by Fix [10] and approximations studied by Patnaik [11].

When the $m_i$ have to be estimated from the data, the limiting noncentral $\chi^2$ distribution still holds, with a reduced number of degrees of freedom. A rigorous proof is obtained from Wald's derivation of the limiting distribution of the likelihood ratio test criterion [12], which becomes equivalent to $X^2$ in large samples.

**6. Conditional $X^2$ tests.** As has been mentioned, additional homogeneous linear restrictions imposed on the deviations $(x_i - m_i)$ have the effect of reducing the number of degrees of freedom attributed to $\chi^2$ in the limiting distribution of $X^2$. These restrictions may arise in the process of fitting, or by the nature of the data. They may also be deliberately imposed by the statistician in the device known as a conditional test. This device is illustrated by the $2 \times 2$ contingency table, in which it has created some stimulating discussion [13].

The data are classified according to two different axes, $A$ and $B$.

|         | $B_1$    | $B_2$    | Totals |
|---------|----------|----------|--------|
| $A_1$   | $x_{11}$ | $x_{12}$ | $r_1$  |
| $A_2$   | $x_{21}$ | $x_{22}$ | $r_2$  |
| Totals......... | $c_1$ | $c_2$ | $n$ |

Data of this kind occur in at least three distinct experimental situations.

($i$) We select a random sample of $n$ from some population and classify every observation into one of the four cells. The symbol $x_{ij}$ denotes the observed number falling in class $A_iB_j$, while $p_{ij}$ denotes the corresponding probability of falling in this class, where the sum of the four $p$'s is unity. The null hypothesis that the two classifications are *independent* amounts to the relation

$$(8) \qquad p_{11}/p_{12} = p_{21}/p_{22}.$$

The joint probability of this group of observations is the usual multinomial

$$(9) \qquad \frac{n!}{x_{11}!\,x_{12}!\,x_{21}!\,x_{22}!}\, p_{11}^{x_{11}}\, p_{12}^{x_{12}}\, p_{21}^{x_{21}}\, p_{22}^{x_{22}}.$$

Only two of the $p_{ij}$ need to be estimated from the data, because of equation (8) and the fact that the $p_{ij}$ add to 1. Thus $X^2$ has $(4 - 2 - 1)$ or 1 degree of freedom.

($ii$) We take a random sample of size $r_1$ from a population denoted by $A_1$, and an *independent* random sample of size $r_2$ from another population denoted by $A_2$. The null hypothesis states that the probability $p$ of an observation falling in $B_1$ is the same in both populations $A_1$ and $A_2$. Given the null hypothesis, the probability of the sample is the product of the two binomials

$$(10) \qquad \left\{ \frac{r_1!}{x_{11}!\,x_{12}!}\, p^{x_{11}} q^{x_{12}} \right\} \left\{ \frac{r_2!}{x_{21}!\,x_{22}!}\, p^{x_{21}} q^{x_{22}} \right\}.$$

This is not the same as the multinomial (9). Given data of type ($i$), however, let us arbitrarily impose the restriction that in repeated sampling we will consider only those tables which have the same marginal totals $r_1$, $r_2$ as our data. Then (9) must be replaced by the conditional distribution of the $x_{ij}$, given $r_1$ and $r_2$. This conditional distribution is easily seen to be the same as (10). For, starting with (9), the distribution of $r_1$ (and hence $r_2$) is the binomial

$$(11) \qquad \frac{n!}{r_1!\,r_2!}\,(p_{11} + p_{12})^{r_1}(p_{21} + p_{22})^{r_2}.$$

To obtain the conditional distribution, we divide (9) by (11). The quotient reduces to (10) if we note that from (8),

$$\frac{p_{11}}{p_{11} + p_{12}} = \frac{p_{21}}{p_{21} + p_{22}} = p \quad (\text{say}).$$

(*iii*) A third case is obtained if *both* sets of marginal totals are regarded as fixed in repeated sampling. Fisher's tea-tasting experiment [14] is an example. The $A$ classification tells whether the milk or the tea was added first, and the $B$ classification tells whether the lady guessed that the milk or the tea was added first. In Fisher's original experiment, he recommended that the lady be informed how many cups were of each kind, and pointed out that she would presumably match her guesses to those two numbers. Thus in repeated trials it is natural to regard both sets of margins as fixed.

The appropriate basic distribution of the $x_{ij}$ is the conditional distribution which develops from (10) if we keep $c_1$ and $c_2$ fixed. This is found to be

$$(12) \qquad\qquad \frac{r_1!\, r_2!\, c_1!\, c_2!}{n!\, x_{11}!\, x_{12}!\, x_{21}!\, x_{22}!}.$$

Case (*i*) has 2 unknown parameters and 1 linear restriction on the $x_{ij}$ ; case (*ii*) has 1 unknown parameter and 2 restrictions, while case (*iii*) has no unknown parameters and 3 restrictions.

Is the same $X^2$ test to be used for all cases? In large samples there is no conflict, because $X^2$ has the same limiting distribution however the linear restrictions arise. This is not so in small samples, where the distribution of $X^2$ differs in the three cases. Fisher [15] recommends that the distribution of $X^2$ obtained in case (*iii*) be taken as the exact small-sample distribution for all three types of data. Questions have been raised about this recommendation.

Originally, part of the objection came perhaps from a feeling that there is something improper in keeping the marginal totals fixed in cases (*i*) and (*ii*), because if we actually drew repeated samples of the same size by the same methods, the margins would not all remain fixed. For a rational appraisal, however, the only relevant factors are the effects of the marginal restrictions on the significance probabilities (or type I errors) and on the power (or type II errors). As regards type I errors, Fisher's recommendation has the great advantage that in case (*iii*) the significance probabilities can be computed exactly, whereas in cases (*i*) and (*ii*) the distribution of $X^2$ involves nuisance parameters, so that "exact" probabilities are not available.

The issue thus reduces to the question whether any loss of power occurs if the case (*iii*) test is employed with the first two types of data. For case (*ii*) data, Barnard [13] proposed a different test which in some circumstances appeared to give a small increase in power. More recently K. D. Tocher [16], has proved the remarkable result that a modification of Fisher's test is the most powerful, in the sense of Neyman and Pearson, for one-tailed tests with any of the three types of data. The modification is necessary to make the problem amenable to Neyman and Pearson's techniques.

The modification may be illustrated by the example which Tocher presents.

### TABLE 1
#### *Tocher's illustration*

| Original table | | | | More extreme cases | | | | | |
|---|---|---|---|---|---|---|---|---|---|
| 2 | 5 | 7 | | 1 | 6 | 7 | 0 | 7 | 7 |
| 3 | 2 | 5 | | 4 | 1 | 5 | 5 | 0 | 5 |
| 5 | 7 | 12 | | 5 | 7 | 12 | 5 | 7 | 12 |

Given the data on the left, we wish to make a one-tailed test at the 5% level. The two possible sets of data which deviate more from the null hypothesis are shown on the right. In Fisher's exact test, we add the probabilities of the three tables as computed by formula (12). This gives

$$0.26515 + 0.04399 + 0.00126 = 0.31040.$$

This value is regarded as the significance probability.

In Tocher's modification, we also compute the total probability of all more extreme cases, that is,

$$0.04399 + 0.00126 = 0.04525.$$

If these numbers, 0.31040 and 0.04525, are both *below* the stated significance level, 0.05, we reject the hypothesis. If they are both *above* 0.05, we accept. If one is above and one is below, as in the present example, we calculate the ratio

$$\frac{0.05 - 0.04525}{0.26515} = 0.01791.$$

Now draw a random number between 0 and 1. If this number is less than 0.01791, we reject; if greater, we accept.

Although this procedure may appear somewhat startling at first sight, the idea is basically simple. Consider how we can obtain a one-tailed test at the 0.05 level from the 2 $\times$ 2 table in this example. If the null hypothesis is rejected only when the two most extreme cases on the right of Table 1 occur, the significance level is actually 0.04525. The third most extreme case, represented by the data on the left of Table 1, occurs with probability 0.26515. Consequently, by the computation above, we obtain a test at the 0.05 level if we also declare as significant a fraction 0.01791 of the cases in which the data on the left are encountered. Tocher selects this fraction by a table of random numbers. There seems no other logical basis for deciding which particular fraction to select.

PART II. SOME ASPECTS OF THE PRACTICAL USE OF THE TEST

**7. The minimum expectation.** Since $\chi^2$ has been established as the limiting distribution of $X^2$ in large samples, it is customary to recommend, in applications of the test, that the smallest expected number in any class should be 10 or (with some writers) 5. If this requirement is not met in the original classification, combination of neighboring classes until the rule is satisfied is recommended. This topic has recently been subject to vigorous discussion among the psychologists [17], [18]. The numbers 10 and 5 appear to have been arbitrarily chosen. A few investigations throw some light on the appropriateness of the rule. The approach has been to examine the exact distribution of $X^2$, when some or all expectations are small, either by mathematical methods or from sampling experiments.

The investigations are scanty and narrow in scope, as is to be expected since work of this type is time-consuming. Thus the recommendations given below may require modification when new evidence becomes available.

To digress for a moment, the problem of investigating the behavior of $X^2$ when expectations are small is an example of a whole class of problems that are relevant to applied statistics. In applications it is an everyday occurrence to use the results of a body of theory in situations where we know, or strongly suspect, that some of the assumptions in the theory are invalid. Thus the literature contains investigations of the $t$-distribution when the parent population is nonnormal, and of the performance of linear regression estimates when the regression in the population is actually nonlinear. Fortunately for applications, the results of theory sometimes remain substantially true even when some assumptions fail to hold. This fact tends to make statistics a more confusing subject than pure mathematics, in which a result is usually either right or wrong.

In any problem of this kind, it is important to define what is meant by saying the results remain "substantially true." I stress this point because a reader who becomes interested in a specific problem and tries to summarize the available knowledge may encounter considerable difficulty. Definitions vary from writer to writer and are sometimes entirely subjective, so that the researches may be presented in a form which baffles any attempt to apply a uniform definition. This remark is not intended as a criticism of work on the $X^2$ problem, where the task of summarizing is comparatively easy. However, I believe that the usefulness of this kind of research would be enhanced by careful attention to the questions: (i) how are we going to measure the disturbance caused by a failure in assumptions, and (ii) when are we going to call this disturbance "serious."

In the present instance, my criterion is to compare the exact $P$ and the $P$ from the $\chi^2$ table, when the null hypothesis is true, in the region in which the tabular $P$ lies between 0.05 and 0.01. This criterion is not ideal, but it does appraise the performance of the tabular approximation in the borderline region between statistical significance and nonsignificance. A disturbance is regarded as unimportant if when the $P$ is 0.05 in the $\chi^2$ table, the exact $P$ lies between 0.04 and 0.06, and if when the tabular $P$ is 0.01, the exact $P$ lies between 0.007 and

0.015. These limits are, of course, arbitrary; some would be content with less conservative limits.

The results suggest that four cases need to be considered separately.

(i) *Goodness of fit tests of bell-shaped curves such as the normal distribution.* The distinguishing feature of this case is that usually only one or two expectations at the tails are small, the others being above the conventional limits of 5 or 10. Cochran [19] has shown that there is little disturbance to the 5% level when a *single* expectation is as low as $\frac{1}{2}$.

This is also true for the 1% level if the number of degrees of freedom in $X^2$ exceeds 6. Two expectations as low as 1 may be allowed with negligible disturbance to the 5% level. Since the discrepancy between an observed and a postulated distribution is often most apparent at the tails, the sensitivity of the $X^2$ test is likely to be decreased by an overdose of pooling at the tails. Thus considerations of the power of the test urge us to use cells with as small expectations as we dare from distributional considerations. The inflexible use of minimum expectations of 5 or 10 may be harmful.

(ii) $2 \times 2$ *contingency tables.* This case is the most thoroughly worked out and can be regarded as solved for practical purposes. Fisher [15] has given the method of obtaining an exact solution, which is not too laborious in samples up to size 30. Tables such as Mainland's [20] give the probability levels of the exact distribution for two samples each of size up to $n = 40$, and Yates' table [21] gives almost exact tests based on $X^2$ after correction for continuity.

(iii) *Tests in which all expectations may be small.* This case occurs from time to time, for example, in genetical studies in which a Mendelian ratio is being compared over a number of small families. Results by Neyman and Pearson [22], Cochran [23] and Sukhatme [24] suggest tentatively that the tabular $\chi^2$ is tolerably accurate provided that all expectations are at least 2.

With very scanty data, there is one danger—that only a few different values of $X^2$ are possible, so that the effects of discontinuity become noticeable. For example, consider the $2 \times 4$ contingency table with marginal totals shown below. All expectations are exactly 2.

|  |  |  |  |  | 8 |
|---|---|---|---|---|---|
|  |  |  |  |  | 8 |
| 4 | 4 | 4 | 4 |  | 16 |

If we construct all tables which satisfy these marginal totals, only seven different values of $X^2$ are found. The exact distribution of $X^2$ and the $\chi^2$ approximation (with 3 degrees of freedom) are shown in Table 2. The agreement is not good, the tabular $P$'s being fairly consistently too low.

With such a small number of values of $X^2$, a correction for continuity comes to mind. To apply this for $X^2 = 2$, we read the $\chi^2$ table at $\chi^2 = 1$, this being half way between 2 and 0 (the next largest value of $X^2$). The corrected $P$'s show a considerable improvement in fit.

In practice, a small table of this kind can be handled by computing the exact distribution of $X^2$ in cases of doubt about the adequacy of the $\chi^2$ approximation. For more complex contingency tables, a systematic method of computation has been given by Freeman and Halton [25].

TABLE 2

*Exact distribution of $X^2$ for a 2 × 4 contingency table. $P\{X^2 \geqq X_0^2\}$*

| $X_0^2$ | Exact | $\chi^2$ Table | Corrected |
|---|---|---|---|
| 0 | 1.000 | 1.000 | 1.000 |
| 2 | .899 | .572 | .801 |
| 4 | .362 | .261 | .391 |
| 6 | .243 | .112 | .172 |
| 8 | .064 | .046 | .072 |
| 10 | .030 | .019 | .029 |
| 16 | .0005 | .0011 | .004 |

(iv) *Tests in which all expectations are small and $X^2$ has many degrees of freedom (say $>60$).* Examples occur in genetical research. The data are presented in, say, a 2 × 200 contingency table, with all 400 expectations small.

In this case, the exact distributions of $X^2$ and $\chi^2$ are both approximately normal, since the degrees of freedom are large. However, the two distributions have different variances, and the normal approximation to the exact distribution is sometimes quite different from the normal approximation to $\chi^2$.

This problem has been studied by Haldane [26], [27] who has worked out the exact mean and variance of $X^2$ for several types of data. His results are given below for the two cases that are perhaps most common.

(a) We have $g$ groups, each containing $s$ individuals, classified into one of two classes. The null hypothesis specifies a *known* constant probability $p$ that any individual falls into the first class. If $x_i$ individuals fall into this class in the $i$th group, and if we wish to test against the alternative that $p$ varies from class to class, a familiar extension of the $X^2$ test is to calculate

$$X^2 = \sum_{i=1}^{g} \frac{(x_i - sp)^2}{spq}$$

with $g$ degrees of freedom. Haldane shows that

$$E(X^2) = g,$$

$$V(X^2) = 2g\left(1 + \frac{1 - 6pq}{2spq}\right).$$

(b) Same data, but $X^2$ computed as for a $2 \times g$ contingency table, since $p$ is unknown.

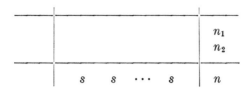

Then

$$E(X^2) = \frac{(g-1)n}{n-1},$$

$$V(X^2) = \frac{2(g-1)n^3 \, (n-g)}{(n-1)^2 \, (n-2)(n-3)} \left(1 - \frac{(n-1)}{n_1 n_2}\right).$$

To take an extreme case, suppose $s = 2$, $g = 160$, $n = 320$, $n_1 = 64$, $n_2 = 256$. The mean and variance of $X^2$ are 159.5 and 159.4 respectively, whereas $\chi^2$ has mean 159 and variance 318, twice as large. The normal approximation to $X^2$ is satisfactory, but $\chi^2$ gives very poor agreement [23]. In practice, such data may be dealt with by use of the normal approximation to the exact distribution, using Haldane's expressions for the mean and variance.

The question remains: where does case (iii) shade into case (iv)? The available data suggest that case (iii) may apply when $X^2$ has less than 15 degrees of freedon, while case (iv) may hold if $X^2$ has more than 60 degrees of freedom. The intervening gap needs investigation.

**8. The correction for continuity.** The exact distribution of $X^2$ is always discontinuous. When all expectations are small, the number of distinct values of $X^2$ may be very limited. In such cases the $\chi^2$ table may give a poor approximation to the exact $P\{X^2 \geqq X_0^2\}$, mainly because the area of a continuous curve is used to approximate the sum of a small number of discrete probabilities. The correction for continuity, introduced by Yates [28], is an attempt to remove this source of error. It amounts to reading the $\chi^2$ table, not at the point $X_0^2$, but at a point halfway between $X_0^2$ and the value of $X^2$ immediately below $X_0^2$ in the discrete series of values.

In practice, the correction is seldom needed except when $X^2$ has 1 degree of freedom, as when testing a single binomial ratio or a $2 \times 2$ contingency table. In the $2 \times 2$ table there are various ways of computing the correction, depending on how one likes to compute $X^2$. My own preference is to find the difference $d$ between $x_i$ and $m_i$, which is the same, apart from sign, in all four cells. The absolute value of $d$ is reduced by $\frac{1}{2}$, and $X^2$ is computed as

$$X^2 = (|d| - \tfrac{1}{2})^2 \sum \frac{1}{m_i}.$$

Note that in the $2 \times 2$ table it is $X$ that is corrected for continuity, not $X^2$, since the successive values of $d$ differ by unity.

When $X^2$ has 1 degree of freedom, a good rule is to apply the correction whenever it produces any appreciable difference in the significance probability. The correction has a tendency to over-correct, changing the tabular $P$ from too small to too large, but is usually an improvement.

If a number of $X^2$ values, each with 1 degree of freedom, are added to form a total $X^2$, the individual $X^2$ values should *not* be corrected for continuity, because the over-correction mounts up in a disconcerting manner [19]. After it has been obtained, the total $X^2$ is corrected, if this is necessary, by the method given in the following paragraph.

Compute the next largest value of $X^2$ which the structure of the data permits. Read the $\chi^2$ table at a point halfway between this value and the observed $X^2$. This procedure was illustrated for a $2 \times 4$ table in the preceding section. Sometimes the next largest value of $X^2$ is not immediately obvious and trial and error is required to find it.

**9. The construction of classes.** When $X^2$ is used to test the hypothesis that the observations follow a *continuous* frequency distribution, the first step is to group the observations into classes. Both the number of classes and the division points between classes are at the disposal of the investigator, and the choices that he makes will affect the sensitivity of the test. I believe that the common practice is to have a moderate number of classes, say between 10 and 25, and to make the class intervals equal. Although information about the best rule for constructing the classes is still meager, the recommendations of those who have looked into this problem are contrary to current practice.

With regard to class intervals, Mann and Wald [29] and Gumbel [30] suggest that these be chosen so that the *expected* number is the same ($= n/k$) in all classes. These authors do not claim that this will increase the power of the test, but merely suggest that it is likely to be a good procedure. Gumbel points out that if this method is used in conjunction with a rule for choosing the value of $k$, much of the arbitrariness that accompanies the construction of class intervals is removed. Under this method, the computational steps are first to estimate the constants (mean, s.d., etc.) which determine the fitted curve, then find the class boundaries which give equal expectations in each class, and finally count the *observed* numbers $x_i$ which fall in the respective classes. The value of $X^2$ is given as

$$X^2 = \frac{k}{n} \sum x_i^2 - n.$$

The paper by Mann and Wald deals with the choice of the number of classes, $k$. The null hypothesis is assumed to specify the distribution completely, and $n$ is assumed large enough so that the limiting $\chi^2$ distribution is applicable.

Some criterion is required to define what is meant by a "best" value of $k$.

It seems natural to try to maximize some property of the power function of the test. The criterion set up by Mann and Wald is a little complex to describe, but this stems from the complexity of the problem.

They define the distance $\Delta$ between the null distribution and any alternative distribution as the maximum difference between the heights of the two cumulative distribution functions. It becomes evident, after some examination of the problem, that there is no hope of choosing $k$ so as to maximize the power function of $X^2$ at all points along its course. They decide to concentrate on maximizing the power at about the point where the power is $\frac{1}{2}$. This is an arbitrary but reasonable choice. The two principal properties possessed by their "best" $k$ are as follows.

(i) For a value of $\Delta$ which they determine, the power of the $X^2$ test is at least $\frac{1}{2}$ for all alternative distributions whose distance from the null distribution is at least $\Delta$. This value $\Delta$ is a simple function of sample size and, as would be expected, it decreases steadily with increasing sample size.

(ii) If any $k$ other than the "best" is chosen, the power of $X^2$ is less than $\frac{1}{2}$ for at least one alternative whose distance from the null distribution exceeds $\Delta$.

The best $k$ is given by the formula

$$k = 4 \left[ \frac{2(n-1)^2}{c^2} \right]^{\frac{1}{5}},$$

where

$$\frac{1}{\sqrt{2\pi}} \int_c^\infty e^{-(x^2/2)} \, dx = \alpha,$$

where $\alpha$ is the significance level. Thus $c = 1.64$ for a test at the 5% level.

The optimum values of $k$ are substantially higher than those customary in practice. For a test at the 5% level, $k$ rises slowly from 31 at $n = 200$ to 78 at $n = 2,000$.

TABLE 3

*Optimum number of classes (Mann and Wald)*

| $n$ | 200 | 400 | 600 | 800 | 1,000 | 1,500 | 2,000 |
|-----|-----|-----|-----|-----|-------|-------|-------|
| $k$ | 31 | 41 | 48 | 54 | 59 | 70 | 78 |

A good exposition and critique of the Mann-Wald paper has been given by Williams [31]. The Mann-Wald method is more tedious to compute than the usual procedure, partly because of the increased number of classes and partly because of the fitting with equal *expected* numbers. Williams shows, however, that the optimum is a broad one, and that the value of $k$ in Table 3 can probably be halved with little loss in sensitivity.

The Mann-Wald paper, although an able performance in a difficult field, is far from a complete investigation of the optimum number of classes. Such an investigation is unlikely to be forthcoming soon. What is the user of the $X^2$ test to conclude from their results? My own reaction has been to put more computational work into $X^2$ tests of continuous distributions, by increasing the number of intervals and using unequal lengths of interval where this is necessary in order to avoid classes with high expected numbers. For sample sizes between 200 and 1000, their recommended expected numbers per class in Table 3 range from 6 to 16. With Williams' modification, the range is from 12 to 32. This does suggest that there is an appreciable loss of power if classes with expectations of more than 50 are commonly used.

**10. Summary recommendations.** The following is an attempt to set down in brief form the recommendations about the computation of $X^2$ which flow from the discussion in this part and from practical experience. The recommendations are not as explicit as I should like. They can, I believe, be made more explicit, but this requires detailed study that goes beyond the scope of the present paper. The total number of observations is $n$.

*I. Attribute data.* The data come to us in grouped form. Pooling of classes is considered undesirable because of loss of power.

(a) *The 2 × 2 table.* Use Fisher's exact test (i) if $n < 20$, (ii) if $20 < n < 40$ and the smallest expectation is less than 5. Mainland's tables [20] are helpful in all such cases. If $n > 40$, use $X^2$, corrected for continuity if the smallest expectation is less than 500.

(b) *Tables with degrees of freedom between 2 and 60 and all expectations less than 5.* If $n$ is so small that Fisher's exact test can be computed without excessive labor, use this. Otherwise use $X^2$, considering whether this needs correction for continuity by finding the next largest value of $X^2$.

(c) *Tables with degrees of freedom greater than 60 and all expectations less than 5.* Try to obtain the exact mean and variance of $X^2$ and use the normal approximation to the exact distribution.

(d) *Tables with more than 1 degree of freedom and some expectations greater than 5.* Use $X^2$ without correction for continuity.

*II. Continuous data.* The data must first be grouped. Use enough cells to keep the expectations down to the levels recommended by Williams (12 per cell for $n = 200$, 20 per cell for $n = 400$, 30 per cell for $n = 1,000$). At the tails, pool (if necessary) so that the minimum expectation is 1.

## PART III. UTILITY OF THE TEST

**11. Criticisms and limitations of the test.** A competent appraisal of the utility of the $X^2$ test would require a sampling survey of scientists and others who try to draw conclusions from data. In such a survey the object would be to discover how frequently these workers have occasion to use a $X^2$ test, and to what extent the application of the test really seems to help them. In fact, such a survey, directed at the use of statistical techniques in general and not merely at the $X^2$ test, might be very illuminating to statisticians if it could be carried out despite the obvious difficulties. Statisticians are, I think, rather quick to jump to conclusions about the kinds of problems which scientists in other fields are supposed to face, and about their presumed uses and misuses of statistical methods and ideas.

In the absence of survey data of this kind, the statistician can give only a personal opinion, based on such contacts as he has had with the users of the $X^2$ test. I will content myself with the cautious statement that since the construction of hypotheses and their continued modification or rejection in the light of new data is one of the standard tools of science, some kind of test of the agreement between theory and data must often be useful. The experiences of Airy and Merriman illustrate the uncomfortable position in which the scientist is placed when he has to state, without the benefit of such a test, whether his observations are in accordance with the predictions of some theory.

On the other hand, a reading of the literature reveals the opinion, expressed by several writers, that the $X^2$ test is of restricted usefulness. The reasons for this critical verdict seem to be diverse. Some of the criticism is directed at the $X^2$ test itself, but some seems to apply to composite, or "general purpose" tests of significance as a whole, and some to *all* tests of significance.

Considering first the criticisms of $X^2$ itself, the name "goodness of fit" is misleading, because the power of the test to detect an underlying disagreement between theory and data is controlled largely by the size of the sample. With a small sample, an alternative hypothesis which departs violently from the null hypothesis may still have a small probability of yielding a significant value of $X^2$. In a very large sample, small and unimportant departures from the null hypothesis are almost certain to be detected. Consequently, when $X^2$ is nonsignificant, the amount by which the null hypothesis has been strengthened depends mainly on the size of sample. This is one of the principal reasons for such misuse of the test as exists. Authors sometimes write as if the validity of their null hypothesis has been greatly strengthened, if not definitely established, by a goodness of fit test made on very scanty data.

Secondly, as Gumbel has pointed out, the $X^2$ test for a continuous frequency distribution is not unique, because of the freedom of choice of number of intervals and end-points of the intervals. Although this is an argument for more standardization in the application of the test, the objection perhaps is minor rather than major. At least, statisticians have not seen any overwhelming advantage in having just one test of a given null hypothesis against a given alternative. In recent years there has been active research in the development of quick, though inefficient, tests for problems in which satisfactory, but less speedy, tests already exist. The tests will give different values of $P$ from the same data, but no serious objections to this situation seem to have been noticed.

There are two available substitute tests which resemble the $X^2$ test in that they are not directed against any specific alternative. One is the $\omega^2$ test (Section 13). This was constructed in order to avoid the grouping of continuous data that is necessary with $X^2$. The other, for data that are in grouped form, is the likelihood ratio test against a completely general alternative hypothesis (Section 14).

One limitation of $X^2$, or of any nonspecific test, is that when the alternative hypotheses can be fairly clearly defined, we may hope to obtain another test, directed against these alternatives, that will be more powerful than $X^2$. An example is Neyman's "smooth" tests (Section 15). These were constructed to detect alternative hypotheses that depart from the null hypothesis in some continuous or smooth fashion. Like $X^2$, the smooth tests are still general, since they do not demand detailed knowledge of the nature of the alternatives. Further down the scale there is a variety of supplementary tests to $X^2$.

Finally, the $X^2$ test is sometimes used when what is needed is not a test of significance of the usual type. There are numerous occasions when the null hypothesis is not expected to be exactly true, but at best approximately true. The argument against $X^2$ in this situation has been developed amusingly by Berkson [32]. He writes "I make the following dogmatic statement, referring for illustration to the normal curve: 'If the normal curve is fitted to a body of data representing any real observations whatever of quantities in the physical world, then if the number of observations is extremely large—for instance, on the order of 200,000—the chi-square $P$ will be small beyond any usual limit of significance.'"

If this statement is granted—and counter-evidence, to put it mildly, is not abundant—then Berkson proceeds to the Socratic conclusion. What is the point of applying a $X^2$ test to a moderate or small sample if we already know that a large sample would show $P$ highly significant?

In his original paper, Karl Pearson was aware of this issue, and did not seem to feel uncomfortable about it. He writes, "Nor again does it appear to follow that if the number be largely increased the same curve will still be a good fit. Roughly the $\chi^2$'s of two samples appear to vary for the same grouping as their total contents. Hence if a curve be a good fit for a large sample, it will be good for a small one, but the converse is not true, and a larger sample may show that our theoretical frequency gives only an approximate law for samples of a certain size. In practice we must attempt to obtain a good fitting frequency for such groupings as are customary or utile. To ascertain the ultimate law of distribution of a population for any groupings, however small, seems a counsel of perfection." Although it is hazardous to try to read another man's mind, his attitude was apparently the defensible one that any theory is at best approximately true, but nevertheless, if we are going to reject a theory, we do so because it does not fit the data that we have, not because it would not fit a much larger sample of data that we do not have.

Nevertheless, I would agree with Berkson that in this situation an ordinary test of significance is not very useful. It is more difficult to say just what we do want. One attack would be to reformulate the null hypothesis so that, instead of testing whether a binomial $p$ equals $p_0$ , we try to construct a test of the null hypothesis that $p$ lies in the specified range $p_0$ , $p_1$ .

As an alternative approach, fiducial or confidence limits seem to be helpful. Suppose that these limits are set up for the difference between two percentages in a $2 \times 2$ contingency table, the ordinary null hypothesis being that the true difference is zero. If the two limits are far from zero, then even when $X^2$ is non-significant we are warned that the data do not establish the null hypothesis as approximately true. If the limits are both near zero, on the other hand, we may be able to conclude that the null hypothesis, although presumably not exactly true, is close enough to the correct hypothesis for all practical purposes.

In testing goodness of fit of a frequency distribution, the extension of this approach is Kolmogorov's method [33] for constructing confidence bounds for the cumulative frequency distribution, given a sample.

To summarize, the $X^2$ test is helpful primarily in the exploratory stages of an investigation, when there is no very clear knowledge of the alternative hypotheses. It is well to remember that the size of sample determines whether the test really is a severe test of the null hypothesis.

**12. Interpretation of high $P$'s.** The question of the interpretation to be placed on very high $P$'s, say those greater than 0.99, has been raised from time to time. In the few instances of this kind that have come my way, my practice has been to give the data further scrutiny before regarding the result as evidence in favor of the null hypothesis.

Events have justified this practice. In nearly every instance, something wrong was discovered, most frequently a numerical mistake or an error in the formula used to compute $X^2$. In one set of data assembled by geneticists, a whole group of $X^2$ showed $P$'s of the order of 0.999. The reason was that these $X^2$ values had been obtained by adding a large number of $X^2$ values, each with 1 degree of freedom, and all the original (1 d.f.) $X^2$ had been corrected for continuity. The over-correction which is a feature of this device had piled up to such an extent that their total $X^2$'s were much smaller than those following the $\chi^2$ distribution. In another case, after discussion with the assistants of the scientist in charge, I surmised that the observations had been influenced by the anticipations of the scientist. Fisher [34] has raised a similar speculation with respect to some of Mendel's results, without any suggestion of improper scientific conduct on the part of Mendel.

## PART IV. TESTS WHICH ARE COMPETITIVE TO $X^2$

**13. The $\omega^2$ test.** Alternatives that have been proposed to the $X^2$ test are of two kinds. Several of the tests, like $X^2$, are "general" tests. Then there is a battery of supplementary tests that are intended for situations where the alternative hypothesis is more definitely known.

The *general* substitute tests that have been proposed have not given $X^2$ very serious competition. This is understandable because of the long history of $X^2$ and of its inclusion as standard doctrine in most elementary courses, and because some of the substitute tests are limited in the type of hypothesis with which they can cope. Moreover, despite the weaknesses of the $X^2$ test discussed in Part III, the advantages of the alternative tests have not yet been clearly enough demonstrated to win many converts. Consequently, Part IV contains only a brief and rather noncommital introduction to these tests.

The first general test, developed by Cramér [35], von Mises [36] and Smirnov [37], was constructed mainly for use with small samples. The null hypothesis completely specifies the frequency distribution followed by the observations. Unlike $X^2$, the $\omega^2$ test requires no grouping of the observations, an obvious advantage with small samples. The test is based on a comparison of the cumulative frequency function $F(x)$ specified by the null hypothesis with an estimate of the cumulative frequency made from the sample. This estimate, $F^*(x)$, is simply $r/n$, where $r$ is the number of observations in the sample which are $\leq x$. The test criterion proposed is the Stieltjes integral

$$\omega^2 = \int_{-\infty}^{\infty} [F(x) - F^*(x)]^2 \, dF(x).$$

If $F(x)$ is continuous, this may be shown to satisfy

$$\omega^2 = \frac{1}{12n^2} + \frac{1}{n} \sum_{r=1}^{n} \left[ F(x_r) - \frac{2r-1}{2n} \right]^2,$$

where the values $x_1, x_2, \cdots, x_n$ are now arranged in increasing order.

The mean and variance of $\omega^2$ are known, and also its limiting distribution (which is nonnormal) as $n \to \infty$. A table of this distribution by Darling and Anderson [38] has appeared recently. Practical use of this test is restricted by the condition that $F(x)$ must be known and by lack of information about the small-sample distribution of $\omega^2$.

**14. The likelihood ratio test.** If the data are presented in grouped form, and if the alternative hypothesis is completely general, it is known that in large samples the $X^2$ test and the likelihood ratio test become equivalent [9]. We start from the usual multinomial

$$Pr = \frac{n!}{x_1! \cdots x_k!} \, p_1^{x_1} p_2^{x_2} \cdots p_k^{x_k}.$$

The likelihood on the null hypothesis is found from $p_i = m_i/n$, where $m_i$ are the expectations estimated by maximum likelihood (unless they are explicitly given). The likelihood on the unrestricted alternative is found from $p_i = x_i/n$. Hence the likelihood ratio becomes

$$\left(\frac{x_1}{m_1}\right)^{x_1} \left(\frac{x_2}{m_2}\right)^{x_2} \cdots \left(\frac{x_k}{m_k}\right)^{x_k}.$$

Its logarithm is

$$L = \sum_{i=1}^{k} x_i \log \left(\frac{x_i}{m_i}\right) = \sum_{i=1}^{k} x_i \log \left\{ 1 + \frac{x_i - m_i}{m_i} \right\}.$$

When this is expanded in a power series in the $(x_i - m_i)$, the leading term is $X^2$ for the maximum likelihood estimates of the parameters.

In view of the equivalence of the two criteria in large samples, there seems no advantage, except one of taste or convenience, in one test over the other.

For small samples, the suggestion has been made from time to time that the likelihood ratio is to be preferred. Examples worked by both tests have been presented and discussed by Neyman and Pearson [22] and Fisher [39]. Since users of statistical methods naturally do not wish to learn more tests than are necessary, a movement to replace $X^2$ by the likelihood ratio seems unlikely to gather momentum unless some definite advantages can be shown to follow. The advantage in computing time is at most small, but there may be an increase in power. The striking way in which many different configurations of the data turn out to give exactly the same value of $X^2$ in small samples suggests an element of coarseness in the $X^2$ test. This coalescence happens to a much reduced extent with the likelihood ratio. However, not enough data about relative power has accumulated to permit a verdict on this issue.

**15. Neyman's smooth tests.** As in the $\omega^2$ test, Neyman [40] postulates that the cumulative frequency $F(x)$ (assumed continuous) is known exactly from the null hypothesis. The first step is to replace the observations $x_i$ by the familiar "probability integral" transforms $y_i$, where

$$y_i = F(x_i).$$

If the null hypothesis is correct, the variates $y_i$ follow a rectangular distribution in the interval $(0, 1)$. The problem, therefore, reduces to that of finding a test for this transformed hypothesis.

Neyman points out that the conceivable alternatives to the null hypothesis fall into two broad classes. The first class, of "smooth" alternatives, contains frequency functions which are continuous and which depart in some gradual and regular manner from the null hypothesis. The second class contains all other alternatives, whose deviation from the null hypothesis is in some respects erratic or discontinuous. The $X^2$ test is not directed specifically at either class, and is to some extent effective against both types of departure from the null hypothesis. Neyman's object is to develop tests sensitive to the first class of alternatives.

If a "smooth" alternative holds, the transforms $y_i$ will no longer follow a rectangular distribution, but will presumably follow a continuous distribution with a limited number of maxima and minima. The proposal is, therefore, to test the $y_i$ for polynomial trends, on the assumption that a polynomial of fairly small degree will satisfactorily represent the smooth alternative. The computations involved and a discussion of the appropriate degree of the polynomial are presented in [40].

### PART V. TESTS WHICH ARE SUPPLEMENTARY TO $X^2$

**16. A supplementary test based on runs.** $X^2$ takes no account of the succession of $+$ and $-$ signs in the deviations $(x_i - m_i)$ between observations and expectations. When a smooth alternative holds, it seems likely that the succession of signs will exhibit some systematic features, such as a long run of $+$'s followed by a run of $-$'s, and this has often been observed in applications of $X^2$. David [41] has adapted the now familiar "run" test as a supplementary test to $X^2$. In the run test, we count the number of runs and refer to a table which shows the significance levels of this quantity, given the total numbers of $+$'s and $-$'s in the series. David has shown that the limiting distribution of the number of runs is independent of that of $X^2$.

The run tests will, of course, be most effective for alternatives which produce few runs, such as a shift in the mean of the distribution. In the reference cited, the test is restricted to the case where the null hypothesis completely specifies the distribution; David states that a test has been developed for the case in which some parameters must be estimated.

**17. Tests based on low moments.** When the null hypothesis postulates that the observations follow a normal, binomial or Poisson distribution, an alternative to $X^2$ that is in fairly common use is to compare the lower moments of the theoretical distribution with estimated moments from the sample. With the normal distribution, the actual values of the mean and variance are rarely given by the null hypothesis, so that a comparison of these moments is not usually possible. Tests of skewness, derived from the third moment, and of kurtosis, derived from the fourth moment, can be made [15].

In the binomial distribution, if $p$ is specified we can compare the sample mean and variance with the theoretical mean and variance. If $p$ is not specified, we can compare variances. Suppose that we have $g$ series of trials, and that each series contains $s$ trials. The number of successes (out of $s$) in the $i$th trial is $x_i$. For a test of the mean proportion of successes, we regard

$$\frac{\left(\dfrac{\bar{x}}{s} - p\right)}{\sqrt{\dfrac{pq}{gs}}}$$

as a normal deviate.

**18. Dispersion tests.** Turning to the variance of $x_i$, if $p$ is specified the estimated variance is $\sum (x_i - sp)^2/g$, while the theoretical variance is $spq$. An appropriate test criterion for the variance is, therefore,

$$\frac{\sum (x_i - sp)^2}{gspq}.$$

If $p$ must be estimated from the data, either because it is unspecified or because the sample estimate disagrees with the postulated $p$, the test criterion becomes

$$\frac{\sum (x_i - \bar{x})^2}{gs\left(\dfrac{\bar{x}}{s}\right)\left(1 - \dfrac{\bar{x}}{s}\right)} = \frac{s \sum (x_i - \bar{x})^2}{g\bar{x}(s - \bar{x})}.$$

As is well known, the related quantities

(13) $$\frac{\sum (x_i - sp)^2}{spq} \quad \text{and} \quad \frac{s \sum (x_i - \bar{x})^2}{\bar{x}(s - \bar{x})}$$

are distributed approximately as $\chi^2$ with $g$ and $(g - 1)$ degrees of freedom, respectively, when the null hypothesis is true.

The variance test can also be made when the number of trials $s_i$ varies from series to series. The test criterion becomes

$$\frac{\sum s_i \left( \frac{x_i}{s_i} - \frac{\bar{x}}{\bar{s}} \right)^2}{\frac{\bar{x}}{\bar{s}} \left( 1 - \frac{\bar{x}}{\bar{s}} \right)} = \frac{\bar{s}^2 \left\{ \sum \frac{x_i^2}{s_i} - \frac{\left( \sum x_i \right)^2}{\sum s_i} \right\}}{\bar{x}(\bar{s} - \bar{x})}.$$

This test criterion can be shown to be identical with $X^2$ as calculated for the $2 \times g$ contingency table.

| $x_1$ | $x_2$ | $\cdots$ | $x_g$ | $\sum x_i$ |
|---|---|---|---|---|
| $s_1 - x_1$ | $s_2 - x_2 \cdots s_g - x_g$ | | | $\sum s_i - \sum x_i$ |
| $s_1$ | $s_2$ | $\cdots$ | $s_g$ | $\sum s_i$ |

It is important to distinguish clearly between this variance $X^2$ test, which is sometimes called a *dispersion* test, and the ordinary goodness of fit $X^2$ test. Suppose that we have 200 families each of size 4, and that every individual belongs to one of two classes $A$ and $a$. The null hypothesis states that the probability $p$ of an $A$ is constant for all families. We may tally the numbers of families that have 0, 1, 2, 3, 4 $A$'s, respectively, and test this frequency distribution against the binomial $(q + p)^4$. This is the ordinary goodness of fit test, which has 3 degrees of freedom if $p$ is unknown. The dispersion test, computed by formula (13), compares the observed variance of this frequency distribution with the theoretical binomial variance.

The dispersion tests frequently prove more sensitive than $X^2$ when the binomial null hypothesis fails because the probability of an $A$ varies from one family to another. The notion of a measure of dispersion of this kind is due to Lexis and antedates the goodness of fit test.

**19. Subdivision of $X^2$ into components.** In the analysis of variance, the subdivision of a sum of squares into single components, or "single degrees of freedom," is a familiar device. If variates $y_i$ are normally and independently distributed with mean 0 and variance $\sigma^2$ on some null hypothesis, these components are obtained by any linear transformation of the form

$$z_i = \sum_{j=1}^{k} l_{ij} y_j \qquad (i = 1, 2, \cdots k)$$

where

$$(14) \qquad \sum_j l_{ij} l_{hj} = \begin{cases} 0, & i \neq h, \\ 1, & i = h. \end{cases}$$

All $z_i$ are normally and independently distributed with mean 0 and variance $\sigma^2$. This transformation enables us to select those $z_i$ that are likely to be sensitive to a particular alternative hypothesis. Often only one or two of the $z_i$ are examined, because it is hard to imagine any feasible alternative that would make the other $z$'s large. Thus the device replaces a "sum of squares" test by a few more specialized tests.

The corresponding subdivision of $X^2$ is easily obtained from Fisher's device of regarding the observed numbers $x_i$ in the cells as following independent Poisson distributions, subject to a single linear restriction. Thus when all expectations are large, we may take

$$y_i = \frac{x_i - m_i}{\sqrt{m_i}}$$

as the set of unit normal deviates. Since these are subject to the linear restriction

$$\sum_{j=1}^{k} (x_j - m_j) = \sum_{j=1}^{k} \sqrt{m_j}\, y_j = 0,$$

we must take

$$Z_1 = \sum \sqrt{m_j}\, y_j / n.$$

Let the remaining $Z_i$ $(i = 2, 3, \cdots k)$ be

$$Z_i = \sum_j l'_{ij} x_j = \sum l'_{ij} \sqrt{m_j}\, y_j + \sum l'_{ij} m_j.$$

If these are to have means zero, we must have

$$(15) \qquad \sum_j l'_{ij} m_j = 0.$$

Note that this relation makes all the remaining $Z_i$ orthogonal with $Z_1$. Since $l_{ij} = l'_{ij} \sqrt{m_j}$, equations (14) become

$$(16) \qquad \sum_j l'_{ij} l'_{hj} m_j = \begin{cases} 0 & i \neq h, \\ 1 & i = h. \end{cases}$$

Any set of $Z_i$ whose coefficients satisfy equations (15) and (16) provide a breakdown of $X^2$ into $(k - 1)$ single components. Then as an algebraic identity,

$$X^2 = \sum_{i=1}^{k} \frac{(x_i - np_i)^2}{np_i} = \sum_{i=1}^{k-1} Z_i^2.$$

As $n$ increases, the individual terms on the right become in the limit independently distributed as $\chi^2$ with 1 degree of freedom. In genetic analysis, where simple interpretations can be attached to the $Z_i$, this tool has proved useful [15].

The application of this breakdown to contingency tables, which requires care, has been elucidated by Lancaster [42] and Irwin [43]. In an $r \times c$ contingency table, $X^2$ can be partitioned into $(r-1)(c-1)$ single components. Each of these represents the usual $X^2$ for a $2 \times 2$ table which is formed by amalgamation of cells in the original table. This breakdown is illustrated below for a $3 \times 3$ table.

*Original table*

| | | | |
|---|---|---|---|
| $x_{11}$ | $x_{12}$ | $x_{13}$ | $r_1$ |
| $x_{21}$ | $x_{22}$ | $x_{23}$ | $r_2$ |
| $x_{31}$ | $x_{32}$ | $x_{33}$ | $r_3$ |
| $c_1$ | $c_2$ | $c_3$ | $n$ |

*Components*

| | | | | | |
|---|---|---|---|---|---|
| $x_{11}$ | $x_{12}$ | $r_{12}$ | $r_{12}$ | $x_{13}$ | $r_1$ |
| $x_{21}$ | $x_{22}$ | $r_{22}$ | $r_{22}$ | $x_{23}$ | $r_2$ |
| $c_{21}$ | $c_{22}$ | $n_{22}$ | $n_{22}$ | $c_{23}$ | $n_{23}$ |
| $c_{21}$ | $c_{22}$ | $n_{22}$ | $n_{22}$ | $c_{23}$ | $n_{23}$ |
| $x_{31}$ | $x_{32}$ | $r_{32}$ | $r_{32}$ | $x_{33}$ | $r_3$ |
| $c_1$ | $c_2$ | $n_{32}$ | $n_{32}$ | $c_3$ | $n$ |

If the $X^2$ are calculated in the usual way for each $2 \times 2$ table, the partition is only approximate, in that in finite samples the individual $X^2$ do not add up to the total $X^2$ for the $3 \times 3$ table. The authors show how to obtain a partition which adds up exactly, that is, which satisfies the sets of equations (15) and (16). It appears that the approximate partition is adequate for most tests of significance; in fact, it has not been shown that the additive partition is really preferable to the approximate partition in small samples.

Another application of the breakdown of $X^2$ is to contingency tables in which numerical scores can be attached to one or both of the classifications. Yates [44] shows how to isolate and compare the regressions of the observations on these scores.

In conclusion, the testing of individual components of $X^2$ is analogous to the use of a set of independent $t$-tests instead of, or in addition to, an $F$-test in the analysis of variance.

I wish to thank T. W. Anderson, E. L. Lehmann and J. W. Tukey for many helpful suggestions.

## REFERENCES

[1] KARL PEARSON, "On the criterion that a given system of deviations from the probable in the case of a correlated system of variables is such that it can be reasonably supposed to have arisen from random sampling," *Philos. Mag. Series 5*, Vol. 50 (1900), pp. 157–172.

[2] H. HOTELLING, "The consistency and ultimate distribution of optimum statistics," *Trans. Am. Math. Soc.*, Vol. 32 (1930), pp. 847–859.

[3] R. A. FISHER, "On the interpretation of chi square from contingency tables, and the calculation of $P$," *Jour. Roy. Stat. Soc.*, Vol. 85 (1922), pp. 87–94.

[4] M. GREENWOOD AND G. U. YULE, "The statistics of anti-typhoid and anti-cholera inoculations and the interpretation of such statistics in general," *Proc. Roy. Soc. Med.*, Vol. 8 (1915), pp. 113–190.

[5] R. A. FISHER, "The conditions under which chi square measures the discrepancy between observation and hypothesis," *Jour. Roy. Stat. Soc.*, Vol. 87 (1924), pp. 442–450.

[6] G. U. YULE, "On the application of the $\chi^2$ method to association and contingency tables, with experimental illustrations," *Jour. Roy. Stat. Soc.*, Vol. 87 (1922), pp. 76–82.

[7] J. BROWNLEE, "Some experiments to test the theory of goodness of fit," *Jour. Roy. Stat. Soc.*, Vol. 87 (1924), pp. 76–82.

[8] H. CRAMÉR, *Mathematical Methods of Statistics*, Princeton University Press, 1946, p. 424.

[9] J. NEYMAN, "Contribution to the theory of the $\chi^2$ test," *Proceedings of the Berkeley Symposium on Mathematical Statistics and Probability*, University of California Press, 1949, pp. 239–273.

[10] E. FIX, "Tables of the Noncentral $\chi^2$," *Univ. California Publ. Stat.*, Vol. 1 (1949), pp. 15–19.

[11] P. B. PATNAIK, "The non-central $\chi^2$- and $F$-distributions and their applications," *Biometrika*, Vol. 36 (1949), pp. 202–232.

[12] A. WALD, "Tests of statistical hypotheses concerning several parameters when the number of observations is large," *Trans. Am. Math. Soc.*, Vol. 54 (1943), pp. 426–482.

[13] G. A. BARNARD, "Significance tests for $2 \times 2$ tables," *Biometrika*, Vol. 34 (1947), pp. 123–138.

[14] R. A. FISHER, *The Design of Experiments*, Oliver and Boyd Ltd., Edinburgh, 1935.

[15] R. A. FISHER, *Statistical Methods for Research Workers*, Edinburgh, 5th and subsequent editions, Oliver and Boyd Ltd., Edinburgh, 1934, Section 21.02.

[16] K. D. TOCHER, "Extension of the Neyman-Pearson theory of tests to discontinuous variates," *Biometrika*, Vol. 37 (1950), pp. 130–144.

[17] D. LEWIS AND C. J. BURKE, "The use and misuse of the chi-square test," *Psych. Bull.*, Vol. 46 (1949), pp. 433–489.

[18] A. L. EDWARDS, "On the use and misuse of the chi-square test—the case of the $2 \times 2$ contingency table," *Psych. Bull.*, Vol. 47 (1950), pp. 341–346.

[19] W. G. COCHRAN, "The $\chi^2$ correction for continuity," *Iowa State Coll. Jour. Sci.*, Vol. 16 (1942), pp. 421–436.

[20] D. MAINLAND, "Statistical methods in medical research," *Canadian Jour. Res., E*, Vol. 26 (1948), pp. 1–166.

[21] R. A. FISHER AND F. YATES, *Statistical Tables for Biological, Agricultural and Medical Research*, Oliver and Boyd, Ltd., Edinburgh, 1938, Table VIII.

[22] J. NEYMAN AND E. S. PEARSON, "Further notes on the $\chi^2$ distribution," *Biometrika*, Vol. 22 (1931), pp. 298–305.

[23] W. G. COCHRAN, "The $\chi^2$ distribution for the binomial and Poisson series, with small expectations," *Annals of Eugenics*, Vol. 7 (1936), pp. 207–217.

[24] P. V. SUKHATME, "On the distribution of $\chi^2$ in small samples of the Poisson series," *Jour. Roy. Stat. Soc. Suppl.*, Vol. 5 (1938), pp. 75–79.

[25] G. H. FREEMAN AND J. H. HALTON, "Note on an exact treatment of contingency, goodness of fit and other problems of significance," *Biometrika*, Vol. 38 (1951), pp. 141–149.

[26] J. B. S. HALDANE, "The exact value of the moments of the distribution of $\chi^2$, used as a test of goodness of fit, when expectations are small," *Biometrika*, Vol. 29 (1937), pp. 133–143.

[27] J. B. S. HALDANE, "The mean and variance of $\chi^2$, when used as a test of homogeneity, when expectations are small," *Biometrika*, Vol. 31 (1939), pp. 346–355.

[28] F. YATES, "Contingency tables involving small numbers and the $\chi^2$ test," *Jour. Roy. Stat. Soc. Suppl.*, Vol. 1 (1934), pp. 217–235.

[29] H. B. MANN AND A. WALD, "On the choice of the number of class intervals in the application of the chi square test," *Annals of Math. Stat.*, Vol. 13 (1942), pp. 306–317.

[30] E. J. GUMBEL, "On the reliability of the classical $\chi^2$ test," *Annals of Math. Stat.*, Vol. 14 (1943), pp. 253–263.

[31] C. ARTHUR WILLIAMS, "On the choice of the number and width of classes for the chi-square test of goodness of fit," *Jour. Am. Stat. Assn.*, Vol. 45 (1950), pp. 77–86.

[32] J. BERKSON, "Some difficulties of interpretation encountered in the application of the chi-square test," *Jour. Am. Stat. Assn.*, Vol. 33 (1938), pp. 526–536.

[33] A. KOLMOGOROV, "Confidence limits for an unknown distribution function," *Annals of Math. Stat.*, Vol. 12 (1941), pp. 461–465.

[34] R. A. FISHER, "Has Mendel's work been re-discovered?" *Annals of Science*, Vol. 1 (1936), pp. 115–137.

[35] H. CRAMÉR, "On the composition of elementary errors," *Skandinavisk Aktuarietidskrift*, Vol. 11 (1928), pp. 13–74, 141–180.

[36] R. VON MISES, *Wahrscheinlichkeitsrechnung und ihre Anwendung in der Statistik und theoretischen Physik*, Deuticke, Leipzig und Wien, 1931, pp. 316–335.

[37] N. SMIRNOV, "Sur la distribution de $\omega^2$," *C. R. Acad. Sci.* Paris, Vol. 202 (1936), p. 449.

[38] D. A. DARLING AND T. W. ANDERSON, "Asymptotic theory of certain goodness of fit criteria based on stochastic processes," *Annals of Math. Stat.*, Vol. 23 (1952), pp. 193–212.

[39] R. A. FISHER, "The significance of deviations from expectation in a Poisson series," *Biometrics*, Vol. 6 (1950), pp. 17–24.

[40] J. NEYMAN, "Smooth test for goodness of fit," *Skandinavisk Aktuarietidskrift*, Vol. 20 (1937), pp. 150–199.

[41] F. N. DAVID, "A $\chi^2$ 'smooth' test for goodness of fit," *Biometrika*, Vol. 34 (1947), pp. 299–310.

[42] H. O. LANCASTER, "The derivation and partition of $\chi^2$ in certain discrete distributions," *Biometrika*, Vol. 36 (1949), pp. 117–129.

[43] J. O. IRWIN, "A note on the subdivision of $\chi^2$ into components," *Biometrika*, Vol. 36 (1949), pp. 130–134.

[44] F. YATES, "The analysis of contingency tables with groupings based on quantitative characters," *Biometrika*, Vol. 35 (1948), pp. 176–181.

# ANALYSIS OF RECORDS WITH A VIEW TO THEIR EVALUATION

## WILLIAM G. COCHRAN

AS Dr. Baehr said, we are all agreed that the problem of trying to find out what effects this Demonstration has produced at the end of the period is going to be difficult. But there are large areas of biological research, where the experimental subjects may be plants or animals or even, in some cases, human beings, and where the scientist does have enough flexibility and control to run the experiment as he pleases. As a temporary escape from the harsh realities of the problems that face us, I would like to consider what precautions the scientist finds it advisable to take, when he does have this degree of control, in order to make the experiment, in his view, a sound one. I shall suppose that the experiment in question has a control group to whom essentially nothing is to be done, and a study group to whom some kind of procedure or treatment is to be applied.

He will first take some steps that are designed to insure the comparability of the two groups. These steps may include pairing or matching of the groups, member by member, if that is thought worth while. They will include some kind of random assignment of the subjects to the groups. After the groups have been made up, some initial measurements or records may be taken in order to verify that the groups really are comparable on characteristics that he thinks are relevant to their probable responsiveness to the treatment.

Secondly, he will take considerable pains to describe what the treatment is. This is necessary for several obvious reasons. It is necessary for communication, in order that other scientists and readers of his work will know what he has done. It is necessary for interpretation, when he and others begin to speculate about the causes behind any of the effects shown by the treatment. It is necessary also for practical application, if the treatment seems to produce

---

effects that are beneficial. People who propose to adopt the treatment in practice must know exactly what the treatment was.

There is a third step, which usually comes early in the planning of the experiment. The scientist will think carefully about the range of effects that the treatment might produce. In an exploratory study he may not be able to do this with much detail: all that he may have is a broad list of possible effects. In other studies he may have narrowed down his field of interest to very few items. Whatever the situation is, he will regard this step as an important one, and there are various devices in the different fields for making sure that this step is not overlooked. Some people like to have what they call a list of hypotheses to be tested: others prefer a list of questions to be answered. With either approach, much care is exercised in constructing and revising this list.

The importance of this step is that it determines what we will attempt to measure and what shall be recorded. The system of records, then, is designed to secure data that will answer the series of questions. When the system of records is being constructed, there should be checking to ensure that the purpose of each record is known, and also that we have not omitted records that will be needed to answer some of the questions.

The process of measurement may require the use of a human observer to a greater or less degree. If the observer is used only incidentally so that it would make no difference to the reading obtained if one observer was suddenly substituted for another, we usually call the measurements *objective*. If the human observer plays a significant role, so that it is doubtful whether another human observer would get the same reading for the same situation, we call them *subjective*. In practice, there is of course a wide range of degrees of subjectivity in measurement.

With any measuring process, the scientist will take precautions to insure that it is unbiased as between the groups—that there isn't some sudden jump in the scale of measurements when he changes from the control to the treated group. He will want to know something about the precision of the measuring device. He will want to know something else that is harder to describe: it might be called the *relevance* of the measurement for his purpose. In other words, he should ask, Does this process really measure what I need to measure? This question of relevance should not be overlooked, because in view of the obvious advantages of objective measurements, it is easy to make the mistake of choosing an objective measurement that isn't relevant in preference to a subjective one that is.

If a human observer is used to a significant degree, the scientist will be innately suspicious and will insist that any observer shall do measurements in each group, that he shall have the same relationship and rapport with the members of each group and that he must not know, when he is measuring a subject, to which group the subject belongs. In some studies the scientist will regard this third precaution as very important, even if the observer is an outside consultant who has no personal interest in obtaining one kind of result rather than another.

After the records are taken, there comes the statistical analysis, which has two general objectives. First, we are aware that the measurements will be affected not only by the treatment, but also by a whole gamut of other influences that are sometimes called the experimental errors. One part of the statistical analysis therefore consists of preliminary computations known as tests of significance, whose purpose is to verify that these experimental errors did not mask the real effects of the treatments. When we get a non-significant result we have reached the disappointing conclusion that whatever real effect of the treatment may have been present, it wasn't large enough to show up convincingly relative to the experimental errors.

The second part of the statistical analysis, which often can be difficult, is an estimation of the sizes of the effects produced by the treatment and an appraisal of the importance of these effects either for practical application or as a contribution to scientific knowledge. This step might seem so obvious as not to need mentioning, but in reading the results of social science studies I have the impression that when the social scientist finds a result that is statistically significant, he sometimes heaves a sigh of relief and says, "Well, that will keep the statisticians quiet," and in his joy he forgets to tell us whether the effect is a large and interesting one or a small and inconsequential one.

Then finally, the last part of the analysis is the real fun—interpretation and discussion.

Had there been time, I would have liked to discuss why all of these precautions are considered worth while, although for most of them this is fairly obvious. Perhaps it is more important to point out that the precautions are not equally necessary, and the extent to which the various precautions are necessary changes from one kind of study to another, and often can be appraised only by judgment.

But in general, the scientist is trying to avoid two types of failure. He may fail to find effects that really are there. He can fail in this way if the measurements are imprecise, if the sizes of the groups are too small, or if he has taken measurements that are not relevant. The second thing to be avoided is bias; that is, something which distorts all the measurements in one group relative to the other. When there is a bias, statistically significant results may be obtained even when the treatment has been ineffective: or the results may be distorted in size and perhaps even in direction.

Bias is particularly to be avoided, because a biased study from a worker with a good reputation may start a period of dispute and discussion that holds up progress for several years.

I would now like to return to reality and consider some of these precautions in the light of this study. If I am not mistaken, there is trouble in varying degrees all along the line, and when I end this recital, I may well be in tears. So I would like to finish by spending a few minutes trying to cheer myself up.

The study does have controls, which were selected by randomization. I was a little shocked to hear an eminent biostatistician like Dr. Fertig express doubts about the effectiveness of this randomization. He may expect to hear more about this deviation from the statistician's party line. With groups as large as 150, I would not be worried about their comparability.

We will, however, be concerned about losses, trying to keep them down, and trying to find out whether they are selective; and if they are at all large, trying to check by such measurements as we can take whether the two groups that are compared in the end are really comparable.

There was also a suggestion that earlier measurements might be made of the controls. A number of proposals were put forward for that purpose yesterday, and these deserve consideration by the team. In this connection the question whether the degree of rapport between the measurer and the subject in the control group can be the same as that in the study group may be important. This question arises whether early measurement of controls is attempted or not. This issue is one argument for having some of the evaluation of effects on both groups done by an outside team, as suggested by Dr. Gruenberg. Such a team cannot measure the study group with the same depth and penetration as the staff of the FHD, but perhaps they can measure some significant variables with the same degree of rapport in both groups.

As to the nature of the treatment, we will have a fairly clear picture in the record of all conferences, including what advice was given, and what the outcome of the advice was. But in the discussions here it has been apparent that the visitors are puzzled as to what the team is trying to do. I am not thinking primarily of such questions as: Is their aim to give therapy for pathology or to stimulate the ability to cope? I don't think that the team need adopt any rigid classification. But everybody will be helped in their understanding of this study if the team can at some time produce a statement as to what their education and their efforts are trying to accomplish. As the team members know, this statement is not easy to construct, but it may be of great importance for the selection of the measurements that are to be taken to assay the effects of the demonstration.

In view of the great difficulty of the problems of measurement in this study, it may be well to remember that records can be of various kinds. The basic record can be a recording of an interview, directive or non-directive, or the answers to a questionnaire. From that, at one stage of summarization there can be an unordered classification which shows simply what changes occurred, taking any classification into which people seem to fall. Then there can be as the next step an ordered classification—much, little, none—and finally, as the last step, there is the metric scale 1, 2, 3, 4, implying an underlying continuum.

I would like to make two general suggestions. First, the less one commits oneself unalterably to the later stages of this classification, the better. This implies that the original records will be kept in an available form so that persons interested can put them together in various ways from different points of view, and that much of the analysis will be done in considerable detail and with fairly primitive classifications. This is not to discourage attempts to construct and utilize metric scales. But the phenomenon under study is very complex, and there is some risk that overall summary scales will contain hidden arbitrary judgments and will have placed together things that are essentially unlike. For example, as Dr. Fertig pointed out, the parts that go into the Individual Evaluation Summary may in the end be more useful than the Summary itself because the implications of the Summary are quite complicated to grasp.

A second maxim is to keep value judgments separate, identified and labeled as far as possible. Naturally, some value judgments are already implied in what the members of the team have decided to try to teach, and this we all appreciate. However, records should be presented so that somebody who doesn't agree with the value judgment of the team can do a different kind of analysis. For instance, the recording of excessive or nonexcessive sibling rivalry is perhaps undesirable in its assumption that we know how much sibling rivalry is good. Recording merely of a scale of sibling rivalry would be safer.

There has been much discussion as to whether some of the changes in emotional health can be measured at all; or at least, whether they can be measured without spending much more time in the process of measurement than is contemplated. I am not competent to discuss that issue and I imagine that the sociologists and psychologists would not agree among themselves about it. If this is the state of affairs, we should not judge harshly those who make a bold attempt and do the best that they can.

It will be extremely difficult to keep the evaluation free from bias. The records taken by the team inevitably will reflect the team's own estimates of what changes are taking place, and an outside group which examines these records for the purpose of evaluation will still be dependent on what the team has put down. Two aspects of the Demonstration have been referred to which may help in this difficulty. The first is the study that Dr. Creedon is proposing on what the people themselves feel about this program, what their expectations were when they came in and how these expectations were changed, which should give a rather independent appraisal of effects produced by the Demonstration. The second device is the use to some extent of an outside team which will construct and take its own records on both groups.

So far as the statistical analysis is concerned, I think that if the other loopholes can be plugged, the statistical analysis will be the least of our worries.

As we have seen, this is a complex area of research and nobody here, I think, would be inclined to promise success in making precise and unbiased measurements of the effects. What can we say to cheer ourselves up?

First, as has been abundantly evident from the discussion, there are a great many facets of interest in this Demonstration, so that despite the difficulties of evaluation, the eggs are by no means all in one basket. Secondly, although this study is unique in a number of ways, it is far from unique in its methodological problems, and particularly in the problem of evaluation. It is, in fact, almost typical of the problem of evaluation that exists in a great deal of current research in public health and sociology. Problems of this kind will probably become more rather than less common. Scientists working in similar field research will be grateful to this project for the constructive ideas that it seems sure to contribute to problems of evaluation, and also perhaps for advice on some pitfalls to avoid.

Then thirdly, a study may be of great value even though the expert looking after the event can point to many things that are unsatisfactory. A reference to Kinsey's studies may be appropriate. Kinsey, as a good quantitative biologist, began to study human sexual behavior. What did he decide to measure? He choose principally two quantities—the earliest age at which a given kind of sexual behavior was first engaged in and the number of orgasms per week in the various kinds of sexual behavior that he distinguished. These records obviously leave much to be desired as the raw material for a penetrating insight into sexual behavior. Further, the count data which he obtained are certainly subject to bias, and it is hard to say how large the biases are. The limitations of these data are evident in the volume on the male, in which many of the most interesting statements are not based on the tabular data, and it is not clear to the reader on what evidence the statements are based.

Nevertheless, in my own opinion, Kinsey's is the best study of sexual behavior in a mass population that has been carried out and is methodologically greatly superior to previous studies. Moreover, his work is a valuable supplement to the more penetrating and thoughtful studies that have been done by some others, but which have had to depend on a few individual cases, whose representativeness is unknown to us.

Finally, I would like to mention one piece of advice which is usually given to Scottish boys shortly after they are weaned, at the critical time when father first puts a golf club in the boy's hands: "Keep your eye on the ball and don't press." In a study of this kind, all sorts of suggestions are made for interesting sidelines that might be explored, but if the team is to get anywhere it must construct a system of priorities as to the tasks that it regards as most essential for its own goals in the study, and it must fight to maintain these priorities against blandishments and outside suggestions, unless and until it has the resources to cope with its main objectives with something to spare for additional studies.

And by "don't press," I mean this: This kind of study is very expensive: it takes a long time and attracts many visitors. The members of the team may come to feel that they are under pressure to produce results and they may begin to worry when some things inevitably go wrong. It is most important to do anything that can be done to lighten this pressure and encourage team members to be more relaxed. I am sure that the visitors have all been greatly impressed by the intelligence, courage, and persistence shown by the team, and I hope that the team members do feel relaxed. They can at least heave a sigh of relief when we leave.

# 51

## STATISTICAL PROBLEMS OF THE KINSEY REPORT*

WILLIAM G. COCHRAN, *Johns Hopkins University*
FREDERICK MOSTELLER, *Harvard University*
JOHN W. TUKEY, *Princeton University*

THIS is the report of a committee appointed by the Commission on Statistical Standards of the American Statistical Association to review the statistical methods used in *Sexual Behavior in the Human Male*. We shall refer both to the book and to its authors (Kinsey, Pomeroy and Martin) as KPM. The committee wishes to emphasize that this report is confined to statistical methodology, and does not concern itself with the appropriateness or the limitations of orgasm as a measure of sexual behavior. The treatment of specific problems has necessitated an examination of some of the statistical and methodological problems of such studies, and the organization of frames of reference in which the statistical methods can be discussed. The committee hopes that both detailed and general considerations will be of service to Dr. Alfred C. Kinsey and his co-workers; to the National Research Council's Committee for Research on Problems of Sex, who requested the appointment of this committee; and to others facing similar statistical or methodological problems.

We have endeavored to write this report in a way that would minimize the possibility of misunderstanding. To do this, it is necessary to

* This article consists of the main text, but not the appendices, of the report of a committee appointed in 1950 by S. S. Wilks as President of the American Statistical Association, to review the statistical methods used by Alfred C. Kinsey, Wardell B. Pomeroy, and Clyde E. Martin in their *Sexual Behavior in the Human Male* (Philadelphia, W. B. Saunders Co., 1948). For further details on the appointment of the committee and its charge, see Section 1, p. 676 below. For an outline of the appendices, as well as of this paper, see Section 3, pp. 678–81. Appendix G, "Principles of Sampling," will appear as an article in the March issue of this JOURNAL. The full report, including both the text given here and the appendices, will be published as a monograph by the American Statistical Association in 1954.

deal with many detailed aspects of the work, one at a time. By judicious selection of topics and attitudes, it would have been possible to write two factually correct reports, one of which would leave the impression with the reader that KPM's work was of the highest quality, the other that the work was of poor quality and that the major issues were evaded. We have not written either of these extreme reports.

Even within the present report, a reader who is trying only to support his own opinions could select sections and topics to buttress either view. In the details of this report the reader will find numerous problems that we feel KPM handled admirably. If he pays attention only to these, he would find support for the opinion that the work is nearly impeccable and that the conclusions must be subtantially correct. There are other problems which we believe KPM failed to handle adequately, in some cases because they did not devote the necessary skill and resources to the problems, in other cases because no solutions for the problems exist at present. The reader who concentrates only on the parts of our report in which such problems are discussed would find support for the opinion that KPM's work is of poor quality.

Our own opinion is that KPM are engaged in a complex program of research involving many problems of measurement and sampling, for some of which there appear at the present to be no satisfactory solutions. While much remains to be done, our overall impression of their work to date is favorable.

Many details are discussed in the body and appendices of this report. The main conclusions are as follows:

1. The statistical and methodological aspects of KPM's work are outstanding in comparison with other leading sex studies. In a comparison with nine other leading sex studies (four supported in part by the same NRC Committee) KPM were superior to all others in the systematic coverage of their material, in the number of items which they covered, in the composition of their sample as regards its age, educational, religious, rural-urban, occupational, and geographic representation, in the number and variety of methodological checks which they employed, and in their statistical analyses. So far as we can judge from our present knowledge, or from the critical evaluations of a number of other qualified specialists, their interviewing was of the best.

2. KPM's interpretations were based in part on tabulated and statistically analyzed data, and in part on data and experience which were not presented because of their nature or because of the limitations of space. Some interpretations appear not to have been based on either of these. We feel that unsubstantiated assertions are not in themselves

inappropriate in a scientific study. The accumulated insight of an experienced worker frequently merits recording when no documentation can be given. However, KPM should have indicated which of their statements were undocumented or undocumentable and should have been more cautious in boldly drawing highly precise conclusions from their limited sample.

3. Many of KPM's findings are subject to question because of a possible bias in the constitution of the sample. This is not a criticism of their work (although it is a criticism of some of their interpretations). No previous sex study of a broad human population known to us, medical, psychiatric, psychological, or sociological, has been able to avoid this difficulty, and we believe that KPM could not have avoided the use of a nonprobability sample at the start of their work. Something may now perhaps be done to study and reduce this possible bias, by a probability sampling program.

In our opinion, no sex study of a broad human population can expect to present incidence data for reported behavior that are *known* to be correct to within a few percentage points. Even with the best available sampling techniques, there will be a certain percentage of the population who refuse to give histories. If the percentage of refusals is 10 per cent or more, then however large the sample, there are no statistical principles which guarantee that the results are correct to within 2 or 3 per cent. The results may actually be correct to within 2 or 3 per cent, but any claim that this is true must be based on the undocumented opinion that the behavior of those who refuse to be interviewed is not very different from that of those who are interviewed. These comments, which are not a criticism of KPM's research, emphasize the difficulty of answering the question: "How accurate are the results?", which is naturally of great interest to any user of the results of a sex study.

4. Many of KPM's findings are subject to question because of possible inaccuracies of memory and report, as are all studies of intimate human behavior among broad segments of the population. No one has proposed any way to remove the dangers of recall (involving both memory and report) and KPM were superior to the nine studies referred to above in their attempts to control and measure these dangers. We have suggested still further expansions of their methodological checks.

Until new methods are found, we believe that no sex study of incidence or frequency in large human populations can hope to measure anything but reported behavior. It may be possible to obtain observed or recorded behavior for certain special groups, but no suggestions have been made by KPM, the critics, or this committee which would make it feasible to study observed or recorded behavior for a large human population. These remarks are intended as a comment on the present status of research techniques in sex studies and not as a criticism of KPM's work.

5. KPM received only limited statistical help, in part because the work was pursued during the War years when such expert help was difficult to find for non-military projects. In view of the limited statistical knowledge which was available to them, as made clear by the failure of their sample size experiment, KPM deserve much credit for the straight thinking which brought them safely by many pitfalls. Their need of adequate statistical assistance continues to be serious. Substantial assistance might come through the development of a statistical clinic at Indiana University, or through the addition of a statistical expert to KPM's own staff. Unfortunately the sort of assistance which might resolve some of their most complex problems would require understanding, background, and techniques that perhaps not more than twenty statisticians in the world possess.

6. A probability sampling program should be seriously considered by KPM. The actual gains from an extensive program are limited, to an extent unknown at present, by refusal rates and indirectly by costs, particularly by the costs of maintaining the present quality of the individual histories by KPM's approach. A step-by-step program, starting with a very small pilot study, is recommended.

7. In addition to proposing a probability sampling program, we have made numerous suggestions in this report for the modification and strengthening of KPM's present approach. The suggestions include expanded methodological checks of their sampling program, a further study of their refusal rate, some modification of their methods of analyses, further comparisons of reported vs. observed behavior, and stricter interpretations of their data. We have been informed by KPM that many of these improvements, including some expansion of their techniques for obtaining data, have already been incorporated in the volume dealing with sexual behavior in the human female.

### CHAPTER I. BACKGROUND AND ORGANIZATION

#### 1. *Organization involved*

This committee, consisting of William G. Cochran, Chairman, Frederick Mosteller, and John W. Tukey, was appointed by President S. S. Wilks in September 1950 as a committee of the Commission on

Statistical Standards of the American Statistical Association. This action was initiated by a request from the Committee for Research on Problems of Sex of the National Research Council, as indicated by the following excerpt from a letter dated May 5, 1950, from Dr. George W. Corner, a member of the NRC Committee, to Dr. Isador Lubin, Chairman of the Commission on Statistical Standards of the American Statistical Association.

> "In accordance with our telephone conversation of yesterday, I am writing to state to you the desire of the Committee for Research in Problems of Sex, of the National Research Council, that the Commission on Standards of the American Statistical Association will provide counsel regarding the research methods of the Institute for Sex Research of Indiana University, led by Dr. Alfred C. Kinsey.
>
> "This Committee has been the major source of financial support of Dr. Kinsey's work, and at its annual meeting on April 27, 1950, again renewed the expression of its confidence in the importance and quality of the work by voting a very substantial grant for the next year.
>
> "Recognizing however that there has been some questioning, in recently published articles, of the validity of the statistical analysis of the results of this investigation, the Committee, as well as Dr. Kinsey's group, is anxious to secure helpful evaluation and advice in order that the second volume of the report, now in preparation, may secure unquestioned acceptance."

Some correspondence ensued, in which Wilks indicated the willingness of the American Statistical Association to provide counsel as requested.

Kinsey, in a letter to Wilks dated August 28, stated that

> "we should make it clear that we deeply appreciate the willingness of the American Statistical Association to undertake such an examination of our statistical methods, that we will give it full cooperation in having access to all of our data as far as the peculiar confidential nature of our data will allow, and that we understand, of course, that the committee shall be free to publish its findings of whatever sort."

In the same letter, Kinsey also made a number of suggestions about the constitution and work of the committee, to the effect that the persons on the committee should be primarily statisticians with experience in human population studies, that they should plan to review the statistical criticisms which have been published about the book on the male, and that they should compare methods used by Kinsey and his associates in their research with methods in other published research in similar fields.

With respect to the research on the human female, Kinsey wrote as follows:

"It should, however, be made clear that all the data that will go into our volume on *Sexual Behavior in the Human Female* are already gathered, that the punch cards have already been set up and most of them punched, and that statistical work is proceeding on that volume now. While the recommendations of the committee may modify further work, it can affect this forthcoming volume only in the form in which the material is presented, the limitations of the conclusions, and the careful description of the limitations of our method and conclusions."

## 2. *Committee procedure*

Although no specific written directive was issued to the committee, the letter quoted earlier from Corner to Lubin sets forth the task assigned to the committee. In one respect the scope was deliberately reduced as compared with that envisaged in the letter. The committee decided not to undertake any examination of the researches and data relating to the human female, in order to avoid disruption of Kinsey's proposed schedule of work.

In October, 1950, the committee spent five days at the Institute for Sex Research of Indiana University, accompanied by Mr. Robert Osborn as assistant. Subsequent meetings of the committee were held at Chicago (December 1950), Princeton (January 1951), Cambridge (May 1951), Baltimore (July 1951) and Princeton (October 1951).

In their review of previous studies of sexual behavior, the committee received major assistance from Dr. W. O. Jenkins, who prepared a series of reports which appear in Appendix B. Mr. A. Kimball Romney prepared a helpful index of the principal criticisms made of the statistical methodology used in the book *Sexual Behavior in the Human Male*.

## 3. *Structure of this report as a whole*

KPM's program of research is a major undertaking, involving more than ten years' work. Any discussion of it which aims at thoroughness must itself be lengthy. In order to keep the main body of our report down to a reasonable length, we have relegated much of the documentation of our conclusions, and all detailed discussion, to the following series of appendices.

A. Discussion of comments by selected technical reviewers.
B. Comparison with other studies.
C. Proposed further work.
D. Probability sampling considerations.
E. The interview and the office as we saw them.
F. Desirable accuracies.
G. Principles of sampling.

Appendix A contains our discussion of the statistical and quantitative methodological content of six of the critical reviews which appeared after the publication of the KPM book. These six were chosen from among the large number of published reviews, because they concentrated their attention on the statistical aspects of the research. Appendix A also includes, where this seems appropriate, discussion of some critical points which were not explicitly raised in the reviews in question.

Appendix B, by W. O. Jenkins, contains a review of the statistical aspects of eight of the major previous sex studies which have been carried out in the United States. Also included are similar reviews of the KPM book and of one more recent study by J. E. Farris. The purpose of this appendix is to provide a basis for comparing the KPM study with the other studies as to comprehensiveness, sampling methods, interviewing methods and statistical analysis.

Appendix C begins by outlining and commenting on suggestions for further work made by the reviewers. It explains the difficulty of estimating the stability of results from a sampling procedure such as KPM's, offers some possible methods for this estimation, and suggests how more appropriate variables for expressing sexual behavior might be developed, and how compound variables might be built on these. It then explores the problem of when to adjust, giving a simple numerical procedure for making the decision, and concludes by summarizing the probability sampling suggestions derived from Appendix D.

Appendix D discusses the problems of analysis and usefulness of probability sampling as a check on a nonprobability sample, particularly when refusal rates are considered; two possible types of probability samples and a probability sampling program which KPM might undertake; and the alternative of studying restricted populations.

Appendix E discusses the interview and the office as we saw them. Appendix F discusses what seems to be known about the accuracy needed in such work as KPM's. Appendix G presents an account of the principles of sampling illustrated with general examples.

Many of the problems faced by KPM occur in most types of sociological investigation. Some are likely to be encountered in almost any kind of scientific investigation. For this reason, we have thought it advisable to present certain of the methodological issues in rather general terms.

The reader is asked to bear in mind that in general our conclusions are not documented in the main body of the report, but in the appendices to which references are given.

## CONTENTS OF FULL REPORT

## 4. *Structure of the main body*

In preparing the main body, we have stressed easy reference and have kept related matters together at the expense of fluency of arrangement and lack of repetition. Thus our main conclusions in a form intended for the general reader take 3 pages in the digest above, while more detailed conclusions, expressed for a more technical audience, take 3 pages in Chapter XI. A particular subject summarized there is also likely to be discussed once in Chapter II, where we try to point out what KPM did, once again in one of Chapters IV to IX, where we assess KPM on an absolute scale, and yet again in Chapter X, where we compare KPM with previous workers in the field. This is repetitive, but we hope that it will permit ready reference and avoid treating subjects out of context.

After this introductory chapter on background structure, the remainder of the main body falls into three parts:

(i) Chapters II and III. In the first of these, we describe, respectively, what choices KPM had to make and what they chose. In Chapter III we outline some essential principles of sampling, which seem not to have been clearly enough formulated or widely enough understood. These chapters are introductory.

(ii) Chapters IV to XI. In the first six of these, we try to compare KPM's work with an absolute standard. The order chosen (interview, sample, methodological checks, analytical techniques, complex examples, interpretation) is that in which the problems arise in an evolving study such as KPM's. Chapter X compares KPM with previous works on the basis of Appendix B, while Chapter XI summarizes the conclusions of this part.

(iii) Chapter XII. This discusses briefly various suggested expenditures of further effort.

## CONTENTS

*Page*

### CHAPTER II. MAJOR AREAS OF CHOICE

#### 5. *What sort of behavior?*

The purpose of Chapter II is to record in summary form the major choices made by KPM.

Certainly the choice of orgasm as the central sort of sexual behavior for study was a major one, leading to consequences whose statistical aspects will be discussed in various places, but this choice is not a matter of *general* quantitative methodology, and hence falls outside the scope of this committee's task.

#### 6. *Whose behavior?*

KPM had to choose the population to which this study should apply. This decision does not seem to have been made clearly. From the basis for the "U. S. Corrections" (p. 105) we should infer it to be "all U. S. white males." If it were the population to which the U. S. Corrected sample actually applies on the average (the *sampled* population, see Section 18), it would be a rather odd white male U. S. Population. It would have age groups, educational status, rural-urban background, marital status and all their combinations according to the 1940 census, but it would have more members in Indiana than in any other state, and it would have been selected to an unknown degree for willingness to volunteer histories of sexual behavior. We do not regard this description of the sampled population as an automatic criticism, as some critics do. We make it here as a factual statement, noting that the careful and wise choice of the sampled population, although difficult, is a relatively free choice of the investigator. More discussion relevant to this point will be found in Chapter II-G (Appendix G).

Further, KPM chose to study the behavior of many (at least 163 in tabular form) segments of this large population, feeling, apparently, both that comparisons among segments would be illuminating and that data for (clinical) application to individuals should come from a reasonably homogeneous segment. KPM's choice of a broad population created many problems, particularly in sampling. Whether they would have been well advised to confine themselves to a more restricted population, e.g., the state of Indiana, is debatable. For our part, we are willing to take their choice as given, and to discuss briefly elsewhere some alternatives for further work (Chapter IX-D).

### 7. *Observed, recorded, or reported behavior*

KPM, interested in actual behavior, had, in principle, the choice of studying observed, recorded, or reported behavior. But since they selected a broad population and orgasm as the type of behavior, their only feasible choice seems to have been *reported* behavior. This situation does not seem likely to change in the foreseeable future.

The choice of reported behavior implies that the question: "On the average, how much difference is there between present reported and past actual behavior?" is seriously involved in any inferences about actual behavior which are attempted from KPM's results. The difference might well be large, leading to a large systematic error in measurement. However, use of observed or recorded behavior in order to avoid this difference does not seem to us a feasible way to measure nationwide incidences and frequencies for KPM's broad population, because it would have produced systematic errors in sampling possibly larger than the error in measurement.

### 8. *Interview or questionnaire, and types thereof*

Having settled on reported behavior, KPM had to decide whether this report should be oral or written, and what methods should be used to elicit it. Their choice was oral, in a face-to-face interview whose flavor was designed to be that of a doctor or family friend. The choice of oral rather than written report:

(1) made it possible to obtain *apparently* satisfactory answers from many more subjects (the percentage of complete illiteracy in the U. S. is small, but the percentage of illiteracy on complex subjects not usually written about is undoubtedly substantial).

(2) permitted and encouraged variation of the form of the questions to suit the subject and the situation.

Those, like some critics, who believe in a repeatable measurement process, regardless of whether or not it measures something that is always relevant, find (2) bad. Those who, like KPM, feel that appropriately flexible wording improves communication and thus improves the quality of report despite the variability resulting from changes in the form of questions, find (2) good.

Given an interview rather than a questionnaire, the remaining choices of KPM follow a consistent pattern. In nearly every case their approach resembled the clinical interview more closely than the psychometric test.

### 9. *Which subjects?*

Here there are various choices, pertaining to:

(1) selection of individuals one at a time or in clusters.
(2) keeping age, education, marital status, etc., segments in the sample proportionate to those in the population or making them of more nearly equal size.
(3) selecting individuals on a catch-as-catch-can basis, a partly randomized basis, or according to a probability sampling plan.

They chose:

(1) to select individuals in clusters.
(2) to keep age, education, marital status, etc., segments more nearly equal in the sample than in the population.
(3) to use no detectable semblance of probability sampling ideas.

The pros and cons will be discussed later.

### 10. *What methodological checks?*

There are choices as to the types of checks and the number of each to be made. The types of checks made by KPM, including

(1) take-retake,
(2) husband-wife,
(3) duplicate recording of interview,
(4) overall comparison of interviews,
(5) others (see Chapter V-A)

seem to cover all those easily thought of. The numbers of checks made are discussed later. Duplicate recording of interviews occurred in an unknown, but presumably small, number of cases. No comparisons from duplicate recordings were reported, perhaps because most occurred in connection with the training of interviewers.

11. *How analyzed and presented?*

In analyzing frequency and incidence of activity, KPM chose to report both raw and "U. S. Corrected" data and to make simple comparisons. Just what was done in general was clearly stated, but the steps involved in detailed computations were not explained. No attempt was made to find helpful scales or composite variables (see Chapters IV-C and V-C).

With the exception of "U. S. Corrections," most of the analysis of the tabular data is confined to straightforward description. Some attention is paid to the problem of sample-population relation in the form of standard errors (presumably underestimated because they were based on the assumption of random sampling). However, this approaches lip service, since many apparent differences are discussed with no attention to significance or nonsignificance. (Again we do not regard this as an automatic criticism, particularly since accurate indication of significance would have been difficult—see Section A-18.)

In analyzing cumulative activity, KPM's main tool was the accumulative incidence curve, a technique which they developed independently.

12. *How interpreted?*

The main choices concerned

(1) extent of warning about possible differences between reported behavior and actual behavior,
(2) extent of warning about possible differences between the sampled population (see Section 18) and the entire U.S. white male population,
(3) extent of warning about sampling fluctuations,
(4) extent of verbal discussion *not* based on evidence presented,
(5) certainty with which conclusions were presented.

Under (1) the emphasis was on methodological checks in order to indicate, as far as they could, how small this difference seemed to KPM to be. Under (2) there was little discussion. Under (3) the warnings were made early, incompletely, but not often. Under (4) the extent of discussion was substantial, most of it aimed at social and legal attitudes about sexual behavior, and descriptions or practices not covered by the tables. Under (5) the conclusions were usually presented with an air of solid certainty.

In general the observations seem to have been interpreted with more fervor than caution, although occasional qualifications may be found.

CHAPTER III. PRINCIPLES OF SAMPLING

13. *Introduction*

It is difficult, if not impossible, to assess the quality of any sample and its analysis without comparing it with a set of principles. This is particularly true of KPM's works. The present chapter endeavors to set down, in compact form, a few of the principles of sampling which are especially relevant to a consideration of KPM's sampling. As we have noted (Section 6), KPM chose to select individuals in groups or clusters, to divide the population into segments and keep segment sizes more nearly equal in the sample than in the population, and to use no semblance of probability sampling ideas. The discussion in this chapter concentrates on these aspects of sampling.

Many readers will, we believe, desire a more connected account of the principles of sampling, with examples and fuller discussion. These are provided in Appendix G. Any reader who finds the statements used in this chapter unclear, or not intuitively acceptable, is urged to turn to Appendix G before proceeding further. Once there, he should read through from the beginning, since argument and exposition there are closely knit and unsuited to piecemeal references.

Whether by biologists, sociologists, engineers, or chemists, sampling is often taken too lightly. In the early years of the present century, it was not uncommon to measure the claws and carapaces of 1000 crabs, or to count the number of veins in each of 1000 leaves, and to attach to the results the "probable error" which would have been appropriate had the 1000 crabs or the 1000 leaves been drawn at random from the population of interest. If the population of interest were all crabs in a wide-spread species, it would be obviously almost impossible to take a simple random sample. But this does not bar us from honestly assessing the likely range of fluctuation of the result. Much effort has been applied in recent years, particularly in sampling human populations, to the development of sampling plans which, *simultaneously,*

(i) are economically feasible,

(ii) give reasonably precise results, and

(iii) show within themselves an honest measure of fluctuation of their results

Any excuse for the practice of treating non-random samples as random ones is now entirely tenuous. Wider knowledge of the principles involved is needed if scientific investigations involving samples (and what such investigation does not involve samples?) are to be solidly based. Additional knowledge of techniques is not so vitally important, though it can lead to substantial economic gains.

14. *Cluster sampling*

A botanist who gathered 10 oak leaves from each of 100 oak trees might feel that he had a fine sample of 1000, and that, if 500 were infected with a certain species of parasites, he had shown that the percentage infection was close to 50%. If he had studied the binomial distribution, he might calculate a standard error according to the usual formula for random samples, $p \pm \sqrt{pq/n}$, which in this case yields $50 \pm 1.6\%$ (since $p = q = .5$ and $n = 1000$). In doing this he would neglect three things:

(i) probable selectivity in selecting trees (favoring large trees, per-haps?

(ii) probable selectivity in choosing leaves from a selected tree (fav-oring well-colored or alternatively, visibly infected leaves per-haps and

(iii) the necessary allowance, in the formula used to compute the standard error, for the fact that he had not selected his leaves individually.

Most scientists are keenly aware of the analogs of (i) and (ii) in their own fields of work, at least as soon as they are pointed out to them. Far fewer seem to realize that, even if the trees were selected at ran-dom from the forest, and 10 leaves were chosen at random from each selected tree, (iii) must still be considered. But if, as might indeed be the case, each tree were either wholly infected or wholly free of infec-tion, then the 1000 leaves tell us *no more* than 100 leaves, one from each tree, since each group of 10 leaves will be all infected or all free of infection. In this event, we should take $n = 100$ in calculating the standard error and find an infection rate of $50 \pm 5\%$. Such an extreme case of increased fluctuation due to sampling in groups or clusters would be detected by almost all scientists, and is not a serious danger. But less extreme cases easily escape detection.

We have just described, as one example of the reasons why the principles of sampling need wider understanding, an example of *cluster sampling*, where the individuals or sampling units are not drawn separately and independently into the sample, but are drawn in clusters, and have tried to make it clear that "individually at ran-dom" formulas do not apply. Cluster sampling is often desirable, but must be analyzed appropriately. KPM's sample was, in the main, a cluster sample, since they built up their sample from groups of people rather than from individuals.

15. *Possibilities of adjustment*

Often the population is divided into segments of known relative size, perhaps from a census. It is sometimes thought that the best method of sampling is to take the same proportion from every segment, so that the sample sizes in the segments match the corresponding population sizes. Such samples do have the advantage of simplifying computations by equalizing weights, and they sometimes lead to a reduction of sampling error. But modern sampling theory shows that optimum allocation of resources usually requires *different* proportions to be sampled from different segments, whether the purpose is to estimate average values over the population or to make analytical comparisons between results in one group of segments and those in another.

When there are disparities in the relative sizes of segments in the sample as compared with the population, whether accidental or planned, these disparities must be taken into account when we attempt to estimate averages over the whole population. One way in which this can be done is by adjustments applied to the segments. Such adjustments proceed as follows. Suppose that we know

  (i)  the true fraction of the population in each segment, and
  (ii) the segment into which each individual in the sample falls.

Then we can weight each individual in the sample by the ratio

$$\frac{\text{fraction of population in that segment}}{\text{fraction of sample in that segment}}.$$

(It is computationally convenient to weight each segment mean with the numerator of this ratio; the result is algebraically identical to that described above.)

The result of adjustment is a new "sampled population"—one such that the relative sizes of its various segments are very nearly correct (according to (i) above). Since the weight is the same for all the sample individuals in a given segment, adjustment does nothing to redress any selectivity which may be present *within* segments. If we adjust in this way, we remove one source of systematic error without affecting other sources at all. The philosophy of such adjustments is discussed further in Section G-12, and it is concluded that they may generally be appropriately made (within the limits discussed in sections C-16 —C-18). Their chief danger is the possible neglect of the possibilities that they may be

(i) entirely too small,

(ii) too large,

(iii) in the wrong direction,

because of unredressed selectivity *within* the segments. When this possibility exists, extreme caution in presenting the results of adjustment is indicated.

## 16. *Probability samples*

When probability samples are used, inferences to the population can be based entirely on statistical principles rather than subject-matter judgment. Moreover, the reliability of the inferences can be judged quantitatively. A probability sample is one in which

(i) each individual (or primary unit) in the sampled population has a known probability of entering the sample,

(ii) the sample is chosen by a process involving one or more steps of automatic randomization consistent with these probabilities, and

(iii) in the analysis of the sample, weights appropriate to the probabilities (i) are used.

Contrary to some opinions, it is *not* necessary, and in fact usually not advisable in a pure probability sample for

(i) all samples to be equally probable, or

(ii) the appearance of one individual in the sample to be unrelated to the appearance of another.

In practice, because some respondents cannot be found or are uncooperative, we usually obtain, at best, approximate probability samples (see Sections A-2 and D-13) and have approximate confidence in our inference.

## 17. *Nonprobability samples*

Samples which are not even approximately probability samples vary widely in both actual and apparent trustworthiness. Their trustworthiness usually increases as they are insulated more and more thoroughly from selective factors which might be related to the quantities being studied. Insulation may be obtained by:

(i) adjustments applied to the segment means in the sample,

(ii) examination of the sample as drawn for signs of selection on a particular factor,

(iii) partial randomization.

Adjustment for segments, as explained in Section 15 above, corrects for any selective factor operation *between* segments, but corrects not at all for selective factors operating *within* segments. If adjustment is to be used, deliberate selectivity between segments may be exercised without danger, *so long as it does not imply selectivity within segments.*

Negative results when the sample is examined for signs for selection on a particular variable are comforting, and strengthen the reliability of the sample. The amount of this strengthening depends very much on the *a priori* importance of the variables checked to what is being studied.

Deliberate (partial) randomization is a step toward a probability sample, and may be very helpful on occasion.

### 18. *Sampled population and target population*

We have found it helpful in our thinking to make a clear distinction between two population concepts. The *target* population is the population of interest, about which we wish to make inferences or draw conclusions. It is the population which we are trying to study. The *sampled* population requires a more careful definition but, speaking popularly, it is the population which we actually succeed in sampling.

The notion of a sampled population can be more clearly described for probability sampling. In order to have probability sampling, we must know the chance that every sampling unit has of entering the sample, and the weight to be attached to the unit in the analysis. The sampled population may be defined as the population generated by repeated application of these chances and these weights. The frequency of occurrence of any particular sampling unit in the sampled population is proportional to the product

(chance of entering the sample) $\times$ (weight used in analysis).

This product is made constant for a probability sample. Thus, with probability sampling, the sampled population consists of all sampling units which have a non-zero chance of selection.

The sampled population is an important concept because by statistical theory we can make quantitative inferential statements, with known chances of error, from sample to sampled population. It must be carefully distinguished from the target population, the population of interest, about which we are tempted to make similar inferential statements.

Even with probability sampling, the sampled and the target population usually differ because of the presence of "refusals," "not-at-homes," "unable to classify," and so on. The consequence of these disturbances is that certain sampling units, although assigned a known chance of selection by the sampling plan, did not in fact have this chance in practice.

With non-probability sampling, the situation is much more obscure. By its definition as given above, the sampled population depends on the existence of a sampling plan (which may be only a vague set of principles in the investigator's head) and on the "chances" that any sampling unit had of being drawn. These chances are not well known—if they were, we should have a probability sample. But in many cases, it is reasonable to behave as if these chances exist and to attempt to estimate them, because they provide the only means of making statistical inferences beyond the non-probability sample to a corresponding "sampled population." The difficulty comes in specifying, or sometimes even thinking about, the nature of the sampled population. It is certain to be a weighted population where, for example, Theodosius Linklater may appear 1.37 times, while Basil Svensson appears only 0.17 times.

Insofar as we make statistical inferences beyond the sample to a larger body of individuals, we make them to the sampled population. The step from sampled population to target population is based on subject-matter knowledge and skill, general information, and intuition —but not on statistical methodology.

CHAPTER IV. THE INTERVIEW AREA

### 19. *Interview vs. questionnaire*

The committee members do not profess authoritative knowledge of interviewing techniques. Nevertheless, the method by which the data were obtained cannot be regarded as outside the scope of the statistical aspects of the research.

For what our opinion is worth, we agree with KPM that a written questionnaire could not have replaced the interview for the broad population contemplated in this study. The questionnaire would not allow flexibility which seems to us necessary in the use of language, in varying the order of questions, in assisting the respondent, in following up particular topics and in dealing with persons of varying degrees of literacy. This is not to imply that the anonymous questionnaire is inherently less accurate than the interview, or that it could not be used fruitfully with certain groups of respondents and certain topics. So far as we are aware, not enough information is available to reach a verdict on these points.

20. *Interviewing technique*

Many investigators have faced the problem of attempting to obtain accurate information about facts which the respondent is thought to be unwilling to report. It is natural to inquire whether KPM, in their interviewing technique, took advantage of accumulated experience as to the best methods for extracting the facts. But it is also well to inquire how much definite experience has been accumulated.

The KPM interview impressed us as an extraordinarily skillful performance. Direct questions are put rapidly in an order which seems to these respondents hard to predict, so that it is difficult to tell what is coming next. Despite the air of briskness, we did not receive the impression that we were being hurried if we wished to reflect before replying, and supplementary questions or information were given if this seemed helpful to the memory. The coded recording of the data was done unobtrusively by the interviewer, so that the interview appeared to be a friendly conversation rather than any kind of an inquisition. These, of course, are personal impressions.

KPM evidently think highly of the virtues of this technique, because it was adopted despite limitations which it imposes on the scope and rate of progress of the study. The technique makes great demands on the interviewer. The long period of training and the personal qualities required have restricted and will continue to restrict the interviewers to a very small number. This limits the speed with which data can be accumulated and also puts restrictions on the type of sampling that can be employed.

The type of interview used by KPM differs markedly from the less directive methods which are sometimes recommended for dealing with taboo subjects. If the subject is likely to feel that his answer to a certain question will affect his prestige in the eyes of the interviewer, a less directive approach would be to conduct the interview in such a way that he gives the desired information without realizing that he is answering the awkward question. The KPM method is the antithesis of this. Research on interviewing techniques has not yet produced any substantial body of evidence as to the superiority of either the less directive methods or the KPM technique.

With regard to specific inaccuracies in the KPM data, we believe that the interview gives an opportunity both for positive and negative bias. The KPM assumption that everyone has engaged in all types of activity seems to some likely to encourage exaggeration by the respondents. (KPM feel (personal communication) that their cross-checks are highly effective in detecting such exaggeration.) On the other hand, our impression from the interview was that a successful denial of certain types of activity would be possible if the subject was prepared to do so, although we do not know the full extent of the KPM cross-checks which would lead them to be suspicious of such a denial. KPM assert (personal communication) that they regard cover-up as a more likely source of bias than exaggeration. Our opinions on this statement are divided.

As KPM point out (p. 48), the subject's willingness to talk about certain types of activity is influenced by the attitudes of the social group to which he belongs. Until evidence to the contrary is presented, the presumption (made by some of the critics) that his final responses will also be influenced is one that cannot be cast aside. The size of these influences is still a matter of opinion. A corresponding element of doubt is present in almost all comparisons between different social levels, both those which provide some of the most interesting comparisons in the book, and those in many other studies.

### CHAPTER V. THE SAMPLING AREA

#### 21. *KPM's sampled population*

As noted above, KPM's sample was deliberately disproportionate, partly in order to cover individual segments defined by age, education, religion, etc., in an adequate manner, partly because of geographical convenience. If the results for individual segments were to be based on samples of at least moderate size, such disproportion was necessary and wise. Its effects on overall results are less clear. It seems impossible to be sure what effect it had on the variability of the final result, and its use is certainly not a demonstrable error as far as variability is concerned.

In their U. S. corrections, KPM provided adjustments for disproportion between segments defined by age, education, and marital status. As noted above (Section 17) we feel that such adjustments are usually appropriate. Due to absence of population data, they did not adjust for religion. The geographical imbalance of their sample was so great that an overall geographic adjustment was not feasible. Thus they compensated for some disproportions, and left others to produce what effects they would.

Their only examination of the sample for signs of selection within segments is their comparison of 100% groups (groups where all members were interviewed) with partial groups (groups where only part of the members were sampled). This gives some insight into the effect of volunteering as a selective factor. Beyond this, KPM report no serious effort to measure the actual effect of volunteering, or to discover what percentage of the population they would be able to persuade to be interviewed.

They made no use of randomization. They might have attempted to sample, say, college seniors from two colleges drawn at random from a large list of colleges, but they are of the opinion (personal communication) that this would have slowed up the work to an unmanageable extent.

All in all, the absence of any orderly sampling plan contrasts strikingly with their usual methodical mode of attack on other problems.

As stated briefly above (Section 6), the "sampled populations" corresponding to

(1) KPM's raw means, and to
(2) KPM's "U.S. corrected" means,

respectively, are startlingly different from the composition of the U. S. white male population. (For example, although these sampled populations have the U. S. average combination of education and rural-urban background, they have half of their members living in Indiana.) Since a complete probability sample seems to have been out of the question at the beginning of the KPM investigation, some such "sampled population" was to be expected, although it might have been somewhat less distorted. Provided that further statistical analyses of the sort indicated in Appendix C, Chapter II-C were made, it would be possible to make adequate rigorous inferences from the sample to this ill-defined "sampled population."

The inference from these vague entities to the U. S. white male population depends on:

(a) the inferrer's view as to what these "sampled populations" are really like, and
(b) the inferrer's judgment as to how (reported) sexual behavior varies within segments.

It is not surprising that experts disagree.

The inference from KPM's sample to the (reported) behavior of all U. S. white males contains a large gap which can be spanned only by expert judgment. This is a common phenomenon in social fields, but is still unfortunate. A considerable bridge across this gap would be furnished by a small probability sample.

## 22. *Could KPM have used probability sampling?*

If probability sampling could have been used, its use would have avoided one of the main gaps in KPM's present chain of inference. We have, therefore, considered this possibility carefully.

The difficulties in applying probability sampling to KPM's study lie in the expenditure of time required to make the contacts necessary to persuade a predesignated man to give a history. By adapting the mechanism of the probability sample to KPM's situation, these difficulties may perhaps be reduced (see Appendix D, Chapter V-D). It would almost certainly have been impractical for KPM to have used a probability sample in the early years of their study. If KPM's apparent "opinions" (p. 39 of KPM) as to the effectiveness of their present techniques of contact are correct, starting a probability sample would have been practical at any time since the appearance of the male volume in 1948.[1] However, KPM (personal communication, 1952) feel that such an interpretation of their written statement is unwarranted.

Since it would not have been feasible for KPM to take a large sample on a probability basis, a reasonable probability sample would be, and would have been, a small one, and its purpose would be:

(1)  to act as a check on the large sample, and
(2)  possibly, to serve as a basis for adjusting the results of the large sample.

A probability sampling program planned to serve these purposes is discussed in Appendix D, Chapter VII-D. Such a program should proceed by stages because of the absence of information on costs and refusal rates.

This conclusion about probability sampling does not excuse KPM from the responsibility for choosing geographical disproportion in order to save travel time and expense. The wisdom or unwisdom of this choice seems to depend on one's view as to the magnitude of geographical differences. Again, it is not surprising that experts disagree.

### CHAPTER VI. METHODOLOGICAL CHECKS

## 23. *Possible checks*

The primary check, if it could be made, is the comparison of *average* actual behavior with *average* reported behavior. *Variability* in the dif-

---

[1] "The number of persons who can provide introductions has continually spread until now, in the present study, we have a network of connections that could put us into almost any group with which we wished to work, anywhere in the country." (P. 39 of KPM.)

ference between actual and reported behavior is secondary in interest, because high variability merely implies the necessity of larger numbers of cases, while large average differences between actual and reported behavior respresent a systematic error that cannot be adjusted without rather complete knowledge. Unfortunately this primary check does not at present seem feasible in studying human sexual behavior as it occurs in our culture.

Of secondary importance are checks of the single actual report with the average actual report, where averages may be taken over fluctuations, time, spouses, and/or interviewers. (See Appendix A, Chapter V-A) In this second category, the following possible comparisons suggest themselves:

1. Reinterviews of the same respondent
2. Comparison of spouses
3. Comparison of interviewers on the same population segment
4. Duplicate interviews by the same interviewer at various times.

### 24. *KPM's checks*

The only comparison of observed and reported behavior which KPM found feasible was the date of appearance of pubic hair, which agreed quite successfully. This is a physical characteristic, different in character and emotional loading from the behavior of main interest. Some subjects may have had to rely upon general information, plus some assistance from the interviewer, in naming a date for themselves. Thus this check furnishes rather weak support.

At the level of rechecks on respondents, some information is available but more is needed. Similarly, comparisons of spouses have been made for a relatively selected group. The checks themselves are encouraging, but more cases are needed.

Some attempts have been made to compare the staff interviewers but since there is some selection in the assignment of cases, these comparisons do not meet the problem as squarely as interviews of the same respondent by different interviewers, or the recorded interview technique.

A comparison of early versus late interviews by Kinsey is given in KPM, but it is hard to tell, for example, whether the 12.4% drop (from 44.9% to 32.5%) in the accumulative incidence for total premarital intercourse at age 19 (single males, education level 13+) from early to late interviews is due to differing groups sampled, instability in the interviewing process, or reasonable sampling variation for cluster sampling (KPM p. 146).

KPM have made serious efforts to check their work in the aspects where checking seems feasible. However, improved and more extensive checking is needed. Although duplicate recording of interviews is mentioned, no data have been published. Even if they must be based on very few cases, such comparisons should be made available.

## CHAPTER VII. ANALYTICAL TECHNIQUES

### 25. *Variables affecting sexual behavior*

After introductory chapters (5 and 6) on early sexual growth and activity, KPM proceed to examine the effects of the following variables:

> Age
> Marital status
> Age of adolescence
> Social level
> Comparison of two generations
> Vertical mobility in the occupational scale
> Rural-urban background
> Religious background

In this chapter we attempt to appraise, in general terms, the analytical techniques used by KPM in their study of these variables.

### 26. *Definition of the variables*

Some of the variables: age of adolescence, social level, occupational level, rural-urban background and religious background, involve problems of definition. These seem to have been in the main thoughtfully handled and presented by KPM. For instance, KPM discuss the relative merits of educational level attained by the subject and of the occupational class of the subject and of his parents as a measure of social level (pp. 330–32). In their opinion, educational level is the most satisfactory criterion and this was adopted for the analysis. In the case of religious affiliation, KPM distinguish between active and inactive profession of religious faith, though the definition of the two terms is not made entirely clear.

The definition which looks least satisfactory is that of age of adolescence (p. 299), where the problem is formidable. The criteria employed by KPM appear difficult for the reader to interpret.

### 27. *Assessing effects of variables*

With a multiplicity of variables which may interact on each other, the task of assessing the importance of each variable individually is not easy. Examination of the variables one by one, ignoring all other variables except the one under scrutiny, may give wrong conclusions, because what appears on the surface to be the effect of one variable may be merely a reflection of the effects of other variables.

A thorough attack on this problem calls for a multiple-variable approach in which all effects are investigated simultaneously. This requires a high degree of statistical maturity and of skill in presentation.

The method utilized by KPM is a compromise. In general, with some exceptions, they regard age, marital status and educational level as basic variables, which are held fixed or compensated for in the investigation of each of the remaining variables. The other variables are disregarded for the moment. Although we have not examined the matter exhaustively, this policy seems to have been justified by events, because KPM claim from their analyses that the other variables, with the exception of age at adolescence, have had relatively minor effects.

## 28. *The measurement of activity*

In the KPM tables, activity is measured by "incidence" (per cent of the population who engage in the activity) as well as by frequency per week. In some tables, both mean and median frequencies are given, and also frequencies for the total and for the active population. There are advantages in presenting various measures. On the other hand, inspection suggests that all these measures are correlated: that is, to some extent they tell the same story. A complex internal analysis would probably show about how many measures are really needed to extract the information in the data and what individual measurements, or combinations of them, are best for this purpose. Perhaps a single one, or at most two, would suffice. As it is, both KPM and the industrious reader have to wade through tables and discussion of a number of different measurements, without being clear whether anything new is learned. Simplification would be pleasant, but is far from essential.

## 29. *Tests of significance*

In the discussion of effects which they regard as real, KPM make little appeal to tests of significance. They often present standard errors attached to the mean frequencies for individual cells. Because sampling was non-random and was by groups, these standard errors, calculated on the assumption of randomness, are under-estimates, perhaps by a substantial amount. The standard errors have a kind of negative virtue, in the sense that if a difference is not significant when judged against these errors, it would not be significant if a valid test could be devised. The problem of devising a realistic estimate of the true standard errors is one of considerable complexity (see Section II-C).

We have been unable to discover from the book the principles by which KPM decide when to regard an effect as real. The size of the effect is one criterion. Size should certainly be taken into account, since an effect may be significant statistically but too small to be of biological or sociological interest. They evidently attach some importance to the consistency with which an effect is exhibited in different parts of a table. As a criterion, consistency is of variable worth. Consistency over different age groups (where age denotes age at the time of the reported activity) is of little worth, since there is inevitably substantial correlation between sampling fluctuations of reported activities at neighboring ages because the same subject appears in neighboring age groups. More weight can be attached to consistency over different educational levels, because different groups of subjects are involved.

To summarize, statements about the data in their tables lie at the level of shrewd descriptive comment, rather than at the level of an attempt to make inferential statements from a sample to a clearly defined population (even though this could not be the U. S. white male population).

We do not propose to discuss the analysis for each variable separately. Two analyses which have attracted much attention will be considered later (Sections 33 to 37).

## 30. U. S. Corrections

In most sampling plans it is necessary to provide a set of weights for the segments of the sampled population to recover accurate estimates for the target population (i.e. the population about which inferences are desired). That such adjustments are usually appropriate, whether probability or nonprobability samples are employed, has already been pointed out (Section 17, see Section II-G).

Since KPM have as their target population U. S. white males, we can reasonably expect them to apply weights in an attempt to correct for disproportionate representation in the sampled population of some segments of the target population.

KPM supply U. S. Corrections (p. 106–9) and use them rather consistently throughout the work. There are no examples given explaining the application of the weights. The critics, and sometimes this committee, have had difficulty in verifying computations where they have been used. Of the 13 tables where corrections could be checked cor pletely, one checked, 10 checked except for one age group each, and two were not checked by the correction mentioned in the text. Apparently the exposition could be improved.

The U. S. Corrections should be used, but it might be possible to make a more effective choice of segments (see A-43 and V-C and II-G).

KPM did not sufficiently warn the reader that U. S. corrected figures are not corrected for selection within segments, and may be seriously biased.

### 31. *The accumulative incidence curve*

KPM have a useful device for summarizing incidence data by age. This accumulative incidence curve gives the percentage of individuals in the sample (reporting for a given age) to whom a particular event has occurred before that age. Although the explanation of the concept of accumulative incidence is not as clear as most of KPM's writing, the computations made are satisfactory. When there are no generation-to-generation changes in the population and no differential recall depending on age at report, this method is particularly justified, because it packs all the incidence data neatly into one grand summary. (For discussion of the critics' comments see A-39.) No better method for overall comparisons seems to be available.

### 32. *Other devices*

1. KPM did some extensive sampling experiments on their data, with a view to discovering the sample size needed for the accuracy they desired. These experiments turned out to be almost valueless because KPM did not take account of the necessary statistical principles (see A-19).

2. The committee had an opportunity to inspect the KPM facilities on a visit to Bloomington, Indiana. We observed that the data sheets were neatly filled out, that the files were well kept, that requests for original data were usually met in a matter of moments, and that the office was well equipped for handling the extensive data with which KPM deal.

3. The KPM volume was written while data were still being collected. Apparently KPM chose to use all the data on hand at the time a particular point was being analyzed (personal communication from KPM). Thus different tables have different totals, a source of annoyance to critics and users of the book. The reasons for this should have been pointed out by KPM. The additional interviewing was deliberately selective with an aim to strengthen weak segments (personal

communication from KPM). It seems to us that, if this strengthening was necessary for later analyses, it would have been worthwhile to add the new material to the early tabulations. This would also have increased comparability and avoided the problems raised by the existence of many different sampled populations.

## CHAPTER VIII. TWO COMPLEX ANALYSES

### 33. *Patterns in successive generations*

In this chapter we discuss briefly two analyses by KPM which have attracted much attention. Our object is to give two specific illustrations of the kind of analysis which they chose to undertake, with comments on their competence.

The first analysis was made by dividing the sample into two groups: those over 33 years of age at the time of interview, with a median age of 43.1 years, and those under 33 years at the time of interview, with a median age of 21.2 years.

Our comments deal with three topics: (i) the statistical methodology employed (ii) KPM's summary of their tables (iii) the general problem of inference from data of this type.

### 34. *Statistical methods*

In the comparisons, educational level and age at the time of the activity are held constant and in nearly all comparisons marital status also. The method used to compare the group means seems satisfactory except for some minor points, discussed in A-25, A-33 and A-43.

It would have been helpful to present classifications of the older and younger groups according to other factors which might influence sexual activity, e.g., rural-urban background, religious affiliation, marital status at age 20 or 25. The two groups would not necessarily agree closely in these break-downs, for there has been a slow drift towards the towns, and perhaps a drift towards "inactive" rather than "active" religious affiliation. For interpretive purposes it is advisable, in any event, to learn as much as possible about the compositions of the older and younger groups. Some critics have claimed that the older generation is "atypical."

### 35. *KPM's summary of their tables*

The data are presented in 8 large tables (98–105). As a statistician learns from experience, a competent summary of a large body of data is not an easy task. KPM give a detailed discussion of the accumulative incidence data for each type of outlet, followed by a similar discussion of the frequency data.

These detailed comments on what the data appear to show seem sound, except that on two occasions where the younger group showed greater sexual activity, KPM ignored or played down the difference between the two groups (Section A-45).

Their general summary statement reads in part as follows:

> "The changes that have occurred in 22 years, as measured by the data given in the present chapter, concern attitudes and minor details of behavior, and nothing that is deeply fundamental in overt activity. There has been nothing as fundamental as the substitution of one type of outlet for another, of masturbation for heterosexual coitus, of coitus for the homosexual, or vice versa. There has not even been a material increase or decrease in the incidences and frequences of most types of activity. . . .
>
> "And the sum total of the measurable effects on American sexual behavior are slight changes in attitudes, some increase in the frequency of masturbation among boys of the lower educational levels, more frequent nocturnal emissions, increased frequencies of premarital petting, earlier coitus for a portion of the male population, and the transferences of a percentage of the pre-marital intercourse from prostitutes to girls who are not prostitutes."

Some critics have objected strongly to this statement, particularly the first paragraph, on the grounds that it gives a biased report by brushing aside the differences in activity, which are almost all in the direction of higher or earlier sexual activity by the younger group. The reporting does appear a little one-sided, in that the reader is encouraged to conclude that the differences are immaterial, although KPM do not state what they mean by a "material" increase. On the other hand, the catalogue of differences, given at the end of the second paragraph above, includes all differences noted either by KPM or the critics, except for an increased homosexual activity in the younger group at educational levels 0–8 and 9–12.

## 36. *Validity of inferences*

Two objections have been made by some critics to any inferences drawn from a comparison of this type. The first is that the groups may not be representative of their generations. KPM have attempted to dispose of this objection, at least in part, by holding educational level and marital status constant. It might be possible to go further and hold other factors constant, or at least examine whether the samples from the two generations differ in these factors. But with non-random sampling the objection is not removed even if a number of factors are held constant, because one or both groups might be biased with respect to some factor whose importance was not realized. Various opinions may be formed as to the strength of the objection, but it can be removed only by the use of probability sampling accompanied by valid tests of significance.

Secondly, in a comparison of this type, the older generation is describing events which involve a much longer period of recall, with a possibility of distortion as events become distant. Further retake studies, if KPM can continue them for a sufficiently long period, may throw some light on the strength of this objection.

The joint effect of these objections is to render the conclusions tentative rather than definitely established.

### 37. *Vertical mobility*

This analysis (pp. 417–47) shows a degree of ingenuity and sophistication which is not too common in quantitative investigations in sociology. The data are arranged in a two-way array according to the occupational class of the subject at the time of interview and the occupational class of the parents. KPM examine whether the pattern of sexual activity of the subject is more strongly associated with the parental occupational class than with that attained by the subject. They conclude (p. 419)

> In general, it will be seen that the sexual history of the individual accords with the pattern of the social group into which he ultimately moves, rather than with the pattern of the social group to which the parent belongs and in which the subject was placed when he lived in the parental home.
>
> The most significant thing shown by these calculations (Tables 107–115) is the evidence that an individual who is ever going to depart from the parental pattern is likely to have done so by the time he has become adolescent.

The amount of data which KPM present in this analysis is worth mention as evidence that they do not shirk work. Tables are given for 7 types of activity. Three age groups are shown in each table. When we classify by occupational level of subject and parent, this leads to 21 two-way tables. Five measures of the type of activity are given, so that a painstaking examination extends over 105 two-way tables.

KPM appear to have paid most attention to the frequency data. Their task is to determine whether this shows a stronger association with the occupational class of the subject or of the parent. In reaching a verdict, they rely on judgment from eye inspection. By a similar eye inspection, we agree with their verdict as a descriptive statement of what the data indicate, although different individuals might disagree as to how definitely their statement holds. Judgments made by one individual for the data on frequencies were that in 7 of the 21 two-way tables, association with subject and parent either was not present at all or looked about equal. In 9 it looked mildly more with the subject and in 5 it looked strongly more with the subject.

It would be of interest to undertake a more objective analysis. Analysis of variance techniques are available for this purpose, although some theoretical problems remain.

So far as interpretation is concerned, the principal disturbing factor is the possibility, which some critics have mentioned, that the subject's reports of his activity are influenced by the social level to which he belongs at the time of interview. KPM maintain that attitudes towards different types of activity are strongly affected by the social level of the subject. Whether they change when he changes his social level would be interesting to discover. Something might be learned by retakes for subjects who had moved in the social scale. To obtain an abundant body of data of this kind will, however, be a slow and difficult process.

### CHAPTER IX. CARE IN INTERPRETATION

#### 38. *Sample and sampled population*

In sample surveys, the inference from sample to sampled population is often relatively straightforward, although not trivial. We can usually set limits so that the statement "the sample agrees with the sampled population within these limits" has approximately the agreed-upon risk. (We may have to work fairly hard to set these limits correctly.) But we have always to remember, and usually must remind the reader steadily, that these limits are not infinitely narrow.

KPM's caution on page 153 (quoted in Appendix A, Section 48) is a caution, but it is not repeated.

In general, their statements about small differences are more forthright than we would care to make.

#### 39. *Sampled population and target population*

When a respectable approximation of a probability sample is involved, the step from sampled population to target population is usually short and the inference strong. Otherwise, the inference is often tortuous and weak. It depends on subject matter knowledge and intuition, and on other barely tangible considerations. These considerations deserve to be brought to the reader's attention, and to be discussed as best the authors may.

This KPM did not do adequately. Their discussion of diversification (p. 92) and 100 per cent samples (p. 93) is only a beginning.

### 40. *Systematic errors of measurement*

Any quantitative study offers the possibility of systematic errors of measurement. It is generally agreed that these possibilities should be placed before the reader and discussed.

In KPM's study these possibilities concentrate on the difference between present reported and past actual behavior. KPM spent Chapter 4 on this question. Their discussion is generally good, except on some questions which arise in connection with generation-to-generation comparison (see Sections A-25 and A-44).

### 41. *Unsupported assertions*

We are convinced that unsubstantiated assertions are not, in themselves, inappropriate in a scientific study. In any complex field, where many questions remain unresolved, the accumulated insight of an experienced worker frequently merits recording when no documentation can be given. However, the author who values his reputation for objectivity will take pains to warn the reader, frequently repetitiously, whenever an unsubstantiated conclusion is being presented, and will choose his words with the greatest care. KPM did not do this.

Many of the most interesting statements in the book are not based on the tabular material presented and it is not made at all clear on what evidence the statements are based. Nevertheless, the statements are presented as if they were well-established conclusions.

### 42. *Some major controversial findings*

Some KPM findings about which much scientific discussion has centered relate to:

(i) stability of sexual patterns,
(ii) homosexuality, and
(iii) the effects of vertical mobility.

In all these areas KPM have made forthright and bold statements. As discussed in more detail in Sections A-45 to A-47 (also see A-25), there are reasons for caution in every one of the three areas.

CHAPTER X. COMPARISON WITH OTHER STUDIES*

## 43. *Interviewing*

Good sex studies have been made using both the personal interview and questionnaire techniques. Given that just one technique is to be employed, KPM's choice of personal interview seems necessary if illiterates or near-illiterates are to be sampled. At present, it is good practice in gathering this type of data to endeavor to have all subjects give information on as many relevant points of the study as possible. No study seems to have done better on this matter than KPM.

Whether it is always good practice to standardize the questions asked is debatable. KPM did not do this and give telling arguments against the practice. Some other studies have standardized the questions, both in personal interview and in self-administered questionnaires, and they have included good arguments in favor of their procedure. In training interviewers KPM seem to have gone to greater lengths (a year of training) in preparing for the *specific* interview used in the study, than any of the other personal interview studies. Information on training of interviewers is fairly hard to come by in all these studies.

Given the choice of personal interview, it is not possible at this writing to be logically certain whether the KPM technique is better or worse than that of the other interview studies, no matter whether one approves or disapproves of the tactics of a diagnostician or medical detective. Some discussion of how the KPM interview appeared to us is given in Appendix E. Numerous cross-checks on frequency and dates of occurrences appear within the KPM interview, while they seem to be lacking in most other studies. Setting aside points on which there is no evidence, KPM's interviewing is as good as or better than that of the other studies reviewed.

---

* The material in this chapter is our inference from the reviews supplied by W. O. Jenkins and presented in Appendix B. We have not personally read all the volumes concerned. The volumes are as follows:

Bromley, Dorothy D., and Britten, Florence H. *Youth and sex*. New York: Harper and Brothers, 1938.

Davis, Katherine B. *Factors in the sex life of twenty-two hundred women*. New York: Harper and Brothers, 1929.

Dickinson, R. L., and Beam, Lura A. *The single woman*. Baltimore: Williams and Wilkins Co., 1934.

Dickinson, R. L., and Beam, Lura A. *A thousand marriages*. Baltimore: Williams and Wilkins Co., 1931

Farris, E. J. *Human fertility and problems of the male*. White Plains, N.Y.: Author's press, 1950.

Hamilton, G. V. *A research in marriage*. New York: A. and C. Boni, 1929.

Kinsey, A. C., Pomeroy, W. B., and Martin, C. E. *Sexual behavior in the human male*. Philadelphia: W. B. Saunders Company, 1948.

Landis, C., et al. *Sex in development*. New York and London: Paul B. Hoeber, 1940.

Landis, C., and Bolles, M. M. *Personality and sexuality of the physically handicapped woman*. New York and London: Paul B. Hoeber, 1942.

Terman, L. M., et al. *Psychological factors in marital happiness*. New York: McGraw-Hill Book Co., 1938

## 44. *Checks*

As for checks on the interviewing process, KPM unquestionably lead the field with 100 per cent samples, retakes, spouse comparisons, early vs. late groups, interviewer comparisons, and the pubic hair study. Some authors mention casual checks with no data supplied. Bromley and Britten compare interview and questionnaire results on different groups. Davis reports a study where 50 subjects were interviewed before and after questionnaire administration, and offers a breakdown by consecutive 100 questionnaires received. Dickinson and Beam's two books speak of comparing verbal reports and physical examination results as a way of verifying the record rather than as a check—no records seem to be published. Farris' comparison of reported vs. personally recorded masturbatory rates omits the critical comparative information. Hamilton finds that different question wordings give different responses, but leaves the matter here. Landis and Bolles use several independent judges for evaluation of scales—but, instead of comparing their results, argue that agreement will be good because of experience and training. They do not compare normal with handicapped subjects. Landis checks with the psychiatric case history as a means of eliminating subjects with discrepancies, and gives data on the agreement of independent judges' ratings. Terman offers spouse comparisons. When KPM's checks are viewed with those of the other leading sex studies in mind, it is clear that a new high level has been established.

## 45. *Sampling*

All studies used volunteer non-probability samples. Some were drawn from more specifiable target populations than others. For example, Bromley and Britten drew exclusively from college volunteers, while Davis used mail-questionnaire respondents from lists of Women's Clubs and college alumnae. Others used well-to-do patients, or clinic groups. Aside from KPM, Bromley and Britten is the only study that seems to have attempted to get nationwide geographic representation (we have omitted M. J. Exner's 1915 study), while Davis has covered the eastern area, and Terman covers part of the California area. Although KPM's sample is heavily charged with college students, a broader representation of social and educational levels is offered than in the other studies. All studies reviewed have special features which make generalizations to specific populations difficult. Certainly KPM's sampling seems never worse and often better than that of the other studies.

## 46. *Analysis*

Most studies confined their analysis to simple descriptive statistics—percentages, means, and medians. A few added ranges, standard deviations, correlation coefficients, and attempted significance tests. About half used two-way breakdowns, usually on background characteristics, as a way of sharpening differences between groups. Three studies offered scales either based on judges' evaluations (Landis, and Landis and Bolles), or scoring of batteries of items (Terman). KPM restricted the use of scales to occupational classification and homosexual-heterosexual rating. They added the accumulative incidence curve, the U. S. corrections, and extensively used fine-grained (high-order) breakdowns. In general, KPM's analysis employed more devices and was more searching than the analyses offered by other studies.

## 47. *Interpretation*

We have already mentioned (33) that KPM are competent at the accurate and understandable verbal description of the meanings of a table whose entries are taken as correct. Some of the other authors have also done well, although the extent of their analysis is usually more limited. In inferring from sampled population to target population, all the studies are weak. The inferences left with the reader (if we are to judge) are much broader than the studies could possibly warrant. Every study has its own precautionary remarks to the effect that the reader must not extend the inferences beyond that of the population studied. Very little attempt is made to describe the target population, to help the reader with the step from sample to sampled population, or to remind him of sampling fluctuations. The precautionary remarks in the opening pages of a study are usually forgotten when the authors come to discuss matters of national policy, morals, legislation, therapy, and psychological and sociological implications toward the end of their book. The reader must then be left with the inference that the findings apply on at least a national scale. Bromley and Britten are more forthright than most. They argue overtly that their volunteer college sample is a representative of all U. S. individuals of college age. Of the 10 studies considered, only two, Davis and Farris, seem to have consistently exercised due caution about generalization from sample to population and warnings to the reader. The last paragraph of the section entitled, "Description of Sample and Sampling Methods" in each review in Appendix B gives one reader's opinion of the generalizations from sample to sampled population intended by the author.

Our reviewer was not asked to gather data that would give us a way of comparing the extent of unsupported statements in the other studies with those of KPM, so this aspect of interpretation remains uncompared by us. It would be very interesting if someone would collect such information, not only in connection with the present work, but with regard to general scientific writing in various fields. This would be no small task.

CHAPTER XI. CONCLUSIONS

## 48. *Interviewing*

(1) The interviewing methods used by KPM may not be ideal, but no substitute has been suggested with evidence that it is an improvement.

(2) The interviewing technique has been subjected to many criticisms (see Section A-11), but on examination the criticisms usually amount to saying "answer is unknown," or "KPM have not demonstrated how good their method is."

These conclusions can be summarized by saying that we need to know more about interviewing in general.

## 49. *Checks*

(1) The types of methodological checks considered by KPM seem to be quite inclusive.

(2) A greater volume of checks—more retakes, etc. is desirable, as is more delicate analysis. (See Sections C-15 and C-18.)

(3) The results of duplicate recording of interviews should be published.

These conclusions can be summarized by saying that KPM's checks were good, but they can afford to supply more.

## 50. *Sampling*

Given U. S. white males as the target population, our conclusions are that:

(1) KPM's starting with a nonprobability sample was justified.

(2) It should perhaps already have been supplemented by at least a small probability sample.

(3) If further general interviewing is contemplated, and perhaps even otherwise, a small probability sample should be planned and taken.

(4) In the absence of a probability-sample benchmark, the present results must be regarded as subject to systematic errors of unknown magnitude due to selective sampling (via volunteering and the like).

## 51. *Analysis*

KPM's analysis is best described as simple and relatively searching. They did not use such techniques as analysis of variance or multiple regression, but they brought out the indications of their data in a work-manlike manner.

In more detail:

(1) their selection of variables for adjustment seemed to be a reasonably effective substitute for more complex analyses,

(2) they gave several measures of activity (giving the reader a choice at the expense of more tables to examine),

(3) they made essentially no use of tests of significance, but cited many standard errors (which were inappropriate for their cluster samples),

(4) they used U. S. Corrections and their (independently developed) accumulative incidence curve. More careful exposition of these devices would have been desirable.

To summarize in another way:

(i) they did not shirk hard work, and

(ii) their summaries were shrewd descriptive comments rather than inferential statements about clearly defined populations.

Their main attempt at inferences was a sample size experiment whose results (i) could have been predicted by statistical theory, (ii) were irrelevant to their cluster sampling.

They continued to add new interviews without redoing earlier tabulations, thus producing an unwarranted effect of sloppiness in the book, although their records were kept carefully and in unusually good shape.

## 52. *Interpretation*

(1) KPM showed competence in accurate and understandable verbal description of the trends and tendencies indicated by their tables. In stating and summarizing what the sample seems to show, they were competent and effective.

(2) Their discussion of the uncertainties in the inferences from the numbers in the tables to the behavior of all U. S. white males was brief, insufficiently repeated, and oftentimes entirely lacking. In instilling due caution about sampling fluctuations and differences between sampled and target populations, they were lax and ineffective.

(3) Their discussion of systematic errors of reporting is careful and detailed (with the exception of some questions bearing on generation comparisons).

(4) Many of their most interesting statements are not based on the tables or any specified evidence, but are nevertheless presented as well-established conclusions. Statements based on data presented, including the most important findings, are made much too boldly and

confidently. In numerous instances their words go substantially beyond the data presented and thereby fall below our standard for good scientific writing.

### 53. *Comparison with other studies*

In comparison with nine other leading sex studies, KPM's work is outstandingly good.

In more detail,

(1) their interviewing ranks with the best,
(2) they have more and better checks,
(3) their geographic and social class representation is broader and better,
(4) their volunteer non-probability sample problem is the same,
(5) they used more varied and searching methods of analysis,
(6) only two of the nine studies (Davis and Farris) were more careful about generalization and warned the reader more thoroughly about its dangers.

Thus, KPM's superiority is marked.

### 54. *The major controversial findings*

It is perhaps fair to regard these four as KPM's major controversial findings:

(1) a high general level of activity, including a high incidence of homosexuality,
(2) a small change from older to younger generations,
(3) a strong relation between activity and socio-economic class,
(4) relations between activity and *changes* of socio-economic class.

All of these KPM set forth as well established conclusions. All are subject to unknown allowances for:

(a) difference between reported and actual behavior,
(b) nonprobability sampling involving volunteering.

While their findings may be substantially correct, it is hard to set any bounds within which the truth is statistically assured to lie (see Appendix A, Section 4.) Once again, we wish to point out that the same difficulties are present in many sociological investigations.

## CHAPTER XII. SUGGESTED EXTENSIONS

### 55. *Probability sampling*

Appendix D discusses the advantages, possibilities and difficulties of probability sampling in some detail.

In brief summary:

(1) Costs and refusal rates together determine the wisdom of extensive probability sampling.
(2) Information on costs and refusal rates is lacking.
(3) Hence probability sampling should begin on a very small scale, say 20 cases.
(4) A step-by-step program, starting at such a scale, seems wise, and is recommended to KPM.

## 56. *Retakes*

While retakes showed high agreement on vital statistics, and moderately high agreement on incidence, the data presented in KPM for frequencies show considerably less agreement. The data do not make clear how much better a retake agrees with a take than with a randomly selected interview for another subject with the same age, religion, social class, etc.

If the agreement is better, then retakes will provide evidence as to non-random agreement—evidence bearing on the much-discussed subject of the constancy of recall. In addition, take-retake differences are clearly so large as to make retakes of two old subjects at least as valuable as a take of one new subject in determining the average behavior of groups (see Section A-24).

If the agreement is no better, then retakes will provide evidence that this was so, and every retake will be as valuable as a new take in determining the average behavior of groups.

In our opinion 500 retakes would help the standing of KPM's data more than 2000 new interviews (selected in the same old way). It would of course be important to determine and report the selective factors which influenced the selection of the retaken subjects.

## 57. *Spouses*

Separate interviews of husband and wife are a useful supplement to retakes, in that they supply the nearest approach to two independent reports of the same action, although the information is restricted for the most part to marital coitus, and is weakened by the possibility of collusion. In the book, KPM present comparisons for 231 pairs of spouses.

In an expansion of this program, various elaborations could be suggested. The first objective should probably be to interview more pairs from the lower educational levels, in order that the agreement between spouses can be examined separately for different educational levels. As in the case of retakes, the data are not wasted so far as the main study is concerned, since they contribute both to the male and female samples.

58. *Presentation*

As the critics point out (Chapters VII-A, I-C), parts of the book are hard to understand because of lack of clarity of presentation. In future editions, the following steps would remove the major ambiguities.

(i) KPM should explain why the numbers of cases change erratically from table to table. In future publication it would be worth substantial effort to avoid these changes.

(ii) Table headings and contents should be critically reviewed as to their lucidity.

(iii) Worked examples of the calculation of U. S. corrections should be given. References under the tables to the variables used for correction should be more precise.

(iv) More discussion should be given, with numerical illustration, of the meaning of accumulative incidence percentages.

(v) More information should be given about the questions asked, with their variations, in the interview. Although this would be extremely laborious to do for the complete interview, one or two blocks of related questions might serve the purpose. For such a block, KPM might describe (a) the variations used in the statement of the questions (b) the variations in the order of questions (c) the reasons for the variations. An illustration of this type would give deeper insight into the logical structure of KPM's interviewing technique and might go far to substantiate their claim (p. 52) that flexibility is one of the strengths of their technique

(vi) Several critics make a strong plea that more information be given about the composition of the sample (see Chapters I-A, I-C). The specific items requested vary with the critic, and some would be a major undertaking both in preparation and publication. A minimum that seems feasible would be to present a multiple classification of the subjects according to the following items *at the time of interview:* age, marital status, occupation, educational status, religious affiliation, place of residence. In addition, more information is needed about the extent to which special groups (e.g., those in penal institutions, homosexual groups) contribute to the tables.

59. *Statistical analyses*

In Appendix C, a number of statistical analyses are outlined which would be a useful contribution to the methodology of studies of this kind. The analyses would require expert statistical direction.

As has been pointed out, the standard errors presented by KPM are invalid, because they were computed on the assumption of random sampling of individuals. A method for calculating standard errors so as to take into account the actual nature of KPM's sampling is given in Chapter II-C. These standard errors would allow a realistic appraisal of the stability of KPM's means. They would indicate by how much the means determined from the present KPM sample are likely to vary from the means of a much larger sample of cases obtained by the KPM methods.

KPM described orgasm rates in terms of per cent incidence and mean or median frequency. However, other mathematical functions of these variables may be more appropriate, leading to simpler statements of the results. Approaches for investigating this question, and the related question of the use of some combination of the variables, are suggested in Chapters III-C and IV-C.

The question of applying adjustments to segment means has already been discussed (Section 17). A technique is presented (Chapter V-C) for reaching practical decisions on the appropriateness of adjustment and on the number of variables for which adjustment should be made.

### 60. *Relative priorities*

We give here our personal collective opinion as to how further effort on the male study might best be spent (we have not tried to evaluate priorities in comparison with the female study, or any other studies which KPM may contemplate).

If the interviewer time which it would require were available, we believe that the effort required for the proposed probability sample would be worthwhile.

So long as it did not interfere with the possibility of a probability sample, available interviewer time should be concentrated:

on retakes when working in or near old areas.
on husband-wife pairs when two interviewers are available.

If the probability sample has already been ruled out, and if fewer interviewer months are available, then an attempt to retake a random sample of previous subjects would be most desirable, whenever possible, husband and wife being taken whenever either is retaken.

Effort in the form of statistical analysis and presentation need not interfere with interviewing, and should be pressed to the extent that experienced and understanding personnel can be found.

*This and the following paper also say "stop, look, and listen" to the researcher in any field of public health who may be less than adequately prepared in statistical method. All three papers are concerned, in general, with the deceptively simple matter of choosing controls, this one marking the pitfalls inherent in pairing or matching.*

# Matching in Analytical Studies*†

## WILLIAM G. COCHRAN

*Professor of Biostatistics, School of Hygiene and Public Health,*
*Johns Hopkins University, Baltimore, Md.*

MOST of the following discussion will be confined to studies in which we compare two populations, which will be called the experimental population and the control population. The experimental population possesses some characteristic (called the experimental factor) the effects of which we wish to investigate: It may consist, for example, of premature infants, of physically handicapped men, of families living in public housing, or of inhabitants of an urban area subject to smoke pollution, the experimental factors being, respectively, prematurity, physical handicaps, public housing, and smoke pollution. I shall suppose that we cannot create the experimental population, but must take it as we find it, except that there may be a choice among several populations that are available for study.

The purpose of the control population is to serve as a standard of comparison by which the effects of the experimental factor are judged. The control population must lack this factor, and ideally it should be similar to the experimental population in anything else that might affect the criterion variables by which

the effects of the factor are measured. Occasionally, an ideal control population can be found, but, more usually, even the most suitable control population will still differ from the experimental population in certain properties which are known or suspected to have some correlation with the criterion variables.

When the control and experimental populations have been determined, the only further resource at our disposal is the selection of the control and experimental *samples* which are to form the basis of the investigation. Sometimes this choice is restricted, because the available experimental population is so small that it is necessary to include all its members, only the control population being sampled.

The problem is to conduct the sampling and the statistical analysis of the results so that any consistent differences which appear between the experimental and the control samples can be ascribed with reasonable confidence to the effects of the factor under investigation.

The first step in any matching process is to select those variables (called the covariables) on which the two samples are to be matched. I shall assume for the moment that this decision has been made; the principles in-

---

* Presented before the Statistics Section of the American Public Health Association at the Eightieth Annual Meeting in Cleveland, Ohio, October 23, 1952.
† Paper No. 288, Department of Biostatistics. Some of the theoretical results used in this paper were obtained under a research contract with the Office of Naval Research.

volved in the decision will be discussed briefly later.

## PAIRING

Matching of the experimental and control samples with respect to the covariables can be accomplished in a number of ways. Conceptually, the simplest is the method of pairing. Each member of the experimental sample is taken in turn, and a partner is sought from the control population which has the same values as the experimental member (within defined limits) for each of the covariables. One way of doing this is to perform a multiple classification of the control population by the variables. We then examine the first member of the experimental sample, pick the cell which contains all control members having the desired set of covariables, and choose as the partner one control member at random from this cell. This procedure is repeated for each member of the experimental sample.

If an occasional cell is found to be empty, it is usually preferable to choose the control partner from a neighboring cell, rather than to omit the experimental member. If numerous cells are found to be empty, this is a danger signal. Either the limits of variation allowed in the covariables are too narrow or the control population is not satisfactory.

The analysis of the results is very simple. The difference (experimental-control) is computed for each pair, and any t-tests are applied directly to this series of differences.

## EFFECTIVENESS OF PAIRING WHEN WE HAVE AN IDEAL CONTROL

It is difficult to discuss the effectiveness of pairing in realistic terms. The advantages of pairing and of covariance analysis are usually demonstrated by means of a linear regression model. I shall present this analysis, but, as will be seen, there is reason to doubt whether the assumptions in the analysis are valid for many of the studies conducted in practice.

Let $y$ denote the variable by which the effects of the experimental factor are measured, and $x$ denote the covariable, assuming for simplicity that there is only one. The model assumes that $y$ has a linear regression on $x$ with the same slope $\beta$ in each population. The equations are as follows:

Experimental population: $y = a + \beta x + d$     (1)
Control population   : $y' = a' + \beta x' + d'$     (2)

The variables $x$ and $d$ are independently distributed and the deviations $d$, $d'$ have means zero in both populations. Further, it is assumed that the means $\overline{X}$, $\overline{X}'$ of $x$ in the two populations are equal, and that $(a - a')$ represents the true effect of the experimental factor, i.e., that no unsuspected biases are present.

In effect, this model postulates that we have been successful in finding an *ideal* control population, since the relation between $y$ and $x$ is the same in both populations and since $x$ has the same average value in both populations.

With this model, the precision given by paired samples can be compared with that given by independent random samples drawn from the two populations. In either method, the effect of the experimnetal factor will be estimated by the difference $(\overline{y} - \overline{y}')$ between the means of the two samples. For the independent samples, each of size $n$, the variance $V_1$ of $(\overline{y} - \overline{y}')$ is

$$V_1 = \frac{2}{n}\sigma_y^2 \qquad (3)$$

assuming for simplicity that $\sigma_y$ is the same in both populations.

With samples paired for $x$, on the

TABLE 1

*Values of (1 — R²)*

| R | 0.2 | 0.3 | 0.4 | 0.5 | 0.6 | 0.7 | 0.8 | 0.9 |
|---|---|---|---|---|---|---|---|---|
| (1 — R²) | 0.96 | 0.91 | 0.84 | 0.75 | 0.64 | 0.51 | 0.36 | 0.19 |

other hand, it follows from equations (1) and (2) that

$$\overline{y} - \overline{y'} = (a - a') + (\overline{d} - \overline{d'})$$

so that the variance $V_p$ of this difference is

$$V_p = \frac{2}{n}\, \sigma^2_d, \qquad (4)$$

This result may be expressed in a more useful form. From (1),

$$\sigma^2_y = \beta^2\, \sigma^2_x + \sigma^2_d, \qquad (5)$$

since $x$ and $d$ are assumed independent. If $\rho$ is the correlation coefficient between $y$ and $x$, then $\beta\sigma_x = \rho\sigma_y$. Thus (5) becomes

$$\sigma^2_y = \rho^2\, \sigma^2_y + \sigma^2_d$$

giving a well-known result in theory,

$$\sigma^2_d = \sigma^2_y\, (1 - \rho^2)$$

Hence, finally, the variance of $(\overline{y} - \overline{y'})$ for the paired samples may be written, from (4)

$$V_p = \frac{2}{n}\, \sigma^2_y\, (1 - \rho^2) \qquad (6)$$

Comparison with (3) shows that the ratio of $V_p$ to $V_i$ is $(1 - \rho^2)$. A more concrete way of expressing this result is as follows. If $n_p$, $n_i$ are the respective sample sizes which make the variances *equal* for paired and independent samples, then $n_p = n_i (1 - \rho^2)$. Thus, the ratio $(1 - \rho^2)$ shows how much we can afford to reduce the sample size, when pairing is employed, without any loss of precision.

If pairing is accomplished for several $x$-variables, all linearly related to $y$, the ratio becomes $(1 - R^2)$ where $R$ is the multiple correlation coefficient between $y$ and the $x$'s. Values of this factor are shown in Table 1.

The reductions in variance are not large until $R$ reaches 0.5. A reduction by one-half, which corresponds to an allowable reduction of the sample size by one-half, requires $R = 0.7$. Unfortunately, although many published papers have discussed the association between morbidity or mortality and such covariables as age, sex, economic status, family size, and degree of overcrowding, the associations tend to be described in general terms, and little information is available about the actual sizes of the correlations that are to be expected in field studies in public health. With some obvious exceptions (e.g., the association between chronic disease and age) my impression is that the multiple correlation coefficient is often below 0.5, so that the gain in precision from pairing is often modest.

SELECTION OF THE COVARIABLES

Table 1 is also relevant in the selection of the covariables on which the pairing is to be based. It is nearly always possible, with a little effort, to produce a substantial number of $x$-variables that *might* have some association with $y$, and this list must be reduced to a few $x$-variables which will actually be used in pairing. It follows from Table 1 that inclusion of a specific $x$-variable in the pairing is worth-while only if it decreases $(1 - R^2)$ by an appreciable amount (say 10 per cent), when this $x$-variable is added to those already selected. Although the practical use of this result requires an intimate knowledge of the relation between $y$ and the $x$'s which is rarely possessed, the result indicates that associations in which the correlation between a covariable and $y$ is of the order of $0.1 - 0.3$ are unlikely to produce useful gains in precision. Thus, in deciding whether to include a covariable, the important question is not "Is there an association?" but "How large is the correlation coefficient?"

### PAIRING VERSUS RANDOM SAMPLES WITH COVARIANCE

If, instead of pairing, we draw random samples of size $n$ from each population and adjust the sample means by covariance, then, on the average,

$$V(\bar{y}_{adj} - \bar{y}'_{adj}) = \frac{2}{n} \sigma^2_y (1 - \rho^2) \left\{ 1 + \frac{1}{2(n-2)} \right\} \qquad (7)*$$

The term in curly brackets represents an increase in variance due to errors in the covariance adjustment. If the covariance adjustment is made for $k$ $x$-variates, by means of a multiple regression, this term becomes

$$1 + \frac{k}{(2n-k-3)}$$

Provided that the sample size $n$ exceeds $10k$, this factor is close to unity, and covariance gives about the same precision as pairing.

In these circumstances there is not much to choose between pairing and covariance. Pairing has the advantage that the computations are simpler, particularly if the samples are paired on several $x$-variables. If the regression is nonlinear, pairing will give higher precision than covariance, unless the presence of nonlinearity is recognized in the covariance analysis and we go to

---

* This result, which is a slightly different form of Dr. Greenberg's result, assumes that the x-variable is normally distributed in the population, but is approximately true even if $x$ is not normal.

the trouble of fitting the appropriate type of regression curve. A difficulty which I have occasionally encountered with covariance is that some scientists have an inborn suspicion of adjustments to the data, and although the adjustments made in the covariance analysis are entirely objective, they may find a rather grudging acceptance.

Pairing has some limitations and disadvantages, the importance of which varies with the type of study. Pairing requires that data on the values of the covariables in the control population be readily accessible; this may not be the case. One disadvantage is the time spent in constructing the pairs. If the experimental sample is small and the control population is large, or if the experimental sample becomes available one member at a time (as with newborn infants or admissions to a hospital) the pairing may be accomplished easily, but it can become tedious if the samples are large and it may impede the progress of the study. A small trial to estimate the time involved in pairing is sometimes advisable. If considerable attrition of the data is expected, as in a long-term follow-up study, the symmetry of pairing is lost. The simplicity of the analysis can be retained only by dropping all partners of "missing" sample members, which involves a loss of information. In order to avoid this loss, it is necessary to use a covariance analysis.

There is one further situation in which pairing may be highly effective. In some studies the experimental population has been drawn from some larger population by a mechanism which operates solely to select certain values of the covariables. For instance, suppose that the experimental population consists of families in public housing in a large city and that these families were selected from a larger population of approved applicants for public housing by some administrative rules. Suppose that the rules give preference to families of veterans and to families of certain sizes (since public housing is built with some preassigned and not necessarily average distribution of family sizes in mind), but are otherwise on a "first come, first served" basis. In this case the approved veteran applicants who are still waiting might constitute a good control population, except that the control and study samples need pairing or matching on family size. (It is not claimed that the selection of entrants to public housing actually operates in this way, the example being intended purely for illustration.)

In a situation of this kind, where $x$ represents family size, the previous regression model might still apply, except that the population mean $\overline{X}, \overline{X}'$ now differ. As a result, some difficulty may be experienced in finding control partners for the experimental sample, but if the pairs can be constructed, equation (6) continues to hold for the variance of $(\overline{y} - \overline{y}')$ in the paired samples.

With the covariance method, the corresponding variance may be shown to be approximately

$$1 + \frac{D^2}{4\sigma_x^2} \qquad (8)$$

times as large as that given by pairing, so that pairing is more efficient than covariance. The increase in precision from pairing relative to covariance is probably not great in practice. For instance, the variance ratio in (8) equals 1.25 when $D = \sigma_x$, that is when the population means $\overline{X}, \overline{X}'$ are distant one standard deviation. This implies a fairly drastic selection operating on the $x$ variable.

If the experimental population involves selection on several $x$ variables, pairing removes the disturbing effects of this selection, provided that *all* the $x$ variables are included in the pairing.

### OTHER METHODS OF MATCHING

As we have seen, pairing is most easily done when one of the populations (usually the control population) is large and the samples are small. If the samples are large, pairing may be time consuming, and if the available populations are not much larger than the desired size of samples, pairing may be impossible. In these circumstances, methods of matching which are less thorough deserve consideration.

$$V(\overline{y}_{adj} - \overline{y}'_{adj}) = \frac{2}{n}\sigma_y^2(1 - \rho^2)\left\{1 + \frac{1}{2(n-2)} + \frac{nD^2}{4(n-2)\sigma_x^2}\right\}$$

where $D = (\overline{X} - \overline{X}')$. The extra term involving $D^2$ appears because the covariance adjustment, $b(\overline{x} - \overline{x}')$, has become larger, since $\overline{x}$ and $\overline{x}'$ no longer have the same population means. The term in $D^2$ is almost independent of the sample size $n$. For $n > 20$, the variance for the covariance method is approximately

In the technic known as *balancing*, we do not pair individually, but select the control sample so that its means agree with those of the experimental sample for each of the covariables. Balancing can usually be carried out more quickly than pairing; it gives the same precision as pairing if the regression of $y$ on the covariables is linear, although balancing is less precise if the regression is nonlinear. As Dr. Greenberg states in his article, balancing requires the use of a covariance analysis in order to perform tests of significance—a point which has often been overlooked.

Another method of obtaining a less rigorous matching is to divide the range of any covariable into three or four classes. Thus, if matching is being done for three covariables, the number of cells produced will lie somewhere between 27 and 64 (some of which may empty). The sample from any cell in the experimental population is drawn at random, and its size is proportional to the number of entries in the cell. The sample from any cell in the control sample is also drawn at random and is made to be the same size as that for the corresponding cell in the experimental sample. If International Business Machines' equipment is used and the samples are large, this method, sometimes called *stratified matching,* is more expeditious than pairing, because it requires a much less detailed breakdown of the population into cells. It is less precise than pairing, since the covariables do not have exactly the same set of values in the control and experimental samples. However, with at least three classes for any covariable, it may be shown by theory that the loss precision is small unless the multiple correlation coefficient exceeds about 0.7. The analysis of the results is slightly more complicated, because we can compare the experimental and control results separately for each cell. In return, this analysis focuses attention on any variation that occurs in the effects of the experimental factor as the levels of the covariables change—in other words, on the interactions of the experimental factor with the covariables. Such information may broaden the results of the study.

There are several variants of this method. For instance, if there is particular interest in interactions, the samples from each cell may be made equal so far as is feasible.

The experimental and control samples need not be the same size. With pairing, we could select $r$ control partners for each member of the experimental sample. The factor $2/n$ in the variance of the mean difference is then reduced to $(r + 1)/rn$. The cost of the study is, of course, increased, but the device may be profitable when the experimental sample is small and the cost of obtaining and processing the control data is not prohibitive.

SITUATION WHEN THE CONTROL
POPULATION IS NOT IDEAL

It is worth reiterating that the previous discussion of the effectiveness of matching or covariance assumes that the control population is ideal, in the sense that the control and experimental populations differ at most through selection on certain covariables which are included in the matching or covariance.

When the two populations differ in other ways, we do not know how effective matching is. It is almost certain to be less effective than when the control population is ideal, and it may be practically ineffective. What is likely to happen is that the regressions of $y$ on the covariables will differ in the two populations. In this event, matching no longer removes all the disturbing effects of the covariables on which we match. Further, there are likely to be other variables with respect to which the populations differ. In so far as these variables are uncorrelated with the matching covariables, their disturbing effects are unchanged by matching. The net result is that the difference between the means of the matched samples is a biased estimate of the effects of the experimental factor. The bias can be expected to be smaller with matched than with unmatched samples, but it may be only slightly smaller.

DISCUSSION AND SUMMARY

The principal conclusion from the preceding discussion is that the selection of the control population is a more crucial step than the selection of a method of matching or of the covariables on which to match. Matching removes the deficiencies of a poor control to only a limited extent.

In turn, this suggests that, whenever feasible, any study should start with a comparison of the experimental and control populations. This is by no means the rule in practice. Frequently, the experimental sample is chosen first and the control population is then searched in order to find the partners. If partners seem hard to find, some misgivings about the control population may arise, but if the pairs can be found, sometimes by selecting an extreme sample from the control population, the study proceeds. This kind of matching may leave us with a control population that does not resemble the study population and a control sample that is a very extreme sample from the control population.

The kind of comparison which I am recommending was conducted by Densen, et al.,[1] as a preliminary to a proposed study on the penicillin treatment of cardiovascular syphilis. From hospital records, it was planned to select a sample of patients who had not received penicillin as a control for an experimental sample of patients who had been treated with penicillin. In a comparison of the available patients from two hospitals, Densen, et al., found many differences between the experimental and control cases and concluded that any matching process would be hard to carry out and that its results would be suspect.

A comparison of this kind does not provide proof that the control population is ideal, since only the covariables can be studied, but not their relations with y. If, however, a number of covariables are included, it is reassur-ing to find agreement, or at most minor disagreement, between the two populations on all these covariables. Where more substantial discrepancies appear, their implications as to the suitability of the control can be considered.

If the control population appears satisfactory, the next step is to select the covariables to be used in matching. At this point any available information about the sizes of the correlations between y and the x's is relevant. Since my impression is that these correlations are frequently low in public health field studies, with some obvious exceptions, my recommendation, in cases of doubt, is to omit an x-variable from the matching rather than to include it. Covariance adjustments can be made should this variable later prove important.

If the samples are small and at least one of the populations is large, so that pairing is not too laborious, pairing has much to recommend it. If the samples are large, stratified matching may be preferable. If the samples are not much smaller than the available populations, random samples should be drawn and covariance adjustments applied.

In conclusion, in observational studies, where it is not feasible to assign a subject at random to the control or experimental sample, we can never be sure that some unsuspected disturbance does not account, in large part, for the observed difference between the two samples. Consequently, the results of tests of significance must be interpreted with more caution in observational studies than in experiments where randomization can be employed. One good practice in observational studies is to check any theory at as many points and in as many ways as ingenuity can devise. An illustration occurs in a study of the relation between inoculation with pertussis vaccine and poliomyelitis, reported by Hill and Knowelden.[2] The experimental sample consisted of children with poliomyelitis and the control sample of children without poliomyelitis, paired for age and sex and living in the same area. The experimental sample showed a marked excess of inoculations during the month preceding onset of poliomyelitis, but no excess of inoculations at intervals larger than a month. Second, during the month preceding onset, paralysis occurred at the site of inoculation in the great majority of cases, but with the interval larger than a month, the site of inoculation was involved in paralysis in only a small minority of cases. The point is that these two independent results greatly strengthen the evidence for a causal relationship. If the same kind of result appears repeatedly when the data are analyzed from widely different points of view, it becomes successively more difficult to imagine any "disturbance" that will explain away *all* the results. Where it can be employed, this technic does much to overcome the handicap under which we all labor in observational studies.

REFERENCES

1. Densen, P. M., et al. Studies in Cardiovascular Syphilis II. Methodologic Problems in the Evaluation of Therapy. *Am. J. Syp., Gonor. & Ven. Dis.* 36, 1:64–76 (Jan.), 1952.
2. Hill, A. Bradford, and Knowelden, J. Inoculation and Poliomyelitis. A Statistical Investigation in England and Wales in 1949. *Brit. M. J.* ii:1 (July), 1950.

# 53

# A SAMPLING INVESTIGATION OF THE EFFICIENCY OF WEIGHTING INVERSELY AS THE ESTIMATED VARIANCE*

William G. Cochran and Sarah Porter Carroll

*Johns Hopkins, Baltimore, Md.*
*and*
*Institute of Statistics, Raleigh, N. C.*

## 1. INTRODUCTION

Suppose that we have a number of estimates $x_i (i = 1, 2, \cdots k)$, normally and independently distributed about the same mean $\mu$ with different variances $\sigma_i^2$. If the values of the $\sigma_i^2$ are known, the best estimate of $\mu$ is generally agreed to be the weighted mean

$$\bar{x}_w = \sum_{i=1}^{k} w_i x_i / w, \qquad \text{where} \qquad w_i = \frac{1}{\sigma_i^2} : w = \sum w_i .$$

If the $\sigma_i^2$ are not known, but we possess estimated variances $s_i^2$, based on $n_i$ degrees of freedom, respectively, analogy suggests the use of a weighted mean with weights inversely proportional to the *estimated* variances. This mean is

$$\bar{x}_{\hat{w}} = \sum_{i=1}^{k} \hat{w}_i x_i / \hat{w}, \qquad \text{where} \qquad \hat{w}_i = \frac{1}{s_i^2} : \hat{w} = \sum \hat{w}_i .$$

Data of this kind may occur when $k$ laboratories make separate determinations $x_i$ of the same physical or chemical quantity, each with an estimated standard error, or when a summary is being made of the results of $k$ replicated experiments, in each of which the difference $x_i$ between a specified pair of treatments has been observed. In practices it cannot be taken for granted that the observations $x_i$ are all estimate, of the *same* mean $\mu$, because personal biases or local conditions of experimentation may render this assumption false. The discussion in this paper is confined to situations in which the assumption holds.

*Research conducted under a contract with the Office of Naval Research.

Some results about the distribution of $\bar{x}_{\hat{w}}$ are known. When the degrees of freedom $n_i$ are all equal to $n$, the limiting distribution of $\bar{x}_{\hat{w}}$, as the number of estimates $k$ tends to infinity, is normal, (1), with mean $\mu$ and variance

$$V(\bar{x}_{\hat{w}}) = \frac{(n-2)}{(n-4)w} \tag{1}$$

The proof requires that $n > 8$ and that the $\sigma_i^2$ are bounded above and below. When the $n_i$ are not equal, the limiting variance takes the more complex form

$$V(\bar{x}_{\hat{w}}) = \frac{\displaystyle\sum_{i=1}^{k} \frac{n_i^2 w_i}{(n_i - 2)(n_i - 4)}}{\left\{ \displaystyle\sum_{i=1}^{k} \left( \frac{n_i w_i}{n_i - 2} \right) \right\}^2} \tag{2}$$

For practical applications, these results may be used as approximations when the number of estimates $k$ that are being combined is large. Until recently, no information has been available as to how well the results apply when $k$ is small. However, Meier (2) has given an approximation to $V(\bar{x}_{\hat{w}})$, valid for any $k$, but neglecting terms of order $1/n_i^2$. His result is

$$V(\bar{x}_{\hat{w}}) \doteq \frac{1}{w} \left[ 1 + \frac{2}{w^2} \sum_{i=1}^{k} \frac{1}{n_i} w_i(w - w_i) \right] \tag{3}$$

Variance formulas (1), (2) and (3) are useful for comparing the precision of $\bar{x}_{\hat{w}}$ with that of other simple estimates of $\mu$—in particular with the unweighted mean of the $x_i$. These formulas cannot be used, however, to attach a standard error to an actual value $\bar{x}_{\hat{w}}$ that has been obtained from a set of data, because the formulas involve the unknown true weights $w_i$. For this purpose, Cochran (1) showed that an unbiased estimate of the limiting variance, when the $n_i$ are equal, is

$$V(\bar{x}_{\hat{w}}) = \frac{n}{(n-4)\hat{w}} \tag{4}$$

Similarly, Meier (2) has shown that an unbiased estimate of (3), neglecting terms of order $1/n_i^2$, is

$$V(\bar{x}_{\hat{w}}) = \frac{1}{\hat{w}} \left[ 1 + \frac{4}{\hat{w}^2} \sum_{i=1}^{k} \frac{1}{n_i} \hat{w}_i(\hat{w} - \hat{w}_i) \right] \tag{5}$$

The present paper gives the results of sampling investigations which were carried out by the junior author (3) in order to learn something about the variance of $\bar{x}_{\hat{w}}$ when $n$ and $k$ are both small. Although the scope of these investigations was restricted by the heavy computation involved, as is often the case with sampling studies, the results provide a partial check on the range of application of Meier's formulas and give some information for values of $n$ and $k$ that are beyond this range.

2. METHOD OF CALCULATION

At first sight, the sampling investigations appeared a formidable task because of the multiplicity of variables. Even confining attention to the case where all $n_i$ are equal to $n$, it was desired to cover rather thoroughly the range of values of both $n$ and $k$ between 2 and 20. Then there was the problem of what sets of variances $\sigma_i^2$ should be investigated.

It appeared, however, that if the variance of $\bar{x}_{\hat{w}}$ was expressed in the form

$$V(\bar{x}_{\hat{w}}) = \frac{f(n, k)}{w} , \qquad (6)$$

the factor $f(n, k)$ would be relatively insensitive to variations in the $\sigma_i^2$.

Several results support this conjecture. From equation (1) it follows that the limiting value of $f(n, k)$, as $k$ tends to infinity, is $(n - 2)/(n - 4)$, for any bounded set of values of $\sigma_i^2$. Further, as $n$ tends to infinity, for any fixed $k$, $f(n, k)$ tends to 1, since the weights then become the correct weights. When $k = 2$, the correct variance can be obtained by numerical integration. Calculations of the variance by this method for a few sets of values of $n_1$ and $n_2$ (Porter, 1947) showed that $f(n, k)$ changed by only a few per cent for $\sigma_1^2/\sigma_2^2$ lying between 0.1 and 10.

Finally, some sampling computations of the variance were made for three different sets of values of $\sigma_i^2$. In the first set, all $\sigma_i^2$ were taken as 1; in the second, the values were 1/2, 1 and 2, each value holding for one-third of the $x_i$'s; in the third, the values were 1/4, 1 and 4. Results are shown in table 1 for $k = 3$, 6, 12 and 15 and for $n = 6$, 10 and 20.

As the values of $\sigma_i^2$ become more unequal, $f(n, k)$ tends to decline. The decreases are small in all cases in table 1, the maximum drop being about 7 per cent for $n = 6$, $k = 12$ and 15. The results suggest that computations made for $\sigma_i^2$ all equal will tend to give values of $f(n, k)$ that are slightly too high but not far in error. Consequently, the principal calculations were made for the case in which all $\sigma_i^2$ are equal to 1.

The procedure was as follows. In samples in which the $s_i^2$ are fixed, and $\sigma_i^2 = 1$, $\bar{x}_w$ is normally distributed with mean $\mu$ and variance

$$\sum_{i=1}^{k} \hat{w}_i^2/w^2 \qquad (7)$$

TABLE 1

Effect of inequality in $\sigma_i^2$ on $f(n, k)$

Values of $f(n, k)$

| $n$ | $\sigma_i^2$ | $k$ = number of estimates | | | |
|---|---|---|---|---|---|
| | | 3 | 6 | 12 | 15 |
| 6 | (1, 1, 1) | 1.22 | 1.39 | 1.54 | 1.59 |
| | (1/2, 1, 2) | 1.22 | 1.39 | 1.52 | 1.54 |
| | (1/4, 1, 4) | 1.18 | 1.32 | 1.43 | 1.47 |
| 10 | (1, 1, 1) | 1.15 | 1.22 | 1.28 | 1.30 |
| | (1/2, 1, 2) | 1.14 | 1.22 | 1.27 | 1.30 |
| | (1/4, 1, 4) | 1.11 | 1.19 | 1.25 | 1.27 |
| 20 | (1, 1, 1) | 1.06 | 1.10 | 1.12 | 1.12 |
| | (1/2, 1, 2) | 1.05 | 1.09 | 1.11 | 1.11 |
| | (1/4, 1, 4) | 1.04 | 1.08 | 1.11 | 1.11 |

The values of $s_i^2$, and hence of $\hat{w}_i = 1/s_i^2$, were obtained by squaring and adding from a table of normal deviates (4). The values of $\hat{w}_i$ were then grouped in sets of $k$, each set yielding one value of the conditional variance of $\bar{x}_w$ by substitution in (7). Enough sets were computed for each $n$ and $k$ so that the mean value of the variance over the group appeared stable (the average coefficient of variation of the mean was 1.9 per cent). Finally, since $w = k$ when all $\sigma_i^2$ are equal to 1, the factor $f(n, k)$ is $k$ times this mean variance, as can be seen from equation (6).

TABLE 2

Values of $f(n, k)$ such that $V(\bar{x}_w) = f(n, k)/w$

| $n$ | $k$ = number of estimates that are being combined | | | | | | | | | | | |
|---|---|---|---|---|---|---|---|---|---|---|---|---|
| | 2* | 2 | 3 | 4 | 5 | 6 | 8 | 10 | 12 | 15 | 20 | ∞† |
| 2 | 1.33 | 1.35 | 1.61 | 1.92 | 2.17 | 2.44 | 2.86 | 3.45 | 3.85 | 4.76 | 5.88 | ∞ |
| 4 | 1.20 | 1.22 | 1.35 | 1.49 | 1.61 | 1.72 | 1.92 | 2.13 | 2.33 | 2.56 | 2.86 | ∞ |
| 6 | 1.14 | 1.15 | 1.22 | 1.30 | 1.35 | 1.39 | 1.45 | 1.49 | 1.54 | 1.59 | 1.64 | 2.00 |
| 8 | 1.11 | 1.11 | 1.20 | 1.23 | 1.28 | 1.28 | 1.33 | 1.37 | 1.39 | 1.43 | 1.45 | 1.50 |
| 10 | 1.09 | 1.09 | 1.15 | 1.18 | 1.20 | 1.22 | 1.25 | 1.27 | 1.28 | 1.30 | 1.32 | 1.33 |
| 12 | 1.08 | 1.09 | 1.12 | 1.15 | 1.16 | 1.18 | 1.18 | 1.20 | 1.20 | 1.23 | 1.23 | 1.25 |
| 15 | 1.06 | 1.09 | 1.10 | 1.12 | 1.15 | 1.15 | 1.16 | 1.18 | 1.19 | 1.20 | 1.20 | 1.18 |
| 20 | 1.05 | 1.05 | 1.06 | 1.09 | 1.10 | 1.10 | 1.11 | 1.11 | 1.12 | 1.12 | 1.12 | 1.12 |

*These values obtained by the formula $f(n, k) = (n + 2)/(n + 1)$

†These values obtained by the formula $f(n, \infty) = (n - 2)/(n - 4)$

### 3. SAMPLING RESULTS FOR THE VARIANCE $\bar{x}_{\hat{w}}$

The values obtained for $f(n, k)$ are shown in table 2. For $k = 2$, with the $\sigma_i^2$ all equal, the exact value of $f(n, k)$ is easily found to be $(n + 2)/(n + 1)$. These exact values appear in the first column of table 2: the corresponding values from the sampling investigation appear in the second column, and indicate good agreement with the exact values.

For $k = \infty$, the values shown in table 2 are obtained from the formula $(n - 2)/(n - 4)w$ for the variance of $\bar{x}_{\hat{w}}$ in the limiting distribution, as given previously in equation (1).

Since the variance of $\bar{x}_{\hat{w}}$ is $1/w$ when the weights are known exactly, the quantity $f(n, k)$ is the factor by which the variance is inflated owing to errors in the estimated weights $\hat{w}_i$. Table 2 indicates that this inflation is less serious when $k$ is small than when $k$ is large. With weights based on 8 degrees of freedom, for instance, the variance is inflated by 50 per cent when many estimates are being combined, but only by 11 per cent when two estimates are being combined.

### 4. COMPARISON WITH THE UNWEIGHTED MEAN

A simple alternative to $\bar{x}_{\hat{w}}$ is the unweighted mean $\bar{x}$. A comparison of the precisions of $\bar{x}$ and $\bar{x}_{\hat{w}}$ is of practical interest, because there is no point in undertaking the extra calculation involved in $\bar{x}_{\hat{w}}$ unless a reasonable gain in precision is anticipated.

The situation most favorable to the unweighted mean is that when the $\sigma_i^2$ are all equal. In this event the unweighted mean is fully efficient. Consequently, the values of $f(n, k)$ in table 2 indicate the maximum inflation in variance that will occur if $\bar{x}_{\hat{w}}$ is used in place of $\bar{x}$. Since in practice we do not know by how much the $\sigma_i^2$ vary, we might be willing to regard this inflation of the variance as a premium paid for insurance against the possibility that the $\sigma_i^2$ vary greatly (in which event $\bar{x}$ would be of low efficiency). Table 2 suggests that if $n$ exceeds 20, the premium is not high, but over most of the table the potential inflation of variance is unfortunately well over 10 per cent.

More generally, the variance of $\bar{x}$ is

$$V(\bar{x}) = \frac{1}{k^2} \sum_{i=1}^{k} \sigma_i^2 = \frac{1}{k^2} \sum_{i=1}^{k} \left( \frac{1}{w_i} \right)$$

as compared with the approximate variance for the weighted mean,

$$V(\bar{x}_{\hat{w}}) = \frac{f(n, k)}{w}$$

By comparing the two variances, working recommendations can be made about the use of the two estimates. The difficulty is, however, to know what amount of variation in the $\sigma_i^2$ is typical of practical conditions. Comparisons will be given for the case $k = 2$. In this case

$$V(\bar{x}) = \frac{w}{4w_1w_2} : V(\bar{x}_{\dot{w}}) = \frac{(n+2)}{(n+1)w},$$

where we have used the approximation to $f(n, 2)$ given in the first column of table 2.

Hence the relative precision of $\bar{x}_{\dot{w}}$ to $\bar{x}$ is

$$\frac{(n+1)}{(n+2)}\frac{w^2}{4w_1w_2} = \frac{(n+1)}{(n+2)}\frac{(1+\varphi)^2}{4\varphi},$$

where $\varphi, = w_1/w_2 = \sigma_2^2/\sigma_1^2$, is the ratio of the variances of the two estimates $x_1$ and $x_2$. Table 3 shows the relative precision (in per cent) for a series of values of $n$ and $\varphi$.

TABLE 3

Relative precision (in per cent) of the weighted to the unweighted mean, for $k = 2$.

| $n$ | \multicolumn{6}{c}{$\varphi = \sigma_2^2/\sigma_1^2$} |
|---|---|---|---|---|---|---|
|  | 1 | 1.5 | 2 | 3 | 4 | 6 |
| 2 | 75 | 78 | 84 | 100 | 117 | 153 |
| 4 | 83 | 87 | 94 | 111 | 130 | 170 |
| 6 | 88 | 91 | 98 | 117 | 137 | 179 |
| 8 | 90 | 94 | 101 | 120 | 141 | 184 |
| 10 | 92 | 96 | 103 | 122 | 143 | 187 |
| 12 | 93 | 97 | 105 | 124 | 145 | 190 |
| 20 | 95 | 100 | 107 | 127 | 149 | 195 |
| ∞ | 100 | 104 | 112 | 133 | 156 | 204 |

If the variance ratio for the two estimates lies between 1 and 2, the maximum possible gain in precision from the weighted mean is at most 12 per cent, and the smaller values of $n$ show a loss in precision. When the variance ratio exceeds 3, on the other hand, the weighted mean is superior, or as good, for all values of $n$ down to 2, and the gains in precision may be substantial.

To summarize, the unweighted mean is preferable if the ratio of the larger (true) variance to the smaller is not more than 2. If the ratio lies between 2 and 3, the unweighted mean appears preferable unless the weights are each based on, say, at least 12 degrees of freedom. If the ratio exceeds 3, the weighted mean is preferable even if only 4 degrees of freedom are available to estimate the weights.

### 5. COMPARISON WITH MEIER'S FORMULA

Table 1 also provides a partial check on Meier's approximate formula (3) for $V(\bar{x}_{\dot{w}})$, subject to the restrictions that the comparison covers only the case where the $\sigma_i^2$ are equal and that the values in table 1 are themselves subject to some sampling error. When all $w_i$ are equal and all $n_i$ are equal, Meier's formula reduces to

$$V(\bar{x}_{\dot{w}}) = \frac{1}{w}\left\{1 + \frac{2(k-1)}{nk}\right\} \tag{8}$$

The ratios of the variances in (8) to those in table 2 are shown in table 4.

TABLE 4
Ratio of variance given by Meier's formula to variance in
Table 2

| $n$ | $k$ = number of estimates | | | | | | | | | | |
|---|---|---|---|---|---|---|---|---|---|---|---|
| | 2 | 3 | 4 | 5 | 6 | 8 | 10 | 12 | 15 | 20 | $\infty$ |
| 2 | 1.13 | 1.04 | .91 | .83 | .75 | .66 | .55 | .50 | .41 | .33 | .00 |
| 4 | 1.04 | .99 | .93 | .87 | .83 | .75 | .68 | .63 | .57 | .52 | .00 |
| 6 | 1.03 | 1.00 | .96 | .94 | .92 | .89 | .87 | .85 | .82 | .80 | .66 |
| 8 | 1.01 | .97 | .97 | .94 | .95 | .91 | .89 | .88 | .86 | .86 | .83 |
| 10 | 1.01 | .98 | .97 | .97 | .96 | .94 | .93 | .92 | .92 | .90 | .90 |
| 12 | 1.00 | .99 | .97 | .97 | .97 | .97 | .96 | .96 | .94 | .94 | .94 |
| 15 | 1.01 | .99 | .98 | .96 | .97 | .97 | .95 | .94 | .93 | .94 | .96 |
| 20 | 1.00 | 1.01 | .99 | .98 | .98 | .98 | .98 | .97 | .97 | .98 | .98 |

From inspection of table 4, Meier's formula appears to underestimate the true variance, the relative underestimation increasing as $k$ increases. If we are willing to regard a 6 per cent underestimation of the variance as tolerable, table 5, derived from table 4, shows the smallest values of $n$ for which Meier's formula is satisfactory in this sense.

When at most 5 estimates are being combined, the sampling investigation suggests that Meier's approximation does remarkably well, being satisfactory for values of $n$ as low as 4 or 6.

The increase in the underestimation by Meier's formula when $k$

TABLE 5
Smallest values of $n$ for which Meier's formula underestimates
by less than 6 per cent.

| Number of estimates, $k$ | 2 | 3 | 4 | 5 | 6 | 8 | $\geq 10$ |
|---|---|---|---|---|---|---|---|
| Smallest no. of d.f., $n$ | 4 | 4 | 6 | 6 | 8 | 10 | 12 |

becomes large can be attributed to the effect of terms in $1/n^2$ and
higher orders. As $k \to \infty$, Meier's formula gives

$$V(\bar{x}_{\hat{w}}) = \frac{1}{w}\left(1 + \frac{2}{n}\right)$$

On the other hand, the correct limiting variance, by formula (1), may be
written

$$V(\bar{x}_{\hat{w}}) = \frac{(n-2)}{(n-4)w} = \frac{1}{w}\left(1 + \frac{2}{n} + \frac{8}{n^2} + \frac{32}{n^3} + \cdots\right)$$

### 6. COMPARISON WITH MEIER'S FORMULA FOR THE ESTIMATED VARIANCE OF $\bar{x}_{\hat{w}}$

The sampling data were also used to investigate the performance of
Meier's formula (5) for the estimated variance of $\bar{x}_{\hat{w}}$. The procedure
was as follows. The formula reads

$$V(\bar{x}_{\hat{w}}) \dot{=} \frac{1}{\hat{w}}\left[1 + \frac{4}{\hat{w}^2}\sum_{i=1}^{k}\frac{1}{n_i}\hat{w}_i(\hat{w} - \hat{w}_i)\right] \tag{5}$$

For any specified $n$ and $k$, a large number of sets of $k$ independent values
of $\hat{w}_i$, each derived from $n$ degrees of freedom, had already been assembled for the determination of $f(n, k)$ as described in section 2. By
substitution in formula (5), each set provided one sample value of
$v(\bar{x}_{\hat{w}})$. The average $\bar{v}(\bar{x}_{\hat{w}})$ of this quantity, taken over all the sets, is an
estimate of the true mean given by Meier's formula. The ratio of
$v(\bar{x}_{\hat{w}})$ to the variance of $\bar{x}_{\hat{w}}$ as found from the same group of sets, i.e.
to $f(n, k)/w$, was then computed. The argument is that if Meier's
formula is unbiased, these ratios should fluctuate about a value close
to 1. As before, the comparison is restricted to the case where the
$\sigma_i^2$ are all equal and the $n_i$ are all equal.

The ratios are shown in table 6. Calculations were made only for
$n \geq 6$, since Meier's formula, which neglects terms of order $1/n^2$, was
not expected to be valid for $n < 6$.

TABLE 6

Ratio of average value of Meier's estimated variance to variance in
Table 1

| $n$ | $k$ = number of estimates | | | | | | | | | |
|---|---|---|---|---|---|---|---|---|---|---|
| | 2 | 3 | 4 | 5 | 6 | 8 | 10 | 12 | 15 | 20 |
| 6 | .96 | .93 | .83 | .83 | .81 | .77 | .75 | .72 | .70 | .68 |
| 8 | 1.01 | .91 | .92 | .87 | .87 | .82 | .81 | .81 | .78 | .76 |
| 10 | 1.01 | .97 | .95 | .93 | .92 | .89 | .89 | .88 | .87 | .86 |
| 12 | .99 | .98 | .97 | .95 | .94 | .94 | .92 | .92 | .89 | .90 |
| 15 | .97 | .98 | .99 | .94 | .93 | .93 | .93 | .92 | .90 | .92 |
| 20 | 1.01 | 1.01 | .98 | .99 | .98 | .96 | .99 | .97 | .99 | .96 |

For $k = 2$, table 6 indicates that Meier's formula does extremely well down to $n = 6$. For higher values of $k$, the formula appears to underestimate to a greater degree than the corresponding formula for the true variance (table 4). If, as in section 5, we accept an underestimation by 6 per cent or less, the smallest values of $n$ for which the formula is satisfactory are shown below for the different values of $k$.

| Number of estimates, $k$ | 2 | 3 | 4 | 5 | 6 | 8 | $\geq 10$ |
|---|---|---|---|---|---|---|---|
| Smallest no. of d.f., $n$ | 6 | 10 | 10 | 12 | 12 | 12 | 20 |

As with the formula for the true variance, the underestimation can be attributed to the effects of neglected terms of higher order in $1/n$. In the limiting distribution when $k \to \infty$, the mean value of Meier's formula (5) can be shown to be

$$\frac{(n-2)}{nw}\left(1' + \frac{4}{n}\right) \tag{9}$$

The first term, $(n-2)/nw$, is the mean value of $1/\hat{w}$ in (5): the second term is the mean value of the expression inside the square brackets in (5).

From equation (1), the correct limiting variance of $\bar{x}_{\hat{w}}$ is $(n-2)/(n-4)w$. For comparison with (9), this may be written

$$\frac{(n-2)}{nw}\left(1 + \frac{4}{n-4}\right) \tag{10}$$

Inspection of (9) and (10) suggests that for large $k$, Meier's formula would be relatively free from bias if the terms in $1/n_i$ were changed to terms in $1/(n_i - 4)$. For $k = 2$, on the other hand, the formula seems excellent as it stands.

As an empirical attempt to improve the performance of the formula, we considered replacing the quantities $n_i$ by quantities $n_i'$, where

$$n_i' = (n_i - 4) + \frac{4}{(k - 1)} = n_i - \frac{4(k - 2)}{(k - 1)} \tag{11}$$

This substitution leaves the formula unchanged when $k = 2$, but gives it the correct mean value in the limiting distribution as $k \to \infty$.

From the sampling data, the average value of the adjusted formula was worked out for each $n$ and $k$, in exactly the same way as for the original formula. The ratios of these average values to the true variance as estimated from the sampling data are shown in table 7. Thus, table 7 presents the same data for the adjusted formula as did table 6 for the original formula.

TABLE 7

Ratio of average value of the adjusted Meier's formula
for the estimated variance to the variance in Table 1.

| $n$ | $k$ = number of estimates | | | | | | | | | |
|---|---|---|---|---|---|---|---|---|---|---|
|  | 2 | 3 | 4 | 5 | 6 | 8 | 10 | 12 | 15 | 20 |
| 6 | .96 | 1.06 | 1.04 | 1.10 | 1.12 | 1.14 | 1.15 | 1.14 | 1.14 | 1.18 |
| 8 | 1.01 | .98 | 1.04 | 1.01 | 1.03 | 1.00 | 1.01 | 1.01 | .99 | .98 |
| 10 | 1.01 | 1.02 | 1.03 | 1.03 | 1.02 | 1.01 | 1.02 | 1.01 | 1.00 | 1.01 |
| 12 | .99 | 1.01 | 1.03 | 1.01 | 1.01 | 1.02 | 1.00 | 1.01 | .99 | 1.00 |
| 15 | .97 | 1.00 | 1.02 | .98 | .98 | .99 | 1.00 | .97 | .96 | .97 |
| 20 | 1.01 | 1.02 | 1.00 | 1.01 | 1.01 | .99 | 1.03 | 1.01 | 1.03 | 1.00 |

The adjusted formula appears very satisfactory down to $n = 8$. For $n = 6$, the adjusted formula works tolerably well for $k \leq 4$, but for larger values of $k$ it gives too high a variance.

### 7. NUMERICAL EXAMPLE

The application of the adjusted formula will be illustrated by the example presented by Meier. The data, from a paper by Snedecor (5), give the percentage of albumin in the plasma protein of normal human subjects, as obtained in 4 different experiments. The relevant figures appear in table 8.

TABLE 8

Illustration of the adjusted formula

| Column | | | | |
|---|---|---|---|---|
| (1) $s_i^2$ | (2) $\hat{w}_i = 1/s_i^2$ | (3) $n_i$ | (4) $\hat{w} - \hat{w}_i$ | (5) $n_i' = n_i - 2.667$ |
| 1.0822 | 0.9241 | 11 | 2.9869 | 8.333 |
| 0.5227 | 1.9133 | 14 | 1.9977 | 11.333 |
| 4.7761 | 0.2094 | 6 | 3.7016 | 3.333 |
| 1.1571 | 0.8643 | 15 | 3.0467 | 12.333 |
| | $\hat{w} = 3.9110$ | | | |

Columns (1)-(3) contain the basic data. Column (4) is formed from column (2). For the $n_i'$, we have from (11)

$$n_i' = n_i - \frac{4(k-2)}{(k-1)} = n_i - \frac{8}{3} = n_i - 2.667.$$

These values appear in column (5). Finally, from the adjusted form of equation (5),

$$v(\bar{x}_{\hat{w}}) = \frac{1}{\hat{w}}\left[1 + \frac{4}{\hat{w}^2}\sum_{i=1}^{k}\frac{1}{n_i'}\hat{w}_i(\hat{w} - \hat{w}_i)\right]$$

$$= 0.2557\left[1 + \frac{4}{(3.9110)^2}\left\{\frac{(0.9241)(2.9869)}{8.333} + \cdots\right.\right.$$

$$\left.\left. + \frac{(0.8643)(3.0467)}{12.333}\right\}\right]$$

$$= 0.3302.$$

This is about 6 per cent higher than Meier's value of 0.3111 as found by the original formula (5). For the approximate number of degrees of freedom to be ascribed to this variance, Meier has suggested

$$f = \frac{\hat{w}^2}{\sum \dfrac{\hat{w}_i^2}{n_i}}$$

This comes out to 38.6 for these data.

Before the publication of Meier's formula, we had constructed an empirical formula for the estimated variance, based on the results of the sampling investigation as follows:

$$v(\bar{x}_{\hat{w}}) = \frac{\bar{n}[k(\bar{n} - 2) + 8]}{\hat{w}(\bar{n} - 2)[k(\bar{n} - 4) + 12]}$$

where $\bar{n}$ is the average number of degrees of freedom in the $k$ estimates. This formula was obtained by fitting a simple algebraic function to the values of $f(n, k)$ which we found. It is subject to the same restriction as the sampling studies, in that it assumes $f(n, k)$ to be independent of the values of the $\sigma_i^2$, whereas Meier's formula has a sounder theoretical basis.

Since Bliss (6), has used this formula (with acknowledgement) in one of his publications, it may be well to remark that down to $n_i = 6$ the formula agrees well enough with the adjusted Meier formula in the cases in which we have checked it, being slightly more conservative. In the present example we have $\bar{n} = 11.5$, $k = 4$, and the formula gives 0.339 for the estimated variance.

SUMMARY

We are given $k$ independent estimates $x_i(i = 1, 2, \cdots k)$ of the same mean $\mu$. The estimates are thought to be of unequal precision, and for the $i$th estimate we have an unbiased estimate $s_i^2$ of its variance $\sigma_i^2$, based on $n_i$ degrees of freedom. This paper describes the results of a sampling investigation undertaken some years ago in order to study the variance of the weighted mean

$$\bar{x}_{\hat{w}} = \sum \hat{w}_i x_i / \sum \hat{w}_i, \qquad \text{where} \qquad \hat{w}_i = 1/s_i^2.$$

The variances were obtained for values of $k$ between 2 and 20, and for values of $n_i$ (assumed all equal) between 2 and 20. The variances found from the sampling investigation were expressed in the form $f(n, k)/w$, where $w = \sum 1/\sigma_i^2$. Since there is reason to believe that the factor $f(n, k)$ is relatively independent of the $\sigma_i^2$, the sampling computations were made for the case in which all $\sigma_i^2$ are equal.

Since $f(n, k) = 1$ when the correct weights $1/\sigma_i^2$ are used, the factor $f(n, k)$ gives a measure of the extent to which the variance of $\bar{x}_{\hat{w}}$ is inflated owing to sampling errors in the weights $\hat{w}_i$. For given $n$, $f(n, k)$ increases steadily as $k$ increases, so that the inflation of variance is smallest when only a few estimates are being combined.

The results for the variance of $\bar{x}_{\hat{w}}$ enable its precision to be compared with that of the unweighted mean $\bar{x}$. When $k = 2$, taking $\sigma_2^2$ as the larger variance, $\bar{x}$ is preferable if $\sigma_2^2/\sigma_1^2 < 2$, while $\bar{x}_{\hat{w}}$ is preferable, for any value of $n$ down to 4, if $\sigma_2^2/\sigma_1^2 > 3$. If this ratio lies between 2 and 3,

$\bar{x}_{\hat{w}}$ appears preferable if the weights are based on at least 12 degrees of freedom each.

The results provide a partial check on approximate formulas recently developed by Meier for the variance and the estimated variance of $\bar{x}_{\hat{w}}$. In these formulas, terms in $1/n_i^2$ are ignored. The comparisons suggest that if 5 or fewer estimates are being combined, Meier's formula for the true variance is satisfactory for values of $n$ down to 6. It is satisfactory for any number of estimates if $n$ is at least 12.

Meier's formula for the estimated variance

$$\frac{1}{\hat{w}} \left\{ 1 + \frac{4}{\hat{w}^2} \sum_{i=1}^{k} \frac{1}{n_i} \hat{w}_i(\hat{w} - \hat{w}_i) \right\}$$

appears adequate down to about $n = 12$, although it tends to be an underestimate. An empirical adjustment, which mades its performance adequate down to $n = 8$, is to replace $n_i$ in the formula by

$$n_i' = n_i - \frac{4(k - 2)}{(k - 1)}$$

### REFERENCES

(1) Cochran, W. G., Problems arising in the analysis of a series of similar experiments. *Jour. Roy. Stat. Soc.*, Supp. *4:* 102–118, 1937.
(2) Meier, P., Variance of a weighted mean. *Biometrics, 9:* 59–73, 1953.
(3) Porter, Sarah, Relative accuracy of weighting inversely as the estimated variance. *M. Sc. Thesis*, North Carolina State College, 1947.
(4) Mahalanobis, P. C., Tables of random samples from a normal population. *Sankhya, 1:* 1–40, 1934.
(5) Snedecor, G. W., The statistical part of the scientific method. *Ann. of the New York Academy of Science, 52:* 742–749, 1950.
(6) Bliss, C. I., *The statistics of bioassay.* Academic Press, Inc., New York, 579, 1952.

# THE PRESENT STRUCTURE OF THE ASSOCIATION*

WILLIAM G. COCHRAN

*The Johns Hopkins University*

FIVE years ago, the Association adopted a new Constitution which was intended to facilitate substantial changes in the nature of the Association. Written Constitutions are not noted for their ability to grip and hold the reader's interest, and I doubt whether many members paid more attention to the new Constitution than was necessary in deciding how to vote on it in 1948. Consequently, I would like to present some impressions of the experience of the Association during the first five years of operation under the new Constitution. I hope that this account will give members a better picture of the present nature of the Association and will lead up to several questions concerning our future development about which I wish to encourage members to do some thinking.

### THE SITUATION AS IT APPEARED IN 1945

Planning for a new Constitution began when the Association was able to resume normal activities towards the end of World War II. In the early discussions about a suitable future pattern for the Association, the committee at work on the new Constitution took note of four developments in the field of statistics that seemed relevant.

1. Statistical techniques had penetrated into a great variety of fields. Up till about 30 years ago, practical statistics dealt mainly with applications to economics, business and government, and the interests of the Association's members tended to reflect this fact. It

---

* Presidential Address at the 113th Annual Meeting of the American Statistical Association, Washington, D. C., December 28, 1953.

is easy to exaggerate the extent to which this was so: the Association has always welcomed statisticians in any field of knowledge and 30 or 40 years ago the *Journal* was publishing important papers on a wide range of topics. But the organized activities of the Association dealt largely with applications in the economic sphere. In the 30's, however, and still more in the early 40's, the increased use of statistical ideas and techniques in such fields as psychology, the various branches of biology, medicine, the social sciences, industrial research and operations, and marketing was a striking phenomenon.

2. During the same period, persons interested in these other developments had founded a number of new societies, among them the Institute of Mathematical Statistics, the Econometric Society, the Psychometric Society and the American Society for Quality Control. All these societies were strongly concerned with statistical techniques, but none of them had any formal relation to the ASA.

3. The membership of the ASA was increasing and might be expected to grow rapidly in the post-war years. In 1945 there were about 3,300 members, at present there are close to 5,000.

4. With the formation of the United Nations, some of its agencies might be expected to foster new developments in international statistics.

In considering the future of the Association in the light of these factors, two principal choices appeared to be open. The Association might continue to give primary attention to applications in economics, leaving applications in other fields to be taken care of by other societies. This would have been a reasonable course of action. Although the Association had received an influx of members whose interests were in other fields, the primary concern of over half the members in 1945 was still with applications to economics or business, as revealed by the 1945 *Directory*.

The second course, the one actually adopted, was to try to give the Association a *central* role with regard to all fields of application of statistics. This decision was advocated by almost all members whose opinions were sought. It was a wise decision from many points of view, particularly when no one knew where important statistical applications might turn up next, when statistical activities were being parcelled out amongst numerous societies and when a strong national body might accomplish much in cooperation with international agencies. We should recognize, however, that the decision involved a real sacrifice, at least for a time, by the members in economics and business, since a relatively homogeneous society catering satisfactorily to them was to be changed into something more amorphous whose future course was harder to predict. These members accepted and encouraged the change with excellent spirit and with, as might be expected, occasional grumbles.

### SOME PROVISIONS OF THE 1948 CONSTITUTION

The decision having been taken, the new Constitution was constructed so as to introduce a number of devices that would make the desired changes easier to accomplish. I would like to describe the purposes, as I understand them, of some of the principal provisions in the 1948 Constitution.

*Associated and affiliated societies.* One of the most difficult questions was: what was to be the relation between the ASA and the other societies dealing with some aspect of statistics that had come into being or might be established in the future? Much thought was given to this question, including a study of various mechanisms that had been adopted by other large central organizations. Finally, it was decided to try two provisions, called *association* and *affiliation.*

Any other society interested in the objects of the ASA may apply to become an Associated or an Affiliated Society. The status of an Associated Society is intended for societies whose interest in statistics is strong: that of an Affiliated Society was intended to cover a looser type of connection, but since this provision was dropped in our recent minor revision of the Constitution, I will not go into detail about it. Proposals for association are examined by our Board of Directors and Council before a decision is taken to grant the status.

Each Associated Society receives the right to appoint two members to the Council of the ASA, one member to the editorial board of *The American Statistician,* and one member to the ASA Committee on Publications for each periodical which it publishes. The ASA is required to offer its publications to the members of Associated Societies on the same basis as to ASA members, and vice versa.

The arrangement involves a slight loss of autonomy by the ASA. In return, it establishes a definite method of liaison, makes our Council more representative of statistical interests as a whole, and puts us in a better position to play the kind of central role that was considered desirable.

*Sections and section committees.* If the ASA is to be a society whose members have a great variety of interests, what can be done to ensure that each of the principal interest-groups within the membership participates to its own satisfaction?

For dealing with this problem, the ASA had a successful precedent in the Biometrics Section, which had been in existence for a number of years. Although only a small fraction of the membership was interested in biometry as such, this Section arranged programs at each annual meeting, held joint sessions at the meetings of a number of the biological societies and published the *Biometrics Bulletin* with financial backing from the ASA.

The 1948 Constitution encouraged the formation of Sections in other broad areas by providing for the establishment of *Section Committees*. The general function of Section Committees is "to further the development of statistics in fields not adequately covered at present by associated or affiliated societies." (Article X, 8). These Committees are represented on the ASA program committee in order to arrange programs in their individual areas. In course of time, a Section Committee may draw up a charter which on approval leads to the formation of a Section. The new Constitution looks still further ahead by providing that when a Section has grown large enough, the Section Committee may take the initiative in organizing an Associated Society.

*Districts and District Committees.* In nation-wide societies that are small, meetings tend to be on a national level. As the society grows in numbers, it becomes feasible to hold regional meetings which give more of the members a chance to participate. In the ASA we have been fortunate in having a long tradition of meetings both at the national level and through our Chapters at the local level. In order to encourage activities and meetings at an intermediate regional level, the Constitution provides for the setting-up of geographical districts. In each, there is a District Committee, with two members from each ASA chapter and from each local unit, if there are any, of any Associated or Affiliated Society. The District Committees thus provide a means for coordinating the activities of the ASA and related societies at both the the local and regional levels.

*Council.* Finally, in order to give the membership a broader representation in the administration of the ASA, the Constitution created a new policy-making body, the Council. This consists of the Board of Directors, the editor of each ASA publication, two representatives from each district and one from each Section Committee with more than 75 members, as well as representatives of Associated Societies and an equal number of representatives-at-large. The Board of Directors, which in former times was the governing body, now serves as the executive committee of the Council. During 1953, the Council had 34 members, as compared with 13 on the Board.

THE ASSOCIATION'S EXPERIENCE UNDER THE 1948 CONSTITUTION

I would now like to describe how the new devices have operated during the past 5 years. In cases where things have not as yet worked quite as actively as was hoped, I do not want to give the impression of washing dirty linen in public, which would be most reprehensible for a President. My defense would be that this linen is not dirty, and it is not being washed, but merely aired.

*Associated Societies.* Up to the present time, only one organization has become linked to us through this provision—the East North American Region of the Biometric Society, which might be regarded as one of our own children grown up, since the Biometric Society is a natural outgrowth of our Biometrics Section.

This modest beginning is not surprising, because no strenuous efforts have been made to bring the provision to the attention of other societies. In my opinion, it is advisable to wait until the ASA has settled down under the new Constitution before exploring with some of the other societies the possibility of a closer relationship, although we have progressed far enough so that any good opportunity for initiating discussions should not be missed. Perhaps the most propitious times will be when cooperation has already arisen about some matter of mutual interest, or when a new society has been launched with the guidance of the ASA. With the older societies, we may also have to recognize and handle tactfully a problem of prestige. Some members of these societies may feel that Association implies in some way a recognition of a lower status. No such status was intended in framing these provisions, under which the ASA sacrifices some autonomy, but the other society does not, as is clearly stated in our Constitution.

*Sections and Section Committees.* Excellent progress has been made in establishing a well-rounded group of Sections. This year, the Section on Social Statistics has been added to those on Biometrics, Business and Economic Statistics .and Training in Statistics. A Committee on Statistics in the Physical Sciences has been at work for 2 years. Jointly these 5 areas appear comprehensive enough to cover the major interests of practically all our members, at least for the time being. Perhaps the largest single group unrepresented by a Section are the members whose primary interest is in statistical theory. So long as the Institute of Mathematical Statistics continues to meet with us, as it has done consistently in the past, such members are unlikely to regard themselves as neglected. In arranging the large number of sessions (currently around 50) which now comprise the program at the annual meeting, the Section representatives have worked most efficiently and amicably, and I believe that we have a smooth mechanism for accomplishing this complicated task. The Section Committees have also been active in varying degrees in other projects, and have been called upon on numerous occasions for advice by the Board and Council.

*Districts and District Committees.* Activity in arranging meetings of something approaching a regional character, which was one of the primary intentions in setting up districts, has proceeded satisfactorily. The initiative, however, has come from different directions on different occasions. The interesting programs at the United Nations headquarters in New York in 1952 and 1953 were a joint venture by several Chapters. The successful series of Institutes at the Universities of of Illinois and Pennsylvania and at the Carnegie Institute of Technology involved cooperative planning among a number of groups, prominent among them being the Business and Economic Statistics Section. The regional meeting to be held in San Francisco in December, 1954, will be the responsibility of the Western District. Thus, what was perhaps the principal object in setting up District Committees is being achieved, although the Committees themselves have not been uniformly active.

*The Council.* In creating the Council, the intent was to give the membership a larger role in the policy-making of the ASA and perhaps also to allow for more deliberation on policy problems. I think it is fair to say that these aims have not been fulfilled thus far. The annual meeting of the Council takes place at the beginning of the new President's term of office, a day or two after the new Council members have been elected. The agenda is a full one, with enough questions calling for immediate decision to leave little time or energy for leisurely discussion of long-range policy problems. The Board members tend to be the more active participants in the discussion, because they are more familiar with the issues than those who are not Board members.

It can be argued, of course, that if affairs are running smoothly without intense Council activity, as they appear to be, there is no point in looking for more work for the Council just to keep them busy. Also, a group with around 30 members is of an awkward size for some types of work and deliberation. The Council can meet at other times and can be polled by mail, so that it stands ready when any important policy matter arises. On the other hand, since the council is our policy-making body, our most representative body, and the body on which nominees from other societies will see us in action, there is a strong case for trying to make it more continuously effective. There are several techniques that would be worth experimentation, and the Board has been considering a plan of action. I am sorry that during my term of office I did not make a beginning.

## THE PRESENT STRUCTURE OF THE ASSOCIATION

As indicated previously, the wording of the 1948 Constitution suggests that the ASA would assume a more definitely central role in statistics by establishing, through association, links with other societies which recognized this role for the ASA. Section Committees were apparently regarded as more of an *interim* mechanism, since the Constitution describes them as applicable to "fields not adequately covered at present by associated or affiliated societies" and regards them as a means for organizing an associated society.

As events have turned out, the formation of Sections and Section Committees has been the predominant feature in the development of the ASA during the past five years, while only a bare beginning has been made in linking ourselves with other societies. This has been a sound order of procedure, in that we have been working hard to try to serve the whole range of statistics, before putting forward claims that we are able to do so. It now looks as if many of our most important activities during the next few years will be in the hands of the Sections. I hope that members of Section committees will realize how important these committees have become. Their useful activity is by no means confined to helping with the program at the Annual Meetings, but may include the planning of more specialized meetings, contributions to the publication program of the ASA and factual studies of problems that confront the content fields.

As the Sections become larger and better established, what will be the next step in the evolution of the ASA? In particular, what will happen if a Section develops into Associated Society or if a society already in existence in the field of the Section becomes associated with us? I do not know the answer, but some recent experiences of the Biometrics Section are worth noting.

After the North American regions of the Biometric Society had been established, the members of the Biometrics Section began a lively discussion of the future of this Section. Some members contended that the Biometrics Section should be dissolved. They claimed that the new regions of the Biometric Society could take care of the welfare of biometry in this country, that their administration would to a large extent be in the hands of ASA members anyway, and that continuation of the Biometrics Section would be an unnecessary duplication of effort.

An opposing view was that for a statistician, membership in the Biometric Society serves a different purpose from membership in the Biometrics Section. At present, about half the members of the Biometric Society are biologists. If this Society is to flourish in its original objectives, it must continue to attract to membership a large number, preferably a majority, of biologists who would not join any statistical association. Thus the Biometric Society gives the statistician the opportunity to talk with *biologists*, learning their problems, working with them, and presenting new techniques for criticism and use. The ASA, on the other hand, is the place where statisticians in biometry can talk with *statisticians* in other content fields, both to find out what new techniques have developed in these fields and to present new ideas in biometry. From this point of view there was a strong argument for continuing the Biometrics Section as a nucleus for attracting future biometricians into the ASA, for cooperating with other Sections and for organizing programs on new or recent discoveries, where the technical level would be too high for most biologists.

After much debate, the decision was taken to continue the Biometrics Section. I do not claim that it was the argument given above which carried the day. Biometricians, like other statisticians, are fond of nice logical distinctions, and each tends to put forward a slightly different reason for advocating the same decision, and to attach great importance to the superiority of his reason over anyone else's, even though to an outsider the reasons are practically indistinguishable. But I hope that the argument will not be overlooked if other Sections blossom into full societies and their members are uncertain whether to continue the Section. If this concept of the purpose of a Section is sound, the greatest benefit will be obtained from the present ASA structure only if there is sustained cooperation among Sections and if members make a habit of attending sessions of several different Sections. There is, of course, nothing to prevent a member from belonging to every Section.

If the opposing view prevails, and if we are to look forward to seeing the Sections disband one by one as Associated Societies are formed (as might happen if there is a general lack of interest in continuing the Sections) then the structure of the ASA will evolve towards something different. A conservative might comment that it would then resemble either a jellyfish or an octopus, depending on how one looks at it. More seriously, I do not mean to suggest that Sections should be kept alive if there is no intrinsic life in them. We should, however, have to re-examine the whole problem of the best type of structure for the ASA under the changed conditions. Actually, some types of organization that did not involve Sections at all were examined in the initial work for the 1948 Constitution, but were rejected as being unsuitable in our present state of growth.

SOME QUESTIONS CONCERNING THE VITALITY OF THE ASSOCIATION

To consider our present structure from a slightly different point of view, I would now like to pose a few broad questions which bear upon what might be called the state of health of the Association.

*Can the ASA maintain the enthusiastic support of its members?* Any large and heterogeneous society is likely to find that it is nobody's darling, because the affections of the members are accorded to some smaller and more homogeneous group in which they feel more at home. As the Association grows larger in its new role, it may be more difficult to give the members a real sense of participation. The *Journal* and *The American Statistician*, as the most tangible benefits from membership, have an important part to play, and it is currently planned to supplement these periodicals from time to time with special monographs and other publications of interest to the members. Meetings of a local or regional character are a beneficial addition to our Annual Meetings as a means of bringing together more of our members. Our Chapters and Sections may accomplish much in giving members a more immediate focus for their interests. Continued joint activity by different Sections will avoid a partitioning into self-contained groups that has occurred in some societies. In addition, I hope that members will continue to agree that statistics needs an all-embracing society, and will appreciate that the Association will inevitably become more diffuse as it succeeds in adopting this role.

*Can the ASA continue to recruit young members?* It is relatively painless for them to enter into membership: students pay only half the regular dues, as do also members under 30 during their first year. The office conducts a continuing campaign to spread information about membership, the groups approached being varied from year to year. As in other societies, our office finds that nothing succeeds so well as a personal approach from a present member, so that it is to our members and to the quality of our publications that we must look mainly for a steady recruitment of young persons.

*Does the structure of the ASA encourage younger members, as they ma-*

*ture, to participate in the running of the ASA?* Since the rapid growth of statistics is recent, we suffer relatively little from government by the grey-haired. Nevertheless, many of our most experienced members are heavily burdened with activities on behalf of scientific societies. For this reason, as well as to keep us supplied with fresh points of view, the talents of younger members should be utilized to the fullest extent. The Chapters and the Section and District Committees provide the first opportunity for younger members to undertake responsible tasks. For service at the national level on the Council or Board, the problem of introducing new blood is more difficult. In the elections, which are by majority vote, my impression is that the candidate who is more widely known (and usually older) is very frequently the winner. Something can be done about this problem both by the Committee on Elections when they nominate candidates and by the President when he appoints committees.

*Is the ASA able to stimulate new developments in statistics?* Some members have expressed the opinion that in the thirties and early forties the ASA missed an opportunity by not playing a more prominent part in the developments which led to the formation of a number of other societies with statistical interests. I am not sure that I would agree. In the Biometric Society, which we did help to establish, I have been slightly disturbed in case the statisticians should play too prominent a role relative to the biologists. In founding this kind of a society, there is something to be said for leaving much of the initiative to the scientists in the subject-matter field, who would not in general be members of the ASA. Nevertheless, our assumption of the role of a central organization with very wide interests does carry more responsibility for helping such developments, rather than leaving them to take place outside the ASA.

Here again we must rely mainly on the Section Committees, particularly when they arrange programs, to be on the lookout for new developments. Inspection of the wide range of our programs in recent years suggests that the committees have been lively and enterprising in this respect. The Board and Council and the office can also help. For a time, the Board felt impelled to adopt a cautious policy owing to our financial difficulties, but fortunately these appear to be well out of the way.

*Is the ASA able to exercise leadership for statistics as a whole?* So far as the use of statistics in government is concerned, our leadership is recognized as a result of a long history of disinterested service to agencies of the government. I believe that international statistical agencies would also join in this recognition.

How do we stand in other areas involving statistical interests? Are we active enough in exercising leadership? These questions are more troublesome. Two areas that have always been of deep concern to the Council and Board are that involving relations with the public and that involving Statistical Standards. A piece of sound and important statistical work may be subjected to unjustified public attack, or a piece of shoddy and unscrupulous work, masking as statistically sound, may threaten to bring discredit on the profession. Should such circumstances arise, I imagine that most members would expect the Association to take corrective action. The problem of doing this effectively raises numerous difficulties. The critical moment for taking action may not be clear: there may be varying opinions about the most appropriate type of action; and the pressure of time may prevent thorough study of the issue before something must be done. For these reasons I am doubtful whether reliance on any standby body, such as the Council or some designated committee, will be adequate. The analogy with a fire brigade is not good, because nobody rings the alarm bell to tell us when to spring into action. The Council and Board have been struggling to consider what program of study might be initiated in order to establish a set of principles and a mode of action for dealing with such emergencies so that we will not be caught unawares. This is a task that needs all the help that members can give. For many of the problems it seems clear that to be fully effective, the ASA must work along with other societies that have statistical interests. Consequently, a program of this kind may be one means of drawing us closer to these societies.

Finally, any account of our present structure must recognize that we are a voluntary organization. Apart from a tiny office staff, everything that we do depends on the voluntary labor of the members. The Association can become what the members want it to be: there is no entrenched bureaucracy to impose its own pattern. Any member with a bright idea will receive an interested hearing (although he may sometimes have to talk a little loudly in order to do so). If his idea is bright enough, he will very likely find himself asked to carry it out as an enterprise of the Association. Secondly, we are a scientific as distinct from a professional society, in the sense that the Association has always worked for the highest statistical standards rather than for the economic interests of its members. As we grow larger, it may be harder to retain this voluntary, scientific character while representing effectively the whole range of statistical activities. For my part, I hope that we can do both.

To summarize, the ASA is in a difficult period of growth in trying to keep up with an extraordinary expansion of statistics which scarcely anyone could have predicted accurately. In particular, the increasing specialization within statistics has set up forces which tend to decrease the amount of common interest amongst members and to split them into separate groups. The task of serving all areas of application in this rapidly-changing environment will require us to be wide-awake, adaptable, and receptive to new ideas and new ventures. My own appraisal would be that during the past five years our Association has made gratifying progress, especially in view of the financial stringencies which inflation imposed upon us. Some of the provisions of the 1948 Constitution have had only modest effects as yet, but these provisions have not proved harmful: they create mechanisms that will increase our flexibility in adapting ourselves to the future growth of the field of statistics. Although much remains to be done, I believe that we now have an organizational pattern that at least for the near future will enable us to take full advantage of our broad, common interests while giving scope also to our more specialized interests.

## PRINCIPLES OF SAMPLING*

WILLIAM G. COCHRAN, *Johns Hopkins University*
FREDERICK MOSTELLER, *Harvard University,*
JOHN W. TUKEY, *Princeton University*

### I. SAMPLES AND THEIR ANALYSES

1. *Introduction*

WHETHER by biologists, sociologists, engineers, or chemists, sampling is all too often taken far too lightly. In the early years of the present century it was not uncommon to measure the claws and carapaces of 1000 crabs, or to count the number of veins in each of 1000 leaves, and then to attach to the results the "probable error" which would have been appropriate had the 1000 crabs or the 1000 leaves been drawn at random from the population of interest. Such actions were unwarranted shotgun marriages between the quantitatively unsophisticated idea of sample as "what you get by grabbing a handful" and the mathematical precise notion of a "simple random sample." In the years between we have learned caution by bitter experience. We insist on some semblance of mechanical (dice, coins, random number tables, etc.) randomization before we treat a sample from an existent population as if it were random. We realize that if someone just "grabs a handful," the individuals in the handful almost always resemble one another (on the average) more than do the members of a simple random sample. Even if the "grabs" are randomly spread around so that every individual has an equal chance of entering the sample, there are difficulties. Since the individuals of grab samples resemble one another *more* than do individuals of random samples, it follows (by a simple mathematical argument) that the means of grab samples resemble one another *less* than the means of random samples of the same size. From a grab sample, therefore, we tend to *under*estimate the variability in the population, although we should have to *over*estimate it in order to obtain valid estimates of variability of grab sample means by substituting such an estimate into the formula for the variability of means of simple random samples. Thus using simple random sample formulas for grab sample means introduces a double

---

* This paper will constitute Appendix G of Cochran, Mosteller, and Tukey, *Statistical Problems of the Kinsey Report*, to be published by the American Statistical Association later this year as a monograph. The main body of this monograph was published in the *Journal* last December (Vol. 48 (1953), pp. 673–716).

bias, both parts of which lead to an unwarranted appearance of higher stability.

Returning to the crabs, we may suppose that the crabs in which we are interested are all the individuals of a wide-ranging species, spread along a few hundred miles of coast. It is obviously impractical to seek to take a simple random sample from the species—no one knows how to give each crab in the species an equal chance of being drawn into the sample (to say nothing of trying to make these chances independent). But this does not bar us from honestly assessing the likely range of fluctuation of the result. Much effort has been applied in recent years, particularly in sampling human populations, to the development of sampling plans which *simultaneously*,

  (i)  are economically feasible
  (ii) give reasonably precise results, and
  (iii) show within themselves an honest measure of fluctuation of their results.

Any excuse for the dangerous practice of treating non-random samples as random ones is now entirely tenuous. Wider knowledge of the principles involved is needed if scientific investigations involving samples (and what such investigation does not?) are to be solidly based. Additional knowledge of techniques is not so vitally important, though it can lead to substantial economic gains.

A botanist who gathered 10 oak leaves from each of 100 oak trees might feel that he had a fine sample of 1000, and that, if 500 were infected with a certain species of parasites, he had shown that the percentage infection was close to 50%. If he had studied the binomial distribution he might calculate a standard error according to the usual formula for random samples, $p \pm \sqrt{pq/n}$, which in this case yields $50 \pm 1.6\%$ (since $p=q=.5$ and $n=1000$). In this doing he would neglect three things:

  (i)  Probable selectivity in selecting trees (favoring large trees, perhaps?),
  (ii) Probable selectivity in choosing leaves from a selected tree (favoring well-colored or, alternatively, visibly infected leaves perhaps), and
  (iii) the necessary allowance, in the formula used to compute the standard error, for the fact that he has not selected his leaves individually at random, as the mathematical model for a simple random sample prescribes.

Most scientists are keenly aware of the analogs of (i) and (ii) in their own fields of work, at least as soon as they are pointed out to them.

Far fewer seem to realize that, even if the trees were selected at random from the forest and the leaves were chosen at random from each selected tree, (iii) must still be considered. But if, as might indeed be the case, each tree were either wholly infected or wholly free of infection, then the 1000 leaves tell us *no more* than 100 leaves, one from each tree. (Each group of 10 leaves will be all infected or all free of infection.) In this case we should take $n = 100$ and find an infection rate of $50 \pm 5\%$.

Such an extreme case of increased fluctuation due to sampling in clusters would be detected by almost all scientists, and is not a serious danger. But less extreme cases easily escape detection and may therefore be very dangerous. This is one example of the reasons why the principles of sampling need wider understanding.

We have just described an example of *cluster sampling*, where the individuals or sampling units are not drawn into the sample independently, but are drawn in clusters, and have tried to make it clear that "individually at random" formulas do not apply. It was not our intention to oppose, by this example, the use of cluster sampling, which is often desirable, but only to speak for proper analysis of its results.

## 2. *Self-weighting probability samples*

There are many ways to draw samples such that each individual or sampling unit in the population has an equal chance of appearing in the sample. Given such a sample, and desiring to estimate the population average of some characteristic, the appropriate procedure is to calculate the (unweighted) mean of all the individual values of that characteristic in the sample. Because weights are equal and require no obvious action, such a sample is *self-weighting*. Because the relative chances of *different* individuals entering the sample are known and compensated for (are, in this case, equal), it is a *probability sample*. (In fact, it would be enough if we knew somewhat less, as is explained in Section 5.)

Such a sample need not be a simple random sample, such as one would obtain by numbering all the individuals in the population, and then using a table of random numbers to select the sample on the basis: one random number, one individual. We illustrate this by giving various examples, some practical and others impractical.

Consider the sample of oak leaves; it might in principle be drawn in the following way. First we list all trees in the forest of interest, recording for each tree its location and the number of leaves it bears. Then we draw a sample of 100 trees, arranging that the probability of a tree's being selected is proportional to the number of leaves which it bears. Then on each selected tree we choose 10 leaves at random. It is easy to verify that each leaf in the forest has an equal chance of being selected. (This is a kind of two-stage sampling with probability proportional to size at the first stage.)

We must emphasize that such terms as "select at random," "choose at random," and the like, always mean that some mechanical device, such as coins, cards, dice, or tables of random numbers, is used.

A more practical way to sample the oak leaves might be to list only the locations of the trees (in some parts of the country this could be done from a single aerial photograph), and then to draw 100 trees in such a way that each tree has an equal chance of being selected. The number of leaves on each tree is now counted and the sample of 1000 is prorated over the 100 trees in proportion to their numbers of leaves. It is again easy to verify that each leaf has an equal chance of appearing in the sample. (This is a kind of two-stage sampling with probability proportional to size at the second stage.)

If the forest is large, and each tree has many leaves, either of these procedures would probably be impractical. A more practical method might involve a four-stage process in which:

(a) the forest is divided into small tracts,

(b) each tract is divided into trees,

(c) each tree is divided into recognizable parts, perhaps limbs, and

(d) each part is divided into leaves.

In drawing a sample, we would begin by drawing a number of tracts, then a number of trees in each tract, then a part or number of parts from each tree, then a number of leaves from each part. This can be done in many ways so that each leaf has an equal chance of appearing in the sample.

A different sort of self-weighting probability sample arises when we draw a sample of names from the Manhattan telephone directory, taking, say, every 17,387th name in alphabetic order starting with one of the first 17,387 names selected at random with equal probability. It is again easy to verify that every name in the book has an equal chance of appearing in the sample (this is a systematic sample with a random start, sometimes referred to as a systematic random sample).

As a final example of this sort, we may consider a national sample of 480 people divided among the 48 states. We cannot divide the 480 cases among the individual states in proportion to population very well, since Nevada would then receive about one-half of a case. If we group the small states into blocks, however, we can arrange for each state or block of states to be large enough so that on a pro rata basis it will have at least 10 cases. Then we can draw samples within each state or block of states in various ways. It is easy to verify that the chances of any two persons entering such a sample (assuming adequate randomness within each state or block of states) are approximately the same, where the approximation arises solely because a whole number of cases has to be assigned to each state or block of states. (This is a rudimentary sort of stratified sample.)

All of these examples were (at least approximately) self-weighting probability samples, and all yield honest estimates of population characteristics. *Each one* requires a *different* formula for assessing the stability of its results! Even if the population characteristic studied is a fraction, almost never will

$$p \pm \sqrt{\frac{\overline{pq}}{n}}$$

be a proper expression for "estimate ± standard error." In every case, a proper formula will require more information from the sample than merely the overall percentage. (Thus, for instance, in the first oak leaf example, the variability from tree to tree of the number infested out of 10 would be needed.)

## 3. *Representativeness*

Another principle which ought not to need recalling is this: By sampling we can learn only about collective properties of populations, not about properties of individuals. We can study the average height, the percentage who wear hats, or the variability in weight of college juniors, or of University of Indiana juniors, or of the juniors belonging to a certain fraternity or club at a certain institution. The population we study may be small or large, but there must be a population—and what we are studying must be a population characteristic. By sampling, we cannot study individuals as particular entities with unique idiosyncrasies; we can study regularities (including typical variabilities as well as typical levels) in a population as exemplified by the individuals in the sample.

Let us return to the self-weighted national sample of 480. Notice that about half of the times that such a sample is drawn, there will be no one in it from Nevada, while almost never will there be anyone from Esmeralda County in that state. Local pride might argue that "this proves that the sample was unrepresentative," but the correct position seems to be this:

(i) the particular persons in the sample are there by accident, and this is appropriate, so far as population characteristics are concerned,

(ii) the sampling plan is representative since each person in the U.S. had *an equal chance* of entering the sample, whether he came from Esmeralda County or Manhattan.

That which can be and should be representative is the *sampling plan,* which includes the manner in which the sample was drawn (essentially a specification of what other samples might have been drawn and what the relative chances of selection were for any two possible samples) *and* how it is to be analyzed.

However great their local pride, the citizens of Esmeralda County, Nevada, are entitled to representation in a national sampling plan only as individual members of the U.S. population. They are *not* entitled to representation as a group, or as particular individuals—only as individual *members* of the U.S. population. The same is true of the citizens of Nevada, who are represented in only half of the actual samples. The citizens of Nevada, as a group, are no more and no less entitled to representation than *any* other group of equal size in the U.S. whether geographical, racial, marital, criminal, selected at random, or selected from those not in a particular national sample.

It is clear that many such groups fail to be represented in any particular sample, yet this is not a criticism of that sample. Representation is not, and should not be, by groups. It is, and should be, by individuals as *members* of the sampled population. Representation is not, and should not be, in any particular sample. It is, and should be, in the sampling *plan.*

## 4. *One method of assessing stability*

Because representativeness is inherent in the sampling plan and not in the particular sample at hand, we can never make adequate use of sample results without some measure of how well the results of this particular sample are likely to agree with the results of other samples which the same sampling plan might have provided. The ability to assess stability fairly is as important as the ability to represent the population fairly. Modern sampling plans concentrate on both.

Such assessment must basically be in terms of sample results, since these are usually our most reliable source of information about the population. There is no reason, however, why assessment should depend only on the sample size and the overall (weighted) sample mean for the characteristic considered. These two suffice when measuring percentages with a simple random sample, but in almost all other cases the situation is more complex.

It would be too bad if, every time such samples were used, the user had to consult a complicated table of alternative formulas, one for each plan, before calculating his standard errors. (These formulas do need to be considered whenever we are trying to do a really good job of maximum stability for minimum cost—considered very carefully in selecting one complex design in preference to another.) Fortunately, however, this complication can often be circumvented.

One of the simplest ways is to build up the sample from a number of independent subsamples, each of which is self-sufficient, though small, and to tabulate the results of interest separately for each subsample. Then variation among separate results gives a simple and honest yardstick for the variability of the result or results obtained by throwing all the samples together. Such a sampling plan involves *interpenetrating replicate subsamples.*

All of us can visualize interpenetrating replicate subsamples when the individuals or sampling units are drawn individually at random. Some examples in more complex cases may be helpful. In the first oak leaf example, we might select randomly, not one sample of 100 trees, but 10 subsamples of 10 trees each. If we then pick 10 leaves at random from each tree, placing them in 10 bags, one for each subsample, and tabulate the results separately, bag by bag, we will have 10 interpenetrating replicate subsamples. Similarly, if we were to pick 10 subsamples out of the Manhattan phone book, with each subsample consisting of every 173,870th name (in alphabetic order) and with the 10 lead names of the 10 subsamples selected at random from the first 173,870 names we would again have 10 interpenetrating replicate subsamples.

We can always analyze 10 results from 10 independent interpenetrating replicate subsamples just as if they were 10 random selected individual measurements and proceed similarly with other numbers of replicate subsamples.

5. *General probability samples*

The types of sample described in the last section are not the only kinds from which we can confidently make inferences from the sample to the population of interest. Besides the trivial cases where the sample amounts to 90% or even 95% of the population, there is a broad class of cases, including those of the last section as special cases. This is the class of *probability samples*, where:

(1) There is a population, the *sampled population*, from which the sample is drawn, and each element of which has some chance of entering the sample.

(2) For each pair of individuals or sampling units which are in the actual sample, the relative chances of their entering the sample are known. (This implies that the sample was selected by a process involving one or more steps of mechanical randomization.)

(3) In the analysis of the actual sample, these relative chances have been compensated for by using relative weights such that

(relative chance) times (relative weight) equals a constant.

(4) For any two possible samples, the sum of the reciprocals of the relative weights of all the individuals in the sample is the same.

(Conditions (3) and (4) can be generalized still further.) In practice of course, we ask only that these four conditions shall hold with a sufficiently high degree of approximation.

We have made the sampling plan representative, not by giving each individual an equal chance to enter the sample and then weighting them equally, but by a more noticeable process of compensation, where those individuals very likely to enter the sample are weighted less, while those unlikely to enter are weighted more when they do appear. The net result is to give each individual an equal chance of affecting the (weighted) sample mean.

Such general probability samples are just as honest and legitimate as the self-weighting probability samples. They often offer substantial advantages in terms of higher stability for lower cost.

We can alter our previous examples, so as to make them examples of general, and not of self-weighting, probability samples. Take first the oak leaf example. We might proceed as follows:

(1) locate all the trees in the forest of interest,

(2) select a sample of trees at random,

(3) for each sampled tree, choose 10 leaves at random and count (or estimate) the total number of leaves,

(4) form the weighted mean by summing the products

(fraction of 10 leaves infested) times
(number of leaves on the tree)

and then divide by the total number of leaves on the 100 trees in the sample.

When we selected trees at random, each tree had an equal probability of selection. When we chose 10 leaves from a tree at random, the chance of getting a particular leaf was

$$\frac{10}{\text{(number of leaves on the tree)}}.$$

Thus the chance of selecting any one leaf was a constant multiple of this and was proportional to the *reciprocal* of the number of leaves of the tree. Hence the correct relative weight is proportional to the number of leaves on the tree, and it is simplest to take it as 1/10 of that number. After all, summing the products

(fraction of 10 infected) *times* (leaves on tree)

or

(1/10) *times* (number out of 10 infected) *times* (leaves on tree)

over all trees in the sample gives the same answer. One-tenth of this answer is given by summing

(1/10) *times* (number out of 1 infected) *times* (leaves on tree)

**or**

$$\text{(number out of 1 infected)} \frac{\text{(leaves on tree)}}{10}$$

which shows that the weighted mean prescribed above is just what would have been obtained with relative weights of (number of leaves on tree)/10.

If in sampling the names in the Manhattan telephone directory, we desired to sample initial letters from P through Z more heavily, we might proceed as follows:

(1) Select one of the first 17,387 names at random with equal probability as the lead name.

(2) Take the lead name, and every 17,387th name in alphabetic order following it, into the sample.

(3) Take every name which begins with $P, Q, R, S, \cdots, Z$ *and* is the 103rd or 207th name after a name selected in step 2 of the sample.

Each name beginning with $A, B, \cdots, N, O$ has a chance of 1/17,387 of entering the sample. Each name beginning with $P, Q, \cdots, Y, Z$ has a chance of 3/17,387 of entering the sample (it enters if any one of *three* names among the first 17,387 is selected as the lead name). Thus the relative weight in the sample of a name beginning with $A, B, \cdots,$ $N, O$ is 3 times that of a name beginning with $P, Q, \cdots, Y, Z$. The weighted mean is found simply as:

$$\frac{3(\text{sum for } A, B, \cdots, N, O\text{'s}) + (\text{sum for } P, Q, \cdots, Y, Z\text{'s})}{3(A, B, \cdots, N, O\text{'s in sample}) + (P, Q, \cdots, Y, Z\text{'s in sample})}.$$

Finally we may wish to distribute our national sample of 480 with 10 in each state. The analysis exactly parallels the oak leaf case, and we have to form the sum of

(mean for state sample) *times* (population of state)

and then to divide by the population of the U.S.

### 6. *Nature and properties of general probability samples*

We can carry over the use of independent interpenetrating replicates to the general case without difficulty. We need only remember that the replicates must be independent. In the oak leaf example, the replicates must come from groups of independently selected trees. In the Manhattan telephone book example, the replicates must be based on independently chosen lead names; in the national sample, the replicates must have members in every state. In every case they must interpenetrate, and do this independently.

It is clear from discussion and examples that general probability samples are inferior to self-weighting probability samples in two ways, for both simplicity of exposition and ease of analysis are decreased! If it were not for compensating advantages, general probability samples would not be used. The main advantages are:

(1) better quality for less cost due to reduction in administrative costs or prelisting cost,

(2) better quality for less cost because of better allocation of effort over strata,

(3) greater possibility of making estimates for individual strata.

All three of these advantages can be illustrated on our examples. In the general oak leaf example, in contrast to the first oak leaf example in Section 2, there is no need to determine the size (number of leaves) of all trees. This is a clear cost reduction, whether in money or time. Suppose that, in the Manhattan telephone book sample, one aim was an opinion study restricted to those of Polish descent. Such persons' names tend to be concentrated in the second part of the alphabet, so that the general sample will bring out more persons of Polish descent and the interviewing effort will be better allocated. In the case of the national sample of 480, the general sample, although probably giving a less stable national result, does permit (rather poor) state-by-state estimates where the self-weighting sample would skip Nevada about half the time.

It is perhaps worth mentioning at this point that, if cost is proportional to the total number of individuals without regard to number of strata or the distribution of interviews among strata, the optimum allocation of interviews is proportional to the product

(size of stratum) *times* (standard deviation within stratum).

In particular, optimum allocation calls for sample strata not in proportion to population strata. If we weight appropriately, disproportionate samples will be better than proportionate ones—if we choose the disproportions wisely.

In specifying the characteristics of a probability sampling at the beginning of this paper, we required that there be a *sampled population*, a population from which the sample comes and each member of which has a chance of entering the sample. We have not said whether or not this is exactly the same population as the population in which we are interested, the *target population*. In practice they are rarely the same, though the difference is frequently small. In human sampling, for example, some persons cannot be found and others refuse to answer. The issues involved in this difference between sampled population and target population are discussed at some length in Part II, and in chapter III-D of Appendix D in our complete report.

### 7. *Stratification and adjustment*

In many cases general probability samples can be thought of in terms of

(1) a subdivision of the population into strata,

(2) a self-weighting probability sample in each stratum, and

(3) combination of the stratum sample means weighted by the size of the stratum.

The general Manhattan telephone book sample can be so regarded. There are two strata, one made up of names beginning in $A$, $B$, $\cdots$, $N$, $O$, and the other made up of names beginning in $P$, $Q$, $\cdots$, $Y$, $Z$. Similarly the general national sample may be thought of as made up of 48 strata, one for each state.

This manner of looking at general probability samples is neat, often helpful, and makes the entire legitimacy of unequal weighting clear in many cases. But it is not general. For in the general oak leaf example, if there were any strata they would be whole trees or parts of trees. And not all trees were sampled. (Still every leaf was fairly represented by its equal chance of affecting the weighted sample mean.) We cannot treat this case as one of simple stratification.

The stratified picture is helpful, but not basic. It must fail as soon as there are more potential strata than sample elements, or as soon as the number of elements entering the sample from a certain stratum is not a constant of the sampling plan. It usually fails sooner. There is no substitute for the relative chances that different individuals or sampling units have of entering the sample. This is the basic thing to consider.

There is another relation of stratification to probability sampling. When sizes of strata are known, there is a possibility of *adjustment*. Consider taking a simple random sample of 100 adults in a tribe where exactly 50% of the adults were known to be males and 50% females. Suppose the sample had 60 males and 40 females. If we followed the pure probability sampling philosophy so far expounded, we should take the equally weighted sample mean as our estimate of the population average. Yet if 59 of the 60 men had herded sheep at some time in their lives, and none of the 40 women, we should be unwise in estimating that 59% of the tribe had herded sheep at some time in their lives. The adjusted mean

$$.50\left(\frac{59}{60}\right) + .50\left(\frac{0}{40}\right) = 49^{+}\%$$

is a far better indicator of what we have learned.

How can adjustment fail? Under some conditions the variability of the adjusted mean is enough greater than that of the unadjusted mean to offset the decrease in bias. It may be a hard choice between adjustment and nonadjustment.

The last example was extreme, and the unwise choice would be made by few. But, again, less extreme cases exist, and the unwise choice, whether it be to adjust or not to adjust, may be made rather easily (and probably has been made many times). A quantitative rule is needed. One is given in chapter V-C of the complete report. In the preceding example the relative sizes of the strata were known exactly. It turns out that inexact knowledge can be included in the computation without great increase in complexity.

An example in Kinsey's area is cited by one critic of the Kinsey report:

> These weighted estimates do not, of course, reflect any population changes since 1940, which introduces some error into the statistics for the present total population. Moreover, on some of the very factors that Kinsey demonstrates to be correlated with sexual behavior, there are no Census data available. For example, religious membership is shown to be a factor affecting sexual behavior, but Census data are lacking and no weights are assigned. While the investigators interviewed members of various religious groups, there is no assurance that each group is proportionately represented, because of the lack of systematic sampling controls. Thus, the proportion of

Jews in Kinsey's sample would seem to be at least 13 per cent whereas their true proportion in the population is of the order of 4 per cent.[1]

Do we know the percentage of Jews well enough to make an adjustment for it? If we can assess the stability of the "4%" figure, the procedure of Chapter V-C will answer this question. Failing this technique, we could translate the question into more direct terms as follows: "In considering Kinsey's results, do we want to have 13 per cent Jews or 4 per cent Jews in the sampled population?" and try to answer with the aid of general knowledge and intuition.

We have discussed the adjustment of a simple random sample. The same considerations apply to the possibility of adjusting any self-weighting or general probability sample. No new complications arise when adjustment is superposed on weighting. The presence of a complication might be suspected in the case where not all segments appear in the sample, and we attempt to use these segments as strata. Careful analysis shows the absence of the complication, as may be illustrated by carrying our example further.

Suppose that the sheep-herding tribe in question contains a known, very small percentage of adults of indeterminate sex, and that none have appeared in our sample. To be sure, their existence affected, albeit slightly, the chances of males and females entering the sample, but it does not affect the thinking which urged us to take the adjusted mean. We still want to adjust, and have only the question "Adjust for what?" to answer.

If the fraction of indeterminate sex is 0.000002, and the remainder are half males and half females, and if our anthropological expert feels that about 1 in 7 of the indeterminate ones has herded sheep, we have a choice between

$$.499999 \left( \frac{59}{60} \right) + .499999 \left( \frac{0}{40} \right) + .000002 \left( \frac{1}{7} \right)$$

which represents adjustment for three strata, one measured subjectively, and

$$.500000 \left( \frac{59}{60} \right) + .500000 \left( \frac{0}{40} \right)$$

which represents adjustment for the two observed strata.

Clearly, in this extreme example, the choice is immaterial. Clearly,

---

[1] Hyman, H. H. and Sheatsley, P. B. "The Kinsey report and survey methodology," *International Journal of Opinion and Attitude Research*, Vol. 2 (1948), 184–85.

also, the estimated accuracy of the anthropologist's judgment must enter. We can again use the methods of Chapter V-C.

## 8. *Upper semiprobability sampling*

Let us be a little more realistic about our botanist and his sample of oak leaves. He might have an aerial photograph, and be willing to select 100 trees at random. But any ladder he takes into the field is likely to be short, and he may not be willing to trust himself in the very top of the tree with lineman's climbing irons. So the sample of 10 leaves that he chooses from each selected tree will not be chosen at random. The lower leaves on the tree are more likely to be chosen than the highest ones.

In the two-stage process of sampling, the first stage has been a probability sample, but the second has not (and may even be entirely unplanned!). These are the characteristic features of an *upper semiprobability sample*. As a consequence, the sampled population agrees with the target population in certain large-scale characteristics, but not in small-scale ones and, usually, not in other large-scale characteristics.

Thus, if in the oak leaf example we use the weights appropriate to different sizes of tree, as we should, the sampled population of leaves will

(1) have the correct relative number of leaves for each tree, but

(2) will have far too many lower leaves and far too few upper leaves.

The large-scale characteristic of being on a particular tree is a matter of agreement between sampled and target populations. The large-scale characteristic of height in the tree (and many small-scale characteristics that the reader can easily set up for himself), is a matter of serious disagreement between sampled and target populations.

The sampled population differs from the target population within each segment, here a tree, although sampled population segments and target population segments are in exact proportion.

If infestation varies between the bottoms and the tops of the trees, this type of sampling will be biased, and, while the inferences from sample to sampled population will be correct, they may be useless or misleading because of the great difference between sampled population and target population.

Such dangers always exist with any kind of nonprobability sampling. Upper semiprobability sampling is no exception. By selecting the trees at random we have stultified biases due to probable selectivity between trees, and this is good. But we have done nothing about almost certain selectivity between leaves on a particular tree—this may be all right, or very bad. It would be nice to always have probability samples, and avoid these difficulties. But this may be impractical. (The conditions under which a nonprobability sample may reasonably be taken are discussed in Part II.)

There is one point which needs to be stressed. The change from probability sampling within segments (in the example, within trees) to some other type of sampling, perhaps even unplanned sampling, shifts a large and sometimes difficult part of the inference from sample to target population—shifts it by moving the sampled population away from the target population toward the sample—shifts it from the shoulders of the statistician to the shoulders of the subject matter "expert." Those who use upper semiprobability samples, or other nonprobability samples, take a heavier load on themselves thereby.

Upper semiprobability samples may be either self-weighting or general. The "quota samples" of the opinion pollers, where interviewers are supposed to meet certain quotas by age, sex, and socioeconomic status, are rather crude forms of upper semiprobability samples, and are often self-weighting. Bias within segments arises, some contribution being due, for example, to the different availability of different 42 year old women of the middle class. The sampled population may contain sexes, ages, and socioeconomic classes in the right ratios, but retiring persons are under-represented (and hermits are almost entirely absent) in comparison with the target population.

Election samples of opinion, although following the same quota pattern, will ordinarily only be self-weighting within states (if we ignore the "who will vote" problem). Predictions are desired for individual states. If Nevada had a mere 100 cases in a self-weighting sample, the total size of a national sample would have to be about 100,000. When national percentages are to be compiled, it would be foolish not to weight each state mean in accordance with the size of the state. No one would favor, we believe, weighting each state equally just because there may be (and probably are) biases within each state.

Disproportionate samples and unequal weights are just as natural and wise a part of upper semiprobability sampling as they are of probability sampling. The difficulties of upper semiprobability sampling do not lie here; instead they lie in the secret and insidious biases due to selectivity within segments.

Our sampling of names from the Manhattan telephone directory might conceivably be drawn by listing the numbers called by subscribers on a certain exchange during a certain time, and then taking into the sample names from each exchange in proportion to the names listed for the exchange. The result would be an upper semiprobability sample with substantial selectivity within the segments, which here are exchanges. The nature of this selectivity would depend on the time of day at which the listing was made.

Whether all segments are represented in an upper semiprobability sample or not, the segments may be used as strata for adjustment. The situation is exactly similar to that for probability sampling. The only difficulty worthy of note is the difficulty of assessing the stability of the various segment means.

Independent interpenetrating replicate subsamples can be used to estimate stabilities of over-all or segment means in upper semiprobability samples without difficulty, if we can obtain a reasonable facsimile of independence in taking the different subsamples. They provide, if really independent, respectable bases for inference from sample to sampled population. We still have a nonprobability sample, however, and there is no reason for the sampled population to agree with the target population. The problem is just reduced to "What was the sampled population?"

What finally is the situation with regard to bias in an upper semiprobability sample? We shall have a weighted mean or an adjusted one. In either case, any bias originally contributed by selectivity between segments will have been substantially removed. *But*, in either case, the contribution to bias due to selectivity within segments will *remain unchanged*. This is an unknown and hence additionally dangerous, sort of bias.

The great danger in weighting or adjusting such samples is not so much that that weighting or adjusting may make the results worse (as it will from time to time) but rather that its use may cause the user to feel that his values are excellent because they are "weighted" or "adjusted" and hence to neglect possible or likely biases within segments. Like all other nonprobability sample results, weighted means from upper semiprobability samples should be *presented and interpreted with caution*.

## 9. *Salvage of unplanned samples*

What can we do for such samples? We can either try to improve the results of their analysis, or try to inquire how good they are anyway. We may try to improve either actual quality, or our belief in that quality. The first has to be by way of manner of weighting or adjustment, the second must involve checking sample characteristics against population characteristics.

Weighting is impossible, since we cannot construct a sampling plan and hence cannot estimate chances of entering the sample in any other manner than by observing the sample itself. So all that we can do under this head is to adjust. We recall the salient points about adjustment, which are the same in a complete salvage operation as they are in any other situation:

(1) The population is divided into segments.
(2) Each individual in the sample can be uniquely assigned to a segment.
(3) The population fraction is either known with inappreciable error or estimated with known stability.
(4) The procedures of Chapter V-C of Appendix C of the complete report are applied to determine whether, or how much, to adjust.

After adjustment, what is the situation as to bias? Even worse than with upper semiprobability sampling, because if we do not adjust, we cannot escape bias by turning to weighting. In summary

(1) whether adjusted or not, the result contains all the effects of all the selectivity exercised *within* segments, while
(2) if adjustment is refused by the methods of Chapter V-C, we face additional biases resulting from selectivity *between* segments of a magnitude comparable with the difference between unadjusted and adjusted mean.

This is, to put it mildly, not a good situation.

Clearly even more caution is needed in presenting and interpreting the results of a salvage operation on an unplanned sample than for any of the other types of sample discussed previously. (If it were not for the psychological danger that adjustment might be regarded as cure, the caution required for results based on the original, unadjusted, unplanned sample would, however, be considerably greater.)

Having adjusted or not as seems best, what else can we do? Only something to make ourselves feel better about the sample. Some other characteristic than that under study can sometimes be compared in the adjusted sample and in the population. A large difference is evidence of substantial bias within segments. Good agreement is comforting, and strengthens the believability of the adjusted mean for the characteristic of interest. The amount of this strengthening depends very much on the *a priori* relation between the two characteristics.

Some would say that an unplanned sample does not deserve adjustment, but the discussion in Part II indicates that if any sort of a summary is to be made, it might as well, in principle, be an adjusted mean.

## II. SYSTEMATIC ERRORS

In order to understand how systematic errors in sampling should be treated, it seems both necessary and desirable to fall back on the analogy with the treatment of systematic errors in measurement. No clear account of the situation for sampling seems to be available in the literature, although understanding of the issues is a prerequisite to the critical assessment of nonprobability samples. On the other hand, one of physical science's greatest and more recurrent problems is the treatment of systematic errors.

### 10. *The presence of systematic errors*

Almost any sort of inquiry that is general and not particular involves both sampling and measurement, whether its aim is to measure the heat conductivity of copper, the uranium content of a hill, the visual acuity of high school boys, the social significance of television or the sexual behavior of the (white) human (U.S.) male. Further, *both* the measurement *and* the sampling will be imperfect in almost every case. We can define away either imperfection in certain cases. But the resulting appearance of perfection is usually only an illusion.

We can define the thermal conductivity of a metal as the average value of the measurements made with a particular sort of apparatus, calibrated and operated in a specified way. If the average is properly specified, then there is no "systematic" error of measurement. Yet even the most operational of physicists would give up this definition when presented with a new type of apparatus, which standard physical theory demonstrated to be less susceptible to error.

We can relate the result of a sampling operation to "the result that would have been obtained if the same persons had applied the same methods to the whole population." But we want to know about the population and not about what we would find by certain methods. In almost all cases, applying the method to the "whole" population would miss certain persons and units.

Recognizing the inevitability of (systematic) error in both measurement and sampling, what are we to do? Clearly, attempt to hold the combined effect of the systematic errors down to a reasonable value. What is reasonable? This must depend on the cost of further reduction and the value of accurate results. How do we *know* that our systematic errors have been reduced sufficiently? We don't! (And neither does the physicist!) We use all the subject-matter knowledge, information and semi-information that we have—we combine it with whatever internal evidence of consistency it seems worthwhile to arrange for the observations to provide. The result is not foolproof. We may learn new things and do better later, but who expects the last words on any subject?

In 1905, a physicist measuring the thermal conductivity of copper would have faced, unknowingly, a very small systematic error due to the heating of his equipment and sample by the absorption of cosmic rays, then unknown to physics. In early 1946, an opinion poller, studying Japanese opinion as to who won the war, would have faced a very small systematic error due to the neglect of the 17 Japanese holdouts, who were discovered later north of Saipan. These cases are entirely parallel. Social, biological and physical scientists all need to remember that they have the same problems, the main difference being the decimal place in which they appear.

If we admit the presence of systematic errors in essentially every case, what then distinguishes good inquiry from bad? Some reasonable criteria would seem to be:

- (1) Reduction of exposure to systematic errors from *either* measurement *or* sampling to a level of unimportance, *if possible and economically feasible,* otherwise
- (1+) Balancing the assignment of available resources to reduction in systematic or variable errors in either measurement or sampling reasonably well, in order to obtain a reasonable amount of information for the "money."
- (2) Careful consideration of possible sources of error and careful examination of the numerical results.
- (3) Presentation of results and inferences in a manner which adequately points out both observed variability and conjectured exposure to systematic error.

In many situations it is easy, and relatively inexpensive, to reduce the systematic errors in sampling to practical unimportance. This is done by using a probability sampling plan, where the chance that any individual or other primary unit shall enter the sample is known, and allowed for, and where adequate randomness is ensured by some scheme of (mechanical) randomization. The systematic errors of such a sample are minimal, and frequently consist of such items as:

- (a) failure of individuals or primary units to appear on the "list" from which selection has been made,
- (b) persons perennially "not at home" or samples "lost,"
- (c) refusals to answer or breakdowns in the measuring device.

These are the hard core of causes of systematic error in sampling. Fortunately, in many situations their effect is small—there a probability sample will remove almost all the systematic error due to sampling.

### 11. *Should a probability sample be taken?*

But this does not mean that it is always good policy to take probability samples. The inquirer may not be able to "afford" the cost in time or money for a probability sample. The opinion pollers do not usually afford a probability sample (instead of designating individuals to be interviewed by a random, mechanical process, they allow their interviewers to select respondents to fill "quotas") and many have criticized them for this. Yet the behavior of the few probability samples in the 1948 election (see pp. 110–112 of *The Pre-election Polls of 1948*, Social Science Research Council Report No. 60) does not make it clear that the opinion pollers should spend their limited resources on probability samples for best results. (Shifts *toward* a probability sample have been promised, and seem likely to be wise.)

The statement "he didn't use a probability sample" is thus *not* a criticism which should end further discussion and doom the inquiry to the cellar. It is always necessary to ask two questions:

(a) Could the inquirer afford a probability sample?
(b) Is the exposure to systematic error from a non-probability sample small enough to be borne?

If the answer is "no" to both, then the inquiry should not be, or have been, made—just as would be the case with a physical inquiry if the systematic errors of all the forms of measurement which the physicist could afford were unbearably large.

If the answer is "yes" to the first question and "no" to the second, then the failure to use a probability sample is very serious, indeed.

If the answer is "yes" to both, then careful consideration of the economic balance is required—however it should be incumbent on the inquirer using a nonprobability sample to show why it gave more information per dollar or per year. (As statisticians, we feel that the onus is on the user of the *non*probability sample. Offhand we know of no expert group who would wish to lift it from his shoulders.)

If the answer is "no" to the first question, and "yes" to the second, then the appropriate reaction would seem to be "lucky man."

Having admitted that the sampling, as well as the measurement, will have some systematic errors, how then do we do our best to make good inferences about the subject of inquiry? Sampling and measurement being on the same footing, we have only to copy, for the sampling area, the procedure which is well established and relatively well understood for measurement. This procedure runs about as follows:

We admit the existence of systematic error—of a difference between the quantity measured (the measured quantity) and the quantity of interest (the target quantity). We ask the observations about the measured quantity. We ask our subject matter knowledge, intuition, and general information about the relation between the measured quantity and the target quantity.

We can repeat this nearly verbatim for sampling:

We admit the existence of systematic error—of a difference between the population sampled (the sampled population) and the population of interest (the target population). We ask the observations about the sampled population. We ask our subject matter knowledge, intuition, and general information about the relation between sampled population and target population.

Notice that the measured quantity is not the raw readings, which usually define a *different* measured quantity, but rather the adjusted values resulting from all the standard corrections appropriate to the method of measurement. (Not the actual gas volume, but the gas volume at standard conditions!) Similarly, the result for the sampled population is not the raw mean of the observations, which usually defines a *different* sampled population, but rather the adjusted or weighted mean, all corrections, weightings and the like appropriate to the method of sampling having been applied. Weighting a sample appropriately is no more fudging the data than is correcting a gas volume for barometric pressure.

The third great virtue of probability sampling is the relative definiteness of the sampled population. It is usually possible to point the finger at most of the groups in the target population who have no chance to enter the sample, who therefore were not in the sampled population; and to point the finger at many of the groups whose chance of entering the sample was less than or more than the chance allotted to them in the computation, who therefore were fractionally or multiply represented in the sampled population. When a nonprobability sample is adjusted and weighted to the best of an expert's ability, on the other hand, it may still be very difficult to say what the sampled population really is. (Selectivity *within* segments cannot be allowed for by weights or adjustments, but it arises to some extent in every nonprobability sample and alters the sampled population.)

## 12. *The value and conditions of adjustment*

Some would say that correcting, adjusting and weighting most nonprobability samples is a waste of time, since you do not *know*, when this process has been completed, to what sampled population the adjusted result refers. This is entirely equivalent to saying that it does not pay to adjust the result of a physical measurement for a known systematic error because there are, undoubtedly, other systematic errors and some of them are likely to be in the other direction. Let us inquire into good practice in the measurement situation, and see what guidance it gives us for the sampling situation.

When will the physicist adjust the principle for the known systematic error? When (i) he has the necessary information and (ii) the adjustment is likely to help. The necessary information includes a theory or empirical formula, and the necessary observations. Empirical formulas and observations are subject to fluctuations, so that adjustment will usually change the magnitude of fluctuations as well as altering the systematic error. The adjustment is likely to help unless the supposed reduction of systematic error coincides with a substantial increase in fluctuations.

If the known systematic error is so small as not to

(1) affect the result by a meaningful amount, or

(2) affect the result by an amount likely to be as large as, or a substantial fraction of, the unknown systematic errors,

then the physicist will report either the adjusted or the unadjusted value. If he reports the unadjusted value, he should state that the adjustment has been examined, and is less than such-and-so. To do this, either he must have calculated the adjustment or he must have had generally applicable and strong evidence that it is small.

In any event, his main care, which he will not always take, must be to warn the reader about the dangers of further systematic errors, perhaps, in some cases, even by saying bluntly that "the adjusted value isn't much better than the raw value," and then provide raw values for those who wish to adjust their own.

If the physicist is aware of systematic errors of serious magnitude and has no basis for adjustment, his practice is to name the measured quantity something, like Brinnell hardness, Charpy impact strength, or if he is a chemist—iodine value, heavy metals as Pb, etc. By analogy, those who feel that the combination of recall and interview technique make Kinsey's results subject to great systematic error might well define "KPM sexual behavior" as a standard term,[2] and work with this.

By analogy then, when should a nonprobability sample be adjusted in principle? (Most probability samples are made to be weighted anyway—this is part of the design and must be carried out.) When (i)

---

[2] The letters KPM stand for Kinsey, Pomeroy and Martin, the authors of *Sexual Behavior in the Human Male*

we have the necessary information and (ii) when the adjustment is likely to help. The necessary information will usually consist of facts or estimates of the true fractions in the population of the various segments.

When is the adjustment likely to help? This problem has usually been a ticklish point requiring technical knowledge and intuition. A quantitative solution is now given in Chapter V–C of Appendix C in the complete report. With this as a guide, it should be possible to make reasonable decisions about the helpfulness of adjustment.

If the decision is to adjust, we should accept the sampled population corresponding to the adjusted mean, and calculate the adjustment. We then report the adjusted value, unless the adjustment is small, when we may report the unadjusted value with the statement that the adjustment alters it by less than such-and-so.

Our main care, which we may not always take, must be to warn the reader about the dangers of further lack of representativeness, perhaps, in some cases, even by saying bluntly that "the adjusted mean isn't much better than the raw mean, even if we took 20 pages to tell you how we did it and six months to do it," and to provide raw means for those who wish to adjust their own.

If we were prepared to report an unadjusted mean, we were clearly inviting inference to some sampled population. Adjustment will give us a sampled population that is usually nearer to the target population. Hence we should adjust.

If we cannot adjust, and must present raw data which we feel *badly* needs adjustment, we may say that this is what we found in these cases—take 'em or leave 'em. Except from the point of view of protecting the reader from over-belief in the results, this would seem to be a counsel of despair. By analogy with the physicist, it seems better to introduce "KPM sexual behavior" and its analogs in such situations.

# 56

# THE ROLE OF A STATISTICAL SOCIETY IN THE NATIONAL SYSTEM OF STATISTICS

By W. G. Cochran, U.S.A. *

RESUMEN

Este artículo considera alguno de los medios por los cuales una Sociedad de Estadística puede ayudar en el mejoramiento del sistema nacional de estadística. En primer lugar, se presenta una clasificación general de los aspectos específicos del sistema nacional que pueden ser de interés particular para la sociedad. Estos aspectos incluyen la cantidad, alcance y calidad de los datos producidos, la organización del sistema nacional, los usos dados a los datos, la sensibilidad del sistema frente a las situaciones cambiantes y la utilización fructífera de las habilidades estadísticas.

Existen varios métodos por medio de los cuales la sociedad puede ejercer su influencia. El método más directo consiste en la creación de comisiones encargadas de estudiar ciertos aspectos del sistema nacional, y hacer representaciones, en nombre de la sociedad, a fin de conseguir que sean puestas en efecto las recomendaciones de las comisiones. Las reuniones de la sociedad proporcionan una oportunidad para revisar, criticar y sugerir cambios en las series estadísticas nacionales, especialmente por medio de sesiones en las que figuren entre los oradores tanto estadísticos gubernamentales como consumidores de sus datos. Las publicaciones de la sociedad pueden desempeñar una función similar y diseminar la información acerca de los nuevos desarrollos y programas en el sistema nacional. La sociedad debe apoyar cualquier oportunidad que surja para permitir a los estadísticos gubernamentales el obtener mayor adiestramiento en estadística. La Real Sociedad de Estadística de Inglaterra, concede certificados y diplomas a aquellos de sus miembros ("Fellows") que pasan un examen prescrito por la Sociedad.

Cuando la sociedad llega a estar bien establecida, los organismos gubernamentales pueden recurrir a ella para su asesoramiento en los problemas estadísticos y para la realización de investigaciones. Algunos ejemplos de este tipo de actividad, tomados de la experiencia de la Asociación Americana de Estadística, se incluyen en este artículo.

The object of this paper is to suggest some ways in which a statistical society can help to improve the national system of statistics. For illustration, I shall use certain experiences of the American Statistical Association (ASA).

* Professor of Biostatistics, School of Hygiene and Public Health, The Johns Hopkins University, Baltimore, Maryland. President of the American Statistical Association 1953. This article was prepared in November 1953.

## Some Aspects of the National System of Statistics

A discussion of this topic gives rise to two different kinds of question. First, what aspects of the national system of statistics should be of concern to the statistical society? Second, what are the various methods by which the statistical society can exert some influence?

With regard to the first question, the following rough classification is intended to suggest some specific features of the national system which may deserve attention.

*1. Quantity and scope of the data produced.* Does the national statistical **series** provide the principal types of basic information that are needed for the sound formulation of national policies? If there are obvious gaps, what constructive proposals can be made to remove them?

*2. Quality of the data.* How trustworthy are the data for the principal uses which might be made of them? Are the necessary definitions of terms clearly given and are adequate warnings issued about known limitations of the data.

*3. Organization of the national system.* Is there overall coordination to insure an integrated program and to prevent duplication of effort or lack of comparability between figures produced by different agencies? Is the collection and publication of data efficiently handled within the separate agencies of the system? Are the agencies free to work objectively without pressure from outside groups with special interests?

*4. Uses of the data.* The statistical society can spread information about the data that are produced by the national system and can point out fruitful uses that might be made of these data by individuals or organizations.

*5. Responsiveness to changing situations.* As times goes on, some series of data become of limited utility and could well be dropped; others require revision; and needs arise for new data. Improved techniques in the collection and tabulation of data are produced. A lively society can do much to bring to light these changing needs and developments.

*6. Fruitful utilization of statistical skills.* Strictly, this item falls under earlier item 3, "Organization," but is given special mention because it is of particular interest to members of a statistical society. There is usually much ignorance among administrators about the qualifications needed for the different types of statistical work and about the functions which a statistician can perform—especially the contributions which he can make when the collection of data is being planned, and the value of the statistical point of view towards many problems that are met at the policy—making levels of government. The society can help in setting up specifications for different types of statistical posts, as has been done, for example, by the IASI Committee on Statistical Education. Attempts by the statistical society to promote greater use of statistical skills must, of course, be conducted with tact; it is important not to create the misleading impression that the society is interested primarily in providing good jobs for statisticians! Nevertheless, if low salary scales are an important factor in hindering the proper use of statistical skills by government, the society should not hesitate to point this out.

The society may also contribute by taking appropriate action when proposals are put forward which would seriously decrease the quantity or quality of statistical work in the government. For instance, if the proposal is to stop an important series of data, the society may bring this proposal to the attention of the principal users of this series, and may wish to express its own opinion on the matter. If statisticians in the government service are being subjected to pressures which make it difficult for them to do objective work of high quality, the society should support the principle that the statistician must be free to seek and report the facts as he finds them. Recently, in the United States, some of the work of the National Bureau of Standards in the design and conduct of experiments was subject to outside criticism, and a number of ill-informed statements received prominence in the public press. The directors of the ASA, after careful consideration of various courses of action, instructed the president of the ASA to take steps to obtain a review of the work of the Bureau of Standards in this field by a body of scientists of the highest competence. Shortly afterwards, a committee to make this review was appointed by the National Academy of Sciences at the request of the Secretary of Commerce and two members of the ASA served on this committee. In my opinion, the noteworthy points about this action by the directors were that the action did not prejudge the issue and that it maintained the principle that the competence and objectivity of scientific work can be judged adequately only by a body of scientists.

I do not think that it would be profitable to attempt to assign any recommended order of priority to the various activities that are suggested by the list above. The practicalities of the situation and the interests of the members will determine what activities they consider most fruitful, and both will change from time to time. I hope that the illustrations given later will indicate some of the possible activities more specifically.

### Direct Activities of the Statistical Society

*Representations in the name of the society.* The most direct type of action is to conduct studies, by means of committees, of certain aspects of the national system of statistics, and to exert pressure, in the name of the society, to put into effect the recommendations from these studies. In the ASA, an example is found as early as 1844, only five years after the founding of the Association. A committee of the Association sent a memorandum to the Congress of the United States, pointing out "various and gross errors" which the committee had discovered in the printed edition of the sixth census of the United States, and concluding as follows: "In view of these facts, the undersigned, in behalf of said association, conceive that such documents ought not to have the sanction of Congress, nor ought they to be regarded as containing true statements relative to the condition of the people and the resources of the United States. They believe it would have been far better to have had no census at all, than such a one as has been published; and they respectfully request your honorable bodies to take such order thereon, and to adopt such measures for the correction of the same, or, if the same cannot be corrected, of discarding and disowning the same, as the good of the country shall require, and as justice and humanity shall demand."

A direct attack of this kind must be used with discretion, particularly if its purpose is to correct deficiencies. There are times when its effectiveness is great, and other times when it arouses a stubborn opposition that in the end results in slower progress than more indirect methods would achieve.

Some scientific societies have a fondness for passing *resolutions* as a means of making their point of view known to the parties whom they wish to influence. In the ASA, this device is used very sparingly, partly because it appears to have had limited effectiveness, and partly because of a prevailing attitude that a resolution in the name of the Association should not be taken unless the great majority of the members are both well-informed about the issue and enthusiastically in favor of the resolution.

*Meetings.* The meetings of a society provide an occasion for reviewing, criticizing, and suggesting changes in the national statistical series. A typical session might have two speakers from a governmental agency and two speakers who are informed users of the data produced by the agency. Such programs give the government statisticians an opportunity to describe recent statistical series and programs, or to ask for advice in the planning of new programs. The users may present useful applications of national data or may raise questions or criticisms about the current programs. Any criticisms are likely to receive careful consideration, since the government statisticians are naturally anxious to stand well in the eyes of their statistical colleagues.

In the 1952 annual meeting of the ASA, the following were some of the sessions devoted to government statistical programs or operations: The new consumers price index; coordination of social statistics at the Federal level; Federal estimates of national savings; Federal sponsorship of data-collecting; activities by non-governmental agencies; labor force and employment trends and projection, 1900-2000; the meaning of statistics classified by industry; new developments in vital statistics.

*Publications.* Like the meetings, the publications of the statistical society can be a vital medium for the exchange of technical information between statisticians in government, industry, and the universities. The *American Statistician,* the news organ of the ASA, contains a section "Federal Statistical Activities" in which new technical developments, programs, and publications of the Federal statistical departments are described. The *Journal* of the ASA is devoted to articles of more permanent interest, and particularly to the results of original research. By means of the *Journal,* government statisticans can learn of new ideas and methods developed in other fields and can find an outlet for the results of their own researches. Reports and recommendations of advisory committees to the Government are usually published either in the *Journal* or the *American Statistician.*

*In-service training.* The society should heartily support any opportunities that arise for enabling government statisticians to obtain further training in statistics, either through short refresher courses or through leaves of absence for more extended study. Administrators sometimes are under the impression that when a statistician has accepted employment, he ought to know everything about his subject and to need no further training. It might be pointed out that statistics is a rapidly developing field in which new techniques of increasing power are coming to light. In the pressure of daily business, it is exceedingly difficult for the government statistician to keep abreast of the latest advances. The best interests of a government agency are served by encouraging its workers to take every opportunity to widen their knowledge.

*Examinations.* Employers of statistcans often find it difficult to appraise the qualifications of applicants for statistical posts because of the lack of standards in the statistical profession. So far as government service is concerned, the statistical society can help, as already mentioned, by giving advice when job descriptions of the different types of statistical post and of the qualifications needed for them are being constructed.

Another step that has been taken by the Royal Statistical Society in Britain is to hold examinations in which the successful candidates receive Certificates or Diplomas, according to the level of the examination. The principal argument in favor of the examinations is that possession of a Certificate or Diploma can be taken by an employer to imply competence in a fairly well-defined area. The examinations might also have the indirect effect of raising the standard of statistical training. There were, however, objections to the holding of examinations from some members of the Royal Economic Society, on the grounds that posts which are described as statistical actually vary greatly in the types of skill and knowledge needed, and that examinations might tend to impose an undesirable degree of uniformity in statistical training. Because of these objections, the examinations are at present restricted to Fellows of the Royal Statistical Society.

## Government Requests for Assistance

As a statistical society becomes well established, government agencies may turn to it for advice and for the conduct of research studies on their statistical problems. Such requests give the society an excellent opportunity to contribute to improvements in the national system. In this connection, I am sure that I need not stress how important it is for the statistical society to earn a reputation for the highest standards of objectivity and to avoid acting in any way as a pressure group for the narrow interests of its members.

Throughout its existence the ASA has maintained close and friendly relations with the major statistical agencies of the government and has frequently been called upon for advice. The usual procedure is for the ASA to appoint a committee of experts to work with the government agency. Unless the task requires an extended period of continuous work, members of these committees customarily serve without pay, except for reimbursement of travel expenses. The ASA maintains the right to publish the report of the committee and insists on freedom of action as to the manner in which the committee shall conduct any investigations needed. The following are some illustrations of ASA committee work done on the request of the Federal Government:

*Committee on Government Statistics and Information Services.* In 1933, this ASA committee was set up at the combined request of the Secretaries of Agriculture, Commerce, Labor, and Interior, "to furnish immediate assistance and advice to the reorganization and improvement of the statistical and informational service of the Federal Government." During the life of this committee, advice was given on a wide range of statistical programs. Committee recommendations covered such fields as employment and payrolls, wages, hours and working conditions, industrial disputes, accident statistics, and the Census of Manufacturers. The committee was also instrumental in the creation of the Central Statistical Board, which later became the Office of Statistical Standards of the Bureau of the Budget. In 1951, the Office of Statistical Standards asked the ASA to set up an advisory committee to make recommendations on important policy issues which arise in the work of the Office. This committee has considered such questions as the following: Should reporting in government surveys be voluntary or compulsory? Should all such reports be confidential, and if so, for what purposes should access to confidential reports be permitted?

*Other ad hoc committees.* In the United States, the Cost of Living Index (later known as the Consumer Price Index), published by the Bureau of Labor Statistics, has become of major importance to millions of workers because of its use during World War II in wage stabilization policy and of numerous agreements made between unions and management in which wage rates depend on the value of the index. In 1943, an ASA advisory committee was established to review the work of the Bureau of Labor Statistics in constructing the index. This careful review by experts led to many improvements in the index and contributed to public acceptance of the index. The confidence placed in the committees work was reflected in a request by the Secretary of Labor in 1949 for another advisory committee on the specific problems involved in the revision of the index. This committee worked closely with personnel of the Bureau of Labor Statistics for almost three years.

Other committees established at government request dealt with morbidity statistics (1942), employment statistics (1945), and the work of the Bureau of Mines (1951).

*Standing committees.* In addition to *ad hoc* committees for specific tasks, the statistical society may establish certain standing committee which are available at any time for providing advice to agencies of the government, or which undertake long-range programs. Examples are the Census Advisory Committee and the Committee on Census Tracts of the ASA. The first works continuously with the appropriate government officials in making and reviewing plans for the collection, tabulation, and publication of census data. The second has done much pioneering work in showing how the Census Tract Statistics published by the Bureau of the Census may be used as basic information in administration and planning by city governments.

*Recruitment of statistical personnel.*   During World War II the ASA cooperated in the establishment of a roster of professional statisticians which aided in making the best use of statistical skills in the war effort. Currently, the ASA is working on a similar project as a phase of the present national defense program.

### Concluding Remarks

Some of the activities which I have mentioned require a well-established society with a large membership. The following are steps which can be taken even by a small society with limited resources:

1.  See that statisticians in government service become members of the society.

2.  Hold meetings at which problems of interest to government statisticians are discussed.

3.  Let administrators know that the society is willing to offer advice to the best of its ability on their statistical problems.

4.  Become informed about the national system of statistics and support desirable additions to that system.

5.  Strive for high standards in all phases of statistical work.

Although this discussion has been confined to those activities of a statistical society that impinge most directly on the national system of statistics, the fact is that almost any step which advances the field of statistics is likely to have a beneficial effect on the national system.

The author wishes to thank Mr. Samuel Weiss for generous help in the preparation of this paper. Any opinions expressed are the author's.

\*     \*     \*

THE RELATION BETWEEN SIMPLE TRANSFORMS AND MAXIMUM LIKELI-
HOOD SOLUTIONS. *By W. G. Cochran.*

Sir Ronald Fisher gives an elegant and compact presentation of the
application of maximum likelihood estimation to analysis of variance
problems that involve binomial data. Since my 1940 paper has been
referred to, I can best contribute to the discussion by summarizing the
argument in that paper, and commenting on it in retrospect. I should
perhaps remark here and now that my 1940 paper was not at all concerned
with *adjusted* transforms such as $\sqrt{x + \frac{1}{2}}$, except for a reference to an
empirical rule suggested by Bartlett for the angular transformation.
The adjective "adjusted" is used in my paper, but it denoted the maximum likelihood solutions.

At the time of my 1940 paper, the square root and angular transformation had been advocated as devices for applying the analysis of
variance to Poisson and binomial data respectively. The object of my

paper, as stated, was "to discuss the theoretical basis for these transformations in more detail, and in particular to examine their relation
to a more exact analysis."

With data of the Poisson type, I first gave the general form of the
maximum likelihood equations, but expressed the opinion that the
numerical solution would be too complicated for frequent use because of
the relatively large number of unknown parameters which usually enter
into the specification of an analysis of variance problem.

I showed, however, that if the effects of treatments, blocks, etc., are
additive on a square-root scale, the maximum likelihood analysis can be
performed in a series of successive approximations, by means of the
method based on working square roots which Fisher exhibits in section
4 of his paper. The approximate $\chi^2$ test of goodness of fit was also
outlined and a numerical example was worked, using the square roots of
the original data as the first step in the approximation.

The same procedure was then developed for binomial data in which
effects are linear on the angular scale. The analysis in terms of working
angles, as given in section 4 of Fisher's paper, was illustrated by a
numerical example.

In short, with Poisson or binomial data, the square root or angular
transformation, respectively can be regarded as the first step in a
maximum likelihood analysis that is valid when effects are additive in
the transformed scale.

In the final sections of the paper, the practical utility of the iterative
process versus that of approximate analysis in the square root or angular
scales was discussed. I pointed out that the iterative processes lead to
maximum likelihood solutions only if (i) effects are additive in the transformed scale and (ii) the residual variation is solely of the Poisson or
binomial type. If either of these conditions fails, the iterative process
loses its claim to be an exact solution. This warning is relevant because
there seems no *a priori* reason for expecting that effects will be additive
in the square root or angular scale, although the assumption of additivity
may be a satisfactory approximation. A more important restriction on
the assumptions, in my own experience, is that biological data which
appear to be of the Poisson or binomial type often contain additional
variation. The $\chi^2$ test is, of course, a safeguard against inappropriate
use of the iterative process.

From experience with several sets of data that had been analyzed by
both methods in preparation for the paper, I also pointed out that the
results of the approximate analysis in the square root or angular scale
seemed to agree well with the analysis based on the final stage of the
iterative process, except for observations giving zero or very small values

in the Poisson case, or zero or 100 per cent in the binomial case. These considerations led me to the following final summary statement, "In practice, these new methods are not recommended to supplant the simple transformations for general use, because it can seldom be assumed that the whole of the experimental error variation follows the Poisson or binomial laws. The more exact analysis may, however, be useful (i) for cases in which the plot yields are very small integers or the ratios of very small integers (ii) in showing how to give proper weight to an occasional zero plot yield."

In retrospect, I am inclined to agree with Fisher that these conclusions are too confident, because they were based on the analysis of only a small number of examples by both methods. I hope that one result of Fisher's paper will be to stimulate the building-up of a substantial body of evidence on the frequency with which the assumptions required in the maximum likelihood analysis hold in practice and on the adequacy of analyses of the unadjusted square roots or angles. My own experience with such data since 1940 is still limited, but as far as it goes it has not led me to modify the conclusions. In the majority of examples that have come my way, the residual variance was considerably greater than the binomial or Poisson variance, so that the iterative procedure in my 1940 paper did not apply. In the remainder, apart from the exceptions noted in my conclusion, analysis of the unadjusted transforms gave results close to the maximum likelihood solutions.

Some further information on this point is supplied by a recent paper (Claringbold, Biggers and Emmens, *Biometrics*, 9, 475, 1953) which compares the angular transformation, using Bartlett's empirical adjustment for zero responses, with two cycles of the maximum likelihood solution. Data from 6 factorial experiments were analyzed by both methods. Summarizing these comparisons, the authors remark that the difference in results is "noticeable, although negligible for practical purposes." They conclude that analysis of the angles "gives a very good basis for the maximum likelihood process and in many practical situations may be sufficiently accurate in itself—the decision, however, will rest with the investigator."

To summarize, an analysis of variance of the simple square roots or angles may be useful in two types of situation. The first concerns heterogeneous data in which the residual variance is greater than the binomial or Poisson variance, but in which we do not feel able to set up a mathematical specification as to the nature of this residual variation. A maximum likelihood solution, which requires this specification, cannot then be made. In using a transformed scale, the hope is that it will bring the data sufficiently close to additivity of effects, normality and equality of variances so that an ordinary analysis of variance in the transformed scale will be reasonably efficient. The second situation, as already mentioned, is one in which the analysis of square roots or angles is regarded as a quick approximation to the maximum likelihood solution, with a saving of time at the expense of some loss of efficiency.

Fisher's attitude, if I do not mistake it, is that this saving in time is often overestimated, and is a poor return for having to use a method that is empirical in place of one that is well-grounded in theory. In this connection, I think that some distinction is appropriate between uses made by the professional statistician and uses made by the biologist. The professional statistician might be expected to give strong preference to a method of analysis for which there is a good theoretical foundation and to find that differences in computing times between approximate and well-grounded analyses are seldom great enough to be a major factor. The biologist, for understandable reasons, is much attracted by quick and simple computational methods, and an additional step or two that is trivial to the professional statistician can appear formidable to the biologist. It is my impression that the speed with which a new technique becomes widely used is considerably influenced by the simplicity or otherwise of the calculations that it requires. Next door to the lecture room in which the probit method is expounded one may still find the laboratory in which the workers compute their LD 50's by the Behrens (Reed-Muench) method. And even the professional statistician is likely to look kindly on methods that may save some time when there are large amounts of routine data to be processed, or when he is in a hurry. Moreover, the element of personal taste seems to enter. We tend, I think, to make exaggerated claims for the computing methods which we happen to like, at the expense of those which we don't like.

The argument in favor of time-saving methods can, of course, be carried too far and can lead to opportunistic and second-rate teaching. But the argument has enough validity to justify the current widespread interest in speedy, if somewhat inefficient, methods of analysis.

With regard to adjusted transforms, I have made some use of the transforms $\log (x + 1)$ and $\sqrt{x + \frac{1}{2}}$ in my own work, again on the empirical grounds that these scales seemed to provide a slightly better discrimination of the treatment effects than the corresponding simple transforms. I agree, however, that emphasis on variance-stabilizing as an end in itself would be misplaced.

# 58

# THE COMBINATION OF ESTIMATES FROM DIFFERENT EXPERIMENTS*

WILLIAM G. COCHRAN

*The Johns Hopkins University,*
*Baltimore, Maryland*

## 1. INTRODUCTION

When we are trying to make the best estimate of some quantity $\mu$ that is available from the research conducted to date, the problem of combining results from different experiments is encountered. The problem is often troublesome, particularly if the individual estimates were made by different workers using different procedures. This paper discusses one of the simpler aspects of the problem, in which there is sufficient uniformity of experimental methods so that the $i$th experiment provides an estimate $x_i$ of $\mu$, and an estimate $s_i$ of the standard error of $x_i$ . The experiments may be, for example, determinations of a physical or astronomical constant by different scientists, or bioassays carried out in different laboratories, or agricultural field experiments laid out in different parts of a region. The quantity $x_i$ may be a simple mean of the observations, as in a physical determination, or the difference between the means of two treatments, as in a comparative experiment, or a median lethal dose, or a regression coefficient.

The problem of making a combined estimate has been discussed previously by Cochran (1937) and Yates and Cochran (1938) for agricultural experiments, and by Bliss (1952) for bioassays in different laboratories. The last two papers give recommendations for the practical worker. My purposes in treating the subject again are to discuss it in more general terms, to take account of some recent theoretical research, and, I hope, to bring the practical recommendations to the attention of some biologists who are not acquainted with the previous papers.

The basic issue with which this paper deals is as follows. The simplest method of combining estimates made in a number of different experiments is to take the arithmetic mean of the estimates. If, however, the experiments vary in size, or appear to be of different precision, the investigator may wonder whether some kind of weighted mean would be more precise. This paper gives recommendations about the kinds of weighted mean that are appropriate, the situations in which they

*Department of Biostatistics, Paper No. 292. This work was assisted by a contract with the Office of Naval Research.

are appropriate, and the circumstances in which the unweighted mean is to be preferred. Methods for obtaining a standard error to be attached to the final estimate are also presented.

The mathematical theory which bears on the problem is complex, and some of the recommendations are based on approximations in theory. Wherever possible, the recommendations are documented by references to published papers. Some theoretical issues are discussed briefly in sections 6 to 9 in cases where the documentation available in the literature does not seem adequate.

### 2. MATHEMATICAL MODELS

As will appear later, the best combined estimate depends on the nature of the data. It is advisable to consider the preliminary question:

Do the values of the $x_i$ agree among themselves within the limits of their experimental errors?

If they do, we may postulate an underlying mathematical model of the form

$$x_i = \mu + e_i , \tag{1}$$

where $e_i$ is the experimental error of $x_i$ .

If the values of the $x_i$ differ by more than can be accounted for by their experimental errors, we require a model of the form

$$x_i = \mu_i + e_i , \tag{2}$$

where $\mu_i$ , which might be called the "true value" in the $i$th experiment, varies from one experiment to another. There are numerous reasons why such variations may exist. They may be the result of differences in the experimental techniques used in the different experiments, of biases that vary in size from one experiment to another, or of real changes in $\mu_i$ due to the environment in which the experiment is conducted. Frequently the investigator is able to predict, from general knowledge or from past experience with the same type of data, whether the $\mu_i$ are likely to vary. In agricultural experiments that are located on farmers' fields throughout an area, for instance, it is quite commonly found that the response to a fertilizer exhibits a real variation from field to field. This variation is often described as an *interaction* of the effect with experiments. Tests of significance for this interaction will be presented later.

If an interaction exists, the type of combined estimate that is wanted requires careful consideration. It is necessary to take into account the purpose for which the combined estimate is to be made and the reasons for the presence of interaction, in so far as these can be discovered.

The following are illustrations of some of the situations that may arise.

(1) In the determination of a physical constant, we might conclude that interaction exists because some of the experiments (e.g. the earlier ones) were done by a technique that is subject to a bias of unknown magnitude, whereas the remainder of the experiments appear to be unbiased. In this event we would presumably discard the results from the biased experiments and consider only a combination of results from the unbiased experiments.

(2) In agricultural experiments the variation in $\mu_i$ may be due mainly to the soil type on which the experiments are conducted. The experiments can then be classified into groups, each of which represents a specific soil type. It may also happen that the number of experiments on a given soil type is not at all proportional to the area of the crop under that soil type in practical farming, perhaps because the experiments were deliberately set up to include some of the rarer types. In this case, if our object is to estimate a mean over some defined farming area, we might adopt the kind of weighted mean that is appropriate to stratified sampling. Thus if $\bar{x}_j$ is the estimated mean for the $j$th soil type, and $A_j$ is the estimated area of the crop under this type in the population, the overall mean is taken as $\sum A_j \bar{x}_j / \sum A_j$.

(3) In the preceding situation we might decide, alternatively, not to estimate the overall mean at all, but to present the individual estimates for the different groups or strata. This practice is advisable where the $\mu_i$ vary so much that different practical recommendations must be given in different strata. Of course, such recommendations are feasible only when the user of the results knows to which stratum he belongs. An example might be experiments on the feeding of chickens, where the results vary with the breed of the chickens.

(4) Occasionally, in laboratory experiments which were thought to be well-controlled, large interactions may appear for which no adequate explanation can be given. In this event it might be best to hand the problem back to the experimenters, on the grounds that there is not much point in attempting a "best" combined estimate until the experimenters can reach better agreement in their results, or at least find out why they disagree.

These illustrations, which do not exhaust the possibilities, bring out the point that the combination of the individual estimates is not a routine matter, but requires clear thinking about both the nature of the data and the function of a combined estimate. However, unless it is decided that no type of combined estimate will serve a useful purpose, we do face the problem of combining at least over certain subgroups of the experiments.

In the remainder of this paper it will be assumed that the experiments which we have decided to combine are a random sample from the population of experiments about which we wish information. This assumption is far from being universally true in practice and should be examined before adopting the methods in this paper, since series of experiments often come into existence in a rather haphazard way.

The discussion will deal only with the combination of a single estimate $x_i$ from each experiment. When each experiment contains more than 2 treatments, we may wish to make a combined analysis of all the experimental results. Some methods for handling this problem are given by Yates and Cochran (1938), Cochran and Cox (1950) and Kempthorne (1952).

### 3. EXPERIMENTS OF THE SAME SIZE AND THE SAME PRECISION

The simplest case is that in which all $k$ experiments are of exactly the same type, with no missing data, and the estimates $x_i$ all have the same error variance $\sigma^2$. In this event the estimated variances $s_i^2$ will each have $n$ degrees of freedom and will each be unbiased estimates of $\sigma^2$. To avoid confusion, note that the symbols $s_i^2$ and $\sigma^2$ refer to the variance of $x_i$, not to the variance per single observation in the experiment.

This case will occur when every experiment has the same precision per observation, and $x_i$ is the same linear function of the observations in the experiment. Thus the variance $\sigma^2$ will be of the form $\sigma_0^2/f$, where $\sigma_0^2$ is the common variance per observation, and $f$ is a divisor which is the same in all experiments. For example, if $x_i$ is an unweighted mean over $r$ replications, $f = r$, and if $x_i$ is the difference between two such means, $f = r/2$. This case would not apply, however, if $x_i$ were the regression of yield on plant number, because the variance of $x_i$ would depend on the distribution of plant numbers in the $i$th experiment.

This case can be handled by familiar and elementary methods, but is included for completeness.

To test whether the $x_i$ are of the same precision we may apply Bartlett's test, in which we compute $\chi^2$, with $(k - 1)$ degrees of freedom, as

$$\chi^2 = \frac{2.303}{C} \left[ nk \log \bar{s}^2 - \sum_{i=1}^{k} n \log s_i^2 \right] \tag{3}$$

where $\bar{s}^2$ is the arithmetic mean of the $s_i^2$ and

$$C = 1 + \frac{(k + 1)}{3nk}$$

Although the investigator can never be sure that the $x_i$ all have the same variance, it is suggested, as a working rule, that the methods in this section are adequate whenever Bartlett's $\chi^2$ is not significant at the 5 per cent level. This opinion is based on the results of a number of sets of data which were worked with and without the assumption of homogeneity. Methods which do not require the assumption are given in section 5.

On the assumption that the $s_i^2$ are homogeneous, the interaction of the $x_i$ with experiments can be tested by means of a standard $F$-test in the analysis of variance (table 1).

TABLE 1
TEST OF THE VARIATION IN $x_i$ FROM EXPERIMENT TO EXPERIMENT

| Source of variation | d.f. | Mean Squares |
|---|---|---|
| Interaction with experiments | $(k - 1)$ | $s_b^2 = \Sigma(x_i - \bar{x})^2/(k - 1)$ |
| Pooled internal error | $nk$ | $\bar{s}^2$ |

*Interactions Absent.*

If there is no interaction ($\mu_i$ all equal to $\mu$), then from equation (1) each $x_i$ is an estimate of $\mu$ with common variance $\sigma^2$. Hence, if the $x_i$ are approximately normally distributed, the recommended estimate of $\mu$ is their unweighted mean $\bar{x}$, with variance $\sigma^2/k$.

To find a sample estimate of the standard error of $\bar{x}$, we may note that the quantities $\bar{s}^2$ and $s_b^2$ are both estimates of $\sigma^2$, with $nk$ and $(k - 1)$ degrees of freedom, respectively. The best estimate of the standard error of $\bar{x}$ is the pooled value

$$\text{s.e.}(\bar{x}) = \sqrt{\frac{nk\bar{s}^2 + (k - 1)s_b^2}{k(nk + k - 1)}} \tag{4}$$

with $(nk + k - 1)$ degrees of freedom. Since $nk$ is usually much greater than $(k - 1)$, the use of $\bar{s}/\sqrt{k}$ as the estimated standard error is not uncommon.

*Interactions Present.*

In this event, the quantity to be estimated is the population mean $\mu$ of the $\mu_i$. Let $\sigma_\mu$ be the standard deviation of the distribution of the $\mu_i$. Then from equation (2),

$$x_i = \mu_i + e_i = \mu + (\mu_i - \mu) + e_i$$

It follows that the estimates $x_i$ vary about $\mu$ with variance $(\sigma_\mu^2 + \sigma^2)$.

Since the $x_i$ are still of equal precision as estimates of $\mu$, the unweighted mean $\bar{x}$ is still the best estimate of $\mu$. However, the variance of $\bar{x}$ is now $(\sigma_\mu^2 + \sigma^2)/k$, and expression (4) cannot be used for the estimated standard error of $\bar{x}$. It is easy to show algebraically that $s_b^2$ in table 1 is an unbiased estimate of $(\sigma_\mu^2 + \sigma^2)$, so that for the standard error of $\bar{x}$ we use

$$\text{s.e.}(\bar{x}) = \frac{s_b}{\sqrt{k}} = \sqrt{\frac{\sum (x_i - \bar{x})^2}{k(k-1)}} \tag{5}$$

To summarize, the only decision that needs to be made is whether we will regard interactions as present or absent. The more conservative procedure is always to regard interactions as present, since estimate (5) for the standard error is valid whether interactions are present or not. If the number of experiments, $k$, is small, however, this estimate has low precision, and one is tempted to use the pooled estimate (4).

A procedure followed by some workers is to pool whenever the $F$ from table 1 is not significant at the 5 per cent level. This has been criticized by others on the grounds that it may underestimate the standard error of $\bar{x}$. The consequences of a rule of this kind have been examined by Bancroft (1944) and Paull (1950) for a series of values of the ratio $I = \sigma_\mu^2/\sigma^2$. Their results show that the rule is somewhat hazardous, in that for moderate values of $I$ (say between 1/4 and 4) it underestimates the standard error and gives too many significant results in a subsequent $t$-test of $\bar{x}$. The alternative rule of pooling only when $F < 2$, suggested by Paull, is safer: its chief defect is that if $I$ is small, it may slightly overestimate the standard error.

### 4. EXPERIMENTS OF DIFFERENT SIZES BUT OF THE SAME PRECISION PER OBSERVATION

Sometimes the experiments that are being combined differ in size and structure, but there is reason to believe that experimental error variances *per observation* are the same in all experiments. If $\sigma_0^2$ denotes this common variance, the variance $\sigma_i^2$ of the estimate $x_i$ will be of the form $\sigma_i^2 = \sigma_0^2/f_i$, where $f_i$ is a factor depending on the type of experiment. For instance, if $x_i$ is a mean over $r_i$ replications, $f_i = r_i$, and if $x_i$ is the difference between two such means, $f_i = r_i/2$. Similarly, if $s_{0i}^2$ denotes the estimated error variance per observation in the $i$th experiment, $s_i^2 = s_{0i}^2/f_i$.

*Example 1.*

This example was obtained by selecting two treatments from a series of experiments on the effectiveness of carbon tetrachloride in killing worms (Nippostrongylus muris) which are parasitic on rats (Whitlock and Bliss, 1943). Each rat was injected with 500 larvae. Eight days later, the rats were treated with varying doses of $CCl_4$ and two days later all rats were killed and the numbers of adult worms were counted for each rat. The treatments to be discussed are the control (no $CCl_4$) and a dose of 0.063 cc per rat. Three experiments included both treatments, with numbers of replications as follows.

| Expt. | Control | 0.063 cc $CCl_4$ |
|---|---|---|
| 1 | 5 | 3 |
| 2 | 5 | 5 |
| 3 | 6 | 7 |

The relevant data are shown in table 2.

TABLE 2

ESTIMATES $(x_i)$ AND VARIANCES PER RAT $(s^2_{0i})$

| Expt. | No. of adult worms | | Difference | | | d.f. |
|---|---|---|---|---|---|---|
| | Control | $CCl_4$ | $x_i$ | $s^2_{0i}$ | $f_i$ | $n_i$ |
| 1 | 290.4 | 204.0 | 86.4 | 3,223 | 1.875 | 10 |
| 2 | 323.2 | 165.2 | 158.0 | 8,370 | 2.500 | 14 |
| 3 | 274.0 | 262.7 | 11.3 | 2,606 | 3.231 | 16 |

The estimates to be combined are the differences $x_i$ between the mean recoveries for the control and the treated rats. The values of $f_i$ may be verified from the numbers of replications already reported. In computing the variances per rat, $s^2_{0i}$, the data from treatments with smaller doses of $CCl_4$ were also used, so that the degrees of freedom are larger than would be provided by the two treatments discussed here.

The first step is to apply Bartlett's $\chi^2$ test to the estimated variances per observation.

$$\chi^2 = \frac{2.303}{C} [n_c \log \bar{s}^2_0 - \sum n_i \log s^2_{0i}]$$

(6)

where

$$n_c = \sum n_i = 40$$

$$\bar{s}_0^2 = \sum n_i s_{0i}^2 / n_c = (191{,}106)/40 = 4{,}777.6$$

$$C = 1 + \frac{1}{3(k-1)} \left\{ \sum \frac{1}{n_i} - \frac{1}{n_c} \right\} = 1.035.$$

The value of $\chi^2$ is 5.59, with $k - 1 = 2$ degrees of freedom. The significance probability is about 0.06, and it is doubtful whether the variances per rat can be considered homogeneous. For the present, this assumption will be made: in section 5 the example will be re-worked without this assumption.

In order to test whether the estimates $x_i$ agree with each other within the limits of their experimental error variances, we carry out a conventional analysis of variance on a single-observation basis (table 3).

TABLE 3
ANALYSIS OF VARIANCE ON A SINGLE-OBSERVATION BASIS

| Source of variation | d.f. | Sum of squares | Mean squares |
|---|---|---|---|
| Interaction with expts. | $(k-1)$ | $\sum f_i(x_i - \bar{x}_w)^2$ | $s_{0b}^2 = \sum f_i(x_i - \bar{x}_w)^2/(k-1)$ |
| Pooled error | $n_c$ | $\sum n_i s_{0i}^2$ | $\bar{s}_0^2 = \sum n_i s_{0i}^2/n_c$ |

Note that the factors $f_i$ are used as weights in computing the sum of squares for the interaction with experiments.

The quantity $\bar{x}_w$ in table 3 is the *weighted* mean

$$\bar{x}_w = \sum f_i x_i / \sum f_i \tag{7}$$

The $F$-ratio, $s_{0b}^2 / \bar{s}_0^2$, gives a test of significance of the presence of interactions.

At this point there are three situations to be considered.

*Interactions absent.*

This case is a familiar one in elementary text-books. We revert to the mathematical model

$$x_i = \mu + e_i$$

so that $x_i$ is an estimate of $\mu$ with variance $\sigma_0^2/f_i$. By least squares theory, the best combined estimate of $\mu$ is the weighted mean $\bar{x}_w$, and its variance is

$$V(\bar{x}_w) = \sigma_0^2 / \sum f_i$$

The most precise combined estimate of $\sigma_0^2$ is obtained by pooling the sums of squares in table 3 to give

$$\hat{s}_0^2 = \frac{\sum f_i(x_i - \bar{x}_w)^2 + \sum n_i s_{0i}^2}{k - 1 + n_c}$$

The standard error of $\bar{x}_w$ is taken as $\hat{s}_0 / \sqrt{\sum f_i}$ , with $(k - 1 + n_c)$ degrees of freedom.

*Interactions large.*

With the more general model

$$x_i = \mu + (\mu_i - \mu) + e_i \ ,$$

the variance of $x_i$ , as an estimate of $\mu$, is

$$V(x_i) = \sigma_\mu^2 + \frac{\sigma_0^2}{f_i}$$

If the values of $\sigma_\mu^2$ and $\sigma_0^2$ were known, the least squares estimate of $\mu$ would be the *semi-weighted* mean

$$\bar{x}_{sw} = \sum W_i x_i / \sum W_i \quad \text{where} \quad W_i = \frac{1}{\sigma_\mu^2 + \frac{\sigma_0^2}{f_i}} \tag{8}$$

The semi-weighted mean (8) includes the weighted mean (7) as a particular case, since it reduces to the weighted mean when $\sigma_\mu = 0$. At the other extreme, when interactions are large, $\sigma_\mu$ is large relative to $\sigma_0$ and the semi-weights are all approximately equal. The semi-weighted mean then differs little from the unweighted mean.

Since it is not profitable to go to the extra trouble of computing the semi-weighted mean unless we are confident that there will be a worth-while gain in precision over the unweighted mean, the precision of the unweighted mean is compared with that of the semi-weighted mean in section 6. The relative precision is found to depend on two factors:

(i) The ratio $I$ of the interaction variance $\sigma_\mu^2$ to the average of the experimental error variances $\sigma_0^2/f_i$ . The higher the value of $I$, the smaller is the loss of precision resulting from the use of the unweighted mean.

(ii) The amount of variation in the factors $f_i$ . As the variation in the $f_i$ increases, the loss of precision resulting from the unweighted mean increases.

The theoretical examination in section 6 leads to the rules given in table 4.

TABLE 4

WORKING RULES FOR THE USE OF THE UNWEIGHTED MEAN

| If ratio of largest to smallest $f_i$ | Use the unweighted mean whenever |
|:---:|:---:|
| $<2$ | $F > 3$ |
| between 2 and 6 | $F > 4$ |
| $>6$ | $F > 5$ |

The rules will be illustrated from example 1. The analysis of variance appears in table 5.

TABLE 5

ANALYSIS OF VARIANCE FOR DATA IN TABLE 2

| Source of variation | d.f. | Sums of squares | Mean squares | $F$ |
|:---|:---:|:---:|:---|:---:|
| Interaction with expts. | 2 | 30,506 | $s_{0b}^2 = 15,253$ | 3.19 |
| Pooled error | 40 | 191,106 | $s_0^2 = 4,778$ | |

The $F$-ratio, 3.19, is almost at the 5 per cent level, indicating a variation in the effectiveness of the $CCl_4$ from experiment to experiment. From table 2 we see that the ratio of the largest to the smallest $f_i$ is less than 2. By the rule in table 4, the unweighted mean of the $x_i$ is recommended since $F$ is over 3. The estimate is

$$\bar{x} = \frac{86.4 + 158.0 + 11.3}{3} = 85.2$$

The standard error of $\bar{x}$ is given by

$$\text{s.e.}_{\bar{x}} = \sqrt{\frac{\sum (x_i - \bar{x})^2}{k(k - 1)}} = \sqrt{\frac{10,763}{(3)(2)}} = 42.3$$

The usual procedure is to attribute $(k - 1)$ or 2 degrees of freedom to this standard error. This is not quite correct, because the $x_i$ are presumably not all of the same precision as estimates of $\mu$, even though we have decided to use their unweighted mean as an overall estimate.

An approximate adjustment which gives a smaller number of degrees of freedom is developed in section 7, although my experience is that it is rarely needed for this application. The adjustment requires some supplementary calculations which are given for another purpose in the subsequent table 6.

In column 2 of table 6, let

$$v_i = s_\mu^2 + \frac{\bar{s}_0^2}{f_i}$$

and let

$$V_1 = \frac{\sum v_i}{k} = 6{,}211 \; : \; V_2 = \frac{\sum v_i^2}{k} = 38{,}773{,}457.$$

The adjusted number of degrees of freedom is

$$n_e = \left\{ \frac{(k-1)^2 V_1^2}{(k-2)V_2 + V_1^2} \right\} = 1.99 \tag{9}$$

The adjustment has a negligible effect.

*Interactions moderate.*

With most sets of data, either the weighted or the unweighted mean will prove to be satisfactory. There remain some cases in which, although $F$ is not large, we believe from the nature of the data that interactions are likely to be present, and are reluctant to rely on either the unweighted or the weighted mean. A sample semi-weighted mean, which is an analogue of the semi-weighted mean in equation (8), may be tried.

In equation (8) the true semi-weights $W_i$ were given as

$$W_i = \frac{1}{\sigma_\mu^2 + \dfrac{\sigma_0^2}{f_i}}$$

The first step is to obtain sample estimates of $\sigma_\mu^2$ and $\sigma_0^2$.

It is easily shown by algebra that the expectations of the two mean squares in the analysis of variance in table 3 are as follows.

$$E(s_{0b}^2) = \sigma_0^2 + \bar{f}' \sigma_\mu^2$$

$$E(\bar{s}_0^2) = \sigma_0^2$$

where

$$\bar{f}' = \frac{1}{(k-1)} \left\{ \sum f_i - \frac{\sum f_i^2}{\sum f_i} \right\}$$

The quantity $\bar{f}'$ is always smaller than the arithmetic mean of the $f_i$. From these results, an unbiased estimate of $\sigma_\mu^2$ is

$$s_\mu^2 = (s_{0b}^2 - \bar{s}_0^2)/\bar{f}' \tag{10}$$

Finally, for the sample semi-weighted mean we take

$$\bar{x}_{sw} = \sum \hat{W}_i x_i / \sum \hat{W}_i$$

where

$$\hat{W}_i = \frac{1}{s_\mu^2 + \dfrac{\bar{s}_0^2}{f_i}}$$

To illustrate the method from example 1, the analysis of variance (table 5) gives

$$s_{0b}^2 = 15{,}253 : \bar{s}_0^2 = 4{,}778.$$

The value of $\bar{f}'$ is found from table 2 to be

$$\bar{f}' = \frac{1}{2}\left\{7.606 - \frac{20.205}{7.606}\right\} = 2.475$$

TABLE 6
CALCULATION OF SEMI-WEIGHTED MEAN

| Expt. | $s_\mu^2 + \dfrac{\bar{s}_0^2}{f_i} = 4{,}232 + \dfrac{4{,}778}{f_i}$ | Reciprocal $\hat{W}_i$ | $x_i$ |
|---|---|---|---|
| 1 | 6,780 | .000147 | 86.4 |
| 2 | 6,143 | .000163 | 158.0 |
| 3 | 5,711 | .000175 | 11.3 |
| Total | | .000485 | |

Hence, from equation (10),

$$s_\mu^2 = \frac{15{,}253 - 4{,}778}{2.475} = 4{,}232.$$

The semi-weights are computed in table 6.

$$\bar{x}_{sw} = \frac{(86.4)(147) + (158.0)(163) + (11.3)(175)}{485} = 83.4$$

This does not differ materially from the unweighted mean, 85.2.

The standard error of $\bar{x}_{sw}$ is, approximately,

$$\text{s.e.}(\bar{x}_{sw}) = \frac{1}{\sqrt{\sum \hat{W}_i}} = \frac{1}{\sqrt{.000485}} = 45.4$$

This formula may give values that are slightly too low, since it ignores the fact that the weights $\hat{W}_i$ are subject to sampling errors. For $t$-tests, it is suggested that $(k - 1)$ degrees of freedom be assigned to the standard error, although the distributions involved have not yet been adequately investigated.

### 5. EXPERIMENTS OF UNEQUAL PRECISION PER OBSERVATION

The methods in this section are to be used when Bartlett's $\chi^2$ is significant or when, for any other reason, the investigator does not wish to assume that the variances per observation are equal. The methods are the same whether the experiments are identical in size and type or not.

As before, the estimate in the $i$th experiment is denoted by $x_i$, and $s_i^2$ is an unbiased estimate of the error variance of $x_i$, based on $n_i$ degrees of freedom.

*The test for interactions.*

The first step is to test for the presence of interactions. One approach is to calculate the ordinary mean square deviation of the $x_i$, i.e.

$$\frac{\sum (x_i - \bar{x})^2}{k - 1}$$

If there are no interactions, this quantity is an unbiased estimate of

$$\bar{\sigma}^2 = \frac{\sum \sigma_i^2}{k}$$

Consequently, an approximate $F$-test of the interactions is made from the ratio

$$F = \frac{\sum (x_i - \bar{x})^2}{(k - 1)\bar{s}^2}$$

where $\bar{s}^2 = \sum s_i^2/k$.

From an elementary point of view, the degrees of freedom might be taken as $(k - 1)$ and $n_c = \sum n_i$. Actually, the calculated $F$ does not follow the tabular $F$-distribution, because the $x_i$ vary in precision. The tabular $F$-distribution might still be used, as an approximation, by reducing the numbers of degrees of freedom ascribed to $F$. Following equation (9), the degrees of freedom in the numerator may be taken as

$$\nu_1 = \frac{(k - 1)^2 V_1^2}{(k - 2) V_2 + V_1^2} \tag{11}$$

where

$$V_1 = \frac{\sum s_i^2}{k} : V_2 = \frac{\sum s_i^4}{k}$$

For the denominator, a familiar rule is

$$\nu_2 = \frac{\left(\sum s_i^2\right)^2}{\sum \dfrac{s_i^4}{n_i}} \qquad (12)$$

There is, however, the further objection that $F$ may be insensitive in detecting the presence of interactions, because experiments with low precision receive the same weight as those with high precision.

An alternative course, less open to these objections, is to base the test of interactions on the *weighted* sum of squares of deviations

$$Q = \sum_{i=1}^{k} w_i(x_i - \bar{x}_w)^2,$$

where

$$w_i = 1/s_i^2$$

$$\bar{x}_w = \sum w_i x_i / \sum w_i$$

If the degrees of freedom $n_i$ are large, $Q$ follows the $\chi^2$ distribution with $(k - 1)$ degrees of freedom. For moderate values of $n_i$, adjustments which transform $Q$ so as to give a better approximation to $\chi^2$ have been worked out by Cochran (1937) for $n_i$ all equal, and by James (1951). Welch (1951) transforms $Q$ so that it may be referred to the $F$-table: his test and that of James are similar in that both neglect terms of order $1/n_i^2$. Although the range of applicability of these tests, which are all approximations, is not yet known, it is suggested that they be used down to $n_i = 6$. To perform Welch's test, we compute the auxiliary quantity

$$a = \sum_{i=1}^{k} \frac{1}{n_i}\left(1 - \frac{w_i}{w}\right)^2$$

where $w = \sum w_i$. Then

$$F_w = \frac{Q}{(k - 1) + \dfrac{2(k - 2)a}{(k + 1)}}$$

with degrees of freedom

$$\nu_1 = (k - 1) : \nu_2 = \frac{(k^2 - 1)}{3a}$$

The $F$ and $F_w$ tests will be illustrated by the data in table 2 for example 1. Although this example was analysed under the assumption that the experiments were of equal precision per observation, the probability value for Bartlett's $\chi^2$, 0.06, casts doubt on this assumption. The first step is to compute the quantities $s_i^2 = s_{0i}^2/f_i$ : these are the estimated variances of the $x_i$ . The remainder of the calculations are arranged in table 7.

TABLE 7
CALCULATIONS FOR THE TEST OF INTERACTIONS

| $x_i$ | $s_i^2$ | $10^2 w_i$ | $w_i x_i$ | $w_i/w$ | $(1 - w_i/w)^2$ | $n_i$ | $(1 - w_i/w)^2/n_i$ |
|---|---|---|---|---|---|---|---|
| 86.4 | 1719 | .0582 | .05028 | .275 | .526 | 10 | .0526 |
| 158.0 | 3348 | .0299 | .04724 | .141 | .738 | 14 | .0527 |
| 11.3 | 807 | .1239 | .01400 | .584 | .173 | 16 | .0108 |
| | 5874 | .2120 | .11152 | 1.000 | | | $a = .1161$ |

*The F-test.*

$$\sum (x_i - \bar{x})^2 = 10{,}763 : \bar{s}^2 = 1{,}958 : F = \frac{5{,}381}{1{,}958} = 2.75$$

The degrees of freedom from an elementary point of view would be 2 and 40. Formulas (11) and (12) will be found to give 1.7 and 30.3 degrees of freedom, respectively. By interpolation between $F(1, 30)$ and $F(2, 30)$, the significance probability comes out at 0.09.

*The $F_w$-test.*

$$Q = \sum w_i x_i^2 - (\sum w_i x_i)^2/w = 11.966 - (.11152)^2/(.002120) = 6.10$$

$$F_w = \frac{6.10}{2 + \dfrac{2(1)(.1161)}{4}} = \frac{6.10}{2.058} = 2.96$$

$$\nu_1 = 2 : \nu_2 = \frac{8}{3(.1161)} = 23.0$$

The significance probability is about 0.07.

The test of interactions is of importance because, as pointed out in section 2, the presence of interactions affects our interpretation of the data and may determine the kind of mean that will be useful. In a borderline case, as in this example, the investigator should take into account both the significance probability and any other knowledge of the data in deciding whether to regard interactions as present or absent. The conservative decision, when in doubt, is to assume interactions present, since the techniques for this situation remain valid even if interactions are absent.

Experience in the application of the $F$- and $F_w$-tests indicates that although the $F_w$-test is more sensitive, the $F$-test, which is simpler to compute, is usually adequate for diagnostic purposes. Consequently, the working rules to be given later are based on the value of $F$.

*Interactions absent.*

If we are willing to assume that interactions are absent, one method of combination is to weight each $x_i$ inversely as its estimated variance $s_i^2$, forming the weighted mean

$$\bar{x}_w = \frac{\sum w_i x_i}{w}, \qquad w_i = 1/s_i^2 : w = \sum w_i$$

The standard error of $\bar{x}_w$ is given approximately by a formula due to Meier (1953), with an adjustment by Cochran and Carroll (1953),

$$\text{s.e.}(\bar{x}_w) = \sqrt{\frac{1}{w}\left\{1 + \frac{4}{w^2}\sum\frac{1}{n_i'}w_i(w - w_i)\right\}} \qquad (13)$$

where

$$n_i' = n_i - \frac{4(k - 2)}{(k - 1)}$$

If the $n_i$ are all equal, formula (13) reduces to the slightly simpler expression

$$\text{s.e.}(\bar{x}_w) = \sqrt{\frac{1}{w}\left\{1 + \frac{4}{n'}\left(1 - \frac{\sum w_i^2}{w^2}\right)\right\}} \qquad (14)$$

The term inside the brackets is an adjustment which takes account of sampling errors in the weights $1/s_i^2$ as estimates of the true weights $1/\sigma_i^2$, and also of the fact that the principal term inside the square root, $1/w$, tends to be an underestimate of the corresponding population expression. These formulas require $n_i \geq 8$: for values of $n_i$ below 8, see section 8.

For the approximate number of degrees of freedom $n_e$ to be attached to this standard error, Meier (1953) suggests

$$n_e = \frac{w^2}{\sum\dfrac{w_i^2}{n_i}} \qquad (15)$$

If the $n_i$ are small, the sampling errors in the weights may be large enough so that the weighted mean is no more precise than the unweighted mean $\bar{x}$, whose standard error is

$$\text{s.e.}(\bar{x}) = \frac{\sqrt{\sum s_i^2}}{k} \tag{16}$$

The approximate number of degrees of freedom $n_e'$ for this s.e. is

$$n_e' = \frac{(\sum s_i^2)^2}{\sum \left(\frac{s_i^4}{n_i}\right)} \tag{17}$$

In seeking some rule which will help in deciding whether to use $\bar{x}$ or $\bar{x}_w$, it is natural to try to base the rule on the value of Bartlett's $\chi^2$, since this will already have been calculated in many cases. Unfortunately, the relation between this $\chi^2$ and the relative precision of $\bar{x}$ to $\bar{x}_w$ is not simple. When the degrees of freedom $n_i$ are large, $\chi^2$ can detect relatively small differences in precision which make $\bar{x}_w$ only slightly more precise than $\bar{x}$. When the $n_i$ are small, on the other hand, $\chi^2$ may sometimes be non-significant even when $\bar{x}_w$ would be substantially better than $\bar{x}$. As a rough guide to the relative precision $R$ of $\bar{x}$ to $\bar{x}_w$, the following formula is suggested.

$$R \dot{=} . \frac{\bar{n}}{(\bar{n} - 2)} e^{-2\chi^2/n_c} \tag{18}$$

where

$$n_c = \sum n_i : \bar{n} = n_c/k$$

This formula was derived as a mathematical approximation and has been checked on a number of sets of data.

Since $\bar{x}$ is preferable on account of its simplicity unless $\bar{x}_w$ brings a worthwhile gain in precision, the investigator will not go far wrong in using $\bar{x}$ unless $R$ is less than 0.9. My experience with actual data has been that often there is little to choose between $\bar{x}$ and $\bar{x}_w$, but occasionally $\bar{x}_w$ wins handsomely.

A warning given by Yates and Cochran (1938) should be repeated. It sometimes happens that there is a correlation between $x_i$ and $s_i^2$, for instance when experiments which have large responses also exhibit high variability. In this event a weighted mean gives too much weight to experiments where the response is low and will be biased.

Example 2 illustrates the rule for choosing between $\bar{x}$ and $\bar{x}_w$.

*Example 2.*

The data in table 8 are the responses in sugar per acre to an applica-

TABLE 8

RESULTS OF 4 EXPERIMENTS ON SUGAR-BEET

| Response to P (cwt) $x_i$ | $s_i^2$ | $w_i = 1/s_i^2$ |
|:---:|:---:|:---:|
| +1.3 | 4.973 | 0.20 |
| +0.4 | 1.416 | 0.71 |
| +0.7 | 6.864 | 0.15 |
| +2.5 | 2.958 | 0.34 |
| Total | 16.211 | 1.40 = w |

tion of superphosphate in 4 experiments on heavy loam soils in the 1936 series of fertilizer trials on sugar-beet in England. Each experiment provided 15 degrees of freedom for error, giving $n_c = 60$.

The value of Bartlett's $\chi^2$ is 9.27, with a probability of about 0.03. In the test for interactions the value of $F$ is less than 1. The response to $P$ had also shown no sign of interactions in several other sets of these sugar-beet experiments, so that the assumption of negligible interactions appeared justifiable.

By formula (18), the crude estimate of $R$ is

$$R^{\cdot} = \cdot \frac{\bar{n}}{(\bar{n} - 2)} e^{-2\chi^2/n_c} = \left(\frac{15}{13}\right) e^{-18.54/60} = 0.85$$

The weighted mean is suggested. Its value is

$$\bar{x}_w = \frac{(1.3)(0.20) + (0.4)(0.71) + (0.7)(0.15) + (2.5)(0.34)}{1.40} = 1.07$$

For the standard error, the simpler form in equation (14) can be used.

$$n' = n - \frac{4(k - 2)}{(k - 1)} = 15 - \frac{(4)(2)}{(3)} = 12.3$$

Hence by equation (14),

$$\text{s.e.}(\bar{x}_w) = \sqrt{\frac{1}{1.40}\left\{1 + \frac{4}{12.3}\left(1 - \frac{.6822}{1.96}\right)\right\}} = 0.93$$

From equation (15), the approximate number of degrees of freedom is

$$n_e = \frac{w^2}{\sum \dfrac{w_i^2}{n_i}} = \frac{n \, w^2}{\sum w_i^2} = \frac{(15)(1.96)}{.6822} = 43$$

The unweighted mean may be verified to be $1.22 \pm 1.01$.

When the numbers of degrees of freedom in the individual experiments are less than 8, the weighted mean will seldom be more precise than the unweighted mean. With the weighted mean, one or two experiments tend to receive very large weights and almost determine the value of the overall mean. If Bartlett's $\chi^2$ is large, the investigator may still feel that some kind of weighting is desirable. A suggested procedure is *partial weighting* (Yates and Cochran, 1938). The same weight is given to all experiments with relatively low values of $s_i^2$, this weight being $\bar{w}_p = 1/\bar{s}_p^2$, where $\bar{s}_p^2$ is the mean of the $s_i^2$ over those experiments that are chosen to have equal weight. Each of the remaining experiments receives its individual weight $w_i = 1/s_i^2$.

The choice of the number of experiments that are to receive equal weight is to some extent arbitrary. A good working rule is to give equal weight to between 1/2 and 2/3 of the experiments (Cochran, 1937). The method prevents an experiment which happens to have a small estimated error from dominating the result, while allowing the less precise experiments to receive lower weights.

*Example 3.*

In studies by the U. S. Public Health Service of observers' abilities to count the number of flies which settle momentarily on a grill, each of 7 observers was shown, for a brief period, grills with known numbers of flies impaled on them and asked to estimate the numbers. For a given grill, each observer made 5 independent estimates. The data in table 9 are for a grill which actually contained 161 flies. Estimated variances are based on 4 degrees of freedom each.

The value of Bartlett's $\chi^2$ was 19.9, with 6 degrees of freedom and a significance probability of less than 0.01. Evidently the observers differ in precision. The $F$-value in the test for interactions was practically 1, giving no indication of any differential bias in observers' error.

The only point of interest in estimating the overall mean is to test whether there is any consistent bias among observers in estimating the 161 flies on the grill. Although inspection of table 9 suggests no such bias, the data will serve to illustrate the application of partial weighting.

It is clear from table 9 that if weighting inversely as $s_i^2$ were employed,

TABLE 9
OBSERVERS' MEAN ESTIMATES AND ERROR VARIANCES

| Observer | Mean estimate $x_i$ | $s_i^2$ | Partial weights |
|----------|---------------------|---------|-----------------|
| 1 | 183.2 | 117.0 | .0129 |
| 2 | 149.0 | 8.1 | .0129 |
| 3 | 154.0 | 235.9 | .0042 |
| 4 | 167.2 | 295.0 | .0034 |
| 5 | 187.2 | 1064.6 | .0009 |
| 6 | 158.0 | 51.2 | .0129 |
| 7 | 143.0 | 134.0 | .0129 |
|  |  |  | $w = .0601$ |

observer 2 would have great influence on the estimate. For partial weighting, we give the same weight to observers 1, 2, 6, and 7. Since $\bar{s}_p^2$ is 77.6 for these observers, $\bar{w}_p = .0129$. The partial weights appear at the right of table 9.

$$\bar{x}_{pw} = \frac{(.0129)(183.2) + \cdots + (.0129)(143.0)}{.0601} = 158.9$$

For the standard error, let

$u$ = no. of experiments given individual weights = 3

$w_u$ = total weight for these $u$ experiments = 0.0085

$p$ = no. of experiments given the same weight = 4

$\bar{n}_p$ = average no. of d.f. for these $p$ experiments = 4

If $\bar{n}_p$ is less than 8,

$$\text{s.e.}(\bar{x}_{pw}) = \frac{1}{w} \sqrt{p\bar{w}_p + \lambda w_u} \tag{19}$$

where $\lambda$ is read as a function of $\bar{n}_p$ and $u$ from table 12 as described in section 8. In this example, with $\bar{n}_p = 4$ and $u = 3$, $\lambda = 1.8$.

$$\text{s.e.}(\bar{x}_{pw}) = \frac{\sqrt{(4)(.0129) + (1.8)(.0085)}}{.0601} = 4.3$$

For $\bar{n}_p$ greater than 8, the standard error may be taken as approximately

$$\text{s.e.}(\bar{x}_{pw}) = \frac{1}{w}\sqrt{p\bar{w}_p + w_u\left\{1 + \frac{4}{w_u^2} \sum \frac{1}{n_i'} w_i(w_u - w_i)\right\}} \tag{20}$$

where the $\sum$ is taken over the $u$ experiments only and

$$n'_i = n_i - \frac{4(u-2)}{(u-1)}$$

Formulas (19) and (20) are revisions of an earlier formula, given by Yates and Cochran (1938), which assumed $p$ and $u$ to be large. In this example, formula (20), although outside of the range of its applicability, agrees well with (19), giving a value of 4.5 for the standard error.

*Interactions present.*

In this case we again have the model

$$x_i = \mu + (\mu_i - \mu) + e_i$$

and the variance of $x_i$ is $(\sigma_\mu^2 + \sigma_i^2)$. The choice of estimate lies between the unweighted mean and the sample semi-weighted mean $\sum \hat{W}_i x_i / \sum \hat{W}_i$, where

$$\hat{W}_i = \frac{1}{s_\mu^2 + s_i^2}$$

The quantity $s_\mu^2$ is computed by formula

$$s_\mu^2 = \frac{\sum (x_i - \bar{x})^2}{(k-1)} - \bar{s}^2$$

where $\bar{s}^2$ is the mean of the $s_i^2$.

As explained in section 4, the relative precision of $\bar{x}$ and $\bar{x}_{sw}$ depends on the size of $\sigma_\mu^2$ and on the amount of variation among the $\sigma_i^2$. Use of $\bar{x}$ when $F$ exceeds 4 is a safe working rule, unless there are extremely large variations in the precisions of the individual $x_i$.

*Example 4.*

Example 1, previously discussed, represents a situation where the data do not indicate very clearly what kind of model and analysis are appropriate. The probability value for Bartlett's $\chi^2$, 0.06, made it doubtful whether equal precision per observation could be postulated. Although this assumption was adopted in the original analysis, tests for interactions without making this assumption were carried out in section 5. The $F$ and $F_w$ values gave probabilities of 0.09 and 0.07, raising the further question whether interactions should be considered as present or absent. Since, however, the value of $F$ was 2.75, the more cautious procedure is to recognize that interactions may be present and use a semi-weighted mean. The subsidiary computations needed are given in table 10.

TABLE 10
COMPUTATIONS FOR THE SEMI-WEIGHTED MEAN

| Expt. | $x_i$ | $s_i^2$ | $s_\mu^2 + s_i^2$ | $10^6 \hat{W}_i$ |
|-------|-------|---------|-------------------|------------------|
| 1     | 86.4  | 1,719   | 5,142             | 194              |
| 2     | 158.0 | 3,348   | 6,771             | 148              |
| 3     | 11.3  | 807     | 4,230             | 236              |
| Totals | 255.7 | 5,874  |                   | 578              |

$$s_\mu^2 = 5{,}381 - 1{,}958 = 3{,}423$$

The calculation proceeds in the right hand columns of table 10. We find

$$\bar{x}_{sw} = 74.1 : \text{s.e.}(\bar{x}_{sw}) = \frac{10^3}{\sqrt{578}} = 41.6$$

The estimate originally made in example 1 was 85.2 ± 42.3. The two estimates do not agree very closely. The difference is due to an apparently fortuitous correlation between $x_i$ and $s_i^2$.

The remaining sections deal with the derivation of some of the appropriate formulas.

### 6. COMPARISON OF THE UNWEIGHTED AND SEMI-WEIGHTED MEANS

Given that interactions are present, the variance of the unweighted mean is

$$V(\bar{x}) = \frac{\sigma_\mu^2 + \bar{\sigma}^2}{k} : \bar{\sigma}^2 = \sum \sigma_i^2 / k \tag{21}$$

If the semi-weights are known exactly, the variance of the semi-weighted mean is

$$V(\bar{x}_{sw}) = \frac{1}{\sum W_i}, \qquad \text{where} \qquad W_i = \frac{1}{\sigma_\mu^2 + \sigma_i^2} \tag{22}$$

Owing to errors in the weights, the variance of the *sample* semi-weighted mean will be greater than (22). Hence the ratio of (21) to (22) gives an upper limit to the relative precision of $\bar{x}_{sw}$ to $\bar{x}$. This ratio is

$$\frac{V(\bar{x})}{V(\bar{x}_{sw})} = \frac{(\sigma_\mu^2 + \bar{\sigma}^2)}{k} \sum \left\{ \frac{1}{\sigma_\mu^2 + \sigma_i^2} \right\} = \lambda \quad \text{(say)} \quad (23)$$

If

$$I = \sigma_\mu^2 / \bar{\sigma}^2$$

is the ratio of the interaction variance to the average error variance, (23) may be written

$$\lambda = \frac{(I + 1)}{k} \sum \left\{ \frac{1}{I + \frac{\sigma_i^2}{\bar{\sigma}^2}} \right\} \quad (24)$$

In the development of a working rule about the choice between $\bar{x}$ and $\bar{x}_{sw}$ , the first step is to find an upper limit to $\lambda$ when we fix the two quantities $I$ and the ratio $r$ of the greatest to the smallest error variances. For the following argument, I am indebted to Dr. Paul Meier.

Let $\sigma_1^2$ be the smallest error variance, and let

$$\sigma_i^2 = r_i \sigma_1^2 \quad (1 \leq r_i \leq r; i > 1)$$

Then

$$k\bar{\sigma}^2 = \sigma_1^2(r_1 + r_2 + \cdots + r_k) = \sigma_1^2 R \quad \text{(say)}$$

Hence (24) becomes

$$\lambda = \frac{(I + 1)}{k} \sum \left\{ \frac{1}{I + \frac{kr_i}{R}} \right\} \quad (25)$$

$$= \frac{(I + 1)}{k^2} \sum \left\{ \frac{R}{(AR + r_i)} \right\} \quad (26)$$

where $A = I/k$.

The argument proceeds by showing that $\lambda$ cannot have a maximum unless every $r_i$ is at one of the ends of its possible range from 1 to $r$. Since $I$ is fixed, we may neglect the term $(I + 1)/k^2$ and consider the quantity

$$\gamma = \sum \left\{ \frac{R}{AR + r_i} \right\} \quad (27)$$

It may be verified that

$$\frac{\partial \gamma}{\partial r_h} = \sum' \left\{ \frac{r_i}{(AR + r_i)^2} \right\} - \frac{R'}{(AR + r_h)^2}$$

$$\frac{\partial^2 \gamma}{\partial r_h^2} = \frac{2R'(A + 1)}{(AR + r_h)^3} - 2A \sum' \left\{ \frac{r_i}{(AR + r_i)^3} \right\}$$

where a prime denotes summation over all terms except that in $r_h$ .

Hence, at any point at which $\partial \gamma / \partial r_h = 0$, we have

$$\frac{\partial^2 \gamma}{\partial r_h^2} = \frac{2(A + 1)}{(AR + r_h)} \sum' \left\{ \frac{r_i}{(AR + r_i)^2} \right\} - 2A \sum' \left\{ \frac{r_i}{(AR + r_i)^3} \right\}$$

$$= \frac{2(A + 1)}{(AR + r_h)} \sum' \frac{r_i}{(AR + r_i)^2} \left\{ 1 - \frac{A}{(A + 1)} \frac{(AR + r_h)}{(AR + r_i)} \right\}$$

The term inside the curly brackets is easily seen to be positive for every $i$. Hence at any point where $\partial \gamma / \partial r_h = 0$, we have the second derivative positive, so that there is no interior maximum.

To find the maximum value of $\lambda$, let $m$ of the $r_i$ be 1, and the remaining $(k - m)$ be $r$. Then from (26)

$$\lambda = \frac{R(I + 1)}{k^2} \left\{ \frac{m}{AR + 1} + \frac{k - m}{AR + r} \right\}$$

where now

$$R = \sum r_i = m + (k - m)r$$

There is no convenient analytic expression for the maximizing value of $m$, but for given $I$ and $r$ the maximum is easily computed numerically. The results in table 11 show the reciprocal of the maximum, i.e. the lower bound to the relative precision of $\bar{x}$ to $\bar{x}_{sw}$ .

TABLE 11
LOWER LIMITS OF RELATIVE PRECISION OF $\bar{x}$ TO $\bar{x}_{sw}$

| $r$ = largest/ smallest error variance | $I$ = ratio of interaction variance to average error variance | | | |
|---|---|---|---|---|
| | 0 | 1 | 2 | 3 |
| 2 | .89 | .97 | .99 | .99 |
| 3 | .75 | .93 | .97 | .98 |
| 4 | .64 | .90 | .95 | .97 |
| 6 | .49 | .85 | .93 | .95 |
| 8 | .40 | .81 | .90 | .94 |
| 16 | .22 | .74 | .86 | .91 |

If an upper limit of 10 per cent in the loss of precision is regarded as tolerable, table 11 shows that the unweighted mean is satisfactory whenever $I$ exceeds 3, or when $I$ is at least 2 and $r$ is 8 or less, or when $I$ is at least 1 and $r$ is 4 or less. Values of $r$ greater than 16 were not included in the table, on the grounds that such cases would represent a very extreme degree of variation in the $\sigma_i^2$ .

The translation of these results into the working rules given in sections 4 and 5 can be made only approximately, since in practice we do not know the value of $I$. In section 5, the numerator of $F$ is an unbiased estimate of $(\sigma_\mu^2 + \bar{\sigma}^2)$, while the denominator is an unbiased estimate of $\bar{\sigma}^2$. Hence, $F$ can be considered as an estimate of $(1 + I)$, although the estimate may be shown to be positively biased. The rule given in section 5, namely to use $\bar{x}$ in general when $F$ exceeds 4, was chosen because with $F > 4$, $I$ is unlikely to be $< 2$, and from table 11 the unweighted mean suffers little loss for $I = 2$ unless the $x_i$ differ greatly in precision.

Similarly, the $F$-ratio in table 3 of section 4 is an estimate of

$$1 + \frac{I\bar{f'}}{k} \sum \left(\frac{1}{f_i}\right)$$

If the range of values of $f_i$ is not too great, this expression is approximately $(1 + I)$, and leads to the rules given in table 4.

These rules are perhaps biased in favor of the semi-weighted mean, because the figures in table 11 are underestimates of the relative precision of $\bar{x}$ to $\bar{x}_{sw}$ .

7. APPROXIMATE NUMBER OF DEGREES OF FREEDOM IN THE STANDARD ERROR
OF $\bar{x}$

In sections 4 and 5 the formula

$$\frac{s_b}{\sqrt{k}} = \sqrt{\frac{\sum (x_i - \bar{x})^2}{k(k - 1)}}$$

has been recommended for the standard error of $\bar{x}$ when interactions are present. Since the $x_i$ do not have equal variances, this standard error is not distributed in the usual way for a root mean square with $(k - 1)$ degrees of freedom.

The distribution of $s_b^2$ will, however, be approximated by a distribution of the usual type for a mean square. The number of degrees of freedom $n_e$ ascribed to this distribution will be chosen so as to give it the correct variance.

Let

$$\theta_i = \sigma_\mu^2 + \sigma_i^2$$

$$\Theta_1 = \frac{\sum \theta_i}{k} \quad : \quad \Theta_2 = \frac{\sum \theta_i^2}{k}$$

If the $x_i$ are normally and independently distributed about $\mu$, the variance of $s_b^2$ is found by algebra to be

$$V(s_b^2) = 2\{(k - 2)\Theta_2 + \Theta_1^2\}/(k - 1)^2 \tag{28}$$

For the typical distribution of a mean square (i.e. that of a multiple of $\chi^2$ with $n_e$ degrees of freedom),

$$V(s_b^2) = \frac{2\{E(s_b^2)\}^2}{n_e} = \frac{2\Theta_1^2}{n_e} \tag{29}$$

Hence, by equating (28) to (29),

$$n_e = \frac{(k - 1)^2 \Theta_1^2}{(k - 2)\Theta_2 + \Theta_1^2} \tag{30}$$

In practice, we must substitute the sample estimates of $\Theta_1$ and $\Theta_2$ .

8.  THE STANDARD ERROR OF THE WEIGHTED MEAN WHEN THE $n_i$ ARE SMALL

Meier's formula (13) or (14) in section 5 is satisfactory for values of $n_i$ down to 8 or down to 6 when $k$ is small. For very small values of $n_i$ , the value of the factor in curly brackets in (13) and (14) has been estimated by experimental sampling, Cochran and Carroll, (1953). Values taken from this paper appear in table 12.

TABLE 12
VALUES OF $\lambda$ FOR WHICH $\lambda/w$ IS AN ESTIMATE OF $V(\bar{x}_w)$

| | Number of Experiments | | | | | | | | | |
|---|---|---|---|---|---|---|---|---|---|---|
| $\bar{n}$ | 2 | 3 | 4 | 5 | 6 | 8 | 10 | 12 | 15 | 20 |
| 2 | 2.0 | 2.9 | 3.9 | 5.1 | 6.1 | 7.9 | 10.6 | 12.6 | 17.1 | 22.8 |
| 4 | 1.5 | 1.8 | 2.2 | 2.5 | 2.7 | 3.2 | 3.7 | 4.1 | 4.7 | 5.4 |
| 6 | 1.3 | 1.5 | 1.7 | 1.8 | 1.9 | 2.0 | 2.1 | 2.2 | 2.3 | 2.4 |
| 8 | 1.2 | 1.5 | 1.5 | 1.6 | 1.6 | 1.7 | 1.8 | 1.8 | 1.9 | 1.9 |

The use of this formula is illustrated in section 9.

## 9. STANDARD ERROR OF THE PARTIALLY WEIGHTED MEAN

Let $\bar{x}_p$ be the mean of the $p$ experiments which each receive equal weight $\bar{w}_p$, and $\bar{x}_u$ be the weighted mean of the remaining $u$ experiments which receive individual weights. For the estimated variance of $\bar{x}_p$ we take the average observed variance in these experiments, divided by $p$; or in other words,

$$s^2(\bar{x}_p) = \frac{1}{p\bar{w}_p}$$

For the estimated variance of $\bar{x}_u$, we use either Meier's formula or, if the $n_i$ are less than 8, the empirical formula in the previous section,

$$s^2(\bar{x}_u) = \frac{\lambda}{w_u}$$

where $\lambda$ is read from table 12 and $w_u$ is the total weight for these $u$ experiments.

The overall mean is

$$\bar{x}_{pw} = \frac{p\bar{w}_p\bar{x}_p + w_u\bar{x}_u}{w}$$

Hence, if $\bar{w}_p$ and $w_u$ can be regarded as free from error,

$$s(\bar{x}_{pw}) = \frac{\sqrt{(p\bar{w}_p)^2 s^2(\bar{x}_p) + w_u^2 s^2(\bar{x}_u)}}{w}$$

$$= \frac{\sqrt{p\bar{w}_p + \lambda w_u}}{w}$$

as given in section 5.

This argument is non-rigorous in several ways. The variance given for $\bar{x}_p$ is too low, since these $p$ experiments are selected because they appear to be precise. Similarly, the variance for $\bar{x}_u$ is too high. Also, errors in the relative weights, $p\bar{w}_p$ and $w_u$, are ignored. However, the formula does reduce to the appropriate values when $p = k$ and when $p = 0$: in intermediate cases my guess is that it may be slightly too low.

## 10. SUMMARY

This paper discusses methods for combining a number of estimates $x_i$ of some quantity $\mu$, made in different experiments. For the $i$th estimate we have an unbiased estimate $s_i^2$ of its variance, based on $n_i$ degrees of freedom.

It is important to find out whether the $x_i$ agree with one another within the limits of their experimental errors. If they do not, i.e. if interactions are present, the type of overall mean that will be useful for future action requires careful consideration. However, in most cases the problem of estimating the mean of the $x_i$ , at least over some subgroup of the experiments, will remain.

If the experiments are of the same type and the $x_i$ are of equal precision, the best estimate in general is the unweighted mean $\bar{x}$, but its standard error differs according as interactions are present or absent.

The second case considered is that in which the experiments are of different types, but the variance $\sigma_0^2$ per observation is the same in all experiments. The variance of $x_i$ is then of the form $\sigma_0^2/f_i$ . If there are no interactions, the best combined estimate is the weighted mean $\sum f_i x_i / \sum f_i$ . If interactions exist, the choice lies between the unweighted mean and a semi-weighted mean. Recommendations for this choice are given. In the semi-weighted mean, the weights $W_i$ are ideally inversely proportional to

$$\sigma_\mu^2 + \frac{\sigma_0^2}{f_i}$$

The semi-weighted mean reduces to the weighted mean when the interaction variance $\sigma_\mu^2 = 0$, and to the unweighted mean when the interaction variance is large. In practice, sample estimates of the weights are used.

Experiments in which the variance per observation is not constant represent perhaps the most common case in practice. In the absence of interactions, possible estimates are the unweighted mean, weighting inversely as $s_i^2$ or, if the $n_i$ are small, a kind of partial weighting. When interactions are present, the unweighted mean or the semi-weighted mean is appropriate. Working rules are given to aid in the selection of an estimate.

In conclusion, the unweighted mean will probably be satisfactory with many sets of data. The principal value in learning about various types of more complex estimates lies in occasional situations in which the unweighted mean would incur a substantial loss of precision, and also in receiving assurance that the unweighted mean is often entirely adequate.

Some approximations in theory that are needed for the practical recommendations are developed in later sections of the paper.

### REFERENCES

Bancroft, T. A. (1944). On biases in estimation due to the use of preliminary tests of significance. *Ann. Math. Stat., 15,* 190–204.
Bliss, C. I. (1952). *The statistics of bioassay.* Academic Press Inc., New York. p. 576.

Cochran, W. G. (1937). Problems arising in the analysis of a series of similar experiments. *Jour. Roy. Stat. Soc., Supp., 4*, 102–118.

Cochran, W. G. and Carroll, S. P. (1953). A sampling investigation of the efficiency of weighting inversely as the estimated variance. *Biometrics, 9*, 447–459.

Cochran, W. G. and Cox, G. M. (1950). *Experimental designs*. John Wiley and Sons, Inc., New York. Chapter 14.

James, G. S. (1951). The comparison of several groups of observations when the ratios of the population variances are unknown. *Biometrika, 38*, 324–329.

Kempthorne, O. (1952). *The design and analysis of experiments*. John Wiley and Sons, Inc., New York. Chapter 28.

Meier, P. (1953). Variance of a weighted mean. *Biometrics, 9*, 59–73.

Paull, A. E. (1950). On a preliminary test for pooling mean squares in the analysis of variance. *Ann. Math. Stat., 21*, 539–556.

Satterthwaite, F. E. (1946). An approximate distribution of estimates of variance components. *Biom. Bull., 2*, 110.

Welch, B. L. (1951). On the comparison of several mean values—an alternative approach. *Biometrika, 38*, 330–336.

Whitlock, J. H. and Bliss, C. I. (1943). A bioassay technique for anthelmintics. *Jour. Parisitology, 29*, 48–58.

Yates, F. and Cochran, W. G. (1938). The analysis of groups of experiments. *Jour. Agr. Sci., 28*, 556–580.

# SOME METHODS FOR STRENGTHENING
## THE COMMON $\chi^2$ TESTS*

WILLIAM G. COCHRAN

*The Johns Hopkins University*

## 1. INTRODUCTION

Since the $\chi^2$ tests of goodness of fit and of association in contingency tables are presented in many courses on statistical methods for beginners in the subject, it is not surprising that $\chi^2$ has become one of the most commonly-used techniques, even by scientists who profess only a smattering of knowledge of statistics. It is also not surprising that the technique is sometimes misused, e.g. by calculating $\chi^2$ from data that are not frequencies or by errors in counting the number of degrees of freedom. A good catalogue of mistakes of this kind has been given by Lewis and Burke (1).

In this paper I want to discuss two kinds of failure to make the best use of $\chi^2$ tests which I have observed from time to time in reading reports of biological research. The first arises because $\chi^2$ tests, as has often been pointed out, are not directed against any specific alternative to the null hypothesis. In the computation of $\chi^2$, the deviations $(f_i - m_i)$ between observed and expected frequencies are squared, divided by $m_i$ in order to equalize the variances (approximately), and added. No attempt is made to detect any particular pattern of deviations $(f_i - m_i)$ that may hold if the null hypothesis is false. One consequence is that the usual $\chi^2$ tests are often insensitive, and do not indicate significant results when the null hypothesis is actually false. Some forethought about the kind of alternative hypothesis that is likely to hold may lead to alternative tests that are more powerful and appropriate. Further, when the ordinary $\chi^2$ test does give a significant result, it does not direct attention to the way in which the null hypothesis disagrees with the data, although the pattern of deviations may be informative and suggestive for future research. The remedy here is to supplement the ordinary test by additional tests that help to reveal the significant type of deviation.

In this paper a number of methods for strengthening or supplementing the most common uses of the ordinary $\chi^2$ test will be presented and illustrated by numerical examples. The principal devices are as follows:

---

*Work assisted by a contract with the Office of Naval Research, U. S. Navy Department. Dept. of Biostatistics paper no. 278.

---

1. Use of small expectations in computing $\chi^2$.
2. Use of a single degree of freedom, or a group of degrees of freedom, from the total $\chi^2$.
3. Use of alternative tests.

Most of the techniques have been available in the literature for some time: indeed, most of them stem from early editions of Fisher's "Statistical Methods for Research Workers." Research which has clarified the problem of subdividing $\chi^2$ in contingency tables is more recent, and still continues.

In the hope of avoiding some confusion, the symbol $X^2$ will be used for the quantity that we calculate from the sample in a chi-square test. The symbol $\chi^2$ itself will refer to a random variate that follows the distribution in the $\chi^2$ tables, and will sometimes be used in the phrase "the $\chi^2$ test."

## 2. USE OF SMALL EXPECTATIONS IN COMPUTING $X^2$

In order to prove that the quantity $X^2 = \sum (f_i - m_i)^2 / m_i$ is distributed as $\chi^2$ when the null hypothesis is true, it is necessary to postulate that the expectations $m_i$ are large: in fact, the proof is strictly valid only as a limiting result when the $m_i$ tend to infinity. For this reason many writers recommend that the $m_i$ be not less than 5 when applying the test in practice, and that neighboring classes be combined if this requirement is not met in the original data. Some writers recommend a lower limit of 10 for the $m_i$ .

It is my opinion that these recommendations are too conservative, and that their application may on occasion result in a substantial loss of power in the test. I give this as an opinion, because not enough research has been done to make the situation quite clear. However, the exact distribution of $\sum (f_i - m_i)^2 / m_i$ , when the expectations are small, has been worked out in a number of particular cases by Sukhatme (2), Neyman and Pearson (3) and Cochran (4), (5). These results indicate that the $\chi^2$ tables give an adequate approximation to the exact distribution even when some $m_i$ are much lower than 5.

Loss of power from following a rule that $m_i \geq 5$ occurs because this rule tends to require grouping of classes at the tails or extremes of the distribution. These are often the places where the difference between the alternative hypothesis and the null hypothesis stands out most clearly, so that the grouping may cover up the most marked difference between the two hypotheses. Information about the extent of the loss of power is unfortunately very scanty, because the power function of $X^2$ is known only as a limiting result when the $m_i$ are large. The following illustration suggests that the loss can be large.

Suppose that we have a sample of $N = 100$. The null hypothesis is that the data follow a Poisson distribution with mean $m$ known to be 1, when actually the data follow the negative binomial distribution $(q - p)^{-n}$, where $n = 2$, $q = 1.5$, $p = 0.5$. What is the chance of rejecting the null hypothesis at the 5% level of significance?

The expected frequencies of 0, 1, 2, 3, and 4 or more occurrences on the two hypotheses are shown in table 1.

TABLE 1.
EXPECTED FREQUENCIES ON THE NULL AND ALTERNATIVE HYPOTHESIS.

|  | Expected frequencies | | | $\dfrac{(m'_i - m_i)^2}{m_i}$ | |
|---|---|---|---|---|---|
|  |  |  |  | Grouped | |
| $i$ | Poisson $m_i$ | Neg. Bin. $m'_i$ | Ungrouped | $m_i \geq 5$ | $m_i \geq 10$ |
| 0 | 36.79 | 44.44 | 1.590 | 1.590 | 1.590 |
| 1 | 36.79 | 29.63 | 1.393 | 1.393 | 1.393 |
| 2 | 18.39 | 14.82 | 0.693 | 0.693 |  |
| 3 | 6.13 | 6.58 | 0.033 | }1.181 | }0.009 |
| 4+ | 1.90 | 4.53 | 3.640 |  |  |
| Totals $N =$ | 100.00 | 100.00 | $\lambda = 7.349$ | $\lambda = 4.857$ | $\lambda = 2.992$ |

If an $m_i$ as low as 1.90 is allowed, we can use 5 classes in the $\chi^2$ test, with 4 degrees of freedom since the $m_i$ are known. To make all $m_i \geq 5$, we must pool the last two classes and have 3 degrees of freedom; and to make all $m_i \geq 10$, we must pool the last three classes and have 2 degrees of freedom. In order to obtain an approximation to the powers of these three ways of applying the $\chi^2$ test, we shall use the asymptotic result for the power function. This is a non-central $\chi^2$ distribution, with parameter $\lambda$ of non-centrality, where

$$\lambda = \sum \frac{(m'_i - m_i)^2}{m_i}$$

The larger the value of $\lambda$, the higher is the power. The contributions to $\lambda$ from each class are shown in the right-hand columns of table 1. For the ungrouped case, note that the extreme class 4+ is much the largest contributor to $\lambda$. Grouping the last two classes considerably reduces this contribution, while grouping the last three classes diminishes it almost to zero. The approximate probabilities of rejecting the null hypothesis may be read from Fix's tables (6) of the non-central $\chi^2$ distribution. These probabilities are 0.56 for the ungrouped case, 0.43 for $m_i \geq 5$ and 0.32 for $m_i \geq 10$. The loss in sensitivity from grouping is evident. Perhaps a more revealing comparison is to compute from Fix's tables the sizes of sample $N$ that would be needed in the two grouped cases in order to bring the probabilities of rejection up to 0.56, the value for the ungrouped case. The results are $N = 136$ when the last two classes are grouped and $N = 191$ when the last three classes are grouped, as against $N = 100$ in the ungrouped case.

This example is only suggestive, and probably favors the ungrouped case slightly, because the computations are based on large-sample results. However, the losses in power from grouping, measured in terms of equivalent sample sizes, are impressive.

### 2.1 Recommendations about minimum expectations

Elsewhere, (7), I have given recommendations about the minimum expectation to be used in $\chi^2$ tests. These working rules may be summarized, in slightly revised form, as follows:

(a) *Goodness of fit tests of unimodal distributions* (*such as the normal or Poisson*). Here the expectations will be small only at one or both tails. Group so that the minimum expectation at each tail is at least 1.

(b) *The 2 $\times$ 2 table.* Use Fisher's exact test (i) if the total $N$ of the table $< 20$, (ii) if $20 < N < 40$ and the smallest expectation is less than 5. Mainland (8) has given useful tables of the exact test for these cases. If $N > 40$ use $X^2$, corrected for continuity.

(c) *Contingency tables with more than 1 d.f.* If relatively few expectations are less than 5 (say in 1 cell out of 5 or more, or 2 cells out of 10 or more), a minimum expectation of 1 is allowable in computing $X^2$.

Contingency tables with most or all expectations below 5 are harder to prescribe for. With very small expectations, the exact distribution of $X^2$ can be calculated without too much labor. Computing methods have been given by Freeman and Halton (9). If $X^2$ has less than 30 degrees of freedom and the minimum expectation is 2 or more, use of the ordinary $\chi^2$ tables is usually adequate. If $X^2$ has more than 30 degrees of freedom, it tends to become normally distributed, but when the expectations are low, the mean and variance are different from those of the tabular $\chi^2$. Expressions for the exact mean and variance have been given by Haldane (10). Compute the exact mean and variance, and treat $X^2$ as normally distributed with that mean and variance.[*]

---

[*]In a previous paper (7) I recommended this procedure only when the degrees of freedom in $X^2$ exceed 60. Some unpublished research suggests that 30 is a better division point.

Further research will presumably change these recommendations, but I do not believe that the recommendations will lead users far astray.

Succeeding sections will deal with some of the common applications of the $\chi^2$ test to goodness of fit problems and to contingency tables. The alternative or supplementary tests to be presented are those that seem most often useful, but they by no means exhaust the possibilities. The important guiding rule is to think about the type of alternative that is likely to hold if the null hypothesis is false, and to select a test that will be sensitive to this kind of alternative.

### 3. THE GOODNESS OF FIT TEST OF THE POISSON DISTRIBUTION

This is the test already referred to in table 1, except that in practice the parameter $m$ must usually be estimated from the data. We have

$$X^2 = \sum \frac{(f_i - m_i)^2}{m_i}$$

where $f_i$ is the observed and $m_i$ the expected frequency of an observation equal to $i(i = 0, 1, 2 \cdots)$.

If the data do not follow the Poisson distribution, two common alternatives are as follows:

(1) The data follow some other single distribution, such as the negative binomial or one of the "contagious" distributions. Another way of discribing this case is to postulate that the individual observations, say $x_i$ , follow Poisson distributions, but with means that vary from observation to observation so as to follow some fixed frequency distribution. For instance, as has been shown, the negative binomial distribution can be produced by assuming that these means follow a Pearson type III distribution.

(2) The means of the observations $x_i$ follow some systematic pattern. With data gathered over several days, the means might be constant within a day, but vary from day to day, or they might exhibit a slow declining trend.

### 3.1 *Test of the variance*

In both cases (1) and (2), a comparison between the observed variance of the observations $x_i$ and the variance predicted from Poisson theory will frequently be more sensitive than the goodness of fit test. The variance test is made by calculating

$$X_?^2 = \sum_{i=1}^{N} \frac{(x_i - \bar{x})^2}{\bar{x}} ,$$

or if the calculation is made from the frequency distribution of the $x_i$ ,

$$X_{\bullet}^2 = \sum_i \frac{f_i (i - m)^2}{m} ,$$

where $m = \bar{x}$ is the sample mean. The quantity $X_{\bullet}^2$ is referred to the $\chi^2$ tables with $(N - 1)$ degrees of freedom. The variance test is an old one: it was introduced by Fisher in the first edition of "Statistical Methods for Research Workers" under the heading "Small samples of the Poisson series", because it can be used when the sample is too small to permit use of the goodness of fit test.

The increased power of the variance test over the goodness of fit test was strikingly shown in some sampling experiments conducted by Berkson (20), in a situation in which the data followed a binomial distribution, which has a *smaller* variance than the Poisson. A rough calculation for the example in table 1 gives 0.76 as the probability of rejecting the Poisson hypothesis by the variance test, as compared with 0.56 for the best of the goodness of fit tests. In practice, I have often found the variance test significant when the goodness of fit test was not. Berkson (21) presents some data that illustrate this point.

### 3.2 *Subdivision of degrees of freedom in the variance test*

If it is suspected that the means of the observations $x_i$ may change in some *systematic* manner, as in case (2) above, more specific tests of significance can be obtained by selecting certain degrees of freedom from $X_{\bullet}^2$ . The ordinary rules of the analysis of variance are followed in subdividing $\sum (x_i - \bar{x})^2$, and the denominator $\bar{x}$ is used to convert the partial sum of squares approximately to $\chi^2$. A few examples will be given. (The formulas presented are intended to make clear the structure of the $\chi^2$ components, but they are not always the speediest formulas to use in computing the components.)

### 3.3 *Test for a change in level*

To test for an abrupt change in the mean of the distribution, occurring after the first $N_1$ observations, we take

$$X^2 = \frac{N_1 N_2}{(N_1 + N_2)} \frac{(\bar{x}_1 - \bar{x}_2)^2}{\bar{x}} \qquad \text{(1 d.f.)}$$

where $N_2 = (N - N_1)$, $\bar{x}_1$ , and $\bar{x}_2$ are the sample means in the two parts of the series and $\bar{x}$ is the overall mean. This test may be extended to compare a group of means.

### 3.4 *Test for a linear trend*

We may anticipate that the means will follow a linear regression on some variate $z_i$ (frequently a time-variable). In this case

$$X^2 = \frac{[\sum (x_i - \bar{x})(z_i - \bar{z})]^2}{\bar{x} \sum (z_i - \bar{z})^2} \qquad \text{(1 d.f.)}$$

### 3.5 *Detecting the point at which a change in level occurs*

This problem has been illustrated by Lancaster (11) in an experiment in which increasing concentrations of disinfectant are poured on a series of suspensions of a bacterial culture, each suspension having a constant amount of bacteria. The observations $x_i$ represent numbers of colonies per plate. The problem is to find the value of $j$ for which the disinfectant is strong enough to begin reducing the number of colonies. To this end we compare each observation with the mean of all previous observations, looking for the first value of $j$ at which $X^2$ becomes large owing to a drop in the number of colonies. We thus obtain a set of independent single degrees of freedom:

$$X_1^2 = \frac{(x_1 - x_2)^2}{2\bar{x}} ; \qquad X_2^2 = \frac{(x_1 + x_2 - 2x_3)^2}{6\bar{x}} ;$$

and in general

$$X_r^2 = \frac{(x_1 + x_2 + \cdots + x_r - rx_{r+1})^2}{r(r+1)\bar{x}}$$

*Note.* The above set of $X^2$ values will add up to the total variance $X_\bullet^2$, but as Lancaster has pointed out, there are other natural ways of subdividing $X_\bullet^2$ in which the separate $X^2$ do not add up to $X_\bullet^2$. When comparing $x_1$ with $x_2$, we might decide to disregard the remainder of the observations and compute $X_1^2$ as

$$X_1'^2 = \frac{(x_1 - x_2)^2}{2\bar{x}_{12}}$$

where $\bar{x}_{12}$ is the mean of $x_1$ and $x_2$. A set of successive $X^2$ values computed in this way will not add up to $X_\bullet^2$, because the denominator changes from term to term, whereas $X_\bullet^2$ carries the denominator $\bar{x}$.

The practice of computing $X^2$ components by using only those parts of the data that are immediately involved has something in its favor (despite the non-additive feature), at least if the total $X_\bullet^2$ has already been shown to be significant. For in that event we have already concluded that the data as a whole do not follow a single Poisson distribution, and the overall mean $\bar{x}$ is of dubious validity as an estimate of the Poisson variance for a part of the data. On the other hand, if the total $X_r^2$ is not significant, but we suspect that some component is, the additive partition is convenient and should be satisfactory in a preliminary examination.

### 3.6 Single degrees of freedom in the goodness of fit test

In the goodness of fit comparison of the observed frequency $f_i$ with the expected frequency $m_i$ of $i$ occurrences, we·can test any linear function of the deviations

$$L = \sum g_i (f_i - m_i)$$

where the $g_i$ are numbers chosen in advance.

In the case in which the mean of the Poisson, and hence the $m_i$, are given in advance, the variance of $L$ is

$$V(L) = \sum g_i^2 m_i - \frac{(\sum g_i m_i)^2}{N} \tag{1}$$

In the more common situation in which the Poisson mean $m$ is estimated from the data,

$$V(L) = \sum g_i^2 m_i - \frac{(\sum g_i m_i)^2}{N} - \frac{[\sum g_i m_i (i - m)]^2}{Nm} \tag{2}$$

where the sums are over the values $0, 1, 2, \cdots$ of $i$. In either case

$$X_1^2 = \frac{L^2}{V(L)} \tag{3}$$

is approximately distributed as $\chi^2$ with 1 d.f. I plan to publish justification for formula (2), which appears to be new. By appropriate choice of the $g_i$, a test specific for a given pattern of deviations is obtained.

In particular, to test any single deviation $(f_i - m_i)$ when $m$ is estimated from the data, we take

$$L = (f_i - m_i) : V(L) = m - \frac{m_i^2}{N} \left\{ 1 + \frac{(i - m)^2}{m} \right\} \tag{4}$$

As an illustration, the data in table 2 are for a sample which gave a satisfactory fit to a Poisson distribution. However, in copying down the frequencies before fitting the Poisson, the frequency of 3 occurrences was erroneously written as 52 instead of 32.

TABLE 2.

GOODNESS OF FIT TEST FOR A SAMPLE WITH A GROSS ERROR.

| $i$ | $f_i$ | $m_i$ | Contribution to $X^2$ |
|------|-------|--------|-------------------------|
| 0    | 52    | 47.65  | 0.40                    |
| 1    | 67    | 77.04  | 1.31                    |
| 2    | 58    | 62.28  | 0.29                    |
| 3    | 52    | 33.56  | 10.13                   |
| 4    | 7     | 13.56  | 3.17                    |
| 5    | 3     | 4.39   | 0.44                    |
| 6+   | 1     | 1.52   | 0.18                    |
| Total | 240  | 240.00 | 15.92                   |

The value of $m$ is 1.6167. The total $X^2$, 15.92 with 5 d.f., is significant at the 1 per cent level. The large contribution to $X^2$ from $i = 3$ excites notice. In order to test this deviation, we take

$$L = f_3 - m_3 = 52 - 33.56 = 18.44$$

$$V(L) = 33.56 - \frac{(33.56)^2}{240}\left\{1 + \frac{(3 - 1.6167)^2}{1.6167}\right\} = 23.31$$

$$X_1^2 = \frac{(18.44)^2}{23.31} = 14.59$$

This comparison accounts for the major part of the total $X^2$. It must be pointed out, however, that the $X_1^2$ test applies only to a deviation picked out before seeing the data. Thus the test can be applied validly, for $i = 0$, say, if we suspect beforehand that the data follow the Poisson distribution for $i \geq 1$; but that the frequency for $i = 0$ may be anomalous. If the test is applied, as here, to a deviation selected because it looks abnormally large, the significance $P$ obtained is too low. I do not have an expression for the correct significance probability when we select the largest deviation. It appears intuitively that the correct probability lies between $P$ and $kP$, where $k$ is the number of classes in the goodness of fit test. Since $P$ is about 0.00013 and $k$ is 7, the upper limit is .00091, which is still highly significant statistically.

#### 4. THE GOODNESS OF FIT TEST OF THE BINOMIAL DISTRIBUTION

For the binomial distribution, there is a series of tests analogous to those given for the Poisson distribution. A typical observation consists of the number of successes $x_j$ out of $n$ independent trials. We have a sample of $N$ such observations. The ordinary goodness of fit test is made by recording the frequency $f_i$ with which $i$ successes occur in the sample, fitting a binomial to these frequencies, and calculating $X^2$ as

$$X^2 = \sum \frac{(f_i - m_i)^2}{m_i},$$

where $m_i$ is the corresponding expected frequency.

As in the Poisson case, departures from the binomial frequently occur either because

(1) the data follow a different frequency distribution, usually with a larger variance (or the probabilities of success $p_j$ show some kind of random variation from observation to observation).

(2) the probabilities $p_j$ are affected by a systematic source of variation.

In both cases, a comparison of the observed and expected variances is likely to be more sensitive than the goodness of fit test.

### 4.1 Test of the variance

The test criteria, all distributed approximately as $\chi^2$, are as follows.

*n constant, p given in advance*

$$X_v^2 = \frac{\sum_i (x_j - np)^2}{npq} = \frac{\sum_i f_i(i - np)^2}{npq}, \qquad N \text{ d.f.}$$

*n constant, p estimated*

$$X_v^2 = \frac{\sum_i (x_j - \bar{x})^2}{n\hat{p}\hat{q}} = \frac{\sum_i f_i(i - n\hat{p})^2}{n\hat{p}\hat{q}}, \qquad (N - 1) \text{ d.f.}$$

where $n\hat{p} = \bar{x}$, and $\hat{q} = 1 - \hat{p}$.

*n varying, p given in advance*

In this case we cannot make a simple goodness of fit test (unless the sample is large enough to be divided into batches, each with $n$ constant, so that the test can be made separately for each batch). If $p_j = x_j/n_j$ is the observed proportion of successes in the $j$th member of the sample,

$$X_v^2 = \frac{\sum n_j(p_j - p)^2}{pq} \qquad N \text{ d.f.}$$

*n varying, p estimated*

$$X_v^2 = \frac{\sum n_i(p_i - \hat{p})^2}{\hat{p}\hat{q}} = \frac{\sum \dfrac{x_i^2}{n_i} - \dfrac{(\sum x_i)^2}{(\sum n_i)}}{\hat{p}\hat{q}}, \qquad (N - 1) \text{ d.f.}$$

where $\hat{p} = (\sum x_i)/(\sum n_i)$ is the estimate of $p$ from the total sample.

There is another way of deriving the variance test. Arrange the data in a $2 \times N$ contingency table, as follows.

| | | | | |
|---|---|---|---|---|
| Successes | $x_1$ | $x_2$ | | $x_N$ |
| Failures | $n_1 - x_1$ | $n_2 - x_2$ | | $n_N - x_N$ |
| Total | $n_1$ | $n_2$ | | $n_N$ |

Then the $X^2$ that is used to test for association in this $2 \times N$ table may be shown to be identical with $X_v^2$.

If $N$ exceeds 30 and the expectations are small, it was pointed out in section 2 that $X^2$ in contingency tables tends to a normal distribution with a mean and variance somewhat different from those of $\chi^2$. Use of Haldane's correct expressions for the mean and variance of $X^2$ was recommended in this case. The same procedure is recommended in the variance test if $N$ exceeds 30 and the average $n_i$ is less than 10. Haldane's expressions are rather complicated when the $n_i$ vary. In the fairly common situation in which $n$ is constant and $p$ is estimated from the data, the following results suffice in almost all cases.

$$E(X_v^2) \doteq (N - 1)\left(1 + \frac{1}{Nn}\right) \qquad (5)$$

$$V(X_v^2) \doteq 2(N - 1)\left(\frac{n - 1}{n}\right)\left(1 - \frac{1 - 7\hat{p}\hat{q}}{Nn\hat{p}\hat{q}}\right) \qquad (6)$$

The important correction term is that in $(n - 1)/n$ in the variance: the terms in $1/Nn$ are usually small. These results will be used in the numerical example which follows.

The data in table 3, taken from a previous paper, (12), illustrate the application of the goodness of fit test and the variance test to the same observations. The original data, due to Dr. J. G. Bald, consisted of 1440 tomato plants in a field having 24 rows with 60 plants in a row. For each plant it was recorded whether the plant was healthy or attached by spotted wilt as of a given date. As one method of examining whether the distribution of diseased plants was random over the field, the plants were divided into 160 groups of 9, each group consisting of 3 plants $\times$ 3 rows. Thus $N = 160, n = 9$. The choice of $n = 9$ was of course arbitrary, and I do not know what would have been the best choice. The obvious alternative to a binomial distribution is that the values of $p_i$ vary from one group of 9 to another, indicating a patchiness in distribution.

TABLE 3.
NUMBERS OF DISEASED PLANTS IN GROUPS OF 9 PLANTS.

| $i$ | $f_i$ | $m_i$ | Contr. to $X^2$ |
|---|---|---|---|
| 0 | 36 | 26.45 | 3.45 |
| 1 | 48 | 52.70 | 0.42 |
| 2 | 38 | 46.67 | 1.61 |
| 3 | 23 | 24.11 | 0.05 |
| 4 | 10 | 8.00 | 0.50 |
| 5 | 3 | | |
| 6 | 1 | 2.05 | 4.25 |
| 7 | 1 | | |
| 8 | 0 | | |
| N = 160 | | 159.98 | 10.28 |

Allowing a minimum expectation of 1, we must pool the last 4 classes in table 3. The value of $X^2$ is 10.28, with 4 d.f., since $p$ is estimated; the significance $P$ is 0.036. (If we pooled the last 5 classes in order to have a minimum expectation of 5, we would obtain a significance $P$ of 0.046.)

For the variance test, we compute $X_v^2$ from the observed frequency distribution in table 3. We have

$$\sum if_i = 261 = Nn\hat{p}, \quad \text{so that } \hat{p} = 0.18125.$$

$$X_v^2 = \frac{\sum i^2 f_i - (\sum if_i)^2/N}{n\hat{p}\hat{q}} = \frac{727 - (261)^2/160}{9(0.18125)(0.81875)}$$

$$= 225.55, \text{ with 159 d.f.}$$

Since $N = 160$ and $n = 9$, we use the normal approximation to the distribution of $X_v^2$, based on the correct mean and variance. In expressions (5) and (6), the terms in $1/Nn$ are negligible ($Nn = 1440$). Hence we take

$$E(X_v^2) = 159 : V(X_v^2) = 2(159)(8)/9 = 282.66$$

The approximate normal deviate is

$$\frac{225.55 - 159}{\sqrt{282.66}} = 3.96$$

This has a significance $P$ less than 0.0001, much lower than that obtained from the goodness of fit tests.

Subdivision of the sum of squares for $X^2_*$, which may be useful in testing for systematic variation of the $p_i$, will be discussed in section 6, which deals with $2 \times N$ contingency tables.

### 4.2 *Single degrees of freedom in the goodness of fit test*

Let

$$L = \sum g_i(f_i - m_i)$$

be a specified linear function of the deviations of observed from expected frequencies. If $p$ is given, formula (1) in section 3.6 holds for the variance of $L$. If $p$ is estimated from the data,

$$V(L) = \sum g_i^2 m_i - \frac{(\sum g_i m_i)^2}{N} - \frac{[\sum g_i m_i (i - n\hat{p})]^2}{N n\hat{p}\hat{q}} \tag{7}$$

For a single deviation, $(f_i - m_i)$, selected in advance,

$$V(L) = m_i - \frac{m_i^2}{N}\left\{1 + \frac{(i - n\hat{p})^2}{n\hat{p}}\right\} \tag{8}$$

Then $L^2/V(L)$ is approximately distributed as $\chi^2$ with 1 d.f.

### 5. THE GOODNESS OF FIT TEST OF THE NORMAL DISTRIBUTION

When the normal distribution is fitted to a body of data, both the mean and the variance are estimated from the sample: consequently, no variance test is possible. However, the variance test is just an application of the general procedure in which we compare the lowest moments (or cumulants) in which the sample can differ from the theoretical distribution that is being fitted. In this sense, the analogue of the variance test is the test for skewness (as given e.g. in Fisher's "Statistical Methods for Research Workers," §14), which will often detect a departure from normality that escapes the goodness of fit test. The test for kurtosis is also useful in this connection.

As with the Poisson and binomial distributions, a specified linear function $L$ of the deviations in the goodness of fit test can be scrutinized. The formula for the variance of $L$ is somewhat more complicated, because the observed frequencies are subject to 3 constraints.   If

$$L = \sum g_i(f_i - m_i)$$

then

$$V(L) = \sum g_i^2 m_i - \frac{\left(\sum g_i m_i\right)^2}{N} - \frac{\left(\sum g_i d_i m_i\right)^2}{Ns^2} - \frac{\left[\sum g_i m_i(d_i^2 - s^2)\right]^2}{2Ns^4}$$

where

$$d_i = \text{(midpoint of } i\text{th class)} - \text{(sample mean)}$$

$$s^2 = \text{sample estimate of variance}$$

In order to apply this test, construct 3 additional columns containing the quantities $g_i m_i$, $d_i$ and $(d_i^2 - s^2)$, respectively.   $V(L)$ is then easily computed.

For testing a *single* deviation, $V(L)$ simplifies to

$$V(L) = m_i - \frac{3m_i^2}{2N} - \frac{m_i^2 d_i^4}{2Ns^4}$$

The series of supplementary tests described above for the Poisson, binomial and normal distributions can be extended to other distributions.   In particular, variance and skewness tests for the negative binomial distribution have been developed by Anscombe (18) and further illustrated by Bliss (19).

### 6. SUBDIVISION OF DEGREES OF FREEDOM IN THE 2 x $N$ CONTINGENCY TABLE

This section describes some useful ways in which the total $X^2$ for a $2 \times N$ contingency table may be subdivided.   The notation, which continues that already used for the binomial distribution, is as follows.

|  | Number of | | | Proportion of Successes |
|---|---|---|---|---|
|  | Successes | Failures | Total |  |
|  | $x_1$ | $n_1 - x_1$ | $n_1$ | $p_1 = x_1/n_1$ |
|  | $x_2$ | $n_2 - x_2$ | $n_2$ | $p_2 = x_2/n_2$ |
|  | $x_N$ | $n_N - x_N$ | $n_N$ | $p_N = x_N/n_N$ |
| Totals | $T_x$ | $T - T_x$ | $T$ | $\hat{p} = T_x/T$ |

For many purposes, a formula due to Brandt and Snedecor is useful in interpreting the total $X^2$. The formula is:

$$X^2 = \frac{\sum_{i=1}^{N} n_i (p_i - \hat{p})^2}{\hat{p}\hat{q}} \tag{9}$$

Thus the total $X^2$ is seen to be a weighted sum of squares of the deviations of the individual proportions of success $p_i$ from their mean, with weights $n_i / \hat{p}\hat{q}$. Consequently, if we subdivide this weighted sum of squares into a set of independent components by the rules of the analysis of variance, we obtain a corresponding subdivision of $X^2$ into independent components.

### 6.1 *Test for a change in the level of p*

In order to test whether the value of $p$ is different in the first $N_1$ rows from that in the subsequent $N_2$ rows ($N = N_1 + N_2$), we may subdivide $X^2$ into the following 3 components.

|                                                        | d.f.        |
|--------------------------------------------------------|-------------|
| Difference between $p$'s in first $N_1$ and last $N_2$ rows | 1           |
| Variation among $p_i$ within the first $N_1$ rows      | $(N_1 - 1)$ |
| Variation among $p_i$ within the last $N_2$ rows       | $(N_2 - 1)$ |

For the following example I am indebted to Dr. Douglas P. Murphy. A group of women known to have cancer of the uterus and a corresponding 'control' group (primarily from a dental clinic and several women's clubs) were selected. From each of a defined set of relatives of the selected person (the proband), data were secured about the presence of cancer. A higher proportion of cancer cases among relatives of the cancer proband would indicate some kind of familial aggregation of the disease, perhaps of genetic origin (13).

In one table, data were presented separately for those relatives who were of the same generation as the proband (e.g. sister) and for those relatives who were one generation earlier (e.g. mother). Some breakdown of this kind is advisable, because cancer attacks mainly in middle or old age, and the 'cancer' and 'control' groups might be found to differ in the proportions of young relatives which they contained. The data, which are a small part of a much more intensive investigation, appear in table 4.

There are several ways of computing the total $X^2$ in a $2 \times N$ table. One form of the Brandt-Snedecor formula is

$$X^2 = \frac{\sum x_i p_i - \hat{p} T_z}{\hat{p}\hat{q}} \tag{10}$$

TABLE 4.
CANCER AMONG RELATIVES OF 'CANCER' AND 'CONTROL' PROBANDS

| | Proband | No. of relatives | | | Proportion with cancer |
| | | With $x_j$ | Without | Total $n_j$ | $p_j$ |
|---|---|---|---|---|---|
| Earlier Generation | Cancer | 86 | 814 | 900 | 0.095556 |
| | Control | 117 | 1038 | 1155 | 0.101299 |
| Same Generation | Cancer | 49 | 1475 | 1524 | 0.032152 |
| | Control | 61 | 1580 | 1641 | 0.037172 |
| | Total | 313 | 4907 | 5220 | $\hat{p} = 0.059962$ $\hat{q} = 0.940038$ $\hat{p}\hat{q} = 0.056367$ |

This expression is useful when there is some question of a systematic variation in the $p_i$ , because we will want to compute the $p_i$ in order to have a look at them. When the $p_i$ have been computed, the numerator of $X^2$ can be obtained from formula (10) directly on the computing machine, without any intermediate writing down. The only disadvantage is that a substantial number of decimal places must be retained in the $p_i$ .

If this method is to be used, first compute $\hat{p}$, $\hat{q}$ and the product $\hat{p}\hat{q}$ (bottom right of table 4). Since $\hat{p}\hat{q}$ is about 1/20, the numerator of $X^2$ must be correct to at least 3 decimal places if we want $X^2$ correct to 2 decimal places. Further, since $T_x$ is 313, we should have 6 decimal places in the $p_i$ . (The symbol $x_j$ should be assigned to the column with the *smaller* numbers: this makes the computations lighter and necessitates fewer decimals in the $p_i$).

From inspection of the $p_i$ in table 4, a large difference in cancer rates between the two generations is evident. Within each generation, the differences in rates between the cancer and control groups appear tiny. To illustrate the methods, the total $X^2$ will be partitioned into the 3 relevant components. All that is necessary is to subdivide the sum of squares in the numerator, and then divide each component by $\hat{p}\hat{q}$. For the numerator of the total $X^2$, we have from formula (10),

$$\text{Total } S.S. = (86)(0.095556) + \cdots + (61)(0.037172) - (313)(0.059962)$$
$$= 5.1447$$

For the comparison of the two generations, we form the auxiliary $2 \times 2$ table.

|  | With $x_j$ | Without | Total $n_j$ | Proportion $p_j$ |
|---|---|---|---|---|
| Earlier generation | 203 | 1852 | 2055 | 0.098783 |
| Same generation | 110 | 3055 | 3165 | 0.034755 |
| Total | 313 | 4907 | 5220 | 0.055962 |

The same formula can be used for this table.

$$\text{Generations } SS = (203)(0.098783) + (110)(0.034755) - $$
$$(313)(0.055962)$$
$$= 5.1079.$$

For the comparisons between cancer and control groups within generations, we have

$$\text{First generation } SS = (86)(0.095556) + (117)(0.101299) - $$
$$(203)(0.098783) = 0.0169.$$

The second generation $SS$, obtained similarly, gives 0.0199. In table 5 the results are summarized and converted to $X^2$ values on division by $\hat{p}\hat{q}$.

TABLE 5.
SUBDIVISION OF $X^2$ INTO COMPONENTS (2 x 4 TABLE)

| Component | d.f. | S.S. | $X^2$ | From 2 x 2 tables |
|---|---|---|---|---|
| First vs. later generation | 1 | 5.1079 | 90.62 | 90.62 |
| Cancer vs. Control: first gen. | 1 | 0.0169 | 0.30 | 0.19 |
| Cancer vs. Control: later gen. | 1 | 0.0199 | 0.35 | 0.59 |
| Total | 3 | 5.1447 | 91.27 | 91.40 |

There is no indication of any difference in cancer rates between the cancer and control groups within either generation. In order to complete this analysis, we should make a combined test of Cancer vs. Control from the two generations. Methods for making tests of this kind are discussed in section 8.

As mentioned previously in connection with the corresponding subdivision for the Poisson distribution, separation of $X^2$ into *additive* components is convenient for a preliminary examination of the data. But if differences in $p$ are found between groups of rows, the $X^2$ values for comparisons made *within* groups need to be recomputed. This is clear in the present example. The additive method requires the assumption that the estimated variance of any $p_i$ is $\hat{p}\hat{q}/n_i$ . However, the huge $X^2$ value of 90.62 between generations shows that the combined $\hat{p}$ cannot be regarded as a valid estimate of the proportion of cancer cases within either of the individual generations.

The procedure is to recompute the two 'within-generation' $X^2$, each from its own $2 \times 2$ table. These values, which no longer add to the original total $X^2$, are given in the right hand column of table 5. The interpretation of the data is not altered. The $X^2$ values computed from individual parts of the table seldom differ greatly from the additive $X^2$, but they can do so in certain circumstances and are worth looking at, as a precaution, in analyses of this type.

In this example the rows were divided into 2 groups. The same methods may be applied to test the variation in $p$ among any number of groups, and also within each group. To obtain an additive separation, we subdivide the numerator of $X^2$ as indicated in the example. Alternatively, for a non-additive separation, we can form an auxiliary table in which each row is a group total and obtain the $X^2$ for this table, and a further $X^2$ for each group considered by itself.

### 6.2 *Test for a linear regression of p.*

In some contingency tables we may expect that the $p_i$ will bear a linear relation to a variate $z_i$ that is defined for each of the $N$ rows of the table. In others, where the rows fall into a natural order, it is not unreasonable to assign scores $z_i$ to the rows, in an attempt to convert the ordering into a continuous scale, with which the $p_i$ may show a linear relation. Since $p_i$ is assigned a weight $n_i/\hat{p}\hat{q}$, the regression coefficient $b$ of $p_i$ on $z_i$ is obtained by the standard formula for weighted regressions:

$$b = \frac{\sum n_i(p_i - \hat{p})(z_i - \bar{z}_w)}{\sum n_i(z_i - \bar{z}_w)^2} \tag{11}$$

where $\bar{z}_w$ is the weighted mean of the $z_i$ .

For computing purposes, the numerator and denominator of $b$ are conveniently expressed as follows (note that $n_i p_i = x_i$):

$$\text{Num.} = \sum x_i z_i - \frac{T_x(\sum n_i z_i)}{T} \tag{12}$$

$$\text{Den.} = \sum n_i z_i^2 - \frac{(\sum n_i z_i)^2}{T} \tag{13}$$

The $X^2$ for regression, with 1 d.f., is

$$X^2 = \frac{(\text{Num.})^2}{\hat{p}\hat{q}(\text{Den.})} \tag{14}$$

As an illustration, the data in table 6, for which I am indebted to the Leonard Wood Memorial, (American Leprosy Foundation) are taken from an experiment on the use of drugs (sulfones and streptomycin) in the treatment of leprosy. The rows denote the *change* in the overall clinical condition of the patient during 48 weeks of treatment: the columns indicate the degree of infiltration (a measure of a certain type of skin damage) present at the beginning of the experiment. The question of interest is whether patients with much initial infiltration progressed differently from those with little infiltration. Patients did not all receive the same drugs, but since no difference in the effects of drugs could be detected, it was thought that the data for different drugs could be combined for this analysis.

TABLE 6.

196 PATIENTS CLASSIFIED ACCORDING TO CHANGE IN CONDITION AND
DEGREE OF INFILTRATION

| Clinical change | | Score $z_j$ | Degree of infiltration | | Total $n_j$ | $p_j = x_j/n_j$ (in %) | $n_j z_j$ |
|---|---|---|---|---|---|---|---|
| | | | $0-7$ | $8-15$ $x_j$ | | | |
| Im-<br>prove-<br>ment | Marked | 3 | 11 | 7 | 18 | 39 | 54 |
| | Moderate | 2 | 27 | 15 | 42 | 36 | 84 |
| | Slight | 1 | 42 | 16 | 58 | 28 | 58 |
| | Stationary | 0 | 53 | 13 | 66 | 20 | 0 |
| | Worse | $-1$ | 11 | 1 | 12 | 8 | $-12$ |
| Total | | | 144 | 52 $T_x$ | 196 $T$ | 0.26531 $\hat{p}$ | 184 |

The total $X^2$ is 6.88, with 4 d.f. ($P$ about 0.16). However, the $p_j$ (the proportions of patients with severe infiltration) decline steadily from 39% in the "markedly improved" class to 8% in the "worse" class. This suggests that a regression of the $p_j$ on the clinical change might furnish a more sensitive test.

The data are typical of many tables in that the rows (clinical changes) are ordered. In order to compute a regression, this ordering must be replaced by a numerical scale. I have supposed that scores 3, 2, 1,

0, $-1$, as shown in table 6, have been assigned to the five classes of clinical change. Such scores are to some extent subjective and arbitrary, and some scientists may feel that the assignment of scores is slightly unscrupulous, or at least they are uncomfortable about it. Actually, any set of scores gives a *valid* test, provided that they are constructed without consulting the results of the experiment. If the set of scores is poor, in that it badly distorts a numerical scale that really does underlie the ordered classification, the test will not be sensitive. The scores should therefore embody the best insight available about the way in which the classification was constructed and used. In the present example, I considered an alternative set 4, 2, 1, 0, $-2$, on the grounds that the doctor seemed to deliberate very carefully before assigning a patient to the "markedly improved" or to the "worse" class, but I decided that the presumption in favor of this scale was not strong enough.

To compute the regression $X^2$, the only supplementary column needed is that of the products $n_i z_i$ , shown on the right of table 6. From equations (12) and (13) for $b$,

$$\text{Num.} = \sum x_i z_i - \frac{T_x(\sum n_i z_i)}{T}$$

$$= (7)(3) + (15)(2) + \cdots + (1)(-1) - \frac{(52)(184)}{196}$$

$$= 17.1837 \tag{15}$$

$$\text{Den.} = \sum n_i z_i^2 - \frac{(\sum n_i z_i)^2}{T}$$

$$= (54)(3) + (84)(2) + \cdots + (-12)(-1) - \frac{(184)^2}{196}$$

$$= 227.2653 \tag{16}$$

$$X^2 \text{ for regression} = \frac{(17.1837)^2}{(227.2653)\hat{p}\hat{q}} = 6.666 \tag{1 d.f.}$$

This is significant at the 1% level. The total $X^2$ has now been subdivided as follows.

| | d.f. | $X^2$ |
|---|---|---|
| Regression of $p_i$ on $z_i$ | 1 | 6.67 |
| Deviations from regression | 3 | 0.21 |
| Total | 4 | 6.88 |

6.3 *Comparison of mean scores.*

There is another way of looking at the relation between degree of infiltration and clinical progress. We might ask whether the degree of improvement is, on the average, better for patients with severe infiltration than for patients with less infiltration. In many applications, including the present one, this is a more natural way of posing the scientific question than by asking whether the proportion of patients with severe infiltration changes from class to class.

Yates (14) has shown how to compute $X^2$, with 1 d.f., for comparing the mean scores, and has proved, in a more general case, that this $X^2$ is identical with the regression $X^2$ which we have just considered. In the 'mean score' approach, it is best to think of the two groups of patients as representing two independent samples. Each patient has been assigned a variate or score $z_i$ , and the $x_i$ now represent the frequencies with which these scores occur in one of the samples. The mean score for a sample is

$$\bar{z} = \frac{\sum x_i z_i}{\sum x_i} = \frac{\sum x_i z_i}{T_x}$$

In table 6, this has the value 0.8194 for patients with 0–7 degrees of infiltration and 1.2692 for patients with 8–15 degrees. The formula for the variance of a mean score is, from Yates (14),

$$V(\bar{z}) = \frac{1}{T_x T} \left[ \sum z_i^2 n_i - \frac{(\sum z_i n_i)^2}{T} \right] \qquad (17)$$

Note that, except for the term in $T_x$ , this formula is exactly the same for each of the two means.

The value of the term in square brackets has already been obtained in equation (16) as 227.2653. Hence, from equation (17) and the data in table 6,

$$V(\bar{z}_1) = \frac{227.2653}{(144)(196)} : V(\bar{z}_2) = \frac{227.2653}{(52)(196)}$$

$$V(\bar{z}_1 - \bar{z}_2) = \frac{227.2653}{196} \left\{ \frac{1}{144} + \frac{1}{52} \right\} = 0.03035$$

The $X^2$ value for the difference in mean scores is

$$X^2 = \frac{(\bar{z}_1 - \bar{z}_2)^2}{V(\bar{z}_1 - \bar{z}_2)} = \frac{(1.2692 - 0.8194)^2}{0.03035} = 6.666$$

in agreement with the regression $X^2$.

6.4 *Step by step comparisons of the $p_i$ .*

Occasionally it is useful to compare $p_1$ with $p_2$ , the weighted mean of $p_1$ and $p_2$ with $p_3$ , and so on. To obtain an additive subdivision of $X^2$, we partition the sum of squares in the numerator of $X^2$ as follows:

$$\frac{n_2\{n_1p_1 - n_1p_2\}^2}{n_1(n_1 + n_2)} ,$$

$$\frac{n_3\{n_1p_1 + n_2p_2 - (n_1 + n_2)p_3\}^2}{(n_1 + n_2)(n_1 + n_2 + n_3)} ,$$

and for the general term,

$$\frac{n_{r+1}\{n_1p_1 + \cdots + n_rp_r - (n_1 + \cdots + n_r)p_{r+1}\}^2}{(n_1 + \cdots + n_r)(n_1 + \cdots + n_{r+1})}$$

Each term is then divided by $\hat{p}\hat{q}$ to convert it to $X^2$ with 1 d.f.

For the corresponding non-additive subdivision, we calculate $X^2$, by the standard methods, for each of the following set of $2 \times 2$ tables.

| $x_1$ | $y_1$ | $n_1$ | $x_1 + x_2$ | $y_1 + y_2$ | $n_1 + n_2$ |
|---|---|---|---|---|---|
| $x_2$ | $y_2$ | $n_2$ | $x_3$ | $y_3$ | $n_3$ |
| $x_1 + x_2$ | $y_1 + y_2$ | | $x_1 + x_2 + x_3$ | $y_1 + y_2 + y_3$ | |

and so on, where we have written $y_i$ for $(n_i - x_i)$.

### 7. THE GENERAL TWO-WAY CONTINGENCY TABLE

For considerations of space, this case will be discussed very briefly. The following notation will be used for the frequencies in the $r \times c$ cells of a table with $c$ columns and $r$ rows.

| | | | | | Total |
|---|---|---|---|---|---|
| | $x_{11}$ | $x_{21}$ | | $x_{c1}$ | $n_1$ |
| | $x_{12}$ | $x_{22}$ | | $x_{c2}$ | $n_2$ |
| | $x_{1r}$ | $x_{2r}$ | | $x_{cr}$ | $n_r$ |
| Total | $T_1$ | $T_2$ | | $T_c$ | $T$ |

One fairly common situation is that the rows can be divided into $g$ groups. We expect that within any group there is no association between rows and columns, but there is reason to examine for association among the group totals. To put it another way, the ratios of the $p_{ij}$ from column to column may be constant within a group but vary from group to group. For this situation, a non-additive separation of the total $X^2$ is obtained by: (1), dividing the contingency table into the appropriate $g$ contingency tables, and calculating the $X^2$ in the ordinary way for each separate table; (2), forming a new $g \times c$ table in which the entries are column totals within each group, and calculating the ordinary $X^2$ for this table.

Lancaster (11) and Irwin (15) have shown how to make a complete (non-additive) separation into single d.f. by subdividing the table into a set of $2 \times 2$ contingency tables, $(r - 1)(c - 1)$ in number, and computing $X^2$ in the ordinary way for each $2 \times 2$ table. The method of forming the $2 \times 2$ tables is easily grasped. The same authors have given general methods for the production of additive subdivisions of the total $X^2$. In the case of the partition into single d.f. just mentioned, Kimball (16) gives an expeditious method of computing the individual $X^2$ values. As in the $2 \times N$ case, the basic difference between additive and non-additive partitions is that in the former the estimated variances used to convert the sums of squares into approximate $\chi^2$ values are obtained from the margins of the whole table, while in the non-additive partition they are obtained separately from the margins of each part.

### 7.1 *Score methods.*

If the rows, say, represent an ordered classification, and scores $z_i$ can be assigned to them, we can compute mean scores

$$\bar{z}_i = \frac{\sum_i z_i x_{ij}}{T_i} = \frac{U_i}{T_i} \qquad (i = 1, 2, \cdots, c)$$

for each column, where $U_i$ may be called the total score for the $i$th column. From equation (17), section 6.3, noting that $T_z$ is now replaced by $T_i$ in our notation, we have

$$V(\bar{z}_i) = \frac{1}{T_i T} \left[ \sum z_i^2 n_i - \frac{(\sum z_i n_i)^2}{T} \right] = \frac{D}{T_i}, \qquad (17)'$$

where the factor $D$ is the same for all columns.

These results provide a test of the hypothesis that there is no difference in mean score from column to column to column. We have

$$X^2 = \frac{\sum T_i (\bar{z}_i - \bar{z})^2}{D} = \frac{\sum \frac{U_i^2}{T_i} - \frac{(\sum U_i)^2}{T}}{D}. \qquad (18)$$

The second form is better for computing. The d.f. are $(c - 1)$. These d.f. may in turn be subdivided for more specific comparisons among the columns. In particular, if scores have also been assigned to each column, we may test the regressions of the column means on these scores. General methods have been given by Yates (14).

### 7.2 Use of the analysis of variance.

For the reader who is familiar with the analysis of variance, but less so with $\chi^2$ tests in contingency tables, it is worth noting that the analyses based on scores may be performed quite satisfactorily by ordinary analysis of variance methods. The approach can be illustrated from the comparison of the mean scores of the two groups of patients which was made from the data in table 6.

Let us regard the data in table 6 as representing two independent samples of data from frequency distributions in which the variates take only the values 3, 2, 1, 0, $-1$. The orthodox analysis of variance, using a pooled estimate of error, is given in table 7.

TABLE 7.
COMPARISON OF MEAN SCORES BY ANALYSIS OF VARIANCE

| Source of variation | d.f. | S.S. | M.S. |
|---|---|---|---|
| Between sample means | 1 | 7.7289 | 7.7289 |
| Within samples | 194 | 219.5364 | 1.1316 |

The $F$-value is 6.830 with 1 and 194 d.f., as compared with the $X^2$ value of 6.666, which we may regard as an $F$-value of 6.666 with 1 and $\infty$ d.f. Clearly, the significance probabilities differ by a negligible amount.

For an $r \times c$ contingency table, the general relation between the $F$ and $X^2$ values for comparing the mean scores for the $c$ columns is

$$F = \frac{X^2}{df} \cdot \frac{(T - c)}{(T - X^2)} \tag{19}$$

where $df$ is the number of degrees of freedom in $X^2$.

This relation may be verified numerically in the present example. Since $X^2 = 6.666$, $df = 1$, $T = 196$, $c = 2$, we obtain 6.830 for the right side of equation (19), in agreement with the actual $F$-value. In most tables, $T$ will be much larger than either $c$ or $X^2$, so that $F$ and $X^2$ will be practically equal. Further, since $F$ usually has a substantial number of d.f. in the denominator, the significance probabilities obtained from $F$ and $X^2$ will usually agree closely.

The $F$-test has one aesthetic objection. As Yates showed, the $X^2$ test gives *exactly* the same result, as it should, for the regression of the $p_i$ on the row scores as for the comparison of the mean scores in the two columns. The $F$-test can also be performed both ways. We can assign a score 0 to one column and a score 1 to the other, and make an $F$-test of the regression of the *row* means on the row scores. This $F$-value, however, is not identical with the $F$-value obtained in table 7. However, the significance levels differ only by trivial amounts except when $T$ is small.

### 8. THE COMBINATION OF 2 x 2 CONTINGENCY TABLES

Suppose that we are comparing the frequencies of some occurrence in two independent samples, and that the whole procedure is repeated a number of times under somewhat differing environmental conditions. The data then consist of a series of $2 \times 2$ tables, and the problem is to make a combined test of significance of the difference in occurrence rates in the two samples. The data obtained in comparing the effectiveness of two agents in dosage-mortality experiments are a typical example, in which the repetitions of the experiment are made under a series of different dosage levels. My concern here, however, is with cases where there is no variate corresponding to dosage level, and no well-established theory of how to combine the data.

One method that is sometimes used is to combine all the data into a single $2 \times 2$ table, for which $X^2$ is computed in the usual way. This procedure is legitimate only if the probability $p$ of an occurrence (on the null hypothesis) can be assumed to be the same in all the individual $2 \times 2$ tables. Consequently, if $p$ obviously varies from table to table, or we suspect that it may vary, this procedure should not be used.

Another favorite technique is to compute the usual $X^2$ separately for each table, and add them, using the fact that the sum of $g$ values of $\chi^2$, each with 1 d.f., is distributed as $\chi^2$ with $g$ d.f. This is a poor method. It takes no account of the signs of the differences $(p_1 - p_2)$ in the two samples, and consequently lacks power in detecting a difference that shows up consistently in the same direction in all or most of the individual tables.

An alternative is to compute the $X$ values, and add them, taking account of the signs of the differences. Since $X$ is approximately normally distributed with mean 0 and unit S.D., the sum of $g$ independent $X$ values is approximately normally distributed with mean 0 and S.D. $\sqrt{g}$. Hence the test criterion,

$$\frac{\sum X}{\sqrt{g}}$$

is referred to the standard normal tables.

This method has much to commend it if the total $N$'s of the individual tables do not differ greatly (say by more than a ratio of 2 to 1) and if the $p$'s are all in the range 20-80%. For the following illustrative data I am indebted to Dr. Martha Rogers. The comparison is between mothers of children in the Baltimore schools who had been referred by their teachers as presenting behavior problems, and mothers of a comparable group of control children who had not been so referred. For each mother, it was recorded whether she had suffered any infant losses (e.g. stillbirths) previous to the birth of the child in the study. The comparison is part of a study of possible associations between behavior problems in children and complications of pregnancy of the mother. Since these loss rates increase with later birth orders, and since the samples might not be comparable in birth orders, the data were examined separately, as a precaution, for 3 birth-order classes (see table 8). The two groups of children are referred to as 'Problems' and 'Controls'.

TABLE 8.
DATA ON NUMBER OF MOTHERS WITH PREVIOUS INFANT LOSSES

| Birth Order | | No. of mothers with | | Total | % Loss |
| --- | --- | --- | --- | --- | --- |
| | | Losses | None | | |
| 2 | Problems | 20 | 82 | 102 | 19.6 |
| | Controls | 10 | 54 | 64 | 15.6 |
| | | 30 | 136 | 166 $= N_1$ | 18.1 |
| 3–4 | Problems | 26 | 41 | 67 | 38.8 |
| | Controls | 16 | 30 | 46 | 34.8 |
| | | 42 | 71 | 113 $= N_2$ | 37.2 |
| 5+ | Problems | 27 | 22 | 49 | 55.1 |
| | Controls | 14 | 23 | 37 | 37.8 |
| | | 41 | 45 | 86 $= N_3$ | 47.7 |

Note that the loss rate is higher in the 'Problems' sample in all 3 tables. Since the $N$'s in the separate tables lie within a 2:1 ratio, and the $p$'s are between 18% and 48%, addition of the $X$ values is indicated. The individual $X$ values are, respectively, 0.650, 0.436, 1.587, all being given the same sign since the difference is in the same direction. For this test, the $X$ values are computed without the correction for continuity. Hence the test criterion is the approximate normal deviate

$$\frac{0.650 + 0.436 + 1.587}{\sqrt{3}} = 1.54$$

The $P$ value is just above 0.10. Addition of the $X^2$ values gives 3.131, with 3 d.f., corresponding to a $P$ of about 0.38.

If the $N$'s and $p$'s do not satisfy the conditions mentioned, addition of the $X$'s tends to lose power. Tables that have very small $N$'s cannot be expected to be of much use in detecting a difference, yet they receive the same weight as tables with large $N$'s. Where differences in the $N$'s are extreme, we need some method of weighting the results from the individual tables. Further, if the $p$'s vary from say 0 to 50%, the difference that we are trying to detect, if present, is unlikely to be constant at all levels of $p$. A large amount of experience suggests that the difference is more likely to be constant on the probit or logit scale. As a further complication, the term $pq$ in the variance of a difference will change from one $2 \times 2$ table to another.

Perhaps the best method for a combined analysis is to transform the data to a probit or logit scale. Examples of this type of analysis are given by Winsor (17) and Dyke and Patterson (22): it is recommended if the data are extensive enough to warrant a searching examination. As an alternative, the following test of significance in the original scale will, I believe, be satisfactory under a wide range of variations in the N's and $p$'s from table to table.

For the $i$th $2 \times 2$ table, let

$n_{i1}, n_{i2}$ = sample sizes
$p_{i1}, p_{i2}$ = observed proportions in the two samples
$\hat{p}_i$ = combined proportion from the margins
$d_i$ = $p_{i1} - p_{i2}$ = observed difference in proportions
$w_i$ = $\dfrac{n_{i1}n_{i2}}{n_{i1} + n_{i2}}$ : $w = \sum w_i$

Then we compute the weighted mean difference

$$\bar{d} = \frac{\sum w_i d_i}{w}$$

TABLE 9.

MORTALITY BY SEX OF DONOR AND SEVERITY OF DISEASE

| Degree of disease | Sex of donor | Number of | | Total | % deaths |
|---|---|---|---|---|---|
| | | Deaths | Surv. | | |
| None | M | 2 | 21 | 23 | 8.7 |
| | F | 0 | 10 | 10 | 0.0 |
| | Total | 2 | 23 | $33 = N_1$ | $6.1 = \hat{p}_1$ |
| Mild | M | 2 | 40 | 42 | 4.8 |
| | F | 0 | 18 | 18 | 0.0 |
| | Total | 2 | 58 | $60 = N_2$ | $3.3 = \hat{p}_2$ |
| Moderate | M | 6 | 33 | 39 | 15.4 |
| | F | 0 | 10 | 10 | 0.0 |
| | Total | 6 | 43 | $49 = N_3$ | $12.2 = \hat{p}_3$ |
| Severe | M | 17 | 16 | 33 | 51.5 |
| | F | 0 | 4 | 4 | 0.0 |
| | Total | 17 | 20 | $37 = N_4$ | $45.9 = \hat{p}_4$ |

This has a standard error

$$\text{S.E.} = \frac{\sqrt{\sum w_i \hat{p}_i \hat{q}_i}}{w}$$

The test criterion is

$$\frac{\bar{d}}{\text{S.E.}} = \frac{\sum w_i d_i}{\sqrt{\sum w_i \hat{p}_i \hat{q}_i}}$$

This is referred to the tables of the normal distribution. As explained in the Appendix, which gives supporting reasons for this criterion, the criterion was constructed so that it would be powerful if the alternative hypothesis implies a constant difference on either the probit or the logit scale. The form of the criterion is not one that I would have selected intuitively, and the reader who feels the same way should consult the Appendix.

The test will be illustrated from data published by Diamond et al. (23). Erythroblastosis foetalis is a disease of newborn infants, sometimes fatal, caused by the presence in the blood of an $Rh+$ baby, of anti-$Rh$ antibody transmitted by his $Rh-$ mother. One form of treatment is an "exchange transfusion," in which as much as possible of the infant's blood is replaced by a donor's blood that is free of anti-$Rh$ antibody. In 179 cases in which this treatment was used in a Boston hospital, the rather startling finding was made that there were no infant deaths out of 42 cases in which a female donor was used, but 27 infant deaths out of 137 cases in which a male donor was used. Since there seemed no *a priori* reason why there should be less hazard with female donors, a statistical investigation was made and reported in the reference.

One possibility was that male donors had been used primarily in the more severe cases. Consequently, the data were classified according to the stage of disease at birth, giving the four 2 × 2 tables shown in table 9.

The $N$'s do not vary greatly, but the $p$'s range from 3 to 46%. The combined test of significance is made from the supplementary data in table 10.

TABLE 10.
COMPUTATIONS FOR THE COMBINED TEST

| Stage | $d_i$ | $\hat{p}_i$ | $\hat{p}_i\hat{q}_i$ | $w_i = \dfrac{n_{i1}n_{i3}}{n_{i1} + n_{i2}}$ |
|---|---|---|---|---|
| None | + 8.7 | 6.1 | 573 | 7.0 |
| Mild | + 4.8 | 3.3 | 319 | 12.6 |
| Moderate | +15.4 | 12.2 | 1071 | 8.0 |
| Severe | +51.5 | 45.9 | 2483 | 3.6 |

$$\frac{\bar{d}}{\text{S.E.}} = \frac{\sum w_i d_i}{\sqrt{\sum w_i \hat{p}_i \hat{q}_i}} = \frac{429.88}{\sqrt{25{,}537}} = 2.69$$

The significance probability is 0.0072. In data of this kind, a nonsignificant result would indicate that the surprising phenomenon can be explained by differences in the selection of cases. A significant result must be interpreted with caution, since conditions were not necessarily comparable for male and female donors even within cases of a given degree of severity. However, a significant result does encourage further study, e.g. by examining results in other hospitals.

The expectations in table 9 are so low that one might well doubt the validity of a normal approximation. At least, I did. However, the exact distribution of $\sum w_i d_i$ can be worked out, with some labor, by writing down the probabilities of all possible configurations for each $2 \times 2$ table. The total numbers of configurations are 3, 3, 7 and 5, respectively, for the four tables, so that the total number of possible samples is 315, of which some have negligible probabilities. The value of the test criterion and the probability was worked out for each sample. The exact significance level was found to be 0.0095, as against the normal approximation of 0.0072. The degree of agreement is reassuring, considering the extreme smallness of the expectations. I am indebted to Mrs. Leah Barron for performing the computations.

### 9. THE EFFECT OF EXTRANEOUS VARIATION

Sometimes count data are subject to extraneous variation of a non-Poisson or non-binomial type, especially when the data are samples from an extensive population about which inferences are to be made. In these circumstances, an answer to the question that is really of interest frequently requires a test of significance which takes account of the extraneous variation. For this purpose, $\chi^2$ tests are inappropriate, because they allow only for Poisson or binomial deviations from the null hypothesis. A few simple examples will be considered.

Suppose that we have a sample of data of the Poisson type and that the sample can be divided on some rational basis into $g$ groups. If we wish to examine whether the mean varies from group to group, it was pointed out (section 3.3) that the variance $X^2$ can be divided into components as follows.

|                | d.f.      |
|----------------|-----------|
| Between groups | $(g - 1)$ |
| Within groups  | $(N - g)$ |

(The latter component may be divided into a contribution from each group.) If the "Within groups" $X^2$ is statistically significant, this is a warning that the "Between groups" $\chi^2$ test may be invalid. Further thought about the situation usually leads us to conclude that we do not want to declare that there are real differences between the true group means unless the observed group means differ by more than can be accounted for from the variation within groups. A more appropriate test is either an $F$-test of the "Between-groups" mean square against the "Within groups" mean square, or; if the observations are small and highly variable, an $F$-test in the square root scale. In fact, if there is reason to expect that the within-group variation will contain a non-Poisson component, use of an $F$-test without troubling to compute $X^2$ is a safer procedure. The same issue may arise in testing whether $p$ varies from group to group within a $2 \times N$ contingency table, where the $\chi^2$ test may have to be replaced by an $F$-test based on the original $p_i$ or on the equivalent angles. If the $n_i$ vary substantially from one proportion to another, the problem of obtaining the most efficient type of $F$-test is discussed in (24).

Another example occurs in the combination of $2 \times 2$ contingency tables discussed in section 8. Suppose that the percentages of success under two procedures are compared repeatedly over a variety of conditions that are planned as a sample of some population of conditions about which we wish to draw conclusions. In this case the mean difference $\bar{d} = \bar{p}_1 - \bar{p}_2$ should be tested against the interaction with replications, instead of by the tests that were presented in section 8.

A thorough discussion of when not to use $\chi^2$ tests in problems of this kind, and of the best alternatives, would be lengthy. It is hoped that the few words of warning given in this section will encourage critical thinking about the suitability of a $\chi^2$ test.

## 10. CONCLUDING REMARKS

In presenting the examples in this paper, I have not attempted to give hard and fast rules as to when the supplementary tests should be used. The most useful principle is to think in advance about the way in which the data seem likely to depart from the null hypothesis. This often leads to the selection of a single test (e.g. the variance test) as the only one that appears appropriate. In other situations we may apply both the variance test (or some breakdown of it) and the goodness of fit test, on the grounds that the latter may reveal some types of departure that would not be discovered by the variance test.

When several tests are applied simultaneously to the same data, the chance that at least one of them will be significant is greater, and sometimes much greater, than the presumed 5% probability. This danger of misleading ourselves about the significance level is now widely recognized, and methods for avoiding it have been produced in some of the simpler problems. Although such methods need further development for the applications discussed in this paper, I believe that an awareness of the problem helps to prevent at least the worst distortions of the significance level.

In conclusion, I wish to thank Dr. Paul Meier for some useful suggestions.

## APPENDIX

The purpose of this Appendix is to give some justification for the method proposed in section 8 for making a combined test of significance of the difference between two proportions in a group of independent $2 \times 2$ tables. In the $i$th table, the observed proportions $p_{i1}$, $p_{i2}$ are based on samples of size $n_{i1}$, $n_{i2}$, respectively, and the observed difference is $d_i = p_{i1} - p_{i2}$. The test criterion proposed is

$$\frac{\sum w_i d_i}{\sqrt{\sum w_i \hat{p}_i \hat{q}_i}} , \qquad \text{where } w_i = \frac{n_{i1} n_{i2}}{n_{i1} + n_{i2}}$$

and

$$\hat{p}_i = \frac{n_{i1} p_{i1} + n_{i2} p_{i2}}{n_{i1} + n_{i2}}$$

is the overall proportion from the margins of the $i$th table.

On the null hypothesis,

$$E(d_i) = 0; \qquad V(d_i) \doteq \hat{p}_i \hat{q}_i \left\{ \frac{1}{n_{i1}} + \frac{1}{n_{i2}} \right\} = \frac{\hat{p}_i \hat{q}_i}{w_i} ,$$

so that the test criterion is approximately normally distributed with mean 0 and s.d. 1. The problem is to show that the test criterion is sensitive in detecting departures from the null hypothesis.

The principal fact to be taken into account is that if the true $p_i$ vary over a wide range, the true difference $\delta_i$ is unlikely to be constant on the observed (proportions) scale, but more likely to be constant on a probit or a logit scale. Suppose, for the moment, that we know the values of the $\delta_i$ on the alternative hypothesis, and consider the more general test criterion

$$Y = \frac{\sum W_i d_i}{\sqrt{\sum W_i^2 \hat{p}_i \hat{q}_i / w_i}}$$

where the $W_i$ are any set of weights. This criterion reduces to the recommended criterion if $W_i = w_i$. It reduces to the sum of the individual $X_i$, divided by the square root of their number, if we put $W_i = 1/\sqrt{\hat{p}_i \hat{q}_i / w_i}$, and can ignore the contribution to the variance of $Y$ arising from variation in the $W_i$. (This restriction is needed because we wish to treat the $W_i$ as fixed weights).

On the null hypothesis, $Y$ is approximately normally distributed with mean 0 and s.d. 1. On the alternative hypothesis,

$$E(\sum W_i d_i) = \sum W_i \delta_i$$

$$V(\sum W_i d_i) = \sum W_i^2 \left\{ \frac{p'_{i1} q'_{i1}}{n_{i1}} + \frac{p'_{i2} p'_{i2}}{n_{i2}} \right\}$$

where $'$ denotes a true proportion. Hence, if we can ignore the contribution to the variance of $Y$ arising from its denominator (as we do in using $\chi^2$ in a $2 \times 2$ table) we have

$$\frac{E(Y)}{\sigma(Y)} = \frac{\sum W_i \delta_i}{\sqrt{\sum W_i^2 \left\{ \dfrac{p'_{i1} q'_{i1}}{n_{i1}} + \dfrac{p'_{i2} q'_{i2}}{n_{i2}} \right\}}}$$

We will maximize the power of $Y$, on the alternative hypothesis, if we choose the $W_i$ so as to maximize the ratio $E(Y)/\sigma(Y)$. By ordinary calculus methods, we find that we must have

$$W_i \propto \frac{\delta_i}{\dfrac{p'_{i1} q'_{i1}}{n_{i1}} + \dfrac{p'_{i2} q'_{i2}}{n_{i2}}}$$

Now if the $\delta_i$ are fairly small, so that $p'_{i1}$ and $p'_{i2}$ are not too far apart, the products $p'_{i1} q'_{i1}$ and $p'_{i2} q'_{i2}$ will both be approximately equal to $p'_i q'_i$, where $p'_i$ is the mean of $p'_{i1}$ and $p'_{i2}$. This gives, as an approximation to the best weights,

$$W_i \propto \frac{\delta_i}{p'_i q'_i \left( \dfrac{1}{n_{i1}} + \dfrac{1}{n_{i2}} \right)} = \frac{\delta_i w_i}{p'_i q'_i}$$

This result is not practically useful, since the $\delta_i$ are unknown. It happens, however, that if the true difference is constant on either the probit or the logit scale, the quantity $\delta_i/p'_i q'_i$ is close to constant over practically the whole range of $p'_i$ from 0 to 100%.

In table 11, which illustrates this result, $p'_{i1}$ has been given a range of values from 0 to 99%. To represent a constant effect on the probit scale, it is assumed that the true probit for the second sample is always 0.5 probit units higher than the probit of $p'_{i1}$. For the logit scale, the true logit for the second sample is taken as 0.8 logit units higher than the logit of $p'_{i1}$.

The sizes of the effects are roughly equal on the two scales for $p'_{i1} = 50\%$, where the alternative hypothesis gives about 69% successes, so that $\delta_i$ is about 19. For both the probit and logit cases, $\delta_i$ varies widely from this value as $p'_{i1}$ varies, but the variation in $\delta_i$ is closely matched by that in $p'_i q'_i$ so that the ratio is practically constant for the logit case and varies only slightly for the probit case. If the $\delta_i$ are made smaller, the constancy of the ratio is improved; if larger, the ratio varies more.

The conclusion from this argument is that if effects are constant, for varying levels of $p'_{i1}$, on either the probit or logit scale, the choice of $W_i = w_i$ gives a test criterion that should be close to the optimum in power.

TABLE 11.

CHECK ON CONSTANCY OF $\delta_i/p'_i q'_i$

| $p'_{i1}(\%)$ | Constant probit effect | | | Constant logit effect | | |
|---|---|---|---|---|---|---|
| | $\delta_i$ | $p'_i q'_i$ | $10^4 \delta_i/p'_i q'_i$ | $\delta_i$ | $p'_i q'_i$ | $10^4 \delta_i/p'_i q'_i$ |
| 1 | 2.38 | 214 | 111 | 1.20 | 157 | 76 |
| 5 | 7.61 | 803 | 95 | 5.49 | 714 | 77 |
| 10 | 11.71 | 1334 | 88 | 9.83 | 1269 | 77 |
| 30 | 19.04 | 2390 | 80 | 18.82 | 2388 | 79 |
| 50 | 19.15 | 2408 | 80 | 19.00 | 2410 | 79 |
| 70 | 14.71 | 1751 | 84 | 13.85 | 1775 | 78 |
| 90 | 6.26 | 640 | 98 | 5.24 | 683 | 77 |
| 95 | 3.40 | 319 | 107 | 2.69 | 352 | 76 |
| 99 | 0.76 | 62 | 122 | 0.55 | 72 | 76 |

REFERENCES

(1) D. Lewis and C. J. Burke, "The use and misuse of chi-square test," *Psych. Bull.*, *46* (1949), 433–498.

(2) P. V. Sukhatme, "On the distribution of $\chi^2$ in small samples of the Poisson series," *Jour. Roy. Stat. Soc. Suppl.*, *5* (1938), 75–79.

(3) J. Neyman and E. S. Pearson, "Further notes on the $\chi^2$ distribution," *Biometrika*, *22* (1931), 298–305.

(4) W. G. Cochran, "The $\chi^2$ correction for continuity," *Iowa State Coll. Jour. Sci.*, *16* (1942), 421–436.

(5) W. G. Cochran, "The $\chi^2$ distribution for the binomial and Poisson series, with small expectations," *Annals of Eugenics*, *7* (1936), 207–217.

(6) E. Fix, "Tables of the Noncentral $\chi^2$," *Univ. California Publ. Stat.*, *1* (1949), 15–19.

(7) W. G. Cochran, "The $\chi^2$ test of goodness of fit," *Annals of Math. Stat.*, *23* (1952), 315–345.

(8) D. Mainland, "Statistical methods in medical research," *Canadian Jour. Res.*, *E, 26* (1948), 1–166.

(9) G. H. Freeman and J. H. Halton, "Note on an exact treatment of contingency, goodness of fit and other problems of significance," *Biometrika, 38* (1951), 141–149.

(10) J. B. S. Haldane, The mean and variance of $\chi^2$ when used as a test of homogeneity, when expectations are small," *Biometrika, 31* (1939), 346–355.

(11) H. O. Lancaster, "The derivation and partition of $\chi^2$ in certain discrete distributions," *Biometrika, 36* (1949), 117–129.

(12) W. G. Cochran, "Statistical analysis of field counts of diseased plants," *Jour. Roy. Stat. Soc. Suppl., 3* (1936), 49–67.

(13) D. P. Murphy, *Heredity in uterine cancer,* Harvard University Press, 1952.

(14) F. Yates, "The analysis of contingency tables with groupings based on quantitative characters," *Biometrika, 35* (1948), 176–181.

(15) J. O. Irwin, "A note on the subdivision of $\chi^2$ into components," *Biometrika, 36* (1949), 130–134.

(16) A. W. Kimball, "Short-cut formulas for the exact partition of $\chi^2$ in contingency tables," *Biometrics, 10* (1954), 452–458.

(17) C. P. Winsor, "Factorial analysis of a multiple dichotomy," *Human Biology, 20* (1948), 195–204.

(18) F. J. Anscombe, "Sampling theory of the negative binomial and logarithmic distributions," *Biometrika, 37* (1950), 358–382.

(19) C. I. Bliss, "Fitting the negative binomial distribution to biological data," *Biometrics, 9* (1953), 176–196.

(20) J. Berkson, "A note on the chi-square test, the Poisson and the binomial," *Jour. Amer. Stat. Ass., 35* (1940), 362–367.

(21) J. Berkson, "Some difficulties of interpretation encountered in the application of the chi-square test," *Jour. Amer. Stat. Ass., 33* (1938), 526–536.

(22) G. V. Dyke and H. D. Patterson, "Analysis of factorial arrangements when the data are proportions," *Biometrics, 8* (1952), 1–12.

(23) Fred H. Allen, Jr., Louis K. Diamond and Joseph B. Watrous, Jr., "Erythroblastosis fetalis. V. The value of blood from female donors for exchange transfusion," *The New England Jour. of Med., 241* (1949), 799–806.

(24) W. G. Cochran, "Analysis of variance for percentages based on unequal numbers," *Jour. Amer. Stat. Assoc., 38* (1943), 287–301.

# 60

## THE RAT POPULATION OF BALTIMORE, 1952

By

ROBERT Z. BROWN,[1] WILLIAM SALLOW,[2] DAVID E. DAVIS[3]
AND WILLIAM G. COCHRAN[4]

(Received for publication September 18, 1954)

One of the more difficult problems in rodent control is an accurate evaluation of changes in the numbers of rodents as a control program proceeds. It may be apparent that a change in rodent population levels has occurred, but short of complete eradication the amount of change is difficult to evaluate. The present paper describes the results of the third of a series of estimates of the number of rats in Baltimore, Md., and introduces some refinements in the method. Previous estimates were made in 1947 (Davis and Fales, 1949) and 1949 (Davis and Fales, 1950). The third estimate was made in the fall of 1952. In addition to evaluating the status of rat population levels in the city, another objective of the work was to develop a method for estimating urban rat populations that might be sufficiently general to permit other cities to adapt it to their local conditions. The species is *Rattus norvegicus.*

This project was made possible by the cooperation of several agencies. The Division of Vertebrate Ecology of The Johns Hopkins University School of Hygiene and Public Health initiated the estimate and served as headquarters. The Communicable Disease Center of the United States Public Health Service at Atlanta, Georgia, loaned Robert Z. Brown who was in direct charge of the training and field work. The Rodent Control Division of the Baltimore City Health Department provided the field workers and the experience of William Sallow. The Bureau of Food Control of the City Health Department, through Ferdinand A. Korff, made available their lists of food establishments. Dr. W. T. Fales, late Director of the Vital Statistics Section, organized the statistical aspects. His untimely death in the middle of the analysis of the data has materially handicapped the preparation of this paper.

## METHODS

For the estimation the city was divided into residential and business areas (see figure 3). The residential areas are divided into census tracts and each block is numbered. The United States Census of Housing 1950 Block Statistics (1950 Housing Report, Volume 5, part 13) was the indispensable guide. Without this report the estimate could not have been made.

The first step in the problem of choosing blocks for sampling was to delimit the area infested by rats. For this purpose the data for 1949 were used to determine which census tracts were known to have had rats in 1949. All census tracts that had rats in 1949 were estimated in 1952. The total number of

[1] Formerly Communicable Disease Center, U. S. Public Health Service, Department of Health, Education and Welfare. Now Colorado College, Colorado Springs, Colo.

[2] Chief, Rodent Control Division, Baltimore City Health Department, Baltimore, Md.

[3] Division of Vertebrate Ecology, The Johns Hopkins University, School of Hygiene and Public Health, Baltimore, Md.

[4] Department of Biostatistics, The Johns Hopkins University, School of Hygiene and Public Health, Baltimore, Md.

TABLE 1

*Summary of census tract data*

| Category | Tracts | | Blocks in tracts | | Blocks surveyed | | Dwelling units | |
|---|---|---|---|---|---|---|---|---|
| | Number | Per cent* | Number | Per cent* | Number | Per cent* | Number | Per cent* |
| Had rats in 1949 | 61 | 36.3 | 2,454 | 37.6 | 301 | 4.6 | 14,048 | 5.05 |
| Tract 17–3 (anomalous) | 1 | .5 | 24 | .4 | 5 | .1 | 300 | .00 |
| "Zero" in 1949 | | | | | | | | |
|   Surveyed 1952 | 7 | 4.2 | 218 | 3.5 | 29 | .4 | 1,493 | .54 |
|   Not surveyed 1952 | 19 | 11.3 | 618 | 9.5 | 0 | — | 28,447 | — |
| Totals | 88 | 52.3 | 3,314 | 51.0 | 335 | 5.1 | 44,288 | 5.59 |
| Total in city | 168 | — | 6,514 | — | — | — | 277,880 | — |

* Per cent refers to the proportion of appropriate total in city.

such tracts estimated was 61 (the difference from the 60 tracts that in 1949 had rats, shown in table 2 of Davis and Fales (1950), is due to division of tracts into parts). In addition, one tract (17–3) was included because in 1949 the blocks estimated happened to be in a slum clearance area and lacked rat signs, although rats were abundant in the tract

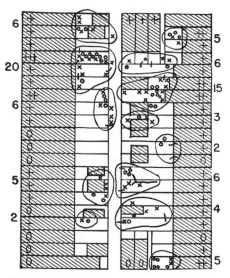

FIGURE 1. Map of rat signs in city block. The circles represent rat holes; the crosses, rat feces; and the lines, runways. Each colony is enclosed by a line and the estimated number written in margin (85 for this block). Block number, date, inspector, etc. are also given.

as a whole in 1949. Further, 7 tracts for which the 1949 estimate was zero were checked in 1952 to indicate the error made by omitting tracts that had zero in 1949. The extent of error involved in this rule will be discussed below.

Within each tract 5 blocks were chosen by the random number table in Snedecor (1946). The numbers of outdoor rats in these blocks were then estimated. In 10 tracts a block was not estimatable because of slum clearance and thus 335 instead of 345 blocks were used. The data for the samples in relation to the totals for the entire city are given in table 1.

The actual estimate of outdoor rats was made essentially as described by Emlen, Stokes and Davis (1949). In brief, the method involves a careful checking of rat signs on a scale map of the block (figure 1). From these signs an estimate of the absolute number of rats is made. It should be noted that this method is much more time consuming and presumably more accurate than the rapid method used in the 1947 and 1949 estimates.

The field work was scheduled to be done in October and November because those months normally have the lowest rainfall and thus would permit field work. Unfortunately, November had almost thrice the normal rainfall and thereby prevented field work for most of the month. The estimating therefore had to be continued through the relatively unfavorable months of December and January. Rainfall interferes with estimating because it obscures the signs of outdoor rats. A reliable estimate cannot be made until two nights have elapsed since rainfall.

The estimates were all made by experienced employees of the Rodent Control Division of the City of Baltimore and were trained for this work by Robert Z. Brown (RZB) for 2 weeks in November. The training consisted of demonstrating the census techniques in infested city blocks and allowing each man to make an independent estimate of the rat population. Several city blocks representing a variety of types of housing and levels of rat populations were utilized. After each block estimate was completed a conference was held in the block and differences in the estimates were analyzed. In some cases the trainees set traps in problem areas to

### TABLE 2a

*Comparison of estimates by various persons at end of training*

| Block | DED | RZB | City Inspectors | | | | | | Average |
|---|---|---|---|---|---|---|---|---|---|
| A |    | 43 | 45 | 38 | 43 | 43 | 36 | 32 | 40 |
| B | 83 | 74 | 96 | 95 | 96 |    |    |    | 89 |
| C | 55 | 40 | 35 | 44 | 38 |    |    |    | 42 |
| D | 80 | 69 | 72 | 93 |    |    |    |    | 79 |
| E | 43 | 50 |    |    |    |    |    |    | 47 |
| F | 25 | 29 |    |    |    |    |    |    | 27 |
| G | 45 |    | 46 | 43 |    |    |    |    | 45 |
| H | 40 |    | 45 | 35 |    |    |    |    | 40 |
| I |    |    | 40 | 27 | 42 | 38 |    |    | 39 |
| J |    |    | 60 | 72 | 106 | 61 |    |    | 75 |

### TABLE 2b

*Comparison of estimates by various persons during estimation*

| Block | Inspector: | | Block | Inspector: | |
|---|---|---|---|---|---|
|  | PJO | City |  | PJO | City |
| a | 10 | 15 | n | 0 | 2 |
| b | 5 | 6 | o | 0 | 2 |
| c | 0 | 13 | p | 0 | 5 |
| d | 40 | 35 | q | 105 | 75 |
| e | 37 | 15 | r | 0 | 6 |
| f | 48 | 27 | s | 100 | 131 |
| g | 5 | 4 | t | 20 | 14 |
| h | 0 | 7 | u | 0 | 14 |
| i | 0 | 25 | v | 4 | 9 |
| j | 4 | 4 | w | 0 | 25 |
| k | 0 | 2 | x | 0 | 8 |
| l | 0 | 8 |  |  |  |
| m | 135 | 82 |  | 513* | 534* |

| Block | Inspector: | | Block | Inspector: | |
|---|---|---|---|---|---|
|  | City | City |  | City | City |
| e | 15 | 7 | ee | 0 | 0 |
| f | 27 | 29 | ff | 2 | 0 |
| g | 4 | 0 | gg | 72 | 95 |
| i | 25 | 6 | hh | 29 | 18 |
| j | 4 | 9 | ii | 19 | 15 |
| aa | 15 | 19 | jj | 12 | 26 |
| bb | 32 | 14 | kk | 8 | 15 |
| cc | 2 | 9 | ll | 0 | 3 |
| dd | 5 | 32 |  |  |  |
|  |  |  |  | 271 | 297 |

* 11 blocks were zero by PJO and City Inspector.

compare with their estimates. Comparisons of the estimates made by the men with the estimates made by RZB and David E. Davis (DED) are given in table 2a. At the end of the census (February, 1953) 46 blocks were chosen randomly and estimated by P. J. Ottenritter (PJO) of the Division of Vertebrate Ecology. His total was 513 rats whereas the estimates by the city men were 534. The close numerical agreement is perhaps deceptive since there was considerable variation (table 2b). However, in

TABLE 3

*Analysis of interviews and inspections*

| "Do you have rats?" | Total interviewed | Actually had rats | Had no rats | Rats present in living quarters | Rats in basement only |
|---|---|---|---|---|---|
| "Substandard" houses | | | | | |
| Yes | 43 | 37 | 6 | 32 | 5 |
| No | 112 | 14 | 98 | 3 | 11 |
| Don't know | 3 | 1 | 2 | 0 | 1 |
| Totals | 158 | 52 | 106 | 35 | 17 |
| "Substandard" apartments | | | | | |
| Yes | 4 | 3 | 1 | 2 | 1 |
| No | 46 | 3 | 43 | 0 | 3 |
| Don't know | 0 | 0 | 0 | 0 | 0 |
| Totals | 50 | 6 | 44 | 2 | 4 |
| "Standard" houses | | | | | |
| Yes | 0 | 0 | 0 | 0 | 0 |
| No | 79 | 0 | 79 | 0 | 0 |
| Don't know | 2 | 0 | 2 | 0 | 0 |
| Totals | 81 | 0 | 81 | 0 | 0 |

some cases a month or more had elapsed between estimates which is sufficient time for considerable change in actual numbers.

A further comparison of estimates is possible because in 17 cases clerical errors allowed a block to be done twice by different men. The sums of the two series were 271 and 297 (table 2b). As above, there may have been considerable true change between the estimates. The problems of bias among the various men will be discussed below.

To estimate the indoor rats a questionnaire to be used by the inspectors was developed. The reason that a questionnaire was used rather than actual inspection is that the questioning required about one third as much time as inspecting. The first step in the de-velopment of the questionnaire was to calibrate the answers into numbers of indoor rats per house. Groups of houses were visited in 3 types of areas: substandard houses, standard houses and substandard apartments. RZB classified 289 dwelling units visited into the following categories according to the occupant's statement: no rats; rat in basement only; rats in 1 room; rats in 2 rooms; rats in 3 or more rooms. Then the occupants' statements were compared with estimates by actual inspections for each house. No rats were found in the 79 standard houses that reported no rats (table 3). In the 112 substandard houses that reported no rats, 14 actually had rats. These rats were mostly in the basement but in 3 cases were also found in rooms. For this reason basement inspections were made for substandard housing in the later survey and no questions were asked about rats in basement.

The inspection yielded the estimates shown in table 4 for the number of rats in relation to the occupants' statements. Actual inspection yielded an estimate of 3 rats per house for houses that had rats only in the basement.

Indoor rats were estimated by the inspectors in the following procedure. In each block that had been selected for estimating the outdoors rats, 6 houses chosen at random were visited, but since about half the people were not home occupants of only about 3 houses per block were questioned. The inspector explained whom he represented and asked the occupant several questions concerning rats. The basic questions were: "Do you have rats? If so, in what rooms?" However, the inspector was permitted to vary the question and to discuss rats with the occupant in order to obtain more truthful (less frightened or resentful) answers. The basement in every substandard dwelling was inspected irrespective of the occupants' answers. This procedure was followed because it was found that occupants of substandard dwellings frequently had little or no knowledge of basement condition.

To get an estimate of the number of rats per dwelling unit the averages derived by RZB's survey were applied

TABLE 4

*Estimate of rats in relation to occupants' statements*

| Occupants' statements | Average rats per house (by inspection) |
|---|---|
| No rats in rooms | 0.08 |
| Rats in 1 room | 2.9 |
| Rats in 2 rooms | 3.2 |
| Rats in 3+ rooms | 4.1 |

TABLE 5

*Summary of approximate labor and costs*

| | Hours | Cost |
|---|---|---|
| Supervisory personnel | 225 | $ 750 |
| City inspectors | 1,672 | 2,875 |
| Miscellaneous expenses (Block maps, transportation, etc.) | | 115 |
| Total | 1,897 | $3,740 |

to the answers obtained by the inspectors during the actual census. For the substandard basements inspectors recorded "presence" or "absence" of rats. The estimate of basement rats was made by reference to the average found during the preliminary survey.

The commercial places were divided into food establishments (licensed) and miscellaneous warehouses, terminals, etc. From the Health Department's alphabetical list (by owner's name) of food establishments, every twentieth retail food store was chosen and inspected. This method is believed to have obtained an unbiased sample because there is no reason to think that a correlation would exist between rat populations and the first letter of the owner's name. The presence or absence of rats in the 592 places visited was determined by men from the Rodent Control Division who had no connection with the Bureau of Food Control.

The miscellaneous commercial areas were visited individually. The number of rats was estimated in 1 produce terminal, 1 race track and 3 large grain elevators.

The number of man-hours and cost of this census can be approximated as shown in table 5.

## RESULTS

The outdoor populations can be estimated from the number of rats per dwelling unit (D.U.) in each tract. The frequency distribution (table 6) has a mean of 0.357 rats per dwelling unit. Since the total D.U. in these 61 tracts is 111,420, the total for outdoor rats is 39,800. Figure 2 shows the density by tracts.

This estimate covers only those tracts that had rats in 1949. Some idea of the extent of omission can be obtained from additional sampling which was done, as mentioned previously, in tracts listed as zero in 1949. Tract 17–3, which by the estimate of 1949 had no rats, was actually surrounded by heavily infested blocks. It was found that the blocks chosen in 1949 all happened to be in a slum-demolition project. The tract in 1952 had 0.5066 rats per D.U., or 700 rats. In another area, 7 additional tracts (which were zero in 1949) in 1952 averaged 0.1073 rats per D.U., or a total of 1,200 rats for the area.

Obviously, rats were present in other tracts also. The number of rats can be approximated in the following manner. The number of D.U. was determined for

TABLE 6

*Numbers of rats per dwelling unit (by tracts)*

| Midpoint rats per D.U. | Tracts |
|---|---|
| 0 | 4 |
| .05 | 11 |
| .15 | 5 |
| .25 | 11 |
| .35 | 12 |
| .45 | 1 |
| .55 | 3 |
| .65 | 4 |
| .75 | 4 |
| .85 | 3 |
| .95 | 1 |
| 1.05 | 1 |
| 1.15 | 0 |
| 1.25 | 1 |
| | — |
| | 61 |

tracts that either (1) were adjacent on at least one side to a tract with 0.30 to 0.59 rats per D.U. or (2) adjacent on at least two sides to tracts that were surveyed (i.e., had rats in 1949). A total of 19 tracts fits these conditions and had 28,447 D.U. Using the value of 0.1073 rats per D.U. (which seems high) gives 3,000 rats. To inspect these 19 tracts would have required one third more work. From the above estimates of the cost, it is apparent that the omission of 3,000 rats would be trivial compared to the cost of securing the data.

For comparability with the 1949 data, the above estimates were made by finding the average number of rats per D.U. and multiplying by the total number of D.U. However, the correlation between the number of rats per block and the number of D.U. per block was low, only about 0.2. In this circumstance a more precise estimate of the total number of rats is obtained by the simpler method of multiplying the average number of rats *per block* by the total number of blocks. For instance, in the 61 inspected tracts, 301 blocks out of 2,454 were surveyed and found to have 3,814 rats, or 12.7 rats per block. Hence the total rats in this area is estimated as $12.7 \times 2,454 = 31,200$ as compared with 39,800 by the "rats per D.U." method. The standard errors attached to these estimates are large, being of the order of 5,500 rats. Part of the difference between the two estimates may, however, represent some unknown positive bias to which the "rats per D.U." method is subject. Estimates made by the "rats per block" method appear in the right hand column of table 7 for the various areas.

The indoor rats in the blocks were estimated by the method described above. In a total of 62 tracts (the 61

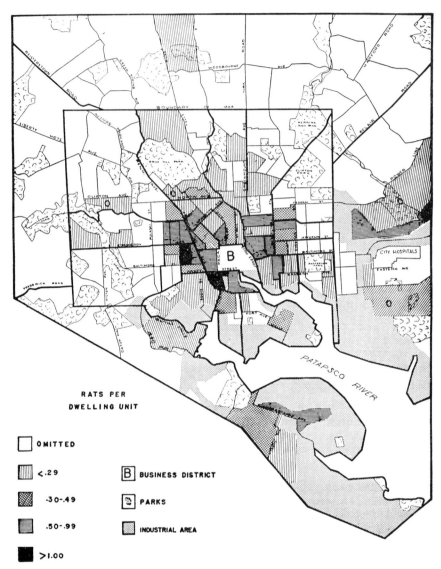

RATS PER
DWELLING UNIT

☐ OMITTED

▨ < .29     B̄ BUSINESS DISTRICT

▩ .30-.49     ▨ PARKS

▤ .50-.99     ▨ INDUSTRIAL AREA

■ >1.00

FIGURE 2. Rats per dwelling unit in 1952.

regular plus the anomalous tract 17–3), 814 D.U. were surveyed. In response to questions a total of 760 occupants said they had no rats. Also, by inspection 75 basements had rats, thus giving about 225 rats. In addition 23 occupants said they had rats in one room ($23 \times 2.9 = 66.7$ rats); 9 occupants said they had rats in 2 rooms ($9 \times 3.2 = 28.8$ rats): and 22 occupants said they had rats in 3 or 4 rooms ($22 \times 4.1 = 90.2$ rats). The grand total is thus **411 rats in 814 D.U.**, or 0.5 rats per D.U. Since there were 112,820 D.U. in these 62 tracts, there were about 56,400 rats.

It will be remembered from the preliminary survey that in 3 out of 112 substandard houses, persons who said they had no rats in rooms actually had a total of 9 rats in rooms. Thus, per-

TABLE 7

*Rat population by type of area*

| Type of place | Population by: | |
| --- | --- | --- |
| | D. U. | Block |
| Outdoors | | |
|     Residential 61 tracts | 39,800 | 31,200 |
|     Residential (17–3) 1 tract | 700 | 700 |
|     Residential (surveyed) 7 tracts | 1,200 | 1,200 |
|     Residential (suspect) 19 tracts | 3,000 | 3,700 |
|       Subtotal | 44,700 | 36,800 |
| Indoors | | |
|     Residential 62 tracts | 56,400 | |
|     Residential (surveyed) 7 tracts | 2,700 | |
|     Residential (suspect) 19 tracts | 6,500 | |
|     Food establishments | 3,000 | 71,200 |
|     Miscellaneous commercial | 1,600 | |
|     Markets | 1,000 | |
| | 115,900 | 108,000 |

sons in substandard houses who said they had no rats should perhaps be charged with $9/112 = 0.08$ rats per inspection. However, this number of rats will be ignored because in all 3 cases rats were also in the basement and hence might be double-counted. Furthermore, one has little confidence in the reliability of an average based on so few cases.

By the same method, the 7 additional tracts gave 0.23 rats per D.U., or 2,700 rats. If we use this figure (0.23) for the 19 tracts that were "suspect" because of location (see above) we get 28,447 D.U. × 0.23 rats per D.U. = 6,500 rats. On this basis the total number of indoor rats is about 66,000, or nearly double the number of outdoor rats.

The numbers of rats in food establishments (mostly retail) were estimated from the data on presence or absence (by inspection) of rats. A total of 22 bakeries and 570 retail stores was in-spected. Only 63 were reported to have rats. To determine the actual number of rats, 14 places were chosen by random numbers from the list of 63. These were inspected by PJO who estimated the number of rats. Actually 8 places had no rats (only old signs), 1 place had about 5 rats and 5 places could not be inspected. Thus, only perhaps half of the 63 places really had rats, or a total of about $30/592 =$ about 5 per cent of the food establishments. There were about 11,500 food establishments; therefore about 600 have rats. If we use a value of 5 rats per place we find 3,000 rats in food establishments.

This infestation rate can be considered from this viewpoint. Suppose that once a year a rat invades a corner grocery, settles down and starts to breed but is detected and eradicated in a month. Weekly inspections for a year would give 4/52, or 8 per cent infested. Consequently, the above 5 per cent infestation rate could be accounted for entirely by invasions and temporary establishments. Thus a figure of 3,000 rats in food establishments is near the minimum that can be attained as long as rats can invade from the surroundings.

Miscellaneous commercial places had a total of about 1,600 rats. In addition, there are a number of municipal markets. Fires in recent years have destroyed 3 old structures, and the others have been greatly improved. Inspection of these markets suggests that 1,000 rats is a generous estimate.

### Results for house mice

While conducting the survey of the indoor rat population, the occupant was asked about the presence of mice. The question was asked in part to avoid confusion of mice with rats and in part to determine the prevalence of mice. In the survey 681 occupants were interviewed and 23.8 per cent answered yes to the question, "Do you have mice?" Assuming that the answer is as reliable for mice as for rats we can conclude that this is a reasonable estimate for these housing areas. No attempt was made to determine the number of mice or to determine prevalence of mice in the food places.

### DISCUSSION

The summation of the various estimates is shown in table 7 and gives a total of about 116,000 rats. Since the 1949 estimate was 58,000 rats (43,000 in residential areas and 15,000 in commercial establishments), it appears that a substantial change occurred from 1949 to 1952, although these figures probably overestimate the true change, for reasons to be explained later.

A comparison of the 1952 map (figure 2) with the 1949 map (figure 3) shows the same distribution in both years. The estimates for particular tracts vary but the distribution and density are similar. A suggestion of a change is present in the southeastern part of the city where a number of tracts were estimated as zero in 1949 but when specially included in 1952 had small numbers of rats. In the 1949 map the triangular white area to the northwest of the business district is the anomalous tract which in 1949 surely had rats but the randomly chosen blocks happened to be in a slum demolition area.

The change estimated in rat population from 1949 to 1952 is almost entirely due to indoor rats. The method used in 1949 did not include inspections of interiors and certainly underestimated the rat population inside houses. This error was suspected and mentioned in the 1949 paper and was the reason for developing the method of indoor estimation used in 1952. Experience with rat populations in Baltimore since 1943 suggests, as stated in 1949, that there has been a true change in the proportion of indoors rats. In retrospect we think it possible that there may have been 40,000 indoor rats in 1949. Since perhaps only half of them were included in the estimate of 43,000 residential rats, this would give figures for 1949 of 23,000 outdoor and 40,000 indoor rats, or 63,000 rats in residential areas. The revised estimate for the city for 1949 would be 78,000 instead of 58,000. Obviously this revision is speculative.

Since a major objective of this estimate was to develop a general method, a statement of the deficiencies and possibilities for improvement is important. In this case a major setback can be attributed to the bad luck of an unusually rainy November. As mentioned above, October and November were chosen because of normally low rainfall. No-

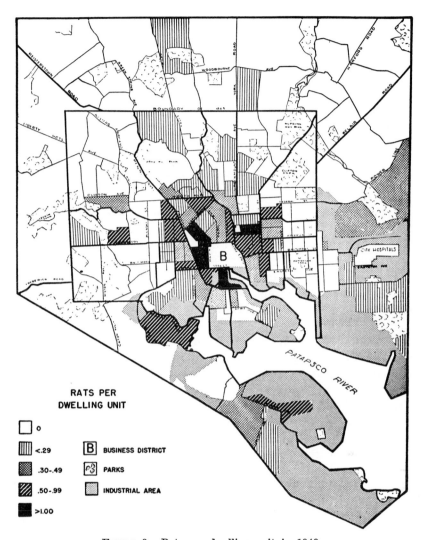

FIGURE 3. Rats per dwelling unit in 1949.

vember, however, had 6.75 inches in contrast to the average of 2.88 inches. The result was that the work dragged out through January instead of being completed in early December. The deficiencies mentioned below were greatly accentuated by this long duration.

The major problem is the actual error of estimate of the rats in a block or house. The difficulty in estimating mammal populations is immense and the extensive literature need not be referred to at this time. Elaborate trapping techniques and statistical analysis still do not reduce the error greatly. Generally speaking, it is rare that a change of less than 50 per cent or even 100 per cent can be detected by trapping techniques. Under these circumstances it is not surprising that the differences from man to man are as great as is shown in table 2. Unfortunately, no scheme is in view that promises to reduce this deficiency.

A number of procedural matters need attention. The maps for the work should be obtained in ample time and all should be ready for the men. Considerable time was lost and confusion resulted from unexpected delays in procuring the maps. Actually the training should not start till all the blocks are selected randomly and the maps are on hand.

Meticulous care in recording the blocks that have been done is essential. In this case 17 blocks were done twice. This defect was in part due to delays in getting maps and in part to the prolongation of the work.

Close supervision and checking are essential. The prolongation through January meant that RZB could not check much of the work since he was available for only 1 month. Therefore, many of the check estimates were not done until some time after the first estimate (accounting in part for the differences of two estimates). The only way that the bookkeeping and supervision can be properly carried out is to have one man working full time on this aspect.

A number of problems exist in the sampling procedure. Some double counting occurred because the estimates of indoor and outdoor rats are essentially separate. The few rats that go in and out of basements will be counted twice.

The standardization of the procedure in respect to basements was not adequate. To economize on time only "substandard" basements were inspected. The types of basements which were included in the "substandard" category were demonstrated to the inspectors during the training period. It was felt that this demonstration plus the experience and reliability of the inspectors was sufficient. In retrospect, however, since no objective written definition of "substandard" was made, the inevitable differences in individual interpretation were magnified. In future programs an objective, written definition of all terms would be most desirable.

Another deficiency was the failure to secure a random subsample of houses for correlation of the answers to questions about indoor rats with actual infestation as ascertained by actual in-inspections. To determine to what extent more than one building was involved in any one infestation all the dwellings in several blocks were inspected. Three types of areas were visited, substandard, standard and substandard apartment building areas (table 3). In addition to this concentrated work a truly random sample of dwellings should have been visited to gain a more accurate estimate for the whole censusing area.

Another defect is that no reinspection was made at houses which were not entered on the first trip. Thus, if no one answered the bell, the house was not included. This error may not be great because there seems little reason to think that persons "not at home" had either more or less rats than persons "at home." The original plan for the estimate provided for correction of all of these defects but the time available did not permit its completion.

The estimates of 1947 and 1949 were based on the number of D.U. in each block. However, the correlation between rats and D.U. is very low $(+ 0.2)$ and hence a per block basis is considered superior. In this paper the estimates on the D.U. basis are given for comparison with the 1947 and 1949 estimates. It is expected in the future to use the block as a basis.

A few comments about the relation of

rat censusing to control programs are pertinent. The rat population in Baltimore has made a substantial increase from 1949 to 1952. Yet the housing rehabilitation and slum clearance programs have been active. Actually, the areas which were rehabilitated in previous years are now degenerating at a rate sufficient to overcome the new work. (Documentation of this statement is reserved for a future publication.) The solution to this situation is improved maintenance of areas freed of rats. Special procedures for reinspection are essential for maintenance of sanitary conditions especially inside houses.

A periodic rat census is important because it provides from time to time during a control program information on the status of rat populations collected *systematically* and *consistently*. By utilizing such a standardized evaluation procedure those in charge of a control program can make valid comparisons of their progress at various stages in the program. "Impressions" gained in the field concerning changes in rat population levels can be substantiated and presented in readily understandable form and hence are of considerable value in public education campaigns or in preparation of future phases of a program.

## Suggestions for a future program

This section contains some suggestions for the design of a future program for estimating the rat population, based on experience accumulated to date.

For outdoor rats, it seems advisable to continue the plan of dividing the residential areas into 3 strata: tracts of high density; "suspect" tracts; tracts of negligible density. As in 1952, the bulk of the effort should be concentrated in the tracts of high density by sampling

4 or 5 blocks per tract. The "suspect" area should be sampled at a lower intensity. In 1952, the "high density" tracts had a standard deviation of 19 rats per block whereas the "suspect" tracts had a standard deviation of 8 rats per block. Statistical theory shows that the best allocation of resources is to sample any stratum with an intensity proportional to the standard deviation in that stratum. Consequently, the 1952 results suggest that the sampling rate in the "suspect" area should not be more than one half that in the "high density" area, and might be as low as one third or one fourth if the "suspect" area is made fairly large in the interests of caution. Tracts considered of negligible density would not be sampled.

For indoor rats the use of initial questioning of the occupant appears to have been successful. To judge from the 1952 data, it would be worth while in the future to inspect all houses at which rats are reported, if permission can be secured. Although such inspections are time-consuming, the total number of houses reporting rats was under 100 in 1952. In addition, a random sample of houses reporting no rats should be inspected, with emphasis on substandard houses if an operationally workable definition can be developed.

The small numbers of rats found in food and other commercial establishments might suggest some reduction in the sampling rate from these establishments. On the other hand, these establishments are of potentially high variability should there be an upsurge in rat density, so that no drastic change in sampling rate is indicated.

In a program of this kind, where estimates are made from droppings and signs, much depends on personal errors and biases of the inspectors. From the data in table 2, a calculation was made of the extent to which such errors might contribute to the variability of the estimate of total number of rats. For the outdoors rats in the high density area, about one third of the variance of the estimate could be attributed to the variability of an inspector's estimates, while the remaining two thirds represented real differences in rat density. Unfortunately, this calculation does not take into account the effect of biases that vary from one inspector to another. An attempt was made to investigate such biases, but the allocation of inspectors to blocks was such that the information provided by the data was too scanty to be trustworthy.

Since the principal use of the estimates is in following time changes, a bias that is common to all inspectors is unimportant if it stays constant throughout time. Biases that vary from inspector to inspector or from time to time are more troublesome, and are not properly taken into account in the standard deviations that can be computed from the data. The necessity of uniform training programs and careful supervision is obvious. In addition, by a careful control of the assignment of inspectors to blocks, it is possible to obtain data that enable us to detect and measure differences in personal biases from one inspector to another. An assignment of this kind should be seriously considered in a future survey.

Some comments are desirable about the cost of an estimate that includes the improvements here suggested. The present program cost $3,700, although the excessive amount of rain made the cost greater than might usually be expected. Let us start with $3,500 as a base and list the cost of each improvement.

| | |
|---|---|
| Base | $3,500 |
| Additional checking of men's work | 250 |
| Better subsample for estimating indoors | 250 |
| Reinspection of places not entered | 500 |
| Inspection of all places answering ''yes'' | 500 |
| | $5,000 |

Included in the above are such miscellaneous costs as longer training, better check on personal bias, etc.

## SUMMARY

The third in a series of estimates of the rat population of Baltimore is presented along with refinements in the methods and suggestions for future programs. The numbers of outdoor and indoor rats were estimated in census tracts that had had rats in 1949. The actual estimating was done by inspectors from the city Rodent Control Division. Outdoor rats were estimated from the rat signs (droppings, tracks, etc.) found during inspection of randomly chosen blocks. Indoor rats were estimated from answers to questions asked of the occupants of randomly chosen dwellings in the census blocks. Commercial places (food establishments) were chosen randomly and inspected and all special problem places (warehouses, terminals, etc.) were visited.

Estimates for outdoor rats were calculated on a rats per dwelling unit basis (44,700 rats) and on a rats per block basis (36,800 rats). The latter estimate seemed more reliable. Indoor rats were estimated at 65,600. Additional rats in food establishments, markets and other commercial places brings the total city estimate to 115,900 on a rats per D.U. basis, or 108,000 on a rats per block basis. Since the rat population in 1949 was about 78,000 (subsequent revision from 1950 publication), there apparently has been a substantial increase.

Deficiencies in the estimating procedures involved an initial delay due to excessive rain, errors in the estimate of rats in a block or a house, delay in obtaining census maps, clerical errors in recording blocks completed and individual differences in interpretation of directions and definitions.

On future programs suggested improvements are additional checking of the inspectors' work to evaluate bias, obtaining a better subsample of houses for estimating indoor rats, reinspection of places not entered (no one at home) and inspection of all places answering "yes" to whether or not they have rats.

REFERENCES

Davis, D. E., and Fales, W. T. The distribution of rats in Baltimore, Maryland. Amer. Jour. Hyg., 1949, *49*: 247–254.

Davis, D. E., and Fales, W. T. The rat population of Baltimore, 1949. Amer. Jour. Hyg., 1950, *52*: 143–146.

Emlen, John T., Stokes, A. W., and Davis, D. E. Methods for estimating populations of brown rats in urban habitats. Ecology, 1949, *30*: 430–442.

Snedecor, George W. Statistical Methods. Ames: Iowa State College Press, 1946. 485 pages.

## RESEARCH TECHNIQUES IN THE STUDY
## OF HUMAN BEINGS

### WILLIAM G. COCHRAN[1]

WHEN Dr. Boudreau invited me to speak tonight, he happened to mention some of the hopes expressed by Dr. Sydenstricker when the Milbank Memorial Fund's division of research was established in 1929. I was struck by one phrase which Dr. Sydenstricker used: he referred to "the possibility of including social data in the domain of scientific research." This phrase set me to trying to sort out my impressions of the quantitative study of human beings as a supposed branch of science. How well is it progressing relative to other branches of science?

Consequently, I would like to present a few of these impressions, with particular reference to the tools of measurement and the general methods of investigation that have been developed.

The claims of the study of social data to be regarded as a branch of science were examined in the 1830's. The occasion was an application made to the British Association for the Advancement of Science to form a section in statistics. In those days, statistical data dealt largely with economic or social matters. The Association appointed a committee to report on the application, and one of its tasks was to consider whether statistics *was* a branch of science. The committee's verdict is interesting. So long as statistics confined itself to the collection, tabulation, and orderly presentation of data, that was science. But if statistics were to concern itself with the interpretation of economic and social data, that would be argumentation, with passions and politics entering, and that was not science. In the picturesque language of the committee, the interpretation of such data could not be allowed as a branch of science, "lest we admit the foul demon of discord into the Eden of philosophy."

The same point of view was maintained a few years later

[1] Professor of Biostatistics, The Johns Hopkins University.

when a statistical society (which became the Royal Statistical Society) was formed in London. The Committee's verdict was in fact embodied in the motto of the new society. This motto consisted of a fat sheaf of wheat, representing the abundant harvest of data that has been collected and tidily arranged. Around the motto was an ornamental ribbon, like the ribbon worn by Miss Atlantic City in the Beauty contests. But in place of the words "Miss Atlantic City" was the Latin motto "Aliis exterendum"—"Let others thrash it out." I am slightly embarrassed that the statisticians should have started their organized career by timidly proclaiming to the world what they will *not* do. The motto is also curious in that the chairman at the early meetings in which statistical organization was discussed was a man well-known to some of you, by the name of Thomas Malthus. It is true that he left much material for others to thrash out, but he did a certain amount of "thrashing out" himself.

Since I shall speak from the viewpoint of the statistician—I can't help it—I must first say a little about statisticians and their relations to scientists. The statistician has long been known as a person who handles data, and the scientist tends to think of seeing a statistician when he has some problem in the analysis of his data. In earlier days, this happened mostly when something had gone wrong with the experiment or survey—or more accurately when the scientist realized that something had gone wrong. As a result, statisticians used to see a sorry collection of the wrecks of research projects.

Now it is a hard fact, which the statistician and the scientist both had to learn, that little could be done to get these wrecks floating again. Usually, some error in the way in which the data were collected made it impossible to draw sound conclusions, manipulate them how you will. This led to two developments. The statistician began to advise scientists to come and see him at the beginning of an investigation—to make him, as it were, an accessory before the fact. Also, the statisticians began to study the process of collecting data in order to learn what procedures and precautions were necessary to ensure that sound conclusions *could* be drawn at the end.

The result is that, at present, the role played by the statistician in the planning of research is often that of verifying that the scientific methodology is sound and sometimes even that of supplying the scientific methodology. Of course, the statistician has other duties. He helps with the arithmetic, tells where the decimal point goes, and he may supply technical formulas from statistical theory, but these are often secondary contributions. Perhaps the statistician's role as a consultant in scientific methods is temporary, because one would expect scientists to perform this function themselves. Indeed, there are signs of a trend in this direction. A few years ago, a conference with a physician about the testing of new drugs on hospital patients seemed to be mainly a matter of trying to wheedle or cajole the physician into taking some precautions that he regarded as a nuisance and as unnecessary. Now he is often found insisting on these precautions himself before the statistician can open his mouth, and the statistician's contribution is to nod his head in agreement at diplomatic intervals. In time, the doctors may decide that they don't need this yes-man.

While it lasts, this role requires close and friendly cooperation between statisticians and research workers. In statistical training centers, something is done to teach young statisticians how to get along with scientists. While in the presence of a number of distinguished scientists, I would like to give a few hints on how to get along with statisticians. In any extensive discussion of a statistical problem, some wag is likely to repeat the old chestnut about the three kinds of lies: "lies, damn lies, and statistics." If you feel an urge to give birth to this witticism, please remember that it is not new, and it was not funny when it was new. The statistician also gets tired of hearing the scientist say "Of course, I am no statistician," in a tone of voice which implies that he is mentioning one of his most sterling virtues. If you *are* no statistician, this fact will probably reveal itself in the course of the conversation, and if you must tell us about it, please do so with an apologetic air. Remember also that the statistician is a poor marriage risk, and may be suffering from marital strains. The reason for this is that the statistician has to cultivate a dislike for imprecise statements, and the person most likely to be making imprecise statements in his vicinity is his unfortunate wife. Statisticians' wives have not, thank goodness, formed an international union, but if they do, the first plank in its platform will be to stop their husbands from being so persnickety.

### Tools of Measurement

The consulting statistician has a fascinating opportunity to learn something about the triumphs and the difficulties of research in different branches of science. He begins to wonder why some branches are forging ahead in an exciting way, while others seem to be creeping. Among the numerous factors that influence the rate of progress of scientific research, two of the most important are the tools of measurement that the research worker has at his disposal and the general methods of investigation available to him.

I shall use the phrase "tools of measurement" in a broad sense, to cover both the range of phenomena that can be measured and the precision of the measuring devices. It can be argued that the available tools of measurement are the most important single factor in determining the rate of progress in a field of research. I do not wish to build up this argument, but one example may be quoted from physics. Towards the end of the last century, the laws of physics seemed to have reached a pinnacle. They were of high accuracy, of immense scope, and were pleasing to common sense. Then improvements in measurement were made that enabled very minute bodies as well as very distant bodies to be more accurately studied. Large cracks appeared in the edifice of physical theory, and to the rescue the physicists had to bring in the revolutionary ideas of quantum mechanics and the theory of relativity. To judge from their difficulties in understanding these concepts and in reconciling them with simple common sense ideas, they must have felt at times as if they had brought in the Marx brothers to repair the building.

The importance that is rightly attached to an improvement in measuring technique is illustrated by the recent award of the Nobel prize in medicine to Dr. Enders and his associates. Their contribution was to grow poliomyelitis virus in tissue culture, and other workers helped to perfect the technique. What does this mean to research in the field? In measuring virus concentrations in specimens from suspected polio cases, as many tests can now be made from the kidneys of a discarded monkey as required 600 monkeys previously. The monkey can be dispensed with entirely, by use of the Hela human cancer cells. Experiments that were impossible can now be done in a week. After a development of this type, any area of research can expect to take great strides forward.

A second impression about measurement techniques is that one never knows where the next advance is coming from. Often it does not come from the field of work that desperately needs the advance. The anthropologists, after measuring skulls from every angle with admirable zeal, have to thank the geneticists for blood group methods that for some purposes are much more reliable. The paleontologists were presented with a new and independent method of dating fossils—radioactive carbon—by the physicists. The electron microscope is a godsend to the manufacturers of paint. And so on.

In the study of human beings many of the problems of measurement are formidable. Not only have we to measure fairly concrete attributes like the state of disease in the individual, (which the doctors will assure us is not easy to measure well) but we need to classify and if possible to measure many things that are hard enough to define in the first place, like motives, morale, intentions, feelings of stress. This means a vast undertaking that has had to start from the ground with rude homemade tools. Thus far, for want of anything better in sight, we have obtained our raw data mainly from what the individual tells us. And the recording instrument has usually been another individual.

We are having to learn about the idiosyncrasies of the human being as a reporter. On the whole, he is surprisingly cooperative, and his good nature in taking up his time to talk to us is heartwarming. He is, in fact, a little too friendly, and will sometimes give the kind of answers which he thinks we would like to have. He is anxious to put on a good front: his statement about the amount he paid for his present car is not entirely to be trusted, and his plans for buying all sorts of expensive gadgets in the future are still less so. On the other hand, he can shut up like a clam. At the end of the war, I helped to gather some data from a carefully selected sample of the German civilian population. According to our results, the Nazi party was one of the world's most exclusive clubs. He is loyal to those whom he likes. The English, in their industrial mortality statistics, were puzzled by the fact that the death rate for the drivers of railway engines (i.e. the locomotive engineers) was above the national average, while that for the man with the apparently less healthy job of stoking the coal furnace was well below. The explanation was that father, after a life of service as stoker, was often posthumously promoted by mother to the position of engine driver on the death certificate.

The recording device—the interviewer—is not perfect either. A quotation from Bertrand Russell, although rather overdrawn, illustrates this point. He is writing about studies of learning in animals. "The animals that have been carefully observed have all displayed the national characteristics of the observer. Animals studied by Americans rush about frantically, with an incredible display of hustle and pep, and at last achieve the desired result by chance. Animals observed by Germans sit still and think, and at last evolve the solution out of their inner consciousness. To the plain man, such as the present writer, this situation is discouraging."

As instances of the amount that has to be learned in order to make the best use of human beings as reporters and recorders, the following are some, but by no means all, of the questions that arise in the planning of morbidity surveys in which the data are obtained by interviews in the home. Over what period of time can the subject remember episodes of illness? What types of illness are easily remembered and accurately reported, and what types are poorly reported? What aids to memory are worth while? How well does the housewife remember and report illnesses of other members of the family? To what extent can the reports be used for a diagnostic classification of the illnesses? How much is gained by checking the reports with physicians who have attended the families? How do lay interviewers compare in effectiveness with public health nurses or medical students? How much information can be picked up at a second visit that was missed at the first? Since a substantial amount of experience has been accumulated for morbidity surveys, at least partial answers can be given to these questions. In other words, something is now known about the precision and the limitations of this type of measuring tool, and about good and bad ways of applying it. Research on more difficult concepts like attitudes and sources of motivation will in time have to answer an analogous list of questions about the interviewer-respondent relationship.

Social scientists are attacking vigorously the fascinating problems involved in devising ways of classifying and measuring what might be called, for want of better words, the strengths and directions of opinions, attitudes, and feelings. They are making surprisingly early use of quantitative scales, with an implied continuum in the background, and have shown ingenuity in constructing methods for testing the internal consistency of the scale and for checking how well the scaled results can be reproduced from a second examination of the same group of people. The criticism has been made, with some justification, that these scales may deceive research workers into thinking that they have measured some rather intangible quantity that they are nowhere in sight of measuring. I don't think that the difficulty arises from the use of quantitative scales themselves: the dangers in pushing this process too fast do not seem to me great. It would be well, however, to be cautious and humble in making claims about what we have measured. Until we are very sure of our ground, use of long Greek names for the things measured might be preferable, rather than claiming to have measured, say, the strength of maternal affection.

## GENERAL METHODS OF INVESTIGATION

Methods of investigation in scientific research can be classified roughly into three types, which may be called chance observations, planned observations, and experiments.

*Chance Observations.* Something unusual strikes the curiosity of an alert scientist, and off he goes into a chain of speculation and then into action. Many of you have heard Sir Alexander Fleming's account of the beginnings of his discovery of penicillin. He happened to notice an unusual contamination from the air of some plates lying in his laboratory. The contribution of chance observations to progress in science must be very great. Last week I was talking to a productive scientist who had had occasion to review carefully his work during the past fifteen years. He remarked that, to his surprise, all his most important discoveries had arisen in unexpected deviations from his main path of research. None of them would have appeared in that anathema of the modern scientist—the "Statement of work to be done during the next fiscal year."

*Planned Observations.* Here the scientist knows what he is after—he knows the questions to which he would like answers —and he maps out a plan of observation which he hopes will provide the answers. Some of the current investigations of the relation between smoking and cancer of the lung are of this type. In the British Medical Research Council's study, all the British doctors were asked three years ago to fill out a questionnaire giving their ages and their recent smoking habits. The rest of the study is just a matter of waiting until a reasonable proportion have died, and then examining whether the death rate and the causes of death are related to smoking habits. Doctors present many advantages for this kind of study: they are likely to cooperate, it is relatively easy to find out if they have died, and when they do die there is reason to believe that the cause of death will be more accurately known than for laymen.

*Experiments.* The word "experiment" has a very broad meaning both in common speech and among scientists. For my present purpose I would like to restrict it to situations in which we are able to *interfere* with nature. In this sense, the essence of an experiment is that we deliberately apply certain chosen procedures for the purpose of measuring their effects. The power of experimentation in speeding up progress in science is tremendous. It has two strong advantages over the observational method. It enables us to select for investigation the factor or factors that will be most informative, whereas with observations, we are restricted to those factors that nature is kind enough to give us the opportunity to observe. The experiment is also the surest method of working out the causal relations that underlie the associations which we observe. With the observational method, the step from correlation to causation is often hazardous and uncertain. For instance, even if several studies in different countries should reach the common conclusion that the death rate from cancer of the lung increases steadily as the amount of smoking increases, the objection will be made (in fact, it has already been made) that this is not a cause and effect relationship, because of the alternative possibility that the kinds of men who smoke heavily are unusually susceptible to cancer of the lung, and would be so even if they did not smoke. Whatever our opinions about the plausibility of this explanation, it is hard to devise an observational study that will clearly support or rebut it. If experimentation were possible, the issue could be cleared up much more easily.

In the study of human beings we are groping our way around among these general methodologies, trying to find which ones pay off best in results. Thus far, observational methods have been used to a large extent, since opportunities for experimentation appear limited.

In particular, we are having to learn how much can be obtained from past data, originally gathered for some other purpose, for example, in connection with the adminstration of a program. Since the data are already there, the method is much speedier than a fresh start would be. In cost, it may mean the difference between $5,000 and $150,000. Although the past data are seldom what we would like to have if we were doing the job anew, yet often there are masses of it, and perhaps it will be possible to select what we need.

Although it is difficult to generalize, experience with past data has been disappointing. It has often given a confused picture from which no clear leads can be drawn, and it has sometimes given leads that turned out to be the wrong ones. The main difficulties appear to be that the definitions used in the data are not rigorous and clear-cut enough for scientific investigations, and that the effects that we wish to study are inextricably tangled up with other effects. Some of my own disappointments with past data remind me of a statement made by Available Jones in the Little Abner comic strip. Available Jones makes his living in part by giving advice. He has two kinds of advice, the 10-cent and the 50-cent kind. Of the 10-cent kind, he says (after some modifications of his spelling): "For 10 cents, I barely listens—in fact I yawns in your face, and the cheap advice you gets will do you more harm than good."

In many human studies, workers are realizing that they must face the long and hard business of planning new observations in order to obtain the 50-cent advice. I do not mean, however, to condemn the use of past data in any outright manner: if a few factors predominate, this should be revealed, and very often, past data are all that we have. Moreover, I owe my first post, in the depression, mainly to the fact that my employers had a large batch of past data which were regarded as a potential mine of information. They hired me to dig it out. I dug furiously, but I doubt whether they received their money's worth. Fortunately, my salary was so low that this moral problem caused me no loss of sleep.

In new studies, we are having to learn how much ground can be covered, that is, how many different questions can be investigated in a single study. At the moment, the lesson seems to be not to be too ambitious. This can be illustrated with respect to one approach to exploratory studies that might be called the method of casting the net widely, if you happen to like it, or the method of shooting blindly in all directions, if you don't happen to like it. Suppose that there is some phenomenon about which not much is known, and we are trying to discover which factors or variables have the most predominant influence on it, or are at least most clearly associated with it. It seems rational to write down all the factors that are likely to have an influence on the phenomenon, include them in the study, and rely on statistical techniques, particularly those of multiple classification or regression, to reveal the most important ones. I know of no one as clever as the social scientist at writing down a ten-page list of factors that might influence any given phenomenon. For the relatively poor results given by this method, the statisticians may be partly to blame, because they may have oversold the power of statistical techniques to unscramble an omelet. If nature mixes things up thoroughly, as she sometimes seems to do, statistical methods will not sort them out very well. Indeed, the more factors that are included in the study, that is, the more painstaking the scientist is, the harder it becomes to disentangle all their effects. Many studies now go to the opposite extreme, concentrating on learning something about a single factor, such as differences between premature and normal children, or between public and slum housing. This means slow progress, and perhaps with more experience some intermediate method will prove rewarding.

Social scientists are having to learn how to observe the same people over a period of years, as in the study of chronic diseases or of the effects of administrative programs. Such studies are expensive and hazardous, because it is difficult to foresee the contingencies that may arise to plague us. For one thing, the human subjects won't stay where they are: off they go to Portland or Honolulu, and if we cannot find means to keep observing them, the group under study dwindles year by year to a remnant consisting of the most settled families. Sometimes it is the scientist who is off to Rangoon or Monte Carlo. My guess would be that we now know how to observe groups for as long as three years, and perhaps for as long as five years: beyond that, there are too few successfully completed studies to be able to say that the technique has been mastered.

In such long-term studies the subjects are sometimes influenced by the fact that they are being studied, in a way that vitiates the purposes of the study. I have heard of farm management studies of poor farmers where the list of questions opened the eyes of some of the farmers to financial opportunities that had never occurred to them. In a few years, these farmers were offering the interviewers jobs. In the British study of smoking and cancer, the Medical Research Council's scientists became alarmed at the number of doctors who replied to the original questionnaire by saying "I have been smoking twenty cigarettes a day, but after reading this questionnaire I have given up smoking for ever."

These long-term studies require, for their direction, a type of scientist who is quite different from the "ivory tower" concept of a scientist. He must be able to assemble a team of workers and to maintain good relations among them: he must obtain the cooperation of various administrators and their agencies, and must handle a considerable amount of paper work. Scientific competence alone does not guarantee success in this type of research: some scientists are too shy, and others too quarrelsome, to meet the requirements.

Social scientists are also having to learn to exercise the kind of ingenuity that is delightful when it comes off. Nature occasionally provides golden opportunities to study some group that will be particularly revealing, as with identical twins who have been reared under different circumstances, or with groups of people who have been long isolated. Ingenuity may also enable us to take the difficult step from correlation to causation. If we have established correlation between two variables A and B in an observational study, we may *think* that A is the cause of B, but nobody saw the murder committed, and the evidence pointing to A as the culprit is only circumstantial. But if by ingenuity we can build up a series of separate pieces of evidence, all pointing to A, it becomes harder and harder to think of an alternative hypothesis that will explain them all away simultaneously. In this connection the social scientist has to use the methods of the detective, the good criminal lawyer, and even the man who is trying to prove that Bacon wrote Shakespeare.

As I have indicated, the use of the more powerful method of experimentation has been small. The obstacles with human subjects are obvious. Yet with persistence and tact, the difficulties can sometimes be overcome, and it may be that experimentation will come to play a more important role than it now does. In medical research on the prevention and cure of disease, some notable successes have been scored by experimentation, and experiments are now being attempted that would, I believe, have been considered impossible a few years ago. The main problem is to secure the tightness of control that is essential for a good experiment, without relaxing the ethical requirement that the welfare of the patient is the paramount consideration.

The trial of the polio vaccine conducted this summer is an example. In some of the states, this trial was made by a method that I would describe as observational, but in others, involving hundreds of thousands of children, the trial was a genuine experiment. The children were divided into two groups at random. Those in one group received, at intervals, three shots of the vaccine. Those in the other group received in the same way three shots of an inert substance that is expected to have no effect. No one in the areas concerned knows which child received vaccine and which control. This, in fact, is known to very few persons, and it will not be revealed until necessary in the final stages of the analysis.

A second example is an international cooperative experiment on drugs for the treatment of leprosy, conducted under the leadership of the Leonard Wood Memorial, that was an organizational masterpiece. The same six drugs were tested at the same time in three different institutions, with uniform methods of measuring and recording the dermatological, neurological and bacteriological progress of the patients, and with random allotments of patients to drugs. One institution was in Japan, one in the Philippines, and one in Pretoria, South Africa. In fact, the chief barrier to progress in this line of research is probably a deficiency in tools of measurement. Since no experimental animal has been found in which leprosy can be studied in the laboratory, it is difficult to obtain clues as to the most promising types of drug to test in the future.

### SUMMARY

The quantitative study of human beings, particularly in their social aspects, is a young field. Because of the multitude of critical problems in human relations facing the world today, research workers are trying to obtain helpful answers on practical questions with rather crude tools of measurement and none too powerful methods of investigation.

In hazarding a few suggestions about the use of resources in this area of research, I should make it clear that I have not surveyed the present use of resources in any adequate way. It may be that my suggestions are already being prosecuted as vigorously as seems worth while.

The field needs to devote ample resources to improving its tools of measurement. This is best done by workers who do not have to produce answers to practical questions at the same time. Raymond Pearl used to urge biologists to stop beating their breasts about the difficulty of doing accurate work in biology. If the biologists would devote as much brains, energy and care to refining their measurements as the physicists do, he claimed that they would obtain as accurate results. Although I think he promised too much, the amount of research that physicists devote to measuring devices as such is impressive, and the returns are equally so. In the social sciences, the work of the psychometricians in the construction of scales is a good beginning. I have heard some hard words about the Rorschach test, but both orthodox and unorthodox methods of measurement should be developed and tried.

Experimentation (in the sense in which I have used it) needs to be exploited as much as possible. The question: "Why can't I do an experiment?" is always worth asking, even if it sounds unrealistic. There may be many opportunities for simple experiments using students as volunteers. A colleague, one of the few men still working on the discouraging task of producing a vaccine for the common cold, finds his volunteer subjects among the convicts.

A balance should be retained between studying what people say they will do and studying what they actually do. Here there is perhaps a contrast between economics and sociology. The economist has kept a close eye on what people do, but has tended to rely on armchair reasoning to uncover the motives for their actions, to the neglect of attempts to study motives independently. The sociologists have been enterprising in tackling the difficult task of studying motives, but they need also to be constantly checking reported motivations against actions.

In addition to scientists engaged in large-scale studies, the field needs a supply of those German animals (in the quotation from Russell) who sit still and think. These might be younger scientists with steady incomes, but with restricted research budgets.

Finally, the field needs to keep strong lines of communication with other branches of science, and particularly with biology, and to recruit some of its research workers from these other branches. For certain research problems, the "interview" method of obtaining data is likely to prove inadequate, and progress may have to await new measuring techniques that are adapted from developments outside of social science. The need for links with biology is obvious: man is biological as well as social; moreover, although biology has access to more powerful and flexible research techniques than social science, many of the problems are the same.

In conclusion, I hope that my comments have not sounded pessimistic. If there is one lesson to be learned from the history of science, it is that the optimists are always right, except that they should have been more optimistic.

# A TEST OF A LINEAR FUNCTION OF THE DEVIATIONS BETWEEN OBSERVED AND EXPECTED NUMBERS*

WILLIAM G. COCHRAN

*The Johns Hopkins University*

## I. INTRODUCTION

As is well known, the $\chi^2$ test of goodness of fit is not directed against any specific pattern of the deviations $(f_i - m_i)$ of the observed frequencies $f_i$ from the expected frequencies $m_i$. For this reason, the $\chi^2$ test is sometimes insensitive in detecting a failure of the null hypothesis. There are, however, a number of alternative or supplementary tests that may be used when it is possible, from the nature of the problem, to predict the type of alternative hypothesis that is most likely to hold if the null hypothesis fails. These tests include a comparison of the variances, or of the third and fourth moments, of the observed and theoretical distributions, and various ways of breaking down $\chi^2$ into components [1].

One additional test of this kind is obtained by selecting any linear function of the deviations,

$$L = \sum g_i(f_i - m_i),$$

where the $g_i$ are numbers, chosen in advance by the person making the test, in such a way that $L$ will be sensitive to the alternative hypothesis that is thought most likely to hold. By suitable assignment of the numbers $g_i$, the criterion $L$ can be made responsive to any anticipated pattern of deviations, either in their signs or in their magnitudes. In particular, if all but one $g_i$ are put equal to zero we obtain the test of an individual deviation. This paper describes how to make an approximate test of significance of the value of $L$. The test is approximate in roughly the same sense in which the goodness of fit $\chi^2$ test is itself approximate, i.e., the test is strictly valid as an asymptotic result when the expectations become large. The theory of the test will be presented first, followed by several illustrative examples.

## II. EXPECTATIONS GIVEN IN ADVANCE

For simplicity, we consider first the case in which the expectations are completely specified by the null hypothesis, so that there are no

* Work assisted by a contract with the Office of Naval Research, Navy Department. Department of Biostatistics paper No. 301.

unknown parameters to be estimated. This is not the common situation, but it does occur, for instance, in testing Mendelian inheritance where the binomial $p$ is given by theory.

Let the known expectations be denoted by the symbols $M_i$ and let

$$N = \sum M_i = \sum f_i$$

be the total size of sample. On the null hypothesis, the observed frequencies $f_i$ follow a multinomial distribution with the following properties:

*Mean:*

$$E(f_i) = M_i \tag{1}$$

*Variance:*

$$V(f_i) = M_i \left\{ 1 - \frac{M_i}{N} \right\} \tag{2}$$

*Covariances:*

$$\text{Cov } (f_i, f_j) = - \frac{M_i M_j}{N} \quad (i \neq j). \tag{3}$$

Then, taking

$$L' = \sum g_i(f_i - M_i),$$

we have, on the null hypothesis,

$$E(L') = 0$$

$$V(L') = \sum_i g_i^2 V(f_i) + 2 \sum_{i<j} g_i g_j \text{ Cov } (f_i, f_j)$$

$$= \sum_i g_i^2 M_i \left\{ 1 - \frac{M_i}{N} \right\} - \frac{2}{N} \sum_{i<j} g_i g_j M_i M_j$$

$$= \sum g_i^2 M_i - \frac{1}{N} (\sum g_i M_i)^2. \tag{4}$$

This is an old result, and is exact for any size of sample [3, §55].

Further, as $N$ becomes large, with fixed $p$'s, so that the $M_i$ become large, the multinomial distribution of the $f_i$ tends to a multivariate normal distribution [2, §30.1], and $L'$ tends to be normally distributed. Hence, the test of significance is made either by treating $L'/\sigma(L')$ as a normal deviate, or by treating

$$\chi_{L'}{}^2 = \frac{L'^2}{V(L')}$$

as $\chi^2$ with 1 degree of freedom.

### III. ONE-PARAMETER ESTIMATION

When the expectations $m_i$ are estimated from the sample, the situation is more complex, and the formula to be given for $V(L)$ is valid only when the expectations are large. Suppose that the $M_i$ are known functions of a single unknown parameter $\theta$, of which the sample estimate is $\hat{\theta}$. Maximum likelihood estimation, or some asymptotically identical method, is assumed. The symbol $L$ will denote the linear function when the expectations are estimated, while $L'$ will be used, as in Section II, when the expectations are known.

We now have

$$L = \sum g_i(f_i - m_i)$$

where the $m_i$ are the values taken by the $M_i$ when $\theta = \hat{\theta}$. We will first find the variances and covariances of the $(f_i - m_i)$, noting that the $m_i$ are now functions of the sample observations.

To save space, two standard results in the theory of maximum likelihood estimation will be quoted. These results, and most results that follow from them, are valid apart from terms that can be neglected when the expectations are large. The symbol $\doteq$ denotes an equation of this type.

The first assumed result is

$$\hat{\theta} - \theta \doteq \frac{1}{I} \sum \frac{(f_i - M_i)}{M_i} \frac{\partial M_i}{\partial \theta}, \tag{5}$$

where

$$I = \sum \frac{1}{M_i} \left( \frac{\partial M_i}{\partial \theta} \right)^2 \tag{6}$$

is Fisher's "amount of information," and is the inverse of the asymptotic variance of $\hat{\theta}$. (In expositions of maximum likelihood theory, the result (5) usually appears in a slightly different form. For instance, in Cramér [2], equation (33.3.4), the right side of (5) has an additional denominator which, as Cramér shows, converges in probability to 1 when the expectations become large.)

The second assumed result is

$$m_i - M_i \doteq (\hat{\theta} - \theta) \frac{\partial M_i}{\partial \theta}, \qquad (7)$$

this being obtained by the first term of a Taylor expansion. Equations (5) and (7) require certain restrictions on the forms of the functions $M_i$ and their derivatives (cf. Cramér, loc. cit.), but I believe that these are satisfied in any of the common applications.

Substitution of the value of $(\hat{\theta} - \theta)$ from (5) into (7) gives

$$m_i - M_i \doteq \frac{1}{I} \frac{\partial M_i}{\partial \theta} \sum_j \frac{(f_j - M_j)}{M_j} \frac{\partial M_j}{\partial \theta}.$$

Hence,

$$f_i - m_i = (f_i - M_i) - (m_i - M_i)$$

$$\doteq (f_i - M_i) - \frac{1}{I} \frac{\partial M_i}{\partial \theta} \sum_j \frac{(f_j - M_j)}{M_j} \frac{\partial M_j}{\partial \theta}. \qquad (8)$$

This is the key equation. It expresses $(f_i - m_i)$ as a linear function of the deviations $(f_j - M_j)$ from the *true* expectations. Since the variances and covariances of these latter deviations are known from equations (2) and (3), we can now find the variance of $(f_i - m_i)$, or of any linear function $L$ of these deviations. Equation (8) also implies that, to the present order of approximation, $E(L) = 0$ when the null hypothesis holds.

Instead of proceeding directly, we shall follow a different route that appears to simplify the algebra. The right side of equation (8) may be interpreted as the deviation of $(f_i - M_i)$ from its linear regression on the variate

$$X = \sum \frac{(f_j - M_j)}{M_j} \frac{\partial M_j}{\partial \theta}. \qquad (9)$$

To see this, we have from (2) and (3)

$$\text{Cov} \{(f_i - M_i), X\} = \frac{\partial M_i}{\partial \theta} - \frac{M_i}{N} \sum_j \frac{\partial M_j}{\partial \theta} = \frac{\partial M_i}{\partial \theta}, \qquad (10)$$

since $\sum M_j = N$, so that its derivative vanishes. Similarly, we find

$$V(X) = \sum \frac{1}{M_i} \left(\frac{\partial M_i}{\partial \theta}\right)^2 = I. \qquad (11)$$

Thus the regression coefficient of $(f_i - M_i)$ on $X$ is

$$b_i = \frac{1}{I} \frac{\partial M_i}{\partial \theta} .$$

Hence, equation (8) may be rewritten as

$$(f_i - m_i) = (f_i - M_i) - b_i X.$$

If

$$L' = \sum g_i(f_i - M_i),$$

it follows that

$$L = \sum g_i(f_i - m_i) = L' - bX,$$

where $b = \sum g_i b_i$ is the regression coefficient of $L'$ on $X$.

Hence the variance of $L$ is equal to the variance of the deviations of $L'$ from its linear regression on $X$, i.e.,

$$V(L) = V(L') - \frac{[\mathrm{Cov}\ (L',\ X)]^2}{V(X)} . \qquad (12)$$

But, from (10) and (11),

$$\mathrm{Cov}\ (L',\ X) = \sum g_i \frac{\partial M_i}{\partial \theta} : V(X) = I.$$

This gives, finally,

$$V(L) \doteqdot ( \sum g_i^2 M_i) - \frac{( \sum g_i M_i)^2}{N} - \frac{1}{I}\left( \sum g_i \frac{\partial M_i}{\partial \theta} \right)^2 . \qquad (13)$$

In practice, we substitute the computed expectations $m_i$ in place of $M_i$. For testing a single deviation, we have

$$\widehat{V}(f_i - m_i) \doteqdot m_i - \frac{m_i^2}{N} - \frac{1}{I}\left( \frac{\partial m_i}{\partial \theta} \right)^2 , \qquad (14)$$

where $\widehat{V}$ denotes an estimated variance.

As in the case where the expectations are given, the test is made either by treating $L/s(L)$ as a normal deviate, or by treating $L^2/\widehat{V}(L)$ as $\chi^2$ with 1 degree of freedom.

Like the goodness of fit test, this test requires some restriction on the smallness of the expectations. The restriction needed will depend on the form of the function $L$. It might be safe to allow some expectations as low as 1 or 2 if these expectations receive relatively small weights $g_i$ in computing $L$. An example of the exact small-sample distribution is presented in the Appendix. So far as it goes, it suggests that the normal approximation may work about as well as does the tabular $\chi^2$ approximation in the ordinary goodness of fit test. Pending further investigation, it seems well to follow the common rule that the minimum expectation should not be less than 5, particularly when a single deviation is being tested. With a single deviation, it is advisable to apply a correction for continuity, by taking the normal deviate as

$$\frac{|f_i - m_i| - \frac{1}{2}}{s(L)}.$$

There is a kind of intuitive interpretation to the result that the variance of $L$ equals the variance of the deviations of $L'$ from its regression on $X$. In Section II, when no parameters were being estimated, the deviations $(f_i - M_i)$ were subject to the single restriction,

$$\sum (f_i - M_i) = 0.$$

It is this constraint that introduces the negative covariance in equation (3) between $f_i$ and $f_j$. When a parameter is being estimated, the maximum likelihood equation imposes a *further* constraint, i.e.,

$$\sum \frac{f_i}{M_i} \frac{\partial M_i}{\partial \theta} = 0,$$

which may be written

$$X = \sum \frac{(f_i - M_i)}{M_i} \frac{\partial M_i}{\partial \theta} = 0.$$

Thus the equation of estimation may be regarded as fixing the value of $X$. This additional restraint leaves the observed frequencies less free to deviate from the theoretical frequencies, and may be expected to diminish the variance of $L$ as compared with that of $L'$. It is not surprising that the appropriate variance is now the variance of the deviations from the regression on the quantity $X$ that is constrained by the equation of estimation.

#### IV. APPLICATION TO THE BINOMIAL AND POISSON DISTRIBUTIONS

Perhaps the most common applications of goodness of fit tests are those to problems in which the null hypothesis specifies either a binomial or a Poisson distribution. Formulas for $V(L)$ appropriate to these cases can be obtained by substituting the appropriate expressions for the $M_i$ and their first derivatives in (13).

The binomial and Poisson distributions have the common property that the equation of estimation makes the sample mean equal to the theoretical mean. By applying the "regression" argument, we can obtain a more general formula for $V(L)$, applicable to any discrete distribution in which the maximum likelihood estimate is the sample mean (or a function of it).

Let $f_i$ denote the frequency of $i$ "successes" $(i=0, 1, 2 \cdots)$. We may take $X$ as

$$X = \sum i(f_i - M_i),$$

since the equation of estimation makes this quantity zero.

Then, it is easy to verify that

$$\text{cov } (L', X) = \sum ig_i M_i - \frac{(\sum g_i M_i)(\sum i M_i)}{N}$$
$$= \sum g_i M_i (i - \mu) \tag{15}$$

where $\mu$ is the mean of the theoretical distribution. Also

$$V(X) = N\sigma^2$$

where $\sigma^2$ is the variance of the theoretical distribution.

Hence, for the estimated variance of deviations from the regression,

$$\widehat{V}(L) \doteq \sum g_i^2 m_i - \frac{(\sum g_i m_i)^2}{N} - \frac{[\sum g_i m_i(i - \widehat{\mu})]^2}{N\widehat{\sigma}^2} \tag{16}$$

where we have substituted sample estimates for the unknown theoretical values involved. For a sample of $N$ from the binomial $(q+p)^n$, we substitute

$$\widehat{\mu} = n\widehat{p} : \widehat{\sigma}^2 = n\widehat{p}\widehat{q}.$$

For a sample of $N$ from the Poisson, we substitute

$$\mu = \widehat{\sigma}^2 = \bar{\imath} = \text{sample mean}.$$

### V. EXAMPLES

*Example 1.* Fisher [3] has analyzed Geissler's data on the distribution of number of boys in 53,680 German families of size 8 (see Table 1). The null hypothesis is that the number of boys follows the binomial

$$(q + p)^8$$

where $p$ is the proportion of boys: $\widehat{p}$, as estimated from the data, is 0.51468.

TABLE 1

NO. OF BOYS IN FAMILIES OF 8

| No. of boys $i$ | No. of families | | | $g_i$ | $ig_i$ |
| | Observed $f_i$ | Expected $m_i$ | Deviations $f_i - m_i$ | | |
|---|---|---|---|---|---|
| 0 | 215 | 165.22 | + 49.78 | +1 | 0 |
| 1 | 1,485 | 1,401.69 | + 83.31 | −1 | −1 |
| 2 | 5,331 | 5,202.65 | +128.35 | +1 | +2 |
| 3 | 10,649 | 11,034.65 | −385.65 | −1 | −3 |
| 4 | 14,959 | 14,627.60 | +331.40 | +1 | +4 |
| 5 | 11,929 | 12,409.87 | −480.87 | −1 | −5 |
| 6 | 6,678 | 6,580.24 | + 97.76 | +1 | +6 |
| 7 | 2,092 | 1,993.78 | + 98.22 | −1 | −7 |
| 8 | 342 | 264.30 | + 77.70 | +1 | +8 |
| | 53,680 | 53,680.00 | | | |

Fisher noted, as is apparent in Table 1, an excess of families with very unequal numbers of boys and girls. He also noted an apparent bias in favor of even numbers of boys. This bias shows up in the central values (2–6 boys). At the extremes, the effect is obscured by the excess of unequally divided families.

To test whether there is an excess of families with even numbers of boys, we may take

$$g_i = + 1 \ (i \text{ even}) : g_i = - 1 \ (i \text{ odd}).$$

We find

$$L = \sum g_i(f_i - m_i) = +1369.98.$$

To apply formula (16) for $V(L)$, we compute

$$\sum g_i^2 m_i = 53,680$$

$$\sum g_i m_i = 0.02$$

$$\sum i g_i m_i = 0.09$$

$$\hat{\mu} = 4.1174$$

$$\hat{\sigma}^2 = 8\hat{p}\hat{q} = 1.9982$$

$$N = 53,680.$$

Now apply formula (16)

$$\widehat{V}(L) \doteq \sum g_i{}^2 m_i - \frac{(\sum g_i m_i)^2}{N} - \frac{[\sum g_i m_i(i - \widehat{u})]^2}{N\widehat{\sigma}^2}.$$

It is clear that the two subtraction terms are entirely negligible, so that

$$\widehat{V}(L) \doteq \sum g_i{}^2 m_i = 53{,}680.$$

The normal deviate is

$$\frac{+1369.98}{\sqrt{53{,}680}} = +5.91$$

indicating a significant excess of families with even numbers of boys.

The reason why the two subtraction terms are negligible is somewhat peculiar to this example. In a binomial with $p = \frac{1}{2}$ and $n$ even, the quantities $\sum g_i m_i$ and $\sum i g_i m_i$ both vanish, as algebraic identities, for this set of $g_i$. Here we have a binomial with $p$ nearly $\frac{1}{2}$. As an exercise in the computations, the reader may try

$$g_i = +1 \ (i \text{ even}) : g_i = 0 \ (i \text{ odd})$$

so as to test the sum of the even deviations. The normal deviate will again be found to be $+5.91$, as would be expected, but the subtraction terms are not negligible.

Although this example serves to illustrate the computations needed in applying the test to binomial data, there are questions about the validity of the application. In working out the frequency distribution of $L$, we have assumed that the coefficients $g_i$ are chosen before seeing the data, whereas the function $L$ was actually constructed for a type of departure from the binomial that was observed in the data. The effect is to make the $L$ test give too many apparently significant results. This point will be discussed further in section VI. Secondly, as already noted, there is an excess of families with very uneven numbers of boys and girls, so that the binomial model used for the null hypothesis probably does not apply exactly. This disturbance will also influence to some degree the frequency distribution of $L$, just as non-normality in the basic data influences the $t$-distribution in a Student's $t$-test.

*Example 2.* This and example 3 are artificial, and are intended to illustrate two properties of the test.

In 100 families of size 2, the frequency with which some attribute occurs is recorded. The binomial $(q+p)^2$ is fitted, the estimate $\hat{p}$ being 0.2.

In this example the goodness of fit $\chi^2$ has 1 $d.f.$ Hence if any one of the single deviations $L = (f_i - m_i)$ is tested by use of equation (16), the $\chi^2$ value for the single deviation should also equal 6.25.

For instance, consider the deviation for $i = 2$. From equation (16),

$$\widehat{V}(f_2 - m_2) = m_2 - \frac{m_2^2}{N} - \frac{m_2(2 - \widehat{\mu})^2}{Nn\widehat{p}\widehat{q}}.$$

Since $\widehat{\mu} = n\widehat{p} = 0.4$, this gives

$$\widehat{V}(f_2 - m_2) = 4 - \frac{16}{100} - \frac{(16)(2.56)}{(100)(0.4)(0.8)} = 2.56.$$

Hence

$$\chi_L^2 = \frac{16}{2.56} = 6.25.$$

TABLE 2

GOODNESS OF FIT TEST OF BINOMIAL DISTRIBUTION

| $i$ | $f_i$ | $m_i$ | Contr. to $\chi^2$ |
|---|---|---|---|
| 0 | 60 | 64 | 0.25 |
| 1 | 40 | 32 | 2.00 |
| 2 | 0 | 4 | 4.00 |
| $N$ = 100 | | 100 | 6.25 (1 $d.f.$) |

The reader may verify that the same value of $\chi_L^2$ is obtained from the deviations for $i = 0$ and $i = 1$. (Corrections for continuity were omitted in this example, since the purpose is to point out an algebraic relationship.)

*Example 3.* A Poisson distribution is fitted to a sample of size 100, in a situation in which the goodness of fit $\chi^2$ again has 1 degree of freedom (Table 3). If a $\chi^2$ value is computed separately for each of the three deviations, we find $\chi_L^2 = 0.365$ for $i = 0$; 0.554 for $i = 1$; and 0.728 for $i = 2$. Thus the "single deviation" tests give *different* results from one another and from the goodness of fit test, in contrast to the result in Example 2.

The discrepancies, which are puzzling at first sight, are a consequence of the grouping of the expectations for $i = 2, 3, 4 \cdots$ which occurs in the 2+ class. On account of this grouping, the value of $\sum im_i$, from Table 3, is 25.743, whereas $\sum if_i$ is 26. Thus the sample value of

$$X = \sum i(f_i - m_i)$$

equals 0.257 instead of 0. The three "single deviation" $\chi^2$ can be brought into much closer agreement with the goodness of fit $\chi^2$ by computing the numerators as

$$[(f_i - m_i) - b_i X]^2,$$

where $X = 0.257$.

On reflection, however, I doubt whether this adjustment is worthwhile, because the goodness of fit $\chi^2$ is also based on the assumption that $\sum if_i = \sum im_i$. In other words, all four values of $\chi^2$ are approximate, and it is not clear that any one of them should be regarded as superior or preferable.

TABLE 3

GOODNESS OF FIT TEST OF POISSON DISTRIBUTION

| $i$ | $f_i$ | $m_i$ | Contr. to $\chi^2$ |
|-----|-------|-------|--------------------|
| 0   | 78    | 77.105 | 0.010 |
| 1   | 18    | 20.047 | 0.209 |
| 2+  | 4     | 2.848  | 0.466 |
|     |       |        | 0.685 (1 d.f.) |

*Example 4.* This example illustrates the test of a single deviation in a frequency distribution that is less familiar than the binomial or Poisson. The figures in Table 4 show the number of adult syphilis patients remaining on the roster of a Baltimore clinic at the beginning of successive two-month periods of observation. All the data refer to the same initial group of 232 patients, so that the successive observations are not independent.

The data are suggestive of an exponential decay curve. Perhaps the simplest mathematical framework that might apply is to suppose that in the hypothetical population of which these data are a sample, there is a constant probability $p$ that any person on the roster at the beginning of a two-month period will drop out during the period. The proportion of the population dropping out in the $i$th period is then $pq^{i-1}$, and the proportion remaining at the end of the 11 periods is $q^{11}$. On

TABLE 4

NUMBER OF PATIENTS REMAINING ON CLINIC ROSTER
AT BEGINNING OF SUCCESSIVE TWO-MONTH PERIODS

| Month following admission | Number of patients on roster |
|:---:|:---:|
| 0 | 232 |
| 2 | 171 |
| 4 | 148 |
| 6 | 134 |
| 8 | 121 |
| 10 | 104 |
| 12 | 90 |
| 14 | 78 |
| 16 | 67 |
| 18 | 61 |
| 20 | 56 |
| 22 | 52 |

this argument, the successive numbers *dropped* in the sample should follow a multinomial distribution with expectations as shown in Table 5. Note that it is necessary to include the number remaining at the end of 11 periods (i.e., the number who would drop out in periods 12+) in order that the numbers add to the original total of 232.

TABLE 5

PATIENTS DROPPED FROM ROSTER

| Period | Number dropped | | $f_i - m_i$ | $x^2$ |
|:---:|:---:|:---:|:---:|:---:|
| | Observed $(f_i)$ | Expected $(m_i)$ | | |
| 1 (0–2) | 61 | 33.09 | +27.91 | 23.54 |
| 2 (2–4) | 23 | 28.37 | − 5.37 | 1.02 |
| 3 (4–6) | 14 | 24.32 | −10.32 | 4.38 |
| 4 (6–8) | 13 | 20.85 | − 7.85 | 2.96 |
| 5 (8–10) | 17 | 17.88 | − 0.88 | 0.04 |
| 6 (10–12) | 14 | 15.33 | − 1.33 | 0.12 |
| 7 (12–14) | 12 | 13.14 | − 1.14 | 0.10 |
| 8 (14–16) | 11 | 11.27 | − 0.27 | 0.01 |
| 9 (16–18) | 6 | 9.66 | − 3.66 | 1.39 |
| 10 (18–20) | 5 | 8.28 | − 3.28 | 1.30 |
| 11 (20–22) | 4 | 7.10 | − 3.10 | 1.35 |
| 12 (22+) | 52 | 42.69 | + 9.31 | 2.03 |
| | 232 | 231.98 | | 38.24 |

There is a simple maximum likelihood estimate of $p$. Let $N$ be the initial number on the roster, $N_i$ the number remaining at the *end* of the $i$th period, and $f_i$ the number dropped during the $i$th period, so that $f_1 = N - N_1 : f_i = N_{i-1} - N_i$, $(i = 2, \cdots 11) : f_{12} = N_{11}$.

The probability of the sample is, apart from factors not involving $p$,

$$P(S) \sim p^{f_1}(pq)^{f_2}(pq^2)^{f_3} \cdots (pq^{10})^{f_{11}}q^{11f_{12}}$$
$$= p^{(f_1+f_2+\cdots+f_{11})}q^{(f_2+2f_3+\cdots+11f_{12})}.$$

But

$$f_1 + f_2 + \cdots + f_{11} = (N - N_1) + (N_1 - N_2) + \cdots + (N_{10} - N_{11})$$
$$= N - N_{11} = D,$$

where $D$ is the total number dropped during the 11 months.

$$f_2 + 2f_3 + \cdots + 11f_{12} = (N_1 - N_2) + 2(N_2 - N_3) + \cdots$$
$$+ 10(N_{10} - N_{11}) + 11N_{11}$$
$$= N_1 + N_2 + \cdots + N_{11} = T - D \text{ (say)}$$

where

$$T = N + N_1 + \cdots + N_{10},$$

is the total of the numbers remaining at the beginning of all periods, excluding the last period. Writing $\Lambda$ for the log of the likelihood, we have

$$\Lambda = \log P(S) \sim D \log p + (T - D) \log q$$
$$\frac{\partial \Lambda}{\partial p} = \frac{D}{p} - \frac{(T - D)}{q}. \tag{17}$$

This gives

$$\hat{p} = \frac{D}{T}.$$

This estimate is a rather natural one. The number dropped in any period, divided by the number present at the beginning of the period, is an unbiased estimate of $p$. The estimate $\hat{p}$ is the total of the numbers dropped, divided by the total of the numbers present.

From Table 4

$$D = 232 - 52 = 180 : T = 232 + 171 + \cdots + 56 = 1262$$
$$\hat{p} = \frac{180}{1262} = 0.14263.$$

The expectations and the deviations are given in Table 5. Clearly, the first period is aberrant, the number dropped being greatly above expectation.

In order to test the deviation for the first period, we use the general formula (14) for its estimated variance, i.e.,

$$\widehat{V}(f_1 - m_1) \doteq m_1 - \frac{m_1^2}{N} - \frac{1}{I}\left(\frac{\partial m_1}{\partial p}\right)^2. \tag{18}$$

Since $M_1 = Np$,

$$\frac{\partial M_1}{\partial p} = N = 232.$$

In a problem of this type, $I$ is usually most easily found by means of the relation

$$I = E\left(-\frac{\partial^2 \Lambda}{\partial p^2}\right).$$

By differentiating equation (17), we have

$$\frac{\partial^2 \Lambda}{\partial p^2} = \frac{-D}{p^2} - \frac{(T - D)}{q^2}.$$

Since

$$E(D) = N(1 - q^{11}),$$
$$E(T - D) = N(q + q^2 + \cdots + q^{11}) = Nq(1 - q^{11})/p,$$

we find

$$I = E\left(\frac{-\partial^2 \Lambda}{\partial p^2}\right) = \frac{N(1 - q^{11})}{p^2 q}.$$

Substitution in formula (18) gives the desired variance (where the last term is written in a form convenient for computation).

$$\widehat{V}(f_1 - m_1) \doteq m_1 - \frac{m_1^2}{N} - \frac{(Np)^2 q}{N(1 - q^{11})}$$

$$= (33.09) - \frac{(33.09)^2}{232} - \frac{(33.09)^2(0.85737)}{189.31} = 23.411.$$

Finally, applying the correction for continuity to $f_1 - m_1 = 27.91$,

$$\chi^2 = \frac{(27.41)^2}{23.411} = 32.09.$$

As would be anticipated, this value is highly significant.

This deviation can be tested, alternatively, by fitting the same model to the data from the second period onwards, starting with the 171 patients left at the end of the first period. The value of the corresponding goodness of fit $\chi^2$ is 6.18, with 9 $d.f.$, as against the original goodness of fit $\chi^2$ of 38.24, with 10 $d.f.$ The difference, 32.06, can be regarded as a test of the fit during the first period. The discrepancy between the values 32.06 and 33.27 (the value given by the $L$ test if no correction for continuity is applied) is presumably due to "small sample" effects. Both methods lead to the conclusion that the model fits satisfactorily after the first period, but that the loss during the first period is too large.

### VI. TESTS SELECTED AFTER INSPECTION OF THE DATA

In the tests described in this paper, the coefficients $g_i$ must be chosen before inspecting the deviations. With the test for a *single* deviation, it is tempting to apply the test to a deviation for which the contribution to $\chi^2$, i.e., $(f_i - m_i)^2 / m_i$, looks suspiciously large. I should not wish to discourage examination of aberrant individual deviations, but the significance probability given by the $L$ test will then be too low. I have not been able to obtain an expression for the significance probability which takes account of the selection after inspection of the deviations. From intuitive reasoning, it appears that this probability will lie between $P$ (that given by the $L$ test) and $kP$, where $k$ is the number of classes in the goodness of fit test. In Example 4, there were reasons for suspecting in advance that the first period would show an abnormal drop, since some patients, having learned that they had syphilis, might shrink from the long course of treatment that was then necessary (the data refer to the 1930's), while others might go elsewhere for treatment. Thus it might have been decided in advance to test the deviation for the first period. On the other hand, if the deviation is picked out purely from inspection, the significance probability appears to lie between $P$ and $12P$. Since $P$ is infinitesimal in this example, there is no doubt about the statistical significance.

The above remarks refer to a single deviation. Suppose now that a linear function $L$ of a number of the deviations is picked out for testing because it looks interestingly large. Then a test that takes account of this selection and errs on the safe side, i.e., gives in general too few significant results, is obtained by referring $L^2/\widehat{V}(L)$ to the $\chi^2$ table with the number of degrees of freedom used in the goodness of fit test.

To show this, we shall show that if the coefficients $g_i$ in $L$ are selected so as to make $L^2/\widehat{V}(L)$ as large as possible, the maximum value of this quantity is equal to the computed value of $\chi^2$ in the ordinary goodness of fit test. In other words, if we always picked out the linear function that would be the "most significant" of all linear functions, we would obtain a valid test of this function by referring $L^2/\widehat{V}(L)$ to the $\chi^2$ table used for the goodness of fit test. Since in practice a linear function that is picked out because it is interesting will not generally give the largest possible normal deviate, we will be making a conservative test of significance if we refer $L^2/\widehat{V}(L)$ to the $\chi^2$ table.

In Example 1 the normal deviate was 5.91. Since $\chi^2$ for these data has seven degrees of freedom, a conservative test is to refer $(5.91)^2$ or 34.93 to the $\chi^2$ table with seven degrees of freedom. The result is still statistically significant.

To prove the needed result, write $\widehat{V}$ for $\widehat{V}(L)$. Then

$$\frac{\partial}{\partial g_i}\left(\frac{L^2}{\widehat{V}}\right) = \frac{2L}{\widehat{V}}\frac{\partial L}{\partial g_i} - \frac{L^2}{\widehat{V}^2}\frac{\partial \widehat{V}}{\partial g_i}.$$

When we set this equal to zero, we obtain the equations

$$\frac{\partial L}{\partial g_i} = \frac{1}{2}\frac{L}{\widehat{V}}\frac{\partial \widehat{V}}{\partial g_i}. \tag{18a}$$

Now from section III,

$$L = \sum g_i(f_i - m_i) : \qquad \frac{\partial L}{\partial g_i} = (f_i - m_i),$$

$$\widehat{V} = \sum g_i^2 m_i - \frac{(\sum g_i m_i)^2}{N} - \frac{1}{I}\left(\sum g_i \frac{\partial m_i}{\partial \theta}\right)^2,$$

$$\frac{1}{2}\frac{\partial \widehat{V}}{\partial g_i} = g_i m_i - \frac{m_i}{N}(\sum g_i m_i) - \frac{1}{I}\frac{\partial m_i}{\partial \theta}\left(\sum g_i \frac{\partial m_i}{\partial \theta}\right).$$

Substitution in (18a) gives, on dividing by $m_i$,

$$\frac{(f_i - m_i)}{m_i} = \frac{L}{\widehat{V}}\left\{g_i - \frac{1}{N}(\sum g_i m_i) - \frac{1}{I}\frac{1}{m_i}\frac{\partial m_i}{\partial \theta}\left(\sum g_i \frac{\partial m_i}{\partial \theta}\right)\right\}.$$

Multiply both sides by $(f_i - m_i)$ and add over all classes. The left side becomes the computed value of $\chi^2$ in the goodness of fit test. On the right side, the first term becomes

$$\frac{L}{\widehat{V}}\left\{\sum g_i(f_i - m_i)\right\} = \frac{L^2}{\widehat{V}},$$

as desired to prove the result. The second and third terms on the right both vanish, the second because $\sum f_i = \sum m_i$, the third because of the maximum likelihood equation of estimation for $\theta$. This completes the proof. The result is valid only asymptotically, since the expression for $\widehat{V}$ is an approximation. When the expectations are given, it is easy to show that the result holds absolutely. Conceptually, this test is of the same kind as that given by Fisher in §64 of *The Design of Experiments*, and later by Scheffé [5], for testing any linear combination of the treatment means in an analysis of variance.

## VII. TWO-PARAMETER ESTIMATION

The "regression" approach extends to the situation where two unknown parameters $\theta_1$ and $\theta_2$ are estimated by maximum likelihood. Since the extension goes smoothly, not all of the details will be presented. We first quote the analogues of the two results (5) and (7) from maximum likelihood theory. The analogue of (5) is the pair of equations

$$\left. \begin{array}{l} I_{11}(\widehat{\theta}_1 - \theta_1) + I_{12}(\widehat{\theta}_2 - \theta_2) \doteq X_1 \\ I_{12}(\widehat{\theta}_1 - \theta_1) + I_{22}(\widehat{\theta}_2 - \theta_2) \doteq X_2 \end{array} \right\}, \tag{19}$$

where

$$I_{uv} = \sum \frac{1}{M_i} \left( \frac{\partial M_i}{\partial \theta_u} \right) \left( \frac{\partial M_i}{\partial \theta_v} \right), \qquad u, v = 1, 2$$

$$X_u = \sum \frac{(f_i - M_i)}{M_i} \frac{\partial M_i}{\partial \theta_u}, \qquad u = 1, 2,$$

these being straightforward extensions of the previous notation.

If $c_{uv}$ is the inverse of the matrix $I_{uv}$, the solutions of these equations are

$$\left. \begin{array}{l} \widehat{\theta}_1 - \theta_1 \doteq c_{11} X_1 + c_{12} X_2 \\ \widehat{\theta}_2 - \theta_2 \doteq c_{12} X_1 + c_{22} X_2 \end{array} \right\}. \tag{20}$$

The analogue of equation (7) is

$$m_i - M_i \doteq (\widehat{\theta}_1 - \theta_1) \frac{\partial M_i}{\partial \theta_1} + (\widehat{\theta}_2 - \theta_2) \frac{\partial M_i}{\partial \theta_2}. \tag{21}$$

Hence

$$(f_i - m_i) = (f_i - M_i) - (\widehat{\theta}_1 - \theta_1) \frac{\partial M_i}{\partial \theta_1} - (\widehat{\theta}_2 - \theta_2) \frac{\partial M_i}{\partial \theta_2}.$$

Substituting from (20) and rearranging, we obtain

$$f_i - m_i \doteq (f_i - M_i) - \left\{ c_{11} \frac{\partial M_i}{\partial \theta_1} + c_{12} \frac{\partial M_i}{\partial \theta_2} \right\} X_1$$

$$- \left\{ c_{21} \frac{\partial M_i}{\partial \theta_1} + c_{22} \frac{\partial M_i}{\partial \theta_2} \right\} X_2.$$

This is the analogue of the key equation (8). It can be shown that the coefficients of $X_1$ and $X_2$ are the regression coefficients of $(f_i - M_i)$ on $X_1$ and $X_2$.

Hence, as before, the variance of $L$ is equal to the residual variance of $L'$ from its multiple regression on $X_1$ and $X_2$. To find this residual variance, we have

$$\text{Cov } (L', X_1) = \sum g_i \frac{\partial M_i}{\partial \theta_1} = S_1, \qquad \text{(say)},$$

$$\text{Cov } (L', X_2) = \sum g_i \frac{\partial M_i}{\partial \theta_2} = S_2, \qquad \text{(say)}.$$

The two regression coefficients (for $L'$ on $X_1$, $X_2$) are

$$b_1 = c_{11}S_1 + c_{12}S_2,$$
$$b_1 = c_{21}S_1 + c_{22}S_2.$$

The reduction in variance due to the regression is

$$b_1S_1 + b_2S_2 = c_{11}S_1{}^2 + 2c_{12}S_1S_2 + c_{22}S_2{}^2.$$

This gives, finally,

$$V(L) \doteq \sum g_i{}^2 M_i - \frac{(\sum g_i M_i)^2}{N} - (c_{11}S_1{}^2 + 2c_{12}S_1S_2 + c_{22}S_2{}^2)$$

as the general formula for two-parameter estimation.

### VIII. APPLICATION TO THE NORMAL DISTRIBUTION

In order to obtain a formula applicable to the normal distribution, we may use the fact that the equations of estimation make the mean and variance of the theoretical distribution equal to those of the sample. Let $d_i$ be the center of the $i$th class. It will simplify the algebra if the origin is placed at the sample mean. Then we take

$$X_1 = \sum d_i(f_i - M_i),$$
$$X_2 = \sum d_i{}^2(f_i - M_i).$$

These choices assume that the sample is large and that the grouping into classes is not too coarse. Because of the grouping, $\sum d_i M_i$ and $\sum d_i^2 M_i$ will not exactly equal the first two moments of the continuous normal distribution and it is assumed here that the discrepancies are not serious. Moreover, in practice we equate the sample *mean square* (dividing by $N-1$) to its expectation, instead of the sum of squares, so that terms in $1/N$ are considered negligible.

It is convenient to have a general formula for the covariance of any two linear functions

$$H = \sum h_i(f_i - M_i) : K = \sum k_i(f_i - M_i).$$

From equations (2) and (3) it is found that

$$\text{Cov } (H, K) = \sum h_i k_i M_i - \frac{(\sum h_i M_i)(\sum k_i M_i)}{N}. \tag{24}$$

The result remains valid when $H$ and $K$ are identical, in which case the covariance becomes a variance.

Hence we obtain the results needed for setting up the regression equations.

$$\text{Cov } (L', X_1) = \sum g_i d_i M_i \tag{25}$$

since $\sum d_i M_i = 0$, because the origin is at the sample mean.

$$\text{Cov } (L', X_2) = \sum g_i d_i^2 M_i - \frac{(\sum g_i M_i)(\sum d_i^2 M_i)}{N}$$

$$\doteq \sum g_i M_i (d_i^2 - s^2), \tag{26}$$

where $s^2$ is the sample variance.

$$V(X_1) = \sum d_i^2 M_i = N \, s^2, \tag{27}$$

$$\text{Cov } (X_1, X_2) = \sum d_i^3 M_i \doteq 0, \tag{28}$$

since the third moment of the theoretical normal distribution is zero. Finally,

$$V(X_2) = \sum d_i^4 M_i - \frac{(\sum d_i^2 M_i)^2}{N} \doteq 2N s^4 \tag{29}$$

since $\mu_4 = 3\mu_2^2$ for the normal distribution.

The five equations (25–29) enable us to set up the regression equations of $L'$ on $X_1$ and $X_2$. For the normal distribution, the equations separate, since $\text{Cov } (X_1, X_2) = 0$. Hence, the estimated variance is approximately

$$\widehat{V}(L) = \sum g_i^2 m_i - \frac{(\sum g_i m_i)^2}{N} - \frac{[\text{Cov } (L, X_1)]^2}{V(X_1)} - \frac{[\text{Cov } (L, X_2)]^2}{V(X_2)}$$

$$= \sum g_i^2 m_i - \frac{(\sum g_i m_i)^2}{N} - \frac{(\sum g_i d_i m_i)^2}{Ns^2} - \frac{[\sum g_i m_i (d_i^2 - s^2)]^2}{2Ns^4}$$

## APPENDIX

### An example of the small-sample distribution of $L'$

The example refers to a multinomial distribution with three classes. The sample size is 10, and the expectations in the three classes are 5, 3 and 2, respectively. This example has been discussed previously, with respect to the ordinary $\chi^2$ test, by Neyman and Pearson [4]. Note that the expectations are fixed. It would be more revealing to work some examples in which the expectations are estimated from the data, but this is considerably more laborious.

The probabilities of each of the 66 possible configurations of the sample were first computed. From these, the exact frequency distributions were worked out for the following two linear functions:

$$L_1' = 6(f_1 - M_1) + 3(f_2 - M_2) + (f_3 - M_3)$$

and

$$L_2' = (f_1 - M_1) + 3(f_2 - M_2) + 6(f_3 - M_3).$$

In $L_1'$, the class with the highest expectation receives the highest weight and the class with the lowest expectation receives the lowest weight. In $L_2'$, these weights are reversed. It was thought that the normal approximation might agree better with the exact distribution for $L_1'$ than for $L_2'$. On the null hypothesis, both $L_1'$ and $L_2'$ have means zero: their standard deviations, as found from equation (4), are 6.3953 and 6.0332, respectively. The exact probabilities and the normal approximations for a two-tailed test are shown in Table 6 for the region in which the exact probability lies between 0.25 and 0.005.

For both $L_1'$ and $L_2'$, the normal approximation tends to underestimate the probability, i.e., to give too many apparently significant results. Table 6 also shows the errors in the normal probabilities as percentages of the true probabilities. The averages of these percentage errors, ignoring sign, are 12 per cent for $L_1'$ and 10 per cent for $L_2'$.

Since the values of the $L''$s proceed by integers, it would be easy to apply a correction for continuity in calculating the normal approximation. This correction removes the tendency of the normal approxima-

TABLE 6

COMPARISON OF EXACT PROBABILITIES AND
NORMAL APPROXIMATIONS

| Deviate | $L_1'$ | | | $L_2'$ | | |
|---|---|---|---|---|---|---|
| | Exact $P$ | Normal approx. | Error in % | Exact $P$ | Normal approx. | Error in % |
| 8 | .257 | .211 | −18 | .213 | .185 | −13 |
| 9 | .172 | .159 | − 8 | .156 | .136 | −13 |
| 10 | .148 | .118 | −20 | .120 | .0903 | −25 |
| 11 | .0997 | .0854 | −14 | .0723 | .0683 | − 6 |
| 12 | .0656 | .0607 | − 7 | .0563 | .0467 | −17 |
| 13 | .0549 | .0422 | −23 | .0303 | .0312 | + 3 |
| 14 | .0289 | .0286 | − 1 | .0210 | .0203 | − 3 |
| 15 | .0185 | .0190 | + 3 | .0138 | .0129 | − 7 |
| 16 | .0130 | .0123 | − 5 | .0084 | .0080 | − 5 |
| 17 | .0064 | .0079 | +23 | .0052 | .0048 | − 8 |

tion to underestimate the true probabilities: in fact the corrected values are mostly overestimates. Apart from this effect, the correction does not improve the approximation. With the correction, the average percentage errors are 15 per cent for $L_1'$ and 12 per cent for $L_2'$.

Contrary to expectations, the normal approximation is not closer for $L_1'$ than for $L_2'$. It is, however, closer in single-tailed tests because the distribution of $L_1'$ is not far from symmetrical, whereas that of $L_2'$ is quite skew.

Although the user must judge for himself whether these approximations are satisfactory for practical use, the agreement seems surprisingly good when one considers that the *largest* expectation is 5.

In conclusion, I wish to thank Dr. W. Kruskal and Dr. P. Meier for some useful suggestions.

### REFERENCES

[1] Cochran, W. G., "Some methods for strengthening the common $\chi^2$ tests," *Biometrics*, 10 (1954), 417–51.
[2] Cramér, H., *Mathematical Methods of Statistics*. Princeton: Princeton University Press, 1946.
[3] Fisher, R. A., *Statistical Methods for Research Workers*. Edinburgh: Oliver and Boyd (1925), §18 in all editions.
[4] Neyman, J. and Pearson, E. S., "Further notes on the $\chi^2$ distribution," *Biometrika*, 22 (1931), 301–2.
[5] Scheffé, H. "A method for judging all contrasts in the analysis of variance," *Biometrika*, 40 (1953), 87–104.

# 63

# CHAPTER 17

# Design and Analysis of Sampling

## by WILLIAM G. COCHRAN

**17.1—Populations.** In the 1908 paper in which he discovered the $t$-test, "Student" opened with the following words: "Any experiment may be regarded as forming an individual of a *population* of experiments which might be performed under the same conditions. A series of experiments is a sample drawn from this population.

"Now any series of experiments is only of value in so far as it enables us to form a judgment as to the statistical constants of the population to which the experiments belong."

From the previous chapters in this book, this way of looking at data should now be familiar. The data obtained in a biological experiment are subject to variation, so that an estimate made from the data is also subject to variation and is, hence, to some degree uncertain. You can visualize, however, that if you could repeat the experiment many times, putting all the results together, the estimate would ultimately settle down to some unchanging value which may be called the true or definitive result of the experiment. The purpose of the statistical analysis of an experiment is to reveal what the data can tell about this true result. The tests of significance and confidence limits which have appeared throughout this book are tools for making statements about the population of experiments of which your data are a sample.

In such problems the sample is concrete, but the population may appear somewhat hypothetical. It is the population of experiments that might be performed, under the same conditions, if you possessed the necessary resources, time, and interest.

In this chapter we turn to situations in which the population is concrete and definite, and the problem is to obtain some desired information about it. Examples are as follows:

| *Population* | *Information Wanted* |
|---|---|
| Ears of corn in a field | Average moisture content |
| Seeds in a large batch | Percentage germination |
| Water in a reservoir | Concentration of certain bacteria |
| Third-grade children in a school | Average weight |

If the population is small, it is sometimes convenient to obtain the information by collecting the data for the whole of the population. More frequently, time and money can be saved by measuring only a sample drawn from the population. When the measurement is destructive, sampling is of course unavoidable.

This chapter presents some methods for selecting a sample and for estimating population characteristics from the data obtained in the sample. During the past 20 years, sampling has come to be relied upon by a great variety of agencies, including government bureaus, market research organizations, and public opinion polls. Concurrently, much has been learned both about the theory and practice of sampling, and a number of books devoted to sample survey methods have appeared (2, 3, 4, 11, 13). In this chapter we explain the general principles of sampling and show how to handle some of the simpler problems that are common in biological work. For more complex problems, references will be given.

**17.2—A simple example.** In the early chapters of this book, you drew samples so as to examine the amount of variation in results from one sample to another and to verify some important results in statistical theory. The same method will illustrate modern ideas about the selection of samples from given populations.

Suppose the population consists of $N = 6$ members, denoted by the letters *a* to *f*. The 6 values of the quantity that is being measured are as follows: *a* 1; b 2; *c* 4; *d* 6; *e* 7; *f* 16. The total for this population is 36. A sample of 3 members is to be drawn in order to estimate this total.

One procedure already familiar to you is to write the letters *a* to *f* on beans or slips of paper, mix them in some container, and draw out 3 letters. In sample survey work, this method of drawing is called *simple random sampling*, or sometimes *random sampling without replacement* (because we do not put a letter back in the receptacle after it has been drawn). Obviously, simple random sampling gives every member an equal chance of being in the sample. It may be shown that the method also gives every combination of three different letters (e.g., *aef* or *cde*) an equal chance of constituting the sample.

How good an estimate of the population total do we obtain by simple random sampling? We are not quite ready to answer this question. Although we know how the sample is to be drawn, we have not yet discussed how the population total is to be estimated from the results of the sample. Since the sample contains 3 members and the population contains 6 members, the simplest procedure is to multiply the sample total by 2, and this is the procedure that will be adopted. You should note that any sampling plan contains two parts—a rule for drawing the sample and a rule for making the estimates from the results of the sample.

We can now write down all possible samples of size 3, make the estimate from each sample, and see how close these estimates lie to the true value of 36. There are 20 possible samples. Their results appear in table

17.2.1, where the successive columns show the composition of the sample, the sample total, the estimated population total, and the error of estimate (estimate *minus* true value).

Some samples, e.g., *abf* and *cde*, do very well, while others like *abc* give poor estimates. Since we do not know in any individual instance whether we will be lucky or unlucky in the choice of a sample, we appraise any sampling plan by looking at its *average* performance.

TABLE 17.2.1

RESULTS FOR ALL POSSIBLE SIMPLE RANDOM SAMPLES OF SIZE 3

| Sample | Sample Total | Estimate of Population Total | Error of Estimate | Sample | Sample Total | Estimate of Population Total | Error of Estimate |
|--------|--------------|------------------------------|-------------------|--------|--------------|------------------------------|-------------------|
| abc | 7 | 14 | −22 | bcd | 12 | 24 | −12 |
| abd | 9 | 18 | −18 | bce | 13 | 26 | −10 |
| abe | 10 | 20 | −16 | bcf | 22 | 44 | + 8 |
| abf | 19 | 38 | + 2 | bde | 15 | 30 | − 6 |
| acd | 11 | 22 | −14 | bdf | 24 | 48 | +12 |
| ace | 12 | 24 | −12 | bef | 25 | 50 | +14 |
| acf | 21 | 42 | + 6 | cde | 17 | 34 | − 2 |
| ade | 14 | 28 | − 8 | cdf | 26 | 52 | +16 |
| adf | 23 | 46 | +10 | cef | 27 | 54 | +18 |
| aef | 24 | 48 | +12 | def | 29 | 58 | +22 |
| | | | | Average | 18 | 36 | 0 |

The average of the errors of estimate, taking account of their signs, is called the *bias* of the estimate (or more generally of the sampling plan). A positive bias implies that the sampling plan gives estimates that are on the whole too high; a negative bias, too low. From table 17.2.1 it is evident that this plan gives unbiased estimates, since the average of the 20 estimates is exactly 36 and consequently the errors of estimate add to zero. With simple random sampling this result holds for any population and any size of sample. Estimates that are unbiased are a desirable feature of a sampling plan. On the other hand, a plan that gives a small bias is not ruled out of consideration if it has other attractive features.

As a measure of the accuracy of the sampling plan we use the variance of the estimates taken about the true population value. This variance is

$$\frac{\Sigma(\text{Error of estimate})^2}{20} = \frac{3504}{20} = 175.2$$

The divisor 20 is used, instead of the divisor 19 with which you have become familiar, this being a common convention among writers on sample surveys. To sum up, this plan is unbiased and has a standard error of estimate of $\sqrt{175.2} = 13.2$. This standard error amounts to 37% of the true population total; evidently the plan is not very accurate for this population.

In simple random sampling the selection of the sample is left to the luck of the draw. No use is made of any knowledge that we may possess about the members of the population. Given such knowledge, we should be able to improve upon simple random sampling by using the knowledge to guide us in the selection of the sample. Much of the recent research on sample survey methods has been directed towards taking advantage of available information about the population to be sampled.

By way of illustration, suppose that before planning the sample we expect that $f$ will give a much higher value than any other member in the population. How can we use this information? It is clear that the estimate from the sample will depend to a considerable extent on whether $f$ falls in the sample or not. This statement can be verified from table 17.2.1; every sample containing $f$ gives an overestimate and every sample without $f$ gives an underestimate.

It seems best, then, to make sure that $f$ appears in every sample. We can do this by dividing the population into two parts or *strata*. Stratum I, which consists of $f$ alone, is completely measured. In stratum II, containing $a$, $b$, $c$, $d$, and $e$, we take a simple random sample of size 2 in order to keep the total sample size equal to 3.

Some forethought is needed in deciding how to estimate the population total. To use twice the sample total, as was done previously, gives too much weight to $f$ and, as already pointed out, will always produce an overestimate of the true total. We can handle this problem by treating the two strata separately. For stratum I we know the total (16) correctly, since we always measure $f$. For stratum II, where 2 members are measured out of 5, the natural procedure is to multiply the sample total in that stratum by 5/2, or 2.5. Hence the appropriate estimate of the population total is

$$16 + 2.5(\text{Sample total in stratum II})$$

TABLE 17.2.2

RESULTS FOR ALL POSSIBLE STRATIFIED RANDOM SAMPLES WITH THE UNEQUAL
SAMPLING FRACTIONS DESCRIBED IN TEXT

| Sample | Sample Total in Stratum II ($T_2$) | Estimate $16 + 2.5\ T_2$ | Error of Estimate |
|--------|------------------------------------|--------------------------|-------------------|
| *abf* | 3 | 23.5 | −12.5 |
| *acf* | 5 | 28.5 | − 7.5 |
| *adf* | 7 | 33.5 | − 2.5 |
| *aef* | 8 | 36.0 | 0.0 |
| *bcf* | 6 | 31.0 | − 5.0 |
| *bdf* | 8 | 36.0 | 0.0 |
| *bef* | 9 | 38.5 | + 2.5 |
| *cdf* | 10 | 41.0 | + 5.0 |
| *cef* | 11 | 43.5 | + 7.5 |
| *def* | 13 | 48.5 | +12.5 |
| Average | | 36.0 | 0.0 |

These estimates are shown for the 10 possible samples in table 17.2.2. Again we note that the estimate is unbiased. Its variance is

$$\frac{\Sigma(\text{Error of estimate})^2}{10} = \frac{487.50}{10} = 48.75$$

The standard error is 7.0 or 19% of the true total. This is a marked improvement over the standard error of 13.2 that was obtained with simple random sampling.

This sampling plan goes by the name of *stratified random sampling with unequal sampling fractions*. The last part of the title denotes the fact that stratum I is completely sampled, whereas stratum II is sampled at a rate of 2 units out of 5, or 40%. Stratification allows us to divide the population into sub-populations or strata that are less variable than the original population, and to sample different parts of the population at different rates when this seems advisable. It is discussed more fully in sections 17.8 and 17.9.

EXAMPLE 17.2.1—In the preceding example, suppose you expect that both $e$ and $f$ will give high values. You decide that the sample shall consist of $e, f$, and one member drawn at random from $a, b, c, d$. Show how to obtain an unbiased estimate of the population total and show that the standard error of this estimate is 7.7. (This sampling plan is not as accurate as the plan in which $f$ alone was placed in a separate stratum, because the actual value for $e$ is not very high.)

EXAMPLE 17.2.2—If previous information suggests that $f$ will be high, $d$ and $e$ moderate, and $a, b$, and $c$ small, we might try stratified sampling with 3 strata. The sample consists of $f$, either $d$ or $e$, and one chosen from $a, b$, and $c$. Work out the unbiased estimate of the population total for each of the 6 possible samples and show that its standard error is 3.9.

**17.3—Probability sampling.** The preceding examples were intended to introduce you to *probability sampling*. This is a general name given to sampling plans in which

(i) every member of the population has a known probability of being included in the sample,

(ii) the sample is drawn by some method of random selection consistent with these probabilities,

(iii) we take account of these probabilities of selection in making the estimates from the sample.

Note that the probability of selection need not be equal for all members of the population: it is sufficient that these probabilities be known. In the first example in the previous section, each member of the population had an equal chance of being in the sample, and each member of the sample received an equal weight in estimating the population total. But in the second example, member $f$ was given a probability 1 of appearing in the sample, as against 2/5 for the rest of the population. This inequality in the probabilities of selection was compensated for by assigning a weight 5/2 to these other members when making the estimate. The use of unequal probabilities produces a substantial gain in precision for some types of population (see section 17.9).

Probability sampling has several advantages. By probability theory it is possible to study the biases and the standard errors of the estimates from different sampling plans. In this way much has been learned about the scope, advantages, and limitations of each plan. This information helps greatly in selecting a suitable plan for a particular sampling job. As will be seen later, most probability sampling plans also enable the standard error of the estimate, and confidence limits for the true population value, to be computed from the results of the sample. Thus, when a probability sample has been taken, we have some idea as to how accurate the estimates are.

Probability sampling is by no means the only way of selecting a sample. An alternative method is to ask someone who has studied the population to point out "average" or "typical" members, and then confine the sample to these members. When the population is highly variable and the sample is small, this method will often give more accurate estimates than probability sampling. Another method is to restrict the sampling to those members that are conveniently accessible. If bales of goods are stacked tightly in a warehouse, it is difficult to get at the inside bales of the pile and one is tempted to confine attention to the outside bales. In many biological problems it is hard to see how a workable probability sample can be devised, as for instance in estimating the number of house flies in a town, or of field mice in a wood, or of plankton in the ocean.

One drawback of these alternative methods is that when the sample has been obtained, there is no way of knowing how accurate the estimate is. Members of the population picked out by an expert as typical may be more or less atypical. Outside bales may or may not be similar to interior bales. Probability sampling formulas for the standard error of the estimate or for confidence limits do not apply to these methods. Consequently, it is wise to use probability sampling unless there is a clear case that this is not feasible or is prohibitively expensive.

**17.4—Listing the population.** In order to apply probability sampling, we must have some way of subdividing the population into units, called *sampling units*, which form the basis for the selection of the sample. The sampling units must be distinct and non-overlapping, and they must together constitute the whole of the population. Further, in order to make some kind of random selection of sampling units, we must be able to number or *list* all the units. As will be seen, we need not always write down the complete list but we must be in a position to construct it. Listing is easily accomplished when the population consists of 5,000 cards neatly arranged in a file, or 300 ears of corn lying on a bench, or the trees in a small orchard. But the subdivision of a population into sampling units that can be listed sometimes presents a difficult practical problem.

Although we have spoken of the population as being concrete and definite, there may be some vagueness about the population which does not become apparent until a sampling is being planned. Before we can come to grips with a population of farms, or of nursing homes, we must define a farm or a nursing home. The definition may require much study and the final decision may have to be partly arbitrary. Two principles to keep in mind are that the definition should be appropriate to the purpose of the sampling and that it should be usable in the field (i.e., the person collecting the information should be able to tell what is in and what is out of the population as defined).

Sometimes the available listings of farms, creameries, or nursing homes are deficient. The list may be out of date, having some members that no longer belong to our population and omitting some that do belong. The list may be based on a definition different from that which we wish to use for our population. These points should be carefully checked before using any list. It often pays to spend considerable effort in revising a list to make it complete and satisfactory, since this may be more economical than to construct a new list. Where a list covers only part of the population, one procedure is to sample this part by means of the list, and construct a separate method of sampling for the unlisted part of the population. Stratified sampling is useful in this situation: all listed members are assigned to one stratum and unlisted members to another.

Preparing a list where none is available may require ingenuity and hard work. To cite an easy example, suppose that we wish to take a number of crop samples, each 2 ft. $\times$ 2 ft., from a plot 200 ft. $\times$ 100 ft. Divide the length of the plot into 100 sections, each 2 ft., and the breadth into 50 sections, each 2 ft. We thus set up a coordinate system that divides the whole plot into 100 $\times$ 50 or 5,000 quadrats, each 2 ft. $\times$ 2 ft. To select a quadrat by simple random sampling, we draw a random number between 1 and 100 and another random number between 1 and 50. These coordinates locate the corner of the quadrat that is farthest from the origin of our system. However, the problem becomes harder if the plot measures 163 ft. $\times$ 100 ft., and much harder if instead of a plot we have an irregularly shaped field. Further, if we have to select a number of areas each 6 in. $\times$ 6 in. from a large field, giving every area an equal chance of selection, the time spent in selecting and locating the sample areas becomes substantial. Partly for this reason, methods of systematic sampling (section 17.7) have come to be favored in routine soil sampling (8).

Another illustration is a method for sampling (for botanical or chemical analysis) the produce of a small plot that is already cut and bulked. The bulk is separated into two parts and a coin is tossed (or a random number drawn) to decide which part shall contain the sample. This part is then separated into two, and the process continues until a sample of about the desired size is obtained. At any stage it is good practice to make the two parts as alike as possible, provided this is done before the coin is tossed. A quicker method, of course, is to grab a handful of about the desired size; this is sometimes satisfactory but sometimes proves to be biased.

In urban sampling in the United States, the city block is often used as a sampling unit, a listing of the blocks being made from a map of the town. For extensive rural sampling, county maps have been divided into areas with boundaries that can be identified in the field and certain of these areas are selected to constitute the sample. The name *area sampling* has come to be associated with these and other methods in which the sampling unit is an area of land. Frequently the principal advantage of area sampling, although not the only one, is that it solves the problem of providing a listing of the population by sampling units.

In many sampling problems there is more than one type or size of sampling unit into which the population can be divided. For instance, in soil sampling in which borings are taken, the size and shape of the borer can be chosen by the sampler. The same is true of the frame used to mark out the area of land that is cut in crop sampling. In a dental survey of the fifth-grade school children in a city, we might regard the child as the sampling unit and select a sample of children from the combined school registers for the city. It would be administratively simpler, however, to take the school as the sampling unit, drawing a sample of schools and examining every fifth-grade child in the selected schools. This approach, in which the sampling unit consists of some natural group (the school) formed from the smaller units in which we are interested (the children), goes by the name of *cluster sampling*.

If you are faced with a choice between different sampling units, the guiding rule is to try to select the one that returns the greatest precision for the available resources. For a fixed size of sample (e.g., 5% of the population), a large sampling unit usually gives less accurate results than a small unit, although there are exceptions. To counterbalance this, it is generally cheaper and easier to take a 5% sample with a large sampling unit than with a small one. A thorough comparison between two units is likely to require a special investigation, in which both sampling errors and costs (or times required) are computed for each unit.

**17.5—Simple random sampling.** In this and later sections, some of the best-known methods for selecting a probability sample will be presented. The goal is to use a sampling plan that will give the highest precision for the resources to be expended, or, equivalently, that will attain a desired degree of precision with the minimum expenditure of resources. It is worth while to become familiar with the principal plans, since they are designed to take advantage of any information that you may have about the structure of the population and about the costs of taking the sample.

In section 17.2 you have already been introduced to *simple random sampling*. This is a method in which the members of the sample are drawn independently with equal probabilities. In order to illustrate the use of a table of random numbers for drawing a random sample, suppose that the population contains $N = 372$ members and that a sample of size $n = 10$ is wanted. Select a 3-digit starting number from table 1.5.1, say the number is 539 in row 11 of columns 80–82. Read down the column and pick out the first ten 3-digit numbers that do not exceed 372. These are 334, 365, 222, 345, 245, 272, 075, 038, 127, and 112. The sample consists of the sampling units that carry these numbers in your listing of the population. If any number appears more than once, ignore it on subsequent appearances and proceed until ten *different* numbers have been found.

If the first digit in $N$ is 1, 2, or 3, this method requires you to skip many numbers in the table because they are too large. (In the above example we had to cover 27 numbers in order to find 10 for the sample.) This does not matter if there are plenty of random numbers. An alternative method is to use all 3-digit numbers up to $2 \times 372 = 744$. Starting at the same place, the first 10 numbers that do not exceed 744 are 539, 334, 615, 736, 365, 222, 345, 660, 431, and 427. Now subtract 372 from all numbers larger than 372. This gives, for the sample, 167, 334, 243, 364, 365, 222, 345, 288, 59, and 55. With $N = 189$, for instance, we can make use of all numbers up to $5 \times 189 = 945$ by this device, subtracting 189 or 378 or 567 or 756 as the case may be.

As mentioned previously, simple random sampling leaves the selection of the sample entirely to chance. It is often a satisfactory method when the population is not highly variable and, in particular, when estimating proportions that are likely to lie between 20% and 80%. On the other hand, if you have any knowledge of the variability in the population, such as that certain segments of it are likely to give higher responses than others, one of the methods to be described later may be more precise.

If $Y_i(i = 1, 2, \ldots N)$ denotes the variable that is being studied, the standard deviation, $\sigma$, of the population is defined as

$$\sigma = \sqrt{\frac{\Sigma(Y_i - \bar{Y})^2}{N - 1}},$$

where $\bar{Y}$ is the population mean of the $Y_i$ and the sum $\Sigma$ is taken over all sampling units in the population.

The standard error of the mean of a simple random sample of size $n$ is:

$$\sigma_{\bar{y}} = \frac{\sigma}{\sqrt{n}}\sqrt{(1 - \phi)},$$

where $\phi = n/N$ is the *sampling fraction*, i.e., the fraction of the population that is included in the sample. (The sampling fraction is commonly denoted by the symbol $f$, but $\phi$ is used here to avoid confusion with our previous use of $f$ for degrees of freedom.)

The term $\sigma/\sqrt{n}$ is already familiar to you: this is the usual formula for the standard error of a sample mean. The second factor, $\sqrt{(1 - \phi)}$, is known as the *finite population correction*. It enters because we are sampling from a population of finite size, $N$, instead of from an infinite population as is assumed in the usual theory. Note that this term makes the standard error zero when $n = N$, as it should do, since we have then measured every unit in the population. In practical applications the finite population correction is close to 1 and can be omitted when $n/N$ is less than 10%, i.e., when the sample comprises less than 10% of the population.

This result is remarkable. For a large population with a fixed amount of variability (a given value of $\sigma$), the standard error of the mean depends mainly on the size of sample and only to a minor extent on the fraction of the population that is sampled. For given $\sigma$, a sample of 100 is almost as precise when the population size is 200,000 as when the population size is 20,000 or 2,000. Intuitively, some people feel that one cannot possibly get accurate results from a sample of 100 out of a population of 200,000, because only a tiny fraction of the population has been measured. Actually, whether the sampling plan is accurate or not depends primarily on the size of $\sigma/\sqrt{n}$. This shows why sampling can bring about a great reduction in the amount of measurement needed.

For the *estimated* standard error of the sample mean we have

$$s_{\bar{y}} = \frac{s}{\sqrt{n}} \sqrt{(1 - \phi)},$$

where $s$ is the standard deviation of the sample, calculated in the usual way.

If the sample is used to estimate the population *total* of the variable under study, the estimate is $N\bar{y}$ and its standard error is

$$s_{N\bar{y}} = \frac{Ns}{\sqrt{n}} \sqrt{(1 - \phi)}$$

In simple random sampling for attributes, where every member of the sample is classified into one of two classes, we take

$$s_p = \sqrt{\frac{pq}{n}} \sqrt{(1 - \phi)}$$

where $p$ is the proportion of the sample that lies in one of the classes. Suppose that 50 families are picked at random from a list of 432 families who possess telephones and that 10 of the families report that they are listening to a certain radio program. Then $p = 0.2$, $q = 0.8$ and

$$s_p = \sqrt{\frac{(0.2)(0.8)}{50}} \sqrt{\left(1 - \frac{50}{432}\right)} = 0.053$$

If we ignore the finite population correction, we find $s_p = 0.057$.

Note that the formula for $s_p$ holds only if each unit is classified as a whole into one of the two classes. If you are using cluster sampling and are classifying individual elements within each cluster, a different formula for $s_p$ must be used. For instance, in estimating the percentage of diseased plants in a field from a sample of 360 plants, the formula above holds if the plants were selected independently and at random. To save time in the field, however, we might have chosen 40 areas, each consisting of 3 plants in each of 3 neighboring rows. With this method the area (a cluster of 9 plants) is the sampling unit. If the distribution of disease in the field were extremely patchy, it might happen that every area had either all 9 plants diseased or no plants diseased. In this event the sample of 40 areas would be no more precise than a sample of 40 independently chosen plants, and we would be deceiving ourselves badly if we thought that we had a binomial sample of 360 plants.

The correct procedure for computing $s_p$ is simple. Calculate $p$ separately for each area (or sampling unit) and apply to these $p$'s the previous formula for continuous variates. That is, if $p_i$ is the percentage diseased in the $i$th area, the sample standard deviation is

$$s = \sqrt{\frac{\Sigma(p_i - p)^2}{(n-1)}},$$

where $n$ is now the number of areas (cluster units). Then

$$s_p = \frac{s}{\sqrt{n}}\sqrt{(1 - \phi)}$$

For instance, suppose that the numbers of diseased plants in the 40 areas were as given in table 17.5.1.

TABLE 17.5.1
NUMBERS OF DISEASED PLANTS (OUT OF 9) IN EACH OF 40 AREAS

| | | | | | | | | | | | | | | | | | | | |
|---|---|---|---|---|---|---|---|---|---|---|---|---|---|---|---|---|---|---|---|
| 2 | 5 | 1 | 1 | 1 | 7 | 0 | 0 | 3 | 2 | 3 | 0 | 0 | 0 | 7 | 0 | 4 | 1 | 2 | 6 |
| 0 | 0 | 1 | 4 | 5 | 0 | 1 | 4 | 2 | 6 | 0 | 2 | 4 | 1 | 7 | 3 | 5 | 0 | 3 | 6 |

Grand total = 99

The standard deviation of this sample is 2.331. Since the *proportions* of diseased plants in the 40 areas are found by dividing the numbers in table 17.5.1 by 9, the standard deviation of the proportions is

$$s = \frac{2.331}{9} = 0.259$$

Hence, (assuming $N$ large),

$$s_p = \frac{s}{\sqrt{n}} = \frac{0.259}{\sqrt{40}} = 0.041$$

For comparison, the result given by the binomial formula will be worked out. From the total in table 17.5.1, $p = 99/360 = 0.275$. The binomial formula is

$$s_p = \sqrt{\frac{pq}{360}} = \sqrt{\frac{(0.275)(0.725)}{360}} = 0.024,$$

giving an overly optimistic notion of the precision of $p$.

Frequently, the clusters are not all of the same size. This happens when the sampling units are areas of land that contain different numbers of the plants that are being classified. Let $m_i$ be the number of elements that are classified in the $i$th unit, and $a_i$ the number that fall into a specified class, so that $p_i = a_i/m_i$. Then $p$, the over-all proportion in the sample, is $(\Sigma a_i)/(\Sigma m_i)$, where each sum is taken over the $n$ cluster units.

The formula for $s$, the standard deviation of the $p_i$, uses the weighted mean square of the $m_i$:

$$s = \sqrt{\frac{1}{(n-1)} \Sigma \left\{ \left(\frac{m_i}{\bar{m}}\right)^2 (p_i - p)^2 \right\}}$$

where $\bar{m} = \Sigma m_i/n$ is the average size of cluster in the sample. This formula is an approximation, no correct expression for $s$ being known in usable form. As before, we have

$$s_p = \frac{s}{\sqrt{n}} \sqrt{(1 - \phi)}$$

For computing purposes, $s$ is better expressed as

$$s = \frac{1}{\bar{m}} \sqrt{\frac{1}{(n-1)} \left\{ \Sigma a_i^2 - 2p\Sigma a_i m_i + p^2 \Sigma m_i^2 \right\}}$$

The sums of squares $\Sigma a_i^2$, $\Sigma m_i^2$ and the sum of products $\Sigma a_i m_i$ are calculated without the usual corrections for the mean. The same value of $s$ is obtained whether the corrections for the mean are applied or not, but it saves time not to apply them.

EXAMPLE 17.5.1—If a sample of 4 from the 16 townships of a county has a standard deviation 45, show that the standard error of the mean is 19.5.

EXAMPLE 17.5.2—In the example presented in section 17.2 we had $N = 6$, $n = 3$, and the values for the 6 members of the population were 1, 2, 4, 6, 7, and 16. The formula for the true standard error of the estimated population total is

$$\sigma_{N\hat{y}} = \frac{N\sigma}{\sqrt{n}} \sqrt{\left(1 - \frac{n}{N}\right)}$$

Verify that this formula agrees with the result, 13.2, which we found by writing down all possible samples.

EXAMPLE 17.5.3—A simple random sample of size 100 is taken in order to estimate some proportion (e.g., the proportion of males) whose value in the population is close to $\frac{1}{2}$. Work out the standard error of the sample proportion $p$ when the size of the population is (i) 200, (ii) 500, (iii) 1,000, (iv) 10,000, (v) 100,000. Note how little the standard error changes for $N$ greater than 1,000.

EXAMPLE 17.5.4—Show that the coefficient of variation of the sample mean is the same as that of the estimated population total.

EXAMPLE 17.5.5—In simple random sampling for attributes, show that the standard error of $p$, for given $N$ and $n$, is greatest when $p$ is 50%, but that the coefficient of variation of $p$ is largest when $p$ is very small.

**17.6—Size of sample.** At an early stage in the design of a sample, the question "How large a sample do I need?" must be considered. Although a precise answer may not be easy to find, for reasons that will appear, there is a rational method of attack on the problem.

Clearly, we want to avoid making the sample so small that the estimate is too inaccurate to be useful. Equally, we want to avoid taking a sample that is too large, in that the estimate is more accurate than we require. Consequently, the first step is to decide how large an error we can tolerate in the estimate. This demands careful thinking about the use to be made of the estimate and about the consequences of a sizeable error. The figure finally reached may be to some extent arbitrary, yet after some thought samplers often find themselves less hesitant about naming a figure than they expected to be.

The next step is to express the allowable error in terms of confidence limits. Suppose that $L$ is the allowable error in the sample mean, and that we are willing to take a 5% chance that the error will exceed $L$. In other words, we want to be reasonably certain that the error will not exceed $L$. Remembering that the 95% confidence limits computed from a sample mean are

$$\bar{y} \pm \frac{2\sigma}{\sqrt{n}},$$

we may put

$$L = \frac{2\sigma}{\sqrt{n}}$$

This gives, for the required sample size,

$$n = \frac{4\sigma^2}{L^2}$$

In order to use this relation, we must have an estimate of the population standard deviation, $\sigma$. Often a good guess can be made from the results of previous samplings of this population or of other similar populations. For example, an experimental sample was taken in 1938 to estimate the yield per acre of wheat in certain districts of North Dakota (7). For a sample of 222 fields, the variance of the yield per acre from field to field was $s^2 = 90.3$ (in bushels$^2$). How many fields are indicated if we wish to estimate the true mean yield within ± 1 bushel, with a 5% risk that the error will exceed 1 bushel? Then

$$n = \frac{4\sigma^2}{L^2} = \frac{4(90.3)}{(1)^2} = 361 \text{ fields}$$

If this estimate were being used to plan a sample in some later year, it would be regarded as tentative, since the variance between fields might change from year to year.

In the absence of data from an earlier sample, we can sometimes estimate $\sigma$ from a knowledge of the range of variation in the population, using the relation between the range and $\sigma$ as described in section 2.2.   If the range is known in a population of size greater than 500, we may take (range)/6 as a crude estimate of $\sigma$.

If the quantity to be estimated is a binomial proportion, the allowable error, $L$, for 95% confidence probability is

$$L = 2\sqrt{\frac{pq}{n}}$$

The sample size required to attain a given limit of error, $L$, is therefore

$$n = \frac{4pq}{L^2}$$

In this formula, $p$, $q$, and $L$ may be expressed either as proportions or as percentages, provided that they are all expressed in the same units.   The result necessitates an advance estimate of $p$.   If $p$ is likely to lie between 35% and 65%, the advance estimate can be quite rough, since the product $pq$ varies little for $p$ lying between these limits.   If, however, $p$ is near zero or 100%, accurate determination of $n$ requires a close guess about the value of $p$.

We have ignored the finite population correction in the formulas presented in this section.   This is satisfactory for the majority of applications. If the computed value of $n$ is found to be more than 10% of the population size, $N$, a revised value $n'$ which takes proper account of the correction is obtained from the relation

$$n' = \frac{n}{1 + \phi}$$

For example, casual inspection of a batch of 480 seedlings indicates that about 15% are diseased.   Suppose we wish to know the size of sample needed to determine $p$, the per cent diseased, to within ± 5%, apart from a 1-in-20 chance.   The formula gives

$$n = \frac{4(15)(85)}{(25)} = 204 \text{ seedlings}$$

At this point we might decide that it would be as quick to classify every seedling as to plan a sample that is a substantial part of the whole batch. If we decide on sampling, we make a revised estimate, $n'$, as

$$n' = \frac{n}{1 + \phi} = \frac{204}{1 + \dfrac{204}{480}} = 143$$

The formulas presented in this section are appropriate to simple random sampling. If some other sampling method is to be used, the general principles for the determination of $n$ remain the same, but the formula for the confidence limits, and hence the formula connecting $L$ with $n$, will change. Formulas applicable to more complex methods of sampling can be obtained in books devoted to the subject, e.g., (2, 4, 11). In practice, the formulas in this section are frequently used to provide a preliminary notion of the value of $n$, even if simple random sampling is not intended to be used. The values of $n$ may be revised later if the proposed method of sampling is markedly different in precision from simple random sampling.

When more than one variable is to be studied, the value of $n$ is first estimated separately for each of the most important variables. If these values do not differ by much, it may be feasible to use the largest of the $n$'s. If the $n$'s differ greatly, one method is to use the largest $n$, but to measure certain items on only a sub-sample of the original sample, e.g., on 200 sampling units out of 1,000. In other situations, great disparity in the $n$'s is an indication that the investigation must be split into two or more separate surveys.

EXAMPLE 17.6.1—A simple random sample of houses is to be taken to estimate the percentage of houses that are unoccupied. The estimate is desired to be correct to within $\pm 1\%$, with 95% confidence. One advance estimate is that the percentage of unoccupied houses will be about 6%; another is that it will be about 4%. What sizes of sample are required on these two forecasts? What size would you recommend?

EXAMPLE 17.6.2—The total number of rats in the residential part of a large city is to be estimated with an error of not more than 20%, apart from a 1-in-20 chance. In a previous survey, the mean number of rats per city block was 9 and the sample standard deviation was 19 (the distribution is extremely skew). Show that a simple random sample of around 450 blocks should suffice.

EXAMPLE 17.6.3—West (12) quotes the following data for 556 full-time farms in Seneca County, New York.

|  | Mean | Standard Deviation Per Farm |
|---|---|---|
| Acres in corn | 8.8 | 9.0 |
| Acres in small grains | 42.0 | 39.5 |
| Acres in hay | 27.9 | 26.9 |

If a coefficient of variation of up to 5% can be tolerated, show that a random sample of about 240 farms is required to estimate the total acreage of each crop in the 556 farms with this degree of precision. (Note that the finite population correction must be used.) This example illustrates a result that has been reached by several different investigators; with small farm populations such as counties, a substantial part of the whole population must be sampled in order to obtain accurate estimates.

**17.7—Systematic sampling.** In order to draw a 10% sample from a list of 730 cards, we might select a random number between 1 and 10, say 3, and pick every 10th card thereafter; i.e., the cards numbered 3, 13, 23, and so on, ending with the card numbered 723. A sample of this kind is known as a *systematic sample*, since the choice of its first member, 3, determines the whole sample.

Systematic sampling has two advantages over simple random sampling. It is easier to draw, since only one random number is required, and it distributes the sample more evenly over the listed population. For this reason systematic sampling often gives more accurate results than simple random sampling. Sometimes the increase in accuracy is large. In routine sampling, systematic selection has become a popular technique.

There are two potential disadvantages. If the population contains a periodic type of variation, and if the interval between successive units in the systematic sample happens to coincide with the wave length (or a multiple of it) we may obtain a sample that is badly biased. To cite extreme instances, a systematic sample of the houses in a city might contain far too many, or too few, corner houses; a systematic sample from a book of names might contain too many, or too few, names listed first on a page, who might be predominantly males, or heads of households, or persons of importance. A systematic sample of the plants in a field might have the selected plants at the same positions along every row. These situations can be avoided by being on the lookout for them and either using some other method of sampling or selecting a new random number frequently. In field sampling, we could select a new random number in each row. Consequently, it is well to know something about the nature of the variability in the population before deciding to use systematic sampling.

The second disadvantage is that from the results of a systematic sample there is no reliable method of estimating the standard error of the sample mean. Textbooks on sampling give various formulas for $s_{\bar{y}}$ that may be tried: each formula is valid for a certain type of population, but a formula can be used with confidence only if we have evidence that the population is of the type to which the formula applies. However, systematic sampling often is a part of a more complex sampling plan in which it is possible to obtain unbiased estimates of the sampling errors.

**17.8—Stratified sampling.** There are three steps in stratified sampling:
    (1) The population is divided into a number of parts, called *strata*.
    (2) A sample is drawn independently in each part.
    (3) As an estimate of the population mean, we use

$$\bar{y}_{st} = \frac{\Sigma N_h \bar{y}_h}{N},$$

where $N_h$ is the total number of sampling units in the $h$th stratum, $\bar{y}_h$ is the sample mean in the $h$th stratum and $N = \Sigma N_h$ is the size of the population. Note that we must know the values of the $N_h$ (i.e., the sizes of the strata) in order to compute this estimate.

There are several reasons why stratification is commonly employed in sampling plans. It can be shown that differences between the strata means in the population do not contribute to the sampling error of the estimate $\bar{y}_{st}$. In other words, the sampling error of $\bar{y}_{st}$ arises solely from variations among sampling units that are in the same stratum. If we can form strata so that a heterogeneous population is divided into parts each of which is fairly homogeneous, we may expect a gain in precision over simple random sampling. In taking 24 soil or crop samples from a rectangular field, we might divide the field into 12 compact plots, and draw 2 samples at random from each plot. Since a small piece of land is usually more homogeneous than a large piece, this stratification will probably bring about an increase in precision, although experience indicates that in this application the increase will be modest rather than spectacular. To estimate total wheat acreage from a sample of farms, we might stratify by size of farm, using any information available for this purpose. In this type of application the gain in precision is frequently large.

In stratified sampling, we can choose the size of sample that is to be taken from any stratum. This freedom of choice gives us scope to do an efficient job of allocating resources to the sampling within strata. In some applications, this is the principal reason for the gain in precision from stratification. Further, when different parts of the population present different problems of listing and sampling, stratification enables these problems to be handled separately. For this reason, hotels and large apartment houses are frequently placed in a separate stratum in a sample of the inhabitants of a city.

We now consider the estimate from stratified sampling and its standard error. For the population mean, the estimate given above may be written

$$\bar{y}_{st} = \frac{1}{N} \Sigma N_h \bar{y}_h = \Sigma W_h \bar{y}_h,$$

where $W_h = N_h/N$ is the relative *weight* attached to the stratum. Note that the sample means, $\bar{y}_h$, in the respective strata are weighted by the sizes, $N_h$, of the strata. The arithmetic mean of the sample observations is no longer the estimate except in one important special case. This occurs with *proportional allocation*, when we sample the same fraction from every stratum. With proportional allocation,

$$\frac{n_1}{N_1} = \frac{n_2}{N_2} = \ldots = \frac{n_h}{N_h} = \frac{n}{N}$$

It follows that

$$W_h = \frac{N_h}{N} = \frac{n_h}{n}$$

Hence

$$\bar{y}_{st} = \Sigma W_h \bar{y}_h = \frac{\Sigma n_h \bar{y}_h}{n} = \bar{y},$$

since $\Sigma n_h \bar{y}_h$ is the total of all observations in the sample. With proportional allocation, we are saved the trouble of computing a weighted mean: the sample is *self-weighting*.

The standard error of $\bar{y}_{st}$ is

$$s(\bar{y}_{st}) = \sqrt{\Sigma W_h^2 \frac{s_h^2}{n_h}}$$

where $s_h^2$ is the sample variance in the $h$th stratum, i.e.,

$$s_h^2 = \frac{\Sigma(Y_{hi} - \bar{y}_h)^2}{n_h - 1},$$

where $Y_{hi}$ is the $i$th member of the sample from the $h$th stratum. The formula for the standard error of $\bar{y}_{st}$ assumes that simple random sampling is used within each stratum and does not include the finite population correction. If the sampling fractions $\phi_h$ exceed 10% in some of the strata, we use the more general formula

$$s(\bar{y}_{st}) = \sqrt{\Sigma \frac{W_h^2 s_h^2}{n_h}(1 - \phi_h)}$$

With proportional allocation the sampling fractions $\phi_h$ are all equal and the general formula simplifies to

$$s(\bar{y}_{st}) = \sqrt{\frac{\Sigma W_h s_h^2}{n}} \cdot \sqrt{(1 - \phi)}$$

If, further, the population variances are the same in all strata (a reasonable assumption in some agricultural applications), we obtain an additional simplification to

$$s(\bar{y}_{st}) = \frac{s_w}{\sqrt{n}}\sqrt{(1 - \phi)}$$

This result is the same as that for the standard error of the mean with simple random sampling, except that $s_w$, the pooled standard deviation *within strata*, appears in place of the sample standard deviation, $s$. In practice, $s_w$ is computed from an analysis of variance of the data.

As an example of proportional allocation, the data in table 17.8.1 come from an early investigation by Clapham (1) of the feasibility of sampling for estimating the yields of small cereal plots. A rectangular plot of barley was divided transversely into 3 equal strata. Ten samples, each

TABLE 17.8.1

ANALYSIS OF VARIANCE OF A STRATIFIED RANDOM SAMPLE
Wheat grain yields — gm. per meter

| Source of Variation | Degrees of Freedom | Sum of Squares | Mean Square |
|---|---|---|---|
| Total | 29 | 8,564 | 295.3 |
| Between strata | 2 | 2,073 | 1,036.5 |
| Within strata | 27 | 6,491 | 240.4 |

a meter length of a single row, were chosen by simple random sampling from each stratum. The problem is to compute the standard error of the estimated mean yield per meter of row.

In this example, $s_w = \sqrt{240.4} = 15.5$, and $n = 30$. Since the sample is only a negligible part of the whole plot, $n/N$ is negligible and

$$s(\bar{y}_{st}) = \frac{s_w}{\sqrt{n}} = \frac{15.5}{\sqrt{30}} = 2.83 \text{ gm.}$$

How effective was the stratification? From the analysis of variance it is seen that the mean square between strata is over four times as large as that within strata. This is an indication of real differences in level of yield from stratum to stratum. It is possible to go further, and estimate what the standard error of the mean would have been if simple random sampling had been used without any stratification. With simple random sampling, the corresponding formula for the standard error of the mean is

$$s(\bar{y}) = \frac{s}{\sqrt{n}},$$

where $s$ is the ordinary sample standard deviation. In the sample under discussion, $s$ is $\sqrt{295.3}$ (from the *total* mean square in table 17.8.1). Hence, as an estimate of the standard error of the mean under simp'e random sampling, we might take

$$s(\bar{y}) = \frac{\sqrt{295.3}}{\sqrt{30}} = 3.14 \text{ gm.,}$$

as compared with 2.83 gms. for stratified random sampling. Stratification has reduced the standard error by about 10%.

This comparison is not quite correct, for the rather subtle reason that the value of $s$ was calculated from the results of a stratified sample and not, as it should have been, from the results of a simple random sample. Valid methods of making the comparison are described for all types of stratified sampling in (2). The approximate method which we used is close enough when the stratification is proportional and at least 10 sampling units are drawn from every stratum.

**17.9—Choice of sample sizes in the individual strata.** It is sometimes thought that in stratified sampling we should sample the same fraction from every stratum; i.e., we should make $n_h/N_h$ the same in all strata, using proportional allocation. A more thorough analysis of the problem shows, however, that the *optimum* allocation is to take $n_h$ proportional to $N_h \sigma_h/\sqrt{c_h}$, where $\sigma_h$ is the standard deviation of the sampling units in the $h$th stratum, and $c_h$ is the cost of sampling per unit in the $h$th stratum. This method of allocation gives the smallest standard error of the estimated mean $\bar{y}_{st}$ for a given total cost of taking the sample. The rule tells us to take a larger sample, as compared with proportional allocation, in a stratum that is unusually variable ($\sigma_h$ large), and a smaller sample in a stratum where sampling is unusually expensive ($c_h$ large). Looked at in this way, the rule is consistent with common sense, as statistical rules always are if we think about them carefully. The rule reduces to proportional allocation when the standard deviation and the cost per unit are the same in all strata.

In order to apply the rule, advance estimates are needed both of the relative standard deviations and of the relative costs in different strata. These estimates need not be highly accurate; rough estimates often give results satisfactorily near to the optimum allocation. When a population is sampled repeatedly, the estimates can be obtained from the results of previous samplings. Even when a population is sampled for the first time, it is sometimes obvious that some strata are much more accessible to sampling than others. In this event it usually pays to hazard a guess about the differences in costs. In other situations we are unable to predict with any confidence which strata will be more variable or more costly, or we think that any such differences will be small. Proportional allocation is then used.

There is one common situation in which disproportionate sampling pays large dividends. This occurs when the principal variable that is being measured has a highly skewed or asymmetrical distribution. Usually, such populations contain a few sampling units that have large values for this variable and many units that have small values. Variables that are related to the sizes of economic institutions are often of this type, for instance, the total sales of grocery stores, the number of patients per hospital, the amounts of butter produced by creameries, and (for certain types of farming) farm income.

With populations of this type, stratification by size of institution is usually highly effective, and the optimum allocation is likely to be much better than proportional allocation. As an illustration, table 17.9.1 shows data for the number of students per institution in a population consisting of the 1,019 senior colleges and universities in the United States. The data, which apply mostly to the 1952–53 academic year, might be used as background information for planning a sample designed to give a quick estimate of total registration in some future year. The institutions are arranged in 4 strata according to size.

TABLE 17.9.1

DATA FOR TOTAL REGISTRATIONS PER SENIOR COLLEGE OR UNIVERSITY,
ARRANGED IN 4 STRATA

| Stratum: Number of Students Per Institution | Number of Institutions $N_h$ | Total Registration for the Stratum | Mean Per Institution $\bar{Y}_h$ | Standard Deviation Per Institution $\sigma_h$ |
|---|---|---|---|---|
| Less than 1,000 | 661 | 292,671 | 443 | 236 |
| 1,000–3,000 | 205 | 345,302 | 1,684 | 625 |
| 3,000–10,000 | 122 | 672,728 | 5,514 | 2,008 |
| Over 10,000 | 31 | 573,693 | 18,506 | 10,023 |
| Total | 1,019 | 1,884,394 | | |

Note that the 31 largest universities, about 3% in number, have 30% of the students, while the smallest group, which contains 65% of the institutions, contributes only 15% of the students. Note also that the within-stratum standard deviation, $\sigma_h$, increases rapidly with increasing size of institution.

Table 17.9.2 shows the calculations needed for choosing the optimun sample sizes within strata. We are assuming equal costs per unit within all strata. The products, $N_h\sigma_h$, are formed and added over all strata. Then the relative sample sizes, $N_h\sigma_h/\Sigma N_h\sigma_h$, are computed. These ratios, when multiplied by the intended sample size $n$, give the sample sizes in the individual strata.

TABLE 17.9.2

CALCULATIONS FOR OBTAINING THE OPTIMUM SAMPLE SIZES IN INDIVIDUAL STRATA

| Stratum: Number of Students | Number of Institutions $N_h$ | $N_h\sigma_h$ | Relative Sample Sizes $N_h\sigma_h/\Sigma N_h\sigma_h$ | Actual Sample Sizes | Sampling Rate (%) |
|---|---|---|---|---|---|
| Less than 1,000 | 661 | 155,996 | .1857 | 65 | 10 |
| 1,000–3,000 | 205 | 128,125 | .1526 | 53 | 26 |
| 3,000–10,000 | 122 | 244,976 | .2917 | 101 | 83 |
| Over 10,000 | 31 | 310,713 | .3700 | 31 | 100 |
| Total | 1,019 | 839,810 | 1.0000 | 250 | |

As a consequence of the large standard deviation in the stratum with the largest universities, the rule requires 37% of the sample to be taken from this stratum. Suppose we are aiming at a total sample size of 250. The rule then calls for (.3700)(250) or 92 universities from this stratum although the stratum contains only 31 universities in all. With highly skewed populations, as here, it is sometimes found that the optimum allocation demands 100% sampling, or even more than this, of the largest institutions. When this situation occurs, the best procedure is to take 100% of the "large" stratum, and employ the rule to distribute the remainder of the sample over the other strata. Following this procedure, we include in the sample all 31 largest institutions, leaving 219 to be distributed among the first 3 strata. In the first stratum, the size of sample is

$$219 \left\{ \frac{.1858}{.1858 + .1526 + .2917} \right\} = 65$$

The allocations, shown in the second column from the right of table 17.9.2, call for over 80% sampling in the second largest group of institutions (101 out of 122), but only a 10% sample of the small colleges. In practice we might decide, for administrative convenience, to take a 100% sample in the second largest group.

It is worth while to ask: Is the optimun allocation much superior to proportional allocation? If not, there is little value in going to the extra trouble of calculating and using the optimun allocation. We cannot, of course, answer this question for a future sample that is not yet taken, but we can compare the two methods of allocation for the 1952–53 registrations for which we have data. To do this, we use the data in tables 17.9.1 and 17.9.2 and the standard error formulas in section 17.8 to compute the standard errors of the estimated population totals by the two methods. These standard errors are found to be 26,000 for the optimum allocation, as against 107,000 for proportional allocation. If simple random sampling had been used, with no stratification, a similar calculation shows that the corresponding standard error would have been 216,000. The reduction in the standard error due to stratification, and the additional reduction due to the optimum allocation, are both striking. In an actual future sampling based on this stratification, the gains in precision would presumably be slightly less than these figures indicate.

**17.10—Stratified sampling for attributes.** If an attribute is being sampled, the estimate appropriate to stratified sampling is

$$p_{st} = \Sigma W_h p_h$$

where $p_h$ is the sample proportion in stratum $h$ and $W_h = N_h/N$ is the stratum weight. To find the standard error of $p_{st}$ we substitute $p_h q_h$ for $s_h^2$ in the formulas previously given in section 17.8.

As an example, consider a sample of 692 families in Iowa to determine, among other things, how many had vegetable gardens in 1943. The families were arranged in 3 strata—urban, rural non-farm, and farm— because it was anticipated that the three groups might show differences in the frequency and size of vegetable gardens. The data are given in table 17.10.1.

TABLE 17.10.1

NUMBERS OF VEGETABLE GARDENS AMONG IOWA FAMILIES, ARRANGED IN 3 STRATA

| Stratum | Number of Families $N_h$ | Weight $W_h$ | Number in Sample $n_h$ | Number With Gardens | Percentage With Gardens |
|---|---|---|---|---|---|
| Urban | 312,393 | 0.445 | 300 | 218 | 72.7 |
| Rural non-farm | 161,077 | 0.230 | 155 | 147 | 94.8 |
| Farm | 228,354 | 0.325 | 237 | 229 | 96.6 |
| | 701,824 | 1.000 | 692 | 594 | |

The numbers of families were taken from the 1940 census. The sample was allotted roughly in proportion to the number of families per stratum, a sample of 1 per 1,000 being aimed at.

The weighted mean percentage of Iowa families having gardens was estimated as

$$\Sigma W_h p_h = (0.445)(72.7) + (0.230)(94.8) + (0.325)(96.6) = 85.6\%$$

This is practically the same as the sample mean percentage, 594/692 or 85.8%, because allocation was so close to proportional.

For the variance of the estimated mean, we have

$$\Sigma W_h^2 p_h q_h / n_h = (0.445)^2(72.7)(27.3)/300 + \text{etc.} = 1.62$$

The standard error, then, is 1.27%.

With a sample of this size, the estimated mean will be approximately normally distributed: the confidence limits may be set as

$$85.6 \pm (2)(1.27) : 83.1\% \text{ and } 88.1\%$$

For the optimum choice of the sample sizes within strata, we should take $n_h$ proportional to $N_h\sqrt{p_h q_h / c_h}$. If the cost of sampling is about the same in all strata, as is true in many surveys, this implies that the fraction sampled, $n_h/N_h$, should be proportional to $\sqrt{p_h q_h}$. Now the quantity $\sqrt{pq}$ changes relatively little as $p$ ranges from 25% to 75%. Consequently, proportional allocation is often highly efficient in stratified sampling for attributes. The optimum allocation will produce a substantial reduction in the standard error, as compared with proportional allocation, only when some of the $p_h$ are close to zero or 100%, or when there are differential costs.

The example on vegetable gardens departs from the strict principles of stratified sampling in that the strata sizes and weights were not known exactly, being obtained from census data three years previously. Errors in the strata weights reduce the gain in precision from stratification and make the standard formulas inapplicable. It is believed that in this example these disturbances are of negligible importance.   Discussions of stratification when there are errors in the weights are given in (2) and (10).

EXAMPLE 17.10.1—The purpose of this example is to compare simple random sampling and systematic sampling of a small population.  The following data are the weights of maize (in 10-gm. units) for 40 successive hills lying in a single row: 104, 38, 105, 86, 63, 32, 47, 0, 80, 42, 37, 48, 85, 66, 110, 0, 73, 65, 101, 47, 0, 36, 16, 33, 22, 32, 33, 0, 35, 82, 37, 45, 30, 76, 45, 70, 70, 63, 83, 34.  To save you time, the population standard deviation is given as 30.1.  Compute the standard deviation of the mean of a simple random sample of 4 hills.  A systematic sample of 4 hills can be taken by choosing a random number between 1 and 10 and taking every 10th hill thereafter. Find the mean $\bar{y}_{sv}$ for each of the 10 possible systematic samples and compute the standard deviation of these means about the true mean $\bar{Y}$ of the population.  Note that the formula is

$$\sigma(\bar{y}_{sv}) = \sqrt{\frac{\Sigma(\bar{y}_{sv} - \bar{Y})^2}{10}}$$

Verify that the standard deviation of the estimate is about 8% lower with systematic sampling.  To what do you think this difference is due?

EXAMPLE 17.10.2—In the example of stratified sampling given in section 17.2, show that the estimate which we used for the population total was $N\bar{y}_{st}$.  From the general formula for the variance of $\bar{y}_{st}$, verify that the variance of the estimated population total is 48.75, as found directly in section 17.2. (Note that stratum I makes no contribution to this variance because $n_h = N_h$ in that stratum.)

EXAMPLE 17.10.3—In stratified sampling for attributes, the optimum sample distribution, with equal costs per unit in all strata, follows from taking $n_h$ proportional to $N_h\sqrt{p_hq_h}$.  It follows that the actual value of $n_h$ is

$$n_h = n\left\{\frac{N_h\sqrt{p_hq_h}}{\Sigma N_h\sqrt{p_hq_h}}\right\}$$

In the Iowa vegetable garden survey, suppose that the $p_h$ values found in the sample can be assumed to be the same as those in the population.  Show that the optimum sample distribution gives sample sizes of 445, 115, and 132 in the respective strata, and that the standard error of the estimated percentage with gardens would then be 1.17%, as compared with 1.27% in the sample itself.

EXAMPLE 17.10.4—For the population of colleges and universities discussed in section 17.9 it was stated that a stratified sample of 250 institutions, with proportional allocation, would have a standard error of 107,000 for the estimated total registration in all 1,019 institutions.  Verify this statement from the data in table 17.9.1.  Note that the standard error of the estimated population total, with proportional allocation, is

$$N\sqrt{\frac{\Sigma W_h\sigma_h^2}{n}}\sqrt{\left(1 - \frac{n}{N}\right)}$$

**17.11—Sampling in two stages.**  Consider the following miscellaneous group of sampling problems: (1) a study of the vitamin A content of butter produced by creameries, (2) a study of the protein content of wheat in the wheat fields in an area, (3) a study of red blood cell counts in a population of men aged 20–30, (4) a study of insect infestation of the leaves of the trees in an orchard, and (5) a study of the number of defective teeth in third-grade children in the schools of a large city.  What do these investigations have in common?  First, in each study an appropriate sampling unit suggests itself naturally—the creamery, the field of wheat, the individual man, the tree, and the school.  Secondly, and this is the important point, in each study the chosen sampling units can be *sub-sampled* instead of being measured completely.  Indeed,  sub-sampling is essential in the first three studies.  No one is going to allow us to take *all* the butter produced by a creamery in order to determine vitamin A content, or all the wheat in a field for the protein determination, or all the blood in a man in order to make a complete count of his red cells.  In the insect infestation study, it might be feasible, although tedious, to examine *all* leaves on any selected tree.  If the insect distribution is spotty, however, we would probably decide to take only a small sample of leaves from any selected tree in order to include more trees.  In the dental study we could take all the third-grade children in any selected school or we could cover a larger sample of schools by examining only a sample of children from the third grade in each selected school.

This type of sampling is called *sampling in two stages*, or sometimes *sub-sampling*.  The first stage is the selection of a sample of *primary sampling units*—the creameries, wheat fields, and so on.  The second stage is the taking of a *sub-sample* of *second-stage units*, or *sub-units*, from each selected primary unit.

As illustrated by these examples, the two-stage method is sometimes the only practicable way in which the sampling can be done.  Even when there is a choice between sub-sampling the units and measuring them completely, two-stage sampling gives the sampler greater scope, since he can choose both the size of the sample of primary units and the size of the sample that is taken from a primary unit.  In some applications an important advantage of two-stage sampling is that it facilitates the problem of listing the population.  Often it is relatively easy to obtain a list of the primary units, but difficult or expensive to list all the sub-units.  To list the trees in an orchard and draw a sample of them is usually simple, but the problem of making a random selection of the leaves on a tree may be very troublesome.  With two-stage sampling this problem is faced only for those trees that are in the sample.  No complete listing of all leaves in the orchard is required.

In the discussion of two-stage sampling we shall assume that the primary units are of approximately the same size.  A simple random sample of $n_1$ primary units is drawn, and the same number $n_2$ of sub-units is selected from each primary unit in the sample.  The estimated standard error of the sample mean $\bar{y}$ *per sub-unit* is then given by the formula

$$s_{\bar{y}} = \frac{1}{\sqrt{n_1}} \sqrt{\frac{\Sigma(\bar{y}_i - \bar{y})^2}{n_1 - 1}},$$

where $\bar{y}_i$ is the mean per sub-unit in the *i*th primary unit.  This formula does not include the finite population correction, but is reliable enough provided that the sample contains less than 10% of all primary units.  Note that the formula makes no use of the individual observations on the sub-units, but only of the primary unit means $\bar{y}_i$.  If the sub-samples are taken for a chemical analysis, a common practice is to composite the sub-sample and make one chemical determination for each primary unit.  With data of this kind we can still calculate $s_{\bar{y}}$.

In section 10.12 you learned about the "components of variance" technique, and applied it to a problem in two-stage sampling. The data were concentrations of calcium in turnip greens, 4 determinations being made for each of 3 leaves. The leaf can be regarded as the primary sampling unit, and the individual determination as the sub-unit. By applying the components of variance technique, you were able to see how the variance of the sample mean was affected by variation between determinations on the same leaf and by variation from leaf to leaf. You could also predict how the variance of the sample mean would change with different numbers of leaves and of determinations per leaf in the experiment.

Since this technique is of wide utility in two-stage sampling, we shall repeat some of the results. The observation on any sub-unit is considered to be the sum of two independent terms. One term, associated with the primary unit, has the same value for all second-stage units in the primary unit, and varies from one primary unit to another with variance $\sigma_1^2$. The second term, which serves to measure differences between second-stage units, varies independently from one sub-unit to another with variance $\sigma_2^2$. Suppose that a sample consists of $n_1$ primary units, from each of which $n_2$ sub-units are drawn. Then the sample as a whole contains $n_1$ independent values of the first term, whereas it contains $n_1 n_2$ independent values of the second term. Hence the variance of the sample mean $\bar{y}$ per sub-unit is

$$\sigma_{\bar{y}}^2 = \frac{\sigma_1^2}{n_1} + \frac{\sigma_2^2}{n_1 n_2}$$

The two components of variance, $\sigma_1^2$ and $\sigma_2^2$, can be estimated from the analysis of variance of a two-stage sample that has been taken. Table 17.11.1 gives the analysis of variance for a study by Immer (6), whose object was to develop a sampling technique for the determination of the sugar percentage in field experiments on sugar beets. Ten beets were chosen from each of 100 plots in a uniformity trial, the plots being the primary units. The sugar percentage was obtained separately for each beet. In order to simulate conditions in field experiments, the Between Plots mean square was computed as the mean square between plots within blocks of 5 plots. This mean square gives the experimental error variance that would apply in a randomized blocks experiment with 5 treatments.

TABLE 17.11.1

ANALYSIS OF VARIANCE OF SUGAR PERCENTAGE OF BEETS (ON A SINGLE-BEET BASIS)

| Source of Variation | Degrees of Freedom | Mean Square | Parameters Estimated |
|---|---|---|---|
| Between plots (primary units) | 80 | 2.9254 | $\sigma_2^2 + 10\,\sigma_1^2$ |
| Between beets (sub-units) within plots | 900 | 2.1374 | $\sigma_2^2$ |

The estimate of $\sigma_1^2$, the Between Plots component of variance, is

$$s_1^2 = \frac{2.9254 - 2.1374}{10} = 0.0788,$$

the divisor 10 being the number of beets (sub-units) taken per plot. As an estimate of $\sigma_2^2$, the within-plots component, we have

$$s_2^2 = 2.1374$$

Hence, if a new experiment is to consist of $n_1$ replications, with $n_2$ beets sampled from each plot, the predicted variance of a treatment mean is

$$s_{\bar{y}}^2 = \frac{0.0788}{n_1} + \frac{2.1374}{n_1 n_2}$$

We shall illustrate a few of the questions that can be answered from these data. How accurate are the treatment means in an experiment with 6 replications and 5 beets per plot? For this experiment we would expect

$$s_{\bar{y}} = \sqrt{\left(\frac{0.0788}{6} + \frac{2.1374}{30}\right)} = 0.29\%$$

Since the sample is large, the sugar percentage figure for a treatment mean would be correct to within $\pm$ (2) (0.29) or 0.58%, with 95% confidence.

If the standard error of a treatment mean is not to exceed 0.2%, what combinations of $n_1$ and $n_2$ are allowable? We must have

$$\frac{0.0788}{n_1} + \frac{2.1374}{n_1 n_2} = (0.2)^2 = 0.04$$

Since $n_1$ and $n_2$ are whole numbers, they will not satisfy this equation exactly: we must make sure that the left side of the equation does not exceed 0.04. You can verify that with 4 replications ($n_1 = 4$), there must be 27 beets per plot; with 8 replications, 9 beets per plot are sufficient; and with 10 replications, 7 beets per plot. As one would expect, the intensity of sub-sampling decreases as the intensity of sampling is increased. The total size of sample also decreases from 108 beets when $n_1 = 4$ to 70 beets when $n_1 = 10$.

**17.12—The allocation of resources in two-stage sampling.** The last example illustrates a general property of two-stage samples. The same standard error can be attained for the sample mean by using various combinations of values of $n_1$ and $n_2$. Which of these choices is the best? The answer depends, naturally, on the cost of adding an extra primary unit to the sample (in this case an extra replication) relative to that of adding an extra sub-unit in each primary unit (in this case an extra beet in each plot). Similarly, in the turnip greens example (section 10.12) the best sampling plan depends on the relative costs of taking an extra leaf and of making an extra determination per leaf. Obviously, if it is cheap to add primary units to the sample but expensive to add sub-units, the most economical plan will be to have many primary units and few (perhaps only one) sub-units per primary unit. For a general solution to this problem, however, we require a more exact formulation of the costs of various alternative plans.

The nature of the costs will vary from one application to another. In many sub-sampling studies the cost of the sample (apart from fixed overhead costs) can be approximated by a relation of the form

$$\text{cost} = c_1 n_1 + c_2 n_1 n_2$$

The factor $c_1$ is the average cost per primary unit of those elements of cost that depend solely on the number of primary units and not on the amount of sub-sampling. Thus in the sugar beet example $c_1$ includes the average cost of land and field operations required to produce one plot. If it were intended in actual experiments to composite all beets from a plot and make a *single* sugar determination for each plot, $c_1$ would also include the cost of making this determination. The factor $c_2$, on the other hand, is the average cost per sub-unit of those constituents of cost that are directly proportional to the total number of sub-units. This includes the average cost per beet of locating and picking the sample of beets. The cost of grinding and mixing the beets prior to the sugar analysis is also assigned to $c_2$, assuming that this cost is roughly proportional to the number of beets taken.

If advance estimates of these constituents of cost are made from some preliminary study, an efficient job of selecting the best amounts of sampling and sub-sampling can be done. The problem may be posed in two different ways. In some studies we specify the desired variance $V$ for the sample mean, and would like to attain this as cheaply as possible. In other applications the total cost $C$ that must not be exceeded is imposed upon us, and we want to get as small a value of $V$ as we can for this outlay. These two problems have basically the same solution. In either case we want to minimize the product

$$VC = \left( \frac{s_1^2}{n_1} + \frac{s_2^2}{n_1 n_2} \right) (c_1 n_1 + c_2 n_1 n_2)$$

Upon expansion, this becomes

$$VC = (s_1^2 c_1 + s_2^2 c_2) + n_2 s_1^2 c_2 + \frac{s_2^2 c_1}{n_2}$$

It can be shown that this expression has its smallest value when

$$n_2 = \sqrt{\frac{c_1 s_2^2}{c_2 s_1^2}}$$

This result gives an estimate of the best number of sub-units (beets) per primary unit (plot). The value of $n_1$ is found by solving either the cost equation or the variance equation for $n_1$, depending on whether cost or variance has been preassigned.

In the sugar beet example we had $s_1^2 = 0.0788$, $s_2^2 = 2.1374$, from which

$$n_2 = \sqrt{\frac{2.1374}{0.0788}} \sqrt{\frac{c_1}{c_2}} = 5.2 \sqrt{\frac{c_1}{c_2}}$$

In this study, cost data were not reported, although it is likely that $c_1$, the cost per plot, would be much greater than $c_2$. Evidently a fairly large number of beets per plot would be advisable. In practice, factors other than the sugar percentage determinations must also be taken into account in deciding on the number of replications in sugar beet experiments.

In the turnip greens example (section 10.12), $n_1$ is the number of leaves and $n_2$ the number of determinations of calcium concentration per leaf. Also, in the present notation,

$$s_1^2 = s_t^2 = 0.0724$$
$$s_2^2 = s^2 = 0.0066$$

Hence the most economical number of determinations per leaf is estimated to be

$$n_2 = \sqrt{\frac{c_1 s_2^2}{c_2 s_1^2}} = \sqrt{\frac{0.0066}{0.0724}} \sqrt{\frac{c_1}{c_2}} = 0.30 \sqrt{\frac{c_1}{c_2}}$$

In practice, $n_2$ must be a whole number, and the smallest value it can have is 1. This equation shows that $n_2 = 1$, i.e., one determination per leaf, unless $c_1$ is many times greater than $c_2$. Actually, since $c_2$ includes the cost of the chemical determinations, it is likely to be greater than $c_1$. The relatively large variation among leaves and the cost considerations both point to the choice of one determination per leaf. This example also illustrates that a choice of $n_2$ can often be made from the equation even when information about relative costs is not too definite. This is because the equation often leads to the same value of $n_2$ for a wide range of ratios of $c_1$ to $c_2$. The values of $n_2$ are subject to sampling errors; for a discussion, see (2).

In section 10.14 you studied an example of *three-stage sampling* of turnip green plants The first stage was represented by plants, the second by leaves within plants, and the third by determinations within a leaf. In the notation of the present section, the variance of the sample mean is

$$s_{\bar{y}}^2 = \frac{s_1^2}{n_1} + \frac{s_2^2}{n_1 n_2} + \frac{s_3^2}{n_1 n_2 n_3}$$

Copying the equation given in section 10.14, we have

$$s_{\bar{y}}^2 = \frac{0.3652}{n_1} + \frac{0.1610}{n_1 n_2} + \frac{0.0067}{n_1 n_2 n_3}$$

To find the most economical values of $n_1$, $n_2$, and $n_3$, we set up a cost equation of the form

$$\text{cost} = c_1 n_1 + c_2 n_1 n_2 + c_3 n_1 n_2 n_3$$

and proceed to minimize the product of the variance and the cost as before. The solutions are

$$n_2 = \sqrt{\frac{c_1 s_2^2}{c_2 s_1^2}}, \qquad n_3 = \sqrt{\frac{c_2 s_3^2}{c_3 s_2^2}},$$

while $n_1$ is found by solving either the cost or the variance equation. Note that the formula for $n_2$ is the same in three-stage as in two-stage sampling, and that the formula for $n_3$ is the natural extension of that for $n_2$. Putting in the numerical values of the variance components, we obtain

$$n_2 = \sqrt{\frac{c_1(0.1610)}{c_2(0.3652)}} = 0.66\sqrt{\frac{c_1}{c_2}}, \quad n_3 = \sqrt{\frac{c_2(0.0067)}{c_3(0.1610)}} = 0.20\sqrt{\frac{c_2}{c_3}}$$

Since the computed value of $n_3$ would be less than 1 for any likely value of $c_2/c_3$, more than one determination per leaf is uneconomical. The optimum number $n_2$ of leaves per plant depends on the ratio $c_1/c_2$. This will vary with the conditions of experimentation. If many plants are being grown for some other purpose, so that there are always ample numbers available for sampling, $c_1$ includes only the extra costs involved in collecting a sample from many plants instead of a few plants. In this event the optimum $n_2$ might also turn out to be 1. If the cost of growing extra plants is to be included in $c_1$, the optimum $n_2$ might be higher than 1.

The device of sub-sampling is widely used in crop, soil, and ecological studies which cover an extensive geographic area; in field experiments where chemical determinations or estimates of soil and crop insects are made; and in laboratory investigations. Research workers who have not previously taken stock of the components of variation and the cost factors that apply to their data may find it rewarding to do so, particularly when some steps in the process are expensive.

If the primary units differ substantially in size from one another, some modification is advisable in the methods discussed here. One technique that has proved efficient is to select primary units with probabilities proportional to their sizes, and to select the same number of sub-sampling units from each primary unit in the sample. For a discussion of the various available methods, see (4, 11, 13).

EXAMPLE 17.12.1—This is the analysis of variance, on a single sub-sample basis, for wheat yield and percentage of protein from data collected in a wheat sampling survey in Kansas in 1939.

| Source of Variation | Yield (bushels per acre) | | Protein (%) | |
|---|---|---|---|---|
| | Degrees of Freedom | Mean Square | Degrees of Freedom | Mean Square |
| Fields | 659 | 434.52 | 659 | 21.388 |
| Samples within fields | 660 | 67.54 | 609 | 2.870 |

Two sub-samples were taken at random from each of 660 fields. Calculate the components of variance for yield. Ans. $s_1^2 = 183.49$, $s_2^2 = 67.54$.

EXAMPLE 17.12.2—For yield, estimate the variance of the sample mean for samples consisting of (i) 1 sub-sample from each of 800 fields, (ii) 2 sub-samples from each of 400 fields, (iii) 8 samples from each of 100 fields. Ans. (i) 0.313, (ii) 0.543, (iii) 1.919.

EXAMPLE 17.12.3—With 2 sub-samples per field, it is desired to take enough fields so that the standard error of the mean yield will be not more than $\frac{1}{2}$ bushel, and at the same time the standard error of the mean protein percentage will be not more than $\frac{1}{8}\%$. How many fields are required? Ans. About 870.

EXAMPLE 17.12.4—Suppose that it takes on the average 1 man-hour to locate and pace a field that is to be sampled. A single protein determination is to be made on the bulked sub-samples from any field. The cost of a determination is equivalent to 1 man-hour. It takes 15 minutes to locate, cut, and tie a sub-sample. From these data and the analysis of variance for protein percentage (example 17.12.1), compute the variance-cost product, $VC$, for each value of $n_2$ from 1 to 5. What is the most economical number of sub-samples per field? Ans. 2. How much more does it cost, for the same $V$, if 4 sub-samples per field are used? Ans. 12%.

**17.13—Ratio and regression estimates.** The *ratio estimate* is a different way of estimating population totals (or means) that is useful in a number of sampling problems. Suppose that you have taken a sample in order to estimate the population total of a variable, $Y$, and that a complete count of the population was made on some previous occasion. Let $X$ denote the value of the variable on the previous occasion. You might then compute the ratio

$$R = \frac{\Sigma Y}{\Sigma X},$$

where the sums are taken over the sample. This ratio is an estimate of the present level of the variate relative to that on the previous occasion. On multiplying the ratio by the known population total on the previous occasion (i.e., by the population total of $X$), you obtain the ratio estimate of the population total of $Y$. Clearly, if the relative change is about the same on all sampling units, the ratio $R$ will be accurate and the estimate of the population total will be a good one.

The ratio estimate can also be used when $X$ is some other kind of supplementary variable. The conditions for a successful application of this estimate are that the ratio $Y/X$ should be relatively constant over the population and that the population total of $X$ should be known. Consider an estimate of the total amount of a crop, just after harvest, made from a sample of farms in some region. For each farm in the sample we record the total yield, $Y$, and the total acreage, $X$, of that crop. In this case the ratio, $R = \Sigma Y/\Sigma X$, is the sample estimate of the mean yield per acre. This is multiplied by the total acreage of the crop in the region, which would have to be known accurately from some other source. This estimate will be precise if the mean yield per acre varies little from farm to farm.

The estimated standard error of the ratio estimate $\hat{Y}_R$ of the population total from a simple random sample of size $n$ is, approximately,

$$s(\hat{Y}_R) = N\sqrt{\frac{\Sigma(Y - RX)^2}{n(n-1)}}$$

The ratio estimate is not always more precise than the simpler estimate $N\bar{y}$ (number of units in population times sample mean). It has been shown that the ratio estimate is more precise only if $\rho$, the correlation coefficient between $Y$ and $X$, exceeds $C_X/2C_Y$, where the $C$'s are the coefficients of variation. Consequently, ratio estimates must not be used indiscriminately, although in appropriate circumstances they produce large gains in precision.

Sometimes the purpose of the sampling is to estimate a ratio, e.g., ratio of dry weight to total weight or ratio of clean wool to total wool. The estimated standard error of the estimate is then

$$s(R) = \frac{1}{\bar{X}}\sqrt{\frac{\Sigma(Y - RX)^2}{n(n-1)}}$$

This formula has already been given (in a different notation) at the end of section 17.5, where the estimation of proportions from cluster sampling was discussed.

In chapter 6 the linear regression,

$$\hat{Y} = a + bx,$$

was discussed. With an auxiliary variable, $X$, you may find that when you plot $Y$ against $X$ from the sample data, the points appear to lie close to a straight line, but the line does not go through the origin. This implies that the ratio $Y/X$ is not constant over the sample. As pointed out in section 6.14, it is then advisable to use a linear regression estimate instead of the ratio estimate. For the population total of $Y$, the linear regression estimate is

$$N\hat{Y} = N\{\bar{y} + b\,(\bar{X} - \bar{x})\},$$

where $\bar{X}$ is the population mean of $X$. The term inside the brackets is the sample mean, $\bar{y}$, adjusted for regression. To see this, suppose that you have taken a sample in which $\bar{y} = 2.35$, $\bar{x} = 1.70$, $\bar{X} = 1.92$, $b = +0.4$. Your first estimate of the population mean would be $\bar{y} = 2.35$. But in the sample the mean value of $X$ is too low by an amount $(1.92 - 1.70) = 0.22$. Further, the value of $b$ tells you that unit increase in $X$ is accompanied, on the average, by $+0.4$ unit increase in $Y$. Hence, to correct for the low value of the mean of $X$, you adjust the sample mean for the regression on $x$, i.e.,

$$2.35 + (+0.4)(0.22) = 2.44 = \bar{y} + b(\bar{X} - \bar{x})$$

To estimate the population total, this value is multiplied by $N$, the number of sampling units in the population.

The standard error of the estimated population total is, approximately,

$$s_{N\hat{Y}} = N s_{y \cdot x} \sqrt{\left( \frac{1}{n} + \frac{(\bar{X} - \bar{x})^2}{\Sigma x^2} \right)}$$

If a finite population correction is required in the standard error formulas presented in this section, insert the factor $\sqrt{(1 - \phi)}$. In finite populations the ratio and regression estimates are both slightly biased, but the bias is seldom important in practice.

**17.14—Sampling methods in biology.** The purpose of this section is to indicate the types of sampling plan currently used in a few of the important applications of sampling in biology.

In making crop and livestock estimates and in studies of farm management or farm economics, we have to deal with a population of farms. In the United States, reliable lists of farms are seldom available. As we have mentioned, a common practice is to take as the sampling unit an area of land, demarcated by easily recognizable boundaries, and containing on the average about 6 farms. Before sampling begins, these areas are marked out on large-scale maps. For national agricultural surveys, maps of this kind have been prepared covering the whole of the country, in what is known as the Master Sample of Agriculture. The areas can be classified into strata in various ways, depending on the purpose of the survey. In some surveys it pays to assemble also a special list of large farms, which are placed in a separate stratum.

For surveys of a region that is not too large, such as a state, the sampling plan involves a stratified sample of these areas. Within a stratum, areas may be selected at random, although frequently a systematic sample is used for convenience. In any area so drawn, the data are obtained for all farms that lie in this area. For national surveys the plan is more complex, multi-stage sampling being employed so as to cut down travel costs and facilitate supervision.

In forestry surveys to estimate the volume of timber of the principal types and age-classes, the first step is to delineate the forest land, either from maps or from aerial photographs. The forest region is then divided into strata according to the type of trees and the density and size of the trees.   This can be done fairly successfully by a careful stereoscopic inspection of recent aerial photographs; results from older photographs tend to be disappointing.   Within each stratum, sample plots are marked out and are measured in detail on the ground by field crews.   Some problems on which research has been done concern the amount of resources that should be devoted to the construction of the strata, relative to that devoted to measurement of sample plots, the best size and shape of plot, and the advantages of optimum versus proportional allocation of the numbers of sample plots per stratum (5).

Studies of wildlife present many complexities, because the members of the population—fish, birds, beavers, field mice—do not stay in the same place and are sometimes difficult to find at all.   Generally speaking, strata can be set up if there is some information about the distribution and habitat of the species in question.   Area sampling units that constitute a random or systematic sample of each stratum can be marked out.

The difficulty is to count the numbers in these areas.   In certain cases a reasonably complete count can be made.   With deer, a field crew may search each sample area thoroughly; fish can be caught by nets or by giving them an electric shock.   Very large animals such as buffalo or antelope can be counted from an airplane in some types of terrain.   One method, used for game birds, is to traverse certain routes by car at the same time of day and the same time of year in successive years, recording the numbers of birds seen.   This approach does not pretend to take a random sample of a stratum, but relies on an assumption that the numbers of birds seen in successive years will be roughly proportional to the total abundances of birds in those years.

With species that cannot be seen easily, various trapping methods are in use.   Sometimes it is possible to catch over 50% of all the animals in a sample area by repeated trapping.   By noting the rate of decline in the catch per trap on successive occasions, one can attempt to estimate the number of animals that have escaped capture and thus make a correction for them.   Where only a relatively small proportion of a large population can be caught, specimens that are caught may be identified by a mark or tag and then released.   An estimate of the size of the population can then be made from the proportion of recaptures found in the traps on a subsequent occasion.   Both of these methods depend on various assumptions about the population—in particular, the assumption that all animals are equally likely to be caught.   There is good reason to believe that this is not so, but little is known at present about the extent of the biases that these methods involve.   Finally, in cases where any visual count is extremely difficult, biologists have tried to construct indices of abundance from indirect signs, such as nests, dens, tracks, or droppings. A general account of methods for wildlife estimation has been given by Scattergood (9).

For the sake of conciseness, some of the difficult practical problems in biological applications of sampling have been omitted. Moreover, sampling methods are subject to continuous evolution, since biologists are working steadily towards improved techniques. The methods described here will doubtless be modified as knowledge accumulates and new aids to sampling are discovered.

REFERENCES

1. A. R. CLAPHAM. Journal of Agricultural Science, 19:214 (1929).

2. W. G. COCHRAN. Sampling Techniques. John Wiley & Sons, New York (1953).

3. W. EDWARDS DEMING. Some Theory of Sampling. John Wiley & Sons, New York (1950).

4. M. H. HANSEN, W. N. HURWITZ, and W. G. MADOW. Sample Survey Methods and Theory. John Wiley & Sons, New York (1953).

5. A. A. HASEL. Chapter 19 in Statistics in Mathematics and Biology. Iowa State University Press (1954).

6. F. R. IMMER. Journal of Agricultural Research, 44:633 (1932).

7. A. J. KING, D. E. McCARTY, and M. McPEAK. U. S. Department of Agriculture Technical Bulletin 814 (1942).

8. J. A. RIGNEY and J. FIELDING REED. Journal of the American Society of Agronomy, 39:26 (1947).

9. L. W. SCATTERGOOD. Chapter 20 in Statistics in Mathematics and Biology. Iowa State University Press (1954).

10. F. F. STEPHAN. Journal of Marketing, 6:38 (1941).

11. P. V. SUKHATME. Sampling Theory of Surveys With Applications. Iowa State University Press (1954).

12. Q. M. WEST. Mimeographed Report, Cornell University Agricultural Experiment Station (1951).

13. F. YATES. Sampling Methods for Censuses and Surveys. Charles Griffin, London (1954).

# 64

## A REJECTION CRITERION BASED UPON THE RANGE†

By C. I. BLISS, W. G. COCHRAN and J. W. TUKEY‡

The experimenter is occasionally faced with an inexplicable, 'aberrant' observation in an otherwise valid set of data. If it is defective and he accepts it—or if it is sound and he rejects it—his results will be biased. By following an objective rule rather than a subjective impression, he can control, and perhaps minimize, his risk of making a wrong decision. The rule proposed here is intended for data consisting of equal-sized sets of replicate measurements, the several sets possibly varying in their means but all being samples from populations with the same variance. It was developed originally to meet the need for a simple rejection criterion for use with several of the bioassays in the *U.S. Pharmacopeia XV* (1955). Other applications could be cited.

For the test which we propose, the range is computed from each set of $n$ measurements, there being $k$ sets in all. The largest range is divided by the sum of all the ranges. The resulting ratio $T$ is compared with the tabular value for the appropriate $n$ and $k$ at the probability level of $P = 0.05$. If the observed ratio exceeds the tabular value (cf. Tables 1 and 2), the set represented by this largest range is assumed to contain an aberrant observation or outlier, which is identified by inspection and rejected. Thus the proposed criterion is closely related to Cochran's test (1941) for the largest variance in a series. The application of this rule controls the probability of biasing a result by failing to reject an outlier.

The test may be illustrated numerically with data from two biological assays that were submitted in collaborative studies sponsored by the U.S. Pharmacopeia. The first is an assay of corticotropin from the concentration of ascorbic acid in the adrenal glands of hypophysectomized rats. Seven rats were assigned at random to each of six dosage groups. Each group, in turn, was injected with one of three dosages of the standard preparation ($S_1$ to $S_3$) and of the test or unknown preparation ($U_1$ to $U_3$). The response $y$ was determined separately from the adrenal glands of each rat. The total for each treatment group is given below, together with its range and sum of squares ($6s^2$):

| Dose | $S_1$ | $S_2$ | $S_3$ | $U_1$ | $U_2$ | $U_3$ |
|---|---|---|---|---|---|---|
| Total response $\Sigma y$ | 28·92 | 24·87 | 20·62 | 26·95 | 25·17 | 20·22 |
| Range ($n = 7$) | 0·86 | 0·79 | 0·37 | 1·49 | 0·80 | 0·76 |
| Variance × 6, $6s^2$ | 0·442 | 0·362 | 0·117 | 1·642 | 0·455 | 0·467 |

The range for $U_1$ (1·49) is the largest and most suspect; when it is divided by the sum of the ranges, we obtain $T = 1·49/5·07 = 0·294$. This exceeds the upper 5 % point of $T = 0·288$ for $k = 6$ and $n = 7$, indicating an outlier. The identification of this group as unusual by our proposed range criterion may be checked against Cochran's test; its critical ratio of

† Prepared, in part, in connexion with research sponsored by the Office of Naval Research.
‡ From The Connecticut Agricultural Experiment Station and Yale University, Johns Hopkins University, and Princeton University, respectively.

$1\cdot642/3\cdot485 = 0\cdot471$ exceeds the upper 1 % point ($0\cdot461$) in the table by Eisenhart, Hastay & Wallis (1947). Among the individual $y$'s in this group ($2\cdot68$, $3\cdot90$, $4\cdot00$, $4\cdot02$, $4\cdot06$, $4\cdot12$ and $4\cdot17$), the smallest response not only falls considerably short of the others, but also contributes to a suspiciously small total response for its dose, when compared with the totals for the other groups.

Our second example is from a turbidimetric assay of vitamin $B_{12}$. Four sets of triplicate tubes were prepared for each of six dosage levels of the reference standard. One set was placed in each of four tube racks, together with triplicate tubes for four dosage levels of a sample or unknown in two of the four racks. Within each rack the tubes were intermingled at random. After incubation overnight, the percentage transmittance of each tube was read in a photometer. The ranges of the $k = 32$ sets of $n = 3$ tubes had the following distribution:

| Observed range of 3 | 0 | 1 | 2 | 3 | 4 | 5 | 6 | 7 | 8 | 13 |
|---|---|---|---|---|---|---|---|---|---|---|
| No. of sets | 1 | 5 | 6 | 8 | 4 | 2 | 3 | 1 | 1 | 1 |

From the total of the ranges ($= 113$), the observed ratio $T = 13/113 = 0\cdot115$. Since $k > 10$, we enter Table 2 with $(k+2)\,T = 34 \times 0\cdot115 = 3\cdot91$, which much exceeds the value of $3\cdot08$, interpolated between $3\cdot03$ and $3\cdot11$ in the column for $n = 3$. On the basis of all 32 ranges, we attribute the range of 13 to an outlier. The three readings comprising this set show transmittances of 24, 34 and 37 %, from which the reading of 24 % would be rejected as aberrant. The test ratio may now be recomputed with the largest of the 31 remaining ranges in the numerator to obtain $T = 8/100 = 0\cdot08$. Since $33 \times 0\cdot08 = 2\cdot64$ is less than the interpolated value ($3\cdot07$) for $n = 3$ and $k = 31$, no other observation would be rejected from the series.

## Distribution

The critical values of the ratio $T$ in Table 1 have been computed for $k = 2$ to 10 ranges, each determined from a set of $n = 2$ to 10 measurements. Table 2 gives critical values of $(k+2)\,T$ for 10 to 50 ranges. Together they should cover the situations that occur most frequently. The original variates are assumed to be distributed normally with a common standard deviation within groups. Since our criterion is homogeneous, we may take $\sigma = 1$ without loss of generality.

Let $w_1, w_2, \ldots, w_k$ be a set of $k$ ranges from normal samples, each of size $n$ and independently and identically distributed. Let $w_*$ be the largest $w$ and $W$ the sum of the $w$'s. Then

$$T_k = \frac{w_*}{W} = \frac{w_*}{(W - w_*) + w_*}$$

$$= \frac{w_*/(W - w_*)}{1 + w_*/(W - w_*)} = \frac{S_{k-1}}{1 + S_{k-1}},$$

where 

$$S_{k-1} = \frac{w_*}{w_1 + w_2 + \ldots + w_k - w_*} = \max\left\{\frac{w_i}{w_1 + \ldots + w_{i-1} + w_{i+1} + \ldots + w_k}\right\}.$$

Thus the distribution of $T_k$ is determined by that of $S_{k-1}$, which is the largest of $k$ identically distributed quantities of the form

$$R_{k-1} = \frac{w_1}{w_2 + w_3 + \ldots + w_k}.$$

The argument used in developing Cochran's test (1941) for the largest variance can be applied to relate the percentage points of $S_{k-1}$ to those of $R_{k-1}$. The value of $S_{k-1}$ for $P = 0.05$ falls between the values of $R_{k-1}$ for

$$P_1 = 0.05/k \quad \text{and} \quad P_2 = 1 - (0.95)^{1/k}.$$

Roughly, $P_2$ is $0.98 P_1$ for all $k$; for $k = 2$, the two values are $P_1 = 0.025$ and $P_2 = 0.02532$, and for $k = 20$, they are $P_1 = 0.0025$ and $P_2 = 0.002558$. Since investigation by one of us suggested that the desired value of $R_{k-1}$ was closer to $P_1$, we have used the compromise probability $P^* = \frac{1}{3}(2P_1 + P_2)$, except where $T_k \geq \frac{1}{2}$ (that is, $S_{k-1} \geq 1$) when $P_1$ is exact.

The calculation of critical levels of $T_k$ for $P = 0.05$ is thus reduced to the calculation of critical values of $R_{k-1}$ for $P = P^*$, from which

$$T_{k,\,0.05} = \frac{R_{k-1,\,P^*}}{1 + R_{k-1,\,P^*}}.$$

Since the numerator and denominator of $R_{k-1}$ are independent, several approximations are available.

Table 1. *Upper 5% points of $T$, the ratio of the largest of $k$ independent normal ranges, each of $n$ observations, to the total of these ranges*

| No. of ranges $k$ | No. of observations $n$ in each range | | | | | | | | |
|---|---|---|---|---|---|---|---|---|---|
| | 2 | 3 | 4 | 5 | 6 | 7 | 8 | 9 | 10 |
| 2 | 0·962 | 0·862 | 0·803 | 0·764 | 0·736 | 0·717 | 0·702 | 0·691 | 0·682 |
| 3 | ·813 | ·667 | ·601 | ·563 | ·539 | ·521 | ·507 | ·498 | ·489 |
| 4 | ·681 | ·538 | ·479 | ·446 | ·425 | ·410 | ·398 | ·389 | ·382 |
| 5 | ·581 | ·451 | ·398 | ·369 | ·351 | ·338 | ·328 | ·320 | ·314 |
| 6 | 0·508 | 0·389 | 0·342 | 0·316 | 0·300 | 0·288 | 0·280 | 0·273 | 0·267 |
| 7 | ·451 | ·342 | ·300 | ·278 | ·263 | ·253 | ·245 | ·239 | ·234 |
| 8 | ·407 | ·305 | ·267 | ·248 | ·234 | ·225 | ·218 | ·213 | ·208 |
| 9 | ·369 | ·276 | ·241 | ·224 | ·211 | ·203 | ·197 | ·192 | ·188 |
| 10 | ·339 | ·253 | ·220 | ·204 | ·193 | ·185 | ·179 | ·174 | ·172 |

Table 2. *Upper 5% points of $(k + 2) T$, where $T$ is the ratio of the largest of $k$ independent normal ranges, each of $n$ observations, to the total of these ranges*

| No. of ranges $k$ | No. of observations $n$ in each range | | | | | | | | |
|---|---|---|---|---|---|---|---|---|---|
| | 2 | 3 | 4 | 5 | 6 | 7 | 8 | 9 | 10 |
| 10 | 4·06 | 3·04 | 2·65 | 2·44$_5$ | 2·30$_5$ | 2·21 | 2·14$_5$ | 2·09 | 2·05 |
| 12 | 4·06 | 3·03 | 2·63$_5$ | 2·42$_5$ | 2·29 | 2·20 | 2·13 | 2·07$_5$ | 2·04 |
| 15 | 4·06 | 3·02 | 2·62$_5$ | 2·41$_5$ | 2·28 | 2·18$_5$ | 2·12 | 2·06$_5$ | 2·02$_5$ |
| 20 | 4·13 | 3·03 | 2·62 | 2·41$_5$ | 2·28 | 2·18$_5$ | 2·11 | 2·05 | 2·01 |
| 50 | 4·26 | 3·11 | 2·67 | 2·44 | 2·29 | 2·19 | 2·11 | 2·06 | 2·01 |

## Approximate $P^*$ values of $R_{k-1}$

The critical values of $R_{k-1}$ were approximated by two methods discussed below. Most of the values in Table 1 were obtained by method A, and most of those in Table 2 by method B. The two agree moderately well, and intermediate tabular values give some weight to both methods.†

A range or total range is often approximated by a multiple of the square root of a $\chi^2$-variate, i.e. by a $\chi$-variate. Equivalent degrees of freedom $\nu$ and scale factors $c$ for a single range are given by Thomson (1953), and for mean ranges by David (1951). From these, the approximate distributions of $w_1$ and of $\overline{w}$ are

$$w_1 \sim c_1 \chi_1 / \sqrt{\nu_1} \quad \text{and} \quad \overline{w} = \frac{w_2 + w_3 + \ldots + w_k}{k-1} \sim c_2 \chi_2 / \sqrt{\nu_2}.$$

When required, more accurate scale factors for mean or total ranges can be found from $c^2 = d_n^2 + V_n/k$ (David, 1951).

Method A consists of the straightforward approximation

$$R_{k-1} = \frac{w_1}{w_2 + w_3 + \ldots + w_k} = \frac{w_1 \sqrt{\nu_2}}{(k-1) c_2 \chi_2} \sim \frac{c_1}{(k-1) c_2} \sqrt{F},$$

where $\sqrt{F} = \dfrac{\chi_1}{\chi_2} \sqrt{\dfrac{\nu_2}{\nu_1}}$ and $F$ has $\nu_1$ and $\nu_2$ degrees of freedom. The required values of $\sqrt{F}$ for critical levels of $P^*$ are obtained by interpolation in a table of $F$.

Method B is less direct and more refined, the $\chi$ approximation appearing only in the denominator of $R_{k-1}$. Let

$$u = \frac{w_2 + w_3 + \ldots + w_k}{(k-1) c_2}.$$

Then the expected mean square of $u$ is $\sigma^2$, and we can regard

$$R_{k-1} = \frac{w_1}{w_2 + w_3 + \ldots + w_k} = \frac{w_1}{(k-1) c_2 u}$$

as the result of studentizing $w_1 / (k-1) c_2$ with the scale estimator $u$. We know the result of studentizing a $\chi$-variate (at probability $P^*$) with the scale-estimator based on $\nu_2$ degrees of freedom. In that case $(F_{\nu_1, \infty})^{\frac{1}{2}}$ is converted into $(F_{\nu_1, \nu_2})^{\frac{1}{2}}$ and the $P^*$ value is increased in the ratio $\sqrt{(F/F_\infty)}$, where $F$ is the $P^*$ value of $F_{\nu_1, \nu_2}$ and $F_\infty$ is the $P^*$ value of $F_{\nu_1, \infty}$. If we take this same studentizing factor as applicable to the studentization of $w_1/(k-1) c_2$ at $P^*$ by $u$, we have, as the method B approximation,

$$R_{k-1} = \frac{w}{(k-1) c_2} \sqrt{\frac{F}{F_\infty}},$$

where $w$ is now the range of $n$ unit normally distributed items at $P^*$ (Pearson, 1942). To improve the accuracy of linear interpolation for non-tabular values of $\nu_1$, $\nu_2$ and $P^*$, selected values of $\sqrt{(F/F_\infty)}$ were first computed from the five-figure tables of Merrington & Thompson (1943), supplemented by the Hald tables (1952) for larger values of $\nu_1$ and for $P = 0 \cdot 001$.

The two approximations differ by the factor

$$\frac{w/c_1}{\sqrt{F_\infty}} = \frac{w/c_1}{\chi_1/\sqrt{\nu_1}},$$

† It is not claimed that the entries in the tables are correct to the last figure given, but we believe that most of them will not differ from the true values by more than a few units in the last place.

which would be unity if the $\chi$-variate approximation to the numerator range were perfect. Method B has the advantage of allowing for idiosyncrasies in the numerator range, rather than hiding them in a $\chi$-variate approximation. For $k > 2$, the total range in the denominator is approximated more closely by the $\chi$-variate than is the simple range in the numerator.†

THE CASE OF $n = 2$

For ranges of 2, $w_1 = \sqrt{2}\,|\,x\,|$, where $x$ is a unit normal deviate. Methods A and B are then identical, since $w_1$ is exactly a $\chi$-variate. The desired critical values are related directly to those of Lord's (1947) statistic

$$u(k-1, 2) = \frac{d_2\,|\,x\,|}{\overline{w}} = \frac{(k-1)\,d_2\,|\,x\,|}{w_2 + w_3 + \ldots + w_k},$$

where $\overline{w}$ is the mean of $k-1$ ranges, and $d_2 = 1{\cdot}1284$ is the average range of a sample of $n = 2$. The critical values of $R_{k-1}$ are thus $\sqrt{2}/d_2(k-1)$ times those of $u(k-1, 2)$. Despite rather substantial interpolation for $P^*$ at some values of $k$, we have used Lord's tables in computing the values for $n = 2$ in Tables 1 and 2.

REFERENCES

COCHRAN, W. G. (1941). The distribution of the largest of a set of estimated variances as a fraction of their total. *Ann. Eugen., Lond.*, **11**, 47.

DAVID, H. A. (1951). Further applications of range to the analysis of variance. *Biometrika*, **38**, 393.

EISENHART, C., HASTAY, M. W. & WALLIS, W. A. (1947). *Selected Techniques of Statistical Analysis.* New York: McGraw-Hill Book Co.

HALD, A. (1952). *Statistical Tables and Formulas.* New York: John Wiley and Sons.

LORD, E. (1947). The use of range in place of standard deviation in the *t*-test. *Biometrika*, **34**, 41.

MERRINGTON, M. & THOMPSON, C. M. (1943). Tables of percentage points of the inverted beta ($F$) distribution. *Biometrika*, **33**, 73.

PEARSON, E. S. (1942). The probability distribution of the range in samples of $n$ observations from a normal population. *Biometrika*, **32**, 301.

THOMSON, G. M. (1953). Scale factors and degrees of freedom for small sample sizes for $\chi$-approximation to the range. *Biometrika*, **40**, 449.

TUKEY, J. W. (1956). Every man his own studentizer. *Memorandum Report* 58. Statistical Res. Grp. Princeton University.

*United States Pharmacopeia XV* (1955). Easton, Pa.: Mack Publishing Co.

† Since the completion of the present calculations method B has been studied further by one of the authors (Tukey, 1956). This investigation indicates that, as used here, method B should give slight, but only slight, overestimates of the critical values.

# 65

## ANALYSIS OF COVARIANCE: ITS NATURE AND USES*

William G. Cochran

*The Johns Hopkins University, Baltimore, Maryland, U.S.A.*

### 1. INTRODUCTION

This paper is intended as an introduction to the subsequent papers in this issue. It discusses the nature and principal uses of the analysis of covariance, and presents the standard methods and tests of significance.

As Fisher [1934] has expressed it, the analysis of covariance "combines the advantages and reconciles the requirements of the two very widely applicable procedures known as regression and analysis of variance." This dual role can be illustrated by a two-way classification in which rows represent treatments, and columns represent blocks or replications. The typical mathematical model appropriate to the analysis of covariance is

$$y_{ij} = \mu + \tau_i + \rho_j + \beta(x_{ij} - x_{..}) + e_{ij} \tag{1}$$

Here $y_{ij}$ is the yield or response, while $x_{ij}$ is an auxiliary variate, sometimes called the *concomitant variate* or *covariate*, on which $y_{ij}$ has a linear regression with regression coefficient $\beta$. The constants $\mu$, $\tau_i$ and $\rho_j$ are the true mean response and the effects of the $i$th treatment and $j$th replication, respectively. The residuals $e_{ij}$ are random variates, assumed in standard theory to be normally and independently distributed with mean zero and common variance.**

From the viewpoint of analysis of variance, equation (1) may be rewritten as

$$y_{ij} - \beta(x_{ij} - x_{..}) = \mu + \tau_i + \rho_j + e_{ij} \tag{2}$$

In this form, (2) is the typical equation for an analysis of variance of the quantities

$$y_{ij} - \beta(x_{ij} - x_{..})$$

---

*Paper No. 319, Department of Biostatistics.
**The symbols $x_{..}$, $y_{..}$ denote overall means, while $x_{i.}$, $y_{i.}$ denote treatment means.

These are the deviations of $y_{ij}$ from its linear regression on $x_{ij}$, or the values of $y_{ij}$ after adjustment for this linear regression. In this setting, $\tau_i$ may be regarded as the true effect of the $i$th treatment on $y_{ij}$, after adjustment for the linear regression on the covariate $x_{ij}$. Thus the technique enables us to remove that part of an observed treatment effect which can be attributed to a linear association with the $x_{ij}$.

When the objective of the analysis is to fit a regression of $y$ on $x$, the parameters $\tau_i$ and $\rho_j$ in equation (1) represent "nuisance" parameters, included in the mathematical specification in order to make it realistic. In this way the analysis of covariance extends the study of regression relationships to data of complex structure in which the nature of the regression is at first sight obscured by structural effects like the $\tau_i$ and $\rho_j$.

## 2. PRINCIPAL USES

These may be grouped under several headings.

2.1 *To increase precision in randomized experiments.* This is probably the most frequent application. The covariate $x$ is a measurement, taken on each experimental unit before the treatments are applied, which is thought to predict to some degree the final response $y$ on that unit. The first illustration of the covariance method in the literature was of this type (Fisher [1932]). The variate $x$ was the yield of tea per plot in a period preceding the start of the experiment, while $y$ was the tea yield at the end of a period of application of treatments (in this illustration, the treatments were "dummy"). Adjustment of the responses $y$ for their regression on $x$ removes the effects of variations in initial yields from the experimental errors, insofar as these effects are measured by the linear regression. In this example these effects might be due either to inherent differences in the tea bushes or to soil fertility differences that were permanent enough to persist during the course of the experiment.

With a linear regression equation, the gain in precision from the covariance adjustment depends primarily on the size of the correlation coefficient $\rho$ between $y$ and $x$ on experimental units (plots) that receive the same treatment. If $\sigma_y^2$ is the experimental error variance when no covariance is employed, the adjustments reduce this variance to a value which is effectively about

$$\sigma_y^2(1 - \rho^2)\left\{1 + \frac{1}{f_e - 2}\right\}$$

where $f_e$ is the number of error d.f. The factor involving $f_e$ is needed to take account of errors in the estimated regression coefficient. If

$\rho$ is less than 0.3 in absolute value, the reduction in variance is inconsequential, but as $\rho$ mounts towards unity, sizeable increases in precision are obtained. In Fisher's example $\rho$ was 0.928, reflecting a high degree of stability in relative yield of a plot from one period to another. The adjustment reduced the error variance roughly to a fraction $\{1 - (0.928)^2\}$, or about one-sixth, of its original value. Some of the most spectacular gains in precision from covariance have occurred in situations like this, in which the covariate represents an initial calibration of the responsiveness of the experimental units.

In this use the function of covariance is the same as that of local control (pairing and blocking). It removes the effects of an environmental source of variation that would otherwise inflate the experimental error. When the relation between $y$ and $x$ is linear, covariance and blocking are about equally effective.* If instead of using covariance we can group the units into replications such that the $x$ values are equal within a replication, this blocking also reduces the error variance to $\sigma_y^2(1 - \rho^2)$.

The potentialities of preliminary measurements as a means of increasing precision have frequently been recognized by experimenters. In animal feeding experiments the response is taken as gain in weight (final-initial weight) rather than final weight itself. Insulin may be assayed from the drop in blood sugar (initial reading—reading 3 hours after injection of insulin) instead of from the 3 hour reading. The weight of a treated muscle on the right side of the body may be taken as a percentage of the weight of the corresponding untreated muscle on the left side. Such adjustments make the best use of the covariate only when the relation between $y$ and $x$ is exactly that implied by the adjustment. In the animal feeding and insulin examples, the assumption is that $\beta = 1$; in the muscles, that $y/x$ is independent of $x$ and has constant variance. If these assumptions do not hold, the adjustment falls short of the optimum and sometimes is worse than no adjustment at all. By a covariance analysis, the experimenter can utilize his knowledge or speculations about the general nature of the relation between $y$ and $x$, but still leave flexibility in the process by including parameters like $\beta$ that are estimated from the data. Incidentally, he can verify from the covariance analysis whether a specific simple adjustment like the use of $(y - x)$ is good enough, as it sometimes is.

In a covariance analysis, the preliminary variate $x$ may be measured on a completely different scale from that of the response $y$,—a situation in which the experimenter would have difficulty in creating a "home-made" method of adjustment. Bartlett [1937] used a visual estimate

---

*See p. 281.

of the degree of saltiness of the soil to adjust cotton yields. Federer and Schlottefeldt [1954] used the serial order (1, 2, $\cdots$ 7) of the plot within a replication as a basis for a quadratic regression adjustment of tobacco data, thereby removing the effects of an unexpected gradient in fertility within the replications. Similarly, the reading performances of children under different methods of instruction may be adjusted for variations in their initial I.Q.'s. Note also that $x$ need not be a direct causal agent of $y$—it may, for instance, merely reflect some characteristic of the environment that also influences $y$.

When covariance is used in this way, it is important to verify that the treatments have had no effect on $x$. This is obviously true when the $x$'s were measured before the treatments were applied. But sometimes the $x$ variates are measured after treatments have been applied, as when plant number shortly before harvest is used to adjust crop yields for uneven growth, or as happened in the index of saltiness used by Bartlett. When the treatments do affect the $x$-values to some extent, the covariance adjustments take on a different meaning. They no longer merely remove a component of experimental error. In addition, they distort the nature of the treatment effect that is being measured. If the higher yields given by superior treatments are due mostly to their effects in increasing numbers of plants, a covariance adjustment, which attempts to measure what yields would be if plant numbers were equal for all treatments, may remove most of the real treatment effect. The $F$-test of treatments against error for the $x$-variate is helpful when there is doubt whether treatments have had some effect on $x$.

2.2. *To remove the effects of disturbing variables in observational studies.* In fields of research in which randomized experiments are not feasible, we may observe two or more groups differing in some characteristic, in the hope of discovering whether there is an association between this characteristic and a response $y$. Examples are differences in heights of urban and rural school children, differences in illness rates between tenants of public and slum housing, and differences in expenditures for luxuries between clerical and manual workers. In observational studies it is widely realized that an observed association, even if statistically significant, may be due wholly or partly to other disturbing variables $x_1$, $x_2$ $\cdots$ in which the groups differ. Where feasible, a common device, analogous to blocking in randomized experiments, is to match the groups for the disturbing variables thought to be most important. In the same way, a covariance adjustment may be tried for $x$-variables that have not been matched.

In a comparison of the heights of children from two different types of school, Greenberg [1953] found that the two groups differed slightly, though not significantly, in mean age. A covariance adjustment for age resulted in a more sensitive comparison of the heights. As a more complex example, Day and Fisher [1937] adjusted the log S. D. of leaf length (used as a measure of within-species variability) for fluctuations in length, breadth and thickness of leaves in comparing populations of *Plantago maritima* from different regions.

In observational studies covariance can perform two distinct functions. One is to remove bias. To illustrate, it follows from model (1) that the *unadjusted* difference $(y_i. - y_j.)$ between the means of two groups is

$$y_i. - y_j. = \tau_i - \tau_j + \beta(x_i. - x_j.) + e_i. - e_j.$$

If the two groups have not been matched for $x$, the difference $(x_i. - x_j.)$ may reflect a real difference in their $x$-distributions, being much larger than can be accounted for by within-group variations. The term $\beta(x_i. - x_j.)$ is then of the nature of a bias which if allowed to remain will render tests of significance and confidence limits invalid. If model (1) applies to the data at hand, a covariance adjustment removes their bias. Most users of covariance in observational studies would, I think, regard coping with bias as its primary function. However, even if there are no real differences between the $x$-distributions in the two groups, so that there is no danger of bias, covariance may still be used to increase the precision of the comparison as in the applications in section 2.1.

Unfortunately, observational studies are subject to difficulties of interpretation from which randomized experiments are free. Although matching and covariance have been skillfully applied, we can never be sure that bias may not be present from some disturbing variable that was overlooked. In randomized experiments, the effects of this variable are distributed among the groups by the randomization in a way that is taken into account in the standard tests of significance. There is no such safeguard in the absence of randomization.

Secondly, when the $x$-variables show real differences among groups—the case in which adjustment is needed most—covariance adjustments involve a greater or less degree of extrapolation. To illustrate by an extreme case, suppose that we were adjusting for differences in parents' income in a comparison of private and public school children, and that the private-school incomes ranged from \$10,000–\$12,000, while the public-school incomes ranged from \$4,000–\$6,000. The covariance would adjust results so that they allegedly applied to a mean income of \$8,000 in each group, although neither group has any observations in which incomes are at or even near this level.

Two consequences of this extrapolation should be noted. Unless the regression equation holds in the region in which observations are lacking, covariance will not remove all the bias, and in practice may remove only a small part of it. Secondly, even if the regression is valid in the no man's land, the standard errors of the adjusted means become large, because the standard error formula in a covariance analysis takes account of the fact that extrapolation is being employed (although it does not allow for errors in the form of the regression equation). Consequently the adjusted differences may become insignificant statistically merely because the adjusted comparisons are of low precision.

When the groups differ widely in $x$, these difficulties imply that the interpretation of an adjusted analysis is speculative rather than soundly based. While there is no sure way out of this difficulty, two precautions are worth observing.

(i) Consider what internal or external evidence exists to indicate whether the regression is valid in the region of extrapolation. Sometimes the fitting of a more complex regression formula serves as a partial check.

(ii) Examine the standard errors of the adjusted group means, particularly when differences become non-significant after adjustment. Confidence limits for the difference in adjusted means will reveal how precise or imprecise the adjusted comparison is.

2.3 *To throw light on the nature of treatment effects.* This application is closely related to the previous one. In a randomized experiment, the effects of several soil fumigants on eelworms, which attack some English farm crops, were compared. After the treatments had been given time to exert their effects, the numbers of eelworm cysts per plot and the yields of the crop, spring oats, were both recorded. Significant effects were produced on both eelworms and oats. It would be of interest to discover whether the reductions in numbers of eelworms were the causal mechanism in producing the observed differences in oats yields. If the treatment effects on the oats ($y$) disappear after adjusting by covariance for differences in the numbers of eelworms ($x$), this suggests, at least at first sight, that the treatment differences are simply a reflection of the differences produced by the fumigants on eelworm numbers. There are numerous instances of this kind in which a concomitant variable might be in part the agent through which the treatments produce their effects on the principal response. A covariance analysis offers the possibility of exploring whether this is so.

Here again, however, there are difficulties which restrict the utility of this ingenious tool. These are analyzed by Fairfield Smith in a later paper in this issue. Since in most of these applications the treatments will have produced significant effects on $x$, there is the problem of extrapolation discussed in the previous section. In addition, as Fairfield Smith illustrates, the interpretation of the adjusted $y$ averages requires careful study. Sometimes these averages have no physical or biological meaning of interest to the investigator, and sometimes they do not have the meaning that is ascribed to them at first glance.

2.4 *To fit regressions in multiple classifications.* The simplest situation, discussed in elementary text books, involves a single classification. By standard techniques we can (i) fit a separate regression of $y$ on $x$ within each class, (ii) test whether the slopes or positions of the lines differ from class to class, and (iii) if advisable, make a combined estimate of a common slope. As an example of a regression in a row $\times$ column classification, the regression of wheat yield on shoot height, number of plants and number of ears was worked out from a series of growth studies on wheat conducted in Britain (Cochran [1938]). Each quartet of observations ($y$, $x_1$, $x_2$, $x_3$) represented the mean over several plots at one station in one year. In order that soil, geographic and seasonal factors would be adequately sampled, data were obtained from 7 stations for each of 5 years, making 35 quartets. Consequently it was necessary to fit constants for the mean yield of each station and the mean yield of each year.

2.5 *To analyze data when some observations are missing.* An interesting by-product of the covariance method, first pointed out by Bartlett (1937), is that it may be used to compute the exact analysis of variance when some observations are missing. To each missing observation we assign any convenient value (*e.g.* 0, 5, or 100) and introduce a dummy $x$-variate that takes the value[1] for the missing unit, and 0 for all other units. The standard covariance computations then give the correct least squares estimates of the treatment means and the exact $F$- and $t$-tests. This method is probably slower than the insertion of a missing value by the Yates formula [1933], but it is useful (i) with unfamiliar classifications where the Yates formula has not been worked out and (ii) where exact $F$- and $t$-tests are important, since the Yates' method gives only approximate tests.

Covariance can also be used (Nair [1939]) to estimate individual yields of a group of plots whose produce has inadvertently been combined, so that only a total over the group is known.

### 3. THE STANDARD COMPUTATIONS

The computations, which will be reviewed briefly, are essentially the same for all mathematical models in which a single regression and a single residual variance are postulated. These cases include the simpler multiple classifications and experimental designs (randomized blocks, cross-over designs, latin squares) as well as balanced and partially balanced incomplete block designs without recovery of inter-block information. Separate discussion is required for hierarchical classifications involving more than one regression equation or more than one residual variance in the specification, as with split-plot designs or incomplete blocks when inter-block information is recovered.

The backbone of the standard procedure is an analysis of sums of squares and products into the Treatments and Error components. Table 1 shows the notation employed

TABLE 1

SUMS OF SQUARES AND PRODUCTS

|            | D. f.          | $(x^2)$   | $(xy)$    | $(y^2)$   |
|------------|----------------|-----------|-----------|-----------|
| Treatments | $(t - 1)$      | $T_{xx}$  | $T_{xy}$  | $T_{yy}$  |
| Error      | $f_e$          | $E_{xx}$  | $E_{xy}$  | $E_{yy}$  |
| Sum        | $t - 1 + f_e$  | $S_{xx}$  | $S_{xy}$  | $S_{yy}$  |

A line is added giving the sums for treatments and error. Thus $S_{xx} = T_{xx} + E_{xx}$, etc.

The error s.s. for $y$ is now divided into two parts: the s.s. for regression on $x$ (1 d.f.) and the s.s. of deviations (Table 2). The same subdivision is made for the Sums.

TABLE 2

PARTITION OF $E_{xx}$ AND $S_{xx}$ INTO COMPONENTS FOR REGRESSION
AND FOR DEVIATIONS

| $(y^2)$ |          | Regression | | Deviations | | |
|---------|----------|------------|-------------|-----------|--------------------------------------|---------|
|         |          | D. f.  S. s. | D. f.     | S. s.     | M. s. |
| Error   | $E_{xx}$ | 1  $E_{xy}^2/E_{xx}$ | $f_e - 1$ | $E_{yy} - E_{xy}^2/E_{xx}$ | $s_e^2$ |
| Sum     | $S_{xx}$ | 1  $S_{xy}^2/S_{xx}$ | $t + f_e - 2$ | $S_{yy} - S_{xy}^2/S_{xx}$ | |
| Treatments (by subtraction) | | | $t - 1$ | $T_{yy} - S_{xy}^2/S_{xx} + E_{xy}^2/E_{xx}$ | $s_t^2$ |

The reduced s.s. for Treatments is obtained by subtracting the deviations s.s. for Error from that for the Sum (last line of Table 2).

The items of information most commonly wanted from this analysis are obtained as follows.

(i) The regression coefficient $\beta$. This is estimated from the Error line; $b = E_{xy}/E_{xx}$. The estimated standard error of $b$ is $s_e/\sqrt{E_{xx}}$, with $(f_e - 1)$ d.f.

(ii) The adjusted estimate of a treatment effect. In the simplest experimental designs, the *unadjusted* estimate for treatment $i$ is simply the mean $y_{i.}$ of all observations having this treatment. The adjusted estimate is

$$y'_{i.} = y_{i.} - b(x_{i.} - x_{..}) \tag{1}$$

(iii) The estimated standard error of any linear function $L = \sum g_i y'_{i.}$ of the adjusted treatment means is

$$s_L = s_e \sqrt{\frac{\sum g_i^2}{r} + \frac{[\sum g_i(x_{i.} - x_{..})]^2}{E_{xx}}} \tag{2}$$

with $(f_e - 1)$ d.f., where $r$ is the number of replications. In particular, the standard error of the difference between two adjusted treatment means is, putting $g_i = 1$, $g_j = -1$, and all other $g$'s $= 0$,

$$s.e.(y'_{i.} - y'_{j.}) = s_e \sqrt{\frac{2}{r} + \frac{(x_{i.} - x_{j.})^2}{E_{xx}}} \tag{3}$$

(iv) For a test of the null hypothesis that all treatment effects are equal, we compute $F = s_t^2/s_e^2$, where $s_t^2$ and $s_e^2$ are the mean squares found in Table 2. This ratio has $(t - 1)$ and $(f_e - 1)$ d.f.

## 4. NATURE OF THE COVARIANCE ADJUSTMENT

The structure of the covariance adjustment, $-b(x_{i.} - x_{..})$, is in accord with common sense: $(x_{i.} - x_{..})$ measures the amount by which the $x$-value for this treatment exceeds the average $x$-value, while $b$ measures the change in $y$ expected to accompany unit change in $x$. In a specific application, the sizes and directions of the adjustments are determined by the data. In this respect a covariance adjustment differs markedly from the type of arbitrary adjustment that has sometimes earned a dubious reputation for the whole process of adjustment. It is not true, however, that a covariance adjustment is entirely objective, since the investigator must choose the type of regression equation (e.g. linear, quadratic, linear in log $x$) from which the adjustment is derived. Moreover, the $x$-variables do not always measure what we would like to think they measure. One occasionally meets extravagant claims for a covariance adjustment, as for example that data have been "adjusted to equalize socio-economic status," when what has actually been done is to adjust for a linear regression on a crude social rating of the father's occupation as reported by anyone who happens to be at home when the interviewer calls.

From equation (3), the estimated variance of the difference between two adjusted means, when averaged over all pairs of means, works out as

$$\frac{2s_e^2}{r}\left\{1 + \frac{t_{xx}}{E_{xx}}\right\} \tag{4}$$

where $t_{xx} = T_{xx}/(t - 1)$ is the treatments *mean square* for $x$. If $t$-tests are to be made between several pairs of means, this expression may be used, as an approximation, for the variance of the difference between any pair (Finney [1946]). This saves the labor of computing a separate standard error for each pair, as is required by the exact formula (3). This device is not recommended when the treatments produce significant effects on $x$, because the variance of the difference may be substantially greater for some pairs than for others, so that the use of a single average variance becomes unsatisfactory.

More generally, the quantity

$$s_e^2\left\{1 + \frac{t_{xx}}{E_{xx}}\right\} \tag{5}$$

may be regarded as the effective error variance *per unit* in a covariance analysis, where the term in brackets is an allowance for sampling errors in $b$. In a completed experiment, the gain in precision from covariance can be estimated by comparing (5) with $s_y^2 = E_{yy}/f_e$, the error mean square for $y$ in its analysis of variance in Table 1. This comparison ignores the loss of 1 d.f. from the error which occurs with a covariance adjustment. The effect of this loss on the sensitivity of the $t$-tests is small even with only 5 d.f. in error.

## 5. THEORY OF THE TECHNIQUE

The theory for the simplest designs will be illustrated by the row $\times$ column classification, with treatments as rows and replications as columns. The model is

$$y_{ij} = \mu + \tau_i + \rho_j + \beta(x_{ij} - x_{..}) + e_{ij} \tag{6}$$

Following the method of least squares, the unknown parameters are estimated by minimizing

$$\sum_{i,j} \{y_{ij} - m - t_i - r_j - b(x_{ij} - x_{..})\}^2 \tag{7}$$

Since $t_i$ need measure only the difference between the effect of the $i$th treatment and the general mean, we may assume

$$\sum_i t_i = 0; \qquad \sum_j r_j = 0.$$

*The estimates.* The algebra involved in finding the estimates can be reduced by introducing the variable

$$x'_{ij} = x_{ij} - x_{i.} - x_{.j} + x_{..} \qquad (8)$$

Those familiar with the analysis of variance will recognize $x'_{ij}$ as the contribution of the $(i, j)$th observation to the error of $x$ in its analysis of variance. The properties of $x'_{ij}$ that will be useful are

$$\sum_i x'_{ij} = 0; \qquad \sum_j x'_{ij} = 0 \qquad (9)$$

$$\sum_{i,j} x'^2_{ij} = E_{xx} ; \qquad \sum_{i,j} x'_{ij} y_{ij} = E_{xy} \qquad (10)$$

The second relation in (10) is less familiar than the others, but is easily verified.

Now the given prediction equation

$$y_{ij} = m + t_i + r_j + b(x_{ij} - x_{..})$$

becomes identical with the prediction equation

$$y_{ij} = m' + t'_i + r'_j + bx'_{ij}$$

if the new estimates $m'$, $t'_i$, $r'_j$ satisfy the relations

$$m = m'; \qquad t_i = t'_i - b(x_{i.} - x_{..}); \qquad r_j = r'_j - b(x_{.j} - x_{..}) \qquad (11)$$

This may be verified by substitution. Further, since $\sum t_i = \sum r_j = 0$, it follows that $\sum t'_i = \sum r'_j = 0$.

Hence, instead of finding $m$, $t_i$, $r_j$ and $b$ so as to minimize (7), we can find $m'$, $t'_i$, $r'_j$ and $b$ to minimize

$$\sum_{i,j} (y_{ij} - m' - t'_i - r'_j - bx'_{ij})^2 \qquad (12)$$

On expansion, (12) becomes

$$\sum_{i,j} (y_{ij} - m' - t'_i - r'_j)^2 - 2b \sum_{i,j} x'_{ij} y_{ij} + b^2 \sum_{i,j} x'^2_{ij}$$

Note that the omitted terms

$$\sum m' x'_{ij} , \qquad \sum t'_i x'_{ij} , \qquad \sum r'_j x'_{ij}$$

vanish because of relations (9).

Using (10), the quantity to be minimized is

$$\sum_{i,j} (y_{ij} - m' - t'_i - r'_j)^2 - 2bE_{xy} + b^2 E_{xx} \tag{13}$$

The advantage of this result is that $b$ is disentangled from the other unknowns. Differentiation with respect to $b$ gives

$$b = E_{xy}/E_{xx} \tag{14}$$

The other unknowns, $m'$, $t'_i$, $r'_j$, must be chosen so as to minimize the first term in (13). But this is exactly the minimization involved in an ordinary analysis of variance of $y$ without covariance. Hence, by the standard results for the analysis of variance,

$$m' = y_{..} ; \qquad t'_i = y_{i.} - y_{..} ; \qquad r'_j = y_{.j} - y_{..}$$

Finally, from (11), the least squares estimates for the covariance analysis are

$$\left. \begin{aligned} m &= y_{..} \\ t_i &= y_{i.} - y_{..} - b(x_{i.} - x_{..}) \\ r_j &= y_{.j} - y_{..} - b(x_{.j} - x_{..}) \end{aligned} \right\} \tag{15}$$

Since the $t_i$ represent deviations from the overall mean, the estimate used in practice is

$$m + t_i = y_{i.} - b(x_{i.} - x_{..}) = y'_{i.} , \quad \text{say.} \tag{16}$$

*Standard errors.* From (13), the residual sum of squares may be written

$$E_{yy} - 2bE_{xy} + b^2 E_{xx} = E_{yy} - \frac{2E^2_{xy}}{E_{xx}} + \frac{E^2_{xy}}{E_{xx}} = E_{yy} - \frac{E^2_{xy}}{E_{xx}} \tag{17}$$

The d.f. are $(f_e - 1)$, since 1 d.f. is subtracted for the regression. Hence $s^2_e$, the residual mean square, is computed as in Table 2.

To find the standard error of $b$, we may write

$$b = \frac{\sum x'_{ij} y_{ij}}{E_{xx}} = \frac{\sum x'_{ij}\{\mu + \tau_i + \rho_j + \beta(x_{ij} - x_{..}) + e_{ij}\}}{E_{xx}}$$

From the properties of the $x'_{ij}$, this reduces to

$$b = \beta + \frac{\sum x'_{ij} e_{ij}}{E_{xx}}$$

This equation expresses $(b - \beta)$ as a linear function of the random residuals $e_{ij}$. Hence

$$\sigma^2_b = E(b - \beta)^2 = \frac{\sigma^2_e \sum x'^2_{ij}}{E^2_{xx}} = \frac{\sigma^2_e}{E_{xx}}$$

In the same way we find, for an adjusted treatment mean,

$$y'_{i.} = m + t_i = \mu + \tau_i + e_{i.} - (x_{i.} - x_{..}) \frac{\sum x'_{ij} e_{ij}}{E_{xx}}$$

For a linear comparison between the adjusted means, it follows that

$$L = \sum g_i y'_{i.} = \sum g_i(\mu + \tau_i) + \sum g_i e_{i.}$$
$$- \{\sum g_i(x_i - x_{..})\} \frac{\sum x'_{ij} e_{ij}}{E_{xx}}$$

By the properties of the $x'_{ij}$, the variate $\sum x'_{ij} e_{ij}$ is uncorrelated with any of the means $e_{i.}$. This gives

$$\sigma_L^2 = \sigma_e^2 \left\{ \frac{\sum g_i^2}{r} + \frac{[\sum g_i(x_{i.} - x_{..})]^2}{E_{xx}} \right\}$$

in agreement with the result in equation (2).

*Note*: The variate $x'_{ij}$ can be used in the same way with all other designs (e.g. latin squares, incomplete blocks) to which the standard covariance theory applies. With an incomplete block design, for example, we define $x'_{ij}$ as

$$x'_{ij} = x_{ij} - m_x - \hat{t}_{ix} - \hat{r}_{jx}$$

where $\hat{t}_{ix}$, $\hat{r}_{jx}$ are the estimates of the treatment and block effects, respectively, in the incomplete block analysis of variance of $x$. The rest of the algebraic development goes through without change.

*Tests of significance.* For the $F$-test, we quote the general theorem on the $F$-test in regression analysis, as applied to this problem. Let

$$D = \sum \{y_{ij} - m - t_i - r_j - b(x_{ij} - x_{..})\}^2$$
$$D'' = \sum \{y_{ij} - m'' - r_j'' - b''(x_{ij} - x_{..})\}^2$$

where the constants are chosen in each case so as to minimize the corresponding sum of squares of deviations. Then if all $\tau_i$ are zero, the quantity

$$\frac{(D'' - D)}{(t - 1)} \bigg/ \frac{D}{(f_e - 1)}$$

is distributed as $F$ with $(t - 1)$ and $(f_e - 1)$ d.f. For proofs see, e.g., Yates [1938], Anderson and Bancroft [1952].

From (17) the denominator $D/(f_e - 1)$ has been shown to be $s_e^2$. By the same approach, $D''$ is the sum of squares of deviations from the "error" regression when only replication effects are eliminated in the analysis of variance. But in that event the "error" will be equal to "treatments + error" in the present analysis. Hence the numerator of $F$, $(D'' - D)/(t - 1)$, is equal to $s_t^2$ as defined in Table 2.

## 6. THE REDUCED SUM OF SQUARES FOR TREATMENTS

The non-mathematical user of the analysis of covariance often finds it easy to rationalize the method by which $b$ is computed and the formula for making the adjustments. The roundabout method by which the numerator of $F$ is computed is, however, apt to appear mysterious. In this section the sum of squares $D'' - D$ in the numerator of $F$ is examined in more detail. Following Fisher, this quantity will be called the *reduced* sum of squares for treatments, $T_{yyR}$ .

Intuitively, one might expect this quantity to be a squared comparison among the adjusted treatment means

$$y_{i.}' = y_{i.} - b(x_{i.} - x_{..})$$

As is well-known, $T_{yyR}$ is not the sum of squares of deviations of these adjusted treatment means. The latter sum of squares, $T_{yyA}$ say, is

$$T_{yyA} = r \sum \{(y_{i.} - y_{..}) - b(x_{i.} - x_{..})\}^2$$

$$= T_{yy} - 2bT_{xy} + b^2 T_{xx} \tag{18}$$

$$= T_{yy} - 2bb_t T_{xx} + b^2 T_{xx} \tag{19}$$

where $b_t = T_{xy}/T_{xx}$ is the regression coefficient as computed from the Treatments line in the analysis of variance in Table 1. From Table 2, on the other hand, the sum of squares $T_{yyR}$ in the numerator of $F$ is

$$T_{yyR} = T_{yy} - \frac{(T_{xy} + E_{xy})^2}{T_{xx} + E_{xx}} + \frac{E_{xy}^2}{E_{xx}} \tag{20}$$

$$= T_{yy} - \frac{(b_t T_{xx} + b E_{xx})^2}{T_{xx} + E_{xx}} + b^2 E_{xx} \tag{21}$$

Subtracting (19) from (21) and taking the common denominator $(T_{xx} + E_{xx})$, we find

$$T_{yyR} = T_{yyA} - \frac{T_{xx}^2 (b_t - b)^2}{T_{xx} + E_{xx}} \tag{22}$$

This result was first given by Yates [1934]. Equation (22) remains valid if $T_{yyR}$ and $T_{yyA}$ are the reduced and adjusted sums of squares for any specific component of the treatments sum of squares, provided that $b_t$ and $T_{xx}$ refer to this component. As Yates showed, the result provides an approximate short-cut method of making $F$-tests when we intend to test several components. The exact test involves going through the procedure in Table 2 separately for each component. We can, however, easily calculate $T_{yyA}$ for each component by relation (18). The $F$-values computed from $T_{yyA}$ will all be too large, but the approximation is usually close if the treatments do not affect $x$, and only a few borderline $F$'s need be recomputed by the exact method. Table 3 compares the approximate and exact $F$'s in an experiment in which the treatments sum of squares was divided into 4 components, (data from Cochran and Cox [1957], s.s. rounded to 100's).

TABLE 3

THE APPROXIMATE F-TEST FROM THE ANALYSIS OF $(y - bx)^2$

|  | D.f. | $x^2$ | $xy$ | $y^2$ | $(y - bx)^2$ | $F_A$ | $F$ | $F_{.05}$ |
|---|---|---|---|---|---|---|---|---|
| | | | Sums of squares | | | | | |
| Treatments | 8 | 292 | $-92$ | 1574 | 2571 | 4.51 | 4.16 | 2.22 |
| Ave. L | 1 | 31 | $-133$ | 572 | 1062 | 14.90 | 14.53 | 4.12 |
| Ave. Q | 1 | 22 | 83 | 311 | 106 | 1.49 | 1.46 | 4.12 |
| Dev. L | 3 | 230 | $-68$ | 434 | 1205 | 5.63 | 5.05 | 2.87 |
| Dev. Q | 3 | 9 | 26 | 257 | 198 | 0.93 | 0.93 | 2.87 |
| Error | 36 | 1214 | 1893 | 5447 | 2495 | | | |

The columns on the right show the approximate ratio $F_A$, the exact $F$ as found from the method in Table 2, and the tabular 5% value. Since none of the $F_A$ values is near the 5% level, no recomputation would be necessary in practice in this example. (The column of mean squares of $(y - bx)^2$ has been omitted.)

The presence of sampling errors in $b$ explains why $T_{yyA}$ is not the correct sum of squares for the numerator of $F$. If the true $\beta$ could be used in $T_{yyA}$, this sum of squares would give an exact $F$-test. The sampling error of $b$ complicates the issue in two ways. Each adjusted treatment mean has a different variance, since

$$V(y'_{i.}) = \sigma_e^2 \left\{ \frac{1}{r} + \frac{(x_{i.} - x_{..})^2}{E_{xx}} \right\} \tag{23}$$

Further, the adjusted means for the $i$th and $j$th treatments are not independent, since $b$ enters into both. Their covariance is

$$\text{cov}(y'_{i.} y'_{j.}) = \frac{\sigma_e^2 (x_{i.} - x_{..})(x_{j.} - x_{..})}{E_{xx}} \tag{24}$$

Consequently, as would be expected from theory, it turns out that the correct numerator $T_{yyR}$ is a quadratic form in the adjusted means,

$$T_{yyR} = \sum_{i,j} a_{ij}(y'_{i.} - y_{..})(y'_{j.} - y_{..})$$

where the matrix $a_{ij}$ is the inverse of the variance-covariance matrix of the $y_{ij}$ (apart from the factor $\sigma_e^2$). To find the $a_{ij}$, we may write, from (22),

$$T_{yyR} = r \sum (y'_i. - y..)^2 - \frac{T_{xx}^2(b_t - b)^2}{T_{xx} + E_{xx}} \tag{25}$$

Now

$$T_{xx}(b_t - b) = T_{xy} - bT_{xx}$$

$$= r \sum_i (x_i. - x..)\{(y_i. - y..) - b(x_i. - x..)\}$$

$$= r \sum_i (x_i. - x..)(y'_i. - y..)$$

Substituting this expression in (25) we find that

$$T_{yyR} = r \sum (y'_i. - y..)^2 - \frac{r^2[\sum (x_i. - x..)(y'_i. - y..)]^2}{T_{xx} + E_{xx}}$$

It follows that

$$a_{ii} = r\left\{1 - \frac{r(x_i. - x..)^2}{T_{xx} + E_{xx}}\right\}$$

$$a_{ij} = -\frac{r^2(x_i. - x..)(x_j. - x..)}{T_{xx} + E_{xx}} \qquad i \neq j \tag{26}$$

The inverse of the matrix $(a_{ij})$ can be shown to be the covariance matrix of the $y'_i$ as given by equations (23) and (24).

It is also instructive to compare $T_{yyA}$ and $T_{yyR}$ with the sum of squares of deviations of the treatment means $y_i.$ from the *treatments* regression on $x$, i.e. with the treatments sum of squares of $(y - b_t x)$. This sum of squares is

$$T_{yy.x} = T_{yy} - \frac{T_{xy}^2}{T_{xx}} = T_{yy} - b_t^2 T_{xx} \tag{27}$$

and has $(t - 2)$ d.f. From (19),

$$T_{yyA} = T_{yy} - 2bb_t T_{xx} + b^2 T_{xx}$$

$$= T_{yy.x} + (b_t - b)^2 T_{xx} \tag{28}$$

using (27). Finally from (22),

$$T_{yyR} = T_{yyA} - \frac{(b_t - b)^2 T_{xx}^2}{T_{xx} + E_{xx}}$$

$$= T_{yy.x} + \frac{(b_t - b)^2 T_{xx} E_{xx}}{T_{xx} + E_{xx}} \tag{29}$$

Equation (29) shows that the reduced sum of squares for treatments separates into two parts: (i) the sum of squares $T_{yy.x}$ with $(t-2)$ d.f. which represents the deviations of the treatment means from their own regression, (ii) a single d.f. which compares the regression coefficients from the treatments and error lines in the analysis of variance. It may be shown that if there are no treatment effects, this sum of squares is distributed as $\chi^2 \sigma_e^2$ with 1 d.f. In this way, equation (29) can be used to prove directly that, on the null hypothesis, $T_{yyR}$ is distributed as $\chi^2 \sigma_e^2$ with $(t-1)$ d.f. In the same vein, equation (28) shows that $T_{yyA}$ consists of $(t-2)$ d.f. that are distributed as $\chi^2 \sigma_e^2$. The remaining 1 d.f. is inflated by the factor $(T_{xx} + E_{xx})/E_{xx}$, as is seen by comparing (28) with (29).

To sum up, the sum of squares $T_{yyR}$ in the numerator of the $F$-test is a quadratic form in the deviations $(y'_i. - y..)$ of the adjusted treatment means. The coefficients $a_{ij}$ in this quadratic form are the inverse of the covariance matrix of the quantities $y'_i.$. As would be expected, $T_{yyR}$ vanishes if all adjusted means are equal. Because of the presence of sampling errors in $b$, $T_{yyR}$ does not equal the usual sum of squares of deviations $T_{yyA}$ of the adjusted treatment means. $T_{yyA}$ always gives too large a sum of squares, but if treatments do not affect $x$ it is a useful approximation to $T_{yyR}$ when several components of the treatments sum of squares are to be tested.

## 7. ASSUMPTIONS REQUIRED FOR THE ANALYSIS OF COVARIANCE

The assumptions required for valid use of the analysis of covariance are the natural extension of those for an analysis of variance, namely,
  (i) Treatment, block and regression effects must be additive as postulated by the model,
  (ii) The residuals $e_{ij}$ must be normally and independently distributed with zero means and the same variance.

Although the effects of failures in these assumptions on the analysis of covariance as such do not appear to have been investigated, much of the related work on the analysis of variance carries over—for instance, that on the effects of non-normality or inhomogeneity of variance in the $e_{ij}$. The general precautions that have been given about the practical use of the analysis of variance should equally be observed in an analysis of covariance, see e.g. Cochran [1947].

Two assumptions that particularly involve the regression term in covariance should be noted. The treatment and the regression effects may not be additive. Bartlett [1937] pointed out that this danger might be present in the cotton-salt example already cited. On plots with a high salt content, the crop might be unable to respond to superior fertilizers. Thus, in an extreme case, the treatment effects may be zero if $x$ lies above a certain value. If this happens, the covariance adjustment may still improve the precision, but (i) the meaning of the adjusted treatment effects become cloudy, and (ii) if covariance is applied in a routine way, the investigator fails to discover the differential nature of the treatment effects—a point that might be important for practical applications.

Secondly, the covariance procedure assumes that the correct form of regression equation has been fitted. Perhaps the most common error to be anticipated is that linear regressions will be used when the true regression is curvilinear. In a randomized experiment in which treatments do not affect $x$, the randomization ensures that the usual interpretations of standard errors and tests of significance are not seriously vitiated, although fitting the correct form of regression would presumably give a larger increase in precision. The danger of misleading results is greater when $x$ shows real differences from treatment to treatment. Later investigations by Fairfield Smith in this issue suggest, however, that this disturbance is serious only in rather extreme situations.

## 8. MULTIPLE COVARIANCE

No new difficulty is presented when there is more than one covariate, although the computations become more lengthy. Only the basic formulae will be given. With two covariates, $x$ and $z$, the regression coefficients are obtained from the equations

$$E_{xx}b_{y.x} + E_{xz}b_{y.z} = E_{xy}$$

$$E_{xz}b_{y.x} + E_{zz}b_{y.z} = E_{zy}$$

If $t$-tests of the adjusted treatment means are wanted, it is advisable to compute the inverse of the $E_{xz}$ matrix, say $(c_{xz})$.

In the simple orthogonal designs like randomized blocks, the adjusted mean of the $i$th treatment is estimated by

$$y'_{i.} = y_{i.} - b_{y.x}(x_{i.} - x_{..}) - b_{y.z}(z_{i.} - z_{..})$$

The variance of the difference between the $i$th and $j$th treatment effects is

$$s_e^2 \left\{ \frac{2}{r} + (x_{i.} - x_{j.})^2 c_{xx} + 2(x_{i.} - x_{j.})(z_{i.} - z_{j.})c_{xz} + (z_{i.} - z_{j.})^2 c_{zz} \right\}$$

The average variance over all pairs of treatments is

$$\bar{s}_{\text{diff}}^2 = \frac{2}{r} s_e^2 (1 + t_{xx}c_{xx} + 2t_{xz}c_{xz} + t_{zz}c_{zz})$$

where $t_{xx}$ is the treatments mean square for $x$, etc. This is the extension of equation (4). If treatments do not affect $x$ or $z$, the square root, $\bar{s}_{\text{diff}}$ , may be used as an approximate standard error of the difference for all $t$-tests.

For the $F$-test, the regression equations from the error line and the (Treatments + Error) line must both be solved. If treatments do not affect $x$ or $z$, an alternative is to begin with an approximate $F$-test based on the analysis of variance of $(y - b_{y.x}x - b_{y.z}z)$. It will be necessary to compute the exact $F$-value only in doubtful cases.

## 9. MORE COMPLEX CLASSIFICATIONS

Three cases that have received attention in the literature are hierarchical classifications, as represented by the split-plot design, incomplete block designs with recovery of interblock information and two-way classifications with unequal numbers in the cells.

In the split-plot design, two independent regression coefficients can be computed, $b_1$ from the whole-plot error line and $b_2$ from the sub-plot error line in the analysis of variance. If the two regression coefficients appear to differ, $b_1$ can be used to adjust whole-plot comparisons and $b_2$ to adjust sub-plot comparisons. The structure of the analysis of variance is given by Kempthorne [1952]. If the suffix $i$ denotes the whole-plot treatments and $j$ the sub-plot treatments, the adjusted means in the two-way table of treatment means can be computed as follows.

$$y'_{ij.} = y_{ij.} - b_1(x_{i..} - x_{...}) - b_2(x_{ij.} - x_{i..}) \tag{30}$$

Although equation (30) looks as if a bivariate regression is being used to make the adjustments, comparisons between whole-plot treatment means $y'_{i..}$ are actually adjusted by $b_1$ alone, since

$$y'_{i..} = y_{i..} - b_1(x_{i..} - x_{...})$$

This form of adjustment avoids a discomforting feature mentioned by Bartlett [1937] and Truett and Fairfield Smith [1956], namely that individual adjusted means may not average to the appropriate whole-plot treatment means. Similarly, it is easy to verify that comparisons between sub-plot treatments and components of the interaction between whole-plot and sub-plot treatments are adjusted by $b_2$ alone. Both regression coefficients enter, however, into certain particular comparisons. For instance, to compare two whole-plot treatments for the same sub-plot treatment, we take, say,

$$y'_{2j.} - y'_{1j.} = y_{2j.} - y_{1j.} - b_1(x_{2..} - x_{1..})$$
$$- b_2(x_{2j.} - x_{1j.} - x_{2..} + x_{1..}) \tag{31}$$

As in the split-plot design without covariance, standard errors of comparisons like this require special investigation. For (31), the estimated variance works out as

$$\frac{2}{r}\frac{s_1^2 + (\gamma - 1)s_2^2}{\gamma} + \frac{(x_{2..} - x_{1..})^2 s_1^2}{E_{1xx}} + \frac{(x_{2j.} - x_{1j.} - x_{2..} + x_{1..})^2 s_2^2}{E_{2xx}}$$

where $\gamma$ is the number of sub-plot treatments, $s_1^2$, $s_2^2$ are the whole- and sub-plot error mean squares, and $E_{1xx}$, $E_{2xx}$ are the whole- and sub-plot error sums of squares for $x$.

If the two regression coefficients can be assumed to be the same, a good working procedure, although not fully efficient, is to adjust all means by the sub-plot coefficient $b_2$. This procedure, originally suggested by Bartlett [1937], has also been recommended after further examination by Truett and Fairfield Smith [1956].

Similar issues arise with incomplete block designs, since separate regressions can be calculated from the Blocks and Intra-block error lines in the analysis of variance. The computations for recovery of inter-block information, using the intra-block error regression to adjust all means, are illustrated for the $9 \times 9$ triple lattice by Cox and Eckhardt [1940] and for the $7 \times 8$ simple rectangular lattice by Robinson and Watson [1949]. The approximations involved in this method need further investigation, particularly for the smaller designs.

Methods for handling covariance analyses in the simpler non-orthogonal multiple classifications are given by Wilks [1938], Hazel [1946] and Das [1953], while Federer [1955] presents methods for lattice squares and changeover designs in which residual effects are to be estimated.

## REFERENCES

Anderson, R. L. and Bancroft, T. A. [1952] *Statistical theory in research*. McGraw-Hill Book Company, New York. Chapter 14.

Bartlett, M. S. [1937] Some examples of statistical methods of research in agriculture. *J. Roy. Stat. Soc. Suppl. 4:* 137–183.

Cochran, W. G. [1938] Crop estimation and its relation to agricultural meteorology. *J. Roy. Stat. Soc. Suppl. 5:* 12–16.

Cochran, W. G. [1947] Some consequences when the assumptions for the analysis of variance are not satisfied. *Biometrics, 3:* 22–38.

Cochran, W. G. and Cox, G. M. [1957] *Experimental designs*. John Wiley and Sons, New York. 2nd Edition.

Cox, G. M. Eckhardt, R. C. and Cochran, W. G. [1940] The analysis of lattice and triple lattice experiments in corn varietal tests. Iowa Agr. Expt. Sta. Res. Bull. 281.

Das, M. N. [1953] Analysis of covariance in two-way classification with disproportionate cell frequencies. *J. Ind. Soc. Agric. Stat. 5:* 161–178.

Day, B. and Fisher, R. A. [1937] The comparison of variability in populations having unequal means. An example of the analysis of covariance with multiple dependent and independent variates. *Ann. Eug. 7:* 333.

Federer, W. T. and Schlottfeldt, C. S. [1954] The use of covariance to control gradients in experiments. *Biometrics 10:* 282–290.

Federer, W. T. [1955] *Experimental design.* Macmillan Company, New York. Chapter 16.

Finney, D. J. [1946] Standard errors of yields adjusted for regression on an independent measurement. *Biometrics 2:* 53.

Fisher, R. A. [1932] *Statistical methods for research workers.* Oliver and Boyd Ltd., Edinburgh. 4th Edition.

Greenberg, B. G. [1953] The use of analysis of covariance and balancing in analytical surveys. *Am. J. of Pub. Health, 43:* 692–699.

Hazel, L. N. [1946] The covariance analysis of multiple classification tables with unequal subclass numbers. *Biometrics 2:* 21–25.

Kempthorne, O. [1952] *The design and analysis of experiments.* John Wiley and Sons, Inc., New York, p. 387.

Nair, K. R. [1939] The application of covariance technique to field experiments with missing or mixed-up yields. *Sankhya 4:* 581–588.

Robinson, H. F. and Watson, G. S. [1949] Analysis of simple and triple rectangular lattice designs. North Carolina Agr. Exp. Sta. Tech. Bull. 88.

Truett, J. Titus, and Smith, H. F. [1956] Adjustment by covariance and consequent tests of significance in split-plot experiments. *Biometrics 12:* 23–39.

Wilks, S. S. [1938] The analysis of variance and covariance in non-orthogonal data. *Metron. 13:* 141–154.

Yates, F. [1933] The analysis of replicated experiments when the field results are incomplete. *Emp. Journal Exp. Agr., 1:* 129–142.

Yates, F. [1934] A complex pig-feeding experiments. *J. Agri. Sci. 24:* 519.

Yates, F. [1938] Orthogonal functions and tests of significance in the analysis of variance. *Jour. Roy. Stat. Soc. Suppl. 5:* 177–180.

---

*(P. 263) For a more thorough comparison of covariance and blocking in this situation, see D. R Cox, Biometrika, 44, 150-158, 1957.

# 66

## THE PHILOSOPHY UNDERLYING THE DESIGN OF EXPERIMENTS

William G. Cochrane
The Johns Hopkins University
Professor of Biostatistics

Introduction. In ordinary speech, the word "experiment" has a broad meaning. It denotes trying out anything new. In our conferences here we shall be using the word in a narrower sense. The essence of an experiment is that we deliberately introduce one or more changes into some process, and take measurements in order to find out the effects of these changes. The changes whose effects are being compared are often called the experimental treatments or more simply the treatments.

The ability to do experiments is one of the most powerful weapons that man has for making advances in his understanding of the world. When conflicting claims or conflicting theories can be put to a crucial test, the workers in a branch of knowledge cannot long remain in error. In fields like economics, sociology and history, on the other hand, where experiments are rarely if ever feasible, it is difficult to get at the true causes behind the events that are observed. Having read that all the economic experts forecast a prolonged rise in the stock market, you may take your meager savings from under the bed and purchase a few good-looking stocks. When the market promptly falls, you may become slightly mad at the experts. Instead, you should be sympathetic and understanding: these men cannot do experiments , and it is hard for them to unravel the complex forces behind the market.

The origin of an experiment usually lies in some information that we would like to have, or in some questions to which we want answers. After carefully phrasing the questions, we select a set of treatments such that comparisons of the effects produced by these treatments will answer the questions. We must then consider the environmental conditions in which these treatments should be applied, and the most suitable kinds of measurement to take. At the end, if the experiment is successful, we find that the results do enable us to answer the questions.

Failures in experimentation. There is much to be learned by considering the causes of failures in experimentation - that is, experiments which do not produce the desired information. Although such failures can be classified in various ways, the following rough grouping will serve my purpose.

1. We may have asked the wrong questions in the beginning. This is perhaps of more frequent occurrence in fundamental research, where the ability to ask the significant questions distinguishes the outstanding from the second-rate scientist. In both fundamental and applied research one can find experiments on questions that have already been answered, since the increasing volume of research makes it harder to keep up with what has been done. And in applied research there are examples where the questions asked were the unimportant rather than the important ones that would have to be faced in putting the results into practice. Before starting to plan an experiment, it is always advisable to ask: "Is this the most informative question to try to answer right now?"

Reproduced with permission from *Proceedings of 1st Conference on the Design of Experiments in Army Research, Development and Testing.* Office of Ordnance Research, U.S. Army, Durham, NC, OOR-1, pages 1–8, 1957.

2. We may have asked the right questions, but the treatments selected are incapable of providing answers to some of the questions. This should not happen in simple experiments with only two or three treatments, but the danger is present in a complex experiment designed to throw light on a number of different questions. It is a hazard particularly associated with experiments that are planned by committees, especially if they contain several strong-minded persons who don't agree with one another. The principal safeguard is to sit down, after the treatments have been selected, and verify for each question the treatment comparisons that will be made in order to answer the questions.

3. In applied research, the conditions under which the experiment is conducted are often strikingly different from those in which results will be applied. Treatment effects that are found in the experiment may not hold up under the conditions of application. There is no entirely safe way out of this difficulty, because much experimentation has to be done with small-scale equipment in a specialized environment, in order to keep down costs and to obtain precise results. In many types of work, the practice is to use the small-scale experiment primarily for screening. Promising candidates from this screening are tested again under conditions that more closely approximate those that will prevail in applications.

I mention this difficulty because it is always well to realize how the conditions of experimentation differ from those of application. Even in pilot experiments it may be possible to include some comparison that help to bridge this gap. Suppose that three different models of some piece of equipment are used in practice and that the properties of these models have already been worked out, so that there is no need to experiment further in comparing them. It may still be advisable to include all three models in experiments designed to test other factors, rather than take the simpler path of confining the experiments to one model. In this way we obtain some check as to whether the other factors perform the same way on every model. Similarly, we may sometimes reject a refinement in experimental technique that otherwise seems attractive, on the grounds that the refinement makes the conditions of the experiment too remote from those of application.

4. We may obtain erroneous results for the effects of the treatments. It has happened several times in medical research that a dramatic cure is found in a first experiment, and perhaps confirmed in a second, causing much excitement in the press. But later and more careful experiments repeatedly fail to find any effect, and after a time medical science reluctantly concludes that the cure doesn't exist. Erroneous results of this type usually happen because some unsuspected bias has crept into the results. Such biases are one of the most frequent causes of failures in experimentation.

5. Finally, the results may be so indecisive as to be useless. This happens when all that we can conclude at the end of the experiment is that the difference in effect between treatment 1 and treatment 2 lies somewhere between a large positive value and a large negative value. In other words, we haven't learned which treatment is superior, nor can we even assume that the two are approximately equal in their effects. Those of you who are new to the design and analysis of experiments may protest that surely we can get

more definite information than this out of an experiment.  Unfortunately,
even with well-conducted experiments, vague conclusions of this kind are
often all that can be drawn.

This type of failure is perhaps less serious than the preceding types.
It arises because the experiment was not precise enough for its purpose,
and it can be remedied by repeating the experiment sufficiently often with
the same treatments.  This, however, is no consolation if decisions have
to be taken before there is time for more experimentation.

In this catalogue of failures in experimentation I have not, of course,
mentioned all types of failures.  I have been told of an agronomist who laid
out an experiment in the semi-arid backlands of Australia, and then as
harvest approached could not remember where he had put it.  There are even
cases where the professor forgot that he had started an experiment until
long after the time when the results should have been recorded.

In the remainder of my remarks, I shall concentrate on failures that
arise from biased results and from indecisive results.  These are the types
of failure in which statistical ideas appear to have been able to help most.

Variability and experimental errors.  One of the most pervasive
features of experimentation is the presence of variability in the results.
The easiest way to find out how much variability you face in your own
results is to apply the same treatment several times and see how well the
results agree in these several repetitions.  A repetition does not mean
just repeating the final readings that are made, but running through from
the beginning the whole process of applying the treatment and taking the
measurements.  In some lines of work, these repetitions agree within one
or two percent.  In this event you may count yourself lucky, in that the
experimental errors are small.  Often, however, the variation in results
from one application of the same treatment to another is much larger:
sometimes it is enormous.  In certain experiments in immunology, for
example, the amount of protective serum that produces a given color-
imetric response is measured.  When a treatment is applied a second time,
about all that we can be sure of is that the dose of serum producing the
same response will lie somewhere between one quarter and four times the
dose that was needed at the first trial.

What causes these variations?  They can enter at any stage in the
conduct of the experiment.  They may be due to lack of uniformity in the
raw material to which treatments are applied.  In experiments in which
the raw material is living, as with animal or human subjects, this is one
of the most important sources of variability.  Uncontrolled changes in
the environment or in the equipment or machinery used, variability in the
human operators and errors in the measuring devices are all contributory
causes.

What do experimental errors do to us?  If we are not careful, we may
finish the experiment with results that are biased and misleading.  If an
experiment with two treatments takes two days to carry through, one way of
doing it that often seems natural to the experimenter is to perform all the
work and take all the measurements on treatment 1 on the first day.

Treatment 2 is handled on the second day. This prevents any danger of mixing up the two treatments, but it is the surest way to invite bias. If the raw material used is somewhat different on the two days, or the equipment is more worn on the second day, or the observers are more careless, all the results obtained for treatment 2 will be subject to a bias. Moreover, the standard methods of statistical analysis give no warning of the presence of bias. These techniques assume, in fact, that no bias is present – a point that is not sufficiently emphasized in introductory courses in statistics.

Some tests were carried out during the last war of three preventatives for sea sickness. Available for the tests were four small ships used for carrying troops. It was proposed to have the ships follow a course a short distance behind one another and to give a specific pill to all the men on a single ship. Administratively, this is the most convenient way to conduct the test. The objection was raised that this procedure might result in a bias if one of the ships proved to be less subject to rolling and pitching than the others. This was not thought likely, because the ships had been built to the same specifications in the same shipyard. However, it was finally agreed to carry out the administratively more difficult plan of giving each pill to one-third of the men on each ship. When the trials were completed it was found that there were adequate amounts of sea sickness on two of the ships. But on the third ship hardly anyone was sick no matter what pill he received, even though one of the pills contained just sugar. Further investigation showed that this ship had given trouble during its seaworthiness trials and that the shipyard had dumped in an extra load of ballast which made it unusually stable.

Even if we are able to avoid biases, experimental errors may result in the kind of indecisive conclusions to which I have already referred. In several repetitions of a test we may find that sometimes treatment 1 wins and sometimes treatment 2. Although treatment 1 wins often enough so that we are convinced of its superiority, it may not be clear how closely we can trust the estimate of the amount of superiority – which is often the important quantity for the practical use of the results.

Naturally, experimental errors are important mainly in relation to the size of difference that the treatments produce, or to the size that we are interested in detecting and studying. The experimenter who faces experimental errors of the order of one or two percent and is dealing with treatments that produce differences of the order of twenty or thirty percent has no problem of this kind. But sooner or later, in most lines of work, there comes a time when we have skimmed off the cream and are no longer working with treatments that produce large differences. When the treatments are producing differences that are of the same order of magnitude as the experimental errors, we have to find some way of coping with these errors.

What can we do about experimental errors? A three-point program might run as follows:

1.  Try to find out the main causes of the experimental errors to which you are subject. Do they lie in the raw material, in equipment that gives erratic performance, in wear or fatigue, in the environment or in errors of the measuring devices? This task may sometimes require an extensive investigation.

2. Having discovered the principal contributors to experimental error, consider for each one what feasible steps, if any, can be taken to reduce or remove its effects. There are many possibilities. I shall discuss a few of them later.

3. After surveying all these proposed steps, select those that will produce the needed amount of reduction in experimental errors most economically and conveniently.

Improvements in technique. One class of methods for cutting down the effects of the principal contributors to experimental errors may be called improvements in technique. If the principal difficulty lies in the variability of the raw material, can we procure more uniform raw material? At one time I was engaged in experiments on the nutrition of pigs in England. We found that our experimental errors were of a size that made precise results difficult to obtain. On the other hand, the rival establishment, Cambridge University, which was doing the same kind of experimentation, had experimental errors low enough so that they had satisfactory precision. A careful comparison of methods revealed only two relevant differences between the two places. We had better statisticians, but Cambridge had better pigs. In Cambridge the pigs had been carefully bred so as to be uniform in their weight gains, while our pigs appeared to have been purchased in a bargain basement and showed a regrettably high degree of variability in their weight gains. Since the only pigs that we could afford were bargain basement pigs, and since an offer to trade a statistician for 20 pigs would probably have been refused by Cambridge, we abandoned this line of experimentation until better resources could be obtained.

Under the same heading come the purchase of better equipment and measuring devices, the standardization of the environment through temperature and humidity controls and so on. Naturally, these facilities cost money and may delay a program of experimentation.

There are three methods of dealing with experimental errors that have been extensively worked upon by the statisticians. These are local control (sometimes called grouping or balancing); randomization; and replication.

Local control. Local control may be illustrated by an experiment with only two treatments, each of which we intend to apply six times in order to get some replication of the results. The general principle is to divide the experiment into six separate little experiments. In each of these we take all precautions that are feasible to ensure that the comparison of treatment 1 and treatment 2 will be an accurate one.

To illustrate, an experiment was conducted in order to find out whether a dose of x-rays might enable a rat to withstand better the effects of a poison gas. There was some reason to believe that this would be so. The experiment contained two groups of rats, one receiving a preliminary dose of x-rays, the other no preliminary treatment. To receive the poison gas the rat was placed under a bell jar into which a steady stream of gas was fed. The time taken for the rat to die was measured.

What are the principal sources of error variation in this experiment?
One is, of course, the rat. Rats vary in their toughness in remaining alive
under doses of the gas. Hence it is important that the two groups contain
equally resistant rats. The resistance of a rat presumably varies with its
sex, its age, its weight and with other factors. The flow of poison gas
into the bell jar might be a second source of variability, since this flow
could not be kept quite uniform from one test to another.

In order to apply local control, therefore, the experimenters selected
for a single trial two rats that were of the same sex and came from the same
litter. This made their genetic backgrounds somewhat similar, which might
affect their ability to resist, and also ensured that they were of the same
age. The two rats in any one trial were both put into the bell jar together,
so that variations in the amount of flow of the gas from one occasion to the
other did not affect the accuracy of the comparison in any single trial.
This is a good example of the use of local control to make sure that a
number of potential sources of experimental error affect each of the treatments equally.

Although the experimenters had evened out the variables that I have
mentioned, they were not able to control weight. The two rats in a pair
differed more or less in weight. The experimenters decided always to give
the x-rays to the lighter rat of the two. They argued that if x-rays
showed a beneficial effect when given to the supposedly weaker rat of the
pair, this would make the final results still more convincing in favor of
x-rays.

Notice the logical confusion in this decision. A series of steps
designed to make the comparison fair and precise is followed by a step that
is designed to make the comparison unfair. The experimenters soon learned
the error of their ways. In each of the first 3 trials, the smaller rat,
the one receiving the x-rays, died first. What conclusion could they draw?
It was time to stop and think.

There are several ways in which the experimenters could have dealt
with the problem presented by variation in weights. What they did was to
toss a coin at each subsequent trial to decide whether the lighter or
heavier rat should receive the x-rays. This is the method of randomization.
It doesn't attempt a complete equilization of the disturbing variable, but
merely ensures that the trial shall be a fair game with respect to this
variable. Randomization is not the best way of handling major sources of
experimental errors, because a careful balancing will take care of them more
adequately. It is very useful, however, for dealing with sources of variation that remain after we have exhausted the resources of balancing. We
hope these sources are minor, but if we are wrong, randomization gives each
treatment the same chance of benefitting from them.

Another method that they could have used was to make the experiment
up in pairs of trials, giving the x-rays to the lighter rat in the first
trial and to the heavier rat in the second trial. This method, based on
2 x 2 latin squares, gives a better balancing out of the weight effect than
randomization. Alternatively they could have recorded the weights and then
at the end adjusted the results so as to equalize weights by an objective

statistical technique known as analysis of covariance.

A friend of mine, after making several attempts to read a book on experimental designs written by Miss Cox and myself, remarked that the subject seemed to be a very complicated one. It is true that the subject abounds with strange names for particular types of designs such as latin squares and graeco-latin squares, and recently with more formidable creatures like partially balanced incomplete blocks, doubly balanced incomplete blocks and so on. Although these designs are unavoidably somewhat complex in detail, their purpose is simple. They are all devices for enabling the experimenter to balance out the effects of the major disturbing variables in a great variety of different situations. It is worthwhile to have many ways for applying the notion of local control, because local control often costs practically nothing to apply, involving merely careful advance thinking about the way in which the experiment should be done.

Replication. Finally, increased precision can always be gained by repeating the experiment enough times, making sure that in each replication the test is independent of the previous replications so that the experimental errors have a chance to average out. In this way good experiments can be done with crude equipment and variable material if we replicate enough times. Of course, replication is not the answer to all our problems because it too costs money and materials.

In this connection there are methods available by which one can make rough estimates, before starting an experiment, of the number of replications needed to detect treatment differences of a given size. More frequent use of these advance estimates would avoid much wastage in experimental work. During the war I had to recommend courses of action on the basis of a summary of the results of the experiments that had been conducted on some scientific question. In a number of these situations the experimental data were practically worthless. Variability was so high and replications so few that the results were too erratic to be relied upon. The point to be emphasized, however, is that in many of these cases it could have been predicted in advance that experiments of the size and type that were done would be almost certain to give indecisive results.

Statistical analysis. By careful technique, local control plus randomization, and use of enough replications we can hope to reduce the effects of experimental errors on the average results for the different treatments to a tolerable amount. In writing our conclusions we must, however, take proper account of the experimental errors that do remain in the estimated treatment effects. The calculations by which this is done may seem mystifying to the beginner, since they derive from the theory of probability. In the standard methods of analysis, each experiment furnishes its own estimate of the magnitude of the experimental errors, making the appropriate allowance for any local control that was employed and for the number of times that the experiment was replicated. The calculations do not allow for biases that have crept into the comparisons, and there seems no way in which this can be done. Constant vigilance against bias should therefore be the watchword of the experimenter.

## Summary

This paper discusses some general principles that should govern controlled experimentation. By way of introduction, some of the main reasons why experiments may fail to provide useful information are outlined, as follows.

1.  The wrong questions were asked in planning the experiment.

2.  The experimental treatments that were selected were incapable of furnishing answers to some of the questions.

3.  The conditions under which the experiment was conducted were too remote from those in which the results were to be applied.

4.  The results obtained were biased.

5.  Although unbiased, the results were so erratic and indecisive as to be useless.

Although the points above are of equal importance, the remainder of the paper concentrates on the last two, on which the statistical viewpoint has the most to contribute.

Since biased and imprecise results arise from uncontrolled variability that affects the results of the experiment, the experimenter should make it his business to find out how large his experimental errors are and what sources of variation are the principal contributors to them. Various methods for reducing the effects of experimental errors and avoiding bias are discussed. These include improvements in technique, local control, randomization and replication. In any given situation, the experimenter is advised to utilize the method that seems to promise the greatest returns.

# 67

# NEWER STATISTICAL METHODS

## INTRODUCTION

THIS paper describes some statistical techniques that may be useful
in human pharmacology and therapeutics, and serves as an introduction to later contributions in this session. Four topics will be presented: (i) the device known as grouping, which forms the basis of
the common types of experimental design such as randomized
blocks, latin squares and incomplete block designs, of which
specific illustrations are given by Drs. Mongar, Dare and Schild,
(ii) sequential experimentation, in which a continuous statistical
analysis of the results is made as they come in, with the hope that
conclusions can be reached more quickly than with the standard
"fixed-size" trial, (iii) suggestions for the study of subject to subject
variability in the relative effectiveness of treatments, (iv) some
methods for the analysis of ranked data of the type obtained when
subjects rate the treatments in order of preference.

Before getting down to business it is worth remarking how much
progress has been made during the past 10 years in experimentation
with human subjects. Planning the experiment is now recognized
to be a major undertaking, necessitating, for a start, having a clear
grasp of the questions it is hoped to answer, and proceeding to
detailed specification of the kind of subjects to be included, the
treatments to be compared and the measurements to be taken. The
points at which different types of trial are particularly vulnerable to
bias are more widely appreciated, as is the value of precautions such
as randomization, "blind" comparisons and the inclusion of
placebos. A growing number of successfully completed trials has
accumulated to serve as guides for future work. Although difficult
decisions still remain, there is more agreement about the circumstances in which ethical considerations justify a well-designed trial
rather than the refusal to experiment.

Human experimentation demands, more than do most research
activities, the effective teamwork of different medical specialists.
Owing to the rarity of many types of disease, co-operation between

Reproduced with permission from *Qualitative Methods in Human Pharmacology and Therapeutics*, pages 119–143. Copyright © 1959, Pergamon Press, Ltd.

different institutions will be increasingly necessary in the future in order to obtain a sufficient number of patients. We are gradually gaining experience in the difficult business of conducting large-scale co-operative trials so that the data can be put together and interpreted as a whole. To cite two examples, the 1954 trial of the Salk polio vaccine involved all schools in over 200 counties scattered throughout the United States. In order to ensure a blind comparison, more than 200,000 children received three separate shots of a saline at prescribed intervals. This number was not the product of megalomania: it was just adequate to provide a reasonably accurate estimate of the effectiveness of the vaccine, and was too small for thorough study of the relation between the children's antibody levels and their clinical reactions. A second example, a masterpiece of organization on an international scale, was a therapeutic trial of five drugs and a placebo for the treatment of leprosy, in which the same experiment was carried out simultaneously, with uniform procedures and records, at four leprosaria, Aisei-en and Komyo-en in Japan, Eversley Childs in the Philippines, and Westfort in South Africa (DOULL 1954).

## GROUPING

In experimental pharmacology a standard recommendation is that whenever feasible, comparisons of treatments should be made *within* subjects rather than *between* subjects. If each subject receives all the treatments, variations in level of response from subject to subject cancel out when treatment averages are compared. The gain in precision is often striking. A within-subject trial is a simple example of grouping. The general principle is to construct groups such that differences between groups represent important sources of variation that may inflate the experimental errors. Then if the experiment can be conducted so that each treatment is represented equally often in every group, differences between groups are automatically eliminated from the comparisons of the treatment averages. The basis for grouping will vary with the type of experiment. The group may be a single subject, as in the within-subject trial, a group of subjects whose final responses are expected to be similar, the order in which treatments are applied or certain operations are done, a particular site on the body, and so on.

In some areas of research, notably agriculture and the physical sciences, plans involving elaborate grouping have been constructed and found useful. Although these more complex arrangements probably have little place in human pharmacology, ingenious applications of the simpler plans of this type have been made in trials utilizing within-subject comparisons. So far as I know, grouping is seldom employed in pharmacological experiments based on between-subject comparisons, sometimes because of difficulties in organization but sometimes, I think, because the opportunities for grouping are overlooked.

Later papers in this session present specific applications of grouping. Some designs based on this principle are illustrated in succeeding sections.

## GROUPING IN WITHIN-SUBJECT COMPARISONS

### Cross-over Designs

Probably the best-known arrangement is the cross-over design for comparing two treatments. Half the subjects receive drug $A$ followed by drug $B$, the remainder drug $B$ followed by drug $A$. This is an example of double grouping, by subjects and by the order in which the treatments are given. The treatment averages are consequently unaffected by variations in level of response from subject to subject or by any systematic difference between the first and second applications. The additional control of the order is not always necessary, but should be included unless experience has shown it to be unimportant.

### Latin Squares

Closely related to the cross-over design is the latin square. This is an arrangement in which each treatment appears in every row and column of a square, the rows and columns representing sources of variation that it is desired to eliminate from the comparison of treatments. Two frequently-quoted examples appear in Fig. 1. In the first, part of a larger experiment, three samples $a$, $b$, $c$ of penicillin sodium, with different percentages of impurities, were injected into three sites (buttocks, triceps, deltoid) on each subject in order to appraise

| Subject | Bu | Site Tr | De | Subject | Order of Treatment 1 | 2 | 3 | 4 |
|---------|-----|-----|-----|---------|-----|-----|-----|-----|
| 1 | a | b | c | 1 | A | B | C | D |
| 2 | b | c | a | 2 | B | A | D | C |
| 3 | c | a | b | 3 | C | D | A | B |
|   |   |   |   | 4 | D | C | B | A |

Fig. 1. Two examples of latin squares

the amount of intramuscular irritation (HERWICK *et al.* 1954). This is a within-subject comparison, but it might not be a good one if the sites differed in irritability under injection. By using 3 subjects to form a latin square, the comparison is balanced with respect to sites (columns) as well as to subjects (rows).

The second experiment (JELLINEK 1946), illustrates the extension of the cross-over principle, in which the columns balance out any consistent effect of the order in which drugs are given. Three analgesics *A*, *B*, *C* and a placebo *D* were tested on patients with recurrent headaches, each pill being tried for a 2 week period. In both experiments the row schedules were followed by a number of subjects rather than a single subject, in order to provide sufficient replication.

*Latin Squares for Estimating Residual Effects*

When different drugs are given in succession to a subject, there may in some situations be residual or carry-over effects of a drug into the next period. In this event the observed effect of a drug *B*, given in the second period, will actually be the sum of the direct effect of *B* and the residual effect of the drug *A* which preceded it in the first period. Comparisons of the average effects of the drugs in the periods in which they were given will therefore be biassed by the presence of residual effects of the other treatments which preceded them. The usual precaution against this disturbance is to allow a sufficient resting time between periods so that residual effects can be assumed negligible. In cases of doubt, however, there are special latin squares from which the sizes of these residual effects can be estimated, with some extra computational labour, and the direct effects freed from this bias (COCHRAN and COX 1957).

Fig. 2 shows a design for three treatments and 6 subjects. The basic plan requires three periods. Its characteristic feature is that over the 6 subjects, each drug is preceded twice by each of the other drugs. This symmetry facilitates the estimation of the residual effects and the adjustment of the direct effects.

If a fourth period can be added in which the drugs given in the third period are repeated, it will be seen from Fig. 2 that over the experiment as a whole each drug is preceded twice by all drugs, including itself. Alternatively, the first period assignment can be repeated in the second period, as is sometimes more convenient. With these arrangements, simpler and more efficient estimates of the direct and residual effects are obtained. In the corresponding design for two treatments, one group of subjects follows the series *ABB*, the other *BAA*. The method of computation, which must be carefully followed, also eliminates differences between subjects and periods from the experimental errors.

| Period | Subject 1 | 2 | 3 | 4 | 5 | 6 |
|---|---|---|---|---|---|---|
| 1 | *A* | *B* | *C* | *A* | *B* | *C* |
| 2 | *B* | *C* | *A* | *C* | *A* | *B* |
| 3 | *C* | *A* | *B* | *B* | *C* | *A* |
| (4) | (*C*) | (*A*) | (*B*) | (*B*) | (*C*) | (*A*) |

Fig. 2. Latin square for estimating residual effects

*Incomplete Block Designs*

In lines of work in which within-subject comparisons have been found to be much more precise than between-subject comparisons, the investigator may wish to compare more treatments than can be given to an individual subject. By means of plans known as *incomplete block designs*, the advantages of within-subject comparisons can be retained in this situation.

| Subject 1 | 2 | 3 | 4 | 5 | 6 | 7 | 8 | 9 | 10 |
|---|---|---|---|---|---|---|---|---|---|
| *a* | *a* | *a* | *a* | *a* | *b* | *b* | *b* | *c* | *c* |
| *b* | *b* | *c* | *d* | *e* | *c* | *d* | *e* | *d* | *d* |
| *c* | *d* | *e* | *f* | *f* | *f* | *e* | *f* | *e* | *f* |

Plan for comparing 6 levels of impurities ($a, \ldots, f$) in penicillin sodium, with 3 injections per subject. HERWICK *et al.* (1954).

| Rabbit | 1 | 2 | 3 | 4 |
|---|---|---|---|---|
| 1st injection | $S_2$ | $S_1$ | $U_2$ | $U_1$ |
| 2nd injection | $U_1$ | $U_2$ | $S_1$ | $S_2$ |

Plan for comparing two levels of the Standard ($S_1, S_2$) and two of the unknown ($U_1, U_2$) in the assay of insulin. (U.S. Pharmacopoeia XV).

Fig. 3. Two examples of incomplete block designs

Fig. 3 shows two examples. The first comes from the comparison previously cited of different samples of penicillin sodium. Actually there were six treatments or samples, i.e. six levels of impurity, to be compared, whereas each subject could test only three samples. These triplets are allotted to subjects in a pattern called a balanced incomplete block design. Its basic property is that every pair of treatments (e.g. $a$, $b$, or $d$, $f$) occurs in two of the subjects. To put it another way, the 5 subjects who received $a$ also received among them each of the other treatments twice.

With an incomplete block design, the simple average of the responses to a treatment cannot be used as an estimate of the effect of that treatment. Treatment $a$, for instance, was tested on subjects 1, 2, 3, 4 and 5, while $b$ was tested on subjects 1, 2, 6, 7 and 8. Thus a comparison of the simple averages for $a$ and $b$ is in part a between-subject comparison. Instead, we may compute the quantity

$D_a$ = (Average response to $a$) − (Average of the 10 other responses given by these 5 subjects.)

This is a within-subject comparison, and by the basic property of the design it is an estimate of the difference between the effect of $a$ and the average effect of the other treatments $b$, $c$, $d$, $e$ and $f$. Now by an algebraic relation,

$$a - \frac{(a+b+c+d+e+f)}{6} = \frac{5}{6}\left\{ a - \frac{(b+c+d+e+f)}{5} \right\}$$

Consequently, $5D_a/6$ is a within-subject estimate of the difference between the effect of $a$ and the average effect of all the treatments. Finally, by adding the general mean of the whole experiment to $5D_a/6$, we obtain an estimate of the average effect of $a$ that provides a within-subject comparison with similar estimates made for the other treatments.

The second example in Fig. 3 is the twin cross-over design in the assay of insulin, now adopted in the U.S. Pharmacopoeia XV. Four treatments (two dosage levels of Standard and two of Unknown) are arranged in blocks of size 2, i.e. two injections per rabbit. In any rabbit, the lower level of one drug is compared with the higher level of the other. By an ingenious but simple calculation, a within-rabbit estimate of the relative potency of the two drugs can be obtained. According to BLISS (1957), this design requires only one-fifth as many blood-sugar determinations as its predecessor in U.S. Pharmacopoeia XIV, and has enabled the Insulin Laboratory in Toronto to reduce its staff by one-half while maintaining the same precision.

## GROUPING IN BETWEEN-SUBJECT COMPARISONS

In order to utilize the simplest type of grouping in between-subject comparisons, we must be able to recognize subjects likely to have similar responses, who can be grouped to form replications. The co-operative leprosy trial provides an illustration. The patients were in the institutions at the start of the trial. Much information about them was available—age, sex, weight, length of time in institution, amount and type of previous drugs, and a classification $(L_1, L_2, L_3)$ by severity of disease. Although it was intended to make dermatological, neurological and bacteriological measurements, an overall rating of clinical progress was considered the most important variable.

Since six drugs were to be compared, the problem was to arrange the patients in groups of six, the first group being the 6 patients with the best clinical prognosis at the start of the trial, and so on until the last group was reached, containing the 6 patients with the poorest initial prognosis. Each treatment was then given to a randomly chosen member of each group. This design is often known as *randomized blocks*, from the terminology used in agriculture. As it happened, the experts were not agreed on the best criteria for predicting clinical progress, and the grouping adopted did not have unanimous support. At the end of the trial, however, it was possible to examine the effectiveness of each factor that might have been used for grouping as well as of the factors that were used, thus furnishing useful information for future trials.

When patients enter a trial over a protracted period, grouping into replications poses organizational problems. Even if we possess good criteria for grouping, we do not know when and if suitable partners for the early patients will appear, and we face the prospect that at the end many patients will be left without partners. While techniques for grouping in this situation appear to have been little studied, two ways of taking advantage of good criteria come to mind.

The first is to use a smaller number of groups, each larger than a single replication. For instance, the prognosis when the patient enters may be classified as "good", "relatively good", "fair", and "poor". Two schemes for assigning treatments are illustrated in Fig. 4, which presents the assignments for the first 6 patients who enter each group. In scheme 1, $A$ and $B$ are balanced over every pair of patients in each group. When the trial ends, the number of $A$'s in a group cannot differ from the number of *B*'s by more than one. A disadvantage, serious for some trials, is that personnel involved in the trial are likely to discover that if patient 3 is an *A*, patient 4 will be a *B*. In scheme 2, along lines suggested by BRADFORD HILL (1955), the *A*'s and *B*'s are balanced only over groups of 6 patients. At the end of the trial there may be more discrepancy between the numbers of *A*'s and *B*'s in a group, but the pattern is harder to guess.

| Patient's number in group | | Prognosis | | |
|:---:|:---:|:---:|:---:|:---:|
| | Good | Rel. good | Fair | Poor |
| | | Scheme 1 | | |
| 1 | *A* | *A* | *B* | *A* |
| 2 | *B* | *B* | *A* | *B* |
| 3 | *A* | *B* | *B* | *B* |
| 4 | *B* | *A* | *A* | *A* |
| 5 | *B* | *B* | *B* | *A* |
| 6 | *A* | *A* | *A* | *B* |
| | | Scheme 2 | | |
| 1 | *B* | *B* | *A* | *A* |
| 2 | *B* | *B* | *A* | *B* |
| 3 | *A* | *A* | *B* | *A* |
| 4 | *A* | *B* | *B* | *A* |
| 5 | *A* | *A* | *A* | *B* |
| 6 | *B* | *A* | *B* | *B* |

Fig. 4. Assignment of treatments to subjects classified in broad groups.

A second possibility is to assign treatments to subjects randomly, ignoring the grouping. At the end of the experiment the results are classified by groups. If there seem to be substantial differences in level from group to group, we can compute estimates of the treatment effects that are balanced with respect to these differences in level. The method, a familiar one in the analysis of non-experimental data, is illustrated in Tables 1 for 2 treatments and three groups. The difference between the treatment means is computed separately for each group. A weighted mean difference is then formed, using weights, derived from the theory of least squares, which give the most efficient estimate of the overall difference. The computation creates, as it were, a within-group comparison of the treatment effects.

*Table 1*

*Within-group estimate of treatment effects (artificial data)*

| Group | Treatment A | | Treatment B | | Diff. $(A-B)$ | Wt. |
|---|---|---|---|---|---|---|
| | No. of subjects | Mean response | No. of subjects | Mean response | | |
| 1 | 6 | 11·2 | 4 | 9·5 | +1·7 | 2·4 |
| 2 | 4 | 8·7 | 4 | 8·0 | +0·7 | 2·0 |
| 3 | 5 | 6·3 | 7 | 4·2 | +2·1 | 2·9 |

$$\text{Weighted mean diff.} = \frac{(2\cdot4)(1\cdot7) + (2\cdot0)(0\cdot7) + (2\cdot9)(2\cdot1)}{2\cdot4 + 2\cdot0 + 2\cdot9}$$

$$= \frac{11\cdot57}{7\cdot3} = 1\cdot58$$

The weight for any group is the product of the numbers of patients for the two treatments, divided by their sum; i.e. for group 1, the weight is (6) (4)/10 = 2·4. The standard error of the weighted mean difference is $s/\sqrt{7\cdot3}$, where $s$ is the standard deviation between subjects in the same group who receive the same treatment, and 7·3 is the sum of the weights.

If the criterion is a continuous measurement, for instance the number of worms excreted by children before treatment in an anthelminthic trial, the procedure illustrated in Table 1 is performed by a technique known as the analysis of covariance. The first step is to compute, by regression methods, the amount of change in the final response that is associated with unit change in the initial excretion level. The final responses to different treatments are then adjusted so as to remove the effects of variations in these initial levels. In most circumstances these techniques achieve almost as much increase in precision as a grouping based on the criterion. The analysis of covariance is a highly flexible tool, usable with any standard design like randomized blocks or latin squares, and may add materially to the precision obtained by the grouping in the design. It may also be employed to investigate how the dose of a drug should be adjusted to the subject's age, weight, height and sex, and has been done for digitoxin by BLISS *et al.* (1953).

An excellent account of designs involving grouping that are

K

suitable for medical research has been given by FINNEY (1955). Detailed plans of the designs appear in COCHRAN and COX (1957).

Since I have been stressing the advantages of grouping, it should be added that the actual gain in precision varies with the circumstances, being sometimes large, sometimes negligible, and, with incomplete block designs, sometimes negative. A grouping may, however, be worth adopting routinely as a precautionary measure, in the sense that although it produces little increase in precision most of the time, there are occasional experiments in which it is highly effective. The gain in precision can be appraised from the results of a completed experiment, as is worth doing if grouping is troublesome. Finally, most of the advantage of grouping is lost if missing observations are numerous. In experiments liable to this nuisance, only the broad grouping illustrated in Fig. 4 is to be recommended.

## SEQUENTIAL TRIALS

In recent years, applications of a technique known as sequential experimentation have begun to appear in pharmacological journals. The usual practice in trials is to decide at the beginning approximately how many subjects will be included. Serious examination of the results is postponed until the final measurements from all these subjects have been gathered, although there may be casual inspection of preliminary data. In a sequential trial, on the other hand, a continuous statistical analysis of the results is made as the data from each subject come in. The trial is stopped as soon as the analysis indicates a clear-cut verdict of statistical significance or non-significance, It seems plausible that this continuous review of the results should often permit conclusions to be reached sooner. We might, for instance, have intended to use 50 subjects per treatment, but the data from the first 30 subjects might already reveal that one treatment was greatly superior, or alternatively might make it clear that no difference of practical importance could emerge.

This reduction in the amount of experimentation is the principal advantage of the sequential method. No single figure as to the amount of the saving can be quoted, because the number of subjects needed in a sequential trial depends on the way in which the results come out. Mathematical study indicates that the saving in numbers of subjects, as compared with a fixed-size trial of the same discriminating power, probably ranges between 10 per cent and 50 per cent, with an average of perhaps 30 per cent, although there may occasionally be a loss.

The sequential method was developed by WALD (1947) during the last war for tests of ammunition and weapons. The characteristics of the situation which make sequential trials especially appropriate may be illustrated by a hypothetical trial designed to compare two shells.

(1) It is natural to conduct the trial sequentially—shell A is fired, then shell B, and so on.

(2) The results of each firing become known before the next firing is made.

(3) The tests are very expensive, so that any reduction in the size of the experiment is a major gain.

(4) The primary object is to pick out the better shell. To put it more exactly, if one shell is superior by an amount of practical importance, the trial should have a high probability of selecting that shell as the winner.

Translating these conditions to human pharmacology, we might say that the sequential method is most likely to prove useful in trials having the following features.

(1) Patients should enter the trial in sequence over a period of time, so that if the trial is stopped abruptly there is a saving both in time and in patients not yet enrolled.

(2) The results of treatment should be known quickly. If not, there may be a substantial number of patients still in process with incomplete records when the statistical analysis indicates that the trial can be terminated. This decreases the savings.

(3) There should be potent reasons for wishing to stop the trial as soon as possible. One is the ethical responsibility of the doctor. Others might be expense, or a risk that if the trial continues for a long time, the drugs tested may be obsolete when the study finishes, on account of the development of newer drugs.

(4) The primary object of the trial should be to perform a test of significance, If, on the other hand, the purpose is to estimate the difference in effect between two treatments with a specified degree of accuracy, the sequential approach, although it may still have advantages, makes no substantial saving in number of patients.

During the past 5 years, a small number of sequential trials have been reported, as summarized in Table 2.

Trials numbered (1), (2), (4) and (6) were short-term trials; trial (5), however, is an application to a comparison of longer duration. Study (3) was a comparison of patient groups rather than a controlled experiment, but is included as a further illustration. I

*Table 2*

*Characteristics of some sequential trials*

| Comparison made | Measurement | Reported by |
|---|---|---|
| (1) Cacl. vs. adrenaline as bronchial dilators | Expiratory flow rate | KILPATRICK and OLDHAM (1954) |
| (2) *n*-Acetyl-*para*-amino-phenol vs. compound codeine as analgesics | Patient's comparative rating on effectiveness | NEWTON and TANNER (1956) |
| (3) Zinc metabolism of cirrhotic and normal patients | Serum zinc concentra-tion | VALLEE *et al.* (1956) |
| (4) Diamorphine vs. phol-codeine vs. placebo as cough suppressants | Patient's preference | SNELL and ARMITAGE (1957) |
| (5) Body-weight regula-tion vs. none during pregnancy | Incidence and severity of pre-eclampsia | BILLEWICZ (1958) |
| (6) Ethoheptazine and as-pirin as analgesics | Individual tests of effec-tiveness | CASS, FREDERICK and BARTHOLOMAY (1958) |

have been informed of a number of trials still in progress, so that additional experience with this technique should accumulate in the near future.

## PLOTTING THE RESULTS

The continuous analysis of results is done by plotting on a chart. The method will be illustrated from two of the trials in Table 2. In the trial reported by NEWTON and TANNER (1956), patients needing regular analgesics were given Tab. Codeine for one week and NAPAP for one week, in randomized order. The procedure was repeated for a second fortnight. As they became available, the results were plotted in Fig. 5, which is taken from their paper. For each patient who rated Tab. Codeine superior, an × was placed to the *right* of the previous cross. For each patient rating NAPAP superior, an × was placed *above* the previous cross. Nothing was plotted for a patient who did not report a consistent preference.

The trial ceases when the succession of crosses reaches a boundary, the verdict appropriate to each boundary being indicated beside it. The boundaries, which are constructed in advance of the trial, are determined by the significance level and the amount of sensitivity desired in the experiment. In this plan there is about a 1 in 10 chance of reaching an outer boundary (i.e. erroneously finding a significant difference) if the drugs are equally effective, and $8\frac{1}{2}$ chances in 10 of reaching the Tab. Codeine boundary if it is preferred, in the long run, by 70 per cent of patients who have a clear preference.

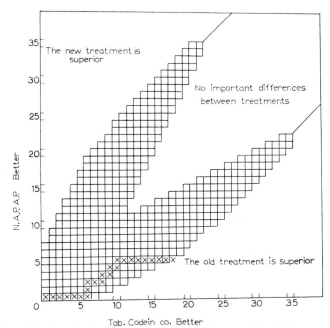

Fig. 5. Chart for plotting the results of a binomial sequential trial. (From NEWTON and TANNER 1956).

Fig. 6 shows one type of chart for a trial in which the measurement $x$ is a continuous variate, in this case the difference between the gain in expiratory flow rate resulting from calcium chloride inhalation and that resulting from adrenaline inhalation in the same subject (KILPATRICK and OLDHAM 1954). The variate computed and plotted as the ordinate is $Sx/\sqrt{Sx^2}$ over all patients whose results have come in. The abscissa is the serial number of the patient.

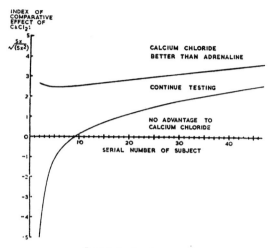

Boundaries for the sequential trial.

Fig. 6. Chart for plotting the results of a sequential trial with a continuous
variable. (From KILPATRICK and OLDHAM 1954).

This figure differs in two respects from Fig. 5. The diagram
provides a one-sided test, the upper boundary signifying a definite
superiority of $CaCl_2$, the lower denoting no important difference. No
boundary was provided to denote a superiority of adrenaline, since
this outcome was considered unlikely. For most purposes a two-
sided diagram, as in Fig. 5, is to be recommended, as the authors
point out in their discussion of the results of this experiment.
Secondly, the boundaries are not closed, and the trial could
theoretically go on for a very long time without reaching either
boundary. All the early sequential diagrams were "open" in this
way. With binomial data, methods for obtaining closed boundaries
were first developed by BROSS (1952): a general procedure has been
given by ARMITAGE (1957). Corresponding schemes for continuous
data have not been produced, so far as I know, although ARMITAGE
(1954) has given a rough working rule for imposing an upper limit
to the number of subjects.

*The Present Status of Sequential Trials*
   Conclusions about the advantages and limitations of the sequential
method in human pharmacology must at present be tentative.

Practical experience with the technique is still scanty, and much mathematical research remains to be done on the construction and properties of sequential trials. The following comments may be helpful to those considering whether to undertake a sequential trial.

For comparative trials, the available plans are restricted to 2 treatments. Either within-subject or between-subject comparisons may be employed. With continuous data, the treatments may be allotted randomly to subjects, or the subjects may be paired. With binomial data (e.g. "success" or "failure") the data *must* be paired, either on a random or on some logical basis, because it has not been possible to work out a scheme with unpaired binomial data. In their experiment, SNELL and ARMITAGE (1957) included three treatments, analysing the data by making comparisons between each of the 3 pairs.

In order to set up a sequential trial, the doctor must decide in advance on 3 quantities:

(1) The significance level at which the test of significance is to be made.

(2) The size of true difference between the treatments that is regarded as *important*, in the sense that if the true difference is of this size or greater, he would like the trial to end by proclaiming the superior treatment as the winner, although he does not greatly mind obtaining an indecisive result if the true difference is less than this amount.

(3) The probability that the trial will detect a true difference of this amount. (This figure is required because it affects the expected length of the trial. A longer trial is needed to make the probability of finding the winner 0·99 instead of 0·9.)

Doctors may find the decisions required in (2) and (3) difficult to make. A natural tendency is to declare that any difference of appreciable size is important to the progress of medicine. However, if the "important difference" is taken to be small, the trial required is likely to be too lengthy to be practicable. By examining the expected lengths of trial for several choices of the "important difference", a workable compromise may emerge. The fact that it forces this kind of decision is not really a disadvantage of the sequential method, because a similar decision should be faced at the start of a fixed-size trial if an intelligent choice of number of subjects is to be made.

A more important point is that the saving in number of patients is sometimes illusory. Consider the comparison of a standard drug with a new drug that will be recommended to replace it if substantially superior. If the trial ends at the boundary marked "new is superior", this means only that the difference in favour of the new drug is statistically significant. It does *not* mean that the difference has been shown to be important in the sense of step (2) above, or that the difference has been accurately estimated. Before deciding what to do next, the actual size of the difference, with its standard error and confidence limits, must be scrutinized. (Correct methods of estimating treatment differences, standard errors and confidence limits at the end of a sequential trial still present mathematical difficulties, although approximations that seem to be adequate are available.) It may be clear that the trial has not estimated the difference in effectiveness of the two drugs accurately enough to permit practical recommendations to be made. Further experimentation, involving more patients and more time, will be needed.

The same problem arises with greater force when several important variables have to be analysed.* When one major variable reaches a boundary, critical examination of the size of the differences for all important variables will be necessary to decide whether to stop or continue. Thus it is hard to estimate what kind of saving sequential trials will make when there are numerous variables.

These remarks probably do not apply if the trial ends as the boundary "Standard is superior" or the boundary "No important difference". In either event we would probably reject the new drug without more ado. This discussion suggests that sequential methods may find their most fruitful applications in screening programmes in which it is desired to reject all new drugs except those likely to be greatly superior to the standard drug. DAVIES (1958) has examined several methods for pharmaceutical screening.

The role of sequential trials in long-term studies remains to be determined. At first sight the method seems unsuitable, because when the plotted points reach a boundary, there may be numerous patients who have started but not completed the trial, so that the saving in numbers of patients is considerably diminished. The saving may still, however be worthwhile, as well as desirable on

---

* There is the additional difficulty that the length of trial required for one variable may be quite different from that required for others. This difficulty is, however, present also in trials of fixed length, in which a sample size that gives satisfactory precision for one important variable may be inadequate for another. In both sequential and fixed-size trials, a compromise solution can sometimes be reached after calculating the length of trial needed separately for each important variable.

ethical grounds. This problem is discussed elsewhere in this volume by Billewicz with respect to a trial in which the treatment, weight regulation during pregnancy, lasted about six months. ARMITAGE (1958) presents a sequential method for comparing two survival-time distributions.

For medical readers, good introductory accounts of the method are given by BROSS (1952) and ARMITAGE (1954), and methods of constructing the boundaries by DAVIES (1956) and RUSHTON (1952) for continuous data and ARMITAGE (1957) for binomial data. Unfortunately, no single source forms a complete and up-to-date manual of the technique. The advice of a statistician should be sought in the planning of a trial.

## SUBJECT TO SUBJECT VARIABILITY IN THE EFFECTS OF TREATMENTS

A criticism of the standard clinical trial is that it estimates only the *average* effects of drugs over all patients, ignoring the possibility that drug $A$, although superior to drug $B$ for most patients, might be the poorer drug for some 20–30 per cent of them. HOGBEN and SIM (1953), who raised this point, have discussed the problems encountered in designing a self-controlled trial for an individual patient.

If important variations in the relative effectiveness of drugs from patient to patient are of frequent occurrence, then, ideally, the objectives of a clinical trial ought to be expanded to provide (i) an investigation of the presence and amount of this variation in effectiveness and (ii) some means for predicting, for practical use, the kinds of patients for whom drug $A$ is highly effective, and those for whom it is less effective relative to $B$. This is an exacting task, and trials that can fulfil these aims will probably seldom be feasible. Two suggestions along these lines will be made.

In situations in which each drug can be given to a subject more than once, the features of the statistical trial and the self-controlled trial for an individual subject can be combined by making, for example, two or three separate comparisons of drugs $A$ and $B$ for every subject in the trial. Because of the small number of replications, conclusions about the performance of the drugs for an individual subject will necessarily be imprecise. But this replication enables us to reinforce the standard clinical trial with an investigation of the variation in performance of the drugs from subject to subject.

With measurements that are on a continuous scale, a standard analysis of variance of the results may be computed, with three components: (i) the average difference between drugs over all subjects (ii) the variation in relative effectiveness of the drugs from subject to subject and (iii) a pooled estimate of experimental error obtained from the internal data for each subject. From components (ii) and (iii) we can make a test of significance for the presence of real differences in performance of the drugs from subject to subject.

It may happen that although there are statistically significant differences, the variation is still small relative to the overall mean difference. In this event it may be safe to recommend drug $A$ for all subjects. If, on the other hand, substantial differences are found, the subjects can be ranked or grouped, at least roughly, according to the relative performance of the two drugs. With the aid of supplementary observations, we may be able to form hypotheses as to why some subjects do well and others poorly with drug $A$. Investigations along these lines, where feasible, might be an illuminating addition to the usual statistical type of trial. One analysis of this type has been reported by NEWTON and TANNER (1956), with further comments by ARMITAGE and HEALY (1957). Since the data were merely the patients' preferences for one drug or the other, the investigation of patient to patient variability was necessarily more limited than can be made with continuous data.

Secondly, in a large clinical trial, any characteristics of the patient that might affect his differential response to the drugs can be recorded. By suitable groupings of the final results, the association of any of these characteristics with the relative performance can be examined. An interesting example is the well-known difference between placebo reactors and placebo non-reactors in their ability to differentiate analgesics.

### ANALYSIS OF PREFERENCES AND RANKINGS

In experimental psychology and sociology, much of the data consists of reports by subjects. A good deal of attention has been given in these fields to investigating the kind of scale on which the subject can most usefully be asked to report, and to the analysis of results expressed in various crude scales. Since some of this research may be relevant to pharmacological experiments in which subjects' ratings are the principal measurements, a few of the simpler methods will be outlined.

*Comparison of Two Treatments.*
Each subject is asked to express his preference, if any, for one of

two treatments $A$ or $B$. With $m$ subjects, the data may be summarized as follows.

|  | Subject's preference $A$ $B$ None | Total |
|---|---|---|
| No. of subjects | $a$        $b$          $c$ | $m$ |

Thus $a$ out of $m$ subjects prefer $A$, while $c$ out of $m$ express no preference for $A$ or $B$.

If subjects who have a firm preference are in the long run equally divided between $A$ and $B$, the numbers $a$ and $b$ should be equal, apart from sampling errors. The null hypothesis that $A$ and $B$ are preferred equally often can therefore be tested by comparing the binomial ratio $a/(a + b)$ with an expected value $1/2$. On the null hypothesis, the standard error of $a/(a + b)$ is

$$\sqrt{\frac{pq}{n}} = \sqrt{\left(\frac{1}{2}\right)\left(\frac{1}{2}\right)\frac{1}{(a + b)}} = \frac{1}{2}\sqrt{\frac{1}{(a + b)}}$$

Hence, for a test of significance, we may take as an approximately normal deviate the quantity

$$\frac{\dfrac{a}{a + b} - \dfrac{1}{2}}{\dfrac{1}{2}\sqrt{\dfrac{1}{a + b}}} = \frac{(a - b)}{\sqrt{a + b}}$$

A correction for continuity is applied by subtracting 1 from the absolute value of $(a - b)$ in the numerator. Sometimes this test is given as a $\chi^2$ test with 1 d.f., where, with the continuity correction,

$$\chi^2 = \frac{(|a - b| - 1)^2}{a + b}$$

Alternatively, an exact test can be made from the binomial tables for $p = \frac{1}{2}$.

If the experiment is of the cross-over design, half the subjects receiving $A$ first and half $B$ first, this test is not quite correct if the order of presentation affects the probability that $A$ or $B$ will be preferred. In this event the standard error given above for $a/(a + b)$ is too large. The overestimation is unlikely to be important in most cases.

*Ranked Comparison of More Than Two Treatments*

Suppose that every subject is asked to compare more than two treatments, ranking them in order of preference. The experiment is of the randomized blocks type, each subject constituting a block. As an example, Table 3 shows a summary of the rankings of four types of chocolate flavouring by each of 32 boys (data from a larger experiment by Bliss *et al.* (1943)). Rank 4 represents the best ranking: 18 boys assigned rank 4 to flavour *A*, 9 assigned rank 3, and so on.

*Table 3.*

*Ranking of 4 chocolate flavours by 32 boys.*

| Flavour | No. of boys giving rank | | | | Sum of ranks | Sum of rankits |
|---|---|---|---|---|---|---|
| | 4 | 3 | 2 | 1 | | |
| *A* | 18 | 9 | 0 | 5 | 104 | 16·09 |
| *B* | 4 | 6 | ·8 | 14 | 64 | −10·90 |
| *C* | 3 | 4 | 20 | 5 | 69 | −6·86 |
| *D* | 7 | 13 | 4 | 8 | 83 | 1·67 |
| Totals | 32 | 32 | 32 | 32 | 320=G | 0·00 |
| Rankits | 1·03 | 0·30 | −0·30 | −1·03 | | |

The row and column labelled "rankits" will be explained later.

Adequate methods for performing a test of the null hypothesis that the mean rankings are the same for all flavours have been developed by Friedman (1940) and Kendall and Babington Smith (1939). From these results it appears that except in very small experiments an ordinary analysis of variance of the data, if used with care, will not lead us far astray in performing *F*- and *t*-tests. With $m = 32$ blocks of $n = 4$ treatments, the relevant computations for setting up an analysis are as follows.

First find the treatment totals and the grand total *G* of the ranks. For flavour *A* the total is

$$(4)(18) + (3)(9) + (2)(0) + (1)(5) = 104.$$

The necessary sums of squares are then obtained.

Correction term: $\quad C = \dfrac{G^2}{mn} = \dfrac{(320)^2}{128} = 800$

Total S.S.: $32(4^2 + 3^2 + 2^2 + 1^2) - C = 960 - 800 = 160$

Treatments S.S.: $\dfrac{(104)^2 + (64)^2 + (69)^2 + (83)^2}{32} - C = 30 \cdot 06.$

The Blocks S.S., i.e. the S.S. between boys, is automatically zero and may be omitted. It is included here to emphasize the relation to a standard analysis of variance for randomized blocks.

*Table 4*

*Analysis of variance of ranks and rankits*

| | d.f. | Ranks S.S. | Ranks m.s. | Rankits S.S. | Rankits m.s. |
|---|---|---|---|---|---|
| Blocks | $m - 1 = 31$ | 0·00 | | 0·00 | |
| Treatments | $n - 1 = 3$ | 30·06 | 10·02 | 13·36 | 4·45 |
| Residual | $(m - 1)(n - 1) = 93$ | 129·94 | $1 \cdot 397 = s^2$ | 60·30 | 0·648 |
| Total | $mn - 1 = 127$ | 160·00 | | 73·66 | |

Whether the Blocks S.S. is included or not, the Residual S.S., from which the estimate of error variation is derived, is assigned $(m - 1)(n - 1)$ d.f.

Kendall and Babington Smith's more exact method uses the same $F$-ratio $10 \cdot 02/1 \cdot 397 = 7 \cdot 17$ as in Table 4, but assigns smaller numbers of d.f. to $F$. For the numerator they propose

$$n_1 = n - 1 - \frac{2}{m} = 3 - \frac{2}{32} = 2 \cdot 94 \text{ d.f.}$$

and for the denominator,

$$n_2 = (m - 1)(n - 1) - \frac{2(m - 1)}{m} = 93 - 1 \cdot 94 = 91 \cdot 06 \text{ d.f.},$$

instead of 3 and 93, respectively. They also recommend a correction for continuity for small experiments. Thus the analysis of variance in Table 4 slightly overrates the statistical significance.

For $t$-tests, the standard error of the difference between two treatments totals in Table 3 is

$$\sqrt{2ms^2} = \sqrt{(2)(32)(1 \cdot 397)} = 9 \cdot 46,$$

this being assigned 93 d.f. in the approximate analysis. Flavour A, an American process chocolate, ranked significantly higher than any of the others, which were different brands of Dutch process. Differences between the Dutch brands were not clearly established.

An alternative method, used by the authors, is to replace the ranks 4, 3, 2, 1 by scores 1·03, 0·30, − 0·30, − 1·03, sometimes called *rankits*. These are the average values of the first, second, third and fourth in order of size of a sample of 4 from a normal population with mean zero and unit standard deviation. FISHER and YATES (1953) give tables of these scores for sample sizes up to 50.

There are two potential advantages of the rankit scores. In some situations it is reasonable to suppose that the responses to the treatments, if they could be measured accurately, would be normally distributed, but that the subject is too crude a measuring instrument to do more than rank the treatments. Under this supposition, rankits should give a more sensitive comparison of the treatments than ranks. Secondly, the standard analysis of variance may be more valid with rankits, since these are in a sense normal deviates. Neither claim has been intensively investigated, although they appear plausible. The principal effect of the rankit scale is that differences between rankings near the top and bottom are increased relative to differences between rankings in the middle. In so far as subjects exercise more care in deciding the top and bottom rankings than the middle rankings, this change in scale seems reasonable.

The analysis of variance of the rankits, shown on the right in Table 4, is carried out in the same way as for ranks. The standard error of the difference between two treatment totals is $\sqrt{(2)(32)(0·648)}$ or 6·44. From the treatment totals (right hand column of Table 3) the same conclusions are found as with ranks.

It is sometimes evident on inspection that some treatments are being ranked more consistently than others, e.g. nearly everyone ranks treatment $D$ first but $A$ and $B$ are found in all positions. If this is so, the pooled estimate of error from the analysis of variance will not apply to the comparison of individual pairs of treatments. In case of doubt, a separate error can be computed for any specific pair, by taking the difference between the ranks or rankits for each subject and performing a $t$-test on the set of differences.

*Ranked Data in Incomplete Block Designs*

Since in many preferential rating studies the subjects can compare only two or three treatments accurately at one time, incomplete block designs have proved popular when the treatments are numerous. In the method of paired comparisons the subject compares two treatments at a time. Over the experiment as a whole every pair is compared equally often.

From the comparisons made between treatments $i$ and $j$ we may compute the proportion of times $p_{ij}$ that treatment $i$ is prefered to treatment $j$. The proportions $p_{ij}$ constitute the basic data. Several methods of analysis have been put forward, the objectives being (i) to test whether the results are consistent, in the sense that they permit a *single* rating score $r_i$ to be assigned to the $i$th treatment, and (ii) if so, to find the rating scores $r_i$.

In the Thurstone–Mosteller method, the underlying hypothesis is that the N.E.D. (or probit) corresponding to $p_{ij}$ is of the form $(r_i - r_j)$. The Bradley–Terry method works with the hypothesis that $\log(p_{ij}/q_{ij})$, called the logit of $p_{ij}$, is of the form $(r_i - r_j)$. If the subject can indicate his degree of preference on a 7-point scale, Scheffé gives a method utilizing the analysis of variance, while Bliss has developed methods based on rankits for both simple preference and degree of preference scales. Reviews of these methods, with numerical illustrations, have been given by BLISS *et al.* (1956) and JACKSON and FLECKENSTEIN (1957). These examples suggest a high degree of agreement between results obtained by different methods, so that the choice of a method may be based mainly on its flexibility and ease of computation.

Not much has been written on the handling of incomplete blocks when more than two treatments are ranked within each block; although DURBIN (1951) gives an overall $F$-test of the null hypothesis that all treatments have the same average rank.

I wish to thank Dr. P. Armitage, Dr. A. F. Bartholomay, Dr. C. I. Bliss and Dr. I. D. J. Bross for helpful suggestions about developments in pharmacological experimentation.

## REFERENCES

ARMITAGE, P. (1954). Sequential tests in prophylactic and therapeutic trials. *Quart. J. Med.*, **23**, 255.

ARMITAGE, P. (1957). Restricted sequential procedures. *Biometrika*, **44**, 9.

ARMITAGE, P. and HEALY, M. J. R. (1957). Interpretation of $\chi^2$ tests. *Biometrics*, **13**, 113.

ARMITAGE, P. (1958). Sequential methods in clinical trials. *Amer. J. Pub. Health*, in press.

BILLEWICZ, W. Z. (1958). Some practical problems in a sequential medical trial. *Proc. Stockholm Conf. Int. Stat. Inst.*, in press.

BLISS, C. I., ANDERSON, E. O. and MARLAND, R. E. (1943). A technique for testing consumer preferences, with special reference to the constituents of ice cream. *Univ. of Conn. Agr. Exp. Sta. Bull.*, 251.

BLISS, C. I., GREINER, T. and GOLD, H. (1953). Estimating the dose of a cardiac glycoside for human subjects. *J. Pharmacol. and exp. Ther.*, **100**, 116.

BLISS, C. I., GREENWOOD, M. L. and WHITE, E. S. (1956). A rankit analysis of paired comparisons for measuring the effect of sprays on flavor. *Biometrics*, **12**, 381–403.

BLISS, C. I. (1957). Some principles of bioassay. *Amer. Scientist*, **45**, 349.

BRADLEY, R. A. and TERRY, M. E. (1952). Rank analysis of incomplete block designs. I. The method of paired comparisons. *Biometrika*, **39**, 324.

BROSS, I. (1952). Sequential medical plans. *Biometrics*, **8**, 188.

CASS, L. J., FREDERIK, W. S. and BARTHOLOMAY, A. F. (1958). Clinical evaluation of ethoheptazine and othoheptazine combined with aspirin. *J. Amer. med. Ass.*, in press.

COCHRAN, W. G. and COX, G. M. (1957). *Experimental designs*. 2nd ed., 133. John Wiley and Sons, New York.

DAVIES, O. L. (Ed.) (1956). *Design and analysis of industrial experiments.* 2nd ed., Chap. 2. Oliver and Boyd, Edinburgh.

DAVIES, O. L. (1958). The design of screening tests in the pharmaceutical industry. *Proc. Stockholm Conf. Int. Stat. Inst.*, in press.

DOULL, J. A. (1954). Studies in lepromatous leprosy, first series. *Int. J. Leprosy*, **22**, 377.

DURBIN, L. (1951). Incomplete blocks in ranking experiments. *Brit. J. Psych.*, **4**, 85.

FINNEY, D. J. (1955). *Experimental design and its statistical basis*. Univ. of Chicago Press.

FISHER, R. A. and YATES, F. (1953). *Statistical tables*. 4th ed. Oliver and Boyd.

FRIEDMAN, M. (1940). The use of ranks to avoid the assumption of normality implicit in the analysis of variance. *J. Amer. stat. Ass.*, **32**, 675.

HERWICK, R. W., WELCH, H., PUTNAM, L. E. and GAMBOA, A. M. (1954). Correlation of the purity of penicillin sodium with intra-muscular irritation in man. *J. Amer. med. Ass.*, **127**, 74.

HILL, A. BRADFORD. (1955). *Principles of medical statistics*. 6th ed., 241. *The Lancet*, London.

HOGBEN, L. and SIM, M. (1953). The self-controlled and self-recorded clinical trial for low-grade morbidity. *Brit. J. prev. and soc. Med.*, **4**, 103.

JACKSON, J. E. and FLECKENSTEIN, M. (1957). An evaluation of some statistical techniques used in the analysis of paired comparison data. *Biometrics*, **13**, 51.

JELLINEK, E. M. (1946). Clinical tests on comparative effectiveness of analgesic drugs. *Biometrics*, **2**, 87.

KENDALL, M. G. and BABINGTON SMITH, B. (1939). The problem of m rankings. *Ann. Math. Stat.*, **10**, 275.

KILPATRICK, G. S. and OLDHAM, P. D. (1954). Calcium chloride and adrenaline as bronchial dilators compared by sequential analysis. *Brit. med. J.*, **2**, 1388.

MOSTELLER, F. (1951). Remarks on the method of paired comparisons. *Psychometrika*, **16**, 3.

NEWTON, D. R. L. and TANNER, J. M. (1956). N-acetyl-para-aminophenol as an analgesic: a controlled clinical trial using the method of sequential analysis. *Brit. med. J.*, **2**, 1096.

RUSHTON, S. (1952). On a two-sided sequential *t*-test. *Biometrika*, **39**, 302.

SCHEFFÉ, H. (1952). An analysis of variance for paired comparisons. *J. Amer. stat. Ass.*, **47**, 381.

SNELL, F. M. and ARMITAGE, P. (1957). Critical comparison of diamorphine and pholcodine as cough suppressants by a new method of sequential analysis. *Lancet*, (i), 860.

VALLEE, B. L., WACKER, W. E. C., BARTHOLOMAY, A. F. and ROBIN, E. D. (1956). Zinc metabolism in hepatic dysfunction. *I. New Eng. J. Med.*, **255**, 403.

WALD, A. (1947). *Sequential analysis*. John Wiley, New York.

# 89

# ESTADO ACTUAL DE LA ENSENANZA ESTADISTICA EN ESTADOS UNIDOS DE AMERICA*

William G. Cochran, E.U.A.**

## SUMMARY

There is no regular instruction in statistics in the high schools (preuniversity level). Recently, however, a course on introductory probability and statistical inference has been tried in over forty schools, using a textbook prepared by the Commission of Mathematics of the College Entrance Examination Board. The course will be tried in additional schools in the near future. The importance attached to undergraduate instruction in statistics varies greatly from one branch of knowledge to another. In business administration and economics in the larger universities, nearly all departments both give and require some instruction in statistics. About 60% of the departments give some statistics in psychology and education, about 40% in sociology, about 25% in biology and engineering and rarely, if at all, in law, physics or chemistry. Required courses in statistics are the rule in schools of public health, but are found in only a minority of schools of medicine. The introductory course usually comprises about 40 hours of lectures or laboratory, and is taught at a low mathematical level. Most of the teaching is decentralized in the individual departments, although a trend towards centralization is noticeable. More advanced courses in statistics are available in the majority of departments or schools of business administration, economics and public health, and in somewhat less than half the department of psychology, education and sociology. Students who wish to take statistics as a major subject for their Bachelor's degree can do so in only a minority of universities. The number has been increasing and the increase may be expected to continue.

The Ph.D. degree with a thesis in statistics is offered in more than 30 institutions. During the last four years, about 30 Ph.D. degrees per year have been given in the United States. (This figure omits some cases where the degree is granted in a subject-matter field, although the training and the thesis were largely statistical in nature.) About 70% of all the Ph.D. graduates in the country came from eight institutions. A number of new major centers of advanced training have, however, been set up during recent years. The graduate curriculum for the degree lasts at least three years and demands course work in mathematics, statistics and sometimes in an applied field, as well as the thesis. Training at the Master's level can be taken in a substantialy greater number of institutions, although exact figures are not available.

For older persons engaged in work of a statistical type, the principal opportunities for training are in concentrated sum-

* Preparado para la I Sesión de la Comisión de Educación Estadística del IASI como documento de trabajo, y para la III Conferencia Interamericana de Estadística, Brasil, junio de 1955, como documento de referencia 3381a. Posteriormente fué actualizado por su autor.
** Profesor de Estadística, Departamento de Estadística, Universidad de Harvard, Cambridge, Massachusetts, U.S.A.

mer-school courses, in evening classes in the large cities and in
short conferences arranged by the professional societies and by
some universities. Government workers in Washington have a
wide choice of courses available through the Graduate School of
the Department of Agriculture.

The demand for trained statisticians continues to exceed
the supply, and the economic rewards in a statistical career are
relatively attractive.

The principal problems are as follows: (a) There is a
shortage of recruits into the field, partly because not enough
young people learn about this career in their undergraduate
work. (b) There is need for more undergraduate instruction of
three types: (1) A general introductory course of a cultural
nature; (2) more technical courses for students in quantitative
fields of study; (3) major undergraduate programs in statistics.
(c) Much groping and experimentation is in progress towards
finding the best way of organizing the teaching of statistics in
a university. There will probably be no single ideal solution.
(d) A good case can be made for some instruction in statistics
in high schools, but no progress can be reported. (e) More could
be done to help persons in positions of responsibility who use
statistical techniques but have inadequate statistical training.

In conclusion, the general picture is heartening, in that a
substantial number of institutions report the strengthening of
one or more aspects of their program in recent years.

## Introducción

Se carece, desgraciadamente, de datos sobre el desarrollo histórico
de la enseñanza estadística en Estados Unidos. Sólo destacaremos algunos
puntos. La enseñanza estadística elemental se ha impartido durante
muchos años en departamentos de economía, administración de negocios
y matemáticas de las universidades. Una encuesta efectuada en 125 esta-
blecimientos universitarios en 1925 por un Comité de la Asociación Estadounidense de Estadística [6], señaló que un curso elemental de estadística
era impartido en el 60% de los departamentos de economía y ciencias
sociales, 42% de los departamentos de administración de negocios, 47%
de los departamentos de matemáticas, 28% de los departamentos de educación, 10% de los departamentos de psicología y 4% de los departamentos
de biología. Cursos más avanzados se impartían, en esa época, en el 30%
de los departamentos de matemáticas, pero en menor extensión que en
cualquier otro departamento.

Durante el período 1935-40, se efectuó un desarrollo importante. En
un pequeño grupo de universidades se crearon puestos para profesores
de estadística que empezaron a preparar programas superiores para la
enseñanza a estadísticos profesionales. Antes de dicho período, un cien-

---

*Nota general*: Los números entre paréntesis cuadrados indican la referencia que se agrega
al final del artículo.

tífico joven que deseara especializarse en estadística podía estudiar sólo en unos pocos departamentos, principalmente en los de matemáticas o economía, en los que algún profesor no obstante estar ocupado como matemático o economista estaba, además, muy interesado en estadística. Los nuevos puestos, fueron emplazados también principalmente en departamentos de matemáticas o estadística. Pero el cambio significativo, fué que el titular designado era considerado como estadístico y se esperaba que se dedicara por entero a trabajar en estadística. De los ocho principales centros de instrucción estadística avanzada de este país, conforme al número de doctorados concedidos, seis empezaron su programa superior después de 1935. La II Guerra Mundial, en la cual se evidenció la importancia de la destreza estadística en muchas fases de los esfuerzos militares, dió impulso tanto a la preparación de estadísticos de carrera, como a la provisión de mayor número de cursos de estadística para estudiantes de otras ramas científicas. La situación actual se presenta a continuación:

## Principales Características de la Enseñanza de la Estadística

### Enseñanza Preuniversitaria

La necesidad de cierta instrucción estadística en la educación secundaria ha sido expresada por varios autores [1], [2], [3], incluyendo al Dr. S. S. Wilks en su discurso presidencial en la Asociación Estadounidense de Estadística. Todos estos autores destacaron la conveniencia de proporcionar al mayor número posible de ciudadanos, alguna instrucción sobre ideas cuantitativas en las ciencias sociales. Según el Dr. Wilks, podría encontrarse un lugar para la instrucción en estadística, omitiendo ciertas partes de la enseñanza secundaria, en álgebra, trigonometría y geometría que son de poca utilidad para los estudiantes, vayan o no a una universidad. Dutka y Kafka [4], presentan esquemas para dos clases de cursos secundarios: Un curso de "Estadística Auxiliar," para estudiantes que después puedan tener a su cargo puestos burocráticos de carácter estadístico y un curso de "Apreciación Estadística," destinado a formar parte de la educación general de todos los estudiantes.

Ultimamente se ha introducido en los liceos el estudio de elementos de probabilidades y de estadística matemática, como resultado de la labor de la Comisión de Matemáticos de la Junta Examinadora de Admisión a los Colegios Superiores. Esta Comisión ha preparado un libro, "Introducción a las Probabilidades e Inferencia Estadística para la Educación Secundaria," y un conjunto de notas y soluciones para los profesores. El contenido de estos libros se enseña a los profesores secundarios de matemáticas durante cursos especiales de verano. El libro ha sido ya utilizado en más de 40 liceos y ha despertado gran interés.

Enseñanza Universitaria

*Número de cursos en estadística*

En Estados Unidos hay alrededor de 1.000 instituciones universitarias donde se dictan cursos para obtener el grado de bachiller. El número de estudiantes matriculados en 1954 fué de 1.900.000. Obtener información fidedigna sobre el estado actual de la enseñanza universitaria, significaría realizar una larga y detallada encuesta, no llevada a cabo aún en este país. El presente informe está basado principalmente en tres fuentes de referencia. La primera es una muestra aleatoria de 44 de las 153 instituciones que tienen más de 3.000 estudiantes cada una. En total, estas 153 instituciones tienen una matrícula de cerca de 1.250.000 estudiantes, o sea, unos dos tercios del total general de 1.900.000 estudiantes. Este grupo comprende la mayor parte de las instituciones más conocidas fuera de los Estados Unidos. Para la encuesta de las 44 instituciones, efectué, con ayuda de mi mujer, un detallado estudio de los prospectos. Este material se mencionará como la muestra de las grandes instituciones.

La segunda fuente es un reciente estudio del Dr. H. H. Chapman y Ruth G. O'Steen, titulado "Estadísticas de las Institutciones Universitarias del Sur; Instrucción, Investigación, Facilidades." El estudio cubrió 193 de las 272 instituciones en los 14 Estados meridionales de los Estados Unidos y fué patrocinado por la Junta Educacional de la Región Meridional. Además del examen de los prospectos, se obtuvo información en un detallado cuestionario llenado por los miembros de las facultades, que enseñan estadística. Le estoy muy reconocido al Dr. Chapman y a la Dra. Leach de la Junta Educacional de la Región Meridional por la gentileza de permitirme utilizar su valioso material, al que se hará referencia como la muestra de las instituciones meridionales.

En comparación con la nación en su conjunto, los Estados meridionales son menos ricos y su economía es más bien agrícola que industrial. Debe mencionarse también que la muestra de las instituciones meridionales cubre toda clase de instituciones.

Además, el Dr. Palmer Johnson [5] ha efectuado un estudio a base de cuestionarios, de la organización, financiamiento, naturaleza y alcance de los programas estadísticos de 21 de las mayores instituciones del medio oeste. Esto ha proporcionado un material adicional útil.

La tabla I muestra la distribución de frecuencias del número de cursos de estadística por institución. En la muestra de las grandes instituciones, cada institución da por lo menos tres cursos, mientras ocho dan 20, con un promedio de 12 cursos por institución. En las instituciones meridionales, el Dr. Chapman realizó gentilmente una división entre las que tenían más de 3.000 estudiantes y las que tenían menos. Cada institución con más de 3.000 estudiantes da, igualmente, por lo menos 3 cursos, con un número promedio algo superior al de la muestra de las grandes instituciones. Para las instituciones meridionales en conjunto, la dispersión es naturalmente mayor, aunque más del 80% de estas instituciones dan, por lo menos, un curso.

TABLA I.  NUMERO DE CURSOS POR INSTITUCION

| Número de cursos | Grandes instituciones (Más de 3.000 alumnos) Frecuencia | Instituciones meridionales | | |
|---|---|---|---|---|
| | | Más de 3.000 alumnos Frecuencia | Menos de 3.000 alumnos Frecuencia | Frecuencia total |
| 0 | — | — | 36 | 36 |
| 1 | — | — | 57 | 57 |
| 2 | — | — | 40 | 40 |
| 3–5 | 10 | 2 | 19 | 21 |
| 6–10 | 13 | 9 | 8 | 17 |
| 11–20 | 13 | 9 | 1 | 10 |
| 21 y más | 8 | 12 | — | 12 |
| | 44 | 32 | 161 | 193 |

Para las grandes instituciones, la tabla II muestra, el porcentaje de departamentos que ofrecen, por lo menos, un curso de estadística para los principales campos de enseñanza. Al efectuar esta tabulación, se consideró que un departamento prometía un curso si aparecía en el curriculum presentado en el respectivo prospecto. En muchos casos el curso se dictaba en el mismo departamento, aunque a veces tenía lugar en el de matemáticas o estadística o alguna otra materia relacionada. "Departamento" se usa aquí como un término general, por ejemplo la universidad puede tener una escuela o una facultad de administración de negocios en lugar de un departamento. Para hacer comparaciones se incluyen algunas cifras del ya mencionado estudio de 1925 [6].

TABLA II.  PORCENTAJE DE DEPARTAMENTOS QUE OFRECEN POR LO MENOS UN CURSO EN ESTADISTICA (DE LA MUESTRA DE LAS GRANDES INSTITUCIONES)

| Departamento | Porcentaje | |
|---|---|---|
| | 1953 | 1925 |
| Administración de Negocios y Economía | 95 | 60 |
| Matemáticas | 77 | 47 |
| Psicología | 64 | 10 |
| Educación | 58 | 28 |
| Sociología | 40 | —* |
| Biología (inclusive Agricultura) | 25 | 4 |
| Ingeniería | 24 | — |
| Derecho | 0 | — |
| Física | 0 | — |
| Química | 0 | — |

* No disponible.

En las grandes instituciones parece que se ha considerado esencial alguna instrucción estadística en administración de negocios y economía. Este no es el caso en psicología y educación, a pesar de que alrededor del 60% de estos departamentos ofrecen dichos cursos, con sociología ligeramente detrás. En biología e ingeniería sólo la cuarta parte de los departamentos ofrece alguna instrucción estadística. En física, química y derecho, tal instrucción estaba completamente ausente. El lugar prominente del departamento de matemáticas en la lista no es sorprendente. No sólo se dictan habitualmente cursos de teoría estadística en los departamentos de matemáticas, sino que en los pequeños colegios se dictan también cursos introductorios a un nivel matemático más bien bajo, proporcionados por las facultades de matemática. Por la tabla II se puede subestimar la extensión en que algunos departamentos proporcionan enseñanza estadística a sus estudiantes. Por ejemplo, un departamento de psicología puede, en la práctica, recomendar a sus estudiantes que tomen un curso introductorio de estadística dado por el departamento de educación, a pesar de que en el prospecto de la institución no se deje constancia de este hecho.

Las cifras dadas para biología en la tabla II no cubren escuelas de medicina o de salud pública. Sólo una pequeña parte de las escuelas de medicina dicta un curso formal sobre métodos estadísticos. En varias escuelas se dan algunas clases como parte de su instrucción en salud pública o medicina preventiva. Las escuelas de salud pública, por el contrario, exigen uniformemente un curso en bioestadística, y muchas de ellas proporcionan varios cursos más avanzados o especializados.

La tabla III, que contiene el número total de cursos ofrecidos por los departamentos, presenta mayor información sobre la popularidad de la instrucción superior de estadística en varios campos de la enseñanza. En general, los resultados coinciden con los que se podrían haber anticipado de la tabla II. Los resultados de las muestras de las grandes instituciones y las instituciones meridionales coinciden bastante. La principal diferencia

TABLA III.  NUMERO DE CURSOS OFRECIDOS POR LOS DEPARTAMENTOS

| Departamento | Grandes Instituciones | Instituciones Meridionales |
|---|---|---|
| Administración de Negocios y Economía | 169 | 256 |
| Matemáticas | 144 | 192 |
| Estadística | 96 | 179 |
| Psicología | 52 | 73 |
| Educación | 58 | 63 |
| Sociología | 26 | 50 |
| Biología (incluyendo Agricultura) | 19 | 79 |
| Ingeniería | 9 | 34 |
| Física | 0 | 5 |
| Química | 0 | 3 |

es que la biología ocupa un lugar más prominente en la lista de cursos en las instituciones meridionales que en la de grandes instituciones, probablemente debido a que la muestra entre las instituciones meridionales contienen una proporción substancialmente mayor de escuelas de agricultura. Prácticamente toda la instrucción estadística encontrada en biología, está asociada con agricultura o silvicultura.

En ambos grupos de instituciones, los cursos impartidos por los departamentos de estadística ocupan el tercer lugar en la tabla III. Só una minoría de instituciones tiene algún tipo de departamento estadístico, pero cuando existe, el número de cursos dictados es relativamente grande. A pesar de que la enseñanza es conducida actualmente, en la mayoría de las instituciones, sobre base descentralizada, muchas de ellas están en el proceso de formar departamentos estadísticos. Puede esperarse así que el rol de los departamentos de estadística en la enseñanza se incremente en forma material en los próximos cinco años.

No es fácil obtener una idea precisa de la frecuencia con que los departamentos de las diversas especializaciones consideran la enseñanza estadística como esencial para el otorgamiento del grado de bachiller. Varía enormemente en la extensión en la que los departamentos prescriben cursos obligatorios. De los prospectos se deduce que al menos un curso en estadística es considerado siempre como esencial para la graduación en administración de negocios y economía. En psicología, educación y sociología, un número importante de los departamentos, aunque siempre en minoría, requieren algún curso en estadística, lo cual es aún raro en ingeniería y en biología. Con escasas excepciones los requisitos estipulados sobre estadística se limitan a un solo curso y esto ocurre en todos los campos específicos.

*Cursos introductorios: nivel, duración y contenido*

Muchos de los cursos en estadística pueden ser considerados *introductorios*, en el sentido de que no se exigen conocimientos previos en Estadística para ser admitidos a ellos. Algunos departamentos especifican algún trabajo en el campo específico como prerrequisito para un primer curso de estadística pero, en general, el único prerrequisito establecido es en matemáticas. Para unos 300 cursos introductorios dados en las escuelas meridionales, los requisitos matemáticos son los siguientes:

Ninguno o no declarado ...................... 136
Permiso del instructor ....................... 11
Algebra (a veces con trigonometría y/o geometría
    analítica) ............................... 107
Cálculo ..................................... 45

Como lo sugieren estos requisitos, los cursos introductorios se dan generalmente a un nivel matemático bajo.

En la gran mayoría de las veces, el curso introductorio se da en el tercer o cuarto año, de un currículo cuatrienal para el grado de bachiller. Fuera de los 278 cursos para pregraduados dictados en las instituciones meridionales, el curso estaba en primer año sólo en dos casos, en segundo año en 35 casos, y en clase mixta de primero y segundo año en cuatro casos. Estos cursos son ligeramente más frecuentes en tercer año que en cuarto año.

La duración más frecuente de los cursos era entre 40 y 50 horas, durante medio año académico (es decir, un semestre). Una escasa minoría de cursos eran algo más breves. En el 60% de los cursos se incluían las clases y los laboratorios a que asistían los estudiantes, mientras que el resto incluía sólo las clases.

El contenido de los cursos, a juzgar por los programas, varía mucho de una especialización a otra, y en un menor grado dentro de una misma especialización. Una indicación del tipo de contenido puede darse brevemente, citando los textos básicos más usados en las diferentes especializaciones. Incidentalmente, dado que más de 70 textos diferentes se usan en los cursos introductorios de las instituciones meridionales, en apariencia ningún texto básico falta:

Administración de negocios y Economía:
1. Croxton & Cowden, *Practical Business Statistics*
2. Simpson & Kafka, *Basic Statistics*
3. Nieswanger, *Elementary Statistical Methods*

Matemáticas
1. Mode, *Elements of Statistics*
2. Hoel, *Introduction to Mathematical Statistics*

Psicología y Educación
1. Garret, *Statistics in Psychology and Education*
2. Edwards, *Statistical Analysis*

Biología:
1. Snedecor, *Statistical Methods*

En Sociología e Ingeniería no existe un libro claramente predominante.

*Cursos de aplicación más avanzados*

En la mayoría de los departamentos de administración de negocios, economía y matemáticas de las grandes instituciones se dicta más de un curso en los que se enseña estadística. En psicología, educación y sociología, lo anterior es válido para algo menos de la mitad de los respectivos departamentos, y en ingeniería y biología para una proporción aún menor. Muchos de estos cursos avanzados son continuación de cursos introductorios, aunque en algunos casos se consideran tópicos más especializados. En administración de negocios y economía, las materias especiales más comumente enseñadas son econometria, estadísticas industriales y control de calidad, muestreo y análisis de series cronológicas; en psicología encontramos la construcción de pruebas, teoría del diseño de experimentos y análisis factorial. "Métodos de investigación en las ciencias sociales" es un título frecuente en los departamentos de sociología, "diseño de experimentos" en los departamentos de agricultura, y "control de calidad" en los cursos de ingeniería.

Al nivel de graduado, la estadística representa un campo mínimo para los estudiantes que aspiran a los grados de "Master" o de Doctor. Esto puede expresar el punto de vista de que la estadística es necesaria para aquellos estudiantes de especializaciones que intentan dedicarse a la investigación, pero es menos esencial para los que no tendrán tal dedicación. Un mínimo en estadística implica generalmente de tres a cinco cursos de la materia. Pocos departamentos especifican matemáticas a nivel del cálculo para que el currículo mínimo pueda comprender un curso introductorio de teoría estadística matemática.

### Programas de especialización en estadística para pregraduados

La amplitud con que la estadística pueda tomarse como materia de especialización para el grado de bachiller,* es un factor de considerable interés, debido a que estos programas son las fuentes más seguras para reclutar futuros estadísticos de carrera. Estos cursos de especialización son ofrecidos en 13 de las 44 grandes instituciones y en 18 de las 193 instituciones meridionales, es decir, que estos programas están aún confinados a una minoría definida de instituciones. La mayoría de estos programas están en las escuelas o departamentos de administración de negocios, y los restantes en los departamentos de matemática o en los departamentos de estadística, separados.

La duración de los cursos de estadística varía de 12 a 24 horas-semestrales, en otras palabras, de 3 a 6 horas semanales durante dos años académicos. El nivel matemático del currículo es generalmente bajo. El requisito común es un año de matemáticas universitarias con algo de introducción al cálculo.

### PREPARACIÓN DE FUTUROS PROFESORES DE ESTADÍSTICA Y DE ESTADÍSTICOS

### Número de graduados

Para los estudiantes que intentan ser estadísticos profesionales, el curso habitual de educación es el que conduce al grado de doctor en filosofía (Ph.D.) Los títulos de las tesis doctorales aprobados de los candidatos al Ph.D. o Sc.D (doctor en ciencias) han sido publicados anualmente en los *Annals of Mathematical Statistics* desde 1950. Estas listas cubren todos los principales centros de enseñanza en estadística

---

* *Nota del Editor*: El grado de bachiller (bachelor) corresponde al primer ciclo de los estudios universitarios.

matemática, pero omiten algunos casos en los cuales el programa de graduación en estadística está dado por un departamento de especialización tal como el de administración de negocios. (Desgraciadamente, no me ha sido posible obtener el número de graduados de este tipo.) Durante el período 1954-7, hubo 189 doctorados de estadística, alrededor de 47 por año.

Los 189 graduados pertenecían a más de 30 instituciones. Los 5 centros principales, Carolina del Norte con 47 doctorados, California con 18, Stanford con 17, Princeton con 14 y el Instituto Politécnico de Virginia con 13, reúnen más de la mitad del total. Puede esperarse que en un futuro próximo se observe un incremento en el número de instituciones que puedan otorgar el título de doctor, así como en la extensión de la enseñanza. Varias instituciones han empezado recien a ofrecer el currículo, mientras que otras que hasta el momento han contribuido con un número insignificante de graduados, han reforzado seriamente su programa en los últimos años.

## Currículo

Muchos de los estudiantes que cursan estadísticas durante su preparación para el doctorado, han hecho estudios preliminares en matemáticas, aunque ocasionalmente los estudiantes de otras especializaciones, como biología y sociología se han interesado en las aplicaciones estadísticas e intentan concentrar la preparación en estadística durante su trabajo de graduación. Muchos de los estudiantes han tenido también algunos cursos previos en Estadística, pero estos cursos varían tanto en cantidad y calidad, que los principales centros de enseñanza estadística avanzada planifican sus cursos de manera que el estudiante pueda iniciarse sin conocimientos previos de la materia. En esta forma, el estudio para graduarse en estadística, contrariamente a lo que ocurre en otras disciplinas viejas, parte desde el comienzo, en lugar de hacerlo desde un nivel relativamente avanzado. Esta situación sin duda continuará hasta que la instrucción secundaria en estadística esté mucho más extendida que lo que está en la actualidad.

El currículo para el doctorado en filosofía, requiere un mínimo de tres años de estudio posterior al grado de bachiller, y frecuentemente toma cuatro o más años. En general, el currículo está dividido entre matemáticas, estadística teórica y aplicada, trabajos en campos específicos, y la preparación de una tesis. Las diferentes instituciones varían en el grado en que dan énfasis a cada uno de estos cuatro tipos de estudios: un número substancial de ellas no requiere, por ejemplo, trabajo fuera del campo de las matemáticas y de la estadística.

En matemáticas, los requerimientos mínimos más comunes son: Algebra (incluyendo matrices y formas cuadráticas) cálculo superior y teoría de funciones de variable real. Otras materias altamente recomendadas en algunos centros que se concentran en estadística matemática, son: funciones de variable compleja, teoría de la medida y de conjuntos de puntos, topología.

En estadística, los cursos básicos más comunes son: teoría de probabilidades y de la distribución, estimación, prueba de hipótesis, mínimos, cuadrados y análisis de la varianza, análisis de múltiples variables, teoría y práctica del muestreo y diseño de experimentos.

También se dictan algunos cursos más especializados, por ejemplo, análisis sucesivo, procesos estocásticos, teoría de decisiones, estimación no paramétrica, operaciones de investigación y análisis.

El estudio de la especialización comprende usualmente tres cursos. Se trata de dar al estudiante alguna idea de los problemas y métodos de investigación de la rama científica en que pretende hacer su carrera. Algunas agencias federales y estatales ofrecen recibir a estadísticos jóvenes como internos durante su trabajo práctico; sin embargo, el número de participantes en estos internados es aún pequeño. Incluso, existen diferencias de opiniones entre los profesores de estadística, respecto de la magnitud de la especialización práctica que es necesaria. Para muchos tipos de consultas, la efectividad de un estadístico es severamente limitada, a menos que tenga un profundo conocimiento de la rama en que deberá informar a sus compañeros de trabajo. Por otra parte, los estudiantes tienen a menudo mucho éxito al aplicar estadística en alguna rama científica muy distinta de aquella en la que esperaban trabajar. Un estadístico despierto puede trabajar fructíferamente en diferentes campos.

El horario de los cursos descritos, ocupa generalmente la mayor parte de los primeros dos años del programa de estudios. Una práctica común es la de someter al estudiante a un examen preliminar, tanto oral como escrito, al cabo del segundo año. Este examen cubre las áreas de estudio del alumno y además proporciona la oportunidad de una discusión preliminar sobre su futura tésis de investigación. El estudiante debe aprobar el examen antes de ser aceptado como candidato al doctorado en filosofía. Algunas veces el examen revela deficiencias en la enseñanza, que pueden ser subsanadas mediante cursos adicionales. Posteriormente, el estudiante se concentra en su tesis, la que requiere de uno a dos años de trabajo hasta quedar completa. La gran mayoría de los problemas de investigación considerados para la tesis, son problemas de estadística matemática.

Aunque el grado de doctor es el requisito más usual para el estudiante que desea un puesto universitario en estadística, el grado de "Master" es a menudo tomado como educación suficiente para iniciar la carrera en la administración fiscal o en los negocios, empezando desde el nivel profesional elemental (junior). De las instituciones meridionales, 15 ofrecen el grado de "Master," contra 10 que ofrecen el de Doctor. En el lustro pasado, estas instituciones concedieron 124 grados de "Master," con mención en estadística, contra 53 grados de Doctor. Aunque no se tienen datos para todo el país, parece que el número de instituciones que ofrecen grados y el número de graduados, son substancialmente mayores en lo que se refiere a "Masters."

Este grado implica un programa de graduación de dos años, con primer año y parte del segundo, dedicado a clases. El nivel matemático es, en general, menor que el del doctorado, y los estudios menos intensos. El programa incluye habitualmente una tesis, en la que a veces se requiere una investigación original, pero otras es suficiente una monografía sobre alguna rama de la estadística.

### Otros tipos de Enseñanza

Para aquellos que no pueden asistir a la universidad como estudiantes de tiempo completo, existen algunas oportunidades disponibles tanto en la instrucción estadística elemental como en aquella suficiente para mantenerlo al corriente de los últimos adelantos. En la mayoría de las grandes ciudades existen cursos universitarios vespertinos. Estos son, generalmente, cursos introductorios de más o menos el mismo nivel y cobertura que los cursos introductorios ofrecidos a los estudiantes regulares de la Universidad. Naturalmente, el número de cursos ofrecidos en cualquier ciudad durante un solo año es pequeño. La Escuela de Graduados del Departamento de Agricultura tiene en Washington, para los empleados del gobierno, un programa que iguala en rango y calidad a los de las mejores universidades. Su prospecto para el año 1954-55 incluía 22 cursos. Esta escuela tiene también convenios con dos universidades para el dictado de cursos vespertinos a empleados federales en Nueva York y Boston.

Fuera de las grandes ciudades hay pocos cursos vespertinos y los estudiantes deben depender principalmente de la lectura individual de libros de texto. Es posible atender a estos grupos con cursos por correspondencia, pero aparentemente esto no se ha hecho en gran escala en el país.

Algunos de los principales centros estadísticos han desarrollado atractivos programas de Escuelas de Verano. Un curso de este tipo dura generalmente seis semanas con enseñanza concentrada durante cinco días de cada semana. Además de los cursos introductorios, se puede dar instrucción más avanzada en ciertos campos específicos, por ejemplo, diseño de experimentos o técnicas de encuestas muestrales. Los empleados que trabajan durante el año pueden, a menudo, arreglarse con sus empleadores para asistir a los cursos, a veces con un sacrificio parcial o total de sus vacaciones habituales. Estos cursos han sido un medio para reclutar varios jóvenes capaces en el campo estadístico.

Tanto las universidades como las sociedades profesionales han hecho experimentos mediante intentos para proporcionar instrucción o la discusión de problemas estadísticos durante periodos muy cortos. Por ejemplo, la Universidad de Carolina del Norte ha realizado, con éxito, varias conferencias de una semana, cada una dedicada a determinada aplicación de la estadística, como agronomía, cultivos vegetales y química, Estos cursos están dirigidos a empleados que tienen ya una buena base de estadística aplicada y que emplean habitualmente técnicas estadísticas en su trabajo. El principal objetivo de la conferencia es el de explicarles algunos nuevos desarrollos y de darles la oportunidad de la discusión mutua de sus problemas estadísticos más apremiantes. Durante los años pasados, la Sociedad Estadounidense de Química ha dedicado una semana de su Conferencia Gordon anual, a aplicaciones estadísticas en química. En las reuniones de sociedades profesionales, que duran generalmente de tres a cuatro días, es práctica habitual establecer tres o cuatro sesiones relacionadas para dar una introducción en alguna rama de la estadística, por ejemplo, el análisis de la varianza. La Sociedad Estadounidense de Control de la Calidad utiliza a menudo este procedimiento en sus reuniones nacionales y regionales. En estas reuniones, las conferencias proporcionan tanto una introducción a la teoría básica de la técnica como a ilustración de aplicaciones dadas por ingenieros que han encontrado útiles tales técnicas.

En Estados Unidos existen muchas oportunidades para realizar estudios e investigaciones de postgraduados. Los principales centros de estadística de las universidades acogen muy favorablemente a tales estudiantes. Dado que la investigación estadística recibe fuertes subvenciones financieras del gobierno, mediante donaciones o contratos, existen generalmente fondos disponibles para pagar honorarios a becados ya graduados. Un buen número de importantes fundaciones privadas financian también estudios avanzados de este tipo. Además, existen programas, tanto federales como de las fundaciones, que patrocinan estudios en el exterior. Las oportunidades para estudios especializados no se han aprovechado suficientemente, debido a que los graduados en estadística, con grado de doctor, tienen abiertas atractivas ventajas en su carrera.

## Cambios desde 1945

De los estudios de las instituciones meridionales y del medio oeste ya citadas en este artículo, se obtuvo información sobre los cambios que habían tenido lugar en los programas de enseñanza estadística desde 1945. El cuadro general es de substanciales aumentos en la cantidad de enseñanza y en los recursos para la misma. De las 62 instituciones meridionales que informaron sobre este punto, 40 mencionaron la adición de por lo menos un curso desde 1945. (Las instituciones meridionales que no contestaron al respecto, probablemente muestren una porción mucho menor en la introducción de nuevos cursos.) De las 15 instituciones del medio oeste que informaron, 13 habían aumentado la cantidad de enseñanza, mientras que 8 habían tenido mayor desarrollo: ya sea la iniciación de un programa tendiente a otorgar el grado de doctor, o el establecimiento de un departamento o de un laboratorio de estadística.

## El Empleo de Estadísticos

Durante los 10 años pasados, la demanda de graduados de "Master" o Doctor, ha excedido la oferta. El graduado idóneo en estadística tiene a su elección por lo menos 3 puestos, cada uno de los cuales su profesor considerará como más adecuado para una persona joven. Las universidades, el comercio y el gobierno están empleando estadísticos en número cada vez mayor. Una tabulación según el tipo de empleador para los 117 graduados con doctorados en el período 1950-53 da el siguiente resultado:

| Tipo de empleador | Número de graduados | % |
|---|---|---|
| Universidades | 73 | 62 |
| Empresas | 21 | 18 |
| Gobierno | 16 | 14 |
| Extranjero | 2 | 2 |
| Ignorado | 5 | 4 |
| | —— | —— |
| | 117 | 100 |

El rol predominante de las universidades como empleador de los doctorados es talvez un reflejo de la forma en que las universidades están extendiendo su cuerpo docente estadístico. Como consecuencias de la escasez, los salarios ofrecidos a jóvenes estadísticos son generalmente más elevados que los que se pagan a graduados en otras ramas de la ciencia que requieren una preparación de similar extensión y complejidad, y los ascensos de los estadísticos jóvenes capaces tienden a ser rápidos. Algunos empleadores, incluso, están dispuestos a ofrecer puestos estadísticos a solicitantes que tengan sólo un mínimo de preparación en la especialización, pero que muestren aptitudes.

### Necesidades y Problemas de la Educación Estadística

Este tema ha sido objeto de enérgicas discusiones entre los profesores de estadística. Durante 1947, aparecieron sobre este asunto informes de tres comisiones diferentes — de la Comisión de Estadística Matemática Aplicada del Consejo Nacional de Investigación, de la Comisión para la Enseñanza de Estadística del Instituto de Estadística Matemática, y (en Inglaterra) de la Comisión para la Enseñanza Estadística de la Real Sociedad de Estadística. Estas tres comisiones estuvieron fundamentalmente de acuerdo en los problemas principales. Estos problemas son aún los más importantes, a pesar de que se ha registrado algún progreso en su solución en los últimos años.

a. Existe una escasez de estadísticos profesionales jóvenes que inicien la carrera. La dificultad no está en la carencia de instituciones en las que se pueda obtener enseñanza avanzada. Actualmente el grado de Doctor en estadística matemática se puede obtener en más de 30 universidades, y un número mayor proporciona instrucción hasta el grado de Master. Aunque los programas del doctorado en algunas de estas instituciones son aún de alcance limitado, los programas están siendo reforzados en un número substancial de instituciones. Las recompensas económicas y las oportunidades son suficientemente grandes para competir favorablemente con cualquier otra rama de la ciencia. La dificultad parece consistir en que los jóvenes, no oyen hablar suficientemente de estadística como una carrera potencial, durante su iniciación universitaria.

b. Hay necesidad de mayor instrucción estadística en la educación secundaria. En lo que se refiere a la oferta de estadísticos de carrera, el paso más útil sería el incremento en el número de departamentos universitarios en los que pudiera tomarse estadística como una materia de especialización para el grado de bachiller. El estudiante secundario que ha tomado varios cursos en estadística como materia de especialización, está en mejores condiciones de ser informado acerca de este campo como futura carrera, que el estudiante que solamente ha tenido un curso introductorio en estadística como complemento a su trabajo en otras ramas científicas. Más aún, a medida que los programas de especialización en estadística se popularicen para los pregraduados, será posible que los correspondientes a la enseñanza para graduación empiecen a un nivel moderadamente avanzado, en lugar de hacerlo desde un comienzo como ocurre en la actualidad.

Existe también la necesidad de mayor instrucción estadística en aquellas ramas científicas en las que la técnica estadística es particularmente útil. Como ya se ha indicado, la existencia de cursos de estadística es actualmente muy común en administración de negocios, economía, psicología, educación y sociología. Las instituciones que proveen este entrenamiento en biología, medicina e ingeniería están aún en minoría y en física y química es raro encontrar enseñanza estadística. Sin embargo, creo que puede decirse con confianza, que la situación está mejorando. Muchos departamentos informan sobre notables adelantos en la disponibilidad de cursos estadísticos en el período siguiente a la terminación de la guerra

Las tres comisiones recomiendan también, un curso introductorio general en estadística que sería tomado por todos los estudiantes de ciencias sociales y naturales y, talvez, incluso por todos los estudiantes universitarios, sobre la base de que, el entrenamiento en el pensamiento cuantitativo acerca de los datos, es esencial para cualquier adulto ilustrado en el mundo, hoy día. De las fuentes citadas en este informe, se desprende que se ha tenido poco éxito en este aspecto. Parece más bien que el progreso será lento, puesto que habrá dificultades para introducir un nuevo curso en el siempre repleto programa escolar.

c. Las tres comisiones discuten también el mejor método para organizar la enseñanza dentro de una universidad. Aunque considerar que este problema debe quedar sujeto a mayores exprimentaciones, tienden a favorecer algún grado de centralización a través de un laboratorio o departamento de estadística. Las ventajas mencionadas en favor de un departamento central, son de que dan a la estadística un estado más definido en el programa docente de la universidad, proporciona la oportunidad del plan coordinado que es necesario en un campo que se expande rápidamente, conduce a una mejor calidad en la enseñanza, y ayuda a prevenir la innecesaria duplicación de cursos, que son generalmente idénticos en su contenido. Según las comisiones, el departamento central o laboratorio debería ser responsable de la enseñanza de los estudiantes que se especialicen en estadística y debería enseñar los cursos introductorios básicos. Cursos más avanzados o apropiados para las especializaciones, podrían ser dictados en el departamento central o en los departamentos especializados. Los profesores que dictaran estos cursos en los departamentos especializados deberían, sin embargo, ser también miembros del departamento de estadística, de manera que el currículo pueda ser conformado sobre una base coherente e integrada.

A pesar de que la enseñanza de estadística está aún descentralizada en la mayoría de las instituciones, ha habido una tendencia hacia la centralización. Desde el final de la guerra, algunas instituciones grandes han organizado departamentos, o laboratorios de estadística, algunas con la intención posterior de dar educación en estadística hasta el nivel de Doctor. En las universidades más pequeñas, el número limitado de miembros de las facultades pesan en favor de cierta centralización de la enseñanza, la que a menudo tiene lugar bajo el alero del departamento de matemáticas.

Los profesores de estadística discuten mucho respecto a si los cursos estadísticos se enseñan mejor como curso general en un departamento de estadística o matemáticas, o si deberían ser dictados en las especializaciones. Los que están de acuerdo con el curso general alegan que evita duplicaciones y proporciona mejor calidad, puesto que el curso es dictado por alguien para quien la estadística es una carrera y ha sido elegido por su competencia en el campo, mientras que el curso en un departamento de especialización es enseñado a menudo por un profesor para el cual la estadística, en sí misma, es sólo de interés secundario y puede tener capacidad inadecuada en el tema. Por el otro lado, los profesores en la especialización discuten a menudo que los estudiantes tienen más interés en un curso que trabaja con problemas y ejemplos extraídos de su propia especialidad, y que los estudiantes encuentran dificultades en adaptar técnicas estadísticas aprendidas en un curso general a las aplicaciones de un campo particular. Existen argumentos sólidos en ambos aspectos de la discusión.

d. Un buen enfoque es el de dar alguna educación estadística en los liceos. Ningún progreso puede mencionarse, sin embargo, debido a que las proposiciones recién están empezando a ser discutidas y llevadas adelante.

e. Como consecuencia del creciente empleo de los métodos estadísticos, muchas personas que ocupan puestos de responsabilidad utilizan actualmente técnicas estadísticas, a pesar de que tienen escasa o ninguna instrucción en la materia. Las principales oportunidades para recibir instrucción es la asistencia a cursos de verano de seis semanas, en clases vespertinas dictadas en las grandes ciudades o en conferencias especiales esporádicas, de corta duración. Muchas personas no pueden aprovechar tales oportunidades. Deberían efectuarse mayores estudios sobre estos grupos y sus necesidades. Una manera de ayudarles podrían ser los cursos por correspondencia, no obstante ciertas desventajas del método de enseñar en esta forma una materia como estadística. El gran avance en la disponibilidad de textos estadísticos, que ha sido un hecho notable en la pasada década, ha facilitado la tarea de aprender mediante el estudio personal.

REFERENCIAS

[1]  National Research Council Committee on Applied Mathematical Statistics. "Personnel and Training Problems Created by the Recent Growth of Applied Statistics in the United States." *Reprint and Circular Series*, No. 128, 1947.

[2]  Deming, W. Edwards and Scates, D.E. "The Need for Statistical Education in High School and College." *The Educational Record*, January 1948, p. 72.

[3]  Wilks, S.S. "Undergraduate Statistical Education." *Journal of the Statistical Association*, Vol. 46, March 1951, p. 1-18.

[4]  Dutka, S. and Kafka, F. "Statistical Training below the College Level." *The American Statistician*, Vol. 4, February 1950, p. 6-7.

[5]  Johnson, Palmer O. "Development of Statistical Programs in Mid-American Universities and Colleges." *The American Statistician*, Vol. 7, October 1953, p. 7-15.

[6]  Glover, James W. "Statistical Teaching in American Colleges and Universities." *Journal of the American Statistical Association*. Vol. 21, 1926, p. 419-424.

*    *    *

# 69

*Eighteen*

## THE DESIGN OF EXPERIMENTS

### WILLIAM G. COCHRAN

---

## INTRODUCTION

In ordinary speech, the word "experiment" has a broad meaning, covering anything that involves trying out something new. In the subject that has come to be known as the design of experiments, the word is used in a narrower sense. The essence of an experiment is that we deliberately apply two or more procedures, called the *treatments*, to some process or operation for the purpose of measuring and comparing their effects. In the preparation of alloys, for instance, the objective of an experiment might be to compare the effects of different concentrations of carbon, or different firing temperatures, or different types of furnaces, or different types of molds into which the molten effluent is poured, on the important qualities of the resulting alloy. The comparative experiment is one of the most powerful weapons available to the scientist. If successful, it provides trustworthy factual information which advances our understanding and contributes to sound practical decisions.

Systematic study of the principles involved in the planning and execution of experiments was initiated over thirty years ago by R. A. (now Sir Ronald) Fisher, with particular reference to agricultural experimentation. There is now a large literature, most of it concerned with the design of a *single* experiment. More recently, attention has shifted to the problem of planning an experimental program in exploratory or developmental research. In this, a series of experiments is to be conducted to discover which of a number of factors or variables have important effects on some response, and to learn something about the nature of these effects.

This chapter contains two main parts. The first presents a brief summary of some principles that should guide the conduct of a single experiment in which the treatments to be compared have already been chosen. In the second, several strategies that have been proposed for carrying out an experimental program will be discussed.

## PART I. THE PLANNING OF A SINGLE EXPERIMENT

*Variability and Its Effects*

An old recipe for an ideal experiment is to keep everything constant except the imposed differences in treatments. If this could be done, any differences in results could be ascribed unequivocally to the differences in treatment. In most lines of experimentation, however, it is either impossible or prohibitively expensive to keep all other factors constant that might influence the results. Variability will be present in the raw materials used, in the state of the environment, in the operations required for the application of the treatments, and in the process of measurement. Thus the *observed* effect of a treatment is subject to an experimental error which represents the joint effect of all these other sources of variability.

Experimental errors have two undesirable consequences. They may produce biases or systematic errors in the estimates of the treatment effects. With two treatments, for example, suppose that each treatment is to be tried six times: in other words, to use a common technical term, the experiment is to have six *replications*. If six applications of a treatment can be processed and the results measured in a single day, one way of doing the experiment is to complete the six replications of treatment 1 on Monday, and to carry out those for treatment 2 on Tuesday. There might, however, be an unanticipated shift in the level of the responses from Monday to Tuesday, due to the use of a new batch of raw material, to a change in humidity, or to a zero error in a reading instrument. If this occurred, all six results from treatment 2 might be found to lie above those for treatment 1, suggesting a consistent superiority of treatment 2, although in fact the differences were caused by the diurnal shift in level.

When a bias of this type is present, the standard methods for the statistical analysis of the results of experiments are incapable of detecting or of correcting for it. Tests of significance and confidence limits become misleading. Biases confuse experimenter and statistician alike.

Experimental errors also produce fluctuations of a more erratic or "random" nature in the results. If several replications are made, these fluctuating errors cause the observed differences in effect between two treatments to vary from one replication to another. Looking over the experiment as a whole, we may not be clear as to what the true size of difference is, nor even whether the superiority of one treatment has been established.

Fluctuating errors are less serious than systematic errors in two respects. Since they do not consistently favor any specific treatment, their effects tend to cancel out over a series of replications. Further, the standard methods of statistical analysis measure the extent to which the estimated treatment effects are influenced by fluctuating errors, so that the experimenter is made aware of the degree of uncertainty in the results. This is small consolation, however, if fluctuating errors in an experiment are so large that no definite conclusions can be drawn.

*Some General Principles*

Three principles which flow obviously from this analysis of the effects of variability may be stated as follows:

1. To take any precautions necessary to reduce systematic errors to a negligible size.

2. To ensure that, despite the presence of fluctuating errors, the treatment effects are estimated with precision adequate for the purpose at hand.

3. To measure the amount of uncertainty that exists in the estimated treatment effects because of experimental errors, and to take account of this uncertainty when drawing conclusions from the results.

In biological research, in which experimental errors are often large, these principles have long been regarded as essential to sound experimentation. In the industrial field, experimental errors tend to be lower, and it has sometimes been thought that we need not worry about them, because any treatment effects that are large enough to be of practical interest will show up clearly despite the errors. While this may be true in certain experiments, there are important segments of industrial research in which experimental errors are just as high as in biological research.

In the next two sections, these principles will be discussed in more detail.

### Precautions Against Systematic Errors

As we have seen, a systematic error is created when some source of variation tends to favor or handicap a specific treatment persistently in all replications. The first step in guarding against such errors is to check over the operations involved in the practical conduct of the experiment, in order to detect whether bias is likely to enter at any stage. While the opportunities for bias vary with the type of experiment, there are several vulnerable stages.

> 1. In the assignment of raw material to the treatments. For instance, if the raw material required for the replications of treatment 1 is allotted first, then that for treatment 2, and so on, it may happen that treatment 1 receives the most responsive batch of raw material, if the experimenter, without realizing it, picks out responsive material first.
> 2. In the order in which operations are carried out on the treatments. Unsuspected time trends are sometimes present in laboratory and factory procedures.
> 3. In the spatial allotment of treatments. There may be systematic differences between different shelves in an oven or different parts of a water bath.
> 4. In the process of measurement, particularly if this is wholly or partly subjective.

With regard to the first three sources of bias, a useful precaution is to make the allotment in question by the use of a table of random numbers. The unsystematic nature of random numbers guards against the persistent repetition of favorable conditions for any one treatment. In other words, randomization helps to ensure that the experimental errors will be fluctuating rather than systematic. A second useful device is that known as blocking or grouping (p. 513). This may be more effective than randomization, in that it may cause certain systematic sources of error to cancel out entirely. In references (2) and (3) numerous combinations of randomization and blocking, appropriate for different experimental situations, are described.

Where measurements are subjective, a standard precaution is to prevent the person making the measurements from knowing which treatment is being measured at any given time. Presentation of treatments in a randomized order is also advisable, in case the measurer has a drift or trend in his measurement errors.

## Methods for Increasing Precision

The standard deviation of the experimental results, computed by the usual methods, provides a measure of the size of the fluctuating errors to which the results are subject. If the standard deviation is undesirably large, we should try to discover at what stages in the experiment the major sources of variability arise. Sometimes this requires a separate and time-consuming investigation of the variability of the raw material or of the precision of measuring devices.

The principal devices for increasing the precision of the final results are outlined below.

*Refinements of technique.* This covers such actions as the procurement of more uniform raw material or more accurate instruments of measurement, standardization of the environment, e.g., by humidity or temperature controls, and the development of small-scale methods of experimentation that can be done with high precision. Sometimes experimentation cannot proceed with any hope of getting results unless such refinements are made. Their disadvantages are that the refinements may be costly, or that, as in the case of small-scale research, the conditions of experimentation come to differ markedly from those to which it is hoped to apply the final results. For this reason, as is well known, small-scale research is often used primarily to discard bad ideas and screen out a few promising treatments that will be tested later under more realistic conditions.

*Replication.* Even with a high standard deviation, satisfactory precision in the treatment means can be obtained if the treatments are replicated often enough, with precautions such as randomization to avoid systematic errors. When experiments must be done under "practical" conditions, so that refinements of technique cannot be introduced, replication is sometimes the most useful weapon. Its disadvantage is the cost involved: since the standard error of the average result for a treatment is $\sigma/\sqrt{r}$, where $r$ is the number of replications and $\sigma$ the standard deviation, the standard error decreases rather slowly as $r$ is increased.

*Auxiliary measurements.* In comparing different commercial feeds for chickens or hogs, the amount by which an animal increases in weight during a given period is usually correlated with its initial weight. The initial weight is an auxiliary measurement that predicts to some extent the final response (i.e., weight) of an animal. By a statistical technique known as the analysis of covariance, these auxiliary measurements can increase the precision of the final responses, the idea being to adjust the final weights so as to remove the effects of variations from animal to animal in initial weights. In general terms, covariance removes the error arising from some source of environmental variation that can be measured by the auxiliary variate but not conveniently controlled.

*Blocking.* By a careful arrangement of the way in which the experiment is done, potential sources of error can often be balanced out so that they affect all treatments equally. Two illustrations will be given. McGehee and Gardner (1) conducted an experiment to measure the effect of factory music programs on the productivity of women engaged in a repetitive operation in rug manufacturing. Four types of musical program (*A, B, C, D*) were compared with no music (*E*), a different program being tested each day. The schedule for the experiment was as follows:

| Week | Mon. | Tues. | Wed. | Thurs. | Fri. |
|------|------|-------|------|--------|------|
| 1    | A    | B     | C    | D      | E    |
| 2    | B    | C     | D    | E      | A    |
| 3    | C    | A     | E    | B      | D    |
| 4    | D    | E     | A    | C      | B    |
| 5    | E    | D     | B    | A      | C    |

The features of interest are that each program appears once on each day of the week, and once during any given week, the arrangement of letters being known as a 5 x 5 latin square. Thus, if productivity is consistently higher on certain days of the week than on others, these differences will cancel when two programs are compared. The same remarks apply to consistent differences in productivity from week to week. A grouping of this kind, such that each treatment occurs equally often in the group, is often called a *block*, the term being borrowed from agricultural experimentation. This latin square employs double blocking, the blocks being respectively days of the week and weeks.

The following slightly more complex arrangement (Table 1) might be used to compare the durabilities of, say, types of leather or rubber on a machine that produces artificial wearing. The machine takes only two specimens in a single run.

Table 1. Incomplete Block Design Used to Compare Five Types of Material on a Wear Machine

|  | Run number | | | | | | | | | |
|  | 1 | 2 | 3 | 4 | 5 | 6 | 7 | 8 | 9 | 10 |
|---|---|---|---|---|---|---|---|---|---|---|
| Right | $A$ | $B$ | $C$ | $D$ | $E$ | $A$ | $B$ | $C$ | $D$ | $E$ |
| Left | $B$ | $C$ | $D$ | $E$ | $A$ | $C$ | $D$ | $E$ | $A$ | $B$ |

Two possible sources of error are that there may be a consistent difference between the amounts of wear on the right and left sides of the machine, and that there may be consistent differences in wear from run to run. The arrangement balances out the "right-left" difference by having each type ($A$, $B$, $C$, $D$, and $E$) run twice on the right side and twice on the left. The differences between runs cannot be balanced in this way, because each run accommodates only two specimens. It may be verified, however, that every pair of letters occurs just once in the same run. It follows that the difference between the average results for type $A$ and the average results of its partners in the same runs (i.e., runs 1, 5, 6, and 9) is an estimate of the average performance of $A$ relative to that of the four other types. From these figures it is easy to obtain comparable estimates of the average performances of the five types, from which any consistent run to run differences have been eliminated. The arrangement is one of a class known as incomplete block designs, the "run" being the incomplete block.[1]

The chief advantage of blocking is that frequently it involves little or no extra cost in the conduct of the experiment, so that even a modest increase in precision from blocking is all to the good. A considerable variety of designs utilizing blocking will be found in references (2) and (3).

Finally, none of the techniques described above has any inherent superiority over the others. In a specific situation the experimenter should utilize those that promise the best return for a given expendi-

[1] For this example the author is indebted to Dr. W. J. Youden.

ture of his resources. His knowledge about the primary sources of variability will, of course, be helpful in making this choice.

With regard to the third principle, i.e., that account must be taken of the experimental errors when drawing conclusions from the results, the standard methods of statistical analysis are now well developed to cope with this problem, provided that systematic errors are absent.


## PART II. THE DESIGN
## OF AN EXPERIMENTAL PROGRAM

Under discussion next are some of the strategies that have been proposed for investigating the effects of a number of different factors or variables on one or more responses. The number of factors may vary from two or three to as high as twenty or more. The factors may be quantitative, like temperature or time, or qualitative, like type of leather or shape of mold. For each quantitative factor there will be a known range within which the factor may be varied in the experiments.

In this type of problem the demands of fundamental research are usually more taxing than those of applied research. In fundamental research the scientist is likely to wish to learn about the interrelationships among the effects of the factors throughout the whole of the range of each quantitative factor and for all variants of each qualitative factor. In applied research the objective is often the more restricted one of finding the levels at which the factors must be set in order to obtain a high response. With this goal we do not need to spend time investigating regions in which the response is low, and we can discard quickly variables that have only minor effects.


### Factorial Experimentation

Perhaps the simplest method of attack is to run a separate experiment for each factor, having as many experiments as there are factors. As an alternative, Fisher pointed out that if all factors can be studied simultaneously in a single factorial experiment, this makes an economical use of resources and provides appropriate data for investigating the interrelationships among the effects of the factors. The point can be illustrated by a program which contains three factors, $A$, $B$ and $C$, each to be investigated at only two levels, say low and high.

To give physical meaning to the explanation, we may consider as an example the three factors, $A$, $B$, and $C$, as the nitrogen, phosphorus, and potassium content of a fertilizer, each comprising 5 per cent of the fertilizer when at the low level and 10 per cent at the high. Each observation for a particular combination of factors would be a measure of the yield of a plot fertilized with a specified weight of fertilizer and quantity of seed.

In the experimental plans (Tables 2 and 3), the presence of a letter signifies that the corresponding factor is held at its high level; the absence of the letter denotes the low level. Thus, the symbol $ac$ implies that factors $A$ and $C$ are being held at their high levels, while $B$ is held at its low level. The treatment combination in which all three factors are at their low levels is usually denoted by the symbol (1). In the example, then, combination (1) is a 5-5-5 fertilizer, while at the other extreme, $abc$ denotes the 10-10-10 mix.

Assuming that four replications are considered advisable, Table 2 shows how the three experiments might be conducted by the "one factor at a time" method. Note that in the first experiment, which is devoted to factor $A$, some decision must be made about the levels at which $B$ and $C$ are fixed during this experiment. We have supposed that $B$ is held at its high level and $C$ at its low level, so that Experiment I compares the treatment combination $b$ with $ab$. Similar choices must be made in the second and third experiments. It is assumed that the high level of $A$ was superior in Experiment I and that this level was used in Experiment II, $ab$ being used in Experiment III for the same reasons. In Table 2, each letter is repeated four times to signify fourfold replication.

Table 2.  Illustration of Single-Factor Experiments

| Exp. I | (Factor $A$) | Exp. II | (Factor $B$) | Exp. III | (Factor $C$) |
|--------|--------------|---------|--------------|----------|--------------|
| $b$ | $ab$ | $a$ | $ab$ | $ab$ | $abc$ |
| $b$ | $ab$ | $a$ | $ab$ | $ab$ | $abc$ |
| $b$ | $ab$ | $a$ | $ab$ | $ab$ | $abc$ |
| $b$ | $ab$ | $a$ | $ab$ | $ab$ | $abc$ |

This series of experiments requires twenty-four treatment combinations to be tested. Its information about relations between the effects of the factors is weak. If someone asks, "Is the effect of $A$ the same at low and high levels of $C$?", the series provides no data to answer the question, since $A$ was tested only at the low level of $C$ in Experiment I.

In factoral experimentation, all treatment combinations that can be made up from the low and high levels of each factor are compared simultaneously in a single experiment, which contains $2^3$ or 8 combinations. Table 3 shows these combinations and illustrates a standard method of analyzing the results.

Table 3. A $2^3$ Factorial Experiment

| Factorial effect | Treatment combination | | | | | | | | Divisor |
|---|---|---|---|---|---|---|---|---|---|
| | (1) | $a$ | $b$ | $ab$ | $c$ | $ac$ | $bc$ | $abc$ | |
| $A$ | − | + | − | + | − | + | − | + | 4 |
| $B$ | − | − | + | + | − | − | + | + | 4 |
| $C$ | − | − | − | − | + | + | + | + | 4 |
| $AB$ | + | − | − | + | + | − | − | + | 4 |
| $AC$ | + | − | + | − | − | + | − | + | 4 |
| $BC$ | + | + | − | − | − | − | + | + | 4 |
| $ABC$ | − | + | + | − | + | − | − | + | 4 |

The eight treatment combinations provide four little experiments on the effect of factor $A$, namely (1) vs. $a$; $b$ vs. $ab$; $c$ vs. $ac$; and $bc$ vs. $abc$. The average of the results of these four comparisons gives an estimate of the effect of $A$ (high level vs. low level) based on four replications. The succession of − and + signs required to compute this average is shown in row $A$ of Table 3.

Similarly, the factorial experiment provides four separate tests of the effect of $B$; (1) vs. $b$; $a$ vs. $ab$; $c$ vs. $bc$; and $ac$ vs. $abc$. The succession of signs needed to compute this effect, with divisor 4 to obtain the average, appears in row $B$. In the same way the experiment gives four replicates for the average effect of $C$. These average effects are called the *main effects* of the factors.

Thus, by testing eight combinations in a single factorial experiment, we obtain fourfold replication for the average effects of each factor, whereas twenty-four combinations had to be tested by the single-factor approach. The secret of the high efficiency of the factorial method is that every treatment combination in the experiment supplies information about the effects of all three factors.

To revert to the question "Is the effect of $A$ the same at low and high levels of $C$?", the factorial experiment throws some light on the answer. The comparisons

$$abc - bc \quad \text{and} \quad ac - c$$

are estimates of the effect of $A$ at the high level of $C$, while the comparisons

$$ab - b \quad \text{and} \quad a - (1)$$

are estimates of the effect of $A$ at the low level of $C$. Consequently the quantity

$$\left\{ \frac{abc - bc + ac - c}{2} \right\} - \left\{ \frac{ab - b + a - (1)}{2} \right\}$$

estimates the difference between the average effect of $A$ for high $C$ and the average effect of $A$ for low $C$. This expression is called the $AC$ interaction. (By a convention which will not be explained here, a divisor 4 is generally used instead of 2.) The series of $-$ and $+$ signs for computing the $AC$ interaction is shown in the row labeled $AC$ in Table 3. By the same process, $AB$ and $BC$ interactions may be computed. We can go further and examine whether the AC interaction is the same at low and high levels of $B$, since the experiment furnishes separate estimates of the $AC$ interaction at these two levels. The difference between these two estimates, with a suitable divisor, is called the $ABC$ three-factor interaction, which is computed as shown in the last line of Table 3.

Briefly, then, the principal advantages of factorial experimentation are that it provides (for a much smaller total amount of experimentation) equally precise estimates of the average effects of the factors as the single-factor method, and it also supplies a convenient body of data for examining relationships between the effects of different factors.

## Additivity of Factorial Effects

The meaning of the various kinds of interactions and the rules for calculating them are explained in the standard textbooks. In many types of work, most of the interactions turn out to be small, particularly those among three or more factors. This is the reason why the analysis of the results of factorial experiments in terms of main effects and interactions has been found useful: frequently it is necessary to report and study only the main effects plus a few interactions between pairs of factors.

If all interactions are negligible, we have the simple situation in which any factor produces the same effect for all combinations of levels of the other factors. The factors are then said to be *additive* in their effects. The term is appropriate because if there are no interactions, the difference between the response to *abc* and the response to the combination (1) in which all three factors are at their low levels will be given by the sum of the main effects of *A*, *B*, and *C*. When effects are additive, the summary of results is particularly simple, since the main effects tell the whole story.

Whether effects are additive or not depends primarily on the way in which nature behaves. The experimenter can, however, exert an influence through the choice of factors, the choice of scale in which results are analyzed, and the choice of the range through which each factor is varied. For instance, if two factors are limiting, in the sense that the response will be poor unless both are in ample supply, we may expect to find an interaction. If the factors produce multiplicative rather than additive effects, analysis of the logs of the responses will restore additivity. As regards the widths of the ranges through which factors are varied, the assumption of additivity is likely to be a closer approximation to the true situation when the ranges are small, so that the effects of the factors are modest rather than large.

Additivity is stressed here because without it, especially if numerous interactions are large, the results of a factorial experiment are apt to be confusing both to interpret and to use. The search for types of factors and scales of analysis which produce approximately additive effects is therefore an important task in multifactorial research.

## Limitations of Factorial Experimentation

Although the factorial experiment is a major contribution to the experimental sciences, there are limitations on its utility. The most obvious is that factorial experiments become prohibitively large and complex if the factors are numerous. Even if only two levels are included for each factor, an experiment with seven factors contains $2^7$

or **128** combinations to be tested, and this number doubles for each additional factor. With quantitative factors whose effects are expected to be curvilinear, more than two levels are desirable, and the size of the experiment mounts still more rapidly. To test five different levels of each of six factors, the number of treatment combinations is $5^6$ or 15,625.

In applied research in which high-yielding combinations of levels of the factors are the primary objective, a factorial experiment may devote too much experimentation to low-yielding combinations. In a $2^7$ experiment, 64 of the tests are made at the low level of $A$. If this level gives uniformly poor responses, an alternative approach that revealed this fact quickly would save unproductive work.

As illustrated previously, an attractive feature of the factorial experiment is that it provides internal replication for estimating the main effects. This replication is extremely useful in lines of work in which experimental errors are high. But if experimental errors are low relative to the amount of change in the response as the factor levels are changed, a high degree of replication may not be needed for finding a good combination of factor levels. In a $2^7$ factorial, any main effect is based on 64 replications: perhaps two or three would suffice.

The length of time required to obtain the results of an experiment will influence research strategy. When there is a long delay in obtaining results, as in experiments on farm crops or in tests of weathering in which treated surfaces are exposed for several months, it seems wise to obtain a large amount of information in a single complex experiment. When results become known quickly, on the other hand, the tendency will be to use smaller experiments of restricted scope, each planned so as to capitalize on the results of the preceding experiments.

### Factorial Experiments in Fractional Replication

If a complete factorial experiment is too large, and the experimenter is unwilling to decrease the number of factors or the number of levels of each factor, he must accept some sacrifice in the quality of information obtained in order to reduce the size of the experiment.

By a selection of part of the treatment combinations in a complete factorial, it is possible to obtain useful information, though with some risk of being misled by the results.

Suppose, for example, that in a $2^3$ factorial, we test only the four combinations

$$a, b, c, abc.$$

The main effect of $A$ is estimated as

$$\frac{(abc) + (a) - (b) - (c)}{2}.$$

Note that this comparison is balanced for $b$ and $c$: i.e., each level of $b$ is represented both on the $+$ and the $-$ sides of the comparison. The same holds for $c$. Thus, this estimate of $A$ is not affected by the size of the $B$ and $C$ effects.

However, if we attempt to compute the $BC$ interaction from the four observations, Table 3 shows that it is given as

$$\frac{(abc) + (a) - (b) - (c)}{2}.$$

That is, the $BC$ interaction and $A$ are identical. $A$ and $BC$ are called *aliases*. It follows that if $B$ and $C$ have a positive interaction, this shows up as an effect of $A$. Consequently, in an experiment in which $A$ has actually no effect, but $B$ and $C$ have effects and a positive interaction, we would erroneously report an effect of $A$. This is the type of hazard present in the use of fractional factorials.

With a $2^4$ factorial, a possible half-replicate is the set

$$(1), ab, ac, ad, bc, bd, cd, abcd.$$

In this case it will be found that all four main effects are independent of two-factor interactions, although the two-factor interactions fall into alias pairs; $AB = CD$; $AC = BD$; $AD = BC$.

With a $2^5$ factorial, 16 of the 32 treatment combinations can be picked to give independent estimates of all five main effects and all ten two-factor interactions. The hazard in this plan is that if some three-factor interactions are not negligible, the estimates of certain of the two-factor interactions are biased. The presence of four-factor interactions produces biased estimates of the main effects.

If all interactions can be assumed negligible, a greater reduction in the size of the experiment can be achieved. Table 4 shows the plan for an experiment with seven factors each at two levels, in which all the main effects are estimated from only eight tests ($\frac{1}{16}$th replicate). The eight treatment combinations appear in the second column. The remaining columns give the series of $-$ and $+$ signs used to estimate the seven main effects.

Table 4. A Fractional Factorial Experiment to Screen Seven Factors in Eight Tests

| Test No. | Treatment | Factorial Effect | | | | | | |
|---|---|---|---|---|---|---|---|---|
| | | $A$ | $B$ | $C$ | $D$ | $E$ | $F$ | $G$ |
| 1 | (1) | − | − | − | − | − | − | − |
| 2 | abcd | + | + | + | + | − | − | − |
| 3 | abef | + | + | − | − | + | + | − |
| 4 | acfg | + | − | + | − | − | + | + |
| 5 | adeg | + | − | − | + | + | − | + |
| 6 | bceg | − | + | + | − | + | − | + |
| 7 | bdfg | − | + | − | + | − | + | + |
| 8 | cdef | − | − | + | + | + | + | − |

To illustrate the basic feature of the plan, consider the four combinations, abcd, abef, acfg and adeg, in which A is at the high level. Note that each of the other letters, b, c, d, e, f and g, appears twice in the four combinations. The same property holds for the four combinations in which a is absent. It follows that in the average effect of A, as estimated from

$$\left\{\frac{abcd + abef + acfg + adeg}{4}\right\} - \left\{\frac{bceg + bdfg + cdef + (1)}{4}\right\},$$

the effects of the six other factors cancel out. Thus an unbiased estimate of the average effect of A, and similarly of all other factors, is obtained with fourfold replication. If any interactions are present, some of these estimates are biased, this being the risk taken in order to obtain a drastic reduction in the size of the experiment.

We could, alternatively, obtain some information about the effects of all seven factors by testing the eight combinations (1), a, b, c, d, e, f and g. With this experiment, however, the estimates of the factorial effects are based on single replications and are positively correlated, since the response to the combination (1) appears with a negative sign in all the estimates $\{a - (1)\}$, $\{b - (1)\}$, etc. The advantages of the fractional replication plan in Table 4 are that we obtain estimates based on four replicates and free from this algebraic correlation.

A number of useful plans with fractional replication have been worked out for the $2^n$ series of factorials, (2), (3), (4), and have found numerous applications in industrial research. If interactions are negligible, the main effects of up to fifteen factors can be estimated in only sixteen tests. In general, the plans are appropriate for experimental situations with the following features.

(i) Experimental errors are high enough so that the internal replication provided by fractional factorials is needed.

(ii) Only two levels of each factor are necessary. This implies either that the factors are qualitative, or, if quantitative, that their main effects are expected to be approximately linear within the range covered.

(iii) Enough is known about the process so that most of the interactions can be assumed negligible or at least small relative to main effects.

These conditions may apply in a screening test in which only a few of the factors are expected to have important effects, but it is desired to include a substantial number of factors in case some factors thought to be minor should turn out to have major effects.

A good account of the uses of fractional factorials in a food research laboratory has been given by Carroll and Dykstra in their recent book, *Experimental Designs in Industry* (13).

With factors at more than two levels, the possibilities are limited. Practical designs have been developed for the $3^4$ factorial in $\frac{1}{3}$ replicate and for combinations like the $4 \times 2^4$, $4^2 \times 2^3$. Horton (14) describes uses of the $4 \times 2^4$, the $4 \times 4 \times 8 \times 8$ and the $3^6 \times 2 \times 2$, with a careful account of the initial planning that led to the adoption of these designs.

When experiments can be completed quickly, a sequence of fractional factorials may form an efficient strategy. The first experiment is small, intended to estimate main effects and perhaps a few two-factor interactions. It may be clear from the results that several factors can be dropped, having only minor effects. The second experiment is then designed to estimate the effects of the principal factors more precisely and to delve somewhat deeper into the interrelations among these effects. A third experiment might be a complete factorial involving the two or three most important factors. Alternatively, results of the first experiment may suggest that all factors should be retained, but that their interactions require more elucidation. This can be done by making the second experiment a further fraction of the same replication. Care is required in constructing the second experiment if the desired information is to be obtained. The problem is discussed by Davies and Hay (5), who first suggested this use of fractional factorials, and in references (2) and (3).

In summary, fractional factorials have become one of the principal research weapons in numerous industrial problems.

### Random Balance Designs

These designs, proposed by Satterthwaite (16), have the same general objective as fractional factorials. The difference is that the treatment combinations are constructed by taking for each factor an independent random selection of the levels of that factor which it has been decided to investigate. If three levels of the first factor are to be included, a series of random selections from the numbers 1,2,3 is made, for example 3,1,1,2,2,1. In the first treatment combination this factor appears at level 3, in the second at level 1, and so on. The restriction is usually made that over the whole experiment, levels 1, 2 and 3 will appear equally often. If the second factor has two levels, repeated random selections of the numbers 1, 2 are made to determine its levels in the successive treatment combinations.

Several advantages are claimed for this type of arrangement. The treatment combinations are easy to set up. There is complete flexibility as to numbers of factors, numbers of levels of each factor and number of combinations tested. In particular, designs can be written down for problems (12 factors at 2 levels and 5 factors at 3 levels in 24 tests) for which no fractional factorial design has been constructed. The experiment can be stopped and analyzed after any reasonable number of tests has been completed.

Random balance designs are, of course, another type of fractional factorial design, since they test only a fraction of the treatment combinations that constitute a complete factorial. The contrast between random balance designs and the older fractional factorials is that in the latter the treatment combinations are carefully selected so that those effects considered most important (usually the main effects of the factors) will be estimated independently of each other. With random balance, in its simplest form as described here, the tested combinations are a random selection from all combinations. For this reason, random balance designs are less efficient, in the sense that for a given size of experiment the principal effects are estimated with higher standard errors, and are harder to analyze and interpret clearly. Satterthwaite has stated (16) that he does not recommend random balance, except where its simplicity of construction is a preponderating consideration, for situations in which a balanced fractional factorial is already available. Critics have also objected that the inclusion of large numbers of factors in small experiments leads to biased and confusing results. An account of the pros and cons of random balance designs will be found in the papers and discussion following reference (16).

It seems likely that in practical applications the distinction between random balance and the older fractional factorial designs, and the controversy about their respective roles, will diminish. Proponents of random balance incorporate fractional factorials into their designs. For instance, if eight of the factors are to appear at two levels, this part of the design will be a balanced fraction of a $2^8$ factorial rather than a random selection of levels. The development of random balance designs has stimulated research on extending the balanced type of fractional factorial so as to produce new designs for problems that are beyond the present scope of these designs.

*Second-Order Designs*

When the factors are quantitative it is natural to think of the response $y$ as a mathematical function of the levels $x_1, x_2, \ldots, x_k$ of the factors. The purpose of the experiment is to investigate the nature of the response surface

$$y = \phi(x_1, x_2, \ldots, x_k).$$

Published research on this problem is confined thus far mainly to situations in which the function $\phi$ can be represented by a plane or by a quadratic function of the $x$'s, although work is in progress on other functions.

The $2^k$ factorial, or fractional factorial, is a convenient design for fitting a plane surface. For instance, with a $2^3$ factorial, the equation

$$y = b_1 x_1 + b_2 x_2 + b_3 x_3$$

may be fitted easily. If the experiment is replicated, or the experimenter has an external estimate of his experimental error, the remaining four degrees of freedom for treatments may be used to judge whether the planar fit is satisfactory within the limits of experimental error.

For fitting a quadratic surface the $3^k$ factorial is convenient, but requires a large experiment if $k$ exceeds 2. Box and Hunter (9) have developed a series of designs, called rotatable second-order designs, specifically for the fitting of quadratic response surfaces. Denote the lower level of any $x$ by $-1$ and the upper level by $+1$, so that the "center" of the design is the point with coordinates $(0, 0, 0, \ldots, 0)$. These designs consist of three parts:

(i) A $2^k$ factorial or fractional factorial.

(ii) A "star," with tests made at the combinations $(\alpha, 0, 0, \ldots, 0)$; $(-\alpha, 0, 0, \ldots, 0)$; $(0, \alpha, 0, 0, \ldots, 0)$; $(0, -\alpha, 0, 0, \ldots, 0)$; etc.

(iii) One or more tests at the center $(0, 0, \ldots, 0)$.

With suitable choices of $\alpha$ and of the number of points at the center, this design will estimate a quadratic response surface with a standard error that is practically constant at all points within a distance 1 from the center. De Baun and Schneider (15) discuss chemical applications of these designs.

The strategies to be described in subsequent sections were developed for the more restricted problem of finding the combination of factor levels that gives a maximum response, and of mapping the general nature of the response in the neighborhood of this optimum set of levels. Since all the strategies leave much scope for variations in the way in which they are applied, they cannot be presented as a series of automatic steps. Only the central idea will be outlined. These methods are probably primarily of use when there is a single response, or only one response of predominating interest. With several responses of equal interest, the fractional factorials and second-order designs mentioned previously may prove more suitable.

## The Single-Factor Method

This is the method already illustrated at the beginning of Part II. A good description is given by Friedman and Savage (7). We shall discuss the method for quantitative factors, although it applies also to qualitative factors. The experimenter first makes a guess at the optimum factor levels, say $x_{11}, x_{21}, \ldots, x_{k1}$. In the first experiment, several levels of the first factor (say 3 to 5) are compared, the levels of all other factors being kept fixed at the initial guesses $x_{i1}$. From the results of the first experiment, the level $x_{12}$ of the first factor that gives the highest response is estimated. The second experiment tests several levels of the second factor, the other factors being fixed at the levels $x_{12}, x_{31}, \ldots, x_{k1}$, and so on. At the end of the first round we have reached a new approximation $x_{12}, x_{22}, \ldots, x_{k2}$ to the optimum combination.

We proceed with a second round, and so on. Various modifications may be introduced. The ranges may be narrowed in subsequent rounds, or some factors may be dropped. More than one factor may be varied simultaneously, as in the steepest ascent method to be described later. The process terminates when the response seems to be capable of little or no further improvement.

The amount of experimentation will depend on (i) the number of factors, (ii) how close the initial guess is to the optimum, (iii) the amount of replication needed in each single-factor experiment, and (iv) the magnitudes of the interactions between the effects of the factors. The method works best when experimental errors are small, so that a single replication suffices in each experiment, and when interactions are absent, i.e., the effects of the factors are additive. If these two conditions hold, a point close to the optimum should be reached at the end of the first round. Interactions slow up progress, because the best value of $x_i$ when the other factors are held at their initial levels may be quite different from its best value when the other factors are held at new levels, so that several rounds are needed for the process to come near to the optimum.

At the beginning of Part II the single-factor method was criticized as being inefficient relative to a complete factorial. Some readers may wonder why it is now presented as a promising method. The reason is that the conditions of the comparison have changed. Suppose that there are three factors, and that five levels of each are considered necessary to locate the optimum level for a factor. A round of the single-factor method, using only one replication, takes 15 tests, whereas a complete replication of the $5^3$ factorial requires 125 tests. The complete factorial supplies much internal replication and enables us to investigate interactions if we want to. But if experimental errors are small the replication is not needed, and if interactions are negligible they will not weaken the performance of the single-factor method in locating the optimum quickly.

*The Method of Steepest Ascent*

This method, introduced by Box and Wilson (8), starts from an initial guess about the position of the optimum. In the subsequent experiments, all factors are varied simultaneously in the hope of proceeding more directly towards the optimum than with the single-factor method. The first experiment is conducted in the neighborhood of the initial guess, and may be a fraction of a $2^n$ factorial. Its purposes are (i) to fit a linear approximation to the response surface, i.e.,

$$\hat{Y} = Y_0 + b_1 x_1 + b_2 x_2 \cdots + b_k x_k, \tag{1}$$

where $\hat{Y}$ is the estimated response, and $x_i$ the level of the $i$th factor, and (ii) to provide data for a test of significance of the adequacy of the linear fit.

If the linear equation appears to fit adequately, a direction of steepest ascent is estimated from its results. Suppose that the origin of the $x$-space is taken as the initially guessed set of levels. Consider all points equidistant from this origin, i.e., lying on the sphere

$$x_1^2 + x_2^2 + \cdots + x_k^2 = r^2.$$

From equation (1) it is easy to show by calculus that the point on the sphere having the highest expected response has coordinates

$$x_i = r b_i / \sqrt{b_1^2 + b_2^2 + \cdots + b_k^2}.$$

Thus, the direction of steepest response is such that $x_i$ must be changed by an amount proportional to $b_i$.

Having chosen a value of $r$, the experimenter makes a new test at this point on the path of steepest ascent. He continues tests on this path so long as the observed improvement in response agrees with the improvement predicted from the linear equation. When the observed improvement begins to fall short, a new linear approximation is fitted, a new path of steepest ascent is computed, and further tests are made on this path.

This process continues until either (i) the linear approximation still holds but the $b_i$ are all small, suggesting that a plateau has been reached, or (ii) the linear equation no longer fits. In the latter event a design to which a second-degree equation can be fitted is used. For this purpose the rotatable second-order designs may be used. The second-degree equation may succeed in locating the position of the optimum, or, if the situation is more complex, analysis of its results suggests the direction of further experimentation.

The method is more sophisticated and complex in operation than the single-factor method. The experimenter's judgment must be exercised in choosing the range through which each factor is varied in the initial $2^k$ experiments. Although this may not be obvious at first sight, these choices have a marked effect on the direction which the calculations will indicate to be that of steepest ascent, and in which further tests are made. The sizes $r$ of subsequent jumps must also be chosen by judgment. Mistakes in these choices may, however, be corrected later in the process.

The method has a high degree of flexibility for coping with different shapes of response surface, and in making advances in response in the face of complex interactions which impede progress by the single-factor method. Advantage is taken also of the internal replication associated with factorial experiments: this should help when experimental errors are sizable.

### *Use of Random Balance Designs*

When the number of factors is large and little is known about the shape of the response surface, a random balance design may be a useful first step (6).

What can be done with the results of a series of such tests? The simplest procedure is to pick the winner, i.e., the test that has given the highest observed response, in the hope that the combination of levels used in this test will be somewhere near the optimum. If experimental errors are negligible, a simple and interesting result holds for the winner. Consider a subregion in the factor space made up of all points at which the true response exceeds some specified value. Then the probability that the winner lies in this subregion of high response is

$$P = 1 - (1 - f)^p$$

where $p$ is the number of random tests made and $f$ is the ratio of the volume of the subregion to the total volume of the factor space.

Table 5. Probability $P$ that the Winner Falls in a Region of High Response

|  | | Number of Tests | |
|---|---|---|---|
| $f$ | 10 | 20 | 30 |
| .05 | .40 | .64 | .87 |
| .10 | .65 | .88 | .99 |
| .20 | .89 | .99 | .99+ |

Table 5 shows that with 20 tests, the probability is 0.88 that the winner falls in the subregion containing the highest 10 per cent of all responses. With 30 tests, this probability jumps to 0.99, and the probability is 0.87 that the winner is in the subregion containing the highest 5 per cent of responses.

This result holds irrespective of the number of factors or of the shape of the response surface. It would apply, for instance, to 20 tests made on 28 or 34 factors. When there are many factors and experimental errors are negligible, the random method appears an inexpensive way of bringing us quickly into a subregion of high response. If the optimum is flat, any point in the best 5 per cent or 10 per cent subregion may give a response satisfactorily near the optimum.

At the end of the first set of tests, something more may be learned by studying the other combinations of levels that give responses close to that of the winner and by plotting the response against the levels of individual factors or small groups of factors. It may be possible to detect and eliminate factors whose levels have a minor influence on the response. A second set of tests can then be started on a smaller number of factors, with levels closer to that of the estimated optimum.

When experimental errors are present, the probabilities in Table 5 are decreased to an extent which will, of course, depend on the sizes of the errors but has not yet been investigated. Errors will also make it difficult to obtain clear-cut conclusions about the importance and nature of the effects of individual factors. Much further research on the utility of the method is needed.

## Comparisons Between the Methods

It seems certain that no single method is best under all conditions (10). The relative advantages will depend on the number of factors, the shape of the response surface, and the sizes of the experimental errors. When the errors are small, the random method looks promising as a starter if many factors (say more than 10) must be included, or if multiple peaks are likely, because the other methods may converge on a peak that is not the highest one. The single-factor method should do well when experimental errors are small and the effects of the factors are approximately additive. The steepest ascent method provides more internal replication and has been claimed to be particularly effective with the type of "rising ridge" surface encountered in chemical experiments, but becomes complex if the factors are numerous.

Comparisons under actual operating conditions will seldom be feasible. In fact, since each strategy allows flexibility and has numerous variants, it is difficult to define the strategies specifically enough to permit comparisons. What can be done is to construct different types of response surfaces mathematically and test the performance of some variants of each method on each surface. Brooks (11) compared several variants of the random, the single-factor, the steepest ascent, and the complete factorial methods on four different surfaces in two variables (i.e., with two factors). Setting each strategy the same task, and taking the most successful variant of each, he found the steepest ascent method best on all four surfaces. The single-factor was practically as good on three of the surfaces but somewhat poorer on a fourth. The complete factorial usually placed third and the random method last. The methods fell in the same order whether experimental error was introduced or not.

Analogues of response surfaces in five variables have been constructed by means of electrical resistance networks (12). These are housed in little black boxes. One contains five dials on which the level of each factor is set. A second box, connected to the first, has a dial by which an experimental error may be added to the true response, and a voltmeter on which the observed response is read.

In the report given by McArthur and Heigl (12), the boxes were used to discover how research teams attack such problems rather than to compare specifically defined strategies. (The authors point out the difficulty already mentioned of defining strategies in sufficient detail for a direct comparison.) The element of cost was introduced into the problem by charging a certain sum for each test and crediting the method with another amount for each 1 per cent gain in the true response.

The two methods used most frequently by the teams were the single-factor and the $\frac{1}{4}$ replicate of a $2^5$ factorial, followed perhaps by further tests. The fractional factorial did better, on the whole, on two of the three surfaces tried, the two strategies being about equally effective in the third, although more tests would be needed for definitive conclusions. Not enough use was made of the steepest ascent or the random methods to permit comparisons.

The results also furnish interesting suggestions about common mistakes in research strategy. These include lack of boldness, staying in a rut, being deceived by experimental errors, failing to eliminate unimportant variables, and not knowing when to stop. As the authors point out, however, no good positive rule about when to stop has emerged as yet from experience with the boxes.

In conclusion, it must be emphasized that knowledge in this area is rudimentary. Statements made in this chapter about the strengths and weaknesses of different strategies should be regarded as highly tentative. In view of the interest which the problem has aroused, there is reason to anticipate a useful volume of soundly based research in the near future.

## REFERENCES

(1) McGehee, W., and Gardner, J. E. "Music in a Complex Industrial Job," *Personnel Psychology,* Vol. 2 (1949), 405-17.

(2) Davies, O. L. (ed.). *Design and Analysis of Industrial Experiments.* New York: Stechert-Hafner Co., 1956, 2nd edition.

(3) Cochran, W. G., and Cox, G. M. *Experimental Designs.* New York: John Wiley and Sons, 1957, 2nd edition.

(4) "Fractional Factorial Experiment Designs for Factors at Two Levels," *Applied Mathematics Series, 48,* National Bureau of Standards, 1957.

(5) Davies, O. L., and Hay, W. A. "The Construction and Uses of Fractional Factorial Designs in Industrial Research," *Biometrics,* Vol. 6 (1950), 233-49.

(6) Anderson, R. L. "Recent Advances in Finding Best Operating Conditions," *Journal of the American Statistical Association,* Vol. 48 (1953), 789-98.

(7) Friedman, M., and Savage, L. J. "Planning Experiments Seeking Maxima," *Techniques of Statistical Analysis.* New York: McGraw-Hill, 1947.

(8) Box, G. E. P., and Wilson, K. W. "On the Experimental Attainment of Optimum Conditions," *Journal of the Royal Statistical Society,* Series B, Vol. 13 (1951), 1-45.

(9) Box, G. E. P., and Hunter, J. S. "Multifactor Experimental Designs for Exploring Response Surfaces," *Annals of Mathematical Statistics,* Vol. 28 (1957), 198-240.

(10) Box, G. E. P. "Integration of Techniques in Process Development," *Transactions of the 11th Convention of the American Society for Quality Control* (1957), pp. 687-702.

(11) Brooks, S. "Comparisons of Methods for Estimating the Optimal Factor Combination." Unpublished Sc.D. thesis, The Johns Hopkins University, 1955.

(12) McArthur, D. S., and Heigl, J. J. "Strategy in Research," *Transactions of the 11th Convention of the American Society for Quality Control* (1957), pp. 1-18.

(13) Carroll, M. B., and Dykstra, C., Jr. "The Application of Fractional Factorials in a Food Research Laboratory," *Experimental Designs in Industry.* New York: John Wiley and Sons, 1958.

(14) Horton, W. H. "Experiences with Fractional Factorials," *Experimental Designs in Industry*. New York: John Wiley and Sons, 1958.

(15) De Baun, R. M., and Schneider, A. M. "Experiences with Response Surface Designs," *Experimental Designs in Industry*. New York: John Wiley and Sons, 1958.

(16) Satterthwaite, F. E. "Random Balance Experimentation." *Technometrics*, Vol. I (1959), 111-193.

W. G. *Cochran*

## Designing Clinical Trials

THE planning and conduct of a clinical trial does not involve any difficult or esoteric intellectual principles. It is mainly a matter of hard work and attention to detail. I shall describe some of the steps that must be gone through in setting up a trial. If a planning group finds itself unable to complete or agree upon one or more of these steps, this is probably a sign that they are not ready to conduct a trial.

First, as Mr. Jablon mentioned, rules must be constructed to tell what kind of patient will be accepted into the trial. These are, in effect, a diagnostic definition of the disease entity for the purposes of this trial.

The definition should be written down, and should be thoroughly understood and agreed upon by all who have major responsibility for the trial. When the symptoms differ widely from patient to patient, the definition may require considerable work. For instance, it may be necessary to define major and minor manifestations, and to require that an accepted patient must exhibit, say, at least two major and at least three minor manifestations.

Numerous special problems may arise. A well-constructed definition may tend to exclude mild cases, whose symptoms are not definite enough to meet the criteria at the time when they are considered for admission. If it is desirable to include mild cases in the trial, the definition must be appropriately chosen, even though this means that some of the accepted cases will later be rejected from the study because a final diagnosis shows that they do not have the disease in question.

If the disease is chronic, some potential patients may already have been receiving one or more of the therapies to be tested. Should this be taken into account in deciding whether to accept them? If the disease has a wide range of severity, with numerous different manifestations, should the definition be constructed so as to accept all kinds of patients, from the mildest to the most severe, or should we attempt to work in some narrower range? There are pros and cons to both questions.

Mr. Jablon stressed that clinical trials are comparative. Their purpose is to compare the effects of two or more different ways of treating patients. The nature of each therapy must be agreed upon and written down in detail, including amount, frequency, and duration. The problem is simplest when it is reasonable to make each therapy exactly the same for all patients subjected to it. But sometimes good practice indicates that the amount or frequency should be varied from patient to patient, being the maximum amount tolerable or being dependent on his rate of progress. There is nothing logically objectionable in defining a therapy so as to allow variation from patient to patient. However, the definition requires extra care, and it is important to report clearly exactly how the therapy was applied, otherwise users of the results of the trial may be misled as to what the therapy actually turned out to be.

The next question is, how many patients do we need? Statistical theory helps at this point, although unfortunately it does not provide an automatic answer to the question. The type of information that theory supplies can be illustrated from a simple situation. Suppose that at the end of the trial in a chronic disease the patients will be rated as "improved" or "not improved." The current standard therapy gives improvement in about 60 per cent of patients and a new therapy has been produced which holds some promise of giving a higher percentage of improvement. Let us ask: what can we expect to learn from a trial with two hundred patients, one hundred on each therapy?

At the end of the trial we make a test of significance of the difference in percentage of improvement under the new and current therapy. Unless the new therapy proves significantly superior, in the statistical sense, I shall suppose that we will drop it from consideration (this is an appropiate decision in many though not in all situations). We do not mind dropping a new therapy that in the long run would be only slightly superior to the current one, but we do not want to drop one that would be greatly superior.

Statistical theory tells us that if the new therapy shows 80 per cent improvement, in the long run, the chance of finding it significantly superior in a trial with two hundred patients is better than 9 out of 10. If its success rate is 75 per cent, or 15 per cent better than the standard, the chance of a significant result is about 3 out of 4; while if its long-run improvement rate is 70 per cent, the chance of a significant result is less than 1 out of 2. In other words, if an increase in the percentage improved from 60 per cent to 70 per cent represents an important medical advance, or is about as much as the new therapy is likely to achieve, two hundred patients are far from sufficient, since the trial is more likely than not to result in the conclusion "no significant difference." On the other hand, if we are interested only in a major improvement from 60 per cent to 80 per cent or better, two hundred patients are ample.

Calculations of this type are useful in giving some idea of the comparative power of the trial. Sometimes they settle the issue, either by revealing that there is no hope of getting the required number of patients, or by showing that the trial is well within our resources. Often, as the example illustrates, they leave us with a judgment decision that must be to some extent a gamble.

Except with some chronic diseases that are institutionalized, relatively few patients are usually available at the start of the trial. They will be obtained as they later enter the hospital or clinic for treatment. Consequently a calculation of number of patients needed is usually translated into the time and number of institutions that will be required to assemble the requisite number. In this connection, initial estimates of the number of suitable patients who will enter an institution during the next year are often overoptimistic, and it is well to be conservative when making this estimate.

It may be clear at this point that there is no hope of completing the trial in a reasonable time in a single institution. If the job is to be done, the coöperation of several institutions must be enlisted. Time does not permit me to discuss the special aspects of coöperative trials, except to say that they involve numerous problems in administration and personal relations, particularly in securing a uniformly high quality of work. They are not to be undertaken lightly.

One aspect of the "sample size" problem may cause confusion. Sometimes the patients are a rather heterogeneous group, and we suspect that the therapy which will be best for one subgroup will not be best for another. It may be thought that this causes no difficulty, because in the analysis and presentation of results the two therapies can be compared separately for each subgroup. If this is to be done, however, the sample size in each subgroup must be large enough so that the comparisons in the individual subgroups will be of satisfactory precision. For a breakdown into two subgroups, the total sample size must usually be at least twice as large as if only a single over-all comparison were being made. If the *differential* effect of the therapies in the two subgroups is important, the increase needed may be substantially greater than this.

We are now ready to allocate therapies to patients.

In experimental design in other fields, ingenious methods have been developed for doing this and found useful, but in clinical trials we are usually content with the simplest procedures. One guiding rule is that the method of allocation should help to preserve blindness in the sense in which Mr. Jablon used this term. The decision as to whether a patient enters a trial or not should always be made before it is known, or is possible to guess, what therapy the patient will receive if he enters. The safest way to ensure this is to allocate patients by means of a list of random numbers that are not available to the person who makes the decision about entering. One slight disadvantage of strict randomization is that sometimes it results in rather an uneven division of patients, for example, 25 on therapy $A$ as against 15 on therapy $B$ out of 40. Simple modifications can be introduced into the randomization which guarantee a more even division without making it easy to guess what the next patient will receive.

Matching has been little used in clinical trials, probably because it seems to involve more administrative trouble than it is likely to be worth. It is feasible with a chronic disease in which most of the patients are available in an institution at the start of the trial, and where numerous measurements have already been made on them as to severity, duration of disease, and prognosis. In this circumstance matched groups may give a more precise comparison than simple randomization by ensuring that each therapy has equal numbers of patients with good, moderate, doubtful, and poor prognosis. Much of the benefit of matching is likely to be lost, however, if a substantial proportion of the patients will have to be dropped from the final analysis of results.

With some chronic diseases, for example recurrent headaches, each patient can receive all the therapies in turn, thus providing an intra-patient comparison. In some cases this arrangement is much more precise than an inter-patient comparison, particularly when the measurement of success depends primarily on subjective reports by the patient. A second potential advantage, unfortunately seldom practicable at present, is that it may be possible to give the patient each therapy a sufficient number of times so that he represents a self-contained experiment. With such data, we are able to examine whether therapy $A$ is superior for all patients and, if not, to discover with reasonable accuracy those patients for whom $A$ is better, those who respond better to $B$, and those who are indifferent.

Finally, the measurements to be used in describing the effects of the therapies must be specified. This is sometimes the principal hurdle. There is naturally a preference for measurements that are objective, since they are likely to be more precise and less vulnerable to personal biases. However, an objective measurement that is not relevant should not be preferred to a subjective one that is. With subjective measurements, the principle of blindness becomes essential.

In conclusion, I have said nothing about the record-keeping side of a trial, but I hope I have brought to your attention some of the principal decisions that must be faced.

# 71

# SOME CLASSIFICATION PROBLEMS
# WITH MULTIVARIATE QUALITATIVE DATA[1]

WILLIAM G. COCHRAN

*Department of Statistics, Harvard University,*
*Cambridge, Massachusetts, U.S.A.*

AND

CARL E. HOPKINS

*Department of Biostatistics, University of Oregon Medical School,*
*Portland, Oregon, U.S.A.*

## 1. INTRODUCTION

Since 1935, when Fisher's discriminant function appeared in the literature, methods for classifying specimens into one of a set of universes, given a series of measurements made on each specimen, have been extensively developed for the case in which the measurements are continuous variates. This paper considers some aspects of the classification problem when the data are qualitative, each measurement taking only a finite (and usually small) number of distinct values, which we shall call *states*. Our interest in the problem arose from discussions about the possible use of discriminant analysis in medical diagnosis. Some diagnostic measurements, particularly those from laboratory tests, give results of the form: $-$, $+$ (2 states); or $-$, doubtful, $+$ (3 states); or (with a liquid), clear, milky, brownish, dark (4 states).

With qualitative data of this type an optimum rule for classification can be obtained as a particular case of the general rule (Rao, [1952], Anderson, [1958]). The rule is exceedingly simple to apply (Section 2). In practice, qualititative data are frequently ordered, as with $-$, doubtful, $+$. The classification rule discussed in this paper takes no explicit advantage of the ordering, as might be done, for instance, by assigning scores to the different states so as to produce quasi-continuous data. The best method of handling ordered qualitative data is a subject worth future investigation.

---

[1]This work was assisted by a contract between the Office of Naval Research, Navy Department, and the Department of Statistics, Harvard University by a Special Research Fellowship of the National Institutes of Health, and by a gift from the Lilly Research Laboratories to the Department of Biostatistics, University of Oregon Medical School. Reproduction in whole or in part is permitted for any purpose of the United States Government.

---

This paper seeks answers to three problems that arise in the use of the proposed rule.

(1) *The effect of the initial sample sizes on the performance of the proposed classification rule.*

In order to construct a rule, we must have preliminary data on some specimens known to be classified correctly, since the rule depends on the joint frequency distributions of the measurements within each universe. Standard classification theory assumes that these distributions are known exactly, although in practice it is sometimes difficult to obtain adequate samples from which to estimate them. The consequences of constructing a rule from preliminary samples of finite sizes are discussed in Sections 3, 4, and 5.

(2) *The relative discriminating power of qualitative and continuous variates.*

Classification is simpler with qualitative than with continuous measurements. For this reason, if a few of the available measurements are continuous while the rest are qualitative, we may be inclined to transform each continuous measurement into a qualitative one by partitioning its frequency distribution, provided that this does not result in too much loss of discriminating power. This consideration led us to investigate the questions: if a normally distributed variate is transformed into a qualitative one with $s$ states, how much discriminating power is lost, and what are the best points of partition of the curve from the point of view of retaining maximum discriminating power (Sections 6 and 7)?

(3) *Use of classification experience for improvement of the rule.*

The optimum rule depends on the relative frequencies $\pi_1$, $\pi_2$, $\pi_3$ etc. with which specimens from the different universes present themselves for classification. The initial estimates of these frequencies, made from previous data or by judgement, may be biased. After a number of specimens have been classified by the rule, it is possible to re-estimate the frequencies $\pi_u$ from the data for these specimens, so that the classification rule may be improved (Section 8).

## 2. AN OPTIMUM RULE FOR CLASSIFICATION

Ideally, the setting up of an optimum rule demands three different kinds of preliminary information. Firstly, we require the joint frequency distribution of the measurements in each universe under consideration. With continuous variates the most common assumption about these distributions is that they are multivariate normal with the same dispersion matrix in each universe. With qualitative variates, suppose that the $j$th measurement takes $s_j$ distinct states. When this measurement is made on a specimen, we learn into which state it falls. A series of qualitative measurements, e.g. one with 2 states, one with 3 states and one with 4 states, defines 24 cells or multivariate states. For any specimen, these three measurements tell us which of the 24 states the specimen occupies. Thus, in general, a set of $k$ qualitative measurements classifies the specimen into one out of a number $S = s_1 s_2 \cdots s_k$ multivariate states. Consequently the joint frequency distribution of the measurements is completely specified for the $u$th universe if we know the probabilities $p_{ui}$ that the specimen falls into the $i$th multivariate state where, for each $u$,

$$\sum_{i=1}^{S} p_{ui} = 1.$$

Secondly, practical use of a classification rule is likely to result in some mistakes in classification. In some applications certain types of mistakes are more serious than others. If the relative costs of different kinds of mistakes can be estimated, the rule should take them into account. Let $c_{vu}$ be the cost incurred when a specimen which actually belongs to universe $V$ is classified as belonging to universe $U$.

Thirdly, when the rule is put into use, the specimens to be classified may not come with equal frequency from the universes. For instance, in a diagnostic screening test, most of the subjects who present themselves may be free from the disease in question. Let $\pi_u$ denote the relative frequency with which specimens come from the $u$th universe, where

$$\sum_u \pi_u = 1.$$

As might be expected, the optimum rule depends on these frequencies.

With this background we now construct an optimum rule. For a specimen which falls into the $i$th multivariate state, we select the universe to which it is to be assigned by minimizing the expected cost of mistakes in classification. Suppose that the specimen is assigned to the $u$th universe. For specimens that actually come from this universe, no mistake is made. Specimens from the $v$th universe will present themselves with relative frequency $\pi_v$, and will fall into the $i$th state with relative frequency $\pi_v p_{vi}$. The expected cost of misclassifying these specimens into the $u$th universe is therefore $\pi_v p_{vi} c_{vu}$. Hence the total expected cost of mistakes in classification is

$$\sum_{v \neq u} \pi_v p_{vi} c_{vu} \, .$$

To apply the optimum rule we compute this quantity for every $u$ and assign specimens to the universe for which the quantity is a minimum.

There are several particular cases of the rule. If $c_{vu} = c_v$, i.e. the cost of misclassifying depends only on the universe from which the specimen comes and not on that into which it is misclassified, the total expected cost over all $m$ universes is

$$\sum_{v \neq u} \pi_v p_{vi} c_v = \sum_{v=1}^{m} \pi_v p_{vi} c_v - \pi_u p_{ui} c_u \, .$$

Since the first term on the right does not depend on $u$, specimens falling into the $i$th state are assigned to the universe for which the triple product $\pi_u p_{ui} c_u$ is greatest.

If the relative costs $c_u$ are taken as equal, the rule minimizes the expected frequency of mistakes in classification. If, further, the relative frequencies $\pi_u$ are also equal, any specimen found in the $i$th state is assigned to the universe $u$ for which the conditional probability $p_{ui}$ is greatest.

For a numerical example we are indebted to Dr. Leslie Kish. The data come from a large study of voting behavior conducted by the Survey Research Center, University of Michigan. (Stokes *et. al.* [1958]). By open-end questionnaires taken during the 1952 and 1956 elections, voters were rated as Democrats $(D)$, Independents $(I)$ or Republicans $(R)$ and also as Unfavorable $(U)$, Neutral $(N)$ or Strongly Favorable $(F)$ to Eisenhower's personality. These are the two predictor measurements, each with 3 states, making 9 bivariate states. The voters were also asked whether they voted for Stevenson or Eisenhower, these being the two universes. The sample sizes were 1003 and 1439.

For illustrative purposes we have set up a rule classifying the subjects as Stevenson or Eisenhower voters by means of the predictor variates. The results of the classification are then compared with the actual voting behavior.

Table 1 shows the relative frequencies $p_{1i}$ and $p_{2i}$ of the $i$th state for Stevenson and Eisenhower voters respectively. We assumed that $c_1 = c_2$ and that $\pi_1 = \pi_2 = \frac{1}{2}$. The resulting classification rule is given on the right in Table 1. Persons falling into the states $DU$, $DN$, $DF$ and $IU$ are classified by the rule as Stevenson voters, since in these states $p_{1i} > p_{2i}$ .

The estimated probabilities of misclassification are easily obtained. Any Eisenhower voter is misclassified if he falls into the states $DU$, $DN$,

*DF* or *IU*. Hence the frequency of misclassification for Eisenhower voters is

$$.033 + .079 + .090 + .017 = .219.$$

The figure for Stevenson voters is 0.133 and the average is 0.176.

As with continuous measurements, it is possible to examine whether a measurement contributes to the classification to a worthwhile extent. If political affiliation alone (*D, I, R*) is recorded, the probabilities in the three states are given in the lower part of Table 1, being obtained of course by addition from the upper part of the table. The probabilities of misclassification are now 0.201 for Stevenson voters and 0.202 for Eisenhower voters. Thus the addition of the attitude measurement reduces the average number of mistakes by about 13 percent.

The use of qualitative variates in attempting to date certain of the

TABLE 1

ILLUSTRATION OF THE CLASSIFICATION RULE FOR TWO MEASUREMENTS

| State (i) | $p_{1i}$ Stevenson | $p_{2i}$ Eisenhower | Resulting classification rule |
|---|---|---|---|
| DU | .347 | .033 | S |
| DN | .359 | .079 | S |
| DF | .094 | .090 | S |
| IU | .068 | .017 | S |
| IN | .078 | .107 | E |
| IF | .026 | .154 | E |
| RU | .002 | .017 | E |
| RN | .023 | .176 | E |
| RF | .004 | .327 | E |
| | 1.001 | 1.000 | |

PROBABILITIES USING POLITICAL AFFILIATION ALONE

| State | Stevenson | Eisenhower | Classification |
|---|---|---|---|
| D | .800 | .202 | S |
| I | .172 | .278 | E |
| R | .029 | .520 | E |

works of Plato (i.e. to arrange a number of universes in a time sequence) has been discussed by Cox and Brandwood [1959].

In the rest of this paper discussion will be confined unless otherwise mentioned to the case of two universes, with $c_1 = c_2 = 1$ and $\pi_1 = \pi_2 = \frac{1}{2}$.

### 3. EFFECTS OF THE FINITE SIZES OF THE INITIAL SAMPLES

The initial sample from each universe serves two purposes. It is used to set up the classification rule and to estimate the probabilities of misclassification for specimens from the two universes, so that we can decide whether the classification is accurate enough to be satisfactory.

Some notation will be needed. To avoid double subscripts, let $U$, $U'$ denote the two universes, and $p_i$, $p_i'$ the true probabilities that a specimen will fall into the $i$th multivariate state. Independent samples of sizes $n$, $n'$ are drawn from the universes. The numbers of specimens falling into the $i$th state are $r_i$, $r_i'$, and the corresponding estimated probabilities are $\hat{p}_i = r_i/n$, $\hat{p}_i' = r_i'/n'$. Since the true $p_i$, $p_i'$ are unknown, the actual classification rule places a specimen in $U$ if $\hat{p}_i > \hat{p}_i'$, in $U'$ if $\hat{p}_i < \hat{p}_i'$. If $\hat{p}_i = \hat{p}_i'$, the decision is made by tossing a fair coin.

For given sample sizes $n$, $n'$ the values of $\hat{p}_i$ and $\hat{p}_i'$ will vary from sample to sample. Thus the actual classification rule and its performance may change from sample to sample. In describing the consequences of finite sample sizes, we usually present the average results over all initial samples of given sizes $n$, $n'$.

The principal consequences of the finite sample sizes are as follows.

1. The probability of misclassification as estimated from the samples is, of course, subject to a sampling error that in moderately large samples may be shown to be approximately of the binomial type.

2. On the average, taken over all pairs of samples from a given pair of universes, the actual probability of misclassification is *greater* than the theoretical optimum: i.e. than the probability that would hold if the true $p_i$, $p_i'$ were used to construct the classification rule.

3. The average estimated probability of misclassification is *less* than the theoretical optimum probability. From 2, it is also less than the average actual probability.

In other words, finite sample sizes involve two types of penalty. Owing to sampling errors in the $\hat{p}_i$, $\hat{p}_i'$, the rule obtained may not be the theoretical optimum, and the probability of misclassification is underestimated from the sample data.

These results may be illustrated, in exaggerated form, by considering samples of size $n = 1$ from universes having two states with the following true probabilities of falling into the states. With the optimum rule the probability of misclassification is 0.1 for specimens from $U$, 0.05 for specimens from $U'$, the average being 0.075.

| State | $U$ | $U'$ | Optimum classification |
|-------|-----|------|------------------------|
| 1 | 0.9 | 0.05 | $U$ |
| 2 | 0.1 | 0.95 | $U'$ |

With $n = 1$, only four types of sample result are possible. Table 2 shows for each type the classification rule that would be set up, the estimated probability of misclassification, and the actual probability that would pertain if that rule were used. A ? means that classification is a toss-up.

To explain the entries in Table 2, consider the last type of result. The wrong classification is made in both states. Nevertheless, if we

TABLE 2

THE CLASSIFICATION RULE FOR SAMPLES OF SIZE 1

| Sample result | | | | Frequency of occurrence | Probability of misclassification Estimated | Actual |
|---|---|---|---|---|---|---|
| State | $U$ | $U'$ | Rule | | | |
| 1 | 1 | 0 | $U$ | $(0.9)(0.95)$ | | |
| 2 | 0 | 1 | $U'$ | $=0.855$ | 0.0 | 0.075 |
| 1 | 1 | 1 | ? | $(0.9)(0.05)$ | | |
| 2 | 0 | 0 | ? | $= 0.045$ | 0.5 | 0.500 |
| 1 | 0 | 0 | ? | $(0.1)(0.95)$ | | |
| 2 | 1 | 1 | ? | $= 0.095$ | 0.5 | 0.500 |
| 1 | 0 | 1 | $U'$ | $(0.1)(0.05)$ | | |
| 2 | 1 | 0 | $U$ | $= 0.005$ | 0.0 | 0.925 |
| | | | | Average | 0.07 | 0.13875 |

believe the samples, we estimate a zero probability of misclassification. The actual probabilities are 0.9 for $U$ and 0.95 for $U'$, giving an average of 0.925.

The overall average actual probability of misclassification is 0.13875 as compared with a theoretical optimum of 0.075 and an average estimated value of 0.07.

The fact that the average actual probability exceeds the theoretical optimum arises, of course, because when $p_i > p'_i$, there will be some pairs of samples in which $\hat{p}_i \leq \hat{p}'_i$. Whenever this occurs, the actual rule for this state is not as good as the optimum.

The fact that the average estimated probability lies below the theoretical optimum probability may be shown as follows. If $p_i > p'_i$, the contribution of this state to the overall theoretical probability of misclassification is $\frac{1}{2}p'_i$. Since $r'_i/n'$ is an unbiased estimate of $p'_i$, the contribution may be written $\frac{1}{2}E(r'_i/n')$, where $E$ denotes a mean value. The estimated contribution, on the other hand, is $\frac{1}{2}$ the average of the *smaller* of $r_i/n$, $r'_i/n'$, which will always be less than $\frac{1}{2}E(r'_i/n')$.

The bias in the estimated probability and the inferiority of the actual to the optimum rule are due primarily to states in which $p_i$ differs little from $p'_i$. These difficulties can be largely avoided by withholding a decision, i.e. making no classification, in such states. For instance, suppose that preliminary samples of size 64 give the following results for a variate with four states.

|  | No. of specimens | |
|---|---|---|
| State | $U$ | $U'$ |
| 1 | 51 | 1 |
| 2 | 8 | 5 |
| 3 | 3 | 5 |
| 4 | 2 | 53 |
|  | 64 | 64 |

States 1 and 4 provide good discrimination, but in states 2 and 3 the estimated probabilities of a correct classification are only about 0.6. Consider a recommended rule to withhold judgement when a specimen occurs in state 2 or 3. From the samples, we have an unbiased estimate 21/128, or 16 percent, of the proportion of specimens for which the rule makes no classification. This estimate is binomial, with standard error $\sqrt{(0.16)(0.84)/128}$. When a decision is made, the probability of misclassification is estimated as 3/107, or about 3 percent. This estimate will be almost exactly a binomial variate, with negligible bias.

If, on the contrary, a decision is made in all states, we would like to obtain from the initial samples a reasonably unbiased estimate of the probability of misclassification for the actual rule developed from the samples. An estimate of the difference between the actual and the theoretical optimum probability of misclassification is also of interest, particularly when it is not too costly to increase the size of the initial samples. These topics are discussed in Sections 4 and 5.

## 4. ESTIMATION OF THE THEORETICAL OPTIMUM PROBABILITY

As shown in Section 3, the sample probability of misclassification is a negatively biased estimate of the theoretical optimum probability. For a given value of $(p_i + p_i')$, it appears intuitively that the underestimation will be greatest when $p_i = p_i'$. In this case the bias in the $i$th state is in absolute value,

$$\tfrac{1}{2}p_i - \{\tfrac{1}{2}E \text{ min. } (\hat{p}_i , \hat{p}_i')\}$$

where $\hat{p}_i$, $\hat{p}_i'$ are independent binomial estimates of $p_i$. An algebraic expression for this quantity as a function of $p_i$, can be developed from the binomial distribution, but is unwieldy unless $n, n'$ are small. An approximation to the bias may be obtained by regarding $\hat{p}_i$, $\hat{p}_i'$ as normally distributed. Since the average value of the smaller of two normal deviates is known to be $-0.56$, the bias in the estimate of $p_i$ when both samples are of size $n$ is approximately

$$-0.56 \; \sqrt{p_i q_i / n}.$$

With $p_i = p_i' = 0.05, n = 50$, for instance, this estimate gives $-0.0173$ as against the correct value of $-0.0170$.

Since this problem will occur mainly in states with low frequencies (unless the classification rule is poor), a relatively crude sample adjustment for bias should be adequate for estimating the overall theoretical optimum probability of misclassification. One suggestion is to use $(\hat{p}_i + \hat{p}_i')/4$ instead of one-half the smaller of $\hat{p}_i$, $\hat{p}_i'$ when estimating the probability of misclassification, this adjustment being made in states in which $\hat{p}_i$, $\hat{p}_i'$ do not differ significantly at, say, the 20 percent level of significance. In the numerical example above, the adjustment is made in states 2 and 3. It amounts to estimating the theoretical optimum probability as

$$\frac{1 + 6.5 + 4 + 2}{128} = \frac{13.5}{128} = 10.5\%$$

as compared with the unadjusted estimate of 11/128 or 8.6 percent.

Examination of individual cases indicates that this adjustment removes most of the bias in states in which $p_i = p_i'$ . When $p_i \neq p_i'$ the adjustment tends to over-correct, and on the whole is likely to be conservative.

### 5. THE AVERAGE INCREASE IN THE ACTUAL PROBABILITY OF MISCLASSIFICATION

The average increase in the actual over the theoretical probability of misclassification may be expressed as a function of the $p_i$ , $p_i'$ and the sample sizes $n$ (assumed equal). In the $i$th state, if $p_i > p_i'$ , the optimum rule gives no misclassification for specimens from $U$, but misclassifies the proportion $p_i'$ of specimens from $U'$ that are in this state. The average frequency of misclassification is therefore $\frac{1}{2}p_i'$ . With the samples, there is no increase in this frequency provided that $r_i > r_i'$ . If $r_i = r_i'$ , we toss a coin. This gives an increase in frequency of

$$\tfrac{1}{4}(p_i + p_i') - \tfrac{1}{2}p_i' = \tfrac{1}{4}(p_i - p_i').$$

If $r_i < r_i'$ , we make the wrong decision, with an increase in frequency of mistakes of

$$\tfrac{1}{2}p_i - \tfrac{1}{2}p_i' = \tfrac{1}{2}(p_i - p_i').$$

Hence, over all states for which $p_i > p_i'$ , the total expected increase in frequency of misclassification may be written

$$\tfrac{1}{2} \sum (p_i - p_i')\{\text{Pr.} (r_i < r_i') + \tfrac{1}{2} \text{Pr.} (r_i = r_i')\}, \qquad (5.1)$$

with an analogous expression for the states in which $p_i < p_i'$ .

The probability inside the curly brackets may be obtained from tables of the binomial distribution. The normal approximation to this quantity is surprisingly good, even for moderate samples and small $p_i$ , $p_i'$ . With this approximation, the estimate is the probability that a normal deviate is less than

$$-\frac{\sqrt{n} \mid p_i - p_i' \mid}{\sqrt{p_i q_i + p_i' q_i'}} \qquad (5.2)$$

No correction for continuity is applied, because of the presence of the term $\frac{1}{2}Pr. (r_i = r_i')$ in the exact expression. As an example of the accuracy, with $p_i = 0.03$, $p_i' = 0.01$, $n = 100$, the normal approximation gives 0.1555 as compared with the exact value 0.1567.

Use of the normal approximation is illustrated in Table 3. From the $\hat{p}_i$ , $\hat{p}_i'$ for each state the normal deviate is computed from (5.2) and the probability is read from the normal tables. The difference

TABLE 3

ESTIMATION OF ACTUAL MINUS OPTIMUM PROBABILITY
OF MISCLASSIFICATION

| State | No. of specimens | | N.D. | Probability | $\|\hat{p}_i - \hat{p}'_i\|$ |
| | $U$ | $U'$ | | | |
|---|---|---|---|---|---|
| 1 | 51 | 1 | Large | 0.0 | — |
| 2 | 8 | 5 | −0.880 | 0.189 | 0.0469 |
| 3 | 3 | 5 | −0.732 | 0.232 | 0.0312 |
| 4 | 2 | 53 | Large | 0.0 | — |
| Total | 64 | 64 | | | |

between the average actual and the theoretical probability of misclassification is estimated as

$$\tfrac{1}{2}\{(0.189)(0.0469) + (0.232)(0.0312)\} = 0.0080 = 0.8\%.$$

At the end of Section 4 the theoretical probability of misclassification
was estimated as 10.5 percent. Adding 0.8 percent, we estimate the
expected actual probability as 11.3 percent.

In order to obtain a general impression of the effect of sample size,
some study was made of expression (5.1) for the average increase of
the actual over the theoretical optimum probability, as a function of
$p_i$, $p'_i$, $n$, and $S$, the number of states. The difficulty in studying this
function lies in the substantial number of parameters involved when
the states are numerous. A preliminary search was made for a smaller
number of parameters which might dominate the performance of the
function. It became evident that the function depends on the overall
optimum probabilities of misclassification, i.e. on the quantities

$$P = \sum_{p_i < p'_i} p_i, \qquad P' = \sum_{p'_i < p_i} p'_i.$$

As $P$, $P'$ diminish, i.e. as the optimum rule becomes better, the average
increase also diminishes, i.e. the actual rule is less likely to differ from
the theoretical optimum. Even with the quantities $P$, $P'$, $n$ and $S$ fixed,
the average increase may vary from almost zero up to a maximum,
depending on the way in which $p_i$, $p'_i$ distribute themselves among the
states. We decided to tabulate the maximum increase for given $P$,
$P'$, $n$ and $S$, although aware that this might produce an unduly pessimistic view of the effect of finite sample size. The worst possible
distributions of the $p_i$, $p'_i$ were found by trial and error. In most
situations, though not in all, the worst case occurs when the probability

$P'$ is distributed equally among all but one of the states for which $p_i > p'_i$. As an example, for 8 states, with $P = P' = 0.1$, the worst distribution in the first four states is as follows.

| $p_i$ | $p'_i$ |
|---|---|
| .69 | .0 |
| .07 | .0333 |
| .07 | .0333 |
| .07 | .0333 |

Table 4 shows the maximum average increases for $n = 50, 100$; $S = 4, 8, 16, 32; P = P' = 0.05. 0.1, 0.2, 0.3, 0.4$.

TABLE 4

MAXIMUM AVERAGE INCREASE IN PROBABILITY OF MISCLASSIFICATION

| Opt. Prob. $P = P'$ | $n = 50$ Number of states | | | | $n = 100$ Number of states | | | |
|---|---|---|---|---|---|---|---|---|
| | 4 | 8 | 16 | 32 | 4 | 8 | 16 | 32 |
| 0.05 | .009 | .018 | .032 | .059 | .006 | .011 | .020 | .034 |
| 0.1 | .011 | .022 | .039 | .067 | .008 | .015 | .025 | .041 |
| 0.2 | .014 | .027 | .049 | .082 | .010 | .020 | .033 | .052 |
| 0.3 | .016 | .034 | .057 | .098 | .011 | .023 | .038 | .062 |
| 0.4 | .025 | .037 | .061 | .071 | .012 | .026 | .041 | .061 |

As mentioned previously, the average increase is smallest when $P, P'$ are small. It becomes larger as the number of states increases, being somewhat less than doubled for each doubling of the number of states. The average increases for $n = 100$ are roughly $1/\sqrt{2}$ times those for $n = 50$. Some irregularities in the table will be noticed for $P = 0.4$. The explanation is that the actual rule always gives at least 50 percent correct classifications. This imposes an upper limit to the maximum average increase for $P = 0.4$.

It is difficult to give general recommendations from this analysis. The following comments may be appropriate.

1. Preliminary samples of around 50 specimens from each universe should give a good indication of those multivariate states in which the classification will be accurate, and of those in which it will be dubious.

2. If a classification is to be made for all multivariate states, the estimated probabilities of misclassification should be increased by the two adjustments for bias.

3. If the number of multivariate states does not exceed 8, Table 4 suggests that samples of size 50 should make the actual rule satisfactorily close to the optimum in most situations. With 8 states and $n = 50$, the actual probabilities in the worst cases are 6.8, 12.2, 22.7 and 33.4 percent, for theoretical optimum probabilities of 5, 10, 20 and 30 percent, respectively. For larger numbers of states, substantially greater samples are indicated.

## 6. THE EFFECT OF REPLACING A NORMAL VARIATE BY A QUALITATIVE VARIATE

When some variates are continuous and some qualitative, Rao's general rule for classification still applies. However, it presents difficulties in application, as Linhart [1959] has pointed out, since a different continuous discriminant is required for each multivariate state defined by the qualitative variates.

In this section the loss of discriminating power when normally distributed variates are partitioned into qualitative variates is examined. It is assumed that the variates are independent. This assumption is unrealistic for applications, but we have not been able to reach general results with correlated variates. With independent variates, simple asymptotic results can be obtained when the number of variates becomes large, and can be checked against the exact results for a small number of variates.

Let the $v$th continuous variate be denoted by $y_v$, $(v = 1, \cdots, k)$, and the two universes by $A$ and $B$. The variate $y_v$ is normal, with s.d. unity in each universe, and with mean 0 in $A$ and $\delta_v$ in $B$, the quantity $\delta_v$ being "the distance apart" of the two populations. The larger $\delta_v$ is, the better $y_v$ is for classification.

Given $k$ independent variates, it is easy to show that the best continuous discriminator, when divided so that its standard deviation is 1, is the quantity

$$R_c = \sum \delta_v y_v / \sqrt{\sum \delta_v^2} . \qquad (6.1)$$

For this discriminator the distance between the two populations is $\sqrt{\sum \delta_v^2}$. As $k$ becomes large with all $\delta_v > 0$, the distance becomes so great that classification is almost perfect. In order to discuss the situation in which some mistakes in classification are made, we assume that as $k$ becomes large the $\delta_v$ tend to zero in such a way that $\sqrt{\sum \delta_v^2}$ remains finite.

The continuous classification rule assigns a specimen to $B$ if $R_c \geq \frac{1}{2} \sqrt{\sum \delta_v^2}$. Since $R_c$ is normally distributed, the probability of misclassification, which is the same for both universes, is the probability that a normal deviate is greater than $\frac{1}{2} \sqrt{\sum \delta_v^2}$.

Let each continuous variate be partitioned into a qualitative variate with six states. Results for fewer than six states are derived as particular cases of the result for six states. For the $v$th variate, let Pr. $(j_v \mid A)$, Pr. $(j_v \mid B)$ be the probabilities that a specimen from $A$ and $B$ respectively falls into the $j_v$th state. The $6^k$ multivariate states into which the specimen can fall may be specified by recording the states $j_1$, $j_2$, $\cdots$, $j_k$ for the individual variates. Since the variates are independent, the probabilities that a specimen falls into a given multivariate state are

$$\prod_{v=1}^{k} \text{Pr.} \ (j_v \mid A) \quad \text{and} \quad \prod_{v=1}^{k} \text{Pr.} \ (j_v \mid B).$$

Consequently the optimum discrete rule places a specimen in $A$ if

$$\prod_{v=1}^{k} \text{Pr.} \ (j_v \mid A) > \prod_{v=1}^{k} \text{Pr.} \ (j_v \mid B),$$

i.e. if

$$\sum_{v=1}^{k} \{ \log \text{Pr.} \ (j_v \mid A) - \log \text{Pr.} \ (j_v \mid B) \} > 0. \qquad (6.2)$$

This may be written

$$\sum x_v > 0 \qquad (6.2)'$$

where

$$x_v = \log \text{Pr.} \ (j_v \mid A) - \log \text{Pr.} \ (j_v \mid B).$$

The quantity $\sum x_v$ is a discrete random variable with $6^k$ possible values, since its value depends on the multivariate state into which the specimen falls. It will be shown that as $k$ becomes large, this quantity tends to be normally distributed.

The way in which an individual variate is partitioned is shown in Figure 1. The six states are labeled, for mnemonic reasons, $A$, $a$, $\alpha$, $\beta$, $b$, $B$: they might be called "strongly, moderately, slightly favorable to $A$", etc. The partition involves two disposable numbers $u_1$, $u_2$, which determine the sizes of the regions $\alpha$, $\beta$ and $a$, $b$. (Four numbers might be used, but symmetry suggests that two will give the best partition.) Strictly, the numbers should be denoted by $u_{1v}$, $u_{2v}$, since it may be profitable to vary them with $v$, but for simplicity the subscript $v$ will be omitted from $u_1$, $u_2$ and $\delta$ at present.

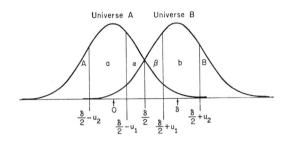

FIGURE 1

PARTITION OF A NORMAL VARIATE INTO SIX STATES

For an individual variate $v$ Table 5 shows the notation used for the probabilities of falling into the respective states.

The probabilities $p_i$ are integrals over the parts of the normal curve indicated in Figure 1. Note that the Pr. $(\mid B)$ values are the same as the Pr. $(\mid A)$ values in reverse order.

TABLE 5

PROBABILITIES THAT A VARIATE WILL FALL INTO THE 6 STATES

|  | State | | | | | |
|---|---|---|---|---|---|---|
|  | $A$ | $a$ | $\alpha$ | $\beta$ | $b$ | $B$ |
| Pr. $(\mid A)$ | $p_1$ | $p_2$ | $p_3$ | $p_4$ | $p_5$ | $p_6$ |
| Pr. $(\mid B)$ | $p_5$ | $p_6$ | $p_4$ | $p_3$ | $p_2$ | $p_1$ |
| $x$ | $w_1$ | $w_2$ | $w_3$ | $-w_3$ | $-w_2$ | $-w_1$ |

The last line of Table 5 shows the six possible values taken by the random variable

$$x = \log \text{Pr.} \, (\mid A) - \log \text{Pr.} \, (\mid B)$$

where

$$w_1 = \log \, (p_1/p_6), \qquad w_2 = \log \, (p_2/p_5), \qquad w_3 = \log \, (p_3/p_4).$$

It follows from Table 5 that

$$E(x \mid A) = w_1(p_1 - p_6) + w_2(p_2 - p_5) + w_3(p_3 - p_4)$$

while $E(x \mid B)$ has the same value with sign reversed. The variance has the same value for both universes, namely

$$V(x) = w_1^2(p_1 + p_6) + w_2^2(p_2 + p_5) + w_3^2(p_3 + p_4) - [E(x \mid A)]^2.$$

Reverting to inequality (6.2)′, its left side is the sum of a large number of independent variables $x$, and hence tends to become normally distributed as $k$ becomes large.

It remains to obtain the values of $E(x \mid A)$ and $V(x)$ when $\delta$ tends to zero. The leading terms in the expressions for the $p_i$ and the $w_i$ are found to be as follows.

Let

$$I_1 = \int_{u_1}^{\infty} z(t)\, dt, \qquad I_2 = \int_{u_2}^{\infty} z(t)\, dt$$

where $z(t)$ is the normal curve, and let $z_0$, $z_1$, $z_2$ be the ordinates of this curve at $0$, $u_1$, $u_2$, respectively. Then, from Figure 1,

$$p_1 = \int_{-\infty}^{\frac{1}{2}\delta - u_2} z(t)\, dt \doteq I_2 + \tfrac{1}{2}\delta z_2 \,,$$

$$p_2 = \int_{\frac{1}{2}\delta - u_2}^{\frac{1}{2}\delta - u_1} z(t)\, dt \doteq I_1 - I_2 + \tfrac{1}{2}\delta(z_1 - z_2).$$

Similarly,

$$p_3 \doteq \tfrac{1}{2} - I_1 + \tfrac{1}{2}\delta(z_0 - z_1), \qquad p_4 \doteq \tfrac{1}{2} - I_1 - \tfrac{1}{2}\delta(z_0 - z_1),$$

$$p_5 \doteq I_1 - I_2 - \tfrac{1}{2}\delta(z_1 - z_2), \qquad p_6 \doteq I_2 - \tfrac{1}{2}\delta z_2 \,,$$

$$w_1 \doteq \frac{\delta z_2}{I_2}, \qquad w_2 \doteq \frac{\delta(z_1 - z_2)}{I_1 - I_2}, \qquad w_3 \doteq \frac{\delta(z_0 - z_1)}{\tfrac{1}{2} - I_1}.$$

Substituting these values in $E(x \mid A)$ and $V(x)$, retaining only the terms of order $\delta^2$, we find

$$E(x \mid A) = \delta^2 \left\{ \frac{z_2^2}{I_2} + \frac{(z_1 - z_2)^2}{I_1 - I_2} + \frac{(z_0 - z_1)^2}{\tfrac{1}{2} - I_1} \right\}. \qquad (6.3)$$

$$V(x) = 2E(x \mid A) = 2\,\delta^2 f(u_1, u_2), \qquad (6.4)$$

writing $f(u_1, u_2)$ for the expression in brackets in (6.3).

We now reintroduce the subscript $v$ in order to sum over the $k$ variates. From inequality (6.2)′ the probability of misclassifying a specimen from $A$ is the probability that the random variable $\sum (x_v \mid A)$ is negative. From (6.3) and (6.4) this variable is approximately normally distributed with

$$\text{Mean} = \sum_{v=1}^{k} \delta_v^2 f(u_{1v}, u_{2v}) : \quad \text{Variance} = 2 \sum_{v=1}^{k} \delta_v^2 f(u_{1v}, u_{2v}).$$

Hence the probability of misclassifying an $A$ is approximately the probability that a normal deviate exceeds

$$\sqrt{\sum \delta_v^2 f(u_{1v}, u_{2v})/2}. \qquad (6.5)$$

In order to minimize the probability we must maximize (6.5). Since $f(u_1, u_{2r})$ does not depend on $\delta_r$, expression (6.5) shows that the maximum is attained by substituting the *same* values of $u_{1r}$, $u_{2r}$ in every term, i.e. the values which maximize $f(u_1, u_2)$. This reduces (6.5) to

$$\sqrt{\text{max. } f(u_1, u_2)} \; \sqrt{\sum \delta_r^2/2}. \tag{6.6}$$

With the untransformed normal variate it was shown earlier that the best rule gives a probability of misclassification equal to the probability that a normal deviate exceeds $\frac{1}{2}\sqrt{\sum \delta_r^2}$. Comparing this result with (6.6), the relative discriminating power of the best qualitative to the best continuous rule is

$$2 \text{ max. } f(u_1, u_2) = 2 \text{ max. } \left\{ \frac{z_2^2}{I_2} + \frac{(z_1 - z_2)^2}{I_1 - I_2} + \frac{(z_0 - z_1)^2}{\frac{1}{2} - I_1} \right\}. \tag{6.7}$$

This result holds in the sense that if a randomly chosen fraction $2 \text{ max. } f(u_1, u_2)$ of the original normal variates is retained, discarding the rest, the best continuous rule becomes equivalent to the best qualitative rule based on all the variates.

The result for five states is obtained by ignoring the distinction between the states $\alpha$ and $\beta$, calling the combined state $D$ (doubtful). The effect is that the last term on the right of (6.7) disappears. For four states we let $u_1 = 0$, the states becoming $A$, $a$, $b$, $B$. For three states, the states $a$ and $b$ are combined into a $D$ region. Finally, for two states we put $u_2 = 0$.

The form of the expression $2f(u_1, u_2)$, the values of $u_1$, $u_2$ giving the best partition of the normal curve, and the relative discriminating power of the qualitative variates appear in Table 7 for 2, 3, 4, 5 and 6 states.

Since the maxima of $f(u_1, u_2)$ are flat, the best values of $u_1$ and $u_2$ are given only to one decimal. For two states, the relative power is $2/\pi$, which will be familiar as the asymptotic power of the sign test. Use of five or six states retains over 90 percent of the power.

The function $2f(u_1, u_2)$ has already appeared in two other problems involving the replacement of normal curve methods by less efficient methods. Ogawa [1951] obtained this function as the efficiency of estimation of the population mean from suitably chosen order statistics in large samples. D. R. Cox [1957] found the same expression for the relative amount of information retained by grouping the normal curve when the group boundaries are chosen to minimize the quantity $E\{x - \mu(x)\}^2/\sigma^2$, where $\mu(x)$ is the mean of the group to which $x$ is assigned. So far as we can see, the three problems are mathematically

TABLE 7

The Asymptotic Relative Discriminating Power of Qualitative to Continuous Normal Variates

| No. of states | $2f(u_1, u_2)$ | Best values of $u_1$ | $u_2$ | Relative power |
|---|---|---|---|---|
| 2 | $2/\pi$ | — | — | 0.636 |
| 3 | $2z_2^2/I_2$ | — | 0.6 | 0.810 |
| 4 | $2\left\{\dfrac{z_2^2}{I_2} + \dfrac{(z_0 - z_2)^2}{\frac{1}{2} - I_2}\right\}$ | — | 1.0 | 0.882 |
| 5 | $2\left\{\dfrac{z_2^2}{I_2} + \dfrac{(z_1 - z_2)^2}{I_1 - I_2}\right\}$ | 0.4 | 1.2 | 0.920 |
| 6 | $2\left\{\dfrac{z_2^2}{I_2} + \dfrac{(z_1 - z_2)^2}{I_1 - I_2} + \dfrac{(z_0 - z_1)^2}{\frac{1}{2} - I_1}\right\}$ | 0.7 | 1.4 | 0.942 |

different, and the identity of results, as Cox has pointed out, seems to depend on particular properties of the normal curve.

## 7. COMPARISONS FOR SMALL NUMBERS OF VARIATES

As a check on the utility of the asymptotic results for practical situations, the best points of partition and the resulting probabilities of misclassification were computed directly for small numbers of variates. For simplicity the distance apart $\delta$ was assumed to be the same for all variates.

Having calculated the probability of misclassification for a given $\delta$ and number of variates $k$, we can compute the number $k'$ of continuous normal variates that are needed to give the same probability of misclassification. For these values of $\delta$ and $k$, the ratio $k'/k$ gives the relative discriminating power of the qualitative to the continuous normal variates. Unlike the asymptotic case, this ratio is of course dependent on the values of $\delta$ and $k$. For fixed $k$, however, the ratio turns out to be almost constant to two decimal places over the range of probabilities of misclassification that are of practical interest.

In the top half of Table 8, the relative discriminatory power with two variates is compared with the asymptotic power from Table 7

TABLE 8

RELATIVE DISCRIMINATING POWER OF QUALITATIVE
TO CONTINUOUS NORMAL VARIATES
FOR SMALL NUMBERS OF VARIATES

| 2 states | | 3 states | | 4 states | | 5 states | |
|---|---|---|---|---|---|---|---|
| $k = 2$ | $k = \infty$ | $k = 2$ | $k = \infty$ | $k = 2$ | $k = \infty$ | $k = 2$ | $k = \infty$ |
| .50 | .64 | .74 | .81 | .84 | .88 | .89 | .92 |

| 2 states $k$ | | | | | | | 3 states $k$ | | | | |
|---|---|---|---|---|---|---|---|---|---|---|---|
| 2 | 3 | 4 | 5 | 6 | 7 | $\infty$ | 2 | 3 | 4 | 5 | $\infty$ |
| .50 | .74 | .56 | .70 | .58 | .68 | .64 | .74 | .76 | .77 | .78 | .81 |

for 2, 3, 4 and 5 states. The "small-sample" powers are all lower than the asymptotic values.

The lower half of Table 8 shows the results with two states and from 2 to 7 variates. The results have one curious feature. With an odd number of variates the relative powers are higher than the asymptotic values. The explanation may be that none of the decisions requires the toss of a coin when the number of variates is odd. In line with this result is the fact that with two states and only one variate ($k = 1$) the classification rule is the same as that obtained from the original normal variate, so that the relative power is unity. With 3 states (Table 8) this peculiarity does not appear, the relative powers increasing steadily with $k$ towards the asymptotic value.

The optimum values of $u_1$ and $u_2$ for small numbers of variates were consistently close to the asymptotic optima. This result held throughout a wide range of values of $\delta \sqrt{k}$ from 0.5 (corresponding to a probability of misclassification of over 40 percent) to 4. Since, moreover, the optima are flat, the asymptotic values of $u_1$ and $u_2$ may be used safely with only two variates.

A few comparisons have been made between the qualitative and continuous discriminators for the case of two correlated variates of equal discriminating power. Suppose that the continuous scales are arranged so that the variates have standard deviations unity, with means zero in population $A$ and $\delta$ in population $B$. If they are positively correlated, as appears to be the more common situation in applications, results obtained for 2 and 3 states indicate that the qualitative discriminator has higher relative power than when the variates are independent. The reverse seems to hold with negative correlation. Consequently our results for independent variates cannot be assumed to be valid for correlated variates. More investigation is needed.

In order to transform a continuous into a qualitative variate in an application we compute the means $\bar{y}_A$, $\bar{y}_B$ of the continuous variate in the initial samples from the two universes and the pooled within-universes standard deviation $s$. Let $\bar{y} = (\bar{y}_A + \bar{y}_B)/2$. With, for instance, five states the best asymptotic values of $u_1$, $u_2$ are $u_1 = 0.4$, $u_2 = 1.2$. The states are constructed as follows.

| Value of $y$ | State |
|---|---|
| less than $\bar{y} - 1.2s$ | $A$ |
| from $\bar{y} - 1.2s$ to $\bar{y} - 0.4s$ | $a$ |
| from $\bar{y} - 0.4s$ to $\bar{y} + 0.4s$ | $D$ |
| from $\bar{y} + 0.4s$ to $\bar{y} + 1.2s$ | $b$ |
| greater than $\bar{y} + 1.2s$ | $B$ |

## 8. ESTIMATION OF THE RELATIVE FREQUENCY WITH WHICH SPECIMENS COME FROM THE TWO UNIVERSES

If specimens present themselves with unequal frequencies $\pi$, $\pi'$ from the two universes, the optimum rule is to assign a specimen to $U$ if $\pi p_i > \pi' p_i'$. This rule requires an initial estimate or guess about the value of $\pi$. With qualitative variates the accuracy of this figure will often be not critical, because the same optimum rule holds over a range of values of $\pi$, although the range becomes shorter as the number of states increases. Table 9 gives an illustration for 4 states. The horizontal line marks the boundary of the best decision region.

TABLE 9

THE OPTIMUM RULE FOR DIFFERENT VALUES OF $\pi$

| State | Range of $\pi$ .011 − .4 | | Range of $\pi$ .4 − .64 | | Range of $\pi$ .64 − .99 | |
|---|---|---|---|---|---|---|
| | $p_i$ | $p_i'$ | $p_i$ | $p_i'$ | $p_i$ | $p_i'$ |
| 1 | .92 | .01 | .92 | .01 | .92 | .01 |
| 2 | .06 | .04 | .06 | .04 | .06 | .04 |
| 3 | .04 | .07 | .04 | .07 | .04 | .07 |
| 4 | .00 | .88 | .00 | .88 | .00 | .88 |

If we guessed $\pi = \pi' = 0.5$, we would use the middle classification rule in Table 9. This remains the best rule so long as $\pi$ lies between 0.4 and 0.64. Even with $\pi$ outside these limits, this rule may be close to the best. For instance, if $\pi$ is 0.7, the average frequency of misclassification using our rule is $(.7)(.04) + (.3)(.05)$, or .043, as against $(.3)(.12)$, or .036 with the optimum rule. With $\pi = 0.8$ the corresponding figures are .042 and .024, a more serious difference.

When $m$ specimens have been classified, the data give the numbers of specimens $r_i$ found in each state, where $\sum r_i = m$. If the $p_i$, $p_i'$ are known, the $r_i$ follow a multinomial distribution with probability

$$P_i = \pi p_i + \pi' p_i'$$

in the $i$th state. The log likelihood is

$$L = \sum_{1}^{s} r_i \log (\pi p_i + \pi' p_i'), \qquad \frac{\partial L}{\partial \pi} = \sum \frac{r_i(p_i - p_i')}{\pi p_i + \pi' p_i'} \tag{8.1}$$

$$E \frac{\partial^2 L}{\partial \pi^2} = -E\left\{ \sum \frac{r_i(p_i - p_i')^2}{(\pi p_i + \pi' p_i')^2} \right\} = -m \sum \frac{(p_i - p_i')^2}{\pi p_i - \pi' p_i'},$$

so that the estimated variance of $\hat{\pi}$ is

$$V(\hat{\pi}) = 1 \Big/ m \sum \frac{(p_i - p_i')^2}{\hat{\pi} p_i + \hat{\pi}' p_i'}. \tag{8.2}$$

The maximum likelihood estimate $\hat{\pi}$ obtained by setting $\partial L/\partial \pi = 0$ in (8.1) can be found by trial, although this is tedious if the states are numerous. One way of obtaining a first trial value of $\pi$ is to note that each state furnishes an unbiased estimate. Since

$$E(r_i) = mP_i = m(\pi p_i + \pi' p_i'),$$

the estimate $\hat{\pi}_i$ from the $i$th state is

$$\hat{\pi}_i = [(r_i/m) - p_i']/(p_i - p_i') : V(\hat{\pi}_i) = P_i Q_i/m(p_i - p_i')^2.$$

These estimates are likely to differ markedly in precision. A fairly good method for a first trial $\hat{\pi}$ is to combine the states for which $p_i > p_i'$, making a single estimate from these states. Table 10 shows a numerical example for a sample of 100 specimens.

From the first two states a trial value is computed as

$$\hat{\pi}_1 = \frac{0.67 - 0.2}{0.6} = 0.78.$$

The values of $r_i(p_i - p_i')$, $P_i$ and $r_i(p_i - p_i')/P_i$ are next computed (Table 10). The total of the latter values gives $\partial L/\partial \pi$ as $-5.2$. Since

TABLE 10

EXAMPLE OF THE M. L. ESTIMATION OF $\pi$ (COMPUTED FOR $\hat{\pi} = .78$)

| $p_i$ | $p_i'$ | $r_i$ | $r_i(p_i - p_i')$ | $P_i$ | $r_i(p_i - p_i')/P_i$ |
|---|---|---|---|---|---|
| .5 | .1 | 37 | 14.8 | .412 | 35.9 |
| .3 | .1 | 30 | 6.0 | .256 | 23.4 |
| .1 | .2 | 9 | −0.9 | .122 | −7.4 |
| .1 | .6 | 24 | −12.0 | .210 | −57.1 |
| | | 100 | | | −5.2 |

$\partial L/\partial \pi$ is a decreasing function of $\pi$ the second trial must be lower. The estimate $\hat{\pi}_2 = 0.75$ gave $+0.5$ for the first derivative. From (8.2) the estimated standard error of $\hat{\pi}_2$ was found to be $\pm 0.075$.

Examination of the formula for the standard error shows that the estimate $\hat{\pi}$ cannot be expected to be precise: the standard error is always larger than the value $\sqrt{\pi(1 - \pi)/m}$ that would apply to a binomial estimate of $\pi$. This suggests that it will be worthwhile to estimate $\pi$ only after a substantial number of specimens have been classified.

## 9. SUMMARY

This paper deals with the problem of assigning specimens to one of two or more universes when the measurements on each specimen are qualitative, each taking a small number of states. After presenting the optimum rule for classifying the specimens, three problems are considered.

The construction of the rule requires initial data on a number of specimens known to be classified correctly. Standard classification theory assumes that these initial samples are infinite in size, although in practice they may be only moderate. The principal effects of the finite sizes of the initial samples are that the probability of misclassification of the rule derived from them is underestimated and that this rule may be inferior to the theoretical optimum rule that we could construct if we had infinite samples. Methods are proposed for obtaining reasonably unbiased estimates of the performance of rules derived from finite samples and for estimating the difference between the actual and the theoretical optimum probability of misclassification. It appears that initial samples of size 50 from each of two universes should be adequate if there are not more than 8 multivariate states. With greater numbers of states, larger sample sizes are needed to ensure that the actual rule will be almost as good as the theoretical optimum.

If most of the variates are qualitative but a few are continuous, one possibility is to transform the continuous variates into qualitative ones, particularly since classification is easier with qualitative than with continuous variates. Asymptotic results are obtained for the best points of partition and the probabilities of misclassification when a large number of independent normal variates are partitioned to form qualitative variates. For qualitative variates with 2, 3, 4, 5 and 6 states the relative efficiencies are 64, 81, 88, 92 and 94 percent respectively. Computations for small numbers of variates show that the asymptotic points of partition remain satisfactory although the relative efficiencies are in general lower.

The optimum rule depends on the relative frequencies with which specimens to be classified present themselves from different universes. Initial estimates of these frequencies must be made in order to set up the rule. With two universes, maximum likelihood estimates of the frequencies from the data for specimens that have been classified by the rule are given. These estimates enable the rule to be improved if the initial estimates differ from the frequencies that apply when the rule is being used.

## REFERENCES

Anderson, T. W. [1958]. *Introduction to multivariate statistical analysis.* John Wiley & Sons, New York, 142–7.

Cox, D. R. [1957]. Note on grouping. *Jour. Amer. Stat. Assn.* 52, 543–7.

Cox, D. R. and Brandwood, L. [1959]. On a discriminatory problem connected with the works of Plato. *Jour. Roy. Stat. Soc. B,* 21, 195–200.

Linhart, H. [1959]. Techniques for discriminant analysis with discrete variables. *Metrika* 2, 138–49.

Ogawa, J. [1951]. Contributions to the theory of systematic statistics. *Osaka Math. Jour.* 4, 175–213.

Rao, C. R. [1952]. *Advanced statistical methods in biometric research.* John Wiley & Sons, New York, Chapter 8.

Stokes, D. E., Campbell, A. and Miller, M. E. [1958]. Components of electoral decision. *Amer. Pol. Sci. Rev. 52,* 367–87.

# COMPARISON OF METHODS FOR DETERMINING STRATUM BOUNDARIES*

by

WILLIAM G. COCHRAN

*Department of Statistics, Harvard University, U.S.A.*

## 1. Introduction

In recent years some interesting work has been done on the problem of dividing a frequency function $f(y)$ into $L$ strata when the purpose is to estimate the population mean of $y$ from a stratified random sample. In particular, Dalenius (1950) found the equations for the stratum boundaries $y_h$ which minimize the variance of the estimated population mean under optimum allocation for fixed total sample size ($n_h = \alpha N_h \sigma_h$, where $n_h$, $N_h$ are the sample and population numbers of sampling units in stratum $h$ and $\sigma_h$ is the within-stratum s.d.). He pointed out that these equations are troublesome to solve. A number of quicker approximate methods of estimating the optimum $y_h$ have been developed by Dalenius and other workers and tested to some extent on theoretical frequency distributions.

In this paper four of the simplest methods will be compared on actual as distinct from theoretical populations. The distributions chosen are skew with a long positive tail and are intended to be roughly representative of those encountered in the sampling of institutions of the type in which there are many small institutions and few large ones. The eight variables are as follows, the range covered by each variable and the size of the population being given in parentheses.

Adjusted gross income per tax return in the U.S. in 1951 ($1,000—$300,000. $N=47,799,000$).

Number of students in four-year U.S. colleges in 1952-3 (200—10,000. $N=677$).
Population of U.S. cities in 1940 (10,000—200,000. $N=1,038$).

Resources of large commercial U.S. banks (70—1,000 million dollars. $N=357$).
Number of farms per area sampling unit in Seneca County, N.Y. (1—18. $N=556$).

Proportion of gross bank loans devoted to (i) commercial and industrial loans (ii) urban real estate loans (iii) agricultural loans—3 frequency distributions (0—100%. $N=13,435$ in each case).

* This work was assisted by a contract between the Office of Naval Research, Navy Department and the Department of Statistics, Harvard University.

The sources of the data are given at the end of this paper.

Two methods of determining the sample sizes in the strata are considered: (i) optimum allocation for fixed sample size as mentioned above and (ii) equal size of sample in all strata. The latter method is included because stratification with one unit per stratum is often used in national surveys conducted by the U.S. Bureau of the Census and has been found convenient in some of the surveys taken by the Indian Statistical Institute, (Mahalanobis, 1954). Stevens (1952) has also discussed the advantages of an equal number of units per stratum.

In practice, of course, the construction of strata cannot be based on the frequency distribution of the variate $y$ that is to be measured in a future survey. At best, we can use the frequency distribution of this variate from a recent census or of some other variate highly correlated with $y$. It was felt, however, that a comparison of the methods using the frequency distribution of $y$ itself would provide a more critical test of their performances. Three cases in which the strata are constructed from the frequency distribution of $y$ at a previous time are included at the end of Section 5.

The comparisons are made for subdivisions into 2, 3 or 4 strata. Restriction to small numbers of strata puts a strain on the methods, most of which depend more or less on the assumption that the number of strata $L$ is reasonably large, so that within a stratum the function $f(y)$ can be assumed to be rectangular. As will be seen, however, some of the methods do very satisfactorily even with small numbers of strata. Although the number of strata used in practice will vary with the conditions of the survey, there are practical advantages in keeping this number small.

## 2. The approximate methods

Let $y_0, y_1, \cdots y_L$ be the stratum boundaries, the strata being numbered $1, 2, \cdots L$, and let $\mu_h$, $\sigma_h$ be the mean and s.d. in stratum $h$ and $W_h = N_h/N$ be the ratio of the number of sampling units in stratum $h$ to the total number in the population. With stratified sampling the usual estimate of the population mean is

$$\bar{y}_{st} = \sum W_h \bar{y}_h$$

where $\bar{y}_h$ is the sample mean in stratum $h$. Its variance is

$$(1) \qquad V(\bar{y}_{st}) = \sum_{h=1}^{L} W_h^2 \sigma_h^2 \left( \frac{1}{n_h} - \frac{1}{N_h} \right).$$

For a fixed total size of sample $n$ it is well known that $V(\bar{y}_{st})$ is minimized by taking $n_h = n W_h \sigma_h \big/ \sum W_h \sigma_h$. The minimum variance is

$$(2) \qquad V_{min}(\bar{y}_{st}) = \frac{1}{n} \left( \sum W_h \sigma_h \right)^2 - \sum \frac{W_h^2 \sigma_h^2}{N_h}.$$

If the sampling fractions $n_h/N_h$ are small in all strata, the last term in (2) can be ignored. The problem is to choose the boundaries $\bar{y}_h$ so as to minimize

$$\sum W_h \sigma_h .$$

By standard calculus methods, Dalenius (1950) showed that the optimum $y_h$ satisfy the equations

(3)
$$\sigma_h + \frac{(y_h - \mu_h)^2}{\sigma_h} = \sigma_{h+1} + \frac{(y_h - \mu_{h+1})^2}{\sigma_{h+1}}$$
$$h = 1, 2, \ldots, L-1$$

Since the $\mu_h$ and $\sigma_h$ cannot be computed until the $y_h$ are known, these equations are not particularly helpful, being scarcely more expeditious than minimizing $\sum W_h \sigma_h$ directly by trial and error.

The justifications for the approximate methods have been given in the literature and will be indicated here only briefly.

I. In the method proposed by Dalenius and Hodges (1957), the values of $\sqrt{f(y)}$ are cumulated. If the cumulated total over the whole range from $y_0$ to $y_L$ is $H$, the approximate stratum boundaries are

$$y_1 = \frac{H}{L} \; : \; y_2 = \frac{2H}{L} \; : \cdots \cdots : y_h = \frac{hH}{L} \; .$$

In other words, the cumulative value of $\sqrt{f(y)}$ is made constant within all strata.

A non-rigorous justification given by Dalenius and Hodges (1959) is as follows. Let

$$z(y) = \int_{y_0}^{y} \sqrt{f(t)} \; dt \; .$$

If the strata are numerous and narrow, the value of $f(y)$ will be approximately constant within a given stratum. Hence

$$W_h = \int_{y_{h-1}}^{y_h} f(t) \; dt \doteq f_h (y_h - y_{h-1}) : \quad \sigma_h \doteq \frac{1}{\sqrt{12}} \Big( y_h - y_{h-1} \Big) :$$

$$z_h - z_{h-1} = \int_{y_{h-1}}^{y_h} \sqrt{f(t)} \; dt \doteq \sqrt{f_h} \, (y_h - y_{h-1}) \; ,$$

where $f_h$ is the "constant" value of $f(y)$ in stratum $h$. By substituting these approximations we find

(4)
$$\sqrt{12} \sum W_h \sigma_h \doteq \sum f_h (y_h - y_{h-1})^2 \doteq \sum (z_h - z_{h-1})^2 \; .$$

Since $(z_L - z_0)$ is fixed, it is easy to see that the sum on the right is minimized by making $z_h - z_{h-1}$ constant. These authors have also developed a second approximation by which the results of this method may be adjusted to bring them closer to the exact solution. In this paper we are concerned only with the simplest approximations.

II. Equation (4) implies that making $z_h - z_{h-1}$ constant is approximately the same as making $W_h \sigma_h$ constant. The result that $W_h \sigma_h$ would be approximately constant for the best choice of boundaries had been conjectured earlier by Dalenius and Gurney (1951). Since it involves the computation of $\sigma_h$, however, this method is not convenient in practice.

III. In the sampling of institutions the variate $y$ is often a measure of the "output" of the sampling unit, e.g. number of inhabitants of a town, number of beds in a hospital. (The word "output" is used rather than "size" to avoid an ambiguous meaning of the term "size of a stratum".) A working rule found useful by Mahalanobis (1952) and Hansen, Hurwitz and Madow (1953) is to construct strata so that the aggregate output is the same in all strata. In our notation this rule makes $W_h \mu_h$ constant in all strata.

As pointed out by Dalenius and Hodges (1957), this rule cannot be of general validity, since the boundaries which it gives are changed if the origin of the scale is changed, whereas the correct solution to the problem is invariant under change of origin. The relation of this rule to the exact solution is not as yet clear. It can be derived from method II by the pragmatic observation that with near-optimum boundaries the coefficients of variation $\mu_h/\sigma_h$ are often found to be approximately the same in all strata. When this holds, the relation $W_h\sigma_h=$constant implies $W_h\mu_k$ =constant. Further analysis of this rule has been given by Kitagawa (1955).

IV. Ekman (1959) suggested making

$$W_h(y_h - y_{h-1}) = \text{constant}.$$

This rule is obtained from the exact rule by showing that if the distribution $f(y)$ is not highly skewed,

$$\left[\frac{\sigma_h^2 + (y_h - \mu_h)^2}{\sigma_h}\right]^2 \doteq \frac{4W_h(y_h - y_{h-1})}{3f(y_h)}.$$

V. Finally, Durbin (1948) proposed a method in a review of Dalenius' doctoral thesis. Let $F(y)$ be the cumulative of $f(y)$. Form a rectangular distribution

$$r(y) = \frac{F(y_L)}{y_L - y_0}$$

over the same range. The stratum boundaries are obtained by taking equal intervals on the cumulative of

$$\frac{1}{2}[r(y) + f(y)].$$

This rule amounts to forming the strata by taking equal areas under a frequency distribution with density half-way between the original distribution and a rectangular distribution. Durbin developed the rule by considering, for $L=2$, the simplest departure from a rectangular distribution, namely a linear function $f(y)$ between $f(0)=1-a$ and $f(1)=1+a$, where $a$ is small.

The rules to be compared are as follows.

I.  Equal intervals on the cumulative of $\sqrt{f(y)}$.

III. Equal aggregate "outputs": $W_h\mu_h=$constant.

IV. $W_h(y_h - y_{h-1})=$constant.

V.  Equal intervals on the cumulative of $\{r(y)+f(y)\}$.

Although at first sight the rules may appear dissimilar, they operate in the same general way with a positive highly skew variate. If strata of equal size are formed, i.e. taking $W_h=$constant or equivalently taking equal intervals on the cumulative of $f(y)$, the strata at the lower end are too narrow and those at the upper end too wide for optimum estimation. On the other hand, if strata of equal length on the $y$ scale are used, i.e. taking $(y_h - y_{h-1})=$constant or equivalently taking equal intervals on the cumulative rectangular distribution, the strata at the lower end are too wide and those at the upper end too narrow. All the rules have the effect of constructing a stratification with boundaries lying between these extremes.

## 3. The shapes of the frequency distributions

Table 1 gives an impression of the shapes of the eight frequency distributions used in this study. It was obtained by dividing the range between $y_0$ and $y_L$ in tenths and finding the percentage of the total frequency lying within each tenth.

Since the upper half of the range usually contained only a small percentage of the total frequency, the last five classes have been combined into one.

Table 1. *Percentage of total frequency falling in successive tenths of the range of y.*

| Range % | Agric. loans | Real estate loans | Ind. loans | Bank res. | Coll. stu. | City pops. | Farms per s.u. | Gross income |
|---|---|---|---|---|---|---|---|---|
| 0–10 | 30.4 | 26.3 | 44.5 | 57.9 | 67.4 | 70.2 | 70.1 | 99.52 |
| 10–20 | 11.4 | 20.7 | 27.8 | 16.2 | 14.4 | 14.1 | 16.3 | 0.36 |
| 20–30 | 10.0 | 15.7 | 14.1 | 8.8 | 6.5 | 5.9 | 7.9 | 0.07 |
| 30–40 | 9.7 | 12.8 | 6.6 | 4.9 | 2.7 | 3.8 | 3.5 | 0.02 |
| 40–50 | 8.8 | 9.7 | 3.5 | 4.8 | 1.9 | 2.1 | 1.3 | 0.02 |
| 50 + | 29.5 | 14.8 | 3.5 | 7.4 | 7.1 | 3.9 | 0.9 | 0.01 |

The distributions are arranged, from left to right, roughly in order of increasing skewness or of increasing departure from a rectangular distribution. The first distribution (percentage of bank loans devoted to agricultural loans) is not far from a rectangular except for an excess frequency in the lowest tenth of the range of $y$. Thereafter the departures become more drastic, ending with the extremely skew distribution of taxable income. (In case it may appear that U.S. taxpayers either have no income or do not report it, we should remark that owing to the high upper limit chosen for the range ($300,000), the lowest tenth of the range contains incomes up to $30,900.)

In a study of this kind it is advisable to have a substantial number of classes in the original frequency distribution. If this number is too low, the choice of stratum boundaries for $L=4$ becomes restricted, the true optimum stratification may be missed and calculation of the within-stratum standard deviations becomes affected by grouping errors. Except for number of farms per sampling unit, a discrete distribution with only 18 values, all other distributions had at least 25 classes available.

## 4. Application of the rules to a numerical example

The method of applying the rules to determine stratum boundaries will be illustrated for the frequency distribution of the percentages of bank loans devoted to commercial loans (Table 2). The $y$ scale ranges from 0 to 100% in 20 equal class intervals. In the original data the two lowest classes were each subdivided into 5 subclasses: for compactness this extra subdivision is not shown here. The boundaries will be for $L=3$.

I. *Equal intervals on* cum. $\sqrt{f}$. The values of $\sqrt{f}$ are cumulated as in column (4). Since the grand total is 389.5, each stratum should contain $389.5/3=129.8$ on the cum. $\sqrt{f}$ scale. The best that can be done is as follows.

|  | Stratum | | |
|---|---|---|---|
|  | 1 | 2 | 3 |
|  | 0–10% | 10–30% | 30–100% |
| cum. $\sqrt{f}$ | 109.1 | 147.3 | 133.1 |

Table 2.  *Calculation of the stratum boundaries by
the four methods for L=3.*

Variable : Percentage of gross loans devoted to industrial loans.

| (1) % | (2) Coded $\mu_h$ | (3) $f_h$ | (4) Cum. $\sqrt{f_h}$ | (5) Cum. $f_h \mu_h$ | (6) Cum. $f_h$ | (7) Cum. $r_h$ |
|---|---|---|---|---|---|---|
| 0–5 | 0.67 | 3,464 | 58.9 | 2,321 | 3,464 | 672 |
| 5–10 | 3 | 2,516 | 109.1 | 9,869 | 5,980 | 1,344 |
| 10–15 | 5 | 2,157 | 155.5 | 20,654 | 8,137 | 2,015 |
| 15–20 | 7 | 1,581 | 195.3 | 31,721 | 9,718 | 2,687 |
| 20–25 | 9 | 1,142 | 229.1 | 41,999 | 10,860 | 3,359 |
| 25–30 | 11 | 746 | 256.4 | 50,205 | 11,606 | 4,030 |
| 30–35 | 13 | 512 | 279.0 | 56,861 | 12,118 | 4,702 |
| 35–40 | 15 | 376 | 298.4 | 62,501 | 12,494 | 5,374 |
| 40–45 | 17 | 265 | 314.7 | 67,006 | 12,759 | 6,046 |
| 45–50 | 19 | 207 | 329.1 | 70,939 | 12,966 | 6,718 |
| 50–55 | 21 | 126 | 340.3 | 73,585 | 13,092 | 7,389 |
| 55–60 | 23 | 107 | 350.6 | 76,046 | 13,199 | 8,061 |
| 60–65 | 25 | 82 | 359.7 | 78,096 | 13,281 | 8,733 |
| 65–70 | 27 | 50 | 366.8 | 79,446 | 13,331 | 9,404 |
| 70–75 | 29 | 39 | 373.0 | 80,577 | 13,370 | 10,076 |
| 75–80 | 31 | 25 | 378.0 | 81,352 | 13,395 | 10,748 |
| 80–85 | 33 | 16 | 382.0 | 81,880 | 13,411 | 11,420 |
| 85–90 | 35 | 19 | 386.4 | 82,545 | 13,430 | 12,092 |
| 90–95 | 37 | 2 | 387.8 | 82,619 | 13,432 | 12,763 |
| 95–100 | 39 | 3 | 389.5 | 82,736 | 13,435 | 13,435 |

With a greater number of class intervals, a more equal partition on the cum. $\sqrt{f}$ scale could of course be obtained.

Sometimes the original frequency distribution contains intervals of unequal length, the intervals becoming wider towards the upper end of the $y$ scale. When the interval changes from one of length $d$ to one of length $ud$, the value of $\sqrt{f}$ in the longer interval must be multiplied by $\sqrt{u}$ when forming cum. $\sqrt{f}$.

III. *Equal aggregate output.* In general, the centers of the original class intervals will be used as estimates of the original $\mu_h$. For this distribution these centers are 2.5%, 7.5%, 12.5% etc., which may be coded as 1, 3, 5, etc. as in column (2). For the first two classes, however, we have more detailed information from the subdivisions into fifths already mentioned. These subdivisions were used to compute more accurate values of $\mu_1$ and $\mu_2$. As it happened, $\mu_2$ came out as 7.50% and was coded as 3 : $\mu_1$ was 1.67%, coded as 0.67.

To apply this method, the values of $f\mu$ are cumulated, column (5). The aggregate output in each stratum should be 82,736/3, or 27,579. For the first stratum, the nearest value of cum. $f_\mu$ is 31,721 for the interval 0-20%. Since $2 \times 27,579 = 55,158$, the second stratum should extened to 35%, at which cum. $f_\mu$ is 56,861.

|          | Stratum |         |          |
|----------|---------|---------|----------|
|          | 1       | 2       | 3        |
|          | 0–20%   | 20–35%  | 35–100%  |
| Output   | 31,721  | 25,140  | 25,875.  |

IV. $W_h(y_h-y_{h-1})=constant$. In this method we equalize the product of the cumulative frequency within the stratum and the width of the stratum. The first step is to cumulate the $f$ values, column (6). The method is a little troublesome to apply, because the value of

$$\sum_h W_h(y_h-y_{h-1})$$

is not constant, depending both on $L$ and on the positions of the boundaries. Hence, for given $L$, it is not at first clear to what figure we should try to equate $W_h(y_h-y_{h-1})$ in the individual strata. A rough guide suggested by Ekman is to compute the product for $L=1$, the value in this case being

$$Q=(\text{cum.}f)\,(y_L-y_0)=(13,435)\,(100)=1,343,500$$

For $L$ strata, the constant value per stratum is approximately $Q/L^2$, or in this example $1,343,500/9\doteq149,000$. This relationship is exact for a rectangular distribution, but tends to give too high a result for skew distributions. As a trial value for the first stratum, we take the interval $0-15\%$, for which $W_h(y_h-y_{h-1})=(8,137)\,(15)$ $=122,055$. Further examination shows that the best partition is to extend the second stratum up to $35\%$, with the following results.

|                        | Stratum  |          |          |
|------------------------|----------|----------|----------|
|                        | 1        | 2        | 3        |
|                        | 0–15%    | 15–35%   | 35–100%  |
| $W_h(y_h-y_{h-1})$     | 122,055  | 79,620   | 85,605.  |

V. *Equal intervals on* cum.$(r+f)$. For this we need the cumulative of a rectangular distribution, column (7). Each class is given a frequency $13,435/20=671.75$. Since cum.$(r+f)$ over all classes is 26,870, each stratum for $L=3$ should contain 8,957. Although a separate column of values of cum.$(r+f)$ can be written down if desired, the strata are found quickly by adding columns (6) and (7) at likely boundary points. The best partition is shown below.

|                | Stratum |         |          |
|----------------|---------|---------|----------|
|                | 1       | 2       | 3        |
|                | 0–15%   | 15–45%  | 45–100%  |
| Cum.$(r+f)$    | 10,152  | 8,653   | 8,065    |

It sometimes happens that two different partitions appear to satisfy the criterion about equally well. In the present instance the strata $0-10\%$, $10-35\%$, $35-100\%$ give $7,324$; $9,496$; $10,050$ for cum.$(r+f)$. In such cases some working rule must be used to make a choice, e.g. taking the partition with the smallest range or the smallest mean deviation.

All four methods locate the boundaries quickly, any difference in time of application being too small to be a major factor. In this example the four methods gave different boundaries, although frequently two methods will agree.

## 5. *Relative precisions of the rules*

For each stratum given by each rule, the value of

$$nV(\bar{y}_{st})=\left(\sum W_h\,\sigma_h\right)^2$$

was computed for $L=2, 3, 4$. In addition, the boundaries that minimize this variance were found by trial. The ratios of $V/V_{min.}$ appear in Table 3. The eight sets of data are arranged from top to bottom roughly in order of increasing departure from a rectangular distribution.

It is clear from Table 3 that the cum. $\sqrt{f}$ rule and the $W_h(y_h-y_{h-1})$ rules have performed consistently well. The cum. $\sqrt{f}$ rule coincided with the actual minimum in 12 of the 24 cases, and in all other cases except one ($L=2$ for income) it gave a variance only trivially higher than the minimum variance. The $W_h(y_h-y_{h-1})$ rule found the minimum 6 times out of the 24, its worst result being a variance ratio 1.16 for $L=2$ with Number of Farms per Sampling Unit. The average of the variance ratios over all 24 cases was close to 1.03 with both rules.

The cum.$(r+f)$ rule did fairly well except on the two most skew populations. The average of $V/V_{min}$ was 1.08 for the first six frequency distributions, but with Number of Farms and Gross Income this method gave some poor boundaries, with an average variance ratio of 1.55 over these six cases.

The "equal aggregate output" rule was relatively unsuccessful on the three "loan" distributions that exhibit the least skewness. This is not surprising, since this rule is not designed to work well for a rectangular distribution with lower end at zero. Over the other distributions it performed erratically, doing well for Bank

Table 3. *Ratios of $V(\bar{y}_{st})$ to $V_{min}(\bar{y}_{st})$ for the different rules.*

| | Agricultural loans | | | | Real estate loans | | | |
|---|---|---|---|---|---|---|---|---|
| $L$ | Cum. $\sqrt{f}$ | $W_h\,\mu_h$ | $W_h(y_h-y_{h-1})$ | Cum. $(r+f)$ | Cum. $\sqrt{f}$ | $W_h\,\mu_h$ | $W_h(y_h-y_{h-1})$ | Cum. $(r+f)$ |
| 2 | 1.03 | 1.36 | 1.03 | 1.03 | 1* | 1.19 | 1.06 | 1.06 |
| 3 | 1.03 | 1.89 | 1.05 | 1* | 1* | 1.28 | 1* | 1.07 |
| 4 | 1* | 1.96 | 1* | 1* | 1* | 1.39 | 1.05 | 1.10 |
| | Industrial loans | | | | Bank resources | | | |
| 2 | 1* | 1.07 | 1.07 | 1.07 | 1.03 | 1.01 | 1.03 | 1.03 |
| 3 | 1* | 1.35 | 1.07 | 1.14 | 1* | 1.07 | 1.04 | 1.12 |
| 4 | 1.01 | 1.42 | 1.01 | 1.21 | 1.09 | 1.05 | 1.09 | 1.22 |
| | College Students | | | | City populations | | | |
| 2 | 1.02 | 1.06 | 1.02 | 1.02 | 1* | 1.03 | 1* | 1* |
| 3 | 1.08 | 1.18 | 1.00 | 1.08 | 1* | 1* | 1.01 | 1.16 |
| 4 | 1.08 | 1.16 | 1.02 | 1.10 | 1.03 | 1.02 | 1.00 | 1* |
| | No. of Farms per sampling unit | | | | Gross income | | | |
| 2 | 1* | 1* | 1.16 | 1.16 | 1.25 | 1.70 | 1* | 1* |
| 3 | 1* | 1.22 | 1* | 1.45 | 1.02 | 1.81 | 1.08 | 2.76 |
| 4 | 1.02 | 1* | 1* | 1.39 | 1* | 1.69 | 1.05 | 1.52 |

A * indicates that the boundaries given by the rule were those that make the variance a minimum.

Resources, City Population and (except for $L=3$) for Farms per Sampling Unit, and poorly for Family Incomes.

In Section 2 this rule was related to the exact solution by (i) noting that with the optimum boundaries, the products $W_h \sigma_h$ might be expected to be approximately constant, and (ii) using an observation that with good strata the ratio $\sigma_h / \mu_h$ is often approximately constant for distributions found in practice. In the examples worked here, the values of $W_h \sigma_h$ in the optimum strata do not diverge greatly, but the coefficients of variation are far from constant.

Since the rate at which $V(\bar{y}_{st})$ is decreased as the number of strata are increased is of interest, Table 4 shows the successive ratios $V_2/V_1$, $V_3/V_2$, $V_4/V_3$ for each distribution, where the suffix denotes the number of strata. These ratios were calculated for the boundaries giving the minimum values of $V$. The distributions are arranged in order of increasing skewness from left to right.

As an approximation, valid for the rectangular distribution, Dalenius (1957) has suggested that $V_L / V_{L-1} \doteq (L-1)^2 / L^2$. This formula gives 0.250, 0.444 and 0.562 for $L=2, 3, 4$ respectively. The averages in Table 4 lie close to these values, but slightly below them. So far as it goes, the table does not indicate any relation between the degree of skewness and the rate of reduction of variance.

Table 4. *Ratio of $\cdot V_{\min}$ for $L$ strata to $V_{\min}$ for $(L-1)$ strata.*

| $L$ | Frequency distribution | | | | | | | | Ave. | $(L-1)^2/L^2$ |
|---|---|---|---|---|---|---|---|---|---|---|
| | 1 | 2 | 3 | 4 | 5 | 6 | 7 | 8 | | |
| 2 | .225 | .266 | .288 | .205 | .178 | .200 | .281 | .214 | .232 | .250 |
| 3 | .402 | .459 | .431 | .410 | .426 | .434 | .377 | .457 | .424 | .444 |
| 4 | .539 | .555 | .537 | .530 | .623 | .550 | .444 | .572 | .544 | .562 |

With two of the distributions (College Students, 1952-3 and City Populations, 1940) it was possible to study the effectiveness of the stratifications set up from these data on the precision of later surveys. For the colleges, bivariate distributions were worked out for 1952-3, 1955-6 enrollments and for the 1952-3, 1958-9 enrollments, simulating surveys with strata constructed from data 3 and 6 years old. Similarly, the 1940 City Population strata were tested on 1950 City Population data.

Optimum allocation of the sample sizes $n_h$ was based on the s.d.'s at the earlier time. If $\sigma_h$, $S_h$ denote the s.d.'s in stratum $h$ at the earlier and later times, respectively, we have

$$\frac{n_h}{n} = \frac{W_h \sigma_h}{\sum W_h \sigma_h}$$

and for the variance in the later survey

$$V(\bar{y}_{st}) = \sum \frac{W_h^2 S_h^2}{n_h} = \frac{1}{n} \left( \sum \frac{W_h S_h^2}{\sigma_n} \right) \left( \sum W_h \sigma_h \right).$$

As often happens, the $S_h$ and $\sigma_h$ were close enough to proportionality so that the increases in variance caused by the difference between the earlier and later optimum allocations were unimportant.

For each rule, Table 5 shows the average values of $V/V_{\min}$ in the earlier and later periods. The data for the earlier period are the simple averages over the six relevent cases in Table 3. For the later period, the $V/V_{\min}$ ratios for 1955-6 and 1958-9 College enrollments were first averaged, these means being then averaged with the City Population figures so as to give each variable equal weight. Note that the $V_{\min}$ values for the later period are obtained from the best boundaries that could be constructed from the *earlier* data.

Table 5.   *Relative effectiveness of the rules at earlier and later periods. Values of $V/V_{\min}$.*

| Distribution | Rule | | | |
|---|---|---|---|---|
| | Cum. $\sqrt{f}$ | $W_h\,\mu_h$ | $W_h(y_h - y_{h-1})$ | Cum. $(r+f)$ |
| Earlier | 1.03 | 1.09 | 1.01 | 1.06 |
| Later | 1.05 | 1.10 | 1.04 | 1.06 |

For these two distributions, all rules did well when applied to the current data and almost equally well when used some years later.

The successive ratios $V_L/V_{L-1}$ are also of interest. When computed from later data these ratios would be expected to be higher and to approach 1 as $L$ becomes larger, because of the imperfect correlation between the $y$ values at the two times. For $L=2, 3, 4$, the average ratios over the two distributions were 0.230, 0.556 and 0.732 as against 0.189, 0.430 and 0.586, respectively, at the earlier time.

## 6. Stratification with equal number of units per stratum

In the previous sections the optimum boundaries were studied under the condition that the stratum sample sizes $n_h$ are chosen to minimize the variance for fixed total sample size. In this section we consider the situation in which, for administrative convenience or other reasons, it is decided to have the same sample size in all strata. It can be anticipated that this restriction will lead to somewhat different optimum stratum boundaries and to higher values of $V(\bar{y}_{st})$. By the methods given by Dalenius (1950), the equations satisfied by the best boundary points $y_h$ are easily found to be

$$W_h\{\sigma_h^2 + (y_h - \mu_h)^2\} = W_{h+1}\{\sigma^2_{h+1} + (y_h - \mu_{h+1})^2\}$$

These are, of course, different from the equations (3) in Section 2 for optimum boundaries under optimum choice of the $n_h$.

On the other hand, it was pointed out in Section 2 that one method of approximating the optimum boundaries with optimum choice of the $n_h$ is to construct the strata so that

$$W_h\,\sigma_h = \text{constant.}$$

If this result actually holds for the optimum boundaries, it follows that the optimum $n_h = n\,W_h\,\sigma_h \Big/ \sum W_h\,\sigma_h$ reduce to

$$n_h = \frac{n}{L} = \text{constant.}$$

This possibility suggests three related questions that can be examined from our data.

(i). For the best boundaries found for optimum choice of $n_h$, are the values of $W_h \sigma_h$ approximately constant?

(ii). If the best boundaries for optimum $n_h$ are used, is there much loss of precision if the sample is drawn with $n_h$ constant?

(iii). Do the best boundaries for constant $n_h$ differ much from those for optimum $n_h$?

With regard to question (i), the coefficient of variation of $W_h \sigma_h$ was computed for each of the 24 sets of optimum boundaries given by the 8 distributions for $L=$ 2, 3, 4. These coefficients averaged about 0.1 for $L=2$, 0.2 for $L=3$ and 0.3 for $L=4$. Thus the values of $W_h \sigma_h$, although not close to constant in a mathematical sense, do not diverge too widely. In the majority of the sets of boundaries there was some evidence of a downward trend in $W_h \sigma_h$ from the lowest to the highest stratum.

Question (ii) is more directly relevant to sample survey practice. For each of the 24 optimum strata for optimum $n_h$, the value of $V(\bar{y}_{st})$ was computed with $n_h = n/L$ and with the optimum $n_h$. Except for one case which requires special treatment, the ratios

$$V_{\text{equal } n_h} / V_{\text{opt. } n_h}$$

were only slightly above 1, the average being 1.04. The exception was the discrete distribution Number of Farms per sampling unit for $L=4$, which gave a ratio 1.36. The lowest of the optimum strata consisted of all sampling units with one farm. Thus the variance within this stratum was zero, and with the optimum choice of $n_h$ this stratum would not be sampled at all. In practice, part of the sample is allocated as a precaution to a stratum (e.g. blocks with no houses) that seems to have zero variance in the advance calculations. Excluding this case, the use of equal $n_h$ for administrative convenience brings about only a trivial increase in $V(\bar{y}_{st})$.

Coming to question (iii), the boundaries that are optimum for equal $n_h$ were found to differ only slightly from those that are optimum for optimum $n_h$, the two sets being the same in the majority of the 24 cases.

A fourth question is: how well do the approximate rules perform if the objective is to obtain the best boundaries for equal $n_h$? The four rules were found to rank themselves in the same order, the cum. $\sqrt{f}$ and $W_h(y_h - y_{h-1})$ being best, cum. $(r+f)$ next and equal output least successful. In general, the ratios $V/V_{\min}$ (corresponding to those given in Table 3) tended to run a little higer. Apparently the minimum of $V$ for varying choice of boundaries is not as flat with equal $n_h$ as with optimum $n_h$.

For the bulk of the computations I wish to thank Joseph Sedransk, Theodore Ingalls Jr. and Ethan Jacob.

## 7. Summary

This paper compares four working rules that have been proposed for approximating the best stratum boundaries when a frequency function $f(y)$ is to be subdivided into $L$ strata. The best boundaries are defined as those that give the smallest variance $V(\bar{y}_{st})$ for the estimated population mean from a stratified sample of size $n$, with optimum choices of the sample sizes $n_h$ in the individual strata.

The four rules are: (i) take equal intervals on the cumulative of $\sqrt{f}$: (ii) choose the boundaries so that $W_h \mu_h = $ constant, where $W_h$ is the total frequency in stratum $h$ : (iii) choose the boundaries so that $W_h(y_h - y_{h-1})$ is constant, where $y_{h-1}$, $y_h$ are the lower and upper bounds of stratum $h$ : (iv) take equal intervals on the cumulative of $(r+f)$, where $r$ is a rectangular distribution with the same total frequency as $f$.

These rules are compared by applying them for $L=2$, 3, 4 to eight skew frequency distributions intended to be somewhat representative of those that occur in practice. In order of increasing skewness, the distributions are as follows: proportion of total bank loans devoted to agricultural, real estate and industrial loans (3 distributions), resources of large banks, numbers of students in four-year colleges, populations of cities, number of farms per area sampling unit and gross incomes from tax returns, all data being from U.S. sources.

All four rules were found to construct the stratum boundaries quickly, with little ambiguity. As a measure of their performance, the ratio $V/V_{min}$ was used, where $V$ is the variance of $\bar{y}_{st}$ given by the strata which the rule sets up, and $V_{min}$ is the corresponding variance given by the best stratum boundaries, found by trial. Rule (i)—equal interval on cum. $\sqrt{f}$—and rule (3)—$W_h(y_h - y_{h-1})=$constant—performed consistently well over the 24 sets of strata, the average of $V/V_{min}$ for each rule being 1.03. Rule (4)—equal intervals on cum. $(r+f)$—was satisfactory except on the two extremely skew distributions (number of farms and gross income). Rule (2) was satisfactory on only four of the eight distributions.

With the best boundaries, the rate at which $V(\bar{y}_{st})$ decreases with increasing numbers of strata agreed well with the formula $V_L/V_{L-1}=(L-1)^2/L^2$ suggested by Dalenius, the figures from the distributions being 0.232, 0.424 and 0.544 for $L=2, 3, 4$ as compared with 0.250, 0.444 and 0.562 given by the approximate formula.

One interesting result is that with the optimum boundaries the values of $W_h \sigma_h$ were sufficiently close to constant so that surveys taken with equal $n_h$ in all strata were almost as precise as surveys with optimum $n_h$. Further, the optimum boundaries for stratification with equal $n_h$ differed little if at all from those for stratification with optimum $n_h$, and the two best rules gave very satisfactory boundaries for use in surveys in which it is intended to make $n_h$ constant. For administrative reasons, stratification with equal $n_h$ is often convenient.

To simulate practical conditions more closely, three bivariate distributions were included in which strata set up from earlier data were used in surveys 3, 6 and 10 years later. In all three cases the strata constructed by the rules were nearly as good as the best strata that could be set up from the earlier data. As would be expected, however, the variances declined less rapidly with increasing numbers of strata.

### References

Dalenius, T. (1950):    The problem of optimum stratification. *Skandinavisk Aktuarietidskrift*, 203-213.

—————— (1957): *Sampling in Sweden*. Chapter 8. Almqvist & Wiksell, Stockholm.

Dalenius, T. and Gurney, M. (1951): The problem of optimum stratification II. *Skandinavisk Aktuarietidskrift*, 133-148.

Dalenius, T. and Hodges, J. L. (1957): The choice of stratification points. *Skandinavisk Aktuarietidskrift*, 198-203.

Dalenius, T. and Hodges, J. L. (1959): Minimum variance stratification. *Journal of the*

*American Statistical Association.* 54, 88–101.

Durbin, J. (1959): Review of "Sampling in Sweden". *Journal of the Royal Statistical Society.* A. 122, 246–248.

Ekman, G. (1959): An approximation useful in univariate stratification, *The Annals of Mathematical Statistics*, 30, 219–229.

Hansen, M. H., Hurwitz, W. and Madow, W. G. 1953: *Sample survey methods and theory.* Vol. 1, John Wiley & Sons, New York.

Kitagawa, T. (1955): Some contributions to the design of sample surveys. *Sankhya* 14, 317–362.

Mahalanobis, P. C. (1952): Some aspects of the design of sample surveys. *Sankhya* 12, 1. 7.

———— (1954). The national sample survey: some aspects of the sample design. *Sankhya* 16, 265–316.

Stevens, W. L. (1952): Samples with the same number in each stratum. *Biometrika* 39, 414–416.

## Sources of data

U.S. Treasury Department: *Statistics of income*, 1951, (adjusted gross incomes in tax returns).

*Warld Almanac.* 1954, 1957, 1960. The World Telegram, New York, (numbers of college students, 1952-3, 1955-6, 1958-9. City populations, 1940, 1950.

*International Bankers Directory.* Rand-McNally Co., New York, 1957, (resources of large commercial banks).

West, Q. M. (1951): *The results of applying a simple random sampling process to farm management data.* Agricultural Experimental Station, Cornell University, (number of farms per sampling unit).

McEvoy, R. H. (1956): Variation in bank asset portfolios. *Journal of Finance.* 11, 463–473, (ratio of agricultural, commercial and industrial, and urban real estate loans to gross loans and discounts).

# RÉSUMÉ

Etant donné une fonction de frequence $f(y)$, $y_0 \le y \le y_L$, le problème est de choisir les limites $y_1, y_2, \cdots y_{L-1}$ qui divisent l'intervalle de $y_0$ a $y_L$ en $L$ strates pour l'échantillonage stratifié. Les meilleures limits sont celles pour lesquelles la variance de l'estimation de la moyenne de la population est minimum, la taille de l'échantillon en strate $h$ étant choisie par répartition au sens de Neyman.

Dalenius (1950) a trouvé les équations qui determinent les limites $y_i$, mais ces équations sont difficiles à appliquer en pratique. Quatre méthodes d'approximation pour determiner les meilleures limits ont été developpées, comme suit.

1. Construire des intervalles egaux sur la frequence cumulative de la distribution de $\sqrt{f(y)}$. (Dalenius and Hodges, 1957)

2. Prendre $W_h \mu_h =$ constante (Mahalanobis, 1952; Hansen, Hurwitz and Madow, 1953), où $W_h$ est la grandeur et $\mu_h$ la moyenne de la strate $h$.

3. Prendre $W_h(y_h - y_{h-1}) =$ constante (Ekman, 1959).

4. Construire des intervalles égaux sur la frequence cumulative de la fonction $f(y) + r(y)$, ou $r(y)$ est une distribution rectangulaire. (Durbin, 1959).

Ces méthodes sont comparées en les appliquant, avec $L = 2, 3, 4$, a huit distributions de frequences asymétriques, qui sont plus ou moins typiques de celles rencontrées dans l'échantillonage des institutions. Les distributions sont:

Proportion des prêts bancaires consacrés aux prêts (i) agricoles (ii) industriels

(iii) fonciers—3 distributions.

Fonds des grandes banques.

Population des villes des Etats-Unis.

Nombre d'étudiants dans les colleges americains.

Nombre de fermes par unité d'échantillon.

Revenu brut imposable des contribuables.

On a trouvé, pour chaque méthode, la variance $V$ de l'estimation de la moyenne de la population et aussi les limites qui donnent la variance minimum $V_{min}$.

Par l'une quelconque de ces quatres méthodes, les limites sont calculées rapidement. Les méthodes 1 et 3 ont donné toujours de bons résultats, le rapport moyen $V/V_{min}$ etant 1,03 par les deux méthodes. La méthode 4 fut satisfaisante, sauf pour les deux distributions les plus asymétriques, mais les résultats de la méthode 2 furent irréguliers.

Avec les limites optimum, le degré de décroissance de la variance, en fonction de $L$, fut d'accord avec la règle d'approximation donnée par Dalenius.

Un résultat interessant est que, pour les limites optimum, la répartition au sens de Neyman a donné pratiquement les mêmes tailles d'échantillon dans toutes les strates.

# Amphetamine Sulfate and Performance

## A Critique

*William R. Pierson, Ph.D., Los Angeles*

IN THE COURSE of reviewing literature for some recent research,[1,2] I noticed particularly a series of papers presented by Smith and Beecher.[3-5] These studies have been widely cited, and many popular sports magazines have mentioned them in connection with the "prevalence" of drug usage among athletes. A critical evaluation, particularly of the first of these studies, reveals serious defects in the collection of the data, naivete and errors in statistical techniques and interpretation, and conclusions which are not substantiated by the data.

Much of the data collected by Smith and Beecher are, by their own admission, unreliable. This critique will not concern itself with those data, nor with the data collected in the "fatigue" aspect of swimmers. It is doubtful if many swimmers begin a race suffering from acute fatigue.

The problem of the first study, as stated by Smith and Beecher, does not concern itself as to whether *any* change results, only whether an *improvement* results. In this instance, the 1-tailed test of significance of differences is the appropriate test. They used the 2-tailed test throughout their study; consequently, none of the probability figures they give for the t test are appropriate. Further, they used various probability levels, with 0.20 as the maximum for indicating statistical significance. This means that any difference which could have occurred by pure chance 20 times out of a 100 is a basis for establishing a "fact." Henry[6] has pointed out the dangers of using a "floating" level of significance, and McNemar[7] states that the 0.05 level is a rather low level for announcing something as "fact." He suggests a criterion of 0.01 probability for statistical significance.

In their first study Smith and Beecher present what they consider to be reliable data on the following groups: 15 swimmers, 9 indoor runners, 9 marathon runners, 6 weight throwers (35 lb.), 4 shot-putters (16 lb.), and 14 swimmers motivated by a reward of steak dinners worth up to $15. They have

Assistant Director, Biokinetics Research Laboratory, Los Angeles County Osteopathic Hospital.

discussed these groups under 6 experiments, and this critique will do the same.

*Experiment 1.*—In this experiment the 15 swimmers were tested. Measurement was by means of a chronoscope which was activated by the starting gun and at least one stop watch. The chronoscope and one stop watch were used on approximately one-half of the races. Although the times were longer for the chronoscope than for the stop watch, in "coordinating" the data this difference was taken into account. Just how this is possible is not explained, but all time information was examined for each race, and the *median* was accepted as accurate. Yet the statistical technique used for evaluating the differences between amphetamine and placebo performances was the t test for correlated *means*. If the median is more accurate in this instance than the mean, there is little justification for using the cited technique; the median test as described by Mood[8] would be more appropriate. Although one would expect greater differences between swimming events than between 100- and 200-yard sprints, the authors divided the swimmers into a 100-yard group, which included free-style and butterfly events, and a 200-yard group, which included free-style, back stroke, and breast stroke events. The scores of the swimmers, as a group, demonstrated a significant difference between amphetamine and placebo performances.

*Experiment 2.*—The subjects of this experiment were 9 indoor runners, 3 each in 600-yard, 1,000-yard, and 1-mile events. The data on one test session are unreliable because of the poor condition of the track. There were no statistically significant differences between placebo and amphetamine performance.

*Experiment 3.*—The design of this experiment was to have been for 3 subjects to run a mile on an outdoor wooden track, 3 subjects to run 600 yards each, and 2 to run 440 yards. However, the data were considered unreliable because the subjects were uncooperative; the track was torn up mid-way through the testing period; there were scheduling difficulties, temperatures varied from 25° F. to 69° F., nd wind velocities were from 8 to 23 miles per our. However, these data were not rejected. None f the differences were significant, even under the mith and Beecher definition, so the data were ombined with Experiments 2 and 4, and they state, In all three experiments using runners as subjects, he groups ran faster under the influence of mphetamine than under placebo." Experiments 2 nd 3 resulted in no significant differences, and Experiment 4 resulted in differences significant only t the 0.10 level (2-tailed). These results indicate hat the sample differences were due entirely to hance and that no "true" difference exists.

*Experiment 4.*—Nine runners training for the Boston marathon participated, and their runs were rom 4.5 to 12.7 miles each. The runners recorded heir own times, and differences in scores between mphetamine and placebo performances were sig-ificant only at the 0.10 (2-tailed) level. The authors tate, "Weather variables added to the variability f the data and tended to reduce the statistical sig-ificance of the improvement shown under amphet-mine by the group." Here again, "improvement" as not been demonstrated, and it is hard to be-ieve that wind and weather would adversely affect nly amphetamine performance and favor only lacebo performance. The authors further state, There were reasons to question the validity of the ata contributed by all three of the subjects who an faster under influence of placebo than am-hetamine."

*Experiment 5.*—The subjects for this experiment vere 9 weight-throwers (35 lb.) and 4 shot-putters 16 lb.). However, the data on 3 of the weight-hrowers were considered unreliable because 2 of hem did not complete all sessions, and 1 subject hanged his style mid-way through the experiment. t is interesting to note that the first 2 of these were he only subjects whose performances were better nder placebo than under amphetamine. The dis-ances thrown by the subjects were not measured, nly estimated by the experimenter (from about 0 ft. away) and by the subject who walked to the ndentation made by the shot or weight. Estimation vas facilitated by drawing chalk-lines every 5 ft. vithin the throwing range. The authors state that he agreement between 2 estimates was *generally* ess than 1 ft. This does not appear to be a very ccurate measure when one considers that the ifferences in reliable mean scores were 1.9 ft. for he weight-throwers and 1.7 for the shot-putters. Nonetheless, Smith and Beecher report that the robability of the differences being due to chance re 0.02 for the weight-throwers and 0.20 for the hot-putters. The same probability figures result rom analyses of both maximum throw and mean hrowing distance.

*Experiment 6.*—The 16 swimmers of this study vere motivated by rewards of steak dinners in an ttempt to simulate the pressure of actual compe-tition, but the data from 2 of the subjects were con-sidered unreliable. The grouping of the subjects was as in Experiment 1 and so was the method of measurement. The probability associated with the difference between placebo and amphetamine for the 14 subjects was 0.10.

The value of the entire study would have been greatly enhanced had the analysis of variance been used rather than individual t tests. This would have resulted in one score which would have indicated whether or not the performances under amphet-amine sulfate are superior to those under placebo, as was the stated purpose of the study. As it is, Smith and Beecher have reported a total of 168 t scores, and the "principle of indifference" would lead one to expect 34 of these to be "significant" at the 0.20 level by pure chance alone. None of the 6 experi-ments demonstrated differences between placebo and amphetamine performances which could not be attributed to chance, i.e., a probability of 0.01 or less. They have thus presented no evidence that a "true" difference does exist, and conclusions to that effect are not justified by the data.

The second study of Smith and Beecher utilized the data from the first study plus the results of a questionnaire which was administered at the same time as the objective data were collected. The main finding of this study is, "The subjective responses of the athletes indicated that 14 mg. of amphetamine sulfate might improve athletic performance by in-fluencing a variety of factors." There also exists, of course, the possibility that it might not. It should be noted that the dosage was administered on a body weight basis, and the 14 mg. is for a 154-lb. subject, a fact not made clear in the summary or abstracts.

The third study in the series is by far the best prepared. The study concerns itself with the effects of amphetamine sulfate on the judgment of time by swimmers. Objective data were collected in the same manner as in Experiment 1 of the first study, and a questionnaire also was administered. Whether or not the data of Experiment 1 of the first study were used as the objective data here, if they were collected simultaneously, or if this is a separate study using different subjects, is not quite clear. The t test was used to evaluate the objective data and the binomial sign test to evaluate the question-naire data. The analyses of the objective data indi-cated no statistically significant difference between placebo and amphetamine performances, and of the 32 analyses of the questionnaire data, the only one to demonstrate a difference significant at the 0.01 level of probability was an instance where the placebo had greater effect than the amphetamine. Smith and Beecher state that, under the dosages administered, no positive conclusion concerning the effect of amphetamine sulfate on judgment dis-tortion can be drawn.

The data on the effects of secobarbital on athletic performance have not been evaluated in this critique. Although Smith and Beecher have presented such data in all 3 studies, the primary purpose of the studies, as reported by them, was to determine if amphetamine sulfate improved the performances of trained athletes.

1721 Griffin Ave., Los Angeles 31.

### References

1. Rasch, P. J.; Pierson, W. R.; and Brubaker, M.: Effects of Amphetamine Sulfate and Meprobamate on Reaction Time and Movement Time, *Int Z Angew Physiol* 18: 280-284, 1960.

2. Pierson, W. R.; Rasch, P. J.; and Brubaker, M.: Some Psychological Effects of Administration of Amphetamine Sulfate and Meprobamate on Speed of Movement and Reaction Time, *Med Del Sport* 1:61-66 (Feb.) 1961.

3. Smith, G. M., and Beecher, H. K.: Amphetamine Sulfate and Athletic Performance: I. Objective Results, *JAMA* 170:542-557 (May 30) 1959. Corrections appeared in *JAMA* 170:1432-1433 (July 18) 1959.

4. ———, ———: Amphetamine, Secobarbital, and Athletic Performance: II. Subjective Evaluation of Performance, Mood States, and Physical States, *JAMA* 172:1502-15 (April 2) 1960.

5. ———, ———: Amphetamine, Secobarbital, and Athletic Performance: III. Quantitative Effects of Judgment, *JAMA* 172:1623-1629 (April 9) 1960.

6. Henry, F. M.: On Use of Rubber Yardstick in Evaluating Difference, *Res Quart* (no. 2) 30:241 (May) 195.

7. McNemar, Q.: *Psychological Statistics*, New York: John Wiley & Sons, Inc., 1949.

8. Mood, A. M.: *Introduction to Theory of Statistics*, New York: McGraw-Hill Book Company, Inc., 1950.

# A Reply

*William G. Cochran, M.A., Gene M. Smith, Ph.D., and Henry K. Beecher, M.D., Boston*

THE COMMENTS by Dr. Pierson, concerning our work with athletes (as cited previously), represent a remarkable demonstration of misunderstanding and misrepresentation of published findings. Most of his critique is devoted to the question of whether the data reported in the first of our 3 papers justify the conclusion that amphetamine improved the performance of the athletes we studied. His conclusion is that "none of the 6 experiments demonstrated differences between placebo and amphetamine performances which could not be attributed to chance, i.e., a probability of 0.01 or less. They [Smith and Beecher] have thus presented no evidence that a 'true' difference does exist, and conclusions to that effect are not justified by the data."

The facts are that 57 athletes were tested. They performed after receiving placebo and after receiving amphetamine sulfate (14 mg. per 70 kg. of body weight). Their scores are presented in the table. A plus score indicates better performance after amphetamine than after placebo; a minus indicates poorer performance after amphetamine. (With the exception of the scores for Experiment 3, the individual scores in the table were all presented in the original publication. The mean effect in the third experiment was reported and was labeled "not significant.")

In the first experiment, 14 of the 15 swimmers (93%) performed better after amphetamine than

after placebo; this was true also of 19 of the 2 runners (73%) in Experiments 2, 3, and 4; 11 the 13 weight-throwers (85%) in Experiment 5; an

### Comparative Effects of Amphetamine (14 Mg/70 Kg. of Body Weight) and Placebo on Performance of Athletes*

| Experiment | Swimmers 1 | Runners 2 | Runners 3 | Runners 4 | Weight-Throwers 5 | Swimmers 6 |
|---|---|---|---|---|---|---|
| | +0.94 | +0.71 | −1.70 | +1.34 | +0.16 | +1.98 |
| | +1.14 | +0.65 | +3.55 | −0.30 | +3.58 | +1.78 |
| | +1.01 | +2.38 | −0.24 | +0.34 | +8.27 | −0.30 |
| | +1.60 | +3.01 | +2.94 | −1.17 | +9.50 | +1.24 |
| | +0.07 | +3.36 | +3.61 | +2.44 | +2.69 | +1.69 |
| | +4.77 | +5.11 | +3.05 | −1.53 | +2.96 | −0.67 |
| | +0.02 | +5.66 | −6.29 | +3.64 | +9.42 | +0.29 |
| | +0.62 | +4.73 | +2.21 | +1.59 | +2.90 | −0.93 |
| | +1.45 | −5.75 | | +3.28 | +0.75 | +1.07 |
| | +1.97 | | | | +3.67 | +1.53 |
| | +1.05 | | | | −2.22 | −0.54 |
| | −0.12 | | | | −6.81 | +2.29 |
| | +0.90 | | | | +6.23 | +0.38 |
| | +1.61 | | | | | −0.20 |
| | +0.37 | | | | | +0.54 |
| | | | | | | +0.11 |

| | Experiment | Mean Difference/ % | T Value | Significance Level Attained (2-Tailed) | (1-Tailed) |
|---|---|---|---|---|---|
| Individual Experiments | 1 | +1.16 | 3.82 | 0.01 | 0.005 |
| | 2 | +2.21 | 1.91 | 0.10 | 0.05 |
| | 3 | +0.89 | 0.72 | ...† | ...† |
| | 4 | +1.07 | 1.72 | 0.20 | 0.10 |
| | 5 | +3.16 | 2.46 | 0.05 | 0.025 |
| | 6 | +0.64 | 2.49 | 0.05 | 0.025 |
| Combined Experiments | 1-5 | +1.77 | 4.14 | 0.0001 | 0.00005 |
| | 2-6 | +1.61 | 3.78 | 0.0002 | 0.0001 |

* Plus score indicates improvement in performance due to amphetamine; minus indicates impairment. Each subject's score is per cent change score obtained by dividing difference between his amphetamine and placebo scores by his placebo score, and then multiplying that value by 100. All subject scores are based on performance made under rested conditions. Scores for weight-throwers are based on the mean rather than maximum throw of each subject each day.
† Not significant.

From the Department of Statistics, Harvard University (Prof. Cochran); and from the Anaesthesia Laboratory, Harvard Medical School at the Massachusetts General Hospital (Drs. Smith and Beecher).

Since Prof. Cochran served as a statistical adviser on the study of amphetamine and athletic performance, he was asked by Drs. Smith and Beecher to join them in writing this rebuttal.

11 of the 16 swimmers (69%) in Experiment 6. Thirteen of the swimmers appeared in both Experiments 1 and 6; to avoid duplication in presenting over-all results, we have combined Experiments 1-5, and we have combined Experiments 2-6.) In combined Experiments 1-5, we find that 81% of the athletes performed better after amphetamine, and 19% performed better after placebo. In combined Experiments 2-6, 75% did better after amphetamine, and 25% did better after placebo.

The table shows that the t test values for the comparisons between amphetamine and placebo are 4.14 for Experiments 1-5 and 3.78 for Experiments 2-6. Both are beyond the 0.0002 level of significance with a 2-tailed test and beyond the 0.0001 level with a 1-tailed test. Thus, the superiority of the amphetamine performance is beyond reasonable question.

In our original publication we described and analyzed each of the 6 experiments separately. We did not emphasize the obvious fact that the 6 experiments gave mutual support to each other. Similarly, the 2 combined analyses reported in the table of this rejoinder were not presented originally because the cross-verification of the separate experiments seemed self-evident. It apparently was not self-evident to Dr. Pierson; his erroneous conclusion, given previously, was based on an examination (one by one) of the 6 experiments described in our first paper.[1]

Dr. Pierson states that none of our 6 experiments, considered individually, showed a difference between amphetamine and placebo which reached the 0.01 level of significance. The fact is that the first experiment did achieve the level. Perhaps part of Dr. Pierson's confusion about Experiment 1 resulted from a regrettable typesetting error which reversed the labels "rested" and "fatigued" in Table 2, making it appear that the significant ($p<0.01$) amphetamine improvement effect for the 15 swimmers occurred in a "fatigued" rather than a "rested" condition. (We were not able to read the galley proof for the article in which Experiment 1 was reported as there was a publication deadline. Corrections later appeared in THE JOURNAL[2] and also in authors' reprints.) The text, however, clearly stated that the amphetamine improvement effect, which reached the 0.01 level of significance, occurred in the "rested" condition. If Dr. Pierson restricted his attention to Table 2, and ignored the text, one might understand how he could have dismissed our significant (0.01 level) finding of an improvement due to amphetamine under the "rested" condition by saying, "It is doubtful if many swimmers begin a race suffering from acute fatigue." (The collection of data under the fatigued condition was primarily for the purpose of obtaining additional information of interest concerning psychopharmacology, not for the purpose of determining whether amphetamine facilitates athletic performance under normal conditions of athletic activity.) The surprising thing, however, is that in addition to ignoring the text, Dr. Pierson apparently did not notice that Tables 1, 3, and 5 all showed quite clearly that the group improvement due to amphetamine ($N=15$) was greater under the "rested" than under the "fatigued" condition. For Dr. Pierson to throw out as irrelevant the significant data of Experiment 1, he had to ignore, misread, or misunderstand over a page of text and 3 tables.

Concerning the second experiment Pierson says, "There were no statistically significant differences between placebo and amphetamine performance." We presented 4 separate analyses of the data of Experiment 2: 1 was significant at the 0.05 level; 2 reached the 0.10 level; and 1 was not significant.

Dr. Pierson objects to our analysis combining the 3 runner experiments (Experiments 2-4); he concludes, "These results indicate that the sample differences were due entirely to chance and that no 'true' difference exists." His conclusion ignores the fact that the statistical evaluation based on the combined runner experiments yielded a probability value beyond the 0.05 level. His statement also demonstrates that he is unaware of a fundamental statistical principle: ". . . the null hypothesis is one which can be rejected but can never be proved."[2]

In commenting on the results obtained with Experiment 5, Pierson mentions that the probability value for the amphetamine-placebo difference was 0.02 for the weight-throwers (35-lb.) and 0.20 for the shot-putters (16-lb.), but he neglected to mention that the probability for all of these men combined is 0.01. (The weight-throwers also included 3 men whose results were considered less unreliable, for reasons previously discussed.[1] If these men are included in the analysis, the combined data give a significance level of 0.05, as reported in our table.)

Dr. Pierson also employs incomplete reporting to minimize the significance of the data of Experiment 6. We presented 4 statistical evaluations of the data in that experiment; 3 of them reached the 0.05 level, and 1 reached the 0.10 level. Dr. Pierson mentions only the comparison which reached the 0.10 level.

By failing to recognize the fact that the 6 experiments cross-verify each other, by employing the tactic of incomplete reporting, and by superficially and inaccurately examining the text and tables pertaining to Experiment 1, Dr. Pierson has confused himself into believing that data which yield a 2-tailed probability value beyond the 0.0001 level (Experiments 1-5 in table) should be attributed to chance.

Dr. Pierson repeatedly refers to our use of the 2-tailed rather than the 1-tailed test; nowhere, however, does he acknowledge the fact that the

2-tailed test is the more conservative choice. Use of the 1-tailed test would have necessitated the assumption that, in comparison with placebo, amphetamine could not possibly have impaired athletic performance. We did not wish to make a gratuitous assumption of this kind. To avoid any confusion, we have included both 2-tailed and 1-tailed probabilities in the table of this rejoinder. As the table shows, the 1-tailed test simply increases the statistical significance of the comparisons by cutting the p values in half.

Dr. Pierson mentions that in our original report we called attention to the fact that certain of our experiments occurred under less ideally controlled conditions than others, and that certain of our subjects were less cooperative and dependable than others. He uses our acknowledgment of that fact to imply that our experiments were undependable and our conclusions unwarranted. We mentioned, for instance, that variability of wind and weather increased the variability of performance of runners who performed outdoors during the winter, that some subjects were less cooperative and dependable than others about taking medications and reporting for sessions on time, and that one session in Experiment 2 was of questionable value because baseball practice had impaired the smoothness of the track that day.

Anyone who has experimented with human subjects knows that such problems are the hard facts of life in clinical and applied research. Throughout our original report we presented statistical analyses both including and excluding the questionable data. There is no scientifically ethical alternative. Moreover, such sources of uncontrolled variability tend, generally, to increase the difficulty of demonstrating, with statistical significance, experimental effects that are real and present. In our original report, and in the table of this rejoinder, we presented evaluations based on all data, including such potentially masking data. In spite of the undesirably high variability in some cases, which we pointed out, the results indicated an unquestionable improvement due to amphetamine.

We have confined our discussion primarily to the facts pertinent to Dr. Pierson's criticism that we presented "conclusions which are not substantiated by the data." His critique contains several additional criticisms as ungrounded as those already discussed. For example, in criticizing our method of presenting the data for the swimming experiments he says ". . . one would expect greater differences between swimming events than between 100- and 200-yard sprints. . . ." Examination of Tables 1 and 11 in the first report indicates that Dr. Pierson is wrong in this regard. The difference in over-all mean times between the 200- and 100-yard freestyle sprints was 79 seconds in Experiment 1, and 78 seconds in Experiment 2, whereas the greatest mean difference between swimming styles at the same distance (200-yard free style versus 200-yard breast stroke) was 35 seconds in both experiments. As a second example, Dr. Pierson says, "It should be noted that the dosage was administered on a body weight basis, and the 14 mg. is for a 154-lb. subject, a fact not made clear in the summary or abstracts." The second sentence of our summary of the paper he refers to reads: "The medicaments were given on a body weight basis, the doses just mentioned being those given to a subject who weighed 70 kg. (154 lb.)." [3] It seems unnecessary to deal at further length with Dr. Pierson's criticisms.

### References

1. Smith, G. M., and Beecher, H. K.: Amphetamine Sulfate and Athletic Performance, JAMA 170:542-557 (May 30) 1959. Correction, ibid., p. 1432 (July 18) 1959.
2. McNemar, Q.: Psychological Statistics, New York: John Wiley & Sons, 1949, p. 67.
3. Smith, G. M., and Beecher, H. K.: Amphetamine, Secobarbital, and Athletic Performance, JAMA 172:1502-1514 (April 2) 1961.

## SPECIAL ARTICLE

## THE ROLE OF MATHEMATICS IN THE MEDICAL SCIENCES

WILLIAM G. COCHRAN*

BOSTON

HISTORICALLY, the applications of mathematical research to biology and the medical sciences have been sporadic, depending on the interest of a few individual workers, and have not had a wide influence on general biologic thinking. The present symposium suggests that a period of renewed activity may have begun.

Mathematical reasoning can contribute to biology in many ways. It may enable the biologist to obtain quantitative estimates in situations in which his information has previously been only qualitative (for example, in estimates of the number of genes involved in the inheritance of a character). When the data are quantitative, a mathematical model based on reasonable assumptions frequently provides estimates that are of higher precision and freer from systematic error. All applications of the method of least squares, including regression and the analysis of variance, have this objective. A mathematical analysis may reveal which variables are critical and which are of minor importance to the success of some operation. In the testing of poliomyelitis vaccine, an obvious critical variable is the ratio of the amount of virus detectable by the monkey or tissue culture test to the amount that causes illness in a child.

A fruitful mathematical theory will predict the results of experiments not yet carried out — in some cases impossible to carry out. In the intensive studies of the Rhesus factor during the 1940's, Mendelian analysis predicted the existence of two genes not then discovered and the serologic properties of two new antibodies. Epidemic theory can indicate by how much the probability of a major epidemic will be reduced by immunization of any given proportion of the susceptible population in a public-health program. In evolution and the study of inbreeding, the consequences of forces acting over many generations can be worked out.

What are the demands on mathematics? Since all mathematical methods are oversimplifications, the degree and kind of simplifications employed are highly important. Although the logistic theory of population growth fits many actual populations remarkably well, the theory is so gross that it tells little about the interplay of the factors determining the growth rate. On the other hand, an overelaborate theory may also be unhelpful because it is capable of agreeing with almost any kind of biologic result that is likely to turn up. A second major requirement of a mathematical theory is that it should be identifiable with the biologic processes that are at work, the parameters in the mathematical system representing biologic entities.

Although much of the mathematics useful in the medical sciences is relatively simple, as Dr. Bernhard and Dr. Levinthal have pointed out, some stimulating problems are appearing. In a recent paper, Neyman[†] has illustrated a variety of unsolved mathematical problems arising from scientific applications of the theory of stochastic processes. In his opinion, this may be the typical situation in sophisticated applications of mathematics in the future. Indeed, it can be maintained that biomathematics has as much to contribute to mathematics as to biology.

The successful use of mathematical ideas will, of course, demand the closest co-operation and understanding between biologists and mathematicians. Anything that encourages biologists to learn more mathematics, and mathematicians to acquire more of the skills needed for applied work is likely to help. More specifically, the mathematician who produces biomathematical results might well work harder than he usually does at the problem of communicating his results to the biologists. The co-operating biologist should take pains to explain his problem fully, should not be diffident about questioning results that do not agree with his biologic intuition and should try to find some simple part of the mathematical analysis that he can work out by himself so that he has a firmer grasp of the essentials of the mathematical development.

*Professor of Statistics, Harvard University.

†Neyman, J. Indeterminism in science and new demands on statisticians. *J. Am. Statist. A.* **55**:625-639, 1960.

# On a Simple Procedure of Unequal Probability Sampling without Replacement

By J. N. K. Rao, H. O. Hartley

*Iowa State University*

and W. G. Cochran

*Harvard University*

[Received April 1962. Revised June 1962]

## Summary

A simple procedure of unequal probability sampling without replacement is proposed. It leads to an estimator of the population total having a smaller variance than is obtained by sampling with replacement. Other advantages of the present method are simplicity of calculation and the possibility of estimating exactly the variance of the estimator.

## 1. Introduction

Given is a finite population of $N$ units with characteristics $y_t$ ($t = 1, 2, ..., N$) whose total $Y = y_1 + y_2 + ... + y_N$ is to be estimated. If a sample of size $n$ is to be drawn from such a population, it is often advantageous to select the units with unequal probability. For example, such a procedure may be useful when measures of sizes $x_t$ are known for all $N$ units in the population which are positively correlated with the characteristics $y_t$. In such cases, one may utilize the knowledge of the $x_t$ by selecting units with probabilities proportional to sizes $x_t$, although this is, of course, not the only way of using the known $x_t$.

Of the literature on sampling with unequal probabilities and without replacement we mention papers by Horvitz and Thompson (1952), Narain (1951), Yates and Grundy (1953), Des Raj (1956) and Hartley and Rao (1962). There are some limitations, of varying importance, attached to all these methods. Briefly speaking, the method of Horvitz and Thompson (1952) is applicable only under severe restrictions on the prescribed probabilities, the unbiased procedures of Narain (1951), Yates and Grundy (1953) and Des Raj (1956) require a cumbersome evaluation of working probabilities, and Hartley and Rao (1962) give only asymptotic variance formulae for the estimates of $Y$ for large and moderate size populations $N$. The present method is an attempt to avoid all these disadvantages at the expense of a slight loss in efficiency. It has the following properties:

(i) It permits the computation of an estimator of the population total which has always a smaller variance than the standard estimator in sampling with unequal probabilities and with replacement.

(ii) Unlike the unbiased procedures of Narain (1951), Yates and Grundy (1953) and Des Raj (1956), the present method does not entail heavy computations, even for sample size $n > 2$, for drawing the sample or computation of the estimator and its variance estimate.

(iii) It enjoys the advantage of exact variance formulae for any population size $N$ as compared with the asymptotic variance formulae of Hartley and Rao (1962). However, a comparison with the latter method shows that the present method will, in many situations, lead to an estimator with a slightly larger variance.

(iv) An unbiased sample estimator of variance that is always positive is available for any $n, N$.

## 2. Single-stage Designs

### 2.1. *The sampling procedure*

Let $p_t$ be the probability for drawing the $t$th unit in the first draw from the whole population. For example, if we are sampling with probability proportional to size of $x_t$, $p_t = x_t/\Sigma x_t$. The sampling procedure consists of the following two stages:

(a) Split the population *at random* into $n$ groups of sizes $N_1, N_2, ..., N_n$ where $N_1 + N_2 + ... + N_n = N$.

(b) Draw a sample of size one with probabilities proportional to $p_t$ from each of these $n$ groups independently.

If the $t$th unit falls in group $i$, the actual probability that it will be selected is $p_t/\pi_i$ where

$$\pi_i = \sum_{\text{Group } i} p_t.$$

### 2.2. *Variance of the estimator of Y*

The estimator of the population total $Y$ is

$$\hat{Y} = \sum_{i=1}^{n} \frac{y_i}{p_i/\pi_i}, \tag{1}$$

where the suffixes $1, 2, ..., n$ denote the $n$ units selected from the $n$ groups separately. Let $E_2$ denote the expectation for a given split of the population and $E_1$ the expectation over all possible splits of the population into $n$ groups of sizes $N_1, N_2, ..., N_n$. Then, for any given split,

$$E_2(\hat{Y}) = \sum_{i=1}^{n} Y_i = Y$$

where

$$Y_i = \sum_{t=1}^{N_i} y_t.$$

Therefore $E(\hat{Y}) = E_1 E_2(\hat{Y}) = E_1(Y) = Y$. Since $\hat{Y}$ is conditionally unbiased, the variance of $\hat{Y}$ is

$$V(\hat{Y}) = E_1 V_2(\hat{Y}),$$

where $V_2$ denotes the variance for a given split. Now

$$V(\hat{Y}) = E_1\{V_2(\hat{Y})\} = E_1\left\{\sum_{i=1}^{n} V_2\left(\frac{y_i}{p_i/\pi_i}\right)\right\} = \sum_{i=1}^{n} E_1 V_2\left(\frac{y_i}{p_i/\pi_i}\right), \tag{2}$$

where

$$V_2\left(\frac{y_i}{p_i/\pi_i}\right) = \sum_{t=1}^{N_i} \frac{p_t}{\pi_i}\left(\frac{y_t}{p_t/\pi_i} - Y_i\right)^2 = \sum_{t<t'} p_t p_{t'}\left(\frac{y_t}{p_t} - \frac{y_{t'}}{p_{t'}}\right)^2, \tag{3}$$

the latter step in (3) following from an algebraic identity. Since $\{N_i(N_i-1)\}/\{N(N-1)\}$ is the probability, in a random split, that a pair of observations fall into the $i$th group, we have, from (3),

$$E_1 V_2\left(\frac{y_i}{p_i/\pi_i}\right) = \frac{N_i(N_i-1)}{N(N-1)} \sum_{t<t'}^{N} p_t p_{t'} \left(\frac{y_t}{p_t} - \frac{y_{t'}}{p_{t'}}\right)^2$$

$$= \frac{N_i(N_i-1)}{N(N-1)} \left(\sum_{t=1}^{N} \frac{y_t^2}{p_t} - Y^2\right). \tag{4}$$

Therefore, from (2) and (4),

$$V(\hat{Y}) = \frac{n\left(\sum_{i=1}^{n} N_i^2 - N\right)}{N(N-1)} \left(\sum_{t=1}^{N} \frac{y_t^2}{np_t} - \frac{Y^2}{n}\right). \tag{5}$$

Now, the estimator of $Y$ in sampling with replacement is

$$\hat{Y}' = \sum_{n} \frac{y_s}{np_s}, \tag{6}$$

where $\sum_{n}$ denotes the summation over the $n$ units drawn with replacement, with variance

$$V(\hat{Y}') = \sum_{t=1}^{N} \frac{y_t^2}{np_t} - \frac{Y^2}{n}. \tag{7}$$

Therefore

$$V(\hat{Y}) = \frac{n\left(\sum_{i=1}^{n} N_i^2 - N\right)}{N(N-1)} V(\hat{Y}'). \tag{8}$$

From (8) it is seen that the choice $N_1 = N_2 = ... = N_n = N/n$ minimizes $V(\hat{Y})$. Therefore, if $N/n = R$, where $R$ is a positive integer, we choose $N_1 = N_2 = ... = N_n = R$, and then (8) reduces to

$$V(\hat{Y}) = \left(1 - \frac{n-1}{N-1}\right) V(\hat{Y}'), \tag{9}$$

which clearly shows the reduction in the variance as compared to sampling with replacement. If $N$ is not a multiple of $n$, we have $N = nR + k$ where $0 < k < n$ and $R$ is a positive integer. Then we choose

$$N_1 = N_2 = ... = N_k = R+1; \quad N_{k+1} = N_{k+2} = ... = N_n = R,$$

and (8) reduces to

$$V(\hat{Y}) = \left\{1 - \frac{n-1}{N-1} + \frac{k(n-k)}{N(N-1)}\right\} V(\hat{Y}'). \tag{10}$$

For $k = 1$ or $n-1$, (10) reduces to

$$V(\hat{Y}) = \left(1 - \frac{n-1}{N}\right) V(\hat{Y}'). \tag{11}$$

### 2.3. *Estimator of variance*

Since

$$E\left(\frac{y_i^2}{p_i^2/\pi_i}\right) = E_1 E_2\left(\frac{y_i^2}{p_i^2/\pi_i}\right) = E_1\left(\sum_{t=1}^{N_i} \frac{y_t^2}{p_t}\right) = \frac{N_i}{N}\sum_{t=1}^{N} \frac{y_t^2}{p_t},$$

$$E\left(\sum_{i=1}^{n} \frac{y_i^2}{p_i^2/\pi_i}\right) = \sum_{t=1}^{N} \frac{y_t^2}{p_t}. \tag{12}$$

Also, if $v(\hat{Y})$ is an unbiased estimator of $V(\hat{Y})$,

$$E\{\hat{Y}^2 - v(\hat{Y})\} = \{E(\hat{Y})\}^2 = Y^2. \tag{13}$$

By subtracting (13) from (12) and substituting the right-hand side in (5), we find that

$$V(\hat{Y}) = E\{v(\hat{Y})\} = E\left\{\frac{\left(\left(\sum\limits_{i=1}^{n} N_i^2 - N\right)\right)}{N(N-1)}\right\}\left\{\sum_{i=1}^{n} \pi_i \frac{y_i^2}{p_i} - \hat{Y}^2 + v(\hat{Y})\right\}, \tag{14}$$

or

$$v(\hat{Y}) = \frac{\left(\sum\limits_{i=1}^{n} N_i^2 - N\right)}{\left(N^2 - \sum\limits_{i=1}^{n} N_i^2\right)}\left\{\sum_{i=1}^{n} \pi_i\left(\frac{y_i}{p_i} - \hat{Y}\right)^2\right\}. \tag{15}$$

With $N_1 = N_2 = \dots = N_k = R+1$; $N_{k+1} = N_{k+2} = \dots = N_n = R$, equation (15) reduces to

$$v(\hat{Y}) = \frac{N^2 + k(n-k) - Nn}{N^2(n-1) - k(n-k)}\left\{\sum_{i=1}^{n} \pi_i\left(\frac{y_i}{p_i} - \hat{Y}\right)^2\right\}. \tag{16}$$

When $k = 0$,

$$v(\hat{Y}) = \frac{1}{(n-1)}\left(1 - \frac{n}{N}\right)\left\{\sum_{i=1}^{n} \pi_i\left(\frac{y_i}{p_i} - \hat{Y}\right)^2\right\}, \tag{17}$$

and when $k = 1$ or $n-1$,

$$v(\hat{Y}) = \frac{1}{(n-1)}\left(1 - \frac{n}{N+1}\right)\left\{\sum_{i=1}^{n} \pi_i\left(\frac{y_i}{p_i} - \hat{Y}\right)^2\right\}. \tag{18}$$

### 3. COMPARISON WITH HARTLEY'S AND RAO'S PROCEDURE

Hartley and Rao (1962) have considered the following sampling procedure. The $N$ units in the population are listed in a *random* order, their $x_t$ are cumulated and a systematic selection of $n$ units from a random start is then made on the cumulation. A necessary condition is that $np_t \leqslant 1$ where $p_t = x_t/\Sigma x_t$. The estimator of the population total $Y$ is

$$\hat{Y}'' = \sum_{s=1}^{n} \frac{y_s}{P_s}, \tag{19}$$

where $P_s$ is the probability for the $s$th unit to be in the sample. For this method $P_s = np_s$. Therefore, when the $y_t$ are exactly proportional to the $p_t$, $V(\hat{Y}'')$ is zero which suggests that considerable reduction in $V(\hat{Y}'')$ will result when the $y_t$ are approximately proportional to the $x_t$. With the "sub-group" method of the present

25

paper it is easy to verify that the probabilities $P_s$ do not necessarily remain proportional to the $p_s$. However, it retains the above optimality property since the estimator $\hat{Y}$ has zero variance when the $y_t$ are proportional to the $p_t$.

The following asymptotic approach has been used by Hartley and Rao (1962) in developing the formulae for variance and estimated variance. In sampling without replacement, the variance in sampling with replacement, namely (7), will be the leading term in the variance and hence the terms of next lower order of magnitude in $N^{-1}$ will represent the gain in precision due to sampling without replacement. Assuming that $P_s \leqslant cN^{-1}$, where $c$ is a constant, it is shown by Hartley and Rao (1962) that

$$V(\hat{Y}'') = \left( \sum_{t=1}^{N} \frac{y_t^2}{np_t} - \frac{Y^2}{n} \right) - \frac{(n-1)}{n} \sum_{t=1}^{N} p_t^2 \left( \frac{y_t}{p_t} - Y \right)^2; \tag{20}$$

this involves the leading term, $V(\hat{Y}')$, and the term of next lower order in $N^{-1}$. This formula is valid only for fairly large values of $N$ and for values of $n$ relatively small compared to $N$. Now from (9) it is seen that

$$V(\hat{Y}) = \left( \sum_{t=1}^{N} \frac{y_t^2}{np_t} - \frac{Y^2}{n} \right) - \frac{(n-1)}{n} \frac{1}{N} \sum_{t=1}^{N} p_t \left( \frac{y_t}{p_t} - Y \right)^2, \tag{21}$$

by considering the leading term, $V(\hat{Y}')$, and the next lower order term in $N^{-1}$. We are restricting to the case of $N_i = N/n = R$ so that the variance of $\hat{Y}$ is minimized. Therefore, both procedures have the same leading term in the variance formula, namely the variance in sampling with replacement. From (20) and (21) it follows that $V(\hat{Y}'') < V(\hat{Y})$ if

$$\sum_{t=1}^{n} p_t^2 \left( \frac{y_t}{p_t} - Y \right)^2 > \frac{1}{N} \sum_{t=1}^{N} p_t \left( \frac{y_t}{p_t} - Y \right)^2. \tag{22}$$

Since sampling with probabilities proportional to size is used primarily in situations in which $y_t$ is approximately proportional to $p_t$, a simple model that seems relevant is (Cochran, 1953, p. 211)

$$y_t = Yp_t + e_t, \tag{23}$$

where $e_t$ is independent of $p_t$. In arrays in which $p_t$ is fixed, assume that

$$E(e_t) = 0; \quad E(e_t^2) = ap_t^g \quad (g > 0). \tag{24}$$

Then (22) can be written as

$$E(p_t p_t^{g-1}) - E(p_t) E(p_t^{g-1}) > 0. \tag{25}$$

The expression on the left-hand side of (25) is the covariance of the variates $p_t$ and $p_t^{g-1}$, and it is positive if $g > 1$, since $0 < p_t < 1$. In practice, because of the positive correlation that usually exists between elements in the same cluster unit, $g$ is likely to lie between 1 and 2, so that it may be expected that (22) is satisfied. Therefore, in many situations we may expect (20) to be smaller than (21). However, (20) is valid only for fairly large values of $N$ whereas the present procedure gives a simple expression for the variance (equation 10) for all values of $N$. Incidentally, (22) is equivalent to the comparison of the variance of ratio estimate (without the finite population correction) and the variance in sampling with replacement (Cochran, 1953, equations (9.26) and (9.27)).

## 4. EXTENSION TO TWO-STAGE DESIGNS

### 4.1. *Variance of the estimator of Y*

Let there be $N$ primary units with $M_t$ secondaries (second-stage units) in the $t$th primary $(t = 1, 2, ..., N)$. We select $n$ primaries with unequal probabilities and without replacement, using the present sampling procedure. If the $t$th primary is selected in the sample, draw a sample of $m_t$ secondaries with equal probabilities and without replacement from the $M_t$ secondaries. Let

$$\bar{y}_t = \sum_{j=1}^{m_t} y_{tj}/m_t,$$

where $y_{tj}$ is the value of the characteristic attached to the $j$th secondary in the $t$th primary. Then

$$\hat{Y} = \sum_{i=1}^{n} \pi_i \frac{M_i \bar{y}_i}{P_i}, \tag{26}$$

where $\pi_i$ is as defined before, is an unbiased estimator of $Y = \Sigma\Sigma y_{tj}$. The estimator (26) becomes self-weighting by taking $M_i/m_i = p_i/\pi_i$. Now

$$V(\hat{Y}) = E_1 V_2(\hat{Y}), \tag{27}$$

where $E_1$ is the expectation over all possible splits of the population of $N$ primaries and $V_2$ is the variance for a particular split. The between primary component in (27) is obtained simply by replacing $y_t$ by

$$Y_t = \sum_{j=1}^{M_t} y_{tj}$$

in (5), and therefore is

$$\frac{n\left(\sum_{i=1}^{n} N_i^2 - N\right)}{N(N-1)} \left(\sum_{t=1}^{N} \frac{Y_t^2}{np_t} - \frac{Y^2}{n}\right). \tag{28}$$

The within primary component is

$$E_1\left\{\sum_{i=1}^{n} \pi_i \sum_{k=1}^{N_i} \frac{M_k^2}{p_k} \left(\frac{1}{m_k} - \frac{1}{M_k}\right) S_k^2\right\}$$

$$= \sum_{i=1}^{n} E_1\left\{\sum_{k=1}^{N_i} M_k^2\left(\frac{1}{m_k} - \frac{1}{M_k}\right) S_k^2 + \sum_{k \neq k'}^{N_i} \frac{M_k^2 p_{k'}}{p_k}\left(\frac{1}{m_k} - \frac{1}{M_k}\right) S_k^2\right\} \tag{29}$$

since

$$\pi_i = \sum_{k=1}^{N_i} p_k,$$

where $S_k^2$ is the mean square of the $y$'s for the $k$th primary. Taking the average over all splits, we find from (29) that the within primary component is

$$\frac{\left(N^2 - \sum\limits_{i=1}^{n} N_i^2\right)}{N(N-1)} \sum_{t=1}^{N} M_t^2 \left(\frac{1}{m_t} - \frac{1}{M_t}\right) S_t^2 + \frac{\left(\sum\limits_{i=1}^{n} N_i^2 - N\right)}{N(N-1)} \sum_{t=1}^{N} \frac{M_t^2}{p_t} \left(\frac{1}{m_t} - \frac{1}{M_t}\right) S_t^2. \tag{30}$$

Therefore, adding (28) and (30), we have

$$V(\hat{Y}) = \frac{n\left(\sum\limits_{i=1}^{n} N_i^2 - N\right)}{N(N-1)} \left(\sum_{t=1}^{N} \frac{Y_t^2}{np_t} - \frac{Y^2}{n}\right) + \frac{\left(N^2 - \sum\limits_{i=1}^{n} N_i^2\right)}{N(N-1)} \sum_{t=1}^{N} M_t^2 \left(\frac{1}{m_t} - \frac{1}{M_t}\right) S_t^2$$

$$+ \frac{\left(\sum\limits_{i=1}^{n} N_i^2 - N\right)}{N(N-1)} \sum_{t=1}^{N} \frac{M_t^2}{p_t} \left(\frac{1}{m_t} - \frac{1}{M_t}\right) S_t^2. \tag{31}$$

With $N_1 = N_2 = ... = N_k = R+1$; $N_{k+1} = N_{k+2} = ... = N_n = R$, (31) reduces to

$$V(\hat{Y}) = \left\{1 - \frac{n-1}{N-1} + \frac{k(n-k)}{N(N-1)}\right\} \left(\sum_{t=1}^{N} \frac{Y_t^2}{np_t} - \frac{Y^2}{n}\right)$$

$$+ \frac{1}{n} \left\{1 - \frac{n-1}{N-1} + \frac{k(n-k)}{N(N-1)}\right\} \sum_{t=1}^{N} \frac{M_t^2}{p_t} \left(\frac{1}{m_t} - \frac{1}{M_t}\right) S_t^2$$

$$+ \left\{\frac{N(n-1)}{n(N-1)} - \frac{k(n-k)}{nN(N-1)}\right\} \sum_{t=1}^{N} M_t^2 \left(\frac{1}{m_t} - \frac{1}{M_t}\right) S_t^2. \tag{32}$$

When $k = 0$, (32) reduces to

$$V(\hat{Y}) = \left(1 - \frac{n-1}{N-1}\right) \left(\sum_{t=1}^{N} \frac{Y_t^2}{np_t} - \frac{Y^2}{n}\right) + \frac{1}{n} \sum_{t=1}^{N} \frac{M_t^2}{p_t} \left(\frac{1}{m_t} - \frac{1}{M_t}\right) S_t^2$$

$$- \frac{n-1}{n(N-1)} \sum_{t=1}^{N} M_t^2 \left(\frac{1}{m_t} - \frac{1}{M_t}\right) S_t^2 \left(\frac{1}{p_t} - N\right). \tag{33}$$

When the primaries are selected with unequal probabilities with replacement and a sample of $m_s$ secondaries is selected independently with equal probabilities and without replacement at each time the $s$th primary is selected (Cochran, 1953, p. 246), the estimator of $Y$ is

$$\hat{Y}' = \sum_{n} \frac{M_s \bar{y}_s}{np_s} \tag{34}$$

with variance

$$V(\hat{Y}') = \left(\sum_{t=1}^{N} \frac{Y_t^2}{np_t} - \frac{Y^2}{n}\right) + \frac{1}{n} \sum_{t=1}^{N} \frac{M_t^2}{p_t} \left(\frac{1}{m_t} - \frac{1}{M_t}\right) S_t^2. \tag{35}$$

Since the last term on the right-hand side of (33) is of lower order of magnitude than the second term, the within primary component of $V(\hat{Y})$ should be of the same order of magnitude as that in $V(\hat{Y}')$. Thus, $V(\hat{Y})$ should be smaller than $V(\hat{Y}')$ in two-stage sampling, although the increase in precision, which affects only the between primary component, will not be large if variation within primaries constitutes a major part of the total variance.

### 4.2. *Estimator of variance*

The estimator of variance in two-stage sampling is easily obtained by using Durbin's (1953) generalization of Yates's rule, which is applicable, and is as follows. The estimate of variance in multi-stage sampling is the sum of two parts. The first part is equal to the estimate of variance calculated on the assumption that the first-stage units have been measured without error. The second part is calculated as if the first-stage units selected were fixed strata, the contribution from each first-stage unit being multiplied by the probability of that unit's inclusion in the sample. Therefore, on replacing $y_i$ by $M_i \bar{y}_i$ in (15), the first part of the estimator of variance is

$$\frac{\left(\sum\limits_{i=1}^{n} N_i^2 - N\right)}{\left(N^2 - \sum\limits_{i=1}^{n} N_i^2\right)} \sum\limits_{i=1}^{n} \pi_i \left(\frac{M_i \bar{y}_i}{p_i} - \hat{Y}\right)^2. \tag{36}$$

Since the probability for the $i$th primary to be in the sample is $p_i/\pi_i$, the second part of the estimator of variance is

$$\sum\limits_{i=1}^{n} (p_i/\pi_i) \frac{\pi_i^2}{p_i^2} M_i^2 \left(\frac{1}{m_i} - \frac{1}{M_i}\right) s_i^2 = \sum\limits_{i=1}^{n} \pi_i \frac{M_i^2}{p_i} \left(\frac{1}{m_i} - \frac{1}{M_i}\right) s_i^2, \tag{37}$$

where $s_i^2$ is the sample mean square of the $y$'s for the $i$th primary. Therefore, adding (36) and (37), we have

$$v(\hat{Y}) = \frac{\left(\sum\limits_{i=1}^{n} N_i^2 - N\right)}{\left(N^2 - \sum\limits_{i=1}^{n} N_i^2\right)} \sum\limits_{i=1}^{n} \pi_i \left(\frac{M_i \bar{y}_i}{p_i} - \hat{Y}\right)^2 + \sum\limits_{i=1}^{n} \pi_i \frac{M_i^2}{p_i} \left(\frac{1}{m_i} - \frac{1}{M_i}\right) s_i^2. \tag{38}$$

### 5. AN EXAMPLE OF TWO-STAGE SAMPLING

To illustrate the computation of the estimate and its standard error in two-stage sampling we take the data of the 19 counties in Iowa which form Economic Areas 3a and b (U.S. Bureau of Census, 1960). The complete enumeration data of the rural population for each township in these counties are available for 1950 and 1960. Here $x_{tj}$ and $y_{tj}$ denote the rural population in the $j$th township of the $t$th county in 1950 and 1960, respectively. We split the population of counties at random into three groups $N_1 = 6$, $N_2 = 6$ and $N_3 = 7$. Then from each group one county is selected with probabilities proportional to the total rural population in 1950, and within each selected county two townships are selected with equal probabilities and without replacement. Here $p_t = X_t/X$, where

$$X_t = \sum\limits_{j=1}^{M_t} x_{tj} \quad \text{and} \quad X = \sum\limits_{t=1}^{N} X_t.$$

Table 1 gives the relevant data from the sample.

TABLE 1

*Data from a two-stage sample*

| | Group 1 | Group 2 | Group 3 |
|---|---|---|---|
| | $N_1 = 6, \pi_1 = \dfrac{42379}{130925}$ | $N_2 = 6, \pi_2 = \dfrac{42174}{130925}$ | $N_3 = 7, \pi_3 = \dfrac{46372}{130925}$ |
| *For the selected county* | $M_1 = 16, X_1 = 7521$ | $M_2 = 18, X_2 = 7393$ | $M_3 = 16, X_3 = 6131$ |
| | $(y_{11}, y_{12}) = (500, 330)$ | $(y_{21}, y_{22}) = (292, 430)$ | $(y_{31}, y_{32}) = (272, 304)$ |

From formula (26), the estimate of the total rural population in 1960 is

$$\hat{Y} = \frac{42379 \times 16 \times 830}{2 \times 7521} + \frac{42174 \times 18 \times 722}{2 \times 7393} + \frac{46372 \times 16 \times 576}{2 \times 6131} = 109336.$$

Also

$$\frac{\left(\sum\limits_{i=1}^{3} N_i^2 - N\right)}{\left(N^2 - \sum\limits_{i=1}^{3} N_i^2\right)} \sum_{i=1}^{3} \pi_i \left(\frac{M_i \bar{y}_i}{p_i} - \hat{Y}\right)^2 = 27883966,$$

$$\sum_{i=1}^{3} \pi_i \frac{M_i^2}{p_i} \left(\frac{1}{m_i} - \frac{1}{M_i}\right) s_i^2 = 17374970.$$

Therefore, from (38), we have $v(\hat{Y}) = 27883966 + 17374970 = 45258936$ or s.e. $(\hat{Y}) = 6728$.

In this example, increased precision can be obtained by using a ratio-type estimator within each selected county. Incidentally, the true value of the total rural population in 1960 is $Y = 111531$.

ACKNOWLEDGEMENTS

The authors wish to thank the referee for some helpful suggestions. J. N. K. Rao's and H. O. Hartley's work is sponsored by the Office of Ordnance Research, United States Army. W. G. Cochran's work is assisted by a contract from the Office of Naval Research, Navy Department.

REFERENCES

BUREAU OF CENSUS (1960), *Census of Population.*
COCHRAN, W. G. (1953), *Sampling Techniques.* New York: Wiley.
DES RAJ (1956), "A note on the determination of optimum probabilities in sampling without replacement", *Sankhyā*, **17**, 197–200.

DURBIN, J. (1953), "Some results in sampling theory when the units are selected with unequal probabilities", *J. R. statist. Soc.* B, **15**, 262–269.

HARTLEY, H. O. and RAO, J. N. K. (1962), "Sampling with unequal probabilities without replacement", *Ann. math. Statist.*, **33**, 350–374.

HORVITZ, D. G. and THOMPSON, D. J. (1952), "A generalization of sampling without replacement from a finite universe", *J. Amer. statist. Assoc.*, **47**, 663–685.

NARAIN, R. D. (1951), "On sampling without replacement with varying probabilities", *J. Ind. Soc. agric. Statist.*, **3**, 169–174.

YATES, F. and GRUNDY, P. M. (1953), "Selection without replacement from within strata with probability proportional to size", *J. R. statist. Soc.* B, **15**, 235–261.

# 76

## THE POTENTIAL CONTRIBUTION OF ELECTRONIC MACHINES
## TO THE FIELD OF STATISTICS

WILLIAM G. COCHRAN

*Department of Statistics*

### 1. Introduction

The field of statistics deals with techniques for the collection, analysis, and interpretation of data. I should like to discuss the impact of electronic machines under two primary headings: (1) mathematical statistics, (2) analysis of data, ending with a few comments about the impact on training and teaching.

Electronic machines have come along at an opportune time for the statistician. As is typical of a relatively new field, many of the standard results in theoretical statistics were obtained without encountering really difficult mathematics. Probability theory, advanced calculus, and matrices, plus a smattering of complex variables and measure theory, have carried us a long way. Similarly the study of methods for the analysis of data naturally concentrated at first on relatively small bodies of data, in which the analysis could be worked in detail on a desk machine.

This situation seems unlikely to persist. As the body of theory has grown, attention has turned to more complex problems, accompanied by a more searching reexamination of older problems for which simple solutions had previously been supplied. In both cases formidable mathematical obstacles are frequent. On the data-analysis side, the collection of large masses of data by governments, states, business firms, and research agencies is a characteristic of the times. Statisticians have also contributed toward making the analysis of data a more complicated business by stressing the importance of multivariable experimentation and the advantages of multistage design of sample surveys.

### 2. Mathematical Statistics

Any research worker in mathematical statistics could, I believe, present a substantial list of important unsolved statistical problems in which (*i*) the primary difficulty is mathematical and (*ii*) electronic machines can help to overcome the difficulty. As illustrations I shall mention first a few problems that are quite old in the field of statistics, followed by several examples from more recent research.

#### 2.1. OLDER PROBLEMS

*Nonlinear regression.* In the fitting of regression equations that are nonlinear in some of the parameters, for example $y = a_1 e^{-b_1 x} + a_2 e^{-b_2 x}$, little is known about the frequency distributions of nonlinear coefficients like $b_1$ and $b_2$ in samples of sizes that are common in practical applications. Our only results are asymptotic, obtained in effect by fitting a linear approximation to the nonlinear relation. The search for nonlinear equations that are tractable by analytical methods has had practically no success.

*Rejection of observations.* The statistician is expected to have some views on the question of rejecting observations that appear anomalous or suspicious. Although numerous rules for guidance in deciding whether to reject observations have been proposed, the performance characteristics of the rules have been very inadequately studied. Thorough understanding of a rule requires a knowledge of its performance under a variety of sizes and numbers of gross errors and a variety of basic frequency distributions of the observations, because a good rejection rule must distinguish between a real gross error and an observation that is merely out on the tail of the frequency distribution. The primary mathematical difficulty in this research is that of handling nonnormal distributions by analytical methods.

*Discriminatory analysis.* In multivariate analysis, several related problems lead to finding a set of weights $w_i$ such that the ratio

$$R = \sum_i \sum_j w_i w_j a_{ij} \bigg/ \sum_i \sum_j w_i w_j b_{ij}$$

is maximized, where $a_{ij}$ and $b_{ij}$ are two independent sample covariance matrices in $p$ variables $(i,j = 1,2,\ldots,p)$. It is known that the turning values of $R$ depend on the roots $\hat{\theta}$ of the determinantal equation

$$|a_{ij} - \hat{\theta}b_{ij}| = 0.$$

The joint frequency function of the $\hat{\theta}$ was discovered by Fisher[1] and Girshick[2] for the situation in which all the corresponding roots in the population are zero. In many problems to which this approach applies, however, some of the population roots are nonzero. We have some large-sample results for this case, but our knowledge of the situation in samples of moderate size has scarcely advanced in the last 20 years.

*Analysis of variance.* The useful technique known as the analysis of variance is based on mathematical models of the type

$$y_{ij} = \mu + \rho_i + \gamma_j + e_{ij},$$

where $i$ denotes the row and $j$ the column in a two-way classification of the observations $y_{ij}$. The standard properties of the model depend on four assumptions: (i) the row effects $\rho_i$ and the column effects $\gamma_j$ are additive, (ii) the residuals $e_{ij}$ are independently distributed, (iii) the $e_{ij}$ all have the same variance $\sigma^2$, (iv) the $e_{ij}$ are normally distributed.

In the widespread applications of this technique, any one of the assumptions may fail—sometimes several simultaneously. Because of its mathematical complexity, research on the consequences of failures in the assumptions is still very incomplete.

## 2.2. NEWER PROBLEMS

*Stochastic processes.* This branch of probability theory deals with any process whose development through space or time has some random or chance element in its structure. In

231

the simplest example, known as the Poisson process, the probability that a signal arrives in the small time interval $t$, $t + dt$, is $\lambda dt$, irrespective of how many have arrived previously. The probability that exactly $n$ signals arrive in an interval of length $T$ is known to be $(\lambda T)^n e^{-\lambda T}/n!$. More complex processes are involved in statistical studies of the growth of bacterial colonies, of epidemics, of queues, of comets, and of carcinogenesis. Neyman[3] maintains that science has reached a stage at which dynamic phenomena are being widely studied and that, if he is to contribute, the statistician must tackle the many unsolved problems in the theory of stochastic processes.

*Bayesian inference and decision theory.* This approach to the major problems of statistical inference avoids some of the unsatisfactory features of the standard solutions to these problems, and is currently a topic of keen research interest as well as of some controversy. It is already evident, however, that in many problems the Bayesian solution cannot be found analytically because of mathematical difficulties.

*Sequential methods.* In the sequential approach, the results of an experiment are analyzed continuously as they come in, in order that the data collection may be stopped as soon as the conclusions become clear enough for our purpose. Although the method holds promise of decreasing the time and labor in investigational work, the properties of sequential plans involve intricate mathematical problems and are insufficiently studied as yet.

## 3. The Role of Electronic Machines

In some of the examples presented above—for instance, the study of criteria for rejection, the distribution of the roots $\hat{\theta}$ in discriminatory analysis, and the development of posterior distributions in Bayesian inference—the obstacle is the evaluation of a multiple integral that has no closed solution. The contribution of the machine will be to perform the integration numerically. In other types of research, for example, sequential analysis with qualitative data and some problems in stochastic processes, the role of the machines will be to make repeated applications of a recurrence formula with great rapidity. In problems like the study of nonlinear regression, it looks at present as if we will have to rely on Monte Carlo methods. These examples suggest that the specific ways in which the machines are utilized will be generally similar to those found in other branches of applied mathematics.

The use of modern statistical methods has been greatly facilitated by statistical tables which provide the results of probability calculations in a form convenient for application. Electronic machines are already filling some of the major gaps in the available tables. Recent publications are an enlarged table of the probability integral of the range (Harter *et al*,[4]) and a table of the hypergeometric distribution (Lieberman and Owen[5]). Computations of an extensive table of the power function of the $F$-distribution, nearing completion at the National Bureau of Standards, will fill a long-felt want.

To summarize, electronic machines represent a major increase in manipulative power and flexibility in mathematical statistics. They can free us from overdependence on the assumption of normality and from confinement to approximate linear solutions of nonlinear

232

problems. They will not, of course, do our statistical thinking for us. Further advances in the basic problems of statistical inference will depend, as in the past, on a stream of new ideas. Even here, however, machines may speed progress by facilitating investigation of the consequences and implications of new ideas. For example, Barnard (personal communication) has pointed out that computer programs can be constructed to tabulate contours of constant likelihood. This provides the opportunity for a more thorough understanding of the performance of the likelihood function as a central concept in statistical inference.

There is much to learn about the most effective ways of using the machines. Indeed, their limitations can already be seen unless we find out how to overcome them. For example, the specification of a problem may involve from, say, four to ten parameters, each of which varies over some range that is of interest. A useful solution to the problem necessitates mapping a complex function of the parameters throughout the region determined by these ranges. Computation of the function for ten levels of each parameter requires a large, sometimes prohibitive, number of calculations. The presentation of results faces questions of expense and bulk.

One approach is to try to map the function satisfactorily from a much smaller panel of values of the parameters, skillfully chosen. A second is to develop mathematical approximations to the function, valid over parts of the region and simple enough so that the user can be asked to substitute in the approximation to save the expense of tabulation. Both approaches will probably demand a combination of mathematical skill and the use of machine results.

To mention a second limitation, the computation of frequency functions to any high degree of accuracy by experimental sampling (Monte Carlo) requires enormous amounts of work. Hartley[6] has estimated that a sample size of over $10^7$ is required for three-decimal accuracy in the cumulative frequency function. This is alleviated somewhat by the numerous ingenious devices that have been introduced into Monte Carlo computations, such as importance sampling, auxiliary variables highly correlated with the variable under study, and the repeated use of the same computation by reweighting the results. While it is perhaps too early to draw conclusions, my guess is that even with the aid of these techniques many problems will remain outside the range of practicability. Perhaps more is to be gained by computing only the low moments of the frequency distribution, which may require much smaller sample sizes, and then approximating the frequency function from a knowledge of its low moments. In 1908, this approach enabled W. S. Gosset, a pioneer in small-sample methods who published under the pseudonym of Student, to guess correctly the frequency distribution of the correlation coefficient in normal samples in which the population correlation is zero.[7] The raw data for his experimental sampling consisted of the heights and the lengths of the middle fingers of the left hands of 3000 criminals.

Another useful approach, which applies to Monte Carlo methods generally, is to remember that "the only good Monte Carlo's are dead Monte Carlo's—the Monte Carlo's we don't have to do" (Trotter and Tukey[8]). If we can divide a problem into a major part that can be handled by analytical mathematics, or by something like numerical integration, leaving only

233

a minor part to be determined by Monte Carlo methods, the reduction in computation time may make the difference between success and failure. This means that we shall be presented with new mathematical problems while learning how to get the most out of machines.

An obvious consequence is that we are going to have to enlist the brains of pure mathematicians, as well as those of applied mathematicians and statisticians. Anything that induces a competent pure mathematician to become interested in playing with machines is likely to be helpful. In this connection, Dr. Ulam's illustrations of the fascinating growth phenomena that can be produced by simple rules of growth suggests that there should be enough to interest and challenge the pure mathematician.

## 4. Analysis of Data

### 4.1. CENSUSES AND SURVEYS

The contribution of machines to the tabulation of large masses of data is well illustrated by the 1960 Census of Population and Housing. Under a legal requirement, the total population of each state must be reported not later than eight months after the beginning of the field enumeration in the census. In all past censuses the state totals were obtained manually as a separate operation from the rest of the processing, because the data could not be coded, punched, and verified in time to meet the deadline by mechanical tabulation.

In 1960 several changes enabled this count to be mechanized. The 100-percent census was restricted to a minimal set of items—age, sex, race, marital status, and relationship to head of household—that could be coded directly (by checking appropriate boxes) on the questionnaires. All other items were transferred to a 25-percent sample of households, recorded on a separate form. The questionnaires are microfilmed. A new electronic device called FOSDIC (Film Optical Sensing Device for Input to Computers), developed by engineers at the National Bureau of Standards, scans the microfilm frames at the rate of 100–200 per minute, and records the relevant data on magnetic tapes, ready for input to the Univac 1105. These machines, of which four were used, are in turn linked to high-speed printers which produce many of the standard tables in a form suitable for photographic reproduction. The whole process, from the census taker's mark to the printed table, is entirely mechanized.

As a result, state totals were given to the President on November 15, 1960. It is expected that most of the standard census report volumes will be available at least a year earlier than with the 1950 census. Although some headaches in processing have appeared, the analysis and publication program is keeping to the anticipated timetable.

At several centers which regularly engage in sample surveys, work is well advanced on standard computer programs for tabulation and analysis. These will provide not only the tabulations and types of estimates commonly used in surveys whose primary objective is descriptive, but also more sophisticated procedures, such as multiple regressions, required in surveys taken for analytical purposes. As these programs become more readily available and freed from their own teething troubles, computers will enhance two of the principal virtues

234

of the sample survey—its speed in producing results and its power as an analytical tool of research.

To revert to the 1960 census, an important feature is automatic editing and quality control of the processing by the computer. Owing to errors in the field, certain data on the questionnaires will be obviously wrong, such as a wife aged 3, some will be dubious, such as a head of household with two wives living in the same house, and some will be missing. The computer is instructed to edit such entries and to insert values for missing entries, using either data available for other members of the household or data for similar households found elsewhere in the census. A continuous record is kept of the amount of editing and insertions done. The computer can also be instructed to flash warnings or to reject data when individual questionnaires or a series of questionnaires do not meet assigned standards.

The skillful use of automatic editing and quality control (available at present only on relatively advanced machines) is of considerable significance for the analysis of large-scale survey data generally. One disadvantage of the machines is that the worker may lose touch with his data, never having studied it in detail. The setting up of an editing, quality-control, and data-rejection program forces the worker to think seriously about the kinds of mistakes and inaccuracies to which his data may be subject and to specify standards of completeness and internal consistency. Manual editing is of course more flexible and, if highly skilled editors were available, should be more accurate. But as Daly[9] has pointed out, "Present-day electronic computers are large enough and fast enough so that they can apply to the data all the rules that a group of clerks could be instructed to apply in detecting and reconciling inconsistencies and inferring missing information. Moreover, the computer does not get bored by this type of work."

4.2.  COMPLEX ANALYSES

Even when the volume of data gathered is not inordinately large, some phenomena require very substantial amounts of computation for study by statistical methods. An example is spectral analysis, a technique used to study the relations between measurements taken at different times or in different places. The problem is not only to get the calculations done, but also to interpret correctly what the results mean. With their ability to perform large numbers of calculations at great speed, computers are an essential tool in such types of research. Tukey[10] gives a lucid introduction to current methods of computation for spectral analysis, which has had useful applications in aerodynamics, astronomy, meterology, human behavior, oceanography, and economics.

Multivariate analysis—the study of the relations among different measurements—is another area in which the amount of calculation required has impeded the exploitation of techniques that have been developed and, more importantly, our understanding of the capabilities of the techniques. It seems obvious that many phenomena must require multivariate methods for their proper elucidation. Yet my impression of the utility of multivariate techniques as a contribution to the study of the world around us has been disappointing. It may be that we have

235

not yet learned how to think in multivariate terms, that we are not using the techniques to best advantage, that we need different techniques, that we are selecting data unwisely, or that we must await improvements in the precision of measurement. Computers should enable us to gain breadth of experience in applying multivariate techniques.

The early influence of computers in this problem may actually be detrimental. A noticeable tendency at present is to throw every available variable into a multivariate analysis, irrespective of its relevance or precision of measurement, just because the machine can now handle the computations. It is to be hoped that this thoughtless enthusiasm will be followed by an attack on the unsolved problem of deciding which variates are worth including in the analysis.

### 4.3. MORE SEARCHING ANALYSES

Every statistical analysis starts from a probabilistic model, intended to represent the relevant aspects of the structure of the data. All these models are oversimplifications, involving a number of unverifiable assumptions. In general, the more assumptions we make, the easier is the analysis and the more precise the results appear to be. Faced with the same data, two statisticians may reach different conclusions, not because their analyses are faulty, but because they started with different models. This contradiction is sometimes embarrassingly evident when two eminent statisticians appear on opposite sides of a law suit.

For this reason, statistical analysis of data cannot be made a fully objective procedure. With the aid of computers, however, we can, much more easily than in the past, analyze a given body of data simultaneously under different models, employing different assumptions and different points of view. This should lead to a more realistic grasp of the process of data analysis and an understanding of the extent to which it is subjective.

### 4.4. ACCESS TO MACHINES

The extent to which computers will be used fruitfully for analysis of data obviously depends on the terms under which access to computers is available. Experience in running computing services shows repeatedly that for maximum effectiveness the service bureau must supply a good deal of statistical consultation. The client may not fully understand the statistical technique that he proposes to employ; it may soon be clear to an expert that the technique will not do what he wants; or the quality of data may not justify an elaborate analysis without additional safeguards. This situation is particularly common in biology, medicine, and the behavioral sciences, in which clients are often relatively untrained in mathematics, and accurate measurement is sometimes hard to achieve. A first-rate service of this type makes heavy demands on the personnel, who must know computing, programming, statistics, and the substantive field and must also possess the appropriate personality characteristics.

My purpose here is only to call attention to this topic, which deserves extensive discus-

236

sion. There is no easy solution to the problems of financing, of cutting down the bottlenecks, of priorities in access, and of deciding how much consultant programming to make available. Not the least problem is that of finding ways of making the programmer's job sufficiently attractive to enlist and hold competent people.

### 5. Teaching and Training

There is much to be said for insisting that every Ph.D candidate in statistics should learn the rudiments of a programming language and of computing on one machine. Fortunately, this can be done without a major expenditure of time. Students seem to enjoy it. Older people in the academic side of statistics were brought up in an era in which many problems were quietly forgotten, because there seemed no hope of an analytical solution. It is difficult in middle age to free oneself from this restricted viewpoint, which colors what is taught as well as the type of research that is done. Students should not be asked to occupy the same straitjacket. For the same reason, some lectures on the contributions that machines can make to the solution of problems in mathematical statistics should also be included. One intriguing prospect is that students can engage in small pieces of statistical research on the machines very early in their course of training.

I do not feel competent to speculate on the long-term effects of computers on teaching, except for a few fairly obvious remarks. We can gradually discard exposition of computing tricks whose sole purpose is to make computations feasible on desk machines, though the ideal rate at which to discard is not entirely clear. The more complex branches of statistics, such as multivariate analysis and time series, should be influenced most quickly by the new potentialities open to us. The neglected problem of specification, as Fisher[11] called it, that is, of deciding on the model used to start the analysis, may be much more fully discussed as more is learned about it. More generally, insofar as computer research leads to a better understanding of the tools of theoretical statistics and their use in data analysis, our lectures should be more soundly based.

This paper owes much to letters from G. A. Barnard, J. Daly, H. O. Hartley, L. Kish, and J. W. Tukey describing their own experiences with computers.

## REFERENCES

1. R. A. Fisher, "The sampling distribution of some statistics obtained from non-linear equations," *Ann. Eugen.* 9 (1939), 238–249.
2. M. A. Girshick, "On the sampling theory of the roots of determinantal equations," *Ann. Math. Stat.* 10 (1939), 203–224.
3. J. Neyman, "Indeterminism in science and new demands on statisticians," *J. Am. Stat. Assoc.* 55 (1960), 625–639.
4. H. L. Harter *et al.*, *The probability integrals of the range and of the studentized range* (WADC Technical Report 58-484, vol. 2; Office of Technical Services, U. S. Dept. of Commerce, Washington, 1959).

5. G. J. Lieberman and D. B. Owen, *Tables of the hypergeometric probability distribution* (Stanford University Press, Stanford, California, 1961).

6. H. O. Hartley, "Changes in the outlook of statistics brought about by modern computers," *Proceedings of the Third Conference on Design of Experiments in Army Research Development and Testing* (Office of Ordnance Research, U. S. Army, 1958), 345–363.

7. Student, "Probable error of a correlation coefficient," *Biometrika 6* (1908), 302–309.

8. H. F. Trotter and J. W. Tukey, "Conditional Monte Carlo in normal samples," in H. A. Meyer, ed., *Symposium on Monte Carlo methods* (Wiley, New York, 1956), 64–79.

9. J. Daly, "Organizational problems related to large scale statistical computations at Bureau of Census," *Statistician 11* (1957), 10–12.

10. J. W. Tukey, "An introduction to the measurement of spectra," in U. Grenander, ed., *Probability and statistics* (The Harald Cramér Volume; Wiley, New York, 1959), 300–330.

11. R. A. Fisher, "On the mathematical foundations of theoretical statistics," *Phil. Trans. Roy. Soc. (London) A 222* (1922), 309–368.

# ON THE PERFORMANCE
# OF THE LINEAR DISCRIMINANT FUNCTION [1]

## by William G. COCHRAN

*Department of Statistics, Harvard University, U. S. A.*

## 1. *Introduction*

This paper deals with the relation between the discriminating power of the linear discriminant function (LDF) and the discriminating powers of the individual variates used in the function. The study was undertaken for two main reasons.

The extensive literature on the LDF gives little guidance on how to select variates for constructing a discriminant. From his knowledge of the problem, an investigator can sometimes suggest a small number of variates that he thinks will have good discriminating power. To these he can often add a second and larger list of variates that *might* be helpful, though he is less sure of this. A meteorologist, for example, could probably supply two such lists of weather variables if the purpose were to predict whether it would rain or not on the next day. When measurements are made on the second list of variables in the two populations between which we wish to discriminate, it may be found that a number of them show little separation between the two populations. In this event it would be useful to have a simple rule by which such variates can be discarded before computing the LDF, on the grounds that their inclusion is unlikely to produce a material increase in discriminating power.

Secondly, given a knowledge of the discriminating power of each of a set of variates, it is sometimes helpful to be able to make a rough estimate of the discriminating power of the LDF without going to the trouble of computing it. This estimate may show whether the LDF is likely to be satisfactory for our purpose or whether it has little hope of being of practical value, so that it is pointless to compute it.

## 2. *The discriminating power of individual variates*

The usual simplifying assumptions will be made. It is supposed that the objective is to classify specimens into one of two populations. The variates $y_i$ follow a

(1) This work was assisted by a contract between the Office of Naval Research, U.S. Navy Department and the Department of Statistics, Harvard University.

1

Reproduced with permission from the *Bulletin of the International Statistical Institute,* Vol. 39, No. 2, pages 435–447. Copyright © 1962.

multivariate normal distribution with the same covariance matrix in each population. Each variate is scaled so that it has unit standard deviation within each population, and so that its mean is zero in population I and has a *positive* value $\delta_i$ (for the $i$th variate) in population II. The sizes of the samples used to construct the function are assumed large enough so that the distances $\delta_i$ and the correlation coefficients $\rho'_{ij}$ are known with negligible error. Finally, the criterion of performance of the LDF will be the probability that a specimen is misclassified.

If the variate $y_i$ is used alone as the discriminant, specimens being allotted to population I if $y_i < \delta_i/2$ and to population II if $y_i > \delta_i/2$, the probability of misclassification is

$$P = \frac{1}{\sqrt{2\pi}} \int_{\delta_i/2}^{\infty} e^{-x^2/2}\, dx \tag{1}$$

this being the same for specimens from either population.

Consequently a ranking of the $\delta'_i s$ ranks the variates in order of their individual discriminating powers. Some values showing the relation between $P_i$, $\delta_i$ and $\delta_i^2$ are given in Table 1.

TABLE 1

*Distances for given probabilities of misclassification*

| $P_i \%$ | 45 | 40 | 35 | 30 | 25 | 20 | 15 | 10 | 5 | 1 |
|---|---|---|---|---|---|---|---|---|---|---|
| $\delta_i$............... | .251 | .507 | .771 | 1.049 | 1.349 | 1.683 | 2.073 | 2.563 | 3.290 | 4.653 |
| $\delta_i^2$............... | .063 | .257 | .549 | 1.100 | 1.820 | 2.833 | 4.296 | 6.569 | 10.82 | 21.65 |

If the aim is high accuracy in classification, variates with error rates of 45 % or 40 %, i. e. with values of $\delta_i$ below 0.5, would be considered poor discriminators. Some workers have deliberately excluded such variates before computing the function. For example, in using physical measurements to distinguish between men of Japanese and Caucasian stock in Hawaii, Horst and Smith (1950) discarded 7 of 18 original variates. Their criterion was that a $t$-test of the difference between the population means did not reach the 1 % level of significance. In our notation this amounts to rejecting any variate for which $\delta_i$ is less than about 0.52. For a similar reason, Wallace and Travers (1938) used only 5 of more than 20 available variates in differentiating successful from unsuccessful salesmen.

Several others have used the same type of criterion as Horst and Smith, i. e. non-significance of $t$ at some level. With samples of sizes $n_1$, $n_2$ from the two populations, $t = \delta \sqrt{n_1 n_2}/\sqrt{n_1 + n_2}$. It seems more appropriate, however, to base any rule on the value of $\delta$, since $t$ is greatly influenced by the sizes of the samples that we happen to have selected for setting-up the discriminant.

With this background we now consider how to estimate the contribution made by a single variate or a group of variates to the performance of the LDF.

2

## 3. *Independent variates*

The problem would be easy if all $p$ variates were independent. In this case the best combined discriminant is

$$Y = \sum_{i=1}^{p} \delta_i y_i / \sqrt{\Sigma \delta_i^2}$$

where Y has been scaled so that its S.D. is unity. Since the mean value of Y is zero in population I and $\sqrt{\Sigma \delta_i^2}$ in population II, the distance for Y is

$$\delta_Y = \sqrt{\Sigma \delta_i^2}. \tag{2}$$

It follows that if we have two independent variates with distances $\delta_1$, $\delta_2$, $(\delta_1 > \delta_2)$, the first variate is equivalent to $m$ of the second variate, where

$$m = \delta_1^2 / \delta_2^2$$

since by (2) $m$ independent variates with distance $\delta_2$ give the same probability of misclassification as one variate with distance $\delta_1$.

From the last line of Table 1, a single variate with a 1% error rate is worth 2 variates with a 5 % rate, over 7 variates with a 20 % rate, 20 variates with a 30 % rate, 84 variates with a 40 % rate and 344 variates with a 45 % rate. If these relationships held generally, it would be an easy matter to decide in a specific application whether a group of poor discriminators were worth including in the calculations. For example, given two variates with distances 1.6, 1.4 and seven additional variates with distances around 0.2, the two best variates have a combined $\Sigma \delta^2$ of 4.52, with a probability of misclassification of 14.4 %. Inclusion of the seven variates increases $\Sigma \delta^2$ to 4.80 and decreases the probability to 13.7 %. The decrease might not be considered worth the extra complexity of an LDF with nine variables.

The complicating factor is the effect of correlations among the variates. As shown by Cochran and Bliss (1948), a covariate which has no discriminating power when used alone, i.e. with $\delta_i = 0$, may greatly reduce the error rate if highly correlated with a discriminant that has been constructed without the covariate. Moreover, as users of the discriminant function know, the best two-variate discriminator chosen from a group of variates need not contain either of the two best individual discriminators.

## 4. *The effect of correlation - two variates*

Suppose we have two variates $y_1$, $y_2$ with distances $\delta_1$, $\delta_2$ and correlation coefficient $\rho$. The variate $y_1$ may represent either a single measurement or a computed discriminant function to which the addition of another variate $y_2$ is being considered.

Since the sign of $\rho$ is of importance in the development, the meaning of positive and negative $\rho's$ will be noted. The variates have been scaled to have zero means in population I and positive means in population II. In other words, the correlation between the population means of $y_1$ and $y_2$ has been set at $+1$. A positive $\rho$ means

3

that the intra-population correlation has the same sign as the correlation between population means.

In applications I believe that positive correlations are likely to be much more frequent than negative correlations. For instance, many applications involve the use of scores in different examinations to distinguish between those who fail at some future task (population I) and those who succeed (population II). The examinations chosen are usually such that a high score indicates ability to perform the subsequent task well. Hence most examination scores should show a positive $\delta_i$. And since good performance in one examination is often accompanied by good performance in others, we would expect most of the $\rho_{ij}$ to be positive. It is more difficult to suggest applications in which negative correlations are the natural expectation. They do occur, however. In the familiar *Iris* example of Fisher (1936), sepal length and sepal width have a $\rho$ of about $+0.6$, yet *Iris versicolor* has a greater sepal length and a smaller sepal width than *Iris setosa*. In our scaling, one of these variables would have its sign changed to make both $\delta_i$ positive, and $\rho$ would become $-0.6$.

An easy way of finding the increase in the squared distance due to the inclusion of $y_2$ is to note that the variate $(y_2 - \rho y_1)$ is independent of $y_1$ and has distance $(\delta_2 - \rho \delta_1)$ and variance $(1 - \rho^2)$. Hence, inclusion of $y_2$ increases the squared distance by the amount

$$\frac{(\delta_2 - \rho \delta_1)^2}{(1 - \rho^2)}.$$

Since we wish to consider $\delta_2$ as a poorer variate that is being added to a better one or to an existing discriminant, we write $\delta_2 = f\delta_1$, where $0 \leqslant f \leqslant 1$. The increase in the squared distance, measured as a fraction of the original squared distance $\delta_1^2$, then becomes

$$\frac{(f - \rho)^2}{(1 - \rho^2)}. \tag{3}$$

This, of course, reduces to $f^2$ if $\rho = 0$. Correlation can therefore be said to be helpful if

$$\frac{(f - \rho)^2}{1 - \rho^2} > f^2. \tag{4}$$

Some conclusions follow.

(i) Any negative correlation is helpful. Indeed, as $\rho$ moves towards $-1$ there is no mathematical limit to the amount by which the squared distance is increased. This illustrates why no rule based on individual discriminating powers can be mathematically sound in predicting joint discriminating power.

(ii) A positive correlation is harmful unless $\rho$ is high enough in relation to $f$. Inequality (4) is satisfied if

$$\rho > \frac{2f}{1 + f^2}. \tag{5}$$

In general, positive correlations have to be high to be helpful. For instance, $\rho$ must exceed 0.94 if $f = 0.7$; 0.8 if $f = 0.5$; and 0.55 if $f = 0.3$.

As further conclusions we may note;

4

(ii*a*) If $f = 1$, i.e. $y_2$ is as good as $y_1$, all positive correlations are harmful.

(ii*b*) If $f = 0$, i.e. $y_2$ is a covariate with no discriminating power, all positive correlations are helpful (as well as all negative correlations). From (3), the proportional increase in the squared distance due to the inclusion of a covariate is

$$\frac{\rho^2}{1 - \rho^2}.$$

If $y_1$ is only a moderately good discriminator, the decrease in the probability of misclassification due to a covariate $y_2$ is small unless $\rho$ is high. For instance, if $\delta_1 = 1.049$, giving an error rate of 30 %, a correlation of 0.63 is required to reduce the error rate to 25 %, and a correlation of 0.78 to bring the error rate down to 20 %. With an initial error rate of 10 %, however, $\rho = 0.63$ brings the rate down to 5 % and $\rho = 0.78$ to about 2 % (*).

It will be noted that $y_2$ adds nothing to the discrimination if $f = \rho$. It can be verified that this situation arises whenever $y_2$ has a linear regression on $y_1$ that is the same in both populations. In this event, $y_2$, before scaling, is of the form

$$y_2 = \alpha + \beta y_1 + e$$

where $e$ is a random variate, independent of $y_1$, with mean zero and the same variance in both populations. This result is a particular case of a general result given by Rao (1954). He showed, in effect, that if all the variates have linear regressions on a set of $k$ variates, these $k$ variates determine an upper limit that can be attained by the squared distance, no matter how many variates are included in the discriminant. The regressions must be the same in both populations.

## 5. *Three variates*

With more than two variates the number of correlation coefficients multiplies, and the effects of correlation are not describable in simple terms. The situation with three variates will be discussed briefly. When a third variate $y_3$ is added to a discriminant containing $y_1$ and $y_2$, the increase in the squared distance may be expressed as follows:

$$\frac{(1 - \rho_{12}^2) \left| \delta_3 - \frac{\rho_{13} (\delta_1 - \rho_{12} \delta_2)}{1 - \rho_{12}^2} - \frac{\rho_{23} (\delta_2 - \rho_{12} \delta_1)}{1 - \rho_{12}^2} \right|^2}{1 - \rho_{12}^2 - (\rho_{13} - \rho_{23})^2 - 2 \rho_{13} \rho_{23} (1 - \rho_{12})}. \tag{6}$$

This reduces, of course, to $\delta_3^2$ if $\rho_{23}$ and $\rho_{13}$ are both zero. Our interest is in the values of $\rho_{23}, \rho_{13}$ that make the squared distance greater than $\delta_3^2$.

If $\delta_1$ and $\delta_2$ are not very different, the terms $(\delta_1 - \rho_{12} \delta_2)$ and $(\delta_2 - \rho_{12} \delta_1)$ in the numerator of (6) will usually both be positive, since this happens for any value of $\rho_{12}$ between $-1$ and $\delta_2 / \delta_1$, where $\delta_2$ is the smaller of $\delta_1, \delta_2$. If this holds, then

(i) The squared distance (6) is always greater than $\delta_3^2$ if $\rho_{12}$ and $\rho_{13}$ are both negative;

(*) In this discussion it is assumed that the covariate is entered in the calculations in the same way as any other variate, whereas Cochran and Bliss (1948) showed that the best method of handling the covariate is slightly different. However, the two methods become equivalent when initial samples are large, as is assumed here.

5

(ii) If $\rho_{12}$, $\rho_{13}$ are of opposite signs, further examination shows that (6) is sometimes greater than and sometimes less than $\delta_3^2$.

(iii) If $\rho_{12}$, $\rho_{13}$ are both positive, their effects appear to be usually harmful except in some cases when one or both are high.

So far as they go, these results are in line with the effects of correlation between two variables as found in Section 4. However, to confuse the issue, we find that if $(\delta_2 - \rho_{12}\,\delta_1)$ is negative, the most favorable situation for $y_3$ occurs when $\rho_{13}$ is negative and $\rho_{23}$ is positive.

## 6. *More than three variates*

With more than three variates, a few simple results on the effects of correlation can be obtained if all intercorrelations are the same. If $p$ variates with distance $\delta_1$, $\delta_2$, ..., $\delta_p$ have the same correlation $\rho_1$ between any pair, several writers have shown that the squared distance for the linear discriminant function is

$$\frac{\{1+(p-1)\,\rho_1\}\;\Sigma\delta^2 - \rho_1\,(\Sigma\delta)^2}{(1-\rho_1)\,\{1+(p-1)\,\rho_1\}}. \tag{7}$$

If $\Sigma\delta^2$ is subtracted, the difference may be expressed as

$$\frac{-\rho_1[(\Sigma\delta)^2 - \Sigma\delta^2 - (p-1)\rho_1\Sigma\delta^2]}{(1-\rho_1)\,\{1+(p-1)\,\rho_1\}}. \tag{8}$$

Since $(\Sigma\delta)^2 > \Sigma\delta^2$, this difference is always positive when $\rho_1$ is negative. If $\rho_1$ is positive, the difference is negative (i.e. correlation is harmful) unless

$$\rho_1 > \frac{(\Sigma\delta)^2 - \Sigma\delta^2}{(p-1)\,\Sigma\delta^2}. \tag{9}$$

As before, high correlations are needed to satisfy (9) unless the $\delta's$ vary widely. For instance, with $p = 4$, $\delta_i = 4, 3, 2, 1$, $\rho_1$ must exceed $7/9 = 0.78$; with $p = 7$, $\delta_i = 4, 3, 2, 2, 1, 1, 1$, $\rho_1$ must exceed $20/27 = 0.74$.

Suppose now that these variates represent a set of $p$ good discriminators. To these we add a set of $q$ poor discriminators having distances $d_1$, $d_2$, ... $d_q$ and correlation $\rho_2$ between every pair of the $q$ variates. If there is no correlation between any good and any poor variate, the squared distance for the combined discriminator, with $(p + q)$ variates, is

$$\frac{\{1+(p-1)\,\rho_1\}\;\Sigma\delta^2 - \rho_1\,(\Sigma\delta)^2}{(1-\rho_1)\,\{1+(p-1)\,\rho_1\}} + \frac{\{1+(q-1)\,\rho_2\}\;\Sigma d^2 - \rho_2\,(\Sigma d)^2}{(1-\rho_2)\,\{1+(q-1)\,\rho_2\}} \tag{10}$$

By comparison with (7) we see that the second term in (10) represents the increase in squared distance due to the set of $q$ variates. By the analysis given above, negative values and high positive values of $\rho_2$ are helpful to the gain in squared distance.

If the correlation between any poor discriminator and any good discriminator is $\rho_3$, Rao (1949) has shown that in our notation the combined squared distance with

6

$(p + q)$ variates becomes the rather complex expression

$$\frac{\Sigma \delta^2}{1-\rho_1} + \frac{\Sigma d^2}{1-\rho_2} + \frac{[qp_3^2 - \rho_1 \{ 1 + (q-1)\rho_2 \}](\Sigma \delta)^2}{(1-\rho_1)\gamma}$$
$$+ \frac{[pp_3^2 - \rho_2 \{ 1 + (p-1)\rho_1 \}](\Sigma d)^2}{(1-\rho_2)\gamma} - \frac{2\rho_3(\Sigma \delta)(\Sigma d)}{\gamma} \qquad (11)$$

where

$$\gamma = \{ 1 + (p-1)\rho_1 \} \{ 1 + (q-1)\rho_2 \} - pq\,\rho_3^2.$$

The difference (11)-(10) measures the change in the squared distance due to a cross-correlation $\rho_3$ between each poor variate and each good variate. After some simplification the difference becomes

$$\frac{q\,\rho_3^2(\Sigma \delta)^2}{\{ 1 + (p-1)\rho_1 \}\gamma} + \frac{p\,\rho_3^2(\Sigma d)^2}{\{ 1 + (q-1)\rho_2 \}\gamma} - \frac{2\,\rho_3(\Sigma \delta)(\Sigma d)}{\gamma}. \qquad (12)$$

From (12) it is found once again that the cross-correlation $\rho_3$ if helpful if $\rho_3$ is negative, but is harmful if $\rho_3$ is positive, unless

$$\rho_3 > \frac{2(\Sigma \delta)(\Sigma d)}{\dfrac{q(\Sigma \delta)^2}{1 + (p-1)\rho_1} + \dfrac{p(\Sigma d)^2}{1 + (q-1)\rho_2}}. \qquad (13)$$

## 7. *Relation of these results to the rejection of poor variates*

To recapitulate, if a group of poor discriminators with distances $d_i$ were uncorrelated with each other and with the good discriminators, their joint contribution to the squared distance would be $\Sigma c$ . The study of the effects of correlation was undertaken to see under what circumstances this value would remain approximately true or would give an overestimate of the joint contribution. The results suggest, although they by no means prove, the following:

(i) If most of the correlations among the poor variates and between poor and good variates are positive and moderate in size, the joint contribution of the poor variates will probably be less than $\Sigma d^2$.

(ii) The contribution is likely to exceed $\Sigma d^2$ if most of the correlations in (i) are negative. I would expect this case, while not impossible, to be rare, since it is difficult to envisage situations in which negative correlations, as defined here, are likely to predominate.

(iii) If correlations are positive and high, the joint contribution may exceed $\Sigma d^2$. Again I doubt that this situation is frequent. It is, of course, easy to have a pair of variates with a high positive correlation. But mostly this arises because they measure almost the same thing. In this event $d_1$ and $d_2$ are likely to be approximately equal, and a very high positive correlation is needed. If $d_1 = 0.9d_2$, for example, expression (5) shows that a positive $\rho$ must exceed 0.9944 to be beneficial.

(iv) A variate which has negative correlations with most of the good variates may be worth inclusion in the discriminant even if its individual performance is poor.

7

(v) With a covariate, any correlation is helpful, although small ones do not help much.

To summarize, if it is the fact that in practice most correlations are positive and modest in size, the analysis suggests that reliance on the value of $\Sigma\,d^2$ in deciding whether to throw away a group of poor discriminators is unlikely to produce a serious mistake.

It has been assumed that the discriminant is set up from samples of infinite sizes. In practice, with samples of sizes $n_1$, $n_2$ from the two populations, the computed $(\bar{y}_{i2} - \bar{y}_{i1})^2 / s_i^2$ is an overestimate of the scaled $d_i^2$. The bias amounts to

$$\mathrm{E}\left\{\frac{(\bar{y}_{i2} - \bar{y}_{i1})^2}{s_i^2}\right\} - d_i^2 = \frac{\nu}{(\nu - 2)}\left\{\frac{1}{n_1} + \frac{1}{n_2}\right\}$$

where $\nu$ is the number of degrees of freedom in $s_i^2$, which will usually be $(n_1 + n_2 - 2)$.

The correction for bias is worth applying unless $n_1$ and $n_2$ are very large. Suppose that $n_1 = n_2 = 50$ and there are six poor variates with estimated values $\widehat{d_i} = (\bar{y}_{i2} - \bar{y}_{i1}) / s_i = 0.41,\ 0.38,\ 0.36,\ 0.34,\ 0.32,\ 0.31$. The estimated joint contribution to the squared distance is

$$\Sigma\,\widehat{d_i^2} = (0.41)^2 + (0.38)^2 + \ldots + (0.31)^2 = 0.7562.$$

With $\nu = 98$, the bias for a single variate is

$$\frac{98}{96} \cdot \frac{2}{50} = \cdot\,0408.$$

The correction for bias in $\Sigma\,\widehat{d_i^2}$ is $6 \times .0408 = .2448$. This reduces the estimated $\Sigma\,\widehat{d_i^2}$ to 0.5114.

## 8. *The performance of the discriminant function in practice*

The previous discussion has involved some speculation about the kinds of correlations that turn up in applications of the discriminant function. In this section, twelve numerical applications reported in the literature are examined. Questions of interest are:

(i) How does the actual squared distance $\mathrm{D}^2$ obtained by the discriminant compare with the estimate $\Sigma\,\delta_i^2$ made on the assumption of independence?

(ii) How does $\mathrm{D}^2$ compare with the estimate made from expression (7) in Section 6 on the assumption that every pair of variates has a correlation equal to the average correlation $\bar{\rho}$?

(iii) What is the distribution of the correlation coefficients found in practice?

In selecting examples the aim was to find a range of types of data, while including the best-known applications. Numerous additional examples involving exam scores or psychological scores could have been added. All variates were scaled so that positive and negative correlations have the meaning discussed in Section 4. Table 2 indicates the measurements used and the nature of the two populations being distinguished.

8

TABLE 2

*Nature of variables and of populations*

| Authors | Variables | Populations |
|---|---|---|
| 1. BEALL ............ | 4 psychological tests..................... | Men and women. |
| 2. DURAND. ......... | Price of car, down payment, income and length of loan. | Good and bad auto loan risks. |
| 3. HAMILTON ........ | Measures of anxiety, dependence guilt, perfectionism, etc. | Non-ulcer dyspeptics and controls. |
| 4. BATEN and DEWITT.. | B. T. U., volatile matter fixed carbone per cent ash of coal. | Coal from two mines. |
| 5. MARTIN .......... | Measurements on mandibles of skulls. | Male and female. |
| 6. MAHALANOBIS et. al. | Stature, sitting height, head length and breadth, nasal length, etc. | Indian ethnic groups. |
| 7. TINTNER. ......... | Length, amplitude, rate of change, etc. in price cycle. | Consumers' and producers' goods. |
| 8. COX and MARTIN... | ph, N. contents and available P. content of soils. | Soils with and without azotobacter. |
| 9. WALLACE.......... | Psychological and intelligence tests. | Successful anc unsuccessful salesmen. |
| 10. FISHER............ | Sepal and petal length and width. | 2 species of Iris. |
| 11. TRAVERS.......... | Tests of intelligence and coordination. | Engineers and air pilots. |
| 12. BARNARD ......... | Skull measurements. | Regression on time. |

Table 3 shows for each example the values of $D^2$ given by the three methods and the corresponding probabilities of misclassification as computed by equation (1) in Section 2 (with D in place of $\delta$). The applications are arranged in order of decreasing $\bar{\rho}$. The quantity $p$ (col. 3) is the number of variables in the discriminant. In

TABLE 3

*Comparison of actual and predicted preformance of LDF.*

| Author* | $\bar{\rho}$ | p | Value of $D^2$ by | | | Prob. of misclassification | | |
|---|---|---|---|---|---|---|---|---|
| | | | LDF | Ind | Equal $\rho$ | LDF | Ind | Equal $\rho$ |
| 1................... | .43 | 4 | 5.79 | 6.09 | 5.31 | .11 | .11 | .13 |
| 2................... | .43 | 4 | 0.40 | 0.46 | 0.40 | .37 | .37 | .37 |
| 3................... | .35 | 6 | 1.07 | 2.32 | 1.11 | .30 | .22 | .30 |
| 4................... | .24 | 4 | 5.33 | 7.73 | 6.08 | .13 | .08 | .11 |
| 5................... | .23 | 6 | 6.04 | 11.74 | 6.08 | .11 | .04 | .11 |
| 6a.................. | .22 | 8 | 7.69 | 12.31 | 7.68 | .083 | .040 | .083 |
| 7................... | .19 | 4 | 3.90 | 5.33 | 4.43 | .16 | .12 | .15 |
| 6b.................. | .15 | 8 | 10.41 | 13.34 | 9.83 | .053 | .034 | .059 |
| 8................... | .12 | 3 | 6.18 | 7.56 | 6.26 | .11 | .08 | .11 |
| 9................... | .10 | 5 | 2.83 | 3.69 | 2.74 | .20 | .17 | .20 |
| 10.................. | .06 | 4 | 103 | 122 | 109 | $0^6 20^+$ | $0^7 17^+$ | $0^7 90^+$ |
| 11.................. | — .02 | 6 | 6.54 | 6.57 | 7.17 | .10 | .10 | .09 |
| 12.................. | — .04 | 4 | 38.3 | 34.8 | 36.4 | .0010 | .0016 | .0013 |

* For identification see Table 2 col. 1.
+ *i. e.* 00000020.

computing Table 3, values of the $(\bar{y}_{i2} - \bar{y}_{i1})$, $s_{ij}$ and $r_{ij}$ were assumed known without error.

*Comparison of the LDF with $\Sigma \delta_i^2$.* The values of $\bar{\rho}$ are all positive except for two small minus values (nos. 11 and 12). The assumption of independence overestimates the value of $D^2$ and therefore underestimates the probability of misclassification in all cases except no. 12. This result supports the contention that if the value of $\Sigma \delta^2$ is used to decide whether to throw away a group of variates before constructing the LDF, a serious mistake will seldom be made.

In estimating how an LDF will actually perform, the $\Sigma \delta^2$ approximation in less successful. The general order of magnitude of the probability of misclassification is indicated, but the estimated probability lies within 20 % of the actual LDF probability in only four examples (nos. 1, 2, 9, 11).

*Comparison of the LDF with the assumption of equal $\rho$.* The assumption of equal $\rho$ does well as a predictor. It gives a higher $D^2$ than the actual LDF 7 times and a lower one 5 times, with one equal value. Except with the low probabilities in examples 10 and 12, the probability of misclassification is always within 20 % of the LDF value.

TABLE 4

*Distribution of the correlation coefficients in the LDF examples*

| Value of $r$ | f | Value of $r$ | f |
|---|---|---|---|
| $< -.5$ | 1 | 0 to 0.1 | 15 |
| $-.5$ to $-.4$ | 2 | 0.1 to 0.2 | 22 |
| $-.4$ to $-.3$ | 1 | 0.2 to 0.3 | 25 |
| $-.3$ to $-.2$ | 4 | 0.3 to 0.4 | 18 |
| $-.2$ to $-.1$ | 4 | 0.4 to 0.5 | 6 |
| $-.1$ to 0 | 9 | 0.5 to 0.6 | 10 |
| | | Over 0.6 | 5 |
| | | | 122 |

Two sets of results (6a and 6b) are given for the well-known U.P. Anthropometric Study by Mahalanobis et. al. (1949). This study was a comparison of many Indian castes and tribes, using a pooled covariance matrix. All the pooled correlation coefficients are positive. In both examples in Table 3, one population was the group called Brahmin (Basti B1) in the study. The six other groups which are at the greatest distance from the B1 group have a common feature. Their mean measurements are all smaller than those for B1 except for nasal breadth, which is larger. Consequently in our scaling, all correlations of nasal breadth with the other seven variates in the LDF become negative, making $\bar{\rho} = 0.15$ in no. 6b. The paper gives the contribution of nasal breadth to $D^2$ when used alone, and also the contribution when it is added as a fifth variate to a discriminant containing head length, head breadth, bizygomatic breadth and nasal length. In line with the mathematical analysis in Section 5, the contribution of nasal breadth to $D^2$ is greater when it is added to the four-variate LDF than when used alone in all six groups with the negative correlations. For the Oraon (T5) group shown in Table 3, these contributions were 0.424 as against 0.186.

10

Somewhat nearer to the B1 group are some groups in which all mean measurements are smaller, giving $\bar{\rho} = 0.22$. The Rajwar ($T_6$) group were used in no. 6a. For this group the contribution of nasal depth to $D^2$ was 0.197 when used alone, but only 0.029 when added to the four-variate LDF.

Table 4 shows the distribution of the 122 correlation coefficients in these examples. It will be noted that 101 of the 122 coefficients are positive and that most of the negative values lie between $-0.2$ and zero. Of the positive coefficients only 15 exceed 0.5. These results support the expectation that in the usual applications of the LDF, correlations are mostly positive and mostly moderate. Moreover, because of sampling errors in the $r_{ij}$, the observed spread in Table 4 exceeds the spread in the corresponding $\rho_{ij}$.

## 9. *Summary*

A common use of the linear discriminant function LDF is to classify specimens into one of two populations, the objective being to minimize the probability of misclassification. This paper considers the problem of estimating the discriminating power of the LDF from a knowledge of the discriminating powers of the individual variates included in the LDF. Such estimates may enable an investigator (i) to discard a group of poor discriminators before computing the LDF (ii) by informing him in advance how well the LDF is likely to perform, to decide whether it is worth while to compute the LDF.

Prediction of the performance of the LDF would be easy if the variates were independent. Study of the effects of correlation among the variates suggests that a variate which is negatively correlated with the other variates is likely to perform better in the LDF than would be anticipated, while a variate that is positively correlated is likely to perform poorly. Reasons are given for expecting that in practice most correlations will be positive, as defined in this paper.

The actual performance of LDF in 12 numerical examples in the literature is examined. With one exception the LDF gives a higher probability of misclassification than predicted on the assumption of independence. The assumption that every pair of variates has a correlation equal to the average correlation gives very satisfactory predictions of the performance of the LDF. Most of the correlations between pairs of variates are positive and moderate or small in size.

These results suggest that it will usually be safe to reject a group of variates if the value of $\Sigma d^2$ (i.e. their joint contribution to $D^2$ on the assumption of independence) is small. Computation of $\Sigma d^2$ does not, of course, require à knowledge of the correlation coefficients. If the correlations have been computed, a poor variate that is negatively correlated with most of the other variates might be included in the LDF. Prediction of $D^2$ for the LDF on the assumption that all correlations are equal to the average correlation appears to be satisfactory unless the probability of misclassification is very low.

# REFERENCES

(1) BARNARD, M. M. (1935). The secular variation of skull characters in four series of Egyptian skulls. *Ann. Eugen.,* 6, 352-371.

(2) BATEN, W. D. & DEWITT, C. C. (1944). Use of the discriminant function in the comparison of proximate coal analyses. *Indus. Eng. Chem. Anal.,* Ed., 16, 32-34.

(3) BEALL, G. (1945). Approximate methods in calculating discriminant functions. *Psychometrika,* 10, 205-217.

(4) COCHRAN, W. G. & BLISS, C. I. (1948). Discriminant functions with covariance. *Ann. Math. Stat.,* 19, 151-176.

(5) COX, G. M. & MARTIN W. P. (1937). Use of a discriminant function for differentiating soils with different Azotobacter populations. *Iowa St. Coll. Jour. Sci.* 11, 323-332.

(6) DURAND, D. (1941). Risk elements in consumer installment financing. Studies in Consumer Installment Financing, 8, *National Bureau of Economic Research,* 125-142.

(7) FISHER, R. A. (1936). The use of multiple measurements in taxonomic problems. *Ann. Eugen.,* 7, 179-188.

(8) HAMILTON, M. (1950). The personality of dyspeptics. *Brit. Jour. Med. Psych.* 23, 182-198.

(9) HORST, P. & SMITH S. (1950). The discrimination of two racial samples. *Psychometrika,* 15, 271-289.

(10) MAHALANOBIS, P. C., MAJUMDAR, D. N. AND RAO, C. R. (1949). Anthropometric survey of the United Provinces, 1941; A statistical study. *Sankhya,* 9, 89-324.

(11) MARTIN, E. S. (1936). A study of an Egyptian series of mandibles, with special reference to mathematical methods of sexing. *Biometrika,* 28, 149-178.

(12) RAO, C. R. (1949). On some problems arising out of discrimination with multiple characters. *Sankhya,* 9, 343-366.

(13) RAO, C. R. (1954). On the use and interpretation of distance functions in statistics. *Bull. Int. Stat. Inst.* XXXIV, 2nd part, 1-10.

(14) TINTNER, G. (1946). Some applications of multivariate analysis to economic data. *Jour. Amer. Stat. Ass.,* 41, 472-500.

(15) TRAVERS, R. M. W. (1939). The use of a discriminant function in the treatment of psychological group differences. *Psychometrika,* 4, 25-32.

(16) WALLACE, N. AND TRAVERS, R. M. W. (1938). A psychometric sociological study of a group of specialty salesmen. *Ann. Eugen.,* 8, 266-302.

12

# RÉSUMÉ

On emploie souvent la fonction discriminante linéaire (LDF) pour classifier des individus en l'une ou l'autre de deux populations, le but étant de réduire au minimum la probabilité d'un classement erroné. Ici il s'agit du problème d'évaluation du pouvoir discriminatoire de la LDF, étant donné le pouvoir discriminatoire individuel de chaque variable dont la LDF est composée. Une telle évaluation permet à l'utilisateur ($i$) de rejeter, avant de calculer la LDF, un groupe de variables dont le pouvoir discriminatoire total semble être faible ($ii$) de décider s'il vaut la peine de calculer la LDF.

Si toutes les variables étaient indépendantes, l'évaluation du pouvoir discriminatoire de la LDF serait facile. Une étude de l'effet de corrélation entre les variables suggère qu'une variable qui a des corrélations négatives avec les autres variables va probablement apporter à la LDF une contribution plus forte que celle qu'on aurait attendue. Une variable à corrélations positives va généralement apporter une moindre contribution. On donne des raisons pour s'attendre à ce qu'en pratique la plupart des corrélations soient positives.

On examine les performances de la LDF sur douze exemples numériques trouvés dans la littérature statistique. A une seule exception près, la fréquence d'une erreur de classement par la LDF est plus grande que la fréquence prévue en supposant l'indépendance des variables. Supposer que toutes les variables ont la même corrélation prévoit très bien la fréquence des erreurs de classement au moyen de la LDF. Dans les douze exemples examinés la plupart des corrélations sont positives et de relatives faibles valeurs.

Ces résultats mènent à la conclusion suivante. Si la contribution d'un groupe de variables au pouvoir discriminatoire, prédite en supposant l'indépendance, est faible, il est peu probable que la contribution réelle soit importante.

13

# SEQUENTIAL EXPERIMENTS
# FOR ESTIMATING THE MEDIAN LETHAL DOSE

William G. COCHRAN and Miles DAVIS
Department of Statistics — Harvard University

SOMMAIRE

Supposons qu'une drogue soit administrée à m sujets et qu'on veuille estimer la quantité de cette drogue produisant une réaction sur 50 % de ces sujets.

Suivant la méthode séquentielle de Robbins et Munro, nous présentons ici un moyen d'estimer cette quantité. Supposons qu'au $n^{\text{ème}}$ essai, la drogue en quantité $x_n$, produise un effet dans une proportion $p_n$. Alors, au $(n+1)^{\text{ème}}$ essai, on utilisera la quantité :

$$x_{n+1} = x_n - \frac{c}{n} \left( p_n - \frac{1}{2} \right)$$

Différents auteurs ont montré que, lorsque n croît, $x_n$ est distribué normalement autour de la dose médiane effective, avec une variance égale à $v/n$, où v est une constante dépendant de c. La valeur optimale de c a été aussi déterminée. Les avantages de cette méthode sont sa précision et sa simplicité, mais on ne peut l'appliquer que dans les cas où la réponse produite est assez rapide pour que l'expérience soit complétée dans un délai raisonnable.

Les propriétés de cette méthode, lorsqu'on estime la dose médiane mortelle dans les expériences toxicologiques faites sur 10 et 50 animaux, ont été examinées à l'aide de machines électroniques.

Si l'on n'a pas à l'avance de bonnes estimations de la dose médiane mortelle et de l'écart-type des doses "juste fatales", on ne peut se baser sur le plan original. On développe alors un plan séquentiel à deux stages et cela semble donner de bons résultats, pourvu que l'écart-type (sur une longue échelle) soit conçu dans un rapport de 4 à 1.

- - - - - - - - - - - - - - -

(1) This work was supported in part by the Office of Naval Research Contract Nonr 1866 (37). Most of the computations were done at the MIT Computation Center, Cambridge, Mass.

## SUMMARY

This paper is a progress report on a sequential method, due to Robbins and Munro, of estimating the amount of a stimulus that produces a response in 50 % of the subjects to which it is applied. Suppose that at the nth step in the experiment a stimulus of amount $x_n$ produces the response in a proportion $p_n$ out of m subjects. Then the amount of stimulus used at trial (n+1) is :

$$x_{n+1} = x_n - \frac{c}{n} \left( p_n - \frac{1}{2} \right).$$

Previous writers have shown that as n increases $x_n$ tends to be normally distributed about median effective dose with a variance $v/n$, where v is a constant depending of c, and have determined the optimum value of c. The advantages of the method are its precision and simplicity, but the method is restricted to situations in which the response is rapid enough so that the experiment can be completed in a reasonable time.

The properties of the method when applied to estimate the median lethal dose in toxicological experiments involving between 10 and 50 animals are examined by the use of electronic machines. The original plan is not reliable if the median lethal dose and the standard deviation of the "just fatal" doses are poorly estimated in advance. A two-stage sequential plan is developed which appears to work well even when the median lethal dose is poorly known in advance, provided that the standard deviation (on a long scale) is known within a ratio of 4 to 1.

## INTRODUCTION

During the last war, Wald introduced a method of sequential analysis by which the results of an experiment are analysed continuously as they come in, the experiment being stopped as soon as a definite verdict of statistical significance or non-significance is reached. The advantage of the method is that it should often permit conclusions to be reached sooner than if the analysis is postponed until a preassigned number of observations has accumulated. For a test of significance of the same power, the average saving in number of observations appears to be about 30 %.

The Wald sequential method has found only limited application in practical experimentation. For this there may be several reasons. The available plans are restricted thus far to the comparison of two treatments , whereas the trend in experimentation is towards more complex experiments of a multifactorial type. In many experiments it is not convenient to plan the experiment sequentially. Some of the mathematical properties of séquential trials are not yet fully understood. One area which seems unusually suitable for sequential methods is medical experimentation with human subjects. Because of the shortage of suitable patients, many experiments must be confined to the comparison of two treatments. The

182

doctor's ethical responsability is a potent reason for stopping the trial as soon as a clear verdict is reached. Patients usually enter the trial over a period of months. Consequently, sequential methods are being tried in this area. Good expository accounts are given by Armitage (1954) and Hajnal et. al. (1959).

Wald's sequential analysis is designed to perform a test of significance. If, instead, the objective is to estimate the difference between the treatment means with a specified precision, the sequential method does not reduce the number of observations required. In this paper we consider a problem in which the purpose of sequential experimentation is to increase the precision of the estimate made from the experiment. As will be seen, the problem is not fully solved, so that this paper is in the nature of a progress report. The method applies to experiments designed to estimate the amount of a stimulus that produces a response in 50 % of the subjects to which it is given. For those unfamiliar with this type of research, some backgroundis presented in the next section, using toxicology as an exemple.

THE MEDIAN LETHAL DOSE (LD 50).

Suppose that it were possible to determine for each animal in a population the exact dose of a toxic agent that is just sufficient to kill the animal. (This can be done, for instance, with digitalis injected slowly into the heart of the cat, but with most poisons it is impracticable). These "just fatal" doses would form a frequency distribution. Experimental evidence suggest that the frequency distribution of the log of the "just fatal" dose is often less skew than that of the original doses.

The two mathematical forms most frequently used to represent the distribution of x, the log dose, are the normal curve and the logistic curve. The frequency fonctions are, respectively,

(normal) :
$$\frac{\beta}{\sqrt{2\pi}} \, e^{-\frac{1}{2}\beta^2(x-\mu)^2} dx \qquad (1)$$

(logistic) :
$$\frac{\beta \, e^{-\beta(x-\mu)} \, dx}{\left[1 + e^{-\beta(x-\mu)}\right]^2} \qquad (2)$$

In both cases $\mu$ is the mean (and also the median) of the distribution. For the normal curve, the standard deviation $\sigma = 1/\beta$ ; for the logistic, $\sigma = \pi/\sqrt{3}\,\beta$. The two curves are closely similar, and the choice between them may be one of convenience.

In practice, the quantity $\mu$, called the median lethal dose (LD 50 or MLD), is used as a measure of the potency of the poison. The principal objective of toxicological experiments is to estimate $\mu$. With most poisons, the way in which the experiment has to be done is to give a specified dose to an animal or group of animals and observe how many die. From (1) and (2) the probability $P(x)$ that an animal dies when given log dose x is as follows,

183

(normal) : $$P(x) = \frac{1}{\sqrt{2\pi}} \int_{-\infty}^{\beta(x-\mu)} e^{-\frac{y^2}{2}} \, dy \qquad (3)$$

(logistic) : $$P(x) = \frac{1}{1 + e^{-\beta(x-\mu)}} \qquad (4)$$

If P is plotted against x over a wide range, the graph for either curve looks roughly as in figure 1.

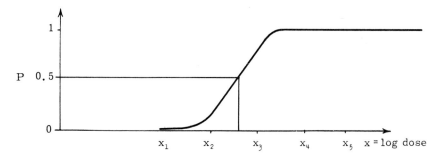

Figure 1 - Graph of proportion killed against log dose.

Since the value of $\mu$ and $\sigma$ are not usually well known in advance of the experiment, the typical non-sequential experiment contains several different dose levels, often equally spaced, as illustrated by the levels $x_1, x_2, \ldots, s_5$, in figure 1. The experimenter hopes that at least two of the levels will fall near the value of $\mu$. Obviously, levels like $x_1$, $x_4$ and $x_5$ which produce no deaths or 100 % deaths give little information about $\mu$. Thus in a non-sequential experiment we expect a certain wastage of effort, in the sense that some of the dose levels tell us little or nothing about the value of the LD 50. The aim in a sequential plan is to choose dosage levels so that most of the testing is done near to $\mu$.

THE UP - AND - DOWN METHOD.

The first sequential plan of this type is the so-called up-and-down method. It was apparently developed originally for testing explosives. Its application to texicology was suggested by Dixon and Mood (1948). To start the process we require an initial guess $x_1$ at the value of $\mu$, and an interval $d$ between successive doses. For the normal curve the best value of $d$ is approximately equal to $\sigma$.

A single animal is given the dose $x_1$. If the animal dies, the second animal receives the dose $(x_1 - d)$. If the first animal survives, the second animal is given the dose $(x_1 + d)$. In the same way, the dose $x_{n+1}$ given to the (n+1)th animal is equal to $(x_n \pm d)$ according as the animal dies or lives at dose $x_n$.

184

It will be clear that even if $x_1$ is far from $\mu$ , the process rapidly brings the test doses near to $\mu$ , and thereafter makes them oscillate above and below $\mu$ . When it is decided to terminate the experiment, a simple method is available for computing an estimate $\hat{\mu}$ of $\mu$ and its standard error, using all or practically all the experimental results. (Brownlee et al. 1953).

Although recognizing that it is difficult to make an entirely fair comparison with a non-sequential experiment, Brownlee et al. concluded from some calculations that the non-sequential experiment required 55 % more observations to estimate $\mu$ with the same precision ; or, in other words, the up-and-down method needs only about 2/3 as many observations. They point out that the principal disadvantage of the up-and-down method is the increased length of time required to complete the experiment, which may be prohibitive unless the response to the stimulus is very rapid. As an alternative they suggest running simultaneously four series of 10 steps each, for example, instead of one series of 40 steps. This alternative gives less precises estimates, but the increase in variance is small if the starting point is within $\pm 4\,\sigma$ of $\mu$ .

THE ROBBINS - MONRO PROCESS.

The present study arose from discussions with a pharmacologist, Dr Denis Hawkins. He was conducting experiments to estimate the LD 50's of a group of quick-acting poisons on mice, using a sequential method suggested by Robbins and Monro (1951). This is a sequential estimation technique of wide applicability, usable in particular to estimate any quantile of a frequency distribution. To apply it to the estimation of the LD 50, suppose that $m$ animals have been tested at dose $x_n$, and that $p_n$ is the proportion dead out of $m$. The Robbins-Monro process states that the next test should be made at level $x_{n+1}$, where :

$$x_{n+1} = x_n - a_n \left( p_n - \frac{1}{2} \right) \tag{5}$$

where the $a_n$ is a sequence of positive numbers. We shall take $a_n = c/n$, where c is a constant.

This method differs from the up-and-down process in three respects. (I) Owing to the factor $n$ in the denominator of $c/n$ the steps become progressively shorter as the trial proceeds. (II) If $m > 1$, the size of step depends also on $p_n$. For instance, if 2 animals die out of 4, $p_n = \frac{1}{2}$ and $x_{n+1} = x_n$. With $m = 1$, $p_n$ is either 0 or 1 and the step is $\pm c/2n$. (III) When the trial stops at step N say, the estimate of the LD 50 is the dose $x_{N+1}$ at which the next test would have been made. There is no reason why this estimate must be used rather than an estimate derived from all the data, but $x_{N+1}$ has the advantage of great simplicity.

185

Results by Robbins and Monro (1951), Chung (1954) and Hodges and Lehmann (1956) showed that as n tends to infinity, $x_n$ becomes normally distributed with mean $\mu$ and variance $v/n$, where v is a constant depending on c and on the distribution of the "just fatal" doses. The following non-rigorous development of v will be given.

If f(y)dy is the frequency function of the quantity $\beta (x - \mu)$, then

$$P(x) = \int_{-\infty}^{\beta(x-u)} f(y)\, dy \qquad (6)$$

To order $1/\sqrt{n}$,

$$P(x) = \frac{1}{2} + \beta (x - \mu) f_o \qquad (7)$$

Where $\mu$ is the median of the distribution and $f_o$ is the ordinate at the median.

From (5), we shall find the variance of $x_{n+1}$ in terms of that of $x_n$, using the lemma,

$$V(x_{n+1}) = V\,[E(x_{n+1}\,|\,x_n)] + E\,[V(x_{n+1}\,|\,x_n)] \qquad (8)$$

Now :

$$E(x_{n+1}\,|\,x_n) = x_n - \frac{c}{n}\left(P_n - \frac{1}{2}\right)$$

$$\doteq \mu + (x_n - \mu)\left[1 - \frac{c\beta f_o}{n}\right]$$

From (7). Hence :

$$V\,[E(x_{n+1}\,|\,x_n)] \doteq V(x_n)\left[1 - \frac{c\beta f_o}{n}\right]^2 \qquad (9)$$

Also, since $p_n$ is binomially distributed about $P_n$, the conditional variance,

$$V(x_{n+1}\,|\,x_n) = \frac{c^2}{n^2}\frac{P_n Q_n}{m}$$

$$\doteq \frac{c^2}{n^2 m}\left[\frac{1}{4} - \beta^2 f_o^2 (x_n - \mu)^2\right] \qquad (10)$$

Hence, from the lemma and (9) and (10),

$$V(x_{n+1}) = V(x_n)\left[1 - \frac{c\beta f_o}{n}\right]^2 + \frac{c^2}{4n^2 m} - \frac{c^2 \beta^2 f_o^2}{n^2 m}\,V(x_n)$$

If $V(x_n) = v/n$, this gives :

186

$$\frac{v}{n+1} \doteq \frac{v}{n} - \frac{v}{n^2} = \frac{v}{n} - \frac{2\,vc\,\beta\,f_o}{n^2} + \frac{c^2}{4\,n^2 m}$$

Where two terms of order $1/n^3$ have been dropped. Hence :

$$v = \frac{c^2}{4m(2c\,\beta\,f_o - 1)} \tag{11}$$

This analysis holds only if $c > 1/2\,\beta f_o$. More penetrating analyses for all positive $c$ are given by Hodges and Lehmann (1956) and Schmetterer (1953).

The value of $c$ minimizes $v$ is $c_o = 1/\beta\,f_o$, the smallest $v$ being $1/4m\beta^2 f_o^2$. For the normal and logistic curves $c_o$ and the limiting variance $v/n$ for $\hat{\mu}$ are as follow (in terms of the standard deviation $\sigma$ of the distribution of $x$.

Normal.    $(f_o = 1/\sqrt{2\,\pi}\;;\;\;\beta = 1/\sigma)$.

$$c_o = \sqrt{2\,\pi}\,\sigma = 2.506\,\sigma \; : \; \frac{v}{n} = \frac{\pi\,\sigma^2}{2\,mn} = \frac{1.571\,\sigma^2}{mn}$$

Logistic.    $(f_o = 1/4 \; : \; \beta = \pi/\sqrt{3}\sigma)$

$$c_o = \frac{4\sqrt{3}\sigma}{\pi} = 2.205\,\sigma \; : \; \frac{v_o}{n} = \frac{12\,c^2}{\pi^2\,mn} = \frac{1.216\,\sigma^2}{mn}\,.$$

## INITIAL CALCULATIONS FOR EXPERIMENTS OF MODERATE SIZE

A  The previous results are asymptotic and do not reveal how large n must be. For his experiments Dr. Hawkins was interested in questions like the following.

(I) Are the results valid for experiments involving a total number of animals (mn) between 10 and 50 ?

(II) Is there a loss of precision if several animals are tested at one time (i. e. taking $m > 1$) so that the number of steps n and the time required to complete the experiment is decreased ? The asymptotic results suggest that the variance of $\hat{\mu}$ depends only on mn.

(III) Initial estimates $x_1$ of $\mu$ and $\hat{\sigma}_1$ of $\sigma$ are required to start the process (since the optimum c depends on $\sigma$). How well does the process work if the initial estimates are poor ?

Calculations of the exact mean $E(\hat{\mu})$ and the mean square error $E(\hat{\mu} - \mu)^2$ of $(\hat{\mu})$ were made for a series of simulated experiments involving 6, 12, 24 and 48 mice. For 6 mice, the computations were done on a Monroe desk machine by Alan Howson ; the remainder were programmed on the IBM 704. The normal law for x was assumed

187

throughout. A range of values of $(x_1 - \mu)/\sigma$ from 0 to 8 (16 in later calculations) was used to show the effect of different starting points. Three values of c were included : $c_o$ (the asymptotic optimum), $2c_o$ and $c_o/2$. The last two cases represent the situation in which the experimenter's guess at $\sigma$ is respectively twice and one-half the correct $\sigma$.

Except in minor details the effects of different values of $x_1$, c and m on MSE($\hat{\mu}$) were the same for mn = 12,24 and 48 animals, so that only selected results will be shown. In figure 2, the value of MSE($\hat{\mu}$)/$\sigma^2$ is plotted against the starting point $x_1$ for c = $c_o/2$, $c_o$ and $2c_o$. The corresponding asymptotic variance 1.571/mn is also shown. The experiment consisted of 12 steps with 1 animal per step.

For c = $c_o$, MSE($\hat{\mu}$) is practically constant (about 10 % above the asymptotic value) if $x_1$ lies within 2 S.D.'s of $\mu$. As $x_1$ moves further away, MSE($\hat{\mu}$) increases rapidly. For c = $c_o/2$, MSE($\hat{\mu}$) remains near asymptotic value only for starting points out to $0.7\sigma$. With double-sized steps, c = $2c_o$, MSE($\hat{\mu}$) runs at first about 30 % higher than with single steps. However, it maintains a practically constant value out to 6 S.D, and is superior to the single step beyond 3 S.D.

The reason for the rapid increases in MSE($\hat{\mu}$) for $c_o = c_o/2$ and $c_o$ is easy to see. If $x_1$ is far from $\mu$ and the number of steps n is small, the results of the shortening of steps through the factor 1/n is that $x_{n+1} = \hat{\mu}$ has not reached $\mu$ by the time the experiment is terminated. Consequently $\hat{\mu}$ is subject to a bias which becomes successively greater as $x_1$ departs from $\mu$.

Now :

$$\text{MSE}(\hat{\mu}) = E(\hat{\mu} - \mu)^2 = V(\hat{\mu}) + (\text{Bias})^2$$

When $x_1$ is far out, the MSE consists mainly of the bias term.

For 24 animals, with c = $c_o$, figure 3 compares MSE($\hat{\mu}$)/$\sigma^2$ for m = 1, n = 24 ; m = 3, n = 8 ; m = 6, n = 4. For $(x_1 - \mu)$ out to about $1.5\sigma$, the higher values of m give slightly smaller mean square errors than m = 1. Beyond this point MSE($\hat{\mu}$) soon becomes intolerably large with m = 3,6 because of the bias term.

To sum up, the original sequential plan with 1 animal per step is satisfactory only if the value of $\sigma$ is known closely, and if the initial start is within 2 S.D. of $\mu$. If these conditions hold, m can be increased without loss of precision. Thus, from figure 3, m = 3, n = 8 does as well as m = 1, n = 24 out to about 2 S.D.

## TWO VARIATIONS

Since many experimenters will not know either $\mu$ or $\sigma$ as well as this in advance, some changes in the process need to be considered. Two variations were investigated. The first is the obvious one of keeping the size of step fixed (i.e. of postponing the introduction of the factor

188

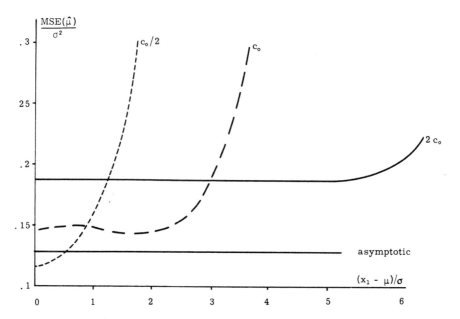

Figure 2 - Relation between $MSE(\hat{\mu})$ and the starting point $x_1 (m = 1, \ n = 2)$.

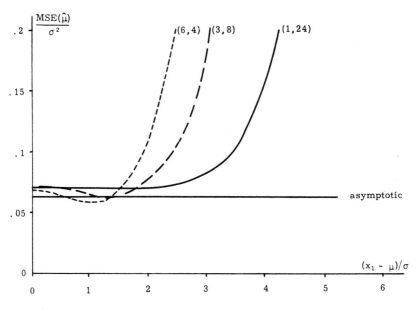

Figure 3 - Relation between $MSE(\hat{\mu})$ and $x_1$ for different $m, n \, (c = c_0)$.

189

$1/n$) as long as the proportion killed $p_n$ is consistently 1 or consistently 0. Typical results for this 'delayed' rule are shown in Table 1, for $c = c_o$.

Table 1

Values of $MSE(\hat{\mu})/\sigma^2$ for the 'delayed' rule ($c = c_o$).

| | | | Starting point $(x_1 - \mu)/\sigma$ | | | | | | |
|---|---|---|---|---|---|---|---|---|---|
| mn | m | n | 0 | 1 | 2 | 3 | 4 | 5 | 6 |
| 12 | 1 | 12 | .150 | .161 | .176 | .191 | .210 | .228 | .254 |
| | 2 | 6 | .144 | .149 | .178 | .232 | .296 | .373 | .566 |
| 24 | 1 | 24 | .071 | .075 | .079 | .082 | .086 | .089 | .093 |
| | 3 | 8 | .070 | .068 | .083 | .111 | .133 | .158 | .193 |
| | 6 | 4 | .069 | .064 | .111 | .231 | .357 | .491 | 1.177 |

The effect of the delayed rule is to make the MSE increase more slowly and steadily as $x_1$ departs from $\mu$. The rule is quite effective for $m = 1$, $n = 12$ and $m = 1$, $n = 24$, the MSE's at 6 S.D. being 0.254 and 0.093, as against 4.53 and 1.63, respectively, with the original rule. For the three cases with $m > 1$, however, the delayed rule still gives undesirably large MSE's when the starting point is beyond 3 S.D.

A second variation was to introduce the factor $n$ more slowly. Instead of the multiplier $c/n$ we used (I) $2c/(n + 1)$, i.e. the series 1, $2/3$, $1/2$, $2/5$, ... and (II) $3c/(n+2)$, i.e. the series 1, $3/4$, $3/5$, $1/2$, ... . The variants had the same general effects as the delayed rule, but were less helpful.

## INITIAL ESTIMATES OF $\mu$ AND $\sigma$

From the preliminary calculations we drew the following conclusions :

(I) Although the delayed rule is helpful, it is not satisfactory for the experimenter whose initial estimate of $\mu$ is poor.

(II) Underestimation of $\sigma$ by a factor of $1/2$ does not result in a loss of precision if the starting point is near $\mu$ (figure 2) but involves great losses otherwise. Overestimation of $\sigma$ by a factor of 2 loses precision even if the start is near $\mu$ (figure 2).

(III) Use of $m > 1$ to shorten the experiment loses little precision if the start is near $\mu$ (figure 3), but is less accurate for starts far removed from $\mu$.

These conclusions bring up the question : now well does the experimenter know $\mu$ and $\sigma$ in advance ? The answers will, of course, depend on the type of experimentation. It appears that in some cases $\mu$ may be known within a range of 10 to 1 (1 unit on the log scale) but in others

190

only within a range of 1,000 to 1 (3 log units). Some information on $\sigma$ is obtainable from reviews by Gaddum (1933) and Bliss and Cattell (1943). They report the estimates $s$ of $\sigma$ in 38 experiments in which death or survival was the criterion (as well as other experiments with different criteria). Note that with $x = \log$ dose, $\sigma$ is dimensionless ; assuming a normal curve, $\sigma$ is log of the ratio of the dose which kills 84.1 % to the dose which kills 50 %.

Thus it is possible to compare values of $s$ from experiments on different drugs and animals. The experimental animals were mostly mice, rats, frogs and cats, although snails and dogs are also represented.

The frequency distribution of $s$ is as follows.

| Class | 0 - .05 | .05 - .1 | .1 - .15 | .15 - .2 | .2 - .3 | .3 - .4 | Over .4 |
|-------|---------|----------|----------|----------|---------|---------|---------|
| Freq. | 1 | 10 | 11 | 7 | 2 | 4 | 3 |

This distribution overestimates the true spread in $\sigma$, since most of the estimates $s$ are subject to a substantial experimental error. About 3/4 of the values of $s$ lie between 0.05 and 0.2, a 4 to 1 ratio, although a few extend as high as 0.5.

## A  TWO-STAGE  PLAN

Suppose the experimenter is confident that on the log scale $\mu$ lies between the limits $\mu_1$ and $\mu_h$, and $\sigma$ between the limits $\sigma_1$ and $\sigma_h$. The results of the preceding section suggest the following two-stage plan.

1/ Take as initial estimates $x_1 = (\mu_1 + \mu_h)/2$,

$$\hat{\sigma}_1 = \sqrt{\sigma_1 \, \sigma_h}$$

2/ Start at $x_1$ with single-mice tests, using a step size of about $(\mu_h - \mu_1)/10$. Even in the worst case, when $\mu$ is at $\mu_1$ or $\mu_h$, this step size should bring the doses near to the LD 50 after 5 or 6 steps. Continue with single mice until a reversal occurs (i.e. a series of deaths followed by a survival or vice versa).

3/ If the step size used does not exceed $1.5\hat{\sigma}_1$, start an ordinary Robbins-Munro process at the mid-point M of the two doses $x_d$ and $x_s$ at which the reversal from death to survival occurs. Take $c = 1.25\hat{\sigma}_1$. More than one mouse per step may be used in the second stage, but the number of steps $n$ should be at least 4, preferably 6. (If the initial step size exceeds $1.5\hat{\sigma}_1$, as will happen when $\mu$ is poorly known and $\hat{\sigma}_1$ is small, test a single mouse at the mid-point of $x_d$ and $x_s$ and use this result in determining M).

The advantage of the first stage is that for a fixed step size, the MSE of M is relatively little influenced by the starting point. If we start far away, more single mice are required to reach M, but its frequency distribution remains about the same.

191

T illustrate the results given by the two-stage process, consider a situation in which $\mu_1 = 1$, $\mu_h = 2.2$ (i.e. $\mu$ is known within a 16 to 1 ratio), $\sigma_1 = 0.05$, $\sigma_h = 0.2$. We take $x_1 = 1.6$, $\hat{\sigma}_1 = \sqrt{(.05)(.2)} = 0.1$, and an initial step size of $1.2/10 = 0.12$.

The results in Table 2 apply to the worst situation with regard to $\mu$, i.e. when $\mu$ actually is at 1.0 or 2.2. Values of $MSE(\hat{\mu})/\sigma^2$ are shown for various second-stage plans involving 12 and 24 mice, and for the true $\sigma = 0.05$, 0.1 and 0.2. The average number of mice required to reach M in the first stage was about 7 in all cases.

Table 2

Values of $MSE(\hat{\mu})$ for a two-stage plan.

| | m,n | $\sigma = 0.05$ | $\sigma = 0.1$ | $\sigma = 0.2$ |
|---|---|---|---|---|
| 12 Mice | 1,12 | .187 | .145 | .162 |
| | 2,6 | .195 | .139 | .171 |
| | 3,4 | .205 | .139 | .191 |
| | 4,3 | .211 | .140 | .212 |
| | asymptotic | .131 | .131 | .131 |
| 24 Mice | 2,12 | .092 | .069 | .103 |
| | 3,8 | .094 | .069 | .117 |
| | 4,6 | .096 | .070 | .129 |
| | 6,4 | .101 | .073 | .156 |
| | asymptotic | .065 | .065 | .065 |
| Ave. no. of mice | | 7.4 | 6.4 | 6.8 |

When $\sigma$ is guessed correctly ($\sigma = 0.1$) the plan works very well. The values of $MSE(\hat{\mu})$ are only slightly above the asymptotic values for 12 and 24 mice and are almost the same for the different values of m.

When $\sigma = 0.05$, i.e. the standard error is overestimated by a factor 2, the MSE is about 40 % above the asymptotic value when n = 12 in the second stage. It increases further, although quite slowly, as $m$ is increased and n decreased.

The same effects appear for $\sigma = 0.2$, except that the increase in the MSE with increasing m is more rapid.

While further examination is needed of the performance of this two-stage plan under various situations, the plan appears to be able to cope with any degree of uncertainty about $\mu$ likely to occur in practice provided that (I) $\sigma$ is known within a range of 4 to 1 and (II) the responses are quick enough so that a total of 16 successive steps is feasible.

192

Variations in the plan which may handle a wider range of $\sigma$ and may improve the results under a 4 to 1 ratio are under consideration.

## REFERENCES

ARMITAGE, P. (1954) - Sequential tests in prophylactic and therapeutic trials. Quart. Jour. Med. 23, 255-274.

BLISS, C.I. and CATTELL, LCK (1943) - Biological Assay. Annual Rev. Phys. 5, 489.

BROWNLEE, K.I., HODGES, J.L. and ROSENBLATT, M. (1953) - The up-and-down method with small samples. Jour. Amer. Stat. Ass. 48, 262-277.

CHUNG, K.L. (1954) - On a stochastic approximation method. Annals Math. Stat. 25, 463-483.

DIXON, W.J. and MOOD, A.M. (1948) - A method for obtaining and analyzing sensitivity data. Jour. Amer. Stat. Ass. 43, 109-126.

GADDUM, J.H. (1933) - Methods of biological assay depending on a quantal response. British Med. Res. Council. Special Report n° 183.

HAJNAL, J., SHARP, J. and POPERT, A.J. (1959) - A method for testing analgesics in rheumatoid arthritis using a sequential prodedure. Annals of Rheumatic Diseases 18, 189-206.

HODGES, J.L. and LEHMANN, E.L. (1956) - Two approximations to the Robbins-Munro process. Proc. Third Berkeley Symposium. Vol. I, 95-104.

ROBBINS, H. and MONRO S. (1951) - A stochastic approximation method. Annals Math. Stat. 22, 400-407.

SCHMETTERER, L. (1953) - Bemerkungen zum Verfahren der Stochastichen Iteration, Osterreich. Ingenieur-Archiv, 7, 111-117.

## DISCUSSION

M. FINNEY : When the number of mice is reasonably large, is there any possibility of making a modified estimation of $\sigma$ after part of the experiment is complete, and using this to improve value of c ?

·Presumably the precision of $\hat{\mu}$ might be seriously disturbed (in any of the experimental designs described) if the distribution of threshold levels were to be more complicated than originally guessed (e.g. bimodal). As in most sequential techniques, the difficulty of building in a check on the validity of initial assumptions remains.

193

13

Pr. COCHRAN : We have given some thought to the suggestion made by Dr. Finney.

Indeed, if some experimenters do not know $\sigma$ in advance within a 4 to 1 ratio, I do not think that any reliable method can be produced without making an estimate of $\sigma$ in the middle of the experiment. In very small experiments, with say less than 30 mice, I doubt whether this estimate of $\sigma$ will be precise enough to be helpful, but with 50 or more mice it might be.

194

## METHODOLOGICAL PROBLEMS IN THE STUDY OF
## HUMAN POPULATIONS

William G. Cochran

*Professor of Statistics, Harvard University, Cambridge, Mass.*

### INTRODUCTION

As is evident from the contents, this publication covers a wide range of health problems. Their common link is that they all require the study of human populations in their natural habitats. This type of research is relatively recent and I think we must anticipate that progress will be in general slow and expensive. In studying human populations we lack much of the power and flexibility of the laboratory scientist, and must often do the kind of study that is feasible rather than the kind that seems most capable of answering our unresolved questions. Many of the phenomena currently under investigation appear to have multiple causes, at least in the present state of our knowledge, so that the search for causative agents against which preventive measures can be developed seems inherently more difficult than with the infectious diseases. Nevertheless, much has already been learned about the conduct of human population studies, and this area of research represents one of the most exciting developments in modern science.

This paper presents some impressions of the current state of methodology in the study of human populations, as seen from the viewpoint of the statistician. My first impression is the rather obvious one that this is preeminently an area in which we learn by doing. The reports of actual experiences in studies directed at particular health impairments, to be given in subsequent papers, are the meat of this monograph.

The first responsibility of the investigator is to have a clear concept of the objectives of his study. In a collaborative study involving investigators with different interests it is particularly helpful to have a statement of the questions that the study is designed to answer and to verify that all collaborators endorse this statement. It is also wise to be sure that one can satisfy the critic who asks "Why is it worth while answering these questions?" In the more intricate areas of research there are times when little progress appears to be made despite well-planned studies. The reason is sometimes that we are asking the wrong questions.

### SAMPLES AND POPULATIONS

The people from whom data are obtained—the sample—are of interest only insofar as the data tell us something about some larger group of people whom statisticians call the population or universe. Further, results obtained from a sample can be extended to a larger population with logical soundness only if the sample is, in a certain technical sense, a probability sample drawn from that population. It follows that in the planning of a study we should start with a careful definition of the target population— the group of people about whom information is wanted. The next step should be to select a method of sampling that gives a probability sample of this population.

The early health studies in human populations were not usually planned in this way. Instead, the investigator tried to find a group of people, often hospital patients or college students, who were convenient to work with and relevant to his purpose. The question "From what population are you sampling?" was seldom raised, and there was little explicit discussion of the extent to which the investigator thought that his results might be generalized. As experience in field investigations has grown, however, it has become increasingly common to tackle the more difficult problem of actually sampling the type of population to which we would like to apply the results. Further, when special groups are studied because they present illuminating opportunities, such as people who have an unusual diet or workers in an industry exposed to a particular health hazard, more pains are taken to describe the ways in which such groups differ from the target population, as a guide to judgment in considering how far the results may be generalized. There is also a tendency to avoid the study of a highly specialized group, no matter how convenient, if it appears to be so atypical that any generalizations to more normal populations will be dubious.

### DESCRIPTIVE AND ANALYTICAL SURVEYS

This growing awareness of the problem of generalization from sample data may be attributed in part to our increasing knowledge of probability sampling. The survey of a well-defined target population by means of a probability sample is one methodological tool in which great advances have been made in recent years. There is now a good supply of simple introductory accounts of sampling, of major texts on the subject, and of reports of actual surveys illustrating the wide scope of the method and the devices used for handling particular problems. This tool should be particularly helpful to local health agencies. If the jurisdiction is a compact area, such as a medium-sized city or a county, the population served by the agency should not be hard to define. A simple type of probability sample requires little in the way of specialized personnel, except where procedures involving medical scientists are included, and may prove to be within the ordinary resources of the agency. At the end of this paper a number of references are given that may be useful to agencies inexperienced in sample surveys.

The report "On the use of sampling in the field of public health," by the Committee on Sampling Techniques, APHA, gives a good account of the potentialities and limitations of sampling, as well as advice on when to call in an expert. It lists the following as possible applications of sampling in the everyday work of public health.

To obtain health information about the persons living in a health department jurisdiction for the planning of a new program.

To test the efficacy of a proposed health department procedure before deciding whether to put it into full-scale use.

To evaluate periodically the results of a health department procedure, such as a health education program.

To survey the environment in a health department jurisdiction, as in sanitation surveys and surveys of housing.

To measure the utilization of health services or the availability of medical care in the jurisdiction.

To obtain information from a large file at less expense and more quickly than would be possible by complete analysis.

To prepare preliminary bulletins on vital statistics in advance of final processing of the records.

To evaluate the reliability or completeness of registration of births and deaths or other record systems.

The most economical type of survey is that in which only overall estimates for the whole population are required. By increasing the size of sample, separate estimates may be obtained for specified subdivisions in the population (e.g., for different age groups or for people in different geographic areas) and comparisons made among these subdivisions. Going a little further, we can investigate whether certain characteristics of individual people, for instance weight, some aspects of diet, or their scores on screening tests, are associated with the aspects of health or behavior that are being studied. In this way the survey becomes a tool for verifying proposed relationships between characteristics of the individual or his environment and his health status or utilization of health services. At a more primitive stage of research, the sample survey may also be used in a search for variables that show associations with health status.

One point sometimes overlooked is that a much larger size of sample is required if precise estimates in each of a number of subdivisions of the population are wanted. Roughly speaking, if the means of each $k$ subdivisions are to be estimated with pre-assigned standard error $\sigma$, the sample must be $k$ times as large as is required if only the overall mean is wanted with standard error $\sigma$. If the difference between any two means is to have standard error $\sigma$, the sample size must be about $2k$ times as large. If *differential* differences are of interest, for instance in inquiring whether a sex difference is the same for two different age groups, the multiplier becomes $4k$. In other words, precise investigation of the interaction between two variables is much more expensive than investigation of the overall effects of a single variable.

Considerable progress has been made in handling the vexing problem of nonresponse. The basic difficulty is not that the size of sample is reduced below our expectations, because this can be allowed for in the planning. But the nonrespondents might have given consistently different replies from the respondents, and if the proportion of nonresponse is substantial, the results from the respondents may give a biased picture of results for the whole population.

Nonresponse comes mainly from three sources. The first, called non-coverage, is a failure to realize that certain households are in the sample. This is an indication of poor field work and should be of negligible proportions in most surveys. The second is a refusal to answer any questions or certain particular questions. A high proportion of refusals is a warning that the questionnaire or method of approach is ineffective. The third arises from the difficulty of finding people at home. Families without children in which husband and wife both work and young males who spend little time in their lodgings are among the more difficult to find.

In household surveys that do not involve objectionable questions or an undue strain on the respondent, a vigorous and intelligent call-back policy usually gives response rates of the order of 93–96 per cent without excessive cost or delay. Success rates below this range should not be regarded as an acceptable performance. The problem is greater when part of a physical examination is to be conducted in the home, as in the taking of blood pressures or of a sample of blood, or when the respondents are asked to come to a medical center for a thorough examination. Earlier attempts to bring a random sample of respondents into a center for examination had response rates of the order of 60 per cent, but I understand that recent work in the National Health Survey has achieved rates of over 80 per cent, with still higher figures when children are being examined. A similar difficulty arises in long-term studies of chronic disease, in which the constant mobility of the American public leads to the wearing away both of the sample and of the staff who are conducting the study. Some years ago I ventured the opinion that while we probably know how to do a good five-year longitudinal study, there were too few examples of successes for longer periods to claim this ability beyond five years. At the time, some English investigators suggested that I was perhaps unduly pessimistic. I shall be interested to see whether this view is confirmed by some of the long-term investigations reported in this monograph.

### Problems of Measurement

I should like to refer to some of the work of the Census Bureau, since this agency has had long experience in the use of probability samples for obtaining data on numbers of people, business and trade, labor, agriculture and housing. In their attempts to improve the quality of their data, problems of measurement now constitute a more serious challenge than problems of sampling. Errors of measurement can be classified roughly into two types: fluctuating errors, that vary from one person to another in the survey, and biases (consistent under- or over-reporting). Fluctuating errors tend to average out with increasing size of sample. Their effects are properly taken into account when standard errors or tests of significance are calculated, so that the statistical analysis still gives a realistic picture of the precision of the results. Fluctuating errors do, of course, decrease the precision, sometimes to the point at which overall results are too imprecise to be useful. Biases, on the other hand, persist in any size of sample. Standard errors, tests of significance, and confidence limits computed by the usual procedures make no allowance for bias and become meaningless if substantial biases are present. The consequence is that the investigator can establish, by the most reputable statistical techniques, results that are simply wrong.

As might be expected, Census experience suggests that measurement errors are most likely to be large when the definition of the measurement is lengthy and intricate or is vague, and when the respondent has to guess about events, particularly those in the past, on which he does not normally keep records. Examples are the distinction between being unemployed and being out of the labor force, the definition of a farm, past illnesses in the family, and the sales of some types of article by a store. Bias is to be expected on matters that the respondent regards as confidential, such as the amount of money kept in savings banks, although the American public often turns out to be less reticent than an outsider would anticipate. Questions on which the answer involves prestige or social acceptability are also liable to bias.

One disturbing result is that interviewers may induce personal biases, apparently quite unconsciously, even on questions to which the answer appears to be almost factual. In a few studies that have managed to estimate the contribution of interviewer biases to the overall inaccuracy of the results, this has been found surprisingly large for some items.

Unfortunately, the study of errors of measurement is difficult. Large numbers of refusals to answer a question or of "Don't know" answers are an obvious indication that the question is a failure. Beyond pointers like this, the principal tools are checks against more accurate outside records, the comparison of different approaches or different interviewers within the same survey, and some types of reinterviews. Usually, record checks are available for only a few items such as days of hospitalization or of sick leave, date of birth, selling price of a house. Sometimes the problem of matching the interview against the record is troublesome. Reinterviews, although highly useful, may be hard to interpret. If the reinterview follows soon after the first interview, the respondent may repeat an earlier erroneous answer because he remembers giving it. If the time interval is widened, the respondent in the second interview may have partly forgotten the circumstances under inquiry. What we really want—a first interview given twice—seems impossible.

In the health field, the problems of measurement are in many respects similar to those encountered by the Census Bureau but on the whole may be somewhat greater. We will want to measure many quantities that are factual and capable of strict definition—for example, the extent to which people utilize specific health services or practise hygienic measures. Then there are a large number of items that involve problems of unambiguous definition, of making the respondent understand what is wanted, or of forgetfulness of past events—for instance, in measuring the prevalence and incidence of particular types of disease. Finally, in certain studies, for example, in trying to discover the reasons for the success or failure of a preventive program, we will want to know why people do, or fail to do, certain acts. This is the area in which least is known about good methods of measurement. More experience and research in this type of measurement are of the highest importance, not merely for the health field, but because better understanding of ourselves is sorely needed in this age of awesome technological power.

What positive advice can be given to a group planning a study? An obvious step is to find out what is already known about the accuracy of the types of measurement contemplated and about alternative measuring instruments for this purpose. Consider some of the choices open to the investigator in a standard household survey.

> Should contact be by mail, telephone, or personal interview or a mixture of these approaches?

> Should the questionnaire be self-administered, a standard list of questions read without deviation by the interviewer, or an 'open end' type of questioning?

> Should the questionnaire be self-coding, or in a form that requires subsequent coding before tabulations can begin?

> Can any available adult answer the questions, or should each respondent answer questions only about himself?

> What quality of education and training are needed in the interviewers?

> If the questions refer to past events, over what periods of time can the respondent report accurately?

> Can the information be obtained in a single visit, or will two or three visits be required?

Information and research on these questions is gradually accumulating in the literature. In view of the large number of population studies undertaken nowadays, frequent reviews of the current state of knowledge on such issues would be valuable. Some references are given at the end of this paper, although I do not know of a comprehensive recent review.

Certain precautions have become a part of good practice. Any definitions to be used in the questionnaire or analysis should be constructed with great care, bearing in mind freedom from ambiguity and relevance to the problem in hand. The questionnaire should be adequately pretested in the field, leaving sufficient time for major revisions if these are necessary.

Similarly, if coding presents any difficulty, independent codings of the pre-test results by different persons should be performed and reconciled. Frequent disagreements are a sign that this item is being poorly measured. If several groups of people are to be compared, each interviewer should work equally in all groups; or, in more complex cases, the interviewers' assignments should be planned so that differences between groups are not affected by overall differences in level obtained by different interviewers. This precaution does not necessarily solve the problem of interviewer biases, since there are situations in which an interviewer shows a bias in one group but not in another, but it greatly helps. Questions about what people would do in hypothetical situations should be avoided. Bradford Hill's "three to one" ratio (1953) is worth bearing in mind. His maxim is that the investigator should ask himself three questions for every one that he proposes to ask the respondents. And one of these three queries should always be "Is this question really necessary?" Multiplication of questions is a common temptation that tends to lower the quality of answers to all questions. Finally, although it is not yet standard practice, the reinterviewing of a subsample of the respondents, whenever feasible, gives some insight into the responses that are highly fluctuating and some clues as to the reasons for inaccuracies.

### Role of Routine Records from a Service Program

When a study is being planned, it is advisable to consider whether some of the data needed are already available. Sometimes it appears at first sight that all the needed data are being collected as part of the records taken in some service program, so that the new study will require only abstraction or copying of the relevant material and some statistical analysis. For those who are inexperienced in the conduct of statistical studies, however, it must be reported that routine records are usually disappointing for research purposes. Often they turn out to be useless, especially when the new study is designed to elucidate cause-and-effect relationships. Research necessitates a series of carefully coordinated questions, involving an amount of detail and specificity that may be irrelevant to the record-keeping of a service program. Research questions must be uniformly and fully answered, whereas routine records may be of mediocre quality, being beset by incompleteness and illegibility. Consequently, when such records appear potentially valuable for avoiding new data-collection, a small but thorough critical study of the performance of the records should be undertaken before committing oneself to rely on them.

On the more positive side, some later papers in this publication will show how records kept for administrative purposes, for instance in pre-insured health plans, can be a fruitful source of information in experienced hands. Further, a routine system can sometimes serve as an economical source of part of the data by making temporary changes in certain of the questions and adding additional personnel to ensure that these questions are answered uniformly and completely for the duration of the study.

## STATISTICAL ANALYSIS

When questionnaires are being constructed, a useful practice is to map out at the same time the type of statistical analysis planned from the results. This should include sketches of all the principal tables, showing the degree of breakdown in each classification and all cross-classifications. It is also helpful to visualize different ways in which the results in these tables might come out, and to note the interpretation that will be made of each type of result. This procedure demands a lively exercise of the imagination and, perhaps for this reason, is often neglected. But there is no surer way of finding out whether all the necessary information is being sought in the questionnaire, whether a number of pointless questions have been included, and of checking whether the questionnaire plus the analysis are capable of meeting the objectives of the study.

Although techniques of statistical analysis have developed greatly during the last 30 years, statistical analysis cannot turn bad data into good data. If the wrong population has been sampled, if measurements have large biases and fluctuating errors, or if many different variables are heavily entangled in the data, the power of elaborate analyses to overcome these deficiencies is feeble. The same is true of electronic machines. There is, indeed, a real danger that the availability of electronic machines will encourage complex analyses of large bodies of routine data that are incompetent to answer the questions posed in research investigations. Experienced investigators, both those in fields where controlled experimentation is possible and those restricted to observational studies, soon learn that the crucial step in any study is to get relevant data of good quality. Fortunately, it usually happens that if the data are wisely planned, the necessary statistical analysis is simple and straightforward.

## TYPES OF STUDY DESIGN

Much of the research in the health field deals with the study of associations between variables, in the hope that these associations will reveal the principal causal agents that produce impairments to health. The best weapon for this purpose is the controlled experiment with randomization. The use of controlled experimentation has now become common in the testing of prophylactic and curative drugs. With ingenuity, there are some opportunities for experiments on topics like the evaluation of health education literature (Greenberg et al., 1953), methods of training health workers or interviewers (e.g., programmed learning versus more traditional instruction) and the comparison of different questionnaires designed to elicit certain information. Much is being learned about motivation and behavior in small experiments with volunteer subjects. My impression is, however, that in most of the researches represented in this monograph the investigator must rely mainly on selected observations rather than on controlled experiments.

Although numerous types of study design are employed in observational studies, the two that appear to be most frequent are (1) analytical surveys of a representative sample from a broad population, as described previously, and (2) studies built around special groups.

The survey has the advantage that the results are generalizable to a known population. It permits the simultaneous study of several potential causal factors, although with the limitation that if these are badly entangled in the sample data, we may be unable to say much about the relative strength of association of each factor with the response variable. As an illustration, obesity, high blood pressure, and a high cholesterol level have all been suggested as factors that increase the risk of heart disease. Suppose for simplicity that each person in the sample has been classified as low or high on each variable. Every person then falls into one of the eight cells.

|  |  | Cholesterol | | | |
|---|---|---|---|---|---|
|  |  | Low | | High | |
|  |  | Low B.P. | High B.P. | Low B.P. | High B.P. |
| Obesity | Low |  |  |  |  |
|  | High |  |  |  |  |

In each cell, the proportion of subjects who have developed heart disease within a given period of time is recorded. For purposes of analysis and interpretation we would like to have approximately equal numbers of subjects in each cell. In this event we can examine conveniently the overall difference in morbidity between the high and the low level of each factor. We can also see whether the difference in morbidity between the low and high obesity groups remains the same or changes as the levels of the other factors change. If nature has been unkind, however, we may find, at one extreme, that practically all the subjects fall either in the high, high, high or in the low, low, low cells. It is then impossible to study the individual effect of each factor. In intermediate situations some cells may be empty and some poorly filled. The individual effects can then be measured only with low precision.

If the population is much larger than the sample, this deficiency can sometimes be remedied by sampling cells that are relatively infrequent in the population at a much higher rate than the easily filled cells. This can be done in such a way that the sample remains a probability sample of the population, although some extra trouble and expense are involved. A good example of the analysis of data of this type, with unequal numbers of observations per cell, has been given by Keyfitz (1952), who used the method to investigate the association between fertility and age of mother at marriage, present age of mother, years of schooling of mother, income level of family, and distance from a city, in Ontario Protestant families.

STUDIES BUILT AROUND SPECIAL GROUPS

These can arise in several ways. One is that we become interested in a single factor, such as a certain complication during pregnancy or a particular home environment, that may be related to the state of health. We seek a group of people—often called the "study group"—in which this factor is concentrated. A second group—the comparison group or "control"—which lacks this factor, but is otherwise similar to the study group, is also required. By making measurements on the two groups we examine whether the suspected differences in health do occur. Alternatively, this type of study is used to learn about possible causes of a given effect. The study group is composed of persons with a relatively rare disease or children with a particular behavior problem and the comparison group of persons free from the defect. The measurements deal with characteristics of the persons or of their past history that may provide clues to agents causing the defects. In some cases there are two study groups, for instance postal clerks and letter carriers, whose differences in work activity offer a contrast that is worth investigating in its relation to health.

The search for a suitable "control" is often a difficult part of this type of research. The prescription usually given is that the study and control groups should be matched on all important "disturbing" variables—that is, on all variables, other than those which we deliberately intend to compare—that may affect the measurements under investigation. In practice, however, the variables on which the two groups are matched usually have to be restricted to two or three, and the matching may have to be a loose stratified matching, Cochran (1953), rather than a person-by-person matching. Individual matching on say four to six variables may be a frustrating task, resulting in the melting away of the study group because no match can be found and requiring an inordinate amount of time. Chapin (1947) quotes a study in which the two groups, initially of sizes 671 and 523, diminished to sizes 23 and 23 after precise matching on six variables. Disturbing variables that are not to be matched should be recorded in each group and their frequency distributions compared before starting the study. Minor inequalities in these frequency distributions (e.g., a control group that is slightly older than the study group) can be removed by adjustments performed in the statistical analysis. With major inequalities, although the formal apparatus of making statistical adjustments can still be applied, I would be inclined to distrust the results, since a hidden extrapolation is often involved.

Sometimes it is reassuring to have more than one control. In studying the incidence of cancer in radiologists a control consisting of another medical specialty yielding about the same age distribution but not subjected to radiation hazards suggests itself. A natural question is: Does it make any difference which specialty is chosen as control?

Since the study group is chosen because it contains a concentration of some variable, it is likely to be rather specialized with respect to some characteristics. Consequently, the criticism can often be made that the study and control groups are not representative samples from any broad population. As I see it, the answer is to admit this criticism but claim that the opportunity to study a single variable in a suitable situation may reveal more about its effects than the attempt to disentangle them from the effects of other variables in a general population survey. Of course, the more highly selective the study and control groups are, the more we are forced to wonder whether the results of the comparison are capable of any generalization. In this dilemma a second control group that is a probability sample of a general population is sometimes helpful. This sample may have to differ from both the first control and the study group in several variables. Agreement in results between the two controls gives some reassurance that the selective factors have not affected the interpretation of the study results. Disagreement may give clues to the effects of the selectivity and warnings against generalization.

In some settings another factor is found to be almost inextricably mixed with the study factor. A comparison of people living in a polluted atmosphere with those in a clean atmosphere may turn out to be also a comparison of poor with well-to-do people, or a comparison of urban with rural living. A certain medical practice whose efficacy we wish to investigate may be given only to patients with a certain diagnostic syndrome, so that no good control group can be found. While there is a temptation to go ahead, arguing that the disturbing factor cannot be important, modern epidemiological practice is to seek another setting in which this disturbance is absent.

This type of study is often simpler to execute and less expensive than the analytical survey. It is, however, a fairly primitive research weapon, especially if the response is affected by a multiplicity of causes. If the factor under study has a predominating effect, this of course should be revealed. But by a single-variable approach it is not possible to learn about the relative importance of different causative variables or about the way in which their effects interact. Moreover there is a tendency, when attention is focused on one factor, to overrate its importance and to find it difficult to think objectively about other possible explanations of our results.

## ASSOCIATION AND CAUSATION

Here we come to the basic weakness of all studies that must depend solely on observations. We may establish associations and show that they remain stable under a wide range of background conditions, but the step to a proof that the association must be due to the causal action of certain specific factors is hard to justify logically. The argument that our groups were not comparable on some variables omitted from the investigation can always be raised.

In facing the situation realistically, about all we can do is to list the alternative explanations of our results, and try to think of the types of observations that would provide a crucial test of the validity of a given alternative. A well-known example occurs in Doll and Hill's (1950) comparison of the rates of smoking in patients with lung cancer and other control patients in the same hospitals. In almost all cases, the interviewers knew whether the patient was a lung cancer patient or not and presumably also knew the general objective of the comparison. Serious consideration was therefore given to the possibility that interviewer biases affected the results, even though each interviewer worked in both groups, by overrating the smoking habits of the lung cancer cases. Fortunately, in 209 patients who were thought at the time of interview to have lung cancer the diagnosis was later found to be erroneous. The rates of smoking for these patients were found to be similar to those of the controls, and significantly smaller than those for the cancer patients. This greatly weakens the alternative explanation that part of the difference in smoking rates was due to interviewer bias. Useful supplementary observations of this type are not often available in the original study. Thus the testing of alternative explanations may require a series of new studies. In some cases the alternatives could be tested rigorously only if controlled experimentation were feasible.

Although this means that we may advance slowly on certain problems, yet if research is prosecuted vigorously it often happens that major technical advances come from unexpected sources. Laboratory studies of the biochemical aspects of disease may dramatically change the picture at any time. So far as their funds permit, the behavioral scientists are also heavily engaged in human population studies and have much experience and sophistication in handling some of the commonest methodological problems. Efforts to secure their participation may be rewarding.

### REFERENCES*

#### Annotated Bibliography

PERROTT, G. ST. J. & K. G. CLARK. 1962. Health studies of human populations: a selected bibliography. Pub. Health Bibliography Series, No. 38. Part II. Methodology. U. S. Govt. Printing Office. Washington, D. C.

This is an up-to-date and highly useful set of references, with notes. It includes the following sections: sample design and selection of controls; defining diagnosis and other study variables; collecting data first hand; nonresponse in surveys; collecting data from existing records; family studies; retrospective and prospective studies; evaluations; inferences and generalizations; and general articles and texts.

#### Descriptive and Analytical Surveys

*Nontechnical introductions.*

Committee on Sampling Techniques. 1954. On the use of sampling in the field of public health. Am. J. Pub. Health. 44: 719–740.

SLONIM, M. J. 1960. Sampling in a Nutshell. Simon and Schuster. New York, N. Y.

COCHRAN, W. G., J. CORNFIELD & M. H. HANSEN. 1951. Modern methods in the sampling of human populations. Am. J. Pub. Health. 41: 1–22.

* I am indebted to Theodore D. Woolsey for some helpful suggestions.

*Books dealing with practical aspects of sampling.*

PARTEN, M. 1950. Surveys, Polls and Samples. Harper & Brothers. New York, N. Y.

MOSER, C. A. 1958. Survey Methods in Social Investigations. Heinemann Ltd. London, England.

*Books presenting theory and practice.*

HANSEN, M. H., W. N. HURWITZ & W. G. MADOW. 1953. Sample Survey Methods and Theory. 2 Vols. John Wiley & Sons. New York, N. Y.

COCHRAN, W. G. 1963. Sampling Techniques. John Wiley & Sons. New York, N. Y. 2nd Ed.

*Articles on special problems.*

HESS, I., D. C. RIEDEL & T. FITZPATRICK. 1961. Probability sampling of hospitals and patients. Univ. of Michigan. Ann Arbor, Mich.

HEMPHILL, F. M. 1952. A sample survey of home injuries. Pub. Health Rep. **67**(10): 1026–1034.

WOOLSEY, T. D. 1956. Sampling methods for a small household survey. Pub. Health Monographs. No. 40.

KISH, L. 1952. A two-stage sample of a city. Am. Soc. Rev. **17**: 761–769.

PATTON, R. 1952. The sampling of records. Pub. Health Rep. **67**(10): 1013–1019.

## Problems of Measurement

PAYNE, S. L. 1951. The Art of Asking Questions. Princeton Univ. Press. Princeton, N. J.

HYMAN, H. H. (Ed.) 1954. Interviewing in Social Research. Univ. of Chicago Press. Chicago, Ill.

NISSELSON, H. & T. D. WOOLSEY. 1959. Some problems of the household interview design for the national health survey. J. Am. Stat. Assoc. **54**: 69–87.

MOONEY, H. W. 1963. Methodology in two California Health Surveys—San Jose (1952) and Statewide (1954–5). Pub. Health Service Monograph No. 70.

WOOLSEY, T. D. & H. NISSELSON. 1956. Some problems in the statistical measurement of chronic disease. American Statistical Association. Washington, D. C.

DORN, H. F. 1951. Methods of measuring incidence and prevalence of chronic disease. Am. J. Pub. Health. **41**: 271–278.

THOMPSON, D. J. & J. TAUBER. 1957. Household survey, individual interview and clinical examination to determine prevalence of heart disease. Am. J. Pub. Health. **47**: 1131–1140.

GRAY, P. G. 1955. The memory factor in social surveys. J. Am. Stat. Assoc. **50**: 344–363.

U. S. National Health Survey. 1958. Concepts and definitions in the health household–interview survey. Health Statistics. Series A–3.

U. S. National Health Survey. 1961. Reporting of hospitalization in the health interview survey. Health Statistics. Series D–No. 4.

U. S. National Health Survey. 1961. Health interview responses compared with medical records. Health Statistics. Series D–No. 5.

## Research Design as Seen by Social Scientists

JAHODA, M., M. DEUTSCH & S. W. COOK. 1951. Research Methods in Social Relations. 2 Vols. The Dryden Press. New York, N. Y.

FESTINGER, L. & D. KATZ. Eds. 1954. Research Methods in the Behavioral Sciences. Staples Press. London, England.

*Other References*

CHAPIN, F. S. 1947. Experimental Designs in Sociological Research. Harper Bros. New York, N. Y.

COCHRAN, W. G. 1953. Matching in analytical studies. Am. J. Pub. Health. **43**: 684–691.

DOLL, R. & A. B. HILL. 1950. Smoking and carcinoma of the lung. Brit. Med. J. **ii**: 739–766.

GREENBERG, B. G., M. E. HARRIS, C. F. MACKINNON & S. S. CHAPMAN. 1953. A method for evaluating the effectiveness of health education literature. Am. J. Pub. Health. **43**: 1147–1155.

HILL, A. B. 1953. Observation and experiment. New Engl. J. Med. **248**: 995–1001.

KEYFITZ, N. 1952. Differential fertility in Ottawa: an application of factorial design to a demographic problem. Population Studies. **6**: 123–134.

# THE BEHRENS-FISHER TEST WHEN THE RANGE OF THE UNKNOWN VARIANCE RATIO IS RESTRICTED*

*By* WILLIAM G. COCHRAN

*Harvard University*

*SUMMARY.* In some applications of the Behrens-Fisher test, it is reasonable to suppose that the unknown variance ratio $\sigma_1^2/\sigma_2^2$ must exceed a known quantity $\lambda_1/\lambda_2$. The effect of this restriction on the significance levels of the test statistic $d$ is examined by computing the probability that $d$ exceeds the tabulated significance level, in the restricted region, for 6, 12, and 24 degrees of freedom in $s_1^2$ and $s_2^2$ and probability levels of 5% and 1%. If the $F$ test made from the data gives strong support to the supposition that $\sigma_1^2/\sigma_2^2 > \lambda_1/\lambda_2$, the disturbance to the Behrens-Fisher significance levels is minor, but otherwise it can be substantial and can lie in either direction. The practical use of these results is discussed.

## 1. INTRODUCTION

In the Behrens-Fisher problem we are given a comparison $x$, normally distributed with mean $\mu$ and variance $(\sigma_1^2 + \sigma_2^2)$. We also have independent estimates $s_1^2$ of $\sigma_1^2$ and $s_2^2$ of $\sigma_2^2$, based on $n_1$ and $n_2$ degrees of freedom. The problem is to test the null hypothesis that $\mu$ has some stated value, usually zero. The test criterion is $d = (x-\mu)/\sqrt{s_1^2 + s_2^2}$.

Although the Behrens-Fisher test is intended to involve no assumptions about the variance ratio $\sigma_1^2/\sigma_2^2$, there are practical problems in which it is reasonable to assume that $\sigma_1^2/\sigma_2^2$ must exceed some known value. To cite an example discussed by Fisher (1941), the mean of a large sample of $(n_1+1)$ crude measurements of some physical quantity may be compared with the mean of a smaller number $(n_2+1)$ of refined measurements, in order to examine whether the two processes are biased relative to one another. If $\sigma_a$, $\sigma_b$ are the standard deviations of the populations of crude and refined measurements, respectively, the assumption $\sigma_a/\sigma_b > 1$ appears justified. In applying the Behrens-Fisher test to this problem, $x$ is the difference between the sample means and

$$s_1^2 = \frac{s_a^2}{(n_1+1)} \quad : \quad s_2^2 = \frac{s_b^2}{(n_2+1)}.$$

Consequently, the restriction $\sigma_a^2/\sigma_b^2 \geqslant 1$ implies that $\sigma_1^2/\sigma_2^2 > (n_2+1)/(n_1+1)$.

* This research was supported by the Contract Nonr 1866(37) with the Office of Naval Research, U. S. Navy Department. The computations were done at the Harvard Computing Center under grant NSF-GP-683 by the National Science Foundation.

This situation occurs also in certain comparisons in the analysis of split-plot or nested experiments. Factor $A$, with $a$ levels, is applied to relatively large plots or experimental units in some standard design. Each unit is divided into $b$ equal subunits, to which are applied the $b$ levels of a second factor $B$. The mathematical model used in the analysis postulates that the error $e_{ijk}$ on the subunit receiving the $i$-th level of $A$, the $j$-th level of $B$ and lying in tne $k$-th replication has mean zero and variance $\sigma^2$. The errors $e_{ijk}$, $e_{i'j'k'}$ on subunits in different units are assumed independent, but errors $e_{ijk}$, $e_{ij'k}$ on two subunits in the same unit have a correlation $\rho$. It follows that the error variance of a unit total, when divided so as to express it on a subunit basis, is

$$\sigma_a^2 = \sigma^2\{1+(b-1)\rho\}. \qquad \qquad \dots \text{(1)}$$

However, the variance of the difference between two subunits in the same unit, also on a single subunit basis, is

$$\sigma_b^2 = \sigma^2(1-\rho). \qquad \qquad \dots \text{(2)}$$

An unbiased estimate $s_a^2$ of $\sigma_a^2$, obtained from the analysis of variance of unit totals, is used for testing the main effects of $A$. The subunit analysis supplies an independent estimate $s_b^2$ of $\sigma_b^2$ for tests involving main effects of $B$ and $AB$ interactions.

If interactions are present and $B$ is a qualitative factor, the experimenter may wish to test comparisons like $(a_2 b_1)-(a_1 b_1)$; that is, the difference between the means for two $A$ levels at the same level of $B$. In terms of the model, the error variance of such comparisons is $2\sigma^2/r$, where $r$ is the number of replications. From equations (1) and (2) an unbiased estimate of this variance is

$$\frac{2\{s_a^2+(b-1)s_b^2\}}{rb}.$$

If $x$ is the mean difference between $(a_2 b_1)$ and $(a_1 b_1)$, the Behrens-Fisher test may be used if we take

$$s_1^2 = \frac{2s_a^2}{rb} \qquad s_2^2 = \frac{2(b-1)s_b^2}{rb} \qquad \qquad \dots \text{(3)}$$

giving to $s_1^2$ the d.f. in $s_a^2$ and to $s_2^2$ those in $s_b^2$.

In many split-plot or nested experiments there is reason to believe that $\rho > 0$. From equations (1) and (2) this implies that $\sigma_a^2/\sigma_b^2 > 1$ and hence in (3) that $\sigma_1^2/\sigma_2^2 > 1/(b-1)$.

The objective of this paper is to make a preliminary exploration of the disturbance produced in the Behrens-Fisher test when tne range of $\sigma_1^2/\sigma_2^2$ is restricted in this way.

## 2. NUMERICAL EXAMPLE

The following notation covers the type of problem illustrated by the two preceding examples. Let $\sigma_a^2$ and $\sigma_b^2$ be the variances in the two populations in which we are prepared to assume that $\sigma_a^2/\sigma_b^2 > 1$. Let $f' = s_a^2/s_b^2$ be the corresponding estimated variance ratio based on $n_1$ and $n_2$ degrees of freedom. The value of $f'$ will be known from the data. Further, let

$$\sigma_1^2 = \lambda_1\sigma_a^2 \qquad \sigma_2^2 = \lambda_2\sigma_b^2 \qquad \qquad \dots \text{(4)}$$

where $\sigma_1^2$ and $\sigma_2^2$ are the variances that enter into the Behrens-Fisher test, $\lambda_1$ and $\lambda_2$ being known numbers that depend on the problem in question. Then the restriction may be written

$$\phi = \frac{\sigma_1^2}{\sigma_2^2} > \frac{\lambda_1}{\lambda_2}. \qquad \dots \ (5)$$

For later calculations it is convenient to rewrite this as

$$\frac{\phi}{f} > \frac{1}{f'} \qquad \dots \ (6)$$

where $f = s_1^2/s_2^2 = \lambda_1 f'/\lambda_2$.

The effect of this additional information about $\phi = \sigma_1^2/\sigma_2^2$ on the significance levels of $d$ may be examined from a result due to Fisher, who showed that the unrestricted significance levels of $d$ can be computed in the following way. The probability that $d$ exceeds a specified value is found first on the assumption that $f = s_1^2/s_2^2$ and $\phi = \sigma_1^2/\sigma_2^2$ are both known. The average value of this probability over all possible values of $\phi$ from 0 to $\infty$ is then calculated by assigning to $\phi/f$ its fiducial distribution for known $f$, this distribution being the tabular $F$ distribution with $n_2$ and $n_1$ d.f.

To obtain the frequency distribution of $d$ when $f$ and $\phi$ are both given, Fisher notes that $x/\sqrt{\sigma_1^2+\sigma_2^2}$ follows the standard normal distribution and that

$$\frac{n_1 s_1^2}{\sigma_1^2} + \frac{n_2 s_2^2}{\sigma_2^2}$$

follows that $\chi^2$ distribution with $(n_1+n_2)$ d.f. Hence a variate that follows Student's distribution with $(n_1+n_2)$ d.f. is

$$
\begin{aligned}
t &= \frac{x\sqrt{n_1+n_2}}{\sqrt{\left(\dfrac{n_1 s_1^2}{\sigma_1^2} + \dfrac{n_2 s_2^2}{\sigma_2^2}\right)(\sigma_1^2+\sigma_2^2)}} \\[2ex]
&= \frac{d\sqrt{s_1^2+s_2^2}\ \sqrt{n_1+n_2}}{\sqrt{\left(\dfrac{n_1 s_1^2}{\sigma_1^2} + \dfrac{n_2 s_2^2}{\sigma_2^2}\right)(\sigma_1^2+\sigma_2^2)}} \\[2ex]
&= \frac{d\sqrt{1+f}\ \sqrt{n_1+n_2}}{\sqrt{\left(n_2 + \dfrac{n_1 f}{\phi}\right)(1+\phi)}}. \qquad \dots \ (7)
\end{aligned}
$$

By means of this relation, the probability that $d$ exceeds any specified value, for given $f$ and $\phi$, is read from the Student $t$-table with $(n_1+n_2)$ degrees of freedom. This probability is then averaged over the fiducial distribution of $\phi/f$ from 0 to $\infty$.

When it is known that $\phi/f > 1/f'$, the natural modification of the Behrens-Fisher technique is to average this probability only over the values of $\phi/f$ that exceed $1/f'$. This is the method that will be adopted in this paper.

From the preceding discussion, writing $u = \phi/f$, the probability that $d$ exceeds the value $d_\alpha$ when we confine ourselves to the restricted region $u > 1/f'$ is given by the expression

$$\int_{1/f'}^{\infty} \frac{u^{\frac{n_2}{2}-1} P\{|t| > d_\alpha g(u)\} du}{(n_1 + n_2 u)^{\frac{n_1 + n_2}{2}}} \bigg/ \int_{1/f'}^{\infty} \frac{u^{\frac{n_2}{2}-1} du}{(n_1 + n_2 u)^{\frac{n_1 + n_2}{2}}} \qquad \ldots \quad (8)$$

where

$$g(u) = \left\{ \frac{u(n_1 + n_2)(1+f)}{(n_1 + n_2 u)(1 + uf)} \right\}^{\frac{1}{2}}$$

and $P\{|t| > d_\alpha g(u)\}$ is the two-tailed probability that Student's $t$ with $(n_1 + n_2)$ degrees of freedom exceeds $d_\alpha g(u)$.

As an illustration, consider a split-plot experiment with $n_1 = 6$, $n_2 = 12$, $b = 2$. These parameters hold if the main units are arranged in a $4 \times 4$ latin square, each unit having two subunits. In the solid line in Figure 1 the probability that $d$ exceeds 2.301 (the 5% Behrens-Fisher value for $n_1 = 6$, $n_2 = 12$, $f = 1$) is plotted against the percentiles of the distribution of $\phi/f$. The average probability (area under the line) is of course 0.05. Throughout most of the range of $\phi/f$ the probability lies below 0.05, this being required to compensate for very high and low values of $\phi$ that make the probability rise steeply towards 1 at both ends. Note that probabilities above 0.05 are mostly contributed by *high* values of $\phi/f$. This happens whenever $n_2 > n_1$.

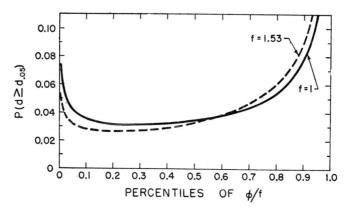

Fig. 1. Probability of exceeding $d_{.05}$ for given $f$ and $\phi/f$ ($n_1 = 6$, $n_2 = 12$).

Suppose that in the split-plot experiment the investigator finds $s_a^2 = s_b^2$, i.e. $f' = 1$. For a split-plot, $f = f'/(b-1)$, so that since $b = 2$, $f = f' = 1$. Hence the restricted region over which the probability must be averaged is $\phi/f > 1$. This is approximately the region to the right of the median 0.5 on the abscissa. Eye inspection suggests that the average probability in the restricted region will exceed 0.05. Numerical integration gives 0.063.

The dotted line in Figure 1 shows the corresponding probabilities for $f = f' = 1.53$. Since this is the 25% significance level of $f'$ it gives slightly greater support to the idea that $\sigma_a^2/\sigma_b^2 > 1$ than does our previous choice of $f' = 1$. For $f' = 1.53$ the restricted region extends from 0.25 to 1 on the abscissa. The average probability in this region is 0.057. As $f'$ increases further, the restricted region increases in size and the average probability moves toward 0.05. For $f'$ lower than 1 the average probability is higher than that for $f' = 1$, being, for example, about 0.084 when $f' = 1/3$. (In each case the probabilities are those of exceeding the 5% value of $d$ for $n_1 = 6$, $n_2 = 12$ and the appropriate $f$.) Similar results hold at the 1% level. The average probability in the curtailed region is 0.015 when $f' = 1$ and 0.012 when $f' = 1.53$.

## 3. RANGE COVERED BY THE COMPUTATIONS

In order to investigate these effects more systematically, a series of calculations were made of the actual probability with which $d$ exceeds $d_{.05}$ and $d_{.01}$, the tabulated Behrens-Fisher significance levels, in the restricted region. The values chosen for $n_1$ and $n_2$ were 6, 12 and 24, giving nine combinations. The quantity $f'$ was taken at its 75%, 50%, 25% and 5% levels. With $f' = s_a^2/s_b^2$ at its 5% level, the data are confirming the investigator's idea that $\sigma_a^2/\sigma_b^2 > 1$, while when $f'$ is at its 75% level, the data are tending to disagree with this apriori assumption. It was not thought necessary to investigate the situations in which $f'$ is at its 1% or 0.1% levels, although such cases might be expected to occur commonly in practice when $\sigma_a^2/\sigma_b^2 > 1$. When $f'$ is at these levels the restricted range is very close to the whole sample space of $\phi/f$, so that the ordinary significance levels of $d$ are unlikely to be much in error.

Expression (8) shows that the probability depends on $f$ as well as on $n_1$, $n_2$ and $f'$. For the two-sample comparison in Section 1,

$$f = (n_2+1)f'/(n_1+1).$$

Thus $f/f'$ is usually close to $n_2/n_1$, lying between this value and unity. In a split-plot experiment $f/f'$ is easily seen to be $n_1/n_2$ if the main units are arranged in a completely randomized design. It is $an_1/(a-1)n_2$ when main units are in randomized blocks and $an_1/(a-2)n_2$ when main units form a latin square, where $a$ is the number of levels of factor $A$. In all three split-plot cases $f/f'$ lies between $n_1/n_2$ and unity.

For such applications it looks as if values of $f/f'$ lying between $n_1/n_2$ and $n_2/n_1$ are primarily of interest. In more complex situations, however, the ratio $f/f'$ need not bear any simple relation to $n_1/n_2$. Consequently the probabilities were computed for $f/f' = 0$, 1/4, 1/2, 1, 2 and $\infty$.

For any specific $n_1$, $n_2$, $f$ and $f'$ the ordinary Behrens-Fisher significance level $d_\alpha$ was first computed on the IBM 7090 by interpolation in the Fisher-Yates tables. Expression (8) was then obtained by numerical integration, using the trigonometric expansion of the integral of the $t$-distribution. I am greatly indebted to Michael Feuer who programmed and conducted the calculations. After debugging, computation of the 432 probability values took 17.9 minutes of machine time.

TABLE 1. PROBABILITY THAT $d$ EXCEEDS THE BEHRENS-FISHER 5% LEVEL IN THE RESTRICTED REGION $\sigma_1^2/\sigma1 > 1/f'$

| $f'$ | $f \to 0$ | $f=f'/4$ | $f=f'/2$ | $f=f'$ | $f=2f'$ | $f \to \infty$ |
|---|---|---|---|---|---|---|
| | | | $n_1=6,\ n_2=6$ | | | |
| 75% | .012 | .023 | .034 | .050 | .070 | .122 |
| 50% | .017 | .029 | .039 | .050 | .062 | .082 |
| 25% | .026 | .037 | .043 | .050 | .055 | .063 |
| 5% | .040 | .046 | .048 | .050 | .051 | .052 |
| | | | $n_1=6,\ n_2=12$ | | | |
| 75% | .024 | .042 | .055 | .073 | .092 | .132 |
| 50% | .030 | .044 | .053 | .064 | .073 | .087 |
| 25% | .036 | .046 | .052 | .057 | .060 | .064 |
| 5% | .045 | .049 | .050 | .051 | .052 | .052 |
| | | | $n_1=6,\ n_2=24$ | | | |
| 75% | .035 | .055 | .069 | .087 | .105 | .139 |
| 50% | .039 | .053 | .062 | .071 | .079 | .090 |
| 25% | .043 | .052 | .056 | .060 | .063 | .065 |
| 5% | .048 | .050 | .051 | .052 | .052 | .053 |
| | | | $n_1=12,\ n_2=6$ | | | |
| 75% | .007 | .013 | .019 | .030 | .045 | .091 |
| 50% | .013 | .021 | .027 | .036 | .047 | .070 |
| 25% | .023 | .030 | .036 | .042 | .048 | .058 |
| 5% | .039 | .044 | .046 | .048 | .050 | .052 |
| | | | $n_1=12,\ n_2=12$ | | | |
| 75% | .019 | .029 | .037 | .050 | .065 | .100 |
| 50% | .025 | .035 | .041 | .050 | .059 | .075 |
| 25% | .033 | .041 | .045 | .050 | .054 | .060 |
| 5% | .044 | .047 | .049 | .050 | .051 | .052 |
| | | | $n_1=12,\ n_2=24$ | | | |
| 75% | .030 | .042 | .051 | .064 | .078 | .107 |
| 50% | .035 | .045 | .051 | .059 | .066 | .078 |
| 25% | .041 | .047 | .051 | .054 | .058 | .062 |
| 5% | .047 | .049 | .050 | .051 | .052 | .052 |
| | | | $n_1=24,\ n_2=6$ | | | |
| 75% | .004 | .008 | .012 | .020 | .032 | .072 |
| 50% | .010 | .016 | .021 | .029 | .038 | .061 |
| 25% | .020 | .027 | .032 | .038 | .044 | .055 |
| 5% | .038 | .042 | .044 | .047 | .048 | .051 |
| | | | $n_1=24,\ n_2=12$ | | | |
| 75% | .014 | .021 | .027 | .037 | .048 | .078 |
| 50% | .022 | .029 | .034 | .041 | .049 | .065 |
| 25% | .031 | .037 | .041 | .045 | .050 | .057 |
| 5% | .043 | .046 | .047 | .049 | .050 | .051 |
| | | | $n_1=24,\ n_2=24$ | | | |
| 75% | .026 | .034 | .041 | .050 | .060 | .084 |
| 50% | .032 | .039 | .044 | .050 | .056 | .068 |
| 25% | .039 | .043 | .047 | .050 | .053 | .058 |
| 5% | .046 | .048 | .049 | .050 | .051 | .052 |

TABLE 2. PROBABILITY THAT $d$ EXCEEDS THE BEHRENS-FISHER 1% LEVEL IN THE RESTRICTED REGION $\sigma_1^2/\sigma_2^2 > 1/f'$

| $f'$ | $f \to 0$ | $f=f'/4$ | $f=f'/2$ | $f=f'$ | $f=2f'$ | $f \to \infty$ |
|---|---|---|---|---|---|---|
| | | | $n_1=6,\ n_2=6$ | | | |
| 75% | .0007 | .0024 | .0048 | .0096 | .0165 | .0322 |
| 50% | .0014 | .0037 | .0062 | .0100 | .0137 | .0186 |
| 25% | .0026 | .0055 | .0078 | .0101 | .0117 | .0131 |
| 5% | .0059 | .0084 | .0094 | .0100 | .0103 | .0105 |
| | | | $n_1=6,\ n_2=12$ | | | |
| 75% | .0030 | .0073 | .0118 | .0182 | .0251 | .0348 |
| 50% | .0040 | .0082 | .0112 | .0147 | .0172 | .0194 |
| 25% | .0055 | .0090 | .0107 | .0121 | .0129 | .0133 |
| 5% | .0080 | .0097 | .0102 | .0104 | .0105 | .0105 |
| | | | $n_1=6,\ n_2=24$ | | | |
| 75% | .0055 | .0118 | .0170 | .0235 | .0294 | .0363 |
| 50% | .0065 | .0112 | .0142 | .0168 | .0185 | .0197 |
| 25% | .0076 | .0107 | .0120 | .0128 | .0131 | .0133 |
| 5% | .0091 | .0102 | .0104 | .0105 | .0105 | .0105 |
| | | | $n_1=12,\ n_2=6$ | | | |
| 75% | .0002 | .0007 | .0014 | .0036 | .0078 | .0234 |
| 50% | .0006 | .0016 | .0028 | .0053 | .0087 | .0160 |
| 25% | .0018 | .0033 | .0050 | .0072 | .0094 | .0123 |
| 5% | .0053 | .0072 | .0083 | .0092 | .0099 | .0104 |
| | | | $n_1=12,\ n_2=12$ | | | |
| 75% | .0018 | .0037 | .0059 | .0098 | .0150 | .0263 |
| 50% | .0028 | .0051 | .0071 | .0100 | .0129 | .0172 |
| 25% | .0045 | .0067 | .0083 | .0100 | .0113 | .0127 |
| 5% | .0075 | .0089 | .0095 | .0100 | .0103 | .0105 |
| | | | $n_1=12,\ n_2=24$ | | | |
| 75% | .0042 | .0075 | .0105 | .0150 | .0198 | .0286 |
| 50% | .0054 | .0083 | .0104 | .0129 | .0151 | .0180 |
| 25% | .0069 | .0090 | .0102 | .0114 | .0122 | .0130 |
| 5% | .0088 | .0097 | .0101 | .0103 | .0105 | .0105 |
| | | | $n_1=24,\ n_2=6$ | | | |
| 75% | .0001 | .0002 | .0006 | .0015 | .0039 | .0172 |
| 50% | .0003 | .0008 | .0015 | .0031 | .0058 | .0135 |
| 25% | .0012 | .0023 | .0035 | .0054 | .0077 | .0115 |
| 5% | .0048 | .0064 | .0075 | .0086 | .0094 | .0103 |
| | | | $n_1=24,\ n_2=12$ | | | |
| 75% | .0010 | .0021 | .0033 | .0057 | .0093 | .0194 |
| 50% | .0020 | .0035 | .0049 | .0071 | .0096 | .0146 |
| 25% | .0038 | .0054 | .0067 | .0083 | .0098 | .0119 |
| 5% | .0072 | .0083 | .0089 | .0096 | .0100 | .0103 |
| | | | $n_1=24,\ n_2=24$ | | | |
| 75% | .0031 | .0051 | .0070 | .0099 | .0135 | .0214 |
| 50% | .0045 | .0064 | .0080 | .0100 | .0120 | .0155 |
| 25% | .0062 | .0078 | .0088 | .0100 | .0110 | .0123 |
| 5% | .0085 | .0093 | .0097 | .0100 | .0102 | .0104 |

## 4. RESULTS

Tables 1 (5% level) and 2 (1% level) show the probability that $d$ exceeds the 5% or 1% level in the Behrens-Fisher table when we make the additional assumption that $\sigma_a^2/\sigma_b^2 > 1$. The results are not too simple to summarize and digest, but the following points emerge.

(1) If one tries to guess the direction of the results intuitively, the easiest case is that in which $f = f'$, so that the restriction becomes $\sigma_1^2/\sigma_2^2 > 1$. It seems natural (at least to me) to guess that this restriction produces the same effect as that of an increase in $s_1^2/s_2^2$ on the ordinary Behrens-Fisher levels, because the additional information suggests that the stability of $d$ will now depend more on the accuracy with which $\sigma_1^2$ is estimated than it does in the unrestricted case. When $n_1 < n_2$, an increase in $s_1^2/s_2^2$ raises the value of $d$ required for 5% significance in the Behrens-Fisher tables for the range of values of $n_1$, $n_2$ considered here. Consequently the restriction should produce probabilities greater than 0.05 or 0.01 in Tables 1 and 2 when $f = f'$ and $n_1 < n_2$. Similarly, the restricted probabilities should be less than 0.05 or 0.01 when $f = f'$ and $n_1 > n_2$. These anticipations are verified in every case in Tables 1 and 2.

When $f = f'$ and $n_1 = n_2$, the same intuition suggests that the probabilities should not be disturbed. To the degree of accuracy shown in Tables 1 and 2 this happens in all 12 cases at the 5% level and in all but four cases at the 1% level. The discrepancies in these four cases are so small that they may be due to rounding errors in the calculations. I have tried to prove by integration that the probability for this set of cases remains exactly at 5% or 1% but have not succeeded, except for the case $f = f' = 1$ in which the result is obvious by symmetry.

(2) As anticipated, when $f'$ is at the 5% level the probabilities remain close to those in the Behrens-Fisher tables except when $f$ is very small or very large. As $f'$ diminishes, the disturbance to the probabilities steadily increases, becoming very substantial when $f'$ is at the 50% and 75% levels.

(3) For given $f'$, $n_1$ and $n_2$ the probability increases monotonically with $f$, as can be verified mathematically.

(4) Although the panel of values of $n_1$ and $n_2$ is not large enough to suggest firm rules of interpolation against these values, it appears in all cases that for fixed $n_1/n_2$, the probability moves towards 0.05 or 0.01 as $(n_1+n_2)$ increases.

### APPLICATION TO DATA

The primary reason for this investigation lies, of course, in the help that it might give to the investigator who wants to apply the Behrens-Fisher test under these restricted conditions. Unfortunately, the amount of help actually supplied is limited. It is clear that for the sample sizes considered, the disturbance to the Behrens-Fisher probabilities can be substantial and can lie in either direction, and no simple rationalization of the whole pattern of results has occurred to me. A more extensive table of the 5% and 1% significance levels of $d$ in the restricted region is perhaps called for, though as a four-variable table it would be inconvenient to use.

The result that the disturbance to the Behrens-Fisher values is minor (except for very small values of $f$ relative to $f'$ ) when $f'$ is at or beyond the 5% level will often be all that the investigator needs to know. In looking for examples that might be typical of the split-plot case, I noticed that in the experiments reported in the well-known books by Snedecor, Federer, Bennett and Franklin, and Cochran and Cox, all the $f'$ values were beyond the 0.5% level ($P < 0.005$) and three of the four were beyond the 0.1% level.

The result that there is either no disturbance or at most a trifling disturbance when $n_1 = n_2$ and $f = f'$ is also useful, since this applies to the comparison of two samples of equal sizes (and to a split-plot with two subunit treatments and main units completely randomized).

For the split-plot experiment with main units completely randomized, in randomized blocks, or in a latin square, we have, as noted previously, $n_1 \leqslant n_2$ and $f/f'$ lying between $n_1/n_2$ and unity. In all such cases in Tables 1 and 2 with $n_1 < n_2$, the Behrens-Fisher significance level is too low for the restricted region, but usually by only a small amount. As an example, the experiment quoted in Goulden's book (1952) has $n_1 = 35$, $n_2 = 42$, $f = f' = 1.17$. This lies at about the 32% level. For $n_1 = 12$, $n_2 = 24$ in Table 1, the probability for $d_{.05}$ and $f = f'$ is 0.054 when $f'$ is at the 25% level and 0.059 when $f'$ is at the 50% level. For $n_1 = 21$, $n_2 = 42$ both probabilities presumably move towards 0.05, with a further move in this direction when $n_1 = 35$, $n_2 = 42$. Hence the probability in the restricted region may be expected to be only slightly above 0.05 in Goulden's example.

In the comparison of two samples of *unequal* sizes, $f/f' = (n_2+1)/(n_1+1)$, which will be approximately $n_2/n_1$ in most practical situations. Two cases may be distinguished. If $n_1 > n_2$, i.e. the less precise sample is larger, Tables 1 and 2 indicate that the probabilities are smaller than 0.05 or 0.01. For instance, in Table 1 with $n_1 = 24$, $n_2 = 6$, $f/f' = 1/4$, the 0.05 probabilities drop to 0.042, 0.027, 0.016 and 0.008. The disturbances can clearly be large if $n_1$ is much greater than $n_2$. The probabilities imply, of course, that a smaller value of $d$ is needed for 5% significance than that given in the Behrens-Fisher tables. If $n_1 < n_2$, on the other hand, the probabilities are higher than the stipulated 0.05 and 0.01, as seen from $n_1 = 6$, $n_2 = 12$, $f/f' = 2$, and from $n_1 = 12$, $n_2 = 24$, $f/f' = 2$.

In conclusion, it is hoped that by using Tables 1 and 2 as illustrated above, the investigator can find out the direction of the disturbance to the Behrens-Fisher significance levels and obtain some idea as to whether the disturbance is likely to be minor or major.

References

Fisher, R. A. (1939) : The comparison of samples with possibly unequal variances. *Ann. Eugen.*, **9**, 174-180.

———— (1941) : The asymptotic approach to Behrens' integral, with further tables for the *d* test of significance. *Ann. Eugen.* **11**, 141–172.

Fisher, R. A. and Yates, F. (1957) : *Statistical Tables,* Oliver and Boyd, Edinburgh, 5th ed., Table VI.

Goulden, C. H. (1952) : *Methods of Statistical Analysis,* John Wiley and Sons, New York, 2nd ed., 225.

*Paper received : August, 1963.*

## APPROXIMATE SIGNIFICANCE LEVELS OF THE BEHRENS-FISHER TEST

William G. Cochran

*Department of Statistics, Harvard University, Cambridge, Mass., U.S.A.*

### INTRODUCTION

Some twenty years ago, not long after the appearance in *Sankhyā* of Sukhatme's tables of the 5% and 1% significance levels of the Behrens-Fisher test (since republished in Fisher and Yates' *Statistical Tables* [1963]), Professor Snedecor asked me to supply an approximation to these significance levels for use in practical work by biologists who did not have access to Sukhatme's tables or who wanted probability levels, such as the 2% or 10%, for which no tables were available. By inspection of Sukhatme's tables I produced an empirical approximation, based on the Student *t*-table. In the intervening period this approximation has found its way into a number of text-books, and I should like consequently to give some account of its accuracy.

### THE APPROXIMATION

In the Behrens-Fisher problem we are concerned with a quantity $x$, normally distributed, with mean zero on the null hypothesis and variance $(\sigma_1^2 + \sigma_2^2)$. We possess independent estimates $s_1^2$ of $\sigma_1^2$ and $s_2^2$ of $\sigma_2^2$, with $n_1$ and $n_2$ degrees of freedom respectively, but do not wish to make any assumption about the value of $\sigma_1^2/\sigma_2^2$. The test criterion tabulated by Sukhatme is $d = x/\sqrt{s_1^2 + s_2^2}$.

Let $t_1$, $t_2$ be the significance levels of Student's $t$ for $n_1$, $n_2$ d.f., respectively, at the desired probability level $\alpha$. The approximate significance level of $d$ is obtained as the weighted mean of $t_1$ and $t_2$ with weights $s_1^2$ and $s_2^2$ :

$$\hat{d}_\alpha = \frac{s_1^2 t_1 + s_2^2 t_2}{s_1^2 + s_2^2}.$$

*Example.* For a certain measurement a sample of 20 psychotic patients gives a variance of 25.1 and a sample of 30 normal persons gives a variance of 12.8. The $F$-ratio 25.1/12.8 is close to the 10 percent (two-tailed) level and since psychotic patients exhibit abnormally large variability on some measurements we may not wish to assume the

Reproduced with permission from *Biometrics*, Vol. 20, No. 1, pages 191–195. Copyright © 1964, Biometric Society.

true variances equal. What is the 5 percent value of $d$ for a comparison of the means of the two samples? Applying the approximation we have

$$s_1^2 = \frac{25.1}{20} = 1.255 \qquad s_2^2 = \frac{12.8}{30} = 0.427$$

$$t_{19} = 2.093 \qquad t_{29} = 2.045$$

$$\hat{d}_{.05} = \frac{(1.255)(2.093) + (0.427)(2.045)}{1.682} = 2.081.$$

To compare with Sukhatme's table, we find $\theta = \tan^{-1}(s_1/s_2) = 59° 45'$. This is close enough to $60°$ to use the entries for that angle, but two-way linear interpolation on $1/n_1$ and $1/n_2$ is required. The interpolated value is 2.076.

In practical work the approximation has the advantage that it need rarely be computed. If $n_1 = n_2$, as happens in many applications, $\hat{d}_\alpha$ reduces to the Student $t$-value for $n_1$ d.f. When $n_1 \neq n_2$, it is often apparent, on looking at $t_1$ and $t_2$, that the value of $d$ computed from the data either exceeds both of them or falls short of both of them. The verdict of the test follows without further computation.

## ACCURACY OF THE APPROXIMATION

As a measure of the accuracy of this approximation, I have compared the actual probability $\alpha'$ that the Behrens-Fisher $d$ exceeds $\hat{d}_\alpha$ with the presumed probability $\alpha$.

When $n_1$ and $n_2$ both become large, it is evident that $\hat{d}_\alpha$ and the correct $d_\alpha$ will agree closely, because both tend to the significance levels of the normal distribution. For any $n_1$ and $n_2$, $\hat{d}_\alpha$ and $d_\alpha$ become identical when $s_1^2/s_2^2$ tends to zero or infinity, since both become the significance level of Student's $t$ for $n_2$ or $n_1$ d.f. These observations suggest that the approximation is likely to be poorest for small or moderate values of $n_1$ and $n_2$ and intermediate values of $s_1^2/s_2^2$.

Consider first $n_1 = n_2 = n$. Inspection of Sukhatme's table and of the more recent table computed by Fisher and Healy [1956] for small, odd degrees of freedom shows that the approximation is in general poorest when $\theta = \tan^{-1}(s_1/s_2) = 45°$, that is, when $s_1 = s_2$, since this is the value at which the significance level of $d$ differs most from Student's $t$ for $n$ d.f. Table 1 gives the $P$ values for $\theta = 45°$, $n = 10, 20, 40$ and two-tailed significance levels of $d = 10\%$, $5\%$ and $1\%$. These were computed from the asymptotic formulae given by Fisher [1941].

The $5\%$ and $10\%$ approximations look satisfactory. At the $1\%$ level the approximation is somewhat too stringent.

With small, odd values of $n_1 = n_2 = n$, Table 2 shows $\alpha'$ for $\theta = 15°$,

TABLE 1

PROBABILITY $\alpha'$ THAT $d$ EXCEEDS THE APPROXIMATE SIGNIFICANCE LEVEL $d_\alpha$ FOR $n_1 = n_2 = n;$ $\theta = 45°$.

| Presumed probability $\alpha$ | $n =$ | | |
|---|---|---|---|
| | 10 | 20 | 40 |
| 10% | .1023 | .1008 | .1004 |
| 5% | .0488 | .0491 | .0494 |
| 1% | .0079 | .0086 | .0092 |

30° and 45°. For this table and subsequent computations for small $n$'s I am indebted to Miles Davis who programmed and conducted the computations. Fisher and Healy's closed expansions of $\alpha'$ in powers of $\sin \theta$ and $\cos \theta$ were used for the computations.

The performance of the approximation for small $n$ is in general similar to that in Table 1. The approximation does well at the 10% and 5% levels. At the 1% level, the approximation tends to underestimate $\alpha$ by 15 to 25 percent for values of $n$ less than 20 and angles between 30° and 60°.

For $n_1$, $n_2$ small and unequal, computations of $\alpha'$ were made at the 10%, 5% and 1% levels for the following sets of values:

$$n_1, n_2 = (3, 5)\ (3, 7)\ (3, 9)\ (5, 7)\ (5, 9)\ (7, 9)$$

$$\theta = 15°, 30°, 45°, 60°, 75°.$$

TABLE 2

VALUES OF $\alpha'$ FOR SMALL VALUES OF $n_1 = n_2 = n$.

| $d$ | $\theta$ | $n =$ | | | |
|---|---|---|---|---|---|
| | | 3 | 5 | 7 | 9 |
| 10% | 15° or 75° | .1031 | .1011 | .1007 | .1005 |
| | 30° or 60° | .1091 | .1043 | .1026 | .1018 |
| | 45° | .1120 | .1059 | .1037 | .1026 |
| 5% | 15° or 75° | .0504 | .0496 | .0495 | .0496 |
| | 30° or 60° | .0518 | .0495 | .0490 | .0490 |
| | 45° | .0527 | .0496 | .0490 | .0488 |
| | 15° or 75° | .0096 | .0094 | .0094 | .0094 |
| 1% | 30° or 60° | .0090 | .0082 | .0082 | .0083 |
| | 45° | .0088 | .0077 | .0076 | .0078 |

Over these 30 values of $\alpha'$, Table 3 shows the average error, (average value of $|\alpha' - \alpha|$) and the maximum error.

TABLE 3

AVERAGE AND MAXIMUM ERRORS FOR $n_1$, $n_2$ SMALL AND UNEQUAL

| $\alpha$ | Average error in $\alpha'$ | Maximum error in $\alpha'$ | Max. occurs at | | |
|---|---|---|---|---|---|
| | | | $n_1$ | $n_2$ | $\theta$ |
| .1000 | .0037 | +.0093 | 3 | 5 | 30° |
| .0500 | .0008 | +.0022 | 3 | 9 | 30° |
| .0100 | .0010 | −.0023 | 7 | 9 | 45° |

As in Tables 1 and 2, all 30 errors were positive at the 10% level and negative at the 1% level. Although most of the errors at the 5% level were negative, a few were positive for the $n_1 = 3$ entries.

As a final check, calculations were made for $n_1 = \infty$, $n_2 = 10$, 20. This case was investigated in detail by Fisher [1941]. It applies approximately to a situation in which two independent estimates of a quantity $\mu$ have been made, one by a large number of imprecise measurements, so that $n_1$ is large, and the second by a small number of precise measurements. The results in Table 4 indicate that the approximation is adequate for this situation.

TABLE 4

VALUES OF $\alpha'$ FOR $(\infty, 10)$ AND $(\infty, 20)$

| $\theta$ | $\alpha = 10\%$ | | $\alpha = 5\%$ | | $\alpha = 1\%$ | |
|---|---|---|---|---|---|---|
| | $(\infty, 10)$ | $(\infty, 20)$ | $(\infty, 10)$ | $(\infty, 20)$ | $(\infty, 10)$ | $(\infty, 20)$ |
| 15° | .1002 | .1000 | .0499 | .0498 | .0099 | .0099 |
| 30° | .1007 | .1002 | .0496 | .0497 | .0095 | .0096 |
| 45° | .1012 | .1004 | .0495 | .0496 | .0090 | .0093 |
| 60° | .1011 | .1004 | .0497 | .0497 | .0090 | .0094 |
| 75° | .1004 | .1002 | .0499 | .0499 | .0096 | .0098 |

## SUMMARY OF RESULTS

For small or moderate $n_1$, $n_2$ the approximation gives slightly too many significant results at the 10% level. The actual probability of exceeding the presumed 10% level averaged about 10.4% over the 42 calculated values represented in Tables 2 and 3. At the 5% level, the approximate test is slightly conservative except for some low values of

$n_1$ or $n_2$ , but gives in general very good agreement with the stipulated 0.05 probability.  The approximation is also conservative at the 1% level.  It is apparently at its worst at this level for $n_1$ , $n_2$ less than 20 and angles near 45°, the significance probability being nearer 0.8% than 1% in the worst cases.  It may be concluded that the approximation is adequate for routine tests of significance made at levels between 1% and 10%, although not for accurate mathematical calculations.

In problems involving linear functions of estimated variances, approximate $t$- and $F$-tests are sometimes made by a method due to Fairfield Smith [1936].  The quantity $(s_1^2 + s_2^2)$ is regarded as having a number of degrees of freedom $n_e$ equal to

$$n_e = \frac{(s_1^2 + s_2^2)^2}{(s_1^4/n_1) + (s_2^4/n_2)}.$$

The quantity $d$ is tested by referring it to Student's $t$-table with $n_e$ d.f. This method does not give a good approximation to the Behrens-Fisher levels: at probabilities between 10% and 1% it produces too many significant results relative to Behrens-Fisher.  For instance, for $n_1 = n_2 = 7$, $\theta = 30°$, the $\alpha'$ values for this method at the 10%, 5% and 1% levels are 12.0%, 6.4% and 1.5% as compared with 10.3%, 4.9% and 0.8% (Table 2) for the approximation in this paper.  This approximation is more nearly related to the Aspin-Welch approach to the problem (Pearson and Hartley [1954]), although I have not checked how closely it agrees with their tables.

## ACKNOWLEDGEMENTS

This work was supported in part by an Office of Naval Research Contract Nonr 1866(37) at Harvard University.  Reproduction in whole or in part is permitted for any purpose of the United States Government.  Most of the computations were done at the M. I. T. Computation Center, Cambridge, Massachusetts.

## REFERENCES

Fisher, R. A. [1941].  The asymptotic approach to Behrens' integral, with further tables for the $d$ test of significance. *Annals of Eugenics 11,* 141–72.

Fisher, R. A., and Healy, M. J. R. [1956].  New tables of Behrens' test of significance. *Jour. Roy. Stat. Soc. B. 18,* 212–16.

Fisher, R. A. and Yates, F. [1963].  *Statistical tables.* Oliver and Boyd, Edinburgh. 6th ed., table VI.

Pearson, E. S. and Hartley, H. O. [1954].  *Biometrika tables for statisticians.* Vol. I, Cambridge University Press, table 11.

Smith, H. F. [1936].  The problem of comparing the results of two experiments with unequal errors. *Jour. Comm. Sci. Ind. Res. (Australia) 9,* 211–12.

Sukhatme, P. V. [1938].  On Fisher and Behrens' test of significance for the difference in means of two normal samples. *Sankhyā 4,* 39–48.

# 82

## WILLIAM G. COCHRAN and MILES DAVIS
## Stochastic Approximation to the Median Effective Dose in Bioassay

### INTRODUCTION

When a stimulus is applied in some types of experimentation, all that we can observe is whether a certain type of response does or does not take place. With explosive powders, an impact applied to the powder may or may not explode it. An insect or animal may die or may survive when given a specific dose of a toxic agent or a virus inoculum. For practical purposes, what we would probably like to know in such experiments is the largest amount of stimulus that can be given without any of the subjects responding, and the smallest amount that will cause all the subjects to respond; e. g. the largest shock that an explosive can stand without any risk of an explosion and the smallest shock that is certain to ignite the powder. But a little reflection, assisted by some attempts to conduct experiments, soon shows that these amounts of stimulus are almost impossible to determine experimentally. Instead, the common practice is to estimate the amount of stimulus that will cause half the subjects to respond. This may be called the median effective dose (ED50) or, in toxicology, the median lethal dose (LD50).

This paper describes some recent work in which the ideas of stochastic approximation are used in the hope of developing experimental methods for estimating the median effective dose that are more economical and convenient in certain situations than the standard methods in current use. Although the mathematical theory of stochastic approximation has some fascinating aspects, the orientation here is applied, and questions of rigor and generality will not be stressed. Further, the possible experimental plans based on stochastic approximation have by no means been thoroughly explored as yet, so that the present account is of the nature of an interim report.

### Mathematical background

Suppose that we could determine for each subject in a population the amount $X$ of the stimulus that would just produce the response in that subject. Over a population of subjects, the variate X would follow a frequency distribution, sometimes called the tolerance distribution. Let $\phi(X)$ be the frequency function. If now a fixed amount x is applied to every subject in the population, the proportion P of subjects who respond is given by

$$P = \int_{-\infty}^{X} \phi(t)\,dt \;.$$

Similarly, if the dose x is applied to a random sample of $\underline{m}$ subjects, the proportion p who respond in the sample will be distributed as a binomial variate with mean P and sample size m, assuming that the population is very much larger than the sample.

Write $y = (t - \mu)/\sigma$, where $\mu$ is the median and $\sigma$ the standard deviation of the tolerance distribution. Suppose that $\phi(t)\,dt$ becomes $f(y)\,dy$. Then

(1)
$$P = \int_{-\infty}^{(x-\mu)/\sigma} f(y)\,dy = \int_{-\infty}^{\beta(x-\mu)} f(y)\,dy$$

where $\beta = 1/\sigma$. In biological assay the quantity $\beta$, called the slope, is a commonly used term. The name arises because a standard method of analyzing the data is to transform the p values to variates z where

(2)
$$p = \int_{-\infty}^{z} f(y)\,dy \;.$$

By comparison with (1) it is clear that if the binomial errors can be neglected, the variate z will plot against x as a straight line with slope $\beta$.

In practice some simplifying assumptions are made about the tolerance distribution $\phi(X)$. By a suitable choice of the scale in which X is recorded (usually the log dose scale in bioassay), it is assumed that $\phi(X)$ is symmetrical about its mean and median, and depends only on the parameters $\mu$ and $\sigma$. Most commonly, $\phi(X)$ is taken to be either the normal or the logistic frequency distribution. These two assumptions usually give closely similar results, the logistic being in some respects easier to work with.

One important consequence of these assumptions may be noted. Suppose that the reduced form of the tolerance frequency function $f(y)$ is the same for two different stimuli and that $\sigma$ is also the same, only the means $\mu$, $\mu'$ being different. It follows that the plots of z against x for the two stimuli should be, apart from binomial

errors, two _parallel_ straight lines.  This fact is used in practice to
provide a test of significance of the null hypothesis that the two $\sigma$'s
are equal.  Further, the dose  x  of one stimulus that produces a
specified proportion  P  of responders is  $\mu + Z\sigma$ , where  Z  is the
transform of  P  as in equation ( 2) .  It follows, and this is the important point, that the difference in dosages required to produce any
specified  P  has the constant value  $\mu - \mu'$  .  If  x  is measured on
a log scale so that  $\mu$  and  $\mu'$  are the logs of the actual dosages,
this means that the ratio of the doses required to produce any specified  P  for the two stimuli is a constant.  This constant, inverted,
is called the relative potency of the two stimuli.  For stimuli that
have similar modes of action, there is a good deal of evidence, although most of it is based on small samples, that the hypothesis of
parallelism ( i. e.  $\sigma = \sigma'$ )  can often be supported.

## Non-sequential experiments

When an experiment is being planned, the investigator does not
of course know the values of  $\mu$  and  $\sigma$  .  As will appear later, his
degree of ignorance about these quantities plays an important role
both in the choice of an experimental strategy and in determining the
precision of the experiment that he decides to conduct.  In practice
the accuracy of initial guesses about  $\mu$  and  $\sigma$  varies greatly from
one situation to another.  In bioassay, a standard agent on which
much previous research has been done is frequently compared with a
new agent thought to have a similar mode of action.  The values of
both  $\mu$  and  $\sigma$  should be fairly well known for the standard.  For the
new agent,  $\sigma'$  will be fairly well known if it can be assumed that
$\sigma'$  is approximately equal to  $\sigma$ ,  but  $\mu$  may be known only poorly.
However, if essentially nothing can be guessed about  $\mu$  for the new
agent, the investigator is likely to start with a small pilot experiment,
perhaps involving only one subject at each dose, from which he makes
a rough initial estimate  $x_0$  of the  ED50 .  In the absence of a standard agent, neither  $\mu$  nor  $\sigma$  may be at all well known initially.  With
inanimate "subjects", such as explosives or plastic water pipe,
Wetherill ( 1963) ,  the material may be more uniform so that  $\mu$  and  $\sigma$
can be guessed fairly successfully in advance from previous results
on similar material.

In the calculations that we have made, the properties of stochastic approximation methods have been investigated for starting values
out to  $\pm 8\sigma$   away from  $\mu$  .  This is probably a wider range than is
needed for any practical application.  The value of  $\sigma$ , with  x
measured on a log scale, is assumed known initially to within a 4 to
1 ratio.  Although this range may be too narrow for some biological
research, it appeared reasonable from inspection of data collected

by Gaddum (1933) and Bliss and Cattell (1943). These workers summarized the estimated values of $\sigma$ found in bioassays on several
different experimental animals with death as the response. The calculations of Brownlee, Hodges and Rosenblatt (1953) and Wetherill
(1963), who also considered this problem, amount to assuming that
$\sigma$ is known within a 2 to 1 ratio.

The usual method of experimentation is illustrated in Figure 1,
which sketches the relation between P and X, for a large range of
values of X, on the assumption of a normal tolerance distribution.
(For a smaller value of $\sigma$, the curve would be steeper in the rising
part.) Given an initial guess at $\mu$, the practice is to test at a number of different levels of the stimulus centered about this initial guess.
The levels are illustrated by the points $x_1, x_2, x_3, x_4$ in Figure 1. The
number of levels may vary from 2 to about 8: they are often equally-
spaced on the x (log dose) scale. The idea is to obtain at least
two levels like $x_2$ and $x_3$ that are in the rising part of the curve
and straddle the ED50. Levels like $x_1$ and $x_4$ which give zero or
100% response contribute little or nothing in the analysis.

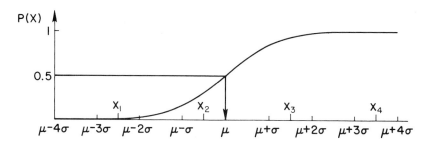

Fig.1. Relation between P(X) and X.

Estimation of the ED50 from the results of such an experiment
has been extensively studied. The more sophisticated methods depend on transforming to the z scale already mentioned, estimating
$\mu$ by fitting the linear regression of z on x . Relatively little advice is available in the literature on the planning of such experiments,
except when $\mu$ and $\sigma$ are well known in advance. The problem of
planning is complex. Decisions have to be made about the number of
levels, the distances between them, and the numbers of subjects to
be tested at the different levels. Part of the problem is that the exact distributions of the maximum likelihood (M. L.) estimate and of
other commonly used estimates of $\mu$ have been worked out only for

the asymptotic case in which the number of animals tested at each
level is large. Bross (1950) and Berkson (1955), who worked the
M. L. estimate for specific small experiments, both encountered a number of troublesome cases in which the estimate either did not exist or
appeared ridiculous.

In attempting to give positive advice for experiments using around
50 subjects, Finney (1951) worked out the effective variances of the
ED50, assuming that the asymptotic formula remains valid, for different numbers and spacings of levels. The calculations were made
for both normal and logistic tolerance distributions. To illustrate, his
recommendation for an experiment with four levels in that equal numbers of subjects be tested at levels expected to give 15%, 35%, 65%
and 85% responses. With a normal tolerance law, this amounts to
selecting the levels $x_0-1.04\sigma_0$, $x_0-0.38\sigma_0$, $x_0+0.38\sigma_0$, $x_0+1.04\sigma_0$,
where $x_0, \sigma_0$ are the preliminary guesses at $\mu$ and $\sigma$ . However,
if $x_0$ is likely to be in error by more than one standard deviation,
this set of levels can scarcely be recommended. For instance, if $x_0$
is actually at $\mu + 1.2\sigma$ while $\sigma_0$ is correct, the lowest level of
stimulus is set at $\mu + 0.16\sigma$ . The expected response at this level
is 56% and the four levels do not straddle the ED50 at all. If $x_0$ is
correct but $\sigma_0 = \sigma/2$, the actual effective variance from the asymptotic formula is about double the value obtained when $\sigma_0$ is correct.
There appears to be no available discussion of satisfactory plans for
experiments of modest size in which the investigator has only a rough
idea initially of the values of $\mu$ and $\sigma$ .

### The Up and Down method

The two stochastic approximation methods that will be considered
differ from the standard plan in that the experiment is conducted sequentially. The Up and Down method, Dixon and Mood (1948), was
first suggested in connection with the testing of explosives by dropping weights on the powder from different heights. In its original form
a single subject is tested at any given time. If at the $\underline{n}$th step in
the process the stimulus is at level $x_n$, the sequential rule is that
the $(n + 1)$th test be made at level

$$
x_{n+1} = \begin{cases} x_n + d & \text{if there is no response at } x_n \\[2em] x_n - d & \text{if there is a response at } x_n , \end{cases}
$$

where $d$ is a constant. When the experiment is terminated at some
chosen value $n = N$, Dixon and Mood give a graph by which the M. L.
estimate of $\mu$ is obtained, assuming a normal tolerance distribution.

From the asymptotic theory they recommend that the step size $\underline{d}$ should be approximately equal to $\sigma$ .

The performance of this process in small samples was studied by Brownlee, Hodges and Rosenblatt (1953) . As a simpler alternative to the M. L. estimate they suggest taking $\hat{\mu}$ as the mean of the levels $x_i$ at which the tests were conducted, omitting the starting level $x_1$ but including the level $x_{N+1}$ at which the next test would have been made had the experiment not stopped at $X_N$ .

The requirement that subjects be tested one at a time severely restricts the utility of this method, since the verdict (response or no response) must be known almost at once if the experiment is not to last an undesirable length of time. In this paper we wish to consider plans that will be feasible if the response is known within an hour or less. This implies that an experiment with $N = 8$ takes up to a day to perform. To meet this situation and still amass a reasonable size of sample, Brownlee, Hodges and Rosenblatt suggest that several processes be run simultaneously with the same starting value $x_1$ . As an estimate $\hat{\bar{\mu}}$ they propose the mean of the estimates $\hat{\mu}_i$ obtained from the different runs. Thus if a sample size of 48 is wanted, six runs with $N = 8$ each might be made. This method is not as accurate as a run of 48 single subjects, because the estimates $\hat{\mu}_i$ are biased unless $x_1$ happens to equal $\mu$, and their average $\hat{\bar{\mu}}$ is subject to the same bias. (The estimate for 48 single subjects is also biased, but the bias is in general smaller.)

Although less efficient, this plan has one possible advantage for estimating the standard error of $\hat{\bar{\mu}}$ . If the bias in $\hat{\bar{\mu}}$ is negligible relative to its standard error, the quantity $\sqrt{\Sigma(\hat{\mu}_i-\hat{\bar{\mu}})^2/k(k-1)}$ is an almost unbiased estimate of s.e. $(\hat{\bar{\mu}})$, where $k$ is the number of repetitions. Dews and Berkson (1954) have produced evidence to suggest that the customary estimates of the standard error of the ED50, computed from the internal results of a single trial, may seriously overrate the precision of the estimate as judged from its stability in repeated experiments of the same type. Their tentative opinion is that the discrepancy is due primarily to an "error of dosage" i.e., to a group of factors that may influence the constancy of the effective administration of the dose. A standard error computed from the observed variability in the $\hat{\mu}_i$ will contain a contribution due to this "error of dosage." On the other hand the number of available degrees of freedom may be scanty.

### The Robbins-Monro process (1951)

This is a general process for estimating the value of X for which a monotonic function $M(X)$ of X attains some specified value $\alpha$ , in situations in which the available estimate $Y(X)$ of $M(X)$ is subject to error. The process has also been adapted by Kiefer and Wolfowitz (1952) to the estimation of the turning value of a regression function. In its application to the estimation of the ED50 the sequential rule may be stated as follows. At the $n$th step let $m_n$ subjects be tested at level $x_n$ and let the proportion dying be $p_n$. Then

$$(3) \qquad x_{n+1} = x_n - a_n(p_n - \tfrac{1}{2})$$

where the $a_n$ are a sequence of positive numbers such that $\Sigma a_n$ diverges and $\Sigma a_n^2$ converges. When the experiment is stopped at the Nth step, the estimate $\hat{\mu}_{RM}$ of the ED50 is simply the level $x_{N+1}$ at which the next test would have taken place.

From consideration of the asymptotic distribution of $x_{N+1}$, Hodges and Lehmann (1956) recommend that $a_n = c/n$, where $c$ is a constant. The rule then becomes

$$(4) \qquad x_{n+1} = x_n - \frac{c}{n}(p_n - \tfrac{1}{2}) \ .$$

Note that, unlike the Up and Down method, the steps become shorter as $n$ increases. The step size also depends on how close the observed proportion of responders $p_n$ is to $1/2$.

With this choice of the $a_n$ it has been shown that for any of the tolerance distributions likely to be encountered in practice, $\hat{\mu}_{RM}$ is asymptotically normally distributed with mean $\mu$. If $m$ subjects are tested at each step, the asymptotic variance after $N$ steps is

$$(5) \qquad V(\hat{\mu}_{RM}) = \frac{c^2}{4mN(2c\beta f_0 - 1)}$$

where $f_0 = f(0)$ is the ordinate of the tolerance distribution at the ED50. This result requires that $c > 1/2\beta f_0$. The value of $c$ that minimizes (5) is $c = 1/\beta f_0$. The minimum variance is

$$(6) \qquad V_{min}(\hat{\mu}_{RM}) = \frac{1}{4mN\beta^2 f_0^2} = \frac{\sigma^2}{4mN f_0^2} \ .$$

For the normal tolerance distribution $f_0 = 1/\sqrt{2\pi}$ and the minimizing value of $c = c_0 = \sqrt{2\pi}\,\sigma = 2.506\sigma$. (Note that the largest step size that can be taken, which happens when all subjects respond or fail to respond at the first test, is $\tfrac{1}{2}c_0$ or $1.25\sigma$). The minimum variance is $\pi\sigma^2/2mN$.

## The small-sample behavior of the

## Up and Down and RM processes

The results to be discussed here are for experiments with $mN = 12$, 24 or 48 subjects, involving not more than 12 steps. We supposed that the investigator who guesses that $\sigma$ lies between known values $\sigma_L$ and $\sigma_H$ will use as his estimate $\sigma_1 = \sqrt{\sigma_L \sigma_H}$ . Thus if $\sigma_H/\sigma_L = 2$ and if the true $\sigma$ actually lies between these limits, the estimate $\sigma_1$ will lie somewhere between $\sigma/\sqrt{2} = 0.71\sigma$ and $\sqrt{2}\sigma = 1.41\sigma$ . If $\sigma_H/\sigma_L = 4$, then $\sigma_1$ lies between $\sigma/2$ and $2\sigma$ .

For the Up and Down process, Brownlee, Hodges and Rosenblatt give recurrence relations by which the mean and mean square error (MSE) of their estimate $\hat{\mu}$, and hence of $\hat{\bar{\mu}}$, can be calculated. From these relations they computed values for $\sigma_1 = 2\sigma/3$, $\sigma$, $3\sigma/2$ and for a range of starting values $x_1$ . For the RM process the mean and MSE of $\hat{\mu}_{RM}$ were programmed for the IBM 704 (later for the 7090) for $\sigma_1 = \sigma/2$, $2\sigma/3$, $\sigma$, $3\sigma/2$, and $2\sigma$ . Both calculations assume a normal tolerance distribution. In both cases it is supposed that the investigator tries to use the recommended optimum step size. That is, he takes $d = \sigma_1$ for the Up and Down process and $c = \sqrt{2\pi}\sigma_1$ for the RM process.

The MSE of the estimate will be used to gauge the accuracy of the process. Our immediate interest is in the values of the MSE for different starting values and guesses at $\sigma$ . Except in minor details, the effects of different values of $x_1$ and $\sigma_1$ on the MSE were the same for the different combinations of m and N that were studied. Figures 2a (Up and Down) and 2b (RM) show the values of $MSE/\sigma^2$ for an experiment with 24 subjects having m = 3, N = 8 . Each figure shows the plot of $MSE/\sigma^2$ against $x_1$ for $\sigma_1 = 2\sigma/3$ , $\sigma$ and $3\sigma/2$ .

In every curve the MSE remains approximately constant for a range of starting values near $\mu$ . Thereafter it begins to rise, soon climbing very steeply. When $\sigma_1 = 2\sigma/3$, the range in which the MSE is constant is quite short, only out to about $0.7\sigma$ . As $\sigma_1$ increases, this 'stable range' also increases in length. On the other hand, if the start is near the ED50, short steps perform better than long steps. In fact for $\sigma_1 = 2\sigma/3$ the constant value of $MSE/\sigma^2$ is about 0.06 for both processes, slightly below the optimum asymptotic value of $\pi/48 = 0.065$.

In general, the RM process is somewhat more accurate than the Up and Down process. For instance, with $\sigma_1 = 3\sigma/2$ the stable value of $MSE/\sigma^2$ is about 0.078 for the RM process as against 0.089 for the Up and Down process. Further, $MSE/\sigma^2$ does not rise above 0.1 for the RM process until $x_1$ is out at $3.9\sigma$, as compared with $2.2\sigma$ for the Up and Down process. The differences in performance, however, are not in any sense sensational.

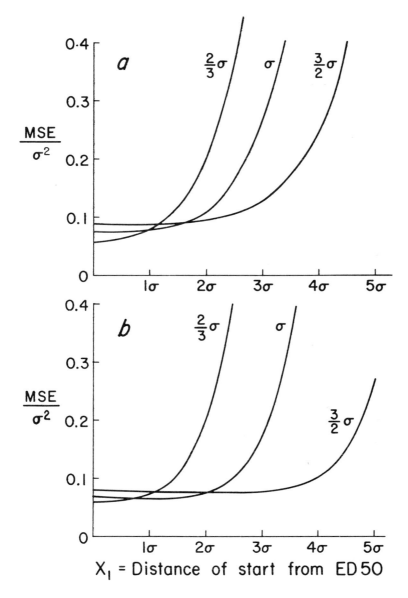

Fig. 2. MSE for Up and Down process *(a)*
and MSE for RM process *(b)*. m = 3, N = 8.

The stable ranges become larger if more steps are used for the same number of subjects, for instance $m = 2$, $N = 12$ as against $m = 3$, $N = 8$, or if $m$ is increased, for instance $m = 6$, $N = 8$ versus $m = 3$, $N = 8$ . But in both cases the increases are usually small, the size of the stable range being determined primarily by the ratio $\sigma_1/\sigma$ .

The explanation of the rapid rise in the MSE is an obvious one. When the starting value is distant from $\mu$, several steps are needed to bring the test levels near to the neighborhood of $\mu$ . By the time the experiment is terminated, the test levels still lie mostly on the same side of $\mu$ as the starting value, so that the estimate is biased. In fact if we start far enough away on the positive side, then with a limited number of steps all subjects are practically certain to respond at all test levels. Thus the variance of the estimate tends to zero and the MSE is composed entirely of bias.

To summarize from the experimenter's point of view, the original Up and Down and RM processes can scarcely be recommended unless either $\mu$ or $\sigma$ is well known in advance. If the starting guess is certain to lie within $\pm\sigma$ of $\mu$ and if $\sigma$ is known within a 2 to 1 ratio, the experimenter might choose $\sigma_1$ slightly higher than $\sqrt{\sigma_L \sigma_H}$ to guard against underestimating $\sigma$ by more than say 20% . This should keep him within the stable range. If $\sigma$ is well known but $\mu$ is not, taking $\sigma_1 = 3\sigma/2$ should keep him in the stable range for starts out to $3\sigma$ . There is, however, need for plans that are more robust than the original processes.

## Some variants

Numerous variants can be suggested in an attempt to make the MSE less dependent on a good initial guess at $\mu$ . For Up and Down runs with single subjects Brownlee, Hodges, and Rosenblatt considered omitting any initial run of levels at which the responses are all the same when computing $\hat{\mu}'$, the mean of the levels at which tests were conducted. For the RM process we investigated a 'delayed' version in which the factor $1/n$ is not introduced so long as $p_n$ remains at 0 or at 1 . We also examined two different choices for the $a_n$ factors, namely $a_n = 2c/(n+1)$ and $a_n = 3c/(n+2)$ . Both changes in the $a_n$ have the effect of slowing down the rate at which the step size is diminished as the experiment proceeds.

As an illustration of the effects of these variants of the RM process, Table 1 shows $MSE/\sigma^2$ for each variant for a range of starting values out to $5\sigma$ . The experiment has $m = 3$, $N = 8$ and $\sigma_1 = \sigma$ .

Table 1.  Values of $MSE/\sigma^2$ for variants of the RM process.

| Starting value $x_1$ | Original process | Delayed process | $2c/(n+1)$ | $3c/(n+2)$ |
|---|---|---|---|---|
| 0 | .070 | .070 | .083 | .104 |
| .62$\sigma$ | .067 | .068 | .083 | .104 |
| 1.25$\sigma$ | .065 | .068 | .082 | .104 |
| 1.88$\sigma$ | .070 | .079 | .083 | .104 |
| 2.51$\sigma$ | .102 | .099 | .085 | .105 |
| 3.13$\sigma$ | .206 | .113 | .096 | .106 |
| 3.76$\sigma$ | .516 | .129 | .132 | .112 |
| 4.39$\sigma$ | 1.242 | .140 | .241 | .123 |
| 5.01$\sigma$ | 2.697 | .159 | .570 | .211 |

The delayed process is the most successful of the three RM variants. It gives about the same results as the original process out to $x_1 = 2.5\sigma$ . Thereafter the $MSE/\sigma^2$ climbs, but much less drastically than with the original process. The two alternative choices of the $a_n$ also extend the stable range as anticipated, but have the disadvantage that they give in general less precision than the original process if $x_1$ happens to be within $\pm 2\sigma$ of $\mu$ .

Summarizing for the whole range of calculations that we have carried out with these variants, it appears that no process of this type is entirely satisfactory for the investigator whose initial $x_1$ may lie far from $\mu$ .

## A two-stage process

These considerations led us to examine a two-stage process. The first stage is conducted with single subjects and a step length of constant size. Its objective is to reach quickly a starting point for the second stage that is reasonably sure to be near the ED50 . The plan has much flexibility, particularly with respect to how the first stage is conducted and how much effort is put into it. For instance, the first stage might be conducted non-sequentially by a simultaneous test of a subject at each of a set of six to ten levels, from which a quick estimate of a starting point for the second stage is made. This is in fact what is commonly done when work is started with a relatively unknown agent.

The type of first stage plan on which we have worked is as follows. A single animal is tested at level $x_0$ . Thereafter the level is moved up or down by steps of length $1.25\sigma_1$ until a reversal in results is obtained. Stage 2 is started at the midpoint $x_1$ of the two levels at which the reversal occurs.

For given step size $1.25\sigma_1$ it turns out that the frequency distribution of $x_1$ depends very little on the original starting point $x_0$, no matter how far away $x_0$ is from $\mu$ . Since an animal started at or beyond $+4\sigma$ is practically certain to respond, it follows that $x_1$ has practically identical distributions for $x_0 = 2.75\sigma$, $4\sigma$, $5.25\sigma$, etc., the only difference being that different numbers of subjects are required to reach a reversal. The distribution of $x_1$ is not the same for $x_0 = 4\sigma$ as for $x_0 = 4.27\sigma$ (say) or $x_0 = 4.82\sigma$, but these distributions have closely similar means, shapes and MSE's . For example, when $\sigma_1 = \sigma$ and $x_0$ is far out, the mean of $x_1$ varies only between $0.11\sigma$ and $0.15\sigma$ and its MSE between $0.59\sigma^2$ and $0.62\sigma^2$

We have already seen that if $\sigma$ is underestimated, it is particularly important to ensure that the second stage RM or UP and Down process starts near $\mu$ . An advantage of the two-stage plan with steps that are some multiple of $\sigma_1$ is that if $\sigma_1$ is too small, the initial steps are short and the first stage gives a somewhat more precise estimate of $\mu$ than it does with long steps. Our choice of $1.25\sigma_1$ for the first stage step size is arbitrary, except that it is suggested by the RM process.

The average number $\bar{n}_1$ of subjects required to reach a reversal is not large. For starting points $x_0$ beyond $\pm 3.75\sigma$, the following equations give the averages approximately.

$$\sigma_1 = \sigma/2 \qquad \bar{n}_1 = 0.9 + 1.6 x_0/\sigma$$

$$\sigma_1 = \sigma \qquad \bar{n}_1 = 1.4 + 0.8 x_0/\sigma$$

$$\sigma_1 = 2\sigma \qquad \bar{n}_1 = 1.8 + 0.4 x_0/\sigma$$

The second stage starts at $x_1$ and consists either of the RM process with $c = \sqrt{2\pi}\ \sigma_1$ or the Up and Down process with step size $\sigma_1$ . Examination of the MSE's at the end of stage 2 showed that while generally satisfactory they were still undesirably inflated by a bad underestimate of $\sigma$, as when $\sigma_1 = \sigma/2$ . To improve this defect, we suggest that the reversal be checked by testing a second subject at each of the two levels. To show how this goes, suppose that $x < x'$ and that the first stage gives 0 at $x$ and 1 at $x'$ , so that $x_1 = (x + x')/2$ . If the two tests give $(0,0)$ at $x$ and $(1,1)$ at $x'$, or $(0,1)$ at $x$ and $(1,0)$ at $x'$, we keep $x_1$ unchanged. But we take $x_1 = x$ if the results are $(0,1)$ and $(1,1)$ , and $x_1 = x'$ if they are $(0,0)$ and $(0,1)$ . This change does not help if $\sigma_1 \geq \sigma$, but is useful insurance against an underestimate.

With this version of stage 1, Table 2 shows $MSE/\sigma^2$ for the estimate of $\mu$ at the end of stage 2 for experiments with 24 subjects requiring 8, 6 or 4 steps in the second stage. (Results for the Up and Down process were not available for $\sigma_1 = \sigma/2$ or $\sigma_1 = 2\sigma$ ).

Table 2.    $MSE/\sigma^2$ for a two-stage plan

| $\sigma_1 =$ | | $\sigma/2$ | $2\sigma/3$ | $\sigma$ | $3\sigma/2$ | $2\sigma$ |
|---|---|---|---|---|---|---|
| k | N | | Second stage:  RM process | | | |
| 3 | 8 | .092 | .070 | .068 | .079 | .094 |
| 4 | 6 | .099 | .070 | .067 | .079 | .096 |
| 6 | 4 | .115 | .075 | .067 | .082 | .101 |
| k | N | | Second stage:  Up and Down process | | | |
| 3 | 8 | --- | .071 | .073 | .089 | --- |
| 4 | 6 | --- | .080 | .070 | .088 | --- |
| 6 | 4 | --- | .090 | .069 | .128 | --- |

Considering the RM process first, the results are satisfactory if N = 6 or 8 steps are used in the second stage and if $\sigma_1$ lies between $2\sigma/3$ and $3\sigma/2$ . The MSE's obtained under these conditions all lie within 20% of the asymptotic value of 0.0655 for 24 subjects. For $\sigma_1 = \sigma/2$ or $\sigma = 2\sigma$ , the MSE's exceed the asymptotic value by 40% to 50% . If only four steps are used at stage 2, results are in general slightly worse than those for N = 6 .

Except when $\sigma_1 = \sigma$ , the Up and Down process does not perform as well as the RM process in the second stage. There is a hint that the Up and Down process is more sensitive to errors in estimating $\sigma$ , and perhaps requires more steps to appear at its best.

## Concluding notes

The extent to which plans of this type become used in practice will of course depend on whether there are situations in which experimenters find the plans convenient and economical of resources. The plans are simple to execute and the estimates of the ED50 are very easily computed. Moreover, as illustrated by the two-stage plan, they offer the prospect of giving estimates whose accuracy is not vitiated by poor initial guesses at the values of $\mu$ and $\sigma$ . Their chief limitation, at least for much experimental work, is that the stepwise method of doing the experiment may not fit the conditions of measurement of the responses. It is for this reason that we have concentrated on plans involving no more than 12 steps.

A second limitation is that we do not at present have a reliable method of computing the standard error of the estimate of $\mu$ . For the two-stage RM process one might contemplate using the asymptotic formula $\pi\sigma^2/2mN$ for the variance of $\hat{\mu}_{RM}$, increased by perhaps 25% in line with the results of Table 2. The trouble is that our estimate of $\sigma^2$ may be poor. For the Up and Down process the procedure already mentioned of calculating the standard error of $\hat{\mu}$ from the observed variability among the $\hat{\mu}_i$ may be used. The disadvantage is that this standard error does not allow for the bias in $\hat{\mu}$ and is rather weak in number of degrees of freedom.

An alternative approach is to analyze the results at the second stage, as if the experiment were non-sequential, by one of the classical methods that provides an estimate of the standard error. Although some preliminary work has been done, this may not be a practical solution except in special circumstances with larger experiments. The chief trouble is that, as previous workers have pointed out, these processes are not designed to give a good estimate of $\sigma$ , so that the estimated s.e. $(\hat{\mu})$, which depends on $\sigma$ , may be rather an unstable quantity. Further, the statistical analysis of the results will become more involved, since we will be using a more complex estimate of $\mu$ . One advantage of research of this type is that we may learn whether the simple estimates used in this paper for the two processes are fully efficient. Wetherill (1963) suggests that the "average level" estimate of $\mu$ for the Up and Down process may be inefficient unless the start is near $\mu$ . Similarly, with the RM process it is natural to ask whether use of the results at levels previous to $x_N$ may not improve the estimate of $\mu$ .

Since a very poor guess at $\sigma$ inflates the MSE even with the two stage process, the question has been asked whether it may be possible to make an estimate of $\sigma$ , say about half-way through the experiment, using this estimate to determine the step size in the last half. With small experiments this proposal is likely to be defeated by the inability of the process to furnish good estimates of $\sigma$ . It might turn out to be feasible in an experiment with 12 steps at the second stage and 96 subjects, although this has not yet been investigated.

The present investigation has assumed a normal tolerance distribution. It will be of considerable interest to discover how much the results are changed under a different tolerance distribution, particularly since both processes are essentially non-parametric. Some work on this question is underway by the second author.

For experiments in which it is feasible to test single subjects and use many steps, the paper by Wetherill (1963) should be consulted. Many forms of the Up and Down process are investigated in order to seek a plan of high efficiency even if $\mu$ and $\sigma$ are not well known.  His recommendation is that the original  Up and Down process be used, with step size  $\sigma_1$ ,  until <u>five</u> changes in response type have occurred.  Thereafter the step size is halved, continuing the process as long as is necessary to attain a desired accuracy in the estimate.  His investigations assume a logistic tolerance distribution and were made by experimental sampling.  In line with this approach, investigation of a longer stage 1 in our process might be fruitful.

For the investigator who could use a sequential plan, although he finds a standard non-sequential plan more convenient, it would be informative to compare these two-stage plans with an alternative in which our first stage is followed by one of the standard plans with, say, four or five fixed levels.  The only comparison in the literature that is partially relevant was made by Brownlee, Hodges and Rosenblatt.  They compared the original  Up and Down  process ( with estimate  $\hat{\mu}$  ) and their variant ( with estimate  $\hat{\mu}'$  )  with a five-level non-sequential plan, assuming that the asymptotic formula for the variance of the latter remains valid in small experiments.  Their results indicated that if the start  $x_1$  is near  $\mu$ ,  the non-sequential plan requires  30% more observations as compared with  $\hat{\mu}'$  and  50% more as compared with  $\hat{\mu}$  for equal accuracy.  For starts out beyond  $2\sigma$  the relative accuracy of the non-sequential plan declines rapidly.

## Summary

This paper discusses the performance of two sequential methods of estimating the median effective dose ( ED50)  in biological assay - the  Up and Down  method of Dixon and Mood and the stochastic approximation method of Robbins and Monro.  The study was made for experiments using from  12  to about  60  subjects, and was restricted to plans that require not more than  12  sequential steps.  This restriction should make the plans operationally feasible for experiments in which the response ( death or survival)  to the stimulus becomes known within an hour or less.  The performance of any plan was judged by the mean square error ( MSE)  of the resulting estimate of the ED50  .

In order to start any experiment for estimating the  ED50,  some initial guess is needed about the value of the  ED50  and about the standard deviation  $\sigma$  of the tolerance distribution.  Our calculations show that the original  Up and Down  or  Robbins-Monro  plans can be relied on to give satisfactory precision only if either the  ED50  or  $\sigma$  is well known in advance.

For situations in which this does not not hold we investigated a two-stage plan.  In the first stage, subjects are tested singly by the Up and Down  method until a reversal occurs.  As a check, another subject is tested at each of the two levels involved in the reversal.

The second stage is started at the estimate $x_1$ of the ED50 given
by the results at these two levels. The frequency distribution of $x_1$
turns out to be practically independent of the original guess $x_0$ at
the ED50 . Further, if at least six steps are used in the second
stage, the MSE of the final estimate with the RM process is close
to the asymptotic minimum provided that $\sigma$ is known within a 2 to 1
ratio (on a log scale) . Even if $\sigma$ is known initially only within a
4 to 1 ratio, use of the RM process in the second stage gives final
estimates whose MSE's are not more than 50% higher than the asymptotic minimum.

## REFERENCES

1.  Berkson, J., (1955). Maximum likelihood and minimum $\chi^2$
    estimates of the logistic function, Jour. Amer. Stat. Assoc.
    Vol. 50. pp. 130-161.

2.  Bliss, C. I. and Cattell, Mc K. (1943). Biological assay. An-
    nual Rev. Phys. 5, 489.

3.  Bross, I. (1950). Estimates of the LD50: a critique. Biomet-
    rics, 6, 413-423.

4.  Brownlee, K.I., Hodges, J. L., and Rosenblatt, M. (1953).
    The up-and-down method with small samples. Jour. Amer. Stat.
    Assoc. 48, 262-277.

5.  Dews, P. B., and Berkson, J. (1954). On the error of bioassay
    with quantal responses. "Statistics and Mathematics in Biology"
    The Iowa State College Press, Ames, Iowa.

6.  Dixon, W. J. and Mood, A. M. (1948). A method for obtaining
    and analyzing sensitivity data. Jour. Amer. Stat. Assoc. 43, 109-
    126.

7.  Finney, D. J. (1951). Statistical Method in Biological Assay.
    Hafner Publishing Company, New York.

8.  Gaddum, J. H. (1933). Methods of biological assay depending
    on a quantal response. British Med. Res. Council. Special
    Report No. 183.

9.  Hodges, J. L. and Lehmann, E. L. (1956). Two approximations
    to the Robbins-Monro process. Proc. Third Berkeley Symposium.
    Vol. I, 95-104. University of California Press.

10. Kiefer, J. and J. Wolfowitz. (1952). Stochastic estimation of the maximum of a regression function. Ann. Math. Stat., Vol. 23, pp. 462-466.

11. Robbins H. and Monro S. (1951). A stochastic approximation method. Ann. Math. Stat. 22, 400-407.

12. Wetherill, G. B. (1963). Sequential estimation of quantal response curves. Jour. Royal Stat. Soc. B, 25, 1-48.

―――――――――――――――――――

This work was assisted by Contract Nonr 1866(37) with the Office of Naval Research, Navy Department. Reproduction in whole or in part is permitted for any purpose of the United States Government. The calculations were done in part at the Computation Center at MIT, Cambridge, Massachusetts.

## DISCUSSION PARTICIPANTS

Dr. Walter T. Federer
   Biological Statistics, Biometrics Unit,
   Cornell University
   Ithaca, New York

Dr. Marvin A. Schneiderman
   Biometry Branch, National Cancer Institute,
   National Institutes of Health
   Bethesda, Maryland

Dr. Donald T. Searls
   Statistical Research Division, Research Triangle Institute
   Box 490, Durham, North Carolina

# DISCUSSION

Walter T. Federer.   What is your advice on the selection of $X_i$ to estimate the median when the active part of the curve is very steep?

In order to increase the precision of the estimated variance why not add two observations at the end of experimentation and locate these two observations at the points of inflection?  Also, it might be advisable to discard the first two observations of your preliminary trial for variance computations.

William G. Cochran.   I'm not sure that I grasp the particular difficulty caused by a very steep curve ( small $\sigma$ ), unless it means that our initial guess $x_1$ is likely to be very far out, or that there is difficulty in giving the exact dose level that the sequential technique calls for.   If $x_1$ is expected to be a poor guess, an approach like that suggested by Dr. Searls may be worthwhile for the first stage.

We have had in mind Dr. Federer's suggestion about additional observations whose purpose is to improve the estimate of s. e. ($\hat{\mu}$) , but have no results on the best number and placement of these observations.

Donald T. Searls.   To avoid wasting a lot of animals for a bad choice of starting point, why not double the distance between levels until a reversal is reached.  At this point the distance between levels could be halved until the original distance between levels was attained.

Example:

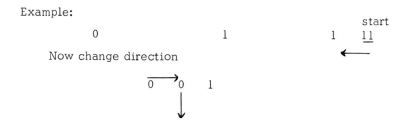

Now change direction

William G. Cochran:   For the first stage of our two-stage
process, this method is certainly an improvement over the use of a
fixed step size when the value of $\sigma$ is underestimated and we start
far out. To cite one calculation, suppose that our guessed value,
$\sigma_1$, is actually $\sigma/2$ and that the start is at $+10\sigma$ . Dr. Searls's
method requires about 9.5 animals to complete his first stage, as
against about 17 with a fixed step size of $1.253\sigma_1$ ( $0.625\sigma$ ) . Also,
somewhat to my surprise, his method gives a smaller MSE in this
case for the estimate $\hat{\mu}$ made at the end of stage 1. With better initial guesses at $\sigma$ and $\mu$ there is less to be gained. With $\sigma_1 = \sigma$
and a start at $+4\sigma$ , for instance, the fixed size takes on the average
only about 4.5 animals for stage 1.

Marvin Schneiderman.   Often we are not interested in the
LD50 -- but rather something like the LD05 (in a therapeutic trial).
What advice have you for handling this problem? Fixed schemes?
Sequential schemes?

William G. Cochran.   Estimation of LD05 is a challenging
problem on which I know of no thorough study. Wetherill (loc. cit.)
made some empirical sampling trials to estimate LD75 and LD95 by
the Robbins-Monro process. He was very disappointed with the results, the estimate being subject to a substantial bias that appeared
to be an inevitable feature of the method. I think his conclusions may
be too pessimistic. His sample sizes were only 15, 25 and 35 .
Since one cannot expect good precision in this problem without large
sample sizes, I would like to see some work done on modifications
of the RM or Up and Down process with big samples.

# COMPARISON OF TWO METHODS OF HANDLING COVARIATES IN DISCRIMINATORY ANALYSIS*

William G. Cochran

## 1. Introduction and summary

The method for constructing a discriminant function from a group of variates, in order to test the difference between the means of two populations or to obtain an index for classifying specimens into one or the other population, is well known. In certain problems of this type, there may be one or more variates, called covariates, that are known to have the same means in the two populations. For instance, in comparing two methods of training for some complex task, initial measurements might be taken of the aptitudes of the subjects for this task and perhaps of their I.Q.'s. The subjects are then divided into two groups by randomization, one group receiving training method A, the other training method B. At the end of the training period, further measures of their abilities in the task, which will be called discriminators, are made. The randomization ensures that the initial aptitudes and I.Q.'s are the same in the two populations. Nevertheless, if these initial measurements are correlated with the discriminators, they may increase the power of the test of significance of the difference in performance given by the two methods.

A technique for handling such covariates in discriminatory analysis was given by Cochran and Bliss [1]. In this approach, which is a natural extension of the ordinary analysis of covariance, the multiple regression of each discriminator on the covariates is first computed from the pooled within-population sums of squares and products. The discriminant function is then constructed from the deviations of the discriminators from these multiple regressions. The procedure leads to a test of significance that is a form of Hotelling's $T^2$ test.

A question that is sometimes asked is : what happens if the covariates are simply included in the discriminant function in exactly the same way as the discriminators ? So far as computing time is concerned, there is

---

* Work done under Contract Nonr 1866 (37) with the Office of Naval Research, U. S. Navy Department.

---

little to choose between this method and the Cochran-Bliss technique. However, the alternative avoids the necessity of learning the special technique for handling the covariates and may be more convenient in routine work. Indeed, Kendall [2] has suggested that the covariates might as well be retained in the discriminant, speculating that little will be lost by this procedure. The present paper compares the two methods of handling covariates.

The following is a summary of the results. For tests of significance the covariance technique is, as expected, more powerful. The two techniques were compared numerically by computing the ratio of the sample size needed with the ordinary technique to that needed with the covariance technique, in order to attain powers in the region 0.5 to 0.9. For a single covariate, this ratio is less than 1.1 only if there are 4 or more discriminators. With 2 covariates the ratio does not drop below 1.1 until there are 8 or more discriminators, and with 3 covariates until there are more than 10 discriminators. For classifying specimens into one of two populations, on the other hand, it appears that the gain from the covariance technique will be trivial provided that the discriminant function is constructed from a reasonably large sample (e.g. at least 100 from each population).

## 2. Numerical example

An example may clarify the difference between the two approaches. The data come from a completely randomized experiment on leprosy, comparing a number of drugs thought to be effective against leprosy bacilli. The results shown here are for one of these drugs $D$ and for an ineffective placebo $P$ (actually, a mild vitamin pill). The covariate $x$ is a total bacteriological score taken over six fixed sites on the body of each patient, while the discriminator $y$ is the same score after a period of daily treatments. A high score indicates the presence of many leprosy bacilli. Table 1 shows the analysis of sums of squares and products and the drug means.

Table 1.  Sums of squares and products

|              | d.f. | $y^2$  | $yx$   | $x^2$  |
|--------------|------|--------|--------|--------|
| Between drugs | 1   | 194.4  | 44.4   | 10.1   |
| Within drugs  | 106 | 4394.0 | 2735.6 | 3166.4 |
| Total         | 107 | 4588.4 | 2780.0 | 3176.5 |

Drug means

|            | Number of patients | $\bar{y}$ | $\bar{x}$ |
|------------|--------------------|-----------|-----------|
| $D$        | $n_1=53$           | 7.2075    | 10.6226   |
| $P$        | $n_2=55$           | 9.8909    | 11.2364   |
| Difference |                    | 2.6834    | 0.6138    |

In this case the Cochran-Bliss technique for testing the difference between the drug means reduces to the usual analysis of covariance. The "within-drugs" regression of $y$ on $x$ is $b=2735.6/3166.4=0.86395$, and the mean square of deviations from the regression is

$$s_{y \cdot x}^2 = \frac{1}{105}\left[ 4394.0 - \frac{(2735.6)^2}{3166.4} \right] = 19.339 \;.$$

If $d_y$, $d_x$ are the mean differences between drugs $P$ and $D$ in $y$ and $x$, respectively, the $F$-test of the difference between the adjusted drug means may be written

$$F = \frac{(d_y - b d_x)^2}{s_{y \cdot x}^2 \left\{ \dfrac{1}{n_1} + \dfrac{1}{n_2} + \dfrac{d_x^2}{W_{xx}} \right\}} \;,$$

where $W_{xx}$ is the within-drugs sum of squares for $x$. Thus,

$$F = \frac{(2.1531)^2}{(19.339)\left\{ \dfrac{1}{53} + \dfrac{1}{55} + \dfrac{(.6138)^2}{3166.4} \right\}} = 6.45$$

with 1 and 105 d.f. The probability level is about 0.013.

If there had been two covariates $x_1$ and $x_2$ it may be noted, for future reference, that the value of $F$ would be

(1)
$$F = \frac{(d_y - b_1 d_{x_1} - b_2 d_{x_2})^2}{s_{y \cdot 12}^2 \left\{ \dfrac{1}{n_1} + \dfrac{1}{n_2} + W^{11} d_{x_1}^2 + 2 W^{12} d_{x_1} d_{x_2} + W^{22} d_{x_2}^2 \right\}}$$

with 1 and 104 d.f., where $W^{ij}$ is the inverse of the matrix $W_{ij}$ of within-drug sums of squares and products for $x_1$ and $x_2$.

If, alternatively, the covariate $x$ is treated exactly like the discriminator $y$, we seek a linear function $L_1 y + L_2 x$ that maximizes the ratio of the sum of squares between drugs to that within drugs. As is well-known, the coefficients $L_i$ are any multiples of the solution of the equations

$$L_1 W_{yy} + L_2 W_{yx} = d_y : \quad 4394.0 L_1 + 2735.6 L_2 = 2.6834$$
$$L_1 W_{yx} + L_2 W_{xx} = d_x : \quad 2735.6 L_1 + 3166.4 L_2 = 0.6138 \;.$$

If the multiple is chosen to make $L_1 = 1$ we find $L_2 = -0.68113$, so that the discriminant function is $y - 0.68113x$. The analysis of variance (table 2) gives an $F$-ratio of 3.41, which by Hotelling's $T^2$ test carries 2 and 105 d.f. The corresponding probability is about 0.04, somewhat higher

Table 2. Analysis of variance of $y - 0.68113x$

| | d.f. | Sums of squares | Mean squares | F |
|---|---|---|---|---|
| Between drugs | 2 | 138.60 | 69.30 | |
| | | | | 3.41 |
| Within drugs | 105 | 2136.42 | 20.35 | |

than the figure 0.013 given by the covariance approach. Another method of computing the $F$ value is to use the fact that the ratio of the sum of squares between drugs to that within drugs for the discriminant function is

$$( 2 )\qquad \frac{n_1 n_2}{(n_1 + n_2)} (W^{yy} d_y^2 + 2 W^{yx} d_y d_x + W^{xx} d_x^2) ,$$

where $W^{yy}$ is the inverse of $W_{yy}$.

## 3. Theory for tests of significance

Suppose that there are $p$ discriminators and $k$ covariates. If the distinction between the two types of variates is ignored, the assumption is that we have a group of $(p+k)$ variates $y_{iuv}$ where $i = 1, 2, \cdots, (p+k)$ denotes the variate, $u = 1, 2$ denotes the population and $v$ denotes the order within the sample. The variates are assumed to follow a multivariate normal distribution with means 0 in the first population and $\delta_i$ in the second, and with the same covariance matrix $\sigma_{ij}$ in both populations. The sample sizes from the two populations are $n_1$ and $n_2$.

If $d_i = \bar{y}_{i2.} - \bar{y}_{i1.}$, the quantities $\sqrt{g}\, d_i$ follow a multivariate normal with means $\sqrt{g}\, \delta_i$ and covariance matrix $\sigma_{ij}$, where $g = n_1 n_2/(n_1 + n_2)$. Further, the within-population sums of squares and products $W_{ij}$ follow a Wishart distribution with d.f. $= n_1 + n_2 - 2 = f$ (say) and with covariance matrix $\sigma_{ij}$. The $d_i$ and the $W_{ij}$ are independent. Hence by the standard theorem leading to Hotelling's $T^2$ test [3], the quantity

$$g \sum_{i, j=1}^{p+k} W^{ij} d_i d_j$$

is distributed as $(p+k)F/(f-p-k+1)$, where $F$ has $(p+k)$ and $(f-p-k+1)$ degrees of freedom and parameter of noncentrality

$$( 3 )\qquad \lambda = g \sum_{i, j=1}^{p+k} \sigma^{ij} \delta_i \delta_j = g \Delta^2 ,$$

where $\sigma^{ij}$ is the inverse of $\sigma_{ij}$ and $\Delta^2$ is Mahalanobis generalized distance between the two populations. This is the generalization of the test in expression (2) in the numerical example.

In the covariance approach let $y_{iuv}$ $(i = 1, 2, \cdots, p)$ denote the discriminators while $x_{\xi uv}$ $(\xi = 1, 2, \cdots, k)$ denote the covariates. The basic assumptions are the same, with the addition that we know that all the $\delta_\xi$ are zero. Detailed proofs for this approach are given in [1] : the following is a sketch of the argument.

Since the generalized distance is invariant under linear transformation, make the transformation

$$y'_{iuv} = y_{iuv} - \sum_{\xi=1}^{k} \beta_{i\xi} x_{\xi uv} ; \quad x'_{iuv} = x_{iuv} ,$$

where $y'_{iuv}$ is the deviation of $y_{iuv}$ from its within-population regression on the $x_{\xi uv}$. In terms of the new variates the covariance matrix becomes

$$\begin{pmatrix} \sigma_{ij.\xi} & 0 \\ 0 & \sigma_{\xi r} \end{pmatrix},$$

where $\sigma_{ij.\xi}$ is the $p \times p$ covariance matrix of the $y'_{iuv}$ and $\sigma_{\xi r}$ is the $k \times k$ covariance matrix of the $x_{\xi uv}$.

Hence the generalized distance in expression (3) may be written

$$\Delta^2 = \sum_{i,j=1}^{p} \sigma^{ij.\xi} \delta_i \delta_j$$

using the fact that every $\delta_\xi = 0$.

The remainder of the argument is conditional on the $x_{iuv}$ remaining fixed from sample to sample. Let $b_{i\xi}$ be the usual regression estimates of the $\beta_{i\xi}$, computed from the within-population sums of squares and products. Thus

$$b_{i\xi} = \sum_{r=1}^{k} W_{ir} W^{\xi r} .$$

Then it can be shown that, conditionally, the quantities

$$d'_i = d_i - \sum_{i=1}^{k} b_{i\xi} d_\xi \qquad (i = 1, 2, \cdots, p)$$

have a multivariate normal distribution with means $\delta_i$ and covariance matrix

$$\sigma_{ij.\xi} \left\{ \frac{1}{n_1} + \frac{1}{n_2} + \sum_{\xi, r=1}^{k} W^{\xi r} d_\xi d_r \right\} = \frac{\sigma_{ij.\xi}}{g'} \quad \text{(say)} .$$

It follows that the quantities $\sqrt{g'} d'_i$ are multivariate normal with means $\sqrt{g'} \delta_i$ and covariance matrix $\sigma_{ij.\xi}$.

Now let the quantities $W_{ij.\xi}$ be the within-population sums of squares and products of the deviations of the $y_{iuv}$ from their estimated regressions on the $x_{\xi uv}$. Computationally,

$$W_{ij.\xi} = W_{ij} - \sum_{\xi=1}^{k} b_{i\xi} W_{j\xi} \, .$$

The $W_{ij.\xi}$ follow a Wishart distribution with $(f-k)$ d.f. and covariance matrix $\sigma_{ij.\xi}$, independently of the $d_i'$.

Hence, finally, by a further application of the theory for Hotelling's $T^2$ test, the quantity

$$g' \sum_{i,j=1}^{p} W^{ij.\xi} d_i' d_j'$$

is distributed as $pF/(f-p-k+1)$, with $p$ and $(f-p-k+1)$ degrees of freedom and parameter of non-centrality

$$\lambda' = g' \sum_{i,j=1}^{p} \sigma^{ij.\xi} \delta_i \delta_j = g' \Delta^2 \, .$$

This result is the general case of the covariance test (1) given in the numerical example.

Thus the ordinary and the covariance approaches both lead to $F$-tests. The $F$-distributions differ in two respects. The numerator has $(p+k)$ degrees of freedom in the ordinary approach as against $p$ in the covariance approach. The parameter of non-centrality in the ordinary approach is

$$(4) \qquad\qquad\qquad \lambda = \frac{n_1 n_2}{(n_1+n_2)} \Delta^2$$

and in the covariance approach is

$$(5) \qquad\qquad \lambda' = g' \Delta^2 = \frac{\dfrac{n_1 n_2}{(n_1+n_2)} \Delta^2}{1 + \dfrac{n_1 n_2}{n_1+n_2} \sum W^{\xi\tau} d_\xi d_\tau} \, .$$

## 4.  Numerical comparison of the two power functions

An immediate problem is that we have to compare an unconditional distribution in the ordinary method with a conditional distribution in the covariance method, in which the parameter of non-centrality in expression (5) depends on the set of $x$'s that appear in the sample. Probably the most satisfactory solution would be to find the unconditional power function for the covariance method, this being a mixture of $F$-distributions with different non-centrality parameters, but this distribution does not seem to be expressible in tractable form.

As an approximation we will use an $F$-distribution that has the average parameter of non-centrality taken over the multivariate normal distribution of the $x_{\xi u v}$. Applying Hotelling's $T^2$ test once again, the quantity

$$\frac{n_1 n_2}{(n_1 + n_2)} \sum_{\xi, \gamma = 1}^{k} W^{\xi \gamma} d_\xi d_\gamma$$

in (5) is distributed as $kF/(f - k + 1)$, with degrees of freedom $k$ and $(f - k + 1)$, where $F$ is now central since all the $\delta_\xi$ are zero. By integration over the $F$-distribution the average value of the non-centrality parameter in (5) is found to be

(6)
$$\bar{\lambda}' = \frac{n_1 n_2}{(n_1 + n_2)} \Delta^2 (f - k + 1)/(f + 1) ,$$

where $f$ is the number of 'within population' degrees of freedom in the data.

The non-central $F$-distribution is still incompletely tabulated. For tests at the 5% and 1% levels, Lehmer [4] has tabulated the values of the non-centrality parameter $\lambda$ needed to give probabilities 0.7 and 0.8 of obtaining a significant result. Charts by Hartley and Pearson [5] and Fox [6] provide values of $\lambda$ for a wider range of powers, though with less accuracy. Our initial approach was as follows. For a given number of discriminators $p$ and covariates $k$, and given denominator degrees of freedom $(f - p - k + 1)$, the values of $\Delta_o^2$ in (4) and $\Delta_c^2$ in (6) needed to attain a specified power $(1 - \beta)$ are obtained. The value of $\Delta_o^2$ was always greater than that of $\Delta_c^2$. Since $\lambda$ in (4) is of dimension unity in $n_1$ and $n_2$, the ratio $\Delta_o^2 / \Delta_c^2$ represents approximately the ratio of the sample size needed with the ordinary technique to that needed with the covariance technique in order to attain the same power. This comparison is slightly unfair to the ordinary technique, because if this technique were applied with a larger sample size there would be more denominator degrees of freedom in $F$. This would increase the power, but the effect is small if $(f - p - k + 1)$ is at all substantial.

Table 3. Ratio of sample sizes needed to obtain specified power $(p = k = 1)$

| Significance level | Power | Denominator | d.f. | | |
|---|---|---|---|---|---|
| | | 20 | 30 | 60 | $\infty$ |
| 5% | 0.8 | 1.236 | 1.234 | 1.230 | 1.228 |
| | 0.7 | 1.256 | 1.251 | 1.249 | 1.248 |
| 1% | 0.8 | 1.211 | 1.202 | 1.194 | 1.186 |
| | 0.7 | 1.224 | 1.216 | 1.208 | 1.201 |

With one discriminator and one covariate, table 3 shows the ratio of sample sizes for powers of 0.7, 0.8 and denominator d.f.=20, 30, 60, ∞, which are obtained conveniently for Lehmer's table.

Three points are noteworthy in table 3: (i) the ratio of sample sizes increases as the power decreases. This increase persists down to the point at which the power equals the significance probability. (ii) the ratio is slightly greater for 5% tests than for 1% tests. (iii) the ratio depends very little on the denominator d.f., although in general it decreases slightly as the denominator d.f. increase. These results held uniformly for all values of $p$, $k$ and $f$ investigated.

Table 4 shows these ratios for a series of values of $p$ and $k$. Since the ratios are almost independent of the denominator d.f., the results in table 4 are given for an infinite number of denominator d.f. Strictly, the results should be presented for several different values of the power. It was thought, however, that the region of most practical interest in tests of significance is that in which the power lies between 0.5 and 0.9. Consequently, the results are the average of the ratios for $(1-\beta)=0.5$, 0.6, 0.7, 0.8, and 0.9. Readers interested in lower values of the power should note that the ratios will be higher than those in table 4.

Table 4. Ratio of sample size needed for the ordinary technique to that for the covariance technique, in order to obtain equal power

| Number of covariates | Number of discriminators | | | | | | | |
|---|---|---|---|---|---|---|---|---|
| | 1 | 2 | 3 | 4 | 5 | 6 | 8 | 10 |
| | Test of significance at 5% level | | | | | | | |
| 1 | 1.26 | 1.14 | 1.10 | 1.08 | 1.07 | 1.06 | 1.04 | 1.04 |
| 2 | 1.42 | 1.26 | 1.19 | 1.15 | 1.13 | 1.11 | 1.09 | 1.07 |
| 3 | 1.58 | 1.36 | 1.27 | 1.21 | 1.18 | 1.16 | 1.12 | 1.10 |
| | Test of significance at 1% level | | | | | | | |
| 1 | 1.20 | 1.12 | 1.09 | 1.07 | 1.06 | 1.05 | 1.04 | 1.03 |
| 2 | 1.35 | 1.22 | 1.17 | 1.13 | 1.11 | 1.10 | 1.07 | 1.07 |
| 3 | 1.47 | 1.31 | 1.23 | 1.19 | 1.16 | 1.14 | 1.11 | 1.10 |

With a single covariate and a single discriminator the sample size must be 26% larger for tests at the 5% level and 20% larger for tests at the 1% level if the covariate is treated like a discriminator instead of being handled by the usual analysis of covariance. If, however, there are 4 or more discriminators and 1 covariate, the gain in precision with the covariance technique becomes quite small. With 2 covariates, the increase in sample size using the ordinary discriminant function technique does not drop below 10% until there are at least 8 discriminators for 5% level tests, and with 3 covariates more than 10 discriminators are needed.

## 5.  Comparison in classification problems

In problems in which new specimens are to be classified into one of the two populations, covariates are likely to appear only infrequently, because there are not many situations in which the investigator can be certain that the means are the same in the two populations. Consider the case in which the covariates are treated like discriminators. We first recapitulate some standard theory. If the necessary parameters were known, the ideal discriminant index for a new specimen having measurements $y_i''$ would be

$$I = \sum_{i=1}^{p+k} L_i y_i'', \qquad \text{where} \quad L_i = \sum_{j=1}^{p+k} \sigma^{ij} \delta_j .$$

This is normally distributed with mean zero for a specimen from the first population and mean

$$(7) \qquad \sum_{i=1}^{p+k} L_i \delta_i = \sum_{i,j=1}^{p+k} \sigma^{ij} \delta_i \delta_j = \varDelta^2$$

in the second population. Its variance is also $\varDelta^2$. Consequently, if a specimen is assigned to the second population whenever $I > \varDelta^2/2$, the probability of misclassification is the normal integral from $\varDelta/2$ to infinity.

When the discriminant is estimated from a sample the index used for classification may be written

$$(8) \qquad \hat{I} = \sum_{i=1}^{p+k} \hat{L}_i \left\{ y_i'' - \frac{1}{2} \left( \bar{y}_{i1.} + \bar{y}_{i2.} \right) \right\} , \quad \text{where} \quad \hat{L}_i = \sum_{j=1}^{p+k} s^{ij} d_j ,$$

a specimen being assigned to the second population when $\hat{I} > 0$.

For given $n_1$, $n_2$, $f$ and number of discriminators, an asymptotic expansion for the probability of misclassification when $\hat{I}$ is used has recently been given by Okamoto [7]. In this expansion the leading term is the normal integral and correction terms in $1/n_1$, $1/n_2$, $1/n_1^2$ and $1/n_2^2$ are added.

It is not clear that the same theory can be applied to the covariance approach. However, an underestimate of the probability of misclassification with this technique can be obtained. In the notation of section 3, suppose that the true regression $\beta_{i\xi}$ of the $y_{iuv}$ on the $x_{\xi uv}$ were known. Then a classification rule analogous to (8) could be constructed from the $p$ variates

$$y_{iuv}' = y_{iuv} - \sum_{\xi=1}^{k} \beta_{i\xi} x_{\xi uv}$$

with

$$\hat{L}_i = \sum_{j=1}^{p} s^{ij.\xi} \{d_j - \sum_{\xi} \beta_{j\xi} d_\xi\} .$$

The probability of misclassification with this rule is given by Okamoto's expansion, noting that there are now $p$ discriminators and that the within-population d.f. are now $(f-k)$.

The fact that the $\beta_{i\xi}$ must be estimated in the covariance approach will presumably increase its probability of misclassification, as will the reduction in d.f. from $f$ to $(f-k)$. Consequently, an overestimate of the superiority of the covariance approach over the ordinary approach is obtained if we compare the probabilities of misclassification with $p$ versus $(p+k)$ discriminators for fixed $n_1$, $n_2$, $f$ and $\Delta$.

For four values of $\Delta$ and $n_1 = n_2 = 100$, $f = 198$, table 5 shows these probabilities for $p = 1$ to 10, computed from [7].

Table 5. Probabilities of misclassification for samples of size 100

| Number of discriminators | $\Delta$ 1 | 2 | 3 | 4 |
|---|---|---|---|---|
| * | .3085 | .1587 | .0668 | .0228 |
| 1 | .3090 | .1593 | .0673 | .0230 |
| 2 | .3112 | .1604 | .0680 | .0234 |
| 3 | .3133 | .1616 | .0687 | .0237 |
| 4 | .3154 | .1628 | .0694 | .0240 |
| 5 | .3175 | .1640 | .0701 | .0244 |
| 6 | .3194 | .1652 | .0708 | .0247 |
| 7 | .3214 | .1664 | .0715 | .0251 |
| 8 | .3232 | .1676 | .0722 | .0254 |
| 9 | .3251 | .1688 | .0729 | .0258 |
| 10 | .3269 | .1700 | .0737 | .0262 |

* Results for the normal integral (all parameters known)

For this size of sample it is clear that an increase by 1, 2 or 3 in the number of discriminators involved causes only a minor increase in the probability of misclassification. With samples of size 50 the increases in the probability as $p$ increases are about double those in table 5. Even in this case the effect of a small increase in $p$ is unlikely to be important for practical purposes.

HARVARD UNIVERSITY

## REFERENCES

[1] W. G. Cochran and C. I. Bliss, "Discriminant functions with covariance," *Ann. Math. Stat.*, 19 (1948), 151-176.

[2] M. G. Kendall, *A course in multivariate analysis*, Charles Griffin, 1957.

[ 3 ] T. W. Anderson, *Introduction to multivariate analysis*, John Wiley and Sons, 1957.

[ 4 ] E. Lehmer, "Inverse tables of probabilities of the second kind," *Ann. Math. Stat.*, 15 (1944), 388-398.

[ 5 ] E. S. Pearson and H. O. Hartley, "Charts of the power function for analysis of variance tests," *Biometrika*, 38 (1951), 112-130.

[ 6 ] M. Fox, "Charts of the power of the *F*-test," *Ann. Math. Stat.*, 27 (1956), 484-497.

[ 7 ] M. Okamoto, "An asymptotic expansion for the distribution of the linear discriminant function," *Ann. Math. Stat.*, 34 (1963), 1286-1301.

# 84

## The Robbins–Monro Method for estimating the Median Lethal Dose

By W. G. Cochran   and   M. Davis

*Harvard University*     *The Johns Hopkins University*

(Received September 1964. Revised October 1964)

### Summary

The Robbins–Monro method provides a sequential plan that gives an attractively simple estimate of the $LD_{50}$ in bioassay and is practicable for agents in which the response (death or survival) is known within an hour. By machine computations, its performance is examined for experiments using 50 or less animals. If the experiment's initial guess at the $LD_{50}$ is within $2\sigma$ of the true $LD_{50}$, a slight variant of the plan (the delayed version) can be recommended. For new agents in which the experimenter's initial knowledge of the $LD_{50}$ and of $\sigma$ may be poor, a two-stage plan is recommended, the first stage using single animals at equally spaced doses, the second stage being an ordinary Robbins–Monro plan. In a preliminary comparison with a standard 4-level non-sequential experiment, the sequential plan gave mean-square errors about 20 per cent lower if $\sigma$ and the $LD_{50}$ are guessed accurately, and much larger gains in accuracy if the $LD_{50}$ is guessed poorly. The problem of attaching a standard error to the Robbins–Monro estimate of the $LD_{50}$ is discussed.

## 1. Introduction

This paper presents the results of some investigations of the performance of the Robbins–Monro (1951) stochastic approximation process, when used in sequential experiments to estimate the median lethal dose ($LD_{50}$). To start the experiment, an initial guess $x_1$ is made at the $LD_{50}$, and $m_1$ animals are given the dose $x_1$. If $k_1$ animals die and $p_1 = k_1/m_1$ is the proportion dead, a second group of $m_2$ animals is tested at the dose level

$$x_2 = x_1 - a_1(p_1 - \tfrac{1}{2}).$$

More generally, the dose level at which the $(n+1)$th group of animals is tested is found from $x_n$ by the equation

$$x_{n+1} = x_n - a_n(p_n - \tfrac{1}{2}),$$

where the $a_n$ are a suitably chosen set of constants. In most of our work we have taken $a_n = c/n$, as recommended by Hodges and Lehmann (1956). When the experimenter decides to terminate the experiment, his estimate $\hat{\mu}$ of the $LD_{50}$ is the dose level at which the next test would have been conducted.

The potential advantages of the method are that fewer animals may be required to estimate the $LD_{50}$ with specified precision than by the usual non-sequential methods and that the simple estimate $\hat{\mu}$ is available at any point at which the investigator decides to terminate the experiment. The practical usefulness of the sequential approach is likely to be confined to situations in which the response to the stimulus becomes known quickly. If the fate of every tested animal is clear in less than an hour, experiments involving 8 steps of the process can be completed in a day. If experiments on a number of agents are to be carried out, agents with responses slower than an hour can be handled without undue delay in obtaining results by the device of starting several experiments at the same time.

The asymptotic properties (*n* large) of the Robbins–Monro process, which can be adapted to estimate any percentile of a cumulative distribution, have attracted much attention. Schmetterer (1961) gives an excellent review. Our interest in the problem arose from questions raised by Dr Denis F. Hawkins, who was trying out the method for measuring the potencies of a group of chemicals on mice. He conducted a number of experiments in order to study the practical performance of the method (Hawkins, 1964). The particular questions that form the subject of this paper are as follows.

(1) Do the asymptotic properties hold in experiments involving less than 50 animals?

(2) To what extent does the precision of the estimate depend on having a good initial guess $x_1$ at the $LD_{50}$?

(3) Asymptotic theory shows that the best choice of *c*, in the factor $c/n$, depends on the standard deviation $\sigma$ of the distribution of just fatal doses (sometimes called the tolerance distribution). Thus an initial guess at $\sigma$ is required in order to conduct the experiment. How sensitive are the results to errors in this initial guess?

(4) With a given number of animals, is there a loss of precision if several animals are tested at each step, thus reducing the number of steps and the time required to complete the experiment? For instance, with 24 animals one might have 24 steps, with 1 animal tested at each step, or 12 steps with 2 animals each, or 8 steps with 3 animals each, and so on.

(5) To put it more generally, can the Robbins–Monro procedure be adapted to give reliable estimates of the $LD_{50}$ even if the initial guesses at $\mu$ and $\sigma$ are very poor?

(6) How does the accuracy of the method compare with that of the standard non-sequential method?

In several respects the present investigation parallels that of Wetherill (1963), who discussed questions (1), (2) and (3) above, as well as numerous topics not included in this paper. Comparison of our results with Wetherill's will be made later. Some of our preliminary results have already been reported (Cochran and Davis, 1963, 1964).

## 2. SOME RESULTS FROM THEORY

As background, some results will be quoted from the theoretical study of the process

$$x_{n+1} = x_n - \frac{c}{n}(p_n - \tfrac{1}{2}). \tag{2.1}$$

We suppose that a constant number *m* of animals is tested at each step and that the proportion dead, $p_n = k_n/m$, is binomially distributed about $P(x_n)$, where

$$P(x_n) = \int_{-\infty}^{x_n} f(y)\,dy. \tag{2.2}$$

The function $f(y)$ is often called the tolerance distribution or the distribution of just fatal doses, since $f(y)\,dy$ is the proportion of animals that a dose between *y* and $(y+dy)$ is just sufficient to kill. The function $P(x)$ is the dosage–response curve.

2

Many results have been obtained relating to the asymptotic distribution of $x_n$ when the number of steps $n$ is large. Those most relevant to this paper are as follows.

(i) The estimate $\hat{\mu}$ becomes normally distributed about $\mu$ with variance

$$c^2/4mn(2cf-1),$$

where $f$ is the ordinate of the tolerance distribution at its median. This result requires $c > 1/2f$.

(ii) It follows that, asymptotically, the best step size is $c = 1/f$, the minimum variance being $1/4mnf^2$. This result suggests that the variance depends only on the total number of animals used, $mn$, and not on $m$ and $n$ individually.

(iii) For the normal tolerance distribution, $f = 1/\sigma\sqrt{(2\pi)}$, so that the optimum step constant $c = \sigma\sqrt{(2\pi)} = 2\cdot506\sigma$. The corresponding variance is $\pi\sigma^2/2mn$. Users of the standard probit method will recognize $2mn/\pi\sigma^2$ as the Fisherian amount of information about $\mu$ supplied by $mn$ tests made exactly at the $LD_{50}$.

(iv) The optimum step constant $c = \sigma\sqrt{(2\pi)}$ requires an initial guess $\sigma_g$ at the value of $\sigma$. If $\sigma_g = r\sigma$ and the guessed optimum step size is used, the effect is to multiply the minimum limiting variance by the factor $r^2/(2r-1)$. This factor reaches $\frac{9}{8}$ when $r = \frac{3}{2}$ or $r = \frac{3}{4}$, and $\frac{4}{3}$ when $r = 2$, suggesting that moderate errors in the step constant are not disastrous.

Expressions for the bias, variance and mean-square error of $\hat{\mu}$ for *finite n* were obtained by Hodges and Lehmann (1956), under the assumption that $P(x)$ can be represented adequately by a linear function of $x$. For fixed $n$, their results show that the mean-square error of $\hat{\mu}$ and the ratio of the bias term to the variance term both increase steadily as the starting point $x_1$ moves away from $\hat{\mu}$. Thus with a poor starting point in an experiment of moderate size, the mean-square error is likely to be large and to consist mainly of bias. Wetherill (1963) considered a more realistic approximation of the linear type and showed that a strictly linear approximation would be likely to underestimate the bias.

To summarize, both the asymptotic results and those of Hodges and Lehmann provide hints about the answers to some of our questions in experiments of practical size, but since it was not clear how far either set of results can be trusted, the first two moments of $\hat{\mu}$ for a panel of values of $m$, $n$, $c$ and $x_1$ were programmed on electronic machines.

### 3. Initial Calculations

In the subsequent calculations the tolerance distribution is assumed normal, with $x = \log$ dose. The values of $m$ (number of animals per step) and $n$ (number of steps) for which calculations were done are shown in Table 3.1. For 6 animals the

TABLE 3.1

*Values of m and n in initial calculations*

| No. of animals | (m, n) combinations |
|---|---|
| 6 | (1, 6) (2, 3) |
| 12 | (1, 12) (2, 6) (3, 4) (4, 3) |
| 24 | (1, 24) (2, 12) (3, 8) (4, 6) (6, 4) |
| 48 | (6, 8) (8, 6) (12, 4) |

distributions of $\hat{\mu}$ were obtained on a desk machine. With larger numbers the mean and mean-square error of $\hat{\mu}$ were programmed on the IBM 709.

Decisions were required about the range of values of $x_1$ (initial guess at $\mu$) and of the step factor $c$ to be covered. In practice the nearness of $x_1$ to $\mu$ varies considerably, depending on the amount of experience that the investigator has had with the drug and animal. Since the computer can easily cover a wide range, computations were made for $x_1$ going from 0 to over $8\sigma$ by jumps of about $0.2\sigma$.

It is more difficult to specify the accuracy with which the investigator can guess $\sigma$. Some information on the values of $\sigma$ that are found in practice is obtainable from reviews by Gaddum (1933, 1953) and Bliss and Cattell (1943). They report estimates $s$ of $\sigma$ from 45 experiments in which the criterion was death or survival. When $x = \log$ dose, the quantity $\sigma$ is dimensionless: assuming a normal curve, $\sigma$ is the log of the ratio of the dose that kills 84·1 per cent to the dose that kills 50 per cent. This makes it possible to present estimates of $\sigma$ for different animals and drugs in the same table. As pointed out by Gaddum, Bliss and Cattell, the value of $\sigma$ obtained in an experiment will depend not only on the type of animal and drug, but also on the experimental design and on the genetic constitution, weight, age and diet of the animals. In the data in Table 3.2 the experimental animals were mostly mice, rats, frogs and cats, although snails and dogs are also represented.

TABLE 3.2

*Frequency distribution of estimates s from
45 experiments with death as the criterion*

| Limits for $s$ | 0–0·05 | 0·05–0·1 | 0·1–0·15 | 0·15–0·2 | 0·2–0·3 | 0·3–0·4 | Over 0·4 |
|---|---|---|---|---|---|---|---|
| Frequency | 2 | 13 | 11 | 8 | 3 | 5 | 3 |

About 70 per cent of the values of $s$ lie between 0·05 and 0·2, a 4 to 1 ratio, although a few extend as high as 0·5. Moreover, the true amount of spread among the $\sigma$'s is overestimated in Table 3.2, since the estimates $s$ are subject to sizeable experimental errors. For instance, in 10 parallel experiments with 35 mice each, reported by Irwin and Cheeseman (1939), the computed values of $s$ ranged from 0·22 to 0·51, although a test of significance showed no sign of real differences among the underlying $\sigma$'s. (Only the average $s$ for these data is included in Table 3.2.)

This analysis suggests that, in deciding on the step size, a worker who has had previous experience with his stock of experimental animals and with the class of drugs under investigation might be able to guess his $\sigma$ in advance to within a 2 to 1 ratio. In a less familiar situation we assumed that the worker can state a lower limit $\sigma_L$ and an upper limit $\sigma_U$ for which $\sigma_U / \sigma_L \leqslant 4$. If these limits are valid and he takes

$$\sigma_g = \surd(\sigma_U \sigma_L)$$

as his guessed value, his step constant $c_g$ will not be in error by a factor of more than 2. To cover this range, our calculations were made for $c/c_0 = \frac{1}{2}, \frac{2}{3}, 1, \frac{3}{2}$, and 2, where $c_0 = \sigma \surd(2\pi)$ is the asymptotically optimum step constant.

These ranges for $x_1$ and $c$ are considerably wider than those used by Wetherill (1963), who took $x_1$ out to $0.83\sigma$ and $c/c_0 = \frac{5}{8}, 1$ and $\frac{11}{8}$. In the application mentioned by Wetherill—the failure of plastic water pipe under impact—the tolerance distribution may be better known initially than is typical in biological assay.

   With $m$ animals per step and $n$ steps, the total number of sequential paths starting from $x_1$ is $(m+1)^n$, which soon mounts appallingly. A method of computing the moments of the estimator $x_{n+1}$ was developed for which the arithmetical labour is proportional to $n(m+1)$. Consider a path in which $r_i$ animals die out of $m$ at the $i$th step $(i = 1, 2, ..., n)$. The successive doses $x_2, x_3, ..., x_n$ at which these tests are made and the resultant estimate $x_{n+1}$ are of course determined by the formula

$$x_{i+1} = x_i - \frac{c}{i}\left(\frac{r_i}{m} - \frac{1}{2}\right).$$

Further, if

$$B(r, x) = \binom{m}{r}\{P(x)\}^r\{1 - P(x)\}^{m-r} \tag{3.1}$$

is the binomial term, the probability of this path is

$$\prod_{i=1}^{n} B(r_i, x_i).$$

Hence the $j$th moment of the estimator $x_{n+1}$ is given by

$$\sum_{r_1=0}^{m}\sum_{r_2=0}^{m}\cdots\sum_{r_n=0}^{m} x_{n+1}^j \prod_{i=1}^{n} B(r_i, x_i). \tag{3.2}$$

This probability can be computed by the following recursive relation. Define

$$\mu_{j,n+1}(x_{n+1}) = x_{n+1}^j, \tag{3.3}$$

$$\mu_{j,i}(x_i) = \sum_{r_i=0}^{m} B(r_i, x_i)\mu_{j\,i+1}\left\{x_i - \frac{c}{i}\left(\frac{r_i}{m} - \frac{1}{2}\right)\right\} \quad (i = 1, 2, ..., n). \tag{3.4}$$

Note that in this recursion the running symbol $i$ works downwards from $n$ to 1. It is not difficult to verify that $\mu_{j,1}(x_1)$ is the $j$th moment of the estimator $x_{n+1}$ as given in formula (3.2).

   With $c_0 = 2\cdot506$ and $\sigma$ taken as 1, a panel of starting values $x_1$ was chosen at intervals of $c_0/12$ from 0 to 16. The programme began by computing and storing the binomial terms in (3.1) and the values of $x$ and $x^2$ for this chosen panel. Using (3.3) and (3.4), the programme then computed for the panel the successive $\mu_{j,i}(x_i)$, down to $\mu_{1,1}(x_1)$ and $\mu_{2,1}(x_1)$, which give the first two moments for a sequence starting at $x_1$. The factor

$$\mu_{j,i+1}\left\{x_i - \frac{c}{i}\left(\frac{r_i}{m} - \frac{1}{2}\right)\right\}$$

in (3.4) required interpolation from the panel of values of $\mu_{j,i+1}(x)$. A seven-point Lagrange formula was used. Clearly this programme can also handle plans in which the number of animals $m$ varies from step to step and will give moments higher than the second if desired.

## 4. RESULTS

   Except in minor details, the effects of different values of $x_1$, $c$ and $m$ on the mean-square error of the estimate, $\text{MSE}(\hat{\mu})$, were the same for $mn = 12, 24$ or $48$

animals. To illustrate the effects of $x_1$ and $c$, Fig. 1 shows $MSE(\hat{\mu})/\sigma^2$ plotted against $(x_1 - \mu)/\sigma$ for the 5 values of $c$ in an experiment with 1 animal per step and 12 steps ($m = 1, n = 12$).

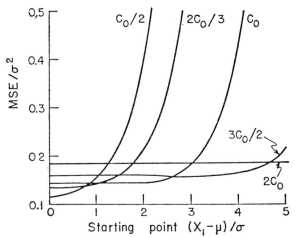

FIG. 1. MSE of the Robbins–Monro estimate plotted against the starting point for 5 different step sizes ($m = 1$, $n = 12$).

For each value of $c$ there is a range of starting values within which the mean-square error remains nearly constant (e.g. out to about $2\cdot4\sigma$ when $c = c_0$). With more distant starting values the MSE rises very rapidly. This rise occurs at the point at which the bias term begins to dominate the MSE. The term "stable range" will denote the range of starting values within which the MSE is approximately constant. The stable range is very short for $c = \frac{1}{2}c_0$ but increases steadily as $c$ increases, extending out to $6\sigma$ when $c$ is twice the asymptotic optimum. The MSE within the stable range also increases steadily with $c$, being below or close to the asymptotic minimum ($0\cdot131$) for $c = \frac{1}{2}c_0$ and $\frac{2}{3}c_0$, but about 43 per cent higher than the asymptotic minimum for $c = 2c_0$.

For an experiment with 24 animals and $c = c_0$, Fig. 2 illustrates the effect of testing 6 and 3 animals per step as compared with 1 animal per step, in order to cut down the number of steps. An increase in $m$ for given $mn$ diminishes the size of the stable range; within the stable range, however, the MSEs are actually slightly smaller when $m$ is increased.

Alternatively, we may examine the effect of an increase in $m$ when $n$ is fixed, e.g. in comparing $m = 6$, $n = 8$ with $m = 3$, $n = 8$. Doubling the value of $m$ in this way reduces the variance of the estimate to almost exactly one-half, but produces only a slight decrease in the bias. Consequently, if $V$, $B$ and $M$ are the variance, bias and MSE when $m = 3, n = 8$ and the start is at $x_1$, the corresponding MSE for $m = 6, n = 8$ may be predicted by the expression $(B^2 + V/2) = (B^2 + M)/2$. This expression over-estimates the MSE, but at most by only a few per cent.

In order to present the results of the calculations for 12 and 24 animals in a compact form, Table 4.1 shows the stable ranges and the MSEs within the stable ranges. For this purpose the stable range, for any given $m$, $n$ and $c$, was taken as the range of starting values within which the MSE is not more than 20 per cent higher than the MSE when the starting value is at the $LD_{50}$. Although the choice of 20 per cent is arbitrary, the results in Table 4.1 change only in minor details if this figure is made 10 per cent or 30 per cent.

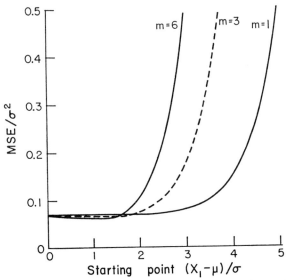

FIG. 2. MSE of the Robbins–Monro estimate plotted against the starting point for $m = 6$, $n = 4$; $m = 3$, $n = 8$; $m = 1$, $n = 24$.

TABLE 4.1

*Values of the stable range (R) and mean-square error (MSE)*
*within the stable range, for $\sigma = 1$*

| Step constant c | $\frac{1}{2}c_0$ | | $\frac{2}{3}c_0$ | | $c_0$ | | $\frac{3}{2}c_0$ | | $2c_0$ | |
|---|---|---|---|---|---|---|---|---|---|---|
| m, n | R | MSE | R | MSE | R | MSE | R | MSE | R | MSE |
| 1, 12 | 0·6 | 0·122 | 1·4 | 0·141 | 2·8 | 0·146 | 4·7 | 0·161 | 6·7 | 0·188 |
| 2, 6 | 0·6 | 0·095 | 1·0 | 0·117 | 2·3 | 0·137 | 4·0 | 0·163 | 5·5 | 0·195 |
| 3, 4 | 0·4 | 0·078 | 0·9 | 0·107 | 2·2 | 0·133 | 3·6 | 0·165 | 5·0 | 0·201 |
| 4, 3 | 0·4 | 0·073 | 0·8 | 0·100 | 2·0 | 0·129 | 3·3 | 0·168 | 4·6 | 0·210 |
| Asy† | | ∞ | | 0·174 | | 0·131 | | 0·147 | | 0·174 |
| 2, 12 | 0·6 | 0·060 | 1·3 | 0·069 | 2·5 | 0·069 | 4·4 | 0·077 | 6·4 | 0·092 |
| 3, 8 | 0·5 | 0·052 | 0·9 | 0·062 | 2·1 | 0·068 | 3·8 | 0·078 | 5·9 | 0·094 |
| 4, 6 | 0·4 | 0·046 | 0·9 | 0·059 | 2·0 | 0·067 | 3·6 | 0·079 | 5·2 | 0·096 |
| 6, 4 | 0·3 | 0·040 | 0·7 | 0·053 | 1·8 | 0·066 | 3·2 | 0·081 | 5·0 | 0·100 |
| Asy† | | ∞ | | 0·087 | | 0·066 | | 0·074 | | 0·087 |

† Asymptotic value of MSE, $\pi r^2 \sigma^2 / 2mn(2r-1)$, where $r = c/c_0$.

The results confirm the indications received from Figs. 1 and 2. The principal conclusions are as follows.

(1) The penalty from underestimation of $\sigma$ (resulting in a short step size) is that the stable range is very restricted, being around $0.8\sigma$ for $c = \frac{2}{3}c_0$ and about $0.5\sigma$ or less for $c = \frac{1}{2}c_0$. On the other hand, if the starting value is within these ranges, the MSE is pleasantly small. If $\sigma$ is overestimated, the stable range is wide-out to over $3\sigma$ for $c = \frac{3}{2}c_0$ and to around $5\sigma$ for $c = 2c_0$, but the stable values of the MSE are from 10 per cent to 20 per cent higher when $c = \frac{3}{2}c_0$ and from 30 per cent to 40 per cent higher when $c = 2c_0$, as compared with those for $c = c_0$.

(2) For a given number of animals, a decrease in the number of steps $(n)$ diminishes the stable range. For starting values that lie within the stable range, a decrease in $n$ gives a smaller MSE when $c \leqslant c_0$, and only a moderate increase when $c > c_0$.

Results for 48 animals (not shown) were in general similar, the MSEs being roughly half of those for 24 animals.

Wetherill (1963) conducted a similar investigation of the MSE of the Robbins–Monro process, except that he assumed a logistic tolerance distribution with 1 animal per step, and used experimental sampling rather than actual determination of the mean and variance of $\hat{\mu}$. He concluded (p. 37) that the process is very efficient for estimation of the $LD_{50}$ and is very robust to errors in the starting value and also to the value of the constant $c$. His results are not in contradiction with ours. His most extreme starting value represents about $0.9\sigma$ on our scale, while his $c$ values correspond roughly to our values $\frac{2}{3}c_0$, $c_0$ and $\frac{3}{2}c_0$. With these values of $c$, Table 4.1 shows that the stable range equals or exceeds $0.9$ except in the cases $m = 4, n = 3$ and $m = 6, n = 4$. Moreover, for a given number of animals, the MSE within the stable range does not vary greatly as $c$ and $m$ vary within these limits. Our own more extensive calculations show, however, that the Robbins–Monro process cannot be described as robust if the starting value and $c$ lie substantially beyond these limits.

For the experimenter the practical indications from this Section appear to be as follows. Suppose first that $\sigma$ is known to within a 2 to 1 ratio, and the guessed $\sigma$ is taken as the geometric mean of the two limits. Then the $c/c_0$ values will lie between $0.707$ and $1.414$. If the starting value is within $\sigma$ of the $LD_{50}$, the Robbins–Monro process with at least 6 steps can be recommended. From Table 4.1 the standard error of the $LD_{50}$ should be about $0.38\sigma$ with 12 animals and $0.27\sigma$ with 24 animals; with 48 animals the standard error is about $0.20\sigma$. If the starting value is known only to within $2\sigma$ of the $LD_{50}$, a safer procedure is to use a value of $\sigma$ close to the upper limit. This guarantees that $c$ will not be much less than $c_0$. The penalty is that $c$ may be almost twice $c_0$. In that event the standard error of $\hat{\mu}$ is increased slightly to about $0.40\sigma$ with 12 animals, $0.31\sigma$ with 24 animals and $0.22\sigma$ with 48 animals.

In less familiar situations, with $\sigma$ known only to within a 4 to 1 ratio, the process gives results within the stable range only if the starting value is within $0.6\sigma$, as is unlikely to be the case.

## 5. THE DELAYED PROCESS AND KESTEN'S PROCESS

If the starting value is poorly known, underestimation of the step size is more damaging than overestimation, because the experiment may terminate before $x_{n+1}$ has reached the neighbourhood of the $LD_{50}$, resulting in a large MSE due to bias. As the experiment proceeds, however, the experimenter is likely to receive warning that he faces this situation by obtaining a succession of results in which all animals survive or all animals die. Hawkins (1964) has illustrated this type of result by running an experiment in which $c$ was deliberately chosen as one-third of his best estimate of $c_0$.

Several variations in the Robbins–Monro process have been suggested in the hope of avoiding these large biases. Perhaps the simplest, which we have called the delayed Robbins–Monro process, is to postpone the introduction of the factor $1/n$ so long as all animals die or all animals survive. With 4 animals per step, if the deaths in the first 4 steps are 4, 4, 4, 3, the step constant is kept at $c$ when determining $x_2$, $x_3$ and $x_4$ from the formula, but changes to $\frac{1}{2}c$ when determining $x_5$, to $\frac{1}{3}c$ when determining $x_6$, and so on. In a second alternative, suggested by Kesten (1958), the step constant is $c$ and $\frac{1}{2}c$ when finding $x_2$ and $x_3$ respectively. Thereafter the step constant is reduced, when computing $x_{i+1}$, only if $(x_i - x_{i-1})$ and $(x_{i-1} - x_{i-2})$ are of opposite sign, suggesting that the test doses are straddling the $LD_{50}$. For instance, if 4, 3, 3 or 4, 3, 2 deaths out of 4 animals occur in the first 3 tests, the step size in calculating $x_4$ remains at $\frac{1}{2}c$, since $(x_2 - x_1)$ is negative and $(x_3 - x_2)$ is either negative or zero. Another possibility, which we have not investigated, would be to increase the step constant $c$ if the results of the first few trials suggest that the doses are all too low or too high.

TABLE 5.1

*Values of $MSE/\sigma^2$ for the ordinary Robbins–Monro process, the delayed Robbins–Monro process and Kesten's process*

| Start $(x_1 - \mu)/\sigma$ | $c = \frac{1}{2}c_0$ | | | $c = c_0$ | | | $c = 2c_0$ | | |
|---|---|---|---|---|---|---|---|---|---|
| | Ord. | Del. | Kes. | Ord. | Del. | Kes. | Ord. | Del. | Kes. |
| *12 steps with 1 animal per step* | | | | | | | | | |
| 0 | 0·116 | 0·136 | 0·106 | 0·142 | 0·147 | 0·159 | 0·184 | 0·184 | 0·263 |
| 1 | 0·151 | 0·186 | 0·135 | 0·142 | 0·156 | 0·153 | 0·185 | 0·186 | 0·251 |
| 2 | 0·375 | 0·265 | 0·237 | 0·143 | 0·173 | 0·167 | 0·185 | 0·190 | 0·244 |
| 3 | 1·402 | 0·313 | 0·390 | 0·177 | 0·187 | 0·199 | 0·184 | 0·197 | 0·256 |
| 4 | 4·278 | 0·360 | 0·639 | 0·418 | 0·204 | 0·240 | 0·183 | 0·205 | 0·273 |
| *6 steps with 2 animals per step* | | | | | | | | | |
| 0 | 0·091 | 0·096 | 0·088 | 0·141 | 0·143 | 0·145 | 0·195 | 0·195 | 0·226 |
| 1 | 0·151 | 0·168 | 0·153 | 0·131 | 0·143 | 0·126 | 0·192 | 0·193 | 0·203 |
| 2 | 0·592 | 0·353 | 0·445 | 0·139 | 0·176 | 0·161 | 0·190 | 0·198 | 0·193 |
| 3 | 2·294 | 0·511 | 1·115 | 0·296 | 0·229 | 0·266 | 0·190 | 0·212 | 0·211 |
| 4 | 6·094 | 0·633 | 1·153 | 1·153 | 0·284 | 0·435 | 0·189 | 0·237 | 0·247 |

Table 5.1 compares the MSEs given by the ordinary, delayed and Kesten processes for experiments with $m = 1$, $n = 12$ and $m = 2$, $n = 6$. Looking first at $c = \frac{1}{2}c_0$, we see that the delayed and Kesten processes are effective in avoiding the great inflation of the MSE that occurs with the ordinary process when the start is out at $3\sigma$ or $4\sigma$. On the other hand, the MSE still increases with both methods as $x_1$ moves away from $\mu$. At $x_1 - \mu = 3\sigma$, the MSE in the delayed process is 2·3 times the MSE at $x_1 = \mu$ when $m = 1$, $n = 12$, and 5·4 times the MSE at $x_1 = \mu$ when $m = 2$, $n = 6$. With the Kesten process, the corresponding ratios are 3·7 and 12·7.

With $c = c_0$ both processes give results similar to those of the ordinary process except at $x_1 - \mu = 4\sigma$, when both are definitely superior to the ordinary process. When $c = 2c_0$, the delayed process becomes almost identical with the original process when the start is within $6\sigma$ of $\mu$. The Kesten process has in general a higher MSE in this case. This happens because the Kesten process sometimes fails to reduce the step size when the last test level is actually near $\mu$, that is, when a reduction would have been beneficial.

To summarize, the delayed and Kesten variants improve the robustness of the Robbins–Monro process over the range from $c = \frac{1}{2}c_0$ to $c = c_0$. In particular, the delayed version can be recommended if the experimenter is confident that his initial guess at $\mu$ is not more than $2\sigma$ away from the true $LD_{50}$.

### 6. Preliminary Trials in a Two-stage Process

With an unfamiliar agent or stock of animals, a common laboratory practice is to try first a series of doses on single animals, in order to get a rough idea of the location of the $LD_{50}$. This practice suggested that we investigate a two-stage experiment in which a fairly crude preliminary trial is followed by the ordinary Robbins–Monro process. The first stage may be either non-sequential or sequential.

With a non-sequential first stage it is supposed that the dose levels are equally spaced on a log scale and that enough levels are tested so that the two lowest doses both give survivals and the two highest doses give deaths. Step sizes of $\sigma_g$ (the guessed value of $\sigma$) and $1.253\sigma_g$ were tried. Since the gain in precision from the small step size more than offset the larger number of animals required, only the results for a step size $\sigma_g$ will be presented. If the dose levels are approximately centred about $\mu$, the number of animals required averages only slightly over 5 when $\sigma_g = 2\sigma$ and about 8 when $\sigma_g = \frac{1}{2}\sigma$. In practice, more will usually be needed since the investigator may not be lucky in having the series of doses centred about $\mu$.

From the results of this preliminary series, the $LD_{50}$ was estimated by the simple Spearman–Kärber method. With doses spaced $\sigma_g$ units apart on the log scale, the estimate is

$$\hat{\mu}_{sk} = x_h + \frac{1}{2}\sigma_g - T\sigma_g,$$

where $x_h$ is the highest dose level used and $T$ is the total number of dead animals. Although the frequency distribution of $\hat{\mu}_{sk}$ depends on the exact positioning of the levels, the MSE varies only slightly with this positioning. A useful feature of the preliminary trials is that a small value of $\sigma_g$ produces a more precise estimate of $\mu$ than a large value. In other words, underestimation of $\sigma$ in the preliminary trials gives more precise estimates $\hat{\mu}_{sk}$ where they will be most needed in the subsequent Robbins–Monro process. The MSEs of $\hat{\mu}_{sk}$ for different step sizes are shown in Table 6.1.

TABLE 6.1

*Values of MSE $(\hat{\mu}_{sk})/\sigma^2$*

| Step size $(\sigma_g)$ | $\frac{1}{2}\sigma$ | $\frac{2}{3}\sigma$ | $\sigma$ | $\frac{3}{2}\sigma$ | $2\sigma$ |
|---|---|---|---|---|---|
| $MSE/\sigma^2$ | 0.282 | 0.376 | 0.564 | 0.849 | 1.178 |

As an illustration of the actual frequency distribution of $\hat{\mu}_{sk}$, the positive half of the distribution, when $\sigma_g = \frac{1}{2}\sigma$ and the levels are centred about $\mu$, is as follows:

| Values of $(\mu_{sk} - \mu)$ | $0.25\sigma$ | $0.75\sigma$ | $1.25\sigma$ | $1.75\sigma$ |
|---|---|---|---|---|
| Probability | 0.3350 | 0.1400 | 0.0235 | 0.0015 |

The distribution is of course symmetrical about $\mu$ in this case. Note that the probability is 0·95 that $\hat{\mu}_{\mathrm{sk}}$ does not differ from $\mu$ by more than 0·75$\sigma$. Thus there is a high probability that the estimate will lie within or fairly close to the stable range of the Robbins–Monro process when $c = \frac{1}{2}c_0$.

Alternatively, the preliminary stage may be conducted sequentially. A simple method is to test a single animal at the guessed value of $\mu$, moving thereafter by steps of size 1·253$\sigma_g$ until a reversal is obtained; e.g. until we get 1, 1, 1, 0 or 0, 0, 1. The estimate $\hat{\mu}_1$ might be taken as the mean of the 2 levels at which the reversal occurs. It was found, however, that this estimate is often poor, since with a decreasing series of doses the first 0 may occur when the lowest level tested is still above $\mu$. A worthwhile precautionary step is to test another animal at each of the levels at which the reversal occurs. If these 2 levels are $x_i$ and $x_{i+1}$, the 4 possible sets of results from the 4 animals that have been tested at these levels are as follows:

*Set*

|         | 1   |   | 2   |   | 3   |   | 4   |   |
|---------|-----|---|-----|---|-----|---|-----|---|
| $x_i$     | 1   | 1 | 1   | 0 | 1   | 0 | 1   | 1 |
| $x_{i+1}$ | 0   | 0 | 0   | 1 | 0   | 0 | 0   | 1 |

If the results shown in either Set 1 or Set 2 are obtained, the estimate $\hat{\mu}_1$ is unchanged at $\frac{1}{2}(x_i + x_{i+1})$. If Set 3 is the result, we take $\hat{\mu}_1 = x_i$, and if Set 4, $\hat{\mu}_1 = x_{i+1}$. Table 6.2 presents the MSEs given by this method.

TABLE 6.2

*Values of MSE $(\hat{\mu}_1)/\sigma^2$ for step size 1·253$\sigma_g$*

| $\sigma_g$ | $\frac{1}{2}\sigma$ | $\frac{2}{3}\sigma$ | $\sigma$ | $\frac{3}{2}\sigma$ | $2\sigma$ |
|------------|---------|---------|---------|---------|---------|
| $MSE/\sigma^2$ | 0·573 | 0·537 | 0·546 | 0·630 | 0·651 |

Comparison with Table 6.1 shows that this method is more precise than the non-sequential method when $\sigma_g > \sigma$, but less precise when $\sigma_g < \sigma$. To improve the precision when $\sigma_g < \sigma$, it would be necessary to add levels below that in which the first 0 appears; in other words, to make the method more similar to the non-sequential one.

The MSEs given in Tables 6.1 and Tables 6.2 are upper limits that apply when the starting value is very far from the LD$_{50}$. Both sets of calculations were made with starting values so far out on the positive side that death was certain. If the experimenter has a starting value that happens to be close to $\mu$, his MSEs in the sequential method will be lower than those shown in Table 6.2. With the non-sequential method, as already mentioned, the starting value, about which the experimenter will presumably centre his levels, makes little difference to the MSE although it does affect the number of animals required to produce survivals at the two lowest doses and deaths at the two highest.

## 7. RESULTS AT THE END OF THE SECOND STAGE

The second stage is an ordinary Robbins–Monro process, starting at $\hat{\mu}_{\mathrm{sk}}$ or $\hat{\mu}_1$ with step constant $c = 2·506\sigma_g$. Since the frequency distributions of $\hat{\mu}_{\mathrm{sk}}$ and $\hat{\mu}_1$ are known, and since the MSE given by a Robbins–Monro process with any specified starting value is obtainable from the machine calculations, the MSEs at the end of the second stage are easily computed. These MSEs appear in Table 7.1 for experiments with 24 and 48 animals in the second stage and with 8, 6 and 4 steps respectively.

TABLE 7.1

*Values of MSE $(\hat{\mu})/\sigma^2$ at the end of Stage 2*

| $m$ | $n$ | $\frac{1}{2}\sigma$ | $\frac{2}{3}\sigma$ | $\sigma$ | $\frac{3}{2}\sigma$ | $2\sigma$ |
|---|---|---|---|---|---|---|
| **24 animals in Stage 2** | | | | | | |
| *Stage 1. Non-sequential* | | | | | | |
| 3 | 8 | 0·064 | 0·065 | 0·068 | 0·078 | 0·094 |
| 4 | 6 | 0·063 | 0·064 | 0·067 | 0·079 | 0·096 |
| 6 | 4 | 0·065 | 0·064 | 0·066 | 0·082 | 0·100 |
| *Stage 1. Sequential* | | | | | | |
| 3 | 8 | 0·092 | 0·070 | 0·068 | 0·079 | 0·094 |
| 4 | 6 | 0·099 | 0·070 | 0·067 | 0·079 | 0·096 |
| 6 | 4 | 0·115 | 0·075 | 0·067 | 0·082 | 0·101 |
| **48 animals in Stage 2** | | | | | | |
| *Stage 1. Non-sequential* | | | | | | |
| 6 | 8 | 0·040 | 0·035 | 0·033 | 0·039 | 0·047 |
| 8 | 6 | 0·042 | 0·036 | 0·033 | 0·039 | 0·048 |
| 12 | 4 | 0·047 | 0·040 | 0·034 | 0·041 | 0·050 |
| *Stage 1. Sequential* | | | | | | |
| 6 | 8 | 0·071 | 0·040 | 0·033 | 0·039 | 0·047 |
| 8 | 6 | 0·076 | 0·044 | 0·034 | 0·039 | 0·048 |
| 12 | 4 | 0·098 | 0·052 | 0·034 | 0·041 | 0·050 |

From the asymptotic formula the corresponding optimum MSEs are $\pi\sigma^2/48 = 0·0654\sigma^2$ (24 animals) and $\pi\sigma^2/96 = 0·0327\sigma^2$ (48 animals).

With a non-sequential first stage the final MSE is almost constant and is close to the asymptotic minimum when $\sigma_g$ lies between $\frac{1}{2}\sigma$ and $\sigma$. When $\sigma_g$ is an overestimate, the MSE increases, being about 18 per cent higher for $\sigma_g = \frac{3}{2}\sigma$ and 40 per cent higher for $\sigma_g = 2\sigma$ if 6 or 8 steps are used, and slightly higher still if only 4 steps are used. As mentioned previously the asymptotic formula predicts increases of 12 per cent and 33 per cent respectively.

With a sequential first stage, the results are almost identical with the above when $\sigma_g$ exceeds $\sigma$ but are poorer when $\sigma_g$ is less than $\sigma$. This is particularly so with 48 animals and $\sigma_g = \frac{1}{2}\sigma$, when the bias still contributes materially to the MSE. In this case, in fact, the MSEs with 48 animals are well over half those with 24 animals. It seems advisable to use the non-sequential version of the first stage unless there are other reasons to the contrary.

To summarize, the two-stage plan, with at least 6 steps in the second stage, estimates the $LD_{50}$ with a standard error (strictly a root-mean-square error) that is at most 20 per cent higher than the asymptotic minimum if the guessed value of $\sigma$ lies between $\frac{1}{2}\sigma$ and $2\sigma$. The 20 per cent value is reached only when $\sigma_g$ is near $2\sigma$.

## 8. COMPARISON OF SEQUENTIAL AND FIXED-LEVEL EXPERIMENTS

Except for investigators who find the Robbins–Monro method especially convenient or are attracted by the simplicity of the estimate, the principal advantage of the method over the standard non-sequential experiment with fixed levels lies in the claim that it gives more precise estimates. A comparison of the precision of sequential and non-sequential experiments is therefore highly relevant. Setting up a comparison that would be regarded as fair by the proponents of both methods, however, appears to require a good deal of preliminary work. The main difficulty is that practically all the available information on the precision of fixed-level trials is based on asymptotic theory. Similarly, the only extensive discussion known to us of the best number and spacings of levels, that of Finney (1964), uses asymptotic formulae.

The present comparison, which is by no means definitive, is confined to an experiment with 24 subjects. The fixed-level plan uses 4 levels with 6 subjects at each level. It might be argued that 3 levels with 8 subjects per level would be preferable, but Finney's calculations (p. 496) suggest that, if $\mu$ and $\sigma$ are well-known in advance, there is little to choose between the best 3-level and 4-level designs, and the 4-level design seemed likely to be more robust for experiments in which $\mu$ and $\sigma$ are not well-known in advance. For a 4-level design with a small number of subjects, Finney recommends (p. 498) that the levels be chosen to give probabilities of dying of 0·15, 0·35, 0·65 and 0·85. With $\mu = 0$, the levels are $-1·036\sigma_g$, $-0·385\sigma_g$, $+0·385\sigma_g$ and $+1·036\sigma_g$, where $\sigma_g$ is the guessed value of $\sigma$. Our first calculations were made with this set of levels, taking $\sigma_g/\sigma$, as in previous work, to be $\frac{1}{2}$, $\frac{2}{3}$, 1, $\frac{3}{2}$ and 2. Two guessed values of $\mu$ were investigated: $\mu_g = 0$ (the correct value) and $\mu_g = \sigma$. These choices produce 10 sets of levels.

For each set, there are $7^4 = 2401$ combinations of numbers dead at the 4 levels. The $LD_{50}$ was estimated by Berkson's minimum normit $\chi^2$ method (Berkson, 1957). This estimate was adopted primarily in order to avoid the iterations required by the maximum likelihood estimator. This decision may have been unfair to the fixed-level experiment. In comparisons of maximum likelihood and minimum normit recently made by Cramer (1964) in experiments with 3 and 5 levels, maximum likelihood gave smaller MSEs for $\hat{\mu}$ except for 3 levels with doses centred close to $\mu$. For cases in which all subjects die or all survive, we followed Berkson's rule of regarding 6 dead out of 6 as $5\frac{1}{2}$ dead out of 6, and 0 dead as $\frac{1}{2}$ dead.

One further issue must be considered. In experiments of this size, cases arise with non-negligible probabilities in which the estimate is indeterminate; for instance, 2 die out of 6 at all 4 levels. There are also cases in which a cautious investigator would have little faith in his estimate, because (i) the slope is negative or (ii) the estimate lies far outside the range of levels actually tested or (iii) the percentages dead at the 4 levels are very irregular or do not straddle 50 per cent (this, of course, is related to (ii)).

Different rules for deleting such cases, of varying degrees of severity, can be constructed. Ideally, we would like a rule that succeeds in deleting cases in which the estimate is very poor but does not reject any more than is clearly advantageous, since deleting presumably means that in practice the experiment would have to be repeated. After examining the performance of several rules, we decided to delete all experiments in which the slope was negative, indeterminate, or positive and less than $\frac{1}{3}\sigma_g$. (The experimenter will of course have had to guess $\sigma_g$ in setting his levels at those recommended by Finney or at any other recommendation.) We believe that this rule was about as favourable to the fixed-level experiments as any that occurred to us. Cramer (1964), who used a somewhat different rule, discusses some of the issues involved.

For the fixed-level experiments, Table 8.1 shows the percentage of cases rejected and the MSE of the retained cases for each of the 10 sets of levels. For later use, the results for $\sigma_g = 3\sigma$ are also included. The table also presents the corresponding MSEs for the Robbins–Monro process with 6 steps and 4 animals per step.

TABLE 8.1

*MSEs for fixed-level and Robbins–Monro experiments*

| | Fixed-level P's guessed at (0·15, 0·35, 0·65, 0·85) | | | | Robbins–Monro $m = 4, n = 6$ | |
| Start | 0 | | $\sigma$ | | 0 | $\sigma$ |
| $\sigma_g/\sigma$ | % cases rejected | $MSE/\sigma^2$ | % cases rejected | $MSE/\sigma^2$ | $MSE/\sigma^2$ | $MSE/\sigma^2$ |
|---|---|---|---|---|---|---|
| $\frac{1}{2}$ | 14·6 | 0·0941 | 27·6 | 0·4051 | 0·0451 | 0·1193 |
| $\frac{2}{3}$ | 8·8 | 0·0919 | 18·7 | 0·4950 | 0·0565 | 0·0776 |
| 1 | 2·4 | 0·0896 | 11·7 | 0·2807 | 0·0692 | 0·0653 |
| $\frac{3}{2}$ | 0·3 | 0·0885 | 3·4 | 0·2161 | 0·0841 | 0·0790 |
| 2 | 0·1 | 0·0906 | 0·7 | 0·1935 | 0·0985 | 0·0959 |
| 3 | 0·0 | 0·0759 | 0·2 | 0·2874 | — | — |

TABLE 8.2

*Ratio of MSE given by the Robbins–Monro method to the MSE given by the fixed-level method*

| | P's guessed at (0·060, 0·282, 0·718, 0·940) start | | P's guessed at (0·15, 0·35, 0·65, 0·85) start | |
| $\sigma_g/\sigma$ | 0 | $\sigma$ | 0 | $\sigma$ |
|---|---|---|---|---|
| $\frac{1}{2}$ | 0·49† | 0·24† | 0·48 | 0·29 |
| $\frac{2}{3}$ | 0·63 | 0·28 | 0·61 | 0·16 |
| 1 | 0·78 | 0·30 | 0·77 | 0·23 |
| $\frac{3}{2}$ | 0·93‡ | 0·41‡ | 0·95 | 0·37 |
| 2 | 1·30 | 0·33 | 1·09 | 0·50 |

† $\sigma_g/\sigma = \frac{4}{9}$ for the fixed-level experiment.
‡ $\sigma_g/\sigma = \frac{4}{3}$ for the fixed-level experiment.

In the fixed-level experiments the frequency of rejected cases is large enough to be annoying in practice when $\sigma_g$ is less than $\sigma$. With the Robbins–Monro method no cases are rejected, so that comparison of the MSEs in Table 8.1 favours the fixed-level experiment unless some penalty for having to repeat the experiment is introduced. The reader may be surprised at the drop in the MSE to 0·0759 for the fixed-level experiment when $\sigma_g = 3\sigma$ and $\mu$ is guessed correctly. The reason is that the set of deaths 0, 0, 6, 6, for which the minimum normit $\chi^2$ estimate is correct, becomes much the most frequent configuration of results. For this reason, a centred set of levels with $\sigma_g = 4\sigma$ gives a still lower MSE of 0·0430, and the MSE approaches zero as the interval is widened further.

In comparing the MSEs given by the two methods, it seemed to us that the fixed-level experiment would show to slightly better overall advantage if the recommended set of levels were those obtained for a centred set when $\sigma_g = \frac{3}{2}\sigma$. These levels give probabilities of dying 0·060, 0·282, 0·718 and 0·940 respectively, spanning a wider range than the original recommendation. Widening the range was recommended by Finney (p. 499) when initial information about $\mu$ and $\sigma$ is scanty. It also reduces the frequency of rejected cases.

Table 8.2 gives the ratios of the Robbins–Monro MSE to the fixed level MSE both for these wider levels and for the original levels. As it turns out, there is very little to choose between the original and the wider levels.

The most striking result from Table 8.2 is the greater robustness of the Robbins–Monro method in coping with under-estimation of $\sigma$ or a relatively poor guess at $\mu$. With $\mu$ and $\sigma$ both guessed accurately, the Robbins–Monro method gives a reduction in MSE of about 22 per cent. If $\mu$ is guessed accurately and $\sigma$ is overestimated, the gain in precision by the sequential plan becomes small, and there may even be a loss, as occurs when $\sigma_g = 2\sigma$. However, when $\sigma_g$ is less than $\sigma$ the Robbins–Monro method is much superior even when $\mu$ is guessed accurately. When $\mu$ is in error by $\sigma$, the Robbins–Monro method uniformly produces large reductions in MSE; in most comparisons made, the MSE by Robbins–Monro is well under half that obtained for the fixed-level experiments.

## 9. THE PROBLEM OF ESTIMATING THE STANDARD ERROR OF THE $LD_{50}$

Being designed to concentrate the test doses near to the $LD_{50}$, the Robbins–Monro process is not well adapted for efficient estimation of $\sigma$, which is required for estimating the standard error of the $LD_{50}$ in a single experiment. For the analogous Up and Down method, Brownlee, Hodges and Rosenblatt (1953) concluded that no reliable estimate of the standard error would be obtainable in small experiments. The same is likely to be true with the Robbins–Monro method. With an experiment involving, say, 40 or more animals, however, the experimenter is likely to ask why he cannot obtain an idea of the precision of his estimate. We have given some thought to the construction of estimators that are related to the sequential nature of the process, but nothing that looks promising has occurred to us. Two suggestions will be presented.

The first is for an experiment in which (i) the two-stage process as described in Section 7 has been followed, and (ii) the experimenter has good reason to believe that his guessed $\sigma$ lies between $\frac{2}{3}\sigma$ and $\frac{3}{2}\sigma$. From Table 7.1, it appears that with 24 animals the mean-square errors are at most $0·082\sigma^2$, and with 48 animals, at most $0·041\sigma^2$. Since $(24)(0·082) \doteq 2$, a rough idea of the root-mean-square error of the $LD_{50}$ is $\sigma_g \sqrt{(2/N)}$, where $N$ is the total number of animals in Stage 2. Obviously, this estimate is very dependent on the accuracy of $\sigma_g$.

A second proposal, made by several workers, is to analyse the results of a Robbins–Monro experiment by one of the standard methods for fixed-level experiments that provides an estimate of the standard error. Wetherill (1963) reports an attempt to use the maximum likelihood estimate in this way, but many iterations were required and the project was dropped. We have conducted a similar investigation, using the Berkson minimum normit $\chi^2$ estimator instead of maximum likelihood. One interesting by-product of this work is that it gives a comparison of the precision of the Robbins–Monro estimator with that of an estimator of the maximum likelihood type that uses all the sequential results more specifically.

The comparison was made for 24 animals, with $n = 4$ steps and $m = 6$ animals per step. Starts were at 0 and $\sigma$, with $\sigma_g/\sigma = \frac{2}{3}$, 1, $\frac{3}{2}$ and 2. We are indebted to Mr Michael Feuer, who programmed and carried out the computations on the IBM 7094. We encountered the problem discussed in Section 8, in that some of the Berkson estimates were indeterminate and some had negative, or only small, positive slopes. Such cases were rejected. Incidentally, the mean-square errors of the Robbins–Monro estimates for the retained cases were practically the same as those for all cases. The rejected cases were not ones in which the Robbins–Monro estimate of the $LD_{50}$ tended to be particularly bad.

The formula for the standard error of the $LD_{50}$ in an individual experiment was the usual asymptotic formula

$$ \mathrm{SE}(\hat{x}_{50}) = \hat{\sigma} \sqrt{\left\{ \frac{1}{\Sigma\, m_i\, w_i} + \frac{(\hat{x}_{50} - \bar{x})^2}{\Sigma\, m_i\, w_i (x_i - \bar{x})^2} \right\}}, \tag{9.1} $$

where $\hat{\sigma}$ is the estimate of $\sigma$, $m_i$ is the number of animals tested at dose level $x_i$ and $w_i$ is the observed normit weight. Table 9.1 shows the percentages of cases rejected, and for the retained cases the actual root MSEs given by the minimum normit and the Robbins–Monro estimators, and the average of the $\mathrm{SE}(\hat{x}_{50})$ as computed by the formula above.

TABLE 9.1

*Comparison of actual $\sqrt{MSE}$ for minimum normit and Robbins–Monro estimators with the average computed SE*

| $\sigma_g/\sigma$ | Start | % cases rejected | Min. normit $\sqrt{MSE}$ | Robbins–Monro $\sqrt{MSE}$ | Ave. $SE(\hat{x}_{50})$ |
|---|---|---|---|---|---|
| $\frac{2}{3}$ | 0 | 8·2 | 0·236 | 0·226 | 0·139 |
|  | $\sigma$ | 19·2 | 0·309 | 0·281 | 0·408 |
| 1 | 0 | 6·9 | 0·266 | 0·259 | 0·175 |
|  | $\sigma$ | 11·6 | 0·305 | 0·247 | 0·361 |
| $\frac{3}{2}$ | 0 | 5·6 | 0·288 | 0·285 | 0·235 |
|  | $\sigma$ | 6·7 | 0·281 | 0·282 | 0·327 |
| 2 | 0 | 4·9 | 0·303 | 0·314 | 0·287 |
|  | $\sigma$ | 3·5 | 0·283 | 0·312 | 0·354 |

The actual $\sqrt{MSE}$ values of the minimum normit and Robbins–Monro estimators agree remarkably well, the only sizeable disagreement occurring when $\sigma_g = \sigma$ and the start is at $\sigma$. There is a hint that Robbins–Monro is slightly more accurate when the step size is short, while the Berkson method may be a trifle better for $\sigma_g = 2\sigma$.

The main purpose of the investigation was to see whether the computed standard errors were reasonably unbiased estimates of the true $\sqrt{MSE}$. As Table 9.1 shows, the average value of the computed SEs (weighting each sample by its probability of occurrence) runs consistently low when the start is at 0. The underestimation is about 40 per cent when $\sigma_g = \frac{2}{3}\sigma$ and decreases steadily, as $\sigma_g$ increases, to about

5 per cent when $\sigma_g = 2\sigma$. For a start at $\sigma$, on the other hand, the average of the computed SEs is about 30 per cent too large when $\sigma_g = \frac{2}{3}\sigma$ and around 20 per cent too large for the other values of $\sigma_g$. It may be that the minimum normit method underestimates $\sigma$ when the start is at zero and overestimates it when the start is at $\sigma$.

Consequently, this investigation must be regarded as inconclusive. The results for the averages of the computed SEs raise doubts as to whether these SEs can be trusted in individual experiments with 24 animals. A useful computation would be to find the confidence limits for $\mu$ from each sample and examine whether the true probabilities of coverage are close to the nominal ones. Further work on larger experiments might also be worth while.

## ACKNOWLEDGEMENTS

The work was done under Contract Nonr 1866 (37) with the Office of Naval Research, U.S. Navy Dept. This research was aided by the National Science Foundation through their grant NSF-GP-683 to the Harvard Computing Center.

## REFERENCES

BERKSON, J. (1957), "Tables for use in estimating the normal distribution function by normit analysis", *Biometrika*, **44**, 411–435.

BLISS, C. I. and CATTELL, M. (1943), "Biological assay", *Ann. Rev. Physiol.*, **5**, 479–539.

BROWNLEE, K. A., HODGES, J. L. and ROSENBLATT, M. (1953), "The Up-and-Down method with small samples", *J. Amer. statist. Ass.*, **48**, 262–277.

COCHRAN, W. G. and DAVIS, M. (1963), "Sequential experiments for estimating the median lethal dose", in *Le Plan d'Expériences*, pp. 181–194. Paris: Centre Nationale de la Recherche Scientifique.

—— —— (1964), "Stochastic approximation to the median effective dose in bioassay", in *Stochastic Models in Medicine and Biology*, ed. J. Gurland, 281–300. Madison: University of Wisconsin Press.

CRAMER, E. M. (1964), "Some comparisons of methods of fitting the dosage response curve for small samples", *J. Amer. statist. Ass.*, **59**, 779–793.

FINNEY, D. J. (1964), *Statistical Method in Biological Assay*, 2nd ed. London: Griffin.

GADDUM, J. H. (1933), "Methods of biological assay depending on a quantal response", *Spec. Rep. Ser. med. Res. Coun. (Lond.)*, No. 183, 12.

—— (1953), "Bioassays and mathematics", *Pharmacol. Rev.*, **5**, 87–134.

HAWKINS, D. F. (1964), "Observations on the application of the Robbins–Monro process to sequential toxicity assays", *Brit. J. Pharmacol. and Chemother.*, **22**, 392–402.

HODGES, J. L. and LEHMANN, E. L. (1956), "Two approximations to the Robbins–Monro process", *Proc. 3rd Berkeley Symp. math. Statist. and Prob.*, **1**, 95–104.

IRWIN, J. O. and CHEESEMAN, E. A. (1939), "On an approximate method of determining the median effective dose and its error in the case of a quantal response", *J. Hyg.*, **39**, 574–580.

KESTEN, H. (1958), "Accelerated stochastic approximation", *Ann. math. Statist.*, **29**, 41–49.

ROBBINS, H. and MONRO, S. (1951), "A stochastic approximation method", *Ann. math. statist.*, **22**, 400–407.

SCHMETTERER, L. (1961), "Stochastic approximation", *Proc. 4th Berkeley Symp. math. Statist. and Prob.*, **1**, 587–609.

WETHERILL, G. B. (1963), "Sequential estimation of quantal response curves", *J. R. statist. Soc. B*, **25**, 1–48.

# 85

## The Planning of Observational Studies of Human Populations

By W. G. Cochran

*Harvard University*

[Read before the Royal Statistical Society on February 17th, 1965,
the President, Mr S. Paul Chambers, C.B., C.I.E., in the Chair]

### 1. Introduction

Since this introduction was written during a period of exposure to the hypnotic effects of election campaign oratory in two countries, I shall not apologize unduly if my title seems to promise more than the paper will attempt to deliver. A more accurate title would be "Comments on some aspects of the planning of certain types of observational studies of human populations". I have in mind studies with two common characteristics:

(i) The objective is to elucidate cause-and-effect relationships, or at least to investigate the relationships between one set of specified variables $x_i$ and a second set $y_i$ in a way that suggests or appraises hypotheses about causation.

(ii) It is not feasible to use controlled experimentation, in the sense of being able to impose the procedures or treatments whose effects it is desired to discover, or to assign subjects at random to different procedures. Some randomization may be employed, however, e.g. in selecting for measurement a random sample from a population that seems suitable for the enquiry at hand.

In recent years such studies have become increasingly common in medicine, public health, education, sociology and psychology. Examples are the studies of the relationship between smoking and health, studies of factors that affect the probability of injuries in motor accidents, studies of the differences in behaviour of school children under permissive and authoritarian régimes, and studies of the effects of new social programmes such as replacing slum housing by public housing.

My experience in this area comes from service on advisory groups that assist in the planning of individual studies and on refereeing teams whose function is to recommend whether the money requested for a proposed study shall be granted. Typical of the latter are the Study Sections of the National Institutes of Health in the United States. In handling applications for statistical studies, a Study Section usually consists of about eight medical specialists and two statisticians. To put it simply, the examination of a proposal boils down to a judgement on two issues. (1) If the investigator succeeds in answering the questions that he proposes to answer, will this be a worthwhile contribution to our knowledge of health and illness? (2) If the investigator does what he proposes to do, is he likely to answer the questions that he proposes to answer? Although the statisticians are presumably placed on the Study Sections in order to help primarily with the second question, the division of labour is informal. In time the statisticians pick up at least some superficial medical expertise by osmosis, and the medical specialists often bring out statistical points that the statisticians have overlooked.

Appraisal of a large number of proposals leaves two general impressions. There is a regular procession of what might be called elementary mistakes. The proposal provides no clue as to what the investigator is trying to find out, if anything. The proposed sample size is too small to offer a reasonable hope of success even under the most optimistic assumptions. The population chosen for the study, although convenient and accessible, is one in which the variables whose effects are to be disentangled and measured just do not vary much. The study is to be based on routine records, found on inspection to be incomplete, frequently illegible and to contain numerous measurements that are either gross errors or wild guesses. The study is liable to a large amount of initial non-response plus loss of subjects as the study proceeds, but the proposal reveals no awareness of this problem. No provision is made for a control or comparison group of subjects, although one is obviously required if any inferences are to be drawn. The two groups of subjects to be compared will obviously differ in some variables that are not under investigation, and the disturbing effects of these variables are likely to be greater than those of the variables that are under investigation. The objective is to discover whether differences in the behaviour of two groups can be explained by certain variables, yet the investigator proposes to match the groups with regard to these same variables. And so on.

Secondly, at the other extreme, there is a different class of proposal. The objective is important, the study will be difficult and costly, and the plan has been carefully thought out. The results will, however, be subject to several sources of bias, and neither the investigator nor the appraising committee can suggest a method of reducing these biases (except that in some cases a completely different type of study might be less vulnerable to bias). The appraisers' judgements about the seriousness of these biases vary widely. This type of proposal leaves the statistician frustrated, because he seems unable to help an obviously capable investigator and because he faces a troublesome decision on whether to recommend support in view of a distinct possibility of misleading results (though he may resolve this by adopting a more general principle that men who are good deserve support, if his terms of reference allow him this latitude).

Drawing on this experience I would like to discuss some of the problems in planning observational studies and some of the current strategies for overcoming them. This type of research, dealing with the acquisition of knowledge that may enable us to enjoy healthier and more harmonious lives, is potentially important, yet I have the impression that it has been somewhat neglected by the statistical profession. (In this remark I am echoing Dr Wold (1956), who stressed the same point when addressing this Society some years ago.) Books on this subject have been written mainly by groups of subject-matter specialists, with the statistician perhaps contributing a chapter on sampling or tests of significance. Good examples from sociology are the books by Jahoda, Deutsch and Cook (1951) and by Festinger and Katz (1953). Similarly, at least in the United States, the statistics departments in universities present courses on the design of experiments and of sample surveys, but instruction on the planning of observational studies, if given, is usually found in the subject-matter departments, often under the title "Research Methods".

It is natural and commendable that subject-matter experts are developing and teaching their own research strategies. One reason is that effective planning of observational studies calls for considerable mastery of the state of research in the subject-matter field, for instance in forming a judgement as to which potential sources of bias are major and must somehow be controlled and which are minor and can be ignored, in knowing whether a proposed method of measurement has been tested for reliability and validity; in appraising which theories of behaviour are consistent with the results of a completed study, or more generally in deciding which type of study it is most fruitful to attempt next. Nevertheless, statisticians have much to contribute —particularly their training and experience in the conditions needed to provide sound inferences and their ability to develop efficient techniques for the analysis of the untidy data that are hard to avoid in observational studies.

Section 2 presents a brief account of the major difficulties in the planning of observational studies. Since detailed discussion of all aspects of planning would be lengthy, later sections are confined to three of these problems—the setting up of comparisons that are to throw light on the causal hypothesis, the handling of disturbing variables, and the step from association to causation—that seem to me to differentiate observational research most clearly from controlled experimentation. In discussing these topics it is relevant to indicate how the problem is tackled in controlled experimentation, because to a large extent, workers in observational research have tried to copy devices that have proved effective in controlled experiments. For instance, Dorn (1953) recommended that the planner of an observational study always ask himself the question, "How would the study be conducted if it were possible to do it by controlled experimentation?".

## 2. MAJOR DIFFICULTIES

In the discussion of Dr Wold's (1956) paper on observational data, Dr Barnard was quoted as having said that a paper of this kind is useful in showing the younger statisticians what difficulties they may be up against. In writing this paper I have been conscious of the danger that a random member of the audience, if asked later for a concise summary of the paper, may quite properly report: "He said that it's all very difficult". A listing of common difficulties is, however, helpful in giving an overall view of the problems that must be overcome if this type of research is to be informative. The following list is, of course personal. Others with a different experience would doubtless construct a different list or at least change the emphasis given to the items on my list.

### 2.1. *Setting Up the Comparisons*

In controlled experimentation the investigator decides on the procedures or treatments whose effects he wishes to compare, and takes steps to apply them. The ability to do this gives him great flexibility and power. With a quantitative variable he can choose the number of levels and the intervals between them that will be most informative. In multi-variable experimentation he can apply combinations of levels, as in factorial design, selected so as to disentangle the effects of the different variables or to map a response surface effectively.

In observational studies the investigator, having decided on the types of comparison that he would like to make, often has to search for some environment in which it may be possible to collect data that provide such comparisons. As examples, this search would probably be the first step if it is desired to study the effects of air pollution on the health of urban dwellers, the type of protection afforded by seat belts under actual accident conditions, differences between the social outlooks of girls who attended co-educational schools and those who attended girls' schools, or the relative effectiveness of surgery and radiation for the treatment of malignant conditions in which ethical considerations forbid randomized experimentation. Often the investigator makes do with comparisons that are far from ideal for his purpose, and sometimes he postpones the study, hoping that later a more suitable environment will be found.

## 2.2. *The Handling of Disturbing Variables*

There is the familiar problem that the response or dependent measurements are usually influenced by many variables other than those under investigation. In controlled experimentation the investigator has three types of weapon at his disposal for handling such disturbing variables: (1) the experiments may be carried out under specialized conditions, e.g. on small plots in agriculture, or in laboratories with temperature and humidity control and highly precise instrumentation, in which some of the principal disturbing variables are absent or have greatly reduced effects; (2) blocking or adjustments made in the analysis can remove the disturbing effects of known major variables; (3) randomization and replication can diminish to a tolerable level the average effects of the remaining disturbing variables, including some whose presence is unknown to the investigator.

In an observational study the research worker can attempt to use the first device by looking for an environment in which some of the most important disturbing variables happen to be absent. This freedom of choice, however, is likely to be limited by the requirement that the environment shall also provide the types of comparison that he wants. If the subjects have been carefully selected so that they are similar as regards the principal disturbing variables, this process may have made them also similar on the variables whose effects we wish to study. Blocking (or matching) and adjustments in the analysis are frequently used. There is, however, an important difference between the demands made on blocking or adjustments in controlled experiments and in observational studies. In controlled experiments, the skilful use of randomization protects against most types of bias arising from disturbing variables. Consequently, the function of blocking or adjustment is to increase precision rather than to guard against bias. In observational studies, in which no random assignment of subjects to comparison groups is possible, blocking and adjustment take on the additional role of protecting against bias. Indeed, this is often their primary role. The extent to which they are capable of doing this will be discussed in Section 3.

## 2.3. *The Step from Association to Causation*

As mentioned, this paper deals with studies whose aim is to elucidate cause and effect relationships. I hope that I shall not be asked to explain exactly what is meant by cause and effect, since writers on the philosophy of science seem unanimously to discard this concept sooner or later as more confusing than helpful in complex situations. But to illustrate situations in which the concept is clear enough, the ultimate goal in applied studies may be to be able to predict the consequences of a new social programme, or of an experience that individual subjects may undergo, or of changes in the subject's living habits. Even in theoretical studies designed mainly to increase our understanding of people's behaviour, the idea of cause and effect is useful in the simpler situations.

In controlled experimentation the investigator who wishes to learn the effects of some procedure can usually go ahead and apply it, if necessary under a variety of other conditions, obtaining a direct answer to the question. A similar approach is sometimes possible in observational research. If a governmental agency has a programme of building new low-cost public housing or a new type of living accommodation for old people in several towns, this may provide the opportunity for studying the effects of the new living conditions on the people who enter them. Indeed, sociologists and economists are sometimes scolded for not being more enterprising in making plans in advance for the direct study of the effects of new public programmes, the co-operative efforts of astronomers and geophysicists in this respect being held out as an example.

For the most part, however, an observational study is a study of the associations between two sets of variables. Attempts to interpret these associations as causal or non-causal must rely heavily on information not supplied by the study, though some information may come from previous studies of a different type. To cite a simple example, suppose that an economist is interested in the question: if families of a certain size and with a certain income received an increase in income, how much of this increase would be spent on food? The data likely to be available or readily collectible relative to this question are a cross-sectional study of the amounts currently spent on food by families with different incomes. The increase in food expenditure per unit increase in income, as computed from this study, may or may not predict the increase that the economist wishes to estimate. In speculating on whether to trust this estimate or how to revise it, he would doubtless use any previous studies or reasonable theories about family spending habits that appeared relevant.

### 2.4. *Inferences from Sample to Population*

In most studies of human beings, the population to which we would like inferences to apply is real, not hypothetical. It is often extensive; the investigator hopes that his conclusions are valid for all males in the country with certain specified characteristics (e.g. of age or marital status). But the population actually sampled is frequently different. It is often narrow in scope, either for financial reasons or because the environment is particularly opportune for providing the appropriate comparisons. Random sampling may not be feasible. Sizeable amounts of non-response may occur. The subjects may be essentially volunteers if the measuring process is troublesome.

Standard statistical methods supply inferences from the sample data to a population of which the data can be regarded as a random sample. This sampled population is often hypothetical and sometimes hard to describe. Judgement as to how far the inferences apply to the target population involves trying to describe the relevant ways in which the sampled and target populations differ, and using any information that gives a clue as to the manner in which these differences will change the inferences. In research in which all sampled populations have to be specialized, a useful safeguard is a series of studies on sampled populations that have different peculiarities. For instance, in the studies comparing the death rates of smokers and non-smokers, the sampled populations were chosen in part because of social forces that facilitated getting good co-operation and accurate data. Fortunately, there are seven large studies, from three countries, all having broad sampled populations. The degree of agreement between studies in the relative death rates of cigarette, cigar, pipe and non-smokers and in the causes of death that show the greatest elevations in the death rates of smokers is impressive. On the other hand, the interpretation of these results is impeded by the fact that five of the studies had sizeable amounts of non-response, while in the remaining two studies no meaningful non-response rates can be calculated. Further, little was done, possibly because the studies were already complex enough in execution, to try to measure the influence of some of the disturbing variables that come to mind.

The deliberate use of sampled populations differing from the target populations is likely to remain a standard practice. This practice is followed also in controlled experimentation, both in agriculture and industry, in which much of the early screening or developmental research is conducted on a small scale that gives precise comparisons and saves money. The most promising results are checked by experiments that more closely approximate the conditions of application. The role of this approach in observation studies has been discussed, but opinion is divided as to the extent to which investigators should attempt to work in the target population itself, despite the extra expense and complexities of execution. Of course, in astronomy and meteorology, and to some extent in clinical medicine, conclusions that seem universal in scope are obtained from highly restricted observational studies. Perhaps in time an increasing body of simple fundamental laws of human behaviour will be uncovered, but in this area there is much to learn about the extent to which, as claimed in the old cockney song "it's the sime the 'ole world over".

### 2.5. *Measurement*

Much of the research on human behaviour and adjustment to life faces formidable problems of measurement. The investigator may want to study concepts like "feelings of ability to cope", "degree of frustration" or "strength of maternal affection". Different investigators develop different measuring instruments (often a series of questions) but it may not be known to what extent they are measuring the same thing, so that the combination or comparison of findings from different studies is rendered uncertain. The problem of varying definitions and measuring instruments is also familiar in clinical medicine. Psychologists and sociologists rightly devote much attention to clarifying the idea of measurement, to sophisticated analyses of the types of bias that may enter with human observers and human reporters and to the study of errors of measurement. In some areas of research, little progress seems in prospect until a substantial improvement in measuring technique is discovered.

The problem of measurement errors also affects the handling of disturbing variables. In many studies it is considered essential that the groups being compared shall be equated or adjusted for differences in socio-economic status. Numerous measures of socio-economic status have been developed, but it is hard to be sure that we have adjusted for the really relevant variable; further, errors of measurement decrease the effectiveness of the adjustment.

### 2.6. *Multiple Variables*

Multiplicity of variables is common, either in the response variables, the potentially causal variables or the important disturbing variables. An example that is far from extreme is the study by Neel and Schull (1956) of the effects of the parents' exposure to atomic radiation in Hiroshima and Nagasaki on the subsequent children. The measure of amount of exposure was a single 4-class variable, except that it was necessary to rate fathers and mothers separately. The indicators of radiation effects on children were the frequencies of stillbirths, neo-natal deaths and gross malformations, the sex ratio, and four measurements of the bodily development of the children. Disturbing variables that were judged important were maternal age, parity and degree of consanguinity in the marriage. Other disturbing variables were carefully considered, although for various reasons no adjustments for them were made. Much larger lists of variables may be present. In studies designed to measure the effects of something on family patterns of life, the response variables may include a health questionnaire, measures of the social activities of the members and of the relations between parents and children, and attempts to assess the degree of satisfaction that the family members derive from their way of living.

The presence of multiple variables raises a number of issues. Despite the advances in multi-variate analysis and in computing aids, summarization of a complex set of tables is still largely an art. Some investigators are appalled to realize how many tables they have to digest. Underestimation of the time and resources required for analysis of results is one of the most frequent features of proposals for research studies. The old maxim that the outlines of the analysis should be carefully sketched as a part of the research plan has lost none of its force. There is also, I believe, useful work to be done by statisticians in learning what some of the newer multivariate techniques really accomplish when applied to data and in explaining this to investigators, many of whom have no clear understanding of the techniques that they are trying to use. We need good expository papers of this type.

Participation in multi-variable studies leaves the impression that a series of lengthy questionnaires weakens the quality of the measurements. One is reminded of Bradford Hill's dictum (1953) that for every question asked of the respondent the investigator should ask himself three questions, one of which should always be: "is this question really necessary?" But when dealing with an imaginative investigator I do not find it easy to determine at what point one should adamantly oppose all further questions, however ingenious and interesting.

### 2.7. *Long-term Studies*

Some studies, e.g. of child growth and development, of chronic disease, or of social programmes whose effects are slow to appear, occupy the full-time energies of research teams for periods of 5, 10 or 15 years. Keeping track of the subjects and persuading them to be measured repeatedly requires much organizational skill, which may be only partially successful. Maintaining the interest of the research team, especially if there are no opportunities for publication for long periods, is another task. In short, this type of study produces a series of administrative and financial problems that are new to most research workers. It also naturally raises the question: is there any quicker way of obtaining useful results? For instance, Kodlin and Thompson (1958) give a useful analysis of the circumstances in which cross-sectional studies conducted at a single time will provide some of the results of long-term studies in growth.

As a final comment, many of these problems arise because investigators are beginning to study a series of new and probably complex phenomena, not because the investigator is restricted to observational methods.

### 3. THE HANDLING OF DISTURBING VARIABLES

Although it may seem to be putting the cart before the horse, some repetition is saved if the discussion of disturbing variables precedes that of the setting up of comparisons. The first step is to construct a list of known disturbing variables. Usually, these are arranged in three classes. (1) Major variables for which some kind of matching or adjustment is considered essential. Their number is kept small in view of the complexities involved in matching or adjusting for many variables simultaneously. (2) Variables for which, ideally, we would like to match or adjust, but content ourselves with some verification that their effects produce little or no bias. (3) Variables whose effects, thought to be minor, are disregarded. I shall consider the comparison of two groups, as occurs in the simplest type of observational study.

In the handling of disturbing variables there are two objectives. We want to protect against bias entering into the estimate of the difference between the two group means for the dependent variable $y$. Secondly, even if there seems no danger of bias, the presence of the disturbing variable may inflate the variance of $\bar{y} - \bar{y}'$ to an extent that makes the comparison imprecise.

### 3.1. *Variables for which no Matching or Adjustment is made*

For a disturbing variable in class (2), it is good practice to measure the variable and check that $\bar{y} - \bar{y}'$ is unlikely to be biased. If the disturbing variable is categorical (e.g. religious affiliation) a condition for absence of bias is that each class has the same frequency in the two populations of which the groups are random samples. The $\chi^2$ test for a $2 \times k$ contingency table is the standard check.

If the disturbing variable $x$ is continuous, let the relations between $y$ and $x$ in the two populations be

$$y = \mu_y + \xi + e; \quad y' = \mu_{y'} + \xi' + e'$$

where the residuals $e$ and $e'$ have zero population means, while $\xi = \phi(x)$ and $\xi' = \phi(x')$ is the regression of $y$ on $x$, assumed the same in both populations. Hence,

$$\bar{y} - \bar{y}' = \mu_y - \mu_{y'} + \bar{\xi} - \bar{\xi}' + \bar{e} - \bar{e}'.$$

A necessary condition for the validity of the usual method of testing the significance of $\bar{y} - \bar{y}'$ or constructing confidence limits for $\mu_y - \mu_{y'}$ is that the population mean of $\bar{\xi} - \bar{\xi}'$ be zero, because the computed variance of $\bar{y} - \bar{y}'$ assumes that $\xi$ acts like a random variable with zero mean.

Assurance that the distribution of $x$ is the same in both populations guarantees that $E(\bar{\xi} - \bar{\xi}') = 0$ for any shape of regression function. A comparison of the frequency distributions of $x$ in the two groups, usually made by the $\chi^2$ test for a $2 \times k$ contingency table, is therefore relevant.

If $\phi(x)$ can be approximated by the polynomial

$$\xi = \phi(x) = \beta_0 + \beta_1 x + \beta_2 (x^2) + \beta_3 (x^3) + \dots$$

the population mean of $\bar{\xi} - \bar{\xi}'$, i.e. the bias in $\bar{y} - \bar{y}'$ due to $x$, is

$$\beta_1 (\mu_1 - \mu_1') + \beta_2 (\mu_2 - \mu_2') + \beta_3 (\mu_3 - \mu_3') + \dots$$

where $\mu_i = E(x^i)$, $\mu_i' = E(x'^i)$ are the $i$th moments of $x$ in the two populations about zero. Thus, verification that the sample means $\bar{x}$, $\bar{x}'$ do not differ by more than sampling error gives assurance only that bias arising from a *linear* regression of $y$ on $x$ is absent or small. A check that the two samples have the same means and variances in $x$ (apart from sampling errors) is assurance against a quadratic regression, and so on.

The polynomial approximation is useful because sometimes the general shape of the regression of $y$ on $x$ is known from previous studies. Many relations are nearly linear or quadratic. Consequently, comparison of the means and variances of $x$ may be more to the point than the $\chi^2$ comparison of the whole frequency distributions. An extreme example is that in which the death rates of two groups of men are being compared and $x$ is age. If $y$ is a (0, 1) variable that denotes survival or death of a man during a year, the regression of $y$ on $x$ in the range 35–80 years is far from linear and is not a polynomial. However, a cubic in which the linear and quadratic terms dominate is often a fair approximation. Thus, unless at least the means and variances of $x$ agree well in the two samples, there is a danger of substantial bias from differences in age. Of course, the relation between death rate and age is so marked at the upper ages that adjustment or matching for age is advisable.

With several $x$-variables, the common practice is to compare the marginal distributions in the two groups for each $x$-variable separately. The above argument makes it clear, however, that if the form of the regression of $y$ on the $x$'s is unknown, identity of the whole multi-variate distribution is required for freedom from bias. Similarly, the polynomial approach indicates that cross-product moments may be involved as well as univariate moments. In view of these extra complexities, it would be useful to know whether a check confined to marginal distributions is in practice likely to give a misleading impression.

Although these checks on the $x$ distribution are usually made by tests of significance, it is not clear what kind of assurance is given by the finding of a non-significant result, nor that a test is the appropriate criterion. An alternative approach will be illustrated for the case in which the regression of $y$ on $x$ is linear, with a residual denoted by $e$. In repeated samples in which $(\bar{x} - \bar{x}')$ is fixed, $\bar{y} - \bar{y}'$ is normally distributed with mean

$$\mu - \mu' + \beta(\bar{x} - \bar{x}')$$

and variance $2\sigma_e^2/n$, this holding whether the mean value of $(\bar{x} - \bar{x}')$ is zero or not. If we assume that there is no bias due to $x$, we regard $\bar{y} - \bar{y}'$ as normally distributed with mean $\mu - \mu'$ and variance $2(\sigma_e^2 + \beta^2 \sigma_x^2)/n$. In large samples, 95 per cent confidence limits for $\mu - \mu'$ are therefore calculated by the formula

$$\bar{y} - \bar{y}' \pm 1 \cdot 96 \sqrt{\left\{ \frac{2}{n} (\sigma_e^2 + \beta^2 \sigma_x^2) \right\}}.$$

From the conditional distribution of $\bar{y} - \bar{y}'$ as given above, the probability that these limits include $\mu - \mu'$ is easily seen to be the probability that a normal deviate lies between the limits

$$-\sqrt{\left(\frac{n}{2}\right)} \frac{\beta(\bar{x} - \bar{x}')}{\sigma_e} \pm 1 \cdot 96 \sqrt{\left\{1 + \frac{\beta^2 \sigma_x^2}{\sigma_e^2}\right\}}.$$

In the preliminary test of significance of $\bar{x} - \bar{x}'$, the test criterion is

$$t = \sqrt{(n/2)} \, (\bar{x} - \bar{x}')/\sigma_x.$$

Hence, the above limits may be written

$$-tv \pm 1 \cdot 96 \sqrt{(1 + v^2)} \tag{3.1}$$

where $v = \beta \sigma_x / \sigma_e$.

Now if variations in $x$ could somehow be removed, the unconditional variance of $y$ would be $\sigma_e^2$. Thus the quantity $v^2$ represents the relative increase in the variance of $y$ due to variations in $x$. This result is a reminder that even if there is no danger of bias from $x$, there is a loss of precision. Placing an $x$ variate in class (2) instead of class (1) implies a judgement that this loss is small, say that $v^2 < 0 \cdot 2$, or $v$ does not exceed $0 \cdot 45$.

Table 3.1 shows the conditional probabilities that the 95 per cent confidence limits actually include $\mu - \mu'$ for $v = 0 \cdot 3$, $0 \cdot 4$, $0 \cdot 5$, $0 \cdot 6$ and $1 \cdot 0$ and $t = 0 \cdot 5$, $1 \cdot 0$, $1 \cdot 5$ and $2 \cdot 0$, computed from (3.1) above.

Note that the value of $t$ is known to the investigator from the preliminary check. If $t < 1$, the conditional probabilities of coverage are greater than the stipulated 95 per cent. If $t = 1.5$, the probabilities are not too far below 95 per cent, provided that the initial guess that $v$ is small was correct. For $t = 2$, its 5 per cent significance level, the coverage is unsatisfactory even if $v$ is small.

TABLE 3.1

*Probability that the 95 per cent limits include $\mu - \mu'$*

| $t$ \ $v$ | 0·3 | 0·4 | 0·5 | 0·6 | 1·0 |
|---|---|---|---|---|---|
| 0·5 | 0·957 | 0·961 | 0·966 | 0·972 | 0·988 |
| 1·0 | 0·950 | 0·950 | 0·951 | 0·953 | 0·963 |
| 1·5 | 0·938 | 0·931 | 0·924 | 0·917 | 0·898 |
| 2·0 | 0·922 | 0·903 | 0·882 | 0·862 | 0·785 |

If a single $x$ variate shows a value of $t$ above 1·5, these results suggest that we have another look at this variate when the values of $y$ become known. At that time we can estimate $v$ and also compare the unadjusted $\bar{y} - \bar{y}'$ with an estimate adjusted for the linear regression, to see whether there is a material difference. If several $x$-variables show $t$ values substantially above 1·5, this raises a question whether the groups are suitable for comparison.

Finally, even equality in the frequency distributions of $x$ does not guarantee absence of bias if the regression function differs in the two populations. This point can be checked when the values of $y$ become known.

### 3.2. Matching and Adjustment

For the major disturbing variables, four methods used in practice will be considered.

1. *Matching.* Pairs are drawn, one from each population, such that $x_i$ and $x_i'$ are identical within some small tolerance. This equates the frequency distributions of $x$ in the two samples. The practical difficulties of matching vary with the situation. If the available populations are much larger than the desired size of sample and the distribution of $x$ differs little in the two populations, matching gives little trouble. If the population reservoirs are limited and show markedly different $x$-distributions, or if several groups or several variables are to be matched, the process can be extremely tedious and may necessitate reducing the desired sample size. A much-quoted example is that of Chapin (1947), who compared the later economic adjustments of boys who completed high school with boys who dropped out. Starting with reservoirs of 671 and 523 boys, he ended with samples of size 23 after matching on six major disturbing variables.

2. *Equal sample sizes within sub-classes.* This procedure attempts to gain most of the advantages of matching with less expenditure of time. Each population is stratified into subclasses by the values of $x$. Within a given sub-class, samples of the same size are drawn from each population, but are not individually matched. To illustrate, suppose that only 100 subjects in group 1 are available, but group 2 has a larger

reservoir, and that there are five sub-classes. To see how things look, 100 subjects are also drawn from population 2. The numbers in each sub-class are found to be as follows.

|  | *Sub-class* | | | | | *Total* |
|---|---|---|---|---|---|---|
|  | 1 | 2 | 3 | 4 | 5 |  |
| Group 1 | 8 | 24 | 30 | 25 | 13 | 100 |
| Group 2 | 20 | 34 | 28 | 15 | 3 | 100 |

Group 2 is to have the same sample size as group 1 in each sub-class. In sub-classes 3, 4 and 5, additional group 2 members will have to be drawn from the population to reach the goals of 30, 25 and 13. There may be difficulty in subclass 5 unless the group 2 reservoir numbers over 400.

In this approach, $\bar{y} - \bar{y}'$ is freed from bias due to differences between means of $x$ in different sub-classes. Some bias may remain from variation of $x$ within sub-classes.

3. *Adjustment for sub-class differences.* In this method the sample sizes are not equalized within sub-classes. However, the estimate of the mean difference is of the form $\Sigma W_j(\bar{y}_j - \bar{y}'_j)$, where $j$ refers to a sub-class. If the variance of $y$ appears constant within sub-classes, $W_j$ may be taken as $n_j n'_j/(n_j + n'_j)$ by the usual least-squares principle. This method and the previous one have approximately the same properties as regards removal of bias, but differ in precision.

4. *Adjustment by regression.* This familiar method, used mostly when $x$ is continuous, consists of computing the within-group regression of $y$ on $x$, and adjusting $(\bar{y} - \bar{y}')$ to remove the effects of this regression. In practice, linear regressions are much the most common.

### 3.2.1. *x categorical or discrete*

The simplest case is that in which $x$ is categorical or discrete and is known without error. Methods 1 and 2 become identical. They are free from bias, provided that the relation between $y$ and $x$ is the same in both populations. With samples of size $n$ in each group, the variance of $\bar{y} - \bar{y}'$ is $2\sigma^2_{y.x}/n$, where $\sigma^2_{y.x}$ is the variance of $y$ in arrays in which $x$ is fixed. Method 3 (adjustment by sub-classification) gives the same protection against bias, but the estimate $\Sigma W_j(\bar{y}_j - \bar{y}'_j)$ has a variance

$$\sigma^2_{u.x}/\Sigma W_j = \sigma^2_{y.x}/n\Sigma p_j p'_j/(p_j + p'_j)$$

where $p_j$ is the proportion of the group 1 sample falling in sub-class $j$. The relative precision of method 3 to methods 1 and 2 is therefore $2\Sigma p_j p'_j/(p_j + p'_j)$. Calculation of this quantity helps in a choice between methods 2 and 3. In the numerical example given with the description of method 2, the initial sub-class sample sizes were 8, 24, 30, 25 and 13 in group 1, and 20, 34, 28, 15 and 3 in group 2. If method 3 is used instead of equating sample sizes within sub-classes, the relative precision works out at 0·92. It might not be considered worth while to go to the extra trouble of equating sample sizes.

If the values of $(\bar{y}_j - \bar{y}'_j)$ differ from sub-class to sub-class or if the residual variance of $y$ is not constant, a different estimate of the mean difference may be adopted, and the result for the relative precision of method 3 will be modified. Adjustment by regression is not possible when $x$ is categorical. Comments on the properties of this technique when $x$ is continuous (next section) apply also when $x$ is discrete.

If $x$ represents an ordered classification (e.g. none, mild, moderate, severe), the remarks in this section do not apply, because $x$ cannot usually be assumed known without error. In such cases it is often more realistic to regard the classification as formed by subdividing the frequency distribution of an underlying continuous variable $u$ into four distinct parts. The situation may indeed not be quite as clear-cut as this, since mistakes in classification may have occurred. As a first approximation I suggest that the results given in section 3.2.2. for $x$ continuous be regarded as applying to ordered classifications also, method 2 corresponding to matching or use of equal sub-class numbers and method 3 to adjustment. This issue has also been discussed by Kihlberg and Narragon (1964).

### 3.2.2. *x continuous*

Individual matching removes any bias arising from $x$, provided that the relation between $y$ and $x$ is the same in both populations and that $x$ is measured without appreciable error. Methods 2 and 3 equate the distributions of $x$ in the two groups only partially, since these distributions may differ within sub-classes. Some residual bias therefore remains in $\bar{y} - \bar{y}'$ and $\Sigma W_j(\bar{y}_j - \bar{y}'_j)$, and the variances of these estimates are increased relative to method 1, since the corresponding functions of $x$ are not zero. Thorough examination of these methods requires an investigation covering different frequency distributions and types of initial bias. The following illustration indicates the general properties of the methods.

Suppose that $x$ is normally distributed in both groups with $\sigma = 1$. In group 1, $\mu = 0$. In group 2, $\mu = -\frac{1}{2}$ and $\mu = -1$ are considered. With $\mu = -\frac{1}{2}$, the initial bias in $x$ in group 2 might be called moderate: a test of $(\bar{x} - \bar{x}')$ in samples of size 100 gives about a 95 per cent chance of finding the difference significant. With $\mu = -1$ the bias is striking, being detectable in samples of size 25.

### TABLE 3.2

*Properties of methods 2 and 3*

| No. of sub-classes | $\mu = -\frac{1}{2}$ Remaining bias | | R.P. of Method 3 | $\mu = -1$ Remaining bias | | R.P. of Method 3 | Remaining variance |
|---|---|---|---|---|---|---|---|
| | Method 2 | Method 3 | | Method 2 | Method 3 | | |
| 2 | 0·184 | 0·190 | 0·96 | 0·382 | 0·430 | 0·87 | 0·39 |
| 3 | 0·105 | 0·109 | 0·95 | 0·218 | 0·259 | 0·84 | 0·23 |
| 4 | 0·071 | 0·076 | 0·95 | 0·147 | 0·183 | 0·82 | 0·15 |
| 5 | 0·052 | 0·055 | 0·95 | 0·109 | 0·135 | 0·81 | 0·10 |

The sub-class boundaries were constructed so that the sub-classes have equal frequencies in population 1. For 2, 3, 4 and 5 subclasses, Table 3.2 gives the relevant results.

The remaining bias in $x$ is $E(\bar{x} - \bar{x}')$ for method 2 and $E\Sigma W_j(\bar{x}_j - \bar{x}'_j)$ for method 3. The original biases were $\frac{1}{2}$ and 1 for the two cases. Method 2 leaves about 36–38 per cent of this bias remaining if there are only two sub-classes and 10–11 per cent remaining with five sub-classes. With moderate bias ($\mu = -\frac{1}{2}$), method 3 has about the same effectiveness in removing bias, and gives only a slightly higher variance, as indicated by the relative precision figures (R.P.) computed as in section 3.2.1. Method 3 is noticeably less effective, when $\mu = -1$, both as regards bias and variance.

The extreme right column of Table 3.2 shows for method 2 the quantity

$$nV(\bar{x} - \bar{x}')/2.$$

Since this quantity would be unity if random samples had been drawn without any sub-classification, it might be described as the proportion $\lambda$ of the original variance of $x$ that remains. The corresponding values for method 3 are only slightly higher and are not shown. How much this variance increases the variance of $\bar{y} - \bar{y}'$ relative to that given by individual matching depends on the correlation between $y$ and $x$. The comparable quantities for $V(\bar{y} - \bar{y}')$ are $(1 - \rho^2 + \lambda\rho^2)$ for methods 2 and 3 and $(1 - \rho^2)$ for method 1. With five sub-classes ($\lambda = 0\cdot10$) the increase in variance with methods 2 and 3 does not reach 10 per cent until $\rho$ exceeds $0\cdot7$. With two sub-classes the loss of precision may be substantial.

Evidently, sub-division into two or three classes has limited effectiveness both in controlling bias and reducing variance. Naturally, investigators prefer to use only a few sub-classes, especially when there are several $x$-variables. Methods 2 and 3 also reduce differences in the higher moments of the distribution of $x$, but illustrations will not be given.

With $\mu = -1$ and five sub-classes, the highest sub-class contains only $3\cdot3$ per cent of population 2 as against 20 per cent of population 1. Method 2 therefore requires a reservoir from population 2 that is around six times the sample from population 1. For individual matching a much larger reservoir would be needed.

Assuming the same linear regression in each population, adjustment by linear regression removes the bias. As regards precision, there is the well-known result

$$V\{\bar{y} - \bar{y}' - b(\bar{x} - \bar{x}')\} = \frac{2\sigma_{y.x}^2}{n}\left\{1 + \frac{n(\bar{x} - \bar{x}')^2}{2\Sigma}\right\}$$

where $\Sigma$, with $2(n-1)$ d.f. is the pooled sum of squares for $x$. In repeated sampling, when $x$ is $N(0, 1)$ and $x'$ is $N(\mu, 1)$,

$$E(\bar{x} - \bar{x}')^2 = \mu^2 + 2/n; \quad E(1/\Sigma) = 1/2(n-2)$$

and the two are independent. Hence the average variance of the adjusted mean is

$$\bar{V} = \frac{2\sigma_{y.x}^2}{n}\left\{1 + \frac{n\mu^2}{4(n-2)} + \frac{1}{2(n-2)}\right\} = \frac{2\sigma_{y.x}^2}{n}\left\{1 + \frac{\mu^2}{4}\right\}$$

when $n$ is large.

With a linear regression, the relative precision of covariance to individual matching is therefore $16/17 = 0\cdot94$ when $\mu = \frac{1}{2}$ and $4/5 = 0\cdot8$ when $\mu = 1$. Covariance is superior to method 3 as regards both bias and variance. It is superior to method 2 in the control of bias. Method 2, with at least five sub-classes, is likely to have a smaller variance unless $\mu$ is small.

If the true regression is quadratic but adjustment is made by a linear regression, the remaining bias in large samples works out as

$$\beta_2\left[(m_2 - m_2') - \frac{(m_1 - m_1')(m_3 + m_3')}{m_2 + m_2'} - \frac{(m_1 - m_1')^2(m_2 - m_2')}{m_2 + m_2'}\right]$$

where $m_i = E(x - \mu_x)^i$ and $m_1 - m_1' = -\mu$ in the above notation. We see again how helpful is equality of the first two moments of the distributions of $x$ and $x'$, since in this event covariance is not needed to remove bias and can concentrate on increasing precision. If the low moments are unequal, methods 2 and 3 would be expected to be more potent than a linear regression in removing non-linear bias, but further investigation is required to appraise whether this superiority is material.

The regression approach has the advantage that separate regressions can be computed in the two populations and used in the adjustment, this being a situation in which bias remains even with individual matching. If the two regressions differed markedly, however, one would be inclined to reconsider whether the groups are suitable for comparison.

If $x$ is subject to appreciable errors of measurement, none of the methods succeeds in complete removal of bias. Suppose that $X = x + d$ is the recorded value of $x$, where $d$ is the error of measurement. If $x$ and $d$ are normally and independently distributed, the expected reduction in $x$ due to a reduction of amount $\mu$ in $X$ is $\mu \sigma_x^2 / (\sigma_x^2 + \sigma_d^2)$. Thus the fraction of the original bias that remains in $X$ after individual matching is $\sigma_d^2 / (\sigma_x^2 + \sigma_d^2)$. Similarly, if $\beta_1$ is the linear regression of $y$ on $x$, the regression of $y$ on $X$ is $\beta_1 \sigma_x^2 / (\sigma_x^2 + \sigma_d^2)$, so that in large samples the fraction of bias remaining in $y$ after regression adjustment is again $\sigma_d^2 / (\sigma_x^2 + \sigma_d^2)$. It is worth remembering that the basic quantity is this ratio. A value of $\sigma_d^2$ that looks large to someone accustomed to highly precise measurements might be small relative to the total variance of $x$.

With several $x$-variables the relative properties of the four methods appear to remain as in the above example. The difficulties of matching and of equating sample sizes mount steadily. Under the adjustment methods, the variance of the adjusted mean difference tends to increase and the analysis becomes more complex. With method 3, a point may be reached at which the investigator wonders whether it is worth adjusting for one or more extra $x$-variables, since the reduction in bias may not compensate for the loss of precision and extra complexity. A criterion for forming a judgement on this question from the sample data has been sketched by Cochran, Mosteller and Tukey (1954), though details need to be worked out.

To summarize, the similarities among the methods are greater than their differences. When feasible, matching is relatively effective. Overall, covariance seems superior to adjustment by sub-classification, though the superiority will seldom be substantial. If the original $x$-distributions diverge widely, none of the methods can be trusted to remove all, or nearly all, the bias. This discussion brings out the importance of finding comparison groups in which the initial differences among the distributions of the disturbing variables are small.

## 4. Setting Up the Comparisons

### 4.1. *The Choice between Different Types of Study*

As mentioned previously, the investigator often has to search for some environment in which a comparison relevant to the causal hypothesis can be made. Sometimes he faces a choice between different types of study. For instance, in the work on the relation between smoking and lung cancer, the crudest approaches were a comparison of the time trends in the lung cancer death rate and in the consumption of tobacco per head within a country, or an examination of the relation between these two figures in different countries at the same time. Then there were numerous studies in which the percentage of smokers among lung-cancer patients was compared with that among patients with other diseases or among the general public. In another approach the lung-cancer death rates of groups of smokers and non-smokers were recorded over a period of years. These comparison groups were obtained either by classifying the members of a large population into different smoking classes by an initial question- naire or by finding a smaller homogeneous group whose members did not smoke, and constructing a comparison group of smokers. An attractive possibiliiy would be to compare twins, preferably identical, of whom one smoked and the other did not, although it seems highly unlikely that enough pairs could be located.

In making a choice between different studies that he might undertake, the investi- gator should consider the resources and time needed and the status of each study with regard to the handling of disturbing variables and to the quality of the measurements. Other relevant factors are: (1) The quantities that can be estimated from the study. Sometimes one study yields only a correlation coefficient while another gives an estimate of the response curve of $y$ to variations in $x$. (2) The range of variation of the suspected causal variable. In general, a study that furnishes a wider range of variation may be expected to give more precise estimates of the effect on $y$. (3) The relation to previous work. One study may be a new approach to the problem, another a repetition of studies done elsewhere. Both have their uses, but preference would normally be given to a new approach, especially if it seems free of some of the biases thought to be present in previous studies. Naturally, all these questions involve judgement. As research by observational methods becomes more widespread and familiar, we should be able to make better appraisals of the productivity of different approaches.

### 4.2. *Some Common Types of Comparison*

In this and the following sections some common types of comparison are presented from the viewpoint of their statistical structure. The simplest plan is a direct compari- son of a few groups (often two) that differ in the hypothetically causal variable. Frequently, only one of the groups is clearly demarcated in advance, and the investi- gator must construct one or more control or comparison groups. For instance, radiologists, particularly the older ones, were formerly exposed to repeated small doses of radiation in their practices. Studies have been made to try to see whether this exposure produced an increase in their probability of dying. For this purpose they must be compared with some other non-exposed group. Ideally, a control, while lacking the suspected causal factor, should have the same distribution as the chosen study group with regard to all major disturbing variables. Sometimes the investigator is not sure whether a specific disturbing variable affects his study group. This situation may require more than one control. In discussing controls for hospitalized lung- cancer patients for a comparison of the proportion of smokers, Mantel and Haenszel (1959) point out that hospital patients in general are known to yield a higher pro- portion of smokers than members of the general public. One possible reason is that smoking histories collected in hospital are more accurate, those obtained from the general public being underestimates. If so, a hospital control is indicated. But if smokers have higher rates of hospitalization, the smoking data being equally accurate, the control should come from the general population. With uncertainties like this, use of both controls is advisable.

4.2.1. *The "before–after" study*

This plan is much used in investigating the effects of new social, economic or medical programmes. If the programme applies to everyone, there is no possibility of finding a control group that does not experience it. At a minimum, the $y$-variables and the principal disturbing $x$-variables are measured before and at appropriate times after the initiation of the programme. This enables us to investigate whether changes in the $y$-variables have occurred over and above those expected from any changes in the $x$-variables. This estimate of the effect of the programme is liable to two types of bias: people's behaviour immediately prior to the start of the programme may be affected by knowledge that the programme is about to begin, and some disturbing variables that affect time changes may be unknown. An estimate that would be free from the first bias is a comparison of the residual $y$ changes during a period after the start of the programme and a period prior to the announcement of the programme, but it is not often feasible to obtain the necessary data.

When the programme (e.g. of new public housing) is available only to certain people, it becomes possible to measure the $y$- and $x$-variables, before and after, both for the group that undertakes the programme and for a control group that does not. The effect of the programme is estimated by the difference between (After–Before) for the programme participants and (After–Before) for the controls, the variable being the residual of $y$ after adjustment for the disturbing variables. One advantage of this design is that we can verify, at the start of the study, whether the two groups are similar in their $y$-variables, instead of having to guess about this from measurements on the $x$-variables alone, as is the usual situation. Despite this, a fully satisfactory control may not come easily. If much initiative is required to get one's family on the eligible list for new housing, families that have shown no such initiative are a dubious control even though they state in a questionnaire that they would like to be in new housing. For certain programmes, e.g. an educational one to improve family health practices, a suggestion is sometimes made of a second control in which $y$ is measured only afterwards, to guard against the possibility that the initial questionnaire alerts the control families to deficiencies in their health practices which they proceed to remedy. My own view is that an educational programme that cannot improve health practices more than can a single questionnaire is not wrongly considered a failure, and that this enlargement of the study is seldom justified.

4.2.2. *The* ex post facto *or retrospective study*

Sometimes an unexpected event has occurred, and the question is "what caused it?" An outbreak of nausea and vomiting follows a picnic meal. If all left-over food has been destroyed, eliminating laboratory analysis, a list of the foods eaten by those who became ill and those who did not, and a description of the symptoms and age and sex distributions of affected and unaffected persons and of the preparation of the dishes served are the main clues to the responsible organism and food. The same approach can be used to investigate why riots have occurred in certain communities while others, at first sight similar, have had no disturbances. The strategy is to set up groups that differ in the $y$-variable, and examine whether they differ in the suspected causal variables.

Because of its relative cheapness and high efficiency under favourable circumstances, this approach is often used in problems in which a direct approach is also feasible. In the smoking–lung cancer relation, numerous studies have been done by both approaches. Another frequent application is a comparison of the successes and failures in some occupation or task, in the hope of discovering the causes, or at least useful predictors, of success and failure.

In the simplest case in which both the cause and the event in question are dichotomous, the two approaches are different ways of sampling the following $2 \times 2$ table, where the $N$'s are the numbers in the population.

| *Postulated cause* | *Event* | | |
|---|---|---|---|
| | *Present* | *Absent* | *Total* |
| *Present* | $N_{11}$ | $N_{12}$ | $N_{1.}$ |
| *Absent* | $N_{21}$ | $N_{22}$ | $N_{2.}$ |
| *Total* | $N_{.1}$ | $N_{.2}$ | $N_{..}$ |

Under the direct approach, a sample is selected from each row and the proportions $N_{11}/N_{1.}$ and $N_{21}/N_{2.}$ are compared. This is expensive if the event is rare, since large samples are needed, or if the event takes years to manifest itself. In the retrospective approach a sample is drawn from each column. If rare, the "event present" column can be sampled at a much higher rate—frequently, all such cases that can be found in the population are taken. Both methods furnish a $\chi^2$ test of the null hypothesis that there is no association. In large samples, if the two methods are to have equal power in detecting a small departure from the null hypothesis, it may be shown that $n_R/n_C = N_{1.}N_{2.}/N_{.1}N_{.2}$, where $n_R, n_C$ are the sample sizes used in the direct and retrospective approaches, respectively. For example, if the event is present in only 1 per cent of the population but the postulated cause occurs in half the population, $n_R/n_C \simeq 25$. Moreover, in the retrospective method the data on the causal variables lie in the past and can be collected without waiting for the effect to develop.

In appraising the size of the effect of the postulated cause, we need to compare the quantities $N_{11}/N_{1.}$ and $N_{21}/N_{2.}$, i.e. the frequencies with which the event occurs when the cause is present and absent, respectively. These are the quantities that are estimated in the direct approach. They can also be estimated by the retrospective approach, provided that

(i) the samples from the columns are random samples and,
(ii) the relative sampling fractions in the two columns are known.

As the retrospective method has been applied in practice, neither condition is usually fulfilled. Cornfield (1951) pointed out that if the event is rare, the ratio of the frequencies, $N_{11}N_{2.}/N_{1.}N_{21}$, will be close to $N_{11}N_{22}/N_{12}N_{21}$. From a retrospective study a consistent estimate of this quantity, called the *relative risk* of the event, is given by the sample cross-product ratio $n_{11}n_{22}/n_{12}n_{21}$, provided that no bias has arisen from non-randomness in sampling. Confidence limits for the relative risk and methods for comparing and combining results from different studies have been given by Cornfield (1956), while Mantel and Haenszel (1959) present methods for estimating an overall relative risk in retrospective studies in which the data have been sub-classified by another variable (e.g. age or location).

The retrospective approach has various weaknesses. The obtainable data about postulated causal factors may be of poor quality, especially if they lie in the distant past or have to be taken from routine records, and the sampling of the columns may be far from random. As more studies in different fields are done by the two methods, the ability of the retrospective approach to estimate relative risk can be more soundly judged. In the smoking–lung cancer relation, the two approaches agreed well on the whole.

### 4.2.3. *Multiple causal variables*

Typical examples are the studies to investigate the roles of measures of blood pressure, obesity and cholesterol levels in the individual as predictors of later heart disease (although these variables would not necessarily be viewed as causal, but perhaps as indicators of the presence of some deeper cause). For the most part, often because there seems no choice, investigators have taken the postulated causal variables as they come in the selected sample, with no deliberate attempt to borrow the idea of factorial design. The disadvantage in this approach is that if the variables are highly correlated it becomes difficult to disentangle their effects. Further, if one of two correlated variables has a high error of measurement while the other does not, the regression coefficient on the first variable is an underestimate and on the second an overestimate.

Exceptions can be cited. In his studies of sexual behaviour, Kinsey's (1948) independent variables (again not necessarily considered causal) were arranged in a multiple classification with something over 300 cells. His announced plan was to obtain a sample of size 300 in each cell, the planned sample size being 100,000. Since, however, he regarded his major problem as that of getting people to tell the truth, his standards for the selection and training of interviewers were exacting, so that his field force was very small. This, plus a haphazard method of sampling, made the realization fall far short of the goal.

The opportunity of using a factorial approach presents itself if there is a population reservoir of size $N$ in which the causal variables have already been measured, and a much smaller sample of size $n$ is to be selected for the actual study. This reservoir might come from some other investigation. Alternatively, if $y$ is expensive to measure, it may be worth while to expend some of the resources on measuring the causal variables in a sample of size $N$, of which $n$ will later be selected for measurement of $y$. Further work on this approach is required. Given the results of the large sample it is not obvious how best to select the sub-sample, with say three or four variables, nor how great a gain in precision over a random sample of size $n$ can be expected. An example of this method was given by Keyfitz (1952) in studying the relation between fertility and five dichotomous demographic variables by sampling from records, although owing to the nature of the variables, he was not able to obtain complete orthogonality in his $2^5$ factorial. Use of an initially larger sample in this way also makes it possible to obtain a better estimate of the response curve of $y$ as the level of a causal variable change.

### 4.2.4. *Population laboratories*

One device that has been tried in a number of large research centres, primarily in public health and sociology, is to select some area, perhaps of 50,000–100,000 persons, that seems appropriate for the type of field research carried on. The background characteristics of the people are measured in an initial census. This is repeated at intervals. Supplementary questions may be added to these censuses to provide reservoirs of data, as in the previous section, for individual studies. The studies, which are mostly carried out on sub-samples, may have widely different foci of interest.

An early example is the Eastern Health District of Baltimore (Fales, 1951; Cochran, 1952), though in this case financial considerations limited the scope of the background data.

A population laboratory of this kind is expensive to maintain. Its advantages are that the background information facilitates the drawing of efficient subsamples and that these can be probability samples from a known population with a broad coverage of urban and rural conditions and of the different social classes.

## 5. THE STEP FROM ASSOCIATION TO CAUSATION

This issue is naturally of great concern to workers in observational research and has received much discussion in individual subject-matter fields. I shall confine myself to a few comments on statistical aspects of the problem.

First, as regards planning. About 20 years ago, when asked in a meeting what can be done in observational studies to clarify the step from association to causation, Sir Ronald Fisher replied: "Make your theories elaborate". The reply puzzled me at first, since by Occam's razor the advice usually given is to make theories as simple as is consistent with the known data. What Sir Ronald meant, as the subsequent discussion showed, was that when constructing a causal hypothesis one should envisage as many *different* consequences of its truth as possible, and plan observational studies to discover whether each of these consequences is found to hold. If a hypothesis predicts that $y$ will increase steadily as the causal variable $z$ increases, a study with at least three levels of $z$ gives a more comprehensive check than one with two levels. A secondary consequence of a hypothesis may be that the relation between $y$ and $z$ changes in a known direction as we move from low to high educational levels. If the study can be made large enough, a verification on this point can be included as well as a determination of the overall relation between $y$ and $z$. The comparisons of the death rates of cigarette smokers and non-smokers are rich in opportunities for this kind of verification. In the largest studies, we can compare the death rate (i) of men who smoked different amounts for the same time, (ii) among smokers of the same amount, of men who had been smoking for different lengths of time, (iii) of ex-smokers and current smokers of the same amount, (iv) among ex-smokers, of those who had previously smoked different amounts, (v) among ex-smokers of the same amount, of those who had stopped recently and those who had stopped for longer periods. The causal hypothesis predicts the direction in which the results should lie for each of these comparisons.

Of course, the number and variety of the consequences depends on the nature of the causal hypothesis, but imaginative thinking will sometimes reveal consequences that were not at first realized, and this multi-phasic attack is one of the most potent weapons in observational studies. In particular, the task of deciding between alternative hypotheses is made easier, since they may agree in predicting some consequences but will differ in others.

Since the initial work on a problem is often done in a restricted population, repetition of the study plan in different environments by different workers has its value, especially in forming a judgement whether results from the sampled population can be extended to a broader target population. Since studies that follow essentially the same plan may be subject to the same biases, an approach with a different plan that escapes some of these biases is highly useful.

When summarizing the results of a study that shows an association consistent with the causal hypothesis, the investigator should always list and discuss all alternative explanations of his results (including different hypotheses and biases in the results) that occur to him. This advice may sound trite, but in practice is often neglected. A model is the section "Validity of the results" by Doll and Hill (1952), in which they present and discuss six alternative explanations of their results in a study. Since this discussion may require use of data extraneous to the study and may be assisted by supplementary observations that can be taken in the study if the investigator realizes the need for them, it is well to anticipate these alternative explanations when the study is being planned.

Similarly, if the study gives an estimate of the size of the effect, possible biases in the estimated size should be discussed. For instance, the responses of crop yields to fertilizers can be estimated by a survey in which farmers report their yields and the amounts of fertilizers used. Clearly, the response will be over-estimated if the better farmers use more fertilizers, but under-estimated if fertilizers are applied primarily to poor soils. If controlled experimentation were impossible in agriculture, so that such surveys were the only means of information about the effects of agricultural practices, we could at least ask supplementary questions to permit adjustment for some known sources of bias and to aid our judgement in discussing remaining biases. As Yates and Boyd (1951) have pointed out, there is no assurance that this process will bring us close to an unbiased estimate. On date of planting of potatoes, surveys over three years gave an average reduction of 0·45 tons per acre for each week's delay, as against 0·5 tons per acre from a limited number of experiments—a good agreement, Boyd (1957). But in these surveys, the response of potatoes to farmyard manure (in the presence of complete fertilizers) averaged 0·1 tons per acre. Statistical adjustments for region, variety, class of seed, date of planting, and acreage grown raised this average to 0·4 tons per acre. Large numbers of controlled experiments give an average of 1·4 tons per acre—more than three times as great. The specialist potato grower has limited available manure but obtains high yields, while from the livestock farmer, who has ample manure, potatoes evidently receive less skilled attention. Although this example is a dampening reminder of the hazards of observational studies, thoughtful judgement about the direction and size of biases is better than no judgement.

In interpreting the results of regression studies on several variables, some postulated as causal and some as disturbing, I believe that more attempts should be made to bring in the investigator's ideas on the directions of causal paths. This type of analysis has been developed to considerable lengths in economics, with much clarification as well as some differences of opinion, and in genetics, under the leadership of Sewall Wright, but has been little tried elsewhere. For those unfamiliar with the technique, illustrations of its potentialities are the discussion by Tukey (1954) of the relations between birth weight, gestation period and litter size in guinea pigs, by Yates (1960) of the relation between health, total income, rent and housing quality, and by Wold (1956) in economics.

The combined evidence on a question that has to be decided mainly from observational studies will usually consist of a heterogeneous collection of results of varying quality, each bearing on some consequence of the causal hypothesis. If some results appear to support the hypothesis, some to contradict it and some are neutral, reaching a verdict demands much skill. Obviously, the investigator should consider whether some revision of his hypothesis will remove the contradictions. In default of this, he cannot avoid an attempt to weigh the evidence for and against, since some results are so vulnerable to bias that they should be given low weight even if supported by routine tests of significance. He should state such judgements forthrightly, remembering his duty to maintain even standards and, if possible, an air of calm detachment. An example that can be recommended is the competent discussion by Cornfield *et al.* (1959) of some of the controversial results in the smoking–lung cancer problem. Even here, some readers may detect a slight shifting of standards that hints at a departure from complete objectivity, though I mention this with trepidation, since the paper has six authors and I am alone. The situation is, of course, more comfortable for the pure scientist, who can always reserve final judgement while stating that there is a strong *prima facie* case, than for the applied scientist who must decide at what point a call for action should be made.

I have written as if the observational statistical studies are the only evidence. Fortunately, there will often be a corpus of laboratory-type research, perhaps using controlled experiments, that endeavours to probe more deeply into the nature of the causal mechanism. In many areas, such research is the primary hope of reaching a full understanding.

In conclusion, much of this paper was written while I enjoyed the hospitality of the Statistics Department, Rothamsted Experimental Station.

REFERENCES

BOYD, D. (1957). "A scrutiny of the British potato crop", *Oper. Res. Quart.*, **8**, 6–21.
CHAPIN, F. S. (1947), *Experimental Designs in Sociological Research.* New York: Harper.
COCHRAN, W. G. (1952), "An appraisal of the repeated population censuses in the Eastern Health District, Baltimore". In *Research in Public Health*, pp. 255–265. New York: Milbank Memorial Fund.
COCHRAN, W. G., MOSTELLER, F. and TUKEY, J. W. (1954), *Statistical Problems of the Kinsey Report*, pp. 246–253. Washington, D.C.: American Statistical Association.
CORNFIELD, J. (1951), "A method of estimating comparative rates from clinical data, applications to cancer of the lung, breast and cervix", *J. Nat. Cancer Inst.*, **11**, 1269–1275.
—— (1956), "A statistical problem arising from retrospective studies", *Proc. Third Berkeley Symp.*, **4**, 135–148.
—— HAENSZEL, W., HAMMOND, E. C., LILIENFELD, A., SHIMKIN, M. B., and WYNDER, E. L. (1959), "Smoking and lung cancer: recent evidence and a discussion of some questions", *J. Nat. Cancer Inst.*, **22**, 173–203.
DOLL, R. and HILL, A. BRADFORD (1952), "A study of the aetiology of carcinoma of the lung", *Brit. Med. J.*, **2**, 1271–1286.
DORN, H. F. (1953), "Philosophy of inferences from retrospective studies", *Amer. J. Public Health*, **43**, 677–683.
FALES, W. T. (1951), "Matched population records in the Eastern Health District, Baltimore, Md.", *Amer. J. Public Health*, **41**, 91.
FESTINGER, L. and KATZ, D. (Eds.) (1953), *Research Methods in the Behavioral Sciences.* New York: Holt, Rinehart & Winston.
HILL, A. BRADFORD (1953), "Observation and experiment", *New England J. Med.*, **248**, 995–1001.
JAHODA, M., DEUTSCH, M. and COOK, S. W. (1951), *Research Methods in Social Relations.* New York: Dryden.
KEYFITZ, N. (1952), "Differential fertility in Ontario. An application of factorial design to a demographic problem", *Population Studies*, **6**, 123–134.
KIHLBERG, J. K. and NARRAGON, E. A. (1964). "A failure of the accident severity classification", *Cornell Aeronautical Lab. Report.* VJ–1823–R8, 62–70.
KINSEY, A. C.; POMEROY, W. B. and MARTIN, C. E. (1948), *Sexual Behavior in the Human Male.* Philadelphia: Saunders.
KODLIN, D. and THOMPSON, D. J. (1958), "An appraisal of the longitudinal approach to studies in growth and development". *Monographs of the Society for Research in Child Development*, No. 67. Lafayette, Indiana.
MANTEL, N. and HAENSZEL, W. (1959), "Statistical aspects of the analysis of data from retrospective studies of disease", *J. Nat. Cancer Inst.*, **22**, 719–748.

NEEL, J. V. and SCHULL, W. J. (1956), *The effect of exposure to the atomic bombs on pregnancy termination in Hiroshima and Nagasaki*. Washington, D.C: Atomic Bomb Casualty Commission.

TUKEY, J. W. (1954), "Causation, regression and path analysis", Ch. 3 in *Statistics and Mathematics in Biology*. Ames: Iowa State College Press.

WOLD, H. (1956), "Causal inference from observational data", *J. R. statist. Soc.* A, **119**, 28–61.

YATES, F. (1960), *Sampling Methods for Censuses and Surveys*, 3rd ed. London: Griffin.

—— and BOYD, D. (1951), "The survey of fertilizer practice: an example of operational research in agriculture", *Brit. Agric. Bull.*, **4**, 206–209.

## DISCUSSION ON PROFESSOR COCHRAN'S PAPER

Sir AUSTIN BRADFORD HILL: I suspect that in our approach to observational studies of the human population, there is only one material difference between Professor Cochran and myself. He, as he points out in an early paragraph of his paper, has (in this situation) largely served as a referee or, at the very least, as a linesman. Over the last 40 years I have had to rush feverishly around the field of play, and in this particular field, unfortunately, most of the missiles are aimed at the players; indeed it is not unknown for the referee to join in. The only comfort I can get is that I suspect that some of those applying for research grants and ruled "offside" by Professor Cochran have probably not quoted the first line of that cockney ditty that he quotes, "it's the sime the 'ole world over". They are more likely to have used the last line!

I agree with him and the random member of the audience whom he quotes that it is all very difficult. I would, however, emphasize—and this arises from Professor Cochran's discussion of the experimental approach—that in this field the difficulties of experiment are no less. That tends sometimes to be overlooked. Take, for instance, the work quoted by Professor Cochran of Neel and Schull on atomic radiation at Hiroshima and Nagasaki and its effect on the human population. If one could make an experiment, surely all the disturbing variables would still have to be recognized? Would they not still have to be considered in setting up the experimental design? Mere randomization of the experimental units would, I believe, be unlikely to give a valuable answer or one that was practically useful.

Later in his paper Professor Cochran rightly quotes with approval the use of a population laboratory. But we are still faced with the problem of how far we can pass from the particular to the general. And in some instances with population laboratories, the particular is *very* particular. The same is true of the clinical trial of a new treatment or a field trial of a vaccine. It is only in quite limited circumstances and with deliberately courted statistical dangers that we can make experiments at all.

How far we can extrapolate from those experiments must always be a matter of concern. And so, as with observations, it is all very difficult. Heaven forbid that in saying so I should appear to denigrate the experimental approach—I have been passionately devoted to it throughout my life; but to statisticians who do not work in this field, and to some who do, I emphasize that it also needs very great care and thought.

With observational studies such as Professor Cochran has described we are, in the last analysis, having to take decisions on circumstantial evidence. Usually, no one sees the murderer slip the arsenic into the teacup and no one sees the *Bacillus typhosus* slink into the tin of corned beef. And so, just as in everyday life, and nearly every day, we take decisions on circumstantial evidence, so must we do in preventive medicine. We cannot escape it.

More often than not the retrospective enquiry, with all the weaknesses that Professor Cochran emphasized, is inevitable. As he himself says, it is only after the victims have appeared that we can start to explore the origin of the epidemic. It is in this way that classical epidemiology has been built up and advanced, but, as Professor Cochran rightly stresses—and this is the important point—along with the results of cognate enquiries of quite other types.

What is essential here is a profound knowledge of the field of work under discussion. Merely to go into the origin of a modern epidemic of typhoid fever from a knowledge of statistical records, their strength and weaknesses, will not take one far. One must have a close familiarity with the habits of the bacillus and of the strength and weaknesses of the laboratory evidence.

Similarly, although this is less well recognized, we will not get far in discussing cancer of the lung and smoking unless we know the whole field, a field with which Professor Cochran became familiar through his membership of the magnificent Advisory Committee to the Surgeon General of the United States Public Health Service. There is indeed a very wide range of data to consider. We have the statistical association in man with all its ramifications in both retrospective and prospective enquiries. We have the histo-pathological evidence of cellular changes in the respiratory mucosa of smokers and non-smokers. We have animal experimentation by the "back-room boys" in the laboratories, using smoke or smoke products, and there is semi-experimental evidence in man when smoking is given up. All this has to be taken into account in reaching a conclusion. It is ignorance of the field, or perhaps neglect of it, that has led some scientists up such curious garden paths that one might almost think that they were random walks!

There comes a time, as Professor Cochran finally observes, when the decision must be made, when the applied scientist must believe that the time for action has arrived—not that I believe that *he personally* should necessarily do anything whatever about it. That time, I believe, may quite rightly and justly be made to vary with the circumstances.

On that subject, I was recently delivered of a presidential address to the Section of Occupational Medicine of the Royal Society of Medicine. In it, I considered nine different ways in which one should study an observed association before passing to causation. In conclusion I said:

"In passing from association to causation, I believe in 'real life' we shall have to consider what flows from that decision. On scientific grounds we should do no such thing. The evidence is there to be judged on its merits and the judgement (in that sense) should be utterly independent of what hangs upon it—or who hangs because of it. But in another and more practical sense we may surely ask what is involved in our decision. That almost inevitably leads us to introduce differential standards before we convict.

"Thus on relatively slight evidence we might decide to restrict the use of a drug for early-morning sickness in pregnant women. If we are wrong in deducing causation from association no great harm will be done. The good lady and the pharmaceutical industry will doubtless survive.

"On fair evidence we might take action on what appears to be an occupational hazard, e.g. we might change from a probably carcinogenic oil to a non-carcinogenic oil in a limited environment and without too much injustice if we are wrong.

"But we should need very strong evidence before we made everyone burn a fuel in their homes that they do not like, stop smoking cigarettes that they do like or eating the fats and sugar that they enjoy. In asking for very strong evidence I would, however, repeat emphatically that this does not imply crossing every 't', and swords with every critic, before we act.

"All scientific work is incomplete, whether it be observational or experimental. All scientific work is liable to be upset or modified by advancing knowledge. That does not confer upon us a freedom to ignore the knowledge we already have nor ever to postpone the action that it appears to demand at a given time."

In the end the decision must turn, I suspect, upon our personalities. Only very limited aid can be sought in those subtle partitions of $\chi^2$ to which the reader of the paper has so eruditely contributed.

We are delighted to have Professor Cochran in this Society again and it gives me a very real and personal pleasure to move this vote of thanks to him.

Mr C. B. WINSTEN: Professor Cochran's paper is a timely one. A large amount of work is being done (and always has been done) on non-experimental data, and statisticians should, of course, make a special contribution. The situation as he sees it in the United States, with "the statistician perhaps contributing a chapter on sampling or tests of significance" (Professor Cochran does not say whether the chapters in different books are all the same) is, on the face of it, depressing. But the picture may look different if we think more about the nature of this type of research.

As R. A. Fisher's enigmatic remark emphasizes, the principal characteristic of non-experimental data is the wealth of possible hypotheses available to explain the patterns of variability which we observe. Even the most innocent-looking scatter diagram, showing an encouraging tendency to correlation, may have a host of different causal schemes which can explain it. Cause may go either way, as we all know. Or there may be some underlying cause influencing both variables. Or an underlying variable may influence one of the variables, and just happen to be correlated with another. Or one variable may have some influence in one part of the range of variation, and another elsewhere.

The patterns of variation may arise from self-selection of the population, as in Professor Cochran's example of the housing list. Or, more subtly, and very important in cross-section studies in economics, the units may adapt, or be forced to adapt, to their own peculiarities. For example, we might expect firms with low productivity to capture less of the market than those with high productivity. If this happens, observed data will over-emphasize the advantages of large units. Again, people presumably adapt to their own peculiarities in choosing or doing a job. We would have to guard against any simple causal scheme in studying the effects of type of work conditions or on those doing the work.

Now such complexity should not be discouraging: it is in fact a challenge. But it does mean, as Professor Cochran says, that research in non-experimental studies should be done by people who have a wide grasp of the field of study and an understanding of the plausibility of different theories. In other words, if statisticians take to this work they must stop being "methods" men and become research workers in the field being studied. To some extent this should be true in experimental studies too, but the need there is not so great. Perhaps, therefore, both teaching and observational studies are best carried out in "subject matter" departments. Perhaps the statisticians interested in such subjects have adapted themselves and migrated to these departments, and this explains Professor Cochran's observations.

Because of the richness of possible explanation, any particular set of data needs far more careful study than experimental data. In experimental data each reading is a sample representing many other readings and therefore often remains anonymous. But in non-experimental data each reading may be a clue to some neglected cause at work. Thus it should be anything but anonymous. At the very least this means that one should identify the individual points in scatter diagrams and get to know why they are there. For example, in studies of variations between regions, or towns, in which I am interested, if one looks at the points which are particularly high or particularly low on the diagram one may be able to pick out particular patterns which suggest a new cause: this may be a useful guide to useful multiple regressions. But such methods mean that one must have, or acquire, detailed knowledge of the towns or regions being considered. Incidentally, it is rather rare to find a linear regression in this type of study.

It is also important to see how much one's measurement of causes might depend on one or two exceptional units. How often, for example, is a study saved from the perils of multi-collinearity by a single point "off the line"? But if this reading is exceptional in this way it may be exceptional in many other ways too.

And one of these other ways may be associated with an unsuspected causal factor. This danger leads to another rule of action for the statistician: always study how far your regressions depend on particular readings: do them several times with and without points where the causal factors are in some way extreme or unusual.

There are dangers, in the light of this, in giving an air of false precision and scientific exactness to this type of work, a precision appropriate to a carefully randomized experiment but inappropriate for data collecting from that untidy laboratory, the real world. It is easy enough to take our data along to a computer, where a regression will be fitted, standard errors will be calculated for the coefficients and, for good measure, a usually inappropriate *t* test will be performed. This will look as though it gives a full measure of the uncertainty of the theory, but it does not, of course, Not only are there nearly always (*pace* Occam, a bad guide, as the quotation from Fisher shows) the possibilities of non-linearities of many different sorts, but also of all sorts of different causal schemes. Standard errors do not measure these sorts of uncertainties. Here I would like to stress how useful the likelihood function is as a mode of passing on to other investigators the objective knowledge and the objective uncertainty found from a particular experiment. The likelihood function can always be extended to include newly conceived hypotheses: it is always clear to what range of hypotheses it applies. What is more, in regression analysis, since the likelihood function is of such a simple form, the sum of squared residuals away from the hypothetical regression, it can even be estimated by eye. Indeed, where a two- or three-dimensional scatter diagram can be used, the eye, when trained, is probably the best computer of likelihood functions. It also has the great advantage of being able to see almost immediately what the computer can only do elaborately, the sensitivity of the likelihood to different shapes of regressions, different points, etc. Unfortunately we cannot use scatter diagrams in four or more dimensions so easily. Here is a field where we must make the computer help to scan the data in the way that is possible without its help in simpler cases.

Professor Cochran has given us a paper on a most important topic. I hope that it will lead to more discussion by statisticians and comparison of their experience in the very wide range of subjects touched upon to-day. I have much pleasure in seconding the vote of thanks.

The vote of thanks was put to the meeting and carried unanimously.

Mr F. D. K. LIDDELL: I should like to add my thanks to Professor Cochran for his paper, which I have found most stimulating, particularly because, in Sir Austin Bradford Hill's terminology, I have just been brought into a particular field of play as a substitute half-way through a game and I have a feeling that the players who have already been taking part will throw bottles, as well as the spectators and the referee.

I should like to say something about the pneumoconiosis field research project carried out by the National Coal Board. I want to do this because there is a belief that if an investigation is on a large enough scale and well enough intentioned, it must provide adequate data.

The population that we are looking at in this research is the complete labour force of a sample of 25 collieries spread throughout Great Britain; at the start of the research 10 years ago, it consisted of 35,000 coal miners. This study falls within Professor Cochran's definitions: the principal objective is to elucidate a cause and effect relationship (it is to discover what level of respirable airborne dust can be tolerated in mining without leading to an unacceptable prevalence of the disease which is called pneumoconiosis—and that means simply dust in the lungs) and it is quite clear that we cannot carry out controlled experimentation.

The response to dust exposure is measured on medical surveys carried out five years apart at each of the 25 collieries. In the interval between the surveys, dust exposure itself is being measured.

The main response variable is change in X-ray appearances but, unfortunately, assessment of such change is subjective and no truly reliable method has yet been found anywhere in the world. We like to think that we are furthest advanced in this country, but even we have a long way to go.

We obtain a second response variable from a questionnaire on respiratory symptoms. This has all the usual difficulties, confused here perhaps even more by many miners' thoughts about compensation when they are answering the questions.

Perhaps the most reliable response we have is that measured in lung function tests. But, unfortunately, lung function deteriorates markedly with the age of the subject—much more than with the degree of pneumoconiosis; further, a more than usually rapid deterioration of lung function may be due not to pneumoconiosis alone but to bronchitis, to smoking, or to some other cause completely unrelated to dust exposure.

Levels of airborne dust in the mining environment vary greatly between jobs and depend on such matters as the mining machines and conditions, so we have no worries that at least we shall find differences in the cause being examined. However, dust levels are highly variable from time to time in the same working place and measurement is difficult. Despite a massive effort in absolute terms, it leads in relative terms to only about one complete shift sampled per man in about 10 years.

The study has to be long-term because the disease takes many years to develop, and the field units are to be congratulated on having solved many of the problems that Professor Cochran mentions on this particular subject. The response rate of the subjects is of the order of 95 per cent of those employed at the time of each medical survey.

Our main concern, however, is with the cohort—that is, the survivors from one survey to the next—and these become only 60 per cent of the initial population. That sounds fine—20,000 subjects still appear an adequate population for study. The real blow is that on the first serial examination of these 20,000 subjects we find only about 800 showing signs of radiological progression and, therefore, of direct concern to us. It is all very fine that the management of the 25 collieries may have been particularly zealous in dust suppression but that does not help us very much from the research point of view, particularly when we bear in mind the many disturbing variables, e.g. smoking habits, obesity, atmospheric pollution.

This material, I am told by physicists and doctors, is better than anything available anywhere else. Nevertheless, the correlations that my colleagues have been able to establish are disappointingly poor.

I have no doubt that in speaking so briefly of the pneumoconiosis field research so soon after having been thrown into its deep end I have distorted considerably, possibly through oversimplification as much as anything else. Nevertheless, I cannot but feel a degree of the frustration that Professor Cochran mentions in his introduction as arising in admittedly rather different circumstances. Here, the basic planning was carried out a decade ago and now we have a mass of data. Quite clearly, we must obtain the maximum benefit from them. I should therefore welcome greatly any suggestions on strategies that Professor Cochran, or anyone else in this distinguished audience, may care to make. Meanwhile, his paper will be most useful in placing the difficulties of our own research in a wider context.

Miss E. M. BROOKE: The Ministry of Health is, naturally, concerned with observations on human populations. In particular, it is concerned with the subject of mental illness, which accounts for nearly half the occupied hospital beds in this country.

Compared with previous speakers, I come almost from the Stone Age because, while it is comparatively easy to talk about miners, about people moving to new towns, or about people with lung cancer, we must at the outset ask who is mentally ill.

Therefore, the population which we want to observe is not at all clearly defined. Except in conditions like general paralysis of the insane, or phenylketonuria, we have no tests. For diagnosis and for detecting the presence of the illness we rely on verbal communication and observation of behaviour. These differ in different cultures and what is normal in one is abnormal in the other.

We are told that mental illness is a deviation from the normal, but no one defines what the normal is. We are also told that we must have tensions if we want to live satisfactory, healthy lives and, therefore, some symptoms of mental illness, such as anxiety, should naturally occur. And so we have to face the problem of when a symptom ceases to be healthy and becomes morbid.

As soon as we have tried to define the population, diagnostic labels are attached to people. Here again we are at a great disadvantage because it is obvious that, even in this country, the diagnostic label must vary very much, as can be seen from the map on p. 149 in *The Geography of Life and Death* by L. Dudley Stamp.

This variation is even greater between countries. In America there are three times as many schizophrenics as depressives. In this country the depressives greatly outweight the schizophrenics. The question is: are the Americans more schizophrenic and are we more depressed or is this a diagnostic variation? A great deal of time has already been devoted to attempts to find the beginnings of an answer to this question.

For an actual investigation of a group of mentally ill people, cases have to be assembled and, therefore, populations should be screened. This is too much for one person and, in order to get a sufficient number of cases, we need an instrument which will enable any number of workers infallibly to find the same case of illness. It is this kind of instrument— a questionnaire which will enable different investigators to arrive at the same kind of case—that has to be developed. So that we are, in fact, at the beginning of shaping our tools.

In the past we have, unfortunately, had to fall back upon hospital populations, but these, of course, are determined by the numbers of beds and by the existence of other facilities. The study of these can lead to considerable fallacies. For example, we notice that the admission rates for single people are much higher than for married and, therefore, the hypothesis was set up that there was a selection against people who were liable to mental illness so that they were not selected as marriage partners. But once one extends one's observations to the people receiving other services, we find that a great many of these missing married people are, in fact, attending the out-patient clinics.

When it comes to cause and effect, we are very much struck in mental illness by the effects which are startling, especially to people who see patients. We are extremely ignorant, I think, of the causes, and there is a great tendency to take coincidences for causes.

This happened, for instance, at the time when a decrease in mental hospital populations coincided with the advent of the tranquillizing drugs. The conclusion straight away appeared to be that one was the effect and the other the cause. In this country, however, the run-down in the mental hospital population started before tranquillizers had gained ground and, at the same time, a great many other factors were coming in, such as the opening of closed wards, bringing the patients into touch with the community and introducing industrial work which would fit them to take their place in society again. Thus we are faced with the problem of discovering which is the causative factor, and in this we have not made very much progress so far.

Retrospective studies are, of course, very popular in this exercise: one sees the problem and one hopes to look back and find the cause. One of the particular joys has been the broken home. This is a quite popular subject and easy to establish, judging by the number of studies in which it features.

People look at the life histories of delinquents or mentally ill people and find the broken home. No one appears to explain why a home that is as broken for Tom as it is for Jim, causes Tom to go into a mental hospital but not Jim. Nor has anyone apparently so far started a longitudinal study to find out how many normal people come from broken homes, how many people from broken homes do not become abnormal, etc.

We are, of course, concerned in the Ministry with the effects of administrative policy and in particular we have adopted the policy that people should, as far as possible, be treated in the community and in general hospitals instead of in large mental hospitals away from the community.

The charge that has been made is that we have not demonstrated the advantage of this. Of course, we cannot demonstrate the advantage until we have worked the system. But it means that we have to organize studies which will enable us to compare the effects of treating people in the general hospital and in the large mental hospital. Here, it is desirable to match groups.

One is then faced with the variables on which to match, and those usually selected are the first suggested: age, sex, marital status, occupation, social class and so on. But it is by no means certain that these are all that are necessary. There is the variation in conditions within the same diagnosis. For example, Leonhard has suggested that schizophrenics can be classified into 20-odd groups and, obviously, even if you compare cases with a slow, insidious onset with those having a sharp onset, considerable variation in their reaction can be observed.

Altogether we are trying to match on some 60 items and we hope to find out by cluster analysis which of these are really significant. At present we are in the position of preparing the tools and hoping that at some future time we may be able to do some useful work with them.

Mr A. S. C. EHRENBERG: Professor Cochran is concerned with "problems . . . that seem . . . to differentiate observational research most clearly from controlled experimentation" and in Section 2 he lists what he calls the major difficulties. Running through these briefly, we have:

In Section 2.1 the specific difficulty mentioned is that the "observational investigator" has to know where to find the phenomena he wants to study; it can, however, be equally helpful if the "controlled experimenter" knows something about his subject-matter.

In Section 2.3 Professor Cochran describes his concept of cause and effect as essentially predictive, but so it would have to be in controlled experimentation.

Under Sampling, in Section 2.4, it is thought that the deliberate use of populations to be sampled which differ from the target populations will remain standard "observational" practice, but this practice is said also to be followed in controlled experimentation.

The "formidable measurement problems" of Section 2.5 must, of course, apply equally well in controlled experimentation, and are to be overcome only by hard work and very little talk.

If in Section 2.6 there is difficulty in telling "imaginative investigators" to stop adding more and more "ingenious" variables to be measured, such investigators can be relegated to their science fiction department irrespective of whether they are "observational" or "controlled".

Again, if in Section 2.7 observational long-term studies are said to lack opportunities for quick publication, we ought to thank God and remember Fisher, who left Rothamsted years before his famously *controlled* long-term experiments were even completed.

Only in Sections 2.2 and 3 dealing with "disturbing variables" is there a suggestion— at first sight—that "observational" and "controlled" studies do differ in the way stated, i.e. that in the latter "the function of blocking or adjustment is to increase precision rather than to guard against bias". However, the "adjustment" procedures given in Section 3 generally do not work. They largely turn on the "dependent" variable $y$ being regressed on the "disturbing" variable $x$, where (i) the distribution of $x$ must differ in the populations to be compared (otherwise there would not be any effective disturbance), and (ii) the regressions are assumed to be the same. This assumption is known to be universally false under condition (i)—a point which was made in Section 5.11(B) in Lindley (1947, p. 234)† and which has been belaboured since in *Applied Statistics* (Vol. XII, 1963 No. 3) to be published this year.

My conclusion is that the problems which Professor Cochran has discussed have nothing to do with any distinction between observational and controlled studies. I am not alone here since Professor Cochran, having led us up the garden path, says exactly the same in his final "comment" at the end of Section 2.

† LINDLEY, D. V. (1947), "Regression lines and the functional relationship", *J. R. Statist Soc.* B, **9**, 218–244.

10

Perhaps I have now prepared the way for the outright rejection of the good statisticians' pseudo-traditional myth which is always implicit and often explicit in Professor Cochran's own paper, namely, that observational studies are at best a second best to so-called controlled experimentation. However, if we survey the work of thousands of scientists over hundreds of years, a conservative estimate might be that at the very least 99·9 per cent of the work has been done by methods which statisticians would never deem fit for their canons of "controlled experimentation". And, of course, the scientists were right, as shown by history and as may be very briefly illustrated by the smoking–lung-cancer problem to which Professor Cochran himself frequently refers in his paper, where I now echo one or two of the remarks made by Sir Austin Bradford Hill.

Suppose that we adopt Professor Cochran's "attractive possibility" (Section 4.1) of comparing twins, and add full-blooded randomization of which twin is to smoke. Then the differing incidence of lung cancer in this "controlled experiment" may be caused by differential rates of contact with cigarette paper or with the noxious metal foil in the packets, by differential exposure to fumes from matches, etc., used for lighting up, by different frequencies of visiting tobacconists, by smokers being either more or less relaxed because they can puff and suck, by non-smokers eating more sweets, etc., and therefore dying early of heart disease, by non-smokers having nothing to do with their hands and getting up to neurotic substitute malpractices, by the well-known fact that smokers "note" cigarette advertisements more and may therefore be impelled to emulate their virile suggestions such as mountain climbing and sailing with ghastly effects on their health such as getting lung cancer (it's the "ads" that did it!), and so on.

In other words, the number of potential "disturbing variables" left in this "perfectly con-trolled experiment" are enough to give a statistician heart failure, especially if he also smokes. Many of the factors can no doubt be, or have already been, eliminated, but this could never be done within the narcissistic confines of the *one* controlled experiment. Controlled experi-mentation is no more than a potentially convenient and just occasionally applicable tactical method for eliminating one or two of the many disturbing variables present in any situation—here people's initial differences, if any, in their predisposition to lung cancer.

Anyway, having found a five times higher incidence of lung cancer amongst non-smokers in this experiment, and for many other good reasons which are well known to *scientists*, the experiment is "repeated" in some sense. Unfortunately then, two or more independent studies of any kind can no longer make up a controlled situation. It follows immediately that controlled experimentation can play no central role in scientific methodology, which concerns itself with the building up of generalized and integrated knowledge. A controlled experiment is necessarily self-contained, and there is room for a *first* study or an *isolated* experiment at most once in any one field of study.

Apart from wasting their own time by attempting to transmute controlled experimen-tation into a philosopher's stone instead of treating it as an occasional convenience, statisticians are—far more importantly—making practical workers feel guilty about "not doing experiments". The stultifying influence of statistical teaching in this particular respect is exceeded in its ill-effects by at least two other aspects of modern statistics, and could therefore be even worse than it is.

Clearly, there is more that one could say, but I conclude still within the general temper of our discussions in this Society by admitting that when I now quote Professor Cochran's own transferable quotation of Professor Barnard's reported comment on Dr Wold's 1956 paper on observational data—namely, that "a paper of this kind is useful in showing the younger statisticians what difficulties they may be up against"—I do feel younger in spirit. Such, we humbly learn, are the workings of cause and effect.

Mr V. SELWYN: As the instigator in the market research world of a large-scale series of observation studies may I in contrast to the previous speakers quote from the hard world of commerce. I should like to contrast observational studies in the market research world, not with controlled experiments, but with verbal techniques.

So much market research work turns on asking people questions and the very act of asking questions produces a series of unknowns. We are dependent upon investigators and upon the words that people use. Therefore, there is a virtue in any technique which avoids questioning.

The exercises with which I have been connected have been concerned with brand consciousness in the motor-oil world. We have taken one measure of brand consciousness, and that is the degree to which the motorist asks for a brand when he buys his motor oil. We will not argue about how far that measures his brand consciousness. We merely take it as a measure.

When we first looked at this particular problem, we realized the cheapest way of finding out was to ask people what they do in their buying situation. But this is not the most desirable way. When we ask we are dependent on respondents' memories, their beliefs, and so on, and memories of occasional purchases can be hazy. Instead, we decided to base a study on observing people at the point of purchase. What does the motorist actually do when he pulls up to a garage forecourt and buys his motor oil?

This study has been repeated over a series of years. Each year 25,000 motorists are observed, of whom about 4,000 buy motor oil.

Now the technique raises a formidable sampling problem. It is the main objection to this technique in market research. We try to base our sample of garages, where we observe, on unknown data—mainly, the degree to which various brands of petrol are sold through various garages. In this garage world, however, there are political factors: i.e. garages are controlled as to what they sell, they are tied to various petrol companies. Some co-operate and some will not. In the final analysis, one is dependent upon a sample of co-operators.

However, we do not depend on the absolute figures of each exercise. By repeating the exercise we get a trend from year to year. On the basis of 4,000 oil purchasers we may find, for example, 71 per cent in one year ask for a brand of oil by name. But instead of depending on that as absolute, we measure what happens the year later and the year following that so as to eliminate the effects of certain variables in the sampling. We may have too many garages of a certain type, small or large, too many on a main road or too many where only pleasure or only business motorists go. These are factors which may affect the numbers asking. Over a period of years we try to minimize as many of these variables as possible.

On the other hand, this operation introduces new variables. Although we try to hold the test on the same week of each year, in June, when we hope to have the same weather, we still have the effects on traffic flows of a variable Whitsun in this country. We may also find that the garage which we visited last year will not co-operate this year, or, alternatively, that the garage changes its policy or the attendants have changed. Instead of finding a lackadaisical attendant who does not mind what brand of oil a motorist has, we might find an aggressive one who tries to force sale.

These points illustrate my contention that one can apply various statistical methods—but everything turns on the raw material. It proves impossible to apply tests to all the variables—for some we do not even know. We try to learn all the possible factors which might affect our answers—but we can only hope to allow for all of them.

Any research technique must have its limitations. In an observation test we can only analyse our motorists by observable factors. We can see whether a motorist is in a saloon or a sports car, or whether he is driving a lorry or a van. But we cannot classify him by age. Investigators only record and do not ask questions. We cannot ask them, for instance, to classify motorists into young, old or middle-aged. For this would prove a subjective assessment on the part of an investigator, on which we do not want to depend.

Yet within all these limitations we have produced very useful results, which, obviously, I cannot go into now. It is a technique which, I feel, is neglected in the market research world, perhaps because of the limitations I have described or perhaps because people have not thought about it sufficiently.

Professor CochRAN subsequently replied in writing as follows:

I am grateful to the speakers for gently correcting an imbalance that my method of presentation may have created and for rounding out the paper with descriptions of the nature of observational research in pneumoconiosis, mental illness and marketing. Sir Austin Bradford Hill and Mr Ehrenberg have reminded us that controlled experiments are faced with much the same set of problems as observational studies and demand an equally high degree of skill in their execution. My reason for bringing controlled experiments into the paper was that the formal study of techniques for planning and analysis has proceeded much further in experimentation than in observational work, and I believe that when grappling with an observational study a knowledge of the lessons learned over the years in experimentation is likely to be useful.

Mr Winsten and Sir Austin have emphasized that the statistician involved in observational studies must become a "subject-matter" expert, because subject-matter knowledge is essential to effective planning and because the information to be summarized on a complex problem is usually widely miscellaneous, ranging from statistical studies to casual but suggestive hints. The degree of attention to detail and imaginative insight that must be used if one is not to be misled in scrutinizing the results of an observational study have been well brought out by Mr Winsten. I am glad that these points were made, because in a statistical paper it is easy to leave an exaggerated impression of the importance of the statistician's standard bunch of tricks.

In the studies described by Mr Liddell, Miss Brooke and Mr Selwyn, I was interested to note that problems of measurement are a major concern. With new tools of measurement, a significant part of the research must be to learn their strengths and weaknesses. This is particularly so when measurement is subjective, for despite Mr Selwyn's reluctance to rely on subjective appraisals, they are often the only ones available. Fortunately, the importance of errors of measurement in the sample surveys taken in many countries is now much better appreciated, with a consequent increase in the research effort devoted to this problem. Although lack of good instruments of measurement is sometimes the chief stumbling-block to progress, there is always room for hope that a major advance may come from an unexpected quarter. The awards of Nobel prizes for developments in tissue culture and paper chromatography are a sign of the importance rightly attached to advances in the science of measurement.

The approach described by Mr Selwyn is an ingenious way of circumventing an erroneous assumption that still persists in some quarters—namely that if one asks people how they would react to a hypothetical situation with which they are not now faced, their replies will be trustworthy for predictive purposes. It is necessary to devise some means of observing how they actually react when faced with the situation.

The areas of research in which Mr Liddell and Miss Brooke are engaged are especially difficult. The list of problems given by Miss Brooke is very familiar to me from my limited acquaintance with the mental illness field, although, alas, I have no good suggestions to offer. I have the impression that the degree of success attained in the long-term follow-up of subjects is greater in this country than in the U.S., perhaps because of the higher mobility in the latter. I was a little puzzled by the phrase "disappointingly poor" applied by Mr Liddell to the correlations obtained. If low correlations are a sure indication that the measurements are inadequate, the word "disappointing" is appropriate. But if they merely imply that the results do not agree with preconceived notions, the adjective hints at an emotional involvement that may obstruct clear thinking.

Since I may already have written too much about controlled experiments, I am reluctant to deal fully with Mr Ehrenberg's comments. His account of their limitations is, I think, much overdrawn. In many countries, large programmes of controlled fertilizer and variety trials are in progress, because past information from such experiments has led to substantial increases in the world's supply of food. The flood of useful new products coming from the research and development sections of industry relies heavily on the skilful use of multi-variable experimentation. A proposal that scientists in these fields should in future confine themselves to observational studies because experimentation is "no more than a potentially convenient and occasionally applicable tactical method" would be considered ridiculous.

With regard to his amusing report on the hypothetical twin-smoking experiment, I know of no one engaged in this field who would regard an identical twin study as more than a small additional link in the evidence, helpful because, as Mr Ehrenberg points out, it removes a genetic difference that is hard to eliminate from most of the other available comparisons. In my own experience I cannot recall instances of statisticians bullying practical workers for not attempting experiments that would have been a waste of time, though they sometimes scold workers for doing bad experiments when they could have done good ones.

Mr Winsten's account of the usefulness of examination of the likelihood function interested me. Now that machines are becoming available for graphical presentation of the results from electronic computers, we may, with some practice, be able to grasp results in more than three dimensions.

On a technical point, Mr Ehrenberg claims that if the distribution of $x$ differs in two populations, the assumption that the regression of $y$ on $x$ is the same in the two populations is "universally false". On the contrary, models in which the assumption holds are easily constructed. We need, for example, merely make $y = \alpha + \beta x + e$ in both populations, where $e$ is a random variable with zero mean for given $x$. His supporting citation from Lindley's well-known paper deals with quite a different problem. Lindley's Section 5.11(B) on p. 234 reads as follows: "The regression line is required for prediction of either true or observed values of one variate from observations of the other *whether or not this latter is in error*: it being understood that the $x_1$ from which $x_2$ is to be predicted comes from the same population as those $x_1$'s used in the estimation of the regression line." Lindley deals with a *single* population in which (i) the regression of $x_2$ on a variate $\xi_1$ is linear, (ii) we cannot measure $\xi_1$ but only $x_1 = \xi_1 + d_1$, where $d_1$ is an error of measurement independent of $\xi_1$. He gives necessary and sufficient conditions for the linearity of the regression of $x_2$ on $x_1$, and shows that if this regression is linear, it should be used for predicting $x_2$ for a new specimen with measured value $x_1$. Lindley's final phrase is a warning that if $x_1$ is measured much more or much less accurately in the new specimen than in the specimens from which the regression was constructed, his results do not apply.

The fact that regressions can be mathematically the same in two populations does not mean that they are the same in applications. Numerous cases can be found in which they are not the same. The appropriate method of adjustment in this situation presents problems, which for limitations of space I did not discuss.

In conclusion, my objectives in writing a mainly descriptive paper on this extensive subject were twofold. I believe, as stated, that statisticians have much to contribute to this challenging field. Secondly, it is a commonplace to find the same problem in study after study. If means can be found for bringing them together for the discussion of tough methodological issues, practical workers in diverse areas of investigation can profit greatly from one another's experience and wisdom, both as regards what to do and what to avoid.

# 86

## ANALYSE DES CLASSIFICATIONS D'ORDRE

### W. G. COCHRAN
Professeur de Statistique, Harvard University, U.S.A.

*Quand les données ont la forme d'une classification par degrés, il est souvent raisonnable de considérer que la classification représente une espèce de groupement d'une échelle de mesurage continue. On peut donc assigner à chaque classe un numéro, qui exprime notre concept de la position de la classe dans l'échelle. Ces numéros créent une variable qui suit une distribution discrète, et permet l'analyse des données au moyen des méthodes développées pour les distributions discrètes et, comme approximation, par les méthodes communes pour les distributions continues.*

*L'avantage de ce procédé est que les méthodes d'analyse des variables discrètes et continues sont plus flexibles et puissantes que les méthodes qu'on a pour les données qualitatives. L'exposé en présente plusieurs illustrations. Un désavantage est que le procédé est plus ou moins subjectif. Mais on trouve que des différences modérées entre les variables assignées par deux investigateurs font généralement peu de différence dans les conclusions.*

*L'étude décrit plusieurs méthodes moins subjectives qui ont été utilisés pour construire les numéros : comparaison avec une mesure continue qu'on considère plus précise, usage d'une population standard dans laquelle la variable associée a une distribution normale, et construction d'une échelle qui aura certaines propriétés désirées et raisonnables.*

*Enfin, il y a deux méthodes objectives – la méthode maximin de Abelson et Tukey, et un $\chi^2$ test de Bartholomew en utilisant le test du rapport des vraisemblances. Cette dernière méthode ne construit pas une variable discrète, mais semble être puissante si le but de l'analyse est un test de signification.*

## 1 - INTRODUCTION

Les méthodes statistiques sont bien adaptées pour l'analyse de deux espèces d'observations. En premier lieu, on trouve les méthodes fondées sur la distribution binomiale, qui servent pour les observations classées dans l'une ou l'autre de deux catégories, comme par exemple les objets approuvés ou rejetés dans une inspection. Pour les caractères quantitatifs, comme la résistance électrique d'un fil, il y a les méthodes développées d'après la loi normale de Gauss-Laplace - l'analyse de variance, la régression, etc.

Quelquefois, cependant, les données ont la forme d'une classification par degrés, peut-être par degrés de gravité ou de préférence. Par exemple, on pourrait décrire les défauts trouvés dans un article comme : mineur, majeur ou critique. Il est évident que dans cette classification, un défaut critique est plus grave qu'un défaut majeur, et un défaut majeur est plus grave qu'un défaut mineur. De même, dans une enquête sur les accidents industriels, on pourrait décrire le degré de préjudice subi par un individu comme léger, modéré, sévère ou fatal. Dans une enquête de marché, une maison de commerce donne deux articles de toilette A et B à chaque ménagère prise dans un échantillon. Après une période d'usage, chaque ménagère exprime sa préférence en choisissant une des cinq classes : préfère nettement A, préfère A, indifférente , préfère B, préfère nettement B.

Jusqu'à présent, les méthodes les plus efficaces pour l'analyse des observations de cette espèce n'ont pas été étudiées à fond. Je veux d'abord exposer une méthode, simple et assez ancienne, qui suffit en pratique pour beaucoup de problèmes. Dans les sections 5 et 6, je décris plusieurs approches alternatives.

Ma première illustration provient d'une expérience sur l'usage des médicaments dans le traitement des lépreux. Au début de l'expérience, il y avait deux types de malades, A et B. Au cours de l'expérience, qui dura presqu'un an, les deux types reçevaient le même traitement médical. A la fin, un expert de la lèpre a examiné soigneusement chaque malade. Il a noté le progrès du malade pendant l'année, et l'a classé dans une des cinq catégories montrées dans la Table 1. La classe la plus favorable est "Amélioration importante", tandis que "Aggravation" représente la situation la moins favorable.

Table 1

Deux groupes de lépreux, classés suivant le degré d'amélioration

| Degré d'amélioration | Groupe A | Groupe B | Total | Distribution A | % | Distribution B |
|---|---|---|---|---|---|---|
| Amélioration importante | 11 | 7 | 7 | 8 | | 14 |
| moyenne | 27 | 15 | 15 | 19 | | 29 |
| faible | 42 | 16 | 16 | 29 | | 31 |
| nulle | 53 | 13 | 13 | 37 | | 25 |
| Aggravation | 11 | 1 | 1 | 8 | | 2 |
| Total | 144 | 52 | 52 | 100 | | 100 |

$$\chi^2 = 6,88 \qquad (4 \text{ d. d. l}) \qquad P = 0,16$$

Conclusion : Pas de différence entre les deux groupes.

La Table 1 à gauche donne les résultats. Le but de l'analyse est de déterminer s'il y a une différence entre les degrés moyens d'amélioration montrés par les deux types de malades.

2 - L'EMPLOI DE $\chi^2$

Le procédé le plus connu pour tester l'hypothèse que les deux dis-

tributions A et B sont les mêmes, à l'exception des erreurs d'échantillonnage, est le test $\chi^2$, ou :

$$\chi^2 = \sum \frac{(\text{Observés - Théoriques})}{\text{Théorique}}$$

La somme s'étend sur les 10 cellules de la Table 1 (5 lignes et
2 colonnes). Les nombres théoriques sont les nombres de malades qu'on
trouverait si les deux distributions étaient réellement les mêmes. Par
exemple, on attendrait $(18)$ $(144)/196 = 13.0$ malades de type A avec
amélioration importante.

La valeur de $\chi^2$ est 6.88, avec 4 d.d.l. Si l'hypothèse nulle est
vraie, la probabilité d'obtenir une valeur de $\chi^2$ aussi grande est 0.16.
Selon le test $\chi^2$, on dirait que les distributions, et par conséquent les
degrés moyens d'amélioration, ne sont pas différents entre A et B.

Mais le $\chi^2$ test n'est pas approprié dans ce cas, car le test ne
spécifie pas une hypothèse alternative particulière, et ne tient pas compte
du fait que nous avons une classification par degrés de gravité. Si on
change l'ordre des classes dans Table 1, la valeur de $\chi^2$ reste la même.
La classification représente une tentative de mesure de la quantité d'amélioration de chaque malade. Notre hypothèse alternative est que, en
moyenne, le degré de progrès diffère entre les deux types A et B. Si
cette hypothèse alternative est vraie, nous nous attendons à trouver que
les malades d'un type montrent une plus grande proportion de cas classés
"Amélioration importante" ou "Amélioration moyenne" et une plus petite
proportion classée "Amélioration nulle" ou "Aggravation".

La Table 1, à droite, donne ces proportions. Notez que le type B
contient 43 % de malades dans les deux meilleures classes, contre 27 %
pour le type A. Dans les deux classes qui ne présentent pas d'amélioration, le type B contient 27 % de malades, et le type A 45 %. Ces résultats suggèrent que les malades de type B ont fait plus de progrès
que ceux de type A.


3 - VARIABLES ASSOCIEES

La méthode que je veux exposer se base sur le point de vue que

1/ si les instruments de mesure étaient assez bons, on pourrait
mesurer le progrès de chaque malade sur une échelle continue,

2/ la classification faite par l'expert représente une espèce de
groupement de cette échelle continue en 5 classes.

De ce point de vue, il semble raisonnable d'assigner à chaque
classe un numéro, qui exprime notre concept de la position de la classe
dans l'échelle. Ces numéros créent une variable, associée à la classification, et qui suit une distribution discrète. Ayant construit la variable
associée, on peut analyser les données au moyen des méthodes développées pour les distributions discrètes. Il n'est pas nécessaire de faire
une hypothèse, peu réaliste, que l'expert a groupé sans erreurs. En
pratique, il y aura des cas où la mesure idéale placerait le malade dans
la deuxième catégorie, mais où l'expert l'a placé dans la première. En
anglais, les numéros qui produisent la variable associée s'appellent généralement "scores".

Comment assigner les numéros ? Ceci dépend de l'information qu'on a concernant la construction des classes, et également des buts de l'enquête. Cette question sera discutée plus tard. Dans l'exemple étudié actuellement, j'ai affecté le numéro 0 à la classe "Amélioration nulle", et les numéros 1 et 2 aux classes "Amélioration faible" et "Amélioration moyenne", puisque l'expert semblait considérer ces classes comme également séparées. Ma première idée était d'assigner le numéro 4 à la classe "Amélioration importante" et le numéro -2 à la classe "Aggravation", parce que l'expert a examiné un malade plus soigneusement et plus longtemps avant de la placer dans une de ces deux classes extrêmes. Mais j'ai décidé que, peut-être, ceci n'était qu'une impression subjective, et j'ai donné les numéros 3 et -1.

Lorsque les numéros sont assignés, on a deux échantillons de la variable associée. On teste la signification de la différence entre les moyennes des deux échantillons par le test t de Student. La Table 2 contient le calcul, le symbole x représentant la variable associée. La méthode est la méthode élémentaire d'application du test t. La valeur de t est 2,616, avec une probabilité un peu moindre qu'une chance sur cent. On peut donc en conclure que les malades de type B ont fait, en moyenne, plus de progrès que ceux de type A.

Table 2

Analyse avec variables associées

| Degré d'amélioration | x | Fréquences A f | Fréquences B f | | A | B |
|---|---|---|---|---|---|---|
| Amélioration importante | 3 | 11 | 7 | $\Sigma f$ | 144 | 52 |
| moyenne | 2 | 27 | 15 | $\Sigma fx$ | 118 | 66 |
| faible | 1 | 42 | 16 | $\bar{x}$ | 0,819 | 1,269 |
| nulle | 0 | 53 | 13 | $\Sigma fx^2$ | 260,0 | 140,0 |
| Aggravation | -1 | 11 | 1 | $(\Sigma fx)^2 / \Sigma f$ | 96,7 | 83,8 |
| | | | | $\Sigma f(x - \bar{x})^2$ | 163,3 | 56,2 |
| | | 144 | 52 | d.d.l. | 143 | 51 |
| | | | | $s^2$ | 1,150 | 1,102 |

$$\bar{\bar{s}}^2 = 219,5/194 = 1,131$$

$$t = \frac{\bar{x}_B - \bar{x}_A}{\sqrt{s^2 \left(\frac{1}{n_A} + \frac{1}{n_B}\right)}} = \frac{0,450}{\sqrt{(1,131) \left(\frac{1}{144} + \frac{1}{52}\right)}} = 2,616$$

$$P = 0,009$$

Conclusion : le Groupe B montre plus d'amélioration, en moyenne, que le Groupe A.

Ce calcul s'étend facilement à la comparaison de plus de deux groupes A, B, C, D, ... On fait d'abord un test F (test de Snedecor) de l'hypothèse nulle que les moyennes de tous les groupes sont égales. Si le test montre qu'il y a des différences, on peut ensuite faire des tests t plus détaillés. Si les groupes représentent des niveaux croissants de quelque ingrédient, on peut obtenir la courbe de régression de la moyenne de chaque groupe sur les niveaux. Par exemple, on pourrait préparer quatre plats de pommes de terre, contenant quatre quantités différentes de sel. Chaque sujet goûte tous les plats et les classe : "Beaucoup trop de sel", "Trop de sel", "Assez de sel" etc. Le but est d'obtenir la relation entre la quantité de sel et la réponse du sujet.

On peut avoir une <u>double</u> classification rangée par degrés. Si on étudie la relation entre les conditions de travail des ouvriers dans une usine et la qualité de leur production, les conditions sont classées par exemple comme supérieure, satisfaisante, mauvaise, et la qualité comme excellente, bonne, pauvre. Dans ce cas on assigne deux variables associées, y et x. L'analyse de données de cette espèce a été présentée par Yates (1948) [8]. Dans certains cas il y a plus d'intérêt à comparer la variabilité des groupes qu'à comparer leurs moyennes. L'exemple dans la Table 3 est tiré des résultats d'une ancienne enquête conduite dans les écoles de Londres. On a essayé de classer selon leur intelligence cent garçons et cent filles pris dans les mêmes écoles.

Table 3

Intelligence des élèves à Londres

|  | Supérieure | au-dessus de la moyenne | Moyenne | au-dessous de la moyenne | Faible | Total |
|---|---|---|---|---|---|---|
| Garçons | 7 | 22 | 48 | 19 | 4 | 100 |
| Filles | 5 | 18 | 58 | 16 | 3 | 100 |
|  | 2 | 1 | 0 | -1 | -2 |  |

$$s_g^2 = 76,9 \qquad s_f^2 = 62,4$$
$$F = 76,9/62,4 = 1,23, \text{d. d. l.} = (99,99)$$

La table ne suggère pas une différence entre le degré moyen d'intelligence des garçons et celui des filles. Mais il semble que les filles sont moins variables entre elles que les garçons. Notez que 58 % des filles ont l'intelligence moyenne, contre 48 % des garçons.

Si on peut assigner une variable associée aux classes d'intelligence, on peut calculer la variance de la mesure d'intelligence pour les garçons et pour les filles. Avec les numéros que j'ai donnés, on trouve $s_g^2 = 76.9$ (garçons) : $s_f^2 = 62.4$ (filles). La valeur de F est $76.9/62.4 = 1,23$, avec 99 et 99 d. d. l. Puisque cette valeur n'est pas significative au niveau 5 % de probabilité, la variabilité supérieure qu'on constate pour les garçons peut être expliquée par les erreurs d'échantillonnage.

## 4 - CRITIQUE DE LA METHODE

Comme nous venons de le voir, la construction d'une variable associée a plusieurs avantages. Elle permet l'emploi de tests qui sont plus puissants que le test $\chi^2$, parce qu'on peut les utiliser pour tester le type particulier d'hypothèse alternative qu'on attend dans les observations. En outre, on peut, par exemple, examiner si la différence moyenne entre le groupe A et le groupe B vaut deux fois la différence moyenne entre le groupe B et le groupe C, ce qui n'est pas possible avec une analyse qualitative. Plus généralement, la méthode est une tentative d'obtenir la plupart des avantages d'une échelle continue de mesure.

La critique principale est, bien entendu, que le procédé est arbitraire. Deux investigateurs peuvent assigner des numéros différents à la même classification. Mais on trouve que dans la majorité des cas, des différences modérées entre les numéros assignés produisent peu de différence entre les conclusions qu'on tire de l'analyse. Par exemple, dans la Table 1 j'ai calculé le test t avec la série de numéros 4, 2, 1, 0, -2 que j'avais d'abord considérée. La valeur de $\underline{t}$ devient 2.46, avec P = 0.015, contre t = 2,62, P = 0.009 produit par la série alternative 3, 2, 1, 0, -1. Les deux analyses mènent à la même conclusion.

Néanmoins, il y a quelquefois de grandes difficultés avec les classes extrêmes. Aux Etats-Unis on a mené beaucoup d'enquêtes sur le degré de dommage souffert par les personnes dans les accidents d'automobiles. Les classes de dommage sont, par exemple, mineur, modéré, sévère, rendant incapable, et fatal. Comment assigner des numéros aux classes : rendant incapable et fatal, afin de les placer dans la même échelle que les autres classes ? De même, un défaut dans une pièce mécanique, tel que la machine ne marche pas, semble être d'un autre degré que celui du défaut qui, quoique gênant, permet à la machine de fonctionner. Dans ces cas, la difficulté peut provenir du fait que la classification ne représente pas une seule échelle. Il faut peut-être deux ou trois variables pour décrire assez exactement la classification.

La variable associée a une distribution discrète non normale : en effet, elle est quelquefois assez dissymétrique. Peut-on vraiment employer les méthodes standard fondées sur la distribution normale, comme le test t et le test F ? Pour examiner cette question dans des cas simples, on peut calculer la distribution exacte de $\underline{t}$ et de $\underline{F}$ par la méthode de "randomization" de Fisher. A mon avis, en pratique, les méthodes normales suffisent généralement, étant entendu que le niveau 5 pour-cent de $\underline{t}$ ou de F trouvé par ces méthodes n'a pas une probabilité exactement égale à 5 %. Il correspond, peut-être, à un niveau entre 4 et 7 pour-cent. Il faut noter aussi deux points :

1/ Les variances peuvent différer d'un groupe à l'autre, spécialement quand les distributions sont biaisées. Si on prépare des pommes de terre avec beaucoup de sel, il peut arriver que tous les sujets les classifient comme "Beaucoup trop de sel", et la variance est nulle, tandis qu'un plat contenant moins de sel peut recevoir tous les degrés de la classification. Aussi, avant de combiner les variances pour obtenir une seule estimation, il est sage de vérifier que les variances ne diffèrent pas, et il faut fréquemment utiliser des méthodes d'analyse qui n'assument pas l'égalité des variances.

2/ Avec des échantillons très petits, c'est-à-dire contenant moins de 20 unités, je suggère l'emploi d'une correction de continuité dans le test t, afin de tenir compte du fait que nous avons une distribution discrète.

## 5 - AUTRES METHODES POUR CONSTRUIRE LES VARIABLES ASSOCIEES

Jusqu'à présent j'ai donné une seule méthode, principalement subjective, pour assigner les numéros. En fait, différents investigateurs ont utilisé plusieurs méthodes, moins subjectives, pour construire les variables associées. Il n'existe pas beaucoup de comparaisons entre les méthodes, de manière qu'on ne peut juger de leurs avantages et désavantages relatifs. Mais, puisque le champ d'application diffère d'une méthode à l'autre, le manque de comparaison n'est pas trop gênant. Je vais décrire brièvement quelques-unes de ces méthodes.

### 5.1 - Comparaison au moyen de mesures plus précises.

Quelquefois il est possible de comparer la classification par degrés de gravité avec une autre mesure du même caractère qu'on considère comme étant plus exacte. Cette possibilité existe quand la classification représente une méthode de mesure rapide et bon marché, tandis que la méthode plus exacte est coûteuse. Par exemple, un observateur qui marche autour d'un arbre peut rapidement classer le dommage causé aux fruits par les insectes, comme nul, léger, modéré, sévère. D'autre part, on peut cueillir le fruit dans un échantillon d'arbres et mesurer le pourcentage de fruits "gâtés ou touchés" sur chaque arbre de l'échantillon par l'inspection des fruits individuels. On peut donc examiner si la classification correspond à un groupement des arbres selon le pourcentage de fruits touchés. Si oui, on peut assigner à chaque classe, comme variable associée, la moyenne des pourcentages pour les arbres qui tombent dans cette classe. S'il y a plusieurs observateurs, il est possible :

1/ de comparer la précision de différents observateurs,

2/ d'examiner si on peut utiliser la même variable associée pour tous les observateurs.

### 5.2 - Usage d'une population standard.

Quand on étudie des échantillons qui proviennent de populations différentes, il peut exister une population standard à laquelle on désire comparer les autres populations. La population standard peut représenter la méthode habituelle utilisée pour fabriquer un article, les autres populations représentant des changements de méthode. Supposons :

1/ que nous avons un grand échantillon tiré de la population standard,

2/ qu'il est raisonnable d'admettre que si le caractère pouvait être mesuré exactement, il aurait approximativement une distribution normale dans la population standard.

Dès lors, dans la population standard on regarde la classification comme un groupement de la distribution normale avec moyenne zéro et écart-type 1. On peut facilement déterminer les bornes de chaque classe dans l'échelle normale. La variable associée est la moyenne de la distribution normale entre ces bornes. Cette méthode est ancienne : le statisticien Karl Pearson l'employait beaucoup.

Pour illustrer les calculs j'ai utilisé les données de la Table 1, en combinant les deux groupes de malades pour construire une "population standard" artificielle. La Table 4 donne les détails.

Table 4

Illustration de la méthode de K. Pearson

| Degré d'amélioration | | % en classes | % cumulés P = F(u) | Borne supérieure de u | Moyenne de u | Variable associée |
|---|---|---|---|---|---|---|
| Aggravation | | 9 | 9 | -1,34 | -1,8 | -0,9 |
| Amélioration | nulle | 21 | 30 | -0,52 | -0,9 | 0 |
| | faible | 30 | 60 | +0,25 | -0,1 | 0,8 |
| | moyenne | 34 | 94 | +1,55 | +0,8 | +1,7 |
| | importante | 6 | 100 | ∞ | +2,0 | +2,9 |

On cumule d'abord les pourcentages dans les classes. Ensuite, on trouve les bornes, en usant la table de la distribution normale cumulative. Par exemple, la table normale montre que 9 % de la distribution reste entre $u = -\infty$ et $u = -1,34$. La moyenne de u entre les bornes a et b est $[Z(a) - Z(b)]/P$, ou $Z(a)$ et $Z(b)$ sont les ordonnées de la distribution pour $u = a$ et $u = b$. Enfin, j'ai ajouté 0.9 à chaque valeur de la variable associée, de manière que la valeur devienne zéro pour la classe "Amélioration nulle". Dans cet exemple, la variable associée diffère peu de la variable initialement utilisée dans la Table 1.

Dans une version alternative de cette méthode, due à Bross (1958), [5] la variable associée est la probabilité cumulative jusqu'au milieu de la classe. Les propriétés de la méthode de Bross, qui évite toute supposition de normalité, sont décrites dans la référence donnée [5].

### 5.3 - Construction d'une échelle qui aura certaines propriétés désirées.

Plusieurs investigateurs ont suggéré cette approche. Dans l'application décrite par Ipsen (1955) [6] une série de niveaux $Z_1$, $Z_2$, $Z_3$... d'un certain traitement ont été appliqués aux sujets, chaque sujet recevant un seul niveau. Plus tard, les sujets sont classés dans des catégories ordonnées qui expriment le degré de succès du traitement. Ayant assigné une variable associée x à la classification, on peut calculer la moyenne $\bar{x}_i$ pour tous les sujets qui ont reçu le niveau $Z_i$. On peut penser que la variable $\bar{x}_i$ doit avoir une relation linéaire avec la variable $Z_i$. Ipsen détermine la valeur $x_j$ qu'on donne à la classe j de manière que le variance due à la régression linéaire entre les $\bar{x}_i$ et les $Z_i$ contienne la proportion maximum de la variance totale de x entre les sujets.

La solution est très simple. La valeur de x assignée à la classe j est la moyenne des valeurs de $Z_i$ pour les sujets qui tombent dans cette classe. Donc, la méthode est presque la même que celle décrite dans la section 5.1, si on considère le variable Z comme la mesure la plus exacte. La Table 5 présente un exemple donné par Ipsen. La variable Z représente le niveau d'immunisation contre le typhus, et les classes A, B, C, D, F sont les degrés de gravité de thyphus. Les moyennes $\bar{Z}$, dans la partie inférieure de la table, sont les valeurs de la variable associée. Dans la dernière ligne, Ipsen a transformé les valeurs de manière que leurs bornes soient 0 et 100.

Table 5

Degré de gravité de la maladie de 71 malades atteints de typhus, classés
selon le niveau d'immunisation

| Niveau d'immunisation (z) | Nombre de malades Degré de gravité du typhus | | | | |
|---|---|---|---|---|---|
| | A | B | C | D | F |
| 1 | | | 5 | 2 | 3 |
| 2 | | 4 | 8 | 4 | 1 |
| 3 | | 2 | 8 | 1 | |
| 4 | | 4 | 3 | | |
| 5 | 11 | 20 | 5 | | |
| f | 11 | 30 | 29 | 7 | 4 |
| $\Sigma$ fz | 55 | 130 | 82 | 13 | 5 |
| $\bar{z}$ | 5,0 | 4,3 | 2,8 | 1,9 | 1,2 |
| valeur associée | 0 | 18 | 58 | 84 | 100 |

Fisher (1947) [7] a décrit une autre application du même type.
Chacun des douze serums a été injecté dans le sang de chacun des douze
sujets. On a marqué les réactions des sujets par les symboles -, ?,
faible, (+), et ++[1], qui forment une classification rangée par degré de
force ou de puissance. On sait que la force de réaction doit différer entre
les sérums et aussi entre les sujets. Dès lors, Fisher a assigné les
valeurs 0, $x_2$, $x_3$, $x_4$, et 1 aux différentes classes, et a déterminé les
valeurs de $x_2$, $x_3$, et $x_4$ qui rendent maximum le rapport de la somme
des carrés entre séra et sujets à la somme totale des carrés dans l'analyse de la variance. Ici, le calcul n'est pas si simple : il faut trouver
les racines caractéristiques d'un déterminant.

# 6 - DEUX METHODES OBJECTIVES

## 6.1 - La méthode "maximin".

Toutes les méthodes précédentes supposent, en effet, que le chercheur connaît un peu plus que l'ordre simple des classes. Par exemple,
avec les lépreux, j'ai jugé que la distance entre les classes "Amélioration
nulle" et "Amélioration faible" est approximativement la même que la
distance entre "Amélioration faible" et "Amélioration moyenne".

Au contraire, la méthode maximin de Abelson et Tukey (1963,
1959) [2] [1] est une méthode objective qui suppose seulement ce qu'on
connaît réellement concernant l'ordre des classes. Ces auteurs supposent qu'il y a une variable associée $x_j$, vraie mais inconnue, pour la-
quelle $x_1 \leqslant x_2 \leqslant x_3 \ldots$ Ils excluent le cas $x_1 = x_2 = x_3 \ldots$ ; c'est-à-dire
qu'il y a au moins une inégalité. Si y est la variable qu'on analyse (par
exemple, proportion de lépreux dans le groupe A), la relation la plus
simple entre $y_j$ et $x_j$ est

$$y = \alpha + \beta x + e$$

---------------

(1) Le symbole (?) indique une réaction probablement faible.

où la variable $\underline{e}$ est une variable de moyenne nulle, de variance $\sigma^2$, la covariance entre $e_i$ et $e_j$ ($i \neq j$) étant nulle. Pour examiner si $y_j$ change d'une façon monotone quand $x_j$ change, il faut estimer $\beta$.

Puisque $x_j$ est inconnu, on construit une variable associée $c_j$. On peut avoir la convention $\Sigma c_j = 0$. Le numérateur du coefficient de régression de $y_j$ par rapport à $c_j$ est $c = \Sigma c_j y_j$. On en déduit :

$$E(c) = \beta \Sigma c_j x_j = \beta \Sigma c_j (x_j - \bar{x})$$

$$V(c) = \sigma^2 \Sigma c_j^2$$

Si on veut tester la quantité c, pour découvrir si $\beta \neq 0$, la puissance du test est déterminée par $E(c)/\sqrt{V(c)}$.

$$\frac{E(c)}{\sqrt{V(c)}} = \frac{\beta \Sigma c_j (x_j - \bar{x})}{\sigma \sqrt{\Sigma c_j^2}} = \left\{ \frac{\beta}{\sigma} \sqrt{\Sigma (x_j - \bar{x})^2} \right\} \left\{ \frac{\Sigma c_j (x_j - \bar{x})}{\sqrt{\Sigma c_j^2 \Sigma (x_j - \bar{x})^2}} \right\}$$

Le premier terme de l'expression à droite ne dépend pas du choix des $c_j$. Le second terme est le coefficient de corrélation r entre $c_j$ et $x_j$. On veut donc rendre r maximum.

Il y a une infinité de valeurs de $x_j$ satisfaisant les relations $x_1 \leqslant x_2 \leqslant x_3 \ldots$ Pour chaque choix des $c_j$, on peut déterminer les $x_j$ qui rendent $\underline{r}$ minimum. Abelson et Tukey recommandent l'utilisation des $c_j$ pour lesquelles le minimum $\underline{r}$ est maximum. Ainsi le nom "maximin".

Pour 2 jusqu'à 8 classes, la Table 6 donne les valeurs des "maximin" $c_j$ (ajustées pour que les valeurs centrales soient -1, 0, et 1.

Table 6

Valeurs des maximins $c_j$ et de $r^2$

| j | n = 2 | n = 3 | n = 4 | n = 5 | n = 6 | n = 7 | n = 8 |
|---|-------|-------|-------|-------|-------|-------|-------|
| 1 | -1 | -1 | -6.5 | -4.4 | -13.0 | -8.1 | -20.8 |
| 2 | 1 | 0 | -1 | -1 | - 3.5 | -2.4 | - 6.4 |
| 3 | | 1 | 1 | 0 | - 1 | -1 | - 3.2 |
| 4 | | | 6.5 | 1 | 1 | 0 | - 1 |
| 5 | | | | 4.4 | 3.5 | 1 | 1 |
| 6 | | | | | 13.0 | 2.4 | 3.2 |
| 7 | | | | | | 8.1 | 6.4 |
| 8 | | | | . | | | 20.8 |
| min $r^2$ | 1.000 | .750 | .651 | .596 | .557 | .530 | .510 |
| min $r^2$ linéaire | 1.000 | .750 | .600 | .500 | .429 | .375 | .333 |
| min $r^2$ lin.-2-4 | 1.000 | .750 | .649 | .588 | .549 | .522 | .493 |

L'échelle $c_j$ n'est pas une fonction linéaire de j. Elle diffère d'une échelle linéaire en ce que les valeurs extrême sont beaucoup plus grandes. Pour n = 8, par exemple, une échelle linéaire aurait les valeurs -7, -5, -3, -1, 1, 3, 5, 7, tandis que les $c_j$ sont -20.8, -6.4, -3.2, -1, 1, 3.2, 6.4, 20.8. En effet, les auteurs notent qu'une échelle qu'ils appellent "linéaire-2-4" est une bonne approximation des maximin c. Pour construire le "linéaire-2-4", on commence avec une échelle linéaire, et on multiplie les valeurs extrêmes par 4, et les valeurs voisines par 2. Pour n = 8, ceci donne -28, -10, -3, 1, 1, 3, 10, 28. La valeur du maximin $r^2$ est 0, 510, et pour le linéaire-2-4 la valeur minimum de $r^2$ est 0, 493, (valeurs données en dessous de la Table 6). Si le lecteur n'a pas la table des $c_j$, les·auteurs remarquent qu'il peut employer l'échelle linéaire-2-4.

Les auteurs ont montré que les valeurs les plus défavorables de la vraie échelle $x_j$ (les valeurs qui donnent les minima $r^2$) sont de l'espèce 0, 1, 1, 1, ... ou 0, 0, 1, 1, 1 : c'est-à-dire les valeurs qui ont une seule inégalité. En dessous de la Table 6 on trouve également les $r^2$ minimum donnés par une variable associée qui est linéaire en j. Pour $n \geqslant 5$, ces $r^2$ sont notablement plus petits que les $r^2$ maximin. Si une échelle comme la linéaire-2-4 semble un peu bizarre au lecteur, il faut remarquer que la méthode de Abelson et Tukey fût développée pour le chercheur qui ne connaît rien de plus que $x_1, \leqslant x_2 \leqslant x_3 \ldots$

La méthode s'étend à d'autres types d'information initiale. Par exemple, les auteurs construisent les $c_j$ dans un exemple avec 6 classes, où on sait que

$$x_1 \leqslant x_2 \leqslant x_3 \leqslant x_5 \leqslant x_6 \qquad \text{et} \qquad x_1 \leqslant x_4 \leqslant x_5$$

mais où on ne connaît pas la relation entre $x_4$ et $x_2$, $x_3$. On peut également utiliser la méthode quand on sait quelque chose sur les distances entre les $x_j$, mais les détails pour ce cas n'ont pas encore été publiés.

### 6.2 - Un $\chi^2$ test alternatif.

Dans cette méthode, due à Bartholomew (1961, 1959) [4], [3], les conditions sont les mêmes que pour la méthode maximin. Pour les membres de l'échantillon qui tombent dans la classe j, la variable qu'on analyse peut être ou une proportion $p_j$ ou la moyenne $\bar{y}_j$ d'une variable continue. On suppose que $E(p_j) = x_j$ (ou $E(\bar{y}_j) = x_j$), les $x_j$ représentant l'échelle vraie mais inconnue. Le problème posé par Bartholomew est de tester l'hypothèse nulle $x_1 = x_2 = x_3 = \ldots$ contre l'alternative $x_1 \leqslant x_2 \leqslant x_3 \leqslant \ldots$

Il construit un test par le critère du rapport des vraisemblances. Pour illustrer cette méthode, dans le cas le plus simple où $\bar{y}_j$ est la moyenne de $n_j$ observations normales, avec $\sigma^2$ connu, nous avons :

$$-2 \sigma^2 \ln L = \Sigma \, n_j \, (\bar{y}_j - x_j)^2$$

où L représente la vraisemblance. Pour l'hypothèse nulle (H.N.), $x_j = x$, et L est maximum lorsque $x = \bar{y} = \Sigma \, n_j \bar{y}_j / \Sigma \, n_j$. Donc :

$$-2 \sigma^2 \ln L_{max} \, (\text{h.N.}) = \Sigma \, n_j \, (\bar{y}_j - \bar{y})^2$$

Pour l'alternative (H.A.), il faut choisir les $x_j$ qui rendent L maximum, compte tenu des relations

$$x_1 \leqslant x_2 \leqslant x_3 \quad \ldots$$

La solution dépend de l'ordre des $\bar{y}_j$. Si

$$\bar{y}_1 \leqslant \bar{y}_2 \leqslant \bar{y}_3 \leqslant \ldots$$

en accord avec l'alternative, la solution est clairement $x_j = \bar{y}_j$, de manière que

$$-2\sigma^2 \ln L_{max}(H.A.) = 0$$

Dans ce cas, le critère pour tester est, ignorant le $2\sigma^2$,

$$[\ln L_{max}(H.A.) - \ln L_{max}(H.N.)] = \Sigma n_j (\bar{y}_j - \bar{y})^2.$$

Mais supposons, avec 3 classes, que

$$\bar{y}_1 \leqslant \bar{y}_3, \quad \bar{y}_2 \leqslant \bar{y}_3, \quad \bar{y}_1 > \bar{y}_2.$$

Bartholomew montre que la solution pour maximiser la H.A. est

$$x_1 = x_2 = \frac{n_1 \bar{y}_1 + n_2 \bar{y}_2}{n_1 + n_2} = \bar{y}_{12} \qquad x_3 = \bar{y}_3$$

Ainsi

$$-2\sigma^2 \ln L_{max}(H.A.) = n_1 (\bar{y}_1 - \bar{y}_{12})^2 + n_2 (\bar{y}_2 - \bar{y}_{12})^2 + n_3 (\bar{y} - \bar{y}_3)^2$$

Dans les cas plus compliqués, il donne les règles simples pour trouver la solution et ensuite le critère. En général, on combine les valeurs des $\bar{y}_j$ voisins pour produire une série de valeurs qui s'accordent avec l'hypothèse alternative. Le test exige des tables spéciales, décrites dans la référence 1959 [3].

Bartholomew (1961) [4] a comparé la puissance de son test avec celle due à la méthode maximin. Bien entendu, sa puissance relative dépend des valeurs des vrais $x_j$. Ni l'une ni l'autre des méthodes n'est supérieure sur toute l'étendue des alternatives. Mais dans les cas les plus défavorables aux maximin $c_j$, le test de Bartholomew gagne matériellement s'il y a plus de quatre classes. La raison semble être qui le critère du rapport des vraisemblances s'adapte à la nature vraie de l'alternative, tandis que les $c_j$ restent fixes pour un nombre donné de classes. La conclusion pratique est que :

1/ si on sait seulement que $x_1 \leqslant x_2 \leqslant x_3 \ldots$ et

2/ si le but est un test de signification, le test de Bartholomew semble être préférable.

Comme la méthode maximin, ce test s'étend :

1/ aux cas dans lesquels l'information sur l'ordre est incomplète ; par exemple, si on sait que $x_4$ est le plus grand, mais l'ordre des $x_1$, $x_2$, $x_3$, est inconnu,

2/ aux cas où l'on a une certaine information imprécise sur les distances entre les $x_j$ (ces résultats ne sont pas encore publiés).

J'ai préparé cet article, à la demande du Professeur Vessereau, pour une des Réunions d'Etudes sur les applications de la statistique dans les entreprises, au temps ou j'étais Professeur à l'Institute de Statistique sur invitation du Professeur Dugué de l'Institut. Je désire remercier le Professeur Dugué, le Professeur Vessereau et leurs collègues pour la généreuse hospitalité avec laquelle ils m'ont accueilli.

# REFERENCES

[1] ABELSON R. P. and TUKEY J. W. (1959) - Efficient conversion of nonmetric information into metric information. Proc. Social Statist. Section, Amer. Statist. Assoc., 226-230.

[2] ABELSON R. P. and TUKEY J. W. (1963) - Efficient utilization of non-numerical information in quantitative analysis. Ann. Math. Statist., 34, 1347-1369.

[3] BARTHOLOMEW D. J. (1959) - A test of homogeneity for ordered alternatives. Biometrika, 46, 36-48 and 328-335.

[4] BARTHOLOMEW D. J. (1961) - A test of homogeneity of means under restricted alternatives. Jour. Roy. Statist. Soc. B, 23, 239-281.

[5] BROSS I. D. (1958) - How to use Ridit analysis. Biometrics, 13, 18-38.

[6] IPSEN J. (1955) - Appropriate scores in bio-assays using death times and survivor symptoms.

[7] FISHER R. A. (1947) - Les Méthodes Statistiques. Presses Universitaires de France. Sections 49.2 et 49.3.

[8] YATES F. (1948) - The analysis of contingency tables with groupings based on quantitative characters.

## Footnote by William G. Cochran

In adding a few notes to Neyman's summary and appraisal of Fisher's contributions, I would like to present an impression of my own about Fisher's outlook, and to give some personal reminiscences of Fisher.

The subject matter of statistics has been defined in various ways. I believe that Fisher thought of statistics as essentially an important part of the mainstream of research in the experimental sciences. His major books, *Statistical Methods for Research Workers*, *Design of Experiments*, and the Fisher-Yates *Statistical Tables* (1) were addressed not to statisticians but to workers in the experimental sciences. The 1925 preface to the first edition of *Statistical Methods* opens as follows (1): "For several years the author has been working in somewhat intimate co-operation with a number of biological research departments; the present book is in every sense the product of

The author is professor of statistics at Harvard University, Cambridge, Massachusetts. This is the text of an address delivered at the December 1966 meeting of the AAAS in Washington.

this circumstance. Daily contact with the statistical problems which present themselves to the laboratory worker has stimulated the purely mathematical researches upon which are based the methods here presented." In those days the principal departments at Rothamsted, from which Fisher was writing, were soil chemistry, soil physics, bacteriology, microbiology, entomology, insecticides, botany, and plant pathology. Except possibly for botany, he contributed to the work of every one of these. His series of papers on distribution theory, which Neyman has described, were undertaken to provide working scientists with a battery of new tools to guide them in analyzing their data.

Neyman has reminded us of the occasion when Fisher was invited to present his ideas before the Royal Statistical Society. He entitled his paper "The logic of inductive inference," but hastened to tell his readers that the title might just as well have been "On making sense of figures"—inserting this homely alternative, I believe, in case his main title might suggest a rather rarified discussion remote from the real task of handling scientific data.

Consistent with this view was his assertion that decision theory was in no sense a generalization of his ideas. He writes (2): " . . . the Natural Sciences can only be successfully conducted by responsible and independent thinkers applying their minds and their imaginations to the detailed interpretation of verifiable observations. The idea that this responsibility can be delegated to a giant computer programmed with Decision Functions belongs to the phantasy of circles rather remote from scientific research." These fighting words might suggest a blanket disapproval of decision theory, but elsewhere he writes of decision theory as the correct approach to a different kind of problem, acceptance sampling (3): "The procedure as a whole is arrived at by minimising the losses due to wrong decisions, or to unnecessary testing, and to frame such a procedure successfully the cost of such faulty decisions must be assessed in advance; equally, also, prior knowledge is required of the expected distribution of the material in supply."

In his own researches he took little interest in the problem of collecting nonexperimental data, as in sample surveys, or in the more perplexing problem of making sound inferences from uncontrolled studies, although such data abound even in the experimental sciences. It is true that his late pamphlet

*Smoking—The Cancer Controversy* dealt with the pitfalls in drawing conclusions about cause and effect from nonexperimental data (4). However, his main contribution in that pamphlet, as I see it, was to claim that there were two alternative hypotheses, both reasonable and neither implicating smoking as the culprit, that might explain the available data on the relation between cigarette smoking and lung cancer death rates. Until data had been gathered that disproved both these alternatives, there was no justification in his view for diatribes or action against cigarette smoking. But as indicative of a rather half-hearted interest in the matter, he presented no detailed analysis to support the claim that these alternatives were in fact consistent with all the observed data, nor did he attempt to outline the types of data that would be needed for a crucial comparison among these hypotheses.

His concept of statistics may explain some of the things that irritated him—for instance, the teaching of a test of significance as a rule for "rejecting" or "accepting" a hypothesis. Like others, I have difficulty in understanding exactly what Fisher meant by a test of significance: he seems to imply different things in different parts of his writings. My general impression is that he regarded it as a piece of evidence that the scientist would somehow weigh, along with all other relevant pieces, in summarizing his current opinion about a hypothesis or in thinking about the nature of the next experiment. A passage in the seventh (1960) edition of *Design of Experiments*, inserted in order to clarify this point, reads as follows (3, p. 25):

In "The Improvement of Natural Knowledge," that is learning by experience, or by planned chains of experiments, conclusions are always provisional and in the nature of progress reports, interpreting and embodying the evidence so far accrued. Convenient as it is to note that a hypothesis is contradicted at some familiar level of significance such as 5% or 2% or 1% we do not, in Inductive Inference, ever need to lose sight of the exact strength which the evidence has in fact reached, or to ignore the fact that with further trial it might come to be stronger, or weaker.

In this connection I wish that Fisher had given more advice on how to appraise "the exact strength of the evidence." I have often wondered, as I suppose does Neyman, why Fisher seems not to have regarded the power of the test as relevant, although he de-

veloped the power functions of most of the common tests of significance.

He was also unhappy, particularly later in life, at seeing statistics taught essentially as mathematics by professors who overelaborated their notation (in order to make their theorems seem difficult, in his opinion) and who gave the impression that they had never seen any data and would hastily leave the room if someone appeared with data.

Although not disagreeing, I perhaps rate Fisher's positive contribution to estimation theory more highly than does Neyman. Given the probabilistic model by which the data were generated, the concept that a specific sample contains a measurable amount of information about a parameter, the delineation of cases in which a sufficient statistic exists, the notion of the efficiency of an estimate and the development of a technique for measuring efficiency—these, although not all original with Fisher, were great steps forward. It was soon evident that these concepts were oversimplifications, applicable to only a limited range of problems, and Fisher, like others after him, struggled hard to find ways of extending their range. But despite the mass of solid and difficult research that has been done since his work, it is noteworthy how often the methods actually used nowadays in data analysis are Fisherian, or are fairly straightforward extensions of his methods.

To turn to some reminiscences, the first is intended to illustrate Fisher's ingenuity in computations, of which he did a great deal with what would now be considered very inferior equipment. When he left Rothamsted in 1933 to become professor of eugenics in London, his assistant, Frank Yates, was appointed head of a one-man department of statistics at Rothamsted, this being the middle of the depression. On the land of the Duke of Bedford at Woburn, barley had been grown in a long-term experiment on the same plots for 50 years. The director of Rothamsted, Sir John Russell, was writing a book on this experiment. In preparation for this, he engaged a young lady to do statistical calculations, but soon afterward she resigned to get married and I was appointed, partly to finish these calculations and partly to assist Yates.

Since the young lady was on her honeymoon, I had first to discover what her calculations were about. For each year she had the barley yield, $Y$, per acre and six variates $x_1 \ldots x_6$ representing the amount and seasonal distribution of rainfall. To each rainfall variate she was fitting a fifth-degree polynomial in time (years), and finding the residuals from this regression. No time trend was being fitted to the barley yields.

Naturally, I plotted each variate against time. The yields had marked trends in time; the rainfall variates showed no sign of any time trend. Her objective was now clear. The time trend in yield was being interpreted as due to slow changes in the soil, not to weather. Consequently, she wanted to remove the effect of this trend in yields by a fifth-degree polynomial on time, before regressing the residuals of yield on the rainfall variates. "But," I said to myself, "she's all mixed up. She solemnly removes the trend from the rainfall variates, which don't have any trend, and doesn't remove it from the yields which do. I shall have to start over again and do it right."

Fortunately, it was then lunchtime. During lunch I asked Yates if he had shown the lady how to set up her calculations or if she had done this herself. The answer was: neither. Before she started, Fisher had given her detailed computing instructions. When I returned from lunch I thought "Perhaps we should not condemn this young woman too hastily." A little algebra showed me the well-known regression result which I should have learned in college. The yields $Y$ have a regression on $x_1 \ldots x_6$ (rainfall) and $x_7 \ldots x_{11}$ (time). If one does not want to invert an $11 \times 11$ matrix, the correct regression coefficients $b_1 \ldots b_6$ can be obtained in two ways: (i) Regress the yields and the rainfall variates on time, and then regress the residuals of yield on the residuals of rainfall. (ii) Regress the rainfall variates on time (whether they have a real trend or not) and regress the direct yields on the residuals of rainfall—this was Fisher's method. Regressing the yields but not the rainfall gives the wrong answer.

Why had he done it this way? Although there was only one set of rainfall variates, there were ten yield variates, from plots with different fertilizer treatments. By Fisher's two-step process, removal of the time trends involved dealing with only a single $5 \times 5$ matrix. Actually, no matrix inversion was required for this step, since Fisher had constructed fifth-degree orthogonal polynomials to do the job.

A second reminiscence might be entitled "Fisher explaining a proof." In one of his lecture courses, he quoted without proof a neat result for what appeared a complex problem. Since all my attempts to prove the result foundered in a maze of algebra, I asked him one day if he would show me how to do the proof. He stated that he had written out a proof, but after opening several file drawers haphazardly, all apparently full of a miscellaneous jumble of papers, he decided that it would be quicker to develop the proof anew. We sat down and he wrote the same equation from which I had started. "The obvious development is in this direction," he said, and wrote an expression two lines in length. After "then I suppose we have to expand this," he produced a three-line equation. I nodded—I had been there too. He scrutinized this expression with obvious distaste and began to stroke his beard. "The only course seems to be this" led to an expression four and one-half lines long. His frown was now thunderous. There was silence, apart from beard stroking, for about 45 seconds. "Well," he said, "the result must come out something like this" and wrote down the compact expression which I had asked him to prove. Class dismissed.

My third experience concerns our joint project. When I went to Cambridge as a student in 1931 my supervisor, Wishart, instructed me, at the request of Fisher, to compute a table of the 1 percent levels of $z = \log_e F$ to seven decimal places for a large panel of different pairs of degrees of freedom. Fisher was doing a corresponding table of the 5 percent levels. For those who think that graduate students nowadays are exploited by their professors, I might mention that Wishart told me he expected me to be working on this table 3 hours a day, 6 days a week, and that the labor was unpaid.

My contacts with Fisher on this project went through three stages. At first, when we met, he would ask about my progress: he had started sooner and was well ahead. Then came a period when he didn't ask, so I would ask him how he was coming along: I was gaining and towards the end of this period I was ahead. The third stage is easily foreseen. I would ask him, and he would hastily change the subject. I can take a hint as well as the next

man. I believe that we last mentioned the project sometime in 1936. The fourth incident exemplifies Fisher as the outraged professor. When he was professor of genetics at Cambridge I called on him one spring morning at his working quarters, Whittinghame Lodge. I was told that he had just received some upsetting news, and was walking in the garden to calm himself. The news was a report from the university committee that was to approve Fisher's proposed teaching program in genetics, to the effect that they had not yet completed their study of his proposal, and there would be a further postponement of a decision, for the seventh time as I recall it, until their next meeting in October. "Cambridge University," said Fisher, "should never appoint a professor who is older than 39. If they do, then by the time his proposal for his teaching program has been approved by the university, he will have reached retirement age."

Finally, there is Fisher, the applied geneticist. We were standing at the corner of Euston Road and Gower Street in London, waiting to cross the road on our way to St. Pancras Station. Traffic was almost continuous and I was worried, because Fisher could scarcely see and I would have to steer him safely across the road. Finally there was a gap, but clearly not large enough to get us across. Before I could stop him he stepped into the stream, crying over his left shoulder "Oh, come on Cochran. A spot of natural selection won't hurt us."

The experience of a period of association with a genius is so exhilarating that I wish every young scientist could have it. I don't know that it helps one to become a better scientist, because relationships and results that we can discern only with great effort, if at all, seem to come in a flash to someone like Fisher. But a glimpse of what the human brain at its best can do encourages a spirit of optimism for the future of *Homo sapiens*.

### References

1. R. A. Fisher, *Statistical Methods for Research Workers* (Oliver and Boyd, Edinburgh, ed. 13, 1963); ———, *Design of Experiments* (Oliver and Boyd, Edinburgh, ed. 7, 1960); R. A. Fisher and F. Yates, *Statistical Tables for Biological, Agricultural and Medical Research Workers* (Hafner, New York, ed. 6, 1963).
2. R. A. Fisher, *Statistical Methods and Scientific Inference* (Hafner, New York, 1956), p. 100.
3. ———, *Design of Experiments*, Sec. 12-1.
4. ———, *Smoking—The Cancer Controversy* (Oliver and Boyd, Edinburgh, 1959).

PLANNING AND ANALYSIS OF NON-EXPERIMENTAL STUDIES*

W. G. Cochran
Harvard University
Cambridge, Massachusetts

1. INTRODUCTION.    During the past 20 years a marked increase
in statistical studies of human populations has taken place.  Several
reasons for this can be suggested.  Successful applications of operations
research during World War II led to an expanded use of this technique in
business and marketing after the war.  Public opinion polls, which proved
interesting and informative as news media, stimulated the growth of
agencies equipped to take sample surveys for clients.  The provision of
increased amounts of money for field research in the social sciences also
contributed.

In many of these studies, the objective is primarily descriptive--to
get the basic facts about some problem.  Examples are the monthly
estimates of numbers of employed and unemployed, or a survey undertaken
in a city to estimate the amount of delinquency among teenage boys according
to some definition of this term.

In other investigations, interest focuses on the study of relationships.
For my purposes, I should like to distinguish two classes within this type,
although they shade into one another.  The first class consists of broad
analytical surveys in which a number of variables are being investigated
simultaneously by multiple classification or multiple regression, or by
setting up models involving systems of equations, as in econometrics.  For
instance, in a recent study organized by the U. S. Office of Education [1],
standard tests were given to school children in grades 1, 3, 6, 9 and 12.
By multiple regression methods, estimates were obtained of the contribution made to the child's performance by various characteristics of the school
attended, by the home environment and parental attributes, and by the child's
aspirations and self-concept.

When these studies are exploratory, the discovery of the relationships
that are present suggests the question:  Why?, leading the investigator to
set up plausible hypotheses about the causal forces at work.  In other
studies, causal hypotheses may already have been proposed, the purpose
of the study being to verify whether the predictions about relationships made
from a casual model are consistent with the results.

*This work was facilitated by a grant from the National Science Foundation
(GS-341).

Reproduced with permission from *Proceedings of the Twelfth Conference on the Design of
Experiments in Army Research and Testing*, U.S. Army Research Office, Durham, NC, ARO-D
Report 67-2, pages 319–336, 1967.

My second class of analytical surveys is narrower in scope and more intimately bound up with the idea of cause and effect. The investigator concentrates on a specific presumed causal agent and tries to measure certain aspects of its effects. Examples are the effects of wearing lap seat belts on the amount and types of injury sustained in auto accidents. the effects of air pollution on illness associated with the respiratory organs, the effect of a new contraceptive device on the birth rate during the next five years, and, to cite a World War II study, the effect of bombing on the morale of the bombed people.

These studies resemble controlled experiments, because we set out to measure the effects of certain 'treatments'--the causal agents. However, in the 'non-experimental' studies with which I am concerned, the investigator is unable, for practical or ethical reasons, to use the two chief weapons of controlled experimentation. He cannot select the subjects who are to receive the causal agent and the subjects from whom it is to be withheld. If the agent is one that may be present in greater or less amount, as with air pollution or bombing, he has no control over these amounts, but must take them as he finds them.

The design and analysis of controlled experiments has become fairly well categorized and standardized. Most university courses on the subject discuss completely randomized, randomized blocks, and latin square plans (sometimes under different names) and go on to factorial experimentation and to techniques for estimating response surfaces. This standardization brings with it the usual benefit of economy of effort: once learned, the techniques of planning and analysis can be applied, often with only minor variations, in widely different areas of research.

With non-experimental studies much less standardization of this type has occurred. There is less cumulative experience with the various types of study plan. In the principal fields in which these plans are used-- sociology, psychology, education, market research, and public health-- workers have only recently begun to learn from one another. Statisticians have shown limited interest in the logical structures of the plans.

While non-experimental studies present many issues that merit discussion, this paper will be confined to three topics, as follows.

> Some preliminary aspects of planning.
> Simple types of study plan.
> Techniques for increasing precision and
>     eliminating bias.

2. PRELIMINARY ASPECTS OF PLANNING. Being unable to apply the causal agent in which he is interested, the investigator in a non-experimental study must first find some locale in which the agent is operating or will operate under conditions suitable for measuring its effects. In this search the following questions must be kept in mind, all of them matters of judgment rather than of black and white.

1. Is the cause operating in sufficient strength? Sometimes, for reasons of convenience or expense, the investigator chooses an environment in which the causal force operates too weakly to allow its effect to be measured in the size of sample that is feasible. For instance, airline pilots might be considered a convenient source from which to study predictors of heart disease, since they receive repeated and thorough medical examinations of which records are kept. On the other hand, one of the criteria by which they are selected is that they are the kind of men who are unlikely to develop heart disease.

2. What other important variables are present whose effects may be confounded with those of the causal variable? How will they be handled? In planning a study of the effects of air pollution, an investigator might look for three residential areas in the same city, one heavily polluted, one moderately, and one relatively free from pollution. But it is likely that the residents of these areas will show a sizeable gradient in socioeconomic levels, which might account for any differences found in respiratory illness. If the investigator confines himself to areas closely similar in socioeconomic level, he may find that the differences in amounts of air pollution are quite small, thus becoming involved in the difficulty mentioned in point 1. Methods for handling confounded variables are discussed later in this paper. If, however, an important variable is too highly correlated with the causal variable, as might be the case in the air pollution example, there may be no way to disentangle their effects.

3. What measurements are to be taken? What is known about the precision and accuracy of the measurements? Many aspects of human life and behavior present formidable problems of measurement: e.g., how does one measure morale? In large studies, the measurement process may be restricted, for reasons of expense, to responses on a printed questionnaire. Substantial biases in measurement can, of course, produce badly misleading results. "Random" errors of measurement of the effects decrease the precision of the results. "Random" errors in measuring the strength of the causal variable (e.g. number of cigarettes smoked per day) will produce an underestimate of the size of the effect. Similarly, "random" errors in measuring a confounded variable decrease the effectiveness of the standard statistical methods for removing the disturbing effects of this variable.

4. If the study is to be made from records already collected by someone else, have the records been checked as to completeness, accuracy, and accessibility? It is always worth considering whether a study can be made from existing records, not only because of cost but because this may be the only way to obtain results in a reasonably short time. Sometimes, investigators construct plans and engage staff for a study on the basis of someone's assurance about the quality of the records that turns out to be greatly over-optimistic, particularly when the records are kept for some legal or administrative purpose but rarely used or examined. A careful pilot survey of the records, designed to reveal any weaknesses for the purpose at hand, is essential before commitments are made.

5. How will the sample size or sizes be determined? In controlled experimentation there are formulas that provide guidance about sample size by calculating the size needed to estimate the effect with a prescribed width of 95% confidence interval, or the size for which some basic test of significance will have a prescribed power. It is advisable to try to use these formulas in non-experimental studies also. However, in order to obtain useful numerical answers from these formulas one must have an estimate of (i) the standard deviation per observation and (ii) the likely size of the effect that is being estimated. In exploratory studies these estimates may be lacking, and the investigator may have to use simply the largest sample size that can be afforded, having speculated that this size is more likely to be too small than too large.

6. If non-response or later melting-away of the sample is anticipated, what are the plans for coping with it? This is a common problem, especially when participation in the study is somewhat of an imposition on the subjects, or when the study extends for several years. Investigators tend to be lax about non-response. The standard call-back or follow-up questionnaire procedures developed in sample surveys are often surprisingly helpful. Sometimes it is feasible to follow people who move within the same metropolitan area even if it is too costly to follow those who leave the area. Sometimes background information about non-respondents is available, or can be obtained by mail, that assists a judgement about the extent to which they bias the conclusions. Speculations about the extent to which non-respondents might bias the results can always be made much more comfortably with a 10% than with a 30% non-response rate.

7. What are the comparisons from which the size of the presumed causal effect will be estimated? Numerous points arise here. In some studies the 'cause present' group is clearly defined, but it is less clear what can be used as a 'cause absent' group for comparable purposes. Often it is important to estimate the causal effect separately in different subgroups of the population (e.g. for people of different ages, for men and women).

The types of adjustment to be made for handling confounded variables are also relevant.

8. Is the environment a 'typical' one from the viewpoint of generalizability of results? Sometimes an ingenious investigator finds a group of people (for instance a special religious sect) among whom the causal force is operating with no important confounded variables. But he may reluctantly decide not to attempt the study in this group, because they seem atypical in so many respects that any generalization of results would appear hazardous.

With some problems of great interest and importance, investigators have to search for a long time before a suitable environment is found. Sometimes none is found: in other cases we are restricted to the type of study that can be done rather than the type we would like to do. Consider the problem of investigating in human subjects the effects of exposure to atomic radiation on illness and death rates. Ideally, the answer would take the form of a dosage-response curve, the rate being expressed as a function of the exposure history (amount and duration).

As pointed out by Seltser and Sartwell [2] , the principal opportunities for investigations in human subjects are confined to the following: (a) the Japanese survivors of the atomic bombs in Hiroshima and Nagasaki, involving a single exposure, (b) groups occupationally exposed to radiation at times when the possible danger from this source was not realized-- radiologists, dentists, and makers of watches with luminous dials, (c) persons who received medical radiation, as in the treatment of some forms of cancer, or infants exposed in utero through pelvic X-rays of the mother in the late stages of pregnancy, and (d) areas of the earth in which natural radioactivity is unusually high.

None of these sources provides more than limited material for constructing a dosage-response curve. To illustrate the types of study that have been undertaken, long-term studies in Hiroshima and Nagasaki were initiated in 1950. In Hiroshima the sample contains about 12,000 people, divided into 4 groups of about 3,000 each, according to their distances from the point of impact of the bomb. The subjects receive regular health examinations, with particular attention to any symptom that might be an after-effect of radiation exposure.

A study of this type is expensive and administratively difficult. Fortunately, the health data also permit many useful investigations of general health questions. From the viewpoint of the dosage-response curve, a weakness is that the dose to which any person was exposed is not known, but has had to be estimated roughly from memory of a person's location and local shielding by buildings at the time when the bomb fell.

Also, the group furthest from the epicenter, who serve as the non-exposed group, differ in some important characteristics from the three exposed groups, and have proved unsatisfactory as a 'control' [3] .

The study by Seltser and Sartwell [2] of the mortality of radiologists is an excellent example of the possibilities from groups occupationally or medically exposed. They chose male members of the Radiological Society of North America. For each member they obtained by a painstaking search the status (dead or alive) as of December 31, 1958, with cause of death and any available information on other factors such as age that might influence duration of life. Research of this type always raises the question: with what are the exposed group to be compared? Ideally, we seek a non-exposed group which is similar to the exposed group with regard to any other variable that is known or suspected to have a material effect on duration of life. (In this example an obviously relevant variable is age.) In an observational study the extent to which this goal can be met is of course dependent on our ability to measure such variables and to find a group that has similar distributions with respect to them.

The authors chose two comparison groups. As the nearest to a non-exposed group they used the American Academy of Ophthalmology and Otolaryngology, whose members rarely have occasion to employ X-radiation. As an intermediate group they also included the American College of Physicians, since some of these members use X-rays, for example, in heart examinations. In such studies the inclusion of a middle group is advantageous in either adding confirmation to the results given by the two extreme groups or in casting doubt upon them. This study, however, again has the weakness that no measures of the doses of radiation experienced by the subjects are available, except as a rough guess for the group as a whole.

3. SIMPLE TYPES OF STUDY PLAN. This section introduces some simpler types of plan, with a brief discussion of their strengths and weaknesses and of the statistical analysis.

3.1 A single group, measured before and after the action of the causal agent. This type is common when the causal agent is of short duration. For example, after complaints about the time taken to go through a cafeteria line, a change in the service is proposed that it is claimed will remove the bottleneck. Before this change is made, the times taken to go through are recorded for a random sample of the users, the same being done after the change is made. In other situations, the causal agent might be DDT spraying of 10 villages, an estimate of the misquito population being made before and after spraying, or a radio and TV appeal which the stations in an area agree to give on a certain day, urging mothers to bring their children into the clinics in a city for immunizations, the number of children appearing for immunization being counted in each clinic during the week before and the week after this appeal.

Unlike the radiologists example, such studies have no comparison group, usually because all members of the population of interest are exposed (at least potentially) to the causal agent. Sometimes, as in the DDT example, a comparison group of unsprayed villages might have been chosen, but is excluded for administrative or financial reasons. Often, a single-group study is the only feasible approach in attempting to learn something about the effects of new governmental programs or laws that apply to everyone.

The absence of a comparison group is, of course, the major weakness. Any other event that produces a change in the level of the variable during the Before-After period has its effects inevitably confounded with those of the causal agent. Campbell and Stanley [4] give a detailed catalogue of these sources of bias in educational research. If the investigator is aware of such other influences he can sometimes ask questions about the reasons for people's change in behavior that help him to judge whether these influences have been important. Knowledge that a change is coming may influence people's behavior immediately before the change, so that the After-Before difference is misleading.

Although the conclusions from studies of this type involve a substantial element of judgment, the studies are, as Campbell and Stanley put it, "worth doing when nothing better can be done". I might express it a little more positively. With new public programs, plans to estimate their effects are often not initiated until some time after the program has been running. By this time it is difficult to get good 'Before' measurements and too late to take precautions or gather supplementary information that might have helped in judging the effects. The question: How can we study the effects of this program? should be raised some time before the program begins.

The statistical analysis usually involves examining the difference between two paired or independent samples. The samples may be sub-classified by another variable, e.g., age of subject, in order to reveal any variation in effect with age.

Sometimes there is reason to expect that the Before measurement will itself influence the subject's behavior. A plan that has been proposed is to have two groups, both exposed to the cause. Whenever feasible, these can be random halves of an initially chosen group. Group 1 is measured 'Before' and 'After', group 2 is measured 'After' only. The idea is that by comparing the two 'After' sets of results, we can test whether the 'Before' measurement influenced the level of the 'After' responses in group 1.

The best method of estimating the size of the causal effect presents a problem involving the pooling of data after performing a test of significance. If the subscripts a and b denote 'After' and 'Before', the difference $(\overline{Y}_{2a} - \overline{Y}_{1b})$ is an unbiased estimate of the causal effect. Assuming a constant variance $\sigma^2$ per subject, this difference has variance $2\sigma^2/n$. The difference $(\overline{Y}_{1a} - \overline{Y}_{1b})$ has variance $2\sigma^2(1-\rho)/n$, where $\rho$ is the correlation between the 'Before' and 'After' measurements for the same subject, but is unbiased only if the 'Before' measurement did not affect the level of the 'After' measurement. The estimates $(\overline{Y}_{2a} - \overline{Y}_{1b})$ and $(\overline{Y}_{1a} - \overline{Y}_{1b})$ are themselves correlated, since $\overline{Y}_{1b}$ appears in both. One approach is to seek a weighted mean of these estimates, with weights determined from the results of the preliminary test of significance of $(\overline{Y}_{1a} - \overline{Y}_{2a})$, that has minimum mean square error subject to a condition that the bias be kept small.

The preceding discussion has been confined to studies in which it is satisfactory to measure the causal effect at a single time after the causal event. In many situations, the causal event may have prolonged effects, or if its effect is likely to die away, the investigator wants to measure this decay curve. For these purposes we need, at a minimum, a series of measurements at intervals of time before the event, followed by a series at intervals after the event. The problem of the model to be used for the analysis of results of this type raises some interesting questions which have been illustrated by Campbell and Stanley [4]. Model-fitting and interpretation are easiest when the 'Before' measurements appear to fluctuate about a constant level; the difficulty increases when the 'Before' and 'After' measurements display trends, particularly those with curvature. The question of serial correlations must also be considered.

3.2 'Cause present' and 'Cause absent' groups. Y measured 'After' only. This is a very common type. The Hiroshima and radiologist studies, investigations of the effectiveness of seat belts in preventing injury in automobile accidents, and the large studies of the death rates of non-smokers and cigarette, cigar, and pipe smokers are examples. As we have seen, there may be several 'cause present' groups, representing different strengths or variations in the causal agent, and more than one 'cause absent' group, particularly where the selection of a control group presents difficulty.

At its simplest, the analysis follow the usual methods for the analysis of one-way classifications or of two-way classifications if pairing or blocking has been employed in forming the groups. Often, however, the analysis of a multiple classification is involved, other variables being introduced in order to diminish the risk of bias, as discussed in section 4, or because the investigator wants to examine interactions of the causal effects with these variables.

An important variant of this method, often called the retrospective method, is much used in epidemiological research. In this, we find a group in which the effect is present and one from which it is absent, and compare the frequency with which the presumed causal agent is found in the two groups. This approach is natural when a group of people show symptoms of food poisoning at a picnic and the cause is being sought. As another example, numerous investigators have selected a group of lung cancer patients and another group of patients in the same hospitals who do not have this disease, comparing the proportions of cigarette smokers in the two groups. With this approach, it is often hard to select the 'effect absent' group and to obtain measurements of high quality. Further, erroneous results may be obtained when there are several causal agents and attention is focussed on one. But with an effect that is rare, this approach may be the only practicable one, and it is often the quickest way of obtaining a preliminary indication for or against a postulated relationship. For a discussion, see [5].

3.3 'Cause present' and 'cause absent' groups. Y measured Before and After. This plan has been used, for example, in studies of the effects of new public housing, as against slum housing, on health and social behavior. When it became known which group of applicants were to move into a new public housing development, a control group of families who would in general remain in slum housing were selected. The basic questionnaires on health and social behavior were obtained both before the move took place and at several times after the successful applicants had moved. In a study of the effects of fluoridation of town water on children's teeth, usually done by a plan of type 3.1, a nearby control town which did not plan to fluoridate could be included if the resources permitted. The state of dental health of a sample of children from both towns would be measured before and some time after the fluoridation in the first town.

With this plan the investigator is in a better position to guard against bias than with plan 3.2. Ideally, the initial distribution of the response variable Y should be the same in the 'cause present' and 'cause absent' groups. Since he has the initial measurements, he can verify whether this seems to be the case. Even if the distributions are somewhat different, it is still possible to compare the amount of change in the two groups during the 'Before-After' period.

A general estimate of the size of the causal effect is

(3.1) 
$$(\overline{Y}_{1a} - \overline{Y}_{2a}) - \beta(\overline{Y}_{1b} - \overline{Y}_{2b}) \; ,$$

where the value of $\beta$ is to be chosen. Suppose that the model is as follows.

Before: $\qquad y_{1bj} = \mu_1 + e_{1bj};$ $\qquad\qquad y_{2bj} = \mu_2 + e_{2bj}$

After: $\qquad y_{1aj} = \mu_1 + \delta + \tau_1 + e_{1aj};$ $\quad y_{2aj} = \mu_2 + \tau_2 + e_{2aj}$ .

Here, $\delta$ represents the causal effect to be estimated; $\tau_1$ and $\tau_2$ represent other time-changes that affect the two groups; and the e's are random variables with means zero. From this model we see that

$$E\{(\overline{Y}_{1a}-\overline{Y}_{2a}) - \beta(\overline{Y}_{1b}-\overline{Y}_{2b})\} = \delta + (\tau_1-\tau_2) + (\mu_1-\mu_2)(1-\beta) .$$

Hence, (i) if $\tau_1 \neq \tau_2$, the plan provides no unbiased estimate of $\delta$: this is, of course, obvious, (ii) if $\tau_1 = \tau_2$ but $\mu_1 \neq \mu_2$ (i.e., the initial levels of the two groups differ), the only unbiased estimate of $\delta$ is given by taking $\beta = 1$. (iii) if $\tau_1 = \tau_2$ and $\mu_1 = \mu_2$, any value of $\beta$ gives an unbiased estimate. Assuming that the e's all have the same variance $\sigma^2$, the estimate (3.1) has variance

$$2\sigma^2 (1-2\beta\rho+\beta^2)/n$$

where $\rho$ is the correlation coefficient between an 'After' and a 'Before' measurement. If the 'After' and 'Before' samples are independent, so that $\rho = 0$, we take $\beta = 0$. If these measurements are paired, the minimum variance is given by $\beta = \rho$. In practice, $\beta$ is estimated in this case by an analysis of covariance of the 'After' on the 'Before' measurements.

4. TECHNIQUES FOR INCREASING PRECISION AND ELIMINATING BIAS. In controlled experiments the investigator relies on randomization, plus other precautions such as 'blindness' in the measurement process, to ensure that biases are kept to a negligible level. As means of increasing precision, blocking and adjustments made by the analysis of covariance are two of the principal weapons.

Devices analogous to blocking and covariance are commonly used in non-experimental studies also. However, since randomization is not available, these devices must perform the double function of eliminating bias and of increasing precision. In fact, since bias is regarded as the chief source of erroneous conclusions, control of bias becomes their principal function.

Suppose that Y is the response or effect variable, and that there is a 'cause present' and a 'cause absent' group. If X is any variable that is related to Y, a bias may arise in $(\overline{Y}_1 - \overline{Y}_2)$, the estimated difference between the means of the two groups, if the distribution of X differs in the two groups. For instance, if the regression of Y on X is linear,

$$Y_{ij} = \mu_i + \beta X_{ij} + e_{ij},$$

where i = 1, 2 denotes the group, and the $e_{ij}$ are residuals with mean zero, then

(4.1) $$E(\overline{Y}_1 - \overline{Y}_2 | X) = \mu_1 - \mu_2 + \beta (\overline{X}_1 - \overline{X}_2) .$$

The term $\beta(\overline{X}_1 - \overline{X}_2)$ is the bias.

In handling these variables the investigator makes a list of the X variables known or thought to be related to Y. These variables are placed in one of the following classes.

(I) Important variables whose effects the investigator will try to remove, either because there seems a danger of bias or because removal will bring a worthwhile increase in precision.

(II) Variables for which the investigator will check whether their distribution is similar in the 'cause present' and 'cause absent' groups. No adjustment will be made for these variables unless the distributions appear sufficiently different so that there seems a danger of bias. This method is employed for variables whose correlation with Y is modest. If Y and X are linearly related, with correlation $\rho$, the fractional reduction in the variance of Y due to elimination of the effect of X cannot exceed $\rho^2$. If $|\rho| \le 0.3$, this reduction is less than 9%: the potential increase in precision is small.

In practice, verification that the distribution of X is similar in the two groups of subjects is often done by forming the frequency distribution of X in each group, with, say, k classes, and making the $\chi^2$ test for a 2 x k contingency table. A low value of $\chi^2$ is taken as assurance that the distributions of X are similar and that there is little risk of bias from the relation between Y and X. This $\chi^2$ test may not be the best procedure. If the regression of Y on X is linear, equation $(4.1)$ shows that comparison of the mean values of X in the two groups is more relevant, since the bias in $(\overline{Y}_1 - \overline{Y}_2)$ comes from the term $(\overline{X}_1 - \overline{X}_2)$. Similarly, if the relation between Y and X is curved and can be approximated by a quadratic regression, comparison of the first two moments of X in the two groups is relevant.

(III) Variables about which nothing will be done, because their relation to Y is judged too tenuous to create trouble. This class also contains X variables which it is not feasible to measure and those of which the investigator is ignorant.

A natural question at this point is: Why not put all the X variables in class I, or at least do so whenever there is any doubt? I don't know the full answer to this, but a partial answer is that the techniques (matching and adjustment) by which we attempt to remove the effects of these X variables become steadily more cumbersome to apply and to interpret as the number of X variables increases. These techniques may be described as follows.

'Ideal' matching. Each member of group 1 has a partner in group 2 who has, within narrow limits, the same value for any X variable for which adjustment is being made. By taking the difference between partners, the effects of these X variables are eliminated, provided that the regression of Y on these X variables is the same in both groups. Clearly, this matching is effective whether the regression is linear or curved.

In practice, the construction of matched pairs often presents difficulty, particularly if matching has to be done on several X variables. Usually, it is necessary to have a large reservoir of subjects for at least one of the two groups; otherwise, it will not be possible to locate partners who agree closely on the values of all the desired X variables. A common experience is that the construction of partners takes much longer than anticipated, that the rules set up about the closeness of the match have to be continually relaxed, and that some subjects have to be omitted because no match is found.

Stratified or frequency matching. This is a looser form of matching which facilitates the construction of partners. The range of each X variable is divided into a number of classes, commonly from 2 to 5 or 6. Thus the X variables create a multiple classification: for instance, with 3 X's and 4 classes per variable there are 64 cells. For a member of the 'cause present' group, any member of the 'cause absent' group who falls in the same cell is an acceptable partner. In the end, what this method amounts to is that in any cell of the multiple classification the two groups have an equal number of subjects. Often, there is no specific designation of partners, since this seems rather pointless.

Stratified matching is the only kind that is feasible for an X that is an ordered classification, such as "mild", "moderate", "severe" or is qualitative, e.g., religious affiliation or urban, suburban, rural.

Adjustment by subclassification. This method is very similar to stratified matching. When selecting the 'cause present' and the 'cause absent' groups we do not attempt any matching. Adjustment for differences in the X distributions in the two groups is accomplished by forming the multiple classification used in stratified matching and making adjustments by a least squares or analysis of variance model.

To illustrate the relation between the two methods, suppose there are X variables and that only 100 subjects are available for the 'cause present' group. To see how the land lies, we classify these subjects, plus 100 from the 'cause absent' group, into 9 cells, assuming that each X variable has 3 cells. In table 1, the numbers of subjects found in each cell are shown, P and A denoting the two groups, Both the P and A sets add to 100.

TABLE 1

Subclassification on two X variables.

| | | $X_1$ | | |
| | | < 20 | 21 - 50 | Over 50 |
|---|---|---|---|---|
| | Mild | P  8 | P  10 | P  19 |
| | | A 23 | A  7 | A  4 |
| $X_2$ | Moderate | P  8 | P  8 | P  16 |
| | | A 26 | A  9 | A  3 |
| | Severe | P  5 | P  11 | P  15 |
| | | A 19 | A  6 | A  3 |

If we are using stratified matching, we select 8 at random out of the 23 A's in the top left cell, discarding the rest. In both the other cells in the top row, we need more A's to reach the desired numbers 10 and 19. Looking the table over, it appears that a reservoir of perhaps 700 or more subjects suitable for the 'cause absent' group would be necessary to build up all the cells to the desired numbers in the P group.

In adjustment by subclassification, as I am using this term, we either accept the A sample as it stands or attempt only to build up cells in which the A sample is very small. The decision depends on the size of the reservoir for the A group, the time and trouble involved in any build up, and the investigator's opinion as to whether the effort is worthwhile.

From the viewpoint of estimation of effects we face a 2 x 3 x 3 table with either stratified matching or adjustment by subclassification. It is assumed that Columns $(X_1)$ and Rows $(X_2)$ both show real effects, and possibly an interaction, since otherwise there would be no need to match or adjust for these X variables.

The simplest situation is that in which there is no interaction of the (P-A) difference with either $X_1$ or $X_2$. In this event the 9 differences $(\overline{P}_{ij} - \overline{A}_{ij})$ are all estimates of the same quantity. It follows that with stratified matching, the difference between the overall sample means $(\overline{P} - \overline{A})$ is free from any confounding with the levels of $X_1$ or $X_2$. The estimate $(\overline{P} - \overline{A})$ has variance $\sigma^2/50$, where $\sigma^2$ is the within-cell variance (assumed constant from cell to cell). If the A sample is accepted as it stands, the corresponding estimate for adjustment by subclassification is a weighted mean of the differences $(\overline{P}_{ij} - \overline{A}_{ij})$, weighting each inversely as its variance. The weights are $n_{1ij}n_{2ij}/(n_{1ij}+n_{2ij})$, where the n's are the sample sizes in the $(i,j)$ cell. For table 1 the variance of this weighted mean difference turns out to be $\sigma^2/36.6$, about 35% larger than with stratified matching. In this situation stratified matching provides a simpler estimate that is more precise.

We may, however, wish to examine whether the (P-A) difference changes with the level of $X_1$ and $X_2$. As Billewicz [6] has pointed out, the ability to examine these interactions is an advantage which these methods hold over 'Ideal' matching. If interactions are found, estimation of the overall difference may become of little interest. The technique needed here is the analysis of multiple classifications with unequal numbers in the cells. While the general least squares theory covering this technique is not new, much remains to be learned about the practical handling and interpretation of such analyses, particularly for investigators who are not expert in statistical methods. The recent paper by Federer and Zelen [7] is a useful contribution.

Adjustment by covariance. Conceptually, this is the same approach as adjustment by subclassification for the case in which the X's are continuous. Covariance may have an advantage and a disadvantage. The grouping of continuous X's into classes in adjustment by subclassification loses some information: covariance avoids this loss. On the other hand, adjustment by subclassification does not involve any assumption that the relation between Y and X is linear. If the investigator follows the common practice of adjusting in covariance only for linear effects of X, covariance is at a disadvantage if the true regression has substantial non-linearity. Of course, this loss can be avoided by adopting a more accurate model in the covariance analysis.

How effective are these techniques? The following comments are based on results quoted in [8] and on some unpublished work. As already mentioned, 'ideal' matching removes bias due to $X_1 \ldots X_k$ under any regression

$$Y_{ij} = \mu_i + \phi(X_{1ij}, \ldots X_{kij}) + e_{ij}, \quad (i = 1, 2)$$

if the regression function $\phi$ is the same in both groups. The variance of $(\overline{Y}_1 - \overline{Y}_2)$ is reduced by the matching to a fraction $(1 - \rho^2)$ of its original value, where $\rho$ is the correlation coefficient between Y and $\phi$. In practice, 'ideal' matching is likely to be at its best when the X's are quantitative and one of the groups has a large reservoir in which matches may be sought, while the other group is small. In this situation, matching should not prove too difficult. Moreover, the other disadvantage of matching--that one cannot examine effectively the interactions of the causal variable with the X variables--scarcely applies when one group is small, since the sample size would probably preclude any precise estimates of interactions.

Covariance adjustment should have about the same effects on bias and precision, with the qualifications that the correct form of the regression equation must be fitted, and that there is some loss of precision from sampling errors in the estimated regression coefficients. If the regression is linear and there happens to be no bias due to the X's, the fraction to which $V(\overline{Y}_1 - \overline{Y}_2)$ is reduced by the covariance adjustment is roughly

(4.2)
$$(1 - \rho^2) \left\{ 1 + \frac{k}{(2n - k - 3)} \right\}.$$

where $\underline{n}$ is the size of sample in each group, so that the regression coefficients are estimated from $2(n-1)$ degrees of freedom. The term in curly brackets will be close to 1 if k is small relative to 2n. However, if there are substantial biases in some of the X's, (4.2) no longer applies, and the corresponding term in curly brackets can be much larger. The performance of this covariance adjustment when the fitted model is of the wrong form deserves further study. Linear covariance adjustments seem to perform surprisingly well when the true regression has a moderate degree of curvature.

The preceding remarks about matching and covariance assume that the X's are measured without appreciable error. Suppose that for an X variable the recorded measurement is x = X + d, where d is a random error of measurement with mean zero, independent of X and of e, the deviation of Y from its regression on X. The effects of these errors of measurement are roughly as follows, where $f = \sigma_d^2/\sigma_x^2 = \sigma_d^2/(\sigma_X^2 + \sigma_d^2)$.

(i)  Matching and covariance remove only a fraction (1-f) of the bias in Y due to X.

(ii)  $V(\overline{Y}_1 - \overline{Y}_2)$ is reduced to the fraction $\{1 - (1-f) \rho^2\}$ of its original value.

While imprecise measurement weakens the performance of these techniques, it is easy to form an exaggerated notion of the size of this effect if some check calculations are not made. For instance, suppose that $\sigma_X$ = 25, nearly all the correct values of X lying between 0 and 125. If we are told that half the observed measurements are wrong by more than 5 units, this seems rather poor quality of measurement. However, a probable error of 5 corresponds to $\sigma_d$ = 7.4, $\sigma_d^2$ = 55, $\sigma_x^2$ = 680, and f = 0.08. Thus, 92% of the bias is still removed.

Now consider stratified matching and adjustment by subclassification as applied to quantitative X's. From the viewpoint of errors of measurement of X , these methods appear crude, since the quantitative scale of an X variable is replaced by a classified variable that takes only the number of distinct values that the number of classes allow. With stratified matching the values of (1-f) are 0.64, 0.79, 0.86, 0.90, and 0.92 for 2, 3, 4, 5, and 6 classes, respectively. Strictly, these values hold only if the regression of Y on X is linear, X is normally distributed, and the classes are of equal size. However, they appear accurate enough as guides to practice when the regression of Y on X is nonlinear, when X has some skewness and kurtosis and when the class sizes depart moderately from equality. The results indicate that at least five or six classes should be used for any X variable which is thought to be a source of substantial bias.

With adjustment by subclassification the preceding (1-f) values apply so far as the removal of bias due to X is concerned. This method suffers an additional loss of precision, as illustrated previously, because of inequalities in the sample sizes of the two groups in the individual cells of the multiple classification.

The situation when $\underline{X}$ is an ordered classification is not so clear. If an ordered classification can be regarded as essentially a grouping of an underlying quantitative $\underline{X}$, the preceding values of $(1-f)$ should be applicable. In practice, however, ordered classifications are often used because no more precise method of measurement is known. If we envisage some accurate measurement X, not yet discovered, it seems reasonable that the ordered classification will contain errors of misclassification as well as grouping errors. These additional errors presumably reduce the values of $(1-f)$, to an extent that does not seem to have been investigated.

Finally, none of the methods can guarantee to remove bias due to an $\underline{X}$ variable that has been omitted from the matching or adjustments. The situation with regard to such omitted variables is interesting. If they happen to have a high correlation with the included $\underline{X}$'s--in other words, if we are lucky--most of their bias will also be removed by the matching or adjustments. This explains, I think, why linear covariance often works well when Y has a quadratic regression on X, since X and $X^2$ have a high correlation in many bodies of data. But one can also meet the opposite situation in which the bias due to omitted X's is inflated by the adjustments. Thus in non-experimental studies there always remains an element of uncertainty in our claims about the size and reality of a presumed causal effect.

# REFERENCES

1. J. S. Coleman et al. Equality of Educational Opportunity. U. S. Office of Education. 1966.

2. R. Seltser and P. E. Sartwell. The influence of occupational exposure to radiation on the mortality of American Radiologists and other medical specialists. Amer. J. Epidemiology, 81, 2-22, 1965.

3. S. Jablon, M. Ishida, and G. W. Beebe. Studies of the Mortality of A-bomb survivors. 2. Radiation Research 21, 423-445, 1964.

4. D. T. Campbell and J. C. Stanley. Experimental and Quasi-Experimental Designs for Research. Rand McNally & Co., Chicago, 1966.

5. B. MacMahon, T. F. Pugh, and J. Ipsen. Epidemiologic Methods. Little, Brown & Co., Boston, 1960.

6. W. Z. Billewicz. The efficiency of matched samples: an empirical investigation. Biometrics, 21, 623-644, 1965.

7. W. T. Federer and M. Zelen. Analysis of multifactor classifications with unequal numbers of observations. Biometrics, 22, 525-551, 1966.

8. W. G. Cochran. The planning of observational studies of human populations. J. Roy. Statist. Soc., A, 128, 234-365, 1965.

# Errors of Measurement in Statistics*

W. G. Cochran

*Harvard University*

In this review of some of the recent work in the study of errors of measurement, attention is centered on the type of mathematical model used to represent errors of measurement, on the extent to which standard techniques of analysis become erroneous and misleading if certain types of errors are present (and the possible remedial procedures), and the techniques that are available for the numerical study of errors of measurement.

## 1. Introduction.

This paper reviews some of the recent work in the study of errors of measurement as they affect data analysis. This is a difficult field, as evidenced by the prolonged struggles of the econometricians to find satisfactory methods for coping with such errors in their investigations of relationships between variables, and by the slow rate of progress that has rewarded major efforts to study errors of measurement in sample surveys. It is also a field of growing importance. For example, the recent rapid increase in data-gathering projects in the social and medical sciences is producing large bodies of data containing variables obviously difficult to measure, such as people's behavior, opinions, feelings, and motivations. Concurrently, there are signs of a rise in research interest in addition to that stimulated by problems in econometrics and sample surveys, and one of the objectives of this review is to encourage research in this area.

Attention will be centered on four aspects of the problem.

(1) The types of mathematical model that have been used to represent errors of measurement.

(2) The extent to which errors of measurement are automatically taken into account in the standard techniques of analysis, and the extent to which these methods become erroneous and misleading if certain types of errors are present.

(3) The amount of harm done by errors of measurement, from the viewpoint of the investigator, in producing unsuspected biases or reduced precision, and the remedial procedures that are available to avoid these undesirable consequences.

(4) Techniques for the study of errors of measurement.

Before looking at mathematical models that attempt to describe the statistical properties of errors of measurement, a reminder about some of the

Received June, 1968.

* This work was facilitated by a grant from the National Science Foundation (GS-341), and by the Office of Naval Research, Contract Nonr 1866(37). This paper is based on the Rietz lecture of the Institute of Mathematical Statistics, Washington, D. C., December, 1967.

different situations in which measurements are made may be useful. The quantity
being measured may be fixed and static through time. It may be fixed at a
point in time, but vary with time. Sometimes we deliberately measure the
wrong quantity, using a substitute measurement that is cheaper or more convenient, as for example in the standard manometer used for measuring blood
pressure by a sleeve wrapped round the arm instead of a direct measurement
of intra-arterial blood pressure. Sometimes we measure the wrong quantity
because we don't know how to measure the right quantity. I once spent some
time trying to dev se a practical method of measuring the strength of maternal
affection in the dairy cow, but did not get far. When the quantity being measured
varies with time, we may in effect be using the wrong measurement if we treat
measurements made some months earlier as if they applied to the situation
today. Sometimes the error of measurement is actually a sampling error. The
amount of insect infestation on a plot in a field experiment may be estimated
by taking from the plot a small sample of plants and recording the mean number
of insects per plant in the sample. As Mahalanobis, 1946, has pointed out,
there may be some margin of physical uncertainty in what is being measured
even at a fixed point in time; the yield of wheat may increase by 4–5% owing
to moisture when the rainy season begins in the parts of India.

To turn to the measuring instrument, measurement and recording may be
completely automatic. Some human action may be involved, but of a simple
and easily checkable type, as in reading a clear, stationary dial. The human
action may involve complex subjective judgment to a greater or less degree.
Several different types of human action may enter. In many sample surveys
it is recognized that errors of measurement can arise from the person being
interviewed, from the interviewer, from the supervisor or leader of a team
of interviewers, and from the processor who transmits the information from
the recorded interview on to the punched cards or tapes that will be analysed.

The preceding examples, though incomplete, suggest that a variety of mathematical models, some simple, some quite complex, may be needed to describe
realistically the types of measurement error relevant to different measurement
problems. Since the introduction of terms representing the presence of errors
of measurement nearly always seems to complicate the mathematical analysis,
two obvious maxims in this field are to use as simple a model as will reasonably
represent the facts and to keep oneself aware of other people's work. The same
model may turn up in very diverse applications and if the consequences of a
certain model have already been worked out, one might as well take advantage
of this. Concurrently, the investigator has a responsibility for producing evidence
that his model does fit adequately. From the literature my impression is that
investigators have needed no urging to use simple models, but are not so well
provided with data that justify the assumptions made in the model, and have
not always worked out the full consequences of the errors of measurement.
For instance, most textbooks on sample survey theory, including my own,
present all the standard theory on the assumption that there are no errors of
measurement, and merely indicate some of the disturbances caused by errors
of measurement briefly in a later chapter.

One common type of error of measurement—the occasional gross error—will not be discussed in this paper. Since computers can print and examine residuals from a fitted model, there is current interest in developing programs by which any likely gross errors are automatically signposted for the investigator to examine and consider. In my opinion, much work of practical interest remains to be done, for instance, more testing of the performance of such methods in the presence of more than one gross error, and further study of the merits of methods like censoring that are less affected by the presence of gross errors.

## PART I. SOME SIMPLE MODELS

### 2. *Continuous variates.*

The subscript $u = 1, 2, \cdots, n$, will denote the member of the sample, while the symbol $Y_u$ refers to the recorded measurement. The symbol $y_u$ denotes the correct or true measurement. For exposition of theory, the symbol $y_u$ is sometimes introduced even when we do not know how to make a correct measurement, though some workers are unhappy with this practice.

The error of the measurement on the $u$th unit is $d_u = Y_u - y_u$. With some measurement processes we can conceive of repeated measurements of the same unit, and with some simple non-destructive processes we can actually carry out such repeated measurements. The subscript $t$ will refer to the $t$th trial or repeated measurement. Thus we write

$$(1) \qquad\qquad Y_{ut} = y_u + d_{ut}$$

In this representation, both $Y_{ut}$ and $d_{ut}$ have a frequency distribution for each member of the population (i.e. for fixed $u$), whereas $y_u$ is assumed fixed for any specific member of the population.

The simplest model is one in which

$$(2.1) \qquad E(d_{ut} \mid u) = 0; \quad E(d_{ut}^2 \mid u) = \sigma^2; \quad E(d_{ut}, d_{ut'}) = 0 \quad (t \neq t');$$
$$E(d_{ut}, d_{vt} \mid u, v) = 0, \qquad u \neq v.$$

In this model the errors are unbiased and have constant variance. They are uncorrelated with the correct values, with one another on different units, and on different trials for the same unit. This model may be expected to apply to some of the simplest measurement processes. It has been used even without evidence that it really applies, because it is sometimes the only model for which the consequences of the errors have been successfully worked out. The additional assumptions that the $d_{ut}$ are normal and independent in the probability sense are often adopted when needed. Sometimes we write $E(d_{ut}^2 \mid u) = \sigma_u^2$, because some units are more difficult to measure precisely than others.

The next stage in the model is to recognize that the measuring instrument may have an overall bias of amount $a$, writing

$$(2) \qquad\qquad Y_{ut} = y_u + a + d_{ut},$$

where assumptions (2.1) about the $d_{ut}$ still apply.

As an introduction to the third stage, I would like to jump back historically to Karl Pearson's 1902 paper on the mathematical theory of errors of measurement. Pearson was interested in the nature of errors of measurement when the quantity being measured is fixed and definite, while the measuring instrument is a human being. He conducted two experiments, each having three persons as measurers. In the first experiment, each person was presented with the same set of lines of unequal lengths drawn on paper, and was asked to bisect each line freehand. The exact middle of each line and the error of measurement was recorded. In the second experiment the task of the measurer was a little harder. A bright line of light slowly traversed a white strip on a black screen. When a bell sounded, each observer drew a mark across a line lying before him on a sheet of paper, to mark the proportional distance of the beam across its strip when the bell sounded. Again, the exact position of the bright line and the error of measurement were carefully noted. The sample sizes were 500 in the bisection series and 520 in the bright line series.

The principal conclusions drawn by Pearson from his analysis of the results were as follows.

(1) In 5 of the 6 cases (3 measurers, 2 experiments), the mean errors differed significantly from zero, the sixth being almost significant at the 5% level. This is the $a$ term in model (2). These overall biases varied in size and direction from one person to another.

(2) For a given measurer, the size of the bias varied throughout the series of trials when the errors were grouped in successive sets of 25. This suggests that with a human measurer the model may need a term $a_u$ representing a bias that varies from one sample member to another.

(3) The errors were not in general normally distributed, but exhibited both some skewness and some kurtosis.

(4) To come to a result that startled Pearson, the errors of two apparently independent observers in measuring the same quantity were positively correlated in 5 cases out of 6. This phenomenon is well known in interview surveys, though the explanations given there do not apply to Pearson's experiments. When two independent interviewers question the same person, the respondent may give an erroneous answer to the first interviewer and simply repeat the same erroneous answer to the second through a conscious or unconscious effect of memory. Alternatively, on a delicate question of opinion, both interviewers may have the same point of view and may induce the same erroneous answer because of the way in which they ask the questions. Pearson's result can have a simple explanation even when the quantity being measured is fixed and objective. If two measurers both tend to underestimate high values and over-estimate low values, these negative correlations between $d_{ut}$ and $y_u$ induce a positive correlation between the errors $d_{ut}$, $d_{ut'}$ of the two measurers. Pearson does not mention this possibility in connection with his results and does not present the results in a way which enables one to examine whether an effect of this type was present. Instead, he uses some uncharacteristically vague language, such as (p. 412) "certain factors affected by the immediate atmosphere seem to be common elements of two or more personalities, and there results from this a tendency in each pair of observers to judge in the same manner", and later (p. 433), "this psychological or organic correlation."

Pearson's results suggest several possible elaborations of model (2). The simplest is to introduce a "variable bias" term $a_u$ , and to make the additional assumption that the $a_u$ are uncorrelated with the correct values $y_u$ ; that is,

$$(3) \qquad Y_{ut} = y_u + a_u + d_{ut} , \qquad \mathrm{Cov} \, (a_u , y_u) = 0.$$

With the same algebra, we may be unwilling to assume that $a_u$ and $y_u$ are uncorrelated, leading to

$$(4) \qquad Y_{ut} = y_u + a_u + d_{ut} , \qquad \mathrm{Cov} \, (a_u , y_u) \neq 0.$$

Further, Pearson's results indicate that intercorrelations in errors of measurement within the sample may occur. In sample surveys the most prominent effect of this kind, and the one most studied, arises from the personal biases of the interviewers, which produce an intra-interviewer covariance term that is positive. Pearson's results also constitute a warning that when we come to study errors of measurement by having two different observers measure the same unit, their errors may not be independent.

The introduction of these correlation terms into the mathematical model complicates the subsequent analysis, and will be postponed until section 17. Much of the work that has been done outside of sample surveys assumes, rightly or wrongly, that errors on different members of the sample are uncorrelated. For the present, therefore, we revert to model (4), with the conditions

$$(2.1) \qquad E(d_{ut} \mid u) = 0; \qquad E(d_{ut}^2 \mid u) = \sigma_u^2 ; \qquad E(d_{ut} , d_{ut'}) = 0, \quad t \neq t';$$

$$E(d_{ut} , d_{vt} \mid u, v) = 0, \quad u \neq v.$$

For some applications it is convenient to amalgamate the terms $y_u$ and $a_u$ by writing $y_u' = y_u + a_u$ , particularly when no feasible method of measuring $y_u$ itself is known. The symbol $y_u'$ might be called the operationally correct value for the $u$th unit. Model (4) then reduces to the simpler form of model (1), i.e.

$$(4)' \qquad Y_{ut} = y_u' + d_{ut} ,$$

though with the above difference in interpretation.

One less general form of model (4) is worth mention, since it has already proved useful. If the relation between the variable bias $a_u$ and the correct value $y_u$ is expressible as a linear regression of $a_u$ on $y_u$ with regression coefficient $\gamma$, we may rewrite model (4) as

$$(4a) \qquad Y_{ut} = \alpha + \beta y_u + a_u + d_{ut} , \qquad E(a_u) = 0, \qquad \mathrm{Cov} \, (a_u , y_u) = 0$$

where $\beta = 1 + \gamma$. In (4a) I have reused the symbol $a_u$ to represent the deviation from this regression, so that the covariance of $a_u$ and $y_u$ is now zero. This model is the basis of Mandel's, 1959, theory of errors of measurement as applied to the analysis of interlaboratory tests. In Mandel's application, equation (4a) applies to a single laboratory, the values of $\alpha$, $\beta$, $\sigma_a^2$ , and $\sigma_u^2$ possibly varying from laboratory to laboratory. This model may also be appropriate when different judges are scaling the same set of subjects on a 0–10 scale. If the scaling involves a subjective element, some judges may be reluctant, while others are willing, to assign very high or very low values. Thus $\beta$ will vary from judge to judge.

We now consider the properties of the recorded sample mean $\bar{Y}$ under model (4). For each sample unit we suppose that $m$ independent trials or measurements are made, the case $m = 1$ being much the most common with human populations. In finding expectations we average first over different trials $t$ for a given unit and then over units. Under conditions (2.1) about the $d_{ut}$, the sample mean is subject to a bias, $a = E(a_u)$. In random sampling from an infinite population or in random sampling with replacement from a finite population, the variance of $\bar{Y}$ for a sample of size $n$ works out as (where $\overline{\sigma_d^2} = E(\sigma_u^2)$)

$$(2.2) \qquad V(\bar{Y}) = \frac{1}{n} \{\sigma_y^2 + \sigma_a^2 + 2 \operatorname{Cov}(y, a)\} + \frac{\overline{\sigma_d^2}}{mn} = \frac{\sigma_{y'}^2}{n} + \frac{\overline{\sigma_d^2}}{mn}.$$

As regards precision, $V(\bar{Y})$ is not necessarily greater than $V(\bar{y})$. If $a_u$ and $y_u$ are uncorrelated or positively correlated, $V(\bar{Y}) > V(\bar{y})$, but the reverse could hold if the $a_u$ and $y_u$ have a strong negative correlation.

For sampling without replacement from a finite population of size $N$, we define $\sigma_{y'}^2$ as $\sum_{u=1}^{N} \{y_u' - E(y_u')\}^2/(N - 1)$. For this case,

$$(2.3) \qquad V(\bar{Y}) = \frac{(N - n)}{N} \frac{\sigma_{y'}^2}{n} + \frac{\overline{\sigma_d^2}}{mn}.$$

With an infinite population, the usual sample estimate of $V(\bar{Y})$ is

$$(2.4) \qquad \hat{V}(\bar{Y}) = \sum_{u=1}^{n} (\bar{Y}_u - \bar{Y})^2/n(n - 1),$$

where $\bar{Y}_u$ denotes the recorded mean of the $m$ measurements from sample unit $u$. The expectation of $\hat{V}(\bar{Y})$ is easily seen to be

$$(2.5) \qquad E\{\hat{V}(\bar{Y})\} = \frac{\sigma_{y'}^2}{n} + \frac{\overline{\sigma_d^2}}{mn}.$$

This result holds also in random sampling with or without replacement from a finite population. By comparison of (2.5) with (2.2), $\hat{V}(\bar{Y})$ remains unbiased under model (4) for an infinite population or for sampling with replacement from a finite population.

In sampling without replacement, $\hat{V}(\bar{Y})$ has a positive bias of amount $\sigma_{y'}^2/N$, (compare (2.5) and (2.3)). If the usual finite population correction is introduced, giving $(N - n)\hat{V}(\bar{Y})/N$, this estimator has a negative bias of amount $\overline{\sigma_d^2}/mN$. Both biases are unimportant when $n/N$ is small.

To sum up, it is reassuring that under model (4) the usual sample estimate of variance takes account of both the variable bias component and the random component of the errors of measurement, except for the slight bias in sampling without replacement. For statistical inference the damaging feature is the overall bias $a$.

## 3. Binomial data

With binomial data in which the correct measurement $y_u$ takes only the values 0 and 1, Hansen, Hurwitz, and Bershad have worked out the consequences of model (4).

$$(4) \qquad Y_{ut} = y_u + a_u + d_{ut},$$

with

$$E(d_{ut} \mid u) = 0, \qquad E(d_{ut}^2 \mid u) = \sigma_u^2, \qquad E(d_{ut}, d_{ut'}) = 0, \qquad t \neq t',$$

$$E(d_{ut}, d_{vt} \mid u, v) = 0, \qquad u \neq v.$$

Consider a unit for which $y_u = 1$. If $Y_{ut} = 1$, there is no error of measurement for this unit on this trial. In order to introduce errors of measurement, we suppose that $Y_{ut} = 0$ on a certain proportion $\theta_u$ of the trials. It follows that $E(Y_{ut} \mid u) = 1 - \theta_u$. Thus, in model (4) with $y_u = 1$, we have $a_u = -\theta_u$, while $d_{ut}$ is a binomial variate with variance $\sigma_u^2 = \theta_u(1 - \theta_u)$. The quantity $\theta_u$ may alternatively be called the *probability of misclassification* for this unit, since a unit with $y_u = 1$, $Y_{ut} = 0$ is being placed in the wrong class.

Similarly, for a unit having $y_v = 0$, we suppose that $Y_{vt}$ takes the value 1 with probability $\phi_v$. This gives $a_v = \phi_v$, $\sigma_v^2 = \phi_v(1 - \phi_v)$. Note that in the binomial case, the variable bias terms are inevitably *negatively* correlated with the correct values.

With a single trial on each unit, what is the unconditional distribution of $Y_{ut}$ in the population? If $y_u = 1$ in a proportion $p$ of the units, and $q = 1 - p$, the probability $P$ that $Y_{ut} = 1$ is

(3.1) $$P = p\{1 - E(\theta_u \mid y_u = 1)\} + qE(\phi_v \mid y_v = 0)$$

(3.2) $$= p(1 - \theta) + q\phi,$$

where $\theta = E(\theta_u \mid y_u = 1)$ is sometimes called the probability of a false negative, while $\phi = E(\phi_v \mid y_v = 0)$ is the probability of a false positive. Further, since $Y_{ut}$ takes only the values 0 and 1, it is binomially distributed. It follows that in random sampling from an infinite population or in sampling with replacement from a finite population, the sample proportion $\hat{P}$ is still distributed like the mean of a binomial sample of size $n$, but with the wrong parameter $P$. This result was first given by Bross, 1954, starting from a model in which $\theta_u = $ constant $= \theta$ and $\phi_v = $ constant $= \phi$.

The net effect of errors of measurement or classification in the binomial case is therefore that the sample proportion $\hat{P}$ remains binomial, but is biased as an estimate of $p$. The bias amounts to $-p\,\theta + q\phi$. In most classifications in practice I see no reason to expect that $\theta$ and $\phi$ are equal. The condition for the absence of bias is the rather curious one that $p\theta = q\phi$. Since $V(\hat{P}) = PQ/n$, while $V(\hat{p}) = pq/n$, errors of measurement cause an increase in variance only if $P$ is nearer $\frac{1}{2}$ than $p$.

In sampling *without replacement* from a finite population, the distribution of $\sum y_u$, the correct number of 1's in the sample, is of course hypergeometric. The distribution of $\sum Y_u$ is not hypergeometric even when all $\theta_u = \theta$ and all $\phi_v = \phi$, but is an untidy expression which will not be given here.

A link between the binomial and the continuous cases can be established in certain simple situations. A binomial variate, such as the proportion of skilled workers over age 55, is often created by starting with an underlying continuous variate (age of skilled workers) and dividing its distribution into two classes. We assume model (1), $Y_{ut} = y_u + d_{ut}$ (or model (4) in the form $Y_{ut} = y_u' + d_{ut}$ in which we are willing to regard $y_u'$ as the operationally correct

TABLE 3.1.

*Values of P, θ, and φ for Given p, $\sqrt{R}$.**

| $\sqrt{R} =$ | .99 | .95 | .90 | .99 | .95 | .90 | .99 | .95 | .90 |
|---|---|---|---|---|---|---|---|---|---|
| $p$ | | $P$ | | | $\theta$ | | | $\phi$ | |
| 0.5 | .500 | .500 | .500 | .045 | .101 | .143 | .045 | .101 | .143 |
| 0.4 | .401 | .405 | .410 | .059 | .119 | .164 | .041 | .087 | .127 |
| 0.3 | .302 | .309 | .319 | .067 | .134 | .183 | .031 | .070 | .105 |
| 0.2 | .202 | .212 | .224 | .081 | .153 | .204 | .023 | .053 | .082 |
| 0.1 | .102 | .112 | .124 | .096 | .179 | .231 | .013 | .033 | .052 |
| 0.05 | .052 | .061 | .069 | .112 | .187 | .251 | .008 | .021 | .034 |
| 0.01 | .011 | .014 | .018 | .128 | .210 | .281 | .002 | .006 | .011 |

* The third decimal in $\theta$ and $\phi$ is not reliable.

measurement and to assume $\sigma_d^2$ constant). On the continuous scale, $\bar{Y}$ is unbiased, but since its variance is $(\sigma_v^2 + \sigma_d^2)/n$, its precision relative to $\bar{y}$ is $\sigma_v^2/(\sigma_v^2 + \sigma_d^2)$. In psychometric scaling, this quantity is sometimes called the coefficient of reliability $R$, since it is the correlation between two independent measurements of the sample units by two different judges or scalers. In terminology used by the Census Bureau, this quantity is $(1 - I)$, where $I = \sigma_d^2/(\sigma_v^2 + \sigma_d^2)$ is the coefficient of inconsistency, (Hansen, Hurwitz, and Pritzker, 1964).

For given $R$ or $I$, and a given value of the boundary $L$ between the two classes, we can compute $p$, $P$, $\theta$, and $\phi$ as follows. Assuming normality, $y_u$ and $Y_{ut}$ follow a bivariate normal with correlation $\sigma_v/\sqrt{\sigma_v^2 + \sigma_d^2} = \sqrt{R}$, while $y_u$ is $N(\mu, \sigma_v^2)$ and $Y_{ut}$ is $N(\mu, \sigma_v^2 + \sigma_d^2)$. Hence, the value of $\theta = P(Y_{ut} < L, y_u > L)/P(y_u > L)$, and similarly that of $\phi$, can be read from tables of the bivariate normal distribution for correlation $\sqrt{R}$. The relations between $\theta$, $\phi$, and $p$ implied by this model can therefore be studied.

As an illustration, table 3.1 shows $P$, $\theta$, and $\phi$ for $\sqrt{R}$, the correlation between $Y_{ut}$ and $y_u$, = 0.99, 0.95, and 0.90, for a series of values of $p$. The value $\sqrt{R} = 0.99$ represents a high quality of measurement, the relative precision of $\bar{Y}$ to $\bar{y}$ being 98%. The two other values of $\sqrt{R}$ correspond to relative precisions of about 90% and 81%.

Only values of $p < 0.5$ are shown. For $p > 0.5$, replace $P$ by $(1 - P)$ and interchange $\theta$ and $\phi$. With $p < 0.5$, $P$ is always positively biased under this model. The bias increases absolutely, and more so relatively, as $p$ moves towards zero, but does not exceed 10%, except for $p \leq 0.1$ when $\sqrt{R} = .95$ and for $p \leq 0.2$ when $\sqrt{R} = .90$. The values of $(\theta + \phi)$ are about 0.1, 0.2, and 0.29 for $\sqrt{R} = .99, .95$, and .90. As $p/q$ diminishes so does $\phi/\theta$, so that the condition for absence of bias, $p\theta = q\phi$, is not too badly disturbed.

## PART II. EFFECTS ON SOME STANDARD STATISTICAL TECHNIQUES.

In this part it is assumed that model (4) holds. The major restriction in this model is its assumption that errors of measurement on different units in the sample are uncorrelated. The validity of this assumption should be carefully considered before applying results from the model to a particular set of data.

## 4. *Estimation of means in analysis of variance.*

The situation here appears to be similar for the standard classifications (one-way, two-way, latin square, incomplete blocks, etc.) and may be illustrated by the two-way classification with fixed effects. In terms of the correct measurements $y_{ij}$ the model for the classification is

$$y_{ij} = \mu + \alpha_i + \beta_j + \epsilon_{ij} \qquad (i = \text{row}, \quad j = \text{column})$$

where $\sum \alpha_i = \sum \beta_j = 0$, and the $\epsilon_{ij}$ are $N(0, \sigma^2)$ and are independent. For the recorded measurement,

$$Y_{ijt} = y_{ij} + a_{ij} + d_{ijt} .$$

The notion of repeated sampling is now a little complex. For fixed $i, j$ we obtain a new $y_{ij}$ by drawing a new residual $\epsilon_{ij}$ . Measurement of this new $y_{ij}$ creates a new variable bias term $a_{ij}$ (because the true value being measured has changed even though $i$ and $j$ are the same) and for each trial of the measurement a new $d_{ijt}$ .

Three cases may arise with respect to the variable biases $a_{ij}$ : (1) they may be correlated with $y_{ij}$ as implied in model (4); (2) they may be uncorrelated with $y_{ij}$ but have a different distribution in different rows or columns, e.g. if different instruments are used in different rows or if the nature of the rows leads to different biases even with the same instrument; (3) they may have the same distribution, uncorrelated with $y_{ij}$ , throughout rows and columns.

In this last case the consequences of the measurement errors are straightforward. We have a row mean,

$$(4.1) \qquad \bar{Y}_{i.t} = \mu + \alpha_i + a + (\bar{a}_{i.} - a) + \bar{d}_{i.t} + \bar{\epsilon}_{i.}.$$

where $a$ is the population mean of the $a_{ij}$ . Since the last three terms in (4.1) all have population means zero in this kind of repeated sampling, all the row means are subject to the same bias $a$. It follows that any contrast or comparison $\sum c_i \bar{Y}_{i.t}$ among the row means, where $\sum c_i = 0$, is unbiased. There is a loss of precision, since the variance of $\bar{Y}_{i.t}$ for a single trial is $(\sigma_a^2 + \sigma_d^2 + \sigma_\epsilon^2)/n$. Further, the means may differ somewhat in precision and there may be some introduction of non-normality if errors of measurement are non-normal.

If the $a_{ij}$ are governed by cases (1) and (2), we face different biases in different row means. Comparisons among estimated row means are biased, and the statistical analysis of the $Y_{ijt}$ provides no clue to the presence of these biases. There is a loss of precision in case (2), and in case (1) also unless the correlation between $a_{ij}$ and $y_{ij}$ is negative and sufficiently large.

While these general remarks can be made from the framework of the model, any real appraisal of the adequacy of the method of measurement in an analysis of variance problem requires planning and collecting data that enable us to examine whether this type of model fits and to study the nature of the bias terms and of the fluctuating terms.

5. *The 2 × 2 table with errors in one classification.*

For data in classifications the 2 × 2 table has been considered by Bross, 1954, and others, the multinomial by Mote and Anderson, 1965, and the general two-way contingency table by Assakul and Proctor, 1965, 1967. Bross discusses the case in which proportions in two populations are being compared from samples. There is no misclassification of the population to which a unit in the data belongs but there is misclassification with respect to the attribute under study.

Quoting (3.2), if $p$ is the correct and $P$ the fallible proportion in a binomial population,

$$(3.2) \qquad P = p(1 - \theta) + q\phi,$$

where $\theta$, $\phi$ are the proportions of false negatives and false positives. If the attribute is the presence of a relatively rare disease (p small), any sizeable value of $\phi$ results in a gross overestimate of $p$ in the recorded population. For example, with $p = 0.01$, $\theta = 0.1$, $\phi = 0.2$,

$$P = (.01)(0.9) + (.99)(0.2) = 0.207,$$

about 20 times as large as $p = 0.01$.

For the difference between the proportions in two populations we have

$$(5.1) \qquad \begin{aligned} P_1 - P_2 &= p_1(1 - \theta_1) + q_1\phi_1 - p_2(1 - \theta_2) - q_2\phi_2 \\ &= p_1(1 - \theta_1 - \phi_1) - p_2(1 - \theta_2 - \phi_2) + \phi_1 - \phi_2 . \end{aligned}$$

If we can assume $\theta_1 = \theta_2$, $\phi_1 = \phi_2$, the results simplify, since (5.1) becomes, as noted by Newell, 1962,

$$(5.2) \qquad P_1 - P_2 = (p_1 - p_2)(1 - \theta - \phi).$$

In this situation the population difference in proportions on the fallible scale is biased downwards. It could even be reversed in sign if $\theta + \phi > 1$.

Since (5.2) implies that $P_1 = P_2$ whenever $p_1 = p_2$, the Type I errors of the test of significance remain correct. Since, however, $|P_1 - P_2| < |p_1 - p_2|$ the power of this test will decrease. Suppose that we have samples of size $n$ from each population when there is no misclassification. With misclassification, let $n'$ be the sample size from each population needed to give the same power. When $n$ is large, the ratio of $n'$ to $n$ is approximately

$$(5.3) \qquad \frac{n'}{n} \doteq \frac{(P_1Q_1 + P_2Q_2)}{(p_1q_1 + p_2q_2)(1 - \theta - \phi)^2}.$$

Strictly, this formula applies as an asymptotic limit when $p_2 - p_1$ is of order $1/\sqrt{n}$. In this event, $n'/n$ can be simplified to $P_1Q_1/p_1q_1(1 - \theta - \phi)^2$.

An interesting tabulation and discussion of $n'/n$ for a range of values of $\theta$ and $\phi$ has been given by Rubin, Rosenbaum, and Cobb, 1956 (using a slightly different formula for $n'/n$). They consider the choice between medical examination of subjects, assumed to classify correctly, and a cheaper interview method that involves mistakes. They estimate that medical examination costs 10 times as much per subject, so that their interest is in the range of values of $p_1$, $p_2$, and $\theta$, $\phi$ with the interview method for which $n'/n$ is substantially less than 10.

The assumption that $\theta_1 = \theta_2$ and $\phi_1 = \phi_2$ should not be made without firm justification. I recall a study in which tubercular persons were being compared with a control sample of healthy people on various characteristics. The field worker warned that her classification of the tubercular persons, whom she knew well, would be much more accurate than that of the unfamiliar control sample. This point has also been stressed by Rubin et al (*loc. cit.*) and by Diamond and Lilinefeld, 1962, who discuss the effect of misclassification on the relative risk ratio $p_1 q_2/p_2 q_1$ .

If $\theta_1 \neq \theta_2$ and $\phi_1 \neq \phi_2$ , equation (5.1) shows that $p_1 = p_2$ does not in general imply $P_1 = P_2$ , so that the Type I errors of the test of significance are increased. There is therefore an increased risk that an entirely spurious relationship is found to be statistically significant, as several writers have warned. More information is needed about the values of the $\theta$'s and $\phi$'s that occur in practice in order to appraise the seriousness of this risk.

Reverting to the link between the binomial and the continuous cases discussed in Section 3, one question is: are errors of measurement more harmful when we convert continuous data to a binomial form for analysis? As regards estimation, the answer seems "yes", because as shown in Section 3, unbiased errors on the continuous scale produce biased estimates of $p$. In tests of significance the reduction in power appears to be the same in the binomial and continuous cases if we can assume normality and have large samples. This follows because on the continuous scale the only difference between the situations with error absent and error present is that the variances of $y$ and $Y$ are different—$\sigma_y^2$ versus $(\sigma_y^2 + \sigma_d^2)$. Hence, the asymptotic relative efficiency of the binomial to the normal tests remains unchanged.

## 6. *Multinomial data.*

We suppose that with correct measuring (no misclassification) the probability that an observation lies in the $i$th class is $p_i$ . The distribution of the sample numbers falling into these classes is multinomial with probabilities $p_i$ . To describe the misclassification in the general case we need a matrix $\theta_{ij}$ , denoting the probability that an observation which correctly belongs to class $i$ is assigned to class $j$, with $\sum_j \theta_{ij} = 1$ for each $i$.

If $P_i$ denotes the probability that an observation is placed in class $i$ under the fallible classification method,

$$(6.1) \qquad\qquad P_i = \sum_i \theta_{ii} p_i .$$

Under random sampling from an infinite population, the distribution of the sample numbers falling into these classes remains multinomial, but with the wrong probabilities $P_i$ . Estimates of the proportions falling into specific classes are therefore biased. Null hypotheses that specify the $p_i$ or specify known relations between the $p_i$ will not hold when the $P_i$ are substituted for the $p_i$ , so that the Type I errors of the standard tests are increased.

Having noted these consequences of the errors in classification, Mote and Anderson, 1965, discuss the prospects of making tests of significance that have correct Type I errors. This can be done if the $\theta_{ij}$ are known. They present their techniques for the case in which the null hypothesis completely specifies the values $p_{0i}$ of the correct $p_i$ (as in the $9:3:3:1$ genetic hypothesis): the extension to hypotheses which specify the $p_{0i}$ as functions of a few parameters that must be estimated from the data appears straightforward in theory, though computations may present more difficulties.

With known $p_{0i}$ and $\theta_{ij}$ their procedure is as follows. The values $P_{0i} = \sum_i \theta_{ij} p_{0j}$ that are consistent with the null hypothesis can be found. The standard $\chi^2$ test is then applied in the form

$$\chi^2 = \sum_i \frac{(n_i - n. P_{0i})^2}{n. P_{0i}}$$

where $n_i$ observations out of $n.$ fall in the $i$th class. In large samples the noncentrality parameter, when the N.H. does not hold, is $\lambda = n. \sum (P_i - P_{0i})^2 / P_{0i}$. By applying Schwarz's inequality, Mote, 1957, has shown that misclassification reduces the value of $\lambda$ and hence the power of the test, as would be expected. (Strictly, the preceding argument requires that $\theta_{ij}$ be a non-singular matrix, since we wish the $p_{0i}$ uniquely determined by the $P_{0i}$.)

In practice, the $\theta_{ij}$ will rarely be known with negligible errors. As increased study of errors of classification is made, a situation more likely to occur is one in which estimates of the $\theta_{ij}$, subject to sampling errors, are available. The extension of the preceding work to this case will then be needed.

With unknown $\theta_{ij}$, there is no hope of estimating them from the $n_i/n.$, there being too many unknown parameters. But if enough is known to express the $\theta_{ij}$ as functions of a few parameters, the number of classes may be large enough to allow estimation of the misclassification matrix. Mote and Anderson illustrate this approach for two models where there are $r$ classes.

|                     (I)                          |                    (II)                            |
| :----------------------------------------------: | :------------------------------------------------: |
| $\theta_{ij} = \theta \quad (j \neq i)$          | $\theta_{ij} = \theta, \qquad |j - k| = 1$         |
| $\theta_{ii} = 1 - (r - 1)\theta$                | $\theta_{ii} = 1 - 2\theta, \quad \text{for} \quad 1 < i < r$ |
|                                                  | $\theta_{11} = \theta_{rr} = 1 - \theta$           |
|                                                  | $\theta_{ij} = 0, \quad \text{otherwise.}$         |

In model I, a unit in class $i$ is equally likely to be misassigned to any of the other classes. In model II, which may apply approximately with ordered classifications, misclassification occurs only between neighboring classes. The $\chi^2$ criterion now takes the form

$$\chi^2 = \sum_{i=1}^{r} \frac{(n_i - n. \hat{P}_{0i})^2}{n. \hat{P}_{0i}},$$

where the sample estimate of $\theta$ has been substituted in $\hat{P}_{0i}$. Estimation by some method of minimizing $\chi^2$ may be easier than by maximum likelihood.

### 7. $r \times c$ Contingency tables.

The work of Assakul and Proctor, 1965, 1967, extends the preceding work in two ways. They allow misclassification in both directions. In an example cited by Roget, 1961, comparing the presence or absence of brain damage at age 7 in a sample of children who had generalized cyanosis in the first week of life and a control sample of healthy births, we might be uncertain both as to the presence of brain injury and as to the possession of cyanosis. Secondly, Assakul and Proctor consider the $r \times c$ contingency table, their principal interest being in the standard $\chi^2$ test.

To describe the misclassification process, let $\theta_{sikj}$ be the probability that an observation which correctly falls in the $(s, k)$ cell is assigned to the $(i, j)$ cell. $\theta_{sikj}$ is now an $rc \times rc$ matrix with the relations

$$(7.1) \qquad \sum_{i=1}^{r} \sum_{j=1}^{c} \theta_{sikj} = 1, \quad \text{for every} \quad s, k.$$

Errors in the $s$, $i$ direction only are represented by making

$$(7.2) \qquad \theta_{sikj} = 0, (k \neq j), \qquad = \theta_{sij} \ (k = j).$$

The following are the principal results. If $p_{ij}$, $P_{ij}$ are the correct and the recorded proportions in the population in the $i$, $j$ cell

$$(7.3) \qquad P_{ij} = \sum_{s=1}^{r} \sum_{k=1}^{c} \theta_{sikj} p_{sk} .$$

The null hypothesis $p_{ij} = p_{i.}p_{.j}$ implies the null hypothesis $P_{ij} = P_{i.}P_{.j}$ and *vice versa*, if and only if the probabilities of misclassification are *independent* in the two directions, that is, if

$$(7.4) \qquad \theta_{sikj} = \rho_{si}\gamma_{kj} .$$

If errors occur in the $s$, $i$ direction only, this condition becomes $\theta_{sikj} = 0$ $(k \neq j)$, $= \theta_{si} \ (k = j)$, a model studied by Mote and Anderson, 1965. With the more general model in (7.2) for errors in one direction, the condition $\theta_{sikj} = \theta_{sij} \ (k = j)$ is the extension of the $2 \times 2$ situation in which $\theta_1 \neq \theta_2$ and $\phi_1 \neq \phi_2$ .

With the independence condition, Type I errors are unchanged and there is no increased risk of rejecting a null hypothesis that actually holds. However, the power of the test in large samples is never increased and nearly always reduced by misclassification. If $p_{ij} = p^0_{.i}.p^0_{.j} + d_{ij}/\sqrt{n}$, where $n$ is the total size of sample, the parameter of non-centrality for the correct $\chi^2$ test is

$$(7.5) \qquad \lambda = \sum_i \sum_j \frac{d_{ij}^2}{p^0_{i.}.p^0_{.j}}.$$

For the test based on misclassified data, the parameter becomes

$$(7.6) \qquad \lambda' = \sum_i \sum_j \frac{\left(\sum_s \sum_k d_{sk}\rho_{si}\gamma_{kj}\right)^2}{\left(\sum_s p^0_{s.}.\rho_{si}\right)\left(\sum_k p^0_{.k}\gamma_{kj}\right)}$$

The ratio $\lambda'/\lambda = n/n'$, where $n$ and $n'$ are the relative sample sizes needed to give equal asymptotic power, and thus measures the deleterious effect of the errors in practical terms.

When the probabilities in rows and columns are not independent, Assakul and Proctor follow the same procedure as Mote and Anderson, showing how to make the appropriate $\chi^2$ test on the $P_{ij}$ when the $\theta_{sikj}$ are known, using an illustration from U. S. Census Bureau research on classification errors. With unknown $\theta_{sikj}$ there is again the possibility that we might know enough to express $\theta_{sikj}$ in terms of a few parameters that can be estimated while fitting $\chi^2$.

## 8. *Linear regression.*

This case has long been studied for its economic applications: an extensive review for the statistician has been given by Madansky, 1959. Here we consider only the effects of errors of measurement on some of the commonest uses of linear regression.

With correct measurements, suppose that $y_u$ has a linear regression on the $k$ variates $x_{iu}$ of the form

$$(8.1) \qquad y_u = \alpha + \sum_{i=1}^{k} \beta_i x_{iu} + \epsilon_u \qquad (u = 1, 2, \cdots, n).$$

The recorded fallible measurements are given by

$$(8.2) \qquad Y_u = y_u + d_u \; ; \qquad X_{iu} = x_{iu} + e_{iu} \, ,$$

where the subscript $t$ has been dropped, assuming a single measurement of each value.

When the errors $e_{iu}$ in $X_{iu}$ are distributed independently of each other and of the $x_{iu}$ , Lindley, 1947, has given the necessary and sufficient conditions that the regression of the $Y_u$ on the $X_{iu}$ remains linear, as follows. Let the cumulant-generating function of the correct measurements be $\psi(t_1 , t_2 , \cdots , t_k)$, defined by the equation

$$\exp \{\psi(t_1 , t_2 , \cdots , t_k)\} = \iint \cdots \int \exp (i \sum t_j x_j) f(x_1 , x_2 , \cdots , x_k) \Pi \, dx_j$$

where $f(x_1 , x_2 , \cdots , x_k)$ is the joint frequency function of the $x_j$ . Let $K(t_1 , t_2 , \cdots , t_k)$ be the corresponding cumulant function of the errors of measurement $e_j$ . Then the necessary and sufficient conditions are

$$(8.3) \qquad \sum_{i=1}^{k} (\beta_i - \beta_i') \frac{\partial \psi}{\partial t_i} = \sum_{i=1}^{k} \beta_i' \frac{\partial K}{\partial t_i} \, ,$$

where the $\beta_i'$ are the regression coefficients of $Y_u$ on the $X_{iu}$ . If the $x_{iu}$ follow a multivariate normal, $\psi$ is a quadratic in the $t_i$ , so that the left side of (8.3), and consequently the right side, is linear in the $t_i$ . It follows in this case that the $e_i$ must be normally distributed if $Y_u$ is to have a linear regression on the $X_{iu}$ .

This basic result still leaves a good deal to be learned for practical applications. Two questions are: (i) what is the situation when the $e_{iu}$ and the $x_{iu}$ are correlated?; (ii) in what way and to what extent is the linearity distorted when condition (8.3) does not hold?

The relations between the $\beta_i'$ and the $\beta_i$ are of interest in studying the effects of errors of measurement on the interpretation of regression coefficients. From (8.3), Lindley expressed these relations in a more familiar form, assuming that the $e_{iu}$ and the $x_{iu}$ are uncorrelated. These results can be extended to include correlations between $e_{iu}$ and $x_{iu}$, assuming no correlation between $e_{iu}$ and $x_{ju}$ or between $e_{iu}$ and $e_{ju}$ ($j \neq i$), as follows.

Although Lindley's conditions (8.3) are a reminder that the regression of $Y_u$ on the $X_{iu}$ may not be strictly linear, we may write, formally,

$$(8.4) \qquad Y_u = \alpha' + \sum \beta_i' X_{iu} + \epsilon_u' ,$$

where $\alpha'$ and the $\beta_i'$ are chosen to minimize

$$E(Y_u - \alpha' - \sum \beta_i' X_{iu})^2,$$

taken over the population, with the consequences that over the population (though not for fixed $X_{iu}$)

$$(8.5) \qquad E(\epsilon_u') = 0; \qquad E(\epsilon_u' X_{iu}) = 0.$$

The expression $(\alpha' + \sum \beta_i' X_{iu})$ might be called the linear component of the regression of $Y$ on the $X_{iu}$. From these relations we can obtain two equivalent expressions for Cov $(YX_i)$. From (8.4),

$$(8.6) \qquad \text{Cov } (YX_i) = \text{Cov } (\alpha' + \sum_i \beta_i' X_i + \epsilon') X_i = \sum_i \beta_i' \sigma_{ij}'$$

where $\sigma_{ij}' = \text{Cov } (X_i' X_j')$. Secondly, from (8.1) and (8.2),

$$(8.7) \qquad \begin{aligned} \text{Cov } (YX_i) &= \text{Cov } (\alpha + \sum_i \beta_i x_i + \epsilon + d)(x_i + e_i) \\ &= \sum_i \beta_i \sigma_{ij} + \beta_j c_j , \end{aligned}$$

where $c_i = \text{Cov } (x_i e_i)$. Hence the equations giving the $\beta_i'$ in terms of the $\beta_i$ are, from (8.6) and (8.7),

$$(8.8) \qquad \sum_i \sigma_{ij}' \beta_i' = \sum_i \sigma_{ij} \beta_i + \beta_j c_j . \qquad (j = 1, 2, \cdots , k)$$

Further, since the population means of $Y_u$ and $y_u$ are the same,

$$(8.9) \qquad \alpha' = \alpha + \sum_i (\beta_i - \beta_i') E(x_i).$$

In (8.8), the relation between $\sigma_{ij}'$ and $\sigma_{ij}$ is as follows.

$$(8.10) \qquad \begin{aligned} \sigma_{ij}' &= \text{Cov } (X_i X_j) = \text{Cov } (x_i + e_i)(x_j + e_j) \\ &= \begin{cases} \sigma_{ij} & (j \neq i) \\ \sigma_{ii} + 2c_i + v_{ii} & (j = i) \end{cases} \end{aligned}$$

where $v_{ii}$ denotes the variance of $e_i$. Thus we need the variances and covariances of the $x_i$ and of the $e_i$ as well as any covariances between $e_i$ and $x_i$ in order to find the $\beta_i'$ from the $\beta_i$. Some practical consequences of the effects of errors on the $\beta_i'$ are discussed in Section 11.

9. *Regression on a single X variable: analysis of covariance.*

In this case (8.8) gives

$$(9.1) \qquad \beta' = \beta \{\sigma_x^2 + \text{Cov } (xe)\} / \{\sigma_x^2 + \sigma_e^2 + 2 \text{Cov } (xe)\}.$$

When $e$ and $x$ are uncorrelated, we have the well-known result,

$$(9.2) \qquad \beta' = \beta \sigma_x^2/(\sigma_x^2 + \sigma_e^2) = \beta R_X ,$$

where $R_X$ is the reliability coefficient of the measurement of $X$. Thus the regression coefficient $\beta$ of $y$ on $x$ is underestimated by the factor $R_X$.

In the analysis of covariance, these errors of measurement hurt us in several ways. Their effects will be presented for the simplest case of the comparison of two independent samples of size $n$, but the effects apply similarly in standard analysis of variance problems. The subscripts 1, 2, denote the two samples. With no error of measurement in either $y$ or $x$, we have

$$(9.3) \qquad \bar{y}_2 - \bar{y}_1 = \alpha_2 - \alpha_1 + \beta(\bar{x}_2 - \bar{x}_1) + \bar{e}_2 - \bar{e}_1$$

We suppose that the objective is to estimate $(\alpha_2 - \alpha_1)$. In randomized experiments the two samples should have the property that $E(\bar{x}_2) = E(\bar{x}_1)$. This gives us the choice of two unbiased estimates: the unadjusted estimate $(\bar{y}_2 - \bar{y}_1)$ with variance $2\sigma_y^2/n$ or the adjusted estimate

$$(9.4) \qquad \bar{y}_2 - \bar{y}_1 - b(\bar{x}_2 - \bar{x}_1),$$

where $b$ is the usual covariance estimate of $\beta$. In large samples the variance of the adjusted estimate is $2\sigma_e^2/n = 2\sigma_y^2(1 - \rho^2)/n$, where $\rho$ is the correlation between $y$ and $x$, since the contribution to the variance of the adjusted estimate arising from the sampling error of $b$ is of order $1/n^2$ and can be ignored.

With errors of measurement in both $Y$ and $X$, we have, in place of (9.3),

$$(9.5) \qquad \bar{Y}_2 - \bar{Y}_1 = \alpha_2' - \alpha_1' + \beta'(\bar{X}_2 - \bar{X}_1) + \bar{e}_2' - \bar{e}_1' .$$

Hence the adjusted estimate

$$(9.6) \qquad \bar{Y}_2 - \bar{Y}_1 - b'(\bar{X}_2 - \bar{X}_1)$$

has expected value

$$(9.7) \qquad \alpha_2' - \alpha_1' = \alpha_2 - \alpha_1 + (\beta - \beta')(\mu_{2x} - \mu_{1x}),$$

from (8.9). Since in a randomized experiment we should have $\mu_{2x} = \mu_{1x}$, the adjusted estimate remains an unbiased estimate of $(\alpha_2 - \alpha_1)$. In large samples its variance is $2\sigma_{e'}^2/n$. In order to examine this quantity we express the $\epsilon'$ in terms of the original residuals $\epsilon$. For any observation,

$$(9.8) \qquad \begin{aligned} \epsilon' &= Y - \alpha' - \beta'X \\ &= y + d - \alpha - (\beta - \beta')\mu_x - \beta'x - \beta'e \end{aligned}$$

from equation (8.9) for $\alpha'$ in terms of $\alpha$. Since $y - \alpha = \beta x + e$,

$$\epsilon' = \epsilon + d + (\beta - \beta')(x - \mu_x) - \beta'e.$$

Assuming $d$ and $e$ uncorrelated with any of the other terms, we have

$$(9.9) \qquad \sigma_{e'}^2 = \sigma_e^2 + \sigma_d^2 + (\beta - \beta')^2\sigma_x^2 + \beta'^2\sigma_e^2 .$$

Using the relations $\beta' = \beta R_X = \beta\sigma_x^2/(\sigma_x^2 + \sigma_e^2)$ and $\beta\sigma_x = \rho\sigma_y$, the terms in $\sigma_x^2$ and $\sigma_e^2$ amalgamate to give

$$(9.10) \qquad \sigma_{e'}^2 = \sigma_e^2 + \sigma_d^2 + (1 - R_X)\rho^2\sigma_y^2 .$$

The effect of errors of measurement in $Y$ is of course to contribute their variance $\sigma_d^2$ to the residual variance, while the effect of errors in $x$ is to put back a fraction $(1 - R_X)$ of the original reduction in variance, $\rho^2 \sigma_y^2$, contributed by the covariance adjustment. Incidentally, if we define a coefficient of reliability $R_Y = \sigma_y^2/(\sigma_y^2 + \sigma_d^2)$ for the measurement of $Y$, (9.10) may be expressed in the neat form

$$(9.11) \qquad \sigma_{\epsilon'}^2 = \sigma_Y^2(1 - \rho^2 R_Y R_X).$$

In non-randomized observational studies, an important use of covariance is to remove a bias in $(\bar{y}_2 - \bar{y}_1)$ as an estimate of $(\alpha_2 - \alpha_1)$. From (9.3), with correct measurements,

$$(9.12) \qquad E(\bar{y}_2 - \bar{y}_1) = \alpha_2 - \alpha_1 + \beta(\mu_{2x} - \mu_{1x}).$$

Suppose that in our study $\mu_{2x} \neq \mu_{1x}$—e.g. the two groups differ in their mean ages, and age is a variable that affects the level of $y$. The unadjusted estimate $(\bar{y}_2 - \bar{y}_1)$ has then a bias of amount $\beta(\mu_{2x} - \mu_{1x})$. The adjusted estimate in (9.4), however, remains unbiased.

With errors of measurement in $Y$ and $X$, we have from (9.6) and (9.7),

$$(9.13) \qquad E\{\bar{Y}_2 - \bar{Y}_1 - b'(\bar{X}_2 - \bar{X}_1)\} = \alpha_2' - \alpha_1'$$

$$= \alpha_2 - \alpha_1 + (\beta - \beta')(\mu_{2x} - \mu_{1x}).$$

Thus the covariance adjustment does not remove all the original bias, but leaves a fraction $(\beta - \beta')/\beta = (1 - R_X)$. Depending on the signs of the terms $(\alpha_2 - \alpha_1)$ and $\beta(\mu_{2x} - \mu_{1x})$, this bias can produce either an overestimate or an underestimate, and will disturb the Type I errors of tests of significance. The bias is, of course, created by the errors in $X$, not those in $Y$.

If there are two independent measurements $X_{iu}$, $X_{iu}'$ ($i = 1, 2$) of each value of $x_{iu}$, Lord, 1960, has given a large-sample method that removes all the bias. The strategy is to replace $b'$ in (9.13) by an unbiased estimate of $\beta$. Since $X_{iu} = x_{iu} + e_{iu}$ and $X_{iu}' = x_{iu} + e_{iu}'$, where $e_{iu}$, $e_{iu}'$ are assumed independent of $x_{iu}$ and of each other, the within-sample pooled covariance of $X_{iu}$ and $X_{iu}'$ ($w$ in Lord's notation) is an unbiased estimate of $\sigma_x^2$. (This fact was noted by Karl Pearson, 1902, and is a key point in Grubbs', 1948, method of studying the sizes of errors of measurement.) Further, the within-sample covariance $v$ of $\bar{X}_{iu}$ and $Y$ is an unbiased estimate of the covariance of $x$ and $y$, where $\bar{X}_{iu}$ is the mean of $X_{iu}$ and $X_{iu}'$. In large enough samples, the quantity $\hat{\beta} = v/w$ can be shown to be an unbiased estimate of $\beta$, so that the estimate

$$(9.14) \qquad \bar{Y}_2 - \bar{Y}_1 - \hat{\beta}(\bar{X}_2 - \bar{X}_1)$$

is free from bias. Actually, $v/w$ has a bias of order $1/n$, and Lord presents an amended estimate $\hat{\beta}^0$ that is freed from the order $1/n$ bias. In his method, the samples need not be equal in size. Further work on this problem has been done by Porter, 1967.

In econometrics there is a substantial body of work dealing with methods of estimation that provide from the $Y$'s and $X$'s consistent estimates of $\beta$ that hopefully will have greatly reduced bias in samples of practical size without requiring specific knowledge of the variances of the $e_i$. The early approach of Wald, 1940, and Bartlett, 1949, through grouping, is well known, and Durbin, 1954, has made numerous useful suggestions. While space considerations decided me against a review of this work, I wish that the effectiveness of some of these methods in moderate-sized samples from different fields of application had been more thoroughly investigated. For some time there has been much interest in investigating the robustness of standard estimation techniques against non-normality and the use of an incorrect model. Obviously, this research should also extend to robustness against the presence of errors of measurement.

## 10. *The Berkson case.*

There are two common situations in which, despite the presence of errors in $X$, $b'$ remains an unbiased estimate of $\beta$. Berkson, 1950, pointed out that in many laboratory experiments, $X$ is set at certain preassigned levels, e.g. $X = 20, 40, 60, 80$ ppm of some ingredient. With errors of measurement present, the actual amounts $x$ of this ingredient will not be exactly 20, 40, 60, 80 ppm, but will vary about these values by means of the relation $x = X + e'$, say, where $e'$ is assumed to have zero mean and to be uncorrelated with $X$. Since we have used the symbol $e$ to denote $X - x$, we can write $e' = -e$. If the relation between $y$ and $x$ is the usual linear one

$$(10.1) \qquad\qquad y = \alpha + \beta x + \epsilon$$

we obtain the relation between $Y$ and $X$ by substituting $Y = y + d, x = X - e$, giving

$$(10.2) \qquad\qquad Y = \alpha + \beta X + \epsilon + d - \beta e.$$

Assuming unbiased measurement and $E(d) = E(e) = 0$, we have $E(Y \mid X) = \alpha + \beta X$, so that the regression of $Y$ on $X$ remains linear with slope $\beta$. However, the residual variance increases from $\sigma_\epsilon^2$ to $(\sigma_\epsilon^2 + \sigma_d^2 + \beta^2 \sigma_e^2)$.

The second case, to which the same argument applies, is that in which the values of $x$ have been grouped and replaced by the midpoints $X$ of the grouping intervals for computing the regression, as is common with large samples. If within each group the values of $x$ are symmetrically distributed about the midpoint of the interval, then $e' = x - X$ has mean zero within each group and is uncorrelated with $X$ (Durbin, 1954). In fact, Durbin gives reasons for suggesting that this type of grouping will often reduce the bias due to errors of measurement in the ungrouped values.

Incidentally, these results are not a contradiction of the general result (9.1) relating $\beta'$ to $\beta$. By (9.1),

$$\beta' = \beta\{\sigma_x^2 + \text{Cov } (xe)\}/\{\sigma_x^2 + \sigma_e^2 + 2\,\text{Cov } (xe)\}.$$

Since $x = X - e$, where $X$ and $e$ are uncorrelated, we have $\text{Cov } (xe) = -\sigma_e^2$. This relation gives $\beta' = \beta$ in (9.1).

In an interesting paper, Box, 1961, extends Berkson's results to multiple linear regression, as would apply in a factorial experiment with preassigned levels of $X$. Given the true relation,

$$(10.3) \qquad\qquad y = \alpha + \sum \beta_i x_i + \epsilon$$

with errors of measurement specified by

(10.4) $\qquad Y = y + d; \qquad x_i = X_i + e'_i \; ; \qquad E(e'_i \mid X_i) = 0.$

we obtain

(10.5) $\qquad Y = \alpha + \sum \beta_i X_i + \epsilon + d + \sum \beta_i e'_i \,.$

Box calls the term $\sum \beta_i e'_i$ the error transmitted from the $x$'s. From (10.5), the estimates of the treatment effects $\beta_i$ are still unbiased. Further, if each component $\epsilon$, $d$, and $\sum \beta_i e'_i$ of the residual has a constant variance throughout the experiment, the experimental error variance is constant throughout the experiment and is estimated without bias by the usual analysis of variance.

Box also examines the situation in non-linear relations. The quadratic model

(10.6) $\qquad y = \alpha + \sum_i \beta_i x_i + \tfrac{1}{2} \sum_i \sum_j \beta_{ij} x_i x_j + \epsilon, \qquad \beta_{ij} = \beta_{ji}$

is of particular interest because of its frequent use in designs for investigating response surfaces. Substituting the fallible measurements from (10.4) gives the rather lengthy expression

(10.7) $\qquad Y = \alpha + \sum \beta_i X_i + \tfrac{1}{2} \sum \sum \beta_{ij} X_i X_j + d + \sum \beta_i e'_i$

$$+ \tfrac{1}{2} \sum \sum \beta_{ij} (X_i e'_j + X_j e'_i) + \tfrac{1}{2} \sum \sum \beta_{ij} e'_i e'_j + \epsilon$$

The term $\tfrac{1}{2} \sum \sum \beta_{ij} e'_i e'_j$ in the residual has mean value $A = \tfrac{1}{2} \sum \sum \beta_{ij} \operatorname{Cov}(e'_i e'_j)$. This is not zero, for even if $e'_i$ and $e'_j$ are uncorrelated (often a reasonable assumption) the mean value is $\tfrac{1}{2} \sum \beta_{ii} \sigma^2_{e_i}$. Consequently, there is a bias in the estimate of $\alpha$, although any contrast or comparison among the treatment means is estimated without bias. Secondly, on account of the term $\tfrac{1}{2} \sum \sum \beta_{ij} \cdot (X_i e'_j + X_j e'_i)$, the residual variance now depends in part on a quadratic function of the $X$'s and will vary from one treatment combination to another. Finally, if the $e'_i$ are normally distributed, the residual term $(\tfrac{1}{2} \sum \sum \beta_{ij} e'_i e'_j - A)$ is distributed as a linear function of $\chi^2$ values, so that the total residual is not normal even if $\epsilon$, $d$, and the $e'_i$ are normal.

If the true response surface is a polynomial of degree $m$ in the $x$'s, the bias term $A$ becomes a polynomial of degree $(m - 2)$ in the $X$'s, while the non-linear fluctuating component of the residual is a polynomial of degree $(2m - 2)$ in the $e'_i$. Box proceeds to consider the question: are the standard methods of estimating the residual variance unbiased in the presence of errors of measurement? With the usual 2-level factorials ·and fractional factorials, the error variance, as computed either from high-order interactions or from the differences between replications of the treatment combinations, remains unbiased. The conditions under which unbiased estimates of error variance are obtainable from partial duplication of the design are also given.

## 11. *Multiple regression and correlation coefficients.*

The effects of errors of measurement on individual regression coefficients in a multiple linear regression are complicated. Even in the case where the regression of $Y$ on the $X_i$ remains linear, the regression coefficient $\beta'_i$ becomes a linear function of $\beta_i$ and of the $\beta_j$ for all $X_j$ that are correlated with $X_i$. We will examine only the case of a regression on two variables.

The correct $x_1$, $x_2$ are assumed to have variances $\sigma_1^2$, $\sigma_2^2$ and correlation $\rho$, while their errors $e_1$, $e_2$ have variances $s_1^2$, $s_2^2$ and are assumed uncorrelated with each other. If $c_1 = \text{Cov}(e_1 x_1)$, $c_2 = \text{Cov}(e_2 x_2)$, the previous relation (8.8) expressing the $\beta_i'$ in terms of the $\beta_i$ gives for $\beta_1'$

$$(11.1) \qquad \beta_1' = [\beta_1\{(\sigma_1^2 + c_1)(\sigma_2^2 + s_2^2 + 2c_2) - \rho^2\sigma_1^2\sigma_2^2\} + \beta_2\rho\sigma_1\sigma_2(s_2^2 + c_2)]/D$$

where

$$D = (\sigma_1^2 + s_1^2 + 2c_1)(\sigma_2^2 + s_2^2 + 2c_2) - \rho^2\sigma_1^2\sigma_2^2 .$$

About the only obvious features of (11.1) are: (i) if $\rho$ is zero, the relation between $\beta_1'$ and $\beta_1$ is the same as in regression on $X_1$ alone, equation (9.1): (ii) if $c_1 = -s_1^2$, $c_2 = -s_2^2$, then $\beta_1' = \beta_1$ (the Berkson case).

If the $e_i$ are uncorrelated with the $x_i$, we may write $\sigma_i^2 = R_i(\sigma_i^2 + s_i^2)$, where $R_1$ and $R_2$ are the reliability coefficients of $X_1$ and $X_2$. Equation (11.1) can then be put in the simpler form

$$(11.2) \qquad \beta_1' = \{\beta_1 R_1(1 - \rho^2 R_2) + \beta_2\beta_{2.1}R_1(1 - R_2)\}/(1 - \rho^2 R_1 R_2)$$

where $\beta_{2.1} = \rho\sigma_2/\sigma_1$ is the regression coefficient of $x_2$ on $x_1$, while

$$(11.3) \qquad \beta_2' = \{\beta_1\beta_{1.2}R_2(1 - R_1) + \beta_2 R_2(1 - \rho^2 R_1)\}/(1 - \rho^2 R_1 R_2).$$

If only $X_1$ is subject to error, i.e. $R_2 = 1$, (11.2) gives $\beta_1' = \beta_1 R_1 f$, where $f$ is the factor $(1 - \rho^2)/(1 - \rho^2 R_1)$, which is essentially less than 1. Hence, $\beta_1'$ is damped down more severely by errors in $x_1$ in bivariate than in univariate regression. Further, with $R_2 = 1$,

$$(11.4) \qquad \beta_2' = \beta_2 + \beta_1\beta_{1.2}(1 - R_1)/(1 - \rho^2 R_1).$$

Thus, even if $x_2$ is measured without error, $\beta_2'$ may either overestimate or underestimate $\beta_2$, this depending on whether the product $\beta_1\beta_{1.2}$ has the same sign as $\beta_2$ or not. With $\beta_1$, $\beta_2$, and $\rho$ all positive (a common situation), it is not hard to produce examples in which $\beta_1 > \beta_2$ but $\beta_1' < \beta_2'$. Thus, interpretation of the relative sizes of different regression coefficients can be severely distorted by errors of measurement.

To illustrate, Table 11.1 shows the values of $\beta_1'$, $\beta_2'$ for $\beta_1 = 2$, $\beta_1 = 1$; $\rho = +0.3$, $-0.3$; and $R_1$, $R_2 = 0.6$, $0.8$, $1.0$. The table assumes the $x$'s scaled so that $\sigma_1 = \sigma_2$, making $\beta_{2.1} = \beta_{1.2} = \rho$, the correlation between $x_1$ and $x_2$.

When $X_1$ and $X_2$ are measured with equal reliability, $R_1 = R_2 = R$, the ratio $\beta_i'/\beta_i$ is somewhat greater than $R$ when $\rho$ is positive and less than $R$ when $\rho$ is negative. With $R = 0.6$, for instance, $\beta_1'/\beta_1 = 0.62$, $\beta_2'/\beta_2 = 0.74$ when $\rho = +0.3$, as against $\beta_1'/\beta_1 = 0.55$, $\beta_2'/\beta_2 = 0.44$ when $\rho = -0.3$. This ratio fluctuates more for the weaker than for the stronger variate, because of the contribution to $\beta_i'$ from the other variate. When $R_1 = 1$ we note, as already mentioned, that $\beta_1'$ exceeds 2 when $\rho = +0.3$ and is less than 2 when $\rho = -0.3$. Lastly, the only case in table 11.1 for which $\beta_1' < \beta_2'$ occurs when $X_1$ is imprecise but $X_2$ is correct ($R_1 = 0.6$, $R_2 = 1$).

For purposes of prediction, interest centers in the effects of these errors on

TABLE 11.1

Values of $\beta_1'$, $\beta_2'$ when $\beta_1 = 2$, $\beta_2 = 1$

| $R_2$ | $R_1 =$ | $\rho = +0.3$ | | | $\rho = -0.3$ | | |
|---|---|---|---|---|---|---|---|
| | | 0.6 | 0.8 | 1.0 | 0.6 | 0.8 | 1.0 |
| 0.6 | $\beta_1'$ | 1.25 | 1.68 | 2.13 | 1.10 | 1.48 | 1.87 |
| | $\beta_2'$ | 0.74 | 0.66 | 0.58 | 0.44 | 0.51 | 0.58 |
| 0.8 | $\beta_1'$ | 1.20 | 1.63 | 2.06 | 1.13 | 1.52 | 1.94 |
| | $\beta_2'$ | 0.99 | 0.89 | 0.78 | 0.59 | 0.69 | 0.78 |
| 1.0 | $\beta_1'$ | 1.15 | 1.57 | 2.00 | 1.15 | 1.57 | 2.00 |
| | $\beta_2'$ | 1.25 | 1.13 | 1.00 | 0.75 | 0.87 | 1.00 |

the square of the multiple correlation coefficient, and on the Mahalanobis squared distance in the related uses of the discriminant function. So far as I know, these effects have not been thoroughly studied. To take the simplest case, suppose that the errors of measurement are uncorrelated with the correct values and with each other. If $\rho_i$ is the simple correlation coefficient between $y$ and $x_i$, the errors reduce this to $\rho_i' = \rho_i \sqrt{R_Y R_i}$. Let $\Gamma$, $\Gamma'$ denote the squared multiple correlation coefficients between $y$ and the $x_i$ and between $Y$ and the $X_i$. If the $x_i$ and $x_j$ were uncorrelated, the effects of the errors would be that $\Gamma'$ equals $\Gamma$ multiplied by

$$(11.1) \qquad R_Y(\sum \rho_i^2 R_i / \sum \rho_i^2).$$

This factor is $R_Y$ multiplied by a weighted mean of the $R_i$, the $R_i$ for the best predictors (with high $\rho_i^2$) receiving the most weight. A similar result holds for the Mahalanobis $D^2$ with $\delta_i^2$ in place of $\rho_i^2$. However, the errors of measurement also reduce any correlation $\rho_{ij}$ between $x_i$ and $x_j$ to $\rho_{ij}' = \rho_{ij} \sqrt{R_i R_j}$ between $X_i$ and $X_j$. In most applications that I have examined, intercorrelations between the $x_i$ are on the whole harmful (i.e. they make $\Gamma$ less than the value it would have if the $x_i$ were uncorrelated). Consequently, this reduction in intercorrelations ($\rho_{ij}' < \rho_{ij}$), mitigates their harmful effect on $\Gamma$. This argument plus some further investigation suggests a rough generalization: if all the $X$'s have reliability $R$, then $\Gamma'$ will usually be slightly greater than $\Gamma R_Y R$. On the other hand, if we are lucky enough to have helpful intercorrelations, $\Gamma'$ appears to be less than $\Gamma R_Y R$.

## PART III. TECHNIQUES FOR THE NUMERICAL STUDY OF ERRORS

### 12. Simultaneous measurement by correct and fallible instruments.

When we try to put numbers into our models, our degree of success depends on the complexity of the model and on the type of data that it is feasible to collect. This section deals with work done on some of the simplest and commonest cases.

Obviously, the most favorable situation is one in which any item can be measured simultaneously by the fallible instrument and by a correct instrument, as Karl Pearson did. If $m$ independent measurements of each of $n$ items are made, consider the model

$$(12.1) \qquad Y_{it} = y_i + a_i + d_{it} \qquad (i = 1, 2, \cdots, n, \quad t = 1, 2, \cdots, m).$$

The mean difference $(\bar{Y}_{..} - \bar{y}_{.})$ provides an estimate of the overall bias $a = E(a_i)$. In the analysis of variance of the individual differences $(Y_{it} - y_i)$, the sum of squares partitions into two components:

|  | d.f. | Mean Square | $E$(M.S.) |
|---|---|---|---|
| Between items | $(n - 1)$ | $s_b^2$ | $\overline{\sigma_d^2} + m\sigma_a^2$ |
| Within items | $n(m - 1)$ | $s_w^2$ | $\overline{\sigma_d^2}$ |

This analysis provides (i) an unbiased estimate $s_w^2$ of the average variance $\overline{\sigma_d^2}$ due to the "random" component of the error, (ii) if normality and constant $\sigma_d^2$ can be assumed, an $F$-test, where $F = s_b^2/s_w^2$, of the null hypothesis $\sigma_a^2 = 0$, namely that there is no variable bias term, (iii) an unbiased estimate $(s_b^2 - s_w^2)/m$ of $\sigma_a^2$ if variable biases are present. The variance of the correct measurements $\sigma_y^2$ is of course estimated directly by $s_y^2$, while the sample covariance of $y_i$ and $(\bar{Y}_{i.} - y_i)$ estimates the covariance (if any) between the $y_i$ and the $a_i$.

Sometimes each item can be measured only once by the fallible instrument $(m = 1)$. In this event the $F$-test of $\sigma_a^2$ cannot be made, and $s_b^2$ estimates $(\overline{\sigma_d^2} + \sigma_a^2)$. If the variable bias term $a_i$ is present but varies from item to item independently of $y_i$, I see no way to detect it with these data, but in some circumstances we might be willing to regard it as "random" and amalgamate it with $d_{it}$. The sample covariance of $(Y_i - y_i)$ on $y_i$ still estimates Cov $(y_i a_i)$, while if $a_i$ has a linear regression $\alpha + \beta y_i$ on $y_i$, the regression coefficient of $(Y_i - y_i)$ on $y_i$ estimates $\beta$.

### 13. *Repeated independent measurements by the fallible instrument.*

More frequently, direct measurement of the $y_i$ is impossible. Sometimes, $m$ independent measurements of each of $n$ items can be made by the same instrument. There is now no prospect of detecting the variable bias terms $a_i$. We must either assume that the model is $Y_{it} = y_i + a + d_{it}$, with $E(d_{it} \mid y_i) = 0$, or regard $y_i' = (y_i + a_i) = E(Y_{it} \mid y_i)$ as the operationally correct measurement. In the following, we assume $Y_{it} = y_i + a + d_{it}$. Although the overall bias $a$ is also undetectable, unbiased estimates of $\sigma_y^2$ and $\overline{\sigma_d^2}$ are obtained from the following analysis of variance of the $Y_{it}$.

|  | d.f. | Means Square | $E$ (M.S.) |
|---|---|---|---|
| Between items | $(n - 1)$ | $s_b^2$ | $\overline{\sigma_d^2} + m\sigma_y^2$ |
| Within items | $n(m - 1)$ | $s_w^2$ | $\overline{\sigma_d^2}$ |

The estimates are: $\hat{\sigma}_y^2 = (s_b^2 - s_w^2)/m$, $\widehat{\overline{\sigma_d^2}} = s_w^2$. With constant $\sigma_d^2$, the coefficient of reliability $R$ is estimated by $\hat{R} = (s_b^2 - s_w^2)/[s_b^2 + (m - 1)s_w^2]$. If the $y_i$ and $d_{it}$ are normal with constant variances, confidence limits for $R$ can be computed, since $s_b^2/s_w^2$ is distributed as a multiple of $F$. The limits work out as follows.

$$(13.1) \qquad R = \frac{m\hat{R} - (F - 1)(1 - \hat{R})}{m + (m - 1)(F - 1)(1 - \hat{R})}.$$

For a two-sided confidence probability $(1 - \alpha)$, substitute the *upper* level $F_{\alpha/2}$ of $F$ with $(n - 1)$ and $n(m - 1)$ d.f. to find the lower limit of $R$. For the upper limit of $R$, substitute $F = 1/F_{\alpha/2}$ with $n(m - 1)$ and $(n - 1)$ d.f. The limits can be computed more quickly from $\phi = s_b^2/s_w^2$ as $R = (\phi - F)/[\phi + (m-1)F]$.

Workers in engineering and physics sometimes prefer to estimate $\sigma_d^2/\sigma_\nu^2 = U = (1 - R)/R$, estimated by $\hat{U} = m s_w^2/(s_b^2 - s_w^2)$. Confidence limits for $U$ are given by

$$(13.2) \qquad U = \frac{mF\hat{U}}{m + (m - 1)(F - 1)\hat{U}}$$

with the same values of $F$ as in (13.1).

14. *Independent measurements by different instruments.*

The basic work on this problem is that of Grubbs, 1948. Consider the model

$$(14.1) \qquad Y_{ij} = y_i + a_j + d_{ij}$$

where $i = 1, 2, \cdots, n$ represents items or specimens, each measured independently by $j = 1, 2, \cdots, m$ instruments. The term $a_j$ is the overall bias of the $j$th instrument. The errors of measurement $d_{ij}$ are assumed uncorrelated with the unknown correct measurements $y_i$, with $E(d_{ij}) = 0$, $E(d_{ij}^2) = \sigma_j^2$, since the instruments may differ in precision.

We start with the usual analysis of variance for a two-way classification, although Grubbs did not use this.

|  | d.f. | Mean Square | E (M.S.) |
|---|---|---|---|
| Between items | $(n - 1)$ | $s_I^2$ | $\bar{\sigma_j^2} + m\sigma_\nu^2$ |
| Between instruments | $(m - 1)$ | $s_M^2$ | $\bar{\sigma_j^2} + n \sum (a_j - \bar{a})^2/(m - 1)$ |
| Interactions | $(n - 1)(m - 1)$ | $s_{IM}^2$ | $\bar{\sigma_j^2}$ |

In the mean squares, the subscripts $I$ and $M$ denote the items and the measurers (instruments), respectively. The relative sizes of $s_M^2$ and $s_{IM}^2$ indicate whether there are differences in the overall biases $a_j$ of the different measuring instruments. The quantity $\hat{\sigma}_\nu^2 = (s_I^2 - s_{IM}^2)/m$, is an unbiased estimate of $\sigma_\nu^2$, while the interactions mean square $s_{IM}^2$ is an unbiased estimate of the average value of $\sigma_j^2$. If the $\sigma_j^2$ can be assumed all equal, say to $\sigma_d^2$, then $s_{IM}^2$ provides an estimate of $\sigma_d^2$. Given normality, equations (13.1) and (13.2) in the preceding section can still be used to find confidence intervals for $R$ and $U$, except that the degrees of freedom in $F$ become $(n - 1)$ and $(n - 1)(m - 1)$, respectively.

This is not all that we want: one of the main reasons for the investigation may be to estimate and compare the $\sigma_j^2$ for individual instruments, in order to find out which is most precise.

Grubbs' approach is by means of the sample variances and covariances for the individual measuring instruments,

$$s_j^2 = \sum_{i=1}^n (Y_{ij} - \bar{Y}_{.j})^2/(n - 1) : s_{jk} = \sum_{i=1}^n (Y_{ij} - \bar{Y}_{.j})(Y_{ik} - \bar{Y}_{.k})/(n - 1).$$

Clearly,

(14.2)                    $E(s_j^2) = \sigma_y^2 + \sigma_j^2 : E(s_{jk}) = \sigma_y^2, \qquad j \neq k.$

As his estimate of $\sigma_y^2$, Grubbs takes the average of the $s_{jk}$, namely

(14.3)                    $\hat{\sigma}_y^2 = 2 \sum\limits_{j=1}^{m} \sum\limits_{k>j}^{m} s_{jk}/m(m-1).$

Gaylor, 1956, showed that this estimate is identical with the analysis of variance estimate, $(s_I^2 - s_{IM}^2)/m$.

As the estimate of $\sigma_j^2$, Grubbs proposes

(14.4)      $\hat{\sigma}_j^2 = s_j^2 - \dfrac{2}{(m-1)} \sum\limits_{k \neq j}^{m} s_{jk} + \dfrac{2}{(m-1)(m-2)} \sum s_{kl}$

where the last sum is over all $(m-1)(m-2)/2$ distinct combinations of the *other* pairs of instruments $(k, l \neq j)$. The particular form of this estimate may puzzle the reader. Since $E(s_j^2) = \sigma_y^2 + \sigma_j^2$, while $E(s_{jk}) = E(s_{kl}) = \sigma_y^2$, any estimator of the form

$$s_j^2 + \sum c_{jk}s_{jk} + \sum c_{kl}s_{kl}, \qquad \sum c_{jk} + \sum c_{kl} = -1$$

is unbiased. Why this one? If $y_i$, $d_{ij}$ are normal and independent,

$$\mathrm{Cov}\,(s_j^2, s_{jk}) = 2\sigma_y^2(\sigma_y^2 + \sigma_j^2)/(n-1) : \mathrm{Cov}\,(s_j^2, s_{kl}) = 2\sigma_y^4/(n-1).$$

If we do not know the relative values of different $\sigma_j^2$, and wish to minimize the variance of $\hat{\sigma}_j^2$ under normality, these results suggest that an estimate of the form

$$s_j^2 - c_1 \sum s_{jk} + c_2 \sum s_{kl}$$

where $c_1 > 0$, $c_2 > 0$, is worth trying. Grubbs' estimate is the only one of this class for which the variance of $\hat{\sigma}_j^2$ does not depend on $\sigma_y^2$ $(m \geq 3)$. As Grubbs showed, assuming normality,

$$(n-1)V(\hat{\sigma}_j^2) = 2\sigma_j^4 + \frac{4\sigma_j^2}{(m-1)} \sum\limits_{k \neq j} \sigma_k^2 + \frac{4}{(m-1)^2(m-2)^2} \sum \sigma_k^2 \sigma_l^2.$$

For most applications, absence of positive terms in $\sigma_y^2$, $\sigma_y^4$ is desirable, since we expect that $\sigma_y^2$ will be substantially greater than any $\sigma_j^2$ unless precision of measurement is very poor. An analysis of variance fan might be tempted to try the alternative estimate $(s_j^2 - \hat{\sigma}_y^2)$, using the $\hat{\sigma}_y^2$ given both by the analysis of variance and by Grubbs. The variance of this estimator involves $\sigma_y^2$ and the estimator does not seem superior to Grubbs' even when $\sigma_y^2$ and the $\sigma_j^2$ are roughly of the same size.

As is usual with estimated components of variance, the estimates $\hat{\sigma}_y^2$ and $\hat{\sigma}_j^2$ can take negative values. For two instruments, Thompson, 1962, has given amended estimates not subject to this disadvantage. Under normality, these estimates are adjusted maximum likelihood estimates. In the two-instrument case the quantity $\hat{R}_1 = s_{12}/s_1^2$ estimates the coefficient of reliability $R_1$ for the first instrument. Since $s_{12}/s_1^2$ is the regression coefficient of $Y_{i2}$ on $Y_{i1}$, a lower confidence limit for $R_1$ is given by

$$(14.5) \qquad R_{1L} = \hat{R}_1 - t_\alpha \sqrt{\frac{(n-2)s_1^4}{s_1^2 s_2^2 - s_{12}^2}}$$

where $t_\alpha$ is the one-sided level with $(n-2)$ d.f. The corresponding upper level $R_{1U}$ is not exact in moderate samples, since $R_{1U}$ can exceed 1 whereas $R$ cannot. Thompson, 1963, has given the corresponding lower level for $\Delta_1 = \sigma_y/\sigma_1 = \sqrt{R_1/(1-R_1)}$.

As Grubbs has remarked, more work is needed on this problem, both on questions of efficient estimation and on the effects of non-normality.

15. *Linear relation between fallible and correct values.*

Continuing the theme of independent measurements by different instruments, the preceding model may be too simple for many applications. Fairfield Smith, 1950, pointed out that the bias of the $j$th instrument may depend on the value $y_i$ that is being measured, the case most studied being that in which this relation is linear. This is in effect the model numbered (4a) in section 2. If $i$ denotes the item being measured and $j$ the measuring instrument, and if $r$ repetitions of each measurement are made, this model may be written

$$Y_{ijt} = \alpha_j + \beta_j(y_i - \bar{y}) + a_{ij} + d_{ijt}$$

where

$$E(a_{ij}) = 0, \quad \text{Cov}\,(a_{ij}, y_i) = 0, \quad E(d_{ijt} \mid ij) = 0, \quad E(d_{ijt}^2 \mid ij) = \sigma_j^2$$

The term $a_{ij}$ measures any failure of this linear relationship to represent fully the systematic errors of measurement. This term might be due to disturbing variables that affect the measurement and have not been controlled. This component of the error cannot be reduced by repeated measurement of the same item by the same instrument.

As already mentioned, this model was used by Mandel, 1959, for the analysis of inter-laboratory round robins, by Mosteller, unpublished memo, when a number of individuals are rated by different judges, and in a slightly different notation by Fairfield Smith, 1950. Although in these applications the $y_i$ are unknown, it will simplify explanation if we assume the $y_i$ known at first.

With the $y_i$ known, the regression for each instrument of $\bar{Y}_{ij.}$ on $y_i$ provides estimates of $\alpha_j$ and $\beta_j$ for that instrument. The residual mean square from this regression is an unbiased estimate of $V(a_{ij}) + \sigma_j^2/r$, while an independent estimate of $\sigma_j^2$ is obtained from the pooled mean square derived from

$$\sum \sum \sum (d_{ijt} - \bar{d}_{ij.})^2.$$

If $\sigma_j^2$, the replication error, seems related to $y_i$, Mandel recommends a preliminary transformation of the data so that $\sigma_j^2$ becomes constant over the range of the data, or at least varies only with $j$, the instrument. If the term $V(a_{ij})$ is present and appears noticeably large for one or two laboratories, Mandel suggests detailed examination of the individual residuals from the regression for these laboratories as a clue to sources of difficulty in them.

When $r = 1$ we cannot isolate $V(a_{ij})$ and have to combine the term $a_{ij}$ with $d_{ij}$. The residual mean square from the regression for instrument $j$ then provides an estimate of the variance of $(a_{ij} + d_{ij})$. In the application considered by Mosteller, with $r = 1$ and $a_{ij} = 0$, the linear component of the model, $\alpha_i + \beta_i(y_i - \bar{y})$, indicates the type of scale used by the $j$th judge. To allow for these difference in breadth of scale, comparisons of the relative precisions of different judges are based on the estimates of $\sigma_i^2/\beta_i^2$ .

Having obtained estimates of the $\alpha_i$ and $\beta_i$ , Mandel proceeds to analyse the between-laboratory variation in terms of $V(\alpha)$ and $V(\beta)$ and to discuss the interpretation of the results. An example with detailed instructions for calculations has been given by Mandel and Lashof, 1959.

Since the $y_i$ are not known in these examples, both Mandel and Mosteller suggest using $\bar{Y}_{i.}$ , the average result over all instruments, in place of $y_i$ as the independent variable in the regressions. Approximate adjustments to take account of this substitution have been worked out by these authors, but more work remains to be done.

## 16. Study of errors with binomial data.

With binomial data a theoretical framework for the study of errors of measurement has been presented in several papers by workers in the U.S. Bureau of the Census, though the development is less extensive than with continuous data. Suppose first that a random sample of $n$ units has been measured both by the correct and by a fallible instrument, with the following results.

Number of units

|  |  | Fallible instrument | | |
|---|---|---|---|---|
|  |  | 1 | 0 | |
| Correct | 1 | $a$ | $b$ | $a + b$ |
| Instrument | 0 | $c$ | $d$ | $c + d$ |
|  |  | $a + c$ | $b + d$ | $n$ |

Reverting to the notation of section 3, the proportion $\theta$ of false negatives is estimated by $b/(a + b)$, the proportion $\phi$ of false positives by $c/(c + d)$, and $p$ by $(a + b)/n$. Alternatively, if we use the notation $Y_i = y_i + \alpha + d_i$ , where $Y_i$ and $y_i$ are $(0, 1)$ variates, the error of measurement, $\alpha + d_i$ , has the value 0 in $(a + d)$ cases, $-1$ in $b$ cases, and $+1$ in $c$ cases. Thus the bias $\alpha$ in the proportion of 1's is estimated by $(c - b)/n$, while the mean square error of $(\alpha + d_i)$ is estimated by $(b + c)/n$. If the fallible measurement is unbiased on the $T(0, 1)$ scale, this quantity also estimates the variance of $d_i$ .

Next consider the case in which we have two independent measurements of each unit by the same fallible instrument, or by different instruments of the same accuracy of measurement. From a $2 \times 2$ table like that above, where rows and columns now refer to the first and second measurements, it is not possible to estimate $p$, $\theta$, and $\phi$. The reason is that $b$ and $c$ are both estimates of $n\{p\theta(1 - \theta) + q\phi(1 - \phi)\}$, so that the table provides only two separate proportions to estimate three quantities. Estimation becomes possible with 3 or more independent measurements.

If $Y_{i1} = y_i + \alpha + d_{i1}$, and $Y_{i2} = y_i + \alpha + d_{i2}$, then $\sigma_d^2$, which equals $E(d_{i1} - d_{i2})^2/2$, is estimated from the $2 \times 2$ table by $(b + c)/2n$. Since $Y_{i1}$ and $Y_{i2}$ are binomial variates, as shown previously, $\sigma_Y^2$ is estimated by either $(a + b)(c + d)/n$ or $(a + c)(b + d)/n$. Hence, the index of inconsistency $I = (1 - R) = \sigma_d^2/\sigma_y^2$ is estimated by

$$\hat{I} = n(b + c)/\{(a + b)(c + d) + (a + c)(b + d)\}.$$

The Bureau of the Census, 1964, has given tables of the values of $\hat{I}$ found by reinterview studies for numerous housing and population characteristics measured in the 1960 U. S. Census. Among characteristics with values of $\hat{I}$ exceeding 0.6 were condition of house, condition of plumbing, and self-employment income. For total income, $\hat{I}$ averaged around 0.45, for years of schooling, around 0.35, and for age (in 5-year classes) around 0.08.

In an important paper, Hansen, Hurwitz, and Pritzker, 1964, discuss methodological aspects of the quantity $g = (b + c)/n$ as an estimator of $2\sigma_d^2$. They show, as would be anticipated, that if the reinterview is by an *improved* measuring instrument, $g$ will underestimate the value of $2\sigma_d^2$ for the original instrument. This was presumably the case in the 1960 Census reinterviews, since the reinterview teams were more expert than the original Census interviewers. Further, correlation between $d_{i1}$ and $d_{i2}$ may be expected if the respondent tends to remember and repeat the same answer at both interviews, or if the second team are instructed to reconcile the answers at the two interviews, a procedure that has obvious appeal in trying to discover the reason for discrepancy between first and second interviews. In the presence of correlation, $g$ estimates $2\sigma_d^2(1 - \rho)$. Thus, both correlation and improved reinterviewing result in estimates $\hat{I}$ that are biased downwards. The authors point out the need for methods of estimating $\rho$, and also the possible presence of what we have called a 'variable bias' term that may itself be correlated between the first and second measurement.

## 17. *Models involving an interviewer effect.*

In sample surveys in which measurements are made by human beings, either by measuring crop yields and acreages or by interviewing respondents, it has long been recognized that the results obtained by the same interviewer on different sampling units may be positively correlated. This correlation may be represented in a form of model (4). If $i$ represents the unit and $j$ the interviewer or human measurer, we write

$$Y_{ijt} = y_i + \alpha_j + a_{ij} + d_{ijt}.$$

Here, $\alpha_j$ is the overall bias and $a_{ij}$ the variable bias of the $j$th interviewer. All variables are assumed uncorrelated with each other. Suppose that in a random sample of size $n$ there are $k$ interviewers, each obtaining $m$ interviews, with $n = km$. Then

(17.1)
$$\sigma_{\bar{Y}}^2 = \frac{1}{n}(\sigma_y^2 + \overline{\sigma_a^2} + \overline{\sigma_d^2}) + \frac{1}{k}\sigma_\alpha^2.$$

Defining the intra-interviewer correlation coefficient as $\rho_w = \text{Cov}(Y_{ij}, Y_{i'j})/\sigma_Y^2 = \sigma_\alpha^2/\sigma_Y^2$, dropping the subscript $t$ when each unit is measured once only, equation (17.1) can be expressed as

$$(17.2) \qquad \sigma_{\bar{Y}}^2 = \frac{1}{n} \sigma_Y^2 \{ 1 + (m-1)\rho_w \}.$$

The consequence of intra-interviewer correlation is that $\sigma_{\bar{Y}}^2$ does not diminish as fast as $1/n$: when $n$ is large, the term $\sigma_a^2/k$ in (17.1) may come to dominate $\sigma_{\bar{Y}}^2$. Hence, there has been much interest in estimating $\rho_w$.

Estimates of $\sigma_a^2/\sigma_Y^2$ by the method of interpenetrating subsamples, due to Mahalanobis, have been investigated by Sukhatme and Seth, 1952, and Kish, 1962, among others. These authors consider the case where $m$ varies from interviewer to interviewer, as is likely to happen in practice, but for simplicity we will keep $m$ fixed. In the simplest situation, each interviewer is assigned to a randomly chosen subsample of $m$ units. This requirement presented no difficulty in the two studies reported by Kish, in which all interviews took place within the same large factory. In the analysis of variance of the results, the mean squares $s_b^2$ (between interviewers) and $s_w^2$ (within interviewers) have the following expectations.

$$E(s_b^2) = \sigma_y^2 + m\sigma_\alpha^2 + \overline{\sigma_a^2} + \overline{\sigma_d^2}$$
$$E(s_w^2) = \sigma_y^2 \qquad\quad + \overline{\sigma_a^2} + \overline{\sigma_d^2}$$

Unbiased estimates of $\sigma_\alpha^2$ and $\sigma_Y^2$ and hence an estimate of $\rho_w$ are obtained from these relations.

Sukhatme and Seth present the analysis of variance under a variety of more complex models. In extensive surveys, complete random assignment to interviewers is not practicable because of the extra travel costs, but it might be feasible to do this for pairs of interviewers within small strata.

In addition to an interest in $\rho_w$, we might wish to estimate $R = \sigma_y^2/\sigma_{\bar{Y}}^2$ for the sample mean, or equivalently $I = 1 - R$, as measuring the relative contribution of errors of measurement to the imprecision of the sample mean. Note that under this model, $I$ for the sample mean depends on $m$, the number of interviews per interviewer, since

$$I_{\bar{Y}} = (m\sigma_i^2 + \overline{\sigma_a^2} + \overline{\sigma_d^2})/(\sigma_y^2 + m\sigma_i^2 + \overline{\sigma_a^2} + \overline{\sigma_d^2}).$$

Sukhatme and Seth present several numerical estimates of $\sigma_\alpha^2$ from Indian agricultural surveys. Kish summarizes the ranges of values of $\rho_w$ obtained in his own and previous studies, including a number of attitudinal variables usually considered ambiguous and difficult to measure. He also notes that in comparisons between the means of different subclasses in the population ('domains of study'), the effects of interviewer biases $\alpha_i$ on precision tend to be much smaller because each interviewer usually works in several different subclasses.

Actually, the analysis of variance above does not allow us to estimate $I$ or $R$. For this, we would need a means of estimating $(\sigma_a^2 + \overline{\sigma_a^2} + \overline{\sigma_d^2})$. If a second independent measurement of the same unit by the same interviewer could be made, the mean square within units would estimate $\overline{\sigma_d^2}$, as Sukhatme and Seth note. However, estimation of $I$ would still not be possible unless the 'variable bias' variance $\overline{\sigma_a^2}$ could be assumed negligible.

In an ingenious further development of this approach, Fellegi, 1964, divides the sample of size $n = mk$ into $k$ subsets of size $m$, say $S_1$, $S_2$, $\cdots$, $S_k$. Suppose we form a $k \times k$ latin square of the letters $S_1$, $S_2$, $\cdots$, $S_k$, randomize it, and take only the first two rows. Assign one interviewer at random to each column. Then, (i) each interviewer collects data from the two subsets to which he has been assigned, and (ii) each subset has been measured twice, by different interviewers. Fellegi describes in detail the different kinds of intercorrelation that may be present in a survey and examines the extent to which relevant variances and covariances can be estimated by this plan, with illustrations from its use in the 1961 Canadian Census.

The presence of intercorrelations of various types both complicates the mathematical models and makes it much more difficult to devise research studies that are capable of estimating the components of error that we would like to estimate. In view of its already undesirable length, this paper has by no means done justice to what has been accomplished, in striving for realistic model-building and in numerical estimation, in the sample survey field.

In conclusion, this review indicates that errors of measurement can sometimes seriously vitiate most standard statistical techniques and at other times have only trivial effects—it depends on the sizes of the relevant variances and covariances. Though further model-building and checking is important, what seems most needed in the present state of development of this area are many studies that permit estimation of these variances and covariances. As Kish, 1962, has emphasized, it is to be hoped that most of these estimation studies can be embedded in ongoing surveys. When an "errors of measurement" study has to be conducted separately, as will sometimes be necessary because of the complexity of such studies, it is always difficult to reproduce the working conditions of an actual survey.

REFERENCES

ASSAKUL, K. and PROCTOR, C. F., 1965. Testing hypotheses with categorical data subject to misclassification. *Institute of Statistics* Mimeograph Series No. 448, North Carolina State University.

ASSAKUL, K. and PROCTOR, C. H., 1967. Testing independence in two-way contingency tables with data subject to misclassification. *Psychometrika 32*, 67–76.

BARTLETT, M. S., 1949. Fitting a straight line when both variables are subject to error. *Biometrics 5*, 207–212.

BERKSON, J., 1950. Are there two regressions? *Jour. Amer. Stat. Assoc. 45*, 164–180.

Box, G. E. P., 1961. The effects of errors in the factor levels and experimental design. *Bull. Int. Stat. Inst. 38*, 339–355.

BROSS, I. D. J., 1954. Misclassification in 2 × 2 tables. *Biometrics 10*, 478–486.

BUREAU OF THE CENSUS, 1964. Evaluation and research program of the U. S. Censuses of Population and Housing, 1960. Series ER60, Nos. 3 and 4.

DIAMOND, E. L. and LILIENFELD, A. M., 1962. Misclassification errors in 2 × 2 tables with one margin fixed: some further comments. *Amer. Jour. Pub. Health 52*, 2106–2110.

DURBIN, J., 1954. Errors in variables. *Review Int. Stat. Inst. 1/3*, 23–32.

FELLEGI, I., 1964. Response variance and its estimation. *Jour. Amer. Stat. Assoc. 59*, 1016–1041.

GAYLOR, D. W., 1956. Equivalence of two measures of product variance. *Jour. Amer. Stat. Assoc. 51*, 451–453.

GRUBBS, F. E., 1948. On estimating precision of measuring instruments and product variability. *Jour. Amer. Stat. Assoc. 43*, 243–264.

HANSEN, M. H., HURWITZ, W. N., and BERSHAD, M., 1961. Measurement errors in censuses and surveys. *Bull. Int. Stat. Inst. 38*, 359–374.

HANSEN, M. H., HURWITZ, W. N., and PRITZKER, L., 1964. The estimation and interpretation of gross differences and the simple response variance. In *Contributions to Statistics*, Statistical Publishing Co., Calcutta.

HANSON, R. H. and MARKS, E. S., 1958. Influence of the interviewer on the accuracy of survey results. *Jour. Amer. Stat. Assoc. 53*, 635–655.

KISH, L., 1962. Studies of interviewer variance for attitudinal variables. *Jour. Amer. Stat. Assoc. 57*, 92–115.

LINDLEY, D. V., 1947. Regression lines and the linear functional relationship. *Jour. Roy. Stat. Soc. B, 9*, 218–224.

LORD, F., 1960. Large-sample covariances analysis when the control variable is fallible. *Jour. Amer. Stat. Assoc. 55*, 307–321.

MADANSKY, A., 1959. The fitting of straight lines when both variables are subject to error. *Jour. Amer. Stat. Assoc. 54*, 173–205.

MAHALANOBIS, P. C., 1946. Recent experiments in statistical sampling in the Indian Statistical Institute. *Jour. Roy. Stat. Soc. 109*, 325–378.

MANDEL, J., 1959. The measuring process. *Technometrics 1*, 251–267.

MANDEL, J., and Lashof, J. W., 1959. The interlaboratory evaluation of testing methods. *A.S.T.M. Bull. 239*, 53–61.

MOTE, V. L., 1957. *Institute of Statistics* Mimeograph Series No. 182, North Carolina State University.

MOTE, V. L. and ANDERSON, R. L., 1965. An investigation of the effect of misclassification on the properties of $\chi^2$ tests in the analysis of categorical data. *Biometrika 52*, 95–109.

NEWELL, D. J., 1962. Errors in the interpretation of errors in epidemiology. *Amer. Jour. Pub. Health 52*, 1925–1928.

PEARSON, K., 1902. On the mathematical theory of errors of judgment. *Phil. Trans. Roy. Soc. Lond. A., 198*, 235–299.

PORTER, A. C., 1967. The effects of using fallible variables in the analysis of covariance. Ph. D. thesis, Wisconsin.

ROGET, E., 1961. A note on measurement errors and detecting real differences. *Jour. Amer. Stat. Assoc. 56*, 314–319.

RUBIN, T., ROSENBAUM, J., and COBB, S., 1956. The use of interview data for the detection of association in field studies. *Jour. Chronic Dis. 4*, 253–266.

SMITH, H. FAIRFIELD, 1950. Estimating precision of measuring instruments. *Jour. Amer. Stat. Assoc. 45*, 447–451.

SUKHATME, P. V. and SETH, G. R., 1952. Non-sampling errors in surveys. *Jour. Ind. Soc. Agr. Stat. 4*, 5–41.

THOMPSON, W. A. JR., 1962. The problem of negative estimates of variance components. *Ann. Math. Stat. 33*, 273–289.

THOMPSON, W. A., JR., 1963. Precision of simultaneous measurement procedures. *Jour. Amer. Stat. Assoc. 58*, 474–479.

WALD, A., 1940. The fitting of straight lines if both variables are subject to error. *Ann. Math. Stat. 11*, 284–300.

# THE EFFECTIVENESS OF ADJUSTMENT BY SUBCLASSIFICATION IN REMOVING BIAS IN OBSERVATIONAL STUDIES

W. G. Cochran

*Harvard University, Cambridge, Mass., U. S. A.*

## SUMMARY

In some investigations, comparison of the means of a variate $y$ in two study groups may be biased because $y$ is related to a variable $x$ whose distribution differs in the two groups. A frequently used device for trying to remove this bias is adjustment by subclassification. The range of $x$ is divided into $c$ subclasses. Weighted means of the subclass means of $y$ are compared, using the same weights for each study group. The effectiveness of this procedure in removing bias depends on several factors, but for monotonic relations between $y$ and $x$, an analytical approach suggests that for $c = 2, 3, 4, 5$, and 6 the percentages of bias removed are roughly 64%, 79%, 86%, 90%, and 92%, respectively. These figures should also serve as a guide when $x$ is an ordered classification (e.g. none, slight, moderate, severe) that can be regarded as a grouping of an underlying continuous variable. The extent to which adjustment reduces the sampling error of the estimated difference between the $y$ means is also examined. An interesting side result is that for $x$ normal, the percentage reduction in the bias of $\bar{x}_2 - \bar{x}_1$ due to adjustment equals the percentage reduction in its variance.

Under a simple mathematical model, errors of measurement in $x$ reduce the amount of bias removed to a fraction $1/(1 + h)$ of its value, where $h$ is the ratio of the variance of the errors of measurement to the variance of the correct measurements. Since ordered classifications are often used because $x$ is difficult to measure, $h$ may be substantial in such cases, though more information is needed on the values of $h$ that are typical in practice.

## 1. INTRODUCTION

Examples of the type of observational study considered in this paper are comparisons of the death rates of men with different smoking habits and Kinsey's comparisons of the frequencies of a specific type of sexual behavior among men of different socioeconomic levels. The investigator studies a response variable $y$ in two or more groups of people who differ with respect to some characteristic (smoking, air pollution, socioeconomic level). He realizes, however, that part or all of the observed differences between the mean values of $y$ in the groups may be due to other variables $x_1$, $x_2$, $\cdots$ in which the groups differ, rather than to the specific characteristic that he is interested in studying.

In planning such studies, it is therefore good practice to note any $x$ variable that is known or suspected to have an important relationship with $y$. Whenever feasible, steps are taken to measure $x$ for the people in each group. The investigator can then examine whether the frequency distributions of $x$ differ in the different groups. If so, he tries to adjust the mean values of $y$ so as to remove any biases that may have arisen because of these differences. Incidentally, $x$ must not be a variable that is itself causally affected by the characteristic (smoking, socioeconomic level) that is under study. If it is, the adjustments remove part of the difference in response that the investigator wants to measure.

In this paper we study the effectiveness of a frequently used method, which will be called adjustment by subclassification. The distribution of $x$ is broken up into 2, 3, or more subclasses. For each group of subjects, the mean value of $y$ is calculated separately within each subclass. Then a weighted mean of these subclass means is calculated for each group, using the same weights for every group. The actual weights employed depend on the judgment of the investigator. Sometimes they are proportional to the numbers of subjects in the subclasses of one group that is regarded as the standard of comparison, sometimes to the combined numbers of subjects in the subclasses over all groups. They may be derived from an extraneous standard population or from least squares theory.

The rationale of this method of adjustment is that when there are numerous subclasses, each with a restricted range of $x$, the within-class distributions of $x$ cannot differ much from one group to another. Comparisons among the subclass means of $y$ for the different groups should therefore be almost free from bias due to $x$. Since the subclass weights are the same for every group, the same remark should apply to the overall adjusted means of $y$.

We are particularly interested in the case in which the number of subclasses is small. Adjustments using 2–5 subclasses per variable are common in practice, especially when there are several $x$ variables for which adjustment is advisable. Further, when $x$ is difficult to measure accurately, the $x$ variable often takes the form of an ordered classification with a limited number of classes, e.g. none, mild, moderate, severe; or for, neutral, against. It is sometimes reasonable to regard an ordered classification as representing a grouping of an underlying continuous $x$ variable. Our results should therefore give an indication of the effectiveness of subclassification when $x$ can be measured only as an ordered classification.

This type of adjustment by subclassification also resembles another procedure known as *frequency matching* or *stratified matching*. In frequency matching with respect to a single variable $x$, the range of $x$ is divided into a number of subclasses. The groups of subjects to be compared are selected in such a way that every group has the same size of sample, say $n_i$, in subclass $i$. It follows that the unweighted means $\bar{y}$ of the groups are all weighted means of the subclass means of $y$, with the same weights $n_i/n$ for every group. Consequently, results about the effectiveness of adjustment by subclassification will apply also to a frequency matching that uses the same set of subclasses.

## 2. AN ILLUSTRATION

The following data were selected from data supplied to the U. S. Surgeon General's Committee from three of the studies in which comparisons of the death rates of men with different smoking habits were made. In each study there are three groups of men: non-smokers, smokers of cigarettes only, and smokers of cigars or pipes or both. The studies are the Canadian study of Best and Walker, with six years experience, the British Doctors' study by Doll and Hill, with five years experience, and the second American Cancer Society study by Hammond, with about 20 months experience. These data were chosen solely for illustration in this paper. They are not the latest reports from these studies, nor do these investigators have any connection with remarks made here.

Table 1 shows for each group the *unadjusted* death rates per 1,000 person-years of experience. These rates were obtained by dividing the total deaths that had occurred in each smoking group by the total number of person-years of exposure ($\times 10^3$) of the men in that group.

These studies in three different countries agree well in the relative death rates of the three groups of smokers. To the naive, the conclusion seems clear. We should urge the cigar and pipe smokers to give up smoking. If they lack the strength of will to do so, they should switch

TABLE 1

DEATH RATES PER 1,000 PERSON-YEARS

| Smoking group | Study | | |
| | Canadian | British | U. S. |
|---|---|---|---|
| Non-smokers | 20.2 | 11.3 | 13.5 |
| Cigarettes only | 20.5 | 14.1 | 13.5 |
| Cigars, pipes | 35.5 | 20.7 | 17.4 |

to cigarettes only. (In the British study the increase in death rates from 11.3 for non-smokers to 14.1 for cigarette smokers is statistically significant by a crude test, but the Canadian and U. S. studies show no corresponding increase.)

Before lending any credence to the comparisons in Table 1, we should ask: are there other variables in which the three groups of smokers may differ, that (i) are related to the probability of dying and (ii) are clearly *not* themselves affected by smoking habits. For men under 40, I suppose the answer is that there are no known variables with a consistent and strong relationship. For men over 40 a variable of this type that becomes of overwhelming importance is of course age. The regression of probability of dying on age for men over 40 is a concave upwards curve, the slope rising more and more steeply as age advances. The mean ages for each group in Table 1 are as follows.

TABLE 2

Mean ages, years

| Smoking group | Study | | |
| | Canadian | British | U. S. |
| --- | --- | --- | --- |
| Non-smokers | 54.9 | 49.1 | 57.0 |
| Cigarettes only | 50.5 | 49.8 | 53.2 |
| Cigars and/or pipe | 65.9 | 55.7 | 59.7 |

The high mean ages for the cigar-pipe smokers are notable and not unexpected. Adjustment to remove the effects of these age differences is essential. Incidentally, even if the mean ages for the three smoking groups had been identical within a study, this would not automatically eliminate age as a possible source of bias in death rates unless the regression of death rate on age were linear with the same slope in each group.

Table 3 shows the adjusted death rates obtained when the age distributions were divided into 2 subclasses, 3 subclasses, and the maximum number of subclasses that the available data allow, these being 12, 9, and 11, respectively. The subclass boundaries were chosen to give subclasses of approximately equal size (this is not necessarily the optimum choice of boundaries). The weights used were those for the non-smokers, so that the non-smoker death rates remain the same for the different numbers of classes.

With the maximum numbers of subclasses, the adjusted cigarette

## TABLE 3
### Adjusted death rates using 2, 3, and 9–11 subclasses

| Number of subclasses | Canadian | | | British | | | U. S. | | |
|---|---|---|---|---|---|---|---|---|---|
| | N. S.* | C.+ | CP' | N. S. | C | CP | N. S. | C | CP |
| 1 | 20.2 | 20.5 | 35.5 | 11.3 | 14.1 | 20.7 | 13.5 | 13.5 | 17.4 |
| 2 | 20.2 | 26.4 | 24.0 | 11.3 | 12.7 | 13.6 | 13.5 | 16.4 | 14.9 |
| 3 | 20.2 | 28.3 | 21.2 | 11.3 | 12.8 | 12.0 | 13.5 | 17.7 | 14.2 |
| 9–11 | 20.2 | 29.5 | 19.8 | 11.3 | 14.8 | 11.0 | 13.5 | 21.2 | 13.7 |

*Non-smokers, +Cigarettes only, 'Cigars, Pipes

death rates now show substantial increases over the non-smoker death rates in all three studies. The adjusted cigar-pipe death rates, on the other hand, exhibit no elevation over those for non-smokers.

Looking down the columns we see that with 2 subclasses substantially over half the effect of the age bias has been removed in most cases, while with 3 subclasses, a little more is removed. The only exception occurs for cigarette smokers in the British study. In this case the mean ages of non-smokers and cigarette smokers agree closely, 49.1 against 49.8 years, so that the adjustments are removing the effects of differences in the shapes of the age distributions rather than in the mean ages.

### 3. TOPICS TO BE DISCUSSED

These topics will be described under three headings, as follows.

1. As already mentioned, results are presented on the effectiveness of adjustment by subclassification in controlling bias when $y$ and $x$ are continuous and the number of subclasses ranges from 2 to 6. In order to introduce some notation, suppose that we are comparing two populations, denoted by the subscripts 1 and 2. Independent random samples have been drawn from each population. Let $u(x)$ represent the population regression of $y$ on $x$. If $y_{1j}$, $y_{2k}$ are random members of the two populations, the model is

$$y_{1j} = \alpha_1 + u(x_{1j}) + e_{1j}, \qquad y_{2k} = \alpha_2 + u(x_{2k}) + e_{2k},$$

where $e_{1j}$, $e_{2k}$ are random residuals with zero means in the respective populations. The quantity to be estimated is $(\alpha_2 - \alpha_1)$. For the unadjusted means of $y$ in the two groups, it follows that

$$E(\bar{y}_1) = \alpha_1 + \bar{u}_1, \qquad E(\bar{y}_2) = \alpha_2 + \bar{u}_2,$$

where

$$\bar{u}_1 = \int u(x)\phi_1(x) \, dx, \qquad \bar{u}_2 = \int u(x)\phi_2(x) \, dx, \qquad (3.1)$$

and $\phi_1(x)$, $\phi_2(x)$ are the frequency functions of $x$ in the two populations. Hence if no adjustment is made, the initial bias due to $x$ is $\bar{u}_2 - \bar{u}_1$ .

As implied above, the regression function $u(x)$ is assumed to be the same in the two populations. In observational studies this assumption does not necessarily hold. When the $u$'s differ in the two populations the meaning of the adjustment and the best method of making it require further study.

In the $i$th subclass, let the boundaries of $x$ be $x_{i-1}$ and $x_i$ and let the sample means of $y$ be $\bar{y}_{1i}$ and $\bar{y}_{2i}$ . We have

$$E(\bar{y}_{1i}) = \alpha_1 + \bar{u}_{1i} ,$$

where

$$\bar{u}_{1i} = \int_{x_{i-1}}^{x_i} u(x)\phi_1(x) \, dx \Big/ \int_{x_{i-1}}^{x_i} \phi_1(x) \, dx. \qquad (3.2)$$

After adjustment, the remaining bias due to $x$ is

$$\sum w_i(\bar{u}_{2i} - \bar{u}_{1i}), \qquad (3.3)$$

where $w_i$ is the weight assigned to subclass $i$. The proportion of the initial bias that is removed by the adjustment is therefore

$$1 - \sum w_i(\bar{u}_{2i} - \bar{u}_{1i})/(\bar{u}_2 - \bar{u}_1). \qquad (3.4)$$

From the nature of expressions (3.1), (3.2), and (3.3), it appears that this proportion will depend on the following quantities: the mathematical form of the regression function $u(x)$, the shapes of the frequency functions $\phi_1(x)$ and $\phi_2(x)$, the number of subclasses, the division points $x_i$ , and the choice of weights $w_i$ .

In view of the numerous variables or functions involved, a thorough investigation may require extensive experimental sampling. Instead, we have attempted to use an analytical approach by restricting the scope of the problem in two ways.

(i) If $u(x)$ is monotone and differentiable, as is presumably the case, for example, in a function designed to represent the regression on age of probability of dying at ages over 40, we can replace $x$ by $u$ in a theoretical investigation, so that the regression becomes linear. The functions $\phi_1(x)$ and $\phi_2(x)$ are replaced by the corresponding frequency functions $f_1(u)$ and $f_2(u)$ into which this transformation converts them. Reverting to $x$, we therefore assume that the regression of $y$ on $x$ is linear and that $x$ has frequency functions $f_1(x)$ and $f_2(x)$ in the two populations.  An important consequence of a linear regression of $y$ on $x$ is that the percentage reduction in the bias of $\bar{y}_2 - \bar{y}_1$ equals that in $\bar{x}_2 - \bar{x}_1$ .

(ii) If $f_1(x)$ and $f_2(x)$ differ only in the value of a single parameter $\theta$ that enters into the specification of each, then, as will be shown, the proportion of the initial bias in $x$ or $y$ that is removed by subclassification can be expressed by a calculus formula when $\theta$ is small.  In particular, if $f_1(x) = f(x)$, $f_2(x) = f(x - \theta)$, so that the distributions differ only by a translation, this proportion becomes

$$\sum_{i=1}^{c} M_i (f_{i-1} - f_i).$$

In this expression, $f_{i-1}$ and $f_i$ are the ordinates of $f(x)$ at the boundaries $x_{i-1}$ and $x_i$ of the $i$th subclass, and $M_i$ is the mean value of $x$ in the $i$th subclass.  Numerical values of these proportions will be presented for $c$ = number of subclasses = $2(1)6$ and for $f(x)$ having the normal, $\chi^2$, and $t$ distributions, as well as for some beta distributions and a distribution simulating the age distribution of males.

2. Some results are given to show the effect of subclassification on the *precision* of the comparison of the two population means, as measured by the ratio of the variance of the adjusted difference $\bar{y}_{2w} - \bar{y}_{1w}$ to that of the initial difference $\bar{y}_2 - \bar{y}_1$ .  Adjustment by subclassification is often used in situations in which the investigator does not suspect any danger of bias, believing that $f_1(x)$ and $f_2(x)$ are closely similar.  Instead, he regards the variance of $y$ as due in part to its regression on $x$.  By controlling on $x$ through the adjustment process, he hopes to reduce the variance of $\bar{y}_{2w} - \bar{y}_{1w}$ .  The objective is the same as in a standard analysis of covariance.

3. The effects of errors of measurement of $x$ on the performance of the adjustments will be examined under a simple mathematical model.  This topic is particularly relevant when $x$ is an ordered classification, since these classifications are often employed because $x$ is difficult to measure accurately, so that errors of measurement are anticipated.

#### 4. PERCENT REDUCTION IN BIAS WITH A LINEAR REGRESSION

Before considering a calculus approach, some preliminary calculations of the percentage reductions in the bias of $\bar{x}_2 - \bar{x}_1$ were made on a computer, using formula (3.4), for the case in which $f_1(x)$ is the normal distribution $N(0, \sigma^2)$, while $f_2(x)$ is $N(\theta, \sigma^2)$.  Equal-sized subclasses in population 1 and equal weights were chosen.  For $\theta/\sigma = 1, \frac{1}{2}, \frac{1}{4}$ and for 2, 3, 4, 5, and 6 subclasses, the percent reductions in the bias of $\bar{x}_2 - \bar{x}_1$ appear in Table 4.

## TABLE 4

PERCENT REDUCTIONS IN BIAS: LINEAR REGRESSION, $x$ NORMAL

| $\theta/\sigma$ | Number of subclasses | | | | |
|---|---|---|---|---|---|
|  | 2 | 3 | 4 | 5 | 6 |
| 1 | 61.8 | 78.2 | 85.3 | 89.1 | 91.5 |
| $\frac{1}{2}$ | 63.2 | 79.1 | 85.9 | 89.6 | 91.8 |
| $\frac{1}{4}$ | 63.6 | 79.3 | 86.0 | 89.7 | 91.9 |

As would be expected, the percentage of the initial bias that is removed by the adjustment increases steadily as the number of sub-classes is increased. Table 4 also indicates that for initial biases which are not too large ($\theta/\sigma \leq \frac{1}{2}$), the percent bias removed may be almost independent of the value of $\theta/\sigma$. This observation suggests that results applicable to any continuous $f(x)$ might be obtainable by a calculus approach in which $\theta/\sigma$ is assumed small. It may be noted that a bias for which $\theta/\sigma = \frac{1}{2}$ is by no means negligible: with two samples of size $n$, the ratio of $\bar{y}_2 - \bar{y}_1$ to its standard error has a bias $\frac{1}{2}\sqrt{n/2}$. With $n = 32$, for example, the probability of finding a significant difference in $\bar{y}_2 - \bar{y}_1$ that is entirely due to this bias is about 0.5 in a two-tailed test at the 5% level.

Let $f(x)$ depend on a parameter $\theta$ that has the value 0 in population 1 and the value $\theta$ in population 2. For the adjustments, the range of $x$ is divided into $c$ subclasses by division points $x_0$, $x_1$, $\cdots$, $x_c$. In the $i$th subclass let $P_i(\theta)$ denote the proportion of the population and $M_i(\theta)$ the mean value of $x$. The weights used may be the $P_i(0)$, the $P_i(\theta)$ or a combination of the two. Since $\theta$ tends to zero in this approach, these different choices of weights become identical. We assume that the $P_i(0)$ are used.

If $M(\theta)$ denotes the overall mean of $x$, the initial bias, $M(\theta) - M(0)$ may be written

$$\sum_{i=1}^{c} [P_i(\theta)M_i(\theta) - P_i(0)M_i(0)] \doteq \theta \sum_{i=1}^{c} \left[ P_i \frac{dM_i}{d\theta} + M_i \frac{dP_i}{d\theta} \right] = \theta \frac{dM}{d\theta}$$

(4.1)

assuming $\theta$ small, where the derivatives are taken at $\theta = 0$. After adjustment, the bias remaining is

$$\sum_{i=1}^{c} P_i(0)[M_i(\theta) - M_i(0)] \doteq \theta \sum_{i=1}^{c} P_i \frac{dM_i}{d\theta}.$$

(4.2)

Consequently, the proportion of the initial bias that is removed is approximately

$$\sum_{i=1}^{c} M_i \frac{dP_i}{d\theta} \Big/ \frac{dM}{d\theta}. \qquad (4.3)$$

The utility of this expression depends, of course, on whether the functions that enter into (4.3) are easily found analytically. If $f_1(x) = f(x)$, $f_2(x) = f(x - \theta)$, the denominator of (4.3) becomes 1, since the initial bias in (4.1) is $\theta$. Further,

$$P_i(\theta) = \int_{x_{i-1}}^{x_i} f(x - \theta) \, dx = \int_{x_{i-1}-\theta}^{x_i-\theta} f(x) \, dx$$

so that at $\theta = 0$,

$$dP_i/d\theta = f(x_{i-1}) - f(x_i). \qquad (4.4)$$

Consequently, the proportional reduction in bias from (4.3) becomes,

$$\sum_{i=1}^{c} M_i(f_{i-1} - f_i), \qquad (4.5)$$

where $f_i = f(x_i)$.

For the unit normal distribution,

$$M_i P_i = \frac{1}{\sqrt{2\pi}} \int_{x_{i-1}}^{x_i} x e^{-\frac{1}{2}x^2} \, dx = f_{i-1} - f_i \qquad (4.6)$$

so that the proportional reduction becomes

$$\sum_{i=1}^{c} (f_{i-1} - f_i)^2/P_i . \qquad (4.7)$$

Expression (4.7) has turned up previously in several papers. In particular, D. R. Cox [1957] obtained it as 1 minus the ratio of the average within-subclass variance of $x$ to the original variance of $x$ when $x$ is normal. The connection between his results and (4.7) will be discussed in section 6. Table 5, taken from his computations, shows the optimum $P_i$ that maximize the percentage reduction in bias, the corresponding reductions, and the reductions obtained when the $P_i$ are all equal.

With the optimum sets of boundaries the central subclass is the largest, the subclasses becoming steadily smaller towards both ends of the range of $x$. Table 5 shows also that the choice of boundaries is not crucial: use of subclasses of equal size diminishes the percent reductions by only about 2% as compared with the optimum boundaries. The percent reductions with equal $P_i$ agree closely with those given in Table 4 for $\theta/\sigma = \frac{1}{4}$ and $\frac{1}{2}$, as had been anticipated.

TABLE 5

OPTIMUM CHOICE OF $P_i$ AND PERCENT REDUCTIONS IN BIAS GIVEN BY
OPTIMUM AND EQUAL $P_i$ WHEN $x$ IS NORMAL

| Number of subclasses, $c$ | Optimum $P_i$ (%) | Maximum reduction | Reduction with equal $P_i$ |
|---|---|---|---|
| 2 | 50%; 50% | 63.7% | 63.7% |
| 3 | 27%; 46%; 27% | 81.0% | 79.3% |
| 4 | 16%; 34%; 34%; 16% | 88.2% | 86.1% |
| 5 | 11%; 24%; 30%; 24%; 11% | 92.0% | 89.7% |
| 6 | 7%; 18%; 25%; 25%; 18%; 7% | 94.2% | 91.9% |

## 5. PERCENT REDUCTION IN BIAS FOR SOME NON-NORMAL DISTRIBUTIONS

Three non-normal distributions were also investigated—$\chi^2$ and $t$ with a range of numbers of degrees of freedom, and some beta distributions. With $\chi^2$ and $t$ it was assumed that population 2 differs from population 1 only in a translation $\theta$, formula (4.5) being used to compute the proportional reduction in bias when $\theta$ is small.

The beta distributions were $B(m, 3)$, for which the frequency function is proportional to $x^{m-1}(1 - x)^2$, where $0 \leq x \leq 1$. The mean of this distribution is $m/(m + 3)$. The sets of pairs $m = (2, 3); (3, 4.5);$ $(4.5, 7);$ and $(7, 12)$ in the two populations give means $(0.4, 0.5);$ $(0.5, 0.6); (0.6, 0.7);$ and $(0.7, 0.8)$, respectively. The beta distribution provides an interesting variant. When $m$ changes, the shape of the distribution changes as well as the mean; further, the distribution has a finite range. The proportional reductions in bias for this distribution were computed directly without using the calculus approximation (4.3).

Two examples simulating situations with non-linear regressions were also worked. The variate $x$ was age, the distribution in population 1 being that of the 1963 U. S. population aged 45–75 by single years. The frequency function is not unlike a trapezoid with a negative slope. The variate $y$ was the death rate, approximately a cubic function of $x$. Two distributions were tried for population 2—a rectangular distribution and the reversed age distribution. Before adjustment for age the mean death rate in population 2 was 22% higher than that in population 1 for the rectangular and 42% higher for the reversed distribution. Table 6 shows the results.

In all these cases the $P_i$ (subclass sizes) were made equal. With $\chi^2$ the percentage reductions differ only trivially from those for the

TABLE 6

PERCENT REDUCTIONS IN BIAS FOR NON-NORMAL DISTRIBUTIONS
WITH EQUAL-SIZED SUBCLASSES

| Distribution of $x$ | | $c$ = number of subclasses | | | | |
|---|---|---|---|---|---|---|
| Pop. 1 | Pop. 2 | 2 | 3 | 4 | 5 | 6 |
| $\chi^2_4$ | $\chi^2_4 + \theta$ | 65.8 | 79.8 | 85.7 | 89.4 | 91.5 |
| $\chi^2_6$ | $\chi^2_6 + \theta$ | 64.9 | 79.6 | 85.9 | 89.4 | 91.6 |
| $\chi^2_{10}$ | $\chi^2_{10} + \theta$ | 64.9 | 79.4 | 85.9 | 89.5 | 91.7 |
| $\chi^2_{20}$ | $\chi^2_{20} + \theta$ | 64.0 | 79.3 | 85.9 | 89.7 | 91.8 |
| $t_3$ | $t_3 + \theta$ | 81.0 | 97.6 | 102.7 | 104.9 | 105.7 |
| $t_5$ | $t_5 + \theta$ | 72.1 | 88.1 | 93.8 | 96.8 | 98.3 |
| $t_{10}$ | $t_{10} + \theta$ | 67.3 | 88.3 | 89.4 | 92.9 | 94.9 |
| $t_{20}$ | $t_{20} + \theta$ | 65.4 | 81.4 | 87.8 | 91.5 | 93.7 |
| $B(2, 3)$ | $B(3, 3)$ | 66.4 | 82.2 | 88.7 | 92.0 | 94.0 |
| $B(3, 3)$ | $B(4.5, 3)$ | 64.5 | 80.5 | 87.3 | 90.9 | 93.1 |
| $B(4.5, 3)$ | $B(7, 3)$ | 62.5 | 78.9 | 85.9 | 89.7 | 92.0 |
| $B(7, 3)$ | $B(12, 3)$ | 60.4 | 77.1 | 84.4 | 88.4 | 90.9 |
| Age | Rect. | 64.1 | 84.5 | 89.4 | 93.0 | 95.0 |
| Age | Reversed | 67.0 | 86.2 | 92.3 | 94.0 | 96.5 |
| $N(0, 1)$ | $N(\theta, 1)$ | 63.7 | 79.3 | 86.1 | 89.7 | 91.9 |

With $\chi^2$ and $t$, the subscripts are the numbers of D.F.

normal distribution even when $\chi^2$ has only 4 D.F. and is strongly skew. With $t$, adjustment by subclassification does substantially better than with the normal distribution when the degrees of freedom are small. The reductions of over 100% that appear when $t$ has 3 D.F. may look erroneous—one cannot remove more bias than there is. What happens is that owing to the nature of the tails of $t$ with small D.F., population 2 has a *lower* mean than population 1 in the highest subclass, the net result being to produce small overall negative biases in place of the initial positive bias.

From occasional calculations with actual populations having a finite range, my impression was that the adjustments perform better than the normal model suggests. The percentages shown in Table 6 for the beta distribution, however, all lie within about ±2% of the corresponding normal values and the averages over the four beta sets are close to the normal values although a little higher for $c \geq 3$. For the two age distribution examples, where the regression of $y$ on $x$ is markedly non-linear, the percentage reductions are consistently higher than with the normal. These increases may be due in part to two factors: equal-sized classes are near the optimum for the trapezoidal age distribution, and the maximum number of classes that the data allow is 30, so that we are comparing 6 subclasses with 30 instead of 6 with an infinite number.

So far as they go, these calculations suggest that the normal values may serve as a working guide to the percentages of bias removed in practice with from 2 to 6 subclasses. There is a hint that the normal values may underestimate the effectiveness of adjustment for some types of distribution of $x$.

## 6. EFFECT OF ADJUSTMENT ON THE PRECISION OF THE COMPARISONS

An extensive study of the relative effectiveness of matching and of adjustment by covariance in increasing the precision of the comparison of the mean values of $y$ in the two groups has been made by Billewicz [1965]. He used a series of experimental samplings designed to simulate the range of models and conditions under which this problem arises in practice. The present analytical approach is a minor supplement to his work.

We assume that $x$ has the same distribution $f(x)$ in the two groups and that the regression of $y$ on $x$ is linear. The formulas are simplest in the case of stratified matching, which is considered first. If independent samples of size $n$ are drawn in the two groups, with no matching or adjustment, the variance of $\bar{y}_2 - \bar{y}_1$ is $V(\bar{y}_2 - \bar{y}_1) = 2\sigma_y^2/n$. Of this, a part $2\beta^2\sigma_x^2/n = 2\rho^2\sigma_y^2/n$ is due to variations in $x$ and a part $2(1-\rho^2)\sigma_y^2/n$ to other sources of variability, where $\beta$ is the regression coefficient of $y$ on $x$ and $\rho$ the correlation between $y$ and $x$.

With stratified matching on $x$, the range of $x$ is divided into $c$ subclasses. In group 1 let $P_i$ be the proportion of the population and $n_i$ the sample number of observations that fall into the $i$th subclass. Group 2 is constrained to have $n_i$ values of $x$ in this subclass also. Consequently, if $\sigma_i^2$ is the variance of $x$ within the $i$th subclass, the variance of $\bar{x}_{2i} - \bar{x}_{1i}$ is $2\sigma_i^2/n_i$. The variance of the overall mean difference $\sum n_i(\bar{x}_{2i} - \bar{x}_{1i})/n$ is $2\sum n_i\sigma_i^2/n^2$. Since $n_i$ is a binomial estimate of $nP_i$, the average value of this variance is $2\sum P_i\sigma_i^2/n$.

The effect of stratified matching is therefore that the contribution of variations in $x$ to $V(\bar{y}_2 - \bar{y}_1)$ is reduced from $2\beta^2\sigma_x^2/n$ to $2\beta^2 \sum P_i\sigma_i^2/n$. The contribution $2(1 - \rho^2)\sigma_y^2/n$ of other sources of variation is unaffected by the matching. If $g = \sum P_i\sigma_i^2/\sigma_x^2$, the net result is to reduce $V(\bar{y}_2 - \bar{y}_1)$ from $2\sigma_y^2/n$ to

$$2\sigma_y^2[g\rho^2 + (1 - \rho^2)]/n = 2\sigma_y^2[1 - (1 - g)\rho^2]/n. \qquad (6.1)$$

A similar formula holds for adjustment by subclassification, except that in moderate-sized samples the factor $g$ is larger than $\sum P_i \sigma_i^2 / \sigma_x^2$ owing to dissimilarities in the sample distributions of $x$ in the two groups. If the samples are of sizes $n_1$, $n_2$, the adjusted mean difference $\sum w_i(\bar{x}_{2i} - \bar{x}_{1i})$ has variance

$$\sum w_i^2 \sigma_i^2 [(1/n_{2i}) + (1/n_{1i})].$$

For given $n_1$, $n_2$, the average value of this variance depends, of course, on the choice of the weights $w_i$. For specified $w_i$ this average can usually be expanded in a series of powers of $1/n_1$ and $1/n_2$, since $n_{1i}$ and $n_{2i}$ are binomial estimates of $n_1 P_i$ and $n_2 P_i$, respectively. For instance, if $w_i = n_{1i}/n$, the two leading terms in the average variance can be shown to be

$$\left(\frac{1}{n_1} + \frac{1}{n_2}\right) \sum P_i \sigma_i^2 + \frac{1}{n_2}\left(\frac{1}{n_1} + \frac{1}{n_2}\right) \sum Q_i \sigma_i^2 , \qquad (6.2)$$

where $Q_i = 1 - P_i$. If the $P_i$ are known and used as weights, the two leading terms are

$$\left(\frac{1}{n_1} + \frac{1}{n_2}\right) \sum P_i \sigma_i^2 + \left(\frac{1}{n_1^2} + \frac{1}{n_2^2}\right) \sum Q_i \sigma_i^2 , \qquad (6.3)$$

where $Q_i = 1 - P_i$. Since the variance of the unadjusted difference $\bar{x}_2 - \bar{x}_1$ is $\sigma_x^2(1/n_1 + 1/n_2)$, the first term in the above expressions leads to the factor $g$. The second term becomes negligible relative to the first as $n_1$ and $n_2$ increase, but can have an appreciable effect in moderate-sized samples. For instance, with $P_i = 1/c$, $Q_i = (c-1)/c$, the second term amounts to increasing the variance by a factor $1 + (c-1)/n_2$ in (6.2) and a factor $1 + (c-1)(n_1^2 + n_2^2)/n_1 n_2(n_1 + n_2)$ in (6.3).

We now consider $g = \sum P_i \sigma_i^2 / \sigma^2$. An alternative expression can be given for $g$. By the well-known result,

$$\sigma^2 = \sum P_i \sigma_i^2 + \sum P_i M_i^2 - \mu^2$$

we have

$$1 - g = \sum P_i M_i^2 / \sigma^2 - \mu^2/\sigma^2. \qquad (6.4)$$

Two incidental results are of interest. For the normal distribution, $N(0, 1)$, $M_i P_i = f_{i-1} - f_i$, by (4.6). Hence, for this distribution, putting $\mu = 0$, $\sigma = 1$ in (6.4),

$$1 - g = \sum M_i(f_{i-1} - f_i). \qquad (6.5)$$

The quantity $1 - g$ is the proportional reduction in the variance of $\bar{x}_2 - \bar{x}_1$ due to adjustment by subclassification. But by (4.5), the same expression holds for the proportional reduction in the bias of $\bar{x}_2 - \bar{x}_1$.

This equivalence of the proportional reductions in variance and bias appears to hold only for the normal distribution.

Secondly, the expression $\sum P_i \sigma_i^2 / n$ is the variance of the mean of a stratified sample of size $n$ with proportional stratification (ignoring the correction for finite population), where $P_i$ and $\sigma_i^2$ refer to the $i$th stratum. Dalenius [1957] has shown that the set of class boundaries $x_i$ which minimize this expression satisfy the simple relations

$$x_i = \tfrac{1}{2}(M_i + M_{i+1}). \tag{6.6}$$

Although these equations still have to be solved by iteration, they are a valuable guide to the iteration that leads to a minimum $g$.

For comparison with the normal, Table 7 shows the percent reductions $100(1 - g)$ with equal-sized classes, for a series of $\chi^2$ and $t$ distributions. In both cases the percent reductions are substantially less than with the normal when the degrees of freedom are small. The reductions for optimum class boundaries (not shown) are somewhat closer to the normal optima but are still inferior.

To come finally to the objective of this section: Table 8 presents the percentage reductions in the variance of $\bar{y}_2 - \bar{y}_1$ that result from stratified matching on $x$, the formula being $100(1 - g)\rho^2$ by (6.1). The main conclusions are: (i) subclassification with a small number of subclasses suffers most, in comparison with an infinite number of classes, when the potential gains in precision are large ($\rho$ high); (ii) on the other hand, with $\rho < 0.7$, as occurs in most applications, the use of at least 4 subclasses realizes nearly all the potential gain.

If $x$ is continuous, we can compare stratified matching with a

TABLE 7

Percent reductions in variance for non-normal distributions, with equal-sized subclasses

| Distribution | $c$ = Number of subclasses | | | | |
|---|---|---|---|---|---|
|  | 2 | 3 | 4 | 5 | 6 |
| $\chi_4^2$ | 55.5 | 71.5 | 79.7 | 83.9 | 86.8 |
| $\chi_6^2$ | 58.2 | 73.8 | 81.4 | 85.7 | 88.2 |
| $\chi_{10}^2$ | 60.3 | 76.3 | 83.5 | 86.8 | 89.6 |
| $\chi_{20}^2$ | 62.0 | 77.4 | 84.7 | 88.4 | 90.6 |
| $t_3$ | 40.5 | 52.3 | 57.8 | 61.7 | 64.4 |
| $t_5$ | 54.0 | 68.9 | 75.4 | 79.8 | 82.7 |
| $t_{10}$ | 59.8 | 75.6 | 82.3 | 86.5 | 89.2 |
| $t_{20}$ | 61.9 | 77.9 | 84.6 | 88.7 | 91.2 |
| Normal | 63.7 | 79.3 | 86.1 | 89.7 | 91.9 |

TABLE 8

PERCENT REDUCTIONS IN $V(\bar{y}_2 - \bar{y}_1)$ RESULTING FROM STRATIFIED
MATCHING ON $x$, WITH EQUAL-SIZED SUBCLASSES AND $x$ NORMAL

| $\rho$ | $c$ = Number of subclasses | | | | | |
|---|---|---|---|---|---|---|
| | 2 | 3 | 4 | 5 | 6 | $\infty$ |
| 0.2 | 2.5 | 3.2 | 3.4 | 3.6 | 3.7 | 4.0 |
| 0.3 | 5.8 | 7.1 | 7.7 | 8.1 | 8.3 | 9.0 |
| 0.4 | 10.2 | 12.7 | 13.8 | 14.4 | 14.7 | 16.0 |
| 0.5 | 15.9 | 19.8 | 21.5 | 22.4 | 23.0 | 25.0 |
| 0.6 | 22.9 | 28.5 | 31.0 | 32.3 | 33.1 | 36.0 |
| 0.7 | 31.2 | 38.9 | 42.2 | 44.0 | 45.0 | 49.0 |
| 0.8 | 40.8 | 50.8 | 55.1 | 57.4 | 58.9 | 64.0 |
| 0.9 | 51.6 | 64.2 | 69.7 | 72.7 | 74.4 | 81.0 |

linear covariance adjustment applied to two independent samples of size $n$. Covariance gives reductions of approximately $[R_\infty - (100 - R_\infty)/2n]$, where $R_\infty$ is the reduction in the column $c = \infty$ in Table 8, the term $(100 - R_\infty)/2n$ being the usual allowance for the sampling error of the regression coefficient. A few calculations will show that for samples of any reasonable size, covariance gives greater gains than stratified matching, as Billewicz also concluded.

Table 8 applies to stratified matching with $x$ normal. Earlier results in this section suggest that reductions will be somewhat less (i) under adjustment by subclassification and (ii) when $x$ is non-normal, at least if the results for $\chi^2$ and $t$ are typical of those with non-normal $x$.

## 7. EFFECT OF ERRORS OF MEASUREMENT IN $x$

We return to the consideration of bias in $y$.

As before, we assume that $y$ has a linear regression on the correct measurement $x$. However, the measurement actually used for adjustments or for the construction of an ordered classification is $X$, which also has a linear regression on $x$ in each population of the form

$$X = \gamma + \beta x + d. \tag{7.1}$$

The term $\gamma$ represents a constant bias in measurement, while if $\beta$ differs from 1, this implies a bias that changes as $x$ changes. The variable $d$ is the random component of the error of measurement, with mean 0, variance $\sigma_d^2$. If $x$ and $d$ are normally and independently distributed, standard bivariate normal theory shows that the regression of $x$

on $X$ is linear, with regression coefficient cov $(x, X)/\sigma_X^2 = \beta\sigma_x^2/(\beta^2\sigma_x^2 + \sigma_d^2)$.

Suppose that $x$ has mean $\mu$ in population 1 and mean $\mu + \theta$ in population 2. On account of relation (7.1), the means of $X$ in the two populations will differ by an amount $\beta\theta$. Since the adjustments are made by means of $X$, their effect on $X$ is to reduce its bias by an amount $\beta\theta\lambda$, where $\lambda$ is the proportional reduction that was shown, as a percentage, in Table 6. Because of the linear regression of $x$ on $X$, this reduction of $\beta\theta\lambda$ in $X$ produces a reduction of amount

$$(\beta\theta\lambda) \cdot \frac{\beta\sigma_x^2}{\beta^2\sigma_x^2 + \sigma_d^2} = \frac{\theta\lambda}{1 + h}$$

in $x$, where $h = \sigma_d^2/\beta^2\sigma_x^2$. Thus the proportional reduction in $x$, which equals that in $y$ on account of the linear regression of $y$ on $x$, is $\lambda/(1 + h)$. The quantity $h$ can be regarded as the ratio of the variance of the random errors of measurement in $x$ to the variance of the correct measurements $x$, because if $X$ is divided by $\beta$ so that a given change in $x$ provides the same expected change in $X$, the random error of measurement becomes $d/\beta$, with variance $\sigma_d^2/\beta^2$.

To sum up for an ordered classification with a linear regression of $y$ on $x$, the adjustment fails to remove all the bias in $y$ for two reasons. Even if there are no random errors of measurement, the ordered classification being constructed from $x$ itself or equivalently from a linear function $\gamma + \beta x$, the proportional reduction in the bias of $y$ is not unity but $\gamma$ as given in Table 6. Secondly, there may be errors of measurement in the variate $X$ by which the ordered classification is made. These errors introduce some misclassifications so far as the correct measurement $x$ is concerned, and reduce $\lambda$ to $\lambda/(1 + h)$.

Consequently, some idea of the value of $h$ is essential in forming a judgment on the effectiveness of adjustment. Although interest in the study of errors of measurement has increased in recent years, not much seems to be known about the values of $h$. In practice, three methods are commonly used to estimate $h$: (i) independent measurements of the same objects by the fallible and the correct method. This approach is feasible only when a correct method exists—usually in areas where the technique of measurement is far advanced; (ii) repeated independent measurements by the fallible ($X$) method or by competing fallible methods; (iii) repeated measurement by a superior but not perfect instrument. With methods (ii) and (iii), $h$ can be estimated for continuous data as shown by Grubbs [1948] and for ordered classifications by unpublished methods reported to me by F. Mosteller. Estimation of $h$ for a binomial variate has been discussed by Hansen et al. [1964].

From such results as I have seen, it is evident that $h$ can vary in practice from a trivial amount to values substantially over 1 in difficult measurements. Thus, when errors of measurement are taken into account, the proportions in Table 6 may be changed only negligibly or may be reduced to figures less than half those shown.

With ordered classifications, the value of $h$ is related to the percentage of wrong classifications by the fallible measurement $X$ in method (i) above, and to the percentage of disagreements in classification by two independent readings $X_1$, $X_2$ of the fallible measurement in method (ii). From the model (7.1), cov $(x, X) = \beta\sigma_x^2$ and cov $(X_1, X_2)$ $= \beta^2\sigma_x^2$, so that $x$ and $X$ follow a bivariate normal with correlation coefficient $1/\sqrt{1 + h}$, while $X_1$ and $X_2$ follow a bivariate normal with correlation coefficient $1/(1 + h)$. Hence, if specimens are being assigned to one of two classes of equal size, the value of $h$ corresponding to a given proportion $p$ of wrong classifications may be read from the tables of the bivariate normal, by noting the value of $p$ needed to make the combined area in the two quadrants $(x > 0, X < 0)$ and $(x < 0, X > 0)$ equal to $p$.

As an illustration, Table 9 shows these percentages for a series of values of $h$. Also shown are the percentage reductions $100h/(1 + h)$ caused by errors in measurement in the amount of bias removed by subclassification, and the approximate percentages $63.7/(1 + h)$ of bias that would actually be removed by adjustment on $X$. For measurements of reasonably high precision with $h < 0.10$, the percentage of wrong classifications agrees closely with the percentage reduction in the bias removed. As the percentage of wrong classifications increases, the effectiveness of adjustment on $X$ drops more sharply. With 15% of wrong classifications, only about half the bias is removed, and with 25%, less than one-third. Disagreements of up to 10% between two fallible measurements correspond to satisfactorily high precision, the reduction in bias removed being at most 5%.

The regression model (7.1) used here is likely to be an oversimplification of the relation between $X$ and $x$ in many applications. The relation may not be linear, and the variance $\sigma_d^2$ may depend on the value of $x$. Some investigation was made of a model in which $d$ is correlated with $x$, but this did not provide any essential generalization, since $d$ can be split into its regression on $x$ and a random component independent of $x$, bringing us back to an ordinary regression model. As already mentioned, what seems most needed in order to appraise the seriousness of errors of measurement are data from which the values of $h$ in different measurement problems can be estimated. Further, this paper has considered only adjustments on a single variable $x$. The effectiveness

TABLE 9

PERCENTAGE OF WRONG CLASSIFICATIONS AND OF DISAGREEMENTS
BETWEEN REPEATED MEASUREMENTS (2 CLASSES) FOR SPECIFIED
RATIOS $h$ OF ERROR VARIANCE TO TRUE VARIANCE

| | Percentage of: | | | |
| $h$ | wrong classifications | disagreements | reduction in bias removed | bias removed |
| --- | --- | --- | --- | --- |
| 0.00 | 0.0 | 0.0 | 0.0 | 63.7 |
| 0.05 | 4.8 | 9.7 | 4.8 | 60.7 |
| 0.10 | 9.5 | 13.6 | 9.1 | 57.9 |
| 0.20 | 13.3 | 18.7 | 16.7 | 53.1 |
| 0.30 | 15.9 | 22.0 | 23.1 | 49.0 |
| 0.50 | 19.5 | 26.8 | 33.3 | 42.5 |
| 1.00 | 25.0 | 33.3 | 50.0 | 31.8 |
| 1.50 | 28.2 | 36.9 | 60.0 | 25.5 |
| 2.00 | 30.4 | 39.2 | 66.7 | 21.2 |

of two-way and three-way classifications used to adjust simultaneously
for two and three $x$-variables also merits investigation.

## ACKNOWLEDGEMENTS

This work was assisted by Contract Nonr 1866(37) with the Office
of Naval Research, Navy Department, and by Grant GS-341 from the
National Science Foundation.

## L'EFFICACITE DE L'AJUSTEMENT PAR SOUS-CLASSIFICATION POUR LA SUPPRESSION DES BIAIS DANS LES ETUDES D'OBSERVATION

### RESUME

Dans certaines recherches la comparaison des moyennes d'une variable $y$ entre
deux groupes étudiés peut être biaisée parce que $y$ est lié à une variable $x$ dont la
distribution diffère dans les deux groupes. Un procédé fréquemment utilisé pour
essayer de supprimer ce biais est l'ajustement par sous-classification. L'étendue $x$
est divisée en $c$ sous-classes. Les moyennes pondérées des moyennes de sous-classes
de $y$ sont comparées en utilisant les mêmes poids pour chaque groupe étudié. L'ef-
ficacité de cette procédure pour supprimer le biais dépend de plusieurs facteurs, mais,
pour des relations monotones entre $y$ et $x$, une approche analytique suggère que
pour $c = 2, 3, 4, 5$, et 6 les pourcentages de biais supprimés sont grossièrement re-
spectivement 64%, 79%, 86%, 90%, et 92%. Ces chiffres devraient également
servir de guide lorsque $x$ est d'une classification ordonnée, (par exemple: aucun,
léger, modéré, grave) qui peut être considérée comme un regroupement d'une variable sous-jacente continue. Le degré avec lequel l'ajustement réduit l'erreur d'échan-

tillonnage de la différence estimée entre les moyennes de $y$ est également examiné. On peut noter au passage un résultat intéressant: pour une variable $x$ normale, la réduction en pourcentage du biais de $\bar{x}_2 - \bar{x}_1$ dû à l'ajustement est égale à la réduction en pourcentage de sa variance.

Sous un modèle mathématique simple, les erreurs de mesure sur $x$ réduisent le montant du biais supprimé d'une fraction $1/(1 + h)$ de sa valeur où $h$ est le rapport de la variance des erreurs de mesure à la variance des mesures correctes. Les classifications ordonnées étant souvent utilisées du fait que $x$ est difficile à mesurer, $h$ peut être assez substantiel dans de tels cas; cependant des renseignements complémentaires sont nécessaires sur les valeurs de $h$ qui sont typiques en pratique.

## REFERENCES

Billewicz, W. Z. [1965]. The efficiency of matched samples: An empirical investigation. *Biometrics 21*, 623–43.

Cox, D. R. [1957]. Note on grouping. *J. Amer. Statist. Ass. 52*, 543–7.

Dalenius, T. [1957]. *Sampling in Sweden*. Almqvist and Wicksell, Stockholm.

Grubbs, F. E. [1948]. On estimating precision of measuring instruments and product variability. *J. Amer. Statist. Ass. 43*, 243–64.

Hansen, M. H., Hurwitz, W. N., and Pritzker, L. [1964]. Measurement errors in censuses and surveys. In *Contributions to Statistics*. Statistical Publ. Company, Calcutta, India.

# Commentary on "Estimation of Error Rates in Discriminant Analysis"

## W. G. Cochran

*Harvard University*

The paper by Lachenbruch and Mickey is interesting and informative on a problem that has awaited solution for some time. Like the authors, I believe that holding out part of the sample for verification will be feasible and convenient in practice only infrequently. Of the remaining methods, the oldest are the $R$ method, in which we use simply the proportion of mistakes made by the sample discriminant on the samples from which it was constructed, and the $D$ or $DS$ methods, in which we read the normal tables at $(-\hat{D}/2)$, where $\hat{D}$ is an estimate of the Mahalanobis distance. It has been suspected that both these methods underestimate the probability of misclassification, as the authors found in their sampling experiments. Nevertheless, the $DS$ method (which uses a reasonably unbiased estimate of $D$) does surprisingly well in these experiments (more on this later).

Previous to this paper, the method I have used is approximately the $OS$ method, an application of Okamoto's expansion of the probability of misclassification in inverse powers of $n_1$, $n_2$ (the sample sizes), and $f$, the degrees of freedom in the estimated covariance matrix. The leading term in the expansion is of course $\phi(-\hat{D}/2)$. I have used this method with some hesitation, because I have not been able to follow the mathematics of Okamoto's expansion. It is reassuring to see how well the $OS$ method does.

Although aware of the $U$ method, I would have regarded it, previous to this paper, as requiring too much computation to be convenient. The authors have made an important contribution in showing how the calculations needed for the $U$ method can be made with only one matrix inversion. The $U$ method can be considered an application of the so-called 'jackknife' technique, originally proposed by Quenouille for reducing the bias in an estimate from order $1/n$ to order $1/n^2$. Incidentally, in an article illustrating the versatility of the jackknife technique, to appear in the forthcoming edition of the "Handbook of Social Psychology," Mosteller and Tukey discuss the use of the jackknife method both for constructing the discriminant and for validating it. The name 'jackknife' seems to me an unfortunate choice, because outside of the U. S., the original meaning of the word 'jackknife' is not well known, so that hearers don't see the point of the name.

A further investigation that might be useful is a comparison of the most promising methods under a non-normal parent population, if one suitable for computer sampling can be found. Intuitively, the $U$ method, the $R$ method, and the Mosteller–Tukey version of the jackknife method appear less sensitive to normality than the others, except that the $R$ method would be discarded because of its poor performance under normality. Secondly, an investigation under more severe test conditions is also relevant. For instance, to cite an extreme case, we might have sample sizes of 30, with 75 initial discriminators, from which a discriminant involving the apparently best 5 discriminators is constructed. Francis (1966), who studied the estimation of the probability of misclassification under these circumstances (small sample size plus selection of variates) used a Bayesian approach with some success, but further testing of this and other methods would be welcome.

I have one statistical quibble with the authors. Their basic measurement is $e$, the absolute difference between the estimated and the true probability—conceptually a continuous variate. In such situations, replacement of the $e$ values by an ordered classification, accompanied by the use of Kolmogoroff–Smirnoff tests, always involves some loss of information which may or may not be important. In an attempt to avoid some of this loss, I compared the average $e$ values for the different methods, assigning scores of 1, 3, 5, 7, and 10 to the five classes (roughly, 1 unit = 0.025 on the $e$ scale). In the most severe tests ($p = 20$), this approach suggests that the $e > .20$ class should have been subdivided, since the poorer methods have more than half their observations in this class.

In the first four columns of the following table, the methods are shown in order of performance as judged by their $\bar{e}$ values for each of the four most difficult tests ($p = 20$ and $p = 8$). These averages were computed from table 3.

| $p = 20$ $m-p-1 \leq 25$ | $p = 20$ $m-p-1 > 25$ | $p = 8$ $m-p-1 \leq 20$ | $p = 8$ $m-p-1 > 20$ | Average scores $p = 20, 8$ | $p = 4, 2$ |
|---|---|---|---|---|---|
| $OS$ 3.2 | $OS$ 3.4 | $OS$ 4.2 | $OS$ 2.7 | $OS$ 3.4 | $DS$ 3.3 |
| $\bar{U}$ 3.6 | $U$ 4.4 | $DS$ 5.2 | $DS$ 3.1 | $\bar{U}$ 4.3 | $\left\{ \begin{array}{l} OS\ 3.5 \\ O\ 3.5 \\ D\ 3.5 \end{array} \right.$ |
| $U$ 4.5 | $\bar{U}$ 4.6 | $\left\{ \begin{array}{l} U\ 5.6 \\ O\ 5.6 \\ \bar{U}\ 5.6 \end{array} \right.$ | $\left\{ \begin{array}{l} \bar{U}\ 3.4 \\ O\ 3.4 \end{array} \right.$ | $U$ 4.6 | |
| $\left\{ \begin{array}{l} DS\ 6.4 \\ O\ 6.4 \end{array} \right.$ | $DS$ 5.5 | | | $DS$ 5.0 | |
| | $O$ 6.1 | | $U$ 3.9 | $O$ 5.4 | $\bar{U}$ 3.6 |
| $D$ 8.1 | $R$ 7.2 | $D$ 6.4 | $D$ 4.3 | $D$ 6.6 | $U$ 4.3 |
| $R$ 8.3 | $D$ 7.7 | $R$ 6.9 | $R$ 5.0 | $R$ 6.8 | $R$ 4.4 |

The $OS$ method ranks first in each of the four hard cases, with the $\bar{U}$ and $U$ methods not far behind. In the two $p = 20$ cases the $DS$ method is somewhat inferior to $OS$, $\bar{U}$, and $U$, but with $p = 8$, $DS$ jumps to second place.

The two right-hand columns of the table show separately the average scores for the $p = 20, 8$ cases and for the $p = 4, 2$ cases. The results confirm those presented by the authors, there being relatively little difference among the methods in the $p = 4, 2$ cases. Overall, the $OS$ and $DS$ methods seem to me to come out a little better than the authors' summary suggests. I was not clear why the $OS$ should do badly, as the authors' state, when $D^2$ is small.

In conclusion, this paper is a good example of the capable use of the computer to attack a problem that has proved too difficult mathematically.

REFERENCE

FRANCIS, I. 1966. Inference in the classification problem. Ph.D. Thesis, Harvard University.

# The Use of Covariance in Observational Studies

By W. G. Cochran

*Harvard University*

## Summary

When two groups of subjects are being compared, one group exposed to some presumed causal force and the other not, adjustment of the difference $(\bar{y}_1 - \bar{y}_2)$ in mean responses by means of a regression on one or more $x$-variables is sometimes employed in order to remove possible biases that arise when subjects are not assigned to groups at random. In such applications, Belson (1956) has suggested that the adjustments be made by means of the regression for the unexposed group only, whereas the routine user of the analysis of covariance employs the pooled regression coefficients from both groups. This note tries to clarify the situations in which Belson's proposal is preferable.

## Introduction

THIS note was stimulated by a paper by Belson (1956) in this journal. His objective was to estimate by means of a written examination what had been learned by the viewers of a series of four B.B.C. television programmes "Bon Voyage", which aimed at teaching French words and phrases useful to a tourist and to remove apprehensions about language difficulties, plus giving some general information. The research plan was of a type sometimes called "After only, with control". It was not feasible to assemble a group of viewers and one of non-viewers (controls) until after the programmes had been shown.

The examination given to both groups contained items designed to test knowledge of French words and phrases presented in "Bon Voyage", knowledge of facts presented and attitude on issues related to visiting France.

Now a viewer with an expert knowledge of French might do well in the test examination without actually having learned anything from the programmes; further, subjects with some knowledge of the French language might be expected to be more likely to choose to view these programmes than subjects ignorant of French. Consequently, the difference $\bar{y}_1 - \bar{y}_2$ between the mean scores of viewers and non-viewers cannot be trusted to measure the effect of the programmes unless we are satisfied that the two groups were comparable in their *initial* knowledge of French and of France. A similar difficulty is encountered whenever attempts are made by this study plan to appraise the effectiveness of a TV or radio programme or of the distribution of informative written material.

Belson was fully aware of this difficulty. For each section of the test examination he sought a number of covariates (which he calls stable correlates) that (i) correlated highly with the $y$ scores for the non-viewers and (ii) were not influenced by the act of viewing. For the examination section—knowledge of French words and phrases presented in "Bon Voyage"—he found two good covariates: $x_1$ was the examination score on a set of control words included in the examination but presumably not in "Bon Voyage", and $x_2$ was a measure of educational level. Together, $x_1$ and $x_2$ gave a multiple correlation of 0·82 with the non-viewers' $y$ scores on the test examination. Incidentally, he points out the importance of having a high multiple correlation coefficient for this purpose. In so far as the multiple correlation falls short of unity, we cannot be sure that any initial bias in the comparison of the two groups will be completely removed by the regression adjustments.

In describing this technique he remarks (p. 197): "It was essential to restrict these analyses to the test results of the non-viewers because the inclusion of those of the other group would, if the programme had in fact produced changes, have led to a misleading attenuation of the correlations." In other words, he insists that in finding the adjusted difference

$$(\bar{y}_1 - \bar{y}_2) - b_1(\bar{x}_{11} - \bar{x}_{12}) - b_2(\bar{x}_{21} - \bar{x}_{22}),$$

the regression coefficients $b_1$ and $b_2$ be estimated from the results for group 2 (the non-viewers) only. Apart from this restriction, the technique is easily recognized as an application of the analysis of covariance. The question to be discussed is: when should $b_1$ and $b_2$ be estimated from non-viewers only, as recommended by Belson, and when from the pooled regressions for viewers and non-viewers, as would be done by a routine user of the analysis of covariance?

### MODELS FOR THE ANALYSIS OF COVARIANCE

The point at issue can be brought out equally well with a single $x$-variate as with two or more. Let $y_{1j}, x_{1j}$ denote the test and covariate scores for the $j$th viewer, and $y_{2j}, x_{2j}$ similarly for the $j$th non-viewer. A mathematical model might be written as follows.

10

*Viewers*

$$y_{1j} = \alpha + \tau_j + \beta x_{1j} + e_{1j},$$

where $\tau_j$ represents the effect of the programme on the $j$th viewer, while $e_{1j}$ is the usual random residual with mean assumed zero.

*Non-viewers*

$$y_{2j} = \alpha + \beta x_{2j} + e_{2j}.$$

With regard to $\tau_j$, three different assumptions seem possible in different applications.

*Case (i)*

$\tau_j = \text{constant} = \tau$. This is the standard model for the analysis of covariance. If we can assume in addition that the residual variance $\sigma^2$ is the same in both groups (this can be checked by an $F$-test), least-squares analysis leads to the use of the usual pooled estimate of $\beta$ from the two groups.

*Case (ii)*

$\tau_j = \tau + \epsilon_j$, where $\tau$ is the average effect of the programme in the population of viewers and $\epsilon_j$ is uncorrelated with $x_{1j}$. Here the effect of the programmes varies from subject to subject, but in a manner unconnected with the value of the covariate. The model for the viewers can be rewritten as

$$y_{1j} = \alpha + \tau + \beta x_j + (e_{1j} + \epsilon_j).$$

It might happen that $\epsilon_j$ is correlated with $e_{1j}$, since characteristics of a viewer that were not measured by the covariate might affect how much was learned from viewing. Whether $\epsilon_j$ is correlated with $e_{1j}$ or not, consequences of this form of the model are:
   (a) The regression coefficients in the two groups should not differ (this can be checked by a $t$-test of $b_1 - b_2$, the difference between the regressions in the two groups) and

   (b)          $$\sigma_1^2 = \sigma^2 + \sigma_\epsilon^2 + 2 \operatorname{cov}(e_{1j}, \epsilon_j) \neq \sigma_2^2.$$

If $\sigma_1^2$ and $\sigma_2^2$ were known, least-squares analysis would lead to a weighted pooled regression of the form

$$\hat{\beta} = \left\{ \frac{\sum y_{1j}(x_{1j} - \bar{x}_1)}{\sigma_1^2} + \frac{\sum y_{2j}(x_{2j} - \bar{x}_2)}{\sigma_2^2} \right\} \bigg/ \left( \frac{\Sigma_1}{\sigma_1^2} + \frac{\Sigma_2}{\sigma_2^2} \right),$$

where $\Sigma_1 = \sum(x_{1j} - \bar{x}_1)^2$ and $\Sigma_2 = \sum(x_{2j} - \bar{x}_2)^2$.

In practice we would not know $\sigma_1^2$ and $\sigma_2^2$, but only the corresponding residual mean squares $s_1^2$ and $s_2^2$ and thus would face the problem of the advisability of weighting inversely as *estimated* variances.

To investigate this matter a little further, let $b_1 = \sum y_{1j}(x_{1j} - \bar{x}_1)/\Sigma_1$ and $b_2 = \sum y_{2j}(x_{2j} - \bar{x}_2)/\Sigma_2$ be the regression coefficients calculated from group 1 and group 2, respectively. Suppose that we choose any weighted mean $w_1 b_1 + w_2 b_2$, where $w_1 + w_2 = 1$, as our estimate $\hat{\beta}$ of $\beta$. The adjusted mean difference

$$\bar{y}_1 - \bar{y}_2 - \hat{\beta}(\bar{x}_1 - \bar{x}_2)$$

may be written

$$\{\bar{y}_1 - \bar{y}_2 - \beta(\bar{x}_1 - \bar{x}_2)\} - \{(\hat{\beta} - \beta)(\bar{x}_1 - \bar{x}_2)\}.$$

By substitution from the model for case (ii), the standard algebra of regression shows that this quantity is an unbiased estimate of $\tau$ with variance

$$\frac{\sigma_1^2}{n_1}+\frac{\sigma_2^2}{n_2}+(\bar{x}_1-\bar{x}_2)^2\left(\frac{w_1^2\sigma_1^2}{\Sigma_1}+\frac{w_2^2\sigma_2^2}{\Sigma_2}\right), \tag{1}$$

where $n_1$, $n_2$ are the sample sizes in groups 1 and 2, and $\sigma_1^2$, $\sigma_2^2$ are as before the two residual variances. The first term above arises from the fact that the regression does not fit perfectly, the second arises from the sampling variance of $\hat{\beta}$. We now study formula (1) for the variance a little more closely.

Consider first the case $E(\bar{x}_1) = E(\bar{x}_2)$. In this situation the covariance adjustment is not needed to remove bias due to $x$, because there is no such bias (though a covariance adjustment might still improve the precision of the comparison between $\bar{y}_1$ and $\bar{y}_2$). In this case

$$E(\bar{x}_1-\bar{x}_2)^2 = \sigma_x^2\left(\frac{1}{n_1}+\frac{1}{n_2}\right).$$

Under normality, the mean value of $1/\Sigma_1$ is known to be $1/(n_1-3)\sigma_x^2$. With a non-normal distribution of $x$, $E(1/\Sigma_1)$ may be expanded in a series of powers of $1/n_1$, of which the leading term is $1/(n_1\sigma_x^2)$. Consequently, in the no-bias case the order of magnitude of the second term in formula (1) for the variance of the adjusted comparison is

$$\left(\frac{1}{n_1}+\frac{1}{n_2}\right)\left(\frac{w_1^2\sigma_1^2}{n_1}+\frac{w_2^2\sigma_2^2}{n_2}\right).$$

For any choice of $w_1$ and $w_2$ and any substantial sample sizes this term is likely to be small relative to the first term

$$\frac{\sigma_1^2}{n_1}+\frac{\sigma_2^2}{n_2}.$$

Thus with, say, $n_1 > 40$, $n_2 > 40$, it does not matter whether we pool or how we pool. In particular, Belson's suggestion, $w_2 = 1$, gives for the second term in the variance an order of magnitude

$$\left(\frac{1}{n_1}+\frac{1}{n_2}\right)\frac{\sigma_2^2}{n_2}$$

and seems satisfactory though not obligatory.

The situation is different when $E(\bar{x}_1) \neq E(\bar{x}_2)$, the case in which covariance is used to help in removing bias due to initial incomparability in the $x$-distributions. If $E(\bar{x}_1-\bar{x}_2) = \lambda\sigma_x$, the leading term in $E(\bar{x}_1-\bar{x}_2)^2$ is $\lambda^2\sigma_x^2$. This is of order unity. In this case the second term in formula (1) becomes of the same order of magnitude as the first term. Reduction of the sampling variance of $\hat{\beta}$ to a minimum therefore becomes essential.

Suppose that the choice is between a regression computed from group 2 only and the usual pooled covariance regression. With a group 2 regression, the multiplier of $(\bar{x}_1-\bar{x}_2)^2\sigma_x^2$ is $\sigma_x^2\sigma_2^2/\Sigma_2$, of order $\sigma_2^2/n_2$. With the usual pooled regression, $w_1 = \Sigma_1/(\Sigma_1+\Sigma_2)$, $w_2 = \Sigma_2/(\Sigma_1+\Sigma_2)$, the multiplier of $(\bar{x}_1-\bar{x}_2)^2\sigma_x^2$ is

$$\sigma_x^2(\Sigma_1\sigma_1^2+\Sigma_2\sigma_2^2)/(\Sigma_1+\Sigma_2)^2, \quad \text{of order} \quad (n_1\sigma_1^2+n_2\sigma_2^2)/(n_1+n_2)^2.$$

10*

Now

$$\frac{\sigma_2^2}{n_2} < \frac{n_1 \sigma_1^2 + n_2 \sigma_2^2}{(n_1 + n_2)^2} \quad \text{only if} \quad \frac{n_1}{n_2} < \frac{\sigma_1^2}{\sigma_2^2} - 2.$$

The advantage in using the group 2 regression therefore is confined to applications in which (i) $\sigma_1^2$ is substantially larger than $\sigma_2^2$, and (ii) $n_2$ is much larger than $n_1$, as may happen if the exposed group is small. For instance, with $n_1 = n_2 = n$, pooling is superior unless $\sigma_1^2 > 3\sigma_2^2$.

If $\sigma_1, \sigma_2$ were known and we could choose the least-squares optimum weights, $w_1 \propto \Sigma_1/\sigma_1^2$, $w_2 \propto \Sigma_2/\sigma_2^2$, the coefficient of $(\bar{x}_1 - \bar{x}_2)^2 \sigma_x^2$ would be

$$\sigma_x^2 \bigg/ \left( \frac{\Sigma_1}{\sigma_1^2} + \frac{\Sigma_2}{\sigma_2^2} \right). \tag{2}$$

This, of course, is always smaller than (or at most equal to) the variance term for the better of the two preceding estimates of $\beta$. What about weighting inversely as the estimated variances, i.e. taking $w_1 \propto \Sigma_1/s_1^2$ and $w_2 \propto \Sigma_2/s_2^2$, where $s_1^2$ and $s_2^2$ are the residual mean squares from the regressions in groups 1 and 2? Neglecting terms of order $1/n_i^2$, Meier (1953) has shown that the effect of the use of estimated instead of correct weights is on the average to multiply the expression (2) by

$$1 + \frac{2w_1(1 - w_1)}{n_1} + \frac{2w_2(1 - w_2)}{n_2}. \tag{3}$$

Under normality, with $w_1 = w_2$ and $n_1 = n_2$, Cochran and Carroll (1953) verified by experimental sampling that Meier's formula holds well for two groups down to $n = 10$. In general, the result (3) suggests that weighting inversely as the estimated variance causes only a negligible loss of precision for $n_1, n_2 > 40$.

*Case (iii)*

$$\tau_j = \gamma + \delta x_{1j} + h_j.$$

This model postulates that the effect of the programme on viewer $j$ is linearly related to the value of the covariate $x_{1j}$ for this subject. The assumption is not unreasonable for this type of application. The programmes might be of a level of difficulty such that a viewer with very poor French ($x_{1j}$ low) would learn little, while one with good French learns a lot. In this case $\delta$ is positive. On the other hand, if very elementary information were given in the programmes, perhaps only these viewers who were initially ignorant ($x_{1j}$ low) would learn much, leading to a negative $\delta$. We could even envisage a quadratic regression, a well-informed viewer already knowing all this stuff and an ignorant viewer being unable to grasp it, with only the moderately informed viewer benefiting from the programmes.

Assuming a linear regression, the model becomes

*Viewers*

$$y_{1j} = \alpha + \gamma + (\beta + \delta) x_{1j} + (e_{1j} + h_j).$$

*Non-viewers*

$$y_{2j} = \alpha + \beta x_{2j} + e_{2j}.$$

A question now arises: what do we mean by the average effect of viewing the programmes? For the sample of viewers, this appears to be

$$\gamma + \delta \bar{x}_1,$$

which becomes the quantity to be estimated. Since the model now postulates a different slope and a different intercept in the two populations, we might as well fit two different lines:

*Viewers*

$$y_{1j} = \alpha_1 + \beta_1 x_{1j} + e_{1j}.$$

*Non-viewers*

$$y_{2j} = \alpha_2 + \beta_2 x_{2j} + e_{2j},$$

where

$$\gamma = \alpha_1 - \alpha_2; \quad \delta = \beta_1 - \beta_2.$$

It follows that the estimate of the average effect of the programmes is

$$
\begin{aligned}
\hat{\gamma} + \hat{\delta}\bar{x}_1 &= (\hat{\alpha}_1 - \hat{\alpha}_2) + (b_1 - b_2)\bar{x}_1 \\
&= \bar{y}_1 - b_1\bar{x}_1 - \bar{y}_2 + b_2\bar{x}_2 + (b_1 - b_2)\bar{x}_1 \\
&= (\bar{y}_1 - \bar{y}_2) - b_2(\bar{x}_1 - x_2),
\end{aligned}
\tag{4}
$$

as recommended by Belson.

In this situation, Belson's suggestion gives an unbiased estimate of the average effect of viewing; the usual covariance estimate is biased.

To sum up, if $b_1$ and $b_2$ differ significantly and if the interpretation given in case (iii) seems a reasonable explanation, Belson's procedure is recommended. In fact, even if the data support the hypothesis that $\beta_1 = \beta_2$, Belson's procedure might be adopted as standard in observational studies when $n_2$, the number of control subjects, is fairly large, provided that the relative contribution of sampling errors in $b_2$ to the variance of Belson's estimate, equation (4), is found to be negligible. This point can be checked when the variance of equation (4) is being calculated. If the contribution is not small, as will happen when there is a sizeable difference $(\bar{x}_1 - \bar{x}_2)$, and if $\beta_1$ and $\beta_2$ are not judged to be different, use of some form of pooled estimate of $b_1$ and $b_2$ is advisable. A more general moral of this note is that before making a routine application of the analysis of covariance in an observational study, the investigator should examine both the adequacy and the implied meaning of his model.

### REFERENCES

BELSON, W. A. (1956). A technique for studying the effects of a television broadcast. *Appl. Statist.*, **V**, 195–202.

COCHRAN, W. G. and CARROLL, S. P. (1953). A sampling investigation of the efficiency of weighting inversely as the estimated variance. *Biometrics*, **9**, 448–459.

MEIER, P. (1953). Variance of a weighted mean. *Biometrics*, **9**, 59–73.

# 93

## Some Effects of Errors of Measurement on Multiple Correlation

W. G. COCHRAN*

In many multiple correlation and regression studies the values of the variables are difficult to measure accurately. This paper attempts to discuss the effects of such errors of measurements on the squared multiple correlation coefficient. With independent errors and a multivariate normal model, the effect is very roughly that $R^2$ becomes reduced to about $R'^2 = R^2 g_v \bar{g}_w$, where $g_v$ is the coefficient of reliability of $y$ and $\bar{g}_w$ is a weighted mean of the coefficients of reliability of the $x$'s. If most intercorrelations among the true $X$'s are detrimental to $R^2$, the value of $R'^2$ may exceed $R^2 g_v \bar{g}_w$ by 10 percent to 20 percent. The situation in which errors are correlated with the true values and the effects of errors on the interpretation of regression coefficients are briefly discussed.

## 1. INTRODUCTION

In recent years there has been an increase in multiple regression studies on problems in which some of the independent variables represent quantities that are obviously difficult to measure and are presumably measured with substantial errors. In the social sciences, for example, these variables may include measures of a person's skills at certain tasks or his attitudes and psychological characteristics, the data being obtained from a questionnaire plus perhaps some kind of examination. Such studies raise the question: to what extent do these errors of measurement weaken or vitiate the uses to which the multiple regression is put? In examining this question, my results fall a good deal short of what is desirable. The only tractable mathematical model is the multivariate normal —simpler than is needed for many applications. Even with this model, the effects of the errors are complex. I have, however, tried to discover approximately what happens in situations representative of at least a substantial number of applications.

This paper deals mainly with the relation between $R^2$, the squared multiple correlation coefficient between $y$ and the $X$'s when these are correctly measured, and $R'^2$, the corresponding value when errors of measurement are present. Multiple regression equations have several different uses. Among them are: (1) To predict the variable $y$. The relevant issue here is the relation between $\sigma_y^2(1-R^2)$ and $\sigma_y^2(1-R'^2)$. (2) To estimate the proportion of the variance of $y$ that is associated with variations in the $X$'s. For this application we usually want to know the value of $R^2$, even if our measurements of the $X$'s are subject To errors. Results given here on the relation between $R'^2$ and $R^2$, plus information on the precision of measurement from pilot or past studies, would enable $R^2$ to be estimated roughly from $R'^2$, but the problem requires further work and

* W. G. Cochran is professor, Department of Statistics, Harvard University. The author is grateful for suggestions received from the editor and a referee. This article was supported by the Office of Naval Research, Contract N00014-67-A-0298-0017.

will not be pursued here. (3) To study and interpret the values of individual regression coefficients $\beta_i$. The nature of the effects of errors of measurement on the values of the $\beta_i$ has been indicated previously [2] and is discussed briefly in Section 8.

## 2. MATHEMATICAL MODEL

Using capital letters to denote correctly measured values, we suppose that in the population the variate $Y_u$ has a linear regression $\alpha + \sum \beta_i X_{iu} + d_u$ on the $k$ $X$'s, where $d$ is the random residual from the regression. Owing to errors of measurement, the variates actually recorded for $Y_u$ and for the $i$th $X$-variate are

$$y_u = Y_u + a + e_u, \qquad x_{iu} = X_{iu} + a_i + e_{iu}, \qquad (2.1)$$

where $a$ and the $a_i$ represent overall constant biases of measurement, while $e_u$ and the $e_{iu}$ are fluctuating components that follow frequency distributions with means zero.

For this type of model Lindley [10] gave the necessary and sufficient relations that must hold between the joint frequency function of the $X_{iu}$ and that of the $e_{iu}$ in order that the regression of $y_u$ on the $x_{iu}$ remains linear. In particular, if $y_u$ and the $X_{iu}$ follow a multivariate normal distribution, the $e_{iu}$ must also follow a multivariate normal. This case is assumed here. Clearly, if $Y_u$, $e_u$, $X_{iu}$, and the $e_{iu}$ jointly follow a multivariate normal, it follows from relations (2.1) that $y_u$ and the $x_{iu}$ also follow a multivariate normal and hence that the regression of $y_u$ on the $x_{iu}$ is linear.

For the present it is assumed further that $e_u$ is independent of $Y_u$ and that any $e_{iu}$ is independent of $X_{iu}$ or any $X_{ju}$ $(j \neq i)$ and of any other $e_{ju}$. These last assumptions are not essential to ensure linearity of the regression of $y_u$ on the $x_{iu}$, and some remarks about the nonindependent case will be made in Section 6. For many applications in which both the $X_{iu}$ and the $e_{iu}$ appear non-normal, it would be desirable to bypass the normality assumptions, but I have no results for this situation.

The bias terms $a$ and $a_i$ in (2.1) affect the constant term in the regression of $y$ on the $x_i$, but do not affect the multiple correlation coefficient between $y$ and the $x_i$, and hence do not enter into the following sections on $R^2$.

## 3. EFFECT ON $R^2$ WHEN $X$'S ARE INDEPENDENT

With $k$ $X$-variates, the following notation will be used for the relevant population parameters:

$\sigma_i^2$ is variance of the correct $X_{iu}$,

$\epsilon^2$, $\epsilon_i^2$ is variance of $e_u$, $e_{iu}$,

$\rho_{ij}$ is correlation coefficient between $X_{iu}$ and $X_{ju}$,

$\delta_i$ is correlation coefficient between $X_{iu}$ and $Y_u$.

The symbol $\delta_i$ is used instead of the more natural $\rho_{iy}$ because this helps to avoid confusion between different kinds of correlation in later discussion. The sign attached to each $X_{iu}$ is assumed chosen so that $\delta_i \geq 0$.

The value of $R^2$, the population squared multiple correlation between $Y$ and the $X_i$, is completely determined by the $\rho_{ij}$ and the $\delta_i$. Primes will denote the corresponding correlations $R'^2$, $\rho'_{ij}$, $\delta_i'$, between the observed $y$ and the $x_i$. From the assumptions we have

$$\rho'_{ij} = \frac{\text{Cov}[(X_i + e_i), (X_j + e_j)]}{\sqrt{(\sigma_i^2 + \epsilon_i^2)(\sigma_j^2 + \epsilon_j^2)}} = \frac{\rho_{ij}\sigma_i\sigma_j}{\sqrt{(\sigma_i^2 + \epsilon_i^2)(\sigma_j^2 + \epsilon_j^2)}}, \quad (3.1)$$

$$\delta_i' = \frac{\text{Cov}[(X_i + e_i), (Y + e)]}{\sqrt{(\sigma_i^2 + \epsilon_i^2)(\sigma_Y^2 + \epsilon^2)}} = \frac{\delta_i\sigma_i\sigma_Y}{\sqrt{(\sigma_i^2 + \epsilon_i^2)(\sigma_Y^2 + \epsilon^2)}}. \quad (3.2)$$

In psychometric writings the quantity $g_i = \sigma_i^2/(\sigma_i^2 + \epsilon_i^2)$ is often called the coefficient of reliability of the measurement $x_i$. Similarly we write $g_y = \sigma_Y^2/(\sigma_Y^2 + \epsilon^2)$ = coefficient of reliability of $y$. Hence, from (3.1) and (3.2),

$$\rho'_{ij} = \rho_{ij}\sqrt{g_i g_j}, \qquad \delta_i' = \delta_i\sqrt{g_i g_y}. \quad (3.3)$$

If the $X$'s are mutually independent, it is well know that

$$R^2 = \sum_{i=1}^{k} \delta_i^2. \quad (3.4)$$

Since our assumptions guarantee that the $x$'s are also independent,

$$R'^2 = \sum_{i=1}^{k} \delta_i'^2 = g_y \sum_{i=1}^{k} \delta_i^2 g_i. \quad (3.5)$$

Hence,

$$R'^2 = R^2 g_y \sum_{i=1}^{k} \delta_i^2 g_i \bigg/ \sum_{i=1}^{k} \delta_i^2 = R^2 g_y \bar{g}_w, \quad (3.6)$$

where $\bar{g}_w$ is a weighted mean of the coefficients of reliability of the $x_i$.

Consider now the residual variance from the regression. With the correct measurements this is $\sigma_Y^2(1 - R^2)$. With the fallible measurements it becomes

$$\sigma_y^2(1 - R'^2) = \sigma_Y^2 + \epsilon^2 - \sigma_y^2 g_y \bar{g}_w R^2 = \sigma_Y^2(1 - R^2\bar{g}_w) + \epsilon^2 \quad (3.7)$$

since $\sigma_y^2 g_y = \sigma_Y^2$. The effect of errors of measurement of $y$ with variance $\epsilon^2$ is simply to increase the residual variance by $\epsilon^2$, as is well known.

Regarding errors of measurement of the $x_i$, two points are worth noting from (3.7) in relation to applications. For a given reliability of measurement, i.e., a given value of $\bar{g}_w$, the deleterious effect on the residual variance increases as $R^2$ increases, being greater when the prediction formula is very good than when it is mediocre. For example, suppose that $\bar{g}_w = 0.5$, representing a poor reliability in measurement of the $x_i$. If $R^2 = 0.9$, the residual variance is increased by errors of measurement of the $x_i$ from $0.1\sigma_Y^2$ to $0.55_Y^2$, over a fivefold increase. With $R^2 = 0.4$, the increase is only from $0.6\sigma_Y^2$ to $0.8\sigma_Y^2$, a 33 percent jump.

Second, as would be expected, the quality of measurement of those $X_i$ that are individually good predictors is much more important than that of poorer predictors. With $k = 2$, $\delta_1 = 0.9$, $\delta_2 = 0.3$, and no errors of measurement, we have $R^2 = 0.90$, $(1 - R^2) = 0.1$. If $g_1 = 0.5$, $g_2 = 1$, this gives $(1 - R'^2) = 0.505$, but with $g_1 = 1$, $g_2 = 0.5$, $(1 - R'^2) = 0.145$, a much smaller increase.

## 4. EFFECT OF CORRELATION BETWEEN X'S: TWO VARIATES

After working several numerical examples, my approach was to try to construct an approximation of the form $R'^2 = R^2 g_y \bar{~}_w f$, where $f$ is a correction factor to allow for the effect of correlations among the $X$'s, being equal to 1 when the $X$'s are independent. With numerous $X$ variables, all intercorrelated, it soon appeared that no simple correction factor was likely to be generally applicable. However, we will continue to study the relation of $R'^2$ to $R^2 g_y \bar{g}_w$. Further, since the effect of errors in $y$ under this present model is always just to introduce the factor $g_y$, this factor will be omitted in what follows so as to concentrate attention on correlations among the $X$'s.

With two $X$-variates having a correlation $\rho$, the values of $R^2$ and $R'^2$ work out as follows:

$$R^2 = (\delta_1^2 + \delta_2^2 - 2\rho\delta_1\delta_2)/(1 - \rho^2), \tag{4.1}$$

$$R'^2 = (g_1\delta_1^2 + g_2\delta_2^2 - 2g_1g_2\rho\delta_1\delta_2)/(1 - g_1g_2\rho^2). \tag{4.2}$$

For given $\delta_1$, $\delta_2$, the correlation $\rho$ lies within the limits $\delta_1\delta_2 \pm \sqrt{(1-\delta_1^2)(1-\delta_2^2)}$; otherwise $R^2$ would exceed 1. Within these limits,

$$R'^2 = R^2 \frac{(g_1\delta_1^2 + g_2\delta_2^2 - 2g_1g_2\rho\delta_1\delta_2)}{(\delta_1^2 + \delta_2^2 - 2\rho\delta_1\delta_2)} \frac{(1 - \rho^2)}{(1 - g_1g_2\rho^2)} \tag{4.3}$$

Inserting $g_w = (g_1\delta_1^2 + g_2\delta_2^2)/(\delta_1^2 + \delta_2^2)$ as a factor, we may write

$$R'^2 = R^2 \bar{g}_w (A)(B), \tag{4.4}$$

where $B$ is the term

$$B = (1 - \rho^2)/(1 - g_1g_2\rho^2). \tag{4.5}$$

For $g_1g_2 < 1$, $\rho \neq 0$, this term is always $< 1$. For fixed $g_1g_2$ it decreases monotonically towards 0 as $\rho$ moves from 0 towards either $+1$ or $-1$.

The remaining factor $A$ takes the form

$$A = \left[1 - \frac{2\rho\delta_1\delta_2}{\dfrac{\delta_1^2}{g_2} + \dfrac{\delta_2^2}{g_1}}\right] \bigg/ \left(1 - \frac{2\rho\delta_1\delta_2}{\delta_1^2 + \delta_2^2}\right). \tag{4.6}$$

For $0 < g_1g_2 < 1$, it follows that $A > 1$ if $\rho$ is positive while $A < 1$ if $\rho$ is negative, since we have postulated that $\delta_1$, $\delta_2$ are both $> 0$.

Hence, if $\rho$ is negative the factor $f = AB$ is always $< 1$, decreasing towards zero as $\rho$ approaches $-1$. If $\rho$ is positive, the situation is not so clear, since $A > 1$ and $B < 1$. However, when $\rho$ is small the factor $A$, which contains only linear terms in $\rho$, tends to dominate $B$ which is quadratic in $\rho$. Thus when $\rho$ is positive, $f = AB$ increases and is greater than 1 when $\rho$ is small, but then decreases as $\rho$ increases further becoming less than 1 if $\rho$ is high enough. The only exception is the case $\delta_1 = \delta_2$. $g_1 = g_2 = g$: $f$ then reduces to $(1+\rho)/1+g\rho)$, which increases from $f = 1$ at $\rho = 0$ to $f = 2/(1+g)$ at $\rho = 1$. Incidentally, when $\delta_1 = \delta_2$, the range of $\rho$ is from $(-1+2\delta_1^2)$ to $+1$.

The size of the product $g_1g_2$ is also relevant to $f$. For given $\rho$, both $A$ and $B$

Table 1. VALUES OF $f = R'^2/R^2\bar{g}_w$ FOR SIX EXAMPLES

| $\rho$ | $\delta_1 = .6,\ \delta_2 = .4$ | | | | $\delta_1 = .7,\ \delta_2 = .2$ | | | |
|---|---|---|---|---|---|---|---|---|
| | $g_i =$ | .9, .7 | .8, .6 | .7, .5 | $g_i =$ | .9, .7 | .8, .6 | .7, .5 |
| | $R^2$ | $f$ | $f$ | $f$ | $R^2$ | $f$ | $f$ | $f$ |
| −.5 | —[a] | — | — | — | .893 | 0.84 | 0.78 | 0.74 |
| −.4 | .848 | 0.87 | 0.82 | 0.78 | .764 | 0.89 | 0.85 | 0.81 |
| −.3 | .730 | 0.91 | 0.88 | 0.85 | .675 | 0.93 | 0.90 | 0.88 |
| −.2 | .642 | 0.95 | 0.92 | 0.90 | .610 | 0.96 | 0.94 | 0.93 |
| −.1 | .574 | 0.98 | 0.97 | 0.96 | .564 | 0.98 | 0.98 | 0.97 |
| 0 | .520 | 1.00 | 1.00 | 1.00 | .530 | 1.00 | 1.00 | 1.00 |
| .1 | .477 | 1.02 | 1.03 | 1.04 | .507 | 1.01 | 1.02 | 1.02 |
| .2 | .442 | 1.04 | 1.06 | 1.07 | .494 | 1.02 | 1.02 | 1.03 |
| .3 | .413 | 1.06 | 1.08 | 1.10 | .490 | 1.02 | 1.02 | 1.03 |
| .4 | .390 | 1.07 | 1.10 | 1.12 | .498 | 1.00 | 1.00 | 1.01 |
| .5 | .373 | 1.08 | 1.11 | 1.14 | .520 | 0.98 | 0.97 | 0.97 |
| .6 | .362 | 1.08 | 1.11 | 1.14 | .566 | 0.94 | 0.91 | 0.90 |
| .7 | .361 | 1.07 | 1.09 | 1.12 | .655 | 0.86 | 0.82 | 0.79 |
| .8 | .378 | 1.03 | 1.03 | 1.06 | .850 | 0.73 | 0.67 | 0.63 |
| .9 | .463 | 0.86 | 0.85 | 0.85 | —[a] | — | — | — |
| | $\bar{g}_w =$ | .838 | .738 | .638 | $\bar{g}_w =$ | .885 | .785 | .685 |

[a] Impossible because $R^2 > 1$.

tend to approach 1 as $g_1 g_2$ increases towards 1. Thus the formula $R'^2 = R^2 \bar{g}_w$ is closer to the truth when $g_1$ and $g_2$ are high.

As an illustration, Table 1 shows the values of $R^2$ and $f$ for $\rho = -0.5(0.1)+0.9$, for six examples. In the first three, $\delta_1 = 0.6$, $\delta_2 = 0.4$, and in the second, $\delta_1 = 0.7$, $\delta_2 = 0.2$. In both sets, $R^2$ is close to 0.5 when $\rho = 0$. The three pairs $g_1, g_2 = (0.9, 0.7)$, $(0.8, 0.6)$, and $(0.7, 0.5)$ are given. The principal difference between the cases $\delta_1 = 0.6$, $\delta_2 = 0.4$ and $\delta_1 = 0.7$, $\delta_2 = 0.2$ is as follows. When $\delta_1$ and $\delta_2$ differ greatly and $\rho$ is positive, $R^2$ begins to increase and $f$ to decrease for quite moderate values of $\rho$ (around 0.3 for $\delta_1 = .7$, $\delta_2 = .2$), while when $\delta_1$ and $\delta_2$ are more nearly equal, $R^2$ decreases and $f$ increases until $\rho$ is closer to 1. The turning value of $f$ is a complicated expression, but is usually close to that of $R^2$.

The complementary sets $g_1, g_2 + (0.7, 0.9)$, $(0.6, 0.8)$, $(0.5, 0.7)$, not shown here, exhibit the same behavior with $f$ lying a little nearer 1, except for high, positive $\rho$ when $f$ becomes less than 1.

The results that $f$ is less than 1 when $\rho$ is negative and that $f$ is usually greater than 1 when $\rho$ is positive and moderate can be rationalized as follows. Suppose we compare the value of $R^2$ from formula (3.1) with the value $R^2_{ind} = \delta_1^2 + \delta_2^2$ that $R^2$ would have if $\rho$ were 0. With $\rho$ negative, $R^2$ increases steadily as $\rho$ departs from 0. With $\rho$ positive, $R^2$ decreases at first but has a minimum at $\rho = \delta_2/\delta_1$, where $\delta_2 \leq \delta_1$, and thereafter increases. It does not reach $R^2_{ind} = (\delta_1^2 + \delta_2^2)$ until $\rho = 2\delta_1\delta_2/(\delta_1^2 + \delta_2^2)$, which is high if $\delta_1$ and $\delta_2$ are not too different. Thus in a bivariate correlation, negative correlation between $x_1$ and $x_2$ increases $R^2$ while positive correlation decreases $R^2$ unless $\rho$ is high enough.

Now the errors of measurement in the $x$'s reduce $\rho$ to $\rho\sqrt{g_1 g_2}$. Thus the errors diminish both a helpful negative $\rho$, making $f<1$ with $\rho$ negative, and a harmful positive $\rho$, making $f>1$ when $\rho$ is positive and moderate.

In these examples $\bar{g}_w$ lies between 0.638 and 0.885. Regarding the crude approximation $R'^2 \cong R^2 \bar{g}_w$, in these examples this is correct to within $\pm 15$ per cent for $\rho$ lying between $-0.3$ and $+0.5$, being much closer than this throughout most of Table 1.

To summarize for two independent variates: when $\rho$ is negative, $f<1$ because the negative correlation $\rho' = \rho \sqrt{g_1 g_2}$ is less helpful to $R'^2$ than the negative correlation $\rho$ is to $R^2$. When $\rho$ is positive and small or modest, $f$ exceeds 1, because the harmful positive correlation between $X_1$ and $X_2$ is decreased by the errors of measurement. If $\rho$ becomes high enough, however, positive correlation becomes helpful and $f$ drops below 1. For given $\rho$, $f$ departs further from 1 as the product $g_1 g_2$ decreases.

With three $X$-variates denoted by the subscripts $i$, $j$, and $k$, the value of $R^2$ may be expressed as

$$R^2 = \frac{\sum_i \delta_i^2(1 - \rho^2_{jk}) - 2 \sum_{j>i} (\rho_{ij} - \rho_{ik}\rho_{jk})\delta_i\delta_j}{1 - \sum_{j>i} \rho^2_{ij} + 2\rho_{12}\rho_{13}\rho_{23}}, \qquad (4.7)$$

while $R'^2$ has the corresponding value found by substituting $\delta_i' = \delta_i \sqrt{g_i}$, $\rho'_{ij} = \rho_{ij} \sqrt{g_i g_j}$. These expressions are discouraging to the prospect of finding an approximation for $f$ that would be valid over a wide range of values of the $g_i$ and the $\rho_{ij}$. With regard to $R^2$ itself, (4.7) suggests that with all $\delta_i > 0$, negative values of $\rho_{ij}$ are likely to be helpful, since the only linear term in the $\rho_{ij}$ is $-2\rho_{ij}\delta_i\delta_j$ in the numerator.

With 3 $X$-variables, a table like Table 1 involves choosing the values of 9 parameters—3 $\delta_i$, 3 $g_i$, and 3 $\rho_{ij}$. The situation rapidly becomes more complex with more than 3 $X$'s. Before proceeding further, we digress to consider the values of the $\rho_{ij}$ and the $g_i$ likely to occur in practice in the hope that a smaller range of $\rho_{ij}$ and $g_i$ might be representative of many practical situations.

## 5. SOME VALUES OF $r_{ij}$ IN PRACTICAL APPLICATIONS

When the sign attached to each $x_i$ is chosen so that $\delta_i \geq 0$, these decisions determine the sign attached to every $\rho_{ij}$. In studying the estimates $r_{ij}$ of the $\rho_{ij}$ found in 12 well-known examples of the discriminant function [1], I noted that most of the $r_{ij}$ are positive and modest in size, while those that are negative are usually small. The same situation appears to hold in many applications of multiple regression. Table 2 shows the distributions of the $r_{ij}$ in (a) the discriminant function examples, (b) the numerical examples of multiple regression given in 12 standard statistical texts, and (c) a single large example—the prediction of verbal ability scores of 12th grade white students in the north of the U. S. from 20 variables representing data on the student, the quality of the school, and the student's home environment [3].

The percentages of $r$'s that are positive are 83, 84, and 74 in the three sets. The negative $r$'s average to between $-0.2$ and 0, the averages of the positive

*Table 2.* DISTRIBUTIONS OF ESTIMATED CORRELATIONS BETWEEN X'S

| $r_{ij}$ | $r_{ij}$ negative | | | $r_{ij}$ | $r_{ij}$ positive | | |
|---|---|---|---|---|---|---|---|
| | D.F. | Texts | Verbal ability | | D.F. | Texts | Verbal ability |
| $< -.5$ | 1 | 1 | 0 | 0 to .1 | 15 | 5 | 58 |
| $-.5$ to $-.4$ | 2 | 0 | 0 | .1 to .2 | 22 | 8 | 41 |
| $-.4$ to $-.3$ | 1 | 0 | 1 | .2 to .3 | 25 | 7 | 25 |
| $-.3$ to $-.2$ | 4 | 2 | 2 | .3 to .4 | 18 | 9 | 7 |
| $-.2$ to $-.1$ | 4 | 2 | 10 | .4 to .5 | 6 | 7 | 3 |
| $-.1$ to 0 | 9 | 5 | 36 | .5 to .6 | 10 | 6 | 6 |
| | | | | .6 to .7 | 4 | 5 | 1 |
| | | | | .7 to .8 | 1 | 4 | 0 |
| | | | | $> .8$ | 0 | 3 | 0 |
| Total | 21 | 10 | 49 | Total | 101 | 54 | 141 |
| $\bar{r}$ | $-0.19$ | $-0.17$ | $-0.09$ | $\bar{r}$ | $+0.30$ | $+0.41$ | $+0.16$ |

$r$'s being a bit higher (0.16 to 0.41). Though some allowance is needed for the sampling errors of the $r_{ij}$, since our interest is in the unknown $\rho_{ij}$, my impression is that most of the degrees of freedom are large enough so that the effects of sampling errors on the average $r$'s should be small. In the discriminant function and text examples there may have been some selection towards more interesting examples, but this would probably affect the sizes of $\delta_i$ rather than the $\rho_{ij}$.

In calculations for three or more $x$'s, these results led me to concentrate on $\rho_{ij}$ less than 0.5, mainly on two cases: (1) all $\rho_{ij}$ positive, and (2) only a minority negative.

## 6. ESTIMATING RELIABILITY

With variables that are hard to measure, the problem of estimating the reliability of measurement is itself formidable. There is an extensive literature dealing with the study of errors of measurement, particularly in sample surveys. However, much of this is not usable for my purpose, because it concentrates on the overall bias in a measuring process, or deals essentially with 0-1 variates, or, when handling continuous variates, arranges them in an ordered classified form and reports only the percentages of cases in which the fallible method was wrong by one or more classes.

Direct estimation of $g$ is possible only when it is feasible to measure both the correct value $X$ and the fallible value $x$ for a sample of items. Data of this type are likely to be available only in the simpler problems of measurement in which a recognized correct method exists. Statements of age, for instance, might be compared with birth certificate records. To cite a few examples, Gray [5] reports $\hat{g} = 0.93$ for verbal reports of number of days sick leave taken during the period July 1 through November 15 as compared with the firm's records, and $\hat{g} = 0.90$ for number of days annual leave taken in the summer months. Data of Kish and Lansing [9] suggest $\hat{g}$ around 0.83 for appraisers' estimates of the selling prices of homes, as compared with actual prices for those homes that had been sold recently. Data collected for the National Center for Health

Statistics [11] provide a $\hat{g}=0.86$ for annual family expenditure on medical care as obtained from a short questionnaire.

Under the Consumer Savings Project, studies have been made of the accuracy of the reporting of financial assets. Ferber [4] presents results of urban and rural studies of time deposits in savings institutions with validation from the records of these institutions. Sample averages gave underestimates of as much as 50 percent or more, because (a) nonrespondents had larger savings than respondents, and (b) among respondents, a substantial number erroneously reported no savings. Respondents who reported savings had generally high accuracy. Thus the distribution of errors of measurement consists of one set of $e_i$ that are small and have mean almost zero, and another set in which $e_i = -X_i$ since $x_i = 0$. Quite apart from the bias, these features make the distribution of errors far from normal and give values of $g$ as low as 0.1 or 0.2.

A more widely used technique for estimating $g$, particularly when no method of making a correct measurement is known, is to attempt to take two *independent* measurements $x_1 = X + e_1$, $x_2 = X + e_2$, where $\text{Cov}(e_1, e_2) = 0$, on a sample of specimens by the fallible process. Assuming further that $X$ is uncorrelated with $e_1$ and $e_2$, the covariance of $x_1$ and $x_2$ estimates $\sigma_X^2$, so that $g$ is estimated by $\text{Cov}(x_1, x_2)/s_x^2$, where $s_x^2$ is the pooled within-sample mean square for $x_1$ and $x_2$.

With independence, this method works even if $x_1$ and $x_2$ are alternative forms of the fallible process that have different $g$'s, since $\text{Cov}(x_1, x_2)/s_{x_1}^2$ estimates $g_1$ for $x_1$ and $\text{Cov}(x_1, x_2)/s_{x_2}^2$ estimates $g_2$ for $x_2$. This method is frequently quoted in the rating of examinations, where $x_1$ and $x_2$ may be split halves of an exam or alternative forms of the exam. Sulzmann, et al. [13] report $\hat{g}$ values averaging 0.81 for test-retest values of standard visual acuity tests. An average $\hat{g}=0.71$ was computed from repetitions of the one-hour blood glucose level (after overnight fasting) in the test for the detection of diabetes [7]. In a review of applications of the Wechsler Intelligence Scale given to normal children aged 5–15, Sells [12] reports $g$ values for 44 subtests in different studies by split-half or test-retest methods. These subtests gave an average $g$ of 0.83, with a range from 0.65 to 0.96, while 16 subtests on retarded or guidance clinic children averaged $\hat{g}=0.89$. Many studies have reported correlations between the Wechsler and the Stanford-Binet scores as alternative measures of facets of intelligence. On normals, the average $g$ was 0.75 (129 subtests) as against an average $g=0.69$ for 23 subtests on mental defectives or retarded children [12].

Numerous complications can arise with this method. The assumption of independence of the two repetitions is vital. With positive correlation $\rho$ between $e_1$ and $e_2$ (e.g., memory of the same wrong answer on two different occasions), the quantity $\text{Cov}(x_1, x_2)/s_x^2$ overestimates $g$. In the simplest correlated case, in which $X$ is uncorrelated with $e_1$ or $e_2$ and $\sigma_{e_1}^2 = \sigma_{e_2}^2$, the quantity $\text{Cov}(x_1, x_2)/s_x^2$ is a consistent estimate of $\{g+\rho(1-g)\}$. The overestimation can easily be substantial. On the other hand, if the correct $X$ varies with time and if the reliability of measurement as of a specific time is wanted, a test-retest correlation may underestimate this $g$, since the retest for a subject is measuring a different $X$ from the test. This effect may account for low $\hat{g}$

values of 0.22 for respiration period, 0.36 for white blood cell count and 0.65 for blood sugar content reported by Guilford [6] from retests one day later on subjects.

In fact, when $X$ varies through time, the relevant variable for inclusion in the multiple correlation may be a complicated function of the levels of $X$ through a period of time, and the appropriate function may not be known. This is presumably the case, for instance, in attempts to relate the amount of cigarette smoking to the death rate. Estimates of the reliability of estimation of $X$ for a specific day or short period of time may be only partially relevant. A similar comment applies to attempts to measure personal traits such as aggressiveness, shyness, or leadership, where the scores on alternative forms of a questionnaire may correlate highly, but neither form may measure the name given by the investigator to the trait.

For these reasons I am uncertain about the range of $g$ values that can be used to describe practical applications in difficult measurement problems. My calculations have been done for the range $g \geq 0.5$. With $g = 0.5$, the variance of the errors of measurement equals the variance of the true measurements, and this seems a rather low standard of measurement. For $g < 0.5$, such as occurs in reports of consumer savings, my approximation for $R^2$ becomes poor. Incidentally, for a given variance of the errors of measurement, the value of $g$ of course depends on $\sigma_X^2$; for the same variable, $g$ can be much higher in a widespread population than in a narrowly restricted one. The low value $\hat{g} = 0.41$ reported by Kinsey [8] for "age at first knowledge of venereal disease" may reflect the fact that the correct $X$ for this variable has a low standard deviation.

## 7. EFFECT OF ERRORS WITH MORE THAN TWO X VARIATES

Returning to the relation between $R'^2$ and $R^2$ with $k$ $X$-variates ($k > 2$), assume first that all $\rho_{ij} > 0$. The simplest situation is that in which $\rho_{ij} = \rho > 0$, $g_i = g$. In this case $R^2$ and $R'^2$ have the values

$$R^2 = \frac{\sum \delta_i^2}{1 + (k-1)\rho}\left[1 + \frac{k\rho \sum (\delta_i - \bar{\delta})^2}{(1-\rho)\sum \delta_i^2}\right], \tag{7.1}$$

$$R'^2 = \frac{g \sum \delta_i^2}{1 + (k-1)g\rho}\left[1 + \frac{gk\rho \sum (\delta_i - \bar{\delta})^2}{(1-g\rho)\sum \delta_i^2}\right]. \tag{7.2}$$

If the $\delta_i$ are approximately equal, i.e., the $X_i$ are individually about equally good, the first terms in (7.1) and (7.2) dominate. Then we have

$$f = \frac{R'^2}{gR^2} \cong \frac{1 + (k-1)\rho}{1 + (k-1)g\rho}. \tag{7.3}$$

For $\rho$ positive, this $f$ exceeds 1 and increases steadily as $\rho$ goes from 0 to 1. For given $g$ and $\rho$, the excess above 1 increases with $k$, the number of variables.

The second terms inside the brackets in (7.1) and (7.2) begin to assume importance when the $\delta_i$ vary substantially. The ratio of the second term in (7.2) to that in (7.1) is $g(1-\rho)/(1-g\rho)$, which is less than 1 and decreases as $\rho$ in-

Table 3. VALUES OF $f = R'^2/gR^2$ FOR THREE EXAMPLES WITH POSITIVE $\rho$

| $k =$ | 3 | | | 5 | | | 10 | | |
|---|---|---|---|---|---|---|---|---|---|
| $\delta_i =$ | .6, .5, .4 | | | .5, $(.4)^2$, .3, .2 | | | .5, .4, $(.3)^2$, $(.2)^3$, $(.1)^3$ | | |
| $g \backslash \rho$ | .2 | .3 | .4 | .2 | .3 | .4 | .2 | .3 | .4 |
| 0.9 | 1.03 | 1.03 | 1.04 | 1.04 | 1.04 | 1.03 | 1.02 | 1.01 | 0.98 |
| 0.8 | 1.06 | 1.07 | 1.08 | 1.08 | 1.08 | 1.07 | 1.05 | 1.02 | 0.98 |
| 0.7 | 1.09 | 1.11 | 1.13 | 1.12 | 1.13 | 1.13 | 1.08 | 1.04 | 0.98 |
| 0.6 | 1.12 | 1.16 | 1.18 | 1.17 | 1.19 | 1.19 | 1.13 | 1.08 | 1.00 |
| 0.5 | 1.15 | 1.21 | 1.25 | 1.23 | 1.26 | 1.26 | 1.18 | 1.12 | 1.03 |

creases. The effect of this term, in conjunction with the first term, is to make the amount by which $f$ rises above 1 smaller for positive and moderate values of $\rho$.

Table 3 shows three examples with positive $\rho_{ij}$ and 3, 5, and 10 $x$-variates, respectively. Values of $f = R'^2/gR^2$ are given for $g = 0.9(0.1)0.5$ and $\rho = .2, .3, .4$.

The examples for three and five variables show rather similar values of $f$. Two offsetting influences are at work: $k = 5$ has more variables, associated with higher $f$'s, but also has more variation among the $\delta_i$, leading to lower $f$'s. The $k = 10$ example, which has still more variation among the $\delta_i$, gives values of $f$ usually nearer 1. With all $\rho$'s positive, the value of $f$ also tends to increase steadily as $g$ diminishes.

Table 4 presents four examples in which a minority of the $\rho_{ij}$ are negative. The values of $k$ are 3, with one $\rho_{ij}$ out of 3 negative; 5, with 4 $\rho_{ij}$ out of 10 negative; and 10, with 9 $\rho_{ij}$ out of 45 negative. As Table 4 shows, with these mixtures of positive and negative $\rho_{ij}$ the values of $f$ stay much closer to unity than when all $\rho_{ij}$ are positive. The main reason is presumably that both the

Table 4. VALUES OF $f = R'^2/gR^2$ FOR EXAMPLES WITH SOME $\rho_{ij}$ NEGATIVE

| $k =$ | 3 | | | 3 | | |
|---|---|---|---|---|---|---|
| $\delta_i =$ | .6, .5, .4 | | | .6, .5, .4 | | |
| $\rho_{ij} =$ | .2, .2, $-.2$ | .3, .3, $-.2$ | .4, .4, $-.2$ | .3, .3, $-.2$ | .3, .3, $-.3$ | .3, .3, $-.4$ |
| $g = 0.9$ | 1.01 | 1.02 | 1.02 | 1.02 | 1.01 | 0.98 |
| 0.8 | 1.02 | 1.04 | 1.05 | 1.04 | 1.02 | 0.98 |
| 0.7 | 1.04 | 1.06 | 1.08 | 1.06 | 1.03 | 0.98 |
| 0.6 | 1.05 | 1.09 | 1.11 | 1.09 | 1.05 | 0.99 |
| 0.5 | 1.07 | 1.11 | 1.15 | 1.11 | 1.07 | 1.00 |

| $k =$ | 5 | | | 10 | | |
|---|---|---|---|---|---|---|
| $\delta_i =$ | .5, .4, .3, .2, .4 | | | .5, .4, $(.3)^2$, $(.2)^3$, $(.1)^3$, .2 | | |
| $\rho_{ij} =$ | .2 $(j \neq 5)$ $-.2$ $(j = 5)$ | .3 $(j \neq 5)$ $-.2$ $(j = 5)$ | .4 $(j \neq 5)$ $-.2$ $(j = 5)$ | .2 $(j \neq 10)$ $-.2$ $(j = 10)$ | .3 $(j \neq 10)$ $-.2$ $(j = 10)$ | .4 $(j \neq 10)$ $-.2$ $(j = 10)$ |
| $g = 0.9$ | 0.99 | 0.99 | 0.99 | 1.01 | 0.98 | 0.99 |
| 0.8 | 0.97 | 0.99 | 0.99 | 1.02 | 1.02 | 0.99 |
| 0.7 | 0.96 | 0.99 | 1.00 | 1.03 | 1.03 | 1.02 |
| 0.6 | 0.96 | 0.99 | 1.00 | 1.06 | 1.06 | 1.02 |
| 0.5 | 0.95 | 1.00 | 1.01 | 1.09 | 1.10 | 1.06 |

helpful negative $\rho_{ij}$ and the harmful positive $\rho_{ij}$ are decreased by the errors in the $x$'s.

When the $g_i$ are very unequal and some $\rho_{ij}$ are negative, $f$ is more erratic, though its behavior still follows the lines indicated above. For instance, if the $g_i$ are such that the harmful correlations remain about the same but the helpful ones are much reduced, $f$ can be substantially less than 1. In the opposite case, the excess of $f$ above 1 can be greater than indicated by Table 4.

So far as they go, the preceding results suggest that the relation $R'^2 = R^2 g_v \bar{g}_w$ may serve as a rough guide to the effect of errors of measurement on the squared multiple correlation coefficient in many applications. The value of $R'^2$ may be up to 10 percent higher than this if most correlations are positive and harmful and the $g$ values exceed 0.7, and up to 25 percent higher if the $g$'s are as low as 0.5. If we are lucky enough to have mainly helpful correlations among the true $X$'s, the errors of measurement tend to reduce them, so that $R'^2$ may be slightly less than $R^2 g_v \bar{g}_w$. These results assume that errors of measurement $e_i$ are independent of one another and of the correct $X$'s.

In the situation in which $e_i$ is correlated with $X_i$ there is a choice of mathematical models. It is not obvious which best describes practical conditions. To specify fully the correlations among the four variables $X_1$, $e_1$, $X_2$, $e_2$ requires six correlation coefficients (with of course some restrictions among their values). If we specify the correlations between $X_1$ and $e_1$, $X_2$ and $e_2$, and $X_1$ and $X_2$, there remain three correlations to be specified. One set of assumptions that seems not unreasonable is that in which

$$e_i = \lambda_i(X_i - \mu_i) + e_i', \tag{7.4}$$

where $\lambda_i$ is the regression of $e_i$ on $X_i$ and the residuals $e_i'$ are uncorrelated with any $X_i$, $X_j$ or $e_j'$ and have variance $\epsilon_i'^2$. In this model the correlations between $e_i$ and $X_j$, $e_j$ and $X_i$, and $e_j$ are only those that arise as reflections of the correlations of $e_i$ with $X_i$ and $e_j$ with $X_j$. Consequences of this model are

$$x_i = X_i + e_i = (1 + \lambda_i)X_i - \lambda_i\mu_i + e_i',$$

$$\delta_i' = \rho_{x_i y} = (1 + \lambda_i)\sigma_i\delta_i / \sqrt{\{(1 + \lambda_i)^2\sigma_i^2 + \epsilon_i'^2\}},$$

$$\rho_{ij}' = (1 + \lambda_i)(1 + \lambda_j)\sigma_i\sigma_j\rho_{ij} / \sqrt{\{(1 + \lambda_i)^2\sigma_i^2 + \epsilon_i'^2\}\{(1 + \lambda_j)^2\sigma_j^2 + \epsilon_j'^2\}}.$$

In the original model with $e_i$ independent of $X_i$ we had $\delta_i' = \sqrt{g_i}\,\delta_i$ and $\rho_{ij}' = \sqrt{g_i g_j}\,\rho_{ij}$. These relations continue to hold in this situation if $g_i$ is redefined as

$$g_i = (1 + \lambda_i)^2\sigma_i^2 / \{(1 + \lambda_i)^2\sigma_i^2 + \epsilon_i'^2\}. \tag{7.5}$$

In this redefinition $g_i$ no longer equals $\sigma_{X_i}^2/\sigma_{x_i}^2$ but $(1+\lambda_i)^2\sigma_{X_i}^2/\sigma_{x_i}^2$. However, if $x_{i1}$, $x_{i2}$ are two repeat estimates of $X_i$ in this model and if $e'_{i1}$ and $e'_{i2}$ are independent, the quantity $\text{Cov}(x_{i1}, x_{i2})/\sigma^2_{x_{i1}}$ still equals this redefined $g_i$ value, so that the usual method of estimation of reliability by repeat measurements still estimates the relevant $g_i$.

For given $\sigma_i^2$, $\epsilon_i^2$, it follows from (7.4) that $\epsilon_i'^2 = \epsilon_i^2 - \lambda_i^2\sigma_i^2$ so that the relevant $g_i$ in (7.5) becomes

$$g_i = (1 + \lambda_i)^2\sigma_i^2 / \{(1 + 2\lambda_i)\sigma_i^2 + \epsilon_i^2\}.$$

This is an increasing function of $\lambda_i$. Thus, under this model positive correlation between $e_i$ and $X_i$ mitigates the effect of errors of measurement, as is not surprising since part of the error $e_i$ is doing the same work as $X_i$. Negative correlation accentuates the effects.

## 8. DISCUSSION

As indicated in the introduction, my interest in this problem was stimulated primarily by two kinds of incidents. Occasionally I see a multiple regression study, based on perhaps 50–80 independent variables involving the measurements of many complex aspects of human behavior, motives and attitudes. The data were obtained by questionnaires filled out in a hurry by apparently disinterested graduate students. The proposal to consign this material at once to the circular file (except that my current wastebasket is rectangular) has some appeal. Second, many multiple regression studies seem to give disappointing results, the discussion in the paper being mainly a search for reasons why $R'^2$ as computed from the data turned out to be so low. Where difficult measurement problems are involved, I have sometimes wondered whether errors of measurement in $y$ or the $x$'s may not supply a substantial part of the explanation.

If it can be trusted to apply to a reasonable number of practical applications, the result that $R'^2$ equals or may somewhat exceed $R^2 g_v \bar{g}_w$ throws some light on both the preceding examples. The result reinforces the need to learn more about the sizes of coefficients of reliability.

Errors of measurement in the $x_i$ also affect other common uses of multiple regressions. For instance, if $x_1$ and $x_2$ are two variables or linear combinations of two sets of variables, each set measuring a distinct aspect of behavior, the investigator sometimes reports (a) the percentage of the variance of $y$ uniquely associated with $x_1$, i.e., the percentage uncorrelated with $x_2$, (b) the percentage uniquely associated with $x_2$, and (c) the percentage that is "common" to $x_1$ and $x_2$. With $\delta_1 = \delta_2 = 0.5$, $\rho = 0.3$, and no errors of measurement, these percentages are (a) 13.5 percent, (b) 13.5 percent, and (c) 11.5 percent, adding to $R^2 = 38.5$ percent. If $g_1 = 0.8$, $g_2 = 1$, the percentages become (a) 10.6 percent, (b) 15.6 percent, and (c) 9.4 percent, and with $g_1 = 0.6$, $g_2 = 1$ they become (a) 7.8 percent, (b) 17.8 percent, and (c) 7.2 percent. In this last case the variable $x_2$ would be reported as having a more important association with $y$ than $x_1$, although the variables are of equal importance if both are measured correctly.

As discussed elsewhere [2], errors of measurement also affect the interpretation of the regression coefficients. In brief, if $\sigma_{ij}$ is the covariance of $X_i$ and $X_j$ and if $\sigma'_{ij} = \sigma_{ij}$ when $i \neq j$ while $\sigma'_{ii} = \sigma_i^2 + \epsilon_i^2$, then in the simplest case when $e_i$ and $X_i$ are independent we have

$$\beta_i' = \beta_i - \sum_{i=1}^{k} \sigma^{ij'} \epsilon_j^2 \beta_j,$$

where $\beta_i'$ is the regression coefficient of $y$ on $x_i$ while $\sigma^{ij'}$ is the inverse of $\sigma'_{ij}$ Separating out the coefficient of $\beta_i$ on the right, we get

$$\beta_i' = \beta_i(1 - \epsilon_i^2 \sigma'_{ii}) - \sum_{j \neq i}^{k} \sigma^{ij'} \epsilon_j^2 \beta_j.$$

Now $\sigma'_{ij} \geq 1/(\sigma_i^2 + \epsilon_i^2)$, so that the coefficient of $\beta_i$ in $\beta_i'$ is at most $\sigma_i^2/(\sigma_i^2 + \epsilon_i^2)$ $= g_i$. Thus the direct effect of an error in $X_i$ on $\beta_i$ is to decrease its absolute value to $g_i \beta_i$ or something less, but $\beta_i'$ also receives contributions from errors of measurement in any other $X_j$ that is correlated with $X_i$. Even if such errors occur in only $X_i$, they can affect the values of all the $\beta_j'$. By working a few examples with varying $g_i$ and $\beta_i$, it becomes evident that interpretation of the $\beta_i'$ as if they were the $\beta_i$ can become quite misleading unless all the $g_i$ are high.

## REFERENCES

[1] Cochran, W. G., "On the Performance of the Linear Discriminant Function," *Bulletin de l'Institut International de Statistique*, 39, 2 (1961), 435–47.

[2] ———, "Errors of Measurement in Statistics," *Technometrics*, 10 (1968) 637–66.

[3] Coleman, J. B., *et al.*, "Equality of Educational Opportunity," Washington: U. S. Government Printing Office, 1966.

[4] Ferber, R., "The Reliability of Consumer Surveys of Financial Holdings: Time Deposits," *Journal of the American Statistical Association*, 60 (1965), 148–63.

[5] Gray, P. G., "The Memory Factor in Social Surveys," *Journal of the American Statistical Association*, 50 (1955), 344–63.

[6] Guilford, J. P., *Personality*, New York: McGraw-Hill, 1959.

[7] Hayner, N. S., *et al.*, "The One-Hour Glucose Tolerance Test," National Center for Health Statistics, Series 2, No. 3, 1963.

[8] Kinsey, A. C., Pomeroy, W. B., and Martin, C. E., *Sexual Behavior in the Human Male*, Philadelphia: W. B. Saunders, 1948.

[9] Kish, L. and Lansing, J. B., "Response Errors in Estimating the Value of Homes," *Journal of the American Statistical Association*, 49 (1954), 520–38.

[10] Lindley, D. V., "Regression Lines and the Linear Functional Relationship," *Journal of the Royal Statistical Society*, B, 9 (1947), 218–24.

[11] National Center for Health Statistics, "Measurement of Personal Health Expenditures," Series 2, No. 2, 1963.

[12] Sells, S. B., "Evaluation of Psychological Measures Used in the Health Examination Survey," National Center for Health Statistics, Series 2, No. 15, 1966.

[13] Sulzmann, J. H., *et al.*, "Data Quoted in Comparison of Two Vision-Testing Devices," National Center for Health Statistics, Series 2, No. 1, 1966.

# *Analysis of Variance*

## W. G. COCHRAN AND C. I. BLISS*

### 1. Introduction

The primary objectives of this chapter are to describe some of the effects of failures in the assumptions that are made in the standard analyses of variance and to give advice on methods of coping with these failures. As a preliminary some examples of the three commonest types of analysis — the one-way, two-way, and Latin-square classifications — will be presented so as to provide a brief review of the technique.

The analysis of variance is a flexible tool with many uses. The writer once heard Fisher describe the analysis-of-variance table as simply a convenient way of doing the bookkeeping for the calculations needed in the preliminary stages of interpreting the results of an experiment.

In the analysis-of-variance table we start with the total sum of squares of deviations of the observations from their mean, $\sum (y - \bar{y})^2$. This quantity is broken down into a number of component sums of squares, SS, one for each of the classifications, plus a residual, or remainder, sum of squares. There is a corresponding breakdown of the total degrees of freedom, $N - 1$, into degrees of freedom, DF, for each of the component sums of squares. When each sum of squares is divided by its degrees of freedom, a column of mean squares, MS, is produced. The most useful outcomes of this calculation are the following:

1. The residual mean square is an unbiased estimate of the error variance per observation in the experiment. Its square root, the standard deviation per observation, is required for computing the standard errors of differences between the treatment means and for testing the significance of these differences.

2. The treatments mean square is a measure of the amount of variation that exists between the treatment means. The ratio of the treatments mean square to the error mean square, usually denoted by $F$, furnishes a test of the null hypothesis that there are no real differences among the effects of the treatments.

3. Often the investigator chooses the treatments so that the experiment will supply answers to a number of separate and distinct questions.

* The final draft of Sections 1 to 9 was prepared by W. G. Cochran; that of Sections 10 to 15, by C. I. Bliss.

In such case the treatments sum of squares can itself be subdivided into a number of parts, each related to one of the questions. Rules for doing this, with illustrations, are presented in Section 5 and applied in Sections 10 and 11.

4. The other classifications in the experiment — for example, animals and dates of testing in a Latin square — are usually employed in the hope that they will remove major sources of variation and thus increase the precision of the comparisons of treatment means. The ratio of the mean square for any such classification to the error mean square indicates the extent to which the classification has been successful in this respect.

## 2. One-way classification

One-way classifications are used in experiments in which the experimental animals or other units have been arranged by randomization (see pp. 4–6) in groups of size $n$, each group receiving a different treatment. Suppose that there are $k$ treatments. The symbol $y_i$ stands for an individual response, where $i$ denotes the treatment ($i$ goes from 1 to $k$). In this one-way classification the treatments are the classes.

The assumptions made as a background to the analysis of variance are usually expressed as follows:

$$y_i = \mu + \alpha_i + e_i \tag{1}$$

The symbol $\mu$ represents the mean response over the whole experiment, and $\alpha_i$ represents the effect of the $i$th treatment. The term $e_i$ stands for the combined effect of all other sources of variation that make one animal differ in response from other animals receiving the same treatment. Thus, $e_i$ represents the contribution of experimental errors. If we can envisage a repetition of the experiment under similar conditions, the terms $\mu$ and $\alpha_i$ remain unchanged, but the $e_i$ vary from one repetition to another. In the standard analysis of variance the $e_i$ are assumed to follow a normal distribution with mean 0 and variance $\sigma^2$ and to be independent of one another.

The analysis-of-variance table depends on an algebraic relation that holds among sums of squares, with a corresponding relation among degrees of freedom. Let $\bar{y}_t$ denote the average of the $n$ observations for a treatment and $\bar{y}$ the overall mean. Then,

$$\text{SS:} \quad \sum (y - \bar{y})^2 = n \sum (\bar{y}_t - \bar{y})^2 + \sum (y - \bar{y}_t)^2 \tag{2}$$

$$\text{DF:} \quad \underset{\text{total}}{kn - 1} = \underset{\text{treatments}}{k - 1} + \underset{\text{residual = error}}{k(n - 1)}$$

Let us consider first the residual sum of squares. From Equation 1 it follows that

$$y_i - \bar{y}_t = e_i - \bar{e}_t$$

since the symbols $\mu$ and $\alpha_i$ cancel out. Hence, within a given treatment we have $\sum (y - \bar{y}_t)^2 = \sum (e_i - \bar{e}_t)^2$. If we divided by $n - 1$, we would get the usual estimate of $\sigma^2$ for a sample of size $n$. Thus, $\sum (y - \bar{y}_t)^2$ estimates $(n - 1)\sigma^2$. When we add over the $k$ treatments, we get a pooled estimate of $k(n - 1)\sigma^2$, so that on dividing by $k(n - 1)$ the residual mean square is an unbiased estimate of $\sigma^2$; it is appropriately called the error mean square.

Now let us consider the treatments mean square. Its average value over repetitions of the experiment may be shown to be

$$\frac{n \sum (\alpha_i - \bar{\alpha})^2}{k - 1} + \sigma^2 \tag{3}$$

This relation is quite informative. If all the $\alpha_i$ are equal (no differences in the effects of the treatments), the treatments mean square, like the

TABLE 1 One-Way Classification for Experiment on Uterine Weights of Immature Rats Four Days after Stilbestrol*

Data, totals, and means:

$$y = \log \text{ weight} - 1.4$$
$$x = \log \text{ dose}$$

| $x$: | 0.30 | 0.45 | 0.60 | 0.75 | 0.90 | |
|---|---|---|---|---|---|---|
| $y$: | −0.038 | 0.083 | 0.219 | 0.261 | 0.308 | |
| | 0.100 | 0.138 | 0.270 | 0.306 | 0.396 | |
| | 0.122 | 0.197 | 0.181 | 0.206 | 0.406 | |
| | 0.139 | 0.197 | 0.237 | 0.341 | 0.456 | |
| | 0.149 | 0.232 | 0.150 | 0.354 | 0.386 | |
| | 0.214 | 0.251 | 0.183 | 0.390 | 0.337 | |
| | 0.144 | 0.173 | 0.223 | 0.309 | 0.354 | |
| Total $T_t$: | 0.830 | 1.271 | 1.463 | 2.167 | 2.643 | $8.374 = T$ |
| Mean $\bar{y}_t$: | 0.119 | 0.182 | 0.209 | 0.310 | 0.378 | $0.239 = \bar{y}$ |

Calculations:

$\sum y^2 = 2.4059$, $C_m = T^2/nk = 8.374^2/35 = 2.0035$

Total SS $= \sum (y - \bar{y})^2 = \sum y^2 - C_m = 0.4024$

Between-doses SS $= \dfrac{\sum T_t^2}{n} - C_m = \dfrac{(0.830^2 + \cdots + 2.643^2)}{7} - 2.0035 = 0.3002$

Analysis of variance:

| Term | DF | SS | MS | F |
|---|---|---|---|---|
| Between doses | $4 = k - 1$ | 0.3002 | 0.0750 | 22.0 ($P < 0.01$) |
| Residual = error | $30 = k(n - 1)$ | 0.1022 | 0.00341 | |
| Total | $34 = kn - 1$ | 0.4024 | | |

* Data of Lee et al. (1942).

error mean square, is an unbiased estimate of $\sigma^2$, so that the ratio $F = $ (treatments MS)/(error MS) should be about 1, apart from sampling fluctuations. As the $\alpha_i$ diverge from one another, the treatments mean square tends to become large because of the first term in Equation 3, causing $F$ to become large.

Table 1 gives a numerical example. The data are the log uterine weights of immature rats after injection with five different doses of stilbestrol. In the analysis-of-variance calculations in the table, Equation 2 for the sums of squares is expressed in a form that is quicker for calculations. First we compute the correction for the mean, $C_m = T^2/kn$, where $T$ is the grand total. Then we rewrite Equation 2 as

$$\underset{\text{total SS}}{\sum y^2 - C_m} = \underset{\text{treatments SS}}{\sum T_t^2/n - C_m} + \underset{\text{error SS}}{\text{(found by subtraction)}}$$

TABLE 2   One-Way Classification (Unequal Sample Sizes) for Experiment on Comb Growth of Capons Injected for Five Days with Androsterone*

Data and calculations:

$$y = \log \text{[length + height (mm)]}$$
$$x = 1 + \log \text{dose}$$

| | | | | | | |
|---|---|---|---|---|---|---|
| $n_t$: | 10 | 10 | 8 | 9 | 9 | $46 = N = \sum n_t$ |
| $n_t x$: | 4.00 | 7.00 | 8.00 | 10.62 | 11.70 | $41.32 = \sum n_t x$ |
| $x$: | 0.40 | 0.70 | 1.00 | 1.18 | 1.30 | |
| $y$: | 0.30 | 0.48 | 0.48 | 0.70 | 0.81 | |
| | 0.30 | 0.54 | 0.60 | 0.78 | 0.93 | |
| | 0.30 | 0.54 | 0.78 | 0.81 | 0.93 | |
| | 0.40 | 0.60 | 0.81 | 0.85 | 0.98 | |
| | 0.48 | 0.65 | 0.81 | 0.90 | 1.00 | |
| | 0.54 | 0.65 | 0.82 | 0.90 | 1.00 | |
| | 0.54 | 0.65 | 0.85 | 0.94 | 1.02 | |
| | 0.60 | 0.70 | 0.85 | 1.00 | 1.02 | |
| | 0.60 | 0.78 | | 1.04 | 1.04 | |
| | 0.74 | 0.81 | | | | |
| $T_t$: | 4.80 | 6.40 | 6.00 | 7.92 | 8.73 | $33.85 = T$ |
| $T_t^2/n_t$: | 2.3040 | 4.0960 | 4.5000 | 6.9696 | 8.4681 | $26.3377 = \sum T_t^2/n_t$ |

$$C_m = T^2/N = 33.85^2/46 = 24.9092$$

Analysis of variance:

| Term | DF | SS | MS | F |
|---|---|---|---|---|
| Doses | 4 | $\sum T_t^2/n_t - C_m = 1.4285$ | 0.3571 | $25.76 \ (P < 0.01)$ |
| Error | 41 | By subtraction $= 0.5684$ | 0.01386 | |
| Total: | 45 | $\sum y^2 - C_m = 1.9969$ | | |

* Data of Emmens (1939b).

Having found the total sum of squares and the treatments sum of squares, we obtain the error sum of squares by subtraction. In the analysis of variance the $F$ value of 22.0 with 4 and 30 DF is highly significant. Actually, in this experiment the value of $F$ is of only mild interest. The objective is to express the uterine weight as a function of the log dose, so we proceed to fit this response curve or, in technical terms, to examine the regression of the treatment means on log dose. Methods of doing this are presented in Section 10.

Other data to be used in Section 10 come from Table 2, an example of a one-way-classification experiment with unequal sample sizes in the classes. The only change required in the calculations is that in finding the treatments sum of squares. The square $T_t^2$ of any treatment total is divided by the sample size $n_t$ for that treatment, the quotient being written down, as shown in the table. The error sum of squares and the degrees of freedom are again found by subtraction in the analysis-of-variance table.

### 3. Two-way classification

Two-way classifications are used in experiments in which the units, say animals, are grouped into replications, and each treatment is applied to a different member within any replication, the member being selected by some randomization process (see pp. 4–9). Usually the purpose in forming replications is to increase the precision of the comparisons among treatments. In a replication one tries to group units expected to show similar responses. For instance, if responses may be expected to be somewhat different in male and female animals and in old and young animals, one would try to have one replication composed of young males, another of young females, and so on.

If $i$ denotes the treatment ($i$ goes from 1 to $k$) and $j$ the replication or group ($j$ goes from 1 to $n$), the model is

$$y_{ij} = \mu + \alpha_i + \beta_j + e_{ij} \tag{4}$$

There are two key assumptions, as follows.

1. Treatment and group effects are *additive*. This implies that the difference between the effects of two treatments is, apart from experimental errors, the same in all groups. There is no natural reason why additivity should obtain. In the first paper in which an analysis of variance appeared in print (Fisher and Mackenzie, 1923) the section introducing this tool was entitled "Analysis of Variation on the Sum Basis." In the experiment the crop was potatoes, the treatments were fertilizers, and the groups were different potato varieties. The authors write: "In order to make clear the method of the analysis of variation,

we will first carry through the process on the assumption that the yield to be expected from a given variety grown with a given manurial treatment is the sum of two quantities, one depending on the variety and the other on the manure. This assumption is evidently an unsatisfactory one." The authors claimed that a product formula, in which the *ratio* of the yields of two treatments was assumed constant from group to group, would be more reasonable. The effects of nonadditivity and methods of handling it are discussed in Sections 6 and 9.

2. As in the one-way classification, the residuals $e_{ij}$ are assumed normally and independently distributed with means 0 and variance $\sigma^2$, that is, NID$(0, \sigma^2)$.

In terms of the observed responses the additive relationship $y$ may be expressed as follows:

$$y = \underset{\text{mean}}{\bar{y}} + \underset{\text{treatment}}{(\bar{y}_t - \bar{y})} + \underset{\text{group}}{(\bar{y}_g - \bar{y})} + \underset{\text{residual = error}}{(y - \bar{y}_t - \bar{y}_g + \bar{y})} \qquad (5)$$

where $\bar{y}_t$ denotes, as before, the average of the $n$ observations for a treatment, $\bar{y}_g$ denotes the average of the $k$ observations in a group or replicate, and $\bar{y}$ denotes the overall mean. Any individual response is expressed as the sum of elements due to the overall mean, the treatment to which it has been subjected, the group in which it lies, and a residual, or error, term. If the additive model, Equation 4, holds, it is easily shown that the residual term above equals $e_{ij} - \bar{e}_t - \bar{e}_g + \bar{e}$ and hence is composed entirely of genuine "error" components. If the model does not hold, the computed residuals are inflated by nonadditivity.

Table 3 describes an experiment of this type, in which the groups are breeds of chicks. The treatments are again increasing doses of stilbestrol, the response being the log of the relative oviduct weight. The table gives the elements from Equation 5. On a desk machine these elements are rather tedious to find and write down, but since electronic computers can easily print all the individual residuals, printouts of residual elements are becoming a routine matter. Inspection of the residual elements for signs of gross errors, nonadditivity, and anything else that looks suspicious is recommended, as will be discussed in Section 6.

With $k$ treatments and $n$ groups the usual method of computing the sums of squares in the analysis of variance of $\sum (y - \bar{y})^2$ is as follows, where $C_m$ is equal to $T^2/kn$, as before.

$$\text{SS:} \qquad \sum \sum y^2 - C_m = (\sum T_t^2/n - C_m) + (\sum T_g^2/k - C_m)$$

$$+ \text{(found by subtraction)} \qquad (6)$$

$$\text{DF:} \qquad \underset{\text{total}}{kn - 1} = \underset{\text{doses}}{(k - 1)} + \underset{\text{breeds}}{(n - 1)} + \underset{\text{error}}{(k - 1)(n - 1)}$$

TABLE 3   Two-Way Classification for Experiment on Oviduct Response of Four Breeds of Chicks to Stilbestrol*

Data, totals, means, and deviations:

$y = \log [100$ oviduct wt. (mg)/body wt. (g)]

| Breed | Dose, µg | | | | Breed total $T_g$ | Breed mean $\bar{y}_g$ | $\bar{y}_g - \bar{y}$ |
|---|---|---|---|---|---|---|---|
| | 50 | 100 | 200 | 400 | | | |
| WL | 0.74 | 0.85 | 1.16 | 1.43 | 4.18 | 1.045 | −0.030 |
| WR | 0.76 | 0.94 | 1.20 | 1.38 | 4.28 | 1.070 | −0.005 |
| RI | 0.68 | 0.92 | 1.24 | 1.48 | 4.32 | 1.080 | 0.005 |
| Hy | 0.64 | 1.01 | 1.32 | 1.45 | 4.42 | 1.105 | 0.030 |
| Treat. total $T_t$: | 2.82 | 3.72 | 4.92 | 5.74 | 17.20 = $T$ | | 0.000 |
| Treat. mean $\bar{y}_t$: | 0.705 | 0.930 | 1.230 | 1.435 | 1.075 = $\bar{y}$ | | |
| $\bar{y}_t - \bar{y}$: | −0.370 | −0.145 | 0.155 | 0.360 | 0.000 | | |

Deviation elements:

| Breed | Dose, µg | | | | Dose, µg | | | |
|---|---|---|---|---|---|---|---|---|
| | 50 | 100 | 200 | 400 | 50 | 100 | 200 | 400 |
| | Element $\bar{y}$ for mean | | | | Element $\bar{y}_t - \bar{y}$ for dose | | | |
| WL | 1.075 | 1.075 | 1.075 | 1.075 | −0.370 | −0.145 | 0.155 | 0.360 |
| WR | 1.075 | 1.075 | 1.075 | 1.075 | −0.370 | −0.145 | 0.155 | 0.360 |
| RI | 1.075 | 1.075 | 1.075 | 1.075 | −0.370 | −0.145 | 0.155 | 0.360 |
| Hy | 1.075 | 1.075 | 1.075 | 1.075 | −0.370 | −0.145 | 0.155 | 0.360 |
| | Element $\bar{y}_g - \bar{y}$ for breeds | | | | Residual $y - \bar{y}_t - \bar{y}_g + \bar{y}$ | | | |
| WL | −0.030 | −0.030 | −0.030 | −0.030 | 0.065 | −0.050 | −0.040 | 0.025 |
| WR | −0.005 | −0.005 | −0.005 | −0.005 | 0.060 | 0.015 | −0.025 | −0.050 |
| RI | 0.005 | 0.005 | 0.005 | 0.005 | −0.030 | −0.015 | 0.005 | 0.040 |
| Hy | 0.030 | 0.030 | 0.030 | 0.030 | −0.095 | 0.050 | 0.060 | −0.015 |

Analysis of variance:

| Term | DF | SS | MS | F |
|---|---|---|---|---|
| Breeds | 3 | 0.0074 | 0.0025 | 0.7 |
| Doses | 3 | 1.2462 | 0.4154 | 110.0 ($P < 0.01$) |
| Error | 9 | 0.0340 | 0.00378 | |
| Total: | 15 | 1.2876 | | |
| $C_m$: | 1 | 18.4900 | | |

* Data of Dorfman and Dorfman (1948).

Alternatively, since we have calculated the elements for doses, breeds, and residuals, the sums of squares for doses, breeds, and error can be found by computing the sums of squares of the elements in the appropriate tables. This calculation may help in understanding both the additive model and the additive sum-of-squares relationship that underlie an analysis of variance. There is no indication of differences in response among breeds ($F = 0.7$), but there are large differences among doses, as might be expected.

## 4. The Latin square

The Latin square is a plan that enables one to group units, say animals, simultaneously in two different ways, so that the treatments are balanced with respect to two potential sources of variability (see pp. 14–16). In the numerical example Table 4, which compares the effects of four doses of insulin on the blood sugar of four rabbits, the first grouping was the individual rabbit, each rabbit serving as his own control, and the second grouping was days, each rabbit being injected with a different dose on four different days.

If $i$ represents the treatment (dose), $j$ the row (date), and $k$ the column (rabbit), the additive model is carried one step further,

$$y_{ijk} = \mu + \alpha_i + \beta_j + \gamma_k + e_{ijk} \tag{7}$$

the $e_{ijk}$ being NID$(0, \sigma^2)$. For our data the additive relation is written as follows:

$$y = \bar{y} + \underset{\text{treatment}}{(\bar{y}_t - \bar{y})} + \underset{\text{row}}{(\bar{y}_r - \bar{y})} + \underset{\text{column}}{(\bar{y}_c - \bar{y})}$$
$$+ \underset{\text{residual = error}}{(y - \bar{y}_t - \bar{y}_c - \bar{y}_r + 2\bar{y})}$$

The elements corresponding to the terms on the right-hand side of this equation are shown in Table 4. In a natural extension of the corresponding result, Equation 6, for the two-way classification, the breakdown of $\sum (y - \bar{y})^2$ in an $n \times n$ Latin square is as follows:

$$\underset{\text{total}}{\sum y^2 - C_m} = \underset{\text{treatments}}{(\sum T_t^2/n - C_m)} + \underset{\text{rows}}{(\sum T_r^2/n - C_m)}$$
$$+ \underset{\text{columns}}{(\sum T_c^2/n - C_m)}$$
$$+ \text{ residual (found by subtraction)}$$

Each sum of squares on the right-hand side can also be found by squaring and adding the elements in the corresponding table of elements. The degrees of freedom for error are $(n - 1)(n - 2)$. In the analysis of variance the mean squares both for dates and for rabbits are substantially greater than the error mean square, indicating that both groupings helped to increase precision, and the doses mean square is large.

TABLE 4  Latin Square for Experiment on Effects of Four Doses of Insulin on Blood Sugar of Rabbits*

Data, totals, and means:

$y$ = mg/(100 cm³ blood sugar) 50 min after injection.

A, B, C, and D denote the four doses of insulin.

Initial variates $y_{ijk}$

| Day of month | Dose and $y$ (rabbit no.) 1 | 2 | 3 | 4 | Row $T_r$ | $\bar{y}_r$ | Treatment Dose† | $T_t$ | $\bar{y}_t$ |
|---|---|---|---|---|---|---|---|---|---|
| 23 | B 24 | C 46 | D 34 | A 48 | 152 | 38 | A S₁ | 224 | 56 |
| 25 | D 33 | A 58 | B 57 | C 60 | 208 | 52 | B S₂ | 160 | 40 |
| 26 | A 57 | D 26 | C 60 | B 45 | 188 | 47 | C U₁ | 212 | 53 |
| 27 | C 46 | B 34 | A 61 | D 47 | 188 | 47 | D U₂ | 140 | 35 |
| $T_c$: | 160 | 164 | 212 | 200 | 736 = $T$ | | | 736 | |
| $\bar{y}_c$: | 40 | 41 | 53 | 50 | | 46 = $\bar{y}$ | | | 46 |

Deviation elements:

| Day of month | Rabbit no. 1 | 2 | 3 | 4 | Row total | Treat. total | Rabbit no. 1 | 2 | 3 | 4 | Row tot. | Treat. total |
|---|---|---|---|---|---|---|---|---|---|---|---|---|
| | Row element $\bar{y}_r - \bar{y}$ | | | | | | Column element $\bar{y}_o - \bar{y}$ | | | | | |
| 23 | −8 | −8 | −8 | −8 | −32 | A 0 | −6 | −5 | 7 | 4 | 0 | A 0 |
| 25 | 6 | 6 | 6 | 6 | 24 | B 0 | −6 | −5 | 7 | 4 | 0 | B 0 |
| 26 | 1 | 1 | 1 | 1 | 4 | C 0 | −6 | −5 | 7 | 4 | 0 | C 0 |
| 27 | 1 | 1 | 1 | 1 | 4 | D 0 | −6 | −5 | 7 | 4 | 0 | D 0 |
| Col. total: | 0 | 0 | 0 | 0 | 0 | | 0 | −24 | −20 | 28 | 16 | 0 | 0 |
| | Treatment element $\bar{y}_t - \bar{y}$ | | | | | | Residual $y - \bar{y}_t - \bar{y}_r - \bar{y}_o + 2\bar{y}$ | | | | | |
| 23 | −6 | 7 | −11 | 10 | 0 | A 40 | −2 | 6 | 0 | −4 | 0 | A 0 |
| 25 | −11 | 10 | −6 | 7 | 0 | B −24 | −2 | 1 | 4 | −3 | 0 | B 0 |
| 26 | 10 | −11 | 7 | −6 | 0 | C 28 | 6 | −5 | −1 | 0 | 0 | C 0 |
| 27 | 7 | −6 | 10 | −11 | 0 | D −44 | −2 | −2 | −3 | 7 | 0 | D 0 |
| Col. total: | 0 | 0 | 0 | 0 | 0 | 0 | 0 | 0 | 0 | 0 | 0 | 0 |

Analysis of variance:

| Term | DF | SS | MS | F |
|---|---|---|---|---|
| Dates (rows) | 3 | 408 | 136.0 | 3.81 |
| Rabbits (columns) | 3 | 504 | 168.0 | 4.71 |
| Doses | 3 | 1,224 | 408.0 | 11.43 ($P < 0.01$) |
| Error | 6 | 214 | 35.7 | |
| Total: | 15 | 2,350 | | |
| $C_m$: | 1 | 33,856 | | |

* Data of Young and Romans (1948).

† S, standard preparation; U, unknown preparation; subscripts 1 and 2 denote low and high dose respectively.

## 5. Comparisons among the treatment means

The analysis of variance provides the value $s^2$ of the experimental error mean square with its degrees of freedom. From $s^2$ we obtain $s$, the standard deviation per observation. When we examine and compare the treatment means in more detail, a few additional rules are useful.

If $\bar{y}_i$ denotes a typical treatment mean, most of the quantities of interest in a summary of the results of an experiment are of the form $\sum c_i \bar{y}_i$, where the $c_i$ are numerical constants whose sum is zero. For example, in the difference between two means, $\bar{y}_1 - \bar{y}_2$, we have $c_1 = 1$ and $c_2 = -1$. If treatments 1 and 2 are of one kind and treatments 3, 4, and 5 are of another kind, one might be interested in a comparison of the average performances of each kind of treatment. This is the quantity

$$L = \frac{\bar{y}_1 + \bar{y}_2}{2} - \frac{\bar{y}_3 + \bar{y}_4 + \bar{y}_5}{3} \qquad (8)$$

Here $c_1 = c_2 = 1/2$ and $c_3 = c_4 = c_5 = -1/3$. A quantity $L = \sum c_i \bar{y}_i$ with $\sum c_i = 0$ is called a *comparison*, or *contrast*, among the treatment means.

*Rule 1.* In a standard design (one-way with equal sample sizes, two-way, or Latin square) the estimated standard error of $L$ is

$$s_L = \frac{s}{\sqrt{n}} \sqrt{\sum c_i^2}$$

where $n$ is the sample size for each treatment mean. The degrees of freedom are those in $s$. In the example above, Equation 8, with four replications, or groups,

$$s_L = \frac{s}{\sqrt{4}} \sqrt{\left(\frac{1}{4} + \frac{1}{4} + \frac{1}{9} + \frac{1}{9} + \frac{1}{9}\right)} = \frac{s}{2}\sqrt{\frac{5}{6}} = 0.456s$$

*Rule 2.* The quantity $nL^2/\sum c_i^2$ is a component of the treatments sum of squares, with 1 DF. The remainder,

$$\text{treatments SS} - nL^2/\sum c_i^2$$

is also a component of the treatments sum of squares, with $k - 2$ DF. The mean square for each component may be tested by an $F$ test, the degrees of freedom being 1 and $f$ for the first component and $k - 2$ and $f$ for the second, where $f$ is the number of error degrees of freedom in the analysis of variance.

Two examples will be given. In the first we suppose the treatments to consist of three agents that are qualitatively similar and a fourth treatment that is a control (no agent). If $\bar{\bar{y}}_3 = (\bar{y}_1 + \bar{y}_2 + \bar{y}_3)/3$ represents the mean of the three agents, the comparison

$$L = \bar{\bar{y}}_3 - \bar{y}_4$$

estimates their *average* effect. Here $\sum c_i^2 = 4/3$; so $3nL^2/4$ is a part of the treatments sum of squares with 1 DF. The remaining part has 2 DF; as might be anticipated, it may be shown to satisfy the following identity:

$$\text{treatments SS} - 3nL^2/4 = n \sum_{i=1}^{3} (\bar{y}_i - \bar{y}_3)^2$$

Thus the remaining part measures differences in effect among the three agents. If the agents are all effective to about the same degree, we may expect to find a significant $F$ for the first component but a nonsignificant $F$ for the second (2 DF) component.

The second example is similar; it occurs when the treatments represent different amounts $x_i$ of some agent. If the relation between $\bar{y}_i$ and $x_i$ appears linear when the two are plotted, we may fit a linear regression of $\bar{y}_i$ on $x_i$. By the usual formula, the regression coefficient is

$$b = \frac{\sum (\bar{y}_i - \bar{y})(x_i - \bar{x})}{\sum (x_i - \bar{x})^2} = \frac{\sum \bar{y}_i(x_i - \bar{x})}{\sum (x_i - \bar{x})^2}$$

since, as may be shown, the term in $\bar{y}$ is 0. Clearly, $b$ is also a comparison among the $\bar{y}_i$, with $c_i = (x_i - \bar{x})/\sum (x_i - \bar{x})^2$ and $\sum c_i^2 = 1/\sum (x_i - \bar{x})^2$. Hence, by Rule 2, the relation

$$nb^2 \sum (x_i - \bar{x})^2 = n[\sum (\bar{y}_i - \bar{y})(x_i - \bar{x})]^2/\sum (x_i - \bar{x})^2$$

is a component of the treatments sum of squares, with 1 DF, representing the slope of the regression of $\bar{y}_i$ on $x_i$.

Now, if the treatment means lay *exactly* on this line, we should have

$$\bar{y}_i = \bar{y} + b(x_i - \bar{x})$$

so that

$$\text{treatments SS} = n \sum (\bar{y}_i - \bar{y})^2 = nb^2 \sum (x_i - \bar{x})^2$$
$$= \text{slope SS (1 DF)}$$

Hence, the *remainder* of the treatments sum of squares, with $k - 2$ DF, must measure the extent to which the treatment means $\bar{y}_i$ deviate from the straight-line relationship. Consequently, the $F$ test of this remainder mean square is used as a test of the linearity of regression; see Section 10.

*Rule 3.* Two comparisons, $L = \sum c_i \bar{y}_i$ and $L' = \sum c_i' \bar{y}_i$, are called *orthogonal* if $\sum c_i c_i' = 0$.

If this relation holds, then $nL^2/\sum c_i^2$ and $nL'^2/\sum c_i'^2$ are *independent* parts of the treatments sum of squares, each with 1 DF. The remainder of the treatments sum of squares, with $k - 3$ DF, is independent of these two components.

It follows that if we construct $k - 1$ mutually orthogonal comparisons, $L, L', L'', \ldots$ (that is, with every pair orthogonal), their combined sum of squares,

$$\frac{nL^2}{\sum c_i^2} + \frac{nL'^2}{\sum c_i'^2} + \frac{nL''^2}{\sum c_i''^2} + \cdots$$

is exactly equal to the treatments sum of squares.

*Note:* In Rules 2 and 3 we often calculate a comparison from the treatment *totals*, that is, $L = \sum c_i T_i$. In this event the sum of squares is $L^2/(n \sum c_i^2)$.

The Latin-square experiment in Table 4 furnishes an example of a complete subdivision of the treatments sum of squares. Since the treatments were two doses of a standard (S) and two doses of an unknown (U) preparation of insulin, an orthogonal set of comparisons that is meaningful is that given in Table 5.

TABLE 5   Orthogonal Subdivision of Treatments Sum of Squares Applied to Results in Table 4 (Initial Variates)

| Treatment totals | A $S_1$ 224 | B $S_2$ 160 | C $U_1$ 212 | D $U_2$ 140 | $L$ | $\sum c_i^2$ | $L^2/(4 \sum c_i^2)$ |
|---|---|---|---|---|---|---|---|
| $U - S, a$ | $-1$ | $-1$ | $+1$ | $+1$ | $-32$ | 4 | 64 |
| Average slope, $b$ | $-1$ | $+1$ | $-1$ | $+1$ | $-136$ | 4 | 1,156 |
| Difference in slopes, $ab$ | $+1$ | $-1$ | $-1$ | $+1$ | $-8$ | 4 | 4 |
| | | | | | | | 1,224* = treatments SS |

* Same as doses sum of squares in Table 4.

Since the error mean square $s^2$ is 35.7 with 6 DF (from Table 4), there is no indication of difference either in average response or in slope between the unknown and the standard. The significant $F$ value in Table 4 can be attributed to the difference in response to the higher and lower doses.

This example illustrates a $2 \times 2$ *factorial* experiment, in which the factors are two preparations each at two dose levels. Further examples of the breakdown of the treatment sum of squares in factorial experiments are given in Section 12.

## 6. Failures in the assumptions

As already noted, most of the assumptions in the analysis of variance concern the residuals. These are assumed to have zero means and constant variance and to be normally and independently distributed.

In addition, the assumption of additivity of treatment and group effects is made in the two-way and Latin-square designs.

In preparation for this paper the writer tried to recall past experiences in which analyses of variance gave badly misleading results and to ask himself, "What was the reason?" In recollection the culprits, in the order of seriousness of mistakes or of confusion that they caused, were as follows (other statisticians might report different lists).

1. The arithmetic was wrong, or the formulas were misunderstood. The remedy is, of course, careful checking of both arithmetic and formulas. In the analysis of variance, when obtained by computer, an excellent check on the sums of squares is agreement between the error sum of squares, found by subtraction as indicated in the preceding sections, and that found as the sum of squares of the residual elements. Mistakes by a factor of $\sqrt{10}$ made in reading square roots also are common.

2. One or more gross errors were present. By "gross error" is meant a recorded value that is badly incorrect for any reason, perhaps because a measurement was calculated, recorded, or copied wrongly, or because something went wrong in applying the treatment on this occasion. A gross error seriously distorts the mean of the treatment that it affects. It can also greatly inflate the error mean square, so that the standard errors computed for unaffected treatment comparisons are too large. General alertness in looking over one's experimental results naturally helps. In the two-way and Latin-square designs it is hard to detect gross errors by eye inspection of the results, since an unusually low individual value might represent a combination of a poor treatment and a poor group. For this reason the residual elements are the place to inspect for gross errors. One method of detection will be presented in Section 7.

3. The error variance $\sigma^2$ was not constant throughout the experiment. The available research suggests that heterogeneity of error variance does not greatly disturb $F$ tests of the treatments sums of squares in the standard designs except sometimes in a one-way classification with markedly unequal sample sizes for different treatments (Scheffé, 1959). However, $t$ tests and confidence limits for individual comparisons among treatment means can be more seriously affected. For instance, if two of the treatments are more stable in their effects than the other treatments, so that the error variance is small for these two treatments, use of the pooled $s$ from the analysis of variance will overestimate the standard error of the difference between these two treatments. Reasons for non-constancy of error variance and methods of handling it are discussed in Section 8.

4. The errors $e_{ij}$ on different units were correlated. This can have serious effects. In a one-way classification, if the terms $e_i$ for different

replicates of a treatment have correlation $\rho$, the variance of the treatment mean $\bar{y}_t$ is $\sigma^2[1 + (n - 1)\rho]/n$ instead of $\sigma^2/n$, while the within-treatment mean square $\sum (y - \bar{y}_t)^2/(n - 1)$ estimates $\sigma^2(1 - \rho)$ instead of $\sigma^2$. If $\rho$ is positive, it follows that the pooled value $s^2/n$ underestimates the true variance of a treatment mean.

Errors that are positively correlated within a treatment can arise in practice for several reasons. If randomization is not used, the different replicates of treatment 1 being applied and processed first for convenience, there may be positive correlations because the most responsive units (the animals) were picked out first or because some other aspect of the conditions of application of the treatment was unusually favorable to a high response on all replicates of that treatment. Examples occur in agricultural crop experiments in which the replicates of treatment 1 are weighed on a day when the crop is quite wet and later treatments are weighed on a dryer day, and in experiments in which the measurement involves human judgment, treatment 1 being scored by one judge and other treatments being scored by other judges with different levels or standards.

If such positive correlations are present in the data, we underestimate the standard errors of treatment means and obtain too many apparently significant results. Since positive correlations are not detected or revealed in the statistical analysis, the principal safeguard is the use of randomization at any stage in the processing at which systematic sources of error may enter. Further, if several measurers are involved, each should measure the same number of replicates of all treatments, and he should not know which treatment he is measuring.

5. In the two-way and more complex classifications the treatment and group effects may not be additive. Roughly speaking, what happens then is that the residual from the additive model in, say, a two-way classification becomes $e_{ij} + h(\alpha_i, \beta_j)$, where $h(\alpha_i, \beta_j)$ represents the nonadditive component needed to make the model correct. The pooled $s^2$ becomes an estimate of $\sigma^2 + \sigma_h^2$. The pooled $s^2/n$ is still approximately unbiased as an estimate of the variance of a treatment mean under the fitted additive model, but $s^2/n$ is inflated in the sense that, if one could transform the data to a scale on which effects were truly additive, the term analogous to $\sigma_h^2$ would disappear, and one would legitimately obtain a smaller error variance. To put it in practical terms, undetected nonadditivity has two consequences: one does not estimate the differences between treatments as precisely as if one fitted the correct model, and one fails to learn that the differences between treatments are not constant from one replicate to another.

In the potato experiment of Fisher and Mackenzie nonadditivity was obvious, because the additive analysis-of-variance model produced *negative* figures for the estimated yields on a number of plots. For less extreme situations a test for nonadditivity that also provides a clue to the kind of transformation needed to attain additivity is presented in Section 9.

6. The reader may be surprised that nothing has been said about non-normality of the residuals. Several reasons can be suggested. First, analysis-of-variance procedures stand up fairly well under moderate amounts of nonnormality. Second, the standard tests for nonnormality — tests of skewness and kurtosis — are not effective in small experiments; thus, we lack powerful weapons for detecting the presence of non-normality. Third, the more harmful types of nonnormality, the so-called long-tailed distributions of errors, tend to produce both extreme observations, which may be picked up in tests for gross errors, and heterogeneity of errors. The latter arises because, as is the case in many nonnormal distributions, the variance is a function of the mean. Thus, in handling gross errors and variance heterogeneity statisticians hope that they are coping also with the more serious effects of nonnormality. It must be admitted, however, that we are not sure that we have the best approach to this general problem.

### 7. A test for outliers

As indicated in Section 6, the first step is to look over the residuals *d* from the fitted model. If the greatest residual appears suspiciously large on eye inspection, a check should be made to see whether any obvious mistake occurred in the experiment in connection with this observation. Sometimes a mistake is discovered such that the correct value can be found and inserted. Sometimes a mistake is discovered but no clue to the correct value; in this event the observation usually is deleted and treated as a missing value, as described in the textbooks (e.g., Bliss, 1967, and Snedecor and Cochran, 1967). Sometimes no mistake is unearthed. Any of three causes may account for a large residual: there actually is a mistake, a gross error, or there is no gross error but the distribution of errors is nonnormal and long-tailed, producing some high residuals, or there is nothing wrong, it is just one of those rare events.

At this point the investigator is likely to ask two questions: "What do I do when no mistake is unearthed?" and, an earlier question, "How do I decide whether the greatest residual looks suspiciously large?" As a guide, several tests for outliers are available. Most of these tests regard an outlier as suspicious if the probability of getting an outlier as large as it under the additive normal model is low, say 5% or less (the tests take account of the fact that the *largest* residual has been picked out).

We shall present a different approach, one due to Anscombe (1960). He regards outlier tests as equivalent to buying insurance against gross errors; since it is assumed that, whenever an outlier is detected, this observation is rejected. Now, any outlier test occasionally leads to rejection of a good observation and, when this happens, the treatment means are not as precisely estimated as they would be if the suspect had been kept. The insurance premium that one pays with Anscombe's test is defined as the percentage increase in the variance of the treatment means through the occasional rejection of good values in the routine use of the test. A 3% premium means that one is willing to incur on the average a 3% increase in the variance between the treatment means through this occasional rejection of good values.

The mathematical solution to this problem is difficult, and the procedure to be given is an approximation due to Anscombe and Tukey (1963). For the three standard designs the residuals $d$ have the following form (there are $k$ treatments and $n$ replicates in the one-way and two-way classifications, and $k$ replicates in the Latin square):

One-way:      $|d| = |y - \bar{y}_t|, \quad f = k(n - 1), \quad N = kn$

Two-way:      $|d| = |y - \bar{y}_t - \bar{y}_g + \bar{y}|,$
                     $f = (k - 1)(n - 1), \quad N = kn$

Latin square:    $|d| = |y - \bar{y}_t - \bar{y}_r - \bar{y}_c + 2\bar{y}|,$
                     $f = (k - 1)(k - 2), \quad N = k^2$

The steps are as follows:

1. Find the one-tailed normal deviate $z$ corresponding to the probability $fP/100N$, where $P$ is the premium *in percent*.

2. Calculate $K = 1.40 + 0.85z$.

3. Calculate $C = K[1 - (K^2 - 2)/4f](f/N)^{1/2}$.

4. Reject the largest residual $d$ if $|d|$ is greater than $Cs$, where $s$ is the standard error per observation.

As an example, let us consider the one-way design in Table 1. The largest deviation is found to come from the observation $-0.038$ for log dose 0.30. This gives

$$d = -0.038 - 0.119 = -0.157, \quad s = \sqrt{0.00341} = 0.0584,$$
$$|d| = 2.69s$$

Further,

$$f = 30 \text{ (error DF)}, \quad N = 35, \quad f/N = 6/7, \quad \sqrt{f/N} = 0.926$$

1. With a 3% premium we have $fP/100N = 0.0257$ and $z = 1.948$.
2. $K = 1.40 + (0.85)(1.948) = 3.06$.
3. $C = (3.06)(1 - 7.36/120)(0.926) = 2.66$.
4. Since $|d| = 2.69s$, we are just in the rejection region.

Two possibilities now present themselves. If our guess is that there was a gross error, we reject the observation and use the missing-value technique. If we think it more likely that this is an indication of a long-tailed distribution of errors, we may, alternatively, replace the $-0.038$ observation with a less extreme one. The latter technique is known as Winsorization in honor of C. P. Winsor, who first suggested its use (see pp. 261, 266 for additional discussion). Finally, it might be remarked that an outlier of which no explanation is found always presents a tricky situation and that research toward improved methods of handling such a situation is continuing.

## 8. Nonconstancy of error variance

Nonconstancy of error variance can occur for several reasons. First, some treatments are more erratic in their effects than others, the error variance changing from treatment to treatment. Second, it is not uncommon to find that the error variance increases as the mean increases; thus, if there are large differences among treatment means, high-yielding treatments may have higher variances then low-yielding ones. Third, nonnormality in the distribution of the errors $e_{ij}$ can also produce a relation between the variance within a treatment and the treatment mean.

If the heterogeneity is caused by erratic behavior of certain treatments, the variance of the comparison $L = \sum c_i \bar{y}_i$ becomes $\sum c_i^2 \sigma_i^2/n$, where $\sigma_i^2$ is the variance within the $i$th treatment. In the one-way classification an unbiased estimate of this variance is given by $\sum c_i^2 s_i^2/n$; $s_i^2$ is the mean square within the $i$th treatment and can be calculated and recorded separately for each treatment. For $t$ tests and confidence intervals an approximate number of degrees of freedom may be assigned to this estimate of variance by the following formula (Satterthwaite, 1946):

$$\text{DF} = (\sum c_i s_i^2)^2 / \sum (c_i^2 s_i^4/f_i)$$

where $f_i$ is the number of degrees of freedom in $s_i^2$.

This approach is not feasible with two-way classifications or Latin squares. With the two-way there are two possibilities. If the treatments that enter into $L = \sum c_i \bar{y}_i$ do not seem to differ much among themselves in their variances, one may calculate a pooled error mean square that applies to them by an analysis of variance in which the other treatments are omitted. If $k'$ treatments enter into $L$, this error mean square has $(n-1)(k'-1)$ DF. Alternatively, one may calculate $L_j = \sum c_i y_{ij}$ separately for each group. Then $L$ is the mean of the $L_j$, and an unbiased estimate of the variance of $L$ is $\sum (L_j - L)^2/n(n-1)$. Unfortunately, this has only $n-1$ DF.

When the heterogeneity takes the form of a smooth relation between the standard deviation $\sigma_i$ within the $i$th treatment and the population mean $\mu_i$ of the $i$th treatment, one should try to express this relation in the form $\sigma_i = \varphi(\mu_i)$, where $\varphi(\mu_i)$ is a simple mathematical function. As an example, let us suppose that the variables $e_{ij}$ have constant variance $\sigma^2$ but that the sources of experimental error operate in a multiplicative rather than an additive way. The replicates of the $i$th treatment then have the values $\mu_i(1 + e_{i1}), \mu_i(1 + e_{i2}), \ldots$, instead of our usual $\mu_i + e_{i1}, \mu_i + e_{i2}, \ldots$. The deviations from the population mean $\mu_i$ of this treatment are $\mu_i e_{i1}, \mu_i e_{i2}, \ldots$, with variance $\sigma_i^2 = \mu_i^2 \sigma^2$. Thus the relation in this example is $\varphi(\mu_i) = \sigma\mu_i$.

Given $\varphi(\mu_i)$, a transformation that puts the data on a scale in which the variance becomes approximately constant is to replace each observation $y$ by the indefinite integral of $dy/\varphi(y)$. Since the integral of $dy/\sigma y$ is $(\log y)/\sigma$, we transform the data to logs before analysis. This variance-

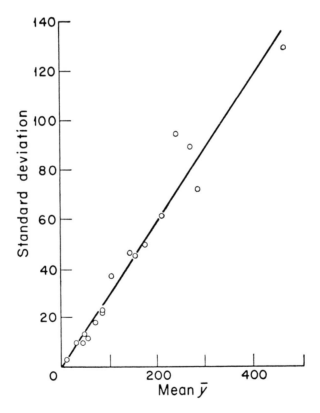

FIGURE 1 Standard deviation versus mean ratio of oviduct to body weight in groups of chicks given different doses of stilbestrol; fitted by a straight line with a zero intercept. From the groups in Table 3 (top) and three additional series.

94.50

stabilizing device is fairly crude, and sometimes the suggested transformation is revised if this seems to stabilize the variance more effectively. Thus, $\log (y + 1)$ is often used instead of $\log y$ for data that cover a wide range. If $\varphi(\mu_i) = a + b\mu_i$, a linear relation that does not go through the origin, the suggested transform is to $\log (y + a/b)$. Sections 13 and 15 illustrate the use of this type of transformation for two well-known nonnormal types of data, binomial percentages and counts (small whole numbers). Section 14 illustrates a different kind of transformation used with qualitative data.

The experiment of Table 3, on the oviduct response of four breeds of chicks, provides an illustration. For each breed and specified amount of stilbestrol a number of chicks, usually about twenty, were used. Consequently, we can calculate the standard deviation and the mean for each breed–amount combination. Figure 1 shows the plot of standard deviation versus $\bar{y}$ *before* the transformation to logs. Since the points lie close to a straight line through the origin, a log transformation is indicated.

## 9. A test for nonadditivity

The rationale of the following test for nonadditivity may be indicated by a consideration of the potato experiment of Fisher and Mackenzie. After criticizing the additive model they proceeded to fit a multiplicative model as more realistic. In our notation the product model is written

$$y_{ij} = \mu(1 + \alpha_i)(1 + \beta_j) + e_{ij}$$

where $\alpha_i$ might be called the deviation of row $i$ from the mean of all rows and $\beta_j$ a similar deviation. Expanding, we have

$$y_{ij} = (\mu + \mu\alpha_i + \mu\beta_j) + \mu\alpha_i\beta_j + e_{ij}$$

Note that the term within the parentheses is an *additive* model. Thus the nonadditive term that appears is $\mu\alpha_i\beta_j$. Apart from the multiplier $\mu$, this is the product of the row deviation and the column deviation. In other words, nonadditivity of this kind is revealed by a linear regression of the residual elements $d$ from the additive model on the product

$$(\bar{y}_t - \bar{y})(\bar{y}_g - \bar{y})$$

Tukey (1949), to whom this method is due, has shown more generally that with any kind of nonadditivity that is removable by a transformation to the scale $y^p$ the residuals $d$ from the additive model in an analysis of $y$ have an approximately linear regression on $x = (\bar{y}_t - \bar{y}) \times (\bar{y}_g - \bar{y})$. The regression coefficient $B_*$ is an estimate of $(1 - p)/\bar{y}$, so that the value of $p$ needed to secure additivity is estimated by $1 - B_* \bar{y}$. In experiments of ordinary size this estimate is very rough; an average value over a series of experiments and the use of judgment are called for.

The technique may be illustrated from the data in Table 3, where the residual elements $d$ and the variates $\bar{y}_t - \bar{y}$ and $\bar{y}_g - \bar{y}$ are shown. In this experiment the variate $y$ is already in the log scale, this transformation being suggested (see Section 8) by the fact that in the original scale $y'$ it was found that $\sigma'$ was proportional to $y'$ (Figure 1). Has additivity also been achieved?

In Figure 2 the residual elements $d$ are plotted against $x$. The scatter about the line is so wide that a line with zero slope would fit almost as well. For a numerical test we compute the numerator and denominator of $B_*$. For the numerator we first compute the sum of products (SP) of the elements of each row in Table 3 with the variates $\bar{y}_t - \bar{y}$. In the first row we have

$$\sum y(\bar{y}_t - \bar{y}) = (0.74)(-0.370) + \cdots + (1.43)(0.360)$$
$$= 0.29755$$

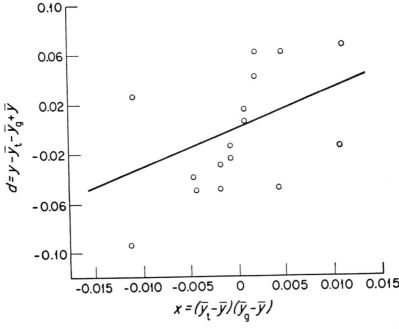

FIGURE 2   Tukey test for nonadditivity after transformation of the data in Table 3 to logarithms. The regression of the residuals in a cross-classification upon the product of the deviations of its treatment and its group mean from the general mean is shown.

The numerator of $B_*$ is the sum of products of these quantities multiplied by the corresponding $\bar{y}_g - \bar{y}$:

| $\sum y(\bar{y}_t - \bar{y})$ | $\bar{y}_g - \bar{y}$ |
|---|---|
| 0.29755 | −0.030 |
| 0.26530 | −0.005 |
| 0.34000 | 0.005 |
| 0.34335 | 0.030 |

Numerator = SP = $1.7475 \times 10^{-3}$

The denominator of $B_*$ is the product of $\sum (\bar{y}_t - \bar{y})^2 = 0.31155$ and $\sum (\bar{y}_g - \bar{y})^2 = 0.00185$, which is $5.7637 \times 10^{-4}$. In the analysis of variance in Table 3 the contribution of nonadditivity, with 1 DF, to the error sum of squares is num.$^2$/den. = 0.0053. The remainder of the error sum of squares is $0.0340 - 0.0053$, with 8 DF, giving a mean square of 0.00359. The $F$ ratio in the test for nonadditivity is 0.0053/0.00359 = 1.47 with 1 and 8 DF, well below the significance level.

Bliss (1967) gives a time-saving way of making this test without finding either the row and column deviations or the residual elements. We let $S_t$ and $S_g$ denote the treatment and group (breed) sums of squares in the analysis of variance. For the denominator we take $S_t S_g$; this is $N$

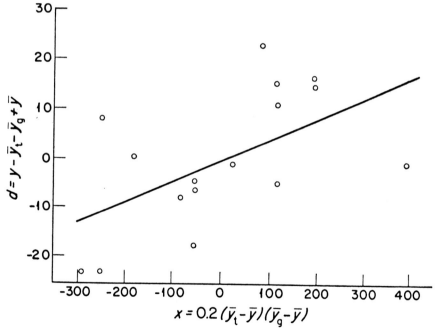

FIGURE 3   Tukey test for nonadditivity before transformation of the data in Table 3 to logarithms.

times the denominator given above. For the numerator we first multiply each row in Table 3 by the corresponding $T_t$, entering these products $\sum (T_t y)$ beside the column totals $T_g$. Now we form the sum of products of the entries in these two columns to get $\sum (T_g T_t y)$. For the numerator we calculate

$$S_* = \sum (T_g T_t y) - T(S_g + S_t + C_m)$$

where $T$ is the grand total and $C_m$ the correction for the mean. Since the quantity $S_*$ is $N$ times the previous numerator, $B_*$ is $S_*/S_t S_g$, and the sum of squares for nonadditivity is $S_*^2/NS_g S_t$ (this calculation is illustrated numerically in Table 9).

The original data $y'$ before transformation to logs gave an $F$ of 3.86 with $P$ about 0.08, as suggested by the plot of $d$ versus $x$ in Figure 3. The value of $1 - B_* \bar{y}'$ was about $-0.2$, suggesting that we replace $y'$ with $(y')^{-0.2}$. A power of $y'$ lying between $-0.2$ and $+0.2$ behaves rather similarly to log $y'$, so that the log transformation that was made is not inconsistent with this suggestion. Sometimes nonadditivity is due to one or two outliers — a point worth keeping in mind.

## 10. Regressions in a one-way design

As described in Sections 2 to 4, an initial step in the analysis of variance partitions the total sum of squares of the deviations of the $N$ responses from their overall mean $\bar{y}$ into categories that can be identified from the design of the experiment. In one-way classifications these are the sums of squares for treatments and for error. When the treatments mean square has more than one degree of freedom, comparisons between treatments may differ widely in significance. The $F$ ratio for treatments tests the null hypothesis that the response under study is the same for all treatments. Even when this aggregate test is not significant statistically, one or more individual contrasts between treatments may be significant.

When treatments differ qualitatively, such as between different test species or between different lots of a hormone, significance testing will depend upon a posteriori comparisons of their respective means (comparisons such as these are considered in Chapter 3). In other experiments the comparisons of interest can be defined a priori and limited in number to the degrees of freedom between treatments. When these comparisons are mutually independent or orthogonal, the total of their separate sums of squares, each with 1 DF, is equal to the treatment sum of squares with $k - 1$ DF. In studies of the response to a hormone or drug the treatments may consist of the different dose levels. Although statistics has sometimes been misapplied to testing which responses from a succession of doses differ significantly, the real experimental objective is to determine an appropriate regression equation and its precision.

As a first step the data are plotted on cross-section paper and fitted with a smooth curve by inspection. From the shape of the curve and the scatter of the points about it scales that will reduce the regression to a straight line or to a simple curve may be selected for the abscissa and ordinate. Such a plot also serves as a check for outliers. Ideally, the response will be equally variable at all dose levels. Especially in border-line cases a plot of this type may suggest what transformation of the response will conform most readily to this ideal.

The treatments sum of squares is subdivided initially into a term, $B^2$, due to the slope of a straight line (1 DF) and a remainder term, $k - 2$ DF, that measures the scatter around the linear regression. The spacings between successive doses, the number of observations at each dose level, or both, may differ from dose to dose. In this most general case we may compute

$$[x^2] = \sum (n_t x^2) - (\sum n_t x)^2/N \tag{9}$$

$$[xy] = \sum (xT_t) - \sum (n_t x) \sum T_t/N \tag{10}$$

$$B^2 = [xy]^2/[x^2] \tag{11}$$

where $n_t$ is the number of responses and $T_t$ is the total of the responses $y$ for each treatment or dose $x$. The square brackets denote a sum of squares or products about a mean. The slope is estimated as

$$b = [xy]/[x^2] \tag{12}$$

with an intercept estimated from the means $\bar{y}$ and $\bar{x}$ as

$$a' = \bar{y} - b\bar{x} \tag{13}$$

The calculation may be illustrated with data on the comb growth of capons in Table 2, where $y = \log$ (length + height) and $x = 1 + \log$ dose. The number of birds ($n_t$) varied from eight to ten between doses, four having been omitted because they were later shown to be regenerating testicular tissue. The analysis of variance in Table 2 has been extended in Table 6 by subdividing the sums of squares for doses into terms for slope and for scatter about the slope. The regression has been plotted in Figure 4. It is evident from the variance ratio for scatter ($F = 0.44$) that a straight line with the equation $Y = 0.2648 + 0.5245x$ adequately describes the effect of androsterone in this experiment.

When the $k$ doses are spaced equally in the preferred units and the frequencies $n_t$ of response are the same at each dose ($n_t = n$), the calculation of the regression can be shortened. It is also easy to test for

TABLE 6   Extension of Analysis of Variance of Androsterone Experiment (Table 2)

| Term | DF | SS | MS | F |
|------|----|----|----|----|
| Slope $B^2$ | 1 | 1.4102 | 1.4102 | 101.7 |
| Scatter | 3 | 0.0183 | 0.0061 | 0.44 |
| Error | 41 | 0.5684 | 0.01386 | |

Here $[x^2] = 5.1255$, Equation 9; $[xy] = 2.6885$, Equation 10: slope $b = 0.5245$, Equation 12; intercept $a' = 0.2648$, Equation 13.

curvature by isolating orthogonally the quadratic term in the parabola

$$Y = \bar{y} + B_1 x_1 + B_2 x_2 \tag{14}$$

For the linear term, or slope $B_1$, successive values of $x$ are coded to $x_1$. When $k$ is an odd number, say 5, then $x_1$ is $-2$, $-1$, 0, 1, and 2. This is extended or shortened for other values of $k$ with an interval between successive doses of $I_1 = 1$. For $k$ even, say 4, $x$ is coded to $x_1$ equal to

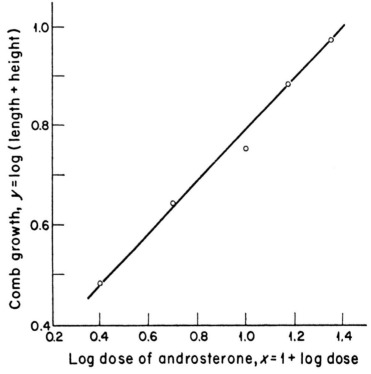

FIGURE 4   Regression of log growth of capon comb upon log dose of androsterone. From Table 6.

$-3$, $-1$, 1, and 3. This is extended similarly for other even values of $k$ in steps of $I_1 = 2$ below and above the mean. In coded units of $x_1$ the slope is determined as

$$B_1 = \sum (x_1 T_t)/(n \sum x_1^2) \tag{15}$$

Coded quadratic polynomials $x_2$, orthogonal with the linear $x_1$, are taken most readily from a table (Fisher and Yates, 1963; Bliss, 1967; Snedecor and Cochran, 1967). These and the intervals $I_2$ are given in Table 7 for

TABLE 7   Orthogonal Polynomials $x_2$ for Quadratic Curvature in Regressions with $k$ Equally Spaced Doses

| No. $k$ of doses | Orthogonal polynomials for dose level | | | | | | | | $\sum x_2^2$ | $I_2$ |
|---|---|---|---|---|---|---|---|---|---|---|
|   | 1 | 2 | 3 | 4 | 5 | 6 | 7 | 8 |   |   |
| 3 | 1 | $-2$ | 1 |   |   |   |   |   | 6 | 3 |
| 4 | 1 | $-1$ | $-1$ | 1 |   |   |   |   | 4 | 1 |
| 5 | 2 | $-1$ | $-2$ | $-1$ | 2 |   |   |   | 14 | 1 |
| 6 | 5 | $-1$ | $-4$ | $-4$ | $-1$ | 5 |   |   | 84 | $\frac{1}{2}$ |
| 7 | 5 | 0 | $-3$ | $-4$ | $-3$ | 0 | 5 |   | 84 | 1 |
| 8 | 7 | 1 | $-3$ | $-5$ | $-5$ | $-3$ | 1 | 7 | 168 | 1 |

$k$ of 3 to 8 dose levels. The regression coefficient for quadratic curvature in terms of $x_2$ is

$$B_2 = \sum (x_2 T_t)/(n \sum x_2^2) \tag{16}$$

The sum of squares for slope and for curvature are computed as

$$B^2 = [\sum (x_1 T_t)]^2/(n \sum x_1^2) \tag{17}$$

$$Q^2 = [\sum (x_2 T_t)]^2/(n \sum x_2^2) \tag{18}$$

each with 1 DF. The remainder with $k - 3$ DF measures the scatter about the parabola, when $Q^2$ is large enough to rule out fitting a straight line.

When the regression is linear, the slope in coded units of $x_1$ may be converted to the original scale of $x$ with

$$b = B_1/i_* \quad \text{if } k \text{ is odd,} \qquad \text{or} \qquad b = 2B_1/i_* \quad \text{if } k \text{ is even} \tag{19}$$

where $i_*$ is the interval between successive values of $x$. If the quadratic term is required, the regression in $x_1$ and $x_2$ can also be transformed to units of $x$ in a parabola. The dose giving a maximal (or minimal) response can be computed directly from the coded coefficients as

$$x_0 = \bar{x} - B_1 I_1 i_*/(2B_2 I_2) \tag{20}$$

The subdivision of the sum of squares for treatments may be illustrated by the uterine weights of immature rats in Table 1 at five dose
levels spaced equally on a log scale with seven responses $y$ (coded) at
each dose. As shown in Section 7, the first log uterine weight ($-0.038$)
at $x = 0.30$ would be judged an outlier by two or more tests, so that the
sample at this dose has been Winsorized (that is, the lowest value,
$-0.038$, is replaced with the next to the lowest value, 0.100, and the
highest value, 0.214, is replaced with the next to the highest value,
0.149, giving 0.903 as the total for the group).

With this adjustment the terms for linear slope and quadratic curvature have been computed (as if there had been no outlier) in Table 8
with the orthogonal coefficients $x_i$, leading to the first two variances,
or mean squares, in the last column. The mean square for scatter with
2 DF and error variance $s^2$ with 28 DF are taken from the revised analysis

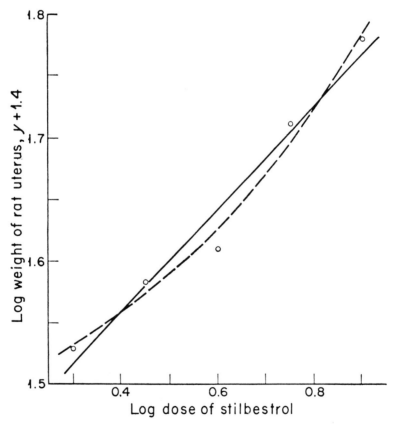

FIGURE 5   Log uterine weight of immature rats as a function of the log dose of
stilbestrol, fitted with a linear regression (solid line) and with a parabola (broken
line). From the data in Table 8.

of variance. Although $F = 2.19$ for quadratic curvature, this was not significant ($P = 0.16$), so the mean values, decoded by adding 1.4 to each mean, are plotted and fitted with a straight line in Figure 5. They have also been fitted with a parabola (broken line). Both equations are given in Table 8.

TABLE 8 Partitioning of Sum of Squares for Response to Stilbestrol (Table 1) after Winsorization of Responses for $x = 0.30$

| Term | Coefficients $x_i$ for $x$ | | | | | $n \sum x_i^2$ | $\sum x_i T_t$ | $\dfrac{(\sum x_i T_t)^2}{n \sum x_i^2}$ |
| | 0.30 | 0.45 | 0.60 | 0.75 | 0.90 | | | |
|---|---|---|---|---|---|---|---|---|
| Slope $x_1$ | $-2$ | $-1$ | 0 | 1 | 2 | 70 | 4.376 | 0.273563 |
| Curvature $x_2$ | 2 | $-1$ | $-2$ | $-1$ | 2 | 98 | 0.728 | 0.005408 |
| $T_t$ | 0.903 | 1.271 | 1.463 | 2.167 | 2.643 | | scatter MS: | 0.002100 |
| $\bar{y}_t$ (decoded) | 1.529 | 1.582 | 1.609 | 1.710 | 1.778 | | error $s^2$: | 0.002467 |

Here $i_* = 0.15$, $B_1 = 4.376/70 = 0.062514$, $b = B_1/i_* = 0.41676$, $B_2 = 0.728/98 = 0.007429$, $\bar{y} = 8.447/35 + 1.4 = 0.24134 + 1.4 = 1.64134$. Straight line: $Y = 1.64134 + 0.41676(x - 0.60)$. Parabola: $Y = 1.64134 + 0.062514x_1 + 0.007429x_2$.

## 11. Regressions in a cross-classification

When a particular treatment effect in a two-way design, such as a slope, accounts for a major part of the treatments sum of squares, it is prudent to test the consistency of this effect between replicate groups by isolating the interaction groups × slope. For a regression with $k$ dose levels the numerator of the regression coefficient is computed separately for each group or replicate directly from the dosage units $x$ or, if coded, from $x_1$ as

$$[xy] = \sum xy - \sum x \sum y/k \qquad \text{or} \qquad [x_1 y] = \sum x_1 y \qquad (21)$$

Their overall total from all replicates will be equal to the numerator of the overall or combined regression for the entire experiment. All will have the same denominator, either of the following:

$$[x^2] = \sum x^2 - (\sum x)^2/k \qquad \text{or} \qquad [x_1^2] = \sum x_1^2 \qquad (22)$$

The sum of squares for differences in slope (or the term for groups × slope) is then computed as either of the following:

$$[B_i^2] = \sum [xy]^2/[x^2] - B^2 \qquad \text{or} \qquad [B_i^2] = \sum [x_1 y]^2/[x_1^2] - B^2 \quad (23)$$

When subtracted from the error sum of squares, or sum of squares of groups × treatments, in the original two-way analysis of variance, the

mean square for the remainder, or groups × scatter, is a new estimate of
the residual, or error, variance $s^2$. If the mean square for groups × slope
is significantly or substantially larger than the new error mean square, it
is the error for the average $B^2$, unless the term for scatter is also significantly larger than the new $s^2$, in which case the error for the slope term
depends upon the mean squares for both scatter and groups × slope.
When the overall quadratic term $Q^2$ from the treatments sum of squares
indicates a significant or nearly significant curvature in the regression,
the interaction groups × quadratic curvature can be separated similarly
from the interaction groups × treatment.

The calculation may be illustrated with the oviduct response of chicks
to stilbestrol, Table 3. The mean response for the four breeds (groups)
at each dose of stilbestrol (on a log scale) is plotted and fitted with a
linear regression in Figure 6. The scatter of points about this line is so
free from trend that separation of a quadratic term is unnecessary.
Because of the even log spacing, $i_* = \log 2 = 0.3010$, the numerator of
the slope has been computed for each breed with the linear polynomials
$x_1$ equal to $-3$, $-1$, 1, and 3 to give the $\sum x_1 y$ in Table 9 and their sum
over the four breeds.

The term for slope in the analysis of variance accounted for all but

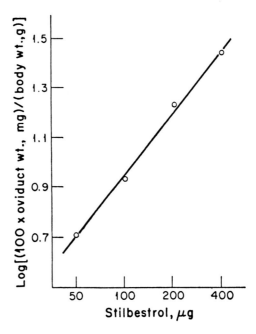

FIGURE 6   Mean log ratio of oviduct to body weight in four breeds of chicks as a
linear function of the log dose of stilbestrol. From the data in Table 9.

TABLE 9  Regression Analysis of Log Oviduct Response of Chicks to Stilbestrol (Table 3)

| Dose, µg | $x_1$ | $T_t$ | Breed | $\sum x_1 y$ | $T_g$ | $\sum (T_t y)$ |
|---|---|---|---|---|---|---|
| 50 | −3 | 2.82 | WL | 2.38 | 4.18 | 19.1642 |
| 100 | −1 | 3.72 | WR | 2.12 | 4.28 | 19.4652 |
| 200 | 1 | 4.92 | RI | 2.72 | 4.32 | 19.9360 |
| 400 | 3 | 5.74 | Hy | 2.74 | 4.42 | 20.3794 |
| Total: | 0 | 17.20 | | 9.96 | 17.20 | 78.9448 |

Calculations:

$B^2 = 9.96^2/(4 \times 20) = 1.2400$, breeds $\times$ slope $= 25.0648/20 - B^2 = 0.0132$, $S_* = 339.617880 - 17.20(19.7436) = 0.027960$, $NS_g S_t = 0.147550$.

Analysis of variance:

| Term | | DF | SS | | MS | $F$ |
|---|---|---|---|---|---|---|
| Breeds | | 3 | 0.0074 | $= S_g$ | 0.0025 | 0.66* |
| Dose, linear $B^2$ | | 1 | 1.2400 ⎫ | | | |
| Dose, scatter | | 2 | 0.0062 ⎭ $= S_t$ | | 0.0031 | 0.82* |
| Breeds $\times$ slope | | .3 | 0.0132 | | 0.0044 | 1.26† |
| Breeds $\times$ scatter | | 6 | 0.0208 | | 0.0035 | |
| | Total: | 15 | 1.2876 | | | |
| | $C_m$: | 1 | 18.4900 | | | |
| Nonadditivity | | 1 | 0.0053 | $= B_*^2$ | 0.0053 | 1.47‡ |
| Remainder | | 8 | 0.0287 | | 0.0036 | |
| Error | | 9 | 0.0340 | | 0.00378 | |

\* Denominator of $F$ ratio is MS error.
† Denominator of $F$ ratio is MS breeds $\times$ scatter.
‡ Denominator of $F$ ratio is MS remainder.

0.5% of the sums of squares for treatments, the remainder being the scatter about the fitted line. By squaring the sums of products $\sum x_1 y$ for the four breeds and dividing their total by $\sum x_1^2 = 20$, the 9 DF for the interaction breeds $\times$ doses has been divided into a sum of squares for breeds $\times$ slope of $1.2532 - 1.2400 = 0.0132$ with 3 DF and a remainder for breeds $\times$ scatter of $0.0340 - 0.0132 = 0.0208$ with 6 DF. From their mean square it is clear that the log oviduct ratio increased similarly with the dose of stilbestrol in all four breeds of chicks.

## 12. Factorial experiments

An important design in which the treatment contrasts can be defined a priori is the factorial experiment (see Chapter 1, Section 2). In its simplest form this is a comparison of two preparations or treatments, each at two or more dose levels, in all possible combinations such as the $2 \times 2$, $2 \times 3$, and $2 \times 4$ designs. These are used quite typically in biological assays for measuring the potency of one preparation, the unknown, relative to a second preparation, the standard (see pp. 106–117). Dosages of both preparations are assigned the same spacing, usually on a log scale, and so far as advance information permits they are expected to produce similar responses at each level.

Each factorial effect is isolated from the sum of squares for treatments by means of orthogonal comparisons, which involve the linear and quadratic coefficients $x_1$ and $x_2$ in Section 10. The number of orthogonal comparisons is equal to the degree of freedom between treatments, and the sum of their respective variances is the same as the sum of squares for treatments.

The $2 \times 2$ factorial design is illustrated by the two-dose insulin assay of Table 4. Its three orthogonal comparisons in Table 5 represent the difference $a$, or $U - S$, in the mean response between the unknown, or test, preparation and the standard, the average, or combined, slope $b$ for the two regressions, and the divergence $ab$ in their slopes when fitted separately, or nonparallelism. Note that the variances in the last column sum to the total sum of squares between treatments in the original analysis of the Latin square. Each may be compared separately with the error variance $s^2 = 35.7$ given in Table 4. These contrasts and their $F$ ratio lead to a test of assay validity and to quantitative estimates of relative potency and of the limits within which it has been determined (see pp. 108–111).

For a $2 \times 3$ factorial design for a parallel-line bioassay of corticotropin the five relevant contrasts are given in Table 10, a one-way

TABLE 10  Corticotropin Assay from Adrenal Ascorbic Acid in Hypophysectomized Rats with Dose Levels of 23.6, 59.0, and 147.5 μg Injected Subcutaneously*

| Row | Factorial coefficients† $x_i$ | | | | | | $\sum x_i^2$ | $T_i = \sum (x_i T_t)$ | $\dfrac{T_i^2}{n \sum x_i^2}$ |
|---|---|---|---|---|---|---|---|---|---|
|  | $S_1$ | $S_2$ | $S_3$ | $U_1$ | $U_2$ | $U_3$ | | | |
| $a$ | −1 | −1 | −1 | 1 | 1 | 1 | 6 | −298 | 2,467 |
| $b$ | −1 | 0 | 1 | −1 | 0 | 1 | 4 | −1,418 | 83,780 |
| $ab$ | 1 | 0 | −1 | −1 | 0 | 1 | 4 | 50 | 104 |
| $q$ | 1 | −2 | 1 | 1 | −2 | 1 | 12 | −240 | 800 |
| $aq$ | −1 | 2 | −1 | 1 | −2 | 1 | 12 | −196 | 534 |
| $T_t$: | 2,536 | 2,180 | 1,802 | 2,379 | 2,146 | 1,695 | | Total: | 87,685 |

* Data of Bliss (1956).
† Here $i = \log 2.5 = 0.3979$, $n = 6$, error $s^2 = 25{,}794/30 = 859.8$ (30 DF).

classification with $n = 6$ rats. The first three contrasts are analogous to those for the $2 \times 2$ factorial design. The remaining two contrasts correspond to quadratic curvature for testing divergence of the standard and unknown from a linear dose–response relation. For details see pp. 111–113.

A $2 \times 4$ factorial experiment compared the motility indices of fowl spermatozoa in two diluents, one with glucose constant at $2\%$ and the other with sodium chloride constant at $0.13\%$, both at pH $= 7.1 \pm 0.1$. Four equally spaced levels of tonicity were tested with each type of solution upon five ejaculates in the two-way classification; see Table 11. The factorial coefficients for isolating the seven orthogonal contrasts between the eight treatments are given in Table 12.

TABLE 11    Motility Indices of Fowl Spermatozoa in Two Diluents at Four Tonicities at pH $= 7.1 \pm 0.1$*

| Ejaculate | Glucose 2.0% | | | | NaCl 0.13% | | | | Total |
| no. | 50 | 100 | 150 | 200 | 50 | 100 | 150 | 200 | $T_g$ |
| --- | --- | --- | --- | --- | --- | --- | --- | --- | --- |
| 1 | 31 | 61 | 67 | 50 | 38 | 65 | 70 | 18 | 400 |
| 2 | 51 | 62 | 45 | 0 | 53 | 55 | 58 | 3 | 327 |
| 3 | 22 | 28 | 32 | 13 | 34 | 42 | 27 | 3 | 201 |
| 4 | 33 | 61 | 71 | 67 | 54 | 67 | 73 | 68 | 494 |
| 5 | 41 | 77 | 76 | 55 | 62 | 70 | 75 | 9 | 465 |
| Total $T_t$: | 178 | 289 | 291 | 185 | 241 | 299 | 303 | 101 | 1887 |

* Data of Wales and White (1958).

The contrast $a$ between the two diluents corresponds to that between standard and unknown in the parallel-line bioassay, and contrasts $b$ and $q$ correspond to the combined linear coefficients and combined quadratic coefficients (Table 7) for $k = 4$ equally spaced dose levels. The coefficients in the first five rows of the analysis parallel those for the $3 \times 2$ factorial parallel-line bioassay of Table 10. The coefficients in the last two rows correspond to scatter about a parabola but have been replaced here with orthogonal coefficients for fitting the cubic term in a polynomial regression. The results of applying the orthogonal coefficients to the $y$'s for each ejaculate and the totals for all ejaculates are given at the right in Table 12. The sum $\sum x_i^2$ in the divisor for each row, 8 or 40, is multiplied by $n = 5$ in computations of the overall, or combined, effect of a treatment and by $n = 1$ in tests of the agreement between ejaculates. For example, the contrast for the linear slope $b$ was $393^2/(5 \times 40) = 772.245$ as the combined linear component (1 DF), and its sum of squares between ejaculates was $[(8^2 + 317^2 + 131^2 + 160^2 + 113^2)/40] - 772.245 = 3,129.830$.

These sums of squares have been entered in the analysis of variance in Table 13; the sum of squares for each treatment effect (1 DF) is in the third column, and that for its interaction with ejaculate (4 DF) is in the

TABLE 12   Factorial Analysis of Motility Indices of Fowl Spermatozoa (Table 11)

| Factorial contrast | Coefficients $x_i$ for four tonicities | | | | | | | | $\sum x_i y$ for ejaculate no. | | | | | $\sum x_i T_i$ | $\sum x_i^2$ |
|---|---|---|---|---|---|---|---|---|---|---|---|---|---|---|---|
| | Glucose 2% | | | | NaCl 0.13% | | | | 1 | 2 | 3 | 4 | 5 | | |
| | 50 | 100 | 150 | 200 | 50 | 100 | 150 | 200 | | | | | | | |
| $a$, diluent | 1 | 1 | 1 | 1 | -1 | -1 | -1 | -1 | 18 | -11 | -11 | -30 | 33 | -1 | 8 |
| $b$, linear slope | -3 | -1 | 1 | 3 | -3 | -1 | 1 | 3 | 8 | -317 | -131 | 160 | -113 | -393 | 40 |
| $ab$, diluent × linear | -3 | -1 | 1 | 3 | 3 | 1 | -1 | -3 | 118 | -23 | 85 | 64 | 195 | 439 | 40 |
| $q$, quadratic term | 1 | -1 | -1 | 1 | 1 | -1 | -1 | 1 | -126 | -113 | -57 | -50 | -131 | -477 | 8 |
| $aq$, diluent × quadratic | 1 | -1 | -1 | 1 | -1 | 1 | 1 | -1 | 32 | 1 | 7 | -14 | 17 | 43 | 8 |
| $c$, cubic term | -1 | 3 | -3 | 1 | -1 | 3 | -3 | 1 | -34 | -59 | -7 | 0 | -51 | -151 | 40 |
| $ac$, diluent × cubic | -1 | 3 | -3 | 1 | 1 | -3 | 3 | -1 | 36 | 59 | -35 | 8 | 85 | 153 | 40 |
| Total $T_i$: | 178 | 289 | 291 | 185 | 241 | 299 | 303 | 101 | | | | | | | |
| Linear for glucose | -3 | -1 | 1 | 3 | 0 | 0 | 0 | 0 | 63 | -170 | -23 | 112 | 41 | 23 | 20 |
| Linear for NaCl | 0 | 0 | 0 | 0 | -3 | -1 | 1 | 3 | -55 | -147 | -108 | 48 | -154 | -416 | 20 |

TABLE 13   Analysis of Variance of Motility Indices of Fowl Spermatozoa (Tables 11 and 12)

| Term | DF | SS | MS | F | Interaction SS, 4 DF |
|---|---|---|---|---|---|
| Ejaculates | 4 | 6,929.650 | 1,732.413 | 19.44 | |
| $a$, diluent | 1 | 0.025 | 0.025 | 0.00 | 319.350 |
| $b$, combined linear | 1 | 772.245 | 772.245 | 0.99* | 3,129.830 |
| $q$, combined quadratic | 1 | 5,688.225 | 5,688.225 | 63.83 | 756.150 |
| $c$, combined cubic | 1 | 114.005 | 114.005 | 1.28 | 68.170 |
| $ab$, diluent × linear | 1 | 963.605 | 963.605 | 10.81 | 631.370 |
| $aq$, diluent × quadratic | 1 | 46.225 | 46.225 | 0.52 | 148.650 |
| $ac$, diluent × cubic | 1 | 117.045 | 117.045 | 1.31 | 215.230 |
| Ejaculate × linear | 4 | 3,129.830 | 782.458 | 8.78 | |
| Other interactions | 24 | 2,138.920 | 89.121 | | |
| Total: | 39 | 19,899.775 | | | 5,268.750 |
| Linear for glucose | 1 | 5.29 | 5.29 | 0.01* | 2,375.860 |
| Linear for NaCl | 1 | 1,730.56 | 1,730.56 | 5.00* | 1,385.340 |

For 2% glucose $Y = 47.15 + 0.23x_1 - 11.925x_2$ and $x_0 = 125 + 0.96 = 125.96$; for 0.13% NaCl $Y = 47.20 - 4.16x_1 - 11.925x_2$ and $x_0 = 125 - 17.44 = 107.56$.

* For these $F$ ratios the appropriate denominator is the corresponding interaction mean square, e.g., $0.99 = 772.245/(3,129.830/4)$.

last column. For a test of the homogeneity of the interactions in the last column the largest sum of squares was divided by the smallest, since all have 4 DF, giving $3,129.83/68.17 = 45.91$ as the maximum $F$ ratio. The expected value of $F$ at $P = 0.05$ for 7 and 4 DF is 33.6 (David, 1952). Accordingly, the five ejaculates differed more in the linear term than would be attributed to chance. This omitted, the next largest $F$ ratio, $756.15/68.17 = 11.09$, fell well below its expected 5% level of 29.5 for 6 DF. These six interaction sums of squares, therefore, may be considered separate estimates of the same error variance, differing from one another only by chance. Since each is based upon only 4 DF, they have been summed to give a common strengthened error of $s^2 = 89.121$ with 24 DF for testing the significance of the corresponding six contrasts between treatments.

Only the quadratic term, $q$, and the interaction diluent × linear, $ab$, were significant. Because of its significant mean square, equal to 782.458 ($F = 8.78$, $P < 0.001$), the interaction ejaculate × linear was the appropriate error for testing the overall effect of slope ($F = 0.99$). Since this interaction was significant, it is of interest to test whether the slope varied more among ejaculates with the glucose than among those with the sodium chloride by subdividing the sum of squares in rows

$b + ab$ as shown in the lower parts of Tables 12 and 13. The variation among ejaculates in the linear term was significantly larger with both diluents than the mean square for other interactions ($F = 6.66$ for glucose and 3.89 for sodium chloride). However, it cancelled out any mean effect for glucose while leading, in the case of sodium chloride, to the second largest factorial sum of squares.

The quadratic term was highly significant and consistent both between diluents and between ejaculates, so that it provided a uniform denominator for computing the tonicity giving the highest motility. Since the linear term for the numerator of this estimate differed between diluents, the tonicity for maximal motility had to be estimated separately with the linear term for each diluent. Substituting in Equation 20 gives $B_1 = 23/100$ for glucose, $B_1 = -416/100$ for sodium chloride, and $B_2 = -477/40$ for their combined quadratic coefficient. The tonicity $x_0$ giving the maximal motility was determined as 125.96 for glucose and as 107.56 for sodium chloride. The two parabolas and their maxima are illustrated in Figure 7. The difference between these two tonicities, $125.96 - 107.56 = 18.40$, is an estimate of the significantly greater tonicity required with 2% glucose in the diluent than with 0.13% sodium chloride.

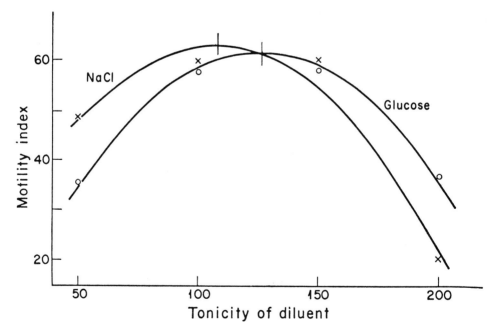

FIGURE 7   Mean motility indices of fowl spermatozoa as a parabolic function of the tonicity in two diluents. From the data in Table 12.

### 13. Analysis of binomial percentages

When the responses in an experiment are the percentages of animals that react, each is potentially a binomial proportion $p$ with a sample variance $pq/n$ that depends upon $p$. A direct analysis of variance of binomial percentages, especially if they were to cover a range wider than 30 to 70%, would not meet the assumption of homoscedasticity but would require weighting of each percentage by the reciprocal of its variance.

When treatments differ quantitatively in a sequence of increasing doses of a drug or hormone, successive percentages of response may represent proportionate areas from an underlying distribution of threshold susceptibilities. On this hypothesis each percentage response, when appropriately transformed, will plot as a straight line against a suitable function of the independent variate, usually its logarithm. If the underlying distribution is normal, the probit is the appropriate transformation. The information in each probit, however, also depends upon the parameter $p$, so that it, too, requires unequal weighting and does not fit the pattern of the analysis of variance.

A scale that is intermediate between no transformation and the probit is the inverse sine of the square root of the percentage. Its variance is constant in large samples when the size $n_t$ of each test group is about the same. Within a range of 7 to 93% the angle is a nearly linear function of the probit. When predictions below 10% or above 90% are not a primary concern and $n_t$ is constant $(n_t = n)$, each observed proportion $\hat{p}$ is transformed to the angle

$$y = \sin^{-1}\sqrt{\hat{p}} \tag{24}$$

As $100p$ increases from 0 to 100%, the angle $y$ for each percentage increases from $0°$ to $90°$ and may be read from a suitable table (Bliss, 1967; Snedecor and Cochran, 1967). The expected variance of $y$ in large samples is

$$\sigma_y^2 = 820.7/n \tag{25}$$

when measured in degrees; it is $1/4n$ when measured in radians.

An approximate adjustment, which reduces the discontinuity near the ends of the scale, is to replace each observed percentage $100a/n = 100\hat{p}$ with an adjusted one,

$$\text{adjusted } \% = \frac{100(a + \frac{1}{4})}{n + \frac{1}{2}} = \frac{50(4a + 1)}{2n + 1} \tag{26}$$

before transformation to angles (Anscombe, 1954). When an analysis of variance has been computed in angles, its agreement with the expected binomial variance in Equation 25 can be tested with the sum of squares for error (with $\nu$ DF) by computing

$$\chi^2 = \nu s^2/\sigma_y^2 \tag{27}$$

Even though $\chi^2$ may be too large or too small for one to consider the test observations homogeneous binomial samples, the transformation is still valid for overdispersion or underdispersion.

The angular transformation may be illustrated with data on the dose–response curve for estrone in female rats that were born in January, spayed in May, and injected with two sequences of four doses of estrone between July and October, 1940. After each dose the response of a rat was called positive if in four vaginal smears on successive days epithelial or squamous cells predominated. Four cages, each of fifteen rats, were tested by means of two $4 \times 4$ Latin squares; the doses and number of positive responses in each test are shown in the upper part of Table 14. The responses with 1 to 4 or 11 to 15 positive were then converted to adjusted percentages, as defined in Equation 26, before being transformed

TABLE 14   All-or-None Dose–Response Curve for Doses A to D of Estrone in Repeated Tests with the Same Rats in Two $4 \times 4$ Latin Squares*

Dose and number $a$ of rats positive in each cage of 15 rats in each test, after doses A = 2.208, B = 2.608, C = 3.008, D = 3.808 μg.

| Cage no. | 1, 7/1–5 | | 2, 7/15–19 | | 3, 7/29–8/1 | | 4, 8/12–16 | | 5, 8/26–31 | | 6, 9/9–13 | | 7, 9/23–27 | | 8, 10/7–11 | | Tot. |
|---|---|---|---|---|---|---|---|---|---|---|---|---|---|---|---|---|---|
| 457 | A | 1 | B | 6 | C | 10 | D | 14 | A | 5 | B | 5 | C | 10 | D | 14 | 65 |
| 458 | B | 3 | A | 4 | D | 13 | C | 7 | B | 7 | A | 3 | D | 13 | C | 9 | 59 |
| 459 | C | 7 | D | 11 | A | 5 | B | 8 | C | 12 | D | 12 | A | 6 | B | 7 | 68 |
| 460 | D | 11 | C | 10 | B | 9 | A | 6 | D | 14 | C | 7 | B | 11 | A | 5 | 73 |

Angular response in degrees from $\% = 100(a + \frac{1}{4})/(15 + \frac{1}{2})$. There may be occasional discrepancies in last digit owing to rounding error.

| Cage no. | 1 | 2 | 3 | 4 | 5 | 6 | 7 | 8 | $T_r$ | Dose x | $T_t$ |
|---|---|---|---|---|---|---|---|---|---|---|---|
| 457 | 16.5 | 39.2 | 54.8 | 73.5 | 35.2 | 35.2 | 54.8 | 73.5 | 382.7 | A −0.222 | 259.4 |
| 458 | 27.3 | 31.6 | 67.6 | 43.1 | 43.1 | 27.3 | 67.6 | 50.8 | 358.4 | B −0.077 | 344.0 |
| 459 | 43.1 | 58.4 | 35.2 | 46.9 | 62.7 | 62.7 | 39.2 | 43.1 | 391.3 | C +0.047 | 407.2 |
| 460 | 58.4 | 54.8 | 50.8 | 39.2 | 73.5 | 43.1 | 58.4 | 35.2 | 413.4 | D +0.252 | 535.2 |
| $T_c$: | 145.3 | 184.0 | 208.4 | 202.7 | 214.5 | 168.3 | 220.0 | 202.6 | 1545.8 | Total: | 1545.8 |

Calculations:
$x = 2(x' - \bar{x}')$, where $x' = \log$ dose; $\sum (xT_t) = 69.9340$, $8 \sum x^2 = 0.967408$, $\bar{x}' = 0.4548$, $\bar{y} = 48.306$, $b = 2 \times 69.9340/0.96741 = 144.580$.

* Data of Pugsley and Morrell (1943).

TABLE 15    Analysis of Variance of Estrone Experiment of Table 14

| Term | DF | SS | MS | F | Cage no. | $\chi^2$; $\nu = 14$ |
|------|-----|------|------|------|------|------|
| Cages | 3 | 193.84 | 64.61 | 3.30 | | |
| Dates or tests | 7 | 1,145.16 | 163.59 | 8.35 | 457 | 16.783 |
| Dose, linear | 1 | 5.055.53 | 5,055.53 | | 458 | 16.471 |
| Dose, scatter | 2 | 7.07 | 3.54 | 0.18 | 459 | 40.588 |
| Error | 18 | 352.66 | 19.59 | | 460 | 17.697 |
| Total: | 31 | 6,754.26 | $\sigma_y^2 = 820.7/15$ | | Total: | 91.539 |
| $C_m$: | 1 | 74,671.80 | $= 54.71$ | | | $P = 0.002$ |

Here $\chi^2 = \nu s^2/\sigma_y^2 = 6.446$; at $P = 0.99$, $\chi^2 = 7.015$ (18 DF).

to angles. The four dose levels in logarithms have been coded to deviations from their mean for computing the analysis of variance and the log dose–response curve in angles.

From $F$ tests in the analysis in Table 15 the mean sensitivity of the rats, as measured from the just effective dose of estrone in each individual, differed between cages and more markedly between tests, when compared with the observed error variance with 18 DF. The dose-response curve was clearly linear in these units (Figure 8) with scatter around the line less than the error mean square. If the variation were binomial, the

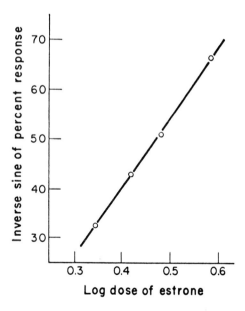

FIGURE 8    Regression of the mean inverse sine of the percentage of positive responses of castrated female rats upon four dose levels of estrone. From the data in Table 14.

observed $s^2 = 19.59$ would be of the same magnitude as its binomial expectation $\sigma_y^2 = 820.7/15 = 54.71$, but it was actually much less. To test the difference, the error sum of squares divided by $\sigma_y^2$ gave $\chi^2 = 6.446$, less than its expectation (7.015) for 18 DF at $P = 0.99$.

Underdispersion such as this may be attributed to balanced sampling. The same individual rats were tested on each occasion. If their inherent sensitivity $p$ to estrone were to differ significantly, the variance in $p$, or $nV(p)$, could reduce the expected variance based upon the means $\overline{pq}$ for the individual rats to

$$E(\sigma^2) = n\overline{pq} - nV(p) \tag{28}$$

For a test of whether $p$ did differ significantly between animals $\chi^2$ was computed, in the right side of Table 15, from the number of positive reactions in the eight tests for each of the fifteen animals in each cage. From the total, $\sum \chi^2 = 91.54$ with 56 DF ($P = 0.002$), we would conclude that the level of susceptibility to estrone did differ, not only between cages but also between rats within cages, and that this reduced $\sigma_y^2$ below its binomial expectation in Equation 25.

## 14. Qualitative responses

Some endocrine responses cannot be measured on an objective scale, but they can be ranked or scored subjectively. To minimize observer's bias with either method, the responses to a series of treatments should be judged by the same observer and the treatments coded or otherwise concealed in a so-called "blind test."

Ranking has the advantage that it is independent of differences between observers in the numerical grade that each might assign to a particular response or to differences within groups in the degree of responsiveness. The responses in a given series might still be ranked in more or less the same sequence of intensities by different observers or by the same observer on different occasions, even though their scores were to differ appreciably. Ranking, however, takes no direct account of the size of the differences between responses, even though it may be easier to rank the extreme values in a series than those in the middle. This subjective experience is reflected in the transformation of ranks to rankits for analysis (Bliss, 1967; Fisher and Yates, 1963), a topic beyond the scope of this chapter.

An experiment that depended upon ranking is a comparison made by van Strik (1961) of the progestational activity of a series of synthetic steroids in immature female rabbits. An analysis in terms both of ranks and of rankits led to essentially the same conclusions.

In tests in which each animal can be used repeatedly scores of 1 or 0 have been proposed by Emmens (1957) and others as an alternative to the angular transformation for an all-or-none, or quantal, response. Each test animal would be scored either as 1 or 0, depending upon whether it reacted or not. In a comparison of intravaginal estrogen and proestrogen at four dose levels in spayed albino mice an analysis based upon 1,536 individual scores had a residual mean square for error one half of that observed between groups or between animals within groups. The $F$ test for treatments (doses), however, was only slightly larger than in an analysis of the same experiment in angles based upon the group responses. In the :c units the error here was of the same magnitude as its binomial expectation, in contrast to the significantly smaller observed error in Table 15.

An extension of the scoring procedure, however, increased the sensitivity of this test. Vaginal smears were taken twice on the third day from the start of the test, there being very few positive smears at other times. Each mouse could be scored in each test as not reacting (0), reacting in only one smear (1), or reacting in both smears (2). The sum over twenty-four spayed mice in each of the eight groups in an $8 \times 8$ design has been analyzed both in angles and in units of this three-term scoring system, with the results shown in Table 16. The increased information from the

TABLE 16   Analyses of Variance Comparing Estrogen (S) and Proestrogen (U) in Mice in Two Units of Response*

| Term | DF | Angles MS | $F$ | Scores MS | $F$ |
|---|---|---|---|---|---|
| Groups | 7 | 74.250 | 2.22 | 22.980 | 1.63 |
| Days | 7 | 145.785 | 4.36 | 59.266 | 4.21 |
| $U - S$ | 1 | 1,092.300 | 32.65 | 848.266 | 60.24 |
| Slope | 1 | 5,016.530 | 149.95 | 2,802.528 | 199.03 |
| Parallels | 1 | 14.365 | 0.43 | 55.278 | 3.93 |
| Curvature | 4 | 34.495 | 1.03 | 9.697 | 0.69 |
| Error | 42 | 33.454† | | 14.081 | |

*Data of Emmens (1957).
†Theoretical $\sigma^2 = 820.7/24 = 34.196$.

double scoring has increased the $F$ ratios appreciably, most markedly in·the test for lack of parallelism, although this was not significant in either unit.

## 15. Response in counts

The response in some endocrine research is recorded in counts, such as the number of mitoses per microscopic field in the vaginal epithelium of the mouse after intravaginal injections of estrogen (Martin and Claringbold, 1960). Of increasing importance are the instrumental counts of radioactivity in radioimmunoassays, numbered usually in the hundreds or thousands. Although commonly converted to percentages on some convenient scale, they potentially may be identified with a basic distributional pattern and analyzed more directly.

If the variation between replicate counts $y'$ were completely random, the variance $V(y')$ between replicates should equal their mean $\bar{y}'$, which would identify the counts as samples from a Poisson distribution. In this case $s^2 = V(y')$ would increase with $\bar{y}'$ in a straight-line relation passing through the origin. The square root of each individual count, $y = \sqrt{y'}$, has a constant variance $V(y) = 1/4$ and is the appropriate unit for computing an analysis of variance or a relative potency. When the original counts $y'$ are small, averaging $y' \le 4$, a better approximation is $y = \sqrt{y' + 0.4}$, or, best of all, the maximum-likelihood solution given by Cochran (1940). A radioimmunoassay for human luteinizing hormone is analyzed in terms of $\sqrt{y'}$ in Chapter 24.

Alternatively, the standard deviation between replicate counts, $s = \sqrt{V(y')}$, may increase linearly with $\bar{y}'$, in which case each count $y'$ is transformed to its logarithm. This relation holds equally in selecting a log transformation for a measurement, as in the oviduct response of chicks to stilbestrol (Figure 6). If the slope $b$ of $s$ upon $\bar{y}'$ has an intercept $a'$ other than zero, the simplest distributional form is the negative binomial (Bliss and Fisher, 1953). The transformation that stabilizes the variance is then $y = \log(y' + a'/b)$ or $y = \log(y' + c)$. In a negative binomial distribution $c$ is an estimate of $\frac{1}{2}k$, where $k$ is its second parameter (Bliss and Owen, 1958). If the rate of decay in a radioimmunoassay were so rapid that its expectation changed appreciably between replicates, a logarithmic transformation of the counts might be preferred to their square roots.

## DISCUSSION

FINNEY   As regards the rejection of observations, I distrust any slick, formal rule which calculates this and calculates that and then, if $A$ is greater than $B$, throws the observation away. I doubt whether Professor Cochran meant anything quite so rigid, but I am afraid that he may have given that impression. I might, as a counter, state my own philosophy. I never like to reject an observation purely on my own responsibility. I want full discussion with the experimenter concerned, a policy that Professor Cochran implied, and I am reluctant to discard any observation *purely* on the basis of calculations. Consider an investigator who says, "Yes, this observation is obviously quite wrong, and I now see what happened. My assistant is notoriously untrustworthy and occasionally makes completely wild measurements." Cochran's rules may have helped to detect the flaw, but the investigator takes the decision to discard. (If an endocrinologist has an assistant who is liable to do such things rather often, possibly it is the assistant who should be rejected and not the observation.) On the other hand, the experimenter concerned may say, "Yes, this observation is obviously wrong. I now recognize that when we are measuring toward the lower end of the scale some important piece of instrumentation becomes unreliable, and we may record an excessively low value." That observation, $-0.038$ in Table 1, possibly has a much higher sampling variance than the rest, in which case a sensible decision may be to reject it and replace it with an expectation based on its being the smallest member of a sample of seven. (I think there are technical difficulties involved here if the experiment is more complicated in design.) Much more troublesome is the situation in which the experimenter says, "This figure looks completely absurd. I have no idea what is wrong with it, but I think it is wrong." One may imagine a table of weights of adult males containing one entry of 25 pounds; this may be a mistake for 250 or for 125 or for something quite different, but obviously it is not the weight of a normal human male. I would reject it, but should I reject 45 pounds, or 65 pounds, or 85 pounds? I don't know, and I don't think it is my job as a statistician to make the decision. If nothing is known about a reason for rejection, and the experimenter is not prepared to say, "This is so absurd that from my biological knowledge I insist on its being rejected," I would prefer to leave it in and not rely on an Anscombe, Tukey, or other rule.

I think something similar applies to transformation. The very interesting presentation of the analysis of variance lacked any mention of the fact that experiments seldom come singly or that, even if they come singly, there may be another one of the same series coming next month. One can have a rule for determining an appropriate transformation, and Professor Cochran stated that usually advocated. However, what is one to do when successive experiments of the same series indicate, as they are likely to do, different transformations? I am unwilling to analyze a relatively small experiment, look at the behavior of the variance, and say, "I think we should use the logarithmic transformation on this." Five weeks later another experiment in the same series may, by this same rule, suggest that the square-root transformation is the best, and a third experiment in the series may suggest another transformation. This probably means that no one experiment is sufficiently precise to indicate any transformation as quite clearly the best. I prefer to sacrifice a little on the quality of the transformation in the interest of having a consistent presentation of the series of experiments. One should also bear in mind that the investigator decides the scale of measurement in which he is interested. If he says that he is interested in the content of some constituent of a body fluid in micrograms per cubic centimeter or in the percentage of animals that show a particular reaction under experimental conditions, I am not doing a very satisfactory job for him if I present my results in terms of the inverse sine of the square root of his values. Yet too little is known, and even less is practised, of methods of inverting a transformation at the end of an analysis in order to return to the required scale.

COCHRAN  Nothing that Professor Finney has said about the rejection of observations is in contradiction to my views. He pointed out the importance of making an investigation before adopting any automatic rejection rule. The reason that there has been a demand for a routine rule is that in many cases such investigations are unhelpful. They fail to uncover information that enables us to substitute a more factually correct value. We then face the question of what to do. Some, like Professor Finney, lean toward a policy of reluctance to reject. Nevertheless, a gross error will make quite a difference in the conclusions. My judgment is that in such situations the routine use of the Anscombe-Tukey technique with a low premium will do more good than harm.

I should mention one other point. There is evidence that any rejection rule that rates a residual by its ratio to the standard error computed from the observations may not do well when there are two or three gross errors or outliers around. If there are two big errors, the standard error may be so inflated by the second gross error that one cannot even detect the first. In this situation, simple plotting of the kind that Dr. Bliss recommended or a test analogous to Daniel's half-normal plot method is likely to do the job better.

BLISS  Rejection of observations was introduced into the Pharmacopeia because on reviewing data derived from various sources we discovered that one statistician had rejected about a quarter of his data whereas another had rejected none. An objective criterion was needed.

SMITH  It was indicated that the rejection of an outlier very seldom leads one astray. In Figure 5 the rejection of the outlier may be misleading. As I understand this example, it represents a growth curve, that is, the effect of a drug, stilbestrol, on the weight of the rat uterus. This function will have an asymptote as the dose increases. It can't just go off helter-skelter. By rejecting that one point in the plot one gets a new value, which is higher than it was originally. Consequently, when one isolates a degree of freedom for a quadratic effect, one is using a curve that, in my view, is essentially incorrect in terms of the underlying model. Thus, there is good reason not to reject it. If it is not rejected, and a quadratic curve is fitted, the quadratic term will not be significant. This is not contrary to Dr. Bliss's position; I think he was merely illustrating the procedure for isolating a single degree of freedom. However, one should first state the mathematical model employed before discussing the rejection of outliers and segregating single degrees of freedom.

BLISS   I should expect this curve to have asymptotes at both the upper and lower extremities. If these doses were nearer the lower asymptote, I should expect curvature in just this direction. That is why I would interpret the rejection of the outlier as an adjustment in the right direction. On the insurance scheme, that first point would be still further off, since it would be replaced with the mean of the remaining observations.

SMITH   I would not do that; such an approach would never occur to me. I would either reject it or retain it. I agree with Professor Finney in that respect.

MCHUGH   I should like to ask Professor Finney two questions regarding transformations. I was interested in the Armitage-Remington example of the effect of a transformation of response in a 2 × 2 factorial experiment (p. 12). In the original scale there was no interaction; when square roots were taken, interaction appeared. Is this an example of what you find disturbing?

My second question concerns the choice of scale. I gather that you view this as not entirely a statistical issue. Do you tend toward the position that the experimenter's theoretical knowledge and experience are the essential criteria?

FINNEY   I would not imply that I never reject or never transform observations. I am pleading for a more critical approach to these problems. The Armitage-Remington illustration does not seem to me particularly relevant. Certainly, if a factorial experiment shows simple additivity on one scale, any nonlinear transformation will distort the addition to some extent. There was nothing atypical about the Armitage-Remington example; the same would almost certainly be true of any simple figures that might have been chosen to illustrate their argument. I would not go so far as to say that the experimenter should be solely responsible for the choice of scale. Perhaps this should be a joint responsibility. The statistician — here I run the risk of being provocative toward ninety per cent of the audience — has the advantage of objectivity. The experimenter has vested interests in the interpretation of his data. He may be tempted to find reasons for a transformation that supports his own hypotheses, and the restraining influence of the statistician may reduce the ill consequences of subjective judgments.

COCHRAN   I have two comments. First, with mild transformations, like the square root for counts and the inverse sine for percentages between 20 and 80, analyses in the original scale or in the transformed scale rarely produce material differences in the important conclusions. This is not so with a more severe transformation like the logarithmic one, which is, in effect, to say, "The world moves on a percentage scale and not on an absolute-difference scale."

Second, what does statistical theory say? It says that one must get one's model right. Transformations as such only enter the theory of estimation (the theory that promises to deliver good estimates) as an approximate means to an end. If one believes that the residuals in the original scale are normal, one does not transform to render a nonadditive model additive. One fits this difficult model with the original data by iteration on a computer. A transformation undertaken to make the model additive is theoretically justified as a good approximation only if it can be shown to retain residuals that are independent and approximately normally distributed. In Fisher's example, in which a multiplication model made more sense to him as a scientist, he did not transform. He analyzed the product model in the original scale.

BLISS   An an example of outliers, in an analysis in angles based upon groups of ten mice the protective action of pretreatment with conditioning injections of *Staphylococcus aureus* and of human or cow's milk was tested against later lethal levels of the same bacterium. From the number of survivors in each group the first working angles gave a chi square from the error term with 83 DF of 121.46 ($P = 0.0061$) or, in other words, one much bigger than its expectation. The residuals were then calculated, the two largest of the 83 DF were discarded, and the analysis was repeated with new working angles. The new chi square showed a probability of $P = 0.303$, in quite satisfactory agreement with the binomial expectation.

EMMENS   I should like to raise the question (which Haldane always used to answer negatively) of whether the majority of biological observations are normally distributed. The middle part may usually be, but many of us find so-called outliers among groups of data in which nothing is wrong, such as the heights and weights of humans. What does one do? Does one get the model right and have new distributions that allow for the nonnormality of many biological observations? Alternatively, does one adhere to the normal distribution and discard observations that don't appear to fit?

COCHRAN   Many of these methods have been developed partly because fitting the model with the correct nonnormal distribution of errors has so far proved too difficult. Computers will ease that process, but on a desk machine transformations, insofar as they increase proximity to normality, are useful.

EMMENS An outstanding feature of the nonnormality of many biological observations is not that they are skewed but that they have an unusual proportion of outliers. These may occur only occasionally, yet with sufficient frequency to cause worry about rejection. If the true distribution includes what would be called outliers in the normal distribution, what should one do?

COCHRAN With a long-tailed distribution of errors I do not recommend the Anscombe-Tukey rejection of outliers. One should either do the hard work of analyzing on a more realistic representation of the error distribution or transform to a scale in which one believes that one comes near enough to normality to permit the application of normal models. Another possibility is the device of Winsorization, though I have had insufficient experience with this technique.

RINARD It was stated earlier that additivity is an assumption necessary for the analysis of variance. From the experimenter's viewpoint nonadditivity — that is, potentiation or inhibition — is often of great interest. Is the analysis of variance useful in detecting these types of interaction? Further, if it is useful and a transformation is undertaken to achieve homogeneity of variance, what happens to such interactions?

COCHRAN The kind of additivity that the analysis of variance requires is additivity of row and column effects, which will usually be treatment and replication, or group, effects. Effects on responses of drug potentiation or inhibition can be studied by response curves fitted to the treatment means or by an appropriate breakdown of the treatments sum of squares. Much of the data that I see are on the effects of inhibitors.

Concerning your second question we should have said more. I would transform to achieve additivity rather than to achieve equal variance. Additivity implies that life is being simply described. If the effect of one factor remains constant over all sorts of variations of the environment or of the levels of other factors, the work is quite easy. Even if that happens in a transformed scale, you have achieved a relatively simple description of a complex phenomenon.

RINARD It may be simple from the statistician's standpoint, but experimenters characteristically think in the original scale of measurement. I should hate to do an experiment, make a transformation to some unaccustomed scale, and thereby fail to detect some interesting hormonal interaction. I want to emphasize that there is danger that this will occur when an investigator looks at such transformed data. Experimenters who use the analysis of variance should be aware of what they are doing when they transform their data.

COCHRAN   That's a good point. However, if interactions are a mere artifact of the scale you are using, they are unlikely to be interesting from the point of view of pure science!

# SOME EFFECTS OF ERRORS OF MEASUREMENT ON LINEAR REGRESSION

W. G. COCHRAN

HARVARD UNIVERSITY

## 1. Introduction

I assume a bivariate distribution of pairs $(y, X)$ in which $y$ has a linear regression on $X$

(1.1) $$y = \beta_0 + \beta_1 X + e,$$

where $e$, $X$ are independently distributed and $E(e|X) = 0$. However, the measurement of $X$ is subject to error. Thus we actually observe pairs $(y, x)$, with $x = X + h$, where $h$ is a random variable representing the error of measurement.

Given a random sample of pairs $(y, x)$, previous writers have discussed various approaches to the problem of making inferences about the line $\beta_0 + \beta_1 X$, sometimes called the *structural* relation between $y$ and $X$. In the present context this line might be called "the regression of $y$ on the correct $X$" to distinguish it from "the regression of $y$ on the fallible $x$." An obviously relevant question is: under assumption (1.1), what is the nature of the regression of $y$ on $x$?

Lindley [5] gave the necessary and sufficient conditions that the regression of $y$ on the fallible $x$ be linear in the narrow sense. This means that $E(y|x)$ is linear in $x$, or equivalently that

(1.2) $$y = \beta_0' + \beta_1' x + e',$$

where $E(e'|x) = 0$. This definition does not require that $e'$ and $x$ be independently distributed. Lindley's proof assumes that the error of measurement $h$ is distributed independently of $X$. His necessary and sufficient conditions are that Fisher's cumulant function (logarithm of the characteristic function) of $h$ be a multiple of that of $X$. Roughly speaking, this implies that $h$ and $X$ belong to the same class of distributions. Thus if $X$ is distributed as $\chi^2\sigma^2$, so is $h$, though the degrees of freedom can differ: if $X$ is normal, $h$ must be normal.

Several writers have discussed the corresponding necessary and sufficient conditions if we demand in addition that the residual $e'$ in (1.2) be distributed independently of $x$. In particular, Fix [3] showed that if the second moment of

This work was supported by the Office of Naval Research through Contract N00014-56A-0298-0017, NR-042-097 with the Department of Statistics, Harvard University.

either $X$ or $h$ exists and $X, h$ are independent as before, the conditions for linearity of regression in this fuller sense are that both $X$ and $h$ be normally distributed.

Thanks partly to computers, numerous regression studies are being done nowadays, particularly in the social sciences and medicine, in which all the variables are difficult to measure, and therefore are presumably measured with sizeable errors. In thinking about mathematical models appropriate to such uses, it seems clear that the forces which determine the nature of the distribution of $h$ (the imperfections of the measuring instrument or process) are quite different from those that determine the nature of the distribution of the correct $X$. Consequently, my opinion is that in such applications even the Lindley conditions will not be satisfied, except perhaps by a fluke or as an approximation (for example, the cumulants of $X$ and $h$ might be similar in the sense that both distributions are close to normal).

This paper considers the regression of $y$ on $x$ when Lindley's conditions are not satisfied. There are at least two reasons for interest in this regression. The objective may be to obtain a consistent estimate of $\beta_1$ for purposes of interpretation or adjustment by covariance (Lord [6]). Secondly, the purpose may be to predict $y$ from the fallible $x$ by the regression technique in which case the shape of this regression is relevant.

The strategy used here is first to construct a straight line relation between $y$ and the fallible $x$ which may be called the *linear component* of the regression of $y$ on $x$. This is the line that we are estimating, in some sense, when we compute a sample linear regression of $y$ on $x$.

The paper then takes a look at the question: what is the nature of the departure from linearity when Lindley's conditions are not satisfied? In particular, does the linear component dominate? If it does, then as Kendall [4] remarks, "A slight departure from linearity will sometimes allow the ordinary theory to be used as an approximation." I have been unable to obtain any general results that are exact, but something can be learned by a combination of an approach *via* moments and the working out of some easy particular cases. These suggest, fortunately, that the linear component often dominates, even with measurements of rather poor reliability, but the issue needs more thorough investigation by someone with greater mathematical power.

## 2. The linear component

As stated, a linear regression of $y$ on the correct $X$ (in the fullest sense) is assumed, namely,

$$(2.1) \qquad\qquad y = \beta_0 + \beta_1 X + e,$$

where $e, X$ are independently distributed and $E(e|X) = 0$. I also assume $X$ scaled so that $E(X) = 0$.

As regards the error of measurement $h$, Lindley's result requires $h$, $X$ to be independently distributed, but this assumption limits the range of applications of the result. Some measuring instruments or methods underestimate high values of $X$ and overestimate low values. I have been unable to obtain conditions analogous to Lindley's when $X$, $h$ follow a general bivariate distribution $\phi(X, h)$ but there is no difficulty in obtaining the linear component in this case. Denote $E(h)$ by $\mu_h$, since measurements may be biased. It is assumed, as seems reasonable for most applications, that $h$ and hence $x$ are distributed independently of $e$. Hence the regression of $y$ on $x$ is, from (2.1),

$$(2.2) \qquad E(y|x) = \beta_0 + \beta_1 E(X|x) = \beta_0 + \beta_1 R(x),$$

say.

Thus we need to find $R(x)$. Let $\phi(X, h)$ be the joint frequency function of $X$, $h$. The marginal distribution of the fallible $x$ is

$$(2.3) \qquad \psi(x) = \int \phi(X, x - X)\, dX,$$

while $R(x) = E(X|x)$ satisfies the equation

$$(2.4) \qquad R(x)\psi(x) = \int X\phi(X, x - X)\, dX.$$

The linear component of $R(x)$ can be defined by fitting the straight line $L(x) = C_0 + C_1 x$ to $R(x)$ by the population analogue of the method of least squares. That is, we choose $C_0$, $C_1$ to minimize

$$(2.5) \qquad \int \{R(x) - C_0 - C_1 x\}^2 \psi(x)\, dx.$$

Clearly,

$$(2.6) \qquad \int R(x)\psi(x)\, dx = \iint X\phi(X, h)\, dX dh = \mu_X = 0,$$

$$(2.7) \qquad \int x R(x)\psi(x)\, dx = \iint (X^2 + Xh)\phi(X, h)\, dX dh = \sigma_X^2 + \sigma_{Xh},$$

where $\sigma_{Xh}$ is the population covariance of $X$ and $h$. Hence the normal equations for $C_0$ and $C_1$ give

$$(2.8) \qquad C_0 = -C_1 \mu_h, \quad C_1 = (\sigma_X^2 + \sigma_{Xh})/(\sigma_X^2 + \sigma_h^2 + 2\sigma_{Xh}).$$

From (2.2), the linear component of the regression of $y$ on $x$ is $L(x) = \beta_0 + \beta_1(C_0 + C_1 x)$. If we write this $\beta_0' + \beta_1' x$, we have

$$(2.9) \qquad \beta_0' = \beta_0 - \beta_1 C_1 \mu_h, \quad \beta_1' = \beta_1 C_1 = \beta_1(\sigma_X^2 + \sigma_{Xh})/(\sigma_X^2 + \sigma_h^2 + 2\sigma_{Xh}).$$

Incidentally, expressions (2.9) for $\beta'_0$, $\beta'_1$ can be obtained directly by noting that our procedure is equivalent to defining $(\beta'_0 + \beta'_1 x)$ as the linear component of the regression of $y$ on $x$ if we write $y = \beta'_0 + \beta'_1 x + e'$ and determine $\beta'_0$, $\beta'_1$ so that the residuals $e'$ satisfy the conditions

$$(2.10) \qquad\qquad E(e') = 0, \qquad \text{Cov } (e', x) = 0.$$

Formulas (2.9) for $\beta'_0$ and $\beta'_1$ agree with the well-known elementary results in the literature, usually obtained on the assumption that $X$, $h$ follow a bivariate normal. Bias in the measurements affects the intercept $\beta'_0$ but not the slope $\beta'_1$. If $h$ and $X$ are uncorrelated, $\beta'_1 = \beta_1 \sigma_X^2/\sigma_x^2$, the factor $\sigma_X^2/\sigma_x^2$ being often called the *reliability* of the measurement $x$. For given $\sigma_X^2$, $\sigma_h^2$, positive correlation of the errors with $X$ makes the underestimation of the slope worse, while negative correlation alleviates it if $\sigma_h^2 < \sigma_X^2$. In the Berkson case [1], the investigator plans to apply preselected amounts $x$ of some agent or treatment in a laboratory experiment, but owing to errors in measuring out this amount, the amount $X$ actually applied is different. Here, Cov $(x, h) = 0$, so that $\sigma_{xh} = -\sigma_h^2$ and (2.9) shows that $\beta'_1 = \beta_1$. This situation also applies when large samples are grouped by their values of $X$ into classes to facilitate the calculation of regression on a desk machine, provided that $x$ is taken as the *mean* of $X$ within each class. The common practice is of course to take $x$ as the midpoint of the class. This makes Cov $(x, h)$ slightly positive for most unimodal distributions of $X$, so that some residual inconsistency in $\beta'_1$ as an estimate of $\beta_1$ remains, though the inconsistency is in general trivial if at least ten classes are used.

Suppose now that $y$ is also subject to an error of measurement $d$. If $Y$ represents the correct value of $y$, we may rewrite the original model (1.1) as

$$(2.11) \qquad\qquad Y = \beta_0 + \beta_1 X + e, \qquad y = Y + d.$$

Hence

$$(2.12) \qquad\qquad E(y|x) = \beta_0 + \beta_1 R(x) + E(d|x).$$

If errors in $y$ are independent of $Y$, $X$ and $h$, then $E(d|x) = \mu_d$, the amount of bias in $d$, and we get the old result that such errors in $y$ do not affect the slope of the regression line. If $d$ is correlated with $Y$, the choice of an appropriate model requires care. Specification of the joint frequency functions $\phi(X, h)$, $\theta(Y, d)$ is not enough to determine $E(d|x)$; we need to know the relation between $d$ and $h$. The following might serve for applications in which the process by which $y$ is measured is independent of that by which $x$ is measured. Noting that $E(X) = 0$, $E(Y) = \beta_0$, write

$$(2.13) \qquad d = \mu_d + \frac{\sigma_{Yd}}{\sigma_Y^2}(Y - \beta_0) + d', \qquad h = \mu_h + \frac{\sigma_{Xh}}{\sigma_X^2} X + h',$$

where $h'$, $d'$, with zero means, are assumed independent of each other and of $X$, $Y$, and $e$. This model does not imply that $d$ and $h$ are independent, since

(2.14)
$$\text{Cov } (dh) = \frac{\sigma_{Yd}\sigma_{Xh}\sigma_{XY}}{\sigma_Y^2\sigma_X^2},$$

but this correlation arises only as a consequence of the $X$, $Y$ correlation.

In some applications there may be further correlation between $d$ and $h$ because the measuring processes are not independent. For instance, an individual pair $(x, y)$ might be estimates of a town population five years apart, where the municipal statisticians use the same techniques in a town, the technique varying from town to town. In general, it will obviously be difficult to know which model to pursue, and to get data for verification of a model.

If (2.13) holds,

(2.15)
$$E(d|x) = \mu_d + \frac{\sigma_{Yd}}{\sigma_Y^2}\beta_1 R(x).$$

From (2.8) and (2.12), we obtain for the linear component of the regression of $y$ on $x$,

(2.16)
$$\left[\beta_0 + \mu_d - \beta_1 C_1 \mu_h \frac{(\sigma_Y^2 + \sigma_{Yd})}{\sigma_Y^2}\right] + \beta_1 x \frac{(\sigma_Y^2 + \sigma_{Yd})(\sigma_X^2 + \sigma_{Xh})}{\sigma_Y^2(\sigma_X^2 + 2\sigma_{Xh} + \sigma_h^2)}.$$

As is obvious from graphical considerations, errors in $y$ that are positively correlated with $Y$ tend to increase the absolute value of $\beta_1'$, whereas errors in $x$ have the opposite effect. With errors in both $y$ and $x$, $\beta_1'$ may be either greater or less than $\beta_1$.

The method of obtaining the linear component extends naturally to a multiple linear regression of $y$ on $x_1, x_2, \cdots, x_k$. Even when the errors of measurement $h_i$ are independent of $X_i$ and of each other—the simplest case—$\beta_i'$ is a linear function of all $\beta_j$ whose corresponding $x_j$ are subject to errors of measurement (Cochran [2]). When the $h_i$ and $X_i$ are correlated, we again meet the problem of specifying the nature of the correlation between $h_i$ and $h_j$.

One objective in working out the relations between $\beta_i'$ and the $\beta_j$ is as a possible means of estimating the coefficients $\beta_j$ of the structural regression by using data from supplementary studies of the errors of measurement. With errors in more than one variate, however, the algebraic results suggest that the information needed about errors of measurement is more than we are likely to be able to obtain.

## 3. Polynomial approach by moments

Now consider the nature of $R(x)$ with errors in $x$ only when Lindley's conditions are not satisfied. Like Lindley, I assume $h$ and $X$ independent, with $\mu_h = 0$. I first chose some simple forms for the frequency functions $f(X)$ of $X$ and $g(h)$ of $h$ for which $R(x)$ can be worked out exactly in closed form. Examples are the $\chi$ distribution with a small number of degrees of freedom, the normal, the uniform, and the exponential types like $e^{-X}$, with $X > 0$, or $\frac{1}{2}e^{-|x|}$, with $-\infty < X < \infty$. Inspection of a few cases indicated that if either $X$ or $h$ follows a skew distribution, the departure of $R(x)$ from linearity, in a region around the mean of $x$, is of the simple type that can be approximated by a quadratic curve (an example will be given in Section 4).

If, however, both $h$ and $X$ are symmetrically distributed about their means 0, then $\psi(x)$ is also symmetrical and $R(-x) = -R(x)$, which suggests a cubic approximation with a zero quadratic term. The equations of the approximating quadratic or cubic, by the least squares method, are obtained easily from the low moments of the distributions of $h$ and $X$. (In the symmetric case it is possible that for some frequency functions a quadratic approximation in $|x|$, with reversal of sign when $x$ is negative, might do better than the cubic, but the fitting requires calculation of some incomplete moments and this has not been pursued.)

In fitting a polynomial approximation of degree $p$ to $R(x)$, we choose the coefficients $C_i$ to minimize

$$(3.1) \qquad \int \{R(x) - \sum_{i=0}^{p} C_i x^i\}^2 \psi(x) \, dx.$$

The $r$th normal equation is

$$(3.2) \qquad \sum_{i=0}^{p} C_i \mu_{r+i, x} = \iint X(X + h)^r f(X) g(h) \, dX dh,$$

where the $\mu$ denote moments about the mean.

Here we are fitting only the simplest nonlinear approximations, say $Q(x)$.

*Case 1.  X or h skew.*

$$(3.3) \qquad Q(x) = C_1 x + C_2 (x^2 - \mu_{2x}),$$

where

$$(3.4) \qquad \Delta = \mu_{4x}\mu_{2x} - \mu_{3x}^2 - \mu_{2x}^3,$$

$$(3.5) \qquad C_1 = (\mu_{4x}\mu_{2X} - \mu_{3x}\mu_{3X} - \mu_{2x}^2\mu_{2X})\Delta^{-1},$$

$$(3.6) \qquad C_2 = (\mu_{2x}\mu_{3X} - \mu_{3x}\mu_{2X})\Delta^{-1}.$$

*Case 2.  X and h both symmetrical.*

$$(3.7) \qquad C(x) = c_1 x + c_3 x^3,$$

where

$$(3.8) \qquad \Delta = \mu_{6x}\mu_{2x} - \mu_{4x}^2,$$

$$(3.9) \qquad c_1 = (\mu_{6x}\mu_{2X} - \mu_{4x}\mu_{4X} - 3\mu_{4x}\mu_{2X}\mu_{2h})\Delta^{-1},$$

$$(3.10) \qquad c_3 = (\mu_{4x}\mu_{2X} - \mu_{2x}\mu_{4X} - 3\mu_{2x}\mu_{2X}\mu_{2h})\Delta^{-1}.$$

There is obvious interest in seeing how well the quadratic and cubic approximations fit $R(x)$. The reduction in the variance of $R(x)$ due to these approximations can be obtained from the normal equations by the usual analysis of variance rule of multiplying the solutions $C_i$ or $c_i$ by the right sides of the normal equations. But I have been unable to obtain an exact expression for the variance of $R(x)$ which does not involve computing an integral, so that I do not have a general result for the closeness of fit in this case, though it can be obtained by numerical integration in specific examples, as discussed in Section 5.

This approach extends also to correlated errors, the expressions for the $C_i$ involving joint moments of $h$ and $X$ which are easily obtained.

## 4. Two examples

As an example of the quadratic approximation I take

$$(4.1) \qquad f(X) = Xe^{-X^2/2}, \qquad X > 0, \qquad g(h) = N(0, \sigma^2).$$

Thus $X$ is skew, essentially a $\chi$ variate with two degrees of freedom, mean $\sqrt{\pi/2}$ and variance $(2 - \pi/2)$, about 0.429, while $h$ is normal. The reliability of the measurement is $(4 - \pi)/(4 - \pi + 2\sigma^2)$, so that measurements with different degrees of reliability from 92 per cent to 54 per cent are represented by taking $\sigma = 0.2(0.1)0.6$. Measurements of lower reliability have been reported, but this range should cover the great majority of applications. Reliability of 50 per cent is far from impressive: the measurement error has as big a variance as the correct measurement. It is not claimed that this example corresponds to any actual situation in practice: although $X$ is essentially positive, $x$ is not.

For this example, $R(x)$ works out as

$$(4.2) \qquad R(x) = \frac{\sigma}{(1 + \sigma^2)^{1/2}} \left\{ \frac{(u^2 + 1)P(u) + uz(u)}{uP(u) + z(u)} \right\}$$

where $u = x/\sigma(1 + \sigma^2)^{1/2}$, $z(u)$, and $P(u)$ are the ordinate and cumulative of $N(0, 1)$. (In this example $X$ was not scaled so that $E(X) = 0$). Figure 1 presents points on $R(x)$ for $\sigma = 0.3$ and $\sigma = 0.6$, plus the line $R(x) = x$ that would apply if there were no error of measurement.

As suggested, the major departure from linearity of the points in Figure 1 near $\mu_x$ is a simple curvature that is well represented by a quadratic curve, though this quadratic would do very badly at points far away from $\mu_x$. For instance, with $x$ and $u$ becoming large and positive, $P(u)$ tends to one and $z(u)$ to zero, so that $R(x)$ behaves like $x/(1 + \sigma^2)$, while for $x$ negative $R(x)$ tends to become asymptotic to the origin. However, the approximation errors in the quadratic $Q(x)$ receive very little weight at these points, since they are far enough away from the mean of $x$ so that $\psi(x)$ is tiny. The mean square error (MSE) of the quadratic approximation, the integral of $[Q(x) - R(x)]^2\psi(x)\,dx$, is only around 0.0003 to 0.0007. The MSE of the linear approximation has the following values.

| $\sigma$ | .2 | .3 | .4 | .5 | .6 |
|---|---|---|---|---|---|
| MSE$[L(x)]$ | .0010 | .0025 | .0041 | .0052 | .0069 |

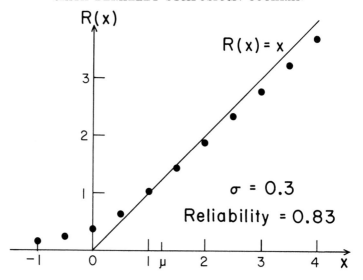

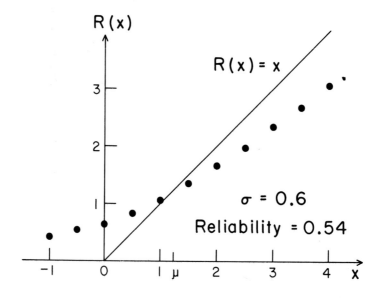

FIGURE 1

Regression $R(x)$ of $X$ (correct) on $x$ (fallible) measurement.
$X$ is $\chi$ (2 d.f.), error of measurement is $N(0, \sigma^2)$.

The example of the symmetrical case has $X = N(0, 1)$, while the measurement error $h$ is uniform between $-L$ and $L$. The reliability is $3/(3 + L^2)$ and with $L = 0.5(0.25)1.5$, it varies between 93 per cent and 57 per cent. Figure 2 shows $R(x)$ for $L = 1$, $L = 1.5$, with reliabilites 0.75 and 0.57. In this example

$$(4.3) \qquad R(x) = [z(x - L) - z(x + L)][C(x + L) - C(x - L)]^{-1}.$$

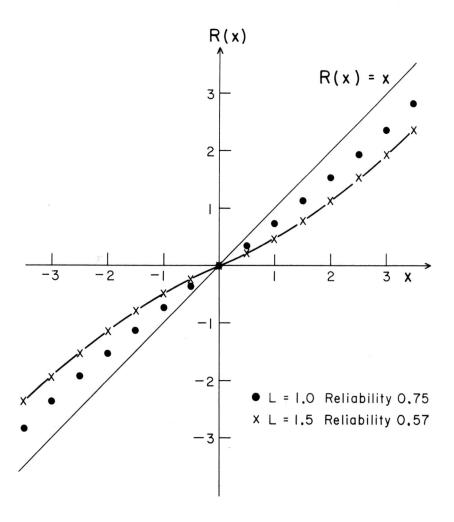

FIGURE 2

Regression of $R(x)$ of $X$ (correct) on $x$ (fallible) measurement. $X$ is $N(0, 1)$, error of measurement is uniform $(-L, L)$.

As $x$ moves far away from its mean at 0, $R(x)$ tends to become $(x - L)$, again linear in $x$ rather than cubic. The MSE of the cubic approximation is very small, the greatest value in the cases worked being 0.00054 at $L = 1.5$. For the linear approximation, the MSE are as follows.

| $L$ | 0.5 | 0.75 | 1 | 1.25 | 1.5 |
|---|---|---|---|---|---|
| $\text{MSE}[L(x)]$ | $.0^45$ | $.0^316$ | $.0011$ | $.0046$ | $.0116$ |

## 5. Adequacy of the linear approximation

In considering the adequacy of the linear approximation, I am not concerned with applications in which the aim is to estimate the structural relationship, but only with those in which (i) the objective is to predict $y$ from $x$, in which case the usual advice is to utilize the observed regression of $y$ on $x$, or (ii) to test the null hypothesis $\beta'_1 = 0$ by a $t$ test, as a means of testing the null hypothesis $\beta_1 = 0$.

First, as Kendall [4] reminded us, Lindley's type of linearity falls short of the assumptions of independence of the residual $e'$ and $x$ and normality of $e'$ that are needed for use of the standard regression formulas. The situation resembles that in the standard texts on sample surveys, in which attempts have been made to construct a theory of linear regression estimators without allotting any specific structure to a relation between $y$ and $x$ except that all values of $y$ and $x$ in the population are bounded. It is known in such cases that the usual sample estimate $\hat{\beta}'_1$ of $\beta'_1$ is biased, the leading term in the bias being $-E(e'x^3)/n\sigma_x^4$, where $n$ is the sample size. The usual formula for the variance of $\hat{\beta}'_1$ holds as a first approximation, but it too has a bias that becomes negligible in large samples, while the numerator and denominator of the $t$ test of $\hat{\beta}'_1$ only become independent asymptotically. At best we can say that the usual methods apply asymptotically.

As regards the effect of errors in $x$ on the precision of regression estimates of $y$, the most important quantity is the variance of the residuals from the regression of $y$ on $x$, or from approximations to this regression that we consider using. It is worth starting with the simplest case in which $X$, $h$ are normal and independent, so that the regression of $y$ on $x$ is linear in both senses. Here $\beta'_1 = \beta_1 \sigma_X^2/\sigma_x^2 = G\beta_1$,

where $G$ is the coefficient of reliability. Further, since

$$(5.1) \qquad \sigma_y^2 = \sigma_e^2 + \beta_1^2\sigma_X^2 = \sigma_{e'}^2 + \beta_1'^2\sigma_x^2 = \sigma_{e'}^2 + G\beta_1^2\sigma_X^2,$$

we have

$$(5.2) \qquad \sigma_{e'}^2 = \sigma_e^2 + (1 - G)\beta_1^2\sigma_X^2 = (1 - \rho^2)\sigma_y^2 + (1 - G)\rho^2\sigma_y^2.$$

This is the familiar result that, even when there is no problem about nonlinearity, errors in $x$ increase the variance of the deviations from the linear prediction model by $(1 - G)\rho^2\sigma_y^2$, an increase that hurts most, for given $G$, when the prediction formula is very good ($\rho$ high).

Returning to the assumptions of this paper, suppose $y$ is predicted from its regression on $x$. Since

(5.3) $$y = \beta_0 + \beta_1 X + e = \beta_0 + \beta_1 R(x) + e',$$

(5.4) $$\sigma_{e'}^2 = \sigma_e^2 + \beta_1^2(\sigma_X^2 - \sigma_R^2) = \sigma_e^2 + \beta_1^2\sigma_X^2(1 - \sigma_R^2/\sigma_X^2).$$

In the examples that I have worked, $\sigma_R^2/\sigma_X^2$ comes numerically to within one or two percentage points of $G = \sigma_X^2/\sigma_x^2$. Consequently, if the population regression of $y$ on $x$ is used for prediction, the loss of precision due to errors in $x$ is about what we expect from the value of the reliability coefficient $G$.

To pass to the additional loss of precision if $L(x)$ instead of $R(x)$ is used in the prediction formula, write

(5.5) $$\beta_0 + \beta_1 R(x) + e' = \beta_0 + \beta_1 L(x) + e'',$$

giving

(5.6) $$\sigma_{e''}^2 = \sigma_{e'}^2 + \beta_1^2(\sigma_R^2 - \sigma_L^2) = \sigma_{e'}^2 + \beta_1^2 \text{ MSE } (L)$$

since the method of fitting $L$ makes Cov $(L, R - L) = 0$. From the little tables of MSE $(L)$ in Section 4, the highest ratios of MSE $(L)$ to $\sigma_X^2$, which occur when reliability is lowest, are around 0.01 to 0.02. If these examples typify what happens in applications, it appears that errors in $x$ create an increase of around $(1 - G)\rho^2\sigma_y^2$ in the variance of residuals, but that even when reliability is low, the additional increase in this variance due to use of $L(x)$ instead of $R(x)$ is unimportant.

Since $e'$ and $x$ are not independent, a look at the conditional distribution of $e'$ for fixed $x$ may be worthwhile. From (5.2),

(5.7) $$e' = e + \beta_1[X - R(x)].$$

By our hypothesis, the term $e$ is independent of $X$, $h$, and hence $x$, so that this term has the same shape of distribution and variance in all arrays with $x$ fixed. The second term is determined by the distribution of $X$ for fixed $x$. In Example 1 this distribution works out as

(5.8) $$f(X|x) = \frac{(1 + \sigma^2)}{\sigma^2\sqrt{2\pi}}\frac{1}{K(x)} X \exp\left\{-\frac{1}{2}\frac{(1 + \sigma^2)}{\sigma^2}\left[X - \frac{x}{1 + \sigma^2}\right]^2\right\}$$

where, with $\mu = x/\sigma(1 + \sigma^2)^{1/2}$,

(5.9) $$K(x) = z(u) + uP(u).$$

This distribution changes in shape as $x$ varies; its variance increases with $x$, the increase being small when the reliability $G$ is high but more marked when $G$ is lower.

In Example 2, $f(X|x)$ is simply the incomplete normal

(5.10) $$f(X|x) = \frac{1}{\sqrt{2\pi}}\frac{1}{K(x)}\exp\{-\tfrac{1}{2}X^2\}; \qquad x - L \leqq X \leqq x + L$$

with $K(x) = P(x + L) - P(x - L)$. When $x$ is at its mean 0, this is symmetrical with its maximum variance, but changes from negative to positive skewness as $x$ changes from negative to positive.

Thus the distribution of $e'$ is a compound of two independent distributions, one unchanging, the other changing with $x$. As we have seen, the average variances of the two components are $\sigma_y^2(1 - \rho^2)$ and approximately $\rho^2\sigma_y^2(1 - G)$. With $G$ high and $\rho$ modest, the unchanging component should dominate, and the assumption that the distribution of $e'$ is the same for different $x$ may be a reasonable approximation, but with say $\rho = 0.8$, $G = 0.5$, the two components have variances $0.36\sigma_y^2$ and $0.32\sigma_y^2$ and are about equally important.

## 6. Summary and discussion

This paper deals with applications in which the standard linear regression model $y = \beta_0 + \beta_1 X + e$, with $e$, $X$ independent and $E(e) = 0$, is assumed to apply to a bivariate sample of pairs $(y, X)$. However, owing to difficulties in measuring the $X$ values, we actually have a bivariate sample $(y, x)$ where $x = X + h$, $h$ being an error of measurement. My opinion is that in applications even Lindley's conditions for linearity of the regression of $y$ on $x$ in the narrow sense will not in general be satisfied.

Facing this situation we can define a linear relation $y = \beta_0' + \beta_1' x$ that may be called the linear component of the regression of $y$ on $x$. The elementary results in the literature for the relations between $\beta_0'$, $\beta_1'$ and $\beta_0$, $\beta_1$ (usually derived on the assumption that $h$, $X$ are normally distributed) hold for this linear component. The linear component can be obtained when $h$ and $X$ are correlated, whereas Lindley assumes $h$, $X$ independent, and extends to errors in $y$ also and to multiple linear regression, subject to problems about specification of the nature of correlated errors.

The next step was to work out the exact regression of $y$ on $x$ for a number of specific examples in which this regression has a closed form. These suggest that the departure from linearity can be approximated by a quadratic in $x$ if either $h$ or $X$ is skew and by a cubic in $x$ if $h$, $X$ are symmetrical. The equations of the approximating quadratic or cubic are easily obtainable from the lower moments of the distributions of $h$ and $X$. Further, in these examples the linear component dominates, in the sense that the mean square deviation of $y$ from the linear component is only slightly larger than that from the exact regression of $y$ on $x$; even for measurements of reliability not much more than 50 per cent.

A further result that holds well in these examples is of interest when the fallible $x$ is used to predict $y$. When $h$ and $X$ are independent and normal, so that the regression of $y$ on $x$ is linear, it is known that the residual variance from the regression of $y$ on $x$ exceeds that for the regression of $y$ on $X$ by $(1 - G)\rho^2\sigma_y^2$, where $G$ is the coefficient of reliability of $x$. This result remains a good approximation when $X$, $h$ are independent but have different distributions so that Lindley's conditions do not hold.

Unfortunately the results here are only suggestive and leave unanswered the important questions. For example, (i) what is the analogous result to Lindley's when $h$ and $X$ are not independent? Some measuring instruments have the property of underestimating high values of $X$ and overestimating low values, and *vice versa*. (ii) How far can the moments approach be trusted? Are there distributions for which the departure from linearity is more complex than a quadratic or cubic? (iii) Lindley's conditions guarantee linearity only in the narrow sense; the deviations $e'$ from the regression of $y$ on $x$ are not independent of $x$. There is reason to believe that in this case linear regression theory can be used asymptotically in large samples, but more needs to be known about the practical importance of the disturbances present in small samples.

## REFERENCES

[1] J. BERKSON, "Are there two regressions?," *J. Amer. Statist. Assoc.*, Vol. 45 (1950), pp. 164–180.

[2] W. G. COCHRAN, "Errors of measurement in statistics," *Technometrics*, Vol. 10 (1968), pp. 637–666.

[3] E. FIX, "Distributions which lead to linear regressions," *Proceedings of the Berkeley Symposium on Mathematical Statistics and Probability*, Berkeley and Los Angeles, University of California Press, 1949, pp. 79–91.

[4] M. G. KENDALL, "Regression, structure and functional relationship." *Biometrika*, Vol. 38 (1951), pp. 11–25.

[5] D. V. LINDLEY, "Regression lines and the linear functional relationship," *J. Roy. Statist. Soc. Ser. B*, Vol. 9 (1947), pp. 218–244.

[6] F. LORD, "Large-sample covariance analysis when the control variable is fallible," *J. Amer. Statist. Assoc.*, Vol. 55 (1960), pp. 307–321.

# 96

DEPARTMENT OF STATISTICS
HARVARD UNIVERSITY

## PERFORMANCE OF A PRELIMINARY TEST OF COMPARABILITY
## IN OBSERVATIONAL STUDIES

by

W. G. COCHRAN·

Technical Report No. 29

February 4, 1970

Contract N00014-67A-0298-0017, NR-042-097

This paper is the result of research supported by the
Office of Naval Research. Reproduction in whole or in
part is permitted by the U. S. Government. Distribution
of this document is unlimited.

Performance of a Preliminary Test of Comparability in
Observational Studies.          W.G.Cochran.

## Summary

This note studies the performance of a preliminary test of
significance sometimes used in observational studies. The means
$\bar{y}_1$, $\bar{y}_2$ of a response variable are being compared in two independent
samples of size $\underline{n}$. It is feared that the comparison may be biased
because a confounding variable x may have different levels (means)
in the two populations. A t-test of $(\bar{x}_1 - \bar{x}_2)$ is made. If $\underline{t}$ is not
significant, $(\bar{y}_1 - \bar{y}_2)$ is used as an estimate of the population difference $\delta_y$; if t is significant, $(\bar{y}_1 - \bar{y}_2)$ is adjusted by an analysis
of covariance. This procedure is found to reduce the remaining
bias in the estimate of $\delta_y$ to a value that does not seriously disturb tests of significance or confidence probabilities based on
this estimate. If the estimation of small values of $\delta_y$ is important,
however, the remaining bias may still be a sizeable percentage of
$\delta_y$, except with large samples.

## 1. Introduction

In an observational study an investigator may plan to compare the mean $\bar{y}$ of a sample of subjects with the mean $\bar{y}_2$ of another sample who differ from the first according to some characteristic or experience (e.g. male-female; drinkers, non-drinkers). He would like to be able to regard the mean difference $(\bar{y}_1 - \bar{y}_2)$ as an unbiased estimate of the difference in effect of the two characteristics. But if he has no control over the sample to which a subject belongs, systematic difference between the samples other than the characteristic in which he is interested may also be present, causing $(\bar{y}_1 - \bar{y}_2)$ to be a biased estimate of the difference in effect of the characteristics. In this situation he is often advised to make a careful appraisal of the comparability of his two samples before proceeding with the study and analysis.

One way of proceeding is to list any other variables (called x-variables here) that are thought to have a substantial influence on y, and if possible to check that each of these variables has a similar distribution in the two samples. A test commonly used for this purpose is the $\chi^2$ test in a 2×k contingency table. The range of x is divided into k classes, and the frequencies $f_{1i}$, $f_{2i}$ are counted for the ith class in the two samples, the $\chi^2$ test being applied to these two sets of frequencies.

The $\chi^2$ test has an obvious intuitive appeal, since if x has a similar distribution in the two samples, there should be little danger of bias arising from any mathematical form of relation between y and x that is the same in the two populations. How much protection is actually obtained from a non-significant $\chi^2$ when the x-distributions differ in the two populations has not been investigated. Further, if the relation between y and x is linear, or approximately so, a test more relevant than $\chi^2$ is a t-test of the difference $(\bar{x}_1 - \bar{x}_2)$. Suppose that the model is as follows:

Pop.1) $\quad y_{ij} = \alpha_1 + \beta x_{ij} + e_{ij}$

Pop.2) $\quad y_{2j} = \alpha_2 + \beta x_{2j} + e_{2j}$

where $(\alpha_1 - \alpha_2)$ is the true difference in effect of the characteristics and the $e_{ij}$ are $N(0, \sigma^2)$ and independent. Hence,

(1.1) $\quad \bar{y}_1 - \bar{y}_2 = \alpha_1 - \alpha_2 + \beta(\bar{x}_1 - \bar{x}_2) + \bar{e}_1 - \bar{e}_2$

(1.2) $\quad E(\bar{y}_1 - \bar{y}_2) = \alpha_1 - \alpha_2 + \beta(\mu_1 - \mu_2)$

where $\mu_1, \mu_2$ are the population means of x. Thus with this model the bias in $(\bar{y}_1 - \bar{y}_2)$ due to differences in the x-distributions in the two populations in $\beta(\mu_1 - \mu_2)$.

Before discussing the t-test of $(\bar{x}_1 - \bar{x}_2)$ we need to mention that there are two other methods of handling possible bias under this model. One is to construct samples matched on the values of x, so that in the matched samples $(\bar{x}_1 - \bar{x}_2)$, and consequently its mean, is forced to be small. The second is to estimate $\beta$ by an analysis of covariance and replace $(\bar{y}_1 - \bar{y}_2)$ by the adjusted estimate

$$(\bar{y}_1 - \bar{y}_2) - \hat{\beta}(\bar{x}_1 - \bar{x}_2)$$

which is an unbiased estimate of $(\alpha_1 - \alpha_2)$ under this model.

Since these techniques are available, one may ask: why bother with tests of comparability? Why not always plan to match or adjust?

4.

Matching or adjustment is to be recommended under this model if the correlation between y and x is high, because whether there is danger of bias or not, variations in x contribute to the variability of y and hence of $(\bar{y}_1-\bar{y}_2)$. In the no-bias situation $(\mu_1=\mu_2)$, it is well known that in a large samples, the covariance adjustment reduces $V(\bar{y}_1-\bar{y}_2)$ by the factor $(1-\rho^2)$ - a worthwhile gain in precision if say $\rho \geq 0.4$ . However, when there are numerous x-variables, matching for all of them may become very troublesome, and multiple covariance also becomes steadily more complex. Consequently, an amended rule of procedure might be as follows. Plan to match or adjust for the 1, 2, or 3 x-variables thought to have high correlations with y, or to contribute major biases. For the remainder, check comparability and match or adjust only if the t-test of $(\bar{x}_1-\bar{x}_2)$ indicates a danger of bias. Finally, there is something to be said for the advice: check comparability for all x-variables. Finding a significant $(\bar{x}_1-\bar{x}_2)$ difference in, say, 3 out of 5 x-variables raises the question whether there are other systematic differences between the samples that we have not thought of, and may make us reconsider whether this is a good pair of samples in which to try to make the comparison $(\bar{y}_1-\bar{y}_2)$. The answer will, of course, depend on how far we are already committed to this locale, and on the other sources of samples that are feasible for us.

The following preliminary procedure will be examined in this note.

1) With two independent samples of size $\underline{n}$, make a two-tailed $\underline{t}$-test

of the quantity

$$t_x = \sqrt{\frac{n}{2}} \frac{(\bar{x}_1 - \bar{x}_2)}{s} \qquad\qquad \text{d.f.} = 2(n-1)$$

where $s^2$ is the pooled within-sample variance of x.

2a)  If $t_x$ is non-significant, take

$$\hat{\alpha}_1 - \hat{\alpha}_2 = \bar{y}_1 - \bar{y}_2$$

as the estimate of $(\alpha_1 - \alpha_2)$; that is, attempt no matching or adjustment.

2b)  If $t_x$ is significant, make a covariance adjustment and take

$$\hat{\alpha}_1 - \hat{\alpha}_2 = (\bar{y}_1 - \bar{y}_2) - \hat{\beta}(\bar{x}_1 - \bar{x}_2)$$

which is known to be unbiased under this model.

For t-tests at the 5%, 10%, and 20% levels; n = 11, 21, 61, and $\infty$, giving 20, 40, 120, and $\infty$ d.f. in $s^2$; and the relevant range of values of $(\mu_1 - \mu_2)$; the remaining bias in $(\bar{y}_1 - \bar{y}_2)$ after applying this procedure has been worked out. If $t_0$ is the significance level of t, the remaining bias is , from (1.1),

$$(1.3) \qquad P(|t_x| \leq t_0) \; \beta E \left\{ (\bar{x}_1 - \bar{x}_2) \; \big| \; |t_x| \leq t_0 \right\}$$

As $(\mu_1 - \mu_2)$ increases from 0, we may expect this remaining bias to be small at first, because although the probability that $\underline{t}$ is non-significant will be high, the conditional mean of $(\bar{x}_1 - \bar{x}_2)$ will be small, since the unconditional mean $(\mu_1 - \mu_2)$ is small.  When $(\mu_1 - \mu_2)$ is large, the probability that $t_x$ is non-significant, and hence the remaining bias, should approach 0.  Presumably the remaining bias reaches a maximum for some intermediate value of $(\mu_1 - \mu_2)$.

## 2. The remaining bias.

When $\underline{n}$ is large so that t becomes nearly normally distributed, the remaining bias can be expressed in a simple form easily calculable from the ordinates and the cumulative probabilities in standard tables of the normal distribution. When $\underline{n}$ is not large, the remaining bias has to be obtained by numerical integration on the computer.

In the large-sample case, $t_x$ is normally distributed with mean $\tau = \sqrt{n}\, (\mu_1 - \mu_2)/\sqrt{2}\,\sigma_x$, while $(\bar{x}_1 - \bar{x}_2) = \sqrt{2}\,\sigma_x\, t_x/\sqrt{n}$ . Hence the remaining bias,

$$P\big(|t_x| \le t_0\big)\beta E\left\{(\bar{x}_1 - \bar{x}_2)\;\middle|\; |t_x| \le t_0\right\} = \sqrt{\frac{2}{n}}\,\beta\sigma_x P\left\{|t_x| \le t_0\right\} E\left\{t_x\;\middle|\; |t| \le t_0\right\}$$

$$= \sqrt{\frac{2}{n}}\,\rho\sigma_y\,\frac{1}{\sqrt{2\pi}}\int_{-t_0}^{t_0} t e^{-\frac{1}{2}(t-\tau)^2}dt$$

If $z(t)$ is the ordinate and $C(t)$ the cumulative probability of the standard normal distribution from $-\infty$ to $t$, evaluation of the integral gives for the remaining bias:

$$\sqrt{\frac{2}{n}}\,\rho\sigma_y\,\left[\,\tau\left\{C(t_0-\tau) - C(-t_0-\tau)\right\} - \left\{z(t_0-\tau) - z(-t_0-\tau)\right\}\right]$$

where $\tau = \sqrt{n}\,(\mu_1-\mu_2)/\sqrt{2}\sigma_x$. This can be calculated for a range of values of $\tau$ from $0$ to $\infty$.

Now consider $\underline{n}$ finite. Instead of using the non-central $\underline{t}$-distribution directly, we write

$$y = \sqrt{\frac{n}{2}}\,\frac{(\bar{x}_1 - \bar{x}_2)}{\sigma_x}\;;\quad u = \frac{s_x}{\sigma_x}\;;\; t = \frac{y}{u}\,,$$

where $s_x$ has $2(n-1)$ d.f. From its definition in (1.3) the remaining bias, apart from the factor $\sqrt{\frac{2}{n}}\,\rho\sigma_y$, is directly expressible as the double integral

$$\frac{\sqrt{2}\ (n-1)^{n-1}}{\sqrt{\pi}\ (n-2)!} \int_0^\infty u^{2n-3}\, e^{-\frac{1}{2}(n-1)u^2}\, du \int_{-t_0 u}^{t_0 u} y e^{-\frac{1}{2}(y-\tau)^2}\, dy$$

## 3. Results

From the preceding results the remaining bias has the form $f\sqrt{\frac{2}{n}}\rho\sigma_y$, where $\underline{f}$ is a function of $\tau = \sqrt{n}(\mu_1-\mu_2)/\sqrt{2}\sigma_x$ and of the d.f. in the $\underline{t}$-test. For the interesting range of values of $\tau$, table I shows $\underline{f}$ and the corresponding probability that $\underline{t}_x$ is non-significant, i.e. that we do not adjust $(\bar{y}_1-\bar{y}_2)$.

## TABLE 1

Values of f and of $P(|t_x| \leq t_0)$ for two samples of size n.  Remaining

bias $= f\sqrt{\frac{2}{n}}\rho\sigma_y$ .

### 5% t-test.

| $\tau$ | f | | | | $P(|t_x| \leq t_0)$ | | | |
|--------|--------|--------|---------|---------|---------|---------|----------|---------|
| | 20 d.f. | 40 d.f. | 120 d.f. | ∞ d.f. | 20 d.f. | 40 d.f. | 120 d.f. | ∞ d.f. |
| 0 | 0 | 0 | 0 | 0 | .95 | .95 | .95 | .95 |
| 0.5 | .355 | .349 | .345 | .342 | .92 | .92 | .92 | .92 |
| 1.0 | .618 | .601 | .589 | .583 | .84 | .84 | .83 | .83 |
| 1.5 | .725 | .691 | .669 | .657 | .70 | .70 | .68 | .68 |
| 2.0 | .666 | .617 | .585 | .570 | .52 | .50 | .49 | .48 |
| 2.5 | .497 | .442 | .408 | .392 | .34 | .32 | .30 | .29 |
| 3.0 | .303 | .256 | .228 | .215 | .19 | .17 | .15 | .15 |

### 10% t-test.

| $\tau$ | f | | | | $P(|t_x| \leq t_0)$ | | | |
|--------|--------|--------|---------|---------|---------|---------|----------|---------|
| 0 | 0 | 0 | 0 | 0 | .90 | .90 | .90 | .90 |
| 0.5 | .274 | .268 | .264 | .262 | .86 | .86 | .86 | .86 |
| 1.0 | .454 | .439 | .429 | .424 | .75 | .74 | .74 | .73 |
| 1.5 | .490 | .467 | .451 | .443 | .58 | .57 | .56 | .56 |
| 2.0 | .405 | .376 | .357 | .348 | .39 | .37 | .37 | .36 |
| 2.5 | .266 | .239 | .222 | .214 | .22 | .21 | .20 | .19 |
| 3.0 | .141 | .121 | .109 | .104 | .11 | .10 | .09 | .09 |

### 20% t-test.

| $\tau$ | f | | | | $P(|t_x| \leq t_0)$ | | | |
|--------|--------|--------|---------|---------|---------|---------|----------|---------|
| 0 | 0 | 0 | 0 | 0 | .80 | .80 | .80 | .80 |
| 0.5 | .170 | .165 | .162 | .160 | .76 | .75 | .75 | .74 |
| 1.0 | .264 | .255 | .249 | .246 | .61 | .60 | .60 | .60 |
| 1.5 | .259 | .247 | .239 | .235 | .42 | .42 | .41 | .41 |
| 2.0 | .189 | .177 | .169 | .165 | .25 | .24 | .24 | .24 |
| 2.5 | .107 | .098 | .092 | .089 | .12 | .12 | .11 | .11 |
| 3.0 | .048 | .04 | .039 | .037 | .05 | .05 | .04 | .04 |

With 5% t-tests the remaining bias is greatest when $\tau$ is around 1.5 for all 4 sample sizes. The maximum value of f is 0.66 in large samples, occurring when the probability of a non-significant t is P = 0.68. At $\tau$ = 1.5 the number of d.f. in t has only a small effect on the performance of the test, the maximum f increasing only to 0.72 when t has only 20 d.f. as compared with $\infty$ d.f., while P increases to 0.70.

In 10% and 20% tests the results are similar, except that the protection against bias is of course greater. In table 1 the greatest observed f in large samples is 0.44 at $\tau$ = 1.5 in 10% tests and f = 0.25 at $\tau$ = 1.0 in 20%. (Actually, the grid of $\tau$ values is too wide to locate the maximum $\tau$ with high accuracy: in both cases the maximum occurs between $\tau$ = 1.0 and $\tau$ = 1.5, but the maximum f values are only trivially different, being 0.45 and 0.25, respectively.) With 20 d.f. in t, the maximum bias is again increased by about 10% as compared with the large-sample value.

## 4. Discussion

We come to the question: is the bias adequately controlled by this preliminary test procedure? This is not easy to answer.

First, the maximum remaining biases that have been reported might occur only at values of $(\mu_1 - \mu_2)$ very unlikely to be encountered in practice. In this event we need not consider these worst possible cases, but some less severe cases more typical of applications. This does not seem to be so. Since $\tau = \sqrt{n} \, (\mu_1 - \mu_2)/\sqrt{2} \, \sigma_x$ and since $\tau$ lies between 1 and 1.5 in the worst cases, these cases have $(\mu_1 - \mu_2)/\sigma_x$ lying between $1.4/\sqrt{n}$ and $2.1/\sqrt{n}$ . For n = 25, these ratios are 0.28 and 0.42, while for n = 100 they are 0.14 and 0.21. It was suggested

10.

earlier that this procedure not be used for x variables from which we suspect major biases, since a decision to match or adjust should then be automatic. A ratio $(\mu_1-\mu_2)/\sigma_x = 0.5$ or higher could be regarded as a major bias in x. At 0.5, for instance, 69% of population 1 lies above the mean of population 2 if x is normal. But at a ratio 0.2, the corresponding figure is 58%. Consequently, it looks as if in fairly large samples the biases $(\mu_1-\mu_2)/\sigma_x$ that make $\tau$ near its maximum could well occur in practice.

The maximum remaining biases may be looked at from two different points of view. We may ask whether the biases are large enough to disturb Type I errors in tests of significance of $(\bar{y}_1-\bar{y}_2)$ or confidence probabilities in confidence interval statements. Secondly, even if the bias in $(\bar{y}_1-\bar{y}_2)$ is large enough to disturb Type I errors badly, it might still be small relative to the size of real difference that is important in practical applications, in other words to the value $\delta_y$ that $E(\bar{y}_1-\bar{y}_2)$ would have in the absence of bias. In many practical judgments about the importance of $\delta_y$, a bias of say 10% or 20% in the estimate of $\delta_y$ might be small enough to permit a correct judgment.

The situation with regard to tests of significance and confidence probabilities is easily examined in samples large enough so that $\underline{t}$ can be regarded as normally distributed. Let P denote $P\{|t_x|\leq t_0\}$, the probability of a non-significant result in the $\underline{t}$-test of $(\bar{x}_1-\bar{x}_2)$. We assume that the investigator sticks to elementary statistical methods. If $(\bar{x}_1-\bar{x}_2)$ is not significant, he assumes that there is no bias and that $\bar{d}_y = (\bar{y}_1-\bar{y}_2)$ is normally distributed with mean $\delta_y$ and variance $2\sigma_y^2/n$. In this case his 95% confidence interval is calculated by assuming that

$$P\left\{|\bar{d}_y - \delta_y| \leq 1.96 \sqrt{\frac{2}{n}}\, \sigma_y\right\} = 0.95 \; .$$

When $|t_x| \leq t_0$, $\bar{d}_y$ is in fact distributed with mean

$$\frac{f}{P}\sqrt{\frac{2}{n}}\, \rho\sigma_y$$

where f is the factor tabulated in table 1 and $P = P\{|t_x| \leq t_0\}$ .
Note that the factor $(f/P)$ appears here instead of the $\underline{f}$ in table 1,
because we now want the conditional mean, not the overall mean.
From the regression model, the conditional variance $V_c$ of $\bar{d}_y$ is

$$V_c(\bar{y}_1 - \bar{y}_2) = \frac{2}{n}\, \sigma_y^{\,2}(1-\rho^2) + \beta^2 V_c(\bar{x}_1 - \bar{x}_2)$$

Since

$$(\bar{x}_1 - \bar{x}_2) = \sqrt{\frac{2}{n}}\, \sigma_x t_x$$

$$V_c(\bar{y}_1 - \bar{y}_2) = \frac{2}{n}\, \sigma_y^{\,2}\left\{(1-\rho^2) + \rho^2 V_c(t_x)\right\}$$

Since $\underline{t}_x$ is assumed $N(\tau,1)$, its conditional variance given that $|t_x| \leq t$
is easily calculated. For $\tau = 1.5$, near the maximum bias, $V_c = 0.4715$
for 5% tests, 0.3844 for 10% tests, and 0.2893 for 20% tests.

A further complication is that conditionally,

$$\bar{y}_1 - \bar{y}_2 = \delta + \beta(\bar{x}_1 - \bar{x}_2) + (\bar{e}_1 - \bar{e}_2)$$

is not normally distributed, being the sum of a normal component
$(\bar{e}_1 - \bar{e}_2)$ and a component $\beta(\bar{x}_1 - \bar{x}_2)$ that is a truncated normal and is
skew when $\tau \neq 0$. For moderate values of $\rho$, however, the variance
of the normal component dominates. Thus for 5% tests, with $\rho = 0.4$,
the variance components are $(1-\rho^2) = 0.84$ and $\rho^2 V_c(t) = (0.16)(0.4715)$
$= 0.11$. Consequently, we have assumed that the assumption that $(\bar{y}_1 - \bar{y}_2)$

is normal will not be badly wrong in these cases.

It follows that the confidence probability $\pi_1$ is

$$\pi_1 = P(\delta_y - 1.96\sqrt{\frac{2}{n}}\,\sigma_y < \bar{d} < \delta_y + 1.96\sqrt{\frac{2}{n}}\,\sigma_y)$$

$$\doteq P\left\{\frac{-1.96 - f\rho/P}{\sqrt{1-(1-V_c)\rho^2}} < z < \frac{+1.96 - f\rho/P}{\sqrt{1-(1-V_c)\rho^2}}\right\}$$

where z is N(0,1).

For $\rho = 0.4$, these probabilities are 0.933, 0.950, and 0.957 for
5%, 10%, and 20% t-tests. Two opposing forces are at work. The
bias tends to lower the confidence probability, but the overestimation of the standard error of $(\bar{y}_1 - \bar{y}_2)$ tends to increase the
probability.

When $t_x$ is significant, the investigator makes a covariance
adjustment which removes the bias, and uses the standard formula
for the variance of an adjusted difference, which in this case is
correct. Consequently his confidence probability $\pi_2$ is also correct.
Hence, the overall confidence probability is

$$\pi = \pi_1\,P + \pi_2(1-P)$$

The values work out as 0.938, 0.950, and 0.954 for 5%, 10%, and 20%
tests. Considering that these are for the worst bias, confidence
probabilities are evidently little disturbed in large samples
(e.g. n > 50).

The relation of the maximum remaining bias to the size of true
difference $\delta_y$ that is of practical importance is difficult to discuss numerically because this value of $\delta_y$ can vary widely. Since
table 1 gives the maximum bias as $f\sqrt{\frac{2}{n}}\,\rho\sigma_y$, it is convenient to

write $|\delta_y| = r\sigma_y$ . Consider a binomial proportion as a simple case. This is unlikely to have a linear relation to a confounding variable x over a wide range, but over a narrower range a linear relation might be a good approximation. If $E(y_{2j}) = 0.5$, then S.D.$(y_{2j}) = 0.5$ . A value of $E(y_{1j})$ to 0.6 might be considered to represent an important advance in some applications. Here $\delta_y = 0.1$ and r = 0.2 . But if $y_{2j}$ represents the presence of a rare defect for which $E(y_{2j}) = 0.01$, a procedure that reduces this value to $E(y_{1j}) = 0.007$ or 0.005 might be a worthwhile discovery. For these cases S.D.$(y_{2j}) = \sqrt{(0.01)(0.99)} = 0.1$ while $\delta_y$ has the values -0.003 or -0.005 In this case, r is only 0.03 or 0.05.

With continuous examination scores, scaled so that the mean score is 500 and the S.D. is 100, a new method of teaching that improved the mean score by 50 points might well attract attention. For this, r = 0.5. While numerous examples of this type could be quoted, the impressions that they leave are:. (i) the value of r that would be considered important varies widely from application to application, (ii) values of r < 0.5 often fall in this category.

Consequently, for r = 0.1(0.1)0.5, it was decided to tabulate the value of n that would reduce the maximum bias to 20% or 10% of $|\delta_y|$, respectively. For example, with 5% t-tests we have in large samples

$$\frac{\text{Max. bias}}{\delta_y} = \frac{0.66\sqrt{2}\ \rho\sigma_y}{\sqrt{n}\ r\ \sigma_y} = \frac{0.93\rho}{\sqrt{n}\ r}$$

Setting this ratio equal to 0.2 or 0.1 gives the required value of n for given $\rho$ . We took $\rho = 0.4$ .

## TABLE 2

Size of each sample needed to **reduce** the maximum remaining bias to 20% or 10% of the population difference $\delta_y$. ($\rho = 0.4$).

Maximum bias = 20%

| $r = \delta_y/\sigma_y =$ | 0.1 | 0.2 | 0.3 | 0.4 | 0.5 |
|---|---|---|---|---|---|
| 5% tests | 346 | 86 | 38 | 22 | <20 |
| 10% tests | 164 | 41 | <20 | <20 | <20 |
| 20% tests | 49 | <20 | <20 | <20 | <20 |

Maximum bias = 10%

| | 0.1 | 0.2 | 0.3 | 0.4 | 0.5 |
|---|---|---|---|---|---|
| 5% tests | 1384 | 346 | 154 | 86 | 55 |
| 10% tests. | 655 | 164 | 73 | 41 | 26 |
| 20% tests | 196 | 49 | <20 | <20 | <20 |

The heavy dependence of the results on the value of $r = \delta_y/\sigma_y$ that is considered important is evident from table 2. For $r \geq 0.2$, a 20% maximum bias requires only modest sample sizes and a 10% bias is attained for sample sizes quite common in practice. The difficult cases are those with $r \leq 0.1$ .

# 6

## *Observational Studies*

WILLIAM G. COCHRAN

---

## 1. Introduction

OBSERVATIONAL STUDIES are a class of statistical studies that have increased in frequency and importance during the past 20 years. In an observational study the investigator is restricted to taking selected observations or measurements on the process under study. For one reason or another he cannot interfere in the process in the way that one does in a controlled laboratory type of experiment.

Observational studies fall roughly into two broad types. The first is often given the name of "analytical surveys." The investigator takes a sample survey of a population of interest and proceeds to conduct statistical analyses of the relations between variables of interest to him. An early example was Kinsey's study (1948) of the relation between the frequencies of certain types of sexual behavior and variables like the age, sex, social level, religious affiliation, rural-urban background, and direction of social mobility of the person involved. Dr. Kinsey gave much thought to the methodological problems that he would face in planning his study. More recently, in what is called the "midtown Manhattan study" (Srole et al., 1962), a team of psychiatrists studied the relation in Manhattan, New York, between age, sex, parental and own social level, ethnic origin, generation in the United States, and religion and nonhospitalized mental illness.

WILLIAM G. COCHRAN is Professor of Statistics, Harvard University, Cambridge, Massachusetts.

This work was assisted by a contract with the Office of Naval Research, Navy Department.

---

The second type of observational study is narrower in scope. The investigator has in mind some agents, procedures, or experiences that may produce certain causal effects (good or bad) on people. These agents are like those the statistician would call *treatments* in a controlled experiment, except that a controlled experiment is not feasible. Examples of this type abound. A simple one structurally is a Cornell study of the effect of wearing a lap seat belt on the amount and type of injury sustained in an automobile collision. This study was done from police and medical records of injuries in automobile accidents. The prospective smoking and health studies (1964) are also a well-known example. These are comparisons of the death rates and causes of death of men and women with different smoking patterns in regard to type and amount. An example known as the "national halothane study" (Bunker et al., 1969) attempted to make a fair comparison of the death rates due to the five leading anesthetics used in hospital operations.

Several factors are probably responsible for the growth in the number of studies of this kind. One is a general increase in funds for research in the social sciences and medicine. A related reason is the growing awareness of social problems. A study known as the "Coleman report" (1966) has attracted much discussion. This was begun because Congress gave the U.S. Office of Education a substantial sum and asked it to conduct a nation-wide survey of elementary schools and high schools to discover to what extent minority-group children in the United States (Blacks, Indians, Puerto Ricans, Mexican-Americans, and Orientals) receive a poorer education than the majority whites. A third reason is the growing area of program evaluation. All over the world, administrative bodies—central, regional, and local—spend the taxpayers' money on new programs intended to benefit some or all of the population or to combat social evils. Similarly, a business organization may institute changes in its operations in the hope of improving the running of the business. The idea is spreading that it might be wise to devote some resources to trying to measure both the intended and the unintended effects of these programs. Such evaluations are difficult to do well, and they make much use of observational studies. Finally, some studies are undertaken to investigate stray reports of unexpected effects that appear from time to time. The halothane study is an example; others are studies of side effects of the contraceptive pill and studies of health effects of air pollution.

This paper is confined mainly to the second, narrower class of observational studies, although some of the problems to be considered are also met in the broader analytical ones.

For this paper I naturally sought a topic that would reflect the outlook and research interests of George Snedecor. In his career activity of helping investigators, he developed a strong interest in the design of experiments, a subject on which numerous texts are now available. The planning of observational studies, in which we would like to do an experiment but cannot, is a closely related topic which cries aloud for George's mature wisdom and the methodological truths that he expounded so clearly.

Succeeding sections will consider some of the common issues that arise in planning.

## 2. The Statement of Objectives

Early in the planning it is helpful to construct and discuss as clear and specific a written statement of the objectives as can be made at that stage. Otherwise it is easy in a study of any complexity to take later decisions that are contrary to the objectives or to find that different team members have conflicting ideas about the purpose of the study. Some investigators prefer a statement in the form of hypotheses to be tested, others in the form of quantities to be estimated or comparisons to be made. An example of the *hypothesis* type comes from a study (Buck et al., 1968), by a Johns Hopkins team, of the effects of coca-chewing by Peruvian Indians. Their hypotheses were stated as follows.

1. Coca, by diminishing the sensation of hunger, has an unfavorable effect on the nutritional state of the habitual chewer. Malnutrition and conditions in which nutritional deficiencies are important disease determinants occur more frequently among chewers than among control subjects.
2. Coca chewing leads to a state of relative indifference which can result in inferior personal hygiene.
3. The work performance of coca chewers is lower than that of comparable non-chewers.

One objection sometimes made to this form of statement is its suggestion that the answers are already known, and thus it hints at personal bias. However, these statements could easily have been put in a neutral form, and the three specific hypotheses about coca were suggested by a previous League of Nations commission. The statements perform the valuable purpose of directing attention to the comparisons and measurements that will be needed.

## 3. The Comparative Structure of the Plan

The statement of objectives should have suggested the type of comparisons on which logical judgments about the effects of treatment would be based. Some of the most common structures are outlined below. First, the study may be restricted to a single group of people, all subject to the same treatment. The timing of the measurements may take several forms.

1. After Only (i.e., after a period during which the treatment should have had time to produce its effects).
2. Before and After (planned comparable measurements both before and after the period of exposure to the agent or treatment).
3. Repeated Before and Repeated After.

In both (1) and (2) there may be a series of After measurements if there is interest in the long-term effects of treatment.

Single-group studies are so weak logically that they should be avoided whenever possible, but in the case of a compulsory law or change in business practice, a comparable group not subject to the treatment may not be available. In an After Only study we can perhaps judge whether or not the situation after the period of treatment was satisfactory but have no basis for judging to what extent, if any, the treatment was a cause, except perhaps by an opinion derived from a subjective impression as to the situation before exposure. Supplementary observations might of course teach something useful about the operation of a law—e.g., that it was widely disobeyed through ignorance or unpopularity with the public or that it was unworkable as too complex for the administrative staff.

In the single-group Before and After study we at least have estimates of the changes that took place during the period of treatment. The problem is to judge the role of the treatment in producing these changes. For this step it is helpful to list and judge any other contributors to the change that can be envisaged. Campbell and Stanley (1966) have provided a useful list with particular reference to the field of education.

Consider a Before-After rise. This might be due to what I vaguely call "external" causes. In an economic study a Before-After rise might accompany a wide variety of "treatments", good or bad, during a period of increasing national employment and prosperity. In educational examinations contributors might be the increasing maturity of the students or familiarity with the tests. In a study of an apparently low group on some variable (e.g., poor at some task) a rise might be due to what is called the regression effect. If a person's score fluctuates from time to time through human variability or measurement error, the "low" group selected is likely to contain persons who were having an unusually bad day or had a negative error of measurement on that day. In the subsequent After measurement, such persons are likely to show a rise in score even under no treatment—either they are having one of their "up" days or the error of measurement is positive on that day. After World War I the French government instituted a wage bonus for civil servants with large families to stimulate an increase in the birthrate and the population of France. I have been told the primary effect was an influx of men with large families into French civil service jobs, creating a Before-After rise that might be interpreted as a success of the "treatment." An English Before-After evaluation of a publicity campaign to encourage people to come into London clinics for needed protective shots obtained a Before-After drop in number of shots given. The clinics, who were asked to keep the records, had persuaded patrons to come in at once if they were known to be intending to have shots (Before), so that these people would be out of the way when the presumed big rush from the campaign started.

A time-series study with repeated measurements Before and After presents interesting problems that of appraising whether the Before-After change during the period of treatment is real in relation to changes that occur from external causes in the Before and After periods and that of deciding what is suggested about the time-response curve to the treatment. Campbell and Ross (1968) give an excellent account of the types of analysis and judgment needed in connection with a study of the Connecticut state law imposing a crackdown on speeding, and Campbell (1969) has discussed the role of this and other techniques in a highly interesting paper on program evaluation.

Single-group studies emphasize a characteristic that is prominent in the analysis of nearly all observational studies—the role of judgment. No matter how well-constructed a mathematical model we have, we cannot expect to plan a statistical analysis that will provide an almost automatic verdict. The statistician who intends to operate in this field must cultivate an ability to judge and weigh the relative importance of different factors whose effects cannot be measured at all accurately.

Reverting to types of structure, we come now to those with more than one group. The simplest is a two-group study of treated and untreated groups (seat-belt wearers and nonwearers). We may also have various treatments or forms of treatment, as in the smoking and health studies (pipes, cigars, cigarettes, different amounts smoked, and ex-smokers who had stopped for different lengths of time and had previously smoked different amounts). Both After Only and Before and After measurements are common. Sometimes both an After Only and a Before-After measurement are recommended for each comparison group if there is interest in studying whether the taking of the Before measurement influenced the After measurement.

Comparison groups bring a great increase in analytical insight. The influence of external causes on both groups will be similar in many types of study and will cancel or be minimized when we compare treatment with no treatment. But such studies raise a new problem—How do we ensure that the groups are comparable? Some relevant statistical techniques are outlined in section 6. In regard to incomparability of the groups the Before and After study is less vulnerable than the After Only since we should be able to judge comparability of the treated and untreated groups on the response variable at a time when they have not been subjected to the difference in treatment. Occasionally, we might even be able to select the two groups by randomization, having a randomized experiment instead of an observational study; but this is not feasible when the groups are self-selected (as in smokers) or selected by some administrative fiat or outside agent (e.g., illness).

## 4. Measurements

The statement of objectives will also have suggested the types of measurements needed; their relevance is obviously important. For instance, early British studies by aerial photographs in World War II were reported to show great damage to German industry. Knowing that early British policy was to bomb the town center and that German factories were often concentrated mainly on the outskirts, Yates (1968) confined his study to the factory areas, with quite a different conclusion which was confirmed when postwar studies could be made. The question of what is considered relevant is particularly important in program evaluation. A program may succeed in its main objectives but have undesirable side effects. The verdict on the program may differ depending on whether or not these side effects are counted in the evaluation.

It is also worth reviewing what is known about the accuracy and precision of proposed measurements. This is especially true in social studies, which often deal with people's attitudes, motivations, opinions, and behavior—factors that are difficult to measure accurately. Since we may have to manage with very imperfect measurements, statisticians need more technical research on the effects of errors of measurement. Three aspects are: (1) more study of the actual distribution of errors of measurement, particularly in multivariate problems, so that we work with realistic models; (2) investigation, from these models, of the effects on the standard types of analysis; (3) study of methods of remedying the situation by different analyses with or without supplementary study of the error distributions. To judge by work to date on the problem of estimating a structural regression, this last problem is formidable.

It is also important to check comparability of measurement in the comparison groups. In a medical study a trained nurse who has worked with one group for years but is a stranger to the other group might elicit different amounts of trustworthy information on sensitive questions. Cancer patients might be better informed about cases of cancer among blood relatives than controls free from cancer.

The scale of the operation may also influence the measuring process. The midtown Manhattan study, for instance, at first planned to use trained psychiatrists for obtaining the key measurements, but they found that only enough psychiatrists could be provided to measure a sample of 100. The analytical aims of the study needed a sample of at least 1,000. In numerous instances the choice seems to lie between doing a study much smaller and narrower in scope than desired but with high quality of measurement, or an extensive study with measurements of dubious quality. I am seldom sure what to advise.

In large studies one occasionally sees a mistake in plans for measurement that is perhaps due to inattention. If two laboratories or judges are needed to measure the responses, an administrator sends all the treatment group to laboratory 1 and the untreated to laboratory 2—it is at least a tidy decision. But any systematic difference between laboratories or judges becomes part of the estimated treatment effect. In such studies there is usually no difficulty in sending half of each group, selected at random, to each judge.

## 5. Observations and Experiments

In the search for techniques that help to ensure comparability in observational studies, it is worth recalling the techniques used in controlled experiments, where the investigator faces similar problems but has more resources to command. In simple terms these techniques might be described as follows.

Identify the major sources of variation (other than the treatments) that affect the response variable. Conduct the experiment and analysis so that the effects of such sources are removed or balanced out. The two principal devices for this purpose are blocking and the analysis of covariance. Blocking is employed at the planning stage of the experiment. With two treatments, for example, the subjects are first grouped into pairs (blocks of size 2) such that the members of a pair are similar with respect to the major anticipated sources of variation. Covariance is used primarily when the response variable $y$ is quantitative and some of the major extraneous sources of variation can also be represented by quantitative variables $x_1, x_2, \ldots$. From a mathematical model expressing $y$ in terms of the treatment effects and the values of the $x_i$, estimates of the treatment effects are obtained that have been adjusted to remove the effects of the $x_i$.

Covariance and blocking may be combined.

For minor and unknown sources of variation, use randomization. Roughly speaking, randomization makes such sources of error equally likely to favor either treatment and ensures that their contribution is included in the standard error of the estimated treatment effect if properly calculated for the plan used.

In general, extraneous sources of variation may influence the estimated treatment effect $\tau$ in two ways. They may create a bias $B$. Instead of estimating the true treatment effect $\tau$, the expected value of $\hat{\tau}$ is $(\tau + B)$, where $B$ is usually unknown. They also increase the variance of $\hat{\tau}$. In experiments a result of randomization and other precautions (e.g., blindness in measurement) is that the investigator usually has little worry about bias. Discussions of the effectiveness of blocking and covariance (e.g., Cox, 1957) are confined to their effect on $V(\hat{\tau})$ and on the power of tests of significance.

In observational studies we cannot use random assignment of subjects, but we can try to use techniques like blocking and covariance. However, in the absence of randomization these techniques have a double task—to remove or reduce bias and to increase precision by decreasing $V(\hat{\tau})$. The reduction of bias should, I think, be regarded as the primary objective—a highly precise estimate of the wrong quantity is not much help.

## 6. Matching and Adjustments

In observational studies as in experiments we start with a list of the most important extraneous sources of variation that affect the response variable. The Cornell study, based on automobile accidents involving seat-belt wearers and nonwearers, listed 12 major variables. The most important was the intensity and direction of the physical force at impact. A head-on collision at 60 mph is a very different matter from a sideswipe at 25 mph. In the smoking–death-rate studies age gradually becomes a predominating variable for men over 55. In the raw data supplied to the Surgeon General's committee by the British and Canadian studies and in a U.S. study cigarette smokers and nonsmokers had about the same death rates. The high death rates occurred among the cigar and pipe smokers. If these data had been believed, television warnings might now be advising cigar and pipe smokers to switch to cigarettes. However, cigar and pipe smokers in these studies were found to be markedly older than nonsmokers, while cigarette smokers were, on the whole, younger. All studies regarded age as a major extraneous variable in the analysis. After adjustment for age differences, death rates for cigar and pipe smokers were close to those for nonsmokers; those for cigarette smokers were consistently higher.

In observational studies three methods are in common use in an attempt to remove bias due to extraneous variables.

*Blocking,* usually known as matching in observational studies. Each member of the treated group has a match or partner in the untreated group. If the $x$ variables are classified, we form the cells created by the multiple classification (e.g., $x_1$ with 3 classes and $x_2$ with 4 classes create 12 cells). A match means a member of the same cell. If $x$ is quantitative (discrete or continuous), a common method is to turn it into a classified variate (e.g., age in 10-year classes). Another method, caliper matching, is to call $x_{11i}$ (in group 1) and $x_{12j}$ (in group 2) matches with respect to $x_1$ if $|x_{11i} - x_{12j}| \le a$.

*Standardization* (adjustment by subclassification). This is the analogue of covariance when the $x$'s are classified and we do not match. Arrange the data from the treated and untreated samples in cells, the $i$th cell containing say $n_{1i}, n_{2i}$ observations with response means $\bar{y}_{1i}, \bar{y}_{2i}$. If the effect $\tau$ of the treatment is the same in every cell, this method depends on the result that for any set of weights $w_i$ with $\Sigma w_i = 1$, the quantity $\hat{\tau} = \Sigma w_i (\bar{y}_{1i} - \bar{y}_{2i})$ is an unbiased estimate of $\tau$ (apart from any within-cell biases). The weights can therefore be chosen to minimize $V(\hat{\tau})$. If it is clear that $\tau$ varies from cell to cell as often happens, the choice of weights becomes more critical, since it determines the quantity $\Sigma w_i \tau_i$ that is being estimated. In vital statistics a common practice is to take the weights from some standard population to which we wish the comparison to apply.

*Covariance* (with $x$'s quantitative), used just as in experiments. The idea of matching is easy to grasp, and the statistical analysis is simple. On the operational side, matching requires a large reservoir in at least one group (treated or untreated) in which to look for matches. The hunt for matches (particularly with caliper matching) may be slow and frustrating, although computers should be able to help if data about the $x$'s can be fed into them. Matching is avoided when the planned sample size is large, there are numerous treatments, subjects become available only slowly through time, and it is not feasible to measure the $x$'s until the samples have already been chosen and $y$ is also being measured.

There has been relatively little study of the effects of these devices on bias and precision, although particular aspects have been discussed by Billewicz (1965), Cochran (1968) and Rubin (1970). If $x$ is classified and two members of the same class are identical in regard to the effect of $x$ on $y$, matching and standardization remove all the bias, while matching should be somewhat superior in regard to precision. I am not sure, however, how often such ideal classifications actually exist. Many classified variables, especially ordered classifications, have an underlying quantitative $x$—e.g., for sex with certain types of response there is a whole gradation from very manly men to very womanly women. This is obviously true for quantitative $x$'s that are deliberately made classified in order to use within-cell matching. In such cases, matching and standardization remove between-cell bias but not within-cell bias. Of an initial bias in means $\mu_{1x} - \mu_{2x}$ they remove about 64%, 80%, 87%, 91%, and 93% with 2, 3, 4, 5, and 6 classes, the actual amount varying a little with the choice of class boundaries and the nature of the $x$ distribution (Cochran, 1968). Caliper matching removes about 76%, 84%, 90%, 95%, and 99% with $a/\sigma_x = 1, 0.8, 0.6, 0.4$, and 0.2. These percentages also apply to $y$ under a linear or nearly linear regression of $y$ on $x$.

With a quantitative $x$, covariance adjustments remove all the initial bias if the correct model is fitted, and they are superior to within-class matching of $x$ when this assumption holds. In practice, covariance nearly always means linear covariance to most users, and some bias remains after covariance adjustment if the $y$, $x$ relation is nonlinear and a linear covariance is fitted. If nonlinearity is of the type that can be approximated by a quadratic curve, results by Rubin (1970) suggest that the residual bias should be small if $\sigma_{1x}^2 = \sigma_{2x}^2$ and $x$ is symmetrical or nearly so in distribution. When $\sigma_{1x}^2/\sigma_{2x}^2$ is 1/2 or 2, the adjustment can either overcorrect or undercorrect to a material extent.

Caliper matching, on the other hand, and even within-class matching do not lean on an assumed linear relation between $y$ and $x$. If $\sigma_{1x}^2/\sigma_{2x}^2$ is near 1 (perhaps between 0.8 and 1.2), the evidence to date suggests, however, that linear covariance is superior to within-class matching in removing bias under a moderately curved $y$, $x$ relation, although more study of this point is needed. Linear covariance applied to even loosely caliper-matched samples should remove nearly all the initial bias in this situation. Billewicz (1965) compared linear covariance and within-class matching (3 or 4 classes) in regard to precision in a model in which $x$ was distributed as $N(0, 1)$ in both populations. For the curved relations $y = 0.4x - 0.1x^2$, $y = 0.8x - 0.14x^2$, and $y = \tanh x$ he found covariance superior in precision on samples of size 40.

Larger studies in which matching becomes impractical present difficult problems in analysis. Protection against bias from numerous $x$ variables is not easy. Further, if there are say four $x$ variables, the treatment effect may change with the levels of $x_2$ and $x_3$. For applications of the conclusions it may be important to find this out. The obvious recourse is to model construction and analysis based on the model, which has been greatly developed, particularly in regression. Nevertheless the Coleman report on education (1966) and the national halothane study (Bunker et al., 1969) illustrate difficulties that remain.

# 7. Further Points on Planning

## Sample Size

Statisticians have developed formulas that provide guidance on the sample size needed in a study. The formulas tend to be harder to use in observational studies than in experiments because less may be known about the likely values of population parameters that appear in the formulas and the formulas assume that bias is negligible. Nevertheless there is frequently something useful to be learned—for instance, that the proposed size looks adequate for estimating a single overall effect of the treatment, but does not if the variation in effect with an $x$ is of major interest.

## Nonresponse

Certain administrative bodies may refuse to cooperate in a study; certain people may be unwilling or unable to answer the questions asked or may not be found at home. In modern studies, standards with regard to the nonresponse problem seem to me to be lax. In both the smoking and Coleman studies nonresponse rates of over 30% were common. The main difficulty with nonresponse is not the reduction in sample size but that nonrespondents may be to some extent different types of people from respondents and give different types of answers, so that results from respondents are biased in this sense. Fortunately, nonresponse can often be reduced materially by hard work during the study, but definite plans for this need to be made in advance.

## Pilot Study

The case for starting with a small pilot study should be considered—for instance, to work out the field procedures and check the understanding and acceptability of the questions and the interviewing methods and time taken. When information is wanted on a new problem, the cheapest and quickest method is to base a study on routine records that already exist. However, such records are often incomplete and have numerous gross errors. A law or administrative rule specifying that records shall be kept does not ensure that the records are usable for research purposes. A good pilot study of the records should reveal the state of affairs. It is worth looking at variances; a suspiciously low variance has sometimes led to detection of the practice of copying previous values instead of making an independent determination.

## Critique

When the draft of plans for a study is prepared, it helps to find a colleague willing to play the role of devil's advocate— to read the plan and to point out any methodological weaknesses that he sees. Since observational studies are vulnerable to such defects, the investigator should of course also be doing this, but it is easy to get in a rut and overlook some aspect. It helps even more if the colleague can suggest ways of removing or reducing these faults.

In the end, however, the best plan that investigator and colleague can devise may still be subject to known weaknesses. In the report of the results these should be discussed in a clearly labeled section, with the investigator's judgment about their impact.

### Sampled and Target Populations

Ideally, the statistician would recommend that a study start with a probability sample of the target population about which the investigator wishes to obtain information. But both in experiments and in observational surveys many factors—feasibility, costs, geography, supply of subjects, opportunity—influence the choice of samples. The population actually sampled may therefore differ in several respects from the target population. In his report the investigator should try to describe the sampled population and relevant target populations and give his opinion as to how any differences might affect the results, although this is admittedly difficult.

One reason why this step is useful is that an administrator in California, say, may want to see the results of a good study on some social issue for policy guidance and may find that the only relevant study was done in Philadelphia or Sweden. He will appreciate help in judging whether to expect the same results in California.

## 8. Judgment about Causality

Techniques of statistical analysis of observational studies have in general employed standard methods and will not be discussed here. When the analysis is completed, there remains the problem of reaching a judgment about causality. On this point I have little to add to a previous discussion (Cochran, 1965). It is well known that evidence of a relationship between $x$ and $y$ is no proof that $x$ causes $y$. The scientific philosophers to whom we might turn for expert guidance on this tricky issue are a disappointment. Almost unanimously and with evident delight they throw the idea of cause and effect overboard. As the statistical study of relationships has become more sophisticated, the statistician might admit, however, that his point of view is not very different, even if he wishes to retain the terms cause and effect.

The probabilistic approach enables us to discard oversimplified deterministic notions that make the idea look ridiculous. We can conceive of a response $y$ having numerous contributory causes, not just one. To say that $x$ is a cause of $y$ does not imply that $x$ is the only cause. With 0, 1 variables we may merely mean that if $x$ is present, the probability that $y$ happens is increased—but not necessarily by much. If $x$ and $y$ are continuous, a causal relation may imply that as $x$ increases, the average value of $y$ increases, or some other feature of its distribution changes. The relation may be affected by the levels of other variables; it may be strengthened or weakened or entirely disappear, depending on these levels. One can see why the idea becomes tortuous. For successful prediction, however, a knowledge of the nature and stability of these relationships is an essential step and this is something that we can try to learn in observational studies.

A claim of proof of cause and effect must carry with it an explanation of the mechanism by which the effect is produced. Except in cases where the mechanism is obvious and undisputed, this may require a completely different type of research from the observational study that is being summarized. Thus in most cases the study ends with an opinion or judgment about causality, not a claim of proof.

Given a specific causal hypothesis that is under investigation, the investigator should think of as many consequences of the hypothesis as he can and in the study try to include response measurements that will verify whether these consequences follow. The cigarette-smoking and death-rate studies are a good example. For causes of death to which smoking is thought to be a leading contributor, we can compare death rates for nonsmokers and for smokers of different amounts, for ex-smokers who have stopped for different lengths of time but used to smoke the same amount, for ex-smokers who have stopped for the same length of time but used to smoke different amounts, and (in later studies) for smokers of filter and nonfilter cigarettes. We can do this separately for men and women and also for causes of death to which, for physiological reasons, smoking should not be a contributor. In each comparison the direction of the difference in death rates and a very rough guess at the relative size can be made from a causal hypothesis and can be put to the test.

The same can be done for any alternative hypotheses that occur to the investigator. It might be possible to include in the study response measurements or supplementary observations for which alternative hypotheses give different predictions. In this way, ingenuity and hard work can produce further relevant data to assist the final judgment. The final report should contain a discussion of the status of the evidence about these alternatives as well as about the main hypothesis under study.

In conclusion, observational studies are an interesting and challenging field which demands a good deal of humility, since we can claim only to be groping toward the truth.

# References

Buck, A. A. et al. 1968. Coca chewing and health. *Am. J. Epidemiol.* 88: 159–77.

Bunker, J. P. et al., eds., 1969. The national halothane study. Washington, D.C.: USGPO.

Billewicz, W. Z. 1965. The efficiency of matched samples. *Biometrics* 21: 623–44.

Campbell, D. T. 1969. Reforms as experiments. *Am. Psychologist* 24: 409–29.

Campbell, D. T., and H. L. Ross. 1968. The Connecticut crackdown on speeding: Time series data in quasi-experimental analysis. *Law and Society Rev.* 3: 33–53.

Campbell, D. T., and J. C. Stanley. 1966. *Experimental and quasi-experimental designs in research.* Chicago: Rand McNally.

Cochran, W. G. 1965. The planning of observational studies. *J. Roy. Statist. Soc. Ser. A,* 128: 234–66.

———. 1968. The effectiveness of adjustment by classification in removing bias in observational studies. *Biometrics* 24: 295–314.

Coleman, J. S. 1966. Equality of educational opportunity. Washington, D.C.: USGPO.

Cox, D. R. 1957. The use of a concomitant variable in selecting an experimental design. *Biometrika* 44: 150–58.

Kinsey, A. C., W. B. Pomeroy, and C. E. Martin. 1948. Sexual behavior in the human male. Philadelphia: Saunders.

Rubin, D. B. 1970. The use of matched sampling and regression adjustment in observational studies. Ph.D. thesis, Harvard Univ., Cambridge.

Srole, L., T. S. Langner, S. T. Michael, M. K. Opler, and T. A. C. Rennie. 1962. Mental health in the metropolis. (The midtown Manhattan study.) New York: McGraw-Hill.

U.S. Surgeon-General's committee. 1964. Smoking and health. Washington, D.C.: USGPO.

Yates, F. 1968. Theory and practice in statistics. *J. Roy. Statist. Soc. Ser. A,* 131: 463–77.

# 98

The ceremony was brought to a close with the following statement by Prof. W. G. Cochran, President of the International Statistical Institute.

Mr. Chairman, ladies and gentlemen, mesdames et messieurs, speaking first for the members of the International Statistical Institute, their accompanying families and all participants in this session, I want to express our warmest thanks to those who have contributed and worked so hard to bring this session about.

Preparation and funding of the scientific and the social programmes in a session of the International Statistical Institute extends over more than two years and involves many people. This session, our 38th, is held under the sponsorship of 14 major scientific and professional organizations in this country. The American Statistical Association has acted on their behalf as organizer of the session, with its President Churchill Eisenhart as Chairman and Pat Riley as Executive Secretary of the Conference Committee. The Social Program Committee is under the able leadership of Mrs. Aryness Joy Wickens.

We are greatly honored by the message from President Nixon which you have just heard.

The list of honorary sponsors of this session greatly broadens the welcome to the ISI, containing members of the United States Cabinet, counselors and advisors to the President, and representatives of many of the leading organizations in this country.

These biennial meetings are unique in bringing together outstanding people in official and academic statistics from many countries under ideal conditions for both concentrated and relaxed discussions and exchange of ideas. Our last Washington session in 1947 was critical in being the first after the end of World War II. Many of us remember gratefully the vision and the tireless efforts of the late Stuart Rice in setting the ISI on its feet again in the new post-war conditions. Since then, the ISI has met in 12 countries throughout the world.

Reproduced with permission from *Proceedings of the 38th Session of the International Statistical Institute,* Vol. 44, Bk. 2, pages 27–30. Copyright © 1972, International Statistical Institute.

At this session I am in the position of one who lives in the host country but has been mercifully spared any of the preparatory labor for the meetings or social events. Please allow me to join our sponsors for a moment in saying how pleased we are to see all who have come to the United States for this session.

On a sad note we shall greatly miss members who have died. Among these, very recent deaths are H. M. Husein (United Arab Republic) and F. F. Stephan (United States).

I pass on merely to remind you briefly of some of the developments in the ISI since the London meeting in 1969.

In addition to our three regular publications - the Bulletin, the Revue, and Statistical Theory and Methods Abstracts - other recent bibliographical activity under ISI auspices is as follows:

The third, enlarged edition of the "Dictionary of Statistical Terms," by M. G. Kendall and W. R. Buckland, was published this spring. A Russian-English and an English-Russian glossary of the terms in this Dictionary, prepared by S. Kotz, will appear soon. Both are published by Oliver and Boyd.

A second edition of the "Bibliography of Index Numbers," prepared under the direction of W. F. Maunder, has been published by the Athlone Press, London.

J. W. Nixon's "Glossary of Terms in Official Statistics" previously translated into Russian, has been translated into Polish by the Central Statistical Office in Warsaw. A Dutch translation is in preparation by the Central Statistical Bureau in The Hague.

In progress are bibliographies on the Design and Analysis of Experiments under the direction of W. T. Federer and on Nonsampling errors under the direction of T. Dalenius. A meeting of the Education Committee to consider a program of further bibliographical work will be held next Monday.

In statistical education, the only training center now under ISI auspices is that in Calcutta, operated jointly with the Indian Statistical Institute. The international training center in Beirut was discontinued in 1970, since the creation of national training courses within the region appeared now to meet the needs of the region. A scholarship fund for students in the Middle East will be administered by the Institute.

In our second Round Table, a small group of experts met in the Netherlands last September to discuss modern techniques in the teaching of statistics, particularly audio-visual aids, use of computers, and programmed learning. Their report, to be published in the Revue, will be of wide interest to teachers of statistics.

At present the ISI has two sections--the International Association for Statistics in the Physical Sciences and the International Association of Municipal Statisticians. Under the vigorous leadership of its President, J. Neyman, the Physical Sciences section, in addition to its meeting here, has arranged six symposia this summer at different places in the U. S., ranging from Boston to Hawaii. One symposium deals with the theory and applications of stochastic point processes, the others with statistical problems in population research, pollution, turbulence, metallurgy and hydrology. In a related area, the ISI will be one of the sponsors of an Advanced Institute on Statistical Ecology, to be held at Pennsylvania State University next summer.

The Municipal Statistics section held its 7th plenary meeting in Strasbourg last year and has collaborated in the preparation of another volume in the series "International Statistical Yearbook of Large Towns". Under its current President, B. Benjamin, this Section is working on expansion and improvement of its programme.

During the London session, a proposal that the ISI should create a new section in the area of sample surveys was made to the General Assembly. This proposal resulted from much preliminary ground work by our Honorary President, Dr. Mahalanobis. In London the General Assembly instructed the Bureau to draft statutes, so that the proposal can be considered in concrete form during the present session. Through a committee headed by I. Fellegi, statutes have now been prepared. With the help of P. Loftus I have also been in touch with the United Nations to ensure that the new section, if established, will have appropriate liaison with the work of the United Nations agencies.

The ISI Bureau has approved many of the recommendations of the Reappraisal Committee, chaired by M. G. Kendall. For those recommendations relating to organizational matters, proposed changes in the Statutes have been prepared and circulated to members for comment.

One recommendation - that the Institute should play a more vigorous and active role with regard to research problems of broad international interest - has been under particular study by a committee headed by A. Stuart. Issues are the type of involvement and the particular research projects most appropriate to a body like ISI, and the question of funding. While much work has been done on these issues, further consideration during this session will be needed.

To pass to a less pleasant theme, I mentioned that the 1947 ISI session was a critical one. The present session will be critical in another sense - that of finance.

Owing to careful management by our Permanent Office under the direction of Mr. Lunenberg, the problem is of fairly recent origin. But steady inflation has put us in a position where we are currently living beyond our income at a rate that cannot be allowed to continue. Steps to meet this problem have been taken by increases in the membership dues, by the creation of corporate memberships and by the introduction of a kind of dues for ex-officio members. I want to thank those ex-officio members who have worked to arrange contributions from their Governments, and members who are helping in the recruitment of Corporate Members. However, despite economies that have also been made, further steps are needed if the present ISI program is to continue. Some hard decisions must be faced.

Before ending, I must express the thanks of ISI members to Mr. Lunenberg and the staff of the Permanent Office for their continued fine service to the Institute. Nobody realizes their vital role more keenly than the President.

Now let me keep you no longer from what I hope will be a pleasant and stimulating session.

# Experiments for Nonlinear Functions
## (R.A. Fisher Memorial Lecture)

### W. G. COCHRAN*

This article reviews work on the planning of experiments with response functions nonlinear in some parameters. Apart from older work on dilution series experiments and quantal bioassays, this field is relatively recent, the mathematical and computing aspects being more complex than for linear responses. An efficient design requires good advance estimates of the parameters. Research has concentrated on (1) optimum estimation of the parameters, under an asymptotic criterion, for both sequential and non-sequential approaches, and (2) tests of the adequacy of the model, discrimination between models and model-building. More work on the small-sample behavior of the designs and on compromise designs that perform well under the typical practical complications would be valuable.

## 1. INTRODUCTION

I would like to discuss some of the work that has been done in designing experiments involving response functions nonlinear in at least one of the parameters. Formally, this excludes the large volume of work on the planning of factorial experiments and on the estimation of multiple regressions, including polynomial response functions, although there are many similarities in both the methods of attack and the results obtained in the linear and nonlinear situations.

This is not an area to which Fisher devoted a great deal of attention. But the first design problem for which he published a solution was nonlinear. This was in his 1922 paper [32] on the mathematical foundations of theoretical statistics, before he had published anything on either the analysis of variance or on randomization and the design of agricultural experiments. The problem is the estimation of the density of small organisms in a liquid by means of a series of dilutions. This problem forms a convenient introduction. It is a one-parameter problem, yet illustrates some of the basic features of nonlinear problems.

In this review I shall try to concentrate on the issues that obviously present themselves, the methods of attack adopted, the progress made thus far, and some problems still awaiting, so far as I know, published research. The area is an exciting one. On the technical side, a high degree of both mathematical and computing skill is required in the more complex problems. On the practical side, there is the important question: is the research producing the kinds of results that assist the investigator in what he regards as his main problems? Equally important and by no means easy, are we able to explain the methods in terms that the experimenter can understand and use?

## 2. DILUTION SERIES EXPERIMENTS

A volume $V$ of a liquid contains $N$ tiny organisms, thoroughly mixed and with no tendency to clumping or mutual rejection. A small volume $x$ is taken out. The probability that this volume contains no organisms is

$$P = \left(1 - \frac{x}{V}\right)^N \cong e^{-Nx/V} = e^{-\theta x}.$$

Here $\theta$, the density per unit volume, is the parameter to be estimated, while $x$ corresponds to the level of a factor which can be chosen by the experimenter. In practice a standard volume is taken out by pipette, a desired $x$ being obtained by diluting the original volume with pure water. The lab test can detect only whether the sample is sterile (contains no organisms) or fertile (contains one or more organisms).

If $n$ samples are drawn for given $x$, the probability that $s$ are sterile is the binomial

$$\frac{n!}{s!(n-s)!} P^s Q^{n-s}.$$

The criterion which Fisher selected can be described in two equivalent ways. One is that he minimized the large-sample formula for the coefficient of variation of the maximum likelihood (ML) estimate of $\theta$. Fisher himself described it as maximizing the sample information about $\log \theta = \theta^2 I(\theta)$, where

$$I(\log \theta) = n(\theta x)^2/(e^{\theta x} - 1). \tag{2.1}$$

He regarded this criterion as the natural one in small as well as large samples, since he used the phrase "without any large-sample approximation" in referring to it.

To maximize $I(\log \theta)$ in (2.1), the quantity $\theta x$ should be set at 1.59, giving $P = 0.20$. To find $x$ such that $x\theta = 1.59$, we need to know $\theta$. This is a standard feature that distinguishes nonlinear from linear problems. In a nonlinear problem, the statistician can say to the experimenter: "You tell me the value of $\theta$ and I promise to

* W.G. Cochran is professor, Department of Statistics, Harvard University, Cambridge, Mass. 02138. This research was conducted under a contract with the Office of Naval Research, Navy Department. The author wishes to thank P. Morse, S.M. Siddik, R.E. Wheeler and the referees for suggestions and information about recent work.

© Journal of the American Statistical Association
December 1973, Volume 68, Number 344
Applications Section

design the best experiment for estimating $\theta$." If the experimenter replies, "Who needs you?," this is natural but not helpful.

What can be done in practice? Three possibilities suggest themselves. With a good initial estimate $\theta_0$ of $\theta$, the experimenter can use Fisher's solution, setting $x = 1.59/\theta_0$, and assuming that he has a good if not an optimum experiment. In Fisher's problem the value of $\theta$ is usually known poorly—perhaps within limits $\theta_L$, $\theta_H$ whose ratio is 100 or 1,000 to 1. The natural first question here is: can the experiment be done sequentially? The first experiment has $x = 1.59/\theta_0$, where $\theta_0$ is perhaps a poor first guess. The second experiment has $x = 1.59/\hat{\theta}_1$, where $\hat{\theta}_1$ is the ML estimate of $\theta$ from the first experiment, and so on, creeping up on the best $\theta x$.

So far as I know, dilution series experiments are routinely done non-sequentially in a single operation. If $\theta$ is thought to lie between $\theta_L$ and $\theta_H$, Fisher's approach was not to optimize anything, but to try to guarantee a specified expected value of $I(\log \theta)$. In a series of twofold dilutions, e.g., the percentage of the total information supplied by different dilutions is shown in the following tabulation:

| $\theta x$ | $I(\%)$ |
|---|---|
| $\geq 8$ | 0.9 |
| 4 | 12.6 |
| 2 | 26.4 |
| 1 | 24.5 |
| $\frac{1}{2}$ | 16.2 |
| $\frac{1}{4}$ | 9.3 |
| $\frac{1}{8}$ | 4.9 |
| $\leq \frac{1}{16}$ | 5.2 |

The five dilutions from $\theta x = 4$ to $\theta x = \frac{1}{4}$ provide 89 percent of the total information. To ensure that these dilutions are covered, we want $x_{\min}\theta_H \leq \frac{1}{4}$ and $x_{\max}\theta_L \geq 4$. This gives $x_{\max}/x_{\min} \geq 16\theta_H/\theta_L$. With $\theta_H/\theta_L = 100$, twelve twofold dilutions suffice to cover this range, and 15 when $\theta_H/\theta_L = 1,000$.

The Rothamsted laboratory which brought the problem to Fisher did 38 dilution series daily, and he observed that daily calculation of the 38 ML estimates would be "exceedingly laborious"—a difficulty that computers can now remove. Even on the desk machines then available, estimating $\theta$ by the method of moments (equating the observed total number of sterile plates to the expected number) could be done in less than five minutes per series by a table which he provided, now Table VIII2 in Fisher and Yates. Further, he showed in 1922 that the method of moments has an asymptotic efficiency of 88 percent. Thus, although one of the principal points in his 1922 paper was the superiority of ML over moments, he recommends moments for this problem for what seemed to him sound practical reasons.

The dilution series example reveals four types of problems that recur throughout nonlinear experiments: (1) setting up one or more criteria by which to judge alternative proposed designs. Often, much weight will be given to getting good estimates of the parameters, (2)

deciding how to proceed when initial estimates of the parameters are dubious. The relative feasibility, cost, and performance of sequential and non-sequential methods become important here, (3) any biometrician, at least, would insist with Fisher that the experiment be capable of providing its own internal estimate of C.V.$(\hat{\theta})$. Dilution series can do this if the model is correct and if large-sample formulas can be trusted in small samples—a point that could stand more checking, (4) checks on the correctness of the model. With twofold dilution, about seven dilutions should provide $P$ values between 5 and 95 percent, giving some data for $\chi^2$ and related checks.

## 3. OTHER WORK BY FISHER

Fisher's remaining work on nonlinear problems mainly involved using the concept of amount of information as helpful in planning data collection, as illustrated in the last chapter of his book [34]. He did much work of this kind, which I will not describe, on the estimation of linkage in humans, animals and plants. In plants, for instance, the amount of linkage between two genes can be estimated by forming a double heterozygote and either crossing it with itself (selfing) or backcrossing it. For estimating close linkage from selfing, he showed that formation of the double heterozygote parent in coupling (AABB × aabb) can be 15 times as efficient as its formation in repulsion (AAbb × aaBB), and is nearly as efficient as backcrossing.

Fisher's first paper on the analysis of variance [35], dealt with a $12 \times 6$ factorial on potatoes. He first presents the standard ANOVA into main effects and interactions. He then remarks that the preceding analysis is given solely for illustration, since the linear model is obviously unsuitable, predicting *negative* expected yields for some of the plots. As more reasonable, he proceeds to fit a nonlinear product model, which can be written

$$E(y_{ij}) = \mu(1 + \alpha_i)(1 + \beta_j).$$

This requires more work but, as anticipated, fits better, the S.S. deviations being 847 against 981. From this paper, I would not have expected Fisher's later ANOVA work to have concentrated so largely on development of the linear model. I am sorry that I never asked him why it did.

## 4. QUANTAL BIOASSAY (NON-SEQUENTIAL)

Another earlier nonlinear problem on which much research for practical experiments has been done is quantal bioassay under a normal or a logit tolerance distribution —a problem again with a 0–1 response. We are comparing a Standard ($S$) with a Test ($T$) preparation thought to contain the same active ingredient and therefore to act like a dilution or concentration of the Standard. Thus if $x$ is log dose, an amount $x$ of $S$ has exactly the same effect as an amount $x - M$ of $T$. Here $M$, the log relative potency of Test to Standard, is the quantity to be estimated.

To illustrate from the normal model, if $n$ subjects are given an amount $x$ of $S$, the proportion responding is binomial with

$$P = \frac{1}{\sigma\sqrt{2\pi}} \int_{-\infty}^{x} \exp\left\{-\tfrac{1}{2}(x - \mu_S)^2/\sigma^2\right\} dx$$

$$= \frac{1}{\sqrt{2\pi}} \int_{-\infty}^{(x-\mu_S)/\sigma} Z(t)dt \quad (4.1)$$

where $Z(t)$ is the ordinate of the Standard normal curve.

For $T$, the formula differs only in that $\mu_T = \mu_S - M$. Thus the problem is a three-parameter one, with one parameter $M$ to be estimated and two nuisance parameters.

For a single agent, Fisher showed that

$$I(\mu) = nZ^2/PQ\sigma^2$$

which is maximized at $P = 0.5$, $x = \mu$. Thus if $\mu_S$, $\mu_T$ and therefore $M$ were known, the optimum experiment would place all subjects at the levels of $S$ and $T$ causing 50 percent response.

Lacking this knowledge, experimenters use two or more levels of each agent (hopefully straddling the 50 percent response) from which the ML estimates of $\mu_S$, $\mu_T$ can be obtained.

If $Y$ is the normal deviate corresponding to $P$ in (4.1)

$$Y = \frac{-\mu}{\sigma} + \frac{x}{\sigma}.$$

For a single agent, Fisher [34] and others—see [30]— showed that ML estimates of $\mu$ and $\sigma$ could be obtained iteratively by a weighted linear regression on $x$ of a transform $y$ (the working transform) of the observed proportion $p = r/n$ of responding subjects. This approach gives two fitted lines

$$\hat{Y}_S = \bar{y}_S + b(x - \bar{x}_S) \quad (4.2)$$

$$\hat{Y}_T = \bar{y}_T + b(x - \bar{x}_T). \quad (4.3)$$

To obtain the same response, $\hat{Y}_S = \hat{Y}_T$, the difference $\hat{M}$ between the required doses $\hat{x}_S$ and $\hat{x}_T$ is, from (4.2) and (4.3),

$$\hat{M} = \bar{x}_S - \bar{x}_T - (\bar{y}_S - \bar{y}_T)/b,$$

where $1/b$ estimates the assumed common $\sigma$. Since $b$ is first estimated separately for Test and Standard, a test of significance of $(b_T - b_S)$ is available and is regarded as an essential check on the basic assumptions before the combined estimate $b$ is made.

Since $\hat{M}$ involves the ratio $(\bar{y}_S - \bar{y}_T)/b$ of two random variables, Finney's criterion [31] for the choice of levels of $x_T$ and $x_S$ and of $n$ is the half-width of Fieller's [29] percent fiducial interval for $M$, which is found to be

$$\frac{1.96}{b(1-g)}\left[(1-g)\left(\frac{1}{s\Sigma nw} + \frac{1}{\tau\Sigma nw}\right) + \frac{(\hat{M} - \bar{x}_S + \bar{x}_T)^2}{S_{xx}}\right]^{\frac{1}{2}} \quad (4.4)$$

where $S_{xx} = \Sigma nw(x - \bar{x})^2$ summed over both agents and $g = (1.96)^2/b^2S_{xx}$ is the square of (1.96 times the coefficient of variation of $b$).

In designing an experiment, the number of levels $k$, their spacing $d$, and the sample size $n$ at each level must be chosen. From previous work on the Standard, good initial estimates of $\beta$ and $\mu_S$ should usually be available and an initial estimate $M_0$ is assumed. The strategy is to make $x_T = x_S - M_0$ at any level. This should make $(\hat{M} - \bar{x}_S + \bar{x}_T)$ in (4.4) small and the corresponding term in (4.4) is often negligible. In this event, with $n$ constant, (4.4) becomes

$$\frac{(1.96)}{b}\left[\frac{2}{n(1 - g)W}\right]^{\frac{1}{2}} \quad (4.5)$$

where $W = \Sigma w$ over the $k$ levels for one agent. Regarding the quantity multiplying (1.96) in (4.5) as a kind of effective standard error of $\hat{M}$, Finney [31, 496–7] tabulates $b^2V_E(\hat{M})$ for $k = 2, 3, 4$, total number of subjects $N = 2kn = 48, 240$, and a range of choices of levels which give $P$ values centered about 50 percent. A similar table is given for the logistic model in which logit $P$ is assumed linear in $x$.

These tables provide estimated optimum spacings and the corresponding $b^2V_E(\hat{M})$ for 2, 3, 4, levels and $N = 48$, 240. Similar tables for other sample sizes and numbers of levels could easily be provided.

The optimum levels assume good initial guesses. The only work that I have seen allowing poor guesses is by Brown [14]. Using the simpler Spearman-Kärber estimates of $\mu_S$, $\mu_T$, he recommends choices of $n$, $d$, $k_S$ and $k_T$ (which he allows to differ), in order to give a desired width of 95 percent confidence interval for $M$. This approach is similar to Fisher's in the dilution series. Naturally, more levels are required to ensure coverage of the 50 percent dose: Brown's worked example gives $k_S = 10$, $k_T = 22$.

Thus, based on the Fieller criterion, available methods furnish:

1. a near-optimum experiment, assuming good initial estimates of $\sigma$, $\mu_S$ and $M$, and using large-sample theory,
2. assuming the model correct, Fieller's limits for the sample data, as a measure of the precision of $\hat{M}$,
3. for more than two levels per agent, tests of the adequacy of the model. The $\chi^2$ for deviations from the model has $(2k - 3)$ d.f. These split into one d.f. for non-parallelism, one d.f. for combined curvature, and $(2k - 5)$ d.f. for other sources. Fortunately, as Finney shows, $k = 4$ does not demand more subjects than $k = 2$.

I know of no intensive study of the robustness of the presumed optima to poor initial guesses at the parameter values. Extensions of Finney's tables to more spacings and more sample sizes would reveal the effects of wrong spacing, through a bad guess at $\sigma$, on $b^2V_E(\hat{M})$. For $r > 1$, it appears from his tables that the effects are more serious if the guess is $\sigma/r$ than if it is $r\sigma$, and more serious with fewer levels, as would be expected. Sample size charts by Healy [36] indicate for $k = 3$ the effects of wrong centering of the doses (through a poor guess at $\mu_S$). More work on robustness and on the small-sample performance of the recommended plans and formulas would be useful.

In the preceding research, the measure of the precision of $\hat{M}$ deals only with "within assay" variation. Several writers have pointed out that even with the same animal and technique, $M$ itself may have substantial variation from one occasion and one laboratory to another. As noted by Dews and Berkson [21], this variation may have serious consequences for the diabetic who is supposed to receive regularly a specified dose of insulin. The planning of a series of assays so as to measure "between assay" variation is a subject somewhat neglected by statisticians.

## 5. QUANTAL BIOASSAY (SEQUENTIAL)

The well-known Up and Down or Staircase method [22, 23, 24] was devised for experiments in which it is convenient to test subjects one at a time, determining the level of the agent for the next subject after seeing the result (0 or 1) for the previous subject. For a given dose spacing $d$, the rule for a single agent (Standard or Test) is the very simple one

$$x_{u+1} = x_u + d \ (\text{if } y_u = 0); \quad x_{u+1} = x_u - d \ (\text{if } y_u = 1).$$

The idea is, of course, to concentrate dose levels in the neighborhood of $\mu$, the median of the response $y$, which the method is designed to estimate. The *nominal* sample size $N$ is defined as the number of trials, beginning with the first pair in which a reversal (0 to 1 or 1 to 0) occurs. The estimate $\hat{\mu}$ of $\mu$ is the mean of the last $N$ values of $x_u$, with an adjustment [23] depending on the numbers of 0's and 1's that were obtained. The mean square error of $\hat{\mu}$ is approximately $2\sigma^2/N$ when $d$ lies between the limits $d = 2\sigma/3$ and $d = 3\sigma/2$, with $d = \sigma$ recommended as the most accurate spacing. This work is based on exact small-sample computations.

A single sequence provides no usable estimate of $\sigma$, which is undesirable if we wish to attach an estimated rms error $\sqrt{2}\hat{\sigma}/\sqrt{N}$ to $\hat{\mu}$. Dixon [23] recommends that the experiment be run in independent sequences with say, $N = 6$ in each sequence. If there are $r$ of these in parallel under the same operating conditions, this speeds up completion of the experiment and allows $V(\hat{\mu})$ to be estimated from $\Sigma(\hat{\mu}_j - \hat{\mu})^2/r(r-1)$. Alternatively, other relevant variables may be changed from one set to another, permitting the effects of these variables on $\mu$ to be investigated by analysis of variance techniques.

For a logistic model, when a single (longer) sequence is being used, Wetherill [44] has proposed a change intended to make the accuracy of $\hat{\mu}$ more robust against a poor initial guess and use of a $d$ too large. After six changes of response type have occurred, estimate $\hat{\mu}$, and restart near $\hat{\mu}$ using half the original spacing. Here there remains the problem of an estimate of $\sigma$ from the data.

Another sequential plan, using the Robbins-Monro stochastic approximation process, attempts to do better

than the Up and Down by steadily shortening the steps as the sequence proceeds. If a group of $n$ subjects are tested at each step, the level of $x$ for the $(u + 1)$th experiment is

$$x_{u+1} = x_u - \frac{c}{u} (p_u - \tfrac{1}{2}).$$

When the experiment is terminated, the estimate $\hat{\mu}$ is the level at which the next experiment would have been conducted [17]. With $g$ steps, the asymptotic formula for $V(\hat{\mu})$ is $\pi\sigma^2/2ng$, the value it would have if all trials could be conducted at the optimum 50 percent level. To guard against a poor initial guess at $\mu$, a 'delayed' version was also suggested in which the step size $c$ remains unchanged until both deaths and survivals have been obtained. A modification with a similar purpose has been proposed by Kesten [38].

For small experiments with $N = ng = 12$, where $T$ is the number of steps, $= 3, 4, 6,$ or $12$, Davis [20] has compared the MSE's of $\hat{\mu}$ for three versions of the Robbins-Monro, two of the Up and Down, and a non-sequential experiment using the Spearman-Kärber estimate, for normal, logistic, uniform and exponential tolerance distributions. This is the first broad comparison of the performances of different plans in small samples. It is reassuring that the recommended step size and the asymptotic formulas for $V(\hat{\mu})$ both perform well for starts within about $1.5\sigma$—about all that can be expected for $N = 12$. Overall, delayed versions of the Up and Down and the Robbins-Monro performed best, both easily beating the non-sequential methods.

## 6. SINGLE CONTINUOUS-VARIABLE RESPONSE— A CRITERION

For the $u$th observation or trial ($u = 1, 2, \cdots, N$) the model now becomes

$$y_u = f(\xi_u; \theta) + \epsilon_u$$
$$= f(\xi_{u1}, \cdots, \xi_{uk}; \theta_1, \cdots, \theta_p) + \epsilon_u. \quad (6.1)$$

Here, $\xi_{ui}$ denotes the level at which the value of the variable $\xi_i$ is set by experimenter in the $u$th trial. There are $k$ such factors or variables, while $p$ is the number of parameters involved in the model. In the simplest models the $\epsilon_u$ are assumed independently $N(0, \sigma^2)$.

The article that provided the impetus to intensive work is that of Box and Lucas [7]. Much related earlier work, dealing primarily with the linear case, had been done by Kiefer [39], Elfving [28], and Chernoff [16], who considered the choice of a criterion and the finding of the design points (levels of the factors $\xi_{ui}$).

The criterion proposed by Box and Lucas assumes interest in all the parameters. It maximizes the generalization of Fisher's amount of information or, equivalently, minimizes the asymptotic formula for Wilks' generalized variance of the ML estimates of the $\theta_j$. From (6.1) the log likelihood is

$$L = -\frac{1}{2\sigma^2} \sum_{u=1}^{N} (y_u - f_u)^2.$$

It follows that the information matrix is

$$E\left(\frac{-\partial^2 L}{\partial\theta_i \partial\theta_j}\right) = \frac{1}{\sigma^2}\sum_{u=1}^{N}\left(\frac{\partial f_u}{\partial\theta_i}\right)\left(\frac{\partial f_u}{\partial\theta_j}\right) = \frac{1}{\sigma^2}(\mathbf{X}'\mathbf{X})$$

where $\mathbf{X}$ is the $N \times p$ matrix

$$(x_{uj}) = \frac{\partial f_u}{\partial\theta_j}.$$

The $x_{uj}$ are known when the factor levels $\xi_{ui}$ and the $\theta_j$ are known. The criterion—choose design points $\xi_{ui}$ to maximize $|\mathbf{X}'\mathbf{X}|$—assumes initial guesses $\theta_j$ for practical use. Other attractive features of this criterion (summarized by M.J. Box and Draper [13]) are as follows:

1. It minimizes the volume of the asymptotic confidence region for the $\theta_j$ [40].
2. From the Bayesian viewpoint, it maximizes the joint posterior probability of the $\theta_j$, given a non-informative prior $\Pi d\theta_j$, for response functions locally linear in the neighborhood of the ML estimates [25].
3. It is invariant under changes of scale of the $\theta_j$.

## 7. FINDING THE DESIGN POINTS—NON-SEQUENTIALLY

Given a criterion, the next step is the complex one of finding design points that satisfy the criterion for a specified $N$ trials. In earlier work, Chernoff [16] considered the case where our interest is in $s \leq p$ of the parameters, the remaining $(p - s)$ being nuisance parameters. His criterion was different—minimizing the average of the asymptotic variances of the $s$ ML estimates. Following Elfving [28], he showed that an optimum design needs at most $s(2p - s + 1)/2$ points, becoming $p(p + 1)/2$ when $s = p$, and $p$ when $s = 1$.

As a start, Box and Lucas [7] assumed initial guesses $\theta_0$ and sought an optimum set of levels when $N = p$, i.e., when there are only as many trials as parameters to be estimated. They point out unappealing features of this decision: no test of the fit of the model, no attempt at robustness against poor initial guesses, to which might be added no data for an experimental estimate of $\sigma^2$.

One advantage with $N = p$ is that $(\mathbf{X}'\mathbf{X})$ is square, so that $|\mathbf{X}'\mathbf{X}| = |\mathbf{X}|^2$ and it suffices to maximize $|\mathbf{X}| = |x_{ui}|$. Illustrative examples worked by Box and Lucas include the exponential growth or decay curve, the Mitscherlich equation, and the two-factor function

$$f(\xi_1, \xi_2, \cdot\theta_1, \theta_2) = \exp(-\theta_1\xi_1 e^{-\theta_2\xi_2}).$$

Depending on the complexity of the problem, methods available for solution are:

1. Geometric or analytic,
2. Calculate $|\mathbf{X}|$ for a grid of values of the $\xi_{ui}$, fit a quadratic to this grid and seek a maximum (with trouble possible if $|\mathbf{X}|$ has more than one turning value),
3. Various computer iterative hill-climbing techniques.

As a simple example with an analytic solution, consider the exponential decay curve

$$f_u = \theta_1 e^{-\theta_2 t_u}$$

where $t_u$ (time) is used for $\xi_{u1}$. The region feasible for experiments is $t(\min) \leq t \leq t(\max)$. For this $f_u$,

$$|\mathbf{X}| = \begin{vmatrix} e^{-\theta_2 t_1} & -t_1\theta_1 e^{-\theta_2 t_1} \\ e^{-\theta_2 t_2} & -t_2\theta_1 e^{-\theta_2 t_2} \end{vmatrix}$$

$$= \theta_1(t_1 - t_2)e^{-\theta_2(t_1+t_2)}.$$

This can be written

$$|\mathbf{X}| = \{\theta_1(t_1 - t_2)e^{-\theta_2(t_1-t_2)}\}\{e^{-2\theta_2 t_2}\}.$$

For given $(t_1 - t_2)$ and with $\theta_2 > 0$, we want $t_2 = t(\min)$. The first curly bracket is maximized when

$$t_1 - t_2 = 1/\theta_2, \text{ giving } t_1 = t(\min) + 1/\theta_2$$

or

$$t_1 = t(\max), \text{ whichever is smaller.}$$

Coming to the case of a single non-sequential experiment with $N > p$, Atkinson and Hunter [2] found in several chemical examples, worked by computer maximizing that with $N$ a multiple of $p$, the optimum plan consisted simply of $N/p$ replications at each of the $p$ optimum sets of levels for the case $N = p$. This result certainly simplifies the finding of optimum plans. Although a counter example showed that the result does not hold in general, they proved, as a sufficient condition, that the result will hold if the region of experimentation lies within a certain ellipsoid in the $x$-space (a point that can be checked by the experimenter).

M.J. Box [8, 10] considered also the case: $N$ not a multiple of $p$. In some problems he found that replications of the $N = p$ solution differing by at most one could be proved to be optimal. In others, while this could not be proved, a computer search was unable to locate anything superior to the near-equal-replication solution. He also considered a one-factor, two-parameter problem with $\xi_{u1}$ = time = $t_u$, where different trials cost different amounts. The problem was to maximize $|\mathbf{X}'\mathbf{X}|$ subject to a fixed cost $C = \Sigma c_u$. The optimum again consisted of experiments at only two times $t_1$, $t_2$, but with the difference that $t_1$ and $t_2$ changed both with $N$ and $C$ and the numbers of replications were no longer near-equal, so that more computing effort was necessary.

The counter-example by Atkinson and Hunter is the linear fitting of a bivariate regression, $f_u = \theta_1\xi_{1u} + \theta_2\xi_{2u}$, with the region of experimentation $0 \leq \xi_{iu} \leq 1$. For $N = p = 2$, the optimum design is at the levels $(1, 0)$ and $(0, 1)$, which gives

$$\mathbf{X}'\mathbf{X} = \begin{pmatrix} 1 & 0 \\ 0 & 1 \end{pmatrix}; \quad |\mathbf{X}'\mathbf{X}| = 1.$$

With $N = 6$, three replications of this plan give

$$\mathbf{X}'\mathbf{X} = \begin{pmatrix} 3 & 0 \\ 0 & 3 \end{pmatrix}; \quad |\mathbf{X}'\mathbf{X}| = 9.$$

But two replications of the three-point plan $(1, 0)$, $(0, 1)$, $(1, 1)$ give

$$\mathbf{X}'\mathbf{X} = \begin{pmatrix} 4 & 2 \\ 2 & 4 \end{pmatrix}; \quad |\mathbf{X}'\mathbf{X}| = 12.$$

The key ellipse in this example is the circle $\xi_1^2 + \xi_2^2 = 1$, and the point $(1, 1)$ in the experimental region lies outside this circle.

The preceding results on the best set of design points are conceptually similar to Fisher's original optimum for the dilution series problem, and assume in effect good initial estimates of the $\theta_j$. With poor initial guesses, the resulting plan will not be optimal in any real sense. I have come across no work analogous to Fisher's, where we start with a wider spread than $p$ points with the object of guaranteeing a specified value of $|\mathbf{X}'\mathbf{X}|$ starting from initial $\theta_j$ assumed known initially only to lie within a certain region.

## 8. FINDING THE DESIGN POINTS SEQUENTIALLY

As would be expected, the methods start with $p$ points, determined by first guesses $\theta_{0j}$, and leading to ML estimates $\hat{\theta}_{1j}$ of all the parameters. Box and Hunter [6] discuss how to add points one at a time. If $(N - 1)$ steps have been completed, so that $\hat{\theta}_{N-1,j}$ are known, then $|\mathbf{X}'\mathbf{X}|$ as a function of the $x$'s for the $N$th point takes the form

$$|\mathbf{X}'\mathbf{X}|_N = \begin{vmatrix} c_{11} + x_{1N}^2 & c_{12} + x_{1N}x_{2N} & \cdots & c_{1p} + x_{1N}x_{pN} \\ c_{12} + x_{1N}x_{2N} & c_{22} + x_{2N}^2 & \cdots & \\ \cdot & \cdot & \cdot & \cdot \\ \cdot & \cdot & \cdot & \cdot \\ c_{1p} + x_{1N}x_{pN} & \cdots & & c_{pp} + x_{pN}^2 \end{vmatrix}$$

where the $c_{ij}$ are known. The criterion is computed for all points of a grid of values of the $\xi_{u1} \cdots \xi_{uk}$ and a quadratic fitted to find the maximizing values.

M.J. Box [10] adds sequential sets of $n = p$ points, each put at the best $p$ design points as estimated from the ML $\hat{\theta}$ obtained from the combined trials conducted to date. After a time, both the ML $\hat{\theta}$ and the indicated set of $p$ design points for the $r$th set begin to change little from those in the $(r - 1)$th set. Box introduces a criterion $R_1$ as a guide to the time when it is no longer worth changing points. A second quantity $R_2$ compares the $|\mathbf{X}'\mathbf{X}|$ value given by all trials conducted to date with the value that $|\mathbf{X}'\mathbf{X}|$ would have if it had been possible to use our current estimate of the best design points in all trials. Thus $R_2$ indicates the amount lost owing to poor initial guesses at the $\theta_j$. In the simulated example (three parameters, two factors), some values of $R_1$ and $R_2$ in sequential plan are listed in the following tabulation:

| Set | $R_1$ | $R_2$ |
|-----|-------|-------|
| 2 | 1.40 | 0.78 |
| 3 | 1.09 | 0.86 |
| 4 | 1.06 | 0.86 |
| 5 | 1.04 | 0.88 |
| 6 | 1.02 | 0.91 |

In order to study the effect on the sequential process of having initial prior information of different amounts about different $\theta_j$, Draper and Hunter [26] took a multinormal prior

$$(2\pi)^{-\frac{1}{2}p} |\mathbf{\Omega}|^{-\frac{1}{2}} \exp\left\{-\tfrac{1}{2}(\theta - \theta_0)'\mathbf{\Omega}^{-1}(\theta - \theta_0)\right\}$$

where $\mathbf{\Omega}$ is the $p \times p$ matrix of variances and covariances and the $\theta_0$ are initial guesses. In the case where $N$ trials had already been completed at chosen levels $\xi_u$, they discussed where to put a further $n$ trials. Their criterion was to maximize the posterior distribution of $\theta$ after $(N + n)$ trials with respect both to $\theta$ and to the values $\xi_u$ $(u = N + 1, \cdots, N + n)$. Assuming $f(\xi_u, \theta)$ to be locally linear, this leads to the approximate criterion: maximize

$$|\mathbf{X}'\mathbf{X} + \sigma^2 \mathbf{\Omega}^{-1}| \tag{8.1}$$

with respect to $\theta$ and $\xi_u$ $(u = N + 1, \cdots, N + n)$. One hurdle is that in (8.1) the values of $\theta$ are hidden in $\mathbf{X}'\mathbf{X}$ and their maximizing values after $(N + n)$ trials depend on observations not yet taken. The natural suggestion is to use $\hat{\theta}_N$ in maximizing (8.1) with respect to the levels $\xi_u$.

The principal value of this type of prior is likely to be the light it throws on how the design would be affected by different amounts of prior information about the different $\theta_j$. As an illustration they work a problem with $\theta_1$, $\theta_2$ independent normals $(0, \sigma_j^2)$, $N = 0$, $n = 2$, and a *single* $\epsilon_{u1}$ $(= t_u,$ a time variable$)$. As $\sigma_1$, $\sigma_2$ vary from 0 to $\infty$, three basic design types predominate: $(t_1, t_2) = (1.2, 6.9)$ for little prior information, $(t_1, t_2) = (1.2, 1.2)$, where the experiment concentrates on estimating $\theta_1$, and $(t_1, t_2) = (6.9, 6.9)$, where the emphasis is on $\theta_2$. Further illustrations of this type would be of interest.

## 9. DESIGNS TO ESTIMATE ONLY A SUBSET OF THE PARAMETERS

Draper and Hunter [25] have noted that the experimenter may intend to estimate only a subset of the parameters, the others being nuisance parameters whose values are of no interest. M.J. Box [12] gives a detailed criterion for constructing such designs. He assumes a noninformative prior $\Pi d\theta_j$, multivariate normal experimental errors, and the local validity of first-term expansions of the response function $f(\xi; \theta)$. Under these assumptions, maximizing the joint posterior density of the subset $\phi$ turns out as might be expected to be equivalent to minimizing the asymptotic generalized variance of the $\hat{\phi}$ at the current best estimates of all parameters.

## 10. TESTS OF FIT OF THE MODEL

As a dividend, the sequential approach might provide some data for a test of fit of the model (at least assuming $\sigma^2$ known), since $y_u$ will have been determined in general at $N > p$ design points. If, however, the successive design points vary over only a restricted part of the experimental region, examination of the residuals may tell us little. Experience with a wider range of nonlinear models may throw more light on this issue. Two approaches to obtaining some check on the model were made by Box and Lucas [7].

First, having computed the combinations of levels of the $\xi_u$ needed to maximize $|\mathbf{X}'\mathbf{X}|$, the experimenter might examine where they occur in the $\xi$ space of interest, and add extra points where he is most worried that the model may be incorrect.

Second, the experimenter may sometimes be reasonably sure that if the model is incorrect, a more general model with, say, one or two extra parameters gives an adequate fit. The example cited is where the model (with a single $\xi$ variable) is

$$f_u = 1 - e^{-\theta_1 \xi_u}$$

where in fact the more general model

$$f_u = \theta_1 \{ e^{-\theta_2 \xi_u} - e^{-\theta_1 \xi_u} \} / (\theta_1 - \theta_2)$$

might be required. The experiment might be planned to estimate $\theta_1$ and $\theta_2$ and test the NH $\theta_2 = 0$, which makes the original model correct.

If the experiment is conducted sequentially, M.J. Box [12] has suggested that the designs might concentrate initially on estimation of all the parameters plus obtaining data to test the NH that the extra parameters have known values (usually 0). Later points could concentrate on more efficient estimation of any simpler model suggested by judgment from the results of these tests.

Atkinson [1] considers designs intended from the beginning to provide tests on the fit of the proposed model. One possibility is to add to the model a low polynomial (e.g., a quadratic) in the $x$ variates. If testing is given high priority, the initial design might concentrate on minimizing the generalized variance of the subset of quadratic terms. In his illustration, however, he notes that the optimum design points with this objective give only 42 percent efficiency in estimating the single parameter in the model and also illustrates a compromise design giving at least 80 percent efficiency. An alternative also illustrated is to add to the model terms in the partial derivatives of the response function with respect to the parameters.

## 11. DISCRIMINATION BETWEEN SPECIFIED MODELS

An approach by Cox [18, 19] for discrimination between two models used a test of significance and was asymmetric: the hypothesis that Model 1 is correct was chosen as the null hypothesis. The test criterion was a modified form of the likelihood ratio, maximizing the asymptotic power against Model 2 as the alternative. Later, in planning an experiment to discriminate between the probit and the logit models from observations confined to three log dosages, Chambers and Cox [15] used a compromise symmetric form of this approach. They first chose a criterion asymptotically powerful against AH = logistic, given NH = probit. For this criterion, they determined the optimum three dosage levels and the proportions of the observations to be put at each level. Then they reversed the procedure, having AH = probit, NH = logistic. Fortunately, the optimum doses did not differ greatly in the two cases, so that good compromise design could be constructed. Unfortunately, as Chambers and Cox note, this plan put the majority of the observations at a high dose level with expected percent killed

over 99.6 percent. Thus the experiment would require large samples, as will not surprise those who have worked with both probits and logits. This approach might end, of course, by rejecting neither model, one specific model, or both models.

An alternative approach, Box and Hill [4], is symmetric and extends to more than two specific models. For two models, the approach supposes that $n$ observations have already been taken (at least enough to estimate any parameters involved) and considers where best to put the $(n + 1)$th for maximum discrimination. At first sight one might be inclined to seek the point (levels of the factors) for which $| \hat{Y}_1 - \hat{Y}_2 |$ is maximized, where $\hat{Y}_1$ and $\hat{Y}_2$ are the model 1 and model 2 estimates from the results for the first $n$ observations. But as Box and Hill note, the precision of estimation of $| \hat{Y}_1 - \hat{Y}_2 |$ is also relevant.

Prior probabilities $\Pi_{i0}$ are first assigned to each model. With two models, the choice might be $\Pi_{i0} = \frac{1}{2}$ and with $m$ models, $\Pi_{i0} = 1/m$. After $n$ runs, the posterior probability for the $i$th model is

$$\Pi_{in} = \Pi_{i,n-1} p_i / \Sigma \Pi_{i,n-1} p_i$$

where $p_i$ is the probability density function of the $n$th observation $y_n$ under model $i$.

The criterion chosen for discrimination uses Shannon's [42] concept of entropy, also known as the Kullback-Liebler information [41]. For $m$ models the entropy is

$$- \sum_{i=1}^{m} \Pi_i \ln \Pi_i.$$

This has its maximum value when $\Pi_i = 1/m$ and becomes steadily smaller as the $\Pi_i$ becomes unequal, i.e., as discrimination improves. Hence the $(n + 1)$th observation is chosen at levels $\xi$ which will maximize the expected decrease in entropy from the $n$th to the $(n + 1)$th experiment.

For two models the resulting discrimination criterion is shown to be

$$D = \Pi_{1n} \Pi_{2n} \left( \int p_1 \ln (p_1 / p_2) dy_{n+1} \right. \\ \left. + \int p_2 \ln (p_2 / p_1) dy_{n+1} \right).$$

If we can further assume that the models are locally linear, with deviations $\xi_u$ that are $N(0, \sigma^2)$, where $\sigma^2$ is known, the criterion becomes, for two models $Y^{(1)}$ and $Y^{(2)}$,

$$D = \frac{1}{2} \Pi_{1n} \Pi_{2n} \left\{ \frac{(\sigma_1^2 - \sigma_2^2)^2}{(\sigma^2 + \sigma_1^2)(\sigma^2 + \sigma_2^2)} \right. \\ \left. + (\hat{Y}_{n+1}^{(1)} - \hat{Y}_{n+1}^{(2)})^2 \left( \frac{1}{\sigma^2 + \sigma_1^2} + \frac{1}{\sigma^2 + \sigma_2^2} \right) \right\},$$

where $\sigma_i^2$ is the approximate variance of $\hat{Y}_{n+1}^{(i)}$. For $m$ criteria, $D$ is the corresponding expression summed over all pairs of models.

One of the examples worked is a simulated example, with four parameters and two factors, to distinguish among the first- to fourth-order reaction curves. The experiment starts with a grid of four points to estimate all four parameters needed. The results proceed as in the table.

### EXAMPLE OF DISCRIMINATION AMONG MODELS

| $n$ | $\xi_1$ | $\xi_2$ | $\Pi_1$ | $\Pi_2$ | $\Pi_3$ | $\Pi_4$ |
|-----|---------|---------|---------|---------|---------|---------|
| 1 | 25 | 575 | | | | |
| 2 | 25 | 475 | | | | |
| 3 | 125 | 575 | | | | |
| 4 | 125 | 475 | .01 | .43 | .50 | .06 |
| 5 | 125 | 600 | .00 | .56 | .43 | .01 |
| 6 | 125 | 600 | .00 | .86 | .13 | .00 |
| 7 | 50 | 450 | .00 | .97 | .02 | .00 |
| 8 | 100 | 600 | .00 | 1.00 | .00 | .00 |

From the beginning, the competition is between the second- and third-order curves, the second-order soon establishing itself as correct.

Box and Hill also work an example in which (1) all models are generalizations of model 1 and (2) model 1 is correct so that *all* models are correct. Here the entropy criterion seems to be given an impossible task, but by their largest $n(15)$, it is tending towards selection of the simplest of the correct models—an admirable performance. However, in more recent examples of this situation in which $n$ was continued to large values, Siddik [43] found that the posterior probability of the simplest correct model rose to a value such as 0.85 to 0.95, but then fluctuated erratically around that value. While it still can be conjectured that the criterion will operate well in practical experiments, its large-sample performance needs further study.

A succeeding paper by Hill, Hunter and Wichern [37] recognizes that the best choice of the $\xi$ levels for discrimination will not in general be those that give the best parameter estimation for the correct model, and seeks to reconcile these conflicting aims. If we knew that model $j$ was the correct model, we would choose the $\xi$ to maximize $\Delta_j = |\mathbf{X}'\mathbf{X}|_j$ for model $j$. Call this value $\Delta_{j,\max}$ and let $\Delta_j$ denote the value of the estimation criterion $\Delta_j$ for any other choice of levels $\xi$. Similarly, let $D_{\max}$ be the maximum expected decrease in entropy, and $D$ the decrease obtained from any other setting of the $\xi$. The criterion which these authors suggest for choosing the $\xi$ levels is

$$C = w_1 D / D_{\max} + w_2 \sum_{j=1}^{m} \Pi_{jn} \Delta_j / \Delta_{j,\max}.$$

The $w_1$ and $w_2$ are weights ($w_1 + w_2 = 1$) which can be changed, as the sequence of runs proceeds, to give increasing weight to good parameter estimation when it becomes clearer that one model is being selected by the discrimination technique. For $w_1$ they suggest, as one possibility,

$$w_1 = \{m(1 - \Pi_{bn})/(m - 1)\}^{\lambda}$$

where $\Pi_{bn}$ is the probability assigned to the best model

before the $(n + 1)$th observation is taken. The quantity $\lambda$ is a positive power that controls the rate of decrease of $w_1$, the weight assigned to the discrimination criterion. Initially, if all $\Pi_{i0} = 1/m$, $w_1$ is unity and all emphasis is given to good discrimination. As $\Pi_{bn}$ approaches one, so does $w_2$, emphasis shifting to estimation for the most likely model.

## 12. MODEL BUILDING

It is more difficult to do justice to the work here, since the strategies will change as the accumulated data suggest new ideas to the experimenter and statistician.

As one approach, Box and Hunter [5, 38] consider the case where the experimenter has at best a tentative model which describes $f(\xi, \theta, t)$. If there are $k$ factors $\xi$ which the experimenter can manipulate, they suggest running the reaction, with measurements of response at certain fixed times, for a $2^k$ factorial or fractional factorial in the levels of the $\xi_i$, widely separated as far as operating restrictions permit. For each combination of the factor levels they estimate each $\theta_j$ and do a standard factorial analysis into main effects and interactions for each $\hat{\theta}_j$. There are two objectives in this procedure: (1) if the model is correct, the $\hat{\theta}_j$ should not change systematically with time or with the changes in the levels of the $\xi_i$, since the $\theta_j$ should be constant, and (2) the way in which the $\hat{\theta}_j$ change may enable the experimenter to specify a vague model more completely, or may suggest relations among the $\theta_j$ and the $\xi_i$ that make sense mechanically.

In their simulated example they use letters $A$, $B$, (the initial concentrations of two reactants), $C$ (the concentrations of a catalyst), and $D$ (the temperature), to denote the factors instead of our $\xi_1, \cdots, \xi_4$. The tentative model was

$$E(y) = \frac{(B)k_1}{k_1 - k_2}(e^{-k_2 t} - e^{-k_1 t}).$$

The initial experiment was a $2^4$ factorial in $A \cdots D$, measured at five times. The nature of the reaction suggested that

$$k_1 = (A)^{p_1}(C)^{q_1}\alpha_1 e^{-\beta_1/T}$$
$$k_2 = (A)^{p_1}(C)^{q_1}\alpha_2 e^{-\beta_1/T}$$

where $T$ is the absolute temperature. (There are now eight parameters to be estimated.) If this suggestion is correct, a factorial analysis of $\ln k_1$ and $\ln k_2$, which was then carried out, should show no effects of $B$ and no interactions involving $A$, $C$, and $D$. The analysis confirmed the model, as did careful examination of the residuals ($y - f$) for the $2^4$ runs at the five times conducted initially. Finally, the eight parameters were estimated from the combined data. The second paper [38] gives further discussion of the examination of residuals, contour diagrams, and plots of the likelihood function as diagnostic aids. The necessity for repeated interchange of ideas between experimenter and statistician is stressed.

An interesting review of approaches and problems in model-building by M.J. Box [19] presents his experiences, with discussion from the audience.

## 13. MORE THAN ONE MEASURED RESPONSE

In some chemical reactions it is possible to measure more than one response $y_{ul}$ ($l = 1, 2, \cdots, L$) which provides information about some or all of the parameters $\theta_j$. The simplest example quoted is the one-parameter exponential

$$y_{1u} = e^{-\theta \xi_u} + \epsilon_{1u}; \quad y_{2u} = 1 - e^{-\theta \xi_u} + \epsilon_{2u}$$

where the single factor $\xi_u$ represents time. Note that $y_{1u}$, $y_{2u}$ do not add to one because of the experimental errors $\epsilon_{lu}$.

In a general approach the model is

$$y_{lu} = f_l(\xi_u; \theta) + \epsilon_{lu}.$$

It has not been considered realistic to assume $\epsilon_{lu}$, $\epsilon_{mu}$ independent. Instead, they are given a multivariate normal distribution with variance-covariance matrix $\sigma_{lm}$.

The first paper on this problem [3] did not assume the $\sigma_{lm}$ known in advance, and merely assigned a 'non-informative' prior distribution to the $\sigma_{lm}$. Later papers took the more tractable problem in which the $\sigma_{lm}$ are assumed known, and will be considered first.

With known $\sigma_{lm}$, Draper and Hunter [25] assigned a Bayesian prior $\Pi d\theta_j$ and followed the method which led to the $|\mathbf{X'X}|$ criterion for a single response, as mentioned in Section 7. It helps to write

$$v_{lm} = \sum_{u=1}^{N} \{y_{lu} - f_{lu}\}\{y_{mu} - f_{mu}\}. \quad (13.1)$$

They find that the posterior probability is

$$p(\theta \mid \mathbf{y}) = c \exp\{-\tfrac{1}{2} \sum_l^L \sum_m^L \sigma^{lm} v_{lm}\}. \quad (13.2)$$

In this approach the $\theta_j$ would be estimated by minimizing

$$\Sigma\Sigma\sigma^{lm}v_{lm}, \quad (13.3)$$

that is, by the natural extension of the method of least squares to the case of multivariate normal deviations.

By an extension of the univariate method, the further assumption that the response functions are approximately linear in the vicinity of the ML estimates leads to the criterion: choose the design points to maximize

$$\Delta = |\sum_{l=1}^{L} \sum_{m=1}^{L} \sigma^{lm} X'_l X_m| \quad (13.4)$$

where for given $l$, $\mathbf{X}_l$ is the $N \times p$ matrix

$$\mathbf{X}_l = (x_{luj}) = \frac{\partial f_l}{\partial \theta_j}(\xi_u; \theta).$$

The matrices $\mathbf{X}_l$ should strictly be evaluated at the ML $\hat{\theta}$ after the experiment has been completed, which cannot be done when the experiment is being planned. If no trials have been conducted, the suggestion is to compute the $\mathbf{X}_l$ for initial guesses $\theta_0$; if $N$ trials have been done and

a further $n$ are being planned, use the $\mathbf{X}_l$ at the ML estimates after $N$ trials.

Illustrations were given for a two-response, one-parameter problem and by M.J. Box [11] for a two response, two parameter and for a two-response, four-parameter problem. Draper and Hunter's interest was to see how the optimum plan and the value of the criterion $\Delta$ in (13.4) varied with $\sigma_{11}$, $\sigma_{22}$, and $\rho$, while Box considered whether replications of the optimum $N = p$ plan were still to be recommended. For $N$ a multiple of $p$, equal replications of this optimum were the best he could find. For $N$ not a multiple of $p$, the best of the near-equal replications was not optimal, but near enough as a good start in a computer search for anything better. M.J. Box comments that this search may not be worth the trouble, though further experience is needed.

Draper and Hunter [27] have also extended to this case the single-response work reported in Section 8 for a multinormal prior. With two responses, for instance, the criterion to be maximized is

$$\Delta = |\sigma^{11}X'_1X_1 + \sigma^{22}X'_2X_2 + \sigma^{12}(X'_1X_2 + X'_2X_1) + \Omega^{-1}|$$

where $\Omega$ is the prior covariance matrix of the $\theta_j$.

Returning to the case of a 'non-informative' prior that leads to the criterion (13.4), M.J. Box [11] considered two practical complications:

1. The response variables may not be measured directly but computed from other prime variables, measured directly, whose values change as the design points change;
2. The factor levels $\xi$ may be themselves subject to error (a familiar problem in experimentation).

Consequences are that the $\sigma_{ij}$ vary with the design points and that it becomes less reasonable to think of 'dependent' variables $y_{lu}$ and 'independent' variables $\xi_{iu}$. Nevertheless, by assuming that the basic measurements are independent, with known variances, he has developed a computer program (essentially involving a known $\sigma_{ij}$ changing with the design points). An example illustrates the application of this technique.

As mentioned, Box and Draper [3] considered the case where the $\sigma_{lm}$ are not known in advance. They assigned the prior $\Pi d\theta_j$ to the parameters and the prior

$$p(\sigma^{lm}) = |\sigma^{lm}|^{-\frac{1}{2}(L+1)}$$

which is the multivariate extension of assigning a uniform prior to (log $\sigma$) in the univariate case. They find the posterior

$$p(\theta \mid \mathbf{y}) = C|v_{lm}|^{-\frac{1}{2}N}$$

where $C$ is a constant. The $\theta_j$ would then be estimated by minimizing $|v_{lm}|$. At first sight this criterion seems rather different from the criterion (13.3): minimize

$$\Sigma\Sigma\sigma^{lm}v_{lm}$$

which emerged when the $\sigma_{lm}$ were assumed known. Box and Draper show, however, that there is a natural resemblance. Let $V_{lm}$ be the cofactor of $v_{lm}$. Now $|v_{lm}|$ can be calculated by multiplying the elements of $v_{lm}$ in any *single* row or column by their cofactors and adding. It follows that

$$|v_{lm}| = \Sigma\Sigma \frac{V_{lm}}{L} v_{lm}. \qquad (13.5)$$

Thus the weights $\sigma^{lm}$ in (13.3) are replaced by weights proportional to the ML estimates of the $\sigma^{lm}$.

The two simulated examples presented both involve only a single $\xi_u$ variate (time). One example has two responses, one parameter, one has 3 responses, 2 parameters. It is now necessary to take $N > p$ in order to obtain estimates of the weights $V_{lm}$. The values chosen were $N = 10$ for the $L = 2$, $p = 1$ example and $N = 12$ for the $L = 3$, $p = 2$ example, no attempt being made to find optimum values of $t$ (design points). From the worked examples a recommendation is made to plot the complete posterior functions for the $\theta_j$ obtained (1) from each individual response function (2) from each pair and (with $L = 3$) from the three combined. These plots indicate the type and amount of information supplied about the respective $\theta_j$ by individual responses and combinations of them. They can also reveal deficiencies in the model, e.g., when the posterior from $y_{1u}$ has little overlap with that for $y_{2u}$.

## 14. COMMENTS

From the work on the $0 - 1$ and the continuous-variable response, we now have a good grasp of the multiple desirable objectives in a nonlinear experiment, a body of techniques that concentrate on a general-purpose criterion for giving good estimates of some or all of the parameters, a method for discriminating among specified models, and an attack on the problem of model-building. Since much of the continuous-variable work is recent, with simulated examples, I would expect a period of digestion by experimenters in industry, with feedback on features that they like and don't like and additional properties desired. Further, as the groups at Wisconsin and I.C.I. warn us, the industrial workers have still harder problems awaiting attack.

It is easy to list much additional related work that would be relevant. To mention a few areas:

(1) Compromise designs, non-optimal by any single criterion, that cope with several different objectives. For instance, a non-sequential plan might deliberately start with $N > p$ distinct points, in order to provide (1) some robustness against poor initial $\theta_0$, (2) either a check on the correctness of the model if an outside estimate of $\sigma^2$ is available, or (3) an internal estimate of $\sigma^2$ if the model can be assumed correct. Something to provide both a check and an estimate of $\sigma^2$ might be possible by a development analogous to Tukey's 1 d.f. for non-additivity under the linear model.

Thus in a paper delivered at these meetings, Wheeler [45] maintains that the experimenter should seek a design that will be reasonably efficient under a variety of

situations which he judges that he may face. Thus for insurance he may want to fit a model more complex than the one that he hopes is correct, he may fear some loss of observations from accidents, and may want at least a specified number of degrees of freedom for estimation of $\sigma^2$. To indicate the inefficiency of any proposed plan, relative to a plan that concentrates solely on efficiency of estimation, Wheeler uses as criterion the relative maximum variance of the predicted response over the experimental region, illustrating how the extensive results on optimum design for linear models, in particular those of Wynn [46], provide computer methods for meeting these goals.

(2) Since the approaches and formulas are to a large extent asymptotic, checks by computer studies on the small sample performance of the 'optimum' plans and formulas.

(3) Extension of the steepest ascent techniques to problems in which there is more than one local optimum, where the objective is to learn something about the global optimum and the effort required to approach it.

(4) Finally, and in no invidious sense, I hope that more people will enter this field, with a resulting broader range of problems attacked, of techniques developed, and of viewpoints. The discussion in the Royal Statistical Society, following Kiefer's [39] presentation of his work on optimum linear plans, revealed doubts about the wisdom of concentrating on optimizing any single criterion. Reasons advanced were that optimizing may require mathematical assumptions or restrictions found unreasonable in many applications, that the experimenter's aims may change when he begins to see some results, and that, in sequential experiments, rules leaving flexibility of judgment to the experimenter and therefore sounding vague to some degree may be better than fixed rules laid down by a statistician's criterion. While this part of the discussion was somewhat negativistic in tone, it suggested that approaches from differing viewpoints have an important role.

[*Received December 1972. Revised May 1973* ]

## REFERENCES

[1] Atkinson, A.C., "Planning Experiments to Detect Inadequate Regression Models," *Biometrika*, 59 (August 1972), 275–93.

[2] ———— and Hunter, W.G., "The Design of Experiments for Parameter Estimation," *Technometrics*, 10 (May 1968), 271–89.

[3] Box, G.E.P. and Draper, N.R., "The Bayesian Estimation of Common Parameters from Several Responses," *Biometrika*, 52 (December 1965), 355–65.

[4] ———— and Hill, W.J., "Discrimination Among Mechanistic Models," *Technometrics*, 9 (February 1967), 57–71.

[5] ———— and Hunter, W.G., "A Useful Method for Model-Building," *Technometrics*, 4 (August 1962), 301–17.

[6] ———— and Hunter, W.G., "Sequential Design of Experiments for Nonlinear Models," *Proceedings of IBM Scientific Computing Symposium in Statistics*, (1965), 113–37.

[7] —— and Lucas, H.L., "Design of Experiments in Nonlinear Situations," *Biometrika*, 46 (June 1959), 77–90.

[8] Box, M.J., "The Occurrence of Replications in Optimal Designs of Experiments to Estimate Parameters in Nonlinear Models," *Journal of the Royal Statistical Society*, Ser. B, 30 (1968), 290–302.

[9] ——, "The Use of Designed Experiments in Nonlinear Model Building," in D.G. Watts, ed., *The Future of Statistics*, New York: Academic Press, 1968, 241–68.

[10] ——, "Some Experiences with a Nonlinear Experimental Design Criterion," *Technometrics*, 12 (August 1970), 569–89.

[11] ——, "Improved Parameter Estimation," *Technometrics*, 12 (May 1970), 219–29.

[12] ——, "An Experimental Design Criterion for Precise Estimation of a Subset of the Parameters in a Nonlinear Model," *Biometrika*, 58 (April 1971), 149–53.

[13] —— and Draper, N.R., "Factorial Designs, the $|X'X|$ Criterion, and Some Related Matters," *Technometrics*, 13 (November 1971), 731–42.

[14] Brown, B.W., Jr., "Planning a Quantal Assay of Potency," *Biometrics*, 22 (June 1966), 322–9.

[15] Chambers, E.A. and Cox, D.R., "Discrimination Between Alternative Binary Response Models," *Biometrika*, 54 (December 1967), 573–8.

[16] Chernoff, H., "Locally Optimum Designs for Estimating Parameters," *Annals of Mathematical Statistics*, 23 (December 1953), 586–602.

[17] Cochran, W.G. and Davis, M., "The Robbins-Monro Method for Estimating the Median Lethal Dose," *Journal of the Royal Statistical Society*, Ser. B, 27 (1965), 28–44.

[18] Cox, D.R., "Tests of Separate Families of Hypotheses," *Proceedings of the 4th Berkeley Symposium*, University of California Press, 1 (1961), 105–23.

[19] ——, "Further Results on Tests of Separate Families of Hypotheses," *Journal of the Royal Statistical Society*, Ser. B, 24 (1962), 406–24.

[20] Davis, M., "Comparison of Sequential Bioassays in Small Samples," *Journal of the Royal Statistical Society*, Ser. B, 33 (1971), 78–87.

[21] Dews, P.B. and Berkson, J., "On the Error of Bioassay with Quantal Response," in O. Kempthorne *et al.*, eds., *Statistics and Mathematics in Biology*, Ames: Iowa State College Press, (1954), 361–70.

[22] Dixon, W.J., "The Up and Down Method in Small Samples," *Journal of the American Statistical Association*, 60 (December 1965), 967–78.

[23] ——, "Quantal-Response Variable Experimentation: The Up-and-Down Method," in J.W. McArthur and T. Colton, eds., *Statistics in Endocrinology*, Cambridge: M.I.T. Press, 1970, 251–68.

[24] —— and Mood, A.M., "A Method for Obtaining and Analyzing Sensitivity Data," *Journal of the American Statistical Association*, 43 (March 1948), 109–26.

[25] Draper, N.R. and Hunter, W.G., "Design of Experiments for Parameter Estimation in Multiresponse Situations," *Biometrika*, 53 (December 1966), 525–33.

[26] —— and Hunter, W.G., "The Use of Prior Distributions in the Design of Experiments for Parameter Estimation in Nonlinear Situations," *Biometrika*, 54 (June 1967), 147–53.

[27] —— and Hunter, W.G., "The Use of Prior Information in the Design of Experiments for Parameter Estimation in Nonlinear Situations: Multiresponse Case," *Biometrika*, 54 (December 1967), 662–5.

[28] Elfving, G., "Optimum Allocation in Linear Regression Theory," *Annals of Mathematical Statistics*, 23 (June 1952), 255–62.

[29] Fieller, E.C., "The Biological Standardization of Insulin," *Journal of the Royal Statistical Society*, Supplement, 7 (1940), 1–64.

[30] Finney, D.J., "The Principles of Biological Assay," *Journal of the Royal Statistical Society*, Supplement, 9 (1947), 46–91.

[31] ——, *Statistical Method in Biological Assay*, 2nd ed., London: Griffin and Co., 1964.

[32] Fisher, R.A., "On the Mathematical Foundations of Theoretical Statistics," *Philosophical Transactions of the Royal Society of London*, Ser. A, 222 (1922), 309–68.

[33] ——, "The Case of Zero Survivors," *Annals of Applied Biology*, 22 (February 1935), 164–65.

[34] ——, "The Design of Experiments," Edinburgh: Oliver and Boyd, Ltd., 1935.

[35] —— and Mackenzie, W.A., "The Manurial Response of Different Potato Varieties," *Journal of Agricultural Science*, 13 (July 1923), 311–20.

[36] Healy, M.J.R., "The Planning of Probit Assays," *Biometrics*, 6 (December 1950), 424–34.

[37] Hill, W.J., Hunter, W.G. and Wichern, D.W., "A Joint Design Criterion for the Dual Problem of Model Discrimination and Parameter Estimation," *Technometrics*, 10 (February 1968), 145–60.

[38] Kesten, H., "Accelerated Stochastic Approximation," *Annals of Mathematical Statistics*, 29 (March 1958), 41–9.

[39] Kiefer, J., "Optimum Experimental Design," *Journal of the Royal Statistical Society*, Ser. B, 21 (1959), 272–319.

[40] ——, "Optimum Designs in Regression Problems II," *Annals of Mathematical Statistics*, 32 (March 1961), 298–325.

[41] Kullback, S. and Liebler, R.A., "On Information and Sufficiency," *Annals of Mathematical Statistics*, 22 (March 1951), 79–86.

[42] Shannon, C.E., "A Mathematical Theory of Communication," *Bell System Technical Journal*, 26 (July and October 1948), 379–423, 623–56.

[43] Siddik, S.M., "Kullback-Liebler Information Function and the Sequential Selection of Experiments to Discriminate Among Several Linear Models," Unpublished Ph.D. thesis, Case Western Reserve University, 1972.

[44] Wetherill, G.B., "Sequential Methods in Statistics," London: Methuen & Co., 1966.

[45] Wheeler, R.E., "Efficient Experimental Design," submitted to *Technometrics*, (1972).

[46] Wynn, H.P., "The Sequential Generation of D-Optimum Experimental Designs," *Annals of Mathematical Statistics*, 41 (October 1970), 1655–64.

# 100

## CONTROLLING BIAS IN OBSERVATIONAL STUDIES : A REVIEW[1]

*By* WILLIAM G. COCHRAN[2]

*Harvard University*

and

DONALD B. RUBIN[3]

*Educational Testing Service* and *Princeton University*

*SUMMARY.* This paper reviews work on the effectiveness of different methods of matched sampling and statistical adjustment, alone and in combination, in reducing bias due to confounding $x$-variables when comparing two populations. The adjustment methods were linear regression adjustment for $x$ continuous and direct standardization for $x$ categorical.

With $x$ continuous, the range of situations examined included linear relations between $y$ and $x$, parallel and non-parallel, monotonic non-linear parallel relations, equal and unequal variances of $x$, and the presence of errors of measurement in $x$.

The percent of initial bias $E(\bar{y}_1 - \bar{y}_2)$ that was removed was used as the criterion. Overall, linear regression adjustment on random samples appeared superior to the matching methods, with linear regression adjustment on matched samples the most robust method. Several different approaches were suggested for the case of multivariate $x$, on which little or no work has been done.

## 1. INTRODUCTION

An observational study differs from an experiment in that the random assignment of treatments (i.e. agents, programs, procedures) to units is absent. As has been pointed out by many writers since Fisher (1925), this randomization is a powerful tool in that many systematic sources of bias are made random. If randomization is absent, it is virtually impossible in many practical circumstances to be convinced that the estimates of the effects of treatments are in fact unbiased. This follows because other variables that affect the dependent variable besides the treatment may be differently distributed across treatment groups, and thus any estimate of the treatment is confounded by these extraneous $x$-variables.

Given the choice between an observational study and an essentially equivalent randomized experiment one would prefer the experiment. Thus in the Report of the President's Commission on Federal Statistics (1971), Light, Mosteller, and Winokur urge greater efforts to use randomized studies in evaluating public programs and in social experimentation, despite the practical difficulties. Often however, random assignment of treatments to units is not feasible, as in the studies of the effects of smoking on health, complications of pregnancy on children, or long-term exposure to

[1]Requests for reprints should be addressed to Donald B. Rubin, Educational Testing Service, Princeton, New Jersey 08540, U.S.A.

[2]Supported by a contract with the Office of Naval Research, Navy Department.

[3]Partially supported by the U.S. Office of Education under contract OEC-0-71-3715.

doses of radiation on uranium mine workers. Also, as in these examples, one might have to wait many years for the results of an experiment while relevant observational data might be at hand. Hence, although inferior to an equivalent experiment, an observational study may be superior to or useful in conjunction with a marginally relevant experiment (e.g. one on the long-term effects of radiation on white rats). In addition, the analysis of data from observational studies can be useful in isolating those treatments that appear to be successful and thus worth further investigation by experimentation, as when studying special teaching methods for underprivileged children.

In dealing with the presence of confounding variables, a basic step in planning an observational study is to list the major confounding variables, design the study to record them, and find some method of removing or reducing the biases that they may cause. In addition, it is useful to speculate about the size and direction of any remaining bias when summarizing the evidence on any differential effects of the treatments.

There are two principal strategies for reducing bias in observational studies. In matching or matched sampling, the samples are drawn from the populations in such a way that the distributions of the confounding variables are similar in some respects in the samples. Alternatively, random samples may be drawn, the estimates of the treatment being adjusted by means of a model relating the dependent variable $y$ to the confounding variable $x$. When $y$ and $x$ are continuous, this model usually involves the regression of $y$ on $x$. A third strategy is to control bias due to the $x$-variables by both matched sampling and statistical adjustment. Notice that the statistical adjustment is performed after all the data are collected, while matched sampling can take place before the dependent variable is recorded.

This paper reviews work on the effectiveness of matching and statistical adjustments in reducing bias in a dependent variable $y$ and two populations $P_1$ and $P_2$ defined by exposure to two treatments. Here, the objective is to estimate the difference $(\tau_1 - \tau_2)$ between the average effects of the treatments on $y$.

Section 2 reviews work on the ability of linear regression adjustment and three matching methods to reduce the bias due to $x$ in the simplest case when both $y$ and $x$ are continuous, there are parallel linear regressions in both populations, and $x$ is the only confounding variable. Section 3 considers complications to this simple case : non-parallel regressions, non-linear regressions, errors of measurement in $x$, and the effect of an omitted confounding variable. Section 4 extends the above cases to include $x$ categorical or made categorical (e.g. low, medium, high). Section 5 presents some multivariate $x$ results which are simple generalizations of the univariate $x$ results. Section 6 considers some multivariate extensions of matching methods. A brief summary of the results and indications for further research are given in Section 7.

## 2. $y, x$ CONTINUOUS : UNIVARIATE PARALLEL LINEAR REGRESSIONS

2.1. *The model.* We begin with the simple case when $y$ and a univariate $x$ are both continuous, and the regressions of $y$ on $x$ are linear and parallel in both populations. For the $j$-th observation from population $i$, the model may be written

$$y_{ij} = \mu_i + \beta(x_{ij} - \eta_i) + e_{ij} \qquad \dots \text{(2.1.1)}$$

with
$$E(e_{ij} | x_{ij}) = 0, \quad E(e_{ij}^2 | x_{ij}) = \sigma_i^2$$

where $\mu_i$ and $\eta_i$ are the means of $y$ and $x$ respectively in population $i$, where $\eta_1 > \eta_2$ without loss of generality. Thus the regressions of $y$ on $x$ differ by the constant

$$E(y_{1j} - y_{2j} | x_{1j} = x_{2j}) = (\mu_1 - \mu_2) - \beta(\eta_1 - \eta_2). \qquad \dots \text{(2.1.2)}$$

If $x$ is the only variable (besides the treatment) that affects $y$ *and* whose distribution differs in the two populations, (2.1.2) equals the difference in the average effects of the treatments, $\tau_1 - \tau_2$. Thus, in this case, the treatment difference in (2.1.2) is constant at any level of $x$.

From (2.1.1) it follows that conditionally on the values of $x_{ij}$ in samples chosen either randomly or solely on $x$,

$$E_c(\bar{y}_1 - \bar{y}_2) = (\mu_1 - \mu_2) + \beta(\bar{x}_1 - \eta_1) - \beta(\bar{x}_2 - \eta_2)$$
$$= \tau_1 - \tau_2 + \beta(\bar{x}_1 - \bar{x}_2). \qquad \dots \text{(2.1.3)}$$

Letting $E_r$ be the expectation over the distribution of variables in random samples,

$$E_r(\bar{y}_1 - \bar{y}_2) = \mu_1 - \mu_2 = \tau_1 - \tau_2 + \beta(\eta_1 - \eta_2) \qquad \dots \text{(2.1.4)}$$

so that the expected bias in $(\bar{y}_1 - \bar{y}_2)$ from random samples is $\beta(\eta_1 - \eta_2)$.

2.2. *Linear regression adjustment.* Since from (2.1.3) $\bar{y}_1 - \bar{y}_2$ is conditionally biased by an amount $\beta(\bar{x}_1 - \bar{x}_2)$ in random and matched samples, it is reasonable to adjust $\bar{y}_1 - \bar{y}_2$ by subtracting an estimate of the bias. The adjusted estimate would then be

$$\hat{\tau}_1 - \hat{\tau}_2 = (\bar{y}_1 - \bar{y}_2) - \hat{\beta}(\bar{x}_1 - \bar{x}_2).$$

In practice, $\hat{\beta}$ is most commonly estimated from the pooled within-sample regressions. With this model, however, $E_c(\hat{\beta}) = \beta$ either for the pooled $\hat{\beta}$ or for $\hat{\beta}$ estimated from sample 1 or sample 2 alone. From (2.1.3) for any of these $\hat{\beta}$,

$$E_c(\hat{\tau}_1 - \hat{\tau}_2) = \mu_1 - \mu_2 - \beta(\eta_1 - \eta_2) = \tau_1 - \tau_2.$$

For this model, the regression adjustment removes all the bias either for random samples or for matched samples selected solely using $x$.

Before using the regression adjusted estimate, the investigator should satisfy himself that the regressions of $y$ on $x$ in the two populations appear linear and parallel. Standard methods of fitting higher order terms in $x$ and separate $\beta$'s in the two samples are appropriate for helping to answer this question.

2.3. *Caliper matching.* In order to construct matched samples of size $n$, the investigator needs initial reservoirs of data of sizes $r_1 n, r_2 n$ from which to seek matches, where $r_i \geqslant 1$ with at least one $r_i > 1$. The work to be reported here is for the case $r_1 = 1$ in which there is a random sample of size $n$ from population 1 to which the sample from population 2 is to be matched from a reservoir of size $rn$ $(r > 1)$. This case is appropriate in studies in which population 1 is of primary interest, population 2 being a control population (untreated or with a standard treatment) with a larger reservoir from which a sample matched to sample 1 is drawn. The case of only one reservoir is a fairly severe test for matching since it is easier to obtain close matches with reservoirs from both populations.

With a random sample from population 1 and some kind of matched sample from population 2 chosen using $x$, relation (2.1.3) gives the expected bias of matched samples as

$$E_m(\bar{y}_1 - \bar{y}) - (\tau_1 - \tau_2) = \beta\{\eta_1 - E_m(\bar{x}_2)\} \qquad \dots \quad (2.3.1)$$

where $E_m$ is the expectation over the distribution of variables in samples from population 2 matched on $x$.

The criterion to be used in judging the effectiveness of matching will be the percentage reduction in bias. From (2.1.4) and (2.3.1) this is

$$\theta = (100) \frac{E_m(\bar{x}_2) - \eta_2}{\eta_1 - \eta_2}.$$

We note that with this model the percentage reduction in expected bias of $(\bar{y}_1 - \bar{y}_2)$ equals that in $(\bar{x}_1 - \bar{x}_2)$.

As a measure of the amount of initial bias in $x$ when appraising methods of matching or adjustment, we chose the quantity

$$B = (\eta_1 - \eta_2) / \left( \frac{\sigma_1^2 + \sigma_2^2}{2} \right)^{1/2}$$

and examined values of $B$ in the range $(0, 1)$. A value of $B = 1$ is considered large. With this bias, the difference $(\bar{x}_1 - \bar{x}_2)$ has about a 90% chance of being detected as significant (5% level) in random samples of 25 when $\sigma_1^2, \sigma_2^2$ are not too unequal. The values of $\sigma_1^2 / \sigma_2^2$ studied were $\frac{1}{2}$, 1, 2.

The first method of matching investigated, often used with $x$ continuous, is paired caliper matching. Each $x_{1j}$ has a partner $x_{2j}$ such that

$$|x_{1j} - x_{2j}| \leqslant c.$$

This method is attractive from two points of view. Although we are assuming at present a *linear* regression of $y$ on $x$, it is clear that a tight caliper matching should remove nearly all the bias in $(\bar{y}_1 - \bar{y}_2)$ under any smooth regression, linear or non-linear, that is the same in both populations. Secondly, at first sight this method provides convenient data for investigating how $E_c(y_{1j} - y_{2j})$ varies with $x$, since $x$ is close to constant for any single pair.

In presenting results on the percent reductions in bias for $x$ normal (Table 2.3.1), we have taken

$$c = a\sqrt{(\sigma_1^2 + \sigma_2^2)/2}$$

where $a = 0.2(0.2)1.0$. Strictly, the results hold for $B < 0.5$ but for $B$ between 0.5 and 1, the percent reductions are only about 1 to $1\frac{1}{2}\%$ lower than the figures shown.

TABLE 2.3.1.  PERCENT REDUCTION IN BIAS OF $x$ FOR CALIPER
MATCHING TO WITHIN $\pm a\sqrt{(\sigma_1^2 + \sigma_2^2)/2}$ WITH $x$ NORMAL

| $a$ | $\sigma_1^2/\sigma_2^2 = \frac{1}{2}$ | $\sigma_1^2/\sigma_2^2 = 1$ | $\sigma_1^2/\sigma_2^2 = 2$ |
|---|---|---|---|
| 0.2 | .99 | .99 | .98 |
| 0.4 | .96 | .95 | .93 |
| 0.6 | .91 | .89 | .86 |
| 0.8 | .86 | .82 | .77 |
| 1.0 | .79 | .74 | .69 |

A tight matching ($a = 0.2$) removes practically all the bias, while a loose matching ($a = 1.0$) removes around 75%. The ratio $\sigma_1^2/\sigma_2^2$ has a minor effect, although performance is somewhat poorer as $\sigma_1^2/\sigma_2^2$ increases.

A disadvantage of caliper matching in practical use is that unless $r$ is quite large there is a non-negligible probability that some of the desired $n$ matches are not found in the reservoir. Nothing seems to be known about the distribution of the number of matches found as a function of $r$, $a$, $(\eta_1 - \eta_2)$ and $\sigma_1^2/\sigma_2^2$. We have not investigated the consequences of incomplete matching as often results in practice. Thus we have no help to give the investigator in estimating the reservoir size needed and the probable percent success in finding caliper matches.

2.4. 'Nearest available' matching. This disadvantage is avoided by a method, (Rubin, 1973a), in which all $n$ pair matches are easily formed by computer. The $n$ values of $x$ from sample 1 and the $rn$ values from reservoir 2 are entered in the computer. In one variant of the method, the sample 1 values of $x$ are first arranged in random order from $x_{11}$ to $x_{1n}$. Starting with $x_{11}$, the computer selects the value $x_{21}$ in reservoir 2 nearest to $x_{11}$ and lays this pair aside. The computer next seeks a 'nearest available' partner for $x_{12}$ from the $(rn-1)$ remaining in reservoir 2, and so on, so that $n$ matches are always found although the value of $a$ is not controlled.

Two other variants of this 'nearest available' method were examined. In these, the members of sample 1 were (i) first ranked from highest to lowest, (ii) first ranked from lowest to highest, before seeking matches from the ranked samples. For $\eta_1 > \eta_2$, Monte Carlo results with $x$ normal showed that for the percent reductions $\theta$ in bias of $(\bar{x}_1 - \bar{x}_2)$, $\theta_{LH} > \theta_{ran} > \theta_{HL}$. If, however, the quality of the matches is judged by the average MSE within pairs, $E_m(x_{1j}-x_{2j})^2$, the order of performance was opposite : $\text{MSE}_{HL} < \text{MSE}_{ran} < \text{MSE}_{LH}$. Both sets of results have rational explanations. The differences in performance were usually small. On balance, random ordering is a reasonable compromise as well as quickest for the computer.

For random ordering, Table 2.4.1 shows the percent reductions in bias of $(\bar{x}_1-\bar{x}_2)$ and hence of $(\bar{y}_1-\bar{y}_2)$ for $r = 2, 3, 4$, $n = 25, 50$ and different combinations of the initial bias $B$ and the $\sigma_1^2/\sigma_2^2$ ratio. Results for $n = 100$ (not shown) differ by at most one or two percentage points from those for $n = 50$, suggesting that the $n = 50$ results hold also for $n > 50$. With this method, the percent reduction in bias decreases steadily as the bias $B$ increases from $1/4$ to $1$, so that results are given separately for the four values of $B$.

As regards the effect of $\sigma_1^2/\sigma_2^2$, matching does best when $\sigma_1^2/\sigma_2^2 = \frac{1}{2}$ and worst when $\sigma_1^2/\sigma_2^2 = 2$. This is not surprising. Since $\eta_1 > \eta_2$ the high values of sample 1 (the ones most likely to cause residual bias) will receive less biased partners when $\sigma_2^2 > \sigma_1^2$.

The investigator planning to use 'nearest available' matching can estimate $B$ and $\sigma_1^2/\sigma_2^2$ from the initial data on $x$. Knowing the value of $r$, he can estimate the expected percent reduction in bias under a linear regression from Table 2.4.1.

TABLE 2.4.1. PERCENT REDUCTION IN BIAS FOR RANDOM ORDER, NEAREST AVAILABLE MATCHING; $x$ NORMAL

| $r \backslash B =$ | $\sigma_1^2/\sigma_2^2 = \frac{1}{2}$ | | | | $\sigma_1^2/\sigma^2 = 1$ | | | | $\sigma_1^2/\sigma_2^2 = 2$ | | | |
|---|---|---|---|---|---|---|---|---|---|---|---|---|
| | $\frac{1}{4}$ | $\frac{1}{2}$ | $\frac{3}{4}$ | $1$ | $\frac{1}{4}$ | $\frac{1}{2}$ | $\frac{3}{4}$ | $1$ | $\frac{1}{4}$ | $\frac{1}{2}$ | $\frac{3}{4}$ | $1$ |
| $n = 25$   2 | 97 | 94 | 89 | 80 | 87 | 82 | 75 | 66 | 63 | 60 | 56 | 48 |
| 3 | 99 | 98 | 97 | 93 | 94 | 91 | 86 | 81 | 77 | 72 | 67 | 61 |
| 4 | 99 | 99 | 99 | 97 | 95 | 95 | 92 | 88 | 81 | 79 | 76 | 68 |
| $n = 50$   2 | 99 | 98 | 93 | 84 | 92 | 87 | 78 | 69 | 66 | 59 | 53 | 51 |
| 3 | 100 | 99 | 99 | 97 | 96 | 95 | 91 | 84 | 79 | 75 | 69 | 63 |
| 4 | 100 | 100 | 100 | 99 | 98 | 97 | 94 | 89 | 86 | 81 | 75 | 71 |

A measure has also been constructed (Rubin, 1973a) of the closeness or quality of the individual pair matches. If pairing were entirely at random, we would have

$$E_m(x_{1j}-x_{2j})^2 = (\sigma_1^2+\sigma_2^2)+(\eta_1-\eta_2)^2$$
$$= (\sigma_1^2+\sigma_2^2)(1+B^2/2).$$

Consequently the quantity

$$100E_m(x_{1j}-x_{2j})^2/(\sigma_1^2+\sigma_2^2)(1+B^2/2)$$

was chosen as the measure. Since results vary little with $n$, only those for $n = 50$ are shown in Table 2.4.2.

TABLE 2.4.2. VALUES OF $100E_m(x_{1j} - x_{2j})^2/(\sigma_1^2 + \sigma_2^2)(1 + B^2/2)$ FOR NEAREST AVAILABLE RANDOM ORDER MATCHING WITH $x$ NORMAL

| $r \backslash B =$ | $\sigma_1^2/\sigma_2^2 = \frac{1}{2}$ | | | | $\sigma_1^2/\sigma_2^2 = 1$ | | | | $\sigma_1^2/\sigma_2^2 = 2$ | | | |
|---|---|---|---|---|---|---|---|---|---|---|---|---|
| | $\frac{1}{4}$ | $\frac{1}{2}$ | $\frac{3}{4}$ | $1$ | $\frac{1}{4}$ | $\frac{1}{2}$ | $\frac{3}{4}$ | $1$ | $\frac{1}{4}$ | $\frac{1}{2}$ | $\frac{3}{4}$ | $1$ |
| 2 | 0 | 1 | 3 | 8 | 1 | 3 | 8 | 15 | 7 | 13 | 20 | 26 |
| 3 | 0 | 0 | 0 | 1 | 0 | 1 | 3 | 6 | 4 | 8 | 12 | 18 |
| 4 | 0 | 0 | 0 | 0 | 0 | 1 | 2 | 4 | 3 | 5 | 9 | 13 |

Except for $\sigma_1^2/\sigma_2^2 = 2$ and $B > \frac{1}{2}$, random ordering gives good quality matches. In fact, since the computer program (Rubin, 1973a) for constructing the matched pairs is very speedy, the investigator can try random, high-low, and low-high ordering. By examining $(\bar{x}_1 - \bar{x}_2)$ and $\Sigma(x_{1j} - x_{2j})^2/n$ for each method, he can select what appears to him the best of the three approaches.

2.5. *Mean matching.* For an investigator who is not interested in pair matching and is confident that the regression is linear, a mean-matching method which concentrates on making $|\bar{x}_1 - \bar{x}_2|$ small has been discussed (Greenberg, 1953). The following simple computer method has been investigated (Rubin, 1973a). Calculate $\bar{x}_1$. Select, from reservoir 2, the $x_{21}$ closest to $\bar{x}_1$, then the $x_{22}$ such that $(x_{21} + x_{22})/2$ is closest to $\bar{x}_1$, and so on until $n$ have been selected. For $n = 50$, Table 2.5.1 shows the percent reductions in bias obtained.

TABLE 2.5.1. PERCENT REDUCTION IN BIAS FOR MEAN MATCHING : $x$ NORMAL

| $r \backslash B =$ | $\sigma_1^2/\sigma_2^2 = \frac{1}{2}$ | | | | $\sigma_1^2/\sigma_2^2 = 1$ | | | | $\sigma_1^2/\sigma_2^2 = 2$ | | | |
|---|---|---|---|---|---|---|---|---|---|---|---|---|
| | $\frac{1}{4}$ | $\frac{1}{2}$ | $\frac{3}{4}$ | $1$ | $\frac{1}{4}$ | $\frac{1}{2}$ | $\frac{3}{4}$ | $1$ | $\frac{1}{4}$ | $\frac{1}{2}$ | $\frac{3}{4}$ | $1$ |
| 2 | 100 | 100 | 98 | 87 | 100 | 99 | 91 | 77 | 100 | 95 | 82 | 67 |
| 3 | 100 | 100 | 100 | 100 | 100 | 100 | 99 | 96 | 100 | 100 | 97 | 84 |
| 4 | 100 | 100 | 100 | 100 | 100 | 100 | 100 | 100 | 100 | 100 | 100 | 95 |

Except in a few difficult cases, particularly $B = 1$, this method of mean matching removes essentially all the bias. So far as we know, mean matching is seldom used, presumably because it relies heavily on the assumption that the regression is linear. With a monotone non-linear regression of $y$ on $x$, one might speculate that mean matching should perform roughly as well as a linear regression adjustment on random samples. But with the regression adjustment, one can examine the relations between $y$ and $x$ in the two samples before deciding whether a linear or non-linear regression adjustment is appropriate, whereas with mean matching performed before $y$ has been observed, one is committed to the assumption of linearity, at least when matching the samples.

## 3. Complications

3.1. *Regressions linear but not parallel.* For $i = 1, 2$, the model becomes

$$y_{ij} = \mu_i + \beta_i(x_{ij} - \eta_i) + e_{ij}. \qquad \text{... (3.1.1)}$$

It follows that for a given level of $x$,

$$E\{(y_{1j} - y_{2j}) \mid x_{1j} = x_{2j} = x\} = \mu_1 - \mu_2 - \beta_1\eta_1 + \beta_2\eta_2 + (\beta_1 - \beta_2)x. \qquad \text{... (3.1.2)}$$

If this quantity is interpreted as measuring the difference in the effects of the two treatments for given $x$, this difference appears to have a linear regression on $x$. At this point the question arises whether a differential treatment effect with $x$ is a reasonable interpretation or whether the $(\beta_1 - \beta_2)$ difference is at least partly due to other characteristics (e.g., effect of omitted $x$-variables) in which the two populations differ. With samples from two populations treated differently, we do not see how this question can be settled on statistical evidence alone. With one study population $P_1$ and two control populations $P_2$, $P_2'$ both subject to $\tau_2$, a finding that $\hat{\beta}_2$ and $\hat{\beta}_2'$ agree closely but differ from $\hat{\beta}_1$ leans in favour of suggesting a differential effect of $(\tau_1 - \tau_2)$.

As it happens, assuming $x$ is the only confounding variable, this issue becomes less crucial if the goal is to estimate the average $(\tau_1 - \tau_2)$ difference over population 1. From (3.1.2) this quantity is

$$E_1(\tau_1 - \tau_2) = (\mu_1 - \mu_2) - \beta_2(\eta_1 - \eta_2). \qquad \text{... (3.1.3)}$$

Since from random samples,

$$E_r(\bar{y}_1 - \bar{y}_2) = \mu_1 - \mu_2, \qquad \text{... (3.1.4)}$$

the initial bias is $\beta_2(\eta_1 - \eta_2)$. With samples matched to a random $\bar{x}_1$,

$$E_m(\bar{y}_1 - \bar{y}_2) = \mu_1 - \mu_2 - \beta_2 E_m(\bar{x}_2) + \beta_2\eta_2,$$

so that the reduction in bias is

$$E_r(\bar{y}_1 - \bar{y}_2) - E_m(\bar{y}_1 - \bar{y}_2) = \beta_2[E_m(\bar{x}_2) - \eta_2].$$

Hence the percent reduction in bias due to matching remains, as before,

$$100[E_m(x_2) - \eta_2]/(\eta_1 - \eta_2)$$

so that previous results for matching apply to non-parallel lines also with this estimand.

As regards regression adjustment, it follows from (3.1.3) and (3.1.4) that

$$E_r[(\bar{y}_1 - \bar{y}_2) - \hat{\beta}_2(\bar{x}_1 - \bar{x}_2)] = (\mu_1 - \mu_2) - \beta_2(\eta_1 - \eta_2) = E_1(\tau_1 - \tau_2).$$

Consequently, in applying the regression adjustment to random samples, use of the regression coefficient calculated from sample 2 provides an unbiased estimate of the desired $E_1(\tau_1-\tau_2)$. This property was noted by Peters (1941), while Belsen (1956) recommended the use of $\hat{\beta}_2$ in comparing listeners $(P_1)$ with non-listeners $(P_2)$ to a BBC television program designed to teach useful French words and phrases to prospective tourists.

With $E_1(\tau_1-\tau_2)$ as the objective, the standard use of the pooled $\hat{\beta}_p$ in the regression adjustment gives biased estimates, though Rubin (1970) has shown that 'nearest available' matching followed by regression adjustment greatly reduces this bias. With matched samples, the standard estimate of $\beta$, following the analysis of covariance in a two-way table, is $\hat{\beta}_d$, the sample regression of matched pair differences, $(y_{1j}-y_{2j})$ on $(x_{1j}-x_{2j})$. Curiously, the Monte Carlo computations show that use of $\hat{\beta}_p$ on matched samples performs better than use of $\hat{\beta}_d$ in this case.

If non-parallelism is interpreted as due to a $(\tau_1-\tau_2)$ difference varying linearly with $x$, the question whether $E_1(\tau_1-\tau_2)$ is the quantity to estimate deserves serious consideration. To take a practice sometimes followed in vital statistics, we might wish to estimate $(\tau_1-\tau_2)$ averaged over a standard population that has mean $\eta_s$ differing from $\eta_1$ and $\eta_2$. The estimand becomes, from (3.1.2)

$$E_s(\tau_1-\tau_2) = \mu_1-\mu_2+\beta_1(\eta_s-\eta_1)-\beta_2(\eta_s-\eta_2).$$

From random samples, an unbiased regression estimate is

$$(\bar{y}_1-\bar{y}_2)+\hat{\beta}_1(\eta_s-\bar{x}_1)-\hat{\beta}_2(\eta_s-\pmb{x}_2) \qquad \ldots \quad (3.1.5)$$

where $\hat{\beta}_1$ and $\hat{\beta}_2$ are the usual least squares estimates from the separate regressions in the two samples.

Alternatively, particularly if $\hat{\beta}_1$ and $\hat{\beta}_2$ differ substantially, no single average of $(\tau_1-\tau_2)$ may be of interest, but rather the values of $(\tau_1-\tau_2)$ at each of a range of values of $x$. As a guide in forming a judgement whether use of a single average difference is adequate for practical application, Rubin (1970) has suggested the following. Suppose that in the range of interest, $x$ lies between $x_L$ and $x_H$. From (3.1.2) the estimated difference in $(\tau_1-\tau_2)$ at these two extremes is

$$(\hat{\beta}_1-\hat{\beta}_2)(x_H-x_L). \qquad \ldots \quad (3.1.6)$$

From (3.1.5), the average $(\tau_1-\tau_2)$ over the range from $x_L$ to $x_H$ is estimated as

$$(\bar{y}_1-\bar{y}_2)+\hat{\beta}_1(\bar{x}-\bar{x}_1)-\hat{\beta}_2(\bar{x}-\bar{x}_2) \quad \text{where} \quad \bar{x} = (x_L+x_H)/2. \qquad \ldots \quad (3.1.7)$$

The ratio of (3.1.6) to (3.1.7) provides some guidance on the proportional error in using simply this average difference.

If it is decided not to use the average difference, the differences $(\tau_1 - \tau_2)$ for specified $x$ can be estimated by standard methods from the separate regressions of $y$ on $x$ in the two samples.

To examine the relation between $(\tau_1 - \tau_2)$ and $x$ from pair-matched samples, it is natural to look at the regression of $(y_{1j} - y_{2j})$ on $\bar{x}._j = (x_{1j} + x_{2j})/2$. However, from the models (3.1.1) it turns out that

$$E\{(y_{1j} - y_{2j})_m \mid \bar{x}._j = x\} = (\mu_1 - \mu_2) - \beta_1 \eta_1 + \beta_2 \eta_2 + (\beta_1 - \beta_2)\bar{x}._j + (\beta_1 + \beta_2)E(d_j \mid \bar{x}._j = x)$$

where $d_j = (x_{1j} - x_{2j})/2$. With $\eta_1 \neq \eta_2$ or $\sigma_1^2 \neq \sigma_2^2$, it appears that $E(d_j \mid \bar{x}._j = x) \neq 0$, so that this method does not estimate the relation (3.1.2) without bias. The bias should be unimportant with tight matching, but would require Monte Carlo investigation.

3.2. *Regression non-linear.* Comparison of the performance of pair-matching with linear regression adjustment is of great interest here, since this is the situation in which, intuitively, pair-matching may be expected to be superior. Use of both weapons—linear regression on matched samples—is also relevant.

Monte Carlo comparisons were made, (Rubin, 1973b), for the monotonic non-linear functions $y = e^{\pm \frac{1}{2}x}$ and $e^{\pm x}$ and the random order nearest available matching method described earlier in Section 2.4. In such studies it is hard to convey to the reader an idea of the amount of non-linearity present. One measure will be quoted. For convenience, the Monte Carlo work was done with $\eta_1 + \eta_2 = 0$ and $(\sigma_1^2 + \sigma_2^2)/2 = 1$. Thus in the average population, $x$ is $N(0, 1)$. In this population the percent of the variance of $y = e^{\pm ax}$ that is attributable to its *linear* component of regression on $x$ is $100a^2/(e^{a^2} - 1)$. For $a = \pm \frac{1}{2}$, $\pm 1$, respectively, 12% and 41% of the variance of $y$ are *not* attributable to the linear component. From this viewpoint, $y = e^{\pm \frac{1}{2}x}$ might be called moderately and $y = e^{\pm x}$ markedly non-linear.

With regression adjustments on random samples, the regression coefficient used in the results presented here is $\hat{\beta}_p$, the pooled within-samples estimate. With regression adjustments on matched samples, the results are for $\hat{\beta}_d$, as would be customary in practice. Rubin (1973b) has investigated use of $\hat{\beta}_1$, $\hat{\beta}_2$, $\hat{\beta}_p$ and $\hat{\beta}_d$ in both situations. He found $\hat{\beta}_p$ in the unmatched case and $\hat{\beta}_d$ in the matched case to be on the whole the best choices.

The results were found to depend markedly on the ratio $\sigma_1^2/\sigma_2^2$. Table 3.2.1 presents percent reductions in bias for $\sigma_1^2/\sigma_2^2 = 1$, the simplest and possibly the most common case. Linear regression on random samples performs admirably, with only a trifling over-adjustment for $y = e^{\pm x}$. Matching is inferior, particularly for $B > \frac{1}{2}$, even with a reservoir of size $4n$ from which to seek matches. Linear regression on matched samples does about as well as linear regression on random samples. Results are for $n = 50$.

Turning to the case $\sigma_1^2/\sigma_2^2 = \frac{1}{2}$ in which better matches can be obtained, note first that linear regression on random samples gives wildly erratic results which call for a rational explanation, sometimes markedly overcorrecting or even (with

$B = \frac{1}{4}$ for $e^x$) greatly increasing the original bias.[4] Matching alone does well, on the average about as well as with a linear relation (Table 2.4.1) when $\sigma_1^2/\sigma_2^2 = \frac{1}{2}$. Linear regression on matched samples is highly effective, being slightly better than matching alone.

TABLE 3.2.1.  PERCENT REDUCTION IN BIAS OF $y(\sigma_1^2/\sigma_2^2 = 1)$; $x$ NORMAL

| method* | $r$ | $B = \frac{1}{4}$ | | | | $B = \frac{1}{2}$ | | | |
|---|---|---|---|---|---|---|---|---|---|
| | | $e^{x/2}$ | $e^{-x/2}$ | $e^x$ | $e^{-x}$ | $e^{x/2}$ | $e^{-x/2}$ | $e^x$ | $e^{-x}$ |
| $R$ | | 100 | 100 | 101 | 101 | 101 | 101 | 102 | 102 |
| $M$ | 2 | 83 | 99 | 70 | 106 | 74 | 94 | 60 | 98 |
| | 3 | 90 | 101 | 79 | 104 | 87 | 98 | 75 | 100 |
| | 4 | 94 | 101 | 87 | 103 | 92 | 99 | 84 | 100 |
| $RM$ | 2 | 99 | 103 | 100 | 108 | 102 | 100 | 106 | 101 |
| | 3 | 100 | 101 | 100 | 103 | 100 | 100 | 102 | 101 |
| | 4 | 100 | 101 | 100 | 102 | 100 | 100 | 101 | 101 |

| method | $r$ | $B = \frac{3}{4}$ | | | | $B = 1$ | | | |
|---|---|---|---|---|---|---|---|---|---|
| | | $e^{x/2}$ | $e^{-x/2}$ | $e^x$ | $e^{-x}$ | $e^{x/2}$ | $e^{-x/2}$ | $e^x$ | $e^{-x}$ |
| $R$ | | 101 | 101 | 104 | 104 | 102 | 102 | 108 | 108 |
| $M$ | 2 | 62 | 87 | 47 | 94 | 53 | 82 | 39 | 91 |
| | 3 | 81 | 96 | 68 | 99 | 70 | 92 | 55 | 97 |
| | 4 | 87 | 98 | 76 | 100 | 79 | 96 | 65 | 99 |
| $RM$ | 2 | 103 | 99 | 110 | 100 | 104 | 99 | 113 | 99 |
| | 3 | 102 | 99 | 105 | 100 | 103 | 100 | 109 | 100 |
| | 4 | 101 | 100 | 103 | 100 | 102 | 100 | 106 | 99 |

*$R$   denotes *linear* regression adjustment on random samples $(\hat{\beta}_p)$.

$M$   denotes 'nearest available' matching.

$RM$   denotes *linear* regression adjustment on matched samples $(\hat{\beta}_d)$.

[4]The most extreme results follow from the nature of the function $e^{\pm ax}$. Consider $e^x$. Its mean value in population $i$ is $e^{(\sigma_i^2/2 + \eta_i)}$. For $B = \frac{1}{4}$, with $\eta_1 = \frac{1}{8}$, $\eta_2 = -\frac{1}{8}$, $\sigma_1^2 = \frac{2}{3}$, $\sigma_2^2 = \frac{4}{3}$, the initial bias in $y$ is *negative*. Since $\eta_1 > \eta_2$ and $\hat{\beta}_p$ is positive, the regression adjustment greatly increases this negative bias, giving $-304\%$ reduction. For $B = \frac{1}{2}$, the initial bias is positive but small, so that regression greatly overcorrects, giving 292% reduction. For $B = \frac{3}{4}$, 1, the initial biases are larger and the overcorrection not so extreme (170%, 139%).

With $\sigma_1^2/\sigma_2^2 = 2$ (Table 3.2.3), linear regression alone performs just as errati-cally as with $\sigma_1^2/\sigma_2^2 = \frac{1}{2}$, the results being in fact the same if $e^{ax}$ is replaced by $e^{-ax}$. As expected from the results in Section 2.3, matching alone is poor. In most cases, regression on matched samples is satisfactory, except for failures with $e^{-x/2}$ and $e^{-x}$ when $B = \frac{1}{4}$ or $\frac{1}{2}$.

TABLE 3.2.2. PERCENT REDUCTION IN BIAS OF $y$
($\sigma_1^2/\sigma_2^2 = \frac{1}{2}$, THE EASIER CASE FOR MATCHING); $x$ NORMAL

| method | $r$ | $B = \dfrac{1}{4}$ | | | | $B = \dfrac{1}{2}$ | | | |
|---|---|---|---|---|---|---|---|---|---|
| | | $e^{x/2}$ | $e^{-x/2}$ | $e^x$ | $e^{-x}$ | $e^{x/2}$ | $e^{-x/2}$ | $e^x$ | $e^{-x}$ |
| R | | 298 | 62 | −304 | 48 | 146 | 80 | 292 | 72 |
| M | 2 | 95 | 99 | 106 | 100 | 96 | 99 | 93 | 99 |
| | 3 | 99 | 100 | 103 | 100 | 98 | 100 | 94 | 100 |
| | 4 | 99 | 100 | 102 | 100 | 99 | 100 | 97 | 100 |
| RM | 2 | 102 | 100 | 96 | 100 | 101 | 100 | 108 | 100 |
| | 3 | 100 | 100 | 100 | 100 | 100 | 100 | 101 | 101 |
| | 4 | 100 | 100 | 100 | 100 | 100 | 100 | 100 | 100 |

| method | $r$ | $B = \dfrac{3}{4}$ | | | | $B = 1$ | | | |
|---|---|---|---|---|---|---|---|---|---|
| | | $e^{x/2}$ | $e^{-x/2}$ | $e^x$ | $e^{-x}$ | $e^{x/2}$ | $e^{-x/2}$ | $e^x$ | $e^{-x}$ |
| R | | 123 | 90 | 170 | 88 | 113 | 96 | 139 | 102 |
| M | 2 | 89 | 96 | 85 | 98 | 76 | 91 | 69 | 96 |
| | 3 | 97 | 100 | 94 | 100 | 94 | 98 | 90 | 99 |
| | 4 | 99 | 100 | 97 | 100 | 97 | 99 | 94 | 100 |
| RM | 2 | 103 | 99 | 113 | 100 | 105 | 99 | 118 | 99 |
| | 3 | 100 | 100 | 102 | 100 | 99 | 99 | 105 | 100 |
| | 4 | 100 | 101 | 101 | 100 | 101 | 100 | 102 | 100 |

3.3. *Regressions parallel but quadratic.* Some further insight into the per-formances of these methods is obtained by considering the model

$$y_{ij} = \tau_i + \beta x_{ij} + \delta x_{ij}^2 + e_{ij}. \qquad \ldots (3.3.1)$$

It follows that

$$E_c(\bar{y}_1 - \bar{y}_2) = (\tau_1 - \tau_2) + \beta(\bar{x}_1 - \bar{x}_2) + \delta(\bar{x}_1^2 - \bar{x}_2^2) + \delta(s_1^2 - s_2^2) \qquad \ldots (3.3.2)$$

where $s_i^2 = \Sigma (x_{ij} - \bar{x}_i)^2/n$. Hence the initial bias in random samples is, uncondi-tionally,

$$(\eta_1 - \eta_2)[\beta + \delta(\eta_1 + \eta_2)] + \delta(\sigma_1^2 - \sigma_2^2) \qquad \ldots (3.3.3)$$

$$= (\eta_1 - \eta_2)\beta + \delta(\sigma_1^2 - \sigma_2^2) \qquad \ldots (3.3.4)$$

TABLE 3.2.3. PERCENT REDUCTION IN BIAS OF $y$
($\sigma_1^2/\sigma_2^2 = 2$, THE HARDER CASE FOR MATCHING); $x$ NORMAL

| method | $r$ | $B = \frac{1}{4}$ | | | | $B = \frac{1}{2}$ | | | |
|--------|-----|-----------|------------|-------|--------|-----------|------------|-------|--------|
| | | $e^{x/2}$ | $e^{-x/2}$ | $e^x$ | $e^{-x}$ | $e^{x/2}$ | $e^{-x/2}$ | $e^x$ | $e^{-x}$ |
| R | | 62 | 298 | 48 | −304 | 80 | 146 | 72 | 292 |
| M | 2 | 48 | 121 | 35 | − 50 | 45 | 81 | 30 | 123 |
| | 3 | 66 | 139 | 51 | − 48 | 60 | 89 | 43 | 118 |
| | 4 | 70 | 121 | 55 | 1 | 65 | 94 | 48 | 126 |
| RM | 2 | 90 | 177 | 90 | − 99 | 100 | 111 | 107 | 171 |
| | 3 | 93 | 149 | 92 | − 29 | 100 | 108 | 105 | 147 |
| | 4 | 95 | 140 | 94 | − 5 | 100 | 107 | 104 | 146 |

| method | $r$ | $B = \frac{3}{4}$ | | | | $B = 1$ | | | |
|--------|-----|-----------|------------|-------|--------|-----------|------------|-------|--------|
| | | $e^{x/2}$ | $e^{-x/2}$ | $e^x$ | $e^{-x}$ | $e^{x/2}$ | $e^{-x/2}$ | $e^x$ | $e^{-x}$ |
| R | | 90 | 123 | 88 | 170 | 96 | 113 | 102 | 139 |
| M | 2 | 38 | 72 | 23 | 90 | 31 | 67 | 16 | 83 |
| | 3 | 55 | 85 | 39 | 98 | 45 | 79 | 28 | 92 |
| | 4 | 60 | 89 | 42 | 100 | 50 | 84 | 29 | 94 |
| RM | 2 | 106 | 102 | 120 | 115 | 109 | 99 | 127 | 104 |
| | 3 | 103 | 102 | 111 | 112 | 106 | 100 | 119 | 104 |
| | 4 | 103 | 101 | 111 | 97 | 105 | 99 | 119 | 102 |

where without loss of generality we have assumed $\eta_1 + \eta_2 = 0$.

Even though $\eta_1 > \eta_2$, if $\delta > 0$ (as appropriate for the positive exponential function) (3.3.4) shows that if $\sigma_1^2 < \sigma_2^2$, the initial bias might be small or even negative. This may indicate why some erratic results appear in the percent reduction in bias with non-linear functions.

From (3.3.2), the remaining bias in matched samples is

$$(\eta_1 - E_m(\bar{x}_2))[\beta + \delta(\eta_1 + E_m(\bar{x}_2))] + \delta\{\sigma_1^2 - E_m(s_2^2)\}. \qquad \ldots \quad (3.3.5)$$

The second term should be minor if the samples are relatively well matched. The first term suggests that in this case the percent reduction in bias should approximate that for parallel-linear regressions if $|\delta/\beta|$ is small. For example, let $\sigma_1^2 = \sigma_2^2 = 1$ and $\theta$ be the percent reduction in bias for $y$ linear. From (3.3.4) and (3.3.5), the percent reduction in bias for $y$ quadratic works out approximately as

$$-(100 - \theta)\, \frac{\delta}{\beta}\, [\eta_1 - E_m(\bar{x}_2)] = \theta \left[ 1 - \frac{\delta}{\beta}\left(1 - \frac{\theta}{100}\right) B \right].$$

For regression adjusted estimates on random samples, $E_c(\hat{\beta}_p)$ may be expressed as

$$E_c(\hat{\beta}_p) = \beta + \delta \left[ \frac{2\bar{x}_1 s_1^2 + 2\bar{x}_2 s_2^2}{s_1^2 + s_2^2} \right] + \frac{\delta(k_{31} + k_{32})}{s_1^2 + s_2^2}$$

where $k_{3i} = \Sigma(x_{ij} - \bar{x}_i)^3/n$ is the sample third moment. From (3.3.2) it follows that the residual bias in the regression adjusted estimate on random samples is conditionally

$$= E_c[(\bar{y}_1 - \bar{y}_2) - \hat{\beta}_p(\bar{x}_1 - \bar{x}_2)] - (\tau_1 - \tau_2)$$

$$= \delta(s_1^2 - s_2^2) + \delta(\bar{x}_1 - \bar{x}_2) \left[ (\bar{x}_1 + \bar{x}_2) - \frac{2(\bar{x}_1 s_1^2 + \bar{x}_2 s_2^2)}{s_1^2 + s_2^2} \right] - \delta(\bar{x}_1 - \bar{x}_2)(k_{31} + k_{32})/(s_1^2 + s_2^2).$$

For a symmetric or near-symmetric distribution of $x$ in both populations the third term becomes unimportant. The first two terms give

$$\delta(s_1^2 - s_2^2)[1 - (\bar{x}_1 - \bar{x}_2)^2/(s_1^2 + s_2^2)].$$

The average residual bias in large random samples after regression adjustment is therefore, for $x$ symmetric and $(\sigma_1^2 + \sigma_2^2)/2 = 1$,

$$\delta(\sigma_1^2 - \sigma_2^2) \left( 1 - \frac{(\eta_1 - \eta_2)^2}{2} \right).$$

This formula suggests, as we found for $e^{\pm ax}$, that with a symmetric $x$ and $\sigma_1^2 = \sigma_2^2$, linear regression adjustment in random samples should remove essentially all the bias when the relation between $y$ and $x$ can be approximated by a quadratic function. The further indication that with $\sigma_1^2 \neq \sigma_2^2$ the residual bias is smaller absolutely as $\eta_1 - \eta_2$ increases towards 1 is at first sight puzzling, but consistent, for example, with the Monte Carlo results for $e^{x/2}$ and $e^x$ when $\sigma_1^2/\sigma_2^2 = 2$ in Table 3.2.3.

To summarize for the exponential and quadratic relationships : If it appears that $\sigma_1^2 \simeq \sigma_2^2$ and $x$ is symmetric (points that can be checked from initial data on $x$) linear regression adjustment on random samples removes all or nearly all the bias. Pair matching alone is inferior. Generally, regression adjustment on pair-matched samples is much the best performer, although sometimes failing in extreme cases. An explanation for this result is given in Rubin (1973b) but is not summarized here because it is quite involved. Further work on adjustment by quadratic regression, on other curvilinear relations, and on the cases $\sigma_1^2/\sigma_2^2 = \frac{3}{4}, \frac{4}{3}$ would be informative.

Before leaving the problem of non-linear regressions, we indicate how the above results can be extended to non-linear response surfaces other than quadratic. Let

$$y_{ij} = \tau_i + g(x_{ij}) + e_{ij}$$

where $g(\cdot)$ is the regression surface. Since $\hat{\beta}_p$ may be written as $\Sigma\Sigma_{i\,j}(y_{ij} - \bar{y}_i)(x_{ij} - \bar{x}_i)/\Sigma\Sigma_{i\,j}(x_{ij} - \bar{x}_i)^2$ the limit of $\hat{\beta}_p$ in large random samples is

$$[\text{cov}_1(x, g(x)) + \text{cov}_2(x, g(x))]/[\text{var}_1(x) + \text{var}_2(x)]$$

where $\text{cov}_i$ and $\text{var}_i$ are the covariances and variances in population $i$. Hence the regression adjusted estimate in large random samples has limiting residual bias

$$E_1(g(x)) - E_2(g(x)) - (\eta_1 - \eta_2)[\text{cov}_1(x, g(x)) + \text{cov}_2(x, g(x))]/[\text{var}_1(x) + \text{var}_2(x)].$$

This quantity can be calculated analytically for many distributions and regression surfaces $g(\,\cdot\,)$, (e.g., normal distributions and exponential $g(\,\cdot\,)$). In addition, if $g$ is expanded in a Taylor series, the residual bias in random or matched samples may be expressed in terms of the moments of $x$ in random and matched samples.

3.4. *Errors of measurement in* $x$. In this section we assume that $y$ has the same linear regression on the correctly measured $x$ (denoted by $X$) in both populations, but that matching or regression adjustment is made with respect to a fallible $x_{ij} = X_{ij} + u_{ij}$, where $u_{ij}$ is an error of measurement. As in Section 2.1 the model is

$$y_{ij} = \mu_i + \beta(X_{ij} - \eta_i) + e_{ij} \qquad \ldots \ (3.4.1)$$

and the expected bias in $(\bar{y}_1 - \bar{y}_2)$ in random samples is as before $\beta(\eta_1 - \eta_2)$.

To cover situations that arise in practice it is desirable to allow (i) $u_{ij}$ and $X_{ij}$ to be correlated, and (ii) $u_{ij}$ to be a biased measurement, with $E_i(u_{ij}) = v_i \neq 0$. A difficulty arises at this point. Even under more restrictive assumptions ($u_{ij}$, $X_{ij}$ independent in a given population and $E_i(u_{ij}) = 0$), Lindley (1947) showed that the regression of $y_{ij}$ on $x_{ij}$ is not linear unless the cumulant generating function of the $u_{ij}$ is a multiple of that of the $X_{ij}$. Lindley's results can be extended to give corresponding conditions when the $u_{ij}$, $X_{ij}$ are correlated. For simplicity we assume that these extended conditions are satisfied.

The linear regressions of $y$ on the fallible $x$ will be written

$$y_{ij} = \mu_i + \beta^*(x_{ij} - \eta_i - v_i) + e_{ij}^*$$

with $E(e_{ij}^* \mid x_{ij}) = 0$. Hence, from (3.4.1),

$$\beta^* = \text{cov}(yx)/\sigma_x^2 = \frac{\beta[\sigma_X^2 + \text{cov}(uX)]}{\sigma_X^2 + \sigma_u^2 + 2\,\text{cov}(uX)}.$$

Unless $\text{cov}(uX) \leqslant -\sigma_u^2$, we have $|\beta^*| < |\beta|$, the slope of the line being damped towards zero. The results in Section 2.2 imply that in random samples or samples matched on $x$ a regression-adjusted estimate $\bar{y}_1 - \bar{y}_2 - \hat{\beta}^*(\bar{x}_1 - \bar{x}_2)$, where $\hat{\beta}^*$ is a least squares estimate of the regression of $y$ on the fallible $x$, changes the initial bias of $\bar{y}_1 - \bar{y}_2$ by the amount

$$-\beta^*(\eta_1 - \eta_2 - v_1 + v_2).$$

Since the initial bias of $\bar{y}_1 - \bar{y}_2$ in random samples is $\beta(\eta_1 - \eta_2)$, the bias of a regression adjusted estimate is

$$(\beta - \beta^*)(\eta_1 - \eta_2) - \beta^*(v_1 - v_2).$$

The last term on the right shows that biased measurements can make an additional contribution ($+$or$-$) to the residual bias. This contribution disappears if the measurement bias is the same in both populations, $v_1 = v_2$. Under this condition the percent reduction in bias due to the regression adjustment is $100\beta^*/\beta$. With the same condition, the percent reduction in bias of $(\bar{y}_1 - \bar{y}_2)$ due to matching on $x$ is easily seen to be

$$\frac{100\beta^*}{\beta} \frac{[E_m(\bar{x}_2) - \eta_2]}{(\eta_1 - \eta_2)}.$$

Thus with this simple model for errors of measurement in $X$, their effects on matching and adjustment are similar—namely to multiply the expected percent reduction in bias by the ratio $\beta^*/\beta$, usually less than 1. With $u$, $X$ uncorrelated, this ratio is the quantity $\sigma_X^2/\sigma_x^2$ often called the reliability of the measurement $x$ (Kendall and Buckland, 1971).

If this reliability, say $(1+a^2)^{-1}$, is known, it can be used to inflate the regression adjustment to have expectation $\beta(\eta_1 - \eta_2)$, (Cochran, 1968b). Thus form the "corrected" regression adjusted estimate

$$\bar{y}_1 - \bar{y}_2 - (1+a^2)\hat{\beta}^*(\bar{x}_1 - \bar{x}_2),$$

which is unbiased for $\tau_1 - \tau_2$ under this model.

In simple examples in which Lindley's conditions are not satisfied, Cochran (1970) found the regression of $y$ on the fallible $x$ to be monotone but curved. A thorough investigation of the effects of errors of measurement would have to attack this case also.

3.5. *Omitted confounding variable.* One of the most common criticisms of the conclusions drawn from an observational study is that they are erroneous because the investigator failed to adjust or match for another confounding variable $z_{ij}$ that affects $y$. He may have been unaware of it, or failed to measure it, or guessed that its effect would be negligible. Even under simple models, however, investigation of the effects of such a variable on the initial bias and on the performance of regression and matching leads to no crisp conclusion that either rebuts or confirms this criticism in any generality.

We assume that $y_{ij}$ has the same linear regression on $x_{ij}$ and $z_{ij}$ in both populations, namely

$$y_{ij} = \mu_i + \beta(x_{ij} - \eta_i) + \gamma(z_{ij} - \nu_i) + e_{ij}. \qquad \dots (3.5.1)$$

Hence, assuming $x$ and $z$ are the only confounding variables,

$$\tau_1 - \tau_2 = E(y_{1j} - y_{2j} \mid x_{1j} = x_{2j}, z_{1j} = z_{2j}) = (\mu_1 - \mu_2) - \beta(\eta_1 - \eta_2) - \gamma(\nu_1 - \nu_2)$$

and the initial bias in $(\bar{y}_1 - \bar{y}_2)$ from random samples is now

$$\beta(\eta_1 - \eta_2) + \gamma(\nu_1 - \nu_2). \qquad \dots (3.5.2)$$

Similarly, the bias in $(\bar{y}_1 - \bar{y}_2)$ from samples matched on $x$ is

$$\beta(\eta_1 - E_m(\bar{x}_2)) + \gamma(\nu_1 - E_m(\bar{z}_2)).$$

Thus, depending on the signs of the parameters involved, the presence of $z_{ij}$ in the model may either increase or decrease (perhaps to an unimportant amount) the previous initial bias $\beta(\eta_1-\eta_2)$. Also, even if $|\eta_1-\eta_2| > |\eta_1-E_m(\bar{x}_2)|$ and $|\nu_1-\nu_2| > |\nu_1-E_m(\bar{z}_2)|$, the bias of $(\bar{y}_1-\bar{y}_2)$ may be greater in matched than random samples.

Suppose now that $z_{ij}$ has linear and parallel regressions on $x_{ij}$ in the two populations :

$$z_{ij} = \nu_i+\lambda(x_{ij}-\eta_i)+\varepsilon_{ij}. \qquad \dots \quad (3.5.3)$$

Then (3.5.1) may be written

$$y_{ij} = \mu_i+(\beta+\gamma\lambda)(x_{ij}-\eta_i)+\varepsilon_{ij}+e_{ij}. \qquad \dots \quad (3.5.4)$$

In (3.5.4) we have returned to the model in Section 2.2 —same linear regression of $y$ on $x$ in both populations. From Section 2.1, the expected change in bias of $(\bar{y}_1-\bar{y}_2)$ due to regression adjustment on $x$ in random samples or samples matched on $x$ is therefore

$$-(\beta+\gamma\lambda)(\eta_1-\eta_2) \qquad \dots \quad (3.5.5)$$

while that due to matching on $x$ is

$$-(\beta+\gamma\lambda)[E_m(\bar{x}_2)-\eta_2]. \qquad \dots \quad (3.5.6)$$

As regards regression, (3.5.2) and (3.5.6) lead to the residual bias

$$\gamma[(\nu_1-\nu_2)-\lambda(\eta_1-\eta_2)]. \qquad \dots \quad (3.5.7)$$

Thus, adjustment on $x$ alone removes the part of the original bias coming from $z$ that is attributable to the linear regression of $z$ on $x$. If $z$ has *identical* linear regressions on $x$ in both populations, so that $(\nu_1-\lambda\eta_1) = (\nu_2-\lambda\eta_2)$, the residual bias is zero as would be expected. With matching in this situation, the residual bias is

$$(\beta+\gamma\lambda)[\eta_1-E_m(\bar{x}_2)]$$

matching being less effective than regression.

With regressions of $z$ on $x$ parallel but not identical, the final bias with either regression or matching could be numerically larger than the initial bias, and no simple statement about the relative merits of regression and matching holds under this model.

If the regressions of $z_{ij}$ on $x_{ij}$ are parallel but non-linear, investigation shows that in large samples, regression and matching remove the part of the bias due to $z$ that is attributable to the linear component of the regression of $z$ on $x$.

## 4. Matching and adjustment by subclassification

4.0. *The two methods.* When the $x$-variable is qualitative, e.g. sex (M, F), it is natural to regard any male from population 1 as a match for any male from population 2 with respect to $x$, or more generally, any two members who fall in the same qualitative class as a match. This method is also used frequently when $x$ is contin-uous, e.g. age. We first divide the range of ages that are of interest into, say, speci-fied 5-year classes 40-44, 45-49, etc. and regard any two persons in the same age class as a match.

In matching to the sample from population 1, let $n_{1j}$ be the number in sample 1 who fall in the $j$-th subclass. From the reservoir from population 2, we seek the same number $n_{2j} = n_{1j}$ in the $j$-th class. The average matched-pair difference, $\Sigma n_{1j}(\bar{y}_{1j} - \bar{y}_{2j})/n$ is of course the difference $(\bar{y}_1 - \bar{y}_2)$ between the two matched sample means, this method being self-weighting.

With random samples from the two populations, the alternative method of adjustment by subclassification starts by classifying both samples into the respective classes. The numbers $n_{1j}$, $n_{2j}$ will now usually differ. However, any weighted mean $\Sigma w_j(\bar{y}_{1j} - \bar{y}_{2j})$, with $\Sigma w_j = 1$, will be subject only to the residual within-class biases insofar as this $x$ is concerned. In practice, different choices of the weights $w_j$ have been used, e.g. sometimes weights directed at minimizing the variance of the weighted difference. For comparison with matching we assume the weights $w_j = n_{1j}/n$.

4.1. *Performance of the two methods.* If sample 1 and reservoir 2 or sample 2 are random samples from their respective populations, as we have been assuming throughout, the $n_{1j}$, $n_{2j}$ who turn up in the final sample are a random sample from those in their population who fall in class $j$ under either method-matching or adjust-ment. Consequently, with the same weights $n_{1j}/n$, the two methods have the same expected residual bias. (An exception is the occasional case of adjustment from initial random samples of equal sizes $n_1 = n_2 = n$, where we find $n_{2j} = 0$ in one or more subclasses, so that subclasses have to be combined to some extent for applica-tion of the 'adjustment by subclassification' method.)

With certain genuinely qualitative classifications it may be reasonable to assume that any two members of the same subclass are identical as regards the effect of this $x$ on $y$. In this event, both matching and adjustment remove all the bias due to $x$, there being no within-class bias. But many qualitative variables like socio-economic status, degree of aggressiveness (mild, moderate, severe), represent an ordered classification of an underlying continuous variable $x$ which at present we are unable to measure accurately. Two members of the same subclass do not have identical values of $x$ in this event. For such cases, and for a variable like age, we assume the model

$$y_{ij} = \tau_i + u(x_{ij}) + e_{ij}, \; i = 1, 2, \; j = 1, 2, ..., c, \qquad \qquad ... \; (4.1.1)$$

the regression of $y$ on $x$ being the same in both populations, with $\tau_1 - \tau_2$ not depending on the value of $x$.

From (4.1.1) the percent reduction in the bias of $y$ due to adjustment by sub-classification of $u$ equals the percent reduction in the bias of $u$. If $u(x) = x$, this also equals the percent reduction in the bias of $x$. If $u(x)$ is a monotone function of $x$, a division of $x$ into classes at the quantiles of $x$ will also be a division of $u$ into classes at the same quantiles of $u$. The percent reductions in bias of $u$ and $x$ will not, however, be equal, since these depend both on the division points and on the frequency distributions, which will differ for $u$ and $x$. The approach adopted by Cochran (1968a) was to start with the case $u(x) = x$, with $x$ normal, and then consider some non-normal distributions of $x$ to throw some light on the situation with $u(x)$ monotone.

In subclassification, the range of $x$ is divided into $c$ classes at division points $x_0, x_1, ..., x_c$. Let $f_i(x)$ be the p.d.f.'s of $x$ in the two populations. The overall means of $x$ are

$$\eta_i = \int x f_i(x)\, dx$$

while in the $j$-th subclass the means are

$$\eta_{ij} = \int_{x_{j-1}}^{x_j} x f_i(x) dx / P_{ij}, \quad \text{where} \quad P_{ij} = \int_{x_{j-1}}^{x_j} f_i(x)\, dx.$$

The initial expected bias in $x$ is $(\eta_1 - \eta_2)$. After matching or adjustment, the weighted mean difference in the two samples is

$$\sum_{j=1}^{c} \frac{n_{1j}}{n} (\bar{x}_{1j} - \bar{x}_{2j}). \qquad \text{... (4.1.2)}$$

Its average value, the expected residual bias, is

$$\sum_{j=1}^{c} P_{1j}(\eta_{1j} - \eta_{2j}). \qquad \text{... (4.1.3)}$$

This expression may be used in calculating the expected percent reduction in bias.

If $f_1(x)$, $f_2(x)$ differ only with respect to a single parameter it is convenient to give it the values 0 and $\Theta$ in populations 1 and 2, respectively. Expression (4.1.3) may be rewritten as

$$\sum_{j=1}^{c} P_j(0)\{\eta_j(0) - \eta_j(\Theta)\}. \qquad \text{... (4.1.4)}$$

A first-term Taylor expansion about 0, assuming $\Theta$ small, seems to work well for biases of practical size, (Cochran, 1968a) and leads to a useful result obtained in a related problem. From (4.1.4) the expected *residual* bias is approximately, expanding about $\Theta = 0$,

$$-\Theta \sum_{j=1}^{c} P_j(0) \frac{d\eta_j(\Theta)}{d\Theta} \qquad \text{... (4.1.5)}$$

the derivative being measured at $\Theta = 0$. On the other hand, the expected *initial* bias is

$$\sum_{j=1}^{c} [P_j(0)\eta_j(0) - P_j(\Theta)\eta_j(\Theta)] \simeq -\Theta \sum_{j=1}^{c} \left[ P_j(0) \frac{d\eta_j(\Theta)}{d\Theta} + \eta_j(0) \frac{dP_j(\Theta)}{d\Theta} \right]. \quad \text{... (4.1.6)}$$

On subtracting (4.1.5) from (4.1.6), the expected proportional reduction in bias is approximately

$$\sum_{j=1}^{c} \eta_j(0) \frac{dP_j(\Theta)}{d\Theta} \Big/ \frac{d\eta(\Theta)}{d\Theta} \qquad \ldots \quad (4.1.7)$$

measured at $\Theta = 0$, where $\eta(\Theta) = \eta_2 = \sum_{j=1}^{c} P_j(\Theta)\eta_j(\Theta)$.

In particular, if $f_1(x) = f(x)$, $f_2(x) = f(x-\Theta)$, the two distributions differing only in their means, we have $\frac{d\eta}{d\Theta} = 1$ and

$$P_j(\Theta) = \int_{x_{j-1}}^{x_j} f(x-\Theta)\, dx = \int_{x_{j-1}-\Theta}^{x_j-\Theta} f(x)\, dx$$

with

$$\frac{dP_j(\Theta)}{d\Theta} = f(x_{j-1}) - f(x_j)$$

at $\Theta = 0$. From (4.1.7), the proportional reduction in bias becomes

$$\sum_{j=1}^{c} \eta_j(0)[f(x_{j-1}) - f(x_j)]. \qquad \ldots \quad (4.1.8)$$

If $f(x)$ is the unit normal distribution, (4.1.8) gives

$$\sum_{j=1}^{c} [f(x_{j-1}) - f(x_j)]^2 / P_j(0) \qquad \ldots \quad (4.1.9)$$

for the proportional reduction in bias. Expression (4.1.9) has been studied in other problems by J. Ogawa (1951) and by D. R. Cox (1957). Cox showed that it is 1 minus the ratio of the average within-class variance to the original variance of $x$ when $x$ is normal. For our purpose, their calculations provide (i) the optimum choices of the $P_{1j}$, (ii) the resulting maximum percent reductions in bias, and (iii) the percent reductions in bias with equal-sized classes $P_{1j} = 1/c$. For $c = 2-10$, the maximum percent reductions are at most about 2% higher than those for equal $P_{1j}$, shown in Table 4.2.1.

TABLE 4.2.1.  PERCENT REDUCTIONS IN BIAS WITH EQUAL-SIZED
CLASS IN POPULATION 1, $x$ NORMAL

| no. of subclasses | 2 | 3 | 4 | 5 | 6 | 8 | 10 |
|---|---|---|---|---|---|---|---|
| % reduction | 64% | 79% | 86% | 90% | 92% | 94% | 96% |

Calculations (Cochran, 1968a) of the percent reductions when $x$ follows $\chi^2$ distributions, $t$ distributions and Beta distributions suggest that the above figures can be used as a rough guide to what to expect in practice when the classification represents an underlying continuous $x$. To remove 80%, 90% and 95% of the initial bias, evidently 3, 5, and 10 classes are required by this method.

## 5. Simple multivariate generalizations

5.1. *Parallel linear regressions.* We now consider the case of many $x$-variables, say $(x^{(1)}, x^{(2)}, ..., x^{(p)})$. Many of the previous results for one $x$ variable have obvious analogues for $p$ $x$-variables, with the $p$-vectors $\boldsymbol{\eta}_i, \boldsymbol{\beta}_i$ and $\boldsymbol{x}_{ij}$ replacing the scalars $\eta_i$, $\beta_i$ and $x_{ij}$. However except in the cases where the adjustment removes all the bias, the conclusions are even less sharp than in the univariate case.

The simplest multivariate case occurs when $y$ has parallel linear regressions on $\boldsymbol{x}$ in both populations

$$y_{ij} = \mu_i + \boldsymbol{\beta}(\boldsymbol{x}_{ij} - \boldsymbol{\eta}_i)' + e_{ij}. \qquad \ldots (5.1.1)$$

The regressions of $y$ on $\boldsymbol{x}$ in the two populations are parallel "planes" with a constant difference of height

$$E(y_{1j} - y_{2j} \mid \boldsymbol{x}_{1j} = \boldsymbol{x}_{2j}) = (\mu_1 - \mu_2) - \boldsymbol{\beta}(\boldsymbol{\eta}_1 - \boldsymbol{\eta}_2)'. \qquad \ldots (5.1.2)$$

If $(x^{(1)}, ..., x^{(2)})$ are the only confounding variables, this constant difference is the treatment difference, $\tau_1 - \tau_2$. From (5.1.1) it follows that conditionally on the values of the $\boldsymbol{x}_{ij}$ in two samples, chosen either randomly or only on the basis of the $x$-variables,

$$E_c(\bar{y}_1 - \bar{y}_2) = \tau_1 - \tau_2 + \boldsymbol{\beta}(\bar{\boldsymbol{x}}_1 - \bar{\boldsymbol{x}}_2)'.$$

The expected bias of $\bar{y}_1 - \bar{y}_2$ in random samples is

$$E_r(\bar{y}_1 - \bar{y}_2) - (\tau_1 - \tau_2) = \boldsymbol{\beta}(\boldsymbol{\eta}_1 - \boldsymbol{\eta}_2)'. \qquad \ldots (5.1.3)$$

Notice that since $\boldsymbol{\beta}$ and $(\boldsymbol{\eta}_1 - \boldsymbol{\eta}_2)$ are vectors the initial bias in $(\bar{y}_1 - \bar{y}_2)$ may be zero even if $\boldsymbol{\beta} \neq \mathbf{0}$ and $(\boldsymbol{\eta}_1 - \boldsymbol{\eta}_2) \neq 0$.

In random $P_1$ and matched $P_2$ samples, the bias is

$$E_m(\bar{y}_1 - \bar{y}_2) - (\tau_1 - \tau_2) = \boldsymbol{\beta}(\boldsymbol{\eta}_1 - E_m(\bar{\boldsymbol{x}}_2))'. \qquad \ldots (5.1.4)$$

Formally, the percent reduction in bias is the natural extension of the univariate result,

$$100\boldsymbol{\beta}(E_m(\bar{\boldsymbol{x}}_2) - \boldsymbol{\eta}_2)' / \boldsymbol{\beta}(\boldsymbol{\eta}_1 - \boldsymbol{\eta}_2)'. \qquad \ldots (5.1.5)$$

But the bias in matched samples may be greater than in random samples even with $E_m(\bar{\boldsymbol{x}}_2)$ closer to $\boldsymbol{\eta}_1$ in all components than $\boldsymbol{\eta}_2$ is to $\boldsymbol{\eta}_1$ (e.g. $(\boldsymbol{\eta}_1 - \boldsymbol{\eta}_2) = (1, -1)$, $(\boldsymbol{\eta}_1 - E_m(\bar{\boldsymbol{x}}_2)) = (\frac{1}{2}, \frac{1}{2})$, $\boldsymbol{\beta} = (1, 1)$), which give initial bias 0 and matched sample bias 1.

The regression adjusted estimate is

$$\hat{\tau}_1 - \hat{\tau}_2 = (\bar{y}_1 - \bar{y}_2) - \hat{\boldsymbol{\beta}}(\bar{\boldsymbol{x}}_1 - \bar{\boldsymbol{x}}_2)' \qquad \ldots (5.1.6)$$

where $\hat{\boldsymbol{\beta}}$ is the vector of estimated regression coefficients of $y$ on $\boldsymbol{x}$. Under this model, $E_c(\hat{\boldsymbol{\beta}}) = \boldsymbol{\beta}$, for $\hat{\boldsymbol{\beta}}_p$, $\hat{\boldsymbol{\beta}}_1$, $\hat{\boldsymbol{\beta}}_2$. Thus, for any of these $\hat{\boldsymbol{\beta}}$ and samples either random or matched on $x$, (5.1.1) and (5.1.6) show that the regression adjusted estimate is unbiased :

$$E_c(\hat{\tau}_1 - \hat{\tau}_2) = \mu_1 - \mu_2 - \boldsymbol{\beta}(\boldsymbol{\eta}_1 - \boldsymbol{\eta}_2)' = \tau_1 - \tau_2.$$

5

5.2. *Non-parallel linear regressions.* As in the univariate case, the regressions of $y$ on $x$ may not be parallel. Assume the objective is to estimate $(\tau_1 - \tau_2)$ averaged over some standard population with mean $x$ vector $\eta_s$ (e.g. $\eta_s = \eta_1$ if $P_1$ is considered the standard). From the multivariate version of (3.1.2), assuming $x$ are the only confounding variables we have

$$E_s(\tau_1 - \tau_2) = \mu_1 - \mu_2 + \beta_1(\eta_s - \eta_1)' - \beta_2(\eta_s - \eta_2)'. \qquad \ldots \ (5.2.1)$$

In random samples $\bar{y}_1 - \bar{y}_2$ has expectation $\mu_1 - \mu_2$ and thus the initial bias is

$$-\beta_1(\eta_s - \eta_1)' + \beta_2(\eta_s - \eta_2)'.$$

If $\eta_s = \eta_1$, this initial bias becomes $\beta_2(\eta_1 - \eta_2)'$.

For random samples or samples selected solely on $x$

$$E_c(\bar{y}_1 - \bar{y}_2) = \mu_1 - \mu_2 + \beta_1(\bar{x}_1 - \eta_1)' - \beta_2(\bar{x}_2 - \eta_2)'$$
$$= E_s(\tau_1 - \tau_2) + \beta_1(\bar{x}_1 - \eta_s)' - \beta_2(\bar{x}_2 - \eta_s)'. \qquad \ldots \ (5.2.2)$$

If $\eta_s = \eta_1$, and sample 1 is a random sample, the bias of $\bar{y}_1 - \bar{y}_2$ is $\beta_2(\eta_1 - E_m(\bar{x}_2))'$ while the initial bias is $\beta_2(\eta_1 - \eta_2)'$. By comparison with (5.1.3) and (5.1.4) it follows that when population 1 is chosen as the standard the effect of matching on bias reduction is the same whether the regressions are parallel or not.

Now consider the regression estimate. Since (5.2.2) gives the conditional bias of $\bar{y}_1 - \bar{y}_2$ it would seem reasonable to estimate this bias using the usual within-sample least squares estimates of $\beta_1$ and $\beta_2$ (and an estimate of $\eta_s$ if necessary) and forming the regression adjusted estimate

$$\bar{y}_1 - \bar{y}_2 - \hat{\beta}_1(\bar{x}_1 - \hat{\eta}_s)' + \hat{\beta}_2(\bar{x}_2 - \hat{\eta}_s)' \qquad \ldots \ (5.2.3)$$

which is an unbiased estimate of $E_s(\tau_1 - \tau_2)$ under the linear regression model. If $\eta_s = \eta_1$ and the first sample is random, this estimate is the natural extension of the univariate result,

$$\bar{y}_1 - \bar{y}_2 - \hat{\beta}_2(\bar{x}_1 - \bar{x}_2)'.$$

If a single summary of the effect of the treatment is not adequate, one could examine the estimated effect at various values of $x$ using (5.2.3) where $\eta_s$ is replaced by the values of $x$ of interest.

5.3. *Non-linear regressions.* If $y$ has non-linear parallel regressions on $x$, expressed by the function $g(x)$, the initial bias, $E_1(g(x)) - E_2(g(x))$, depends on the higher moments of the distributions of $x$ in $P_1$ and $P_2$ (e.g. the covariance matrices $\Sigma_1$ and $\Sigma_2$ if $x$ is normal) as well as the means. The large sample limit of the pooled regression adjusted estimate in random samples is

$$E_1(g(x)) - E_2(g(x)) - (\eta_1 - \eta_2)[\Sigma_1 + \Sigma_2]^{-1} C'$$

where the $k$-th component of the $p$-vector $C$ is $\mathrm{cov}_1(x^{(k)}g(x)) + \mathrm{cov}_2(x^{(k)}g(x))$.

This quantity, as well as similar quantities for the case of parallel, non-linear regressions, can be obtained analytically for many distributions and regression functions. As far as we know, no work has been done on this problem or the more difficult one involving matched samples, in which case the distribution of $x$ in matched samples may not be analytically tractable. Expanding $g(x)$ in a Taylor series would enable one to expand the limiting residual bias in terms of the moments of $x$ in random and matched samples (matched moments for the regression adjusted estimate based on matched pairs).

5.4. *Errors of measurement in x*. Assume that $y$ has parallel linear regressions on the correctly measured matching variables $X$, that $X$ are the only confounding variables, but that matching and regression adjustment are done on the fallible $x = X + u$. Hence

$$y_{ij} = \mu_i + \beta(X_{ij} - \eta_i)' + e_{ij}$$

and the initial bias in random samples is $\beta(\eta_1 - \eta_2)'$. If $y$ has a linear regression on the fallible $x$ (e.g. $y$, $X$, $u$ are multivariate normal), let

$$y_{ij} = \mu_i + \beta^*(x_{ij} - \eta_i - v_i) + e_{ij}^*$$

where $E(e_{ij}^* | x_{ij}) = 0$, $E(u_{ij}) = v_i$, and $\beta^* = \Sigma_x^{-1} \operatorname{cov}(y, x)$, where $\Sigma_x$ is the covariance matrix of the $x$ variables.

A regression adjusted estimate based on random samples or samples matched on $x$ changes the initial bias by the amount $-\beta^*(\eta_1 - \eta_2 - v_1 + v_2)'$, and thus the bias of a regression adjusted estimate is

$$(\beta - \beta^*)(\eta_1 - \eta_2)' - \beta^*(v_1 - v_2)'. \qquad \dots \quad (5.4.1)$$

Some simple results can be obtained for the special case when $v_1 = v_2$ (equally biased measurements in both populations), $X$ and $u$ are uncorrelated, and the covariance matrix of $u$ is proportional to the covariance matrix of $X$, say $a^2 \Sigma_X$. With the latter two conditions $\beta^*$ becomes $(1 + a^2)^{-1} \beta$. This result and $v_1 = v_2$ imply that (5.4.1) becomes

$$\frac{a^2}{1 + a^2} \beta(\eta_1 - \eta_2)'$$

and the percent reduction in bias due to regression adjustment is

$$100/(1 + a^2),$$

as in the univariate case, since $1/(1 + a^2)$ corresponds to the reliability, which we are assuming to be uniform for all variables. Under this same set of special conditions, the percent reduction in bias due to matching on $x$ would be

$$\frac{100}{1 + a^2} \beta(E_m(\bar{x}_2) - \eta_2)' / \beta(\eta_1 - \eta_2)'.$$

Under models different from the above special case, clearcut results appear more difficult to obtain, and the percent reduction in bias for regression adjustment or matching is not necessarily between 0 and 100 percent even with $\nu_1 = \nu_2$, $X$ and $u$ uncorrelated, and all $u_i$ independent.

If one knew $\Sigma_u$, one could form a "corrected" regression-adjusted estimate that is in large samples unbiased for $\tau_1 - \tau_2$. That is, assuming $x$ and $u$ are uncorrelated, form

$$\bar{y}_1 - \bar{y}_2 - \hat{\Sigma}_X^{-1} \, \hat{\Sigma}_x \, \hat{\beta}^*(\bar{x}_1 - \bar{x}_2)' \qquad \ldots \quad (5.4.2)$$

where $\hat{\beta}^*$ is the usual least squares estimate of the regression of $y$ on $x$, $\hat{\Sigma}_x$ is the estimated within group covariance matrix of $x$ and $\hat{\Sigma}_X = \hat{\Sigma}_x - \Sigma_u$. In the special case when $\Sigma_u = a^2 \Sigma_X$, the estimate simplifies to the analogue of the univariate result if $a^2$ is known

$$\bar{y}_1 - \bar{y}_2 - (1 + a^2)\hat{\beta}^*$$

which is unbiased for $\tau_1 - \tau_2$.

5.5. *Omitted confounding variables.* Assume that $y$ has parallel regressions on $(x, z)$ in the populations but that matching and/or adjustment is done on the $x$ variables alone. Also assume that $x$ and $z$ are the only confounding variables. This multivariate case is very similar to the univariate one of Section 3.5 and the multivariate analogs of all the formulas follow in an obvious manner. The basic result is that if $z$ has a linear regression on $x$, $z$ can be decomposed into $z_a$ along $x$ and $z_0$ orthogonal to $x$, and adjustment on $x$ is also adjustment on $z_a$ but does not affect $z_0$.

## 6. Some Multivariate Generalizations of Univariate Matching Methods

6.1. *Caliper matching.* Thus far we have not discussed any specific multivariate matching methods. The obvious extension of caliper matching is to seek in reservoir 2 a match for each $x_{1j}$ such that $|x_{1j}^{(k)} - x_{2j}^{(k)}| < c_k$ for $k = 1, 2, \ldots, p$. This method is used in practice, the difficulty being the large size of reservoir needed to find matches.

The effect of this method on $E_m(\bar{y}_1 - \bar{y}_2)$ could be calculated from univariate results if all $x$ were independently distributed in $P_2$ (this restriction will be relaxed shortly). This follows because selection on $x_{ij}$ from $P_2$ would not affect the other $x$ variables, and so the percent reduction in the bias of the variate $x^{(k)}$ under this method would be the same as that under the univariate caliper matching $|x_{1j}^{(k)} - x_{2j}^{(k)}| < c_k$. From these $p$ percent reductions, the percent reduction in bias could be calculated for any $y$ that is linear in the $x$. For example, with $p = 2$, let $B = 0.5$ and $\sigma_1^2/\sigma_2^2 = \frac{1}{2}$ for $x^{(1)}$, while $B = 0.25$ and $\sigma_1^2/\sigma_2^2 = 2$ for $x^{(2)}$. Then if $c_1 = 0.4\sqrt{(\sigma_1^2 + \sigma_2^2)/2}$ and $c_2 = 0.8\sqrt{(\sigma_1^2 + \sigma_2^2)/2}$, the reductions for $x^{(1)}$ and $x^{(2)}$ from Table 2.3.1 are about 96% and 77%. That for $x^{(1)} + x^{(2)}$, for instance, is about $[(.96)(.5) + (.77)(.25)]/(.75) = 90\%$.

With this approach, an attempt to select the $c_k$ from initial estimates of $\eta_1^{(k)} - \eta_2^{(k)}$ and $\sigma_1^2/\sigma_2^2$ for $x^{(k)}$ so that matching gives the same percent reduction in bias for each $x^{(k)}$ has some appeal, particularly when matching for more than one $y$ or when it is uncertain which $x^{(k)}$ are more important. Whenever this property does not hold, matching can *increase* the bias for some $y$'s linear in $\boldsymbol{x}$. For instance, in the preceding example matching would increase the bias for $x^{(1)} - 2x^{(2)}$, whose bias is initially zero.

In general of course, the matching variables are not independently distributed in $P_2$, but if they are normally distributed (or more generally spherically distributed, Dempster, 1969) there exists a simple linear transformation

$$\boldsymbol{z} = \boldsymbol{x}\boldsymbol{H} \quad \text{where} \quad \boldsymbol{H}'\boldsymbol{H} = \boldsymbol{\Sigma}_2^{-1} \qquad \qquad \dots \quad (6.1.1)$$

such that the $\boldsymbol{z}$ are independently distributed in $P_2$. Hence, assuming (1) $\boldsymbol{x}$ normal in $P_2$, (2) a large sample from $P_2$ so that $\boldsymbol{H}$ is essentially known and all matches can be obtained, and (3) the caliper matching method defined above is used on the $\boldsymbol{z} = \boldsymbol{x}\boldsymbol{H}$ variables, Table 2.3.1 can be used to calculate the percent reduction in bias for each of the $z^{(k)}$. Also, from these $p$ percent reductions in bias, the percent reduction in bias can be calculated for any linear combination of the $z^{(k)}$, such as any $x^{(k)}$ or any $y$ that is linear in $\boldsymbol{x}$.

We consider caliper matching on the transformed variables to be a reasonable generalization of univariate caliper matching to use in practice. Caliper matching on the original $x$ variables defines a fixed $p$-dimensional "rectangular" neighborhood about each $\boldsymbol{x}_{1j}$ in which an acceptable match can be found. If caliper matching is used on the $z$-variables, a neighborhood is defined about each $\boldsymbol{x}_{ij}$ that in general is no longer a simple rectangle with sides perpendicular to the $x$-variables but a $p$-dimensional parallelopiped whose sides are not perpendicular to the $x$-variables but to the $p$ linear combinations of the $x$-variables corresponding to the $\boldsymbol{z}$. Since the original choice of a rectangular neighborhood (e.g. rather than a circular one) was merely for convenience, the neighborhood defined by the $\boldsymbol{z}$ calipers should be just as satisfactory.

6.2. *Categorical matching.* As a second example of a commonly used matching method for which we can apply the univariate results, assume the categorical matching method of Section 4 is used with $c_k$ categories for each matching variable, the final match for each member of the first sample being chosen from the members of the second sample lying in the same categories on all variables. If this matching is performed on the transformed variables $\boldsymbol{z}$ given in (6.1.1), normality is assumed, and the reservoir is large, Table 4.2.1 can be used to calculate the percent reduction in bias of each $z^{(k)}$ in the final matched sample, and thus of each $x^{(k)}$ or any $y$ linear in $\boldsymbol{x}$. Actually Table 4.2.1 requires the ratio of variances to be 1 and $B$ moderate or small but could be extended to include more cases.

By adjusting the number of categories used per matching variable $z^{(k)}$ as a function of $E_1(z^{(k)}) - E_2(z^{(k)})$ and $\mathrm{var}_1(z^{(k)})/\mathrm{var}_2(z^{(k)})$ one can obtain approximately the same percent reduction in bias of any $y$ that is linear in $\boldsymbol{x}$.

6.3. *Discriminant matching.* As a final example of multivariate matching methods for which some of the previous univariate results are applicable, assume the transformation in (6.1.1) will be used with $H$ defined so that $(\eta_1 - \eta_2)H \propto (1, 0, ..., 0)$. Univariate matches are then obtained on $z^{(1)}$, the best linear discriminant with respect to the $\Sigma_2$ inner product, as suggested by Rubin (1970). Note with this method there is no (mean) bias orthogonal to the discriminant (i.e. $E_1 z^{(k)} = E_2 z^{(k)}$, $k = 2, ..., p$); hence, if the $x$ are normal in $P_2$ (so that $z^{(1)}$ and $(z^{(2)}, ..., z^{(P)})$ are independent), the percent reduction in bias for any linear function of the $x$ equals the percent reduction in bias of $z^{(1)}$.

Tables 2.3.1, 2.5.1, 2.4.1, or 4.2.1 can then be used to calculate the percent reduction in bias for each $x^{(k)}$ when univariate caliper, mean, nearest available or categorical matching is used on the discriminant. In using these tables $\sigma_1^2/\sigma_2^2$ is the ratio of the $z^{(1)}$ variances in $P_1$ and $P_2$, $(\bar{x}_1 - \bar{x}_2)\Sigma_2^{-1} \Sigma_1 \Sigma_2^{-1}(\bar{x}_1 - \bar{x}_2)'/(\bar{x}_1 - \bar{x}_2)\Sigma_2^{-1}(\bar{x}_1 - \bar{x}_2)'$, and $B$ is the number of standard deviations between the means of $z^{(1)}$ in $P_1$ and $P_2$, $(\bar{x}_1 - \bar{x}_2)\Sigma_2^{-1}(\bar{x}_1 - \bar{x}_2)'/\sqrt{\frac{1}{2}(\sigma_1^2 + \sigma_2^2)}$. Note that for many matching variables, this $B$ could be quite large even if the means of each matching variable are moderately similar in $P_1$ and $P_2$.

Discriminant matching has several appealing properties :

(1) it is easy to control the sizes of the final matched samples to be exactly of size $n$;

(2) if $x$ is approximately normal in $P_2$ the method should do a good job of reducing bias of any $y$ linear in $x$, even for a modest reservoir; this follows from an examination of Tables 2.4.1 and 2.5.1 ;

(3) if $x$ is approximately normal in both $P_1$ and $P_2$ with $\Sigma_1 \simeq \Sigma_2$, pair matching should do a good job of reducing the bias of any type of regression when the reservoir is large and/or when combined with regression adjustment.

The third point follows from the fact that if $x$ is normal in $P_1$ and $P_2$ with $\Sigma_1 = \Sigma_2$, orthogonal to the discriminant the distributions of the matching variables are identical in $P_1$ and $P_2$ and unaffected by the matching. Hence, for any $y$, all bias is due to the different distributions of the discriminant, and Tables 3.2.1-3.2.3 indicate that with moderate $r$, matching and regression adjustment remove much of this bias; also when $r \to \infty$ the distributions of all matching variables will be the same in the matched samples if nearest available matching is used and $x$ is normal with $\Sigma_1 = \Sigma_2$.

In addition, if one had to choose one linear combination of the $x$ along which a non-linear $y$ is changing most rapidly, and thus on which to obtain close pair matches, the discriminant seems reasonable since the matching variables were presumably chosen not only because their distributions differ in $P_1$ and $P_2$ but also because they are correlated with $y$.

Of course, the joint distributions of matching variables are not assured to be similar in the matched samples, as they would be with pair matches having tight calipers or with a large number of categories using the methods of Sections 6.1 or 6.2. However, the ability to find tight pair matches on all matching variables in a highly multivariate situation seems dubious even with moderately large $r$. The implications of these points require study.

In practice the discriminant is never known exactly. However, symmetry arguments (Rubin, 1973c) show that under normality in $P_2$, matching on the sample-based discriminant still yields the same percent reduction in expected bias for each $x^{(k)}$.

6.4. *Other matching methods.* There are two kinds of problems with the preceding matching methods. First, for those utilizing all the $z$ it is difficult to control the size of the final sample of matches. Thus with the caliper or categorical methods little is known about the actual reservoir size needed to be confident of obtaining a match for each member of the first sample, although an argument suggests that the ratio of reservoir to sample size for $p$ variables i.i.d. in $P_1$ and $P_2$ is roughly the $p$-th power of the ratio for one variable. The use of caliper matching to obtain matched samples in a practical problem is described in Althauser and Rubin (1970).

When using mean and nearest available matching on the discriminant it is easy to control the final matched sample to have size $n$. However using discriminant matching, individual matched pairs are not close on all variables and they rely on specific distributional assumptions to insure that the samples are well-matched, even as $r \to \infty$.

An alternative is to try to define matching methods more analogous to the univariate nearest available matching method using some definition of "distance" between $x_{1j}$ and $x_{2j}$. We might choose the $n$ matches by ordering the $x_{1j}$ in some way (e.g. randomly) and then assigning as a match the nearest $x_{2j}$ as defined by some multivariate distance measure. Such methods will be called nearest available metric matching methods.

A simple class of metrics is defined by an inner product matrix, $D$, so that the distance from $x_{1j}$ to $x_{2j}$ is $(x_{1j}-x_{2j}) D(x_{1j}-x_{2j})'$. Rather obvious choices for $D$ are $\Sigma_1^{-1}$ or $\Sigma_2^{-1}$ yielding the Mahalanobis (1927) distance between $x_{1j}$ and $x_{2j}$ with respect to either inner product. If $\Sigma_1 \propto \Sigma_2$ and $x$ is spherical, symmetry implies that either Mahalanobis distance yields the same percent reduction in bias for each $x^{(k)}$.

More generally, unpublished symmetry arguments (Rubin, 1973c) show that for $x$ spherical and an inner product metric, the same percent reduction in bias is obtained for each $x^{(k)}$ if and only if

(1) The $P_i$ covariance matrices of $x$ orthogonal to the discriminants are proportional :

$$\Sigma_+ = \Sigma_1 - \frac{1}{s_1^2}(\eta_1-\eta_2)'(\eta_1-\eta_2) = c\left[\Sigma_2 - \frac{1}{s_2^2}(\eta_1-\eta_2)'(\eta_1-\eta_2)\right]$$

where $s_i^2 =$ the variance of discriminant in $P_i = (\eta_1 - \eta_2)\Sigma_i^{-1}(\eta_1 - \eta_2)'$. (Note that this implies the discriminants with respect to the $P_1$ and $P_2$ inner products are proportional).

(2) The inner product matrix $D$ used for matching is proportional to $[\Sigma_+ + k(\eta_1 - \eta_2)'(\eta_1 - \eta_2)]^{-1}$ with $k \geqslant 0$ (if $k = 0$ or $\infty$, the inverse is a generalized inverse, Rao, 1973).

The choice of $k = \infty$ yields matching along the discriminant, $k = 0$ yields matching in the space orthogonal to the discriminant, $k = s_1^{-2}$ yields matching using the $P_1$ Mahalanobis distance and $k = cs_2^{-2}$ yields matching using the $P_2$ Mahalanobis distance. Symmetry arguments also show that under normality and condition (1), using the sample estimates of $\Sigma_+$ and $(\eta_1 - \eta_2)$ gives the same percent reduction in bias for each $x^{(k)}$.

There are of course other ways to define distance between $x_{1j}$ and $x_{2j}$, for example by the Minkowski metric

$$\left[ \prod_{k=1}^{p} |x_{1j}^{(k)} - x_{2j}^{(k)}|^{\gamma} \right]^{1/\gamma} \quad \text{for some } \gamma > 0.$$

Nothing seems to be known about the performance of such matching methods.

A final class of methods that has not been explored might be described as sample metric matching. The simplest example would be to minimize distance between the means $\bar{x}_1$ and $\bar{x}_2$ with respect to a metric. More interesting and robust against non-linearity would be to minimize a measure of the difference between the empirical distribution functions.

7.1. *Summary comments.* This review of methods of controlling bias in observational studies has concentrated on the performance of linear regression adjustments and various matching methods in reducing the initial bias of $y$ due to differences in the distribution of confounding variables, $x$, in two populations; this seemed to us the most important aspect in observational studies. We have not considered the effects of these techniques on increasing precision, as becomes the focus of interest in randomized experiments.

If the $x$ variables are the only confounding variables, linear regression adjustment on random samples removes all the initial bias when the $(y, x)$ relations are linear and parallel. With only one $x$ and parallel monotonic curved relations of the types examined, linear adjustment on random samples again removes essentially all the bias if $\sigma_1^2 = \sigma_2^2$ and the distributions of $x$ are symmetric, but may perform very erratically if $\sigma_1^2/\sigma_2^2$ is not near 1, or if the distributions of $x$ are asymmetric.

Except in studies from past records, like the Cornell studies of the effectiveness of seat belts in auto accidents (Kihlberg and Robinson, 1968) matching must usually be performed before $y$ has been measured. A drawback is the time and frustration involved in seeking matches from the available reservoirs, but this will be alleviated if computer methods like the 'nearest available' are extended to more than one $x$. The appeal of matching lies in the simplicity of the concept and the intuitive idea that a tight matching should work well whether the relation between $y$ and $x$ is linear or curved. In our studies with one $x$, however, the matching methods alone did not perform as well as linear regression under either a linear $(y, x)$ relation, or a monotonic non-linear relation with $\sigma_1^2 = \sigma_2^2$ and $x$ symmetric. Regression adjustment on matched samples also removes all the bias in the linear case and is about as effective as regression on random samples in the non-linear case. If the $(y, x)$ relation is non-linear and $\sigma_1^2$ and $\sigma_2^2$ are very different, matching followed by regression adjustment on matched pairs performs best. Monte Carlo results on more moderate $\sigma_1^2/\sigma_2^2$ and asymmetric $x$ would be helpful.

Overall, linear regression adjustment is recommended as superior to matching alone when $x$ is continuous and only a moderate reservoir is available. In a similar comparison with more emphasis on precision, Billewicz (1965) reports that regression was more effective than matching in this respect also. However, it appears that the approach of pair matching *plus* regression adjustment on matched pairs is generally superior to either method alone.

An obvious approach not considered here is to try adjustment by a quadratic regression if this appears to fit well in both samples; there appears to be no work on this problem.

Indeed, this review has indicated numerous topics on which little or no work has been done. Even with univariate $x$ these include research on the sizes of reservoirs needed to obtain caliper or categorical matches, on the effectiveness of the commonly used technique of incomplete matching in which members of sample 1 that lack good matches are discarded, and in methods of relaxing the restrictive assumptions of linearity and normality as suggested in Section 3.3 (and Section 5.3 for the multivariate case). For the case of a dichotomous dependent variable the only work seems to be that of McKinlay (1973).

In Sections 6.1-6.4 we have suggested several multivariate extensions of the matching methods but very little is known about their effectiveness. In this connection a survey of the commonly used methods of control, reservoir sizes and number of variables that occur in applications would be useful in guiding the scope of further research.

REFERENCES

ALTHAUSER, R. P. and RUBIN, D. B. (1970) : The computerized construction of a matched sample. *American Journal of Sociology*, 76, 325-346.

BELSEN, W. A. (1956) : A technique for studying the effects of a television broadcast. *Applied Statistics*, V, 195-202.

BILLEWICZ, W. Z. (1965) : The efficiency of matched samples : an empirical investigation. *Biometrics*, 21, 623-643.

6

Cochran, W. G. (1968a) : The effectiveness of adjustment by sub-classification in removing bias in observational studies. *Biometrics*, 24, 295-313.

——— (1968b) : Errors of measurement in statistics. *Technometrics*, 10, 637-666.

——— (1970) : Some effects of errors of measurement on linear regression. *Proceedings of the 6th Berkeley Symposium*, I, 527-539.

Cox, D. R. (1957) : Note on grouping. *Journal of American Statistical Association*, 52, 543-547.

Dempster, A. P. (1969) : *Elements of Continuous Multivariate Analysis*, Addison Wesley.

Fisher, R. A. (1925) : *Statistical Methods for Research Workers*, 1st edition, Oliver and Boyd.

Greenberg, B. G. (1953) : The use of covariance and balancing in analytical surveys. *American Journal of Public Health*, 43, 692-699.

Kendall, M. G. and Buckland, W. R. (1971) : *A Dictionary of Statistical Terms*, Oliver and Boyd.

Kihlberg, J. K. and Robinson, S. J. (1968) : Seat belt use and injury patterns in automobile accidents. *Cornell Aeronautical Laboratory Report* No. VJ-1823-R30.

Light, R. J., Mosteller, F. and Winokur, H. S. (1971) : Using controlled field studies to improve public policy. *Federal Statistics* (report of the President's Commission) 11, 367-402.

Lindley, D. V. (1947) : Regression lines and the linear functianal relationship. *Journal of the Royal Statistical Society*, B, 9, 218-224.

Mahalanobis, P. C. (1927) : Analysis of race mixture in Bengal. *J. Asiatic Society of Bengal*, 23, 301-333.

McKinlay, S. J. (1973) : An assessment of the relative effectiveness of several measures of association in removing bias from a comparison of qualitative variables. As yet unpublished

Ogawa, J. (1951) : Contributions to the theory of systematic statistics. *Osaka Mathematical Journal*, 4, 175-213.

Peters, C. C. (1941) : A method of matching groups for experiment with no loss of population. *Journal of Educational Research*, 34, 606-612.

Rao, C. R. (1973) : *Linear Statistical Inference and Its Applications*, 2nd edition, Wiley.

Rubin, D. B. (1970) : The Use of Matched Sampling and Regression Adjustment in Observational Studies. (Ph.D. thesis, Harvard University.)

——— (1973a) : Matching to remove bias in observational studies. *Biometrics*, 29, 159-183.

——— (1973b) : The use of matched sampling and regression adjustment to remove bias in observational studies. *Biometrics*, 29, 185-203.

——— (1973c) : Multivariate matching methods that are equal percent bias reducing : Some analytic results. Unpublished manuscript.

*Paper received : May, 1973.*

W. G. Cochran

## 21 The Vital Role of Randomization in Experiments and Surveys

**An Early Agricultural Experiment**

To begin with experimentation, much of this is comparative. The investigator applies a number of different agents or procedures, often called the treatments, to his experimental material—human beings, animals, clouds, chemical reactions, for example—and takes measurements in order to compare the effects of the different treatments.

In planning an experiment so as to obtain accurate comparisons, the natural attitude of the laboratory experimenter is to keep everything constant except the difference in treatments, leaving nothing to chance. Any difference found in the result for treatment $A$ and that for treatment $B$ must then be due to the difference in the effects of the two treatments. But keeping everything constant is rarely feasible even in the laboratory. It is impossible when we are comparing human beings (identical twins differ in many respects), animals, or a seeded with an unseeded cloud system.

For coping with the variability present in the great bulk of experiments, the revolutionary concept introduced by R. A. Fisher was the idea that something *must* be left to chance.[1] In putting forward the device of randomization, he insisted that the final choice of the plan for the experiment be determined by the outcome of a series of coin tossings or by some similar gambling device.

At first sight, the deliberate use of gambling as an essential step in an experiment sounds most unscientific, to say the least. In order to see what randomization does for us, we need to consider the difficulties encountered in drawing conclusions from experiments on variable material, and the techniques developed historically for handling these difficulties.

The difficulties can be illustrated from the results of seven trials conducted in 1764 by the great English experimental agronomist Arthur Young, at the age of 23.[2] He wished to compare the relative profitability of sowing wheat by drilling the seed in rows ("the new husbandry") as against broadcasting the seed ("the old husbandry") on seven fields that he owned. The total area in any single trial was either an acre or a half-acre. Each area was divided into two equal parts—one drilled, one broadcast. Young states: "the soil exactly the same; the time of culture, and, in a word, every circumstance equal in both." (Here, Young is trying to keep everything constant except the differ-

Department of Statistics, Science Center, Harvard University, Cambridge, Massachusetts 02138.

Reproduced with permission from *The Heritage of Copernicus*, J. Neyman (Ed.), pages 445–463. Copyright © 1974, Massachusetts Institute of Technology Press.

ence in methods of sowing.) All expenses for ploughing, seed, sowing, weeding, and harvesting were recorded minutely on each half in pounds, shillings, pence, hapennies, and farthings (the monetary system that the British have only recently abandoned). At harvest (on the same day) a sample from each half was sent to market to determine the selling price.

Table 21.1 shows the differences in profit per acre (drilling minus broadcasting) in the seven trials.

Table 21.1. Difference ($D - B$) in Profit per Acre (pounds sterling).

| Trial | 1 | 2 | 3 | 4 | 5 | 6 | 7 | Average |
|---|---|---|---|---|---|---|---|---|
| $D - B$ | −0.3 | −0.3 | −2.3 | −0.7 | +1.7 | −2.2 | −3.0 | −1.01 |

The range of results, from £1.7 in favor of drilling to £3.0 in favor of broadcasting, is typical of experimentation with variable material. Despite his statement: "the soil exactly the same," Young knew well that two half-acres do not give the same yield even under the same treatment and husbandry. In fact, in his introduction to the long series of experiments that he conducted, he writes "he (the reader) will no where find a connected train of experiments invariably successful enough to create suspicions."

In drawing conclusions from a limited number of trials with discordant results, Young had no standard methods available and had to confine himself to giving his descriptive impressions of the relative performances of the two treatments. He also made two general observations that are as relevant today as in 1764.

The first simply amounts to saying that experimenters are human rather than coldly objective. In reviewing previous experiments, he noted how often the investigator had a "favorite," as he called it, among the treatments. Writing of Jethro Tull, the famous inventor of the seed drill, Young states "Mr. Tull . . . lets nothing escape his pen to destroy his measure (i.e., favorite treatment)."

Young realized further that any conclusions drawn from a series of trials apply only to the type of situation under which the trials were conducted. He warned that his results could not be trusted to hold on a different type of soil, or under different farm management, or in different years with their changing weather. This caution is well illustrated by the seven trials that Young conducted in the next year 1765. This time, drilling won in six trials out of seven, with an average superiority of £0.57 per acre. Young's warning might be expressed as: "Experiments must be capable of being considered a representative sample of the population to which the conclusions are to be applied." Clearly, conducting experiments so that conclusions are widely applicable is not easy.

### Developments in Astronomy and Probability

Attempts to draw conclusions from the results of series of trials of this type suggest the need for two objective techniques: (1) a technique to assist in appraising the strength of a claim that broadcasting was definitely superior to drilling under the 1764 conditions, (2) a measure of the accuracy or reliability of the average gain of £1.01 per acre for broadcasting in these seven trials.

One fascinating aspect of science is that advances needed in one branch may come from quite different branches. In the present case, techniques helpful for summarizing agricultural experiments were developed from the study of errors of measurement in astronomy and from the mathematical theory of probability, mainly during the nineteenth century.

Astronomical measurements are often difficult to make accurately; when the astronomer repeats the measurement, he gets a different result. This variation is ascribed to errors of measurement, since the astronomer is measuring the same quantity on each occasion. The practice arose of repeating the measurement several times, taking the average of the measurements as presumably more accurate than any single measurement. This naturally led to the question: How accurate is the average of 3, 5, or 10 measurements? It is clear that we cannot hope to give a simple definite answer to this question, such as "the average is too large by 0.39," for this would require a knowledge of the quantity being measured. Some other way is needed.

Errors of measurement may be studied by repeating a measurement many times until the average can be regarded as accurate. The difference between an individual measurement and the average is then the size of the individual *error* of measurement. When these errors are classified by their sizes in a frequency distribution, the frequencies can often be well approximated by a bell-shaped mathematical curve, symmetrical about zero error. In this curve, the relative frequency of errors with sizes between $x$ and $(x + dx)$ is

$$\frac{1}{\sigma\sqrt{2\pi}} e^{-\frac{1}{2}\frac{x^2}{\sigma^2}} dx. \tag{21.1}$$

The symbol $\sigma$, the *standard deviation* of the distribution, determines how widely spread the errors are. Strictly, $\sigma^2$ is the average of the squares of the sizes of the errors and is called the *variance* of the distribution.

A property of this curve is that the absolute error (ignoring sign) is equally likely to exceed or be less than $0.675\sigma$. This quantity, the *probable error,* was often used to describe the accuracy of a measuring process. Further, the absolute error has only a 5 percent chance of exceeding $1.96\sigma$. The curve came to be called the Normal Law of Error. In standard tables, the quantity tabulated is $x/\sigma$, which has mean 0, standard deviation 1. The tables give the probability that $x/\sigma$ exceeds any given value.

If $y$ is a measurement of a quantity $\mu$, the error of measurement is of course $x = y - \mu$. Hence, for measurements following the Normal Law of Error, the frequency distribution of the measurements themselves is, from Equation 21.1,

$$\frac{1}{\sigma\sqrt{2\pi}}\, e^{-\frac{1}{2}\frac{(y-\mu)^2}{\sigma^2}}\, dy. \tag{21.2}$$

By the symmetry of this curve, $\mu$ is the mean of the distribution.

Thus far, we have noted that repeated measurements of a quantity $\mu$ are often approximately normally distributed about $\mu$. Combined with this observation, a theorem in mathematical probability provided objective statements about the accuracy of the average of a number of estimates. The theorem goes as follows.

Suppose we have made $n$ measurements $y_1, y_2, \ldots, y_n$ of $\mu$ and can regard them as independent and following the normal law, with mean $\mu$. (The term "independent" means that the error $(y - \mu)$ of a measurement is not influenced in any way by the other measurements made.) The theorem states that their average $\bar{y}$ is also normally distributed with mean $\mu$, and standard deviation $\sigma/\sqrt{n}$. Hence, the quantity $(\bar{y} - \mu)/\sigma/\sqrt{n} = \sqrt{n}(\bar{y} - \mu)/\sigma$ follows the standard normal tables.

From this result, objective statements in probability terms can be made about the accuracy or reliability of the mean $\bar{y}$ of several independent, normally distributed measurements. For example, there is a 50 percent chance that the error in $\bar{y}$ lies between $-0.675\sigma/\sqrt{n}$ and $+0.675\sigma/\sqrt{n}$, a 95 percent chance that the error in $\bar{y}$ lies between $-1.96\sigma/\sqrt{n}$ and $+1.96\sigma/\sqrt{n}$, and so on for any chosen level of probability. It is the variability of the measurements that forces us to use probability statements in describing the accuracy of their averages.

These statements require a knowledge of $\sigma$, which we seldom possess. From $n$ independent measurements, it was known, however, that an estimate of $\sigma^2$ is given by

$$s^2 = \sum_{i=1}^{n} (y_i - \bar{y})^2/(n - 1).$$

Thus in practice, statements about the accuracy of the sample mean $\bar{y}$ have to be based on the quantity $t = \sqrt{n}(\bar{y} - \mu)/s$ rather than on the quantity $\sqrt{n}(\bar{y} - \mu)/\sigma$ that follows the standard normal tables. In an important step forward in 1908, W. S. Gosset, a chemist in the Guinness brewery in Dublin, obtained and tabulated the distribution of $t$ for a series of $n$ independent and normally distributed measurements. By replacing the standard normal tables by the $t$-table, probability statements can be made that take account of the number of measurements on which the estimate of $\sigma$ is actually based. Since Gosset wrote under the pen name of "Student," his table became known as Student's $t$-table.

The possibility of using these ideas in drawing conclusions from the results of experiments in agriculture and biology began to be realized toward the end of the nineteenth century. We can see how we might try to make statements about the accuracy of Young's average difference of £1.01 per acre in favor of broadcasting. He has seven measurements $y_1, \ldots, y_7$ of this difference $\mu$, giving $s = 1.607$ and $t = \sqrt{7}(1.01 - \mu)/(1.607)$. From the $t$-table for seven measurements, there is a 50 percent chance that $t$ lies between $\pm 0.718$, or that $\mu$ lies between $1.01 \pm (1.607)(0.718)/\sqrt{7} = (0.57, 1.45)$. Similarly, limits for $\mu$ corresponding to any desired degree of certainty can be found. We have had to assume, however, something that we do not know, namely that the measurements $y_i$ can be regarded as normally and independently distributed about $\mu$.

## Uniformity Trials in Agriculture

Although experiments in agriculture had been conducted regularly at least since the early 1700s, it is remarkable that serious studies of methods for conducting field experiments did not begin until around 1910. The studies used an ingenious device called *uniformity trials*. (Mercer and Hall,[3] Wood and Stratton,[4] Montgomery.[5]) In these, an area suitable for experiments was divided into a large number of small plots, harvested and weighed separately. The treatment and the farm husbandry were the same on *all* plots. Thus, the variation in yield from plot to plot was a direct measure of the experimental error due to the heterogeneity of the soil.

The goal was to find the best size and shape of plot, number of plots (replications) per treatment, and plan for the layout of the experiment on the field. For any proposed plan, "dummy" treatments $\overline{A}, \overline{B}, \overline{C}, \ldots$ were assigned to the plots to see which plans gave the most accurate comparisons among the dummy treatment means A, B, C, and so on.

Some indications from uniformity trials were as follows.
1. The fertility pattern over a field is complex and varies from field to field.
2. Not surprisingly, plots near one another tend to be more alike in yield than plots far apart.
3. A rising or falling trend in yields along either side of the field is often seen.
4. Since the amount of variation from plot to plot changes with the field and with the crop, each experiment must provide its own estimate $s/\sqrt{n}$ of the standard deviation of the treatment averages $\overline{A}, \overline{B}, \overline{C}, \ldots$

From these indications, a good plan for accurate experimentation, according to Student,[6,7] and Beavan[8] places different treatments $A, B, C$, and so on, on plots *near* one another, so that the comparisons among them are more accurate. For the same reason, replications of the same treatment should be scattered as widely as possible across both the length and breadth of the field, so that the averages for all treatments would be affected in the same way by rising or falling trends parallel to the sides. Various checkerboard plans that seemed to meet these objectives were recommended.

For comparing two varieties of a cereal, a simple and convenient plan was Beavan's Half-Drill strip, which used plots extending the complete length of the field. The order across the breadth was *ABBAABBA . . . ABBA*. The advantage of this order, as is easily verified, is that any straight-line trend across the breadth is completely eliminated from the error of the comparison $(\overline{A} - \overline{B})$.

### Enter Randomization

When Fisher began work on field experiments at the Rothamsted Experimental Station in the early 1920s, he found systematic plans of the preceding type advocated and widely employed. Their purpose was the obviously important one of providing accurate comparisons among the treatment means $\overline{A}, \overline{B}, \overline{C}, \ldots$ He realized also that objective conclusions from the results of an experiment could be stated only in probability terms. These probability statements required that the experiment provide a valid estimate of the variance and hence of the standard deviation of comparisons between the treatment means. A further assumption required was that experimental errors were independent and followed a normal distribution.

Fisher did not at first consider trying to satisfy all these diverse requirements when planning an experiment. He concentrated on the provision of a valid estimate of error. The problem, he noted, was that the *actual* errors of the comparisons between $\overline{A}, \overline{B}, \overline{C}, \ldots$ arose from the differences in fertility between plots treated *differently,* while the estimate of error must come from differences in fertility between plots treated *alike.* How can the experimenter ensure that his estimate of error measures the actual errors? As Fisher saw it,[9]

An estimate of error so derived will only be valid for its purpose if we make sure that, in the plot arrangement, pairs of plots treated alike are not nearer toegther, or further apart than, or in any relevant way distinguishable from pairs of plots treated differently. Systematic plans deliberately placed plots treated differently as near together as possible, while plots treated alike were scattered as far apart as possibile, thus violating these conditions.

Suppose that $k$ treatments, each with $n$ replications, are to be compared on $N = nk$ plots. One way of obtaining a valid estimate of error, noted Fisher, is to arrange the plots *completely at random.* The plot numbers 1 to $N$ can be written on cards and mixed thoroughly. The $n$ plots to receive treatment $A$ are drawn, and so on. Tables of 1 million of the digits 0 to 9 in random order have been published to facilitate this process.[10]

This method clearly satisfies the desired condition. Whether a pair of plots receives the same or different treatments depends only on the result of the gamble, and in no way on the properties or positions of the plots. Algebraically, the method has the following property. Let the plot yields due to soil fertility be any amounts $y_1, y_2, \ldots, y_N$. For any treatment $A$, it can be proved that, if we average over all possible random arrangements, the actual error variance of $\bar{A}$ equals the usual estimate of the error variance of $\bar{A}$, both averages being equal to

$$\sum_{i=1}^{N} (y_i - \bar{y})^2/n(N-1).$$

The importance of this result is that it holds for *any* $y_i$, that is, for any fertility pattern.

The experimenter may object that this method, complete randomization, makes no use of any knowledge or judgment that he has about the nature of the variation in his material, leaving the arrangement of the experiment entirely to chance. On the contrary, Fisher pointed out that the introduction of randomization did not prevent the experimenter from using such knowledge to increase the accuracy of the comparisons between the treatment means.[11] But he must do this in ways that also guaranteed an unbiased estimate of his actual error variance. To compare $k$ treatments, the experimenter could first form groups or blocks of $k$ plots (or units) that he judged to be closely similar in fertility. The plan, a *randomized blocks* design, puts each treatment once in every block, its position in the block being determined by random numbers, independently in every block. For this plan the average actual error variance of a treatment mean over all possible randomizations again equals the average estimated variance. Both averages now equal

(Average variance within blocks)$/n$.

Thus, if the experimenter's judgment was correct that his blocks were highly uniform, he obtained more accurate comparisons. In any event he obtained unbiased estimates of error. Since each treatment is in every block, this plan automatically eliminates all differences in fertility between block means from the errors of the *actual* comparisons. These differences must be eliminated also in estimating the error variance. This is done by a simple calculation known as the analysis of variance, developed by Fisher.

More flexible plans using the device of blocking in one or in two directions at right angles were produced and are given in books on the design of experiments (for example, Fisher[12]). Each plan has its appropriate method of randomization and method of estimating the error variance.

### Randomization and Probability

The act of randomization, as Fisher[13] showed later, has much more basic consequences. It can supply its own foundation for the probability statements made in the conclusions, without requiring any gratuitous assumption by the experimenter that his experimental errors are independent and normally distributed. Two examples will be given.

The first example, Fisher,[14] considers a simple experiment to test the claim of a lady that by tasting a cup of tea she can tell whether the milk or the tea infusion was added first. The plan consists of preparing four cups with milk added first, four with tea added first, and presenting them in random order to the lady for tasting and judging. The lady is told that there will be four cups of each kind.

Suppose we agree to admit the lady's claim if she classifies all eight cups correctly. How likely are we to be wrong? It is known that there are $8!/4!4! = 70$ distinct orders in which the eight cups can be presented. If she has no discriminating ability, she has only a 1 in 70 chance of naming the correct order. What if she is correct on three of the "milk first" cups, wrong on the fourth? We would not then admit her claim, since probability calculation shows that the random order gives her 17 chances out of 70, or nearly 1 in 4, of getting this or a better result entirely by blind guessing. This type of argument, a *test of significance,* is used in appraising a claim that there is a real difference between the effects of two treatments. As a working rule it is common to regard this claim as not established if the observed $(\overline{A} - \overline{B})$ difference or an even larger one could have occurred by chance with a probability greater than 1 in 20, or 5 percent, if the two treatments had identical effects.

Good experimental technique—ensuring that the tea is of the same strength in all cups—makes the "cups of tea" experiment more sensitive, while careless technique—making some cups of China and some of Indian tea—decreases the sensitivity by increasing the "error" variation in taste between cups. It is the randomization that justifies the probability argument used in the test of significance.

As a second example, consider four *ABBA* sandwiches (sixteen plots) in a half-drill strip experiment. This plan assumes that, apart from a possible straight-line trend in yields, the variation from sandwich to sandwich is normal and independent. In effect, systematic plans always assume that Nature does the randomization for us.

However, to take an extreme example, suppose that Nature does not cooperate at all, the trend in yields being quadratic with no other variation. On the "null" hypothesis that $A$ and $B$ have exactly the same effect, the plot yields can be represented by the squares of the numbers from 0 to 15, as in Table 21.2.

In every sandwich, the difference $(\bar{A} - \bar{B})$ is 2. Thus the overall actual

Table 21.2. Simulated Yields of 16 Plots.

|   | Yields | $(A - B)$ In Pairs | $(A - B)$ In Sandwiches |
|---|--------|--------------------|-------------------------|
| $A$ | 0 | | |
| | | $-1$ | |
| $B$ | 1 | | |
| | | | 2 |
| $B$ | 4 | | |
| | | 5 | |
| $A$ | 9 | | |
| $A$ | 16 | | |
| | | $-9$ | |
| $B$ | 25 | | |
| | | | 2 |
| $B$ | 36 | | |
| | | 13 | |
| $A$ | 49 | | |
| $A$ | 64 | | |
| | | $-17$ | |
| $B$ | 81 | | |
| | | | 2 |
| $B$ | 100 | | |
| | | 21 | |
| $A$ | 121 | | |
| $A$ | 144 | | |
| | | $-25$ | |
| $B$ | 169 | | |
| | | | 2 |
| $B$ | 196 | | |
| | | 29 | |
| $A$ | 225 | | |

error of $(\overline{A} - \overline{B})$ is also 2. With a continuation of this quadratic trend, the actual error remains 2 no matter how many sandwiches there are (that is how large the experiment).

What about the estimation of error variance? Student suggested two different methods involving different assumptions.[15] His preferred method was to estimate the error variation from the variation among the $(\overline{A} - \overline{B})$ means for each sandwich. Since every sandwich mean is 2, Student's estimate of the error variance is 0 for any size of experiment, as against an actual error of 2 in $(\overline{A} - \overline{B})$ due to the variation in plot yields. An experimenter might object that this example is unrealistic in assuming a completely systematic pattern of variation. This is true, but the pattern can contain major systematic features of which the experimenter is unaware.

How can a randomized plan handle this extreme situation? With blocks of two neighboring plots, a randomized blocks design is not advisable if a rising or falling fertility trend is suspected. The randomization might by chance give $A$ preceding $B$ in every block, clearly undesirable. It is better to have $A$ precede $B$ in four of the eight blocks, chosen at random, with $B$ preceding $A$ in the other four. This plan, the Cross-Over, is much used in medical research in which drugs $A$ and $B$ are given to each patient in succession to obtain intrapatient comparisons, and an effect of the order in which the drugs are given is feared.

As with the cups of tea, there are 70 possible random choices of the four blocks in which $A$ precedes $B$. If $T$ is the $(A - B)$ total, algebra shows that Student's $t$ always increases whenever $T$ increases, being easily calculated from $T$. Thus, if the true difference in effect between $A$ and $B$ is zero (the null hypothesis), the randomization distribution of Student's $t$ can be obtained by writing down the frequency distribution of the 70 values of $T$. On the null hypothesis, the value $T = \pm56$ is nearest to that required for significance at the 5 percent level, the probability of this or a greater value $(\pm64)$ being $4/70 = 0.057$. Similarly, if this were an actual experiment in which we were estimating the true difference $\mu$ between the effects of $A$ and $B$, we could find limits $\mu_L$, $\mu_H$ such that the statement "$\mu$ lies between $\mu_L$ and $\mu_H$" has a 5.7 percent chance of being wrong.

For practical work, Fisher did not recommend the randomization test as a routine replacement for the use of Student's $t$ table. It is slower, and conclusions usually agree reasonably well by the two methods. In this example, with errors neither random nor normal, Student's table gives $P = 0.046$ corresponding to our $P = 0.057$ for $T = 56$. The randomization test, however, dispenses with the assumption that experimental errors are normal and independent and serves as a fundamental check.

As comparative experimentation spread to a multiplicity of areas other than agriculture, randomized plans were speedily adopted. This is not surprising. The patterns of error variation in such areas were diverse and often poorly known apart from a few obvious features. A randomized plan enabled the experimenter to use his judgment in attempting to increase the precision of his comparisons, and provided a foundation for valid conclusions whether his judgment was correct or not.

Randomization in an experiment by no means removes all the complexities in drawing conclusions from the results. The probability statements from the randomization set apply strictly to the data found in the experiment. They do not remove Young's warning about inferences to broader populations. The psychology students in a randomized experiment in University A are not necessarily representative of psychology students in general, even less so of young adults in general. This point can sometimes be handled by devising an experiment or a series of experiments that are done on a random sample of the members of the population to which we wish our conclusions to apply, though questions of feasibility and cost usually prevent this. Randomization at least sets us on the right road.

### Consequences of Failures to Randomize

The sources of error variation in an experiment fall into three classes: (1) Those controlled by devices like blocking; they are removed from the actual errors of the comparisons and should be removed from the estimated errors. (2) Those randomized; they affect equally the actual and the estimated errors of the comparisons of treatment effects. (3) Those neither controlled nor randomized; these are the sources that can make the probability statements incorrect and misleading in the conclusions. The most frequent consequences are biased comparisons.

For example, in a large experiment in 1930, 10,000 school children in Lanarkshire, Scotland, received 3/4 pint of milk daily while another 10,000 children in the same schools received no milk, as controls in a study of the effects of milk on height and weight. The teachers selected the two groups of children, in certain cases by ballot (presumably random drawing) and in others on an alphabetical system. Student comments:[16] "So far so good, but after invoking the goddess of chance they unfortunately wavered in their adherence to her for we read: 'In any particular school where there was any group to which these methods had given an undue proportion of well-fed or ill-nourished children, others were substituted in order to obtain a more level selection.' "

Student continues: "This is just the sort of after-thought . . . which is apt to spoil the best laid plans." At the start of the experiment the selected controls were definitely superior in height and weight to the "feeders" by an amount equivalent to three months growth in weight and four months growth in height. In "improving" on the random selection, it looks as if some teachers unconsciously tended to assign the milk to those who seemed to need it most.

In this connection, a useful property of randomization is that it protects the experimenter against a tendency to bias of which he is unaware. If, instead of randomizing, he picks out first from the batch the ten rabbits to receive treatment $A$, he may have a tendency to select the larger rabbits first, so that treatment $A$ is given to larger and perhaps sturdier rabbits than treatment $B$. Medical experimenters are so well alerted to the danger of personal bias that if feasible they supplement randomization by another proviso called double blindness—meaning that neither the patient nor the doctor who is measuring the effect of each drug shall know which drug is being measured. A medical researcher who says: "My experiment was randomized and double-blind" may sound to the layman to be groping in the dark, but there are sound reasons for this introduction of chaos.

The consequences of failure to randomize are seen most markedly in the important class of comparative studies in which randomization is not feasible. The illness and death rates of samples of cigarette smokers and nonsmokers are compared in order to measure the effects of cigarette smoking (if any) on health. The two samples are usually found to differ in their age distributions. Smokers and nonsmokers may also differ in their eating and drinking habits, amounts of exercise, and in numerous other variables that can affect their health. About all that the investigator can do is: (1) Take supplementary measurements to discover whether such disturbing differences exist between the two samples. (2) If they do, try to remove their effects on $(\bar{A} - \bar{B})$ by statistical adjustments. Such adjustments again require a mathematical model with assumed random and independent residual variation. (3) Give his opinion about the effects of disturbing variables for which he has been unable to adjust.

Since, as Young observed, investigators often have "favorites" among the treatments, other investigators may give contrary opinions or point out sources of disturbing variation of which the investigator was apparently unaware. The conclusions become a matter of debate, rather than anything firmly established. (This is what happened in the Lanarkshire Milk Experiment.) For this and other reasons, the Report of the President's Commission on Federal Statistics contains a section, by Light, Mosteller, and Winokur, advocating greater efforts to introduce controlled, randomized studies in evaluations of the effects of public policies.[17]

### Sample Surveys
In the second half of the nineteenth century, census bureaus in a number of countries began experimenting with sampling as a tool for estimating the population means and totals of important statistics. Sometimes a sample (perhaps 1 percent, 10 percent, 20 percent) of the latest census returns was being preserved for further analyses. Or certain specific information—demographic, agricultural, or social—was wanted for a geographic area or a large town, and collection from a sample promised to save time and labor.

If results from the sample were to be trustworthy, the sample must obviously be representative of the population from which it was drawn. Two approaches to what was called Representative Sampling were developed.

### Random Selection of Units
The units might be houses, farms, or marriages or tax returns on lists. Random selection gave every unit an equal and independent chance of being included in the sample. This method depended on an even more remarkable property of randomization, the Central Limit Theorem. Averages per unit calculated from large random samples tend to become normally distributed about the population mean, whether the original data are normal or not. This result is proved in probability theory and can be verified in the classroom by experimental sampling. Thus, the normal distribution provided probability statements about the size of the error in any average estimated from the sample.

A further development was stratified random sampling. The units in the population were first arranged in subpopulations or strata, the objective being to have each stratum internally homogeneous. Thus, in an urban study of household incomes and expenditures, the strata might be suburbs, arranged from poorest to wealthiest. An independent random sample of households was taken from each stratum, the number $m_i$ drawn in the $i$th stratum being made proportional to the total number $M_i$ in the stratum. This device ensured that the sample was representative of the different economic levels present in the city.

**Purposive Selection of Groups of Units**

It was realized that labor could be saved by sampling complete groups of units, whole blocks in a town instead of individual households. However, a random sample of 200 blocks out of the 2000 in a town was recognized as less representative than a random sample of 1 household in 10. Purposive selection attempted to overcome this defect. The first step was to choose certain control variables related to the subject matter of the study. If census data were available, the 200 blocks were deliberately selected so that the distributions of the control variables in the sample were similar to those in the whole town. It was thought that this method would make the sample more representative for other variables as well.

As in the planning of experiments there were two rival candidates: random selection and purposive nonrandom selection. After considerable practical experience had been gained with both methods, the International Statistical Institute appointed a commission in the 1920s to report on methods of representative sampling. The summary of the report by Jensen described both methods, stating:[18]

The particular advantage of the random method is that one can always be sure of the degree of accuracy with which one is working, as any precision required can always be attained by including a suitable number of units in the sample. The weak point in this method lies in the difficulties of carrying out in practice the strict rules which are demanded by the application of the law of large numbers.

(The primary difficulties mentioned were the need for an exact definition of the unit of sampling, for example, a household or a farm, the need for strictly independent random sampling, and the fact that failure of certain types of people to answer the questionnaire could bias the results. The latter problem is always present in sample surveys, but cannot be blamed on the random selection.)

The summary continues:

The advantage of the purposive method is that it is capable of being applied to practically every field of research, even when the conditions of selection at random are lacking. Furthermore, a saving in time and labour will often be possible owing to the fact that the units which are included in the sample are selected by groups. Whether it may broadly be said that this method is inferior to the random selection as regards precision is debatable; but at any rate there is this difference, that in the purposive method the precision can generally not be so easily measured by mathematical means and in some cases some essential feature in such measurement cannot be ascertained.

(The last comment recognized the same problem as arose with a systematic experimental design. The precision of a purposive sample could be estimated only by making unverifiable assumptions about the nature of the variation in the population.)

On balance, this 1926 report seemed to favor purposive selection as more widely applicable and often labor saving. Purposive selection was, however, already on the way out. For example, in November 1926 the Italian statisticians Gini and Galvani had to select about a 15 percent sample of the 1921 Italian census, to be preserved for further analyses. The census returns were arranged by communes (8354 in number), grouped into 214 districts. They decided to select 29 districts purposively, in such a way that the sample means for seven control variables approximately equaled their means for the whole country. For a noncontrol variable $y$, they used the data from the 29 districts to construct a linear relation that predicted $y$ from the control variables $x_i$. From this relation, they predicted $y$ for all 214 districts and hence its population mean. They found, however, that their method gave poor estimates of the population means for important noncontrol variables. Their final judgment was that stratified random sampling, with the commune as a sampling unit, would have given much better results, although they noted that it would be difficult to decide on the best choice of strata.

The paper that was most effective in displaying the limitations of purposive selection was published by Neyman in 1934.[19] He showed that even if the Gini and Galvani sample consisted of all 214 districts, their prediction method would give the correct population mean for a noncontrol variable

$y$ only if a particular linear relation between $y$ and the control variables $x$ held throughout the 214 districts. Further, the Gini and Galvani sample estimate was the most precise of its type only if a further relation about the nature of the variability of $y$ in the population also held. Both relations might occasionally hold in practice, but, as he illustrated from survey data, they could not be assumed to hold with any generality. Thus, the user of purposive selection was gambling on knowledge about the population that he did not possess.

In the same paper Neyman also developed the theory needed for the practical use of stratified random sampling to best advantage. This method, with further developments such as unequal probabilities of selection of different units, and the use of ratio estimates to known control variables, became the basic sampling tool in surveys all over the world.

Without the element of random selection, even a very large sample can mislead. The *Literary Digest*'s prediction of a Landon victory over Roosevelt in 1936 was based on a sample of more than 2 million questionnaires. The trouble was that the sample was drawn from lists of telephone and automobile owners, who tended to vote Republican in those days.

**Conclusion**

In both experiments and sample surveys, methods with randomization as an essential component challenged and replaced methods depending on deliberate choice by the investigator. Unlike the Copernican situation, systematic or purposive methods did not have behind them centuries of authoritative support, since the serious study of how to plan either experiments or sample surveys with variable material took place only within the last century. But the idea that in an experiment or survey the investigator must rely to some extent on chance was revolutionary. The natural tendency of investigators was to think that as experts they could surely do better than reliance on blind chance. But this may not be so even in simple problems. In studies of the growth in height of wheat plants, agronomists selected samples after careful inspection of the plants.[20] When the wheat was 2 feet high, they tended to select samples with a positive bias in height. When the wheat was 4 feet high, their selected samples had a negative bias in height.

Randomization is also "pleasing to the mind" in its insistence that data must be collected in a way that justified the probability statements to be used in drawing conclusions from them. As Fisher put it with regard to experiments,[21]

Owing to the fact, however, that the material conduct of an experiment had been regarded as a different business from its statistical interpretation, serious lacunae had been permitted between what had in fact been done, and what was to be assumed for mathematical purposes. . . . It was necessary to treat the question of the field procedure and that of the statistical analysis as but two aspects of a single problem, and an examination of the relationship between the two aspects showed that once the practical field procedure was fixed, only a single method of statistical analysis could be valid, and, what was of more practical importance, that its validity depended on the introduction of a random element in the arrangement of the plots.

# References

1. R. A. Fisher, *Statistical Methods for Research Workers*. First ed., Oliver and Boyd, Edinburgh, 1925.

2. Arthur Young, *A Course of Experimental Agriculture*. Exshaw et al., Dublin, Vol. *1*, 1771, pp. 136–155.

3. W. B. Mercer and A. D. Hall, "The Experimental Error of Field Trials." *J. Agric. Sci.*, Vol. *4*, 1911. pp. 109–132.

4. T. B. Wood and F. J. M. Stratton, "The Interpretation of Experimental Results." *J. Agric. Sci.*, Vol. *3*, 1910, pp. 417–440.

5. E. G. Montgomery, "Variation in Yield and Methods of Arranging Plots to Secure Comparative Results." *Nebs. Agric. Exp. Sta., 25th Ann. Rpt.*, 1911.

6. Student, "Appendix to Mercer and Hall's paper on 'The Experimental Error of Field Trials.'" *J. Agric. Sci.*, Vol. *4*, 1911, pp. 128–132.

7. Student, "On Testing Varieties of Cereals." *Biometrika*, Vol. *15*, 1923, pp. 271–293.

8. E. S. Beavan, "Trials of New Varieties of Cereals." *J. Minist. Agric.*, Vol. *29*, 1922.

9. R. A. Fisher, "The Arrangement of Field Experiments." *J. Minist. Agric.*, Vol. *33*, 1926, pp. 503–513.

10. Rand Corporation, *A Million Random Digits*. Free Press, Glencoe, Ill., 1955.

11. Fisher, *Statistical Methods for Research Workers*, 1925.

12. R. A. Fisher, *The Design of Experiments*. First ed., Oliver and Boyd, Edinburgh, 1935.

13. Ibid.

14. Ibid.

15. Student, "On Testing Varieties of Cereals," 1923.

16. Student, "The Lanarkshire Milk Experiment." *Biometrika*, Vol., *23*, 1931, pp. 398–406.

17. R. J. Light, F. Mosteller, and H. S. Winokur, Jr., "Using Controlled Field Studies to Improve Public Policy." In *Federal Statistics*, Report of the President's Commission, Vol. *II*, 1971, pp. 367–402.

18. A. Jensen, "Report on the Representative Method in Statistics." *Bull. Int. Statist. Inst.*, Vol. *22*, 1926, pp. 359–377.

19. J. Neyman, "On the Two Different Aspects of the Representative Method: The Method of Stratified Sampling and the Method of Purposive Selection." *J. Roy. Statis. Soc.*, Vol. *97*, 1934, pp. 558–606.

20. F. Yates, "Some Examples of Biased Sampling." *Ann. Engrg.*, Vol. *6*, 1935, pp. 202–213.

21. R. A. Fisher, "The Contributions of Rothamsted to the Development of the Science of Statistics." *Rothamsted Exp. Sta. Report*, 1933, pp. 1–8.

# Two Recent Areas of Sample Survey Research

W. G. COCHRAN

*Harvard University, Cambridge, Mass.* 02138, *USA*

## 1. Introduction

Three areas of vigorous activity in sample survey research during the last ten years have been (i) the construction of plans, with their corresponding estimates, for selection of primary sampling units (psu) with unequal probabilities and without replacement, (ii) a search for methods of estimating from the sample the variances of non-linear estimates made in the analysis of complex surveys, and (iii) further development of mathematical models for studying the effects of various types of errors of measurement made in surveys. In view of space considerations, this paper will be confined to two studies in areas (i) and (ii) that present useful comparisons among the principal methods available at present.

## 2. Sampling with unequal probabilities

### 2.1. Recent work

During the 1960's there was a burst of activity in producing methods for selecting samples with unequal probabilities and without replacement. Most of the methods assumed sample size $n = 2$, appropriate to the case of a national stratified sample with $n_h = 2$ in each stratum. A smaller number of methods dealt witn $n$ greater than 2. By the time that Brewer and Hanif (1969) prepared their review of the methods, over 35 methods had appeared in the literature.

In view of this profusion of methods, two papers by J.N.K. Rao and Bayliss (1969, 1970) make a useful contribution by comparing the performances of a group of the methods in single-stage sampling that have the following desirable properties: (a) the variance of the estimator of the population total should always be smaller than that of the usual estimator in sampling with replacement, (b) a non-negative unbiased variance estimator should be available, and (c) computations should not be too difficult.

The methods were compared from three points of view: (i) relative simplicity, (ii) the efficiencies of the estimators $\hat{Y}$ of the population total, as judged by the inverses of their actual variances, and (iii) the stabilities of the sample estimates of the variance of $\hat{Y}$, as judged by the inverses of the variances of these estimators $\hat{V}(\hat{Y})$.

I shall confine myself to a slightly smaller group of methods. The efficiencies of methods are known to depend on the probability $\pi_i$ of inclusion of unit i in the sample and the joint probability $\pi_{ij}$ of inclusion of units i and j both in the sample.

## 2.2. The methods studied

For $n = 2$, there is a set of equivalent methods due to Brewer (1963), Rao (1965), and Durbin (1967) which keep $\pi_i = 2x_i/X$ proportional to the chosen measure of size and use the Horvitz-Thompson estimator. For sample selection, the convenient method of Durbin (1967) will be used as illustration. The first unit is selected with probability $p_i = x_i/X$. If unit i is selected first, the second unit j is selected with probability *proportional* to

$$p_j[(1 - 2p_i)^{-1} + (1 - 2p_j)^{-1}].$$

The total probability of inclusion of unit i can then be shown to be $\pi_i = 2p_i$. This method uses the Horvitz-Thompson estimator of the population total $Y = \Sigma^N y_i$, giving

$$\hat{Y}_B = \sum^n \frac{y_i}{\pi_i} = \left(\frac{y_1}{\pi_1} + \frac{y_2}{\pi_2}\right)$$

for $n = 2$. The estimated variance, due to Yates-Grundy (1953) is for $n = 2$

$$\hat{V}(\hat{Y}_B) = \frac{(\pi_1\pi_2 - \pi_{12})}{\pi_{12}} \left(\frac{y_1}{\pi_1} - \frac{y_2}{\pi_2}\right)^2.$$

Murthy (1957) selects the first unit with probability $p_i$ and the second with probability $p_j/(1 - p_i)$. For general n, his estimator is

$$\hat{Y}_M = \sum^n \frac{P(S|i)}{P(S)} y_i$$

where $P(S|i)$ is the probability of selecting the observed sample, given that unit i is selected first and $P(S)$ is the probability of selecting the observed sample. For $n = 2$ this estimate becomes

$$\hat{Y}_M = \left[(1 - p_2)\frac{y_1}{p_1} + (1 - p_1)\frac{y_2}{p_2}\right]/(2 - p_1 - p_2)$$

with variance estimator

$$\hat{V}(\hat{Y}_M) = \frac{(1 - p_1)(1 - p_2)(1 - p_1 - p_2)}{(2 - p_1 - p_2)^2} \left(\frac{y_1}{p_1} - \frac{y_2}{p_2}\right)^2.$$

In the Rao-Hartley-Cochran (1962) method, the population is split at random into two groups with numbers of units $N_1$, $N_2$, as nearly equal as possible. One unit is selected independently from each group with probability $p_i/P_g$, where $P_g = \Sigma p_i$ over the group. The estimator is

$$\hat{Y}_{RHC} = \sum_g^n P_g \frac{y_i}{p_i} = \left(\frac{y_1}{p_1} P_1 + \frac{y_2}{p_2} P_2\right).$$

The variance estimator is ($n = 2$)

$$\hat{V}(\hat{Y}_{RHC}) = c_0 P_1 P_2 \left(\frac{y_1}{p_1} - \frac{y_2}{p_2}\right)^2$$

where $c_0 = (N - 2)/N$ for $N$ even and $(N - 1)/(N + 1)$ for $N$ odd.

Other methods studied by Rao and Bayliss were those due to Fellegi (1963) and Hanurav (1967), which turned out to be very similar in performance to the Brewer et al. methods. Lahiri's (1951) early method of drawing the sample with probability $\Sigma p_i$ so as to produce an unbiased ratio estimator

$$\hat{Y}_L = (y_1 + y_2)/(p_1 + p_2)$$

performed erratically as regards efficiency of the estimator. Methods by Des Raj (1956), similar to Murthy's but slightly inferior, were also compared.

### 2.3. Comparisons by Rao and Bayliss for $n = 2$

The methods were compared in three circumstances:

(1) on 7 very small ($N = 4, 5, 6$) artificial populations used by writers as illustrations, these will be omitted here,

(2) on 20 natural populations to which these methods might be applied, with $N$ ranging from 9 to 35,

(3) on the much-used super-population model with a linear regression

$$y_i = \beta x_i + e_i$$

where the $e_i$ are uncorrelated, with conditional means 0 given $x_i$, and conditional variances $ax_i^g$ with $g = 1, 1.5, 1.75, 2$.

In presenting results, the Brewer-Rao-Durbin methods were taken as standard, the figures given being the percent gains ($+$) or losses ($-$) in efficiency of the other methods with respect to this method. Table 2.1 shows

Table 2.1

Percent gains in efficiency of the estimators over
the Brewer-Rao-Durbin estimator

| | | Estimator | |
|---|---|---|---|
| | Murthy | R. H. C. | Replacement |
| % gains | | Natural populations | |
| Mean | + 2.7 | + 0.4 | − 7.4 |
| Extremes | (− 2, + 18) | (− 7, + 7) | (− 17, − 1) |
| | | Linear model ($g = 1$) | |
| Mean | + 2.7 | Not given | − 10.0 |
| Extremes | (0, + 12) | Not given | (− 21, − 4) |
| | | Linear model ($g = 1.5$) | |
| Mean | + 2.0 | − 2.6 | − 10.0 |
| Extremes | (0, + 5) | (− 11,0) | (− 21, − 4) |
| | | Linear model ($g = 2$) | |
| Mean | − 0.6 | − 5.3 | − 12.4 |
| Extremes | (− 6, − 0) | (− 23, − 1) | (− 32, − 4) |

results for relative efficiencies of the estimators. These relative efficiencies have a slightly skew distribution, but the arithmetic mean and the lowest and highest extreme values will serve as summary statistics. The results for the natural populations were those of most interest to me. The results for the linear model use the sample $x$-distributions found in the natural populations. Comparisons with the standard method of unequal probability sampling with replacement are also included, with its customary variance estimator

$$\hat{V}(\hat{Y}) = \left(\frac{y_1}{\pi_1} - \frac{y_2}{\pi_2}\right)^2 .$$

In the natural populations, the average gains show very little difference among the three 'without-replacement' methods. The Murthy method does slightly better than the others, the 'with-replacement' method being about 7% poorer. About the same story holds under the linear model with $g = 1$ but the Brewer-Rao-Durbin method improves relatively as $g$ increases, the rank order at $g = 2$ being Brewer, Murthy, RHC, though still with minor differences.

For the relative stabilities of the estimators of variance, the figures are much more skew. I resorted in Table 2.2 to the median, followed by the quartiles and then by the extreme values. The order in both the natural populations and the linear models is now R.H.C., Murthy, Brewer-Rao-Durbin.

Table 2.2

Percent gains in efficiency of the variance estimator
over the Brewer-Rao-Durbin variance estimator

| | Method | | |
| | Murthy | R. H. C. | Replacement |
|---|---|---|---|
| | Natural populations | | |
| Median | 4.5 | 9.0 | − 3.0 |
| Quartiles | (3,16) | (5,27) | (− 7,3) |
| Extremes | (− 3,301) | (− 5,508) | (− 13,322) |
| | Linear model ($g = 1$) | | |
| Median | 8.0 | 16.0 | 4.5 |
| Quartiles | (4,24) | (8,42) | (− 3,16) |
| Extremes | (1,277) | (2,433) | (− 13,284) |
| | Linear model ($g = 1.5$) | | |
| Median | 4.0 | 6.5 | − 2.5 |
| Quartiles | (2,22) | (4,34) | (− 5,7) |
| Extremes | (1,370) | (1,543 ) | (− 14,346) |
| | Linear model ($g = 2$) | | |
| Median | 1.5 | 1.5 | − 8.0 |
| Quartiles | (1,13) | (1,15) | (− 9, − 6) |
| Extremes | (0,406) | (0,457) | (− 14,288) |

As the upper quartiles and the extremes indicate, the R.H.C. and Murthy variance estimators give substantial gains in stability in about $\frac{1}{4}$ of the populations.

## 2.4. Comparisons for $n = 3$ and $n = 4$

For $n > 2$ the Murthy and the RHC methods of selecting the sample and of making the estimate extend easily. For RHC the variance estimator is

$$\hat{V}(\hat{Y}_{\text{RHC}}) = k \sum_{1}^{n} P_g \left( \frac{y_i}{p_i} - \hat{Y}_{RHC} \right)^2$$

where

$$k = \left( \sum^{n} N_g^2 - N \right) \Big/ \left( N^2 - \sum^{n} N_g^2 \right).$$

The procedure that minimizes the variance is to make the group numbers $N_g$ as equal as possible. For Murthy,

$$\hat{V}(\hat{Y}_M) = \sum_{i}^{n} \sum_{j>i}^{n} p_i p_j [P(S)P(S\,|\,i,j) - P(S\,|\,i)P(S\,|\,j)]P(S)^{-2} \left( \frac{y_i}{p_i} - \frac{y_j}{p_j} \right)^2$$

where $P(S|i,j)$ is the conditional probability of drawing $S$ given that units $i, j$ were drawn first.

Extension of the BRD methods so as to keep $\pi_i = np_i$ requires ingenuity. Rao and Bayliss (1970) use Sampford's (1967) extension. In one form of his method, the first unit is drawn with probability $p_i$. All subsequent units are drawn with probabilities proportional to $\lambda_i = p_i/(1 - np_i)$ and with replacement. If all **n** units are different, accept the sample, otherwise reject as soon as a unit appears twice and try again. Sampford shows that for this method, $\pi_i = np_i$, and gives a rule for calculating the $\pi_{ij}$.

Table 2.3 shows the Rao and Bayliss (1970) results for the percent gains in efficiency over Sampford's method for $n = 3, 4$ for the natural populations studied.

Table 2.3

Percent gains in efficiency of the estimator's over Sampford's estimator

| | | Estimator | |
| % gains | Murthy | R. H. C. | Replacement |
|---|---|---|---|
| | $n = 3$ (14 natural populations) | | |
| Mean | $+ 2.2$ | $- 1.4$ | $- 13.6$ |
| Extremes | $(- 2,9)$ | $(- 11,10)$ | $(- 24, - 1)$ |
| | $n = 4$ (10 natural populations) | | |
| Mean | $+ 3.8$ | $- 5.2$ | $- 27.8$ |
| Extremes | $(- 4,33)$ | $(- 24,33)$ | $(- 39, - 3)$ |

As with $n = 2$ the mean differences among the 'without replacement' methods are small, though Murthy's still maintains a slight lead. The RHC method becomes slightly inferior to Sampford's. Consistent with the latter trend is a study by Sampford (1969) of a natural population (farms in Orkney) with $n = 12$, $N = 35$, where RHC suffered a loss of 24% in precision relative to Sampford.

As regards the stability of the variance estimator, however, the RHC method increases its superiority, with Murthy not far behind (Table 2.4).

Since more weight will usually be given to precision of estimation than to stability of variance estimation, I concur with Rao and Bayliss in describing Murthy's method as the best compromise performer in these studies. For $n_h = 2$, however, the most frequent case in stratified sampling, there seems little enough to choose among about six methods mentioned so that the choice may be made on other features, e.g. convenience or ease with which the method applies in rotation sampling.

Table 2.4

Percent gains in stability of the variance
estimators over Sampford's estimator

| % gains | Murthy | RHC | Replacement |
|---|---|---|---|
| | | $n = 3$ (14 populations) | |
| Median | + 9.5 | + 18.5 | + 0.5 |
| Quartiles | (6,15) | (9,22) | (− 9,41) |
| Extremes | (− 5,45) | (− 10,47) | (− 18,22) |
| | | $n = 4$ (10 populations) | |
| Median | + 20.0 | + 29.0 | − 7.5 |
| Quartiles | (11,83) | (12,140) | (− 16,52) |
| Extremes | (4,120) | (2,242) | (− 23,235) |

These studies are for single-stage sampling. In multi-stage sampling, a necessity in extensive surveys, Durbin (1967) has stressed that methods which draw the primary units without replacement have the annoying feature that unbiased variance estimates from the sample require the separate calculation of both a between-psu and a within-psu component. Thus for the Horvitz-Thompson estimator with $n = 2$, in multi-stage sampling,

$$\hat{V}(\hat{Y}_{BRD}) = \left(\frac{\pi_1 \pi_2}{\pi_{12}} - 1\right) \left(\frac{\hat{y}_1}{\pi_1} - \frac{\hat{y}_2}{\pi_2}\right)^2 + \sum_i^2 \pi_i \hat{V}_2(\hat{y}_i)$$

where $\hat{y}_i$ is the unbiased sample estimate of the psu total $y_i$ and $\hat{V}_2(\hat{y}_i)$ an unbiased estimate of the component of $V(\hat{y}_i)$ due to second and further stages.

Durbin has suggested techniques for $n = 2$ designed to simplify the first component and avoid most of the calculation of the second component, including an approximate method in which each $n = 2$ sample necessitates calculation of only a single square for variance estimation, while retaining the advantage of without-replacement psu selection. This method worked well in an application to British election statistics and it would be worthwhile to examine its performance and the stability of the variance estimator in other natural populations.

## 3. Estimates of the sampling variance of non-linear estimators

### 3.1. Introduction

The problem of computing sample estimates of the variances of non-linear statistics has also received increased attention of late. The most familiar example is the ratio estimator, but statistics like partial and multiple regression and correlation coefficients are also involved in the analysis of survey data. Moreover, the methods need to be extended so as to handle stratified sampling, cluster sampling, multi-stage sampling and devices like post stratification and adjustments intended to decrease non-response bias. In making analytical comparisons, statistics like Student's **t** are widely employed.

Three approaches have been tried.

### 3.2. Taylor series approximations

Let $\mathbf{v} = (v_1, v_2, \cdots, v_m)$ be a set of statistics whose expected value is the set $\mathbf{V} = (V_1, V_2, \cdots V_m)$. If the function to be estimated is $\theta = F(\mathbf{V})$, estimated by $\hat{\theta} = F(\mathbf{v})$, we write

$$\hat{\theta} - \theta = F(\mathbf{v}) - F(\mathbf{V}) \approx \sum_{i=1}^{m} (v_i - V_i) \frac{\partial F}{\partial V_i}$$

using the first-term Taylor expansion, where the partial derivatives are to be evaluated at $v_i = V_i$ but in practice usually have to be evaluated at $v_i$. The variance of $\hat{\theta}$ is then approximated by the variance of the linear function

$$\left[ \sum_{i=1}^{m} (v_i - V_i) \frac{\partial F}{\partial V_i} \right]$$

expressible in terms of the variances and covariances of the $v_i$. This is the method which produced the large-sample formulas for variance and estimated variance of a ratio. In extending this approach to more complex estimates and surveys, papers by Keyfitz (1957), Kish (1968), Tepping (1968) and Woodruff (1971) have shown that ingenuity in the method used to compute the variance of the linearized form can considerably simplify the numerical work. Little is known about the performance of this method in moderate-sized samples, though there is growing evidence of substantial underestimation of the variance for the ratio estimate by this method.

### 3.3. Balanced repeated replications (BRR)

The method of repeated replication is related to Mahalanobis' method of interpenetrating subsamples. The method was first employed by the U.S. Census Bureau for estimating variances in their Current Population Survey. Most of the work has been done on national stratified samples with two psu per stratum.

With two units per stratum chosen with equal probability, a half-sample is obtained by selecting at random one unit from each stratum. The statistic $\hat{\theta}$ of interest is calculated for the half-sample. Let $\hat{\theta}_H, \hat{\theta}_C, \hat{\theta}_S$ denote the estimate as computed from the chosen half-sample, the complementary half-sample, and the whole sample. For a strictly linear estimate it is easy to verify that the quantities

$$(\hat{\theta}_H - \hat{\theta}_S)^2 = (\hat{\theta}_C - \hat{\theta}_S)^2 = \tfrac{1}{4}(\hat{\theta}_H - \hat{\theta}_C)^2 \qquad (3.3.1)$$

are all unbiased estimates of $V(\hat{\theta})$, each with 1 degree of freedom, if the finite population correction is negligible. The method amounts to repeating this process until a desired number of replications has been obtained, and, more importantly, in using it for non-linear statistics, though without formal analytical support.

With $L$ strata, the possible number of replications is $2^L$. For linear estimates, McCarthy (1966) proved that an orthogonal subset, using the Plackett and Burman (1946) orthogonal main-effect designs for the $2^n$ factorial, produces the same variance estimate as the complete set, and requires at most $(L + 3)$ half-replicates. If this number is still too large, McCarthy suggests use of a subset of the balanced set. In a later review of this method, McCarthy (1969) found that the three half-sample variance estimators in (3.3.1) agreed well with one another in some non-linear situations—15 combined ratio estimates, 24 partial regression coefficients, and 8 multiple regression coefficients in 27 strata from a Health examination survey by the National Center for Health Statistics.

### 3.4. The Jackknife method

The method often called the Jackknife was first suggested by Quenouille (1949), as a technique for reducing the bias in an estimator like the ratio from $O(1/n)$ to $O(1/n^2)$. It has been applied mostly to simple random samples. Suppose $n = mg$. The sample is divided into **g** groups of size **m**. Let $\hat{\theta}_S$ denote the estimate made from the complete sample $S$ and let $\hat{\theta}_{(S-j)}$ denote the estimate made by omitting the **j**th group. Let

$$\hat{\theta}_j = g\hat{\theta}_S - (g-1)\hat{\theta}_{(S-j)}.$$

Quenouille showed that if $\hat{\theta}_S$ has bias $O(1/n)$, the average of the $\hat{\theta}_j$, say $\hat{\theta}_J$, has bias $O(1/n^2)$. Going further, Tukey (1958) suggested that in many non-linear problems the quantities $\hat{\theta}_j$, which he called pseudo-values, might be used to estimate $V(\hat{\theta}_J)$ by the usual rule for random samples, viz.

$$\hat{V}(\hat{\theta}_J) = \sum_{j=1}^{g} (\hat{\theta}_j - \hat{\theta}_J)^2 / g(g-1).$$

He suggested further that

$$t' = (\hat{\theta}_J - \theta) / \sqrt{\hat{V}(\hat{\theta}_J)}$$

would often be approximately distributed as $t$ with $(g-1)$ d.f. With linear estimates from normal data it is easily verified that $t'$ is identical to $t$ calculated in the usual way.

Some analytic support for this approach has come in the form of asymptotic results. Under broad assumptions, Brillinger (1964) showed that for maximum likelihood estimates, $t'$ tends to a $t_{(g-1)}$ distribution as $m$ tends to infinity. Arvesen (1969) gave a similar result for estimates (of the form known as Hoeffding's $U$-statistics) that are symmetrical with respect to the members of the sample.

### 3.5. Frankel's comparison of the three approaches

Using data from the March 1967 Current Population Survey, Frankel (1971) has made a Monte Carlo comparison of the performance of the three methods in relatively small samples of sizes 12, 24, and 60, 2 units being drawn from each of 6, 12, and 30 strata. The population in this study contained 3240 primary units of average size 14.1 households. For a given number of strata, all strata were of the same size; e.g. with 12 strata, each has 270 psu. Thus the study involves cluster units of unequal sizes and proportional stratification, but not unequal probability selection.

The types of estimates examined were: 8 means (effectively ratio estimates), 12 differences of means, 12 simple correlations, 8 partial regression coefficients, 6 partial correlation coefficients and 2 multiple correlation coefficients. As regards the estimates themselves, the average relative biases $[E(\hat{\theta}) - \theta]/\theta$ were less than 1% for means and differences between means, less than 7% for simple correlations and 5% for partial regression coefficients even with the smallest sample size 12. The average relative biases were somewhat larger for partial and multiple correlations (12% and 16% for $n = 12$). Examination of the ratio of these biases to the standard error of $\hat{\theta}$ showed that in the worst cases, two-sided 95% confidence interval statements might have confidence probability nearer 90% if this bias were the only source of trouble.

Frankel compared four variants of both the BRR and the Jackknife method for estimating $V(\hat{\theta}_S)$, where $S$ denotes the whole sample. We consider here only the variant that appeared to perform best. For the BRR method this was

$$\hat{V}_B(\hat{\theta}_S) = (1-f)\text{Ave}\left[\tfrac{1}{2}(\hat{\theta}_H - \hat{\theta}_S)^2 + \tfrac{1}{2}(\hat{\theta}_C - \hat{\theta}_S)^2\right] \qquad (3.5.1)$$

where $f$ is the finite population correction averaged over the Plackett-Burman orthogonal set of half-samples.

It is not at first sight obvious how to extend the Jackknife method to stratified sampling. Frankel's choice was as follows. Let $\hat{\theta}_{Hi}$ be the estimate obtained by omitting one unit in stratum $i$ and doubling the value given by the other unit. Let

$$\hat{V}_{JH} = (1-f)\sum_{i=1}^{L}(\hat{\theta}_{Hi} - \hat{\theta}_S)^2$$

with the complementary estimator

$$\hat{V}_{JC} = (1-f)\sum_{i=1}^{L}(\hat{\theta}_{Ci} - \hat{\theta}_S)^2.$$

Then

$$\hat{V}_J(\hat{\theta}_S) = \tfrac{1}{2}[\hat{V}_{JH} + \hat{V}_{JC}]. \qquad (3.5.2)$$

The number of independent Monte Carlo drawings was 300 for 6 and 12 strata and 200 for 30 strata.

## 3.6. Results

The average percent relative biases in the variance estimators by the three methods appear in Table 3.1. No method is consistently best in this respect. All methods do satisfactorily for means and differences between means, the average biases being under 5% except for 6.7% with BRR for means when $n = 24$. BRR is the only method that does well for simple correlation coefficients and is superior to the Jackknife for partial $r$'s (the Taylor method was not tried for partial $r$'s and multiple $R$'s because of its complexity in these cases.) On the other hand, both Taylor and the Jackknife are superior to BRR for partial regression coefficients. For the 2 multiple $R$'s, BRR beats the Jackknife but neither does well.

Table 3.2 studies the performance of an approximate **t**-statistic derived from each method in the form

$$t' = [\hat{\theta} - E(\hat{\theta})]/s(\hat{\theta})$$

which is assigned the number of d.f.—6, 12, or 30—that would ordinarily be given for this sampling plan. Frankel tabulated the relative frequency

W. G. COCHRAN

Table 3.1

Average percent relative biases of the estimators
of variance: $100[\hat{V}(\hat{\theta}_s) - V(\hat{\theta}_s)]/V(\hat{\theta}_s)$

| Statistics | No. averaged | Taylor | Method BRR 6 strata ($n = 12$) | J |
|---|---|---|---|---|
| Means | 8 | − 3.8 | + 3.5 | − 1.6 |
| Diffs. | 12 | − 4.4 | + 2.9 | − 2.1 |
| Correlations r | 12 | − 25.1 | + 0.2 | − 12.7 |
| Partial b's | 8 | − 7.2 | + 19.1 | − 0.2 |
| Partial r's | 6 | N.G.* | + 7.9 | − 11.1 |
| Multiple R's | 2 | N.G. | − 1.9 | − 20.0 |
| | | | 12 strata ($n = 24$) | |
| Means | 8 | + 2.5 | + 6.7 | − 3.8 |
| Diffs. | 12 | − 0.0 | + 3.9 | − 1.0 |
| Correlations r | 12 | − 21.7 | − 3.5 | − 12.7 |
| Partial b's | 8 | − 2.3 | + 10.9 | 0.0 |
| Partial r's | 6 | N.G. | − 3.2 | − 11.8 |
| Multiple R's | 2 | N.G. | − 17.6 | − 31.0 |
| | | | 30 strata ($n = 60$) | |
| Means | 8 | − 0.0 | + 1.3 | 0.0 |
| Diffs. | 12 | − 4.1 | − 2.3 | − 3.8 |
| Correlations r | 12 | − 14.9 | − 2.3 | − 9.2 |
| Partial b's | 8 | − 1.8 | + 7.2 | + 2.3 |
| Partial r's | 6 | N.G. | − 4.3 | − 7.3 |
| Multiple R's | 2 | N.G. | − 9.7 | − 23.0 |

*N.G. = not given

with which $t'$ fell within certain fixed numerical values related to the normal distribution (e.g. 2.576, 1.960, 1.645). For this presentation I have chosen the values nearest to the 95% and 90% two-sided limits of **t**—so as to obtain a check on confidence probabilities at or near these levels. The figures in Table 3.2 are [1—(confidence probability)], since errors in the **t**-approximation stand out more clearly in this scale.

Comparison of the observed relative frequencies in the Monte Carlo study with the relevant Student-**t** frequencies places the methods almost uniformly in the order (1) BRR, (2) Jackknife, (3) Taylor. In most cases the observed frequency with which **t**' lies in the tails is greater than the frequency obtained by assuming that **t**' follows Student's *t*-distribution. Is the agreement between the **t**' and **t** tail-frequencies good enough?

Table 3.2

Relative frequency with which $[\hat{\theta} - E(\hat{\theta})]/s(\hat{\theta})$ lies outside certain limits, compared with probabilities $P(t)$ for Student's t

| | 6 strata | | | | | |
| | Limits $\pm$ 2.576: $P(t_6) = .042$ | | | Limits $\pm$ 1.960: $P(t_6) = .098$ | | |
| | Taylor | BRR | J | Taylor | BRR | J |
|---|---|---|---|---|---|---|
| Means | .052 | .044 | .049 | .112 | .096 | .106 |
| Diffs. | .055 | .050 | .054 | .116 | .100 | .106 |
| r's | .084 | .052 | .069 | .163 | .114 | .137 |
| Partial b's | .058 | .034 | .048 | .127 | .085 | .117 |
| Partial r's | N.G.* | .043 | .063 | N.G. | .092 | .132 |
| Mult. R's | N.G. | .065 | .088 | N.G. | .105 | .160 |
| | 12 strata | | | | | |
| | Limits $\pm$ 1.960: $P(t_{12}) = .074$ | | | Limits $\pm$ 1.645: $P(t_{12}) = .126$ | | |
| | Taylor | BRR | J | Taylor | BRR | J |
| Means | .081 | .078 | .080 | .135 | .130 | .134 |
| Diffs. | .093 | .088 | .092 | .148 | .138 | .144 |
| r's | .141 | .103 | .125 | .197 | .156 | .174 |
| Partial b's | .088 | .066 | .084 | .150 | .125 | .146 |
| Partial r's | N.G. | .088 | .112 | N.G. | .131 | .174 |
| Mult. R's | N.G. | .150 | .187 | N.G. | .210 | .262 |
| | 30 strata | | | | | |
| | Limits $\pm$ 1.960: $P(t_{30}) = .059$ | | | Limits $\pm$ 1.645: $P(t_{30}) = .110$ | | |
| | Taylor | BRR | J | Taylor | BRR | J |
| Means | .057 | .056 | .057 | .112 | .109 | .112 |
| Diffs. | .057 | .054 | .057 | .116 | .112 | .116 |
| r's | .102 | .089 | .098 | .164 | .138 | .153 |
| Partial b's | .068 | .062 | .068 | .117 | .110 | .116 |
| Partial r's | N.G. | .103 | .121 | N.G. | .156 | .181 |
| Mult. R's | N.G. | .175 | .208 | N.G. | .265 | .298 |

*N.G = not given

In discussing the routine use of Student's t with well-behaved data in introductory applied courses, a rough working attitude that I have sometimes given is to regard a tabular 5% two-tailed *t* as corresponding to an actual two-tailed frequency not exactly 5%, but lying somewhere between 4% and 7%. Judged by this rule, the best BRR method is highly satisfactory for means, differences between means, and regression coefficients. For simple and partial correlation coefficients it is mostly satisfactory except for the puzzling feature, as Frankel notes, that the agreement worsens as **n** increases in his study. For multiple correlations, where the same puzzling feature appears, the observed **t'** frequency can run over twice as high as the **t**-

frequency. One possibility is that methods like BRR and the Jackknife can run into trouble when dealing with statistics like $r$ and $R$ that are restricted to lie between finite boundaries.

Overall, Frankel's results seem to me encouraging news. For his sample sizes 12, 24, 60, I would hesitate to trust a $t$-approximation to $R$ even in random samples from an infinite population and with normal deviations from the regression.

Since the best BRR method in (3.5.1) requires calculation of both $\hat{\theta}_H$ and $\hat{\theta}_C$, it is noteworthy that use of $\hat{V}(\hat{\theta}_S) = (1 - f)\operatorname{Ave}(\hat{\theta}_H - \hat{\theta}_S)^2$, which does not require calculation of $\hat{\theta}_C$, does almost as well in $t'$ frequencies and might be recommended as speedier. Frankel himself reserves judgment on this question pending more extensive data.

There is obviously room for much further work of this type, dealing with different natural populations, different degrees of complexity in the sample design, different types of non-linear estimate, and different sample sizes.

Though it appears formidable, analytic work may help us to understand why, as I see it, $t'$ performs better in general than I would have anticipated with these sample sizes. Other questions are whether monotonic transformations may help with estimates like $R$, and how best to estimate $\theta$. In regression analysis in multi-stage samples, for instance, the between-psu regression may have a different interpretation and give different $\hat{\beta}_i$ values from within-psu regressions. But Frankel has made a most welcome initial attack.

## References

Arvesen, J. N. (1969). Jackknifing $U$-Statistics. *Ann. Math. Statist.*, **40**, 2076–2100.

Bayliss, D. L. and Rao, J. N. K. (1970). An empirical study of stabilities of estimators and variance estimators in unequal probability sampling ($n = 3$ or 4). *J. Amer. Statist. Ass.*, **65**, 1645–1667.

Brewer, K. R. W. (1963). A model of systematic sampling with unequal probabilities. *Australian J. Statist.*, **5**, 5–13.

Brewer, K. R. W. and Hanif, M. (1969). Sampling without replacement with probability proportional to size. Unpublished manuscript.

Brillinger, D. R. (1964). The asymptotic behaviour of Tukey's general method of setting approximate confidence limits (the Jackknife) when applied to maximum likelihood estimates. *Rev. Int. Statist. Inst.*, **32**, 202–206.

Des Raj (1956). Some estimators in sampling with varying probabilities without replacement. *J. Amer. Statist. Ass.*, **51**, 269–284.

Durbin, J. (1967). Estimation of sampling errors in multi-stage samples. *Appl. Statist.*, **16**, 152–164.

Fellegi, I. P. (1963). Sampling with varying probabilities without replacement: rotating and non-rotating samples. *J. Amer. Statist. Ass.*, **58**, 183–201.

Frankel, M. R. (1971). An empirical investigation of some properties of multivariate statistical estimates from complex samples. Ph.D. thesis, University of Michigan.

Hanurav, T. V. (1967). Optimum utilization of auxiliary information: $\pi$PS sampling of two units from $c$ stratum. *J. Roy. Statist. Soc. Ser. B*, **29**, 374–391.

Keyfitz, N. (1957). Estimates of sampling variance when two units are selected from each stratum. *J. Amer. Statist. Ass.*, **46**, 105–109.

Kish, L. (1968). Standard errors for indexes from complex samples. *J. Amer. Statist. Ass.*, **63**, 512–529.

Lahiri, D. B. (1951). A method for sample selection for providing unbiased ratio estimates. *Bull. Int. Statist. Inst.*, **33**, 133–140.

McCarthy, P. J. (1966). Replication: an approach to the analysis of data from complex surveys. Nat. Center for Health Statist., Washington, D. C., Series 2, No. 31.

McCarthy, P. J. (1969). Pseudo-replication: half samples. *Rev. Int. Statist. Inst.*, **37**, 239–264.

Murthy, M. N. (1957). Ordered and unordered estimators in sampling without replacement. *Sankhya*, **18**, 379–390.

Plackett, R. L. and Burman, J. P. (1946). The design of optimum multifactorial experiments. *Biometrika*, **33**, 305–325.

Quenouille, M. H. (1956). Notes on bias in estimation. *Biometrika*, **43**, 353–360.

Rao, J. N. K. (1965). On two simple schemes of unequal probability sampling without replacement. *J. Indian Statist. Ass.*, **3**, 173–180.

Rao, J. N. K. and Bayliss, D. N. (1969). An empirical study of the stabilities of estimators and variance estimators in unequal probability sampling of two units per stratum. *J. Amer. Statist. Ass.*, **64**, 540–559.

Rao, J. N. K., Hartley, H. O. and Cochran, W. G. (1962). On a simple procedure of unequal probability sampling without replacement. *J. Roy. Statist. Soc. Ser. B*, **24**, 482–491.

Sampford, M. R. (1967). On sampling without replacement with unequal probabilities of selection. *Biometrika*, **54**, 499–513.

Sampford, M. R. (1969). A comparison of some possible methods of sampling for smallist populations, with units of unequal size. In *New Developments in Survey Sampling*, Wiley, New York, 170–187.

Tepping, B. (1968). The estimation of variance in complex surveys. *Proc. Soc. Statist. Sect. Amer. Statist. Ass.*, 11–18.

Tukey, J. W. (1958). Bias and confidence in not quite large samples. *Ann. Math. Statist.* **29**, 614 (Abstract).

Woodruff, R. S. (1971). A simple method for approximating the variance of a complicated estimate. *J. Amer. Statist. Ass.*, **66**, 411–414.

Yates, F. and Grundy, P. M. (1953). Selection without replacement from within strata with probability proportional to size. *J. Roy. Statist. Soc. Ser. B*, **15**, 253–261.

# Bartlett's Work on Agricultural Experiments

## W. G. COCHRAN

**Abstract**

A selection of Bartlett's work as an operating statistician in his first
position as statistician 1934–38 at the I.C.I. agricultural research station
at Jealott's Hill, Berks., is described. This illustrates some of the
methods he used for the efficient detection of treatment effects and for
an appraisal of the suitability of the experimental designs that were
being used.

## 1. Introduction

From 1934 to 1938 Bartlett was statistician at the I.C.I. agricultural
research station at Jealott's Hill, Berks., this being his first full-time job.
His work involved dealing with the design and analysis of randomized
experiments both in the field and laboratory. By 1934 the Fisherian
methods for comparative experimentation had been produced in rapid
succession — randomized blocks, latin squares, factorial design (including the confounding of interactions), with the analysis of variance and
covariance as primary techniques in aiding the interpretation of the
results. At that time we were just learning how to use these tools to best
advantage. A selection of Bartlett's work in this process will show
something of his methods as an operating statistician.

## 2. The efficient detection of treatment effects

In a section with this title, Bartlett (1937) notes that the more statisticians can condense the description of the treatment effects, the more
likely they will be able to make efficient tests of these effects. In the

absence of any available theory of the mode of action of treatments, the statistician has to seek efficient *empirical* reduction methods.

*Isolation of single d.f.* This now standard device is illustrated in a $4 \times 4$ factorial on weights of sugar-beet tops, with 4 degrees of closeness of spacing $(S)$ and 4 levels of complete fertilizer $(M)$. Although the degrees of closeness are not in strictly equal steps, the $S_L$, $M_L$ and $S_L \times M_L$ single d.f. are found to summarize adequately the performance of the 16 treatment combinations, there being little or no advantage from close spacing unless manure is given. Nowadays, standard computer programs for factorial designs usually allow a breakdown of main effects and 2-factor interactions into polynomial components of this type, but may not cover components of the non-polynomial type.

*Observations repeated in time on each unit* (*plot*). With observations taken at $n$ successive times on each unit, the problem of a summary analysis is more difficult. One approach that is often natural is to fit a response curve $y(t)$ to each plot from the $n$ available times, studying how the treatments affect the nature of this response curve or concentrating on its most important aspects in relation to the objectives of the experiment. Two examples, Bartlett (1937), will be cited.

The first came from experiments on the effect of shading and types of nitrogen on the herbage on a lawn in successive weeks. One response variable $y$ was the percentage area covered by clover each week. For this the response curve $y(t)$ on a plot appeared graphically to be a declining non-linear function. Bartlett notes, however, that the rate of change of this percent with time will depend both on the amount of clover available to expand and on the area $(100 - y)$ available to expand into (or, with a negative rate of change, on the area of competing species). These comments suggest as an approximation the equation

$$\frac{dy}{dt} = Ay(100 - y)$$

giving

$$\ln(y/(100 - y)) = Bt + c.$$

This leads to the familiar logit transformation $x = \ln(y/(100-y))$ as a device usually effective in straightening the relation $x(t)$ of $x$ against time. Analyses of variance were made of the linear regression of $x$ on time, to give a compact empirical summary of the treatment effects.

The second example was an experiment (7 treatments, 6 blocks) on germination in peas. For each treatment, 50 seeds per replicate were placed in a pan on 5 May 1936, daily counts being made of the *total* number $y_t$ germinated by day $t$. Here the key economic variables are the overall total number germinating and some measure of the earliness of germination. For the latter, Bartlett suggests a 'rate index', which for the $n$ days in the experiment is

$$\left(\sum_1^n y_t\right)\Big/(ny_n).$$

The choice of this index may appear puzzling. But if $d_j$ seeds germinate on day $j$, then $y_t = \Sigma' d_j$, so that if we omit the divisor $n$, the index becomes

$$\{nd_1 + (n-1)d_2 + \cdots + d_n\}\Big/\left(\sum^n d_j\right).$$

This quantity can be described as the mean earliness of germination, the days being counted backwards, $1, 2, \cdots, n$, from the last date, as Bartlett points out. The quantities $y_n$ (total number germinated) and the rate index then constitute the most important variables for a summary analysis. (In the data the counts were not entirely daily, one day having two counts, but Bartlett uses the index as an empirical measure that serves to denote earliness.)

*An ingenious use of confounding.* A $3^3$ experiment on wheat was conducted in 1934, the factors being varieties, spacings and nitrogen in 4 blocks of 27 plots. In 1935 a $3^2$ experiment on maize was planned on the same site, the treatments being the same 3 spacings and levels of nitrogen. It was desired in addition to measure any residual effects of the 1934 spacings and $N$ levels, comparing them with the direct effects of these factors. By using three of the four 1934 blocks, Bartlett noted that he could construct a plan that was a single replication of a $3^4$ ($S_D, N_D, S_R, N_R$) with 2 d.f. from the 4-factor interactions confounded with blocks, where $D$ denotes the direct and $R$ the residual effects.

The composition of the 1935 blocks (I, II, III) is given by the following diagram.

| | Residual factors | | | | Direct factors | | | | | 1935 blocks | | |
|---|---|---|---|---|---|---|---|---|---|---|---|---|
| | $S_0$ | $S_1$ | $S_2$ | | $S_0$ | $S_1$ | $S_2$ | | | $X$ | $Y$ | $Z$ |
| $N_0$ | $X$ | $Y$ | $Z$ | $N_0$ | $A$ | $B$ | $C$ | $A$ | I | II | III |
| $N_1$ | $Z$ | $X$ | $Y$ | $N_1$ | $C$ | $A$ | $B$ | $B$ | III | I | II |
| $N_2$ | $Y$ | $Z$ | $X$ | $N_2$ | $B$ | $C$ | $A$ | $C$ | II | III | I |

Thus the 1935 block I had the 27 treatment combinations $AX, BY, CZ$ where $A$ denotes the three direct combinations $N_0S_0, N_1S_1, N_2S_2$ and $X$ the corresponding three residual combinations $N_0S_0, N_1S_1, N_2S_2$. The error mean square was estimated from the remaining 4-factor interactions (14 d.f.) and the 3-factor interactions which contained a 2-factor residual interaction (16.d.f). Significant residual effects of $N$ and $S$ were found, but were negligible relative to the large direct effects.

*Salt damage in experiments in Egypt.* Accumulation of salt in the upper layers of the soil in Egypt can be a major source of variation in some experiments, reducing the cotton yield. In three experiments with serious spotty damage, a visual estimate of the degree of saltiness of the plot soil was made on each plot on a 0 to 10 scale, (Crowther and Bartlett (1938)). Since the regression of yield on this index appeared linear, a covariance adjustment for the regression of yield on salt was made. This device reduced the error mean square by 54% and 38% in two experiments though by only 5% in a third.

For many statisticians, analysis of the covariance-adjusted yields would be assumed to have removed the disturbing effects of salt damage. But Bartlett reminds us that covariance may not really *adjust* for salt damage, since severe salt concentrations might decrease the ability of the crop to respond to the treatments. To examine whether this effect of the salt was present, he recalculates the analysis, omitting the 12 plots most severely salty. In the re-analysis the linear response to nitrogen increased by 16%, so that his warning about the interpretation of covariance was justified.

*Recovery of intra-block information.* Bartlett (1938) examined an ingenious method proposed by Papadakis (1937) for handling patchy fields. The technique might be described as the recovery of *intra-block* information about fertility. First calculate for each plot yield $y$ the quantity $y' = (y - \bar{y}_t + \bar{y})$, where $\bar{y}_t$ is the mean of the treatment on the plot. The quantities $y'$ are regarded as estimates of the fertility levels of the plots, though nothing would be gained by a covariance regression of the $y$ on the $y'$. Instead, a covariate $x$ is found by using the fertility $y'$ of neighboring plots to predict the fertility level of the plot. With long narrow plots, $x$ is the mean of the two $y'$ values contiguous to the long side of the plot. With squarish plots, $x$ can be the mean of the $y'$ for the four contiguous plots, with special rules for edge plots. The treatment yields $\bar{y}_t$ are adjusted for the covariance on $x$.

Bartlett's contribution has two parts. In the first, accepting the method at face value, he applies it to two $4 \times 3^2 \times 2$ factorials on cotton in blocks of 72 plots. The standard errors per plot are reduced from 12.0% to 8.6%, and from 8.9% to 7.0% — certainly worthwhile gains. He goes on to investigate the approximate theoretical validity of the technique by examining algebraically the first-order disturbances to the standard covariance formulas. He finds that the assumptions required are likely to be least disturbed when the blocks are large and suggests from his analysis that, when calculating the residual error mean square, 2 d.f. be allotted to the regression on $x$ (instead of the usual 1 d.f.) He concludes that the method is approximately valid with, at my guess, at least 30 plots per block, and may occasionally be useful if an experiment has large blocks found to be internally heterogeneous.

## 3. Analysis as a guide to design

A notable feature of Bartlett's work was the use of supplementary analysis of the completed results of an experiment in forming a judgment as to whether the design used was the most suitable for its purpose, or whether an alternative should be tried for similar experiments in the future. This working habit, natural in the early stages of experimentation, is worth forming in any major program of experimentation. Some examples follow.

*Split-plot experiments.* Split-plot experiments were quite common in the 1920's and 30's, especially if a larger plot was more convenient for some factors than for others. A series of large-scale factorials on cotton in Egypt had two main-plot factors — varieties (3) and watering (2 levels) — and two sub-plot factors — spacing (3 levels) and nitrogen (3 levels). In appraising the suitability of the split-plot plans, Bartlett (1937) estimates, from the main-plot and sub-plot error mean squares in 10 experiments, what the efficiencies of main-plot and sub-plot comparisons would have been with ordinary randomized blocks on the same sites, relative to those obtained from the split-plot plans used. The median relative efficiencies were 3.2 for main-plot comparisons and 0.82 for sub-plot comparisons.

These figures suggest that there will be a net gain, as regards precision of main effects, from reverting to randomized blocks. This was done in two experiments in 1936, which showed satisfactory precision. Similarly, with factorials employing partial confounding of interactions, Bartlett estimates from the results the gains in efficiency for unconfounded effects and gains or losses for confounded effects as a check on the performance of the design versus randomized blocks.

*Fertility diagrams.* A further device used by Bartlett (Crowther and Bartlett (1938)) was the fertility diagram, in combination with the size of the standard error per plot. Construction of fertility diagrams from uniformity trial data was well known in agriculture as a guide to the suitability of the field and to the best method of blocking. Subject to the assumption of additivity of the effects, a fertility diagram that serves the same purposes can be constructed from the quantities $(y - \bar{y}_t + \bar{y})$, where $y$ is the yield on an individual plot and $\bar{y}_t$ the mean yield of the treatment on that plot. As with uniformity data, these diagrams often revealed a patchiness in fertility with no regular pattern, from which geographically compact blocks seem the best protection, although the judgment might be to avoid some fields in future experiments or (with large blocks) to try the Papadakis method already described.

*Experiments on dairy cows.* The cow is an expensive animal for experimentation owing to the high variability of milk yield from cow to cow. Standard errors of 25% per cow in milk yield are not uncommon. One method used in the 1930's to circumvent this difficulty was to make the cow serve as its own control by using a cross-over or latin square design with the treatments given in succession to the cow for short periods during the lactation cycle. Bartlett (1935) notes two disadvantages of this method. The number of treatments is limited to two or three, and it is by no means clear that the performances of treatments in practice over the whole lactation cycle are predictable from their performance during a short period in that cycle. Thus, precision may have been obtained by asking the wrong question.

He takes advantage of an experiment that had no significant treatment effects to regard the data as a uniformity trial. The alternatives to the cross-over design that he wishes to consider are to give a cow uniform treatment during the first $n$ weeks ($n = 1, 2, 3, 5$) and a single treatment during the remaining $(20-n)$ weeks of the lactation period. He finds that a covariance adjustment for the early period yields reduces the s.e. per cow from 25.2% to around 8% for $n = 1, 2, 3, 5$, a level at which the measurement of treatment differences of economic value becomes feasible. This analysis indicates that one week's preliminary period is enough, but Bartlett suggests three weeks as a safer period to protect against temporary fluctuations. This method was used in a 1934 winter feeding trial, resulting in a standard error of 7.0%.

This experiment had one missing cow which ran dry early. He notes (Bartlett (1937)) that the exact least squares analysis of data with a missing value can be obtained by inserting any number $y$ for the missing value and doing a covariance analysis on a dummy $x$-variate which takes the value 1 for the missing value and 0 elsewhere. This approach has recently been adapted (Rubin (1972)) as a computer method of handling analyses of data with missing values.

*A non-linear response.* In insecticide tests under controlled laboratory conditions, the proportion $\hat{p}$ killed out of $n$ by a given dose can often be assumed to be binomial. Under a normal tolerance distribution, the theoretical proportion killed is assumed to be related to the log dose $x$ by the relation

$$p = \frac{1}{\sqrt{(2\pi)}} \int_{-\infty}^{(x-\mu)/\sigma} z(u)\,du,$$

where $z(u)$ is the standard normal ordinate at $u$. This is a two-parameter problem. For the estimation of $\mu$, the median lethal dose, the Fisherian amount of information is well-known to be

$$I(\mu) = \frac{nz^2}{\sigma^2 pq} = \frac{z^2}{\sigma^2 V(\hat{p})}$$

and to be maximized if $x$ can be chosen to produce an expected 50% kill.

Bartlett's (1936) problem arose from insecticide tests in the field, not the laboratory. Leatherjackets (larvae that live in the soil and attack the roots of cereals and grass) are controlled by applying toxic emulsions to the plots, the experiment having 5 treatments, including a control. The surviving leatherjackets are counted by application of a standard emulsion which brings the leatherjackets to the surface, two square foot samples being counted per plot.

Still assuming the normal tolerance distribution, Bartlett notes that the number $n\hat{q}$ surviving in the field will not be binomial (further, $n$ is unknown). The simplest assumption is that $n\hat{q}$ is Poisson, for which $V(n\hat{q}) = nq = m$ (say). But in the field the variance is often found to contain also a component in $m^2$. Consequently he writes

$$V(n\hat{q}) = \alpha m (1 + \beta m) = \alpha m (1 + \lambda q) \propto q (1 + \lambda q).$$

Consequently, for these field tests the information is more accurately expressed as

$$I(\mu) \propto \frac{z^2}{\sigma^2 q(1 + \lambda q)}.$$

He plots the curve of $I(\mu)$ against $p$ (proportion killed) for a series of values of $\lambda$. For $1 \leqq \lambda \leqq 3$, the information is maximized by aiming at a percent kill around 80% instead of the 50% for a binomial model. (The experimental example presented suggests $\lambda \simeq 3$.)

He warns that this is a 'rough and ready' analysis and that other experiments may not follow this model, but gives this calculation as another example of the kinds of questions that should be asked in monitoring the efficiency of experiments.

## 4. Concluding remarks

In some respects the preceding examples do not do justice to Bartlett. They do not illustrate his originality and power in tackling unsolved problems in theory and they mostly concern techniques that now are, or should be, well-known. But they illustrate that he is equally at home in applied statistics, continually keeping in mind the objectives of experimental design and analysis, questions of feasibility and convenience, and ready to employ approximate theoretical methods when they would be adequate for his purpose.

## References

BARTLETT, M. S. (1935) An examination of the value of covariance in dairy cow nutrition experiments. *J. Agric. Sci.* **25**, 238–244.

BARTLETT, M. S. (1936) Some notes on insecticide tests in the laboratory and in the field. *J. R. Statist. Soc. Suppl.* **3**, 185–194.

BARTLETT, M. S. (1937) Some examples of statistical methods of research in agriculture and applied biology. *J. R. Statist. Soc. Suppl.* **4**, 138–183.

BARTLETT, M. S. (1938) The approximate recovery of information from replicated field experiments with large blocks. *J. Agric. Sci.* **28**, 418–427.

CROWTHER, F. AND BARTLETT, M. S. (1938) Experimental and statistical technique of some complex cotton experiments in Egypt. *Emp. J. Exp. Agric.* **6**, 53–68.

PAPADAKIS, J. S. (1937) A statistical method for field experiments. *Bull. Inst. Amél. Plantes à Salonique* **23**, 1–30.

RUBIN, D. R. (1972) A non-iterative algorithm for least squares estimation of missing values in any analysis of variance design. *Appl. Statist.* **21**, 136–141.

# THE ROLE OF STATISTICS IN NATIONAL HEALTH POLICY DECISIONS

WILLIAM G. COCHRAN[1]

## INTRODUCTION

A rational choice between alternative policy decisions must lean heavily on relevant quantitative information, in a word, on statistics. Thus, statistics have a very important role to play in health policy decisions. In many decisions, two kinds of statistics are needed, descriptive statistics about the nature of the situation that requires improvement and the resources that might be used in this effort, and estimates of the effects of different policies for using these resources. The current conflicting views of Senator Edward Kennedy and Representative Paul Rogers on health manpower legislation, for example, both deal with the same problem, a shortage of health services manpower in urban ghettos and rural areas. In correcting this maldistribution, they favor different policies: a period of mandatory service by young doctors, and financial inducements, respectively. A rational decision will presumably require detailed descriptive information about the extent and severity of this maldistribution, and information as well about the effectiveness and acceptability of these two policies and any others that come under serious consideration by physicians and the general public.

This paper will initially briefly consider the two major branches of methodological statistics: sample surveys as a means of providing descriptive statistics, and techniques for studying the effects of different procedures or agents. Some issues that arise in using these techniques as a guide for health policy decisions, such as preparing summaries of the state of the evidence,

[1] Department of Statistics, Harvard University, Cambridge, MA 02138.

the role of the statistician on the policy-making board, and the monitoring and evaluating of policies in operation will be discussed. Finally, the role of statistics in the work of the Bureau of Drugs of the Food and Drug Administration (FDA) will briefly be described, using a published review of this role.

## SAMPLE SURVEY TECHNIQUES

Sample survey techniques constitute one of the most important methodological advances in our field. Comprehensive descriptive information about a large population can be collected by measuring only a small fraction of the population, with major gains in cost and time. A sampling rate of only one in 1240 is used in the Current Population Survey of the Census Bureau which provides the monthly estimates of the number and percentage of unemployed. In fact, many types of descriptive information cannot be collected except by sample surveys. The techniques used today were mostly developed during the period 1925–1960.

By 1960 there had been years of experience in the regular collection of sample data on economic, business, and labor conditions on a national level by the Census Bureau. In health, the forerunner of regular national samples was the report in 1953 of the Fales Subcommittee (1) of the US National Committee on Vital and Health Statistics. This report reviewed the need for morbidity statistics as a guide in administrative planning, in the evaluation of health programs, in the provision of medical and dental services and personnel to meet the health needs of the nation, and in medical research. In light of these needs the report then proposed a national mor-

Reproduced with permission from the *American Journal of Epidemiology*, Vol. 104, No. 4, pages 370–379. Copyright © 1976, American Journal of Epidemiology.

bidity sample on a continuing basis, specifying the type of data to be collected, the frequency and geographic detail, the sampling plan and sample size. A series of special studies for obtaining supplementary data was also recommended.

A bill authorizing a National Health Survey Program was signed in 1956. This program has four parts: 1) a weekly health interview sample of households, 2) a health examination survey (a series of separate surveys), 3) health resources surveys (a series covering hospital, dental and nursing services), and 4) surveys of vital records related to births and deaths. The publication record from these surveys is impressive, though there are complaints of publication delays. The April 1972 compilation of reports available without charge to the public listed 295 publications. A later listing and topical index to the publications for the period 1962–1974 contained 399 publications, of which 69 are out of print. These publications and accompanying sets of data are the most comprehensive body of general information on the nation's health.

As background information for health policy decisions, the national sample surveys can be highly useful when the decision requires national data or data for a few broad regions. Sampling is less effective when the policy problem demands separate information for small subgroups or localities within the population. The sample sizes in these subgroups become too small. In attempting to obtain information on small subgroups from a national survey, something can be done, first by internal analysis of the survey data to study the sizes of the local variations, and then by attempts to utilize the small-sample data from nearby areas or areas judged to be similar.

A difficulty that arises with some important health and economic data is that errors of measurement become sizeable, sometimes larger than sampling errors.

The respondent may fail to answer some questions, the answers to others may be at best crude guesses sometimes distorted by self-interest, and errors occur in recording and coding data. In studying measurement errors, we may not know any way in which to make a correct measurement, but can only make repeated erroneous measurements, frequently correlated. Progress in the study of errors of measurement has therefore been slow and expensive. Often the consequences for decision-making are background data that are biased and of limited accuracy. Furthermore, attaching realistic limits of accuracy to the sample estimates may amount to little more than judgmental guesswork, since with most survey methods the standard errors computed from the sample do not take proper account of measurement errors. This difficulty is not confined to sample surveys. Census data are also subject to biases from measurement errors. A related difficulty may be in definitions. A definition in a survey may fail to count people whom the policy-makers consider relevant and assume are being counted.

The National Health Survey was planned to provide data useful to health-policy planners and others. However, any continuing general-purpose survey may in time become outdated because it is unresponsive to changing needs as perceived by users when new problems arise. Criticism is sometimes directed toward the survey statisticians, who are accused of concentrating on what they believe to be their expertise: planning for the collection of samples and making estimates from them, without either interest or competence in judging the utility of what is being collected. This complaint certainly does not apply to the sampling groups in either the Census Bureau or the National Center for Health Statistics (NCHS).

Both groups have made continued efforts to keep informed of the needs of users, through outside advisory committees, requests for information, and special surveys of users and potential users. A recent report (2) of a Technical Consulting Panel (TCP) of the US National Committee on Vital and Health Statistics reviews NCHS efforts in this direction. These efforts included an indepth questionnaire in 1968 sent to a sample of 952 significant users identified from a preceding larger sample. As topics needing the greatest additional emphasis, the indepth sample ranked collection of data on measures of health, medical economics, utilization of services, and demography in that order.

This TCP consisted of an economist, sociologist, demographer, health planner, former State health statistician, and an NCHS staff statistician. The panel was given a very broad mandate to identify health and demographic statistical problems which warrant intensive analytical studies of NCHS data. In view of the expertise represented in the TCP, however, the panel restricted themselves to methods of increasing the analytical potential of NCHS data for descriptive information about the current status of health care for the population. In their review of NCHS data, needs for relevant subject-matter information were divided into four classes: 1) those that can be met from existing NCHS data, 2) those requiring new questions or the merger of existing data files within the current survey framework, 3) those requiring new survey methods, e.g., cohort studies, and 4) those requiring new surveys of special groups. This type of classification enables the survey organization to estimate the resources, cost, and feasibility of supplying the information. The TCP also recommended that other TCP's be formed to review the utility of NCHS data for other important user needs and proposed new communication channels to existing and potential users.

In summary, the available descriptive data from surveys may not be as detailed nor as accurate as desired for policy decisions. Additional special analyses oriented towards the policy problem may be necessary in order to make the best use of the available data.

## COMPARATIVE STUDIES OF THE EFFECTS OF DIFFERENT PROCEDURES, AGENTS, OR TREATMENTS

Two general strategies can be used. The first and most widely used is often called an observational study. The investigator seeks a population where the different procedures are now or will be at work. In studying the effects of a single procedure he looks for a control group of people not subject to the procedure. The approach varies according to the resources and flexibility of the investigator. Sometimes only a single-group study can be done, by observing a group for a period before and after exposure to the procedure.

When the variables for comparison have been developed, samples are drawn and data collected from which a statistical analysis will provide comparisons of the quantitative effects of the procedures. Usually, an observational study provides the most readily available comparison —sometimes the only kind available. Its weakness, widely recognized, is its vulnerability to bias. For different procedures being compared, the people receiving them, the conditions of application, the measurements taken, and other factors suspected or unsuspected can differ systematically in ways that may lead to biased estimates of the effects. Techniques such as matching and adjustments using regression often reduce these biases substantially but cannot be guaranteed to remove them. A claim that an effect has been proved to be causal is hard to defend.

The net result is that estimates of effects from observational studies usually carry only limited conviction. They are nearly always open to dispute by those who for one reason or another disagree with the estimates, and the objections can seldom be fully refuted.

The second strategy is controlled experimentation. For this, the investigator must possess the power to choose the subjects exposed to each procedure and have sufficient control so that conditions of application and measurement of effects are strictly comparable for different procedures. Devices like blocking, double blindness, and randomization are used to try to ensure that comparisons of effects are unbiased and efficiently made, and that the experiment provides a measure of the accuracy of these comparisons. The principal techniques for planning such studies and their statistical analysis were developed during the period 1925–1955.

The technique began to be used in agriculture, spreading to other branches of biology, psychology, education, and industry. Soon after the end of World War II the simpler types of controlled experiments on human subjects appeared for applications in medicine and public health. There followed its gradual acceptance by the medical profession, including techniques for handling the ethical problems, and the development of the administrative skills for the conduct, monitoring and bookkeeping required in multi-clinic or multi-hospital studies. The results of a controlled, randomized experiment came to be regarded as basically more trustworthy than those of an observational study because of the greater control against bias.

For policy decisions about the future, however, the results of controlled studies and observational studies must both be regarded as vulnerable to bias. It is rarely feasible to conduct these studies, especially controlled experiments, on anything resembling a random sample of the subjects or under the conditions to which the results will be applied for policy decisions. The discrepancy between the sampled and the target population is the potential source of this bias. Statisticians have given little attention to this problem beyond mentioning it and perhaps presenting a judgment about the kind of populations to which the investigator thinks that his results can be safely applied.

Controlled experiments have been a valuable source of estimates of effects in some health policy decisions, as in the approval or rejection of New Drug Applications (NDA) by the Food and Drug Administration (FDA). They have seldom been used in studies of the effects of health and social programs, because of the high degree of control required and the many administrative difficulties, including the problem of using randomization. But in view of the unsatisfactory nature of the conclusions that can be drawn from observational studies, several investigators in recent years have urged the increased use of controlled, randomized experimentation in measuring the effects of social and health programs. In 1971, Light et al. (3) recommended: "Controlled field studies should be used more frequently to improve both new and continuing social, health, and welfare programs." They outline the steps that would help to implement this recommendation.

In the same year, the Social Science Research Council appointed a committee to make an intensive study of the role that controlled experimentation might play in measuring the effects of social programs. The committee members represented a wide range of experience in the methodological aspects of both strategies: observational studies and controlled experiments. This effort resulted in a book entitled *Social Experimentation*, by Riecken and Boruch (4), addressed to policy and decision makers. The book appraises the advantages, limitations and practical possibilities of controlled experimentation in evaluating policies. An appendix gives abstracts of completed randomized experiments on the effects of social programs as well as of a few experiments related to the health field. The latter estimated the effects of several programs for the treatment of mental illness, the rehabilitation of nursing-home patients, and the Taiwan experiment on the acceptability of the intra-uterine device (IUD) by families. Except for the experiment in Taiwan, they were local and on a small scale.

The future role of controlled, randomized experiments on a broad scale to compare the effects of different health policies will depend on a number of factors: learning the administrative, data-collection and analytical skills required to conduct them and obtaining understanding and acceptance by the public and legislators. A favorable situation may occur if a forthcoming decision is anticipated and money is allocated for a pilot study specifically designed to guide a policy decision. At this stage, the experimenter is more likely to be in a position to apply randomization and to take other precautions to ensure comparability.

### REVIEWING THE STATE OF EVIDENCE

An important contribution which can be made by statisticians is to collaborate with health workers and other scientists and prepare summaries of the data relevant to an imminent policy decision. For descriptive background data, a broadening of the approach suggested by the TCP which considered health care, is a natural step. For a specific health problem on which a policy decision may be needed soon, an ad-hoc study committee might be organized to summarize the evidence from all sources, both national and local surveys. This may not be easy. It may require obtaining a combined estimate from surveys which differ to some extent in their definitions, methods of measurement, and degrees of accuracy insofar as these can be judged, and which show substantial differences in their individual estimates. Statisticians must give more attention to this frustrating task if they are to assist in the best use of the data.

A corresponding job is to prepare estimates of effects relevant to health policy decisions, using the combined evidence in observational studies and controlled experiments. Two examples will be given, the first dealing with food supply.

This summary (5), by an agronomist, Crowther, and a statistician, Yates, was carried out at the beginning of World War II. It was directed at providing data for British war-time policy in making the best use of the available supplies of the fertilizers, nitrogen (N), phosphorus (P), and potash (K). Both potash and phosphorus had to be imported. After expressing surprise that no summary of available data had previously been made, they analyzed all of the published results of experiments on the main arable crops in Great Britain since 1900 plus similar experiments in other northern European countries.

Methodologically, this might be considered one of the easier summary analyses apart from the large amount of work involved. They were dealing mostly with randomized experiments. For a specific fertilizer and crop, the response curve was known to be well approximated by an exponential curve rising to an asymptote, and interactions between the effects of fertilizers were small except for those with dung, which contains N, P, and K. Thus for specific fertilizer costs and crop prices, they were able to calculate optimum amounts of N, P, and K, and to estimate the effects of failing to apply to optimum amounts. Calculations were done separately for the absence and presence of dung. From a sample survey they were able to compare these amounts with those actually used by farmers in 1937, drawing attention to major discrepancies. In the event of wartime shortages of fertilizers, they also recommended how reductions below the optima should be allocated between N, P, and K. This kind of summary analysis can be used directly by policymakers.

The study leading to the Surgeon-General's Report (6) on smoking and health is an example of a study specifically commissioned to help a policy recommendation which the Surgeon-General would have to make. The advisory committee reviewed the mass of evidence, almost all from various types of observational studies, and gave its judgments on the effects of smoking on a wide range of diseases. The committee consisted of seven medical specialists, one chemist, one epidemiologist, and one statistician. A personal observation that the medical specialists on the team were spending a good deal of their time analyzing and interpreting the data underscores the importance of the role of statistics.

In some problems that are attracting public concern, for example, various kinds of pollution, it becomes evident that more research must be done on cause and effect relationships before policy-makers can be guided to suitable actions. At this stage a different kind of review is useful: a bibliography of the data available on the health factors and on the distribution of the possible injurious pollutants throughout the population. The objective is to help research, by indicating the kinds of studies of relationships that can be initiated with existing data and the kinds that will require collecting new data before a study can begin. An example is the paper by Goldsmith (7), dealing with the effects of environmental pollution on morbidity and mortality. For each health variable and each pollutant, he rates availability of data on a five-point scale. The highest rating is given where data are well-organized and regularly published: the lowest where a method needs to be developed and validated for obtaining the measurements.

All the preceding summaries take time—usually years. The Crowther-Yates summary, although done in a situation that might require a crisis decision, was not completed until 1941. The administrator like the Surgeon-General who anticipates a decision in time to initiate a study of the relevant statistical evidence can make an important contribution to informed decision-making.

## The statistician's role in policy making

There are strong arguments for the presence on policy-making boards of statisticians experienced in collaborating with health workers. They can help to explain the meaning of the data, to draw attention to the subjects covered and not covered in the data, to avoid statistical fallacies, and to discuss the implications of the inaccuracies inevitable in all data. The statistician is not a policy-maker and has no expertise or experience in policy decisions. However, his training in the logical analysis of a problem and his freedom (if it exists) from emotional involvement with one decision or another may enable him to contribute a useful comment or question in the decision-process, if done tactfully. His goal should be to make himself useful. Further, some policy decisions are forced on very short notice by popular or political demand. Given very limited time, the statisticians can give major help in selecting the small body of data that there will be time to analyze, in making a rough and ready analysis, and in summarizing the results.

A recent international conference of national committees on vital and health statistics puts the role of the statistician in policy decisions a bit more forcibly in its report (8): "The statistician must concern himself with policy-making. He should be a member of any policy-making team; not usurping the right and responsibility of the administrator to decide (and to do his own job to his own satisfaction) but listened to and respected by the administrator as the supplier of the statistical information most relevant to the problem under consideration and as a skilled interpreter of the information, indicating the possible policy choices and the probable outcome of different strategies." I hope we can live up to these lofty goals.

In this connection, this and other reports interested in the fruitful uses of statistics, stress the responsibility of the statistician in learning how to communicate with users. A question which I will not discuss here but which is relevant is whether our current academic training in statistics helps or hinders this communication. There are forces pulling in both directions. The increasing complexity of methodological statistics and supporting mathematics as more difficult problems are attacked has meant that many doctoral students learn very little about any field in which statistics is useful. On the other hand, much of the research money which supports thesis work comes from user agencies, so that some students are working on problems presented by users. Futher, graduate students are in demand as part-time consultants by users. This provides useful training, though from the occasional feedback on the relative merits of the advice offered perhaps they could profit from more supervision.

## POLICY DECISIONS AND THEIR CONSEQUENCES

When health policy decisions on national programs are being made, I judge that in most cases the statistical information will be inadequate in some respects. In particular, forecasts of the future effects of alternative policies will often be derived from data of questionable accuracy and relevance or from little data at all. For instance, there is much current discussion and thought in preparing for a more complete National Health Insurance Plan. I doubt if anyone knows how well the adopted plan will work in giving the people access to health care, how much it will cost, and how it will be received by the medical profession and public. In such situations we must expect that few policies will be completely successful without some alteration, and expect as well that some

will prove failures. In order to obtain an idea of the success record of innovative programs in the social and health field, Gilbert et al. (9) chose 28 innovations whose effects had been evaluated by well-run controlled randomized experiments. Most of the experiments were local, though the innovations could have been applied nationally. They constructed a five-point rating scale for the degree of success or failure of the program and made judgmental ratings from the results of the experiments. They found that about one-fifth of the programs were clear and substantial successes, about one-fifth had small to moderate positive effects, while the remaining three-fifths either had no discernible effects or were mixed in their effects, a few being harmful.

If something like this performance record applies to health policy decisions, some obvious recommendations are in order: 1) keep policies flexible since we may later wish to alter some policies, 2) make plans for monitoring the operation of the policy from a quality-control viewpoint, and 3) set aside resources for an evaluation of the effects of the policy. All three recommendations involve major statistical activities. Statisticians already have some experience in monitoring quality by sample surveys, e.g., of hospital death rates for the more hazardous types of surgery, or of clinical laboratories in reporting on test specimens. But for rare breakdowns with serious consequences, statisticians have only limited expertise on how best to estimate the rates.

As regards evaluation, a strictly descriptive account of events following the policy decision should be a relatively easy task. A before-after comparison should reveal whether the intended improvements are occurring or not, but must be planned before the policy is put into effect or it may be found that the before and after measurements are not comparable. Moreover, this does not provide the evidence as to whether any changes are due to the policy. Measuring the effects of the policy in a causative sense will call for much ingenuity. In a national program, one approach is to plan a number of small evaluative studies in different localities throughout the nation. In a given locality one may wish to compare the health experience of people exposed to the policy and of a control group not exposed to it, or of a group exposed to some variant of the policy that seems to be a possible improvement. Important questions are: Is it feasible to construct such comparison groups? Can some form of randomization be used in forming the groups so that there is a series of controlled, randomized experiments on the same question in a sample of localities? Gilbert et al. (9) recommended that strong efforts be made towards gaining acceptance by administrators and the public of the need for such evaluations and for controlled experiments as the best research tool. In view of the normal institutional inflexibility, a convincing body of data will presumably be needed to induce administrators to stop an ongoing policy or to make changes to it.

In this area, another responsibility of the statisticians in collaboration with other specialists is to come forth with advice when they think that the accumulated evidence strongly suggests serious consideration of a major change in policy, and when they believe that further evaluative data are needed before any change in policy can be made.

## SOME STATISTICAL OPERATIONS IN THE BUREAU OF DRUGS (FDA)

The following notes are taken from a review by Daniel et al. (10) of the statistical work in the Bureau of Drugs (FDA). The purpose of that review was to appraise the quality of this work and make recommendations directed towards improving it. The interest here is in noting some of the statistical problems encountered in an agency which continually makes important health policy decisions and evaluations. The February 1, 1970 reorganization statement for the FDA states that among numerous other functions "the Bureau of Drugs develops standards and medical policy and conducts research on efficacy, reliability, and safety of drugs and devices for man; reviews and evaluates New Drug Applications (NDA's) and claims for investigational drugs, conducts clinical studies on safety and efficacy of drugs and devices; operates an adverse drug reaction reporting system . . . "

As pointed out by Daniel et al. (10), some of the standard statistical techniques involved in the day-to-day work of the Bureau are sampling and sample surveys, bioassay, including the slippery problem of estimation of the "no effect" dose of a toxic agent, design of experiments, quality control, statistical modeling and the resultant fitting of equations, and large-scale data processing. My subsequent comments are confined to three major functions of the Bureau, decisions on NDA's, reports on the efficacy of drugs already on the market, and the monitoring of adverse reactions.

In support of NDA's, the data on drug efficacy submitted by manufacturers often consist of numerous multi-clinic studies of widely varying quality and relevance. Thus the FDA staff have to wade through the least attractive and most time-consuming type of combined analysis of data to reach a summary verdict on efficacy. In an attempt to alleviate this problem, the FDA staff, in cooperation with industry scientists, have begun to issue guidelines on useful experimental designs for testing different types of drugs. The hope is to discourage the collection and submission of miscellaneous tests of poor quality. It has been noted that statisticians in the Bureau of Drugs have begun to participate in the evaluations of NDA's leading to acceptance or rejection.

The 1962 Kefauver-Harris Amendments

to the Food, Drug, and Cosmetic Act require that drugs placed on the market since 1938 be shown by the FDA to be efficacious and safe. This mammoth task was given to 30 review panels selected by the National Academy of Sciences, National Research Council Drug Efficacy Study Policy Advisory Committee. Not surprisingly, many members of these panels expressed concern about the generally poor quality of the evidence of efficacy of the drugs reviewed; in many cases, the available information was insufficient for a full assessment of efficacy (11). If there are drugs in widespread and common use for which information on efficacy is lacking, this situation raises further questions. Should some clinical trials be started at this late date in order to make this assessment? Since resources for this effort will presumably be limited, what criteria should be used in selecting the drugs most in need of further information by such trials?

The monitoring of relatively rare adverse reactions is a formidable task. Ideally, one wants to know not only the overall nature, severity, and rate of a reaction, but whether it is restricted to certain kinds of subjects or patients who can be identified. Neither compulsory nor voluntary reporting of adverse reactions seems to produce satisfactory data for analysis towards these ends. The reports are miscellaneous, incomplete, and lack denominators, and seldom reveal whether the reaction can definitely be regarded as a causal effect of the drug. Large sample surveys will be needed, planned with specific objectives and initiated when the drug comes on the large-scale market. The number of monitoring surveys of this kind that are in effect is not known to me.

## CONCLUDING COMMENTS

Like other reports on the role of statistics in almost any activity, suggestions have been presented in this paper on the need

for more statistical studies of various kinds. Since statistical studies are becoming more costly like everything else, another policy decision that arises is: Which studies should be financed, and which, though sounding useful and well-planned, will be refused financing because of limited resources? At present it seems that such decisions are made on a haphazard basis. Some discussion of what the guiding policy should be, would be worthwhile.

On a much broader level, a very important decision is: What policy should be used in determining which diseases should be allocated health research funds? What kinds of statistics are relevant to this decision and what role should they play?

The preparation of this paper has left me with several impressions. In the education of statisticians, more emphasis should be placed on producing a supply of statisticians interested and experienced in collaborating as members of a team composed of an administrator, physician, and epidemiologist, for producing data to guide decisions. On the technical side there are a number of areas in which statisticians are currently somewhat weak: the combined analysis of data of different structure and quality from different sources; the ability to weigh data which have varying degrees of inaccuracy as a basis in reaching decisions; and conducting randomized experiments to measure the effects of national health policies.

### REFERENCES

1. US National Health Survey: Origin and Program of the U.S. National Health Survey, 1958. Washington DC Department of Health, Education, and Welfare, PHS Publication No 548-A1
2. US National Committee on Vital and Health Statistics: The analytical potential of NCHS data for health care systems. Public Health Service, Health Resources Administration, Washington DC DHEW Publication No (HRA) 76-1454, 1975
3. Light RJ, Mosteller F, Winokur HS, Jr: Using controlled field studies to improve public policy. Federal Statistics (Report of the President's Commission) 2:367–402. Washington DC, US GPO, 1971
4. Riecken HW, Boruch RF: Social Experimenta-

tion. A method for planning and evaluating social intervention. New York, Academic Press, 1974

5. Crowther EM, Yates F: Fertilizer policy in wartime: the fertilizer requirements of arable crops. Empire Journal of Experimental Agriculture 9:77–97, 1941

6. Report of the Advisory Committee to the Surgeon General of the Public Health Service: US Department of Health, Education, and Welfare, Washington DC, US GPO, PHS Publication No 1103, 1964

7. Goldsmith JR: Statistical problems and strategies in environmental epidemiology. *In* Proceedings of the Sixth Berkeley Symposium on Mathematical Statistics and Probability. Edited by LM Lecam, J Neyman, EL Scott. Berkeley, University of California Press, 1972, pp 1–28

8. World Health Organization: New approaches in health statistics. Report of the Second International Conference of National Committees on Vital and Health Statistics. Geneva, WHO Tech Rep Ser 559, 1974, p 23

9. Gilbert JP, Light RJ, Mosteller F: Assessing social innovations: an empirical base for policy. *In:* Evaluation and Experiment: Some Critical Issues in Assessing Social Programs. Edited by C Bennett, A Lumsdaine. New York, Academic Press, 1975 pp 39–193

10. Daniel C, Tufte ER, Kadane JA: Statistics and data analysis in the Food and Drug Administration. Federal Statistics (Report of the President's Commission) 2:65–95. Washington DC, US GPO, 1971

11. Hershel J, Miettinen OS, Shapiro S, et al: Comprehensive Drug Surveillance. JAMA 213:1455–1460, 1970

# 105

1.

EARLY DEVELOPMENT
OF TECHNIQUES IN
COMPARATIVE EXPERIMENTATION

William G. Cochran

Department of Statistics
Harvard University
Cambridge, Massachusetts

The historical development of statistical ideas and techniques as regards their practical application has been relatively little studied.
It seems likely that in time we will come across centuries-old examples
of individual comparative experiments or discussions of the principles
to be followed in comparative experimentation. Thus, Stigler (1973)
has noted that in the eleventh century the Arabic doctor Avicenna laid
down seven rules for medical experimentation on human subjects, including a recommendation for replication and the use of controls and a
warning of the dangers of confounding variables. In 1627, Francis
Bacon published an account of the effects of steeping wheat seeds for
12 hours in nine different concoctions (e.g., water mixed with cow
dung, urine, three different wines), with unsteeped seed as controls,
on the speed of germination and the heartiness of growth. He noted
that these comparisons (made in single replication) were important to
profitability, as most of the steepings were cheap. The results suggested the inadvisability of wasting wine, since the claret treatment
proved inferior to the control, whereas wheat steeped in the other two
wines did not germinate at all. The winner was seed steeped in urine.

As far as I know, however, the study of the practical conduct of
comparative experiments as we know it today was begun and pursued during the eighteenth and nineteenth centuries in agricultural field
experimentation.  I shall start with the great English agronomist,
Arthur Young, who made comparative field experimentation his main
occupation for a number of years.

ARTHUR YOUNG'S CONTRIBUTIONS

In 1763, at the age of 22, Arthur Young inherited a farm in England
from his father.  He decided to devote himself to experiments designed
to discover the most profitable methods of farming.  In succeeding
years he carried out on his farm a large number of field experiments
on the principal crops in his region, publishing their results and
his conclusions in a three-volume book *A Course of Experimental
Agriculture* (1771).

What were Young's ideas on the conduct of field experiments?  In
many ways, they were surprisingly modern.  First, he stressed that
experiments must be *comparative*.  When comparing a new method with a
standard method, *both* must be included in the experiment, even if the
farmer already knows a great deal about the general performance of the
standard method in past years or on different fields.  Young's reason
was that the performance of any method on specific fields in a specific
year was sufficiently affected by soil fertility, drainage, and climate
that only a comparison on the same fields and weather could be trusted.
An example of his method of conducting field experiments is his early
comparison of sowing wheat in rows by the drill (the new husbandry)
compared with broadcasting the seed (the old husbandry).  A single
comparison or trial was conducted on large plots--an acre or a half-
acre in a field split into halves--one drilled, one broadcast.  Of the
two halves, Young (1771) writes:  "the soil exactly the same; the time
of culture, and in a word every circumstance equal in both."  Not
surprisingly, he said nothing about how he decided which half to drill,
which to broadcast.

Second, Young had the idea that because of this variability caused by soil fertility, drainage, insect pests, and other factors, the results of a single trial could not be trusted. The drill-broadcast comparison involved seven trials (replications) in seven fields, in which he used his overall impression of the seven sets of results in drawing conclusions. He went further, noting that his comparative results might change in a different year with differing weather. For this reason, he repeated his experiments on different plots in five successive years, drawing his overall conclusions from a summary table of the annual mean profit per acre for drilling and broadcasting. It was not until the present century that this amount of replication became standard practice.

To revert to Young's insistence on the comparative nature of experiments, he considered bringing in the yields for past crops by the old method (broadcasting) in his overall summary of the results, but he rejected this idea. He wrote "nothing of this sort should be done, even on the same soil, unless the experiments were absolutely comparative; it may be a matter of amusement or curiosity, but of no utility, no authority." This issue persists today. In reviewing the present state of knowledge about the relative merits of two therapies for hospitalized patients, we may find a few well-controlled experiments and a larger number of doctors' observations on their experiences with one or the other therapy. Young would seem to suggest that to consider the latter group is a waste of time.

Young stressed careful measurement. All expenses were recorded on each half in pounds, shillings, pence, hapennies, and farthings and at harvest a sample of wheat from each half was sent to market on the same day to determine the selling price.

Young had no quantitative technique for statistical analysis. But he had remarkable awareness of two problems that still plague us today in experimentation and statistical studies generally. One is the problem of bias in the investigator. Each of Young's books begins with what we would now call a review of literature, and I was surprised at how

many past field experiments (mostly in single replication) he found
to review in the 1760s.  He wrote (1771):

> The many volumes upon agriculture which I have turned over,
> guarded me against a too common delusion, and ever fatal in an
> inquiry after truth; then adopting a favorite notion, and form-
> ing experiments with an eye to confirm it.  There is scarcely a
> modern book on agriculture, but carries marks of this unhappy
> vanity in the author, which must render its authority doubtful
> to every sensible reader.  The design of perusing such works
> was to find practical and experimental directions in doubtful
> points; and my disappointment gave me a disgust at favorite
> hypotheses.

He went on to give examples in which the authors slanted both the
data they presented and their summary remarks to support their favorite.

I imagine that any statistician who does much consulting on the
planning of experiments has had the same experience as I have, of
experimenters, including competent ones, who begin the consultation
by saying "I want to do an experiement to show that ... ."  He knows
the answer.  In experiments with human subjects, special protective
devices such as "double-blindness" have had to be adopted in which,
if feasible, neither the subject nor the person measuring the response
knows which treatment is being measured.  It has been noted that this
difficulty may be present in what is now called social experimentation--
the attempt to use randomized comparative experiments in the evaluation of the effects of social programs.  Some sociologists take the
view that new social programs, undertaken with laudable aims, will
sometimes, perhaps often, be found to have few beneficial effects.
The administrator in charge of the program will not find it easy to
accept and publicize a negative finding about the value of the program
that he has been directing.

Furthermore, Young was well aware of the problem of drawing
conclusions that extend beyond the experimental results to the broader
set of conditions to which we are interested in applying the results.
He stressed the pitfalls in what we now call inferences from sample
to population.  He warned that his conclusions could not be trusted
to apply on a different farm, with different soil and farm management
practices.  On a more subtle issue, he also noted that his results

would not apply as a guide to long-term agricultural policy. The
reason was that drilling made the field easier to weed than did
broadcasting. This might then enable wheat--the principal cash crop--
to be grown more frequently on a field, instead of the older method
of leaving the field fallow in alternate years, or then growing a
coarser but less profitable crop in order to smother out the weeds.

On this question of inferences from the experimental results, a
statistician would insist that, ideally, experiments should be
conducted under a representative or random sample of the broad set of
conditions (the population, as we call it) to which the results are
to be applied. But for reasons of expense and feasibility, this is
a requirement seldom satisfied in practice even today. It can be done
fairly well in very simple experiments--as when I am comparing the
closeness of shaving myself (as the population) by an electric shaver
or a specific type of blade, or the relative time it takes me to drive
home from the office by routes A and B. But in more complex experiments, it is quite hard, in teaching, to find examples in which the
requirement was deliberately satisfied. Often, for example, the
effects of different procedures on the behavior of young adults are
compared with the behavior of the students who happen to be around the
psychology department of University Y at a specific time. As Student
(1926) put it, "in some cases it is only by courtesy that experiments
can be considered to be a random sample of any population." In clinical
trials on ill patients, where it is obviously desirable that superior
therapies be adopted in medical practice, we may not be able to do much
better than describe the relevant characteristics of the patients in
the experiment as a guide to the kind of population to which we judge
the experimental results to be applicable, with supplementary analyses
that may guide speculation about the relevance of the results to
populations of a different type.

THE NINETEENTH CENTURY

If I may digress from my main subject for a moment, this issue of
statistical inference from the results was faced again ca. 1830, when
a proposal was made to form a statistics section of the British

Association for the Advancement of Science.  The Association quite
properly appointed a committee, under the distinguished chairmanship
of Thomas Malthus, and asked it to report on the question:  Is
statistics a branch of science?  The committee readily agreed that
insofar as statistics dealt with the collection and orderly tabulation
of data--that was science.  But on the question "Is the statistical
interpretation of the results scientifically respectable?," a violent
split arose into pros and cons.  The cons won, and they won again a
few years later in 1834 when the Statistical Society of London, later
to become the Royal Statistical Society, was formed.  Their victory
was symbolized in the emblem chosen by the Society.  This was a fat,
neatly bound sheaf of healthy wheat--presumably representing the
abundant data collected, and well-tabulated.  On the binding ribbon
was the Society's motto--the Latin words, *Aliis exterendum,* which
literally means "Let *others* thrash it out."  It seems strange that
statisticians in England should have begun their organization by
timidly proclaiming to the world what they would *not* do.  As a Scot
who has lived and worked among the English, this does not sound like
the English to me.

I do not know the full reason for this attitude.  A contributing
factor may have been that most of the senior statisticians of the time
were heads of official statistical agencies, whose tasks were confined
to collection and tabulation of data.  Interpreters of data bearing on
social, economic, or political matters often disagreed violently as to
the conclusions flowing from the data, a practice that they have
continued to this day.  By 1840, the Society was already beginning
to strain against the limitation, and its meetings have always been
noted for their vigorous discussions of all aspects of statistics as
we regard it.

Returning to experimentation, an illustration of the ideas that
were current in the nineteenth century was James Johnston's book
*Experimental Agriculture* (1849), which was devoted to advice on the
practical conduct of field experiments regarded as a scientific problem.
Among points made by Johnston were the following:

1. He stressed the importance of doing experiments and doing them well. A badly made experiment was not merely time and money wasted, but led to the adoption of incorrect results into standard books, to loss of money in practice by the erroneous advice it gave, and to the neglect of further researches. (I have heard this point made recently with regard to medical experiments on seriously ill patients, where there is often a question for the doctor if it is ethical to conduct an experiment, but from the broader viewpoint a question of whether it is ethical *not* to conduct experiments.)

2. He had observed that plots near one another tended to give similar yields. Hence, he recommended that repetitions of the same treatment be scattered, with the consequence that *different* treatments were placed on *neighboring* plots. A common rule in helping to achieve this was to place repetitions of the same treatment in relation to one another by the Knight's move in chess.

3. He hinted, though without elaboration, at the value of factorial experimentation, by insisting that with two fertilizers a and b, the four treatments--none, a, b, and ab-- should all be compared.

4. He realized the vital need for a quantitative theory of variation, writing "I have elsewhere drawn attention to the importance of this question--'when are results to be considered as identical?' ... As yet we do not possess any such system of mean results, though few things would at present do more to clear up our ideas as to the precise influence of this or that substance on the growth of plants."

As Johnston did not know, considerable progress on the mathematical side toward a quantitative theory of variation had already been made by the early nineteenth century by workers such as De Moivre, Gauss, and Laplace. Results available by about 1820 were the concept of the standard deviation or standard error as a measure of the amount of variation in a population. Then there were the theory of least squares, the distribution of the means of samples from a normal distribution, and some forms of the Central Limit Theorem, with applications in particular to errors of observation in astronomy. Formulas were available for calculating what was called the *probable error* of the average of a sample of independent normal observations. This was a quantity such that the actual error in the mean was equally likely to exceed or fall short of this quantity. I have not come across specific attempts to apply these methods in drawing conclusions from field trials during the nineteenth century, but I would be surprised if such attempts do not exist.

Many will recall, for instance, Charles Darwin's discussion in 1876 of the heights attained by corn plants in his comparison of crossed and self-fertilized corn (*Zea mays*) in pot experiments, which Fisher reproduced in his book, *Design of Experiments* (1935). This experiment was similar to Young's comparison of drilling and broadcasting wheat in format, except that Darwin had 15 replications as against Young's 7. Darwin sent his data to Francis Galton for statistical advice. Galton's discussion makes it clear that he was aware of the probable error and of the result that the averages of independent samples from a normal distribution are themselves normally distributed. But, as I interpret his writing, two gaps caused him to abandon his attempts to apply these mathematical results to Darwin's data. The first--the one noted by Fisher--was that even assuming the 15 differences in height (crossed minus selfed) to be normally distributed, the calculation of the probable error required a good estimate of the standard deviation of the differences, which Galton felt 15 observations were insufficient to provide. The second problem was that 15 observations were also not nearly enough to tell him the actual law of distribution followed by the individual differences in height. As Galton put it: "the real difficulty lies in our ignorance of the precise law followed by the series."

Toward the end of the nineteenth century and into the twentieth, there were two main lines of development in the layout of field experiments. Field experimentation in agriculture has the convenient property that the size and shape of the experimental unit--the plot-- is to a large extent controllable by the experimenter. Investigations of different sizes and shapes of plots led to a realization that for a given total area of land, replicated small plots (e.g., 100 square meters, or 1/40 acre) were a good choice (e.g., Wagner, 1898). These were much smaller than the quarter- or half-acre plots used by Young.

The introduction of many new varieties of farm crops following Mendel's work produced two reasons for even smaller plots with higher numbers of replications. Because the difference in the cost of using variety A rather than variety B in practical farming was small, lying only in the cost of seed, relatively small differences in mean yield became of economic importance, so that experiments of greater discriminating power were needed. Second, with new varieties sometimes only a limited amount of seed was available for experimentation.

This period also saw numerous extensions of Johnston's recommendation that unlike treatments be placed on plots near one another. Highly ingenious systematic plans were produced, the objective of which was to obtain the maximum precision in the comparison of different treatments. Three examples of such plans are shown below.

|   | SONNE |   |   |   | KNIGHT'S MOVE |   |   |   |
|---|---|---|---|---|---|---|---|---|
| \multicolumn{4}{c}{7 treatments} | \multicolumn{5}{c}{5 treatments} |
| \multicolumn{4}{c}{4 replications} | \multicolumn{5}{c}{5 replications} |

<pre>
        SONNE                KNIGHT'S MOVE

     7 treatments            5 treatments
     4 replications          5 replications

     A  F  E  G              A  B  C  D  E

     B  G  D  F              C  D  E  A  B

     C  A  C  E              E  A  B  C  D

     D  B  B  D              B  C  D  E  A

     E  C  A  C              D  E  A  B  C

     F  D  G  B

     G  E  F  A
</pre>

HALF-DRILL STRIP

2 treatments, number of replications flexible

ABBA│ABBA│ABBA│ABBA│ABBA│....

The Sonne plan automatically eliminated systematic differences in fertility between columns. Furthermore, if the rows were numbered from 1 to 7, the mean position of each letter was the same (4), so that any *linear* trend in fertility from row to row was also automatically eliminated from comparisons among the treatment means. The Knight's move is a particular 5 × 5 Latin square with the

additional property that no treatment appears twice on a diagonal--
helpful in the event of fertility gradients parallel to the diagonals.

The half-drill strip, invented by E. S. Beavan, was very convenient
for comparing two varieties of a cereal.  In the drill, the seed boxes
on the left half were sown with variety A, on the right with B.  A
plot (one half-drill wide) extended the length of one side of the
chosen area.  Having sown plots AB, the horse turned at the end, going
back to sow BA and complete the sandwich ABBA.  Any linear trend within
a sandwich was eliminated from the mean yield, even if the trend differed from sandwich to sandwich.  Systematic plans of these types became the recommended methods for use in field experiments.

More detailed information about the methods used in experimentation
from Francis Bacon on can be found in Young's (1771) review of literature and in two reviews, one by Fussell (1935) and one by Crowther
(1937), which I found most helpful.

THE TWENTIETH CENTURY--INITIAL WORK TOWARD
QUANTITATIVE INTERPRETATION OF RESULTS

The first major step in attempting to quantify results of experimentation was Student's 1908 paper, "The probable error of a mean."
'Student' was the pen name of William D. Gosset.  On leaving Oxford,
Gosset went to work in 1899 as a Brewer in Dublin for Messrs. Guinness,
who liked their workers to use pen names if publishing papers.  It
apparently is not known how or when Student became interested in
statistics, but many of the statistical problems that he encountered
at Guinness were small-sample problems, for which the available large-
sample results were at best only a dubious approximation.  As we have
noted, the probable error $0.67\sigma$ of a normally distributed value x--
the result that Galton was trying to use--requires knowledge or a good
estimate of $\sigma$.  In his 1908 paper, Student set out to find the distribution of the amount of error $(\bar{x} - \mu)$ in the sample mean, when divided
by s, where s was the estimate of $\sigma$ from a sample of any known size.
From this distribution of $(\bar{x} - \mu)/s$, the probable error of a mean $\bar{x}$
could be calculated for any size of sample.  He was well aware of the
point that worried Galton--a small sample was insufficient to determine

the form of the distribution of **x**. But he chose the normal distribution for simplicity, giving his opinion: "it appears probable that the deviation from normality must be very severe to lead to serious error." I believe that subsequent work has justified this judgment, if we have independence and no wild outliers.

Although his mathematical analysis of the problem was incomplete, he got the right answer. The steps were as follows. He worked out the first four moments of the distribution of $s^2$, and noted that they were the same as those for a Pearson's Type III curve, which he guessed, correctly, was the distribution followed by $s^2$. He then showed that $s^2$ and $\bar{x}$ are uncorrelated, and assumed (again correctly) that they were independent. From these results, finding the distribution of $(\bar{x} - \mu)/s$ was an easy integral. He constructed a table of the cumulative function for sample sizes from 4 to 10. (In modern notation, his table in a column headed n is that for $t/\sqrt{n-1}$ with $(n-1)$ degrees of freedom.) He also checked that for $n = 10$, his t-table agreed quite well with the corresponding normal table.

This remarkable paper had two further sections. Student did what we would now call two Monte Carlo studies, using the heights and the left middle finger measurements of 3,000 criminals, arranged in 750 random samples of size 4. Both empirical t-distributions based on the 750 calculated values of t agreed well with his table by Karl Pearson's $\chi^2$ test of goodness of fit. A further section gave three illustrations of the use of the table in practice, one to a paired experiment like Young's, but which compared the hours of sleep produced by two soporific drugs, one to pot experiments on wheat, and one to a field experiment on barley.

The t-distribution did not spread like wildfire. In his foreword to Student's *Collected Papers*, (1942), McMullen wrote "For a long time after its discovery and publication the use of this test hardly spread outside Guinness's brewery". Even in September 1922, 14 years later, we find Student writing to Fisher: "I am sending you a copy of Student's Tables as you are the only man that's ever likely to use them!" Young research workers who feel that the world is very slow to appreciate

their results might be heartened by this example. The world is
indeed a little slow at times to realize how brilliant we are.

Soon after the publication of Student's 1908 paper, two papers
at last appeared the objective of which was to explain and use
probability results in planning and analyzing agricultural experiments.
The first was the result of the collaboration by Wood, an agronomist,
and Stratton, an astronomer, and was entitled "The Interpretation of
Experimental Results" (Wood and Stratton, 1910). This paper first
illustrated the kinds of frequency distributions of the measurements
found on small plots--normal, skew, and even trimodal. Then, after
remarking that "no two branches of study could be more widely separated
than Agriculture and Astronomy," they noted that both branches suffered
from variability caused by the weather. They explained the astronomer's
method of estimating the accuracy of his averages by the use of the
probable error. Its calculation and meaning were illustrated by data
on percent dry matter in mangels and further examples. It was used
to perform something like both a one-tailed and a two-tailed test of
significance of the difference between two treatment means. Estimates
were made of the numbers of replications needed both in feeding experiments on animals and yield experiments on crops in order that an
observed difference of a given size between two treatment means would
be significant at the 1 in 30 level. (Thus, although they did not find
the number of replications necessary to control the power of the tested
difference at a given level--this had to wait for the work of Neyman,
et al. (1935)--they were getting close.) For these calculations they
needed, of course, estimates of the probable error for a single animal
or plot, obtained from a survey of replicated experiments.
The second paper on the conduct of field experiments, "The
experimental error of field trials", was written by two agronomists,
Mercer and Hall (1911). They investigated the effects of size and
shape of plot, type of experimental plan, and number of replications
on the standard deviation of the mean, using data for wheat and mangels
all treated alike. They recommended five replications of 40th acre
plots as giving 2% for the standard deviation of a mean. As with the
writers in the nineteenth century, a systematic distribution of the

treatments was recommended, with replications of the same treatments scattered and unlike treatments placed near one another. An appendix by Student proposed an efficient systematic plan, with sandwiches ABBA in two directions, for comparing two treatments. Neither paper used Student's t-table, although Student gave two references to it.

Both papers used the results of what were called *uniformity trials,* in which a large number of small plots, all treated similarly, were harvested. These results provided data from which the standard errors obtained from different sizes and shapes of plots, numbers of replications, and different experimental plans could be estimated. Agriculture is, I think, unique in the amount of effort devoted to the study of the variability with which field experiments had to cope. A catalog prepared by Cochran (1937) contained references to 191 uniformity trials on field crops and 31 trials on tree crop experiments.

To summarize from this incomplete review, it took roughly a full century to accomplish two major steps. These steps were (1) to begin applying the probability theory already available in astronomy so as to provide objective quantitative methods for the interpretation of the results from experiments on variable material, and (2) to establish detailed methods for the efficient practical conduct of field experiments. Further major changes were soon to come, however.

ENTER FISHER

Fisher joined the Rothamsted Experimental Station in 1919. As I have noted elsewhere (Cochran, 1973), his first solution in 1922 to a problem in designing an experiment dealt with an experiment in which the measurement was a binomial variable that was nonlinear in the quantity to be estimated. This quantity was the density of tiny organisms in a liquid, where the test could detect whether a sample contained no organisms or at least one organism, but could not count the number of organisms. Nonlinear problems, which occur also in quantal bioassay, have come to be extensively studied in more complex situations from 1960 onward, but this work stands apart from Fisher's writings on field experimentation.

His initial steps in the development of the analysis of variance,
as applied to a three-way classification, appeared in a paper by
Fisher and Mackenzie (1923), which dealt with the analysis of a
factorial experiment on potatoes. In the same year, in his paper
"On testing varieties of cereals," Student (1923) reproduced the
algebraic instructions for a two-way analysis of variance, which he
had received from Fisher in correspondence. Student applied them to
an experiment on barley with eight varieties in 20 replications. Both
experiments had been laid out systematically. Furthermore, the potato
experiment was actually of the split-plot type, for which Fisher's
analysis was not fully correct. Evidently, he was feeling his way
at that time.

Student's paper had two minor points of interest. He knew that
with only *two* varieties in 20 replications, he could find the variance
of their differences, from the 20 individual differences within
replications. How was he to extend this to *eight* varieties? With
eight varieties, 28 different pairs could be formed, and he realized
that a natural, but unappealing, method was to repeat the preceding
calculation 28 times, finding the average of the 28 variances of
differences. Fisher's analysis of variance table gave the identical
result much more quickly.

Second, Student introduced the term "variance"--the square of
the standard deviation--as useful in studying the contributions of
different sources to the variability of an observation. In a later
footnote he pointed out that he was not the author of this term, as
Fisher had used it since 1918.

Fisher's earliest writings on general strategy in field experiments
were contained in the first edition of his book *Statistical Methods for
Research Workers* (1925) and in a paper of 10-1/2 small pages (1926),
entitled "The arrangement of field experiments." This short paper
presented nearly all his principal ideas on the planning of experiments.

Fisher began with an explanation of the concept of a test of
significance. Manure was applied to an acre of land, whereas a
neighboring acre was left unmanured but was otherwise handled
similarly. The yield on the manured plot was 10% higher. What

**105.16**

reason was there to think that the 10% increase was the result of the manure and not soil heterogeneity? Fisher noted that if the experimenter could say that in 20 years of past experience with the *same* treatment on the two acres the difference had never before reached 10%:

> ... the evidence would have reached a point which may be called the verge of significance; for it is convenient to draw the line at about the level at which we can say 'Either there is something in the treatment or a coincidence has occurred such as does not occur more than once in twenty trials.' This level, which we may call the 5 per cent point, would be indicated, though very roughly, by the greatest chance deviation observed in twenty successive trials.

(Fisher noted that it would take about 500 years of previous experience to determine the 5% significance level reasonably accurately by what we call the Monte Carlo method.) I have given this argument in some detail, because students sometimes ask "How did the 5% significance level or Type I error come to be used as a standard?" I am not sure, but this is the first comment known to me on the choice of 5%. Fisher went on:

> If one in twenty does not seem high enough odds, we may, if we prefer it, draw the line at one in fifty (the 2 per cent point) or one in a hundred (the 1 per cent point). Personally, the writer prefers to set a low standard of significance at the 5 per cent point, and ignore entirely all results which fail to reach this level.

Fisher sounds fairly casual about the choice of 5% for the significance level, as the words 'convenient' and 'prefers' have indicated.

He then remarked that if the experimenter had the actual yields for, say, the 10 past years under uniform treatment, and if he could trust the theory of errors, he could calculate the 5% significance level of the difference by using an estimated standard error and Student's t-table, which Fisher had included in the first edition of *Statistical Methods* a year earlier.

Fisher next pointed out that methods had been devised for obtaining from the results of the experiment itself an estimate of the standard error of the difference between two treatment means-- thereby providing the technique that Johnston was calling for nearly

80 years earlier. Since this estimate had to come from the differences between plots treated alike, the solution had to lie in
replication. However, the recommended systematic plans in current
use, which scattered the replicates and juxtaposed unlike treatments,
deliberately violated a basic assumption made in the mathematical
theory behind the probable error. If the experimenter's judgment
was correct, differences between the scattered replicates would
consistently overestimate the real errors of the differences between
the means of unlike treatments. If his judgment was wrong, the
differences between replicates would underestimate the relevant errors.
The only way to be certain of a valid estimate of error was to ensure
that the relative positions of plots treated alike did not differ in
any systematic or relevant way from those of plots treated differently.

There was one easy solution. Having decided on the numbers of
treatments and replicates, the experimenter could assign treatments
to plots entirely at random, e.g., by drawing numbered balls from a
well-mixed bag. Fisher seemed to regard the truth of this claim as
obvious, giving no mathematical discussion.

Fisher realized that to an experienced experimenter, replacement
of a carefully chosen systematic design by a randomized design might
seem to involve  an undesirable loss of precision as the price of this
valid estimate of error. Not so, he said. By the randomized blocks
design, with compact blocks of neighboring plots, the goal of having
unlike treatments near one another could be combined with independent
randomization *within each block*. He stressed, however, that it was
now necessary to change the method of calculating the standard error.
This plan automatically eliminated consistent differences between
block fertility levels from the actual errors of the differences between treatment means; therefore, they must also be eliminated from
the estimate of error by the analysis of variance method which had
been developed by Fisher. The Latin square plan went a step further,
with each treatment occurring once in each row and each column of
the square. Thus, fertility gradients along both the length and the
breadth of a field could be eliminated from the actual and the

estimated errors. He counted and classified the Latin squares up to
the 6 × 6, offering to send the user squares drawn at random from all
possible squares of the desired size.

The final section of this 1926 paper, entitled "Complex
experimentation" contained Fisher's advocacy of what we now call
factorial design, in which he indulged in a brief general philosophic
statement.

> No aphorism is more frequently repeated in connection with field
> trials, than that we must ask Nature few questions, or, ideally,
> one question, at a time. The writer is convinced that this view
> is wholly mistaken. Nature, he suggests, will best respond to a
> logical and carefully thought out questionnaire; indeed, if we
> ask her a single question, she will often refuse to answer until
> some other topic has been discussed.

He then gave the field plan of a 3 × 2 × 2 factorial experiment
on oats in 8 randomized blocks of 12 plots, the treatments being 3
amounts of nitrogen fertilizer, either in the form of ammonium sulfate
or ammonium chloride, applied early or late in the season, in all
12 combinations.

He stressed the following advantages of this approach.

1. The first was its efficiency. Every plot yield or observation provided some information about the effects of each of
   the three factors, whereas in a single-factor experiment a
   plot yield provided information only on the single factor
   under investigation. The 96 plots in this experiment supplied
   32 replications on the average differences between sulfate
   and chloride, between early and late application, and between
   different amounts of nitrogen. To obtain the same number of
   replications, single-factor experiments would have required
   224 plots (more than twice as many).

2. Many questions about the interrelationships between the effects
   of the factors could be investigated; for instance, was the
   effect of nitrogen the same in early and late application?

3. The advantage that Fisher regarded as the most important was
   that the average effect of any factor was averaged over a
   number of variations in the conditions as regards the other
   factors, giving these results, as he put it, a very much
   wider inductive basis than given by single-question methods.

4. The final paragraph of this paper noted that it would sometimes be advantageous to sacrifice information deliberately
   on certain interrelations (interactions) believed to be

unimportant, confounding them with block differences in order
to reduce the size of block and thus increase the precision
of the more important comparisons. This subject, confounding,
came to be extensively worked out later.

These four techniques--blocking, randomization, factorial design,
and the analysis of variance--explained more fully in *The Design of
Experiments* (Fisher, 1935), came to be cornerstones in comparative
experimentation on variable material.

After 1930, work on the relevant mathematical properties of
randomization sets began to appear. Papers by Eden and Yates (1931),
Tedin (1931), Bartlett (1935), and Welch (1937) investigated numerically or algebraically how well the discrete distribution of F over
the randomization set approximated the tabular distribution for
randomized blocks and Latin squares under the null hypothesis.

Although generalization from this limited number of studies is
risky, these studies indicate that the 5% level in the F table corresponds to a randomization test level somewhere between 3% and 7%.
Under a less restrictive model in which treatments affect individual
plots differentially, Neyman et al. (1935) showed that the error in
randomized blocks remain unbiased over the set for the F test of the
null hypothesis of no average treatment effects, but the error is
subject to some bias in the Latin square, which leans very heavily
on strict additivity.

Consider any standard design (e.g., randomized blocks) with
fixed mean, block and treatment effects $\mu$, $\beta_j$, $\tau_k$, and with
given experimental errors $e_{ij}$ of any structure on each plot. Under
an additive model, the observed yield $y_{ij}$ on the ith plot in the jth
block will be

$$y_{ij} = \mu + \tau + \beta_j + e_{ij}$$

where $\tau$ is the treatment effect that the randomization happens to
assign to that plot. Calculate any comparison $\Sigma L_i \bar{y}_i$ among the
treatment means for each possible outcome of the randomization. Its
variance over the randomization set can be proved to be identically
equal to the average over the set of its estimated variances, found

by using the error mean square and the rules in the analysis of variance. It is in this sense that randomization guarantees a valid estimate of error, irrespective of the nature of the experimental errors on the individual plots. The assumption of strict additivity is, of course, crucial here. For example, with several levels of nitrogen, it is often found that the variance of $N_{\ell in}$ from block to block exceeds that of $N_{quad}$, so that errors must be calculated separately for the two comparisons. This is a counterexample with a relation between $\tau$ and e. But randomization still makes the separate errors valid.

The proponents of systematic designs, including Student, were by no means immediately converted to randomization. Debating papers on both sides of the argument appeared at intervals for about the next 12 years. The protagonists occasionally indulged in derogatory remarks about one another's scientific acumen, a practice that I have heard Fisher describe privately as "just some Billingsgate"-- the name of the London fish market noted for the salty language of its vendors. Claims made by those who favored systematic plans were

1. It was better, scientifically, to have smaller *real* errors when comparing different treatment means, even at the expense of some overestimation in the *estimated* standard errors attached to these means. (An argument on this point can be quite stimulating.)

2. Some specific outcomes of a randomization looked obviously unwise or risky, e.g., getting AB, AB, AB, AB, AB, AB, in a paired experiment laid out in six blocks in a single strip. Sometimes, as in this case, such undesirable plans could be avoided, still meeting Fisher's conditions, by introducing additional blocking into the plan as in a crossover design. However, in more complex cases, attempts to develop general methods for finding randomization sets free from undesirable elements have had only limited success.

3. By the adoption of a more realistic mathematical model applicable to a systematic plan, a method of analysis that provided a sufficiently valid estimate of the standard error might be found for systematic designs. I gather that Neyman (1935) looked into this type of approach more generally in a paper that was published in 1929, which I have not seen. In competent statistical hands this approach might have worked quite satisfactorily. It is sometimes used as a rescue operation when an unfortunate outcome of a randomization is

unnoticed until the responses were measured, e.g., Outhwaite
and Rutherford (1955). As a simple example, Student (1938)
suggested that with Beavan's sandwich design, the sandwich
be regarded as the unit, the error being calculated from
the variation in $(\bar{A} - \bar{B})$ between sandwiches. If this did
not provide enough degrees of freedom, he suggested using
the variation between half-sandwiches, with one degree of
freedom subtracted for the average of the linear effects
which this plan removed.

The debate on randomization had mostly died out by 1938. As
comparative experimentation spread rapidly beyond agriculture, one
circumstance favored the adoption of Fisherian randomized plans. In
agriculture, systematic plans had emerged from over 60 years of
extensive study of the nature of variations in soil fertility. In
other areas of research, investigators often had no comparable knowledge or data about the nature of the variations they faced beyond a
few trends easily handled by blocking. They were less likely to feel:
"I can do better than randomized blocks."

I detect a hint of this attitude in Student's (1931a) discussion
of the Lanarkshire milk experiment that had been carried out in 1930.
This experiment was one of the earlier ones in which something like
randomization was used. Moreover, it was in the difficult area now
called social experimentation. The experiment involved 20,000 school-
children, of whom 5,000 received 3/4 pint daily of raw milk, 5,000
received 3/4 pint daily of pasteurized milk, and 10,000 received no
milk supplement in school. In a given school, only two treatments
were compared, either None vs Raw or None vs Pasteurized. The
assignment of children to treatment in a class was either by ballot
(which I take it implies randomization) or alphabetically.

Student approved this invocation of the goddess of chance.
Unfortunately, the plan that was used might be called improved
randomization. If the original allocation gave an undue proportion
of well- or ill-nourished children to one treatment, the teachers
could make substitutions. It looked as if what enough teachers did
instead was to give the milk to those children who needed it most.
At the start of the experiment the "no milk" children averaged 3
months growth in weight and 4 months growth in height superior to

the "milk" children. Student noted that even with statistical adjustments, these biases made conclusions about the effects of milk doubtful.

Comparing the effects of pasteurized and raw milk, Fisher and S. Bartlett (1931) concluded from these data that pasteurized milk was inferior to raw milk with regard to the increases in both weight and height. Student's view was that since the assignment of raw or pasteurized milk to schools was not random, he would be very chary of drawing any conclusion on this question from these data. It would have been tempting to ask Fisher and Student: "Whose side are you on?"

Student's recommended plan was that if a large-scale experiment was wanted, randomized blocks of two treatments (e.g., raw and pasteurized), to be used in a class, children in the same block being of the same age and sex and of similar height, weight, and physical condition. He also suggested that randomized blocks of 50 pairs of identical twins would probably give more reliable information as well as be much less expensive. In another paper (1931b) in the same year, he recommended fully systematic plans for yield trials in agriculture as superior to Fisher's randomized blocks and Latin squares. He seemed to be saying: "In an unfamiliar area, better stick to randomization."

ACKNOWLEDGMENT

This work was supported by Contract No. N00014-67A-0298-0017 with the Office of Naval Research, Navy Department.

REFERENCES

Bacon, F. (1627), *Sylva Sylvarum, or a Naturall History*, pp. 109-110. William Rawley, London.

Bartlett, M. S. (1935), The effect of non-normality on the t distribution. *Proc. Cambridge Phil. Soc. 31,* 223-231.

Cochran, W. G. (1937), Catalogue cf uniformity trial data. *J. Roy. Statist. Soc. B 4,* 233-253.

Cochran, W. G. (1973), Experiments for non-linear functions. *J. Am. Statist. Assoc. 68,* 771-781.

Crowther, E. M. (1936), The technique of modern field experiments.
    *J. Roy. Ag. Soc. Engl. 97,* 1-28.

Eden, T., and Yates, F. (1931), On the validity of Fisher's z test
    when applied to an actual example of non-normal data. *J. Ag.
    Sci. 23,* 6-17.

Fisher, R. A. (1925), *Statistical Methods for Research Workers,*
    Oliver and Boyd, Edinburg.

Fisher, R. A. (1926), The arrangement of field experiments. *J.
    Ministry Ag. Sept. 33,* 503-513.

Fisher, R. A. (1935), *The Design of Experiments.* Oliver and Boyd,
    Edinburgh.

Fisher, R. A., and Mackenzie, W. A. (1923), The manurial response of
    different potato varieties. *J. Ag. Sci. 13,* 311-320.

Fisher, R. A., and Bartlett, S. (1931), Pasteurized and raw milk.
    *Nature 127,* 591-592.

Fussell, G. E. (1935), The technique of early field experiments.
    *J. Roy. Ag. Soc. Engl. 96,* 78-88.

Johnston, J. F. W. (1849), *Experimental Agriculture, Being the Results
    of Past and Suggestions for Future Experiments in Scientific and
    Practical Agriculture.* W. Blackwood and Sons, Edinburgh.

McMullen, L. (1942), Preface to *Student's Collected Papers.* Cambridge
    Univ. Press, Cambridge.

Mercer, W. B., and Hall, A. D. (1911), The experimental error of
    field trials. *J. Ag. Sci. 4,* 109-132.

Neyman, J., Iwaszkiewicz, K., and Kolodziejczyk, S. (1935), Statistical
    problems in agricultural experimentation. *J. Roy. Statist. Soc.
    Suppl. 2,* 108-180.

Outhwaite, A. D., and Rutherford, A. (1955), Covariance analysis as
    an alternative to stratification in the control of gradients.
    *Biometrics 11,* 431-440.

Stigler, S. M. (1973), Gergonne's 1815 paper on the design and analysis
    of polynomial regression experiments, Tech. Rept. 344, Univ. of
    Wisconsin, Madison.

Student (1908), The probable error of a mean. *Biometrika 6,* 1-24.

Student (1923), On testing varieties of cereals. *Biometrika 15,*
    271-294.

Student (1931a), The Lanarkshire milk experiment. *Biometrika 23,*
    398-407.

Student (1931b), Yield trials. *Balliere's Encyclopedia of Scientific
    Agriculture,* 1342-1360. Balliere and Co., London.

Student (1938), Comparison between balanced and random arrangements
    of field plots. *Biometrika 29,* 363-379.

Tedin, O. (1931), The influence of systematic plot arrangements upon the estimate of error in field experiments. *J. Ag. Sci. 21,* 191-208.

Wagner, P. (1898), *Suggestions for Conducting Reliable Practical Experiments* [English transl.]. Chemical Works, London.

Welch, B. L. (1937), On the z-test in randomized blocks and Latin squares. *Biometrika 29,* 21-52.

Wood, T. B., and Stratton, F. J. M. (1910), The interpretation of experimental results. *J. Ag. Sci. 3,* 417-440.

Young, A. (1771), *A Course of Experimental Agriculture.* Exshaw et al., Dublin.

# 11 106

# Experiments in surgical treatment of duodenal ulcer

William G. Cochran
Persi Diaconis
Allan P. Donner
David C. Hoaglin
Nicholas E. O'Connor
Osler L. Peterson
Victor M. Rosenoer

## 1. INTRODUCTION

This chapter presents a critical survey of randomized clinical trials comparing vagotomy and pyloroplasty with vagotomy and antrectomy as elective treatments for duodenal ulcer.

The second section describes the study group and the iterative process which led to this report. Section 3 sketches the health impact of peptic ulcer disease. Section 4 summarizes the natural history of the disease and briefly describes the two operative treatments.

The study group drafted lists of medical and statistical criteria to be used in comparing the selected studies. These criteria appear in Sections 5 and 6. For the four published studies a detailed consideration (in Section 7) includes tabulations of their critical features, a discussion of the overall findings, and a discussion of their major medical and statistical problems.

Section 8 contains a detailed discussion of principal problem areas common to all studies. The statistical problems are sample size, the need in these investigations for nonoperated controls, and problems of loss to follow-up and of ran-

•Supported in part by grants from the Robert Wood Johnson Foundation through the Harvard Faculty Seminar on Human Experimentation in Health and Medicine during 1973–74.
•Supported in part by Division of Research, Lahey Clinic Foundation, Grant# 0683.
•Dr. Peterson was a fellow at the Center for Advanced Study in the Behavioral Sciences (1974–75) under a program of senior medical fellowships supported by the Kaiser Family Foundation when this report was being written and revised.

domization. Medical problems include tests for completeness of vagotomy (the Hollander test) and evaluation of outcome and side effects according to the Visick scale.

Section 9 contains a summary of the group's findings and general recommendations.

## 2. BACKGROUND TO THE PRESENT STUDY

The Working Group on Protocol Issues prepared this report as part of the Faculty Seminar on Human Experimentation in Health and Medicine held at Harvard University during the 1973-74 academic year and continued as the Faculty Seminar on Analysis of Health and Medical Practices during 1974-75. Focusing on surgical treatment of duodenal ulcer gave the group many concrete examples of problems in preparation, implementation, and assessment of protocols for clinical trials; but attention was not confined to this one area, and many of the findings reported here should be useful in other areas as well.

The decision to study surgical treatment of duodenal ulcer disease was taken at the urging of a physician member of the group, who reported sharp disagreement among surgeons about the relative merits of the several operations used. Surgical treatment at present includes three or four operations which could be reliably characterized by risk. The most common symptom, pain, is subjective and variable, thus posing measurement problems in defining disease severity. The smaller group of patients who suffer severe hemorrhage, perforation, or gastric outflow obstruction provides more objective evidence of disease severity. Following patients over several years to determine early, intermediate, and late results raises interesting practical problems. These complex problems made the subject attractive for critical review.

A major goal of the review was to find possible solutions for some of these problems. Such a review, it was foreseen, might raise a number of policy issues. Was there as much uncertainty about the effectiveness of the several operations as the reported discussion by surgeons suggested? Could an independent review contribute to the resolution of any problems that might be found? In part the group was also testing the utility of a rigorous review using defined standards as a scientific tool in examining policy questions.

To assess the relative merits of operations used in elective surgical treatment of peptic ulcer disease, one turns first to randomized comparative studies because they are substantially more reliable in eliminating biases than are nonrandomized studies. While this position is not universally accepted, the evidence is decidedly in its favor. (Section 8.7 discusses some aspects of randomization.) To narrow and simplify the study, the group limited it to elective gastric surgery, thus eliminating operations for bleeding, perforation, and obstruction. We further decided to concentrate on two operations, vagotomy with antrectomy and vagotomy with drainage. The small number of randomized studies involving comparison of these two widely used operations made it possible to examine each study in considerable detail.

The group drafted sets of medical and statistical criteria to apply to such studies. We had also examined a number of clinical trials in which partial gastrectomy was compared with drainage operations. The clear preference in recent years for vagotomy in combination with less extensive gastric surgery led to the later restriction. The criteria used in the final study were developed during the early exploratory period. A search (based on the Index Medicus) of the English-language literature published since 1948 produced a list of about 75 trials in the problem area. Closer examination (by teams pairing a statistician and an M.D.) eliminated those which did not report randomized comparisons of the two surgical procedures and left a total of four studies: Price et al. (1970), supplemented by Johnson et al. (1970) and Postlethwait (1973), Jordan and Condon (1970), Sawyers and Scott (1971), and Howard et al. (1973). The careful work underlying these four extensive studies proved very valuable as a basis for our work. Without the substantial progress which their authors had already made, we would have been able to learn far less about comparative studies in this problem area.

The full group then read, discussed, and evaluated the published reports of the four studies, applying the medical and statistical criteria. This allowed elaboration of the various criteria as particular features of individual studies and reports were considered. The results of these deliberations are given in the remainder of the present report.

## 3. IMPACT ON HEALTH

Peptic ulcer disease* is a common and recurring cause of morbidity. Its prevalence rate is 22.0 per thousand among males and 12.6 per thousand among females (Wilson, 1974). The highest prevalence rate is found in the age group 45 to 64, where the rates per thousand population are 45 for males and 22.8 for females. Prevalence varies with income, from 22 per thousand among persons with incomes under $5,000 per annum to 14.6 per thousand among persons with incomes above $10,000 (in 1968).

Though peptic ulcer is not one of the most common causes of disability, it is nevertheless significant (Namey and Wilson, 1973). Among the approximately 22 million persons with some activity limitation due to health, approximately 550,000 (or 2.5%) suffer some limitation on their normal activities because of peptic ulcer disease.

A 1969-70 study of hospitalizations in New England (exclusive of Connecticut) found that the average charge per hospital stay was $772 (Jones et al., 1976). Peptic ulcer patients, whose admission rate was 1.9 per thousand population (1.2% of all admissions), incurred substantially higher bills per stay ($1,153 on the average). A calculation based on these data indicates that more than $16 million in charges for hospitalization are incurred by peptic ulcer patients annually, representing 1.85% of the estimated charges for all hospitalization, or nearly $2 per New England resident.

Finally, peptic ulcer, though not among the most common causes of death, is nevertheless a significant one (Namey and Wilson, 1973). In 1968 there were

---

*The term "peptic ulcer" includes esophageal, gastric, duodenal, and anastomotic ulcers. Among these, duodenal ulcers are by far the most common.

about 10,000 deaths in the United States from peptic ulcer disease. This represents about 0.53% of all deaths.

## 4. NATURAL HISTORY AND SURGICAL THERAPY

Most studies and descriptions of duodenal ulcer disease are based upon patients treated in medical centers for particularly complex and intractable problems. It is often difficult, therefore, to use such studies as a basis for describing the natural history of duodenal ulcer disease for the population as a whole. From the more careful clinical studies we judge that the average healing time of a duodenal ulcer is about 6 weeks. A high percentage of patients will have recurrences. However, a large percentage will recover and be completely free of recurrences within five years (45-90%). The probability that patients will be cured and become symptom-free apparently diminishes with disease duration. A substantial number of ulcer patients will require hospitalization (40%) though this is likely to be influenced by health insurance coverage as well as by the disease. A minority of patients with duodenal ulcers will have bleeding complications with their disease (15-20%). A smaller proportion (7-10%) will experience a perforation of their ulcer.

The surgical therapy of duodenal ulcer disease is directed at reducing the amount of acid secreted by the stomach. This plan makes the basic assumption that an increase in acid secretion by the stomach is associated with an increased incidence of duodenal ulcer disease. The vagus nerves, which run alongside the esophagus and innervate the stomach, control the cephalic phase of gastric acid secretion and provide a stimulus to motility. Cutting these nerves thus decreases secretion of acid by the parietal cells in the upper half of the stomach, but it also decreases motility so that emptying of the stomach is impaired. By widening the outlet of the stomach, a pyloroplasty provides adequate emptying or drainage. An antrectomy achieves better drainage by removing the lower portion of the stomach. Since that portion of the stomach contains the cells which secrete the hormone gastrin, a powerful stimulator of acid secretion by the parietal cells, the combination of vagotomy and antrectomy reduces acid secretion in two ways.

## 5. MEDICAL CRITERIA

Initially, the group had an extensive list of medical criteria for evaluating surgical studies. However, the discovery in reviewing the papers that many of the relevant details were not included led to a narrowing of the list.

The basic preoperative information is the patient's age and sex and the history and presence of ulcer and other diseases which may influence risk or outcome. Details of the patient's history should include the length of time the patient has had ulcer symptoms and the duration and severity of those symptoms (if possible, in terms of the amount of work lost, frequency and duration of dieting, medication, adherence to treatment, and number and duration of hospitalizations required).

In establishing the diagnosis of an ulcer, one or more of the following criteria would be acceptable:
1. Ulcer seen on an upper gastrointestinal barium x-ray.
2. Ulcer seen by endoscopy.
3. Confirmation of the ulcer either at surgery or by the pathologist.

Postoperatively, the study should yield the following information:
1. Operative mortality, which includes death up to 30 days following the operation.
2. Length of stay in the hospital following operation.
3. Length of time it takes to return to the patient's usual activity.
4. Length of time the patient was followed-up.
5. Incidence of ulcer symptom recurrence.
6. Recurrence of a proven ulcer (proven by one of the above criteria).
7. Incidence and severity of dumping following the operation.
8. Incidence and severity of diarrhea (consisting of three or more bowel movements per day).
9. Amount of weight loss following surgery.
10. Permanent change in meal frequency and amounts.

## 6. STATISTICAL CRITERIA

The statistical criteria used in this study are as follows:

(1) The specific aims of the study should be clearly stated, and the total sample size should be consistent with these aims.

(2) The patients selected for the trial should be representative of a larger population of patients at risk of having the operation(s) under study. The requirements for eligibility should be listed. If certain kinds of patients were excluded from the trial, this should be stated, along with the reason for exclusion and the number of patients excluded. On the other hand, if there is a bias toward *including* certain kinds of patients, this should also be stated.

(3) Factors affecting the randomization procedure should be specified in detail. If patients were withdrawn or not given the randomly selected operation or omitted from the study after randomization, there should be sufficient information presented to judge the effect this may have had on the comparison of the treatments. Ideally, there should be a control group consisting of patients with characteristics as similar as possible to those in the treatment groups. For a disease as poorly understood and as unpredictable as duodenal ulcer, this is particularly important.

(4) The treatment (surgical) procedures should be described, together with the range of modifications permitted.

(5) The measurements recorded for the patients must be clearly defined and should be objective whenever possible. The processes by which the measurements are taken should be described in detail and should be consistent from patient to patient for all treatments. Follow-up should be "blind" whenever possible.

(6) Frequency distributions of important clinical characteristics such as age, sex, disease severity, complications, and other diseases in the study population

should be given. The treatment groups should be compared with regard to the most important of these so that comparability or its lack can be assessed.

(7) Frequency distributions of modifications in the surgical procedure for each treatment group should be given, as well as distributions for surgical and pathological findings, if any.

(8) Follow-up procedure should be carefully described. If there is substantial loss to follow-up, the authors should indicate how they dealt with the resulting possibility of bias. All missing cases must be accounted for.

(9) The authors should indicate clearly which statistical procedures were used in addition to reporting a *P* value. Wherever they may be of value in judging the size of differences between treatment effects, estimates of variability (such as standard errors) should be given.

(10) The conclusions from the study should be clearly stated. It should be possible to assess the seriousness of any deficiencies in design or execution of the study from the material presented.

With a few modifications these criteria can readily be applied to randomized clinical trials on other problems.

We discuss several general points here and return in Section 8 to sample size, randomization, follow-up, and the possibility of a control group, giving particular attention to how these arise in studying surgical treatment of duodenal ulcer.

From experience we appreciate the difficulties which may arise in the sampling phase of an ideal study (Cochran, Mosteller, and Tukey, 1954). It is seldom feasible to proceed by first defining the target population to which we wish the results of the study to apply and then selecting for the experiment a study group which is a random sample from this population. It remains true, however, that the most useful studies are those in which the study group is evidently representative of a population which presents itself to others who can use the results. If several target populations may be distinguished, the report should identify them and indicate how broadly the results may apply. Overall, clear statements of the criteria for inclusion in the study and information about relevant characteristics of the patients are valuable in documenting the relationship between study group and target population.

Naturally enough, not all measurements can be objective. A common example in the treatment of duodenal ulcer is "patient satisfaction." When asked about satisfaction with the outcome, a patient may feel that some sort of positive answer is expected. Such a subtle bias could be very difficult to assess.

As a general rule, the papers we have examined could well give substantially more detail on characteristics of the study populations. Simple frequency distributions for single characteristics are not commonplace, and such counts for important *combinations* of characteristics seldom appear.

## 7. DISCUSSION OF PAPERS STUDIED

In examining and comparing the four surgical studies, the group applied the medical and statistical criteria presented in Sections 5 and 6. Since these criteria were sharpened during preliminary comparisons of various studies, they cannot be expected to show any one paper in the most favorable light. Thus the discussions of the individual studies often point out possibilities for improvement, and space limitations do not permit us to emphasize their numerous strengths.

To aid the reader in comparing the four papers studied, we use tabular form to present information on attributes, follow-up, and outcomes of each one. Since some entries in the three tables are abbreviated, we explain them before proceeding to the discussion of each paper. ("N.R.," which appears in all three parts, indicates that the information was not reported or was inadequate.)

### Attributes of Study (Table 11-1)

Most of this information is self-explanatory. "Typical age" is used instead of "average age" because some studies reported median age or some other summary value. The information on "surgeons" gives a rough indication of the level of training and experience of those who performed the operations. "Timing of randomization" is discussed further in Section 8.7. In characterizing "diagnosis" we sought to determine how the presence of an ulcer was established; this is particularly important in analyzing recurrence. One of the critical attributes of a randomized clinical trial is definition of the patients accepted as eligible for randomization. The treatment of this study measure was disappointing in these four generally good studies. Only one (Howard et al., 1973) gave a quantitative description of severity of the disease.

### Follow-up (Table 11-2)

A reference to a numbered table (for example, "(Table I)" under Postlethwait) indicates that the information is given in some detail by that table of the particular paper and could not adequately be summarized. Under "timing" the shorthand "1,3,6(6)36(12)" means that patients were seen at 1 month, 3 months, from 6 months to 36 months at intervals of 6 months, and thereafter at intervals of 12 months (the number in parentheses gives the interval length). The entry "blind?" indicates whether the interviewer knew what operation that patient had had, while "independent?" reflects whether the interviewer was someone other than the surgeon(s) involved.

### Outcomes (Table 11-3)

The three major treatment outcomes are early and late deaths and improvement of symptoms. As is customary, "operative mortality" counts deaths up to 30 days after operation. "Total mortality" includes later deaths from causes unrelated to the operation. "Success" gives the percentage of patients whose overall result was judged excellent or good on a Visick-type scale (described in Section 8.6).

**Table 11–1.** Attributes of four randomized surgical studies of duodenal ulcer. (P = vagotomy/pyloroplasty; A = vagotomy/antrectomy)

|  |  | *Jordan/Condon* | *Sawyers/Scott* | *Howard et al.* | *Postlethwait* |
|---|---|---|---|---|---|
| Number of patients | P | 108[a] | 40 | 100 | 337 |
|  | A | 92 | 39 | 73 | 331[b] |
| Percent male | P | 100% | (72.2%) | 100% | 100% |
|  | A | 100% |  | 100% | 100% |
| Typical age (years) | P | 49 | (range = 16 to 76) | 54 | 46 |
|  | A | 46 |  | 53 | 47 |
| Hospital(s) |  | V.A.* | university | V.A. | V.A. |
| Surgeons |  | 95% residents | authors | 90% residents | residents and attending staff |
| Timing of randomization |  | before operation | N.R. | during operation | during operation |
| Diagnosis |  | X-ray | N.R. | X-ray, acid | N.R. |
| Indication for surgery |  | based on "clinical history" | N.R. | duration of pain in years (mostly 5 to 10) | intractability, not defined beyond "usual symptoms" |

*Veterans Administration
[a]The original randomization assigned 98 patients to pyloroplasty and 102 to antrectomy, but the protocol was violated in 10 cases because resection was considered too hazardous.
[b]This study also involved hemigastrectomy and gastric resection. An alternative analysis could combine the results for antrectomy and hemigastrectomy.

**Table 11–2.** Comparison of follow-up in four surgical studies of duodenal ulcer.

|  | *Jordan/Condon* | *Sawyers/Scott* | *Howard et al.* | *Postlethwait* |
|---|---|---|---|---|
| length (mo.) min, mean, max | 24,NR,60 | 12,30,45 | 24,NR,120 | (Table I) |
| timing (mo.) | 1,3,6(6)36(12) | N.R. | N.R. | 6,24,60 |
| % loss | 14.4 (3 yrs.)[c] | N.R. | N.R. | 15.3 (5 yrs.) |
| blind? | no | N.R. | N.R. | yes |
| independent? | at 3 yrs. | N.R. | N.R. | yes |

[c]Table 1 in the paper gives the number of patients expected and those actually studied at yearly intervals up to five years. Because of an unexplained sharp drop from the number studied during the third year to the number expected during the fourth year, we give the loss after three years.

Table 11–3. Comparison of selected outcomes in four surgical studies of duodenal ulcer. (All values in %)

|  |  | *Jordan/Condon* | *Sawyers/Scott* | *Howard et al.* | *Postlethwait* |
|---|---|---|---|---|---|
| operative | P | 1.9 | 0 | 0 | 0.6[d] |
| mortality | A | 0 | 0 | 0 | 0.9 |
| total | P | 8.3 | 0 | 0 | N.R. |
| mortality | A | 4.3 | 0 | 0 |  |
| *recurrence* |  |  |  |  |  |
| rate | P | 7.4 | 0 | 10.0 | 6.6 |
|  | A | 0 | 2.6[e] | 3.9 | 1.7 |
| delay | | (Table 9a) | 17 mo.[e] | (Table VI) | (Table I) |
| proof | | N.R. | x-ray | x-ray, surgery | surgery |
| diarrhea | P | (Table 8) | 17.5 | 14.0 | 20.7 |
|  | A |  | 5.1 | 22.0 | 21.5 |
| dumping | P | (Table 6) | 25.0 | 8.5 | N.R. |
|  | A |  | 20.5 | 27.0 |  |
| weight loss | | (Table 7) | N.R. | (Table V) | N.R. |
| success | P | (Table 13) | 95.0 | N.R. | 83.0 |
| (V-scale) | A |  | 97.4 |  | 89.2 |

[d]If we included related deaths which occurred at 31, 35, 37, and 37 days after operation, these rates would rise to 0.9 and 1.5, respectively.
[e]One patient developed a recurrent ulcer at 17 months.

## Jordan and Condon (1970)

Among the patients in this study initially randomized to vagotomy and resection, ten were judged to be in too hazardous a condition for a resection, and hence those 10 patients had a vagotomy and pyloroplasty and were left in the study. The patients having vagotomy and pyloroplasty had the only operative mortality (2 patients) and the only ulcer recurrences (8 patients). The paper does not state whether any of the deaths or recurrences came in those 10 patients who were initially randomized for resection, so it is not possible to determine whether this departure from the protocol may account for the increased incidence of postoperative mortality and ulcer recurrence in the vagotomy and drainage group. The noticeable age difference between the two groups leaves a similar unanswered question. It is worth mentioning that this study is the only one of the four in which the operative mortality was higher for vagotomy/pyloroplasty than for vagotomy/antrectomy.

A second problem is that patients with emergency indications for ulcer surgery were included in the study, and it is not clear which cases of ulcer recurrence or mortality were patients operated on for emergency indications.

A third difficulty is that the percentage of follow-up, though initially high, declined rapidly: 81% of the patients were followed for 3 years, but only 20%

were followed for a full 5 years. This is too short a period of time to determine the true ulcer recurrence rate for either procedure.

### Sawyers and Scott (1971)

This brief paper has several deficiencies. First, the study sample, with only 40 patients per treatment, is too small to establish statistically significant differences in rates of operative mortality or ulcer recurrence between the two groups. (Section 8.2 discusses this in more detail.) Second, the sample may not have been typical of the larger group of patients with ulcer disease in that the male-to-female ratio in this series was three to one. Furthermore, there was no breakdown of sex distribution by operation. Third, follow-up time was short: the minimum was 12 months, and the mean was 30 months. This is too short a time to make an accurate statement about ulcer recurrence. Fourth, the authors did not clearly outline their criteria for diagnosis of a duodenal ulcer either preoperatively or as a recurrence. (Only one patient seemingly had a recurrence.) Fifth, no data were given on the incidence of recurrent ulcer symptoms following operation. The conclusions reached by the authors are based not only on their own data, but also on data from other published studies.

### Howard, Murphy, and Humphrey (1973)

This is a brief but informative paper. The groups are well outlined and described prior to surgery as well as following surgery. This study demonstrated no important difference in mortality rate or ulcer recurrence for the two groups. One criticism is that the paper does not give the number of patients lost to follow-up. Also, although the maximum length is much better than in the other series, the follow-up is still a bit short to determine the true ulcer recurrence rate. One table (Table VI) shows that patients were still suffering recurrent ulcers as late as 4 and 5 years after their operation. There was a high incidence of recurrence after vagotomy and drainage; however, the numbers of patients in the study were not large enough to make this difference significant.

### Postlethwait (1973)

This paper has to be considered together with earlier reports on this group of patients (Price et al., 1970; Johnson et al., 1970); we emphasize it because it is the most recent in the series. It is, in general, an excellent paper and the only one which describes the patients lost to follow-up and attempts to assess the effects of the losses. There is some confusion, however, in indications for operation on patients in the study. In the 5-year follow-up paper, the patients were described as those undergoing elective surgery, and patients needing emergency surgery were excluded. However, the earlier paper (Price et al., 1970) states clearly that half the patients were operated on for perforation, hemorrhage, or obstruction. Thus it is not clear how directly the results apply to elective operations for duodenal ulcer disease. The length of follow-up, however, is admirable, and the system of randomization (Johnson et al., 1970) is also very good.

Finally this is the only one of the four studies with sample sizes which are large enough to provide a good chance of detecting the small differences in rates of mortality and recurrence which are associated with the two operations (see Section 8.2).

## 8. PROBLEM CONSIDERATIONS

In this section we return to several important considerations and discuss them in more detail. Some, such as the Visick scale and tests for completeness of vagotomy, are particular to treatment of duodenal ulcer but typical of problems which can arise in other surgical studies. Others—randomization, follow-up, aggregation, and sample size—must generally be faced in any study.

### 8.1 Combining the Evidence from Various Studies

For the operations studied, the four papers selected for review tend to show differences in outcomes that are consistent with those reported from a number of nonrandomized series. In general, vagotomy and pyloroplasty was a safer operation than vagotomy and antrectomy. Vagotomy and pyloroplasty, however, was not quite as good an operation as vagotomy and antrectomy when measured by relief of ulcer symptoms. These differences were small and not always consistent; none reached statistical significance. It would be useful if the results of the four studies could be combined. Then small one-sided effects which are not significant individually might yield an overall significant finding. Such a combination seems initially sensible in view of the similarity of the patients and clinics involved, but a closer look shows that any combination without more specific data than are currently available would be unwise.

What if the comparison were limited to recurrence rates following operations on elective patients only? One problem is that three of the studies have apparently pooled emergency and elective patients. While the overall breakdown into emergency and elective is described, no breakdown by operation is given. A second problem arises when we ask which recurrence rate should be studied. Two clear candidates are

$$R_1 = \frac{\text{number of recurrences}}{\text{total in study from start}}$$

and

$$R_2 = \frac{\text{number of recurrences}}{\text{number remaining alive in study}}$$

A potential problem with $R_1$ is that one of the operations might have a higher death rate and higher recurrence rate, but the higher recurrence rate would not be evident because too few patients were exposed to the possibility of recurrence. Similar comments apply to follow-up losses. The second rate $R_2$ represents the fraction of those remaining in the study for a fixed period who developed a recurrent ulcer. Unfortunately, it is not possible to estimate $R_2$ for elective patients from the data currently available. A further difficulty is that the studies have far from uniform lengths of follow-up. Postlethwait (1973) reports a uniform follow-up of 5 years, but the other papers have various mixtures of length of follow-up which are not always well reported. We discuss the special problems of follow-up further in Section 8.5. Our overall conclusion is that, given the variation of reporting, there is no reliable way of combining the findings from the four studies.

We have become aware that peptic ulcer disease presents an unusual problem for the clinical investigator. The difficulty of making accurate and reliable measurements and the requirement for lengthy follow-up may explain why the many clinical studies published have left the value of the various operations used unproven. Occasionally the conclusions of different studies have been contradictory. It would be desirable if one of the professional organizations concerned with gastroenterology were to publish standards to guide investigation and reporting. Standard descriptions of symptoms (such as duration, severity, dieting, medications, and hospitalization), treatment, and follow-up (including intervals, duration, methods of reassessment, and details of losses), if used by all investigators, would provide a basis for combining different studies, much as in multicenter trials.

## 8.2. Sample Size

In the statistical analysis of the results of an experiment that compares two operations, the two groups of patients in the experiment are regarded as samples drawn from the much larger aggregate or population of patients about whom we are trying to learn. For any postulated percentages $P_1$, $P_2$ of occurrence in this population of some outcome of the operations, formulas are available (Cochran and Cox, 1957, pp. 23-27) for calculating the probability of finding a statistically significant difference in the experiment, at any chosen level of significance, for any specified sample sizes $n_1$, $n_2$. These formulas can help the investigator in planning an experiment. In the randomized experiment on the Salk polio vaccine trial (Francis et al., 1957), for instance, the calculations suggested that at least 200,000 children in each group (vaccine and placebo) would be necessary to guarantee a high probability of finding a significant difference if, in fact, the vaccine reduced the risk of paralytic polio by 50% in the population of children in grades 1-3.

Most medical or surgical experiments have numerous outcomes that are measured and compared for the two procedures. Usually the investigator selects from one to three of these outcomes that he regards as most important and makes the calculation of $n_1$, $n_2$ separately for each one. The next problem is to decide on outcome values $P_1$ and $P_2$ that are realistic. This problem is simplified if operation 1 is a standard method on which past data about $P_1$ values have accumulated. The investigator can then name a value $P_2$ such that he regards the difference $(P_2 - P_1)$ as of scientific or practical importance. He can then find the sample size required in the experiment in order to have only a small probability of finding an inconclusive, nonsignificant result if the "true" difference is

$(P_2 - P_1)$ or greater. When there are little past data on the performance of either operation, selection of $P_1$, $P_2$ values may be very difficult and the sample size calculation of limited help.

In comparing vagotomy/pyloroplasty (VP) and vagotomy/antrectomy (VA), the deaths (during the operation and first month post-op) are important outcomes. For these deaths the average death rate $(P_1 + P_2)/2$ in the four studies was about 1%. With these low $P$ values, sample sizes needed to "discriminate" in this sense between $P_1$ and $P_2$ are very high, e.g., $n = 850$ in each group for $P_1 = 0.5\%$ and $P_2 = 2\%$ in order to have an 80% chance of detecting this difference at the 5% level.

The three studies that reported on ulcer recurrences within 5 years showed rates of about 8% for VP and 3% for VA. Some questions remain since the follow-up period was not always fully reported. If, for instance, these values held in the population of patients sampled by these experiments, the probability of finding a significant difference at the 5% level (two-tailed test using inverse sine transformation) is:

| Sample Size $(n_1 = n_2 = n)$ | Probability |
|:---:|:---:|
| 100 | 0.34 |
| 200 | 0.60 |
| 300 | 0.78 |

Thus, sample sizes of somewhat over 300 would be needed to confidently distinguish between $P_1 = 8\%$ for VP and $P_2 = 3\%$ for VA, while this sample size can expect to detect only a marked difference in the death rates.

Analogous sample size calculations are available (Cochran and Cox, 1957) when the outcome is a discrete or continuous variable or when the primary interest is in a good estimate of the difference $(P_1 - P_2)$. (See Chapter 12 for further discussion of sample size and power of tests.)

## 8.3 Nonoperated Controls

In this section we return to the possibility of a control group (Statistical Criterion 3) in which patients would receive further medical therapy instead of a surgical treatment. This can be justified ethically when intractability is the only indication for surgery, because patients may recover, or their disease may become inactive (the more modest description used by gastroenterologists) many years after onset. In some cases it should provide a more accurate basis of information in choosing a therapy for duodenal ulcer disease. For example, when the natural history of the disease may be changing, such a control group would be an important reference point. If, by contrast, medical therapy had been thoroughly demonstrated to be inferior, then the study would focus on comparing surgical treatments, and a nonoperated control group would be of little benefit.

Our search of the literature did not locate any mention of nonoperated control groups. This may be the consequence of surgery being restricted to patients who have had a reasonable trial of medical therapy and who continue to experience disabling symptoms. It would be difficult under these circumstances to randomize such patients into a nonoperated control group to continue ineffective (for them) medical therapy. Nonetheless, in light of the possible strengthened inferences that a control group permits, we feel that this possibility deserves further comment.

## 8.4 Tests for Completeness of Vagotomy

The insulin test introduced in 1946 by Hollander has been extensively used to determine the completeness of vagotomy. Hypoglycemia is a potent cause of gastric acid secretion, the stimulus being mediated through the vagus nerves. In 1942, Jemerin, Hollander, and Weinstein reported that the insulin test was reliable in differentiating innervated (positive result) from denervated (negative result) gastric pouches in dogs. In the clinic, the result was less clear: 48% of the vagotomized patients had positive insulin test 10–14 days postoperatively. Lewis (1951) stressed the contrast between the ease with which complete insulin negativity is produced by vagotomy in the dog, and the frequency with which a positive response to insulin is found in man after a presumably complete operation.

On theoretical grounds we must question the validity of the insulin test. The notion that stimulation of acid secretion by insulin is mediated solely by the vagus in man is an unproven assumption. Even more suspect is the corollary that any response to insulin after vagotomy indicates an incomplete vagotomy, as there is no independent test for completeness of the vagotomy. Certainly, incomplete vagotomy or the recovery or regeneration of injured fibers may contribute to a positive insulin response after vagotomy. However, nonvagal neural mechanisms may become operative after vagotomy as suggested by Kronborg (1973), and nonneural mechanisms such as the hypoglycemic release of epinephrine, which in turn releases gastrin and stimulates acid gastric secretions (Stadil and Rehfeld, 1973), may play an important role.

Re-exploration studies by Fawcett et al. (1969) and Kronborg (1973) indicate that a positive insulin test can occur in the presence of a surgically complete vagotomy. The percentage of patients with a positive Hollander test in whom intact vagi were found was not greater than the percentage with negative insulin tests.

Does the insulin test, then, have any place in assessing the clinical results of vagotomy? With the evidence before us, no definite conclusion can be drawn. Many surgeons feel that it is important for them to have some objective evidence that they have indeed cut the vagi. In this situation the insulin test provides uncertain supportive evidence.

## 8.5. Follow-up

In this section we discuss length, causes of missing information, techniques for reducing losses, and treatment of the resulting data.

The duration of follow-up necessary to encompass all cases of ulcer recurrence following ulcer surgery is not really well established. It appears that, both for vagotomy/pyloroplasty and for vagotomy/hemigastrectomy, the length of follow-up should be at least 8 years. In several published studies (Howard et al., 1973; Goligher et al., 1964 and 1968; Brooks et al., 1975) ulcer recurrences, plotted against time, seem to follow a roughly bimodal distribution with early recurrences coming within the first 3 years and then late recurrences occurring between 5 and 8 years.

Detailed analysis of the follow-up data should include life-table applications to provide estimates of the mean length of time until recurrence or death, as well as the distributions of these events. If the number of patients permits, it is useful to see the effects of factors like age and disease severity on these distributions.

Thorough analysis of follow-up data becomes more difficult when patients are lost to follow-up. There are four principal causes of missing follow-up information: (1) Experiments must be evaluated after a finite length of time. (2) Patients enter the experiment at different times. (3) Patients die from unrelated causes. (4) Patients are lost to follow-up for other causes.

We briefly discuss each of these four problems.

1. Length of follow-up has been mentioned above and in Section 8.1. Studies must continue long enough to allow recurrences to take place. We recommend a minimum length of 8 years. Whatever the length of the study, reports must be completed, and the classical problem of using censored data (that is, data from which a known number of potential observations are missing at one or both extremes) to assess survival curves arises (see the references cited below).

2. Most clinical studies continue for several years during which patients enter the study continuously until an adequate number of subjects has been treated. This seems inevitable in large-scale comparisons of surgical treatments. Among the papers, the study by Howard et al. had follow-up varying from 2 to 10 years for this reason. Jordan and Condon are less definite about the length of follow-up, but their Table 1 (p. 548) indicates new patient entries throughout the 5-year trial. Life-table procedures are designed to handle these types of data.

3. Postoperative death rates have immediate utility in comparing safety of two treatments, and all the studies considered have separate accounting for patients who died postoperatively or died later from "unrelated causes." None, however, makes use of the total mortality in the comparison of the two operations. While it is clear that gastric surgery can be performed with considerable safety, the amount of diarrhea and dumping after operation indicates physiological derangements of sufficient severity to make some excess of deaths among surviving patients quite possible. Careful follow-up of sufficient patients to provide assurance on this point is certainly justified. Deaths from "unrelated causes" therefore cannot be lightly ignored; they may prove in time to be "related causes." Neglecting these patients and analyzing the data as if they were never in the study also bias mean recurrence rates upward since patients who were known to be free of recurrences for some time are omitted from the computation.

4. Patients may be lost to a study for reasons like emigration or lack of cooperation. Price et al., the only one of the four studies to clearly describe losses,

lost over 15% of the patients under study. With recurrence rates as similar as they are for the two operations under study, even a slight difference in losses from each group could render comparisons of observed recurrence rates meaningless. Price et al. performed the useful check of reviewing the records of lost patients to see if there were significant differences in operation, age, condition, or other characteristics and found no differences.

A search produced few references to available techniques for guarding against follow-up losses. There seems to be no substitute for determination. Several Ph.D. dissertations were found at the Harvard School of Public Health in which graduate students taking up a problem 10 to 15 years after a study managed to locate about 95% of the patients previously studied. William Taylor* of the Mayo Clinic writes that they frequently have losses under 5% in studies involving very long-term follow-up. This is achieved by writing letters directly to patients and not going through their doctors; if no reply is forthcoming, the telephone is used. If the patient is not found, a vigorous search is undertaken, including use of bill-collecting agencies, who apparently have experience with similar problems. A loss as high as 15% in a population as well documented as veterans appears unnecessary and wasteful.

There are two extreme approaches to treatment of missing patients, each with its biases. First, it may be argued that if a patient had a recurrence, he would be likely to rejoin the study, so it is assumed that all lost patients are disease-free. This approach clearly results in a bias toward lower estimated recurrence rates. Second, some studies eliminate all data pertaining to patients lost to follow-up. Since such patients were known to be disease-free for a certain length of time, this procedure will result in a bias toward higher recurrence rates. Discussion of similar problems and techniques for handling them will be found in the references below.

A battery of statistical techniques has been developed to try to deal with problems of follow-up (1 through 4 above). Koch et al. (1972) describe methods using techniques developed for contingency tables. The method of Kaplan and Meier (1958) is widely used for estimating survival curves in the face of various types of missing data. Efron (1967) discusses the comparison of two treatments under the conditions discussed in 1 to 4 above. All three papers contain many references to the statistical and medical literature.

## 8.6 Visick Scale

The Visick scale (1948) has been used by Goligher (1970) and others to evaluate the overall symptomatic state of the patient. It is a four-point scale on which the patient is rated from unsatisfactory to excellent following an assessment made by one or more judges. The assessment may take place once (Visick recommends 6 months to 12 years after the operation) or periodically. The purpose in any case is to gauge the severity of the operation's side effects in a manner which is both standardized and objective. However, the reproducibility of the clinical assessment by different physicians is unknown at present, and the Visick scale's

---

*Personal communication to Osler Peterson.

true worth is undocumented. (For further discussion of Visick scale see Chapter 10.)

## 8.7 Randomization

In view of the emphasis given to randomized studies throughout this chapter, some further discussion of randomization and of particular features of the present problem is desirable. This section reviews the concept and implications of randomization, describes its application to ulcer surgery, and examines some implications of a randomization scheme in one of the published studies.

The basic notion of randomization is simple: patients or experimental subjects are assigned to treatments by chance in such a way that each patient has an equal chance of receiving any of the treatments under study. Such familiar chance mechanisms as tossing a coin or drawing a card from a well-shuffled pack could provide the random element, but it is generally preferable to use a table of random digits or some other device whose randomness is reliable and thoroughly understood.

The purpose of randomization is almost equally simple. In any substantial experiment a great many factors may affect the outcome, and differences between the results of treatments are seldom overwhelmingly clear, even after careful analysis. Randomization makes it possible to avoid biases (from expected as well as unexpected sources) which could cause difficulty in deciding whether an apparent difference between treatments is a consequence of real differences between the treatments or is simply an artifact of the way patients were assigned to treatments. Without randomization, conclusions from the comparison are in doubt. It is important to realize that one cannot expect the same benefits from a scheme of haphazard assignment, performed without any *intentional* biases, because the underlying "random" mechanism cannot be well understood, and subtle biases may, in fact, be present. Overall, the role of randomization in evaluating social, medical, and sociomedical programs has been nicely stated by Gilbert, Light, and Mosteller (1975): "randomization, together with careful control and implementation, gives strength and persuasiveness to an evaluation that cannot ordinarily be obtained by other means." (Gilbert, McPeek, and Mosteller elaborate on these questions in Chapter 9.)

In planning and carrying out a randomized controlled trial of (elective) surgical treatments for duodenal ulcer, the basic randomization procedure should be straightforward, but practical problems can arise. Chief among these are the specification of the target population and the timing of the actual randomization. When several surgical procedures are to be compared, it may be that not all patients are suitable candidates for all the operations. In this case, the decision on whether to include the "nonoperable" patients in the population of experimental subjects has considerable impact on the inferences ultimately formulated. Since a patient might often be "nonoperable" because of poorer health, a decision to exclude such patients would naturally reduce the scope of justifiable inferences; conclusions would then apply only to a subpopulation healthier than the general population.

If "nonoperable" patients are not automatically excluded, the timing of the randomization becomes important, and some sort of "escape mechanism" may be required. Three possibilities and their consequences are:

1. *Randomize before the operation begins.* A patient assigned to an unacceptable treatment is then removed from the study. The more severe operations would receive healthier patients, and a bias could favor such operations.

2. *Randomize before the operation begins.* A patient assigned to an unacceptable operation would be treated with a specified "default" operation. As in alternative 1, bias would tend to favor the more severe operations. In the present case, however, it is possible to estimate this bias by comparing patients originally assigned to the default operation with those assigned to it for reasons of poorer health.

3. *Randomize either before the operation or during it* (when complications have just become evident). A patient assigned to an unacceptable operation is rerandomized among the acceptable operations. An unbiased comparison is possible within each subset of operations over which randomization or rerandomization is performed. Further discussion at the end of the section deals with an example.

The central issue in this discussion of timing is the role a patient's condition may play in selecting the operation. It is important to have a clear understanding of the relation between the experimental population and the general population of patients who might be candidates for treatment by one or more of the operations compared in the experiment.

The extensive study reported on by Price et al. (1970) and Postlethwait (1973) provides an interesting example.

"Briefly, the operations selected for application on a random basis were vagotomy and drainage, vagotomy and antrectomy, vagotomy and hemigastrectomy, and gastric resection.... After abdominal exploration, the surgeon determined whether or not the condition of the patient and the pathologic changes in the duodenum would permit any one of the four operations to be performed without endangering the welfare of the patient or compromising the outcome. The suitable patients were randomly assigned by opening the next numbered sealed envelope, and the indicated operation was performed [Postlethwait, p. 387]."

Here the randomization is a variant of possibility 3 above. While the procedure used in this study is certainly a very good one, it is important to realize that comparisons between some operations must be interpreted with care. For example, in using the results for vagotomy/drainage and vagotomy/antrectomy one must keep firmly in mind the selectivity imposed by requiring that the condition of the patient permit the more severe operation of gastric resection. In this important respect the study group is not necessarily representative of the general population of patients who would be suitable candidates for either vagotomy/drainage or vagotomy/antrectomy if those two were the only operations considered.

In this experiment a slightly more elaborate randomization scheme might have provided a basis for broader conclusions. Such an alternative may be valuable when, as may have been the case here, many patients are not eligible for randomization over the full set of operations. Price et al. report that "1,358 patients who met certain criteria were randomly assigned one of [the] four operations. During the same period, 1,630 patients had definitive operations for duodenal ulcer but the type of operation was not randomly assigned" (p. 233). It seems likely that many of the "rejected" patients might have been eligible for randomization over a subset of the four operations. Since the four operations are ordered by extent of resection (and, most likely, by severity), only two such subsets are likely to be sensible: (vagotomy/drainage, vagotomy/antrectomy, vagotomy/hemigastrectomy) and (vagotomy/drainage, vagotomy/antrectomy). One might hope that this approach would yield considerably more information without seriously complicating the administration of the study.

## 9. SUMMARY AND CONCLUSIONS

Elective surgery for intractable duodenal ulcer offers a substantial focus for a broad critical examination of protocol issues. By systematically studying published reports on four randomized studies which compare vagotomy/pyloroplasty with vagotomy/antrectomy, our group of physicians and statisticians has been able to consolidate some quantitative results on the outcomes of these two operations and to identify several possibilities for improvement in reporting and analysis of results from surgical experiments.

While the relation between gastric acid and duodenal ulcer is not well understood, the prevalence of the two treatments described in this chapter makes consideration of their risks and effects mandatory. The exercise of summarizing the four studies (Jordan and Condon, 1970; Sawyers and Scott, 1971; Howard et al., 1973; and Postlethwait, 1973) in a single comparative table (see Section 7) was beneficial because it facilitated evaluations and at the same time emphasized the difficulties of putting studies on a common basis. This was especially true for follow-up because few of the necessary details were available.

In view of this it is not easy to make definite statements about the overall picture of outcome on the basis of these four studies. Since the results would obviously be useful, however, we attempted to do so for five major outcomes: operative mortality, recurrence, weight loss, dumping, and diarrhea. *Operative mortality* is the easiest outcome to assess; for both vagotomy/pyloroplasty and vagotomy/antrectomy it is acceptably low, and none of the four studies shows a significant difference between the two procedures. *Recurrence* rate is higher for vagotomy/pyloroplasty than for vagotomy/antrectomy, and the difference was statistically significant in the Postlethwait study. For this outcome, as well as for others, we must be cautious: loss to follow-up is substantial, and it may affect these comparisons. Length of follow-up may also be too short. We wish the evidence were stronger. *Weight loss* shows no real difference between the two procedures, but only two of the papers include any numerical evidence. The evidence on *dumping* is mixed, and it seems wisest not to attempt a conclusion. Jordan and Condon report year-by-year data on sweating, palpitation, and weakness and find no difference. The difference for Sawyers and Scott is not significant, but Howard et al. found vagotomy/antrectomy significantly worse (at the 5% level). Postlethwait reports finding a significant "linear trend of less dumping in direct relationship to the amount of stomach removed" but gives no data which would permit direct comparison of vagotomy/pyloroplasty and vagotomy/antrectomy. On *diarrhea* the findings are generally in agreement, and we can suggest a conclusion: no difference between the two procedures.

While these conclusions are a natural product of our study, the difficulties encountered in reaching them are more important and more general. It is quite reasonable to try to reach firmer conclusions and a more comprehensive view by combining the results from available studies. This is, at least implicitly, what anyone reviewing the literature would do; one expects evidence to accumulate as research proceeds. Our attempt to do this systematically and quantitatively, however, leads us to emphasize the problems in reaching a satisfactory and defensible synthesis of evidence from several studies. Often, we have found, the published reports make this process unnecessarily difficult.

A prominent example is the definition of intractability. None of the four studies we examined gave a clear and precise description of criteria for intractability. This one reporting flaw is typical of a number of others, but none of them is difficult to remedy. We conclude that much more attention should be given to establishing uniform definitions and methods of reporting and analysis.

Another serious problem is the very considerable percentage of patients lost to follow-up. If these losses are not drastically reduced, there may be biases of unknown size and direction, and the effective number of patients in the study will be substantially reduced.

In making effective use of the patients in a study, careful attention must also be paid to the method of randomization, especially when three or more surgical treatments are being compared (see Section 8.7).

The question of adequate sample size is intimately related to the problems of comparing and combining results from several studies. If results on key outcomes cannot be combined across studies, then it may be quite difficult to justify a trial in which the number of patients is too small to give a high probability of reaching a definitive conclusion.

Overall, these steps would help a number of surgical experiments to produce stronger and more useful results. We look forward to opportunities to put many of them into effect.

### Acknowledgments

The authors are grateful to Paul A. Blake, Pierce Gardner, Bernard Rosner, and James H. Warram, Jr., for valuable discussion at various stages and to John P. Gilbert and Frederick Mosteller for critical reading of earlier drafts.

### References

Brooks JR, Kia D, Membrano AA: Truncal vagotomy and pyloroplasty for duodenal ulcer. Arch Surg 110:822, 1975

Cochran WG, Cox GM: Experimental Designs. Second edition. New York, John Wiley and Sons, 1957

Cochran WG, Mosteller F, Tukey JW: Principles of sampling. J Am Stat Assoc 49: 13, 1954

Efron B: The two sample problem with censored data, Proceedings of the Fifth Berkeley Symposium on Mathematical Statistics and Probability. Vol. 4. Edited by LM LeCam, J Neyman. Berkeley, California, University of California Press, 1967, p 831

Fawcett AN, Johnston D, Duthie HL: Revagotomy for recurrent ulcer after vagotomy and drainage for duodenal ulcer. Br J Surg 56: 111, 1969

Flood CA: The results of medical treatment of peptic ulcer. J Chronic Dis 1: 43, 1955

Francis T, Napier JA, Voight RB, et al: Evaluation of the 1954 Field Trial of Poliomyelitis Vaccine, Final Report. Ann Arbor, Michigan, Edwards Brothers, Inc., 1957

Fry J: Peptic ulcer: a profile. Br Med J 2:809, 1964

Gilbert JP, Light RJ, Mosteller F: Assessing social innovations: An empirical base for policy. Evaluation and Experiment: Some Critical Issues in Assessing Social Programs. Edited by CA Bennett and AA Lumsdaine. New York, Academic Press, 1975

Goligher JC: The comparative results of different operations in the elective treatment of duodenal ulcer. Br J Surg 57: 780, 1970

Goligher JC, Pulvertaft CN, Watkinson G: Controlled trial of vagotomy and gastro-enterostomy, vagotomy and antrectomy, and subtotal gastrectomy in elective treatment of duodenal ulcer: Interim report. Br Med J 1: 455, 1964

Goligher JC, Pulvertaft CN, de Dombal FT, et al: Five- to eight-year results of Leeds/York controlled trial of elective surgery for duodenal ulcer. Br Med J 2:781, 1968

Hogan MD: A Study of Postoperative Complications Following Surgery for Duodenal Ulcer. Ph.D. thesis. Chapel Hill, North Carolina, Department of Biostatistics, University of North Carolina, 1969

Howard RJ, Murphy WR, Humphrey EW: A prospective randomized study of the elective surgical treatment for duodenal ulcer: Two- to ten-year follow-up study. Surgery 73:256, 1973

Johnson WD, Grizzle JE, Postlethwait RW: Veterans Administration cooperative study of surgery for duodenal ulcer—I: description and evaluation of method of randomization. Arch Surg 101:391, 1970

Jones SH, Carr J, Peterson OL: A comparison of hospitalizations in New England before and after Medicare. Social Security Bulletin (in press), 1976

Jordan PH, Condon RE: A prospective evaluation of vagotomy-pyloroplasty and vagotomy-antrectomy for treatment of duodenal ulcer. Ann Surg 172: 547, 1970

Kaplan EL, Meier P: Nonparametric estimation from incomplete observations. J Am Stat Assoc 53:457, 1958

Kennedy T: Which vagotomy? Which drainage? Proc R Soc Med 67:3, 1974

Koch GG, Johnson WD, Tolley HD: A linear models approach to the analysis of survival and extent of disease in multidimensional contingency tables. J Am Stat Assoc 67:783, 1972

Kronborg O: The discriminatory ability of gastric acid secretion tests in the diagnosis of recurrence after truncal vagotomy and drainage for duodenal ulcer. Scand J Gastroenterol 8:483, 1973

Lewis FJ: The effect of parasympathetic or sympathetic denervation on total stomach pouch secretion in dogs. Surgery 30:578, 1951

Lipetz S: Gastro-intestinal ulceration and non-ulcerative dyspepsia in an urban practice. Br Med J 2:172,1955

Namey C, Wilson RW: Age patterns in medical care, illness and disability, U.S. 1968–69. Data from the National Health Survey, Series 10, No 70 (DHEW Publication No [HSM] 73–1026). Washington, DC, Government Printing Office, 1973

Postlethwait RW: Five year follow-up results of operations for duodenal ulcer. Surg Gynecol Obstet 137:387, 1973

Price WE, Grizzle JE, Postlethwait RW, et al: Results of operation for duodenal ulcer. Surg Gynecol Obstet 131:233, 1970

Sawyers JL, Scott WH: Selective gastric vagotomy with antrectomy or pyloroplasty. Ann Surg 174:541, 1971

Stadil F, Rehfeld JF: Release of gastrin by epinephrine in man. Gastroenterology 65:210, 1973

Visick AH: A study of the failures after gastrectomy. Ann R Coll Surg Engl 3: 266, 1948

Vital Statistics of the United States, 1968. Vol. 2, part A. DHEW, Public Health Service, National Center for Health Statistics

Wilder CS: Limitation of Activity and Mobility due to Chronic Conditions, U.S., 1972. Data from the National Health Survey, Series 10, No 96 (DHEW Publication No [HRA] 75–1523). Washington, DC, Government Printing Office, 1975

Wilson RW: Prevalence of Selected Chronic Digestive Conditions, U.S., July-December 1968. Data from the National Health Survey, Series 10, No 83 (DHEW Publication No [HRA] 75–1510). Washington, DC, Government Printing Office, 1974

# 107

# Laplace's Ratio Estimator

*William G. Cochran*

HARVARD UNIVERSITY

This paper describes the sample survey from which Laplace estimated the population of France as of September 22, 1802 by means of a ratio estimator (ratio of population to births during the preceding year). His analysis of the distribution of the error in his estimate used a Bayesian approach with a uniform prior for the ratio of births to population. He assumed that the data for his sample and for France itself were both drawn at random from an infinite superpopulation. He found the large-sample distribution of his error of estimate to be approximately normal, with a small bias and a variance that he calculates. He ends by calculating the probability that his estimate of the population of France is in error by more than half a million. To my knowledge this is the first time that an asymptotic distribution of the ratio estimate has been worked out.

## 1. INTRODUCTION

Our friend HO Hartley has contributed to so many important areas in statistics that authors of this Festschrift should not find it difficult to select a topic related to some of HO's work. Assuming that we are not expected to rival his originality and brilliance, I would like to write about an early use of the ratio estimator in a sample survey. This is the well-known survey from which Laplace estimated the population of France as of September 22, 1802.

This survey interests me for at least two reasons. It is a survey taken under the direction of the French government at the request of a private citizen.

*Key words and phrases* Laplace, ratio estimator, sample survey, error of estimate, superpopulation.

However, as Dr. Stephen Stigler has reminded me, by 1802 Laplace already had previous government service and political prestige having been Minister of the Interior for six weeks in 1799, President of the Senate in 1801, and been made a Grand Officer of the Legion of Honor by Napoleon in 1802. Thus the sample survey may have had a more official origin than I at first thought.

Secondly, Laplace not only gave the formula for the ratio estimate, but worked out the large-sample distribution of the error in his estimate, showing this distribution to be normal with a bias and a variance that he gives, and noting that in large samples the effect of the bias becomes negligible. He ends by calculating the probability that his estimate of the population of France is in error by more than half a million. His work is the first known to me in which the sampling distribution of the ratio estimate has been tackled.

His survey is described with the arithmetical estimates but without any details of the error theory in Chapter 8 of the English translation: "A philosophical essay on probabilities," (Laplace, 1820, English translation, 1951). His assumptions and mathematical strategy plus the resulting normal approximation are given in Chapter 6 of the "Théorie Analytique des Probabilités." For the details of the analysis the reader is referred to an earlier chapter in which Laplace worked on a similar problem.

## 2. THE SURVEY AND THE ESTIMATE

In France, registration of births was required, and Laplace assumed without comment that he could obtain the numbers of births in a year for France as a whole or for any administrative subdivision. In this he may have been a little naïve. The fact that some action is compulsory in a country, even in a dictatorship, does not ensure that everyone takes this action. Actually, Laplace's theoretical assumptions do not demand completeness of birth registration, but something less—namely, that any percent incompleteness should be constant throughout France, apart from a binomial sampling error.

From such data as he could get his hands on, Laplace had noticed that the ratio of population to births during the preceding year was relatively stable. He persuaded the French government to take a sample of the small administrative districts known as communes, and count the total population $y$ in the sample communes on September 22, 1802. From the known total number of registered births during the preceding year in these communes, $x$, and in the whole country $X$ the ratio estimate $\hat{Y}_R = Xy/x$ of the population of France could be calculated.

The method—registered births × estimated ratio of population to births—was not original with Laplace. It dates back at least to John Graunt in 1662. In his "Observations on the bills of mortality," Graunt suggested it as one method of estimating the population of a country in which the number of annual births was known. In fact, Graunt went one better than Laplace, and estimated the key ratio $y/x$ without any sample at all, by what we might call crude guesswork. He argued that since women of childbearing age seemed to be having a baby about every second year, the number of women of childbearing age could be estimated as $2X$, where $X$ was the number of annual births. Then he guessed that the number of women aged between 16 and 76 was about twice the number of women of childbearing age and was therefore $2 \times 2 \times X$. (The number of live women over 76 was, I suppose, negligible.) He regarded every woman over 16 as belonging to a family, and, by a step which I do not fully understand, counted $2 \times 2 \times X$ as the number of families. The final step was to estimate the average size of a family as 8—husband, wife, three children, and three servants or lodgers. Thus to estimate population from births, multiply by $2 \times 2 \times 8 = 32$.

Commenting on this arithmetic, Greenwood (1941) thought that Graunt had made two big errors. As an average, a baby every two years was much too often; Greenwood thought the first multiplier should have been 4 instead of 2. On the other hand, 8 looked much too large as the average size of a family. Since these two errors were in opposite directions, perhaps Graunt was lucky. For what it is worth, Greenwood quotes data from the British Census of 1851, nearly 200 years after Graunt, which gave a ratio of population to births as 31.47.

Laplace's use of the ratio estimate based on a sample was not even original in France, two other Frenchmen having used the method in the late eighteenth century, as Stephan (1948) has noted.

Laplace's sample was what we now call a two-stage sample. He first selected 30 departments—the large administrative areas—chosen to range over the different climates in France. In each sample department a number of communes were selected for complete population counts on September 22, 1802. The criterion for selection was that the communes should have zealous and intelligent mayors: Laplace wanted accurate population counts. The combined population of the sample communes as of September 22, 1802 was 2,037,615 (I guess about a 7% sample).

At this point Laplace did two things that puzzle me slightly. As regards births, he totaled the sample births for the three-year period September 22, 1799 to September 22, 1802, finding a value 215,599 so that his sample $x$ is 215,599/3. No explanation is given for the use of the three-year period. One might guess that the objective was to obtain higher precision from a larger sample with little effort since the births were already on record. But the device might not accomplish this if the birth rate varied from year to year. In his theoretical development of the sampling distribution of the error in $\hat{Y}_R$, he acts as if $x$ was counted for only the year preceding September 22, 1802. In his numerical estimate, however, the sample ratio $y/x$ was taken as $3(2,037,615)/(215,599) = 28.852345$.

Secondly, having been responsible for this important survey, he might be expected to present and use as $X$ the actual number of registered births in France in the year preceding September 22, 1802. Instead, he writes casually: "supposing that the number of annual births in France is one million, which is nearly correct, we find, on multiplying by the preceding ratio $(y/x)$, the population of France to be 28,352,845 persons."

## 3. THE SAMPLING ERROR: STANDARD METHODS

Suppose that we had the population and birth data $x_{ij}$, $y_{ij}$ for the $j$th sample commune in the $i$th department and wanted to try to estimate the sampling error of Laplace's $\hat{Y}_R$. The sample was a two-stage sample, but with no use of probability sampling. We might first judge whether the sample of departments could be regarded as a random sample, or perhaps better as a geographically stratified random sample. Secondly, we might judge whether the distribution of selected communes with zealous and intelligent mayors within a department seemed to approximate a random selection. Depending on these judgments we might decide to use the textbook formula for the large-sample variance of the ratio estimator in two-stage sampling. In this formula, $V(\hat{Y}_R)$ depends on the variances and the covariance of the $y_{ij}$, $x_{ij}$ between departments and between communes in the same department.

Laplace simplified his mathematical problem by a sweeping assumption. He considered an infinite urn consisting of white and black balls and representing a superpopulation of French citizens alive on September 22, 1802. The white balls represented those born in the preceding year and registered. The ratio $p$ (proportion of white balls) is unknown. He regarded the ratio $x/y$ from the sample of communes as a binomial estimate of $p$. Choice of this model seems at odds with his deliberate choice of sample departments scattered throughout France, which presupposes a judgment that the birth rate $p$ varies from department to department.

If we adopt Laplace's assumption and assume the sampling fraction $y/Y$ small, we write $\hat{Y}_R = Xy/x = X/\hat{p}$, where $\hat{p}$ is a binomial estimator of $p$ based on a sample of $y$ persons. By Taylor's approximation,

$$\hat{Y}_R = \frac{X}{\hat{p}} \cong \frac{X}{p} - \frac{X(\hat{p} - p)}{p^2} + \frac{X(\hat{p} - p)^2}{p^3}. \qquad (1)$$

With $Y = X/p$, it follows that with binomial sampling $\hat{Y}_R$ has a bias whose leading term is $Xpq/yp^3 = Xq/yp^2$. On substituting $\hat{p} = x/y$, $\hat{q} = (y - x)/y$, the sample estimate of this bias is $X(y - x)/x^2$.

From (1), the approximate variance of $\hat{Y}_R$ is

$$V(\hat{Y}_R) \cong X^2 V(\hat{p})/p^4 = X^2 pq/yp^4 = X^2 q/yp^3. \tag{2}$$

The ratio of the bias to the s.e. of $\hat{Y}_R$ is approximately $(q/yp)^{1/2}$, becoming negligible for $y$ large.

If we drop the assumption that $y/Y$ is small, and use the hypergeometric, the variance (2) is multiplied by the factor $(Y - y)/(Y - 1)$.

The sample estimate of the variance (2) is

$$\tilde{v}(\hat{Y}_R) = X^2(y - x)y^3/y^2x^3 = X^2 y(y - x)/x^3. \tag{3}$$

The preceding analysis assumes in effect that $y$ and $E(x) = yp$ are both large.

Laplace's analysis finds that $\hat{Y}_R$ is approximately normally distributed in large samples, with the same leading term $X(y - x)/x^2$ in the bias, but with a variance slightly larger than the estimate in (3), namely

$$v(\hat{Y}_R) = X(X + x)y(y - x)/x^3. \tag{4}$$

There is also the difference that since his approach was Bayesian, he would regard (4) as the correct asymptotic variance given the sample information rather than as a sample estimate of the variance. If $y/Y$ is small, the difference between (3) and (4) is minor, since $x/X$ will nearly always be small so that $X(X + x) \cong X^2$.

## 4. LAPLACE'S ANALYSIS OF THE SAMPLING ERROR

When we examine Laplace's analysis more carefully, we note that he assumes that France itself and the sample communes are *both* binomial samples from the infinite superpopulation or urn with unknown $p$ (ratio of births to population). Thus he regards $X$ as a random binomial variable from a sample of unknown size $Y$, the population of France. So far as I know, this use of an infinite superpopulation in studying the properties of sampling methods was not reintroduced into sample survey theory until 1963, when Brewer (1963), followed by Royall (1970), applied it to the ratio estimator with the following model in the superpopulation

$$y = \beta x + \varepsilon \qquad \varepsilon \sim (0, \lambda x), \tag{5}$$

this being the model under which the ratio estimator is expected to perform well.

If we follow Brewer and Royall in obtaining results conditional on the known value of $X$, we write $\hat{Y}_R = X/p_y$, $Y = X/p_Y$, where $p_y$ and $p_Y$ are estimates of $p$ obtained from binomial samples of sizes $y$, $Y$. Then

$$\hat{Y}_R - Y = X\left(\frac{1}{p_y} - \frac{1}{p_Y}\right) \cong X(p_Y - p_y)/p^2. \tag{6}$$

Averaging over repeated selections of France itself and of the random subsample of communes drawn from France, we get

$$E(\hat{Y}_R - Y)^2 = EV(\hat{Y}_R) \cong \frac{X^2 pq}{p^4}\left(\frac{1}{y} - \frac{1}{Y}\right) \tag{7}$$

$$= \frac{X^2 q}{yp^3}\frac{(Y - y)}{Y}. \tag{8}$$

This is practically the same result for $V(\hat{Y}_R)$ as we obtained when we replaced the binomial by the hypergeometric in (2).

The reason why Laplace's variance (4) differs from (8) and from the sample estimate of (8) does not lie in his Bayesian approach, but in his assumption that France and the sample of communes were *independent* samples, whereas the latter is a subsample drawn from France.

To proceed with Laplace's analysis, he begins by assuming that the unknown $p$ follows a uniform prior $dp$ ($0 \le p \le 1$). A comment here is that, assuming almost no advance knowledge, Laplace might have tightened his prior by assuming, say, $p < 0.2$. For even if the women of childbearing age work overtime, the denominator of $p$ contains many women of nonchildbearing ages as well as men.

Given the binomial sample data from the communes ($x$ successes out of $y$ trials), the posterior distribution of $p$ is then

$$p^x(1 - p)^{y-x}\bigg/\int_0^1 p^x(1 - p)^{y-x}\,dp. \tag{9}$$

Then he assumes that he has a second independent binomial sample, France itself, which gave $X$ successes (births) out of an unknown number of trials $Y$ (the population of France). He writes $Y = \hat{Y}_R + z$ and concentrates on the distribution of $z$, the error in the ratio estimate of the population of France. The probability of $X$ successes in $(Xy/x) + z$ trials is

$$\frac{\left(\dfrac{Xy}{x} + z\right)!}{X!\left[\dfrac{X(y - x)}{x} + z\right]!}p^X(1 - p)^{[X(y-x)/x]+z} \tag{10}$$

In order to find the ordinate of the distribution of $z$ in terms of the known quantities $X$, $y$, $z$, Laplace first multiplies (10) by (9). This is where he assumes that his two samples, France and the communes, are independent. He then integrates this product with respect to $p$ from 0 to 1. He multiplies the resulting expression by $dz$ and divides by its integral with respect to $z$ from $-\infty$ to $\infty$ in order to obtain a distribution function $f(z)$ whose integral is unity. He proceeds to find an approximate form of this distribution when $x$, $X$, and $y$ are all large.

Multiplication of (10) by (9) followed by integration with respect to $p$ yields

$$P(z\,|\,X, y, x) \propto \frac{\left(\dfrac{Xy}{x} + z\right)!}{X! \left[\dfrac{X(y-x)}{x} + z\right]!} \cdot \frac{\displaystyle\int_0^1 p^{X+x}(1-p)^{[(X+x)(y-x)/x]+z}\,dp}{\displaystyle\int_0^1 p^x(1-p)^{y-x}\,dp} \qquad (11)$$

$$= \frac{(y+1)!\,(X+x)!\left(\dfrac{Xy}{x} + z\right)!\left[\dfrac{(X+x)(y-x)}{x} + z\right]!}{X!\,x!\,(y-x)!\left[\dfrac{X(y-x)}{x} + z\right]!\left[\dfrac{(X+x)y}{x} + 1 + z\right]!}. \qquad (12)$$

In finding an approximation to this multiple of the distribution of $z$, we may discard all factorials not involving $z$ which will cancel when (12) is divided by its integral over $z$. In the four large factorials in (12) involving $z$, Laplace first applied Stirling's approximation to $n!$, then expanded in a Taylor series up to terms in $z^2$. Ignoring the terms in $\sqrt{2\pi}$ we have

$$\ln[(\phi + z)!] \cong (\phi + z + \tfrac{1}{2}) \ln \phi + (\phi + z + \tfrac{1}{2}) \ln[1 + (z/\phi)] - \phi - z. \qquad (13)$$

Discarding terms in (13) not involving $z$ and using Taylor's expansion and noting that all four $\phi$ terms in (12) are large, we have

$$\ln(\phi + z)! \cong z \ln \phi + \left(\phi + z + \frac{1}{2}\right)\left(\frac{z}{\phi} - \frac{z^2}{2\phi^2}\right) - z \qquad (14)$$

$$\cong z \ln \phi + \frac{z^2}{2\phi} + \frac{z}{2\phi}. \qquad (15)$$

In (12) there are four $\phi$ terms, $Xy/x$, etc. The leading term in the sum of the four $\ln \phi$ terms in (15) works out as $-\ln[1 + x/(X + x)y] \cong -x/(X + x)y$. The leading term in the sum of the four $(1/\phi)$ terms in (15) is

$-x^3/X(X + x)y(y - x)$. Hence, from (12) and (15), if $f(z)$ is the frequency distribution of $z$,

$$\ln f(z) \cong \frac{-xz}{(X + x)y} - \frac{x^3(z^2 + z)}{2X(X + x)y(y - x)} \tag{16}$$

$$f(z) \cong \exp\left\{-\frac{1}{2}\frac{x^3}{X(X + x)y(y - x)}\left[z^2 + z - \frac{2X(y - x)}{x^2}z\right]\right\} \tag{17}$$

(apart from the multiplier needed to make the integral unity).

Thus Laplace's analysis showed the large-sample distribution of the error $z$ in the ratio estimate to be normal, with a bias whose leading term is $X(y - x)/x^2$ if $x/X$ is negligible, and a variance $X(X + x)y(y - x)/x^3$.

Given Laplace's binomial assumption and his data, the standard error of $z$ is calculated to be 107,550. This makes the odds about 300,000 to 1 against an error of more than half a million in his estimate.

It is unfortunate that Laplace should have made a mistake in probability in a book on the theory of probabilities. In his application, however, the mistake was of little consequence. His working out of the large-sample distribution of the ratio estimator and his concept of the superpopulation as a tool in studying estimates from samples are pioneering achievements.

## REFERENCES

BREWER, K. W. R. (1963). Ratio estimation in finite populations: Some results deducible from the assumption of an underlying stochastic process. *Austral. J. Statist.* **5** 93–105.

GRAUNT, J. (1662). Natural and political observations made upon the bills of mortality. Reprinted by Johns Hopkins Press, Baltimore, Maryland, 1939.

GREENWOOD, M. (1941). Medical statistics from Graunt to Farr. I. *Biometrika* **32** 101–127.

LAPLACE, P. S. (1820). A philosophical essay on probabilities. English translation, Dover, N.Y., 1951.

ROYALL, R. M. (1970). On finite population sampling theory under certain linear regression models. *Biometrika* **57** 377–387.

STEPHAN, F. F. (1948). History of the uses of modern sampling procedures. *J. Amer. Statist. Assoc.* **43** 12–39.

# 108

## *Some Reflections*

WILLIAM G. COCHRAN

According to the poet, coming events are supposed to cast their shadows before. But when Gertrude and I were on the faculty of the Statistical Laboratory at Iowa State College during the years 1938–40, I had no precognition of what, or how much, she would accomplish. She was then keeping an eye on the staff (under Mary Clem) who did the calculations—we didn't call it computing in those days—and was setting up the statistical analyses needed by Professor Snedecor's and her own extensive consultations. Her research was primarily on experimental design. In 1940 she had a paper in the *Annals* on the construction of balanced incomplete blocks, and we had a joint paper with R. C. Eckhardt on lattices and triple lattices.

She was also assembling her series of typed, mimeographed notes on standard designs. Each note indicated the kind of experiment for which the design was suitable and gave an example of its use, with the arithmetic needed to do the analysis and summarize the results. These notes led to the book *Experimental Designs,* published in 1950. I don't remember when or how we decided to write a book—the suggestion may have come from Walter Shewhart, who was then advising Wiley's on a series of books on statistics. Our division of labor was easily agreed upon. Since Gertrude disliked the job of writing even more than I did, she took responsibility for the experimental plans, and I took primary responsibility for the descriptive text and any hidden theory. Misprints and errors in the Cochran text have kept me busy at times for quite a number of years. In over 25 years, one misprint in one treatment in one plan has been brought to my attention. This performance was typical of Gertrude's work. The letter pointing out this misprint cheered me up no end.

I must not attempt to write the paper by Anderson, Monroe and Nelson that follows this note. But in trying to understand how Gertrude succeeded in doing so much for our profession, I think that she realized earlier and more keenly than almost all of us how useful statistics and the statistical point of view can be in human affairs. Her writings in the 1930's can fairly be described as "spreading the word"—explaining new techniques in language that potential users could understand. She also realized that a great deal of difficult organizational, administrative, and money-raising work would have to be done if institutions that could supply statistical education and expertise were to become available. From my contacts with Gertrude I have had no reason to think that she enjoyed this work—rather, she gritted her teeth and set herself to it. I hope, and I believe, that later she came to enjoy finding herself so good at it. Some of her proposals might sound too ambitious and difficult to attain—she had high standards—but if the proposals made sense to her she was never deterred by difficulties. Her job was to overcome them. For instance, of the faculty that she engaged at North Carolina in statistics, five members, including Gertrude herself, have been elected to the National Academy of Sciences—Bose, Cochran, Cox, Hoeffding, and Hotelling.

For handling difficulties she was gifted with unusual stamina, to which she added courage and determination. I remember her telling me, on one of her quick trips from Raleigh to Ames, how annoyed she was with herself. After driving all the way from Raleigh without rest, she found when about 20 minutes from Ames that she could not face driving another mile. She had to phone home and ask friends to come out and drive her home. Her organization work was also helped, I believe, by a keen natural interest in other people which she nurtured. In her annual Xmas letters I was always surprised at the large number of people she mentioned. She knew all their names, where they lived, and what their interests were. Her letters always wanted to know about our children and, later, our grandchildren. When Betty and I visited the Pacific and Australia for the ISI meetings, we needed a guardian for our younger daughter in case something went wrong. We asked Gertrude—that was the kind of responsibility for which Gertrude was the obvious choice—and she agreed as we knew she would.

She derived a great deal of pleasure from having many friends, from meetings with them all over the world, and from her many activities as a hostess. In 1976 she wrote: "A big treat, January 31—February 4, El Mahdy and Zinet Said were my houseguests", and later: "More houseguests, Mo Kantabutra and Dr. Charlotte Young, with talking and fun." When I visited her in Duke Hospital in November 1977, her chief concern was that the doctors might not let her home in time to be hostess to a Japanese couple who were coming to stay with her. Her professional instincts were not entirely at rest during that visit. Apparently she had not had a spell of hospitalization previously. She recited to me a list of the ways in which the running of the hospital could be improved; why hadn't these changes already been made? My explanation: "But Gertrude, you don't realize that's the way hospitals are" got nowhere.

Her professional life was rounded out very effectively by participating and delighting in all the tourism that could be attached to her world travels, by all sorts of little collections of things that interested her in her travels, by her many hobbies, by planning and building houses, and by her love of flowers. At age 76, among other travels, she went by bus from Fairbanks, Alaska, to Dawson City, Yukon. The bus broke down 80 miles from the next village. "The experiences of hitchiking were exciting, especially for the girls who rode 80 miles in a transport truck." Then on by train from Dawson City to Skagway and by boat to Vancouver. The 1976 Xmas letter ends: "All is fine, I'm just slower."

# 109

## Detecting Systematic Errors in Multi-Clinic Observational Data

NANNY WERMUTH

Johannes Gutenberg Universität, Mainz, Federal Republic of Germany

W. G. COCHRAN

Statistics Department, Harvard University, Cambridge, Massachusetts 02138, U. S. A

### Summary

*In multi-clinic studies it is hard to maintain a uniformly high quality of measurement and coding. Systematic errors almost always occur, in spite of the best of intentions and the most rigid protocols. It is the statistician's responsibility to plan for the detection of these errors, as well as to try to avoid them and not be misled by them. The practice of examining the univariate and multivariate sample frequency distributions of the variables under study, with an eye open for anything that looks puzzling, can be very helpful in detecting and trying to correct systematic errors that would bias the analysis. Examples are given from a 21-clinic study on pregnancy and child development.*

### 1. Systematic Errors

In research on variables that may be related to morbidity rates, cooperative observational studies among a number of clinics are often necessary in order to attain a desired sample size and to sample a more extensive population. A constant problem in multi-clinic studies is that of maintaining a consistently high quality of measurement and coding from clinic to clinic and of avoiding systematic errors. Misconceptions by the statistical analyst about what has been observed, measured, or defined, can have serious effects on the interpretation of the results.

### 2. Use of Frequency Distributions

As a precaution against misunderstandings of the data and undetected systematic errors, examination of the frequency distributions of all variables relevant to the relationships or hypotheses under study is often recommended before any analysis is undertaken. In particular we recommend the following steps:

(1) Examine all univariate distributions for the sample data.

(2) If results are available also for a broader population close to the target population, compare the frequency distributions in the sample and the broader population. Discrepancies may suggest something wrong with the sample measurements that needs investigation, or may give clues as to how well the sampled and target populations agree.

---

*Key Words:* Frequency distributions; Missing observations; Multi-clinic studies; Observational studies; Systematic errors.

---

(3) Compare the sample frequencies for each clinic, looking for clinics that appear to be outliers.

(4) If a substantial proportion of the measurements are missing for any variable $x_i$, investigate the reasons why. In this connection it may help to compare the distributions of other variables in the subsample in which $x_i$ was measured and in the whole sample, as will be illustrated (Example 7).

(5) Proceed similarly for two-way distributions for the main variables under study.

In examining these frequency distributions, look particularly for anything puzzling or surprising. Is the explanation a systematic error in measurement, or something wrong with a definition? Can this error be corrected for the data, or will new measurements be necessary? It is at this stage that a pilot study can be most useful. Some pilot work is always advisable in multi-clinic studies, especially in checking the clarity and understanding of questions, but its size and extent are matters of judgment. A small study may not reveal discrepancies and a large one is costly and slows up the main study.

## 3. Examples

The following examples illustrate the use of frequency distributions in seeking out systematic errors. The examples were taken from a collaborative study, in 21 clinics, of the relations between events in pregnancy and child development (DFG-Forschungsbericht 1976). The study began in 1964 in the Federal Republic of Germany. All women entered the study during their first trimester of pregnancy, returned for repeated examinations, and kept diaries on the course of their pregnancies and on the development of their children up to the age of three years whenever possible. For a report of the study, 478 classified variables for 7,870 women were available for analyses. The frequency distribution checks in the following examples are classified into univariate checks, bivariate checks, and checks involving missing values.

### 3.1. Univariate Checks

*Example 1.* In the case of a $(0,1)$ variable, the sample frequency distribution reduces to a single proportion unless we are interested in the order in which 0's and 1's appear. Two things were puzzling about the reported frequency of women who had nephritis during the year prior to the onset of pregnancy. First, there were no missing values, an answer being recorded for every woman in the sample. Did all women in fact know whether they had nephritis during the past year? Secondly, the proportion of nephritis cases in the sample was much lower than the prevalence rates reported in other studies.

The explanation here lay in the question from which the data were coded. This question did not ask directly about nephritis, being instead: "Which illnesses did you have during the past year?" Direct questions about the specific diseases, with possible answers "Yes," "No," "Don't know," are necessary in attempts to estimate prevalence rates. This is the type of issue that can be picked up in an initial pilot study before the form of the questions is fixed.

*Example 2.* The sample frequency distribution of the weights of the mothers prior to pregnancy, recorded in kg, was bimodal, showing one region of high frequencies between 50 and 70 kg, and another between 100 and 140 kg. It was clear that many women had stated their weights in lbs, contrary to instructions. Detection and correction of the weights that were recorded in lbs was not difficult, because in doubtful cases there were other weight records available from the repeated examinations during pregnancy.

*Example 3.* Each mother was asked at what age her child started to speak two-word sentences. This question, if posed correctly, meant "At what age did the child not only speak single words but combine at least two words, like 'Mama, drink,' meaning 'Mama, I want a drink.' " When the responses by clinics were compared, one of the 21 clinics appeared to be an outlier. In this clinic the ages for two-word sentences were two years or more, as against values below two years in all other clinics. Investigation showed that in this clinic the question had been interpreted as asking for the age at which the child began to speak grammatically correct subject-predicate sentences. The answers of this clinic were subsequently omitted.

### 3.2. Bivariate Checks

*Example 4.* One part of this study was a comparison of congenital defects in the child for mothers who had a subclinical infection of rubella (German measles) during the first trimester and mothers who had not. To obtain a criterion for the rubella status, titer values the rubella at the beginning and end of the first trimester were recorded on a scale: $0, 2^2, 2^3, 2^4,$ ..., $2^{12}$. A mother was coded as having had rubella if the titer value had increased by at least two steps from the beginning to the end of the first trimester.

However, a bivariate classification of the mother's initial titer value against her rubella status (yes, no) caused this definition to be reconsidered. A substantial number of women had initial rubella titer values so high that an increase by two or more units on the scale was either impossible or very unlikely. Moreover, a high initial titer was an indication of a very recent rubella infection. Mothers with such high initial titers were treated in some analyses as a third comparison group.

*Example 5.* Another bivariate table revealed a puzzling error in coding. Pregnancies were classified by duration and also by whether the pregnancy ended in an abortion or not. For a number of women with more than 260 days pregnancy, the outcome was coded as an abortion. If the code for duration was correct, such cases should have been coded as stillbirths: after 190 days of pregnancy, a fetus is generally regarded as capable of living. Since only a few cases were involved, correct codes were obtained by going back to the individual case records.

### 3.3. Checks Related to Missing Data

*Example 6.* Starting in 1968 all newborns in the study were to be examined for antibodies to rubella, the results being stated as titer-values. Examination of the titer-values for individual clinics showed that (1) some clinics had high missing-value rates for this variable, and (2) in these clinics the recorded values were in general greater than in other clinics. Why? Investigation showed that in these clinics the antibody status of the newborn was determined only when the physician saw a specific reason for it, such as a typical malformation in the body, or a high titer-value or observed infection in the mother. Inclusion of the incomplete data from these clinics in studies of factors related to the child's antibody status might result in biases. Consequently, the data were not used for this purpose but instead for investigating when a high titer-value of the mother were transferred to the child. For this analysis it was judged that the biases would be smaller.

*Example 7.* One statistical analysis dealt with the relation between the still birth rate and presence or absence of the symptoms edema, hypertension, and proteinuria in the mother at the time of birth.

However, no proteinuria was recorded for a number of mothers. On comparing the subsample in which all three study variables were measured with the complete sample, the relative frequencies of edema and hypertension, both singly and together, were found to be much lower in the subsample. Evidently, proteinuria had not been measured in a number of mothers with definite edema and hypertension symptoms. What had happened was this: When a woman in poor condition came to the clinic close to the time of birth showing either or both edema and hypertension, some physicians decide to induce labor immediately and not waste time by determining the amount of protein in the urine. In order to avoid these missing values, the symptom value for a variable at the last examination prior to birth was used if the value at birth was not available.

In conclusion, the preceding errors are all elementary, and might scarcely be worth writing about except that they can easily be overlooked by the consulting statistician who believes without checking what he or she reads or is told about the way in which the data have been measured. Moreover, visiting each clinic to inspect the measurement process in action, though highly desirable as a check, is often not feasible. Scrutiny and simple analyses of frequency distributions, as illustrated, will help to detect and correct some systematic errors in the original data.

*Résumé*

*Dans la recherche collaborative entre cliniques, il est difficile de s'assurer que les mesures qu'on va analyser sont de bonne qualité et ont le même sens dans toutes les cliniques. A ce but, il est utile à examiner, avant l'analyse, les distributions de fréquence des variables principes, en cherchant des erreurs de mesurage ou de classification qui pourraient biaiser les résultats. Le papier donne des exemples de ce pratique, priées d'une recherche entre vingt et un cliniques, sur la grosseuse et le développement de l'enfant.*

*Reference*

DFG-Forschungsbericht (1976): *Schwangerschaftsverlauf und Kindesentwicklung*, Harald Boldt Verlag, Boppard.

*Received August* 1978; *Revised April* 1979

# 110

## EXPERIMENTAL DESIGN

I. THE DESIGN OF EXPERIMENTS     *William G. Cochran*

I

### THE DESIGN OF EXPERIMENTS

In scientific research, the word "experiment" often denotes the type of study in which the investigator deliberately introduces certain changes into a process and makes observations or measurements in order to evaluate and compare the effects of different changes. These changes are called the *treatments*. Common examples of treatments are different kinds of stimuli presented to human subjects or animals or different kinds of situations with which the investigator faces them, in order to see how they respond. In exploratory work, the objective may be simply to discover whether the stimuli produce any measurable responses, while at a later stage in research the purpose may be to verify or disprove certain hypotheses that have been put forward about the directions and sizes of the responses to treatments. In applied work, measurement of the size of the response is often important, since this may determine whether a new treatment is practically useful.

A distinction is often made between a controlled experiment and an uncontrolled observational study. In the latter, the investigator does not interfere in the process, except in deciding which phenomena to observe or measure. Suppose that it is desired to assess the effectiveness of a new teaching machine that has been much discussed. An observational study might consist in comparing the achievement of students in those schools that have adopted the new technique with the achievement of students in schools that have not. If the schools that adopt the new technique show higher achievement, the objection may be raised that this increase is not necessarily caused by the machine, as the schools that have tried a new method are likely to be more enterprising and successful and may have students who are more competent and better prepared. Examination of previous records of the schools may support these criticisms. In a proper experiment on the same question, the investigator decides which students are to be taught by the new machine and which by the standard technique. It is his responsibility to ensure that the two techniques are compared on students of equal ability and degree of preparation, so that these criticisms no longer have validity.

The advantage of the proper experiment over the observational study lies in this increased ability to elucidate cause-and-effect relationships. Both types of study can establish associations between a stimulus and a response; but when the investigator is limited to observations, it is hard to find a situation in which there is only one explanation of the association. If the investigator can show by repeated experiments that the same stimulus is always followed by the same response and if he has designed the experiments so that other factors that might produce this response are absent, he is in a much stronger position to claim that the stimulus causes the response. (However, there are many social science fields where true experimentation is not possible and careful observational investigations are the only source of information.) [See, *for example*, EXPERIMENTAL DESIGN, *article on* QUASI-EXPERIMENTAL DESIGN; SURVEY ANALYSIS. *See also* Becker 1968; Wax 1968.]

Briefly, the principal steps in the planning of a controlled experiment are as follows. The treatments must be selected and defined and must be relevant to the questions originally posed. The *experimental units* to which the treatments are to be applied must be chosen. In the social sciences, the experimental unit is frequently a single animal or human subject. The unit may, however, be a group of subjects, for instance, a class in comparisons of teaching methods. An important point is that the choice of subjects and of the environmental conditions of the experiment determine the range of validity of the results.

The next step is to determine the size of the sample—the number of subjects or of classes. In general, the precision of the experiment increases as the sample size increases, but usually a balance must be struck between the precision desired and the costs involved. The method for allocating treatments to subjects must be specified, as must the detailed conduct of the experiment. Other factors that might influence the outcome must be controlled (by *blocking* or *randomization*, as discussed later) so that they favor each treatment equally. The responses or criteria by which the treatments will be rated must be defined. These may be simple classifications or measurements on a discrete or continuous scale. Like the treatments, the responses must be relevant to the questions originally posed.

When data from the experiment become available, the statistical analysis of these data and the preparation of a report on the results of the experiment are final steps. This report faces a number of questions. Have differences in the effects of different treatments been clearly shown? What can be said about the sizes of these differences? To what types of experimental unit, for example, of human subjects, can the reader safely apply the reported results?

**History.** The early history of ideas on the planning of experiments appears to have been but little studied (Boring 1954). Modern concepts of experimental design are due primarily to R. A. Fisher, who developed them from 1919 to 1930 in the planning of agricultural field experiments at the Rothamsted Experimental Station in England. The main features of Fisher's approach are as follows (*randomization, blocking,* and *factorial experimentation* will be discussed later):

(1) The requirement that an experiment itself furnish a meaningful estimate of the underlying variability to which the measurements of the responses to treatments are subject.

(2) The use of randomization to provide these estimates of variability.

(3) The use of blocking in order to balance out known extraneous sources of variation.

(4) The principle that the statistical analysis of the results is determined by the way in which the experiment was conducted.

(5) The concept of factorial experimentation, which stresses the advantages of investigating the effects of different factors or variables in a single complex experiment, instead of devoting a separate experiment to each factor.

These ideas were stated very concisely by Fisher in 1925 and 1926 but more completely in 1935.

## Experimental error

**Some sources of experimental error.** A major problem in experimentation is that the responses of the experimental units are influenced by many sources of variation other than the treatments. For example, subjects differ in their ability to perform a task under standard conditions: a treatment that is allotted to an unusually capable group of subjects will appear to do well; the instruments by which the responses are measured may be liable to errors of measurement; both the applied treatment and the environment may lack uniformity from one occasion to another.

In some experiments, the effects of subject-to-subject variation are avoided by giving every treatment to each subject in succession, so that comparisons are made within subjects. Even then, however, learning, fatigue, or delayed consequences of previously applied treatments may influence the response actually measured after a particular treatment.

The primary consequence of extraneous sources of variation, called *experimental errors,* is a masking of the effects of the treatments. The observed difference between the effects of two treatments is the sum of the true difference and a contribution due to these errors. If the errors are large, the experimenter obtains a poor estimate of the true difference; then the experiment is said to be of low precision.

*Bias.* It is useful to distinguish between random error and error due to bias. A bias, or systematic error, affects alike all subjects who receive a specific treatment. Random error varies from subject to subject. In a child growth study in which children were weighed in their clothes, a bias would arise if the final weights of all children receiving one treatment were taken on a cold day, on which heavy clothing was worn, while the children receiving a second treatment were weighed on a mild day, on which lighter clothing was worn. In general, bias cannot be detected in the analysis of the results, so that the conclusions drawn by statistical methods about the true effects of the treatments are misleading.

It follows that constant vigilance against bias is one of the requisites of good experimentation. The devices of randomization and blocking, if used intelligently, do much to guard against bias. Additional precautions are necessary in certain types of experiments. If the measurements are subjective evaluations or clinical judgments, the expectations and prejudices of the judges and subjects may influence the results if it is known which treatment any of the subjects received. Consequently, it is important to ensure, whenever it is feasible, that neither the subject nor the person taking the measurement knows which treatment the subject is receiving; this is called a "double blind" experiment. For example, in experiments that compare different drugs taken as pills all the pills should look alike and be administered in the same way. If there is a no-drug treatment, it is common practice to administer an inert pill, called a *placebo,* in order to achieve this concealment.

▶ Even in well-planned experiments, bias may enter in more subtle forms. In education, a new method of teaching may be highly successful in early experiments as compared with standard methods, partly because it is a welcome change from a standard routine. It may be much less successful when it has been widely adopted and is no longer a novelty. Many experiments on human behavior and resistance to stress and pain are carried out on subjects who are volunteers in one sense or another; because they may behave or react differently from nonvolunteers, conclusions from the experiments have limited applicability. In their book on the volunteer subject, Rosenthal and Rosnow (1975) quote McNemar's earlier statement (1946) "The existing science of human behavior is largely the science of the behavior of sophomores."

**Methods for reducing experimental error.** Several devices are used to remove or decrease bias and random errors due to extraneous sources of variation that are thought to be substantial. One group of devices may be called refinements of technique. If the response is the skill of the subject in performing an unfamiliar task, a major source of error may be that subjects learn this task at different rates. An obvious precaution is to give each subject enough practice to reach his plateau of skill before starting the experiment. The explanation of the task to the subjects must be clear; otherwise, some subjects may be uncertain what they are supposed to do. Removal from an environment that is noisy and subject to distractions may produce more uniform performance. The tasks assigned to the subjects may be too easy or too hard so that all perform well or poorly under any treatment, making discrimination between the treatments impossible. The reduction of errors in measurement of the response often requires prolonged research. In psychometrics, much of the work on scaling is directed toward finding superior instruments of measurement [see SCALING].

*Blocking.* In many experiments involving comparisons between subjects, the investigator knows that the response will vary widely from subject to subject, even under the same treatment. Often it is possible to obtain beforehand a measurement that is a good predictor of the response of the subject. A child's average score on previous tests in arithmetic may predict well how he will perform on an arithmetic test given at the end of a teaching experiment. Such initial data can be used to increase the precision of the experiment by forming blocks consisting of children of approximately equal ability. If there are three teaching methods, the first block contains the three children with the best initial scores. Each child in this block is assigned to a different teaching method. The second block contains the three next best children, and so on. The purpose of the blocking is to guarantee that each teaching method is tried on an equal number of good, moderate, and poor performers in arithmetic. The resulting gain in precision may be striking.

The term "block" comes from agricultural experimentation in which the block is a compact piece of land. With human subjects, an arrangement of this kind is sometimes called a *matched pairs* design (with two treatments) or a *matched groups* design (with more than two treatments).

A single blocking can help to balance out the effects of several different sources of variation. In a two-treatment experiment on rats, a block comprising littermates of the same sex equalizes the two treatments for age and sex and to some extent for genetic inheritance and weight also. If the conditions of the experiment are subject to uncontrolled time trends, the two rats in a block can be tested at approximately the same time.

*Adjustments in the statistical analysis.* Given an initial predictor, $x$, of the final response, $y$, an alternative to blocking is to make adjustments in the statistical analysis in the hope of removing the influence of variations in $x$. If $x$ and $y$ represent initial and final scores in a test of some type of skill, the simplest adjustment is to replace $y$ by $y - x$, the improvement in score, as the measure of response. This change does not always increase precision. The error variance of $y - x$ for a subject may be written $\sigma_y^2 + \sigma_x^2 - 2\rho\sigma_y\sigma_x$, where $\rho$ is the correlation between $y$ and $x$. This is less than $\sigma_y^2$ only if $\rho$ exceeds $\sigma_x/2\sigma_y$.

A more accurate method of adjustment is given by the analysis of covariance. In this approach, the measure of response is $y - bx$. The quantity $b$, computed from the results of the experiment, is an estimate of the average change in $y$ per unit increase in $x$. The adjustment accords with common sense. If the average $x$ value is three units higher for treatment $A$ than for treatment $B$, and if $b$ is found to be $\frac{2}{3}$, the adjustment reduces the difference between the average $y$ values by two units.

If the relation between $y$ and $x$ is linear, the use of a predictor, $x$, to form blocks gives about the same increase in precision as its use in a covariance analysis. For a more detailed comparison in small experiments, see Cox (1957). Blocking by means of $x$ may be superior if the relation between $y$ and $x$ is not linear. Thus, a covariance adjustment on $x$ is helpful mainly when blocking has been used to balance out some other variable or when blocking by means of $x$ is, for some reason, not feasible. One disadvantage of the covariance adjustment is that it requires considerable extra computation. A simpler adjustment such as $y - x$ is sometimes preferred even at some loss of precision.

*Randomization.* Randomization requires the use of a table of random numbers, or an equivalent device to decide some step in the experiment, most frequently the allotment of treatments to subjects [see RANDOM NUMBERS].

Suppose that three treatments—A, B, C—are to be assigned to 90 subjects without blocking. The subjects are numbered from 1 to 90. In a two-digit column of random numbers, the numbers 01 to 09 represent subjects 1 to 9, respectively; the numbers 10 to 19 represent subjects 10 to 19, respectively, and so on. The numbers from 91 to 99 and the number 00 are ignored. The 30 subjects whose numbers are drawn first from the table are assigned to treatment A, the next 30 to B, and the remaining 30 to C.

In the simplest kind of blocking, the subjects or experimental units are arranged in 30 blocks of three subjects each. One in each block is to receive A, one B, and one C. This decision is made by randomization, numbering the subjects in any block from 1 to 3 and using a single column of random digits for the draw.

Unlike blocking, which attempts to eliminate the effects of an extraneous source of variation, randomization merely ensures that each treatment has an equal chance of being favored or handicapped by the extraneous source. In the blocked experiment above, randomization might assign the best subject in every block to treatment A. The probability that this happens is, however, only 1 in $3^{30}$. Whenever possible, blocking should be used for all major sources of variation, randomization being confined to the minor sources. The use of randomization is not limited to the allotment of treatments to subjects. For example, if time trends are suspected at some stage in the experiment, the order in which the subjects within a block are processed may be randomized. Of course, if time trends are likely to be large, blocking should be used for them as well as randomization, as illustrated later in this article by the crossover design.

In his *Design of Experiments*, Fisher illustrated how the act of randomization often allows the investigator to carry out valid tests for the treatment means without assuming the form of the frequency distribution of the data (1935). The calculations, although tedious in large experiments, enable the experimenter to free himself from the assumptions required in the standard analysis of variance. Indeed, one method of justifying the standard methods for the statistical analysis of experimental results is to show that these methods usually give serviceable approximations to the results of randomization theory [see NONPARAMETRIC STATISTICS; see also Kempthorne 1952].

*Size of experiment.*   An important practical decision is that affecting the number of subjects or experimental units to be included in an experiment. For comparing a pair of treatments there

are two common approaches to this problem. One approach is to specify that the observed difference between the treatment means be correct to within some amount ±d chosen by the investigator. The other approach is to specify the power of the test of significance of this difference.

Consider first the case in which the response is measured on a continuous scale. If $\sigma$ is the standard deviation per unit of the experimental errors and if each treatment is allotted to $n$ units, the standard error of the observed difference between two treatment means is $\sqrt{2}\,\sigma/\sqrt{n}$ for the simpler types of experimental design. Assuming that this difference is approximately normally distributed, the probability that the difference is in error by more than $d = 1.96\,\sqrt{2}\,\sigma/\sqrt{n}$ is about 0.05 (from the normal tables). The probability becomes 0.01 if $d$ is increased to $2.58\,\sqrt{2}\,\sigma/\sqrt{n}$. Thus, although there is no finite $n$ such that the error is certain to be less than $d$, nevertheless, from the normal tables, a value of $n$ can be computed to reduce the probability that the error exceeds $d$ to some small quantity $\alpha$ such as 0.05. Taking $\alpha = 0.05$ gives $n = 7.7\sigma^2/d^2 \cong 8\sigma^2/d^2$. The value of $\sigma$ is usually estimated from previous experiments or preliminary work on this experiment.

If the criterion is the proportion of units that fall into some class (for instance, the proportion of subjects who complete a task successfully), the corresponding formula for $n$, with $\alpha = 0.05$, is

$$n \cong 4[p_1(1 - p_1) + p_2(1 - p_2)]/d^2,$$

where $p_1$, $p_2$ are the true proportions of success for the two treatments and $d$ is the maximum tolerable error in the observed difference in proportions. Use of this formula requires advance estimates of $p_1$ and $p_2$. Fortunately, if these lie between 0.3 and 0.7 the quantity $p(1 - p)$ varies only between 0.21 and 0.25.

The choice of the value of $d$ should, of course, depend on the use to be made of the results, but an element of judgment often enters into the decision.

The second approach (specifying the power) is appropriate, for instance, when a new treatment is being compared with a standard treatment and when the investigator intends to discard the new treatment unless the test of significance shows that it is superior to the standard. He does not mind discarding the new treatment if its true superiority is slight. But if the true difference (new − standard) exceeds some amount, $\Delta$, he wants the probability of finding a significant difference to have some high value, $\beta$ (perhaps 0.95, 0.9, or 0.8).

With continuous data, the required value of $n$

is approximately

$$n \cong 2\sigma^2(\xi_\alpha + \xi_{1-\beta})^2/\Delta^2,$$

where

$\xi_\alpha$ = normal deviate corresponding to the significance level, $\alpha$, used in the test of significance,

and

$\xi_{1-\beta}$ = normal deviate for a *one-tailed* probability $1 - \beta$.

For instance, if the test of significance is a one-tailed test at the 5% level and $\beta$ is 0.9, so that $\xi_\alpha = 1.64$ and $\xi_{1-\beta} = 1.28$, then $n \cong 17\sigma^2/\Delta^2$. The values of $\Delta$, $\alpha$, and $\beta$ are chosen by the investigator With proportions, an approximate formula is

$$n \cong 2(\xi_\alpha + \xi_{1-\beta})^2 \bar{p}\bar{q}/(p_2 - p_1)^2,$$

where $\bar{p} = (p_1 + p_2)/2$ and $\bar{q} = 1 - \bar{p}$ and $p_2 - p_1$ is the size of difference to be detected. One lesson that this formula teaches is that large samples are needed to detect small or moderate differences between two proportions. For instance, with $p_1 = 0.3$, $p_2 = 0.4$, $\alpha = 0.05$ (two-tailed), and $\beta = 0.8$, the formula gives $n = 357$ in each sample, or a total of 714 subjects.

More accurate tables for $n$, with proportions and continuous data, are given in Cochran and Cox (1950) and a fuller discussion of the sample size problem in Cox (1958).

If the investigator is uncertain about the best values to choose for $\Delta$, it is instructive to compute the value of $\Delta$ that will be detected, say with probability 80% or 90%, for an experiment of the size that is feasible. Some experiments, especially with proportions, are almost doomed to failure, in the sense that they have little chance of detecting a true difference of the size that a new treatment is likely to produce. It is well to know this before doing the experiment.

*Controls.* Some experiments require a *control*, or comparison, treatment. For a discussion of the different meanings of the word "control" and an account of the history of this device, see Boring (1954). In a group of families having a prepaid medical care plan, it is proposed to examine the effects of providing, over a period of time, additional free psychiatric consultation. An intensive initial study is made of the mental health and social adjustment of the families who are to receive this extra service, followed by a similar inventory at the end. In order to appraise whether the differences (final − initial) can be attributed to the psychiatric guidance, it is necessary to include a control group of families, measured at the beginning and at the end, who do not receive this service. An argument might also be made for a second control group that does not receive the service and is measured only at the end. The reason is that the initial psychiatric appraisal may cause some families in the first control group to seek psychiatric guidance on their own, thus diluting the treatment effect that is to be studied. Whether such disturbances are important enough to warrant a second control is usually a matter of judgment.

The families in the control groups, like those in the treated group, must be selected by randomization from the total set of families available for the experiment. This type of evaluatory study presents other problems. It is difficult to conceal the treatment group to which a family belongs from the research workers who make the final measurements, so that any preconceptions of these workers may vitiate the results. Second, the exact nature of the extra psychiatric guidance can only be discovered as the experiment proceeds. It is important to keep detailed records of the services rendered and of the persons to whom they were given.

### Factorial experimentation

In many programs of research, the investigator intends to examine the effects of several different types of variables on some response (for example, in an experiment on the accuracy of tracking, the effect of speed of the object, the type of motion of the object, and the type of handle used by the human tracker). In factorial designs, these variables are investigated simultaneously in the same experiment. The advantages of this approach are that it makes economical use of resources and provides convenient data for studying the interrelationships of the effects of different variables.

These points may be illustrated by an experiment with three factors or variables, $A$, $B$, and $C$, each at two levels (that is, two speeds of the object, etc.). Denote the two levels of $A$ by $a_1$ and $a_2$, and similarly for $B$ and $C$. The treatments consist of all possible combinations of the levels of the factors. There are eight combinations:

(1) $a_1 b_1 c_1$    (3) $a_1 b_2 c_1$    (5) $a_1 b_1 c_2$    (7) $a_1 b_2 c_2$
(2) $a_2 b_1 c_1$    (4) $a_2 b_2 c_1$    (6) $a_2 b_1 c_2$    (8) $a_2 b_2 c_2$

Suppose that one observation is taken on each of the eight combinations. What information do these give on factor $A$? The comparison $(2) - (1)$, that is, the difference between the observations for combinations $(2)$ and $(1)$, is clearly an estimate of the difference in response, $a_2 - a_1$, since the factors $B$ and $C$ are held fixed at their lower levels. Similarly, $(4) - (3)$ gives an estimate of $a_2 - a_1$, with $B$ held at its higher level and $C$ at its lower level. The differences $(6) - (5)$ and $(8) - (7)$ supply two further estimates of $a_2 - a_1$. The average of these four differences provides a comparison of $a_2$ with $a_1$ based on two samples of size four and is called the *main effect* of $A$.

Turning to $B$, it may be verified that $(3) - (1)$, $(4) - (2)$, $(7) - (5)$, and $(8) - (6)$ are four comparisons of $b_2$ with $b_1$. Their average is the main effect of $B$. Similarly, $(5) - (1)$, $(6) - (2)$, $(7) - (3)$, and $(8) - (4)$ provide four comparisons of $c_2$ with $c_1$.

Thus the testing of eight treatment combinations in the factorial experiment gives estimates of the effects of each of the factors $A$, $B$, and $C$ based on samples of size four. If a separate experiment were devoted to each factor, as in the "one variable at a time" approach, 24 combinations would have to be tested (eight in each experiment) in order to furnish estimates based on samples of size four. The economy in the factorial approach is achieved because every observation contributes information on all factors.

In many areas of research, it is important to study the relations between the effects of different factors. Consider the following question: Is the difference in response between $a_2$ and $a_1$ affected by the level of $B$? The comparison

$$(a_2b_2 - a_1b_2) - (a_2b_1 - a_1b_1),$$

where each quantity has been averaged over the two levels of $C$, measures the difference between the response to $A$ when $B$ is at its higher level and the response to $A$ when $B$ is at its lower level. This quantity might be called the effect of $B$ on the response to $A$. The same expression rearranged as follows,

$$(a_2b_2 - a_2b_1) - (a_1b_2 - a_1b_1),$$

also measures the effect of $A$ on the response to $B$. It is called the *AB two-factor interaction*. (Some writers introduce a multiplier, $\frac{1}{2}$, for conventional reasons.) The $AC$ and $BC$ interactions are computed similarly.

The analysis can be carried further. The $AB$ interaction can be estimated separately for the two levels of $C$. The difference between these quantities is the effect of $C$ on the $AB$ interaction. The same expression is found to measure the effect of $A$ on the $BC$ interaction and the effect of $B$ on the $AC$ interaction. It is called the *ABC three-factor interaction*.

The extent to which different factors exhibit interactions depends mostly on the way in which nature behaves. Absence of interaction implies that the effects of the different factors are mutually additive. In some fields of application, main effects are usually large relative to two-factor interactions, and two-factor interactions are large relative to three-factor interactions, which are often negligible. Sometimes a transformation of the scale in which the data are analyzed removes most of the interactions [see STATISTICAL ANALYSIS, SPECIAL PROBLEMS OF, *article on* TRANSFORMATIONS OF DATA]. There are, however, many experiments in which the nature and the sizes of the interactions are of primary interest.

The factorial experiment is a powerful weapon for investigating responses affected by many stimuli. The number of levels of a factor is not restricted to two and is often three or four. The chief limitation is that the experiment may become too large and unwieldy to be conducted successfully. Fortunately, the supply of rats and university students is large enough so that factorial experiments are widely used in research on learning, motivation, personality, and human engineering (see, for example, Cattell 1968).

Several developments mitigate this problem of expanding size. If most interactions may safely be assumed to be negligible, good estimates of the main effects and of the interactions considered likely to be important can be obtained from an experiment in which a wisely chosen fraction (say $\frac{1}{2}$ or $\frac{1}{4}$) of the totality of treatment combinations is tested. The device of *confounding* (see Cochran & Cox 1950, chapter 6, esp. pp. 183–186; Cox 1958, sec. 12.3) enables the investigator to use a relatively small sized block in order to increase precision, at the expense of a sacrifice of information on certain interactions that are expected to be negligible. If all the factors represent continuous variables $(x_1, x_2, \cdots)$ and the objective is to map the *response surface* that expresses the response, $y$, as a function of $x_1, x_2, \cdots$, then one of the designs specially adapted for this purpose may be used. [*For discussion of these topics, see* EXPERIMENTAL DESIGN, *article on* RESPONSE SURFACES; *see also* Cox 1958; Davies 1954.]

In the remainder of this article, some of the commonest types of experimental design are outlined.

*Randomized groups.* The randomized group arrangement, also called the one-way layout, the simple randomized design, and the completely randomized design, is the simplest type of plan. Treatments are allotted to experimental units at random, as described in the discussion of "Randomization."

above. No blocking is used at any stage of the experiment; and, since any number of treatments and any number of units per treatment may be employed, the design has great flexibility. If mishaps cause certain of the responses to be missing, the statistical analysis is only slightly complicated. Since, however, the design takes no advantage of blocking, it is used primarily when no criteria for blocking are available, when criteria previously used for blocking have proved ineffective, or when the response is not highly variable from unit to unit.

*Randomized blocks.* If there are $v$ treatments and the units can be grouped into blocks of size $v$, such that units in the same block are expected to give about the same final response under uniform treatment, then a randomized blocks design is appropriate. Each treatment is allotted at random to one of the units in any block. This design is, in general, more precise than randomized groups and is very extensively used.

Sometimes the blocks are formed by assessing or scoring the subjects on an initial variable related to the final response. It may be of interest to examine whether the comparative effects of the treatments are the same for subjects with high scores as for those with low scores. This can be done by an extension of the analysis of variance appropriate to the randomized blocks design. For example, with four treatments, sixty subjects, and fifteen blocks, the blocks might be classified into three levels, *high*, *medium*, or *low*, there being five blocks in each class. A useful partition of the degrees of freedom ($df$) in the analysis of variance of this "treatments × levels" design is as follows:

|  | df |
|---|---|
| Between levels | 2 |
| Between blocks at the same level | 12 |
| Treatments | 3 |
| Treatments × levels interactions | 6 |
| Treatments × blocks within levels | 36 |
| Total | 59 |

The mean square for interaction is tested, against the mean square for treatments × blocks within levels, by the usual $F$-test. Methods for constructing the levels and the problem of testing the overall effects of treatments in different experimental situations are discussed in Lindquist (1953). [*See* LINEAR HYPOTHESES, *article on* ANALYSIS OF VARIANCE.]

*The crossover design.* The crossover design is suitable for within-subject comparisons in which each subject receives all the treatments in succes-

sion. With three treatments, for example, a plan in which every subject receives the treatments in the order *ABC* is liable to bias if there happen to be systematic differences between the first, second, and third positions, due to time trends, learning, or fatigue. One design that mitigates this difficulty is the following: a third of the subjects, selected at random, get the treatments in the order *ABC*, a third get *BCA*, and the remaining third get *CAB*. The analysis of variance resembles that for randomized blocks except that the sum of squares representing the differences between the over-all means for the three positions is subtracted from the error sum of squares.

*The Latin square.* A square array of letters (treatments) such that each letter appears once in every row and column is called a Latin square. The following are two 4 × 4 squares.

| (1) | | | | | (2) | | | |
|---|---|---|---|---|---|---|---|---|
| C | A | B | D | | A | B | C | D |
| A | B | D | C | | B | C | D | A |
| B | D | C | A | | D | A | B | C |
| D | C | A | B | | C | D | A | B |

This layout permits simultaneous blocking in two directions. The rows and columns often represent extraneous sources of variation to be balanced out. In an experiment that compared the effects of five types of music programs on the output of factory workers doing a monotonous job, a 5 × 5 Latin square was used. The columns denoted days of the week and the rows denoted weeks. When there are numerous subjects, the design used is frequently a group of Latin squares.

For within-subject comparisons, the possibility of a residual or carry-over effect from one period to the next may be suspected. If such effects are present (and if one conventionally lets columns in the above squares correspond to subjects and rows correspond to order of treatment) then square (1) is bad, since each treatment is always preceded by the same treatment (A by C, etc.). By the use of square (2), in which every treatment is preceded once by each of the other treatments, the residual effects can be estimated and unbiased estimates obtained of the direct effects (see Cochran & Cox 1950, sec. 4.6a; Edwards 1950, pp. 274–275). If there is strong interest in the residual effects, a more suitable design is the *extra-period Latin square*. This is a design like square (2). in which the treatments C, D, A, B in the fourth period are given again in a fifth period.

*Balanced incomplete blocks.* When the number of treatments, $v$, exceeds the size of block, $k$, that appears suitable, a balanced incomplete blocks design is often appropriate. In examining the taste preferences of adults for seven flavors of ice cream in a within-subject test, it is likely that a subject can make an accurate comparison among only three flavors before his discrimination becomes insensitive. Thus $v = 7$, $k = 3$. In a comparison of three methods of teaching high school students, the class may be the experimental unit and the school a suitable block. In a school district, it may be possible to find twelve high schools each having two classes at the appropriate level. Thus $v = 3$, $k = 2$.

Balanced incomplete blocks (BIB) are an extension of randomized blocks that enable differences among blocks to be eliminated from the experimental errors by simple adjustments performed in the statistical analysis. Examples for $v = 7$, $k = 3$ and for $v = 3$, $k = 2$ are as follows (columns are blocks):

|  | | $v = 7$, $k = 3$ | | | | | $v = 3$, $k = 2$ | |
|---|---|---|---|---|---|---|---|---|---|
| A | B | C | D | E | F | G | A | B | C |
| B | C | D | E | F | G | A | B | C | A |
| D | E | F | G | A | B | C | | | |

The basic property of the design is that each pair of treatments occurs together (in the same block) equally often.

In both plans shown, it happens that each row contains every treatment. This is not generally true of BIB designs, but this extra property can sometimes be used to advantage. With $v = 7$, for instance, if the row specifies the order in which the types of ice cream are tasted, the experiment is also balanced against any consistent order effect. This extension of the BIB is known as an *incomplete Latin square* or a *Youden square*. In the high schools experiment, the plan for $v = 3$ would be repeated four times, since there are twelve schools.

**Comparisons between and within subjects.** Certain factorial experiments are conducted so that some comparisons are made within subjects and others are made between subjects. Suppose that the criterion is the performance of the subjects on an easy task, $T_1$, and a difficult task, $T_2$, each subject attempting both tasks. This part of the experiment is a standard crossover design. Suppose further that these tasks are explained to half the subjects in a discouraging manner, $S_1$, and to the other half in a supportive manner, $S_2$. It is of interest to discover whether these preliminary suggestions, $S$, have an effect on performance and whether this effect differs for easy and hard tasks. The basic plan, requiring four subjects, is shown in the first three lines of Table 1, where $O$ denotes the order in which the tasks are performed.

**Table 1**

| Subject | 1 | | 2 | | 3 | | 4 | |
|---|---|---|---|---|---|---|---|---|
| Order | $O_1$ | $O_2$ | $O_1$ | $O_2$ | $O_1$ | $O_2$ | $O_1$ | $O_2$ |
| Treatment | $T_1S_1$ | $T_2S_1$ | $T_2S_1$ | $T_1S_1$ | $T_1S_2$ | $T_2S_2$ | $T_2S_2$ | $T_1S_2$ |
| $(T_2 - T_1)$ | − | + | + | − | − | + | + | − |
| $(S_2 - S_1)$ | − | − | − | − | + | + | + | + |
| $TS$ | + | − | − | + | − | + | + | − |
| $TO$ | + | + | − | − | + | + | − | − |

The comparison $T_2 - T_1$, which gives the main effect of $T$, is shown under the treatments line. This is clearly a within-subject comparison since each subject carries a + and a −. The main effect of suggestion, $S_2 - S_1$, is a between-subject comparison: subjects 3 and 4 carry + signs while subjects 1 and 2 carry − signs. The $TS$ interaction, measured by $T_2S_2 - T_1S_2 - T_2S_1 + T_1S_1$, is seen to be a within-subject comparison.

Since within-subject comparisons are usually more precise than between-subject comparisons, an important property of this design is that it gives relatively high precision on the $T$ and $TS$ effects at the expense of lower precision on $S$. The design is particularly effective for studying interactions. Sometimes the between-subject factors involve a classification of the subjects. For instance, the subjects might be classified into three levels of anxiety, $A$, by a preliminary rating, with equal numbers of males and females of each degree included. In this situation, the factorial effects $A$, $S$ (for sex), and $AS$ are between-subject comparisons. Their interactions with $T$ are within-subject comparisons.

The example may present another complication. Subjects who tackle the hard task after doing the easy task may perform better than those who tackle the hard task first. This effect is measured by a $TO$ interaction, shown in the last line in Table 1. Note that the $TO$ interaction turns out to be a between-subject comparison. The same is true of the $TSO$ three-factor interaction.

In designs of this type, known in agriculture as *split-plot* designs, separate estimates of error are calculated for between-subject and within-subject

**Table 2**

| Source | df |
|---|---|
| **Between subjects** | |
| S | 1 |
| TO | 1 |
| TSO | 1 |
| Error b | $4(n - 1)$ |
| **Within subjects** | |
| O | 1 |
| T | 1 |
| TS | 1 |
| SO | 1 |
| Error w | $4(n - 1)$ |

comparisons. With $4n$ subjects, the partition of degrees of freedom in the example is shown in Table 2 (if it is also desired to examine the $TO$ and $TSO$ interactions).

Plans and computing instructions for all the common types of design are given in Cochran and Cox (1950); and Lindquist (1953), Edwards (1950), and Winer (1962) are good texts on experimentation in psychology and education.

WILLIAM G. COCHRAN

[Directly related are the articles under LINEAR HYPOTHESES.]

### BIBLIOGRAPHY

►BECKER, HOWARD S. 1968 Observation: I. Social Observation and Social Case Studies. Volume 11, pages 232–238 in International Encyclopedia of the Social Sciences. Edited by David L. Sills. New York: Macmillan and Free Press.

BORING, EDWIN G. 1954 The Nature and History of Experimental Control. American Journal of Psychology 67:573–589.

CAMPBELL, DONALD T.; and STANLEY, JULIAN C. 1963 Experimental and Quasi-experimental Designs for Research on Teaching. Pages 171–246 in Nathaniel L. Gage (editor), Handbook of Research on Teaching. Chicago: Rand McNally.

►CATTELL, RAYMOND B. 1968 Traits. Volume 16, pages 123–128 in International Encyclopedia of the Social Sciences. Edited by David L. Sills. New York: Macmillan and Free Press.

COCHRAN, WILLIAM G.; and COX, GERTRUDE M. (1950) 1957 Experimental Designs. 2d ed. New York: Wiley.

COX, D. R. 1957 The Use of a Concomitant Variable in Selecting an Experimental Design. Biometrika 44: 150–158.

COX, D. R. 1958 Planning of Experiments. New York: Wiley.

DAVIES, OWEN L. (editor) (1954) 1956 The Design and Analysis of Industrial Experiments. 2d ed., rev. Edinburgh: Oliver & Boyd; New York: Hafner.

EDWARDS, ALLEN (1950) 1960 Experimental Design in Psychological Research. Rev. ed. New York: Holt.

○FISHER, R. A. (1925) 1970 Statistical Methods for Research Workers. 14th ed., rev. & enl. New York: Hafner; Edinburgh: Oliver & Boyd.

FISHER, R. A. (1926) 1950 The Arrangement of Field Experiments. Pages 17 502a–17.513 in R. A. Fisher, Contributions to Mathematical Statistics. New York: Wiley. → First published in Volume 33 of the Journal of the Ministry of Agriculture.

○FISHER, R. A. (1935) 1971 The Design of Experiments. 9th ed. New York: Hafner; Edinburgh: Oliver & Boyd.

KEMPTHORNE, OSCAR 1952 The Design and Analysis of Experiments. New York: Wiley.

LINDQUIST, EVERET F. 1953 Design and Analysis of Experiments in Psychology and Education. Boston: Houghton Mifflin.

►McNEMAR, QUINN 1946 Opinion Attitude Methodology. Psychological Bulletin 43:289–374.

►ROSENTHAL, ROBERT; and ROSNOW, R. L. 1975 The Volunteer Subject. New York: Wiley.

►WAX, ROSALIE HANKEY 1968 Observation: II. Participant Observation. Volume 11, pages 238–241 in International Encyclopedia of the Social Sciences. Edited by David L. Sills. New York: Macmillan and Free Press.

WINER, B. J. 1962 Statistical Principles in Experimental Design. New York: McGraw-Hill.

## Postscript

**Social experimentation.** When a social program that is intended to solve or alleviate a social, economic, or health problem has been in progress for some time, an effort to measure the effects of the program as a guide to a policy decision about its future is a natural step. These efforts are important because many programs are costly, and experience has suggested that only a minority of new programs are successful (Light et al. 1971). The traditional method of evaluating the effects of the programs has been a post hoc observational study. But owing to such factors as poor retrospective measurement of the state of the problem before the program started, the absence of any comparable control group of people who do not receive the program, and vested interests in the continuation of the program, the evaluations were often badly biased.

Since the mid-1960s a group of social scientists have been stressing two points. Wherever feasible, an evaluation of the effects of a new social program should be conducted as a study in itself before a policy decision is made whether to put the program into effect on a large scale. Second, much more strenuous efforts should be made to use randomized experimentation with a control group in this evaluation. Experimentation on the effects of social programs involves many difficulties—a few are mentioned in the experiment on the effects of family psychiatric care described in the main article. Problems of planning, measurement, record keeping, managerial resources, cost, and interpretation of results are encountered, as is a delay—usually of some years—of the decision whether to institute the program itself on a large scale. In 1971 a committee of the Social Science Research Council decided to prepare a book, Social Experimentation (Riecken & Boruch 1974), on the available knowledge about the use of randomized experiments in planning and evaluating social programs. This book also contains examples of well-planned social experiments that have been completed—for example, on fertility control programs, on types of training for police recruits, on mental health rehabilitation programs, on teaching self-care to nursing home patients, on negative taxation, and on the absence of a bond as a requisite for bail. A companion volume, *Experimental Testing of Public Policy* (Social Science Research Council Conference on Social Experiments 1976), has also appeared. The future of this development will be important and interesting to observe.

WILLIAM G. COCHRAN

[*See also* EVALUATION RESEARCH; PUBLIC POLICY AND STATISTICS; *and the biography of* YOUDEN.]

ADDITIONAL BIBLIOGRAPHY

LIGHT, RICHARD J.; MOSTELLER, FREDERICK; and WINOKUR, H. S. JR. 1971 Using Controlled Field Studies to Improve Public Policy. Volume 2, chapter 6 in U.S. President's Commission on Federal Statistics, *Federal Statistics: Report.* Washington: Government Printing Office.

RIECKEN, HENRY W.; and BORUCH, ROBERT F. 1974 *Social Experimentation: A Method for Planning and Evaluating Social Intervention.* New York: Academic Press.

SOCIAL SCIENCE RESEARCH COUNCIL CONFERENCE ON SOCIAL EXPERIMENTS, BOULDER, 1974 1976 *Experimental Testing of Public Policy: Proceedings.* Edited by Robert F. Boruch and Henry W. Riecken. Boulder, Colo.: Westview.

This article is a revised version of one that was published in *The International Encyclopedia of Social Sciences* in 1968. New paragraphs and references are noted with → and a rewritten paragraph with a ◯. The postscript and additional bibliography are also new.

FISHER AND THE ANALYSIS OF VARIANCE

William G. Cochran


## 1. Introduction

The development of the analysis of variance and many of its applications is one of the main evidences of Fisher's genius. In this lecture I have described some of Fisher's papers on analysis of variance that particularly interested me. The first paper on this topic (with W.A. Mackenzie) appeared in 1923 [CP 32]. Two aspects of this paper are of historical interest. At that time Fisher did not fully understand the rules of the analysis of variance -- his analysis is wrong -- nor the role of randomization. Secondly, although the analysis of variance is closely tied to additive models, Fisher rejects the additive model in his first analysis of variance, proceeding to a multiplicative model as more reasonable.

Three years later Fisher had a fairly complete mastery of the main ramifications of the analysis of variance, including tests of significance, the role of randomization in providing estimates of error from the results of an experiment, the use of blocking, the analysis of split-plot experiments, factorial experimentation and confounding, as well as the null distribution of $z = \frac{1}{2} \ln F$ and its relation to the $\chi^2$, t, and normal distributions.

Other works referred to in this lecture are a fairly intricate early example of confounding (the experiment was conducted in 1927), the analysis of covariance, work on the analysis of variance of Poisson or binomial data, and an attempt by Fisher to derive the distribution theory for a non-linear regression problem. These examples give an indication of Fisher's point of view and of the variety of problems that he tackled.

## 2. 1923: The First Paper on the Analysis of Variance [CP 32]

The first published paper with an analysis of variance was the analysis by Fisher and Mackenzie of the results of a 2×12×3 factorial experiment on potatoes, which appeared in 1923. The factors were: (1) farmyard manure (dung) -- absent

and present, (2) twelve standard varieties of potatoes, and (3) no potash, potassium sulphate, and potassium chloride. The experiment was laid out in a nested or
split-split plot design.

The field was divided into two large pieces or blocks of land, one of which
was given dung, the other not. Thus no estimate of the standard error for the
average effect of dung was available, since this factor appeared in only single
replication. Each block had 36 plots, containing the 12 varieties in three-fold
replication, except that in the undunged block the variety Kitchener of Khartoum
appeared only twice. This arrangement provides 47 degrees of freedom for testing
the mean square for Varieties and the Varieties × Dung interactions. Fisher indicates particular interest in testing Variety × Manures interactions, since only in
their absence could varietal comparisons made on a single manure, or manurial
comparisons made on a single variety, be trusted.

No randomization was used in this layout. Following the procedure recommended
at that time, the layout apparently attempts to minimize the errors of the differences between treatment means by using a chessboard arrangement that places different treatments near one another so far as is feasible. This arrangement utilizes
the discovery from uniformity trials that plots near one another in a field tend
to give closely similar yields. A consequence is, of course, that the analysis of
variance estimate of the error variance per plot, which is derived from differences
in yield between plots receiving the same treatment, will tend to overestimate,
since plots treated alike are farther apart than plots receiving different treatments. Fisher does not comment on the absence of randomization or on the chessboard design. Apparently in 1923 he had not begun to think about the conditions
necessary for an experiment to supply an unbiased estimate of error.

As regards the third factor, each plot contained three rows, one receiving
potassium sulphate, one potassium chloride, and one no potassium. Thus the K
comparisons and the VK, DK, and VDK interactions were made on a sub-plot basis.
The analysis of variance that would now be considered appropriate for this experiment is shown in Table 1.

Table 1: Analysis of Variance for the Split-Split Plot Experiment

| Source of Variation | d.f. |
|---|---|
| Blocks (D) | 1 |
| V | 11 |
| VD | 11 |
| Between-plot error | 47 |
| K | 2 |
| VK | 22 |
| DK | 2 |
| VDK | 22 |
| Within-plot error | 94 |
| Total | 212 |

The Fisher-Mackenzie analysis of variance (Table 2) differs from Table 1 in two respects. They did not realize that the split-split feature necessitated two separate estimates of error. They combined the two estimates, giving 94 + 47 = 141 d.f. Also they did not distinguish D and K as separate factors, reporting 3 × 2 = 6 manures.

Table 2: Fisher and Mackenzie's Analysis of Variance

| Source of Variation | d.f. | Mean Square | F |
|---|---|---|---|
| Manuring | 5 | 1231.6 | 98.77 |
| Variety | 11 | 258.5 | 20.73 |
| Deviations from summation formula | 55 | 17.84 | 1.43 |
| Variation between parallel plots | 141 | 12.47 | |
| Total | 212 | – | |

This section of their paper is described as "analysis of variation on the sum basis" and in Table 2, the line "Deviations from summation formula", with 55 d.f., represents the Varieties × Manures interactions.

At that time Fisher had not published or tabulated the null distribution of F. For the Varieties × Manures interactions, his analysis gives F = 17.84/12.47 = 1.43 with 55 and 141 d.f. In testing this value, he uses an approximation to the null F-distribution that works well enough if the degrees of freedom exceed 20. He assumes that $\ln s_i$ is normal with mean $\ln \sigma_i$ and S.E. $= \sqrt{1/2v_i}$, where $v_i$ is the number of degrees of freedom for $s_i^2$. It follows that on the null hypothesis

$c_1 = c_2$, $\frac{1}{2} \ln F = 0.1788$ has S.E. approximately $(1/110 + 1/282)^{\frac{1}{2}} = 0.1124$. The normal deviate $0.1788/0.1124 = 1.591$, giving $P = 0.056$ in a one-tailed test. On this test the interactions mean square is not quite significant, although the authors note that the effects of both varieties and manuring are clearly significant.

Having concluded this section, the authors continue:

> The above test is only given as an illustration of the method; the summation formulae for combining the effects of variety and manurial treatment is evidently quite unsuitable for the purpose. No one would expect to obtain from a low yielding variety the same actual increase in yield which a high yielding variety would give; the falsity of such an assumption is emphasized by the fact that the expected values $(a + b - \bar{x})$ calculated on such an assumption, are often negative in the unmanured series. A far more natural assumption is that the yield should be the product of two factors, one depending on the variety and the other on the manure.

The remark about negative expected values may be an exaggeration. In this experiment, I found only one, under the additive model, that for the undunged rows receiving no potash, for which the expectation is $(6.03 + 4.47 - 11.76) = -1.26$. The logical point that a summation formula can lead to negative expected values is, however, sound.

### 3. The Product Formula -- Multiplicative Effects

If $x_{ij}$ is the mean yield of the ith manure and the jth variety, the model for the product formula is

$$x_{ij} = a_i b_j + \epsilon_{ij}.$$

The normal equations for estimating the $a_i$ and $b_j$ by the method of least squares are

$$\sum_j b_j x_{ij} = a_i \sum_j b_j^2, \tag{1}$$

$$\sum_i a_i x_{ij} = b_j \sum_i a_i^2. \tag{2}$$

Fisher notes that these equations may be solved by iteration. From first approximations to the $a_i$, equation (2) gives first approximations to the $b_j$, then equation (1) gives second approximations to the $a_i$, and so on until the solutions converge.

By this method he obtains a residual sum of squares of 847 on a single-observation basis.

As an alternative route, he notes that equations (1) and (2) give

$$\sum_i \sum_j a_i b_j x_{ij} = (\sum_i a_i^2)(\sum_j b_j^2) = \lambda. \tag{3}$$

From equation (3), it follows that the residual sum of squares is

$$\sum_i \sum_j (x_{ij} - a_i b_j)^2 = \sum_i \sum_j x_{ij}^2 - \lambda,$$

so that the value of $\lambda$ is the reduction in sum of squares due to the product model and leads to the residual sum of squares.

Further, if equation (1), with k instead of j as the running subscript, is used to eliminate the $a_i$ from equation (2), we get

$$\sum_i b_i (\sum_i x_{ij} x_{ik}) = b_j \sum_i a_i^2 \sum_k b_k^2 = b_j \lambda. \tag{4}$$

Hence if

$$c_{jk} = \sum_i x_{ij} x_{ik},$$

equation (4) shows that $\lambda$ is a root of the determinantal equation

$$|c_{jk} - \lambda I| = 0.$$

By solving this equation Fisher finds the residual sum of squares to be 846.3, confirming the iterative solutions. The residual mean square is $846.3/55 = 15.39$, slightly better than the residual mean square 17.8 from the additive model, as Fisher seems to have expected. Rather surprisingly, practically all of Fisher's later work on the analysis of variance uses the additive model. Later papers give no indication as to why the product model was dropped. Perhaps Fisher found, as I did, that the additive model is a good approximation unless main effects are large, as well as being simpler to handle than the product model.

## 4. The Distribution of $z = \frac{1}{2} \ln F$

In the next year, 1924, Fisher presented the null distribution of $z = \frac{1}{2} \ln F$ in an informative paper at the International Mathematical Congress at Toronto [CP 36]. This paper showed the relation between the z-distribution and the $\chi^2$, normal, and t-distributions. I understand that the Toronto paper was not published until 1928.

Fisher does not give an analytical proof of the distribution, but remarks that he found it first when studying the distribution of the intraclass correlation coefficient. With n classes of size g, the covariance between members of the same class was estimated as

$$2 \sum_{i=1}^{n} \sum_{j=1}^{g} \sum_{k>g}^{g} (x_{ij}-\bar{x})(x_{ik}-\bar{x})/ng(g-1), \tag{6}$$

the number of products being $ng(g-1)/2$. With the variance of the $x_{ij}$ estimated as

$$s^2 = \sum_{1}^{n} \sum_{1}^{g} (x_{ij}-\bar{x})^2/ng, \tag{7}$$

we get for the intraclass correlation coefficient

$$r = \frac{2 \sum_{i=1}^{n} \sum_{j=1}^{g} \sum_{k>j}^{g} (x_{ij}-\bar{x})(x_{ik}-\bar{x})}{(g-1) \sum_{i=1}^{n} \sum_{j=1}^{g} (x_{ij}-\bar{x})^2} . \tag{8}$$

Harris (1913) noted that with g large, the number of cross-product terms made r in equation (8) tedious to calculate. In finding a quicker method of calculating r, he established the link between r and the analysis of variance by proving that if SSB, SSW denote the sums of squares between and within groups,

$$\frac{SSB}{SSW} = \frac{(n-1)}{n(g-1)} F = \frac{\nu_1}{\nu_2} F = \frac{1 + (g-1)r}{(g-1)(1-r)} . \tag{9}$$

Equation (9) gives the relation between F and r. The quantities $\nu_1$ and $\nu_2$ are the degrees of freedom between and within classes, $\nu_1 = (n-1)$, $\nu_2 = n(g-1)$, so that $(g-1) = \nu_2/(\nu_1+1)$.

If c is a constant of integration, the density of r in the null case, which Fisher [CP 14] found in 1921, is

$$f(r) = c(1-r)^{(g-1)n/2 - 1} [1 + (g-1)r]^{(n-3)/2} \qquad (10)$$

Substitution of $F = e^{2z}$ for r from expression (9), plus some tidying, gives the null distribution of z:

$$c'e^{\nu_1 z} dz(\nu_2 + \nu_1 e^{2z})^{-(\nu_1+\nu_2)/2} \qquad (11)$$

This method of finding the distribution of z probably explains why the discussion of the analysis of variance in Statistical Methods for Research Workers [SMRW], which appeared in 1925, immediately follows intra-class correlation, coming in the same chapter.

### 5.  1925-26:  Further Development of Analysis of Variance -- Randomization, Blocking, Confounding

When Statistical Methods for Research Workers [SMSW] appeared in 1925, it was evident that Fisher understood the function of randomization in experimentation. He writes in [SMSW, p. 248]: "The first requirement which governs all well-planned experiments is that the experiment should yield not only a comparison of different manures, treatments, varieties, etc. but also a means of testing the significance of such differences as are observed". And later "For our test of significance to be valid the differences in fertility between plots chosen as parallels must be truly representative of the differences between plots with different treatment; and we cannot assume that this is the case if our plots have been chosen in any way according to a prearranged system".

He notes that valid tests of significance are obtained if treatments are assigned to plots wholly at random (e.g., by shuffling or by a table of random numbers). Going further, he observes that the device of blocking, as in randomized blocks and the Latin square, combined with appropriate randomization, can give increased accuracy in the comparison of different treatments without sacrificing valid tests of significance.

It is also clear that by 1925 Fisher's earlier confusion about the nature
of the experimental errors in split-plot experiments had been removed.  In
Statistical Methods for Research Workers [SMRW] he analyzes the dunged half of
the 1923 2×12×3 split-split-plot experiment as a split-plot experiment with
varieties in the plots and potash dressings in the sub-plots (rows), presenting
the usual two errors -- one for comparison among plots and one for comparisons
among sub-plots in the same plot.

A remarkable paper of 10½ small pages appeared in 1926 [CP 48], containing
many of Fisher's ideas on the layout of field experiments.  The paper, titled
"The Arrangement of Field Experiments", opens by noting the recent interest of
agriculturalists in the errors of field experiments, aided by study of the results
of uniformity trials.  It then raises the questions:  What is meant by a valid
estimate of error?  When is a result statistically significant?  The paper is,
I think, the first in which Fisher puts forward his preference for a low standard
of significance at the 5% level, noting that others may prefer a different level
(2% or 1%).  He then notes the vital role of replication in providing data
from which to estimate errors, and repeats in more detail the argument for the
importance of randomization in providing a valid estimate of error, including a
description of randomized blocks and Latin square designs as arrangements that
can give increased accuracy in the comparison of different treatments.

Then follows the advocacy of factorial experimentation, including the famous
passage: "No aphorism is more frequently repeated in connection with field
trials, than that we must ask Nature few questions, or, ideally, one question, at
a time.  The writer is convinced that this view is wholly mistaken.  Nature, he
suggests, will best respond to a logical and carefully thought out questionnaire;
indeed, if we ask her a single question, she will often refuse to answer until
some other topic has been discussed."

After giving an example of a factorial experiment, he ends this remarkable
paper by pointing out that it will sometimes be advantageous to sacrifice information on some treatment comparisons by confounding them with certain elements
of soil heterogeneity, in order to increase accuracy or other treatment comparisons
considered more important.

## 6. An Early Example of Confounding

In 1927 Rothamsted conducted a 4×3×2 factorial experiment on barley, partially confounded in blocks of 12 plots, that presented problems both in the estimation of effects and in the analysis of variance. Fisher used this example twice: in an expository monograph [CP 90] with J. Wishart on analysis of the results of field experiments, and in a revised form in section 52 of his book Design of Experiments [DOE].

In a standard 4×3×2 factorial, with for example, 4 levels of K, 3 levels of N, and 2 levels of P, all 24 treatment combinations are distinct. In blocks of 12 plots we can include all 12 KN combinations and all 6 NP combinations, so that the KN and NP interactions need not be confounded with blocks. However, since there are 8 KP treatment combinations, the KP and KNP interactions must be partially confounded in blocks of 12 plots. In the best plan, one example of the two blocks in a replicate is as follows:

Block Ia

| KNP | KNP | KNP | KNP |
|-----|-----|-----|-----|
| 000 | 100 | 201 | 301 |
| 011 | 111 | 210 | 310 |
| 021 | 121 | 220 | 320 |

Block Ib

| KNP | KNP | KNP | KNP |
|-----|-----|-----|-----|
| 001 | 101 | 200 | 300 |
| 010 | 110 | 211 | 311 |
| 020 | 120 | 221 | 321 |

With any level of K, P must appear twice at the 1 level and once at the 0 level, or vice versa. In block Ia, P appears twice at the 1 level with K0 and K1, and twice at the 0 level with K2 and K3. The result is that in the KP interaction the comparison $(K3 + K2 - K1 - K0)(P1 - P0)$ is partially confounded, while the two independent comparisons $(K3 - K2 + K1 - K0)(P1 - P0)$ and $(K3 - K2 - K1 + K0)(P1 - P0)$ are unconfounded with blocks Ia and Ib. The partially confounded comparison in KP can be estimated clear of blocks from the data in block Ia as

$$\tfrac{1}{4}[301 - \frac{(310+320)}{2} + 201 - \frac{(210+220)}{2} - \frac{(111+121)}{2} + 110 - \frac{(011+021)}{2} + 000].$$

The variance of this comparison from block Ia is therefore $6\sigma^2/16 = 3\sigma^2/8$, as against $2\sigma^2/6 = \sigma^2/3$ for the two components of KP that are unconfounded. In Fisher's terminology the units of information on the 3 comparisons in KP from block Ia are therefore $(3 + 3 + 8/3)/\sigma^2 = 26/3\sigma^2$ as against $9/\sigma^2 = 27/3\sigma^2$ if KP were unconfounded. Hence in this plan KP is confounded only to the extent 1/27. Similarly, the KNP interaction may be shown to have relative information 23/27, being confounded to the extent 4/27 or about 15%.

The extra complication in the 1927 Rothamsted barley experiment is that the factors were 4 types or qualities of N fertilizer -- sulphate of ammonia, chloride of ammonia, cyananide, and urea, 3 levels of N (0, 1, 2), and 2 levels of P (0, 1). Clearly, there is no difference among the four qualities of N at the lowest level of N where no N is given. Thus there are only 18 distinct treatment combinations in this example.

The partial confounding and the estimation of PQ and QNP must be reconsidered. In making quality comparisons we obviously want to omit the 0 level of N. The simplest method is to make them from the sum of the N1 and N2 levels, though, as we shall see, there is an argument for making them from the (N1 + 2N2) results, giving double weight to the N2 level.

The plan previously presented in blocks Ia and Ib is the one used in this experiment in two replicates, replacing K by Q. With the (N1+N2) levels, it looks at first sight as if the comparison $(Q3 + Q2 - Q1 - Q0)(P1 - P0)$ is completely confounded. When we calculate this comparison from block Ia we get

$$S = -(310 + 320 + 210 + 220 + 111 + 121 + 011 + 021)/8,$$

all terms being - with none +. But we can get an unconfounded intra-block estimate of the relevant part of this QP comparison from

$$[S + 2(000 + 100 + 201 + 301)]/8.$$

In this way blocks Ia and Ib give an intra-block estimate of the comparison $(Q3 + Q2 - Q1 - Q0)(P1 - P0)$ in QP. The variance of this estimate is $2(8 + 16)\sigma^2/64 = 3\sigma^2/4$. The two unconfounded components of QP give estimates from blocks Ia and Ib with variance $\sigma^2/4$. Fisher notes that the partially confounded component of QP has a variance 3 times that of the unconfounded components and is relatively poorly measured. He also notes that this component is the largest of the three numerically, which at first sight suggests that there was an

111.26

unlucky choice of the component to confound. However, in the final analysis of variance, QP was not significant. The overall information on all three components from blocks Ia and Ib is $(4 + 4 + 4/3)/\sigma^2 = 28/3\sigma^2$, as against $36/3\sigma^2$ with no confounding. The relative information on QP is 7/9 -- not as good as 26/27 obtained when all 24 treatment combinations were distinct.

In experiments with qualitative and quantitative factors, another question of interest is: How should quality differences be measured? It is reasonable that the difference between two qualities at the N2 level should be twice that at the N1 level. In this event the comparison (N2-N1) will show no QN interaction, making the main effects of Q larger when measured as (2N2 + N1) rather than as (N2 + N1). Fisher tried both methods, as shown in Table 3.

Table 3:  Two Methods of Measuring Quality Differences in N

| N Level | Totals (8 replicates) | | | |
|---|---|---|---|---|
| | Sulphate | Chloride | Cyanamide | Urea |
| 1 | 1524 | 1618 | 1615 | 1469 |
| 2 | 1693 | 2110 | 1607 | 1965 |

Analysis of Variance

| | Using (N2 + N1) | | Using (2N2 + N1) | |
|---|---|---|---|---|
| | d.f. | Sum of Squares | d.f. | Sum of Squares |
| Q | 3 | 21,739 | 3 | 33,032 |
| QN | 3 | 23,332 | 3 | 12,039 |
| Sum | 6 | 45,071 | 6 | 45,071 |

Evidently (2N2 + N1) works better, giving larger Q main effects and smaller QN interactions.

The analysis of variance for the whole experiment is shown in Table 4. The linear response to N is large (74%), while the effects of P and the quality differences are both significant at the 1% level. None of the overall interactions is significant.

Table 4: Analysis of Variance of the 4×3×2 Experiment

| Source of Variation | d.f. | Sum of Squares | Mean Squares | F |
|---|---|---|---|---|
| Blocks | 3 | 12,216 | | |
| N linear | 1 | 308,505 | 308,505 | 138.1** |
| N quadratic | 1 | 7,420 | 7,420 | 4.9* |
| P | 1 | 18,881 | 18,881 | 12.5** |
| Q | 3 | 33,032 | 11,011 | 7.3** |
| QN | 3 | 12,039 | 4,013 | 2.7 |
| QP | 3 | 7,870 | 2,623 | 1.7 |
| QNP | 3 | 6,634 | 2,211 | 1.5 |
| Error | 27 | 40,763 | 1,510 | - |

## 7. Analysis of Covariance

Fisher introduced the analysis of covariance in the fourth edition of Statistical Methods [SMRW] in 1932. The application was not to an actual experiment but to uniformity trial data Y of 16 plots of tea bushes, arranged in a 4×4 square. The covariate X consisted of the yields of the same 16 plots in the preceding growth period. If the experiment is laid out in a 4×4 Latin square, the error mean square is found to be 97.2. On the other hand, if the previous yields X are used to form four randomized blocks, maximizing the previous differences among blocks, the error mean square in the current period (Y data) is reduced to 40.7. Use of the Latin square plan plus the covariance adjustments on previous yields does still better, reducing the error mean square to 15.1, a very marked gain in accuracy over the original Latin square.

Since there are no treatments in this example, a description of the F-test of the adjusted treatment means was postponed until the 5th edition of Statistical Methods [SMRW, 1934].

## 8. Analysis of Variance for Poisson and Binomial Data

In 1940 Cochran published a paper on the analysis of variance for data that follow the Poisson or binomial laws. With Poisson data, the method recommended at that time by Bartlett (1936) was to transform the data to square roots, and perform the analysis of variance in the square root scale. The idea was that in the square root scale the error variance should be approximately constant, so that the pooled error in the analysis of variance could be used, and there was a little evidence that the square roots were not far from normally distributed. With binomial data, Bliss (1937) recommended an analysis in the angular arcsine $\sqrt{p}$ scale, for similar reasons. Later, various adjustments to these transforms were introduced that were claimed to give a more constant variance on the transformed scale.

With Poisson data, Cochran's interest was as follows. Suppose one could assume that effects were additive in the square root scale, but wanted maximum likelihood estimates of the treatment effects, how would these differ from the estimates obtained in the recommended analysis of variance in the square root scale? With randomized blocks data, for example, let $\mu_{ij} = E(x_{ij})$. With an additive model in the square root scale, we have

$$\eta_{ij} = \mu_{ij}^{\frac{1}{2}} = \mu + \alpha_i + \beta_j. \tag{12}$$

Also, with Poisson data the log likelihood is

$$L = \sum_i \sum_j x_{ij} \ln \mu_{ij} - \mu_{ij}; \quad \frac{\partial L}{\partial \mu_{ij}} = \sum_i \sum_j \frac{(x_{ij} - \mu_{ij})}{\mu_{ij}}. \tag{13}$$

Hence, the normal equation $\partial L / \partial \alpha_i = 0$ for the parameter $\alpha_i$ is

$$\sum_j \frac{2}{\sqrt{\mu_{ij}}} (x_{ij} - \mu_{ij}) = 0. \tag{14}$$

Let $y_{ij} = \sqrt{x_{ij}}$. By a first-term Taylor expansion,

$$x_{ij} - \mu_{ij} \cong 2\sqrt{\mu_{ij}} (y_{ij} - \eta_{ij}). \tag{15}$$

Substituting from (15) back into expression (14) gives for the normal equation for $\alpha_i$ in the square root scale (with ML estimation),

$$4 \sum_j (y_{ij} - \eta_{ij}) = 0. \tag{16}$$

But expression (16) is the standard equation of estimation for $\alpha_i$ in an analysis of variance in the square root ($y_{ij}$) scale. Equation (16) is only approximate, depending as it does on a first-term Taylor expansion. Fisher's device, which Cochran used, is to turn this approach into a series of successive approximations to the ML estimates. Define an adjusted transform, $y'_{ij}$, so that equation (15) becomes correct. That is, let

$$y'_{ij} = \hat{n}_{ij} + \frac{(x_{ij}-\hat{\mu}_{ij})}{2\sqrt{\hat{\mu}_{ij}}} = \hat{n}_{ij} + \frac{(x_{ij}-\hat{n}_{ij}^2)}{2\hat{n}_{ij}} \; . \tag{17}$$

Equation (17) is used iteratively. From an analysis of variance in square roots, estimate the $\hat{n}_{ij}$. From (17), get a second approximation to the $y'_{ij}$. Analyze these $y'_{ij}$ and get third approximations to the $y'_{ij}$ from (17), continuing until the $\hat{n}_{ij}$ converge to ML estimates.

From a number of examples with Poisson or binomial data in which I found ML estimates of the treatment effects in this way, my summary impressions were: (i) the ML estimates usually differed little from those given by an analysis of variance in the square root or angular scale, and (ii) the adjustments to the transforms needed to give ML estimates were rather similar to those recommended to give more nearly constant variance in the transformed scale.

To illustrate what I mean, Table 5 gives results from the analysis of a 5×5 Latin square with Poisson data. The table shows (i) the observations, (ii) the square roots of the observations, (iii) the adjusted transforms that give the ML estimates by Fisher's method. Where the same observation, e.g., 4 occurred several times in the experiment, the mean of the adjusted transforms is shown in Table 5, (iv) the values $(\sqrt{x} + \sqrt{x+\frac{1}{2}})/2$. This adjusted transform is similar to the adjusted transform $(\sqrt{x} + \sqrt{x+1})$ recommended for better variance stabilizing by Freeman and Tukey (see Mosteller and Youtz (1961)), except that the adjustment in (iv) is somewhat milder. Note how closely the ML adjusted transforms in (iii) agree with the values $(\sqrt{x} + \sqrt{x+\frac{1}{2}})/2$ suggested in (iv) for improved variance stabilization.

In working these examples I also noticed on several occasions that data which appeared at first sight to be Poisson or binomial had extraneous variation present, as revealed by a $\chi^2$ test. With such data I formed the judgment that analysis of the transforms or adjusted transforms might be justified as a working approxima-tion, but Fisher's method could not be regarded as a more exact analysis, since the data were not Poisson or binomial. I formed the opinion that adjusted trans-forms might serve as a substitute for Fisher's ML solutions.

Table 5:   Two Sets of Adjusted Square Roots with Poisson Data

| (i) | x | 0 | 1 | 2 | 3 | 4 |
|---|---|---|---|---|---|---|
| (ii) | $\sqrt{x}$ | 0 | 1 | 1.41 | 1.73 | 2.00 |
| (iii) | ML adj. | 0.54 | 1.05 | 1.48 | 1.77 | 2.05 |
| (iv) | $(\sqrt{x} + \sqrt{x+\frac{1}{2}})/2$ | 0.35 | 1.12 | 1.50 | 1.80 | 2.06 |
| (i) | x | 5 | 6 | 8 | 9 | 17 |
| (ii) | $\sqrt{x}$ | 2.24 | 2.43 | 2.82 | 3.00 | 4.12 |
| (iii) | ML adj. | 2.25 | 2.47 | 2.86 | 3.04 | 4.17 |
| (iv) | $(\sqrt{x} + \sqrt{x+\frac{1}{2}})/2$ | 2.29 | 2.50 | 2.87 | 3.04 | 4.15 |

In 1954, Fisher [CP 254] presented a paper in which he repeated his method of finding maximum likelihood estimates in the analysis of non-normal data by transforming the data.  He then remarked that some authors seemed to think erroneously that the objective of the transformation was to find a constant error variance in the transformed scale.  He noted that with Poisson data, Bartlett had proposed the adjusted scale $\sqrt{x+\frac{1}{2}}$ while Anscombe had proposed $\sqrt{x+3/8}$, neither author realizing that in Fisher's view these adjustments were a waste of time.  He brought Cochran into the act by noting that while Cochran presented the ML approach, he did not "totally disavow" the analysis of transformed data adjusted for variance stabilization.

In retrospect I agree that Fisher's approach is superior.  It specifies a definite mathematical model, and uses maximum likelihood estimation, recognized as preferable to least squares estimation with non-normal data.  My results were that analysis of variance on the transformed scale, with or without variance-stabilizing adjustments, agreed closely with the ML estimates, and was a good working method, particularly when extraneous variation is present so that the assumptions leading to Fisher's ML solutions do not apply.

## 9.  An Example of Non-Linear Regression

With regressions non-linear in some of the parameters, examples in which something approaching the null distribution of F can be found are rare.  An example
given by Fisher [CP 163] is the harmonic regression

$$y_x = \mu + \alpha \cos(\theta x) + \beta \sin(\theta x) + \epsilon \tag{18}$$

for $x = 1,2,\ldots,n = 2s + 1$.  This regression is linear in $\mu$, $\alpha$, $\beta$, but non-linear
in $\theta$.  With a linear regression the parameter $\mu$ is estimated by the mean $\bar{y}$, and in
the null case the three parameters $\alpha$, $\beta$, and $\theta$ would be expected to account for
about a fraction of 3/2s of the sum of squares $\Sigma(y-\bar{y})^2$.

Fisher notes that the $n = 2s + 1$ observations may be expressed as (2s+1)
orthogonal linear functions, namely the mean $\bar{y}$ and the harmonic functions $\Sigma a_x y_x$,
$\Sigma b_x y_x$, where

$$a_x = \sqrt{\frac{2}{2s+1}} \cos \left(\frac{2\pi px}{2s+1}\right) \qquad b_x = \sqrt{\frac{2}{2s+1}} \sin \left(\frac{2\pi px}{2s+1}\right), \tag{19}$$

$p = 1,2,\ldots,s$.  For example, with $n = 5$, $s = 2$ the four cos and sin terms are shown
in Table 6 (without the multiplier $\sqrt{2/(2s+1)}$.

Table 6:  Harmonic Terms (n = 5, s = 2)

| p \ x | | 1 | 2 | 3 | 4 | 5 |
|---|---|---|---|---|---|---|
| 1 | cos | .309 | −.809 | −.809 | .309 | 1 |
| 1 | sin | .951 | .588 | −.588 | −.951 | 0 |
| 2 | cos | −.809 | .309 | .309 | −.809 | 1 |
| 2 | sin | .588 | −.951 | .951 | −.588 | 0 |

These terms are mutually orthogonal and are orthogonal to the mean $\bar{y}$.
Therefore if $\alpha = \beta = 0$, so that the data have no harmonic trend, the quantities

$$ss_p = [\Sigma_x (a_x y_x)]^2 + [\Sigma_x (b_x y_x)]^2 \tag{20}$$

are independently distributed as $\chi^2 \sigma^2$ with 2 d.f. for any fixed value of $p =$
$1,2,\ldots,s$.  The sum of squares (20) is the reduction in $\Sigma(y-\bar{y})^2$ due to fitting
the model (18) with any given p and accompanying $a_x$ and $b_x$.  Of the s values of p,
suppose we select the value that gives the greatest reduction in $\Sigma(y-\bar{y})^2$.  Fisher

[CP 75] had shown previously that for this value of p, the ratio $g = ss_p/\Sigma(y-\bar{y})^2$ is distributed so that the probability of exceeding any value $g$ is

$$P = s(1-g)^{s-1} - \frac{s(s-1)}{2}(1-2g)^{s-1} + \ldots + (-)^{k-1}\frac{s!}{(s-k)!k!}(1-kg)^{s-1},$$

the series stopping at the largest integer k less than $1/g$. The upper 5% and 1% levels of the distribution of g have been tabulated by Eisenhart, Hastay, and Wallis (1947).

Fisher notes that if we fit the optimum $\theta$ in the model (18), it will give a reduction in the sum of squares at least as great as the best of the p's. Hence the table of g can be used to test the null hypothesis $\theta = 0$ under the model (18). The verdict of the test can be trusted if it finds a significant harmonic component, but if it fails to reject the null hypothesis, there might still be a value of $\theta$, better than the best of the p's, that would give a significant result.

Fisher's work also shows that this is an example in which linear theory cannot be trusted. As s increases, the ratio $g = \max ss_p/\Sigma(y-\bar{y})^2$ does not tend when s is large to $3/2s$, but to the larger value $(\gamma + \ln s)/s$ when s is large, where $\gamma$ is Euler's constant. Similarly, the use of the F-ratio for 3 and (2s-3) d.f. for a joint test of $\hat{\alpha}$, $\hat{\beta}$, and $\hat{\theta}$, as linear theory would suggest, can be badly wrong. Some 5% F values for 3 and (2s-3) d.f. are shown below, along with the 5% F values from the g table, which are themselves a little low for testing the optimum $\hat{\theta}$.

Table 7:  Comparison of F values

| n | 11 | 21 | 41 | 61 |
|---|----|----|----|----|
| $F_{3,2s-3}$ | 4.35 | 3.20 | 2.86 | 2.52 |
| F from g-distribution | 5.04 | 4.54 | 4.57 | 4.69 |

The linear approximation becomes badly wrong as n increases.

-34-

## References

Bartlett, M.S. (1936). "The Square Root Transformation in the Analysis of Variance," Journal of the Royal Statistical Society, 3, 68-78.

Bliss, C.I. (1937). "The Analysis of Field Experimental Data Expressed in Percentages," Plant Protection (Leningrad), 67-77.

Cochran, W.G. (1940). "The Analysis of Variance when Experimental Errors Follow the Poisson or Binomial Laws," The Annals of Mathematical Statistics, 11, 335-347.

Eisenhart, C., Hastay, M.W. and W.S. Wallis (1947). Techniques of Statistical Analysis. New York: McGraw-Hill.

Harris, J.A. (1913). "On the Calculation of Intra-Class and Inter-Class Coefficients of Correlation from Class Moments when the Number of Possible Combinations is Large," Biometrika, 9, 446-472.

Mosteller, F. and C. Youtz (1961). "Tables of the Freeman-Tukey Transformations for the Binomial and Poisson Distributions," Biometrika, 48, 433-440.

# 112

SUMMARIZING THE RESULTS OF A SERIES OF EXPERIMENTS

William G. Cochran

Professor of Statistics Emeritus, Harvard University

I first met this problem in the thirties in agriculture. I wrote a
paper on it (1), and later a more ambitious paper with Yates (2), in which
a number of examples were worked. We tried to see in what respects the
analysis of a group of experiments resembled and in what respects it differed
from the analysis of a single experiment.

The need to summarize results of a series of experiments on the same
treatments arises in two types of application. The first type may be described as exploratory; a number of experiments on the relative performance
of something or of two treatments have been carried out, and we are trying
to answer the question; what is the present state of knowledge about the
relative merits of the two treatments? For instance, the recent academy
study of saccharin started with the experiments in which large doses were
given to rats; these were the prime experiments. To cite a second example,
Yates and Crowther realized at the beginning of World War II that Britain
would have to import most of her fertilizers during the war and would be
short of fertilizers. Accordingly, they summarized the experiments (4)
about the responses of the common farm crops to fertilizers in order to
answer the question: What is the present state of knowledge about the
effects of fertilizers and to provide material for an intelligent rationing
system for fertilizers?

As another example, I was in a group that studied two common methods of
surgery for duodenal ulcer--vagotomy (cutting the vagus nerves) plus a radi-
cal antrectomy (which removes the lower portion of the stomach) versus
vagotomy plus the milder pyloroplasty (which widens the outlet of the stomach
to provide better drainage).

Reproduced with permission from *Proceedings of the 25th Conference on the Design of Experiments in Army Research Development and Testing,* U.S. Army Research Office, Durham, NC,
ARO Report 80-2, pages 21–33, 1980.

We found four experiments that appeared to have been carefully done and properly randomized. We could have come across a number of comparisons that were well done but not randomized--the type sometimes called observational studies. Since often we cannot use randomization and have to make a comparison without it, I would have been interested in including the observational studies so as to learn whether they agreed with the randomized studies and if not, why not? But the medical members of our team had been too well brought up by statisticians, and refused to look at anything but randomized experiments. In this type of surgery, we may expect the experiment to be of different designs and perphaps differing numbers of replications.

The second type of application occurs commonly in agriculture. It differs from the first in two ways. It is known that the relative performance of a treatment (variety of a crop or fertilizer) is likely to vary both from field to field within a year and from year to year. Thus experiments are likely to be repeated in different fields and for a number of years. Secondly, there is a better chance that the experiments, being jointly planned, are of the same design and number of replications. For instance, when the growing of sugarbeets was introduced into Britain after World War I, the government conducted 3×3×3 factorials (ultimately 30 per year) at the leading centers for a number of years.

The objective of the experiments may be a series of decisions as to which varieties of a crop look promising and should be kept for further testing, which varieties should be discarded, and which varieties having been fully tested, should be part of an approved list and have their seed made available to farmers. As an example, Patterson and Silvey (5) have described the trials of varieties of cereals that Britain has conducted in recent years, the designs being incomplete blocks. This kind of screening program is not confined to agriculture. It may be used in seeking the best drugs or vaccines for some purpose in medicine, or in seeking persons best capable of doing some task. In 1963, Federer gave a bibliography of some 500 papers on screening programs.

## 2. Miscellaneous Experiments in Exploratory Work

I'll start with exploratory experiments done by different people at different places and times. Since these experiments were not planned as a coordinated series, we must expect them to differ in designs, and in numbers of replications. First we must think of the question: of what population, if any, can these experiments be considered something approaching a random sample? Is this population relevant to future applications of any conclusions that we draw? In some cases we may reluctantly conclude that the experiments do not sample any population of interest to us, and decide not to prepare any summary. In some cases the experiments are so variable that some must be thrown out before any summary is attempted. The way in which the experiments were done also affects the nature of the population that they sample. The nature of the experiments also affects the kind of population that they sample. In the National Academy study of saccharin to which I referred, the doses in the laboratory experiments were so large that the estimates of the effects of more normal doses depended to a substantial extent on the kind of model used in extrapolating the experimental results. In experiments comparing two methods of surgery, the experiments may be confined, for ethical or logical reasons, to the kind of patients whose doctors state beforehand that they can safely take either method of surgery. Otherwise, it is difficult to interpret the results of the experiments. This restriction affects the character of the population to whom the conclusions apply.

In agriculture, as I have stated, we have to contend with variations in both space and time: But in other fields of application there may be no strong reasons to consider time as a separate source of variation, even though the experiments will have presumably been done at different times. So in considering a summary of miscellaneous exploratory experiments, I shall combine time and space and speak of treatments × places.

use Bartlett's test, or if the data seem nonnormal and we want a more robust test, we can use Levene's test, based on the absolute value of the deviations that lead to the $s^2$, that appears to be less affected by nonnormality. If the $s^2$ seem markedly heterogeneous, the F-test of the interactions against the pooled error is not exact, but assingning a number of df to the pooled error by Satterthwaite's approximation should provide an approximate test.

The next step is to reach one of three decisions about the Treatments × experiments interactions. (i) that it is negligible, (ii) that it is not negligible but has no discernable structure. By this I mean that although the effects of the treatments vary from experiment to experiment, we have no information for making different predictions in different parts of the population and must draw single overall conclusions about the effects of the treatments.(iii)The third case is that in which the interaction is of a nature that we think we understand, and is large enough so that different treatments win in different parts of the population that can be described. In this case we expect to recommend different treatments for different parts of the population.

Consider first case (i) in which we judge that the treatments × experiments interactions are negligible. If the experiments differ in number of replications and in their error variances, a question to be considered is: Should the treatment means in individual experiments be weighted in forming the overall means, so as to give more influence to the more accurate experiments? If so, what should the weights be? If the error variances $\sigma_j^2$ were known, the weights should presumably be $w_j = \sigma_j^2/n_j$, but the variances are only estimated, unless the $\sigma_j^2$ appear to be equal so that weights $n_j$ can be experiments interactions for the ith Treatment in the jth experiment. We may also expect experiments to have different variances $\sigma_j^2$ per observation and to differ in number of replications $n_j$.

For the jth experiment, a model that seems reasonable with a quantitative response is that the mean of the ith Treatment in the jth experiment is

$$\bar{y}_{ij} = \mu + t_i + \gamma_{ij} + \bar{e}_{ij}$$

where $\gamma_{ij}$ is the treatments × experiments interaction and the variance of the error term $e_{ij}$ is $\sigma_j^2/n_{ij}$ ($i = 1,2,\ldots,t$; $j = 1,2,\ldots,k$).

In a combined analysis of these means, a reasonable first step is to form a two-way treatments × experiment table of these means. If all treatments are present in all experiments, an analysis of variance into the following components should be easy.

|  | df |
|---|---|
| Experiments | $(k-1)$ |
| Treatments | $(t-1)$ |
| Treatments × Experiments | $(t-1)(k-1)$ |
| Pooled error | |

The purpose is to test the interaction. If some treatments are missing from some experiments, a least squares analysis appropriate to missing data is used. In this case the Treatments line is Treatments, adjusted for experiments.

The pooled error in the analysis of variance of the treatment means is $(1/k)(s^2/n_{ij})$, or if the $s_j^2$ seem to be homogeneous, $s^2(1/n_j)$. We will want to examine whether the $s_j^2$ appear to be heterogeneous, since this affects the F-test of the ratio treatments × experiments/pooled error. For this we can used for the treatment in the jth experiment. Various authors have worked on this problem of weighting with fallible weights.

The first step is to find out if there is much gain in accuracy from the use of weighted means. If the $s_j^2$ appear to be homogeneous, and the weights are the known values $n_j/\bar{s}_j^2$, this can be done, because the ratio of the variance of the weighted to the unweighted mean of the $y_{ij}$ is $(\sum w_j)(\sum 1/w_j)/k^2$. For instance, if one-third of the experiments each have $n_j$ with relative values 1, 1/2, 1/4, the relative value of the variance of the weighted to the unweighted mean is $36/49 = 0.73$. The situation is less favorable to the weighting if the $s_j^2$ differ, so that we have to use something like estimated weights $n_j/s_j^2$. Under normality, the maximum likelihood estimate of the overall mean $\mu_i$ is

$$\sum_j \frac{n_j(f_j - 1)}{f_j s_j^2 + n_j(\bar{y}_{ij} - \hat{\mu}_i)^2} (\bar{y}_{ij} - \hat{\mu}_i) = 0 .$$

This has to be found iteratively. In this type of estimate, an experiment with low $s_j^2$ and apparently high precision is prevented from dominating the overall mean if it disagrees markedly from the value suggested by the other experiments, since the term $n_j(\bar{y}_{ij} - \hat{\mu}_i)^2$ will be large, and will decrease the weight given to this experiment.

Some years ago, C. R. Rao (7) brought out a new method of estimating variances and variance components called the MINQUE (minimum norm quadratic unbiased estimator). Since I have been interested in this problem for over 40 years, I asked J.N.K. Rao of Carleton University and P.S.R.S. Rao of the University of Rochester if the MINQUE method would lead to improved estimates of the weighted mean. Both men looked into the problem--J.N.K. Rao in the case with no treatments $\times$ experiments interactions which is now being considered and P.S. Rao in the case in which we assume a random treatments $\times$ experiments interaction with variance $\sigma_\gamma^2$, which also has to be estimated. Both men discovered what I had suspected in working with MINQUE--that if one is trying to produce an improved method of estimating variance components, it may not be wise to make the estimates unbiased. With unbiased methods one may get variance component estimates that sometimes take negative values and have large variances. Both men produced adjustments to MINQUE that are essentially positive. J.N.K. Rao's method (8) uses non-iterative weights rather similar to the maximum likelihood weights. The weights are

$$w_j = n_j(f_j + 1)/[f_j s_j^2 + n_j(\bar{y}_{ij} - \bar{y}_i)^2] \, ,$$

where $\bar{y}_i$ is the unweighted mean of the $\bar{y}_{ij}$. Some limited Monte Carlo studies have shown that the weighting does better than the maximum likelihood estimates of the treatments means except when differences in the error variances are extreme. This estimate also does better than MINQUE and better than the simple weights $w_j = n_j/s_j^2$ and is probably the best found thus far.

For estimating the gain in accuracy from the use of erroneous weights like these, the previous figures for the relative accuracy of weighted to unweighted means must be reduced, because of sampling error in the weights. The dampening factor depends both on the average df with which $s_j^2$ are estimated, and on the amount of heterogeneity in the weights. For the previous example with weights proportional to 1, 1/2 and 1/3 in thirds, and 1.36 if the weights are known, the dampening factor is approximately $(\bar{f} + 6)/(\bar{f} + 8)$, where $\bar{f}$ is the average number of df in $s_j^2$.

Thus if the $s_j^2$ have 6 df on the average, the relative efficiency of weighted to unweighted means is estimated as $(12)(1.36)/(14) = 1.18$--a rather modest gain from weighting. Before resorting to weighting, check also that weighted means apply to the same population as unweighted means. For example, if the weights tend to be high when the mean yields of the experiments are also high, we may conclude that the results for weighted means apply to a population having a higher mean yield than our actual population and decide not to use weighted means.

For comparison between the estimated means of the treatments, we need standard errors. With unweighted means, the estimate of their standard error is $\sqrt{\overline{\Sigma(s_j^2/n_j)}/k}$. With the experimental error variances of the individual experiments taken as homogeneous, the estimated variance of the mean weighted as $n_j$ is $s^2/\Sigma n_j$. For Rao's estimate with fallible weights, Rao (8) has given a rough estimate of the variance of this weighted mean, which also implies a dampening factor for the fact that fallible weights are being used. The jacknife estimate is another possibility.

When the treatments × experiments interaction is significant, we need to see if we can understand the nature of the interaction. For this, a two-way treatments × experiments table of residuals is helpful. Sometimes there is no winner; different treatments appear to win in different parts of the population, but either we do not fully understand the interaction or do not wish to use it in a recommendation. Sometimes there are two distinguishable parts of the population in which the ranking of the treatment is different, and we understand why. Student (10) cites an example. After a long series of experiments, the Irish Department of Agriculture introduced Spratt-Archer barley as the best suited to the country. In one county the farmers refused to grow it, claiming that their native barley was superior. In order to convince these farmers, the Department of Agriculture made some special comparisons in this county of the native barley versus Spratt-Archer. To their surprise, the native barley was superior. The reason also became clear. This barley is a quick-starting variety. Now in this county, farming is rather lackadaisical, so that the weeds flourish. The weeds tended to smother the Spratt-Archer barley, which starts slowly, but the native barley, starting quickly, could smother the weeds. Another maxim from this example is make sure the experiments sample the population to which their results will be applied.

If there are two parts which have $k_1$ and $k_2$ experiments, the following breakdown of the interaction is relevant

|            |                       | df               |
|------------|-----------------------|------------------|
| Treatments | (Part I - Part II)    | $(t-1)$          |
| Treatments | Part I experiments    | $(t-1)(k_1-1)$   |
| Treatments | Part II experiments   | $(t-1)(k_2-1)$   |

In this breakdown, we expect the first term to be large and the other parts small. In addition, we need to analyze parts I and II separately, in order to see if there are definite treatments differences in each part.

If the interaction is significant and is assumed to be random, the variance of a treatment mean in an individual experiment is $(\sigma_\gamma^2 + \sigma^2/n_j)$, which moves nearer equality because of the term $\sigma_\gamma^2$ but also means that an extra parameter has to be estimated if weighted means are contemplated. In a Monte Carlo study by P.S.R.S. Rao, Kaplan, and Cochran (9) several types of weighted means including a revised MINQUE were included but the unweighted mean proved very hard to beat, as might be expected, unless $\sigma_Y^2$ is small and the variation in the $\sigma_j^2$ is extreme. Use of the unweighted mean has the advantage that an unbiased estimate mean of the variance of the overall mean of a treatment is $\Sigma(y_{1j} - \bar{y})^2 / k(k-1)$.

If the original observations are in proportions, remember that a decision, e.g. whether a single overall mean has enough advantage overall over the other means to recommend it, or whether two means should be recommended for different parts of the population, must be made in proportions. If the combined analysis is made in some other scale, such as angles or logits, because it is thought nearer to normality or in some ways more suitable, remember that means in the original proportional scale will be slightly biased when we transfer back. Quenouille (11) has given approximate corrections for this bias, which do not appear to be well known. Let $s^2$ be our estimate of the variance of $\bar{z}$ (where z denotes the transformed scale), that is, the mean in the transformed scale. If an angular transformation is used, Quenouille's correction for bias in the transformed mean is to increase $\sin^2 \bar{z}$ by $\frac{1}{2}(1 - e^{-2\bar{z}})\cos(2\bar{z})$. If logits are used with equal weights, the usual procedure is to take $p = e^z / (1 + e^z)$ when transforming back to p. Quenouille's correction for bias is to add $(n-1)s^2/2n$ to $\bar{z}$ before taking $e^z / (1 + e^z)$.

3. Variations in Both Time and Space

This situation is likely to occur primarily in agriculture. Since the experiments are likely to be jointly planned, they may have the same designs and number of replications, the same experiment being repeated at the same place for four or five years. As mentioned, the number of years will commonly be limited to at most four or five, since a larger number slows up any recommendations. But the experiments may not have the same numbers of replication - more may have been added in later years. In varietal trials, a new variety may be added in the second or third years, so that different treatments may have different numbers of years at any given time. However, unless the numbers of replications differ greatly, a preliminary analysis of the treatment means will usually be adequate and is fairly easy, although there are extra complications and full least squares may have to be used if some treatments are only present in the later years.

It will usually be necessary to treat the treatments $\times$ years variation as random, with variance $\sigma^2_{ty}$, even if it does not act like a random variate. A good deal is known about the influence of weather on crops, and we may have found, for instance, that in a good year the best treatments have a greater advantage, so that the treatments $\times$ years interaction is definitely not random. But a superior treatment before recommendation, must be superior, on the average, over a span of years, taking $\sigma^2_{ty}$ into account, since we cannot recommend a different treatment for different years.

The preliminary analysis of variance and the expected values of the means squares are shown below. I have treated the treatments $\times$ places interactions as random as well as treatments $\times$ years, since this is usually the assumption that has to be made if it is a question of recommending the overall use of one treatment.

| | df | Expected value of mean squares |
|---|---|---|
| Treatments | (t-1) | $\bar{\sigma}^2 + n\sigma^2_{tpy} + np\sigma^2_{ty} + ny\sigma^2_{tp} + npy\sigma^2_t$ |
| T × Y | (t-1)(y-1) | $\bar{\sigma}^2 + n\sigma^2_{tpy} + np\sigma^2_{ty}$ |
| T × P | (t-1)(p-1) | $\bar{\sigma}^2 + n\sigma^2_{tpy} + ny\sigma^2_{tp}$ |
| T × P × Y | (t-1)(p-1)(y-1) | $\bar{\sigma}^2 + n\sigma^2_{tpy}$ |
| Pooled error | | $\bar{\sigma}^2$ |

In presenting the expected values, I have taken the simplest case, in which all experiments are of the same size and design, the symbols n, t, p, and y standing for number of replications, number of treatments, number of places, and number of years. The symbol $\bar{\sigma}^2$ is, of course, the true pooled error variance. The $MS_{tpy}$ is tested against error, and if F is about 1, this mean square may be combined with the pooled error. The expected values are written as if treatments are also random, with variance $\sigma^2_t$. If the effects of treatments are fixed, as they usually are, replace $\sigma^2_t$ by what is usually called $s^2_t = \Sigma(t-\bar{t})^2/(n-1)$.

From the expected values it is clear that the treatments × years and treatments × places interactions are tested by an appropriate F test (approximate if $\sigma^2_j$ varies from experiment to experiment) against the mean square for the tpy three-factor interaction, and that an unbiased estimate of $\sigma^2_{ty}$ is $(MS_{ty} - MS_{tpy})/np$. For the main effects of Treatments, no single line in the analysis of variance is a proper error. An unbiased estimate of the error variance for the error of a treatment mean, if interactions are present and random, is

$$MS_{tp} + MS_{ty} - MS_{tpy}$$

and an approximate F test of the treatments mean square may be made by taking $F = MS_t/(MS_{tp} + MS_{ty} - MS_{tpy})$, with Satterthwaite's approximation used to ascribe a number of df to the denominator. However, in a small Monte Carlo study of experiments, Hudson and Krutchkoff (13) found, somewhat surprisingly, that a rival $F = (MS_t + MS_{tpy})/(MS_{tp} + MS_{ty})$ using Satterthwaite, had somewhat better power and recommended it, although it did not approximate the 5% and 1% levels of F when the null hypothesis was true.

Since whether we recommend one treatment, two treatments or suspend judgement for some reason depends mainly on how the treatments vary in effects from place to place, the two-way table of treatments and places deserves careful study. The treatments x places interaction is sometimes heterogeneous; some comparisons of some treatments have a higher mean square interaction than others. Subdivisions of the treatments and places and the treatments x places sum of squares should be tried.

Thus, as we have seen, the summary of a series of experiments calls mainly for experience in the analysis of variance, which we now have. It is well to adopt something of the attitude in exploratory analysis and be on the lookout for anything unexpected, since the nature of the tp interaction is often a hard thing to puzzle out.

## REFERENCES

(1) W. G. Cochran (1937). Problem arising in the analysis of a series of similar experiments. Supp. J. Roy. Statist. Soc., 4, 102-118.

(2) F. Yates and W. G. Cochran (1938). The analysis of groups of experiments, J. Agric. Sci., 28, 550-580.

(3) W. G. Cochran et al. (1977). Experiments in surgical treatment of duodenal ulcer. In Costs, Risks and Benefits of Surgery, Bunker, Barnes and Mosteller (eds.), Oxford University Press.

(4) F. Yates and E. M. Crowther (1941). Fertilizer policy in wartime. The fertilizer requirements of arable crops, Emp. J. Exp. Agric., 9, 77-97.

(5) H. D. Patterson and V. Silvey (1979). Statutory and recommended list trials of crop varieties in The United Kingdoms, J. Roy. Stat. Soc. (in press).

(6) F. E. Satterthwaite (1946). An approximate distribution of estimates of variance components, Biometrics Bull., 2, 110-114.

(7) C. R. Rao (1970). Estimation of heteroscedastic variances in linear models, J. Amer. Statist. Assoc., 65, 161-72.

(8) J. N. K. Rao (1979). Estimating the common mean of possibly different normal populations. A simulation study, unpublished manuscript.

(9) P. S. R. S. Rao, J. Kaplan and W. G. Cochran (1979). Estimators for the one-way random effects model with unequal error variances, unpublished manuscript.

(10) Student (1931). Agriculture field experiments, Nature, 14 March, 404.

(11) M. H. Quenouille (1953). The design and analysis of experiments. Charles Griffin and Co. Ltd., London

(12) W. T. Federer (1963). Procedures and designs useful for screening material in selection and allocation, with a bibliography, Biometrics, 19, 553-587.

(13) J. D. Hudson and R. G. Krutchkoff. A Monte Carlo investigation of the size and power of tests employing Satterthwaite's synthetic mean squares. Biometrika, 55, 431-433.

# 113

## Discussion

*William G. Cochran*


I am glad that Dr. Hilts has given a good deal of attention to the second half of the nineteenth century. My own limited reading in epidemic theory jumped directly from Farr to Hamer, and I knew nothing of the contributions of Ransome and Whitelegge. I don't think, however, that the similar jump in Dr. Serfling's historical review is to be criticized. He was writing a compact review of the historical development of epidemic theory, and naturally concentrated on people like Hamer who actually constructed a theory, working out the mathematical consequences of assumptions about the way in which disease spreads from one person to another.

What struck me from Dr. Hilts' account of the period 1840-1900 was how well the epidemiologists were doing in getting ready for attempts to construct epidemic theories. Through both compulsory and voluntary notification acts, they were collecting regular reports of cases of different kinds of infectious diseases. They were thinking about the nature of any epidemic waves or cycles that they seemed to observe. They were naming and discussing major factors that might explain these features and might be introduced into a theory; in particular, the density of susceptibles in relation to the supply of infectives, changes in the virulence of the infecting organism or in the infectivity of the remaining susceptibles as the epidemic proceeded, the rather vague miasma factor and other meteorological or geographical factors.

This kind of activity impresses me because the period 1850-1900 seems to have been rather quiet and unexciting, with few major breakthroughs, as regards the general development of statistical theory and applications, though I should caution that the period has not been extensively studied. What was happening in statistics? We had Galton's idea of the regression line in 1889, but we had to wait over 30 years until Fisher provided the equipment to put it to full use. The most prolific mathematical statistician of that time, Edgeworth, made some contributions to tests of significance, tests of goodness of fit, asymptotic theory, and estimation, including maximum likelihood. But he created no school, and his name is hardly ever mentioned to students in statistics classes. In the 1890s the heads of Census Bureaus in Western Europe were just beginning to try out sampling methods as a means of saving time and money. The most advanced methodologically of the sampling statisticians was, I think, the Norwegian A.N. Kiaer, who was using and advocating what we now call proportional stratification. The application in which the design of experiments was most advanced at that time was to agricultural field experiments. The agronomists had done limited studies of soil fertility patterns as a guide to the layout of field experiments. The most consistent feature that they found was that adjacent plots tended to give similar yields. Hence, in the layout of comparative trials, the recommendation was that different treatments be placed next to one another. The field of econometrics was further ahead than epidemiology in methods for describing economic waves and cycles, but not, I think, in explanatory theory. My memory may be faulty, but I can recall no work on what I would call causative or explanatory model-building in biology at that time.

The epidemiologists, however, were trying to come to grips with a very complex model-building job. When one thinks of the steps by which infection passes from one person to another with different diseases, it seems clear that construction of a realistic epidemic theory must be a very difficult business indeed. I don't think I was invited here to teach you epidemiology, but let me indicate some of the difficulties in epidemic theory as seen by M.S. Bartlett, the leading mathematical statistician who has worked on epidemic theory. I'll quote in slightly condensed form from his major paper in 1956.[1] He writes:

In spite of the brilliant pioneering work of Farr, Hamer, and Ross, and of important later studies by Soper, Greenwood, McKendrick, E.B. Wilson and others, . . . a quantitative theory of epidemics in any complete sense is still a very long way off. The well-known complexity of most epidemiological phenomena is hardly surprising, for not only does it depend on the interactions between "hosts" and infecting organisms, each individual interaction itself usually a complicated and fluctuating biological process, but it is also. . . . a struggle between opposing populations, the size of which may play a vital role. This last aspect . . . can only be discussed in terms of statistical concepts . . . From the time of Ross at least, the importance of studying the nature, density, and mode of transmission of the infecting agent has been recognized, although reliable information of this kind is often comparatively meagre. . . . The virus or bacterial populations may be in a continuous genetic or other biological state of flux. One need merely recall, for example, the existence of different strains of influenza virus, or the evidence for strains of different virulence in experimental epidemiological studies.

I will add only that the theory must be probabilistic or stochastic, not deterministic. A given set of initial conditions may lead to no epidemic, or to epidemics of different degrees of severity and duration, the outcome depending partly on chance mechanisms. Moreover, because of their mathematical complexity, stochastic processes in time and space began to be tackled relatively late by mathematical statisticians. Proposing conferences is not one of my hobbies—but a case can be made for a short conference of epidemiologists and mathematical statisticians on the future outlook for epidemic theory. The subject offers plenty that is challenging to the mathematical statistician. If the problem of communication can be overcome, the epidemiologists might indicate the kinds of questions of interest to them to which results from theory might help to provide answers.

Given these difficulties, one might wonder why the epidemiologists around 1900 hoped to get anywhere in epidemic theory with the meager tools that they had. Sir William Hamer may have had the right attitude, to judge from his comments on Soper's paper in 1929. He quotes Professor Boycott's advice to take advantage of any specially easy examples that Providence pushes under our noses. He cites measles as such an example, because of the insistent invariability of the measles organism, coupled with the lasting protection afforded by one attack of the disease. He commends Ransome for singling out measles for study, to be followed by Whitelegge, Campbell, Munro and others.

As regards attempts to construct an epidemic theory, Sir William goes on to quote Professor Boycott's warning of the danger of cultivating thoughts and reading books to which we are not equal. But he believed that the Professor would not entirely disapprove of putting our crude ideas about the spread of a disease into some kind of a model, as we would now express it, and seeing how well the model fitted the facts that we know.

Coming to the period after Hamer's 1906 paper, I am not sure whether Dr. Hilts implied that Karl Pearson and Brownlee held up the development of epidemic theory after Hamer. Personally, I doubt whether they or the biometric approach did so. For one thing, I don't think that after 1906 statisticians had the combination of methodology and some knowledge of epidemiology to be able to extend Hamer's approach for quite some time. When I first read Soper's 1929 paper around 1932, I regarded it as a pioneering paper with no obvious forerunners on the mathematical statistical side.

Soper's paper was very well received when he presented it to the Royal Statistical Society. But the discussants at the meeting indicated the complexity of the problem by noting that the size of the community, the locale—urban or rural—the amount of crowding, and the time of year at which measles start all had an effect on the likely nature of the epidemic. Others noted, as Soper had done, that under-notification and mis-diagnosis made it difficult to check theory against the available data, and still others called for extension of the theory to different diseases, for example, polio.

It is true that Brownlee, as a medical man who was chief statistician of the Medical Research Council after 1914, and as the most prolific writer on epidemics, was in a strong position of influence. However, his insistence that an explanation must fit what was known about the shapes of epidemic curves, now that these could be studied more thoroughly by biometric methods, was reasonable logically. His weaknesses were, I suppose, that he did not think hard enough as an epidemiologist, he did not realize that his biometric data were not very accurate and he was unaware of the complexities of epidemic theory, for which he is hardly to be blamed.

Brownlee did not influence Sir Ronald Ross, who published[2] his mathematical theory of the epidemiology of malaria in 1911 in the second edition of his book. This was a real epidemiologist's theory, based on what Ross knew about the process by which malaria spreads from one person to another. In a 1915 paper[3] Ross was already questioning Brownlee's pet hypothesis that changes in the virulence of the infecting organism were the major factor in explaining the shapes of epidemic curves. In some brilliant work, Ross later developed mathematics that extended his theory. This extension could, I believe, take into account both the density of susceptibles and changes in infectivity over time. Ross pointed out that his theory might also be applicable both to other diseases and to crowd phenomena outside of epidemiology.

In conclusion, the layman can find two situations when reading about a period of earlier scientific history. There may be one or two investigators whom I would describe as being just ahead of their time.

They produced fruitful ideas, now taken for granted, that had no visible impact on the development of scientific thought, and were rediscovered years later. On the other hand, as I read Dr. Hilts' account of the period 1840-1900, it seems to show, as he has noted, an orderly and useful progression of ideas.

## Notes

1. M. S. Bartlett, "Deterministic and stochastic models for recurrent epidemics," *Proc. Third Berkeley Symp.*, 1956, *4:* 81-109.

2. R. Ross, *The Prevention of Malaria*, 2nd ed. (London: Murray, 1911).

3. R. Ross, "Some a priori pathometric equations," *Brit. Med. J.*, March 27, 1915.

# Estimators for the One-Way Random Effects Model With Unequal Error Variances

PODURI S.R.S. RAO, JACK KAPLAN, and WILLIAM G. COCHRAN*

---

Using the random effects model, $y_{ij} = \mu + \alpha_i + \epsilon_{ij}$, ($i = 1, \ldots, k; j = 1, \ldots, n_i$), where $\alpha_i$ and $\epsilon_{ij}$ are normal with means zero and variances $\sigma_\alpha^2$ and $\sigma_i^2$, this article considers eight methods of estimating $\sigma_i^2$, $\sigma_\alpha^2$ and thirteen corresponding procedures of estimating $\mu$. Biases and mean squared errors (MSE's) of these procedures are examined for variations in the magnitudes of the unknown parameters, the sample sizes, and the number of groups.

KEY WORDS: Combining the results of experiments; Unequal variances and unequal sample sizes; Estimators for the variances and the mean; ANOVA and MINQUE type of estimators; Two-stage procedure; Biases and mean squared errors.

## 1. INTRODUCTION

This article assumes a one-way random effects model

$$y_{ij} = \mu + \alpha_i + \epsilon_{ij} \tag{1}$$

($i = 1, \ldots, k; j = 1, \ldots, n_i$), in which $\alpha_i$ and $\epsilon_{ij}$ are normally and independently distributed with means zero and variances $\sigma_\alpha^2$, $\sigma_i^2$. The problem is to estimate $\mu$, $\sigma_\alpha^2$, and the $\sigma_i^2$ from the available data. With regard to the estimation of $\mu$, the results can be applied to estimate the average difference between two treatment means over a series of randomized experiments in any of the standard designs, conducted at a set of different locations or at different times. This problem arises when the combined results of a set of experiments in agriculture, biology, medicine, and industry are being reported.

Cochran (1937) discussed the estimation of $\mu$ when the experiments have the same design and number of replications. With randomized blocks, for instance, $y_{ij}$ in (1) could represent the difference between two treatments in the $j$th block of the $i$th experiment. The mean difference $\bar{y}_i$ for this experiment has variance $\sigma_\alpha^2 + (\sigma_i^2/n_i)$ as an estimator of $\mu$. The relevant statistics are $\bar{y}_i$, its estimated variance $s_i^2/n_i$, and the number of degrees of freedom in $s_i^2$. For estimators of $\mu$, Cochran (1937) considered the unweighted mean, the weighted and semiweighted means of the $\bar{y}_i$, and the maximum likelihood estimator. Yates and Cochran (1938) applied these estimators to three sets of data: randomized blocks experiments comparing different varieties of barley that were repeated at different locations and in different years; 13 latin squares that estimated the effects of fertilizer on sugar beet; and a further series on sugar beet with differing numbers of replications.

The special case of the model (1) in which $\sigma_\alpha^2$ can be assumed to be zero, that is, $E(y_{ij}) = \mu$, $V(y_{ij}) = \sigma_i^2$, has been considered by Cochran and Carroll (1953), Meier (1953), Bement and Williams (1969), Levy (1970), and others. For estimating $\mu$, these authors examined the relative merits of the grand mean, the weighted least squares estimator with weights estimated from the samples, and the maximum likelihood estimator (MLE). Also, J.N.K. Rao and Subrahmaniam (1971), J.N.K. Rao (1973), and Chaubey and P.S.R.S. Rao (1976) examined the advantages of estimating the $\sigma_i^2$ by the principle of the MINQUE (minimum norm quadratic unbiased estimator) suggested by C.R. Rao (1970, 1971), and then estimating $\mu$ by weighted least squares. Further work on this approach using modifications of the MINQUE principle has been done by Fuller and J.N.K. Rao (1978) and P.S.R.S. Rao and Chaubey (1978).

For the present model in (1), the weighted least squares estimator (WLS) of $\mu$ is

$$\hat{\mu} = \frac{\sum t_i \bar{y}_i}{\sum t_i}, \tag{2}$$

where $t_i = n_i/(\sigma_i^2 + n_i\sigma_\alpha^2)$ and $\bar{y}_i$ is the mean of the $i$th group. In this article we present the results of our investigation on the efficiencies of different types of estimators for the variance and the mean, without assuming that $\sigma_\alpha^2 = 0$. For $\sigma_i^2$, $\sigma_\alpha^2$, and $\mu$, we consider 11 estimators based on the analysis of variance (ANOVA) and MINQUE procedures and their modifications. For $\mu$, we also consider the grand mean (GM) $\bar{y}$ and the mean of the group means (MGM), $\sum \bar{y}_i/k$, which do not require the prior estimation of $\sigma_i^2$ and $\sigma_\alpha^2$.

The estimators are described in the next section, and the variances of the estimators are derived in Section 3. Throughout this article, the summations over the sub-

* William G. Cochran was Professor (Emeritus), Department of Statistics, Harvard University, Cambridge, MA 02138. Jack Kaplan is Assistant Professor, Department of Biometry, Case Western Reserve University, Cleveland, OH 44106. Poduri S.R.S. Rao is Professor, University of Rochester, Rochester, NY 14627. The authors would like to thank the editor, the associate editor, and the referees for their valuable suggestions. Thanks also to Uchila N. Umesh for his assistance with the computations and to Lorraine Ziegenfuss for her excellent typing of the manuscript.

© Journal of the American Statistical Association
March 1981, Volume 76, Number 373
Theory and Methods Section

script $i$ range from 1 to $k$, and those over $j$ range from 1 to $n_i$. The sampling experiment is described in Section 4. In Section 5 we present the results for some cases of the model. The conclusions from the entire study regarding the relative merits of the estimators are presented in Section 6.

## 2. ESTIMATORS FOR THE VARIANCES AND THE MEAN

### 2.1 Grand Mean and the Mean of the Group Means

If all the $\sigma_i^2$ are equal and $\sigma_\alpha^2 = 0$, $(\sigma_i^2 + n_i\sigma_\alpha^2)$ is a constant and the estimator in (2) becomes the GM

$$\bar{y} = \frac{\sum n_i \bar{y}_i}{n}, \tag{3}$$

where $n = \sum n_i$. On the other hand, when $(\sigma_i^2/n_i)$ do not vary, $\hat{\mu}$ in (2) becomes

$$\bar{y}^* = \frac{1}{k}\sum \bar{y}_i, \tag{4}$$

which is the MGM. The variances of these two estimators are equal to $\sum (n_i^2 v_i/n^2)$ and $\sum (v_i/k^2)$, respectively, where $v_i = \sigma_\alpha^2 + (\sigma_i^2/n_i)$.

Clearly there is no need to estimate $\sigma_i^2$ and $\sigma_\alpha^2$ for the GM and MGM. For all the following procedures in Sections 2.2 through 2.7, however, $\sigma_i^2$ and $\sigma_\alpha^2$ are estimated first and then $\mu$ is obtained from (2).

### 2.2 ANOVA and Unweighted Sums of Squares Estimators

With the between and within classes sums of squares, the ANOVA type of unbiased estimators for $\sigma_i^2$ and $\sigma_\alpha^2$ is obtained from

$$(n_i - 1)\hat{\sigma}_i^2 = \sum_j (y_{ij} - \bar{y}_i)^2 \tag{5}$$

and

$$\left(n - \frac{\sum n_i^2}{n}\right)\hat{\sigma}_\alpha^2 + \sum_i\left(1 - \frac{n_i}{n}\right)\hat{\sigma}_i^2 \tag{6}$$

$$= \sum n_i(\bar{y}_i - \bar{y})^2.$$

For the procedure with the unweighted sums of squares (USS), $\hat{\sigma}_i^2$ is obtained from (5) and then $\sigma_\alpha^2$ from

$$(k - 1)\hat{\sigma}_\alpha^2 + \frac{k - 1}{k}\sum\frac{\hat{\sigma}_i^2}{n_i} = \sum (\bar{y}_i - \bar{y}^*)^2, \tag{7}$$

where $\bar{y}^*$ is as defined in (4). This procedure also results in unbiased estimators for $\sigma_i^2$ and $\sigma_\alpha^2$.

### 2.3 The MINQUE With A Priori Weights

In matrix notation the model in (1) can be written as

$$Y = X\beta + \epsilon, \tag{8}$$

where $Y$ and $\epsilon$ are $(n \times 1)$ column vectors. All the elements of the column vector $X$ are equal to unity and $\beta$

$= \mu$. Further,

$$\epsilon = U_\alpha\xi_\alpha + U_1\xi_1 + \cdots + U_k\xi_k, \tag{9}$$

where $U_i$ and $\xi_i$ are defined as follows. Let $I_i$ denote the identity matrix of order $(n_i \times n_i)$. Then, $U'_i = (0 : I_i : 0)$ and $\xi_i$ is an $(n_i \times 1)$ vector with mean zero and dispersion $\sigma_i^2 I_i$. The columns of $U_\alpha$ can be written as $e_1, \ldots, e_k$. The first $n_1$ elements of $e_1$ are unities and the rest are all equal to zero. Similar notation holds for $e_2, \ldots, e_k$. The vector $\xi_\alpha$ has mean zero and dispersion $\sigma_\alpha^2 I$. Thus the dispersion of $\epsilon$ is

$$D(\epsilon) = \sigma_\alpha^2 V_\alpha + \sigma_1^2 V_1 + \cdots + \sigma_k^2 V_k, \tag{10}$$

where $V_\alpha = U_\alpha U'_\alpha$ and $V_i = U_i U'_i$, $(i = 1, \ldots, k)$. In some situations prior information about the unknown variances may be available. Let $\gamma_\alpha^2, \gamma_1^2, \ldots, \gamma_k^2$ denote the a priori values of $\sigma_\alpha^2, \sigma_1^2, \ldots, \sigma_k^2$. To estimate $p_\alpha\sigma_\alpha^2 + p_1\sigma_1^2 + \cdots + p_k\sigma_k^2$ for chosen constants $(p_\alpha, p_1, \ldots, p_k)$, the MINQUE with a priori values is given by the quadratic form $Y'AY$. The matrix $A$ is obtained by minimizing $\text{tr}AV_*AV_*$, where

$$V_* = \gamma_\alpha^2 V_\alpha + \gamma_1^2 V_1 + \cdots + \gamma_k^2 V_k, \tag{11}$$

subject to the condition for invariance, $AX = 0$, and the conditions for unbiasedness, $\text{tr}AV_\alpha = p_\alpha$ and $\text{tr}AV_i = p_i$. The solution to the preceding principle is given by C.R. Rao (1971, 1972) as

$$A = \lambda_\alpha RV_\alpha R + \sum_{i=1}^k \lambda_i RV_i R, \tag{12}$$

where $W = V_*^{-1}$, $P = X(X'WX)^{-1}X'W$, $Q = I - P$, and $R = WQ = Q'W$. The coefficients $\lambda_i$ in (12) are obtained through the conditions for unbiasedness.

For the present model in (1), simplifying the solution in (12), we obtain the MINQUE for $p_\alpha\sigma_\alpha^2 + p_1\sigma_1^2 + \cdots + p_k\sigma_k^2$ as

$$Y'AY = \lambda_\alpha \sum_1^k w_i^2(\bar{y}_i - \bar{y})^2 \tag{13}$$

$$+ \sum_1^k \lambda_i[(n_i - 1)s_i^2/r_i^2 + (w_i^2/n_i)(\bar{y}_i - \bar{y})^2],$$

where $n_i\bar{y}_i = \sum y_{ij}$, $(n_i - 1)s_i^2 = \sum (y_{ij} - \bar{y}_i)^2$, $w_i = n_i/(n_i + r_i)$, $\bar{y} = \sum w_i\bar{y}_i/\sum w_i$, and $r_i = (\gamma_i^2/\gamma_\alpha^2)$.

The estimators of $\sigma_\alpha^2$ and $\sigma_i^2$ are obtained from (13) by using the conditions for unbiasedness. Let

$$T_1 = \sum w_i, \quad T_2 = \sum w_i^2,$$

$$P_i = 1 - 2(w_i/T_1) + (T_2/T_1^2),$$

$$a_i = 1 - 2(w_i/T_1) + n_i^2(n_i - 1)/r_i^2 w_i^2,$$

$$T_3 = \sum (w_i^2/a_i), \quad T_4 = \sum w_i^2 P_i, \quad T_5 = \sum (w_i^2 P_i/a_i),$$

$$T_6 = \sum (w_i^2 P_i^2/a_i), \quad T_7 = \sum (n_i/a_i),$$

$$T_8 = (\sum n_i P_i/a_i) \quad \text{and} \quad T_9 = T_1^2 + T_3. \tag{14}$$

For estimating $\sigma_\alpha^2$, $\lambda_\alpha$ and $\lambda_i$ $(i = 1, \ldots, k)$ are obtained from

$$(T_5^2/T_9 - T_6 + T_4)\lambda_\alpha = 1 \tag{15}$$

and

$$\lambda_i = (T_5/T_9 - P_i)(n_i/a_i)\lambda_\alpha . \tag{16}$$

Similarly, for estimating $\sigma_i^2$ $(i = 1, \ldots, k)$, $\lambda_\alpha$, $\lambda_i$, and $\lambda_j$, $(j = i)$ are obtained from

$$(T_4 + T_5^2/T_9 - T_6)\lambda_\alpha = (T_5/T_9 - P_i)(n_i/a_i) , \tag{17}$$

$$\lambda_i = (n_i^2/w_i^2 a_i) - (n_i/a_i)^2/T_9$$
$$+ \lambda_\alpha(n_i/a_i)(T_5/T_9 - P_i) , \tag{18}$$

and

$$\lambda_j = -(n_i/a_i)(n_j/a_j)/T_9 + \lambda_\alpha(n_j/a_j)(T_5/T_9 - P_j) . \tag{18a}$$

Kaplan (1979) used (13) to obtain the estimates of $\sigma_\alpha^2$ and $\sigma_i^2$ for the simulations of this study. Swallow and Searle (1978) derive the MINQUE's of $\sigma_\alpha^2$ and $\sigma_i^2$ when the latter are equal and compare their merits with the ANOVA estimators.

It can be shown that when all the $\sigma_i^2$ are equal to $\sigma^2$ and all the $n_i$ are equal to $m$, the MINQUE's of $\sigma^2$ and $\sigma_\alpha^2$ are the same as the ANOVA estimators

$$\hat\sigma^2 = \frac{\sum\sum (y_{ij} - \bar y_i)^2}{n - k} . \tag{19}$$

and

$$\hat\sigma_\sigma^2 = \frac{\sum (\bar y_i - \bar y)^2}{k - 1} - \frac{\hat\sigma^2}{m} . \tag{20}$$

It can also be seen from (13) that the MINQUE of $\sigma_\alpha^2$ or $\sigma_i^2$ does not depend on the a priori values $\gamma_i^2$ and $\gamma_\alpha^2$, but it depends on the relative values $r_i = (\gamma_i^2/\gamma_\alpha^2)$. For applications in which there is no prior information about inequalities in $\sigma_i^2$, we examined the merits of the MINQUE method in which $r_i$ $(i = 1, \ldots, k)$ are assumed to be equal and to have the values 2, 1, $(\frac{1}{2})$, and $(\frac{1}{10})$.

## 2.4 A Two-Stage Procedure

Since the assumption that the $r_i$ are all equal may be unrealistic, at the first stage we estimated $\sigma_i^2$ and $\sigma_\alpha^2$ from (13) with $r_i = 1$. Now $r_i$ is obtained from these estimators, and the variances are estimated at the second stage, again using (13).

## 2.5 A Nonnegative Estimator

For estimating the unequal variances of a linear model, the Euclidean norm of the MINQUE is minimized without the condition for unbiasedness in P.S.R.S. Rao and Chaubey (1978). The procedure for the general case of unequal variances and covariances is described in P.S.R.S. Rao (1977). For the present model, with all the $r_i$ equal to unity, this modified procedure provides the estimators

$$k\hat\sigma_\alpha^2 = \sum l_i^2(\bar y_i - \bar{\bar y})^2 \tag{21}$$

and

$$n_i\hat\sigma_i^2 = \sum_j (y_{ij} - \bar y_i)^2 + (l_i^2/n_i)(\bar y_i - \bar{\bar y})^2 , \tag{22}$$

where $l_i = n_i/(n_i + 1)$ and $\bar{\bar y} = (\sum l_i\bar y_i)/(\sum l_i)$.

These estimators may be called the average of the squared residuals (ASR) type of estimators, or the MINQE type, and they are nonnegative. We note that the estimators based on the ANOVA, unweighted means, and MINQUE procedures may take negative values.

## 2.6 ASR-SSE Type of Estimators

In this procedure $\hat\sigma_i^2 = \sum (y_{ij} - \bar y_i)^2/(n_i - 1)$ and $\hat\sigma_\alpha^2$ is obtained from (21). Clearly this hybrid procedure provides unbiased estimators for $\sigma_i^2$, but a biased one for $\sigma_\alpha^2$.

## 2.7 ANOVA and MINQUE Estimators With the Assumption That All the $\sigma_i^2$ Are Equal

For the random components model, the assumption $\sigma_\alpha^2 = 0$ considered in Section 2.1 for the GM may not be appropriate. Without making this assumption, but assuming that $\sigma_i^2$ are equal, we may obtain the ANOVA type of estimators for the common variance $\sigma^2$ and for $\sigma_\alpha^2$ from

$$(n - k)\hat\sigma^2 = \sum\sum (y_{ij} - \bar y_i)^2 \tag{23}$$

and

$$\left(n - \frac{\sum n_i^2}{n}\right)\hat\sigma_\alpha^2 + (k - 1)\hat\sigma^2 = \sum n_i(\bar y_i - \bar y)^2 . \tag{24}$$

In this article these estimators are called ANOVA EQUAL.

With the same assumptions, we have also estimated $\sigma^2$ and $\sigma_\alpha^2$ through the MINQUE method with the common value of the relative a priori weight $r$ equal to unity. These estimators are called MINQUE EQUAL.

## 2.8 The Maximum Likelihood Estimator

From (1), the logarithm of the likelihood function $L$ can be written as

$$-2lnL = 2nln\sqrt{2\pi} + \sum (n_i - 1)ln\sigma_i^2$$
$$+ \sum ln(n_iv_i) + \sum (n_i - 1)(s_i^2/\sigma_i^2) \tag{25}$$
$$+ \sum t_i(\bar y_i - \mu)^2 ,$$

where $v_i = \sigma_\alpha^2 + (\sigma_i^2/n_i)$ and $t_i = 1/v_i$, as defined earlier. The estimating equations can be shown to be

$$-\frac{n_i(n_i - 1)}{\sigma_i^2} + \frac{n_i(n_i - 1)s_i^2}{\sigma_i^4} - t_i + t_i^2(\bar y_i - \mu)^2 = 0 ,$$

$$-\sum t_i + \sum t_i^2(\bar y_i - \mu)^2 = 0$$

and

$$\sum t_i(\bar y_i - \mu) = 0 . \tag{26}$$

We note that the MLE is also discussed in Cochran (1937).

The asymptotic variances of $\hat{\sigma}_\alpha^2$ and $\hat{\mu}$ for the maximum likelihood procedure can be shown to be

$$V(\hat{\sigma}_\alpha^2) = 2/(\sum t_i^2) \tag{27}$$

and

$$V(\hat{\mu}) = 1/(\sum t_i) . \tag{28}$$

## 2.9 Comments on the Choice of the Estimation Procedures

As described in Sections 2.1 through 2.7, we have considered the following 13 estimators for $\mu$: GM, MGM, ANOVA, USS, MINQUE with the four a priori weights, TWO STAGE, ASR, ASR-SSE, ANOVA EQUAL, and MINQUE EQUAL.

Our attempts to obtain the MLE for $\sigma_i^2$, $\sigma_\alpha^2$, and $\mu$ have not resulted in much success, since the iterations did not converge and the procedure proved to be computationally expensive. In Tables 3 and 5, we present the asymptotic variances in (27) and (28) for some illustrative cases. Some caution is needed in comparing them with the variances and the MSE's of the remaining estimators, which are presented in those tables for finite values of $k$ and $n_i$. The reader is referred to Hemmerle and Hartley (1973) and J.N.K. Rao (1980) for implementing the maximum likelihood procedure.

In addition to the estimators considered in this article, other types of estimators may be considered. For instance, the estimators of Henderson's (1953) type reviewed by Searle (1968) and the restricted maximum likelihood estimator reviewed by Harville (1977). Our investigation in this article is concerned with estimators of the ANOVA and MINQUE type, which can be obtained only with a little amount of computation, and GM and MGM, which do not require the prior estimation of the unknown variances.

## 3. VARIANCES OF THE ESTIMATORS OF $\sigma_\alpha^2$

### 3.1 A General Result

Let

$$Q = \sum_1^k c_i(\bar{y}_i - \bar{y})^2 , \tag{29}$$

where $\bar{y} = \sum m_i \bar{y}_i$, and $m_i$ and $c_i$ are chosen constants. Let $a_i$ denote the column vector with the $i$th element equal to $(1 - m_i)$ and the rest of the elements equal to $-m_j$ ($j \neq i$). Denote $a_i a'_i$ by $A_i$ and $\sum c_i A_i$ by $B$. Let $D$ denote the diagonal matrix with elements $v_i = \sigma_\alpha^2 + (\sigma_i^2/n_i)$. Now the expectation and variance of $Q$ can be shown to be

$$E(Q) = \sum c_i(a'_i D a_i) = \sum c_i \, \text{tr} D A_i = \text{tr}(BD)$$
$$= \sum c_i(1 - 2m_i)v_i + (\sum c_i)(\sum m_i^2 v_i) \tag{30}$$

and

$$(\tfrac{1}{2})V(Q) = \text{tr}(BDBD) = \sum_i c_i^2(a'_i D a_i)^2$$

$$+ \sum\sum_{i \neq j} c_i c_j(a'_i D a_j)^2$$

$$= \sum c_i^2(1 - 2m_i)^2 v_i^2 + (\sum c_i^2)(\sum m_i^2 v_i)^2$$

$$+ 2[\sum c_i^2(1 - 2m_i)v_i](\sum m_i^2 v_i)$$

$$+ [(\sum c_i)^2 - \sum c_i^2](\sum m_i^2 v_i)^2$$

$$+ 2(\sum c_i m_i^2 v_i^2)(\sum c_i)$$

$$+ 2(\sum c_i m_i v_i)^2 - 4(\sum c_i^2 m_i^2 v_i^2)$$

$$- 4(\sum m_i^2 v_i)[(\sum c_i m_i v_i)(\sum c_i)]$$

$$- \sum c_i^2 m_i v_i] . \tag{31}$$

### 3.2 Variances of the ANOVA and USS Estimators

From the preceding results or otherwise, the variances of $\hat{\sigma}_\alpha^2$ derived in (6) and (7) are respectively given by

$$(\tfrac{1}{2})(n - \sum n_i^2/n)^2 V(\hat{\sigma}_\alpha^2) = \sum (1 - 2n_i/n)n_i^2 v_i^2$$
$$+ (\sum n_i^2 v_i/n)^2 + \sum [(1 - n_i/n)^2 \sigma_i^4/(n_i - 1)] \tag{32}$$

and

$$(\tfrac{1}{2})(k - 1)^2 V(\hat{\sigma}_\alpha^2) = [(k - 2)/k]\sum v_i^2 + (1/k^2)(\sum v_i)^2$$
$$+ [(k - 1)^2/k^2]\sum [\sigma_i^4/n_i^2(n_i - 1)] . \tag{33}$$

### 3.3 Variance of the MINQUE

From (13), the MINQUE can be written as

$$\hat{\sigma}_\alpha^2 = \sum c_i(\bar{y}_i - \dot{y})^2 + \sum d_i(n_i - 1)s_i^2 , \tag{34}$$

where $c_i = \lambda_\alpha + (\lambda_i/n_i)w_i^2$, $m_i = w_i/(\sum w_i)$, $\dot{y} = \sum m_i \bar{y}_i$, and $d_i = (\lambda_i/r_i^2)$. The values of $\lambda_\alpha$ and $\lambda_i$ are obtained from (14) through (16). Thus

$$V(\hat{\sigma}_\alpha^2) = V(Q) + 2\sum (n_i - 1)d_i^2 \sigma_i^4 . \tag{35}$$

We note that $V(Q)$ is obtained from (31) with the above value for $c_i$.

### 3.4 Expected Value and Variance of the ASR

The ASR in (21) can be written as

$$k\hat{\sigma}_\alpha^2 = \sum c_i(\bar{y}_i - \bar{y})^2 , \tag{36}$$

where $c_i = n_i^2/(n_i + 1)^2$, $m_i = [n_i/(n_i + 1)]/\sum [n_i/(n_i + 1)]$. With this $c_i$ and $m_i$, the expected value and variance of ASR are obtained from (30) and (31).

### 3.5 Variances of the Remaining Estimators

Clearly variances of the ANOVA EQUAL and MINQUE EQUAL can be obtained from the results of Sections 3.2 and 3.3. We note that ASR and ASR-SSE procedures result in the same estimator for $\sigma_\alpha^2$. The two-stage estimator and its MSE depend on the estimated value of $r_i$. The asymptotic variance of the MLE is presented in (28).

## 4. THE SAMPLING EXPERIMENT

### 4.1 Selection of the Parameters and the Sample Sizes

Cochran (1937) and Cochran and Carroll (1953) have considered a number of issues while selecting the values of the parameters for their investigation. With a similar motivation, we have selected several sets of the parameters for this study. The different patterns of the parameters are intended for studying the effects of variations in $\sigma_i^2$, $n_i$, $(\sigma_\alpha^2/\sigma_i^2)$, and $k$ on the performances of the estimators.

For the number of groups $k$, we considered 3, 6, and 9. With each one of the chosen patterns of $\sigma_i^2$ and $n_i$, we considered the values of 0, 5, 10, 20, and 100 for $\sigma_\alpha^2$. Overall we considered a total of about 300 different patterns of the parameters, most of which are presented in Table 1. The variation among the $\sigma_i^2$ ranged from zero to 262 and the variation among the $n_i$ from 2.7 to 288. In Table 1 the case of the sample sizes (3,3,5,5,7,7) and variances (20,10,5) refer to $k = 6$ groups with variances (20,20,10,10,5,5) with the indicated sample sizes.

With each one of the specified patterns, we generated samples from the normal distribution through the procedure of Marsaglia, Ananthanarayanan, and Paul (1976) and computed the estimators of $\sigma_i^2$, $\sigma_\alpha^2$, and $\mu$. Of course, the simulation is not needed for the GM and MGM or for the estimation of $\sigma_i^2$ and $\sigma_\alpha^2$ through the ASR. The final results are obtained by repeating the process several times—a minimum of 400 and a maximum of 2,000 repetitions for each case of the model.

### 4.2 Adjustments for Negativeness

Our computations indicate that for a given pattern of sample sizes and parameters the probability of any one of the estimators of $\sigma_\alpha^2$, other than the ASR, taking a negative value is almost the same except for the ANOVA

EQUAL and MINQUE EQUAL; for the latter two estimators, this probability is much higher for some patterns and lower for the others. As we would expect, the probability of negative estimates is higher when $\sigma_\alpha^2$ is small. For instance, when $\sigma_\alpha^2 = 0$ and $\sigma_i^2 = \sigma^2 = 10$, for most of the procedures our estimate of the probability is about 50 to 60 percent; but when $\sigma_\alpha^2 = 100$, this figure reduces to three percent or less. Clearly the estimator of $\sigma_i^2$ is nonnegative for the ANOVA type and also the ASR, but it may take negative values for the MINQUE.

When an estimate of the variance becomes negative, usually a small quantity or zero is substituted for the estimate. Thompson and Anderson (1975) have successfully tried other types of truncations. In this study we required $(n_i\sigma_\alpha^2 + \sigma_i^2)$ to be positive and considered the following three types of adjustments. If the estimates of $\sigma_\alpha^2$ and $(n_i\sigma_\alpha^2 + \sigma_i^2)$ become negative, set $\sigma_\alpha^2$ to be zero and estimate $\sigma_i^2$ by the sample variance $s_i^2$. If the estimates of $\sigma_i^2$ and $(n_i\sigma_\alpha + \sigma_i^2)$ are found to be negative, set $\sigma_\alpha^2$ to be zero and assume all the $\sigma_i^2$ to be equal. If the estimate of $(n_i\sigma_\alpha^2 + \sigma_i^2)$ is positive but that of $\sigma_i^2$ or $\sigma_\alpha^2$ is negative, do not make any adjustments, since this situation may merely indicate the smallness of the latter parameters.

## 5. RESULTS FOR SOME CASES OF THE MODEL

The conclusions in the next section are based on the results of the *entire* investigation through the different sets of the variances and sample sizes described in Table 1.

For the sake of illustration, we present here the results for eight cases of the model, described in Table 2. Space does not permit us to present the results for more cases. The values of $M = \sigma_\alpha^2/(\sum n_i\sigma_i^2/n)$ and $Q = \max v_i/\min v_i$ are also given in that table.

In Table 3, we present the actual variances of the ANOVA, USS, the MINQUE with the a priori weights (2, 1, .5, .1), and also the bias, variance, and MSE of the ASR, along with the asymptotic variance of the MLE. It can be seen that when the a priori values are correct, the variance of the MINQUE is smaller than the vari-

## Table 1. Patterns of the Variances and Sample Sizes. X and Y Are the Numbers of the Patterns of the Sample Sizes and the Variances

| Sample Sizes: $n_i$ | | | | Variances: $\sigma_i^2$ | | Combinations Used in the Study: (X; Y) |
|---|---|---|---|---|---|---|
| X | Pattern | X | Pattern | Y | Pattern | |
| 1 | 3, 3, 9 | 11 | 5(2), 15(2), 25(2) | 1 | 10, 10, 10 | (1; 1,2,3), (2; 1,2,3), |
| 2 | 3, 5, 7 | 12 | 10(2), 15(2), 20(2) | 2 | 20, 10, 5 | (3; 1,5,6), (4; 1,2,3), |
| 3 | 3, 3, 24 | 13 | 3(3), 3(3), 9(3) | 3 | 40, 10, 2.5 | (5; 1,2,3), (6; 1,6,7), |
| 4 | 3, 10, 17 | 14 | 3(3), 5(3), 7(3) | 4 | 4, 10, 20 | (8; 1,2,3), (9; 1,2,3), |
| 5 | 3, 15, 27 | 15 | 3(3), 3(4), 12(2) | 5 | 2.5, 10, 40 | (10; 1,3,4), (13; 1,6,7), |
| 6 | 3, 3, 39 | 16 | 5(3), 10(3), 15(3) | 6 | 10, 5, 20 | (14; 1,6,7), (15; 1,6,7), |
| 7 | 15, 30, 45 | 17 | 3(3), 10(3), 17(3) | 7 | 10, 2.5, 40 | (16; 1,3,4,8), (17; 1,3,5,7), |
| 8 | 3(2), 5(2), 7(2) | 18 | 8(3), 10(3), 12(3) | 8 | 5, 10, 15 | (11; 8), (12; 8), (18; 7), |
| 9 | 3(2), 3(3), 15 | 19 | 10(3), 20(3), 30(3) | | | (19; 8), (20; 8) |
| 10 | 3(2), 3(2), 9(2) | 20 | 10(3), 5(3), 15(3) | | | |

NOTE: As an illustration, the X pattern numbered 8 refers to $k = 6$ with sample sizes (3,3,5,5,7,7). In the (X; Y) combinations, as an illustration, (1; 1,2,3) refers to three cases in which $k = 3$ with sample sizes (3,3,9) in each of them and variances equal to (10,10,10), (20,10,5), and (40,10,2.5).

Table 2. Values of $\sigma_i^2$ and $n_i$ of the Illustrated Cases and the Values of M and Q When $\sigma_\alpha^2$ Takes the Values 0, 5, 10, 20, and 100

| Case | $\sigma_i^2$ and $n_i$ | M and Q | $\sigma_\alpha^2$ | | | | |
|------|------|------|------|------|------|------|------|
| | | | 0 | 5 | 10 | 20 | 100 |
| a | 10,10,10 | 0 | | .5 | 1.0 | 2.0 | 10.0 |
| | 3,10,17 | 5.67 | | 1.49 | 1.26 | 1.13 | 1.03 |
| b | 20,10,5 | 0 | | .61 | 1.22 | 2.5 | 12.25 |
| | 3,10,17 | 22.67 | | 2.20 | 1.62 | 1.31 | 1.06 |
| c | 40,10,25 | 0 | | .57 | 1.14 | 2.29 | 11.43 |
| | 3,10,17 | 90.7 | | 3.56 | 2.30 | 1.65 | 1.13 |
| d | 10,10,10 | 0 | | .5 | 1.0 | 2.0 | 10.0 |
| | 3, 3,24 | 8.0 | | 1.54 | 1.28 | 1.14 | 1.03 |
| e | 5,10,20 | 0 | | .29 | .57 | 1.14 | 5.71 |
| | 3, 3,24 | 4.0 | | 1.43 | 1.23 | 1.12 | 1.02 |
| f | 2.5,10,40 | 0 | | .15 | .3 | .6 | 3.0 |
| | 3, 3, 24 | 4.0 | | 1.43 | 1.23 | 1.12 | 1.02 |
| g | 10,10,10,10,10,10, 10,10,10 | 0 | | .5 | 1.0 | 2.0 | 10.0 |
| | 5,5,5,10,10,10, 15,15,15 | 3 | | 1.24 | 1.13 | 1.06 | 1.01 |
| h | 5,5,5,10,10,10, 15,15,15 | 0 | | .46 | .92 | 1.85 | 9.23 |
| | 10,10,10,5,5,5, 15,15,15 | 4.0 | | 1.27 | 1.14 | 1.07 | 1.01 |

ances of the remaining unbiased estimators (see cases (a) and (d) of Table 3, for instance). We have not presented the variances of the ANOVA EQUAL and MINQUE EQUAL here and in the remaining tables since we found them to increase substantially when the implied assumption of the equality of the $\sigma_i^2$ is not satisfied. Notice that the two-stage estimator requires the estimation of the a priori values first. The values in this table are obtained from the expressions in Section 3.

Although all the estimators of $\sigma_\alpha^2$ except the ASR are unbiased, they become biased because of the truncation described in Section 4. In Table 4 we present the MSE's of the estimators for the eight illustrated cases. We have found negligible differences between the MINQUE's with the four different a priori weights, and in Table 4 we present the MSE for the weight $r_i$ equal to one.

The actual variances of the unbiased estimators GM and MGM and the asymptotic variance of MLE for the illustrated cases are presented in Table 5. We have found that the estimators of $\mu$ of the remaining procedures, including for the ASR, have negligible bias. The MSE's of the estimators of these procedures obtained through the simulated experiment are given in Table 6. We have found the MSE's of these estimators to be almost the same as the variance of MGM (given in Table 5) when $\sigma_\alpha^2$ is 20 or higher. We have not presented the MSE of the ASR-SSE in this table since we did not find it performed better than ASR.

## 6. RELATIVE MERITS OF THE ESTIMATORS

As described in Section 5, the following conclusions are obtained from all the results of our investigation.

## 6.1 Variances and MSE's of the Estimators of $\sigma_\alpha^2$

1. The expressions for the expectation and variance of the ASR are given in Section 3. For an illustration, when all the $\sigma_i^2$ are equal to $\sigma^2$ and the $n_i$ are equal to $m$, the expectation and variance of $\dot{\sigma}_\alpha^2$ for this procedure are

$$E(\dot{\sigma}_\alpha^2) = [m/(m + 1)]^2(k - 1)v/k \tag{37}$$

and

$$V(\dot{\sigma}_\alpha^2) = 2[m/(m + 1)]^4(k - 1)v^2/k^2 , \tag{38}$$

where $v = \sigma_\alpha^2 + (\sigma^2/m)$. Thus the variance of ASR is smaller than the asympototic variance $2v^2/k$ of the MLE. Further, if $\sigma_\alpha^2 = 0$, the ratio of the MSE of ASR to the latter variance is equal to $[m/(m + 1)]^4(k - 1)(k + 1)/2k$. Thus, for instance, the ASR is superior to the MLE if the sample size is smaller than 14, 6 and 5 when $k = 3$, 4, and 5, respectively.

We have found that when $\sigma_\alpha^2 > 0$, the MSE of ASR is smaller than the MSE's of the rest of the estimators and the asymptotic variance of the MLE for all the different cases of the model considered in this investigation, even when $\sigma_i^2$ and $n_i$ vary. For this case of $\sigma_\alpha^2$, the MINQUE ranks next to ASR; the TWO STAGE and USS are close to the MINQUE. For designs with a small number of observations, there may be a considerable amount of reduction in the MSE obtained by replacing an unbiased estimator by a suitable biased one (see La Motte 1976).

If $(\sigma_i^2/n_i)$ is small, $\bar{y}_i$ has variance $\sigma_\alpha^2$, the biased estimator $\sum (\bar{y}_i - \bar{y})^2/(k + 1)$, where $\bar{y} = (\sum \bar{y}_i/k)$, has MSE $2\sigma_\alpha^4/(k + 1)$, and the ASR is close to this estimator. Another plausible reason for the good performance of the ASR is that it is unaffected by the estimator of $\sigma_i^2$, unlike the remaining estimators.

We have found the bias of ASR to be nonnegative when $\sigma_\alpha^2 = 0$ and negative when $\sigma_\alpha^2 > 0$. This bias increases with $\sigma_\alpha^2$ but decreases as $k$ and $n_i$ increase.

2. When $\sigma_\alpha^2 = 0$, ANOVA has, in general, smaller MSE than ASR, MINQUE, TWO STAGE, and USS, provided that variation among $n_i$ is small.

3. The MSE's of all the estimators increase as the variation among $\sigma_i^2$ increases for given $n_i$, as the variation among $n_i$ increases for given $\sigma_i^2$, as the variation among $(\sigma_i^2/n_i)$ increases, and as $\sigma_\alpha^2$ increases for given $\sigma_i^2$ and $n_i$. An increase in $k$ decreases the MSE's. The difference in the MSE's of ANOVA, USS, MINQUE, and TWO STAGE becomes small as $\sigma_\alpha^2$ becomes large.

4. All the MSE's increase as $M$ increases and $Q$ decreases.

## 6.2 Variances and MSE's of the Estimators of $\mu$

1. When $\sigma_\alpha^2 > 0$, we found the variance of MGM to be smaller than the variance of GM and the MSE's of the remaining estimators, the ASR ranks second, and the rest of the estimators have almost the same MSE.

Table 3. Actual Variances of the Estimators of $\sigma_\alpha^2$ When Its True Values Are 0, 5, 10, 20, and 100. Bias of ASR and Its MSE Are Also Presented. Asymptotic Variance of the MLE Is Denoted by MLE*. Letters a–h Refer to the Eight Cases of Table 2

| | $\sigma_\alpha^2$ | | | | | $\sigma_\alpha^2$ | | | | |
|---|---|---|---|---|---|---|---|---|---|---|
| | 0 | 5 | 10 | 20 | 100 | 0 | 5 | 10 | 20 | 100 |
| | | | a | | | | | b | | |
| ANOVA | 1.76 | 45.5 | 152.8 | 558.1 | 12,955.2 | 4.17 | 48.7 | 156.8 | 563.7 | 12,973.3 |
| USS | 4.69 | 46.1 | 137.5 | 470.3 | 10,332.8 | 16.07 | 67.6 | 169.2 | 522.2 | 10,546.8 |
| MINQUE (2) | 2.84 | 42.7 | 136.2 | 484.5 | 11,006.5 | 8.76 | 53.1 | 151.2 | 508.6 | 11,103.3 |
| MINQUE (1) | 3.56 | 43.3 | 134.3 | 469.7 | 10,522.8 | 11.66 | 58.2 | 156.0 | 505.0 | 10,667.0 |
| MINQUE (.5) | 4.06 | 44.4 | 135.0 | 467.2 | 10,372.6 | 13.65 | 62.3 | 161.2 | 510.0 | 10,548.4 |
| MINQUE (.1) | 4.56 | 45.7 | 136.8 | 469.1 | 10,329.8 | 15.55 | 66.4 | 167.3 | 519.1 | 10,535.6 |
| ASR VAR | .67 | 10.6 | 33.3 | 117.1 | 2,625.6 | 2.06 | 13.7 | 38.2 | 125.4 | 2,661.8 |
| ASR BIAS | .76 | −1.74 | −4.24 | −9.24 | −49.2 | 1.18 | −1.32 | −3.82 | −8.82 | −48.82 |
| ASR MSE | 1.24 | 13.65 | 51.3 | 202.5 | 5,050.1 | 3.45 | 15.5 | 52.8 | 203.2 | 5,044.9 |
| MLE* | .50 | 26.95 | 87.7 | 309.4 | 6,884.3 | .16 | 28.25 | 93.9 | 327.7 | 7,009.3 |
| | | | c | | | | | d | | |
| ANOVA | 14.59 | 63.5 | 176.0 | 591.7 | 13,071.5 | 5.41 | 54.3 | 161.6 | 551.7 | 12,093.1 |
| USS | 61.23 | 134.5 | 257.8 | 654.3 | 11,026.6 | 8.99 | 57.6 | 156.2 | 503.4 | 10,481.2 |
| MINQUE (2) | 32.68 | 88.2 | 197.5 | 577.2 | 11,350.5 | 6.30 | 53.3 | 154.1 | 517.5 | 11,184.4 |
| MINQUE (1) | 44.05 | 106.0 | 219.1 | 598.9 | 11,007.1 | 7.28 | 54.0 | 152.1 | 502.2 | 10,696.5 |
| MINQUE (.5) | 51.79 | 118.6 | 235.8 | 621.2 | 10,951.5 | 8.01 | 55.3 | 152.9 | 499.5 | 10,528.8 |
| MINQUE (.1) | 59.19 | 131.1 | 252.9 | 646.7 | 10,998.9 | 8.77 | 57.0 | 155.3 | 501.9 | 10,477.4 |
| ASR VAR | 7.59 | 23.2 | 51.6 | 146.6 | 2,745.7 | 1.04 | 10.3 | 29.6 | 98.7 | 2,108.6 |
| ASR BIAS | 2.10 | −.4 | −2.9 | −7.7 | −47.9 | .98 | −1.8 | −4.57 | −10.1 | −54.6 |
| ASR MSE | 12.01 | 23.4 | 60.0 | 209.0 | 5,039.5 | 2.00 | 13.5 | 50.5 | 201.3 | 5,086.4 |
| MLE* | .042 | 29.2 | 100.9 | 355.2 | 7,257.1 | .337 | 31.81 | 97.7 | 329.4 | 6,981.4 |
| | | | e | | | | | f | | |
| ANOVA | 4.18 | 51.6 | 157.5 | 544.7 | 12,063.0 | 5.31 | 56.1 | 165.5 | 559.5 | 12,132.8 |
| USS | 5.87 | 50.3 | 144.8 | 483.6 | 10,394.8 | 5.67 | 50.1 | 144.5 | 483.4 | 10,394.5 |
| MINQUE (2) | 4.57 | 49.4 | 148.1 | 507.1 | 11,139.3 | 5.29 | 52.4 | 153.4 | 417.1 | 11,186.5 |
| MINQUE (1) | 5.03 | 48.8 | 144.0 | 488.3 | 10,635.9 | 5.37 | 50.6 | 147.1 | 494.1 | 10,663.6 |
| MINQUE (.5) | 5.38 | 49.2 | 143.4 | 483.0 | 10,457.0 | 5.48 | 50.03 | 145.0 | 486.1 | 10,472.0 |
| MINQUE (.1) | 5.76 | 50.0 | 144.3 | 482.9 | 10,394.2 | 5.62 | 50.02 | 144.5 | 483.4 | 10,397.4 |
| ASR VAR | .74 | 9.4 | 28.1 | 96.1 | 2,096.6 | .84 | 9.75 | 28.8 | 97.2 | 2,101.8 |
| ASR BIAS | .83 | −1.95 | −4.73 | −10.3 | −54.7 | .85 | −1.92 | −4.7 | −10.3 | −54.7 |
| ASR MSE | 1.42 | 13.2 | 50.5 | 201.8 | 5,091.4 | 1.57 | 13.44 | 50.9 | 202.4 | 5,093.5 |
| MLE* | 1.06 | 30.2 | 93.1 | 318.9 | 6,926.3 | 1.06 | 30.17 | 93.1 | 318.9 | 6,926.3 |
| | | | g | | | | | h | | |
| ANOVA | .296 | 10.1 | 34.3 | 126.0 | 2,939.5 | .326 | 10.2 | 34.6 | 126.6 | 2,942.4 |
| USS | .528 | 9.8 | 31.6 | 112.8 | 2,561.6 | .507 | 9.7 | 31.3 | 112.2 | 2,558.8 |
| MINQUE (2) | .401 | 9.5 | 31.5 | 114.0 | 2,624.4 | .390 | 9.4 | 31.3 | 113.7 | 2,622.6 |
| MINQUE (1) | .451 | 9.6 | 31.3 | 112.7 | 2,579.2 | .435 | 9.5 | 31.1 | 112.2 | 2,576.9 |
| MINQUE (.5) | .519 | 9.8 | 31.6 | 112.6 | 2,561.4 | .467 | 9.5 | 31.1 | 111.9 | 2,562.8 |
| MINQUE (.1) | .486 | 9.7 | 31.4 | 112.5 | 2,565.3 | .498 | 9.6 | 31.3 | 112.1 | 2,558.6 |
| ASR VAR | .194 | 4.8 | 15.8 | 56.8 | 1,301.2 | .187 | 4.7 | 15.7 | 56.6 | 1,299.9 |
| ASR BIAS | .834 | −.61 | −2.06 | −4.96 | −28.12 | .798 | −.65 | −2.10 | −5.0 | −28.2 |
| ASR MSE | .889 | 5.2 | 20.0 | 81.4 | 2,092.1 | .823 | 5.2 | 20.1 | 81.5 | 2,093.0 |
| MLE* | .191 | 8.4 | 27.8 | 99.9 | 2,276.7 | .127 | 8.2 | 27.5 | 99.3 | 2,274.1 |

2. When $\sigma_\alpha^2 = 0$, the variance of GM is smaller than the variance of MGM and the MSE's of the rest of the estimators, provided $k$ is small. When $k$ is large, the remaining estimators are usually superior to GM; especially, the MSE of ASR and the variance of MGM are smaller than the variance of GM.

3. As in the case of estimating $\sigma_\alpha^2$, an increase in the variations among $\sigma_i^2$, $n_i$, $(\sigma_i^2/n_i)$ results in an increase in the MSE's of all the estimators. An increase in $k$, even without increasing the total sample size, decreases the MSE's of the estimators. When $\sigma_\alpha^2$ is large, the difference in the MSE's of the estimators becomes small.

4. We note that GM and MGM are unbiased and the variance of GM coincides with the asymptotic variance of the MLE, provided $\sigma_\alpha^2 = 0$ and $\sigma_i^2$ are equal. All the procedures have negligible bias, and their MSE's approach the asymptotic variance of the MLE as $\sigma_\alpha^2$ becomes large; the variance of MGM also approaches the latter when $\sigma_\alpha^2$ is large or as $k$ is large.

5. When $M > 0$, MGM and ASR have smaller MSE's than the rest. Specifically, when $M$ is between zero and one, both MGM and ASR are superior to the remaining estimators. If $M$ exceeds one, the variance of MGM is smaller than the MSE's of the rest of the estimators.

Table 4. MSE's of the Estimators of $\sigma_\alpha^2$ From the Simulation Study

| | $\sigma_\alpha^2$ | | | | | $\sigma_\alpha^2$ | | | | |
|---|---|---|---|---|---|---|---|---|---|---|
| | 0 | 5 | 10 | 20 | 100 | 0 | 5 | 10 | 20 | 100 |
| | | | *a* | | | | | *b* | | |
| ANOVA | 1.07 | 37.8 | 159.8 | 557.0 | 13,550 | 2.84 | 40.4 | 164.0 | 683.0 | 11,980 |
| USS | 2.45 | 38.7 | 135.6 | 481.1 | 10,360 | 10.97 | 49.5 | 144.5 | 626.3 | 9,606 |
| MINQUE | 1.89 | 36.1 | 134.8 | 477.8 | 10,770 | 7.96 | 44.3 | 142.3 | 608.0 | 9,810 |
| TWO STAGE | 2.60 | 37.3 | 132.5 | 471.4 | 10,540 | 7.93 | 49.9 | 139.7 | 609.0 | 9,478 |
| ASR | 1.24 | 13.7 | 51.3 | 202.5 | 5,050 | 3.45 | 15.5 | 52.8 | 203.2 | 5,045 |
| | | | *c* | | | | | *d* | | |
| ANOVA | 14.21 | 58.1 | 142.1 | 556.2 | 10,450 | 3.99 | 74.0 | 110.1 | 457.3 | 12,860 |
| USS | 60.20 | 77.5 | 210.9 | 580.5 | 9,760 | 6.93 | 60.3 | 114.9 | 444.3 | 9,706 |
| MINQUE | 43.28 | 67.7 | 178.4 | 543.5 | 9,409 | 5.51 | 63.7 | 108.6 | 430.6 | 10,500 |
| TWO STAGE | 32.91 | 73.0 | 189.1 | 538.3 | 9,505 | 7.27 | 60.5 | 108.6 | 434.4 | 9,774 |
| ASR | 12.01 | 23.4 | 60.0 | 209.0 | 5,040 | 2.00 | 13.5 | 50.5 | 201.3 | 5,086 |
| | | | *e* | | | | | *f* | | |
| ANOVA | 2.56 | 43.5 | 179.6 | 590.3 | 12,920 | 3.37 | 49.3 | 206.8 | 588.8 | 12,610 |
| USS | 3.27 | 45.2 | 167.2 | 490.6 | 12,310 | 3.08 | 47.2 | 156.2 | 474.0 | 10,620 |
| MINQUE | 2.86 | 42.6 | 165.3 | 510.3 | 12,080 | 3.08 | 46.4 | 168.9 | 499.2 | 10,890 |
| TWO STAGE | 3.32 | 43.2 | 163.9 | 480.0 | 12,290 | 3.10 | 46.8 | 155.6 | 469.3 | 10,640 |
| ASR | 1.42 | 13.2 | 50.5 | 201.8 | 5,091 | 1.57 | 13.4 | 50.9 | 202.4 | 5,094 |
| | | | *g* | | | | | *h* | | |
| ANOVA | .18 | 9.36 | 34.1 | 125.9 | 2,904 | .26 | 10.1 | 37.7 | 121.8 | 3,094 |
| USS | .32 | 9.33 | 32.0 | 118.7 | 2,648 | .33 | 9.6 | 34.6 | 113.9 | 2,582 |
| MINQUE | .27 | 9.06 | 31.6 | 117.2 | 2,640 | .28 | 9.4 | 34.3 | 112.0 | 2,627 |
| TWO STAGE | 1.38 | 8.79 | 30.9 | 116.3 | 2,621 | .40 | 9.1 | 33.4 | 111.3 | 2,563 |
| ASR | .89 | 5.20 | 20.0 | 81.4 | 2,092 | .823 | 5.2 | 20.1 | 81.5 | 2,093 |

NOTE: For the sake of comparison, the exact MSE of ASR (rounded off value) is reproduced here from Table 3.

Table 5. Variances of GM and MGM and the Asymptotic Variance of MLE of $\mu$ When $\sigma_\alpha^2 = 0, 5, 10, 20,$ and 100

| | $\sigma_\alpha^2$ | | | | | $\sigma_\alpha^2$ | | | | |
|---|---|---|---|---|---|---|---|---|---|---|
| | 0 | 5 | 10 | 20 | 100 | 0 | 5 | 10 | 20 | 100 |
| | | | *a* | | | | | *b* | | |
| GM | .333 | 2.54 | 4.76 | 9.18 | 44.56 | .272 | 2.48 | 4.69 | 9.12 | 44.49 |
| MGM | .547 | 2.21 | 3.88 | 7.21 | 33.88 | .885 | 2.55 | 4.22 | 7.55 | 34.22 |
| MLE* | .333 | 2.15 | 3.84 | 7.19 | 33.88 | .220 | 2.27 | 4.03 | 7.44 | 34.19 |
| | | | *c* | | | | | *d* | | |
| GM | .292 | 2.50 | 4.71 | 9.14 | 44.51 | .333 | 3.63 | 6.93 | 13.53 | 66.53 |
| MGM | 1.509 | 3.28 | 4.94 | 8.28 | 34.94 | .787 | 2.45 | 4.12 | 7.45 | 34.12 |
| MLE* | .127 | 2.41 | 4.30 | 7.86 | 34.83 | .333 | 2.36 | 4.07 | 7.42 | 34.11 |
| | | | *e* | | | | | *f* | | |
| GM | .583 | 3.88 | 7.18 | 13.78 | 66.58 | 1.108 | 4.41 | 7.71 | 14.31 | 67.11 |
| MGM | .648 | 2.31 | 3.98 | 7.31 | 33.98 | .648 | 2.31 | 3.98 | 7.31 | 33.98 |
| MLE* | .476 | 2.27 | 3.95 | 7.30 | 33.98 | .476 | 2.27 | 3.95 | 7.30 | 33.98 |
| | | | *g* | | | | | *h* | | |
| GM | .111 | .76 | 1.41 | 2.70 | 13.07 | .120 | .769 | 1.42 | 2.71 | 13.08 |
| MGM | .136 | .69 | 1.25 | 2.36 | 11.25 | .130 | .685 | 1.24 | 2.35 | 11.24 |
| MLE* | .111 | .686 | 1.24 | 2.36 | 11.25 | .095 | .678 | 1.24 | 2.35 | 11.24 |

Table 6. MSE's of the Estimators of $\mu$ From the Simulation Study When $\sigma_\alpha^2 = 0, 5, and 10$

| | $\sigma_\alpha^2$ | | | $\sigma_\alpha^2$ | | |
|---|---|---|---|---|---|---|
| | 0 | 5 | 10 | 0 | 5 | 10 |
| | | a | | | b | |
| ANOVA | .458 | 2.28 | 4.09 | .447 | 2.61 | 4.42 |
| USS | .453 | 2.29 | 4.07 | .469 | 2.63 | 4.44 |
| MINQUE | .446 | 2.28 | 4.07 | .454 | 2.63 | 4.43 |
| TWO STAGE | .435 | 2.29 | 4.07 | .446 | 2.63 | 4.41 |
| ASR | .460 | 2.25 | 4.04 | .491 | 2.54 | 4.36 |
| | | c | | | d | |
| ANOVA | .625 | 2.70 | 4.92 | .694 | 2.56 | 4.57 |
| USS | .800 | 2.86 | 4.98 | .655 | 2.63 | 4.58 |
| MINQUE | .739 | 2.77 | 4.95 | .655 | 2.60 | 4.55 |
| TWO STAGE | .635 | 2.74 | 4.97 | .649 | 2.60 | 4.59 |
| ASR | .786 | 2.59 | 4.77 | .725 | 2.47 | 4.50 |
| | | e | | | f | |
| ANOVA | .719 | 2.52 | 4.00 | .635 | 2.61 | 4.25 |
| USS | .698 | 2.52 | 4.02 | .633 | 2.60 | 4.26 |
| MINQUE | .698 | 2.49 | 3.99 | .629 | 2.59 | 4.25 |
| TWO STAGE | .645 | 2.48 | 3.99 | .597 | 2.56 | 4.25 |
| ASR | .655 | 2.42 | 3.97 | .584 | 2.54 | 4.23 |
| | | g | | | h | |
| ANOVA | .156 | .72 | 1.25 | .138 | .68 | 1.22 |
| USS | .157 | .72 | 1.25 | .136 | .68 | 1.22 |
| MINQUE | .153 | .71 | 1.25 | .132 | .68 | 1.22 |
| TWO STAGE | .148 | .71 | 1.25 | .120 | .68 | 1.22 |
| ASR | .126 | .71 | 1.25 | .119 | .68 | 1.22 |

6. We found that MGM is preferable to the rest when $Q$ is between 1.0 and 1.5, both MGM and ASR when $Q$ is between 1.5 and 2.5, and GM when $Q$ is more than 2.5. In two cases, when $Q$ was 1.16 and 1.45, Yates and Cochran (1938) preferred MGM to other estimators.

[*Received August 1979. Revised June 1980.*]

## REFERENCES

BEMENT, T.R., and WILLIAMS, J.S. (1969), "Variance of Weighted Regression Estimators When Sampling Errors Are Independent and Heteroscedastic," *Journal of the American Statistical Association*, 64, 1369–1382.

CHAUBEY, Y.P., and RAO, P.S.R.S. (1976), "Efficiencies of Five Estimators of Two Linear Models With Unequal Variances," *Sankhyā*, Ser. B, 38, 364–370.

COCHRAN, W.G. (1937), "Problems Arising in the Analysis of a Series of Similar Experiments," *Journal of the Royal Statistical Society*, Supplement 4, 102–118.

COCHRAN, W.G., and CARROLL, S.P. (1953), "A Sampling Investigation of the Efficiency of Weighting Inversely As the Estimated Variance," *Biometrics*, 9, 447–459.

FULLER, W.A., and RAO, J.N.K. (1978), "Estimation for a Linear Regression Model With Unknown Diagonal Covariance Matrix," *Annals of Statistics*, 6, 1149–1158.

HARVILLE, D.A. (1977), "Maximum Likelihood Approaches to Variance Component Estimation and to Related Problems," *Journal of the American Statistical Association*, 72, 320–338.

HEMMERLE, W.J., and HARTLEY, H.O. (1973), "Computing Maximum Likelihood Estimation for the Mixed AOV Model Using the W-transformation," *Technometrics*, 15, 819–831.

HENDERSON, C.A. (1953), "Estimation of Variance and Covariance Components," *Biometrics*, 9, 226–252.

KAPLAN, J. (1979), "Some Results on Estimating Variance Components in Unbalanced Designs," PhD thesis, University of Rochester.

LA MOTTE, L.R. (1976), "Invariant Quadratic Estimators in the Random, One-Way ANOVA Model," *Biometrics*, 32, 793–804.

LEVY, P. (1970), "Combining Independent Estimators—An Empirical Sampling Study," *Technometrics*, 12, 162–165.

MARSAGLIA, G., ANANTHANARAYANAN, K. and PAUL, N.J. (1976), "Improvements on Fast Methods for Generating Normal Random Variables," *Inf. Processing Letters*, 5, 27–30.

MEIER, P. (1953), "Variance of a Weighted Mean," *Biometrics*, 9, 59–73.

RAO, C.R. (1970), "Estimation of Heteroscedastic Variances in Linear Models," *Journal of the American Statistical Association*, 65, 161–172.

——— (1971), "Estimation of Variance and Covariance Components—MINQUE Theory," *Journal of Multivariate Analysis*, 3, 257–275.

——— (1972), "Estimation of Variance and Covariance Components in Linear Models," *Journal of the American Statistical Association*, 67, 112–115.

RAO, J.N.K. (1973), "On the Estimation of Heteroscedastic Variances," *Biometrics*, 29, 11–24.

——— (1980), "Estimating the Common Mean of Possibly Different Normal Populations—A Simulation Study," *Journal of the American Statistical Association*, 75, 447–453.

RAO, J.N.K., and SUBRAHAMANIAM, K. (1971), "Combining Independent Estimators and Estimation in Linear Regression With Unequal Variances," *Biometrics*, 27, 971–990.

RAO, P.S.R.S. (1977), "Theory of the MINQUE—A Review," *Sankhyā*, Ser. A, 201–210.

RAO, P.S.R.S., and CHAUBEY, Y.P. (1978), "Three Modifications of the Principle of the MINQUE," *Communications in Statistics*, A7, 767–778.

SEARLE, S.R. (1968), "Another Look at Henderson's Methods of Estimating Variance Components," *Biometrics*, 24, 749–787.

SWALLOW, W.H., and SEARLE, S.R. (1978), "Minimum Variance Quadratic Unbiased Estimation (MINQUE) of Variance Components," *Technometrics*, 20, 265–272.

THOMPSON, W.O., and ANDERSON, R.L. (1975), "A Comparison of Designs and Estimators for the Two-Stage Nested Random Model," *Technometrics*, 17, 37–44.

YATES, F., and COCHRAN, W.G. (1938), "The Analysis of Groups of Experiments," *Journal of Agricultural Sciences*, 28, 556–580.

# 115

## The Horvitz-Thompson Estimator

The objectives of this paper, Horvitz and Thompson (1952), are to give an unbiased estimator of the population total when sampling with known unequal probabilities. The sampling variance of this estimator is also given, as well as an unbiased estimator of the sampling variance. It is well known that no unbiased estimator with minimum variance can be given, except by a fluke, because if the probability $\pi_i$ is proportional to $y_i$ (usually unknown) the estimator $\sum y_i/\pi_i$ is known to have zero variance when $\pi_i = y_i/Y_i$, where $Y_i = \sum y_i$ is the population total.

If $\underline{n}$ sampling units are drawn independently from a population with probabilities $\pi_i$, equal or unequal, the Horvitz-Thompson unbiased estimator of the population total is

$$\hat{Y} = \sum_{i=1}^{n} \frac{Y_i}{\pi_i}$$

where $Y_i$ is the total on the ith unit, and $\pi_i$ is assumed non-zero for any method of sampling. Horvitz and Thompson showed that the variance of the estimator is

$$V(\hat{Y}) = \sum_{i=1}^{N} \frac{(1-\pi_i)}{\pi_i} Y_i^2 + 2 \sum_{i \neq j}^{N} \frac{(\pi_{ij} - \pi_i \pi_j)}{\pi_i \pi_j} Y_i Y_j$$

where $\pi_{ij}$ is the probability (assumed non-zero for any pair) that units $\underline{i}$ and $\underline{j}$ are both in the sample. They also showed that an unbiased sample estimator of variance is

$$\hat{V}(\hat{Y}) = \sum_{i=1}^{n} \frac{(1-\pi_i)}{\pi_i^2} Y_i^2 + 2 \sum_{i=1}^{N} \sum_{j \neq i}^{N} \frac{(\pi_{ij} - \pi_i \pi_j)}{\pi_{ij}} \frac{}{\pi_i \pi_j} Y_i Y_j \quad .$$

An alternative sample estimate of variance, given by Yates and Grundy (1953), is

$$V(Y) = \sum_{i=1}^{n} \sum_{j\neq i}^{n} \frac{(\pi_{ij} - \pi_i \pi_j)}{\pi_{ij}} \left(\frac{Y_i}{\pi_i} - \frac{Y_j}{\pi_j}\right)^2$$

taken over every pair of units in the sample, with the same restrictions that $\pi_{ij} \neq 0$ for any pair of units in the sample. Both of the estimators of variance can be negative and are therefore rather unstable quantities. Rao and Singh (1973) compared the two variance estimators in samples with $\underline{n} = 2$ from 34 small natural populations using a method of sample selection due to Brewer (1963) which will be described later. The Yates and Grundy estimator of the variance was found to be considerably more stable.

In connection with the Yates-Grundy estimator of variance, Vijayan (19. noted that if $\underline{n} = 2$ a necessary and sufficient condition that their estimate of variance be non-negative is $(\pi_i \pi_j - \pi_{ij}) \geq 0$ for all i,j, but by a counter-example he pointed out that if n > 2 and a non-negative estimator of variance exists, this condition is sufficient but not necessary for all estimators.

If the sample is drawn with equal probabilities, the HT estimator becomes the estimator $\hat{Y} = NY/n$ commonly used, since $\pi_i = n/N$. But as Horvi and Thompson remark, there may be an advantage in accuracy in drawing the sample with unequal probabilities. As an example, they take a population c sisting of 20 city blocks in Ames, Iowa; the variate $Y_i$ being the estimated number of households in the block. For sampling, they divide the populatio into two strata by the number of houses and consider a sample with n = 2, one unit being drawn from each stratum. When units are drawn with equal probability the variance of the estimated total is 7,873, but when units ar drawn with probability proportional to $Y_i$, the variance of the estimated po lation total falls to 3,934.

With n > 1 when sampling without replacement, one must be careful in order to attain the desired probability, since they always move towards equality if we are careless. The probability that the ith unit is drawn at either the first or the second draw, if we follow the natural method and draw the second unit with probability $\pi_i/(1-\pi_j)$, where unit $\underline{j}$ was drawn first, is

$$\pi_i + \sum_j \frac{\pi_i \pi_j}{1-\pi_j} = \pi_i(1 + \sum_{j \neq i}^N \frac{\pi_j}{1-\pi_j}) = \pi_i(1 + A - \frac{\pi_i}{1-\pi_i})$$

where

$$A = \sum_{j=1}^N \frac{\pi_i}{1-\pi_i} \quad .$$

Dividing these values by 2, Yates and Grundy (1953) found that with original probabilities $\pi_i$ = 0.4, 0.3, 0.2, and 0.1, and n = 2, the average probabilities of drawing unit $\underline{i}$ on either the first or second draw become 0.1173, 0.2206, 0.3042, and 0.3579. Thus some care is necessary in order to attain any desired $\pi$'s. For example, Brewer (1963), Rao (1965), and Durbin (1967) all gave methods that have the desired $\pi_i$ and $\pi_{ij}$. Brewer draws the first unit with probability proportional to $\pi_i(1-\pi_i)/(1-2\pi_i)$ and the second with probability equal to $\pi_i/(1-\pi_j)$, where unit $\underline{j}$ was the unit drawn first. The divisor needed to convert the terms for the probability of being drawn first into probabilities is of course their sum

$$D = \sum_{i=1}^N \frac{\pi_i(1-\pi_i)}{1-2\pi_i} = \frac{1}{2}(1 + \sum_i \frac{\pi_i}{1-2\pi_i}) \quad .$$

With n = 2 the probability that a unit was drawn is the sum of the probabilities that the unit was drawn first and that it was drawn second. Thus this probability is

$$\frac{\pi_i(1-\pi_i)}{D(1-2\pi_i)} + \frac{1}{D}\sum_{j\neq i}^{N}\frac{\pi_j(1-\pi_j)\pi_i}{(1-2\pi_j)(1-\pi_j)} = \frac{\pi_i}{D}(1 + \sum_{j=1}^{N}\frac{\pi_j}{1-2\pi_j}) \ .$$

But this equals $2\pi_i$ by definition of D.

Durbin draws the first unit with probability $\pi_i$ and the second unit with probability proportional to

$$\pi_j[\frac{1}{1-2\pi_i} + \frac{1}{1-2\pi_j}]$$

where unit $\underline{j}$ was drawn first. It turns out that the probability that the ith unit was drawn either first or second is $2\pi_i$. Thus Durbin's method has the property, which Brewer's does not, that $\pi_i$ is the probability of being drawn at either the first or second draw, which is useful if one wants to replace a panel of people.

For systematic sampling Madow (1949) has given a simple method which keeps $\pi_i \propto nX_i$, where the units are cluster units of size $X_i$. As usual with systematic sampling, no unbiased sample estimate of variance is available. Sometimes one wants to draw the sample with probability proportional to $\sum x_i$ where $x_i$ is an auxiliary variable with an approximately constant ratio to the variable $Y_i$. If we draw the first unit with probability $x_i$, and subsequent units with equal probabilities, the probability of a sample is proportional to $\sum x_i$, as shown by Midzuno (1951).

Later writers examined further properties of the estimator: Hanurau (1967) investigated good sampling strategies with $\pi_i \propto X_i$ and the use of the Horvitz-Thompson estimator. He considered that $\pi_{ij}/\pi_i\pi_j$ should be positive, less than 1, but not too small, otherwise the Yates-Grundy estimator has a large variance. He produced a method of drawing the sample that has these properties. Hege (1967), following Ajagaonkar, stated that if an estimator is unbiased and has the smallest variance of any unbiased estimator in a class, it is called the "necessary best estimator" in that class. He showed that the Horvitz-Thompson estimator is a necessary best estimator among the class of linear estimates of Y, with arbitrary probabilities of selection without replacement at each draw. T. J. Rao (1971) examined several good sampling strategies, with use of the Horvitz-Thompson estimator and fixed cost of sampling.

## References

Brewer, K.W.R. (1963). A model of systematic sampling with unequal probabilities, Australian Jour. Stat., 5, 5-13.

Durbin, J. (1967). Design of multistage surveys for the estimation of sampling errors, App. Stat., 16, 152-164.

Hanurau, T.V. (1975). Optimum utilization of auxiliary information: πps sampling of two units from a stratum, Jour. Roy. Stat. Soc. B, 29, 374-391.

Hege, V.S. (1967). An optimum property of the Horvitz-Thompson estimate, Jour. Amer. Stat. Assoc., 62, 1013-1017.

Horvitz, D.G. and D.J. Thompson (1952). A generalization of sampling without replacement from a finite universe, Jour. Amer. Stat. Assoc., 47, 663-685.

Madow, W.G. (1949). On the theory of systematic sampling II, Ann. Math. Stat., 20, 333-354.

Midzuno, H. (1951). On the sampling system with probability to sum of sizes, Ann. Inst. Stat. Math., 2, 99-108.

Rao, J.N.K. (1965). On two simple schemes of unequal probability sampling without replacement, Jour. Ind. Stat. Assoc., 3, 173-180.

Rao, J.N.K. and M.P. Singh (1973). On the choice of estimator in survey sampling, Australian Jour. Stat., 15, 95-104.

Rao, T.J. (1971). πPS sampling designs and the Horvitz-Thompson estimator, Jour. Amer. Stat. Assoc., 66, 872-875.

Yates, F. and P.M. Grundy (1953). Selection without replacement from within strata with probability proportional to size, *Jour*. *Roy*. *Stat*. *Soc*. B, 15, 253-261.

## CHAPTER I - INTRODUCTION

1.1  <u>Incomplete data</u>.--In censuses and in almost all sample surveys, a number of different variables are measured on all selected units. The incomplete or missing data with which this report deals are generally of two types.  Sometimes, none of the variables is measured for a unit or subunit--the mail questionnaire was not returned, the interviewer could not find anyone at home, and so on.  This type of miss will be called a <u>unit nonresponse</u>.  Alternatively, most of the questions for a unit are answered, but for certain questions either no answer is given or the answer is judged to contain a gross error and is deleted during editing.  Usually, such questions are ones that are sensitive, e.g., income, or ones for which the respondent does not have the information.  This type of miss is called an <u>item nonresponse</u>.

The most obvious consequence of nonresponse is that a smaller sample of data is available for making the estimates than was originally planned.  In addition, there is often reason to believe that as a group nonrespondents differ systematically from respondents.  As Deming (1950) expressed it with regard to human population:  "The people who do not respond are in some ways and in varying degrees different from them  that do."  Thus, in making estimates from the available respondent data, the sampler may face biases that are essentially unknown in size and direction.

A paper for the Panel on Incomplete Data, Committee on National Statistics, National Academy of Sciences, to be published by Academic Press, Inc., New York. Reproduced with the permission of Academic Press, Inc., New York.

1.2  <u>Some early work on unit nonresponses</u>.--Discussions of the problem of nonresponse and of some possible methods for handling it
seem to have first appeared in the 1930's and 1940's--the same
period during which many of the standard techniques and results in
probability sampling were developed.

The method of quota sampling, discussed in Chapter II, is an early
example of a group of methods whose objective is to cut down or
eliminate nonresponse when the sample data are being collected in
the field.  Quota sampling was developed by Cherington, Roper, Gallup,
and Crossley as a means of saving field costs in taking stratified
samples  for public opinion surveys.  This method formally avoids
unit nonresponse by not requiring that any specific respondents be
questioned.  Instead, the interviewer is told merely how many persons
who fall in each of a set of strata must be interviewed.  The strata
are usually cells of a multiple classification intended to be recognizable in the field by the interviewer--often a classification by
variables like sex, color, geographic residence, plus age and income
by broad classes.  The method may involve some waste time in the
field in partial interviews of people until the interviewer learns
that they do not fall into any stratum in which respondents are
still needed, and, of course, the method abandons random selection
within strata.

Incomplete data are encountered in randomized experiments as well
as in sample surveys, and the development of methods for making

estimates of the effects of the treatment from experiments with missing data also began in the 1930's. The technique most widely used
for this purpose in field experiments is worth noting. This technique, due to Yates (1933), applies the standard Least Squares
model for the analysis of the complete results of field experiments
to the incomplete data that are present. Yates does this by "filling
in" or imputing values for all the missing observations by formulas
such that the standard analysis of the completed data gives the same
estimates as the Least Squares analysis of the incomplete data.
The advantage of this approach is that investigators who know only
how to analyze complete data can use Yates' method, while others
may prefer to apply the Least Squares analysis directly to the
incomplete data that are present. The same choice may be made in
a number of sample survey problems. Thus, Hartley (1958) shows
how to adapt the "fill-in" technique to finding maximum likelihood
estimates of parameters from incomplete random samples of data for
discrete distributions.

Since this approach might be able to handle many problems with
incomplete data, the question: when is the approach valid, is
important. The approach does not require that the units (plots,
subjects, etc.) for which data are missing are a random subsample
of the units in the experiment. For every missing unit it requires
a knowledge of the treatment and block, or more generally of the

values of all background variables that enter into the prediction of the yield on the unit in the Least Squares equations. Furthermore, the fact that an observation is missing must be known to be unrelated to the size of the experimental error that would have arisen on this unit. In other words, the "miss" must be attributable to outside causes and not in any way to the yield on the unit. For instance, this assumption is satisfied for a corner plot in a field experiment that is in a rather exposed place and is trampled by a cow. It is not satisfied when two replicates of variety 3 are battered to the ground in a thunderstorm and are missing (not harvested), if variety 3 is known to have weak stalks. Finally, note that the results of the analysis of the incomplete data are conditional on the set of background variables for the missing observations.

A realization that standard methods can be applied to the incomplete data that we have does not, however, dispose of the statistical problems created by missing data in such cases. In randomized experiments, missing observations almost always make the analysis of the results more complicated for the experimentor. A number of papers have been written to present convenient methods of performing the statistical analyses, at first on desk machines and later for computer programs, given various configurations of missing data. With certain configurations, it has been found that some treatment contrasts that are of interest cannot be estimated from the incomplete data. New problems in statistical theory were encountered, for example, in finding out what can be estimated when data are missing, and in the extension of covariance adjustment to this case.

1.3 <u>Methods using call-backs</u>.--With human populations, one method of cutting down nonresponse rates is to make repeated attempts to contact each sample respondent and have the questions answered. Thus Deming (1945) writes: "Call-backs on a subsample of those people not at home on the first call must be made until results are obtained. In continuing surveys covering identical units, some method of substitution must be worked out for units that cease to exist. If a mailed questionnaire is used, there must be unrelenting follow-up by mail, telephone, and telegraph; and, finally, if necessary, by personal interview of a subsample of those unmoved by less expensive devices." Deming notes that the idea of call-backs on a subsample of initial nonrespondents was presented by W. N. Hurwitz in 1943 at a seminar of the Graduate School of the Department of Agriculture.

In 1946, Hansen and Hurwitz (1946) published the theory covering the subsampling of nonrespondents for a random sample. They had in mind surveys in which the first contact was made by mail, followed by an interview subsampling of the nonrespondents (with, I suppose, some rule about the required number of call-backs). Describing their paper they write: "Personal interviews generally elicit a substantially complete response, but the cost per schedule is, of course, considerably higher than it would be for the mail question-naire method. The purpose of this paper is to indicate a technique which combines the advantages of both procedures."

Later in the same paper, Deming (1945) writes: "Studies in the removal of the biases that arise in cheap methods of collecting data...are worthy of the best efforts of statisticians. However complicated the theory of sampling may appear to be, the problem of predicting biases is by comparison much more complicated and less developed. There are as yet no theories of bias in any way comparable to the ones of sampling. There is an open field here for the joint efforts of statistician and psychologist."

In this connection it was realized around this time that examination of the numbers of completed questionnaires and of the mean responses obtained in successive calls could give valuable information about the effectiveness of call-backs and about the sizes and directions of the biases produced by failure to call back. Several papers pub-lished during 1939-44 on nonresponse bias and the use of call-backs may be cited--Stanton (1939), Rollins (1940), Saudet and Wibon (1940), Hilgard and Payne (1944). In their discussion of the use of call-backs, Clausen and Ford (1947) give an example of a mailed question-naire with two follow-ups. The response variable was the percentage of veterans in school or in training for post-war jobs. The initial sample size was 14,606, the response rate after three mailings being

88%. Table 1 shows the mean responses (percent of veterans in school or training) at each mailing, the percentage of the initial sample who replied to each mailing, and the cumulative percents.

Table 1

Percent of veterans in school or training in
three successive mailings

|  | First mailing | Second mailing | Third mailing | Estimate for nonrespondents |
|---|---|---|---|---|
| % in school or training............ | 42 | 25 | 29 | (25) |
| % of initial sample who replied........ | 54 | 23 | 11 | 12 |
| Cumulative % response | 54 | 77 | 88 | 100 |

The percentage in school or training declined steadily in the first three waves, so that reliance on early returns would have overestimated this percentage. From Table 1 an estimate of the percent in school or training can be made for the final 12% of nonrespondents, by plotting the response data against the cumulative percent response and making the estimate by judgment or by some kind of regression method. Clausen and Ford's estimate for this final group is 25 percent in school or training. In turn, this leads to a weighted estimate of the population mean that attempts to correct for nonresponse bias, namely: $[(.54)(42)+(.23)(25)+(.11)(29)+(.12)(35)] = 35$ percent in school or training. Hendricks (1948, 1949) has discussed this technique.

On a related topic, the recent book by the psychologists Rosenthal and Rosnow (1975) summarizes the results of many investigations of the characteristics of those who volunteer, as compared with those

who decline to volunteer, as subjects in research studies on human
behavior and learning. The book estimates how the use of volunteers
biases conclusions drawn from the studies.

From analyses of call-back data, samplers learn about the sizes
and directions of biases with different questions and different
types of populations. It is usually found that respondents in a
population are more intelligent, better educated, more interested
in the questions, and are less embarrassed or threatened by the
questions. A sampler might be able to predict from such data roughly
the biases to be expected and the costs per completed interview
at successive calls, particularly in repetitive surveys of the same
population for the same items. These data may permit the construction
of models for the effects of different calls that lead to the
appraisal of different call-back policies from a cost-benefit viewpoint. For instance, Deming (1953) gives methods (based on a model)
that provide a specified mean square error in the estimated population total at minimum cost. With a binomial response variate, Birnbaum and Sirken (1950) use the fact that the error in an estimated
proportion is always bounded to determine the initial sample size
and number of call-backs that make the absolute error of estimate
in the sample proportion $\leq \delta$ with probability $(1-\alpha)$ at minimum expected
cost.

The major reviews in 1946 by Yates (1946) and Mahalanobis (1946)
of their work on the development of sample surveys did not discuss

the handling of nonresponse problems, perhaps because nonresponse
is rare in crop-cutting surveys. Yates did, however, criticize
quota sampling, stating that "such a method may very easily introduce serious biases." In the discussion following Yates' paper,
Hartley proposed another method for reducing biases due to missing
data. He began by calling attention to the biases caused by the
omission of "not at home" households in probability samples: "Now,
it is well known that if these 'nobody in' households are left out,
the remainder of the original sample is seriously biased. For instance, there will be an unduly low proportion of housewives with
part- or full-time work, an unduly high proportion of young children,
an unduly low proportion of 'queuers,' and many other misrepresentations.
A classic example of this bias is mentioned by Professor Bradford Hill
in his book, Principle of Medical Statistics. The example is the
Ministry of Health report on the influenza pandemic of 1918. In
this enquiry houses which were found closed at the time of the visit
had to be ignored."

Hartley noted that repeated call-backs at "not at home" households
would cut down the nonresponses, but commented: "whatever the success
of the method, it is bound to be very laborious." He therefore proposed for consideration a cheaper alternative. Make a single call,
but ask additional questions whose purpose is to find out, for the
"at home" households, the proportion of the time of interviewing during which an adult will be found at home. The sample responses are

then stratified by these proportions of time at home. The stratum
weights are estimated from the sample proportions, and the form of
estimate appropriate to stratified sampling is used. This proposal
does not produce data for the not-at-homes, but the stratified estimate might prove to have less bias than the sample mean, and the
method should indicate those variables for which the responses are
correlated, and those for which they are not correlated, with the proportion of time spent at home. Hartley's proposal led to the Politz-
Simmons methods (1949, 1950).

In his reply, Yates (1946) doubted whether Hartley's methods, though
ingenious, would be very satisfactory in practice. As an alternative
that would also reduce the amount of calling back, Yates proposed sub-
sampling "the houses for which say more than one call-back is required,
selecting say 50 percent of such houses for further calls, subsequently
doubling the contribution of this 50 percent to the total sample." This
is, of course, one version of the method described by Hansen and Hurwitz
(1946).

1.4  Item nonresponse.--In all but the most recent work on data with
item nonresponses, the approach has been to analyze the available incomplete data. As with randomized experiments, this analysis does not
require the assumption that units with item nonresponses are a random
subsample of the original data. The probability that an item is
missing may depend on the values of other items for the unit.

But given these values the probability must be constant. For instance, the probability that income is unreported in a survey may depend on the sex, age, and educational level of the respondent. But if for males of a given age and educational level, those with unusually high incomes are those who do not report incomes for one reason or another, analysis of the incomplete data will be subject to bias.

In the earliest paper on item nonresponses, Wilks (1932) considers a random sample from a bivariate normal distribution. In $n_{12}$ observations, $X_1$ and $X_2$ are both measured, in $n_{10}$ observations, $X_1$ is measured but $X_2$ is missing, while in $n_{02}$ observations, $X_2$ is measured but $X_1$ is missing. Wilks finds maximum likelihood estimates of the parameters $\mu_1$, $\mu_2$, $\sigma_1$, $\sigma_2$, and $\rho$, and gives the large-sample formulas for the variances and covariance of $\hat{\mu}_1$ and $\hat{\mu}_2$. He notes that his estimates are not particularly simple, and devotes a substantial part of the paper to studying the efficiencies of simpler estimates of the five parameters. For instance, if the $n_{02}$ units in which $X_1$ was missing were a random sub-sample, we might use the mean of all the $(n_{12} + n_{10})$ measured values of $X_1$ as an estimate of $\mu_1$.

A more general concept developed by Wilks is that of the joint efficiency of a set of estimates. He defines this as the reciprocal of the determinant of the matrix of the large-sample variances and covariances of the estimates, following the approach by Fisher. He proves that this quantity is a maximum when the estimates are maximum likelihood estimates. This concept of efficiency is useful in Wilks' work in that he can calculate the relative efficiencies of the sets of simpler estimates that he and the reader might consider using, so that the reader knows what penalty is being paid for simplicity.

No extensions of Wilks' work appeared until the 1950's. For more than 2 multivariate normal variates, Mathai (1951) showed how to write down the equations leading to the maximum likelihood estimates of the population means and to the information matrix, but did not discuss practical methods of solution.

Lord (1955) considers a trivariate normal sample in which $X_1$ is missing in a subsample of size $n_{023}$ while $X_2$ is missing in a sample of size $n_{103}$. The original sample size is $n = (n_{023} + n_{103})$, so that $X_3$ is measured for all such data, leaving eight parameters $\mu_1$, $\mu_2$, $\mu_3$, $\sigma_1$, $\sigma_2$, $\sigma_3$, $\rho_{13}$, and $\rho_{23}$ to be estimated. Lord obtains explicit estimates. A device that he uses is to estimate first $\sigma_3$, $\sigma_{1 \cdot 3}$, $\sigma_{2 \cdot 3}$, $\beta_{1 \cdot 3}$, and $\beta_{2 \cdot 3}$, which uniquely determine $\sigma_1$, $\sigma_2$, $\sigma_3$, $\rho_{13}$ and $\rho_{23}$, where the notation $\sigma_{1 \cdot 3}$, $\beta_{1 \cdot 3}$, etc., is the usual one for linear regressions on one independent variate.

Edgett (1956) develops the maximum likelihood estimates of the nine parameters in a trivariate normal in which $X_1$ is missing in part of the sample.

Anderson (1957) presents an approach that produces the maximum likelihood estimates for any number of variates, given certain patterns of missing data. In the simplest pattern of this type, the variates can be arranged in order so that for any unit in which $X_2$ is measured, so is $X_1$ and $X_2$, and so on. The approach depends on the fact that the likelihood

factorizes into the marginal distribution of $X_1$; the conditional

distribution of $X_2$ given $X_1$; the conditional distribution of $X_3$

given $X_1$ and $X_2$; and so on. It follows that $\mu_1$ and $\sigma_1$ are estimated

from the univariate distribution of $X_1$ in the whole sample since $X_1$

has no misses, while $\mu_2$, $\sigma_2$, and $\rho_{12}$ are estimated from the part of

the sample in which $X_1$ and $X_2$ are measured and the part in which $X_1$

alone is measured, using the same technique as in double sampling

with regression, and so on. As illustrations of this approach,

Anderson works out the solutions given by Lord and Edgett for their

trivariate problems. (Wilks' problem does not have Anderson's

pattern of missing data.)

A different method proposed by Buck (1960) is also of interest. This

method employs multiple regression rather than maximum likelihood, and

uses the set of observations for which all items are measured to

substitute predicted or imputed values for all missing items on any

units. For instance, if $X_1$ and $X_4$ are missing on a unit, the predicted

value $X_1$ for this unit is obtained from the regression of $X_1$ on $X_2$,

$X_3$, $X_5$,...,$X_k$ in the sample in which all items are measured, by sub-

stituting the known values of $X_2$, $X_3$, $X_5$...for the unit in question.

With a random sample the estimated population mean for the item is

the mean of the observed and predicted values of this item for all

the sample members. The matrix of sample mean squares and products, calculated from observed and predicted values, requires some adjustment in order to remove biases. This matrix is needed for regression calculations applicable to the whole sample.

1.5 <u>Contents of volume</u>.--We now indicate the contents of the rest of this volume, which contains three parts, each with a number of chapters. Part I - methods of data collection - describes a number of techniques, used in the field, whose objective is to cut down the amount of missing data and so make the problem less acute. Part II - data processing and analysis - deals with methods for making estimates and carrying out further analyses in the presence of nonresponse, including methods of weighting when estimating population totals or means, and methods of imputation that fill in estimated values for the misses and complete the data for further analyses. In Part III - nonresponse in the context of total survey error - the total error in a survey is broken down into the principal components that affect it with different methods of imputation.

1.6 <u>Part I - Methods of data collection</u>.--Three of these techniques have been mentioned. For quota sampling, studies of its performance relative to probability sampling have been made in the book by Stephan and McCarthy (1959). A combination of probability sampling and quota samplings has been developed by Sudman (1967), in which probability sampling is used down to the block level in the selection of the sample, and the quota method is introduced only in selecting people within blocks.

Chapter 8 of Part I gives a critique of the quota sampling method and examples of its current use by two sampling organizations.

Chapter 6 describes double sampling, gives the theory and an illustration, and includes the more realistic methods of Srinath and J.N.K. Rao of determining the size of the subsample. Included also are the extension of the method to more than one subsample and to the ratio and regression estimators. Bayesian work on double sampling is described in Chapter 7. This involves its application to the estimation of population proportions, sampling methods that draw two samples, one pilot and one main, from the respondents and two from the nonrespondents, and the Bayesian applications to the estimation of regression coefficients.

On call-backs, data on the number of respondents found at successive calls are given from several sources, and Durbin and Stuart's (1954) data on the relative costs of obtaining a completed interview at successive calls - this in Chapter 2. The model and strategy from which to make cost-benefit studies of different call-back policies, developed by Deming with further work by P.S.R.S. Rao, is described with an illustration.

For households surveys another strategy is to expand the number of respondents who can be asked to supply the information desired if the selected respondent is unavailable. Various rules that have been used to define acceptable respondents are described in Chapter 3. A critique is given of the quality of the data obtained in studies from different types of respondent. Instead of a proxy who gives information about the selected respondent, some surveys allow substitution of a different respondent if the intended respondent cannot be contacted or refuses. This substitution is of course a departure from strict probability sampling and usually the substitution rule is designed in the hope that the substitute will be similar to the respondent wanted. Chapter 4 describes substitution rules and gives a number of surveys in which the substitute data could be compared with that from the original respondents. Conclusions are that substitutions may reduce but do not eliminate respondent bias, and probably work best when the sampling units are small in number and such that a good deal of information about them is known with which to find good substitutes. No theory about substitutes seems to have been developed.

The use of proxy respondents to supply desired information is more fully developed in network sampling (Chapter 9). The linkage rule that defines the proxy respondents may be based on kinship, friendship, or proximity to the household. With these rules, the number of sources that can give information about a selected sample member, and hence the probability of being a respondent, vary from member to member, so that weighted estimates of population totals are required. Chapter 9 describes the kind of intended respondents for whom network sampling was produced, presents the theory, including costs, and gives its application to the problem of coverage (data missing because the sampling units in question are not in the frame).

Some surveys have used small amounts of money or modest gifts with the aim of increasing response rates, particularly in mail surveys...

Chapter 5 presents some studies that have tried this method, including
studies giving results on the effect of the financial incentives on
response rates. Incentives seem to help response rates in mail surveys:
more studies are needed to indicate whether they help in interview surveys. In another area, Part II will also include an account of the
randomized response method. In this method, designed to encourage
responses to sensitive questions, one of two questions is given to
each respondents, and the interviewer does not know whether the sensitive
question or an innocent one is being answered. It is hoped that this
information encourages respondents to answer, since they are not revealing sensitive information to the respondent.

1.7 Part II. Data processing and analysis.--Chapter 1 presents the
concepts and notation that will be used in considering nonresponse later
in this part . At the beginning, the chapter describes the randomization and Bayesian approaches to sample survey theory in the absence of
nonresponse. In the randomization approach used in most textbooks,
the values of the responses in the population are regarded as fixed
numbers with no specific frequency distribution. The frequency distribution of estimates of the population total is that generated by
the set of random samples and their resulting estimates that the sampling
plan can provide. This approach avoids the assumption of any models that
the population values follow and may hope to obtain robust results. But
in samples with missing data, some models must be constructed about
the missing data in order to make estimates. The Bayesian approach

to survey sampling, described in Chapter 1, is model-oriented.  The
responses are assumed drawn at random from some superpopulation model,
while the nonresponses are also drawn from a superpopulation model
that may be the same or different from the model for the responses.
It is noted that in large samples the inferences are approximately the
same in the randomization and Bayesian approaches if the distribution
of the population data is normal under the Bayesian approach.  The
presence of nonresponse creates difficulties under both systems,
since the assumptions that must be made about the distribution of the
nonresponses are unverifiable.  The Bayesian system may be in better
shape for this task, since it also is accustomed to making assumptions
about the distribution of the responses, but the practical details
of its application may be complicated.

A compromise approach to the problem of making inferences about the
population total is to use Bayesian methods to make a series of
plausible assumptions about the distributions of the responses that nonrespondents would have given had they responded.  By a drawing from
any model for nonrespondents we can impute values for the missing data,
complete the data, and hence calculate 95% confidence intervals for the
population mean.  A series of these imputations may be of two types--
repeated drawings from the same model and drawings from different models
Multiple imputations of this type will reveal how variation in drawing
from the same model and variation in drawing from different models
about nonresponse will affect the 95% intervals, or more generally
analyses made on the incomplete sample data.  This idea of multiple

imputations, discussed in Chapter 6 of Part III, can make us more fully aware of the kinds of conclusions that can safely be drawn from analyses of incomplete sample data and the extent to which they will vary because of incompleteness.

The second chapter in Part III discusses weighting adjustments for nonresponse. We give first some introductory material. In one group of methods widely used in survey practice, the sample of respondents is partitioned into classes or strata by variables thought to be related to the probability of unit nonresponse so that units in the same stratum have about the same probability of being a nonrespondent. Then the assumption is made that this probability is actually constant for all members of the same stratum.

In stratum h, let $\Pi_h$ be the probability that any unit is selected, and $\phi_h$ the probability that the unit responds if selected. If $\hat{Y}_{hr}$ is the sample total of the respondents in stratum h, the estimated population total for this group of methods is

$$\sum_h \frac{1}{\Pi_h \phi_h} \hat{Y}_{hr} = \sum_h W_h \hat{Y}_{hr}$$

where

$$W_h = 1/\Pi_h \phi_h$$

is a _weighting_ or _inflation_ factor.

In different applications of this method, the best estimates of $W_h$ that the data provide are used. The simplest case is that of simple random sampling in wich the population totals $N_h$ for the strata are known.

Here, $\Pi_h = n_h/N_h$, and we take $n_{hr}/n_h$ as our estimate of $\phi_h$, so that

$$\Pi_h \phi_h = n_{hr}/N_h$$

and the estimated population total is

$$\hat{Y} = \sum_h N_h \hat{Y}_{hr}/n_{hr} = \sum_h N_h \bar{y}_{hr} \quad .$$

Variants of the method depend on the nature of the classification. We will discuss several in this section under the assumption that the original sample was a simple random one.

In the technique known as post-stratification, the classification is made after the data have been collected. In such cases the $N_h$ are rarely known.* If the original sample sizes $n_h$ are known, we may estim $N_h$ by $Nn_h/n$, since in post-stratification the sample distributes itself approximately proportionally in the strata. This gives

$$\hat{Y} = \frac{N}{n} \sum_h \left( \frac{n_h}{n_{hr}} \hat{Y}_{hr} \right) = \frac{N}{n} \cdot \sum_h n_h \bar{y}_{hr}$$

For example, the estimator by Hansen and Hurwitz mentioned in Section 1.3 is of this type. There are two post-strata. Stratum 1 consists of those who respond to the first (mail) call, with sample size $n_1$. Stratum 2 has original sample size $n_2 = (n-n_1)$, though the interview sample actually taken is of size $n_2/k$, with sample mean $\bar{y}_2$. The estimator is

$$\hat{Y} = \frac{N}{n} \left( n_1 \bar{y}_1 + n_2 \bar{y}_2 \right)$$

---

*When the $N_h$ are unknown, this method also goes by the name of a "weighting class" adjustment.

In the method of Bartholomew (1961), calls are made on <u>all</u> $n_2$ in

stratum 2. Any nonrespondents at the second call are assumed to be

a random subsample of the $n_2$, so that the estimator is the same as

that of Hansen and Hurwitz except that $\bar{y}_2$ is replaced by $\bar{y}_{2r}$, the mean

for the respondents at the second call.

Unless the $N_h$ are known, post-stratification suffers a loss of pre-

cision because the strata weights are estimated from sample data.

In some applications, the post-strata are formed by cross-classifica-

tions on two variables. The $N_{ij}$ in stratum $(i,j)$ are not known, but

one or both of the marginal totals $N_{i.}$ of $N_{.j}$ are known. In this event

we may try to improve on the usual post-stratification estimator

$$\hat{Y} = N \sum_{ij} \frac{n_{ij}}{n} \bar{y}_{ijr}$$

replacing it by the weighted estimator

$$\hat{Y} = N \sum_{ij} w_{ij} \cdot \bar{y}_{ijr}$$

where the $w_{ij}$ are estimated from both the $n_{ij}$ and the known marginal

proportions $N_{i.}/N$ and $N_{.j}/N$. This method of estimation is known as

<u>raking</u>. Some of its potential applications will be described.

Other information may be used to estimate the probability of response

$\phi_h$. In the Politz-Simmons method, only a single call is made. Inter-

viewing takes place on the six evenings Monday to Saturday. Any

respondent found at home is asked on how many of the other five weeknights

he or she was at home at this time.  If the answer is h nights, the
quantity (h+1)/6 is taken as an estimate of $\phi_h$.  The sample results
are post-stratified into six strata (h=0,1,...5).  This method, of
course, omits from consideration those out of the total sample of
size n who are never found at home; hence, instead of an estimate of
the form

$$\hat{Y} = \sum_h \frac{1}{\Pi_h \phi_h} \hat{Y}_{hr}$$

the method uses a ratio-type estimator

$$\hat{Y} = N \frac{\Sigma \hat{Y}_{hr}/\hat{\phi}_h}{\Sigma 1/\hat{\phi}_h}$$

In the chapter on weighting adjustments in Part II, the emphasis has been
placed on comparisons among alternative estimators of the population total
in the presence of missing data.  In particular, the case where there is
no adjustment for nonresponse is compared to estimators based on three
different assumptions about the probability of a response, namely a constant
probability, probability proportional to a know value of $x_i$, and probability
proportional to an unknown $x_i$.  These assumptions are usually applied within
strata in a post-stratified estimate.  Sometimes the units can be grouped
into a two-way classification such that the probability of response is proportional to $u_h v_k$ for a unit in the (h,k) cell of the classification.  Estimates of the population total are made by raking, using the information
available about the $u_h$ and the $v_k$.  Some recommendations for future study
are given.  An appendix gives adjustments to the Hurwitz-Thompson estimate
of the population corresponding to the three assumptions given above about
the probability of a response.  The biases, variances, and estimated varianc

of these estimates are given, with a numerical example at the end.
With weighting methods, estimates of population totals are made without filling-in the answers that nonrespondents are assumed to have
given if they had responded. For the simpler weighting methods, however, one can construct a method of imputing a value for each missing
unit such that the population total, estimated from the completed
sample data, agrees with the estimation by the weighting method. For
example, suppose that in a stratum the probability of response is
assumed constant. If $n_r$ respond out of n selected, with sample total $\hat{Y}_r$
for the respondents, the usual weighting estimate of the population
total is $N\hat{\bar{Y}}_r/n_r$. This estimate can be obtained by imputing the value
of $\bar{y}$ for each missing value. Chapter 3 in Part II gives an account
of one of the most common imputation methods, known as the "hot deck"
method. In a hot deck procedure, a sample reported value is substituted
for the missing value: in a "cold deck" procedure, the value substituted
is not a sample value. Usually, the sample is first classified by
background variables thought to be related to the probability of a
miss, so that substitution of a member of the same class will introduce
little bias. In a sequential hot deck method, the class is first put
in some kind of order, and the last reported sample value is substituted, or occasionally something more complex. The hot deck methods
in three large surveys are described. Some work on the variance of the
methods is given. With a simple random sample of size n, suppose that
m units are missing and (n-m) report. If the m are missing at random,
the variance of the mean of the reporters is $\sigma^2/(n-m)$. If a sample

member is substituted for each miss, the variance of the sample mean

is (with replacement)

$$V(\bar{y}_h) = \frac{\sigma^2}{n}\left[1+ \frac{m(2n-m-1)}{(n-m)n}\right]$$

This variance is always greater than or equal to $\sigma^2/(n-m)$. The variance is slightly different in a sequential hot deck method. The chapter concludes by describing five studies in which the hot deck method has been compared with alternatives, such as weighting, ratio or regression, for estimating the population total or mean.

In Chapters 5 and 6 of Part III, item and unit nonresponse are considered in the context of sampling from an infinite superpopulation. In Chapter 5, items or whole units are assumed to be missing at random. After preliminaries, methods of finding maximum likelihood estimates are discussed, including the most general patterns of missing data to which Anderson's method already described are applicable, iterative methods, including the EM algorithm, for handling more general patterns of missing data, and two-way contingency tables and multinomial distribution with missing values.

In Chapter 6, the superpopulation model is extended to cover cases in which the responding units are not a random sample of the sampling units. As might be expected, the cases that can be covered are those in which the probability of nonresponse is a function of y, though it may also depend on other variables $y_2$ and $y_3$. Some examples of ml estimates are given in such cases; also, a numerical illustration is shown, in which the population mean is estimated by ml and by the sample mean.

1.8  Part III.--Totals and variances are provided for five methods of imputation--called zero substitution, weighting, duplication, historical substitution, and the Hurwitz-Thompson method.  Variance formulas are extended to cover two-stage sampling, of which a numerical illustration is given.